# GENETICS

## ANALYSIS OF GENES AND GENOMES

SEVENTH EDITION

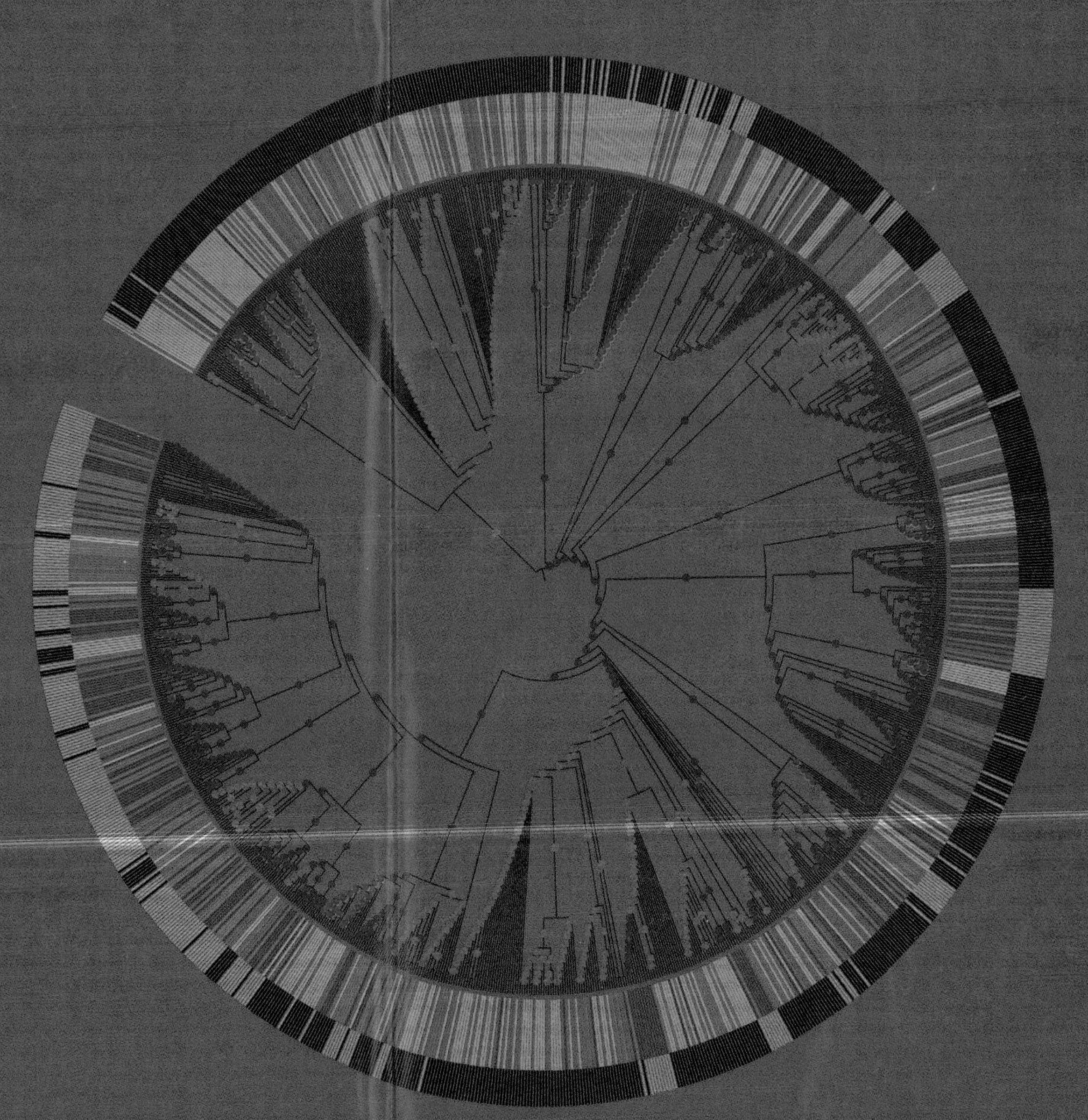

Courtesy of Lawrence David, Dana Hunt, Eric Alm, and Martin Polz.

DANIEL L. HARTL AND ELIZABETH W. JONES

JONES AND BARTLETT PUBLISHERS
*Sudbury, Massachusetts*
BOSTON TORONTO LONDON SINGAPORE

*World Headquarters*
Jones and Bartlett Publishers
40 Tall Pine Drive
Sudbury, MA 01776
978-443-5000
info@jbpub.com
www.jbpub.com

Jones and Bartlett Publishers
Canada
6339 Ormindale Way
Mississauga, Ontario L5V 1J2
Canada

Jones and Bartlett Publishers
International
Barb House, Barb Mews
London W6 7PA
United Kingdom

Jones and Bartlett's books and products are available through most bookstores and online booksellers. To contact Jones and Bartlett Publishers directly, call 800-832-0034, fax 978-443-8000, or visit our website www.jbpub.com.

Substantial discounts on bulk quantities of Jones and Bartlett's publications are available to corporations, professional associations, and other qualified organizations. For details and specific discount information, contact the special sales department at Jones and Bartlett via the above contact information or send an email to specialsales@jbpub.com.



Production Credits
Chief Executive Officer: Clayton Jones
Chief Operating Officer: Don W. Jones, Jr.
President, Higher Education and Professional Publishing: Robert W. Holland, Jr.
V.P., Sales and Marketing: William J. Kane
V.P., Design and Production: Anne Spencer
V.P., Manufacturing and Inventory Control: Therese Connell
Publisher, Higher Education: Cathleen Sether
Acquisitions Editor, Science: Shoshanna Goldberg
Managing Editor, Science: Dean W. DeChambeau
Associate Editor, Science: Molly Steinbach
Editorial Assistant, Science: Caroline Perry
Production Editor: Rachel Rossi
Senior Marketing Manager: Andrea DeFronzo
Photo Research Manager/Photographer: Kimberly Potvin
Text Design: Anne Spencer
Cover Design: Anne Spencer

Cover and Title Page Image: Courtesy of Lawrence David, Dana Hunt, Eric Alm, and Martin Polz.
Printing and Binding: Courier Kendallville
Cover Printing: Courier Kendallville

ISBN: 978-0-7637-7215-4

Library of Congress Cataloging-in-Publication Data
Hartl, Daniel L.
Genetics : analysis of genes and genomes / Daniel L. Hartl and Elizabeth W. Jones. -- 7th ed.
p. ; cm.
Includes bibliographical references and index.
ISBN 978-0-7637-5868-4 (alk. paper)
1. Genetics. I. Jones, Elizabeth W. II. Title.
[DNLM: 1. Genetics. 2. Genomics. QU 450 H331g2009]

QH430.H3733 2009
576.5--dc22 2008001012

6048
Printed in the United States of America
12 11 10 09 08 10 9 8 7 6 5 4 3 2

To the best teachers we ever had—Our parents and our students

# BRIEF CONTENTS

1 Genes, Genomes, and Genetic Analysis . . . . . 1

2 DNA Structure and Genetic Variation . . . . . 38

3 Transmission Genetics: The Principle of Segregation . . . . . 77

4 Chromosomes and Sex-Chromosome Inheritance . . . . . 114

5 Genetic Linkage and Chromosome Mapping . . . . . 150

6 Molecular Biology of DNA Replication and Recombination . . . . . 190

7 Molecular Organization of Chromosomes . . . . . 221

8 Human Karyotypes and Chromosome Behavior . . . . . 251

9 Genetics of Bacteria and Their Viruses . . . . . 295

10 Molecular Biology of Gene Expression . . . . . 341

11 Molecular Mechanisms of Gene Regulation . . . . . 380

12 Genomics, Proteomics, and Transgenics . . . . . 431

13 Genetic Control of Development . . . . . 471

14 Molecular Mechanisms of Mutation and DNA Repair . . . . . 509

15 Molecular Genetics of the Cell Cycle and Cancer . . . . . 551

16 Mitochondrial DNA and Extranuclear Inheritance . . . . . 587

17 Molecular Evolution and Population Genetics . . . . . 611

18 The Genetic Basis of Complex Traits . . . . . 651

# CONTENTS

**Preface** xv
**Acknowledgments** xxiii
**About the Authors** xxvi
**In Memoriam** xxvii
**About the Cover** xxviii

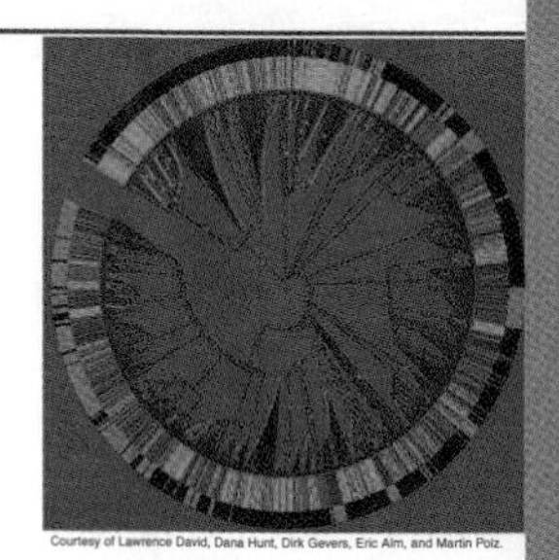

Courtesy of Lawrence David, Dana Hunt, Dirk Gevers, Eric Alm, and Martin Polz.

## 1 Genes, Genomes, and Genetic Analysis 1

**1.1 DNA: The Genetic Material 2**
Experimental Proof of the Genetic Function of DNA 3
Genetic Role of DNA in Bacteriophage 4
**1.2 DNA Structure and Replication 6**
An Overview of DNA Replication 9
**1.3 Genes and Proteins 10**
Inborn Errors of Metabolism as a Cause of Hereditary Disease 10
**1.4 Genetic Analysis 14**
Mutant Genes and Defective Proteins 15
Complementation Test for Mutations in the Same Gene 17
Analysis of Complementation Data 20
Other Applications of Genetic Analysis 20
**1.5 Gene Expression: The Central Dogma 21**
Transcription 23
Translation 24
The Genetic Code 25
**1.6 Mutation 26**
**1.7 Genes and Environment 28**
**1.8 The Molecular Unity of Life 29**
Prokaryotes and Eukaryotes 30
Evolutionary Relationships Among Eukaryotes 31
Genomes and Proteomes 32

## 2 DNA Structure and Genetic Variation 38

**2.1 Genetic Differences Among Individuals 39**
DNA Markers as Landmarks in Chromosomes 39
**2.2 The Molecular Structure of DNA 40**
Polynucleotide Chains 41
Base Pairing and Base Stacking 43
Antiparallel Strands 45
DNA Structure as Related to Function 46
**2.3 The Separation and Identification of Genomic DNA Fragments 47**
Restriction Enzymes and Site-Specific DNA Cleavage 48
Gel Electrophoresis 50
Nucleic Acid Hybridization 52
The Southern Blot 54
**2.4 Selective Replication of Genomic DNA Fragments 55**
Constraints on DNA Replication: Primers and 5′-to-3′ Strand Elongation 56
The Polymerase Chain Reaction 57
**2.5 The Terminology of Genetic Analysis 60**
**2.6 Types of DNA Markers Present in Genomic DNA 62**
Single-Nucleotide Polymorphisms (SNPs) 62
Restriction Fragment Length Polymorphisms (RFLPs) 63
Tandem Repeat Polymorphisms 66
Copy-Number Polymorphisms (CNPs) 68
**2.7 Applications of DNA Markers 68**
Genetic Markers, Genetic Mapping, and "Disease Genes" 68
Other Uses for DNA Markers 69

# 3 Transmission Genetics: The Principle of Segregation 77

**3.1 Morphological and Molecular Phenotypes 78**

**3.2 Segregation of a Single Gene 80**
Phenotypic Ratios in the $F_2$ Generation 82
The Principle of Segregation 83
Verification of Segregation 85
The Testcross and the Backcross 86

**3.3 Segregation of Two or More Genes 87**
The Principle of Independent Assortment 88
The Testcross with Unlinked Genes 88
Three or More Genes 89

**3.4 Probability in Genetic Analysis 90**
Elementary Outcomes and Events 92
Probability of the Union of Events 93
Probability of the Intersection of Events 93
Conditional Probability 94
Bayes' Theorem 96

**3.5 Human Pedigree Analysis 97**
Characteristics of Dominant and Recessive Inheritance 97
Most Human Genetic Variation is Not "Bad" 99
Molecular Markers in Human Pedigrees 100

**3.6 Incomplete Dominance and Epistasis 101**
Multiple Alleles 102
Human ABO Blood Groups 103
Epistasis 105

# 4 Chromosomes and Sex-Chromosome Inheritance 114

**4.1 The Stability of Chromosome Complements 115**

**4.2 Mitosis 116**

**4.3 Meiosis 119**
The First Meiotic Division: Reduction 124
The Second Meiotic Division: Equation 128

**4.4 Sex-Chromosome Inheritance 128**
Chromosomal Determination of Sex 128
X-Linked Inheritance 129
Pedigree Characteristics of Human X-linked Inheritance 133
Heterogametic Females 133
Nondisjunction as Proof of the Chromosome Theory of Heredity 134
Sex Determination in *Drosophila* 135

**4.5 Probability in the Prediction of Progeny Distributions 136**
Using the Binomial Distribution in Genetics 136
Meaning of the Binomial Coefficient 138

**4.6 Testing Goodness of Fit to a Genetic Hypothesis 139**
The Chi-Square Method 139
Are Mendel's Data Too Good to Be True? 143

# 5 Genetic Linkage and Chromosome Mapping 150

**5.1 Linkage and Recombination of Genes in a Chromosome 151**
Coupling Versus Repulsion of Syntenic Alleles 153
The Chi-Square Test for Linkage 153
Each Pair of Linked Genes Has a Characteristic Frequency of Recombination 154
Recombination in Females Versus Males 155

**5.2 Genetic Mapping 155**
Map Distance and Frequency of Recombination 156
Crossing over 159
Recombination Between Genes Results from a Physical Exchange Between Chromosomes 159
Crossing over Takes Place at the Four-Strand Stage of Meiosis 161
Multiple Crossovers 164

**5.3 Genetic Mapping in a Three-Point Testcross 166**
Chromosome Interference in Double Crossovers 167
Genetic Mapping Functions 168
Genetic Map Distance and Physical Distance 169

5.4 **Genetic Mapping in Human Pedigrees 170**
Maximum Likelihood and Lod Scores 170

5.5 **Mapping by Tetrad Analysis 173**
Analysis of Unordered Tetrads 173
Genetic Mapping with Unordered Tetrads 175
Analysis of Ordered Tetrads 177

5.6 **Special Features of Recombination 180**
Recombination Within Genes 180
Mitotic Recombination 181

## 6 Molecular Biology of DNA Replication and Recombination 190

6.1 **Problems of Initiation, Elongation, and Incorporation Error 191**

6.2 **Semiconservative Replication of Double-Stranded DNA 192**
The Meselson–Stahl Experiment 192
Semiconservative Replication of DNA in Chromosomes 194
Theta Replication of Circular DNA Molecules 196
Rolling-Circle Replication 198
Multiple Origins and Bidirectional Replication in Eukaryotes 198

6.3 **Unwinding, Stabilization, and Stress Relief 200**

6.4 **Initiation by a Primosome Complex 201**

6.5 **Chain Elongation and Proofreading 202**

6.6 **Discontinuous Replication of the Lagging Strand 204**
Fragments in the Replication Fork 204
The Joining of Precursor Fragments 205

6.7 **Terminator Sequencing of DNA 207**
Sanger Sequencing 208
Massively Parallel Sequencing 209

6.8 **Molecular Mechanisms of Recombination 210**
Gene Conversion and Mismatch Repair 211
Double-Strand Break and Repair Model 212

## 7 Molecular Organization of Chromosomes 221

7.1 **Genome Size and Evolutionary Complexity: The C-Value Paradox 222**

7.2 **The Supercoiling of DNA 223**
Topoisomerase Enzymes 224

7.3 **The Structure of Bacterial Chromosomes 226**

7.4 **The Structure of Eukaryotic Chromosomes 226**
The Nucleosome: The Structural Unit of Chromatin 226
The Nucleosome Core Particle 226
Chromosome Territories in the Nucleus 229
Chromosome Condensation 230

7.5 **Polytene Chromosomes 232**

7.6 **Repetitive Nucleotide Sequences in Eukaryotic Genomes 232**
Kinetics of DNA Renaturation 233
Analysis of Genome Size and Repetitive Sequences by Renaturation Kinetics 235

7.7 **Unique and Repetitive Sequences in Eukaryotic Genomes 237**
Unique Sequences 238
Highly Repetitive Sequences 238
Middle-Repetitive Sequences 239

7.8 **Molecular Structure of the Centromere 239**

7.9 **Molecular Structure of the Telomere 241**

## 8 Human Karyotypes and Chromosome Behavior 251

8.1 **The Human Karyotype 252**
Standard Karyotypes 252
The Centromere and Chromosome Stability 255
Dosage Compensation of X-Linked Genes 256
The Calico Cat 259
Pseudoautosomal Inheritance 259
Active Genes in the "Inactive" X Chromosome 260
Gene Content and Evolution of the Y Chromosome 260
Tracing Human History Through the Y Chromosome 262

**8.2 Chromosome Abnormalities in Human Pregnancies 264**
Down Syndrome and Other Viable Trisomie 266
Trisomic Segregation 267
Sex-Chromosome Abnormalities 268
Environmental Effects on Nondisjunction 269
**8.3 Chromosomal Deletions and Duplications 269**
Deletions 270
Deletion Mapping 271
Duplications 273
Unequal Crossing Over in Red–Green Color Blindness 273
**8.4 Genetics of Chromosomal Inversions 275**
Paracentric Inversion (Not Including the Centromere) 275
Pericentric Inversion (Including the Centromere) 277
**8.5 Chromosomal Translocations 277**
Reciprocal Translocations 278
Genetic Mapping of a Translocation Breakpoint 278
Robertsonian Translocations 280
Translocations and Trisomy 21 281
Translocation Complexes in *Oenothera* 282
**8.6 Genomic Position Effects on Gene Expression 282**
**8.7 Polyploidy in Plant Evolution 284**
Sexual Versus Asexual Polyploidization 285
Autopolyploids and Allopolyploids 286
Monoploid Organisms 288
**8.8 Genome Evolution in the Grass Family (Gramineae) 289**

## 9 Genetics of Bacteria and Their Viruses 295

**9.1 Mobile DNA 296**
Plasmids 296
The F Plasmid: A Conjugative Plasmid 297
Insertion Sequences and Transposons 298
Mobilization of Nonconjugative Plasmids 299
Integrons and Antibiotic-Resistance Cassettes 300
Pathogenicity Islands 301
Multiple-Antibiotic-Resistant Bacteria 302
**9.2 Bacterial Genetics 302**
Mutant Phenotypes 302
Mechanisms of Genetic Exchange 303
**9.3 DNA-Mediated Transformation 303**
**9.4 Conjugation 305**
Cointegrate Formation and Hfr Cells 305
Time-of-Entry Mapping 306
F′ Plasmids 311
**9.5 Transduction 312**
The Phage Lytic Cycle 312
Generalized Transduction 313
**9.6 Bacteriophage Genetics 316**
Plaque Formation and Phage Mutants 317
Genetic Recombination in the Lytic Cycle 318
Genetic and Physical Maps of Phage T4 319
Fine Structure of the *rII* Gene in Bacteriophage T4 321
**9.7 Lysogeny and Specialized Transduction 325**
Site-Specific Recombination and Lysogeny 326
Specialized Transduction 332

## 10 Molecular Biology of Gene Expression 341

**10.1 Amino Acids, Polypeptides, and Proteins 342**
**10.2 Colinearity Between Coding Sequences and Polypeptides 344**
**10.3 Transcription 344**
Overview of RNA Synthesis 344
Types of RNA Polymerase 346
Promoter Recognition 347
Mechanism of Transcription 348
Genetic Evidence for Promoters and Terminators 350
**10.4 Messenger RNA 350**
**10.5 RNA Processing in Eukaryotes 352**
5′ Capping and 3′ Polyadenylation 352
Splicing of Intervening Sequences 352
Characteristics of Human Transcripts 353
Coupling of Transcription and RNA Processing 354
Mechanism of RNA Splicing 354
Effects of Intron Mutations 357
Exon Shuffle in the Origin of New Genes 357
**10.6 Translation 358**
Nonsense-Mediated Decay 359

Initiation by mRNA Scanning 359
Elongation 361
Release 362
Protein Folding and Chaperones 364

**10.7 Complex Translation Units 366**
Polysomes 366
Polycistronic mRNA 366

**10.8 The Standard Genetic Code 368**
Genetic Evidence for a Triplet Code 368
How the Code Was Cracked 370
Features of the Standard Code 371
Transfer RNA and Aminoacyl-tRNA Synthetase Enzymes 371
Redundancy and Wobble 372
Nonsense Suppression 373

## 11 Molecular Mechanisms of Gene Regulation 380

**11.1 Transcriptional Regulation in Prokaryotes 381**
Inducible and Repressible Systems of Negative Regulation 381
Positive Regulation 382

**11.2 The Operon System of Gene Regulation 383**
$Lac^-$ Mutants 383
Inducible and Constitutive Synthesis and Repression 384
The Repressor 384
The Operator Region 385
The Promoter Region 385
The Operon System of Transcriptional Regulation 386
Positive Regulation of the Lactose Operon 387
Regulation of the Tryptophan Operon 390

**11.3 Regulation Through Transcription Termination 392**
Attenuation 392
Riboswitches 394

**11.4 Regulation in Bacteriophage Lambda 396**

**11.5 Transcriptional Regulation in Eukaryotes 398**
Galactose Metabolism in Yeast 398
Transcriptional Activator Proteins 401
Transcriptional Enhancers and Transcriptional Silencers 401
Deletion Scanning 402
The Eukaryotic Transcription Complex 404
Chromatin-Remodeling Complexes 407
Alternative Promoters 407

**11.6 Epigenetic Mechanims of Transcriptional Regulation 409**
Cytosine Methylation 411
Methylation and Transcriptional Inactivation 411
Genomic Imprinting in the Female and Male Germ Lines 412

**11.7 Regulation Through RNA Processing and Decay 413**
Alternative Splicing 413
Messenger RNA Stability 414

**11.8 RNA Interference 414**

**11.9 Translational Control 417**
Small Regulatory RNAs Controlling Translation 417

**11.10 Programmed DNA Rearrangements 419**
Gene Amplification 419
Antibody and T-Cell Receptor Variability 419
Mating-Type Interconversion 422
Transcriptional Control of Mating Type 423

## 12 Genomics, Proteomics, and Transgenics 431

**12.1 Site-Specific DNA Cleavage and Cloning Vectors 432**
Production of DNA Fragments with Defined Ends 432
Recombinant DNA Molecules 434
Plasmid, Lambda, and Cosmid Vectors 434

**12.2 Cloning Strategies 436**
Joining DNA Fragments 436
Insertion of a Particular DNA Molecule into a Vector 437
The Use of Reverse Transcriptase: cDNA and RT–PCR 439

**12.3 Detection of Recombinant Molecules 440**
Gene Inactivation in the Vector Molecule 440
Cloning of Large DNA Fragments 442
Screening for Particular Recombinants 442

12.4 **Genomics and Proteomics 443**
Genomic Sequencing 444
Genome Annotation 444
Comparative Genomics 445
Transcriptional Profiling 447
Two-Hybrid Analysis of Protein Interactions 452

12.5 **Transgenic Organisms 454**
Germ-Line Transformation in Animals 455
Genetic Engineering in Plants 458
Transformation Rescue 459
Site-Directed Mutagenesis and Knockout Mutations 460

12.6 **Some Applications of Genetic Engineering 461**
Giant Salmon with Engineered Growth Hormone 461
Nutritionally Engineered Rice 462
Production of Useful Proteins 462
Genetic Engineering with Animal Viruses 463

## 13 Genetic Control of Development 471

13.1 **Genetic Determinants of Development 472**

13.2 **Early Embryonic Development in Animals 472**
Autonomous Development and Intercellular Signaling 473
Composition and Organization of Oocytes 475
Early Development and Activation of the Zygotic Genome 476

13.3 **Genetic Analysis of Development in the Nematode 477**
Analysis of Cell Lineages 478
Mutations Affecting Cell Lineages 478
Programmed Cell Death 478
Loss-of-Function and Gain-of-Function Alleles 480
Epistasis in the Analysis of Developmental Switches 483

13.4 **Genetic Control of Development in *Drosophila* 485**
Maternal-Effect Genes and Zygotic Genes 488
Genetic Basis of Pattern Formation in Early Development 489
Coordinate Genes 491
Gap Genes 493
Pair-Rule Genes 493
Segment-Polarity Genes 494
Interactions in the Regulatory Hierarchy 494
Metamorphosis of the Adult Fly 496
Homeotic Genes 498
*HOX* Genes in Evolution 498

13.5 **Genetic Control of Development in Higher Plants 500**
Flower Development in *Arabidopsis* 500
Combinatorial Determination of the Floral Organs 502

## 14 Molecular Mechanisms of Mutation and DNA Repair 509

14.1 **Types of Mutations 510**
Germ-Line and Somatic Mutations 510
Conditional Mutations 510
Classification by Function 511

14.2 **The Molecular Basis of Mutation 512**
Nucleotide Substitutions 512
Missense Mutations: The Example of Sickle-Cell Anemia 513
Insertions, Deletions, and Frameshift Mutations 515
Dynamic Mutation of Trinucleotide Repeats 516
Cytosine Methylation and Gene Inactivation 518

14.3 **Transposable Elements 520**
Molecular Mechanisms of Transposition 520
Transposable Elements as Agents of Mutation 523
Transposable Elements in the Human Genome 524
RIP: A Defense Against Transposons 525

14.4 **Spontaneous Mutation 525**
The Nonadaptive Nature of Mutation 525
Estimation of Mutation Rates 527
Hot Spots of Mutation 527

14.5 **Mutagens 529**
Depurination 529
Oxidation 530
Base-Analog Mutagens 531
Chemical Agents That Modify DNA 531
Intercalating Agents 532
Ultraviolet Irradiation 532
Ionizing Radiation 533

Genetic Effects of the Chernobyl Nuclear Accident 536

**14.6 Mechanisms of DNA Repair 537**
Mismatch Repair 538
Base Excision Repair 540
AP Repair 541
Nucleotide Excision Repair 541
Photoreactivation 542
DNA Damage Bypass 542
The SOS Repair System 543

**14.7 Reverse Mutations and Suppressor Mutations 543**
Intragenic Suppression 543
Intergenic Suppression 544
The Ames Test for Mutagen/Carcinogen Detection 545

## 15 Molecular Genetics of the Cell Cycle and Cancer 551

**15.1 The Cell Cycle 552**
Key Events in the Cell Cycle 552
Transcriptional Program of the Cell Cycle 553

**15.2 Genetic Analysis of the Cell Cycle 554**
Mutations Affecting Progression Through the Cell Cycle 554

**15.3 Progression Through the Cell Cycle 557**
Cyclins and Cyclin-Dependent Protein Kinases 557
Targets of the Cyclin–CDK Complexes 559
Triggers for the $G_1$/S and $G_2$/M Transitions 559
Protein Degradation Helps Regulate the Cell Cycle 561

**15.4 Checkpoints in the Cell Cycle 562**
The DNA Damage Checkpoint 562
The Centrosome Duplication Checkpoint 567
The Spindle Assembly Checkpoint 567
The Spindle Position Checkpoint 568

**15.5 Cancer Cells 569**
Oncogenes and Proto-Oncogenes 570
Tumor-Suppressor Genes 573

**15.6 Hereditary Cancer Syndromes 575**
Defects in Cell-Cycle Regulation and Checkpoints 575
Defects in DNA Repair 578

**15.7 Genetics of the Acute Leukemias 578**

## 16 Mitochondrial DNA and Extranuclear Inheritance 587

**16.1 Patterns of Extranuclear Inheritance 588**
Mitochondrial Genetic Diseases 588
Heteroplasmy 591
Maternal Inheritance and Maternal Effects 592
Tracing Population History Through Mitochondrial DNA 592

**16.2 Organelle Heredity 592**
RNA Editing 593
The Genetic Codes of Organelles 593
Leaf Variegation in Four-O'Clock Plants 594
Drug Resistance in *Chlamydomonas* 595
Respiration-Defective Mitochondrial Mutants 598
Cytoplasmic Male Sterility in Plants 598

**16.3 The Evolutionary Origin of Organelles 600**

**16.4 Cytoplasmic Transmission of Symbionts 601**

**16.5 Maternal Effect in Snail Shell Coiling 603**

## 17 Molecular Evolution and Population Genetics 611

**17.1 Molecular Evolution 612**
Gene Trees 612
Bootstrapping 614
Gene Trees and Species Trees 615
Rates of Protein Evolution 615
Rates of DNA Evolution 616
Origins of New Genes: Orthologs and Paralogs 617

**17.2 Population Genetics 618**
Allele Frequencies and Genotype Frequencies 619
Random Mating and the Hardy–Weinberg Principle 620
Implications of the Hardy–Weinberg Principle 622
A Test for Random Mating 622

Frequency of Heterozygous Genotypes 623
Multiple Alleles 623
DNA Typing 624
X-Linked Genes 627

**17.3 Inbreeding 627**
The Inbreeding Coefficient 628
Allelic Identity by Descent 629
Calculation of the Inbreeding Coefficient from Pedigrees 630
Effects of Inbreeding 632

**17.4 Genetics and Evolution 632**

**17.5 Mutation and Migration 633**
Irreversible Mutation 633
Reversible Mutation 634

**17.6 Natural Selection 634**
Selection in a Laboratory Experiment 635
Selection in Diploid Organisms 636
Components of Fitness 637
Selection–Mutation Balance 637
Heterozygote Superiority 638

**17.7 Random Genetic Drift 639**

**17.8 Tracing Human History Through Mitochondrial DNA 643**

## 18 The Genetic Basis of Complex Traits 651

**18.1 Complex Traits 652**
Continuous, Categorical, and Threshold Traits 652
The Normal Distribution 653

**18.2 Causes of Variation 655**
Genotypic Variation 656
Environmental Variation 658
Genetics and Environment Combined 658
Genotype-by-Environment Interaction and Association 660

**18.3 Genetic Analysis of Complex Traits 660**
The Number of Genes Affecting Complex Traits 661
Broad-Sense Heritability 662
Twin Studies 663

**18.4 Artificial Selection 663**
Narrow-Sense Heritability 664
Phenotypic Change with Individual Selection: A Prediction Equation 665
Long-Term Artificial Selection 666
Inbreeding Depression and Heterosis 666

**18.5 Correlation Between Relatives 667**
Covariance and Correlation 667
The Geometrical Meaning of a Correlation 667
Estimation of Narrow-Sense Heritability 668

**18.6 Heritabilities of Threshold Traits 669**

**18.7 Identification of Genes Affecting Complex Traits 670**
Linkage Analysis in the Genetic Mapping of Quantitative Trait Loci 670
The Number and Nature of QTLs 672
Candidate Genes for Complex Traits 674

**Answers to Even-Numbered Problems 683**
**Further Reading 701**
**Word Roots 705**
**Concise Dictionary of Genetics and Genomics 709**
**Index 735**

# PREFACE

**Today's students are enthusiastic** to learn about genetics and genomics. Their curiosity is refreshed by almost daily reports in the media of new discoveries dealing with genetic differences in drug responses, genetic risk factors for disease, and genetic evidence for human origins and history. There are also ethical controversies: Should genetic manipulation be used on patients for the treatment of disease? Should human embryonic stem cells be used in research? Should human beings be cloned? And social controversies: Should there be laws governing genetic privacy? Who should have access to genetic testing records, and for what purpose?

For the teacher of genetics, the challenges are:

- To sustain the students' enthusiasm;
- To help motivate a desire to understand the principles of genetics in a comprehensive and rigorous way;
- To guide students in gaining an understanding that genetics is not only a set of principles but also an experimental approach to solving a wide range of biological problems; and
- To help students learn to think about genetic problems and about the wider social and ethical issues arising from genetics and genomics.

**Genetics: Analysis of Genes and Genomes, Seventh Edition**, addresses these challenges while also showing the beauty, logical clarity, and unity of the subject. Our pedagogical approach is to treat transmission genetics, molecular genetics, and evolutionary genetics as fully integrated subjects. This appeals to most modern geneticists, who recognize that the distinctions between the subfields are artificial. The chapters have been arranged to match the organization of the overwhelming majority of college courses in genetics, including our own.

Our aim has been to provide a clear, comprehensive, rigorous, and balanced introduction to genetics and genomics at the college level. It is our belief that a good course should maintain the right balance between two important aspects of genetics. The first aspect is that genetics is a body of knowledge pertaining to genetic transmission, function, and mutation; the second is that genetics is an experimental approach, or a kit of "tools," for the study of biological processes such as development or behavior. The rationale of the book is that any student claiming a knowledge of genetics must:

- Understand the basic processes of gene transmission, mutation, expression, and regulation;
- Be familiar with the principal experimental methods that geneticists and molecular biologists use in their studies, and recognize the advantages and limitations of these approaches;
- Be able to think like a geneticist at the elementary level of being able to formulate genetic hypotheses, work out their consequences, and test the results against observed data;
- Be able to state genetic principles in his or her own words and to recognize the key terms of genetics in context;
- Be able to solve problems of several types, including single-concept exercises that require application of definitions or the basic principles of genetics, problems in genetic analysis in which several concepts must be applied in logical order, and problems in quantitative analysis that call for some numerical calculation;
- Gain some sense of the social and historical context in which genetics and genomics has developed and is continuing to develop; and
- Acquire a basic familiarity with the genetic resources and information that are available through the Internet.

## Special Features

We have included many special features to help students achieve these learning goals. The text is clearly and concisely written in a somewhat relaxed prose style without being chummy or excessively familiar. Each chapter contains two or three **Connections** in which the text material is connected to excerpts of classic papers that report key experiments in genetics or raise important social, ethical, or legal issues in genetics. Each Connection has a brief introduction of its own, explaining the importance of the experiment and the historical context in which it was carried out. At the end of each chapter is a **Chapter Summary** in the form of bullet points, a **Review the Basics** feature in which the major concepts are reviewed in the form of questions for discussion, a **Guide to Problem Solving** in which typical types of problems that apply the principles are worked in full, a large number of **Analysis and Applications** problems for solution (with answers to even-numbered problems at the end of the book), and a special set of **Challenge Problems** for students wanting to stretch. The chapter-end material also includes instructions for accessing the book's Web site, where there are complete chapter summaries, key terms, many more problems and exercises of various types, and a feature called **Genetics on the Web** to acquaint students with key Internet sites containing resources in genetics and genomics. At the end of the book are **Answers to Even-Numbered Problems** in which the logic of each answer is explained in full, a list of **Further Reading** for students who want additional depth, a compilation of **Word Roots** to help students master the essential terminology of genetics and genomics, and a **Concise Dictionary of Genetics and Genomics**.

## What's New in the Seventh Edition?

This edition has been completely revised and updated. Each chapter has been thoroughly reworked. Nonessential or outdated material has been deleted, new methods and results have been added, and several of the chapters have been reorganized to allow seamless integration of the new material. Several new Connections have been added, and there are more than 70 new or updated illustrations and tables. Hundreds of new, classroom-tested problems have been added, and the Guide to Problem Solving has been expanded.

### Chapter Organization

In order to help students keep track of the main issues and avoid being distracted by details, each chapter begins with an Outline showing the road map of the territory ahead. An opening paragraph gives an overview of the chapter, illustrates the subject with some specific examples, and shows how the material is connected to genetics as a whole. The text makes liberal use of numbered lists and "bullets" in order to help students organize their learning, as well as summary statements set off in special type in order to emphasize important principles.

## Contents and Organization

The organization of the chapters is that favored by the majority of instructors who teach genetics. It is the organization we use in our own courses. An important feature is the presence of two introductory chapters providing a broad overview of DNA, genes, and genomes—what they are, how they function, how they change by mutation, and how they evolve through time. Today, most students learn about DNA in grade school or high school; in our teaching, we have found it artificial to pretend that DNA does not exist until the middle of the term. The introductory chapters serve to connect the more advanced concepts that students are about to learn with what they already know. They also serve to provide each student with a solid framework for integrating the material that comes later.

Throughout each chapter, there is a balance between observation and theory, between principle and concrete example, and between challenge and motivation. Molecular, classical, and evolutionary genetics are completely integrated. Throughout the book are frequent references to human genetics. The book is also liberally populated with applications to other animals and plants, including the key model organisms used in genetics and genomics.

A number of points related to organization and coverage should be noted:

- **Chapter 1** is an overview of genetics designed to bring students with disparate backgrounds to a common level of understanding. This chapter enables classical, molecular, and evolutionary genetics to be integrated in the rest of the book. Included in Chapter 1 are the basic concepts of molecular genetics: DNA structure, replication, expression, and mutation. New to this edition is a thorough introduction to genetic analysis, focusing on the classical experiments of Beadle and Tatum but also summarizing applications of genetic analysis that have led to six Nobel Prizes.
- **Chapter 2** emphasizes that the primary tools of the modern geneticist derive from the experimental manipulation of DNA. It includes a more detailed look at DNA structure, and it introduces the principal methods of DNA manipulation including restriction enzymes, electrophoresis, DNA hybridization, Southern blotting, and the polymerase chain reaction. We also introduce single-nucleotide polymorphisms (SNPs) and copy-number polymorphisms (CNPs), and discuss how these types of genetic markers can be assayed on a genome-wide scale using oligonucleotide microarrays (DNA chips).
- **Chapters 3 through 5** are the core of Mendelian genetics, including segregation and independent assortment, the chromosome theory of heredity, mitosis and meiosis, linkage and chromosome mapping, and tetrad analysis in fungi. Also included is the basic probability framework of Mendelian genetics and the testing of genetic models by means of the chi-square test. Unique in Chapter 3 is the integration of molecular genetics with Mendel's experiments. We describe the molecular basis of the wrinkled mutation and show how a modern geneticist would carry out Mendel's study, examining the molecular phenotypes on the one hand and the morphological phenotypes on the other. This pedagogy provides a solid basis for understanding not only Mendel's experiments as he actually performed and interpreted them, but also for understanding how modern molecular approaches are used in genetic analysis. Molecular markers are also integrated into human genetic analysis, which includes a new discussion of the molecular genetics of the taster polymorphism and its relation to one's dislike of broccoli.
- **Chapters 6 and 7** deal with the molecular structure and replication of DNA and the molecular organization of chromosomes. Chapter 6 also covers the molecular mechanisms of recombination, and Chapter 7 includes a discussion of repetitive DNA sequences in eukaryotic genomes and the molecular structures of centromeres and telomeres. New to this edition is an expanded discussion of DNA sequencing, including the newest massively parallel sequencing technologies that can yield a billion base pairs of DNA sequence within a few hours.

- **Chapter 8** covers the principles of chromosome mechanics with special reference to human chromosome number and structure and the types of aberrations that are found in human chromosomes. The genetic implications of chromosome abnormalities—duplications, deficiencies, inversions, and translocations—are also discussed. New to this edition are data on the incomplete inactivation of genes in the "inactive" X Chromosome, updated information on the molecular structure of the Y Chromosome, and the evidence from genomic sequencing for ancient polyploidy in a number of key model organisms.
- **Chapter 9** deals with the principles of genetics in bacteria, beginning with the genetics of mobile DNA, plasmids, and integrons, and their relationships to the evolution of multiple antibiotic resistance. It examines mechanisms of genetic recombination in microbes including transformation, conjugation, and transduction, as well as temperate and virulent bacteriophages. This edition also includes a discussion of the mosaic structure of bacterial chromosomes as evidenced by genomic and pathogenicity islands of DNA introduced by horizontal transmission.
- **Chapters 10 and 11** deal with molecular genetics in the strict sense. These chapters include the classical principles of gene expression and gene regulation. They include broad aspects of gene regulation that are topics of much current research: chromatin remodeling complexes, imprinting and other epigenetic modifications of gene expression, riboswitches, and the mechanisms of gene regulation through siRNA and microRNA. For the first time we examine the eukaryotic transcription complex in detail, and include an expanded discussion of regulation of the classical *GAL* system in yeast.
- **Chapter 12** focuses on recombinant DNA and genomics. Included are the use of restriction enzymes and vectors in recombinant DNA, cloning strategies, site-directed mutagenesis, the production of genetically defined transgenic animals and plants, and applications of genetic engineering. We emphasize the extraordinary number and variety of genomes that have been sequenced and summarize the principles that have emerged, while acknowledging that it will probably require decades to fully understand all that the genomic sequences imply. This edition includes an introduction to comparative genomics and computational genomics. Functional genomics is introduced by examining how DNA microarrays are used to study global patterns of coordinated gene expression, and proteomics is exemplified by a discussion of two-hybrid analysis and its application.
- **Chapter 13** deals with the genetic control of development, focusing on genetic analysis of development in nematodes (*Caenorhabditis elegans*) and *Drosophila,* and also includes a thorough examination of the genetic basis of floral development in *Arabidopsis thaliana.* For the first time we examine how the principle of epistasis is used in the genetic analysis of developmental mutants to identify the regulatory switches that control developmental processes.
- **Chapter 14** deals with the molecular basis of mutation. A thorough discussion of different types of mutation also includes genetically unstable sequences, such as the trinucleotide repeats whose amplification causes the fragile–X syndrome of mental retardation. The molecular action of various types of mutagens is examined, along with the genetic effects of the Chernobyl nuclear accident. Chapter 14 also covers the important field of DNA repair, with updated and expanded discussions of ectopic recombination, nucleotide excision repair, and DNA-damage bypass repair.
- **Chapter 15** stresses cancer from the standpoint of the genetic control of cell division, with emphasis on the checkpoints that, in normal cells, result either in inhibition of cell division or in programmed cell death (apoptosis). Cancer results from a series of successive mutations, usually in somatic cells, that overcome the normal checkpoints that control cellular proliferation. In this edition we include an expanded discussion of the characteristics of cancer cells, the role of telomerase, the poison-subunit mutants of p53, and the role of microRNA in cell-cycle regulation.
- **Chapter 16** covers organelle genetics, including genetic defects in human mitochondrial DNA. The history of domestication of the dog is elucidated by an examination of

the mitochondrial DNA sequences of domestic breeds of dogs in comparison with their wild relatives.

- **Chapters 17 and 18** deal with population and evolutionary genetics and include an overview of the principles of molecular evolution. The discussion also includes DNA typing in criminal investigations and paternity testing, the effects of inbreeding, and the evolutionary mechanisms that drive changes in allele frequency. How mitochondrial DNA sequences inform us about human history and migration is also examined. The approach to quantitative genetics includes a discussion of how particular genes influencing quantitative traits (QTLs, or quantitative-trait loci) may be identified and mapped by linkage analysis. New in this edition is discussion of the emerging principles from studies of QTLs affecting human traits. There is also a section on the genetic determinants of human behavior with emphasis on a depression-related polymorphism in the human serotonin transporter gene and its relation to the response to "club drugs" such as Ecstasy.

## Connections

A unique feature of this book is found in boxes called Connections. Each chapter has two or three of these boxes. They are our way of connecting genetics to the world outside the classroom. All of the Connections include short excerpts from the original literature of genetics, usually papers, each introduced with a short explanatory passage. Many of the Connections are excerpts from classic papers, such as Mendel's paper, but by no means are all "classic" papers old papers. Many of them are very recent.

The pieces are called Connections because each connects the material to something that broadens or enriches implications. Some of the Connections raise issues of ethics in the application of genetic knowledge or social issues that need to addressed. Some of the pieces were published originally in French, others in German. These appear in English translation. In papers that use outmoded or unfamiliar terminology, or that use archaic gene symbols, we have substituted the modern equivalent because the use of a consistent terminology in the text and in the Connections makes the material more accessible to the student.

**Genetics on the Web** features links that introduce students to the vast repository of information about genetics and genomics that is available on the Internet. Genetics is a dynamic science, and through this venue we can introduce the newest discoveries as soon as they appear to keep the textbook up to date. The Internet sites are accessed through the use of key words that are highlighted in the book's web site. The URLs are maintained as hot links at the publisher's Web site:

**http://biology.jbpub.com/book/genetics**

and are kept constantly up to date, tracking the address of each site if it should change.

## Problems

Each chapter provides about 50 problems, graded in difficulty, for the students to test their understanding of the material. The problems are of several different types:

*Review the Basics* problems ask for genetic principles to be restated in the student's own words; some are matters of definition or call for the application of elementary principles.

*Guide to Problem Solving* demonstrates problems worked in full. The concepts needed to solve the problem, and the reasoning behind the answer, are explained in detail. This feature serves as a review of the important concepts used in working problems. It also highlights some of the most common mistakes made by beginning students and gives pointers on how the student can avoid falling into these conceptual traps.

*Analysis and Applications* problems are more traditional types of genetic problems in which several concepts must be applied in logical order and often require some numerical calculation. The level of mathematics is that of arithmetic and elementary probability as it pertains to genetics. None of the problems uses mathematics beyond elementary algebra.

*Challenge Problems* are similar to those in Analysis and Applications, but they are a degree more challenging, often because they require a more extensive analysis of data before the question can be answered.

## Solutions

The answers to even-numbered *Analysis and Applications* problems are included in the answer section at the end of the book. The answers are complete. They explain the logical foundation of the solution and lay out the methods. The answers to the rest of the *Analysis and Applications* and *Challenge Problems* are available in the *Student Solutions Manual and Supplemental Problems* book.

We find that many of our students, like students everywhere, often sneak a look at the answer before attempting to solve a problem. This is a pity. Working backward from the answer should be a last resort. This is because problems are valuable opportunities to learn. Problems that the student cannot solve are usually more important than the ones that can be solved, because the sticklers usually identify trouble spots, areas of confusion, or gaps in understanding. We therefore urge our students to try answering each question before looking at the answer.

## Word Roots and Concise Dictionary

At the end of the book we include a short list of *Word Roots* that are used in many of the technical terms of genetics. The *Word Roots* are useful in helping students to interpret and to remember the meaning of key terms. There is also a *Concise Dictionary of Genetics* for students to check their understanding of the key words or look up any technical terms they may have forgotten. The *Dictionary* includes not only the key words but also genetic terms that students are likely to encounter in exploring the Internet or in their further reading.

## Further Reading

This feature includes recommendations for *Further Reading* for the student who either wants more information or who needs an alternative explanation for the material presented in the book. Some additional "classic" papers and historical perspectives are included.

## Illustrations

The art program is spectacular and a learning aid itself. Every chapter is richly illustrated with beautiful graphics in which color is used functionally to enhance the value of each illustration. The illustrations are also heavily annotated with explanatory boxes explaining step-by-step what is happening at each level of the illustration. These labels make the art inviting as well as informative. They also allow the illustrations to stand relatively independently of the text, enabling the student to review material without rereading the whole chapter.

The art program is used not only for its visual appeal, but also to increase the pedagogical value of the book:

- Characteristic colors and shapes have been used consistently throughout the book to indicate different types of molecules—DNA, mRNA, tRNA, and so forth. For example, DNA is illustrated in any one of a number of ways, depending on the level of resolution necessary for the illustration, and each time a particular level of resolution is depicted, the DNA is shown in the same way. It avoids a great deal of potential confusion that DNA, RNA, and proteins are represented in the same manner in every chapter.
- There are numerous full-color photographs of molecular models in three dimensions; these give a strong visual reinforcement of the concept of macromolecules as physical entities with defined three-dimensional shapes and charge distributions that serve as the basis of interaction with other macromolecules.
- The page design is clean, crisp, and uncluttered. As a result, the book is pleasant to look at and easy to read.

## Flexibility

There is no necessary reason to start at the beginning and proceed straight to the end. Each chapter is a self-contained unit that stands on its own. This feature gives the book the flexibility to be used in a variety of course formats. Throughout the book, we have integrated molecular and classical principles, so you can begin a course with almost any of the chapters. Most teachers will prefer starting with the overview in Chapter 1 because it brings every student to the same basic level of understanding. Chapter 2 introduces the basic experimental manipulations used in modern genetics and serves to integrate molecular and classical genetics in the discussion of Mendel in Chapter 3. Teachers preferring the Mendel-early format should start with Chapter 1, continue with Chapter 3, then backtrack to Chapter 2. Some teachers are partial to a chromosomes-early format, which would suggest the order chapter 1, 4, 2, 3, and 5. A novel approach would be a genomes-first format, which could be implemented by beginning with chapters 1, 2, and 12. The writing and illustration program was designed to accommodate a variety of formats, and we encourage teachers to take advantage of this flexibility in order to meet their own special needs.

## Supplements

Jones and Bartlett Publishers offers a suite of traditional and interactive multimedia supplements to assist instructors and aid students in mastering genetics. Additional information and review copies of any of the following items are available through your Jones and Bartlett Sales Representative.

### Instructor's ToolKit CD-ROM includes:

- A Test Bank containing over a thousand questions and complete answers. The questions, authored by Elena Lozovsky of Harvard University and Teri Shors of University of Wisconsin – Oshkosh, are a mix of factual, descriptive, and quantitative types. A typical chapter contains multiple-choice, fill-in-the-blank, and short answer questions. The Test Bank is provided as Rich Text files.
- The PowerPoint® Image Bank is an easy-to-use multimedia tool that provides all of the illustrations and photos from the text (to which Jones and Bartlett holds the rights to reproduce electronically) for use in classroom presentation. You can select images you need or easily generate your own slide shows, print the files for transparency creation. Many images have already been inserted into the PowerPoint Lecture Outline presentations for ease of use.
- A PowerPoint presentation containing the detailed outline for each chapter of *Genetics: Analysis of Genes and Genomes, Seventh Edition* is included. The presentation, designed to mirror the text, is constructed flexibly in order to meet your lecture's organization. The outline is open, allowing you to provide the elements you deem necessary, whether it be new text or more images from the Image Bank.
- Microsoft Word files of the *Student Solutions Manual and Supplemental Problems* are provided. This manual is also available to your students as a printed supplemental text. It provides detailed answers for all of the problems at the end of chapters. It also contains supplemental problems with explanations to help give your students further practice with difficult genetics concepts.

### Genetics on the Web
### http://biology.jbpub.com/book/genetics

Updated and expanded for this edition, the student Web site Genetics on the Web offers support to students as they further explore course content and the field of genetics. Students will find Animated Flashcards that reinforce important terms and Crossword Puzzles that provide an exciting way to review course material. Research and Reference links point students to valuable information on genetics-related organizations and news outlets. For students' convenience, an Interactive Online Glossary is available for quick reference.

### Student Solutions Manual and Supplemental Problems

This completely revised supplemental text, available to all students, is comprised of two parts. First, it contains a complete set of solutions for all of the end-of-chapter problems in the text.

Second, it contains hundreds of supplemental problems with complete solutions, as well as a chapter summary and list of key terms for each chapter of the text. The problems, written by Elena Lozovsky of Harvard University, are a tremendous resource for students seeking an extra self-assessment of understanding. The problems are designed to test their knowledge of each chapter's content, but particularly to practice critical thinking and working through some of the more difficult concepts in genetics.

### *The Gist of Genetics: Guide to Learning and Review*

Written by Rowland H. Davis and Stephen G. Weller of the University of California, Irvine, this study aid uses illustrations, tables, and text outlines to review all of the fundamental elements of genetics. It includes extensive practice problems and review questions with solutions for self-check. *The Gist* helps students formulate appropriate questions and generate hypotheses that can be tested with classical principles and modern genetic techniques.

# ACKNOWLEDGMENTS

We are indebted to many colleagues whose advice, thoughtful reviews, and comments have helped in the preparation of this and previous editions. Their expert recommendations are reflected in the content, organization, and presentation of the material.

Laura Adamkewicz, George Mason University, Fairfax VA
Jeremy C. Ahouse, Brandeis University, Waltham MA
Mary Alleman, Duquesne University, Pittsburgh PA
Peter D. Ayling, University of Hull, Hull, United Kingdom
John C. Bauer, Stratagene, Inc., La Jolla CA
Anna W. Berkovitz, Purdue University, West Lafayette IN
Mary K. B. Berlyn, Yale University, New Haven CT
Thomas A. Bobik, University of Florida, Gainesville FL
David Botstein, Princeton University, Princeton NJ
Colin G. Brooks, The Medical School, Newcastle, United Kingdom
Pierre Carol, Université Joseph Fourier, Grenoble, France
Domenico Carputo, University of Naples, Naples, Italy
Sean Carroll, University of Wisconsin, Madison WI
Chris Caton, Univeristy of Birmingham, Birmingham AL
John Celenza, Boston University, Boston MA
Alan C. Christensen, University of Nebraska–Lincoln, Lincoln NE
Christoph Cremer, University of Heidelberg, Heidelberg, Germany
Marion Cremer, Ludwig Maximilians University, Munich, Germany
Thomas Cremer, Ludwig Maximilians University, Munich, Germany
Leslie Dendy, University of New Mexico, Los Alamos NM
Stephen J. DíSurney, University of Mississippi, University MS
John W. Drake, National Institute of Environmental Health Sciences, Research Triangle Park NC
Kathleen Dunn, Boston College, Boston MA
Chris Easton, State University of New York, Binghamton NY
Wolfgang Epstein, University of Chicago, Chicago IL
Brian E. Fee, Manhattan College, Riverdale NY
Gyula Ficsor, Western Michigan University, Kalamazoo MI
Robert G. Fowler, San Jose State University, San Jose CA
David W. Francis, University of Delaware, Newark DE
Gail Gasparich, Towson University, Towson MD

Elliott S. Goldstein, Arizona State University, Tempe AZ
Patrick Guilfoile, Bemidji State University, Bemidji MN
Jeffrey C. Hall, Brandeis University, Waltham MA
Mark L. Hammond, Campbell University, Buies Creek NC
Steven Henikoff, Fred Hutchinson Cancer Research Center, Seattle WA
Charles Hoffman, Boston College, Boston MA
Ivan Huber, Fairleigh Dickenson University, Madison NJ
Kerry Hull, Bishop's University, Quebec, Canada
Lynn A. Hunter, University of Pittsburgh, Pittsburgh PA
Richard Imberski, University of Maryland, College Park MD
Joyce Katich, Monsanto, Inc., St. Louis MO
Jeane M. Kennedy, Monsanto, Inc., St. Louis MO
Jeffrey King, University of Berne, Switzerland
Anita Klein, University of New Hampshire, Durham, NH
Tobias A. Knoch, German Cancer Research Center, Heidelberg, Germany
Yan B. Linhart, University of Colorado, Boulder CO
K. Brooks Low, Yale University, New Haven CT
Lauren McIntyre, University of Florida, Gainesville, FL
Sally A. MacKenzie, Purdue University, West Lafayette IN
Gustavo Maroni, University of North Carolina, Chapel Hill NC
Jeffrey Mitton, University of Colorado, Boulder CO
Robert K. Mortimer, University of California, Berkeley CA
Gisela Mosig, Vanderbilt University, Nashville TN
Steve O'Brien, National Cancer Institute, Frederick MD
Kevin O'Hare, Imperial College, London, United Kingdom
Ronald L. Phillips, University of Minnesota, St. Paul MN
Robert Pruitt, Purdue University, West Lafayette IN
Pamela Reinagel, California Institute of Technology, Pasadena CA
Susanne Renner, University of Missouri, St. Louis MO
Andrew J. Roger, Dalhousie University, Halifax, Nova Scotia, Canada
Kenneth E. Rudd, National Library of Medicine, Bethesda MD
Thomas F. Savage, Oregon State University, Corvallis OR
Joseph Schlammandinger, University of Debrecen, Hungary
David Shepard, University of Delaware, Newark DE
Alastair G.B. Simpson, Dalhousie University, Halifax, Nova Scotia, Canada
Leslie Smith, National Institute of Environmental Health Sciences, Research Triangle Park NC
Charles Staben, University of Kentucky, Lexington KY
Johan H. Stuy, Florida State University, Talahassee FL
David T. Sullivan, Syracuse University, Syracuse NY
Jeanne Sullivan, West Virginia Wesleyan College, Buckhannon WV
Millard Susman, University of Wisconsin, Madison WI
Barbara Taylor, Oregon State University, Corvallis OR
Irwin Tessman, Purdue University, West Lafayette IN
Micheal Tully, University of Bath, Bath, United Kingdom
David Ussery, The Technical University of Denmark, Lyngby, Denmark
George von Dassow, Friday Harbor Laboratories, Friday Harbor WA
Denise Wallack, Muhlenberg College, Allentown PA
Kenneth E. Weber, University of Southern Maine, Gorham ME
Tamara Western, Okanagan University College, Kelowna, British Columbia, Canada

We especially want to acknowledge Elena R. Lozovsky of Harvard University for her work on the *Student Solutions Manual* and *Supplemental Problems*.

We also wish to acknowledge the superb editorial and production staff who helped make this book possible: Cathleen Sether, Dean DeChambeau, Molly Steinbach, Anne Spencer, Rachel Rossi, and Kimberly Potvin of Jones and Bartlett. Much of the credit for the attractiveness and readability of the book should go to them. We also thank Jones and Bartlett for their ongoing commitment to high quality in book production. We are also grateful to the many people, acknowledged in the legends of the figures, who contributed photographs, drawings, and micrographs from their own research and publications. Every effort has been made to obtain permission to use copyrighted material and to make full disclosure of its source. We are grateful to the authors, journal editors, and publishers for their cooperation. Any errors or omissions are wholly inadvertent and will be corrected at the first opportunity.

# ABOUT THE AUTHORS

**Daniel L. Hartl** is Higgins Professor of Biology at Harvard University and a member of the National Academy of Sciences and the American Academy of Arts and Sciences. He received his B.S. degree and Ph.D. from the University of Wisconsin and carried out postdoctoral research at the University of California at Berkeley. His research interests include molecular genetics, genomics, molecular evolution, and population genetics.

**Elizabeth W. Jones** was the Dr. Frederick A. Schwertz Distinguished Professor of Life Sciences and Head of the Biological Sciences Department at Carnegie Mellon University. She received a B.S. degree in chemistry and a Ph.D. in genetics from the University of Washington in Seattle. Her research focused on gene regulation and the genetic control of cellular form.

# IN MEMORIAM

Elizabeth W. Jones passed away unexpectedly on June 11, 2008.

Beth, as she was known to her friends, served as the Dr. Frederick A. Schwertz Distinguished Professor of Life Sciences and Head of the Biological Sciences Department at Carnegie Mellon University in Pittsburgh, Pennsylvania. Throughout her career she published more than 80 papers and edited more than 20 books and monographs in genetics and cell biology. She was a Fellow of the American Association for the Advancement of Science, served as president of the Genetics Society of America, and was the Editor-in-Chief of GENETICS for 12 years.

A leader in the field, she was known for her efficiency, her vision, her rigor, and her gracious nature. Her wonderful sense of humor could make you laugh till tears. Beth was an outstanding editor, and the consummate geneticist, both in her own research and in her teaching. Her contributions to teaching, as reflected in the seven editions of the *Genetics* and *Essential Genetics* textbooks she co-authored with Daniel Hartl, have inspired generations of students.

We honor her memory as a beloved friend and colleague, and as a leading geneticist and scientific educator.

# ABOUT THE COVER

The cover, graphic courtesy of Dana Hunt, Lawrence David, Dirk Gevers, Sarah Preheim, Eric Alm, and Martin Polz at the Massachusetts Institute of Technology, illustrates data on 1024 isolates of *Vibrio* bacteria from ocean samples. *Vibrio* is an important group of bacteria because some species are associated with cholera and food poisoning.

The outside band indicates time of collection, gray for Spring and black for Fall.

The second ring indicates how the bacteria were collected.

Each water sample was successively run through 63, 5, 1, and 0.22 μm filters. It is hypothesized that ocean *Vibrio* have adapted to living on different sized particles and marine plankton.

In the second ring, blue denotes cells captured on the 0.22 μm filter, green those on the 1 μm filter, yellow those on the 5 μm filter, and red those on the 63 μm filter.

The pedigree in the middle depicts the inferred ancestral relationships among the isolates based on the similarity of their DNA sequences; more closely-related *Vibrio* are more proximate on the diagram. Ancestral *Vibrio* (the interior dots) share identical colors if their descendants share similar genetic sequences, as well as collection times and particle associations. Note the clustering of isolates with respect to sequence similarity, time of collection, and particle size.

# CHAPTER 1

# Genes, Genomes and Genetic Analysis

The filamentous fungus *Neurospora crassa* figures prominently in the history of genetic analysis. This photograph shows sexual spores of the organism. Each group of eight spores forms within a tubelike structure. In this case some of the spores are green because they express a genetically engineered green-fluorescent protein.



## CHAPTER OUTLINE


**1.1 DNA: The Genetic Material**
- Experimental Proof of the Genetic Function of DNA
- Genetic Role of DNA in Bacteriophage

**1.2 DNA Structure and Replication**
- An Overview of DNA Replication

**1.3 Genes and Proteins**
- Inborn Errors of Metabolism as a Cause of Hereditary Disease
- Mutant Genes and Defective Proteins

**1.4 Genetic Analysis**
- Complementation Test for Mutations in the Same Gene
- Analysis of Complementation Data
- Other Applications of Genetic Analysis

**1.5 Gene Expression: The Central Dogma**
- Transcription
- Translation
- The Genetic Code

**1.6 Mutation**

**1.7 Genes and Environment**

**1.8 The Molecular Unity of Life**
- Prokaryotes and Eukaryotes
- Evolutionary Relationships Among Eukaryotes
- Genomes and Proteomes

CONNECTION The Black Urine Disease
Archibald E. Garrod 1908
*Inborn Errors of Metabolism*

CONNECTION One Gene, One Enzyme
George W. Beadle and Edward L. Tatum
*Genetic Control of Biochemical Reactions in Neurospora*

Each species of living organism has a unique set of inherited characteristics that makes it different from every other species. Each species has its own developmental plan—often described as a sort of "blueprint" for building the organism—which is encoded in the DNA molecules present in its cells. This developmental plan determines the characteristics that are inherited. Because organisms in the same species share the same developmental plan, organisms that are members of the same species usually resemble one another. For example, it is easy to distinguish a human being from a chimpanzee or a gorilla. A human being habitually stands upright and has long legs, relatively little body hair, a large brain, and a flat face with a prominent nose, jutting chin, distinct lips, and small teeth. All of these traits are inherited—part of our developmental plan—and help set us apart as *Homo sapiens.*

But human beings are by no means identical. Many traits, or observable characteristics, differ from one person to another. In addition to notable differences between males and females, there is a great deal of variation in hair color, eye color, skin color, height, weight, personality traits, and other characteristics. Some human traits (such as sex) are transmitted biologically, others culturally. The color of our eyes results from biological inheritance, but the native language we learned as a child results from cultural inheritance. Many traits are influenced jointly by biological inheritance and environmental factors. For example, weight is determined in part by inheritance but also in part by environment: how much food we eat, its nutritional content, our exercise regimen, and so forth.

**Genetics** is the study of biologically inherited traits, including traits that are influenced in part by the environment. **Genomics** is the study of all the genes in an organism to understand their molecular organization, function, interaction, and evolutionary history. The fundamental concept of genetics and genomics is that:

> Inherited traits are determined by the elements of heredity that are transmitted from parents to offspring in reproduction; these elements of heredity are called **genes**.

The existence of genes and the rules governing their transmission from generation to generation were first articulated by Gregor Mendel in 1866 (Chapter 3). Mendel's formulation of inheritance was in terms of the abstract rules by which hereditary elements (he called them "factors") are transmitted from parents to offspring. His objects of study were garden peas, with variable traits like pea color and plant height. At one time genetics could be studied only through the progeny produced from matings. Genetic differences between species were impossible to define, because organisms of different species usually do not mate, or they produce hybrid progeny that die or are sterile. The approach to the study of genetics through the analysis of the offspring from matings is often referred to as *classical* genetics, or organismic or morphological genetics. Given the advances of *molecular,* or modern, genetics, it is possible to study differences between species through the comparison and analysis of the DNA itself. There is no fundamental distinction between classical and molecular genetics. They are different and complementary ways of studying the same thing: the function of the genetic material. In this book we include many examples showing how molecular and classical genetics can be used in combination to enhance the power of genetic analysis.

The foundation of genetics as a molecular science dates back to 1869, just three years after Mendel reported his experiments. It was in 1869 that Friedrich Miescher discovered a new type of weak acid, abundant in the nuclei of white blood cells. Miescher's weak acid turned out to be the chemical substance we now call **DNA (deoxyribonucleic acid).** For many years the biological function of DNA was unknown, and no role in heredity was ascribed to it. This first section shows how DNA was eventually isolated and identified as the material that genes are made of.

## 1.1 DNA: The Genetic Material

That the cell nucleus plays a key role in inheritance was recognized in the 1870s by the observation that the nuclei of male and female reproductive cells undergo fusion in the process of fertilization. Soon thereafter, **chromosomes** were first observed inside the nucleus as threadlike objects that become visible in the light microscope when the cell is stained with certain dyes. Chromosomes were found to exhibit a characteristic "splitting" behavior in which each daughter cell formed by cell division receives an identical complement of chromosomes (Chapter 4). Further evidence for the importance of chromosomes was provided by the observation that, whereas the number of chromosomes in each cell may differ among biological species, the number of chromosomes is nearly always constant within the cells of any particular species. These features of chromosomes were well understood by about 1900, and they made it seem likely that chromosomes were the carriers of the genes.

By the 1920s, several lines of indirect evidence began to suggest a close relationship between chromosomes and DNA. Microscopic studies with special stains showed that DNA is present in chromosomes. Chromosomes also contain various types of proteins, but the amount and kinds of chromosomal proteins differ greatly from one cell type to another, whereas the amount of DNA per cell is constant. Furthermore, nearly all of the DNA present in cells of higher organisms is present in the chromosomes. These arguments for DNA as the genetic material were unconvincing, however, because crude chemical analyses had suggested (erroneously, as it turned out) that DNA lacks the chemical diversity needed in a genetic substance. The favored candidate for the genetic material was protein, because proteins were known to be an exceedingly diverse collection of molecules. Proteins therefore became widely accepted as the genetic material, and DNA was assumed to function merely as the structural framework of the chromosomes. The experiments described below finally demonstrated that DNA is the genetic material.

## ■ Experimental Proof of the Genetic Function of DNA

An important first step was taken by Frederick Griffith in 1928 when he demonstrated that the genetic material can be transferred from one bacterial cell to another. He was working with two strains of the bacterium *Streptococcus pneumoniae* identified as S and R. When a bacterial cell is grown on solid medium, it undergoes repeated cell divisions to form a visible clump of cells called a **colony.** The S type of *S. pneumoniae* synthesizes a gelatinous capsule composed of complex carbohydrate (polysaccharide). The enveloping capsule makes each colony large and gives it a glistening or smooth (S) appearance (**FIGURE 1.1**). This capsule also enables the bacterium to cause pneumonia by protecting it from the defense mechanisms of an infected animal. The R strains of *S. pneumoniae* are unable to synthesize the capsular polysaccharide; they form small colonies that have a rough (R) surface (Figure 1.1). The R strain of the bacterium does not cause pneumonia, because without the capsule the bacteria are inactivated by the immune system of the host. Both types of bacteria "breed true" in the sense that the progeny formed by cell division have the capsular type of the parent, either S or R.

Mice injected with living S cells get pneumonia. Mice injected either with living R cells or with heat-killed S cells remain healthy. Here is Griffith's critical finding: mice injected with a *mixture* of living R cells and heat-killed S cells contract the disease and often die of pneumonia (**FIGURE 1.2**). Bacteria isolated from blood samples of these dead mice produce S cultures with a capsule typical of the injected S cells, even though the injected S cells had been killed by heat. Evidently, the injected material from the dead S cells includes a substance that can be transferred to living R cells that enables them to synthesize the S-type cap-

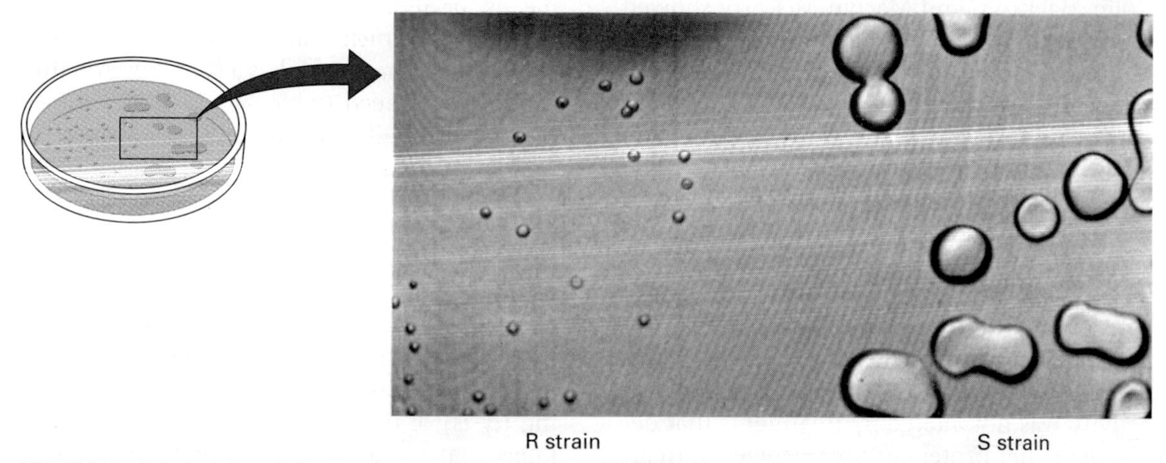


**FIGURE 1.1** Colonies of rough (R, the small colonies) and smooth (S, the large colonies) strains of *Streptococcus pneumoniae.* The S colonies are larger because of the gelatinous capsule on the S cells. [Photograph reproduced from *Journal of Experimental Medicine* by O.T. Avery, et al. Copyright 1994 by Rockefeller University Press. Reproduced with permission of Rockefeller University Press in the format Textbook via Copyright Clearance Center.]

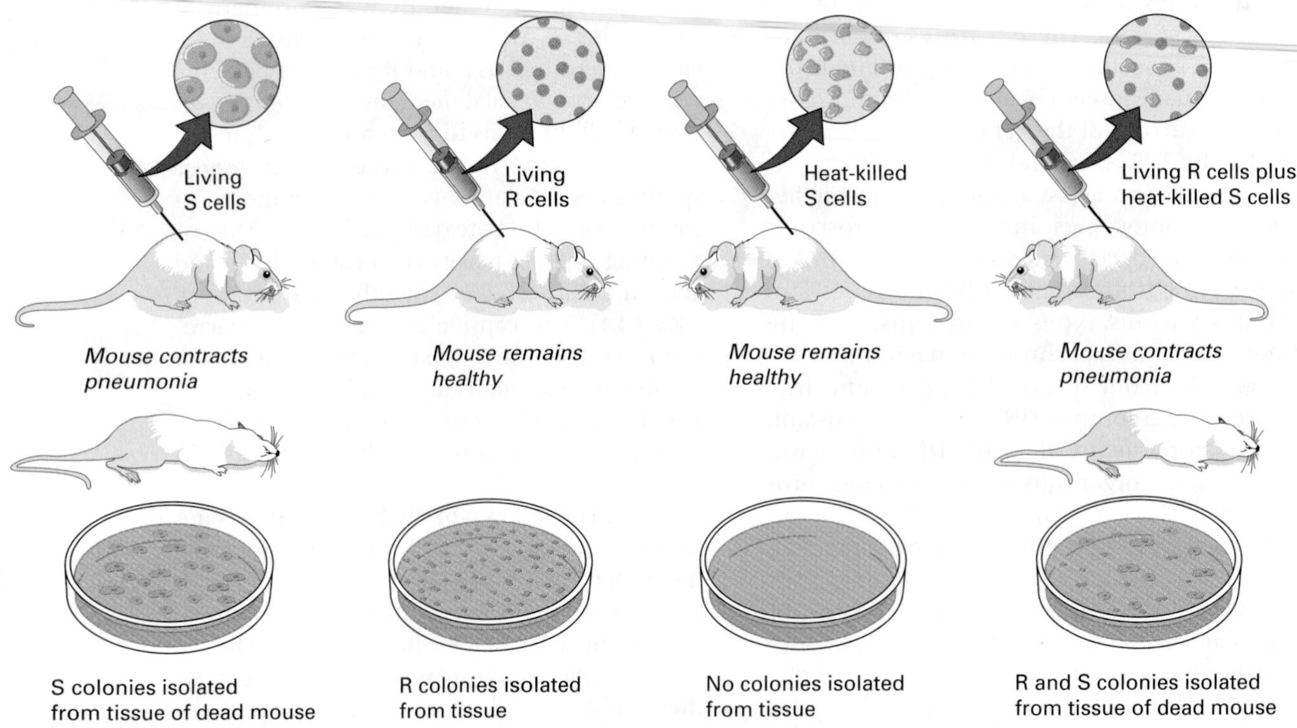


**FIGURE 1.2** The Griffith's experiment demonstrating bacterial transformation. A mouse remains healthy if injected with either the nonvirulent R strain of *S. pneumoniae* or heat-killed cell fragments of the usually virulent S strain. R cells in the presence of heat-killed S cells are transformed into the virulent S strain, causing pneumonia in the mouse.

sule. In other words, the R bacteria can be changed—or undergo **transformation**—into S bacteria. Furthermore, the ability to synthesize the capsule is inherited by descendants of the transformed bacteria.

Transformation in *Streptococcus* was originally discovered in 1928, but it was not until 1944 that the chemical substance responsible for changing the R cells into S cells was identified. In a milestone experiment, Oswald Avery, Colin MacLeod, and Maclyn McCarty showed that the substance causing the transformation of R cells into S cells was DNA. In doing these experiments, they first needed to develop chemical procedures for isolating almost pure DNA from cells, which had never been done before. When they added DNA isolated from S to growing cultures of R cells, they observed transformation—that is, a few cells of type S were produced. Although the DNA preparations contained traces of protein and RNA (ribonucleic acid, an abundant cellular macromolecule chemically related to DNA), the transforming activity was not altered by treatments that destroyed either protein or RNA. However, treatments that destroyed DNA eliminated the transforming activity (**FIGURE 1.3**). These experiments implied that the substance responsible for genetic transformation was the DNA of the cell—hence that DNA is the genetic material.

## ■ Genetic Role of DNA in Bacteriophage

Another pivotal finding was reported by Alfred Hershey and Martha Chase in 1952. They studied cells of the intestinal bacterium *Escherichia coli* after infection by the virus T2. A virus that attacks bacterial cells is called a **bacteriophage,** a term often shortened to **phage.** *Bacteriophage* means "bacteria-eater." The structure of a bacteriophage T2 particle is illustrated in **FIGURE 1.4**. It is exceedingly small, yet it has a complex structure composed of head (which contains the phage DNA), collar, tail, and tail fibers. (The head of a human sperm is about 30–50 times larger in both length and width than the head of T2.) Hershey and Chase were already aware that T2 infection proceeds via the attachment of a phage particle by the tip of its tail to the bacterial cell wall, entry of phage material into the cell, multiplication of this material to form a hundred or more progeny phage, and release of the progeny phage by bursting (lysis) of the bacterial host cell. They also knew that T2 particles were composed of DNA and protein in approximately equal amounts.

Because DNA contains phosphorus but no sulfur, whereas most proteins contain sulfur but

(A) The transforming activity in S cells is not destroyed by heat.

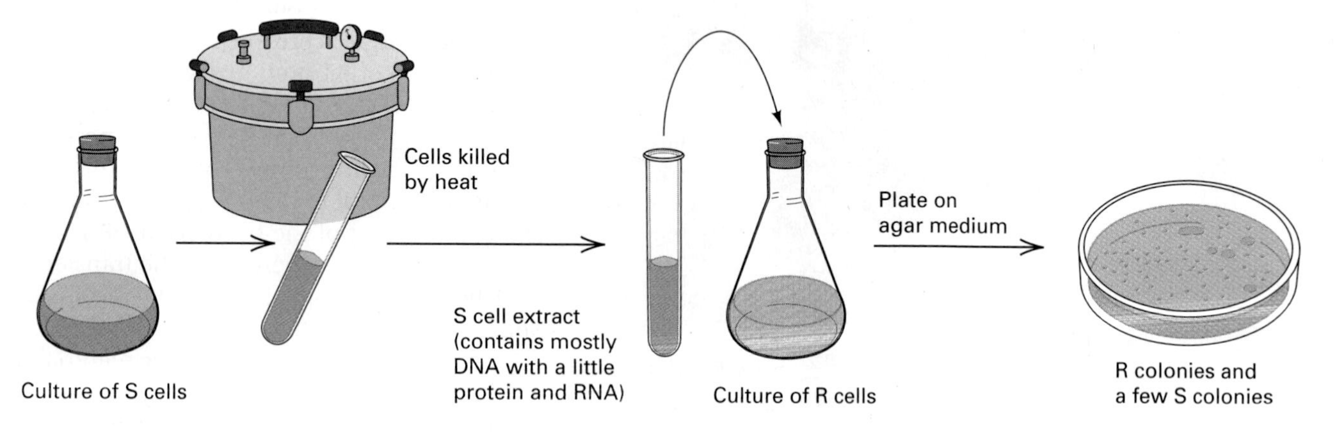


(B) The transforming activity is not destroyed by either protease or RNase.

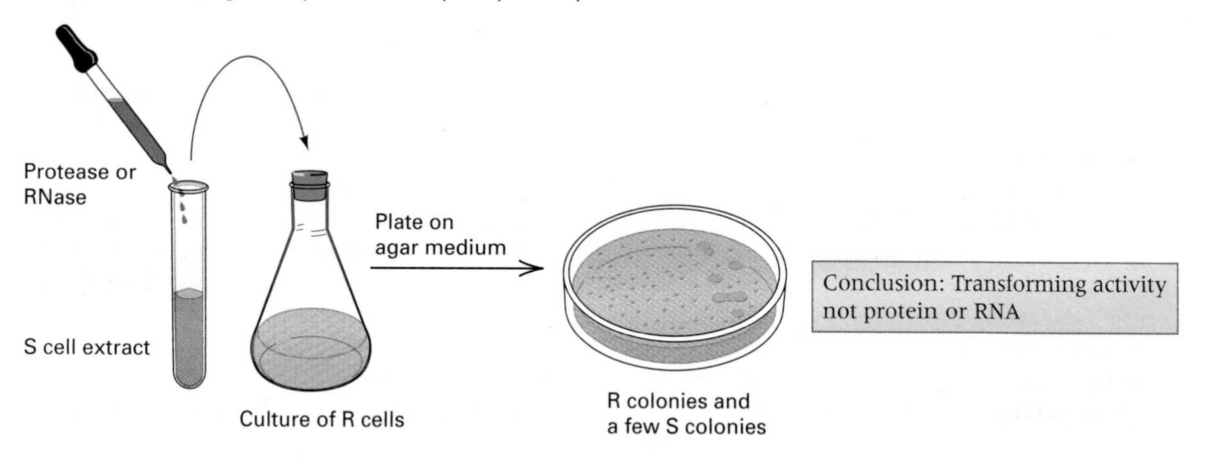


(C) The transforming activity is destroyed by DNase.

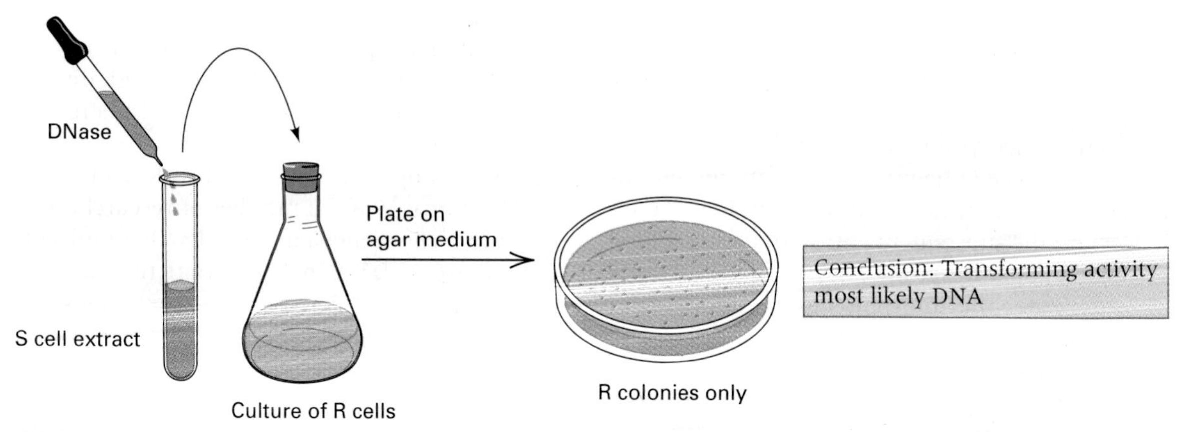


**FIGURE 1.3** A diagram of the Avery–MacLeod–McCarty experiment demonstrating that DNA is the active material in bacterial transformation. (A) Purified DNA extracted from heat-killed S cells can convert some living R cells into S cells, but the material may still contain undetectable traces of protein and/or RNA. (B) The transforming activity is not destroyed by either protease or RNase. (C) The transforming activity is destroyed by DNase and so probably consists of DNA.

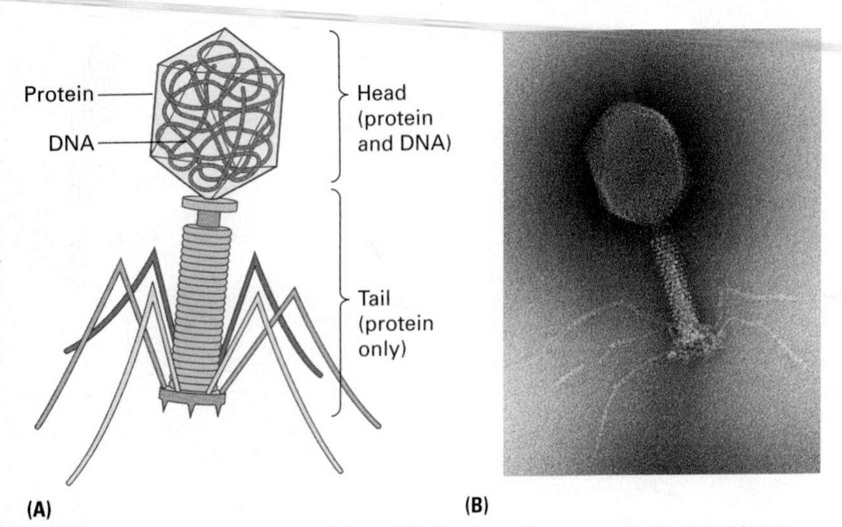


**FIGURE 1.4** (A) Drawing of *E. coli* phage T2, showing various components. The DNA is confined to the interior of the head. (B) An electron micrograph of phage T4, a closely related phage. [Electron micrograph courtesy of Robert Duda, University of Pittsburgh.]

no phosphorus, it is possible to label DNA and proteins differentially by using radioactive isotopes of the two elements. Hershey and Chase produced particles containing radioactive DNA by infecting *E. coli* cells that had been grown for several generations in a medium that included $^{32}$P (a radioactive isotope of phosphorus) and then collecting the phage progeny. Other particles containing labeled proteins were obtained in the same way, by using medium that included $^{35}$S (a radioactive isotope of sulfur).

In the experiments summarized in **FIGURE 1.5**, nonradioactive *E. coli* cells were infected with phage labeled with *either* $^{32}$P (part A) *or* $^{35}$S (part B) in order to follow the DNA and proteins individually. Infected cells were separated from unattached phage particles by centrifugation, resuspended in fresh medium, and then swirled violently in a kitchen blender to shear attached phage material from the cell surfaces. This treatment was found to have no effect on the subsequent course of the infection, which implies that the phage genetic material must enter the infected cells very soon after phage attachment. The kitchen blender turned out to be the critical piece of equipment. Other methods had been tried to tear the phage heads from the bacterial cell surface, but nothing had worked reliably. Hershey later explained, "We tried various grinding arrangements, with results that weren't very encouraging. When Margaret McDonald loaned us her kitchen blender, the experiment promptly succeeded."

After the phage heads were removed by the blender treatment, the infected bacteria were examined. Most of the radioactivity from $^{32}$P-labeled phage was found to be associated with the bacteria, whereas only a small fraction of the $^{35}$S radioactivity was present in the infected cells. The retention of most of the labeled DNA, contrasted with the loss of most of the labeled protein, implied that a T2 phage transfers most of its DNA, but very little of its protein, to the cell it infects. The critical finding (Figure 1.5) was that about 50 percent of the transferred $^{32}$P-labeled DNA, but less than 1 percent of the transferred $^{35}$S-labeled protein, was inherited by the *progeny* phage particles. Hershey and Chase interpreted this result to mean that the genetic material in T2 phage is DNA.

The experiments of Avery, MacLeod, and McCarty and those of Hershey and Chase are regarded as classics in the demonstration that genes consist of DNA. At the present time, the equivalent of the transformation experiment is carried out daily in many research laboratories throughout the world, usually with bacteria, yeast, or animal or plant cells grown in culture. These experiments indicate that DNA is the genetic material in these organisms as well as in phage T2. Although there are no known exceptions to the generalization that DNA is the genetic material in all cellular organisms and many viruses, in several types of viruses the genetic material consists of RNA. These RNA-containing viruses include the human immunodeficiency virus HIV-1 that causes AIDS (aquired immune deficiency disease).

## 1.2 DNA Structure and Replication

The inference that DNA is the genetic material still left many questions unanswered. How is the DNA in a gene duplicated when a cell divides? How does the DNA in a gene control a hereditary trait? What happens to the DNA when a mutation (a change in the DNA) takes place in a gene? In the early 1950s, a number of researchers began to try to understand the detailed molecular structure of DNA in hopes that the structure alone would suggest answers to these questions. In 1953 James Watson and Francis Crick at Cambridge University proposed the first essentially correct three-dimensional structure of the DNA molecule. The structure was dazzling in its elegance and revolutionary in suggesting how DNA duplicates itself, controls hereditary traits, and undergoes mutation. Even while their tin-and-wire model of the DNA molecule was still incomplete, Crick would visit his favorite pub and exclaim "we have discovered the secret of life."

In the Watson–Crick structure, DNA consists of two long chains of subunits, each twisted

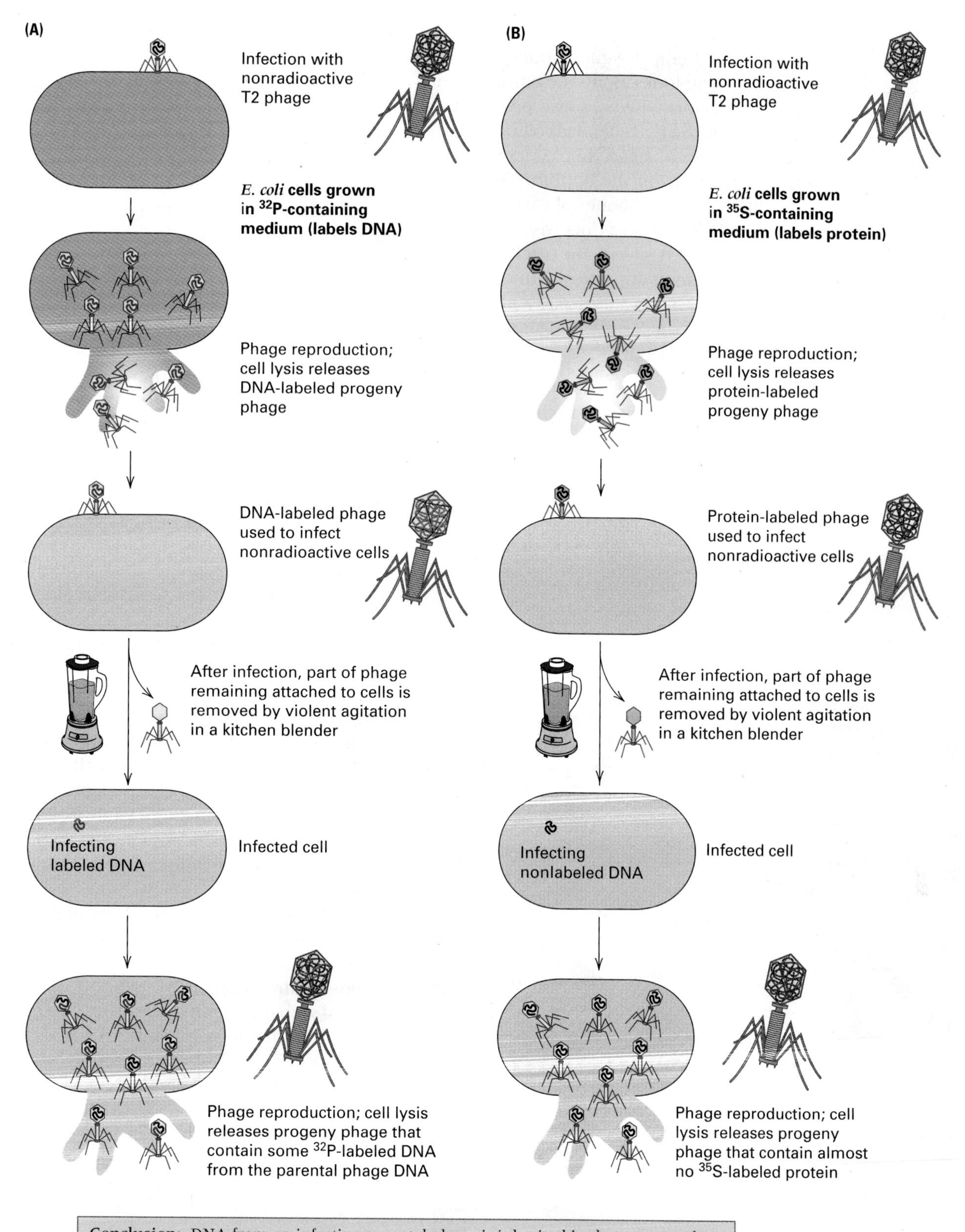
(A)
Infection with nonradioactive T2 phage
E. coli cells grown in ³²P-containing medium (labels DNA)
Phage reproduction; cell lysis releases DNA-labeled progeny phage
DNA-labeled phage used to infect nonradioactive cells
After infection, part of phage remaining attached to cells is removed by violent agitation in a kitchen blender
Infecting labeled DNA
Infected cell
Phage reproduction; cell lysis releases progeny phage that contain some ³²P-labeled DNA from the parental phage DNA
(B)
Infection with nonradioactive T2 phage
E. coli cells grown in ³⁵S-containing medium (labels protein)
Phage reproduction; cell lysis releases protein-labeled progeny phage
Protein-labeled phage used to infect nonradioactive cells
After infection, part of phage remaining attached to cells is removed by violent agitation in a kitchen blender
Infecting nonlabeled DNA
Infected cell
Phage reproduction; cell lysis releases progeny phage that contain almost no ³⁵S-labeled protein
Conclusion: DNA from an infecting parental phage is inherited in the progeny phage

**FIGURE 1.5** The Hershey–Chase ("blender") experiment demonstrating that DNA, not protein, is responsible for directing the reproduction of phage T2 in infected *E. coli* cells. (A) Radioactive DNA is transmitted to progeny phage in substantial amounts. (B) Radioactive protein is transmitted to progeny phage in negligible amounts.

around the other to form a double-stranded helix. The double helix is right-handed, which means that as one looks along the barrel, each chain follows a clockwise path as it progresses. You can visualize the right-handed coiling in part A of **FIGURE 1.6** if you imagine yourself looking up into the structure from the bottom. The dark spheres outline the "backbone" of each individual strand, and they coil in a clockwise direction. The subunits of each strand are **nucleotides,** each of which contains any one of four chemical constituents called **bases** attached to a phosphorylated molecule of the 5-carbon sugar **deoxyribose.** The four bases in DNA are:

- **Adenine (A)**
- **Thymine (T)**
- **Guanine (G)**
- **Cytosine (C)**

The chemical structures of the nucleotides and bases need not concern us at this time. They are examined in Chapter 2. A key point for our present purposes is that the bases in the double helix are paired as shown in Figure 1.6B. That is:

> At any position on the paired strands of a DNA molecule, if one strand has an A, then the partner strand has a T; and if one strand has a G, then the partner strand has a C.

The pairing between A and T and between G and C is said to be **complementary;** the complement of A is T, and the complement of G is C. The complementary pairing means that each base along one strand of the DNA is matched with a base in the opposite position on the other strand. Furthermore:

> Nothing restricts the sequence of bases in a single strand, so any sequence could be present along one strand.

This principle explains how only four bases in DNA can code for the huge amount of information needed to make an organism. It is the *sequence* of bases along the DNA that encodes the genetic information, and the sequence is completely unrestricted.

The complementary pairing is also called **Watson–Crick pairing.** In the three-dimensional structure in Figure 1.6A, the base pairs are represented by the lighter spheres filling the interior of the double helix. The base pairs lie almost flat, stacked on top of one another perpendicular to the long axis of the double helix, like pennies in a roll. When discussing a DNA molecule, biologists frequently refer to the individual strands as **single-stranded DNA** and to the double helix as **double-stranded DNA** or **duplex DNA.**

Each DNA strand has a **polarity,** or directionality, like a chain of circus elephants linked trunk to tail. In this analogy, each elephant corresponds to one nucleotide along the DNA strand. The polarity is determined by the direction in which the nucleotides are pointing. The "trunk" end of the strand is called the *3′ end* of the strand, and the "tail" end is called the *5′ end.* In double-stranded DNA, the paired strands are oriented in opposite directions, the 5′ end of one strand aligned with the 3′ end of the other. The molecular basis of the polarity, and the reason for the opposite orientation of the strands in duplex DNA, are explained in Chapter 2. In illustrating DNA molecules in this book, we use an arrow-like ribbon to represent the backbone, and we use tabs jutting off the ribbon to represent the nucleotides. The polarity of a DNA strand is indicated by the direction of the arrow-like ribbon. The tail of the arrow represents the 5′ end of the DNA strand, the head the 3′ end.

Beyond the most optimistic hopes, knowledge of the structure of DNA immediately gave clues to its function:

1. The sequence of bases in DNA could be copied by using each of the separate

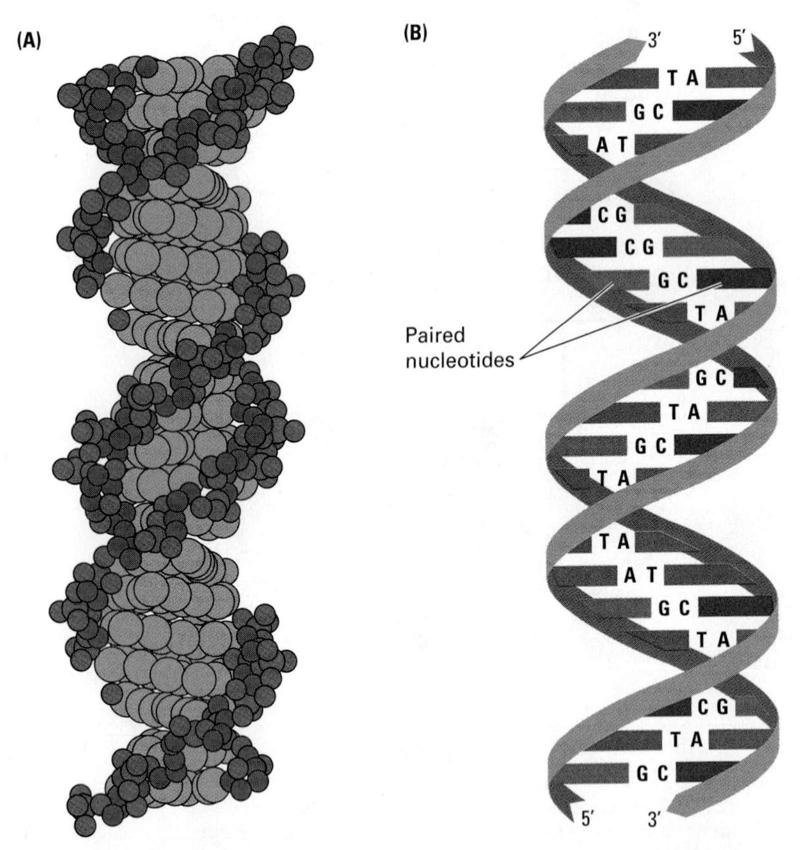


**FIGURE 1.6** Molecular structure of the DNA double helix in the standard "B form." (A) A space-filling model, in which each atom is depicted as a sphere. (B) A diagram highlighting the helical strands around the outside of the molecule and the A–T and G–C base pairs inside.

"partner" strands as a pattern for the creation of a new partner strand with a complementary sequence of bases.
2. The DNA could contain genetic information in coded form in the sequence of bases, analogous to letters printed on a strip of paper.
3. Changes in genetic information (mutations) could result from errors in copying in which the base sequence of the DNA became altered.

In the remainder of this chapter, we discuss some of the implications of these clues.

## An Overview of DNA Replication

Watson and Crick noted that the structure of DNA itself suggested a mechanism for its replication. "It has not escaped our notice," they wrote, "that the specific base pairing we have postulated immediately suggests a copying mechanism." The copying process in which a single DNA molecule gives rise to two identical molecules is called **replication.** The replication mechanism that Watson and Crick had in mind is illustrated in **FIGURE 1.7**.

As shown in part A of Figure 1.7, the strands of the original (parent) duplex sepa-

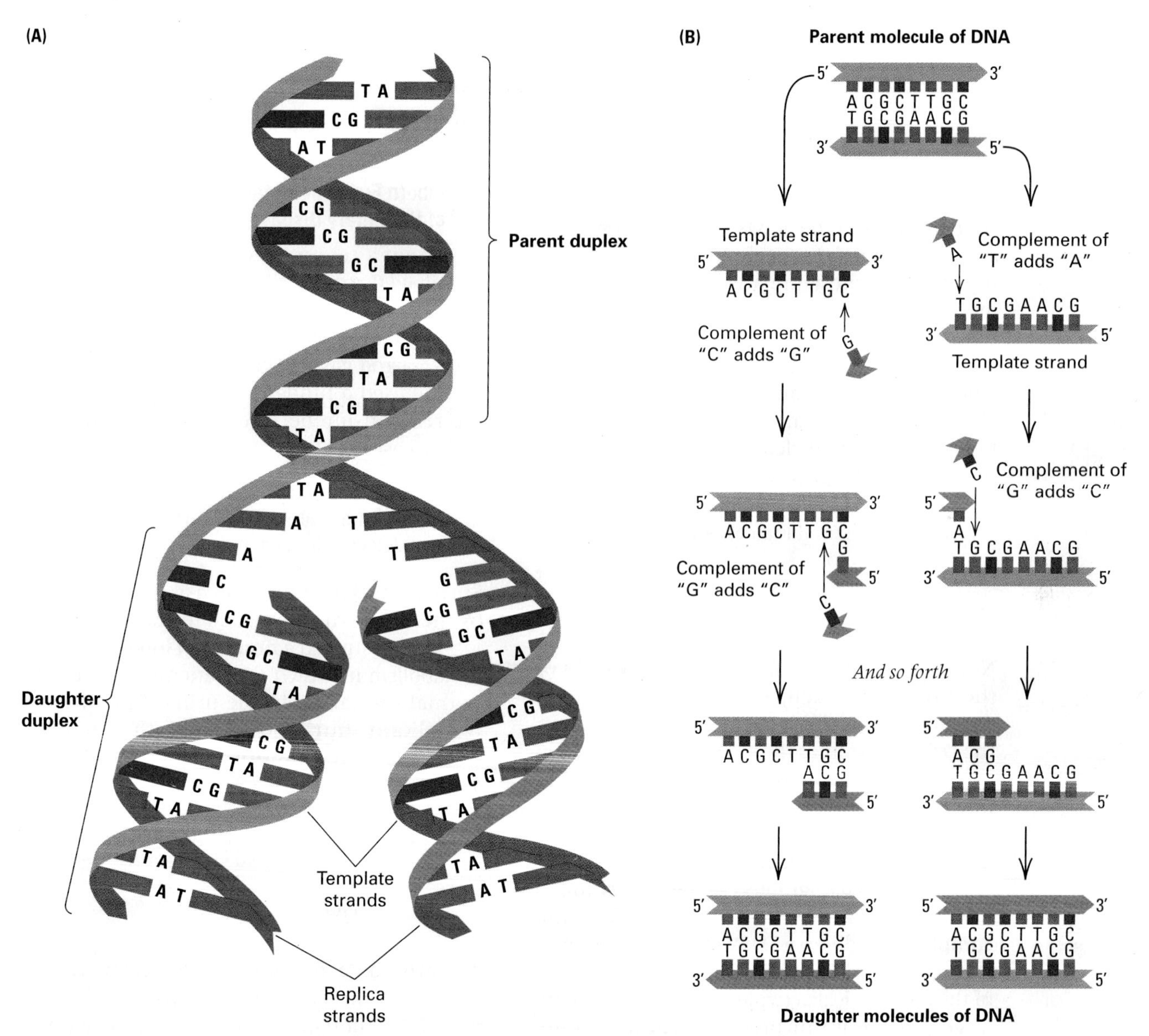


**FIGURE 1.7** Replication of DNA. (A) Replication of a DNA duplex as originally envisioned by Watson and Crick. As the parental strands separate, each parental strand serves as a template for the formation of a new daughter strand by means of A–T and G–C base pairing. (B) Greater detail showing how each of the parental strands serves as a template for the production of a complementary daughter strand, which grows in length by the successive addition of single nucleotides to the 3′ end.

rate, and each individual strand serves as a pattern, or **template,** for the synthesis of a new strand (replica). The replica strands are synthesized by the addition of successive nucleotides in such a way that each base in the replica is complementary (in the Watson–Crick pairing sense) to the base across the way in the template strand (Figure 1.7B). Although the mechanism in Figure 1.7 is simple in principle, it is a complex process that is fraught with geometrical problems and requires a variety of enzymes and other proteins. The details are examined in Chapter 6. The end result of replication is that a single double-stranded molecule becomes replicated into two copies with identical sequences:

```
            5'-ACGCTTGC-3'
            3'-TGCGAACG-5'
               ↙      ↘
5'-ACGCTTGC-3'          5'-ACGCTTGC-3'
3'-TGCGAACG-5'          3'-TGCGAACG-5'
```

Here the bases in the newly synthesized strands are shown in red. In the duplex on the left, the top strand is the template from the parental molecule and the bottom strand is newly synthesized; in the duplex on the right, the bottom strand is the template from the parental molecule and the top strand is newly synthesized. Note in Figure 1.7B that in the synthesis of each new strand, new nucleotides are added only to the 3′ end of the growing chain:

> The obligatory elongation of a DNA strand only at the 3′ end is an essential feature of DNA replication.

## 1.3 Genes and Proteins

Now that we have some basic understanding of the structural makeup of the genetic material, we can ask how the sequence of nucleotides in the DNA determines the biochemical characteristics of cells and organisms. If the sequence of nucleotides along the DNA is thought of as a string of letters on a sheet of paper, then the genes are made up of distinct words that form sentences and paragraphs that give meaning to the pattern of letters. What is created from the complex and diverse DNA codes is protein, a class of macromolecules that carries out most of the biochemical activities in the cell. Cells are largely made up of proteins. These include structural proteins that give the cell rigidity and mobility, proteins that form pores in the cell membrane to control the traffic of small molecules into and out of the cell, and receptor proteins that regulate cellular activities in response to molecular signals from the growth medium or from other cells. Proteins are also responsible for most of the metabolic activities of cells. They are essential for the synthesis and breakdown of organic molecules and for generating the chemical energy needed for cellular activities. In 1878 the term **enzyme** was introduced to refer to the biological catalysts that accelerate biochemical reactions in cells. By 1900, thanks largely to the work of the German biochemist Emil Fischer, enzymes were shown to be proteins. As often happens in science, nature's "mistakes" provide clues as to how things work. Such was the case in establishing a relationship between genes and disease, because a "mistake" in a gene (a mutation) can result in a "mistake" (lack of function) in the corresponding protein. This provided a fruitful avenue of research for the study of genetics.

### Inborn Errors of Metabolism as a Cause of Hereditary Disease

About a century ago, the British physician Archibald Garrod realized that certain heritable diseases followed the rules of transmission that Mendel had described for his garden peas. In 1908 Garrod gave a series of lectures in which he proposed a fundamental hypothesis about the relationship between heredity, enzymes, and disease:

> Any hereditary disease in which cellular metabolism is abnormal results from an inherited defect in an enzyme.

Such diseases became known as **inborn errors of metabolism,** a term still in use today.

Garrod studied a number of inborn errors of metabolism in which the patients excreted abnormal substances in the urine. One of these was **alkaptonuria.** In this case, the abnormal substance excreted is **homogentisic acid:**

OH, O, $CH_2$–C, OH, HO

An early name for homogentisic acid was *alkapton,* hence the name *alkaptonuria* for this disease. Even though alkaptonuria is rare, with an incidence of about one in 200,000 people, it was well known even before Garrod studied it. The disease itself is relatively mild, but it has one striking symptom: The urine of the patient turns

FIGURE 1.8 Urine from a person with alkaptonuria turns black because of the oxidation of the homogentisic acid that it contains. [Courtesy of Daniel De Aguiar.]

black because of the oxidation of homogentisic acid (FIGURE 1.8). This is why alkaptonuria is also called *black urine disease.* An early case was described in the year 1649:

> The patient was a boy who passed black urine and who, at the age of fourteen years, was submitted to a drastic course of treatment that had for its aim the subduing of the fiery heat of his viscera, which was supposed to bring about the condition in question by charring and blackening his bile. Among the measures prescribed were bleedings, purgation, baths, a cold and watery diet, and drugs galore. None of these had any obvious effect, and eventually the patient, who tired of the futile and superfluous therapy, resolved to let things take their natural course. None of the predicted evils ensued. He married, begat a large family, and lived a long and healthy life, always passing urine black as ink. (Recounted by Garrod, 1908.)

Garrod was primarily interested in the biochemistry of alkaptonuria, but he took note of family studies that indicated that the disease was inherited as though it were due to a defect in a single gene. As to the biochemistry, he deduced that the problem in alkaptonuria was the patients' inability to break down the phenyl ring of six carbons that is present in homogentisic acid. Where does this ring come from? Most animals obtain it from foods in their diet. Garrod proposed that homogentisic acid originates as a breakdown product of two amino acids, phenylalanine and tyrosine, which also contain a phenyl ring. An **amino acid** is one of the "building blocks" from which proteins are made. Phenylalanine and tyrosine are constituents of normal proteins. The scheme that illustrates the relationship between the molecules is shown in FIGURE 1.9. Any such sequence of biochemical reactions is called a **biochemi-**

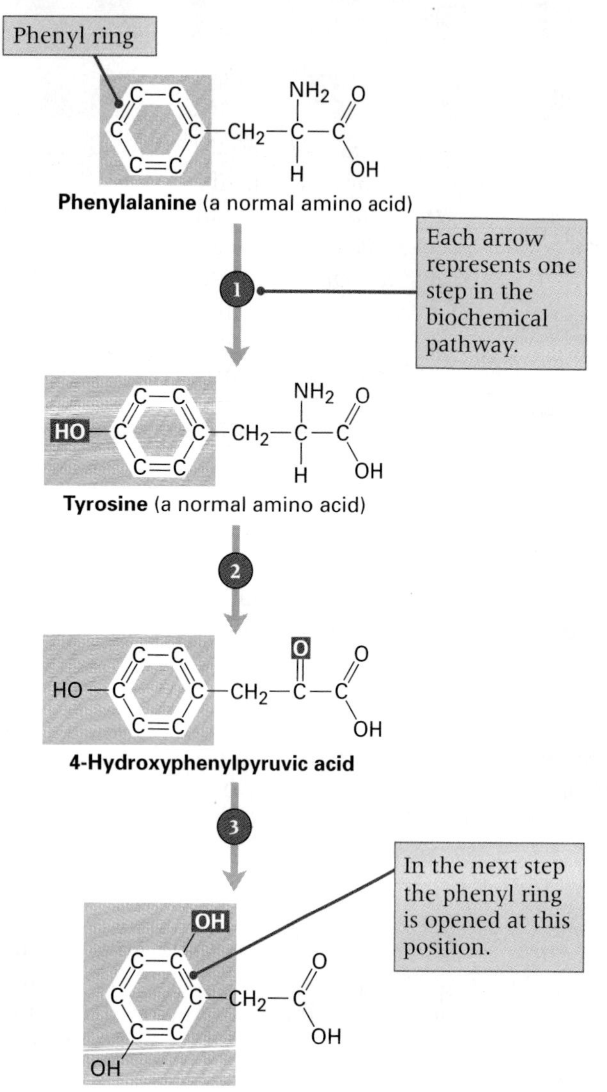


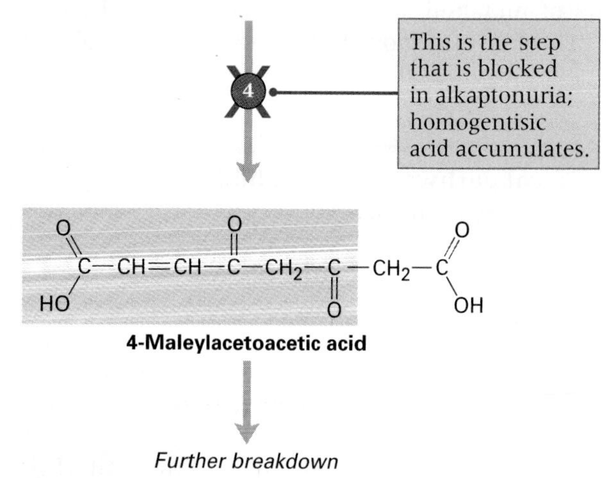


FIGURE 1.9 Metabolic pathway for the breakdown of phenylalanine and tyrosine. Each step in the pathway, represented by an arrow, requires a specific enzyme to catalyze the reaction. The key step in the breakdown of homogentisic acid is the breaking open of the phenyl ring.

## connection
### The Black Urine Disease

**Archibald E. Garrod 1908**
St. Bartholomew's Hospital,
London, England
*Inborn Errors of Metabolism*

*Although he was a distinguished physician, Garrod's lectures on the relationship between heredity and congenital defects in metabolism had no impact when they were delivered. The important concept that one gene corresponds to one enzyme (the "one gene–one enzyme hypothesis") was developed independently in the 1940s by George W. Beadle and Edward L. Tatum, who used the bread mold* Neurospora crassa *as their experimental organism. When Beadle finally became aware of* Inborn Errors of Metabolism, *he was generous in praising it. This excerpt shows Garrod at his best, interweaving history, clinical medicine, heredity, and biochemistry in his account of alkaptonuria. The excerpt also illustrates how the severity of a genetic disease depends on its social context. Garrod writes as though alkaptonuria were a harmless curiosity. This is indeed largely true when the life expectancy is short. With today's longer life span, alkaptonuria patients accumulate the dark pigment in their cartilage and joints and can eventually develop arthritis.*

To students of heredity the inborn errors of metabolism offer a promising field of investigation. . . . It was pointed out [by others] that the mode of incidence of alkaptonuria finds a ready explanation if the anomaly be regarded as a rare recessive character in the Mendelian sense. . . . Of the cases of alkaptonuria a very large proportion have been in the children of first cousin marriages. . . . It is also noteworthy that, if one takes families with five or more children [with both parents normal and at least one child affected with alkaptonuria], the totals work out in strict conformity to Mendel's law, i.e. 57 [normal children] : 19 [affected children] in the proportions 3 : 1. . . . Of inborn errors of metabolism, alkaptonuria is that of which we know most. In itself it is a trifling matter, inconvenient rather than harmful. . . . Indications of the anomaly may be detected in early medical writings, such as that in 1584 of a schoolboy who, although he enjoyed good health, continuously excreted black urine; and that in 1609 of a monk who exhibited a similar peculiarity and stated that he had done so all his life. . . . There are no sufficient grounds [for doubting that the blackening substance in the urine originally called alkapton] is homogentisic acid, the excretion of which is the essential feature of the alkaptonuric. . . . Homogentisic acid is a product of normal metabolism. . . . The most likely sources of the phenyl ring in homogentisic acid are phenylalanine and tyrosine, [because when these amino acids are administered to an alkaptonuric] they cause a very conspicuous increase in the output of homogentisic acid. . . . Where the alkaptonuric differs from the normal individual is in having no power of destroying homogentisic acid when formed—in other words of breaking up the phenyl ring of that compound. . . . We may further conceive that the splitting of the phenyl ring in normal metabolism is the work of a special enzyme and that in congenital alkaptonuria this enzyme is wanting.

***We may further conceive that the splitting of the phenyl ring in normal metabolism is the work of a special enzyme and that in congenital alkaptonuria this enzyme is wanting.***

Source: H. Harris. *Garrod's Inborn Errors of Metabolism, Second edition.* Oxford University Press (1963). Originally published in London, England by the Oxford University Press.

**cal pathway** or a **metabolic pathway.** Each arrow in the pathway represents a single step depicting the transition from the "input" or **substrate molecule,** shown at the head of the arrow, to the "output" or **product molecule,** shown at the tip. Biochemical pathways are usually oriented either vertically with the arrows pointing down, as in Figure 1.9, or horizontally, with the arrows pointing from left to right. Garrod did not know all of the details of the pathway in Figure 1.9, but he did understand that the key step in the breakdown of homogentisic acid is the breaking open of the phenyl ring and that the phenyl ring in homogentisic acid comes from dietary phenylalanine and tyrosine.

What allows each step in a biochemical pathway to occur? Garrod correctly surmised that each step requires a specific enzyme to catalyze the reaction for the chemical transformation. Persons with an inborn error of metabolism, such as alkaptonuria, have a defect in a single step of a metabolic pathway because they lack a functional enzyme for that step. When an enzyme in a pathway is defective, the pathway is said to have a **block** at that step. One frequent result of a blocked pathway is that the substrate of the defective enzyme accumulates. Observing the accumulation of homogentisic acid in patients with alkaptonuria, Garrod proposed that there must be an enzyme whose function is to open the phenyl ring of homogentisic acid and that this enzyme is missing in these patients. Isolation of the enzyme that opens the phenyl ring of homogentisic acid was not actually achieved until 50 years after Garrod's lectures. In normal people it is found in cells of the liver, and just as Garrod had predicted, the enzyme is defective in patients with alkaptonuria.

The pathway for the breakdown of phenylalanine and tyrosine, as it is understood today, is shown in **FIGURE 1.10**. In this figure the emphasis is on the enzymes rather than on the structures of the **metabolites,** or small molecules, on which the enzymes act. Each step in the pathway requires the presence of a particular enzyme that catalyzes that step. Although Garrod knew only about alkaptonuria, in which the defective enzyme is homogentisic acid 1,2 dioxygenase, we now know the clinical consequences of defects in the other enzymes. Unlike alkaptonuria, which is a relatively benign inherited disease, the others are very serious. The condition known as **phenylketonuria (PKU)** results from the absence of (or a defect in) the enzyme **phenylalanine hydroxylase (PAH).** When this step in the pathway is blocked, phenylalanine accumulates. The excess phenylalanine is broken down into harmful metabolites that cause defects in myelin formation that damage a child's developing nervous system and lead to severe mental retardation.

However, if PKU is diagnosed in children soon enough after birth, they can be placed on a specially formulated diet low in phenylalanine. The child is allowed only as much phenylalanine as can be used in the synthesis of proteins, so excess phenylalanine does not accumulate. The special diet is very strict. It excludes meat, poultry, fish, eggs, milk and milk products, legumes, nuts, and bakery goods manufactured with regular flour. These foods are replaced by an expensive synthetic formula. With the special diet, however, the detrimental effects of excess phenylalanine on mental development can largely be avoided, although in adult women with PKU who are pregnant, the fetus is at risk. In many countries, including the United States, all newborn babies have their blood tested for chemical signs of PKU. Routine screening is cost-effective because PKU is relatively common. In the United States, the incidence is about 1 in 8000 among Caucasian births. The disease is less common in other ethnic groups.

In the metabolic pathway in Figure 1.10, defects in the breakdown of tyrosine or of 4-hydroxyphenylpyruvic acid lead to types of tyrosinemia. These are also severe diseases. Type II

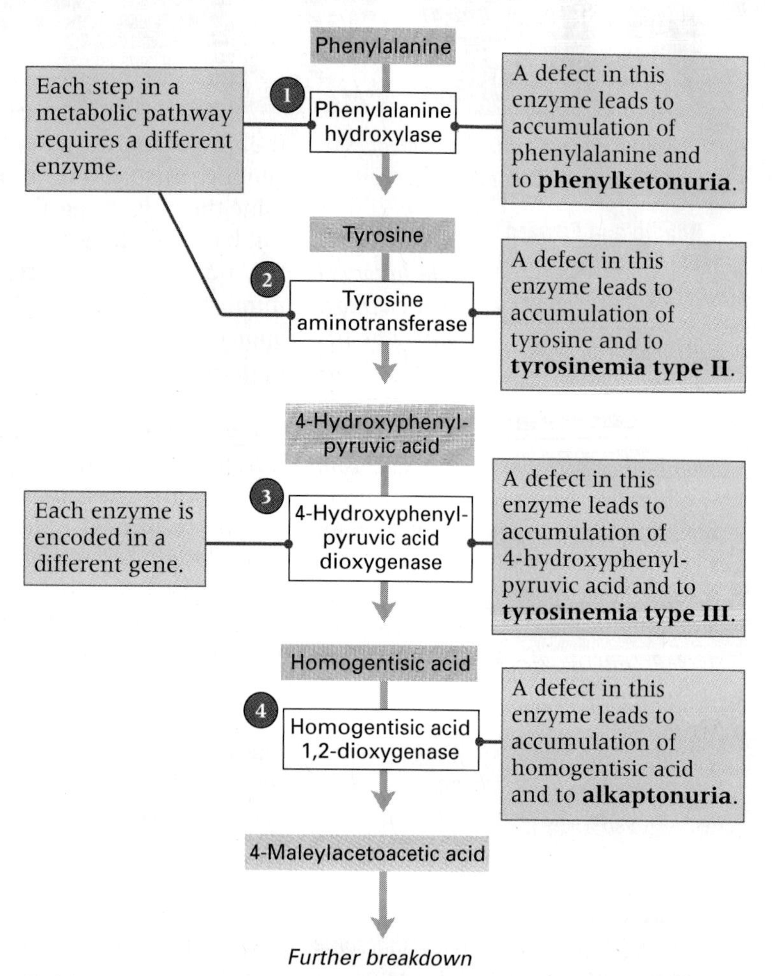


**FIGURE 1.10** Inborn errors of metabolism that affect the breakdown of phenylalanine and tyrosine. An inherited disease results when any of the enzymes is missing or defective. Alkaptonuria results from a mutant homogentisic acid 1,2 dioxygenase; phenylketonuria results from a mutant phenylalanine hydroxylase.

is associated with skin lesions and mental retardation, Type III with severe liver dysfunction.

The genes for the enzymes in the biochemical pathway in Figure 1.10 have all been identified and the nucleotide sequence of the DNA determined. In the following list, and throughout this book, we use the standard typographical convention that *genes* are written in *italic* type, whereas gene products are not printed in italics. This convention is convenient, because it means that the protein product of a gene can be represented with the same symbol as the gene itself, but whereas the gene symbol is in italics, the protein symbol is not. In Figure 1.10 the numbers correspond to the following genes and enzymes.

1. The gene *PAH* on the long arm of chromosome 12 encodes phenylalanine hydroxylase (PAH).
2. The gene *TAT* on the long arm of chromosome 16 encodes tyrosine aminotransferase (TAT).

## connection

## One Gene, One Enzyme

**George W. Beadle and Edward L. Tatum 1941**
*Stanford University, Stanford, California*
*Genetic Control of Biochemical Reactions in* Neurospora

*How do genes control metabolic processes? The suggestion that genes control enzymes was made very early in the history of genetics, most notably by the British physician Archibald Garrod in his 1908 book Inborn Errors of Metabolism. But the precise relationship between genes and enzymes was still uncertain. Perhaps each enzyme is controlled by more than one gene, or perhaps each gene contributes to the control of several enzymes. The classic experiments of Beadle and Tatum showed that the relationship is usually remarkably simple: One gene codes for one enzyme. The pioneering experiments united genetics and biochemistry, and for the "one gene, one enzyme" concept, Beadle and Tatum were awarded a Nobel Prize in 1958 (Joshua Lederberg shared the prize for his contributions to microbial genetics). Because we now know that some enzymes contain polypeptide chains encoded by two (or occasionally more) different genes, a more accurate statement of the principle is "one gene, one polypeptide." Beadle and Tatum's experiments also demonstrate the importance of choosing the right organism. Neurospora had been introduced as a genetic organism only a few years earlier, and Beadle and Tatum realized that they could take advantage of the ability of this organism to grow on a simple medium composed of known substances.*

From the standpoint of physiological genetics the development and functioning of an organism consist essentially of an integrated system of chemical reactions controlled in some manner by genes. . . . In investigating the roles of genes, the physiological geneticist usually attempts to determine the physiological and biochemical bases of already known hereditary traits. . . . There are, however, a number of limitations inherent in this approach. Perhaps the most serious of these is that the investigator must in general confine himself to the study of non-lethal heritable characters. Such characters are likely to involve more or less non-essential so-called "terminal" reactions. . . . A second difficulty is that the standard approach to the problem implies the use of characters with visible manifestations. Many such characters involve morphological variations, and these are likely to be based on systems of biochemical reactions so complex as to make analysis exceedingly difficult. . . . Considerations such as those just outlined have led us to investigate the general problem of the genetic control of development and metabolic reactions by reversing the ordinary procedure and, instead of attempting to work out the chemical bases of known genetic characters, to set out to determine if and how genes control known biochemical reactions. The ascomycete *Neurospora* offers many advantages for such an approach and is well suited to genetic studies. Accordingly, our program has been built around this organism. The procedure is based on the assumption that x-ray treatment will induce mutations in genes concerned with the control of known specific chemical reactions. If the organism must be able to carry out a certain chemical reaction to survive on a given medium, a mutant unable to do this will obviously be lethal on this medium. Such a mutant can be maintained and studied, however, if it will grow on a medium to which has been added the essential product of the genetically blocked reaction. . . . Among approximately 2000 strains [derived from single cells after x-ray treatment], three mutants have been found that grow essentially normally on the complete medium and scarcely at all on the minimal medium. One of these strains proved to be unable to synthesize vitamin $B_6$ (pyridoxine). A second strain turned out to be unable to synthesize vitamin $B_1$ (thiamine). A third strain has been found to be unable to synthesize para-aminobenzoic acid. . . . These preliminary results appear to us to indicate that the approach may offer considerable promise as a method of learning more about how genes regulate development and function. For example, it should be possible, by finding a number of mutants unable to carry out a particular step in a given synthesis, to determine whether only one gene is ordinarily concerned with the immediate regulation of a given specific chemical reaction.

***These preliminary results appear to us to indicate that the approach may offer considerable promise as a method of learning more about how genes regulate development and function.***

Source: G. W. Beadle and E. L. Tatum, *Proc. Natl. Acad. Sci. USA* 27 (1941): 499-506.

3. The gene *HPD* on the long arm of chromosome 12 encodes 4-hydroxyphenylpyruvic acid dioxygenase (HPD).
4. The gene *HGD* on the long arm of chromosome 3 encodes homogentisic acid 1,2 dioxygenase (HGD).

## 1.4 Genetic Analysis

The genetic implications of Garrod's research on inborn errors of metabolism were not widely appreciated, most likely because his writings focused primarily on biochemical pathways rather than inheritance. The definitive connection between genes and enzymes came from studies carried out in the 1940s by George W. Beadle and Edward L. Tatum using a filamentous fungus *Neurospora crassa,* commonly called red bread mold, an organism they chose because both genetic and biochemical analysis could be done with ease. In these experiments, they identified new mutations that each caused a block in the metabolic pathway for the synthesis of some needed nutrient, and showed that each of these blocks corresponded to a defective enzyme needed for one step in the pathway. The research was important not only because it solidified the link between genetics and biochemistry, but also because the experimental approach, now called **genetic analysis,** has been successfully applied to understanding a wide range of biological processes from the genetic control of cell cycle and cancer to that of development and behavior. We shall therefore examine the Beadle-Tatum experiments in some detail.

## Mutant Genes and Defective Proteins

*N. crassa* grows in the form of filaments on a great variety of substrates including laboratory medium containing only inorganic salts, a sugar, and the vitamin biotin. Such a medium is known as a **minimal medium** because it contains only the nutrients that are essential for growth of the organism. The filaments consist of a mass of branched threads separated into interconnected, multinucleate compartments allowing free interchange of nuclei and cytoplasm. Each nucleus contains a single set of seven chromosomes. Beadle and Tatum recognized that the ability of *Neurospora* to grow in minimal medium implied that the organism must be able to synthesize all metabolic components other than biotin. If the biosynthetic pathways needed for growth are controlled by genes, then a mutation in a gene responsible for synthesizing an essential nutrient would be expected to render a strain unable to grow unless the strain was provided with the nutrient.

These ideas were tested in the following way. Spores of nonmutant *Neurospora* were irradiated with either x-rays or ultraviolet light to produce mutant strains with various nutritional requirements. (Why these treatments cause mutations is discussed in Chapter 14.) The isolation of a set of mutants affecting any biological process, in this case metabolism, is called a **mutant screen.** In the initial step for identifying mutants, summarized in **FIGURE 1.11**, the irradiated spores (pur-

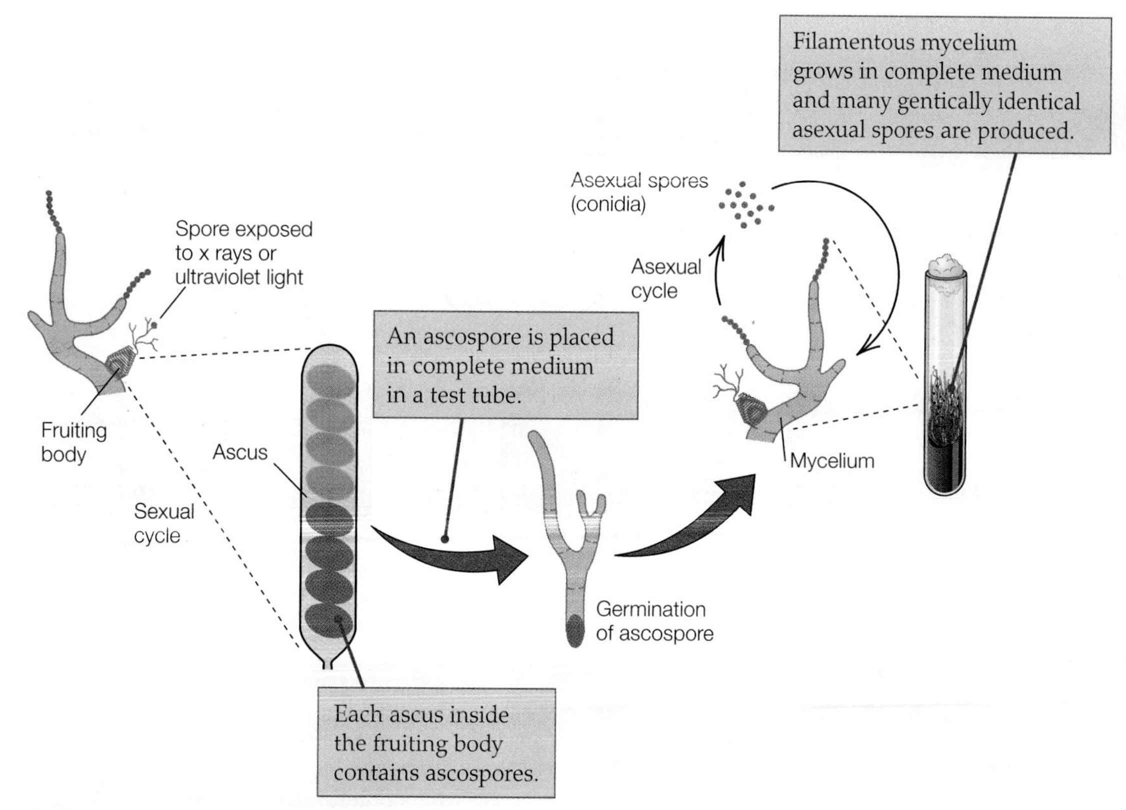


**FIGURE 1.11** Beadle and Tatum obtained mutants of the filamentous fungus *Neurospora crassa* by exposing asexual spores to x-rays or ultraviolet light. The treated spores were used to start the sexual cycle in fruiting bodies. After any pair of cells and their nuclei undergo fusion, meiosis takes place almost immediately and results in eight sexual spores (ascospores) included in a single ascus. These are removed individually and cultured in complete medium. Ascospores that carry new nutritional mutants are identified later by their inability to grow in minimal medium.

ple) were used in crosses with an untreated strain (green). Ascospores produced by the sexual cycle in fruiting bodies were individually germinated in **complete medium,** a complex medium enriched with a variety of amino acids, vitamins, and other substances expected to be essential metabolites whose synthesis could be blocked by a mutation. Even those ascospores containing a new mutation affecting synthesis of an essential nutrient would be expected to germinate and grow in complete medium.

To identify which of the irradiated ascospores contained a new mutation affecting the synthesis of an essential nutrient, conidia from each culture were transferred to minimal medium (**FIGURE 1.12A**). The vast majority of cultures could grow on minimal medium; these were discarded because they lacked any new mutation of the desired type. The retained cultures were the small number unable to grow in minimal medium, because they contained a new mutation blocking the synthesis of some essential nutrient.

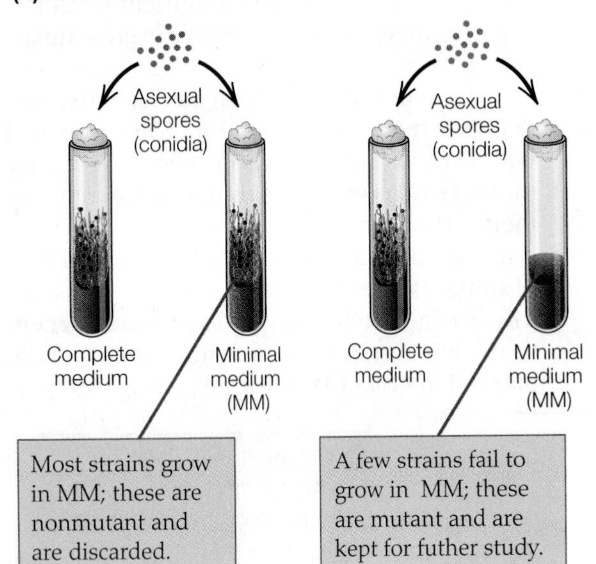
(A) Test for nutritional mutants
Asexual spores (conidia)
Complete medium
Minimal medium (MM)
Asexual spores (conidia)
Complete medium
Minimal medium (MM)
Most strains grow in MM; these are nonmutant and are discarded.
A few strains fail to grow in MM; these are mutant and are kept for futher study.

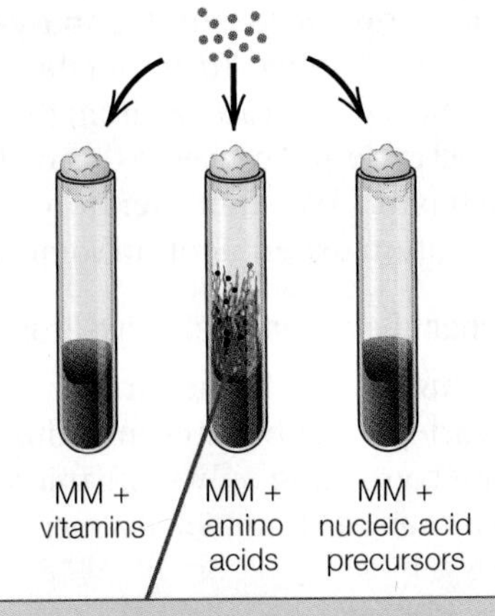
(B) Test for required nutrient
MM + vitamins
MM + amino acids
MM + nucleic acid precursors
Mutant strains are tested on MM supplemented with mixtures of nutrients. The mutant strain here requires amino acids to grow.

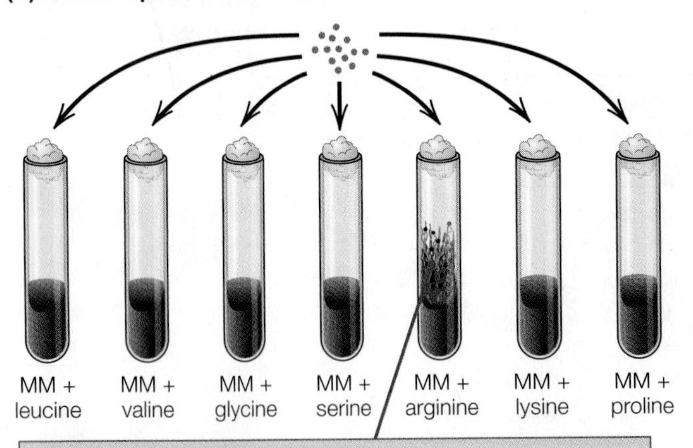
(C) Test for specific amino acid
MM + leucine
MM + valine
MM + glycine
MM + serine
MM + arginine
MM + lysine
MM + proline
Mutant strains requiring amino acids are tested on MM supplemented with each of the amino acids in turn. The mutant strain here requires arginine to grow.

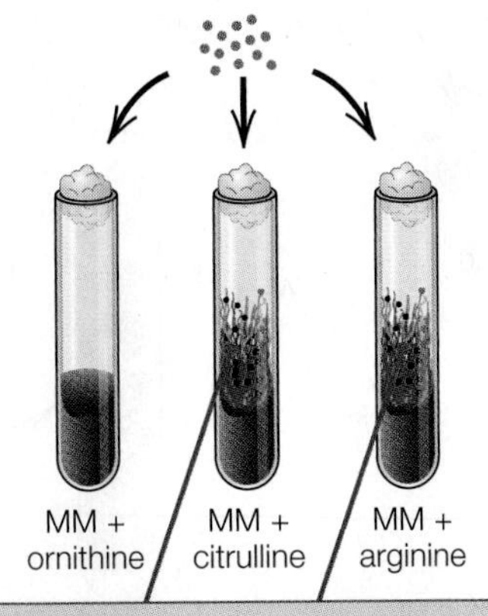
(D) Test of arginine precursors
MM + ornithine
MM + citrulline
MM + arginine
Mutant strains requiring arginine are tested on MM supplemented with possible precursors of arginine. This mutant strain grows on citrulline or arginine, but not ornithine.

**FIGURE 1.12** (A) Mutant spores can grow in complete medium but not in minimal medium. (B) Each new mutant is tested for growth in minimal medium supplemented with a mixture of nutrients. (C) Mutants that can grow on minimal medium supplemented with amino acid are tested with each amino acid individually. (D) Mutants unable to grow in the absence of arginine are tested with likely precursors of arginine.

Samples from a portion of each mutant culture were then transferred to a series of media to determine whether the mutation results in a requirement for a vitamin, an amino acid, or some other substance (**FIGURE 1.12B**). In the example illustrated, the mutant strain requires one (or possibly more than one) amino acid, because a mixture of all amino acids added to the minimal medium allows growth. Because the proportion of irradiated cultures with new mutations was very small, only a negligible number of cultures contained two or more new mutations that had occurred simultaneously.

For nutritional mutants requiring amino acids, further experiments testing each of the amino acid individually usually revealed that only one amino acid was required to be added to minimal medium in order to support growth. In **FIGURE 1.12C**, the mutant strain requires the amino acid arginine. Even in the 1940s some of the possible intermediates in amino acid biosynthesis had been identified. These were recognized by their chemical resemblance to the amino acid and by being present at low levels in the cells of organisms. In the case of arginine, two candidates were ornithine and citrulline. All mutants requiring arginine therefore were tested in medium supplemented with either ornithine alone or citrulline alone (**FIGURE 1.12D**). One class of arginine-requiring mutants, designated Class I, was able to grow in minimal medium supplemented with either ornithine, citrulline, or arginine. Other mutants, designated Class II, were able to grow in minimal medium supplemented with either citrulline or arginine, but not ornithine. A third class, Class III, was able to grow only in minimal medium supplemented with arginine.

The types of arginine-requiring mutants illustrate the logic of genetic analysis as applied to metabolic pathways. The logic is easiest to see in the context of the metabolic pathway, which is shown in **FIGURE 1.13**, where arginine is the end product of a linear metabolic pathway starting with some precursor metabolite, and ornithine and citrulline are intermediates in the pathway. The mutants support this structure of the pathway because:

- Mutants able to grow in the presence of either ornithine, citrulline, or arginine must have a metabolic block between the precursor metabolite and both of the intermediates.
- Mutants able to grow only in the presence of arginine must have a metabolic block in the pathway between arginine and the most "downstream" of the intermediates.
- Mutants able to grow in the presence of citrulline or arginine but not ornithine imply that ornithine is "upstream" from citrulline in the metabolic pathway.

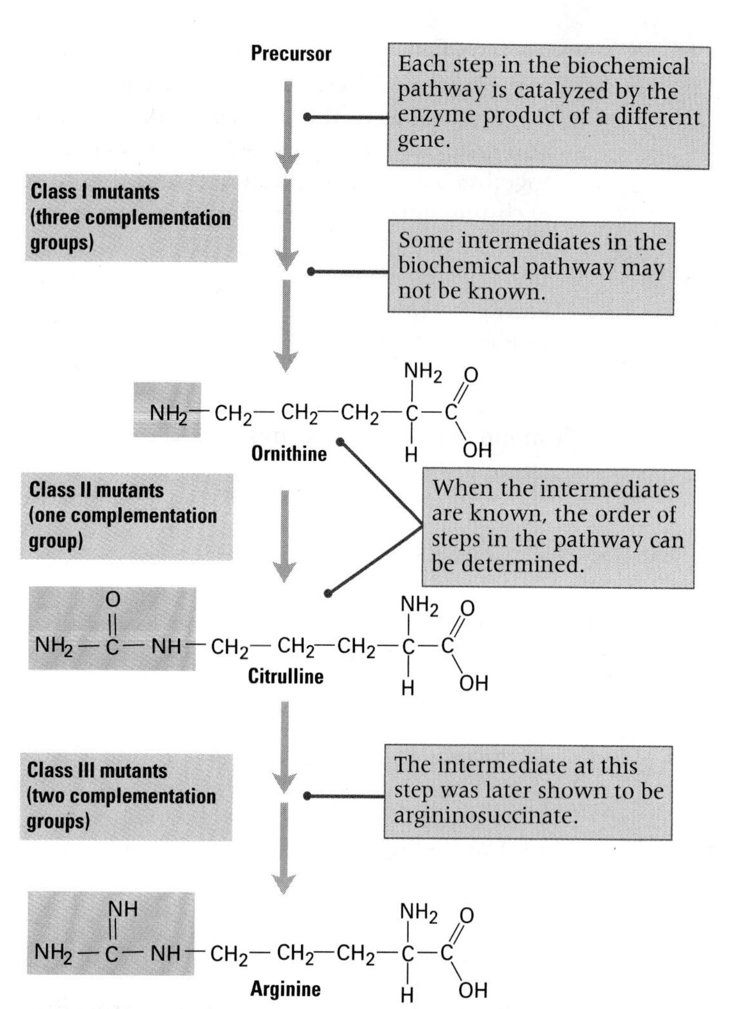


**FIGURE 1.13** Metabolic pathway for arginine biosynthesis inferred from genetic analysis of *Neurospora* mutants.

The structure of the pathway was further confirmed by the observations that Class III mutants accumulate citrulline and Class II mutants accumulate ornithine. Finally, it was shown by direct biochemical experiment that the inferred enzymes were actually present in nonmutant strains but defective in mutant strains.

## Complementation Test for Mutations in the Same Gene

Beadle and Tatum were fortunate to study metabolic pathways in a relatively simple organism in which one gene corresponds to one enzyme. In such a situation, genetic analysis of the mutants reveals a great deal more about the metabolic pathway than merely the order of the intermediates. By classifying each mutation according to the particular gene it is in, and grouping all the mutations in each gene together, each set of mutations, and therefore each individual gene, corresponds to one enzymatic step in the metabolic pathway. In Figure 1.13, for example, the mutants in Class I each correspond to one of three genes, which implies that there are three steps in the pathway between the precursor and ornithine. Similarly, the mutants in Class III each correspond to one of two genes, which implies that there are two steps in the pathway between the citrulline and arginine. However, all of the mutants in Class II have mutations in the same gene, which implies only one step in the pathway between ornithine and citrulline.

Mutations that have defects in the same gene are identified by means of a **complementation test,** in which two mutations are brought together into the same cell. In most multicellular organisms and even some sexual unicellular organisms, the usual way to do this is by means of a mating. When two parents, each carrying one of the two mutations, are crossed, fertilization beings the reproductive cells containing the two mutations together, and through ordinary cell division each cell in the resulting offspring carries one copy of each mutant gene. In *Neurospora* this procedure does not work because nuclear fusion is followed almost immediately by the formation of ascospores, each of which has only one set of chromosomes.

Complementation tests are nevertheless possible in *Neurospora* owing to the multinucleate nature of the filaments. Certain strains, including those studied by Beadle and Tatum, have the property that when the filaments from two mutant (or nonmutant) organisms come into physical contact, the filaments fuse and the new filament contains multiple nuclei from each of the participating partners. This sort of hybrid filament is called a **heterokaryon** and it contains mutant forms of both genes. The word roots of the term *heterokaryon* mean "different nuclei." (A list of the most common word roots used in genetics can be found at the end of this book.)

When a heterokaryon formed from two nutritional mutants is inoculated into minimal medium, it may grow or it may fail to grow. If it grows in minimal medium, the mutant genes are said to undergo **complementation**, and this result indicates that the mutations are in different genes. On the contrary, if the heterokaryon fails to grow in minimal medium, the result indicates **noncomplementation,** and the two mutations are inferred to be in the same gene.

The inferences from complementation or noncomplementation emerge from the logic illustrated in **FIGURE 1.14**. Here the multinucleate filament is shown, and the mutant nuclei are color coded according to which of two different genes (red or purple) is mutant. The red and purple squiggles represent the proteins encoded in the mutant nuclei, and a "burst" represents a defect in the protein at that position resulting from a mutation in the corresponding gene.

Part A depicts the situation when the mutant strains have mutations in different genes. In the heterokaryon, the red nuclei produce mutant forms of the red protein and normal forms of the purple protein, whereas the purple nuclei produce mutant forms of the purple protein and normal forms of the red protein. The result is that the red/purple heterokaryon has normal forms of both the red and purple protein. It also has mutant forms of both proteins, but these do not matter. What matters is that the normal forms allow the heterokaryon to grow on minimal medium because all needed nutrients can be synthesized. In other words, the normal purple gene in the red nucleus complements the defective purple gene in the purple nucleus, and the other way around. The logic of complementation is captured in the ancient nursery rhyme, "Jack Sprat could eat no fat / His wife could eat no lean / And so between the two of them / They licked the platter clean," because each partner makes up for the defect in the other.

Part B in Figure 1.14 shows a heterokaryon formed between mutants with defects in the same gene, in this case purple. Both of the purple nuclei encode a normal form of the red protein, but each purple nucleus encodes a defective purple protein. When the nuclei are together, two different mutant forms of the purple protein are produced, and so the biosynthetic pathway that requires the purple protein is still blocked and the heterokaryon is unable to grow in minimal medium. In other words, the mutants 2 and 3 in Figure 1.14 fail to complement, and so they are judged to have mutations in the same gene.

The following principle underlies the complementation test.

> **The Principle of Complementation:** A complementation test brings two mutant genes together in the same cell or organism. If this cell or organism is nonmutant, the muta-

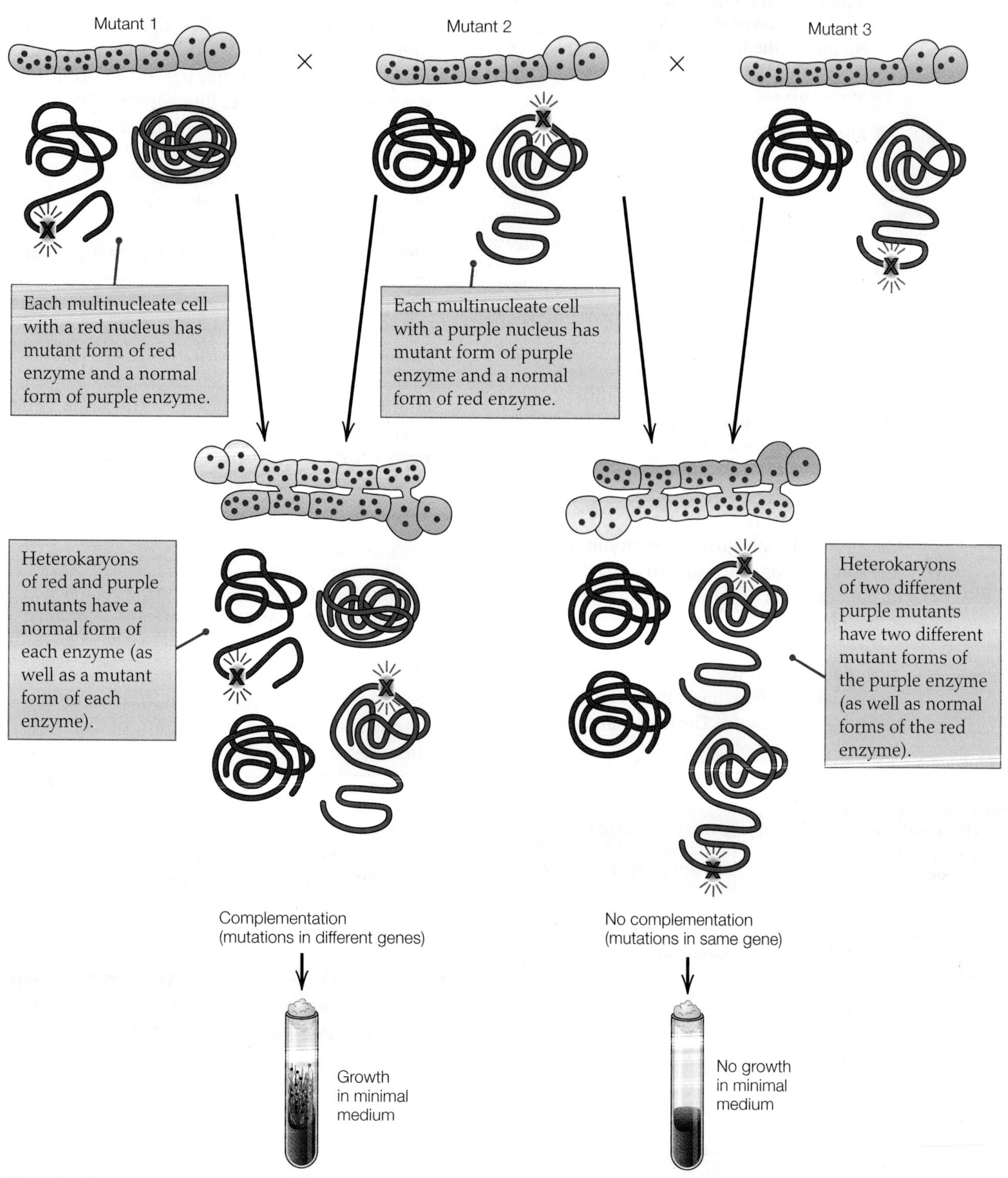

(A) Heterokaryon formation
(B) Heterokaryon formation
Mutant 1
Mutant 2
Mutant 3
Each multinucleate cell with a red nucleus has mutant form of red enzyme and a normal form of purple enzyme.
Each multinucleate cell with a purple nucleus has mutant form of purple enzyme and a normal form of red enzyme.
Heterokaryons of red and purple mutants have a normal form of each enzyme (as well as a mutant form of each enzyme).
Heterokaryons of two different purple mutants have two different mutant forms of the purple enzyme (as well as normal forms of the red enzyme).
Complementation (mutations in different genes)
No complementation (mutations in same gene)
Growth in minimal medium
No growth in minimal medium

**FIGURE 1.14** Molecular interpretation of a complementation test using heterokaryons to determine whether two mutant strains have mutations in different genes (A) or mutations in the same gene (B). In (A) each nucleus contributes a nonmutant form of one or the other polypeptide chain, and so the heterokaryon is able to grow in minimal medium. In (B) both nuclei contribute a mutant form of the same polypeptide chain; hence, no nonmutant form of that polypeptide can be synthesized and the heterokaryon is unable to grow in minimal medium.

tions are said to complement one another and it means that the parental strains have mutations in different genes. If the cell or organism is mutant, the mutations fail to complement one another, and it means that the parental mutations are in the same gene.

## ■ Analysis of Complementation Data

In the mutant screen for *Neurospora* mutants requiring arginine, Beadle and Tatum found that mutants in different classes (Class I, Class II, and Class III in Figure 1.13) always complemented one another. This result makes sense, because the genes in each class encode enzymes that act at different levels between the known intermediates. However, some of the mutants in Class I failed to complement others in Class I, and some in Class III failed to complement others in Class III. These results allow the number of genes in each class to be identified.

To illustrate this aspect of genetic analysis, we will consider six mutant strains in Class III. These strains were taken in pairs to form heterokaryons and their growth on minimal medium assessed. The data are shown in **FIGURE 1.15A**. The mutant genes in the six strains are denoted *x1*, *x2*, and so forth, and the data are presented in the form of a matrix in which + indicates growth in minimal medium (complementation) and − indicates lack of growth in minimal medium (lack of complementation). The diagonal entries are all −, which reflects the fact that two copies of the identical mutation cannot show complementation. The pattern of + and − signs in the matrix indicate that mutations *x1* and *x5* fail to complement one another; hence *x1* and *x5* are mutations in the same gene. Likewise, mutations *x2*, *x3*, *x4*, and *x6* fail to complement one another in all pairwise combinations; hence *x2*, *x3*, *x4*, and *x6* are all mutations in the same gene, but a different gene from that represented by *x1* and *x5*.

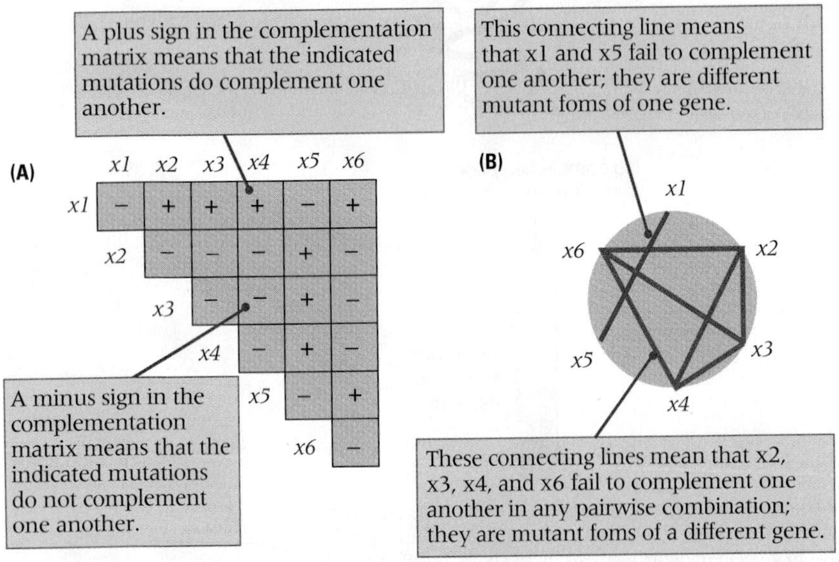

Data in a complementation matrix conveniently can be analyzed by arranging the mutant genes in the form of a circle as shown in **FIGURE 1.15B**. Then, for each possible pair of mutations, connect the pair by a straight line if the mutations *fail* to complement (− signs in part A). According to the principle of complementation, these lines connect mutations that are in the same gene. Each of the groups of noncomplementing mutations is called a **complementation group.** As we have seen, each complementation group defines a gene, so the complementation test actually provides the geneticist's operational definition:

> A *gene* is defined experimentally as a set of mutations that make up a single complementation group. Any pair of mutations within a complementation group fail to complement one another.

The mutations in Figure 1.15 therefore represent two genes, and mutation of any one of them results in the inability of the strain to convert citrulline to arginine. On the assumption that one gene encodes one enzyme, which is largely true for metabolic enzymes in *Neurospora*, the pathway from citrulline to arginine in Figure 1.13 must comprise two steps with an unknown intermediate in between. This intermediate was later found to be argininosuccinate. Likewise, Class I mutants defined three complementation groups and hence there are three enzymatic steps from the precursor to ornithine. These intermediates also were soon identified. Finally, Class II mutations all failed to complement one another, and the finding of only one complementation group means that there is but a single enzymatic step that converts ornithine to citrulline.

A plus sign in the complementation matrix means that the indicated mutations do complement one another.

This connecting line means that x1 and x5 fail to complement one another; they are different mutant foms of one gene.

(A)

|  | x1 | x2 | x3 | x4 | x5 | x6 |
|---|---|---|---|---|---|---|
| x1 | − | + | + | + | − | + |
| x2 |  | − | − | − | + | − |
| x3 |  |  | − | − | + | − |
| x4 |  |  |  | − | + | − |
| x5 |  |  |  |  | − | + |
| x6 |  |  |  |  |  | − |

A minus sign in the complementation matrix means that the indicated mutations do not complement one another.

(B)

x1
x6
x2
x5
x3
x4

These connecting lines mean that x2, x3, x4, and x6 fail to complement one another in any pairwise combination; they are mutant foms of a different gene.

**FIGURE 1.15** A method for interpreting the results of complementation tests. (A) Arrange the mutations in a circle. (B) Connect by a straight line any pair of mutations that fail to complement—that is, that yield a mutant heterokaryon. Any pair of mutations connected by a straight line are mutations in the same gene, and are more than likely mutations at different nucleotide sites in the gene. This example shows two complementation groups, each of which represents a single gene needed for arginine biosynthesis.

## ■ Other Applications of Genetic Analysis

The type of genetic analysis pioneered by Beadle and Tatum is immensely powerful for identifying the genetic control of complex biological processes. Their approach lays out a systematic path, a cookbook if you will, for gene discovery. First, decide what process you want to study. Next figure out what characteristics mutant organisms with a disruption in that process would display. Then do a mutant screen for mutants showing these characteristics. Carry out complementation tests to find out how many different genes you have identified. Finally, find out what the products of those genes are, what they do, how they interact with each other, and in what order they function.

Beadle and Tatum themselves analyzed many metabolic pathways for a wide variety of essential nutrients, but their experiments were especially important in deciphering pathways of amino acid biosynthesis. Their findings over just a few years are said to have "contributed more knowledge of amino acid biosynthetic pathways than had been accumulated during decades of traditional study." They were awarded the 1958 Nobel Prize in Physiology or Medicine for their research, and in the intervening years at least five more Nobel Prizes in Physiology or Medicine were awarded in which genetic analysis carried out along the lines of Beadle and Tatum played a significant role. Here is a list, with quotations from the official citations of the Nobel Foundation:

- 1958: George Beadle and Edward Tatum "for their discovery that genes act by regulating definite chemical events," shared with Joshua Lederberg "for his discoveries concerning genetic recombination and the organization of the genetic material of bacteria" (Chapter 9).
- 1965: François Jacob, André Lwoff, and Jacques Monod "for their discoveries concerning genetic control of enzyme and virus synthesis" (Chapter 11).
- 1995: Edward B. Lewis, Christiane Nüsslein-Volhard, and Eric F. Wieschaus "for their discoveries concerning the genetic control of early embryonic development" (Chapter 13).
- 2001: Leland H. Hartwell, Tim Hunt, and Sir Paul Nurse "for their discoveries of key regulators of the cell cycle" (Chapter 15).
- 2002: Sydney Brenner, H. Robert Horvitz, and John E. Sulston "for their discoveries concerning genetic regulation of organ development and programmed cell death" (Chapter 13).
- 2007: Mario R. Capecchi, Martin J. Evans, and Oliver Smithies 'for their discoveries of principles for introducing specific gene modifications in mice by the use of embryopnic stem cells' (Chapter 12).

The Beadle and Tatum experiments established that a defective enzyme results from a mutant gene, but how? For all they knew, genes *were* enzymes. This would have been a logical hypothesis at the time. We now know that the relationship between genes and enzymes is somewhat indirect. With a few exceptions, each enzyme is *encoded* in a particular sequence of nucleotides present in a region of DNA. The DNA region that codes for the enzyme, as well as adjacent regions that regulate when and in which cells the enzyme is produced, make up the "gene" that encodes the enzyme. Next we turn to the issue of *how* genes code for enzymes and other proteins.

## 1.5 Gene Expression: The Central Dogma

Watson and Crick were correct in proposing that the genetic information in DNA is contained in the sequence of bases in a manner analogous to letters printed on a strip of paper. In a region of DNA that directs the synthesis of a protein, the genetic code for the protein is contained in only one strand, and it is decoded in a linear order. A typical protein is made up of one or more polypeptide chains; each **polypeptide chain** consists of a linear sequence of amino acids connected end to end. For example, the enzyme PAH consists of four identical polypeptide chains, each 452 amino acids in length. In the decoding of DNA, each successive "code word" in the DNA specifies the next amino acid to be added to the polypeptide chain as it is being made. The amount of DNA required to code for the polypeptide chain of PAH is therefore $452 \times 3 = 1356$ nucleotide pairs. The entire gene is very much longer—about 90,000 nucleotide pairs. Only 1.5 percent of the gene is devoted to coding for the amino acids. The noncoding part includes some sequences that control the activity of the gene, but it is not known how much of the gene is involved in regulation.

There are 20 different amino acids. Only four bases code for these 20 amino acids, with each "word" in the genetic code consisting of three adjacent bases. For example, the base sequence ATG specifies the amino acid methionine (Met), TCC specifies serine (Ser), ACT specifies threonine (Thr), and GCG specifies alanine (Ala). There are 64 possible three-base combinations but only 20 amino acids because some combinations specify the same amino acid. For example, TCT, TCC, TCA, TCG, AGT, and AGC all code for serine (Ser), and CTT,

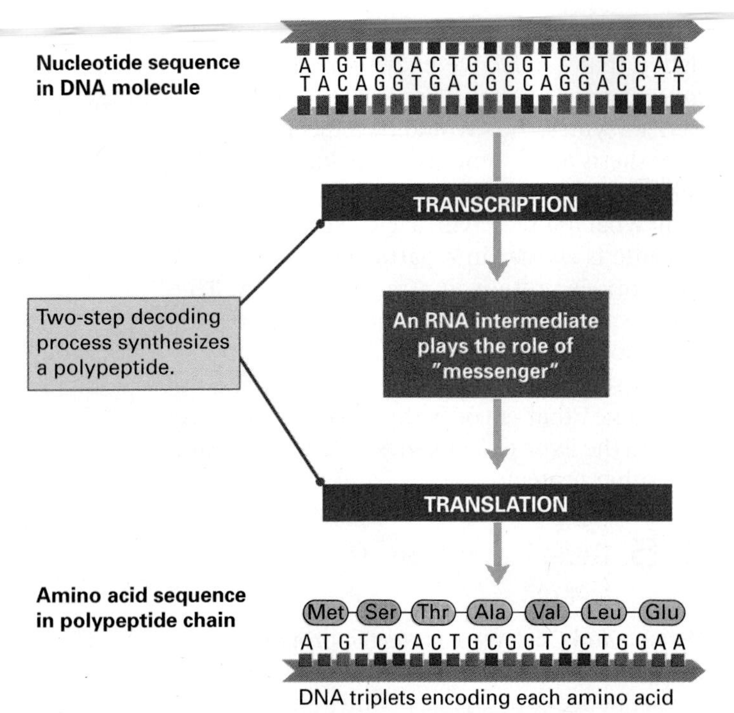


**FIGURE 1.16** DNA sequence coding for the first seven amino acids in a polypeptide chain. The DNA sequence specifies the amino acid sequence through a molecule of RNA that serves as an intermediary "messenger." Although the decoding process is indirect, the net result is that each amino acid in the polypeptide chain is specified by a group of three adjacent bases in the DNA. In this example, the polypeptide chain is that of phenylalanine hydroxylase (PAH).

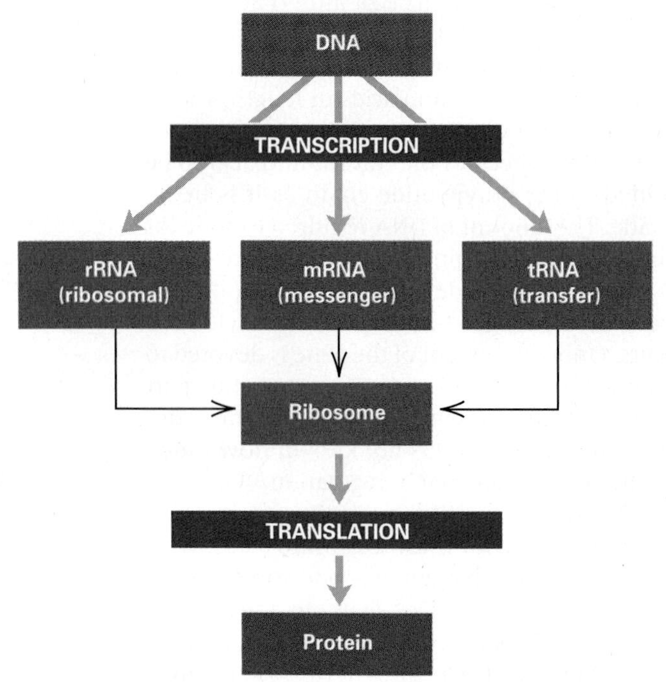


**FIGURE 1.17** The "central dogma" of molecular genetics: DNA codes for RNA, and RNA codes for protein. The DNA → RNA step is transcription, and the RNA → protein step is translation.

CTC, CTA, CTG, TTA, and TTG all code for leucine (Leu). An example of the relationship between the nucleotide sequence in a DNA duplex and the amino acid sequence of the corresponding protein is shown in **FIGURE 1.16**. This particular DNA duplex is the human sequence that codes for the first seven amino acids in the polypeptide chain of PAH.

The scheme outlined in Figure 1.16 indicates that DNA codes for protein not directly but indirectly through the processes of *transcription* and *translation*. The indirect route of information transfer,

$$\text{DNA} \rightarrow \text{RNA} \rightarrow \text{Protein}$$

is known as the **central dogma** of molecular genetics. The term *dogma* means "set of beliefs"; it dates from the time the idea was put forward first as a theory. Since then the "dogma" has been confirmed experimentally, but the term persists. The central dogma is shown in **FIGURE 1.17**. The main concept in the central dogma is that DNA does not code for protein directly but rather acts through an intermediary molecule of **ribonucleic acid (RNA).** The structure of RNA is similar to, but not identical with, that of DNA. There is a difference in the sugar (RNA contains the sugar **ribose** instead of deoxyribose), RNA is usually single-stranded (not a duplex), and RNA contains the base **uracil (U)** instead of thymine (T), which is present in DNA. Actually, three types of RNA take part in the synthesis of proteins:

- A molecule of **messenger RNA (mRNA),** which carries the genetic information from DNA and is used as a template for polypeptide synthesis. In most mRNA molecules, there is a high proportion of nucleotides that actually code for amino acids. For example, the mRNA for PAH is 2400 nucleotides in length and codes for a polypeptide of 452 amino acids; in this case, more than 50 percent of the length of the mRNA codes for amino acids.
- Four types of **ribosomal RNA (rRNA),** which are major constituents of the cellular particles called **ribosomes** on which polypeptide synthesis takes place.
- A set of about 45 **transfer RNA (tRNA)** molecules, each of which carries a particular amino acid as well as a three-base recognition region that base-pairs with a group of three adjacent bases in the mRNA. As each tRNA participates in translation, its amino acid becomes the terminal subunit added to the length of the growing

polypeptide chain. A tRNA that carries methionine is denoted $tRNA^{Met}$, one that carries serine is denoted $tRNA^{Ser}$, and so forth. (Because there are more than 20 different tRNAs but only 20 amino acids, some amino acids correspond to more than one tRNA.)

The central dogma is the fundamental principle of molecular genetics because it summarizes how the genetic information in DNA becomes expressed in the amino acid sequence in a polypeptide chain:

> The sequence of nucleotides in a gene specifies the sequence of nucleotides in a molecule of messenger RNA; in turn, the sequence of nucleotides in the messenger RNA specifies the sequence of amino acids in the polypeptide chain.

Given a process as conceptually simple as DNA coding for protein, what might account for the additional complexity of RNA intermediaries? One possible reason is that an RNA intermediate gives another level for control, for example, by degrading the mRNA for an unneeded protein. Another possible reason may be historical. RNA structure is unique in having both an informational content present in its sequence of bases and a complex, folded three-dimensional structure that endows some RNA molecules with catalytic activities. Many scientists believe that in the earliest forms of life, RNA served both for genetic information and catalysis. As evolu tion proceeded, the informational role was transferred to DNA and the catalytic role to protein. However, RNA became locked into its central location as a go-between in the processes of information transfer and protein synthesis. This hypothesis implies that the participation of RNA in protein synthesis is a relic of the earliest stages of evolution—a "molecular fossil." The hypothesis is supported by a variety of observations. For example, (1) DNA replication requires an RNA molecule in order to get started (Chapter 6), (2) an RNA molecule is essential in the synthesis of the tips of the chromosomes (Chapter 7), and (3) some RNA molecules act to catalyze key reactions in protein synthesis (Chapter 10).

## ■ Transcription

The manner in which genetic information is transferred from DNA to RNA is shown in **FIGURE 1.18**. The DNA opens up, and one of the strands is used as a template for the synthesis of a complementary strand of RNA. (How the template strand is chosen is discussed in Chapter 10.) The process of making an RNA strand from a DNA template is **transcription,** and the RNA molecule that is made is the **transcript.** The base sequence in the RNA is complementary (in the Watson–Crick pairing sense) to that in the DNA template, except that U (which pairs with A) is present in the RNA in place of T. The rules of base pairing between DNA and RNA are summarized in **FIGURE 1.19**. Each RNA strand has a polarity—a 5′ end and a 3′ end—and, as in the synthesis of DNA, nucleotides are added only to the 3′ end of a growing RNA strand. Hence, the 5′ end of the RNA transcript is synthesized first, and transcrip-

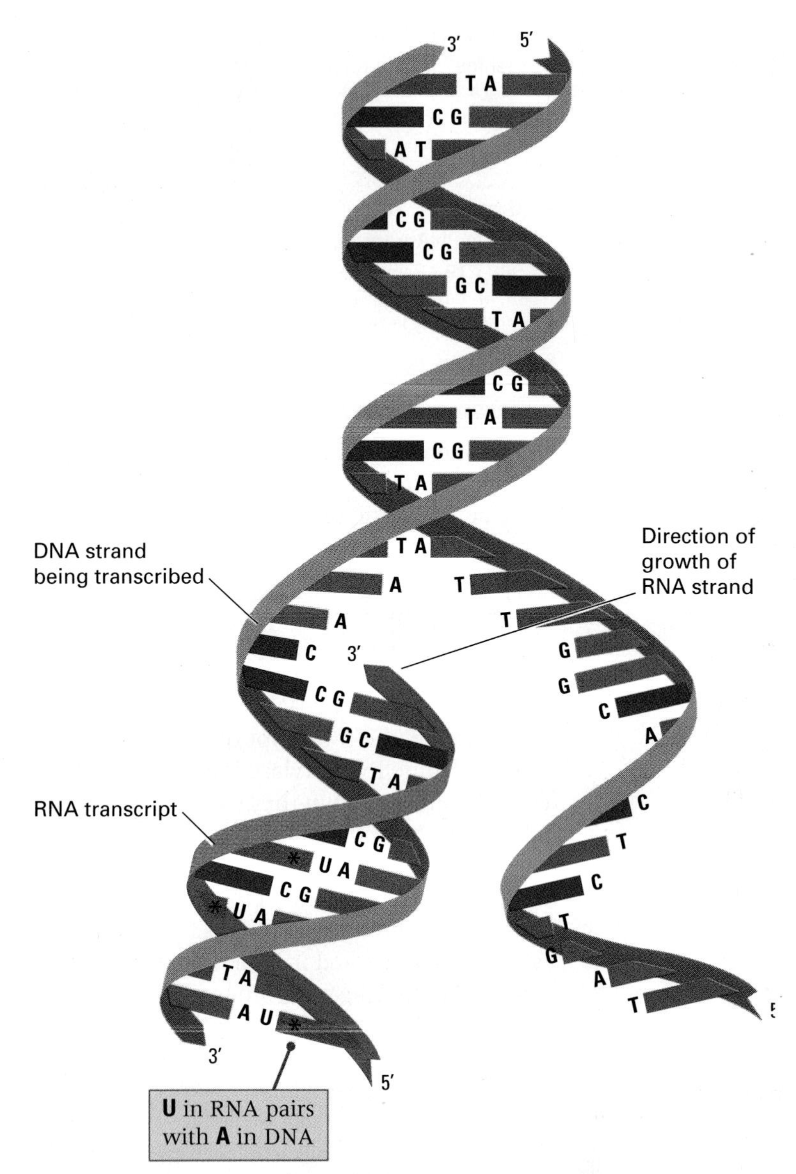


**FIGURE 1.18** Transcription is the production of an RNA strand that is complementary in base sequence to a DNA strand. In this example, the DNA strand at the bottom is being transcribed into a strand of RNA. Note that in an RNA molecule, the base U (uracil) plays the role of T (thymine) in that it pairs with A (adenine). Each A—U pair is marked.

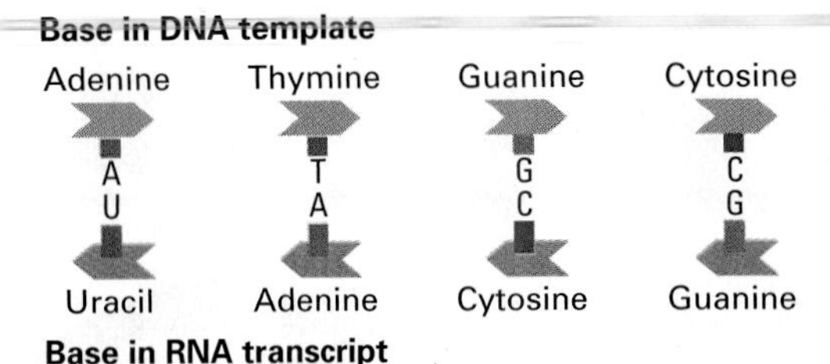
Base in DNA template
Adenine
Thymine
Guanine
Cytosine
A
U
T
A
G
C
C
G
Uracil
Adenine
Cytosine
Guanine
Base in RNA transcript

**FIGURE 1.19** Pairing between bases in DNA and in RNA. The DNA bases A, T, G, and C pair with the RNA bases U, A, C, and G, respectively.

tion proceeds along the template DNA strand in the 3′-to-5′ direction. Each gene includes nucleotide sequences that initiate and terminate transcription. The RNA transcript made from any gene begins at the initiation site in the template strand, which is located "upstream" from the amino acid–coding region, and ends at the termination site, which is located "downstream" from the amino acid–coding region. For any gene, the length of the RNA transcript is very much smaller than the length of the DNA in the chromosome. For example, the transcript of the *PAH* gene for phenylalanine hydroxylase is about 90,000 nucleotides in length, but the DNA in chromosome 12 is about 130,000,000 nucleotide pairs. In this case, the length of the *PAH* transcript is less than 0.1 percent of the length of the DNA in the chromosome. A different gene in chromosome 12 would be transcribed from a different region of the DNA molecule in chromosome 12, and perhaps from the opposite strand, but the transcribed region would again be small in comparison with the total length of the DNA in the chromosome.

## Translation

The synthesis of a polypeptide under the direction of an mRNA molecule is known as **translation.** Although the sequence of bases in the mRNA codes for the sequence of amino acids in a polypeptide, the molecules that actually do the "translating" are the tRNA molecules. The mRNA molecule is translated in nonoverlapping groups of three bases called **codons.** For each codon in the mRNA that specifies an amino acid, there is one tRNA molecule containing a complementary group of three adjacent bases that can pair with the codon. The correct amino acid is attached to one end of the tRNA, and when the tRNA comes into line, the amino acid to which it is attached becomes the most recent addition to the growing end of the polypeptide chain.

The role of tRNA in translation is illustrated in **FIGURE 1.20** and can be described as follows:

> The mRNA is read codon by codon. Each codon that specifies an amino acid matches with a complementary group of three adjacent bases in a single tRNA molecule. One end of the tRNA is attached to the correct amino acid, so the correct amino acid is brought into line.

The tRNA molecules used in translation do not line up along the mRNA simultaneously as shown in Figure 1.20. The process of translation takes place on a ribosome, which combines with

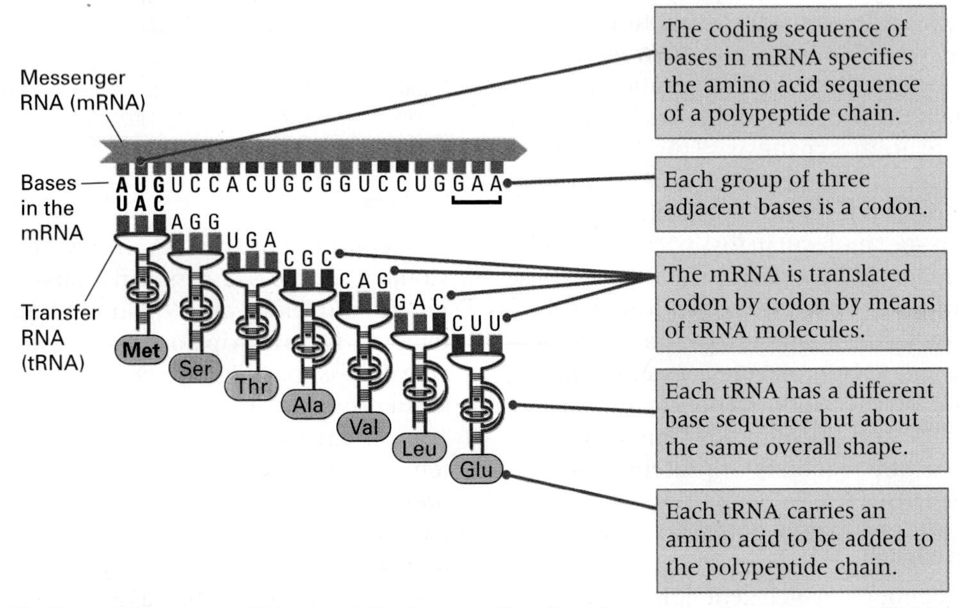
Messenger
RNA (mRNA)
Bases in the mRNA
AUGUCCACUGCGGUCCUGGAA
UAC
AGG
UGA
CGC
CAG
GAC
CUU
Transfer
RNA
(tRNA)
Met
Ser
Thr
Ala
Val
Leu
Glu
The coding sequence of bases in mRNA specifies the amino acid sequence of a polypeptide chain.
Each group of three adjacent bases is a codon.
The mRNA is translated codon by codon by means of tRNA molecules.
Each tRNA has a different base sequence but about the same overall shape.
Each tRNA carries an amino acid to be added to the polypeptide chain.

**FIGURE 1.20** The role of messenger RNA in translation is to carry the information contained in a sequence of DNA bases to a ribosome, where it is translated into a polypeptide chain. Translation is mediated by transfer RNA (tRNA) molecules, each of which can base-pair with a group of three adjacent bases in the mRNA. Each tRNA also carries an amino acid. As each tRNA, in turn, is brought to the ribosome, the growing polypeptide chain is elongated.

a single mRNA and moves along it from one end to the other in steps, three nucleotides at a time (codon by codon). As each new codon comes into place, the next tRNA binds with the ribosome. Then the growing end of the polypeptide chain becomes attached to the amino acid on the tRNA. In this way, each tRNA in turn serves temporarily to hold the polypeptide chain as it is being synthesized. As the polypeptide chain is transferred from each tRNA to the next in line, the tRNA that previously held the polypeptide is released from the ribosome. The polypeptide chain elongates one amino acid at each step until any one of three particular codons specifying "stop" is encountered. At this point, synthesis of the chain of amino acids is finished, and the polypeptide chain is released from the ribosome. (This brief description of translation glosses over many of the details that are presented in Chapter 10.)

### The Genetic Code

Figure 1.20 indicates that the mRNA codon AUG specifies methionine (Met) in the polypeptide chain, UCC specifies Ser (serine), ACU specifies Thr (threonine), and so on. The complete decoding table is called the **genetic code,** and it is shown in **TABLE 1.1**. For any codon, the column on the left corresponds to the first nucleotide in the codon (reading from the 5′ end), the row across the top corresponds to the second nucleotide, and the column on the right corresponds to the third nucleotide. The complete codon is given in the body of the table, along with the amino acid (or translational "stop") that the codon specifies. Each amino acid is designated by its full name and by a three-letter abbreviation as well as a single-letter abbreviation. Both types of abbreviations are used in molecular genetics. The code in Table 1.1 is the "standard" genetic code used in translation in the cells of nearly all organisms. In Chapter 10 we examine general features of the standard genetic code and the minor differences found in the genetic codes of certain organisms and cellular organelles. At this point, we are interested mainly in understanding how the genetic code is used to translate the codons in mRNA into the amino acids in a polypeptide chain.

In addition to the 61 codons that code only for amino acids, there are four codons that have specialized functions:

- The codon AUG, which specifies Met (methionine), is also the "start" codon for polypeptide synthesis. The positioning of a $tRNA^{Met}$ bound to AUG is one of the first

**Table 1.1 The standard genetic code**

| First nucleotide in codon (5′ end) | Second nucleotide in codon: U | C | A | G | Third nucleotide in codon (3′ end) |
|---|---|---|---|---|---|
| U | UUU Phe F *Phenylalanine* | UCU Ser S *Serine* | UAU Tyr Y *Tyrosine* | UGU Cys C *Cysteine* | U |
| U | UUC Phe F *Phenylalanine* | UCC Ser S *Serine* | UAC Tyr Y *Tyrosine* | UGC Cys C *Cysteine* | C |
| U | UUA Leu L *Leucine* | UCA Ser S *Serine* | UAA Termination | UGA Termination | A |
| U | UUG Leu L *Leucine* | UCG Ser S *Serine* | UAG Termination | UGG Trp W *Tryptophan* | G |
| C | CUU Leu L *Leucine* | CCU Pro P *Proline* | CAU His H *Histidine* | CGU Arg R *Arginine* | U |
| C | CUC Leu L *Leucine* | CCC Pro P *Proline* | CAC His H *Histidine* | CGC Arg R *Arginine* | C |
| C | CUA Leu L *Leucine* | CCA Pro P *Proline* | CAA Gln Q *Glutamine* | CGA Arg R *Arginine* | A |
| C | CUG Leu L *Leucine* | CCG Pro P *Proline* | CAG Gln Q *Glutamine* | CGG Arg R *Arginine* | G |
| A | AUU Ile I *Isoleucine* | ACU Thr T *Threonine* | AAU Asn N *Asparagine* | AGU Ser S *Serine* | U |
| A | AUC Ile I *Isoleucine* | ACC Thr T *Threonine* | AAC Asn N *Asparagine* | AGC Ser S *Serine* | C |
| A | AUA Ile I *Isoleucine* | ACA Thr T *Threonine* | AAA Lys K *Lysine* | AGA Arg R *Arginine* | A |
| A | AUG Met M *Methionine* | ACG Thr T *Threonine* | AAG Lys K *Lysine* | AGG Arg R *Arginine* | G |
| G | GUU Val V *Valine* | GCU Ala A *Alanine* | GAU Asp D *Aspartic acid* | GGU Gly G *Glycine* | U |
| G | GUC Val V *Valine* | GCC Ala A *Alanine* | GAC Asp D *Aspartic acid* | GGC Gly G *Glycine* | C |
| G | GUA Val V *Valine* | GCA Ala A *Alanine* | GAA Glu E *Glutamic acid* | GGA Gly G *Glycine* | A |
| G | GUG Val V *Valine* | GCG Ala A *Alanine* | GAG Glu E *Glutamic acid* | GGG Gly G *Glycine* | G |

Codon — Three-letter and single-letter abbreviations

steps in the initiation of polypeptide synthesis, so all polypeptide chains begin with Met. (Many polypeptides have the initial Met cleaved off after translation is complete.) In most organisms, the $tRNA^{Met}$ used for initiation of translation is the same $tRNA^{Met}$ used to specify methionine at internal positions in a polypeptide chain.
- The codons UAA, UAG, and UGA are each a "stop" that specifies the termination of translation and results in release of the completed polypeptide chain from the ribosome. These codons do not have tRNA molecules that recognize them but are instead recognized by protein factors that terminate translation.

How the genetic code table is used to infer the amino acid sequence of a polypeptide chain can be illustrated by using PAH again, in particular the DNA sequence coding for amino acids 1 through 7. The DNA sequence is

5′-ATGTCCACTGCGGTCCTGGAA-3′
3′-TACAGGTGACGCCAGGACCTT-5′

This region is transcribed into RNA in a left-to-right direction, and because RNA grows by the addition of successive nucleotides to the 3′ end

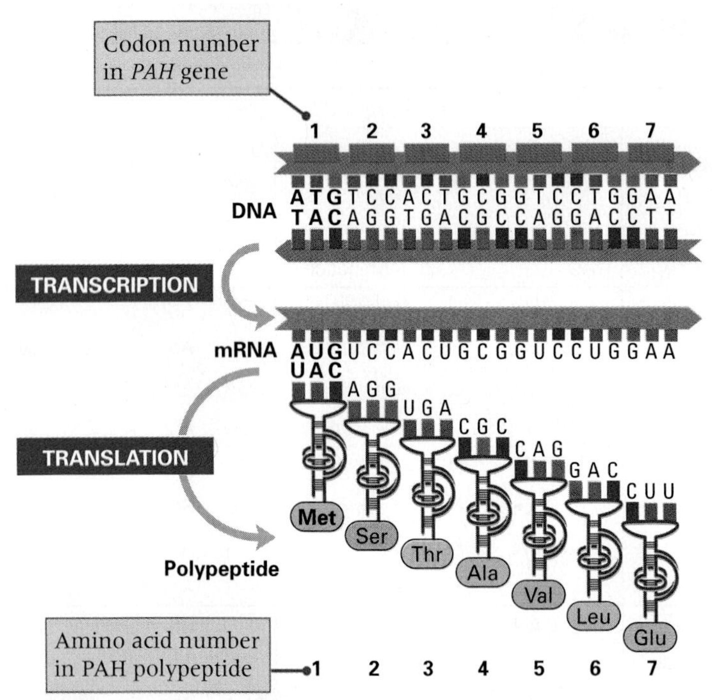


**FIGURE 1.21** The central dogma in action. The DNA that encodes PAH serves as a template for the production of a messenger RNA, and the mRNA serves to specify the sequence of amino acids in the PAH polypeptide chain through interactions with the ribosome and tRNA molecules. The total number of amino acids in the PAH polypeptide chain is 452, but only the first 7 are shown.

(Figure 1.18), it is the bottom strand that is transcribed. The nucleotide sequence of the RNA is that of the top strand of the DNA, except that U replaces T, so the mRNA for amino acids 1 through 7 is

5′-AUGUCCACUGCGGUCCUGGAA-3′

The codons are read from left to right according to the genetic code shown in Table 1.1. Codon AUG codes for Met (methionine), UCC codes for Ser (serine), and so on. Altogether, the amino acid sequence of this region of the polypeptide is

| 5′-AUG | UCC | ACU | GCG | GUC | CUG | GAA-3′ |
|---|---|---|---|---|---|---|
| Met | Ser | Thr | Ala | Val | Leu | Glu |

or, in terms of the single-letter abbreviations,

| 5′-AUG | UCC | ACU | GCG | GUC | CUG | GAA-3′ |
|---|---|---|---|---|---|---|
| M | S | T | A | V | L | E |

The full decoding operation for this region of the *PAH* gene is shown in **FIGURE 1.21**. In this figure, the initiation codon AUG is highlighted because some patients with PKU have a mutation in this particular codon. As might be expected from the fact that methionine is the initiation codon for polypeptide synthesis, cells in patients with this particular mutation fail to produce any of the PAH polypeptide. Mutation and its consequences are considered next.

## 1.6 Mutation

The term **mutation** refers to any heritable change in a gene (or, more generally, in the genetic material) or to the process by which such a change takes place. One type of mutation results in a change in the sequence of bases in DNA. The change may be simple, such as the substitution of one pair of bases in a duplex molecule for a different pair of bases. For example, a C—G pair in a duplex molecule may mutate to T—A, A—T, or G—C. The change in nucleotide sequence may also be more complex, such as the deletion or addition of base pairs. These and other types of mutations are considered in Chapter 7. Geneticists also use the term **mutant,** which refers to the result of a mutation. A mutation yields a mutant gene, which in turn produces a mutant mRNA, a mutant protein, and finally a mutant organism that exhibits the effects of the mutation—for example, an inborn error of metabolism.

DNA from patients from all over the world who have phenylketonuria has been studied to determine what types of mutations are responsible for the inborn error. There are a large va-

**THE WOMEN IN THE WEDDING** photograph are sisters. Both have two copies of the same mutant *PAH* gene. The bride is the younger of the two. She was diagnosed just three days after birth and put on the PKU diet soon after. Her older sister, the maid of honor, was diagnosed too late to begin the diet and is mentally retarded. The two-year old pictured in the photo at the right is the daughter of the married couple. They planned the pregnancy: dietary control was strict from conception to delivery to avoid the hazards of excess phenylalanine harming the fetus. Their daughter has passed all developmental milestones with distinction. [Courtesy of Charles R. Scriver, Montreal Children's Hospital Research Institute, McGill University.]

riety of mutant types. More than 400 different mutations have been described in the gene for PAH. In some cases part of the gene is missing, so the genetic information to make a complete PAH enzyme is absent. In other cases the genetic defect is more subtle, but the result is still either the failure to produce a PAH protein or the production of a PAH protein that is inactive. In the mutation shown in **FIGURE 1.22**, substitution of a G—C base pair for the normal A—T base pair at the very first position in the coding sequence changes the normal codon AUG (Met) used for the initiation of translation into the codon GUG, which normally specifies valine (Val) and cannot be used as a "start" codon. The result is that translation of the PAH mRNA cannot occur, and so no PAH polypeptide is made. This mutant is designated M1V because the codon for M (methionine) at amino acid position 1 in the PAH polypeptide has been changed to a codon for V (valine). Although the M1V mutant is quite rare worldwide, it is common in some localities, such as Québec Province in Canada.

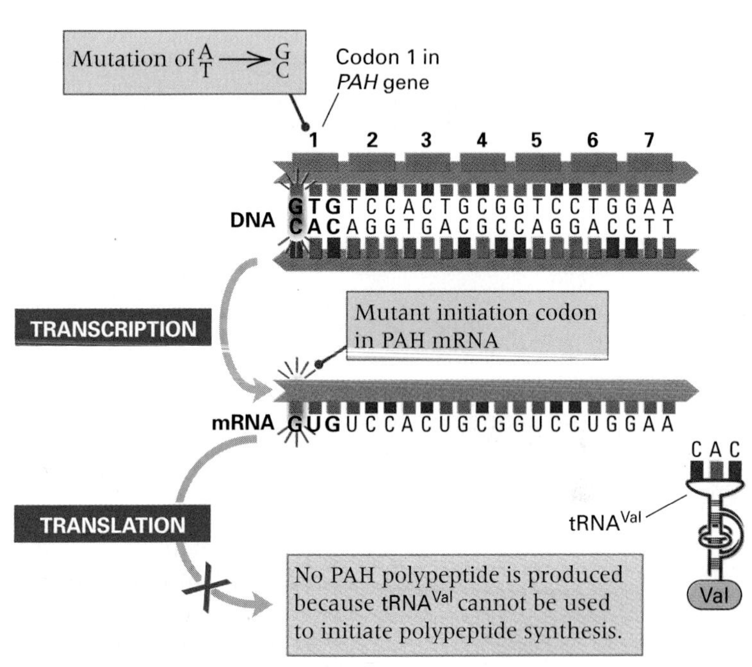


**FIGURE 1.22** The M1V mutant in the *PAH* gene. The methionine codon needed for initiation mutates into a codon for valine. Translation cannot be initiated, and no PAH polypeptide is produced.

One PAH mutant that is quite common is designated R408W, which means that codon 408 in the PAH polypeptide chain has been changed from one coding for arginine (R) to one coding for tryptophan (W). This mutation is one of the four most common among European Caucasians with PKU. The molecular basis of the mutant is shown in **FIGURE 1.23**. In this case, the first base pair in codon 408 is changed from a C—G base pair into a T—A base pair. The result is that the PAH mRNA has a mutant codon at position 408; specifically, it has UGG instead of CGG. Translation does occur in this mutant because everything else about the mRNA is normal, but the result is that the mutant PAH carries a tryptophan (Trp) instead of an arginine (Arg) at position 408 in the polypeptide chain. The consequence of the seemingly minor change of one amino acid is very drastic. Although the R408W polypeptide is complete, the enzyme has less than 3 percent of the activity of the normal enzyme.

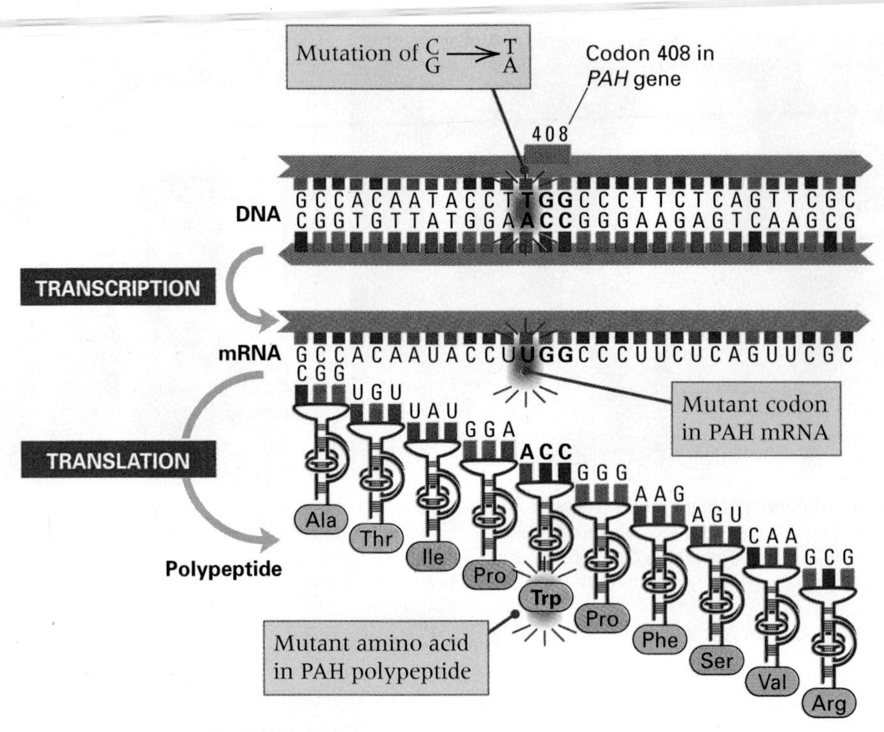


**FIGURE 1.23** The R408W mutant in the *PAH* gene. Codon 408 for arginine (R) is mutated into a codon for tryptophan (W). The result is that position 408 in the mutant PAH polypeptide is occupied by tryptophan rather than by arginine. The mutant protein has only a low level of PAH enzyme activity.

## 1.7 Genes and Environment

Inborn errors of metabolism illustrate the general principle that genes code for proteins and that mutant genes code for mutant proteins. In cases such as PKU, mutant proteins cause such a drastic change in metabolism that a severe genetic defect results. But biology is not necessarily destiny. Organisms are also affected by the environment. PKU serves as an example of this principle, because patients who adhere to a diet restricted in the amount of phenylalanine develop mental capacities within the normal range. What is true in this example is true in general. Most traits are determined by the interaction of genes and environment.

It is also true that most traits are affected by multiple genes. No one knows how many genes are involved in the development and maturation of the brain and nervous system, but the number must be in the thousands. This number is in addition to the genes that are required in all cells to carry out metabolism and other basic life functions. It is easy to lose sight of the multiplicity of genes when considering extreme examples, such as PKU, in which a single mutation can have such a drastic effect on mental development. The situation is the same as that with any complex machine. An airplane can function properly provided that thousands of parts work together in harmony, but only one defective part, if that part affects a vital system, can bring it down. Likewise, the development and functioning of every trait require a large number of genes working in harmony, but in some cases a single mutant gene can have catastrophic consequences.

In other words, the relationship between a gene and a trait is not necessarily a simple one. The biochemistry of organisms is a complex branching network in which different enzymes may share substrates, yield the same products, or be responsive to the same regulatory elements. The result is that most visible traits of organisms are the net result of many genes acting together and in combination with environmental factors. PKU affords examples of each of three principles governing these interactions:

1. ***One gene can affect more than one trait.*** Children with extreme forms of PKU often have blond hair and reduced body pigment. This is because the absence of PAH is a metabolic block that prevents conversion of phenylalanine into tyrosine, which is the precursor of the pigment melanin. The relationship between severe mental retardation and decreased pigmentation in PKU makes sense only if one knows the metabolic connections among phenylalanine, tyrosine,

**FIGURE 1.24** Among cats with white fur and blue eyes, about 40 percent are born deaf. We do not know why there is defective hearing nor why it is so often associated with coat and eye color. This form of deafness can be regarded as a pleiotropic effect of white fur and blue eyes.

and melanin. If these connections were not known, the traits would seem completely unrelated. PKU is not unusual in this regard. Many mutant genes affect multiple traits through their secondary or indirect effects. The various, sometimes seemingly unrelated, effects of a mutant gene are called **pleiotropic effects,** and the phenomenon itself is known as **pleiotropy. FIGURE 1.24** shows a cat with white fur and blue eyes, a pattern of pigmentation that is often (about 40 percent of the time) associated with deafness. Hence deafness can be regarded as a pleiotropic effect of white coat and blue eye color. The developmental basis of this pleiotropy is unknown.

2. ***Any trait can be affected by more than one gene.*** We discussed this principle earlier in connection with the large number of genes that are required for the normal development and functioning of the brain and nervous system. Among these are genes that affect the function of the blood–brain barrier, which consists of specialized glial cells wrapped around tight capillary walls in the brain, forming an impediment to the passage of most water-soluble molecules from the blood to the brain. The blood–brain barrier therefore affects the extent to which excess free phenylalanine in the blood can enter the brain itself. Because the effectiveness of the blood–brain barrier differs among individuals, PKU patients with very similar levels of blood phenylalanine can have dramatically different levels of cognitive development. This also explains in part why adherence to a controlled-phenylalanine diet is critically important in children but less so in adults; the blood–brain barrier is less well developed in children and is therefore less effective in blocking the excess phenylalanine.

   Multiple genes affect even simpler metabolic traits. Phenylalanine breakdown and excretion serve as a convenient example. The metabolic pathway is illustrated in Figure 1.10. Four enzymes in the pathway are indicated, but even more enzymes are involved at the stage labeled "further breakdown." Because differences in the activity of any of these enzymes can affect the rate at which phenylalanine can be broken down and excreted, all of the enzymes in the pathway are important in determining the amount of excess phenylalanine in the blood of patients with PKU.

3. ***Most traits are affected by environmental factors as well as by genes.*** Here we come back to the low-phenylalanine diet. Children with PKU are not doomed to severe mental deficiency. Their capabilities can be brought into the normal range by dietary treatment. PKU serves as an example of what motivates geneticists to try to discover the molecular basis of inherited disease. The hope is that knowing the metabolic basis of the disease will eventually make it possible to develop methods for clinical intervention through diet, medication, or other treatments that will ameliorate the severity of the disease.

## 1.8 The Molecular Unity of Life

The pathway for the breakdown and excretion of phenylalanine is by no means unique to human beings. One of the remarkable generalizations to have emerged from molecular genetics is that organisms that are very distinct—for example, plants and animals—share many features in their genetics and biochemistry. These similarities indicate a fundamental "unity of life":

> All creatures on Earth share many features of the genetic apparatus, including genetic information encoded in the sequence of bases in DNA, transcription into RNA, and transla-

Dr. Jeremy Burgess/Science Photo Library/Photo Researchers, Inc.

**MEMBRANE-BOUND MITOCHONDRIA,** similar in form and function, are present in both plant (top) and animal (right) cells. Despite the fact that these cells come from two very different organisms, their similarities speak to a fundamental unity of life. Taking up much of each cell is the nucleus, where the genetic material resides.

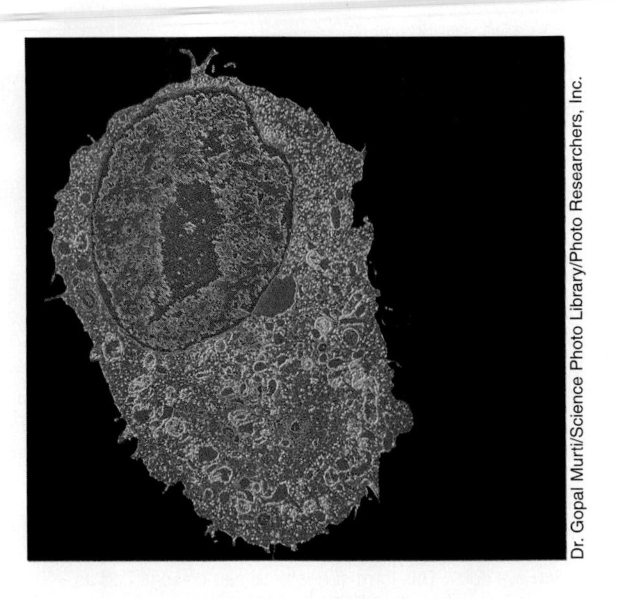

Dr. Gopal Murti/Science Photo Library/Photo Researchers, Inc.

tion into protein on ribosomes with the use of transfer RNAs. All creatures also share certain characteristics in their biochemistry, including many enzymes and other proteins that are similar in amino acid sequence, three-dimensional structure, and function.

## Prokaryotes and Eukaryotes

Why do organisms share a common set of similar genes and proteins? Because all creatures share a common origin. The process of **evolution** takes place when a population of organisms descended from a common ancestor gradually changes in genetic composition through time. From an evolutionary perspective, the unity of fundamental molecular processes is derived by inheritance from a distant common ancestor in which the molecular mechanisms were already in place.

Not only the unity of life, but also many other features of living organisms become comprehensible from an evolutionary perspective. For example, the interposition of an RNA intermediate in the basic flow of genetic information from DNA to RNA to protein makes sense if the earliest forms of life used RNA for both genetic information and enzyme catalysis. The importance of the evolutionary perspective in understanding aspects of biology that seem pointless or needlessly complex is summed up in a famous aphorism of the evolutionary biologist Theodosius Dobzhansky: "Nothing in biology makes sense except in the light of evolution."

One indication of the common ancestry among Earth's creatures is illustrated in **FIGURE 1.25**. The tree of relationships was inferred from similarities in nucleotide sequence in an RNA molecule found in the small subunit of the ribosome. Three major kingdoms of organisms are distinguished:

1. ***Bacteria*** This group includes most bacteria and cyanobacteria (formerly called blue-green algae). Cells of these organisms lack a membrane-bounded nucleus and mitochondria, are surrounded by a cell wall, and divide by binary fission.
2. ***Archaea*** This group was initially discovered among microorganisms that produce methane gas or that live in extreme environments, such as hot springs or high salt concentrations. They are widely distributed in more normal environments as well. Like those of bacteria, the cells of archaeans lack internal membranes. DNA sequence analysis indicates that the machinery for DNA replication and transcription in archaeans resembles that of eukaryans, whereas their metabolism strongly resembles that of bacteria. About half of the genes found in the kingdom Archaea are unique to this group.
3. ***Eukarya*** This group includes all organisms whose cells contain an elaborate network of internal membranes, a membrane-bounded nucleus, and mitochondria. Their DNA is organized into true chromosomes, and cell division takes place by means of mitosis (discussed in Chapter 4). The eukaryotes include plants and animals as well as fungi and many single-celled organisms, such as amoebae and ciliated protozoa.

The members of the groups Bacteria and Archaea are often grouped together into a larger

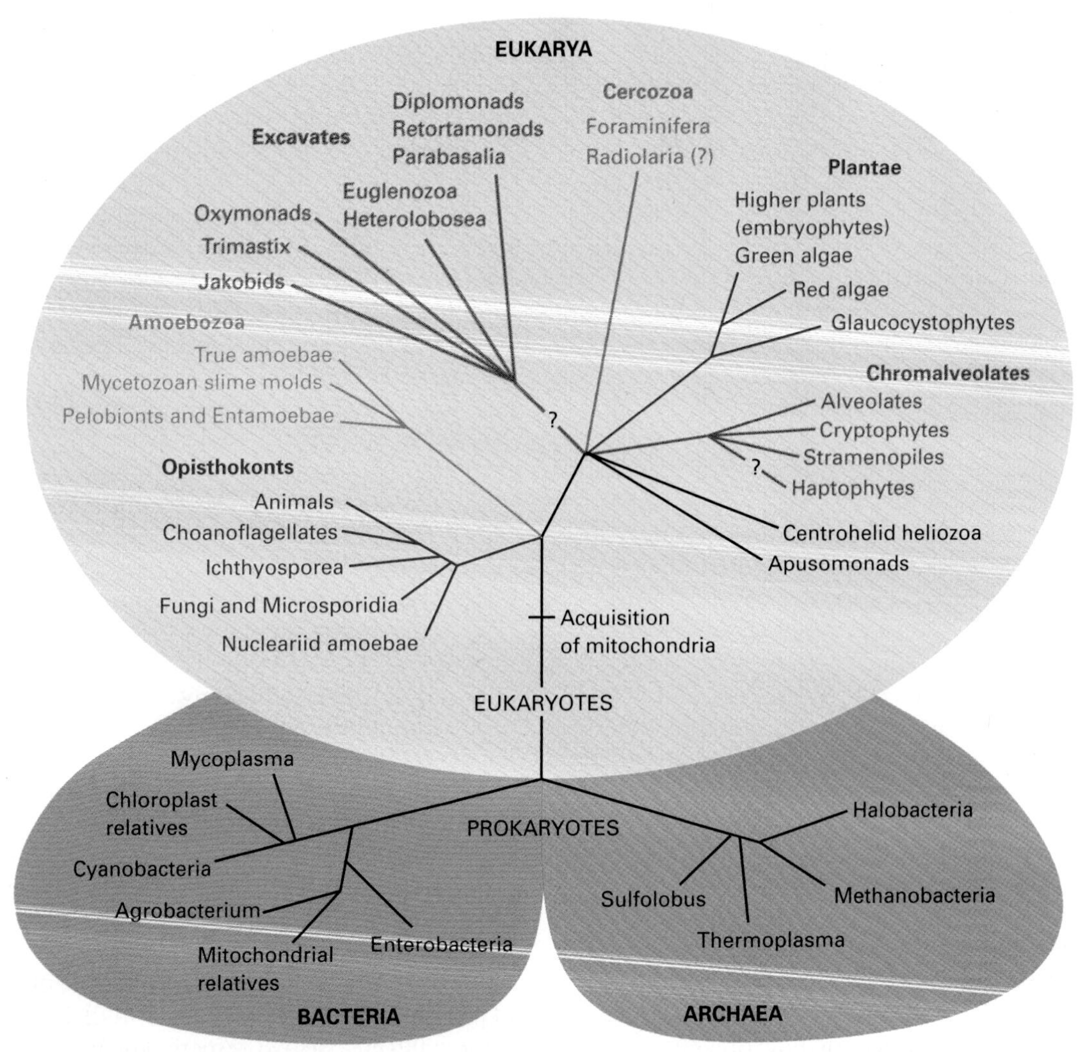


**FIGURE 1.25** Evolutionary relationships among the major life forms as inferred from similarities in DNA sequence. The three major groups—Bacteria, Archaea, and Eukarya—are evident. Although the great diversity among eukaryotic forms is illustrated, there is also much more diversity among bacterial and archaeal forms than indicated here. [Courtesy of Andrew J. Roger, Alastair B. Simpson, and Mitchell L. Sogin.]

assemblage called **prokaryotes,** which literally means "before [the evolution of] the nucleus." This terminology is convenient for designating prokaryotes as a group in contrast with **eukaryotes,** which literally means "good [well-formed] nucleus."

## ■ Evolutionary Relationships Among Eukaryotes

In Figure 1.25, the treelike relationship within each of the major groups (Eukarya, Bacteria, and Archaea) indicates that each major group evolved from a single common ancestor. The trees for bacteria and archaea are greatly simplified in order to emphasize the bacterial origins of mitochondria and chloroplasts, as well as the unusual environments in which many archaeans live. The tree for eukaryotes is more detailed, but still excludes some important groups whose positions in the tree are unclear.

Until quite recently, the eukaryotic tree was based primarily upon RNA sequences present in the small subunit of the ribosome. These data suggested a eukaryotic evolutionary tree having, at its top, a "crown" of more complex eukaryotes including animals, fungi, plants, and most algae. Along the "trunk" of the tree was a series of branches corresponding to less complex organisms with sometimes bizarre features or life cycles. At the base of the tree were a number of groups of unicellular organisms lacking mitochondria.

Figure 1.25 shows how dramatically this picture has changed. This tree for the eukaryotes is based not only on ribosomal RNA, but also the sequences of many protein-coding genes, mitochondrial DNA, and especially on the presence or absence of a unique gene fusion that joins dihydrofolate reductase with thymidylate synthase. These data imply that unicellular eukaryotes lacking mitochondria are not the descendants of the earliest eukaryotes that existed prior to the acquisition of mitochondria. These organisms are in fact derived from eukaryotes that originally possessed mitochondria, but the mitochondria have been lost. This scenario is also supported by the finding that the known genomic sequences of organisms lacking mitochondria show that genes derived from mitochondrial DNA are present in their nucleus, as is also found in the genomes of other eukaryotes.

Many of the groups of eukaryotic organisms in Figure 1.25 have odd and unfamiliar names—even to most professional biologists. Some of the defining features of these organisms are given in the dictionary at the end of the book. These organisms demonstrate the astonishing diversity even among eukaryotes. The known eukaryotic organisms, for the most part, fall into six major groups. Most human beings would not think of themselves as belonging to the *opisthokonts,* although this group is the longest recognized and includes animals, fungi, and numerous unicellular relatives. Most typical amoebae and slime molds are included in the *amoebozoa.* Multicellular green plants are familiar to everyone, but the larger grouping *plantae* includes not only land plants and green algae, but also red algae and many obscure forms. The group *chromalveolates* unites most other eukaryotic algae with dinoflagellates, ciliates, apicomplexans (such as the malaria parasite), and brown algae. The grouping referred to as *cercozoa* includes flagellates and atypical amoebae. The grouping designated *excavates* is more controversial (indicated by the question mark); it includes many flagellated forms without mitochondria that were previously thought to be near the base of the tree. The question marks in Figure 1.21 serve to emphasize that understanding the true evolutionary tree of eukaryotes is an ongoing process that periodically requires updating as new data become available.

## ■ Genomes and Proteomes

The totality of DNA in a single cell is called the *genome* of the organism. In sexual organisms, the genome is usually regarded as the DNA present in a reproductive cell. Modern methods for sequencing DNA are so rapid and efficient that the complete DNA sequence is known for more than 1000 genomes and large DNA molecules. These include the genomes of at least two species each of mammals (including the human genome and the mouse genome), fish, fruit flies, nematodes, yeast and higher plants, as well as hundreds of bacteria, archaea, cellular organelles, viruses and viroids, and large plasmids. (A *viroid* is a type of infectious circular RNA molecule associated with certain plant diseases, and a *plasmid* is a type of accessory DNA molecule found in cells.)

**TABLE 1.2** gives some examples of organisms whose genomes have been completely sequenced. Genome size is given in megabases (Mb), or millions of base pairs. The genome of the bacterium *Hemophilus influenzae,* like that of most bacteria, is very compact in that most of it codes for proteins. A high density of genes, relative to the amount of DNA, is also found in the budding yeast (5538 genes), the nematode worm (18,250 genes), the fruit fly (13,350 genes), and the diminutive flowering plant *Arabidopsis thaliana* (about 25,000 genes). The human genome, by contrast, contains large amounts of noncoding DNA. Comparison with the nematode is illuminating. Whereas the human genome is about 30 times larger than that of the worm, the number of genes is less than 2 times larger. This discrepancy reflects the fact that only about 1.5 percent of the human genome sequence codes for protein. (About 27 percent of the human genome is present in genes, but much of the DNA sequence present in genes is not protein-coding.)

The complete set of proteins encoded in the genome is known as the **proteome.** In less complex genomes, such as the yeast, worm, and fruit fly, the number of proteins in an organism's proteome is approximately the same as the number of genes. However, as we shall see in Chapter 11, some genes encode two or more proteins through a process called *alternative splicing* in which segments of the original RNA transcript are joined together in a variety of combinations to produce different messenger RNAs. Alternative splicing is especially prevalent in the human genome. At least one third of human genes, and possibly as many as two thirds, undergo alternative splicing, and among the genes that undergo alternative splicing, the number of different messenger RNAs per gene ranges from 2 to 7. Hence, with its seemingly limited repertoire of 25,000 genes, the human genome can create approximately 60,000–90,000 different mRNAs. The widespread use of alternative splicing to multiply the coding capacity of genes is one source of human genetic complexity.

Most eukaryotic organisms contain *families* of related proteins that can be grouped according to similarities in their amino acid sequence. These families exist because the evolution of a new gene function is typically preceded by the duplication of an already existing gene, followed by changes in nucleotide sequence in one of the

**Table 1.2 Comparisons of genomes and proteomes**

| Organism | Genome size, Mb[a] (approximate) | Number of genes (approximate) | Number of distinct proteins in proteome[b] (approximate) | Shared protein families |
|---|---|---|---|---|
| *Hemophilus influenzae* (causes bacterial meningitis) | 1.9 | 1700 | 1400 | |
| *Saccharomyces cerevisiae* (budding yeast) | 13 | 6000 | 4400 | |
| *Caenorhabditis elegans* (soil nematode) | 100 | 20,000 | 9500 | 3000 |
| *Drosophila melanogaster* (fruit fly) | 120[c] | 16,000 | 8000 | 5000 |
| *Mus musculus* (laboratory mouse) | 2500 | 25,000 | 10,000 | 7000[d] |
| *Homo sapiens* (human being) | 2900[e] | 25,000 | 10,000 | 9900[f] |

[a]Millions of base pairs.
[b]Excludes "families" of proteins with similar sequences (and hence related functions).
[c]Excludes 60 Mb of specialized DNA ("heterochromatin") that has a very low content of genes.
[d]Based on similarity with sequences in messenger RNA (mRNA).
[e]For convenience, this estimate is rounded to 3000 Mb elsewhere in this book.
[f]Based on the observation that only about 1% of mouse genes lack a similar gene in the human genome, and *vice versa*.

copies that gives rise to the new function. The new function is usually similar to the previous one (for example, a change in the substrate specificity of a transporter protein), so that the new protein retains enough similarity in amino acid sequence to the original that their common ancestry can be recognized.

The molecular unity of life can be seen in the similarity of proteins in the proteome among diverse types of organisms. Such comparisons are shown in the right-hand column in Table 1.2. In this tabulation, each family of related proteins is counted only once, in order to estimate the number of proteins in the proteome that are "distinct" in the sense that their sequences are dissimilar. In yeast, worms, and flies, the number of distinct proteins is approximately 4400, 9500, and 8000, respectively. The brackets in Table 1.2 indicate the number of distinct proteins that share sequence similarity between species. From these comparisons, it appears that most multicellular animals share 5,000–10,000 proteins that are similar in sequence and function. Approximately 3000 of these are shared with eukaryotes as distantly related as yeast, and approximately 1000 with prokaryotes as distantly related as bacteria. What these comparisons among proteomes imply is that biological systems are based on protein components numbering in the thousands. This is a challenging level of complexity to understand, but the challenge is much less intimidating than it appeared to be at an earlier time when human cells were thought to produce as many as a million different proteins.

## CHAPTER SUMMARY

- Inherited traits are affected by genes.
- Genes are composed of the chemical deoxyribonucleic acid (DNA).
- DNA replicates to form copies of itself that are identical (except for rare mutations).
- DNA contains a genetic code specifying what types of enzymes and other proteins are made in cells.
- DNA occasionally mutates, and the mutant forms specify altered proteins that have reduced activity or stability.
- A mutant enzyme is an "inborn error of metabolism" that blocks one step in a biochemical pathway for the metabolism of small molecules.
- Traits are affected by environment as well as by genes.
- All living organisms share certain features of cellular physiology and metabolism owing to descent from common ancestors.
- The molecular unity of life is seen in comparisons of genomes and proteomes.

## REVIEW THE BASICS

- What were the key experiments showing that DNA is the genetic material?
- How did understanding the molecular structure of DNA give clues to its ability to replicate, to code for proteins, and to undergo mutation?
- Why is *the* pairing of complementary bases a key feature of DNA replication?
- What is the process of transcription and in what ways does it differ from DNA replication?
- What three types of RNA participate in protein synthesis, and what is the role of each type of RNA?
- What is the "genetic code," and how is it relevant to the translation of a polypeptide chain from a molecule of messenger RNA?
- What is an inborn error of metabolism? How did this concept serve as a bridge between genetics and biochemistry?
- How does the "central dogma" explain Garrod's discovery that nonfunctional enzymes result from mutant genes?
- Explain why many mutant forms of phenylalanine hydroxylase have a simple amino acid replacement, yet the mutant polypeptide chains are absent or present in very small amounts.
- What is a pleiotropic effect of a gene mutation? Give an example.
- What are some of the major differences in cellular organization among Bacteria, Archaea, and Eukarya?

## GUIDE TO PROBLEM SOLVING

**Problem 1** In the human gene for the protein huntingtin (so named because it is associated with Huntington disease), the first 21 nucleotides in the amino-acid-coding region are

3′-TACCCACCGTTATAAGAGAGT-5′

What is the sequence of a partner strand?

**Answer** The base pairing between the strands is A with T and C with G, but it is equally important that the strands in a DNA duplex have opposite polarity. The complementary strand is therefore oriented with its 5′-end at the left. The base sequence of the partner strand is

5′-ATGGGTGGCAATATTCTCTCA-3′

**Problem 2** If the DNA duplex for huntingtin in Problem 1 were transcribed from left to right, deduce the base sequence of the RNA in this coding region.

**Answer** To deduce the RNA sequence, we must apply three concepts. First, the base pairing is such so that an A, T, C and G in DNA template strand is transcribed as U, A, G and C, respectively, in the RNA. Second, the DNA template strand and RNA transcript have the opposite polarity. Third (and critically for this problem), the RNA strand is always transcribed in the 5′-to-3′ direction. Because we are told that transcription takes place from left to right, we can deduce that the transcribed strand is that given in Problem 1. The RNA transcript therefore has the base sequence

5′-AUGGGUGGCAAUAUUCUCUCA-3′

**Problem 3** Given the RNA sequence coding for part of human huntingtin deduced in Problem 2, what is the amino acid sequence in this part of huntingtin?

**Answer** The polypeptide chain is translated in successive groups of three nucleotides (codons), starting at 5′ end of the coding sequence and moving in the 5′-to-3′ direction. The amino acid corresponding to each codon can be find in the genetic code table. The first seven amino acids in the polypeptide chain are

```
5′-AUG GGU GGC AAU AUU CUC UCA-3′
   Met Gly Gly Asn Ile Leu Ser
```

**Problem 4** Suppose that a mutation in the human huntingtin gene occurs in the DNA sequence shown in Problem 1. In this mutation, the red G is replaced with a T. What is the nucleotide sequence of this region of the DNA duplex (both strands), and that of the messenger RNA, and what is amino acid sequence of the mutant huntingtin?

**Answer** The DNA, RNA, and polypeptide chain have the sequences as follows, where the differences from the nonmutant gene are in red. The mutation results in stop codon and premature termination of huntingtin synthesis.

```
DNA (transcribed strand)
        3′-TAC CCA CCG TTA TAA GAG ATT-5′
DNA (nontranscribed strand)
        5′-ATG GGT GGC AAT ATT CTC TAA-3′
RNA coding region
        5′-AUG GGU GGC AAU AUU CUC UAA-3′
Polypeptide chain
           Met Gly Gly Asn Ile Leu STOP
```

## ANALYSIS AND APPLICATIONS

**1.1** Classify each of the following statements as true or false.

**(a)** Each gene is responsible for only one visible trait.

**(b)** Every trait is potentially affected by many genes.

**(c)** The sequence of nucleotides in a gene specifies the sequence of amino acids in a protein encoded by the gene.

**(d)** There is one-to-one correspondence between the set of codons in the genetic code and the set of amino acids encoded.

**1.2** From their examination of the structure of DNA, what were Watson and Crick able to infer about the probable mechanisms of DNA replication, coding capability, and mutation?

**1.3** What does it mean to say that each strand of a duplex DNA molecule has a polarity? What does it mean to say that the paired strands in a duplex molecule have opposite polarity?

**1.4** What important observation about S and R strains of *Streptococcus pneumoniae* prompted Avery, MacLeod, and McCarty to study this organism?

**1.5** In the transformation experiments of Avery, MacLeod, and McCarty, what was the strongest evidence that the substance responsible for the transformation was DNA rather than protein?

**1.6** A chemical called phenol (carbonic acid) destroys proteins but not nucleic acids, and strong alkali such as sodium hydroxide destroys both proteins and nucleic acids. In the transformation experiments with *Streptococcus pneumoniae,* what result would be expected if the S-strain extract had been treated with phenol? If it had been treated with strong alkali?

**1.7** What feature of the physical organization of bacteriophage T2 made it suitable for use in the Hershey-Chase experiments?

**1.8** Like DNA, molecules of RNA contain large amounts of phosphorus. When Hershey and Chase grew their T2 phage in bacterial cells in the presence of radioactive phosphorus, the RNA must also have incorporated the labeled phosphorus, and yet the experimental result was not compromised. Why not?

**1.9** The DNA extracted from a bacteriophage contains 16 percent A, 16 percent T, 34 percent G, and 34 percent C. What can you conclude about the structure of this DNA molecule?

**1.10** The DNA extracted from a bacteriophage consists of 20 percent A, 34 percent T, 35 percent G, and 11 percent C. What is unusual about this DNA? What can you conclude about its structure?

**1.11** A region along a DNA strand that is transcribed contains no A. What base will be missing in the corresponding region of the RNA?

**1.12** If one strand of a DNA duplex has the sequence 5′-ATCAG-3′, what is the sequence of the complementary strand? (Write the answer with the 5′ end at the left.)

**1.13** Suppose that a double-stranded DNA molecule is separated into its constituent strands, and the strands are purified using a high-speed centrifuge. In one of the strands the base composition is 25 percent A, 18 percent T, 20 percent G, and 37 percent C. What is the base composition of the other strand?

**1.14** Consider a region along one strand of a double-stranded DNA molecule consists of tandem repeats of the trinucleotide 5′-GTA-3′, so that the sequence in this strand is 5′-GTAGTAGTAGT...-3′ What is the sequence in the other strand? (Write the answer with the 5′ end at the left.)

**1.15** If the *template* strand of a DNA duplex has the sequence 5′-TCAG-3′, what is the sequence of the RNA transcript? (Write the answer with the 5′ end at the left.)

**1.16** If the *nontemplate* strand of a DNA duplex has the sequence 5′-ATCAG-3′, what is the sequence of the RNA transcript across this region? (Write the answer with the 5′ end at the left).

**1.17** A duplex DNA molecule contains a random sequence of the four nucleotides with equal proportions of each. What is the average spacing between consecutive occurrences of the sequence 5′-ATGC-3′? Between consecutive occurrences of the sequence 5′-TACGGC-3′?

**1.18** An RNA molecule folds back upon itself to form a "hairpin"structure held together by a region of base pairing. One segment of the molecule in the paired region has the base sequence 5′-UAUCGUAU-3′. What is the base sequence with which this segment is paired?

**1.19** A synthetic mRNA molecule consists of the repeating base sequence

5′-AAAAAAAA...-3′

When this molecule is translated *in vitro* using ribosomes, transfer RNAs, and other necessary constituents from *E. coli,* the result is a polypeptide chain consisting of the repeating amino acid Lys – Lys – Lys - ... If you assume that the genetic code is a triplet code, what does this result imply about the codon for lysine (Lys)?

**1.20** A synthetic mRNA molecule consisting of the repeating base sequence

5′-UUUUUUUUUUUU...-3′

is terminated by the addition, to the right-hand end, of a single nucleotide bearing A. When translated *in vitro,* the resulting polypeptide consists of a repeating sequence of phenylalanines terminated by a single leucine. What does this result imply about the codon for leucine?

**1.21** With *in vitro* translation of an RNA into a polypeptide chain, the translation can begin anywhere along the RNA molecule. A synthetic RNA molecule has the sequence

5′-CGCUUACCACAUGUCGCGAAC-3′

How many reading frames are possible if this molecule is translated *in vitro*? How many reading frames are possible if this molecule is translated *in vivo,* in which translation starts with the codon AUG?

**1.22** You have sequenced both strands of a double-stranded DNA molecule. To evaluate the potential of this molecule for coding amino acids, you conceptually transcribe it into RNA and then conceptually translate the RNA into a polypeptide chain. How many reading frames would you have to examine?

**1.23** A synthetic mRNA molecule consists of the repeating base sequence

5′-ACACACACACACAC...-3′

When this molecule is translated *in vitro,* the result is a polypeptide chain consisting of the alternating amino acids Thr – His – Thr – His – Thr – His... Why do the amino acids alternate? What does this result imply about the codons for threonine (Thr) and histidine (His)?

**1.24** A synthetic mRNA molecule consists of the repeating base sequence

5′-AUCAUCAUCAUCAUC...-3′

When this molecule is translated *in vitro,* the result is a mixture of three different polypeptide chains. One consists of repeating isoleucines (Ile – Ile – Ile – Ile ...), another of repeating serines (Ser – Ser – Ser - Ser...), and the third of repeating histidines (His – His – His – His...). What does this result imply about the manner in which an mRNA is translated?

**1.25** How is it possible for a gene with a mutation in the coding region to encode a polypeptide with the same amino acid sequence as the nonmutant gene?

**1.26** A polymer is made that has a random sequence consisting of 75 percent Gs and 25 percent Us. Among the amino acids in the polypeptide chains resulting from *in vitro* translation, what is the expected frequency of Trp? of Val? of Phe?

**1.27** Results of complementation tests of the six independent recessive mutations *a–f* are shown in the accompanying matrix, where + indicates complementation and – indicates lack of complementation. Classify the mutations into complementation groups, and write the names of the mutations in each complementation group in the blanks provided below. (Some of the blanks may remain empty.)

| | *b* | *c* | *d* | *e* | *f* |
|---|---|---|---|---|---|
| *a* | + | + | – | – | + |
| *b* | | + | + | + | + |
| *c* | | | + | + | + |
| *d* | | | | – | + |
| *e* | | | | | + |

Mutations in complementation group 1: ___ ___ ___ ___ ___
Mutations in complementation group 2: ___ ___ ___ ___ ___
Mutations in complementation group 3: ___ ___ ___ ___ ___
Mutations in complementation group 4: ___ ___ ___ ___ ___

**1.28** A mutant screen is carried out in *Neurospora crassa* for mutants that are unable to synthesize an amino acid, that we shall symbolize as G. A number of mutants are isolated and classified into four groups according to their ability to grow (+) or not grow (–) in minimal medium supplemented with possible precursors D, E, and F. The data are shown in the accompanying table.

| | D | E | F | G |
|---|---|---|---|---|
| Class 1 | + | + | + | + |
| Class 2 | – | – | – | + |
| Class 3 | – | – | + | + |
| Class 4 | + | – | + | + |

Complete the diagram shown below. Each arrow indicates one or more biochemical reactions. Within each circle write the class of mutants (1–4) whose products contribute to the reactions symbolized by the arrow, and in the squares write the name of the amino acid or precursor (D – G) at that position in the pathway.

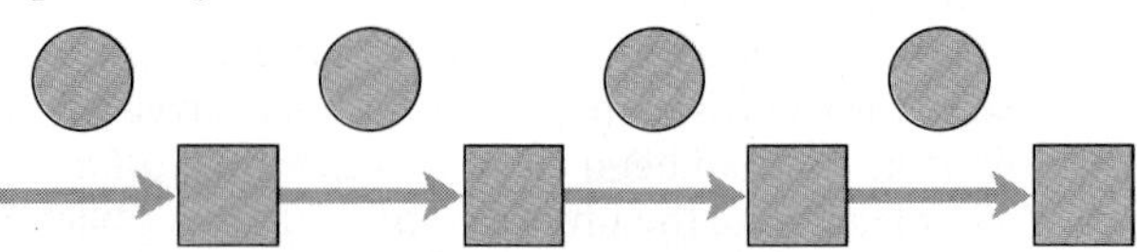

**1.29** The coding sequence in the messenger RNA for amino acids 1 through 10 of human phenylalanine hydroxylase is

5′-AUGUCCACUGCGGUCCUGGAAAACCCAGGC-3′

**(a)** What are the first 10 amino acids?
**(b)** What sequence would result from a mutant RNA in which the red A was changed to G?
**(c)** What sequence would result from a mutant RNA in which the red C was changed to G?
**(d)** What sequence would result from a mutant RNA in which the red U was changed to C?
**(e)** What sequence would result from a mutant RNA in which the red G was changed to U?

**1.30** A "frameshift" mutation is a mutation in which some number of base pairs, other than a multiple of three, is inserted into or deleted from a coding region of DNA. The result is that, at the point of the frameshift mutation, the reading frame of protein translation is shifted with respect to the nonmutant gene. To see the consequence of a frameshift mutation, consider that the coding sequence in the messenger RNA for the first 10 amino acids in human beta hemoglobin (part of the oxygen carrying protein in the blood) is

5′-AUG GUG CAC CUG ACU CCU GAG GAG AAG UCU...-3′

**(a)** What is the amino acid sequence in this part of the polypeptide chain?
**(b)** What would be the consequence of a frameshift mutation resulting in an RNA missing the red U?
**(c)** What would be the consequence of a frameshift mutation resulting in an RNA with a G inserted immediately in front of the red U?

**Challenge Problem 1** With regard to the wildtype and mutant RNA molecules described in Problem 1.30, deduce the base sequence in both strands of the corresponding double-stranded DNA for:

**(a)** The wildtype sequence
**(b)** The single-base deletion
**(c)** The single-base insertion

**Challenge Problem 2** You are carrying out Beadle-Tatum type experiments to analyze a metabolic pathway in *Neurospora*. You know that the precursor in the pathway is a molecule symbolized as P and that the product is a vitamin symbolized as Z. You are sure that the pathway from P to Z is linear and that the molecules W, X, and Y are intermediates. However, there may be other intermediates not yet identified. You obtain 10 independent mutations that cannot grow on minimal medium supplemented with P but can grow on minimal medium supplemented with Z. The 10 mutants fall into four classes that can grow (+) or cannot grow (–) on minimal medium supplemented with the nutrients W, X, or Y. The data are shown in the accompanying table.

| | P | W | X | Y | Z |
|---|---|---|---|---|---|
| Class I (mutant 5) | – | – | – | – | + |
| Class II (mutants 1, 2, 3, 4, 5, 6, 7) | – | + | + | + | + |
| Class III (mutant 2) | – | + | – | + | + |
| Class IV (mutants 8, 9, 10) | – | – | – | + | + |

Draw a linear metabolic pathway with P on the left and Z on the right in which each of the intermediates W, X, and Y shown in the order in which they occur in the metabolic pathway in the synthesis of Z from P.

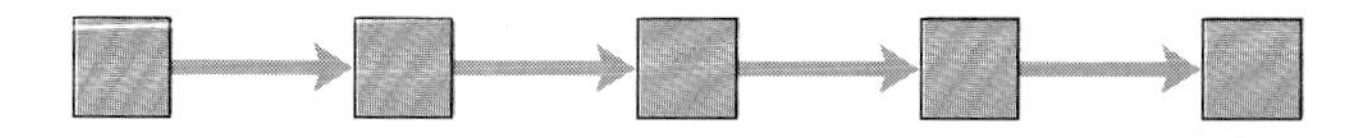

**Challenge Problem 3** The 10 mutants in the Challenge Problem 2 were tested for complementation in all pairwise combinations using heterokaryons. The results are shown in the matrix, in which + indicates the ability of the heterokaryon to grow in minimal medium and – indicates inability to grow in minimal medium.

| | 2 | 3 | 4 | 5 | 6 | 7 | 8 | 9 | 10 |
|---|---|---|---|---|---|---|---|---|---|
| Mutant 1 | + | + | – | + | + | + | + | + | + |
| 2 | | + | + | + | + | + | + | + | + |
| 3 | | | + | + | + | – | + | + | + |
| 4 | | | | + | + | + | + | + | + |
| 5 | | | | | + | + | + | + | + |
| 6 | | | | | | + | + | + | + |
| 7 | | | | | | | + | + | + |
| 8 | | | | | | | | + | + |
| 9 | | | | | | | | | – |

Assume that each complementation group defines a different gene, and assume further that each gene encodes an enzyme that catalyzes a single step in the metabolic pathway, that converts one molecule of substrate into one molecule of product.

**(a)** Redraw the metabolic pathway deduced from Challenge Problem 2. Use a right arrow to indicate each enzymatic step in the pathway, and label each arrow with the mutant number 1–10 that blocks the enzymatic step. In some cases, you will not be able to specify the order in which the enzymes occur in the pathway, so you may write them in any order you wish.

**(b)** In the metabolic pathway above that you have deduced from the data, how many UNKNOWN intermediates are there between the precursor P and the vitamin Z?

## GENETICS *on the web*

GeNETics on the Web will introduce you to some of the most important sites for finding genetics information on the Internet. To explore these sites, visit the Jones and Bartlett companion site to accompany *Genetics: Analysis of Genes and Genomes, Seventh Edition* at http://biology.jbpub.com/book/genetics.

There you will find a chapter-by-chapter list of highlighted keywords. When you select one of the keywords, you will be linked to a Web site containing information related to that keyword.

# CHAPTER 2

# DNA Structure and Genetic Variation

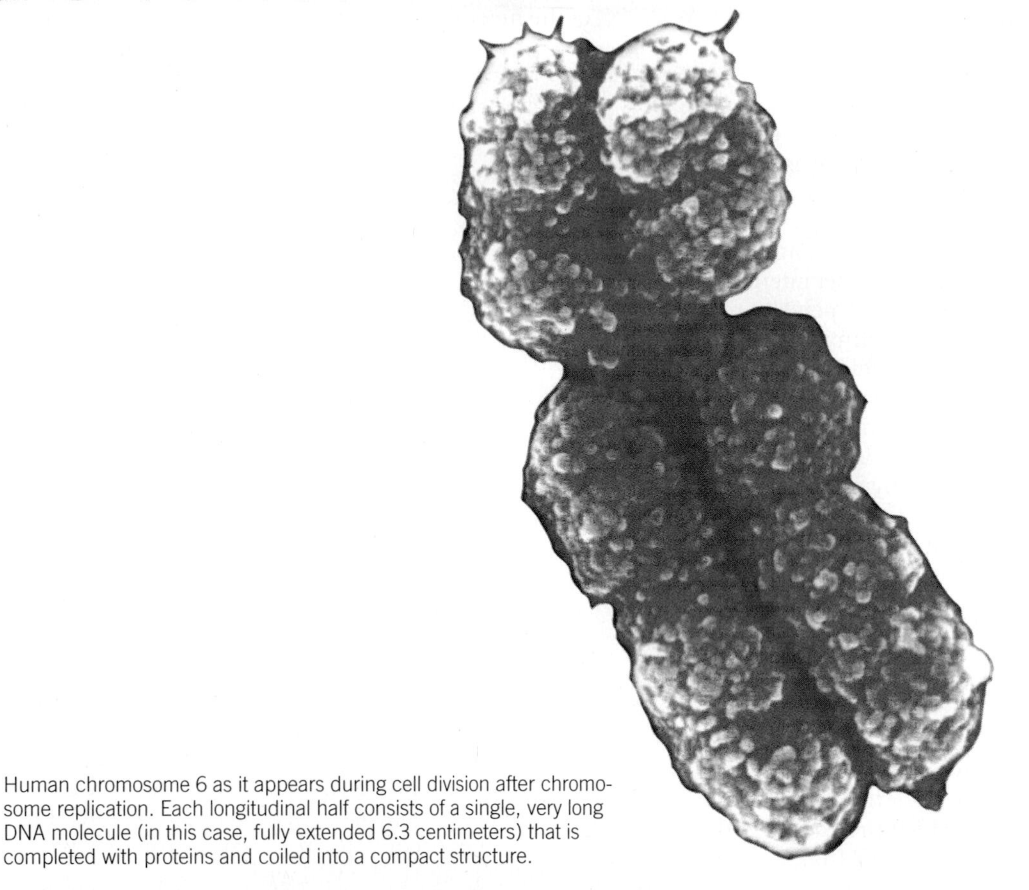

Human chromosome 6 as it appears during cell division after chromosome replication. Each longitudinal half consists of a single, very long DNA molecule (in this case, fully extended 6.3 centimeters) that is completed with proteins and coiled into a compact structure.

## CHAPTER OUTLINE


**2.1** Genetic Differences Among Individuals
- DNA Markers as Landmarks in Chromosomes

**2.2** The Molecular Structure of DNA
- Polynucleotide Chains
- Base Pairing and Base Stacking
- Antiparallel Strands
- DNA Structure as Related to Function

**2.3** The Separation and Identification of Genomic DNA Fragments
- Restriction Enzymes and Site-Specific DNA Cleavage
- Gel Electrophoresis
- Nucleic Acid Hybridization
- The Southern Blot

**2.4** Selective Replication of Genomic DNA Fragments
- Constraints on DNA Replication: Primers and 5′-to-3′ Strand Elongation
- The Polymerase Chain Reaction

**2.5** The Terminology of Genetic Analysis

**2.6** Types of DNA Markers Present in Genomic DNA
- Single-Nucleotide Polymorphisms (SNPs)
- Restriction Fragment Length Polymorphisms (RFLPs)
- Tandem Repeat Polymorphisms
- Copy-Number Polymorphisms (CNPs)

**2.7** Applications of DNA Markers
- Genetic Markers, Genetic Mapping, and "Disease Genes"
- Other Uses for DNA Markers

CONNECTION The Double Helix
James D. Watson and Francis H. C. Crick 1953
*A Structure for Deoxyribose Nucleic Acid*

CONNECTION Origin of the Human Genetic Linkage Map
David Botstein, Raymond L. White, Mark Skolnick, and Ronald W. Davis 1980
*Construction of a Genetic Linkage Map in Man Using Restriction Fragment Length Polymorphisms*

Since the mid-1970s, studies in genetics have undergone a revolution based on the use of increasingly sophisticated ways to isolate and identify specific fragments of DNA. The culmination of these techniques was large-scale genomic sequencing—the ability to determine the correct sequence of the base pairs that make up the DNA in an entire genome and to identify the sequences associated with genes. Because many of the laboratory organisms used in genetics experiments have relatively small genomes, these sequences were completed first. The techniques used to sequence these simpler genomes were then scaled up to sequence much larger genomes, including the human genome.

## 2.1 Genetic Differences Among Individuals

The human genome in a reproductive cell consists of approximately 3 billion base pairs organized into 23 distinct chromosomes (each chromosome contains a single molecule of duplex DNA). A typical chromosome can contain several hundred to several thousand genes, arranged in linear order along the DNA molecule present in the chromosome. The sequences that make up the protein-coding part of these genes actually account for only about 1.3 percent of the entire genome. The other 98.7 percent of the sequences do not code for proteins. Some noncoding sequences are genetic "chaff" that gets separated from the protein-coding "wheat" when genes are transcribed and the RNA transcript is processed into messenger RNA. Other noncoding sequences are relatively short sequences that are found in hundreds of thousands of copies scattered throughout the genome. Still other noncoding sequences are remnants of genes called *pseudogenes*. As might be expected, identifying the protein-coding genes from among the large background of noncoding DNA in the human genome is a challenge in itself.

Geneticists often speak of the nucleotide sequence of "the" human genome because corresponding DNA sequences from any two individuals are identical at about 99.9% of their nucleotide sites. These shared sequences are our evolutionary legacy; they contain the genetic information that makes us human beings. In reality, however, there are many different human genomes. Geneticists have great interest in the 0.1 percent of the human DNA sequence—3 million base pairs—that differs from one genome to the next, because these differences include the mutations that are responsible for genetic diseases such as phenylketonuria and other inborn errors of metabolism, as well as the mutations that increase the risk of more complex diseases such as heart disease, breast cancer, and diabetes.

Fortunately, only a small proportion of all differences in DNA sequence are associated with disease. Some of the others are associated with inherited differences in height, weight, hair color, eye color, facial features, and other traits. Most of the genetic differences between people are completely harmless. Many have no detectable effects on appearance or health. Such differences can be studied only through direct examination of the DNA itself. These harmless differences are nevertheless important, because they serve as genetic markers.

### DNA Markers as Landmarks in Chromosomes

In genetics, a **genetic marker** is any difference in DNA, no matter how it is detected, whose pattern of transmission from generation to generation can be tracked. Each individual who carries the marker also carries a length of chromosome on either side of it, so that the marker *marks* a particular region of the genome. Any difference in DNA sequence between two individuals can serve as a genetic marker. Although genetic markers are often harmless in themselves, they allow the positions of disease genes to be located along the chromosomes and their DNA to be isolated, identified, and studied.

Genetic markers that are detected by direct analysis of the DNA are often called **DNA markers.** DNA markers are important in genetics because they serve as landmarks in long DNA molecules, such as those found in chromosomes, which allow genetic differences among individuals to be tracked. They are like signposts along a highway. Using DNA markers as landmarks, the geneticist can identify the positions of normal genes, mutant genes, breaks in chromosomes, and other features important in genetic analysis (**FIGURE 2.1**).

The detection of DNA markers usually requires that the **genomic DNA** (the total DNA extracted from cells of an organism) be fragmented into pieces of manageable size (usually a few thousand nucleotide pairs) that can be manipulated in laboratory experiments. In the following sections, we examine some of the principal ways in which DNA is manipulated to reveal genetic differences among individuals,

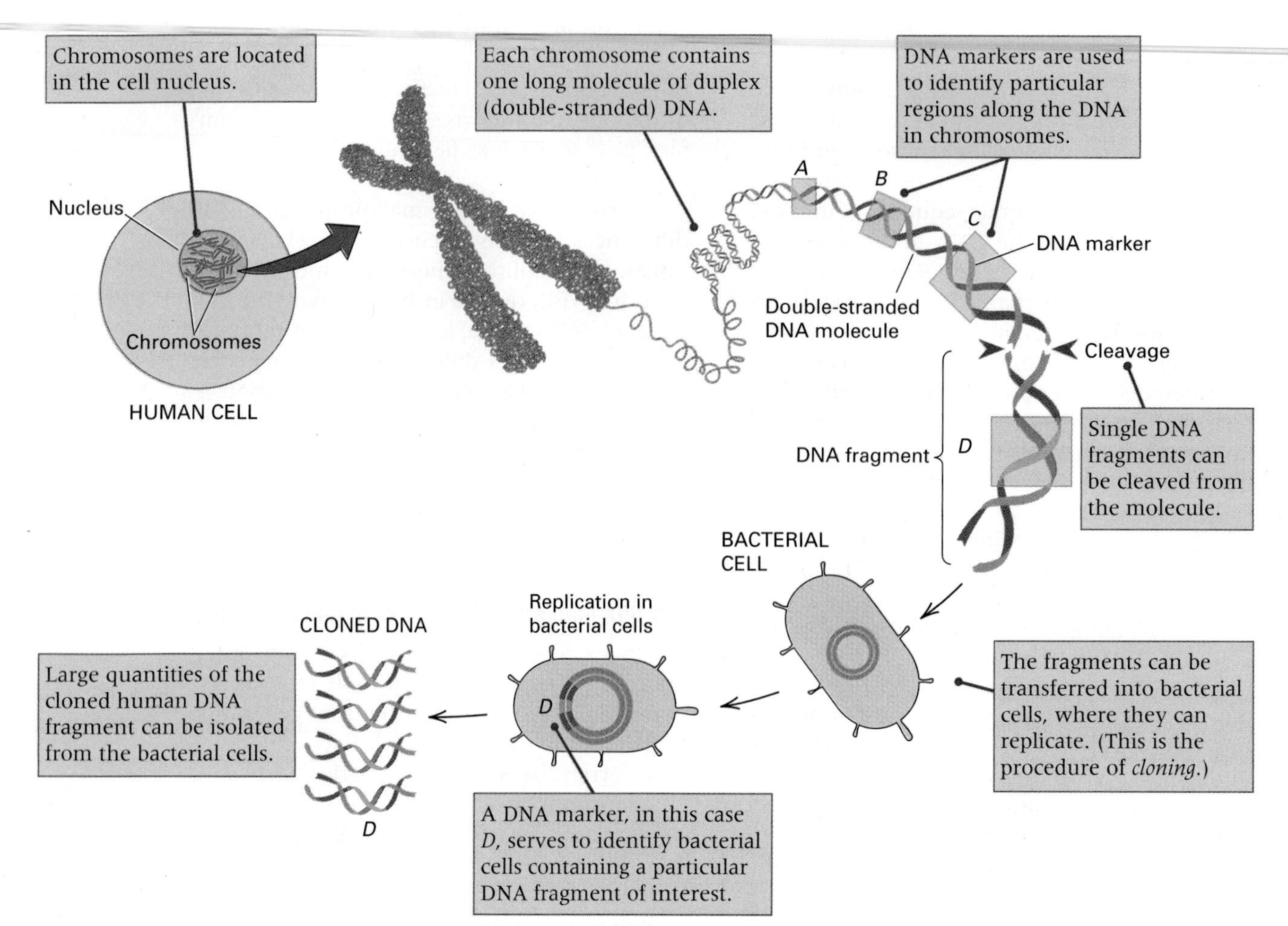


**FIGURE 2.1** DNA markers serve as landmarks that identify physical positions along a DNA molecule, such as DNA from a chromosome. As shown above, a DNA marker can also be used to identify bacterial cells into which a particular fragment of DNA has been introduced. The procedure of DNA cloning is not quite as simple as indicated here; it is discussed further in Chapter 12.

whether or not these differences find outward expression. Use of these methods broadens the scope of genetics, making it possible to carry out genetic analysis in *any* organism. This means that detailed genetic analysis is no longer restricted to human beings, domesticated animals, cultivated plants, and the relatively small number of model organisms favorable for genetic studies. Direct study of DNA eliminates the need for prior identification of genetic differences between individuals; it even eliminates the need for controlled crosses. The methods of molecular analysis discussed in this chapter have transformed genetics:

> The manipulation of DNA is the basic experimental operation in modern genetics.

The methods described in this chapter are the principal techniques used in virtually every modern genetics laboratory.

## 2.2 The Molecular Structure of DNA

Modern experimental methods for the manipulation and analysis of DNA grew out of a detailed understanding of its molecular structure and replication. Therefore, to understand these methods, one needs to know something about the molecular structure of DNA. We saw in Chapter 1 that DNA is a helix of two paired, complementary strands, each composed of an ordered string of *nucleotides*, each bearing one of the bases A (adenine), T (thymine), G (guanine), or cytosine (C). Watson–Crick base pairing between A and T and between G and C in the complementary strands holds the strands together. The complementary strands also hold the key to replication, because each strand can serve as a template for the synthesis of a new complementary strand. We will now take a closer look at DNA structure and at the key features of its replication.

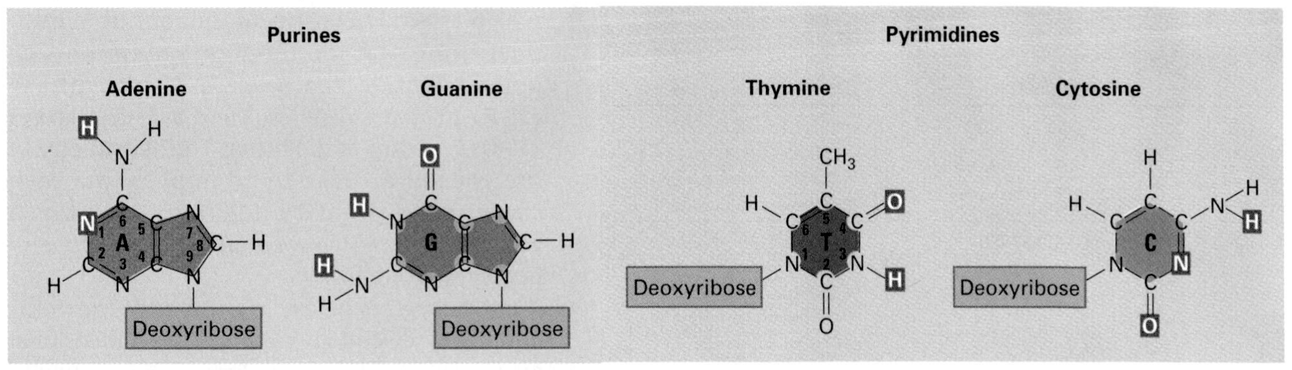


**FIGURE 2.2** Chemical structures of the four nitrogen-containing bases in DNA: adenine, thymine, guanine, and cytosine. The nitrogen atom linked to the deoxyribose sugar is indicated. The atoms shown in red participate in hydrogen bonding between the DNA base pairs.

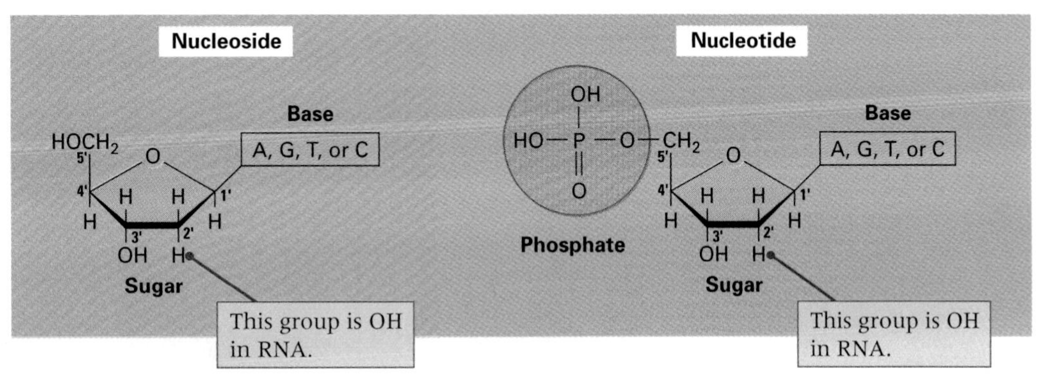


**FIGURE 2.3** A typical nucleotide, showing the three major components (phosphate, sugar, and base), the difference between DNA and RNA, and the distinction between a nucleoside (no phosphate group) and a nucleotide (with phosphate). Nucleotides are monophosphates (with one phosphate group). Nucleoside diphosphates contain two phosphate groups, and nucleoside triphosphates contain three.

## Polynucleotide Chains

In terms of biochemistry, a DNA strand is a polymer—a large molecule built from repeating units. The units in DNA are composed of 2′-deoxyribose (a five-carbon sugar), phosphoric acid, and the four nitrogen-containing bases denoted A, T, G, and C. The chemical structures of the bases are shown in **FIGURE 2.2**. Note that two of the bases have a double-ring structure; these are called **purines.** The other two bases have a single-ring structure; these are called **pyrimidines.**

- The purine bases are adenine (A) and guanine (G).
- The pyrimidine bases are thymine (T) and cytosine (C).

In DNA, each base is chemically linked to one molecule of the sugar deoxyribose, forming a compound called a **nucleoside.** When a phosphate group is also attached to the sugar, the nucleoside becomes a **nucleotide** (**FIGURE 2.3**). Thus, a nucleotide is a nucleoside plus a phosphate. In the conventional numbering of the carbon atoms in the sugar in Figure 2.3, the carbon atom to which the base is attached is the 1′ carbon. (The atoms in the sugar are given primed numbers to distinguish them from atoms in the bases.) The nomenclature of the nucleoside and nucleotide derivatives of the DNA bases is summarized in **TABLE 2.1**. Most of these terms are not needed in this book; they are included because they are likely to be encountered in further reading.

In nucleic acids, such as DNA and RNA, the nucleotides are joined to form a **polynucleotide chain,** in which the phosphate attached to the 5′ carbon of one sugar is linked to the hydroxyl group attached to the 3′ carbon of the next sugar in line (**FIGURE 2.4**). The chemical bonds by which the sugar components of adjacent nucleotides are linked through the phosphate groups are called **phosphodiester bonds.** The 5′–3′–5′–3′ orientation of these linkages continues throughout the chain,

## Table 2.1 DNA nomenclature

| Base | Nucleoside | Nucleotide |
|---|---|---|
| Adenine (A) | Deoxyadenosine | Deoxyadenosine-5′ monophosphate (dAMP) diphosphate (dADP) triphosphate (dATP) |
| Guanine (G) | Deoxyguanosine | Deoxyguanosine-5′ monophosphate (dGMP) diphosphate (dGDP) triphosphate (dGTP) |
| Thymine (T) | Deoxythymidine | Deoxythymidine-5′ monophosphate (dTMP) diphosphate (dTDP) triphosphate (dTTP) |
| Cytosine (C) | Deoxycytidine | Deoxycytidine-5′ monophosphate (dCMP) diphosphate (dCDP) triphosphate (dCTP) |

which typically consists of millions of nucleotides. Note that the terminal groups of each polynucleotide chain are a **5'-phosphate (5'-P) group** at one end and a **3'-hydroxyl (3'-OH) group** at the other. The asymmetry of the ends of a DNA strand implies that each strand has a **polarity** determined by which end bears the 5′ phosphate and which end bears the 3′ hydroxyl.

A few years before Watson and Crick proposed their essentially correct three-dimensional structure of DNA as a double helix, Erwin Chargaff developed a chemical technique to measure the amount of each base present in DNA. As we describe this technique, we denote the molar concentration of any base by the symbol for the base enclosed in square brackets; for

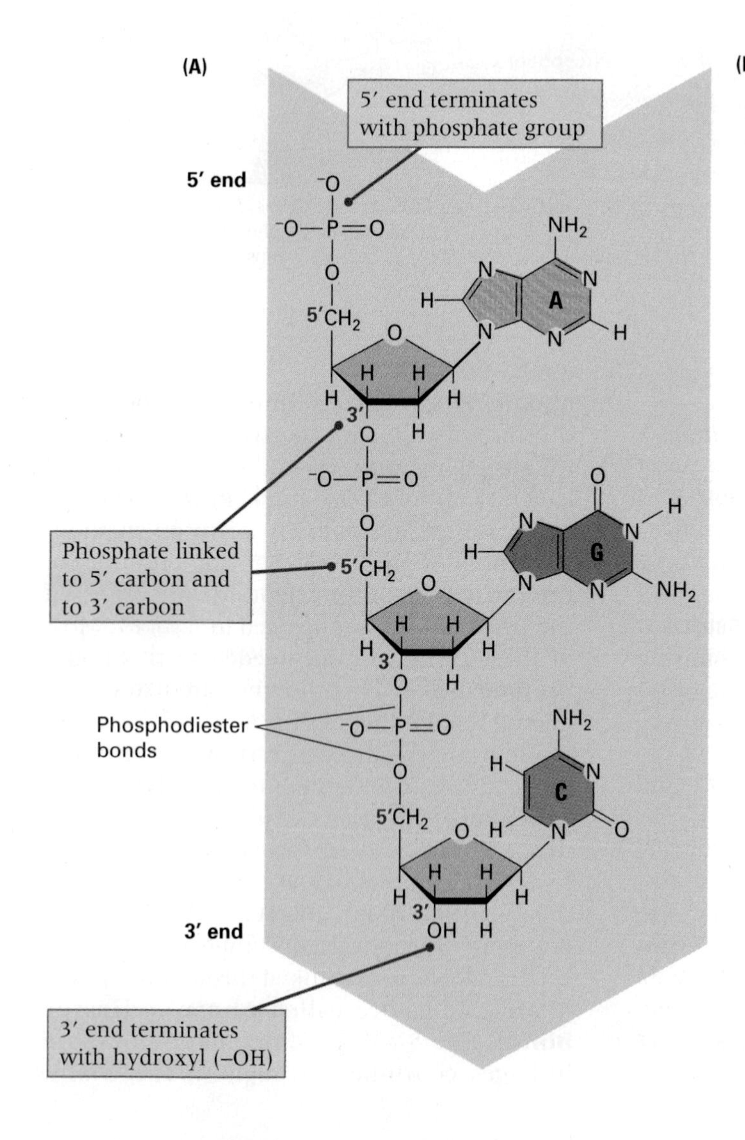


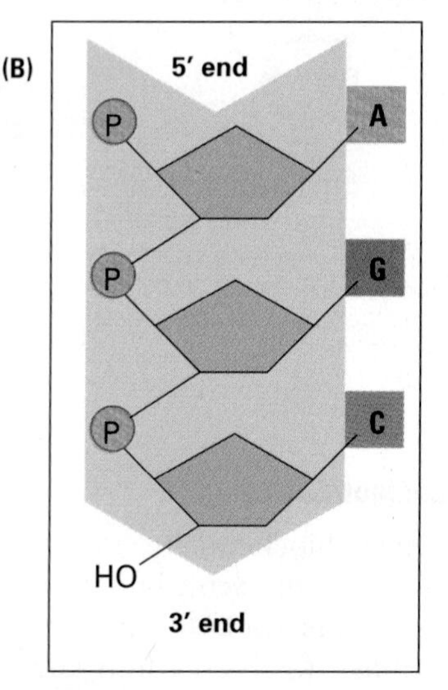


**FIGURE 2.4** Three nucleotides at the 5′ end of a single polynucleotide strand. (A) The chemical structure of the sugar–phosphate linkages, showing the 5′-to-3′ orientation of the strand (the red numbers are those assigned to the carbon atoms). (B) A common schematic way to depict a polynucleotide strand.

**Table 2.2** Base composition of DNA from different organisms

| Organism | Base (and percentage of total bases) Adenine | Thymine | Guanine | Cytosine | Base composition (percent G + C) |
|---|---|---|---|---|---|
| Bacteriophage T7 | 26.0 | 26.0 | 24.0 | 24.0 | 48.0 |
| Bacteria | | | | | |
| *Clostridium perfringens* | 36.9 | 36.3 | 14.0 | 12.8 | 26.8 |
| *Streptococcus pneumoniae* | 30.2 | 29.5 | 21.6 | 18.7 | 40.3 |
| *Escherichia coli* | 24.7 | 23.6 | 26.0 | 25.7 | 51.7 |
| *Sarcina lutea* | 13.4 | 12.4 | 37.1 | 37.1 | 74.2 |
| Fungi | | | | | |
| *Saccharomyces cerevisiae* | 31.7 | 32.6 | 18.3 | 17.4 | 35.7 |
| *Neurospora crassa* | 23.0 | 22.3 | 27.1 | 27.6 | 54.7 |
| Higher plants | | | | | |
| Wheat | 27.3 | 27.2 | 22.7 | 22.8* | 45.5 |
| Maize | 26.8 | 27.2 | 22.8 | 23.2* | 46.0 |
| Animals | | | | | |
| *Drosophila melanagaster* | 30.8 | 29.4 | 19.6 | 20.2 | 39.8 |
| Pig | 29.4 | 29.6 | 20.5 | 20.5 | 41.0 |
| Salmon | 29.7 | 29.1 | 20.8 | 20.4 | 41.2 |
| Human being | 29.8 | 31.8 | 20.2 | 18.2 | 38.4 |

*Includes one-fourth 5-methylcytosine, a modified form of cytosine found in most plants more complex than algae and in many animals

example, [A] denotes the molar concentration of adenine. Chargaff used his technique to measure the [A], [T], [G], and [C] content of the DNA from a variety of sources. He found that the **base composition** of the DNA, defined as the **percent G + C**, differs among species but is constant in all cells of an organism and within a species. Data on the base composition of DNA from a variety of organisms are given in **TABLE 2.2**.

Chargaff also observed certain regular relationships among the molar concentrations of the different bases. These relationships are now called **Chargaff's rules**:

- The amount of adenine equals that of thymine: [A] = [T].
- The amount of guanine equals that of cytosine: [G] = [C].
- The amount of the purine bases equals that of the pyrimidine bases:

  [A] + [G] = [T] + [C].

Although the chemical basis of these observations was not known at the time, one of the appealing features of the Watson–Crick structure of paired complementary strands was that it explained Chargaff's rules. Because A is always paired with T in double-stranded DNA, it must follow that [A] = [T]. Similarly, because G is paired with C, we know that [G] = [C]. The third rule follows by addition of the other two: [A] + [G] = [T] + [C]. In the next section, we examine the molecular basis of base pairing in more detail.

## Base Pairing and Base Stacking

In the three-dimensional structure of the DNA molecule proposed in 1953 by Watson and Crick, the molecule consists of two polynucleotide chains twisted around one another to form a double-stranded helix in which adenine and thymine, and guanine and cytosine, are paired in opposite strands (**FIGURE 2.5**). In the standard structure, which is called the **B form of DNA,** each chain makes one complete turn every 34 Å. The helix is right-handed, which means that as one looks down the barrel, each chain follows a clockwise path as it progresses. The bases are spaced at 3.4 Å, so there are ten bases per helical turn in each strand and ten base pairs per turn of the double helix.

The strands feature **base pairing,** in which each base is paired to a complementary base in the other strand by hydrogen bonds. (A **hydrogen bond** is a weak bond in which two participating atoms share a hydrogen atom between them.) The hydrogen bonds provide one type of force holding the strands together. In Watson–Crick base pairing, adenine (A) pairs with thymine (T), and guanine (G) pairs with cytosine (C). The hydrogen bonds that form in the

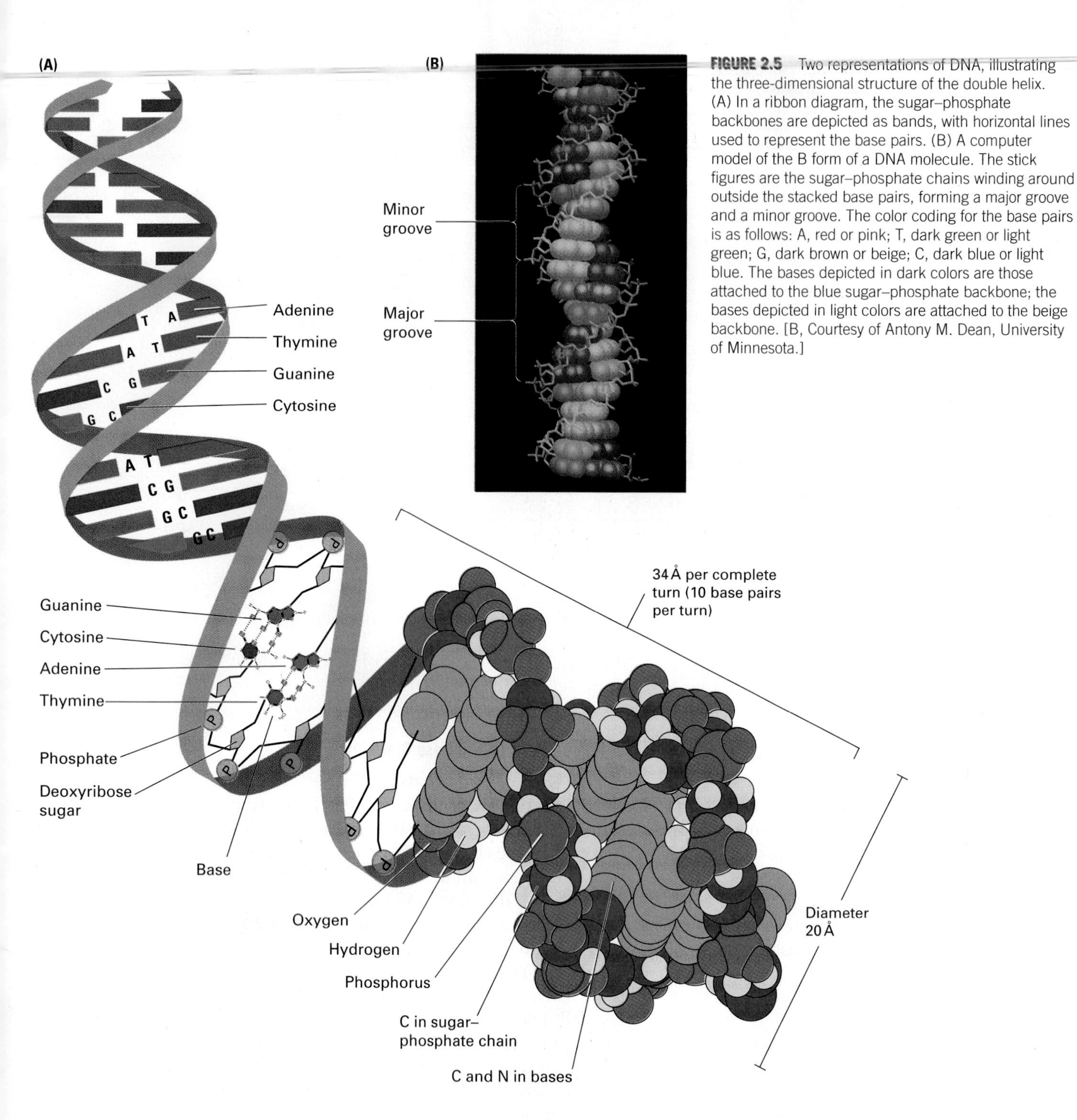


**FIGURE 2.5** Two representations of DNA, illustrating the three-dimensional structure of the double helix. (A) In a ribbon diagram, the sugar–phosphate backbones are depicted as bands, with horizontal lines used to represent the base pairs. (B) A computer model of the B form of a DNA molecule. The stick figures are the sugar–phosphate chains winding around outside the stacked base pairs, forming a major groove and a minor groove. The color coding for the base pairs is as follows: A, red or pink; T, dark green or light green; G, dark brown or beige; C, dark blue or light blue. The bases depicted in dark colors are those attached to the blue sugar–phosphate backbone; the bases depicted in light colors are attached to the beige backbone. [B, Courtesy of Antony M. Dean, University of Minnesota.]

adenine–thymine base pair and in the guanine–cytosine pair are illustrated in **FIGURE 2.6**. Note that an A—T pair (Figure 2.6A and B) has two hydrogen bonds and that a G—C pair (Figure 2.6C and D) has three hydrogen bonds. This means that the hydrogen bonding between G and C is stronger in the sense that it requires more energy to break; for example, the amount of heat required to separate the paired strands in a DNA duplex increases with the percent of G + C. Because nothing restricts the sequence of bases in a single strand, any sequence could be present along one strand. This explains Chargaff's observation that DNA from different organisms may differ in base composition. However, because the strands in duplex DNA are complementary, Chargaff's rules of [A] = [T] and [G] = [C] are true whatever the base composition.

In the B form of DNA, the paired bases are stacked on top of one another like pennies in a roll. The upper and lower faces of each nitroge-

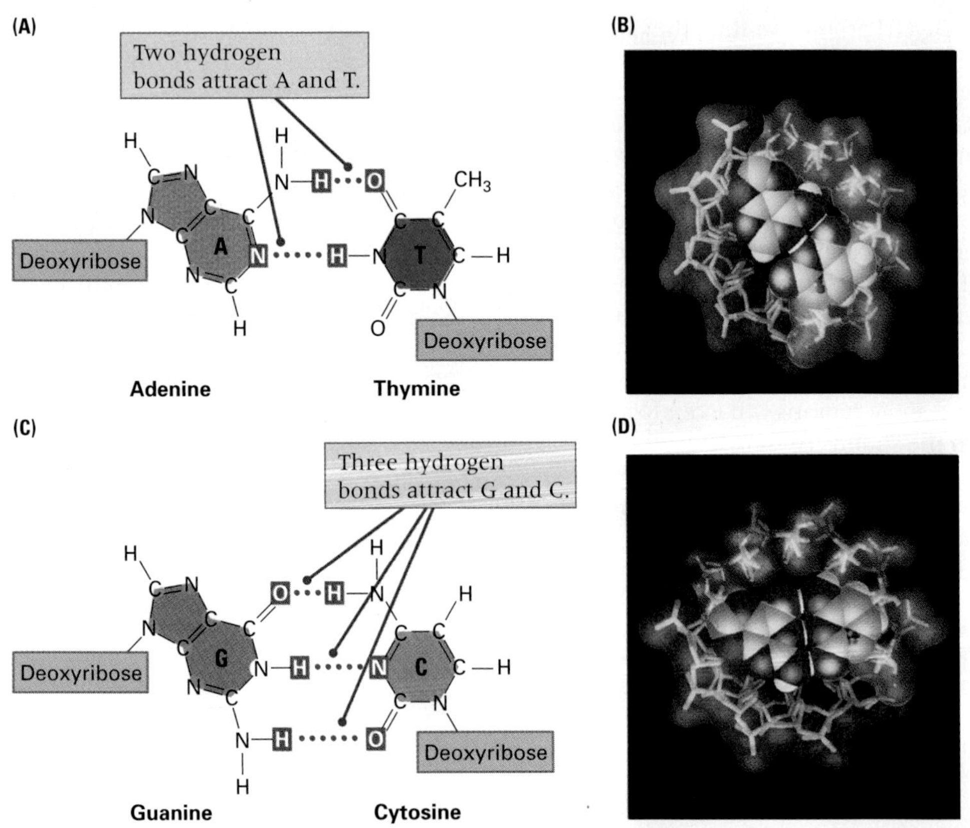


**FIGURE 2.6** Normal base pairs in DNA. On the left, the hydrogen bonds (dotted lines) with the joined atoms are shown in red. (A and B) A–T base pairing. (C and D) G–C base pairing. In the space-filling models (B and D), the colors are as follows: C, gray; N, blue; O, red; and H (shown in the bases only), white. Each hydrogen bond is depicted as a white disk squeezed between the atoms sharing the hydrogen. The stick figures on the outside represent the backbones winding around the stacked base pairs. [Space-filling models courtesy of Antony M. Dean, University of Minnesota.]

nous base are relatively flat and nonpolar (uncharged). These surfaces are said to be *hydrophobic* because they bind poorly to water molecules, which are very polar. (The polarity refers to the asymmetrical distribution of charge across the V-shaped water molecule; the oxygen at the base of the V tends to be quite negative, whereas the hydrogens at the tips are quite positive). Owing to their repulsion of water molecules, the paired nitrogenous bases tend to stack on top of one another in such a way as to exclude the maximum amount of water from the interior of the double helix. This feature of double-stranded DNA is known as **base stacking.** A double-stranded DNA molecule therefore has a hydrophobic core composed of stacked bases, and it is the energy of base stacking that provides double-stranded DNA with much of its chemical stability.

When discussing a DNA molecule, molecular biologists frequently refer to the individual strands as single strands or as single-stranded DNA; they refer to the double helix as double-stranded DNA or as a *duplex* molecule. The two grooves spiraling along outside of the double helix are not symmetrical; one groove, called the **major groove,** is larger than the other, which is called the **minor groove.** Proteins that interact with double-stranded DNA often have regions that make contact with the base pairs by fitting into the major groove, into the minor groove, or into both grooves (Figure 2.5B).

## Antiparallel Strands

Each backbone in a double helix consists of deoxyribose sugars alternating with phosphate groups that link the 3′ carbon atom of one sugar to the 5′ carbon of the next in line (Figure 2.4). The two polynucleotide strands of the double helix have opposite polarity in the sense that the 5′ end of one strand is paired with the 3′ end of the other strand. Strands with such an arrangement are said to be **antiparallel.** One implication of antiparallel strands in duplex DNA is that in each pair of bases, one base is attached to a sugar that lies above the plane of pairing, and the other base is attached to a sugar that lies below the plane of pairing. Another implication is that each terminus of the double helix possesses one 5′-P group (on one strand)

and one 3′-OH group (on the other strand), as shown in **FIGURE 2.7**.

The diagram of the DNA duplex in Figure 2.5 is static and so somewhat misleading. DNA is a dynamic molecule, constantly in motion. In some regions, the strands can separate briefly and then come together again in the same conformation or in a different one. The right-handed double helix in Figure 2.5 is the standard B form, but depending on conditions, DNA can actually form more than 20 slightly different variants of a right-handed helix, and some regions can even form helices in which the strands twist to the left (called the *Z form of DNA*). If there are complementary stretches of nucleotides in the same strand, then a single strand, separated from its partner, can fold back upon itself like a hairpin. Even

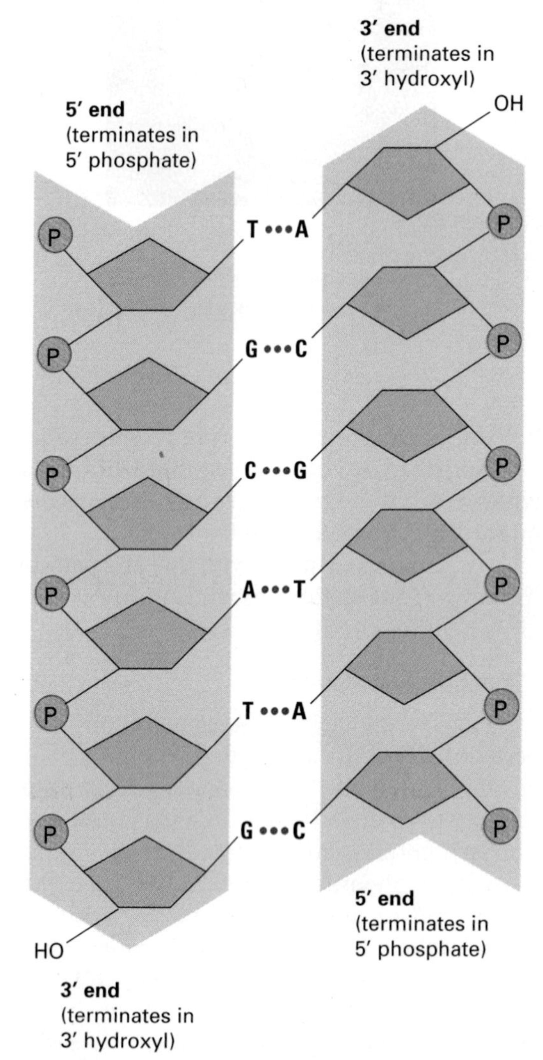


**FIGURE 2.7** A segment of a DNA molecule, showing the antiparallel orientation of the complementary strands. The shaded blue arrows indicate the 5′-to-3′ direction of each strand. The phosphates (P) join the 3′ carbon atom of one deoxyribose (horizontal line) to the 5′ carbon atom of the adjacent deoxyribose.

triple helices consisting of three strands can form in regions of DNA that contain suitable base sequences.

## DNA Structure as Related to Function

In the structure of the DNA molecule, we can see how three essential requirements of a genetic material are met.

1. Any genetic material must be able to be replicated accurately, so that the information it contains will be precisely replicated and inherited by daughter cells. The basis for exact duplication of a DNA molecule is the pairing of A with T and of G with C in the two polynucleotide chains. Unwinding and separation of the strands, with each free strand being copied, results in the formation of two identical double helices (see Figure 1.7).
2. A genetic material must also have the capacity to carry all of the information needed to direct the organization and metabolic activities of the cell. As we saw in Chapter 1, the product of most genes is a protein molecule—a polymer composed of repeating units of amino acids. The sequence of amino acids in the protein determines its chemical and physical properties. A gene is expressed when its protein product is synthesized, and one requirement of the genetic material is that it direct the sequence in which amino acid units are added to the end of a growing protein molecule. In DNA, this is done by means of a genetic code in which groups of three bases specify amino acids. Because the four bases in a DNA molecule can be arranged in any sequence, and because the sequence can vary from one part of the molecule to another and from organism to organism, DNA can contain a great many unique regions, each of which can be a distinct gene.
3. A genetic material must also be capable of undergoing occasional mutations in which the information it carries is altered. Furthermore, so that mutations will be heritable, the mutant molecules must be capable of being replicated as faithfully as the parental molecule. This feature is necessary to account for the evolution of diverse organisms through the slow accumulation of favorable mutations. Watson and Crick suggested that heritable mutations might be possible in DNA by rare mispairing of the

**connection**

## The Double Helix

**James D. Watson and Francis H. C. Crick 1953**
Cavendish Laboratory, Cambridge, England
*A Structure for Deoxyribose Nucleic Acid*

*This is one of the watershed papers of twentieth-century biology. Oddly enough, the paper received only a lukewarm reception. Biochemists studying DNA were interested primarily in its physical properties rather than its structure, and geneticists were preoccupied with understanding the action of mutagens such as x-rays. There was great interest in discovering the mechanism of protein synthesis, but it was not clear how (or even if) DNA was involved in this process. Only around 1960, when the critical role of RNA in protein synthesis had become clear, did the importance of the Watson–Crick structure begin to be appreciated. Watson and Crick benefited tremendously in knowing that their structure was consistent with the unpublished structural studies of Maurice Wilkins and Rosalind Franklin. The same issue of* Nature *that included the Watson and Crick paper also included, back to back, a paper from the Wilkins group and one from the Franklin group detailing their data and the consistency of their data with the proposed structure. It has been said that Franklin was poised a mere two half-steps from making the discovery herself, alone. In any event, Watson and Crick and Wilkins were awarded the 1962 Nobel Prize for their discovery of DNA structure. Rosalind Franklin, tragically, died of cancer in 1958 at the age of 38.*

We wish to suggest a structure for the salt of deoxyribose nucleic acid (DNA). . . . The structure has two helical chains each coiled round the same axis. . . . Both chains follow right-handed helices, but the two chains run in opposite directions. . . . The bases are on the inside of the helix and the phosphates on the outside. . . . There is a residue on each chain every 3.4 Å and the structure repeats after 10 residues. . . . The novel feature of the structure is the manner in which the two chains are held together by the purine and pyrimidine bases. The planes of the bases are perpendicular to the fiber axis. They are joined together in pairs, a single base from one chain being hydrogen-bonded to a single base from the other chain, so that the two lie side by side. One of the pair must be a purine and the other a pyrimidine for bonding to occur. . . . Only specific pairs of bases can bond together. These pairs are adenine (purine) with thymine (pyrimidine), and guanine (purine) with cytosine (pyrimidine). In other words, if an adenine forms one member of a pair, on either chain, then on these assumptions the other member must be thymine; similarly for guanine and cytosine. The sequence of bases on a single chain does not appear to be restricted in any way. However, if only specific pairs of bases can be formed, it follows that if the sequence of bases on one chain is given, then the sequence on the other chain is automatically determined. . . . It has not escaped our notice that the specific pairing we have postulated immediately suggests a plausible copying mechanism for the genetic material. . . . We are much indebted to Dr. Jerry Donohue for constant advice and criticism, especially on interatomic distances. We have also been stimulated by a knowledge of the general nature of the unpublished experimental results and ideas of Dr. Maurice H. F. Wilkins, Dr. Rosalind Franklin and their co-workers at King's College, London.

***If only specific pairs of bases can be formed, it follows that if the sequence of bases on one chain is given, then the sequence on the other chain is automatically determined.***

Source: J.D. Watson and F.H.C. Crick, *Nature* 171(1953): 737–738.

bases, with the result that an incorrect nucleotide becomes incorporated into a replicating DNA strand.

## 2.3 The Separation and Identification of Genomic DNA Fragments

The following sections show how an understanding of DNA structure and replication has been put to practical use in the development of procedures for the separation and identification of particular DNA fragments. These methods are used primarily either to identify DNA markers or to aid in the isolation of particular DNA fragments that are of genetic interest. For example, consider a pedigree of familial breast cancer in which a particular DNA fragment serves as a marker for a region of chromosome that also includes the mutant gene responsible for the increased risk. In this case the ability to identify the presence or absence of the marker for each woman in the pedigree is critically important for predicting her risk of breast cancer. To take another example, suppose one is testing the hypothesis that a mutation causing a genetic disease is present in a particular DNA fragment. In this situation it is important to be able to pinpoint this fragment using genetic markers in order to isolate the fragment from affected individuals, verify whether the hypothesis is true, and identify the nature of the mutation.

Most procedures for the separation and identification of DNA fragments can be grouped into two general categories:

1. Those that identify a specific DNA fragment present in genomic DNA by making use of the fact that complementary single-stranded DNA sequences can, under the proper conditions, form a duplex molecule. These procedures rely on *nucleic acid hybridization.*
2. Those that use prior knowledge of the sequence at the ends of a DNA fragment to specifically and repeatedly replicate this one fragment from genomic DNA. These procedures rely on selective DNA replication *(amplification)* by means of the *polymerase chain reaction.*

The major difference between these approaches is that the first (relying on nucleic acid hybridization) identifies fragments that are present in the genomic DNA itself, whereas the second (relying on DNA amplification) identifies experimentally manufactured *replicas* of fragments whose original templates (but not the replicas) were present in the genomic DNA. This difference has practical implications:

- Hybridization methods require a greater amount of genomic DNA for the experimental procedures, but relatively large fragments can be identified, and no prior knowledge of the DNA sequence is necessary.
- Amplification methods require extremely small amounts of genomic DNA for the experimental procedures, but the amplification is usually restricted to relatively small fragments, and some prior knowledge of DNA sequence is necessary.

The following sections discuss both types of approaches and give examples of how they are used. In methods that use nucleic acid hybridization to identify particular fragments present in genomic DNA, the first step is usually cutting the genomic DNA into fragments of experimentally manageable size. This procedure is discussed next.

## Restriction Enzymes and Site-Specific DNA Cleavage

Procedures for chemical isolation of DNA usually lead to random breakage of double-stranded molecules into an average length of about 50,000 base pairs. This length is denoted 50 kb, where kb stands for **kilobases** (1 kb = 1000 base pairs). A length of 50 kb is close to the length of double-stranded DNA present in the bacteriophage λ that infects *E. coli.* The 50-kb fragments can be made shorter by vigorous shearing forces, but one of the problems with breaking large DNA molecules into smaller fragments by random shearing is that the fragments containing a particular gene, or part of a gene, will be of different sizes. In other words, with random shearing, it is not possible to isolate and identify a *particular* DNA fragment on the basis of its size and sequence content, because each randomly sheared molecule that contains the desired sequence somewhere within it differs in size from all other molecules that contain the sequence. In this section we describe an important enzymatic technique that can be used for cleaving DNA molecules at specific sites. This method ensures that all DNA fragments that contain a particular sequence have the same size; furthermore, each fragment that contains the desired sequence has the sequence located at exactly the same position within the fragment.

The cleavage method makes use of an important class of DNA-cleaving enzymes isolated primarily from bacteria. The enzymes are called **restriction endonucleases** or **restriction enzymes,** and they are able to cleave DNA molecules at the positions at which particular, short sequences of bases are present. These naturally occurring enzymes serve to protect the bacterial cell by disabling the DNA of bacteriophages that attack it. Their discovery earned Werner Arber of Switzerland a Nobel Prize in 1978. Technically, the enzymes are known as *type II restriction endonucleases.* The restriction enzyme *Bam*HI is one example; it recognizes the double-stranded sequence

5′-GGATCC-3′
3′-CCTAGG-5′

and cleaves each strand between the G-bearing nucleotides shown in red. **FIGURE 2.8** shows how

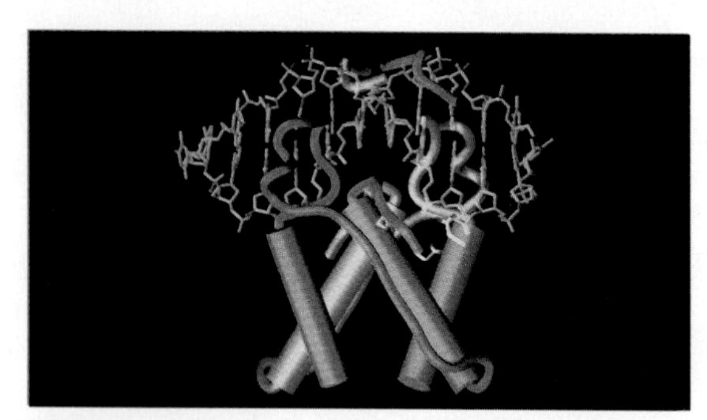

**FIGURE 2.8** Structure of the part of the restriction enzyme *Bam*HI that comes into contact with its recognition site in the DNA (blue). The pink and green cylinders represent regions of the enzyme in which the amino acid chain is twisted in the form of a right-handed helix. [Reproduced from T. Newman, et al., *Science* 269 (1995): 656-663. Reprinted with permission from AAAS.]

the regions that make up the active site of *Bam*HI contact the recognition site (blue) just prior to cleavage, and the cleavage reaction is indicated in **FIGURE 2.9**.

**TABLE 2.3** lists nine of the several hundred restriction enzymes that are known. Most restriction enzymes are named after the species in which they were found. *Bam*HI, for example, was isolated from the bacterium *Bacillus amyloliquefaciens* strain H, and it is the first (I) restriction enzyme isolated from this organism. Because the first three letters in the name of each restriction enzyme stand for the bacterial species of origin, these letters are printed in italics; the rest of the symbols in the name are not italicized. Most restriction enzymes recog-

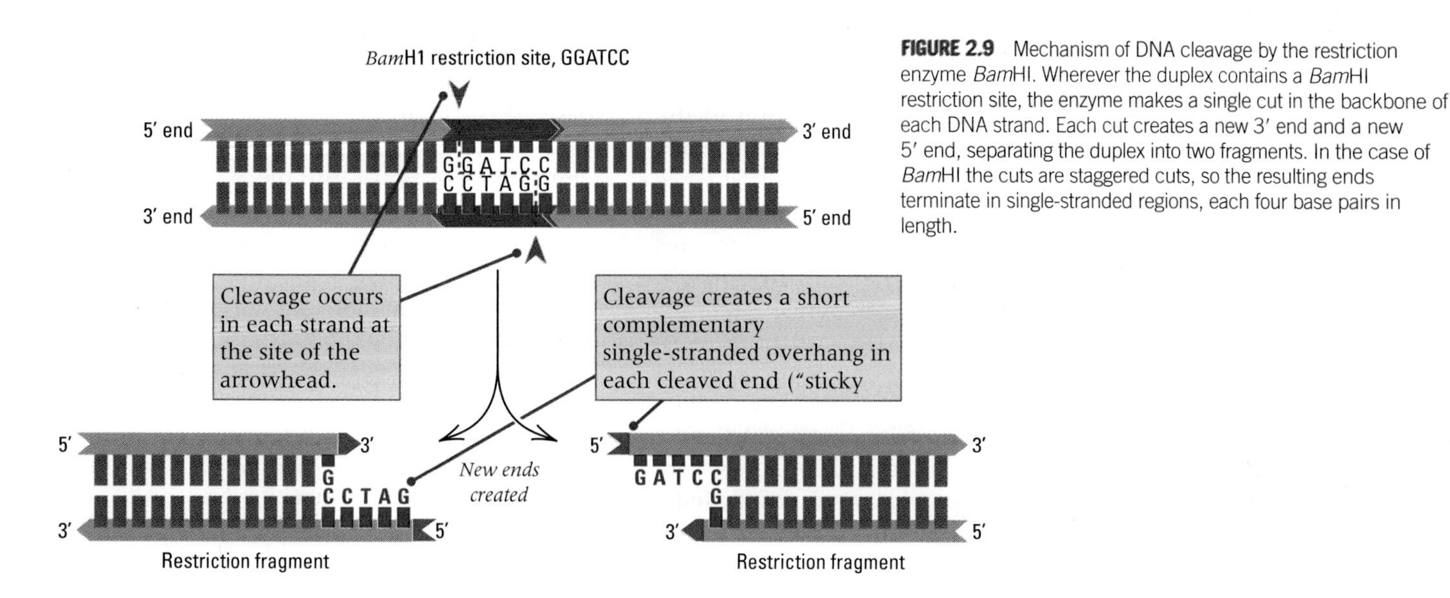


**FIGURE 2.9** Mechanism of DNA cleavage by the restriction enzyme *Bam*HI. Wherever the duplex contains a *Bam*HI restriction site, the enzyme makes a single cut in the backbone of each DNA strand. Each cut creates a new 3′ end and a new 5′ end, separating the duplex into two fragments. In the case of *Bam*HI the cuts are staggered cuts, so the resulting ends terminate in single-stranded regions, each four base pairs in length.

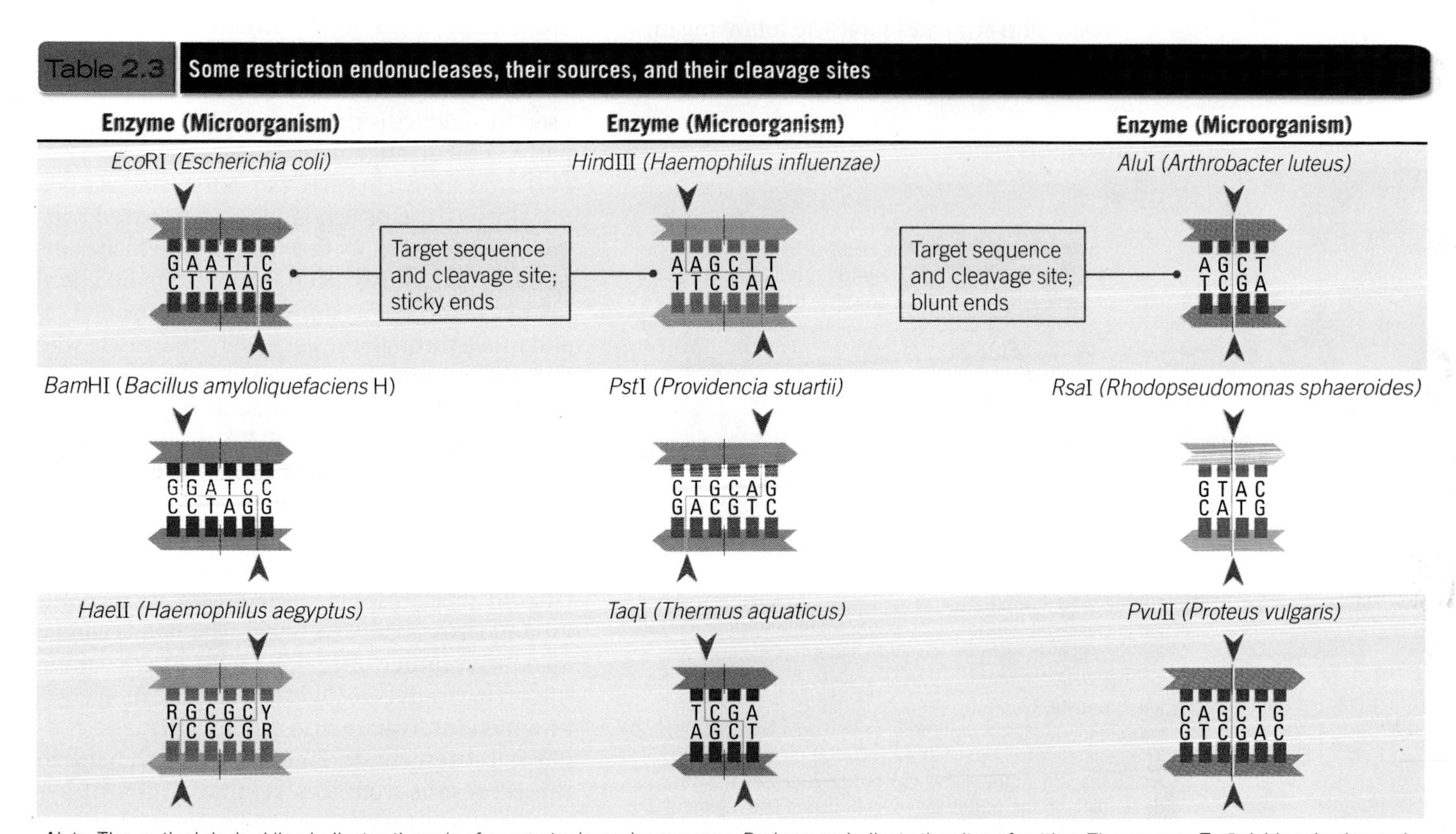

**Table 2.3** Some restriction endonucleases, their sources, and their cleavage sites

| Enzyme (Microorganism) | Enzyme (Microorganism) | Enzyme (Microorganism) |
|---|---|---|
| *Eco*RI (*Escherichia coli*)<br>GAATTC<br>CTTAAG | *Hind*III (*Haemophilus influenzae*)<br>AAGCTT<br>TTCGAA | *Alu*I (*Arthrobacter luteus*)<br>AGCT<br>TCGA |
| *Bam*HI (*Bacillus amyloliquefaciens* H)<br>GGATCC<br>CCTAGG | *Pst*I (*Providencia stuartii*)<br>CTGCAG<br>GACGTC | *Rsa*I (*Rhodopseudomonas sphaeroides*)<br>GTAC<br>CATG |
| *Hae*II (*Haemophilus aegyptus*)<br>RGCGCY<br>YCGCGR | *Taq*I (*Thermus aquaticus*)<br>TCGA<br>AGCT | *Pvu*II (*Proteus vulgaris*)<br>CAGCTG<br>GTCGAC |



*Note:* The vertical dashed line indicates the axis of symmetry in each sequence. Red arrows indicate the sites of cutting. The enzyme *Taq*I yields cohesive ends consisting of two nucleotides, whereas the cohesive ends produced by the other enzymes contain four nucleotides. R and Y refer respectively to any complementary purines and pyrimidines.

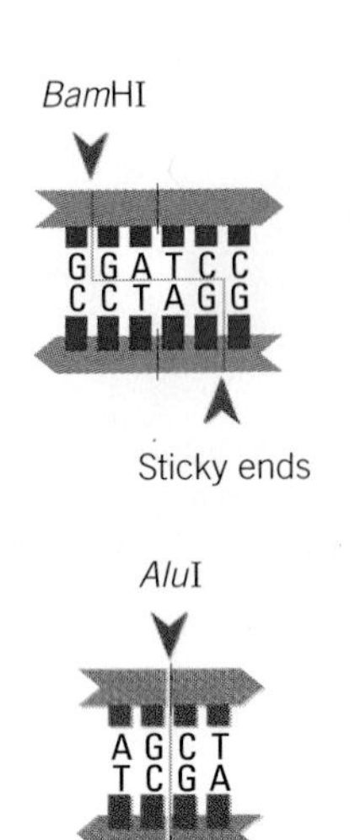


nize only one short base sequence, usually four or six nucleotide pairs. The enzyme binds with the DNA at these sites and makes a break in each strand of the DNA molecule, producing free 3′-OH and 5′-P groups at each position. The nucleotide sequence recognized for cleavage by a restriction enzyme is called the **restriction site** of the enzyme. The examples in Table 2.3 show that some restriction enzymes cleave their restriction site asymmetrically (at different sites in the two DNA strands), but other restriction enzymes cleave symmetrically (at the same site in both strands). The former leave **sticky ends** because each end of the cleaved site has a small, single-stranded overhang that is complementary in base sequence to the other end (Figure 2.9). In contrast, enzymes that have symmetrical cleavage sites yield DNA fragments that have **blunt ends.** In virtually all cases, the restriction site of a restriction enzyme reads the same on both strands, provided that the opposite polarity of the strands is taken into account; for example, each strand in the restriction site of *Bam*HI reads 5′-GGATCC-3′ (Figure 2.9). A DNA sequence with this type of symmetry is called a **palindrome.** (In ordinary English, a palindrome is a word or phrase that reads the same forwards and backwards, such as "madam.")

Restriction enzymes have the following important characteristics:

- Most restriction enzymes recognize a single restriction site.
- The restriction site is recognized without regard to the source of the DNA.
- Because most restriction enzymes recognize a unique restriction-site sequence, the number of cuts in the DNA from a particular organism is determined by the number of restriction sites present.

The DNA fragment produced by a pair of adjacent cuts in a DNA molecule is called a **restriction fragment.** A large DNA molecule typically will be cut into many restriction fragments of different sizes. For example, an *E. coli* DNA molecule, which contains $4.6 \times 10^6$ base pairs, is cut into several hundred to several thousand fragments, and mammalian genomic DNA is cut into more than a million fragments.

## Gel Electrophoresis

The DNA fragments produced by a restriction enzyme can be separated by size using the fact that DNA is negatively charged and moves in response to an electric field. If the terminals of an electrical power source are connected to the opposite ends of a horizontal tube containing a DNA solution, then the DNA molecules will move toward the positive end of the tube at a rate that depends on the electric field strength and on the shape and size of the molecules. The movement of charged molecules in an electric field is called *electrophoresis.*

The type of electrophoresis most commonly used in genetics is **gel electrophoresis.** An experimental arrangement for gel electrophoresis of DNA is shown in **FIGURE 2.10**. A thin slab of a gel, usually agarose or acrylamide, is prepared containing small slots (called *wells*) into which samples are placed. An electric field is applied, and the negatively charged DNA molecules penetrate and move through the gel toward the anode (the positively charged electrode). A gel is a complex molecular network that contains narrow, tortuous passages, so smaller DNA molecules pass through more easily; hence the rate of movement increases as the size of the DNA fragment decreases. **FIGURE 2.11** shows the result of electrophoresis of a set of double-stranded DNA molecules in an agarose gel. Each discrete region containing DNA is called a **band.** The bands can be visualized under ultraviolet light after soaking the gel in the dye *ethidium bromide,* the molecules of which intercalate into duplex DNA and render it fluorescent. In Figure 2.11, each band in the gel results from the fact that all DNA fragments of a given size have migrated to the same position in the gel. To produce a visible

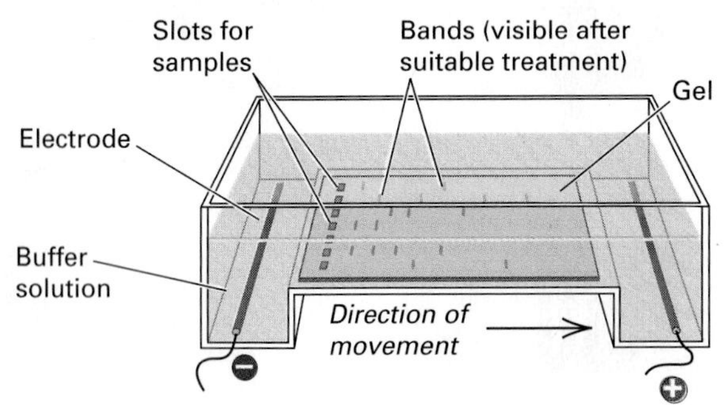


**FIGURE 2.10** Apparatus for gel electrophoresis. Liquid gel is allowed to harden with an appropriately shaped mold in place to form slots for the samples. After electrophoresis, the DNA fragments, located at various positions in the gel, are made visible by immersing the gel in a solution containing a reagent that binds to or reacts with DNA. The separated fragments in a sample appear as bands, which may be either visibly colored or fluorescent, depending on the particular reagent used. The region of a gel in which the fragments in one sample can move is called a *lane.*

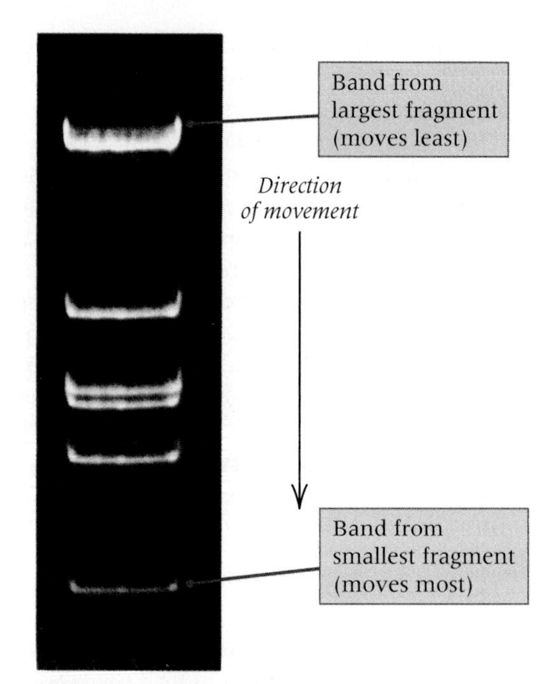


**FIGURE 2.11** Gel electrophoresis of DNA. Fragments of different sizes were mixed and placed in a well. Electrophoresis was in the downward direction. The DNA has been made visible by the addition of a dye (ethidium bromide) that binds to DNA and that fluoresces when the gel is illuminated with short-wavelength ultraviolet light.

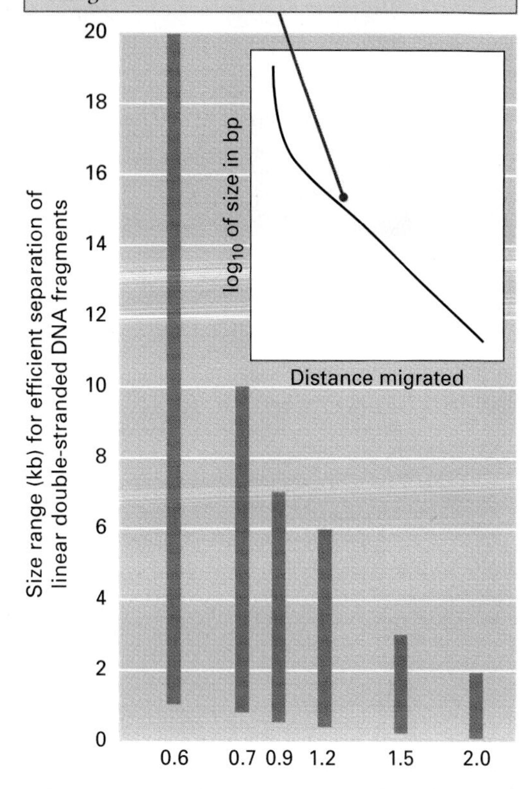


**FIGURE 2.12** In agarose gels, the concentration of agarose is an important factor in determining the size range of DNA fragments that can be separated.

band, a minimum of about $5 \times 10^{-9}$ grams of DNA is required, which for a fragment of size 3 kb works out to about $10^9$ molecules. The point is that a very large number of copies of any particular DNA fragment must be present in order to yield a visible band in an electrophoresis gel.

A linear double-stranded DNA fragment has an electrophoretic mobility that decreases in proportion to the logarithm of its length in base pairs—the longer the fragment, the slower it moves—but the proportionality constant depends on the agarose concentration, the composition of the buffering solution, and the electrophoretic conditions. This means that different concentrations of agarose allow efficient separation of different size ranges of DNA fragments (see **FIGURE 2.12**). Less dense gels, such as 0.6 percent agarose, are used to separate larger fragments; whereas more dense gels, such as 2 percent agarose, are used to separate smaller fragments. The inset in Figure 2.12 shows the dependence of electrophoretic mobility on the logarithm of fragment size. It also indicates that the linear relationship breaks down for the largest fragments that can be resolved under a given set of conditions.

Because of the sequence specificity of cleavage, *a particular restriction enzyme produces a unique set of fragments for a particular DNA molecule.* Another enzyme will produce a different set of fragments from the same DNA molecule. In **FIGURE 2.13**, this principle is illustrated for the digestion of *E. coli* phage λ DNA by either *Eco*RI or *Bam*HI (see part B). The locations of the cleavage sites for these enzymes in λ DNA are shown in Figures 2.13A and C. A diagram showing sites of cleavage along a DNA molecule is called a **restriction map.** Particular DNA fragments can be isolated by cutting out the small region of the gel that contains the fragment and removing the DNA from the gel. One important use of isolated restriction fragments employs the enzyme *DNA ligase* to insert them into self-replicating molecules such as bacteriophage, plasmids, or even small artificial chromosomes (Figure 2.1). These procedures constitute **DNA cloning** and are the

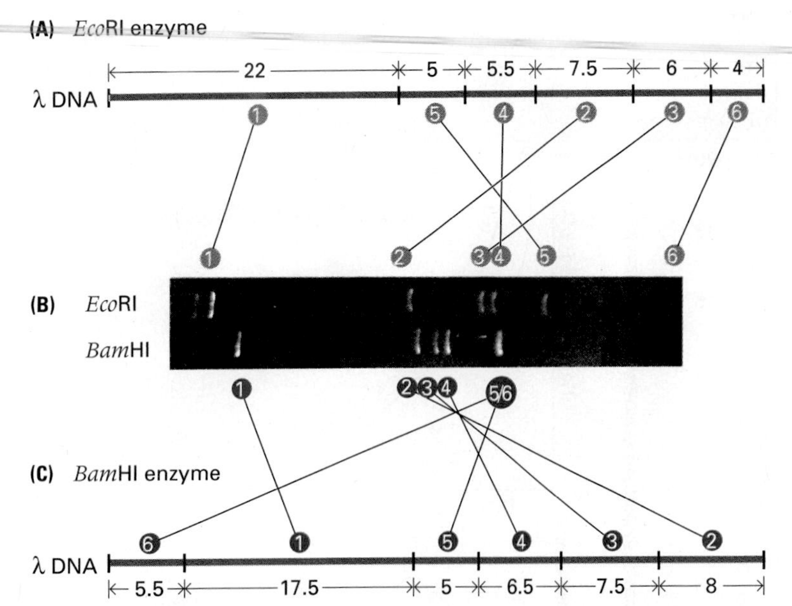


**FIGURE 2.13** Restriction maps of λ DNA for the restriction enzymes (A) *Eco*RI and (C) *Bam*HI. The vertical bars indicate the sites of cutting. The numbers within the arrows are the approximate lengths of the fragments in kilobase pairs (kb). (B) An electrophoresis gel of *Bam*HI and *Eco*RI enzyme digests of λ DNA. Numbers indicate fragments in order from largest (1) to smallest (6); the circled numbers on the maps correspond to the numbers beside the gel. The DNA has not undergone electrophoresis long enough to separate bands 5 and 6 of the *Bam*HI digest.

basis of one form of *genetic engineering*, discussed further in Chapter 12.

## ■ Nucleic Acid Hybridization

Most genomes are sufficiently large and complex that digestion with a restriction enzyme produces many bands that are the same or similar in size. Identifying a particular DNA fragment in a background of many other fragments of similar size presents a needle-in-a-haystack problem. Suppose, for example, that we are interested in a particular 3.0 *Bam*HI fragment from the human genome that serves as a marker indicating the presence of a genetic risk factor toward breast cancer among women in a particular pedigree. On the basis of size alone, this fragment of 3.0 kb is indistinguishable from fragments ranging from about 2.9 to 3.1 kb. How many fragments in this size range are expected? When human genomic DNA is cleaved with *Bam*HI, the average length of a restriction fragment is $4^6$ = 4096 base pairs, and the expected total number of *Bam*HI fragments is about 730,000; in the size range 2.9–3.1 kb, the expected number of fragments is about 17,000. What this means is that even though we know that the fragment we are interested in is 3 kb in length, it is only one of 17,000 fragments that are so similar in size that the one of interest cannot be distinguished from the others by length alone.

This identification task is actually harder than finding a needle in a haystack because haystacks are usually dry. A more accurate analogy would be looking for a needle in a haystack that had been pitched into a swimming pool full of water. This analogy is more relevant because gels, even though they contain a supporting matrix to make them semisolid, are primarily composed of water, and each DNA molecule within a gel is surrounded entirely by water. Clearly, we need some method by which the molecules in a gel can be immobilized and our *specific* fragment identified.

The DNA fragments in a gel are usually immobilized by transferring them onto a sheet of special filter paper consisting of nitrocellulose, to which DNA can be permanently (covalently) bound. How this is done is described in the next section. In this section we examine how the two strands in a double helix can be separated to form single strands and how, under the proper conditions, two single strands that are complementary or nearly complementary in sequence can come back together to form a different double helix. The separation of the strands is called **denaturation,** and the coming together of complementary strands is called **nucleic acid hybridization** or **renaturation.** (The term *hybridization* is appropriate because the two strands that anneal to form a DNA duplex may not be exactly the same strands that were paired prior to denaturation.) The practical applications of nucleic acid hybridization are many:

- A small part of a DNA fragment can be hybridized with a much larger DNA fragment. This principle is used in identifying specific DNA fragments in a complex mixture, such as the 3-kb *Bam*HI marker for breast cancer that we have been considering. Applications of this type include the tracking of genetic markers in pedigrees and the isolation of fragments containing a particular mutant gene.
- A DNA fragment from one gene can be hybridized with similar fragments from other genes in the same genome; this principle is used to identify different members of *families* of genes that are similar, but not identical, in sequence and that have related functions.

- A DNA fragment from one species can be hybridized with similar sequences from other species. This allows the isolation of genes that have the same or related functions in multiple species. It is used to study aspects of molecular evolution, such as how differences in sequence are correlated with differences in function, and the patterns and rates of change in gene sequences as they evolve.

As we saw in Section 2.2, the double-stranded helical structure of DNA is maintained by base stacking and by hydrogen bonding between the complementary base pairs. When solutions containing DNA fragments are raised to temperatures in the range 85°C–100°C, or to the high pH of strong alkaline solutions, the paired strands begin to separate. Unwinding of the helix happens in less than a few minutes (the time depends on the length of the molecule). A common way to detect denaturation is by measuring the capacity of DNA in solution to absorb ultraviolet light of wavelength 260 nm, because the absorption at 260 nm ($A_{260}$) of a solution of single-stranded molecules is 37 percent higher than the absorption of the double-stranded molecules at the same concentration. As shown in **FIGURE 2.14**, the progress of denaturation can be followed by slowly heating a solution of double-stranded DNA and recording the value of $A_{260}$ at various temperatures. The temperature required for denaturation increases with G + C content, not only because G–C base pairs have three hydrogen bonds and A–T base pairs two, but because consecutive G–C base pairs have stronger base stacking.

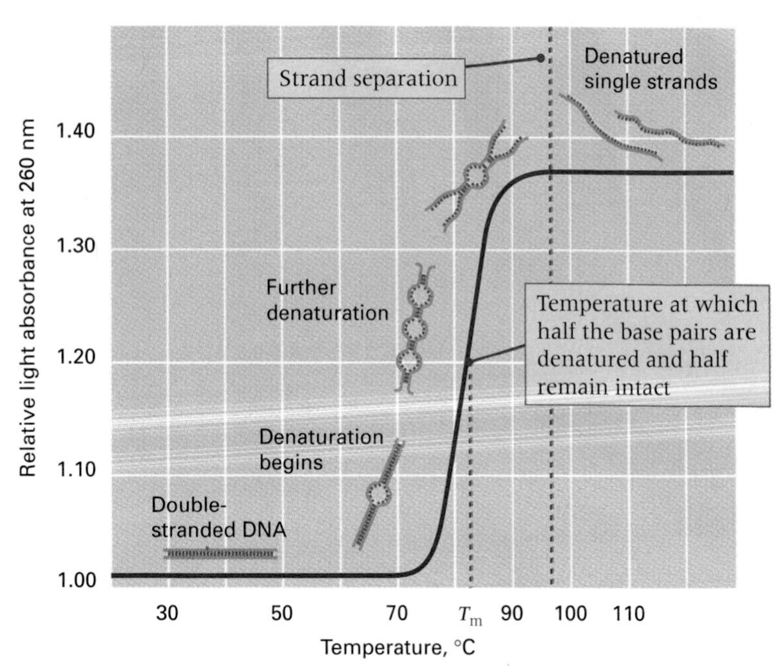


**FIGURE 2.14** Mechanism of denaturation of DNA by heat. The temperature at which 50 percent of the base pairs are denatured is the *melting temperature*, symbolized as $T_m$.

In order for denatured DNA strands to be able to undergo renaturation, two requirements must be met:

1. The salt concentration must be high (>0.25M) to neutralize the negative charges of the phosphate groups, which would otherwise cause the complementary strands to repel one another.
2. The temperature must be high enough to disrupt hydrogen bonds that form at random between short sequences of bases within the same strand, but not so high that stable base pairs between the complementary strands are disrupted.

The initial phase of renaturation is a slow process because the rate is limited by the random chance that a region of two complementary strands will come together to form a short sequence of correct base pairs. This initial pairing step is followed by a rapid pairing of the remaining complementary bases and rewinding of the helix. Rewinding is accomplished in a matter of seconds, and its rate is independent of DNA concentration because the complementary strands have already found each other.

The example of nucleic acid hybridization in **FIGURE 2.15** will enable us to understand some of the molecular details and also to see how hybridization is used to "tag" and identify a particular DNA fragment. Shown in part A is a solution of denatured DNA, called the **probe,** in which each molecule has been labeled with either radioactive atoms or light-emitting molecules. Probe DNA is typically obtained from a clone, and the labeled probe usually contains denatured forms of both strands present in the original duplex molecule. (This has led to some confusing terminology. Geneticists say that probe DNA hybridizes with DNA fragments containing sequences that are *similar* to the probe, rather than *complementary*. What actually occurs is that one strand of the probe undergoes hybridization with a complementary sequence in the fragment. But because the probe usually contains both strands, hybridization takes place with any fragment that contains a similar sequence, each strand in the probe undergoing hybridization with the complementary sequence in the fragment.)

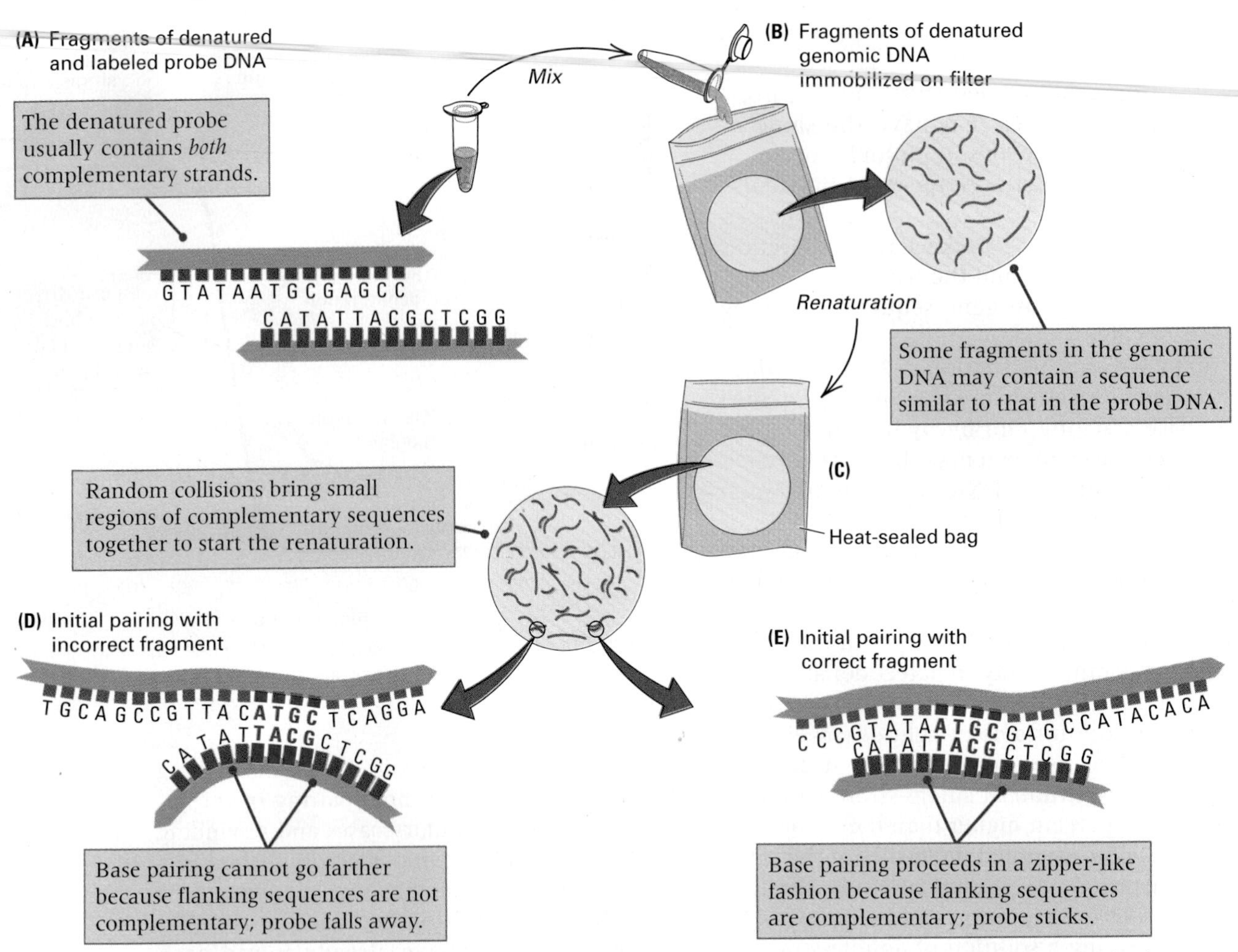
(A) Fragments of denatured and labeled probe DNA
Mix
(B) Fragments of denatured genomic DNA immobilized on filter
The denatured probe usually contains *both* complementary strands.
GTATAATGCGAGCC
CATATTACGCTCGG
Renaturation
Some fragments in the genomic DNA may contain a sequence similar to that in the probe DNA.
(C)
Random collisions bring small regions of complementary sequences together to start the renaturation.
Heat-sealed bag
(D) Initial pairing with incorrect fragment
TGCAGCCGTTACATGCTCAGGA
CATATTACGCTCGG
(E) Initial pairing with correct fragment
CCCGTATAATGCGAGCCATACACA
CATATTACGCTCGG
Base pairing cannot go farther because flanking sequences are not complementary; probe falls away.
Base pairing proceeds in a zipper-like fashion because flanking sequences are complementary; probe sticks.

**FIGURE 2.15** Nucleic acid hybridization. (A) Duplex molecules of probe DNA (obtained from a clone) are denatured and (B) placed in contact with a filter to which are attached denatured strands of genomic DNA. (C) Under the proper conditions of salt concentration and temperature, short complementary stretches come together by random collision. (D) If the sequences flanking the paired region are not complementary, then the pairing is unstable and the strands come apart again. (E) If the sequences flanking the paired region are complementary, then further base pairing stabilizes the renatured duplex.

Part B in Figure 2.15 is a diagram of genomic DNA fragments that have been immobilized on a nitrocellulose filter. When the probe is mixed with the genomic fragments (part C), random collisions bring short, complementary stretches together. If the region of complementary sequence is short (part D), then random collision cannot initiate renaturation because the flanking sequences cannot pair; in this case the probe falls off almost immediately. If, however, a collision brings short sequences together in the correct register (part E), then this initiates renaturation, because the pairing proceeds zipper-like from the initial contact. The main point is that DNA fragments are able to hybridize only if the length of the region in which they can pair is sufficiently long. Some mismatches in the paired region can be tolerated. How many mismatches are allowed is determined by the conditions of the experiment. The lower the temperature at which the hybridization is carried out, and the higher the salt concentration, the greater the proportion of mismatches that are tolerated.

## The Southern Blot

The ability to renature DNA in the manner outlined in Figure 12.15 means that solution containing a small fragment of denatured DNA, if it is suitably labeled (for example, with radioactive $^{32}P$), can be combined with a complex mixture of denatured DNA fragments, and upon renaturation the small fragment will "tag" with radioactivity any molecules in the complex mixture with which it can hybridize. The radioactive tag allows these molecules to be identified.

The methods of DNA cleavage, electrophoresis, transfer to nitrocellulose, and hybridization with a probe are all combined in the **Southern blot,** named after its inventor Edwin

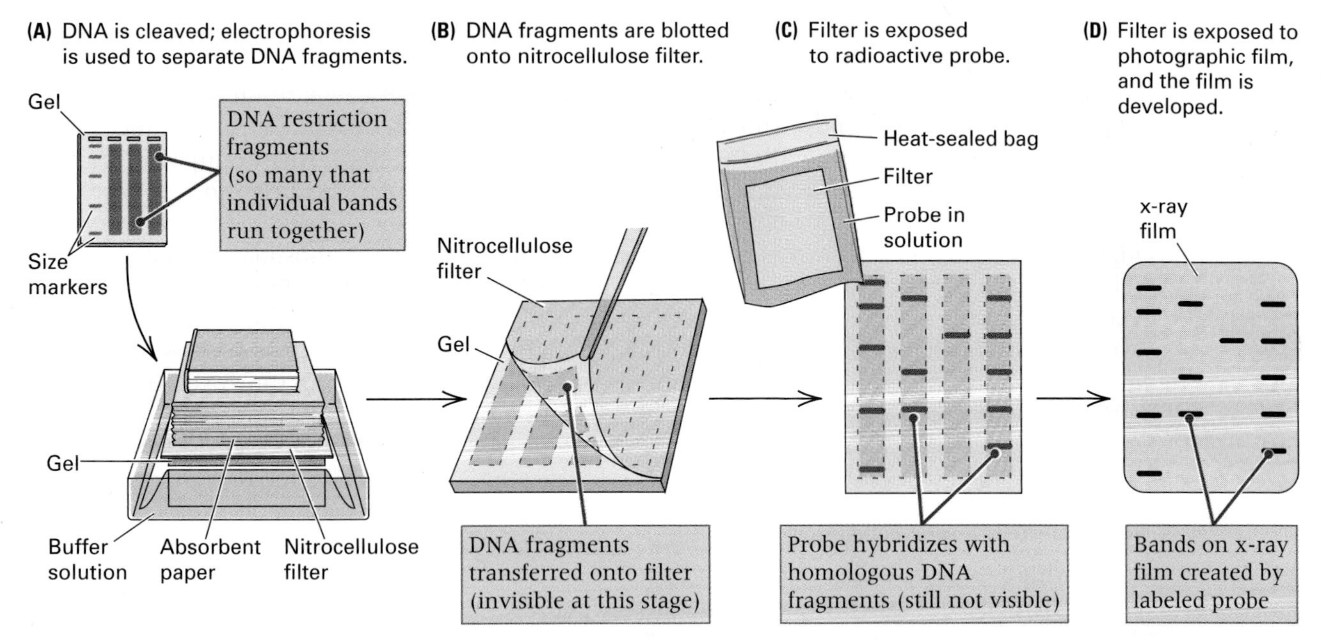
(A) DNA is cleaved; electrophoresis is used to separate DNA fragments.
(B) DNA fragments are blotted onto nitrocellulose filter.
(C) Filter is exposed to radioactive probe.
(D) Filter is exposed to photographic film, and the film is developed.
Gel
DNA restriction fragments (so many that individual bands run together)
Size markers
Gel
Buffer solution
Absorbent paper
Nitrocellulose filter
Nitrocellulose filter
Gel
DNA fragments transferred onto filter (invisible at this stage)
Heat-sealed bag
Filter
Probe in solution
Probe hybridizes with homologous DNA fragments (still not visible)
x-ray film
Bands on x-ray film created by labeled probe

**FIGURE 2.16** The Southern blot: An experimental procedure for identifying the position of a specific DNA fragment in a gel.

Southern. In this procedure, a gel in which DNA molecules have been separated by electrophoresis is treated with alkali to denature the DNA and render it single-stranded (**FIGURE 2.16**). Then the DNA is transferred to a sheet of nitrocellulose filter in such a way that the relative positions of the DNA fragments are maintained. The transfer is accomplished by overlaying the nitrocellulose onto the gel and stacking many layers of absorbent paper on top; the absorbent paper sucks water molecules from the gel and through the nitrocellulose, to which the DNA fragments adhere. (This step is the "blot" component of the Southern blot, parts A and B.) Then the filter is treated so that the single-stranded DNA becomes permanently bound. The treated filter is mixed with a solution containing denatured probe (DNA or RNA) under conditions that allow complementary strands to hybridize to form duplex molecules (part C). Radioactive or other label present in the probe becomes stably bound to the filter, and therefore resistant to removal by washing, only at positions at which base sequences complementary to the probe are already present on the filter, so that the probe can form duplex molecules. The label is located by placing the paper in contact with x-ray film. After development of the film, blackened regions indicate positions of bands containing the radioactive or light-emitting label (part D).

The procedure in Figure 2.16 solves the wet-haystack problem by transferring and immobilizing the genomic DNA fragments to a filter and identifying, by hybridization, the ones that are of interest. Practical applications of Southern blotting center on identifying DNA fragments that contain sequences similar to the probe DNA or RNA, where the proportion of mismatched nucleotides allowed is determined by the conditions of hybridization. The advantages of the Southern blot are convenience and sensitivity. The sensitivity comes from the fact that both hybridization with a labeled probe and the use of photographic film amplify the signal; under typical conditions, a band can be observed on the film with only $5 \times 10^{-12}$ grams of DNA—a thousand times less DNA than the amount required to produce a visible band in the gel itself.

## 2.4 Selective Replication of Genomic DNA Fragments

Although nucleic acid hybridization allows a particular DNA fragment to be identified when present in a complex mixture of fragments, it does not enable the fragment to be separated from the others and purified. Obtaining the fragment in purified form requires cloning, which is straightforward but time-consuming. (Cloning methods are discussed in Chapter 12.) However, if the fragment of interest is not too long, and if the nucleotide sequence at each end is known, then it becomes possible to obtain large quantities of the fragment merely by selective replication. This process is called **amplification.** How would one know the sequence of the ends? Let us return to our example of the 3.0-kb *Bam*HI fragment that serves to mark a risk factor for breast cancer in certain pedigrees. Suppose that this fragment is cloned and sequenced from one affected individual, and it is found that, relative to the normal genomic sequence in this region, the *Bam*HI fragment is missing a region of 500 base pairs. At this point the sequences at the ends of the fragment are known, and we can also infer that amplification of genomic DNA from individuals with the risk factor will yield a band of 3.0 kb, whereas amplification from genomic DNA of noncarriers will yield a band of 3.5 kb. This difference allows every person in the pedigree to be diagnosed as a carrier or noncarrier merely by means of DNA amplification. To understand how amplification works, it is first necessary to examine a few key features of DNA replication.

## Constraints on DNA Replication: Primers and 5′-to-3′ Strand Elongation

As with most metabolic reactions in living cells, nucleic acids are synthesized in chemical reactions controlled by enzymes. An enzyme that forms the sugar–phosphate bond (the phosphodiester bond) between adjacent nucleotides in a nucleic acid chain is called a **DNA polymerase.** A variety of DNA polymerases have been purified, and for amplification of a DNA fragment, the DNA synthesis is carried out *in vitro* by combining purified cellular components in a test tube under precisely defined conditions. (*In vitro*, literally "in glass," implies the absence of living cells.)

In order for DNA polymerase to catalyze synthesis of a new DNA strand, preexisting single-stranded DNA must be present. Each single-stranded DNA molecule present in the reaction mix can serve as a template upon which a new partner strand is created by the DNA polymerase. For DNA replication to take place, the 5′-triphosphates of the four deoxynucleosides must also be present. This requirement is rather obvious, because the nucleoside triphosphates are the precursors from which new DNA strands are created. The triphosphates needed are the compounds denoted in Table 2.1 as dATP, dGTP, dTTP, and dCTP, which contain the bases adenine, guanine, thymine, and cytosine, respectively. Details of the structures of dCTP and dGTP are shown in **FIGURE 2.17**, in which the phosphate groups cleaved off during DNA synthesis are indicated. DNA synthesis requires all four nucleoside 5′-triphosphates and does not take place if any of them is omitted.

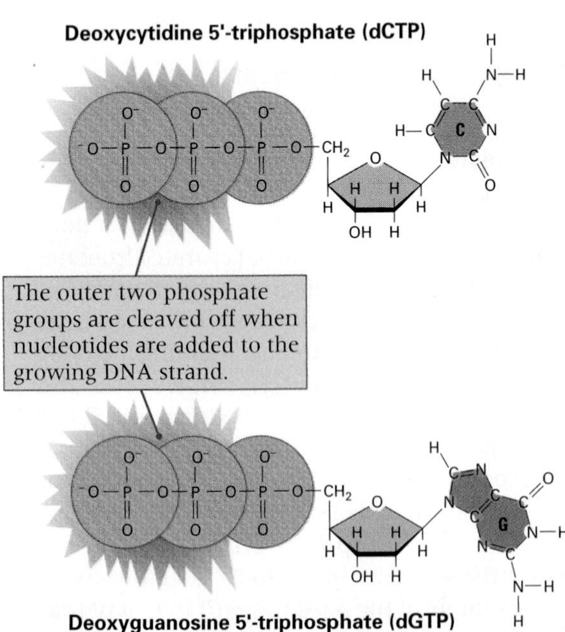

Deoxycytidine 5'-triphosphate (dCTP)
The outer two phosphate groups are cleaved off when nucleotides are added to the growing DNA strand.
Deoxyguanosine 5'-triphosphate (dGTP)

**FIGURE 2.17** Two deoxynucleoside triphosphates used in DNA synthesis. The outer two phosphate groups are removed during synthesis.

A feature found in all DNA polymerases is that:

> A DNA polymerase can only *elongate* a DNA strand. It is not possible for DNA polymerase to *initiate* synthesis of a new strand, even when a template molecule is present.

One important implication of this principle is that DNA synthesis requires a preexisting segment of nucleic acid that is hydrogen-bonded to the template strand. This segment is called a **primer.** Because the primer molecule can be very short, it is an **oligonucleotide,** which literally means "few nucleotides." As we shall see in Chapter 6, in living cells the primer is a short segment of RNA, but in DNA amplification *in vitro*, the primer employed is usually DNA.

It is the 3′ end of the primer that is essential, because, as emphasized in Chapter 1:

> DNA synthesis proceeds only by addition of successive nucleotides to the 3′ end of the growing strand. In other words, chain elongation always takes place in the 5′-to-3′ direction (5′ → 3′).

The reason for the 5′ → 3′ direction of chain elongation is illustrated in **FIGURE 2.18**. It is a consequence of the fact that the reaction catalyzed by DNA polymerase is the formation of a phosphodiester bond between the free 3′-OH group of the chain being extended and the innermost phosphorus atom of the nucleoside triphosphate being incorporated at the 3′ end. Recognition of the appropriate incoming nucleoside triphosphate in replication depends on base pairing with the opposite nucleotide in the template strand. DNA polymerase will usually catalyze the polymerization reaction that incorporates the new nucleotide at the primer terminus only when the correct base pair is present. The same DNA polymerase is used to add each of the four

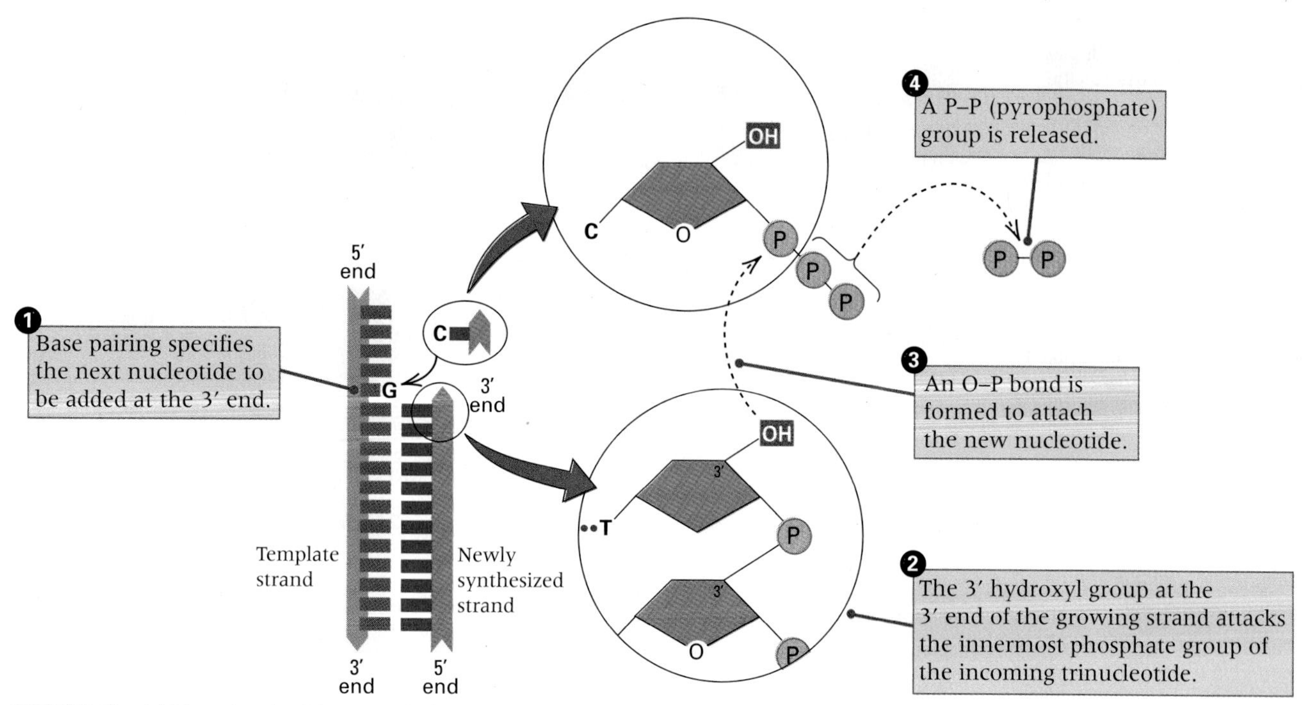

5′ end
Base pairing specifies the next nucleotide to be added at the 3′ end.
G
C
3′ end
Template strand
Newly synthesized strand
3′ end
5′ end
OH
C
O
P
P
P
A P–P (pyrophosphate) group is released.
P
P
An O–P bond is formed to attach the new nucleotide.
OH
T
P
O
P
The 3′ hydroxyl group at the 3′ end of the growing strand attacks the innermost phosphate group of the incoming trinucleotide.

**FIGURE 2.18** Addition of nucleotides to the 3′-OH terminus of a growing strand. The recognition step is shown as the formation of hydrogen bonds between the A and the T. The chemical reaction is that the 3′-OH group of the 3′ end of the growing chain attacks the innermost phosphate group of the incoming trinucleotide.

deoxynucleoside phosphates to the 3′-OH terminus of the growing strand.

## The Polymerase Chain Reaction

The requirement for an oligonucleotide primer, and the constraint that chain elongation must always occur in the 5′ → 3′ direction, make it possible to obtain large quantities of a particular DNA sequence by selective amplification *in vitro*. The method for selective amplification is called the **polymerase chain reaction (PCR).** For its invention, Californian Kary B. Mullis was awarded a Nobel Prize in 1993. PCR amplification uses DNA polymerase and a *pair* of short, synthetic oligonucleotide primers, usually 18–22 nucleotides in length, that are complementary in sequence to the ends of the DNA sequence to be amplified. **FIGURE 2.19** gives an example in which the primer oligonucleotides (green) are 9-mers. These are too short for most practical purposes, but they will serve for illustration. The original duplex molecule (part A) is shown in blue. This duplex is mixed with a vast excess of primer molecules, DNA polymerase, and all four nucleoside triphosphates. When the temperature is raised, the strands of the duplex denature and become separated. When the temperature is lowered again to allow renaturation, the primers, because they are in great excess, hybridize (or *anneal*) to the separated template strands (part B). Note that the primer sequences are different from each other but complementary to sequences present in opposite strands of the original DNA duplex and flanking the region to be amplified. The primers are oriented with their 3′ ends pointing in the direction of the region to be amplified, because each DNA strand elongates only at the 3′ end. After the primers have annealed, each is elongated by DNA polymerase using the original strand as a template, and the newly synthesized DNA strands (red) grow toward each other as synthesis proceeds (part C). Note that:

> A region of duplex DNA present in the original reaction mix can be PCR-amplified only if the region is flanked by the primer oligonucleotides.

To start a second cycle of PCR amplification, the temperature is raised again to denature the duplex DNA. Upon lowering of the temperature, the original parental strands anneal with the primers and are replicated as shown in Figure 2.19B and C. The daughter strands produced in the first round of amplification also anneal with primers and are replicated, as shown in part D. In this case, although the daughter duplex molecules are identical in sequence to the original parental molecule, they consist entirely of primer oligonucleotides and nonparental DNA that was synthesized in either the first or the second cycle of PCR. As successive cycles of denaturation, primer annealing,

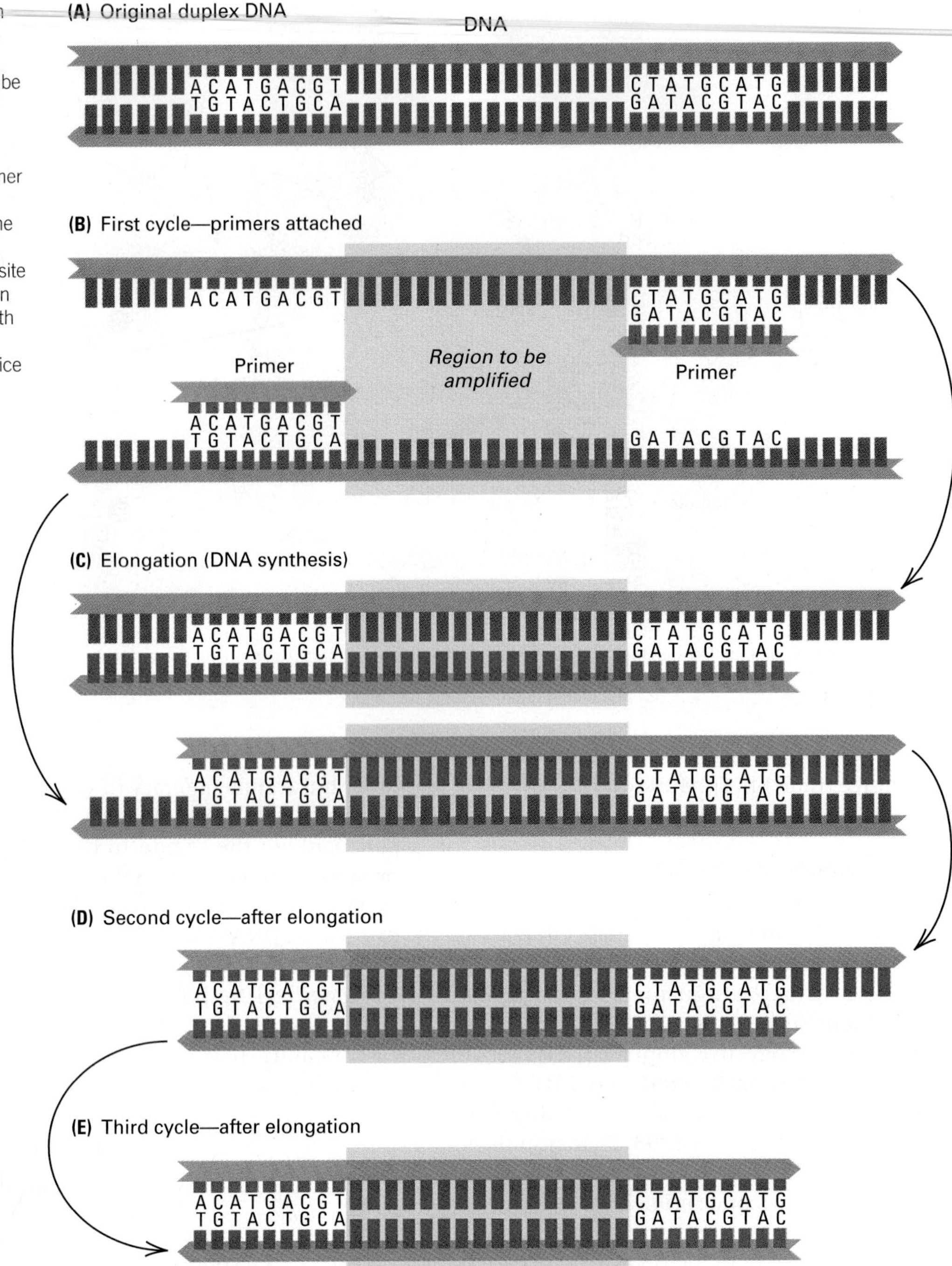


**FIGURE 2.19** Role of primer sequences in PCR amplification. (A) Target DNA duplex (blue), showing sequences chosen as the primer-binding sites flanking the region to be amplified. (B) Primer (green) bound to denatured strands of target DNA. (C) First round of amplification. Newly synthesized DNA is shown in pink. Note that each primer is extended *beyond* the other primer site. (D) Second round of amplification (only one strand shown); in this round, the newly synthesized strand terminates at the opposite primer site. (E) Third round of amplification (only one strand shown); in this round, both strands are truncated at the primer sites. Primer sequences are normally at least twice as long as shown here.

and elongation occur, the original parental strands are diluted out by the proliferation of new daughter strands until eventually, virtually every molecule produced in the PCR has the structure shown in part D.

The power of PCR amplification is that the number of copies of the template strand increases in exponential progression: 1, 2, 4, 8, 16, 32, 64, 128, 256, 512, 1024, and so forth, doubling with each cycle of replication. Starting with a mixture containing as little as one molecule of the fragment of interest, repeated rounds of DNA replication increase the number of amplified molecules exponentially. For example, starting with a single molecule, 25 rounds of DNA replication will result in $2^{25} = 3.4 \times 10^7$ molecules. This number of molecules of the amplified fragment is so much greater than that of the other unamplified molecules in the original mixture that the amplified DNA can often be used without further purification. For example, a single fragment of 3 kb in *E. coli* accounts for only 0.06 percent of the DNA in this organism. However, if this single fragment were replicated through 25 rounds of replication, then 99.995 percent of the resulting mixture would consist of the amplified sequence. A 3-kb fragment of human DNA constitutes only 0.0001 percent of the total genome size. Amplification of a 3-kb fragment of human DNA to 99.995 percent purity would require about 34 cycles of PCR.

An overview of the polymerase chain reaction is shown in **FIGURE 2.20**. The DNA sequence to be amplified is again shown in blue and the

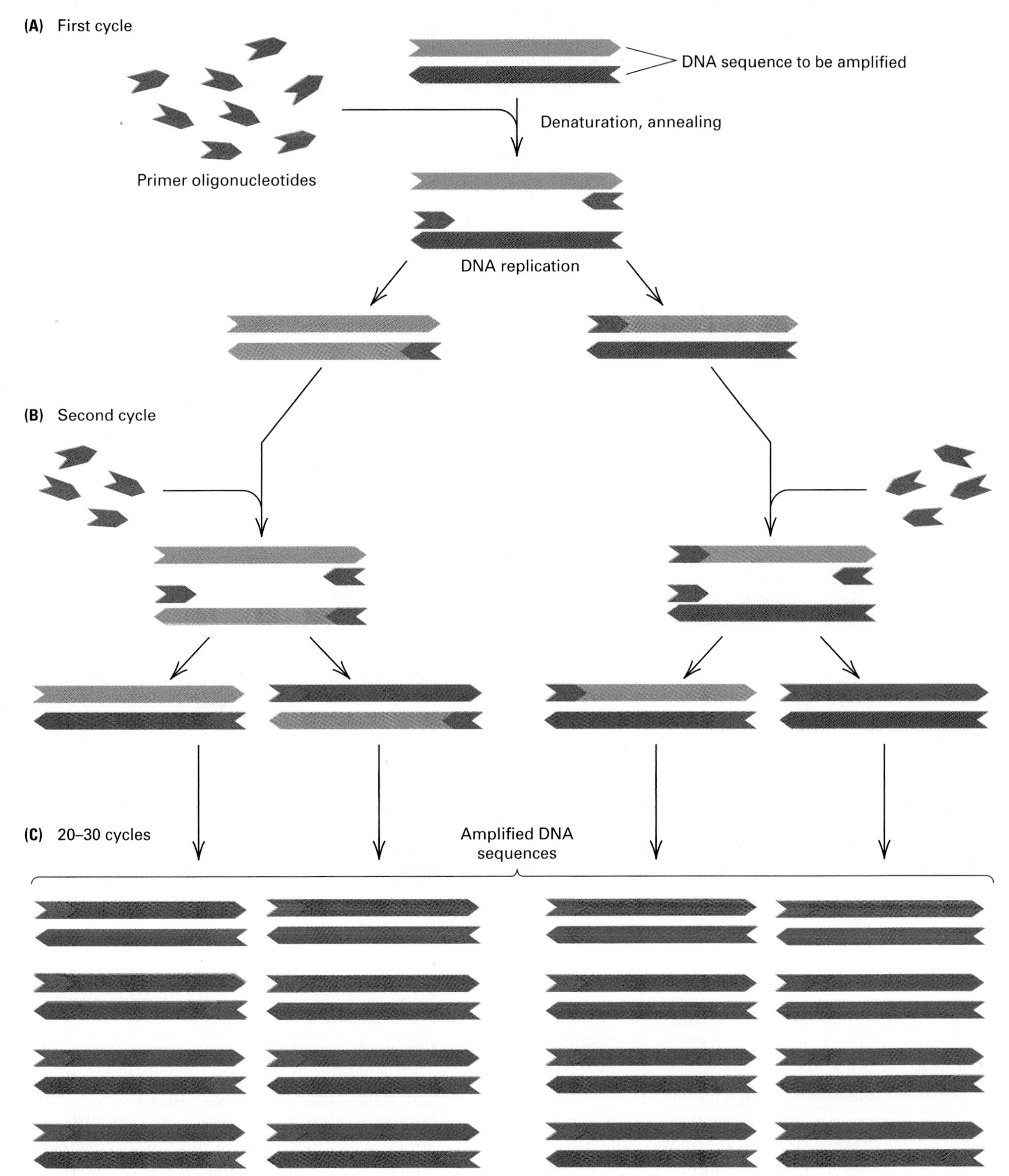


**FIGURE 2.20** Polymerase chain reaction (PCR) for amplification of particular DNA sequences. Only the region to be amplified is shown. Oligonucleotide primers (green) that are complementary to the ends of the target sequence (blue) are used in repeated rounds of denaturation, annealing, and DNA replication. Newly replicated DNA is shown in pink. The number of copies of the target sequence doubles in each round of replication, eventually overwhelming any other sequences that may be present.

oligonucleotide primers in green. The oligonucleotides anneal to the ends of the sequence to be amplified and become the substrates for chain elongation by DNA polymerase. In the first cycle of PCR amplification, the DNA is denatured to separate the strands. The denaturation temperature is usually around 95°C. Then the temperature is decreased to allow annealing in the presence of a vast excess of the primer oligonucleotides. The annealing temperature is typically in the range of 50°C–60°C, depending largely on the G + C content of the oligonucleotide primers. To complete the cycle, the temperature is raised slightly, to about 70°C, for the elongation of each primer. The steps of denaturation, renaturation, and replication are repeated from 20–30 times, and in each cycle the number of molecules of the amplified sequence is doubled.

Implementation of PCR with conventional DNA polymerases is not practical, because at the high temperature necessary for denaturation, the polymerase is itself irreversibly unfolded and becomes inactive. However, DNA polymerase isolated from certain organisms is heat stable because the organisms normally live in hot springs at temperatures well above 90°C, such as are found in Yellowstone National Park. Such organisms are said to be **thermophiles.** The most widely used heat-stable DNA polymerase is called *Taq* polymerase, because it was originally isolated from the thermophilic bacterium *Thermus aquaticus.*

PCR amplification is very useful for generating large quantities of a specific DNA sequence. The principal limitation of the technique is that the DNA sequences at the ends of the region to be amplified must be known so that primer oligonucleotides can be synthesized. In addition, sequences longer than about 5000 base pairs cannot be replicated efficiently by conventional PCR procedures, although there are modifications of PCR that allow longer fragments to be amplified. On the other hand, many applications require amplification of relatively small fragments. The major advantage of PCR amplification is that it requires only trace amounts of template DNA. Theoretically only one template molecule is required, but in practice the amplification of a single molecule may fail because the molecule may, by chance, be broken or damaged. But amplification is usually reliable with as few as 10–100 template molecules, which makes PCR amplification 10,000–100,000 times more sensitive than detection via nucleic acid hybridization.

The exquisite sensitivity of PCR amplification has led to its use in DNA typing for criminal cases in which a minuscule amount of biological material has been left behind by the perpetrator (skin cells on a cigarette butt or hair-root cells on a single hair can yield enough template DNA for amplification). In research, PCR is widely used in the study of independent mutations in a gene whose sequence is known in order to identify the molecular basis of each mutation, to study DNA sequence variation among alternative forms of a gene that may be present in natural populations, or to examine differences among genes with the same function in different species. The PCR procedure has also come into widespread use in clinical laboratories for diagnosis. To take just one very important example, the presence of the human immunodeficiency virus (HIV), which causes acquired immune deficiency syndrome (AIDS), can be detected in trace quantities in blood banks via PCR by using primers complementary to sequences in the viral genetic material. These and other applications of PCR are facilitated by the fact that the procedure lends itself to automation by the use of mechanical robots to set up and run the reactions.

## 2.5 The Terminology of Genetic Analysis

In order to discuss the types of DNA markers that modern geneticists commonly use in genetic analysis, we must first introduce some key terms that provide the essential vocabulary of genetics. These terms can be understood with reference to **FIGURE 2.21**. In Chapter 1 we defined a **gene** as an element of heredity, transmitted from parents to offspring in reproduction, that influences one or more hereditary traits. Chemically, a gene is a sequence of nucleotides along a DNA molecule. In a population of organisms, not all copies of a gene may have exactly the same nucleotide sequence. For example, whereas one form of a gene may have the codon GCA at a certain position, another form of the same gene may have the codon GCG. Both codons specify alanine. Hence, the two forms of the gene encode the same sequence of amino acids yet differ in DNA sequence. The alternative forms of a gene are called **alleles** of the gene. Different alleles may also code for different amino acid sequences, sometimes with drastic effects. Recall the example of the *PAH* gene for phenylalanine hydroxylase in Chapter 1, in which a change in codon 408 from CGG (arginine) to TGG (tryptophan) results in an inactive enzyme that becomes expressed as the inborn error of metabolism phenylketonuria.

Within a cell, genes are arranged in linear order along microscopic thread-like bodies called **chromosomes,** which we will examine in detail in Chapters 4 and 7. Each human reproductive cell contains one complete set of 23 chromosomes containing $3 \times 10^9$ base pairs of DNA. A typical chromosome contains several hundred to several thousand genes. In humans the average is approximately 1000 genes per chromosome. Each chromosome contains a single molecule of duplex DNA along its length, complexed with proteins and very tightly coiled. The DNA in the average human chromosome, when fully extended, has relative dimensions comparable to those of a wet spaghetti noodle 25 miles long; when the DNA is coiled in the form of a chromosome, its physical compaction is comparable to that of the same noodle coiled and packed into an 18-foot canoe.

The physical position of a gene along a chromosome is called the **locus** of the gene. In most higher organisms, including human beings, each cell other than a sperm or egg contains two copies of each type of chromosome—one inherited from the mother and one from the father. Each member of such a pair of chromosomes is said to be **homologous** to the other. (The chromosomes that determine sex are an important exception, discussed in Chapter 4, which we will ignore for now.) At any locus, therefore, each individual carries two alleles, because one allele is present at a corresponding position in each of the homologous maternal and paternal chromosomes (Figure 2.21).

The genetic constitution of an individual is called its **genotype.** For a particular gene, if the two alleles at the locus in an individual are indistinguishable from each other, then for this gene the genotype of the individual is said to be **homozygous** for the allele that is present. If the two alleles at the locus are different from each other, then for this gene the genotype of the individual is said to be **heterozygous** for the alleles that are present. Typographically, genes are indicated in italics, and alleles are typically distinguished by uppercase or lowercase letters (*A* versus *a*), subscripts ($A_1$ versus $A_2$), superscripts ($a^+$ versus $a^-$), or sometimes just + and −. Using these symbols, homozygous genes would be portrayed by any of these formulas: *AA*, *aa*, $A_1A_1$, $A_2A_2$, $a^+a^+$, $a^-a^-$, +/+, or −/−. As in the last two examples, the slash is sometimes used to separate alleles present in homologous chromosomes to avoid ambiguity. Heterozygous genes would be portrayed by any of the formulas *Aa*, $A_1A_2$, $a^+a^-$, or +/−. In Figure 2.21, the genotype *Bb* is heterozygous because the *B* and *b* alleles are distinguishable (which is why they are assigned different symbols), whereas the genotype *CC* is homozygous. These genotypes could also be written as *B*/*b* and *C*/*C*, respectively.

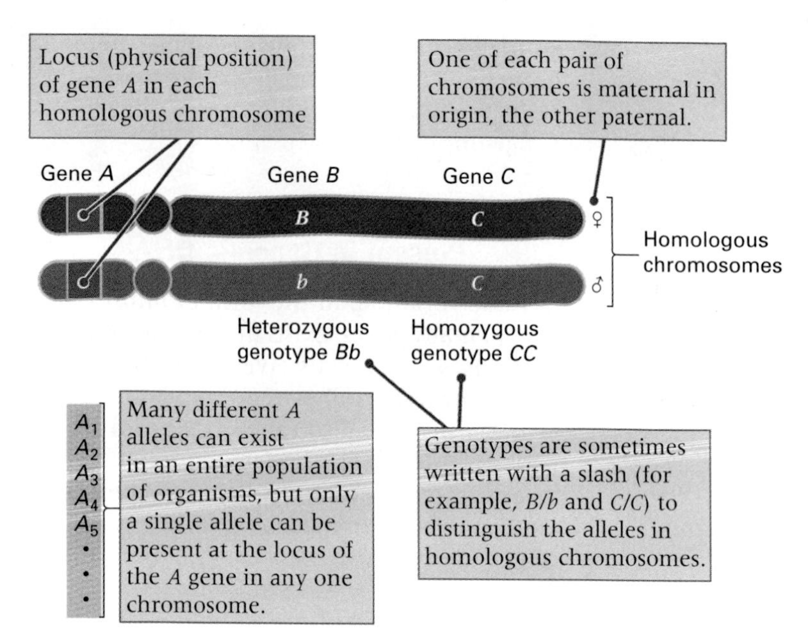


**FIGURE 2.21** Key concepts and terms used in modern genetics. Note that a single gene can have any number of alleles in the population as a whole, but no more than two alleles can be present in any one individual.

Whereas the alleles that are present in an individual constitute its genotype, the physical or biochemical expression of the genotype is called the **phenotype.** To put it as simply as possible, the distinction is that the genotype of an individual is what is on the *inside* (the alleles in the DNA), whereas the phenotype is what is on the *outside* (the observable traits, including biochemical traits, behavioral traits, and so forth). The distinction between genotype and phenotype is critically important because there usually is not a one-to-one correspondence between genes and traits. Most complex traits, such as hair color, skin color, height, weight, behavior, life span, and susceptibility to disease, are influenced by many genes. Most traits are also influenced more or less strongly by environment. This means that the same genotype can result in different phenotypes, depending on the environment. Compare, for example, two people with a genetic risk for lung cancer; if one smokes and the other does not, the smoker is much more likely to develop the disease. Environmental effects also imply that the same phenotype can result from more than one genotype; smoking again provides an example, because most smokers who are not genetically at risk can also develop lung cancer.

## 2.6 Types of DNA Markers Present in Genomic DNA

Genetic variation, in the form of multiple alleles of many genes, exists in most natural populations of organisms. We have called such genetic differences between individuals *DNA markers*; they are also called **DNA polymorphisms.** (The term *polymorphism* literally means "multiple forms.") The methods of DNA manipulation examined in Sections 2.3 and 2.4 can be used in a variety of combinations to detect differences among individuals. Anyone who reads the literature in modern genetics will encounter a bewildering variety of acronyms referring to different ways in which genetic polymorphisms are detected. The different approaches are in use because no single method is ideal for all applications, each method has its own advantages and limitations, and new methods are continually being developed. In this section we examine some of the principal methods for detecting DNA polymorphisms among individuals.

### Single-Nucleotide Polymorphisms (SNPs)

A **single-nucleotide polymorphism,** or **SNP** (pronounced "snip"), is present at a particular nucleotide site if the DNA molecules in the population often differ in the identity of the nucleotide pair that occupies the site. For example, some DNA molecules may have a T—A base pair at a particular nucleotide site, whereas other DNA molecules in the same population may have a C—G base pair at the same site. This difference constitutes a SNP. The SNP defines two "alleles" for which there could be three genotypes among individuals in the population: homozygous with T—A at the corresponding site in both homologous chromosomes, homozygous with C—G at the corresponding site in both homologous chromosomes, or heterozygous with T—A in one chromosome and C—G in the homologous chromosome. The word *allele* is in quotation marks above because the SNP need not be in a coding sequence, or even in a gene. In the human genome, any two randomly chosen DNA molecules are likely to differ at one SNP site about every 1000 bp in noncoding DNA and at about one SNP site every 3000 bp in protein-coding DNA. Note, in the definition of a SNP, the stipulation that DNA molecules must differ at the nucleotide site "often." This provision excludes rare genetic variation of the sort found in less than 1 percent of the DNA molecules in a population. The reason for the exclusion is that genetic variants that are too rare are not generally as useful in genetic analysis as the more common variants. SNPs are the most common form of genetic differences among people. About 3 million SNPs have been identified that are relatively common in the human population, and about 1 million of these are typically used in a search for SNPs that might be associated with complex diseases such as diabetes or high blood pressure.

Identifying the particular nucleotide present at each of a million SNPs is made possible through the use of **DNA microarrays** composed of about 20 million infinitesimal spots on a glass slide about the size of a postage stamp. Each tiny spot contains a unique DNA oligonucleotide sequence present in millions of copies synthesized by microchemistry when the microarray is manufactured. Each oligonucleotide sequence is designed to hybridize specifically with small fragments of genomic DNA that include one or the other of the nucleotide pairs present in a SNP. The microarrays also include numerous controls for each hybridization. The controls consist of oligonucleotides containing deliberate mismatches that are intended to guard against being misled by particular nucleotide sequences that are particularly "sticky" and hybridize too readily with genomic fragments and other sequences that form structures that hybridize poorly or not at all. Such microarrays, sometimes called **SNP chips**, enable the SNP genotype of an individual to be determined with nearly 100 percent accuracy.

The principles behind oligonucleotide hybridization are illustrated in **FIGURE 2.22**. Here the length of each oligonucleotide is seven nucleotides, but in practice this is too short. Typical SNP chips consist of oligonucleotides at least 25 nucleotides in length. Part A shows the two types of DNA duplexes that might form a SNP. In this example, some chromosomes carry a DNA molecule with a TA base pair at the position shown in red, whereas the DNA molecular in other chromosomes has a CG base pair at the corresponding position. Short fragments of genomic DNA are labeled with a fluorescent tag and then the single strands hybridized with a SNP chip containing the complementary oligonucleotides as well as the numerous controls. The duplex containing the TA will hybridize only with the two oligonucleotides on the left, and that containing the CG will hy-

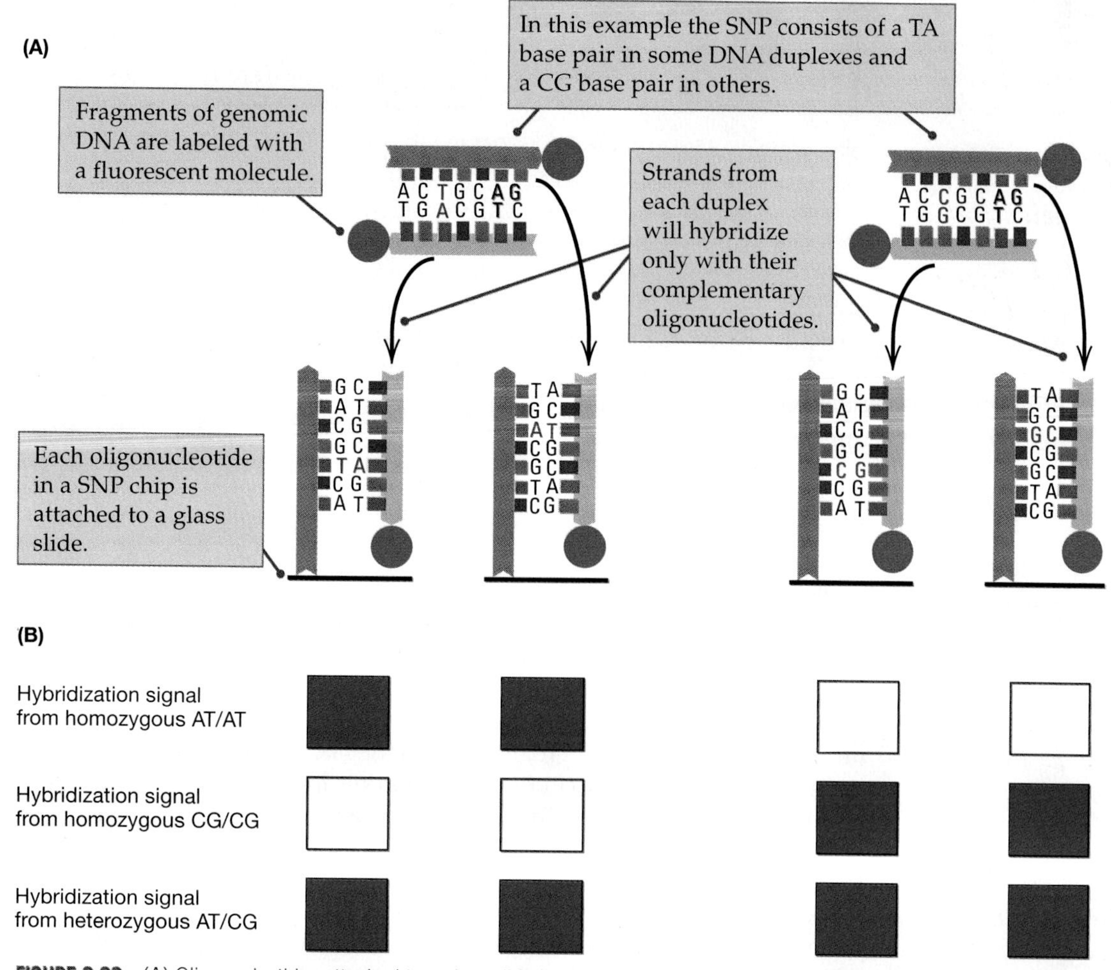
(A)
In this example the SNP consists of a TA base pair in some DNA duplexes and a CG base pair in others.
Fragments of genomic DNA are labeled with a fluorescent molecule.
ACTGCAG
TGACGTC
Strands from each duplex will hybridize only with their complementary oligonucleotides.
ACCGCAG
TGGCGTC
Each oligonucleotide in a SNP chip is attached to a glass slide.
(B)
Hybridization signal from homozygous AT/AT
Hybridization signal from homozygous CG/CG
Hybridization signal from heterozygous AT/CG

**FIGURE 2.22** (A) Oligonucleotides attached to a glass slide in a SNP chip can be used to identify duplex DNA molecules containing alternative base pairs for a SNP, in this example a TA base pair versus a CG base pair. (B) The SNP genotype of an individual can be determined by hybridization because DNA from genotypes that are homozygous TA/TA, homozygous CG/CG, or heterozygous TA/CG each gives a different pattern of fluorescence.

bridize only with the two oligonucleotides on the right.

After the hybridization takes place, the SNP chip is examined with fluorescence microscopy to detect the spots that fluoresce due to the tag on the genomic DNA. The possible patterns are shown in part B. Genomic DNA from an individual whose chromosomes contain two copies of the TA form of the duplex (homozygous TA/TA) will cause the two leftmost spots to fluoresce but the two rightmost spots will remain unlabeled. Similarly, genomic DNA from a homozygous CG/CG individual will cause the two rightmost spots to fluoresce but not the two on the left. Finally, genomic DNA from a heterozygous TA/CG individual will cause fluorescence of all four spots because the TA duplex labels the two leftmost spots and the CG duplex labels the two rightmost spots.

Use of SNP chips or other available technologies for high-throughput genotyping of millions of SNPs in thousands of individuals allows genetic risk factors for disease to be identified. A typical study compares the genotypes of patients with particular diseases with healthy people matched with the patients for such factors as sex, age, and ethnic group. Comparing the SNP genotypes among these groups often reveals which SNPs in the genome mark the location of the genetic risk factors, as will be explained in greater detail in Section 2.7.

### Restriction Fragment Length Polymorphisms (RFLPs)

Although most SNPs require DNA sequencing to be studied, those that happen to be located within a restriction site can be analyzed with a Southern blot. An example of this situation is shown in **FIGURE 2.23**, where the SNP consists of a T–A nucleotide pair in some molecules and a C–G pair in others. In this example, the polymorphic nucleotide site is included in a cleavage site for the restriction enzyme *Eco*RI (5′-GAATTC-3′). The two nearest flanking *Eco*RI

## connection

# Origin of the Human Genetic Linkage Map

**David Botstein,[1] Raymond L. White,[2] Mark Skolnick,[3] and Ronald W. Davis[4] 1980**

[1]Massachusetts Institute of Technology, Cambridge, Massachusetts

[2]University of Massachusetts Medical Center, Worcester, Massachusetts

[3]University of Utah, Salt Lake City, Utah

[4]Stanford University, Stanford, California

*Construction of a Genetic Linkage Map in Man Using Restriction Fragment Length Polymorphisms*

*This historic paper stimulated a major international effort to establish a genetic linkage map of the human genome based on DNA polymorphisms. Pedigree studies using these genetic markers soon led to the chromosomal localization and identification of mutant genes for hundreds of human diseases. A more ambitious goal, still only partly achieved, is to understand the genetic and environmental interactions involved in complex traits such as heart disease and cancer. The "small set of large pedigrees" called for in the excerpt was soon established by the Centre d'Etude du Polymorphisme Humain (CEPH) in Paris, France, and made available to investigators worldwide for genetic linkage studies. Today the CEPH maintains a database on the individuals in these pedigrees that comprises approximately 15,000 polymorphic DNA markers and 3 million genotypes.*

No method of systematically mapping human genes has been devised, largely because of the paucity of highly polymorphic marker loci. The advent of recombinant DNA technology has suggested a theoretically possible way to define an arbitrarily large number of arbitrarily polymorphic marker loci. . . . A subset of such polymorphisms can readily be detected as differences in the length of DNA fragments after digestion with DNA sequence-specific restriction endonucleases. These restriction fragment length polymorphisms (RFLPs) can be easily assayed in individuals, facilitating large population studies. . . . [Genetic mapping] of many DNA marker loci should allow the establishment of a set of well-spaced, highly polymorphic genetic markers covering the entire human genome [and enabling] any trait caused wholly or partially by a major locus segregating in a pedigree to be mapped. Such a procedure would not require any knowledge of the biochemical nature of the trait or of the nature of the alterations in the DNA responsible for the trait. . . . The most efficient procedure will be to study a small set of large pedigrees which have been genotyped for all known polymorphic markers. . . . The resolution of genetic and environmental components of disease . . . must involve unraveling the underlying genetic predisposition, understanding the environmental contributions, and understanding the variability of expression of the phenotype. In principle, linked marker loci can allow one to establish, with high certainty, the genotype of an individual and, consequently, assess much more precisely the contribution of modifying factors such as secondary genes, likelihood of expression of the phenotype, and environment.

***In principle, linked marker loci can allow one to establish, with high certainty, the genotype of an individual.***

Source: D. Botstein, et al., *Am. J. Hum. Genet.* 32(1980): 314–331.

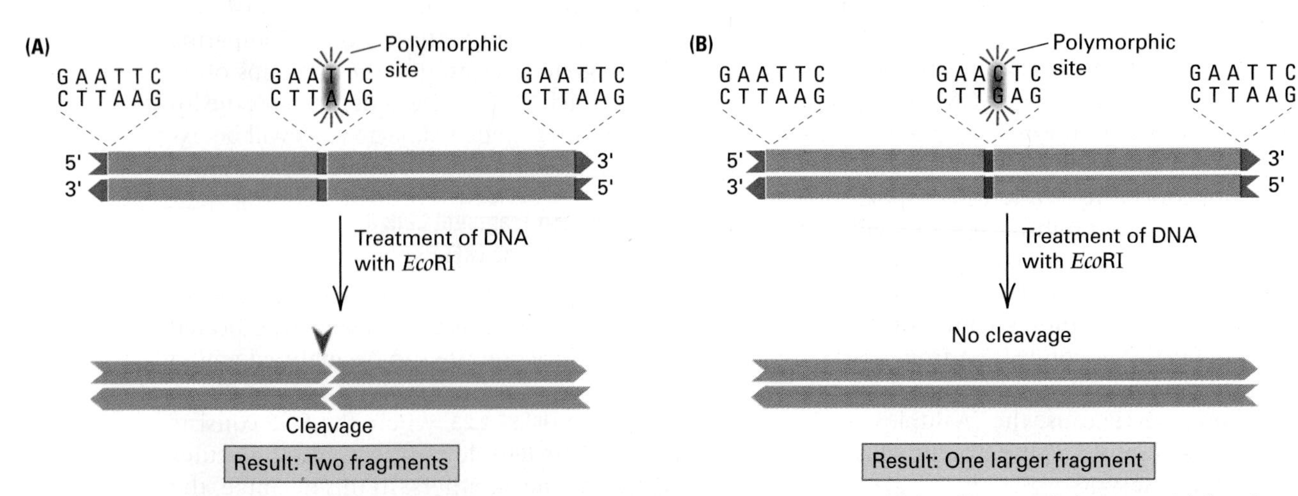


**FIGURE 2.23** A difference in the DNA sequence of two molecules can be detected if the difference eliminates a restriction site. (A) This molecule contains three restriction sites for *Eco*RI, including one at each end. It is cleaved into two fragments by the enzyme. (B) This molecule has an altered *Eco*RI site in the middle, in which 5′-GAATTC-3′ becomes 5′-GAACTC-3′. The altered site cannot be cleaved by *Eco*RI, so treatment of this molecule with *Eco*RI results in one larger fragment.

sites are also shown. In this kind of situation, DNA molecules with T–A at the SNP will be cleaved at both flanking sites and also at the middle site, yielding two *Eco*RI restriction fragments. Alternatively, DNA molecules with C–G at the SNP will be cleaved at both flanking sites but not at the middle site (because the presence of C–G destroys the *Eco*RI restriction site) and so will yield only one larger restriction fragment. A SNP that eliminates a restriction site is known as a **restriction fragment length polymorphism,** or **RFLP** (pronounced either as "riflip" or by spelling it out).

Because RFLPs change the number and size of DNA fragments produced by digestion with a restriction enzyme, they can be detected by the Southern blotting procedure discussed in Section 2.3. An example appears in **FIGURE 2.24**. In this case the labeled probe DNA hybridizes near the restriction site at the far left and identifies the position of this restriction fragment in the electrophoresis gel. The duplex molecule labeled "allele *A*" has a restriction site in the middle and, when cleaved and subjected to electrophoresis, yields a small band that contains sequences homologous to the probe DNA. The duplex molecule labeled "allele *a*" lacks the middle restriction site and yields a larger band. In this situation there can be three genotypes—*AA, Aa,* or *aa,* depending on which alleles are present in the homologous chromosomes—and all three genotypes can be distinguished as

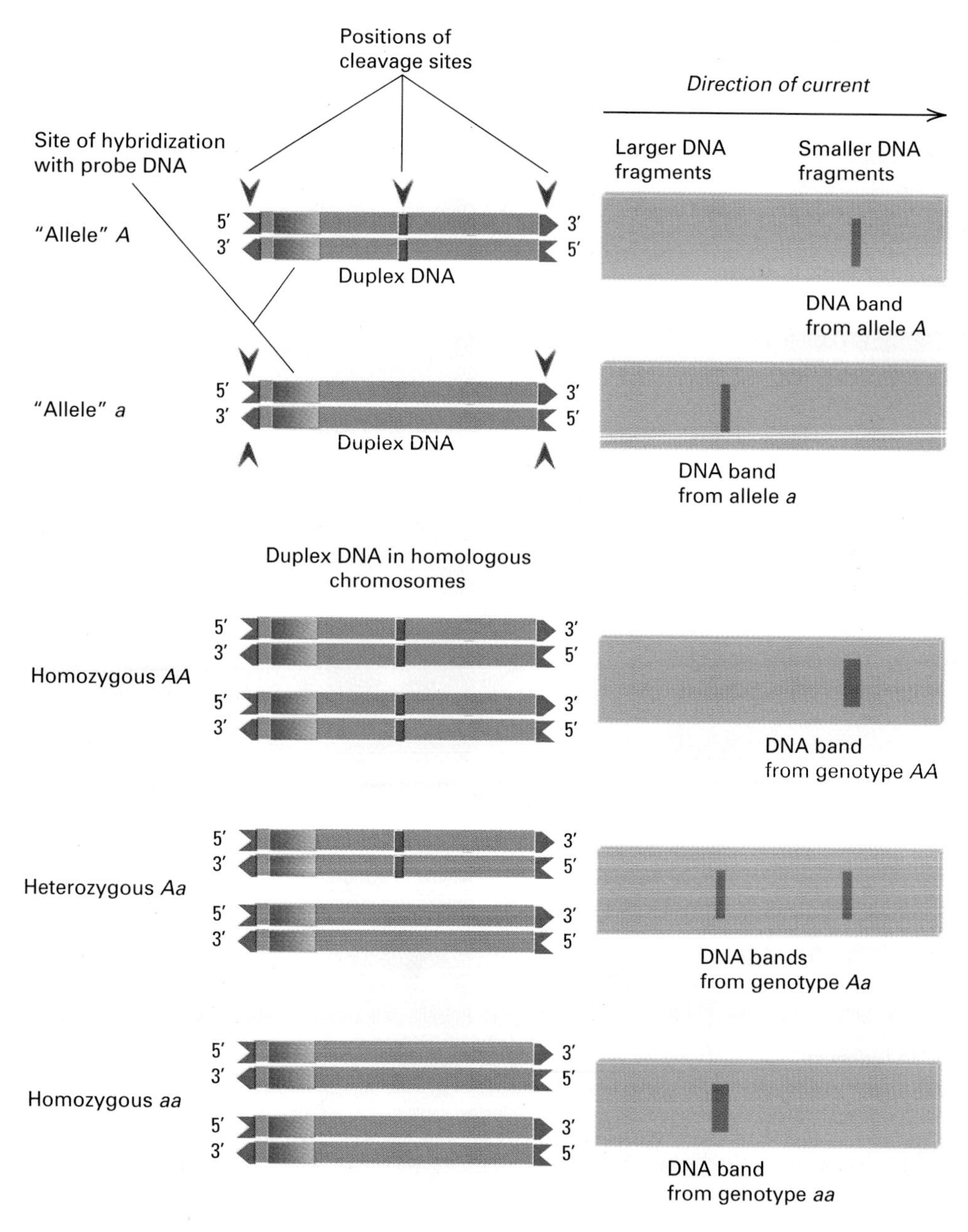


**FIGURE 2.24** In a restriction fragment length polymorphism (RFLP), alleles may differ in the presence or absence of a cleavage site in the DNA. In this example, the *a* allele lacks a restriction site that is present in the DNA of the *A* allele. The difference in fragment length can be detected by Southern blotting. RFLP alleles are codominant, which means (as shown at the bottom) that DNA from the heterozygous *Aa* genotype yields each of the single bands observed in DNA from homozygous *AA* and *aa* genotypes.

shown in the Figure 2.24. Homozygous *AA* yields only a small fragment, homozygous *aa* yields only a large fragment, and heterozygous *Aa* yields both a small and a large fragment. Because the presence of both the *A* and *a* alleles can be detected in heterozygous *Aa* genotypes, *A* and *a* are said to be **codominant.** In Figure 2.23, the bands from *AA* and *aa* have been shown as somewhat thicker than those from *Aa*, because each *AA* genotype has two copies of the *A* allele and each *aa* genotype has two copies of the *a* allele, compared with only one copy of each allele in the heterozygous genotype *Aa*.

## ■ Tandem Repeat Polymorphisms

An important type of DNA polymorphism results from differences in the number of copies of a DNA sequence that may be repeated many times in tandem at a particular site in a chromosome. Any particular chromosome may have any number of copies of the tandem repeat, typically ranging from ten to a few hundred. **FIGURE 2.25** illustrates DNA molecules that differ

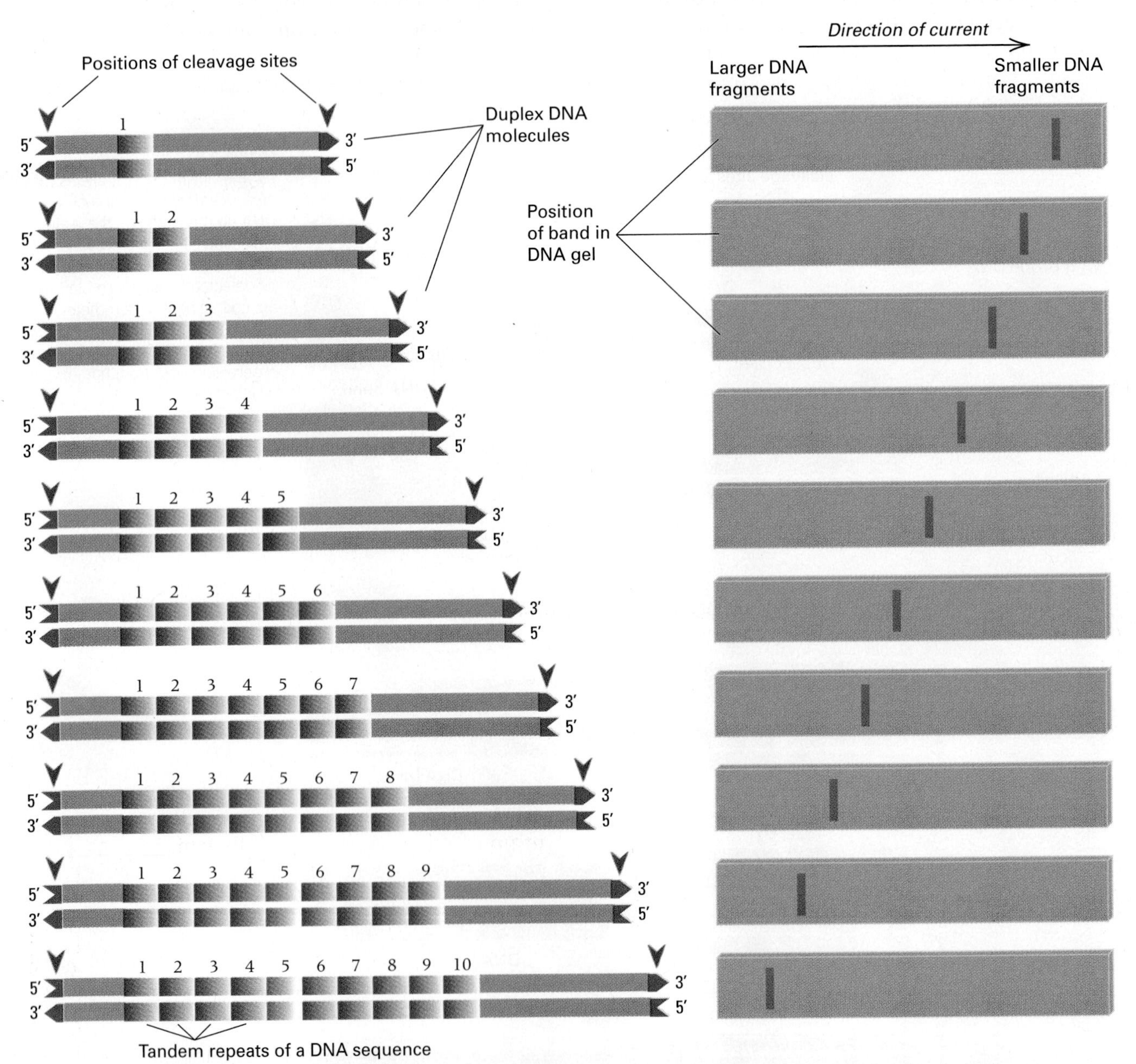


**FIGURE 2.25** A genetic polymorphism in which the alleles in a population differ in the number of copies of a DNA sequence (typically 2–60 bp) that is repeated in tandem along the chromosome. This example shows alleles in which the repeat number varies from 1 to 10. Cleavage at restriction sites flanking the repeat yields a unique fragment length for each allele. The alleles can also be distinguished by the size of the fragment amplified by PCR using primers that flank the repeat.

in the number of tandem repeats. In this case the number of repeats varies from 1 to 10. When any of the DNA molecules is cleaved with a restriction endonuclease that cleaves at sites flanking the tandem repeat, the size of the resulting restriction fragment is determined by the number of repeats that it contains. A molecule of duplex DNA containing the repeats can also be amplified by means of the polymerase chain reaction (PCR) using primers flanking the tandem repeat. Whether obtained by endonuclease digestion or PCR, the resulting DNA fragments increase in size according to the number of repeats they contain. Therefore, as shown at the right in Figure 2.25, two DNA molecules that differ in the number of copies of the tandem repeat can be distinguished because each will produce a different sized DNA fragment that can be separated by means of electrophoresis.

The acronym **SSR** stands for **simple sequence repeat.** An example of an SSR is the repeating sequence 5'-...ACACACAC...-3'. Two reasons make SSRs important in human genetic analysis:

1. SSRs are abundant in the human genome.
2. SSRs are often highly polymorphic in human populations.

As to their abundance in the human genome, the most common SSRs are tabulated in **TABLE 2.4**. There are many more dinucleotide repeats than trinucleotide repeats, but there are great differences in abundance within each class. The most common dinucleotide repeat is 5'-...ACACACAC...-3', which is present at more than 80,000 locations throughout the human genome. On average, the human genome has one SSR every 2 kb of human DNA, or about 1.5 million SSRs altogether.

**Table 2.4** Some simple sequence repeats in the human genome

| SSR repeat unit | Number of SSRs in the human genome |
|---|---|
| 5′-AC-3′ | 80,330 |
| 5′-AT-3′ | 56,260 |
| 5′-AG-3′ | 23,780 |
| 5′-GC-3′ | 290 |
| 5′-AAT-3′ | 11,890 |
| 5′-AAC-3′ | 7,540 |
| 5′-AGG-3′ | 4,350 |
| 5′-AAG-3′ | 4,060 |
| 5′-ATG-3′ | 2,030 |
| 5′-CGG-3′ | 1,740 |
| 5′-ACC-3′ | 1,160 |
| 5′-AGC-3′ | 870 |
| 5′-ACT-3′ | 580 |

Source: Data from International Human Genome Sequencing Consortium, *Nature* 409(2001): 860-921.

The tendency of SSRs to be polymorphic in copy number makes them very useful as genetic markers. The typical SSR has a repeat length of 2–9 nucleotides. A polymorphism in the number of repeats when the repeating unit is longer (in the range 10–60 nucleotides) is called a **variable number of tandem repeats (VNTR)**. Because the repeat unit is longer in VNTRs, DNA fragments that differ in the number of copies are more easily distinguished in electrophoresis gels than are restriction fragments from SSRPs (Figure 2.25). For this reason VNTRs have been useful in **DNA typing** (sometimes called **DNA fingerprinting**) for individual identification and for assessing the degree of genetic relatedness between individuals.

Not only do tandem repeats of DNA sequences tend to be polymorphic, but most polymorphisms tend to have a large number of alleles. Each "allele" corresponds to a DNA molecule with a different number of copies of the repeating unit, as illustrated in Figure 2.25. In other words, a polymorphism in the number of tandem repeats usually has **multiple alleles** in the population. Even with multiple alleles, however, any particular chromosome must carry only one of the alleles defined by the number of tandem repeats, and any individual genotype can carry at most two different alleles. Nevertheless, a large number of alleles implies an even larger number of genotypes. For example, even with only 10 alleles in a population, there could be 10 different homozygous genotypes and 45 different heterozygous genotypes.

More generally, with $n$ alleles there are a total of $n(n + 1)/2$ possible genotypes of which $n$ are homozygous and $n(n - 1)/2$ are heterozygous. Consequently, if an SSRP or VNTR has a relatively large number of alleles in a population, but no one allele is exceptionally common, then each of the many genotypes in the population will have a relatively low frequency. If the genotypes at 6–8 highly polymorphic loci are considered simultaneously, then each possible multiple-locus genotype is exceedingly rare. The very low frequency of any multiple-locus genotype is what gives tandem-repeat polymorphisms their utility in DNA typing. The use of DNA typing in criminal investigation is exemplified in Problems 2.29 and Challenge Problem 2 at the end of this chapter and discussed further in Chapter 17.

### ■ Copy-Number Polymorphisms (CNPs)

In addition to the small-scale variation in copy number represented by such genomic features as tandem repeats of short sequences (SSRPs and VNTRs), a substantial portion of the human genome can be duplicated or deleted in much larger but still submicroscopic chunks ranging from 1 kb to 1 Mb (Mb stands for *megabase pairs*, or one million base pairs). This type of variation is known as **copy-number polymorphism** or **CNP**. The extra or missing copies of the genome in CNPs can be detected by means of hybridization with oligonucleotides in DNA microarrays. Because each spot on the microarray consists of millions of identical copies of a particular oligonucleotide sequence, the number of these that undergo hybridization depends on how many copies of the complementary sequence are present in genomic DNA. A typical region is present in two copies (one inherited from the mother and the other from the father). If an individual has an extra copy of the region, the ratio of hybridization and therefore fluorescence intensity will be 3:2, whereas if an individual has a missing copy, the ratio of hybridization and therefore fluorescence intensity will be 1:2. These differences can readily be detected with DNA microarrays. Moreoever, because CNPs are relatively large, a microarray typically will have many different oligonucleotides that are complementary to sequences at intervals across the CNP; hence, the CNP will result in a increase or decrease in signal intensity of all the oligonucleotides included in the CNP. Current SNP chips also include about a million oligonucleotide probes designed to detect known CNPs.

CNPs by definition exceed 1 kb in size, but many are much larger. In one study of about 300 individuals with their ancestry in Africa, Europe, or Asia, approximately 1500 CNPs were discovered by hybridization with microarrays. These averaged 200–300 kb in length. In the aggregate, the CNPs included 300–450 million base pairs, or 10–15 percent of the nucleotides in the entire genome. Many of the CNPs were located in regions near known mutant genes associated with hereditary diseases. CNPs in alpha and beta hemoglobin genes are known to be associated with resistance to malaria, and CNPs in an HIV-1 receptor gene *CCL3* are associated with resistance to AIDS. The prevalence of CNPs has prompted the inclusion of CNP oligonucleotides onto DNA microarrays in order to be able to assess what effects CNPs may have on the risk of complex diseases such as diabetes or Alzheimer disease.

## 2.7 Applications of DNA Markers

Why are geneticists interested in DNA markers and DNA polymorphisms? Their interest can be justified on any number of grounds. In this section we consider the reasons most often cited.

### ■ Genetic Markers, Genetic Mapping, and "Disease Genes"

Perhaps the key goal in studying DNA polymorphisms in human genetics is to identify the chromosomal location of mutant genes associated with hereditary diseases. In the context of disorders caused by the interaction of multiple genetic and environmental factors, such as heart disease, cancer, diabetes, depression, and so forth, it is important to think of a harmful allele as a **risk factor** for the disease, which increases the probability of occurrence of the disease, rather than as a sole causative agent. This needs to be emphasized, especially because genetic risk factors are often called **disease genes**. For example, the major "disease gene" for breast cancer in women is the gene *BRCA1*. For women who carry a mutant allele of *BRCA1*, the lifetime risk of breast cancer is about 36 percent, and hence most women with this genetic risk factor do *not* develop breast cancer. On the other hand, among women who are not carriers, the lifetime risk of breast cancer is about 12 percent, and hence many women without the genetic risk factor *do* develop breast cancer. Indeed, *BRCA1* mutations are found in only 16 percent of affected women who have a family history of breast cancer. The importance of a genetic risk factor can be expressed quantitatively as the **relative risk,** which equals the risk of the disease in persons who carry the risk factor as compared to the risk in persons who do not. The relative risk for *BRCA1* equals 3.0 (calculated as 36 percent/12 percent).

The utility of DNA polymorphisms in locating and identifying disease genes results from **genetic linkage,** the tendency for genes that are sufficiently close together in a chromosome to be inherited together. Genetic linkage will be discussed in detail in Chapter 5, but the key concepts are summarized in **FIGURE 2.26**, which shows the location of many DNA polymorphisms along a chromosome that also carries a genetic risk factor denoted *D* (for disease gene). Each DNA polymorphism serves as a genetic marker for its own location in the chromosome. The importance of genetic linkage is that DNA markers that are sufficiently close to the disease gene will tend to be inherited together with the disease gene in pedigrees—and the closer the markers, the stronger this association. Hence, the initial approach

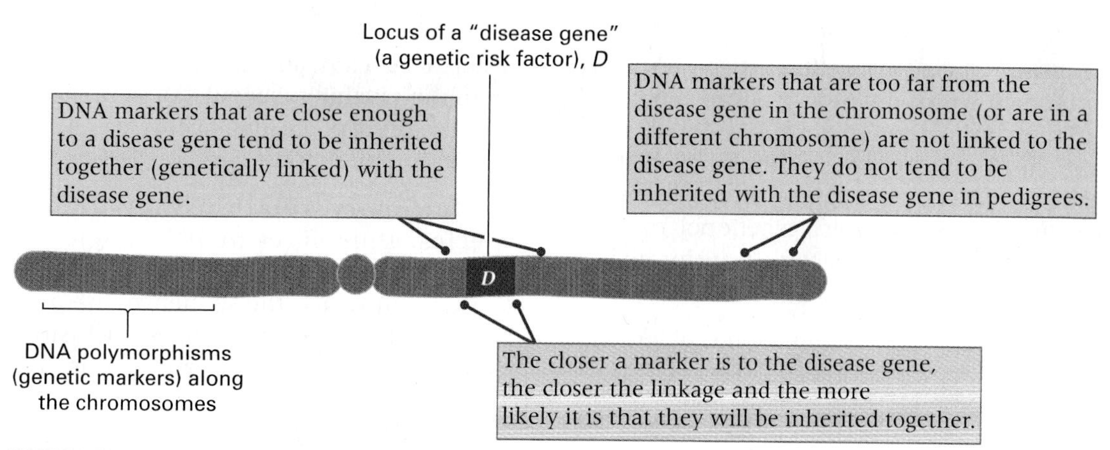
Locus of a "disease gene"
(a genetic risk factor), D
DNA markers that are close enough to a disease gene tend to be inherited together (genetically linked) with the disease gene.
DNA markers that are too far from the disease gene in the chromosome (or are in a different chromosome) are not linked to the disease gene. They do not tend to be inherited with the disease gene in pedigrees.
D
DNA polymorphisms (genetic markers) along the chromosomes
The closer a marker is to the disease gene, the closer the linkage and the more likely it is that they will be inherited together.

**FIGURE 2.26** Concepts in genetic localization of genetic risk factors for disease. Polymorphic DNA markers (indicated by the vertical lines) that are close to a genetic risk factor (*D*) in the chromosome tend to be inherited together with the disease itself. The genomic location of the risk factor is determined by examining the known genomic locations of the DNA polymorphisms that are linked with it.

to the identification of a disease gene is to find DNA markers that are genetically linked with the disease gene in order to identify its chromosomal location, a procedure known as **genetic mapping**. Once the chromosomal position is known, other methods can be used to pinpoint the disease gene itself and to study its functions.

If genetic linkage seems a roundabout way to identify disease genes, consider the alternative. The human genome contains approximately 30,000 genes. If genetic linkage did not exist, then we would have to examine 30,000 DNA polymorphisms, one in each gene, in order to identify a disease gene. But the human genome has only 23 pairs of chromosomes, and because of genetic linkage and the power of genetic mapping, it actually requires only a few hundred DNA polymorphisms to identify the chromosome and approximate location of a genetic risk factor.

## Other Uses for DNA Markers

DNA polymorphisms are widely used in all aspects of modern genetics because they provide a large number of easily accessed genetic markers for genetic mapping and other purposes. Among the other uses of DNA polymorphisms are the following.

**Individual identification.** We have already mentioned that DNA polymorphisms have application as a means of DNA typing (DNA fingerprinting) to identify different individuals in a population. DNA typing in other organisms is used to determine individual animals in endangered species and to identify the degree of genetic relatedness among individual organisms that live in packs or herds. For example, DNA typing in wild horses has shown that the wild stallion in charge of a harem of mares actually sires fewer than one-third of the foals.

**Epidemiology and food safety science.** DNA typing also has important applications in tracking the spread of viral and bacterial epidemic diseases, as well as in identifying the source of contamination in contaminated foods.

**Human population history.** DNA polymorphisms provide important information in anthropology to reconstruct the evolutionary origin, global expansion, and diversification of the human population.

**Improvement of domesticated plants and animals.** Plant and animal breeders have turned to DNA polymorphisms as genetic markers in pedigree studies to identify, by genetic mapping, genes that are associated with favorable traits in order to incorporate these genes into currently used varieties of plants and breeds of animals.

**History of domestication.** Plant and animal breeders also study genetic polymorphisms to identify the wild ancestors of cultivated plants and domesticated animals, as well as to infer the practices of artificial selection that led to genetic changes in these species during domestication.

**DNA polymorphisms as ecological indicators.** DNA polymorphisms are being evaluated as biological indicators of genetic diversity in key indicator species present in biological communities exposed to chemical, biological, or physical stress. They are also used to monitor genetic diversity in endangered species and species bred in captivity.

**Evolutionary genetics.** DNA polymorphisms are studied in an effort to describe the patterns in which different types of genetic variation occur throughout the genome, to infer the evolutionary mechanisms by which genetic variation is maintained, and to illuminate the processes by which genetic polymorphisms within species become transformed into genetic differences between species.

**Population studies.** Population ecologists employ DNA polymorphisms to assess the level of genetic variation in diverse populations of organisms that differ in genetic organization (prokaryotes, eukaryotes, organelles), population size, breeding structure, or life-history characters, and they use genetic polymorphisms within subpopulations of a species as indicators of population history, patterns of migration, and so forth.

**Evolutionary relationships among species.** Differences in DNA sequences between species is the basis of *molecular phylogenetics,* in which the sequences are analyzed to determine the ancestral history (phylogeny) of the species and to trace the origin of morphological, behavioral, and other types of adaptations that have arisen in the course of evolution.

## CHAPTER SUMMARY

- A DNA strand is a polymer of A, T, G and C deoxyribonucleotides joined 3′-to-5′ by phosphodiester bonds.
- The two DNA strands in a duplex are held together by hydrogen bonding between the A–T and G–C base pairs and by base stacking of the paired bases.
- Each type of restriction endonuclease enzyme cleaves double-stranded DNA at a particular sequence of bases usually four or six nucleotides in length.
- The DNA fragments produced by a restriction enzyme can be separated by electrophoresis, isolated, sequenced, and manipulated in other ways.
- Separated strands of DNA or RNA that are complementary in nucleotide sequence can come together (hybridize) spontaneously to form duplexes.
- DNA replication takes place only by elongation of the growing strand in the 5′-to-3′ direction through the addition of successive nucleotides to the 3′ end.
- In the polymerase chain reaction, short oligonucleotide primers are used in successive cycles of DNA replication to amplify selectively a particular region of a DNA duplex.
- Genetic markers in DNA provide a large number of easily accessed sites in the genome that can be used to identify the chromosomal locations of disease genes, for DNA typing in individual identification, for the genetic improvement of cultivated plants and domesticated animals, and for many other applications.

## REVIEW THE BASICS

- What four bases are commonly found in the nucleotides in DNA? Which form base pairs?
- Which chemical groups are present at the extreme 3′ and 5′ ends of a single polynucleotide strand?
- What does it mean to say that a single strand of DNA strand has a polarity? What does it mean to say that the DNA strands in a duplex molecule are antiparallel?
- What are restriction enzymes and why are they important in the study of particular DNA fragments? What does it mean to say that most restriction sites are palindromes?
- Describe how a Southern blot is carried out. Explain what it used for. What is the role of the probe?
- How does the polymerase chain reaction work? What is it used for? What information about the target sequence must be known in advance? What is the role of the oligonucleotide primers?
- What is a DNA marker? Explain how harmless DNA markers can serve as aids in identifying disease genes through genetic mapping.
- Define and given an example of each of the following key genetic terms: locus, allele, genotype, heterozygous, homozygous, phenotype.

## GUIDE TO PROBLEM SOLVING

**Problem 1** Linear double-stranded DNA from a bacteriophage is isolated, labeled at both 5′ ends with $^{32}P$, and digested with restriction enzymes. *Mlu*I produces fragments of 5, 7, and 13 kb. A Southern blot of this digest shows the radioactive label in the bands at 5 and 7 kb. *Hind*III cleaves the same molecule into 3, 7 and 15 kb fragments, and in this case the 3 and 7 kb bands contain the $^{32}P$. Digestion with both enzymes produces fragments of sizes 2, 3, 4, 5, and 11 kb.

**(a)** Draw a diagram of the linear DNA showing the relative positions of the *Mlu*I and *Hind*III sites and the distances in kb between them.

**(b)** A probe labeled with $^{32}P$ made from one of the bacteriophage genes hybridizes with the 5 kb fragment from the *Mlu*I digest and the 3 kb and 15 kb fragments from the *Hind*III digest.

Show the approximate location of the gene on the restriction map.

**Answer**

**(a)** The best way to start is to create separate maps for the *Mlu*I and *Hind*III sites. The 13 kb fragment in the *Mlu*I digest is not labeled. This implies that this fragment is in the middle of DNA molecule.
The restriction map for *Mlu*I digest is therefore

**I**

In the *Hind*III digest the 15 kb fragment is not labeled and, therefore, this fragment is in the middle. There are two ways to orient the restriction map of the *Hind*III digest relative to the *Mlu*I map.

This is one way

**II**

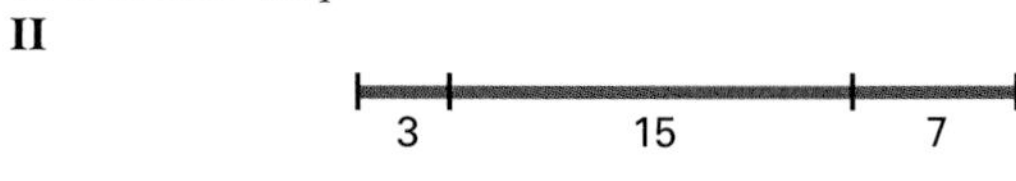


This is the other way

**III**

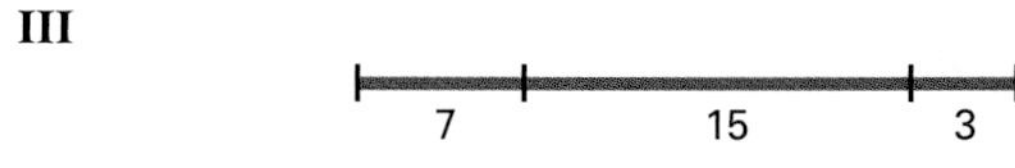


When we combine maps I and II into map IV, the predicted double-digest fragments are 2, 3, 7 and 13 kb, and this prediction is different from the result observed. Hence, map IV and therefore map II are wrong.

**IV**

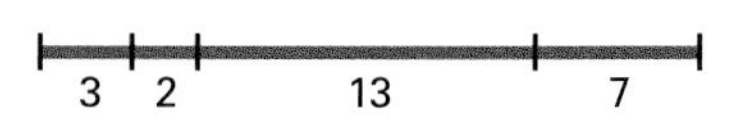


On the other hand, combining maps I and III yields the outcome shown here as map V, which does explain the observations.

**V**

**(b)** The labeled probe from the bacteriophage gene hybridizes with the 5 kb fragment from the *Mlu*I digest and the 3 kb and 15 kb fragments from the *Hind*III digest. These observations allow one to conclude that the 2 kb fragment in the correct restriction map V contains sequences corresponding to the gene, and the 5′ and 3′ ends of the gene overlap across the 5 and 11 kb fragments. The approximation location of the gene in the restriction map is shown by the horizontal bar in the accompanying diagram.

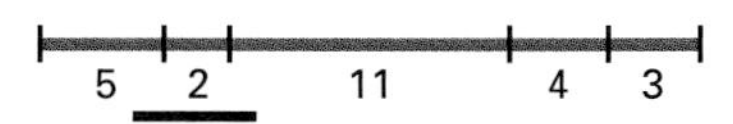


**Problem** A geneticist plans to use the polymerase chain reaction (PCR) to amplify part of the DNA sequence shown below, using oligonucleotide primers that hybridize in the regions marked in red. (These are illustrative only and too short to be used in practice.) Specify the sequence of the primers that should be used, including the polarity, and deduce the sequence of the amplified DNA fragment.

5′-GCGAAACGATCCTCATCCTGTCTCTTGATCAGAGCTTGATCCCCTG-3′
3′-CGCTTTGCTAGGAGTAGGACAGAGAACTAGTCTCGAACTAGGGGAC-5′

**Answer** The primers should be able to pair with the chosen primer sites and must be oriented with their 3′ ends facing each other. Thus the "forward" primer (the one that is elongated in left-to-right direction) should have the sequence 5′-ACGAT-3′, and the "reverse" primer (the one that is elongated in right-to-left direction) should have the sequence 3′-ACTAG-5′. *Because* the convention for writing nucleic acid sequences is to put 5′ end on the left, the reverse primer is 5′-GATCA-3′.

**Problem** The genetic material of the bacteriophage M13 is a single-stranded DNA molecule of 6407 nucleotides, whose complete sequence can be found in publicly accessible databases such as GenBank. Upon M13 DNA being introduced into a bacterial cell, a complementary DNA strand is synthesized by bacterial enzymes, resulting in a double-stranded replicative form of the phage genome. How would you determine whether or not both strands of the replicative form are transcribed within an infected cell?

**Answer** To determine whether both DNA strands are templates for RNA synthesis, one could isolate the replicative form of the phage DNA, separate the two strands, and test each strand separately for the ability to hybridize with mRNA found in M13-infected cells. Alternatively (and less definitively), one could examine the genome sequence of M13 in GenBank and search both DNA strands for open reading frames that could code for proteins. As it happens, only one strand of M13 contains such open reading frames, and hence only one DNA strand is likely to be transcribed.

**Problem** A plasmid of 6200 base pairs (bp) includes at least three complete copies of a viral gene arranged in tandem. Cleavage of the plasmid with the restriction enzyme *Hind*III results in fragments of 900, 1300, and 4000 bp. The tandem copies of the gene are contained within only one of the *Hind*III fragments. The gene encodes a protein of 405 amino acids. Which *Hind*III fragment is likely to contain the gene copies?

**Answer** To encode a protein of 405 amino acids requires at least 1215 bp. Three tandem copies of the gene would therefore require 3645 bp, and four copies 4860 bp. The smallest *Hind*III fragment is not long enough to contain even one copy of the gene. The 1300 bp fragment could contain one copy but not two or more. Only the largest fragment could include three copies of the gene, and it could not include more than three copies.

**2.1** What chemical groups are present at the 3′ and 5′ ends of a single polynucleotide molecule?

**2.2** In the deoxyribonucleotide shown below, which carbon atom carries the phosphate group and which carries the 3′ hydroxyl group?

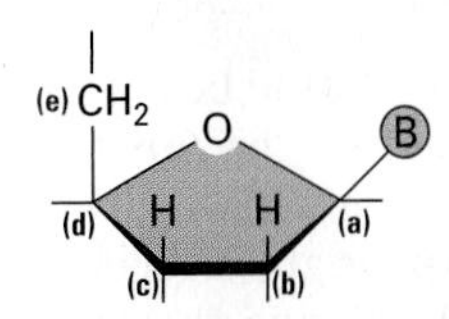


**2.3** Many restriction enzymes produce restriction fragments that have "sticky ends." What does this mean?

**2.4** Which of the following sequences are palindromes and which are not? Explain your answer.

**(a)** 5′-CCGG-3′
**(b)** 5′-TTTT-3′
**(c)** 5′-GCTAGC-3′
**(d)** 5′-CCGCTC-3′
**(e)** 5′-AAGGTT-3′

**2.5** The list below gives half of each of a set of palindromic restriction sites. Replace the Ns to complete the sequence of each restriction site.

**(a)** 5′-AGNN-3′
**(b)** 5′-ATGNNN-3′
**(c)** 5′-ATTNNN-3′
**(d)** 5′-NNNAGC-3′

**2.6** Apart from nucleotide sequence, what is different about the ends of restriction fragments produced by the following restriction enzymes? (The downward arrow represents the site of cleavage in each strand.)

**(a)** *Sca*I (5′-AGT↓ACT-3′)
**(b)** *Nhe*I (5′-G↓CTAG-3′)
**(c)** *Cfo*I (5′-GCG↓C-3′)

**2.7** A solution contains double-stranded DNA fragments of size 4 kb, 8 kb, 10 kb, and 13 kb that are separated in an electrophoresis gel. In the accompanying diagram of the gel, match the fragments sizes with the correct bands.

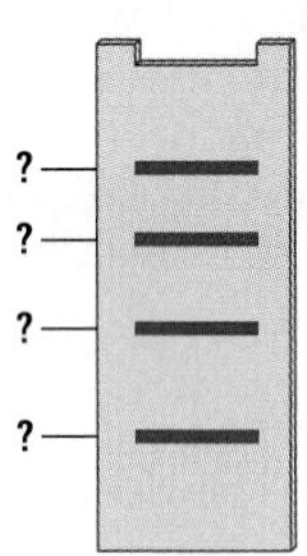


**2.8** The linear DNA fragment shown here has cleavage sites for *Aat*II (*A*) and *Xho*I (*X*). In the accompanying diagram of an electrophoresis gel, indicate the positions at which bands would be found after digestion with

**(a)** *Aat*II alone
**(b)** *Xho*I alone
**(c)** *Aat*II and *Xho*I together

The dashed lines on the right indicate the positions to which bands of 1–12 kb would migrate.

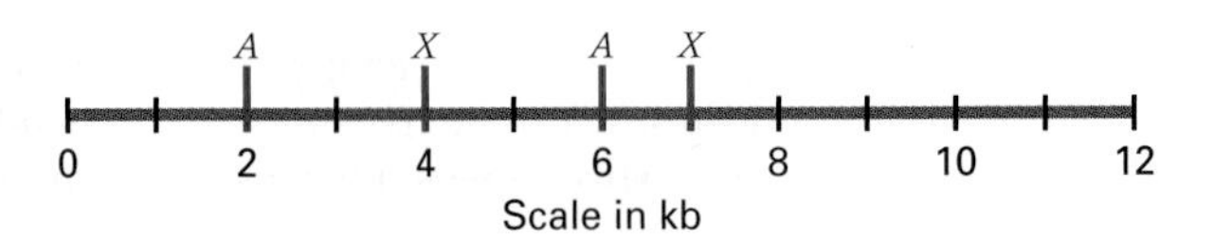


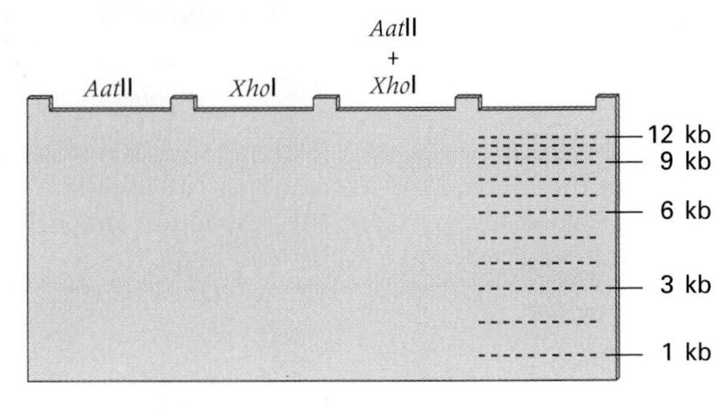


**2.9.** The circular DNA molecule shown here has cleavage sites for *Aat*II and *Xho*I. In the accompanying diagram of an electrophoresis gel, indicate the positions at which bands would be found after digestion with

**(a)** *Aat*II alone
**(b)** *Xho*I alone
**(c)** *Aat*II and *Xho*I together

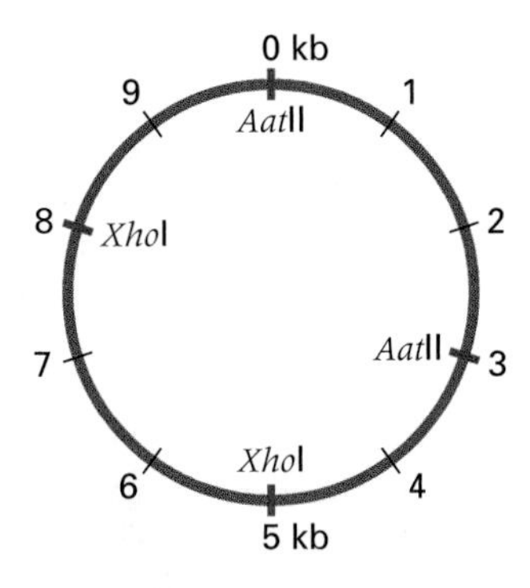


The dashed lines indicate the positions to which bands of 1–12 kb would migrate.

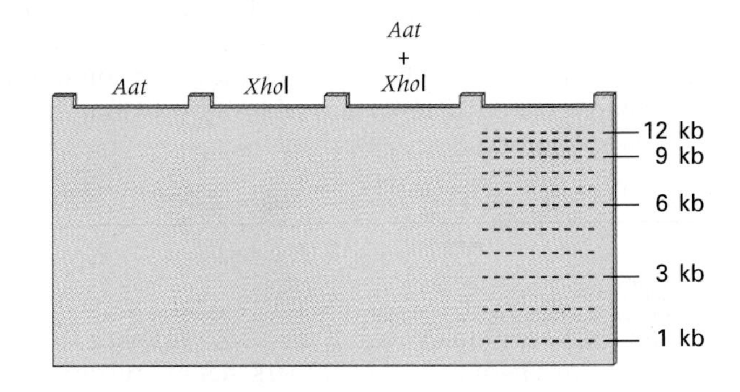


**2.10** Consider the accompanying diagram of a region of duplex DNA, in which the Bs represent bases in Watson-Crick pairs. Specify as precisely as possible the identity of:

**(a)** $B_5$, assuming that $B_1$ = A.
**(b)** $B_6$, assuming that $B_2$ = C.

**(c)** $B_7$, assuming that $B_3$ = Any purine.
**(d)** $B_8$, assuming that $B_4$ = A or T

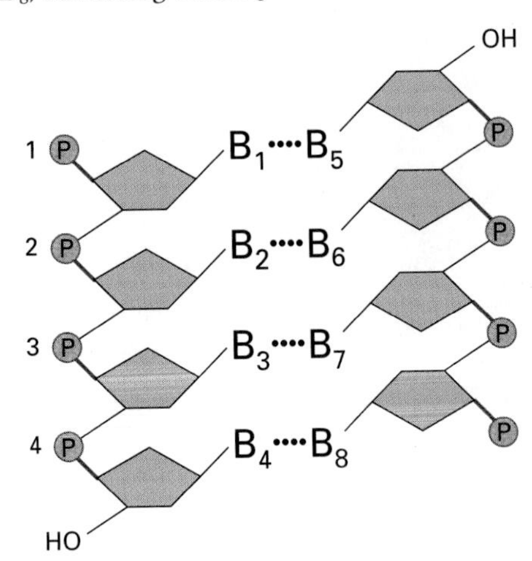


**2.11** In the precursor nucleotides of the DNA duplex diagrammed in Problem 2.10, with which base was each of the phosphate groups 1–4 associated with prior to its incorporation into the polynucleotide strand?

**2.12** In a random sequence consisting of equal proportions of all four nucleotides, what is probability that a particular short sequence of nucleotides matches a restriction site for:
**(a)** A restriction enzyme with a 4-base cleavage site?
**(b)** A restriction enzyme with a 6-base cleavage site?
**(c)** A restriction enzyme with an 8-base cleavage site?

**2.13** In a random sequence consisting of equal proportions of all four nucleotides, what is the average distance between restriction sites for:
**(a)** A restriction enzyme with a 4-base cleavage site?
**(b)** A restriction enzyme with a 6-base cleavage site?
**(c)** A restriction enzyme with an 8-base cleavage site?

**2.14** If *Haemophilus influenzae* DNA were essentially a random sequence of $4.6 \times 10^6$ bp with equal proportions of all four nucleotides (this is an oversimplification), approximately how many restriction fragments would be expected from cleavage with
**(a)** A "4-cutter" restriction enzyme?
**(b)** A "6-cutter" restriction enzyme?
**(c)** An "8-cutter" restriction enzyme?

**2.15** Consider the restriction enzymes *Asi*SI (cleavage site 5′-GCGAT↓CGC-3′) and *Bsi*EI (cleavage site 5′-CGAT↓CG-3′), where the downward arrow denotes the site of cleavage in each strand. Is every *Asi*SI site a *Bsi*EI site? Is every *Bsi*EI site a *Asi*SI site? Explain your answer.

**2.16** A DNA duplex with the sequence shown below is cleaved with *Kas*I (cleavage site 5′-GG↓CGCC-3′), where the arrow denotes the site of cleavage in each strand. If the resulting fragments were brought together in the same order as in the original duplex and the breaks in the backbones sealed, what possible DNA duplexes would be expected?

5′-CTGGGGCGCCCTCGTCAGCGAGGGGGCGCCGAT-3′
3′-GACCCCGCGGGAGCAGTCGCTCCCCCGCGGCTA-5′

**2.17** The restriction enzymes *Pst*I, *Pvu*II, and *Mlu*I have the following restriction sites, where the arrow indicates the site of cleavage in each strand:

*Pst*I 5′-CTGCA↓G-3′
*Pvu*II 5′-CAG↓CTG-3′
*Mlu*I 5′-A↓CGCGT-3′

A DNA duplex with the following sequence is digested

5′-ATGCCCTGCAGTACCATGACGCGTTACGCAGCTGATCGAAACGCGTATATATGCC-3′
3′-TACGGGACGTCATGGTACTGCGCAATGCGTCGACTAGCTTTGCGCATATATACGG-5′

What fragments would result from cleavage with:
**(a)** *Pst*I?
**(b)** *Pvu*II?
**(c)** *Mlu*I?

**2.18** Consider the sequence:

5′-CTGCAGGTG-3′
3′-GACGTCCAC-5′

If this sequence were cleaved with *Pst*I (5′-CTGCAG-3′), could it still be cleaved with *Pvu*II (5′-CAGCTG-3′)? If it were cleaved with *Pvu*II, could it still be cleaved with *Pst*I? Explain your answer.

**2.19** A circular DNA molecule is cleaved with *Afe*I, *Nhe*I, or both restriction enzymes together. The accompanying diagram shows the resulting electrophoresis gel, with the band sizes indicated. Draw a diagram of the circular DNA showing the relative positions of the *Afe*I and *Nhe*I sites.

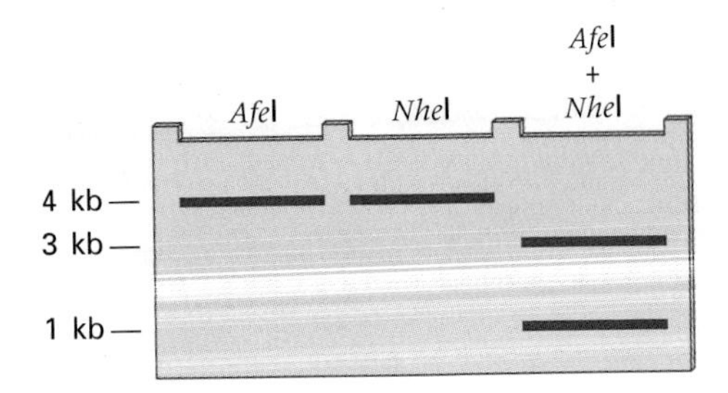


**2.20** In the diagrams of DNA fragments shown here, the tick marks indicate the positions of restriction sites for a particular restriction enzyme. A mixture of the two types of molecules is digested and analyzed with a Southern blot using either probe A or probe B, which hybridize to the fragments at the positions shown by the rectangles. In the accompanying gel diagram, indicate the bands that would result from each of these probes. (The scale on the right shows the expected positions of fragments from 1–12 kb.)

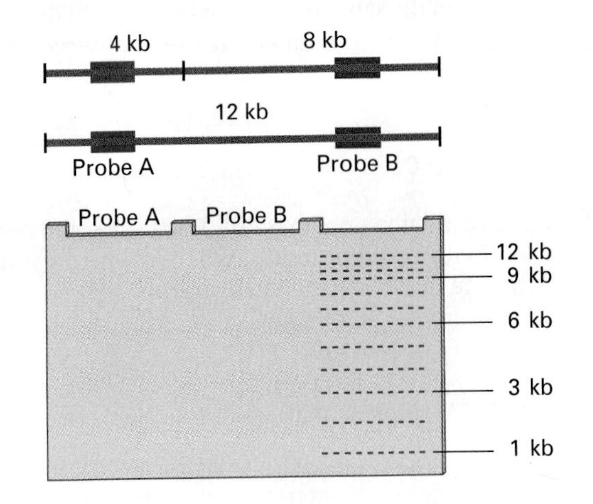


**2.21** In the accompanying diagram, the tick marks indicate the positions of restriction sites in two alternative DNA fragments that can be present at the a locus in a human chromosome. An RFLP analysis is carried out, using probe DNA that hybridizes with the fragments at the position shown by the rectangle. With respect to this RFLP, how many genotypes are possible? (Use the symbol $A_1$ to refer to the allele that yields the upper DNA fragment, and $A_2$ to refer to the allele that yields the lower DNA fragment.) In the accompanying gel diagram, indicate the genotypes across the top and the phenotype (band position or positions) expected of each genotype. (The scale on the right shows the expected positions of fragments from 1–12 kb.)

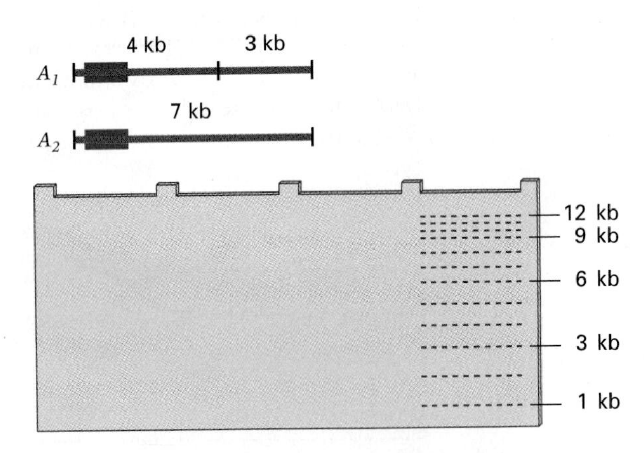


**2.22** The accompanying diagram shows the DNA fragments associated with an RFLP in the human genome revealed by a probe that hybridizes at the position shown by the rectangle. The tick marks are cleavage sites for the restriction enzyme used in the RFLP analysis. How many alleles would this restriction enzyme and this probe detect? How many genotypes would be possible? In the accompanying gel diagram, indicate the phenotype (pattern of bands) expected of each genotype. (The scale on the right shows the expected positions of fragments from 1–12 kb.)

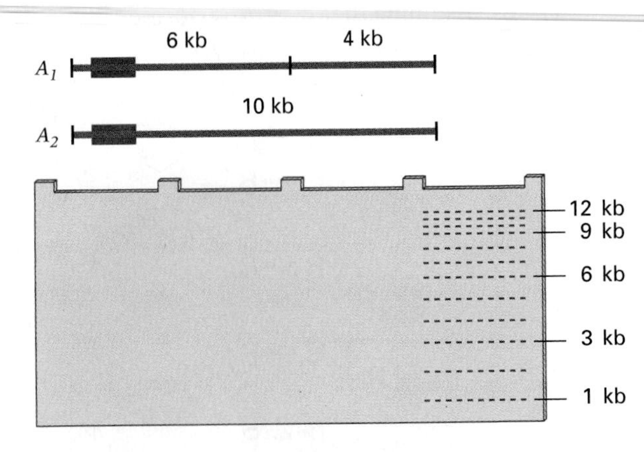


**2.23** The thick horizontal lines shown here represent alternative DNA molecules at a particular locus in a human chromosome. The tick marks indicate the positions of restriction sites for a particular restriction enzyme. Genomic DNA from a sample of people is digested and analyzed by a Southern blot using a probe DNA that hybridizes at the position shown by the rectangle. How many possible RFLP alleles would be observed in the sample? How many genotypes?

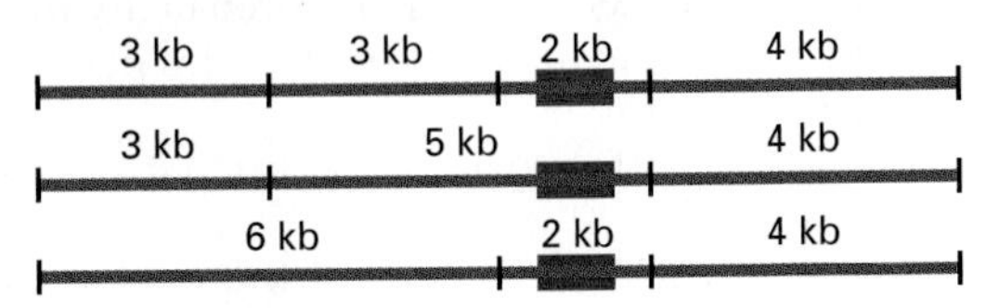


**2.24** The RFLP described in Problem 2.23 is analyzed with the same restriction enzyme but a different probe, which hybridizes at the site indicated here by the rectangle. How many RFLP alleles would be found in this case? How many genotypes? (Use the symbols $A_1$, $A_2$, etc. to indicate the alleles.) In the accompanying gel diagram, indicate the genotypes across the top and the phenotype (band position or positions) expected of each genotype. (The scale on the right shows the expected positions of fragments from 1–12 kb.)

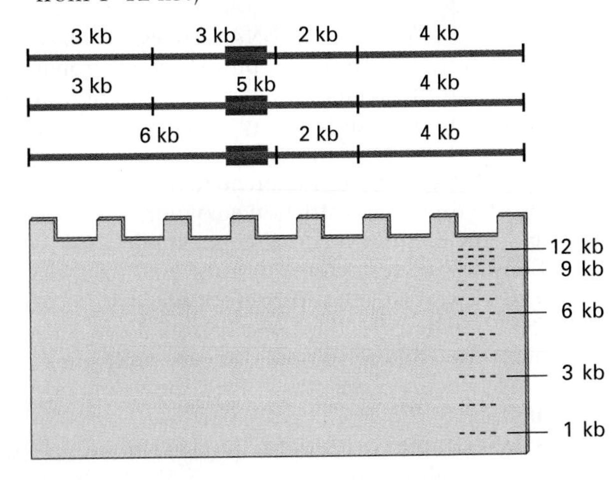

**2.25** If pentamers were long enough to serve as specific oligonucleotide primers for PCR (in practice they are too short), what DNA fragment would be amplified using the "forward" primer 5′-AATGC-3′ and the "reverse" primer 3′-GCATG-5′ acting on the double-stranded DNA molecule shown here?

5′-GATTACCGGTAAATGCCGGATTAACCCGGGTTATCAGGCCACGTACAACTGGAGTCC-3′
3′-CTAATGGCCATTTACGGCCTAATTGGGCCCAATAGTCCGGTGCATGTTGACCTCAGG-5′

**2.26** Would the primer pairs 5′-AATGC-3′ and 3′-GCATG-5′ amplify the same fragment described in the Problem 2.25? Explain your answer.

**2.27** Suppose that a fragment of human DNA of length 3 kb is to be amplified by PCR. The total genome size is $3 \times 10^6$ kb, which equals $3 \times 10^9$ base pairs.
**(a)** Prior to amplification, what fraction of the total DNA does the target sequence constitute?
**(b)** What fraction does it constitute after 10 cycles of PCR?
**(c)** After 20 cycles of PCR?
**(d)** After 30 cycles of PCR?

**2.28** Some polymorphisms can be identified using oligonucleotides with randomly chosen sequences as primers. These are known and RAPD polymorphisms, where RAPD stands for *randomly amplified polymorphic DNA*. Typically, the primer oliginucleotides are sufficiently short that they will hybridize at multiple sites in the genome, and thus serve to amplify multiple fragments. An amplified fragment is regarded as polymorphic if it can be amplified in some individuals but not others. The accompanying gel shows a typical pattern observed with randomly amplified DNA from four individuals. In this case, which bands are the RAPD polymorphisms?

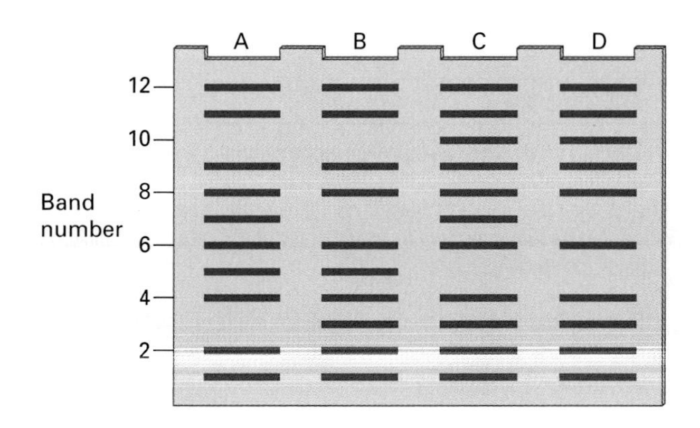


**2.29** A cigarette butt found at the scene of a robbery is found to have a sufficient number of epithelial cells stuck to the paper that the DNA can be extracted and analyzed by DNA typing. Shown here are the results of typing for three probes (locus 1–locus 3) of the evidence (X) and cells from seven suspects (A–G). Which of the suspects can be excluded? Which cannot be excluded? Can you identify the perpetrator? Explain you reasoning.

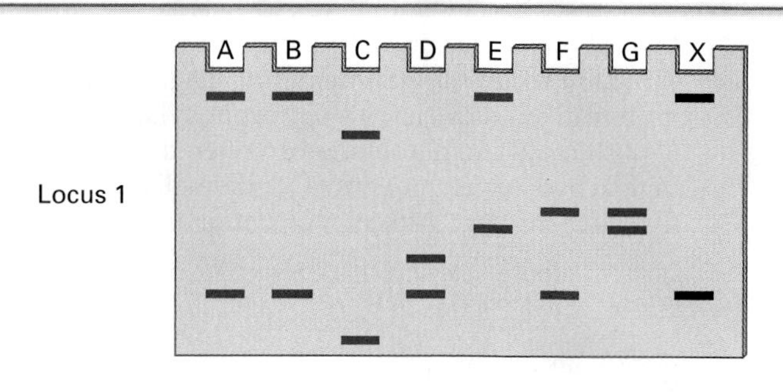


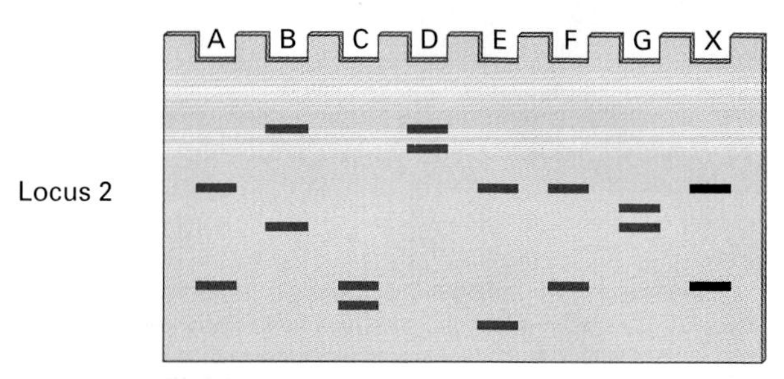


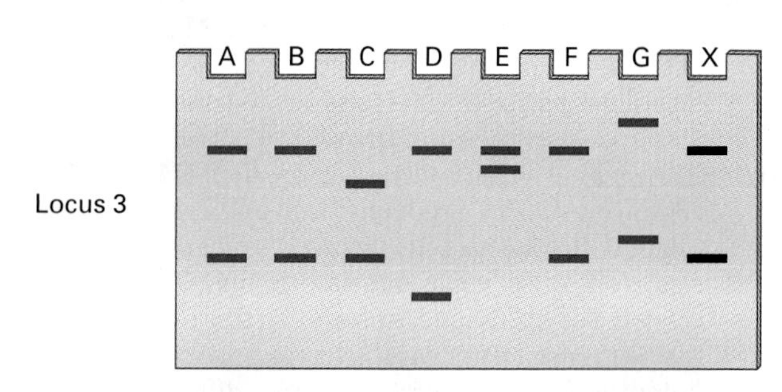


**2.30** A woman is uncertain which of two men is the father of her child. DNA typing is carried out on blood from the child (C), the mother (M), and on the two males (A and B), using probes for a highly polymorphic DNA marker on two different chromosomes ("locus 1" and "locus 2"). The result is shown in the accompanying diagram. Does either or both of the tested loci rule out any of the males as being the possible father? Explain your reasoning.

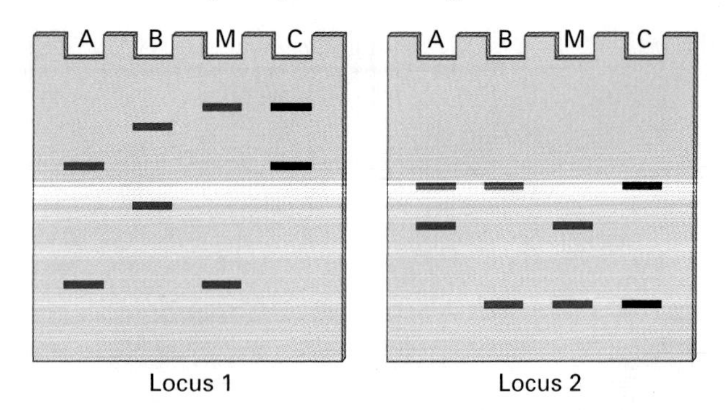

**2.31** Snake venom diesterase cleaves the chemical bonds shown in red in the accompanying diagram, leaving mononucleotides that are phosphorylated in the 3′ position. If the phosphates numbered 2 and 4 are radioactive, which mononucleotides will be radioactive after cleavage with snake venom diesterase?

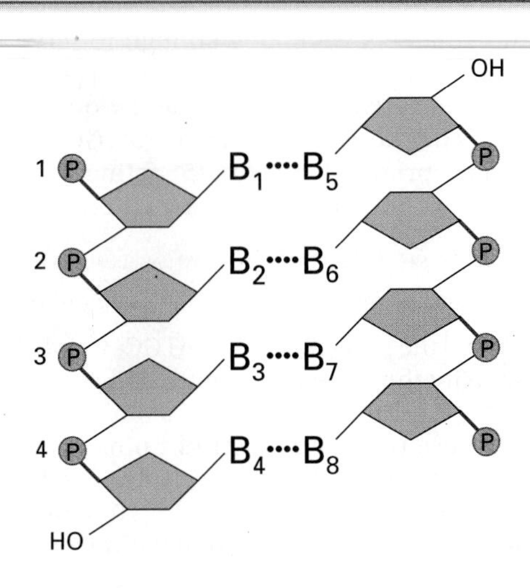


## CHALLENGE PROBLEMS

**Challenge Problem 1** The genome of *Drosophila melanogaster* is $180 \times 10^6$ bp and a fragment of size 1.8 kb is to be amplified by PCR. How many cycles of PCR are necessary for the amplified target sequence to constitute at least 99 percent of the total DNA?

**Challenge Problem 2** A young victim of murder is found in an advanced state of decomposition and cannot be identified. Police suspect the victim is a one of five persons reported by their parents as missing. DNA typing is carried out on tissues from the victim (X) and on the five sets of parents (A–E), using probes for a highly polymorphic DNA marker on two different chromosomes ("locus 1" and "locus 2"). The result is shown in the accompanying diagram. How do you interpret the fact that genomic DNA from each individual yields two bands? Can you identify the parents of the victim? Explain your reasoning.

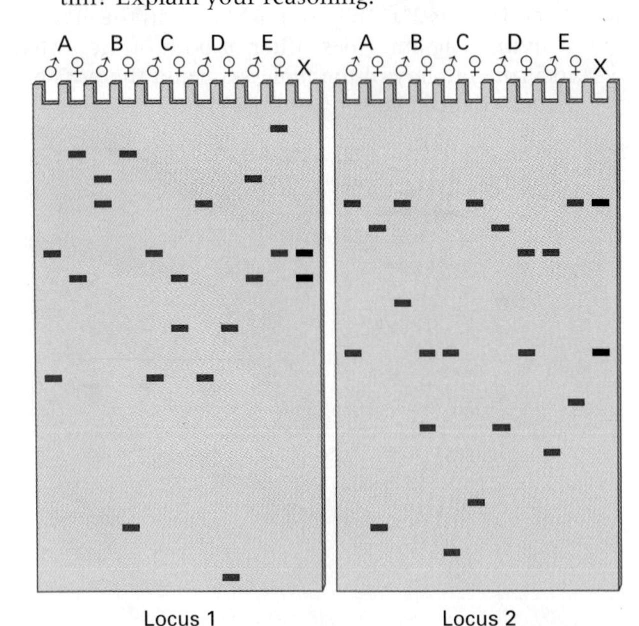


**Challenge Problem 3** The snake venom diesterase enzyme described in the Problem 2.31 was originally used in a procedure called "nearest neighbor" analysis. In this procedure, a DNA strand is synthesized in the presence of all four trinucleotides, one of which carries a radioactive phosphate in the (innermost) position. Then the DNA is digested to completion with snake venom diesterase, and the resulting mononucleotides are separated and assayed for radioactivity. Examine the diagram in Problem 2.31 and explain:

**(a)** How this procedure reveals the "nearest neighbors" of the radioactive nucleotide.

**(b)** Whether the "nearest neighbor" is on the 5′ or the 3′ side of the labeled nucleotide.

### GENETICS *on the web*

GeNETics on the Web will introduce you to some of the most important sites for finding genetics information on the Internet. To explore these sites, visit the Jones and Bartlett companion site to accompany *Genetics: Analysis of Genes and Genomes, Seventh Edition* at http://biology.jbpub.com/book/genetics.

There you will find a chapter-by-chapter list of highlighted keywords. When you select one of the keywords, you will be linked to a Web site containing information related to that keyword.

# CHAPTER 3

# Transmission Genetics: The Principle of Segregation

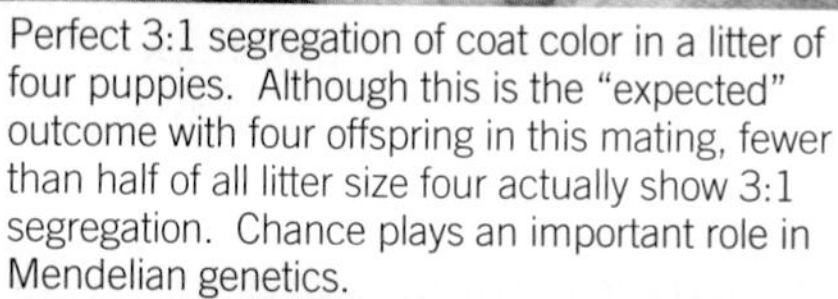

Perfect 3:1 segregation of coat color in a litter of four puppies. Although this is the "expected" outcome with four offspring in this mating, fewer than half of all litter size four actually show 3:1 segregation. Chance plays an important role in Mendelian genetics.

## CHAPTER OUTLINE


**3.1** Morphological and Molecular Phenotypes

**3.2** Segregation of a Single Gene
- Phenotypic Ratios in the $F_2$ Generation
- The Principle of Segregation
- Verification of Segregation
- The Testcross and the Backcross

**3.3** Segregation of Two or More Genes
- The Principle of Independent Assortment
- The Testcross with Unlinked Genes
- Three or More Genes

**3.4** Probability in Genetic Analysis
- Elementary Outcomes and Events
- Probability of the Union of Events
- Probability of the Intersection of Events
- Conditional Probability
- Bayes Theorem

**3.5** Human Pedigree Analysis
- Characteristics of Dominant and Recessive Inheritance
- Most Human Genetic Variation is Not "Bad"
- Molecular Markers in Human Pedigrees

**3.6** Incomplete Dominance and Epistasis
- Multiple Alleles
- Human ABO Blood Groups
- Epistasis

CONNECTION What Did Gregor Mendel Think He Discovered?
Gregor Mendel 1866
*Experiments on Plant Hybrids*

CONNECTION This Land is Your Land
The Huntington's Disease Collaborative Research Group 1993
*A Novel Gene Containing a Trinucleotide Repeat That Is Expanded and Unstable on Huntington's Disease Chromosomes*

In this chapter we consider how genes are transmitted from parents to offspring and how this determines the distribution of genotypes and phenotypes among related individuals. The study of the inheritance of traits constitutes **transmission genetics.** This subject is also called **Mendelian genetics** because the underlying principles were first deduced from experiments in garden peas (*Pisum sativum*) carried out in the years 1856–1863 by Gregor Mendel, a monk at the monastery of St. Thomas in the town of Brno (Brünn), in the Czech Republic. He reported his experiments to a local natural history society, published the results and his interpretation in its scientific journal in 1866, and began exchanging letters with one of the leading botanists of the time. His experiments were careful and exceptionally well documented, and his paper contains the first clear exposition of the statistical rules governing the transmission of genes from generation to generation. Nevertheless, Mendel's paper was ignored for 34 years until its significance was finally appreciated.

## 3.1 Morphological and Molecular Phenotypes

Until the advent of molecular genetics, geneticists dealt mainly with morphological traits, in which the differences between organisms can be expressed in terms of color, shape, or size. Mendel studied seven morphological traits contrasting in seed shape, seed color, flower color, pod shape, and so forth (**FIGURE 3.1**). Perhaps the most widely known example of a contrasting Mendelian trait is round versus wrinkled seeds. As pea seeds dry, they lose water and shrink. Round seeds are round because they shrink uniformly; wrinkled seeds are wrinkled because they shrink irregularly. The wrinkled phenotype is due to the absence of a branched-chain form of starch known as *amylopectin,* which is not synthesized in wrinkled seeds owing to a defect in the enzyme starch-branching enzyme I (SBEI).

The nonmutant, or **wildtype,** allele of the gene for SBEI is designated *W* and the mutant allele *w*. Seeds that are heterozygous *Ww* have only half as much SBEI enzyme as wildtype homozygous *WW* seeds, but this half the normal amount of enzyme produces enough amylopectin for the heterozygous *Ww* seeds to shrink uniformly and remain phenotypically round. Hence, with respect to seed shape, genotypes *WW* and *Ww* have the same phenotype (round). The *W* allele is called the **dominant allele** and *w* is called the **recessive allele.**

The molecular basis of the wrinkled mutation is that the *SBEI* gene has become interrupted by the insertion, into the gene, of a DNA

**IN THIS SMALL** garden plot adjacent to the monastery of St. Thomas, Gregor Mendel grew more than 33,500 pea plants over a period of eight years, including more than 6400 plants in one year alone. He received some help from two fellow monks who assisted in the experiments.

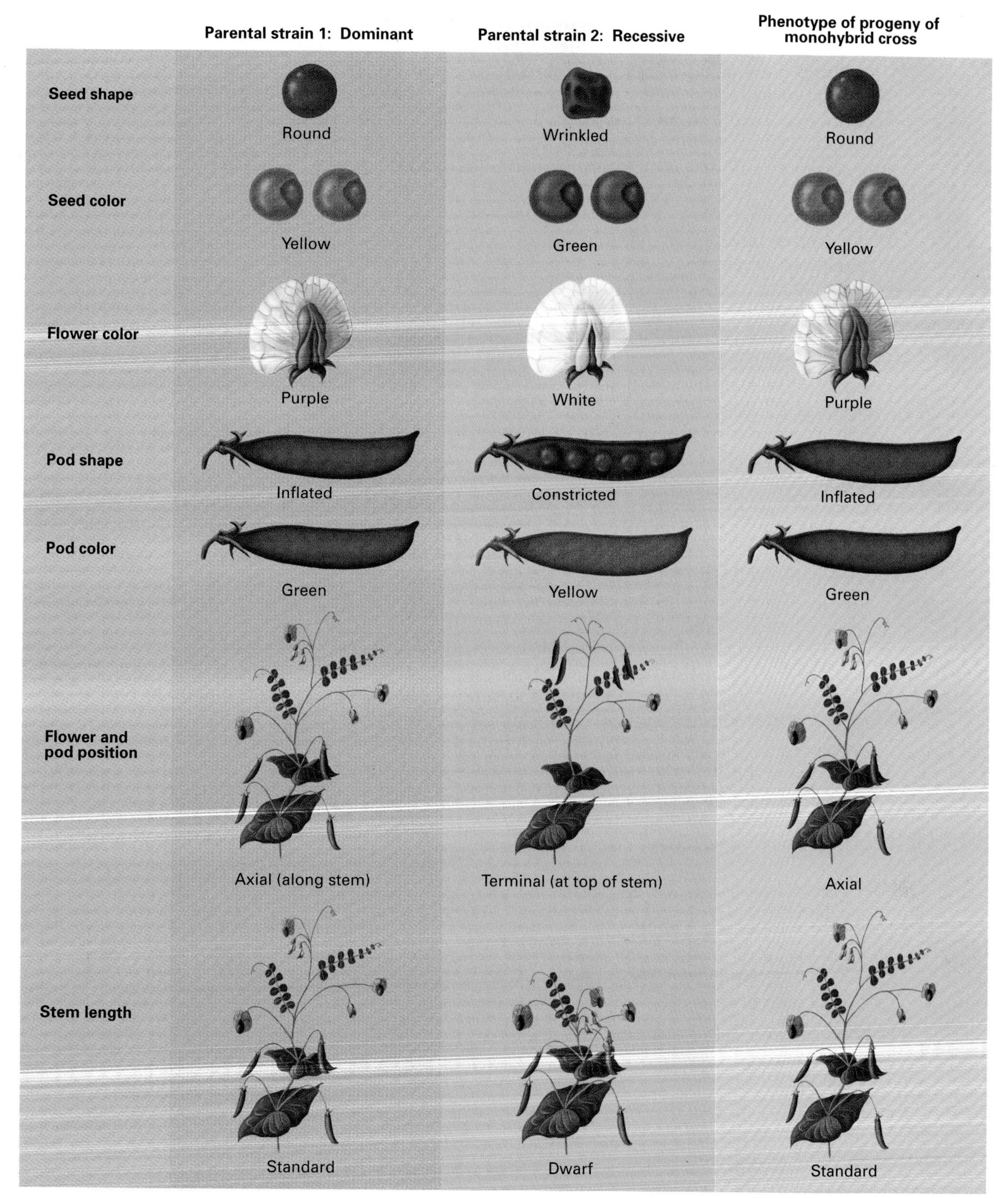


**FIGURE 3.1** The seven different traits in peas studied by Mendel. The phenotype shown at the far right is the dominant trait, which appears in the hybrid produced by crossing.

sequence called a **transposable element.** Such DNA sequences are capable of moving (*transposition*) from one location to another within a chromosome or between chromosomes. The molecular mechanism of transposition is discussed in Chapter 14, but for our present purposes it is necessary to know only that transposable elements are present in most genomes, especially the large genomes of eukaryotes, and that many spontaneous mutations result from the insertion of transposable elements into a gene.

**FIGURE 3.2** includes a diagram of the DNA structure of the wildtype *W* and mutant *w* alleles and shows the DNA insertion that interrupts the *w* allele. Highlighted are two *Eco*RI restriction sites, present in both alleles, that flank the site of the insertion. The diagram in part C indicates the pattern of bands that would be expected if one were to carry out a Southern blot (Section 2.3) in which genomic DNA was digested to completion with *Eco*RI and then the resulting fragments were separated by electrophoresis and hybridized with a labeled probe complementary to a region shared between the *W* and *w* alleles. The *Eco*RI fragment from the *W* allele would be smaller than that of the *w* allele, because of the inserted DNA in the *w* allele, and thus it would migrate faster than the corresponding fragment from the *w* allele and move to a position closer to the bottom of the gel. Genomic DNA from homozygous *WW* would yield a single, faster-migrating band; that from homozygous *ww* a single, slower-migrating band; and that from heterozygous *Ww* two bands with the same electrophoretic mobilities as those observed in the homozygous genotypes. In Figure 3.2C, the band in the homozygous genotypes is shown as somewhat thicker than those from the heterozygous genotype, because the single band in each homozygous genotype comes from the two copies of whichever allele is homozygous, and thus it contains more DNA than the corresponding DNA in the heterozygous genotype, in which only one copy of each allele is present.

Hence, as illustrated in Figure 3.2C, the RFLP analysis clearly distinguishes between the genotypes *WW, Ww*, and *ww*, because the heterozygous *Ww* genotype exhibits both of the bands observed in the homozygous genotypes. This situation is described by saying that the *W* and *w* alleles are **codominant** with respect to the molecular phenotype. However, as indicated by the seed shapes in Figure 3.2C, *W* is dominant over *w* with respect to morphological phenotype. In the discussion that follows, we use the RFLP analysis of the *W* and *w* alleles to emphasize the importance of molecular phenotypes in modern genetics and to demonstrate experimentally the principles of genetic transmission.

## 3.2 Segregation of a Single Gene

Mendel selected peas for his experiments for two primary reasons. First, he had access to varieties that differed in contrasting traits, such as round versus wrinkled seeds and yellow versus green seeds. Second, his earlier studies had indicated that peas usually reproduce by self-pollination, in which pollen produced in a flower is used to fertilize the eggs in the same flower. Producing hybrids by cross-pollination (**outcrossing**), re-

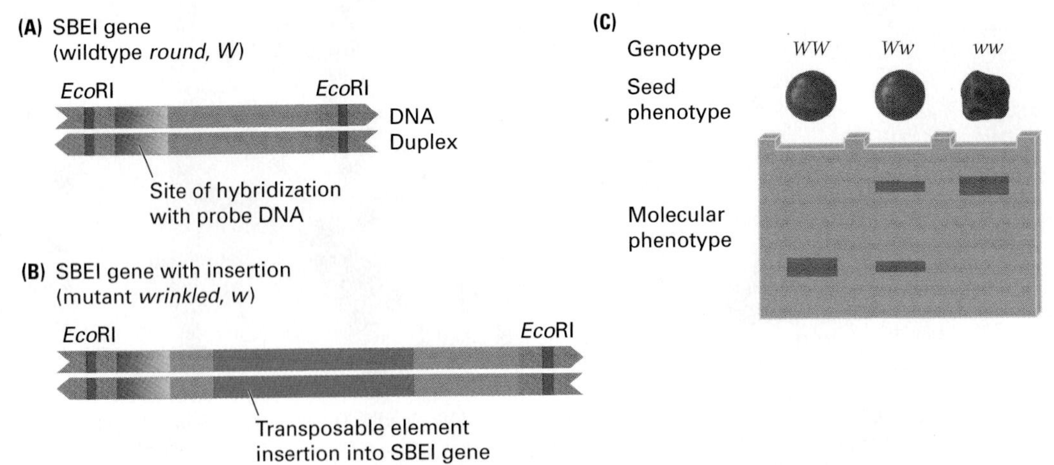


**FIGURE 3.2** (A) *W* (round) is an allele of a gene that specifies the amino acid sequence of starch branching enzyme I (SBEI). (B) *w* (wrinkled) is an allele that encodes an inactive form of the enzyme because its DNA sequence is interrupted by the insertion of a transposable element. (C) At the level of the morphological phenotype, *W* is dominant to *w*: Genotypes *WW* and *Ww* have round seeds, whereas genotype *ww* has wrinkled seeds. The molecular difference between the alleles can be detected as a restriction fragment length polymorphism (RFLP) using the enzyme *Eco*RI and a probe that hybridizes at the site shown. At the molecular level, the alleles are codominant: DNA from each genotype yields a different molecular phenotype—a single band differing in size for homozygous *WW* and *ww*, and both bands for heterozygous *Ww*.

quired painstaking surgery on immature flowers to reveal the receptive female structures, excise and discard the undeveloped male pollen-producing structures, and expose the female structures to mature pollen from a a different plant. After the cross-fertilization, each dissected flower was enclosed in a fine mesh bag to prevent stray pollen from accessing the female structures inside. The relatively small space needed to grow each plant, and the relatively large number of progeny that could be obtained, gave him the opportunity, as he says in his paper, to "determine the number of different forms in which hybrid progeny appear" and to "ascertain their numerical interrelationships."

The fact that garden peas are normally self-fertilizing means that in the absence of deliberate outcrossing, plants with contrasting traits are usually homozygous for alternative alleles of a gene affecting the trait; for example, plants with round seeds have genotype *WW* and those with wrinkled seeds genotype *ww*. The homozygous genotypes are indicated experimentally by the observation that hereditary traits in each variety are **true-breeding**, which means that plants produce only progeny like themselves when allowed to self-pollinate normally. Outcrossing between plants that differ in one or more traits creates a **hybrid**. If the parents differ in one, two, or three traits, the hybrid is a *monohybrid, dihybrid,* or *trihybrid.* In keeping track of parents and their hybrid progeny, we say that the parents constitute the **$P_1$ generation** and their hybrid progeny the **$F_1$ generation**.

Because garden peas are sexual organisms, each cross can be performed in two ways, depending on which phenotype is present in the female parent and which in the male parent. With round versus wrinkled, for example, the female parent can be round and the male wrinkled, or the reverse. These are called **reciprocal crosses**. Mendel was the first to demonstrate the following important principle:

> The outcome of a genetic cross does not depend on which trait is present in the male and which is present in the female; reciprocal crosses yield the same result.

This principle is illustrated for round and wrinkled seeds in **FIGURE 3.3**. The gel icons show the

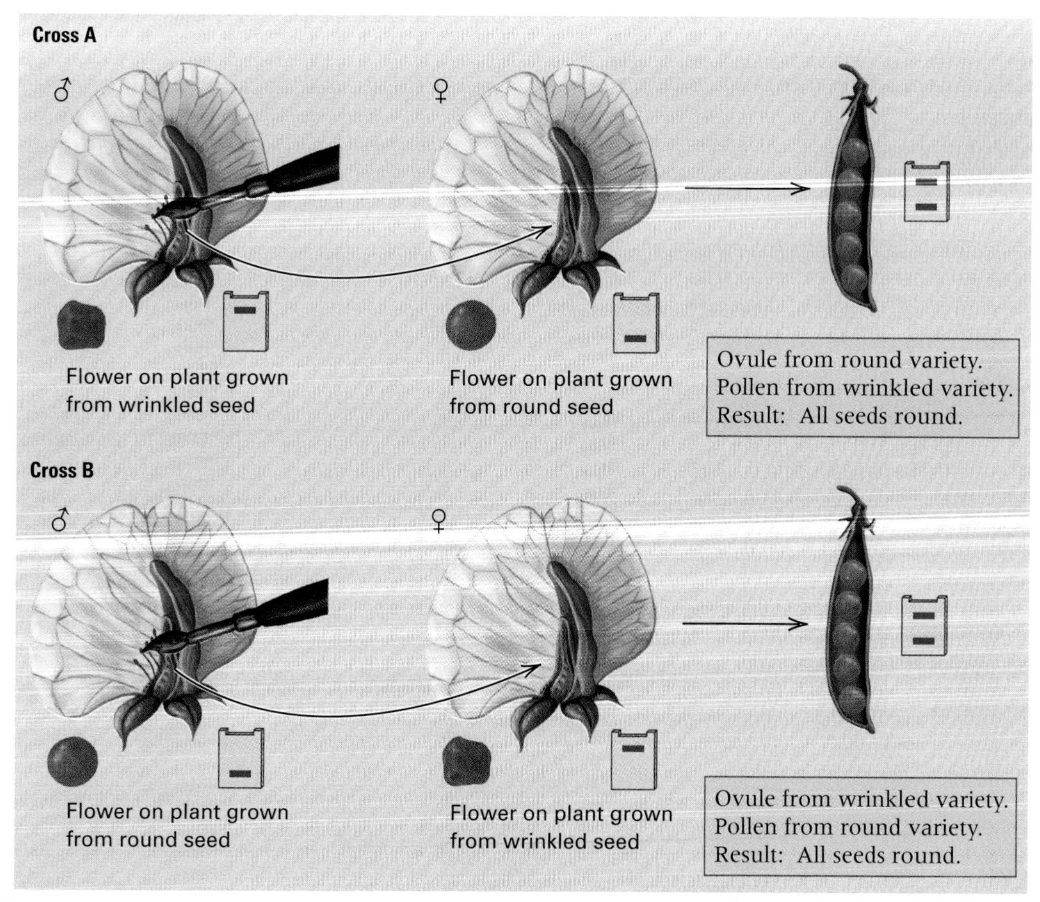


**FIGURE 3.3** Morphological and molecular phenotypes showing the equivalence of reciprocal crosses. In this example, the hybrid seeds are round and yield an RFLP pattern with two bands, irrespective of the direction of the cross.

RFLP bands that genomic DNA from each type of seed in these crosses would yield. The genotypes of the crosses and their progeny are as follows:

Cross A: *WW* ♀ × *ww* ♂ → *Ww*
Cross B: *ww* ♀ × *WW* ♂ → *Ww*

In both reciprocal crosses, the progeny have the morphological phenotype of round seeds, but as shown by the RFLP diagrams on the right, all progeny genotypes are actually heterozygous *Ww* and therefore different from either parent. The genetic equivalence of reciprocal crosses illustrated in Figure 3.3 is a principle that is quite general in its applicability, but there is an important exception, having to do with sex chromosomes, that will be discussed in Chapter 4.

In the following section we examine a few of Mendel's original experiments in the context of RFLP analysis in order to relate the morphological phenotypes and their ratios to the molecular phenotypes that would be expected.

## ■ Phenotypic Ratios in the $F_2$ Generation

Although the progeny of the crosses in Figure 3.3 have the dominant phenotype of round seeds, the RFLP analysis shows that they are actually heterozygous. The *w* allele is hidden with respect to the morphological phenotype because *w* is recessive to *W*. Nevertheless, the wrinkled phenotype reappears in the next generation when the hybrid progeny are allowed to undergo self-fertilization. For example, if the round $F_1$ seeds from Cross A in Figure 3.3 were grown into sexually mature plants and allowed to undergo self-fertilization, some of the resulting seeds would be round and others wrinkled. The progeny seeds produced by self-fertilization of the $F_1$ generation constitute the **$F_2$ generation.** When Mendel carried out this experiment, in the $F_2$ generation he observed the results shown in the following diagram:

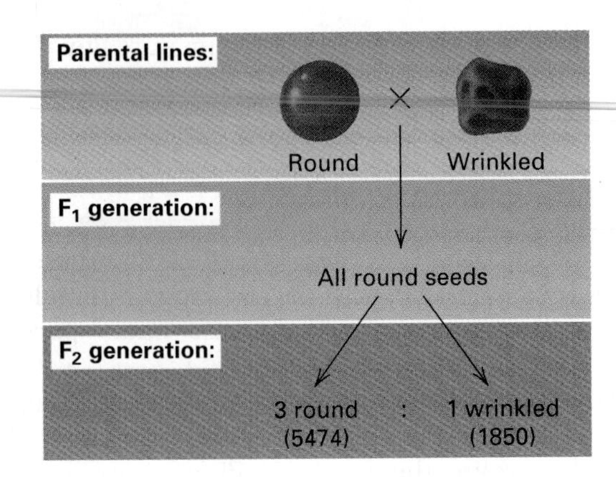


Note that the ratio 5474 : 1850 is approximately 3 : 1.

A 3 : 1 ratio of dominant : recessive forms in the $F_2$ progeny is characteristic of simple Mendelian inheritance. Mendel's data demonstrating this point are shown in **TABLE 3.1**. Note that the first two traits in the table (round versus wrinkled seeds and yellow versus green seeds) have many more observations than any of the others. The reason is that these traits can be classified directly in the seeds, whereas the others can be classified only in the mature plants, and Mendel could analyze many more seeds than he could mature plants. The principal observations from the data in Table 3.1 are:

- The $F_1$ hybrids express only the dominant trait (because the $F_1$ progeny are heterozygous—for example, *Ww*)
- In the $F_2$ generation, plants with either the dominant or the recessive trait are present (which means that some $F_2$ progeny are homozygous—for example, *ww*).
- In the $F_2$ generation, there are approximately three times as many plants with the dominant phenotype as plants with the recessive phenotype.

The 3 : 1 ratio observed in the $F_2$ generation is the key to understanding the mechanism of genetic transmission. In the next section we use RFLP analysis of the *W* and *w* alleles to explain why this ratio is produced.

**Table 3.1 Results of Mendel's monohybrid experiments**

| Parental traits | $F_1$ trait | Number of $F_2$ progeny | $F_2$ ratio |
|---|---|---|---|
| Round × wrinkled (seeds) | Round | 5474 round, 1850 wrinkled | 2.96 : 1 |
| Yellow × green (seeds) | Yellow | 6022 yellow, 2001 green | 3.01 : 1 |
| Purple × white (flowers) | Purple | 705 purple, 224 white | 3.15 : 1 |
| Inflated × constricted (pods) | Inflated | 882 inflated, 299 constricted | 2.95 : 1 |
| Green × yellow (unripe pods) | Green | 428 green, 152 yellow | 2.82 : 1 |
| Axial × terminal (flower position) | Axial | 651 axial, 207 terminal | 3.14 : 1 |
| Long × short (stems) | Long | 787 long, 277 short | 2.84 : 1 |

## ■ The Principle of Segregation

The 3 : 1 ratio can be explained with reference to **FIGURE 3.4**. This is the heart of Mendelian genetics. You should master it and be able to use it to deduce the progeny types produced in crosses. The diagram illustrates these key features of single-gene inheritance:

1. Genes come in pairs, which means that a cell or individual has *two* copies (alleles) of each gene.
2. For each pair of genes, the alleles may be identical (homozygous *WW* or homozygous *ww*), or they may be different (heterozygous *Ww*).
3. Each reproductive cell (**gamete**) produced by an individual contains only *one* allele of each gene (that is, either *W* or *w*).
4. In the formation of gametes, any particular gamete is equally likely to include either allele (hence, from a heterozygous *Ww* genotype, half the gametes contain *W* and the other half contain *w*).
5. The union of male and female reproductive cells is a random process that reunites the alleles in pairs.

The essential feature of transmission genetics is the separation, technically called **segregation**, in unaltered form, of the two alleles in an individual during the formation of its reproductive cells. Segregation corresponds to points 3 and 4 in the foregoing list. The prin-

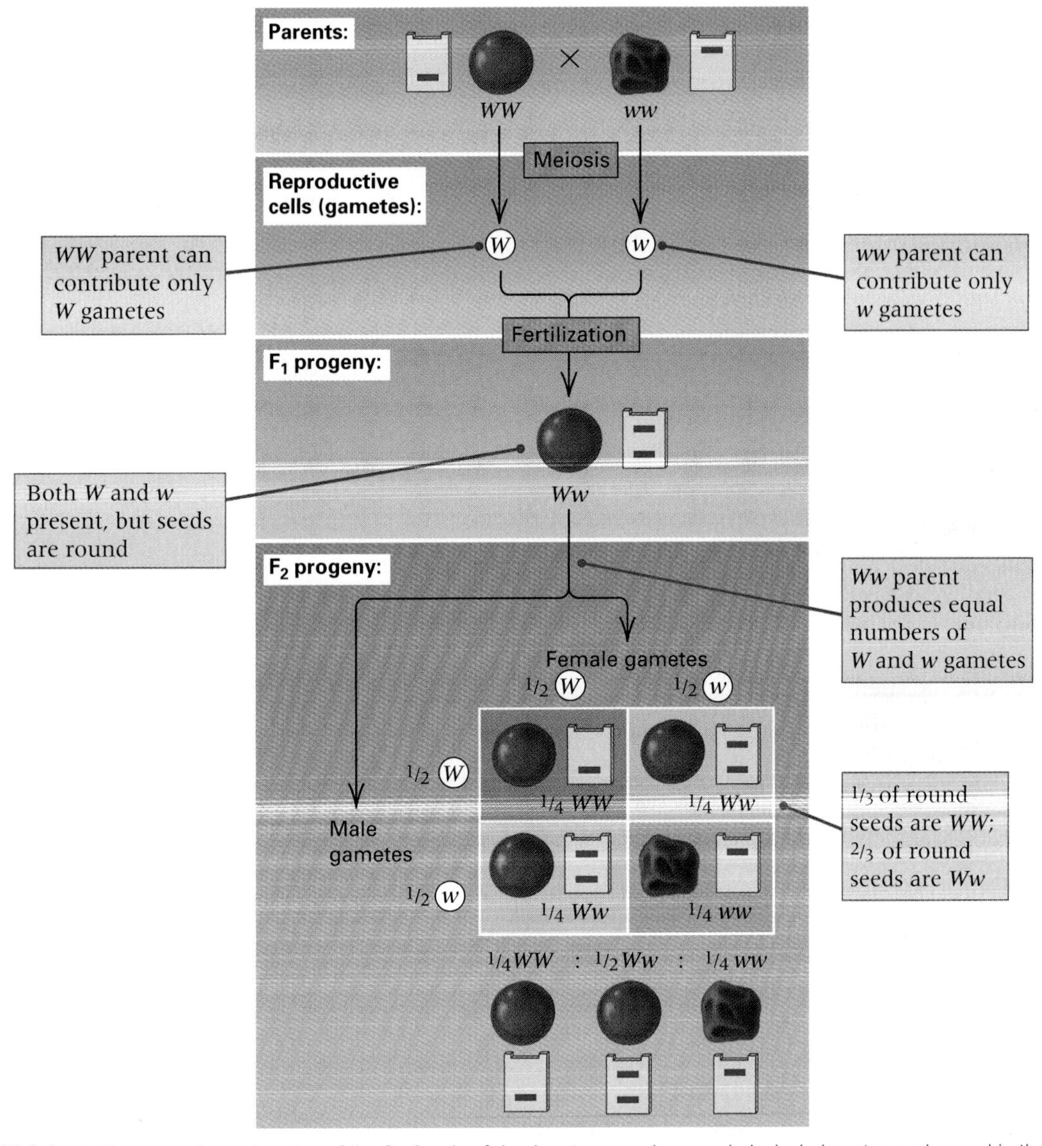


**FIGURE 3.4** A diagrammatic explanation of the 3 : 1 ratio of dominant : recessive morphological phenotypes observed in the $F_2$ generation of a monohybrid cross. The 3 : 1 ratio is observed because of dominance. Note that the ratio of *WW* : *Ww* : *ww* genotypes in the $F_2$ generation is 1 : 2 : 1, as can be seen from the restriction fragment phenotypes.

# connection

## What Did Gregor Mendel Think He Discovered?

**Gregor Mendel 1866**
Monastery of St. Thomas, Brno [then Brünn], Czech Republic
*Experiments on Plant Hybrids (original in German)*

*Mendel's paper is remarkable for its precision and clarity. It is worth reading in its entirety for this reason alone. Although the most important discovery attributed to Mendel is segregation, he never uses this term. His description of segregation is found in the first passage in italics in the excerpt. (All of the italics are reproduced from the original.) In his description of the process, he takes us carefully through the separation of* A *and* a *in gametes and their coming together again at random in fertilization. One flaw in the description is Mendel's occasional confusion between genotype and phenotype, which is illustrated by his writing* A *instead of* AA *and* a *instead of* aa *in the display toward the end of the passage. Most early geneticists made no consistent distinction between genotype and phenotype until 1909, when the terms themselves were coined.*

Artificial fertilization undertaken on ornamental plants to obtain new color variants initiated the experiments reported here. The striking regularity with which the same hybrid forms always reappeared whenever fertilization between like species took place suggested further experiments whose task it was to follow that development of hybrids in their progeny. . . . This paper discusses the attempt at such a detailed experiment. . . . Whether the plan by which the individual experiments were set up and carried out was adequate to the assigned task should be decided by a benevolent judgment. . . . [Here the experimental results are described in detail.] Thus experimentation also justifies the assumption *that pea hybrids form germinal and pollen cells that in their composition correspond in equal numbers to all the constant forms resulting from the combination of traits united through fertilization.* The difference of forms among the progeny of hybrids, as well as the ratios in which they are observed, find an adequate explanation in the principle [of segregation] just deduced. The simplest case is given by the series for *one pair of differing traits.* It is shown that this series is described by the expression: $A + 2Aa + a$, in which $A$ and $a$ signify the forms with constant differing traits, and $Aa$ the form hybrid for both. The series contains four individuals in three different terms. In their production, pollen and germinal cells of form $A$ and $a$ participate, on the average, equally in fertilization; therefore each form manifests itself twice, since four individuals are produced. Participating in fertilization are thus:

**Whether the plan by which the individual experiments were set up and carried out was adequate to the assigned task should be decided by a benevolent judgment.**

| | |
|---|---|
| Pollen cells | $A + A + a + a$ |
| Germinal cells | $A + A + a + a$ |

It is entirely a matter of chance which of the two kinds of pollen combines with each single germinal cell. However, according to the laws of probability, in an average of many cases it will always happen that every pollen form $A$ and $a$ will unite equally often with every germinal-cell form $A$ and $a$; therefore, in fertilization, one of the two pollen cells $A$ will meet a germinal cell $A$, the other a germinal cell $a$, and equally, one pollen cell $a$ will become associated with a germinal cell $A$, and the other $a$.

| | | | | |
|---|---|---|---|---|
| Pollen cells | $A$ | $A$ | $a$ | $a$ |
| | ↓ | ↘ | ↙ | ↓ |
| Germinal cells | $A$ | $A$ | $a$ | $a$ |

The result of fertilization can be visualized by writing the designations for associated germinal and pollen cells in the form of fractions, pollen cells above the line, germinal cells below. In the case under discussion one obtains

$$\frac{A}{A} + \frac{A}{a} + \frac{a}{A} + \frac{a}{a}$$

In the first and fourth terms germinal and pollen cells are alike; therefore the products of their association must be constant, namely $A$ and $a$; in the second and third, however, a union of the two differing parental traits takes place again, therefore the forms arising from such fertilizations are absolutely identical with the hybrid from which they derive. *Thus, repeated hybridization takes place.* The striking phenomenon, that hybrids are able to produce, in addition to the two parental types, progeny that resemble themselves is thus explained: $Aa$ and $aA$ both give the same association, $Aa$, since, as mentioned earlier, it makes no difference to the consequence of fertilization which of the two traits belongs to the pollen and which to the germinal cell. Therefore

$$\frac{A}{A} + \frac{A}{a} + \frac{a}{A} + \frac{a}{a} = A + 2Aa + a$$

This represents the *average* course of self-fertilization of hybrids when two differing traits are associated in them. In individual flowers and individual plants, however, the ratio in which the members of the series are formed may be subject to not insignificant deviations. . . . Thus it was proven experimentally that, in *Pisum,* hybrids form *different kinds* of germinal and pollen cells and that this is the reason for the variability of their offspring.

Source: G. Mendel, *Verhandlungen des naturforschenden den Vereines in Brünn* 4(1866): 3–47.

ciple of segregation is sometimes called *Mendel's first law.*

> **The Principle of Segregation:** In the formation of gametes, the paired hereditary determinants separate (segregate) in such a way that each gamete is equally likely to contain either member of the pair.

Another key feature of transmission genetics is that the hereditary determinants are present as pairs in both the parental organisms and the progeny organisms but as single copies in the reproductive cells. This feature corresponds to points 1 and 5 in the foregoing list.

Figure 3.4 illustrates the biological mechanism underlying the important Mendelian ratios in the $F_2$ generation of 3 : 1 for phenotypes and 1 : 2 : 1 for genotypes. To understand these ratios, consider first the parental generation in which the original cross is *WW* × *ww*. The sex of the parents is not stated because reciprocal crosses yield identical results. (There is, however, a convention in genetics that unless otherwise specified, crosses are given with the female parent listed first.) In the original cross, the *WW* parent produces only *W*-containing gametes, whereas the *ww* parent produces only *w*-containing gametes. Segregation still takes place in the homozygous genotypes as well as in the heterozygous genotype, even though all of the gametes carry the same type of allele (*W* from homozygous *WW* and *w* from homozygous *ww*). When the *W*-bearing and *w*-bearing gametes come together in fertilization, the hybrid genotype is heterozygous *Ww*, which is shown by the bands in the gel icon next to the $F_1$ progeny. With regard to seed shape, the hybrid *Ww* seeds are round because *W* is dominant over *w*.

When the heterozygous $F_1$ progeny form gametes, segregation implies that half the gametes will contain the *W* allele and the other half will contain the *w* allele. These gametes come together at random when an $F_1$ individual is self-fertilized or when two $F_1$ individuals are crossed. The result of random fertilization can be deduced from the sort of cross-multiplication square shown at the bottom of the figure, in which the female gametes and their frequencies are arrayed across the top margin and the male gametes and their frequencies along the left-hand margin. This calculating device is widely used in genetics and is called a **Punnett square** after its inventor Reginald C. Punnett (1875–1967). The Punnett square in Figure 3.4 shows that random combinations of the $F_1$ gametes result in an $F_2$ generation with the genotypic composition 1/4 *WW*, 1/2 *Ww*, and 1/4 *ww*. This can be confirmed by the RFLP banding patterns in the gel icons because of the codominance of *W* and *w* with respect to the molecular phenotype. But because *W* is dominant over *w* with respect to the morphological phenotoype, the *WW* and *Ww* genotypes have round seeds and the *ww* genotypes have wrinkled seeds, yielding the phenotypic ratio of round : wrinkled seeds of 3 : 1. Hence, it is a combination of segregation, random union of gametes, and dominance that results in the 3 : 1 ratio.

The ratio of $F_2$ genotypes is as important as the ratio of $F_2$ phenotypes. The Punnett square in Figure 3.4 also shows that the ratio of *WW* : *Ww* : *ww* genotypes is 1 : 2 : 1, which can be confirmed directly by the RFLP analysis.

## Verification of Segregation

The round seeds in Figure 3.4 conceal a genotypic ratio of 1 *WW* : 2 *Ww*. To say the same thing in another way, among the $F_2$ seeds that are round (or, more generally, among organisms that show the dominant morphological phenotype), 1/3 are homozygous (in this example, *WW*) and 2/3 are heterozygous (in this example, *Ww*). This conclusion is obvious from the RFLP patterns in Figure 3.4, but it is not at all obvious from the morphological phenotypes. Unless you knew something about genetics already, it would be a very bold hypothesis, because it implies that two organisms with the same morphological phenotype (in this case round seeds) might nevertheless differ in molecular phenotype and in genotype.

Yet this is exactly what Mendel proposed. But how could this hypothesis be tested experimentally? He realized that it could be tested via self-fertilization of the $F_2$ plants. With self-fertilization, plants grown from the homozygous *WW* genotypes should be true-breeding for round seeds, whereas those from the heterozygous *Ww* genotypes should yield round and wrinkled seeds in the ratio of 3 : 1. On the other hand, the plants grown from wrinkled seeds should be true-breeding for wrinkled because these plants are homozygous *ww*. The results Mendel obtained are summarized in **FIGURE 3.5**. As predicted from the genetic hypothesis, plants grown from $F_2$ wrinkled seeds were true breeding for wrinkled seeds, yielding only wrinkled seeds in the $F_3$ generation. But some of the plants grown from round seeds showed evidence of segregation. Among 565 plants grown from round $F_2$ seeds, 372 plants produced both round and wrinkled seeds in a proportion very close to 3 : 1, whereas the re-

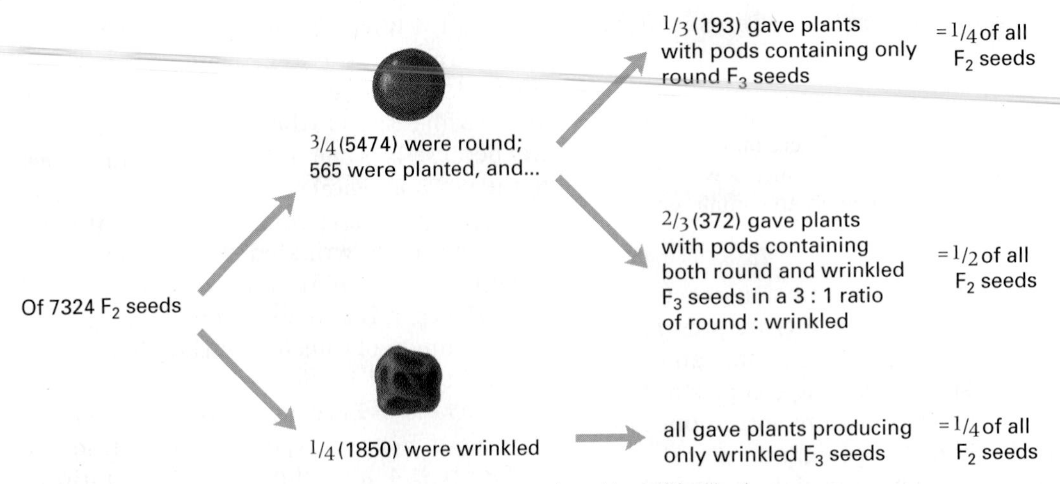
1/3 (193) gave plants with pods containing only round F3 seeds
= 1/4 of all F2 seeds
3/4 (5474) were round; 565 were planted, and...
2/3 (372) gave plants with pods containing both round and wrinkled F3 seeds in a 3 : 1 ratio of round : wrinkled
= 1/2 of all F2 seeds
Of 7324 F2 seeds
1/4 (1850) were wrinkled
all gave plants producing only wrinkled F3 seeds
= 1/4 of all F2 seeds

**FIGURE 3.5** Summary of $F_2$ phenotypes and the progeny produced by self-fertilization.

maining 193 plants produced only round seeds in the $F_3$ generation. The ratio 193 : 372 equals 1 : 1.93, which is very close to the ratio 1 : 2 of *WW* : *Ww* genotypes predicted from the genetic hypothesis.

An important feature of the homozygous round and homozygous wrinkled seeds produced in the $F_2$ and $F_3$ generations is that the phenotypes are exactly the same as those observed in the original parents in the $P_1$ generation. This makes sense in terms of DNA, because the DNA of each allele remains unaltered unless a new mutation happens to occur. Mendel described this result in a letter by saying that in the progeny of crosses, "the two parental traits appear, separated and unchanged, and there is nothing to indicate that one of them has either inherited or taken over anything from the other." From this finding, he concluded that the hereditary determinants for the traits in the parental lines were transmitted as two different elements that retain their purity in the hybrids. In other words, the hereditary determinants do not "mix" or "contaminate" each other. In modern terminology, this means that, with rare but important exceptions, genes are transmitted unchanged from generation to generation.

## The Testcross and the Backcross

Another straightforward way of testing the genetic hypothesis in Figure 3.4 is by means of a **testcross,** a cross between an organism that is heterozygous for one or more genes (for example, *Ww*) and an organism that is homozygous for the recessive allele (for example, *ww*). The result of such a testcross is shown in **FIGURE 3.6**. Because the heterozygous parent is expected to produce *W* and *w* gametes in equal numbers, whereas the homozygous recessive produces only *w* gametes, the expected progeny are 1/2 with the genotype *Ww* and 1/2 with the genotype *ww*. The former have the dominant phenotype because *W* is dominant over *w*, and the latter have the recessive phenotype. A testcross is often extremely useful in genetic analysis:

> In a testcross, the phenotypes of the progeny reveal the relative frequencies of the different gametes produced by the heterozygous parent, because the recessive parent contributes only recessive alleles.

Mendel carried out a series of testcrosses with various traits. The results are shown in **TABLE 3.2**. In all cases, the ratio of phenotypes among the test-

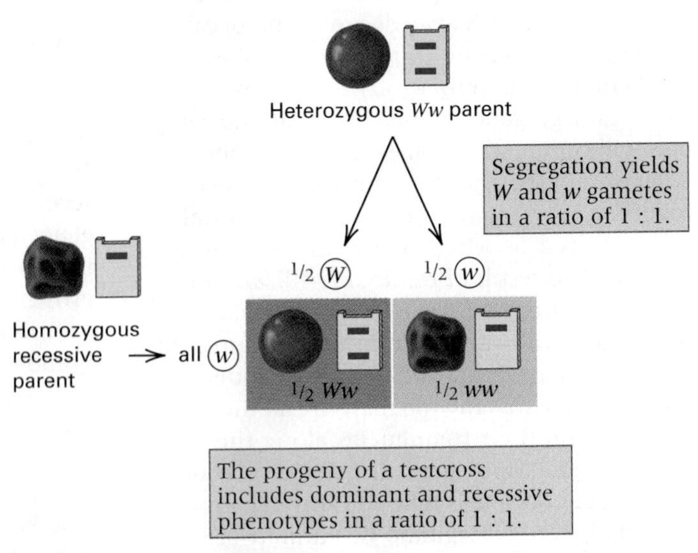
Heterozygous *Ww* parent
Segregation yields *W* and *w* gametes in a ratio of 1 : 1.
1/2 W
1/2 w
Homozygous recessive parent
all w
1/2 Ww
1/2 ww
The progeny of a testcross includes dominant and recessive phenotypes in a ratio of 1 : 1.

**FIGURE 3.6** In a testcross of a *Ww* heterozygous parent with a *ww* homozygous recessive, the progeny are *Ww* and *ww* in the ratio of 1 : 1. A testcross shows the result of segregation.

Table 3.2 **Mendel's testcross results**

| Testcross ($F_1$ heterozygote × homozygous recessive) | Progeny from testcross | Ratio |
|---|---|---|
| Round × wrinkled seeds | 193 round, 192 wrinkled | 1.01 : 1 |
| Yellow × green seeds | 196 yellow, 189 green | 1.04 : 1 |
| Purple × white flowers | 85 purple, 81 white | 1.05 : 1 |
| Long × short stems | 87 long, 79 short | 1.10 : 1 |

cross progeny is very close to the 1 : 1 ratio expected from segregation of the alleles in the heterozygous parent.

Another valuable type of cross is a **backcross**, in which hybrid organisms are crossed with one of the parental genotypes. Backcrosses are commonly used by geneticists and by plant and animal breeders, as we will see in later chapters. Note that the testcrosses in Table 3.2 are also backcrosses, because in each case, the $F_1$ heterozygous parent came from a cross between the homozygous dominant and the homozygous recessive.

## 3.3 Segregation of Two or More Genes

The results of many genetic crosses depend on the segregation of the alleles of two or more genes. The genes may be in different chromosomes or in the same chromosome. Although in this section we consider the case of genes that are in two different chromosomes, the same principles apply to genes that are in the same chromosome but are so far distant from each other that they segregate independently. The case of *linkage* of genes in the same chromosome is examined in Chapter 5.

To illustrate the principles, we consider again a cross between homozygous genotypes, but in this case homozygous for the alleles of two genes. A specific example is a true-breeding variety of garden peas with seeds that are wrinkled and green (genotype *ww gg*) versus a variety with seeds that are round and yellow (genotype *WW GG*). As suggested by the use of uppercase and lowercase symbols for the alleles, the dominant alleles are *W* and *G*, the recessive alleles *w* and *g*. The mutant gene responsible for Mendel's green seeds has been identified. It is an inborn error in metabolism that blocks the metabolic pathway for breaking down the green pigment chlorophyll. Homozygous mutant seeds are unable to break down their chlorophyll and therefore they stay green, whereas wildtype seeds do break down their chlorophyll and turn yellow like the leaves of certain trees in the autumn. The gene is officially named *staygreen*.

A cross of *WW GG* plants with *ww gg* plants yields $F_1$ seeds with the genotype *Ww Gg*, which are phenotypically round and yellow because of the dominance relations. When the $F_1$ seeds are grown into mature plants and self-fertilized, the $F_2$ progeny show the result of simultaneous segregation of the *W*, *w* allele pair and the *G*, *g* allele pair. When Mendel perfomed this cross, he obtained the following numbers of $F_2$ seeds:

| | |
|---|---|
| Round, yellow | 315 |
| Round, green | 108 |
| Wrinkled, yellow | 101 |
| Wrinkled, green | 32 |
| Total | 556 |

In these data, the first thing to be noted is the expected 3 : 1 ratio for each trait considered separately. The ratio of round : wrinkled (pooling across yellow and green) is

$$\begin{aligned}&(315 + 108) : (101 + 32)\\&= 423 : 133\\&= 3.18 : 1\end{aligned}$$

And the ratio of yellow : green (pooling across round and wrinkled) is

$$\begin{aligned}&(315 + 101) : (108 + 32)\\&= 416 : 140\\&= 2.97 : 1\end{aligned}$$

Both of these ratios are in satisfactory agreement with 3 : 1. (Testing for goodness of fit to a predicted ratio is described in Chapter 4.) Furthermore, in the $F_2$ progeny of the dihybrid cross, the separate 3 : 1 ratios for the two traits were combined at random. With random combinations, as shown in **FIGURE 3.7**, among the 3/4 of the progeny that are round, 3/4 will be yellow and 1/4 green; similarly, among the 1/4 of the progeny that are wrinkled, 3/4 will be yellow and 1/4 green. The overall proportions of round yellow to round green to wrinkled yellow to wrinkled green are therefore expected to be

$$\begin{aligned}&3/4 \times 3/4 : 3/4 \times 1/4 : 1/4 \times 3/4 : 1/4 \times 1/4\\&= 9/16 : 3/16 : 3/16 : 1/16\end{aligned}$$

The observed ratio of 315 : 108 : 101 : 32 equals 9.84 : 3.38 : 3.16 : 1, which is a satisfactory fit

FIGURE 3.7 The 3 : 1 ratio of round : wrinkled, when combined at random with the 3 : 1 ratio of yellow : green, yields the 9 : 3 : 3 : 1 ratio observed in the $F_2$ progeny of the dihybrid cross.

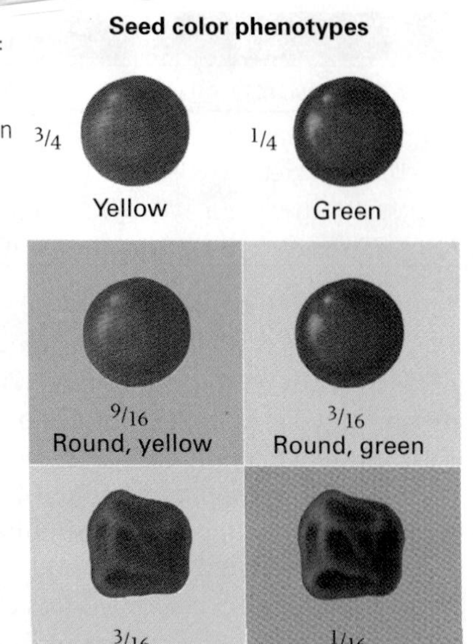


Ratio of phenotypes in the $F_2$ progeny of a dihybrid cross is 9 : 3 : 3 : 1.

FIGURE 3.8 Independent segregation of the *W*, *w* and *G*, *g* allele pairs means that among each of the *W* and *w* gametic classes, the ratio of *G* : *g* is 1 : 1. Likewise, among each of the *G* and *g* gametic classes, the ratio of *W* : *w* is 1 : 1.

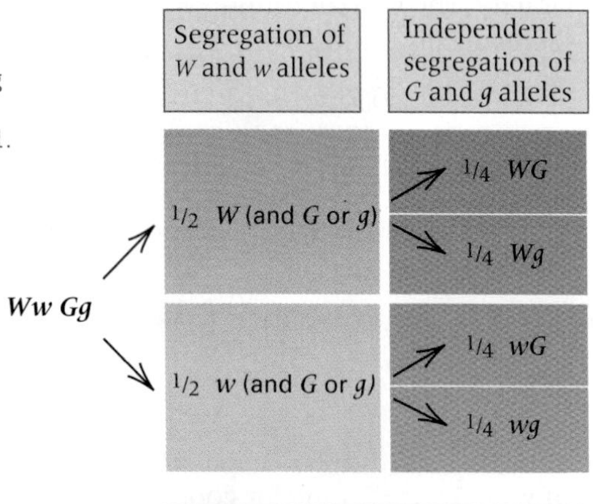


Result : An equal frequency of all four possible types of gametes

to the 9 : 3 : 3 : 1 ratio expected from the Punnett square in Figure 3.7.

## ■ The Principle of Independent Assortment

The independent segregation of the *W*, *w* and *G*, *g* allele pairs is illustrated in **FIGURE 3.8**. What independence means is that if a gamete contains *W*, it is equally likely to contain *G* or *g*; and if a gamete contains *w*, it is equally likely to contain *G* or *g*. The implication is that the four gametes are formed in equal frequencies:

$$1/4\ WG \quad 1/4\ Wg \quad 1/4\ wG \quad 1/4\ wg$$

The result of independent assortment when the four types of gametes combine at random to form the zygotes of the next generation is shown in **FIGURE 3.9**. Note that the expected ratio of phenotypes among the $F_2$ progeny is

$$9 : 3 : 3 : 1$$

However, as the Punnett square also shows, the ratio of genotypes in the $F_2$ generation is more complex; it is

$$1 : 2 : 1 : 2 : 4 : 2 : 1 : 2 : 1$$

The reason for this ratio is shown in **FIGURE 3.10**. Among seeds that have the *WW* genotype, the ratio of

$$GG : Gg : gg \text{ equals } 1 : 2 : 1$$

Among seeds that have the *Ww* genotype, the ratio is

$$2 : 4 : 2$$

(which is a 1 : 2 : 1 ratio multiplied by 2 because there are twice as many *Ww* genotypes as either *WW* or *ww*). And among seeds that have the *ww* genotype, the ratio of

$$GG : Gg : gg \quad \text{equals} \quad 1 : 2 : 1$$

The phenotypes of the seeds are shown beneath the genotypes, and the combined phenotypic ratio is

$$9 : 3 : 3 : 1$$

The principle of independent segregation of two pairs of alleles in different chromosomes (or located sufficiently far apart in the same chromosome) has come to be known as the principle of independent assortment. It is also sometimes referred to as *Mendel's second law.*

> **The Principle of Independent Assortment:** Segregation of the members of any pair of alleles is independent of the segregation of other pairs in the formation of reproductive cells.

Although the principle of independent assortment is fundamental in Mendelian genetics, the phenomenon of linkage, caused by proximity of genes in the same chromosome, is an important exception.

## ■ The Testcross with Unlinked Genes

Genes that show independent assortment are said to be **unlinked.** The hypothesis of independent assortment can be tested directly in a testcross with the double homozygous recessive:

$$Ww\ Gg \times ww\ gg$$

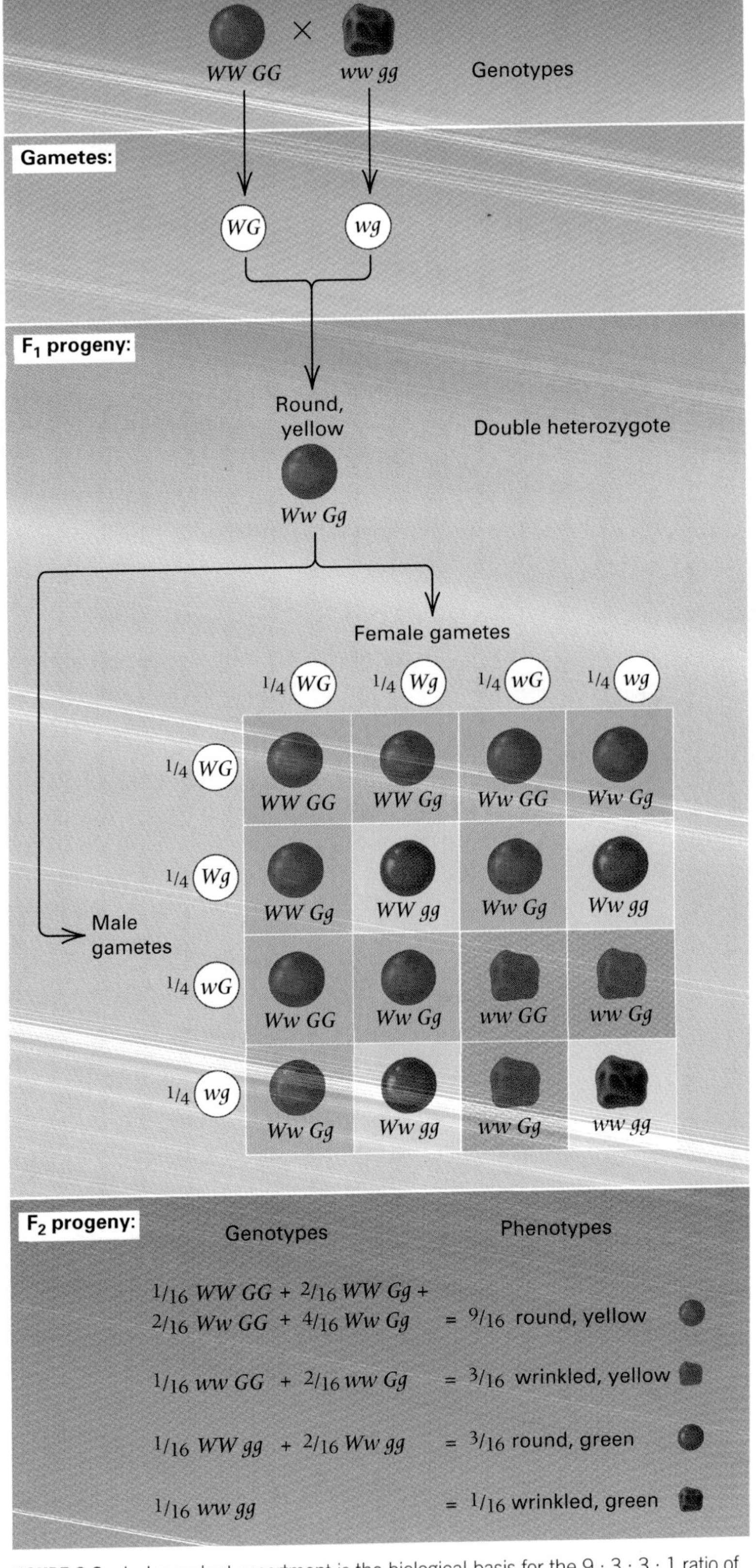


**FIGURE 3.9** Independent assortment is the biological basis for the 9 : 3 : 3 : 1 ratio of $F_2$ phenotypes resulting from a dihybrid cross.

The result of the testcross is shown in **FIGURE 3.11**. Because plants with doubly heterozygous genotypes produce four types of gametes—*W G, W g, w G,* and *w g*—in equal frequencies, whereas the plants with *ww gg* genotypes produce only *w g* gametes, the possible progeny genotypes are *Ww Gg, Ww gg, ww Gg,* and *ww gg,* and these are expected in equal frequencies. Because of the dominance relations—*W* over *w* and *G* over *g*—the progeny phenotypes are expected to be round yellow, round green, wrinkled yellow, and wrinkled green in a ratio of

$$1 : 1 : 1 : 1$$

As with the one-gene testcross, in a two-gene testcross the ratio of progeny phenotypes is a direct demonstration of the ratio of gametes produced by the doubly heterozygous parent. In the actual cross, Mendel obtained 55 round yellow, 51 round green, 49 wrinkled yellow, and 53 wrinkled green, which is in good agreement with the predicted 1 : 1 : 1 : 1 ratio.

## Three or More Genes

A Punnett square of the type in Figure 3.9 is a nice way to show the logic behind the genotype and phenotype frequencies for two genes that undergo independent assortment. As a method for solving problems, however, a Punnett square is not very efficient. In working a problem, especially when time is limited as during an exam, drawing and filling out the whole square takes too long. Another approach is less graphical but quicker. For example, the expected frequency of any genotype with independent assortment for two genes can be obtained by picking out the corresponding term in the expression

$$(\tfrac{1}{4}\ WW + \tfrac{1}{2}\ Ww + \tfrac{1}{4}\ ww) \times (\tfrac{1}{4}\ GG + \tfrac{1}{2}\ Gg + \tfrac{1}{4}\ gg)$$

Likewise, the expected frequency of any phenotype is given by the corresponding term in the expression

$$(\tfrac{3}{4}\text{ round} + \tfrac{1}{4}\text{ wrinkled}) \times (\tfrac{3}{4}\text{ yellow} + \tfrac{1}{4}\text{ green})$$

For example, the expected frequency of *Ww gg* genoypes in the $F_2$ generation is given by $\tfrac{1}{2} \times \tfrac{1}{4} = \tfrac{1}{8}$, and the expected frequency of round, green phenotypes is given by $\tfrac{3}{4} \times \tfrac{1}{4} = \tfrac{3}{16}$.

The time saved by use of the algebraic expressions is small when the cross has only two genes, but it is already considerable with three genes that show independent assortment. Consider an example of a trihybrid cross with the allele pairs (*W, w*), (*G, g*), and (*P, p*), where *P*

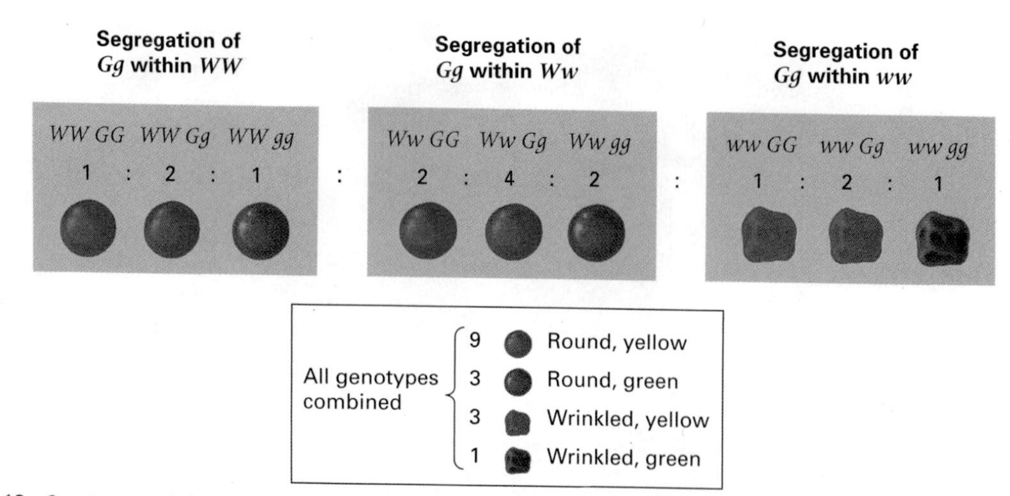


**FIGURE 3.10** Genotypes and phenotypes of the $F_2$ progeny of the dihybrid cross for seed shape and seed color.

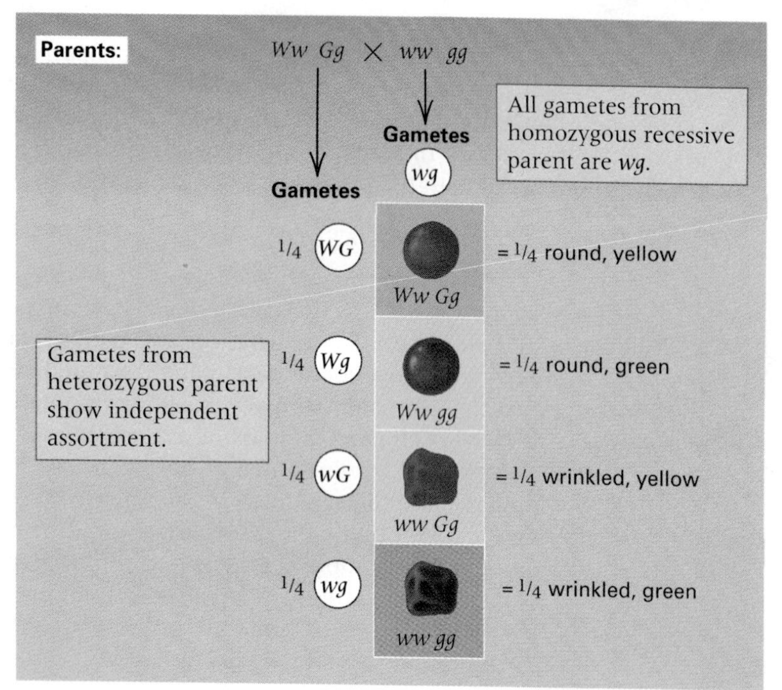


**FIGURE 3.11** Genotypes and phenotypes resulting from a testcross of the *Ww Gg* double heterozygote.

is a dominant allele for purple flowers and *p* is a recessive mutation for white flowers. In this case, the Punnett square is a cube containing 64 cells that is tricky to draw and tedious to complete. The genotype frequencies nevertheless correspond to terms in the expression

$$(\tfrac{1}{4}\ WW + \tfrac{1}{2}\ Ww + \tfrac{1}{4}\ ww) \times (\tfrac{1}{4}\ GG + \tfrac{1}{2}\ Gg + \tfrac{1}{4}\ gg) \times (\tfrac{1}{4}\ PP + \tfrac{1}{2}\ Pp + \tfrac{1}{4}\ pp)$$

and the phenotype frequencies to terms in

$$(\tfrac{3}{4}\text{ round} + \tfrac{1}{4}\text{ wrinkled}) \times (\tfrac{3}{4}\text{ yellow} + \tfrac{1}{4}\text{ green}) \times (\tfrac{3}{4}\text{ purple} + \tfrac{1}{4}\text{ white})$$

If you wish to know the probability that an offspring genotype is *Ww Gg Pp*, for example, this can be obtained by multiplying the frequencies $\frac{1}{2} \times \frac{1}{2} \times \frac{1}{2} = \frac{1}{8}$. Likewise, if you wish to know the probability that offspring phenotype is round, yellow, purple, this can be calculated as $\frac{3}{4} \times \frac{3}{4} \times \frac{3}{4} = \frac{24}{64}$.

Mendel actually carried out this three-gene cross. The results for the phenotypes are shown in **FIGURE 3.12**. They conform well to the ratio of 27 : 9 : 9 : 9 : 3 : 3 : 3 : 1 expected from independent assortment. The various progeny types represent $3^3 = 27$ different genotypes. Mendel did a testcross for each offspring having the dominant phenotype for one or more traits in order to determine whether its genotype was homozygous or heterozygous for each of the genes. This effort alone required 632 testcrosses. Little wonder that he complained that "of all the experiments, [this one] required the most time and effort."

Now that we have emphasized the efficiency and utility of a more abstract approach to genetic calculations than the Punnett square, we are in a position to explain a little more formally the foundations of probability as they relate to Mendelian genetics and to take such reasoning an important step further.

## 3.4 Probability in Genetic Analysis

As you already know, Mendel's laws of genetic transmission are fundamentally laws of chance (probability). He surpassed any of his contemporaries in understanding that his principles of inheritance accounted for the different types of progeny he observed as well as for the ratios in which they were found. No discoveries in genetics made since Mendel's time have undermined the fundamental role of chance in heredity that he was the first to recognize.

To fully understand Mendelian genetics, we therefore need to understand the elementary

$(^3/_4 W- + ^1/_4 ww) \times (^3/_4 G- + ^1/_4 gg) \times (^3/_4 P- + ^1/_4 pp)$

| | | | Observed number | Expected number |
|---|---|---|---|---|
| $^{27}/_{64}$ | $W-\ G-\ P-$ | Round, yellow, purple | 269 | 270 |
| $^{9}/_{64}$ | $W-\ G-\ pp$ | Round, yellow, white | 98 | 90 |
| $^{9}/_{64}$ | $W-\ gg\ P-$ | Round, green, purple | 86 | 90 |
| $^{9}/_{64}$ | $ww G-\ P-$ | Wrinkled, yellow, purple | 88 | 90 |
| $^{3}/_{64}$ | $W-\ gg\ pp$ | Round, green, white | 27 | 30 |
| $^{3}/_{64}$ | $ww G-\ pp$ | Wrinkled, yellow, white | 34 | 30 |
| $^{3}/_{64}$ | $ww\ gg\ P-$ | Wrinkled, green, purple | 30 | 30 |
| $^{1}/_{64}$ | $ww\ gg\ pp$ | Wrinkled, green, white | 7 | 10 |

- For any one gene, the ratio of phenotypes is 48 : 16 = 3 : 1
- For any pair of genes, the ratio of phenotypes is 36 : 12 : 12 : 4 = 9 : 3 : 3 : 1

**FIGURE 3.12** With independent assortment, the expected ratio of phenotypes in a trihybrid cross is obtained by multiplying the three independent 3 : 1 ratios of he dominant and recessive phenotypes. A dash used in a genotype symbol indicates that either the dominant or the recessive allele is present; for example, *W*– refers collectively to the genotypes *WW* and *Ww*. (The expected numbers total 640 rather than 639 because of round-off error.)

principles of probability. Every problem in probability begins with an experiment, which may be real or imaginary. In genetics, the experiment is typically a cross. Associated with the experiment is a set of possible outcomes of the experiment. In genetics the possible outcomes are typically genotypes or phenotypes. The possible outcomes are called *elementary outcomes*. They are elementary outcomes in the sense that none of them can be reduced to combinations of the others. In our applications of probability, the number of elementary outcomes is often relatively small, or in any case can be enumerated. The principles of probability also can deal with conceptual experiments in which there are an infinite number of elementary outcomes, but this requires some technicalities that are beyond the scope of this book.

Each elementary outcome is assigned a *probability* that is proportional to its likelihood of occurrence. In principle the probabilities can be assigned arbitrarily. There are only two rules. First, the probability of each elementary outcome must be a nonnegative number between 0 and 1, and may actually equal 0 or 1. An elementary outcome with a probability of 0 cannot occur, and one with a probability of 1 *must* occur. The second rule is that the sum of the probabilities of all the elementary outcomes must equal 1. This rule assures that some one of the elementary outcome must occur. These two rules also handle an annoying question often asked in regard to a coin toss: What happens if it lands on its edge? The answer is that this elementary outcome is assigned a probability of 0, and so we need not bother with it.

Here it will be helpful to consider a specific example. Consider the conceptual experiment of self-fertilization of the $F_1$ progeny of a cross between pea plants homozygous for the round *W* and yellow *G* alleles with plant homozygous for *w* and *g*. Recall that the *W*, *w* and *G*, *g* allele pairs undergo independent assortment, and so there are 16 elementary outcomes. These are shown in **FIGURE 3.13A**. Each of the elementary outcomes is equally likely, and so the probability of each elementary outcome is assigned a value of $^1/_{16}$. Note that the progeny genotype *Ww Gg* is listed four times. This is because there are four possible combinations of parental gametes that can yield the progeny genotype *Ww Gg* (see Figure 3.9).

The enumeration of the elementary outcomes and their probabilities constitute what is often called the *sample space* of the probability problem. For the progeny from self-fertilization of *Ww Gg* the sample space is shown in Figure 3.13A. This is the sample space in which all probability considerations regarding this conceptual experiment take place.

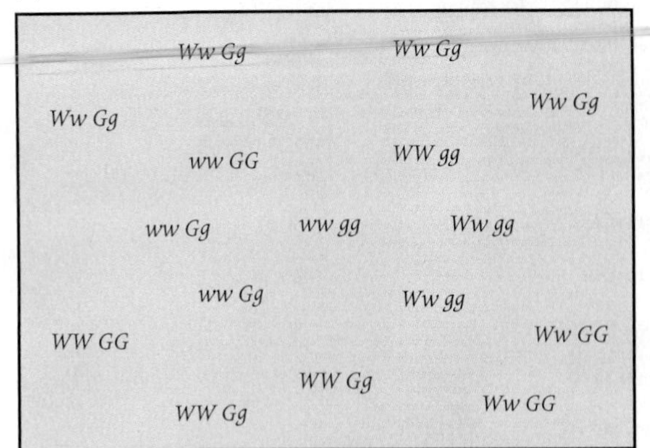


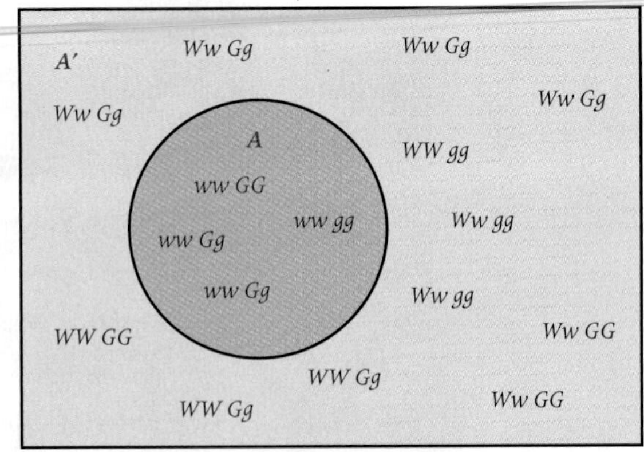




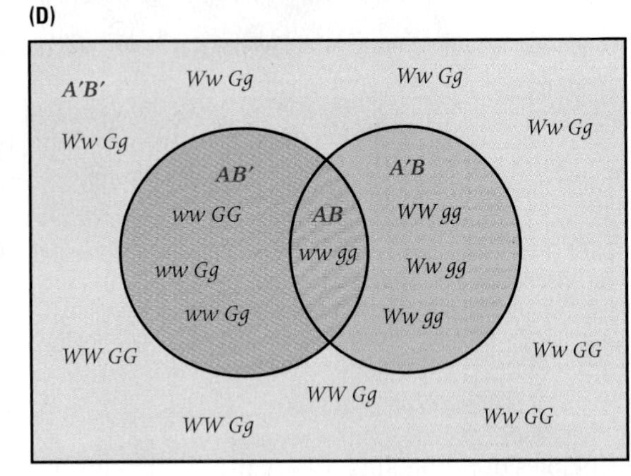


**FIGURE 3.13** (A) Sample space for the possible progeny in a cross for the allele pairs *W*, *w* for round versus wrinkled seeds and *G*, *g* for yellow versus green seeds. (B) The event *A* includes all genotypes whose phenotype is wrinkled. (C) The event *B* includes all genotypes whose phenotype is yellow. (D) The union and intersection (yellow) of *A* and *B*.

## ■ Elementary Outcomes and Events

Any combination of elementary outcomes constitutes an **event** in the sample space. In Figure 3.13B, the circle labeled *A* includes four elementary outcomes defining the event *A* as "the offspring genotype is *ww*." The event A could be defined equivalently as "the offspring phenotype is wrinkled." The alternative ways of defining *A* in words shows how subsets of elementary outcomes can relate genotypes and phenotypes. Corresponding to every event is a *probability* of that event, in this case symbolized Pr{*A*}. A fundamental principle of probability is that:

> The **probability** of any event equals the sum of the probabilities of all the elementary outcomes included in the event.

In the example in Figure 3.13B, therefore, Pr{*A*} = $^4/_{16}$ because the event *A* consists of four elementary outcomes (possible progeny genotypes), each of which has probability of $^1/_{16}$. An event may include no elementary outcomes, in which case its probability is 0; or it may include all elementary outcomes, in which case its probability is 1.

The elementary outcomes outside the circle defining event *A* in Figure 3.13B define another event that we have denoted *A*′. This event consists of all elementary outcomes not present in *A*, and so in other words it consists of all progeny whose genotype is not *ww*, or equivalently it consists of progeny whose phenotype is round. The event *A*′ is called the *complement* of *A*, and it is also variously denoted as $A^C$, *A*, or not-*A*. The probability of *A*′ again equals the sum of the probabilities of the elementary outcomes that constitute *A*′, and so in this case Pr{*A*′} = $^{12}/_{16}$.

Events also can be composed of other events. To see how, consider the event denoted *B* in Figure 3.13C. Event *B* is defined as all progeny from the cross whose genotype is *gg*, which could also be defined as all progeny whose seeds are green. Because *B* includes four elementary outcomes, its probability is Pr{*B*} = $^4/_{16}$. Again there is a complementary event *B*′ defined as all elementary outcomes not included in *B*; or to define *B*′ in another way, all progeny whose genotype is either *GG* or *Gg*. There are 12 elementary outcomes in *B*′, and so Pr{*B*′} = $^{12}/_{16}$.

Look again at Figures 3.13B and C and consider the event consisting of elementary outcomes that are included in $A$ or included in $B$ or both. This event is called the **union** of $A$ and $B$. We will denote the union of $A$ and $B$ as $A + B$, but in many probability textbooks you may find it symbolized as $A \cup B$, where the symbol $\cup$ is pronounced "cup." The event $A + B$ consists of seven elementary outcomes in which the progeny genotype is either *ww*, *gg*, or both, and so $Pr\{A + B\} = 7/16$.

Another important way in which two events $A$ and $B$ can be combined is called the **intersection** of $A$ and $B$, and it consists of all the elementary outcomes that are included in both $A$ and $B$. An example is shown in Figure 3.13D, where the intersection of $A$ and $B$ is shaded yellow. We will denote the intersection of $A$ and $B$ as $AB$, but in probability textbooks the intersection is often represented as $A \cap B$, where the symbol $\cap$ is pronounced "cap."

## Probability of the Union of Events

As with all events, the probability of the union of events $A + B$ equals the sum of the probabilities of the elementary outcomes that are included in $A$ or $B$ or both. Likewise, the probability of the intersection of events $AB$ equals the sum of the probabilities of the elementary outcomes that are included in both $A$ and $B$. In general, an equation for the probability of the union of $A$ and $B$ is

$$\Pr\{A + B\} = \Pr\{A\} + \Pr\{B\} - \Pr\{AB\} \qquad (1)$$

The reason for subtracting $\Pr\{AB\}$ will become evident by looking at Figure 3.13D. Because the progeny genotype *ww gg* is included in both $A$ and $B$, when the probabilities of the elementary outcomes in $A$ are added to those in $B$, the genotype *ww gg* is included twice, and to correct for the overcounting the probability of this outcome must be subtracted from the total. Because the elementary outcomes counted twice are exactly those that are present in both $A$ and $B$, they constitute the intersection of $A$ and $B$. Hence, $\Pr\{AB\}$ is the quantity that is subtracted in Equation 1.

An important special case of Equation 1 pertains when $A$ and $B$ do not overlap, that is, when they have no elementary outcomes in common. In this case $A$ and $B$ are said to be **mutually exclusive** (or *disjoint*), and so $\Pr\{AB\} = 0$. Therefore, when $A$ and $B$ are mutually exclusive, Equation 1 becomes

$$\Pr\{A + B\} = \Pr\{A\} + \Pr\{B\} \qquad (2)$$

This equation is sometimes called the **addition rule** for mutually exclusive events. Events are mutually exclusive when the occurrence of one event excludes the occurrence of the other. In other words, mutually exclusive events are mutually incompatible in the sense that no elementary outcome can be present in both. An example, again based on the experiment of crossing *Ww Gg* × *Ww Gg*, is shown in **FIGURE 3.14**. Here we have defined events $A^*$ and $B^*$ in such a way that they are modified versions of $A$ and $B$ that do not overlap. In words, the event $A^*$ may be defined as "the progeny phenotype is wrinkled but not green," and the event $B^*$ may be defined as "the progeny phenotype is green but not wrinkled." Defined in this way, $A^*$ and $B^*$ are mutually exclusive, and therefore Equation 2 has the implication that $\Pr\{A^* + B^*\} = \Pr\{A^*\} + \Pr\{B^*\} = 6/16$.

## Probability of the Intersection of Events

Returning again to events $A$ and $B$ in Figure 3.13D, consider the probability of the intersection $AB$. This event consists of all elementary outcomes that are included in both $A$ and $B$, which in this example consists of one genotype only, namely *ww gg*. The probability of $AB$ is therefore $\Pr\{AB\} = 1/16$. This is a special case in which the events are **independent**, which means that knowing whether or not the event $A$ occurs tells you nothing about whether or not the event $B$ occurs. The probability of the joint occurrence of independent events has the special property that it is proportional to the product of the probabilities of the individual events. In the example in Figure 3.13D we have already seen that $\Pr\{A\} = 4/16$ and that $\Pr\{B\} = 4/16$, and so in this case, $\Pr\{AB\} = 4/16 \times 4/16 = 16/256 = 1/16$. The events $A$ and $B$ in this example illustrate an important general principle: When the events $A$ and $B$ are independent, their probability is given by

$$\Pr\{AB\} = \Pr\{A\}\Pr\{B\} \qquad (3)$$

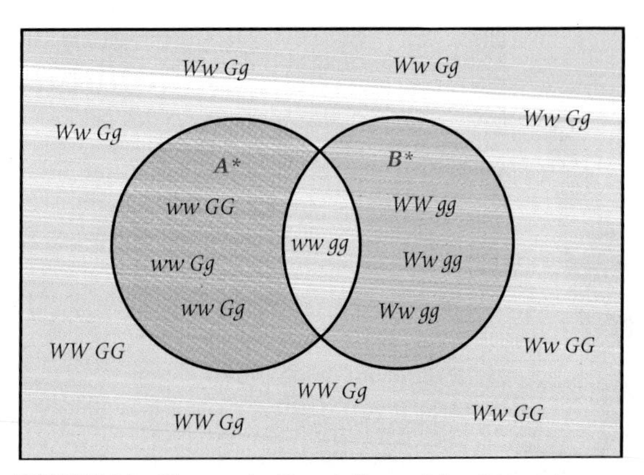


**FIGURE 3.14** The events $A^*$ and $B^*$ are defined in such a way that they each exclude genotypes whose phenotype is wrinkled and green. $A^*$ and $B^*$ do not overlap and hence are mutually exclusive.

This equation is sometimes called the **multiplication rule** for independent events. The choice of terms for "independent events" and "independent assortment" is not fortuitous, because independent assortment means that knowing the genotype of an offspring for the *W*, *w* allele pair tells you nothing about the genotype for the *G*, *g* allele pair. This principle is illustrated in **FIGURE 3.15A.** Another situation in genetics in which independence is the rule is shown in part B, which deals with successive offspring from a cross. Successive offspring are independent because the genotypes of early progeny have no influence on the probabilities in later progeny. The independence of successive offspring contradicts the widespread belief that in each human familiy, the ratio of girls to boys must "even out" at approximately 1:1. According to this reasoning, a family with four girls would be more likely to have a boy the next time around. But this belief is supported neither by theory nor by actual data. The data indicate that human families are equally likely to have a girl or a boy on any birth, irrespective of the sex distribution in previous births. Although statistics guarantees that the sex ratio will balance out when averaged over a very large number of cases, this does not imply that it will equalize in any individual case. To be concrete, among families in which there are five children, those consisting of five boys balance those consisting of five girls, yielding an overall sex ratio of 1 : 1; nevertheless, both of these sex ratios are unusual.

In anticipation of thinking about non-independent events, consider that every elementary outcome in event *B* in Figure 3.13D must either be in *A* or in *A′*. In particular, the offspring genotype *ww gg* that is included in *B* is also in *A*, whereas all the other genotypes included in *B* are in *A′*. This principle is true in general. Any event *B* can always be written as the union of two other events, *BA* and *BA′*. Furthermore, *BA* and *BA′* are mutually exclusive, because no elementary outcome can be in both *A* and *A′*. Equation 2 therefore implies that

$$\Pr\{B\} = \Pr\{BA\} + \Pr\{BA'\} \qquad (4)$$

Since $\Pr\{BA\} = \frac{1}{16}$ and $\Pr\{BA'\} = \frac{3}{16}$, it follows that $\Pr\{B\} = \frac{1}{16} + \frac{3}{16} = \frac{4}{16}$. As we shall see in the next section, Equation 4 has many important applications.

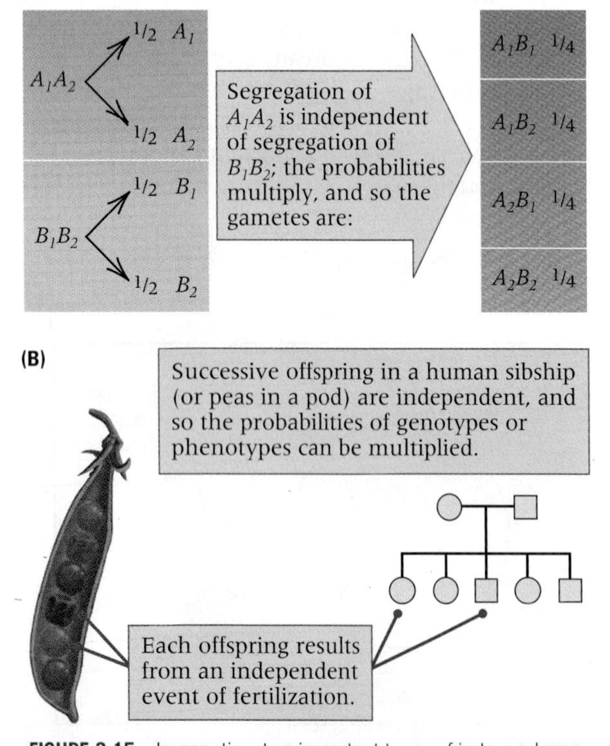


**FIGURE 3.15** In genetics, two important types of independence are (A) independent segregation of alleles that show independent assortment and (B) independent fertilizations resulting in successive offspring.

## ■ Conditional Probability

In some situations there may be partial information about the outcome of an experiment that can affect the chance that an event has occurred. The known information is called *prior information*, and it is taken into account by means of the concept of conditional probability. The *conditional probability* that an event *A* has occurred, given prior information that another event *B* has occurred, is denoted by the symbol $\Pr\{A \mid B\}$, where the event $A \mid B$ is pronounced as "*A* given *B*." To take a specific example, consider the events *A* and *B* in **FIGURE 3.16.** Here the event *A* can be described in words as "an offspring is wrinked," and event *B* as "an offspring is either wrinkled or green or both." In this example, if we had prior knowledge that *B* had occurred it would change our assessment of the probability that *A* had occurred. The reason is simple: given that *B* has occurred eliminates nine elementary outcomes from the sample space. The prior information that *B* has occurred therefore defines a new experiment and sample space because elementary outcomes that are included in *B′*, the complement of *B*, are not allowed. Given that *B* has occurred, the probability of any event is equal to its relative likelihood within *B*. Because *B* includes seven elementary outcomes and each is equally likely, each of these elementary outcomes has a probability of $\frac{1}{16}$. Hence

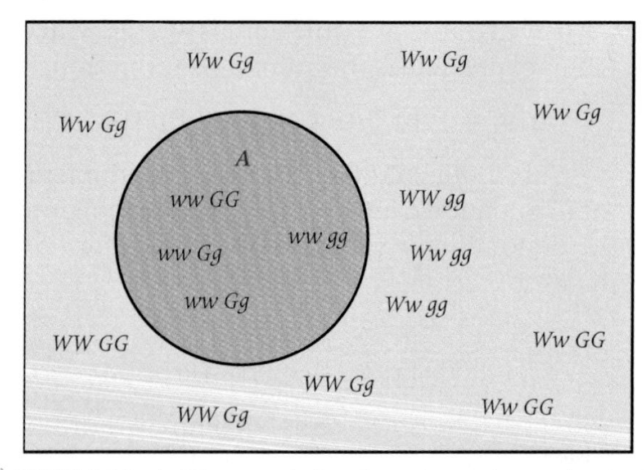


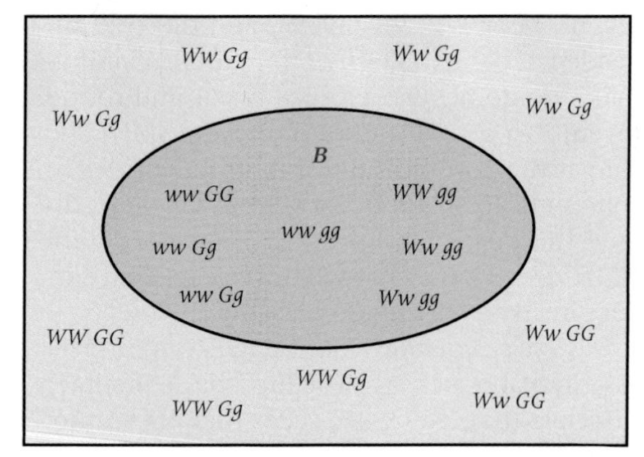


**FIGURE 3.16** In this example the elementary outcomes in event *A* are completely included among those in event *B*. Knowing that the event *B* has occurred gives important information about whether the event *A* has occurred.

the probability that *A* occurs, given that *B* has occurred, is $\Pr\{A \mid B\} = 3/7$.

These ideas may become clearer in the technical definition of conditional probability. The **conditional probability** $\Pr\{A \mid B\}$ is *defined* as

$$\Pr\{A \mid B\} = \frac{\Pr\{AB\}}{\Pr\{B\}} \qquad (5)$$

provided that $\Pr\{B\} \neq 0$. The denominator in this equation automatically adjusts the sample space for the fact that only elementary outcomes included in *B* are allowed. For the events in Figure 3.16, for example, $\Pr\{AB\} = 4/16$ and $\Pr\{B\} = 7/16$, and so Equation 5 says that $\Pr\{A \mid B\} = (4/16)/(7/16) = 4/7$, which we calculated earlier simply by redefining the probability of each of the six elementary outcomes in *B* as $1/7$.

To illustrate a more complex application of Equation 5, consider the cross involving the allele pairs (*W, w*), (*G, g*), and (*P, p*) whose $F_2$ phenotypes are shown in Figure 3.12. Suppose you were asked, "What is the probability that a plant has genotype *Ww Gg Pp*, given that it arose from a round yellow seed and has purple flowers?" Let *A* be the event "the offspring genotype is *Ww Gg Pp*" and *B* be the event "the offspring phenotype is round, yellow, and purple." In the earlier discussion of this cross, we showed that $\Pr\{A\} = 1/2 \times 1/2 \times 1/2 = 1/8$. In this case $\Pr\{AB\} = \Pr\{A\}$, because a plant of genotype *Ww Gg Pp* must have phenotype corresponding to *B*, namely, round, yellow, and purple. Likewise we showed earlier that $\Pr\{B\} = 3/4 \times 3/4 \times 3/4 = 27/64$. Hence, according to Equation 5, the probability that an offspring genotype is *Ww Gg Pp* (event *A*) given that its phenotype is round, yellow, and purple (event *B*) is given by $\Pr\{AB\}/\Pr\{B\} = (1/8)/(27/64) = 8/27$. This answer also can be gotten by mean of a Punnett square for three segregating genes, but this approach requires very much more time and effort.

Equation 5 becomes a particularly powerful tool in probability when combined with Equation 4, in which case its form becomes

$$\Pr\{A \mid B\} = \frac{\Pr\{AB\}}{\Pr\{BA\} + \Pr\{BA'\}} \qquad (6)$$

Note that $\Pr\{AB\} = \Pr\{BA\}$ because the events *AB* and *BA* consist of exactly the same elementary outcomes. To illustrate the utility of Equation 6, consider a conceptual experiment in which we choose one round, green seed from the sample space in Figure 3.13A. What is the probability that the genotype of the seed is *Ww*? To use Equation 6, let *B* be the event that "a randomly chosen green seed is round" and *A* be the event that "a randomly chosen green seed has genotype *Ww*." Then $BA = AB$ is the event that "a randomly chosen green seed has genotype *Ww* and is round" and $BA'$ is the event that "a randomly chosen green seed has genotype *WW* and is round." Mendel's principle of segregation tells us that $\Pr\{BA\} = \Pr\{AB\} = 1/2$, and $\Pr\{BA'\} = 1/4$. Putting all this together in the form of Equation 6 yields

$$\Pr\{A \mid B\} = \frac{1/2}{1/2 + 1/4} = 2/3$$

If you are averse to algebraic equations, application of Equation 6 also can be carried out by means of a table laid out as that shown below, again taking as an example the probability that a randomly chosen green, round seed has genotype *Ww*:

| | **B** |
|---|---|
| ***A*** 1/2 | 1/2 |
| ***A'*** 1/4 | 1/4 |
| Sum | 3/4 |

On the left are the probabilities that a randomly chosen green seed has genotype *Ww* (event $A$) or $WW$ (event $A'$), and the column headed $B$ contains entries for the probability that a randomly chosen seed is $Ww$ and round (event $BA$) or $WW$ and round (event $BA'$). The sum at the bottom is the sum of the entries in column $B$. To obtain $\Pr\{A \mid B\}$ from the table, divide the entry for $BA$ by the sum to obtain $(\frac{1}{2})/(\frac{3}{4}) = \frac{2}{3}$, which is what we calculated earlier directly from Equation 6.

These calculations may seem like an unnecessarily elaborate way to deduce the elementary principle that, among $F_2$ seeds that are round, $\frac{2}{3}$ are $Ww$ and $\frac{1}{3}$ are $WW$, but we shall see in the next section that this type of formal reasoning allows much more complex types of probability problems to be solved.

## Bayes Theorem

The wonderful insight of conditional probability is credited to the mind of Thomas Bayes (1702–1761), a Presbyterian minister in Tunbridge Wells, England. His ideas were summarized in his *Essay Towards Solving a Problem in the Doctrine of Chances*, published in 1763, two years after his death. What is now known as **Bayes theorem** can be written in several forms, including the forms in Equations 5 and 6. Many sources give Bayes theorem in another form in which the numerator and denominator in Equation 6 are expressed more fully. This can be done using the expressions in **TABLE 3.3**. These expressions yield a more complete form of Bayes theorem, but they results in an equation that appears much more formidable than Equation 6. Note that the layout of Table 3.3 is the same as that used earlier, and in particular the $B$ column consists of the entries $\Pr\{BA\}$ and $\Pr\{BA'\}$. These entries have also been converted into statements of conditional probability using Equation 5, which can be written as $\Pr\{AB\} = \Pr\{B\}\Pr\{A \mid B\}$ merely by multiplying both sides by $\Pr\{B\}$. The roles of $A$ and $B$ in this expression can be interchanged by relabeling $A$ as $B$ and $B$ as $A$. It then follows that $\Pr\{BA\} = \Pr\{A\}\Pr\{B \mid A\}$ and also that $\Pr\{BA'\} = \Pr\{A'\}\Pr\{B \mid A'\}$. These expressions correspond to those on the right hand side of Table 3.3. The sum of column $B$ in Table 3.3 still equals $\Pr\{B\}$, which when expressed as the sum of its terms equals

$$\Pr\{B\} = \Pr\{A\}\Pr\{B \mid A\} + \Pr\{A'\}\Pr\{B \mid A'\}$$

Substituting $\Pr\{A\}\Pr\{B \mid A\}$ into the numerator of Equation 6 and the above expression into the denominator yields an alternative form of Bayes theorem:

$$\Pr\{A \mid B\} = \frac{\Pr\{A\}\Pr\{B \mid A\}}{\Pr\{A\}\Pr\{B \mid A\} + \Pr\{A'\}\Pr\{B \mid A'\}} \quad (7)$$

Equation 7 looks extremely complex, but it is merely another way of writing Equation 5 or Equation 6.

We will illustrate application of Equation (7) using the tabular layout of Table 3.3 in connection with the situation shown in **FIGURE 3.17**. In this example, suppose that a round seed (seed 1) was chosen at random from among the green seeds in Figure 3.5. Suppose that we wish to know whether its genotype is $Ww$ or $WW$ and so carry out a testcross. We therefore grow a plant from this seed and fertilize the ovules with pollen from a plant grown from the wrinkled green seed (seed 2). Expecting a large number of progeny seeds, we hope to be able to specify the genotype of seed 1 because if it is $Ww$ there will be $\frac{1}{4}$ wrinkled seeds among the progeny. Unfortunately, the plant grown from seed 1 was attacked by vicious pea weevils leaving one soli-

**Table 3.3 Tabular setup for applying Bayes theorem**

| | $B$ |
|---|---|
| $A$ $\Pr\{A\}$ | $\Pr\{BA\} = \Pr\{A\}\Pr\{B \mid A\}$ |
| $A'$ $\Pr\{A'\}$ | $\Pr\{BA'\} = \Pr\{A'\}\Pr\{B \mid A'\}$ |
| Sum | $\Pr\{B\}$ |

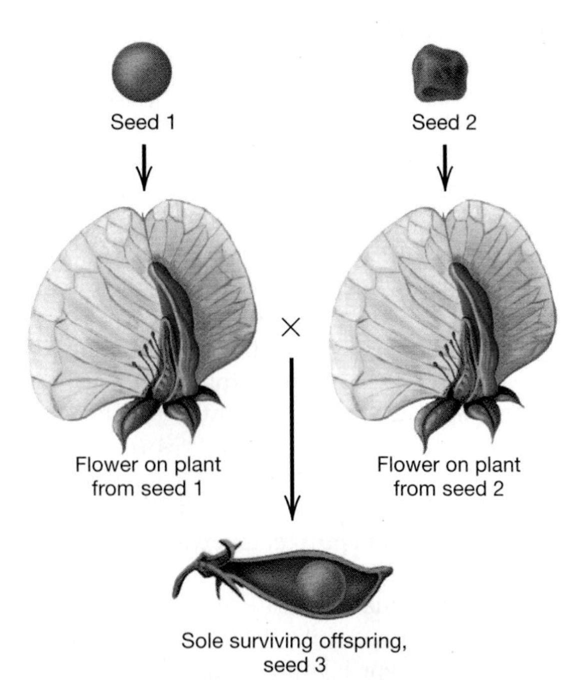


**FIGURE 3.17** A testcross of a plant grown from a single round, green seed taken from the $F_2$ progeny of a dihybrid cross. Only one offspring seed from the textcross survives, but this contributes data that bear on the probability that seed 1 has genotype $WW$ versus $Ww$.

tary, stunted pod containing only one forlorn offspring, namely seed 3, which proved to be round. Based on this finding, what is the probability that the seed 1 was of genotype *Ww*?

We can solve this problem using Bayes theorem in the form of Table 3.3. The relevant calculations are shown below. The event *A* is "the seed 1 has genotype *Ww*" and event *A'* is "the seed 1 has genotype *WW*." We have already seen that the probability that a seed known to be round had genotype *Ww* is $\frac{2}{3}$ and the probability that it has genotype *WW* is $\frac{1}{3}$. These are the entries in the

| | *B* |
|---|---|
| *A* 2/3 | (2/3)(1/2) |
| *A'* 1/3 | (1/3)(1) |
| Sum | 2/3 |

left-hand column. In column *B*, the entry on the top is $\Pr\{A\}\{\Pr\{B \mid A\}$, where *B* is the event "the seed 3 is round." Likewise the entry $(\frac{1}{3})(1)$ is $\Pr\{A'\}\{\Pr\{B \mid A'\}$. The sum at the bottom is $\Pr\{B\}$, and hence the required conditional probability $\Pr\{A \mid B\} = [(\frac{2}{3})(\frac{1}{2})]/(\frac{2}{3}) = \frac{1}{2}$. Note that this is quite different from the probability $\frac{2}{3}$ that we would have deduced if the seed 3 had not survived.

## 3.5 Human Pedigree Analysis

Large deviations from expected genetic ratios are often found in individual human families and in domesticated large animals because of the relatively small number of progeny. The effects of segregation are nevertheless evident upon examination of the phenotypes among several generations of related individuals. A diagram of a family tree showing the phenotype of each individual among a group of relatives is a **pedigree.** In this section we introduce basic concepts in pedigree analysis.

### Characteristics of Dominant and Recessive Inheritance

**FIGURE 3.18** defines the standard symbols used in depicting human pedigrees. Females are represented by circles and males by squares. (A diamond is used if the sex is unknown—as, for example, in a miscarriage.) Persons with the phenotype of interest are indicated by colored or shaded symbols. For recessive alleles, heterozygous carriers are sometimes depicted with half-filled symbols. A mating between a female and a male is indicated by joining their symbols with a horizontal line, which is connected vertically to a second horizontal line, below, that connects the symbols for their offspring. The offspring within a sibship, called **siblings** or **sibs,** are represented from left to right in order of their birth.

A pedigree for the trait *Huntington disease,* caused by a dominant mutation, is shown in **FIGURE 3.19**. The numbers in the pedigree are for convenience in referring to particular persons. The successive generations are designated by Roman numerals. Within any generation, all of the persons are numbered consecutively from left to right. The pedigree starts with the

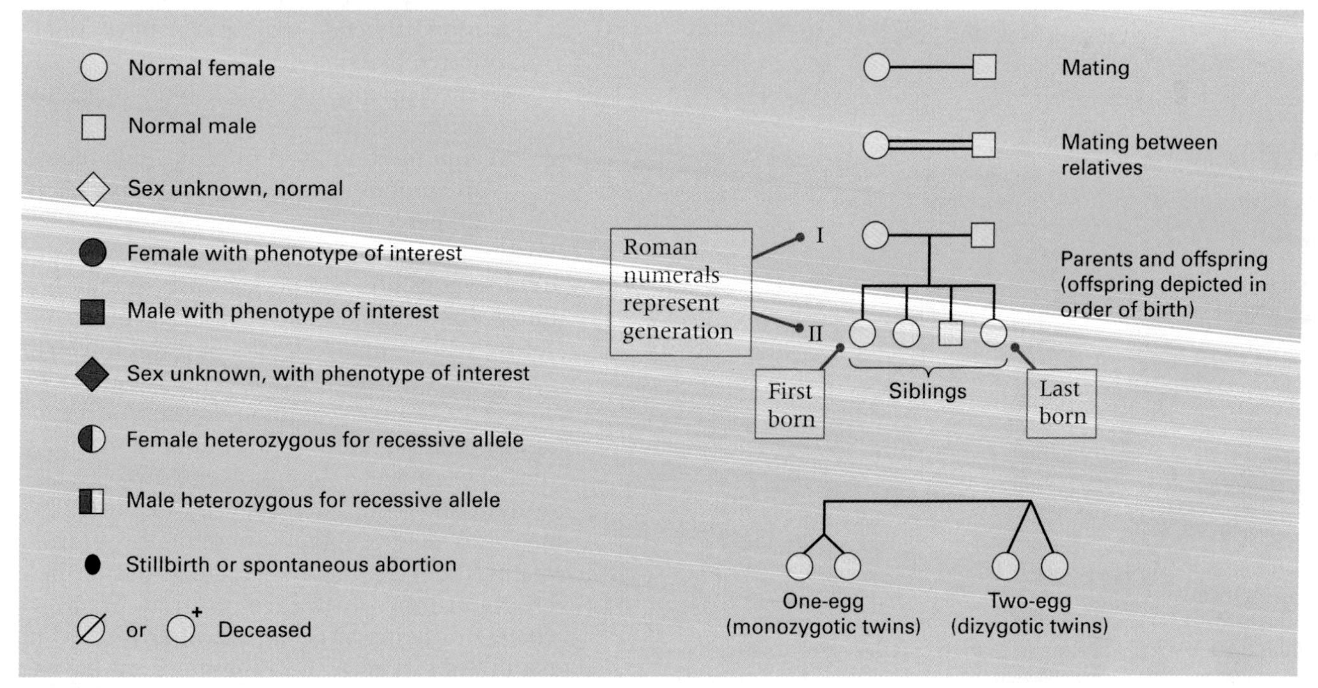


**FIGURE 3.18** Conventional symbols used in depicting human pedigrees.

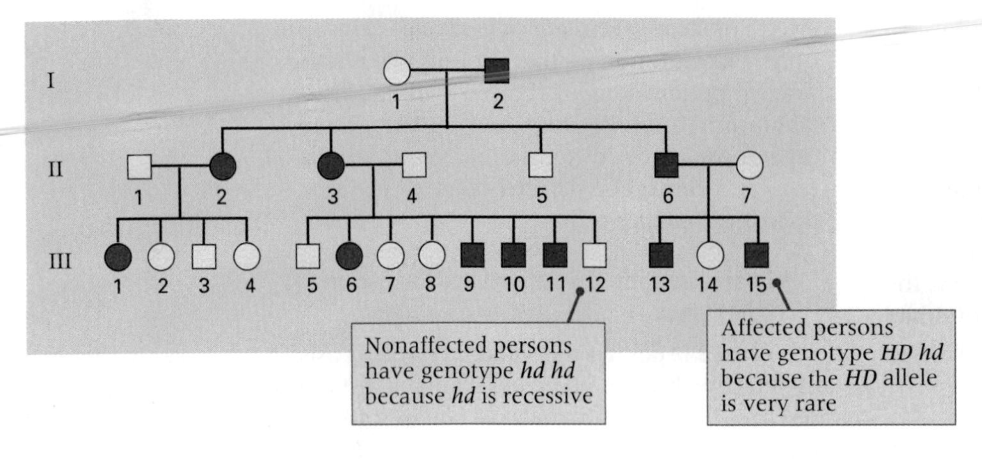
I
II
III
Nonaffected persons have genotype *hd hd* because *hd* is recessive
Affected persons have genotype *HD hd* because the *HD* allele is very rare

**FIGURE 3.19** Pedigree of a human family showing the inheritance of the dominant gene for Huntington disease. Females and males are represented by circles and squares, respectively. Red symbols indicate persons affected with the disease.

woman I-1 and the man I-2. The man has Huntington disease, which is a progressive nerve degeneration that usually begins about middle age. It results in severe physical and mental disability and then death. The dominant allele, *HD*, that causes Huntington disease is rare. All affected persons in the pedigree have the heterozygous genotype *HD hd*, whereas nonaffected persons have the homozygous normal genotype *hd hd*. The disease has complete penetrance. The **penetrance** of a genetic disorder is the proportion of individuals with the at-risk genotype who actually express the trait; *complete penetrance* means the trait is expressed in 100 percent of persons with that genotype. The pedigree demonstrates the following characteristic features of inheritance due to a rare dominant allele with complete penetrance.

1. Females and males are equally likely to be affected.
2. Affected offspring have one affected parent (except for rare new mutations), and the affected parent is equally likely to be the mother or the father.
3. On average, half of the individuals in sibships with an affected parent are affected.

A pedigree for a trait due to a homozygous recessive allele is shown in **FIGURE 3.20**. The trait is *albinism,* absence of pigment in the skin, hair, and iris of the eyes. This pedigree illustrates characteristics of inheritance due to a rare recessive allele with complete penetrance:

1. Females and males are equally likely to be affected.
2. Affected individuals, if they reproduce, usually have unaffected progeny.
3. Most affected individuals have unaffected parents.
4. The parents of affected individuals are often relatives.
5. Among siblings of affected individuals, the proportion affected is approximately 25 percent.

With rare recessive inheritance, the mates of homozygous affected persons are usually homozygous for the normal allele, so all of the offspring will be heterozygous and not affected. Heterozygous **carriers** of the mutant allele are considerably more common than homozygous affected individuals, because it is more likely that a person will inherit only one copy of a rare mutant allele than two copies. Most homozygous recessive genotypes therefore result from matings between carriers (heterozygous × heterozygous), in which each offspring has a 1/4 chance of being affected. Another important feature of rare recessive inheritance is that the parents of affected individuals are often related

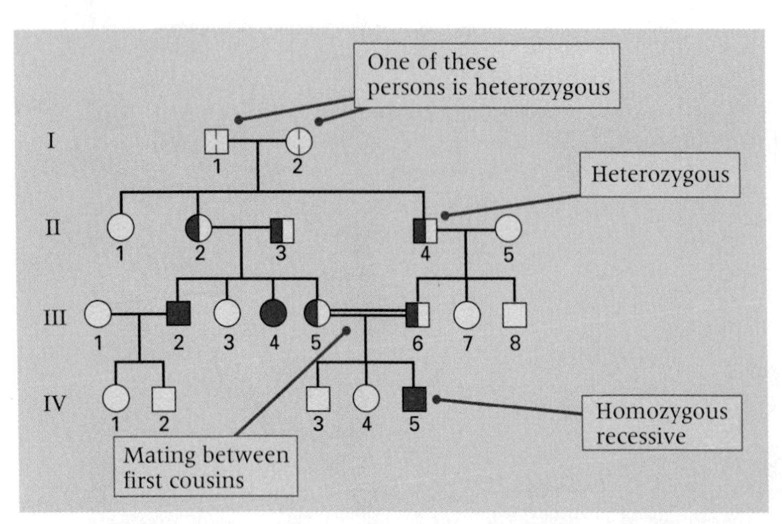
One of these persons is heterozygous
I
Heterozygous
II
III
IV
Homozygous recessive
Mating between first cousins

**FIGURE 3.20** Pedigree of albinism. With recessive inheritance, affected persons (filled symbols) often have unaffected parents. The double horizontal line indicates a mating between relatives—in this case, first cousins.

(**consanguineous**). A mating between relatives is indicated with a double line connecting the partners, as for the first-cousin mating in Figure 3.20. Mating between relatives is important for recessive alleles to become homozygous, because when a recessive allele is rare, it is more likely to become homozygous through inheritance from a common ancestor than from parents who are completely unrelated. The reason is that the carrier of a rare allele may have many descendants who are also carriers. If two of these carriers should mate with each other (for example, in a first-cousin mating), then the hidden recessive allele can become homozygous with a probability of 1/4. Mating between relatives constitutes *inbreeding*, and the consequences of inbreeding are discussed further in Chapter 17. Because an affected individual indicates that the parents are heterozygous carriers, the expected proportion of affected individuals among the siblings is approximately 25 percent, but the exact value depends on the details of how affected individuals are identified and included in the database.

### Most Human Genetic Variation is Not "Bad"

Before the advent of molecular methods, there were many practical obstacles to the study of human genetics. With the exception of traits such as the ABO and other blood groups, few traits showing simple Mendelian inheritance were known. Most of these were associated with genetic diseases, and these presented special challenges:

- Most genes that cause simple Mendelian genetic diseases are rare, so they are observed in only a small number of families.
- Many genes for simple Mendelian diseases are recessive, so they are not detected in heterozygous genotypes.
- The number of offspring per human family is relatively small, so segregation cannot usually be detected in single sibships.
- The human geneticist cannot perform testcrosses, backcrosses, or other experimental matings.

This book includes numerous examples of how molecular genetics has revolutionized the study of human genetics. For example, in Chapter 2 we discussed the prevalence of single-nucleotide polymorphisms (SNPs) and copy-number polymorphisms (CNPs) in the human genome. Only a very small number of these common polymorphisms are associated with diseases, and even those serve as genetic risk factors that interact with other risk factors, including other genes and the environment. Although most SNPs and CNPs have no adverse effects, both types of variation show simple Mendelian inheritance, which is why human geneticists have come to rely on them for family and population studies.

But a few, seemingly harmless, simple-Mendelian traits had been detected in the premolecular era. One of the best known was associated with the ability to taste a chemical substance known as **phenylthiocarbamide (PTC)**, which has the molecular structure shown here.

$NH_2$
C=S
N
H

PTC is an artificial chemical created by an industrial chemist in the early 1930s. The taste polymorphism was discovered one day when he carelessly released a cloud of powdered PTC into the air. The PTC powder didn't bother the chemist at all, but his lab mate loudly complained about the bitter taste it left in his mouth. Out of curiosity, the chemist started to test family and friends for their ability to taste PTC, and he recruited a geneticist who began to study the situation. It was soon shown from family studies that the ability to taste PTC is a trait inherited as a simple Mendelian dominant. In European populations about 70 percent of the people are tasters and 30 percent are nontasters, but these proportions differ greatly among ethnic groups. Among people of African or Asian origin the frequency of tasters is about 90 percent, whereas among Australian aborigines it is only about 50 percent.

The ability to taste PTC is quantitative, however. The most sensitive tasters can taste concentrations as low as 0.001 millimolar (mM) whereas the most insensitive nontasters fail to detect concentrations as high as 10 mM. For classifying individuals as "tasters" or "nontasters" an arbitrary cutoff is employed, typically at a concentration of 0.2 mM PTC. Most of the variation in tasting ability between tasters and nontasters is due to the major taster polymorphism, but there are also differences due to other genes, gender, and probably environmental factors. The result of the other variables is that about 5 percent of the heterozygous tasters get classified as "nontaster" and at least 5 percent of the homozygous nontasters become classified as "tasters."

The molecular basis of the taster polymorphism is now known to reside in a taste receptor protein known as hTAS2R38. There are several alleles of the gene, but the most common forms of the protein differ by three amino acids at scattered positions along the protein. The allelic forms are known as *PAV* and *AVI*, because the three key amino acids in the PAV protein are proline, alanine, and valine, whereas these three positions in the AVI protein are occupied by alanine, valine, and isoleucine. The PAV form is the one that confers the ability to taste PTC.

When you think about it, a polymorphism in PTC tasting makes no sense. PTC is a completely artificial chemical synthesized in the laboratory, so why should there be a polymorphism in the ability to taste it? A clue comes from the observation that the chemical structure of PTC resembles a large and heterogeneous class of molecules called *glucosinolates*. These are compounds synthesized by some plants, including some human food plants, and their synthesis likely evolved as a chemical defense against plant-eating insects. Among the plants that produce glucosinolates is one singled out by former President George H. W. Bush, who in 1989 removed broccoli from the White House menu, proclaiming: "I do not like broccoli. And I haven't liked it since I was a little kid and my mother made me eat it. And I'm President of the United States and I'm not going to eat any more broccoli!" In good-humored protest, broccoli growers throughout the country sent him tons of the stuff. Ironically, 17 years after Bush Sr.'s broccoli boycott, new studies showed that individuals carrying the PAV form of the hTAS2R38 taste receptor do, in fact, find broccoli to be significantly more bitter tasting than individuals homozygous for the allele encoding AVI form. Tasters also report a greater perceived bitterness for collard greens, turnips, rutabagas, and horseradish.

## Molecular Markers in Human Pedigrees

Because techniques for manipulating DNA allow direct access to the DNA, modern genetic studies of human pedigrees are carried out primarily using genetic markers present in the DNA itself, rather than through the phenotypes produced by mutant genes. Various types of DNA polymorphisms were discussed in Chapter 2, along with the methods by which they are detected and studied. An example of a DNA polymorphism segregating in a three-generation human pedigree is shown in **FIGURE 3.21**. The type of polymorphism is a *simple sequence repeat polymorphism (SSRP)*, in which each allele differs in size according to the number of copies it contains of a short DNA sequence repeated in tandem. The differences in size are detected by electrophoresis after amplification of the region by PCR. SSRPs usually have many codominant alleles, and the majority of individuals are heterozygous for two different alleles. In the example in Figure 3.21, each of the parents is heterozygous, as are all of the children.

Six alleles are depicted in Figure 3.21, denoted by $A_1$ through $A_6$. In the gel, the numbers of the bands correspond to the subscripts of the alleles. The mating in generation II is between two heterozygous genotypes: $A_4A_6 \times A_1A_3$. Because of segregation in each parent, four genotypes are possible among the offspring ($A_4A_1$, $A_4A_3$, $A_6A_1$, and $A_6A_3$); these would conventionally be written with the smaller subscript first, as $A_1A_4$, $A_3A_4$, $A_1A_6$, and $A_3A_6$. With random fertilization the offspring genotypes are equally likely, as may be verified from a Punnett square for the mating. Figure 3.21 illustrates some of the principal advantages of multiple, codominant alleles for human pedigree analysis:

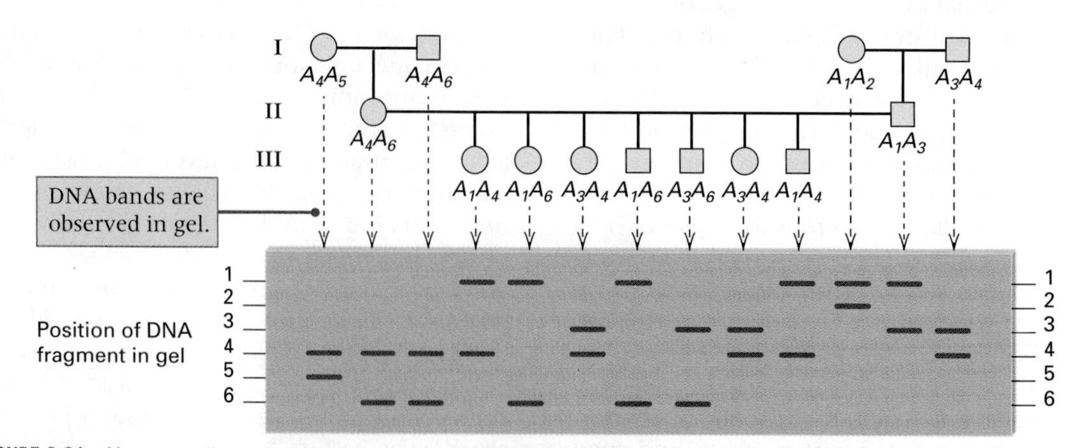


**FIGURE 3.21** Human pedigree showing segregation of SSRP alleles. Six alleles ($A_1$–$A_6$) are present in the pedigree, but any one person can have only one allele (if homozygous) or two alleles (if heterozygous).

**connection**

## This Land is Your Land

**The Huntington's Disease Collaborative Research Group 1993**
Comprising 58 authors among 9 institutions
*A Novel Gene Containing a Trinucleotide Repeat That Is Expanded and Unstable on Huntington's Disease Chromosomes*

*Modern genetic research is sometimes carried out by large collaborative groups in a number of research institutions scattered across several countries. This approach is exemplified by the search for the gene responsible for Huntington disease. The search was highly publicized because of the severity of the disease, the late age of onset, and the dominant inheritance. Famed folk singer Woody Guthrie, who wrote "This Land Is Your Land" and other well-known tunes, died of the disease in 1967. When the gene was identified, it turned out to encode a protein (now called huntingtin) of unknown function that is expressed in many cell types throughout the body and not, as expected, exclusively in nervous tissue. Within the coding sequence of this gene is a trinucleotide repeat (5′-CAG-3′) that is repeated in tandem a number of times according to the general formula (5′-CAG-3′)$_n$. Among normal alleles, the number n of repeats ranges from 11 to 34 with an average of 18; among mutant alleles, the number of repeats ranges from 40 to 86. This tandem repeat is genetically unstable in that it can, by some unknown mechanism, increase in copy number ("expand"). In two cases in which a new mutant allele was analyzed, one had increased in repeat number from 36 to 44 and the other from 33 to 49. This is a mutational mechanism that is quite common in some human genetic diseases. The excerpt cites several other examples. The authors also emphasize that their discovery raises important ethical issues, including genetic testing, confidentiality, and informed consent.*

Huntington disease (HD) is a progressive neurodegenerative disorder characterized by motor disturbance, cognitive loss, and psychiatric manifestations. It is inherited in an autosomal dominant fashion and affects approximately 1 in 10,000 individuals in most populations of European origin. The hallmark of HD is a distinctive choreic [jerky] movement disorder that typically has a subtle, insidious onset in the fourth to fifth decade of life and gradually worsens over a course of 10 to 20 years until death. . . .The genetic defect causing HD was assigned to chromosome 4 in one of the first successful linkage analyses using DNA markers in humans. Since that time, we have pursued an approach to isolating and characterizing the HD gene based on progressively refining its localization. . . . [We have found that a] 500-kb segment is the most likely site of the genetic defect. [The abbreviation kb stands for kilobase pairs; 1 kb equals 1000 base pairs.] Within this region, we have identified a large gene, spanning approximately 210 kb, that encodes a previously undescribed protein. The reading frame contains a polymorphic (CAG)$_n$ trinucleotide repeat with at least 17 alleles in the normal population, varying from 11 to 34 CAG copies. On HD chromosomes, the length of the trinucleotide repeat is substantially increased. . . . Elongation of a trinucleotide repeat sequence has been implicated previously as the cause of three quite different human disorders, the fragile-X syndrome, myotonic dystrophy, and spino-bulbar muscular atrophy. . . . It can be expected that the capacity to monitor directly the size of the trinucleotide repeat in individuals "at risk" for HD will revolutionize testing for the disorder. . . . We consider it of the utmost importance that the current internationally accepted guidelines and counseling protocols for testing people at risk continue to be observed, and that samples from unaffected relatives should not be tested inadvertently or without full consent. . . . With the mystery of the genetic basis of HD apparently solved, [it opens] the next challenges in the effort to understand and to treat this devastating disorder.

***We consider it of the utmost importance that the current internationally accepted guidelines and counseling protocols for testing people at risk continue to be observed, and that samples from unaffected relatives should not be tested inadvertently or without full consent.***

Source: The Huntington's Disease Collaborative Research Group, *Cell* 72(1993): 971–983

1. Heterozygous genotypes can be distinguished from homozygous genotypes.
2. Many individuals in the population are heterozygous, and so many matings are informative in regard to segregation.
3. Each segregating genetic marker yields up to four distinguishable offspring genotypes.

## 3.6 Incomplete Dominance and Epistasis

Dominance and codominance are not the only possibilities for pairs of alleles. There are situations of **incomplete dominance,** in which the phenotype of the heterozygous genotype is intermediate between the phenotypes of the ho-

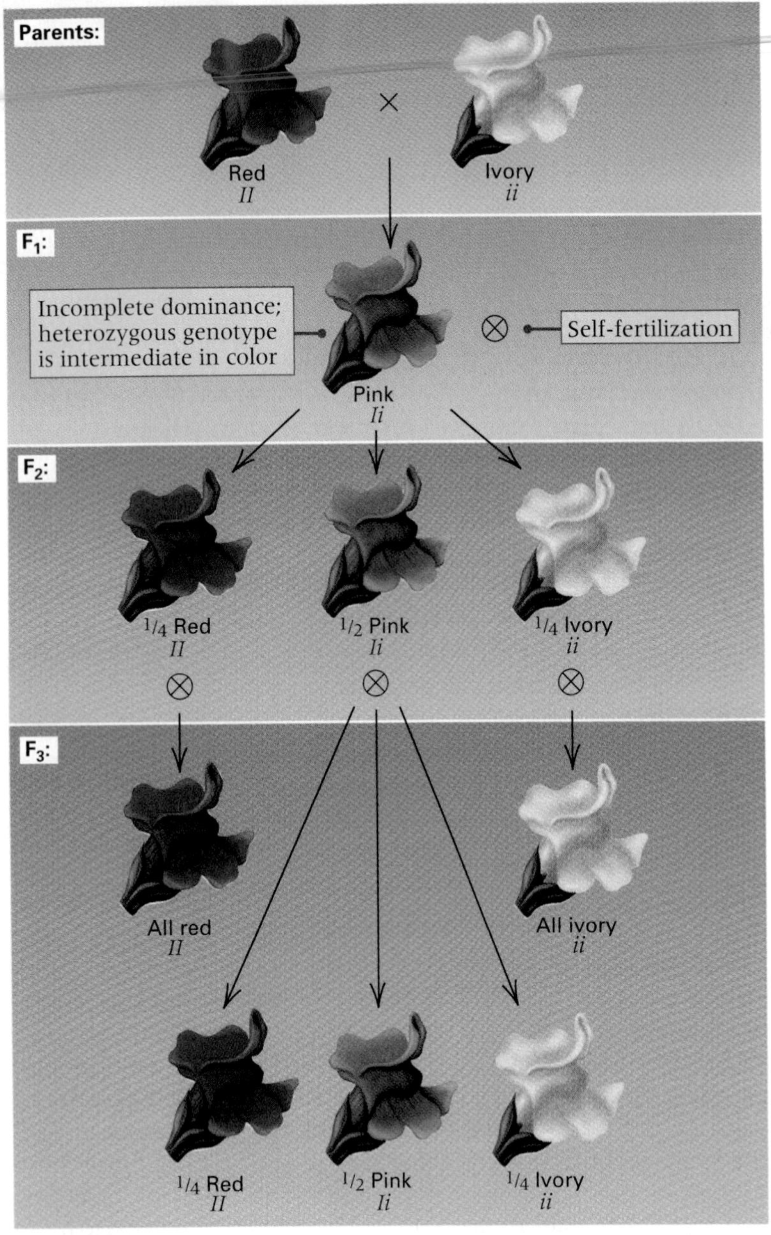


**FIGURE 3.22** Red versus white flower color in snapdragons shows incomplete dominance.

mozygous genotypes. A classic example of incomplete dominance concerns flower color in the snapdragon *Antirrhinum majus* (**FIGURE 3.22**). In wildtype flowers, a red type of anthocyanin pigment is formed by a sequence of enzymatic reactions. A wildtype enzyme, encoded by the *I* allele, is limiting to the rate of the overall reaction, so the amount of red pigment is determined by the amount of enzyme that the *I* allele produces. The alternative *i* allele codes for an inactive enzyme, and *ii* flowers are ivory in color. Because the amount of the critical enzyme is reduced in *Ii* heterozygotes, the amount of red pigment in the flowers is reduced also, and the effect of the dilution is to make the flowers pink. A cross between plants differing in flower color therefore gives direct phenotypic evidence of segregation (Figure 3.22). The cross *II* (red) × *ii* (ivory) yields $F_1$ plants with genotype *Ii* and pink flowers. In the $F_2$ progeny obtained by self-pollination of the $F_1$ hybrids, one experiment resulted in 22 plants with red flowers, 52 with pink flowers, and 23 with ivory flowers, which fits the expected ratio of 1 : 2 : 1.

## ■ Multiple Alleles

The occurrence of **multiple alleles** is exemplified by the alleles $A_1$–$A_6$ of the SSRP marker in the human pedigree in Figure 3.21. Multiple alleles are relatively common in natural populations and, as in this example, can be detected most easily by molecular methods. In the DNA of a gene, each nucleotide can be A, T, G, or C, so a gene of $n$ nucleotides can theoretically mutate at any of the positions to any of the three other nucleotides. The number of possible single-nucleotide differences in a gene of length $n$ is therefore $3 \times n$. If $n = 5000$, for example, there are potentially 15,000 alleles (not counting any of the possibilities with more than one nucleotide substitution). Most of the potential alleles do not actually exist at any one time. Some are absent because they did not occur, others did occur but were eliminated by chance or because they were harmful, and still others are present but at such a low frequency that they remain undetected. Nevertheless, at the level of DNA sequence, most genes in most natural populations have multiple alleles, all of which can be considered "wildtype." Multiple wildtype alleles are useful in such applications as DNA typing because two unrelated people are unlikely to have the same genotype, especially if several different loci, each with multiple alleles, are examined. Many harmful mutations also exist in multiple forms. Recall from Chapter 1 that more than 400 mutant forms of the phenylalanine hydroxylase gene have been identified in patients with phenylketonuria. The alleles $A_1$–$A_6$ in Figure 3.21 also illustrate that although a *population* of organisms may contain any number of alleles, any particular organism or cell may carry no more than two, and any gamete may carry no more than one.

In some cases, the multiple alleles of a gene exist merely by chance and reflect the history of mutations that have taken place in the population and the dissemination of these mutations among population subgroups by migration and interbreeding. In other cases, there are biological mechanisms that favor the maintenance of a large number of alleles. For example, genes that control self-sterility in certain flowering plants can have large numbers of allelic types. This type of self-sterility is found in species of red clover that grow wild in many pastures. The self-sterility genes prevent self-fertilization because a pollen grain can undergo pollen tube growth and fertilization only if it contains a self-sterility allele different from either of the alleles present in the flower on which it lands. In other words, a pollen grain containing an allele already present in a flower will not function on that flower. Because all pollen grains produced by a plant must contain one of the self-sterility alleles present in the plant, pollen cannot function on the same plant that produced it, and self-fertilization cannot take place. Under these conditions, any plant with a new allele has a selective advantage, because pollen that contains the new allele can fertilize all flowers except those on the same plant. Through evolution, populations of red clover have accumulated hundreds of alleles of the self-sterility gene, many of which have been isolated and their DNA sequences determined. Many of the alleles differ at multiple nucleotide sites, which implies that the alleles in the population are very old.

### Human ABO Blood Groups

In a multiple allelic series, there may be different dominance relationships between different pairs of alleles. An example is found in the human ABO blood groups, which are determined by three alleles denoted $I^A$, $I^B$, and $I^O$. (Actually, there are two slightly different variants of the $I^A$ allele.) The blood group of any person may be A, B, AB, or O, depending on the type of polysaccharide (polymer of sugars) present on the surface of red blood cells. One of two different polysaccharides, A or B, can be formed from a precursor molecule that is modified by the enzyme product of the $I^A$ or the $I^B$ allele. The gene product is a glycosyl transferase enzyme that attaches a sugar unit to the precursor (**FIGURE 3.23**). The $I^A$ or the $I^B$ alleles encode different forms of the enzyme with replacements at four amino acid sites; these alter

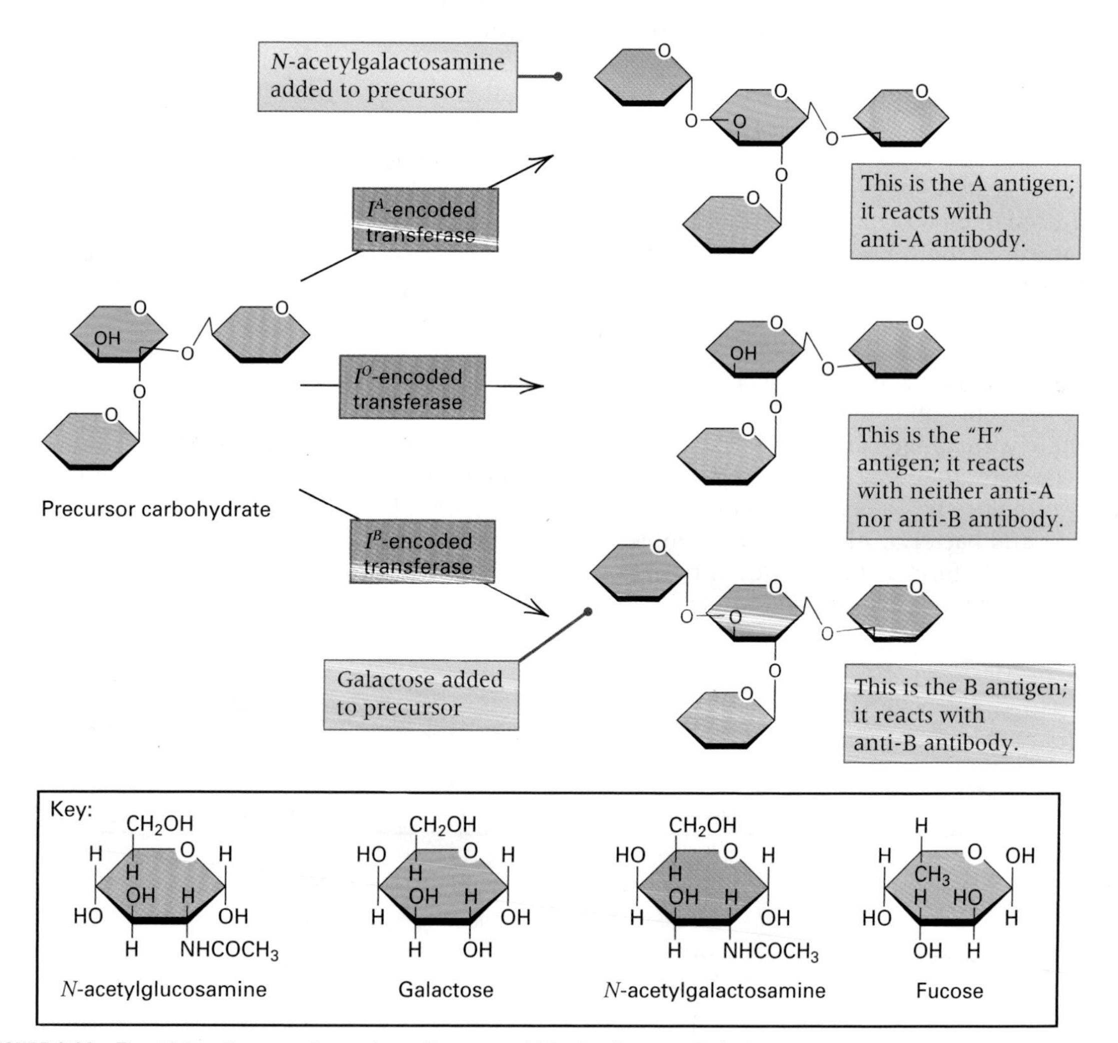


**FIGURE 3.23** The ABO antigens on the surface of human red blood cells are carbohydrates.

the substrate specificity so that each enzyme attaches a different sugar. People of genotype $I^AI^A$ produce red blood cells having only the A polysaccharide and are said to have blood type A. Those of genotype $I^BI^B$ have red blood cells with only the B polysaccharide and have blood type B. Heterozygous $I^AI^B$ people have red cells with both the A and the B polysaccharides and have blood type AB. The third allele, $I^O$, encodes an enzymatically inactive protein that leaves the precursor unchanged; neither the A nor the B type of polysaccharide is produced. Homozygous $I^OI^O$ persons therefore lack both the A and the B polysaccharides and are said to have blood type O.

In this multiple allelic series, the $I^A$ and $I^B$ alleles are codominant: The heterozygous genotype has the characteristics of both homozygous genotypes—the presence of both the A and the B carbohydrate on the red blood cells. On the other hand, the $I^O$ allele is recessive to both $I^A$ and $I^B$. Hence, heterozygous $I^AI^O$ genotypes produce the A polysaccharide and have blood type A, and heterozygous $I^BI^O$ genotypes produce the B polysaccharide and have blood type B. The genotypes and phenotypes of the ABO blood group system are summarized in the first three columns of **TABLE 3.4**.

ABO blood groups are important in medicine because of the need for blood transfusions. A crucial feature of the ABO system is that most human blood contains antibodies to either the A or the B polysaccharide. An **antibody** is a protein made by the immune system in response to a stimulating molecule called an **antigen** and is capable of binding to the antigen. An antibody is usually specific in that it recognizes only one antigen. Some antibodies combine with antigen and form large molecular aggregates that may precipitate.

Antibodies act to defend against invading viruses and bacteria. Although antibodies do not normally form without prior stimulation by the antigen, people capable of producing anti-A and anti-B antibodies do produce them. Production of these antibodies may be stimulated by antigens that are similar to polysaccharides A and B and that are present on the surfaces of many common bacteria. However, a mechanism called *tolerance* prevents an organism from producing antibodies against its own antigens. This mechanism ensures that A antigen or B antigen elicits antibody production only in people whose own red blood cells do not contain A or B, respectively. The end result:

> People of blood type O make both anti-A and anti-B antibodies, those of blood type A make anti-B antibodies, those of blood type B make anti-A antibodies, and those of blood type AB make neither type of antibody.

The antibodies found in the blood fluid of people with each of the ABO blood types are shown in the fourth column in Table 3.4. The clinical significance of the ABO blood groups is that transfusion of blood containing A or B red-cell antigens into persons who make antibodies against them results in an agglutination reaction in which the donor red blood cells are clumped. In this reaction, the anti-A antibody agglutinates red blood cells of either blood type A or blood type AB, because both carry the A antigen. Similarly, anti-B antibody agglutinates red blood cells of either blood type B or blood type AB. When the blood cells agglutinate, many blood vessels are blocked, and the recipient of the transfusion goes into shock and may die. Incompatibility in the other direction, in which the donor blood contains antibodies against the recipient's red blood cells, is usually acceptable because the donor's antibodies are diluted so rapidly that clumping is avoided. The types of compatible blood transfusions are shown in the last two columns of Table 3.4. Note that a person of blood type AB can receive blood from a person of any other ABO type; type AB is called

**Table 3.4 Genetic control of the human ABO blood groups**

| Genotype | Antigens present on red blood cells | ABO blood group phenotype | Antibodies present in blood fluid | Blood types that can be tolerated in transfusion | Blood types that can accept blood for transfusion |
|---|---|---|---|---|---|
| $I^AI^A$ | A | Type A | Anti-B | A & O | A & AB |
| $I^AI^O$ | A | Type A | Anti-B | A & O | A & AB |
| $I^BI^B$ | B | Type B | Anti-A | B & O | B & AB |
| $I^BI^O$ | B | Type B | Anti-A | B & O | B & AB |
| $I^AI^B$ | A & B | Type AB | Neither anti-A nor anti-B | A, B, AB & O | AB only |
| $I^OI^O$ | Neither A nor B | Type O | Anti-A & anti-B | O only | A, B, AB & O |

a *universal recipient.* Conversely, a person of blood type O can donate blood to a person of any ABO type; type O is called a *universal donor.*

## Epistasis

In Chapter 1 we saw how the products of several genes may be necessary to carry out all the steps in a biochemical pathway. In genetic crosses in which two mutations that affect different steps in a single pathway are both segregating, the typical $F_2$ ratio of 9 : 3 : 3 : 1 is not observed. Gene interaction that perturbs the normal Mendelian ratios is known as **epistasis.** One type of epistasis is illustrated by the interaction of the *C, c* and *P, p* allele pairs affecting flower coloration in peas. These genes encode enzymes in the biochemical pathway for the synthesis of anthocyanin pigment, and the production of anthocyanin requires the presence of at least one wildtype dominant allele of each gene. The proper way to represent this situation genetically is to write the required genotype as

$$C- \; P-$$

where each dash is a "blank" that may be filled with either allele of the gene. Hence *C*— comprises the genotypes *CC* and *Cc*, and likewise *P*— comprises the genotypes *PP* and *Pp*. All four genotypes are included in the symbol *C*— *P*—, and only these genotypes, have purple flowers.

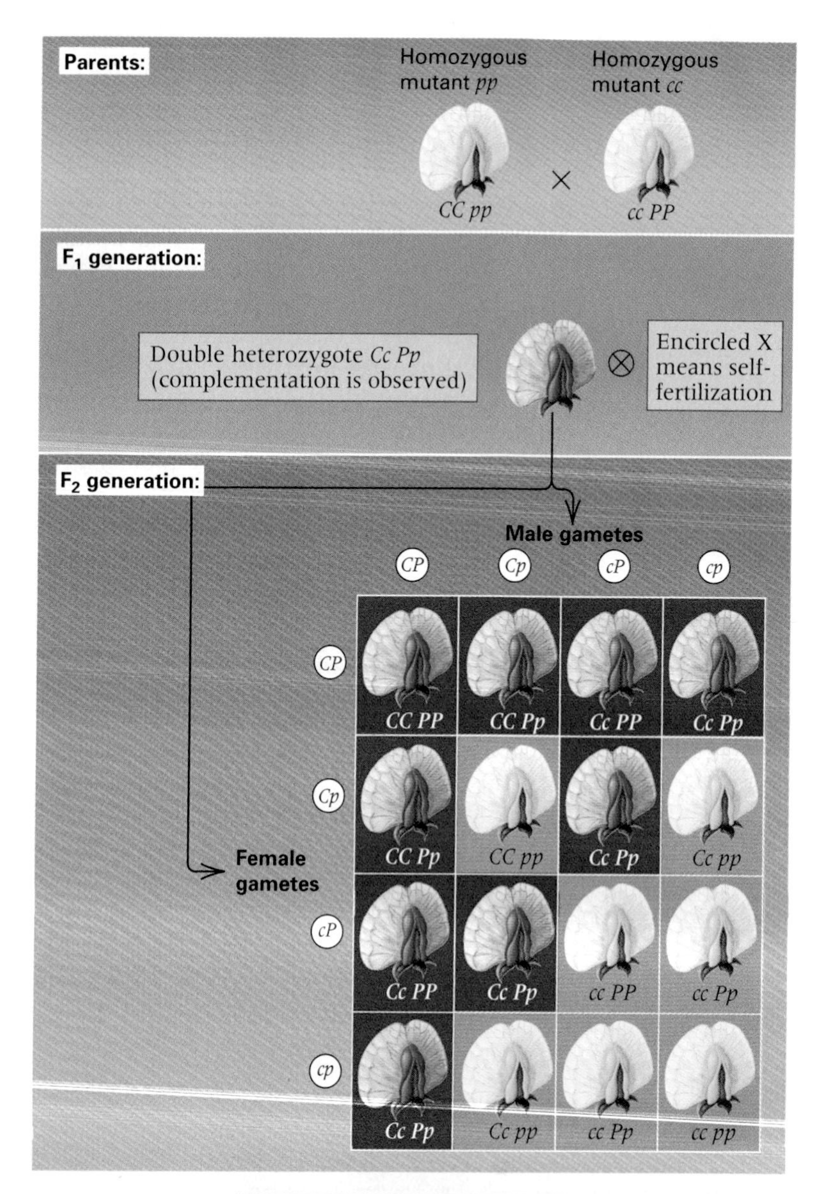


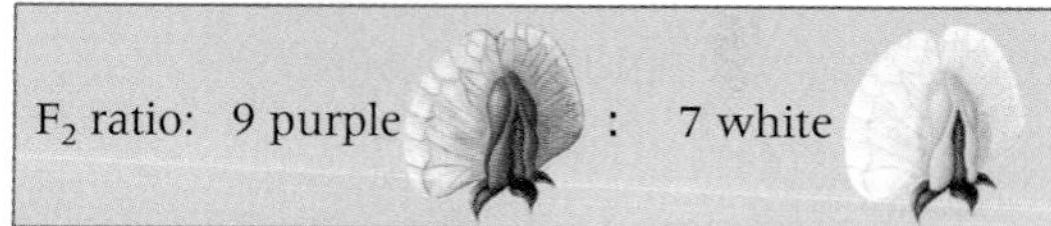


**FIGURE 3.24** Epistasis in the determination of flower color in peas. Formation of the purple pigment requires the dominant allele of both the *C* and *P* genes. With this type of epistasis, the dihybrid $F_2$ ratio is modified to 9 purple : 7 white.

**FIGURE 3.24** shows a cross between the homozygous recessive genotypes *CC pp* and *cc PP*. The phenotype of the flowers in the $F_1$ generation is the wildtype purple. Why? Because the *C* allele is dominant to *c* and the *P* allele is dominant to *p*, the $F_1$ plant is a double heterozygote of genotype *Cc Pp* and therefore has purple flowers. The result is nevertheless strange, because the original cross involved two homozygous recessive mutants, each of which had white flowers. Once we have been told that the mutant *c* and the mutant *p* alleles are in different genes, the finding of wildtype flowers in the $F_1$ generation is logical. But what if we did not know whether the mutant alleles were in different genes? What if both mutants were picked up in a mutant screen, or discovered in different laboratories, and we did not know whether the *c* allele and the *p* allele were alleles of the same gene or alleles of different genes? The answer is that the phenotype of the $F_1$ progeny in a cross like that in Figure 3.24 tells you. If the phenotype of the $F_1$ progeny is wildtype, as in Figure 3.24, this observation tells you that *c* and *p* are alleles of *different* genes. On the other hand, if the phenotype of the $F_1$ progeny is mutant (in this case a plant with white flowers), this result tells you that *c* and *p* are alleles of the *same* gene. In the latter case it would be best to change their names in such a way as to signify their allelism, such as $c = c_1$ and $p = c_2$, because then the genotype of the $F_1$ plant would be written as $c_1/c_2$ and it would be obvious from the genotype symbol that the phenotype is mutant, because each allele of the gene is mutant.

This discussion will hopefully remind you of the Beadle-Tatum experiments in Chapter 1

(Section 1.4) and in particular their use of the *complementation test* to determine whether two nutritional mutants in *Neurospora* were, or were not, mutants of the same gene. The principle in Figure 3.24 is exactly the same. The difference is that Beadle and Tatum used *Neurospora* heterokaryons, in which mutant alleles are brought together by forming hybrid filaments with two types of haploid mutant nuclei, whereas in diploid organisms like the peas in Figure 3.24 the hybrid nuclei are created by crossing two homozygous recessive mutants. The main point is that complementation tests are also used in sexual diploid organisms to identify which recessive mutations are alleles and which are not, and the phenotype of the $F_1$ generation in Figure 3.24 illustrates the principle.

Whereas the $F_1$ generation in Figure 3.24 illustrates complementation, the $F_2$ generation illustrates one type of epistasis. Suppose that plants of the $F_1$ generation are self-fertilized (indicated by the encircled cross sign), and assume that the (*C*, *c*) and (*P*, *p*) allele pairs undergo independent assortment. The Punnett square in Figure 3.24 gives the genotypes and phenotypes of the $F_2$ generation. Because only the *C*− *P*− progeny have purple flowers, the ratio of purple flowers to white flowers in the $F_2$ generation is 9 : 7. The epistasis does not change the result of independent segregation, it merely conceals the fact that the underlying ratio of the genotypes *C*− *P*− : *C*− *pp* : *cc P*− : *cc pp* is 9 : 3 : 3 : 1.

For a trait determined by the interaction of two genes, each with a dominant allele, there are only a limited number of ways in which the 9 : 3 : 3 : 1 dihybrid ratio can be modified. The possibilities are illustrated in **FIGURE 3.25**. In the absence of epistasis, the $F_2$ ratio of phenotypes is 9 : 3 : 3 : 1. In each row, different colors indicate different phenotypes. For example, in the modified ratio at the bottom, the phenotypes of the "3 : 3 : 1" classes are indistinguishable, resulting in a 9 : 7 ratio. This is the ratio observed in the segregation of the *C*, *c* and *P*, *p* alleles in Figure 3.24, with its 9 : 7 ratio of purple to white flowers. Taking all the possible modified ratios in Figure 3.25 together, there are nine possible dihybrid ratios when both genes show complete dominance. Examples of each of the modified ratios are known. The types of epistasis that result in these modified ratios are illustrated in the following examples, which are taken from a variety of organisms. Other examples can be found in the problems at the end of the chapter.

**9 : 7** is observed when a homozygous recessive mutation in either or both of two different genes results in the same mutant phenotype, as in Figure 3.24.

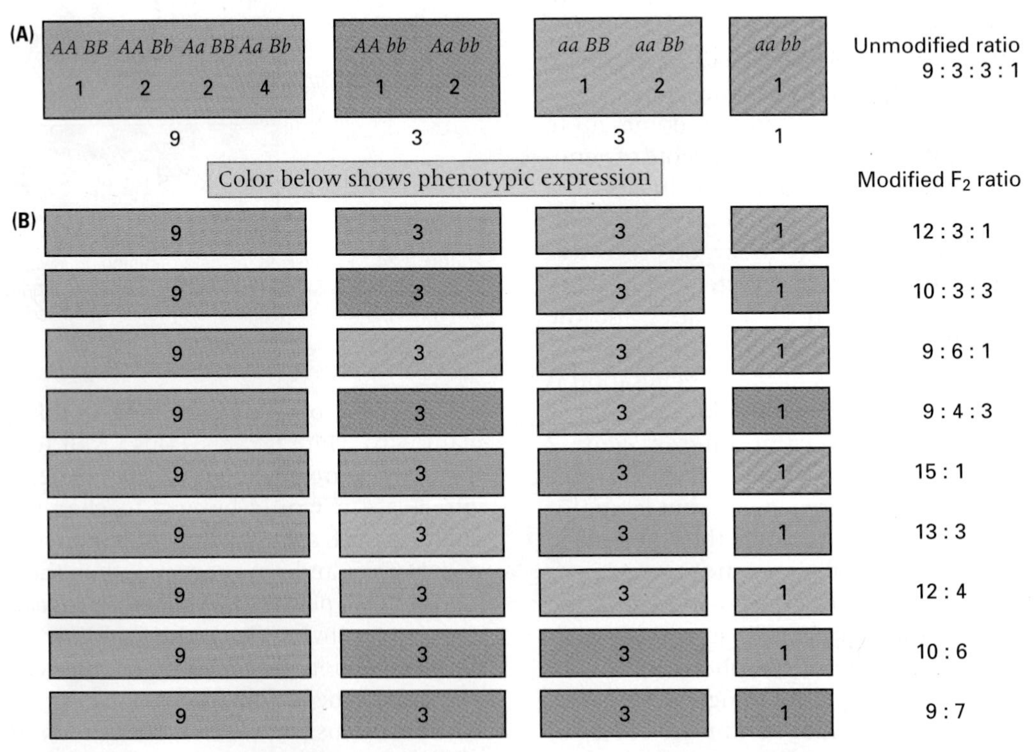


**FIGURE 3.25** Modified $F_2$ dihybrid ratios. In each row, different colors indicate different phenotypes.

*12 : 3 : 1* results when the presence of a dominant allele at one locus masks the genotype at a different locus, such as the *A*− genotype rendering the *B*− and *bb* genotypes indistinguishable. For example, in genetic study of the color of the hull in oat seeds, a variety with white hulls was crossed with a variety with black hulls. The $F_1$ hybrid seeds had black hulls. Among 560 progeny in the $F_2$ generation, the hull phenotypes observed were 418 black, 106 gray, and 36 white. The ratio of phenotypes is 11.6 : 3.9 : 1, or very nearly 12 : 3 : 1. A genetic hypothesis to explain these results is that the black-hull phenotype is due to the presence of a dominant allele *A* and the gray-hull phenotype is due to another dominant allele *B* whose effect is apparent only in *aa* genotypes. On the basis of this hypothesis, the original varieties had genotypes *aa bb* (white) and *AA BB* (black). The $F_1$ has genotype *Aa Bb* (black). If the *A, a* allele pair and the *B, b* allele pair undergo independent assortment, then the $F_2$ generation is expected to have the genotypic and phenotypic composition 9/16 *A*− *B*− (black hull), 3/16 *A*− *bb* (black hull), 3/16 *aa B*− (gray hull), 1/16 *aa bb* (white hull), or 12 : 3 : 1.

*13 : 3* is illustrated by the difference between White Leghorn chickens (genotype *CC II*) and White Wyandotte chickens (genotype *cc ii*). Both breeds have white feathers because the *C* allele is necessary for colored feathers, but the *I* allele in White Leghorns is a dominant inhibitor of feather coloration. The $F_1$ generation of a dihybrid cross between these breeds has the genotype *Cc Ii*, which is expressed as white feathers because of the inhibitory effects of the *I* allele. In the $F_2$ generation, only the *C*− *ii* genotype has colored feathers, so there is a 13 : 3 ratio of white : colored.

*9 : 4 : 3* is observed when homozygosity for a recessive allele with respect to one gene masks the expression of the genotype of a different gene. For example, if the *aa* genotype has the same phenotype regardless of whether the genotype is *B*− or *bb*, then the 9 : 4 : 3 ratio results. As an example, in mice the grayish "agouti" coat color results from a horizontal band of yellow pigment just beneath the tip of each hair. The agouti pattern is due to a dominant allele *A*, and in *aa* animals the coat color is black. A second dominant allele, *C*, is necessary for the formation of hair pigments of any kind, and *cc* animals are albino (white). In a cross of *AA CC* (agouti) × *aa cc* (albino), the $F_1$ progeny are *Aa Cc* and phenotypically agouti. Crosses between $F_1$ males and females produce $F_2$ progeny in the proportions 9/16 *A*− *C*− (agouti), 3/16 *A*− *cc* (albino), 3/16 *aa C*− (black), 1/16 *aa cc* (albino), or 9 agouti : 4 albino : 3 black.

*9 : 6 : 1* implies that homozygosity for either of two recessive alleles yields the same phenotype but that the phenotype of the double homozygote is different. In Duroc–Jersey pigs, red coat color requires the presence of two dominant alleles *R* and *S*. Pigs of genotype *R*− *ss* and *rr S*− have sandy-colored coats, and *rr ss* pigs are white. The $F_2$ ratio is therefore 9/16 *R*− *S*− (red), 3/16 *R*− *ss* (sandy), 3/16 *rr S*− (sandy), 1/16 *rr ss* (white), or 9 red : 6 sandy : 1 white.

## CHAPTER SUMMARY

- Inherited traits are determined by the genes present in the reproductive cells united in fertilization.
- Genes are usually inherited in pairs: one from the mother and one from the father.
- The genes in a pair may differ in DNA sequence and in their effect on the expression of a particular inherited trait.
- The maternally and paternally inherited genes are not changed by being together in the same organism.
- In the formation of reproductive cells, the paired genes separate again into different cells.
- Random combinations of reproductive cells containing different genes result in Mendel's ratios of traits appearing among the progeny.
- The ratios actually observed for any trait are determined by the types of dominance and gene interaction.

## REVIEW THE BASICS

- What is the principle of segregation, and how is this principle demonstrated in the results of a single-gene (monohybrid) cross?
- What is the principle of independent assortment, and how is this principle demonstrated in the results of a two-gene (dihybrid) cross?
- Explain why random union of male and female gametes is necessary for Mendelian segregation and independent assortment to be observed in the progeny of a cross.
- What is the difference between mutually exclusive events and independent events? How are the probabilities of these two types of events combined? Give two examples of genetic events that are mutually exclusive and two examples of genetic events that are independent.
- When two pairs of alleles show independent assortment, under what conditions will a 9 : 3 : 3 : 1 ratio of phenotypes in the $F_2$ generation *not* be observed?
- Explain the following statement: "Among the $F_2$ progeny of a dihybrid cross, the ratio of genotypes is 1 : 2 : 1, but among the progeny that express the dominant phenotype, the ratio of genotypes is 1 : 2."
- What are the principal features of human pedigrees in which a rare dominant allele is segregating? In which a rare recessive allele is segregating?
- What is a mutant screen and how is it used in genetic analysis?
- Explain the statement: "In genetics, a gene is identified experimentally by a set of mutant alleles that fail to show complementation." What is complementation? How does a complementation test enable a geneticist to determine whether two different mutations are or are not mutations in the same gene?
- What does it mean to say that epistasis results in a "modified dihybrid $F_2$ ratio?" Give two examples of a modified dihybrid $F_2$ ratio, and explain the gene interactions that result in the modified ratio.

## GUIDE TO PROBLEM SOLVING

**Problem 1** Complete the table by inserting 0, 1/4, 1/2, or 1 for the probability of each genotype of the progeny from each type of mating. For which mating are the parents identical in genotype but the progeny as variable in the genotype as they can be for a single locus? For which mating are the parents as different as they can be for a single locus but the progeny identical to each other and different from either parent?

| | Progeny genotypes | | |
|---|---|---|---|
| **Mating** | ***AA*** | ***Aa*** | ***aa*** |
| *AA* × *AA* | | | |
| *AA* × *Aa* | | | |
| *AA* × *aa* | | | |
| *Aa* × *Aa* | | | |
| *Aa* × *aa* | | | |
| *aa* × *aa* | | | |

**Answer** This kind of table is fundamental to being able to solve almost any type of quantitative problem in transmission genetics.

| | Progeny genotypes | | |
|---|---|---|---|
| **Mating** | ***AA*** | ***Aa*** | ***aa*** |
| *AA* × *AA* | 1 | 0 | 0 |
| *AA* × *Aa* | 1/2 | 1/2 | 0 |
| *AA* × *aa* | 0 | 1 | 0 |
| *Aa* × *Aa* | 1/4 | 1/2 | 1/4 |
| *Aa* × *aa* | 0 | 1/2 | 1/2 |
| *aa* × *aa* | 0 | 0 | 1 |

The mating for which the parents are identical in genotype, but the progeny are as variable in genotype as they can be for a single locus, is *Aa* × *Aa*. The mating for which the parents are as different as they can be for a single locus, but the progeny identical to each other and different from either parent, is *AA* × *aa*

**Problem 2** The pedigree below shows the inheritance of a rare, simple Mendelian dominant mutation with penetrance equal to 1/3. (A penetrance of 1/3 means that 1/3 of the individuals with the mutant genotype actually express the mutant phenotype.) The woman I-1 necessarily has the genotype *Mm*, where *M* is the mutant allele. Because the mutant allele is rare, you may assume that the male I-2 has the genotype *mm*. Individual II-1 is not affected. What is the probability that II-1 has the genotype *Mm*?

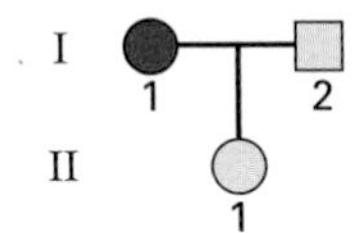


**Answer** This is a typical problem that makes use of Bayes' theorem. Let *A* be the event that individual II-1 has genotype *Mm*, and let *B* be the event that individual II-1 is not affected. Then event *A'* is therefore the event that individual II-1 has the genotype *mm*. Because I-2 has genotype *Mm*, then Pr{A} = 1/2 and also Pr{*A'*} = 1/2. Now we can apply Bayes' theorem:

$$\Pr\{A \mid B\} = \frac{\Pr\{B \mid A\}\Pr\{A\}}{\Pr\{B \mid A\}\Pr\{A\} + \Pr\{B \mid A'\}\Pr\{A'\}}$$

where Pr{*B* | *A*} is the probability that an individual of genotype *Mm* is not affected and Pr{*B* | *A'*} is the probability that an individual of genotype *mm* is not affected. Because the penetrance is 1/3, then Pr{*B* | *A*} = 2/3. Because the genotype *mm* is never affected, P{*B* | *A'*} = 1. Putting all this together, we obtain the answer we were seeking:

$$\Pr\{A \mid B\} = \frac{(2/3)(1/2)}{(2/3)(1/2) + (1)(1/2)} = 2/5$$

Another approach avoids the machinery of Bayes' theorem. Because individual I-1 has genotype *Mm*, the possible offspring of I-1 are of three types: (1) *Mm* and affected with probability 1/2 × 1/3 = 1/6, (2) *Mm* and not affected with probability 1/2 × 2/3 = 1/3, and (3) *mm* and not affected with probability 1/2. Because we know that II-1 is not affected, possibility (1) can be ruled out, and so the probability that II-1 has genotype *Mm*, given that she is not affected, is given by (1/3)/[(1/3 + 1/2)] = 2/5, which agrees with the answer above.

**Problem** In domesticated chickens, a dominant allele *C* is required for colored feathers, but a dominant allele *I* of an unlinked gene is an inhibitor of color that overrides the effects of *C*. White Leghorns have genotype *CC II* whereas White Wyandottes have genotype *cc ii*. Both breeds are white, but for different reasons. In the $F_2$ generation of a cross between White Leghorns and White Wyandottes:

**(a)** What is the phenotypic ratio of white : colored?

**(b)** Among $F_2$ chicks that are white, what is the proportion of *Cc Ii*?

**Answer**

**(a)** The initial cross of *CC II* × *cc ii* will yield $F_1$ chickens with genotype *Cc Ii*. Because the genes are not linked, the offspring in the $F_2$ generation will have a ratio of the genotypes 9 *C– I–* : 3 *C– ii* : 3 *cc I–* : 1 *cc ii*. Only the chickens with genotype *C– ii* will be colored because *C* is required for colored feathers but *I* inhibits the effect of *C*. This type of interaction of genes, or *epistasis*, will lead to a phenotypic ratio of 13 white : 3 colored.

**(b)** In any dihybrid cross with unlinked genes, the probability that an $F_2$ progeny is heterozygous for both genes equals (1/2) × (1/2) = 1/4. Because only 13/16 of the chicks in this example are white, the proportion of *Cc Ii* among the white chicks is (1/4)/(13/16) = 4/13. You can also get this result by counting squares in the Punnett square for a dihybrid cross: A total of 13 genotypes yield white chicks, and among these four are heterozygous for both genes.

**Problem** In Duroc-Jersey pigs, animals with genotype *R– S–* are red, those with genotype *R– ss* or *rr S–* are sandy colored, and those with the genotype *rr ss* are white. The genes show independent assortment. In the $F_2$ generation of a cross between the genotypes *RR ss* and *rr SS*:

**(a)** What is the phenotypic ratio of red : nonred?

**(b)** Among $F_2$ piglets that are red, what is the proportion of *Rr Ss*?

**Answer**

**(a)** The initial cross *RR ss* × *rr SS* will yield $F_1$ pigs with genotype *Rr Ss*. The progeny of the $F_2$ generation will have the genotypic ratio 9 *R– S–* : 3 *R– ss* : 3 *rr S–* : 1 *rr ss*. Only the pigs with genotype *R– S–* will be red, while all the others will be nonred (some sandy and some white, in the ratio of 6 : 1). This type of interaction of two genes (epistasis) will lead to a phenotypic ratio of 9 red : 7 nonred.

**(b)** 4/9. The proportion of pigs with the genotype *Rr Ss* is the frequency of those heterozygous for both genes, which equals 1/2 × 1/2 = 1/4. Because only 9/16 of the pigs are red, the proportion of *Rr Ss* among them is (1/2 × 1/2)/(9/16) = 4/9. You can also get this answer by counting squares in the Punnett square for a dihybrid cross: A total of 9 genotypes yield red pigs, among which 4 are heterozygous for both genes.

## ANALYSIS AND APPLICATIONS

**3.1** In the cross *Aa Bb Cc Dd* × *Aa Bb Cc Dd*, in which all genes undergo independent assortment, what proportion of offspring are expected to be heterozygous for all four genes?

**3.2** Consider a gene with four alleles, $A_1$, $A_2$, $A_3$, and $A_4$. In a cross between $A_1A_2$ and $A_3A_4$, what is the probability that a particular offspring inherits either $A_1$ or $A_3$ or both?

**3.3** Assuming equal numbers of boys and girls, if a mating has already produced a girl, what is the probability that the next child will be a boy? If a mating has already produced two girls, what is the probability that the next child will be a boy? On what type of probability argument do you base your answers?

**3.4** A cross is carried out between genotypes *Aa BB Cc dd Ee* and *Aa Bb cc DD Ee*. How many genotypes of progeny are possible?

**3.5** An individual of genotype *AA Bb Cc DD Ee* is testcrossed. Assuming that the loci undergo independent assortment, what fraction of the progeny are expected to have the genotype *Aa Bb Cc Dd Ee*?

**3.6** Homozygous pea plants with round seeds are crossed to homozygous pea plants with wrinkled seeds. The $F_1$ progeny undergo self-fertilization. A single round seed is chosen at random from the $F_2$ progeny, and its DNA examined by electrophoresis as described in the text. What is the probability that the observed gel pattern will be B?

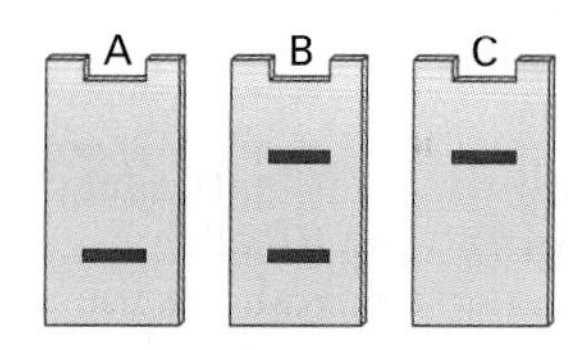


**3.7** Assume that the trihybrid cross *MM SS tt* × *mm ss TT* is made in a plant species in which *M* and *S* are dominant but there is no dominance between *T* and *t*. Consider the $F_2$ progeny from this cross, and assume independent assortment.

**(a)** How many phenotypic classes are expected?

**(b)** What is the probability of genotype *MM SS tt*?

**(c)** What proportion would be expected to be homozygous for all genes?

**3.8** Shown here is a pedigree and a gel diagram indicating the clinical phenotypes with respect to phenylketonuria and the molecular phenotypes with respect to an RFLP that overlaps the *PAH* gene for phenylalanine hydroxylase. The individual II-1 is affected.

**(a)** Indicate the expected molecular phenotype of II-1.

**(b)** Indicate the possible molecular phenotypes of II-2.

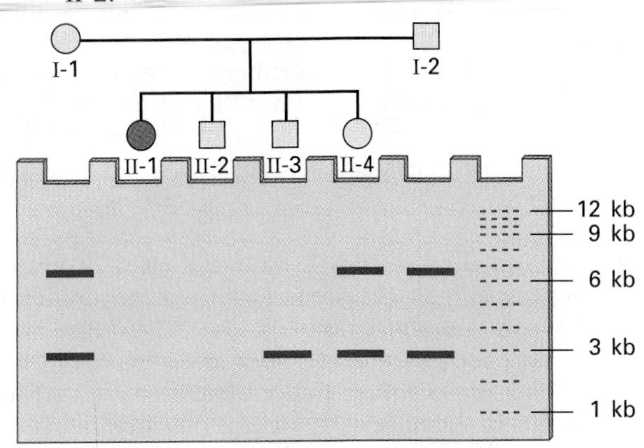


**3.9** What mode of inheritance is suggested by the following pedigree? Based on this hypothesis, and assuming that the trait is rare and has complete penetrance, what are the possible genotypes of all individuals in this pedigree?

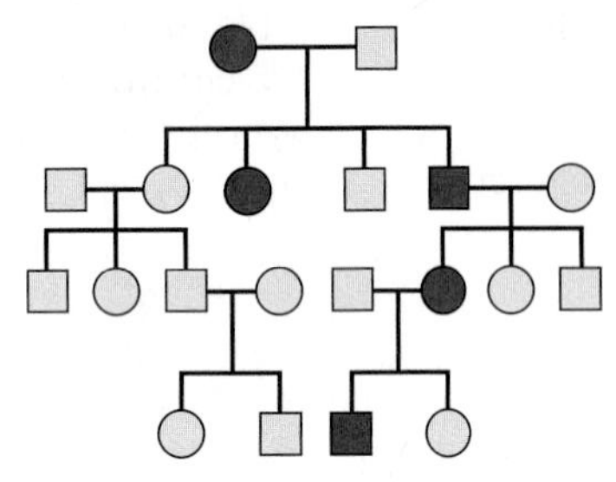

**3.10** Huntington disease is a rare neurodegenerative human disease determined by a dominant allele, *HD*. The disorder is usually manifested after the age of 45. A young man has learned that his father has developed the disease.

**(a)** What is the probability that the young man will later develop the disorder?

**(b)** What is the probability that a child of the young man carries the *HD* allele?

**3.11** The gel diagram below shows the banding patterns observed among the progeny of a cross between double heterozygous genotypes shown at the left (P1 and P2). On the right are shown the patterns of bands observed among the progeny. The bands are numbered 1–4.

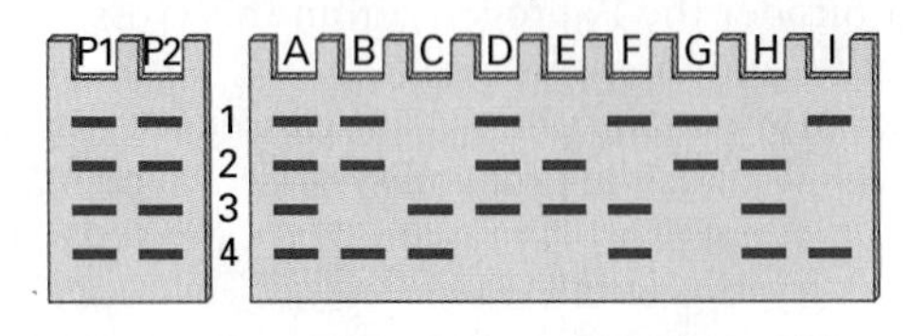


**(a)** Identify which pairs of numbered bands correspond to the two segregating pairs of alleles.

**(b)** Assuming that the two pairs of alleles undergo independent assortment, what is the probability that an offspring of the cross shows the banding pattern in lane D?

**3.12** Assume that the trait in the accompanying pedigree is due to simple Mendelian inheritance.

**(a)** Is it likely to be due to a dominant allele or a recessive allele? Explain.

**(b)** What is the meaning of the double horizontal line connecting III-1 with III-2?

**(c)** What is the biological relationship between III-1 and III-2?

**(d)** If the allele responsible for the condition is rare, what are the most likely genotypes of all of the persons in the pedigree in generations I, II, and III? (Use *A* and *a* for the dominant and recessive alleles, respectively.)

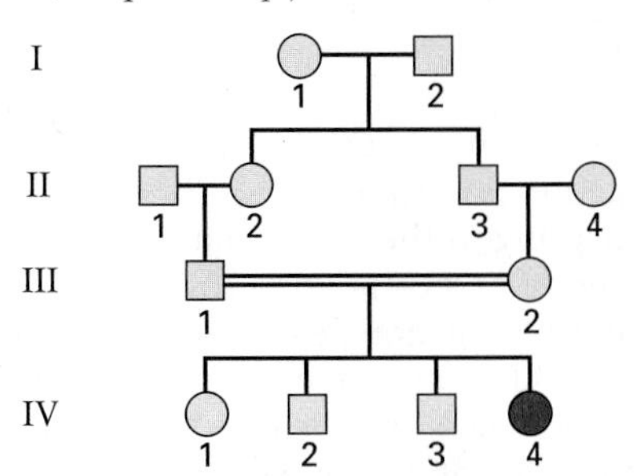


**3.13** In *Drosophila*, the mutant allele $bw^{dts}$ causing brown eyes (normal eyes are red) is temperature sensitive. In flies reared at 29°C the mutant allele is dominant, but in flies reared at 22°C the mutant allele is recessive. In a cross of $bw^{dts}/+ \times bw^{dts}/+$, where the + sign denotes the wildtype allele of $bw^{dts}$, what is the expected ratio of brown-eyed flies to red-eyed flies if the progeny are reared at 29°C? At 22°C?

**3.14** Pedigree analysis tells you that a particular parent may have the genotype *AA BB* or *AA Bb*, each with the same probability. Assuming independent assortment, what is the probability of this parent to produce an *A b* gamete? What is the probability of the parent to produce an *A B* gamete?

**3.15** The pedigree shown here is for a rare autosomal recessive trait with complete penetrance. You may assume that no one the pedigree has the recessive allele unless that person inherits it from either I-1 or II-4 or both.

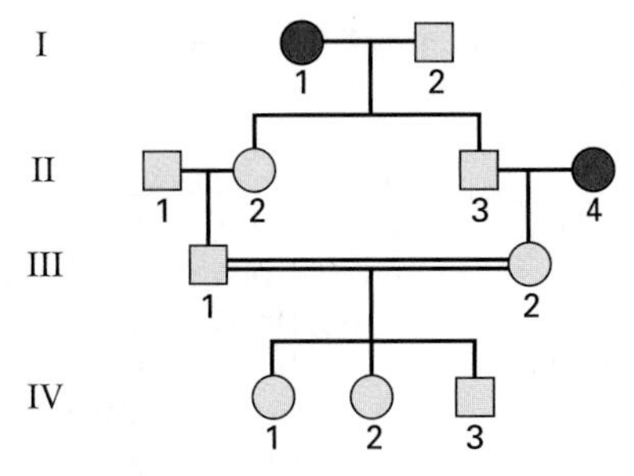


**(a)** Ignoring the sibship in generation IV, what is the probability that both parents in the first cousin mating are heterozygous?

**(b)** Taking the sibship in generation IV into account, what is the probability that both parents in the first cousin mating are heterozygous?

**3.16** A clinical test for a certain disease is 100 percent accurate in individuals who are affected but also yields a false positive result in 0.2 percent of healthy individuals. If the proportion of affected individuals in a population is 0.002, what is overall probability that an individual selected at random will yield a positive test result?

**3.17** The accompanying pedigree and gel diagram shows the phenotypes of the parents for an RFLP that has multiple alleles. What are the possible phenotypes of the progeny, and in what proportions are they expected?

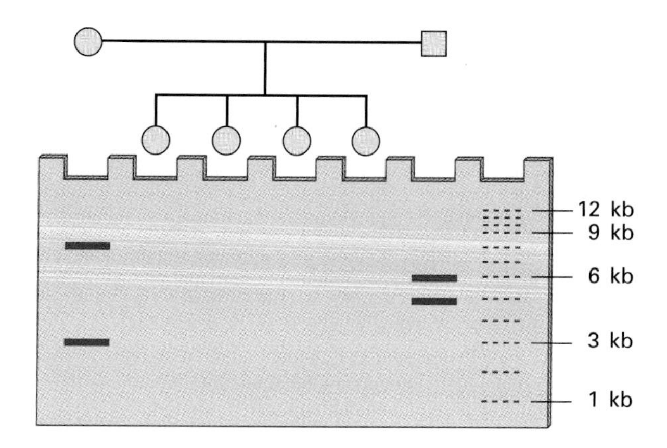


**3.18** The pollen from round pea seeds obtained from the $F_2$ generation of a cross between a true-breeding strain with round seeds and a true-breeding strain with wrinkled seeds was collected and used *en masse* to fertilize plants from the true-breeding wrinkled strain. What fraction of the progeny is expected to have wrinkled seeds? (Assume equal fertility among all genotypes.)

**3.19** Heterozygous *Cp cp* chickens express a condition called creeper, in which the leg and wing bones are shorter than normal (*cp cp*). The dominant *Cp* allele is lethal when homozygous. Two alleles of an independently segregating gene determine white (*W*−) versus yellow (*ww*) skin color. From matings between chickens heterozygous for both of these genes, what phenotypic classes will be represented among the viable progeny, and what are their expected relative frequencies?

**3.20** The $F_2$ progeny from a particular cross exhibit a modified dihybrid ratio of 9 : 7 (instead of 9 : 3 : 3 : 1). What phenotypic ratio would be expected from a testcross of the $F_1$ progeny?

**3.21** In the mating *Aa* × *Aa*, what is the smallest number of offspring, *n*, for which the probability of at least one *aa* offspring exceeds 95 percent?

**3.22** A woman is affected with a trait due to a dominant mutant allele that shows 50% penetrance. If she has a child, what is the probability that it will be affected?

**3.23** The pattern of coat coloration in dogs is determined by the alleles of a single gene, with *S* (solid) being dominant over *s* (spotted). Black coat color is determined by the dominant allele *A* of a second gene, and homozygous recessive *aa* animals are tan. A female having a solid tan coat is mated with a male having a solid black coat and produces a litter of six pups. The phenotypes of the pups are 2 solid tan, 2 solid black, 1 spotted tan, and 1 spotted black. What are the genotypes of the parents?

**3.24** Consider a phenotype for which the allele *N* is dominant to the allele *n*. A mating *Nn* × *Nn* is carried out, and one individual with the dominant phenotype is chosen at random. This individual is testcrossed and the mating yields four offspring, each with the dominant phenotype. What is the probability that the parent with the dominant phenotype has the genotype *Nn*?

**3.25** Some polymorphisms can be identified using oligonucleotides with randomly chosen sequences as primers. These are known and RAPD polymorphisms, where RAPD stands for *randomly amplified polymorphic DNA*. Typically, a RAPD polymorphism results from a site in which only some of the chromosomes in a population will bind with the primers and yield a band, whereas the other chromosomes will not bind with the primers and so not yield a band. DNA from an individual who is heterozygous for the site will yield a band. The gel diagram shown here includes the phenotype of two parents (X and Y) with respect to two RAPD polymorphisms corresponding to different sites in the genome. Each parent is homozygous for the site associated with the RAPD band its DNA exhibits. The two RAPD polymorphisms result from amplification of different sites in the genome that undergo independent assortment. In the gel diagram, indicate the expected phenotype of the $F_1$ progeny as well as all possible phenotypes of the $F_2$ progeny along with their expected proportions.

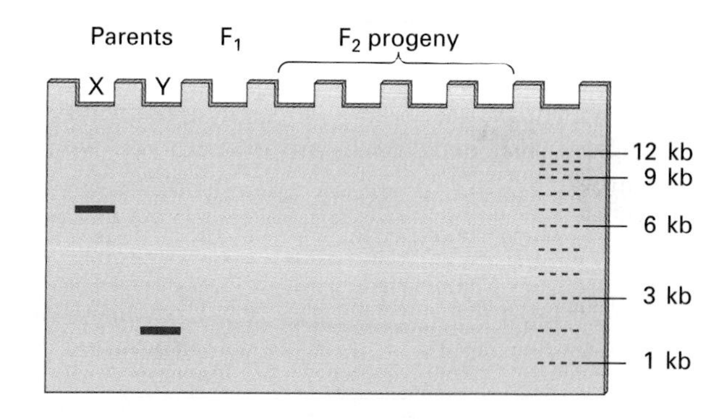


**3.26** The accompanying gel diagram shows the phenotype of two parents (X and Y), each homozygous for two RFLPs that undergo independent assortment. Parent X has genotype $A_1A_1\ B_1B_1$, where the $A_1$ allele yields a band of 4 kb and the $B_1$ allele yields a band of 6 kb, and parent Y has genotype $A_2A_2\ B_2B_2$, where the $A_2$ allele yields a band of 8 kb and the $B_2$ allele yields a band of 2 kb. Show the expected phenotype of the $F_1$ progeny as well as all possible phenotypes of the $F_2$ progeny along with their expected proportions.

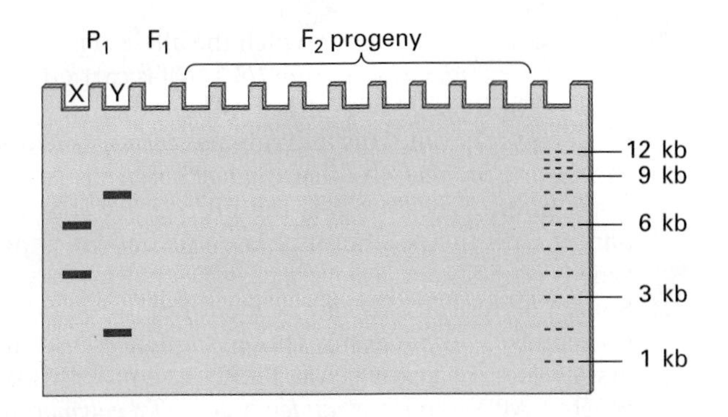


**3.27** Complementation tests of the recessive mutant genes *a* through *f* produced the data in the accompanying matrix. The circles represent missing data. Assuming that all of the missing mutant combinations would yield data consistent with the entries that are known, complete the table by filling each circle with a + or − as needed.

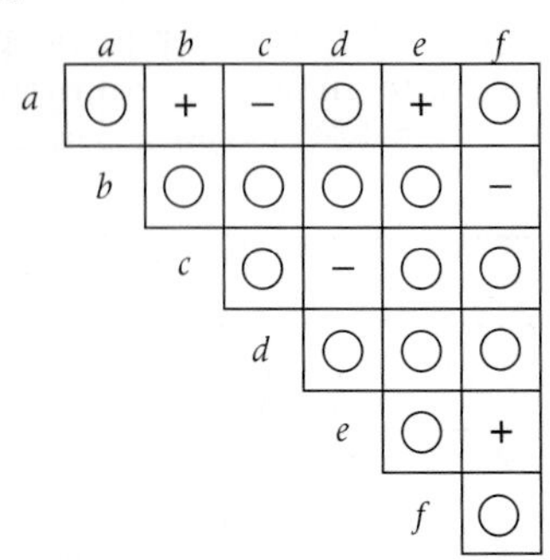

| | *a* | *b* | *c* | *d* | *e* | *f* |
|---|---|---|---|---|---|---|
| *a* | O | + | − | O | + | O |
| *b* | | O | O | O | O | − |
| *c* | | | O | − | O | O |
| *d* | | | | O | O | O |
| *e* | | | | | O | + |
| *f* | | | | | | O |

**3.28** In plants, certain mutant genes are known that affect the ability of gametes to participate in fertilization. Suppose that an allele *A* is such a mutation and that pollen cells bearing the *A* allele are only half as likely to survive and participate in fertilization as pollen cells bearing the *a* allele. Complete the Punnett square for the $F_2$ generation in a monohybrid cross. What is the expected ratio of *AA* : *Aa* : *aa* plants in the $F_2$ generation?

## CHALLENGE PROBLEMS

**Challenge Problem 1** Diagrammed here is DNA from a wildtype gene (top) and a mutant allele (bottom) that has an insertion of a transposable element that inactivates the gene. The transposable element is present in many copies scattered throughout the genome. The symbols *B* and *E* represent the positions of restriction sites for *Bam*HI and *Eco*RI, respectively, and the rectangles show sites of hybridization with each of three probes (A, B, and C) that are available. The dots at the left indicate that the nearest site of either *Bam*HI or *Eco*RI cleavage is very far to the left of the region shown. Explain which probe and which single restriction enzyme you would use for RFLP analysis to identify both alleles. Also, explain why any other choices would be unsuitable.

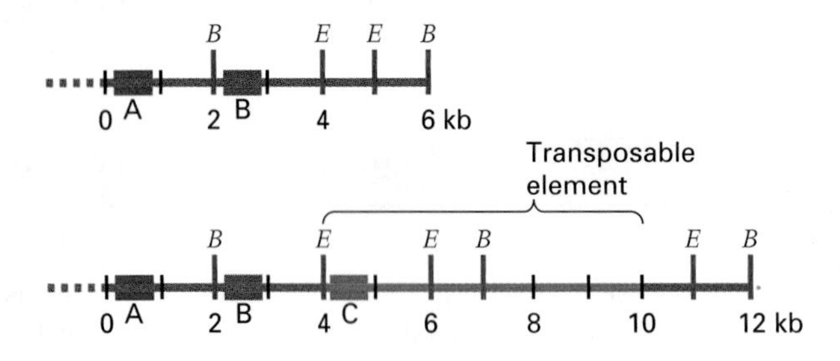


**Challenge Problem 2** Meiotic drive is an unusual phenomenon in which two alleles do not show Mendelian segregation from the heterozygous genotype. Examples are known from mammals, insects, fungi, and other organisms. The usual mechanism is one in which both types of gametes are formed, but one of them fails to function normally. The excess of the driving allele over the other can range from a small amount to nearly 100 percent. Suppose that *D* is an allele showing meiotic drive against its alternative allele *d*, and suppose that *Dd* heterozygotes produce functional *D*-bearing and *d*-bearing gametes in the proportions 3/4 : 1/4. In the mating *Dd* × *Dd*,

**(a)** What are the expected proportions of *DD*, *Dd*, and *dd* genotypes?

**(b)** If *D* is dominant, what are the expected proportions of *D*− and *dd* phenotypes?

**(c)** Among the *D*− phenotypes, what is the ratio of *DD* : *Dd*?

**(d)** Answer parts (a) through (c), assuming that the meiotic drive takes place in only one sex.

Challenge Problem 3 The accompanying table summarizes the effect of inherited tissue antigens on the acceptance or rejection of transplanted tissues, such as skin grafts, in mammals. The tissue antigens are determined in a codominant fashion, so that tissue taken from a donor of genotype *Aa* carries both the A and the a antigen. In the table, the + sign means that a graft of donor tissue is accepted by the recipient, and the − sign means that a graft of donor tissue is rejected by the recipient. The rule is: *Any graft will be rejected whenever the donor tissue contains an antigen not present in the recipient.* In other words, any transplant will be accepted if, and only if, the donor tissue does *not* contain an antigen different from any already present in the recipient.

| | | Donor | | |
|---|---|---|---|---|
| | | *AA* | *Aa* | *aa* |
| Recipient | *AA* | + | − | − |
| | *Aa* | + | + | + |
| | *aa* | − | − | + |

The diagram illustrated here shows all possible skin grafts between inbred (homozygous) strains of mice ($P_1$ and $P_2$) and their $F_1$ and $F_2$ progeny. Assume that the inbred lines $P_1$ and $P_2$ differ in only one tissue-compatibility gene. For each of the arrows, what is the probability of acceptance of a graft in which the donor is an animal chosen at random from the population shown at the base of the arrow and the recipient is an animal chosen at random from the population indicated by the arrowhead?

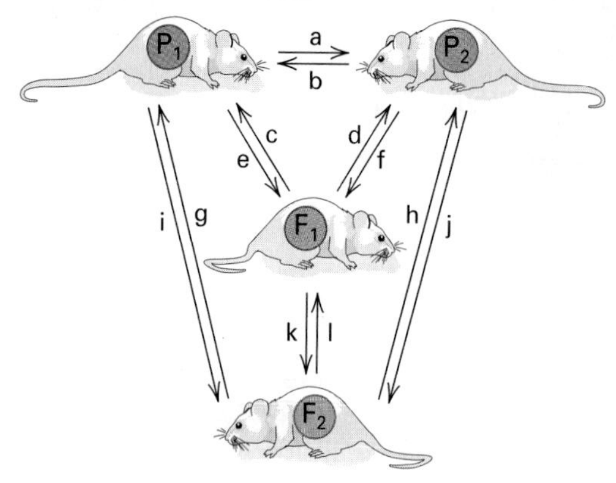


GENETICS on the web

GeNETics on the Web will introduce you to some of the most important sites for finding genetics information on the Internet. To explore these sites, visit the Jones and Bartlett companion site to accompany *Genetics: Analysis of Genes and Genomes, Seventh Edition* at http://biology.jbpub.com/book/genetics.

There you will find a chapter-by-chapter list of highlighted keywords. When you select one of the keywords, you will be linked to a Web site containing information related to that keyword.

# CHAPTER 4

# Chromosomes and Sex-Chromosome Inheritance

Fluorescence micrograph showing anaphase of mitosis in a lung cell of the rough-skinned newt *Taricha granulosa*. The cell is stained for DNA (blue), microtubules (green), and intermediate filaments (red).



## CHAPTER OUTLINE


**4.1** The Stability of Chromosome Complements

**4.2** Mitosis

**4.3** Meiosis
- The First Meiotic Division: Reduction
- The Second Meiotic Division: Equation

**4.4** Sex-Chromosome Inheritance
- Chromosomal Determination of Sex
- X-Linked Inheritance
- Pedigree Characteristics of Human X-Linked Inheritance
- Heterogametic Females
- Nondisjunction as Proof of the Chromosome Theory of Heredity
- Sex Determination in *Drosophila*

**4.5** Probability in the Prediction of Progeny Distributions
- Using the Binomial Distribution in Genetics
- Meaning of the Binomial Coefficient

**4.6** Testing Goodness of Fit to a Genetic Hypothesis
- The Chi-Square Method
- Are Mendel's Data Too Good to Be True?

CONNECTION Grasshopper, Grasshopper
E. Eleanor Carothers 1913
*The Mendelian Ratio in Relation to Certain Orthopteran Chromosomes*

CONNECTION The White-Eyed Male
Thomas Hunt Morgan 1910
*Sex-Limited Inheritance in* Drosophila

CONNECTION Seeds of Doubt
Ronald Aylmer Fisher 1936
*Has Mendel's Work Been Rediscovered?*

It came as no great revelation that genes are located in chromosomes. The parallel between their properties made this quite obvious:

1. Genes come in pairs; chromosomes come in pairs.
2. Alleles of a gene segregate; homologous chromosomes segregate.
3. Unlinked genes undergo independent assortment; nonhomologous chromosomes undergo independent assortment.

These parallels were first pointed out in 1903, and after that time there was little doubt that chromosomes are the cellular carriers of the genes. But parallels do not constitute scientific proof, nor does widespread agreement among scientists. In this chapter we shall examine some of the experimental evidence that was at the time—and still is—regarded as sufficient to prove the chromosome theory of heredity.

## 4.1 The Stability of Chromosome Complements

The cell nucleus was first discovered in 1831, but not until the late 1860s was it understood that nuclear division nearly always accompanies cell division. The importance of the nucleus in inheritance was reinforced by the nearly simultaneous discovery that the nuclei of two gametes fuse in the process of fertilization. The next major advance came in the 1880s with the discovery of **chromosomes**, easily visualized in the light microscope with the use of certain dyes. A few years later, chromosomes were found to segregate by an orderly process into the daughter cells formed by cell division as well as into the gametes formed by the division of reproductive cells. Three important regularities were observed about the **chromosome complement** (the complete set of chromosomes) of plants and animals.

1. The nucleus of each **somatic cell** (a cell of the body, in contrast to a **germ cell,** or gamete) contains a fixed number of chromosomes typical of the particular species. This number varies tremendously among species, and chromosome number bears little relation to the complexity of the organism (**TABLE 4.1**).
2. The chromosomes in the nuclei of somatic cells are usually present in pairs. For example, the 46 chromosomes of human beings consist of 23 pairs (**FIGURE 4.1**). Cells with nuclei that contain two similar sets of chromosomes are called **diploid**. A diploid individual carries two alleleic copies of each gene present in each pair of chromosomes. The chromosomes occur in pairs because one chromosome of each pair derives from the maternal parent and the other from the paternal parent of the organism.
3. The gametes that unite in fertilization to produce the diploid somatic cells have nuclei that contain only one set of chromosomes, consisting of one member of each pair. The gametic nuclei are **haploid**.

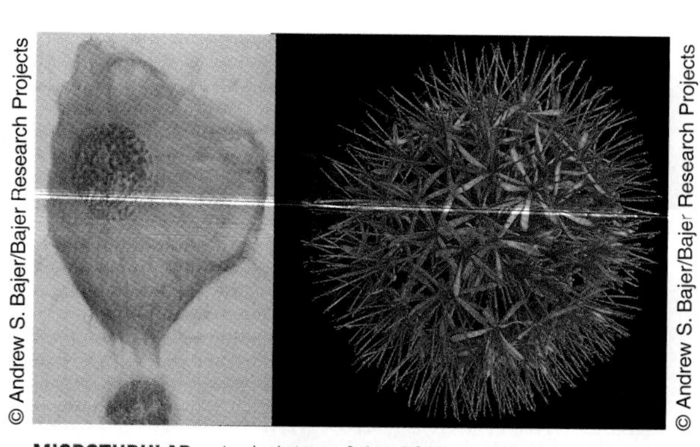


**MICROTUBULAR** cytoskeleton of the African globe amaryllus (*Scadoxus*), which becomes transformed into the spindle in mitosis, as seen through a light microscope.

**Table 4.1 Somatic (diploid) chromosome numbers of some plant and animal species**

| Organism | Chromosome number | Organism | Chromosome number |
|---|---|---|---|
| Field horsetail | 216 | Yeast (*Saccharomyces cerevisiae*) | 32 |
| Bracken fern | 116 | Fruit fly (*Drosophilia melanogaster*) | 8 |
| Giant sequoia | 22 | Nematode (*Caenorhabditis elegans*) | 11 ♂, 12 ♀ |
| Macaroni wheat | 28 | House fly | 12 |
| Bread wheat | 42 | Scorpion | 4 |
| Fava bean | 12 | Geometrid moth | 224 |
| Garden pea | 14 | Common toad | 22 |
| Mustard cress (*Arabidopsis thaliana*) | 10 | Chicken | 78 |
| Corn (*Zea mays*) | 20 | Mouse | 40 |
| Lily | 24 | Gibbon | 44 |
| Snapdragon | 16 | Human being | 46 |

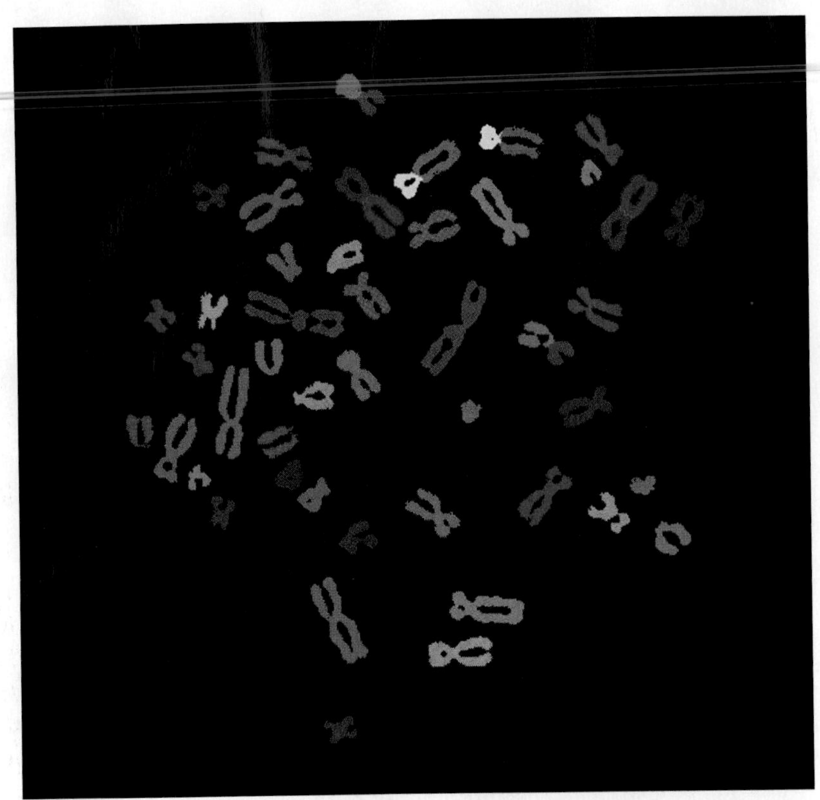

**FIGURE 4.1** Chromosome complement of a human male. There are 46 chromosomes, present in 23 pairs. At the stage of the division cycle in which these chromosomes were observed, each chromosome consists of two identical halves lying side by side longitudinally. Except for the members of one chromosome pair (the pair that determines sex), the members of all the chromosome pairs are the same color because they contain DNA molecules that were labeled with the same mixture of fluorescent dyes. The colors differ from one pair to the next because the dye mixtures differ in color. In some cases, the long and the short arm have been labeled with a different color. [Courtesy of Michael R. Speicher, Institute of Genome Genetics, Medical University of Graz.]

In multicellular organisms that develop from single cells, the presence of the diploid chromosome number in somatic cells and of the haploid chromosome number in germ cells indicates that there are *two* processes of nuclear division that differ in their outcome. One of these (mitosis) maintains the chromosome number; the other (meiosis) reduces the number by half. These two processes are examined in the following sections.

## 4.2 Mitosis

**Mitosis** is a process of nuclear division that ensures that each of two daughter cells receives a diploid complement of chromosomes identical to the diploid complement of the parent cell. Mitosis is usually accompanied by **cytokinesis,** the process in which the cell itself divides to yield two daughter cells. The basic process of mitosis is remarkably uniform in all organisms:

1. Each chromosome is already present as a duplicated structure at the beginning of nuclear division. (The duplication of each chromosome coincides with the replication of the DNA molecule it contains.)
2. Each chromosome divides longitudinally into identical halves that separate from each other.
3. The separated chromosome halves move in opposite directions, and each becomes included in one of the two daughter nuclei that are formed.

In a cell not undergoing mitosis, the chromosomes are invisible with a light microscope. Each consists of an elongated thread too thin to be seen. This stage of the cell cycle is called **interphase.** In preparation for mitosis, the DNA in the chromosomes is replicated during a period of interphase called **S** (**FIGURE 4.2**), which stands for *synthesis* of DNA. DNA replication is accompanied by chromosome duplication. Before and after S, there are periods, called $\mathbf{G_1}$ and $\mathbf{G_2}$, respectively, in which DNA replication does not take place. The **cell cycle** (the life cycle of a cell) is commonly described in terms of these three interphase periods followed by mitosis, **M**. The order of events is therefore $G_1 \rightarrow S \rightarrow G_2 \rightarrow M$, as shown in Figure 4.2. In this representation, the M period also includes the division of the cytoplasm (cytokinesis) into two approximately equal parts, each containing one daughter nucleus. The length of time required for a complete life cycle varies with cell type; it is 18–24 hours for the majority of cells in higher eukaryotes. The relative duration of the different periods in the cycle also varies considerably with cell type. Mitosis is usually the shortest period, requiring 0.5–2 hours.

The cell cycle is an active, regulated process controlled by mechanisms that are essentially identical in all eukaryotes. These are discussed in detail in Chapter 15. The transitions from $G_1$ into S and from $G_2$ into M are called **checkpoints** because the transitions are delayed unless key processes have been completed (Figure 4.2). For example, at the $G_1$/S checkpoint, either sufficient time must have elapsed since the preceding mitosis (in some cell types) or (in other cell types) the cell must have attained sufficient size for DNA replication to be initiated. Similarly, the $G_2$/M checkpoint requires that DNA replication and repair of any DNA damage be completed for the M phase to commence.

Illustrated in **FIGURE 4.3** are the essential features of chromosome behavior in mitosis. Mitosis is conventionally divided into four stages: **prophase, metaphase, anaphase,** and **telophase**. (If you have trouble remembering the order, you can jog your memory with "*p*eas *m*ake *a*wful *t*arts.") The stages have the following characteristics:

FIGURE 4.2 The cell cycle of a typical mammalian cell growing in tissue culture with a generation time of 24 hours.

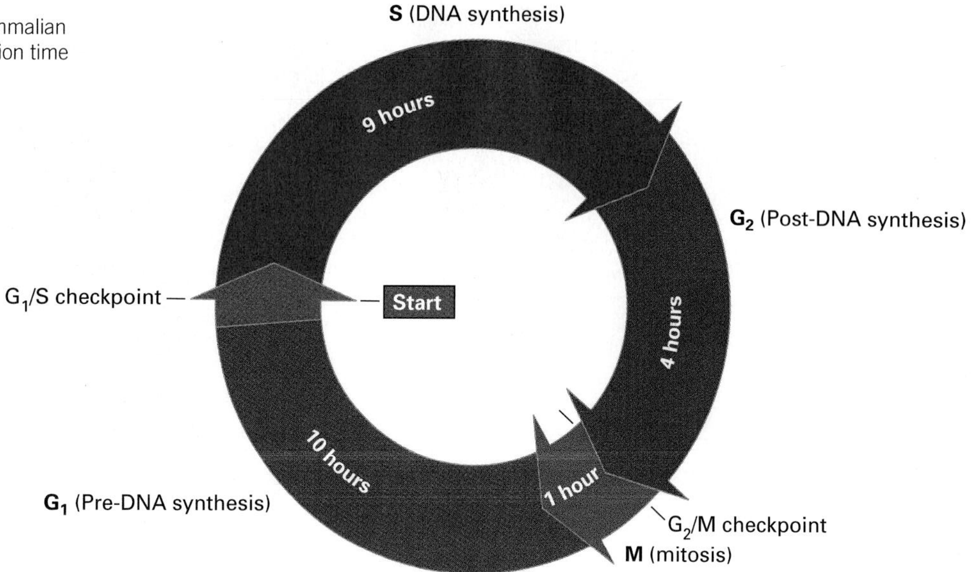


**1.** ***Prophase*** In interphase, the chromosomes have the form of extended filaments and cannot be seen with a light microscope as discrete bodies. Except for the presence of one or more conspicuous dark bodies (**nucleoli**), the nucleus has a diffuse, granular appearance. The beginning of prophase is marked by the condensation of chromosomes to form visibly distinct threads within the nucleus. Chromatin condensation is brought about by large protein complexes known as *condensins*. At this stage each chromosome is already longitudinally double, consisting of two closely associated subunits called **chromatids**. The longitudinally bipartite nature of each chromosome is readily seen later in prophase. Each pair of chromatids is the product of the duplication of one chromosome in the S period of interphase. The chromatids in a pair are held together at a specific region of the chromosome called the **centromere.** As prophase progresses, the chromosomes become shorter and thicker as a result of intricate coiling (Chapter 7). At the end of prophase, the nucleoli disappear and the nuclear envelope, a membrane surrounding the nucleus, abruptly disintegrates.

**2.** ***Metaphase*** At the beginning of metaphase, the **mitotic spindle** forms. The spindle is an elongated, football-shaped array of spindle fibers consisting primarily of microtubules formed by polymerization of the protein tubulin. Many other proteins and at least one RNA-protein complex regulate tubulin polymerization and microtubule organization. The ends or *poles* of the spindle, where the microtubules converge, mark the locations of the *centrosomes*, which are the microtubule organizing centers where tubulin polymerization is initiated. Each pair of centrosomes results from the duplication of a single centrosome that takes place in interphase, followed by migration of the daughter centrosomes to opposite sides of the nuclear envelope. (These processes are discussed in more detail in Chapter 15.)

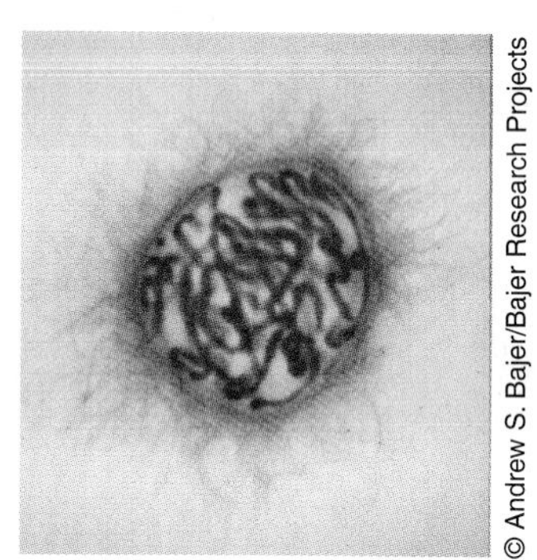


**PROPHASE** of *Scadoxus.*

The spindle features three types of microtubules: (1) those that anchor the centrosome to the cell membrane, (2) those that arch between the centrosomes, and (3) those that become attached to the chromosomes. The manner in which these are established exemplifies an important organizing principle of biology that may be called *exploration and stabilization.* As microtubules are formed, new tubulin subunits are added to the growing end of the polymer, but the growing end can also become unstable and initiate a processes of depolymerization in which the microtubule shrinks. Several proteins help regulate the balance between polymerization and depolymerization. In spindle formation, microtubules grow out from the spindle poles in essentially random directions (this is the exploration part of the process), but unless something happens to stabilize the growing end, each polymer will ultimately undergo depolymerization and disappear. For the microtubules that become attached to the chromosomes, the event that stabilizes the growing end is contact with a structure technically known as the **kinetochore**, which coincides with the position of the centromere. The process of random

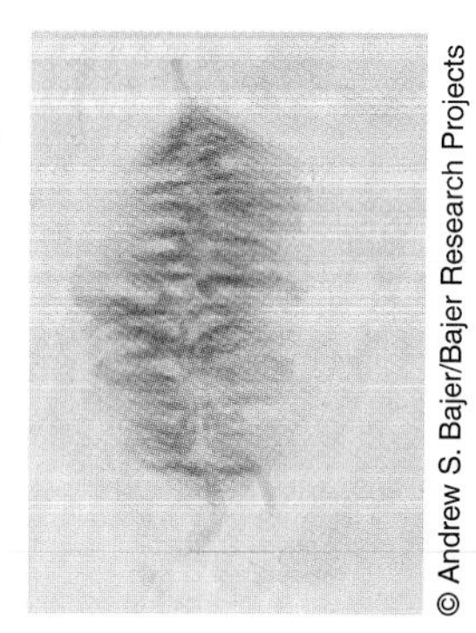


**METAPHASE** of *Scadoxus.*

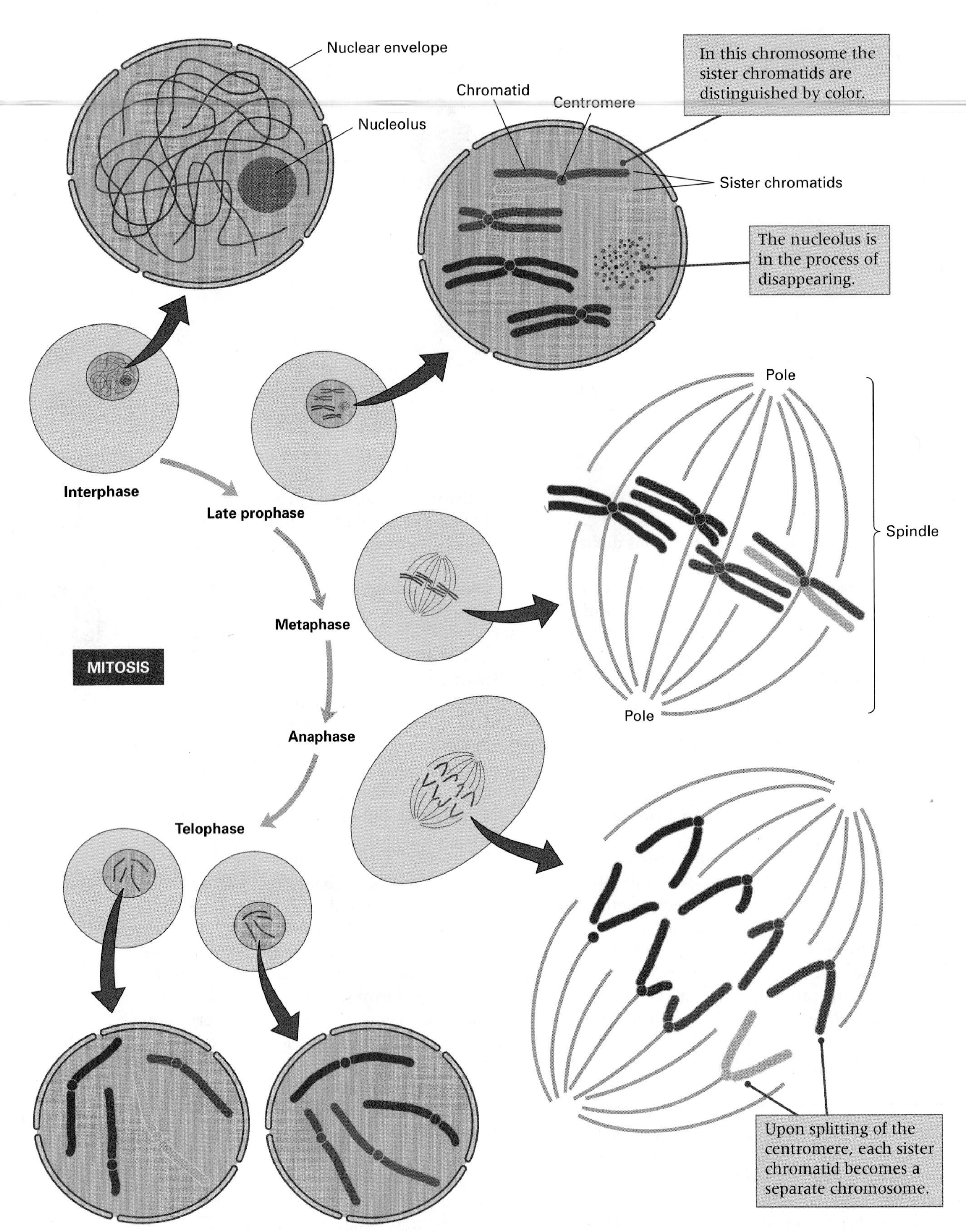


**FIGURE 4.3** Chromosome behavior during mitosis in an organism with two pairs of chromosomes (red/rose versus green/blue). At each stage, the smaller inner diagram represents the entire cell, and the larger diagram is an exploded view showing the chromosomes at that stage.

exploration and stabilization thereby results in a situation in which only those chromosomal microtubules that make contact with a kinetochore become stabilized and the others depolymerize. Analogous types of stabilization likely also account for the microtubules that attach to the cell membrane and those that arch between the centrosomes.

After the spindle fibers have become attached to the chromosomes, each chromosome is moved to a position near the center of the cell where its kinetochore lies on an imaginary plane approximately equidistant from the spindle poles. This imaginary plane is called the **metaphase plate**. Aligned on the metaphase plate, the chromosomes reach their maximum condensation and are easiest to count and examine for differences in shape and appearance.

Proper chromosome alignment is an important cell cycle control checkpoint at metaphase in both mitosis and meiosis. In a cell in which a chromosome is attached to only one pole of the spindle, the completion of metaphase is delayed. The signal for proper chromosome alignment comes from the kinetochore, and the chemical nature of the signal seems to be the dephosphorylation of certain kinetochore-associated proteins. Through the signaling mechanism, when all of the kinetochores are under tension and aligned on the metaphase plate, the metaphase checkpoint is passed and the cell continues the process of division.

**3. *Anaphase*** In anaphase, the proteins holding the chromatids together are dissolved. The centromeres become separate, and the two **sister chromatids** of each chromosome move toward opposite poles of the spindle. Once the centromeres separate, each sister chromatid is regarded as a separate chromosome in its own right. Chromosome movement results in part from progressive shortening of the spindle fibers attached to the centromeres, which pulls the chromosomes in opposite directions toward the poles, and often also from a temporary elongation of the dividing cell in a direction paralleling the spindle. At the completion of anaphase, the chromosomes lie in two groups near opposite poles of the spindle. Each group contains the same number of chromosomes that was present in the original interphase nucleus.

**4. *Telophase*** In telophase, a nuclear envelope forms around each compact group of chromosomes, nucleoli are formed, and the spindle disappears. The chromosomes undergo a reversal of condensation until they are no longer visible as discrete entities. The two daughter nuclei slowly assume a typical interphase appearance as the cytoplasm of the cell divides into two by means of a gradually deepening furrow around the periphery. (In plants, a new cell wall is synthesized between the daughter cells and separates them.)

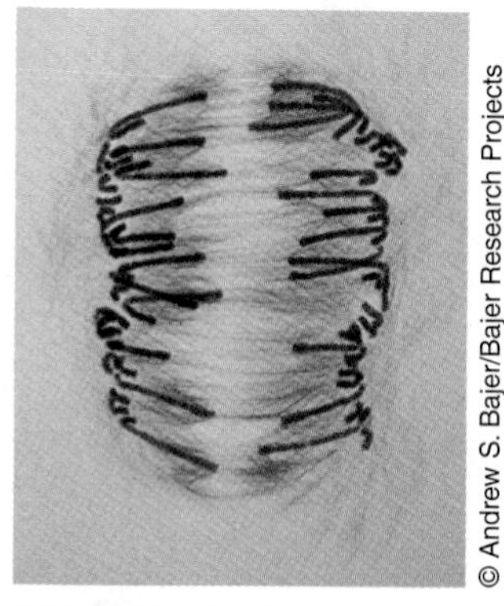


**ANAPHASE** of *Scadoxus*.

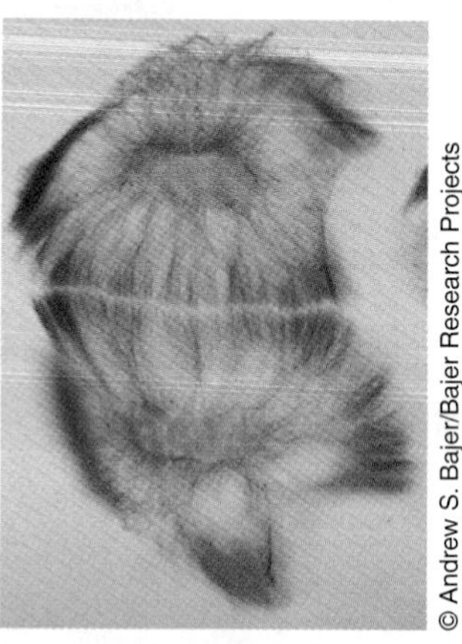


**TELOPHASE** of *Scadoxus*.

## 4.3 Meiosis

**Meiosis** is the mode of cell division that results in haploid daughter cells containing only one member of each pair of chromosomes. The process generates genetic diversity because each daughter cell contains a different set of alleles. Meiosis consists of two successive nuclear divisions. An overview of the chromosome behavior is outlined in **FIGURE 4.4**.

1. Prior to the first nuclear division, the members of each pair of homologous chromosomes become closely associated along their length (Figure 4.4). Each member of the pair is already replicated and consists of two sister chromatids joined at the centromere. The pairing of the homologous chromosomes therefore produces a four-stranded structure.
2. In the first nuclear division, the homologous chromosomes are separated from each other, the members of each pair going to opposite poles of the spindle (Figure 4.4B). Each daughter chromosome consists of two chromatids attached to a common centromere (Figure 4.4C), so both of the two nuclei that are formed contain a haploid set of chromosomes. (Chromosomes are enumerated by counting the number of centromeres, not the number of chromatids.)
3. The second nuclear division loosely resembles a mitotic division, *but there is no DNA replication.* At metaphase, the chromosomes align on the metaphase plate; and at anaphase, the chromatids of each chromosome are separated into opposite daughter nuclei (Figure 4.4D). The net effect of the two divisions in meiosis is the creation of four haploid daughter nuclei, each containing the equivalent of a single sister chromatid

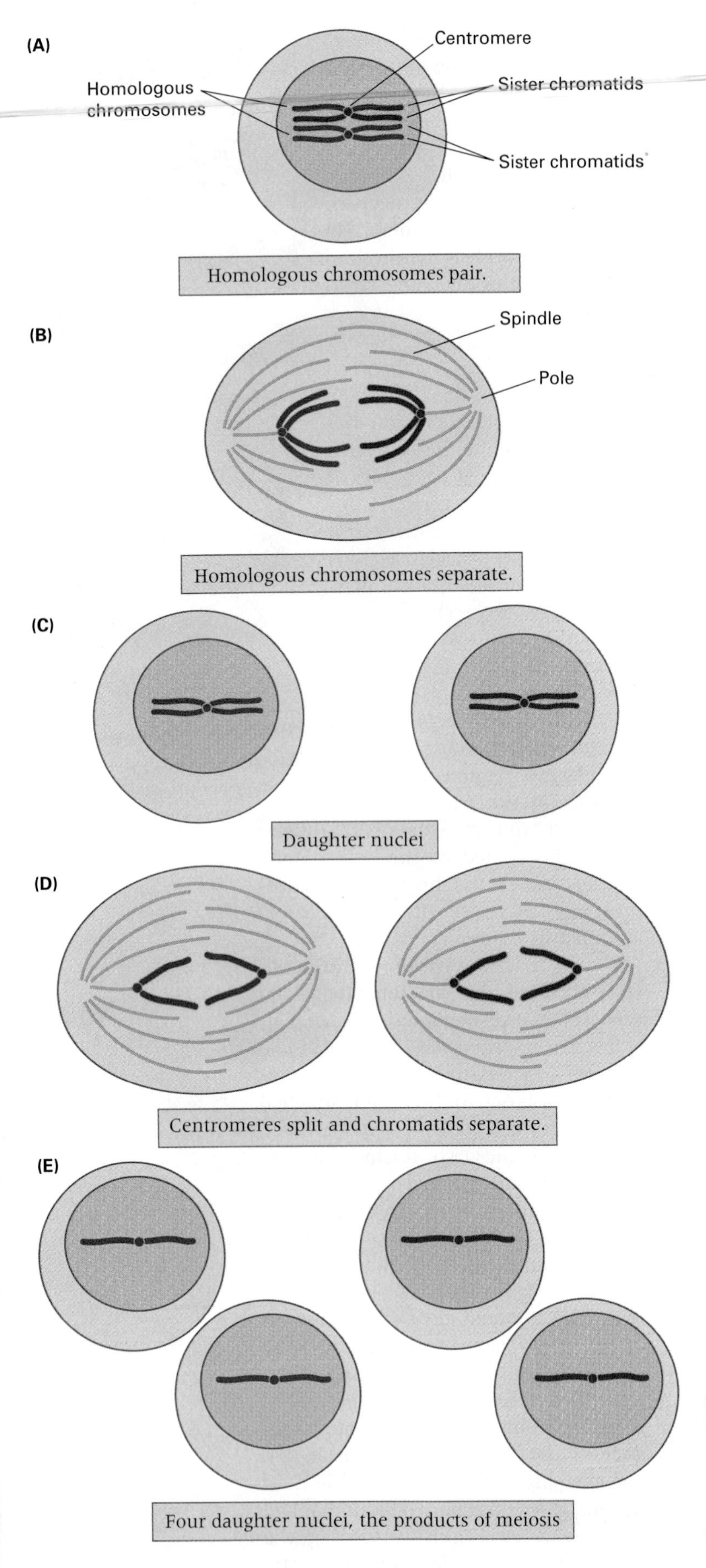


from each pair of homologous chromosomes (Figure 4.4E).

Figure 4.4 does not show that the paired homologous chromosomes can exchange genes. The exchanges result in the formation of chromosomes that consist of segments from one homologous chromosome intermixed with segments from the other. In Figure 4.4, the exchanged chromosomes would be depicted as segments of alternating color. The exchange process is one of the critical features of meiosis, and it will be examined in the next section.

In animals, meiosis takes place in specific cells called *meiocytes,* a general term for the primary oocytes and spermatocytes in the gamete-forming tissues (**FIGURE 4.5**). The *oocytes* form egg cells, and the *spermatocytes* form sperm cells. Although the process of meiosis is similar in all sexually reproducing organisms, in the female of both animals and plants, only one of the four products develops into a functional cell while the other three disintegrate. In animals, the products of meiosis form either sperm or eggs. In plants, the situation is slightly more complicated:

1. The products of meiosis typically form *spores,* which undergo one or more mitotic divisions to produce a haploid *gametophyte* organism. The gametophyte produces gametes by mitotic division of a haploid nucleus (**FIGURE 4.6**).
2. Fusion of haploid gametes creates a diploid zygote that develops into the *sporophyte* plant, which undergoes meiosis to produce spores and so restarts the cycle.

Meiosis is a more complex and considerably longer process than mitosis and usually requires days or even weeks. The entire process of meiosis is illustrated in its cellular context in **FIGURE 4.7**. The essence is that *meiosis consists of two divisions of the nucleus but only one replication of the chromosomes.* The nuclear divisions—called the *first meiotic division* and the *second meiotic division*—can be separated into a sequence of stages similar to those used to describe mitosis. The distinctive events of this important process occur during the first division of the nucleus. These events are described in the following section.

**FIGURE 4.4** Overview of the behavior of a single pair of homologous chromosomes in meiosis. The key differences from mitosis are the pairing of homologous chromosomes (A) and the two successive nuclear divisions (B and D) that reduce the chromosome number by half. For clarity, this diagram does not incorporate crossing over, an interchange of chromosome segments that takes place at the stage depicted in part A.

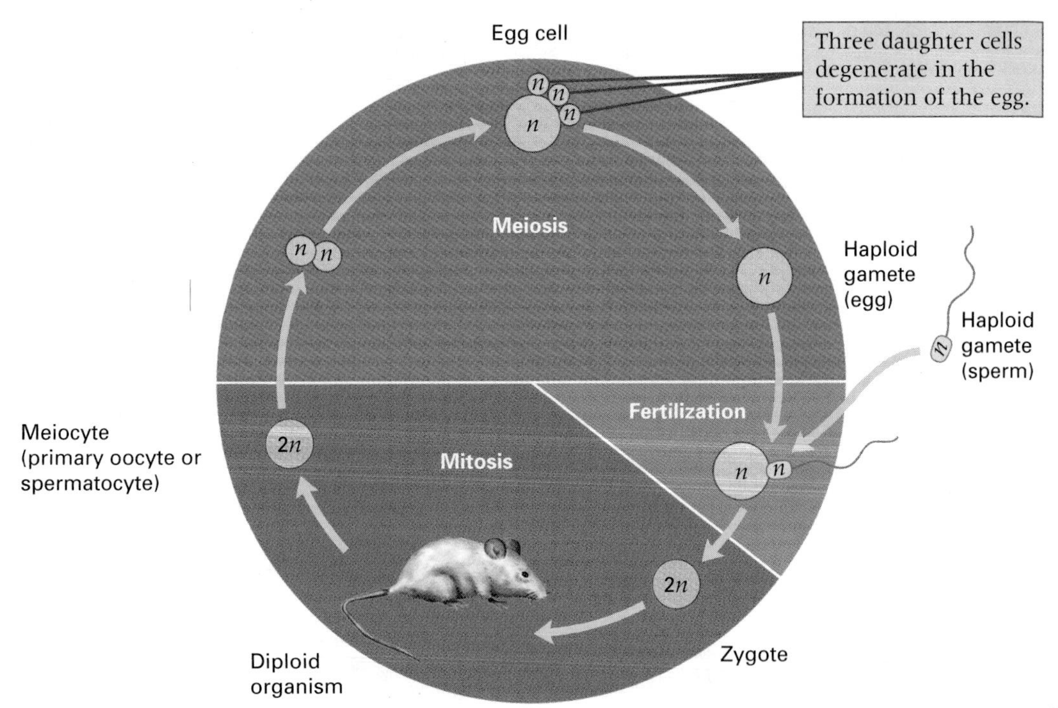


**FIGURE 4.5** The life cycle of a typical animal. The number *n* is the number of chromosomes in the haploid chromosome complement. In males, the four products of meiosis develop into functional sperm; in females, only one of the four products develops into an egg.

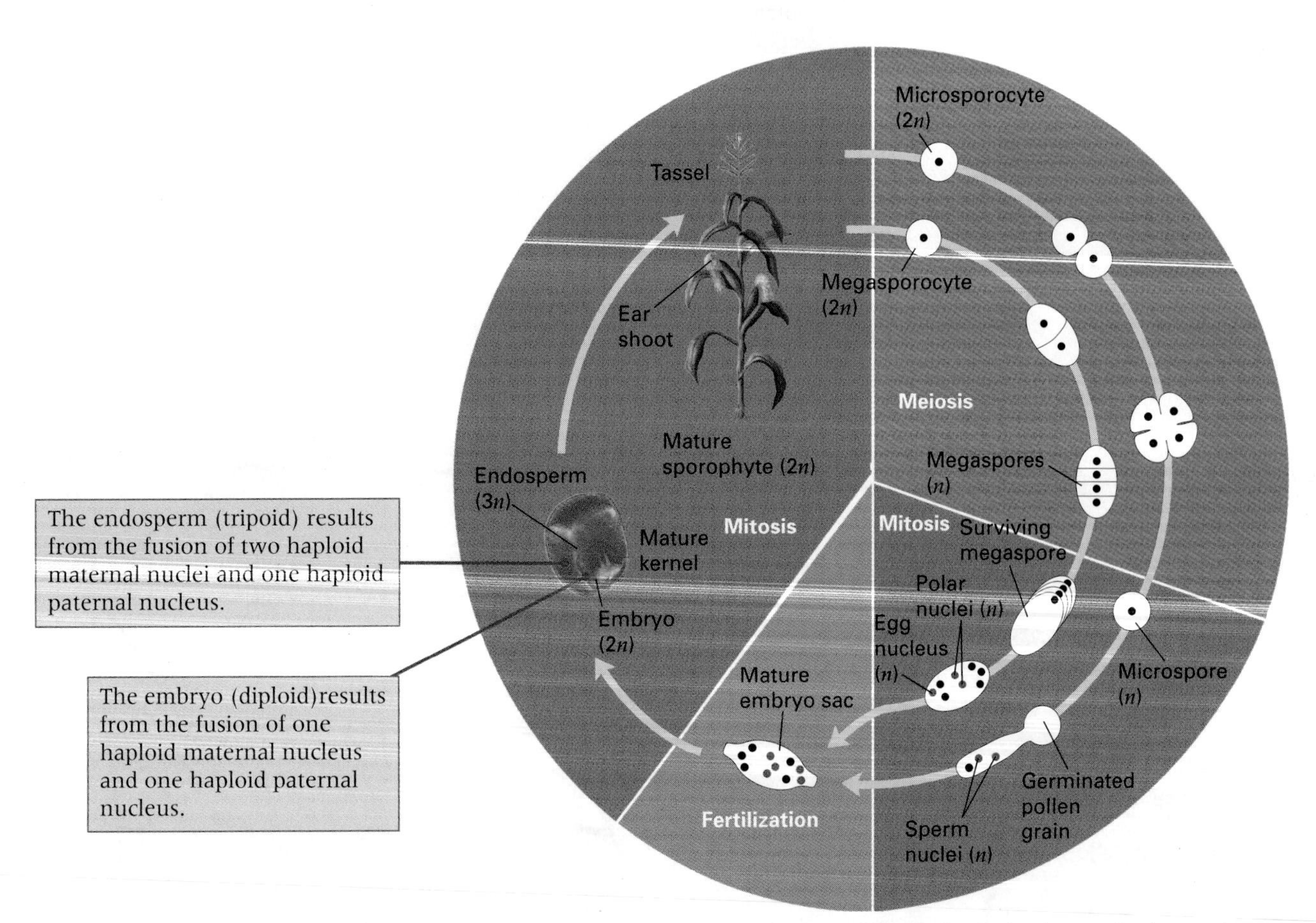


**FIGURE 4.6** The life cycle of corn, *Zea mays*. As is typical in higher plants, the diploid spore-producing (sporophyte) generation is conspicuous, whereas the gamete-producing (gametophyte) generation is microscopic. The egg-producing spore is the *megaspore*, and the sperm-producing spore is the *microspore*. Nuclei participating in meiosis and fertilization are shown in yellow and green, respectively.

**FIGURE 4.7** Chromosome behavior during meiosis in an organism with two pairs of homologous chromosomes (red/rose and green/blue). At each stage, the small diagram represents the entire cell and the larger diagram is an expanded view of the chromosomes at that stage.

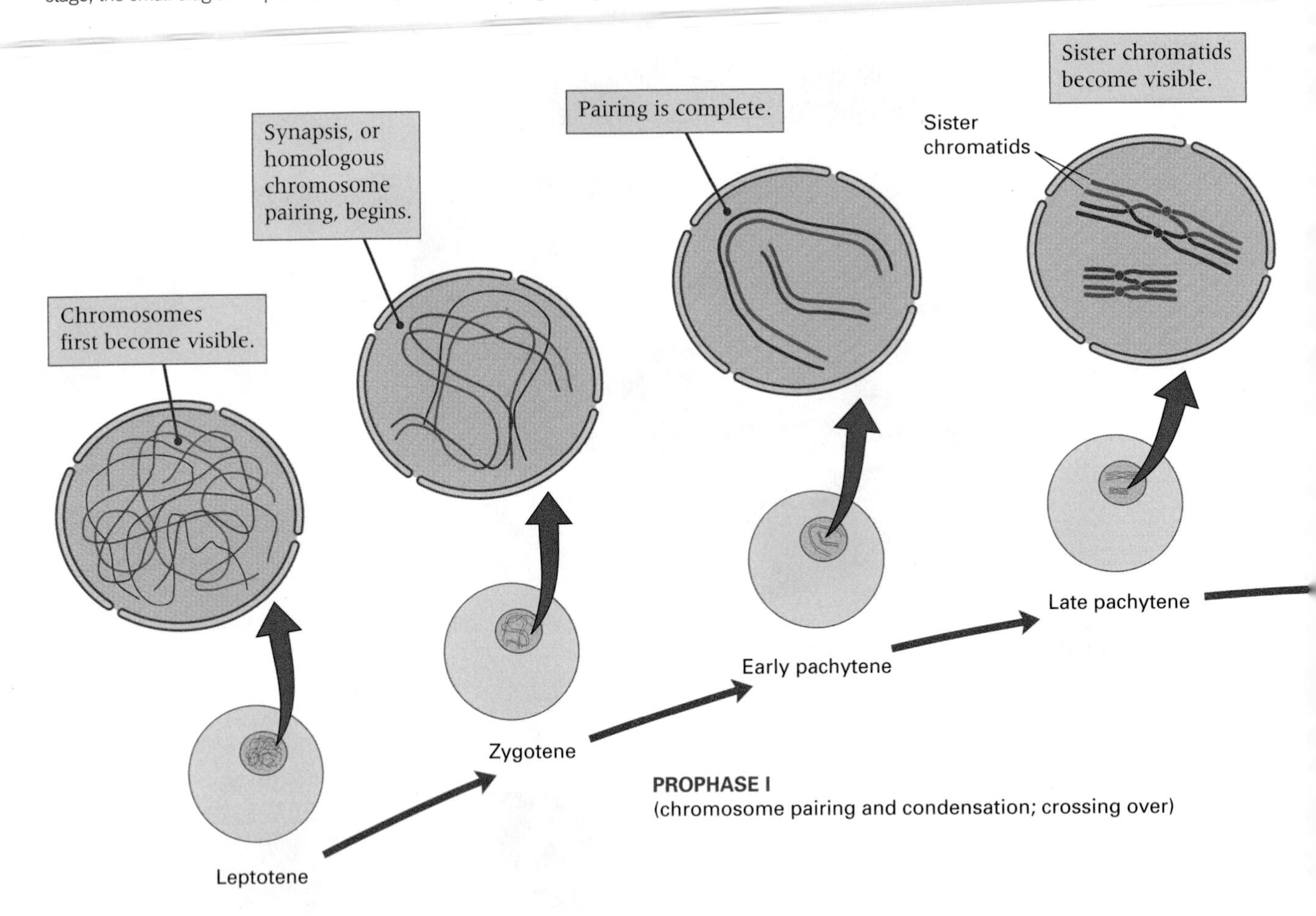


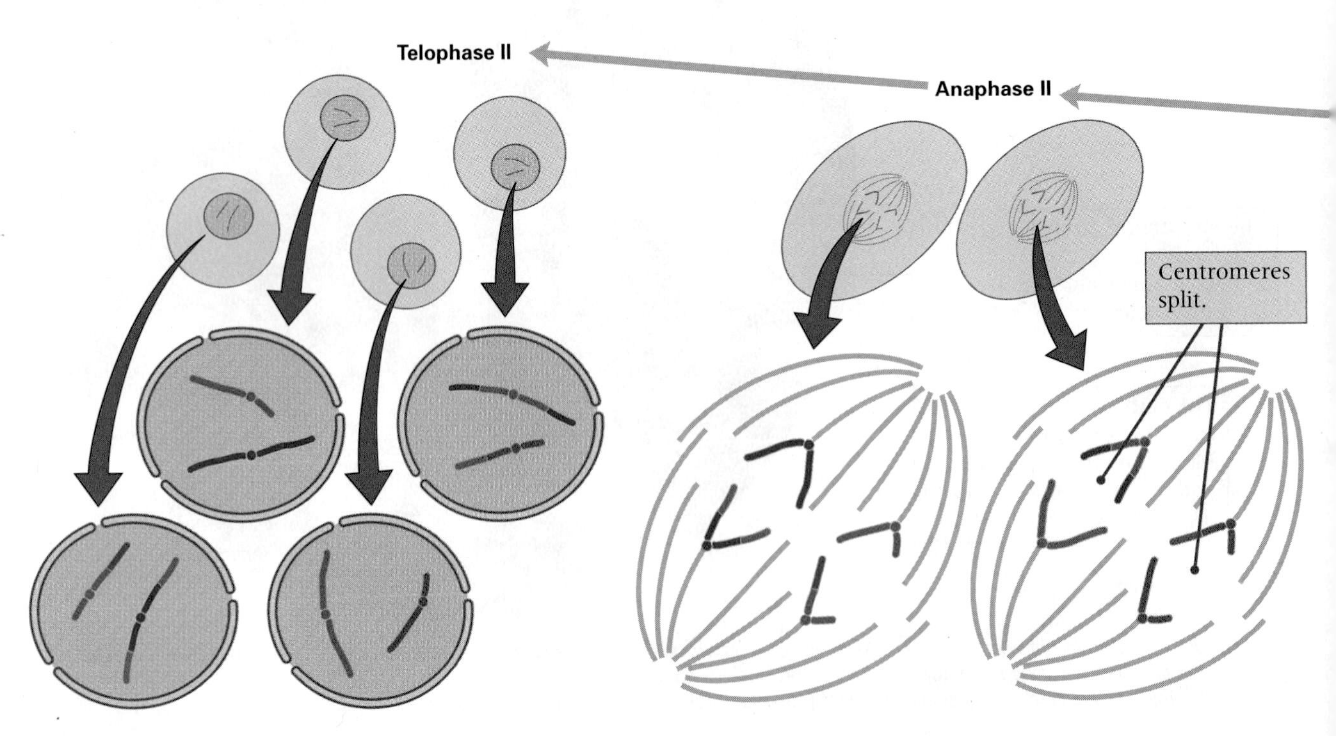

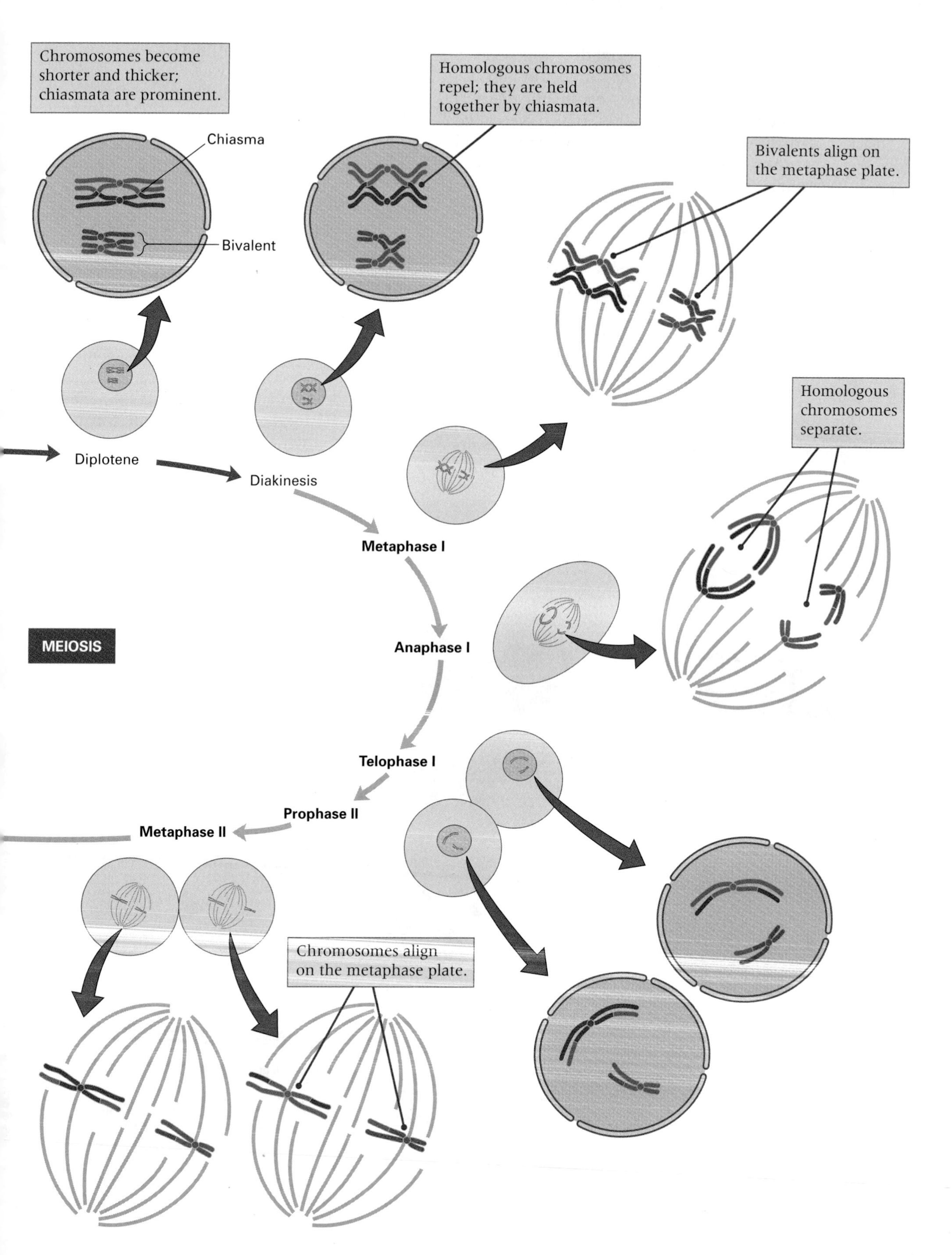

Chromosomes become shorter and thicker; chiasmata are prominent.
Chiasma
Bivalent
Homologous chromosomes repel; they are held together by chiasmata.
Bivalents align on the metaphase plate.
Homologous chromosomes separate.
Diplotene
Diakinesis
Metaphase I
MEIOSIS
Anaphase I
Telophase I
Prophase II
Metaphase II
Chromosomes align on the metaphase plate.

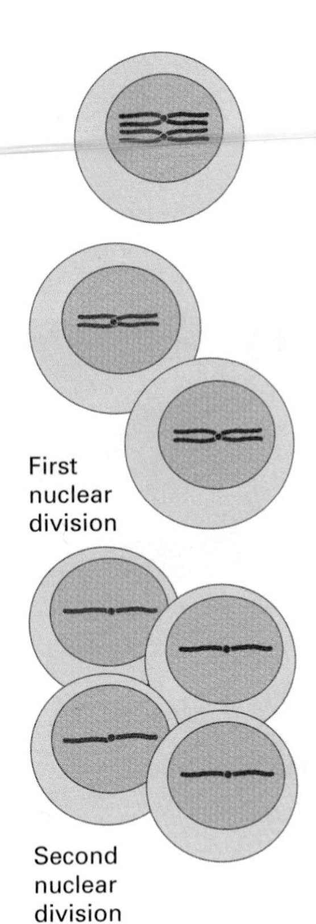


## ■ The First Meiotic Division: Reduction

The first meiotic division (meiosis I) is sometimes called the **reductional division** because it divides the chromosome number in half. By analogy with mitosis, the first meiotic division can be split into four stages, which are called **prophase I**, **metaphase I**, **anaphase I**, and **telophase I**. These stages are generally more complex than their counterparts in mitosis. The stages and substages can be visualized with reference to **FIGURES 4.7 AND 4.8**.

**1.** ***Prophase I*** This long stage lasts several days in most higher organisms. It is commonly divided into five substages: *leptotene, zygotene, pachytene, diplotene,* and *diakinesis*. These terms describe the appearance of the chromosomes at each substage.

In **leptotene**, which literally means "thin thread," the chromosomes first become visible as long, threadlike structures. The pairs of sister chromatids can be distinguished by electron microscopy. In this initial phase of chromosome condensation, numerous dense granules appear at irregular intervals along their length. These localized contractions, called *chromomeres,* have a characteristic number, size, and position in a given chromosome (Figure 4.8A).

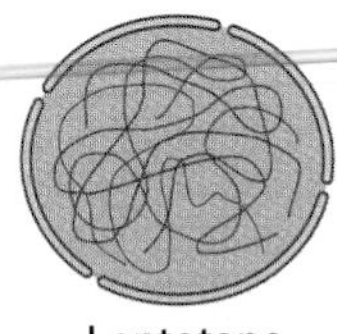


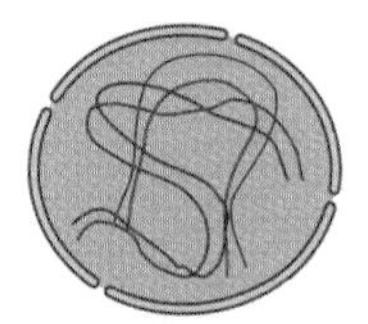


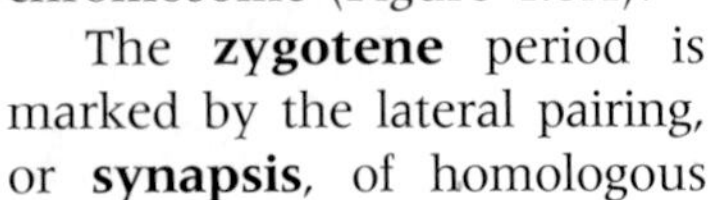

The **zygotene** period is marked by the lateral pairing, or **synapsis**, of homologous chromosomes, beginning at the chromosome tips. (The term *zygotene* means "paired threads.") As the pairing proceeds in zipper-like fashion along the length of the chromosomes, it results in a precise chromomere-by-chromomere association (Figure 4.8B and F). Synapsis is facilitated by the **synaptonemal complex**, a protein structure that helps hold the aligned homologous chromosomes together. Each

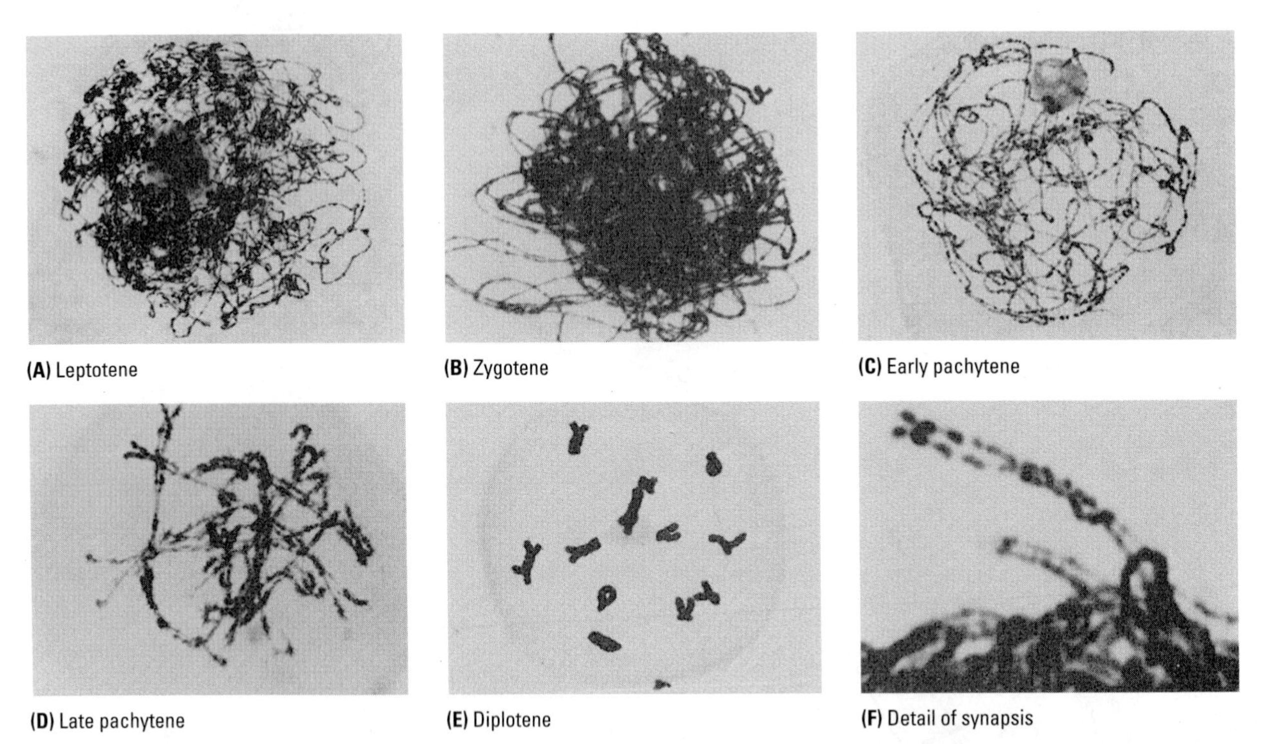


**FIGURE 4.8** Substages of prophase I in microsporocytes of the lily (*Lilium longiflorum*). (A) Leptotene: condensation of the chromosomes is initiated and beadlike chromomeres are visible along the length of the chromosomes. (B) Zygotene: pairing (synapsis) of homologous chromosomes occurs (paired and unpaired regions can be seen particularly at the lower left in this photograph). (C) Early pachytene: synapsis is completed and crossing over between homologous chromosomes occurs. (D) Late pachytene: continuation of the shortening and thickening of the chromosomes. (E) Diplotene: mutual repulsion of the paired homologous chromosomes, which remain held together at one or more cross points (chiasmata) along their length. Diakinesis follows (not shown): the chromosomes reach their maximum contraction. (F) Zygotene (at higher magnification in another cell) showing paired homologs and matching of chromomeres during synapsis. [Parts A, B, C, E, and F courtesy of Marta Walters and Santa Barbara Botanic Gardens, Santa Barbara, California. Part D courtesy of Herbert Stern. Used with permission of Ruth Stern.]

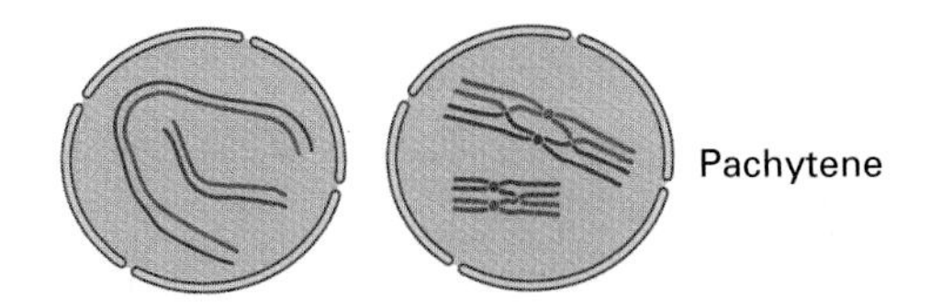


pair of synapsed homologous chromosomes is referred to as a **bivalent.**

Throughout **pachytene** (Figure 4.8C and D), which literally means "thick thread," the chromosomes continue to shorten and thicken (Figure 4.7). By late pachytene, it can sometimes be seen that each bivalent (that is, each set of paired chromosomes) actually consists of a **tetrad** of four chromatids, but the two sister chromatids of each chromosome are usually juxtaposed very tightly. Genetic exchange by means of **crossing over** takes place during pachytene. In Figure 4.7, the sites of exchange are indicated by the points where chromatids of different colors cross over each other.

At the onset of **diplotene**, the synaptonemal complex breaks down and the synapsed chromosomes begin to separate. Diplotene means "double thread," and the diplotene chromosomes are now clearly double (Figure 4.8E). The pairs of homologous chromosomes remain held together at intervals by cross-connections resulting from crossing over. Each cross-connection is called a **chiasma** (plural, chiasmata) and is formed by a breakage and rejoining between nonsister chromatids. As shown in the chromosome and diagram in **FIGURE 4.9**, *a chiasma results from physical exchange between chromatids of homologous chromosomes.* In normal meiosis, each bivalent usually has at least one chiasma, and bivalents of long chromosomes often have three or more chiasmata.

The final period of prophase I is **diakinesis**, in which the homologous chromosomes seem to repel each other and the segments not connected by chiasmata move apart. (Diakinesis means "moving apart.") It is at this substage of prophase I that the chromosomes attain their maximum condensation. The homologous chromosomes in each bivalent remain connected by at least one chiasma, which persists until the first meiotic anaphase. Near the end of diakinesis, the formation of a spindle is initiated, and the nuclear envelope breaks down.

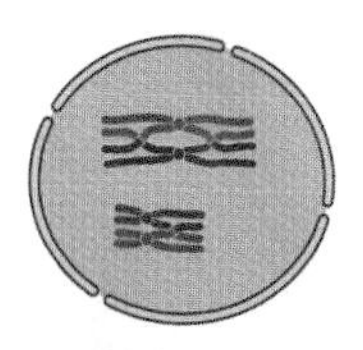

**2. *Metaphase I*** Each bivalent is maneuvered into a position straddling the metaphase plate with the centromeres of the homologous chromosomes oriented to opposite poles of the spindle (**FIGURE 4.10**A). The orientation of the centromeres determines which member of each bivalent will subsequently move to each pole, and whether the maternal or the paternal centromere is oriented toward a particular pole is completely a matter of chance. As shown in **FIGURE 4.11**, the bivalents formed from nonhomologous pairs of chromosomes can be oriented on the metaphase plate in either of two ways. If each of the nonhomologous chromosomes is heterozygous for a pair of alleles, then one type of alignment results in *A B* and *a b* gametes, whereas the other type results in *A b* and *a B* gametes (Figure 4.11). Because the metaphase alignment takes place at random, the two types of alignment—and therefore the four types of gametes—are equally frequent. The ratio of the four types of gametes is 1 : 1 : 1 : 1, which means that the *A, a* and *B, b* pairs of alleles undergo independent assortment. That is,

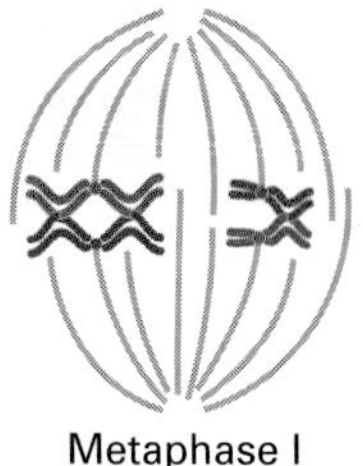


> Genes on different chromosomes undergo independent assortment because nonhomologous chromosomes align at random on the metaphase plate in meiosis I.

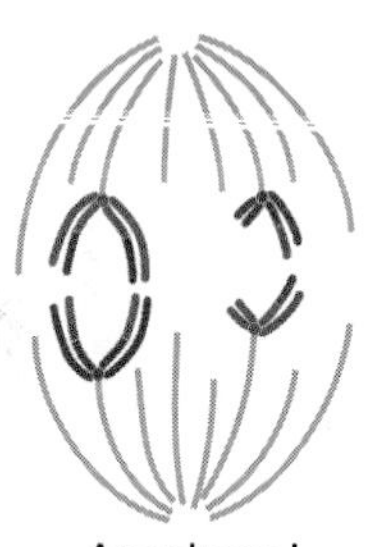


**3. *Anaphase I*** In this stage, homologous chromosomes, each composed of two

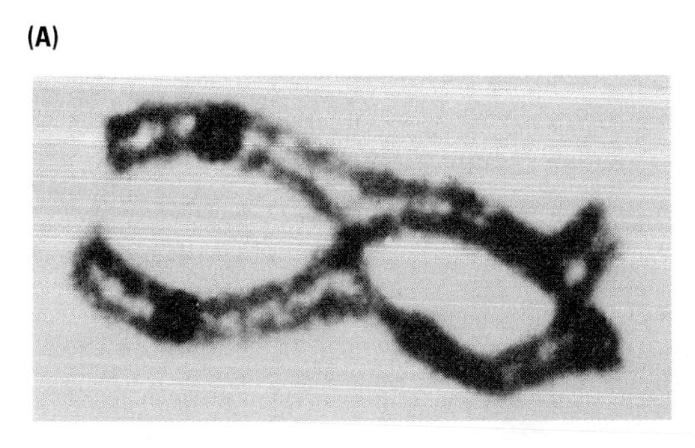


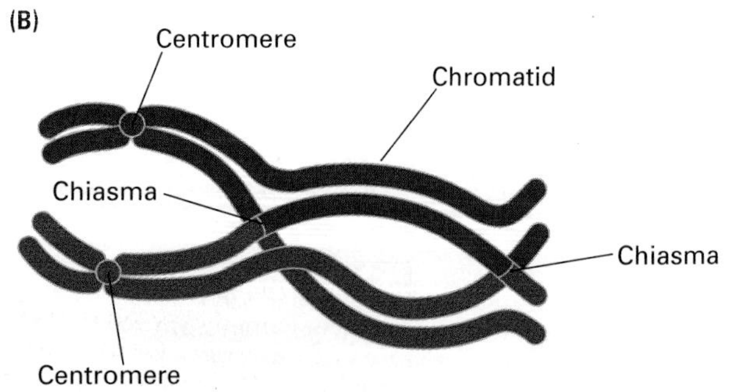


**FIGURE 4.9** Light micrograph (A) and interpretive drawing (B) of a bivalent consisting of a pair of homologous chromosomes. This bivalent was photographed at late diplotene in a spermatocyte of the salamander *Oedipina poelzi*. It shows two chiasmata where the chromatids of the homologous chromosomes appear to exchange pairing partners. [Reproduced from *The Mechanics of Inheritance* by Franklin W. Stahl. Copyright © 1964 by Prentice-Hall, Inc. Reprinted by permission of Pearson Education, Inc.]

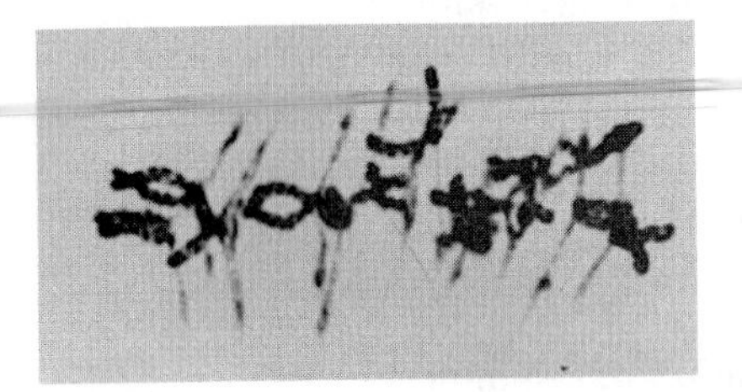

**(A)** Metaphase I

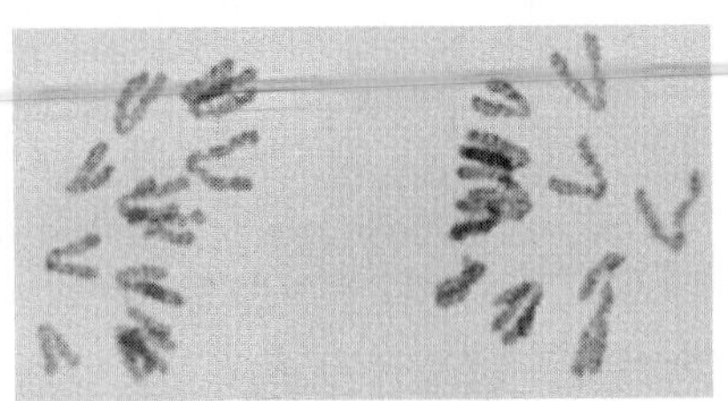

**(B)** Anaphase I

**FIGURE 4.10** Later meiotic stages in microsporocytes of the lily *Lilium longiflorum:* (A) metaphase I; (B) anaphase I; (C) metaphase II; (D) anaphase II; (E) telophase II. Cell walls have begun to form in telophase, which will lead to the formation of four pollen grains. [Courtesy of Herbert Stern. Used with permission of Ruth Stern.]

**(C)** Metaphase II (telophase I and propase II not shown)

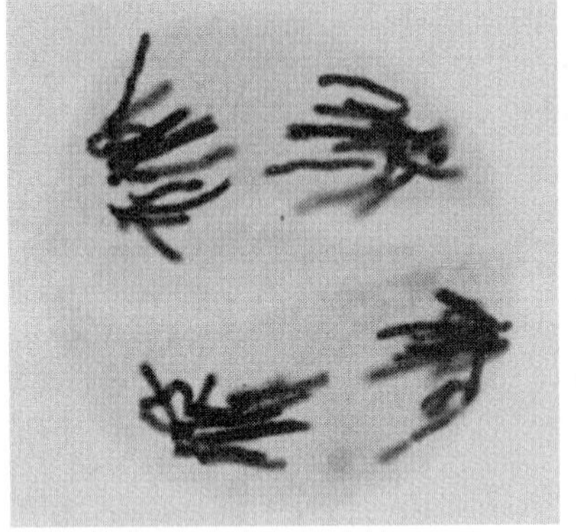

**(D)** Anaphase II

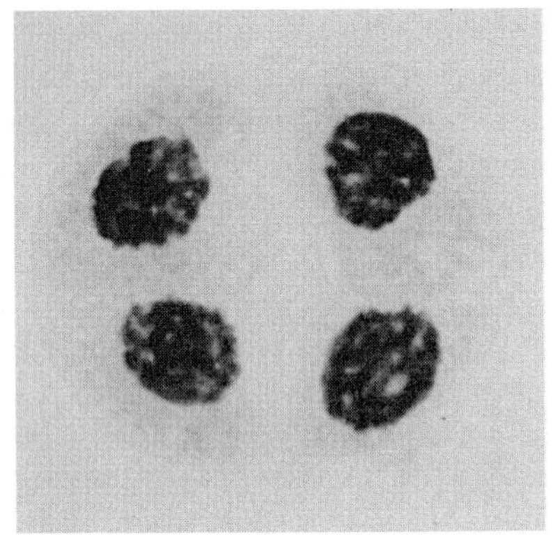

**(E)** Telophase II

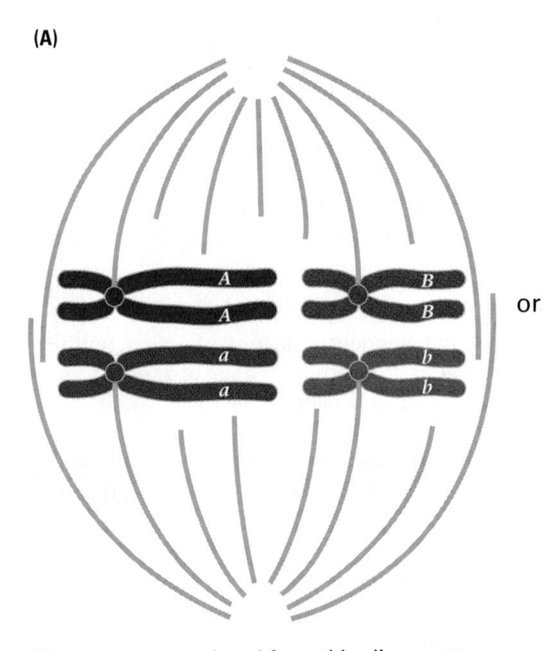
(A)
A
A
a
a
B
B
b
b
or
(B)
A
A
a
a
b
b
B
B
The gametes produced from this alignment are
$AB : AB : ab : ab$
The gametes produced from this alignment are
$Ab : Ab : aB : aB$
Because the alignments are equally likely, the overall ratio of gametes is
$AB : Ab : aB : ab = 1 : 1 : 1 : 1$
This ratio is characteristic of independent assortment.

**FIGURE 4.11** Independent assortment of genes (A or B) on nonhomologous chromosomes results from random alignment of nonhomologous chromosomes at metaphase I.

## connection
## Grasshopper, Grasshopper

*As an undergraduate researcher, Carothers showed that nonhomologous chromosomes undergo independent assortment in meiosis. For this purpose she studied a grasshopper in which one pair of homologous chromosomes had members of unequal length. At the first anaphase of meiosis in males, she could determine by observation whether the longer or the shorter chromosome went in the same direction as the X chromosome. As detailed in this paper, she found 154 of the former and 146 of the latter, a result in very close agreement with the 1 : 1 ratio expected from independent assortment. There is no mention of the Y chromosome because in the grasshopper she studied, the females have the sex chromosome constitution XX, whereas the males have the sex chromosome constitution X. In the males she examined, therefore, the X chromosome did not have a pairing partner. The instrument referred to as a camera lucida was at that time in widespread use for studying chromosomes and other microscopic objects. It is an optical instrument containing a prism or an arrangement of mirrors that, when mounted on a microscope, reflects an image of the microscopic object onto a piece of paper where it can be traced.*

**E. Eleanor Carothers 1913**
University of Kansas,
Lawrence, Kansas
*The Mendelian Ratio in Relation to Certain Orthopteran Chromosomes*

The aim of this paper is to describe the behavior of an unequal bivalent in the primary spermatocytes of certain grasshoppers. The distribution of the chromosomes of this bivalent, in relation to the X chromosome, follows the laws of chance; and, therefore, affords direct cytological support of Mendel's laws. This distribution is easily traced on account of a very distinct difference in size of the homologous chromosomes. Thus another link is added to the already long chain of evidence that the chromosomes are distinct morphological individuals continuous from generation to generation, and, as such, are the bearers of the hereditary qualities. . . . This work is based chiefly on *Brachystola magna* [a short-horned grasshopper]. . . . The entire complex of chromosomes can be separated into two groups, one containing six small chromosomes and the other seventeen larger ones. [One of the larger ones is the X chromosome.] Examination shows that this group of six small chromosomes is composed of five of about equal size and one decidedly larger. [One of the small ones is the homolog of the decidedly larger one, making this pair of chromosomes unequal in size.] . . . In early metaphases the chromosomes appear as twelve separate individuals [the bivalents]. Side views show the X chromosome in its characteristic position near one pole. . . . Three hundred cells were drawn under the camera lucida to determine the distribution of the chromosomes in the asymmetrical bivalent in relation to the X chromosome. . . . In the 300 cells drawn, the smaller chromosome went to the same nucleus as the X chromosome 146 times, or in 48.7 percent of the cases; and the larger one, 154 times, or in 51.3 percent of the cases. . . . A consideration of the limited number of chromosomes and the large number of characters in any animal or plant will make it evident that each chromosome must control numerous different characters. . . . Since the rediscovery of Mendel's laws, increased knowledge has been constantly bringing into line facts that at first seemed utterly incompatible with them. There is no cytological explanation of any other form of inheritance. . . . It seems to me probable that all inheritance is, in reality, Mendelian.

***Another link is added to the already long chain of evidence that the chromosomes are distinct morphological individuals continuous from generation to generation, and, as such, are the bearers of the hereditary qualities.***

Source: E. E. Carothers, *J. Morphol.* 24 (1913): 487-511.

chromatids joined at an undivided centromere, separate from one another and move to opposite poles of the spindle (Figure 4.10B). Chromosome separation at anaphase is the cellular basis of the segregation of alleles:

> The physical separation of homologous chromosomes in anaphase is the physical basis of Mendel's principle of segregation.

Note, however, that the centromeres of the sister chromatids are tightly stuck together and behave as a single unit. A specific protein acts as a glue holding the sister centromeres together. This protein appears in the centromeres and adjacent chromosome arms during S phase and persists throughout meiosis I. It disappears only at anaphase II, when sister-centromere cohesion is lost and the sister centromeres separate.

**4. *Telophase I*** At the completion of anaphase I, a haploid set of chromosomes consisting of one homolog from each bivalent is located near each pole of the spindle (Figure 4.6). In telophase, the spindle breaks down, and, depending on the species, either a nuclear envelope briefly forms around each group of chromosomes or the chromosomes enter the second meiotic division after only a limited uncoiling.

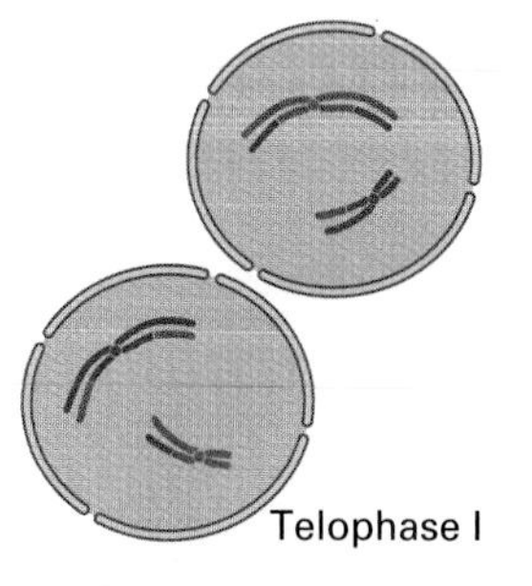

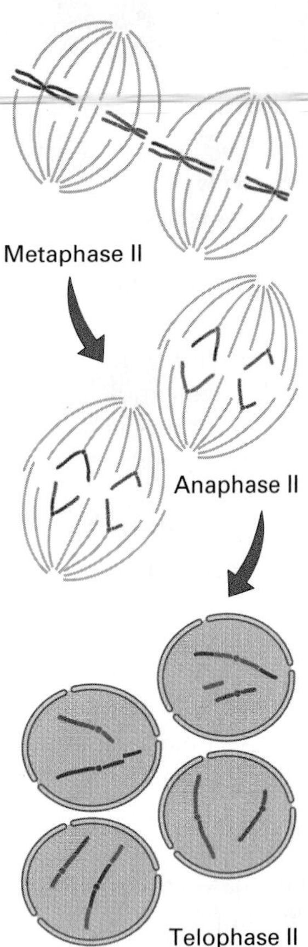


### ■ The Second Meiotic Division: Equation

The second meiotic division (meiosis II) is sometimes called the **equational division** because the chromosome number remains the same in each cell before and after the second division. In some species, the chromosomes pass directly from telophase I to **prophase II** without loss of condensation; in others, there is a brief pause between the two meiotic divisions and the chromosomes may "decondense" (uncoil) somewhat. *Chromosome replication never takes place between the two divisions;* the chromosomes present at the beginning of the second division are identical to those present at the end of the first.

After a short prophase (prophase II) and the formation of second-division spindles, the centromeres of the chromosomes in each nucleus become aligned on the central plane of the spindle at **metaphase II** (Figure 4.10C). In **anaphase II** the protein holding the sister centromeres together breaks down. As a result, the sister centromeres appear to split longitudinally, and the chromatids of each chromosome move to opposite poles of the spindle (Figure 4.10D). Once the centromeres split at anaphase II, each chromatid is considered to be a separate chromosome.

**Telophase II** (Figure 4.10E) is marked by a transition to the interphase condition of the chromosomes in the four haploid nuclei, accompanied by division of the cytoplasm. Thus, the second meiotic division superficially resembles a mitotic division. However, there is an important difference:

> The chromatids of a chromosome are usually not genetically identical along their entire length because of crossing over associated with the formation of chiasmata during prophase of the first division.

## 4.4 Sex-Chromosome Inheritance

The first rigorous experimental proof that genes are parts of chromosomes was obtained in experiments concerned with the pattern of transmission of the **sex chromosomes,** the chromosomes responsible for determination of the separate sexes in some plants and in nearly all animals. We will examine these results in this section.

### ■ Chromosomal Determination of Sex

The sex chromosomes are an exception to the rule that all chromosomes of diploid organisms are present in pairs of morphologically similar homologs. As early as 1891, microscopic analysis had shown that one of the chromosomes in males of some insect species does not have a homolog. This unpaired chromosome was called the **X chromosome,** and it was present in all somatic cells of the males but in only half the sperm cells. The biological significance of these observations became clear when females of the same species were shown to have two X chromosomes.

In other species in which the females have two X chromosomes, the male has one X chromosome along with a morphologically unmatched chromosome. The unmatched chromosome is referred to as the **Y chromosome,** and it pairs with the X chromosome during meiosis in males, usually along only part of its length because of a limited region of homology. The difference in chromosomal constitution between males and females is a chromosomal mechanism for determining sex at the time of fertilization. Whereas every egg cell contains an X chromosome, half of the sperm cells contain an X chromosome and the rest contain a Y chromosome. Fertilization of an X-bearing egg by an X-bearing sperm results in an XX zygote, which normally develops into a female; and fertilization by a Y-bearing sperm results in an XY zygote, which normally develops into a male (**FIGURE 4.12**). The result is a criss-cross pattern of inheritance of the X chromosome in which a male receives his X chromosome from his mother and transmits it only to his daughters.

The XX–XY type of chromosomal sex determination is found in mammals, including human beings, in many insects, and in other animals, as well as in some flowering plants. The female is called the **homogametic** sex because only one type of gamete (X-bearing) is produced, and the male is called the **heterogametic** sex because two different types of gametes (X-bearing and Y-bearing) are produced. When the union of gametes in fertilization is random, a sex ratio at fertilization of 1 : 1 is expected because males produce equal numbers of X-bearing and Y-bearing sperm.

The X and Y chromosomes together constitute the sex chromosomes; this term distinguishes them from other pairs of chromosomes, which are called **autosomes.** Although the sex chromosomes control the developmental switch that determines the earliest stages of female or male development, the developmental process itself requires many genes scattered throughout the genome, including genes on the autosomes. The X chromosome also contains many genes

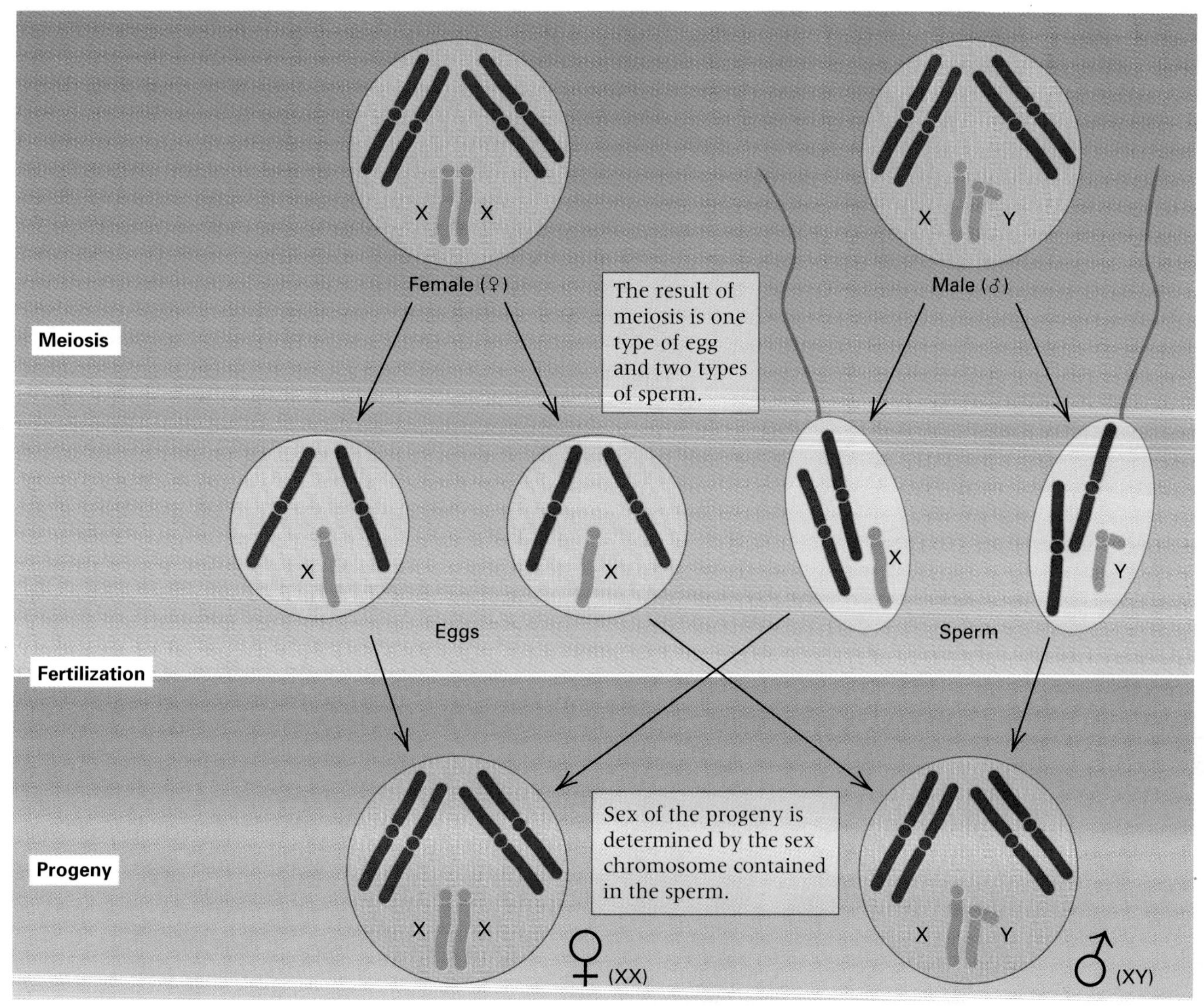


**FIGURE 4.12** The chromosomal basis of sex determination in mammals, many insects, and other animals.

with functions unrelated to sexual differentiation, as we will see in the next section. In most organisms, including human beings, the Y chromosome carries few genes other than those related to male determination.

## ■ X-Linked Inheritance

The compelling evidence that genes are located in chromosomes came from the study of a *Drosophila* gene for white eyes, which proved to be present in the X chromosome. Recall that in Mendel's crosses, it did not matter which trait was present in the male parent and which in the female parent. Reciprocal crosses gave the same result. One of the earliest exceptions to this rule was found by Thomas Hunt Morgan in 1910 in an early study of a mutant fruit fly that had white eyes. The wildtype eye color is a brick-red combination of red and brown pigments (**FIGURE 4.13**). Although white eyes can result from certain combinations of autosomal genes that eliminate the pigments individually, the white-eye mutation that Morgan studied results in a metabolic block that knocks out both pigments simultaneously. (The gene codes for a transmembrane protein that is necessary to transport the eye pigment precursors into the pigment cells.)

Morgan's study started with a single male with white eyes that appeared in a wildtype laboratory population that had been maintained for many generations. In a mating of this male with wildtype females, all of the $F_1$ progeny of both sexes had red eyes, which showed that the allele for white eyes is recessive. In the $F_2$ progeny from the mating of $F_1$ males with females, Morgan observed 2459 red-eyed females, 1011 red-eyed males, and 782 white-eyed males. It was clear that the white-eyed phenotype was somehow connected with sex, because all of the white-eyed flies were males.

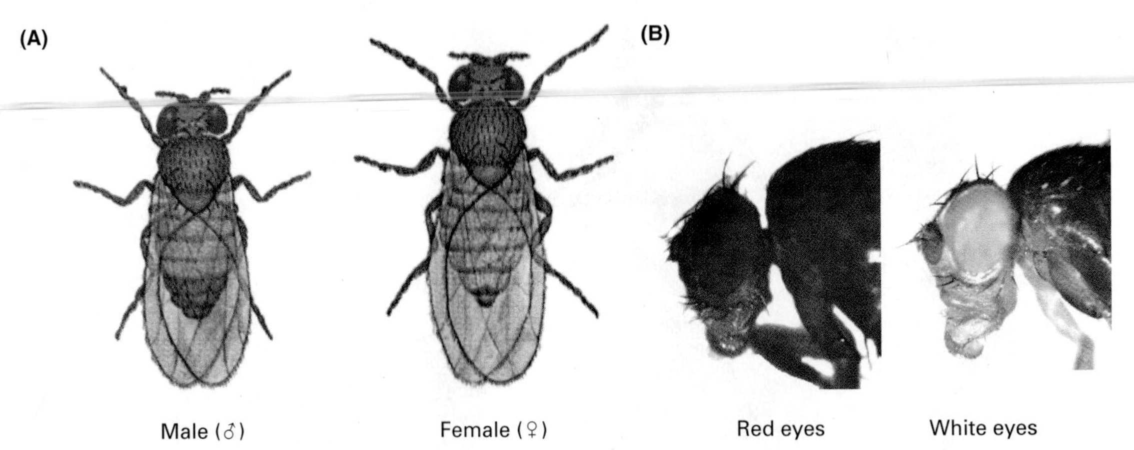


**FIGURE 4.13** Drawings (A) of a male and a female fruit fly, *Drosophila melanogaster.* The photographs (B) show the eyes of a wildtype red-eyed male and a mutant white-eyed male. [Illustrations © Carolina Biological Supply Company. Used with permission. Photographs courtesy of E. R. Lozovsky.]

**MORGAN'S GROUP** partying in the fly room at Columbia University in 1919, to celebrate A. H. Sturtevant's return from military service in World War I. T. H. Morgan is on the far right in the back row, with H. J. Muller beside him. Sturtevant is in the foreground, leaning back in his chair, and C. B. Bridges is in shirtsleeves sitting next to an apelike creature dubbed "Pithecanthropus," a dummy made up for the occasion. [Courtesy of the Archives, California Institute of Technology.]

On the other hand, white eyes were not restricted to males. For example, when red-eyed $F_1$ females from the cross of wildtype ♀ × white ♂ were backcrossed with their white-eyed fathers, the progeny consisted of both red-eyed and white-eyed females and red-eyed and white-eyed males in approximately equal numbers.

A key observation came from the mating of white-eyed females with wildtype males. All the female progeny had wildtype eyes, but all the male progeny had white eyes. This is the reciprocal of the original cross of wildtype ♀ × white ♂, which had given only wildtype females and wildtype males, so the reciprocal crosses gave different results.

Morgan realized that reciprocal crosses would yield different results if the allele for white eyes were present in the X chromosome. A gene on the X chromosome is said to be **X-linked.** The normal chromosome complement of *Drosophila melanogaster* is shown in **FIGURE 4.14**. Females have an XX chromosome comple-

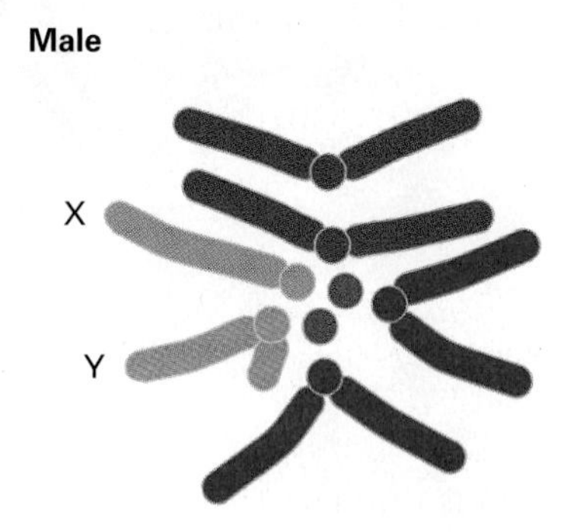


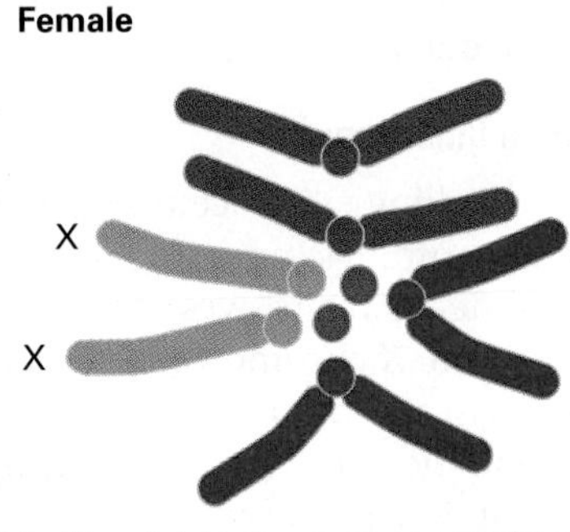


**FIGURE 4.14** The diploid chromosome complements of a male and a female *Drosophila melanogaster.* The centromere of the X chromosome is nearly terminal, but that of the Y chromosome divides the chromosome into two unequal arms. The large autosomes (chromosomes 2 and 3, shown in blue and green) are not easily distinguishable in these types of cells. The tiny autosome (chromosome 4, shown in yellow) appears as a dot.

## connection

### The White-Eyed Male

*Morgan's genetic analysis of the white-eye mutation marks the beginning of* Drosophila *genetics. It is in the nature of science that as knowledge increases, the terms used to describe things change also. This paper affords an example, because the term* sex-limited inheritance *is used today to mean something completely different from Morgan's usage. What Morgan was referring to is now called X-linked inheritance or sex-linked inheritance. To avoid confusion, we have taken the liberty of substituting the modern equivalent wherever appropriate. Morgan was also unaware that* Drosophila *males had a Y chromosome. He thought that females were XX and males X, as in grasshoppers (see the Carothers paper). We have also supplied the missing Y chromosome. On the other hand, Morgan's gene symbols have been retained as in the original. He uses* R *for the wild-type allele for red eyes and* W *for the recessive allele for white eyes. This is a curious departure from the convention, already introduced by Mendel, that dominant and recessive alleles should be represented by the same symbol. Today we use* w *for the recessive allele and* $w^+$ *for the dominant allele.*

**Thomas Hunt Morgan 1910**
Columbia University,
New York, New York
*Sex-Limited Inheritance in* Drosophila

In a pedigree culture of *Drosophila* that had been running for nearly a year through a considerable number of generations, a male appeared with white eyes. The normal flies have brilliant red eyes. The white-eyed male, bred to his red-eyed sisters, produced 1,237 red-eyed offspring. . . . The $F_1$ hybrids, inbred, produced

| | |
|---|---|
| 2,459 | red-eyed females |
| 1,011 | red-eyed males |
| 782 | white-eyed males |

*No white-eyed females appeared.* The new character showed itself to be sex-linked in the sense that it was transmitted only to the grandsons. But that the character is not incompatible with femaleness is shown by the following experiment. The white-eyed male (mutant) was later crossed with some of his daughters ($F_1$), and produced

| | |
|---|---|
| 129 | red-eyed females |
| 132 | red-eyed males |
| 88 | white-eyed females |
| 86 | white-eyed males |

The results show that the new character, white eyes, can be carried over to the females by a suitable cross, and is in consequence in this sense not limited to one sex. It will be noted that the four classes of individuals occur in approximately equal numbers (25 percent). . . . The results just described can be accounted for by the following hypothesis. Assume that all of the spermatozoa of the white-eyed male carry the "factor" for white eyes "W"; that half of the spermatozoa carry a sex factor "X," [and] the other half lack it, *i.e.*, the male is heterozygous for sex. [The male is actually XY.] Thus, the symbol for the male is "WXY", and for his two kinds of spermatozoa WX—Y. Assume that all of the eggs of the red-eyed female carry the red-eyed "factor" R; and that all of the eggs (after meiosis) carry one X each, the symbol for the red-eyed female will be therefore RRXX and that for her eggs will be RX. . . . The hypothesis just utilized to explain these results first obtained can be tested in several ways. [There follow four types of crosses, each yielding the expected result.] . . . In order to obtain these results it is necessary to assume that, when the two classes of spermatozoa are formed in the RXY male, R and X go together. . . . The fact is that this R and X are combined and have never existed apart.

***No white-eyed females appeared.***

Source: T. H. Morgan, *Science* 32 (1910): 120–122.

ment, whereas males are XY. Morgan's hypothesis was that an X chromosome contains either a wildtype $w^+$ allele or a mutant $w$ allele and that the Y chromosome does not contain a counterpart of the *white* gene. Using the *white* allele present in an X chromosome to represent the entire X chromosome, we can write the genotype of a white-eyed male as $w$Y and that of a wildtype male as $w^+$Y. Because the $w$ allele is recessive, white-eyed females are of genotype $ww$ and wildtype females are either heterozygous $w^+w$ or homozygous $w^+w^+$. The implications of this model for reciprocal crosses are shown in **FIGURE 4.15**. The mating wildtype ♀ × white ♂ is Cross A, and that of white ♀ × wildtype ♂ is Cross B.

The X-linked mode of inheritance does account for the different phenotypic ratios observed in the $F_1$ and $F_2$ progeny from the crosses. The characteristics of X-linked inheritance can be summarized as follows:

1. Reciprocal crosses resulting in different phenotypic ratios in the sexes often indicate X-linked inheritance. In the case of white eyes in *Drosophila*, the cross of a red-eyed female with a white-eyed male yields all red-eyed progeny (Figure 4.15, Cross A), whereas the cross of a white-eyed female with a red-eyed male yields red-eyed female progeny and white-eyed male progeny (Figure 4.15, Cross B).

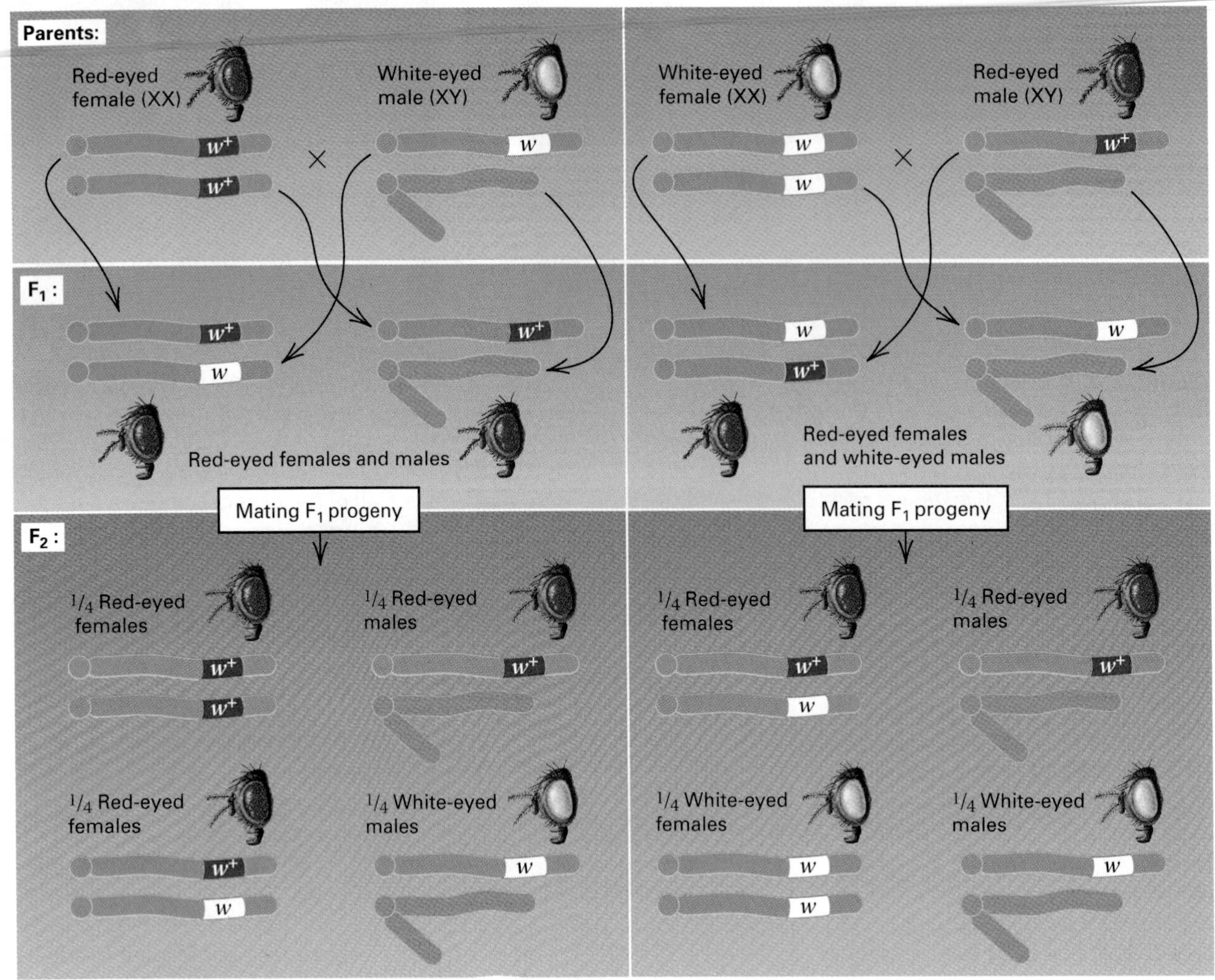


**FIGURE 4.15** Chromosomal interpretation of the results obtained in $F_1$ and $F_2$ progenies in crosses of *Drosophila*. Cross A is a mating of a wildtype (red-eyed) female with a white-eyed male. Cross B is the reciprocal mating of a white-eyed female with a red-eyed male. In the X chromosome, the wildtype $w^+$ allele is shown in red, the mutant $w$ allele in white. The Y chromosome does not carry either allele of the $w$ gene.

2. Heterozygous females transmit each X-linked allele to approximately half of their daughters and half of their sons; this is illustrated in the $F_2$ generation of Cross B in Figure 4.15.
3. Males that inherit an X-linked recessive allele exhibit the recessive trait because the Y chromosome does not contain a wildtype counterpart of the gene. Affected males transmit the recessive allele to all of their daughters but none of their sons; this principle is illustrated in the $F_1$ generation of Cross A in Figure 4.15. Any male that is not affected must carry the wildtype allele in his X chromosome.

The essence of X-linked inheritance is captured in the Punnett square in **FIGURE 4.16**: A male transmits his X chromosome only to his daugh-

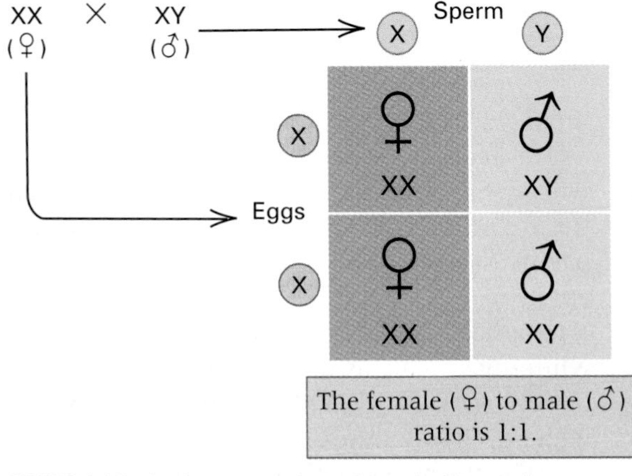


**FIGURE 4.16** In chromosomal sex determination, each son gets his X chromosome from his mother and his Y chromosome from his father.

ters, whereas a female transmits an X chromosome to offspring of both sexes.

## Pedigree Characteristics of Human X-Linked Inheritance

An example of a human trait with an X-linked pattern of inheritance is *hemophilia A,* a severe disorder of blood clotting determined by a recessive allele. Affected persons lack a blood-clotting protein called factor VIII that is needed for normal clotting, and they suffer excessive, often life-threatening bleeding after injury. A famous pedigree of hemophilia starts with Queen Victoria of England (**FIGURE 4.17**). One of her sons, Leopold, was hemophilic, and two of her daughters were heterozygous carriers of the gene. Two of Victoria's granddaughters were also carriers, and by marriage they introduced the gene into the royal families of Russia and Spain. The heir to the Russian throne of the Romanoffs, Tsarevich Alexis, was afflicted with the condition. He inherited the gene from his mother, the Tsarina Alexandra, one of Victoria's granddaughters. The Tsar, the Tsarina, Alexis, and his four sisters were all executed by the Russian Bolsheviks in the 1918 revolution. Ironically, the present royal family of England is descended from a normal son of Victoria and is free of the disease.

X-linked inheritance in human pedigrees shows several characteristics that distinguish it from other modes of genetic transmission.

1. For any rare trait due to an X-linked recessive allele, the affected individuals are exclusively, or almost exclusively, male. There is an excess of males because females carrying the rare X-linked recessive are almost exclusively heterozygous and so do not express the mutant phenotype.
2. Affected males who reproduce have normal sons. This follows from the fact that a male transmits his X chromosome only to his daughters.
3. A woman whose father was affected has normal sons and affected sons in the ratio 1 : 1. This is true because any daughter of an affected male must be heterozygous for the recessive allele.

## Heterogametic Females

In some organisms, the homogametic and heterogametic sexes are reversed; that is, the males are XX and the females are XY. This type of sex determination is found in birds, in some reptiles and fish, and in moths and butterflies. The reversal of XX and XY in the sexes results in

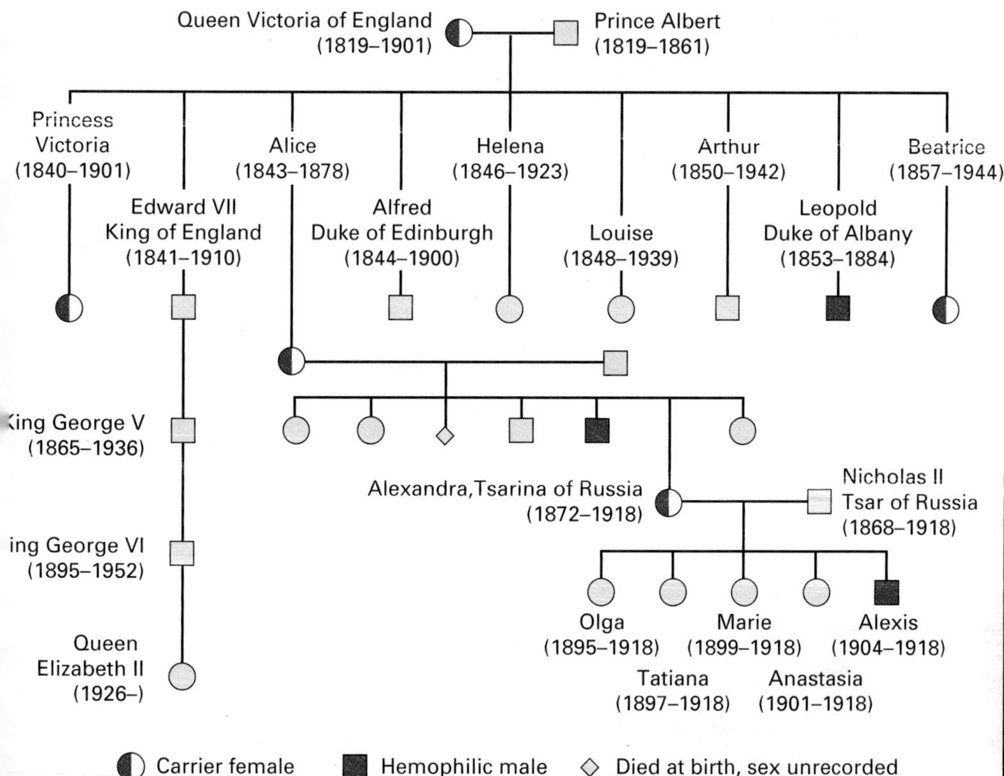


Courtesy of the Boston Public Library, Print Department.

GURE 4.17 Genetic transmission of hemophilia A among the descendants of Queen Victoria of England, cluding her granddaughter, Tsarina Alexandra of Russia, and Alexandra's five children. The photograph is that of ar Nicholas II, Tsarina Alexandra, and their children. Tsarevich Alexis was afflicted with hemophilia. [*Source:* ulver Pictures.]

an opposite pattern of nonreciprocal inheritance of X-linked genes. To distinguish sex determination in these organisms from the usual XX–XY mechanism, the sex chromosome constitution in the homogametic sex is designated ZZ and that in the heterogametic sex as WZ. Hence in organisms with heterogametic females, the chromosomal constitution of the females is designated WZ and that of the male ZZ.

A specific example of Z-linked inheritance in chickens is shown in **FIGURE 4.18**. A few breeds of chickens have feathers with alternating transverse bands of light and dark color, resulting in a phenotype known as barred. In other breeds the feathers are uniformly colored and nonbarred. Reciprocal crosses between true-breeding barred and true-breeding nonbarred breeds give the results shown, which indicate that the gene that determines barring must be dominant and must be located in the Z chromosome.

**FIGURE 4.18** Barred feathers in chickens, a classic example showing that chromosomal sex determination in birds is the reverse of that in mammals. In birds, females are the heterogametic sex. The Z chromosome carrying the dominant barred mutation is indicated in red.

## Nondisjunction as Proof of the Chromosome Theory of Heredity

The parallel between the inheritance of the *Drosophila white* mutation and the genetic transmission of the X chromosome supported the chromosome theory of heredity that genes are parts of chromosomes. Other experiments with *Drosophila* provided the definitive proof.

One of Morgan's students, Calvin Bridges, discovered rare exceptions to the expected pattern of inheritance in crosses with several X-linked genes. For example, when white-eyed *Drosophila* females were mated with red-eyed males, most of the progeny consisted of the expected red-eyed females and white-eyed males. However, about 1 in every 2000 $F_1$ flies was an exception, either a white-eyed female or a red-eyed male. Bridges showed that these rare exceptional offspring resulted from occasional failure of the two X chromosomes in the mother to separate from each other during meiosis—a phenomenon called **nondisjunction.** The consequence of nondisjunction of the X chromosome is the formation of some eggs with two X chromosomes and others with none. Four classes of zygotes are expected from the fertilization of these abnormal eggs (**FIGURE 4.19**). Animals with no X chromosome are not detected because embryos that lack an X are not viable; likewise, most progeny with three X chromosomes die early in development. Microscopic examination of the chromosomes of the exceptional progeny from the cross white ♀ × wildtype ♂ showed that the exceptional white-eyed females had two X chromosomes *plus* a Y chromosome, and the exceptional red-eyed males had a single X but were *lacking* a Y. The latter, with a sex-chromosome constitution denoted XO, were sterile males.

These and related experiments demonstrated conclusively the validity of the chromosome theory of heredity.

> **Chromosome theory of heredity:** Genes are physically located within chromosomes.

Bridges's evidence for the chromosome theory was that exceptional behavior on the part of chromosomes is precisely paralleled by exceptional inheritance of their genes. This proof of the

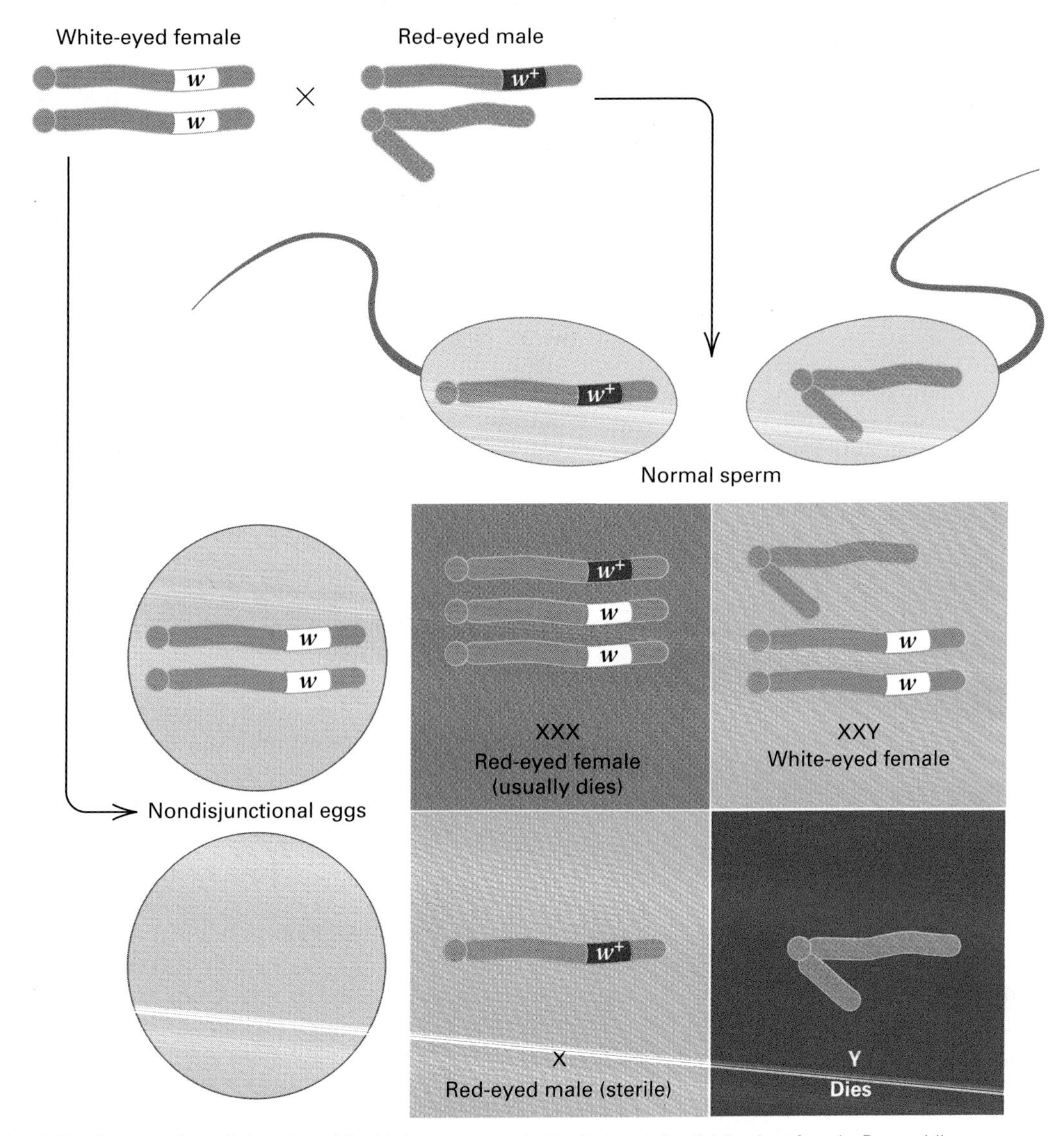


**FIGURE 4.19** Results of nondisjunction of the X chromosomes in the first meiotic division in a female *Drosophila*.

chromosome theory ranks among the most important and elegant experiments in genetics.

## Sex Determination in *Drosophila*

In the XX–XY mechanism of sex determination, the Y chromosome is associated with the male. In some organisms, including human beings, this association occurs because the presence of the Y chromosome triggers events in embryonic development that result in the male sexual characteristics (Chapter 8). *Drosophila* is unusual among organisms with an XX–XY type of sex determination because the Y chromosome, although associated with maleness, is not male-determining. This is demonstrated by the finding, shown in **FIGURE 4.20**, that in *Drosophila*, XXY embryos develop into morphologically normal, fertile females, whereas XO embryos develop into morphologically normal, but sterile, males. (The O is written in the formula XO to emphasize that a sex chromosome is missing.) The sterility of XO males shows that the Y chromosome, though not necessary for male development, is essential for male fertility; in fact, the *Drosophila* Y chromosome contains six genes required for the formation of normal sperm.

The genetic determination of sex in *Drosophila* depends on the number of X chromosomes present in an individual fly compared with the number of sets of autosomes. In *Drosophila*, a haploid set of autosomes consists of one copy each of chromosomes 2, 3, and 4 (the autosomes). Normal diploid flies have two haploid sets of autosomes (a homologous pair each of chromosomes 2, 3, and 4) plus either two X chromosomes in a female or one X and one Y chromosome in a male (Figure 4.14). If

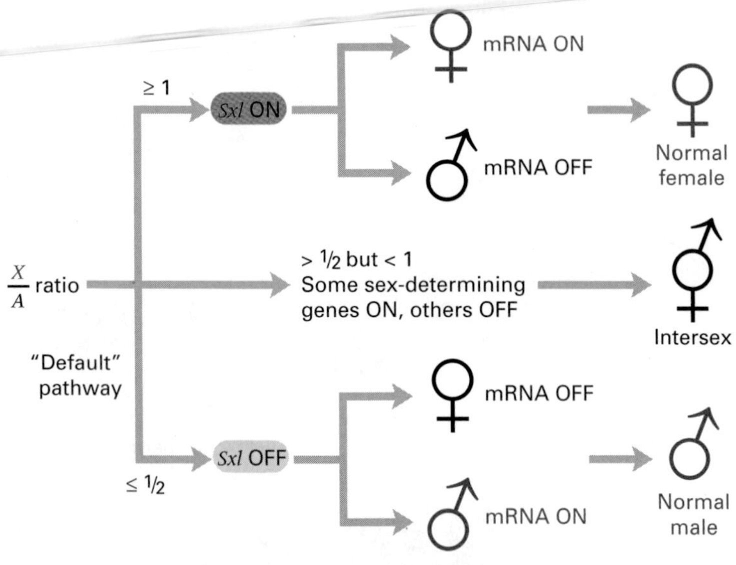


**FIGURE 4.20** Early steps in the genetic control of sex determination in *Drosophila* through the activity of the Sex-lethal gene *Sxl*.

we use A to represent a complete haploid complement of autosomes, then

$$A = \text{chromosome 2} + \text{chromosome 3} + \text{chromosome 4}$$

In these terms, a normal male has the chromosomal complement XYAA, and the ratio of X chromosomes to sets of autosomes (the X/A ratio) equals 1X : 2A, or 1 : 2. Normal females have the chromosomal complement XXAA, so the female X/A ratio is 2X : 2A, or 1 : 1. Flies with X/A ratios smaller than 1 : 2 (for example, XAAA, which means one X chromosome and three sets of autosomes) are male; those with X/A ratios greater than 1 : 1 (for example, XXXAA, which implies three X chromosomes and two sets of autosomes) are female. Intermediate X/A ratios such as 2 : 3 (for example, XXAAA, with two X chromosomes and three sets of autosomes) develop as intersexes with some characteristics of each sex.

Sexual differentiation in *Drosophila* is controlled by a gene called *Sex-lethal (Sxl)*. The *Sxl* gene codes for two somewhat different proteins, depending on whether a male-specific coding region is included in the messenger RNA. Furthermore, the amount of Sxl protein present in the early embryo regulates the expression of the *Sxl* gene by a feedback mechanism. At low levels of Sxl protein, the male-specific form of the protein is made and shuts off further expression of the gene. At higher levels of the Sxl protein, the female-specific form of the protein is made and the gene continues to be expressed. An outline of the genetic control of sex determination in *Drosophila* is shown in Figure 4.20.

## 4.5 Probability in the Prediction of Progeny Distributions

Genetic transmission includes a large component of chance. A particular gamete from an *Aa* organism might or might not include the *A* allele, depending on chance. A particular gamete from an *Aa Bb* organism might or might not include both the *A* and *B* alleles, depending on the chance orientation of the chromosomes on the metaphase I plate. Genetic ratios result not only from the chance assortment of genes into gametes, but also from the chance combination of gametes into zygotes. Although exact predictions are not possible for any particular event, it is possible to determine the probability that a particular event will be realized, as we saw in Chapter 3. In this section, we consider some additional probability methods used in interpreting genetic data.

### ■ Using the Binomial Distribution in Genetics

The addition rule of probability deals with outcomes of a genetic cross that are mutually exclusive. Outcomes are "mutually exclusive" if they are incompatible in the sense that they cannot occur at the same time. For example, the possible sex distributions among three children consist of four mutually exclusive outcomes: zero, one, two, or three girls. These outcomes have probabilities 1/8, 3/8, 3/8, and 1/8, respectively. The addition rule states that the overall probability of any combination of mutually exclusive events is equal to the sum of the probabilities of the events taken separately. For example, the probability that a sibship of size three contains *at least one girl* includes the outcomes one, two, and three girls, so the overall probability of at least one girl equals $3/8 + 3/8 + 1/8 = 7/8$.

The multiplication rule of probability deals with outcomes of a genetic cross that are independent. Any two possible outcomes are independent if the knowledge that one outcome is actually realized provides no information about whether the other is realized also. For example, in a sequence of births, the sex of any particular child is not affected by the sex of any sibling born earlier and has no influence whatsoever on the distribution of sexes of any siblings born later. Each successive birth is independent of all the others. When possible outcomes are independent, the multiplication rule states that the probability of any combination of outcomes being realized equals the product of the probabilities of all the individual outcomes taken separately. For example, the probability that a sibship of three children will consist of three girls equals $1/2 \times 1/2 \times 1/2$, because the probability of each birth resulting in a girl is 1/2, and the successive births are independent.

Probability calculations in genetics frequently use the addition and multiplication rules together. For example, the probability that all three children in a family will be of the same sex uses both the addition and the multiplication rules. The probability that all three will be girls is $(1/2)(1/2)(1/2) = 1/8$, and the probability that all three will be boys is also 1/8. Because these outcomes are mutually exclusive (a sibship of size three cannot include three boys *and* three girls), the probability of either three girls or three boys is the sum of the two probabilities, or $1/8 + 1/8 = 1/4$. The other possible outcomes for sibships of size three are that two of the children will be girls and the other a boy, and that two will be boys and the other a girl. For each of these outcomes, three different orders of birth are possible—for example, GGB, GBG, and BGG—each having a probability of $1/2 \times 1/2 \times 1/2 = 1/8$. The probability of two girls and a boy, disregarding birth order, is the sum of the probabilities for the three possible orders, or 3/8; likewise, the probability of two boys and a girl is also 3/8. Therefore, the distribution of probabilities for the sex ratio in families with three children is

| GGG | GGB | GBB | BBB |
|---|---|---|---|
| | GBG | BGB | |
| | BGG | BBG | |

$$(1/2)^3 + 3(1/2)^2(1/2) + 3(1/2)(1/2)^2 + (1/2)^3 =$$
$$1/8 + 3/8 + 3/8 + 1/8 = 1$$

The sex ratio information in this display can be obtained more directly by expanding the binomial expression $(p + q)^n$, in which $p$ is the probability of the birth of a girl (1/2), $q$ is the probability of the birth of a boy (1/2), and $n$ is the number of children. In the present example,

$$(p + q)^3 = 1p^3 + 3p^2q + 3pq^2 + 1q^3$$

in which the red numerals are the possible number of birth orders for each sex distribution. Similarly, the binomial distribution of probabilities for the sex ratios in families of five children is

$$(p + q)^5 = 1p^5 + 5p^4q + 10p^3q^2 + 10p^2q^3 + 5pq^4 + 1q^5$$

Each term tells us the probability of a particular combination. For example, the third term is the probability of three girls ($p^3$) and two boys ($q^2$) in a family that has five children:

$$10(1/2)^3(1/2)^2 = 10/32 = 5/16$$

There are $n + 1$ terms in a binomial expansion to the power $n$. The exponents of $p$ decrease by one from $n$ in the first term to 0 in the last term, and the exponents of $q$ increase by one from 0 in the first term to $n$ in the last term. The coefficients generated by successive values of $n$ can be arranged in a regular triangle known as **Pascal's triangle,** shown for $n = 0$ to 10 in **FIGURE 4.21**. The horizontal rows of the

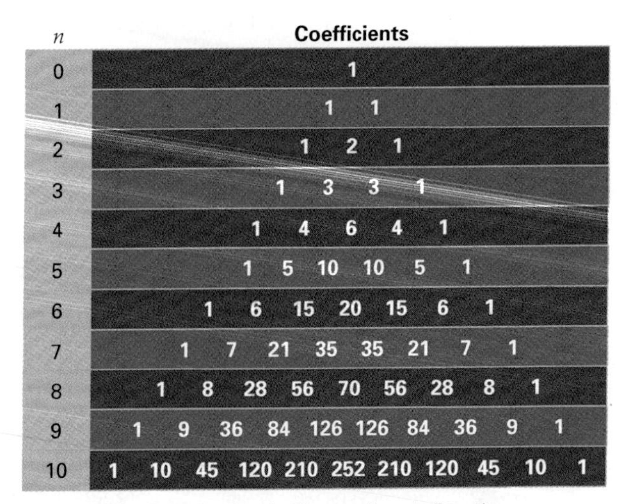


**FIGURE 4.21** Pascal's triangle. The numbers in the $n$th row are the coefficients of the terms in the expansion of the polynomial $(p + q)^n$.

triangle are symmetrical; each row begins and ends with a 1, and each other entry can be obtained as the sum of the two numbers on either side of it in the row above.

To generalize just a bit, if the probability of event A is $p$, that of event B is $q$, and the two events are independent and mutually exclusive, then the probability that A will be realized four times and B two times (in a specified order) is, by the multiplication rule, $p^4q^2$. Usually we are interested in a combination of events regardless of their order, such as "four A and two B." In this case, we multiply the probability that the combination 4A : 2B will be realized in any specified order by the number of possible orders. The number of different combinations of six events, four of type A and two of type B, is given by the coefficient of $p^4q^2$ in the expansion of $(p\text{ A} + q\text{ B})^6$. This coefficient can be found in the row for $n = 6$ in Pascal's triangle, as the fifth entry from the left. (It is the fifth because the successive entries are the coefficients of $p^0q^6$, $p^1q^5$, $p^2q^4$, $p^3q^3$, $p^4q^2$, $p^5q^1$, and $p^6q^0$.) Hence, the overall probability of four realizations of A and two realizations of B in six trials is given by $15p^4q^2$.

The general rule for repeated trials of events with constant probabilities is as follows:

> If the probability of event A is $p$ and the probability of the alternative event B is $q$, the probability that, in $n$ trials, event A is realized $s$ times and event B is realized $t$ times is

$$\frac{n!}{s!t!}p^sq^t \qquad (1)$$

in which $s + t = n$ and $p + q = 1$. The symbol $n!$ is read as "$n$ factorial," and it stands for the product of all positive integers from 1 through $n$ (that is, $n! = 1 \times 2 \times 3 \times \cdots \times n$). Values $n$ factorial from $n = 0$ through $n = 15$ are given in **TABLE 4.2**. The value $0! = 1$ is defined arbitrarily to generalize its use in mathematical formulas. The magnitude of $n!$ increases very rapidly; 15! is more than a trillion.

**Table 4.2 Factorials**

| $n$ | $n!$ | $n$ | $n!$ |
|---|---|---|---|
| 0 | 1 | 8 | 40,320 |
| 1 | 1 | 9 | 362,880 |
| 2 | 2 | 10 | 3,628,800 |
| 3 | 6 | 11 | 39,916,800 |
| 4 | 24 | 12 | 479,001,600 |
| 5 | 120 | 13 | 6,227,020,800 |
| 6 | 720 | 14 | 87,178,291,200 |
| 7 | 5040 | 15 | 1,307,674,368,000 |

Equation (1) applies even when either $s$ or $t$ equals 0 because $0! = 1$. (Remember also that any number raised to the zero power equals 1.) Any individual term in the expansion of the binomial $(p + q)^n$ is given by Equation (1) for the appropriate values of $s$ and $t$. In Pascal's triangle, successive entries in the $n$th row are the values of $n!/(s!\,t!)$ for $s = 0, 1, 2, \ldots, n$.

Let us consider a specific application of Equation (1), in which we calculate the probability that a mating between two heterozygous parents will yield exactly the expected 3 : 1 ratio of the dominant and recessive traits among sibships of a particular size. The probability $p$ of a child showing the dominant trait is 3/4, and the probability $q$ of a child showing the recessive trait is 1/4. Suppose we wanted to know how often families with eight children contain exactly six children with the dominant phenotype and two with the recessive phenotype. This is the "expected" Mendelian ratio. In this case, $n = 8, s = 6, t = 2$, and the probability of this combination of events is

$$\frac{8!}{6!\,2!}p^6q^2 = \frac{8!}{6! \times 2!}(3/4)^6(1/4)^2 = 0.31$$

That is, in only 31 percent of the families with eight children would the offspring exhibit the expected 3 : 1 phenotypic ratio; the other sibships would deviate in one direction or the other because of chance variation. The importance of this example is in demonstrating that although a 3 : 1 ratio is the "expected" outcome (and is also the single most likely outcome), the majority of the families (69 percent) actually have a distribution of offspring different from 3 : 1.

## Meaning of the Binomial Coefficient

The factorial part of the binomial expansion in Equation (1), which equals $n!/(s!\,t!)$, is called the *binomial coefficient*. As we have noted, this ratio enumerates all possible ways in which $s$ elements of one kind and $t$ elements of another kind can be arranged in order, provided that the $s$ elements and the $t$ elements are not distinguished among themselves. A specific example might include $s$ yellow peas and $t$ green peas. Although the yellow peas and the green peas can be distinguished from each other because they have different colors, the yellow peas are not distinguishable from one another (because they are all yellow), nor are the green peas (because they are all green).

The reasoning behind the factorial formula begins with the observation that the total number of elements is $s + t = n$. Given $n$ elements, each distinct from the next, the number of different ways in which they can be arranged is:

$$n \times (n-1) \times (n-2) \times \cdots \times 3 \times 2 \times 1$$

Why? Because the first element can be chosen in $n$ ways, and once it is chosen, the next can be chosen in $n - 1$ ways (because only $n - 1$ are left to choose from), and once the first two are chosen, the third can be chosen in $n - 2$ ways, and so on. Finally, once $n - 1$ elements have been chosen, there is only one way to choose the last element. The $s + t$ elements can be arranged in $n!$ ways, provided that the elements are all distinguished among themselves. However, applying again the argument we just used, each of the $n!$ particular arrangements must include $s!$ different arrangements of the $s$ elements and $t!$ different arrangements of the $t$ elements, or $s! \times t!$ altogether. Dividing $n!$ by $s! \times t!$ therefore yields the binomial coefficient for the number of ways in which the $s$ elements and the $t$ elements can be arranged when the elements of each type are not distinguished among themselves.

## 4.6 Testing Goodness of Fit to a Genetic Hypothesis

Geneticists often need to decide whether an observed ratio is in satisfactory agreement with a theoretical prediction. Mere inspection of the data is unsatisfactory because different investigators may disagree. Suppose, for example, that we crossed a plant having purple flowers with a plant having white flowers and, among the progeny, observed 14 plants with purple flowers and 6 with white flowers. Is this result close enough to be accepted as a 1 : 1 ratio? What if we observed 15 plants with purple flowers and 5 with white flowers? Is this result consistent with a 1 : 1 ratio? There is bound to be statistical variation in the observed results from one experiment to the next. Who is to say what results are consistent with a particular genetic hypothesis? In this section, we describe a test of whether observed results deviate too far from a theoretical expectation. The test is called a test for **goodness of fit,** where the word *fit* means how closely the observed results "fit," or agree with, the expected results.

### The Chi-Square Method

A conventional measure of goodness of fit is a value called **chi-square** (its symbol is $\chi^2$), which is calculated from the number of progeny observed in each of various classes, compared with the number expected in each of the classes on the basis of some genetic hypothesis. For example, in a cross between plants with purple flowers and those with white flowers, we may be interested in testing the hypothesis that the parent with purple flowers is heterozygous for one pair of alleles determining flower color and that the parent with white flowers is homozygous recessive. Suppose further that we examine 20 progeny plants from the mating and find that 14 are purple and 6 are white. The procedure for testing this genetic hypothesis (or any other genetic hypothesis) by means of the chi-square method is as follows:

1. ***State the genetic hypothesis in detail, specifying the genotypes and phenotypes of the parents and the possible progeny.*** In the example using flower color, the genetic hypothesis implies that the genotypes in the cross purple × white could be symbolized as *Pp* × *pp*. The possible progeny genotypes are *Pp* and *pp*.
2. ***Use the rules of probability to make explicit predictions of the types and proportions of progeny that should be observed if the genetic hypothesis is true. Convert the proportions to numbers of progeny (percentages are not allowed in a $\chi^2$ test).*** If the hypothesis about the flower-color cross is true, then we should expect the progeny genotypes *Pp* and *pp* to occur in a ratio of 1 : 1. Because the hypothesis is that *Pp* flowers are purple and *pp* flowers are white, we expect the phenotypes of the progeny to be purple or white in the ratio 1 : 1. Among 20 progeny, the expected numbers are 10 purple and 10 white.
3. ***For each class of progeny in turn, subtract the expected number from the observed number. Square this difference and divide the result by the expected number.*** In our example, the calculation for the purple progeny is $(14 - 10)^2/10 = 1.6$, and that for the white progeny is $(6 - 10)^2/10 = 1.6$.
4. ***Sum the result of the numbers calculated in step 3 for all classes of progeny. The summation is the value of $\chi^2$ for these data.*** The sum for the purple and white classes of progeny is $1.6 + 1.6 = 3.2$, and this is the value of $\chi^2$ for the experiment, calculated on the assumption that our genetic hypothesis is correct.

In symbols, the calculation of $\chi^2$ can be represented by the expression

$$\chi^2 = \sum \frac{(\text{Observed} - \text{Expected})^2}{\text{Expected}}$$

in which $\Sigma$ means the summation over all the classes of progeny. Note that $\chi^2$ is calculated using the observed and expected *numbers,* not the proportions, ratios, or percentages. Using something other than the actual numbers is the most common beginner's mistake in applying the $\chi^2$ method. The $\chi^2$ value is reasonable as a measure of goodness of fit, because the closer the observed numbers are to the expected numbers, the smaller the value of $\chi^2$. A value of $\chi^2 = 0$ means that the observed numbers fit the expected numbers perfectly.

As another example of the calculation of $\chi^2$, suppose that the progeny of an $F_1 \times F_1$ cross includes two contrasting phenotypes observed in the numbers 99 and 45. In this case the genetic hypothesis might be that the trait is determined by a pair of alleles of a single gene, in which case the expected ratio of dominant : recessive phenotypes among the $F_2$ progeny is 3 : 1. Considering the data, the question is whether the observed ratio of 99 : 45 is in satisfactory agreement with the expected 3 : 1. Calculation of the value of $\chi^2$ is illustrated in **TABLE 4.3**. The total number of progeny is $99 + 45 = 144$. The *expected* numbers in the two classes, on the basis of the genetic hypothesis that the true ratio is 3 : 1, are calculated as $(3/4) \times 144 = 108$ and $(1/4) \times 144 = 36$. Because there are two classes of data, there are two terms in the $\chi^2$ calculation:

$$\begin{aligned} \chi^2 &= \frac{(99 - 108)^2}{108} + \frac{(45 - 36)^2}{36} \\ &= 0.75 + 2.25 \\ &= 3.00 \end{aligned}$$

Once the $\chi^2$ value has been calculated, the next step is to interpret whether this value represents a good fit or a bad fit to the expected numbers. This assessment is done with the aid of the graphs in **FIGURE 4.22**. The $x$-axis gives the $\chi^2$ values that reflect goodness of fit, and the $y$-axis gives the probability $P$ that a worse fit (or one equally bad) would be obtained by chance, assuming that the genetic hypothesis is true. If the genetic hypothesis is true, then the observed numbers should be reasonably close to the expected numbers. Suppose that the observed $\chi^2$ is so large that the probability of a fit as bad or worse is very small. Then the observed results do *not* fit the theoretical expectations. This means that the genetic hypothesis used to calculate the expected numbers of progeny must be rejected, because the observed numbers of progeny deviate too much from the expected numbers.

In practice, the critical values of $P$ are conventionally chosen as 0.05 (the 5 percent level) and 0.01 (the 1 percent level). For $P$ values ranging from 0.01 to 0.05, the probability that chance alone would lead to a fit as bad or worse is between 1 in 20 experiments and between 1 in 100, respectively. This is the middle region in Figure 4.22; if the $P$ value falls in this range, the correctness of the genetic hypothesis is considered very doubtful. The result is said to be **statistically significant** at the 5 percent level. For $P$ values smaller than 0.01, the probability that chance alone would lead to a fit as bad or worse is less than 1 in 100 experiments. This is the lower region in Figure 4.22; in this case, the result is said to be **highly significant** at the 1 percent level, and the genetic hypothesis is rejected outright. If the terminology of statistical significance seems backward, it is because the term *significant* refers to the magnitude of the deviation between the observed and the expected numbers; in a result that is statistically significant, there is a large ("significant") difference between what is observed and what is expected.

To use Figure 4.22 to determine the $P$ value corresponding to a calculated $\chi^2$, we need the number of **degrees of freedom** of the particular $\chi^2$ test. For the type of $\chi^2$ test illustrated in Table 4.3, the number of degrees of freedom equals the number of classes of data minus 1. Table 4.3 contains two classes of data (wildtype and mutant), so the number of degrees of freedom is $2 - 1 = 1$. The reason for subtracting 1 is that, in calculating the expected numbers

**Table 4.3 Calculation of $\chi^2$ for a monohybrid ratio**

| Phenotype (class) | Observed number | Expected number | Deviation from expected | (Deviation)² / expected number |
|---|---|---|---|---|
| Wildtype | 99 | 108 | −9 | 0.75 |
| Mutant | 45 | 36 | 19 | 2.25 |
| Total | 144 | 144 | | $\chi^2 = 3.00$ |

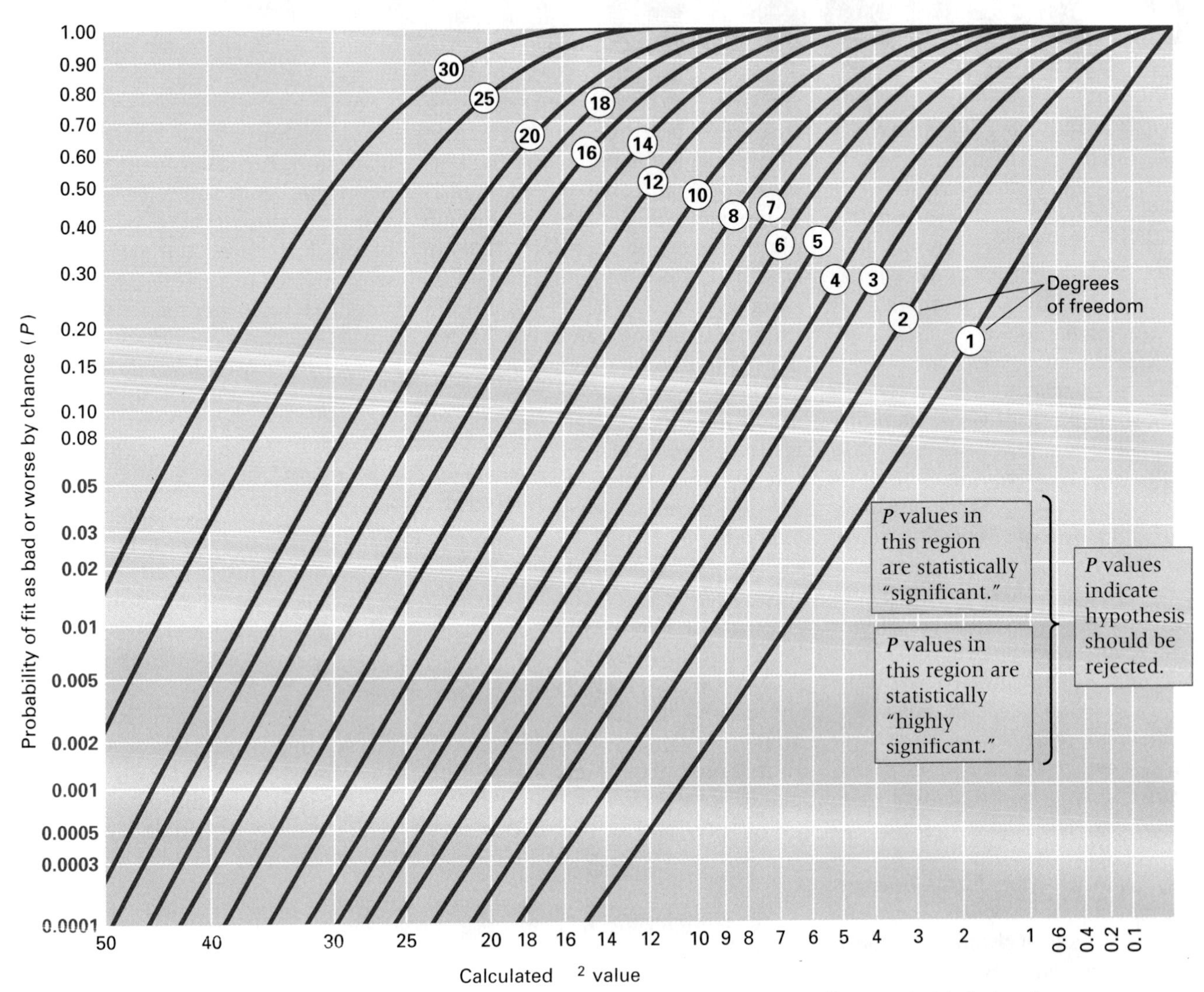


**FIGURE 4.22** Graphs for interpreting goodness of fit to genetic predictions using the chi-square test. For any calculated value of $\chi^2$ along the *x*-axis, the *y*-axis gives the probability *P* that chance alone would produce a fit as bad as or worse than that actually observed, when the genetic predictions are correct. Tests with *P* in the pink region (less than 5 percent) or in the green region (less than 1 percent) are regarded as statistically significant and normally require rejection of the genetic hypothesis that led to the prediction.

of progeny, we make sure that the total number of progeny is the same as that actually observed. For this reason, one of the classes of data is not really "free" to contain any number we might specify; because the expected number in one class must be adjusted to make the total come out correctly, one "degree of freedom" is lost. Analogous $\chi^2$ tests with three classes of data have 2 degrees of freedom, and those with four classes of data have 3 degrees of freedom.

Once we have determined the appropriate number of degrees of freedom, we can interpret the $\chi^2$ value in Table 4.3. Refer to Figure 4.22, and observe that each curve is labeled with its degrees of freedom. To determine the *P* value for the data in Table 4.3, in which the $\chi^2$ value is 3.00, first find the location of $\chi^2 = 3.00$ along the *x*-axis in Figure 4.22. Trace vertically on this line until you intersect the curve with 1 degree of freedom. Then trace horizontally to the left until you intersect the *y*-axis, and read the *P* value; in this case, $P = 0.08$. This means that chance alone would produce a $\chi^2$ value as great as or greater than 3.00 in about 8 percent of experiments of the type in Table 4.3; and, because the *P* value is within the upper region, the goodness of fit to the hypothesis of a 3 : 1 ratio of wildtype : mutant is judged to be satisfactory.

As a second illustration of the $\chi^2$ test, we will determine the goodness of fit of Mendel's round-versus-wrinkled data to the expected 3 : 1 ratio. Among the 7324 seeds that he observed, 5474 were round and 1850 were wrinkled. The expected numbers are $(3/4) \times 7324 = 5493$

# connection
## Seeds of Doubt

*R. A. Fisher, one of the founders of modern statistics, was also interested in genetics. He gave Mendel's data a thorough going over and made an "abominable discovery." Fisher's unpleasant discovery was that some of Mendel's experiments yielded a better fit to the wrong expected values than they did to the right expected values. At issue are two series of experiments consisting of progeny tests in which $F_2$ plants with the dominant phenotype were self-fertilized and their progeny examined for segregation to ascertain whether each parent was heterozygous or homozygous. In the first series of experiments, Mendel explicitly states that he cultivated 10 seeds from each plant. What Mendel did not realize, apparently, is that inferring the genotype of the parent on the basis of the phenotypes of 10 progeny introduces a slight bias. The reason is shown in the accompanying illustration. Because a fraction $(3/4)^{10}$ of all progenies from a heterozygous parent will not exhibit segregation, purely as a result of chance, this proportion of Aa parents gets misclassified as AA. The expected proportion of "apparent" AA plants is $(1/3) + (2/3)(3/4)^{10}$ and that of Aa plants is $(2/3)[1 - (3/4)^{10}]$, for a ratio of 0.37 : 0.63. In the first series of experiments, among 600 plants tested, Mendel reports a ratio of 0.335 : 0.665, which is in better agreement with the incorrect expectation of 0.33 : 0.67 than with 0.37 : 0.63. In the second series of experiments, among 473 progeny, Mendel reports a ratio of 0.32 : 0.68, which is again in better agreement with 0.33 : 0.67 than with 0.37 : 0.63. Modern scholars have concluded that Fisher most likely misinterpreted Mendel's description of the second series of experiments, and that the results of Mendel's first series of experiments actually fit Fisher's expectation very well.*

**Ronald Aylmer Fisher 1936**
University College, London, England
*Has Mendel's Work Been Rediscovered?*

In connection with these tests of homozygosity by examining ten offspring formed by self-fertilization, it is disconcerting to find that the proportion of plants misclassified by this test is not inappreciable. Between 5 and 6 percent of the heterozygous plants will be classified as homozygous. . . . Now among 600 plants tested by Mendel 201 were classified as homozygous and 399 as heterozygous. . . . The deviation [from the true expected values of 222 and 378] is one to be taken seriously. . . . A deviation as fortunate as Mendel's is to be expected once in twenty-nine trials. . . . [In the second series of experiments], a total deviation of the magnitude observed, and in the right direction, is only to be expected once in 444 trials; there is therefore a serious discrepancy. . . . If we could suppose that larger progenies, say fifteen plants, were grown on this occasion, the greater part of the discrepancy would be removed. . . . Such an explanation, however, could not explain the discrepancy observed in the first group of experiments, in which the procedure is specified, without the occurrence of a coincidence of considerable improbability. . . . The reconstruction [of Mendel's experiments] gives no doubt whatever that his report is to be taken entirely literally, and that his experiments were carried out in just the way and much in the order that they are recounted. The detailed reconstruction of his programme on this assumption leads to no discrepancy whatsoever. A serious and almost inexplicable discrepancy has, however, appeared, in that in two series of results the numbers

***The reconstruction [of Mendel's experiments] gives no doubt whatever that his report is to be taken entirely literally, and that his experiments were carried out in just the way and much in the order that they are recounted.***

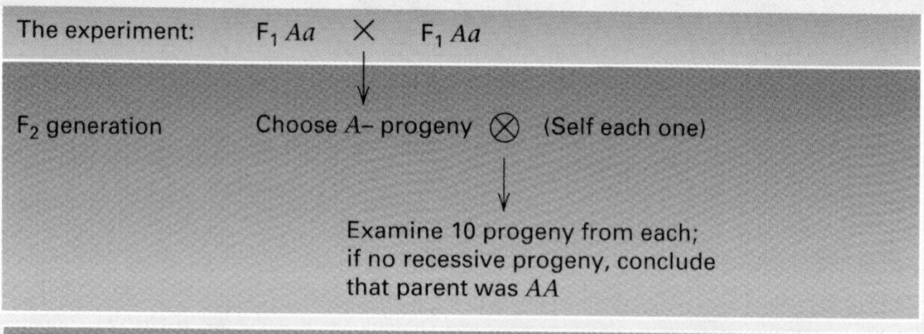


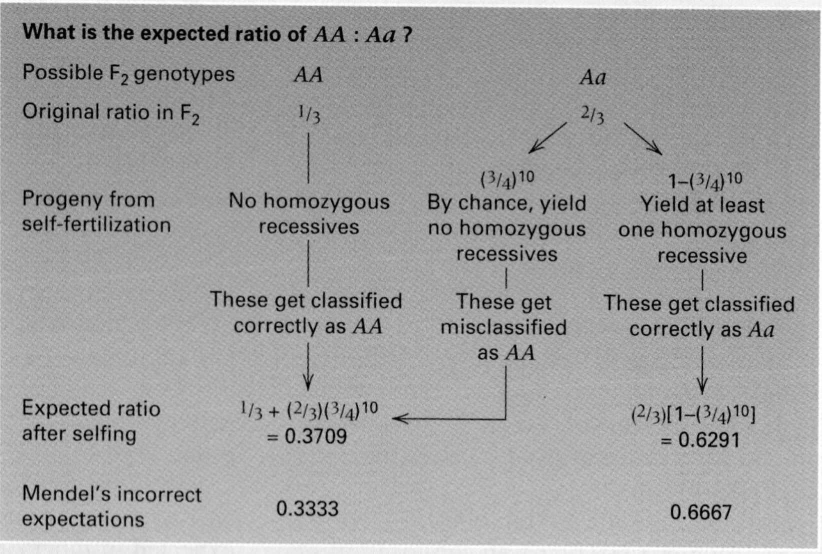

observed agree excellently with the two to one ratio, which Mendel himself expected, but differ significantly from what should have been expected had his theory been corrected to allow for the small size of his test progenies. . . . Although no explanation can be expected to be satisfactory, it remains a possibility among others that Mendel was deceived by some assistant who knew too well what was expected.

Source: R. A. Fisher, *Ann. Sci.* 1 (1936): 115-137.

round and $(1/4) \times 7324 = 1831$ wrinkled. The $\chi^2$ value is calculated as:

$$\chi^2 = \frac{(5474 - 5493)^2}{5493} + \frac{(1850 - 1831)^2}{1831}$$
$$= 0.26$$

The fact that the $\chi^2$ is less than 1 already implies that the fit is very good. To find out how good, note that the number of degrees of freedom equals $2 - 1 = 1$ because there are two classes of data (round and wrinkled). From Figure 4.22, the $P$ value for $\chi^2 = 0.26$ with 1 degree of freedom is approximately 0.65. This means that in about 65 percent of all experiments of this type, a fit as bad or worse would be expected simply because of chance. Only about 35 percent of all experiments would yield a better fit.

## Are Mendel's Data Too Good to Be True?

Many of Mendel's experimental results are very close to the expected values. For the ratios listed in Table 3.1 in Chapter 3, the $\chi^2$ values are 0.26 (round versus wrinkled seeds), 0.01 (yellow versus green seeds), 0.39 (purple versus white flowers), 0.06 (inflated versus constricted pods), 0.45 (green versus yellow pods), 0.35 (axial versus terminal flowers), and 0.61 (long versus short stems). (As an exercise in $\chi^2$, you should confirm these calculations for yourself.) All of the $\chi^2$ tests have $P$ values of 0.45 or greater (Figure 4.22), which means that the reported results are in excellent agreement with the theoretical expectations.

The statistician Ronald Fisher pointed out in 1936 that Mendel's results are *suspiciously* close to the theoretical expectations. In a large number of experiments, some experiments can be expected to yield fits that appear doubtful simply because of chance variation from one experiment to the next. In Mendel's data, the doubtful values that are to be expected appear to be missing. **FIGURE 4.23** shows the observed deviations in Mendel's experiments compared with the deviations expected by chance. (The measure of deviation is the square root of the $\chi^2$ value, assigned either a plus or a minus sign according to whether the dominant or the recessive phenotypic class was in excess of the expected number.) For each magnitude of deviation, the height of the bar on the right gives the number of experiments that Mendel observed with such a magnitude of deviation, and that of the bar on the left gives the number of experiments expected to deviate by this amount

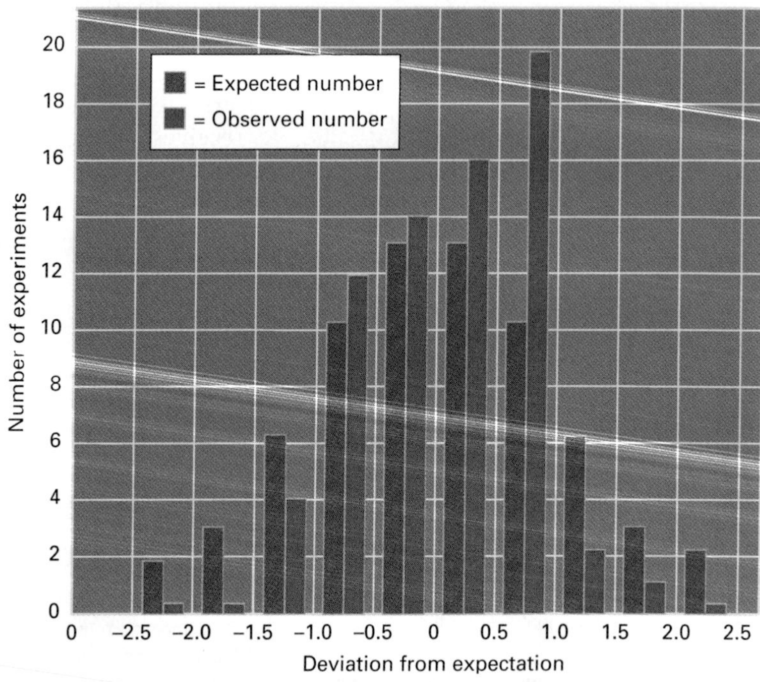


**FIGURE 4.23** Distribution of deviations observed in 69 of Mendel's experiments (yellow bars) compared with expected values (orange bars). There is no suggestion that the data in the middle have been adjusted to improve the fit, but there are fewer experiments with large deviations than might be expected. Several experiments with large deviations may have been discarded or repeated.

as a result of chance alone. There are clearly too few experiments with deviations smaller than −1 or larger than +1. This type of discrepancy could be explained if Mendel discarded or repeated a few experiments with large deviations that made him suspect that the results were not to be trusted.

Did Mendel cheat? Did he deliberately falsify his data to make them appear better? Mendel's paper reports extremely deviant ratios from individual plants, as well as experiments repeated when the first results were doubtful. These are not the kinds of things a dishonest person would admit. Only a small bias is necessary to explain the excessive goodness of fit in Figure 4.23. In a count of seeds or individual plants, only about 2 phenotypes per 1000 would need to be assigned to the wrong category to account for the bias in the 91 percent of the data generated by the testing of monohybrid ratios. The excessive fit could also be explained if three or four entire experiments were discarded or repeated because deviant results were attributed to pollen contamination or other accident. After careful reexamination of Mendel's data in 1966, the evolutionary geneticist Sewall Wright concluded,

> Mendel was the first to count segregants at all. It is rather too much to expect that he would be aware of the precautions now known to be necessary for completely objective data. . . . Checking of counts that one does not like, but not of others, can lead to systematic bias toward agreement. I doubt whether there are many geneticists even now whose data, if extensive, would stand up wholly satisfactorily under the $\chi^2$ text. . . . Taking everything into account, I am confident that there was no deliberate effort at falsification.

Mendel's data are some of the most extensive and complete "raw data" ever published in genetics. Additional examinations of the data will surely be carried out as new statistical approaches are developed. However, the principal point to be emphasized is that up to the present time, no reputable statistician has alleged that Mendel knowingly and deliberately adjusted his data in favor of the theoretical expectation.

## CHAPTER SUMMARY

- Chromosomes in eukaryotic cells are usually present in pairs.
- The chromosomes of each pair separate in meiosis, one going to each gamete.
- In meiosis, the chromosomes of different pairs undergo independent assortment because nonhomologous chromosomes move independently.
- In many animals, sex is determined by a special pair of chromosomes, the X and Y.
- The "criss-cross" pattern of inheritance of X-linked genes is determined by the fact that a male receives his X chromosome only from his mother and transmits it only to his daughters.
- Irregularities in the inheritance of an X-linked gene in *Drosophila* gave experimental proof of the chromosomal theory of heredity.
- The progeny of genetic crosses follow the binomial probability formula.
- The chi-square statistical test is used to determine how well the observed genetic data agree with the expectations derived from a hypothesis.

## REVIEW THE BASICS

- What is the genetic significance of the fact that gametes contain half the chromosome complement of somatic cells?
- The term *mitosis* derives from the Greek *mitos*, which means "thread." The term *meiosis* derives from the Greek *meioun*, which means "to make smaller." What feature, or features, of these types of nuclear division might have led to the choice of these terms?
- Explain the meaning of the terms *reductional division* and *equational division*. What is reduced or kept equal? To which nuclear divisions do the terms refer?
- Explain the following statement: "Independent alignment of nonhomologous chromosomes at metaphase I of meiosis is the physical basis of independent assortment of genes in different chromosomes."
- What are some of the important differences between the first meiotic division and the second meiotic division?
- Draw a diagram of a bivalent and label the following parts: centromere, sister chromatids, nonsister chromatids, homologous chromosomes, chiasma.
- What are the principal characteristics of human pedigrees in which a rare X-linked recessive allele is segregating?
- In what ways is the inheritance of a Y-linked gene different from that of an X-linked gene?
- How did nondisjunction "prove" the chromosome theory of heredity?
- Why is a statistical test necessary to determine whether an observed set of data yields an acceptable fit to the result expected from a particular genetic hypothesis?
- What statistical test is often used for this purpose?
- What are the conventional *P* values for significant and highly significant, and what do these numbers mean?

**Problem 1** A recessive mutation in an X-linked gene results in hemophilia, marked by a prolonged increase in the time needed for blood to clot. Suppose that two phenotypically normal parents produce three normal daughters and a son affected with hemophilia.

**(a)** What is the probability that all of the daughters are heterozygous carriers?

**(b)** If one of the daughters mates with a normal male and produces a son, what is the probability that the son will be affected?

**Answer**

**(a)** Because the phenotypically normal parents have an affected son who gets his only X chromosome from his mother, the mother must be a carrier of the mutation. Therefore, the probability that any particular daughter is a carrier is $\frac{1}{2}$. The probability that all three daughters are carriers is $(\frac{1}{2})^3$ because their births are independent events.

**(b)** If the daughter is not a carrier, the probability of an affected son is 0; and if the daughter is a carrier, the probability of an affected son is $\frac{1}{2}$. Because the probability of the daughter being a carrier is $\frac{1}{2}$ (part **a**), the overall probability of an affected son is $(\frac{1}{2}) \times 0 + (\frac{1}{2}) \times (\frac{1}{2}) = \frac{1}{4}$.

**Problem 2** In a sibship with seven children, what is the probability that it includes four boys and three girls or three boys and four girls? Assume that each child has an equal likelihood of being a boy or a girl.

**Answer** The probability that the sibship consist of four boys and three girls, in any order of birth, equals $[7!/(4!3!)](\frac{1}{2})^4(\frac{1}{2})^3$, where the factor in square brackets is the number of possible birth orders of four boys and three girls. This probability works out to $\frac{35}{128}$. Likewise, the probability that the sibship consists of three boys and four girls is $[7!/(3!4!)](\frac{1}{2})^3(\frac{1}{2})^4$, which also equals $\frac{35}{128}$. The question asks for the probability of either four boys and three girls or three boys and four girls, and these events are mutually exclusive; hence the overall probability is $\frac{35}{128} + \frac{35}{128} = \frac{35}{64}$, or about 55%.

**Problem 3** A geneticist carries out a cross between two strains of fruit flies that are heterozygous for a recessive allele of each of two genes, *st* (scarlet) and *bw* (brown), affecting eye color. Flies homozygous for *st* have bright red (scarlet) eyes, whereas $st^+/st^+$ and $st^+/st$ genotypes have wildtype brick-red eyes. Flies homozygous for *bw* have brown eyes, whereas $bw^+/bw^+$ and $bw^+/b$ genotypes have wildtype brick-red eyes. The double homozygous genotype *st/st; bw/bw* results in white eyes. In a test of independent assortment of these two genes, a geneticist crosses $st^+/st$; $bw^+/bw$ females with $st^+/st$; $bw^+/bw$ males. Among 240 progeny there are 150 flies with wildtype eyes, 36 with scarlet eyes, 46 with brown eyes, and 8 with white eyes.

**(a)** Under the null hypothesis that these genes undergo independent assortment, what are the expected numbers in each phenotypic class?

**(b)** What is the value of the chi-square in a test of goodness of fit between the observed values and the expected values based on the null hypothesis of independent assortment?

**(c)** How many degrees of freedom does this chi-square value have?

**(d)** What is the *P*-value for the chi-square value in this goodness-of-fit test? Does this *P*-value support the null hypothesis of independent assortment or should the hypothesis be rejected?

**(e)** Would a greater chi-square value increase or decrease the *P*-value?

**Answer**

**(a)** Because this is a dihybrid cross $st^+/st$; $bw^+/bw \times st^+/st$; $bw^+/bw$, the ratio of 9 : 3 : 3 : 1 of the offspring phenotypes should be expected if the genes assort independently. We can calculate the expected number of flies in each class of the progeny:

| Phenotype | Progeny | Expected number |
|---|---|---|
| wildtype | $st^+/-$; $bw^+/-$ | (9/16) × 240 = 135 |
| scarlet | $st/st$; $bw^+/-$ | (3/16) × 240 = 45 |
| brown | $st^+/-$; $bw/bw$ | (3/16) × 240 = 45 |
| white | $st/st$; $bw/bw$ | (1/16) × 240 = 15 |

**(b)** The chi-square is calculated as

$$\chi^2 = \sum \frac{(\text{observed} - \text{expected})^2}{\text{expected}}$$

where the sum is over all classes of progeny. In this case,

$$\chi^2 = \frac{(150-135)^2}{135} + \frac{(36-45)^2}{45} + \frac{(46-45)^2}{45} + \frac{(8-15)^2}{15} = 6.76$$

**(c)** The number of degrees of freedom equals the number of classes of data minus 1. In this case there are four classes of data; thus there are three degrees of freedom.

**(d)** The *P*-value for a chi-square value of 6.76 with 3 degrees of freedom equals 0.08. This *P*-value is greater than 0.05. Therefore, we should not reject the null hypothesis of independent assortment.

**(e)** A greater chi-square value would decrease the *P*-value.

**Problem 4** The mutation for bar-shaped eyes in *Drosophila melanogaster* shows the following features of genetic transmission:

**(a)** The mating of bar-eyed males with wildtype females produces wildtype sons and bar-eyed daughters.

**(b)** The bar-eyed females from the mating in part **a**, when mated with wildtype males, yield a 1 : 1 ratio of bar : wildtype sons and a 1 : 1 ratio of bar : wildtype daughters.

What mode of inheritance do these characteristics suggest?

**Answer** Because the sexes are affected unequally in the progeny of the mating in **a**, some association with the sex chromosomes is suggested. The progeny from mating **a** provide the important clue. Because a male receives his X-chromosome from his mother, then the observation that all males are wildtype suggests that the gene for bar eyes is on the X-chromosome. The fact that all daughters are affected is also consistent with X-linkage, provided that the *bar* mutation is dominant. Mating **b** confirms the hypothesis of a dominant X-linked gene, because the females from a mating **a** would have the genotype *bar*/+ and so would produce the observed progeny. The data are therefore consistent with the *bar* mutation being an X-linked dominant.

## ANALYSIS AND APPLICATIONS

**4.1** The diagrams below illustrate a pair of homologous chromosomes in prophase I of meiosis. Which diagram corresponds to each stage: leptotene, zygotene, pachytene, diplotene, and diakinesis?

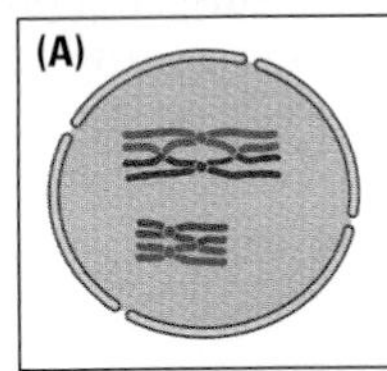

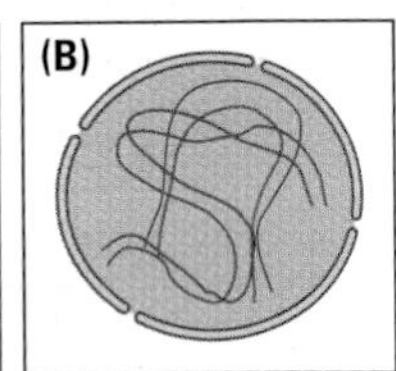

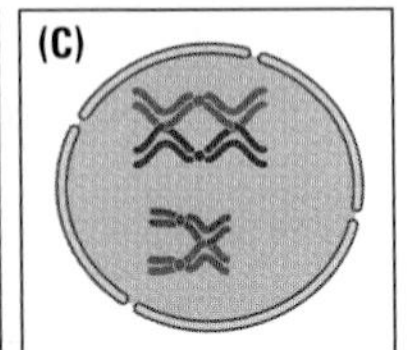

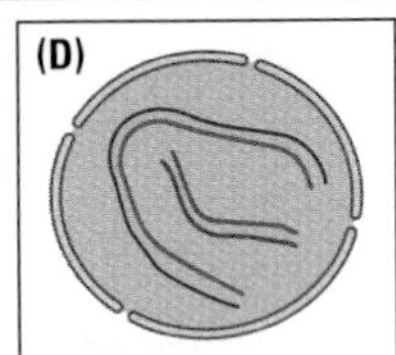

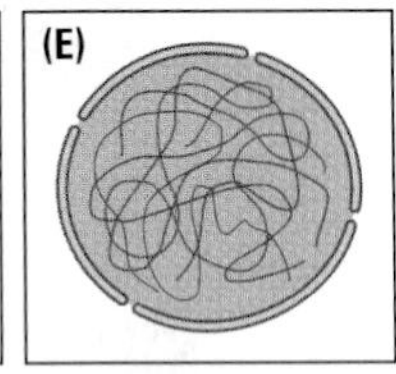


**4.2** The diagrams below illustrate anaphase in an organism that has two pairs of homologous chromosomes. Identify the stages as mitosis, meiosis I, or meiosis II.

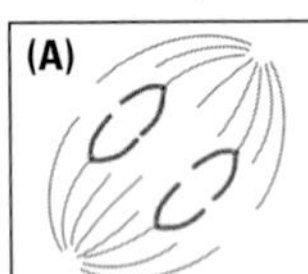

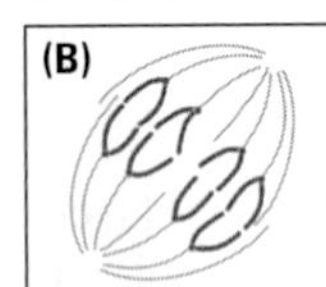

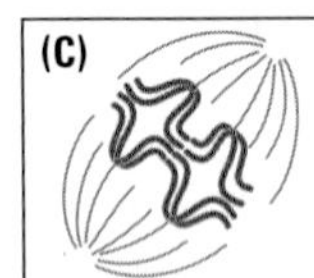


**4.3** What is the probability that four offspring from the mating *Aa* × *Aa* consist of exactly 3 *A*− and 1 *aa*?

**4.4** A woman who is heterozygous for both a phenylketonuria mutation and an X-linked hemophilia mutation has a child with a phenotypically normal man who is also heterozygous for a phenylketonuria mutation. What is the probability that the child will be affected by both diseases? (Assume that they are equally likely to have a boy or a girl.)

**4.5** In the accompanying pedigree of X-linked Duchenne muscular dystrophy, what is the probability that the woman III-3 is a heterozygous carrier?

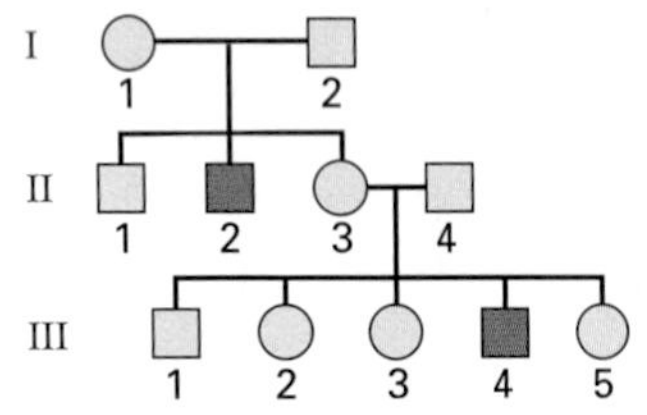


**4.6** A chromosomally normal woman and a chromosomally normal man have a son whose sex-chromosome constitution is XYY. In which parent, and in which meiotic division, did the nondisjunction take place?

**4.7** *Drosophila virilis* is a diploid organism with 6 pairs of chromosomes (12 chromosomes altogether). How many chromatids and chromosomes are present in the following stages of cell division:
(a) Metaphase of mitosis?
(b) Metaphase I of meiosis?
(c) Metaphase II of meiosis?

**4.8** In a trihybrid cross with genes that undergo independent assortment:
(a) What is the expected proportion of triply heterozygous offspring in the $F_2$ generation?
(b) What is the expected proportion of triply homozygous offspring in the $F_2$ generation?

**4.9** The diploid gekkonid lizard *Gonatodes taniae* from Venezuela has a somatic chromosome number of 16. If the centromeres of the 8 homologous pairs are designated as *Aa, Bb, Cc, Dd, Ee, Ff, Gg,* and *Hh*:
(a) How many different combinations of centromeres could be produced during meiosis?
(b) What is the probability that a gamete will contain only those centromeres designated by capital letters?

**4.10** Among sibships consisting of six children, and assuming a sex ratio of 1 : 1:
(a) What is the proportion with no girls?
(b) What is the proportion with exactly one girl?
(c) What is the proportion with exactly two girls?
(d) What is the proportion with exactly three girls?
(e) What is the proportion with three or more boys?

**4.11** A litter of cats includes eight kittens. What is the probability that it contains an even number each of males and females? Assume an equal likelihood of male and female kittens, and for purposes of this problem regard 0 as an even number. Does the answer surprise you? Why?

**4.12** A horticulturalist crossed a true-breeding onion plant with red bulbs to a true-breeding onion plant with white bulbs. All of the $F_1$ plants had white bulbs. When seeds resulting from self-fertilization of the $F_1$ plants were grown, onion bulbs were recovered in the ratio of 12 white bulbs : 3 red bulbs : 1 yellow bulb. Propose a hypothesis to explain these observations.

**4.13** Among 160 progeny in the $F_2$ generation of a dihybrid cross, a geneticist observes four distinct phenotypes in the ratio 91: 21 : 37 : 11. She believes this result may be consistent with a ratio of 9 : 3 : 3 : 1. To test this hypothesis, she calculates the chi-square value. Does the test support her hypothesis, or should she reject it?

**4.14** What is the value of the chi-square that tests goodness of fit between the observed numbers 40 : 60 as compared with the expected numbers 50 : 50?

**4.15** What is the probability that a sibship of seven children includes at least one boy and at least one girl? Assume that a sex ratio is 1 : 1.

**4.16** How many genotypes are possible for an autosomal gene with six alleles? How many genotypes are possible with an X-linked gene with six alleles?

**4.17** The growth habit of the Virginia groundnut *Arachis hypogaea* can be "runner" (spreading) or "bunch" (compact). Two pure-breeding strains of groundnuts with the bunch growth habit are crossed. The $F_1$ plants have the runner growth habit. If the plants are allowed to self-fertilize, the $F_2$ progeny ratio is 9 runner: 7 bunch. What genetic hypothesis can account for these observations?

**4.18** In some human pedigrees, the blue/brown eye color variation segregates like a single-gene difference with the brown allele dominant to the blue allele. Two brown-eyed individuals each of whom had a blue-eyed parent mate. Assuming that the trait segregates like a single-gene difference in this pedigree:

**(a)** What is the probability that the first child will have brown eyes?

**(b)** If a brown-eyed child results, what is the probability that the child will be heterozygous?

**(c)** What is the probability that this couple will have three children with blue eyes? That none of the three children will have blue eyes?

**4.19** In the pedigree below, the male I-2 is affected with red-green color blindness owing to an X-linked mutation. What is the probability that male IV-1 is color blind? (Assume that the only possible source of the color blindness mutation in the pedigree is that from male I-2.)

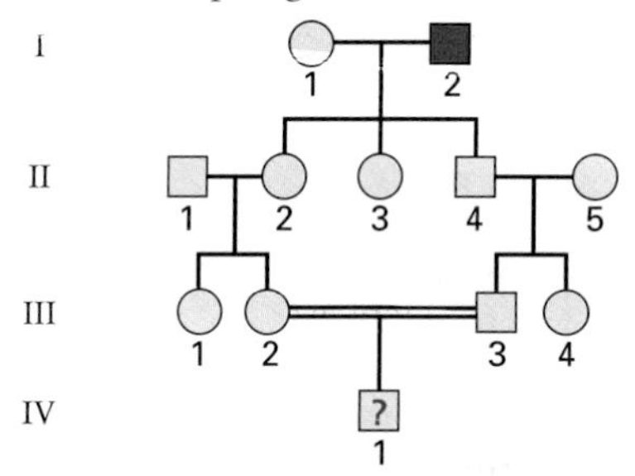


**4.20** The male II-1 in the pedigree shown here is affected with a trait due to a rare X-linked recessive allele. Use the principles of conditional probability to calculate the probability that the female III-1 is a heterozygous carrier.

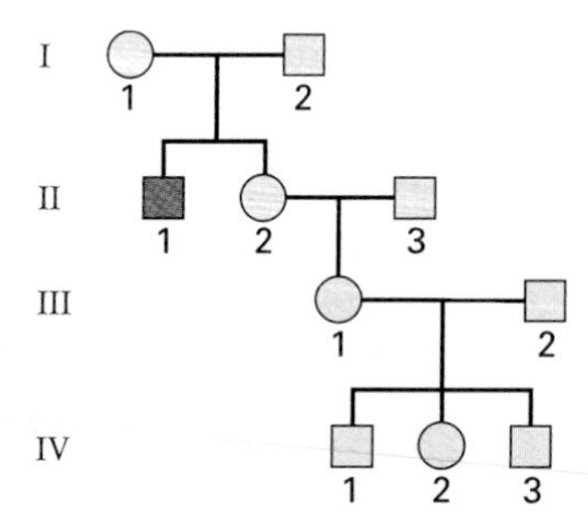


**4.21** The pedigree shown below is the same as that in Problem 4.20 but with additional information based on molecular analysis. The male II-1 is again affected with a trait due to a rare X-linked recessive allele. The bands in the gel are restriction fragments that identify the wildtype allele (the shorter fragment S corresponding to the band near the bottom of the gel) or the mutant allele (the longer fragment L corresponding to the band near the top of the gel). From the information given in the pedigree and the gels, calculate the probability that the female III-1 is a heterozygous carrier.

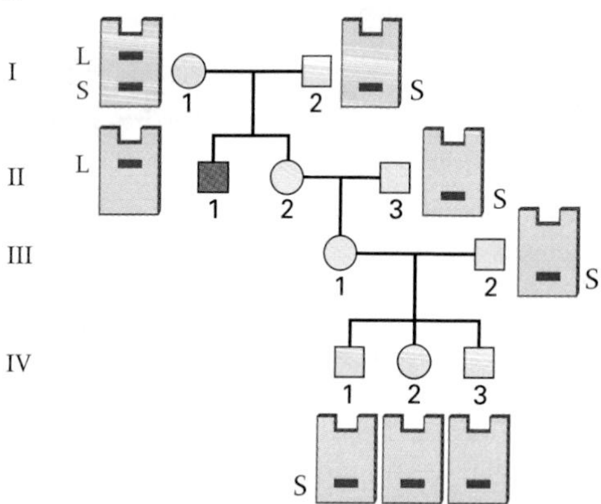


**4.22** In the pedigree shown here, the males I-2 and III-2 are affected with a trait due to a common X-linked recessive allele. What is the probability that individual III-1 is heterozygous?

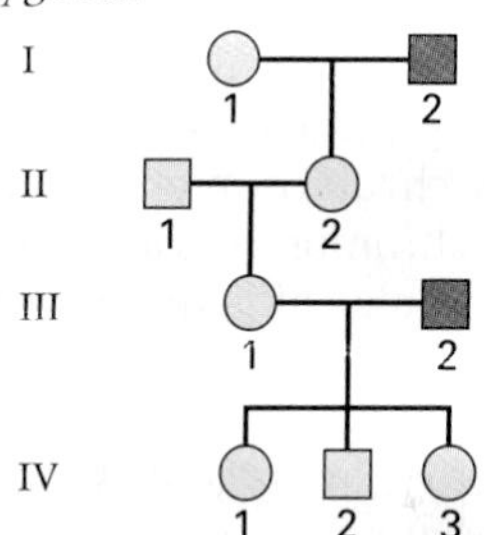


**4.23** Yellow body color in *Drosophila* is determined by the recessive allele *y* of an X-linked gene, and the wildtype gray body color is determined by the $y^+$ allele. What genotype and phenotype ratios would be expected from the following crosses?

**(a)** yellow male × wildtype female

**(b)** yellow female × wildtype male

**(c)** daughter from mating in part **a** × wildtype male

**(d)** daughter from mating in part **a** × yellow male

**4.24** The accompanying pedigree and gel diagram show the molecular phenotypes for an RFLP with two alleles. What mode of inheritance does the pedigree suggest? Based on this hypothesis, and using $A_1$ to represent the RFLP allele associated with the 3-kb band and $A_2$ to represent that associated with the 8-kb band, deduce the genotype of each individual in the pedigree.

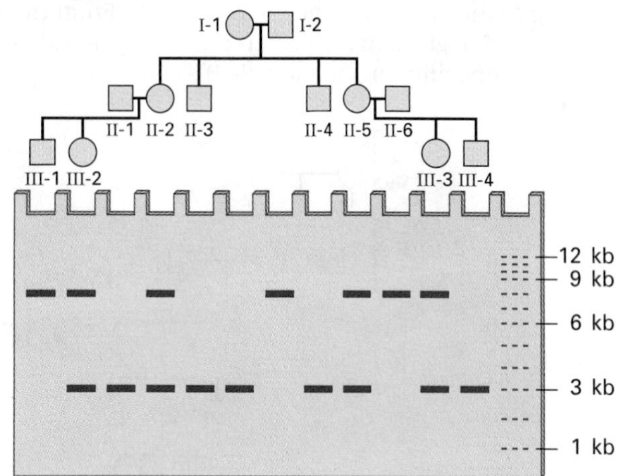


**4.25** Tall, yellow-flowered crocus is mated with short, white-flowered crocus. Both varieties are true breeding. All the $F_1$ plants are backcrossed with the short, white-flowered variety. This backcross yielded 800 progeny in the proportions 234 tall yellow, 203 tall white, 175 short yellow, and 188 short white plants. Does the observed result fit the genetic hypothesis of 1 : 1 : 1 : 1 segregation as assessed by a $\chi^2$ test?

**4.26** Attached-X chromosomes in *Drosophila* are formed from two X chromosomes attached to a common centromere. Females of genotype C(1)RM/Y, in which C(1)RM denotes the attached-X chromosomes, produce C(1)RM-bearing and Y-bearing gametes in equal proportions. What progeny is expected to result from the mating between a male carrying the X-linked allele, *y*, for yellow body and an attached-X female with wildtype body? How does this result differ from the typical pattern of X-linked inheritance? (Note: *Drosophila* zygotes containing three X chromosomes or no X chromosomes do not survive.)

**4.27** A rare dominant autosomal mutation *W* results in wooly, curled hair in some European pedigrees. A woman with wooly hair with blood group O marries a man with straight hair (*ww*) with blood group AB. The genes are in different chromosomes.

**(a)** What are the chances that the mating will produce a wooly-haired group B child?

**(b)** What are the chances that the mating will produce a straight-haired group B child?

**(c)** If three straight-haired children with blood group A children are born to these parents, what are the chances that the next child born will be wooly-haired and blood group B?

**4.28** Ducklings of the domesticated mallard *Anas platyrhynchos* may exhibit dark brown dorsal plumage known as mallard, almost black dorsal plumage known as dusky, or a pattern known as restricted in which the black is confined to patches on the head and tail. These phenotypes result from the action of three alleles of a single autosomal gene. Three types of crosses were made, with the following results:

1. Restricted × mallard: all $F_1$ are restricted; crosses of $F_1 \times F_1$ yield an $F_2$ generation with the ratio 3 restricted : 1 mallard.
2. Mallard × dusky: all $F_1$ are mallard; crosses of $F_1 \times F_1$ produce an $F_2$ generation with the ratio 3 mallard : 1 dusky.
3. Restricted × dusky: all $F_1$ are restricted; crosses of $F_1 \times F_1$ produce an $F_2$ generation with the ratio 3 restricted : 1 dusky.

**(a)** Assume that an $F_1$ male from cross 1 is mated with an $F_1$ female from cross 2. List the phenotypes and their expected frequencies among the progeny of this cross.

**(b)** Assume that an $F_1$ male from cross 3 is mated with an $F_1$ female from cross 2. List the phenotypes and their expected frequencies among the progeny of this cross.

**4.29** In cattle, an allele for the absence of horns shows complete dominance: *HH* and *Hh* are hornless or polled, and *hh* is horned. On the other hand, the effect of the allele producing red coat (*R*) shows incomplete dominance with the allele producing white coat color (*r*). The heterozygous genotype *Rr* is roan colored, an intermediate color in which white hairs are intermixed with red hairs. *H* and *R* undergo independent assortment.

**(a)** What would be the phenotype of an $F_1$ offspring from the mating *RR HH* × *rr hh*?

**(b)** What would be the phenotypes and their expected proportions in the $F_2$ progeny of the cross $F_1 \times F_1$ from part **a**?

**(c)** What would be the phenotypes and their expected proportions among the progeny derived from crossing $F_1$ individuals from part **a** to the original horned white stock?

**4.30** Familial Mediterranean fever (FMF) is a hereditary inflammatory disorder that affects groups of patients originating from around the Mediterranean Sea, especially Armenian and non-Ashkenazi Jewish populations, Anatolian Turks, and Levantine Arabs. FMF is inherited as an autosomal recessive. It is not known whether the gene causing the disease is the same in these different populations. To investigate this possibility, linkage of the allele to a series of VNTR (variable number of tandem repeat) polymorphisms was examined. One VNTR in chromosome 16 was especially informative. For this VNTR, two extended Armenian families from the same geographic region gave the results shown in the accompanying pedigrees and gels.

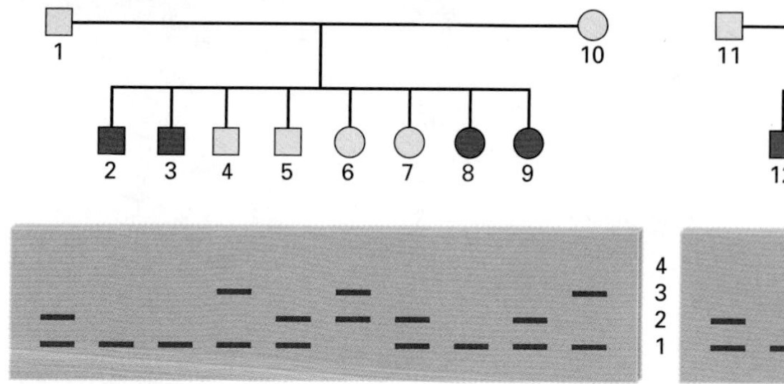


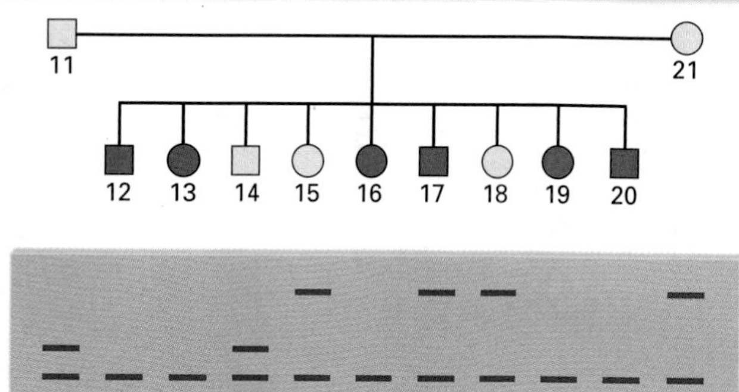


The numbers between the gels correspond to the VNTR alleles.

**(a)** Is there evidence for linkage of the allele for FMF to any VNTR allele?

**(b)** Do any of the data conflict with your explanation? If so, how do you account for it?

## CHALLENGE PROBLEMS

**Challenge Problem 1** The accompanying pedigree and gel diagram pertain to a morphological phenotype (shaded symbols in the pedigree) and a molecular phenotype (a two-allele RFLP). What mode of inheritance do these data suggest for each trait?

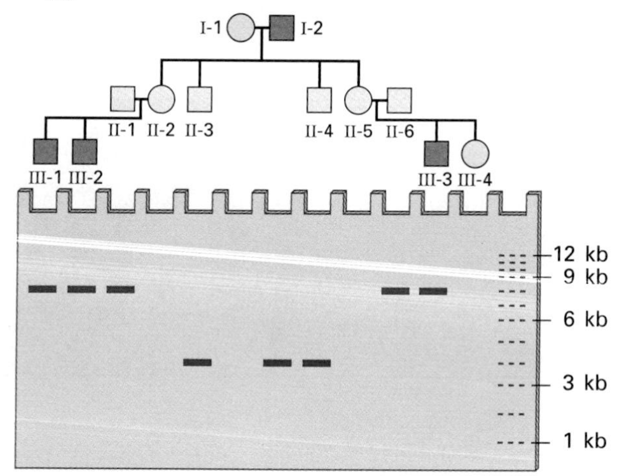


**Challenge Problem 2** In mice, the dominant $T$ allele results in a short tail, but the homozygous $T/T$ genotype is lethal. A cross between two short-tailed mice produces a litter of five pups. What is the expected distribution of tailless to tailed mice? Why are the two most likely outcomes equally probable?

**Challenge Problem 3** X-linked hemophilia is present in the pedigree illustrated here. What is the probability that the woman III-5 is a carrier, taking into account the information that she has had two normal sons? What is the probability that her next son will be affected?

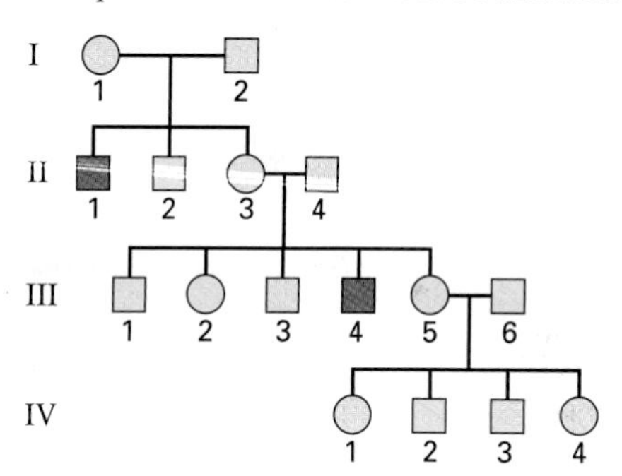


## GENETICS *on the web*

GeNETics on the Web will introduce you to some of the most important sites for finding genetics information on the Internet. To explore these sites, visit the Jones and Bartlett companion site to accompany *Genetics: Analysis of Genes and Genomes, Seventh Edition* at http://biology.jbpub.com/book/genetics.

There you will find a chapter-by-chapter list of highlighted keywords. When you select one of the keywords, you will be linked to a Web site containing information related to that keyword.

# CHAPTER 5

# Genetic Linkage and Chromosome Mapping

Genetic mapping identifies the relative positions of genetic markers along a chromosome based on the frequency of recombination between the genetic markers. In modern genetic maps, most of the genetic markers are molecular markers such as single-nucleotide polymorphisms, restriction-fragment length polymorphisms, or simple-sequence repeat polymorphisms.



## CHAPTER OUTLINE

**5.1 Linkage and Recombination of Genes in a Chromosome**
- Coupling Versus Repulsion of Syntenic Alleles
- The Chi-Square Test for Linkage
- Each Pair of Linked Genes Has a Characteristic Frequency of Recombination
- Recombination in Females Versus Males

**5.2 Genetic Mapping**
- Map Distance and Frequency of Recombination
- Crossing Over
- Recombination Between Genes Results from a Physical Exchange Between Chromosomes
- Crossing Over Takes Place at the Four-Strand Stage of Meiosis
- Multiple Crossovers

**5.3 Genetic Mapping in a Three-Point Testcross**
- Chromosome Interference in Double Crossovers
- Genetic Mapping Functions
- Genetic Map Distance and Physical Distance

**5.4 Genetic Mapping in Human Pedigrees**
- Maximum Likelihood and Lod Scores

**5.5 Mapping by Tetrad Analysis**
- Analysis of Unordered Tetrads
- Genetic Mapping with Unordered Tetrads
- Analysis of Ordered Tetrads

**5.6 Special Features of Recombination**
- Recombination Within Genes
- Mitotic Recombination

CONNECTION **Genes All in a Row**
Alfred H. Sturtevant 1913
*The Linear Arrangement of Six Sex-Linked Factors in* Drosophila, *as Shown by Their Mode of Association*

CONNECTION **Human Gene Map**
Jeffrey C. Murry and 26 other investigators 1994
*A Comprehensive Human Linkage Map with Centimorgan Density*

Because homologous chromosomes behave as units when they undergo meiosis, one might expect that genes located in the same chromosome would not undergo the same pattern of independent assortment observed in nonhomologous chromosomes. Genes that are always transmitted together are said to show complete **linkage.** Pioneering studies of this issue were carried out by Thomas Hunt Morgan, who studied mutations located in the X chromosome of *Drosophila*. He did observe linkage, but what he observed was an incomplete form of linkage, evidenced by progeny who exhibited a combination of traits not exhibited in the parents.

Linkage between genes is usually incomplete because homologous chromosomes can undergo an exchange of segments when they are paired. So, while the alleles present in each chromosome do tend to remain together in inheritance, some chromosomes can be transmitted that have new combinations of alleles. An exchange event (crossing over) between homologous chromosomes results in the **recombination** of genes, yielding daughter chromosomes that carry combinations of alleles not present in the parental chromosomes. What we learn in this chapter is that the probability of crossing over between any two genes serves as a measure of genetic distance between the genes. This makes it possible to construct a *genetic map*, a diagram of a chromosome showing the relative positions of the genes. The genetic mapping of linked genes is an important research tool in genetics because it enables a new gene to be assigned to a chromosome and even to a precise position relative to other genes in the same chromosome. Genetic mapping is often a first step in the identification and isolation of a new gene. Genetic mapping is essential in human genetics for the identification of genes associated with hereditary diseases, such as the genes whose presence predisposes women carriers to the development of breast cancer.

## 5.1 Linkage and Recombination of Genes in a Chromosome

As we saw in Chapter 4, a direct test of independent assortment is to carry out a testcross between an $F_1$ double heterozygote (*Aa Bb*) and the double recessive homozygote (*aa bb*). Shown in **FIGURE 5.1** are the expected gametes from the *Aa Bb* parent when the genes are in different chromosomes. Because the possible metaphase I alignments of homologous chromosomes are equally likely, the double heterozygote produces all four possible types of gametes in equal proportions. The alleles present in a gamete are typically written in a row with a space between each gene; in this case the four

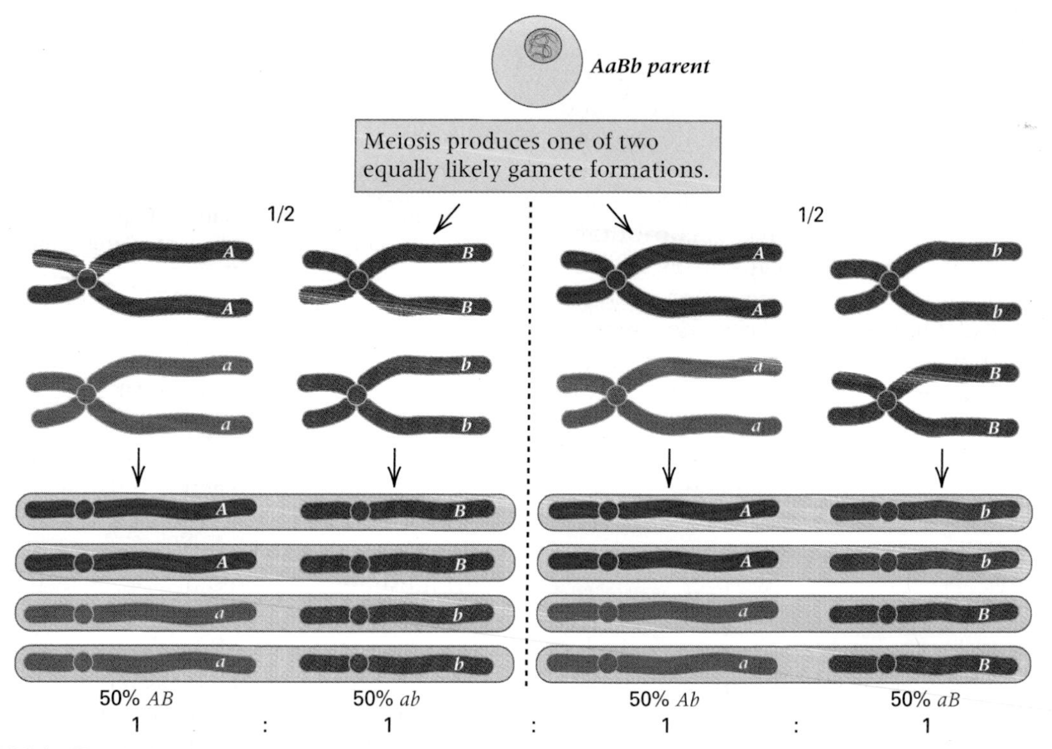


**FIGURE 5.1** The physical basis of the independent assortment of genes in different chromosomes is that nonhomologous chromosomes orient independently at metaphase I of meiosis. In this example, the orientation on the left results in *A B* and *a b* gametes, whereas that on the right results in *A b* and *a B* gametes.

equally likely types of gametes are *A B*, *A b*, *a B*, and *a b*.

In Morgan's early experiments with *Drosophila*, he examined the segregation of the X-linked alleles *w* versus $w^+$, which determine white versus normal red eyes, and the alleles *m* versus $m^+$, which determine miniature versus normal wing size. When genes are linked, it is important to keep track of which alleles are present together in the same chromosome. This is generally done by using a slash (also called a virgule) to separate the alleles present in homologous chromosomes. Morgan's initial cross was between females with white eyes and normal wings and males with red eyes and miniature wings. Using the slash, we would write this cross as:

$$w\ m^+/w\ m^+\ ♀♀ \times w^+\ m/\text{Y}\ ♂♂$$

The resulting $F_1$ progeny consisted of wildtype females (genotype $w\ m^+/w^+\ m$) and white-eyed, nonminiature males (genotype $w\ m^+/\text{Y}$). Then the $F_1$ progeny were crossed:

$$w\ m^+/w^+\ m\ ♀♀ \times w\ m^+/\text{Y}\ ♂♂$$

In this cross, X chromosomes carrying either the allele combination $w\ m^+$ or the allele combination $w^+\ m$ are called **parental types** of chromosomes; this terminology reflects the fact that parents involved in the cross carry the alleles in these configurations. If the *w* and *m* genes were completely linked, then all of the offspring would inherit one or the other of the parental types of chromosome. On the other hand, if recombination between the *w* and *m* genes takes place, then some of the progeny will inherit an X chromosome with either the allele combination $w^+\ m^+$ or the allele combination *w m*. These are known as **recombinant types** of chromosomes because they contain allele combinations that are different from those present in the parental chromosomes.

When the cross displayed above was carried out, the female progeny were all nonminiature with a 1 : 1 ratio of red : white eyes (if you do not understand why, write out the Punnett square), and the male progeny were as follows:

| | | |
|---|---|---|
| white eye, normal wing ($w\ m^+/\text{Y}$) | 226 | 66.5 percent parental types |
| red eye, miniature wing ($w^+\ m/\text{Y}$) | 202 | |
| red eye, normal wing ($w^+\ m^+/\text{Y}$) | 114 | 33.5 percent recombinant types |
| white eye, miniature wing ($w\ m/\text{Y}$) | 102 | |
| | 644 | |

Because each male receives his X chromosome from his mother, a male's phenotype reveals the genotype of the X chromosome that he inherited. The results of the experiment show a great departure from the 1 : 1 : 1 : 1 ratio of the four male phenotypes expected with independent assortment, yet the ratios of $w : w^+$ and of $m : m^+$ are both approximately 1 : 1. This type of deviation from independent assortment shows that the parental combination of the alleles for the *w* and *m* genes tend to remain together in inheritance, but they are not completely linked. In this experiment, the combinations of alleles in the parental chromosomes were present in 428/644 = 66.5 percent of the $F_2$ males, and recombinant (nonparental) chromosomes were present in 216/644 = 33.5 percent of the $F_2$ males. The 33.5 percent value for the recombinant chromosomes is called the **frequency of recombination**; it can also be written as 0.335. This frequency of recombination should be contrasted with both the 50 percent recombination expected with independent assortment and the 0 percent recombination expected with complete linkage.

The recombinant X chromosomes $w^+m^+$ and *w m* result from crossing over in meiosis in $F_1$ females. In this example, the frequency of recombination between the linked *w* and *m* genes was 33.5 percent, but with other pairs of linked genes it ranges from near 0 to 50 percent. Even genes in the same chromosome, if they are sufficiently far apart, can undergo independent assortment, which means that their frequency of recombination equals 50 percent. This implies the following principle:

> Genes with recombination frequencies less than 50 percent are present in the same chromosome (linked). Two genes that undergo independent assortment, indicated by a recombination frequency equal to 50 percent, either are in nonhomologous chromosomes or are located far apart in a single chromosome.

The possibility that two genes in the same chromosome may be unlinked creates a problem in terminology. We cannot say all genes in the same chromosome are "linked," because linkage means a recombination frequency of less than 50 percent, and genes that are sufficiently far apart are actually unlinked. The term that is used instead is *synteny:* Two genes are said to be **syntenic** if they are in the same chromosome, regardless of whether they show independent assortment or linkage.

## Coupling Versus Repulsion of Syntenic Alleles

Geneticists use a slash notation for linked genes that has the general form $w^+ \, m/w \, m^+$. This notation is a simplification of the following more descriptive but cumbersome form:

$$\frac{w^+ \qquad m}{w \qquad m^+}$$

In this form of notation, the horizontal line replaces the slash in separating the two homologous chromosomes. For consistency, the linked genes are always written in the same linear order. This convention makes it possible to indicate the wildtype allele of a gene with a plus sign in the appropriate position. For example, the genotype $w \, m^+/w^+ \, m$ could also be written without ambiguity as $w +/+ m$. This system of gene notation is used for *Drosophila* and a number of other organisms.

A genotype that is heterozygous for each of two linked genes can have the alleles in either of two possible configurations, as shown in **FIGURE 5.2** for the *w* and *m* genes. In one configuration, called the ***trans***, or **repulsion**, configuration, the mutant alleles are in opposite chromosomes, and the genotype is written as $w +/+ m$ (Figure 5.2A). In the alternative configuration, called the ***cis***, or **coupling**, configuration, the mutant alleles are present in the same chromosome, and the genotype is written as $w \, m/+ +$.

Morgan's study of linkage between the *white* and *miniature* alleles began with the *trans* configuration. He also studied progeny from the *cis* configuration of the *w* and *m* alleles, which results from the cross of white, miniature females with red, nonminiature males. This cross is shown in the "+" notation as:

$$\frac{w \, m}{w \, m} ♀♀ \times \frac{+ \, +}{Y} ♂♂$$

In this case, the $F_1$ females were phenotypically wildtype double heterozygotes, and the males had white eyes and miniature wings. A cross of the $F_1$ progeny is written, again using the "+" notation, as:

$$\frac{w \, m}{+ +} ♀♀ \times \frac{w \, m}{Y} ♂♂$$

This cross produced the following progeny:

| | | |
|---|---|---|
| red eye, normal wing ($++/w \, m$ ♀♀ and $++/Y$ ♂♂) | 395 | 62.3 percent parental types |
| white eye, miniature wing ($w \, m/w \, m$ ♀♀ and $w \, m/Y$ ♂♂) | 382 | |
| white eye, normal wing ($w +/w \, m$ ♀♀ and $w +/Y$ ♂♂) | 223 | 37.7 percent recombinant types |
| red eye, miniature wing ($+ m/w \, m$ ♀♀ and $+ m/Y$ ♂♂) | 247 | |
| | 1247 | |

This result demonstrates that the frequency of recombination with *w* and *m* in the *cis*, or coupling, configuration is the same as that observed with *w* and *m* in the *trans*, or repulsion, configuration: 37.7 percent versus 33.5 percent. The difference is within the range expected from sampling variation among experiments. However, in this case, the chromosomes constituting the parental and recombinant classes of offspring are reversed. They are reversed because the original parents of the $F_1$ female were different: repulsion ($w +/+ m$) in one case and coupling ($w \, m/+ +$) in the other. The repeated finding of equal recombination frequencies in experiments of this kind leads to the following conclusion:

> Recombination between linked genes takes place with the same frequency whether the alleles of the genes are in the *trans* (repulsion) configuration or in the *cis* (coupling) configuration; it is the same no matter how the alleles are arranged.

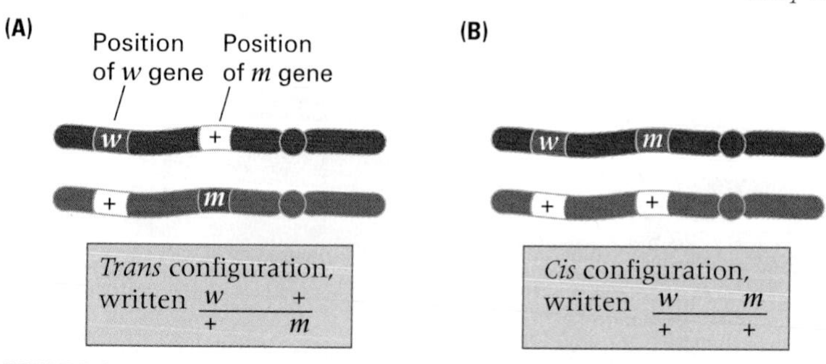


**FIGURE 5.2** Possible configurations of the mutant alleles in a genotype heterozygous for two mutations. (A) The *trans*, or repulsion, configuration has the mutant alleles on opposite chromosomes. (B) The *cis*, or coupling, configuration has the mutant alleles on the same chromosome.

## The Chi-Square Test for Linkage

Any estimated value of the frequency of recombination ($r$) is subject to sampling variation from one experiment to the next. How can the statistical significance of an observed value of $r < 1/2$ be tested against the alternative hypothesis that $r = 0.5$? To take a specific example, how can we be sure that the value $r = 0.377$ observed in the test of $w \, m$ in the coupling configuration is not a statistical accident? Perhaps the true value is $r = 0.50$ and the genes, although syntenic, are nevertheless unlinked. Statistical significance for linkage can be tested on the basis of the observed numbers by means of the chi-square test examined in Chapter 4. As an example, we will use the *w m* experiment yielding the estimate $r = 0.377$. The overall chi-square value in the testcross data is given in the top row of **TABLE 5.1**. With four classes of progeny there are $4 - 1 = 3$ degrees of freedom. But only one of these is relevant to the hypothesis of linkage. The others pertain to the segregation of *w* versus $w^+$ and of *m* versus $m^+$. This point is shown by the chi-square values in the bottom rows of the table. The test of 1 : 1 segregation of *w* versus $w^+$ entails pooling the *m* and $m^+$ data to obtain the observed ratio:

$$w : w^+ = 382 + 223 : 395 + 247$$
$$= 605 : 642$$

To test for 1 : 1 of the *m* versus $m^+$ segregation, we pool the *w* and $w^+$ data:

$$m : m^+ = 382 + 247 : 395 + 223$$
$$= 629 : 618$$

The hypothesis of no linkage ($r = 1/2$) predicts a 1 : 1 ratio of recombinant : nonrecombinant progeny, so in this case we pool the data as:

Recombinant : nonrecombinant

$$= 223 + 247 : 395 + 382$$
$$= 470 : 777$$

The three individual chi-square values, each with 1 degree of freedom (d.f.), are shown in Table 5.1. The overall chi-square with 3 d.f. has a *P* value that can be calculated by numerical methods as $1.5 \times 10^{-16}$; this leaves no doubt that the underlying ratio of the four classes of progeny differs from 1 : 1 : 1 : 1. To learn the source of the discrepancy, we examine the individual chi-square values that have 1 d. f. Neither the *w* versus $w^+$ nor the *m* versus $m^+$ segregation gives any evidence for departure from 1 : 1, the *P* values being 0.295 and 0.755, respectively. However, the chi-square value for recombinants versus nonrecombinants is highly significant, with a *P* value that can be calculated as $3.5 \times 10^{-18}$. From this test we can safely conclude that the cause of the discrepancy is linkage, and we have already estimated the frequency of recombination as $r = 470/1247 = 0.377$.

Note also in Table 5.1 that the 1-d.f. chi-square values sum to the overall chi-square of 76.775. The additivity means that the three hypotheses being tested are independent in the sense that the truth or falsehood of any of them has no bearing on the validity of the others. In practice, we are not often interested in such a detailed analysis of segregation as is outlined in Table 5.1. Usually we want to test only for linkage, and this test is based on the chi-square value for recombinants versus nonrecombinants:

> The chi-square test for linkage is a test for an equal number (1 : 1 ratio) of recombinant versus nonrecombinant progeny, which has 1 degree of freedom.

Testing the individual 1 : 1 segregations is usually unnecessary.

## Each Pair of Linked Genes Has a Characteristic Frequency of Recombination

The recessive allele *y* of another X-linked gene in *Drosophila* results in yellow body color instead of the usual gray color determined by the $y^+$ allele. In one experiment, white-eyed females were mated with males having yellow bodies, and then the wildtype $F_1$ females were testcrossed with yellow-bodied, white-eyed males:

$$+\,w/+\,w \;♀♀ \times y\,+/Y \;♂♂$$
$$\downarrow$$
$$+\,w/y\,+ \;♀♀ \times yw/Y \;♂♂$$

**Table 5.1 Testing for linkage with chi-square**

| Chi-square test | Calculation | Degrees of freedom | | *P* value |
|---|---|---|---|---|
| Overall chi square | $(395 - 311.75)^2/311.75 +$ $(382 - 311.75)^2/311.75 +$ $(223 - 311.75)^2/311.75 +$ $(247 - 311.75)^2/311.75 =$ | 76.775 | 3 | $1.5 \times 10^{-16}$ |
| Segregation of *w* vs. $w^+$ | $(605 - 623.5)^2/623.5 +$ $(642 - 623.5)^2/623.5 =$ | 1.098 | 1 | 0.295 |
| Segregation of *m* vs. $m^+$ | $(629 - 623.5)^2/623.5 +$ $(618 - 623.5)^2/623.5 =$ | 0.097 | 1 | 0.755 |
| Linkage (recombinant vs. nonrecombinant) | $(470 - 623.5)^2/623.5 +$ $(777 - 623.5)^2/623.5 =$ | 75.580 | 1 | $3.5 \times 10^{-18}$ |
| Sum of 1-d.f. chi-square values | | 76.775 | 3 | |

The progeny were

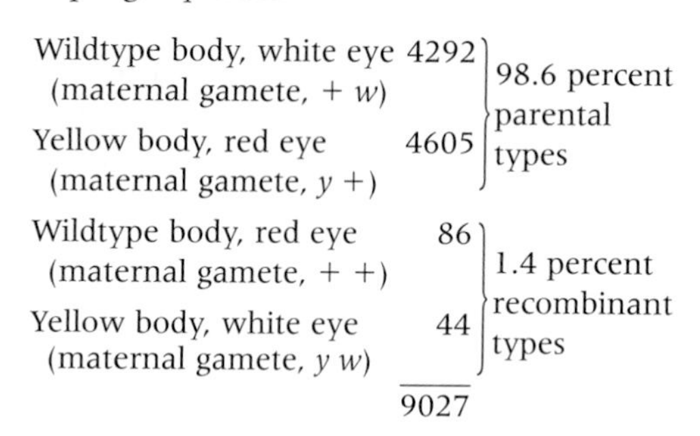

| | | |
|---|---|---|
| Wildtype body, white eye (maternal gamete, + *w*) | 4292 | 98.6 percent parental types |
| Yellow body, red eye (maternal gamete, *y* +) | 4605 | |
| Wildtype body, red eye (maternal gamete, + +) | 86 | 1.4 percent recombinant types |
| Yellow body, white eye (maternal gamete, *y w*) | 44 | |
| | 9027 | |

In a second experiment, yellow-bodied, white-eyed females were crossed with wildtype males, and then the $F_1$ wildtype females and $F_1$ yellow-bodied, white-eyed males were intercrossed:

$$y\,w/y\,w ♀♀ \times ++/\mathrm{Y} ♂♂$$
$$\downarrow$$
$$+\,+/y\,w ♀♀ \times y\,w/\mathrm{Y} ♂♂$$

In this case, 98.6 percent of the $F_2$ progeny had parental chromosomes and 1.3 percent had recombinant chromosomes. The parental and recombinant types were reversed in the reciprocal crosses, but the recombination frequency was virtually the same. Females with the *trans* genotype *y* +/+ *w* produced about 1.4 percent recombinant progeny, carrying either of the recombinant chromosomes *y w* or + +; similarly, females with the *cis* genotype *y w*/+ + produced about 1.3 percent recombinant progeny, carrying either of the recombinant chromosomes *y* + or + *w*. However, the recombination frequency between the genes for yellow body and white eyes was much lower than that between the genes for white eyes and miniature wings (1.4 percent versus about 35 percent). These and other experiments reinforce the following conclusions:

- The recombination frequency is a characteristic of a particular pair of genes.
- Recombination frequencies are the same in *cis* (coupling) and *trans* (repulsion) heterozygotes.

### Recombination in Females Versus Males

*Drosophila* is unusual in that recombination does not take place in males. Although it is not known how (or why) crossing over is prevented in males, the result of the absence of recombination in *Drosophila* males is that all syntenic genes (those located in the same chromosome) show complete linkage in the male. For example, the genes *cn* (cinnabar eyes) and *bw* (brown eyes) are both in chromosome 2 but are so far apart that, in females, there is 50 percent recombination. Thus the cross

$$\frac{cn\ bw}{+\,+} ♀♀ \times \frac{cn\ bw}{cn\ bw} ♂♂$$

yields progeny of genotype + +/*cn bw* and *cn bw*/*cn bw* (the nonrecombinant types) as well as *cn* +/*cn bw* and + *bw*/*cn bw* (the recombinant types) in the proportions 1 : 1 : 1 : 1. However, because there is no crossing over in males, the reciprocal cross

$$\frac{cn\ bw}{cn\ bw} ♀♀ \times \frac{cn\ bw}{+\,+} ♂♂$$

yields progeny only of the nonrecombinant genotypes ++/*cn bw* and *cn bw*/*cn bw* in equal proportions. The absence of recombination in *Drosophila* males is a convenience often made use of in experimental design. As shown in the case of *cn* and *bw*, all the alleles present in any chromosome in a male must be transmitted as a group, without being recombined with alleles present in the homologous chromosome.

The complete absence of crossing over in *Drosophila* males is atypical, but it is not unusual for frequencies of recombination to differ between the sexes. For example, the human genome also exhibits more recombination in females than in males, but the difference is not nearly so extreme as that in *Drosophila*. On the average, across the entire human genome, the recombination frequency between genetic markers in males is approximately 60 percent of the value observed between the same genetic markers in females.

## 5.2 Genetic Mapping

The linkage of the genes in a chromosome can be represented in the form of a **genetic map**, which shows the linear order of the genes along the chromosome with the distances between adjacent genes proportional to the frequency of recombination between them. A genetic map is also called a **linkage map** or a **chromosome map**. The concept of genetic mapping was first developed by Morgan's student Alfred H. Sturtevant in 1913. The early geneticists understood that recombination between genes takes place by an exchange of segments between homologous chromosomes in the process now called crossing over. Each crossover is manifested physically as a chiasma, or cross-shaped configuration, between homologous chromosomes; chiasmata are observed in prophase I of meiosis (Chapter 4). Each chiasma results from the breaking and rejoining of nonsister

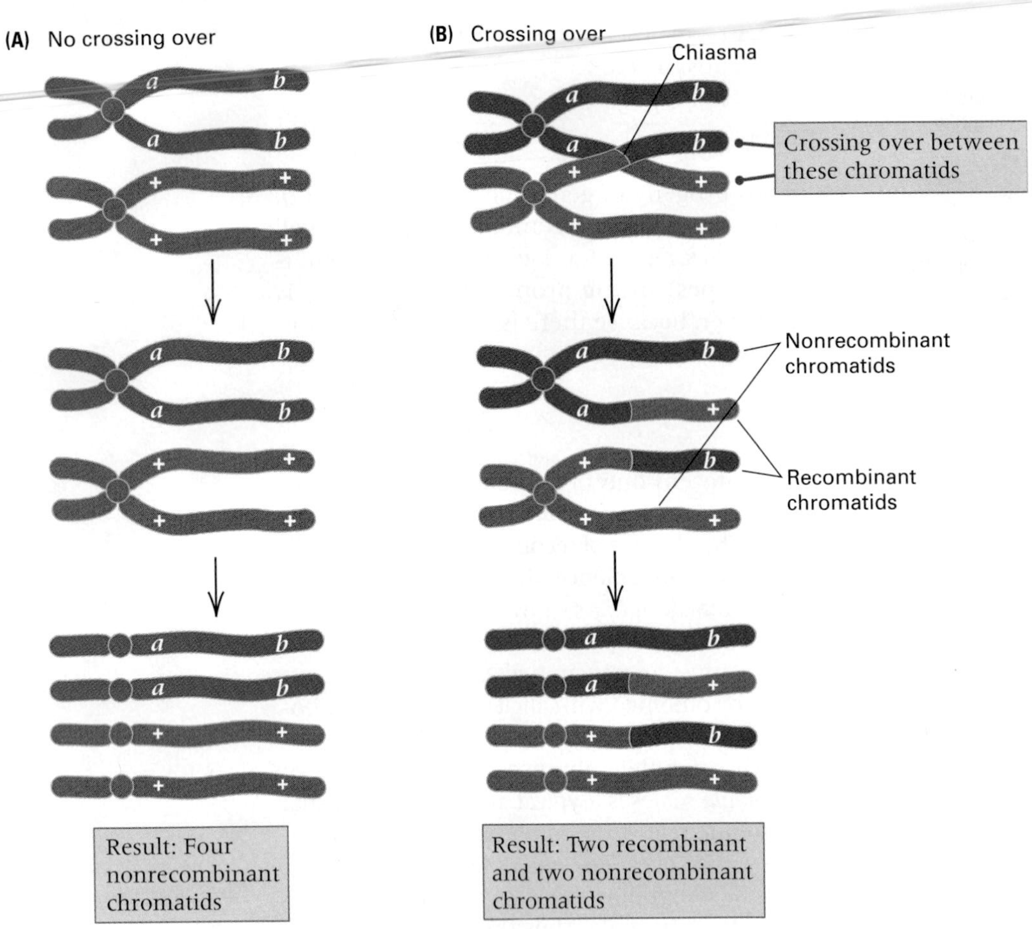


**FIGURE 5.3** (A) Alleles in the same chromosome remain together when there is no crossing over between them. (B) When there is crossing over, the result is two recombinant and two nonrecombinant products, because the exchange is between only two of the four chromatids.

chromatids, with the result that there is an exchange of corresponding segments between them. The theory of crossing over is that each chiasma results in a new association of genetic markers. This process is illustrated in **FIGURE 5.3**. When there is no crossing over (Figure 5.3A), the alleles present in each homologous chromosome remain in the same combination. When crossing over does take place (Figure 5.3B), the outermost alleles in two of the chromatids are interchanged (recombined).

## Map Distance and Frequency of Recombination

The unit of distance in a genetic map is called a **map unit**. Over short intervals, 1 map unit equals 1 percent recombination. For example, two genes that recombine with a frequency of 3.5 percent are said to be located 3.5 map units apart. One map unit is also called a **centimorgan** (abbreviated cM) in honor of T. H. Morgan. A distance of 3.5 map units therefore equals 3.5 centimorgans and indicates 3.5 percent recombination between the genes. For ease of reference, we list the four completely equivalent ways in which a genetic distance between two genes may be represented.

- As a *frequency of recombination* (in the foregoing example, 0.035)
- As a *percent recombination* (here 3.5 percent)
- As a *map distance* in map units (in this case, 3.5 map units)
- As a map distance in *centimorgans* (here 3.5 centimorgans, abbreviated 3.5 cM)

Physically, 1 map unit can be defined as the length of the chromosome in which, on the average, 1 crossover is formed in every 50 cells undergoing meiosis. This principle is illustrated in **FIGURE 5.4**. If 1 meiotic cell in 50 has a crossover, the frequency of *crossing over* equals 1/50, or 2 percent. Yet the frequency of *recombination* between the genes is 1 percent. The correspondence of 1 percent recombination with 2 percent crossing over is a little confusing until one realizes that each crossover results in 2 recombinant chromatids and 2 nonrecombinant chromatids (Figure 5.4). A frequency of crossing over of 2 percent means that, among the 200 chromosomes that result from meiosis in 50 cells, exactly 2 chromosomes (those involved in the exchange) are recombinant for genetic markers flanking the crossover. To put the matter in another way, 2 percent crossing over corresponds to 1 percent recombination because only half the chromatids in each cell with an exchange are actually recombinant.

In situations in which there are genetic markers along the chromosome, such as the *A*, *a* and *B*, *b* pairs of alleles in Figure 5.4, recombination between the marker genes takes place only when crossing over occurs *between* the genes. **FIGURE 5.5** illustrates a case in which a crossover takes place between the gene *A* and the centromere, rather than between the genes *A* and *B*. The crossover does result in the physical exchange of segments between the innermost chromatids. But because the exchange is located outside the region between *A* and *B*, all of the resulting gametes

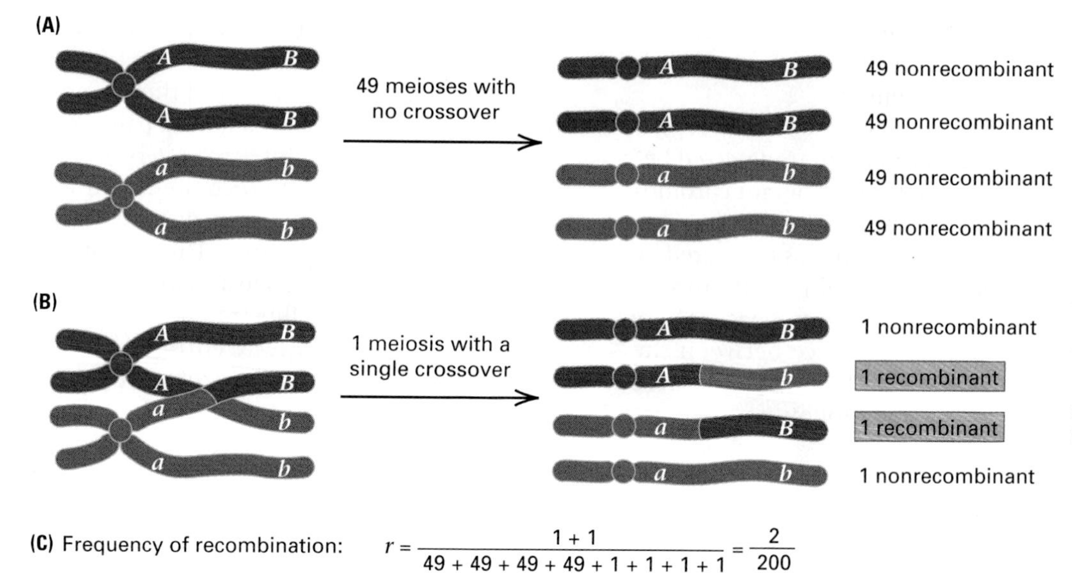


**FIGURE 5.4** Chromosomal configurations in 50 meiotic cells, in which 1 cell has a crossover between two genes. (A) The 49 cells without a crossover result in 98 *A B* and 98 *a b* chromosomes; these are all nonrecombinant. (B) The cell with a crossover yields chromosomes that are *A B*, *A b*, *a B*, and *a b*, of which the middle two types are recombinant chromosomes. (C) The recombination frequency equals 2/200, or 1 percent (also called 1 map unit or 1 cM). Therefore, 1 percent recombination means that 1 meiotic cell in 50 has a crossover in the region between the genes.

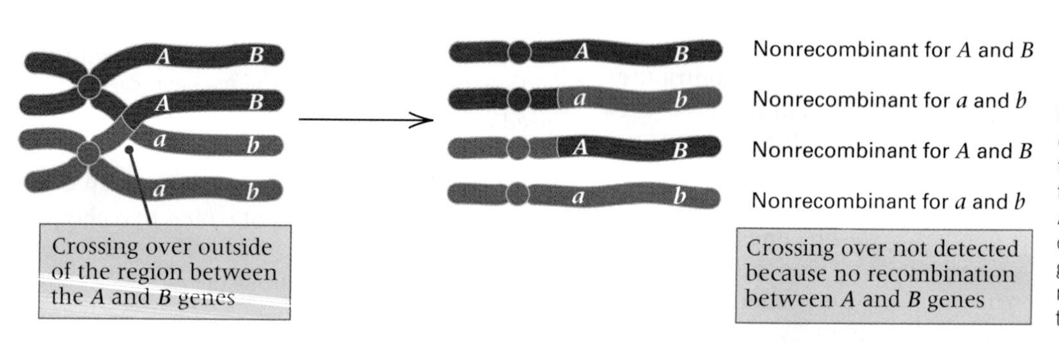


**FIGURE 5.5** Crossing over outside of the region between two genes is not detectable through recombination. Although a segment of chromosome is exchanged, the genetic markers stay in the nonrecombinant configurations, in this case *A B* and *a b*.

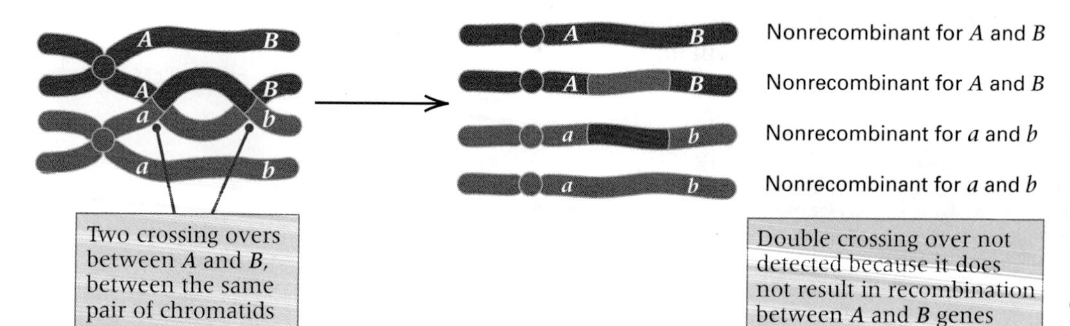


**FIGURE 5.6** If two crossovers take place between marker genes, and both involve the same pair of chromatids, then neither crossover is detected because all of the resulting chromosomes are nonrecombinant *A B* or *a b*.

must carry either the *A B* or the *a b* allele combination. These are nonrecombinant chromosomes. The presence of the crossover is undetected because it is not in the region between the genetic markers.

In some cases, the region between genetic markers is large enough that two (or even more) crossovers can be formed in a single meiotic cell. One possible configuration for two crossovers is shown in **FIGURE 5.6**. In this example, both crossovers are between the same pair of chromatids. The result is that there is a physical exchange of a segment of chromosome between the marker genes, but the double crossover remains undetected because the markers themselves are not recombined. The absence of recombination results from the fact that insofar as recombinat on between *A* and *B* is concerned, the second crossover reverses the effect of the first. The resulting chromosomes are either *A B* or *a b*, both of which are nonrecombinant.

Given that double crossing over in a region between two genes can remain undetected because it does not result in recombinant chromosomes, there is an important distinction between the distance between two genes as measured by the *recombination frequency* and that distance as measured in *map units*. Map units measure how much crossing over takes place between the genes. For any two genes, the **map distance** between them is defined as follows:

$$\text{Map distance (cM)} = (1/2) \times \begin{array}{c}\text{average number} \\ \text{of crossovers in} \\ \text{the region per} \\ \text{meiotic cell}\end{array} \times 100 \qquad (1)$$

The recombination frequency, on the other hand, reflects how much recombination is actually *observed* in a particular experiment. Double crossovers that do not yield recombinant gametes, such as the one in Figure 5.6, do contribute to the map distance but do not contribute to the recombination frequency. The distinction is important only when the region in question is large enough that double crossing over can occur. If the region between the genes is so short that no more than one crossover can be formed in the region in any one meiosis, then map units and recombination frequencies are the same (because there are no multiple crossovers that can undo each other). This is the basis for saying that, for short distances, 1 map unit equals 1 percent recombination. Over an interval so short as to yield 1 percent observed recombination, multiple crossovers are usually precluded, so the map distance equals the recombination frequency in this case.

For a series of linked genes, if the chromosomal region between each adjacent pair is sufficiently short that multiple crossovers are not formed in the region, the recombination frequencies (and hence the map distances) between the genes are additive. This important feature of recombination, and also the logic used in genetic mapping, is illustrated in **FIGURE 5.7**. The genes are all in the X chromosome of *Drosophila*: *y* (yellow body), *rb* (ruby eye color), and *cv* (shortened wing crossvein). The recombination frequency between genes *y* and *rb* is 7.5 percent, and that between *rb* and *cv* is 6.2 percent. The genetic map might be any one of three possibilities, depending on whether *y*, *cv*, or *rb* is in the middle. Map A, which has *y* in the middle, can be excluded because it implies that the recombination frequency between *rb* and *cv* should be greater than that between *rb* and *y*, which contradicts the observed data.

Maps B and C are both consistent with the recombination frequencies. They differ in their predictions regarding the recombination frequency between *y* and *cv*. In map B the predicted distance is 1.3 map units, whereas in map C the predicted distance is 13.3 map units. In reality, the observed recombination frequency between *y* and *cv* is 13.3 percent. Map C is therefore correct.

There are actually two genetic maps corresponding to map C. They differ only in whether *y* is placed at the left or at the right. One map is:

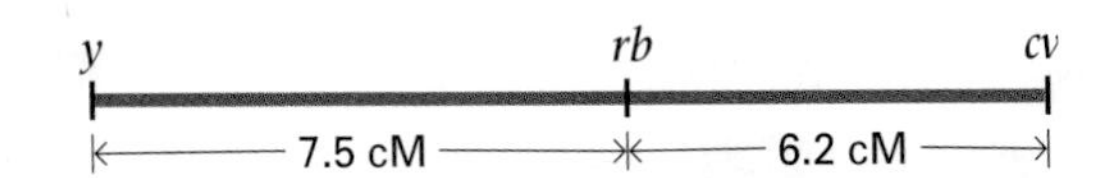


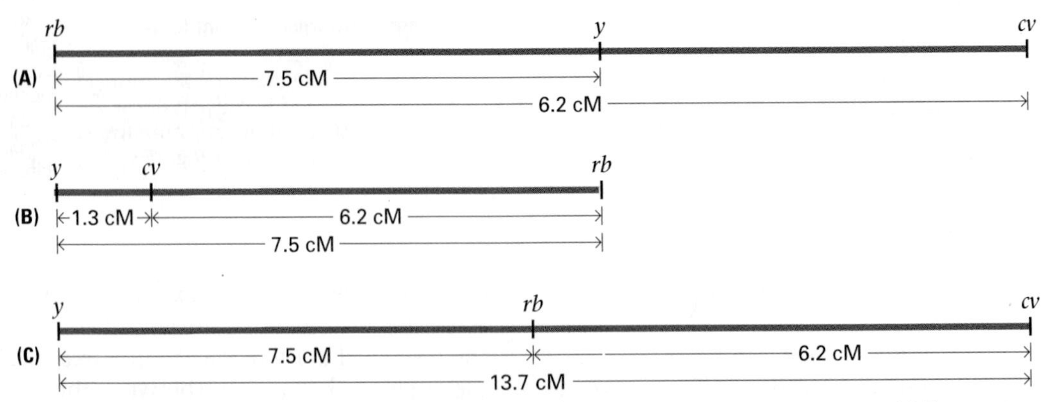


**FIGURE 5.7** In *Drosophila*, the genes *y* (yellow body) and *rb* (ruby eyes) have a recombination frequency of 7.5 percent, and *rb* and *cv* (shortened wing crossvein) have a recombination frequency of 6.2 percent. There are three possible genetic maps, depending on whether *y* is in the middle (A), *cv* is in the middle (B), or *rb* is in the middle (C). Map A can be excluded because it implies that *rb* and *y* are closer than *rb* and *cv*, whereas the observed recombination frequency between *rb* and *y* is larger than that between *rb* and *cv*. Maps B and C are compatible with the data given.

The other map is:

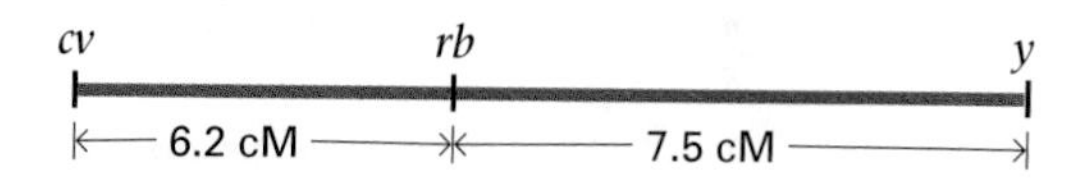


These two ways of depicting the genetic map are completely equivalent.

By this type of reasoning, all the known genes in a chromosome can be assigned a position in a genetic map of the chromosome by considering successive subsets of three markers. Each set of syntenic genes constitutes a **linkage group.** The number of linkage groups is the same as the haploid number of chromosomes of the species. For example, cultivated corn *(Zea mays)* has ten pairs of chromosomes and ten linkage groups. A partial genetic map of chromosome 10 is shown in **FIGURE 5.8**, along with the phenotypes shown by some of the mutants. The ears of corn in Figure 5.8C and 5.8F demonstrate the result of Mendelian segregation. Figure 5.8C shows a 3 : 1 segregation of yellow : orange kernels produced by the recessive *orange pericarp*-2 (*orp*-2)

Courtesy of M. G. Neuffer, College of Agriculture, Food, and Natural Resources, University of Missouri.

allele in a cross between two heterozygous genotypes, and Figure 5.8F shows a 1 : 1 segregation of marbled : white kernels produced by the dominant allele *R1-mb* in a cross between a heterozygous genotype and a homozygous normal.

Courtesy of M. G. Neuffer, College of Agriculture, Food, and Natural Resources, University of Missouri.

## Crossing over

The orderly arrangement of genes represented by a genetic map is consistent with the conclusion that each gene occupies a well-defined site (*locus*) in the chromosome. In a heterozygote, the alleles occupy corresponding locations in the homologous chromosomes. Crossing over is brought about by a physical exchange of segments resulting in a new association of genes in the same chromosome. This process has the following features:

1. The exchange of segments between parental chromatids takes place in the first meiotic prophase, *after the chromosomes have duplicated.* At this stage the four chromatids (strands) of a pair of homologous chromosomes are closely synapsed. Crossing over is a physical exchange between chromatids in a pair of homologous chromosomes.
2. The exchange process consists of the breaking and rejoining of the two chromatids, resulting in the *reciprocal* exchange of equal and corresponding segments between them (see Figure 5.3).
3. The sites of crossing over are more or less random along the length of a chromosome pair. Hence, the probability of crossing over between two genes increases as the physical distance between the genes along the chromosome becomes larger. This principle is the basis of genetic mapping.

## Recombination Between Genes Results from a Physical Exchange Between Chromosomes

A classic demonstration that recombination is associated with a physical exchange of segments between homologous chromosomes made use of two structurally altered X chromosomes of *Drosophila* that permitted parental and recombinant chromosomes to be recognized in a microscope. This experiment, performed in 1936 by Curt Stern, is outlined in **FIGURE 5.9**. One of the abnormal X chromosomes was missing a segment that had become attached to chromosome 4; this X chromosome could be identified by its missing terminal segment. The second aberrant X chromosome could be identified because it had a small second arm consisting of a piece of the Y chromosome. The mutant alleles *car* (a recessive allele resulting in carnation eye color instead of wildtype red) and *B* (a dominant allele resulting in bar-shaped eyes instead of round) were present in the first abnormal X chromosome, and the wildtype alleles of these genes were in the second abnormal X. Females with the two structurally and genetically marked X chromosomes were mated with males having a normal X that carried the recessive alleles of the genes (Figure 5.9). Among the progeny from this cross, flies

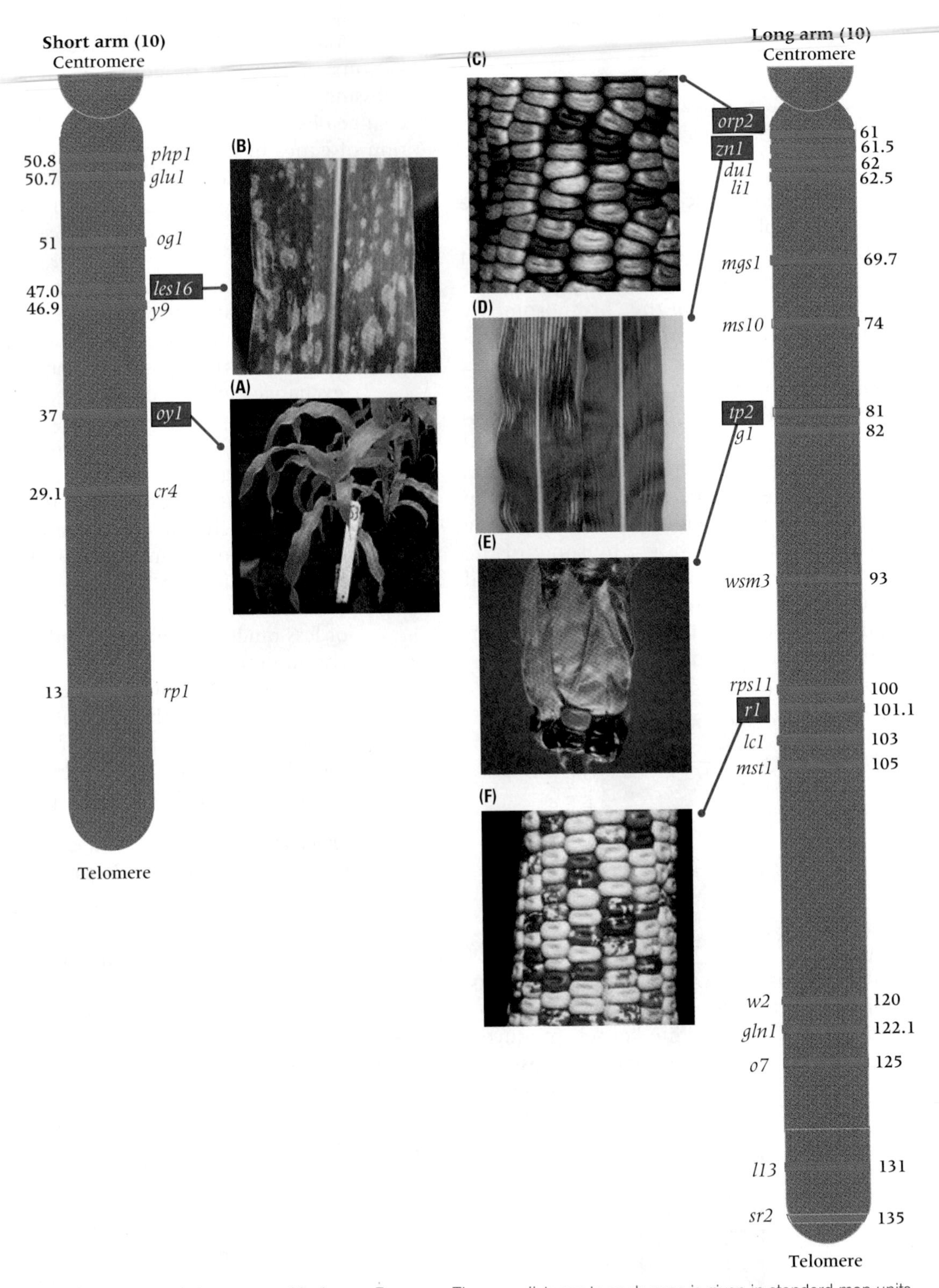


**FIGURE 5.8** Genetic map of chromosome 10 of corn, *Zea mays*. The map distance to each gene is given in standard map units (centimorgans) relative to a position 0 for the telomere of the short arm (lower left). (A) Mutations in the gene *oil yellow-1 (oy1)* result in a yellow-green plant. The plant in the foreground is heterozygous for the dominant allele *Oy1*; behind is a normal plant. (B) Mutations in the gene *lesion-16 (les16)* result in many small to medium-sized, irregularly spaced, discolored spots on the leaf blade and sheath. The photograph shows the phenotype of a heterozygote for *Les16*, a dominant allele. (C) The *orp2* allele is a recessive expressed as orange pericarp, a maternal tissue that surrounds the kernels. The ear shows the segregation of *orp2* in a cross between two heterozygous genotypes, yielding a 3 : 1 ratio of yellow : orange seeds. (D) The gene *zn1* is *zebra necrotic-1*, in which dying tissue appears in longitudinal leaf bands. The leaf on the left is homozygous *zn1*, that on the the right is wildtype. (E) Mutations in the gene *teopod-2 (tp2)* result in many small, partially podded ears and a simple tassel. An ear from a plant heterozygous for the dominant allele *Tp2* is shown. (F) The mutation *R1-mb* is an allele of the *r1* gene, resulting in red or purple color in the aleurone layer of the seed. Note the marbled color in kernels of an ear segregating for *R1-mb*. [Adapted from an illustration by E. H. Coe. Photographs courtesy of M. G. Neuffer, College of Agriculture, Food, and Natural Resources, University of Missouri.]

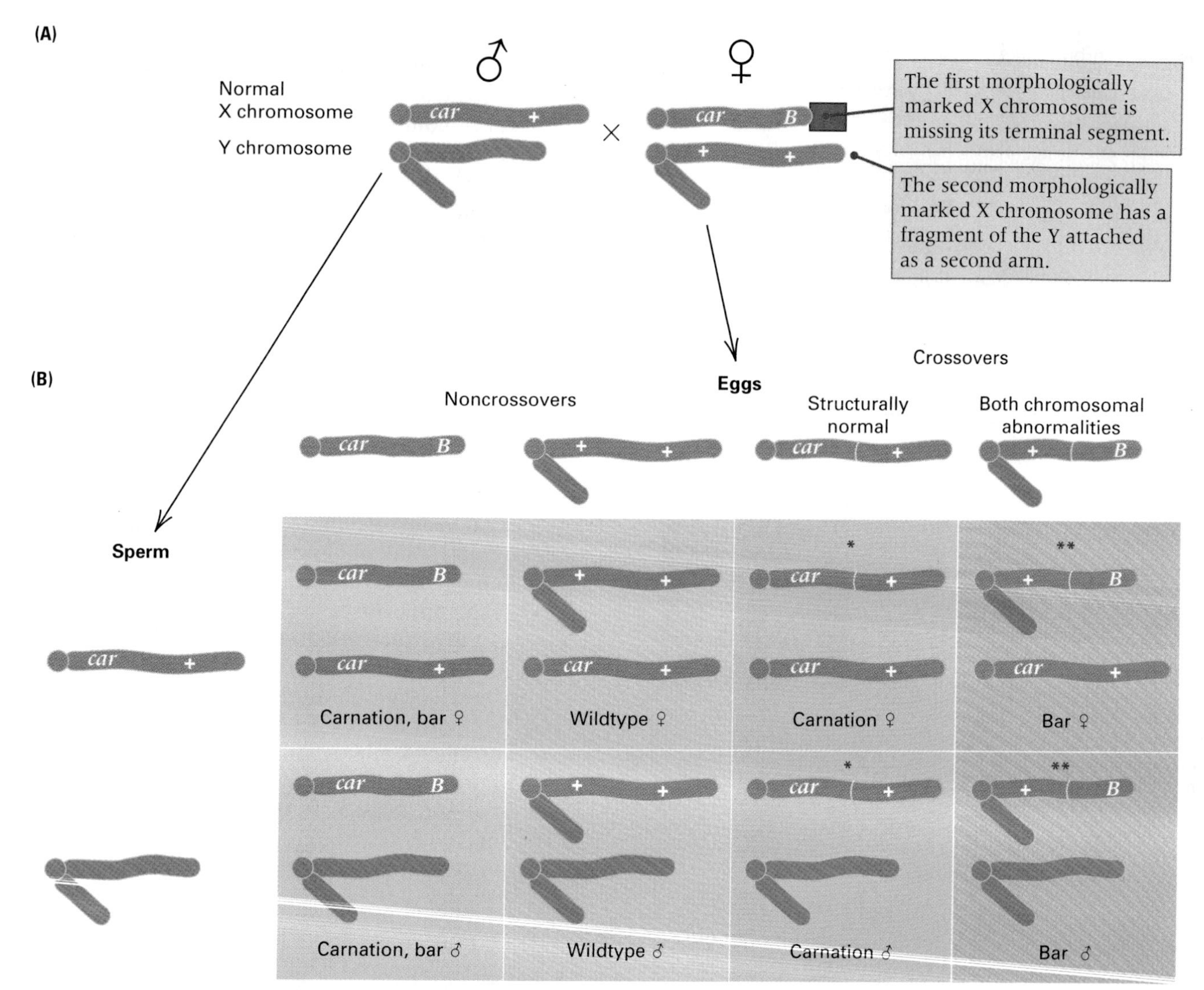


**FIGURE 5.9** (A) Diagram of a cross in which the two X chromosomes in a *Drosophila* female are morphologically distinguishable from each other and from a normal X chromosome. One X chromosome has a missing terminal segment, and the other has a second arm consisting of a fragment of the Y chromosome. (B) Result of the cross. The offspring that are carnation but not bar (single asterisks) contain a structurally normal X chromosome, and the offspring that are bar but not carnation (double asterisks) contain an X chromosome with both morphological markers. The result demonstrates that genetic recombination between marker genes is associated with physical exchange between homologous chromosomes. Segregation of the missing terminal segment of the X chromosome, which is attached to chromosome 4, is not shown.

with parental or recombinant chromosomes could be recognized by their eye color and shape. In the genetically recombinant progeny from the cross, the X chromosome had the appearance that would be expected if recombination of the genes were accompanied by an exchange that recombined the chromosome markers. In particular, the progeny with wildtype, bar-shaped eyes had an X chromosome with a missing terminal segment and the attached Y arm; and the progeny with carnation-colored, round eyes had a structurally normal X chromosome with no missing terminus and no Y arm. As expected, the nonrecombinant progeny were found to have an X chromosome structurally identical with one of the X chromosomes in the mother.

## Crossing Over Takes Place at the Four-Strand Stage of Meiosis

So far we have asserted, without citing experimental evidence, that crossing over takes place in meiosis after the chromosomes have duplicated, at the stage when each bivalent has four chromatid strands. One experimental proof that crossing over takes place after the chromosomes have duplicated came from a study of laboratory stocks of *D. melanogaster* in which the two X chromosomes in a female are joined to a common centromere to form an aberrant chromosome called an **attached-X**, or **compound-X**, chromosome. The normal X chromosome in *Drosophila* has a centromere almost at the end of the chromosome. The

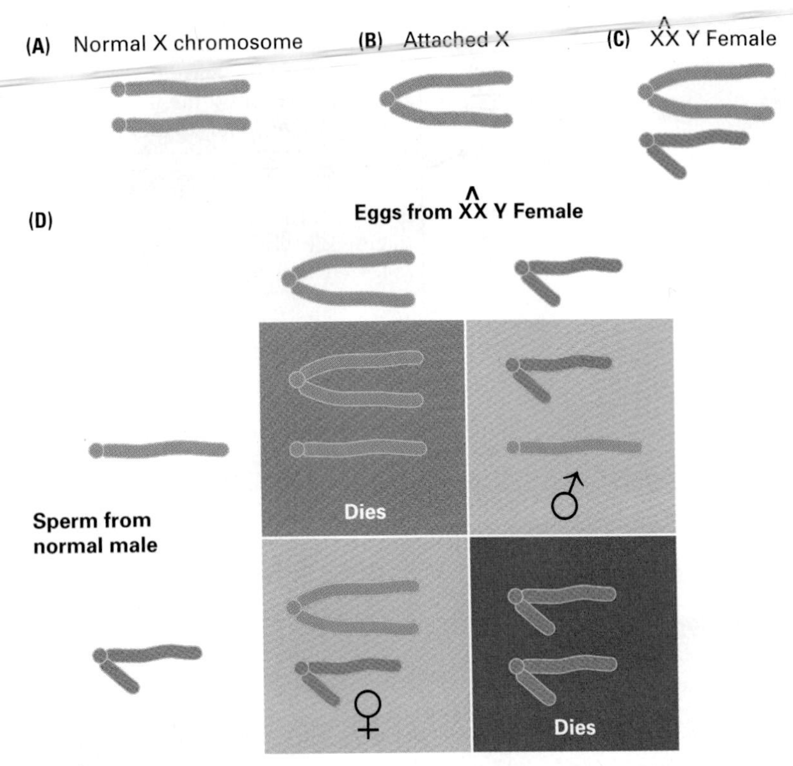


**FIGURE 5.10** Attached-X (compound-X) chromosomes in *Drosophila*. (A) A structurally normal X chromosome in a female. (B) An attached-X chromosome, with the long arms of two normal X chromosomes attached to a common centromere. (C) Typical attached-X females also contain a Y chromosome. (D) Outcome of a cross between an attached-X female and a normal male. Genotypes with either three X chromosomes or no X chromosomes are lethal. Note that in this case, a male fly receives its X chromosome from its father and its Y chromosome from its mother—the opposite of the usual situation in *Drosophila*.

attachment of two of these chromosomes to a single centromere results in a chromosome with two equal arms, each consisting of a virtually complete X. Females with a compound-X chromosome usually contain a Y chromosome as well, and they produce two classes of viable offspring: females who have the maternal compound-X chromosome along with a paternal Y chromosome, and males with the maternal Y chromosome along with a paternal X chromosome (**FIGURE 5.10**). Attached-X chromosomes are frequently used to study X-linked genes in *Drosophila*, because when an attached-X female is mated with a male carrying any X-linked mutation, the sons receive the mutant X chromosome and the daughters receive the attached-X chromosome. In matings with attached-X females, therefore, the inheritance of an X-linked gene in the male passes from father to son to grandson, and so forth, which is the opposite of usual X-linked inheritance.

In an attached-X chromosome in which one X carries a recessive allele and the other carries the wildtype nonmutant allele, crossing over between the X-chromosome arms can yield attached-X products in which the recessive allele is present in both arms of the attached-X chromosome (**FIGURE 5.11**). Hence, attached-X females that are heterozygous can produce some female progeny that are homozygous for the recessive

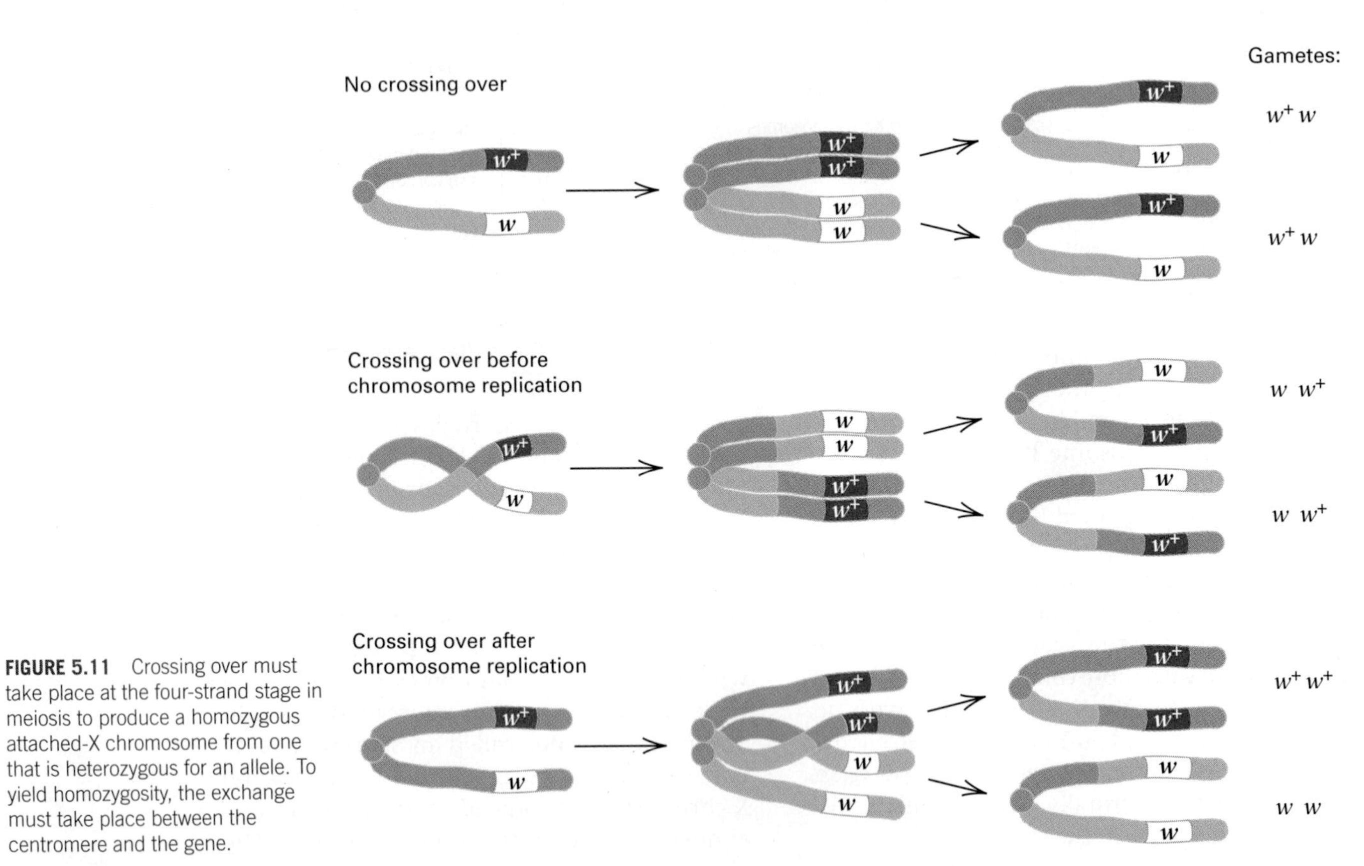


**FIGURE 5.11** Crossing over must take place at the four-strand stage in meiosis to produce a homozygous attached-X chromosome from one that is heterozygous for an allele. To yield homozygosity, the exchange must take place between the centromere and the gene.

## connection

### Genes All in a Row

*Genetic mapping remains the cornerstone of genetic analysis. It is the principal technique used in modern human genetics to identify the chromosomal location of mutant genes associated with inherited diseases. The genetic markers used in human genetics are often single-nucleotide polymorphisms (SNPs) that differ from one person to the next, but the basic principles of genetic mapping are the same as those originally enunciated by Sturtevant. In this excerpt, we have substituted the symbols presently in use for the genes,* y *(yellow body),* w *(white eyes),* v *(vermilion eyes),* m *(miniature wings), and* r *(rudimentary wings). (The sixth gene mentioned is another mutant allele of white, now called white-eosin.) In this paper, Sturtevant uses the term* crossing over *instead of* recombination *and* crossovers *instead of* recombinant chromosomes. *We have retained his original terms but, in a few cases, have put the modern equivalent in brackets.*

**Alfred H. Sturtevant 1913**
Columbia University,
New York, New York
*The Linear Arrangement of Six Sex-Linked Factors in* Drosophila*, as Shown by Their Mode of Association*

Morgan, by crossing white eyed, long winged flies to those with red eyes and rudimentary wings (the new sex-linked character), obtained, in $F_2$, white eyed rudimentary winged flies. This could happen only if "crossing-over" [recombination] is possible; which means, on the assumption that both of these factors are in the X chromosome, that an interchange of materials between homologous chromosomes occurs (in the female only, since the male has only one X chromosome). A point not noticed at this time came out later in connection with other sex-linked factors in *Drosophila*. It became evident that some of the sex-linked factors are associated, i.e., that crossing-over does not occur freely between some factors, as shown by the fact that the combinations present in the $F_1$ flies are much more frequent in the $F_2$ than are new combinations of the same characters. This means, on the chromosome view, that the chromosomes, or at least certain segments of them, are much more likely to remain intact during meiosis than they are to interchange materials. . . . It would seem, if this hypothesis be correct, that the proportion of "crossovers" [recombinant chromosomes] could be used as an index of the distance between any two factors. Then by determining the distances (in the above sense) between A and B and between B and C, one should be able to predict AC. . . . Just how far our theory stands the test is shown by the data below, giving observed percent of crossovers [recombinant chromosomes] and the distances calculated [from the summation of shorter intervals].

***These results form a new argument in favor of the chromosome view of inheritance, since they strongly indicate that the factors investigated are arranged in a linear series.***

| Factors | Calculated distance | Observed percentage of crossovers |
|---|---|---|
| *y*—*v* | 30.7 | 32.2 |
| *y*—*m* | 33.7 | 35.5 |
| *y*—*r* | 57.6 | 37.6 |
| *w*—*m* | 32.7 | 33.7 |
| *w*—*r* | 56.6 | 45.2 |

It will be noticed at once that the longer distances, *y*—*r* and *w*—*r*, give smaller per cent of crossovers, than the calculation calls for. This is a point which was to be expected and is probably due to the occurrence of two breaks in the same chromosome, or "double crossing-over." But in the case of the shorter distances the correspondence with expectation is perhaps as close as was to be expected with the small numbers that are available. . . . It has been found possible to arrange six sex-linked factors in *Drosophila* in a linear series, using the number of crossovers per 100 cases [the frequency of recombination] as an index of the distance between any two factors. A source of error in predicting the strength of association between untried factors is found in double crossing-over. The occurrence of this phenomenon is demonstrated. . . . These results form a new argument in favor of the chromosome view of inheritance, since they strongly indicate that the factors investigated are arranged in a linear series.

Source: Z. H. Sturtevant, *J. Exp. Zool.* 14 (2005): 43-59.

allele. The frequency with which homozygosity is observed increases with increasing map distance of the gene from the centromere. From the diagrams in Figure 5.11, it is clear that homozygosity can result only if the crossover between the gene and the centromere takes place after the chromosome has duplicated. The finding of homozygous attached-X female progeny therefore implies that crossing over takes place at the *four-strand* stage of meiosis. If this were not the case, and crossing over happened before duplication of the chromosome (at the *two-strand* stage), it would result only in a swap of the alleles between the chromosome arms and would never yield the homozygous products that are actually observed.

## ■ Multiple Crossovers

When two genes are located far apart along a chromosome, more than one crossover can occur between them in a single meiosis, and the probability of multiple crossovers increases with the distance between the genes. Multiple crossing over complicates the interpretation of recombination data. The problem is that some of the multiple exchanges between genes do not result in recombination and hence are not detected. As we saw in Figure 5.6, the effect of one crossover can be canceled by another crossover farther along the way. If two exchanges between the same two chromatids take place between the genes *A* and *B*, then their net effect will be that all chromosomes are nonrecombinant, either *A B* or *a b*. Two of the products of this meiosis have an interchange of their middle segments, but the chromosomes are not recombinant for the genetic markers, so they are genetically indistinguishable from noncrossover chromosomes. The possibility of such canceling means that the true map distance, defined in Equation (1), is *underestimated* by setting it equal to the observed frequency of recombination.

In higher organisms, double crossing over is effectively precluded in chromosome segments that are sufficiently short. Therefore, by using recombination data for closely linked markers to build up genetic linkage maps, we can avoid multiple crossovers that cancel each other's effects. For genes that are distant along a chromosome, the true map distance is estimated by summing the recombination frequencies across shorter subintervals within which multiple crossovers do not occur.

The *minimum* recombination frequency between two genes is 0. The recombination frequency also has a maximum:

> No matter how far apart two genes may be, the maximum frequency of recombination between any two genes is 50 percent.

Fifty percent recombination is the same value that would be observed if the genes were on nonhomologous chromosomes and assorted independently. The maximum frequency of recombination is observed when the genes are so far apart in the chromosome that at least one crossover is nearly always formed between them. Figure 5.3B shows that a single exchange in every meiosis would result in half of the products having parental combinations and the other half having recombinant combinations of the genes. Two exchanges between two genes have the same effect, as shown in **FIGURE 5.12**. Figure 5.12A shows a *two-strand double crossover,* in which the same chromatids participate in both exchanges; no recombination of the marker genes is detectable. When the two exchanges have one chromatid in common (*three-strand double crossover,* Figure 5.12B and C), the result is indistinguishable from that of a single exchange: Two products with parental combinations and two with recombinant combinations are produced. Note that there are two types of three-strand doubles, depending on which three chromatids participate. The final possibility is that the second exchange involves the chromatids that did not participate in the first exchange (*four-strand double crossover,* Figure 5.12D), in which case all four products are recombinant.

In most organisms, when double crossovers are formed, the chromatids that take part in the two exchange events are selected at random. In this case, the expected proportions of the three types of double exchanges are 1/4 two-strand doubles, 1/2 three-strand doubles, and 1/4 four-strand doubles. This means that, on the average,

$$(1/4)(0) + (1/2)(2) + (1/4)(4) = 2$$

recombinant chromatids will be found among the four chromatids produced from meioses with two exchanges between a pair of genes. This is the same proportion obtained with a single exchange between the genes. Moreover, a maximum of 50 percent recombination is obtained for any number of exchanges.

In the discussion of Figure 5.12, we emphasized that in most organisms, the chromatids taking part in double-exchange events are selected at random. Then the maximum frequency of recombination is 50 percent. When there is a *nonrandom* choice of chromatids in successive crossovers, the phenomenon is called **chromatid interference**. It can be seen in Figure 5.12 that, relative to a random choice of chromatids, an excess of four-strand double crossing over (*positive chromatid interference*) results in a maximum frequency of recombination greater than 50 percent; likewise, an excess of two-strand double crossing-over (*negative chromatid interference*) results in a maximum frequency of recombination less than 50 percent. Therefore, the finding that the maximum frequency of recombination between two genes in the same chromosome is not 50 percent can be regarded as evidence for chromatid interference. Positive chromatid interference has not yet been ob-

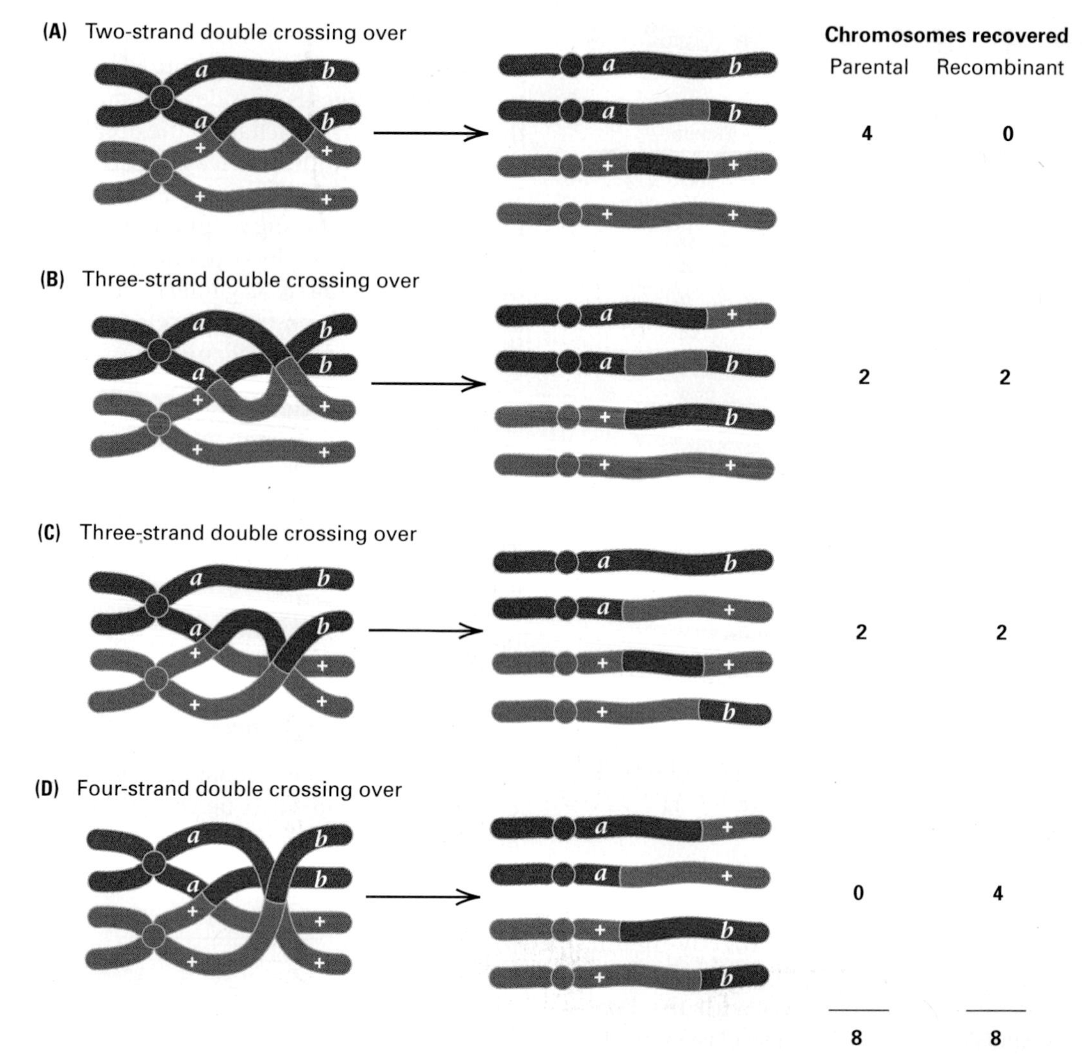


**FIGURE 5.12** The result of two crossovers in the interval between two genes is indistinguishable from independent assortment of the genes, provided that the chromatids participate at random in the exchanges. (A) A two-strand double crossover. (B, C) The two types of three-strand double crossovers. (D) A four-strand double crossover.

served in any organism; negative chromatid interference has been reported in some fungi.

Double crossing over is detectable in recombination experiments that employ **three-point crosses**, which include three pairs of alleles. If a third pair of alleles, $c^+$ and $c$, is located between the two with which we have been concerned (the outermost genetic markers), then double exchanges in the region can be detected when the crossovers flank the $c$ gene (**FIGURE 5.13**). The two crossovers, which in this example take place between the same pair of chromatids, would result in a reciprocal exchange of the $c^+$ and $c$ alleles between the chromatids. A three-point cross is an efficient way to obtain recombination data; it is also a simple method for determining the order of the three genes, as we shall see in the next section.

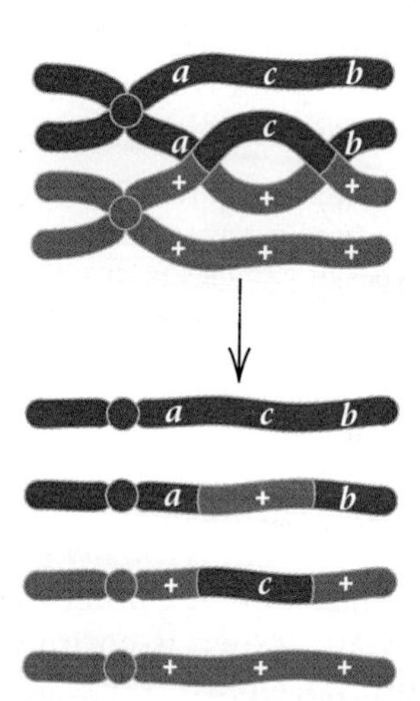


**FIGURE 5.13** A two-strand double crossover that spans the middle pair of alleles in a triple heterozygote results in a reciprocal exchange of the middle pair of alleles between the two participating chromatids.

**Table 5.2** Progeny from a three-point testcross in corn

| Phenotype of testcross progeny | Genotype of gamete from hybrid parent | Number |
|---|---|---|
| Normal (wildtype) | *Lz Gl Su* | 286 |
| Lazy | *lz Gl Su* | 33 |
| Glossy | *Lz gl Su* | 59 |
| Sugary | *Lz Gl su* | 4 |
| Lazy, glossy | *lz gl Su* | 2 |
| Lazy, sugary | *lz Gl su* | 44 |
| Glossy, sugary | *Lz gl su* | 40 |
| Lazy, glossy, sugary | *lz gl su* | 272 |

## 5.3 Genetic Mapping in a Three-Point Testcross

The data in **TABLE 5.2** result from a testcross in corn with three linked genes. We will use these to illustrate the analysis of a three-point cross. The recessive alleles are *lz* (for lazy or prostrate growth habit), *gl* (for glossy leaf), and *su* (for sugary endosperm), and the multiply heterozygous parent in the cross had the genotype:

$$\frac{Lz\ Gl\ Su}{lz\ gl\ su}$$

Therefore, the two classes of progeny that inherit noncrossover (parental-type) gametes are the normal plants and plants with the lazy-glossy-sugary phenotype. These classes are far larger than any of the crossover classes. If the combination of dominant and recessive alleles in the chromosomes of the heterozygous parent were unknown, then we could deduce from their relative frequencies among the progeny that the noncrossover gametes were *Lz Gl Su* and *lz gl su*. This point is important enough to state as a general principle:

> In any genetic cross involving linked genes, no matter how complex, the two most frequent types of gametes with respect to any pair of genes are *nonrecombinant;* these provide the linkage phase (*cis* versus *trans*) of the alleles of the genes in the multiply heterozygous parent.

In mapping experiments, the order of genes along the chromosome is usually not known in advance. In Table 5.2, the order in which the three genes are shown is entirely arbitrary. There is, however, an easy way to determine the correct order from three-point data. Simply identify the genotypes of the double-crossover gametes produced by the heterozygous parent and compare them with the nonrecombinant gametes. Because the probability of two simultaneous exchanges is considerably smaller than that of either single exchange, the double-crossover gametes will be the least frequent types. Table 5.2 shows that the classes composed of four plants with the sugary phenotype and two plants with the lazy-glossy phenotype (products of the *Lz Gl su* and *lz gl Su* gametes, respectively) are the least frequent and therefore constitute the double-crossover progeny.

> The effect of double crossing over, as Figure 5.13 shows, is to interchange the members of the *middle* pair of alleles between the chromosomes.

This means that if the parental chromosomes are

*Lz Gl Su* and *lz gl su*

and the double-crossover chromosomes are

*Lz Gl su* and *lz gl Su*

then *Su* and *su* are interchanged by the double crossing over and must be the middle pair of alleles. Therefore, the genotype of the heterozygous parent in the cross should be written as

$$\frac{Lz\quad Su\quad Gl}{lz\quad su\quad gl}$$

which is now diagrammed correctly with respect to both the order of the genes and the configuration of alleles in the homologous chromosomes. A two-strand double crossover between chromatids of these parental types is diagrammed below, and the products can be seen to correspond to the two types of gametes identified in the data as the double crossovers.

$$\frac{Lz\quad Su\quad Gl}{lz\quad su\quad gl} \rightarrow \frac{Lz\quad su\quad Gl}{lz\quad Su\quad gl}$$

From this diagram, it can also be seen that the reciprocal products of a single crossover between *lz* and *su* would be *Lz su gl* and *lz Su Gl* and that the products of a single exchange between *su* and *gl* would be *Lz Su gl* and *lz su Gl*.

We can now summarize the data in a more informative way, writing the genes in correct order and identifying the numbers of the different chromosome types produced by the heterozygous parent that are present in the progeny.

| | | | | |
|---|---|---|---|---|
| Lz | Su | Gl | 286 | Parental types |
| lz | su | gl | 272 | |
| Lz | su | gl | 40 | Single crossovers between *lz* and *su* |
| lz | Su | Gl | 33 | |
| Lz | Su | gl | 59 | Single crossovers between *su* and *gl* |
| lz | su | Gl | 44 | |
| Lz | su | Gl | 4 | Double-crossover types |
| lz | Su | gl | 2 | |
| | | | 740 | |

Note that each class of single recombinants includes two reciprocal products and that these are found in approximately equal frequencies (40 versus 33, and 59 versus 44). This observation illustrates an important principle:

> The two reciprocal products that result from any crossover, or any combination of crossovers, are expected to appear in approximately equal frequencies among the progeny.

In calculating the frequency of recombination from the data, remember that the double-recombinant chromosomes result from *two* exchanges, one in each of the chromosome regions defined by the three genes. Therefore, chromosomes that are recombinant between *lz* and *su* are represented by the following chromosome types:

| | | | |
|---|---|---|---|
| Lz | su | gl | 40 |
| lz | Su | Gl | 33 |
| Lz | su | Gl | 4 |
| lz | Su | gl | 2 |
| | | | 79 |

Because 79/740, or 10.7 percent, of the chromosomes recovered in the progeny are recombinant between the *lz* and *su* genes, the map distance between these genes is estimated as 10.7 map units, or 10.7 centimorgans. Similarly, the chromosomes that are recombinant between *su* and *gl* are represented by:

| | | | |
|---|---|---|---|
| Lz | Su | gl | 59 |
| lz | su | Gl | 44 |
| Lz | su | Gl | 4 |
| lz | Su | gl | 2 |
| | | | 109 |

The recombination frequency between the *su* and *gl* genes is 109/740, or 14.7 percent, so the map distance between them is estimated as 14.7 map units, or 14.7 centimorgans. The genetic map of the chromosome segment in which the three genes are located is therefore:

*lz* ——— *su* ——— *gl*

|←10.7 map units→|←—— 14.7 map units ——→|

One error that many students make in analyzing three-point crosses is to forget to include the double recombinants when calculating the recombination frequency between adjacent genes. You can keep from falling into this trap by remembering that the double-recombinant chromosomes have single recombination in *both* regions.

## Chromosome Interference in Double Crossovers

The ability to detect double crossovers makes it possible to determine whether exchanges in two different regions are formed independently. Using the information from the example with corn, we know from the recombination frequencies that the probability of recombination is 0.107 between *lz* and *su* and 0.147 between *su* and *gl*. If crossing over is independent in the two regions (which means that the formation of one exchange does not alter the probability of the second exchange), then the probability of an exchange in both regions is the product of these separate probabilities, or $0.107 \times 0.147 = 0.0157$. This implies that in a sample of 740 gametes, the expected number of double crossovers would be $740 \times 0.0157 = 11.6$, whereas the number actually observed is only 6. Such deficiencies in the observed number of double crossovers are common and identify a phenomenon called **chromosome interference**, in which crossing over in one region of a chromosome reduces the probability of a second crossover in a nearby region. Because chromosome interference is nearly universal, whereas chromatid interference is virtually unknown, the term *interference*, when used without qualification, nearly always refers to chromosome interference.

The **coefficient of coincidence** is the observed number of double-recombinant chromosomes divided by the expected number. Its value provides a quantitative measure of the degree of interference, defined as:

$$\begin{aligned} i &= \text{interference} \\ &= 1 - \text{coefficient of coincidence} \end{aligned}$$

From the data in our example, the coefficient of coincidence is $6/11.6 = 0.51$, which means that the observed number of double crossovers was only 51 percent of the number we would expect to observe if crossing over in the two regions were independent. The value of the interference depends on the distance between the genetic markers and on the species. In some species, the interference increases as the distance between the two outside markers becomes smaller, until a point is reached at which double crossing over is eliminated; that is, no double crossovers are found, and the coefficient of coincidence equals 0 (or, to say the same thing, the interference equals 1). In *Drosophila*, this distance is about 10 map units. In most organisms, when the total distance between the genetic markers is greater than about 30 map units, interference essentially disappears and the coefficient of coincidence approaches 1.

## Genetic Mapping Functions

The effect of interference on the relationship between genetic map distance and the frequency of recombination is illustrated in **FIGURE 5.14**. Each curve is an example of a **mapping function**, which is the mathematical relation between the genetic distance across an interval in map units (centimorgans) and the observed frequency of recombination across the interval. In other words, a mapping function converts a *map distance* between genetic markers into a *recombination frequency* between the markers. As we have seen, when the map distance between the markers is small, the recombination frequency equals the map distance. This principle is reflected in the curves in Figure 5.14 in the region in which the map distance is smaller than about 15 cM. At less than this distance, all of the curves are nearly straight lines, which means that map distance and recombination frequency are equal: 1 map unit equals 1 percent recombination, and 10 map units equal 10 percent recombination.

For distances greater than 15 map units, the curves differ. The upper curve is based on the assumption of complete interference *i*, so that $i = 1$. With this mapping function, the linear relation holds all the way to a map distance of 50 cM. Then the curve terminates abruptly at a map distance of 50 map units. It terminates because complete interference means that a chromosome can have no more than one crossover along its length; because one crossover corresponds to 50 map units, this is the maximum length of the chromosome.

The other two curves in Figure 5.14 are for different patterns of interference along the chromosome. The bottom curve is usually called *Haldane's mapping function* after its inventor. It assumes no interference ($i = 0$), and the mathematical form of the function is $r = (1/2)(1 - e^{-d/50})$, where $d$ is the map distance in centimorgans. Any mapping function for which $i$ is between 0 and 1 must lie in the interval between the top and bottom curves. The particular example shown is *Kosambi's mapping function*, in which the interference is assumed to decrease as a linear function of distance according to $i = 1 - 2r$. Although simple in its underlying assumptions, the formula for Kosambi's function is not simple.

Most mapping functions are almost linear near the origin, as are those in Figure 5.14. This near linearity implies that for map distances

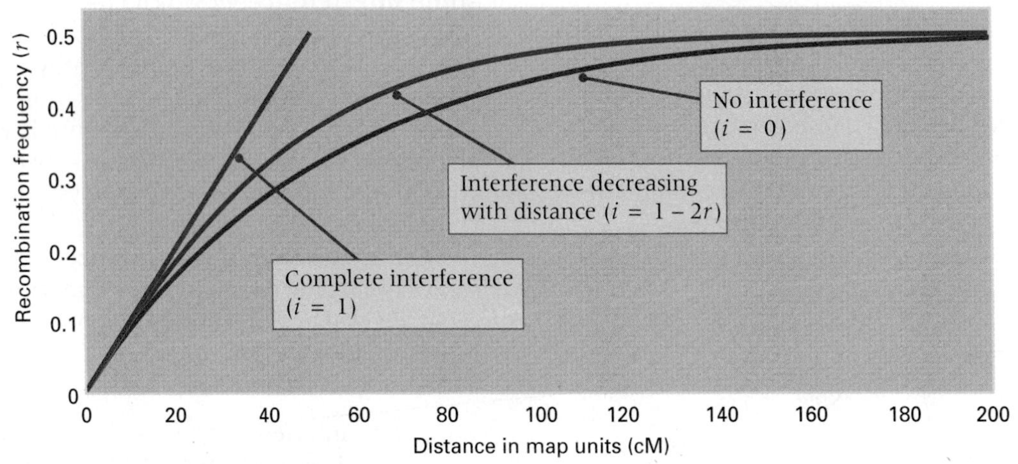


**FIGURE 5.14** A mapping function is the relation between genetic map distance across an interval and the observed frequency of recombination across the interval. Map distance is defined as one-half the average number of crossovers converted into a percentage. The three mapping functions correspond to different assumptions about interference, *i*. In the first curve, $i = 1$ (complete interference); in this case the curve terminates abruptly because there can be at most one crossover per chromosome. In the third curve, $i = 0$ (no interference). The mapping function in the middle is based on the assumption that *i* decreases as a linear function of distance.

smaller than about 15 cM, whatever the pattern of chromosome interference, there are so few double recombinants that the recombination frequency in percent essentially equals the map distance. Hence, the map distance between two widely separated genetic markers can be estimated with some confidence by summing the map distances across smaller segments between the markers, provided that each of the smaller segments is less than about 15 map units in length.

## Genetic Map Distance and Physical Distance

Generally speaking, the greater the physical separation between genes along a chromosome, the greater the map distance between them. Physical distance and genetic map distance are usually correlated because a greater distance between genetic markers affords a greater chance for crossing over to take place.

On the other hand, the general correlation between physical distance and genetic map distance is by no means absolute. We have already noted that the frequency of recombination between genes may differ in males and females. An unequal frequency of recombination means that the sexes can have different map distances in their genetic maps, although the physical chromosomes of the two sexes are the same and the genes must have the same linear order. An extreme example is that of *Drosophila* males, in which no crossing over occurs. In *Drosophila* males, the map distance between any pair of genes located in the same chromosome is 0; nevertheless, genes in different chromosomes do undergo independent assortment.

The general correlation between physical distance and genetic map distance can break down even in a single chromosome. For example, crossing over is much less frequent in some regions of the chromosome than in other regions. The term **heterochromatin** refers to certain regions of the chromosome that have a dense, compact structure in interphase; these regions take up many of the standard dyes used to make chromosomes visible. The rest of the chromatin, which becomes visible only after chromosome condensation in mitosis or meiosis, is called **euchromatin**. In most organisms, the major heterochromatic regions are adjacent to the centromere; smaller blocks are present at the ends of the chromosome arms (the telomeres) and interspersed with the euchromatin. In general, crossing over is much less frequent in regions of heterochromatin than in regions of euchromatin.

Because there is less crossing over in heterochromatin, a given length of heterochromatin will appear much shorter in the genetic map than an equal length of euchromatin. In heterochromatic regions, therefore, the genetic map gives a distorted picture of the physical map. An example of such distortion appears in **FIGURE 5.15**, which compares the physical map and the genetic map of chromosome 2 in *Drosophila*. The physical map is depicted as the chromosome appears in metaphase of mitosis. Two genes near the tips and two near the euchromatin–heterochromatin junction are indicated in the genetic map. The map distances across the euchromatic arms are 54.5 and 49.5 map units, respectively, for a total euchromatic map distance of 104.0 map units. However, the heterochromatin, which constitutes approximately 25 percent of the entire chromosome, has a genetic length in map units of only 3.0. The distorted length of the heterochromatin in the genetic map results from the reduced frequency of crossing over in the heterochromatin. In spite of the distortion of the genetic map across the heterochromatin, in the regions of euchromatin there is a good correlation between the physical distance between genes and their distance in map units in the genetic map.

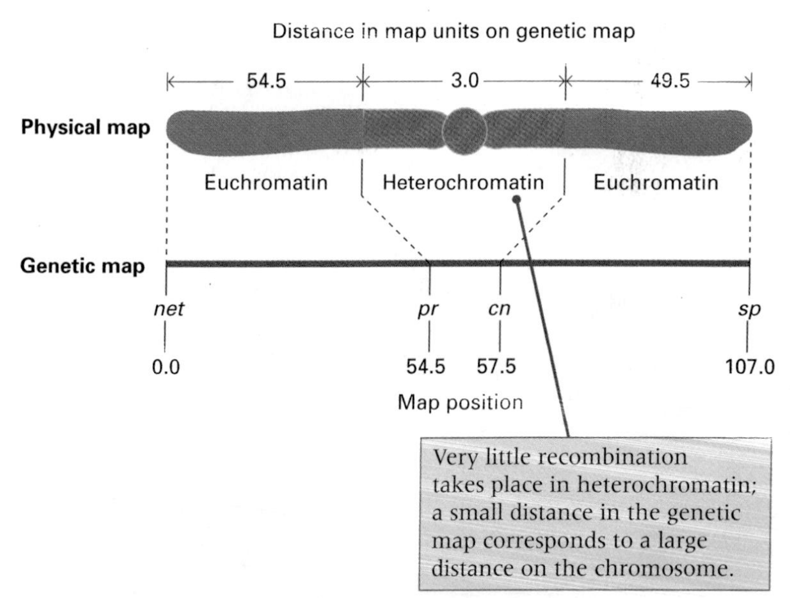


**FIGURE 5.15** Chromosome 2 in *Drosophila* as it appears in metaphase of mitosis (physical map, top) and in the genetic map (bottom). Heterochromatin and euchromatin are in contrasting colors. The genes are *net* (net wing veins), *pr* (purple eye color), *cn* (cinnabar eye color), and *sp* (speck of wing pigment). The total map length is 54.5 + 49.5 + 3.0 = 107.0 map units. The heterochromatin accounts for 3.0/107.0 = 2.8 percent of the total map length but constitutes approximately 25 percent of the physical length of the metaphase chromosome.

## 5.4 Genetic Mapping in Human Pedigrees

In Chapter 2 we emphasized the utility of DNA polymorphisms as genetic markers in the human genome for mapping genetic risk factors ("disease genes"). Precise genetic mapping is often the first step in cloning a genetic risk factor. The present human genetic map is based on studies of a standard set of large pedigrees established by the Centre d'Etude du Polymorphisme Humain (CEPH) in Paris. Each genetic marker is at a location that is known both in map position (centimorgans) and in physical position (location along the DNA sequence). There is about 60 percent more recombination in females than in males, so the female and male genetic maps differ in length. The female map is about 4400 cM; the male map about 2700 cM. Averaged over both sexes, the length of the human genetic map for all 23 pairs of chromosomes is about 3500 cM. Because the total DNA content per haploid set of chromosomes is 2.9 billion base pairs, there is, very roughly, 1.2 cM per million base pairs in the human genome.

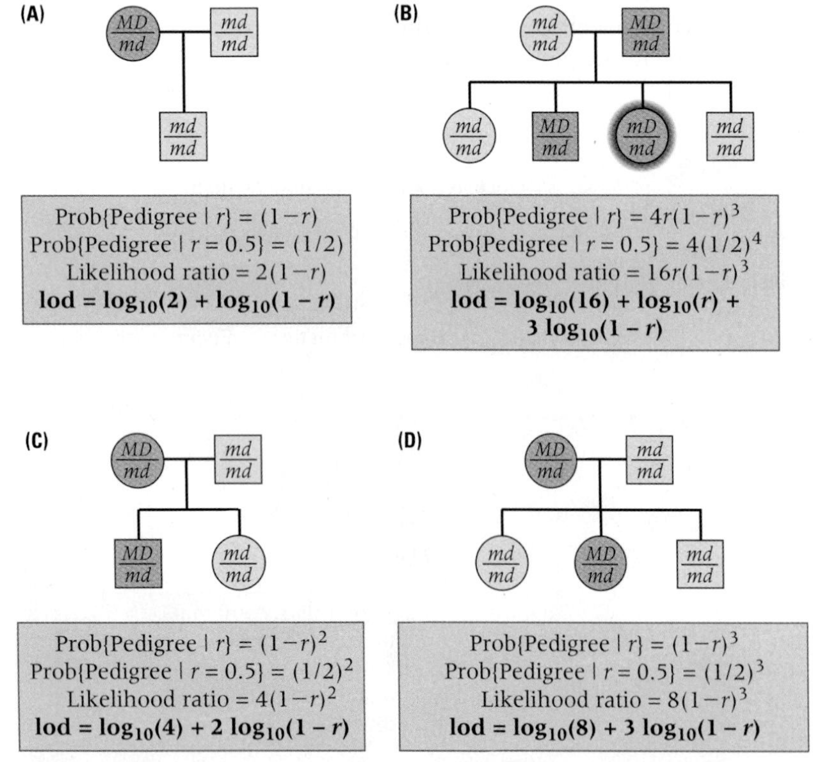


**FIGURE 5.16** Pedigrees showing genetic linkage of a molecular marker *M* to a disease gene *D*. For each pedigree in turn, each progeny is classified as either recombinant or nonrecombinant, and the probability of the particular offspring distribution is calculated for recombination frequency *r* using the binomial distribution. This probability is divided by that for the special case $r = 1/2$ (no linkage) to obtain the likelihood ratio. (Note that the binomial coefficients cancel.) For any value of *r*, the likelihood ratio is the relative probability of the pedigree assuming linkage to that assuming no linkage. The logarithm (base 10) of the likelihood ratio constitutes the lod score for the pedigree.

### ■ Maximum Likelihood and Lod Scores

The methods used for mapping in human genetics are somewhat different from those for experimental organisms, because

- Matings cannot be arranged for experimental convenience.
- Coupling ($A\ B/a\ b$) versus repulsion ($A\ b/a\ B$) heterozygotes cannot usually be distinguished.
- Progeny numbers are small.

In this section we examine a method commonly used for linkage analysis in human genetics, which allows data from multiple pedigrees to be pooled.

**FIGURE 5.16** depicts four pedigrees segregating for a polymorphic DNA marker and a trait determined by a dominant allele *D*. For simplicity, we assume complete penetrance of the dominant allele. The symbols *M* and *m* represent alleles of the molecular marker, and for the sake of illustration, we assume that the linkage phase of the double-heterozygous parents is known to be $M\ D/m\ d$ in each pedigree. For linkage studies, each offspring is classified as to whether it is or is not recombinant. In this example, there is only 1 recombinant progeny, indicated by the arrow in pedigree B, and 9 nonrecombinant progeny. Using the same analysis as in Section 5.1 and other examples in this chapter, we would estimate the frequency of recombination as $r = 1/(1 + 9) = 0.10$. The problem is that the numbers are so small that a chi-square test for statistical significance is inappropriate.

How can we decide whether the pedigrees in Figure 5.16 yield significant evidence for linkage? The first step is to deduce the probability of the offspring from each mating, using the binomial distribution discussed in Chapter 4, in terms of the number of recombinants and nonrecombinants. For simplicity, in this example we classify each progeny only with respect to whether it is recombinant or nonrecombinant and ignore sex, affected status, and other traits. For example, the sibship in pedigree B consists of 1 recombinant and 3 nonrecombinants, and so the binomial probability of this sibship is $(4!/1!3!)r^1(1 - r)^3 = 4r(1 - r)^3$. This and the other probabilities are shown immediately below each pedigree. The next step is to calculate, for each pedigree, the probability of its distribution of recombinant and nonrecombinant progeny under the assumption that $r = 0.5$ (independent assortment), which assumes that the genes are either in different chromosomes or widely separated in the same chromosome. These probabilities are given in the second line below each

## connection

### Human Gene Map

*Assessed by the distance between genetic markers, the human genetic map is one of the most dense genetic maps of any organism. At present, the human map includes more than 5000 genetic markers.*

**Jeffrey C. Murry and 26 other investigators 1994**
University of Iowa and 9 other research institutions
*A Comprehensive Human Linkage Map with Centimorgan Density*

*The assembly of the human genetic map represents a major achievement in international science. It was pieced together by hundreds of investigators working in many countries throughout the world. This paper is one example of a progress report illustrating several important features. First, the human genetic map was made possible by researchers studying a single, very extensive set of families assembled by CEPH (Centre d'Étude du Polymorphisme Humain, Paris) so that their data could be collated. Second, a very large proportion of the markers in the human map were detected by means of the polymerase chain reaction (PCR) to amplify short tandem repeats (STRs) that are abundant and highly polymorphic in the human genome. Third, the data are so extensive that they are not presented in the paper itself but rather in a large wall chart and (more conveniently) at a site on the Internet. Fourth, the authors point out the utility of the genetic map in determining the chromosomal locations of mutations that cause disease, using the methods of this chapter, as a first step in identifying the genes themselves. Finally, as is clear from the closing passages of the excerpt, the authors are well aware of the implications of modern human genetics for ethics, law, and social policy.*

For the first time, humans have been presented with the capability of understanding their own genetic makeup and how it contributes to morbidity of the individual and the species. Rapid scientific advances have made this possible, and developments in molecular biology, genetics, and computing, coupled with a cooperative and interactive biomedical community, have accelerated the progress of investigation into human inherited disorders. A primary engine driving these advances has been the development and use of human gene maps that allow the rapid positional assignment of an inherited disorder as a starting point for gene identification and characterization. . . . The genetic maps were based on genotypes generated from DNA samples obtained from the CEPH reference pedigree set. Since this material is publicly available, individual research groups can add their own genetic markers in the future. . . . Although only markers genotyped on CEPH reference families are included in the maps, the list is extensive. The CEPH database contains blood group markers and protein polymorphisms as well as a variety of DNA-based markers including RFLPs and short tandem repeat polymorphisms (STRPs). There has been an emphasis on PCR-based markers. . . . The 3617 STRP markers alone provide about one marker every $1 \times 10^6$ base pairs (bp) throughout the genome. Once an initial localization [of a new gene] is identified by means of a set of genome-wide genetic maps, subsequent steps can be undertaken to increasingly narrow the region to be searched and eventually to select a specific interval on which studies of physical reagents, such as cloned DNA fragments, can be carried out. . . . Once the physical reagents are in hand, gene and mutation searches are possible with a number of strategies, . . . but considerable problems and difficulties still remain. . . . Public access to databases is an especially important feature of the Human Genome Project. They are easy to use and facilitate rapid communication of new findings and can be updated efficiently. . . . Finally, it should be emphasized that the new opportunities and challenges for biomedical research provided by these marker-dense maps also create an urgency to face the parallel challenges in the areas of ethics, law, and social policy. Our ability to distinguish individuals for forensic purposes, identify genetic predispositions for rare and common inherited disorders, and to characterize, if present, the underlying nature of the genetic components of normal trait variability such as height, intelligence, sexual preference, or personality type, has never been greater. Although technically feasible, whether these maps should be used for these ends should be resolved after open dialogue to review those implications and devise policy to deal with the as yet unpredictable outcomes.

**It should be emphasized that the new opportunities and challenges for biomedical research provided by these marker-dense maps also create an urgency to face the parallel challenges in the areas of ethics, law, and social policy.**

Source: J. C. Murray, et al., *Science* 265 (1994): 2049-2054.

pedigree. For each pedigree, the ratio of the probability of the pedigree given an arbitrary value of $r$, symbolized Prob{Pedigree | $r$}, to that with $r = 0.5$, symbolized Prob{Pedigree | $r = 0.5$}, is called the **likelihood ratio**; this is given in the third line. (Note that, in calculating the likelihood ratio, the binomial coefficients in the numerator and denominator cancel.) From the likelihood ratios we obtain the **lod score** for each pedigree, which is the logarithm (base 10) of the likelihood ratio. These are given in the bottom line under each pedigree. (The term *lod* stands for *log-odds*, because the likelihood ratio is sometimes called the *odds*.)

Because the pedigrees are independent, the probabilities multiply across pedigrees, which means that the likelihood ratios also multiply. Hence, the lod scores are additive across pedigrees, because $\log_{10}(xy) = \log_{10}(x) + \log_{10}(y)$. For the pedigrees in Figure 5.16, therefore, we can add the lod scores to obtain the overall lod:

$$\text{Lod} = \log_{10}(2 \times 16 \times 4 \times 8) + \log_{10}(r) + 9\log_{10}(1 - r)$$

The frequency of recombination is estimated as the value of $r$ that maximizes the lod score for all the pedigrees. The curve of lod against $r$ is plotted in **FIGURE 5.17**. In this example the maximum lod is at $r = 0.10$. This estimate makes intuitive sense because 1 progeny, among the total of 10, carries a recombinant chromosome. To assess the strength of the evidence for linkage, the convention in human genetics is to regard a lod score greater than 3 as being statistically significant. The rationale is that for two genes chosen at random, the odds against linkage in the human genome are about 50 : 1 (modern genetic maps put it closer to 66 : 1), and conventional statistical significance at the 5 percent level corresponds to odds of 20 : 1. Hence, for pedigree data to override the *a priori* odds against linkage at a 5 percent level requires odds of $(50 \times 20) : (1 \times 1)$, or 1000 : 1; this corresponds to a lod score of 3 because $\log_{10}(1000) = 3$. On the other side, a lod score $< -2$ is generally regarded as significant evidence *against* linkage. Any value $-2 < \text{lod} < 3$ is considered uninformative in regard to linkage, and additional data are required. In the example we are considering, the maximum lod at $r = 0.10$ is lod $= 1.60$, so the evidence for linkage is not regarded as statistically significant. There are too few total progeny.

Although 10 progeny are not sufficient to make a strong case for linkage in this example, 20 progeny would be sufficient. Suppose that the size of each sibship in Figure 5.16 were doubled and that 2 of the total of 20 progeny carried a recombinant chromosome. In this case the overall lod score would be:

$$\text{Lod} = \log_{10}(2^{20}) + 2\log_{10}(r) + 18\log_{10}(1 - r)$$

This curve is plotted in Figure 5.17 also. The maximum is again at $r = 0.10$, but in this case the maximum lod equals 3.20. Because the maximum lod is greater than 3, the evidence for linkage is regarded as statistically significant.

The analysis in Figure 5.17 illustrates the underlying principles of linkage analysis in human pedigrees. In practice, the genetic mapping of disease genes presents many complications not dealt with here. For example, if the linkage phase is not known, the probability for each pedigree is calculated as a weighted average, assuming an equal likelihood of coupling and repulsion. There are also problems with incomplete penetrance. For example, the major genetic risk factor for human breast cancer *BRCA1* has a penetrance of 37 percent by age 40, 66 percent by age 55, and 85 percent by age 80, so the ages of women in breast cancer pedigrees must be taken into account. There are also different genetic as well as environmental causes of breast cancer, yielding risks for women who do not carry *BRCA1* of 0.4 percent by age 40, 3 percent by age 55, and 8 percent by age 80. Sorting out complexities such as these requires

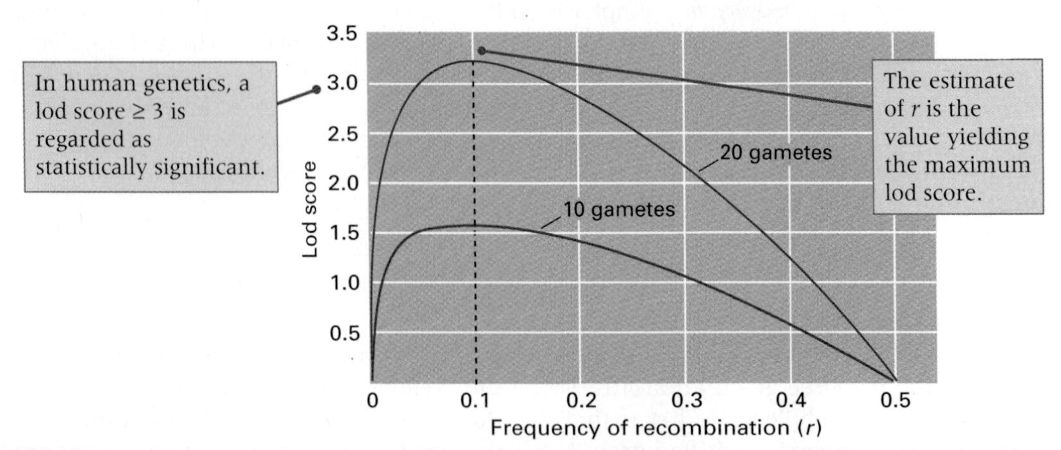


**FIGURE 5.17** Overall lod score for the pedigrees in Figure 5.16 showing maximum lod at $r = 0.10$. The lod based on 10 gametes is not statistically significant; that based on 20 gametes is statistically significant (lod $\geq 3$).

considerably more data than for simple Mendelian inheritance. For example, for a gene that doubles the risk of a complex disease, localizing the gene to within 1 cM ($\approx 10^6$ bp) through comparisons of marker genotypes in pairs of affected siblings requires a median of 700 sib pairs.

## 5.5 Mapping by Tetrad Analysis

Some species of fungi are especially favorable for genetic analysis. They are haploid, so recessive mutations are expressed, and they produce very large numbers of progeny. Most important, each meiotic tetrad is contained in a saclike structure called an **ascus** and can be recovered as an intact group. Each product of meiosis is included in a separate reproductive cell called an **ascospore**, and all of the ascospores formed from one meiotic cell remain together in the ascus (**FIGURE 5.18**). The advantage of using these organisms to study recombination is the potential for analyzing all of the products from each meiotic division. Certain aberrant asci showing segregation ratios other than 1 : 1 provide critical information about the molecular mechanism of recombination. This will be discussed in Chapter 6. In this section we show how the analysis of meiotic products in asci (*tetrad analysis*) can be used to study genetic linkage.

In these organisms, the life cycle is relatively short. In budding yeast, the diploid is induced to undergo meiosis by growth in medium that is deficient in its nutritional composition. The resulting haploid meiotic products, which form the ascospores, germinate to yield the vegetative stage of the organism (**FIGURE 5.19**). In some species, each of the four products of meiosis subsequently undergoes a mitotic division, with the result that each member of the tetrad yields a *pair* of genetically identical ascospores. In most of the organisms, the meiotic products, or their derivatives, are not arranged in any particular order in the ascus. However, bread molds of the genus *Neurospora* and related species have the useful characteristic that the meiotic products are arranged in a definite order directly related to the planes of the meiotic divisions. We will examine the ordered system after first looking at unordered tetrads.

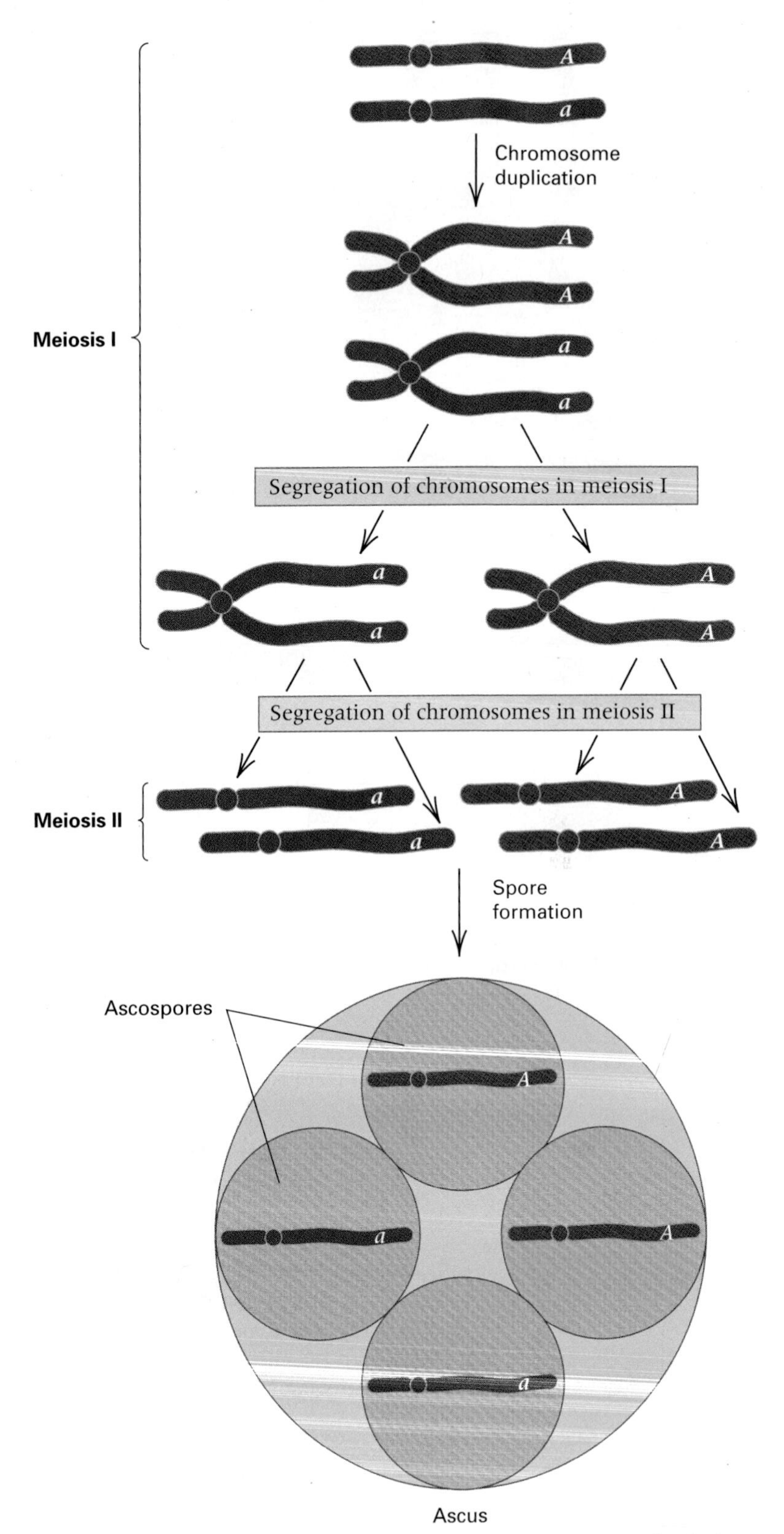


**FIGURE 5.18** Formation of an ascus containing all of the four products of a single meiosis. Each product of meiosis forms a reproductive cell called an ascospore; these cells are held together in the ascus. Segregation of one chromosome pair is shown.

### ■ Analysis of Unordered Tetrads

In the tetrads, when two pairs of alleles are segregating, three patterns of segregation are possible. For example, in the cross

$$A\,B \times a\,b$$

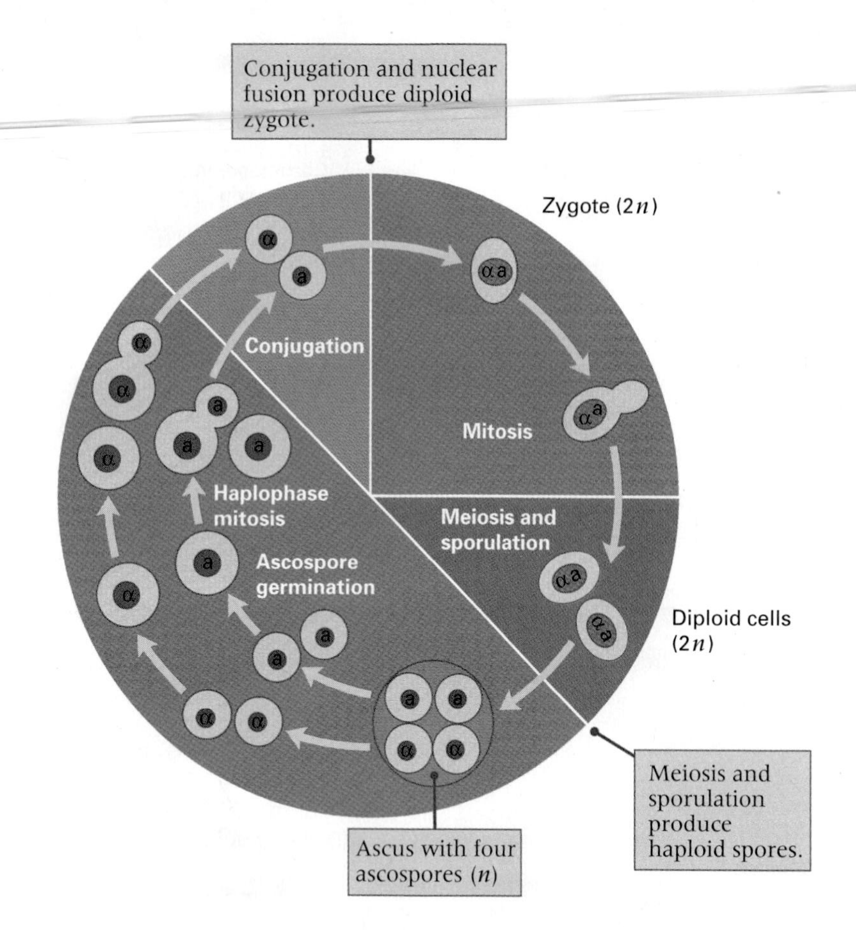


**FIGURE 5.19** Life cycle of the budding yeast *Saccharomyces cerevisiae*. Mating type is determined by the alleles **a** and α. Both haploid and diploid cells normally multiply by mitosis (budding). Depletion of nutrients in the growth medium induces meiosis and sporulation of cells in the diploid state. Diploid nuclei are orange; haploid nuclei are yellow.

the three types of tetrads are:

*AB AB ab ab* — Referred to as a **parental ditype (PD) tetrad.** Only two genotypes are represented, and the alleles have the same combinations found in the parents.

*Ab Ab aB aB* — Referred to as a **nonparental ditype (NPD) tetrad.** Only two genotypes are represented, but the alleles have nonparental combinations.

*AB Ab aB ab* — Referred to as a **tetratype (TT) tetrad.** All four possible genotypes are present.

The formation of tetrads, each containing all four products of a single meiosis, provides experimental proof of two fundamental principles of transmission genetics. The first principle is that of Mendelian segregation. Observe that each type of tetrad has a ratio of 2 *A* : 2 *a* and 2 *B* : 2 *b*. These ratios imply that Mendelian segregation is not merely a statistical average of ratios across a large number of meiotic divisions. Rather:

> Mendelian segregation of alleles takes place in each individual meiotic division and results in exactly two daughter cells carrying one allele and exactly two daughter cells carrying the alternative allele.

The second principle relates to crossing over. Observe that tetratype tetrads are composed of the allele configurations *A B, A b, a B* and *a b*. In other words, two of the products are parental types and two are recombinant types. The equality of parental and recombinant types implies that:

> Crossing over takes place at the four-strand stage of meiosis and results in two chromosomes that are recombinant and two chromosomes that are nonrecombinant.

Tetrads have played an important role in the genetics of fungi because linkage between genes is easily identified by the analysis of tetrads. The detection of linkage is based on the relative numbers of parental ditype (PD) and nonparental ditype (NPD) tetrads:

> When genes are *unlinked,* the parental-ditype tetrads and nonparental-ditype tetrads are expected in equal frequencies (PD = NPD). *Linkage* is indicated when nonparental-ditype tetrads appear with a much lower frequency than parental-ditype tetrads (NPD << PD).

The reason for the equality PD = NPD for unlinked genes is shown in **FIGURE 5.20** (part A) for two pairs of alleles located in different chromosomes. In the absence of crossing over between either gene and its centromere, the two chromosomal configurations are equally likely at metaphase I, so PD = NPD. When there is crossing over between either gene and its centromere (Figure 5.20B), a tetratype tetrad results, but this does not change the fact that PD = NPD.

In contrast, when genes are linked, parental ditypes are far more frequent than nonparental ditypes. To see why, assume that the genes are linked and consider the events required for the production of the three types of tetrads. **FIGURE 5.21** shows that when no crossing over takes place between the genes, a PD tetrad is formed. Single crossing over between the genes results in a TT tetrad. The formation of a two-strand, three-strand, or four-strand double crossover results in a PD, TT, or NPD tetrad, respectively. With linked genes, meiotic cells with no crossovers will always outnumber those with four-strand double crossovers. This results in the inequality NPD << PD that is characteristic of linkage.

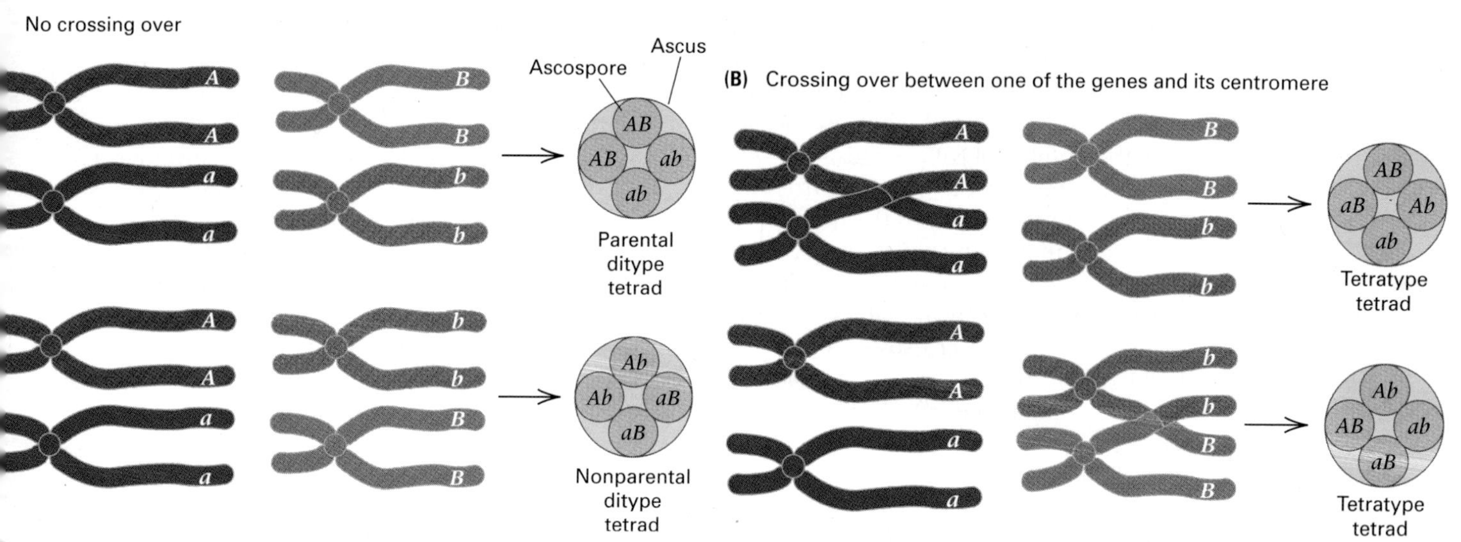


URE 5.20 Types of unordered asci produced with two genes in different chromosomes. (A) In the absence of crossing over, dom arrangement of chromosome pairs at metaphase I results in two different combinations of chromatids, one yielding PD ads and the other NPD tetrads. (B) When crossing over takes place between one gene and its centromere, the two chromo- ne arrangements yield TT tetrads. If both genes are closely linked to their centromeres (so that crossing over between each e and its centromere is rare), few TT tetrads are produced.

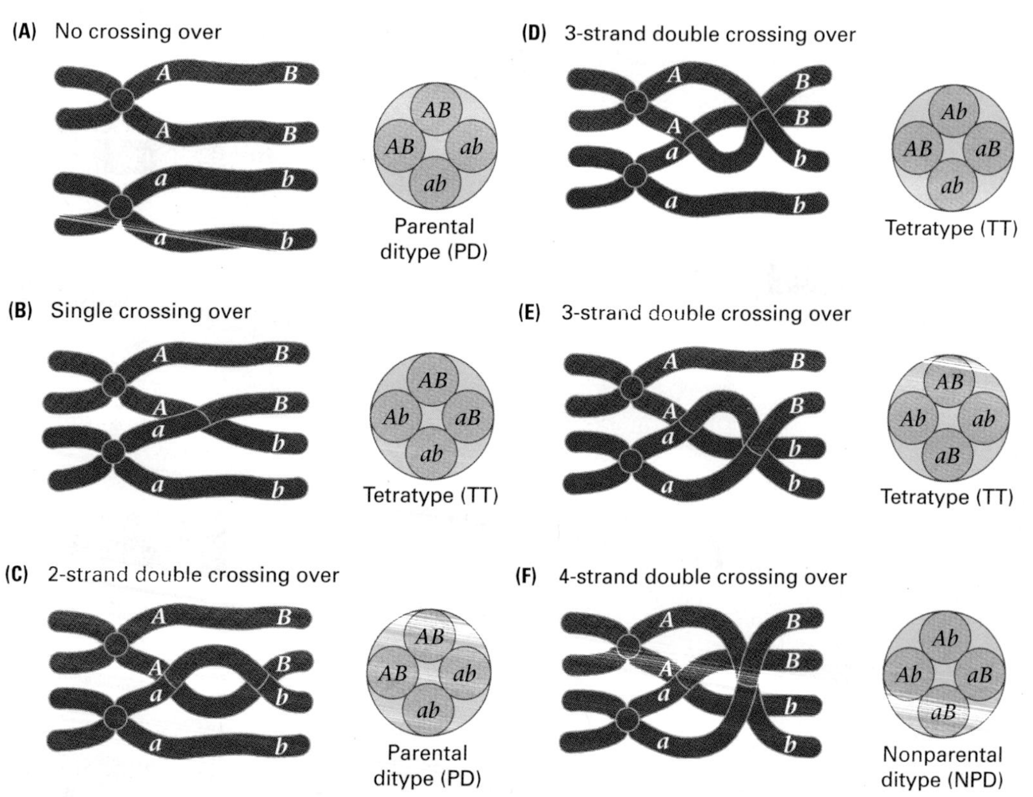


**FIGURE 5.21** Types of tetrads produced with two linked genes. In the absence of crossing over (A), a PD tetrad is produced. With a single crossover between the genes (B), a TT tetrad is produced. Among the four possible types of double crossovers between the genes (C, D, E, and F), only the four-strand double crossover in part F yields an NPD tetrad.

## Genetic Mapping with Unordered Tetrads

The relative frequencies of the different types of tetrads can be used to determine the map distance between two linked genes, assuming that the genes are sufficiently close that triple crossovers and higher levels of crossing over can be neglected. To determine the map distance, we first use the number of NPD tetrads to estimate the frequency of double crossovers. The four types of double crossovers in Figure 5.21 are expected in equal frequency. Thus the number of tetrads resulting from double crossovers (parts C, D, E, and F of Figure 5.21) should equal four times the number of NPD tetrads (part F), or 4 × [NPD], where the square brackets denote the observed number of NPD tetrads. At the same time, the number of TT tetrads resulting from three-strand double crossovers (parts D and E) should equal twice the number of NPD tetrads, or 2 × [NPD] (part F). Subtracting 2 × [NPD] from the total number of tetratypes yields the number of tetratype tetrads that originate from single crossovers, or [TT] − 2[NPD]. By definition (Equation 1), the map distance between two genes equals half the frequency of single-crossover tetrads ([TT] − 2[NPD]) plus the frequency of double-crossover tetrads (4[NPD]) times 100. Thus,

$$\text{Map distance} = \frac{(1/2)([TT] - 2[NPD]) + 4[NPD]}{\text{Total number of tetrads}} \times 100$$

$$= \frac{(1/2)[TT] + 3[NPD]}{\text{Total number of tetrads}} \times 100 \qquad (2)$$

The 1/2 in these expressions corrects for the fact that only half of the meiotic products in a TT tetrad produced by a single crossover are recombinant. In the special case where the linked genes are close enough that [NPD] = 0 (no double crossovers), the map distance equals the percentage of TT tetrads divided by 2.

A systematic format for analyzing linkage relationships for allelic pairs in unordered tetrads is presented in the form of a logical flow diagram in **FIGURE 5.22**. All of the equations are based on the assumption that there is no chromatid interference. The right-hand side summarizes the equa-

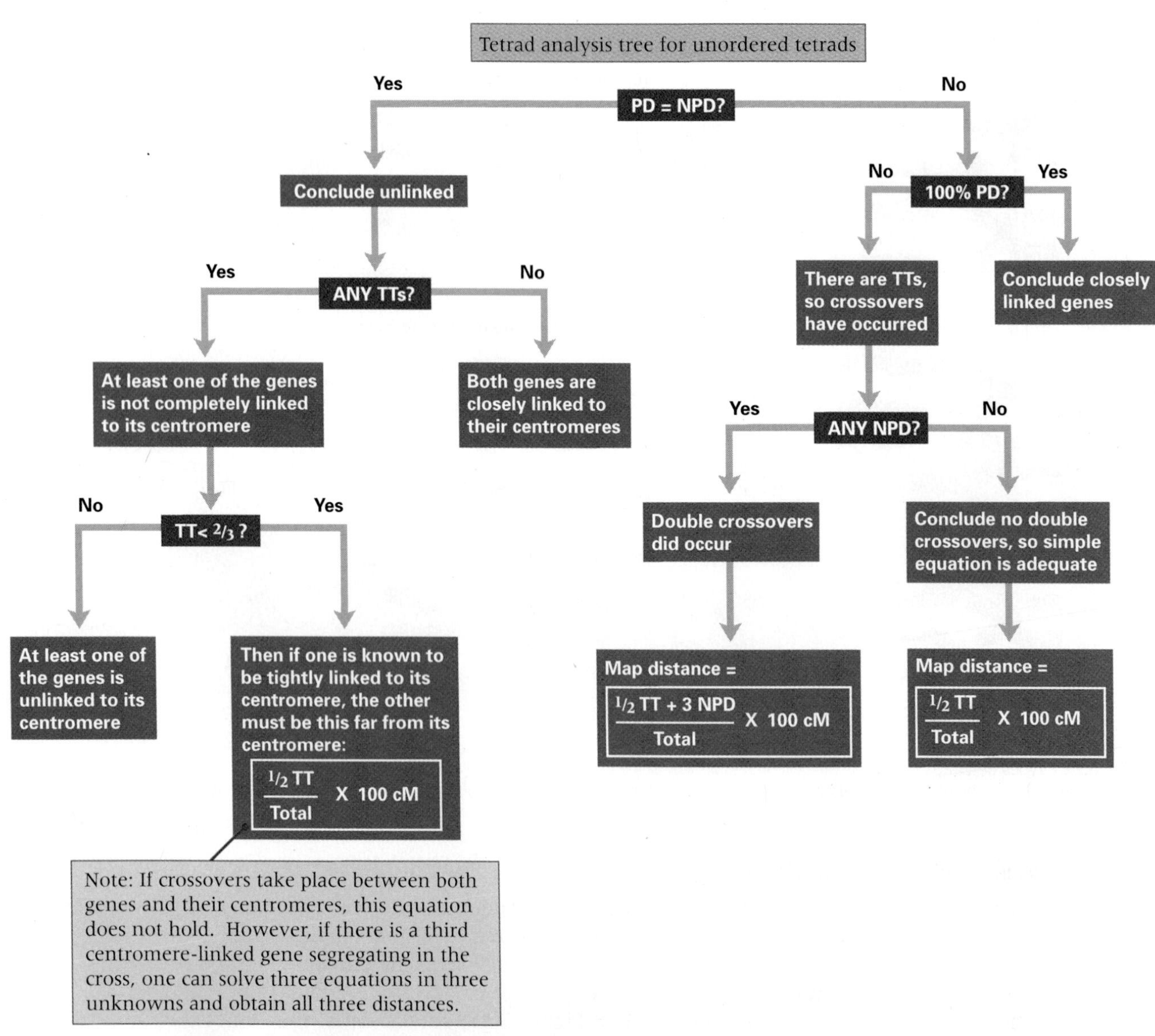


**FIGURE 5.22** Tree diagram for analyzing linkage relations from unordered tetrads.

tions for linked genes that we derived on the basis of Figure 5.21. The left-hand side pertains to unlinked genes, and it shows how the frequency of tetratypes can be used to make inferences about the distance of each gene from its own centromere.

As an example of the calculations for linked genes, consider a two-factor cross that yields 112 PD, 4 NPD, and 24 TT tetrads. The fact that NPD $<<$ PD (that is, $4 << 112$) indicates that the two genes are linked and leads us down the right branch of the tree. Because both TT and NPD tetrads are recovered, we take the first left fork, and because there are NPD tetrads, we also take the next left fork. The appropriate equation is therefore Equation (2). Substitution of the values yields a map distance of:

$$\text{Map distance} = \frac{[(1/2) \times 24] + (3 \times 4)}{112 + 4 + 24} \times 100$$
$$= 17.1 \text{ cM}$$

Note that this mapping procedure differs from that presented earlier in the chapter in that recombination frequencies are not calculated directly from the number of recombinant and nonrecombinant chromatids but, rather, from the inferred types of crossovers.

As Figure 5.22 indicates, tetrad analysis yields a great deal of information about the linkage relationship between genetic markers and the distance of unlinked markers to their respective centromeres. However, it is not necessary to carry out a full tetrad analysis for estimating linkage. The alternative is to examine spores chosen at random after allowing the tetrads to break open and disseminate their spores. This procedure is called **random-spore analysis,** and the linkage relationships are determined exactly as described earlier for *Drosophila* and *Zea mays*. In particular, the frequency of recombination equals the number of spores that are recombinant for the genetic markers divided by the total number of spores.

## Analysis of Ordered Tetrads

*Neurospora crassa* is celebrated as the organism in which Beadle and Tatum carried out their Nobel Prize experiments on the relation between genes and enzymes (Chapter 1). Here we look at some of the special genetic features that the organism has to offer. In *Neurospora*, the products of meiosis are contained in an *ordered* array of ascospores (**FIGURE 5.23**). A zygote nucleus contained in a sac-like ascus undergoes meiosis almost immediately after it is formed. The four nuclei produced by meiosis are in a linear, ordered sequence in the ascus, and each of them undergoes a mitotic division to form two genetically identical and adjacent ascospores. Each mature ascus contains eight ascospores arranged in four pairs, each pair derived from one of the products of meiosis. The ascospores can be removed one by one from an ascus and each germinated in a culture tube to determine its genotypes.

Ordered asci also can be classified as PD, NPD, or TT with respect to two pairs of alleles; hence the tree diagram in Figure 5.22 can be used to analyze the linkage data. In addition, the ordered arrangement of meiotic products makes it possible to determine the recombination frequency between any particular gene and its centromere. The logic of the mapping technique is based on the feature of meiosis shown in **FIGURE 5.24**:

> Homologous centromeres of parental chromosomes separate at the first meiotic division; the centromeres of sister chromatids separate at the second meiotic division.

Thus, in the absence of crossing over between a gene and its centromere, the alleles of the gene (for example, *A* and *a*) must separate in the first meiotic division; this separation is called **first-division segregation.** If, instead, a crossover is formed between the gene and its centromere, the *A* and *a* alleles do not become separated until the second meiotic division; this separation is called **second-division segregation.** The distinction between first-division and second-division segregation is shown in Figure 5.24. As shown in part A, first-division segregation is indicated by either of two possible arrangements of the spores: *A A a a* and *a a A A*. However, as shown in part B, second-division segregation is indicated by any of four patterns: *A a A a*, *a A a A*, *A a a A*, and *a A A a*.

The percentage of asci with second-division segregation can be used to map a gene with respect to its centromere. For example, let us assume that 30 percent of a sample of asci from a cross exhibit second-division segregation for the *A* and *a* alleles. This means that 30 percent of the cells undergoing meiosis had a crossover between the *A* gene and its centromere. Furthermore, in each cell in which crossing over takes place, two of the chromatids are recombinant and two are nonrecombinant. In other words, a frequency of crossing over of 30 percent corresponds to a recombination frequency of 15 percent. By convention, map distance refers to the frequency of recombinant meiotic

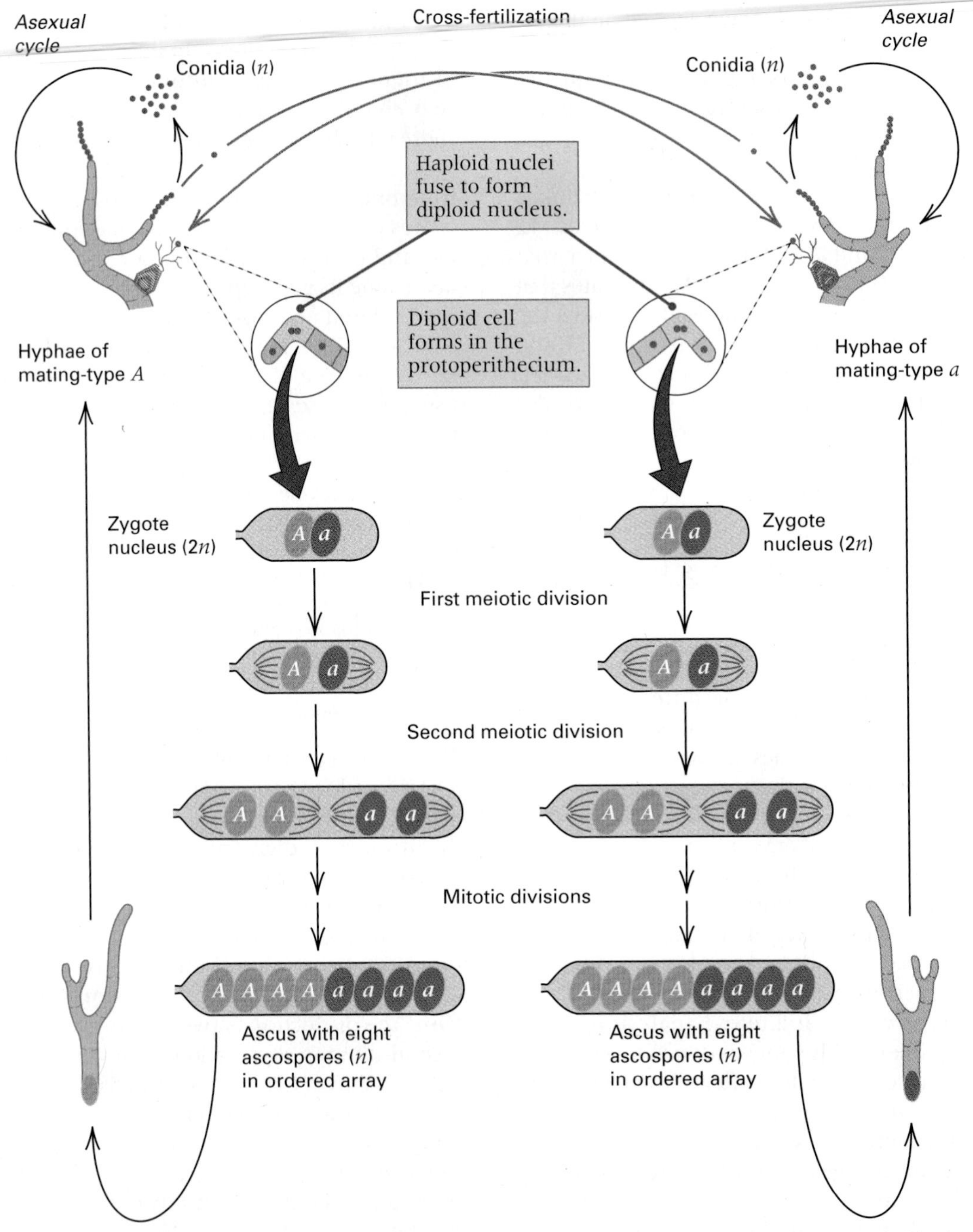


**FIGURE 5.23** The life cycle of *Neurospora crassa.* The vegetative body consists of partly segmented filaments called hyphae. Conidia are asexual spores that function in the fertilization of organisms of the opposite mating type. A protoperithecium develops into a structure in which numerous cells undergo meiosis.

products rather than to the frequency of cells with crossovers. Therefore, the map distance between a gene and its centromere is given by

$$\frac{(1/2)(\text{Asci with second-division segregation patterns})}{\text{Total number of asci}} \times 100 \qquad (3)$$

This equation is valid as long as the gene is close enough to the centromere for multiple crossovers to be neglected. Reliable linkage values are best determined for genes that are near the centromere. The location of more distant genes is accomplished by mapping them relative to other genes nearer the centromere.

If a gene is far from its centromere, crossing over between the gene and its centromere is so frequent that the *A* and *a* alleles become randomized with respect to the four chromatids. The result is that the six possible spore arrangements shown in Figure 5.24 are all equally likely. This equal likelihood reflects the patterns that can result from choosing randomly among 2 *A* and

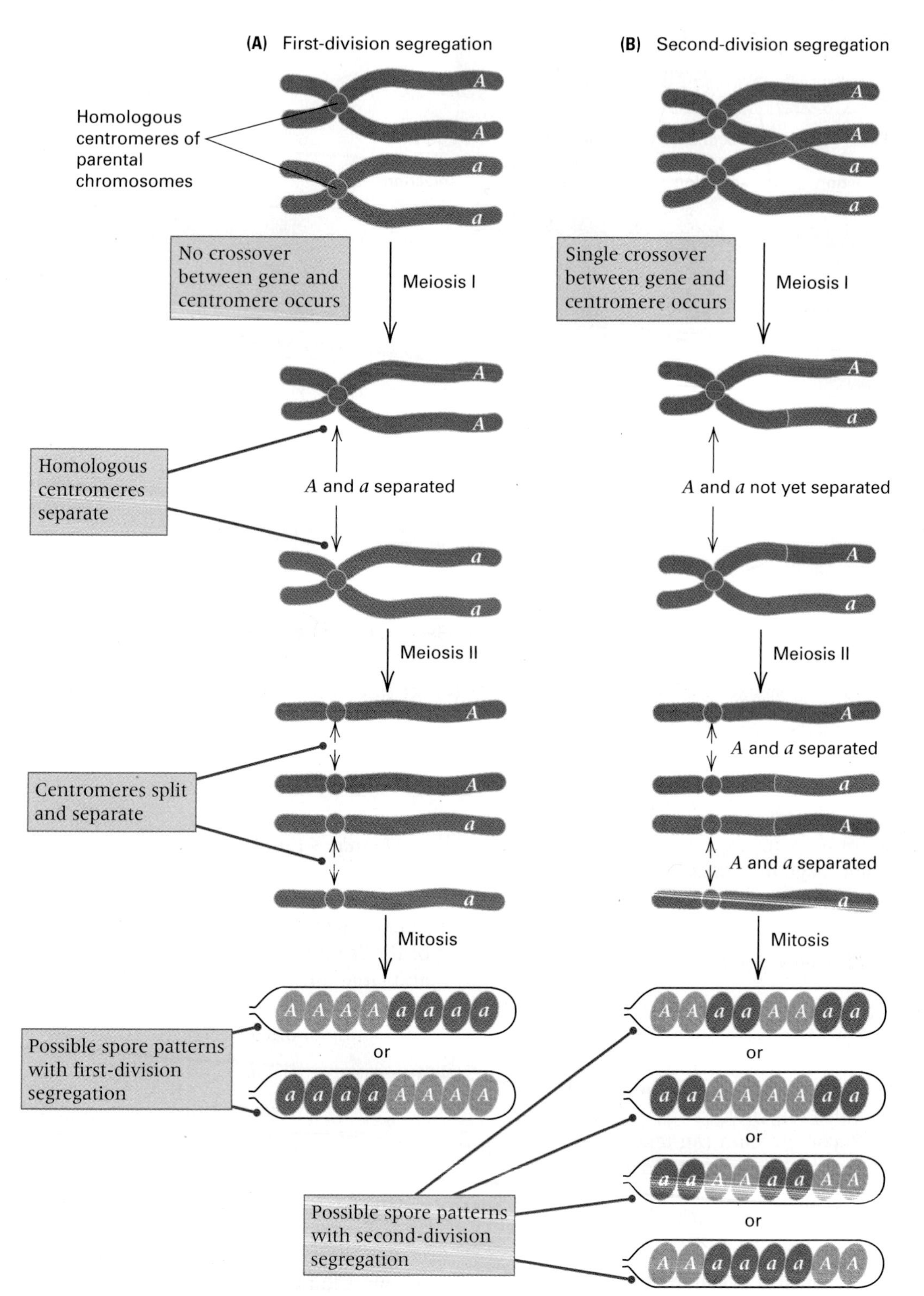


**FIGURE 5.24** First- and second-division segregation in *Neurospora.* (A) First-division segregation patterns are found in the ascus when crossing over between the gene and centromere does not take place. The alleles separate (segregate) in meiosis I. Two spore patterns are possible, depending on the orientation of the pair of chromosomes on the first-division spindle. (B) Second-division segregation patterns are found in the ascus when crossing over between the gene and the centromere delays separation of *A* from *a* until meiosis II. Four patterns of spores are possible, depending on the orientation of the pair of chromosomes on the first-division spindle and that of the chromatids of each chromosome on the second-division spindle.

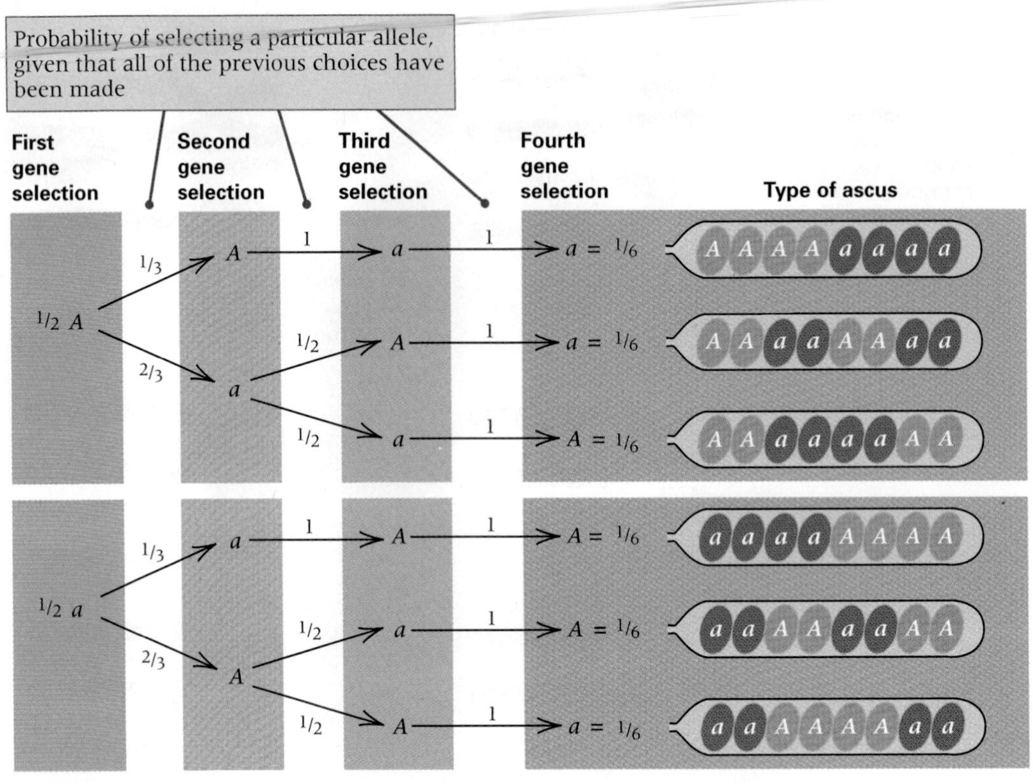


**FIGURE 5.25** Diagram showing the result of free recombination between an allelic pair, *A* and *a*, and the centromere. The frequency of second-division segregation equals 2/3. This is the maximum frequency of second-division segregation, provided there is no chromatid interference.

2 *a* spore pairs in the ascus, as shown by the branching diagram in **FIGURE 5.25**. Therefore, in the absence of chromatid interference,

> The maximum frequency of second-division segregation asci is 2/3.

## 5.6 Special Features of Recombination

In this section we examine two classic experiments indicating that:

1. Recombination can take place between the nucleotides in a single gene.
2. Recombination between genes can occur during mitosis.

### Recombination Within Genes

Initial evidence for recombination within genes came from investigations of a gene in the X chromosome of *D. melanogaster* known as *lozenge* (*lz*). Mutant alleles of *lozenge* are recessive, and their phenotypic effects include a disturbed arrangement of the facets of the compound eye and a reduction in the eye pigments. Numerous *lz* alleles with distinguishable phenotypes are known.

Females heterozygous for two different *lz* alleles, one in each homologous chromosome, have lozenge eyes. When such heterozygous females were crossed with males that had one of the *lz* mutant alleles in the X chromosome, and large numbers of progeny were examined, flies with normal eyes were occasionally found. For example, one cross carried out was

$$\frac{X \quad + \quad lz^{BS}+ \quad +}{X \quad ct \quad +lz^{g} \quad v} \times \frac{+ \quad +lz^{g} \quad v \quad X}{Y}$$

in which $lz^{BS}$ and $lz^{g}$ are mutant alleles of *lozenge*, and *ct* (cut wing) and *v* (vermilion eye color) are genetic markers that map 7.7 units to the left and 5.3 units to the right, respectively, of the *lozenge* locus. From this cross, 134 males and females with wildtype eyes were found among more than 16,000 progeny, for a frequency of $8 \times 10^{-3}$. These exceptional progeny might have resulted from a reverse mutation of one or of the other of the mutant *lozenge* alleles to $lz^{+}$, but the observed frequency, though small, was much greater than the known frequencies of reverse mutations. The genetic constitution of the maternally derived

X chromosomes in the normal-eyed offspring was also inconsistent with such an explanation. The male offspring had cut wings and the females were *ct*/+ heterozygotes. That is, all of the rare nonlozenge progeny had an X chromosome with the constitution

$$\underline{ct \quad ++ \quad +}$$

which could be accounted for by recombination between the two *lozenge* alleles. Proof of this conclusion came from the detection of the reciprocal recombinant chromosome

$$\underline{+ \quad lz^{BS}lz^{g} \quad v}$$

in five male progeny that had a lozenge phenotype distinctly different from the phenotype resulting from the presence of either a $lz^{BS}$ or $lz^{g}$ allele alone. The exceptional males also had vermilion eyes, as expected from reciprocal recombination. The observation of intragenic recombination indicated that genes have a subunit structure, now known to be a sequence of nucleotides, and that the multiplicity of allelic forms might be due to mutations at different sites (nucleotides) in the gene.

## Mitotic Recombination

Genetic exchange can also take place in mitosis, although at a frequency about a thousandfold lower than in meiosis and probably by a somewhat different mechanism than meiotic recombination. The first evidence for mitotic recombination was obtained in experiments with *Drosophila*, but the phenomenon has been studied most carefully in fungi such as yeast and *Aspergillus*, in which the frequencies are higher than in most other organisms. Genetic maps can be constructed from mitotic recombination frequencies. In some organisms in which a sexual cycle is unknown, mitotic recombination is the only method of obtaining linkage data. In organisms in which both meiotic recombination and mitotic recombination are found, mitotic recombination is always at a much lower frequency. In mitotic recombination, the relative map distances between particular genes sometimes correspond to those based on meiotic recombination frequencies, but for unknown reasons, the distances are often markedly different. The discrepancies may reflect different mechanisms of exchange and perhaps a nonrandom distribution of potential sites of exchange.

The genetic implications of mitotic recombination are illustrated in **FIGURE 5.26**. Each of the homologous chromosomes has two genetic markers distal to the site of the breakage and reunion. At the following anaphase, each centromere splits and the daughter cells receive one centromere of each color, along with the chromatid attached to it. The disjunction can happen

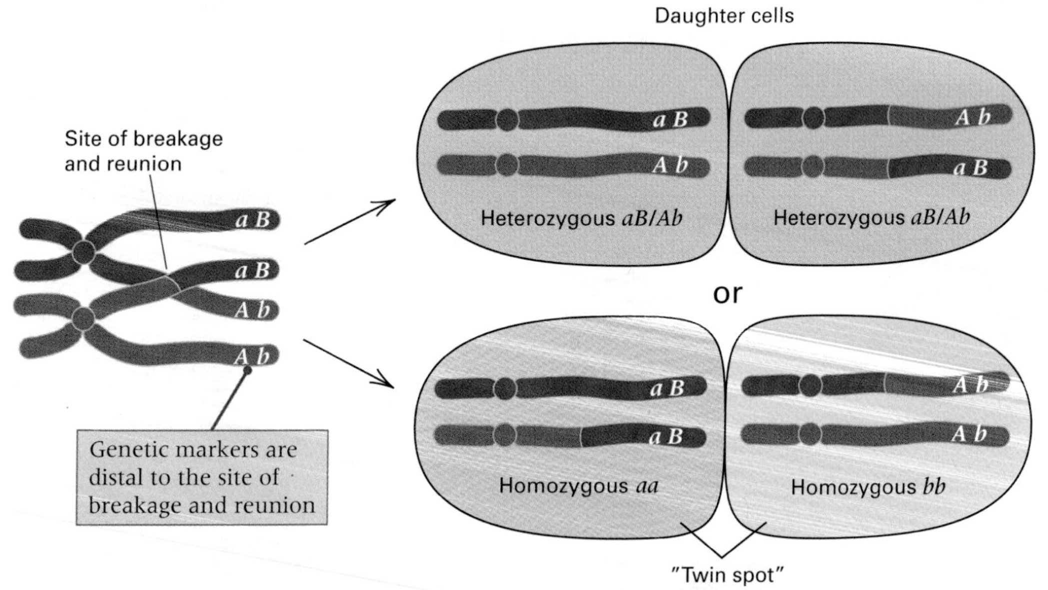


**FIGURE 5.26** Mitotic recombination. Homologous chromosomes in prophase of mitosis are shown in blue and green. At anaphase, each centromere splits, and each daughter cell receives one centromere of each color and its attached chromatid. When a rare reciprocal exchange takes place between nonsister chromatids of homologous chromosomes, the daughter cells can be either of two types. If the nonexchanged chromatids and the exchanged chromatids go together (top right), both cells are genetically like the parent—in this example, *A b*/*a B*. If one nonexchanged and one exchanged chromatid go to each cell together (bottom right), the result is that alleles distal to the point of exchange become homozygous. In this example, one daughter cell is *a B*/*a B* and the other *A b*/*A b*. Adjacent cell lineages are therefore homozygous for either *aa* or *bb* and can be detected phenotypically as a "twin spot" of *aa* somatic tissue adjacent to *bb* somatic tissue.

in two ways, as shown on the right. If the two nonexchanged and the two exchanged chromatids go together (top), the result of the exchange is not detectable genetically. However, if each nonexchanged chromatid goes with one of the exchanged chromatids, then the result is that the region of the chromosome distal to the site of the exchange becomes homozygous. With an appropriate configuration of genetic markers, the exchange is detectable as a *twin spot* in which cells of one homozygous genotype (in this example, an *a B* / *a B* spot) are adjacent to cells of a different homozygous genotype (in this example, an *A b* / *A b* spot). The original experiments made use of the X-linked markers $A = y^+$ and $a = y$ (*y* is *yellow* body color) along with $B = sn^+$ and $b = sn$ (*sn* is *singed* bristles). In females, the result of mitotic recombination was observed as a twin spot in which a patch of cuticle with a yellow color and normal bristles was adjacent to a patch of cuticle with normal color and singed bristles.

The rate of mitotic recombination can be increased substantially through x-ray treatment and the use of certain mutations. Mitotic recombination is a useful tool in genetics because it results in the production of **somatic mosaics**, organisms that contain two or more genetically different types of tissue. For example, the mosaic *Drosophila* females discussed in the previous paragraph consist of predominantly wildtype tissue but include some patches of $y^+ \, sn/y^+ \, sn$ and some patches of $y \, sn^+/y \, sn^+$. Each twin spot of mutant tissue derives from repeated division of a single cell that became homozygous through a mitotic exchange. In modern developmental genetics, the movement and differentiation of particular cell lineages are tracked by the study of patches of genetically marked tissue resulting from mitotic recombination.

## CHAPTER SUMMARY

- Genes that are located sufficiently close together in a chromosome do not undergo independent assortment; they are said to be linked.
- The alleles of linked genes tend to be inherited as a group.
- Crossing over between homologous chromosomes results in recombination that breaks up combinations of linked alleles.
- A genetic map depicts the relative positions of genes along a chromosome.
- The map distance between genes in a genetic map is related to the rate of recombination between the genes.
- Physical distance along a chromosome is often (but not always) correlated with map distance.
- Tetrads are sensitive indicators of linkage because each contains all the products of a single meiosis.
- Recombination can take place between nucleotides within a single gene.

## REVIEW THE BASICS

- In genetic analysis, why is it important to know the map position of a gene along a chromosome?
- How is the recombination frequency estimated? What does it mean to say that two genes have a recombination frequency of 10 percent? What is the maximum frequency of recombination between two genes? Is there a maximum map distance between genes?
- Why is the frequency of recombination over a long interval of a chromosome always smaller than the map distance over the same interval?
- For a region between two genes in which no more than one crossover can take place in each cell undergoing meiosis, the frequency of recombination is equal to one half the frequency of meiotic cells in which a chiasma occurs in the region. Explain why the factor of one half occurs when we compare recombination frequency to chiasma frequency.
- What is a double crossover? How many different kinds of double crossovers are possible? Diagram each kind of double crossover in a region between two heterozygous genes in a bivalent, along with the resulting chromatids.
- What is interference? In a region of high interference, would you observe more or fewer double crossovers than expected? Explain.
- What is special about the inheritance of an attached-X chromosome, compared with the inheritance of unattached X chromosomes?
- Explain how in fungi, the occurrence of tetratype tetrads for two linked genes implies that crossing over must take place at the four-strand stage of meiosis.
- Explain how the analysis of fungal tetrads for two linked genes allows double crossovers to be detected in a two-point cross (a cross in which only two genes are segregating).
- Explain how the recovery of ordered tetrads allows any gene to be mapped relative to the centromere of the chromosome on which the gene is located.

**Problem 1** Two genes in chromosome 7 of corn are identified by the recessive alleles *gl* (glossy), determining glossy leaves, and *ra* (ramosa), determining branching of ears. When a plant heterozygous for each of these alleles was crossed with a homozygous recessive plant, the progeny consisted of the following genotypes with the numbers of each indicated:

| | | | |
|---|---|---|---|
| *Gl ra / gl ra* | 93 | *gl ra / gl ra* | 4 |
| *Gl Ra / gl ra* | 6 | *gl Ra / gl ra* | 105 |

What is the genotype of the doubly heterozygous parent? Estimate the frequency of recombination between these genes.

**Answer** The most frequent gametes from the doubly heterozygous parent are the nonrecombinant gametes, in this case *Gl ra* and *gl Ra*. The genotype of a parent was therefore *Gl ra* / *gl Ra*. The recombinant gametes are *gl ra* and *Gl Ra*, and the frequency of recombination is estimated as $(6 + 4)/(93 + 105 + 6 + 4) = 4.8$ percent.

**Problem 2** A cross in corn is carried out with three linked genes in a triply heterozygous genotype for the allele pairs (*A*, *a*), (*D*, *d*), and (*R*, *r*). The dominant *A* allele results in red leaves; *aa* plants have green leaves. The dominant *D* allele results in tall plants; *dd* plants are dwarfed. The dominant *R* allele results in ragged leaf margin; *rr* plants have smooth leaf margins. The triply heterozygote is crossed with homozygous recessives for all three alleles, and the following phenotypes of progeny are observed. Assume that interference in each interval is complete, so that the map distance in centimorgans (cM) corresponds to the percent recombination,

| | |
|---|---|
| red, tall, ragged | 265 |
| red, tall, smooth | 24 |
| red, dwarf, ragged | 120 |
| red, dwarf, smooth | 90 |
| green, tall, ragged | 70 |
| green, tall, smooth | 140 |
| green, dwarf, ragged | 16 |
| green, dwarf, smooth | 275 |
| Total | 1000 |

**(a)** What is the order of the genes?
**(b)** What is the genotype of the triply heterozygous parent with the genes written in correct order?
**(c)** What is the shortest map distance between adjacent genes?
**(d)** What is the longest map distance between adjacent genes?
**(e)** What is the value of the interference across this interval?

**Answer**

**(a)** To solve this type of problem, you should first deduce the genotype of the gamete contributed by the triply heterozygous parent to each class of progeny. In this case, the gamete types are, from top to bottom, *A D R*, *A D r*, *A d R*, *A d r*, *a D R*, *a D r*, *a d R*, and *a d r*. To determine the order of the genes, you must identify which gene is in the middle. You can do this by comparing the parental type gametes with the double crossovers. The two parental types are *A D R* and *a d r*, because these are the most frequent, and the double crossover types are *A D r* and *a d R*, as these are the least frequent. Comparing the parental types with the double crossover types reveals that *R* and *r* are interchanged relative to the alleles of the other two genes. Hence, the order of the genes is *A R D* or equivalently *D R A*. For concreteness we will use the order *A R D*.
**(b)** The triply heterozygous parent therefore has the genotype *A R D*/*a r d*.
**(c)** The distance between the genes *R* and *D* equals the number of single crossovers plus the number of double crossovers in this region divided by the total number of progeny or $(90 + 70 + 24 + 16)/1000 = 20$ cM. (The double crossovers must be included, because they include a crossover in the R—D interval.)
**(d)** The distance between the genes *A* and *R* equals the number of single crossovers plus the number of double crossovers in this region divided by the total number of the progeny or $(120 + 140 + 24 + 16)/1000 = 30$ cM.
**(e)** The expected number of double crossovers equals $0.2 \times 0.3 \times 1000 = 60$, whereas the observed number of double crossovers is $24 + 16 = 40$. The interference *i* therefore equals $1 - (40/60) = 1/3$.

**Problem 3** In the accompanying human pedigree, the parents I-1 and I-2 are both heterozygous for a recessive allele for cystic fibrosis (*cf*) and also heterozygous for a restriction fragment length polymorphism (RFLP) whose phenotypes are shown in the diagram of the electrophoresis gel. DNA fragments *A* and *a* are the result of a restriction-enzyme cleavage site that differs between alleles *A* and *a*. The RFLP is linked to the *CF* gene. II-1 and II-2 are first cousins, and a molecular analysis of their parents and siblings indicates that both II-1 and II-2 have the genotype *CF A*/ *cf a*. The frequency of recombination between the genes is 12 percent. None of the children II-1 through II-5 is affected with cystic fibrosis.

**(a)** What is the probability that the child II-6 is affected?
**(b)** What is the probability that the child II-1 is a carrier?

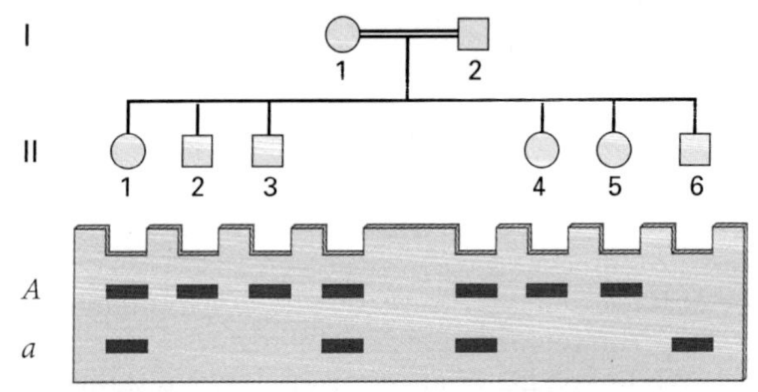


**Answer**

**(a)** For II-6 to be affected, his genotype would have to be *cf a*/ *cf a*, so both chromosomes would have to be nonrecombinant *cf a*. The probability of a nonrecombinant *cf a* chromosome from the mother is $(1 - 0.12)/2 = 0.44$ and from the father is also $(1 - 0.12)/2 = 0.44$.

The probability that II-6 is affected is therefore 0.44 × 0.44 = 0.194.

**(b)** For II-1 to be a carrier, her genotype would have to be *CF A/cf a* or *CF a/cf A*. The *CF A/cf a* genotype has either a *CF A* nonrecombinant chromosome from the mother and a *cf a* nonrecombinant chromosome from the father, or a *cf a* nonrecombinant chromosome from the mother and a *CF A* nonrecombinant chromosome from the father; the overall probability of *CF A/cf a* is therefore $[(1 - 0.12)/2] \times [(1 - 0.12)/2] + [(1 - 0.12)/2] \times [(1 - 0.12)/2] = 0.3872$. The *CF a/cf A* genotype has either a *CF a* recombinant chromosome from the mother and a *cf A* recombinant chromosome from the father, or a *cf A* recombinant chromosome from the mother and a *CF a* recombinant chromosome from the father; the probability is therefore $0.12/2 \times 0.12/2 + 0.12/2 \times 0.12/2 = 0.0072$. The overall probability that II-1 is a carrier therefore equals $0.3872 + 0.0072 = 0.3944$. (Note that the possibility of two recombinant chromosomes is much less likely than that of two nonrecombinant chromosomes.)

**Problem** The recessive mutations *b* (black body color), *st* (scarlet eye color), and *hk* (hooked bristles) identify three autosomal genes in *D. melanogaster*. The following progeny were obtained from a testcross of females heterozygous for all three genes. In each class of progeny, only the mutant phenotypes are designated.

| | |
|---|---|
| black, scarlet | 243 |
| black | 241 |
| black, hooked | 15 |
| black, hooked, scarlet | 10 |
| hooked | 235 |
| hooked, scarlet | 226 |
| scarlet | 12 |
| wildtype | 18 |

What conclusions are possible concerning the linkage relations of these three genes? Calculate any appropriate map distances.

**Answer** For problems of this type, the best strategy is first to consider the genes in pairs, ignoring the third gene. For the *b, st* pair, the relevant comparisons are black, scarlet (243 + 10), black (241 + 15), scarlet (226 + 12), and nonblack, nonscarlet (235 + 18). The sums are 253, 256, 238, and 253, respectively, the approximate equality indicating that *b* and *st* are unlinked. Similarly, for the *hk, st* gene pair, the relevant comparisons are hooked, scarlet (10 + 226), hooked (15 + 235), scarlet (243 + 12), and nonhooked nonscarlet (241 + 18). In this case the sums are 236, 250, 255, and 259, respectively, with approximate equality in this case indicating independent assortment of *hk* and *st*. For the remaining allele pair *b, hk*, the relevant comparisons are black hooked (15 + 10), black (243 + 241), hooked (235 + 226), and nonblack nonhooked (12 + 18). The sums are 25, 484, 461, and 30, respectively, yielding an estimated frequency of recombination of $(25 + 30)/(25 + 484 + 461 + 30) = 0.055$.

## ANALYSIS AND APPLICATIONS

**5.1** What gametes, and in what ratios, are produced by male and female *Drosophila* of genotype *A b / a B* when the genes are present in the same chromosome and the frequency of recombination between them is 3 percent? (Note: The answer differs for males and females.)

**5.2** Which of the following statements are true of two genes that show the maximum frequency of recombination of 50%?
- **(a)** The genes may be in different chromosomes.
- **(b)** The genes may be quite far apart but in the same chromosome.
- **(c)** The genes undergo independent assortment.
- **(d)** If the genes are syntenic, more than one crossover occurs between the genes in almost every meiosis.
- **(e)** The genes always undergo first division segregation.
- **(f)** The genes are located in heterochromatin.
- **(g)** The coefficient of coincidence equals zero.

**5.3** Two linked genes are separated by a distance such that exactly 4 percent of the cells undergoing meiosis have one crossover (chiasma) between the genes and 96 percent have no crossover. What is the percent recombination between the genes?

**5.4** A double heterozygote has the coupling configuration *A B/a b* of two genes that have a frequency of recombination of 0.05. If one gamete is chosen at random, what is the probability that it is a nonrecombinant gamete?

**5.5** In *Neurospora*, a fungus with ordered tetrads, a gene is located at a distance of 8 map units from the centromere. What is the expected frequency of second-division segregation of the gene?

**5.6** If the chromatids involved in each of two crossovers are independent, what is the expected ratio of two-strand : three-strand : four-strand doubles?

**5.7** For two linked genes in an organism with unordered tetrads, such as budding yeast, what type of tetrad results from a four-strand double crossover?

**5.8** Consider a hypothetical organism in which all double crossovers are four-strand double crossovers. Two genes are separated by a physical distance such that each meiotic cell has a double crossover in the region between the genes. What would be the frequency of recombination between these genes? How does this differ from the situation in real organisms?

**5.9** What is the map distance between two genes if 90 percent of the meiotic cells have no crossover in the region between them, 6 percent have one, 3 percent have two, and 1 percent has three?

**5.10** In an organism such as budding yeast that produces ordered tetrads, if two pairs of alleles are segregating, how many different types of tetrads are possible?

**5.11** Consider an organism heterozygous for two genes located in the same chromosome in the coupling configuration *A B* / *a b*. If a single crossover occurs between the genes in every cell undergoing meiosis, and no multiple crossovers occur between the genes, what is the recombination frequency between the genes? What effect would multiple crossovers have on the frequency of recombination?

**5.12** The accompanying diagram shows the linkage map of the three genes *a, b,* and *c,* where the distances are given in centimorgans (map units).

*a* ——— 10 ——— *b* ——— 15 ——— *c*

Assume that double crossovers across these intervals show a coefficient coincidence of 0.40. Among 1000 gametes from an individual of genotype *A B C*/*a b c,* what are the expected numbers of each of the possible allele combinations?

**5.13** The dominant allele for Huntington disease is segregating in the accompanying pedigree. Also shown is the pattern of DNA bands observed for a restriction fragment length polymorphism (RFLP). The bands are labeled as 1 and 2. Using *HD* and *hd* for the mutant and nonmutant Huntington alleles, and $A_1$ and $A_2$ for the alleles yielding the RFLP bands 1 and 2, respectively:

**(a)** Deduce the genotype of each person in the pedigree with respect to the Huntington alleles and to the RFLP alleles.

**(b)** Explain why there are no persons who have only the smaller DNA fragment (fragment 2, the lower band).

**(c)** Indicate whether you find any evidence of linkage between the Huntington gene and the RFLP.

**(d)** If you do see evidence of linkage, list each person who carries a recombinant chromosome according to the numbers across the bottom of the gel.

**(e)** Estimate the frequency of recombination between the Huntington gene and the RFLP.

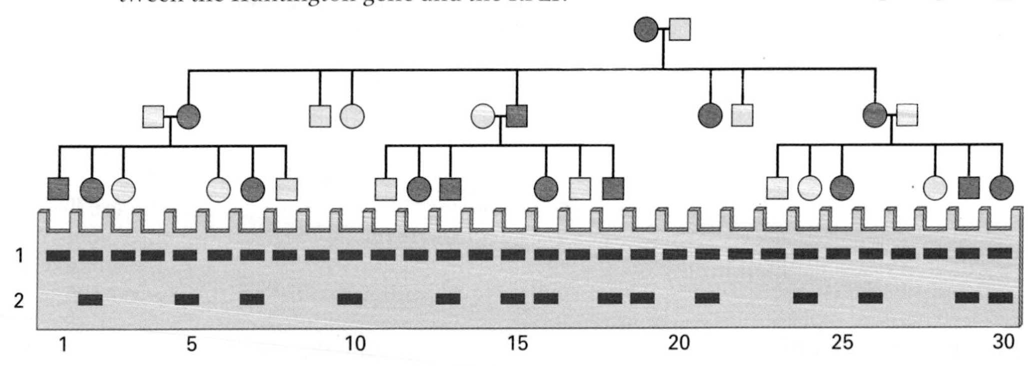


**5.14** Individual plants homozygous for various combinations of linked genes *R* and *T* (i.e., *RR TT, RR tt, rr TT,* or *rr tt*) were crossed in pairs to produce $F_1$ plants, which were then self-fertilized to produce an $F_2$ generation. *R* is dominant to *r,* and *T* is dominant to *t.* What is the chromosomal constitution (e.g., *R T* / *r t*) for each $F_1$ parent of the following $F_2$ phenotypic distributions:

| Self fertilized $F_1$ plants | $F_2$ phenotypes | | | |
|---|---|---|---|---|
| | *R T* | *R t* | *r T* | *r t* |
| **(a)** | 106 | 34 | 0 | 0 |
| **(b)** | 0 | 0 | 104 | 36 |
| **(c)** | 99 | 0 | 31 | 0 |
| **(d)** | 64 | 10 | 8 | 18 |

**5.15** The genes for *vermilion* (*v*) eyes, *sable* (*s*) body color, and *forked* (*f*) bristles are on the X-chromosome of *Drosophila melanogaster*. They map in the order *v s f,* with 10 map units between *v* and *s* and 14 map units between *s* and *f.* Suppose that you cross a wildtype male with a *v s f* female and then testcross the $F_1$ females.

**(a)** Indicate the progeny classes and their expected numbers among 2000 total progeny, assuming that the interference is $i = 1$.

**(b)** Make the same calculation assuming an interference value of $i = 0.5$.

**(c)** Assume again that $i = 1$, and suppose that you cross a homozygous wildtype female to a *v s f* male and then mate the progeny among themselves. What are the possible classes of progeny and their expected numbers among 2000 progeny animals?

**5.16** Consider the following cross in budding yeast *Saccharomyces cerevisiae,* which produces unordered asci:

*leu2 trp1* + × + + *met14*

The resulting diploid was induced to undergo meiosis, and the asci were dissected and each spore analyzed genetically. The following tetrad types were found. The spore genotypes in each tetrad are listed vertically, with the observed number of each type of tetrad at the bottom. For convenience, *leu2*, *trp1*, and *met14* are abbreviated *l*, *t*, and *m*, respectively.

| | | | | | |
|---|---|---|---|---|---|
| *l t m* | *l t* + | + *t* + | + *t m* | + *t m* | + *t* + |
| *l t m* | *l t* + | + *t* + | + *t m* | *l t m* | *l t* + |
| + + + | + + *m* | *l* + *m* | *l* + + | + + + | + + *m* |
| + + + | + + *m* | *l* + *m* | *l* + + | *l* + + | *l* + *m* |
| 230 | 235 | 215 | 220 | 54 | 46 |

Analyze these data as fully as you can with respect to the linkage of each gene with every other gene and with respect to its centromere. Where a genetic linkage is indicated, calculate the map distance.

**5.17** The diagram below shows bands corresponding to restriction-fragment length polymorphisms in each of two genes, one with alleles called *A* and *a* and the other with alleles called *B* and *b*. The banding patterns of the parents $P_1$ and $P_2$ are shown on the left, and the number of each possible type of banding pattern observed among 200 offspring are shown at the right.

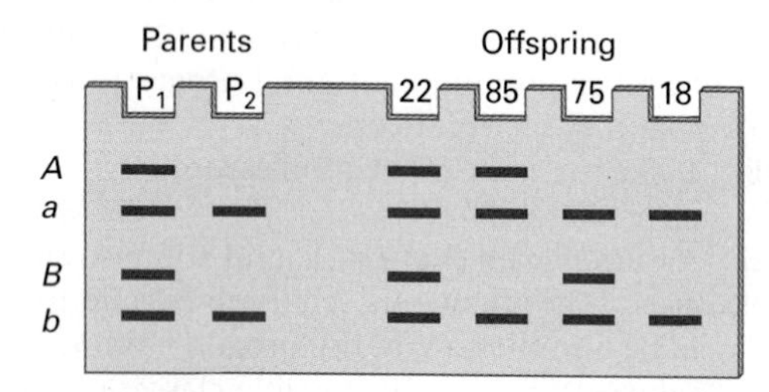


**(a)** What is the genotype of the parent $P_1$?

**(b)** What is the frequency of recombination between *A* and *B*?

**5.18** The data in the accompanying table show 1000 gametes observed from a triply heterozygous parent in a three-point testcross to determine the genetic map of three linked genes. Neither the parental genotype of the heterozygous parent nor the order of the genes is known.

| | |
|---|---|
| *F G H* | 4 |
| *F G h* | 41 |
| *F g H* | 393 |
| *F g h* | 50 |
| *f G H* | 64 |
| *f G h* | 413 |
| *f g H* | 33 |
| *f g h* | 2 |

**(a)** What is the genotype of the heterozygous parent?

**(b)** What is the correct linear order of the genes?

**(c)** What is the map distance between the middle gene and its nearest neighbor?

**(d)** What is the map distance between the middle gene and its farthest neighbor?

**(e)** What is the expected number of double crossovers?

**(f)** What is the coincidence *c* and the interference *i*?

**5.19** The diagram that follows shows bands corresponding to codominant molecular markers in each of three linked autosomal genes, with the alleles denoted (*A*, *a*), (*E*, *e*) and (*H*, *h*) for the three loci. The banding patterns of the parents $P_1$ and $P_2$ are shown on the left, and the number of each possible type of banding pattern observed among 1000 offspring is shown at the right.

**(a)** Fill each blank with the observed number of the gametic type.

| Gamete | Number |
|---|---|
| *A E H* | ____ |
| *A E h* | ____ |
| *A e H* | ____ |
| *A e h* | ____ |
| *a E H* | ____ |
| *a E h* | ____ |
| *a e H* | ____ |
| *a e h* | ____ |

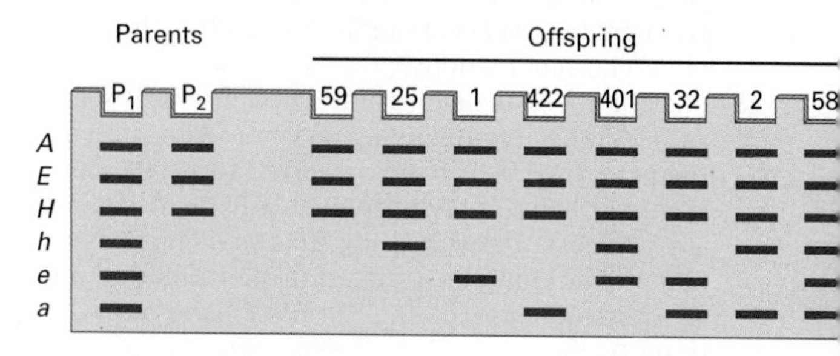


**(b)** Which of the three genes is in the middle?

**(c)** What is the genotype of the $P_1$ parent?

**(d)** What is the distance in map units between the middle gene and its nearest neighbor?

**(e)** What is the distance in map units between the middle gene and its farthest neighbor?

**(f)** What is the expected number of double crossovers?

**(g)** Calculate the coincidence (*c*) and the interference (*i*) for these data.

**5.20** In the nematode *Caenorhabditis elegans*, the mutations *dpy*-21 (dumpy) and *unc*-34 (uncoordinated) identify linked genes that affect body conformation and coordination of movement. The frequency of recombination between the genes is 24 percent. If the heterozygote *dpy*-21 +/+ *unc*-34 undergoes self-fertilization (the normal mode of reproduction in this organism), what fraction of the progeny is expected to be both dumpy and uncoordinated?

**5.21** A strain of mice is homozygous for certain alleles of five genes: *AA bb cc DD ee*. A second strain has the genotype *aa BB CC dd EE*. Genes *D* and *E* are linked to each other, being 20 cM apart. All other genes assort independently. The two strains are crossed to generate $F_1$ individuals, and the $F_1$ females are testcrossed.

**(a)** Considering only the genes *A*, *B*, and *C*, what is the probability of obtaining an offspring from the testcross that has the same phenotype as the $F_1$ female parent?

**(b)** Considering all the genes, what is the probability that an offspring from the testcross has the same phenotype as the $F_1$ female parent?

**5.22** A *Drosophila* geneticist exposes flies to a mutagenic chemical and obtains nine mutations in the X chromosome that are lethal when homozygous. The mutations are tested in pairs in complementation tests, with the results shown in the accompanying table. A + indicates complementation (that is, flies carrying both mutations survive), and a – indicates noncomplementation (flies carrying both mutations die). How many genes (complementation groups) are represented by the mutations, and which mutations belong to each complementation group?

| | 1 | 2 | 3 | 4 | 5 | 6 | 7 | 8 | 9 |
|---|---|---|---|---|---|---|---|---|---|
| **1** | – | + | – | + | + | + | – | + | + |
| **2** | | – | + | + | + | + | + | + | + |
| **3** | | | – | + | + | + | – | + | + |
| **4** | | | | – | + | – | + | + | + |
| **5** | | | | | – | + | + | + | – |
| **6** | | | | | | – | + | + | + |
| **7** | | | | | | | – | + | + |
| **8** | | | | | | | | – | + |
| **9** | | | | | | | | | – |

**5.23** A cross is carried out between the homozygous genotypes *A B C* and *a b c*. The triply heterozygous $F_1$ offspring are then crossed with homozygous *a b c*. The resulting progeny and their numbers are:

| | |
|---|---|
| *A B C/a b c* | 177 |
| *A B c/a b c* | 89 |
| *A b C/a b c* | 81 |
| *A b c/a b c* | 180 |
| *a B C/a b c* | 173 |
| *a B c/a b c* | 71 |
| *a b C/a b c* | 68 |
| *a b c/a b c* | 161 |
| Total | 1000 |

**(a)** What is the value of the chi-square test for linkage between the *A* and *B* genes?

**(b)** Does the observed result fit the genetic hypothesis that the genes are unlinked?

**5.24** Suppose you make a cross of two strains of budding yeast with the following genotypes: *cys3 vps8 ade1* × + + +. The diploid undergoes meiosis and forms asci. The following tetrad types in the numbers shown were obtained. For convenience, *c*, *v* and *a* designate *cys3*, *vps8* and *ade1*, respectively. Although yeast tetrads are unordered, one can order them by including a heterozygous marker that is tightly linked to its centromere. Assume that such a marker was included in this cross and that the tetrad genotypes are presented as ordered tetrads.

| | | | | |
|---|---|---|---|---|
| *c v a* | *c v a* | *c v a* | *c v a* | + *v a* |
| *c v a* | + *v a* | + + *a* | *c v* + | *c v* + |
| + + + | *c* + + | *c v* + | + + *a* | + + *a* |
| + + + | + + + | + + + | + + + | *c* + + |
| 159 | 83 | 41 | 116 | 1 |

**(a)** Evaluate these data for linkage between each gene and its centromere.

**(b)** For any gene linked to its centromere calculate the map distance from the data given.

**5.25** The following classes and frequencies of ordered tetrads were obtained from the cross $a^+ \, b^+ \times a \, b$ in *Neurospora crassa*. (Only one member of each pair of spores is shown.)

| Spore pair | | | | Number of asci |
|---|---|---|---|---|
| **1–2** | **3–4** | **5–6** | **7–8** | |
| $a^+b^+$ | $a^+b^+$ | $a\ b$ | $a\,b$ | 1766 |
| $a^+b^+$ | $a\ b$ | $a^+b^+$ | $a\,b$ | 220 |
| $a^+b^+$ | $a\ b^+$ | $a^+b$ | $a\,b$ | 14 |

What is the order of the genes in relation to the centromere?

**5.26** In the domestic cat, the genes for yellow coat color (actually, a sort of orange) and for black coat color are alleles of an X-linked gene. Heterozygous females show patches of yellow and patches of black fur, a pattern called *calico*. The pedigree illustrated here shows the distribution of coat color in a number of litters of kittens, along with the patterns of DNA bands observed for two restriction fragment polymorphisms. Bands 1 and 2 result from DNA fragments produced by alleles $A_1$ and $A_2$; bands 3 and 4 result from DNA fragments produced by alleles $B_3$ and $B_4$ of a different gene. Examine the pedigree and assign complete genotypes to as many animals possible.

**(a)** Is there any evidence that the coat-color gene and the $A_1$, $A_2$ allelic pair are linked?

**(b)** Is there any evidence that the coat-color gene and the $B_3$, $B_4$ allelic pair are linked?

**(c)** If there is evidence of linkage, identify which animals a carry recombinant chromosome.

**(d)** Estimate the frequency of recombination between the linked genes. (Hint: All of the doubly heterozygous females in generations 1 and 2 have the same *cis-trans* configuration of alleles.)

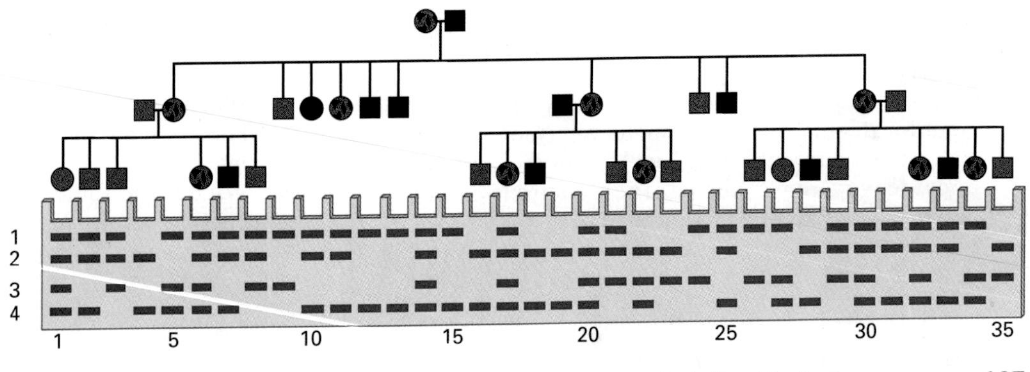

**5.27** A genetic map of a region with three linked genes *A*, *B*, and *R* is shown below.

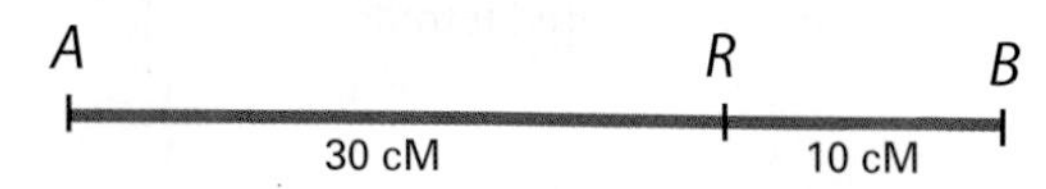


Suppose that you cross a pure-breeding *A R B* strain with a pure-breeding *a r b* strain, and then testcross the $F_1$ progeny.

**(a)** What percentage of the testcross progeny would you expect to be phenotypically *A r*?

**(b)** What percentage of the testcross progeny do you expect to be phenotypically *A R B*? You may assume that the interference across the region is $i = 0$.

**5.28** The filled symbols in the pedigree below depict a trait due to homozygosity for an autosomal recessive allele designated *a*. The gel patterns for a linked restriction-fragment length polymorphism with alleles $M_1$ and $M_2$ are also shown. Individuals I-1 and II-2 are not affected, but individuals I-2 and II-1 are affected. Amniocentesis has been carried out on II-3 and the gel pattern is as shown; however the phenotype of II-3 is not known (indicated by the question mark). The frequency of recombination between the restriction-fragment length polymorphism and the autosomal recessive is 25 percent. It is not known whether the genotype of individual I-1 is $M_1A/M_2\ a$ (we may call this the coupling configuration) or $M_1\ a/M_2\ A$ (repulsion), but you may assume that in the population as a whole these possibilities are equally likely.

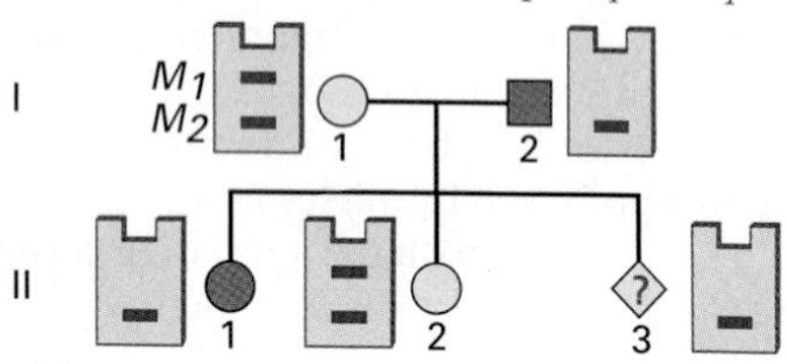


**(a)** Let *X* stand for the event that II-1 is affected and II-2 is not affected. What is the probability that individual I-1 has the coupling configuration (*C*), given *X*?

**(b)** What is the probability that individual I-1 has the repulsion configuration (*R*), given *X*.

**(c)** What is the probability of the event *Y* that individual II-3 has genotype *aa*, taking into account the gel pattern shown in the pedigree?

**5.29** In the pedigree shown here, the male I-2 is affected with recessive X-linked hemophilia A. The gel patterns show a restriction-fragment length polymorphism (RFLP) of a DNA fragment that is also X-linked and located at a distance showing 30 percent recombination with the hemophilia locus. The RFLP alleles are designated *s*, which yields the slow-moving band near the top of the gel; and *f*, which yields the fast-moving band near the bottom of the gel. Let *A* and *a* represent the nonmutant and mutant forms of the hemophilia gene, respectively, and assume that the male I-2 is the only possible source of the hemophilia mutation in this pedigree.

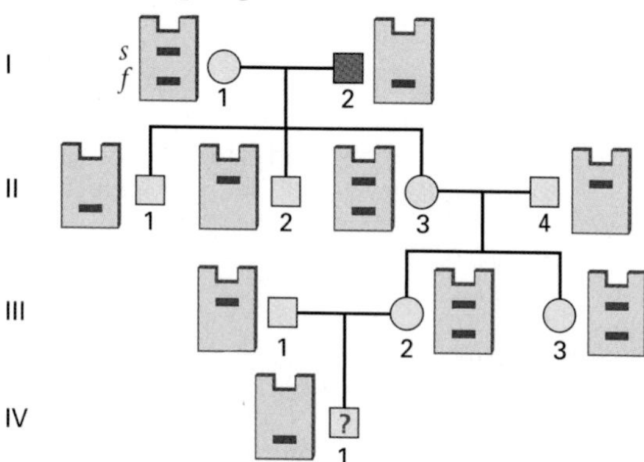


**(a)** What is the probability that the woman II-3 is heterozygous for the hemophilia mutation?

**(b)** What is the probability that the woman III-2 is heterozygous for the hemophilia mutation?

**(c)** What is the probability that IV-1 is affected with hemophilia?

## CHALLENGE PROBLEMS

**Challenge Problem 1** Analyze the data in Problem 5.24 to determine the linkage between each pair of genes. For any pair of linked genes, calculate the map distance.

**Challenge Problem 2** The linkage relationships of four genes were studied in the flowering plant *Arabidopsis thaliana*. The recessive allele *cnx* confers resistance to chlorate, *ga3* results in gibberellin-responsive dwarfism, *sex1* causes overproduction of starch, and *th2* creates a requirement for thiamine. A plant homozygous for all four genetic markers was crossed with wildtype, and the $F_1$ progeny were testcrossed. The gametes from the heterozygous parent were found to be:

| | | | | | | | | | |
|---|---|---|---|---|---|---|---|---|---|
| cnx | ga3 | sex1 | th2 | 185 | cnx | ga3 | sex1 | + | 29 |
| cnx | ga3 | + | th2 | 205 | cnx | ga3 | + | + | 28 |
| cnx | + | sex1 | th2 | 2 | cnx | + | sex1 | + | 26 |
| cnx | + | + | th2 | 3 | cnx | + | + | + | 19 |
| + | ga3 | sex1 | th2 | 24 | + | ga3 | sex1 | + | 1 |
| + | ga3 | + | th2 | 21 | + | ga3 | + | + | 4 |
| + | + | sex1 | th2 | 26 | + | + | sex1 | + | 198 |
| + | + | + | th2 | 31 | + | + | + | + | 198 |

**(a)** Which of the genes, if any, are linked?

**(b)** For any genes that are linked, construct a genetic map.

**(c)** What is the value of the chromosome interference across any region of linked genes?

**Challenge Problem 3** Tissue antigens, such as those determining the success or failure of skin grafts, are determined in a codominant fashion. A graft will be rejected if, and only if, the donor tissue contains an antigen not present in the recipient. The diagram illustrated here shows all possible skin grafts between two strains of mice and their progeny. The allele pairs *A*, *a* and *B*, *b* each determine a different tissue antigen, and the genes are linked with a frequency of recombination of 20 percent. For each of the arrows, what is the probability of acceptance of a graft in which the donor is an animal chosen at random from the population shown at the base of the arrow and the recipient is an animal chosen at random from the population indicated by the arrowhead?

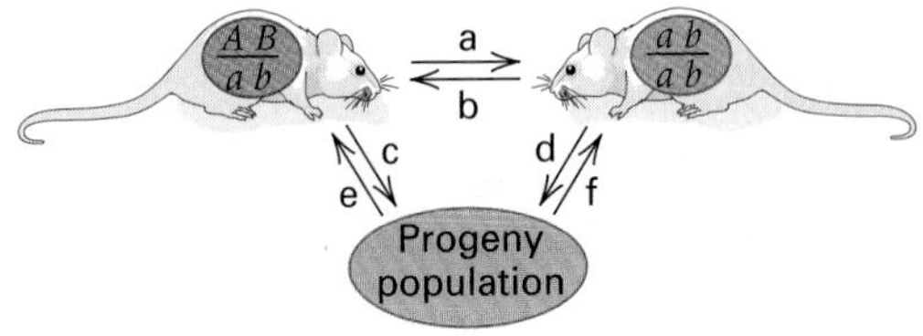

**Challenge Problem 4** The Christmas Grinch has a genetic mapping function given by the formula

$$r = 50[1-(0.2)^{d/100}]$$

where $r$ is the frequency of recombination (in percent) and $d$ is the map distance (in centimorgans).

**(a)** What frequency of recombination corresponds to 10 map units? To 20 map units? To 50 map units? To 100 map units?

**(b)** What map distance corresponds to 1 percent recombination? To 5 percent recombination? To 25 percent recombination? To 40 percent recombination? What happens when $r = 50$ percent?

**(c)** Draw a graph of frequency of recombination against map distance using this mapping function.

## GENETICS *on the web*

GeNETics on the Web will introduce you to some of the most important sites for finding genetics information on the Internet. To explore these sites, visit the Jones and Bartlett companion site to accompany *Genetics: Analysis of Genes and Genomes, Seventh Edition* at http://biology.jbpub.com/book/genetics.

There you will find a chapter-by-chapter list of highlighted keywords. When you select one of the keywords, you will be linked to a Web site containing information related to that keyword.

# CHAPTER 6

# Molecular Biology of DNA Replication and Recombination

Data from a high-throughput massively parallel DNA sequencing machine. Each of the four colors (red, blue, green, yellow) corresponds to a different base (A, T, G, or C) along the sequence of the molecule. [Courtesy of the DOE Joint Genome Institute.]

## CHAPTER OUTLINE


**6.1** Problems of Initiation, Elongation, and Incorporation Error

**6.2** Semiconservative Replication of Double-Stranded DNA
- The Meselson–Stahl Experiment
- Semiconservative Replication of DNA in Chromosomes
- Theta Replication of Circular DNA Molecules
- Rolling-Circle Replication
- Multiple Origins and Bidirectional Replication in Eukaryotes

**6.3** Unwinding, Stabilization, and Stress Relief

**6.4** Initiation by a Primosome Complex

**6.5** Chain Elongation and Proofreading

**6.6** Discontinuous Replication of the Lagging Strand
- Fragments in the Replication Fork
- The Joining of Precursor Fragments

**6.7** Terminator Sequencing of DNA
- Sanger Sequencing
- Massively Parallel Sequencing

**6.8** Molecular Mechanisms of Recombination
- Gene Conversion and Mismatch Repair
- Double-Strand Break and Repair Model

CONNECTION Replication by Halves
Matthew Meselson and Franklin W. Stahl 1958
*The Replication of DNA in* Escherichia coli

CONNECTION Happy Holliday
Robin Holliday 1964
*A Mechanism for Gene Conversion in Fungi*

## connection
### Happy Holliday

**Robin Holliday 1964**
John Innes Institute,
Bayfordbury, Hertfordshire,
United Kingdom
*A Mechanism for Gene Conversion in Fungi*

*When Holliday wrote this paper, gene conversion was a well-known but rather mysterious phenomenon. The favored explanation was the "copy-choice" model, in which, during DNA replication, the growing point of a daughter strand could switch its template between DNA strands of the same polarity in the parental DNA duplexes. A 3 : 1 conversion ratio could therefore be explained by postulating that there had been a* template *switch in the region of the conversion. The copy-choice model predicted that DNA replicates conservatively rather than semiconservatively, that chromosome pairing must take place prior to DNA replication, and that multiple crossovers should always involve the same pair of chromatids. None of these predictions was supported by the experimental evidence, but no satisfactory alternative hypothesis was available. The Holliday model, although it is now known to be too simplistic, nevertheless solved the mystery in proposing, as the mechanism of gene conversion, the cross-pairing of DNA strands (rather than template switching) to create heteroduplex regions in which any mismatched bases could be enzymatically repaired.*

Tetrad analysis [in many fungi] has demonstrated that any heterozygous locus normally gives the expected 2 : 2 Mendelian ratio; but that aberrant tetrads also occasionally occur, where one allele is represented three times and the other once. . . . In spite of its difficulties the copy-choice hypothesis is still favored by a number of authors. This is so because it is assumed that conversion is impossible without some genetic replication. . . . In this paper an origin for conversion events which does not involve genetic replication will be discussed. . . . The model uses the complementarity of the two strands of DNA in order to explain specific chromatid pairing at the molecular level. After genetic replication and general genetic pairing, at certain points DNA molecules from opposite homologous chromatids unravel to form single strands, and these then anneal or coil up with the complementary strands from the other chromatid. Thus, specific or effective pairing over short regions could occur. There are a number of possibilities of varying complexity as to how the DNA might get into this configuration. . . . If the annealed region spans a point of heterozygosity—a mutant site—mispairing of bases will occur. . . . Repair mechanisms may operate, and these, by adjusting the base sequences in order to restore normal base pairing, would bring about gene conversion in the absence of any genetic replication. The model indicates how a precise breakage and rejoining of chromatids could occur in the vicinity of the conversion, so that conversion would frequently be accompanied by the recombination of outside markers.

***The model indicates how a precise breakage and rejoining of chromatids could occur.***

Source: R. Holliday, *Genet. Res.* 5 (1964): 282–304.

bivalent and are important in the proper alignment of the bivalent at the metaphase plate in preparation for anaphase I. Bivalents that lack chiasmata to help hold them together are prone to undergo nondisjunction.

The currently favored model for homologous recombination is the **double-strand break and repair model** outlined in **FIGURE 6.30**. In this model, recombination between DNA strands is initiated by a double-stranded break in a DNA duplex (Figure 6.30A). After a double-stranded break, the size of the gap is usually increased by nuclease digestion of the broken ends, with greater degradation of the 5′ ends leaving overhanging 3′ ends as shown in the illustration. These gaps are repaired using the unbroken homologous DNA molecule as a template, but in meiosis the repair process can result in crossovers that yield chiasmata between nonsister chromatids. These crossovers are also the physical basis of what is observed genetically as recombination. In some organisms, including humans and other mammals, the DNA breaks are much more likely to occur at certain positions in the genome than others. Crossovers resulting in recombination are much more likely to occur at these positions, and so they are referred to as *hot spots* of recombination.

A double-stranded break does not necessarily result in a crossover, however. Repair of the double-stranded break by the noncrossover pathway is illustrated in Figure 6.30B. The first step in repair is that a broken 3′ end invades the homologous unbroken DNA duplex, forming a short heteroduplex region with one strand and a looped-out region of the other strand called a **D loop**. (Specific proteins are required to mediate strand invasion; in *E. coli* the strand-invasion

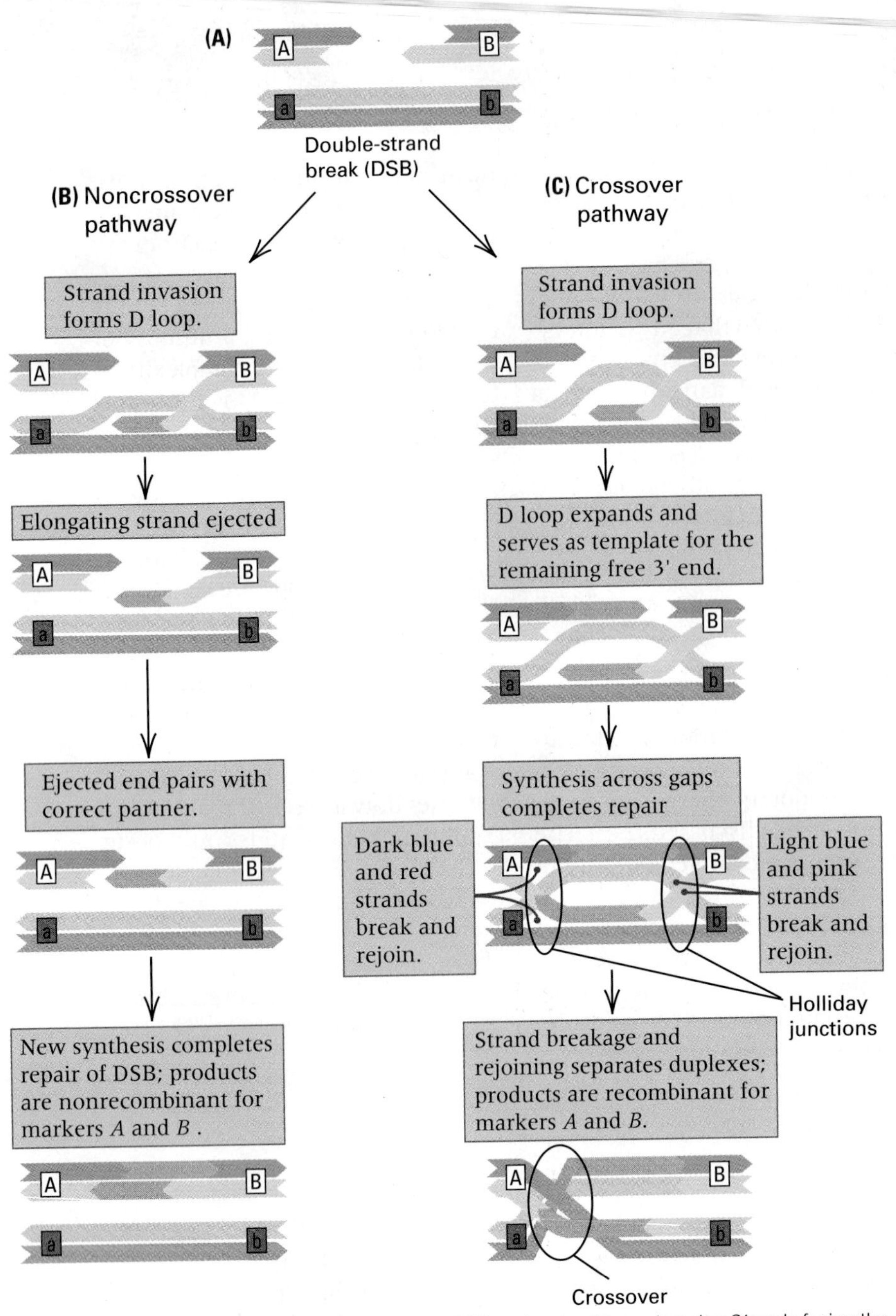
(A)
A
B
a
b
Double-strand break (DSB)
(B) Noncrossover pathway
(C) Crossover pathway
Strand invasion forms D loop.
Strand invasion forms D loop.
Elongating strand ejected
D loop expands and serves as template for the remaining free 3' end.
Ejected end pairs with correct partner.
Synthesis across gaps completes repair
Dark blue and red strands break and rejoin.
Light blue and pink strands break and rejoin.
Holliday junctions
New synthesis completes repair of DSB; products are nonrecombinant for markers A and B .
Strand breakage and rejoining separates duplexes; products are recombinant for markers A and B.
Crossover

**FIGURE 6.30** (A) Double-strand break in a duplex DNA molecule with overhanging 3′ ends facing the gap. Repair of the break makes use of the nonbroken homologous duplex. (B) Repair pathway that does not result in crossing over, although the heteroduplex regions can undergo gene conversion. (C) Repair pathway that does result in crossing over, also with possible gene conversion in heteroduplex regions. [Adapted from D. K. Bishop and D. Zickler, *Cell* 117 (2004): 9–15.]

protein is known as RecA.) In the illustration, the heteroduplex region is the region where the light blue strand is paired with the red strand. Because it is a heteroduplex, any base-pair mismatches in this region could be corrected by mismatch repair in such a way as to result in gene conversion. Such heteroduplex regions are typically only a few hundred base pairs in length. They are much shorter than a gene and vastly shorter than a chromosome, and so gene conversions are rare events except for short regions very near the site of a double-stranded break.

At one end of the heteroduplex, the free 3′ end of the broken DNA strand is extended (brown), but after a time it is ejected from the template, and the strands of the unbroken duplex are able come together again. At this point, the extension of the 3′ end is long enough that pairing can take place with the complementary strand in the broken duplex. At the same time, this pairing provides a template for the 3′ end of the other broken strand. Extension of the 3′ ends across the remaining gaps completes the repair of the double stranded break. Note that although gene conversion can occur in the noncrossover pathway, the resulting duplex DNA molecules are nonrecombinant.

The crossover pathway for repairing a double-stranded break is illustrated in Figure 6.30C. Again, invasion of the unbroken duplex forms a D loop and a short heteroduplex region in which gene conversion can occur. As in the noncrossover pathway, the free 3′ end of the broken DNA strand is extended (brown), but in this case it continues until it displaces the partner strand (pink) of the template strand (red). The displaced strand can then serve as a template for the elongation of the 3′ end of the other broken strand. Eventually, the extensions of the broken strands become long enough that they can be attached to the broken 5′ ends. This completes the repair of the double-stranded break, but note that the resulting structure includes two places where the strands have exchanged pairing partners. Each of the structures where pairing partners are switched is called a **Holliday junction**, named after Robin Holliday, who first predicted that such structures would be involved in recombination.

The problem with Holliday junctions is that they are places where DNA strands from different duplex molecules are interconnected. How the strands are interconnected is shown for the DNA double helices in **FIGURE 6.31**, part A. Resolution of the Holliday junctions is necessary for the DNA molecules to become free of one another. This requires breakage and rejoining of one pair of DNA strands at each Holliday junction. The breakage and rejoining is an enzymatic function carried out by an enzyme called the **Holliday junction-resolving enzyme**.

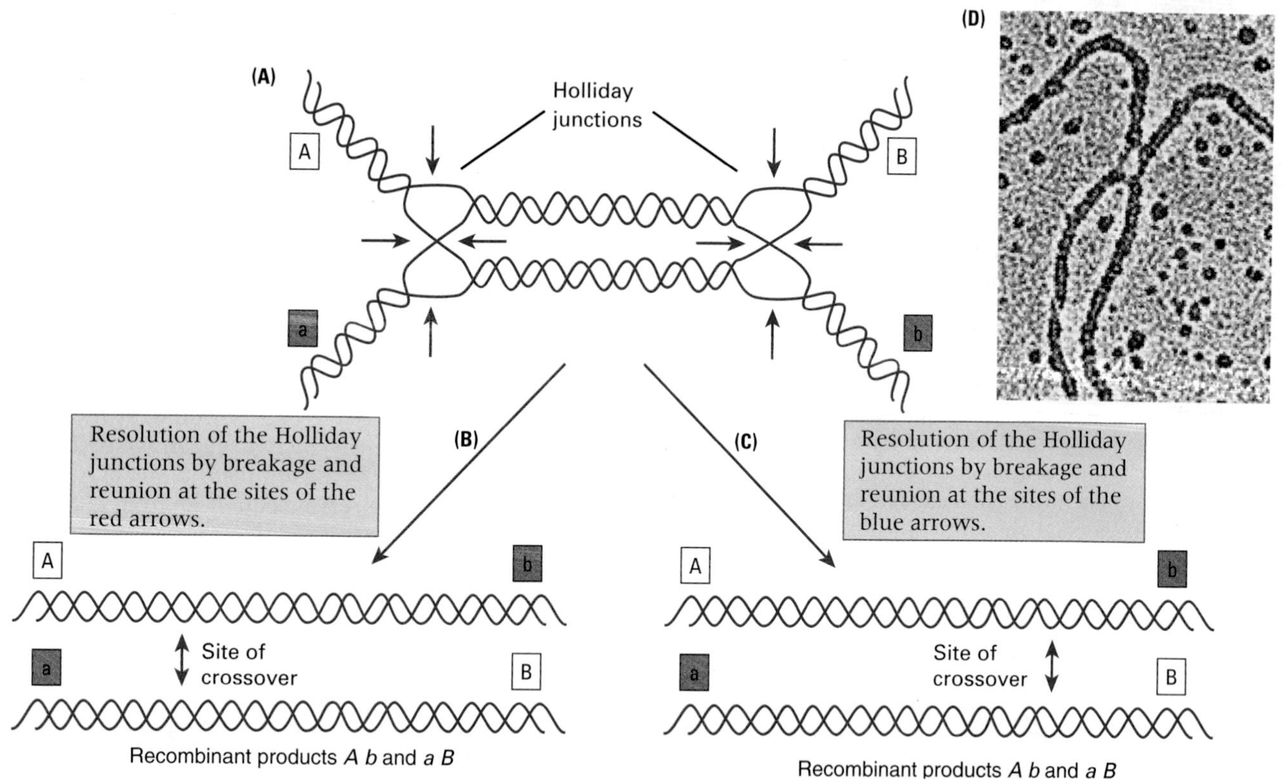


**FIGURE 6.31** (A) Two Holliday junctions in a pair of DNA molecules undergoing recombination; (B and C) two modes of resolution depending which strands are broken and rejoined. Part D is an electron micrograph showing a single Holliday junction between a pair of DNA molecules.

Parts B and C in Figure 6.31 show two ways in which the Holliday structures can be resolved. Breakage and rejoining of the strands indicated by the red arrows results in a crossover at the site of the left-hand Holliday junction, whereas breakage and rejoining of the strands indicated by the blue arrows results in a crossover at the site of the right-hand Holliday junction. In both cases the resulting DNA molecules have a crossover that yields reciprocal recombinant *A b* and *a B* products. (In principle, resolution also could take place at the red arrows in one Holliday junction and the blue arrows in the other, but these resolutions result in noncrossover products. It is unclear how often these noncrossover types of resolution take place.)

## CHAPTER SUMMARY

- DNA replication is semiconservative; the parental strands remain intact and serve as templates for daughter strand synthesis.
- New polynucleotide strands are initiated by a primosome containing an RNA polymerase that synthesizes a short RNA primer complementary to a region of the template strand.
- Each RNA primer is elongated by DNA polymerase, which adds successive deoxyribonucleotides to the 3′ end of the growing chain. The leading strand, whose 3′ end faces the replication fork, is synthesized continuously; the lagging strand is synthesized in relatively short precursor fragments (Okazaki fragments).
- The major DNA polymerase has a 3′-to-5′ exonuclease function that serves for proofreading, in which the last-added nucleotide is removed if it contains an incorrect base.
- Dideoxy sequencing is a chain-termination method of DNA sequencing in which the nucleotide sequence of a growing strand is deduced from the lengths of successive fragments whose elongation was terminated by the incorporation of a dideoxyribonucleotide lacking a 3′-OH group.
- At the DNA level, recombination is initiated by a double-stranded break in a DNA molecule. Use of the homologous DNA molecule as a template for repair can result in a crossover, in which both strands of the participating DNA molecules are broken and rejoined.

## GUIDE TO PROBLEM SOLVING

**Problem 1** Chromosome 1 of the chicken (*Gallus gallus*) contains $2 \times 10^8$ nucleotide pairs. How long would it take to replicate this molecule if replication started at one end and proceeded to the other, assuming a typical eukaryotic rate of DNA synthesis is 50 nucleotide pairs per second?

**Answer** Replication would take $(2 \times 10^8/50) = 4 \times 10^6$ sec or 46.3 days. The answer explains why DNA molecules in eukaryotes must have many origins of replication.

**Problem 2** The illustrated DNA sequence in the gel shown was obtained by the dideoxy sequencing method. The color code is black (G), green (A), red (T), and purple (C). What is nucleotide sequence of the daughter strand synthesized in the sequencing reaction in part A? In part B? What is the nucleotide sequence of the complementary template strand in each case? Label each end as 5′ or 3′. Can you see any special relationship between the sequences in A and B?

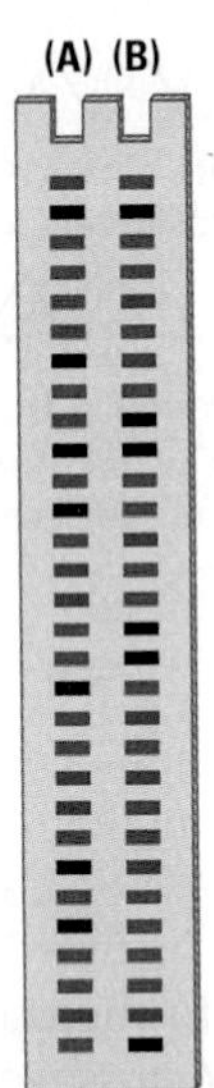


**Answer** The bands in the gel result from incomplete chains which synthesis was terminated by the incorporation of the dideoxynucleotide. The fragments are ordered by size with the smallest at the bottom of the gel. Because DNA strands elongate by addition of the new nucleotides to the 3′ end, the strand synthesized in the sequencing reaction A has the sequence, from bottom to top, of 5′-ACTAGAG ACCATGATCCTGTGATGAATAGC-3′. The template strand is complementary and antiparallel, and so its sequence is 3′-TGATCTCTGGTACTAGGACACTACTTATCG-5′. Similarly, in part B, the daughter strand synthesized in the sequencing reaction is 5′-GCTATTCATCACAGGATCATG GTCTCTAGT-3′ and its template strand is 3′-CGATAAGT AGTGTCCTAGTACCAGAGATCA-5′. If you write the template strand in B with the 5′ end at the left, you will see that it is complementary to the template strand in A. That is, the same region of DNA was sequenced from opposite strands.

**Problem 3** In budding yeast *Saccharomyces cerevisiae*, the wildtype gene *ADE-2* and one of its mutant *ade-2* alleles differ by a single base pair in the DNA; where the wildtype allele has an A–T base pair, the mutant allele has a G–C base pair. If the region in which the alleles differ is included in a heteroduplex, and both mismatches are repaired, how would one explain an ascus containing:

**(a)** 3 *ADE-2* : 1 *ade-2* spores.
**(b)** 1 *ADE-2* : 3 *ade-2* spores.
**(c)** 2 *ADE-2* : 2 *ade-2* spores.

**Answer** **a.** C→T mismatch repair in one heteroduplex, G→A in the other. **b.** T→C mismatch repair in one heteroduplex, A→G in the other. **c.** Either a T→C mismatch repair in one heteroduplex and a G→A in the other, or a C→T mismatch repair in one heteroduplex and a A→G in the other.

## ANALYSIS AND APPLICATIONS

**6.1** In both prokaryotes and eukaryotes, primers are needed for the initiation of DNA replication by the major DNA polymerases. What are the differences between prokaryotic and eukaryotic primers?

**6.2** What is the function of the 3′-to-5′ exonuclease activity associated with DNA polymerase, and what are the consequences for the cell if this function is inactivated by mutation?

**6.3** What chemical groups are joined by DNA ligase? By DNA polymerase?

**6.4** What protein is responsible for initiating the unwinding of double-stranded DNA by separating the parental DNA strands?

**6.5** What protein eliminates the torsional stress caused by unwinding of double-stranded DNA?

**6.6** Identify the chemical group that is at the indicated terminus of the daughter strand in the accompanying diagram of rolling circle replication.

**6.7** For the replication bubble illustrated here, indicate the leading strand and the lagging strand at each replication fork and identify the ends as 3′ or 5′.

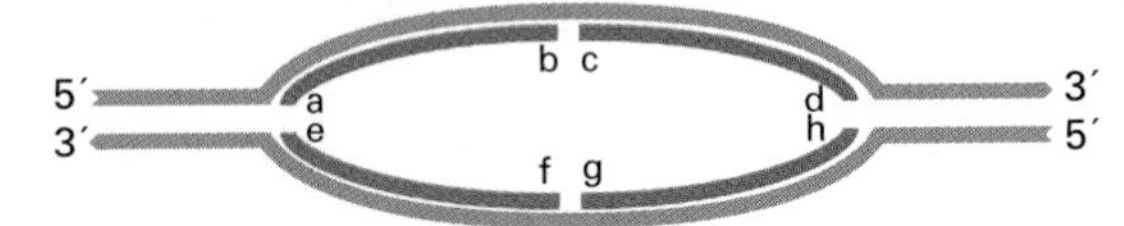


**6.8** In their experiment on DNA replication, Meselson and Stahl separated the parental and daughter DNA molecules using

**(a)** autoradiography
**(b)** starch-gel electrophoresis
**(c)** density-gradient centrifugation
**(d)** crystallography

**6.9** In the Meselson-Stahl experiment on DNA replication, what fraction of the DNA was composed of one light strand and one heavy strand ("hybrid") after one generation of growth in medium containing $^{14}N$? After two generations of growth in medium containing $^{14}N$? What fraction of hybrid DNA is expected after *n* generations of growth in medium containing $^{14}N$?

**6.10** Identify the atoms or chemical groups *a–d* attached to each sugar carbon atom in the diagram of deoxyribose shown here.

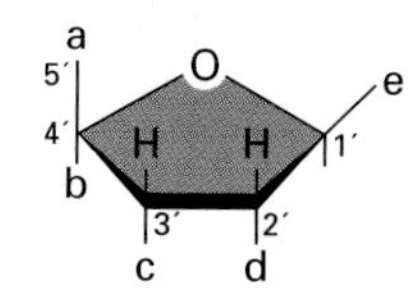


**6.11** Classify the following structure as that of a nucleoside or a nucleotide. Identify its components: What is the sugar? What is the base? Is there a phosphate? Is the base shown here a purine or a pyrimidine?

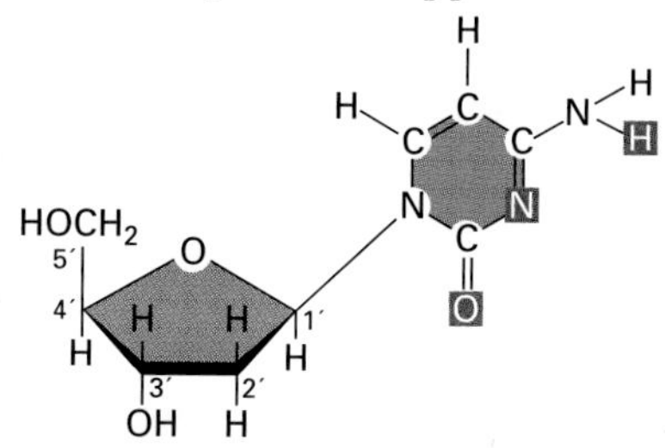


**6.12** Shown here are two duplex DNA molecules that single-stranded at each end. (The single-stranded ends are called "overhanging ends."). Suppose we put these two DNA molecules into a test tube with all the components necessary for DNA replication and incubate the mixture. Will there be any polymerization reaction? If so, show the resulting products. If no reaction takes place, explain why not.

```
5'-ATGGATCCTTATAAC   -3'  5'-    CTAGTACTGGTGC-3'
3'-  CCTAGGAATATTGAT-5'  3'-GAAGATCATGACCA   -5'
```

**6.13** Below is the structure of a derivative of a deoxynucleotide.

**(a)** What makes it different from a normal deoxynucleotide?

**(b)** Is this nucleotide a purine or a pyrimidine? Identify the base.

**(c)** Assuming correct base pairing, could DNA polymerase add this nucleotide to a DNA strand during DNA replication? Explain.

**(d)** If DNA polymerase were to add this nucleotide to a new DNA strand, what happens to DNA synthesis? Explain.

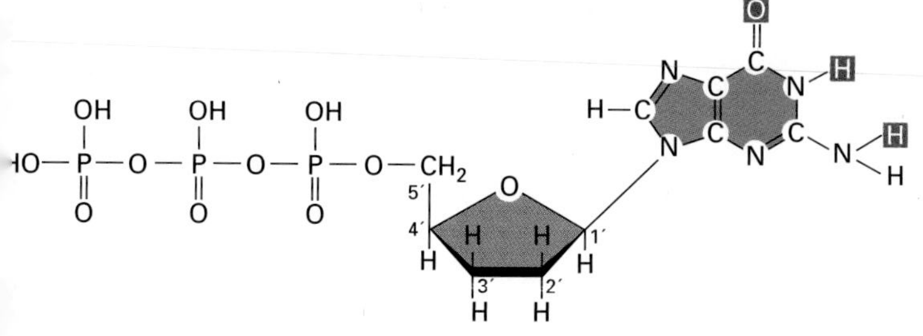


**6.14** In the Meselson-Stahl experiment, $^{15}N$-labeled cells were shifted to $^{14}N$ medium. For the semiconservative and conservative modes of replication, what proportion of $^{15}N$–$^{15}N$, $^{15}N$–$^{14}N$, and $^{14}N$–$^{14}N$ DNA would you expect to find after 1, 2, and 3 rounds of replication?

**6.15** An asteroid probe brings back a bacterial species that has DNA as its genetic material. You perform a Meselson-Stahl experiment and show that, after one round of replication in $^{14}N$ medium, half of the daughter DNA duplexes have $^{15}N$ in both strands whereas the other half have $^{14}N$ in both strands. Interpret these data.

**6.16** A reference strain of *E. coli* has a circular genome consisting of $4.3 \times 10^6$ base pairs of DNA. Replication starts at one location in the genome and proceeds in both directions. Under one particular set of growth conditions, 800 base pairs of DNA are formed per second at each replication fork.

**(a)** At how many revolutions per minute (rpm) is each replication fork rotating?

**(b)** How long does it take to replicate the entire *E. coli* genome under these conditions?

**6.17** Match the enzymes **1–5** below with their functions **a–e**:

| | |
|---|---|
| **1.** DNA ligase | **(a)** Unwinds the DNA double helix during replication |
| **2.** DNA polymerase | **(b)** Cleaves double-stranded DNA at specific, short sequences |
| **3.** DNA helicase | **(c)** Copies template DNA into RNA |
| **4.** RNA polymerase | **(d)** Replicates DNA |
| **5.** Restriction enzyme | **(e)** Joins broken DNA strands |

**6.18** If the percentage of guanine in the double-stranded DNA of a certain species decreased by 5 percent over evolutionary time, what would you expect to have happened to the percentage of adenine in that DNA?

**6.19** A biochemist isolates and purifies what she thinks are all cellular components needed for DNA replication. But when she adds template DNA, although replication occurs, each daughter DNA molecule consists of a normal strand paired with numerous segments of DNA a few hundred nucleotides long. What has she probably left out of the mixture?

**6.20** What would happen to DNA replication in a mutant strain of *E. coli* that had a mutation in the gene for DNA ligase, such that the enzyme was functional at 37°C but nonfunctional at 42°C?

**6.21** In the early years of DNA study, an elegant combined chemical and enzymatic technique enabled scientists to identify "nearest neighbor" nucleotides (adjacent nucleotides in a DNA strand). For example, if the single-stranded tetranucleotide 5′-AGTC-3′ were treated in this way, the nearest neighbors would be AG, GT, and TC. (Unless otherwise specified, nucleotide sequences are always written with the 5′ terminus at the left). Prior to the time that techniques became available for

DNA sequencing, nearest-neighbor analysis was used to determine sequence relations. Nearest-neighbor analysis indicated that complementary DNA strands are antiparallel, a phenomenon that you are asked to examine in this problem by predicting some nearest-neighbor frequencies. Assume that you have determined the frequencies of the following nearest neighbors: AG, 0.15; GT, 0.03; GA, 0.08; TT, 0.10.

**(a)** What are the nearest-neighbor frequencies of CT, AC, TC, and AA?

**(b)** If DNA had a parallel (rather than an antiparallel) structure, what nearest-neighbor frequencies could you deduce from the observed values?

**6.22** DNA isolated from two different species has the same base ratios (A + T/G + C). Does this mean that they have the same nearest neighbor frequencies? (If you are unfamiliar with nearest neighbors, please see Problem 21.)

**6.23** In contrast with the question in Problem 22, would organisms with similar nearest-neighbor frequencies have similar base ratios (A + T/G + C)?

**6.24** In rolling circle replication, after the initial single-stranded nick is produced, which strand of the circular DNA becomes the leading strand and which becomes the lagging strand?

**6.25** In rolling circle replication, do both parental DNA strands replicate simultaneously, or is one of them replicated first?

**6.26** Suppose that a "light" ($^{14}$N-labeled) DNA duplex forms a Holliday structure with a "heavy" ($^{15}$N-labeled) DNA duplex with the Holliday junction exactly in the middle. You allow the Holliday junction to be resolved, and then centrifuge the resulting molecules in a CsCl gradient. What would you expect the density of the resulting molecules to be after resolution of the Holliday junction?

**6.27** The following sequence of nucleotides is present along one strand of a DNA duplex at one of the replication forks in a replication bubble. Synthesis of an RNA primer on this template begins by copying the base in red.

3′-...AGTAGATACGTCCAG...-5′

**(a)** If the RNA primer consists of eight nucleotides, what is its base sequence?

**(b)** In the intact RNA primer, which nucleotide has a free hydroxyl (–OH) terminus?

**(c)** What is the chemical group is attached to the 5′ end of the primer?

**(d)** If the opposite strand of the parental DNA duplex is replicated as the leading strand, is the replication fork moving from left-to-right or right-to-left?

**6.28** The diagrams below depict the fluorescence patterns of chromosomes in mitotic metaphase after one and two rounds of DNA replication in the presence of BUdR, and the dotted lines represent the DNA strands in the DNA duplex present in each chromatid. Depict the BUdR labeling of each chromatid by (1) making the line solid if the strand is labeled with BUdR or (2) leaving it dashed if it is not labeled with BUdR.

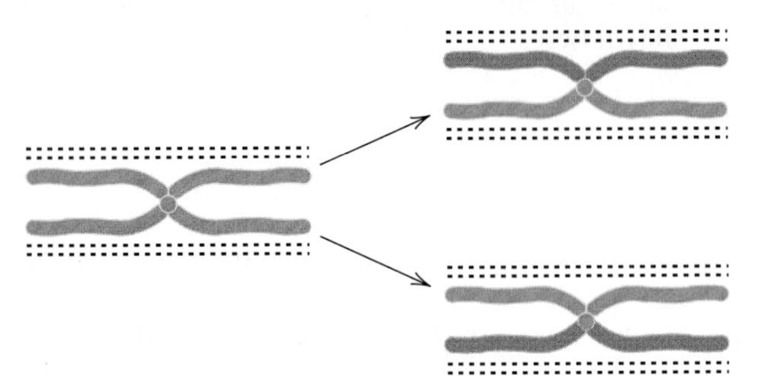

**6.29** Which of the following statements are true for double-stranded DNA? Explain your reasoning. The symbol R stands for purine (A or G), and Y stands for pyrimidine (C or T).

**(a)** $A + C = G + T$

**(b)** $A + G = T + C$

**(c)** $R = Y$

**6.30** How do the DNA sequences shown in the accompanying gel differ from each other? The color code is black (G), green (A), red (T), and purple (C).

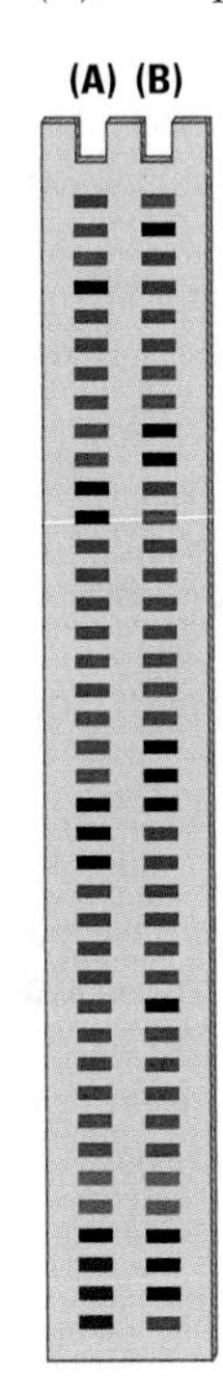

## CHALLENGE PROBLEMS

**Challenge Problem 1** The mitotic chromosome illustrated here was observed after two rounds of DNA replication in the presence of BUdR. The dotted lines represent the DNA strands in the DNA duplex present in each chromatid. Explain what process might have resulted in such a pattern of fluorescence. Depict the BUdR labeling of each chromatid by (1) making the line solid if the strand is labeled with BUdR or (2) leaving it dashed if it is not labeled with BUdR.

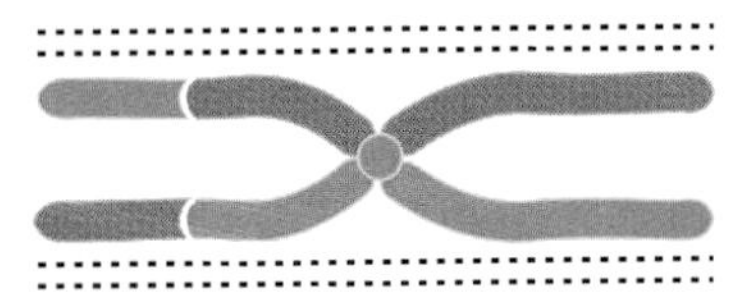

**Challenge Problem 2** A cross involving four linked genes (*A B C D* × *a b c d*) is made in budding yeast *Saccharomyces cerevisae*. The genes *B* and *C* are extremely close together in the DNA. Among 200 asci, one is found to contain the spores *A b c d, a b c D, A B C D,* and *a b c d*.

**(a)** What is unusual about this ascus?
**(b)** Can it be explained by a single event?
**(c)** What does the type of tetrad with respect to the *A, a* and *D, d* allele pairs indicate about this event?
**(d)** Could an ascus containing *A b c D, a b c d, A B C D,* and *a b c d* have the same explanation?

**Challenge Problem 3** Tetrads are analyzed from a cross involving three linked genes in *Neurospora crassa*, $his^+$ $co^+$ $leu^+$ × *his co leu*. A small percentage of asci showed 3 : 1 segregation of each of the markers. Explain the following results.

**(a)** Among those with 3 : 1 segregation of *his*, the *co–leu* pair showed PD >> NPD (linkage).
**(b)** Among those with 3 : 1 segregation of *co*, the *his–leu* pair showed PD = NPD (no linkage!).
**(c)** Among those with 3 : 1 segregation of *leu*, the *his–co* pair showed PD >> NPD (linkage).

### GENETICS on the web

GeNETics on the Web will introduce you to some of the most important sites for finding genetics information on the Internet. To explore these sites, visit the Jones and Bartlett companion site to accompany *Genetics: Analysis of Genes and Genomes, Seventh Edition* at http://biology.jbpub.com/book/genetics.

There you will find a chapter-by-chapter list of highlighted keywords. When you select one of the keywords, you will be linked to a Web site containing information related to that keyword.

# CHAPTER 7

# Molecular Organization of Chromosomes

The Hessian fly *Mayetiola destructor* is a major insect pest of cultivated wheat. It has the smallest genome size known among insects, a mere 50% of that of *Drosophila*. The largest insect genome is 200 times greater than that of the Hessian fly.



## CHAPTER OUTLINE


**7.1** Genome Size and Evolutionary Complexity: The C-Value Paradox

**7.2** The Supercoiling of DNA
- Topoisomerase Enzymes

**7.3** The Structure of Bacterial Chromosomes

**7.4** The Structure of Eukaryotic Chromosomes
- The Nucleosome: The Structural Unit of Chromatin
- The Nucleosome Core Particle
- Chromosome Territories in the Nucleus
- Chromosome Condensation

**7.5** Polytene Chromosomes

**7.6** Repetitive Nucleotide Sequences in Eukaryotic Genomes
- Kinetics of DNA Renaturation
- Analysis of Genome Size and Repetitive Sequences by Renaturation Kinetics

**7.7** Unique and Repetitive Sequences in Eukaryotic Genomes
- Unique Sequences
- Highly Repetitive Sequences
- Middle-Repetitive Sequences

**7.8** Molecular Structure of the Centromere

**7.9** Molecular Structure of the Telomere

CONNECTION Post-Genomics Genetics
Shawn Ahmed and Jonathan Hodgkin, 2000
*MRT-2 Checkpoint Protein Is Required for Germline Immortality and Telomere Replication in* C. elegans

CONNECTION Telomeres: The Beginning of the End
Carol W. Greider and Elizabeth H. Blackburn 1987
*The Telomere Terminal Transferase of* Tetrahymena *Is a Ribonucleoprotein Enzyme with Two Kinds of Primer Specificity*

In this chapter we examine the molecular structures of prokaryotic genomes and of eukaryotic chromosomes. Genome size and organization differ profoundly between prokaryotes and eukaryotes. Prokaryotes are unicellular organisms with relatively small genomes consisting of one or more DNA molecules that contain the essential genetic information. Usually the genome consists of just one molecule, but in some prokaryotes the essential genes are divided between two or more molecules. The cells may also contain one or more copies of accessory DNA molecules (*plasmids*) that contain nonessential information. The genome and the plasmids are often circular DNA molecules, but some prokaryotic genomes and plasmids are linear. Eukarotyic cells usually have larger genomes in the form of linear DNA molecules packaged with characteristic types of proteins (*histones*) into chromosomes, which are gathered together inside a nucleus enclosed in a nuclear envelope. (Circular chromosomes are occasionally found, but they are aberrations.) The DNA molecules in mitochondria and chloroplasts are also circular. The genetics of these organelles is the subject of Chapter 16.

## 7.1 Genome Size and Evolutionary Complexity: The C-Value Paradox

A summary of a small sample of genome sizes is shown in **TABLE 7.1**. Bacteriophage MS2 is one of the smallest viruses; it has only four genes in a single-stranded RNA molecule containing 3569 nucleotides. SV40 virus, which infects monkey and human cells, has a genetic complement of five genes in a circular double-stranded DNA molecule of about 5 kb (5000 nucleotide pairs). The more complex phages and animal viruses have as many as 250 genes and DNA molecules ranging from 50 to 300 kb. Prokaryotic genomes are substantially larger. Archaeal genomes (for example, *Methanococcus jannaschii*) are generally similar in size to bacterial genomes. Some prokaryotic genomes consist of linear DNA, others circular DNA. For example, the chromosome of the spirochete *Borrelia burgdorferi*, the agent of

**Table 7.1 Genome size of some representative viral, bacterial, and eukaryotic genomes**

| Genome | Approximate genome size in thousands of nucleotides | Form |
|---|---|---|
| Viruses | | |
| MS2 | 4 | Single-stranded RNA |
| Human immunodeficiency virus (HIV) | 9 | Single-stranded RNA |
| Colorado tick fever virus | 29 | Linear double-stranded RNA |
| SV40 | 5 | Circular double-stranded DNA |
| $\phi$X174 | 5 | Circular single-stranded DNA; double-stranded replicative form |
| $\lambda$ | 50 | Linear double-stranded DNA |
| Herpes simplex | 152 | Linear double-stranded DNA |
| T2,T4,T6 | 165 | Linear double-stranded DNA |
| Smallpox | 267 | Linear double-stranded DNA |
| Prokaryotes | | |
| *Methanococcus jannaschii* | 1,600 | Circular double-stranded DNA |
| *Escherichia coli K12* | 4,600 | Circular double-stranded DNA |
| *Borrelia burgdorferi* | 910 | Linear double-stranded DNA |
| Eukaryotes | | Haploid chromosome number |
| *Saccharomyces cerevisiae* (yeast) | 13,000 | 16 |
| *Caenorhabditis elegans* (nematode) | 97,000 | 6 |
| *Arabidopsis thaliana* (wall cress) | 100,000 | 5 |
| *Drosophila melanogaster* (fruit fly) | 180,000 | 4 |
| *Takifugu rubripes* (fish) | 400,000 | 22 |
| *Homo sapiens* (human being) | 3,000,000 | 23 |
| *Zea mays* (maize) | 4,500,000 | 10 |
| *Amphiuma means* (salamander) | 90,000,000 | 14 |

Lyme disease, is a linear DNA molecule of about 910 kb, and that of *Escherichia coli* strain K12 is a circular DNA molecule of 4600 kb. The genomes of unicellular eukaryotes are even larger. The genome size of budding yeast, *Saccharomyces cerevisiae*, is 13,000 kb. The units of length of nucleic acids in which genome sizes are typically expressed are as follows:

1. ***kilobase (kb)*** $10^3$ nucleotide subunits
2. ***megabase (Mb)*** $10^6$ nucleotide subunits
3. ***gigabase (Gb)*** $10^9$ nucleotide subunits

In these terms, viral genomes are typically in the range 100–1000 kb, bacterial genomes typically in the range 1–10 Mb, and eukaryotic genomes typically in the range 100–1000 Mb. (The smallest eukaryotic genomes are about 10 Mb.)

Among multicellular eukaryotes, however, genome size often differs enormously among species with the same level of metabolic, developmental, and behavioral complexity. This lack of correlation is known as the **C-value paradox.** The differences are often quite incredible. Differences in genome size among species of protozoa are 5800-fold, among arthropods 250-fold, fish 350-fold, algae 5000-fold, and angiosperms thousandfold (**FIGURE 7.1**). The term *paradox* is amply justified by observing that the genome of *Amphiuma* salamanders is 30 times larger than the size of the human genome.

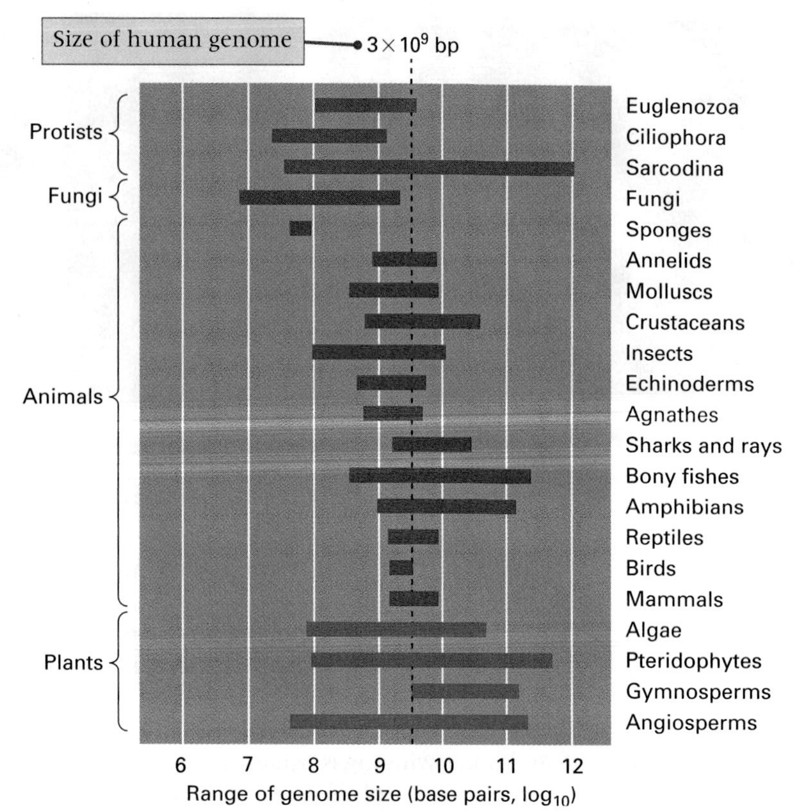


**FIGURE 7.1** Genome size ranges over several orders of magnitude in some groups of organisms, and genome size is not correlated with developmental, metabolic, or behavioral complexity.

The C-value paradox is not due to differences in the number of genes. For example, rice and maize have about the same number of genes (transcripts and proteins), but the maize genome at 2.5 Gb is about six times larger than that of rice at 0.4 Gb. In nearly all higher animals and plants, the actual number of genes is much less than the theoretical maximum. The reason for the discrepancy is that in higher organisms, much of the DNA has functions other than coding for the amino acid sequence of proteins (as we shall see in Chapter 10), and most of the rest consists of transposable elements and other types of repetitive sequences discussed in Section 7.6.

## 7.2 The Supercoiling of DNA

The genome of *E. coli* is a circular DNA molecule about 1500 $\mu$m long compacted inside a cell only about 2 $\mu$m long and 1 $\mu$m in diameter. The DNA molecule must be compacted; otherwise it would not fit inside the cell. At the same time, the genes must be available for transcription as their expression is needed. Part of the compaction of the circular chromosome is due to its being **supercoiled**, which means that segments of double-stranded DNA are twisted around one another, in a manner analogous to the way a telephone cord can be twisted around itself. The geometry of supercoiling can be illustrated by a simple example (**FIGURE 7.2**). Consider first a linear duplex DNA molecule whose ends are joined in such a way that each strand forms a continuous circle. The covalently closed circular DNA molecule is said to be **relaxed** if no twisting is present other than the helical twisting (Figure 7.2A). The individual polynucleotide strands of a relaxed circle form the usual right-handed (positive) helical structure with ten nucleotide pairs per turn of the helix. Suppose you were to cut one strand in a relaxed circle and unwind it one complete rotation of 360° so as to undo one complete turn of the double helix. When the ends were rejoined again, the result would be a circular helix that is "underwound." Because a DNA molecule has a strong tendency to maintain its standard helical form with ten nucleotide pairs per turn, the circular molecule would respond to the underwinding in one of two ways: (1) by forming regions with "bubbles" in which the bases are unpaired (Figure 7.2B) or (2) by twisting the circular molecule in the opposite

**AN EXAMPLE** of the C-value paradox. The Japanese pufferfish *Takifugu rubripes* has a genome size of 400 Mb; the two-toed salamander *Amphiuma means* one of 90,000 Mb. The latter is no more complex than the former. [Left photo © Ken Lucas/Visuals Unlimited; Right photo © Phil Dotson/Photo Researchers, Inc.]

sense from the direction of underwinding. This twisting is the supercoiling, and a molecule with right-handed twists is **negatively supercoiled** (Figure 7.2C). Examples of supercoiled molecules are shown in Figure 7.2C and D. The two responses to underwinding are not independent, and underwinding is usually accommodated by a combination of the two processes: An underwound molecule contains some bubbles of unpaired bases and some supercoiling, although supercoiling predominates.

While *E. coli* and most other bacteria have negatively supercoiled DNA, most thermophiles and other members of the kingdom Archaea have positively supercoiled DNA, in which the double helix is relatively overwound and the twists in the supercoiled DNA are left-handed. What about supercoiling in eukaryotes? The DNA in eukaryotic chromosomes typically has only a few regions that are supercoiled, and these are usually near genes that are actively being transcribed.

## ■ Topoisomerase Enzymes

The supercoiling of natural DNA molecules is produced by DNA **topoisomerase** enzymes. Enzymes of the class denoted *topoisomerase I* act by wrapping themselves completely around a DNA duplex and causing a single-stranded nick by breaking a phosphodiester bond in the backbone of one of the DNA strands. The nicked ends are swiveled around the intact strand, changing the helical twist (number of helical turns). Then the nick is resealed and

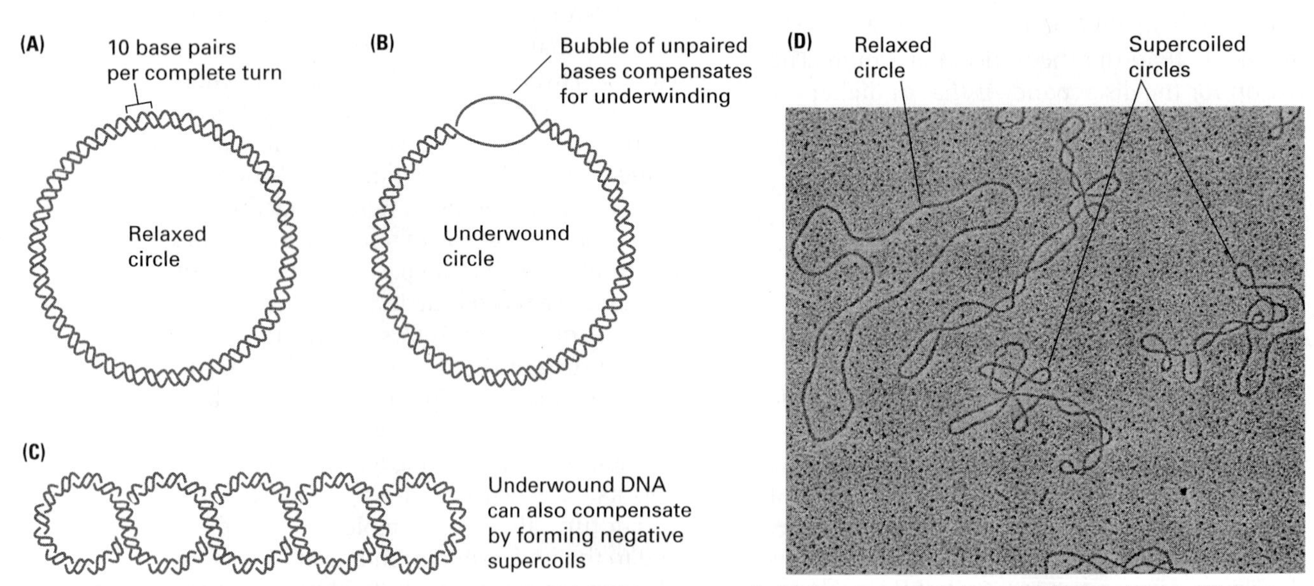
(A)
10 base pairs per complete turn
Relaxed circle
(B)
Bubble of unpaired bases compensates for underwinding
Underwound circle
(C)
Underwound DNA can also compensate by forming negative supercoils
(D)
Relaxed circle
Supercoiled circles

**FIGURE 7.2** Different states of a covalent circle. (A) A nonsupercoiled (relaxed) covalent circle with 36 helical turns. (B) An underwound covalent circle with only 32 helical turns. (C) The molecule in part B, but with four twists to eliminate the underwinding. (D) Electron micrograph showing nicked circular and supercoiled DNA of phage PM2. Note that no bases are unpaired in part C. In solution, parts B and C would be in equilibrium. [Part D, courtesy of K. Gopal Murti, St. Jude Children's Research Hospital, Memphis, TN.]

the duplex released from the enzyme. Depending on conditions, topoisomerase I enzymes can either increase or decrease the amount of supercoiling.

A second class of topoisomerase enzymes is called *topoisomerase II*. These enzymes work by producing a double-stranded gap in one molecule through which another double-stranded molecule is passed. In this way topoisomerase II enzymes are able to pass one DNA duplex entirely through another or to separate two circular DNA molecules that are interlocked. The mechanism of action of topoisomerase II from budding yeast *Saccharomyces cerevisiae* is illustrated in **FIGURE 7.3**. The molecular structure of the enzyme includes two sets of "jaws" set approximately at right angles (Figure 7.3A). The inner set clamps one of the duplexes, the outer set the other (Figure 7.3B and C). To allow the outer DNA molecule to pass completely through the inner one, the inner duplex is first cleaved, and then the outer duplex is passed through the gap (Figure 7.3D and E). After passage, the gap is repaired and both molecules are released (Figure 7.3F).

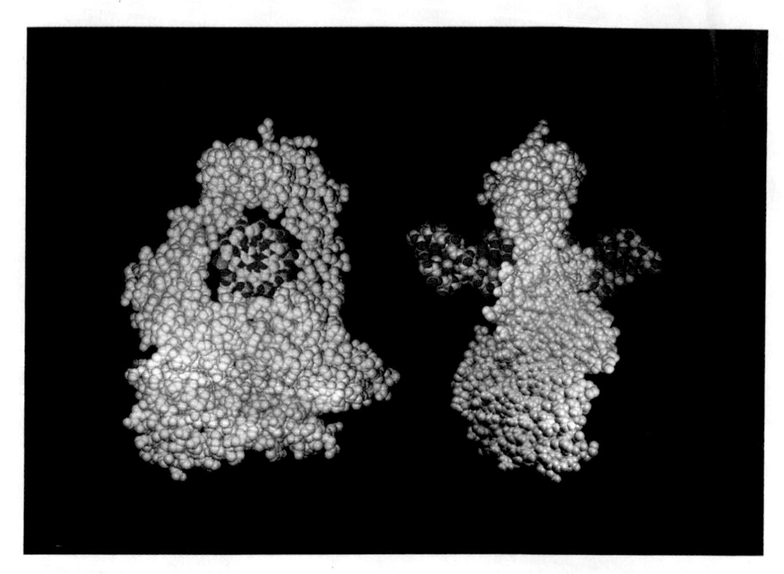

**PROPOSED STRUCTURE** of an association between duplex DNA and topoisomerase I in which the DNA passes through a hole formed in the enzyme. The two views are perpendicular to each other. The structure of the enzyme includes a pair of jaw-like projections that can open and close. The intermediate shown here is created when the enzyme closes its jaws around a DNA molecule, completely enclosing it. What happens then is that one DNA strand is cleaved, swiveled, and reconnected, whereupon the jaws open and the DNA molecule is released. [Reprinted by permission from Macmillan Publishers Ltd: C. D. Lima, J. C. Wang, and A. Mondragón, *Nature* 367 (1994): 138-146, http://www.nature.com/nature.]

In a supercoiled DNA molecule free of proteins that maintain the supercoiling, any nick eliminates all supercoiling because the strain of underwinding is relaxed by free rotation of the intact strand about the sugar–phosphate bond opposite the break. Therefore, any treatment that nicks DNA relaxes the supercoiling. Single-stranded nicks can be produced by any of a variety of enzymes, such as **deoxyribonuclease (DNase)**, that cleave sugar–phosphate bonds.

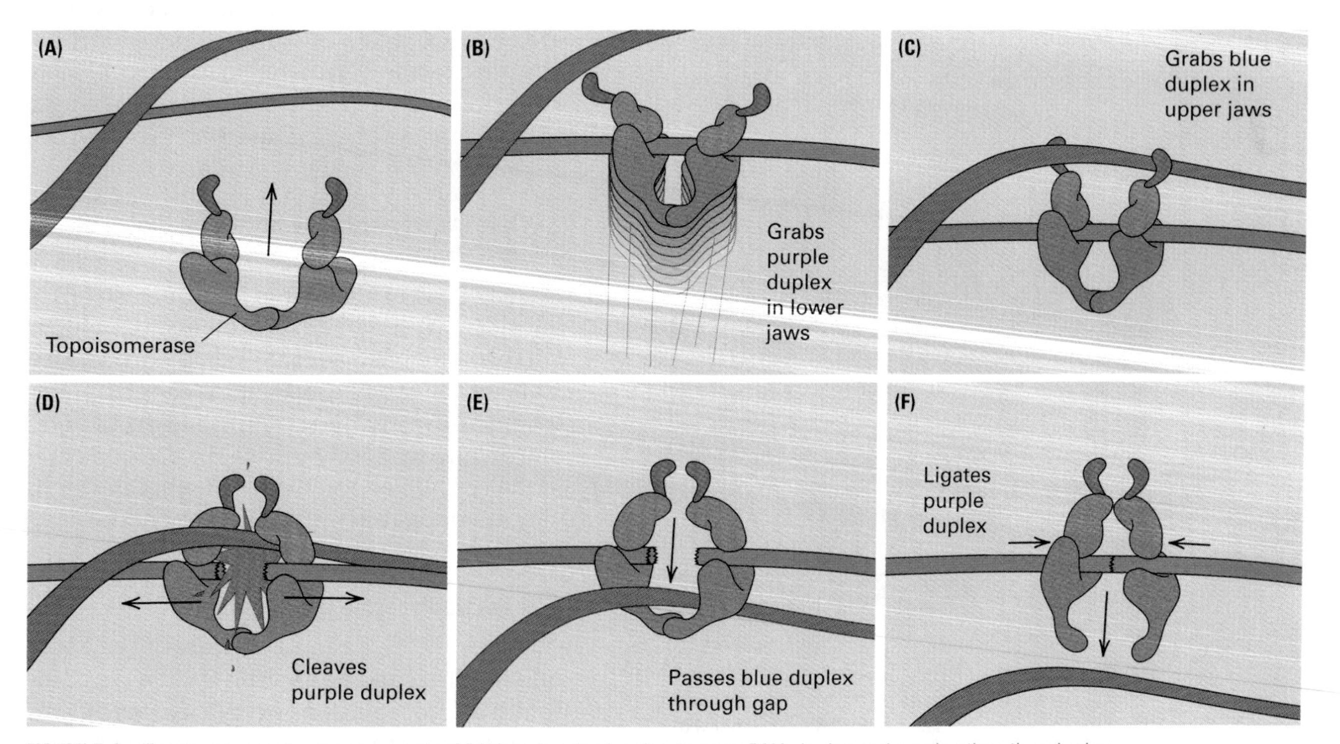


**FIGURE 7.3** Topoisomerase II untangles a pair of DNA molecules by cleaving one DNA duplex and passing the other duplex through the gap. [Adapted from J. M. Berger, et al., *Nature* 379 (1996): 225–232.]

## 7.3 The Structure of Bacterial Chromosomes

One part of the compactness of DNA in a bacterial cell is due to supercoiling. Another part is due to proteins that bind with DNA to wrap, bend, or compact the molecule. Four proteins account for much of this level of packaging (HU for wrapping, FIS and IHF for bending, and HNS for compaction), and together they constitute approximately 10 percent of the total protein in the cell. The result is that the chromosome of *E. coli* is a condensed unit called a folded chromosome or **nucleoid**. The term *chromosome* is technically a misnomer for this structure, because it is not a true chromosome in the sense of a eukaryotic chromosome. Nevertheless, the term is widely used. The most striking feature of the bacterial nucleoid is that the DNA is organized into a set of looped domains (**FIGURE 7.4**). As isolated from bacterial cells, the nucleoid contains small amounts of several proteins that are thought to be responsible for the multiply looped structure. The degree of condensation of the isolated nucleoid (that is, its physical dimensions) is affected by a variety of factors, and some controversy exists about the state of the nucleoid within the cell.

Figure 7.4 also shows loops of DNA in the *E. coli* chromosome that are supercoiled. Those loops that are not supercoiled result from the action of DNases during isolation. The presence of both supercoiled and relaxed loops indicates that the loops are in some way independent of one another. In the preceding section, we stated that supercoiling is generally eliminated in a DNA molecule by one single-strand break. However, such a break in the *E. coli* chromosome does not eliminate all supercoiling. If nucleoids, all of whose loops are supercoiled, are treated with a DNase and examined at various times after treatment, it is observed that supercoiling is relaxed in one loop at a time, not in all loops at once (**FIGURE 7.5**). The loops must be isolated from one another in such a way that rotation in one loop is not transmitted to other loops. The independence is probably the result of proteins that bind to the DNA in a way that prevents rotation of the helix.

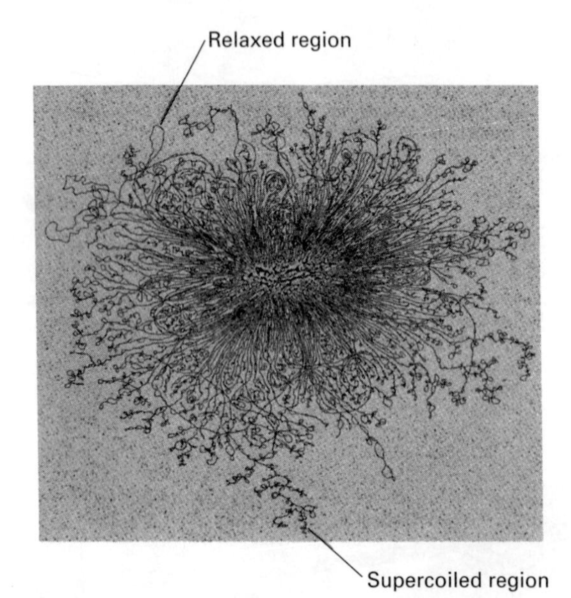


**FIGURE 7.4** Electron micrograph of an *E. coli* chromosome, showing the multiple loops emerging from a central region. [Courtesy of Bruno Zimm and Ruth Kavenoff. Used with permission of Georgianna Zimm, University of California, San Diego.]

## 7.4 The Structure of Eukaryotic Chromosomes

A remarkable feature of the genetic apparatus of eukaryotes is the feat of packaging in which an enormous length of genetic material is condensed into a relatively few small chromosomes. The largest of the 23 human chromosomes contains a DNA molecule that is 82 mm ($8.2 \times 10^4$ $\mu$m) in length. However, at metaphase of a mitotic division, this DNA molecule is condensed into a compact structure about 10 $\mu$m long and less than 1 $\mu$m in diameter. An analogy may make it easier to appreciate the prodigious feat of packaging that such chromosome condensation represents. If the DNA molecule in human chromosome 1 (the largest chromosome) were a cooked spaghetti noodle 1 mm in diameter, it would stretch for 25 miles; in chromosome condensation, this noodle is gathered together, coil upon coil, until at metaphase it is a canoe-sized tangle of spaghetti 16 feet long and 2 feet wide. After cell division, the noodle is unwound again.

### ■ The Nucleosome: The Structural Unit of Chromatin

The DNA of all eukaryotic chromosomes is associated with numerous protein molecules in a stable, ordered aggregate called **chromatin**. Some of the proteins present in chromatin determine chromosome structure and the changes in structure that occur during the division cycle of the cell. Other chromatin proteins appear to play important roles in regulating chromosome functions.

### ■ The Nucleosome Core Particle

The simplest form of chromatin is present in nondividing eukaryotic cells, when chromosomes are not sufficiently condensed to be visible by light microscopy. Chromatin isolated from such cells is a complex aggregate of DNA and

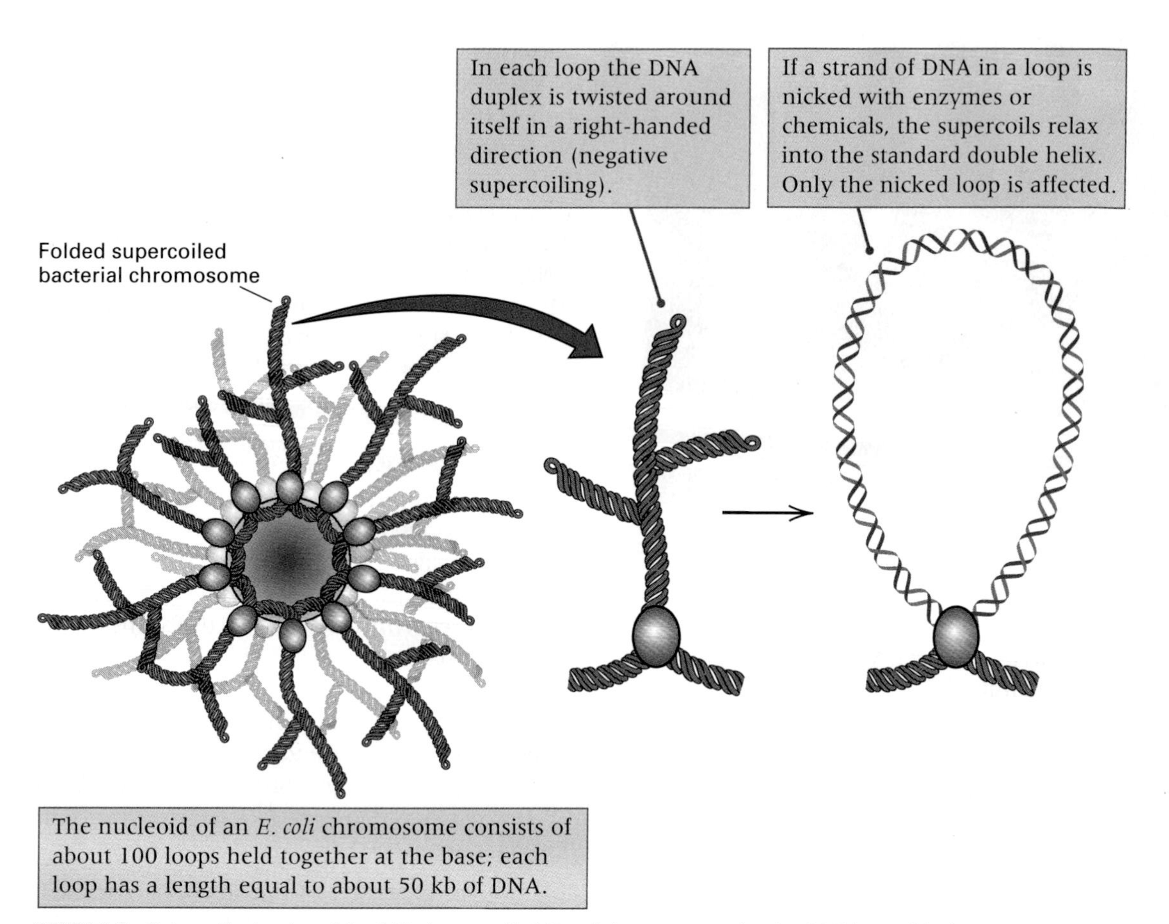

In each loop the DNA duplex is twisted around itself in a right-handed direction (negative supercoiling).
If a strand of DNA in a loop is nicked with enzymes or chemicals, the supercoils relax into the standard double helix. Only the nicked loop is affected.
Folded supercoiled bacterial chromosome
The nucleoid of an *E. coli* chromosome consists of about 100 loops held together at the base; each loop has a length equal to about 50 kb of DNA.


**FIGURE 7.5** Schematic drawing of the folded supercoiled *E. coli* chromosome, showing DNA loops (blue) attached to a protein core (orange) and the opening of loops by nicks. [Adapted from N. J. Trun and J. F. Marko, *ASM News* 64 (1998): 276–283.]

proteins. The major class of proteins comprises the **histone** proteins. Histones are largely responsible for the structure of chromatin. Five major types of histones—H1, H2A, H2B, H3, and H4—are present in the chromatin of nearly all eukaryotes in amounts about equal in mass to that of the DNA. Histones are small proteins (100–200 amino acids) that differ from most other proteins in that from 20 to 30 percent of the amino acids are lysine and arginine, both of which have a positive charge. (Only a few percent of the amino acids of a typical protein are lysine and arginine.) The positive charges enable histone molecules to bind to DNA, primarily by electrostatic attraction to the negatively charged phosphate groups in the sugar–phosphate backbone of DNA. Placing chromatin in a solution with a high salt concentration (for example, 2 molar NaCl) to eliminate the electrostatic attraction causes the histones to dissociate from the DNA. Histones also bind tightly to each other; both DNA–histone and histone–histone binding are important for chromatin structure.

The histone molecules from different organisms are remarkably similar, with the exception of H1. In fact, the amino acid sequences of H3 molecules from widely different species are almost identical. For example, the sequences of H3 of cow chromatin and pea chromatin differ by only 4 of 135 amino acids. The H4 proteins of all organisms also are quite similar; cow and pea H4 differ by only 2 of 102 amino acids. There are few other proteins whose amino acid sequences vary so little from one species to the next. When the variation between organisms is very small, we say that the sequence is highly **conserved**. The extraordinary conservation in histone composition through hundreds of millions of years of evolutionary divergence is consistent with the important role of these proteins in the structural organization of eukaryotic chromosomes.

In the electron microscope, chromatin resembles a regularly beaded thread (**FIGURE 7.6**). The bead-like units in chromatin are called **nucleosomes**. The organization of the nucleosomes in chromatin is illustrated in **FIGURE 7.7**, part A. Each unit has a definite composition, consisting of two molecules each of H2A, H2B, H3, and H4, a segment of DNA containing about 200 nucleotide pairs, and one molecule of histone H1. The complex of two subunits each of H2A, H2B, H3, and

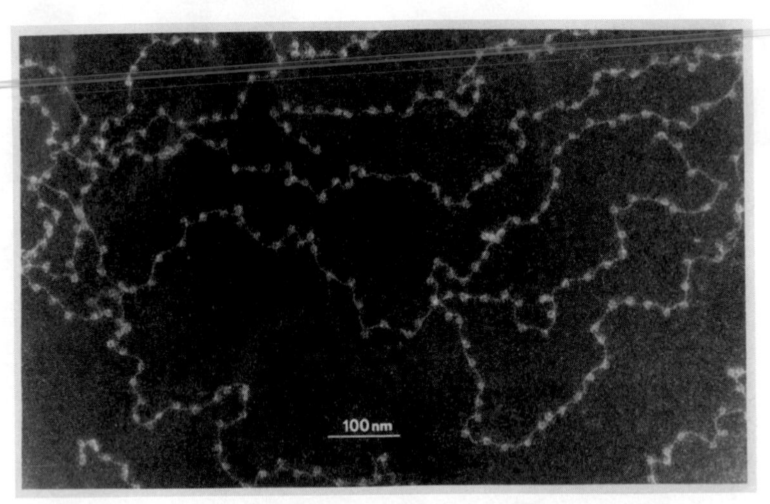


**FIGURE 7.6** Dark-field electron micrograph of chromatin, showing the beaded structure at low salt concentration. The beads have diameters of about 100 Å. [Courtesy of Ada L. Olins and Donald E. Olins, Bowdoin College.]

H4, as well as part of the DNA, forms each "bead," and the remaining DNA and histone H1 bridges between the beads.

Brief treatment of chromatin with certain DNases (for example, *micrococcal nuclease* from the bacterium *Staphylococcus aureus*) yields a collection of small particles of quite uniform size consisting only of histones and DNA. The DNA fragments in these particles are of lengths equal to about 200 nucleotide pairs or small multiples of that unit size (the precise size varies with species and tissue). These particles result from cleavage of the linker DNA segments between the beads (Figure 7.7B). More extensive treatment with DNase results in loss of the H1 histone and digestion of all the DNA except that protected by the histones in the bead. The resulting structure is called a **core particle**, the detailed structure of which is shown in **FIGURE 7.8**. It consists of two molecules each of H2A (yellow), H2B (red), H3 (blue), and H4 (green), around which is wound a segment of DNA of about 145 nucleotide pairs. Each nucleosome is composed of a core particle, the linker DNA between adjacent core particles (which is removed by extensive nuclease digestion), and one molecule of H1; the H1 binds to the histone octamer and to the linker DNA, causing the linkers that extend from both sides of the core particle to cross and draw nearer to the octamer, though some of the linker DNA

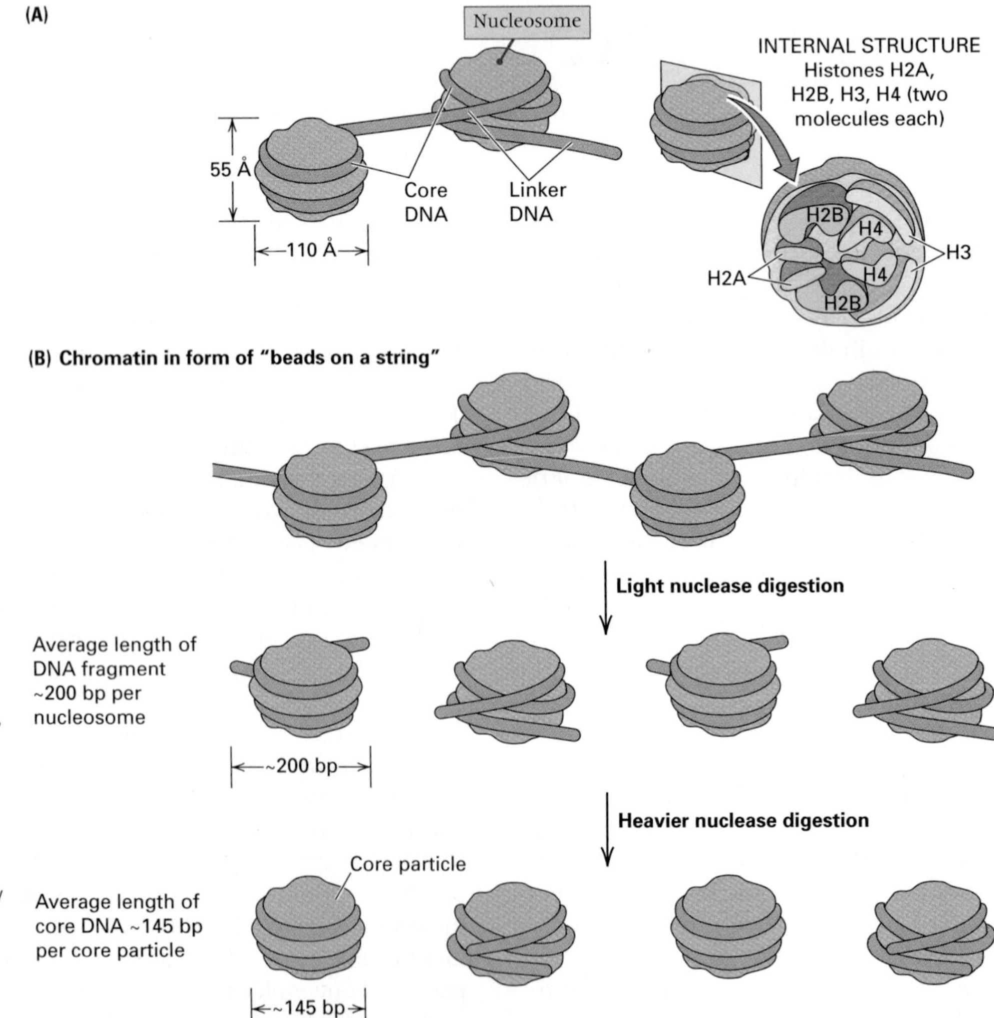


**FIGURE 7.7** (A) Organization of nucleosomes. The DNA molecule is wound one and three-fourths turns around a histone octamer called the core particle. If H1 were present, it would bind to the octamer surface and to the linkers, causing the linkers to cross. (B) Effect of treatment with micrococcal nuclease. Brief treatment cleaves the DNA between the nucleosomes and results in core particles associated with histone H1 and approximately 200 base pairs of DNA. More extensive treatment results in loss of H1 and digestion of all but 145 base pairs of DNA in intimate contact with each core particle.

does not come into contact with any histones. The size of the linker ranges from 20 to 100 nucleotide pairs for different species and varies even in different cell types in the same organism (55 nucleotide pairs is usually considered an average size). Little is known about the structure of the linker DNA or about whether it has a special genetic function, and the cause of the variation in its length is also unknown.

The amino ends of the histone proteins, which constitute about 25 percent of the total length, are known as **histone tails** because they are accessible to enzymes that modify particular amino acid residues such as by the addition of one or more acetyl ($-COCH_3$), methyl ($-CH_3$), or phosphate [$-OP(=O)(OH)_2$] groups to the amino acid. The modifications resulting from these histone acetylase, methylase, or phosphorylase enzymes are reversible by the corresponding histone deacetylase, demethylase, or dephosphorylase enzymes. The histone modifications, especially of histones H3 and H4, are important to gene activity. Acetylated histones tend to bind DNA more loosely and usually render chromatin more accessible to transcription, whereas methylated histones can either promote or impede transcription depending on the particular histone residue that is modified. Histone modifications are thought to be important features of *chromatin remodeling* that takes place in the regulation of gene activity, which will be examined in greater detail in Chapter 11.

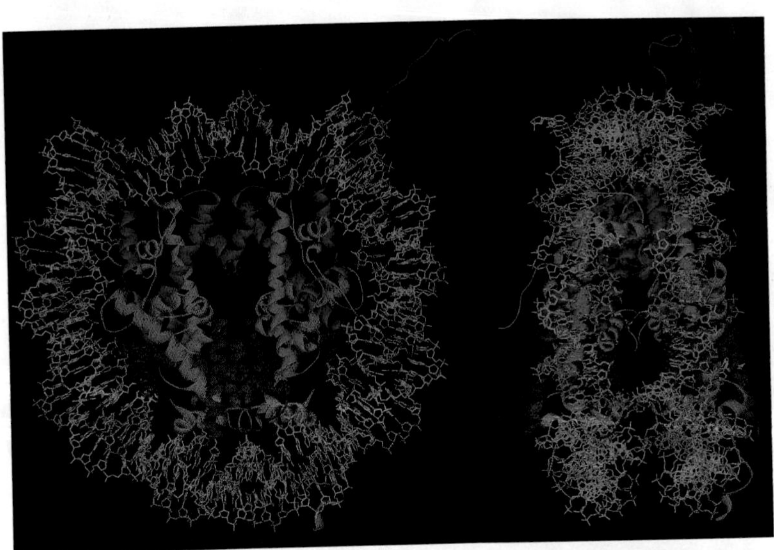

**FIGURE 7.8** Subunit structure of the nucleosome core particle with the DNA duplex wrapped around it. The color coding for the histone monomers is H2A, yellow; H2B, red; H3, blue; and H4, green. [Protein Data Bank 1AOI. K. Luger, et al., *Nature* 389 (1997): 251–260.]

## Chromosome Territories in the Nucleus

At the salt concentration present in living cells, the nucleosome fiber compacts into a shorter, thicker fiber with an average diameter ranging from 300 to 350 Å; this is called the **30-nm chromatin fiber** (**FIGURE 7.9**, part A). In forming the 30-nm chromatin fiber, the string of nucleosomes forms a series of stacked right-handed coils (part B) in which each nucleosome is attached to its neighbor by linker DNA that stretches nearly linearly across to the opposite side of the coil. Looking down at the 30-nm fiber from the top (part C), one can trace the path of the linker DNA as it travels down the length of the fiber. In each revolution around the fiber axis, the path of the linker DNA closely approximates the shape of a seven-pointed star.

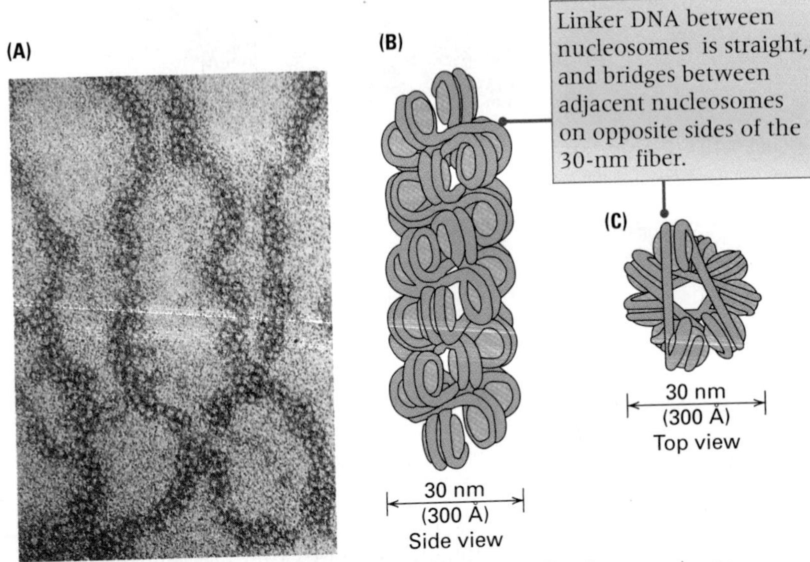
(A)
(B)
Linker DNA between nucleosomes is straight, and bridges between adjacent nucleosomes on opposite sides of the 30-nm fiber.
(C)
30 nm
(300 Å)
Top view
30 nm
(300 Å)
Side view

**FIGURE 7.9** (A) Electron micrograph of the 30-nm chromatin fiber in mouse chromosomes. (B and C) A model of the chromatin fiber in which the DNA (blue-gray) is wound around each nucleosome. [Part A, courtesy of Barbara A. Hamkalo, University of California, Irvine; Part B, adapted from J. T. Finch and A. Klug, *Proc. Natl. Acad. Sci. USA* 73 (1976): 1897–1901.]

In the nucleus of a nondividing cell, the 30-nm chromatin fiber is organized into higher order structures that can be visualized using modern methods of optical sectioning and image reconstruction. **FIGURE 7.10** shows a computer-generated image of 30-nm chromatin fibers within the nucleus of a nondividing cell. The chromatin fibers are folded into small *chromatin loops* with a DNA content of approximately 100 kb each, and these are further organized into *chromatin domains* with a DNA content of approximately 1 Mb each. Each chromosome arm occupies a discrete **chromosome territory**, denoted by the different colors. In cells cycling through mitosis, the chromosome territories are disrupted when the chromosomes condense and the cell divides, but they are reconstituted again in the next interphase. However, the

**FIGURE 7.10** Computer-generated image of chromosome territories formed by 30-nm chromatin fibers within the nucleus of a nondividing cell. [Courtesy of Tobias A. Knoch, Erasmus MC, Rotterdam, and Kirchhoff-Institute for Physics, Ruperto–Carola University, Heidelberg: TA. Knoch@taknoch.org].

chromosome territories may differ in position in different cell types as well as in the same cell type at different times in development.

Chromosome territories are correlated with gene densities. The territories of chromatin domains containing relatively few genes tend to be located near the periphery of the nucleus or near the nucleolus, whereas the territories of chromosome domains that are relatively gene rich tend to be located toward the interior of the nucleus. For example, human chromosome 18 (85 Mb in size) is relatively gene poor whereas chromosome 19 (67 Mb in size) is relatively gene rich. In the nucleus, chromosome 18 territories tend to be at the nuclear periphery whereas those of chromosome 19 tend to be in the interior.

The spaces between the chromatin domains form a network of channels, like the holes permeating through a sponge. The channels are large enough to allow passage of the molecular machinery for replication, transcription, and RNA processing. Evidence suggests that these molecules gain access to chromatin by means of passive diffusion. Replication, transcription, and RNA processing are all ordered processes. DNA replication takes place in small, discrete regions that exhibit a reproducible temporal and spatial pattern, and transcription takes place in a few hundred discrete locations. However, many important details are still unknown about the organization of chromatin in the nucleus and how chromatin territories function in the coordination of the central molecular processes of replication, transcription, and RNA processing.

## ■ Chromosome Condensation

The hierarchical nature of chromosome structure is illustrated in **FIGURE 7.11**. Assembly of DNA and histones is the first level, resulting in a sevenfold reduction in length of the DNA and the formation of a beaded flexible fiber 110 Å (11 nm) wide (Figure 7.11B), roughly five times the width of free DNA (Figure 7.11A). The structure of chromatin varies with the concentration of salts, and the 110-Å fiber is present only when the salt concentration is quite low. In the living cell this is usually compacted into the 30-nm chromatin fiber (Figure 7.11C), which in the interphase nucleus is folded into 100-kb chromatin loops that are organized into 1-Mb chromatin domains that form the chromosome territories.

In cells cycling through mitosis, the interphase chromatin organization is replaced by a more compact organization in which the 30-nm chromatin fiber condenses into a chromatid of the metaphase chromosome (Figure 7.11D through F). Chromosome condensation is an ordered, energy-consuming process orchestrated by a protein complex called *condensin* that works together with topoisomerase II to actively coil the chromatin. The condensin complex is composed of at least five proteins including two types of proteins denoted SMC (stands for *structural maintenance of chromosomes*), specifically, SMC2 and SMC4. Histone H3 phosphorylation also plays a role in chromatin condensation. Mammals have at least two kinds of condensins, one of which is specific for regions near the kinetochore. Although the structures of some condensin proteins are known, the details of chromatin condensation are still largely unknown, and there is no strong evidence supporting any of the particular coiled structures greater than the 30-nm chromatin fiber depicted in Figure 7.11.

In electron micrographs of isolated metaphase chromosomes from which histones have been removed, the partly unfolded DNA has the form of an enormous number of loops that seem to extend from a central core or *scaffold* (**FIGURE 7.12**). This chromosomal **scaffold** is

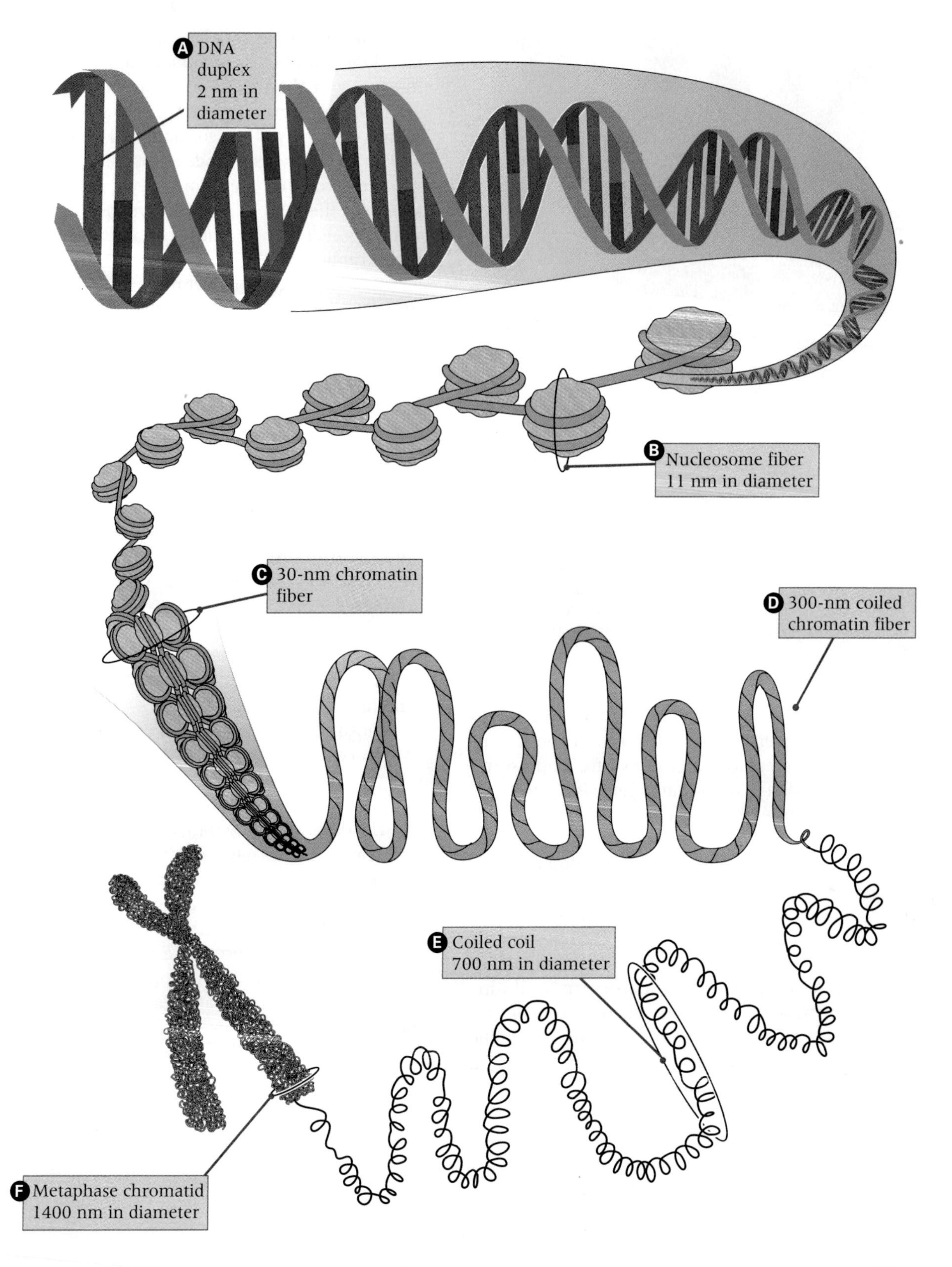


**FIGURE 7.11** Various stages in the condensation of DNA (A) and chromatin (B through E) in forming a metaphase chromosome (F). The dimensions indicate known sizes of intermediates, but the detailed structures in D-F are hypothetical.

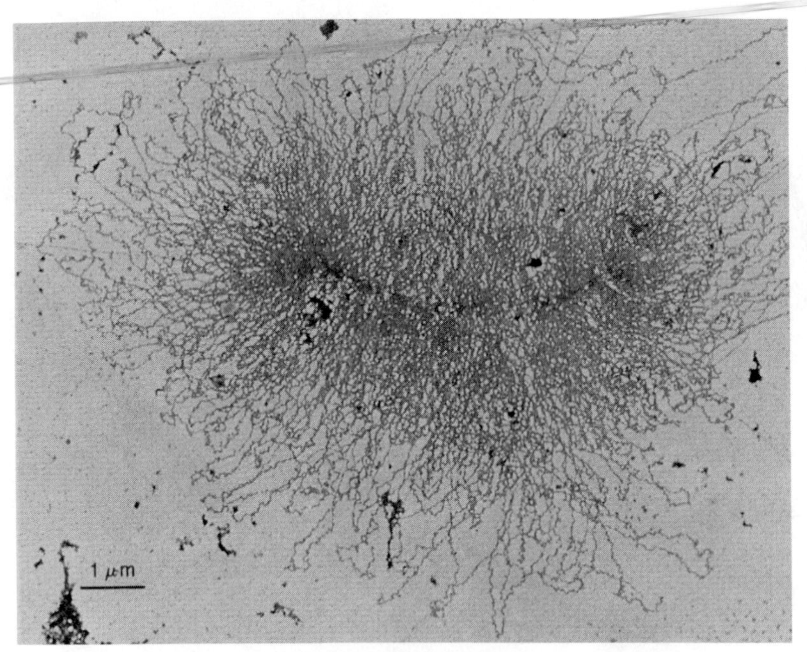


**FIGURE 7.12** Electron micrograph of a partially disrupted anaphase chromosome of the milkweed bug *Oncopeltus fasciatus*, showing multiple loops of 30-nm chromatin at the periphery. [Reproduced from *Insect Ultrastructure*, vol. 1, 1982, p. 222, "Morphological Analysis of Transcription in Insect Embryos," V. Voe, et al., © Plenum Publications Corporation with kind permission from Springer Science and Business Media. Photo courtesy of Victoria Foe and Charles Laird, University of Washington, and Hugh Forrest, University of Texas.]

composed of a number of nonhistone chromosomal proteins including a member of the SMC2 family different from that found in condensin. Electron microscopic studies of chromosome condensation in mitosis and meiosis suggest that the scaffold extends along the chromatid and that the 30-nm fiber becomes arranged into a series of loops radiating from the scaffold. Details are not known about the additional folding that is required of the fiber in each loop to produce the fully condensed metaphase chromosome.

The genetic significance of the compaction of DNA and protein into chromatin and ultimately into the metaphase chromosome is that it greatly facilitates the movement of the genetic material during nuclear division. Relative to a fully extended DNA molecule, the length of a metaphase chromosome is reduced by a factor of approximately $10^4$. Without chromosome condensation, the chromosomes would become so entangled that there would be many more abnormalities in the distribution of genetic material into daughter cells.

## 7.5 Polytene Chromosomes

Giant chromosomes called **polytene chromosomes** are found in the nuclei of cells of the salivary glands and certain other tissues of the larvae of *Drosophila* and other two-winged (dipteran) flies. These chromosomes contain about 1000 identical DNA molecules laterally aligned. Each has a length and cross-sectional diameter many times greater than those of the corresponding chromosome at mitotic metaphase in ordinary somatic cells, as well as a constant and distinctive pattern of transverse banding (**FIGURE 7.13**). The polytene structures are formed by repeated replication of the DNA in a closely synapsed pair of homologous chromosomes without separation of the replicated chromatin strands. Polytene chromosomes are atypical chromosomes and are formed in "terminal" tissues of the larva that are eliminated in the formation of the pupa. Although they do not contribute to the adult fly, the polytene tissues of larvae have been especially valuable in the genetics of *Drosophila*, as will become apparent in Chapter 8.

In polytene nuclei of *D. melanogaster* and other dipteran species, large blocks of heterochromatin adjacent to the centromeres are aggregated into a single compact mass called the **chromocenter**. Because the two largest chromosomes in *Drosophila* (chromosomes numbered 2 and 3) have centrally located centromeres, the chromosomes appear in the configuration shown in **FIGURE 7.14**. The paired X chromosomes (in a female), the left and right arms of chromosomes 2 and 3, and a short chromosome (chromosome 4) project from the chromocenter. In a male, the Y chromosome, which consists almost entirely of heterochromatin, is incorporated in the chromocenter.

The darkly staining transverse bands in polytene chromosomes have about a tenfold range in width. These bands result from the side-by-side alignment of tightly folded regions of the individual chromatin strands that are often visible in mitotic and meiotic prophase chromosomes as chromomeres. More DNA is present within the bands than in the interband (lightly stained) regions. About 5000 bands have been identified in the *D. melanogaster* polytene chromosomes. This linear array of bands, which has a pattern that is constant and characteristic for each species, provides a finely detailed **cytological map** of the chromosomes. The banding pattern is such that observers with sufficient training and experience can identify short regions in any of the chromosomes (Figure 7.14).

## 7.6 Repetitive Nucleotide Sequences in Eukaryotic Genomes

In bacterial genomes the variation in average base composition from one region to another is usually quite small. In eukaryotes, by contrast,

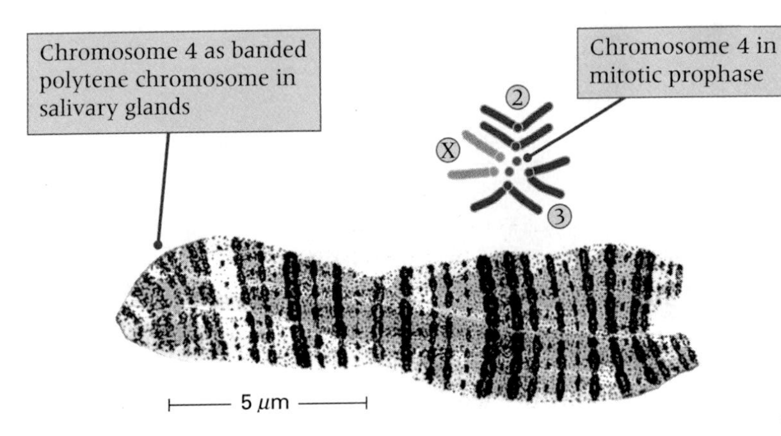


**FIGURE 7.13** The polytene fourth chromosome of *Drosophila melanogaster.* The somatic chromosomes of *Drosophila*, drawn to scale with respect to the polytene fourth chromosome, are shown at the upper right as they appear in mitotic prophase. [Reproduced from C. Bridges, "Salivary chromosome maps...", *J. Hered.* 26 (1935): 60–64, by permission of Oxford University Press.]

some components of the genome have a base composition that is quite different from the average. For example, one component of crab DNA is only 3 percent G + C, compared to an average of approximately 50 percent G + C for the rest of the genome. The components with unusually low or unusually high G + C contents often consist of fairly short nucleotide sequences that may be tandemly repeated as many as a million times in a haploid genome. Such repeated sequences are known as **satellite DNA.** In the mouse, satellite DNA accounts for about 10 percent of the genome. Other **repetitive sequences** also are present in eukaryotic DNA. Because repetitive DNA consists of many highly similar or identical sequences, fragments of repetitive DNA renature more readily than fragments of nonrepetitive DNA. Information about the size of repeated sequences and the number of copies of a particular sequence can be obtained through studies of the rate of renaturation. The quantitative analysis of DNA renaturation is considered next.

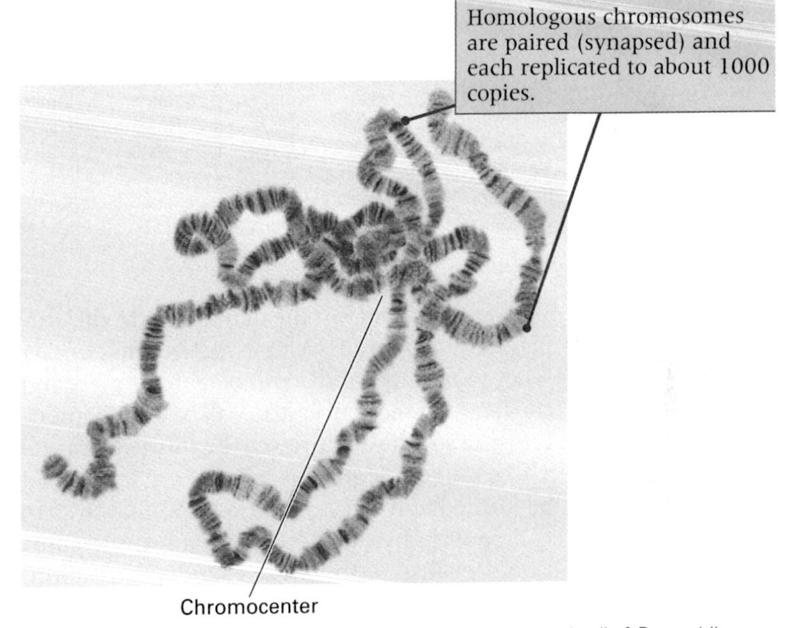

Homologous chromosomes are paired (synapsed) and each replicated to about 1000 copies.
Chromocenter

**FIGURE 7.14** Polytene chromosomes from a larval salivary gland cell of *Drosophila melanogaster.* The chromocenter is the central region in which the centromeric regions of all chromosomes are united. [© Andrew Syred/Photo Researchers, Inc.]

## ■ Kinetics of DNA Renaturation

In Chapter 2 we emphasized that complementary single strands of DNA denatured by heat can undergo renaturation at a lower temperature to form duplex molecules in solution. The initial step is the random collision of short complementary sequences. If the flanking sequences are not complementary, then even at the lower temperature thermal motion will disrupt the chance pairing and the individual strands will go their separate ways again. If the flanking sequences are complementary, however, the initial short duplex region will very quickly increase in size by additional base pairing on either side. Eventually a long duplex molecule can be formed. The base pairing need not be perfect, but the mismatches need to be sufficiently rare that the duplex molecule can resist thermal denaturation. The initial collision between pairs of complementary single strands is the rate-limiting step in renaturation. Because the chance of initial collision is concentration-dependent, the rate of reassociation increases with DNA concentration (**FIGURE 7.15**, part A). Any increase in DNA concentration results in a corresponding increase in the number of potential pairing partners for a given strand.

Overall DNA concentration is only one factor affecting the rate of renaturation. The overall complexity of the sequences also matters. By the *complexity* of a DNA sequence we mean the number of possible pairing partners for any DNA strand present in a given quantity of DNA. For example, consider a 10-kb duplex of synthetic DNA consisting of poly dA in one strand paired with poly dU in the other, and compare this with a 10-kb duplex consisting of a random sequence of nucleotides. For a given concentration, strands from the dA•dU duplex will renature much faster than the random sequence.

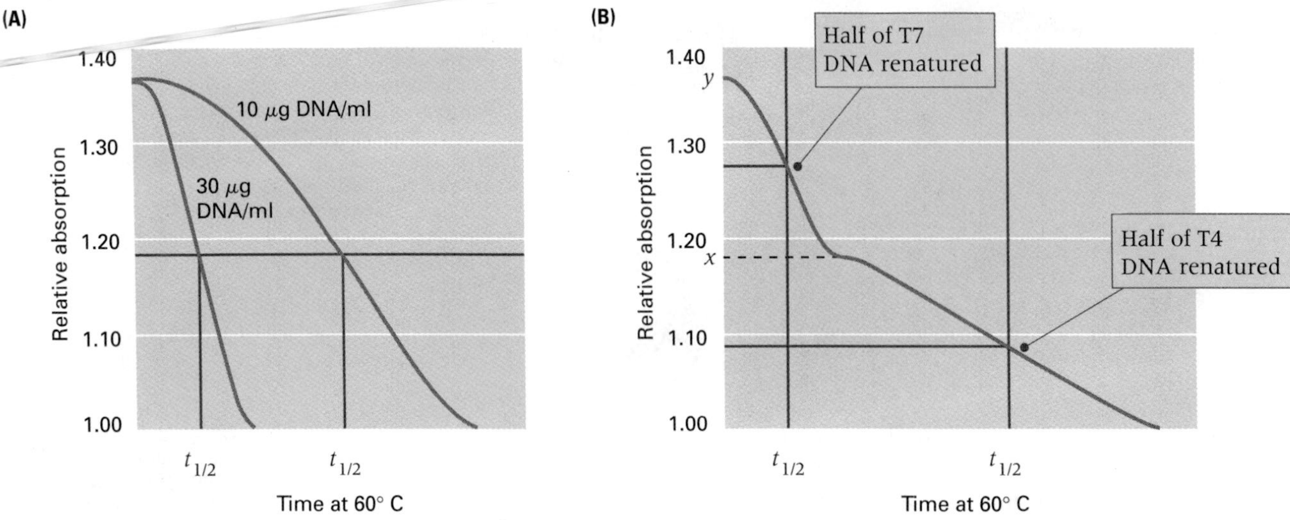


**FIGURE 7.15** (A) Dependence of renaturation time on the concentration of phage T7 DNA. After a period at 90° C to separate the strands, the DNA was cooled to 60° C. Renaturation is complete when the relative absorption reaches 1. (B) Renaturation of a mixture of T4 and T7 DNA, each at the same temperature. Extrapolation (black dashed line) yields the early portion of the T4 curve; the ratio of the absorptions at points *x* and *y* yields the fraction of the total DNA that is T4 DNA. The times required for half-completion of renaturation, $t_{1/2}$, are obtained by drawing the red horizontal lines, which divide each curve equally in the vertical direction, and then extending the red vertical lines to the time axis.

The reason is that, for the denatured dA•dU duplex, any region of dA that comes into contact with any region of dU will be able to pair, and the flanking sequences also will be complementary and so allow a duplex to be formed. For the 10-kb random sequence, however, a duplex can be formed only if a strand by chance happens to collide with its exact complement in the solution. Because of the more rapid renaturation, we say that the dA•dU duplex is less complex than the random sequence.

Sequence complexity also depends on length. For example, compare the 10-kb random sequence with a 1000-kb random sequence. The 10-kb sequence will renature more rapidly because, for a given concentration of DNA nucleotides, there will be many more copies of the 10-kb sequence than copies of the 1000-kb sequence, and therefore a strand of the 10-kb sequence is much more likely to collide with a complementary partner. The 10-kb random sequence is said to be less complex than the 1000-kb random sequence. The difference in rate of renaturation is found even when the duplex molecules are fragmented into many components of equal size, as is normally done in actual experiments. Breaking the molecules does not matter because the molar concentration of nucleotides is the same for the broken molecules as for the unbroken molecules.

The implications of sequence complexity are shown in a real example in Figure 7.15B. This experiment shows the renaturation of a mixture of DNA molecules from the bacteriophages T7 and T4. These molecules have no common nucleotide sequences, and they also differ in molecular mass because the T7 genome is about 1/4 the size of that of the T4 genome (39.9 kb for T7 versus 168.9 kb for T4). Therefore, in solutions in which the number of grams of DNA per milliliter is the same for each bacteriophage genome, as in this experiment, the molar concentration of T7 is about four times greater than that of T4. Because the bacteriophage genomes are not complementary, strands from T7 can pair only with other strands from T7, and likewise for strands from T4. Because the molar concentration is greater for T7, however, the T7 strands will begin to renature first. The renaturation curve for the mixture is shown in 7.15B. In this case the curve consists of two steps: one for the more rapidly renaturing T7 molecules and the other for the more slowly renaturing T4 molecules. Each step in the curve accounts for half of the change in the absorption of the solution at 260 nanometers (*A*260) because the initial concentration (in μg of DNA per milliliter, which is proportional to *A*260) of each type of molecule is the same.

As is apparent in Figure 7.15, the renaturation curve of a molecule that contains no repeating base sequences consists of only a single step, and the rate of renaturation is a function of the size of the molecule. Steps in a renaturation curve are not produced merely by fragmenting molecules into pieces. Steps result from renaturation of a mixture of sequences (or a genome) that includes both a unique component and multiple copies of different repeated sequences, because *fragments containing the more numerous repeated sequences will renature more rapidly than the fragments containing portions of the unique sequence.* For example, consider a molecule containing 50,000 base pairs that consists of 100 tandem copies of a repeated sequence of 500 base pairs. If the molecules are broken into about 100 fragments of roughly equal size, each fragment will be about 500 base pairs. Although the renaturation curve for the fragments will have a single step, the rate of renaturation will be characteristic of molecules 500 nucleotides in length. If a genome contains multiple families of repeated sequences whose abundances differ, the renaturation curve will have steps: one step for each repeating sequence. This principle is the basis of the analysis of renaturation kinetics.

Renaturation kinetics can be described in a simple mathematical form, because the reaction is one in which the rate-limiting step is the initial collision of two molecules. In such a case, the fraction of single strands remaining denatured at a time $t$ after the start of renaturation is given by the expression

$$\frac{C}{C_0} = \frac{1}{[1 + kC_0t]} \qquad (1)$$

in which $C$ is the concentration of single-stranded DNA in moles of nucleotide per liter, $C_0$ is the initial concentration, and $k$ is a constant. The expression $C_0t$ is commonly called **Cot**, and a plot of $C/C_0$ versus $C_0t$ is called a **Cot curve**. The units of $C_0t$ are the product of the DNA concentration (moles per liter) and time (seconds), which is written as $\frac{\text{moles}}{\text{liter}} \bullet \text{sec.}$

When renaturation is half completed ($t = t_{1/2}$), then $C/C_0 = 1/2$. Substitution of these values into Equation (1) leads, after some simplification, to:

$$C_0t_{1/2} = 1/k \qquad (2)$$

The value of $1/k$ depends on experimental conditions, but for a particular set of conditions, the value is proportional to the number of nucleotides in the renaturing sequences. The longer the sequence, the greater the time to achieve half-complete renaturation for a particular starting concentration (because the number of molecules is smaller). Equations (1) and (2) apply to a single molecular species, a point to which we will return. If a molecule consists of several subsequences, then one needs to know $C_0$ for each subsequence, and a set of values of $1/k$ will be obtained (one for each step in the renaturation curve), each value depending on the length of the subsequence. What is meant by the "length" of the sequence that determines the rate of renaturation is best described by example. A DNA molecule containing only adenine in one strand (and thymine in the other) has a repeating length of 1. The repeating tetranucleotide . . . GACTGACT . . . has a repeating length of 4. A nonrepeating DNA molecule containing $n$ nucleotide pairs has a unique length of $n$.

Experimentally, the number of nucleotides per repeating unit is not determined directly. Generally, renaturation curves for a series of molecules of known molecular mass *with no repeating elements in their sequences* (and hence yielding one-step renaturation curves) serve as standards. Molecules composed of short repeating sequences are also occasionally used. A set of curves of this kind is shown in **FIGURE 7.16**. Note that two of these simple curves represent the entire genomes of *E. coli* and phage T4. With the standard conditions for Cot analysis used to obtain this set of curves, the sequence length $n$ (in base pairs) that yields a particular value of $C_0t_{1/2}$ is

$$n = (5 \times 10^5)C_0t_{1/2}$$

in which $t$ is in seconds, $C_0$ is in nucleotides per liter, and $5 \times 10^5$ is a constant dependent on the conditions of renaturation. Through this formula, the experimentally determined value of Cot ($C_0t_{1/2}$) yields an estimate of the repeat length, $n$, of a repetitive sequence. Note again that $C_0$ is not the overall DNA concentration but the concentration of the individual sequence producing a particular step in a curve. How one obtains the necessary value of $C_0$ will become clear when we analyze a Cot curve. Such an analysis begins by first noting the number of steps in the curve (each step of which represents a sequence or class of sequences of a particular length) and the fraction of the material represented by each step. The observed value of $C_0t_{1/2}$ for each step must be corrected by first inferring the value of $C_0$ for each sequence class. The lengths of the sequences are then determined from these corrected values by comparison to standards, and the sequence lengths and sequence abundances, as a proportion of the total, are compared to obtain the number of copies of each sequence. This analysis is best understood by looking at an example, which we will do next.

## Analysis of Genome Size and Repetitive Sequences by Renaturation Kinetics

**FIGURE 7.17** shows a Cot curve typical of those obtained in analyses of the renaturation kinetics of eukaryotic genomes. Three discrete steps are evident: 50 percent of the DNA has $C_0t_{1/2} = 10^3$

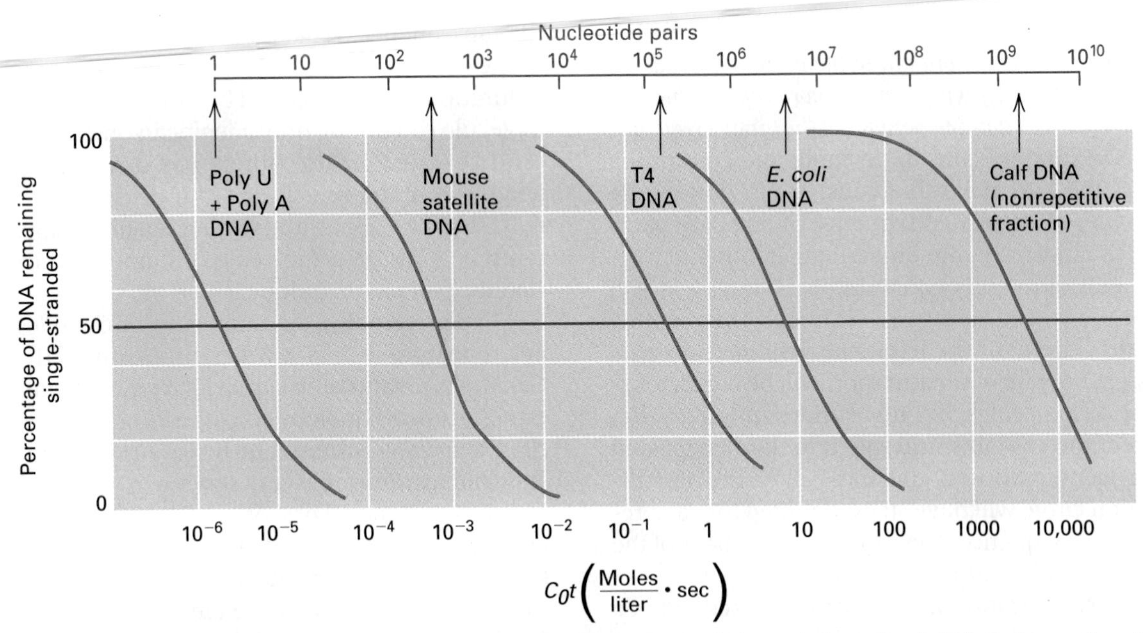


**FIGURE 7.16** A set of Cot curves for various DNA samples. The black arrows pointing up to the red scale indicate the number of nucleotide pairs for each sample; they align with the intersection of each curve with the horizontal red line (the point of half-renaturation, or $C_0t_{1/2}$). The *y*-axis on this graph can be related to that on Figure 7.15 by noting that maximum absorption represents totally single-stranded DNA and minimum absorption represents totally double-stranded DNA.

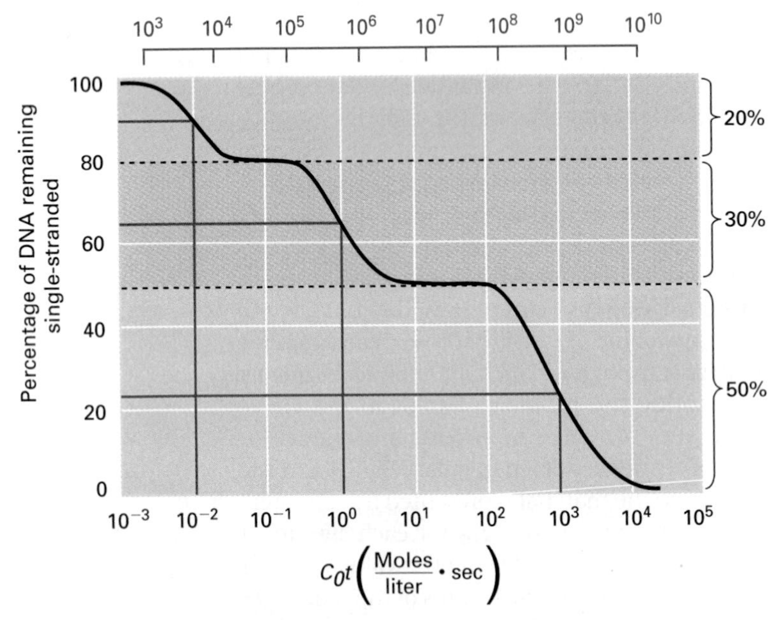


**FIGURE 7.17** The Cot curve analyzed in the text. The scale at the top is the same as that in Figure 7.16. The black dashed lines indicate the fractional contribution of each class of molecules to the total DNA.

(blue), 30 percent has $C_0t_{1/2} = 10^0 = 1$ (beige), and 20 percent has $C_0t_{1/2} = 10^{-2}$ (green). The scale at the top of the figure was obtained from Cot analysis of molecules that have unique sequences of known lengths, as in Figure 7.16. The sequence sizes cannot be determined directly from the observed $C_0t_{1/2}$ values, because each value of $C_0$ used in plotting the horizontal axis in the body of the figure is the total DNA concentration in the renaturation mixture. Multiplying each $C_0t_{1/2}$ value by the fraction of the total DNA that it represents yields the necessary corrected $C_0t_{1/2}$ values (that is, $0.50 \times 10^3$, $0.30 \times 1$, and $0.20 \times 10^{-2}$). From the size scale at the top of Figure 7.17, the corresponding sequence sizes (repeat lengths) are approximately $3.0 \times 10^8$, $2.2 \times 10^5$, and $1 \times 10^3$ nucleotide pairs, respectively.

To determine the approximate number of copies of each sequence, we make use of the fact that the number of copies of a sequence having a particular renaturation rate is inversely proportional to $t_{1/2}$ and hence to the observed (uncorrected) $C_0t_{1/2}$ values for each class. Thus, if the haploid genome contains only one copy of the longest sequence ($3.0 \times 10^8$ base pairs), it contains $10^3$ copies of the sequence (or sequences) of length $2.2 \times 10^5$ base pairs and $10^5$ copies of the sequence (or sequences) of length $1 \times 10^3$ base pairs. An estimate of the total number of base pairs per genome would in this case be:

$$(3.0 \times 10^8) + (10^3 \times 2.2 \times 10^5) + (10^5 \times 1 \times 10^3) = 6.2 \times 10^8 \text{ nucleotide pairs}$$

The different sequence components of a eukaryotic DNA molecule can be isolated by procedures that recover the double-stranded molecules formed at different times during a renaturation reaction. The method used is to allow renaturation to proceed only to a particular $C_0t_{1/2}$ value. The reassociated molecules present at that point are then separated from the remaining single-stranded molecules, usually

## connection

### Post-Genomics Genetics

*This paper demonstrates that formal genetic analysis—mutation and mapping—still provides a uniquely powerful approach to understanding the biology of an organism, even when, as in this case, the complete DNA sequence of the organism's genome is known. The reasons are, first, that sequence does not always reveal function and, second, that many genes have two or more functions, only one of which is recognized. In this example, analysis of a mutation with the insidious phenotype of gradual germ-line extinction revealed a gene with an unexpected dual function in telomere replication and response to DNA damage. Of what use was the genomic sequence? Once the new mutation was genetically mapped, a likely candidate gene could immediately be identified by inspection of all the coding sequences in the region. The suspect was then quickly shown to be the culprit by sequencing the candidate gene in the mutant.*

**Shawn Ahmed and Jonathan Hodgkin 2000**
Medical Research Council Laboratory of Molecular Biology
Cambridge, United Kingdom
*MRT-2 Checkpoint Protein Is Required for Germline Immortality and Telomere Replication in* C. elegans

In most higher organisms, the germ line is responsible for perpetuation of the species . . . and can be thought of as an immortal cell lineage. . . . To investigate how the germ line achieves immortality, a screen was conducted for *Caenorhabditis elegans* mutants with mortal germ lines. . . . Sixteen independent *mortal germ-line mutants* were identified [among 400 mutagenized lines] that became effectively sterile between generations $F_4$ and $F_{16}$. . . . The unexpectedly large number of *mrt* mutants that were identified in our small pilot screen suggests a conservative estimate of ~50 genes that are specifically required for germ-line immortality in *C. elegans*. . . . For one of these mutants, *mrt-2*, the generation at sterility varies widely, from generation $F_{10}$ to generation $F_{28}$. . . . These results suggest that *mrt-2* accumulates some kind of damage that is inherited. . . . The *mrt-2* germ line is hypersensitive to x-rays, which can damage DNA by inducing double-stranded breaks. . . . To gain further insight into the genome stability defect, the chromosomes were examined. Early generation *mrt-2* worms contained the six pairs of chromosomes expected for wildtype, whereas late-generation *mrt-2* worms contained only three, four, or five chromosomes, suggesting the occurrence of chromosome fusions. . . . All chromosome fusions tested were homozygous viable, indicating that they were not missing any essential genes and were therefore probably end-to-end fusions. . . . The late-onset end-to-end chromosome fusion phenotype suggested a defect in telomere replication. . . . [Molecular studies showed that] the telomeres of *mrt-2* worms shorten at the rate of about 12 base pairs per cell division (about 125 base pairs per generation). . . . The *mrt-2* gene is a *C. elegans* homolog of the *Schizosaccharomyces pombe rad1*$^+$ and *Saccharomyces cerevisiae RAD17* checkpoint genes that are conserved from yeast to mammals. In yeast, these genes are required to delay cell-cycle progression in response to DNA damage or in response to a block in DNA replication. . . . The checkpoint protein described here, MRT-2, is required both for telomere replication and for response to DNA damage.

***The late-onset end-to-end chromosome fusion phenotype suggested a defect in telomere replication***

Source: S. Ahmed and J. Hodgkin, *Nature* 403(2000): 159–164.

by passing the solution through a tube filled with a form of calcium phosphate crystal (hydroxyapatite) that preferentially binds double-stranded DNA.

## 7.7 Unique and Repetitive Sequences in Eukaryotic Genomes

The nucleotide sequence composition of many eukaryotic genomes has been examined via analysis of the renaturation kinetics of DNA. The principal finding is that eukaryotic organisms differ widely in the proportion of the genome that consists of repetitive DNA sequences and in the types of repetitive sequences that are present. In most eukaryotic genomes, the DNA consists of three major components:

1. **Unique, or single-copy, sequences** This is usually the major component, typically constituting 30–75 percent of the chromosomal DNA in most organisms.
2. **Highly repetitive sequences** This component usually constitutes 5–45 percent of the genome, depending on the species. Some of these sequences are the satellite DNA referred to earlier. The sequences in this class are typically 5–300 nucleotide pairs per repeat and present in as many as $10^5$ copies per genome.
3. **Middle-repetitive sequences** This component typically constitutes 1–30

percent of a eukaryotic genome; it includes sequences that are usually present in 10–1000 copies per genome.

These different components can be identified not only by the kinetics of DNA reassociation, but also by the number of bands that appear in Southern blots with the use of appropriate probes and by other methods. It should be clear from the preceding discussion of DNA reassociation that the dividing line between many middle-repetitive sequences and highly repetitive sequences is arbitrary.

### ■ Unique Sequences

Most gene sequences and the adjacent nucleotide sequences required for their expression are contained in the unique-sequence component. With minor exceptions (for example, the repetition of one or a few genes), the genomes of viruses and prokaryotes are composed entirely of single-copy sequences; in contrast, such sequences constitute only about 40 percent of the total genome in some sea urchin species, a little more than 50 percent of the human genome, and about 70 percent of the *D. melanogaster* genome.

### ■ Highly Repetitive Sequences

Many highly repetitive sequences are localized in blocks of tandem repeats, whereas others are dispersed throughout the genome. Dispersed highly repetitive sequences include some SINE and LINE retrotransposable elements that are found at high copy number in certain genomes, including the human genome (Chapter 14). Among the localized highly repetitive sequences, most are fairly short. They can differ greatly in abundance from one genome to the next: 6 percent in the human genome, 18 percent in the *D. melanogaster* genome, but 45 percent in the *D. virilis* genome. One of the simplest possible repetitive sequences makes up 25 percent of the genomes of certain species of land crabs; it is composed of an alternating AT sequence (· · · ATATATAT · · ·) with about 3 percent G + C interspersed. In the *D. virilis* genome, the major components of the localized highly repetitive class are three different but related sequences of seven base pairs rich in A + T:

```
5′-ACAAACT-3′
5′-ATAAACT-3′
5′-ACAAATT-3′
```

Blocks of satellite (highly repetitive) sequences in the genomes of several organisms have been located by *in situ* hybridization with metaphase chromosomes. The satellite sequences located by this method have been found to be in the regions of the chromosomes called **heterochromatin**. These are regions that condense earlier in prophase than the rest of the chromosome and are darkly stainable by many standard dyes used to make chromosomes visible (**FIGURE 7.18**); sometimes the heterochromatin remains highly condensed throughout the cell cycle. The **euchromatin**, which makes up most of the genome, is visible only in the mitotic cycle. The major heterochromatic regions are adjacent to the centromere; smaller blocks are present at the ends of the chromosome arms (the telomeres) and interspersed with the euchromatin. In many species, an entire chromosome, such as the Y chromosome in *D. melanogaster* and in mammals, is almost completely heterochromatic. Different highly repetitive sequences have been purified from *D. melanogaster*, and *in situ* hybridization has shown that each sequence

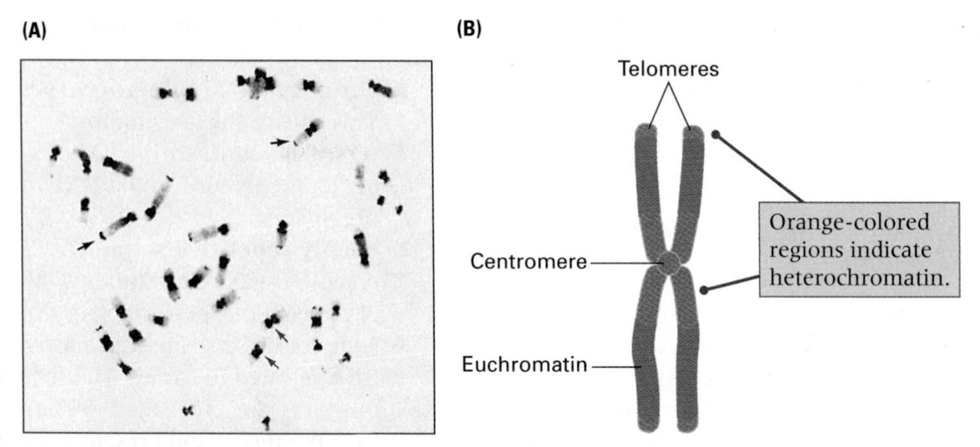


**FIGURE 7.18** (A) Metaphase chromosomes of the ground squirrel *Ammospermophilus harrissi*, stained to show the heterochromatic regions near the centromere of most chromosomes (red arrows) and the telomeres of some chromosomes (black arrows). (B) An interpretive drawing. [Courtesy of T. C. Hsu, Ph.D., and used with permission of Sen Pathak, Ph.D., Anderson Cancer Center, University of Texas.]

has its own distinctive distribution among the chromosomes.

The genetic content of heterochromatin is summarized in the following generalization:

> The number of genes located in heterochromatin is small relative to the number in euchromatin.

The relatively small number of genes means that many large blocks of heterochromatin are genetically almost inert, or devoid of function. Indeed, heterochromatic blocks can often be rearranged in the genome, duplicated, or even deleted without major phenotypic consequences.

### ■ Middle-Repetitive Sequences

Middle-repetitive sequences constitute about 12 percent of the *D. melanogaster* genome and 40 percent or more of the human and other eukaryotic genomes. These sequences differ greatly in the number of copies and in their locations in the genome. They comprise many families of related sequences and include several groups of genes. For example, the two major types of ribosomal RNA are transcribed from a tandem pair of genes that is repeated several hundred times in most eukaryotic genomes. The genes for tRNA molecules as well as those for the histone proteins are repeated in the genomes of all eukaryotes. The copies of the tRNA genes are scattered, whereas those for the histones are usually clustered in tandem repeats. Each histone gene is repeated about 10 times per genome in chickens, 20 times in mammals, about 100 times in *Drosophila,* and as many as 600 times in some species of sea urchins.

Most of the dispersed middle-repetitive DNA in the *D. melanogaster* genome consists of transposable elements (Chapter 14). There are about 50 different families of such elements—with names such as *P, hobo, copia, jockey, Helena, roo, 412, 297,* and *mdg*—each of which is present in 10–50 copies that occur at widely scattered locations throughout the chromosomes.

## 7.8 Molecular Structure of the Centromere

The centromere is a specific region of the eukaryotic chromosome that becomes visible as a narrow constriction along the condensed chromosome. It serves as a central component of the *kinetochore,* the complex of DNA and proteins to which the spindle fibers attach in order to move the chromosomes in both mitosis and meiosis. The kinetochore is also the site at which the spindle fibers shorten, causing the chromosomes to move toward the poles. Electron microscopic analysis has shown that in some organisms—for example, the budding yeast *S. cerevisiae*—a single spindle-protein fiber is attached to centromeric chromatin. Most other organisms have multiple spindle fibers attached to each centromeric region.

Centromeres exhibit considerable structural variation among species. At one extreme are **holocentric chromosomes**, which appear to have centromeric sequences spread throughout their length (constituting what is called a *diffuse centromere*). The nematode *Caenorhabditis elegans* has holocentric chromosomes. In a holocentric chromosome, microtubules attach along the entire length of each chromatid. If a holocentric chromosome is broken into fragments by x-rays, each fragment behaves as a separate, smaller holocentric chromosome that moves normally in cell division. Diffuse centromeres are very poorly understood and will not be discussed further.

The conventional type of centromere is the **localized centromere**, in which microtubules attach to a single region of the chromosome (the kinetochore). Localized centromeres fall into two categories, *point centromeres* and *regional centromeres.* Point centromeres are found in a number of different yeasts, including *S. cerevisiae,* and they are relatively small in terms of their DNA content. Other eukaryotes, including higher eukaryotes, have regional centromeres that may contain hundreds of kilobases of DNA. These regions contain repetitive DNA sequences packaged in nucleosomes that feature a centromere-specific form of histone H3.

The chromatin segment of the centromeres of *S. cerevisiae* has a unique structure that is exceedingly resistant to the action of various DNases; it has been isolated as a protein–DNA complex containing from 220 to 250 nucleotide pairs. The nucleosomal constitution and DNA sequences of all of the yeast centromeres are known. Several common features of the sequences are shown in **FIGURE 7.19**. All yeast centromeres have sequences highly similar to those indicated for regions CDE1, CDE2, and CDE3, but the sequence of region CDE4 varies from one centromere to another. Region 2 is noteworthy in that approximately 90 percent of the base pairs are A–T pairs. The centromeric DNA is contained in a structure designated *the centromeric core particle* that is larger and contains more DNA than the 160 nucleotide pairs in a typical yeast nucleosome core particle. This structure is responsible for the resistance of centromeric DNA to DNase. The spindle fiber is believed to be attached directly to this particle (Figure 7.19B).

The centromeres of budding yeast are unusually small and simple. In higher eukaryotes, each centromeric region encompasses a million base pairs or more, to which numerous spindle fibers become attached. **FIGURE 7.20** illustrates the organization of DNA sequences in the centromeric region of the short arm of human chromosome 2. The organization is typical in that it includes a patchwork of DNA sequences derived from duplicated regions of euchromatin from different chromosomes. How these patchworks are put together is not clear, but the duplications occur at an estimated rate of six to seven events per million years. Chromosome 2 is also typical in containing a large fraction of repetitive DNA sequences. The region at the left (nearest the centromere) consists of tandem repeats of a family of related 170-bp DNA sequences called **alphoid DNA**. Most human chromosomes contain 100–1000 copies of alphoid DNA (**FIGURE 7.21**). DNA

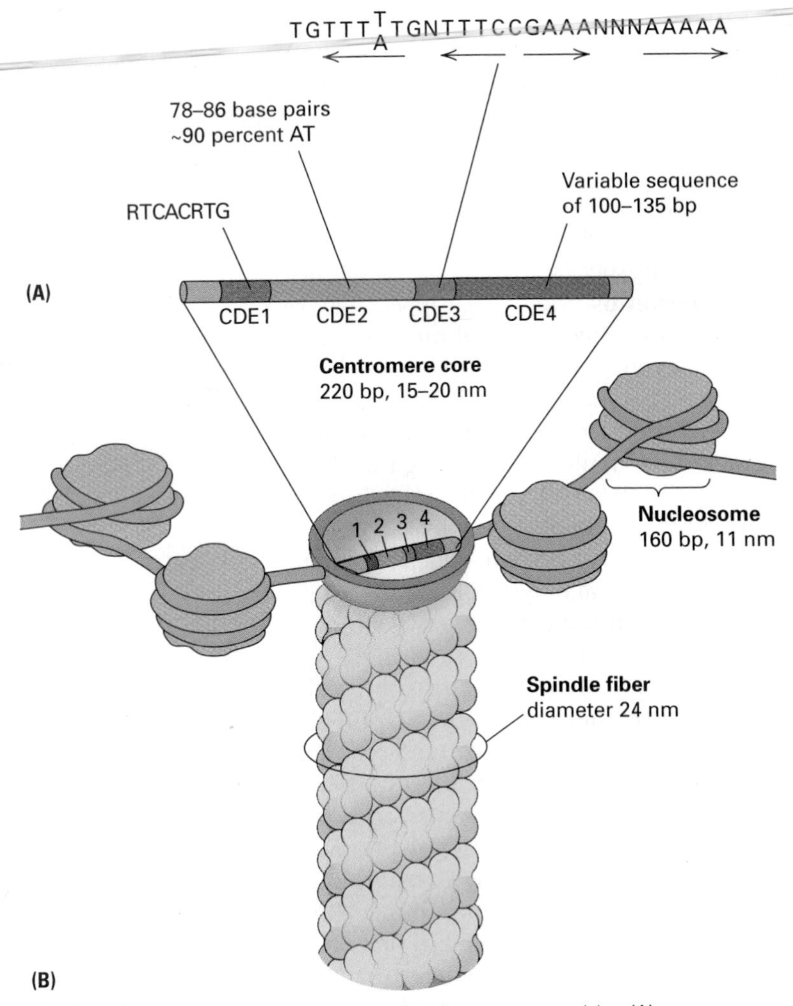

TGTTT T/A TGNTTTCCGAAANNNAAAAA
78–86 base pairs
~90 percent AT
Variable sequence
of 100–135 bp
RTCACRTG
(A)
CDE1
CDE2
CDE3
CDE4
Centromere core
220 bp, 15–20 nm
1 2 3 4
Nucleosome
160 bp, 11 nm
Spindle fiber
diameter 24 nm
(B)

**FIGURE 7.19** A centromere from the yeast *Saccharomyces cerevisiae.* (A) Diagram of centromeric DNA, showing the major regions (CDE1 through CDE4) common to all yeast centromeres. The letter R stands for any purine (A or G), and the letter N indicates any nucleotide. Inverted-repeat segments in region 3 are indicated by arrows. The sequence of region CDE4 varies from one centromere to the next. (B) Positions of the centromere core and the nucleosomes on the DNA. The DNA is wrapped around histones in the nucleosomes, but the detailed organization and composition of the centromere core are unknown. [Adapted from K. S. Bloom, M. Fitzgerald-Hayes, and J. Carbon, *Cold Spring Harb. Symp. Quant. Biol.* 47 (1982): 1175.]

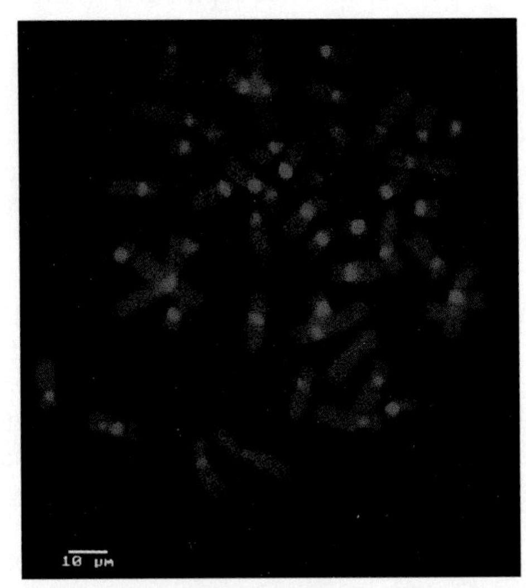

10 μm

**FIGURE 7.21** Hybridization of human metaphase chromosomes (red) with alpha-satellite DNA. The yellow areas result from hybridization with the labeled DNA. The sites of hybridization of the alpha satellite coincide with the centromeric regions of all 46 chromosomes. [Courtesy of Paula Coelho and Claudio E. Sunkel Cariola, IBMC.]

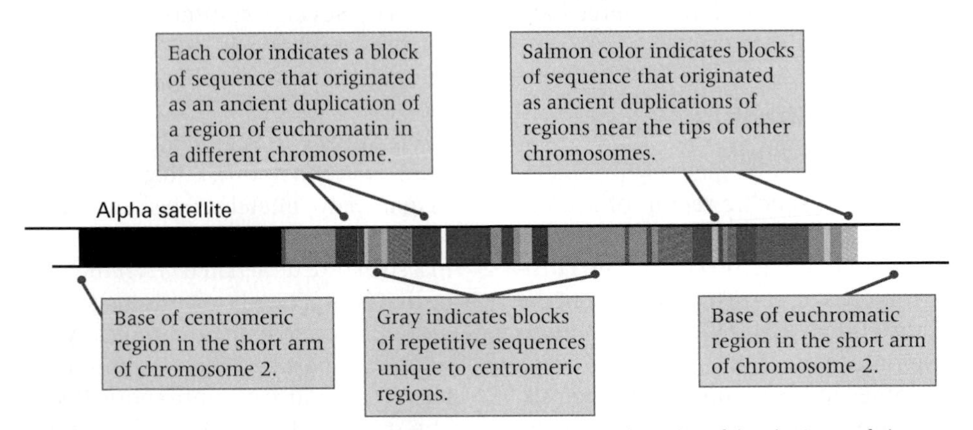

Each color indicates a block of sequence that originated as an ancient duplication of a region of euchromatin in a different chromosome.
Salmon color indicates blocks of sequence that originated as ancient duplications of regions near the tips of other chromosomes.
Alpha satellite
Base of centromeric region in the short arm of chromosome 2.
Gray indicates blocks of repetitive sequences unique to centromeric regions.
Base of euchromatic region in the short arm of chromosome 2.

**FIGURE 7.20** Mosaic ancestry of duplicated sequences present in the centromeric region of the short arm of chromosome 2. The total length of the region shown is approximately 750 kb. [Adapted from X. She, et al., *Nature* 430 (2004): 857–864.]

strands, and when the primer is removed, the complementary strand has a short (8–12 nucleotides) single-stranded overhang at the 3′ end. The single-stranded overhang is subject to degradation by exonucleases. Without some mechanism to restore the digested end, the DNA molecule in a chromosome would become slightly shorter with each replication. There is a mechanism for end restoration, and in mutant cells in which the mechanism is defective, each chromosome end does become shorter in each replication until, eventually, there is so much degradation that the cell dies.

The mechanism of restoring the ends of a DNA molecule in a chromosome relies on an enzyme called **telomerase**. This enzyme works by adding tandem repeats of a simple sequence to the 3′ end of a DNA strand. In the ciliated protozoan *Tetrahymena,* in which the enzyme was first discovered, the simple repeating sequence is 5′-TTGGGG-3′, and in humans and other vertebrate organisms it is 5′-TTAGGG-3′. (*Drosophila* is unusual even among insects in having specialized non-LTR retrotransposable elements at the chromosome tips instead of simple repeats.) The tandem repeats of the simple sequence, along with associated proteins, constitute the telomere. As the repeating telomere sequence is being elongated, DNA replication occurs to synthesize a partner strand, and so, for example, the telomere sequence of the right-hand end of any *Tetrahymena* chromosome would be a DNA duplex of the form

```
5′...TTGGGGTTGGGGTTGGGGTTGGGGTTGGGGTTGGGG-3′
3′...AACCCCAACCCCAACCCC-5′
```

This DNA duplex has a single-stranded overhang at the 3′ end that can be elongated further by the telomerase. The role of telomerase in the replication of chromosomal DNA is illustrated in **FIGURE 7.22**. Part A represents the duplex DNA in a chromosome; the telomere sequences are shown in red. Because DNA replication cannot start precisely at the 3′ end, the 5′ end of each daughter strand in part B is a little shorter than the template strand from which it was replicated. The unreplicated part of the telomere sequence is subject to degradation by nucleases. The 3′ end of each daughter molecule also has a shortened telomere, because this end is replicated from the underhanging 5′ end of the telomere in the parental strand. The shortened telomere remaining at each 3′ end is the substrate of the telomerase, which elongates each 3′ end by the addition of more repeating telomere units—in the case of *Tetrahymena,* 5′-TTGGGG-3′. Telomere elongation restores the structure of the original parental chromosome in which each end has a larger number of telomere repeats at the 3′ end and a smaller number of repeats at the 5′ end.

Relatively few copies of the telomere repeat are necessary to prime the telomerase to add more copies and form a telomere. Remarkably, the telomerase enzyme incorporates an essential RNA molecule, called a **guide RNA**, that contains sequences complementary to the telomere repeat and that serves as a template for telomere synthesis and elongation. For example, the *Tetrahymena* guide RNA contains the sequence 3′-AACCCCAAC-5′. The guide RNA undergoes base pairing with the telomere repeat and serves as a template for telomere elongation by the addition of more repeating units (**FIGURE 7.23**). The complementary DNA strand of the telomere is synthesized exactly like the lagging strand in ordinary DNA replication (Chapter 6). In the telomeric regions of most eukaryotic chromosomes, there are also longer, moderately repetitive DNA sequences immediately preceding the terminal repeats. These sequences differ among organisms and even among different chromosomes in the same organism. Telomeric regions are also associated with proteins that inhibit gene expression in the region, which is known as *telomeric silencing.*

What limits the length of a telomere? In yeast, molecules of a protein called Rap1 bind to the telomere sequence as it is being elongated until about 17 Rap1 molecules have accumulated. At this point, telomere elongation stops, possibly because Rap1 itself inhibits telomerase activity, and then other proteins (including some proteins implicated in DNA repair) join the complex and bind to the extreme end of the telomere. Because each Rap1 molecule binds to approximately 18 base pairs of the telomere, the predicted length of a yeast telomere is $17 \times 18 = 306$ base pairs, which is very close to the value observed. Additional evidence for the role of Rap1 comes from mutations in the *RAP1* gene that produce a Rap1 protein that cannot bind to telomere sequences; in these mutant strains, massive telomere elongation is observed.

Telomeric sequences found in a variety of organisms are shown in **TABLE 7.2**. Each telomere sequence in Table 7.2 is written with the elongated 3′ end at the far right. For example, the *Tetrahymena* formula reads $C_4A_2/T_2G_4$, which means the telomere sequence is

```
3′-AACCCC-5′
5′-TTGGGG-3′ →
```

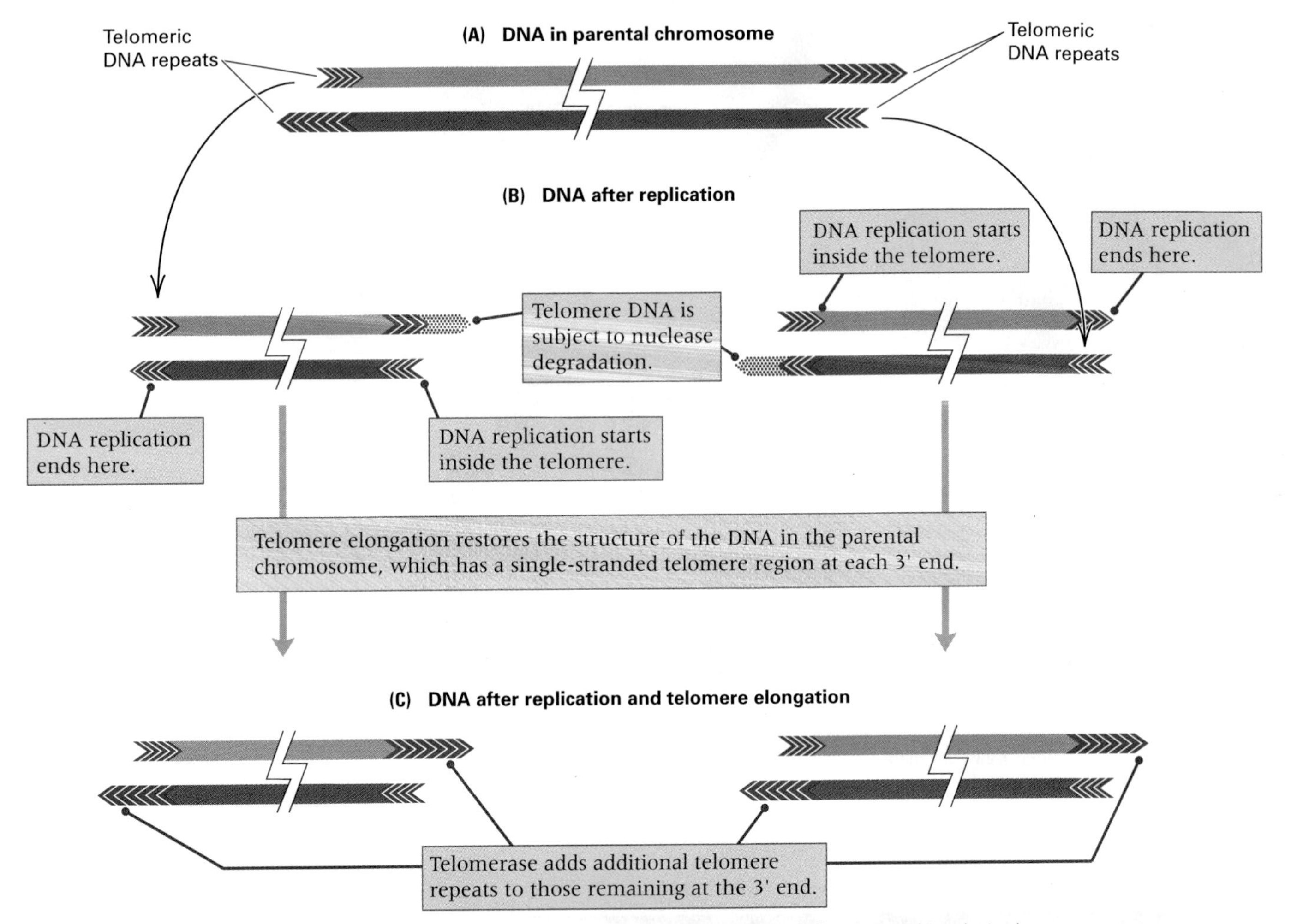


**FIGURE 7.22** The function of telomerase. (A) Chromosomal DNA is double-stranded. Each end of each strand terminates in a set of telomere repeats, but the 3′ end of each strand is longer as a result of telomerase action after the previous replication. (B) In the replication of each parental DNA strand, the new daughter DNA strand is initiated within the telomere repeat at the 3′ end; the telomere at the 5′ end in the new strand is shorter than that in the parental strand. The unreplicated 3′ end of the parental strand is vulnerable to digestion by nucleases. (C) In the daughter DNA duplexes formed by replication, the 3′ strand at each end is elongated by the addition of telomere repeats by the telomerase. The length and 3′ overhang of the telomeres are restored to the state that was present in the original parental molecule.

where the arrow indicates elongation by telomerase (compare with Figure 7.23). Similarly, the formula for the vertebrate telomere is $C_3TA_2/T_2AG_3$, which means the telomere sequence is

```
3′-AATCCC-5′
5′-TTAGGG-3′ →
```

In many organisms the length of a run varies; for example, $C_{1-8}$ means variation ranging from as few as one C to as many as eight Cs from one repeat to the next. There is also occasional variation in which nucleotide occupies a given position; this is indicated by small stacked letters.

There is a strong tendency for the elongated strand of the telomere repeats in Table 7.2 to be rich in guanine nucleotides. Guanine bases are special in that they have the capacity to hydrogen-bond to one another in the form of *G-quartets* (**FIGURE 7.24**). A protein has been isolated from the ciliated protozoan *Oxytricha* that binds specifically to the telomeric DNA of linear chromosomes. The $\beta$ subunit of this telomere-binding protein promotes G-quartet formation by the telomeric DNA of *Oxytricha:*

```
···5′-TTTTGGGGTTTTGGGGT-3′···
```

or *Tetrahymena:*

```
···5′-TTGGGGTTGGGGT-3′···
```

It is hypothesized that within telomeres, telomeric DNA may be organized in special three-dimensional conformations that include such G-quartets. Models showing the possible positions of G-quartets in *Oxytricha* and *Tetrahymena* are shown in Figure 7.24.

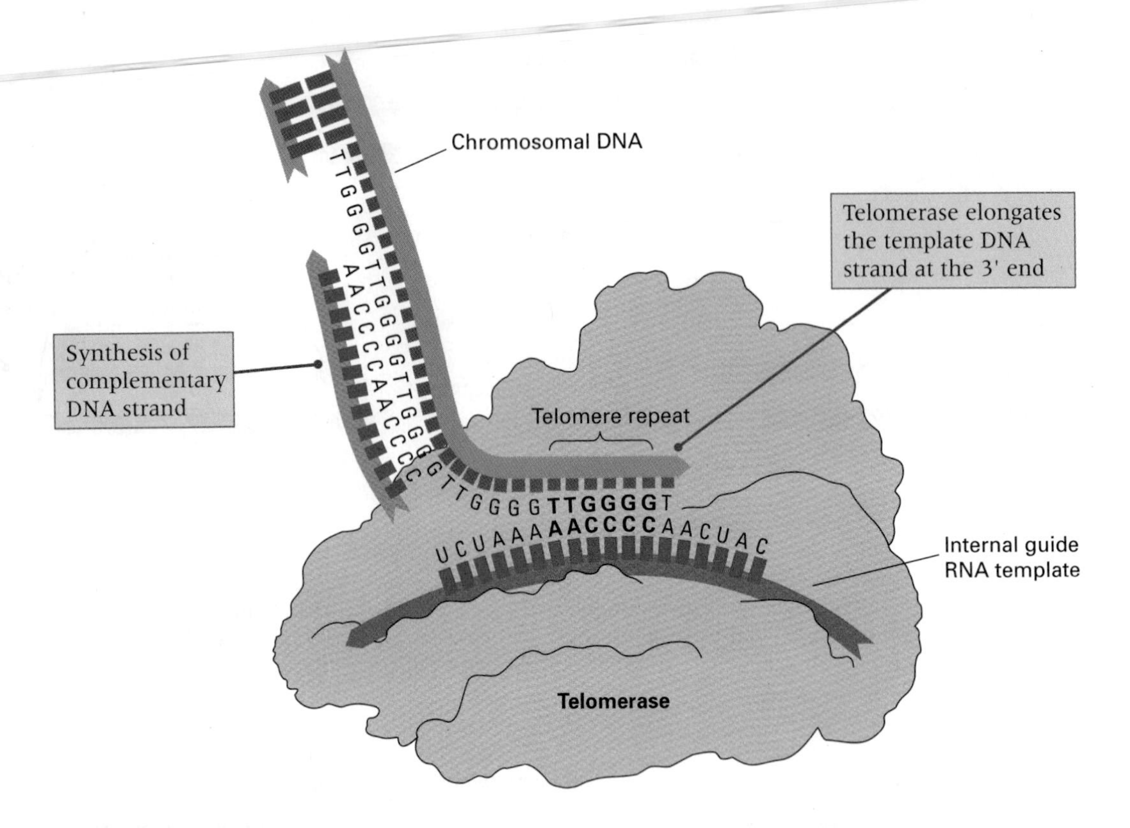


**FIGURE 7.23** Telomere formation in *Tetrahymena*. The telomerase enzyme contains an internal RNA with a sequence complementary to the telomere repeat. The RNA undergoes base pairing with the telomere repeat and serves as a template for telomere elongation. The newly forming DNA strand is produced by DNA polymerase.

**Table 7.2 Sequences of telomeric DNAs**

| Organism | Sequence |
|---|---|
| Protozoa | |
| *Tetrahymena* | $C_4A_2/T_2G_4$ |
| *Paramecium* | $C_3{}^{C}_{A}A_2/T_2{}^{G}_{T}G_3$ |
| *Plasmodium* | $C_3T{}^{A}_{G}A_2/T_2{}^{T}_{C}AG_3$ |
| *Trypanosoma* | $C_3TA_2/T_2AG_3$ |
| *Giardia* | $C_3TA/TAG_3$ |
| Slime molds | |
| *Physarum* | $C_3TA_2/T_2AG_3$ |
| *Dictyostelium* | $C_{1-8}T/AG_{1-8}$ |
| Fungi | |
| *Saccharomyces* | $C_{2-3}ACA_{1-6}/T_{1-6}GTG_{2-3}$ |
| *Candida* | $ACAC_2A_2GA_2GT_2AGACATC_2GT/ACG_2ATGTCTA_2CT_2CT_2G_2TGT$ |
| *Schizosaccharomyces* | $C_{1-6}G_{0-1}T_{0-1}GTA_{1-2}/T_{1-2}ACA_{0-1}C_{0-1}G_{1-6}$ |
| *Neurospora* | $C_3TA_2/T_2AG_3$ |
| Invertebrates | |
| *Caenorhabditis* | $GC_2TA_2/T_2AG_2C$ |
| *Bombyx*, some other insects | $C_2TA_2/T_2AG_2$ |
| Vertebrates | $C_3TA_2/T_2AG_3$ |
| Plants | |
| *Chlamydomonas* | $C_3TA_4/T_4AG_3$ |
| *Arabidopsis* | $C_3TA_3/T_3AG_3$ |
| Tomato | $C_3A{}^{T}_{A}T_2/A_2{}^{A}_{T}TG_3$ |

Source: Reproduced from V. A. Zakian, *Science* 270 (1995): 1601–1607. Reprinted with permission from AAAS.

**FIGURE 7.24** Models of telomere structure in *Oxytricha* (A) and *Tetrahymena* (B) that incorporate the G-quartet structure. The arrowhead indicates the 3′ end of the DNA strand. The indicated G bases participate in an unusual type of base pairing called Hoogsteen base pairing, and the resulting G-quartets are drawn as green squares. There are two G-quartets postulated in the *Oxytricha* telomere, three in *Tetrahymena*. [Adapted from J. R. Williamson, M. K. Raghuraman, and T. R. Cech, *Cell* 59 (1989): 871–880.]

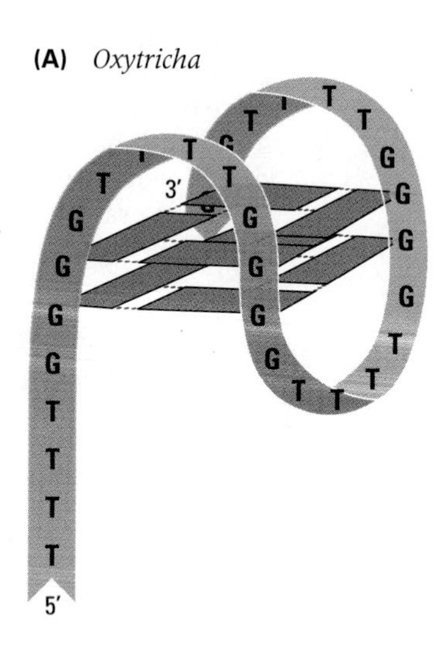


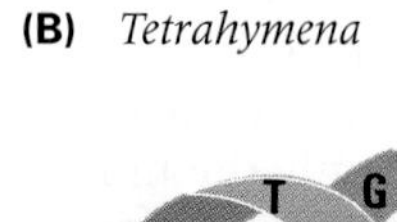


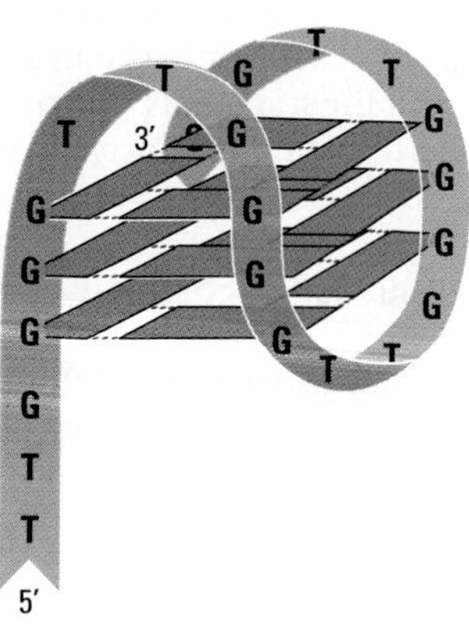


## CHAPTER SUMMARY

- Most organisms have a genome composed of one or more molecules of double-stranded DNA. In eukaryotes the nuclear DNA molecules are usually linear, but in prokaryotes some DNA molecules are circular and others linear. Mitochondrial and chloroplast DNA is also circular. Circular DNA molecules are invariably supercoiled.
- Genome size varies greatly among organisms and sometimes even among closely related organisms. Among metazoans, there is no correlation between the genome size of an organism and its evolutionary or metabolic complexity; this is known as the C-value paradox.
- DNA is rarely in a fully extended state but rather is folded in an intricate way to reduce its effective volume. In bacteria, circular chromosomal DNA is bound with certain proteins and folded to form a multiply looped structure of independently supercoiled domains called a nucleoid.
- Each eukaryotic chromosome contains a single, usually very long, DNA molecule. This molecule is combined with histone proteins to form chromatin. The elementary unit of chromatin is the nucleosomal core particle, an octamer of histone proteins around which is wrapped a length of duplex DNA.
- Most eukaryotic genomes, in addition to the "unique" DNA sequences that include the majority of genes, also contain DNA sequences that are highly repetitive or moderately repetitive.
- The centromere is a specialized DNA structure that functions as the center of chromosome movement in nuclear division.
- The telomere is another specialized DNA structure that serves to stabilize the chromosome tips from shortening through progressive loss of DNA.

## GUIDE TO PROBLEM SOLVING

**Problem 1** Genomic DNA is prepared from a sample of normal tissue and cleaved with a restriction enzyme that has a cleavage site in a single-copy sequence present in the subtelomeric DNA of a particular chromosome. A Southern blot is carried out using a probe for this single-copy subtelomeric sequence. As a control, a sample of the cleaved DNA is also hybridized with a probe homologous to a sequence known to be in the interior of the same chromosome. The result is shown in the accompanying gel diagram. Explain why the band produced by the probe for the internal sequence is very sharp and well defined, whereas that produced by the probe for the subtelomeric sequence is diffuse and spread over a range of sizes.

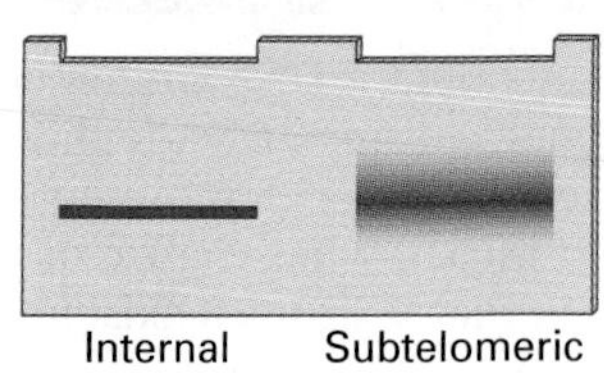

**Answer** The internal sequence gives a sharp band because all the fragments containing the sequence were produced by cleavage of each chromosomal DNA molecule at the nearest restriction site on each side; the fragments are therefore all of the same size. The subtelomeric sequence gives a diffuse band because only the unique restriction site is common to all the fragments; the other end terminates in the telomeric sequence, which can differ in length from one cell to the next.

**Problem 2** G–C base pairs add more stability to a DNA duplex than do A–T base pairs, not only because they have an extra hydrogen bond, but also because of their stronger base stacking. Because of base stacking, the temperature required for denaturation increases not only with the G–C content but also with the length of continuous G–C tracts in the molecule.

**(a)** Which of the two illustrated DNA molecules (1 or 2) would have the lower temperature for strand separation? Why?

**(b)** Which process—denaturation or renaturation—is dependent on the concentration of DNA? Explain your answer.

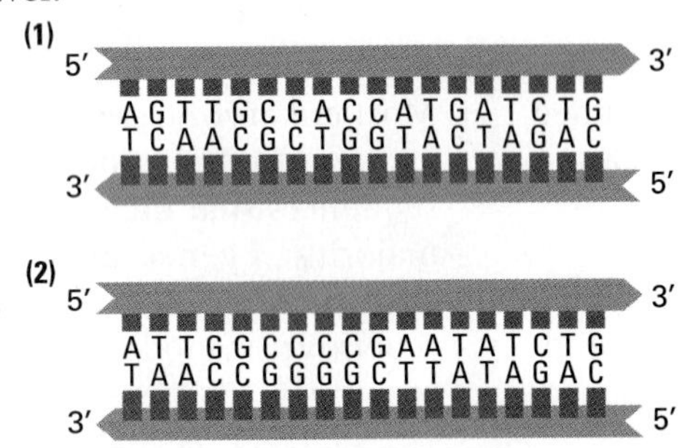


**Answer**

**(a)** Here, the lengths of the G–C tracts are important. Molecule 2 has a long G–C segment, which will be the last region to separate. Hence, molecule 1 has the lower temperature for strand separation.

**(b)** Renaturation is concentration-dependent. The explanation is that renaturation requires complementary molecules to collide *by chance* before the formation of base pairs can occur.

**Problem 3** Consider the schematic $C_0t$ curve for genomic DNA shown here. It shows that the genome of the organism contains two classes of DNA sequence, repetitive and nonrepetitive (unique).

**(a)** What fraction of the DNA is contained in sequences that are repetitive?

**(b)** What fraction of the DNA is contained in sequences that are unique?

**(c)** If the unique sequences are present once per haploid genome, how many copies per haploid genome are there of the repetitive classes?

**(d)** Assuming that the average sequence length in each class of sequences is given by $n = (5 \times 10^5)C_0t_{1/2}$, what is the average sequence length of each class of sequences?

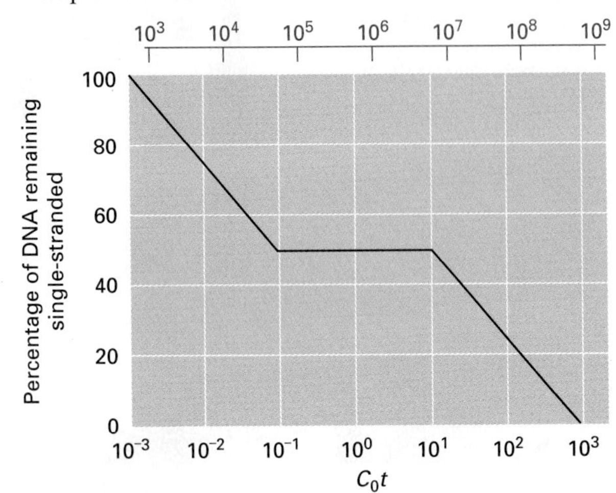


**Answer**

**(a)** The percentages are read off the *y* axis, subtracting the value for the percent DNA remaining single-stranded after each repetitive component of the DNA has renatured from the value prior to the renaturation of this component. The $C_0t$ values are read off the *x*-axis corresponding to the midpoint of the renaturation curve of each component. Hence 100 – 50 = 50 percent consists of repetitive DNA with a $C_0t_{1/2}$ of $1 \times 10^{-2}$.

**(b)** Since the $C_0t$ curve shows only two types of sequences, and 50 percent are repetitive, the remaining and 50 percent consists of unique-sequence DNA with a $C_0t_{1/2}$ of $1 \times 10^2$.

**(c)** The relative abundances of the sequence classes are inversely proportional to their $C_0t_{1/2}$ values; in this example, the relative abundances of repetitive and unique-sequence DNA are $1/(1 \times 10^{-2}) = 100$ and $1/(1 \times 10^2) = 0.01$. Assuming that the unique sequences are present once per haploid genome, then the relative abundances of the repetitive and unique sequences classes per haploid genome are 10,000 : 1.

**(d)** The average sequence length of the repetitive fraction is $n = (5 \times 10^5)(1 \times 10^{-2}) = 5 \times 10^3$ base pairs, and that of the unique-sequence fraction is given by $(5 \times 10^5)(1 \times 10^2) = 5 \times 10^7$ base pairs. These numbers could also be estimated from the scale across the top.

## ANALYSIS AND APPLICATIONS

**7.1** Why is it that telomeric DNA cannot be replicated by the normal mechanism of DNA replication?

**7.2** If you had never seen a chromosome in a microscope but had seen nuclei, what feature of DNA structure would tell you that DNA must exist in a highly coiled state within cells?

**7.3** Distinguish between a nucleosome and a nucleosome core particle. What protein molecules does each contain, and what length of DNA?

**7.4** Consider a length of duplex DNA packaged into nucleosomes with linkers of 55 base pairs (bp) between nucleosomes. How many nucleosomes would be expected in a region of 1 kb? In a region of 50 kb?

**7.5** What would be expected to happen to the telomeres of a cell through successive mitotic divisions if the telomerase were defective?

**7.6** Is there an analog of a telomere in the *E. coli* chromosome?

**7.7** Recall that nearest neighbors in DNA refer to dinucleotides of the form 5′–NN–3′, where NN are the nearest-neighbor nucleotides. Human telomeric DNA consists of long double-stranded tracts of the repeating sequence 5′–TTAGGG–3′, terminated by a 3′ overhang consisting of 5′–TTAGGG–3′ repeats. Suppose that human telomeres are isolated and digested with a 3′-to-5′ exonuclease, which degrades the single-stranded overhangs. What are the types and relative frequencies of nearest neighbors in the remaining double-stranded telomeric DNA?

**7.8** What would be expected to happen to the chromosomes of a cell if the protein Rap1 were defective and could not bind to the telomere?

**7.9** What causes the bands in a polytene chromosome? Do polytene chromosomes undergo division?

**7.10** Do the banded regions of the *Drosophila* polytene chromosomes consist of euchromatin or heterochromatin? Explain your answer.

**7.11** The polytene chromosomes of *Drosophila melanogaster* contain about 5000 bands. The size of the euchromatic genome is $120 \times 10^6$ base pairs, and there are about 15,000 genes. Assume that the amount of DNA between the bands is negligible.

**(a)** What is the average amount of DNA per band?

**(b)** What is the average number of genes per band?

**7.12** Which of the DNA molecules shown here would require a higher temperature for denaturation?

**(a)**
```
5'-ACTGTCATAGAT-3'
3'-TGACAGTATCTA-5'
```

**(b)**
```
5'-GTCACGGCTAGC-3'
3'-CAGTGCCGATCG-5'
```

**7.13** Explain why DNA fragments containing the repeating sequence ATATAT would renature much more rapidly than DNA fragments of the same length containing a random sequence of As and Ts.

**7.14** Suppose you uniformly increased the size of duplex DNA by a factor of 500,000. The diameter would then be 1 millimeter, or about the diameter of a thin spaghetti noodle. What is the actual length of a real DNA molecule, which, if scaled up 500,000 times, would extend the length of the route of the Boston Marathon (about 42,000 meters).

**7.15** Many transposable elements contain a pair of inverted repeat sequences at the ends. What effect would the inverted repeats have on the rapidity of renaturation of such molecules in solution?

**7.16** The tip of the *Drosophila* X chromosome is broken by x-rays near the locus of the wildtype allele of the gene for white eyes. The fragment bearing the telomere is lost, but the wildtype allele $w^+$ of the gene remains. Although it lacks a telomere (*Drosophila* telomeres are unusual in being composed of specialized types of transposable elements rather than simple repeating sequences), the chromosome is stable enough to be transmitted through successive generations. However, many males with white eyes are recovered; these result from a high rate of spontaneous mutation of $w^+$ to $w$. Explain why this result might be expected, and specify the nature of the $w$ mutations.

**7.17** Telomerase extends the 3′ end of duplex DNA, resulting in a 3′ overhang at each end. In vertebrates, the 3′ overhang consists of many repeats of the sequence 5′–TTAGGG–3′. In a vertebrate chromosome, are the 3′ overhangs at opposite ends compatible in the sense that they could undergo base pairing to form a duplex molecule?

**7.18** In higher primates, the telomere repeat sequence is give by

```
5'-TTAGGGTTAGGGTTAGGG-3'
3'-AATCCCAATCCC-5'
```

In the human lineage, chromosome 2 was formed by end-to-end fusion of the telomeres of two separate chromosomes. The fusion preserved the perfect six-nucleotide repeat of both telomeres, so that no nucleotides were duplicated or deleted. Write the sequence of human chromosome 2 at the junction where the telomere repeats fused. Also show one telomere repeat on each side of the junction.

**7.19** Endonuclease S1 can break single-stranded DNA but does not break double-stranded linear DNA. However, S1 can cleave supercoiled DNA, usually making a single-strand break. Why does this occur?

**7.20** Consider a long linear DNA molecule, one end of which is rotated four times with respect to the other end in the unwinding direction.

**(a)** If the two ends are joined to keep the molecule in the underwound state, how many base pairs will be broken?

**(b)** If the underwound molecule is allowed to form a supercoil, how many twists will be present?

**7.21** DNA from species A, labeled with $^{14}N$ and randomly fragmented, is renatured with an equal concentration of DNA from species B, labeled with $^{15}N$ and randomly fragmented, and then centrifuged to equilibrium in CsCl. Five percent of the total renatured DNA has a hybrid density. What fraction of the base sequences is common to the two species?

**7.22** The temperature at which half of the base pairs in a sample of double-stranded DNA are denatured is called the *melting temperature* and designated $T_m$. Over a wide range of base compositions, the $T_m$ is approximately linear in the percentage of G + C base pairs, sybolized $g$. Deduce the approximate relationship between $T_m$ and $g$, given that DNA with $g = 22$ percent G + C has $T_m = 78°C$ and that DNA with $g = 73$ percent G + C has $T_m = 98.9°C$.

**7.23** Use the relationship in Problem 7.22 to find the approximate melting temperature for a bacterial species whose genome is:

**(a)** A + T rich (70 percent A–T).

**(b)** Composed of equal proportions of A + T and G + C base pairs.

**(c)** G + C rich (70 percent G–C).

**7.24** Use the relationship in Problem 7.22 to find the G + C content of:

**(a)** The human genome, with $T_m = 84.7°C$.
**(b)** The rice genome, with $T_m = 87.0°C$.

**7.25** Under standard conditions of renaturation, the average sequence length of a class of sequences $n = (5 \times 10^5)\ C_0t_{1/2}$, where $n$ may be interpreted as a measure of sequence complexity. Under these standard conditions, what $C_0t_{1/2}$ is expected for DNA corresponding to the following genomes of animal viruses. You may assume that none of the viruses contain repetitive sequences.

**(a)** Polyoma virus, genome size 5 kb.
**(b)** Adenovirus, genome size 35 kb.
**(c)** Herpes virus, genome size 230 kb.

**7.26** Suppose you carry out a renaturation experiment with bacteriophage lambda DNA under standard conditions and observe a $C_0t_{1/2} = 10^{-1}$, corresponding to a sequence complexity of 50 kb. Then you carry out the same experiment, but with twice the DNA concentration. What is the new value of $C_0t_{1/2}$? What does this value imply about the sequence complexity of the DNA? Are the two values for sequence complexity in conflict? Explain.

**7.27** The accompanying diagram shows two types of repeated sequences that can occur at the ends of a transposable element. Part (A) shows direct repeats, part (B) shows inverted repeats. Suppose that the repeats undergo pairing and recombination. What are the genetic consequences if:

**(a)** The sequences are direct repeats?
**(b)** The sequences are inverted repeats?

Draw diagrams to support your answers.

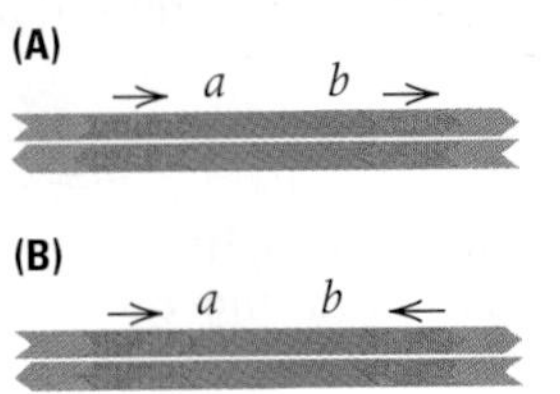


**7.28** A relaxed covalently closed circular DNA molecule has all bases paired. A protein molecule binds to the circle, creates a nick, winds the nicked strand one complete turn around the unnicked strand in a clockwise direction, seals the nick, and then detaches from the circle.

**(a)** Is the helix underwound or overwound?
**(b)** Will the molecule undergo supercoiling to relieve the stress?
**(c)** Is the supercoiling positive or negative?

**7.29** A covalently closed circular DNA molecule with no supercoiling has 10 disrupted base pairs. If it is allowed to supercoil, how many turns of supercoiling are necessary to restore complete base pairing?

**7.30** Draw a schematic of a $C_0t$ curve of the type shown in Worked Problem 3 for genomic DNA that consists of 30 percent repetitive DNA with a $C_0t_{1/2}$ of $10^{-2}$, 30 percent repetitive DNA with a $C_0t_{1/2}$ of $10^{-1}$, and 40 percent nonrepetitive DNA with a $C_0t_{1/2}$ of $10^2$.

## CHALLENGE PROBLEMS

**Challenge Problem 1** A sample of identical DNA molecules, each containing about 3000 base pairs per molecule, is mixed with histone octamers under conditions that allow formation of chromatin. The reconstituted chromatin is then treated with a nuclease, and enzymatic digestion is allowed to occur. The histones are removed, and the positions of the cuts in the DNA are identified by sequencing the fragments. It is found that the breaks have been made at random positions and, as expected, at about 200-base-pair intervals. The experiment is then repeated with a single variation. A protein known to bind to DNA is added to the DNA sample before addition of the histone octamers. Again, reconstituted chromatin is formed and digested with nuclease. In this experiment, it is found that the breaks are again at intervals of about 200 base pairs, but they are localized at particular positions in the base sequence. At each position, the site of breakage can vary over only a 2- to 3-base range. However, examination of the base sequences in which the breaks have occurred does not indicate that breakage occurs in a particular sequence; in other words, each 2- to 3-base region in which cutting occurs has a different sequence. Explain the difference between the two experiments.

**Challenge Problem 2** A sequence of middle-repetitive DNA from *Drosophila* is isolated and purified. It is used as a template in a polymerization reaction with DNA polymerase and radioactive substrates, and highly radioactive probe DNA is prepared. The probe DNA is then used in an *in situ* hybridization experiment with cells containing polytene chromosomes obtained from ten different flies of the same species. Autoradiography indicates that the radioactive material is localized to about 20 sites in the genome, but they are in different sites in each fly examined. What does this observation suggest about the DNA sequence being studied?

Challenge Problem 3 Consider the Cot curve shown in the figure.

**(a)** What fraction of the DNA is contained in components that are unique? In components that are repetitive?

**(b)** If the unique sequences are present once per haploid genome, how many copies per haploid genome are there of the other two classes?

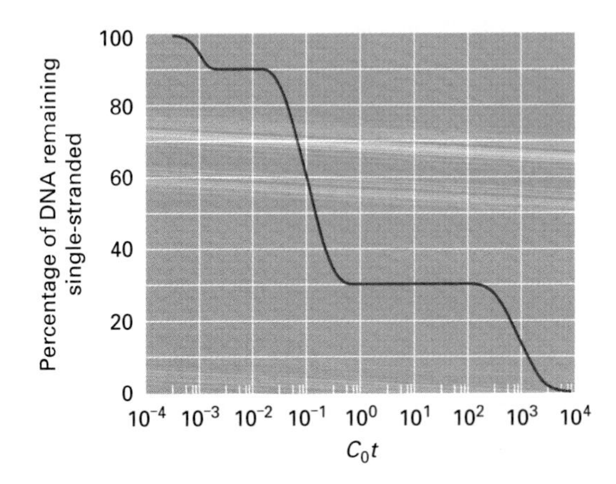


## GENETICS on the web

GeNETics on the Web will introduce you to some of the most important sites for finding genetics information on the Internet. To explore these sites, visit the Jones and Bartlett companion site to accompany *Genetics: Analysis of Genes and Genomes, Seventh Edition* at http://biology.jbpub.com/book/genetics.

There you will find a chapter-by-chapter list of highlighted keywords. When you select one of the keywords, you will be linked to a Web site containing information related to that keyword.

# CHAPTER 8

# Human Karyotypes and Chromosome Behavior

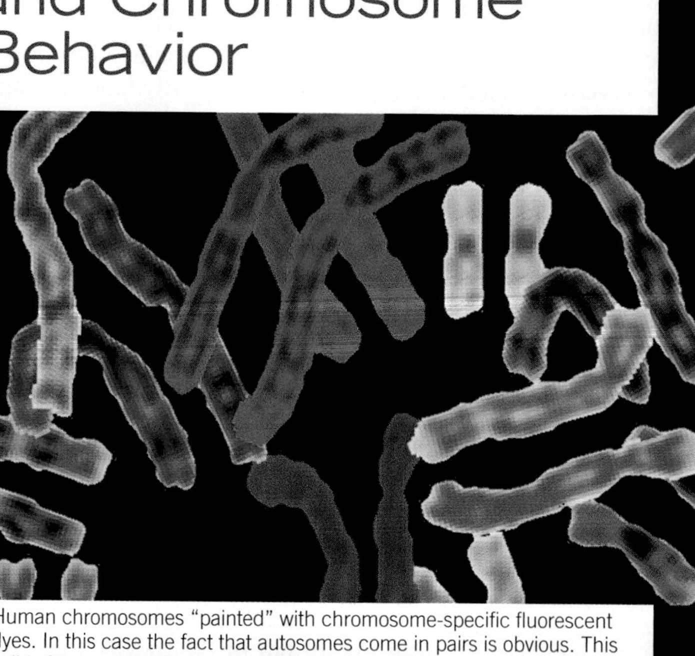

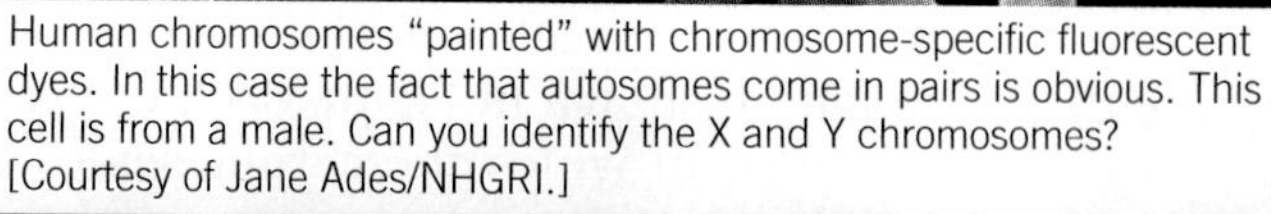

Human chromosomes "painted" with chromosome-specific fluorescent dyes. In this case the fact that autosomes come in pairs is obvious. This cell is from a male. Can you identify the X and Y chromosomes? [Courtesy of Jane Ades/NHGRI.]

## CHAPTER OUTLINE


**8.1 The Human Karyotype**
- Standard Karyotypes
- The Centromere and Chromosome Stability
- Dosage Compensation of X-Linked Genes
- The Calico Cat
- Pseudoautosomal Inheritance
- Active Genes in the "Inactive" X chromosome
- Gene Content and Evolution of the Y Chromosome
- Tracing Human History Through the Y Chromosome

**8.2 Chromosome Abnormalities in Human Pregnancies**
- Down Syndrome and Other Viable Trisomies
- Trisomic Segregation
- Sex-Chromosome Abnormalities
- Environmental Effects on Nondisjunction

**8.3 Chromosomal Deletions and Duplications**
- Deletions
- Deletion Mapping
- Duplications
- Unequal Crossing Over in Red–Green Color Blindness

**8.4 Genetics of Chromosomal Inversions**
- Paracentric Inversion (Not Including the Centromere)
- Pericentric Inversion (Including the Centromere)

**8.5 Chromosomal Translocations**
- Reciprocal Translocations
- Genetic Mapping of a Translocation Breakpoint
- Robertsonian Translocations
- Translocations and Trisomy 21
- Translocation Complexes in *Oenothera*

**8.6 Genomic Position Effects on Gene Expression**

**8.7 Polyploidy in Plant Evolution**
- Sexual Versus Asexual Polyploidization
- Autopolyploids and Allopolyploids
- Monoploid Oganisms

**8.8 Genome Evolution in the Grass Family (Gramineae)**

CONNECTION **Lyonization of an X Chromosome**
Mary F. Lyon 1961
*Gene Action in the X Chromosome of the Mouse* (Mus musculus L.)

CONNECTION **The First Human Chromosomal Disorder Identified**
Jerome Lejeune, Marthe Gautier, and Raymond Turpin 1959
*Study of the Somatic Chromosomes of Nine Down Syndrome Children* [original in French]

In most species, organisms with an extra chromosome or a missing chromosome usually have developmental or other types of abnormalities. The abnormalities result from the increase or decrease in copy number (*dosage*) of the genes in that chromosome. Some organisms, usually rare, are found to have a variation in chromosome structure. The abnormal chromosome may have a particular segment missing, duplicated, reversed in orientation, or attached to a different chromosome. Each of these structural abnormalities has different genetic implications. In this chapter we consider the human chromosome complement and some of the major chromosomal abnormalities encountered in human populations. We also examine chromosome abnormalities in other organisms. Generally speaking, animals are much less tolerant of chromosomal changes than are plants. As we shall see, especially in plants, the acquisition of entire extra sets of chromosomes is not necessarily harmful. In some lineages of plants, the reduplication of entire chromosome sets has been a major process in genome evolution and the origin of species.

## 8.1 The Human Karyotype

The normal chromosome complement of a cell in mitotic metaphase from a human male is illustrated in **FIGURE 8.1**. The chromosomes have been labeled via a technique called **chromosome painting,** in which different colors are "painted" on each chromosome by hybridization with DNA strands labeled with different fluorescent dyes. Individual chromosomes are first isolated by any of a variety of techniques (for example, microdissection followed by DNA amplification with PCR); then the chromosome-specific DNA samples are labeled with fluorescence. A mixture of differently labeled strands from all the chromosomes is used in hybridization with metaphase chromosomes squashed onto a glass slide, allowing the fluorescent strands to hybridize with complementary strands present in the chromosomes. Unhybridized DNA is washed from the slide, and the preparation is examined through a confocal microscope to read the fluorescent signals for conversion into visible colors. (A confocal microscope produces images of a single region in a single focal plane, because it is able to reject scattered and extraneous light.)

### Standard Karyotypes

Chromosome painting dramatically identifies the pairs of homologous chromosomes. The presentation shown in Figure 8.1A is a *metaphase spread,* in which the chromosomes are arranged as they appear in the cytological preparation. A more conventional representation, called a **karyotype**, is shown in Figure 8.1B. In a karyotype, the autosomes in the metaphase spread are rearranged systematically in pairs, from longest to shortest, and numbered from 1 (the longest) through 22. In this example the sex chromosomes are set off at the bottom right. The single X and Y chromosomes are evi-

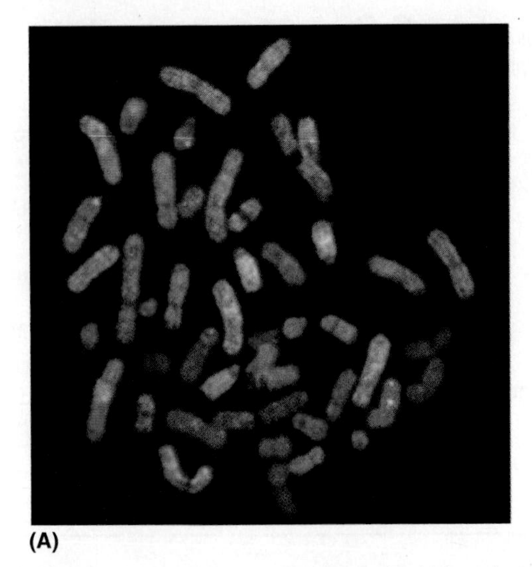

(A)

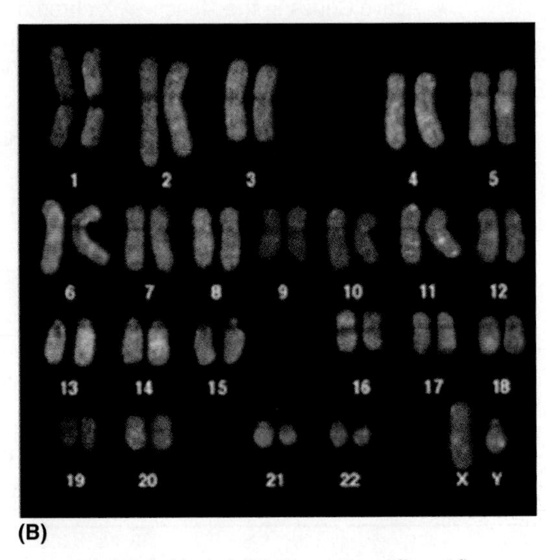


(B)

**FIGURE 8.1** Human chromosome painting, in which each pair of chromosomes is labeled by hybridization with a different fluorescent probe. (A) Metaphase spread showing the chromosomes in a random arrangement as they were squashed onto the slide. (B) A karyotype, in which the chomosomes have been grouped in pairs and arranged in conventional order. Chromosomes 1–20 are arranged in order of decreasing size, but for historical reasons, chromosome 21 precedes chromosome 22, even though chromosome 21 is smaller. [Courtesy of Johannes Wienberg, Ludwig-Maximillians-University, and Thomas Ried, National Institutes of Health.]

dent. The karyotype of a normal human female has a pair of X chromosomes, instead of an X and a Y, in addition to the 22 pairs of autosomes. Chromosome painting is of considerable utility in human cytogenetics because even complex chromosome rearrangements can be detected rapidly and easily.

Another, less colorful metaphase spread and karyotype are shown in **FIGURE 8.2**. In this case the chromosomes have been treated with a staining reagent called Giemsa, which causes the chromosomes to exhibit transverse bands (**G-bands**) that are specific for each pair of homologs. These bands permit smaller segments of each chromosome arm to be identified. The G-bands replicate in mid to late S phase and are relatively gene poor, whereas the interbands (called *R-bands*) replicate early in S phase and are relatively gene rich.

The chromosomes are grouped into seven sets denoted by the letters A through G. (The X chromosome is included in group C, the Y in group G.) These conventional groupings date from a time prior to G-banding and chromosome painting, when the chromosomes could be sorted only by size and centromere position.

The nomenclature of the banding patterns in human chromosomes is shown in **FIGURE 8.3**, where the red letter beneath each chromosome indicates its group. For each chromosome, the short arm is designated with the letter *p*, which stands for "petite," and the long arm by the letter *q*, which stands for "not-p." Within each arm the regions are numbered. The first digit is the major region, numbered consecutively proceeding from the centromere toward the telomere; within each region, the second number indicates the next-smaller division, again numbered outward from the centromere. For example, the designation 1p34 indicates chromosome 1, short arm, division 34. Some divisions can be subdivided still further by the bands and interbands within them, which are numbered consecutively and indicated by a digit placed after a decimal point following the main division; for example, 1p36.2 means the second band in 1p36. Some familiar genetic landmarks in the human genome are the Rh (Rhesus) blood-group locus, which is in the region between 1p34 and 1p36.2; the ABO blood-group locus at 9q34; the red–green color blindness genes at Xq28; and the male-determining gene on the Y chromosome, called *SRY* (sex-determining region, Y) at Yp11.3.

**FIGURE 8.4** shows a chromosome painting of a human chromosome complement at metaphase

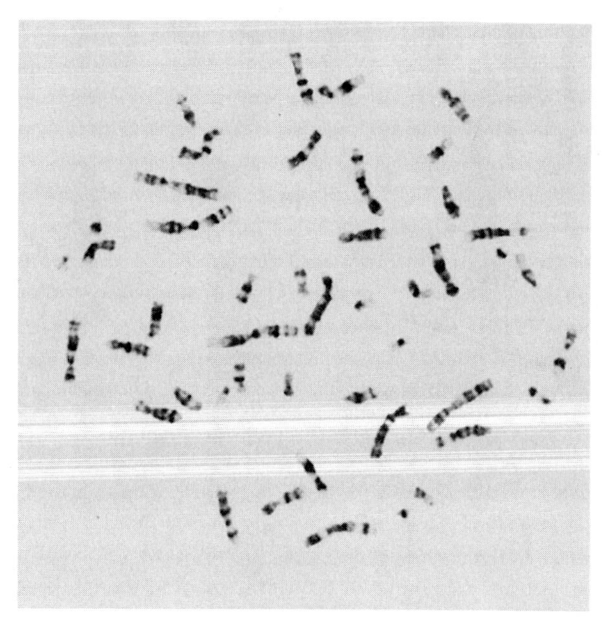

**(A)** Photograph of metaphase chromosomes

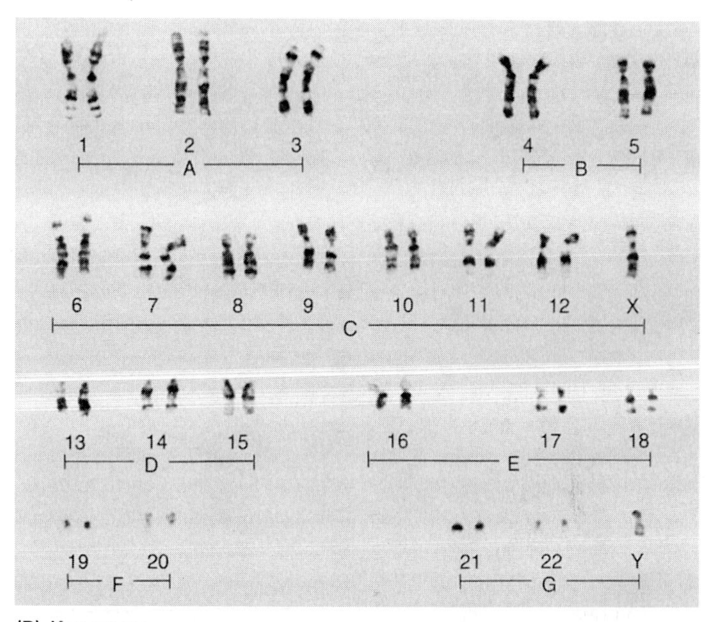


**(B)** Karyotype

**FIGURE 8.2** A karyotype of a normal human male. Blood cells arrested in metaphase were stained with Giemsa and photographed with a microscope. (A) The chromosomes as seen in the cell by microscopy. (B) The chromosomes have been cut out of the photograph and paired with their homologs. [Courtesy of Patricia A. Jacobs, Wessex Regional Genetics Laboratory, Salisbury District Hospital.]

of mitosis. Only one of each pair of homologous autosomes is shown, along with the X and Y chromosomes. Below each chromosome is the amount of DNA in the chromosome in megabase pairs (Mb), the estimated number of genes from the human genome sequence, and the approximate gene density. Gene density is not highly correlated with chromosome size, and it can differ greatly from one chromosome to the

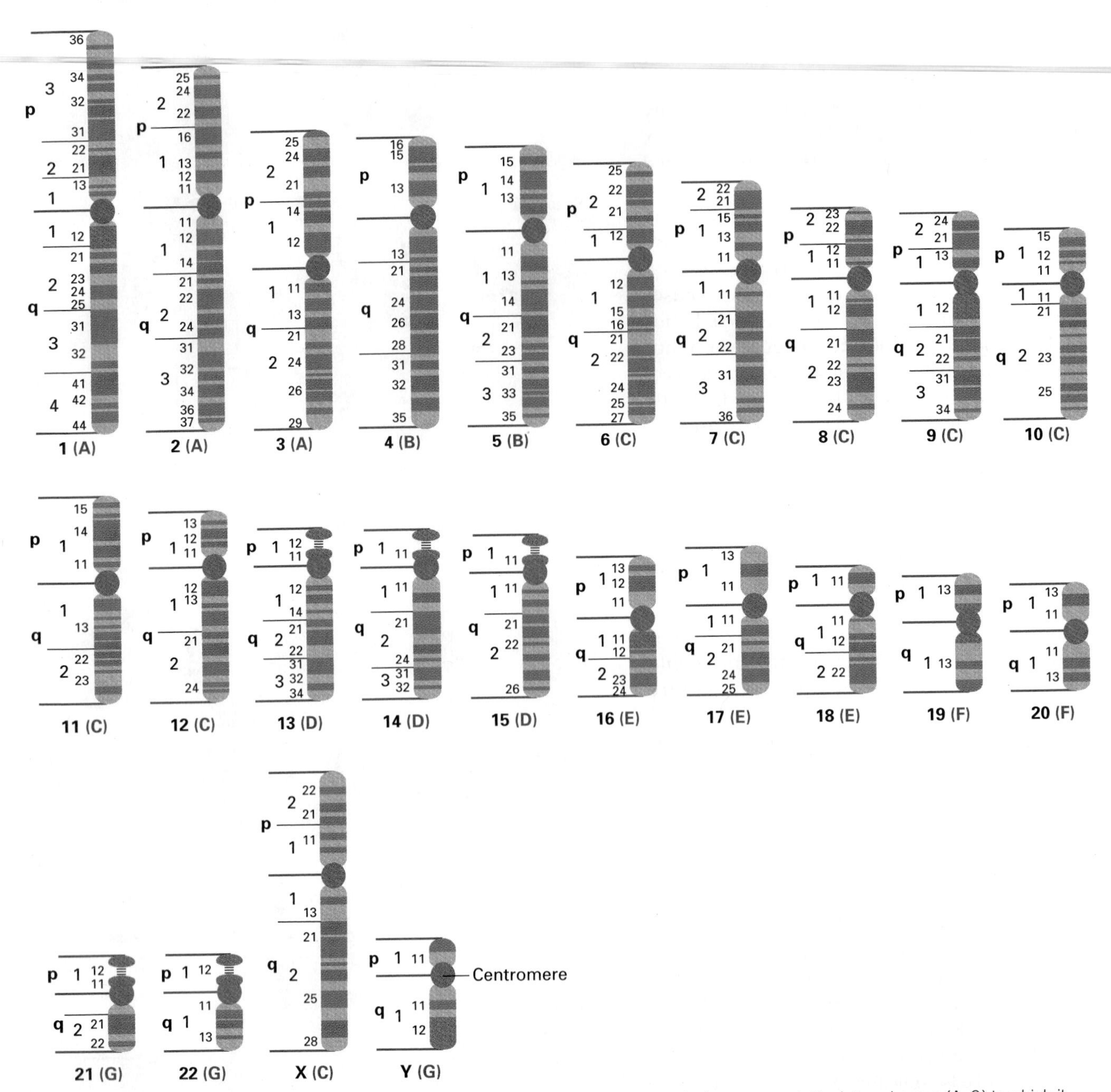


**FIGURE 8.3** Designations of the bands and interbands in the human karyotype. Beneath each chromosome is the lettered group (A–G) to which it belongs.

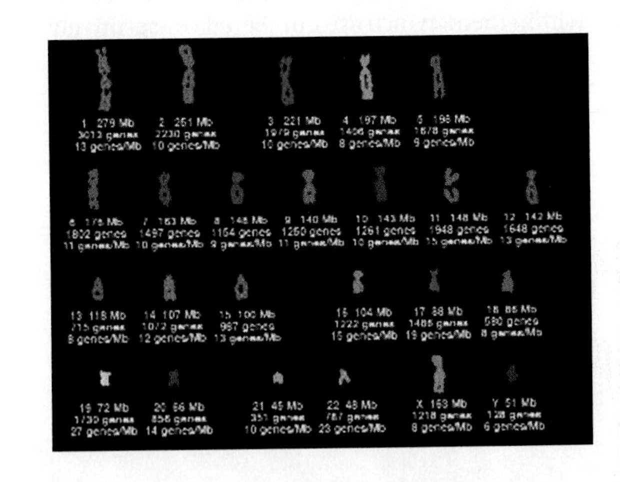

**FIGURE 8.4** The human chromosome complement at metaphase of mitosis showing the amount of DNA in each chromosome, the estimated number of genes, and the approximate gene density. For the autosomes, only one of each homologous pair is shown. [Sequence data from International Human Genome Sequencing Consortium, *Nature* 409 (2001): 860–921, and J. C. Venter, et al., *Science* 291 (2001): 1304–1351. Chromosome image courtesy of Michael R. Speicher, Institute of Human Genetics, Medical University of Graz.]

next. Two of the smallest chromosomes, 19 and 22, have the highest gene densities (27 and 23 genes per Mb, respectively), and two of the largest chromosomes (4 and 5) have among the smallest gene densities (8 and 9 genes per Mb, respectively).

For ease of reference, some of the principal conventions used in describing human chromosomes and human chromosome abnormalities are listed in **TABLE 8.1**. The suffix *ter* is used for "terminal"; hence pter and qter refer to the terminal parts of the short and long arms, respectively. The + and − symbols refer to extra or missing chromosomal material. Preceding a chromosome symbol, they mean the entire chromosome is affected; following it, they refer to only part of the chromosome. For example, +21 means the presence of an extra copy of chromosome 21 (characteristic of Down syndrome, also called Down's syndrome), whereas 21q22ter+ means the presence of an extra copy of part of chromosome 21 extending from band q22 through to the telomere. The designation *mos* means **mosaic**, an individual composed of two or more genetically distinct types of cells. The other terms in Table 8.1 are self-explanatory; many of these chromosome abnormalities are discussed in detail later in this chapter.

### The Centromere and Chromosome Stability

As is true in nearly all eukaryotic organisms, each human chromosome is linear and has a single centromere. (Rarely, a *ring* chromosome is found, which results from breakage and rejoining of the telomeres of a linear chromosome.) Linear monocentric chromosomes are often classified according to the relative position of their centromeres, which determines the appearance of the daughter chromosomes as they separate from each other in anaphase (**FIGURE 8.5**). A chromosome with its centromere about in the middle is a **metacentric chromosome**; the arms are of approximately equal length, and each daughter chromosome forms a V shape at anaphase. When the centromere is somewhat off-center, the chromosome is a **submetacentric chromosome**, and each daughter chromosome forms a J shape at anaphase. A chromosome with the centromere very close to one end appears I-shaped at anaphase because the arms are grossly unequal in length; such a chromosome is **acrocentric**.

Chromosomes with a single centromere are usually the only ones that are reliably transmitted from parental cells to daughter cells or from parental organisms to their progeny. When a cell divides, spindle fibers attach to the centromere of each chromosome and pull the sister chromatids to opposite poles. A chromosome that lacks a centromere is an **acentric** chromosome. Acentric chromosomes are genetically unstable because they cannot be maneuvered properly during cell division and are lost. Occasionally, a chromosome arises that has two centromeres and is said to be **dicentric**. A dicentric chromosome is also genetically unstable because it is not transmitted in a predictable fashion. The dicentric chromosome is frequently lost from a cell when the two centromeres proceed to opposite poles in the course of cell division; in this case, the chromosome is stretched and forms a *bridge* between the daughter cells. This bridge may not be included in either daughter nucleus, or it may break, with the result that each daughter nucleus receives a broken chromosome. We will consider one mechanism by which dicentric and acentric chromosomes are formed when we discuss inversions.

Although most dicentric chromosomes are genetically unstable, if the two centromeres are close enough together, they can frequently behave as a single unit and be transmitted normally. This principle was important in the evolution of human chromosome 2. Among higher

**Table 8.1 Conventional karyotype symbols used in human genetics**

| Symbol | Meaning |
|---|---|
| A–G | Chromosome groups |
| 1–22 | Autosome designations |
| X, Y | Sex-chromosome designations |
| p | Short arm of chromosome |
| q | Long arm of chromosome |
| ter | Terminal portion: pter refers to terminal portion of short arm, qter to terminal portion of long arm |
| + | Preceding a chromosome designation, indicates that the chromosome or arm is extra; following a designation, indicates that the chromosome or arm is larger than normal |
| − | Preceding a chromosome designation, indicates that the chromosome or arm is missing; following a designation, indicates that the chromosome or arm is smaller than normal |
| mos | Mosaic |
| / | Separates karyotypes of clones in mosaics—e.g., 47, XXX/45,X |
| dup | Duplication |
| dir dup | Direct duplication |
| inv dup | Inverted duplication |
| del | Deletion |
| inv | Inversion |
| t | Translocation |
| rcp | Reciprocal translocation |
| rob | Robertsonian translocation |
| r | Ring chromosome |
| i | Isochromosome (two identical arms attached to a single centromere, like an attached-X chromosome in *Drosophila*) |

primates, chimpanzees and human beings have 23 pairs of chromosomes that are similar in morphology and G-banding pattern, but chimpanzees have no obvious homolog of human chromosome 2, a large metacentric chromosome. Instead, chimpanzees have two medium-sized acrocentric chromosomes not found in the human genome. The cause of this situation is shown in **FIGURE 8.6**. The G-banding patterns indicate that human chromosome 2 was formed by fusion of the telomeres between the short arms of two acrocentric chromosomes that, in chimpanzees, remain acrocentrics. The fusion created a dicentric chromosome with its two centromeres close together. The new chromosome must have been sufficiently stable to be retained in the human lineage. The present-day human chromosome 2 has only a single functional centromere. The chromosome fusion reduced the chromosome number in the human lineage from 48, which is characteristic of the great apes (chimpanzee, gorilla, and orangutan), to the number 46.

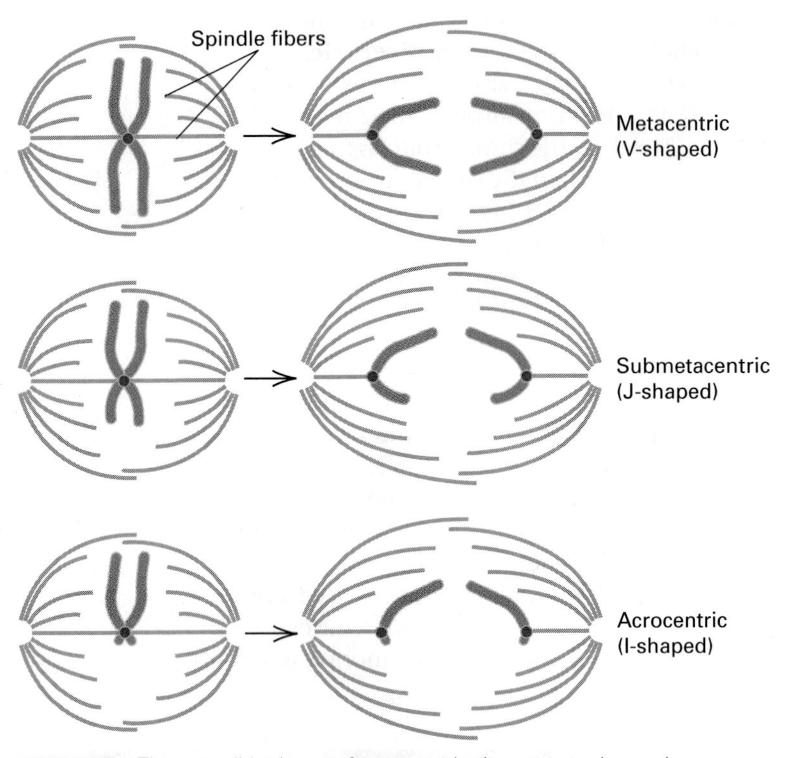


**FIGURE 8.5** Three possible shapes of monocentric chromosomes in anaphase as determined by the position of the centromere (shown in dark blue).

## ■ Dosage Compensation of X-Linked Genes

For all organisms with XX–XY sex determination, there is a problem of the dosage of genes on the X and Y chromosome, because females have two copies of this chromosome whereas males have only one. (There is less of a problem with Y-linked genes, because the Y chromosome is largely heterochromatic and carries relatively few genes.) In most organisms, a mechanism of **dosage compensation** has evolved in which the unequal dosage in the sexes is corrected either by increasing the activity of genes in the X

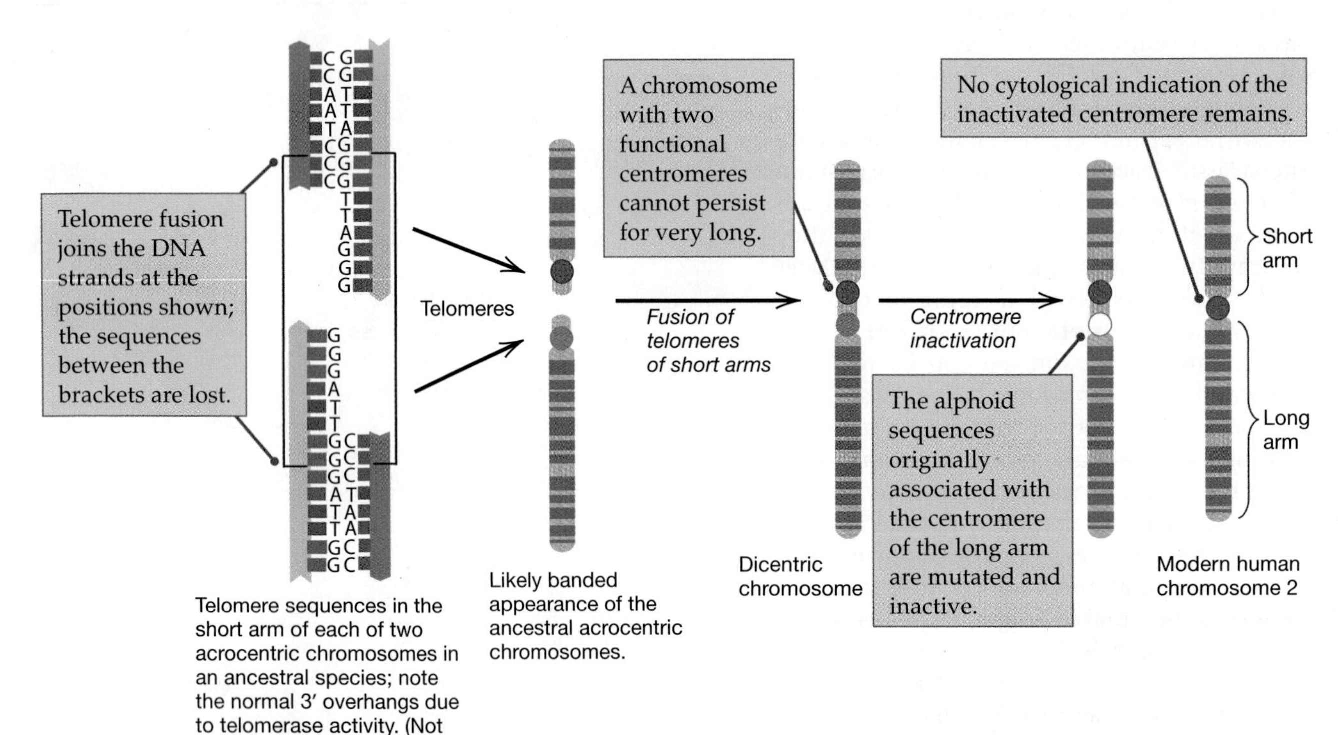

Telomere fusion joins the DNA strands at the positions shown; the sequences between the brackets are lost.
Telomeres
A chromosome with two functional centromeres cannot persist for very long.
Fusion of telomeres of short arms
Centromere inactivation
No cytological indication of the inactivated centromere remains.
The alphoid sequences originally associated with the centromere of the long arm are mutated and inactive.
Short arm
Long arm
Telomere sequences in the short arm of each of two acrocentric chromosomes in an ancestral species; note the normal 3′ overhangs due to telomerase activity. (Not drawn to scale.)
Likely banded appearance of the ancestral acrocentric chromosomes.
Dicentric chromosome
Modern human chromosome 2

**FIGURE 8.6** Human ancestors had 24 pairs of chromosomes rather than 23. In the evolution of the human genome, two acrocentric chromosomes fused to create human chromosome 2.

## connection

### Telomeres: The Beginning of the End

**Carol W. Greider and Elizabeth H. Blackburn 1987**
University of California, Berkeley, California
*The Telomere Terminal Transferase of* Tetrahymena *Is a Ribonucleoprotein Enzyme with Two Kinds of Primer Specificity*

*What a wonderful surprise that an RNA is a key ingredient in the formation of telomeres! Two limitations of DNA polymerase are that it requires a primer oligonucleotide and that it can elongate a DNA strand only at the 3′ end. The limitations imply that the 3′ end of a DNA strand should become progressively shorter with each round of replication, owing to the need for a primer at that extremity. (Chromosomes without proper telomeres do, in fact, become progressively shorter.) The organism used in this study,* Tetrahymena, *is a ciliated protozoan. Each cell has a specialized type of nucleus called a macronucleus that contains many hundreds of small chromosomes. The level of telomerase activity is high because each of these tiny chromosomes needs a pair of telomeres. The convenience of using* Tetrahymena *for the study of telomere function illustrates a principle that runs through the history of genetics: Breakthroughs often come from choosing just the right organism to study. The authors speculate about the possible presence of a "guide" RNA in the telomerase. They were exactly right.*

Precise recognition of nucleic acids is often carried out by enzymes that contain both RNA and protein components. For some of these ribonucleoproteins (RNPs), the RNA components provide specificity to the reaction by base-pairing with the substrate. The recognition that RNA can act catalytically has led to the increase in the number of known RNP-catalyzed reactions. We report here that an RNP is involved in synthesizing the telomeric sequences found at the ends of *Tetrahymena* macronuclear chromosomes. . . . We have previously reported the identification of an activity in *Tetrahymena* cell extracts that adds telomeric repeats onto appropriate telomeric sequence primers in a nontemplated manner. . . . Repeats of the *Tetrahymena* telomeric sequence TTGGGG are added, 1 nucleotide at a time, onto the 3′ end of the input primer. . . . We have begun characterizing and purifying the telomerase enzyme in order to investigate the mechanisms controlling the specificity of the reaction. . . . We propose that the RNA component(s) of telomerase may play a role in specifying the sequence of the added telomeric TTGGGG repeats, in recognizing the structure of the G-rich telomeric sequence primers, or both. . . . In the course of purifying the telomerase from crude extracts, we noted a marked sensitivity to salts [in which high] concentrations inactivated the telomeric elongation activity. The salt sensitivity and the large size of the enzyme suggested that the telomerase may be a complex containing a nucleic acid component. To test whether the telomerase contained an essential nucleic acid, we treated active fractions with either micrococcal nuclease or RNase A. The nuclease activity of each of these enzymes abolished the telomeric elongation activity. . . . These experiments suggest that the telomerase contains an essential RNA component. . . . The RNA of telomerase may simply provide a scaffold for the assembly of proteins in the active enzyme complex; however, . . . it is tempting to speculate that the RNA component of the telomerase might be involved in determining the sequence of the telomeric repeats that are synthesized and/or the specific primer recognition. If the RNA of telomerase contains the sequence CCCCAA, this sequence could act as an internal guide sequence.

***. . . it is tempting to speculate that the RNA component of the telomerase might be involved in determining the sequence of the telomeric repeats that are synthesized . . .***

Source: C. W. Greider and E. H. Blackburn, *Cell* 51(1987): 887–898.

sequences needed for spindle fiber attachment are interspersed among the alphoid repeats, and the alphoid repeats themselves appear to contribute to centromere activity.

## 7.9 Molecular Structure of the Telomere

Each end of a linear chromosome is composed of a special DNA–protein structure called a **telomere.** Genetic and microscopic observations first indicated that telomeres are essential for chromosome stability. In *Drosophila,* Hermann J. Muller found that chromosomes without ends could not be recovered after chromosomes were broken by treatment with x-rays. In maize, Barbara McClintock observed that broken chromosomes frequently fuse with one another and form new chromosomes with abnormal structures often having two centromeres. As we saw in Chapter 6, DNA polymerase cannot initiate DNA synthesis by itself, but instead requires an RNA primer. At the end of a chromosome, the primer terminates the 5′ end of one of the DNA

chromosome in males or by reducing the activity of genes in the X chromosome in females. Opposite extremes of dosage compensation are illustrated by the fruit fly *Drosophila melanogaster* and the nematode worm *Caenorhabditis elegans*:

- In males of *D. melanogaster*, a complex including at least one male-specific protein is recruited to the X chromosome, which alters its chromatin structure, most likely by acetylation of a specific lysine in histone H4. This results in an increase in the transcriptional activity of the single X chromosome in males to a level equal to the sum of that of both X chromosomes in females.
- In females of *C. elegans*, a protein complex is recruited to the X chromosomes, which decreases the level of transcription of each X chromosome by half. This results in a decrease in the level of transcription of each X chromosome in females, so that both X chromosomes together have the same transcriptional activity as that of the single X chromosome in males.

The mechanism of dosage compensation in mammals is seemingly simple. In the early cleavage divisions of the embryo, at roughly the 64–128 cell stage, one and only one X chromosome in each cell, chosen at random, remains genetically active, and all the other X chromosomes that may be present in the cell undergo a process of **X inactivation**. Any X chromosome that is inactivated in a particular somatic cell remains inactive in all the descendants of that cell (**FIGURE 8.7**); hence the inactive state of an X chromosome is inherited from parental cell to daughter cell.

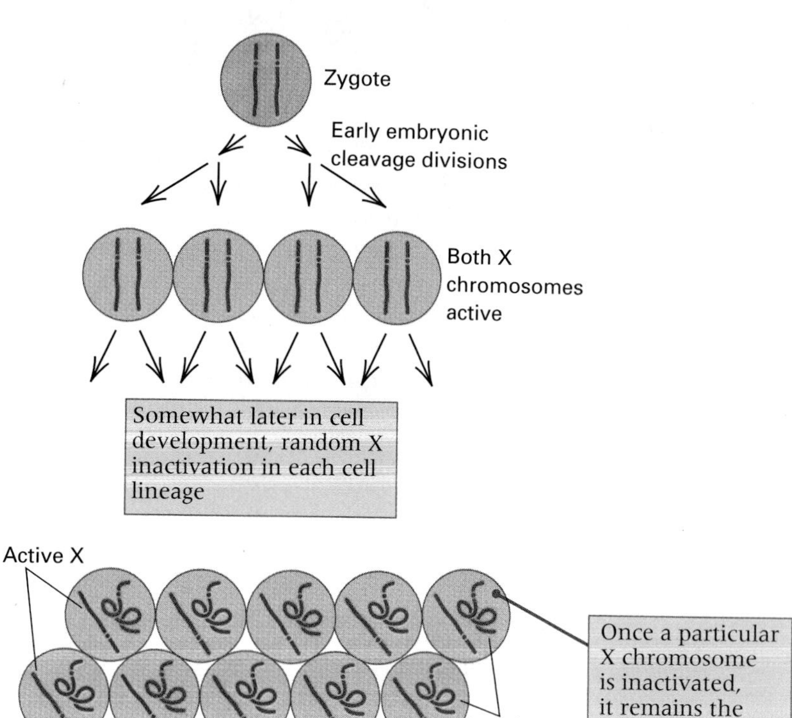


**FIGURE 8.7** Schematic diagram of somatic cells of a normal female, showing that the female is a mosaic for X-linked genes. The two X chromosomes are shown in red and blue. An active X is depicted as a straight chromosome, an inactive X as a tangle. Each cell has just one active X, but the particular X that remains active is a matter of chance. In human beings, the inactivation includes all but a few genes in the tip of the short arm.

The process of X inactivation takes place in all embryos with two or more X chromosomes, including normal XX females. The inactivation process is one of chromosome condensation initiated at a site called *XIC* (for *X-inactivation center*) near the centromere on the long arm between Xq11.2 and Xq21.1. Cytologically, this site marks a "bend" in the chromosome in mitotic metaphase, which is thought to be a visible manifestation of the X inactivation. The finding that the Xq11.2–Xq21.1 region is present in all structurally abnormal X chromosomes is taken as one line of evidence for its essential function in X inactivation, because structurally abnormal X chromosomes lacking this region would not be inactivated and therefore cause lethality and not be observed. Another line of evidence is that the *XIC* includes a transcribed region designated *Xist* (for *X-inactivation–specific transcript*). Transcription of *Xist* is the earliest event observed in X inactivation, and *Xist* is the only gene known to be transcribed only from the inactive X chromosome. Remarkably, the spliced transcript of *Xist* does not contain an open reading frame encoding a protein. It appears to function as a structural RNA, and as transcription of *Xist* continues, the spliced transcript progressively coats the inactive X chromosome, spreading outward from the *XIC*. Thereafter, other molecular changes take place along the inactive X chromosome that are typically associated with gene silencing, including heavy cytosine methylation in 5′ regulatory regions, deacetylation and methylation of histones, and aggregation of heterochromatin-specific DNA-binding proteins. In mouse embryos in which the *Xist* homolog has been disrupted, X inactivation does not take place, which demonstrates that *Xist* is essential for inactivation.

X-chromosome inactivation has two consequences. First, it results in dosage compensation. It equalizes the number of active copies of X-linked genes in females and males. Although a female has two X chromosomes and a male has only one, because of inactivation of one X

## connection

## Lyonization of an X Chromosome

**Mary F. Lyon 1961**
Medical Research Council,
Harwell, England
*Gene Action in the X Chromosome of the Mouse* (Mus musculus L.)

*How do organisms solve the problem that females possess two X chromosomes whereas males have only one? Unless there were some type of correction (called dosage compensation), the unequal number would mean that for all the genes in the X chromosome, cells in females would have twice as much gene product as cells in males. It would be difficult for the developing organism to cope with such a large difference in dosage for so many genes. The problem of dosage compensation has been solved by different organisms in different ways. The hypothesis put forward in this paper is that in the mouse (and, by inference, in other mammals), the mechanism is very simple: One of the X chromosomes, chosen at random in each cell lineage early in development, becomes inactivated and remains inactivated in all descendant cells in the lineage. In certain cells, the inactive X chromosome becomes visible in interphase as a deeply staining "sex-chromatin body." We now know that there are a few genes in the short arm of the X chromosome that are not inactivated. There is also good evidence that the inactivation of the rest of the X chromosome takes place sequentially from an "X-inactivation center." We also know that in marsupial mammals, such as the kangaroo, it is always the paternal X chromosome that is inactivated.*

It has been suggested that the so-called sex-chromatin body is composed of one heteropyknotic [that is, deeply staining during interphase] X chromosome. . . . The present communication suggests that evidence of mouse genetics indicates: (1) that the heteropyknotic X chromosome can be either paternal or maternal in origin, in different cells of the same animal; (2) that it is genetically inactivated. The evidence has two main parts. First, the normal phenotype of XO females in the mouse shows that only one active X chromosome is necessary for normal development, including sexual development. The second piece of evidence concerns the mosaic phenotype of female mice heterozygous for some sex-linked [X-linked] mutants. All sex-linked mutants so far known affecting coat colour cause a "mottled" or "dappled" phenotype, with patches of normal and mutant colour. . . . It is here suggested that this mosaic phenotype is due to the inactivation of one or other X chromosome early in embryonic development. If this is true, pigment cells descended from the cells in which the chromosome carrying the mutant gene was inactivated will give rise to a normal-coloured patch and those in which the chromosome carrying the normal gene was inactivated will give rise to a mutant-coloured patch. . . . Thus this hypothesis predicts that for all sex-linked genes of the mouse in which the phenotype is due to localized gene action the heterozygote will have a mosaic appearance. . . . The genetic evidence does not indicate at what early stage of embryonic development the inactivation of the one X chromosome occurs. . . . The sex-chromatin body is thought to be formed from one X chromosome in the rat and in the opossum. If this should prove to be the case in all mammals, then all female mammals heterozygous for sex-linked mutant genes would be expected to show the same phenomena as those in the mouse. The coat of the tortoiseshell cat, being a mosaic of the black and yellow colours of the two homozygous genotypes, fulfills this expectation.

***The coat of the tortoiseshell cat, being a mosaic of the black and yellow colours of the two homozygous genotypes, fulfills this expectation.***

Source: M. F. Lyon, *Nature* 190 (1961): 372.

chromosome in each of the somatic cells of a female, both sexes have the same number of active X chromosomes. The mechanism of dosage compensation by means of X inactivation was originally proposed by Mary Lyon and is called the **single-active-X principle**.

The second consequence of X-chromosome inactivation is that a normal female is a sort of mosaic for X-linked genes. That is, each somatic cell expresses the genes in only one X chromosome, but the X chromosome that is active genetically differs from one cell to the next. This mosaicism can be observed directly in females that are heterozygous for X-linked alleles that determine different forms of an enzyme, A and B: When cells from a heterozygous female are individually cultured in the laboratory, half of the clones are found to produce only the A form of the enzyme and the other half to produce only the B form. Mosaicism can also be observed directly in women who are heterozygous for an X-linked recessive mutation that results in the absence of sweat glands; these women exhibit large patches of skin in which sweat glands are present (these patches are derived from embryonic cells in which the normal X chromosome remained active and the mutant X was inactivated) and other large patches of skin in which sweat glands are absent (these patches are derived from embryonic cells in which the normal X chromosome was inactivated and the mutant X remained active).

Although all mammals use X-chromosome inactivation for dosage compensation in females, the choice of which chromosome to inactivate is not always random. In marsupial mammals, which include the kangaroo, the koala, and the wombat, the X chromosome that is inactivated is always the one contributed by the father. The result is that female marsupials are not genetic mosaics of paternal and maternal X-linked genes. A female marsupial expresses the X-linked genes that she inherited from her mother.

**FIGURE 8.8** This cat's name is Ginger. She is a female that is heterozygous for the orange and black coat color alleles and shows the classic "calico" pattern of patches of orange and black fur.

## The Calico Cat

In nonmarsupial mammals, the result of random X inactivation in females can sometimes be observed in the external phenotype. One example is the "calico" pattern of coat coloration in female cats. Two alleles affecting coat color are present in the X chromosome in cats. One allele results in an orange coat color (sometimes referred to as yellow), the other in a black coat color. Because he has only one X chromosome, a normal male has either the orange or the black allele. A female can be heterozygous for orange and black, and in this case the coat color is "calico"—a mosaic of orange and black patches. The orange and black patches result from X-chromosome inactivation. In cell lineages in which the X chromosome bearing the orange allele is inactivated, the X chromosome with the black allele is active and so the fur is black. In cell lineages in which the X chromosome with the black allele is inactivated, the orange allele in the active X chromosome results in orange fur (**FIGURE 8.8**).

The white patches have a completely different explanation. The white patches are due to an autosomal gene *S* for white spotting, which prevents pigment formation in the cell lineages in which it is expressed. Why the *S* gene is expressed in some cell lineages and not others is not known. Homozygous *SS* cats have more white than heterozygous *Ss* cats. Ginger, the female cat in the Figure 8.8, is homozygous *SS*.

## Pseudoautosomal Inheritance

The silencing of genes in the inactive X chromosome evolved gradually as the X and Y chromosomes progressively diverged from their ancestral chromosomes and the Y chromosome began to lose the function of most of its genes. The gene inactivation in the inactive X chromosome therefore affects individual genes and blocks of genes, and some genes in the inactive X are not silenced. Some of the genes that escape X inactivation have functional homologs in the Y chromosome, whereas others do not.

Two continuous regions that escape X inactivation are found at the tips of the long and short arms. These are regions in which the Y chromosome does retain functional homologous genes. These regions of homology enable the X and Y chromosomes to synapse in spermatogenesis, and a crossover takes place that holds the chromosomes together to ensure their proper segregation during anaphase I. The regions of shared X–Y homology define the **pseudoautosomal regions:** *PARp* is a 2.7-Mb region at the terminus of the short arms and *PARq* is a 0.3-Mb region at the terminus of the long arms. Because crossing over regularly takes place at least in *PARp*, the rate of recombination per nucleotide pair is at least twentyfold greater *PARp* in than in the autosomes.

The pedigree patterns of inheritance of genes in the pseudoautosomal regions are indistinguishable from patterns characteristic of autosomal inheritance. The reason is that a mutant allele in a pseudoautosomal region is neither completely X-linked nor completely Y-linked, but can move back and forth between the X and Y chromosomes because of recombination in the pseudoautosomal region. A gene that shows an autosome-like pattern of inheritance, but that is known from molecular studies to reside in a pseudoautosomal region, is said to show *pseudoautosomal inheritance*.

## ■ Active Genes in the "Inactive" X Chromosome

In addition to genes in the pseudoautosomal regions *PARp* and *PARq*, some other genes in the "inactive" X chromosome also show transcriptional activity. In particular, a study of the level of transcription of 624 genes along the human X chromosome has shown that about 15 percent of the X-linked genes escape inactivation to some degree. The "inactive X" is therefore not completely silenced. The location of the partially active genes along the chromosome is shown in **FIGURE 8.9**. The transcribed genes occur in large blocks that tend to be located in the distal portions of the arms, especially the short arm, a pattern suggesting that escape from X inactivation may be correlated with distance from *Xist*.

In any event, most of the genes that escape inactivation have levels of transcription that range from 15–50 percent of those observed for their homologs in the active X chromosome; hence the level of activity is robust. The number of genes that escape complete X inactivation, and their levels of transcription, could readily account for some of the differences in expression of traits between males and females, for phenotypic variation among females heterozygous for such X-linked mutations as hemophilia A, and for phenotypic variation among individuals with abnormal numbers of X chromosomes.

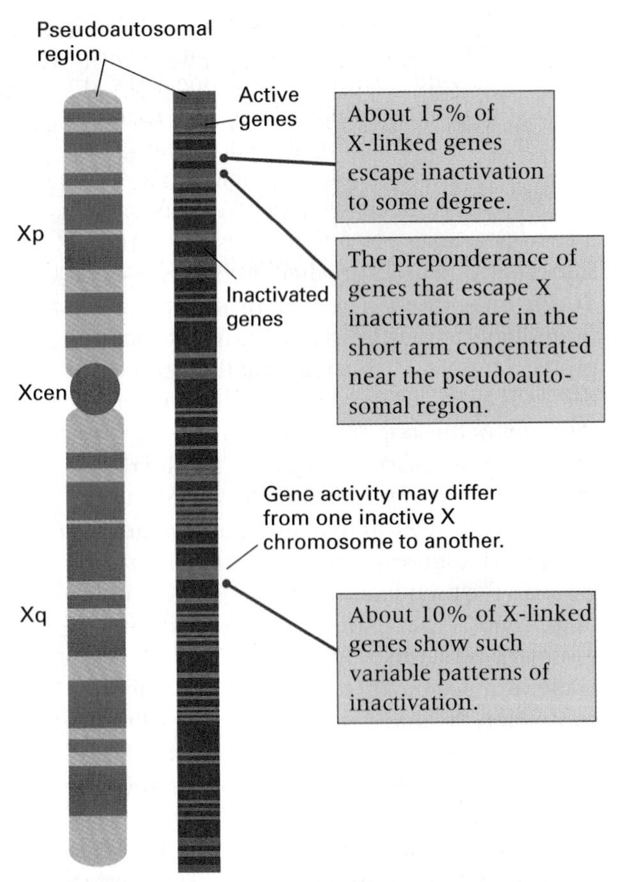
Pseudoautosomal region
Active genes
About 15% of X-linked genes escape inactivation to some degree.
Xp
The preponderance of genes that escape X inactivation are in the short arm concentrated near the pseudoautosomal region.
Inactivated genes
Xcen
Gene activity may differ from one inactive X chromosome to another.
Xq
About 10% of X-linked genes show such variable patterns of inactivation.

**FIGURE 8.9** Distribution along the human X chromosome of the approximately 15 percent of genes that escape complete transcriptional silencing in the inactive X chromosome. [Data from L. Carrel and H. F. Willard, *Nature* 434 (2005): 400–404.]

## ■ Gene Content and Evolution of the Y Chromosome

Comparative cytogenetic and molecular studies suggest that the X and Y chromosomes began their existence as a pair of ordinary autosomes in the common ancestor of modern mammals and birds. They started to diverge in DNA sequence and gene content at about the same time that the evolutionary lineage of mammals diverged from that of birds, some 300–350 million years ago (MY). One must assume that prior to this time, recombination took place at normal levels throughout the entire proto-X and proto-Y chromosomes and that their gene contents were identical.

In the human genome as it exists today, the Y chromosome includes genes that encode about 26 distinct proteins, many of which are important for male fertility. The Y chromosome is also populated with many repeat sequences, some extremely long, which can undergo gene conversion and serve as sites for homologous recombination. One of the key genes in the Y chromosome is the master sex-determining gene *SRY* at position Yp11.3. This band is in the short arm of the Y, near but not included in the pseudoautosomal region. The gene *SRY* codes for a protein transcription factor, the **testis-determining factor (TDF)**, which directs embryonic development toward the male sex by inducing the undifferentiated embryonic genital ridge, the precursor of the gonad, to develop as a testis. A transcription factor is a protein that stimulates transcription of its target genes.

Once *SRY* had evolved as a sex-determining mechanism, the Y chromosome began to diverge in DNA sequence from the X chromosome, and the region of possible X–Y recombination became progressively restricted to the telomeric regions. In regions with no X–Y recombination, there is a steady selection pressure for genes in the Y chromosome to undergo mutational degeneration into nonfunctional states. This results from the forced heterozygosity of the Y chromosome, which allows multiple deleterious mutations to accumulate through time because there is no opportunity for recombination to regenerate Y chromosomes that are free of deleterious mutations. Hence any Y-linked gene whose function is nonessential will tend to degenerate gradually because of the accumulation of mutations, and at the same time there will be selection pressure for dosage compensation of the homologous gene in the X chromosome. Eventually, only the dosage-compensated X-linked gene will remain functional.

Apparently blocks of genes were removed from the region of X–Y recombination in large chunks. Molecular evidence for this conclusion is summarized in **FIGURE 8.10**. Shown at the left are the locations of some protein-coding sequences in the short arm of the modern X chromosome from band Xp11 to the telomere. All of these genes have homologous sequences in the modern Y chromosome, as shown at the right. The amount of sequence divergence between the X and Y homologs shows a remarkable pattern. In the synonymous positions of the codons, where a nucleotide substitution can occur without changing the encoded amino acid, the sequence divergence (measured, roughly speaking, as the proportion of synonymous sites that differ between the X and Y homologs) for the genes *GYG2–AMELX* is 0.07–0.11, for *TB4X–UTX* it is 0.23–0.36, for *SMCX* it is 0.52, and for other genes outside the region shown it is > 0.94. Because the evolutionary rate of nucleotide substitutions at synonymous sites in mammalian genes is known, we can say that these levels of divergence correspond to divergence times of 30–50 million years (MY) for *GYG2–AMELX*, 80–130 MY for *TB4X–UTX*, 130–170 MY for *SMCX*, and 300–350 MY for other genes outside the region shown. The simplest explanation is that these were the times when successive blocks of genes were removed from the region of X–Y recombination by chromosome rearrangements, namely, inversions, in which a region somewhere in the interior of the Y chromosome becomes reversed in orientation. One way in which such an inversion can take place is by means of homologous recombination between two repeated DNA sequences that are present in opposite orientations at different locations in the Y chromosome, a mechanism that is examined in greater detail in Section 8.4. Whether the inversions actually took place via

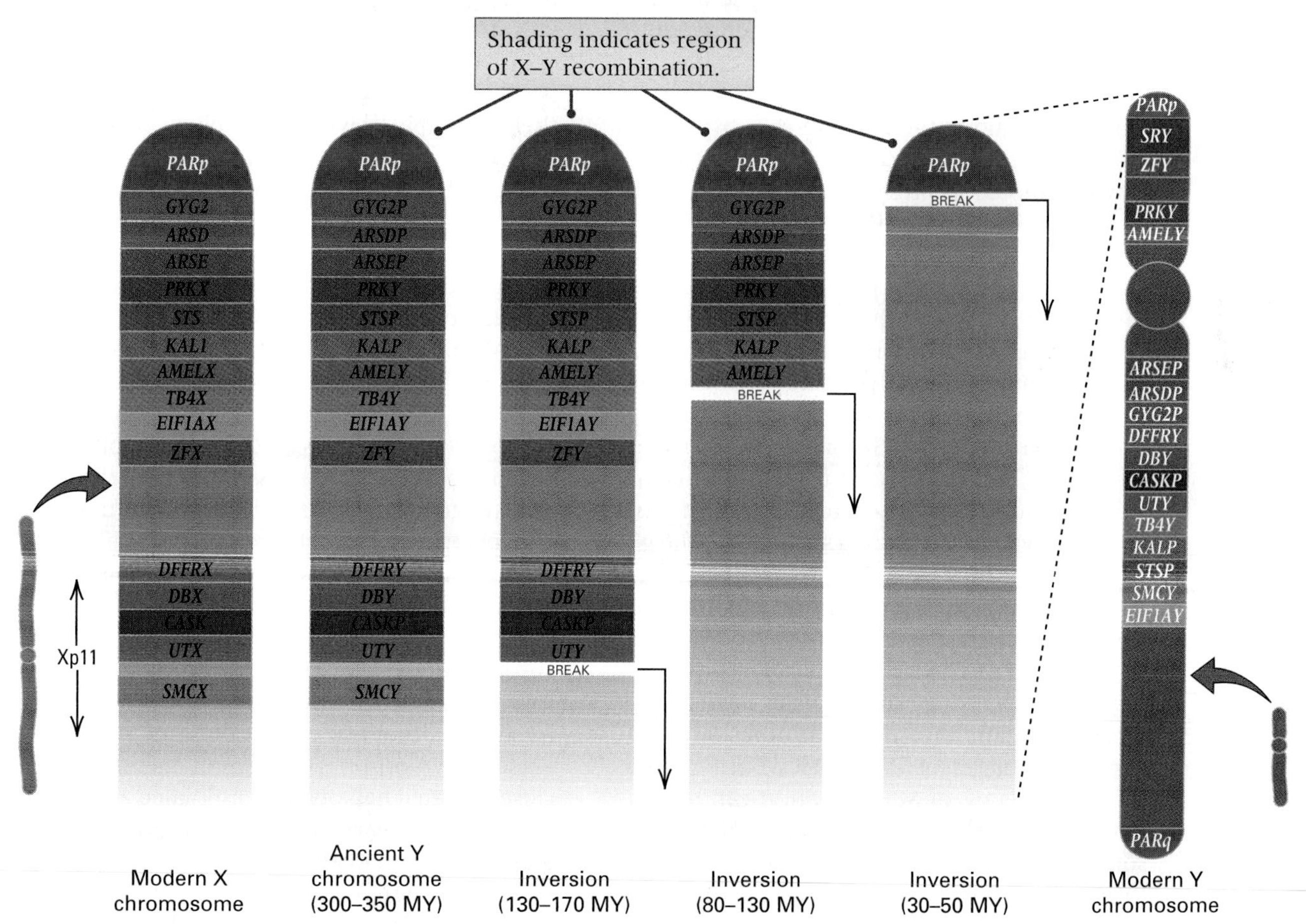


**FIGURE 8.10** Progressive shortening of the mammalian X–Y pseudoautosomal region through time due to inversions in the Y chromosome, inferred from DNA sequence data. The arrows denote the distal (nearest the telemere) breakpoint of each successive inversion interrupting the pseudoautosomal region. [Data from B. T. Lahn and D. C. Page, *Science* 286 (1999): 964–967.]

homologous recombination or were generated by some other process of breakage and reunion is not known. However, the inversions happened, the evolutionary reconstruction in Figure 8.10 shows that the 130–170 MY inversion breakpoint was adjacent to *UTY*, the 80–130 MY inversion breakpoint adjacent to *AMELY*, and the 30–50 MY year breakpoint adjacent to the present-day pseudoautosomal region on the short arm. As each of these inversions was fixed in the evolving Y chromosome, it removed the corresponding block of genes from the region of X–Y recombination, so that the rate of sequence divergence between the X and Y homologs accelerated. Other rearrangements in the Y chromosome, not able to be traced from these data, led to some additional scrambling of the gene order in the Y chromosome.

## ■ Tracing Human History Through the Y Chromosome

Because the Y chromosome does not undergo recombination along most of its length, genetic markers in the Y chromosome are completely linked and so remain together as the chromosome is transmitted from generation to generation. The genetic relation between Y chromosomes can therefore be traced, because chromosomes that are closely related will share more alleles along their length than will more distantly related chromosomes. The set of alleles at two or more loci present in a particular chromosome is called a **haplotype**. For many genealogical studies of the Y chromosome, simple-sequence repeat (SSR) polymorphisms are convenient because of their relatively high rate of mutation due to replication errors and the large number of alleles. The logic is that Y chromosomes with haplotypes that share alleles at each of 20–30 SSRs across the chromosome must have descended from the same ancestral Y chromosome in the very recent past. For haplotypes differing at a single locus the genetic relationship is less close, for those differing at two loci it is still less close, and so forth. This simple logic is the basis of tracing population history through Y-chromosome polymorphisms. Haplotypes that share many alleles have a more recent common ancestral Y chromosome than haplotypes that share fewer alleles. Furthermore, because the rate of SSR mutation can be estimated, the time at which the ancestral chromosome existed can be deduced. This reasoning forms the basis of the estimate that the most recent common ancestor of all extant human Y chromosomes existed 50,000–150,000 years ago. Such estimates are not highly precise, and there are many assumptions that must be made. Much can be learned about human population history through studies of the Y chromosome. The following discussion highlights three specific examples pertaining to Genghis Khan, the Cohanim, and the origin of European Gypsies.

**A Legacy of Genghis Khan.** At its maximum extent stretching from China to Russia through to the Middle East and then into Eastern Europe, the Mongol Empire of the 13th century comprised the largest land empire that history has known. The founder was born Temujin around 1162. As a young man he organized a confederation of tribes, who around 1200 took to their small Mongolian ponies equipped with high wooden saddles and stirrups, and armed with bow and arrow began to conquer their neighbors. Soon thereafter, Temujin adopted the name Genghis Khan, which means Universal Ruler. He was often merciless, exterminating the men and boys of rebellious cities and kidnapping the women and girls. In answer to a question about the source of happiness, he is reputed to have said, "The greatest happiness is to vanquish your enemies, to chase them before you, to rob them of their wealth, to see those dear to them bathed in tears, to clasp to your bosom their wives and daughters." Through their multiple wives, concubines, and innumerable unrecorded sexual conquests, Genghis Khan and his descendants were very prolific. His eldest son Tushi had 40 acknowledged sons, and his grandson Kubilai Khan (under whom the Mongol Empire reached its maximum extent) had 22 acknowledged sons.

Although the legacy of Genghis Khan is well recorded in history, it was hardly expected that it would show up in studies of the Y chromosome. But genotyping studies of 32 markers along the Y chromosome of 2123 men sampled from throughout a large region of Asia yielded the remarkable result in **FIGURE 8.11**. Each circle represents a population sample, with its area proportional to the sample size. The red sectors denote the relative frequency of a group of nearly identical Y-chromosome haplotypes, whereas the white sectors represent the relative frequency of other haplotypes that are genetically much more diverse. The most recent common ancestor of the closely related haplotypes is estimated as existing 1000 ± 300 years ago. Furthermore, the geographical region in which the closely related haplotypes cluster is included largely within the Mongol Empire (shading). The sole exception is population 10, composed of the ethnic Hazara of Pakistan. This provides a clue to the origin of the

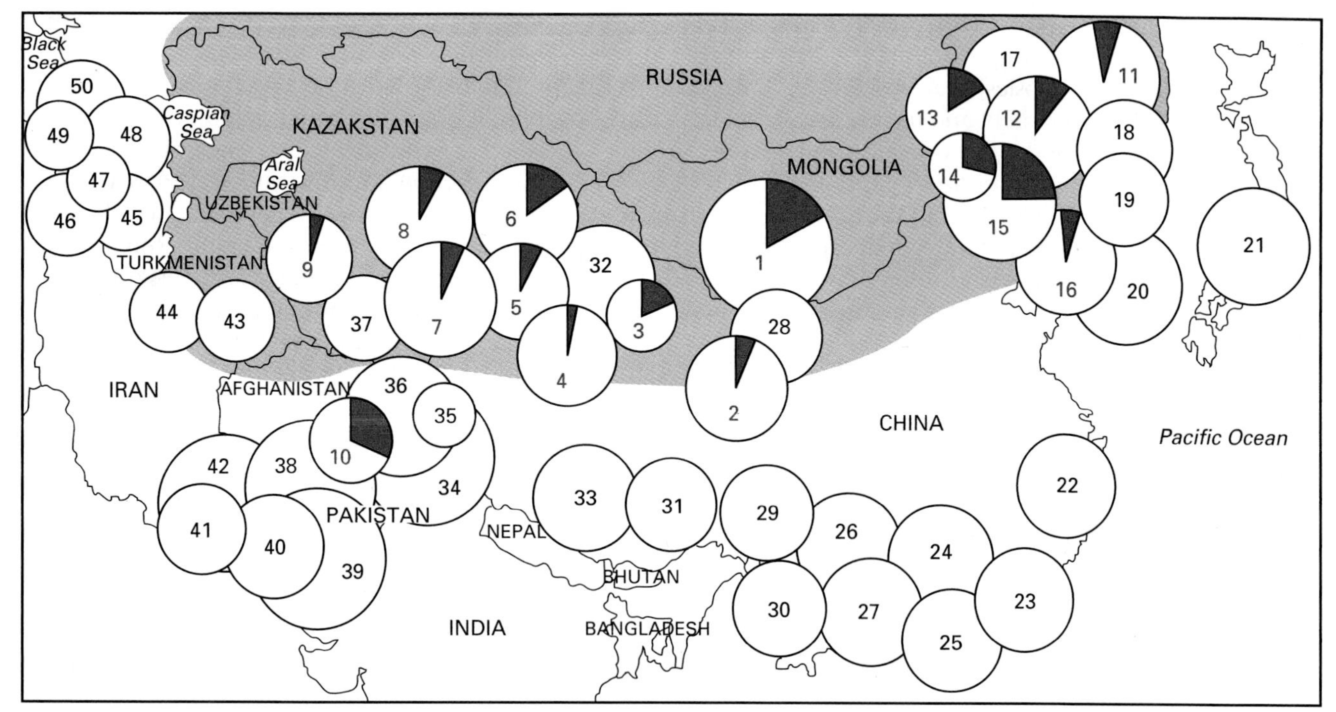


**FIGURE 8.11** Distribution of Y-chromosome haplotypes (red), presumed to have descended from Genghis Khan or his close male relatives, among populations near and bordering the ancient Mongol Empire. The specific population groups are: (1) Mongolian, (2) Han [Gansu], (3) Chinese Kazak, (4) Han [Xinjiang], (5) Xibe, (6) Uyghur, (7) Kyrgyz, (8) Kazak, (9) Uzbek, (10) Hazara, (11) Hezhe, (12) Daur, (13) Ewenki, (14) Han [Inner Mongolian], (15) Inner Mongolian, (16) Manchu, (17) Oroqen, (18) Han [Heilongjiang], (19) Chinese Korean, (20) Korean, (21) Japanese, (22) Shezu, (23) Han [Guangdong], (24) Yaozu [Liannan], (25) Lizu, (26) Buyi, (27) Yaozu [Bama], (28) Huizu, (29) Han [Sichuan], (30) Hani, (31) Qiangzu, (32) Chinese Uyghur, (33) Tibetan, (34) Burusho, (35) Balti, (36) Kalash, (37) Tajik, (38) Baloch, (39) Parsi, (40) Makrani Negroid, (41) Makrani Baloch, (42) Brahui, (43) Turkmen, (44) Kurd, (45) Azeni, (46) Armenian, (47) Lezgi, (48) Georgian, (49) Ossetian, (50) Svan. [Adapted from T. Zerjal, *Am. J. Hum. Genet.* 72 (2003): 717–721.]

closely related Y chromosomes, because the Hazara consider themselves to be of Mongol origin, and many claim to be direct male-line descendants of Genghis Khan. Whatever their origin, the closely related Y chromosomes are found in about 8% of the males throughout a large region of Asia (populations 1–16). Direct proof of the connection with Genghis Khan in principle could be obtained by determining the haplotype of the Y chromosome in material recovered from his grave. He died in 1227 from injuries sustained in a fall from a horse, and he is said to have been buried in secret at a site near his birthplace.

**A Legacy of the Cohanim.** The Lemba are a group of about 50,000 Bantu-speaking people living predominantly in South Africa and Zimbabwe. They drew attention about 100 years ago because of their vaguely Jewish customs including dietary restrictions and male circumcision, and especially because of their oral history, which holds that their ancestors arrived by boat from a city called Sena, variously placed in Yemen, Judea, Egypt, or Ethiopia. Studies of 12 polymorphic Y-chromosome markers among 136 Lemba males from six clans have shed some light on the situation. The Y chromosomes from the Lemba derive from one of two lineages. One is closely related to the Bantu and the other is clearly Semitic. About 50% of the Y chromosomes of one Lemba clan (the Buba) have haplotypes closely related to a haplotype of Judaic origin called the *Cohen modal haplotype* because it occurs primarily in the Cohanim (the plural of Cohen), the priestly lineage said to be descended from Moses's brother Aaron. Although the Cohen model haplotype has a frequency at least 50% in the Cohanim, it is rare in other Semitic groups. This finding affords some support for the Lemba's oral history, and the estimated time for the most recent common ancestor of the Lemba and Cohanim Y chromosomes is roughly 3000–5000 years. The earliest of these dates would be consistent with the time that the Assyrian King Shalmaneser V sent the ten tribes of Israel into exile. Sometimes known as the "black Jews of south Africa," the Lemba are technically not Jewish. Judaism is transmitted through the maternal lineage, and Lemba tradition holds that only men survived the perilous voyage from Sena.

**Origin of European Gypsies.** Arriving in Eastern Europe about 1000 years ago, the Roma (Gypsies) were persecuted for centuries. They were held and bartered as slaves until the 1860s, and they were the only ethnic group besides Jews to be singled out for extermination in the Nazi death camps. Today they number more than twelve million people located in many countries around the world. Their origin has been disputed. The term "Gypsy" reflects a legend that they originated in Egypt, but their language (Romanes) has some similarities to languages of the Indian subcontinent.

Studies of the Y chromosome have clarified this situation, too. A group of closely related haplotypes was found among men in all of 14 Romani populations studied and accounted for 44.8 percent of all the Romani Y chromosomes. Elsewhere in the world this haplotype is found in the Indian subcontinent. In this study, mitochondrial DNA haplotypes were also examined. Mitochondrial DNA is also convenient for tracing population history because it does not undergo recombination and is transmitted through the female (Chapter 16). A particular group of mitochondrial DNA haplotypes was found in 26.5 percent of the female lineages among the Romani populations. This haplotype derives from the Indian subcontinent, too. The origin of the Y chromosomal and mitochondrial DNA haplotypes and the relatively high frequency of a small number of haplotypes among the Roma is consistent with a small group of founders originating in the Indian subcontinent. Given the time of their appearance in Eastern Europe, it has been suggested that their migration was actually a flight from the invading armies of Mahmud of Ghazni from what is now Afghanistan.

## 8.2 Chromosome Abnormalities in Human Pregnancies

Approximately 15 percent of all recognized pregnancies in human beings terminate in spontaneous abortion, and in about half of all spontaneous abortions, the fetus has a major chromosome abnormality. **TABLE 8.2** summarizes the average rates of chromosome abnormality found per 100,000 recognized pregnancies in several studies. The term **trisomy** refers to an otherwise diploid organism that has an extra copy of an individual chromosome. Many of the spontaneously aborted fetuses have trisomy of one of the autosomes. Triploids, which have three sets of chromosomes (total count 69), and tetraploids, which have four sets of chromosomes (total count 92), are also common in spontaneous abortions. Triploids and tetraploids are examples of **euploid** conditions, because they have the same relative gene dosage as found in the diploid. In contrast, relative gene dosage is upset in a trisomic, because three copies of the genes located in the trisomic chromosome are present, whereas two copies of the genes in the other chromosomes are present. Such unbalanced chromosome complements are said to be **aneuploid**. Although it is not apparent in the data in Table 8.2, in most organisms, euploid abnormalities generally have less severe phenotypic effects than aneuploid abnormalities. In Table 8.2 the term *balanced translocation* refers to a euploid condition in which nonhomologous chromosomes have an interchange of part, but all of the parts are present; the term *unbalanced translocation* refers to an aneuploid condition in which some part of the genome is missing. The much greater survivorship of the balanced translocation indicates that a euploid chromosomal abnormality is generally less harmful than an aneuploid chromosome abnormality.

**Table 8.2** Chromosome abnormalities per 100,000 recognized human pregnancies

| | | 15,000 spontaneous abortions | 85,000 live births |
|---|---|---|---|
| Trisomy | | | |
| A: | 1 | 0 | 0 |
| | 2 | 159 | 0 |
| | 3 | 53 | 0 |
| B: | 4 | 95 | 0 |
| | 5 | 0 | 0 |
| C: | 6–12 | 561 | 0 |
| D: | 13 | 128 | 17 |
| | 14 | 275 | 0 |
| | 15 | 318 | 0 |
| E: | 16 | 1229 | 0 |
| | 17 | 10 | 0 |
| | 18 | 223 | 13 |
| F: | 19–20 | 52 | 0 |
| G: | 21 | 350 | 113 |
| | 22 | 424 | 0 |
| Sex chromosomes | | | |
| XYY | | 4 | 46 |
| XXY | | 4 | 44 |
| XO | | 1350 | 8 |
| XXX | | 21 | 44 |
| Translocations | | | |
| Balanced | | 14 | 164 |
| Unbalanced | | 225 | 52 |
| Polyploid | | | |
| Triploid | | 1275 | 0 |
| Tetraploid | | 450 | 0 |
| Other (mosaics, etc.) | | 280 | 49 |
| **Total** | | **7500** | **550** |

connection

# The First Human Chromosomal Disorder Identified


**-rome Lejeune, Marthe -utier, and Raymond Turpin**
**-59**
-ional Center for Scientific -earch, Paris, France
*-dy of the Somatic -romosomes of Nine Down -drome Children*
-ginal in French]


*Down syndrome had been one of the greatest mysteries in human genetics. One of the most common forms of mental retardation, the syndrome did not follow any pattern of Mendelian inheritance. Yet some families had two or more children with Down syndrome. (Many of these cases are now known to be due to a translocation involving chromosome 21.) This paper marked a turning point in human genetics by demonstrating that Down syndrome actually results from the presence of an extra chromosome. It was the first identified chromosomal disorder. The excerpt uses the term* telocentric, *which means a chromosome that has its centromere very near one end. In the human genome, the smallest chromosomes are three very small telocentric chromosomes. These are chromosomes 21, 22, and the Y. A normal male has five small telocentrics (21, 21, 22, 22, and Y); a normal female has four (21, 21, 22, and 22). (The X is a medium-sized chromosome with its centromere somewhat off center.) In the table that follows, note the variation in chromosome counts in the "doubtful" cells. The methods for counting chromosomes were then very difficult, and many errors were made either by counting two nearby chromosomes as one or by including in the count of one nucleus a chromosome that actually belonged to a nearby nucleus. Lejeune and collaborators wisely chose to ignore these doubtful counts and based their conclusion only on the "perfect" cells. Sometimes good science is a matter of knowing which data to ignore.*

The culture of fibroblast cells from nine Down syndrome children reveals the presence of 47 chromosomes, the supernumerary chromosome being a small telocentric one. The hypothesis of the chromosomal determination of Down syndrome is considered. . . . The observations made in these nine cases (five boys and four girls) are recorded in the [accompanying] table.

The number of cells counted in each case may seem relatively small. This is due to the fact that only the pictures [of the spread chromosomes] that claim a minimum of interpretation have been retained in this table. The apparent variation in the chromosome number in the "doubtful" cells, that is to say, cells in which each chromosome cannot be noted individually with certainty, has been pointed out by several authors. It does not seem to us that this phenomenon represents a cytological reality, but merely reflects the difficulties of a delicate technique. It therefore seems logical to prefer a small number of absolutely certain counts ("perfect" cells in the table) to a mass of doubtful observations, the statistical variance of which rests solely on the lack of precision of the observations. Analysis of the chromosome set of the "perfect" cells reveals the presence in Down syndrome boys of 6 small telocentric chromosomes (instead of 5 in the normal man) and 5 small telocentric ones in Down syndrome girls (instead of 4 in the normal woman). . . . It therefore seems legitimate to conclude that there exists in Down syndrome children a small supernumerary telocentric chromosome, accounting for the abnormal figure of 47. To explain these observations, the hypothesis of nondisjunction of a pair of small telocentric chromosomes at the time of meiosis can be considered. . . . It is, however, not possible to say that the supernumerary small telocentric chromosome is indeed a normal chromosome and at the present time the possibility cannot be discarded that a fragment resulting from another type of aberration is involved.


***Analysis of the chromosome set of the "perfect" cells reveals the presence in Down syndrome boys of 6 small telocentric chromosomes (instead of 5 in the normal man) and 5 small telocentric ones in Down syndrome girls (instead of 4 in the normal woman).***


**Number of chromosomes**

| | | "Doubtful" cells | | | "Perfect" cells | | |
|---|---|---|---|---|---|---|---|
| | | **46** | **47** | **48** | **46** | **47** | **48** |
| Boys | 1 | 6 | 10 | 2 | — | 11 | — |
| | 2 | — | 2 | 1 | — | 9 | — |
| | 3 | — | 1 | 1 | — | 7 | — |
| | 4 | — | 3 | — | — | 1 | — |
| | 5 | — | — | — | — | 8 | — |
| Girls | 1 | 1 | 6 | 1 | — | 5 | — |
| | 2 | 1 | 2 | — | — | 8 | — |
| | 3 | 1 | 2 | 1 | — | 4 | — |
| | 4 | 1 | 1 | 2 | — | 4 | — |

Source: J. Lejeune, M. Cautier, and R. Turpin, *Comptes Rendus Hebd. Seances Acad. Sci.* 248 (1959): 1721–1722.

When an otherwise diploid organism has a missing copy of an individual chromosome, the condition is known as *monosomy.* In most organisms, chromosome loss (resulting in monosomy) is a more frequent event than chromosome gain (resulting in trisomy). However, monosomies are conspicuously absent in the data on spontaneous abortions in Table 8.2. Their absence is undoubtedly due to another feature of monosomy:

> A missing copy of a chromosome (monosomy) usually results in more harmful effects than an extra copy of the same chromosome (trisomy).

In human fertilizations, monosomic zygotes are probably created in even greater numbers than trisomic zygotes, but monosomy is not found among aborted fetuses in Table 8.2 because the abortions take place so early in development that the pregnancy goes unrecognized by the mother. Data relevant to very early abortions come from medical records of women attempting to become pregnant who while trying to conceive undergo a pregnancy hormone test every day. The hormone assayed is human chorionic gonadotropin, a glycoprotein first produced by the embryo soon after conception at about the time of implantation in the uterine wall. The results are that most such women conceive every month, but that in 50–60 percent of the cases, implantation fails to occur or the embryo undergoes spontaneous abortion shortly thereafter. Given the high level of chromosomal abnormalities in the late spontaneous abortions in Table 8.2, it seems likely that the majority of these early spontaneous abortions also have chromosomal abnormalities, primarily monosomy. These data imply a huge fetal wastage, but this serves the important biological function of eliminating many fetuses that would be grossly abnormal in their physical and behavioral development because of major chromosomal abnormalities.

**DOWN SYNDROME** stacking blocks perfectly. [© PhotoCreate/ShutterStock, Inc.]

### Down Syndrome and Other Viable Trisomies

Table 8.2 demonstrates that monosomy or trisomy of most human autosomes is incompatible with life. There are three exceptions: trisomy 13, trisomy 18, and trisomy 21. The first two are rare conditions associated with major developmental abnormalities, and the affected infants can survive for only a few days or weeks.

Trisomy 21 is **Down syndrome** (or *Down's syndrome*), which occurs in about 1 in 750 liveborn children. Its major symptom is mental retardation, but there can also be multiple physical abnormalities, such as heart defects. Affected children are small in stature because of delayed maturation of the skeletal system; their muscle tone is poor, resulting in a characteristic facial appearance; and they have a shortened life span of usually less than 50 years. Nevertheless, for a major chromosomal abnormality, the symptoms are relatively mild, and most children with Down syndrome can relate well to other people.

> Children with Down syndrome usually take great pleasure in their surroundings, their families, their toys, their playmates. Happiness comes easily, and throughout life they usually maintain a childlike good humor. They are not burdened with the grown-up cares that come to most people with adolescence and adulthood. Life is simpler and less complex. The emotions that others feel seem to be less intense for them. They are sometimes sad, happy, angry, or irritable, like everyone else, but their moods are generally not so profound and they blow away more quickly. . . . Children with Down syndrome, though slow, are still very responsive to their environment, to those around them, and to the affection and encouragement they receive from others. [Quoted from D. W. Smith and A. A. Wilson. *The Child with Down's Syndrome.* (Philadelphia: Saunders, 1973.)]

Most cases of Down syndrome are caused by nondisjunction, which means the failure of homologous chromosomes to separate in meiosis, as explained in Chapter 4. The result of chromosome-21 nondisjunction is one gamete that contains two copies of chromosome 21 and one that contains none. If the gamete with two copies participates in fertilization, then a zygote with trisomy 21 is produced. The gamete with no copy may also participate in fertilization, but zygotes with monosomy 21 do not survive even through the first few days or weeks of pregnancy. About three-fourths of the trisomy-21 fetuses also undergo spontaneous abortion (Table 8.2). If this were not the case, and all trisomy-21 fetuses survived to birth, the incidence of Down syndrome would rise to 1 in 250, approximately a threefold increase over the incidence actually observed.

Chromosome 21 is a small chromosome and therefore is somewhat less likely to undergo meiotic crossing over than a longer one. Noncrossover bivalents sometimes have difficulty aligning at the metaphase plate because they lack a chiasma to hold them together, so there is an increased risk of nondisjunction. Among the events of nondisjunction that result in Down syndrome, about 40 percent are derived from such nonexchange bivalents.

For unknown reasons, nondisjunction of chromosome 21 is more likely to happen in oogenesis than in spermatogenesis, and so the abnormal gamete in Down syndrome is usually the egg. Furthermore, the risk of nondisjunction of chromosome 21 increases dramatically with the age of the mother, resulting in a risk of Down syndrome that reaches 6 percent in mothers age 45 and older (**FIGURE 8.12**). For this reason, many physicians recommend that older women who are pregnant have cells from the fetus tested in order to detect Down syndrome prenatally. This can be done 15–16 weeks after fertilization by **amniocentesis**, in which cells of the developing fetus are obtained by insertion of a fine needle through the wall of the uterus and into the sac of fluid (the *amnion*) that contains the fetus, or it can be done at 10–11 weeks after fertilization by a procedure called *chorionic villus sampling* (*CVS*), which uses cells from a zygote-derived embryonic membrane (the *chorion*) associated with the placenta. Early diagnosis is desirable, but CVS has about a threefold higher risk of inducing a miscarriage than does amniocentesis.

About 3 percent of all cases of Down syndrome are due not to simple nondisjunction, but to an abnormality in chromosome structure. In these cases the risk of recurrence of the syndrome in subsequent children is very high—up to 20 percent of births. The high risk is caused by a chromosomal translocation in one of the parents, in which chromosome 21 has been broken and become attached to another chromosome. This situation is considered in Section 8.5.

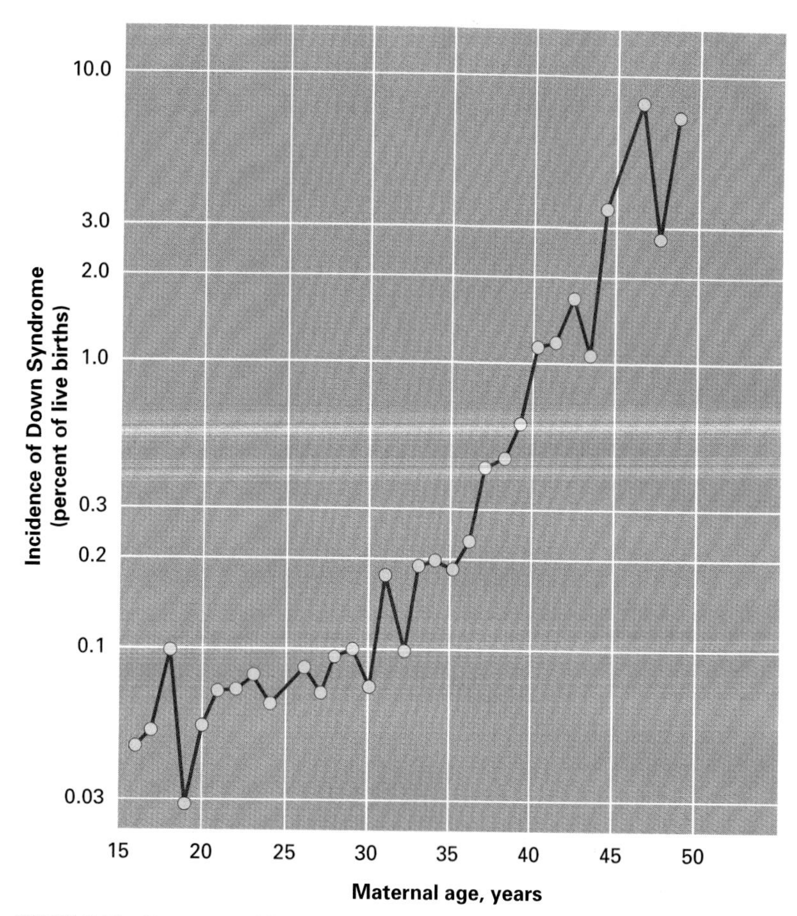


**FIGURE 8.12** Frequency of Down syndrome (number of cases per 100 live births) related to age of mother. The graph is based on 438 Down syndrome births (among 330,859 total births) in Sweden in the period 1968 to 1970. [Data from E. B. Hook and A. Lindsjö, *Am. J. Hum. Genet.* 30 (1978): 19–27.]

## Trisomic Segregation

In a trisomic organism, the segregation of chromosomes in meiosis is upset because the trisomic chromosome has two pairing partners instead of one. The behavior of the chromosomes in meiosis depends on the manner in which the homologous chromosome arms pair and on the chiasmata formed between them. In some cells, the three chromosomes form a **trivalent** in which distinct parts of one chromosome are paired with homologous parts of each of the others (**FIGURE 8.13**, part A). In metaphase, the trivalent is usually oriented with two centromeres pointing toward one pole and the other centromere pointing toward the other pole. The result is that at the end of both meiotic divisions, one pair of gametes contains two copies of the trisomic chromosome, and the other pair of gametes contains only a single copy. Alternatively, the trisomic chromosome can form one normal bivalent and one **univalent**, or unpaired chromosome, as shown in Figure 8.13B. In anaphase I, the bivalent

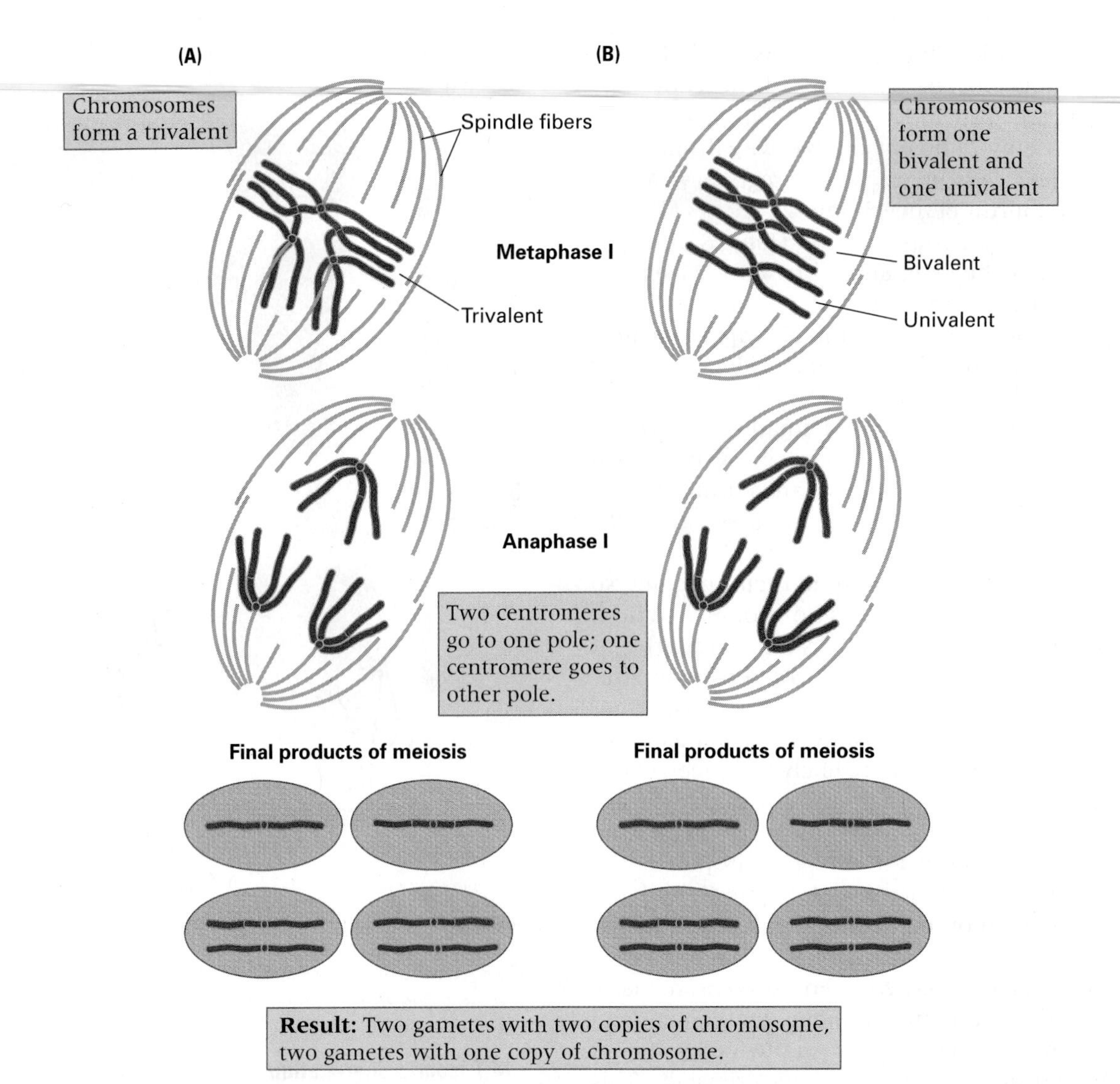
(A)
(B)
Chromosomes form a trivalent
Spindle fibers
Chromosomes form one bivalent and one univalent
Metaphase I
Bivalent
Trivalent
Univalent
Anaphase I
Two centromeres go to one pole; one centromere goes to other pole.
Final products of meiosis
Final products of meiosis
Result: Two gametes with two copies of chromosome, two gametes with one copy of chromosome.

**FIGURE 8.13** Meiotic pairing in a trisomic. (A) Formation of a trivalent. (B) Formation of a univalent and a bivalent. Both types of pairing result in one pair of gametes containing two copies of the trisomic chromosome and the other pair of gametes containing one copy of the trisomic chromosome.

disjoins normally and the univalent usually proceeds randomly to one pole or the other. Again, the end result is the formation of two products of meiosis that contain two copies of the trisomic chromosome and two products of meiosis that contain one copy. To state the matter in another way, a trisomic organism with three copies of a chromosome (say, C C C) will produce gametes among which half contain two copies (C C) and half contain one (C). When mated with a chromosomally normal individual, a trisomic is therefore expected to produce trisomic or normal progeny in a ratio of 1 : 1. This theoretical expectation is borne out in experimental organisms by matings of trisomics and in human beings by the finding that, in the few known cases when a person with Down syndrome reproduces, the frequency of Down syndrome in the offspring is approximately 50 percent.

## ■ Sex-Chromosome Abnormalities

Sex-chromosome trisomies, as a group, are even more frequent among newborns than is trisomy 21 (Table 8.2). There are two reasons why extra sex chromosomes have phenotypic effects that are relatively mild compared with those of autosomal trisomies. First, the single-active-X principle results in the silencing of most X-linked genes in all but one X chromosome in each somatic cell. Second, the Y chromosome contains relatively few functional genes.

The four most common types of sex-chromosome abnormalities are described below. The karyotypes are given in the conventional fashion, with the total number of chromosomes listed first, followed by the sex chromosomes that are present. For example, in the designation 47,XXX the number 47 refers to the total number of chromosomes, and

XXX indicates that the person has three X chromosomes.

- *47,XXX* This condition is often called simply **trisomy-X**. People with the karyotype 47,XXX are female. Most such women are mentally and phenotypically normal and give no physical or behavioral indication that an extra X chromosome is present. Mild mental impairment may be somewhat more frequent than among 46,XX females.
- *47,XYY* This condition is often called the **double-Y** karyotype. These people are male and tend to be taller than average, but they are otherwise phenotypically normal. At one time it was thought that 47,XYY males developed severe personality disorders and were at a high risk of committing crimes of violence, a belief based on the finding of 47,XYY among violent criminals. Further study indicated that the frequency of 47,XYY males in the general population was much higher than had been supposed. Some 47,XYY males have slightly impaired mental function. Although their rate of criminality is higher than that of normal males, the crimes are mainly nonviolent crimes such as petty theft. The majority of 47,XYY males are phenotypically and psychologically normal, have mental capabilities in the normal range, and have no criminal convictions.
- *47,XXY* This condition is called **Klinefelter syndrome**. Affected persons are male. They tend to be tall, do not undergo normal sexual maturation, are sterile, and in some cases have enlargement of the breasts. Mild mental impairment is common.
- *45,X* Monosomy of the X chromosome in females is called **Turner syndrome**. Affected persons are phenotypically female, but short in stature and without sexual maturation. Mental abilities are typically within the normal range. Note in Table 8.2 that more than 99 percent of 45,X fetuses undergo spontaneous abortion. This indicates that the condition usually has profound deleterious effects on development. Table 8.2 also shows that monosomy of the X chromosome is much more frequent than trisomy, which is almost certainly true of autosomes as well, in spite of the absence of monosomics in recognized pregnancies.

#### Environmental Effects on Nondisjunction

Because a large fraction of aneuploid zygotes result in miscarriage or in live births with congenital defects or mental retardation, the identification of environmental hazards that may increase the incidence of meiotic errors is of great importance. Environmental risk factors that have been suggested include radiation, smoking, alcohol consumption, oral contraceptives, fertility drugs, environmental pollutants, pesticides, and so forth. When significant effects have been found, they are usually small and not always reproducible, in part due to confounding effects of other factors such as maternal age. In view of the maternal-age effect, the female sex hormone estrogen and molecules resembling estrogen have long been under suspicion. In view of this background, it was no great surprise to learn that modest concentrations of a common estrogen mimic known as bisphenol A [technical name 2,2-(4,4-dihydroxy-diphenol)propane] caused about an eightfold increase in the incidence of aneuploidy in mice.

Bisphenol A is the basic subunit of polycarbonate plastic products widely used as a can liner in the food and beverage industry. In its polymerized form it may be completely harmless, but the monomers can leach out of plastic products that are damaged. The effect of bisphenol A was discovered by chance in studies of aneuploidy in the mouse. In the course of microscopic studies, the researchers noted a sudden increase in the incidence of meiotic cells in which one or more chromosomes failed to align properly on the metaphase II plate. At the same time the incidence of aneuploid oocytes increased from 1.4 percent to 11.6 percent.

The investigators succeeded in tracing the cause to the inadvertent use of a harsh alkaline detergent to wash the polycarbonate cages and water bottles. They suspected that the detergent was causing bisphenol A monomers to leach out of the plastic, and follow-up experiments with bisphenol A monomers demonstrated its effect on aneuploidy directly. Although the studies were carried out in female mice, the similarity of the meiotic process in female humans is a matter of concern. More generally the study raises a warning flag about environmental agents that may increase the rate of aneuploidy, particularly agents that have estrogen-like activity.

## 8.3 Chromosomal Deletions and Duplications

We now turn to the genetic implications of aberrations in *chromosome structure*. In the history of genetics, chromosome abnormalities were originally discovered through their genetic effects, which, though confusing at first, eventually came to be understood and were later confirmed directly by microscopic observations. There are several principal types of structural aberrations—deletions, duplications, inversions, and translocations—each of which has characteristic genetic effects.

## Deletions

A chromosome in which a segment is missing is said to have a **deletion** or a **deficiency**. Deletions are generally harmful to the organism, and the usual rule is the larger the deletion, the greater the harm. Very large deletions are usually lethal, even when heterozygous with a normal chromosome. Small deletions are often viable when they are heterozygous with a structurally normal homolog, because the normal homolog supplies gene products that are necessary for survival. However, even small deletions are usually homozygous-lethal (that is, when both members of a pair of homologous chromosomes carry the deletion).

Among the *copy-number polymorphisms (CNPs)* observed in the human genome (Chapter 2), deletions account for a significant proportion. Some of these have important phenotypic consequences. Examples include *autism spectrum disorders (ASDs)*, which are characterized by communication deficits, social impairment, and repetitive behaviors occurring prior to the age of 3, usually requiring extensive family support and medical intervention. ASDs affect an estimated 1/166 individuals with a ratio of affected males to affected females of about 5 : 1. An important genetic component is indicated by the observation that, if one member of a pair of genetically identical twins has ASD, there is a 70–90 percent chance that the other twin is also affected. In addition, genetic mapping studies have implicated genes in 20 different chromosomes. However, copy-number polymorphisms are associated with ASD in about 10 percent of the cases in which only one child in a sibship is affected. Most but not all of the ASD-related CNPs are deletions.

Deletions can be formed in two major ways. One is by chromosome breakage and reunion. Chromosome breaks result from double-stranded breaks in the DNA backbone. Chromosome breaks occur spontaneously at a low rate, but they also can be induced by x-rays and certain chemicals that cause double-stranded breaks in DNA. A deletion is created when a chromosome arm is broken in two places, when the broken ends bearing the centromere and the telomere fuse and the part left out remains as an acentric fragment that is lost.

Deletions can also be created by homologous recombination between repeated DNA sequences present at different sites along the DNA, a process known as **ectopic recombination**. An example is shown in **FIGURE 8.14**, part A. In this case each copy of the repeated DNA sequence is indicated by a color gradient. Note that the gradient runs from left to right in both copies, which indicates that both copies of the repeated DNA sequence have the same orientation along the DNA, a configuration known as **direct repeats**. If the direct repeats undergo pairing and homologous recombination, the result is a deletion of the material between the direct repeats, because the small circular acentric fragment containing this material is lost.

Examples of deletions caused by ectopic recombination in the human Y chromosome are shown in Figure 8.14B. This figure depicts a region of the Y chromosome that includes several large repeated sequences, shown here as gradients. The red and blue gradients indicate direct repeats, whereas the yellow gradients indicate **inverted repeats**, in which the repeated sequences are in reverse orientation. The yellow inverted repeats include genes important for male fertility. To give an idea of scale, the red repeats are each 229 kb in length, and the region between the repeats is 3.5 Mb. As shown in the diagram, homologous recombination in the red repeats results in loss of both sets of copies of the male-fertility genes. About one in 4000 males has this deletion, which causes complete sterility. A deletion with less drastic effects results from recombination within the blue repeats, which yields a deletion of only one set of the fertility genes. Although loss of these genes does not result in complete sterility, it does impair spermatogenesis. Nevertheless about 1 percent of males have a Y chromosome with this deletion.

In laboratory organisms, deletions can be detected genetically by making use of the fact that a chromosome with a deletion no longer carries the wildtype alleles of the genes that are missing. For example, in *Drosophila*, many deletions of a gene called *Notch*, which results in a notched wing margin, are large enough to remove the nearby wildtype allele of *white*. When these deleted chromosomes are heterozygous with a structurally normal chromosome carrying the recessive *w* allele, the fly has white eyes

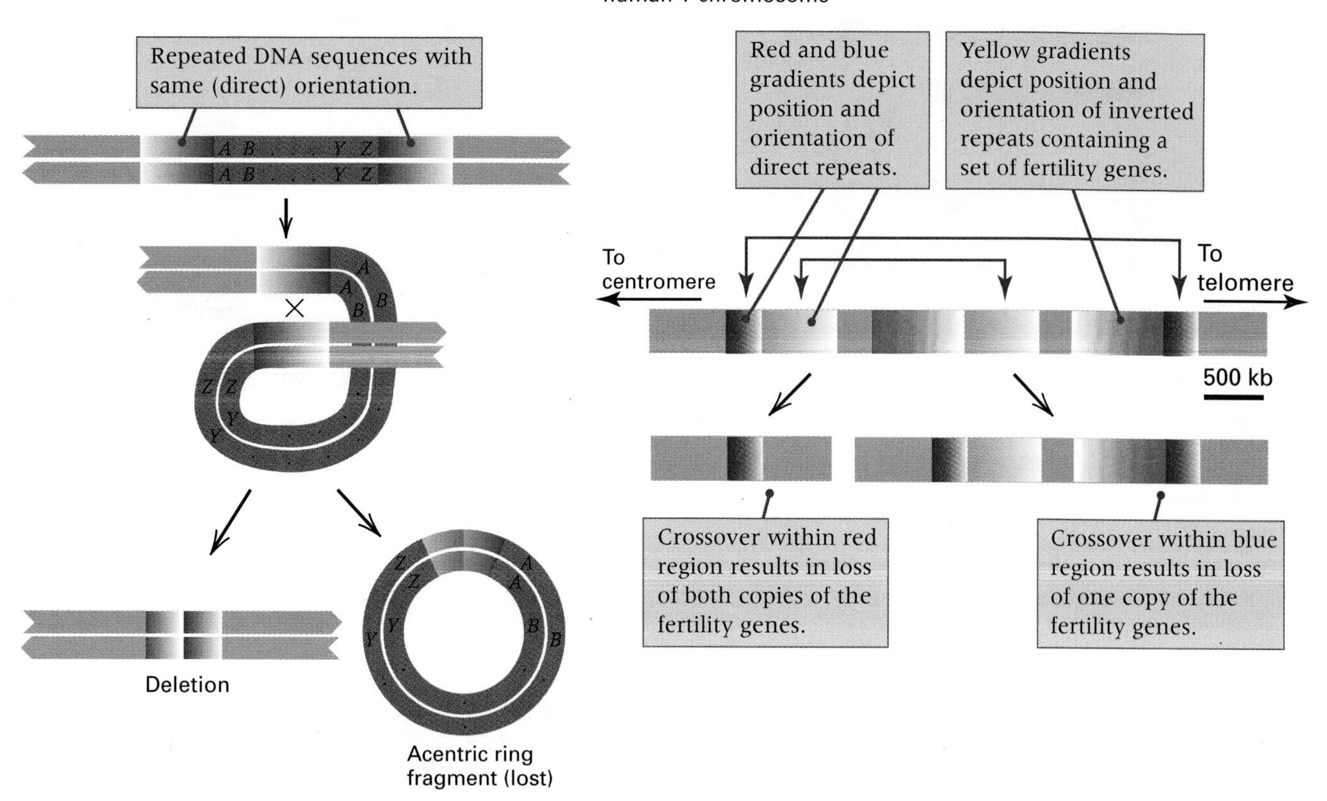
(A) Ectopic recombination
Repeated DNA sequences with same (direct) orientation.
Deletion
Acentric ring fragment (lost)
(B) AZFc region of human Y chromosome
Red and blue gradients depict position and orientation of direct repeats.
Yellow gradients depict position and orientation of inverted repeats containing a set of fertility genes.
To centromere
To telomere
500 kb
Crossover within red region results in loss of both copies of the fertility genes.
Crossover within blue region results in loss of one copy of the fertility genes.

**FIGURE 8.14** Ectopic recombination between direct repeats in the same DNA molecule results in deletion of the material between the repeats. (A) The process as it occurs at the level of the DNA molecule. (B) An example showing the consequences of ectopic recombination in the human Y chromosome.

because the wildtype $w^+$ allele is no longer present in the deleted *Notch* chromosome. This **uncovering** of the recessive allele implies that the corresponding wildtype allele of *white* has also been deleted. Once a deletion has been identified, its size can be assessed genetically by determining which recessive mutations in the region are uncovered by the deletion. This method is illustrated in **FIGURE 8.15**.

## Deletion Mapping

With the banded polytene chromosomes in *Drosophila* salivary glands, it is possible to study deletions and other chromosome aberrations physically. For example, all the *Notch* deletions cause particular bands to be missing in the salivary chromosomes. Physical mapping of deletions also allows individual genes, otherwise known only from genetic studies, to be assigned to specific bands or regions in the salivary chromosomes.

Physical mapping of genes in part of the *Drosophila* X chromosome is illustrated in **FIGURE 8.16**. The banded chromosome is shown near the top, along with the numbering system used to refer to specific bands. Each chromosome is divided into numbered sections (the X chromosome comprises sections 1 through 20), and each of the sections is divided into subdivisions designated A through F. Within each lettered subdivision, the bands are numbered in order, and so, for example, 3A6 is the sixth band in subdivision A of section 3 (Figure 8.16). On the average, each band contains about 20 kb of DNA, but there is considerable variation in DNA content from band to band.

In Figure 8.16, the mutant X chromosomes labeled I through VI have deletions. The deleted part of each chromosome is shown in red. These deletions define regions along the chromosome, some of which correspond to specific bands. For example, the deleted region in both chromosome I and chromosome II that is present in all the other chromosomes consists of band 3A3. In crosses, only deletions I and II uncover the mutation *zeste (z)*, so the *z* gene must be in band 3A3, as indicated at the top. Similarly, the recessive-lethal mutation

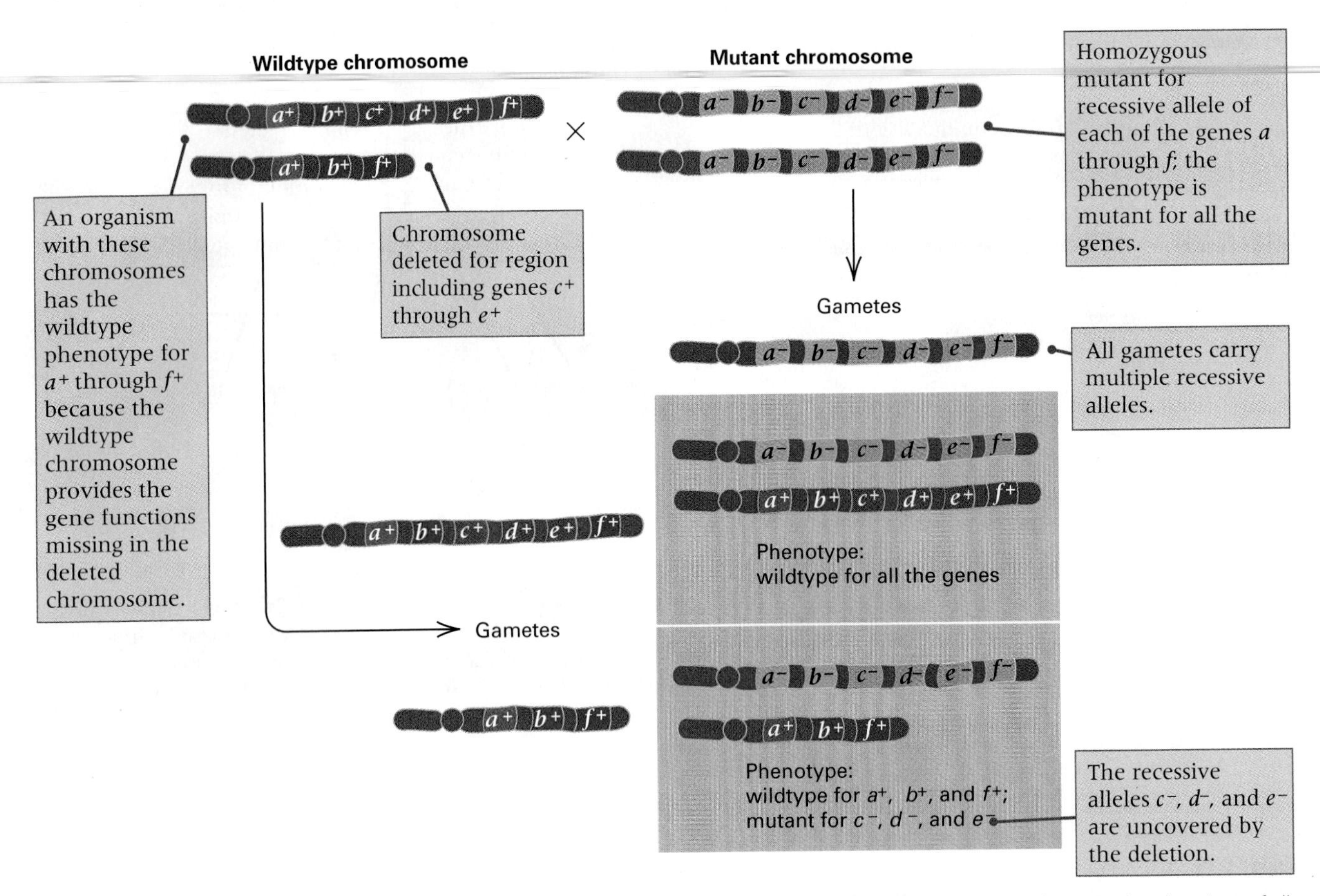


**FIGURE 8.15** Mapping of a deletion by testcrosses. The $F_1$ heterozygotes with the deletion express the recessive phenotype of all deleted genes. The expressed recessive alleles are said to be uncovered by the deletion.

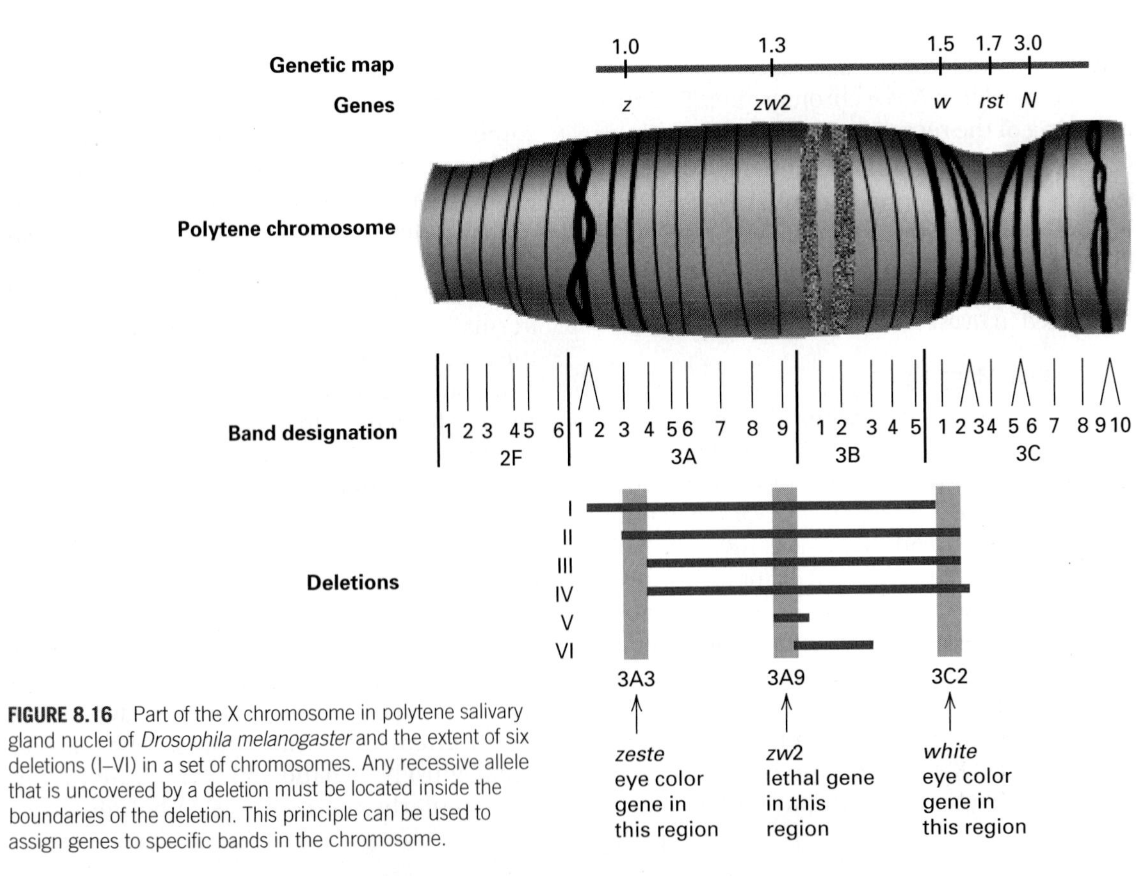


**FIGURE 8.16** Part of the X chromosome in polytene salivary gland nuclei of *Drosophila melanogaster* and the extent of six deletions (I–VI) in a set of chromosomes. Any recessive allele that is uncovered by a deletion must be located inside the boundaries of the deletion. This principle can be used to assign genes to specific bands in the chromosome.

*zw2* is uncovered by all deletions except VI; therefore, the *zw2* gene must be in band 3A9. As a final example, the *w* mutation is uncovered only by deletions II, III, and IV; thus the *w* gene must be in band 3C2. The *rst* (rough eye texture) and *N* (notched wing margin) genes are not uncovered by any of the deletions. These genes were localized by a similar analysis of overlapping deletions in regions 3C5 to 3C10.

## Duplications

Some abnormal chromosomes have a region that is present twice, which is a **duplication**. Certain duplications have phenotypic effects of their own. An example is the *Bar* duplication in *Drosophila,* which is a tandem duplication of about 1 Mb comprising bands 16A1 through 16A7 in the X chromosome. A **tandem duplication** is one in which the duplicated segment is present in the same orientation immediately adjacent to the normal region in the chromosome. In the case of *Bar,* the tandem duplication produces a dominant phenotype of bar-shaped eyes.

Tandem duplications are able to produce even more copies of the duplicated region by means of a process called **unequal crossing over**, which is actually a type of ectopic recombination. **FIGURE 8.17**, part A, illustrates the chromosomes in meiosis of an organism that is homozygous for a tandem duplication (brown region). When they undergo synapsis, these chromosomes can mispair with each other, as illustrated in Figure 8.17B. A crossover within the mispaired part of the duplication (Figure 8.17C) will thereby produce a chromatid carrying a triplication and a reciprocal product (labeled "single copy" in Figure 8.17D) that has lost the duplication. For the *Bar* region, the triplication can be recognized because it produces an even greater reduction in eye size than the duplication.

The most frequent effect of a duplication is a reduction in the probability of survival (reduced viability). In general, survival decreases with increasing size of the duplication. However, deletions are usually more harmful than duplications of comparable size.

## Unequal Crossing Over in Human Red–Green Color Blindness

Human color vision is mediated by three light-sensitive protein pigments present in the cone cells of the retina. Each of the pigments is related to *rhodopsin,* the pigment that is found in the rod cells and mediates vision in dim light. The light sensitivities of the cone pigments are toward blue, red, and green. These are the pure colors that our visual pigments detect. We perceive all other colors as mixtures of blue, red, and green. The gene for the blue-sensitive pigment resides somewhere in 7q22-qter (which means 7q22 to the terminus), and the genes for the red and green pigments are on the X chromosome at Xq28 separated by less than 5 cM (approximately 5 Mb of DNA). Because the red and green pigments arose from the duplication of a single ancestral pigment gene and are still 96 percent identical in amino acid sequence, the genes are similar enough that they can pair and undergo unequal crossing over. The process of unequal crossing over is the genetic basis of red-green color blindness.

Almost everyone is familiar with **red–green color blindness**; it is one of the most common inherited conditions in human beings.

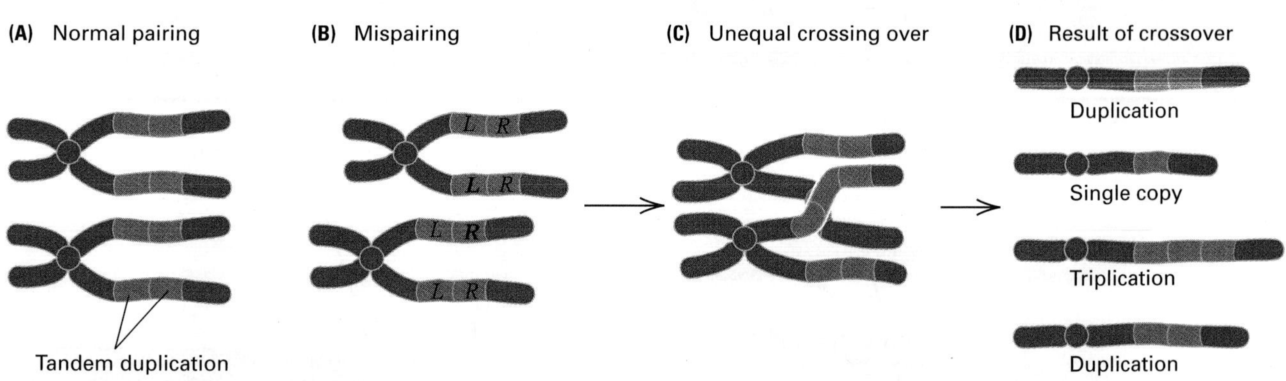


**FIGURE 8.17** An increase in the number of copies of a chromosome segment resulting from unequal crossing over of tandem duplications (brown). (A) Normal synapsis of chromosomes with a tandem duplication. (B) Mispairing. The right-hand element of the lower chromosome is paired with the left-hand element of the upper chromosome. (C) Crossing-over within the mispaired duplication, which is called unequal crossing over. (D) The outcome of unequal crossing over. One product contains a single copy of the duplicated region, another chromosome contains a triplication, and the two strands that do not participate in the crossover retain the duplication.

Approximately 5 percent of males have some form of red–green color blindness. The preponderance of affected males immediately suggests X-linked inheritance, which is confirmed by pedigree studies (Section 4.4). Affected males have normal sons and carrier daughters, and the carrier daughters have 50 percent affected sons and 50 percent carrier daughters.

Actually, there are several distinct varieties of red–green color blindness. Defects in red vision include *protanopia,* an inability to perceive red, and *protanomaly,* an impaired ability to perceive red. The comparable defects in green perception are called *deuteranopia* and *deuteranomaly,* respectively. Isolation of the red-pigment and green-pigment genes and study of their organization in people with normal and defective color vision have indicated quite clearly how the "-opias" and "-omalies" differ, and have also explained why the frequency of color blindness is relatively so high.

The organization of the red-pigment and green-pigment genes in men with normal vision is illustrated in **FIGURE 8.18**, part A. Unexpectedly, a significant proportion of normal X chromosomes contain two or three green-pigment genes. How these arise by unequal crossing over is shown in Figure 8.18B. The red-pigment and green-pigment genes pair, and a crossover takes place in the region of homology between the genes. The result is a duplication of the green-pigment gene in one chromosome and a deletion of the green-pigment gene in the other.

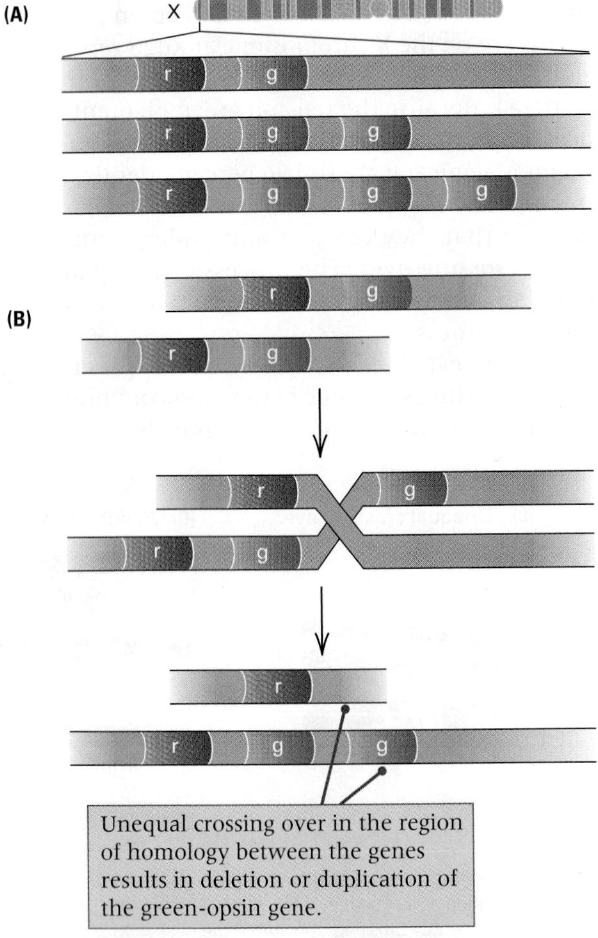
(A)
X
r
g
(B)
Unequal crossing over in the region of homology between the genes results in deletion or duplication of the green-opsin gene.

**FIGURE 8.18** (A) Organization of red-pigment and green-pigment genes in three wildtype X chromosomes. (B) Origin of multiple green-pigment genes by unequal crossing over.

The recombinational origins of the defects in color vision are illustrated in **FIGURE 8.19**. The top chromosome in Figure 8.19A is the result of deletion of the green-pigment gene shown earlier in Figure 8.18B. Males with such an X chromosome have deuteranopia, or "green-blindness." Other types of abnormal pigments result when crossing over takes place within mispaired red-pigment and green-pigment genes. Crossing over between the genes yields a **chimeric gene**, which is a composite gene: part of one joined with part of the other. The chimeric gene in Figure 8.19A joins the 5′ end of the green-pigment gene with the 3′ end of the red-pigment gene. If the crossover point is nearer the 5′ end (toward the left in the figure), then the resulting chimeric gene is mostly "red" in sequence, so the chromosome causes deuteranopia, or "green-blindness." However, if the crossover point is nearer the 3′ end (toward the right in the figure), then most of the green-pigment gene remains intact, and the chromosome causes deuteranomaly.

Chromosomes associated with defects in red vision are illustrated in Figure 8.19B. The chimeric genes are the reciprocal products of the unequal crossovers that yield defects in green vision. In this case, the chimeric gene consists of the red-pigment gene at the 5′ end and the green-pigment gene at the 3′ end. If

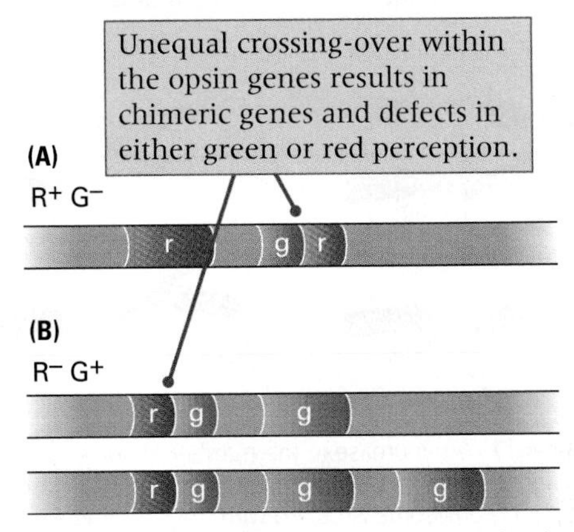
Unequal crossing-over within the opsin genes results in chimeric genes and defects in either green or red perception.
(A)
R+ G−
r
g
(B)
R− G+

**FIGURE 8.19** Genetic basis of absent or impaired red–green color vision. (A) Defects in green vision. (B) Defects in red vision.

the crossover point is nearer the 5′ end, most of the red-pigment gene is replaced with the green-pigment gene. The result is protanopia, or "red-blindness." The same is true of the other chromosome indicated in Figure 8.19B. However, if the crossover point is nearer the 3′ end, then most of the red-pigment gene remains intact, and the result is protanomaly.

## 8.4 Genetics of Chromosomal Inversions

Another important type of chromosome abnormality is an **inversion**, a chromosome in which the linear order of a group of genes is the reverse of the normal order. An inversion can be formed by a two-break event in a chromosome in which the middle segment is reversed in orientation before the breaks are healed. An inversion also can be created by ectopic recombination between DNA sequences that are inverted repeats, as illustrated for the DNA duplexes in **FIGURE 8.20**. In this diagram the differently colored gradients represent the inverted repeats, and the letters represent the order of genes in the region between the inverted repeats. Ectopic recombination between the repeats results in a chromosome with an inversion in the order of the genes between the repeats.

In an organism that is heterozygous for an inversion, one chromosome is structurally normal (wildtype), and the other carries an inversion. These chromosomes pass through mitosis without difficulty, because each chromosome duplicates and its chromatids are separated into the daughter cells without regard to the other chromosome.

Although a heterozygous inversion causes no problems in mitosis, there can be problems in meiosis. These result from homologous recombination in the region that is inverted. To understand why, recall that in prophase I of meiosis the homologous chromosomes are attracted gene for gene in the process of *synapsis*. The pairing of homologous chromosomes is shown for a heterozygous inversion in **FIGURE 8.21**. In this diagram the gradients represent the orientation of the DNA sequences along the homologous chromosomes. The region in blue is inverted in one homolog but not in the other, and so the inversion is heterozygous. In an inversion heterozygote, in order for gene-for-gene pairing to take place everywhere along the length of the chromosome, one or the other of the chromosomes must twist into a loop in the region in which the gene order is inverted. In Figure 8.21, it is the structurally normal chromosome that is shown as looped, but in other cells it may be the inverted chromosome that is looped. In either case, the loop is called an **inversion loop**.

### ■ Paracentric Inversion (Not Including the Centromere)

The inversion loop itself does not create a problem. The looping apparently takes place without difficulty and can be observed through

Repeated DNA sequences present in reversed orientation (inverted repeats).

A B . . . Y Z

Inversion

RE 8.20 Ectopic recombination between inverted repeats results ı inversion.

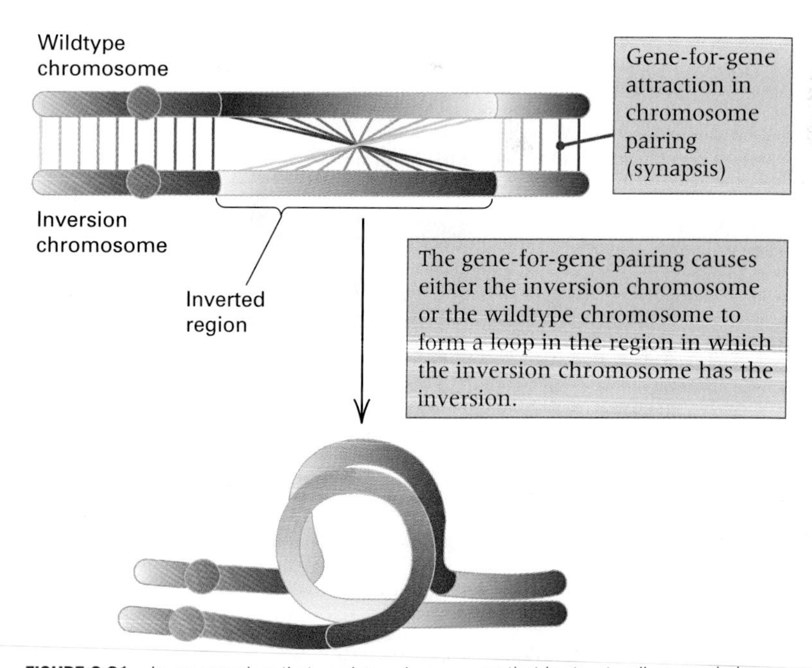


**FIGURE 8.21** In an organism that carries a chromosome that is structurally normal along with a homologous chromosome bearing an inversion, the gene-for-gene attraction between the chromosomes during synapsis causes one of the chromosomes to form into a loop in the region in which the gene order is inverted. In this example, the structurally normal chromosome forms the loop. Only two of the four chromatids are shown.

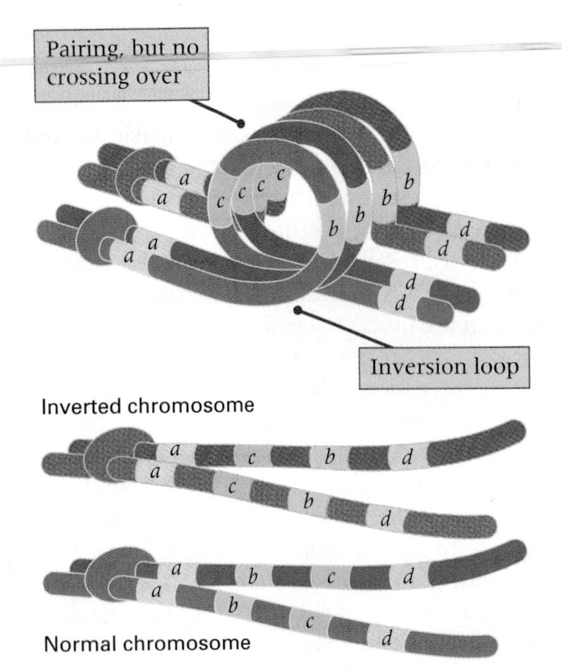


**FIGURE 8.22** In an inversion heterozygote, if there is no crossing over within the inversion loop, then the homologous chromosomes disjoin without problems. Two of the resulting gametes carry the inverted chromosome, and two carry the structurally normal chromosome.

the microscope. As long as there is no crossing over within the inversion, the homologous chromosomes can separate normally at anaphase I, as illustrated in **FIGURE 8.22**. On the other hand, when there is crossing over within the inversion loop, the chromatids involved in the crossing over become physically joined, and the result is the formation of chromosomes containing large duplications and deletions. The products of the crossing over can be deduced from **FIGURE 8.23**, part A, by tracing along the chromatids. The outer chromatids are the ones not participating in the crossover. One of these contains the inverted sequence and the other the normal sequence, as shown in Figure 8.23B. Because of the crossover, the inner chromatids, which did participate in the crossover, are connected.

The inversion shown in Figure 8.23, in which the centromere is not included in the inverted region, is known as a **paracentric inversion**, which means inverted "beside" (*para-*) the centromere. As the figure shows, the products of crossing over include a dicentric and an acentric chromosome. Neither the dicentric chromosome nor the acentric chromosome can be included in a normal gamete. The acentric chromosome is usually lost because it lacks a centromere and, in any case, has a deletion of the *a* region and a duplication of the *d* region. The dicentric chromosome is also often lost because it is held on the meiotic spindle by the chromatid bridging between the centromeres; in any case, this chromosome is deleted for the *d* region and duplicated for the *a* region. Hence, when there is a crossover in the inversion loop, the only chromatids that can be recovered in the gametes are the chromatids that did not participate in the crossover. One of these carries the inversion, and the other does not. It is for this reason that inversions prevent the recovery of crossover products. In the early years of genetics, before their identity as inversions was discovered, inversion-bearing chromosomes were known as "crossover suppressors."

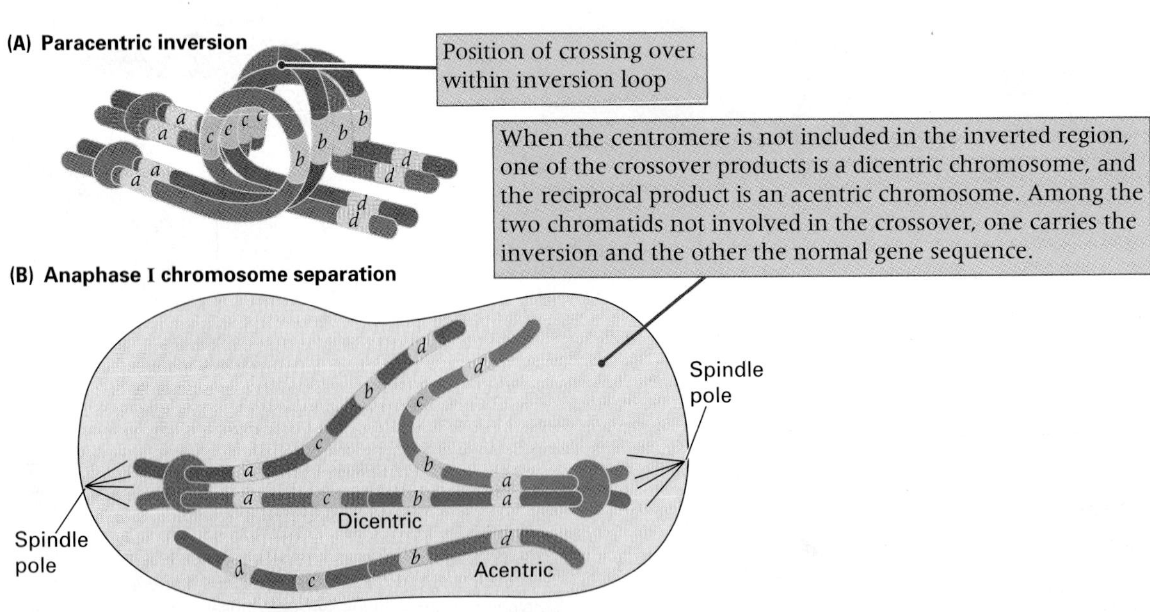


**FIGURE 8.23** (A) Synapsis between homologous chromosomes, one of which contains an inversion. There is a crossover within the inversion loop. (B) Anaphase I configuration resulting from the crossover.

### Pericentric Inversion (Including the Centromere)

When the inversion does include the centromere, it is called a **pericentric inversion**, which means "around" (*peri-*) the centromere. Chromatids with duplications and deficiencies are also created by crossing over within the inversion loop of a pericentric inversion, but in this case the crossover products are monocentric. The situation is illustrated in **FIGURE 8.24**, part A. The diagram is identical to that in Figure 8.23 except for the position of the centromere. The products of crossing over can again be deduced by tracing the chromatids. In this case, both products of the crossover are monocentric, but one chromatid carries a duplication of *a* and a deletion of *d*, and the other carries a duplication of *d* and a deletion of *a* (Figure 8.24B). Although either of these chromosomes could be included in a gamete, the duplication and deficiency usually cause inviability. Thus, as with the paracentric inversion, the products of recombination are not recovered, but for a different reason. Among the chromatids not participating in the crossover in Figure 8.24A, one carries the pericentric inversion, and the other has the normal sequence.

## 8.5 Chromosomal Translocations

A chromosomal aberration resulting from the interchange of parts between nonhomologous chromosomes is called a **translocation**. Translocations can be formed by an interchange of parts between two broken chromosomes or by ectopic recombination between copies of repeated DNA sequences present in two nonhomologous chromosomes. In **FIGURE 8.25**, organism A is homozygous for two pairs of structurally normal chromosomes. Organism B contains one structurally normal pair of chromosomes and another pair of chromosomes that have undergone an interchange of terminal parts. The organism is said to be *heterozygous* for the translocation.

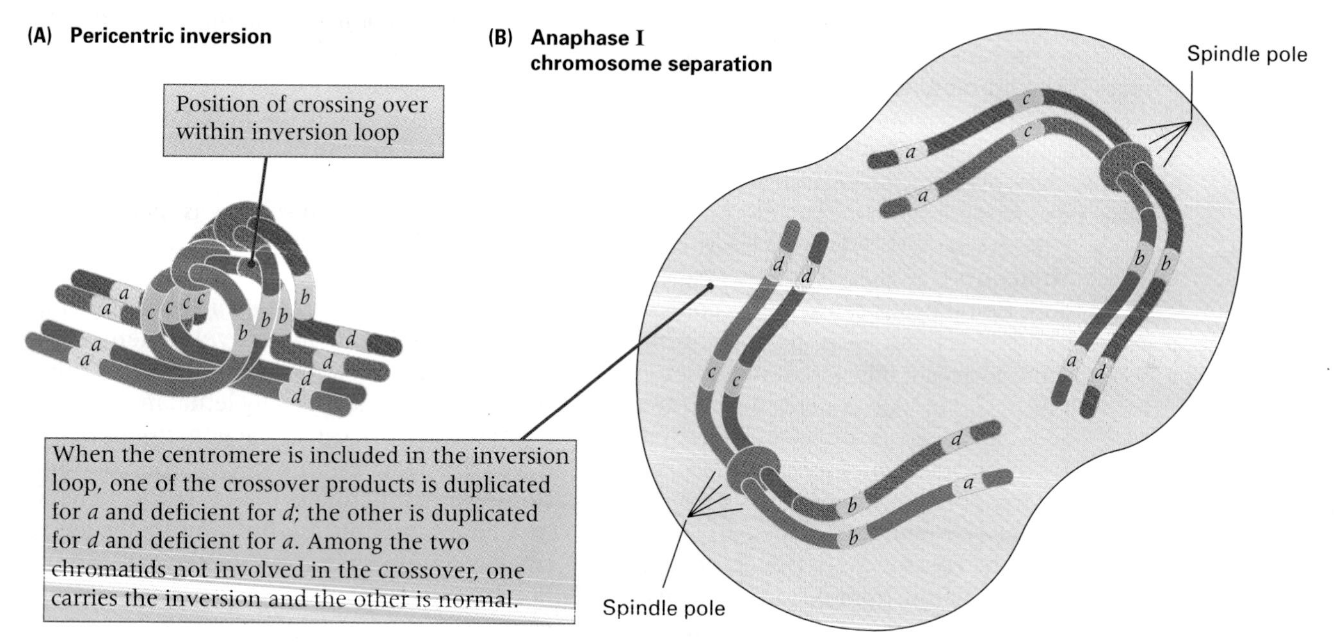
(A) Pericentric inversion
Position of crossing over within inversion loop
a
c
b
d
When the centromere is included in the inversion loop, one of the crossover products is duplicated for a and deficient for d; the other is duplicated for d and deficient for a. Among the two chromatids not involved in the crossover, one carries the inversion and the other is normal.
(B) Anaphase I chromosome separation
Spindle pole
Spindle pole

**FIGURE 8.24** (A) Synapsis between homologous chromosomes, one of which carries a pericentric inversion. A crossover within the inversion loop is shown. (B) Anaphase I configuration resulting from the crossover.

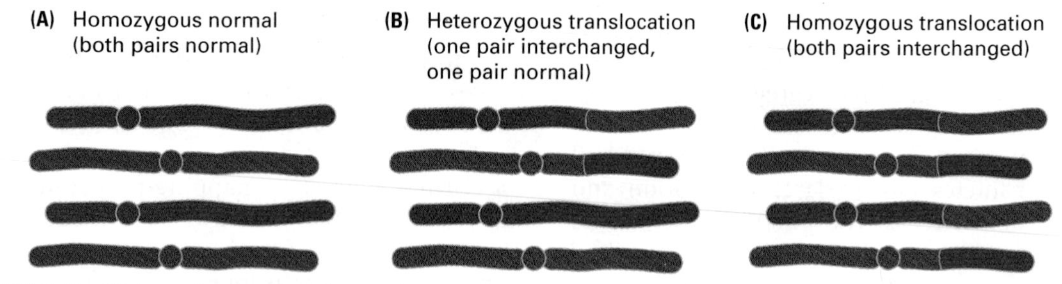
(A) Homozygous normal (both pairs normal)
(B) Heterozygous translocation (one pair interchanged, one pair normal)
(C) Homozygous translocation (both pairs interchanged)

**FIGURE 8.25** (A) Two pairs of nonhomologous chromosomes in a diploid organism. (B) Heterozygous reciprocal translocation, in which two nonhomologous chromosomes (the two at the top) have interchanged terminal segments. (C) Homozygous reciprocal translocation.

## Reciprocal Translocations

The translocation in Figure 8.25 is properly called a **reciprocal translocation** because it consists of two reciprocally interchanged parts. As indicated in Figure 8.25C, an organism can also be homozygous for the translocation if both pairs of homologous chromosomes have undergone an interchange of parts.

An organism that is heterozygous for a reciprocal translocation usually produces only about half as many offspring as normal—a condition called **semisterility**. The reason for the semisterility is production of unbalanced gametes in meiosis. When meiosis takes place in a translocation heterozygote, the normal and translocated chromosomes must undergo synapsis, as shown in **FIGURE 8.26**. Ordinarily, there would also be chiasmata between nonsister chromatids in the arms of the homologous chromosomes, but these are not shown, as though the translocation were present in an organism with no crossing over, such as a male *Drosophila*. Segregation from this configuration can take place in any of three ways. In the list that follows, the symbol 1 + 2 ⟷ 3 + 4 means that at the first meiotic anaphase, the chromosomes in Figure 8.26 labeled 1 and 2 go to one pole and those labeled 3 and 4 go to the opposite pole. The red numbers indicate the two parts of the reciprocal translocation. The three types of segregation are as follows:

- *1 + 2 ⟷ 3 + 4* This mode is called **adjacent-1** or *nondisjunctional* segregation. Homologous centromeres go to opposite poles, but each normal chromosome goes with one part of the reciprocal translocation. All gametes formed from adjacent-1 segregation have a large duplication and deficiency for the distal part of the translocated chromosomes. (The *distal* part of a chromosome is the part farthest from the centromere.) The pair of gametes originating from the 1 + 2 pole are duplicated for the distal part of the blue chromosome and deficient for the distal part of the red chromosome; the pair of gametes from the 3 + 4 pole have the reciprocal deficiency and duplication.
- *1 + 3 ⟷ 2 + 4* This mode is **adjacent-2** or *disjunctional* segregation, in which homologous centromeres go to the same pole at anaphase I. In this case, all gametes have a large duplication and deficiency of the proximal part of the translocated chromosome. (The *proximal* part of a chromosome is the part closest to the centromere.) The pair of gametes from the 1 + 3 pole have a duplication of the proximal part of the red chromosome and a deficiency of the proximal part of the blue chromosome; the pair of gametes from the 2 + 4 pole have the reciprocal deficiency and duplication.
- *1 + 4 ⟷ 2 + 3* In this type of segregation, which is called **alternate** segregation, the gametes are all balanced (euploid), which means that none has a duplication or deficiency. The gametes from the 1 + 4 pole have both parts of the reciprocal translocation; those from the 2 + 3 pole have both normal chromosomes.

The semisterility of genotypes that are heterozygous for a reciprocal translocation results from lethality due to the duplication and deficiency gametes produced by adjacent-1 and adjacent-2 segregation. Although the expected frequencies of adjacent-1 : adjacent-2 : alternate segregation are $\frac{1}{4}:\frac{1}{4}:\frac{1}{2}$, in practice the frequencies with which these types of segregation take place are strongly influenced by the position of the translocation breakpoints, by the number and distribution of chiasmata in the interstitial region between the centromere and each breakpoint, and by whether the quadrivalent tends to open out into a ring-shaped structure on the metaphase plate.

Translocation semisterility is manifested in different stages in the life histories of plants and animals. Plants have an elaborate gametophyte phase of the life cycle, a haploid phase in which complex metabolic and developmental processes are necessary. In plants, large duplications and deficiencies are usually lethal in the gametophyte stage. Because the gametophyte produces the gametes, in higher plants the semisterility is manifested as pollen or seed lethality. In animals, by contrast, only minimal gene activity is necessary in the gametes, which function in spite of very large duplications and deficiencies. In animals, therefore, the semisterility is usually manifested as zygotic lethality.

## Genetic Mapping of a Translocation Breakpoint

In species in which translocation heterozygotes exhibit semisterility, the semisterility can be used as the phenotype to map the breakpoint of the translocation just as though it were a normal gene. The mapping procedure can be made clear by means of an example. Translocation *TB-10L1* is a translocation with one breakpoint in the long arm of chromosome 10 in maize, and it results in semisterility when heterozygous. A cross is made between a trans-

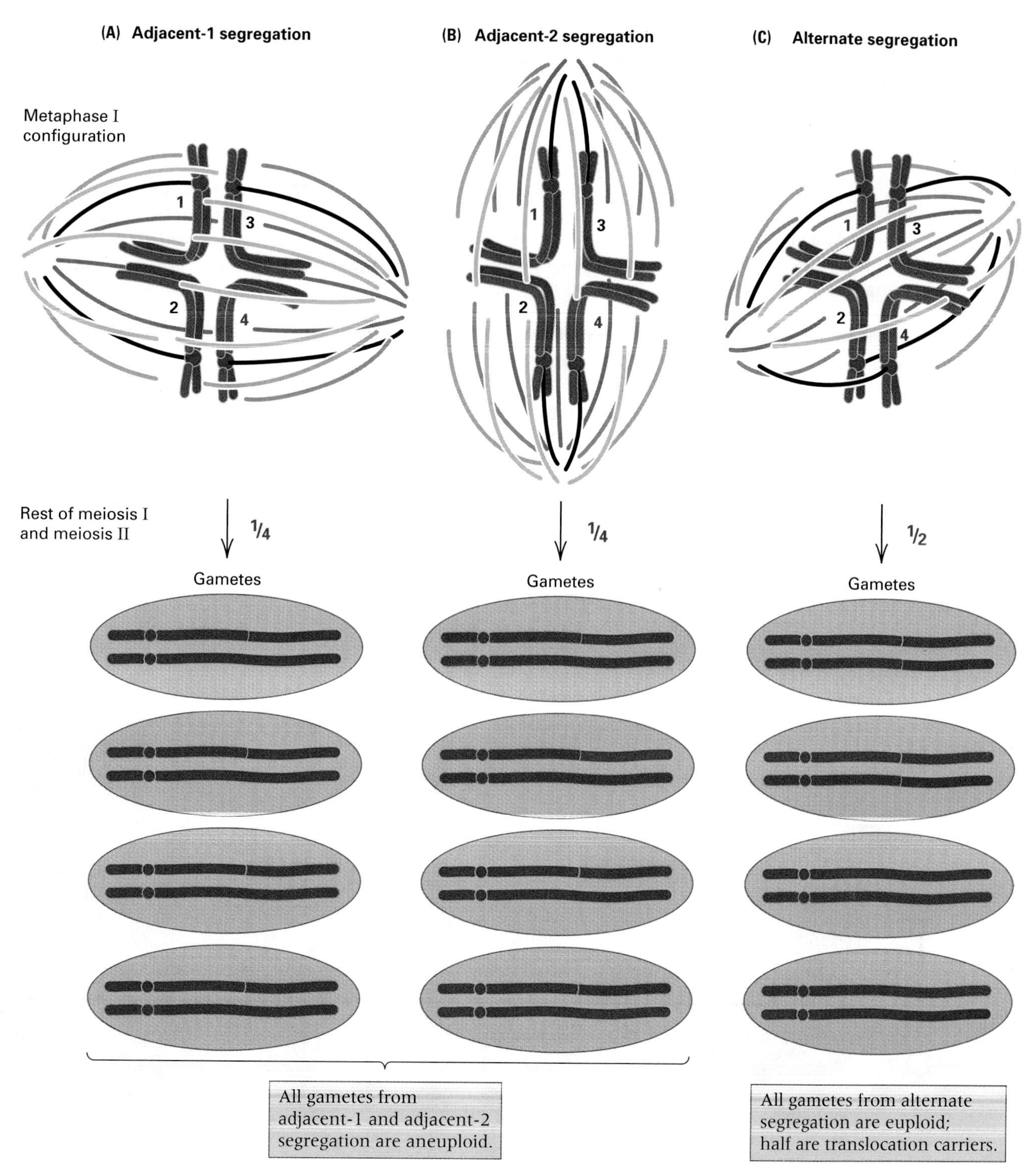


**FIGURE 8.26** Quadrivalents formed in the synapsis of a heterozygous reciprocal translocation and their expected frequencies. The translocated chromosomes are numbered in red, their normal homologs in black. No chiasmata are shown. (A) Adjacent-1 segregation: homologous centromeres separate at anaphase I; all of the resulting gametes have a duplication of one terminal segment and a deficiency of the other. (B) Adjacent-2 segregation: homologous centromeres go together at anaphase I; all of the resulting gametes have a duplication of one proximal segment and a deficiency of the other. (C) Alternate segregation: half of the gametes receive both parts of the reciprocal translocation and the other half receive both normal chromosomes.

location heterozygote and a genotype homozygous for both *zn1* (*zebra necrotic 1*) and *tp2* (*teopod 2*), and semisterile progeny are testcrossed with *zn1 tp2* homozygotes. (The phenotypes of *zn1* and *tp2* are shown in Figure 5.8.) The parental genotype is therefore *Zn1 Tp2 TB-10L1*/*zn1 tp2* +, where the + denotes the position of the translocation breakpoint in the homologous chromosome. The progeny phenotypes are as follows:

| | | | |
|---|---|---|---|
| nonzebrastripe | nonteopod | semisterile | 392 |
| nonzebrastripe | nonteopod | fertile | 3 |
| nonzebrastripe | teopod | semisterile | 42 |
| nonzebrastripe | teopod | fertile | 73 |
| zebrastripe | nonteopod | semisterile | 83 |
| zebrastripe | nonteopod | fertile | 34 |
| zebrastripe | teopod | semisterile | 1 |
| zebrastripe | teopod | fertile | 372 |

These data are analyzed exactly like the three-point crosses in Section 5.3. The double recombinants are present in the rarest classes of progeny and differ from the parental genotypes in exchanging the gene that is in the middle of the three. The double recombinants are *Zn1 Tp2* + and *zn1 tp2 TB-10L1*, which means that the translocation breakpoint lies between *zn1* and *tp2*. Hence the map distance between *zn1* and the breakpoint is

$$(3 + 73 + 83 + 1)/1000 = 16 \text{ cM}$$

and that between the translocation breakpoint and *tp2* is

$$(3 + 42 + 34 + 1)/1000 = 8 \text{ cM}$$

The genetic map of the region containing the *TB-10L1* translocation breakpoint is

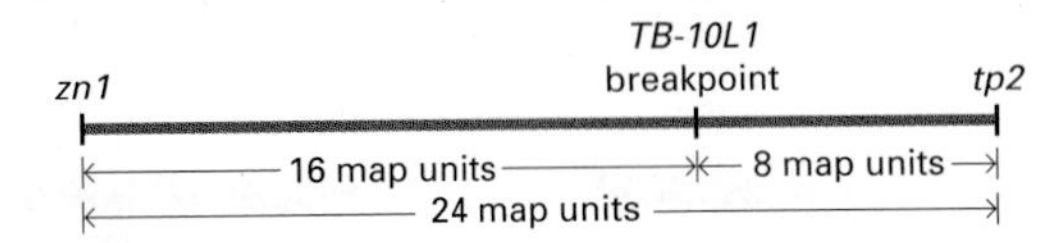


## Robertsonian Translocations

A special type of *non*reciprocal translocation is a **Robertsonian translocation**, in which two nonhomologous acrocentric chromosomes undergo fusion of their short arms yielding a chromosome with a single functional centromere (**FIGURE 8.27**). Robertsonian translocations are important in human genetics, especially as a risk factor to be considered in Down syndrome. When chromosome 21 is one of the ac-

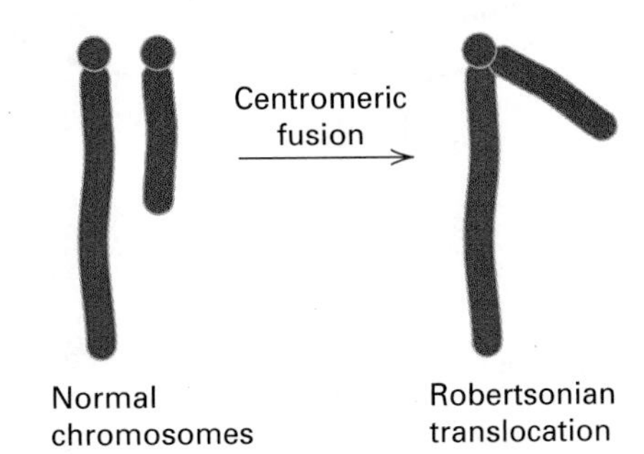


**FIGURE 8.27** Formation of a Robertsonian translocation by fusion of two acrocentric chromosomes in the centromeric region.

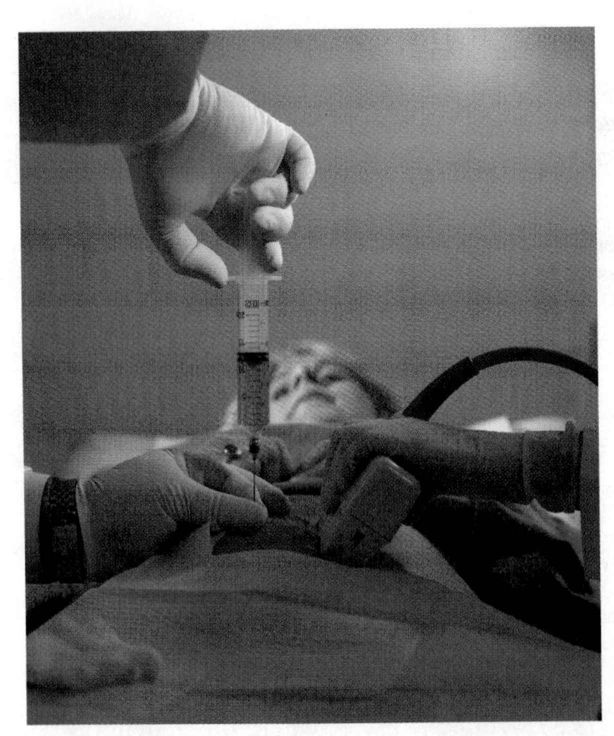

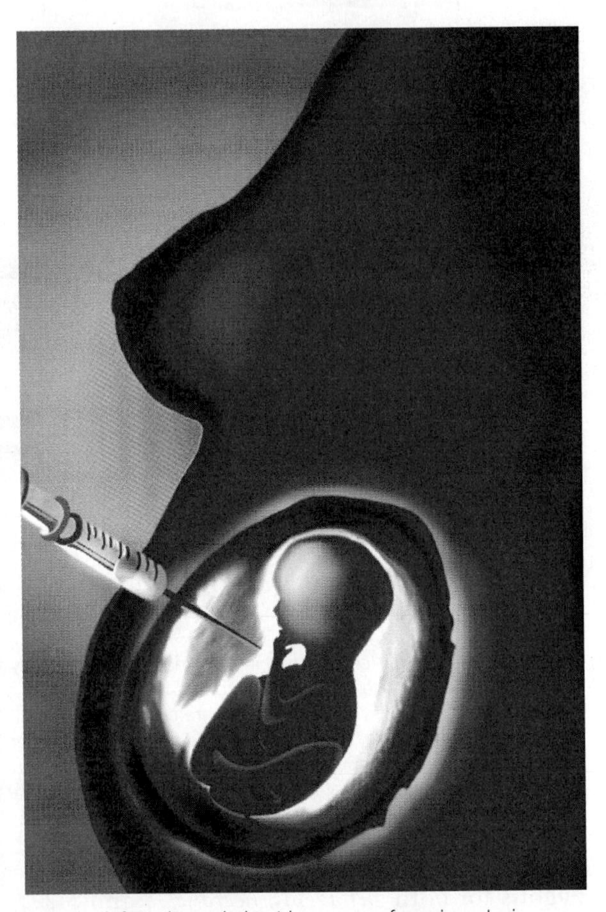

Prenatal diagnosis of chromosomal abnormalities and some other genetic defects is carried out by means of amniocentesis performed at 15–18 weeks of pregnancy. Using ultrasound images as a guide, a needle is inserted through the membranes surrounding the fetus, and a small sample of fluid containing fetal cells is removed. [Left: © Yoav Levy/Phototake/Alamy Images. Right: © VEM/Photo Researchers, Inc.]

rocentrics in a Robertsonian translocation, the rearrangement leads to a familial type of Down syndrome in which the risk of recurrence is very high. Approximately 3 percent of children with Down syndrome are found to have one parent with such a translocation.

## Translocations and Trisomy 21

Heterozygous carriers of a Robertsonian translocation are phenotypically normal, but a high risk of Down syndrome results from aberrant segregation in meiosis. The possible modes of segregation are shown in **FIGURE 8.28**. A

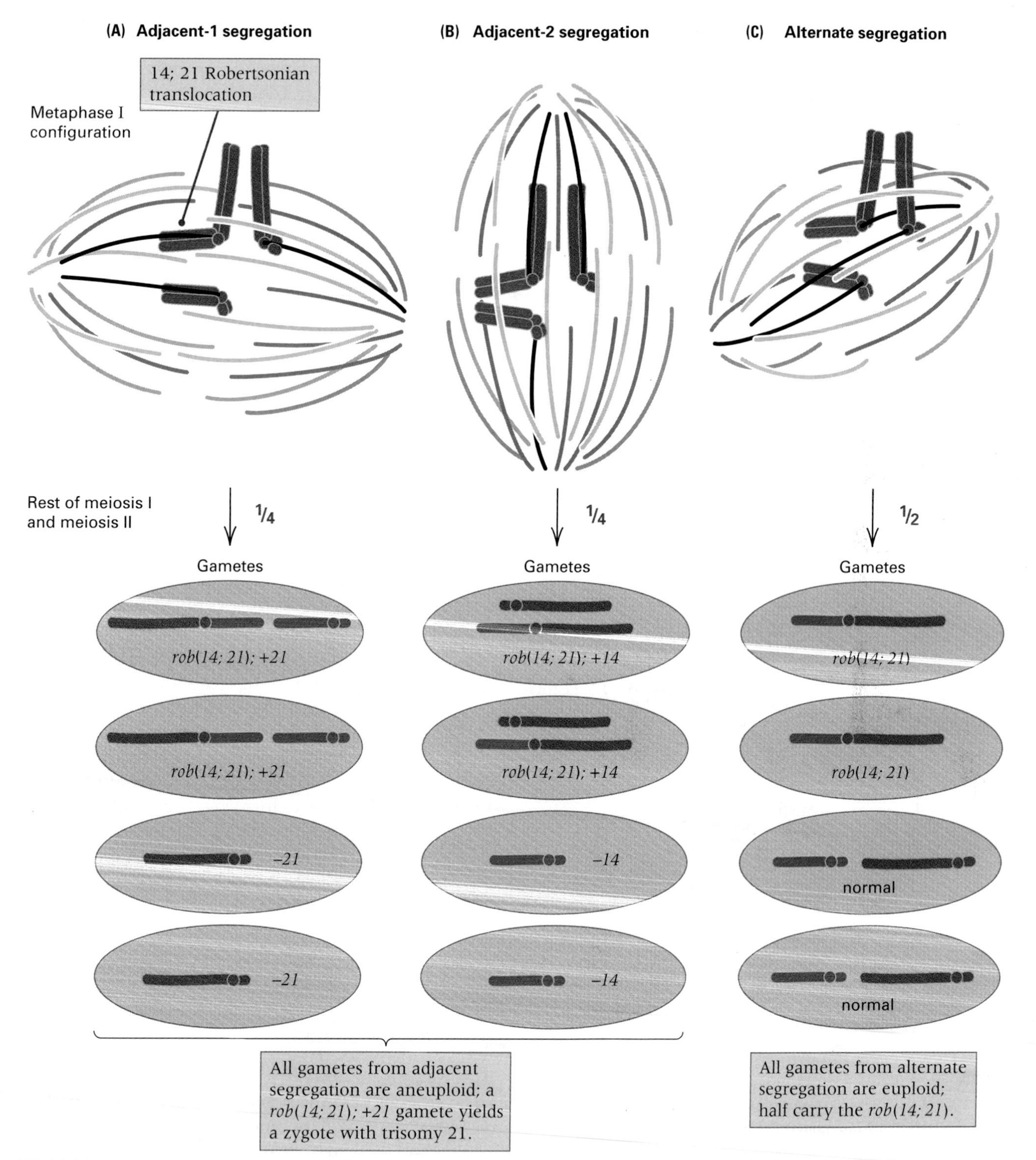


**FIGURE 8.28** Segregation of Robertsonian translocation between chromosomes 14 and 21 and the expected frequencies. (A) Adjacent-1 segregation: in which the gametes formed from the pole at the top have, effectively, an extra copy of chromosome 21. (B) Adjacent-2 segregation: The gametes are either duplicated or deficient for chromosome 14. (C) Alternate segregation: Half of the gametes give rise to phenotypically normal children who are carriers of the Robertsonian translocation.

Robertsonian translocation can join chromosome 21 with any other acrocentric chromosome, but for concreteness we use chromosome 14. The symbol *rob* refers to the translocation, and the + sign or − sign preceding a chromosome number designates an extra copy, or a missing copy, of the entire chromosome.

Among the several possible types of gametes that can arise, one contains a normal chromosome 21 along with the 14;21 Robertsonian translocation (Figure 8.28A). If this aberrant gamete is used in fertilization, then the fetus will contain two copies of the normal chromosome 21 plus the 14;21 translocation. In effect, the fetus contains three copies of chromosome 21 and hence has Down syndrome. The other abnormal gametes that result from adjacent-1 or adjacent-2 segregation either are missing chromosome 21 or chromosome 14 or contain effectively two copies of chromosome 14 (Figure 8.28A and B). If these gametes participate in fertilization, the result is monosomy 21, monosomy 14, or trisomy 14, respectively. The monosomic embryos undergo very early spontaneous abortion; the trisomy-14 fetus undergoes spontaneous abortion later in pregnancy. Hence, families with a high risk of translocation Down syndrome also have a high risk of spontaneous abortion due to other chromosome abnormalities. Alternate segregation of a Robertsonian translocation yields gametes that carry either the translocation or both normal chromosomes (Figure 8.28C). Because these gametes derive from reciprocal products of meiosis, the nonaffected children have a 50 percent risk of carrying the translocation.

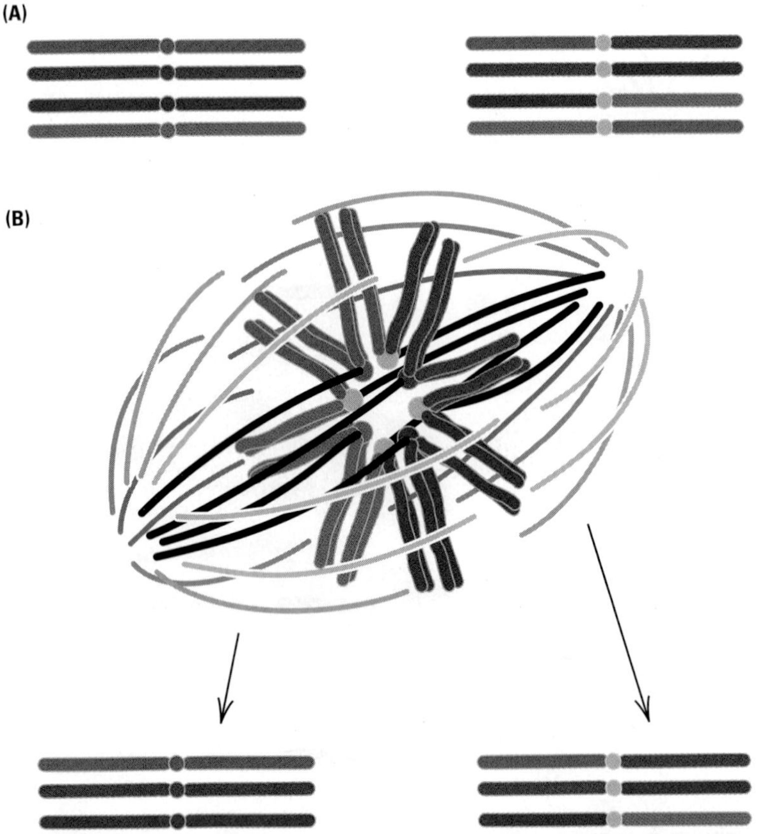

(A)

(B)

**FIGURE 8.29** (A) Complex translocation heterozygote of the type found in some species of *Oenothera*. The chromosomes at the top left are not rearranged. Those at the top right are connected by a chain of translocations, each chromosome having exchanged an arm with the next chromosome in line. (B) At metaphase I in meiosis, the pairing configuration of the translocation heterozygote is a ring of chromosomes in which each arm is paired with its proper partner; note that each chromosome consists of two chromatids. Alternate segregation from the metaphase ring yields, after the second meiotic division, two types of gametes: those containing all normal chromosomes and those containing all translocated chromosomes.

### ■ Translocation Complexes in *Oenothera*

Certain groups of plants have reciprocal translocations present in natural populations without the semisterility usually expected. Among these are the evening primroses in the genus *Oenothera,* of which there are about 100 wild species native to North America and many cultivated varieties. *Oenothera* escapes the translocation semisterility because segregation is always in the alternate mode. In some species of *Oenothera,* entire sets of chromosomes are interconnected through a chain of reciprocal translocations of the type illustrated in **FIGURE 8.29**, part A. The chromosomes at the top left are all normal; those at the top right are translocated in such a way that each chromosome has exchanged an arm with the next in line. When the chromosomes in the complex translocation heterozygote undergo synapsis, the result is a ring of chromosomes (Figure 8.29B). The astonishing feature of meiosis in such *Oenothera* heterozygotes is that the segregation is exclusively of the alternate type, so the only gametes formed contain either the entire set of normal chromosomes or the entire set of translocated chromosomes. This is not the end of the surprises. In *Oenothera,* one of the gametic types is inviable in the pollen and the other is inviable in the ovule, so fertilization restores the karyotype of the complex translocation heterozygote!

## 8.6 Genomic Position Effects on Gene Expression

Genes near the breakpoints of chromosomal rearrangement become repositioned in the genome and flanked by new neighboring genes. In many cases, the repositioning of a gene affects its level of expression or, in some cases, its ability to function; this is called **position effect**. Such effects have been studied extensively in *Drosophila* and also in yeast. In *Drosophila*, the most common type of position effect results in a mottled (mosaic) phenotype that is observed as interspersed patches of wildtype cells, in which the wildtype gene is expressed, and mutant cells, in which the wildtype gene is inactivated. The phenotype is said to show *variegation*, and the phenomenon is called **position-effect variegation (PEV)**. In the older literature, PEV is often referred to as variegated or V-type position effect.

PEV usually results from a chromosome aberration that moves a wildtype gene from a position in euchromatin to a new position in or near heterochromatin (**FIGURE 8.30**). (Euchromatin and heterochromatin are discussed in Chapter 7.) **FIGURE 8.31** illustrates some of the patterns of wildtype (red) and mutant (white) facets that are observed in male flies that carry a rearranged X chromosome in which an inversion repositions the wildtype $w^+$ allele into heterochromatin. The same types of patterns are found in females heterozygous for the rearranged X chromosome and an X chromosome carrying the *w* allele. The patterns of $w^+$ expression coincide with the clonal lineages in the eye; that is, all of the red cells in a particular patch derive from a single ancestral cell in the embryo in which the $w^+$ allele was activated. In contrast with the pattern shown in Figure 8.31, other chromosome rearrangements with PEV yield a salt-and-pepper pattern of mosaicism, which consists of numerous very small patches of wildtype tissue, and still others yield a combination of many small and a few large patches. These patterns imply that gene activation can be very late in development as well as very early.

Although the mechanism of PEV is not understood in detail, it is thought to result from the unusual chromatin structure of heterochromatin interfering with gene activation. The determination of gene expression or nonexpression is thought to take place when the boundary between condensed heterochromatin and euchromatin is established. Where heterochromatin is juxtaposed with euchromatin, the chromatin condensation characteristic of heterochromatin may spread into the adjacent euchromatin, inactivating euchromatic genes in the cell and all of its descendants. A similar inactivation phenomenon takes place in cells of female mammals when euchromatic genes translocated to the X

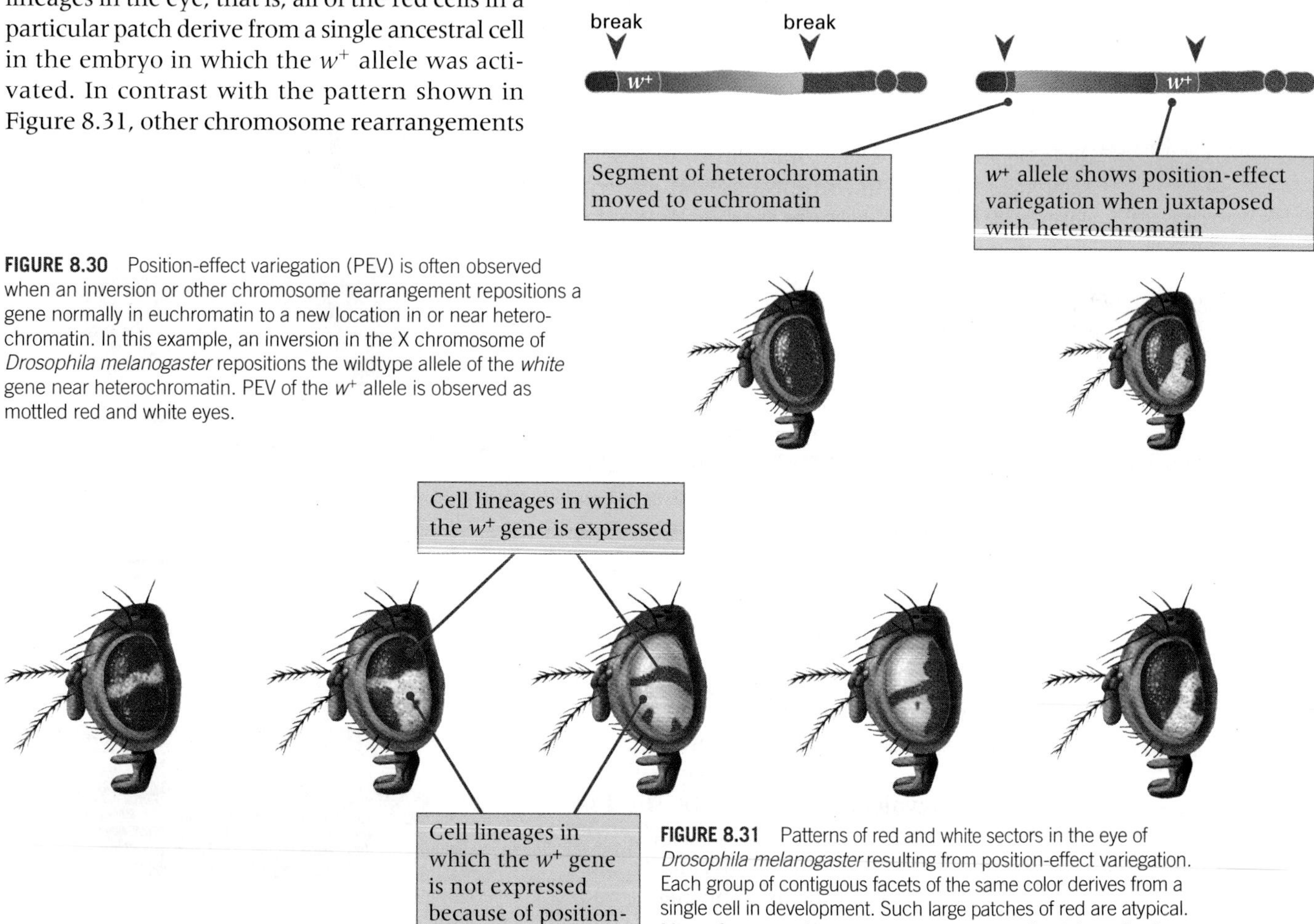

(A) Normal X chromosome
break
break
w+
(B) Inverted X chromosome
w+
Segment of heterochromatin moved to euchromatin
w+ allele shows position-effect variegation when juxtaposed with heterochromatin
Cell lineages in which the w+ gene is expressed
Cell lineages in which the w+ gene is not expressed because of position-effect variegation

**FIGURE 8.30** Position-effect variegation (PEV) is often observed when an inversion or other chromosome rearrangement repositions a gene normally in euchromatin to a new location in or near heterochromatin. In this example, an inversion in the X chromosome of *Drosophila melanogaster* repositions the wildtype allele of the *white* gene near heterochromatin. PEV of the $w^+$ allele is observed as mottled red and white eyes.

**FIGURE 8.31** Patterns of red and white sectors in the eye of *Drosophila melanogaster* resulting from position-effect variegation. Each group of contiguous facets of the same color derives from a single cell in development. Such large patches of red are atypical. More often, one observes numerous very small patches of red or a mixture of many small and a few large patches.

FIGURE 8.32 Chromosome numbers in diploid and polyploid species of *Chrysanthemum*. Each set of homologous chromosomes is depicted in a different color.

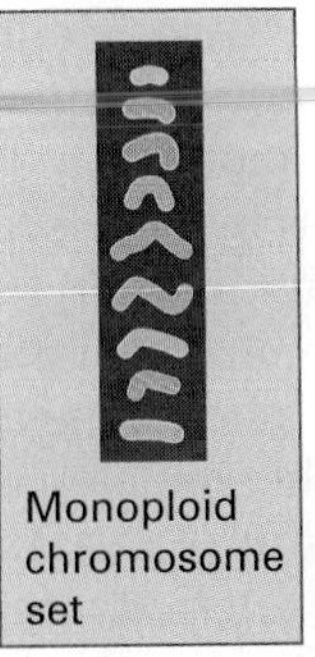


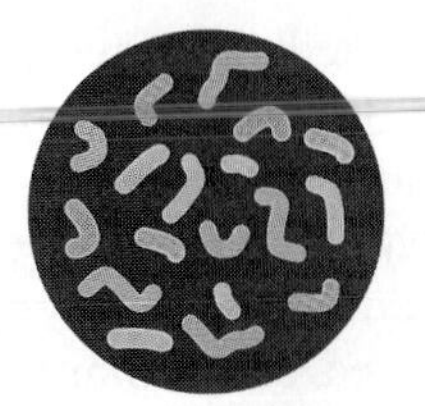


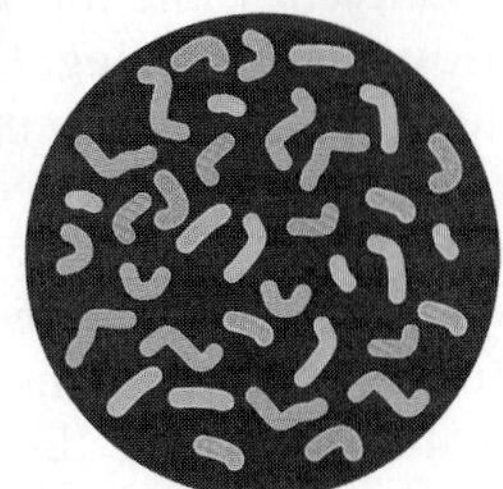


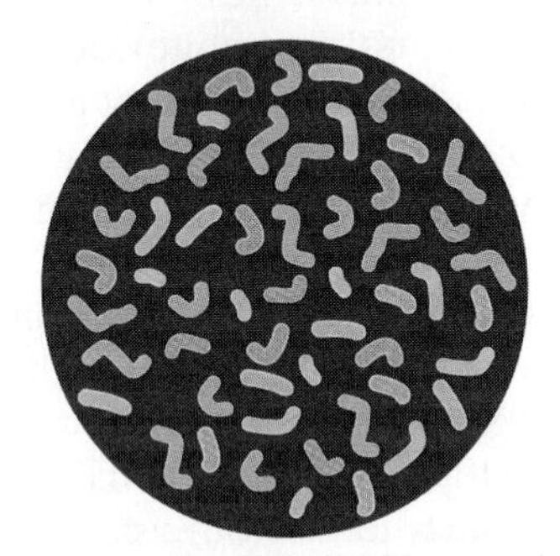


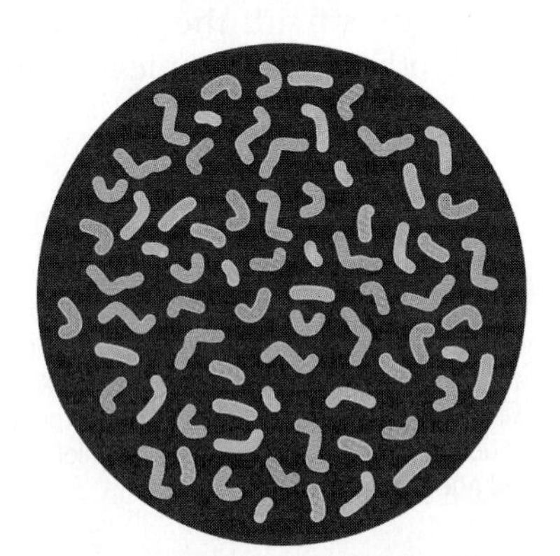


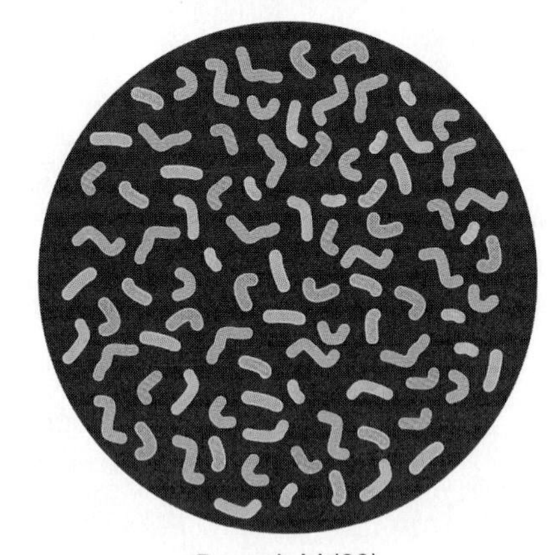


chromosome become heterochromatic and inactive. The length of the euchromatic region that is inactivated ranges from 1 to 50 bands in the *Drosophila* polytene chromosomes, depending on the particular chromosome abnormality. At the molecular level, this range is approximately 20–1000 kb. The term *spreading*, with its implication of smooth continuity, should not be taken literally, however, because in both *Drosophila* and mammals, chromosomal regions affected by PEV are known to contain active genes interspersed with inactive ones. It is likely that heterochromatic regions associate to form a relatively compact "compartment" in the nucleus, and the presence of heterochromatin near a euchromatic gene may silence the gene simply by attracting it into the heterochromatic compartment.

## 8.7 Polyploidy in Plant Evolution

The genus *Chrysanthemum* illustrates **polyploidy**, an important phenomenon, found often in higher plants, in which a species has a genome composed of multiple complete sets of chromosomes. One *Chrysanthemum* species, a diploid species, has 18 chromosomes. A closely related species has 36 chromosomes. However, comparison of chromosome morphology indicates that the 36-chromosome species has two complete sets of the chromosomes found in the 18-chromosome species (FIGURE 8.32). The basic chromosome set in the group, from which all the other genomes are formed, is called the **monoploid** chromosome set. In *Chrysanthemum*, the monoploid chromosome number is 9. The diploid species has two complete copies of the monoploid set, or 18 chromosomes altogether. The 36-chromosome species has four copies of the monoploid set ($4 \times 9 = 36$) and is a **tetraploid**. Other species of *Chrysanthemum* have 54 chromosomes ($6 \times 9$, constituting the *hexaploid*), 72 chromosomes ($8 \times 9$, constituting the *octoploid*), and 90 chromosomes ($10 \times 9$, constituting the *decaploid*).

In meiosis, the chromosomes of all *Chrysanthemum* species synapse normally in pairs to form bivalents (Section 4.3). The 18-chromosome species forms 9 bivalents, the 36-chromosome species forms 18 bivalents, the 54-chromosome species forms 27 bivalents, and so forth. Gametes receive one chromosome from each bivalent, so the number of chromosomes in the gametes of any species is exactly half the number of chromosomes in its somatic cells. The chromosomes present in the gametes of a species constitute the **haploid** set of chromosomes. In the species of *Chrysanthemum* with 90 chromosomes, for example, the haploid chromosome number is 45; in meiosis, 45 bivalents are formed, so each gamete contains 45 chromosomes. When two such gametes come together in fertilization, the complete set of 90 chromosomes in the species is restored. Thus, the gametes of a polyploid organism are not always monoploid, as they are in a diploid organism; for example, a tetraploid organism has diploid gametes.

The distinction between the terms *monoploid* and *haploid* is subtle:

- The *monoploid* chromosome set is the basic set of chromosomes that is multiplied in a polyploid series of species, such as *Chrysanthemum.*
- The *haploid* chromosome set is the set of chromosomes present in a gamete, irrespective of the chromosome number in the species.

The potential confusion arises because of diploid organisms, in which the monoploid chromosome set and the haploid chromosome set are the same. Considering the tetraploid helps to clarify the difference; it contains four monoploid chromosome sets, and the haploid gametes are diploid.

Polyploidy is widespread in certain plant groups. Among angiosperms (flowering plants), 30–80 percent of existing species are thought to have originated as some form of polyploid. In some groups the frequency is even higher; for example, in pteridophytes (a group that includes ferns, horsetails, and clubmosses), the frequency of polyploid species may be as high as 95 percent. The fern *Ophioglossum pycnostichum* is an 84-ploid with a chromosome number of 1260, a record among plants.

Valuable agricultural crops that are polyploid include wheat, corn, cotton, potatoes, bananas, coffee, sugar cane, peanuts, and apples. Polyploidy often leads to an increase in the size of individual cells, and polyploid plants are often larger and more vigorous than their diploid ancestors; however, there are many exceptions to these generalizations. Polyploidy is rare in vertebrate animals, but it is found in a few groups of invertebrates. One reason why polyploidy is rare in animals is the difficulty in regular segregation of the sex chromosomes. For example, a tetraploid animal with XXXX females and XXYY males would produce XX eggs and XY sperm (if all chromosomes paired to form bivalents), so the progeny would be exclusively XXXY and thus unlike either of the parents.

Polyploid plants found in nature nearly always have an even number of sets of chromosomes, because organisms that have an odd number have low fertility. Organisms with three monoploid sets of chromosomes are known as **triploids**. As far as growth is concerned, a triploid is quite normal because the triploid condition does not interfere with mitosis; in mitosis in triploids (or any other type of polyploid), each chromosome replicates and divides just as in a diploid. However, because each chromosome has more than one pairing partner, chromosome segregation is severely upset in meiosis, and most gametes are defective (Figure 8.13). Unless the organism can perpetuate itself by means of asexual reproduction, it will eventually become extinct. The infertility of triploids is sometimes of commercial benefit. For example, the seeds of "seedless" watermelons are small and edible because the plant is triploid and most of the seeds fail to develop to full size. In Florida and in certain other states, weed control in waterways is aided by the release of weed-eating fish (the grass carp), which do not become overpopulated because the released fish are sterile triploids.

### Sexual Versus Asexual Polyploidization

Polyploid organisms can be produced in two principal ways, which are illustrated in **FIGURE 8.33** for the example of tetraploidy. In the mechanism known as **sexual polyploidization**, the increase in chromosome number takes place in *meiosis* through the formation of *unreduced gametes* that have double the normal complement of chromosomes. Unreduced gametes are formed in many species at frequencies of 1–40 percent, and the frequency can be under genetic control. For example, in the potato, a single recessive mutation that acts during pollen formation causes the first-division and second-division meiotic spindles to be oriented in the same direction (rather than being at right angles to each other as in nonmutant cells), with the result that a pollen nucleus forms around each of the two adjacent groups of telophase II chromosomes, yielding unreduced gametes. Also in the potato, a different recessive mutation acts to eliminate the second meiotic division during the formation of female gametes, again resulting in

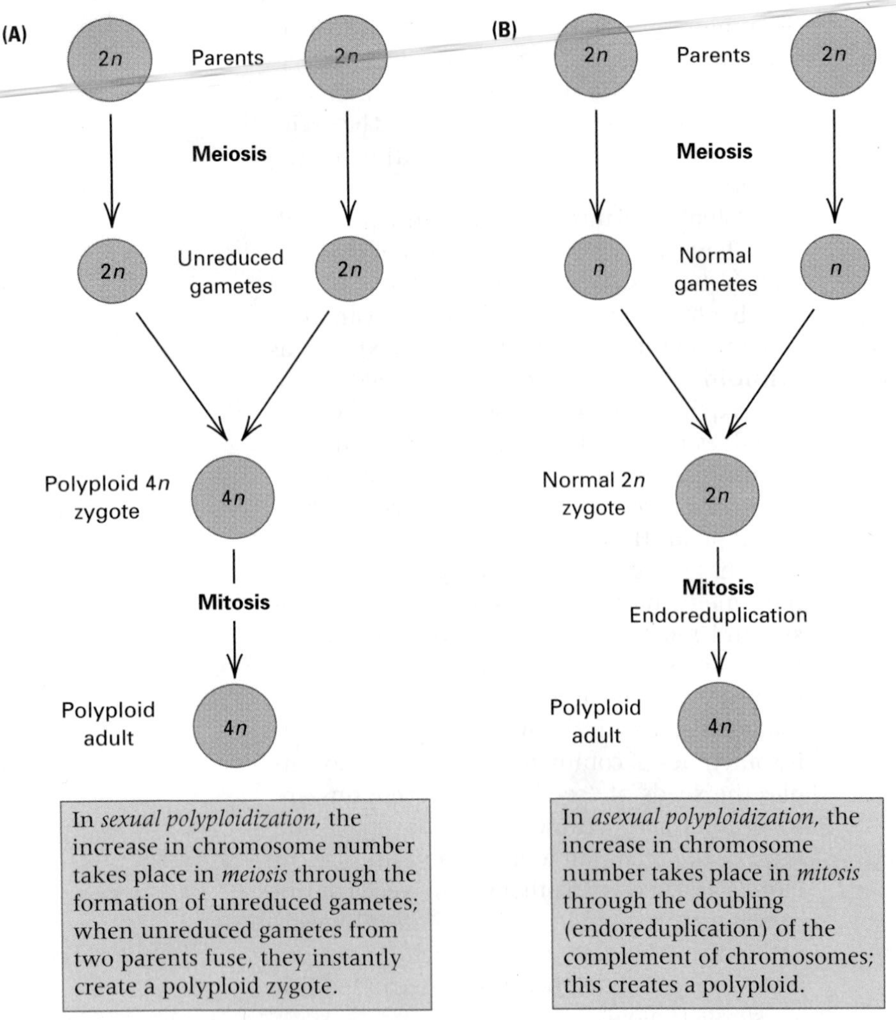
(A)
2n
Parents
2n
Meiosis
2n
Unreduced gametes
2n
Polyploid 4n zygote
4n
Mitosis
Polyploid adult
4n
In sexual polyploidization, the increase in chromosome number takes place in meiosis through the formation of unreduced gametes; when unreduced gametes from two parents fuse, they instantly create a polyploid zygote.
(B)
2n
Parents
2n
Meiosis
n
Normal gametes
n
Normal 2n zygote
2n
Mitosis
Endoreduplication
Polyploid adult
4n
In asexual polyploidization, the increase in chromosome number takes place in mitosis through the doubling (endoreduplication) of the complement of chromosomes; this creates a polyploid.

**FIGURE 8.33** Formation of a tetraploid organism by (A) sexual polyploidization and (B) asexual polyploidization. The symbol $n$ stands for the monoploid chromosome number.

unreduced gametes. Figure 8.33A shows two unreduced $2n$ gametes yielding a $4n$ tetraploid, but there are many other possibilities. For example, union of an unreduced $2n$ gamete with a normal $n$ gamete yields a $3n$ triploid.

The other principal mechanism of polyploid formation is **asexual polyploidization** (Figure 8.33B), in which the doubling of the chromosome number takes place in *mitosis*. Chromosome doubling through an abortive mitotic division is called **endoreduplication**. In a plant species that can undergo self-fertilization, endoreduplication creates a new, genetically stable species, because if the chromosomes in the tetraploid can pair two-by-two in meiosis, they can segregate regularly and yield gametes with a full complement of chromosomes. Self-fertilization of such a tetraploid restores the chromosome number, so the tetraploid condition can be perpetuated.

The genetics of tetraploid species, and that of other polyploids, is more complex than that of diploid species because the organism carries more than two alleles of any gene. With two alleles in a diploid, only three genotypes are possible: *AA, Aa,* and *aa*. In a tetraploid, by contrast, five genotypes are possible: *AAAA, AAAa, AAaa, Aaaa,* and *aaaa*. Among these genotypes, the middle three represent different types of tetraploid heterozygotes.

An octoploid species (eight sets of chromosomes) can be generated by sexual or asexual polyploidization of a tetraploid. Again, if only bivalents form in meiosis, then an octoploid organism can be perpetuated sexually by self-fertilization or through crosses with other octoploids. Furthermore, cross-fertilization between an octoploid and a tetraploid results in a hexaploid (6 sets of chromosomes). Repeated episodes of polyploidization and cross-fertilization may ultimately produce an entire polyploid series of closely related organisms that differ in chromosome number, as exemplified in *Chrysanthemum*.

## Autopolyploids and Allopolyploids

*Chrysanthemum* represents a type of polyploidy, known as **autopolyploidy**, in which all chromosomes in the polyploid species derive from a single diploid ancestral species. In many cases of polyploidy, the polyploid species have complete sets of chromosomes from two or more *different* ancestral species. Such polyploids are known as **allopolyploids**. They derive from occasional hybridization between different diploid species when pollen from one species germinates on the stigma of another species and sexually fertilizes the ovule. The pollen may be carried to the wrong flower by wind, insects, or other pollinators. **FIGURE 8.34** illustrates hybridization between species A and B in which polyploidization leads to the formation of an allopolyploid (in this case, an *allotetraploid*), which carries a complete diploid genome from each of its two ancestral species. The formation of allopolyploids through hybridization is an extremely important process in plant evolution and plant breeding. At least half of all naturally occurring polyploids are allopolyploids. Wheat provides an excellent example of allopolyploidy. Cultivated bread wheat is a hexaploid with 42 chromosomes constituting a complete diploid genome of 14 chromosomes from each of three ancestral species. The 42-chromosome allopolyploid is thought to have originated by the series of hybridizations outlined in **FIGURE 8.35**.

An example of chromosome painting to reveal the ancestral origin of the chromosome sets

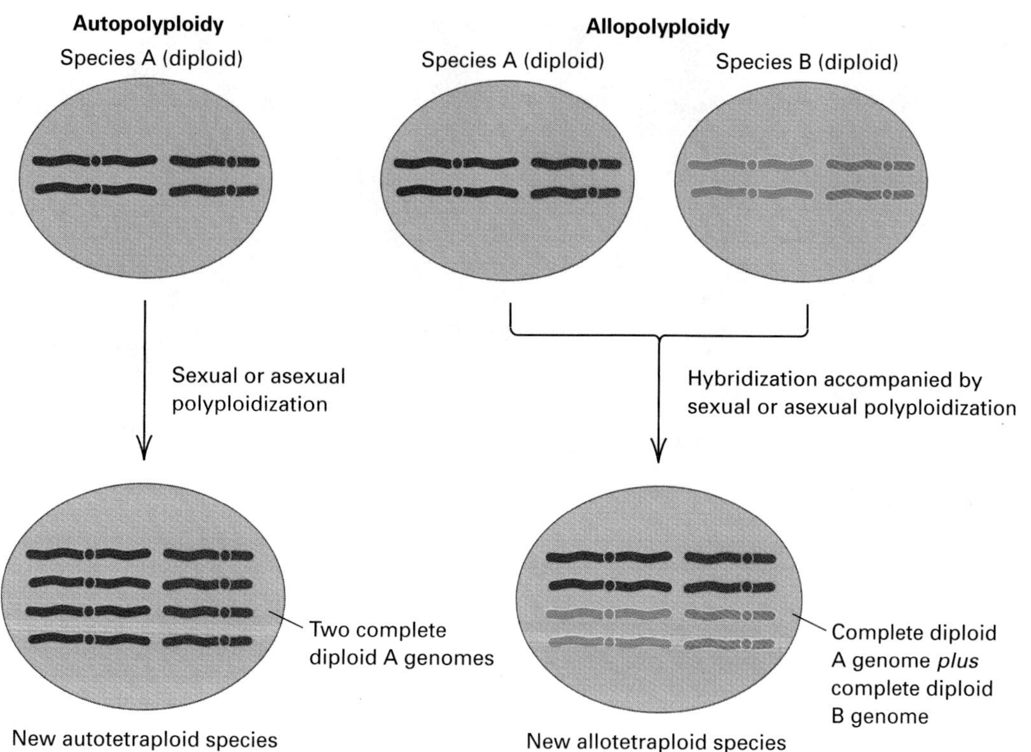


**FIGURE 8.34** Autopolypoids have chromosome sets from a single species; allopolypoids have chromosome sets from different species.

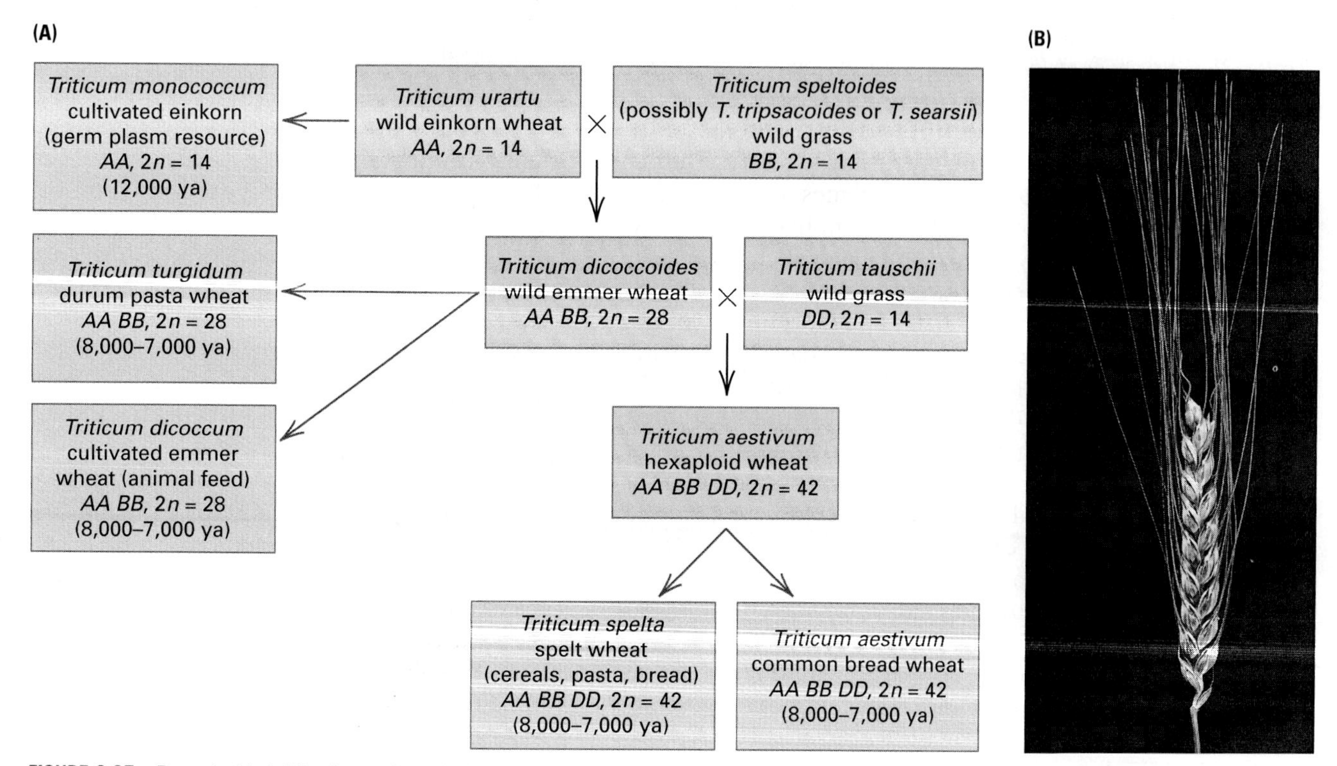


**FIGURE 8.35** Repeated hybridization and polyploidization in the origin of wheat. (A) Each of the A, B, and D genomes has seven chromosomes, and $2n$ is the total chromosome number for each species. Wild species are in green boxes and domesticated species are in yellow boxes along with the approximate time of domestication (ya = years ago). (B) The spike of *T. turgidum*, one of the earliest cultivated wheats. [Photo courtesy of Gordon Kimber, Department of Agronomy, University of Missouri.]

in an allopolyploid is shown in **FIGURE 8.36**. The flower is from a variety of crocus called Golden Yellow. Its genome contains seven pairs of chromosomes, which are shown painted in yellow and green. Golden Yellow was thought to be an allopolyploid formed by hybridization of two closely related species followed by endoreduplication of the chromosomes in the hybrid. The putative ancestral species are *Crocus flavus*, which has four pairs of chromosomes, and *Crocus*

**FIGURE 8.36** Flower of the crocus variety Golden Yellow and chromosome painting that reveals its origin as an allopolyploid. Its seven pairs of chromosomes are shown at the right. The chromosomes in green hybridized with DNA from *C. angustifolius*, which has three pairs of chromosomes, and those in yellow hybridized with DNA from *C. flavus*, which has four pairs of chromosomes. [Part A © Elena Elisseeva/ShutterStock, Inc.; Part B reproduced from M. Orgaard, N. Jacobsen, and J. S. Heslop–Harrison, "The hybrid origin of two cultivars of crocus . . . ," *Ann. Bot.* 76 (1995): 253–262, by permission of Oxford University Press.]

*angustifolius,* which has three pairs of chromosomes. To paint the chromosomes of Golden Yellow, DNA from *C. flavus* was isolated and labeled with a fluorescent green dye, and that from *C. angustifolius* was isolated and labeled with a fluorescent yellow dye. The result of the chromosome painting is very clear: Three pairs of chromosomes hybridize with the green-labeled DNA from *C. flavus,* and four pairs of chromosomes hybridize with the yellow-labeled DNA from *C. angustifolius.*

Recently formed polyploids such as those in wheat and crocus can be detected easily because, in wheat, the size and morphology of the chromosomes in each genome have been maintained, and, in crocus, the DNA sequences retain enough similarity to their ancestral sequences to undergo genome-specific DNA hybridization in chromosome paints. Ancient polyploids are more difficult to detect, but complete genomic sequencing can be informative because DNA sequences can be recognized as sharing a common ancestor long after they have become so different that DNA hybridization can no longer occur. Genomic sequencing has revealed some surprising examples of ancient polyploidy, including budding yeast *(Saccharomyces cerevisiae),* the flowering plant *Arabidopsis thaliana,* and rice *(Oryza sativa).* In each of these cases, the ancient polyploidization yielded a tetraploid, and the event took place so long ago that many of the duplicated genes have been deleted and lost, and many others have undergone so much divergence that their common origin through polyploidization is obscured except by detailed sequence and functional comparisons. Genetically and functionally, the ancient polyploids behave as conventional diploids, as if each gene were a single-copy gene.

## Monoploid Organisms

An organism is monoploid if it develops from a cell with a single set of chromosomes. Meiosis cannot take place normally in the germ cells of a monoploid, because each chromosome lacks a pairing partner, and hence monoploids are usually sterile. Monoploid organisms are quite rare, but they occur naturally in certain insect species (ants, bees) in which males are derived from unfertilized eggs. These monoploid males are fertile because the gametes are produced by a modified meiosis in which chromosomes do not separate in meiosis I.

Monoploids are important in plant breeding because, in the selection of diploid organisms with desired properties, favorable recessive alleles may be masked by heterozygosity. This problem can be avoided by studying monoploids, provided that their sterility can be overcome. In many plants, the production of monoploids capable of reproducing can be stimulated by conditions that yield aberrant cell divisions. Two techniques make this possible.

With some diploid plants, monoploids can be derived from cells in the anthers (the pollen-bearing structures). Extreme chilling of the anthers causes some of the haploid cells destined to become pollen grains to begin to divide. These cells are monoploid as well as haploid. If the cold-shocked cells are placed on an agar surface containing suitable nutrients and certain plant hormones, then a small dividing mass of cells called an *embryoid* forms. A subsequent change of plant hormones in the growth medium causes the embryoid to form a small plant with roots and leaves that can be potted in soil and allowed to grow normally. Because monoploid cells have only a single set of chromosomes, their genotypes can be identified without regard to the dominance or recessiveness of individual alleles. A plant breeder can then select a monoploid plant with the desired traits. In some cases, the desired genes are present in the original diploid plant and are merely sorted out and selected in the monoploids. In other cases, the anthers are treated with mutagenic agents in the hope of producing the desired traits.

When a desired mutation is isolated in a monoploid, it is necessary to convert the monoploid into a homozygous diploid because the monoploid plant is sterile and does not produce seeds. Converting the monoploid into a diploid is possible by treatment of the meristematic tissue (the growing point of a stem or branch) with the substance **colchicine**. This chemical is an inhibitor of the formation of the mitotic spindle. When the treated cells in the monoploid meristem begin mitosis, the chromosomes replicate normally; however, the colchicine blocks metaphase and anaphase, so the result is endoreduplication (doubling of each chromosome in a given cell). Many of the cells are killed by colchicine, but, fortunately for the plant breeder, a few of the monoploid cells are converted into the diploid state (**FIGURE 8.37**). The colchicine is removed to allow continued cell multiplication, and many of the now-diploid cells multiply to form a small sector of tissue that can be recognized microscopically. If placed on a nutrient–agar surface, this tissue will develop into a complete plant. Such plants, which are completely homozygous, are fertile and produce normal seeds.

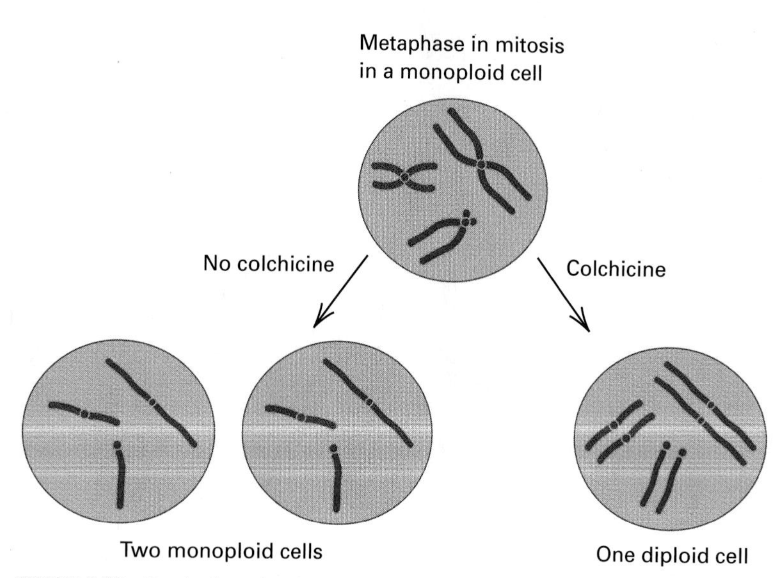


**FIGURE 8.37** Production of a diploid from a monoploid by treatment with colchicine. The colchicine disrupts the spindle and thereby prevents separation of the chromatids after the centromeres (circles) divide.

## 8.8 Genome Evolution in the Grass Family (Gramineae)

The cereal grasses are our most important crop plants. They include rice, wheat, maize, millet, sugar cane, sorghum, and other cereals. The genomes of grass species vary enormously in size. The smallest, at 400 Mb, is found in rice; the largest, at 16,000 Mb, is found in wheat. Although some of the difference in genome size results from the fact that wheat is an allohexaploid, a far more important factor is the large variation from one species to the next in types and amount of repetitive DNA sequences present. Each chromosome in wheat contains approximately 25 times as much DNA as each chromosome in rice. For comparison, maize has a genome size of 2500 Mb; it is intermediate in size among the grasses and approximately the same size as the human genome.

Despite the large variation in chromosome number and genome size in the grass family, there are a number of genetic and physical linkages between single-copy genes that are remarkably conserved amid a background of very rapidly evolving repetitive DNA sequences. In particular, each of the conserved regions can be identified in all the grasses and referred to a similar region in the rice genome. The situation is as depicted in **FIGURE 8.38**. The rice chromosome pairs are numbered R1 through R12, and the conserved regions within each chromosome are indicated by lowercase letters—for example, R1a and R1b. In each of the other species, each chromosome pair is diagrammed according to the arrangement of segments of the rice genome that contain single-copy DNA sequences homologous to those in the corresponding region of the chromosome of the species in question. For example, the wheat monoploid chromosome set is designated W1 through W7. One region of W1 contains single-copy sequences that are homologous to those in rice segment R5a, another contains single-copy sequences that are homologous to those in rice segment R10, and still another contains single-copy sequences that are homologous to those in rice segment R5b. The genomes of the other grass species can be aligned

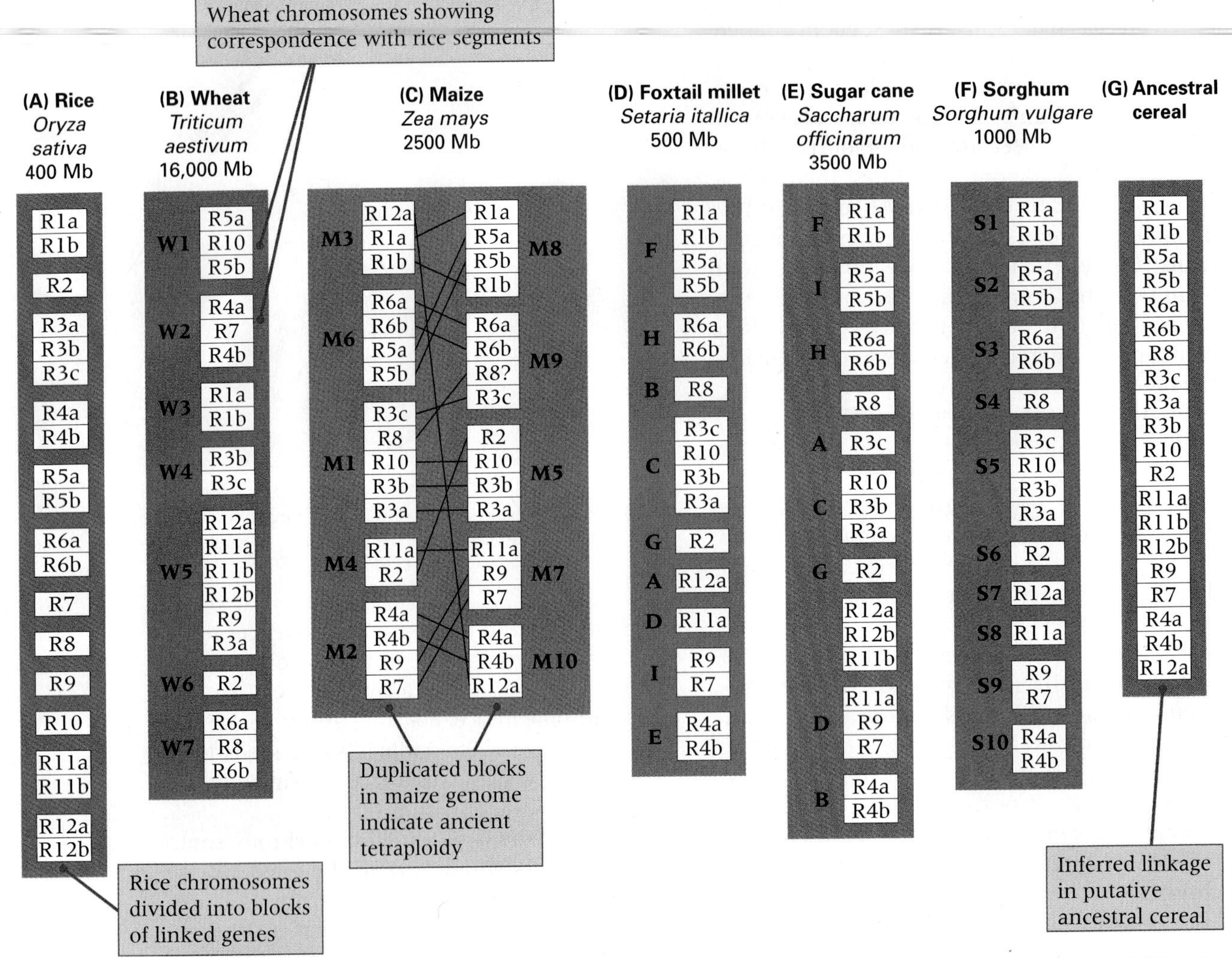


**FIGURE 8.38** Conserved linkages (synteny groups) between the rice genome (A) and that of other grass species: wheat (B), maize (C), foxtail millet (D), sugar cane (E), and sorghum (F). Genome sizes are given in millions of base pairs (Mb). Part (G) depicts the inferred order of segments in a hypothetical ancestral cereal genome. [Data from G. Moore, *Curr. Opin. Genet. Dev.* 5 (1995): 717–724.]

with those of rice as shown. Each of such conserved genetic and physical linkages is called a **synteny group**.

Synteny groups are found in other species comparisons as well. For example, the human and mouse genomes share about 180 synteny groups owing to about an equal number of chromosome rearrangements that took place in the approximately 80 million years since these species last shared a common ancestor. These synteny groups are often useful in identifying the mouse homolog of a human gene.

## CHAPTER SUMMARY

- The standard human karyotype consists of 22 pairs of autosomes plus one pair of sex chromosomes.
- Chromosome abnormalities are a major factor in human spontaneous abortions and an important cause of genetic disorders, such as trisomy 21 (Down syndrome).
- Dosage compensation in mammals results from genetic inactivation (silencing) of X chromosomes at an early stage in embryonic development, except that one X chromosome, chosen at random, remains active in each cell lineage (the single-active-X principle). Some genes in the "inactive" X chromosomes are not silenced.
- Genetically unbalanced (aneuploid) chromosomal complements due to an extra or missing chromosome often have more severe effects on phenotype than does addition of a genetically

balanced (euploid) entire extra set of chromosomes.

- By a process of mispairing and unequal crossing over, genes that are duplicated in tandem along the chromosome can give rise to chromosomes with even more copies.
- Gene-for-gene pairing between a wildtype chromosome and one that contains an inversion of a segment of genes results in the formation of a loop in one of the chromosomes; crossing over within the "inversion loop" leads to chromosomal abnormalities.
- Reciprocal translocations result in abnormal gametes because they upset segregation.
- Duplication of the entire chromosome complement present in a species—or in a hybrid between species—is a major process in the evolution of higher plants.

## GUIDE TO PROBLEM SOLVING

**Problem 1** Six bands in a salivary gland chromosome of *Drosophila* are shown in the accompanying figure, along with the extent of five deletions (Del1–Del5).

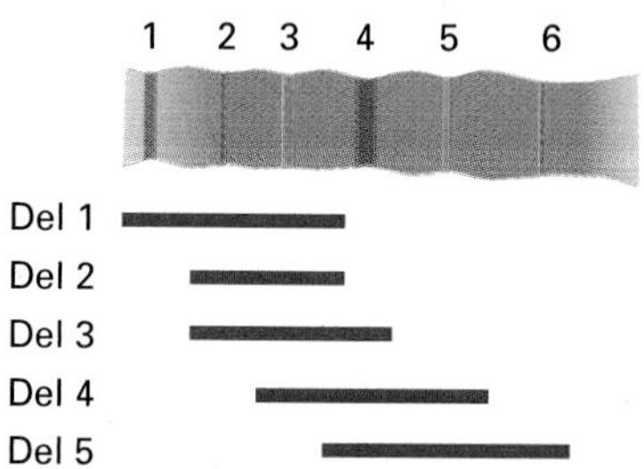


Recessive alleles *a, b, c, d, e,* and *f* are known to be in the region, but their order is unknown. When the deletions are heterozygous with each allele, the following results are obtained:

| | *a* | *b* | *c* | *d* | *e* | *f* |
|---|---|---|---|---|---|---|
| Del 1 | – | – | – | + | + | + |
| Del 2 | – | + | – | + | + | + |
| Del 3 | – | + | – | + | – | + |
| Del 4 | + | + | – | – | – | + |
| Del 5 | + | + | + | – | – | – |

In this table, the – means that the deletion is missing the corresponding wildtype allele (the deletion uncovers the recessive allele) and + means that the corresponding wildtype allele is still present. Use these data to infer the position of each gene relative to the salivary gland chromosome bands.

**Answer** Genes *b, a, c, e, d,* and *f* are located in bands 1, 2, 3, 4, 5, and 6, respectively. The reasoning is as follows. Because deletion 1 uncovers any gene in band 1 but the other deletions do not, the pattern – + + + + observed for gene *b* puts *b* in band 1. Deletions 1–3 uncover band 2 but deletions 4–5 do not, and so the pattern – – – + + observed for gene *a* localizes gene *a* to band 2. Genes in band 3 are uncovered by deletion 1–4 but not 5, and so the pattern – – – – + implies that gene *c* is in band 3. Band 4 is uncovered by deletions 3–5 but not 1–2, which means that gene *e,* with pattern + + – – –, is in band 4. Genes in band 5 are uncovered by deletions 4–5 but not 1–3, and so the pattern + + + – – observed for gene *d* places it in band 5. Genes in band 6 are uncovered by deletion 5 but not 1–4, and so the pattern + + + + – puts gene *f* in band 6.

**Problem 2** The accompanying illustration is a simplified version of pairing between a chromosome with an inversion and its normal homolog, which is extremely useful when the question of interest concerns the consequences of crossing-over within the inversion loop. In the simplified version, only the inverted region is shown paired, the normal regions unpaired. In the case of a paracentric inversion, this means that the centromeres are on opposite sides. (In real cells, of course, the inverted region forms a loop.) In this diagram it is easy to see that a crossover at the position of the arrow results in a dicentric chromosome and an acentric fragment (in addition to the normal and inverted chromosomes that do not participate in the crossover). Deduce the consequence of a two-strand double crossover when the first one takes place as shown and the second one takes place between the genes *C* and *D*.

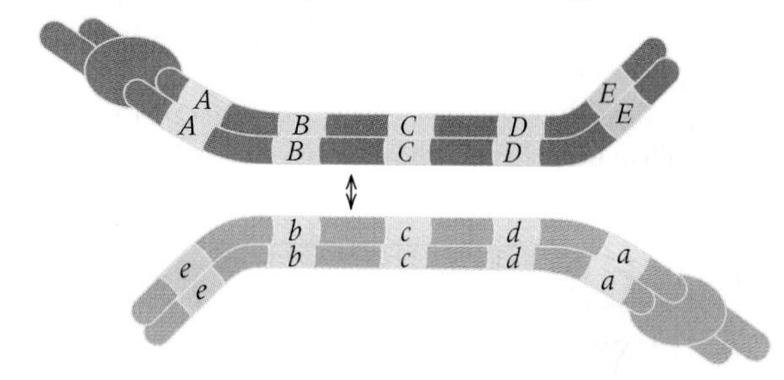


**Answer** The result is one normal chromosome and one inverted chromosome (these derive from chromatids not involved in either crossover), a normal chromosome carrying the alleles *A B c D E,* and an inverted chromosome carrying the alleles *a b C d e*. Note that the two-strand double crossover has transferred an allele between the normal and the inverted chromosomes.

**Problem 3** Indicate what types of gametes would be expected, and in what proportions, from an organism that is:

**(a)** Trisomic for a single chromosome.

**(b)** Monosomic for a single chromosome.

**Answer**

**(a)** In a trisomic organism, either two of the chromosomes will form a pair and the other remain unpaired, or all three chromosomes will come together, each chromosome pairing along part of its length with both of the others. These situations are shown in the illustration. The anaphase I segregation determines the chromosomes included in the gametes. In either type of pairing, two gametes receive two copies of the chromosome and two gametes receive one copy. Therefore, trisomic segregation leads to disomic or monosomic gametes in an expected ratio of 1 : 1.

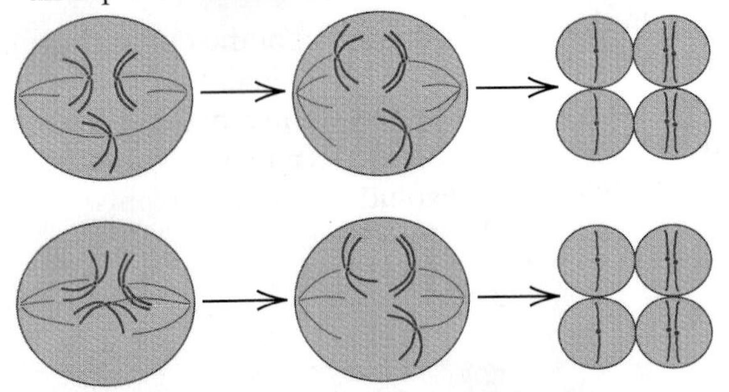

**(b)** In a monosomic organism, the monosomic chromosome must remain unpaired. At anaphase I, it goes to one pole or the other. The end result is two gametes containing no copies of the chromosome and two gametes containing one copy. Therefore, monosomic segregation leads to nullisomic (missing the chromosome) gametes or monosomic gametes in an expected ratio of 1 : 1.

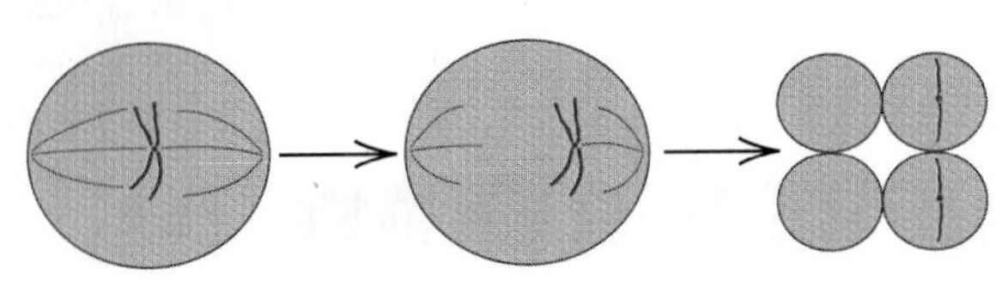

## ANALYSIS AND APPLICATIONS

**8.1** What is the biological function of the obligatory crossover that occurs between the X and Y chromosome during meiosis in the pseudoautosomal region?

**8.2** What is the genetic consequence of the obligatory crossover that occurs between the X and Y chromosome during meiosis in the pseudoautosomal region?

**8.3** In certain types of white blood cells, the inactive X chromosome in females can be observed microscopically as a densely staining heterochromatic body in the nucleus of the cells during interphase. This densely staining blob is called a *Barr body*. In cells with more than one inactivated X chromosome, the number of Barr bodies equals the number of X chromosomes that are inactivated. How many Barr bodies would be present in white blood cells of an individual of karyotype 49, XXXYY?

**8.4** How many Barr bodies would be present in each of the following human conditions?
- **(a)** Klinefelter syndrome
- **(b)** Turner syndrome
- **(c)** Down syndrome
- **(d)** 47, XYY
- **(e)** 47, XXX

**8.5** Why is it not completely correct to refer to the X chromosome that forms a Barr body as "inactive" in a normal human female?

**8.6** A spontaneously aborted human fetus is found to have 45 chromosomes. What is the most probable karyotype? Had the fetus survived, what genetic disorder would it have had?

**8.7** A spontaneously aborted human fetus was found to have the karyotype 92, XXYY. If the zygote resulted from the fusion of normal X-bearing and Y-bearing gametes, what might have happened to the chromosomes in the zygote to result in this karyotype?

**8.8** Color blindness in human beings is an X-linked trait. A man who is color blind has a 45, X (Turner syndrome) daughter who is also color blind. Did the nondisjunction that led to the 45, X child occur in the mother or the father? Explain the evidence supporting your answer.

**8.9** The long arm of the normal human Y chromosome contains a large region that is duplicated. In some mutant Y chromosomes the region between the duplications is missing, and only one copy of the duplication is present. How can this observation be explained? Does the observation tell you anything about how the duplications are oriented in the chromosome?

**8.10** A chromosome has the gene sequence *A B C D E F G*. What is the sequence following a *C*-through-*E* inversion? Following a *C*-through-*E* deletion? Two chromosomes with the sequences *A B C D E F G* and *M N O P Q R S T U V* undergo a reciprocal translocation after breaks in *E F* and *S T*. What are the possible products? Which products are genetically stable?

**8.11** A recessive mutation in the human genome results in a condition called anhidrotic ectodermal dysplasia, which is associated with an absence of sweat glands. The condition can be detected by studies of the electrical conductivity of the skin, because skin without sweat glands has a lower electrical conductivity (higher resistance) than normal skin. In kinships in which the recessive allele is segregating, affected males are found to show low conductance uniformly across their skin surface, as do affected females. However, heterozygous females show a mosaic pattern with normal conductance in some patches of skin and low conductance in others. The pattern of tissue lacking sweat glands is different for each mosaic female examined. How could this pattern of gene expression be explained?

**8.12** A phenotypically normal woman has a child with Down syndrome. The woman is found to have 45 chromosomes. What kind of chromosome abnormality can account for these observations? How many chromosomes does the affected child have? How does this differ from the usual chromosome number and karyotype of a child with Down syndrome?

**8.13** The vast majority of progeny from either 47, XXX females or 47, XYY males are karyotypically normal 46, XX or 46, XY. Is this finding the one expected?

**8.14** Genes *a, b, c, d, e,* and *f* are closely linked in a chromosome, but their order is unknown. Three deletions in the region are found to uncover recessive alleles of the genes as follows: Deletion 1 uncovers *a, b,* and *d.* Deletion 2 uncovers *a, d, c,* and *e.* Deletion 3 uncovers *e* and *f.* What is the order of the genes? In this problem, you will see that there is enough information to order most, but not all, of the genes. Suggest what experiments you might carry out to complete the ordering.

**8.15** Recessive genes *a, b, c, d, e,* and *f* are closely linked in a chromosome, but their order is unknown. Four deletions in the region are examined. One deletion uncovers *a, c,* and *f;* another uncovers *b* and *c;* the third uncovers *d* and *f;* and the fourth uncovers *d, e,* and *f.* What is the order of the genes?

**8.16** Recessive genes *a, b, c, d, e,* and *f* are closely linked in a chromosome, but their order is unknown. Three deletions in the region are examined. One deletion uncovers *a, d,* and *e;* another uncovers *c, d,* and *f;* and the third uncovers *b* and *c.* What is the order of the genes?

**8.17** Four strains of *Drosophila melanogaster* are isolated from different localities. The banding patterns of a particular region of salivary chromosome 2 have the following configurations (each letter denotes a band).

**(a)** *a b f e d c g h i j* **(c)** *a b f e h g i d c j*
**(b)** *a b c d e f g h i j* **(d)** *a b f e h g c d i j*

Assuming that part **c** represents the ancestral sequence, deduce the evolutionary ancestry of the other chromosomes.

**8.18** Two species of Australian grasshoppers coexist side by side. In meiosis, each has 5 pairs of chromosomes. When the species are crossed, the chromosomes in the hybrid form 3 pairs of chromosomes and one group of 4. What alteration in chromosome structure could account for these results?

**8.19** Why are translocation heterozygotes semisterile? Why are translocation homozygotes fully fertile? If a translocation homozygote is crossed with an individual that has normal chromosomes, what fraction of the $F_1$ progeny is expected to be semisterile?

**8.20** Asiatic wild cotton *(Gossypium arboreum)* has 13 pairs of chromosomes, and an American wild species *(G. thurberi)* also has 13 pairs of chromosomes. Interspecific crosses between the species are sterile because of highly irregular chromosome pairing in meiosis. The American cultivated cotton *(G. hirsutum)* has 26 pairs of chromosomes and is fully fertile. Crosses between *G. arboreum* and *G. hirsutum* and crosses between *G. thurberi* and *G. hirsutum* produce plants that, in meiosis, exhibit 13 pairs of chromosomes (bivalents) and 13 chromosomes with no pairing partner (univalents). What do these cytological data tell us about the genetic origin of present-day American cultivated cotton?

**8.21** From how many possible pathways of hybridization and endoreduplication could the diploid species with genomes *AA, BB,* and *CC* yield an allohexaploid with the genome *AABBCC*? Each of *A, B,* and *C* stands for one monoploid set of chromosomes. (One such pathway, for example, is (*AA* × *BB*) × *CC,* which means that *A* and *B* underwent hybridization and polyploidization to yield the allotetraploid *AABB,* which then underwent hybridization and polyploidization with *CC* to yield the allohexaploid *AABBCC.*)

**8.22** What types of hybridization and endoreduplication could account for a polyploid species with the genome composition *AABBBBCC,* where the diploid ancestors have genome compositions *AA, BB,* and *CC,* respectively? (This type of polyploidy is known as *segmental allopolyploidy.*)

**8.23** An autopolyploid series similar to *Chrysanthemum* consists of five species. The basic monoploid chromosome number in the group is 7. What chromosome numbers would be expected among the species?

**8.24** Yellow body ( *y*) is a recessive mutation near the tip of the X chromosome of *Drosophila.* A wildtype male was irradiated with x rays and crossed with *yy* females, and one $y^+$ son was observed. This male was mated with *yy* females, and the offspring were

| | |
|---|---|
| yellow females | 242 |
| yellow males | 0 |
| wildtype females | 0 |
| wildtype males | 260 |

The yellow females were found to be chromosomally normal, and the $y^+$ males were found to breed in the same manner as their father. What type of chromosome abnormality could account for these results?

**8.25** Curly wings (*Cy*) is a dominant mutation in the second chromosome of *Drosophila.* A *Cy*/+ female was irradiated with x-rays and crossed with normal males; the *Cy*/+ sons were then mated individually with normal females. From one cross, the progeny were:

| | |
|---|---|
| curly males | 0 |
| curly females | 157 |
| wildtype males | 142 |
| wildtype females | 0 |

What abnormality in chromosome structure is the most likely explanation for these results? (Remember that crossing-over does not take place in male *Drosophila.*)

**8.26** Semisterile maize plants heterozygous for a reciprocal translocation between chromosomes 1 and 2 were crossed with chromosomally normal plants homozygous for a recessive allele in chromosome 2 that results in virescent leaves. When semisterile $F_1$ plants were testcrossed, the following data were obtained:

| | |
|---|---|
| semisterile virescent | 36 |
| fertile virescent | 240 |
| semisterile nonvirescent | 282 |
| fertile nonveriscent | 42 |

**(a)** What is the frequency of recombination between the gene for virescent leaves and the translocation breakpoint in chromosome 2?

**(b)** What phenotypic ratio in the testcross progeny would be expected if the gene for virescent leaves had not been in one of the chromosomes involved in the translocation?

**8.27** A strain of semisterile maize heterozygous for a reciprocal translocation between chromosomes 1 and 2 was crossed with chromosomally normal plants homozygous for the recessive mutations *brachytic* and *fine-stripe* on chromosome 1. When semisterile $F_1$ plants were crossed with plants of the *brachytic, fine-stripe* parental strain, the following phenotypes were found in a total of 1000 $F_2$ progeny.

| | Semisterile | Fertile |
|---|---|---|
| wildtype | 434 | 4 |
| brachytic | 30 | 20 |
| fine-stripe | 17 | 35 |
| brachytic, fine-stripe | 6 | 454 |

What are the recombination frequencies between *brachytic* and the translocation breakpoint and between *fine-stripe* and the translocation breakpoint?

**8.28** In *Drosophila*, ruby eye color (*rb*) results from a recessive allele in the X chromosome, and dumpy wings (*dp*) results from a recessive allele in chromosome 2. Among the progeny of flies exposed to x-rays, a phenotypically normal male carrying a reciprocal translocation between the X chromosome and chromosome 2 was obtained. The genetic and chromosomal constitution of this male is shown in the illustration. What phenotypes and what relative frequencies would be expected among the progeny from the mating of this male to a female with ruby eyes and dumpy wings? (Remember that there is no crossing over in *Drosophila* males.)

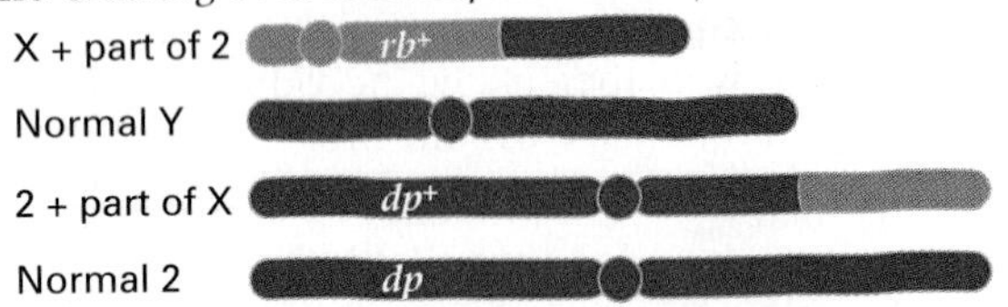


**8.29** In the homologous chromosomes shown here, the red region represents an inverted segment of chromosome. An individual of this genotype was crossed with an *a b c d e* homozygote. Most of the offspring were either *A B C D E* or *a b c d e*, but a few rare offspring were obtained that were *A B c D E*. What events occurring in meiosis in the inversion heterozygote can explain these rare progeny? Is the gene sequence in the rare progeny normal or inverted?

**8.30** An unusual strain of yeast, thought to be a normal haploid, was crossed with a normal haploid strain carrying the mutation *his7*. The *his7* mutation is located in chromosome 2 and is an allele of a gene normally required for synthesis of the amino acid histidine. Among 9 tetrads analyzed from this cross, the following types of segregation were observed:

| | |
|---|---|
| 4 wildtype : 0 *his7* | 4 tetrads |
| 2 wildtype : 2 *his7* | 1 tetrad |
| 3 wildtype : 1 *his7* | 4 tetrads |

However, when the unusual strain was crossed with haploid strains with recessive markers on other chromosomes, segregation in the tetrads was always 2 : 2. What type of chromosome abnormality in the wildtype strain might account for the unexpected segregation when the strain is mated with *his7*?

## CHALLENGE PROBLEMS

**Challenge Problem 1** Use the "trick" given in Worked Problem 2 for drawing the paired configuration between an inversion-bearing chromosome and its homolog to deduce the consequences of the following types of double crossovers, where the first is at the position shown in the illustration in Worked Problem 2 and the second is between the genes *C* and *D*.

**(a)** Two-strand double crossover.
**(b)** Three-strand double crossover (one normal homolog not included).
**(c)** Three-strand double crossover (one inverted homolog not included).
**(d)** Four-strand double crossover.

**Challenge Problem 2** Females heterozygous for the reciprocal translocation in Problem 8.28 and carrying normal X and 2 chromosomes containing *rb* and *dp*, respectively, are crossed with ruby, dumpy males. What genotypes and phenotypes are expected among 1000 progeny if the recombination frequency between *rb* and the translocation breakpoint is 10 percent and that between *dp* and the translocation breakpoint is 30 percent. Assume there is no interference.

**Challenge Problem 3** An otherwise diploid strain of yeast is trisomic for one chromosome. The strain carries a recessive allele *a* in one copy of the trisomic chromosome and the dominant allele *A* in each of the other copies, yielding the genotype *AAa*. The strain is sporulated and tetrads are classified for *A* and *a*. Three types of tetrads are found: (1) 4 *A* : 0 *a*, (2) 2 *A* : 2 *a*, and (3) 3 *A* : 1 *a*. Assume that the trisomic chromosome forms a bivalent and a univalent at random. Show that: **(a)** if *A* and *a* are completely linked to the centromere, then the expected ratio of 4 : 0 to 2 : 0 to 3 : 1 tetrads is 1 : 2 : 0. **(b)** if *A* and *a* are unlinked to the centromere ($r = 0.5$), then the expected ratio of 4 : 0 to 2 : 0 to 3 : 1 tetrads is 4 : 1 : 4.

### GENETICS *on the web*

GeNETics on the Web will introduce you to some of the most important sites for finding genetics information on the Internet. To explore these sites, visit the Jones and Bartlett companion site to accompany *Genetics: Analysis of Genes and Genomes, Seventh Edition* at http://biology.jbpub.com/book/genetics.

There you will find a chapter-by-chapter list of highlighted keywords. When you select one of the keywords, you will be linked to a Web site containing information related to that keyword.

# CHAPTER 9

# Genetics of Bacteria and Their Viruses

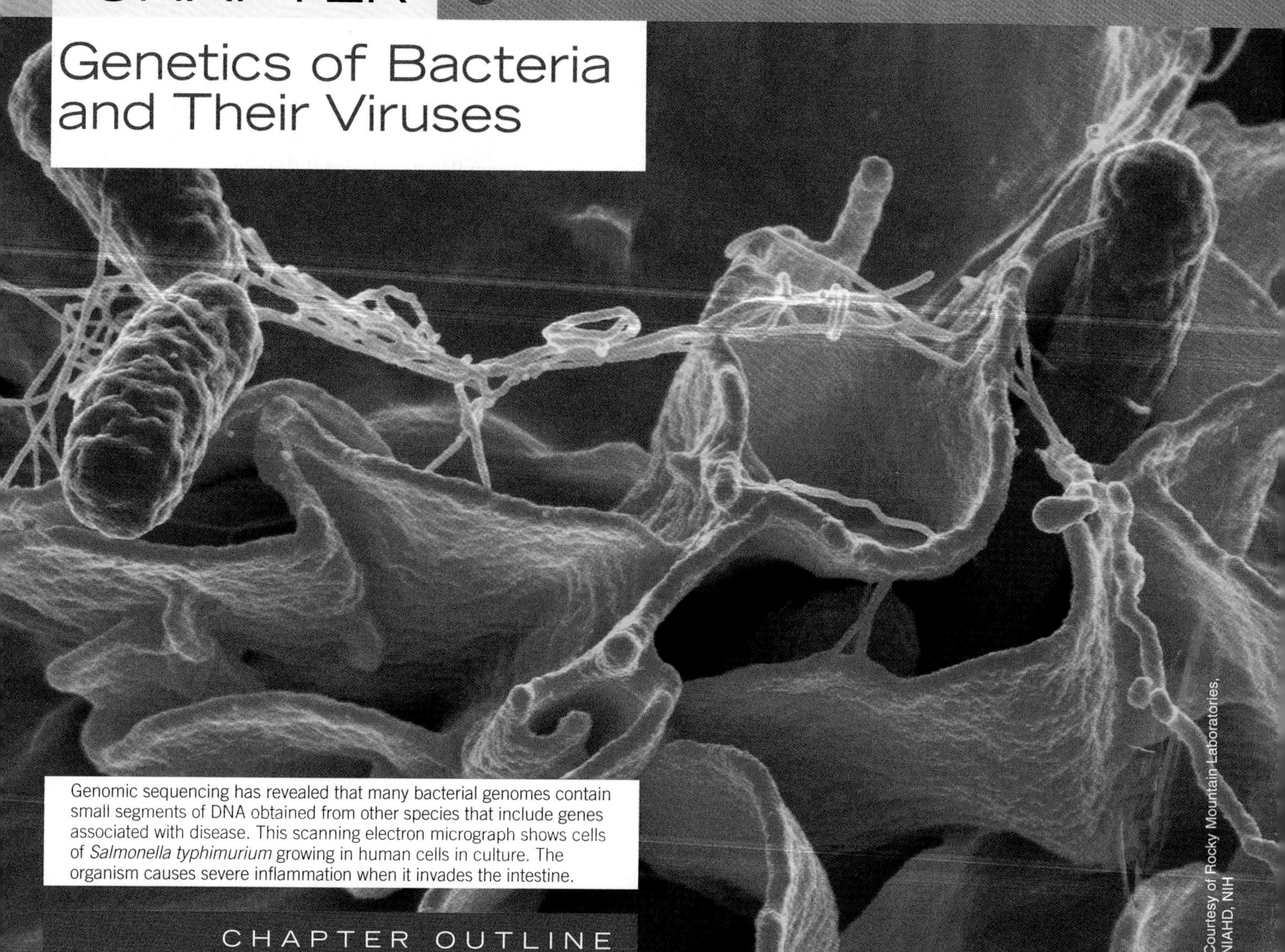


Genomic sequencing has revealed that many bacterial genomes contain small segments of DNA obtained from other species that include genes associated with disease. This scanning electron micrograph shows cells of *Salmonella typhimurium* growing in human cells in culture. The organism causes severe inflammation when it invades the intestine.

## CHAPTER OUTLINE


**9.1 Mobile DNA**
- Plasmids
- The F Plasmid: A Conjugative Plasmid
- Insertion Sequences and Transposons
- Mobilization of Nonconjugative Plasmids
- Integrons and Antibiotic-Resistance Cassettes
- Pathogenicity Islands
- Multiple-Antibiotic-Resistant Bacteria

**9.2 Bacterial Genetics**
- Mutant Phenotypes
- Mechanisms of Genetic Exchange

**9.3 DNA-Mediated Transformation**

**9.4 Conjugation**
- Cointegrate Formation and Hfr Cells
- Time-of-Entry Mapping
- F′ Plasmids

**9.5 Transduction**
- The Phage Lytic Cycle
- Generalized Transduction

**9.6 Bacteriophage Genetics**
- Plaque Formation and Phage Mutants
- Genetic Recombination in the Lytic Cycle
- Genetic and Physical Maps of Phage T4
- Fine Structure of the *rII* Gene in Bacteriophage T4

**9.7 Lysogeny and Specialized Transduction**
- Site-Specific Recombination and Lysogeny
- Specialized Transduction

CONNECTION The Sex Life of Bacteria
Joshua Lederberg and Edward L. Tatum 1946
*Gene Recombination in* Escherichia coli

CONNECTION Origin of Phage Genetics
Alfred D. Hershey and Raquel Rotman 1948
*Genetic Recombination Between Host-Range and Plaque-Type Mutants of Bacteriophage in Single Bacterial Cells*

CONNECTION Artoo
Seymour Benzer 1955
*Fine Structure of a Genetic Region in Bacteriophage*

Bacteria and their viruses (bacteriophage) have unique reproductive systems with multiple and novel mechanisms of genetic exchange. Some bacterial DNA sequences can become mobile by any of a variety of mechanisms. This feature enables them to become widely disseminated within a bacterial population and even to spread between species. In this chapter we discuss the genetic systems of bacteria and bacteriophage. We begin by examining **mobile DNA:** sequences that can be transferred between DNA molecules and from one cell to another. The ability to share genes in this manner, even among different bacterial species, is a unique feature of bacterial genetic systems.

## 9.1 Mobile DNA

A high percentage of bacteria isolated from clinical infections are resistant to one or more antibiotics. Most of them are resistant to multiple antibiotics. Some are resistant to all antibiotics in routine use. The problem has become so severe that many of the antibiotics that were at one time most effective and had the fewest side effects are now virtually useless. The widespread antibiotic-resistance genes almost never originate from new mutations in the bacterial genome. They are acquired, usually several at a time, in various forms of mobile DNA.

### Plasmids

**Plasmids** are nonessential DNA molecules that exist inside bacterial cells. They replicate independently of the bacterial genome and segregate to the progeny when a bacterial cell divides, so they can be maintained indefinitely in a bacterial lineage. Many plasmids are circular DNA molecules, but others are linear. The number of copies of a particular plasmid in a cell varies from one plasmid to the next, depending on the mechanism by which replication is regulated. High-copy-number plasmids are found in as many as 50 copies per host cell, whereas low-copy-number plasmids are present in 1–2 copies per cell. Plasmids range in size from a few kilobases (kb) to a few hundred kilobases (**FIGURE 9.1**) and are found in most bacterial species that have been studied. In *E. coli,* most plasmids are either quite small (up to about 10 kb) or quite large (greater than 40 kb). A typical *E. coli* isolate contains three different small plasmids, each present in multiple copies per cell, and one large plasmid present in a single copy per cell. The presence of plasmids can be detected physically by electron microscopy, as in Figure 9.1, or by gel electrophoresis of DNA samples. Some plasmids can be detected because of phenotypic characteristics that they confer on the host cell. The phenotype most commonly studied is antibiotic resistance. For example, a plasmid containing a tetracycline-resistance gene (*tet-r*) will enable the host bacterial cell to form colonies on medium containing tetracycline.

Plasmids rely on the DNA-replication enzymes of the host cell for their reproduction, but the initiation of replication is controlled by plasmid genes. In high-copy-number plasmids,

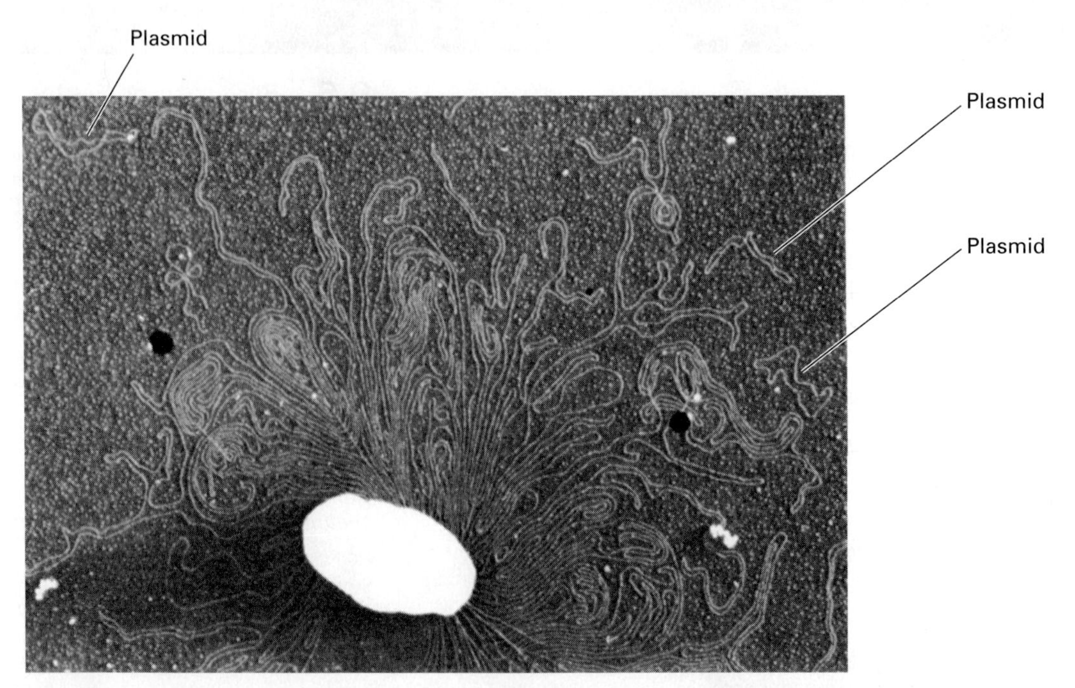


**FIGURE 9.1** Electron micrograph of a ruptured *E. coli* cell, showing released chromosomal DNA and several plasmid molecules. [Courtesy of David Dressler and Huntington Potter. Used with permission of Huntington Potter, Johnnie B. Byrd Alzheimer's Center & Researching Institute.]

replication is initiated multiple times during replication of the host genome, but in low-copy-number plasmids, replication is initiated only once per round of replication of the host genome. All types of plasmids contain sequences that promote their segregation into both daughter cells produced by fission of the host cell, so spontaneous loss of plasmids is uncommon.

### The F Plasmid: A Conjugative Plasmid

Many large plasmids contain genes that enable the plasmid to be transferred between cells. An example found in the bacterium *E. coli* is a large plasmid called the **F factor**. (The F stands for *fertility*.) Cells containing the F factor are designated $F^+$ ("F plus"); those lacking the F factor are designated $F^-$ ("F minus"). Transfer of the F plasmid between cells is mediated by a tube-like structure called a *pilus* (plural *pili*) that makes contact between the $F^+$ and $F^-$ cells (**FIGURE 9.2**). The joining of bacterial cells in the transfer process is called **conjugation**, and the plasmids that can be transferred in this manner are called **conjugative plasmids**. Not all plasmids are conjugative plasmids. Most small plasmids are nonconjugative; they can be maintained in a bacterial lineage as the cells divide, but they do not contain the approximately 20 genes necessary for pilus assembly or those for DNA transfer. Hence, they are unable to be transferred on their own. As we shall see later, however, they are able to employ the genetic trickery of recombination in order to tag along with conjugative plasmids, and in this way nonconjugative plasmids can be mobilized for cell-to-cell transfer.

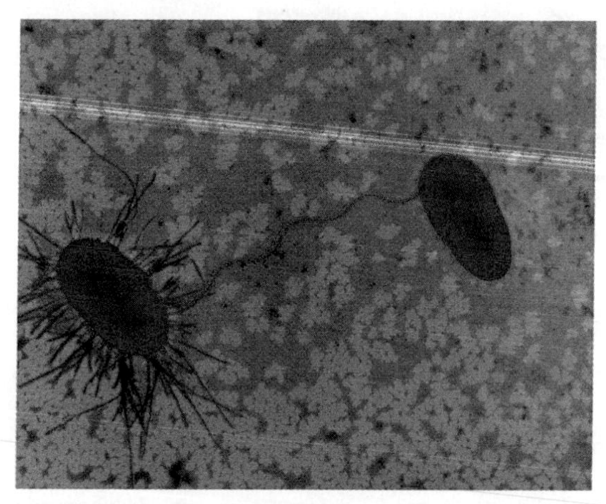

**FIGURE 9.2** Pilus connecting two cells of *E. coli*. The cell with the multiple appendages (used for colonizing the intestine) contains an F plasmid that encodes the proteins necessary to produce the pilus. Prior to transfer of the plasmid DNA, the pilus shortens and draws the cells closer together. For ease of visualization, this pilus is coated with a bacteriophage that targets the F pilus. [© Dennis Kunkel/Phototake, Inc./Alamy Images.]

The F plasmid is a low-copy-number plasmid, present in one to two copies per cell. It replicates once per cell cycle and segregates to both daughter cells in cell division. The F factor is approximately 100 kb in length and contains many genes that govern its maintenance in the cell and its transmission between cells.

Conjugation begins with physical contact between a donor cell and a recipient cell, as shown in Figure 9.2. Conjugation is mediated by a multisubunit protein complex that is assembled within the cytoplasm and inner membrane of the $F^+$ bacterium. This macromolecular structure penetrates through the outer membrane and secretes the F pilus. The major protein of the pilus (called *pilin*), as well as the protein components of the secretion apparatus, are encoded in the F factor. Once the pilus contacts an $F^-$ cell, the pilus retracts and the cell membranes of the donor and recipient are brought into close proximity. The secretion apparatus forms a *conjugation junction* containing a pore through which DNA passes from the $F^+$ to the $F^-$ cell. Prior to DNA transfer, a complex of proteins called a *relaxosome* binds to the F-factor origin of transfer, *oriT*. An enzyme called *relaxase*, encoded in the F-factor gene *TraI*, nicks the F-factor DNA at a unique site called *nic*, binds covalently to the 5′ end of the nicked strand, and unwinds about 200 base pairs of DNA. It is this protein-capped single strand of DNA that is transferred through the pore into the $F^-$ cell.

DNA transfer is always accompanied by replication of the plasmid. Rolling-circle replication (explained in Section 6.2) is initiated at the 3′ end of the nicked strand and replaces the transferred single strand. Lagging strand DNA synthesis in the recipient converts the transferred single strand into double-stranded DNA (**FIGURE 9.3**). When transfer is complete, the linear F strand becomes circular again in the recipient cell. Note that because one replica remains in the donor while the other is transferred to the recipient, after transfer both cells contain F and can function as donors. The $F^-$ cell has been converted into an $F^+$ cell.

The transfer of the F plasmid requires only a few minutes. In laboratory cultures, if a small number of donor cells is mixed with an excess of recipient cells, F spreads throughout the population in a few hours, and all cells ultimately become $F^+$. Transfer is not so efficient under natural

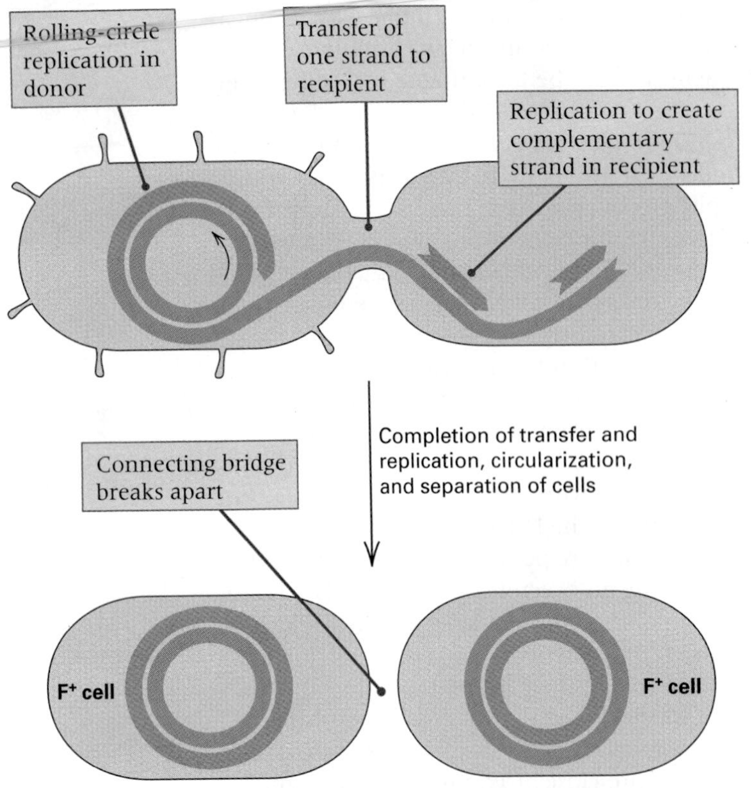


**FIGURE 9.3** Transfer of F from an $F^+$ to an $F^-$ cell. Pairing of the cells triggers rolling-circle replication. Rose represents DNA synthesized during pairing. For clarity, the bacterial chromosome is not shown, and the plasmid is drawn overly large; the plasmid is in fact much smaller than a bacterial chromosome.

conditions, and only about 10 percent of naturally occurring *E. coli* cells contain the F factor.

## Insertion Sequences and Transposons

Transposable elements are the key agents of gene mobilization. These DNA sequences will be discussed in Chapter 14 in terms of their molecular mechanisms of transposition and their role as mutagenic agents. Here we focus on their role in bacterial genetics, particularly as agents through which antibiotic-resistance genes can be mobilized.

Bacteria contain a wide variety of transposable elements. The smallest and simplest are **insertion sequences**, or **IS elements**, which are typically 1–3 kb in length and usually encode only the transposase protein required for transposition and one or more additional proteins that regulate the rate of transposition. Like many transposable elements in eukaryotes, they possess inverted-repeat sequences at their termini, which are used by the transposase for recognizing and mobilizing the IS element. Upon insertion, they create a short, direct duplication of the target sequence at each end of the inserted element. The DNA organization of the insertion sequence IS*50* is diagrammed in **FIGURE 9.4**, part A.

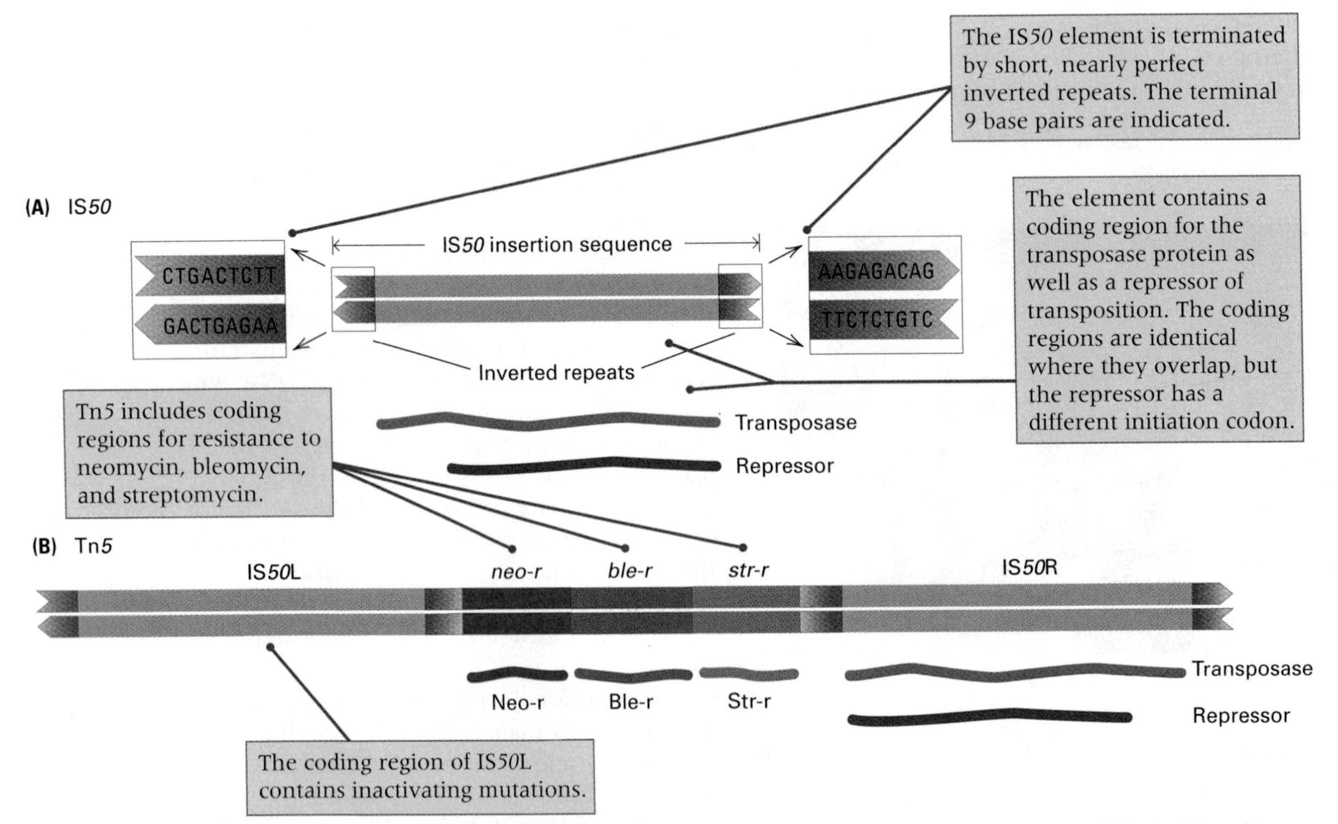


**FIGURE 9.4** Transposable elements in bacteria. (A) Insertion sequence IS*50*. (B) Composite transposon Tn*5* consisting of two IS*50* elements flanking a set of antibiotic-resistance genes.

Other transposable elements in bacteria contain one or more genes unrelated to transposition that can be mobilized along with the transposable element; this type of element is called a **transposon**. The length of a typical transposon is several kilobases, but a few are much longer. Much of the widespread antibiotic resistance among bacteria is due to the spread of transposons that include one or more (usually multiple) antibiotic-resistance genes. When a transposon mobilizes and inserts into a conjugative plasmid, it can be widely disseminated among different bacterial hosts by means of conjugation.

Some transposons have composite structures with antibiotic-resistance genes sandwiched between insertion sequences, as is the case with the Tn*5* element illustrated in Figure 9.4B, which terminates in two IS*50* elements in inverted orientation. Transposons are usually designated by the abbreviation Tn followed by an italicized number (for example, Tn*5*). When it is necessary to refer to genes carried in such an element, the usual designations for the genes are used. For example, Tn*5* (*neo-r ble-r str-r*) contains genes for resistance to three different antibiotics: neomycin, bleomycin, and streptomycin.

## Mobilization of Nonconjugative Plasmids

Nonconjugative and conjugative plasmids typically coexist in the same cell along with host genomic DNA, and when a transposable element is mobilized, all of the DNA molecules present are potential targets for insertion. In time, many of the plasmids in a bacterial lineage can acquire copies of a transposable element present in the host DNA, and the host DNA can acquire copies of a transposable element present in the plasmids. In this manner, transposable elements become disseminated among independently replicating DNA molecules. The result is that most bacteria contain multiple copies of different types of transposable elements, some in the host genome, some in plasmids, and some in both. In *E. coli*, for example, natural isolates contain an average of one to six genomic copies of each of six naturally occurring IS elements, and among the cells that contain a particular IS element, 20–60 percent also contain copies in one or more plasmids.

The result is that many nonconjugative and conjugative plasmids present in a bacterial cell come to carry one or more copies of the same transposable element. Because these copies are homologous DNA sequences, they can serve as substrates for recombination by the processes discussed in Section 6.8. What happens when two plasmids undergo recombination in a region of homology? The result is shown in **FIGURE 9.5**. The recombination forms a composite plasmid called a **cointegrate**. If one of the participating plasmids is nonconjugative and the other is conjugative, then the cointegrate is also a conjugative plasmid and so can be transferred in conjugation. After conjugation, the nonconjugative plasmid can become free of the cointegrate by recombination between the same sequences that created it. By the mechanism of cointegrate formation,

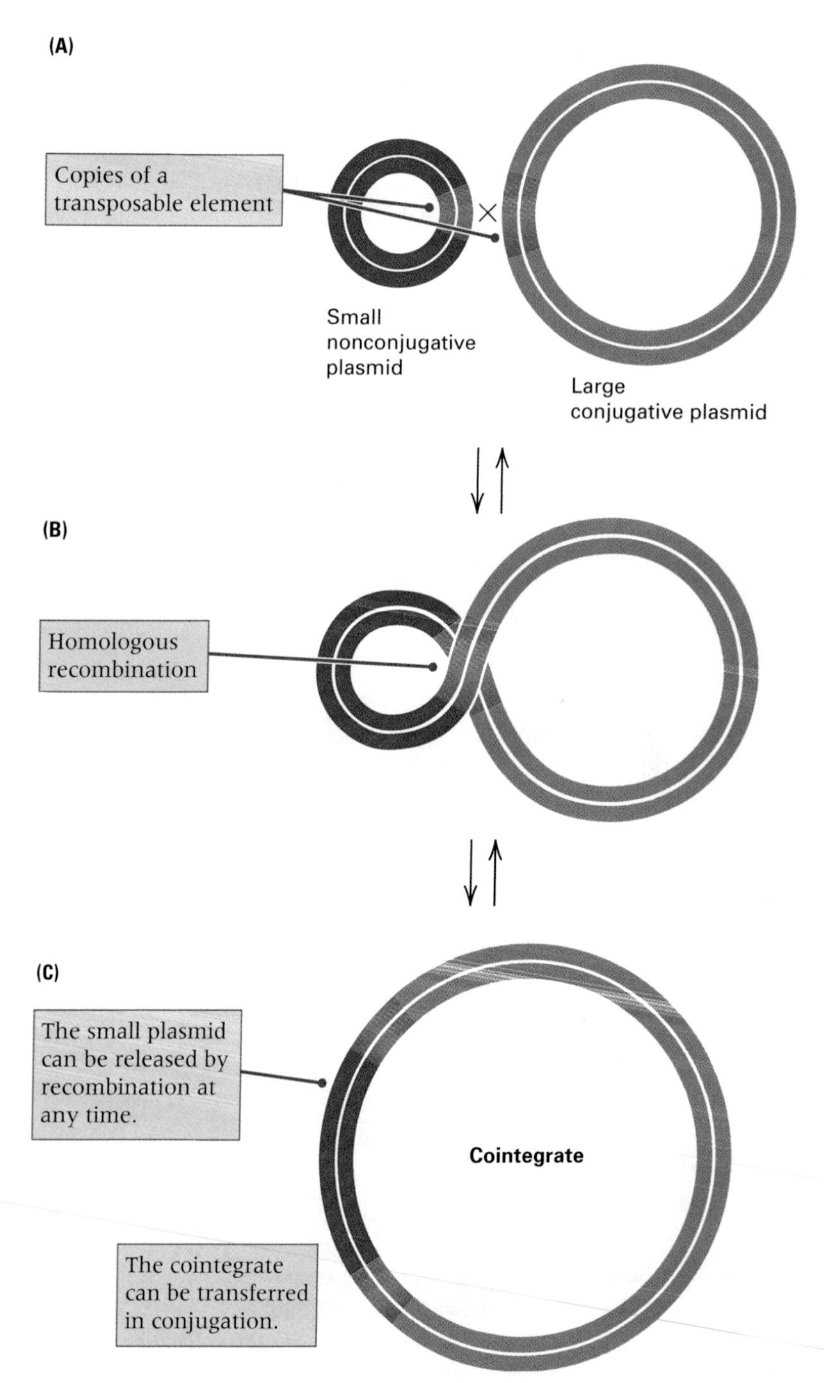


**FIGURE 9.5** Cointegrate formed between two plasmids by recombination between homologous sequences (for example, copies of a transposable element) present in both plasmids.

therefore, nonconjugative plasmids can temporarily ride along with conjugative plasmids and be transferred from cell to cell.

## Integrons and Antibiotic-Resistance Cassettes

In the evolution of multiple antibiotic resistance, bacteria have also made liberal use of a set of enzymes known as *site-specific recombinases,* which were present in bacterial populations long before the antibiotic era and functioned in the evolution of other traits. Each type of **site-specific recombinase** binds with a specific nucleotide sequence in duplex DNA. When the site is present in each of two duplex DNA molecules, the recombinase brings the sites together and catalyzes a reciprocal exchange between the duplexes. An example is shown in **FIGURE 9.6**, where the site-specific recombinase joins a circular DNA molecule with a linear DNA molecule. Note that the reaction can proceed in the reverse direction, too, and free the circle from the cointegrate.

An example of a site-specific recombinase is an enzyme called the *Cre recombinase,* which is encoded in a gene in the *E. coli* bacteriophage P1. The Cre recognition sequence is called *loxP;* it is the 34-bp sequence shown in Figure 9.6. Although the terminal 13 base pairs at each end of the *loxP* sequence form a perfect inverted repeat, the *loxP* sequence nevertheless has a directionality because the central region is asymmetrical. This kind of partial symmetry is typical of site-specific recombinases. Recombination between two *loxP* sites preserves the *loxP* sequences because the participating sites are identical, and hence the recombination reaction is reversible. Some site-specific recombinases favor the reaction that brings two molecules together into a cointegrate. Others (including Cre) favor the reaction that splits a cointegrate into two separate molecules. Some site-specific recombinases bring together and recombine sites that are similar but not identical; in these cases the recombination does not preserve the recognition sites, and so the reaction is not necessarily reversible.

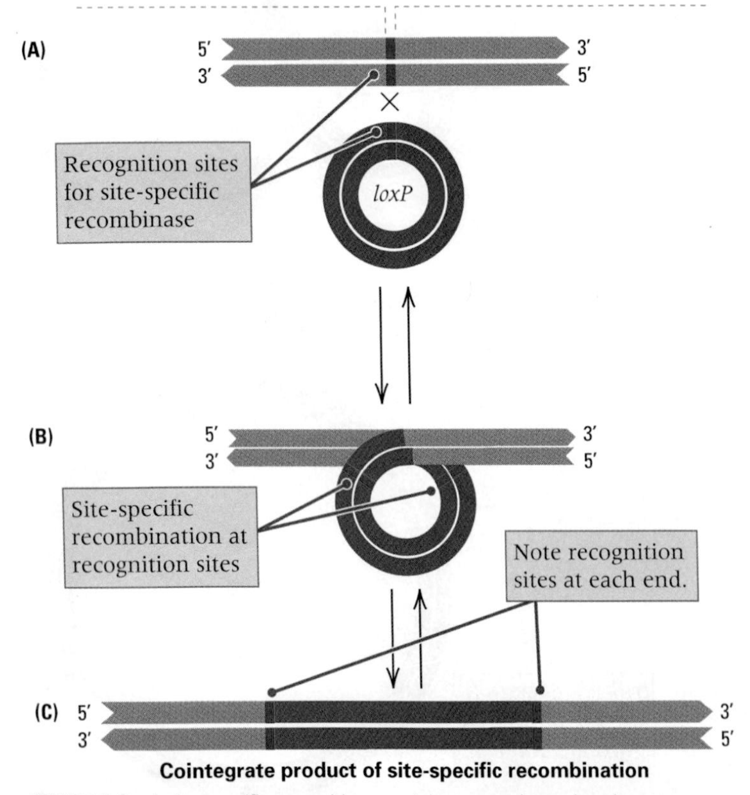


**FIGURE 9.6** A site-specific recombinase catalyzes a reciprocal exchange between two specific sequences. No other sequences can serve as substrates. The recognition site for the Cre recombinase (*loxP*) is shown at the top.

Site-specific recombinases are used in the assembly of multiple-antibiotic-resistance units called *integrons.* An **integron** is a DNA element that encodes a site-specific recombinase as well as a recognition region that allows other sequences with similar recognition regions to be incorporated into the integron by recombination. The elements that integrons acquire are known as *cassettes.* In the context of integrons, a **cassette** is a circular antibiotic-resistance-coding region flanked by a recognition region for an integron. Because the site-specific recombinase integrates cassettes, the integron recombinase is usually called an **integrase**.

Several different types of integrons have been characterized. The best known of these are the Class 1 integrons, which include a site-specific recombinase denoted Int1 and invariably a coding region (*sul1*) that confers resistance to sulphonamide antibiotics. The molecular structure of a Class 1 integron is shown in **FIGURE 9.7**, part A. Also shown is the mechanism by which antibiotic-resistance cassettes are sequentially acquired. The Int1 integrase catalyzes a site-specific recombination between a sequence denoted *attI* present in the integron and a similar sequence denoted *attC* (also called the *59-base element*) in the cassette. All *attC* regions are similar, but no two are identical.

Figure 9.7A shows how a cassette is captured by site-specific recombination between *attI* and *attC*. In general, antibiotic-resistance cassettes contain protein-coding regions but lack promoter sequences of their own. They can be transcribed only by read-through transcription from an adjacent promoter. The integron

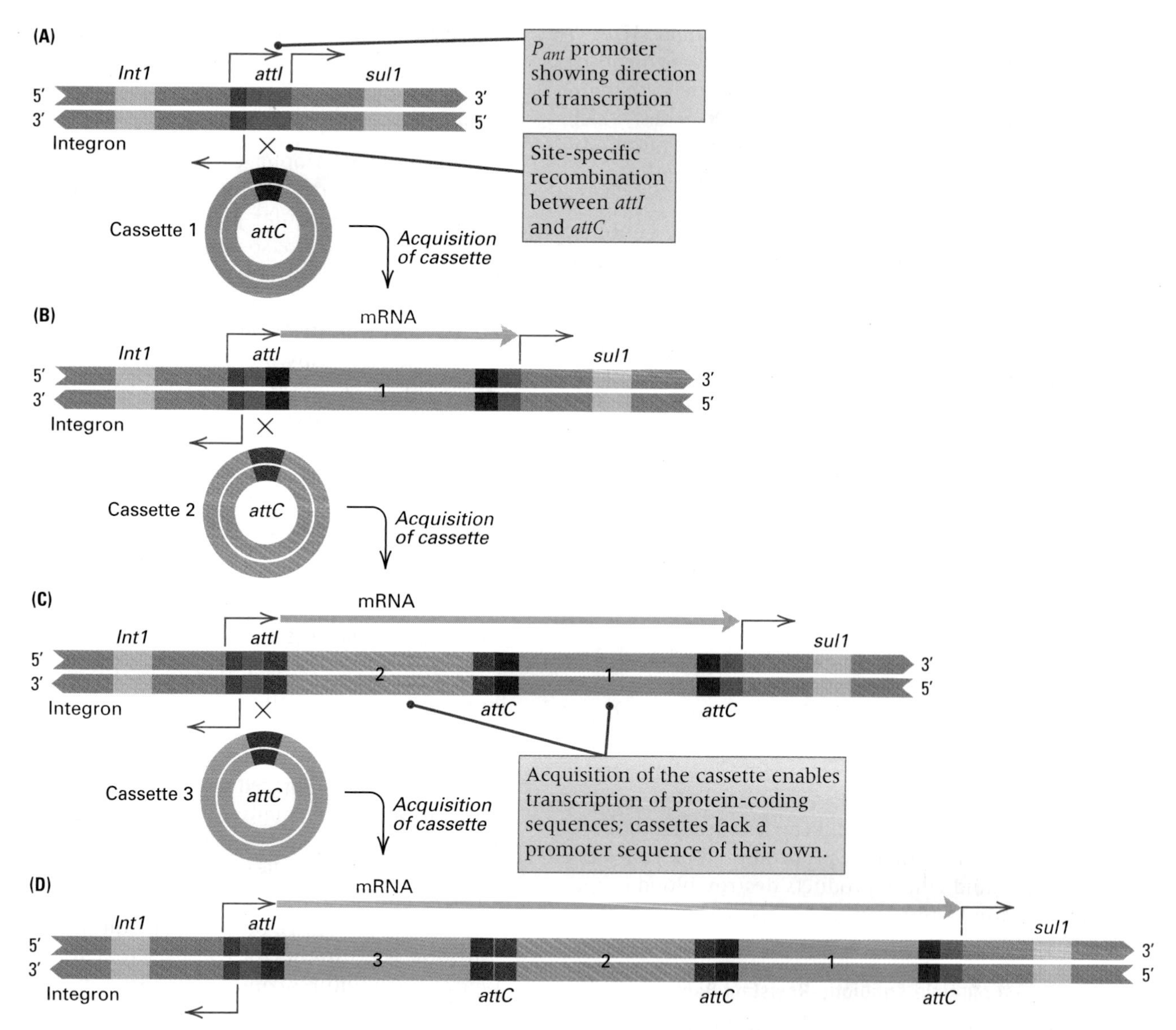


**FIGURE 9.7** Mechanism by which an integron sequentially captures cassettes by site-specific recombination between the *attI* site in the integron and the *attC* site in the cassette.

provides the needed promoter, called $P_{ant}$, at a position upstream from the *attI* site, so that when a cassette is captured, the coding sequence can be expressed. More than 40 different promoterless cassettes that encode proteins for resistance to antibiotics including $\beta$-lactams, aminoglycosides, chloramphenicol, trimethoprim, and streptothricin have been identified.

Once one cassette is in place, as shown in Figure 9.7B, a second can be captured using the same *attI* site and the *attC* present in the new cassette. Note that the new cassette is integrated immediately adjacent to the *attI* site and that the mRNA produced from the $P_{ant}$ promoter includes the coding sequences for both cassettes. In Figure 9.7C, a third cassette is added to the integron and the mRNA from $P_{ant}$ becomes even longer. When there are multiple cassettes, as shown here, all of them are cotranscribed from $P_{ant}$, but the downstream coding sequences are transcribed less frequently because there is a greater chance that transcription will terminate before reaching them. This constraint sets a practical limit on the number of cassettes that can be transcribed efficiently, but integrons with up to 10 antibiotic-resistance cassettes have been found.

## Pathogenicity Islands

All the comings and goings of plasmids and genes jumping about because of transposons and integrons suggest that the bacterial genome is a patchwork of segments of diverse origin inserted into a core set of genes. A patchwork model first gained strong support from genome sequencing of several independent isolates (strains) of *E. coli*. The range of genome sizes among the strains was more than 10 percent—4.63 Mb in strain MG1655 versus 5.23 Mb in strain CF073—which constitutes more than 1000 genes. All sequenced strains share a common set of about 3800 genes, which probably represents the core set of genes inherited from the original ancestor of what we now call *E. coli*.

The interspersed regions of the genome, present in some strains but not others, are due to *genomic islands* of DNA containing multiple genes that were acquired from other bacteria. When these islands of acquired DNA contain genes that cause disease, they are called **pathogenicity islands**. Examples of pathogenicity islands are found in *E. coli* strain O157:H7. This is a pathogenic strain typically spread through contaminated food or water that causes bloody diarrhea and sometimes kidney failure. The strain sickens about 100,000 people per year in the United States alone, among whom about 100 die. The O157:H7 strain contains about 1400 genes not present in the laboratory strain *E. coli* K12. These acquired genes include a pathogenicity island encoding factors that allow the cells to stick to the intestinal wall and to secrete specific proteins into the host cells. Kidney failure is promoted by a toxin encoded in an integrated bacteriophage that inhibits protein synthesis and causes vascular damage. The bloody diarrhea is promoted by genes in a plasmid whose products destroy blood coagulation factors and cause the destruction of red blood cells.

### ■ Multiple-Antibiotic-Resistant Bacteria

In nature, a conjugative plasmid can, through time, accumulate different transposons containing multiple independent antibiotic-resistance genes, or transposons containing integrons that have acquired multiple-antibiotic-resistance cassettes, with the result that the plasmid confers resistance to a large number of completely unrelated antibiotics. These multiple-resistance plasmids are called **R plasmids**. Some R plasmids are closely related to the F plasmid and clearly evolved from the F factor. The evolution of R plasmids is promoted by the use (and overuse) of antibiotics, which selects for resistant cells because, in the presence of antibiotics, resistant cells have a growth advantage over sensitive cells. The presence of multiple antibiotics in the environment selects for multiple-drug resistance. Serious clinical complications result when plasmids resistant to multiple drugs are transferred to bacterial pathogens, or agents of disease. Infections with some pathogens that contain R factors are extremely difficult to treat, because the pathogen may be resistant to most or all antibiotics currently in use.

## 9.2 Bacterial Genetics

Bacteria and bacteriophage bring to traditional types of genetic analysis four important advantages over multicellular plants and animals. First, they are haploid, so dominance or recessiveness of alleles is not a complication in identifying genotype. Second, a new generation is produced in minutes rather than in weeks or months, which vastly increases the rate of accumulation of genetic data. Third, they are easy to grow in enormous numbers under controlled laboratory conditions, which facilitates molecular studies and the analysis of rare genetic events. Fourth, the individual members of these large populations are genetically identical; that is, each laboratory population is a *clone* of genetically identical cells.

### ■ Mutant Phenotypes

Many species of bacteria can be grown in liquid medium or on the surface of a semisolid growth medium gelled with agar. Bacteria used in genetic analysis are usually grown on agar. A single bacterial cell placed on an agar medium will grow and divide many times, forming a visible cluster of cells called a *colony* (**FIGURE 9.8**). The number of bacterial cells present in a liquid culture can be determined by spreading a known volume of the culture on a solid medium and counting the number of colonies that form. Typical *E. coli* cultures contain up to $10^9$ cells/ml. The appearance of colonies or the ability or inability to form colonies on particular media can in some cases be used to identify the genotypes of bacterial cells.

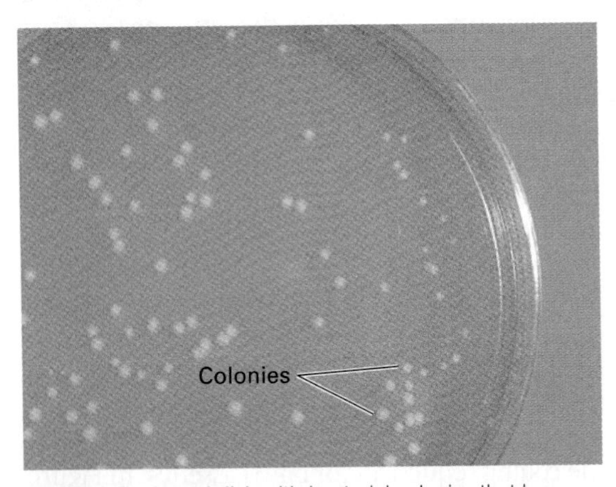


**FIGURE 9.8** A petri dish with bacterial colonies that have formed on a solid medium. The heavy streaks of growth result from colonies so densely packed that there is no space between them. [Courtesy of Dr. Jim Feeley/CDC.]

As we have seen in earlier chapters, genetic analysis requires mutants. With bacteria, three types of mutants are particularly useful:

1. ***Antibiotic-resistant mutants*** These mutants are able to grow in the presence of an antibiotic, such as streptomycin (Str) or tetracycline (Tet). For example, streptomycin-sensitive (Str-s) cells have the wildtype phenotype and fail to form colonies on medium that contains streptomycin, but streptomycin-resistant (Str-r) mutants can form colonies on such a medium.
2. ***Nutritional mutants*** Wildtype bacteria can synthesize most of the complex nutrients they need from simple molecules present in the growth medium. The wildtype cells are said to be **prototrophs**. The ability to grow in simple medium can be lost by mutations that disable the enzymes used in synthesizing the complex nutrients. Mutant cells are unable to synthesize an essential nutrient and cannot grow unless the required nutrient is supplied in the medium. Such a mutant bacterium is said to be an **auxotroph** for the particular nutrient. For example, a methionine auxotroph cannot grow on a **minimal medium** that contains only inorganic salts and a source of energy and carbon atoms (such as glucose), but the methionine auxotroph *is* able to grow if the minimal medium is supplemented with methionine.
3. ***Carbon-source mutants*** Such mutant cells cannot utilize particular substances as sources of carbon atoms or of energy. For example, $Lac^-$ mutants cannot utilize the sugar lactose for growth and are unable to form colonies on minimal medium that contains only lactose as the carbon source.

A medium on which all cells can form colonies is called a **nonselective medium**. Mutants and wildtype cells may or may not be distinguishable by growth on a nonselective medium. If the medium allows growth of only one type of cell (either wildtype or mutant), then it is said to be **selective**. For example, a medium containing streptomycin is selective for the Str-r phenotype and selective against the Str-s phenotype; similarly, minimal medium containing lactose as the sole carbon source is selective for $Lac^+$ cells and against $Lac^-$ cells.

In bacterial genetics, phenotype and genotype are designated in the following way. A phenotype is designated by three letters, the first of which is capitalized, with a superscript + or − to denote presence or absence of the designated phenotype, and with s or r for sensitivity or resistance. A genotype is designated by lowercase italicized letters. Thus, a cell unable to grow without a supplement of leucine (a leucine auxotroph) has a $Leu^-$ phenotype; this would usually result from a $leu^-$ mutation. Often the − superscript is omitted, but using it prevents ambiguity.

### Mechanisms of Genetic Exchange

A bacterial chromosomal DNA molecule almost never encounters another complete DNA molecule from a different bacterial cell. Instead, genetic exchange is usually between a chromosomal fragment from one cell and intact chromosomal DNA from another cell. Furthermore, a clear donor–recipient relationship exists; the donor cell is the source of a DNA fragment, which is transferred to the recipient cell by one of several mechanisms, and exchange of genetic material takes place in the recipient by means of reciprocal recombination between homologous DNA sequences. When the donor DNA is a linear molecule, incorporation of donor DNA into the recipient chromosome requires at least two exchange events, one at each end. Because the donor molecule contains only a fraction of a complete genome, only an even number of exchanges results in a viable product. The usual outcome of these events is the recovery of only one of the crossover products. However, in some situations, the transferred DNA is circular, and a single exchange results in total incorporation of the circular donor DNA into the chromosome of the recipient.

Three major types of genetic transfer are found in bacteria: *transformation*, in which a DNA molecule is taken up from the external environment and incorporated into the genome; *conjugation*, in which donor DNA is transferred from one bacterial cell to another by direct contact; and *transduction*, in which DNA is transferred from one bacterial cell to another by a bacterial virus. Recombination frequencies that result from these processes are used to produce genetic maps of bacteria. Although the maps are exceedingly useful, they differ in major respects from the types of maps obtained from crosses in eukaryotes, because genetic maps in eukaryotes are based on frequencies of exchange events between chromatids in meiosis.

## 9.3 DNA-Mediated Transformation

**Transformation** is a process by which recipient cells acquire genes from free DNA molecules in the surrounding medium. Transformation with purified DNA was the first experimental proof that DNA is the genetic material (Chapter 1). In these experiments, a rough-colony phenotype of *Streptococcus pneumoniae* was changed to a smooth-colony phenotype by exposure of the cells to DNA from a smooth-colony strain. In the laboratory, donor DNA is usually isolated from donor cells and then added to a suspension of recipient cells. In natural settings, such as soil, free DNA can become available by spontaneous breakage (lysis) of donor cells.

Transformation begins with a recipient cell's uptake of a DNA fragment from the surrounding medium and terminates with *one strand* of donor DNA replacing the homologous segment in the recipient DNA. Most bacterial species are probably capable of the recombination step, but many species have only a very limited ability to take up free DNA efficiently. Even in a species capable of transformation, DNA is able to penetrate only some of the cells in a growing population. However, many bacterial species can be made competent to take up DNA, provided that the cells are subjected to an appropriate chemical treatment (for example, treatment with $CaCl_2$).

Transformation is a convenient technique for gene mapping in some species. When DNA is isolated from a donor bacterium (**FIGURE 9.9**), it

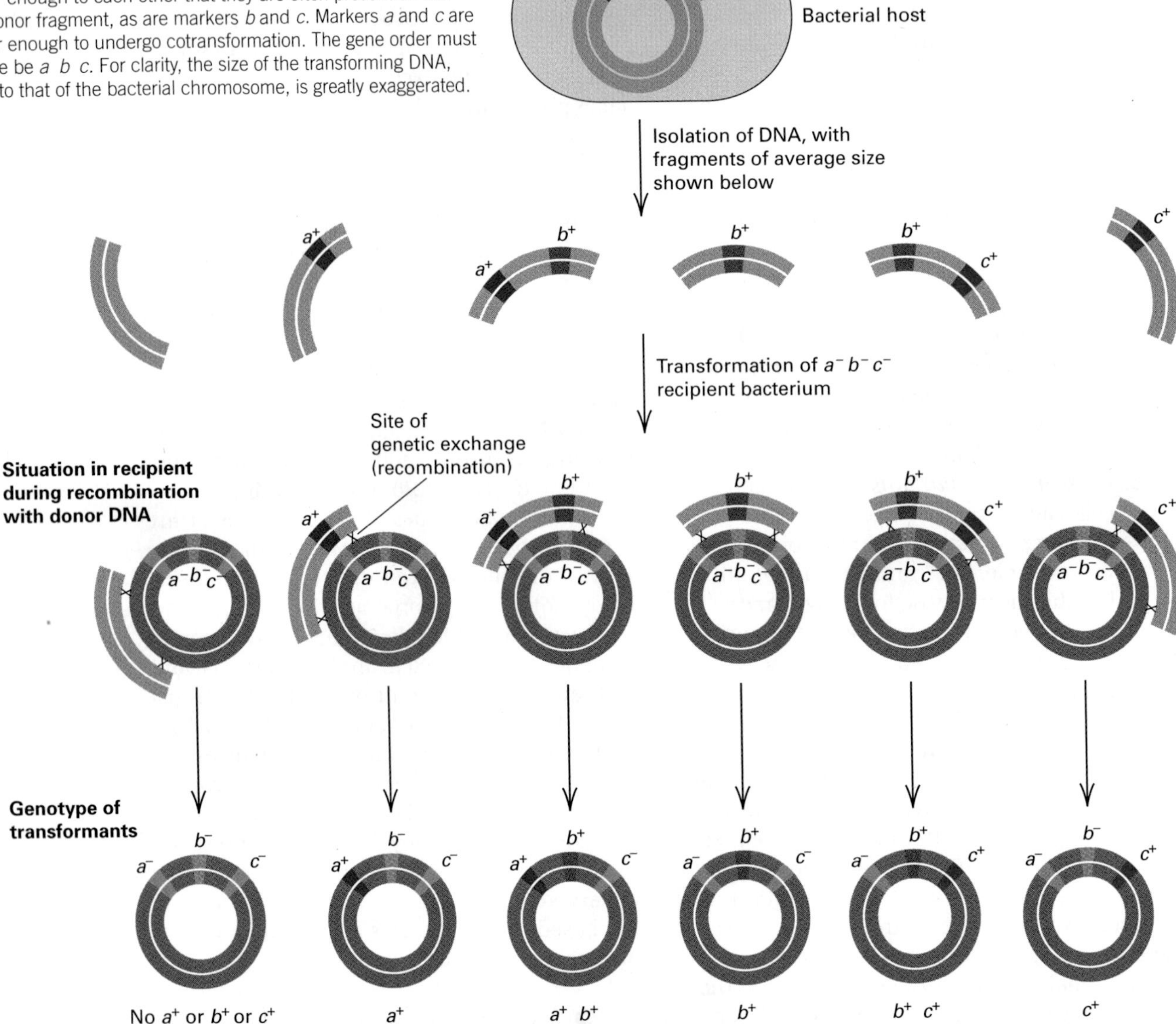


**FIGURE 9.9** Cotransformation of linked markers. Markers *a* and *b* are near enough to each other that they are often present on the same donor fragment, as are markers *b* and *c*. Markers *a* and *c* are not near enough to undergo cotransformation. The gene order must therefore be *a b c*. For clarity, the size of the transforming DNA, relative to that of the bacterial chromosome, is greatly exaggerated.

is invariably broken into small fragments. In most species, with suitable recipient cells and excess external DNA, transformation takes place at a frequency of about 1 transformed cell per $10^3$ cells. If two genes, *a* and *b*, used as genetic markers, are so widely separated in the donor chromosome that they are always contained in two different DNA fragments, then the probability of simultaneous transformation (**cotransformation**) of an $a^-b^-$ recipient into wildtype is the product of the probabilities of transformation of each genetic marker, or roughly $10^{-3} \times 10^{-3}$, which equals one $a^+b^+$ transformant per $10^6$ recipient cells. However, if the two genes are so near one another that they are often present in a single donor fragment, then the frequency of cotransformation is nearly the same as the frequency of single-gene transformation, or one wildtype transformant per $10^3$ recipients. The general principle is as follows:

> Cotransformation of two genes at a frequency substantially greater than the product of the single-gene transformations implies that the two genes are close together in the bacterial chromosome.

Studies of the ability of various pairs of genes to be cotransformed also yield gene order. For example, if genes *a* and *b* can be cotransformed, and genes *b* and *c* can be cotransformed, but genes *a* and *c* cannot, then the gene order must be *a b c* (Figure 9.9). Note that cotransformation frequencies are not equivalent to the recombination frequencies used in mapping in eukaryotes, because they are determined by the size distribution of donor fragments and the likelihood of recombination between bacterial DNA molecules rather than by the occurrence of chiasmata in synapsed homologous chromosomes (Chapter 5).

## 9.4 Conjugation

*Conjugation* is a process in which DNA is transferred from a bacterial donor cell to a recipient cell by cell-to-cell contact. We have already examined this process in the context of conjugative plasmids. In this section we shall see how the same process can transfer genes present in the bacterial chromosome.

### Cointegrate Formation and Hfr Cells

Transfer of chromosomal genes between *E. coli* cells was first observed by Joshua Lederberg in 1951. Although it was not known at the time, the exchange took place because the donor cells were $F^+$, and in a few cells the F factor had become integrated into the bacterial chromosome (**FIGURE 9.10**). These are known as **Hfr cells**. Hfr stands for **high frequency of recombination**, which refers to the relatively high frequency with which donor genes are transferred to the recipient. The integration process is essentially the same as the formation of a cointegrate between two plasmids illustrated in Figure 9.5. Insertion sequences (Section 9.1) are instrumental in the origin of Hfr bacteria from $F^+$ cells, because the F plasmid normally integrates through genetic exchange between an IS element present in F and a homologous copy that has transposed to an essentially random site in the bacterial chromosome. Because the F factor can exist either separate from the chromosome or incorporated into it, it qualifies as an **episome**—a genetic element that can exist free in the cell or as a segment of DNA integrated into the chromosome.

In an Hfr cell (Figure 9.10), the bacterial chromosome remains circular, though enlarged about 2 percent by the integrated F-factor DNA. Integration of F is an infrequent event, but single cells containing integrated F can be isolated and cultured. When an Hfr cell undergoes

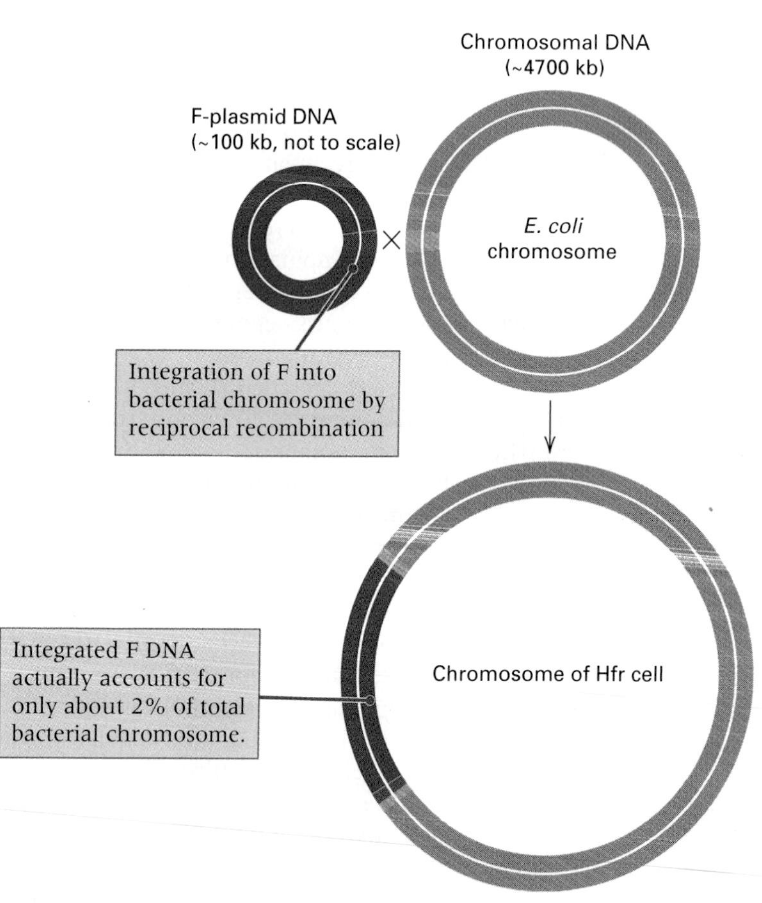


**FIGURE 9.10** Integration of F (blue circle) by recombination between a nucleotide sequence in F and a homologous sequence (usually an insertion sequence) in the bacterial chromosome. The F plasmid DNA is shown greatly enlarged relative to the size of the bacterial chromosome.

conjugation, the process of transfer of the F factor is initiated in the same manner as in an $F^+$ cell. However, because the F factor is part of the bacterial chromosome, transfer from an Hfr cell also includes DNA from the bacterial chromosome.

The Hfr × $F^-$ conjugation process is illustrated in **FIGURE 9.11**. The stages of transfer are much like those by which F is transferred to $F^-$ cells: coming together of donor and recipient cells, rolling-circle replication in the donor cell, and conversion of the transferred single-stranded DNA into double-stranded DNA by lagging-strand synthesis in the recipient. However, in the case of Hfr matings, the transferred DNA does not become circular and is not capable of further replication in the recipient, because the transferred F factor is not complete. The replication and associated transfer of the chromosomal DNA are controlled by the F factor and are initiated in the Hfr chromosome at the origin of transfer, a site in F denoted *oriT,* the same position at which replication and transfer begin in a free F plasmid. A part of F is the first DNA transferred, chromosomal genes are transferred next, and the remaining part of F is the last DNA to enter the recipient. Because the conjugating cells usually break apart long before the entire bacterial chromosome is transferred, the final part of F is almost never transferred into the recipient. Incorporation of donor DNA into the recipient chromosome requires homologous recombination in the recipient. Two recombination events result in the replacement of a segment of recipient DNA with a homologous segment of donor DNA, and the segment of recipient DNA that has been replaced is degraded (Figure 9.11).

Several differences between F transfer and Hfr transfer are notable.

- It takes 100 minutes under the usual conditions for an entire bacterial chromosome to be transferred, in contrast with about 2 minutes for the transfer of F. The difference in time is a result of the relative sizes of F and the chromosome (100 kb for F versus 4600 kb for *E.coli* K12).
- During transfer of Hfr DNA into a recipient cell, the mating pair usually breaks apart before the entire chromosome is transferred. Under usual conditions, several hundred genes are transferred before the cells separate.
- In a mating between Hfr and $F^-$ cells, the $F^-$ recipient remains $F^-$ because cell separation usually takes place before the final segment of F is transferred.
- In Hfr transfer, some regions in the transferred DNA fragment become incorporated into the recipient chromosome. The incorporated regions replace homologous regions in the recipient chromosome. The result is that some $F^-$ cells become recombinants, containing one or more genes from the Hfr donor cell. For example, in a mating between Hfr *leu*$^+$ and $F^-$ *leu*$^-$, some $F^-$ *leu*$^+$ cells arise. However, *the genotype of the donor Hfr cell remains unchanged.*

Genetic analysis requires that recombinant recipients be identified. Because the recombinants derive from recipient cells, a method is needed to eliminate the donor cells. The usual procedure is to employ an $F^-$ recipient containing an allele that can be selected. Genes that confer antibiotic resistance are especially useful for this purpose. For instance, after a mating between Hfr *leu*$^+$*str-s* and $F^-$ *leu*$^-$*str-r* cells, the Hfr Str-s cells can be selectively killed by plating the mating mixture on medium containing streptomycin. A selective medium that lacks leucine can then be used to distinguish between the nonrecombinant and the recombinant recipients. The $F^-$ *leu*$^-$ parent cannot grow in medium that lacks leucine, but recombinant $F^-$ *leu*$^+$ cells can grow because they possess a *leu*$^+$ gene. Only recombinant recipients—that is, cells having the genotype *leu*$^+$*str-r*—form colonies on a selective medium containing streptomycin and lacking leucine.

When a mating is done in this way, the transferred marker that is selected by the growth conditions (*leu*$^+$ in this case) is called a **selected marker**, and the marker used to prevent growth of the donor (*str-s* in this case) is called the **counterselected marker**. The counterselected marker should be located at such a place in the chromosome that most mating cells will have broken apart before it is transferred, and the selected marker must not be present in the $F^-$ cells. The selective agents can then be used to select the recombinant $F^-$ cells and eliminate the Hfr donors. Selection and counterselection are necessary in bacterial matings because recombinants constitute only a small proportion of the entire population of cells (in spite of the name "high frequency of recombination").

## Time-of-Entry Mapping

Genes can be mapped by Hfr × $F^-$ matings. However, the genetic map is quite different from all maps that we have seen so far in that it is not a linkage map but a transfer-order map. It is obtained by deliberate interruption of DNA

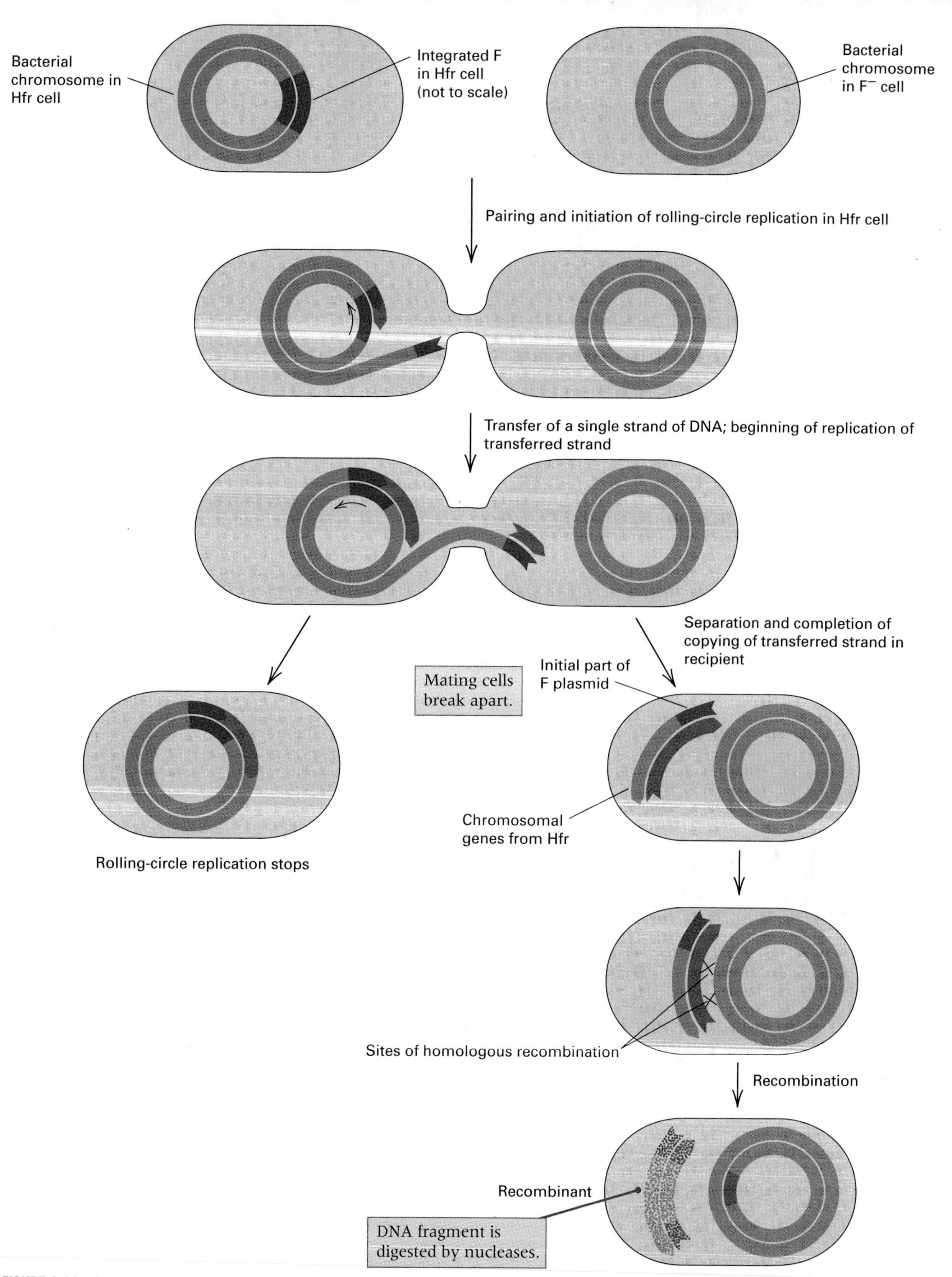


**FIGURE 9.11** Stages in the transfer and production of recombinants in an Hfr × $F^-$ mating. Pairing initiates rolling-circle replication within the F sequence in the Hfr cell, resulting in the transfer of a single strand of DNA. The single strand is converted into double-stranded DNA in the recipient. The mating cells usually break apart before the entire chromosome is transferred. Recombination takes place between the Hfr fragment and the $F^-$ chromosome and leads to recombinants containing genes from the Hfr chromosome. Note that only a part of F is transferred. This part of F is not incorporated into the recipient chromosome. The recipient remains $F^-$.

transfer in the course of mating—for example, by violent agitation of the suspension of mating cells in a kitchen blender. The time at which a particular gene is transferred can be determined by breaking the mating cells apart at various times and noting the earliest time at which breakage no longer prevents recombinants from appearing. This procedure is called the **interrupted-mating technique**. When this is done with Hfr $\times$ F$^-$ matings, the number of recombinants of any particular allele increases with the time during which the cells are in contact. This phenomenon is illustrated in **TABLE 9.1**. The reason for the increase is that different Hfr $\times$ F$^-$ pairs initiate conjugation and chromosome transfer at slightly different times.

A greater understanding of the transfer process can be obtained by observing the results of a mating with several genetic markers. For example, consider the mating

$$\text{Hfr } a^+\ b^+\ c^+\ d^+\ e^+ str\text{-}s \times \quad \text{F}^-\ a^-\ b^-\ c^-\ d^-\ e^- str\text{-}r$$

in which $a^-$ cells require nutrient A, $b^-$ cells require nutrient B, and so forth. At various times after mixing of the cells, samples are agitated violently and then plated on a series of media that contain streptomycin and different combinations of the five substances A through E (in each medium, one of the five is left out). Colonies that form on the medium lacking A are $a^+str\text{-}r$, those growing without B are $b^+str\text{-}r$, and so forth. All of these data can be plotted on a single graph to give a set of curves, as shown in part A of **FIGURE 9.12**. Four features of this set of curves are notable.

1. The number of recombinants in each curve increases with length of time of mating.
2. For each marker, there is a time (the **time of entry**) before which no recombinants are detected.
3. Each curve has a linear region that can be extrapolated back to the time axis, defining the time of entry of each gene $a^+, b^+, \ldots, e^+$.
4. The number of recombinants of each type reaches a maximum, the value of which decreases with successive times of entry.

The explanation for the time-of-entry phenomenon is the following: Not all donor cells start transferring DNA at the same time, so the number of recombinants increases with time. Transfer begins at the replication origin of F in the Hfr chromosome. Genes are transferred in linear order to the recipient, and the time of entry of a gene is the time at which that gene first enters a recipient in the population. Separation of a mating pair prevents further transfer and limits the number of recombinants seen at a particular time.

The times of entry of the genes used in the mating just described can be placed on a map, as shown in Figure 9.12B. The numbers on this map and the others are genetic distances between the markers. In this context, "genetic distance" is measured as minutes between the times of entry. Mating with another F$^-$ with genotype $b^-\ e^-\ f^-\ g^-\ h^-\ str\text{-}r$ could be used to locate the three genes *f*, *g*, and *h*. Data for the second recipient would yield a map such as that shown in Figure 9.12C. Because genes *b* and *e* are common to both maps, the two maps can be combined to form a more comprehensive map, as shown in Figure 9.12D.

Studies with different Hfr strains (Figure 9.12E) also are informative. It is usually

**Table 9.1** Data showing the production of Leu$^+$ Str-r recombinants in a cross between Hfr *leu$^+$ str-s* and F$^-$ *leu$^-$ str-r* cells when mating is interrupted at various times

| Minutes after mating | Number of Leu$^+$ Str-r recombinants per 100 Hfr cells |
|---|---|
| 0 | 0 |
| 3 | 0 |
| 6 | 6 |
| 9 | 15 |
| 12 | 24 |
| 15 | 33 |
| 18 | 42 |
| 21 | 43 |
| 24 | 43 |
| 27 | 43 |

*Note:* Minutes after mating means minutes after the Hfr and F$^-$ cell suspensions are mixed. Extrapolation of the recombination data to a value of zero recombinants indicates that the earliest time of entry of the *leu$^+$* marker is 4 minutes.

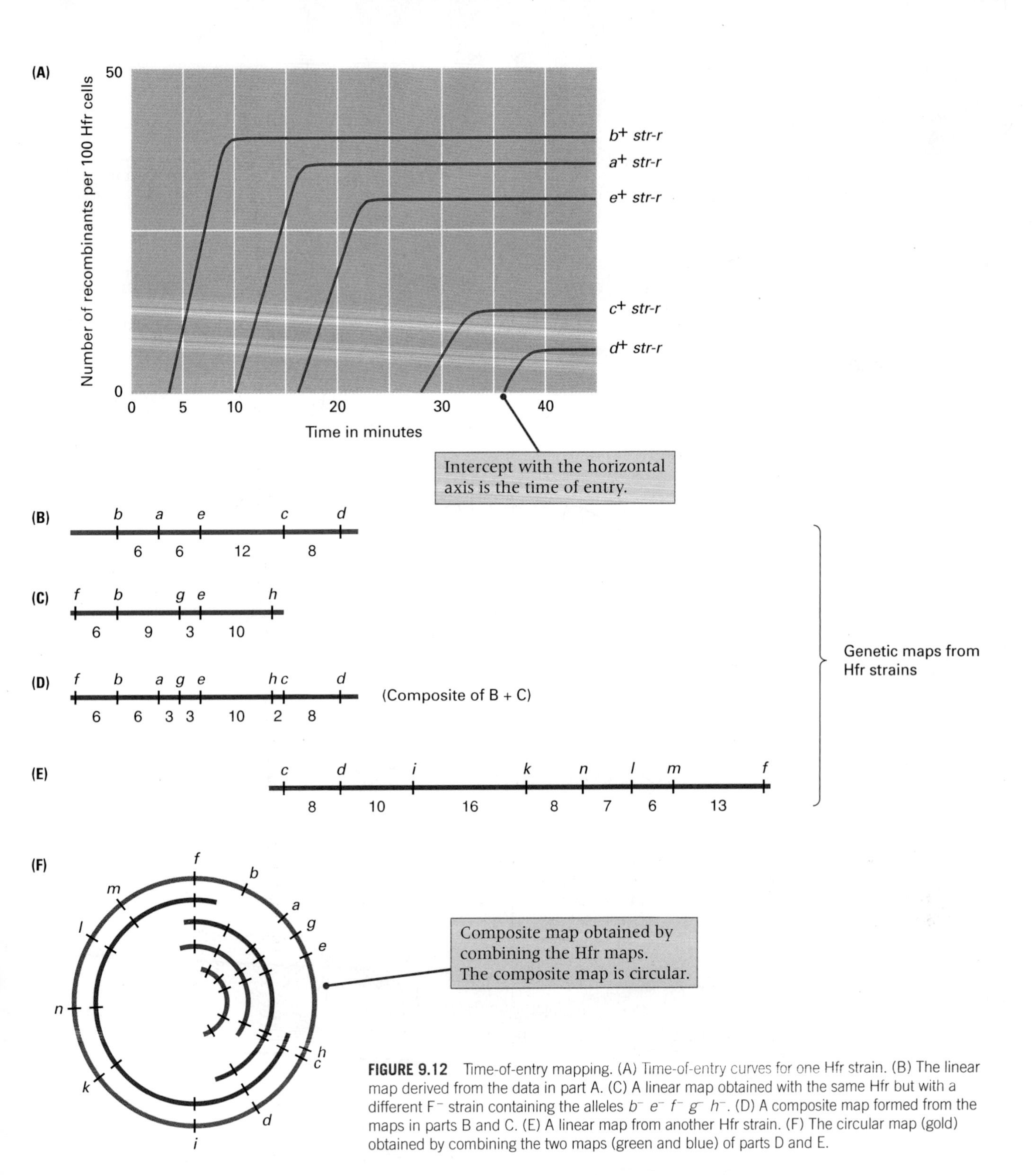


**FIGURE 9.12** Time-of-entry mapping. (A) Time-of-entry curves for one Hfr strain. (B) The linear map derived from the data in part A. (C) A linear map obtained with the same Hfr but with a different F$^-$ strain containing the alleles $b^-$ $e^-$ $f^-$ $g^-$ $h^-$. (D) A composite map formed from the maps in parts B and C. (E) A linear map from another Hfr strain. (F) The circular map (gold) obtained by combining the two maps (green and blue) of parts D and E.

found that different Hfr strains are distinguishable by their origins and directions of transfer, which indicates that F can integrate at numerous sites in the chromosome and in both possible orientations. Combining the maps obtained with different Hfr strains yields a composite map that is *circular*, as illustrated in Figure 9.12F. The circularity of the map is a result of the circularity of the *E. coli* chromosome in F$^-$ cells and the multiple points of integration of the F plasmid; if F could integrate at only one site and in one orientation, the map would be linear.

A great many such mapping experiments have been carried out, and the data have been combined to provide an accurate genetic map of the entire *E. coli* chromosome (**FIGURE 9.13**). Both the DNA molecule and the genetic map are circular. The entire chromosome requires 100

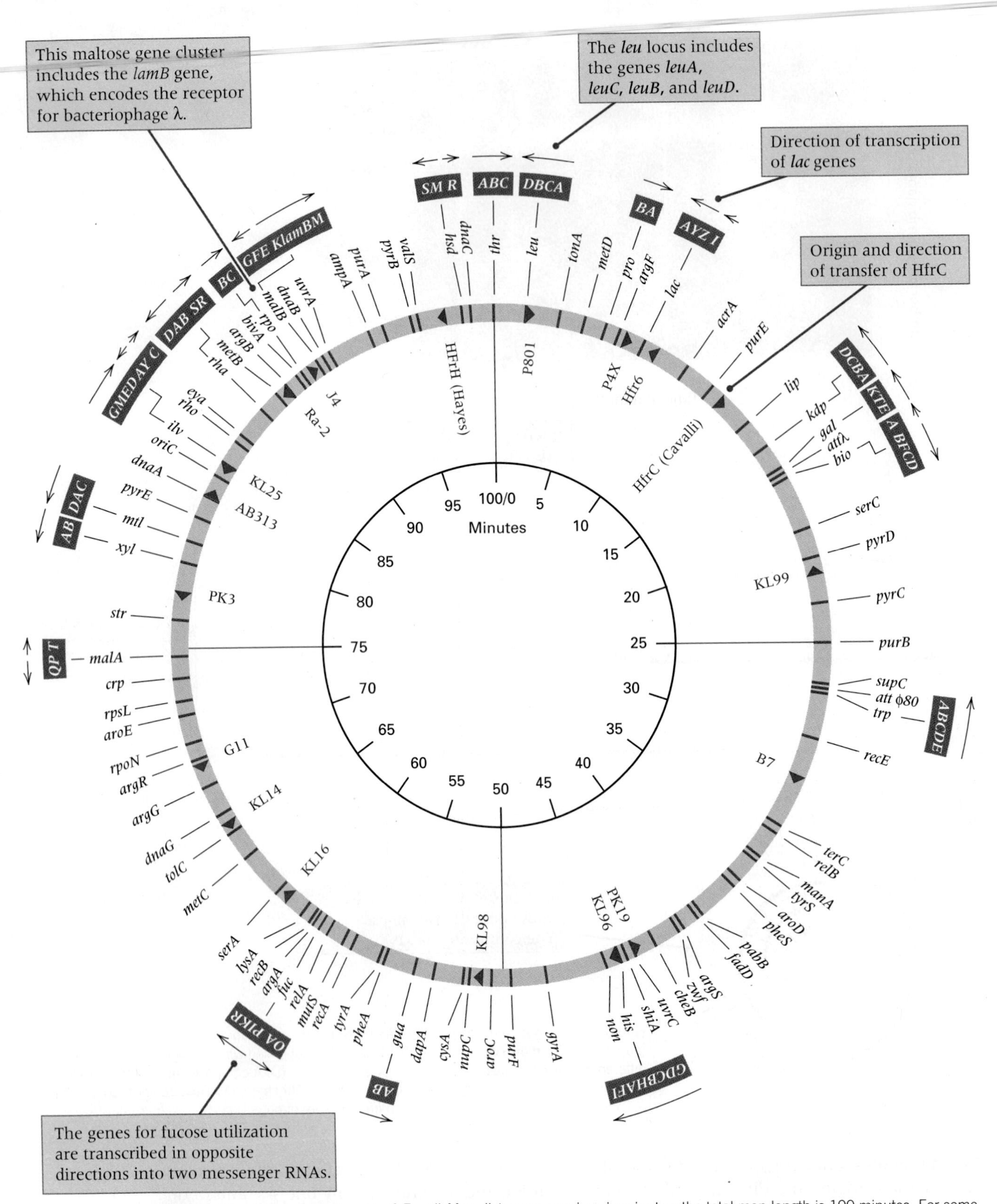

This maltose gene cluster includes the lamB gene, which encodes the receptor for bacteriophage λ.
The leu locus includes the genes leuA, leuC, leuB, and leuD.
Direction of transcription of lac genes
Origin and direction of transfer of HfrC
The genes for fucose utilization are transcribed in opposite directions into two messenger RNAs.
100/0
Minutes
HfrH (Hayes)
HfrC (Cavalli)

**FIGURE 9.13** Circular genetic map of *E. coli*. Map distances are given in minutes; the total map length is 100 minutes. For some of the loci that encode functionally related gene products, the map order of the clustered genes is shown, along with the direction of transcription and length of transcript (black arrows). The purple arrowheads show the origin and direction of transfer of a number of Hfr strains. For example, HfrH transfers *thr* very early, followed by *leu* and other genes in a clockwise direction.

minutes to be transferred (it usually breaks first), so the total map length is 100 minutes. In the outer circle in Figure 9.13, the arrows indicate the direction of transcription and the coding regions included in each transcript. The purple arrowheads show the origin and direction of transfer of a number of Hfr strains. Transfer from HfrC, for example, goes counterclockwise, starting with *purE acrA lac.*

The region between minutes 84 and 86 of the *E. coli* map is shown in greater detail in **FIGURE 9.14**. Genes and gene clusters transcribed as a unit are indicated by the arrows pointing in the direction of transcription, shown to the right of the blue line. To the left of this line are either regions that are not transcribed or genes in which the direction of transcription is unknown. The symbol *oriC* represents the origin of DNA replication of the *E. coli* chromosome.



**FIGURE 9.14** A more dense genetic map, showing genes in the region between 84 and 86 minutes of the *E. coli* chromosome. Arrows on the right indicate the direction of transcription and the length of the messenger RNA for each gene or gene cluster. On the left, *oriC* is the origin of replication of the chromosome. [Reproduced from F. C. Neidhardt, et. al. (eds). *Escherichia coli and Salmonella: Cellular and Molecular Biology, Second edition.* ASM Press, Washington, D.C. (1996).]

## ■ F′ Plasmids

Occasionally, F is excised from Hfr DNA by an exchange between the same sequences used in the integration event. However, in some

A ruptured cell of *Escherichia coli* with its DNA content pouring out. The molecule is obviously much longer than the dimensions of the cell. The cell dimensions are approximately 1 $\mu$m $\times$ 2 $\mu$m whereas the circumference of the circular DNA molecule is approximately 1600 $\mu$m. [One micrometer ($\mu$m) = $10^{-3}$ millimeter = 3.9 x $10^{-5}$ inches.] [© Dr. Gopal Murti/Photo Researchers, Inc.]

cases, the excision process is not a precise reversal of integration. Instead, breakage and reunion take place between nonhomologous sequences at the boundary of F and nearby chromosomal DNA (**FIGURE 9.15**). Aberrant excision creates a plasmid containing a fragment of chromosomal DNA, which is called an **F′ plasmid** ("F prime"). By the use of Hfr strains having different origins of transfer, F′ plasmids with chromosomal segments from many regions of the chromosome have been isolated. These elements are extremely useful because they render any recipient cell diploid for the region of the chromosome carried by the plasmid. These diploid regions make possible dominance tests and gene-dosage tests (studies of the effects on gene expression of increasing the number of copies of a gene). Because only a part of the genome is diploid, cells containing an F′ plasmid are **partial diploids**, also called **merodiploids**. We will examine genetic analysis using F′ plasmids in Chapter 11 when we discuss the *E. coli lac* genes.

## 9.5 Transduction

In the process of **transduction**, a bacterial DNA fragment is transferred from one bacterial cell to another by a phage particle containing the bacterial DNA. Such a particle is called a **transducing phage**.

### The Phage Lytic Cycle

In the **lytic cycle** of a bacteriophage, a phage particle attaches to a host bacterium and injects its nucleic acid into the cell. The phage

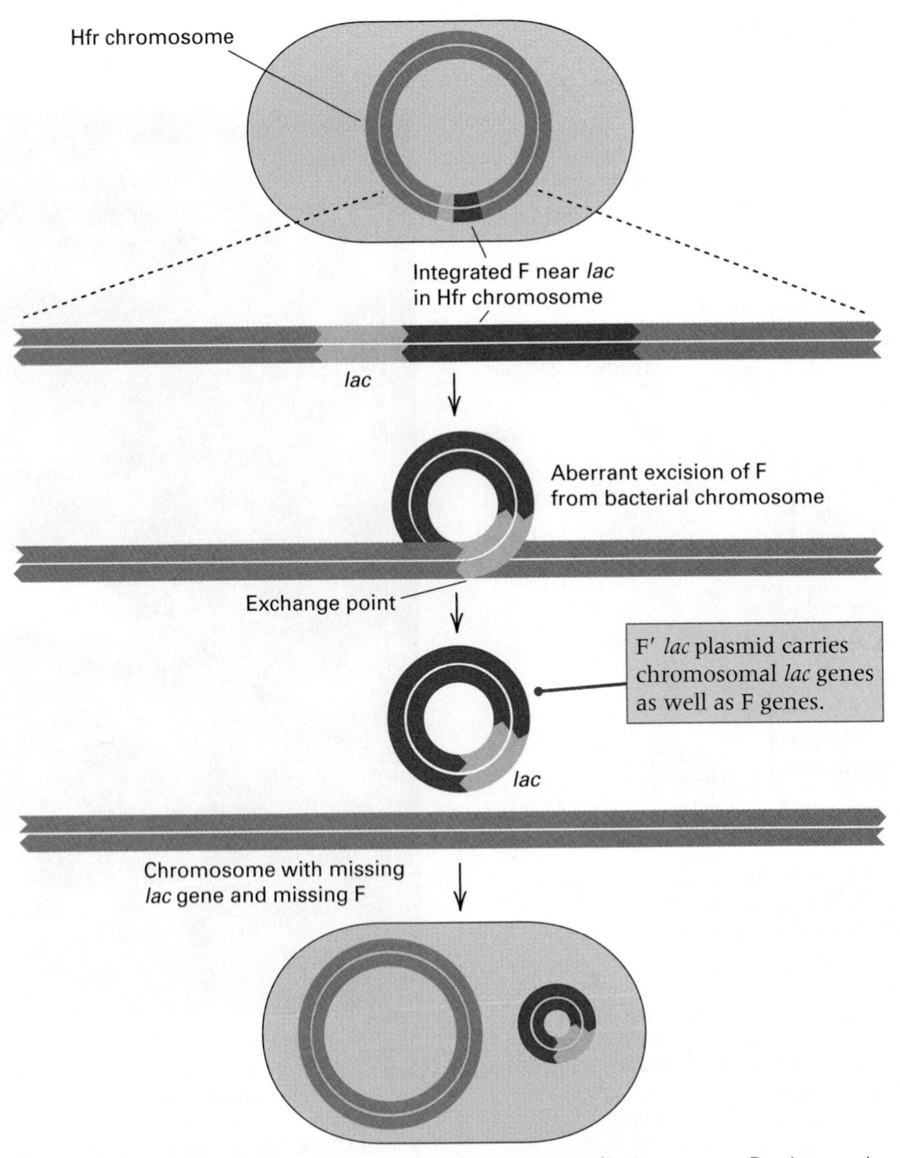


**FIGURE 9.15** Formation of an F′ *lac* plasmid by aberrant excision of F from an Hfr chromosome. Breakage and reunion are between nonhomologous regions.

## connection

### The Sex Life of Bacteria

**Joshua Lederberg and Edward L. Tatum 1946**
Yale University, New Haven, Connecticut
*Gene Recombination in* Escherichia coli

*After their discovery in the nineteenth century, bacteria were considered "things apart"—unlike other organisms in fundamental ways. Lederberg and Tatum's discovery of what at first appeared to be a conventional sexual cycle was a sensation, completely unexpected. It meant that bacteria could be considered "genetic organisms" along with yeast,* Neurospora, Drosophila, *and other genetic favorites. For this and related discoveries, Lederberg and Tatum were awarded the 1958 Nobel Prize along with George W. Beadle. In this excerpt, you will note that the authors discuss bacterial recombination as requiring a cell fusion that would bring both parental genomes together. This interpretation shows that it is possible to make exactly the right observation, and realize its significance, but not quite grasp what is really going on. The conclusion that bacterial recombination involved unidirectional transfer was reached much later, after the discovery of Hfr strains and the development of the interrupted-mating technique.*

Analysis of mixed cultures of nutritional mutants has revealed the presence of new types which strongly suggest the occurrence of a sexual process in the bacterium *Escherichia coli.* The mutants consist of strains which differ from their parent wildtype strain K-12 in lacking the ability to synthesize growth factors. As a result of these deficiencies they will only grow in media supplemented with their specific nutritional requirements. In these mutants single nutritional requirements are established as single mutational steps under the influence of x-rays or ultraviolet light. By successive treatments, strains with several requirements have been obtained. In the recombination studies here reported, two triple mutants have been used, one requiring threonine, leucine and thiamin, the other requiring biotin, phenylalanine and cystine. The strains were grown in mixed culture in complete medium. The cells were washed with sterile water and inoculated heavily into synthetic agar medium, to which various supplements had been added to allow the growth of colonies of various nutritional types. This procedure readily allows the detection of very small numbers of cell types different from the parental forms. The only new types found in "pure" cultures of the individual mutants were occasional forms which had reverted for a single factor, giving strains which required only two of the original three substances. In mixed cultures, however, a variety of types has been found. These include wildtype strains with no growth-factor deficiencies and single mutant types requiring only thiamin or phenylalanine. . . . These types can most reasonably be interpreted as instances of the assortment of genes in new combinations. In order that various genes may have the opportunity to recombine, a cell fusion would be required. . . . The fusion presumably occurs only rarely, since in the cultures investigated only one cell in a million can be classified as a recombinant type. . . . These experiments imply the occurrence of a sexual process in the bacterium *Escherichia coli.*

**These types can most reasonably be interpreted as instances of the assortment of genes in new combinations.**

Source: J. Lederberg and E. L. Tatum, *Nature* 158(1946): 558.

nucleic acid is transcribed and translated into phage proteins, and it replicates many times. Finally, newly synthesized nucleic acid molecules are packaged into protein shells (forming progeny phage), and then the particles are released from the cell. We have already examined the lytic cycle of phage T2 (see Figure 1.5 on page 7). All types of bacteriophage can undergo a lytic cycle; a phage capable *only* of lytic growth is called **virulent**. T2 is a virulent phage.

#### ■ Generalized Transduction

**Generalized transduction** is a process in which a bacteriophage particle can carry a small piece of any part of the host bacterial genome. Types of phage able to carry out this process are known as *generalized transducing phage.* In the lytic cycle, some particles that contain only DNA obtained from the host bacterium, rather than phage DNA are produced. The term *general* is used to emphasize that the bacterial DNA fragment can be derived from *any* part of the bacterial chromosome.

An example of a generalized transducing phage is the *E. coli* phage P1. During the lytic cycle of phage P1, illustrated in **FIGURE 9.16**, the overwhelming majority of progeny particles contain only phage DNA. However, as part of the process of infection, the phage makes a nuclease that cleaves the bacterial DNA into fragments. Occasionally, single fragments of bacterial DNA comparable in size to P1 DNA are

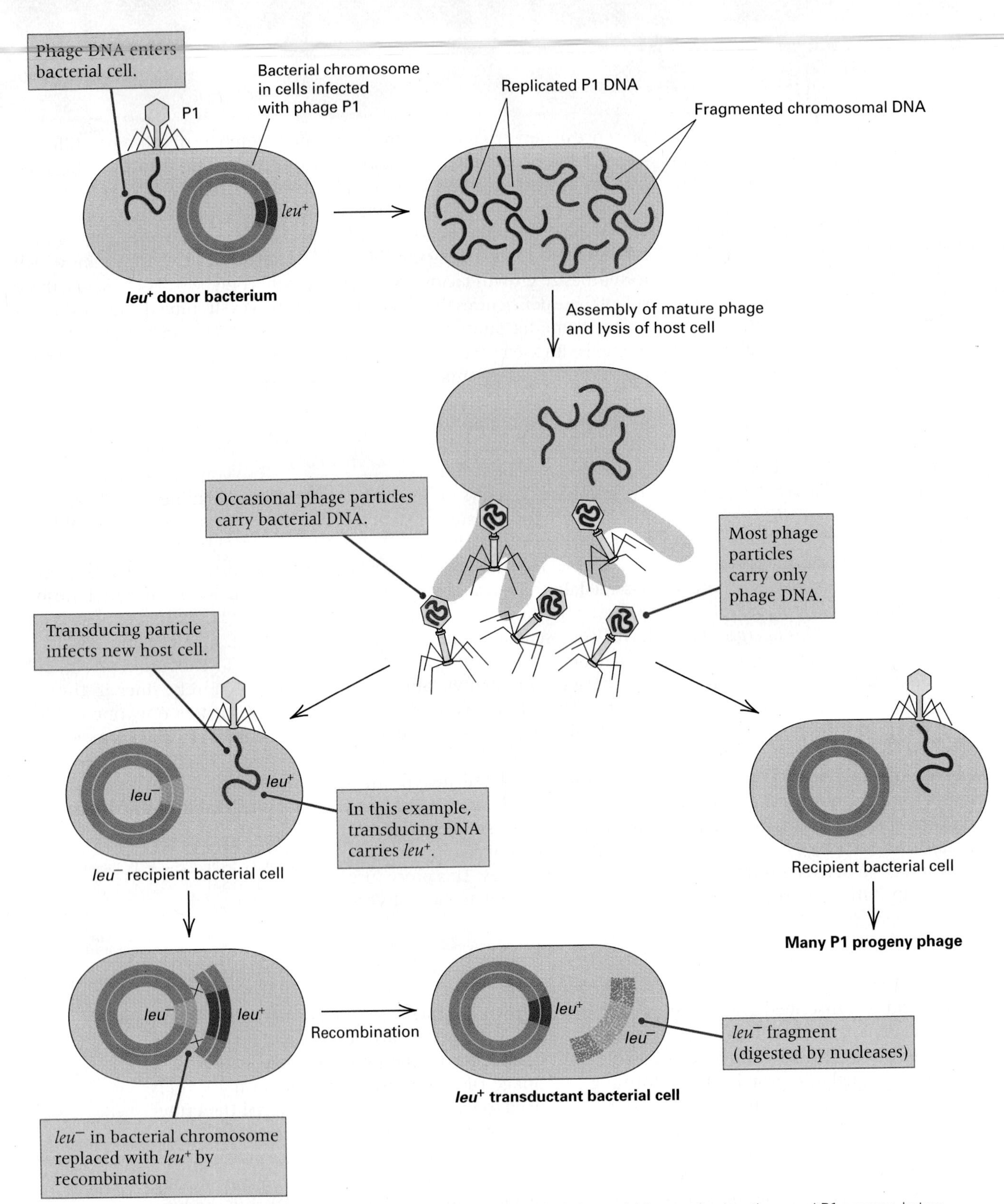


**FIGURE 9.16** Generalized transduction. Phage P1 infects a *leu*$^+$ donor, yielding predominantly normal P1 progeny, but an occasional particle carries *leu*$^+$ (or any other small segment of bacterial DNA) instead of phage DNA. When the progeny population infects a bacterial culture, the transducing particle can yield a transductant. Note that the recombination step requires two exchanges. For clarity, double-stranded phage DNA is drawn as a single line.

packaged into phage particles in place of P1 DNA (Figure 9.16). The positions of the nuclease cuts in the host chromosome are random, so a transducing particle may contain a fragment derived from any region of the host DNA. A large population of P1 phages will contain a few particles carrying any bacterial gene. On the average, any particular gene is present in roughly one transducing particle per $10^6$ progeny phages. When a transducing particle adsorbs to a bacterium, the bacterial DNA contained in the phage head is injected into the cell and becomes available for recombination with the homologous region of the host chromosome. A typical P1 transducing particle contains 100–115 kb of bacterial DNA.

Let us now examine the events that follow infection of a bacterium by a generalized transducing particle obtained by growth of P1 on wildtype *E. coli* containing a *leu*$^+$ gene. This is shown at the bottom in Figure 9.16. When such a particle adsorbs to a bacterial cell of *leu*$^-$ genotype and injects the DNA that it contains into the cell, the cell survives because the phage head contained only bacterial genes and no phage genes. A recombination event exchanging the *leu*$^+$ allele carried by the phage for the *leu*$^-$ allele carried by the host converts the genotype of the host cell from *leu*$^-$ into *leu*$^+$. In such an experiment, typically about 1 *leu*$^-$ cell in $10^6$ becomes *leu*$^+$. Such frequencies are easily detected on selective growth medium. For example, if the infected cell is placed on solid medium that lacks leucine, it is able to multiply and a *leu*$^+$ colony forms. A colony does not form unless recombination inserted the *leu*$^+$ allele.

The fragment of bacterial DNA contained in a transducing particle is large enough to include about 50 genes, so transduction provides a valuable tool for genetic linkage studies of short regions of the bacterial genome. Consider a population of P1 prepared from a *leu*$^+$ *gal*$^+$ *bio*$^+$ bacterium. This sample contains particles able to transfer any of these alleles to another cell; that is, a *leu*$^+$ particle can transduce a *leu*$^-$ cell to *leu*$^+$, or a *gal*$^+$ particle can transduce a *gal*$^-$ cell to *gal*$^+$. Furthermore, if a *leu*$^-$ *gal*$^-$ culture is infected with phage, then both *leu*$^+$ *gal*$^-$ and *leu*$^-$ *gal*$^+$ bacteria are produced. However, *leu*$^+$ *gal*$^+$ colonies do not arise because the *leu* and *gal* genes are too far apart to be included in the same DNA fragment (**FIGURE 9.17**, part A).

The situation is quite different with a recipient cell with genotype *gal*$^-$ *bio*$^-$, because the *gal* and *bio* genes are so closely linked that both genes are sometimes present in a single DNA fragment carried in a transducing particle—namely, a *gal*$^+$–*bio*$^+$ particle (Figure 9.17B). However, not all *gal*$^+$ transducing particles also include *bio*$^+$, nor do all *bio*$^+$ particles include *gal*$^+$. The probability of both markers being in a single particle—and hence the probability of simultaneous transduction of both markers (**cotransduction**)—depends on how close to each other the genes are. The closer they are, the greater the frequency of cotransduction.

Cotransduction of the *gal*$^+$–*bio*$^+$ pair can be detected by plating infected cells on the appropriate growth medium. If *bio*$^+$ transductants are selected by spreading the infected cells on a glucose-containing medium that lacks biotin, then both *gal*$^+$ *bio*$^+$ and *gal*$^-$ *bio*$^+$ colonies will grow. If these colonies are tested for the *gal* marker, then 42 percent are found to be *gal*$^+$ *bio*$^+$ and the rest *gal*$^-$ *bio*$^+$; similarly, if *gal*$^+$ transductants are selected, then about 42 percent are found to be *gal*$^+$ *bio*$^+$. In other words, the *frequency of cotransduction* of *gal* and *bio* is 42 percent, which means that 42 percent of all transducing particles that contain one gene also include the other.

Studies of cotransduction can be used to map closely linked genetic markers by means of three-factor crosses analogous to those described in Chapter 5. That is, P1 is grown on wildtype bacteria and used to transduce cells that carry a mutation of each of three closely linked genes. Cotransductants containing various pairs of wildtype alleles are examined. The gene located in the middle can be identified because its wildtype allele is nearly always cotransduced with the wildtype alleles of the genes that flank it. For example, in Figure 9.17B, a genetic marker located between *gal*$^+$ and *bio*$^+$ will almost always be present in *gal*$^+$ *bio*$^+$ transductants.

How is the frequency of cotransduction between genes related to their map distance in minutes in the standard *E. coli* map? The theoretical relationship depends on the size of a molecule in a transducing phage relative to the size of the entire chromosome. For bacteriophage P1, the distance in minutes between markers on a conjugative map is related to the frequency of cotransduction approximately as follows:

$$\text{Map distance in minutes} = 2 - 2 \times (\text{Cotransduction frequency})^{1/3} \quad (1)$$

This equation implies that the frequency of cotransduction falls off very rapidly as the distance between the markers increases. In the formula, the shortest distance for which the cotransduction frequency equals 0 is 2 minutes; hence any markers separated by more than

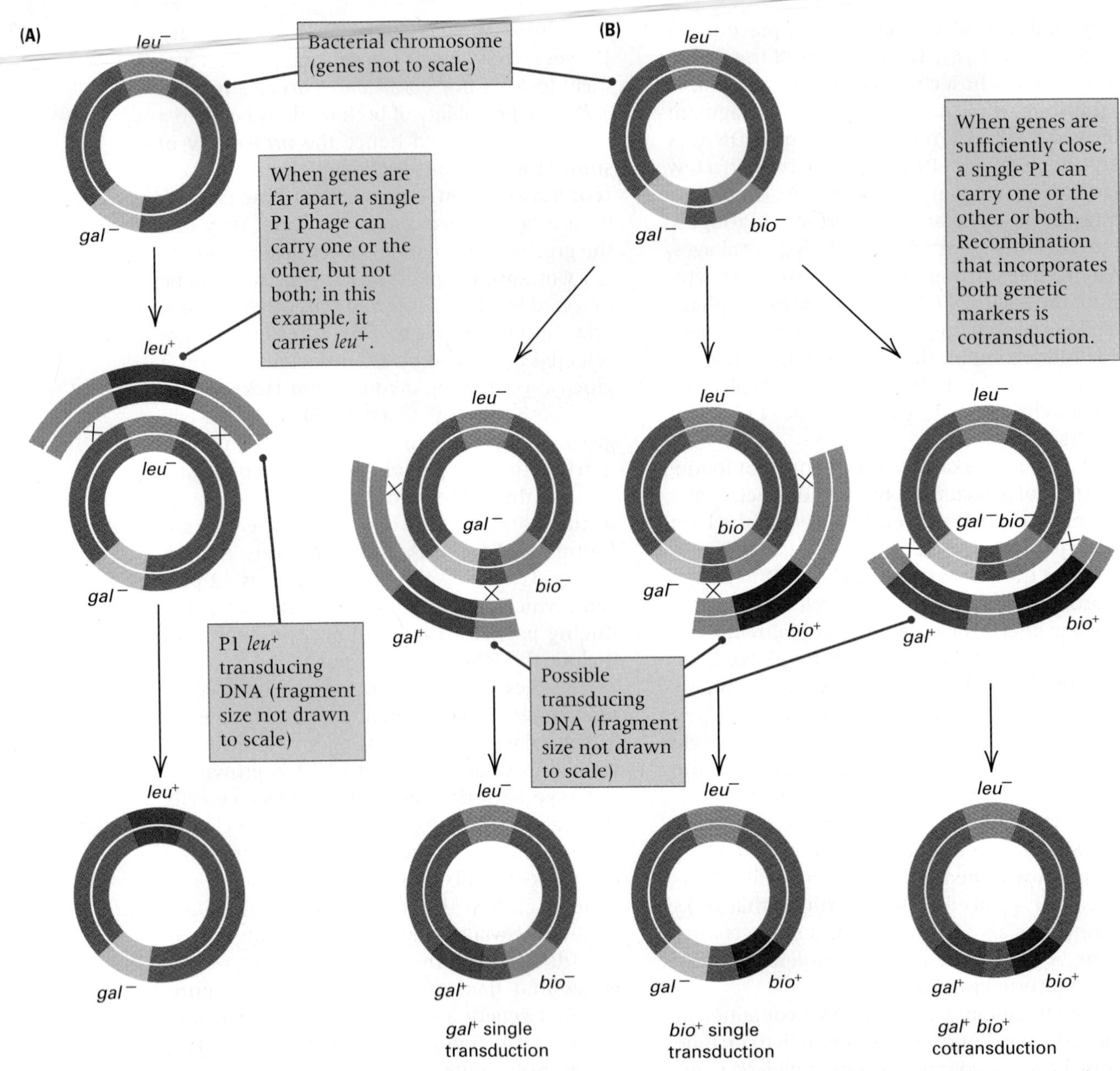


**FIGURE 9.17** Demonstration of linkage of the *gal* and *bio* genes by cotransduction. (A) A P1 transducing particle carrying the $leu^+$ allele can convert a $leu^-$ $gal^-$ cell into a $leu^+$ $gal^-$ genotype (but cannot produce a $leu^+$ $gal^+$ genotype). (B) The transductants that could be formed by three possible types of transducing particles: one carrying $gal^+$, one carrying $bio^+$, and one carrying the linked alleles $gal^+$ $bio^+$. The third type results in cotransduction. For clarity, the size of the DNA fragment in a transducing particle, relative to the size of the bacterial chromosome, is not drawn to scale.

2 minutes will not be cotransduced by phage P1. For *gal* and *bio*, which have a cotransduction frequency of about 42 percent, the distance equals 0.5 minute. (The *gal* and *bio* genes are located at 17.0 and 17.5 minutes, respectively, in the genetic map in Figure 9.13.)

## 9.6 Bacteriophage Genetics

Except for newly arising mutants, phage progeny from a bacterium infected by a single phage have the same genotype as the parental phage. However, if two phage particles with *different* genotypes infect a single bacterial cell, new genotypes can arise by genetic recombination. This process differs significantly from genetic recombination in eukaryotes in two ways:

1. The number of participating DNA molecules differs from one cell to the next.
2. Reciprocal recombinants are not always recovered in equal frequencies from a single infected cell.

Several phage particles can infect a single bacterium, and the DNA of each particle can replicate. In a multiply infected cell in which a lytic cycle is under way, there are genetic exchanges between

## connection

## Origin of Phage Genetics

*Alfred D. Hershey and Raquel Rotman 1948*
*Washington University, St. Louis, Missouri*
*Genetic Recombination Between Host-Range and Plaque-Type Mutants of Bacteriophage in Single Bacterial Cells*

*Even though bacteriophages require a bacterial host to survive and reproduce, their reproductive cycle conforms in all respects to that of cellular organisms, including DNA as their genetic material, semiconservative replication of DNA, heritable mutations, and, as shown first in this report, a process of recombination. Hershey and Rotman were quite unsure whether phage recombination resulted from physical exchange of DNA molecules and whether, when two molecules recombined, both reciprocal products of the exchange were produced. Later research showed that the exchanges are reciprocal and that they are physical exchanges.*

We have shown that any two of several mutants of the bacterial virus T2 interact with each other, in bacterial cells infected with both, to give rise to wildtype and double mutant genetic recombinants. . . . In principle, the experimental technique we have to describe is very similar to genetic crossing. One starts with a pair of mutants, each corresponding to a mutant haploid germ cell differing from wildtype by a different unit change. Bacterial cells are infected with both members of the pair, and during viral growth the pair interact to produce viral progeny corresponding to germ cells of a new generation, but now including some individuals differing from wildtype by both unit changes, and other individuals differing from wildtype not at all. The analogy to other genetic recombination is obvious, and it is natural to look for a common mechanism. . . . The crosses between *h* and *r* mutants yield the linkage system shown below.

***The analogy to other genetic recombination is obvious, and it is natural to look for a common mechanism.***

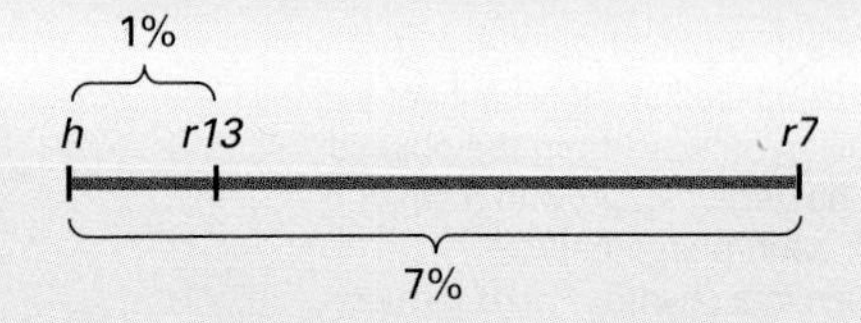


In the diagram, *h* refers to bacterial host range and *r13* and *r7* to two different, rapidly lysing mutants. The data show that the *h* locus is very closely linked to *r13* (less than 1 percent of wildtype), and that the linkage relation between *r13* and *r7* is 6 percent. . . . The results further show that the two recombinants [wildtype and double mutant] appear in equal numbers in any one cross, and that pairs of reverse crosses yield equal numbers of recombinants. It is these relations that increase the resemblance to simple types of Mendelian segregation. . . . A hypothesis is proposed according to which one visualizes genetic interaction not between two viral particles, but between two sets of independently multiplying chromosome-like structures. Genetic exchange occurs either by reassortment of these structures, or by something like crossing over between homologous pairs, depending on the structural relation between the genetic factors concerned. The interpretation made brings the linkage relations into superficial agreement with the requirements of linear structure, but there is little evidence that the genetic exchanges are reciprocal, and accordingly little evidence that they are material exchanges.

Source: A. D. Hershey and R. Rotman, *Genetics* 34(1949): 44–71.

phage DNA molecules. If the infecting particles carry different mutations as genetic markers, then recombinant phages result. Recombination can take place anywhere along two homologous DNA molecules.

### Plaque Formation and Phage Mutants

Phages are easily detected because, in a lytic cycle, an infected cell breaks open (a process called **lysis**) and releases phage particles to the growth medium. The formation of plaques can be observed as outlined in part A of **FIGURE 9.18**. A large number of bacteria (about $10^8$) are placed on a solid medium. After a period of growth, a continuous turbid layer of bacteria results. If a phage is present at the time the bacteria are placed on the medium, it adsorbs to a cell, and shortly afterward the infected cell lyses and releases many phages. Each of these progeny adsorbs to a nearby bacterium, and after another lytic cycle, these bacteria in turn release phage that can infect still other bacteria in the vicinity. (Progeny phage remain close to their site of origin because their size prevents diffusion in agar medium.) These cycles of infection continue, and after several hours the phage destroy all of the bacteria in a localized area, giving rise to a clear, transparent region—a **plaque**—in the otherwise turbid layer of confluent bacterial growth. Phage can multiply only in growing bacterial cells, so exhaustion of nutrients in the growth medium limits phage multiplication and

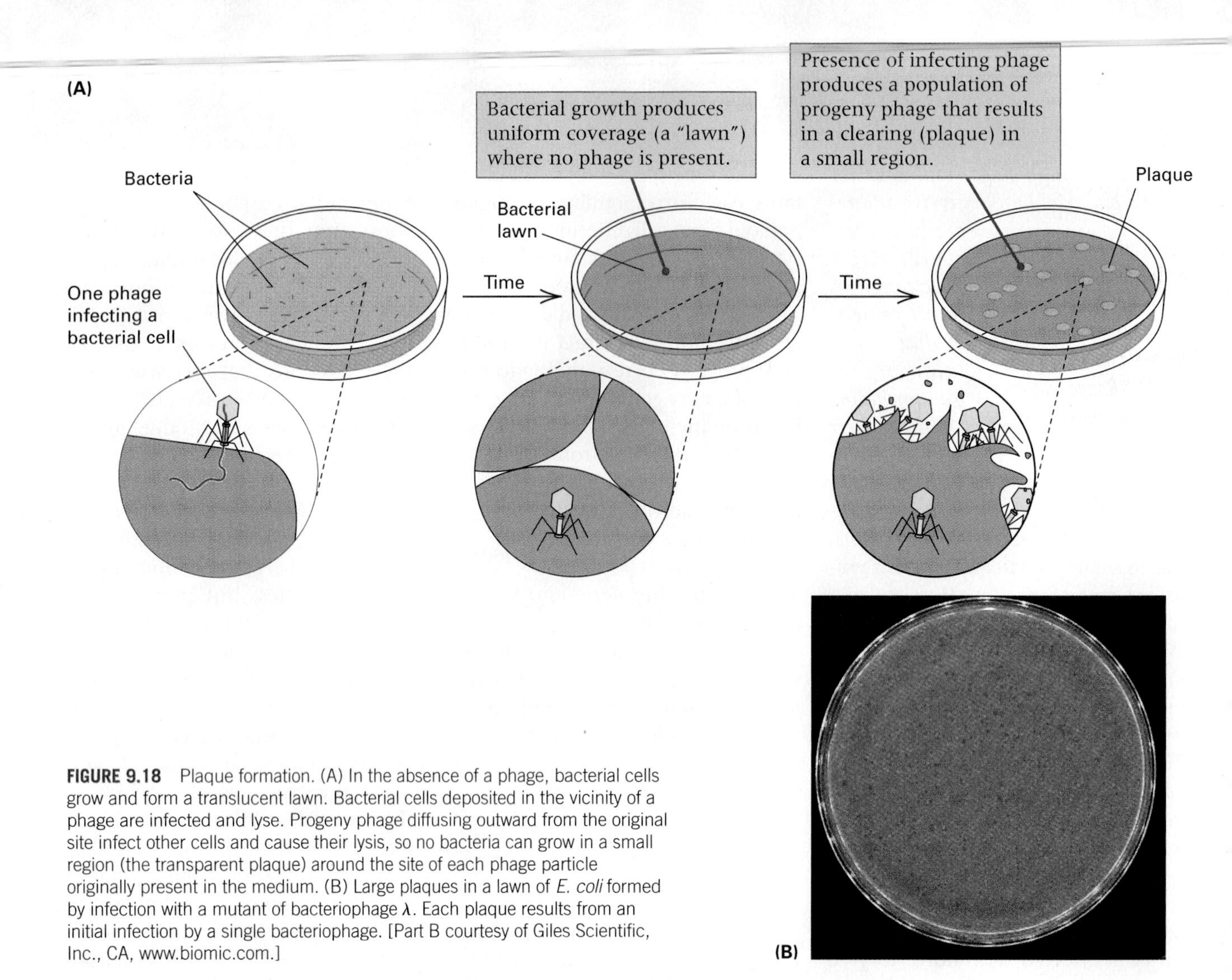
(A)
Bacterial growth produces uniform coverage (a "lawn") where no phage is present.
Presence of infecting phage produces a population of progeny phage that results in a clearing (plaque) in a small region.
Bacteria
Plaque
Bacterial lawn
One phage infecting a bacterial cell
Time
Time
(B)

**FIGURE 9.18** Plaque formation. (A) In the absence of a phage, bacterial cells grow and form a translucent lawn. Bacterial cells deposited in the vicinity of a phage are infected and lyse. Progeny phage diffusing outward from the original site infect other cells and cause their lysis, so no bacteria can grow in a small region (the transparent plaque) around the site of each phage particle originally present in the medium. (B) Large plaques in a lawn of *E. coli* formed by infection with a mutant of bacteriophage λ. Each plaque results from an initial infection by a single bacteriophage. [Part B courtesy of Giles Scientific, Inc., CA, www.biomic.com.]

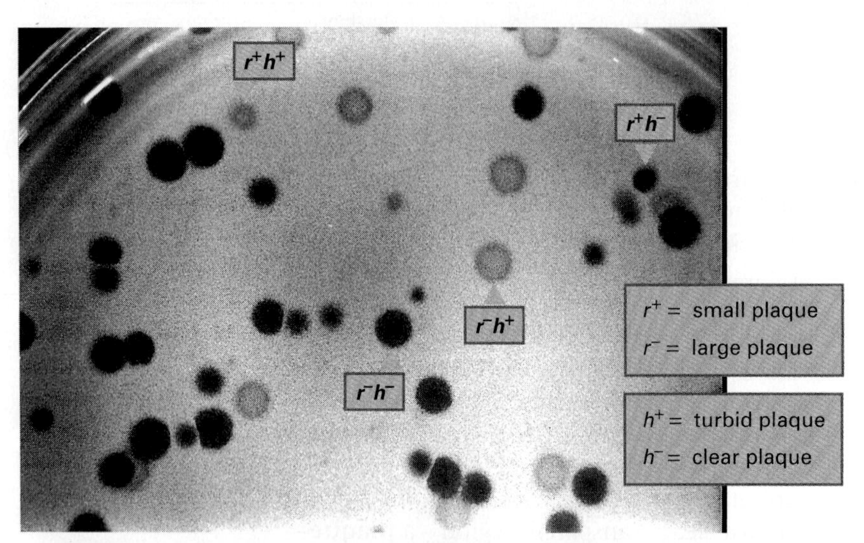
r+h+
r+h−
r−h+
r−h−
r+ = small plaque
r− = large plaque
h+ = turbid plaque
h− = clear plaque

**FIGURE 9.19** A phage cross is performed by infecting host cells with both parental types of phage simultaneously. This example shows the progeny of a cross between T4 phage of genotypes $r^- h^+$ and $r^+ h^-$ when both parental phages infect cells of *E. coli*. The $r^+h^+$ and $r^-h^-$ genotypes are recombinants. [Courtesy of Leslie Smith and John W. Drake, National Institutes of Health.]

the size of the plaque. Because a plaque is a result of an initial infection by one phage particle, the number of individual phages originally present on the medium can be counted.

The genotypes of phage mutants can be determined by studying the plaques. In some cases, the appearance of the plaque is sufficient. For example, phage mutations that decrease the number of phage progeny from infected cells often yield smaller plaques. Large plaques can be produced by mutants that cause premature lysis of infected cells, so that each round of infection proceeds more quickly (Figure 9.18B).

## Genetic Recombination in the Lytic Cycle

If two phage particles with different genotypes infect a single bacterium, then some phage progeny are genetically recombinant. **FIGURE 9.19** shows plaques from progeny of a mixed infection with *E. coli* phage T4 mutants. The $r^-$ (*rapid lysis*) allele results in large plaques, and the $h^-$

(*host range*) allele results in clear plaques. The cross is written as:

$r^- h^+$ (*large turbid plaque*) $\times$

$r^+ h^-$ (*small clear plaque*)

Four plaque types can be seen in Figure 9.19. Two—the large turbid plaque and the small clear plaque—correspond to the phenotypes of the parental phages. The other two phenotypes—the large clear plaque and the small turbid plaque—are recombinants that correspond to the genotypes $r^- h^-$ and $r^+ h^+$, respectively. When many bacteria are infected, approximately equal numbers of reciprocal recombinant types are usually found among the progeny phage. In an experiment like that shown in Figure 9.19, in which each of the four genotypes yields a different phenotype of plaque morphology, the number of each of the genotypes can be counted by examining each of the plaques that is formed. The recombination frequency, expressed as a percentage, is defined as:

$$\text{Recombination frequency} = \frac{\text{Number of recombinant phage}}{\text{Total number of phage}} \times 100$$

Recombination frequencies can be used to estimate map distances, just as they are in eukaryotes.

## Genetic and Physical Maps of Phage T4

Early mapping experiments indicated that mutations in T4 mapped in three separate clusters. However, all three clusters showed linkage to one another. In elegant experiments with three-point crosses, George Streisinger and colleagues demonstrated in 1964 that the genetic map for T4 phage is actually circular.

In each cross, they mapped three or four genetic markers with respect to one another and proceeded systematically through the entire T4 genome, eventually demonstrating all of the linkages shown in **FIGURE 9.20**. Many additional genes were identified and mapped later by other researchers (**FIGURE 9.21**), and the results were fully consistent with the circular map. In Figure 9.21, the regions indicated in the innermost circle are the three clusters of T4 markers that had initially been identified and mapped. The outermost circle in Figure 9.21 presents a much larger set of markers establishing the overall circularity of the

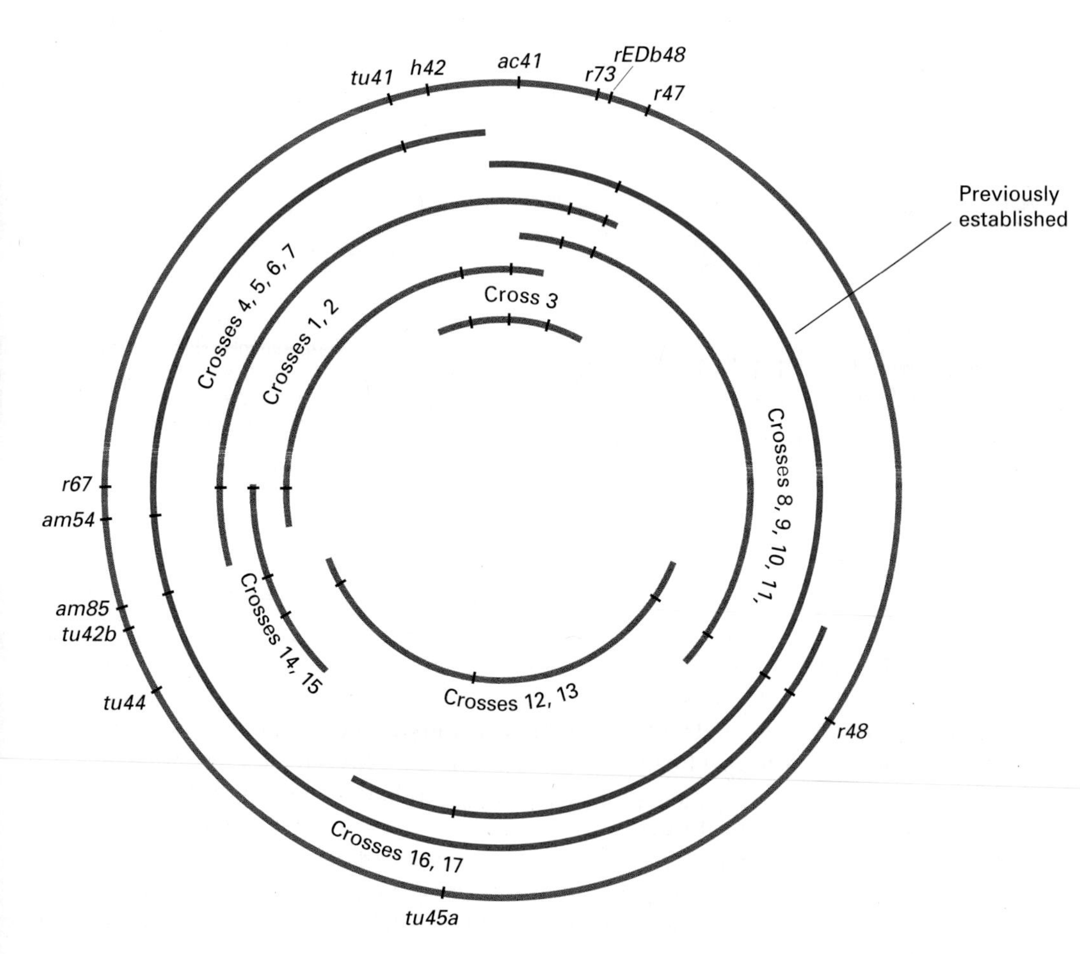


**FIGURE 9.20** Circular genetic map of phage T4. Each arc connects three or four markers that were mapped in the crosses indicated. The circular map results when all data are considered together. [Adapted from G. Streisinger, R. S. Edgar, and G. H. Denhardt. *Proc. Natl. Acad. Sci. USA* 51(1964): 775–779.]

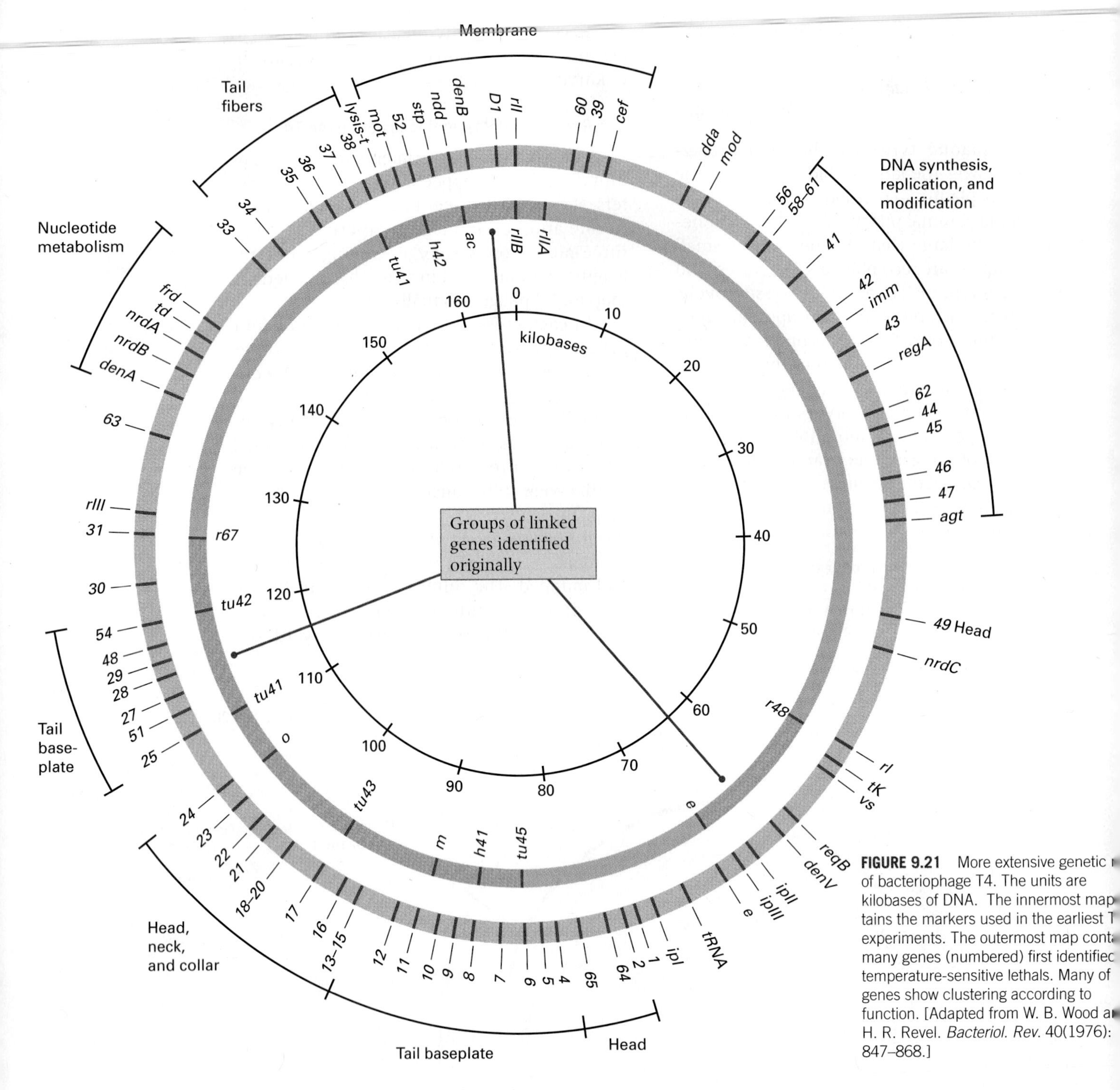


**FIGURE 9.21** More extensive genetic of bacteriophage T4. The units are kilobases of DNA. The innermost map tains the markers used in the earliest T experiments. The outermost map cont many genes (numbered) first identified temperature-sensitive lethals. Many of genes show clustering according to function. [Adapted from W. B. Wood a H. R. Revel. *Bacteriol. Rev.* 40(1976): 847–868.]

genetic map beyond doubt. The T4 genetic map in Figure 9.21 also indicates that genes in T4 show extensive clustering according to the function for which they are required. For example, there is a large cluster of genes for DNA replication in the upper right quadrant, and there is a cluster of genes for phage head components near the bottom.

In view of the circular nature of the T4 genetic map, it came as quite a surprise to find that the DNA molecule in a T4 phage particle is a single *linear* molecule. This discovery was completely unexpected and at first seemed inconsistent with the genetic data. However, the discrepancy was resolved by the finding that the very ends of the phage T4 DNA are duplicated, or have **terminal redundancy**. Because of this redundancy, each molecule is about 2 percent longer than would be expected. As the DNA is replicating inside the cell, recombination between each of the duplicated ends of one T4 genome with homologous sequences in other T4 genomes results in products much longer than can be contained in a T4 head (**FIGURE 9.22**). Such *concatemeric* molecules are

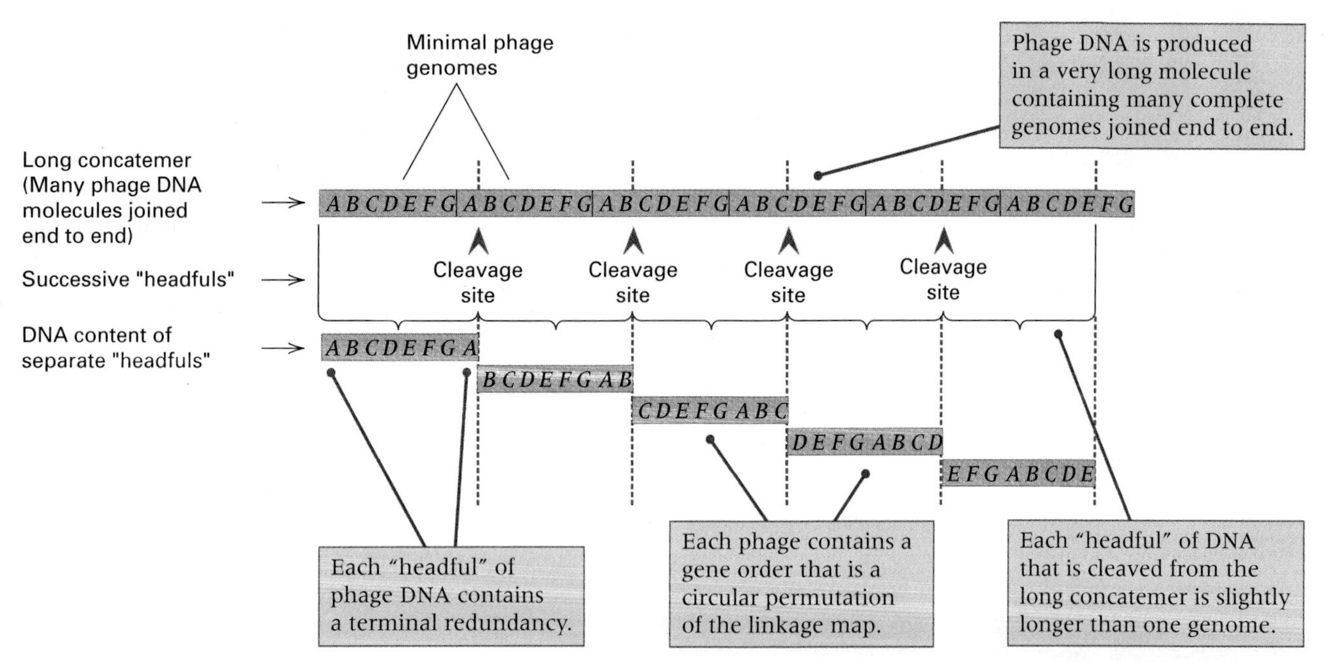


**FIGURE 9.22** Terminal redundancy and circular permutation of the phage T4 linear DNA molecule. A long *concatemer*, or series of molecules linked together, is formed inside the cell by genetic recombination. Each successive headful of DNA is slightly longer than unit size and contains terminal redundancy. Except for the terminal redundancy, the molecules are also circular permutations of each other.

formed because recombination in the T4 genome is very frequent, averaging about 20 recombination events per chromosome. When the DNA is packaged, it is enzymatically cleaved into "headful" packages consisting of about 102 percent of the minimum length of the T4 genome; hence the duplication at the ends. Because of the headful packaging mechanism, each T4 DNA molecule is terminally redundant. Moreover, in a whole population of DNA molecules present in a phage sample, except for the short terminal redundancy, each of the molecules is also a **circular permutation** of the others. (A molecule is a circular permutation of another if the other sequence can be created by joining its ends and cleaving at an internal position.) Because each T4 molecule is a circular permutation, different molecules begin at different points in the DNA sequence, but they always incorporate a little more than one complete genome. The properties of terminal redundancy and circular permutation account for the circular genetic map.

## Fine Structure of the *rII* Gene in Bacteriophage T4

In Section 5.6, we described experiments with *Drosophila* that showed that recombination within a gene could take place. A genotype heterozygous for two different mutant alleles of the same gene could, at low frequency, generate recombinant chromosomes carrying either both mutations or neither mutation. These experiments gave the first indication of intragenic recombination and the first evidence that genes have a fine subunit structure. Other studies of genes were performed in the years that followed this experiment, but none could equal the fine-structure mapping of the *rII* gene in bacteriophage T4 carried out by Seymour Benzer. Using novel genetic mapping techniques that reduced the number of required crosses from more than half a million to several thousand, Benzer succeeded in mapping 2400 independent mutations in the *rII* locus of phage T4.

Wildtype T4 bacteriophage is able to multiply in *E. coli* strains B and K12(λ) and gives small ragged plaques. The designation K12(λ) signifies a K12 strain that is lysogenic for bacteriophage λ (described in Section 9.7). Mutations in the *rII* gene of T4 result in large round plaques on strain B but completely prevent T4 from propagating in strain K12(λ). If *E. coli* cells of strain B are all infected with two different *rII* mutants, then recombination between the mutations can be detected, even if the frequency is very low, by taking advantage of the inability of *rII* mutants to grow on K12(λ). Plating the progeny phages on K12(λ) selects for growth of the $rII^+$ recombinant progeny, because only these recombinants can grow. Furthermore, because very large numbers of progeny phage can be examined (numbers of $10^{10}$ bacteriophage/ml are not unusual), even very low frequencies of recombination can be detected. Typical results for three different *rII* mutations might yield a map like the following:

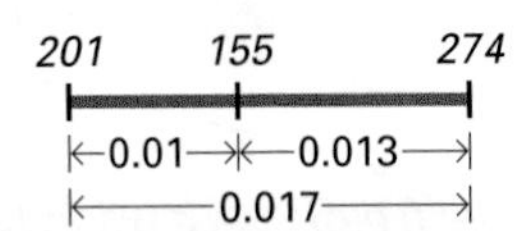


where the numbers above the line designate *rII* alleles and the numbers below are map distances given as two times the percent of $rII^+$, namely:

$$\text{Map distance} = \frac{2 \times \left[\begin{array}{c}\text{Number of}\\ \text{plaques per phage}\\ \text{on strain K12}(\lambda)\end{array}\right]}{\text{Number of plaques per phage on strain B}} \times 100$$

(In this case the ratio is multiplied by 2 because only the $rII^+$ recombinants are detected, rather than both reciprocal products of recombination.)

Some mutations failed to recombine with several mutations, each of which recombined with the others. These were interpreted to be deletion mutations, because they prevented recombination with two or more "point" mutations known to be at different sites in the gene. Each deletion eliminated a part of the bacteriophage genome, including a region of the *rII* gene. The use of deletions greatly simplified the ordering and mapping of thousands of mutations.

**FIGURE 9.23** depicts the array of deletion mutations used for mapping. Deletion mapping is

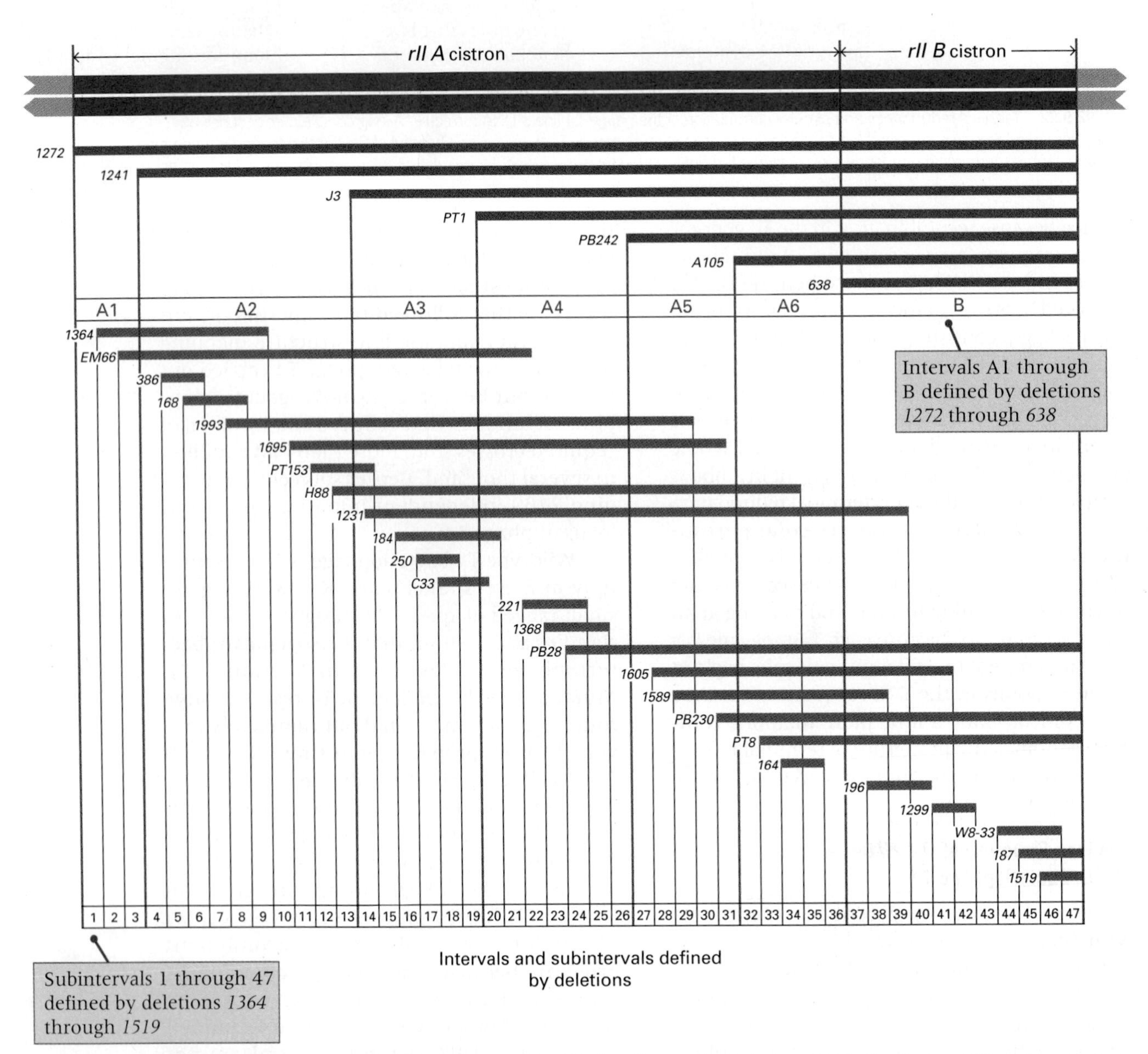


**FIGURE 9.23** The array of deletion mutations used to divide the *rII* locus of bacteriophage T4 into 7 regions and 47 smaller subregions. The extent of each deletion is indicated by a horizontal orange or dark blue bar. Any deletion endpoint used to establish a boundary between regions or subregions is indicated with a vertical line. [Adapted from S. Benzer, *Proc. Natl. Acad. Sci. USA* 47(1961): 403–426.]

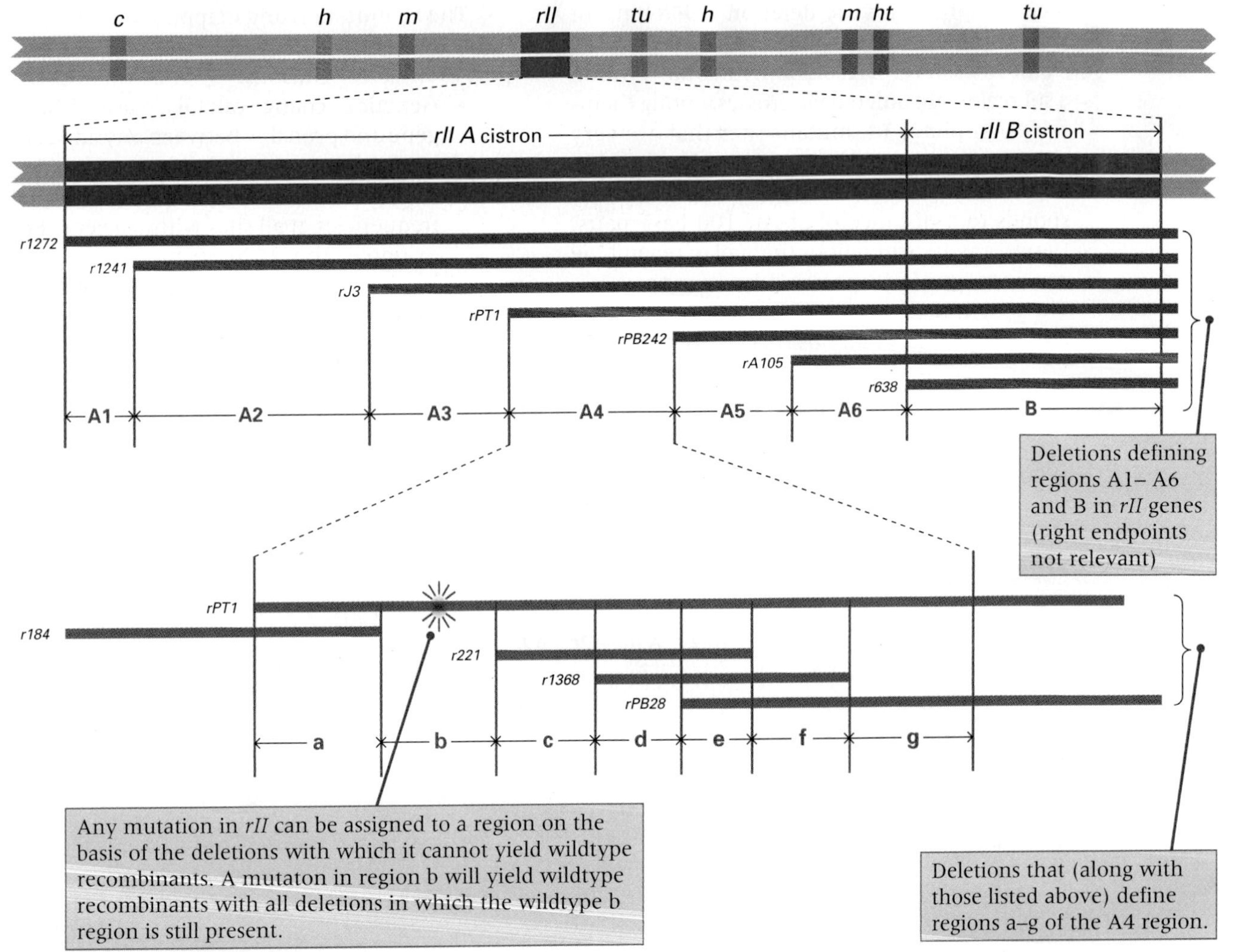


**FIGURE 9.24** The location of the *rII* locus (green) in relation to other genetic markers. Major subdivisions of the *rII* region are defined by the left ends of seven deletion mutations (deleted DNA indicated in orange). All seven deletions extend through the right end of the *rII* region. Further subdivision of the A4 region is made possible by the additional deletions shown at the bottom of the figure (deleted DNA indicated in dark blue). For simplicity, only the deleted regions are shown.

based on the presence or absence of recombinants; each cross yields a yes-or-no answer and so avoids many of the ambiguities of genetic maps based on frequencies of recombination. In any cross between an unknown "point" mutation (for example, a simple nucleotide substitution) and one of the deletions, the presence of wildtype progeny means that the point mutation is outside of the region missing in the deletion. (The reciprocal product of recombination, which carries the deletion plus the point mutation, is not detected in these experiments.) On the other hand, if the point mutation is present in the region missing in the deletion, then wildtype recombinant progeny cannot be produced. Because each cross clearly reveals whether a particular mutation is within the region missing in the deletion, deletion mapping also substantially reduces the amount of work needed to map a large number of mutations.

The series of crosses that would be made in order to map a particular *rII* mutation is presented in **FIGURE 9.24**, in which the large intervals A1 through A6 plus B are the same as those in Figure 9.23. To illustrate the method, suppose that a particular mutation being examined is located in the region denoted A4. This mutation would fail to yield wildtype recombinants in crosses with the large deletions *r1272, r1241, rJ3,* and *rPT1,* but it would yield wildtype recombinant progeny in crosses with the large deletions *rPB242, rA105,* and *r638.* Conversely, any mutation that yielded the same pattern of outcomes in crosses with the large deletions would be assigned to region A4. Still finer resolution of the genetic map within region A4 is made possible by the set of deletions shown at the bottom of Figure 9.24, the endpoints of which define seven subregions (a through g) within A4. For example, a mutation in region A4 that yields wildtype recombinants with the deletion *r1368* but not with *r221* would be assigned to the c subregion. At the finest level of resolution, mutations within a subregion are ordered via crosses among themselves. In phage T4, mutant sites that are very close can be separated by recombination, because on the average, 1 percent recombination corresponds to a distance of about 100 base pairs. Hence, any two mutations that fail to recombine can be assigned to the same site within the gene. The genetic map generated for a large number of independent *rII* mutations is given in **FIGURE 9.25**.

The *rII* mutation and mapping studies were important because they gave experimental support to these conclusions:

- Genetic exchange can take place within a gene and probably between any adjacent base pairs.
- Mutations are not produced at equal frequencies at all sites within a gene. For example, the 2400 *rII* mutations were located at only 304 sites. One of these sites accounted for 474 mutations (Figure 9.25); a site that shows such a high fre-

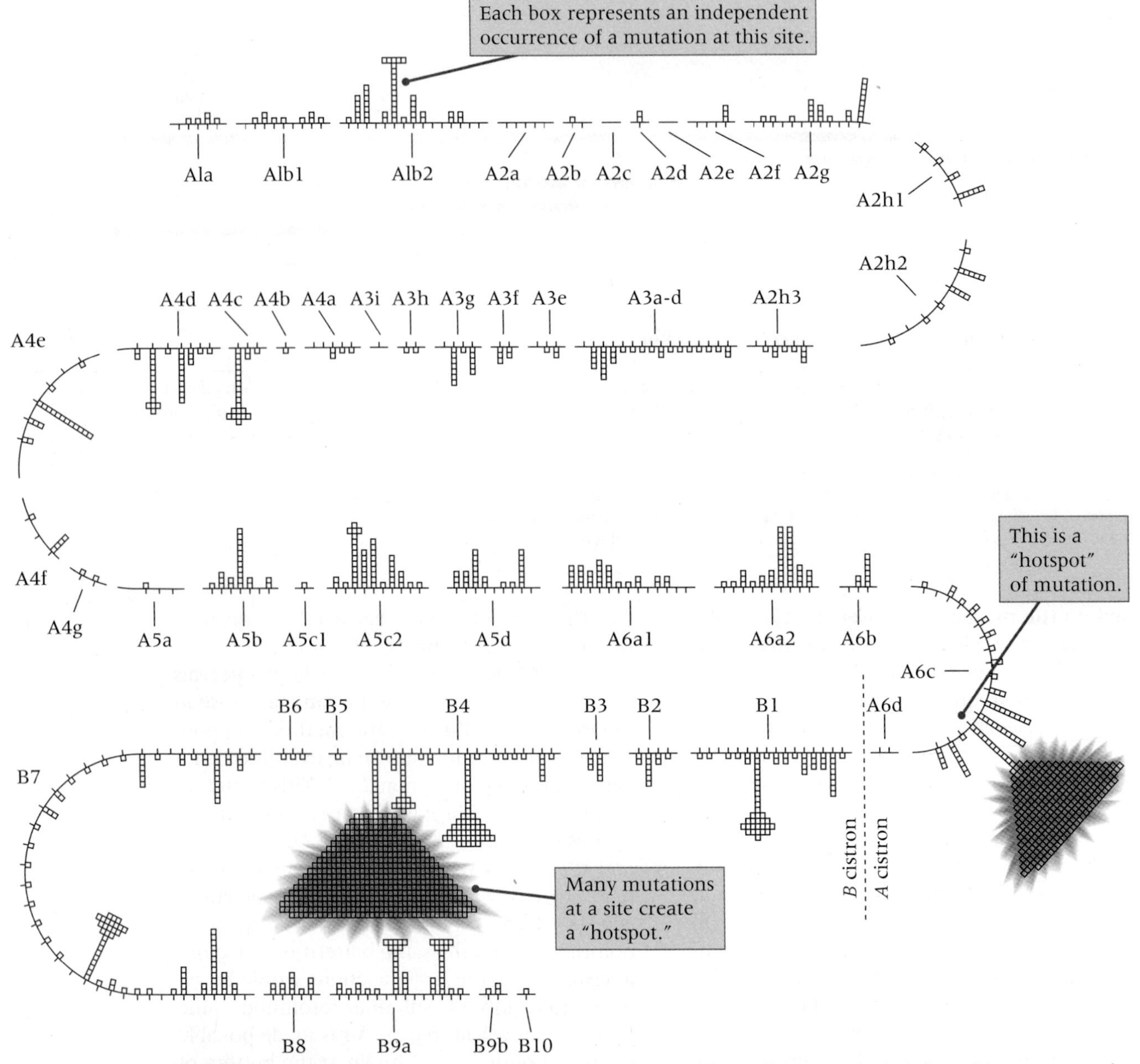


**FIGURE 9.25** Genetic map of the *rII* locus of phage T4. Each small square indicates a separate, independent occurrence of a mutation at the indicated site. The arrangement of sites within each A or B segment is arbitrary. [Adapted from S. Benzer, *Proc. Natl. Acad. Sci. USA* 47(1961): 403–426.]

quency of mutation is called a **hotspot** of mutation. At sites other than hotspots, mutations were recovered only once or a few times.

The *rII* analysis was also important because it helped to distinguish experimentally between three distinct meanings of the word *gene*. Most commonly, the word *gene* refers to a unit of function. Physically, this corresponds to a protein-coding segment of DNA. Benzer assigned the term **cistron** to this unit of function, and the term is still occasionally used. The unit of function is normally defined experimentally by a complementation test (see Section 3.7), and indeed, two units of function, the *rIIA* cistron and the *rIIB* cistron, were defined by complementation. The limits of *rIIA* and *rIIB* are shown in Figures 9.23 and 9.25. The complementation between *rIIA* and *rIIB* is observed when two types of T4 phage, one with a mutation in *rIIA* and the other with a mutation in *rIIB*, are used simultaneously to infect *E. coli* strain K12(λ). The mutations complement; multiply infected cells produce normal numbers of phage progeny. In contrast, when K12(λ) is simultaneously infected with two phages having different *rIIA* mutations or with two phages having different *rIIB* mutations, no progeny phages are produced.

Besides the meaning of function, clarified by use of the term *cistron*, the term *gene* has two other distinct meanings: (1) the unit of genetic transmission that participates in recombination and (2) the unit of genetic change or mutation. Physically, both the recombinational and the mutational units correspond to the individual base pairs in a gene. Despite potential ambiguity, the term *gene* is still the most important word in genetics, and in most cases, the shade of meaning intended is clear from the context.

## 9.7 Lysogeny and Specialized Transduction

A genetic map of the *E. coli* phage λ based on recombination in the lytic cycle is shown in **FIGURE 9.26**. The map is linear, and the DNA molecule in the λ phage particle is also linear.

The complete DNA sequence of the genome of λ phage is known, and many of the genes and gene products and their functions have been identified. The map of genes in Figure 9.26 shows where each gene is located along the DNA molecule, scaled in kilobase pairs (kb), rather than in terms of its position in the genetic map. However, a genetic map with distances in centimorgans (cM), or percent recombination,

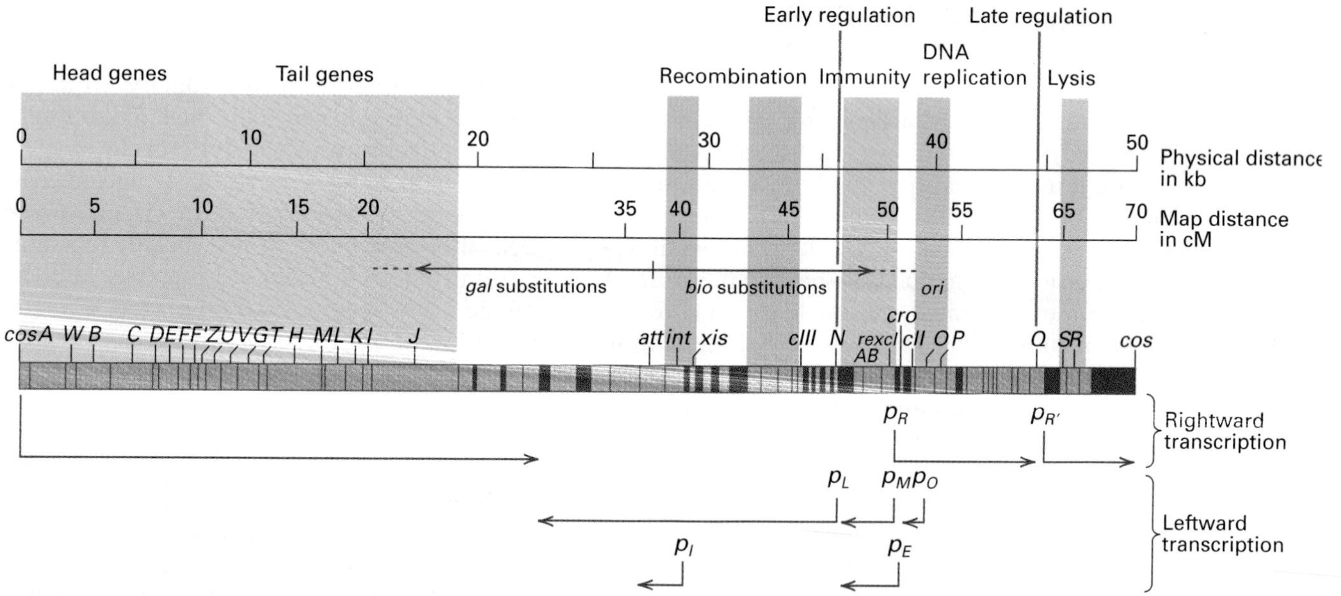


**FIGURE 9.26** Molecular and lytic-cycle genetic map of bacteriophage λ. The molecular map is scaled in kilobase pairs (kb), the genetic map in centimorgans (cM), a unit equivalent in this case to percent recombination. Clusters of genes with related functions are indicated by the pastel-colored boxes. The positions of the regulatory genes *N* and *Q* are indicated by vertical lines at about 35 and 45 kb, respectively. Promoters are indicated by the letter *p*. The direction of transcription and length of transcript are indicated by arrows. Light brown regions are coding sequences; dark brown regions are unlikely to code for proteins. The *att* (attachment site) is the site of recombination for integration of the λ prophage. The origin of replication (*ori*) is the region denoted *O*. The extents of possible substitutions of λ DNA with *E. coli* DNA in the specialized transducing phages λ*dgal* and λ*dbio* are indicated by arrows just above the gene map; such transducing phages are discussed in the section on specialized transduction.

has been placed directly below the molecular scale. Comparing the scales reveals that the frequency of recombination is not uniform along the molecule. For example, the 5 map units between genes *H* and *I* span about 3.1 kb, whereas the 5 map units between the genes *int* and *cIII* span about 4.8 kb.

The λ map is also interesting for what it implies about the evolution of the phage genome. The genes in λ show extensive clustering by function. The left half of the map consists entirely of genes whose products (head and tail proteins) are required for assembly of the phage structure, and within this region the head genes and the tail genes themselves form subclusters. The right half of the λ genome also shows several gene clusters, which include genes for DNA replication, recombination, and lysis. The genes are clustered not only by function but also according to the time at which their products are synthesized. For example, the *N* gene acts early; genes *O* and *P* are active later; and genes *Q, S,* and *R* and the head-tail cluster are expressed last. The transcription patterns for mRNA synthesis are thus very simple and efficient. There are only two rightward transcripts, and all late genes except for *Q* are transcribed into the same mRNA.

## ■ Site-Specific Recombination and Lysogeny

Infection with the phage λ can have an outcome other than the lytic cycle, however. The phage λ is an example of a **temperate phage**, which means that it can undergo a process called *lysogeny.* In a lysogenic infection, the phage does not multiply, the bacterial cell does not lyse, and no progeny phage are produced. The bacterial cell is seemingly unaffected, and no phage are apparent within it. In fact, in **lysogeny**, what happens is that the phage DNA becomes integrated into the bacterial chromosome and is replicated passively along with it. As a result, a phage DNA molecule is transmitted to each bacterial daughter cell.

Integration of the phage DNA into the bacterial chromosome is a function of a site-specific recombinase (integrase) encoded in the phage. Each type of temperate phage has its own integration site in the bacterial chromosome. For phage λ the integration site (called *att*λ) is at 17.3 minutes on the genetic map, between *gal* and *bio,* and for phage ϕ80 (phi 80) the integration site (*att*ϕ80) is at 28.2 minutes; both sites are shown in the map in Figure 9.13. The integrated phage DNA is called a **prophage**, and the bacterial cell carrying the prophage is called a **lysogen**. A strain lysogenic for λ has the symbol (λ) appended to its name. For example, the strain *E. coli* K12(λ) that made possible the fine-structure analysis of the *rII* region of phage T4 was a K12 strain that had become lysogenic for λ.

Lysogeny comes about when the λ integrase catalyzes a site-specific recombination between *att*λ in the bacterial chromosome and a site called *att* in the middle of the lytic phage map (Figure 9.26). For this mechanism to work, the phage DNA must be in the form of a circle. But in the mature phage, the DNA is a linear molecule, so how is the circular molecule formed? As shown in **FIGURE 9.27**, the circular molecule is formed inside the bacterial cell immediately after infection. The extreme ends of phage λ DNA are single stranded, each with 12 unpaired bases, which are complementary in sequence so that they can undergo base pairing with each other to form the circular molecule. The single-stranded ends are called **cohesive ends** to indicate their ability to undergo base pairing. The packaging of DNA in phage λ does not follow a headful mechanism, like T4. Rather, the λ packaging process recognizes specific sequences that are cleaved to produce the cohesive ends. Upon entering the cell, the complementary ends anneal to form a nicked circle, and ligation seals the nicks (**FIGURE 9.28**). Circularization, which takes place early in both the lytic and lysogenic cycles, is a necessary event in both cycles—for DNA replication in the lytic mode, for prophage integration in lysogeny. In about 75 percent of infected cells, the circular molecule replicates, and the lytic cycle ensues. However, in about 25 percent of infected cells, the circular λ molecule and the circular *E. coli* DNA molecule undergo site-specific recombination catalyzed by the λ integrase, and the phage DNA becomes incorporated into the bacterial chromosome. Because λ can exist either as an autonomous genetic element (in the lytic cycle) or as an integrated element in the chromosome (in lysogeny), λ, like the F factor, is classified as an episome.

The bacterial and phage attachment sites each consist of three segments. The central segment of 15 base pairs has the same nucleotide sequence in both attachment sites and is the region in which the recombination event takes place. The phage attachment site is denoted *POP′* (*P* for *phage*), and the bacterial attachment site is denoted *BOB′* (*B* for *bacteria*). *POP′* is much the larger of the two *att* sites, because *P* and *P′* extend 105 nucleotides beyond the core *O* region of 15 base pairs, whereas *B* and *B′* extend only 4 nucleotides beyond the core *O* region (**FIGURE 9.29**). The

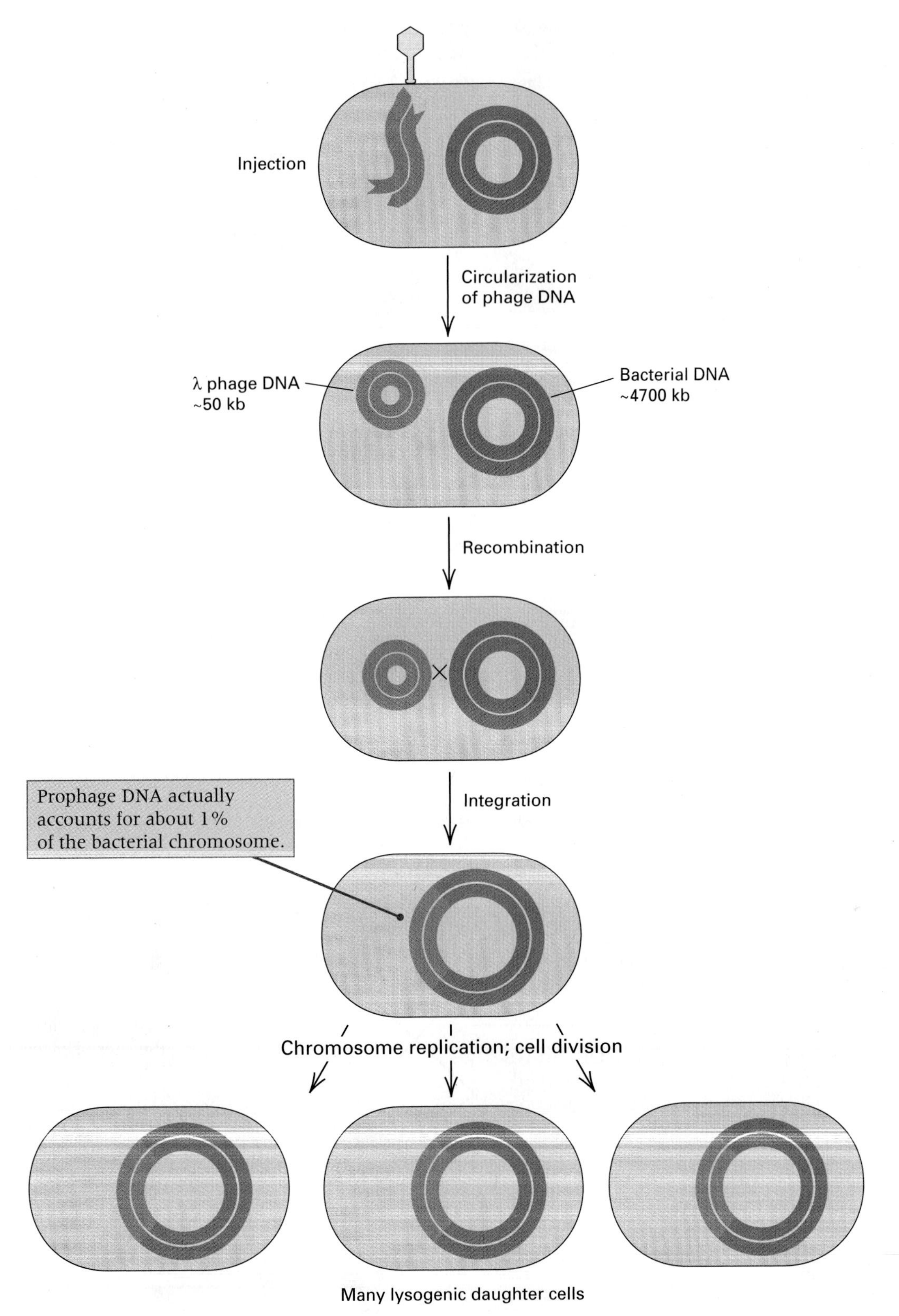


**FIGURE 9.27** Integration of phage DNA into the bacterial chromosome during lysogeny. Genes needed to establish lysogeny are expressed early in infection and then repressed. The integrated region in brown represents the prophage DNA. For clarity, the phage DNA is drawn much larger than to scale; the size of phage λ DNA is actually about 1 percent of the size of the *E. coli* genome.

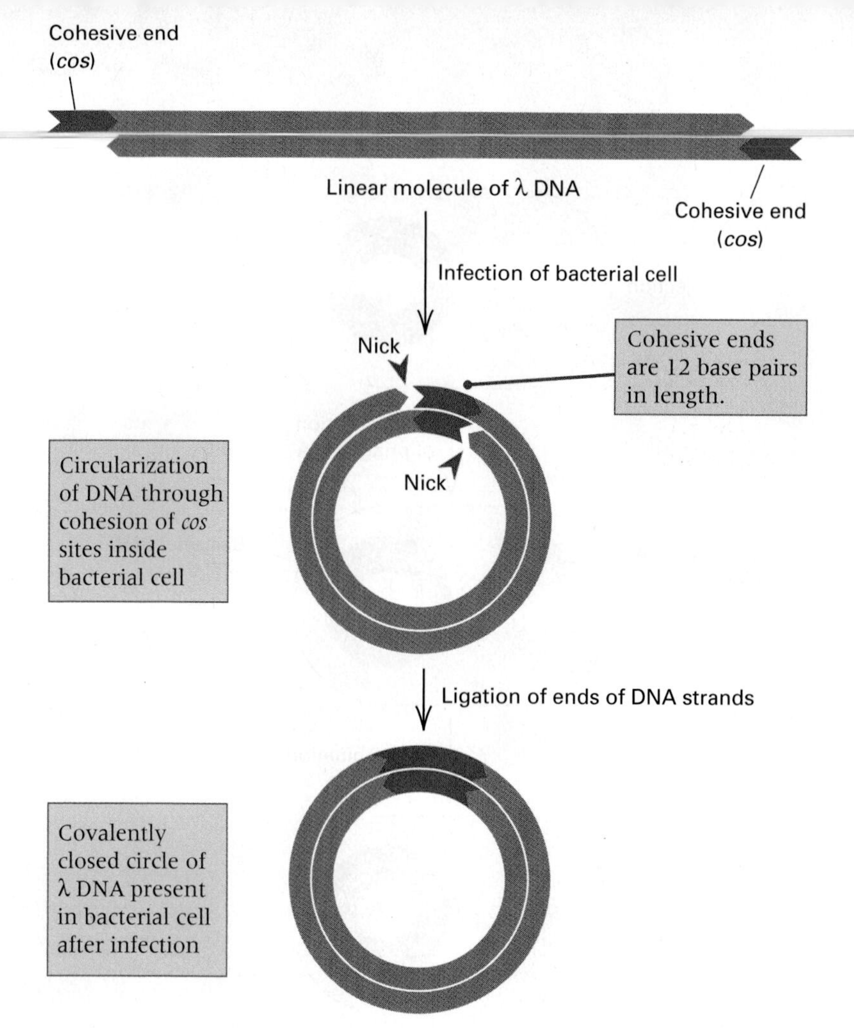

Cohesive end
(*cos*)
Linear molecule of λ DNA
Cohesive end
(*cos*)
Infection of bacterial cell
Nick
Cohesive ends are 12 base pairs in length.
Circularization of DNA through cohesion of *cos* sites inside bacterial cell
Nick
Ligation of ends of DNA strands
Covalently closed circle of λ DNA present in bacterial cell after infection

**FIGURE 9.28** A diagram of a linear λ DNA molecule, showing the cohesive ends (complementary single-stranded ends). Circularization by means of base pairing between the cohesive ends forms an open (nicked) circle, which is converted into a covalently closed (uninterrupted) circle by sealing (ligation) of the single-strand breaks. The length of the cohesive ends is 12 base pairs in a total molecule of approximately 50 kb.

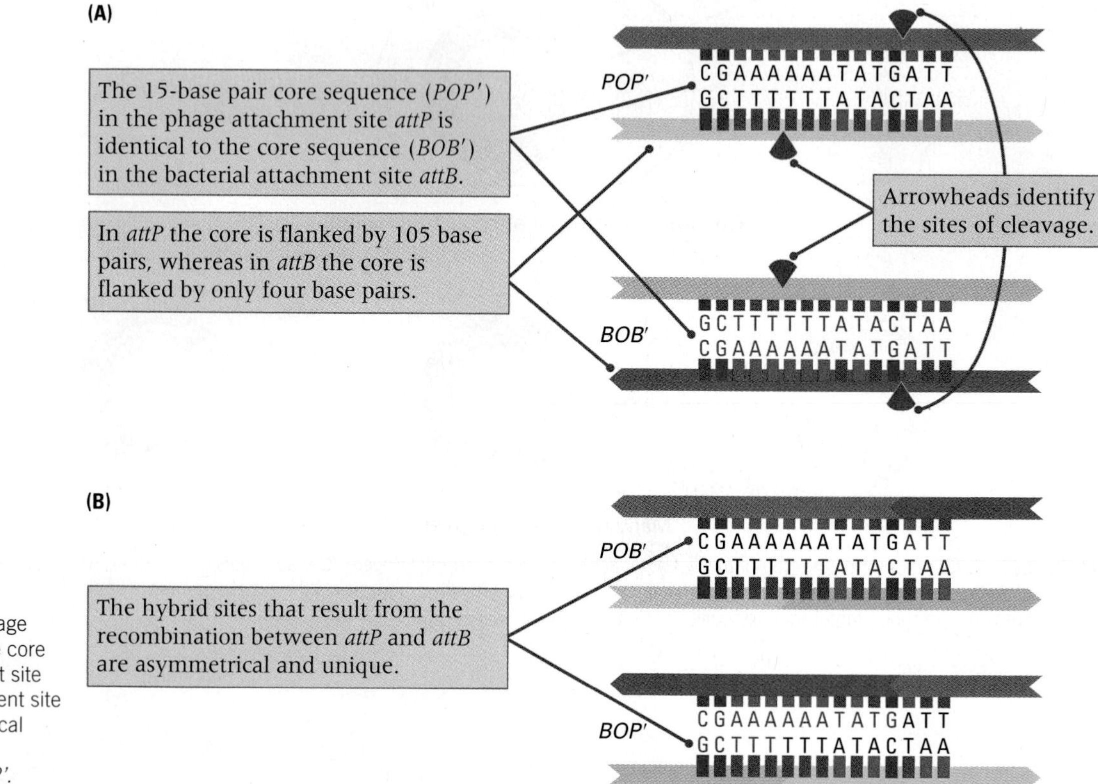

(A)
The 15-base pair core sequence (*POP′*) in the phage attachment site *attP* is identical to the core sequence (*BOB′*) in the bacterial attachment site *attB*.
In *attP* the core is flanked by 105 base pairs, whereas in *attB* the core is flanked by only four base pairs.
POP′
CGAAAAAATATGATT
GCTTTTTTATACTAA
Arrowheads identify the sites of cleavage.
BOB′
GCTTTTTTATACTAA
CGAAAAAATATGATT
(B)
POB′
CGAAAAAATATGATT
GCTTTTTTATACTAA
The hybrid sites that result from the recombination between *attP* and *attB* are asymmetrical and unique.
BOP′
CGAAAAAATATGATT
GCTTTTTTATACTAA

**FIGURE 9.29** (A) Sites of cleavage resulting from λ integrase in the core regions of the phage attachment site *POP′* and the bacterial attachment site *BOB′*. (B) The results of reciprocal recombination are the hybrid attachment sites *POB′* and *BOP′*.

## connection

### Artoo

**Seymour Benzer 1955**
Purdue University,
West Lafayette, Indiana
*Fine Structure of a Genetic Region in Bacteriophage*

*Just as every fan of "Star Wars" can identify the lovable robot Artoo-Detoo (a.k.a. R2-D2), every geneticist can identify the gene* rII. *The* rII *gene in bacteriophage T4 was the first experimental example of genetic fine structure. Benzer used the special property that* rII *mutants cannot grow on* E. coli *strain K but can grow on strain B to examine recombination between different nucleotides within the* rII *gene. He demonstrated that the* rII *gene was divisible by recombination. (It is now known that, in principle, recombination can take place between any adjacent nucleotides.) But if the gene can be subdivided by recombination, then what is a gene, anyway? If two different mutations can undergo recombination whether or not they are in the same gene, then how can one decide, experimentally, whether two different mutations are or are not alleles? Benzer realized that the key experimental operation in the definition of allelism was not recombination but rather the complementation test. This is a rare paper with two great ideas in it: recombination within a gene, and the use of the complementation test to determine experimentally whether two different mutations are or are not alleles of the same gene.*

The phenomenon of genetic recombination provides a powerful tool for separating mutations and discerning their positions along a chromosome. When it comes to very close neighboring mutations, a difficulty arises, since the closer two mutations lie to one another, the smaller is the probability that recombination between them will occur. Therefore, failure to observe recombinant types ordinarily does not justify the conclusion that the two mutations are inseparable. . . . A high degree of resolution can best be achieved if there is available a selective feature for the detection of small proportions of recombinants. Such a feature is offered by the case of the *rII* mutants of T4 bacteriophage described in this paper. The wildtype phage produces plaques on either of two bacterial hosts, *Escherichia coli* strain B or strain K, while a mutant of the *rII* group produces plaques only on B. Therefore, if a cross is made between two different *rII* mutants any wildtype recombinants which arise, even in proportions as low as $10^{-8}$, can be detected by plating on strain K. . . . In this way, a series of eight *rII* mutants of T4 have been crossed with each other. The results of these crosses are given in the figure on the bottom of this page.

***The functional relatedness of two closely linked mutations causing similar defects may be tested by constructing diploid heterozygotes containing the two mutations.***

The distances are only roughly additive; there is some systematic deviation in the sense that a long distance tends to be smaller than the sum of its component shorter ones, [which is accounted for by multiple recombination events]. . . . Thus, while all *rII* mutants in this set fall into a small portion of the phage linkage map, it is possible to seriate [order] them unambiguously, and their positions *within* the region are well scattered. . . . *Test for allelism.* The functional relatedness of two closely linked mutations causing similar defects may be tested by constructing diploid heterozygotes containing the two mutations. . . . The *trans* form, containing one of the mutations in each chromosome, may or may not produce the wild phenotype. If it does [complementation], it is concluded that the two mutations in question are located in separate functional units. [They are alleles of different genes.] . . . In order to characterize a unit of genetic "function," it is necessary to define what function is meant. . . . On the basis of phenotype tests of *trans* configuration heterozygotes [complementation tests], the *rII* region can be subdivided into two functionally separable segments. Each segment may have the "function" of specifying the sequence of amino acids in a polypeptide chain.

Source: S. Benzer, *Proc. Natl. Acad. Sci. USA* 41(1955): 344–354.

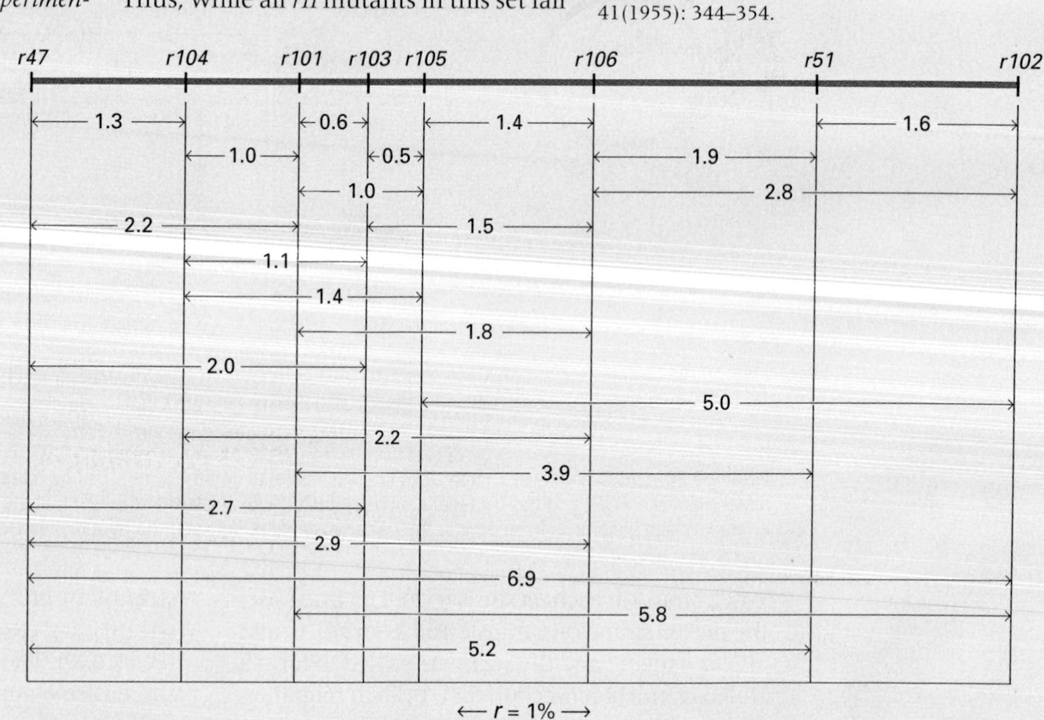

Electron micrograph of two particles of bacteriophage λ. The head geometry is icosahedral, which means that it consists of 20 identical triangular faces and 12 corners. Inside the head is a double-stranded DNA molecule of approximately 50,000 nucleotide pairs, terminating in complementary single-stranded regions of 12 nucleotides at each 5′ end. The tail consists of a stack of 35 protein disks forming a hollow tube through which the DNA passes into the infected cell. [© Dr. G. Murti/Visuals Unlimited]

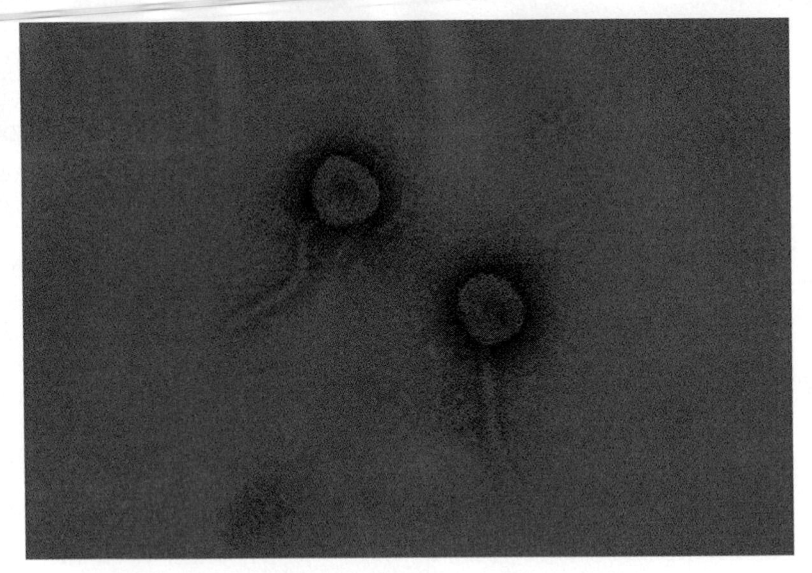

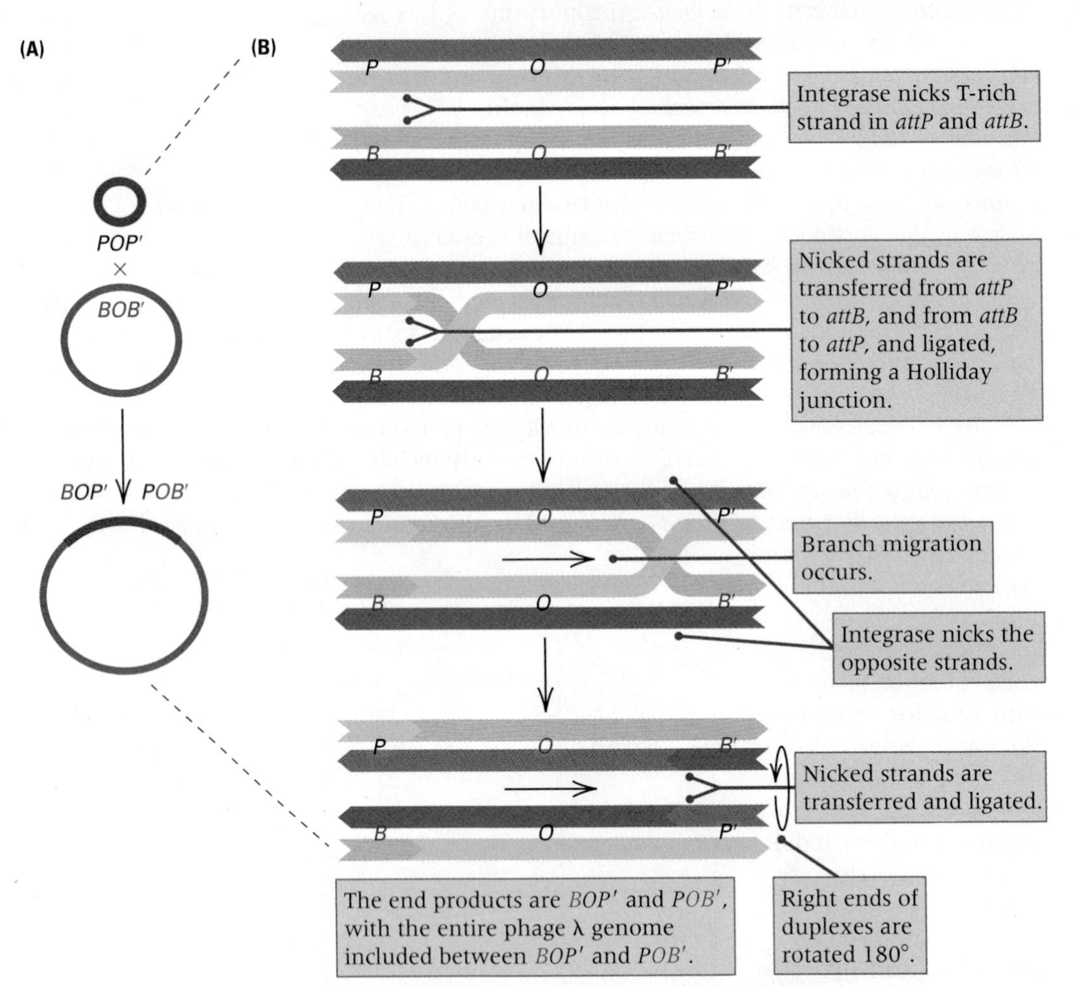


**FIGURE 9.30** (A) Overview of λ integration into the bacterial chromosome. (B) Mechanism of recombination between the phage attachment site *POP′* and the bacterial attachment site *BOB′*. Note that the result of the reciprocal recombination is that the entire λ phage genome is inserted into the bacterial genome between the hybrid sites *BOP′* and *POB′*.

T-rich strand in each *att* site is nicked by integrase, the nicked strands are transferred from *attP* to *attB* and *attB* to *attP,* and the nicks are sealed to form a Holliday junction (Section 6.8). Branch migration occurs rightward through the core, the opposite strands in *attP* and *attB* are nicked, the nicked strands are exchanged, and the nicks resealed (**FIGURE 9.30**). The products are *POB′* and *BOP′*, with the entire λ genome included between the two hybrid *att* sites.

The geometry of phage integration is shown in **FIGURE 9.31**. As a result of the recombination event, the genetic map of the prophage is not the same as the map of the phage but is, rather, a circular permutation of it that arises from the central location of *att* (the *POP′* site) in the phage lytic map and the circularization of the phage DNA.

The correct model for prophage integration was first suggested by Allan Campbell in 1962. The model was confirmed via bacterial crosses of lysogens with nonlysogens, as well as transduction by phage P1. Campbell found that the order of genes in the integrated prophage was

*N* *mi R A m6* *J*

where *m6* is a mutation affecting head formation and *mi* is one affecting lysis. In contrast, the order of genes in the free phage as determined by general recombination is

*A m6* *J att N* *mi R*

The prophage map is thus a circular permutation of the map of the lytic phage. The prophage is inserted into the *E. coli* chromosome between the genes *gal* and *bio*, as indicated in **FIGURE 9.32**. An additional finding that confirmed the physical insertion of λ was that the integrated prophage increased the distance between *gal* and *bio*. In a nonlysogen, the cotransduction frequency between *gal* and *bio* is about 42 percent, which from Equation (1) corresponds to a map distance of 0.5 minute. In a λ lysogen, the *gal–bio* cotransduction frequency is 1.6 percent, corresponding to a map distance of 1.5 minutes. The 1-minute difference in map distance corresponds to 46 kb, because the complete chromosome of 4600 kb is transferred in 100 minutes; this value is within experimental error of the physical size of λ (50 kb).

When a temperate phage undergoes lysogeny, the phage genes become part of the bacterial chromosome, so it might be expected that the phenotype of the bacterium would change. But most phage genes in a prophage are kept in an inactive state by a **phage repressor** protein

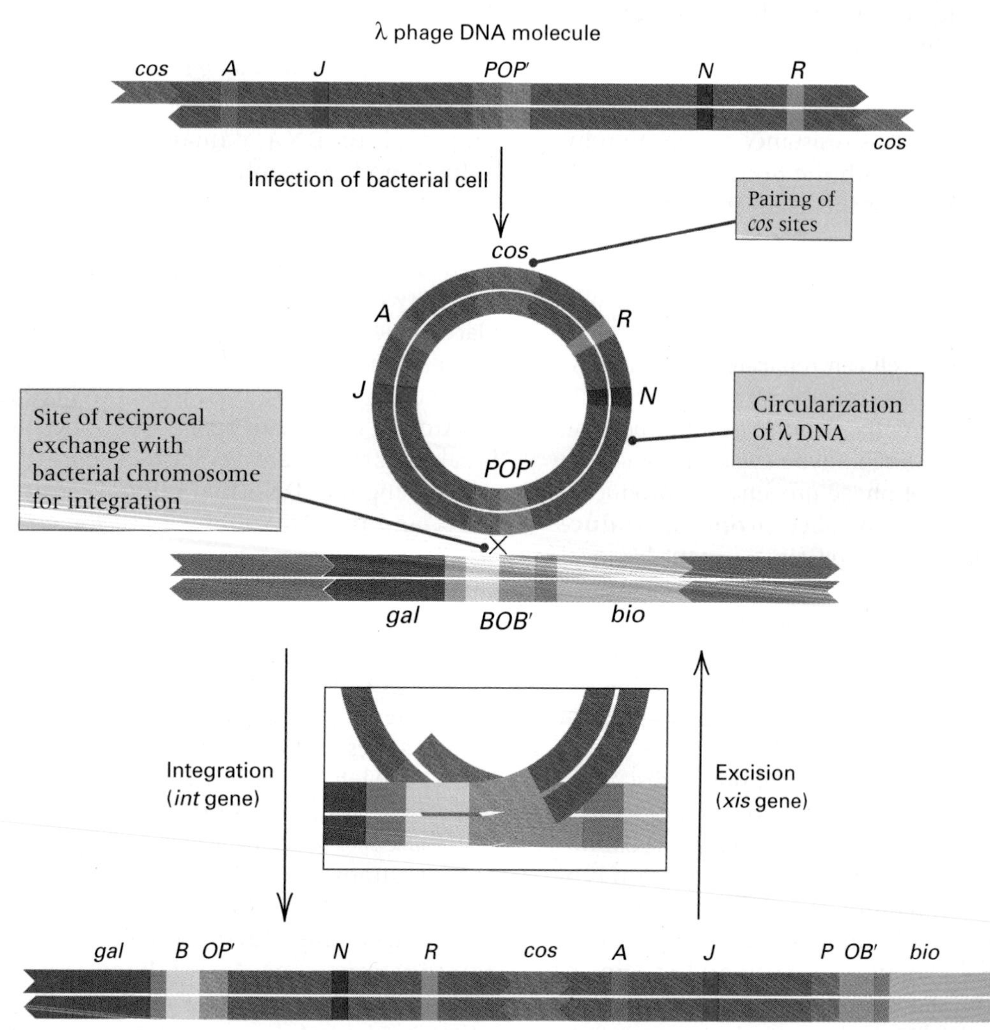


**FIGURE 9.31** The geometry of integration and excision of phage λ. The phage attachment site is *POP′*. The bacterial attachment site is *BOB′*. The prophage is flanked by two hybrid attachment sites denoted *BOP′* and *POB′*.

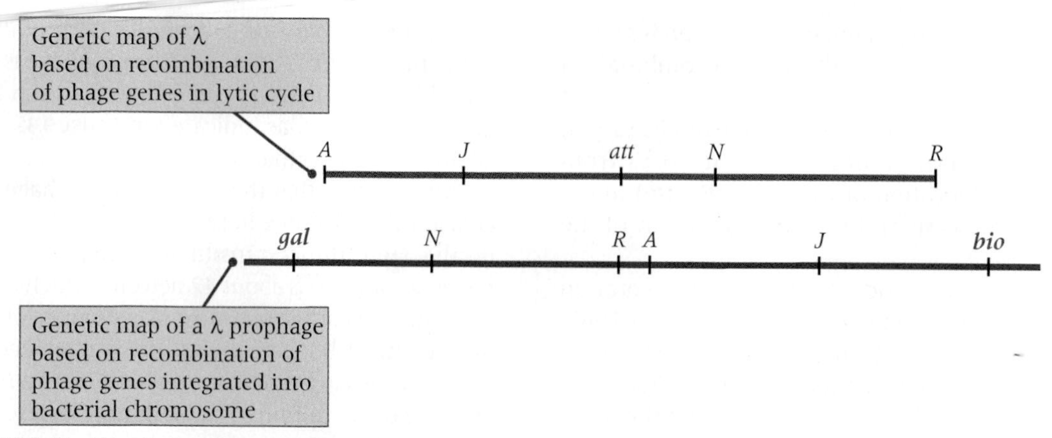


**FIGURE 9.32** The lytic genetic map of phage λ as determined by recombination in the lytic cycle (gold) compared with the prophage genetic map. Only a few genes are shown in order to provide reference markers. The *gal* and *bio* genes (red) are bacterial genes flanking the *att*λ in the *E. coli* chromosome.

encoded in one of the phage genes. The repressor protein is synthesized initially by the infecting phage and then continually by the prophage. The gene that codes for the repressor is frequently the only prophage gene that is expressed in lysogens. If a lysogen is infected with a phage of the same type as the prophage—for example, λ infecting a λ lysogen—then the repressor present within the cell from the prophage prevents expression of the genes of the infecting phage. This resistance to infection by a phage identical with the prophage, which is called **immunity**, is the usual criterion for determining whether a bacterial cell contains a particular prophage. For example, λ will not form plaques on bacteria that contain a λ prophage.

A lysogenic cell can replicate nearly indefinitely without the release of phage progeny. However, the prophage can sometimes become activated to undergo a lytic cycle in which the usual number of phage progeny are produced. This phenomenon is called **prophage induction**, and it is initiated by damage to the bacterial DNA. The damage sometimes happens spontaneously but is more often caused by some environmental agent, such as chemicals or radiation. The ability to be induced is advantageous for the phage because the phage DNA can escape from a damaged cell. The biochemical mechanism of induction is complex and will not be discussed, but the excision of the phage is straightforward.

Excision is another site-specific recombination event that reverses the integration process. Excision requires the phage enzyme integrase plus an additional phage protein called *excisionase*. Genetic evidence and studies of physical binding of purified excisionase, integrase, and λ DNA indicate that excisionase binds to integrase and thereby enables the latter to recognize the prophage attachment sites *BOP′* and *POB′*; once bound to these sites, integrase makes cuts in the *O* sequence and recreates the *BOB′* and *POP′* sites. This reverses the integration reaction, causing excision of the prophage (Figure 9.31).

## ■ Specialized Transduction

When a bacterium lysogenic for phage λ is subjected to DNA damage that leads to induction, the prophage is usually excised from the chromosome precisely. However, in about 1 cell per $10^6$ to $10^7$ cells, an excision error is made (**FIGURE 9.33**), and a chance breakage in two nonhomologous sequences takes place—one break within the prophage and the other in the bacterial DNA. The free ends of the excised DNA are then joined to produce a DNA circle capable of replication. The sites of breakage are not always located so as to produce a length of DNA that can fit in a λ phage head, and the DNA may be too large or too small. Sometimes, however, a molecule that can replicate and be packaged forms. In λ lysogens, the prophage lies between the *gal* and *bio* genes, and because the aberrant cut in the host DNA can be either to the right or to the left of the prophage, particles that carry either the *bio* genes (cut at the right) or the *gal* genes (cut at the left) can arise. The resulting phages are called λ*bio* and λ*gal* transducing particles. These are *specialized transducing phages* because they can transduce only certain bacterial genes (*gal* or *bio*), in contrast with the P1-type of generalized transducing phage, which can transduce any gene. **Specialized transduction** is therefore a process in which the DNA of an excised prophage can carry a small piece

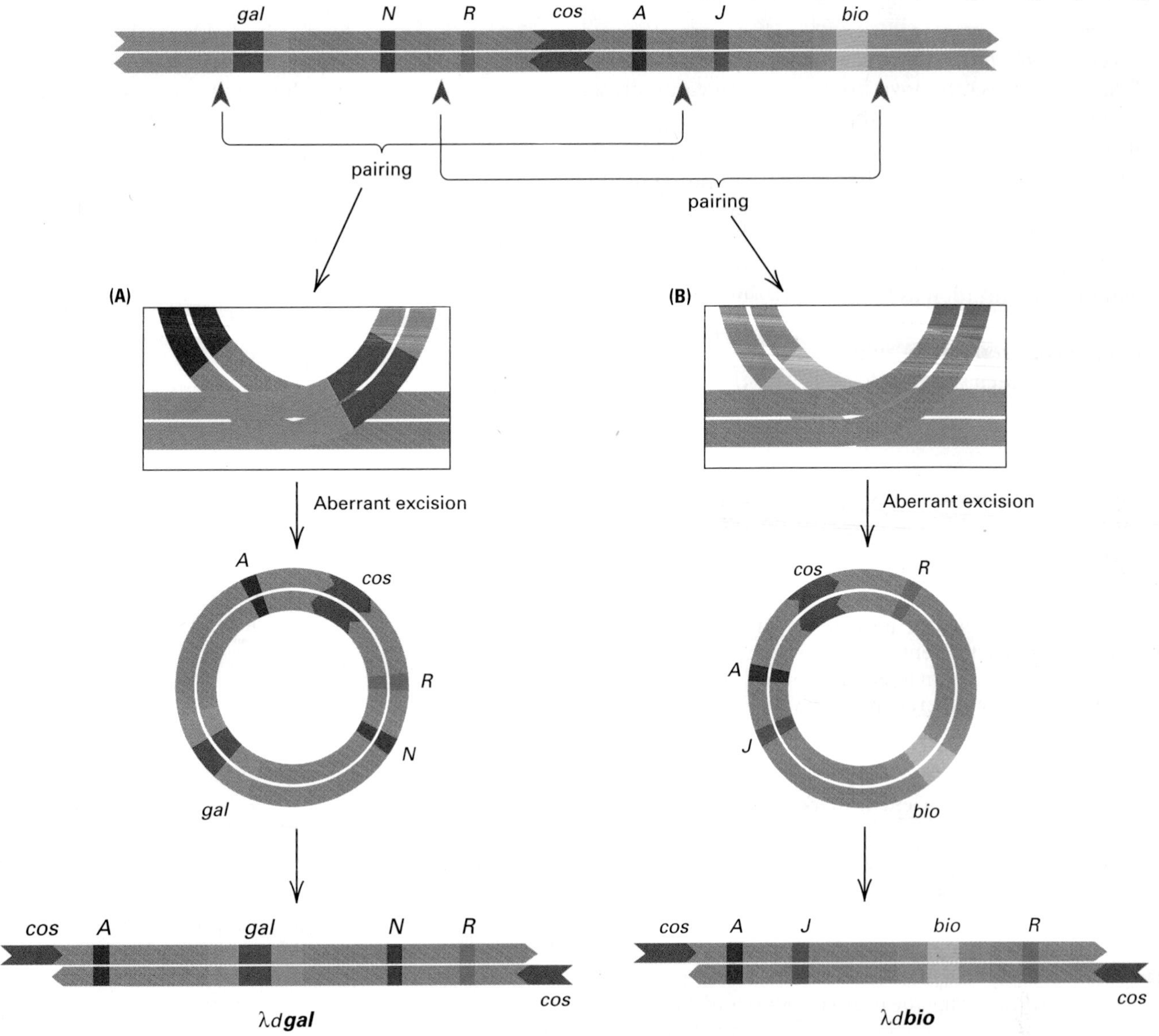


**FIGURE 9.33** Aberrant excision leading to the production of specialized λ transducing phages. (A) Formation of a *gal*-transducing phage (λ*dgal*). (B) Formation of a *bio*-transducing phage (λ*dbio*).

of the host bacterial genome adjacent to the phage attachment site.

Often the specialized transducing phages are defective, because essential λ genes are replaced by bacterial genes during formation of the $gal^+$-bearing or $bio^+$-bearing λ molecules. The phages are thus called λ*dgal* (defective, $gal^+$-transducing) and λ*dbio* (defective, $bio^+$-transducing), respectively. They are unable by themselves to infect *E. coli* productively. Addition of a wildtype helper phage both allows production of a lysate rich in λ*dgal* or λ*dbio* transducing phage and formation of a double lysogen. Presumably, integration of the wildtype λ provides hybrid attachment sites *POB′* and *BOP′* that allow integration of λ*dbio* or λ*dgal* by homologous recombination between identical hybrid attachment sites.

## CHAPTER SUMMARY

- Bacteria take advantage of several mechanisms by which DNA sequences can move from one DNA molecule to another, from one cell to another, or even from one bacterial species to another.
- These mechanisms of genetic exchange can result in a mosaic genome organization as found in *E. coli*, in which each strain shares a set of core genes with all other strains, but each strain also contains multiple genomic islands of inserted DNA that distinguish it from other strains. Some of these islands include genes that can cause disease
- In transformation, a bacterial cell takes up DNA from the surrounding medium and incorporates the DNA into its genome by homologous recombination, replacing some host genes in the process.
- In *E. coli*, the F (fertility) plasmid can mobilize the chromosome for transfer to another cell in the process of conjugation.
- Some types of bacteriophages can incorporate bacterial genes and transfer them into new host cells in the process of transduction.
- DNA molecules from related bacteriophages that are present in the same host cell can undergo genetic recombination.
- Some viruses are able to integrate their DNA into that of the host cell, where it replicates along with the host DNA and is transmitted to progeny cells.

## GUIDE TO PROBLEM SOLVING

**Problem 1** Three independent integrations of the F factor into the chromosome of a bacterial species related to *E. coli* yielded four different Hfr derivatives (HfrX, HfrY, and HfrZ), each with a different origin and possibly a different direction of transfer of markers. These were examined in interrupted-mating experiments and were found to transfer chromosomal genes at the times shown in the accompanying table.

| | Genetic marker | | | | | |
|---|---|---|---|---|---|---|
| Hfr | *ala* | *arg* | *his* | *ser* | *thr* | *met* |
| X | — | 24 | 12 | — | — | 4 |
| Y | 9 | 25 | — | — | 3 | — |
| Z | — | — | — | 11 | — | 29 |

**(a)** Draw a diagram of the origin and direction of transfer of each Hfr, showing the relative positions of the chromosomal markers and the distances between them in minutes.

**(b)** Draw a composite genetic map using the data from all three of the Hfr strains.

**Answer**

**(a)** Consider each Hfr in turn and, starting with the earliest-entering gene, write down the name of each gene as it enters. The difference in time of entry between adjacent genes is the distance in minutes between the genes.

**(b)** Arrange the maps of the Hfr strains so that any shared genes between two or more Hfr strains coincide. Then draw a composite map based on the overlaps between the individual Hfr strains. The individual Hfr maps and the composite map are shown below. Maps that are the mirror image of those shown are also correct.

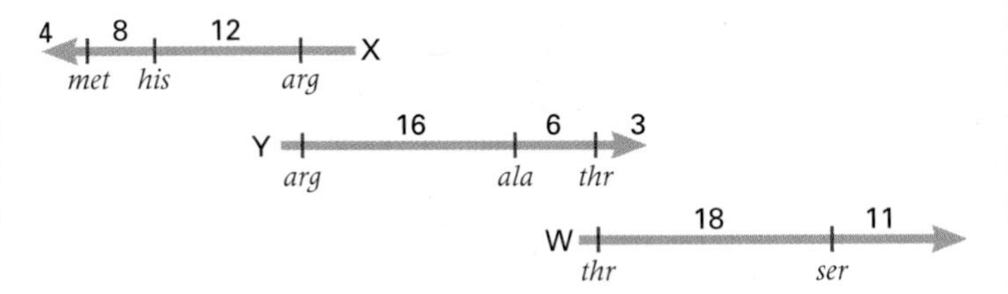


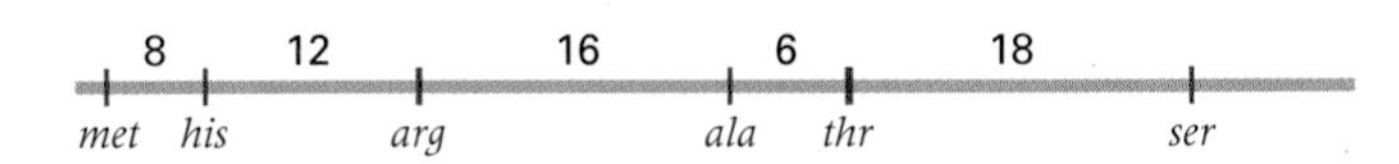


**Problem 2** Cotransduction experiments were carried out to determine the order of three closely linked genes in *E. coli*. P1 phage grown on a strain of genotype $prp^+$ $dad^+$ $uxa^+$ were used to transduce a recipient strain of genotype $prp^-$ $dad^-$ $uxa^-$. The key observations were:

**(1)** When $prp^+$ transductants were selected, many of these were $uxa^+$ but very few were $dad^+$.

**(2)** When $uxa^+$ transductants were selected, many were also either $prp^+$ or $dad^+$, though usually not both. Which gene is in the middle?

**Answer** The three possible genetic maps are shown here. Map (**a**) is consistent with observation (1) provided that dad is farther from *prp* than *uxa* is. However, map (**a**) is not consistent with observation (2), because this map implies that *uxa–dad* contransduction should be much rarer than *uxa–prp* cotransduction. Map (**b**) can be rejected because it predicts that *prp–uxa* cotransduction should be much less frequent than *prp–dad* cotransduction; this is inconsistent with observation (1). Map (**c**) is the only map that is consistent with both observation (1) and observation (2).

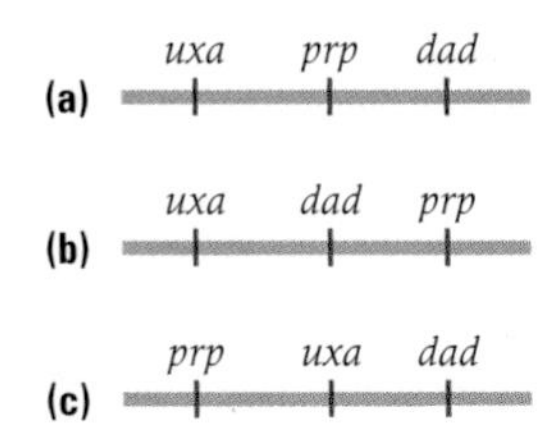

**Problem** Numbers of phage or bacteria in a suspension are usually so large that single colonies or plaques would be impossible to observe without suitable dilution. The usual dilutions are 100-fold, in which 0.1 ml of the suspension is mixed with 9.9 ml of dilution buffer; or 10-fold, in which 1 ml of suspension is mixed with 9 ml of suspension buffer. Usually *serial dilutions* are necessary, in which the suspension is diluted once into dilution buffer, the resulting suspension mixed thoroughly and diluted a second time, the resulting suspension mixed thoroughly and diluted a third time, and so forth. The serial dilution factors multiply, so, for example, two 100-fold and two 10-fold dilutions yield an overall dilution factor of $100 \times 100 \times 10 \times 10 = 10^6$. As an example of serial dilution, suppose that a suspension of bacteriophage is serially diluted through four dilutions of 100-fold each and one dilution of 10-fold. From the final dilution, a volume of 0.1 ml is mixed with a great excess of growing bacteria and then spread over nutrient agar in a petri dish and incubated overnight. The next day, 35 plaques are visible. Estimate the number of viable phage per milliliter (the phage *titer*) in the original undiluted suspension.

**Answer** There two straightforward ways to estimate the original titer. The first is to work through the dilutions backward, starting with the number of plaques observed, in this case 35. The original titer must then have been $35 \times 10$ (correction for plating only 0.1 ml of the final dilution) $\times 10$ (for the tenfold dilution) $\times 100 \times 100 \times 100 \times 100$ (for the four hundredfold dilutions), which equals $35 \times 10^{10}$ phage per ml in the original solution. Alternatively, one could let $t$ be the original phage titer, which is reduced to $t \times 10^{-9}$ by the serial dilutions and again by $10^{-1}$ by plating only 0.1 ml of the final dilution, yielding an expected number of plaques of $t \times 10^{-10}$. Setting $t \times 10^{-10} = 35$ and solving for $t$ yields $t = 35 \times 10^{10}$.

## ANALYSIS AND APPLICATIONS

**9.1** A suspension of *E. coli* is serially diluted through two dilutions of 100-fold each and two dilutions of 10-fold each. From the final dilution, a volume of 0.1 ml is spread over nutrient agar in a petri dish and incubated overnight. The next day, 42 colonies are visible. Estimate the number of viable bacteria per milliliter in the original undiluted suspension.

**9.2** You are given a suspension of bacteria and told that it contains $7 \times 10^8$ viable cells per milliliter. What combination of 100-fold and 10-fold serial dilutions would you carry out so that 0.1 ml of the final dilution would contain approximately 70 viable cells?

**9.3** Time-of-entry experiments indicate that a gene for resistance to kanamycin, *kan-r*, is closely linked to a gene for resistance to bacteriophage T5, *T5-r*. A geneticist has strains of two genotypes available, *kan-r T5-s* (resistant to kanamycin, sensitive to phage T5) and *kan-s T5-r* (sensitive to kanamycin, resistant to phage T5). In an experiment using phage P1 to learn whether the genes are close enough for cotransduction, how would the experiment be done:

**(a)** Which strain would you infect with the P1 to obtain the transducing phage?

**(b)** Which strain would you infect with the transducing phage?

**(c)** Which marker would you select to obtain transductants?

**(d)** How would you determine whether the genes were cotransduced?

**9.4** Phage T2 (related to phage T4) normally forms small, clear plaques on a lawn of *E. coli* strain B. A mutant of strain B called B/2 ("B bar 2") is unable to adsorb T2 phage particles, so no plaques are formed. T2*h* is a phage with a host-range mutation that is capable of adsorbing to both strains B and B/2, and it forms normal-looking plaques. If *E. coli* strains B and B/2 are mixed in equal proportions and used to generate a lawn, what will be the appearance of plaques made by T2 and T2*h*?

**9.5** A standard laboratory strain of *E. coli* has a genome size of about 5 Mb (megabases) and contains about 3500 genes. The generalized transducing phage P1 can transfer a segment of DNA of about 100 kb. On the average, how many genes are in a segment of this size?

**9.6** Given that bacteriophage lambda has a genome size of 50 kilobase pairs, what is the approximate genetic length of the prophage in minutes when inserted into an *E coli* strain with a genome size of 5 Mb? (*Hint*: There are 100 minutes in the entire *E. coli* genetic map.)

**9.7** In the specialized transduction of a $gal^-$ $bio^-$ strain of *E. coli* using bacteriophage lambda from a $gal^+$ $bio^+$ lysogen, what medium would select for $gal^+$ transductants without selecting for $bio^+$?

**9.8** If $leu^+$ *str-r* recombinants are desired from the cross Hfr $leu^+$ *str-s* $\times$ $F^-$ $leu^-$ *str-r*, on what kind of medium should the matings pairs be plated? Which are the selected and which the counterselected markers?

**9.9** Naturally occurring *E. coli* strains range in genome size from about 4.5 Mb to about 5.5 Mb, owing to the presence of genomic islands in some lineages but not in others. If a strain with a genome size of 5 Mb has a genetic map length of 100 minutes, what is the map length of a strain with a genome size of 4.5 Mb? Of a strain with a genome size of 5.5 Mb?

**9.10** For generalized transduction with phage P1, the theoretical relationship between map distance in minutes and frequency of cotransduction is

$$\textit{map distance} = 2 - 2 \times (\textit{cotransduction frequency})^{1/3}$$

**(a)** What map distance in minutes corresponds to 75 percent cotransduction? to 50 percent cotransduction? to 25 percent cotransduction?

**(b)** What is the cotransduction frequency between genetic markers separated by half a minute? one minute? two minutes? greater than two minutes?

**(c)** If genes *a*, *b*, and *c* are in alphabetical order and the cotransduction frequencies are 30 percent for *a b* and 10 percent for *b c*, what is the expected cotransduction frequency between *a* and *c*?

**9.11** Bacterial cells of genotype *pur*$^-$ *pro*$^+$ *his*$^+$ were transduced with P1 bacteriophage grown on bacteria of genotype *pur*$^+$ *pro*$^-$ *his*$^-$. Transductants containing *pur*$^+$ were selected and tested for the unselected markers *pro* and *his*. The numbers of *pur* colonies with each of four genotypes are as follows:

| | |
|---|---|
| *pro*$^+$ *his*$^+$ | 102 |
| *pro*$^-$ *his*$^+$ | 25 |
| *pro*$^+$ *his*$^-$ | 160 |
| *pro*$^-$ *his*$^-$ | 1 |

What is the gene order?

**9.12** Five genes in a bacterial strain are being studied. The gene *tau-r* confers resistance to the bacteriophage *tau*, the gene *top-r* confers resistance to the antibiotic topomycin, the gene *arg*$^+$ is necessary for arginine biosynthesis, the gene *suc*$^+$ is necessary for growth on sucrose, and the gene *mot*$^+$ is necessary for cell motility. The accompanying table shows the results of cotransduction experiments. A + means that the genes can be cotransduced, a – means they they cannot be contransduced, and NT means not tested. What is the order of the genes?

| | *top* | *arg* | *suc* | *mot* |
|---|---|---|---|---|
| *tau* | – | + | + | – |
| *top* | | NT | + | + |
| *arg* | | | – | NT |
| *suc* | | | | NT |

**9.13** A strain of *Escherichia coli*, which was originally isolated from a single colony and had been stored for 50 years in agar in a tightly sealed glass vial, was taken out of the agar and streaked onto nutrient agar to purify single colonies. Each colony represented a separate lineage descended from the original strain. Genomic DNA from 12 colonies was extracted and cleaved with *Eco*RI, the fragments were separated by electrophoresis and transferred to a membrane filter, and the filter was hybridized with a probe for the transposable element known as insertion sequence IS*5*. (There are no *Eco*RI cleavage sites within the sequence of IS*5*.) The result is shown in the accompanying diagram.

**(a)** Explain why some bands are identical across all strains.
**(b)** Explain why some bands are different.
**(c)** Estimate the rate of IS*5* transposition per lineage per year.
**(d)** Estimate the rate of IS*5* transposition per element per lineage per year.

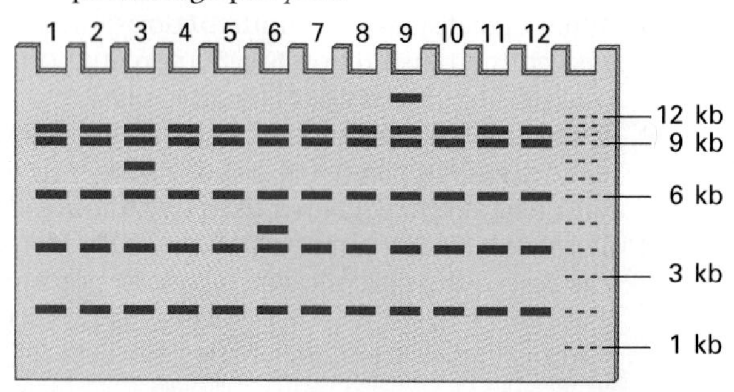


**9.14** In *Escherichia coli*, the genes *lac*$^+$, *gal*$^+$, *trp*$^+$, and *his*$^+$ are necessary for utilization of lactose, utilization of galactose, synthesis of tryptophan, and synthesis of histidine, respectively, and *str-r* and *str-s* are alleles for streptomycin resistance and sensitivity, respectively. You wish to carry out the cross Hfr *lac*$^+$ *gal*$^+$ *trp*$^+$ *his*$^+$ *str-s* × F$^-$ *lac*$^-$ *gal*$^-$ *trp*$^-$ *his*$^-$ *str-r*, using each of the markers in turn as selected markers and streptomycin resistance as a counterselected marker. What medium should you use for each of the following purposes?

**(a)** To select for *lac*$^+$ and counterselect for *str-r*.
**(b)** To select for *gal*$^+$ and counterselect for *str-r*.
**(c)** To select for *trp*$^+$ and counterselect for *str-r*.
**(d)** To select for *his*$^+$ and counterselect for *str-r*.

**9.15** In the previous problem, what medium should you use to determine each of the following?

**(a)** Which of the *lac*$^+$ *str-r* colonies are *gal*$^+$.
**(b)** Which of the *lac*$^+$ *str-r* colonies are *trp*$^+$.
**(c)** Which of the *lac*$^+$ *str-r* colonies are *his*$^+$.
**(d)** Which of the *lac*$^+$ *str-r* colonies are *trp*$^+$ and *his*$^+$.

**9.16** An Hfr donor of genotype *a*$^+$ *b*$^+$ *c*$^+$ *str-s* is mated with an F$^-$ recipient of genotype *a*$^-$ *b*$^-$ *c*$^-$ *str-r*; the order of transfer is *a b c*. None of the genes is transferred early, the distance between *a* and *b* is the same as that between *b* and *c*, and none of the markers is near *str*. Recombinants are selected as usual by plating on medium that lacks particular nutrients and contains streptomycin. Which of the following statements are true? Explain. (More than one are true.)

**(a)** The number of *a*$^+$ *str-r* colonies will be larger than the number of *c*$^+$ *str-r* colonies.
**(b)** The number of *b*$^+$ *str-r* colonies will be smaller than the number of *c*$^+$ *str-r* colonies.
**(c)** The number of *a*$^+$ *b*$^+$ *str-r* colonies will be smaller than the number of *b*$^+$ *str-r* colonies.
**(d)** The number of *a*$^+$ *b*$^+$ *str-r* colonies will be about equal to the number of *b*$^+$ *str-r* colonies.
**(e)** Most *a*$^+$ *c*$^+$ *str-r* colonies will also be *b*$^+$.
**(f)** Most *b*$^-$ *c*$^+$ *str-r* colonies will also be *a*$^-$.
**(g)** The number of *a*$^+$ *b*$^+$ *c*$^-$ *str-r* colonies will be smaller than the number of *a*$^+$ *b*$^-$ *c*$^-$ *str-r* colonies.

**9.17** *Salmonella enterica* is closely related to *Escherichia coli*. It can be infected with the F plasmid, which can integrate into the chromosome to produce Hfr strains. These can be mated with F$^-$ *E. coli* to study the order and time of entry of genetic markers. The data shown here pertain to times of entry of four genetic markers in crosses of *E. coli* Hfr × *E. coli* F$^-$ and *S. enterica* Hfr × *E. coli* F$^-$.

| | *Ile* | *met* | *pro* | *arg* |
|---|---|---|---|---|
| *E. coli* HFr × *E. coli* F$^-$ | 28 | 20 | 6 | 22 |
| *S. enterica* Hfr × *E. coli* F$^-$ | 4 | 22 | 47 | 18 |

**(a)** How do the genetic maps of *E. coli* and *S. enterica* compare with respect to these genes?
**(b)** What are the origin and direction of transfer in each Hfr?

**9.18** An Hfr strain transfers genes in alphabetical order. When tetracycline sensitivity is used for counterselection, the number of $h^+$ *tet-r* colonies is 1000-fold lower than the number of $h^+$ *str-r* colonies found when streptomycin sensitivity is used for counterselection. Suggest an explanation for the difference.

**9.19** An Hfr strain transfers genes in the order *a b c*. In an Hfr $a^+$ $b^+$ $c^+$ *str-s* × F⁻ $a^-$ $b^-$ $c^-$ *str-r* mating, do all $b^+$ *str-r* recombinants receive the $a^+$ allele? Are all $b^+$ *str-r* recombinants also $a^+$? Explain your answers.

**9.20** If the genes in a bacterial chromosome are in alphabetical order and an Hfr cell transfers genes in the order *g h i . . . d e f*, what types of F′ plasmids could be derived from the Hfr?

**9.21** In an Hfr $lac^+$ $met^+$ *str-s* × F⁻ $lac^-$ $met^-$ *str-r* mating, $met^+$ is transferred much later than $lac^+$. If cells are plated on minimal medium containing glucose and streptomycin, what fraction of the cells are expected to be $lac^+$? If cells are plated on minimal medium containing lactose, methionine, and streptomycin, what fraction of the cells are expected to be $met^+$?

**9.22** In the mating

Hfr $met^-$ $his^+$ $leu^+$ $trp^+$ × F⁻ $met^+$ $his^-$ $leu^-$ $trp^-$

the $met^-$ marker is known to be transferred very late. After a short time after the Hfr and F⁻ cells are mixed, the mating cells are interrupted, and the cell suspension is plated on four different growth media. The amino acids in the growth media and the number of colonies observed on each are as follows:

| | |
|---|---|
| histidine + tryptophan | 250 colonies |
| histidine + leucine | 50 colonies |
| leucine + tryptophan | 500 colonies |
| histidine | 10 colonies |

In this experiment, what is the purpose of the $met^-$ mutation in the Hfr strain? What is the order of transfer of the genes? Why is the number of colonies so small for the medium containing only histidine?

**9.23** A bacteriophage has a genome consisting of a small linear molecule of double-stranded DNA that contains a unique (one per genome) cleavage site for a restriction enzyme in a gene that is essential for phage maturation and lysis. In most strains (type A) of the host bacterium, the phage forms plaques with nearly 100% efficiency (one plaque per infecting phage). However, in a few strains (type B), infection results in almost no plaques. When phage are isolated from the few plaques on type B, they subsequently form plaques with nearly 100% efficiency on type B. DNA analysis of these plaque-forming phage DNA reveals that a cytosine in the unique restriction site has been methylated. These phage also form plaques on type A, but afterwards the progeny phage again have low plaque-forming ability on type B. Explain these results.

**9.24** An Hfr *str-s* strain transfers genes in alphabetical order

$a^+$ $b^+$ $c^+$ . . . $x^+$ $y^+$ $z^+$

with $a^+$ transferred early and $z^+$ a terminal marker. This Hfr strain is mated with F⁻ $z^-$ *str-r* cells. The mating mixture is agitated violently 15 minutes after mixing to break apart conjugating cells and then plated on a medium that lacks nutrient Z and contains streptomycin. The $z^+$ gene is far from the *str-r* gene. The yield of $z^+$ *str-r* colonies is about one per $10^7$ Hfr cells. What are two possible modes of origin and genotypes of such a colony? How could the two possibilities be distinguished?

**9.25** A time-of-entry experiment is carried out in the mating

Hfr $a^+$ $b^+$ $c^+$ $d^+$ *str-s* × F⁻ $a^-$ $b^-$ $c^-$ $d^-$ *str-r*

where the spacing between the genes is equal. The data in the accompanying table are obtained.

| Time of mating in minutes | Number of recombinants of indicated genotype per 200 Hfr | | | |
|---|---|---|---|---|
| | $a^+$ *str-r* | $b^+$ *str-r* | $c^+$ *str-r* | $d^+$ *str-r* |
| 0 | 0.01 | 0.006 | 0.008 | 0.0001 |
| 10 | 5 | 0.1 | 0.01 | 0.0004 |
| 15 | 50 | 3 | 0.1 | 0.001 |
| 20 | 100 | 35 | 2 | 0.001 |
| 25 | 105 | 80 | 20 | 0.1 |
| 30 | 110 | 82 | 43 | 0.2 |
| 40 | 105 | 80 | 40 | 0.3 |
| 50 | 105 | 80 | 40 | 0.4 |
| 60 | 105 | 81 | 42 | 0.4 |
| 70 | 103 | 80 | 41 | 0.4 |

What are the times of entry of each gene? Suggest one possible reason for the low frequency of $d^+$ *str-r* recombinants.

**9.26** Four independent integrations of the F factor into the chromosome of an unusual strain of *E. coli* yielded four different Hfr derivatives of the strain (HfrW, HfrX, HfrY, and HfrZ), each with a different origin and possibly a different direction of transfer of markers. These were examined in interrupted-mating experiments and were found to transfer chromosomal genes at the times shown in the accompanying table.

**(a)** Draw a circular genetic map, with position 0 (also 100) minutes at the top, showing the order of the chromosomal genes and the distance (in minutes) between adjacent genes. (The marker *leu* is near 2 minutes on the standard map.) Annotate the genetic map with arrows indicating the origin and direction of transfer of each Hfr and the distance (in minutes) from the origin of transfer to the first marker transferred.

**(b)** How does the genetic map of this particular strain compare with that of the standard *E. coli* strain in Figure 9.13? Suggest an explanation of any discrepancy between the genetic maps.

| | **Genetic marker** | | | | | | | | | |
|---|---|---|---|---|---|---|---|---|---|---|
| **Hfr** | ***his*** | ***lac*** | ***leu*** | ***lip*** | ***pheS*** | ***pro*** | ***pyrD*** | ***terC*** | ***tonA*** | ***trp*** |
| W | – | 20 | – | 11 | – | – | 3 | – | – | – |
| X | – | – | 18 | – | – | – | 31 | – | 20 | 25 |
| Y | – | 13 | – | 22 | – | 6 | – | 2 | – | – |
| Z | 19 | – | – | – | 12 | 4 | – | 8 | – | – |

**9.27** You have reason to believe that a number of *rII* mutants of bacteriophage T4 are deletion mutations. You cross them in all possible combinations in *E. coli* strain B and plate them on *E. coli* strain K12(λ) to determine whether $r^+$ recombinants are formed. The formation of $r^+$ recombinants indicates that the mutations can recombine and so, if they are deletions, they must be nonoverlapping. The results are given in the accompanying table, in which *a* through *f* indicate an *rII* mutation and + indicates the formation of $r^+$ recombinant progeny in the cross. Assemble a deletion map for these mutations, using a line to indicate the DNA segment that is deleted in each mutant.

| | *a* | *b* | *c* | *d* | *e* | *f* |
|---|---|---|---|---|---|---|
| *a* | – | – | – | – | – | – |
| *b* | | – | – | + | + | – |
| *c* | | | – | + | – | + |
| *d* | | | | – | – | + |
| *e* | | | | | – | + |
| *f* | | | | | | – |

**9.28** Seven *rII* mutants, *t* through *z*, thought to carry point mutations, are crossed to the six deletion mutants in Problem 9.27 and scored for the ability to produce $r^+$ recombinants. The results are given in the accompanying table.

| | *a* | *b* | *c* | *d* | *e* | *f* |
|---|---|---|---|---|---|---|
| *t* | – | – | – | + | + | + |
| *u* | – | + | – | + | + | + |
| *v* | + | + | + | – | + | + |
| *w* | + | + | + | + | + | – |
| *x* | – | + | + | + | + | – |
| *y* | – | – | + | + | + | – |
| *z* | – | + | + | – | – | + |

Using the deletion end points to define genetic intervals along the *rII* gene, position each point mutation within an interval. Refine the deletion end points if required by the data.

**9.29** You cross the *rII* point mutations *t* through *z* in Problem 9.28 in all possible combinations in K12(λ) to assess complementation. The results are shown in the accompanying complementation matrix.

| | *t* | *u* | *v* | *w* | *x* | *y* | *z* |
|---|---|---|---|---|---|---|---|
| *t* | – | – | + | – | – | – | + |
| *u* | | – | + | – | – | – | + |
| *v* | | | – | + | + | + | – |
| *w* | | | | – | – | – | + |
| *x* | | | | | – | – | + |
| *y* | | | | | | – | + |
| *z* | | | | | | | – |

The mutations *v* and *z* complement deletion *c* but not deletion *e*. The mutations *v* and *z* also fail to complement a mutant known to be defective for *rIIB* function but not for *rIIA* function. Which mutants are in the *rIIA* cistron and which in the *rIIB* cistron? Where does the boundary between the two genes lie? Assemble a complementation matrix for the deletion mutants, filling in all the squares if possible.

**9.30** P2 and P4 are bacteriophages of *E. coli*. They have the following properties: (1) When P2 phage infects a bacterium, the bacterium eventually bursts, yielding about 100 P2 progeny. (2) When P4 infects a bacterium, the bacterium survives because P4 is a defective phage. (3) When P2 phage and P4 phage coinfect the same bacterium, lysis of the bacterium yields 100 P4 progeny and no P2 progeny, because the presence of P4 inhibits the growth of P2. Suppose $3 \times 10^8$ P2 and $2 \times 10^8$ P4 are added to $10^8$ bacterial cells.

**(a)** How many bacteria will not be infected with either phage?
**(b)** How many bacteria will survive?
**(c)** How many bacteria will produce P2 progeny?
**(d)** How many bacteria will produce P4 progeny?
(*Hint:* If the average number of phage per cell in a mixed suspension is *x*, then the probability that a cell remains uninfected is $e^{-x}$.)

## CHALLENGE PROBLEMS

**Challenge Problem 1** Bacteriophage 363 can be used for generalized transduction of genetic markers in *E. coli*. François Jacob used phage 363 to demonstrate that the lysogenic state of bacteriophage can be genetically transferred from one strain to another so long as the experiment is carried out at 20°C, a temperature at which phage do not multiply. Jacob studied four strains of bacteria:

A: $thr^-$ $lac^-$ $gal^-$
B: $thr^-$ $lac^-$ $gal^-$ (λ)
C: $thr^+$ $lac^+$ $gal^+$
D: $thr^+$ $lac^+$ $gal^+$ (λ)

In these designations, the symbol (λ) means that the strain is lysogenic for phage lambda. Jacob made two kinds of crosses. Phage grown on strain D were used to transduce strain A, and phage grown on strain C were used to transduce strain B. The results are given in the table below. Give a genetic explanation of these results. What genetic element is being transduced to give or to remove lysogeny?

| | Thr$^+$ | | Lac$^+$ | | Gal$^+$ | |
|---|---|---|---|---|---|---|
| Recipient strain | Colonies tested | Colonies lysogenic | Colonies tested | Colonies lysogenic | Colonies tested | Colonies lysogenic |
| A | 400 | 0 | 400 | 0 | 400 | 24 |
| B | 400 | 400 | 400 | 400 | 400 | 368 |

**Challenge Problem 2** In studying a novel type of antibiotic resistance, a geneticist carries out crosses between three Hfr derivatives of a natural isolate of *Escherichia coli* and an F$^-$ laboratory strain carrying the mutant alleles shown in the illustration. The natural isolate carries a duplication of the wildtype allele of the gene *C,* in which the duplicate copies are on opposite sides of the chromosome. The origin, extent, and direction of transfer in three Hfr crosses are shown by the arrows. Draw the genetic map that will result if the geneticist assumes that only one copy of each wildtype allele was present in the Hfr strains.

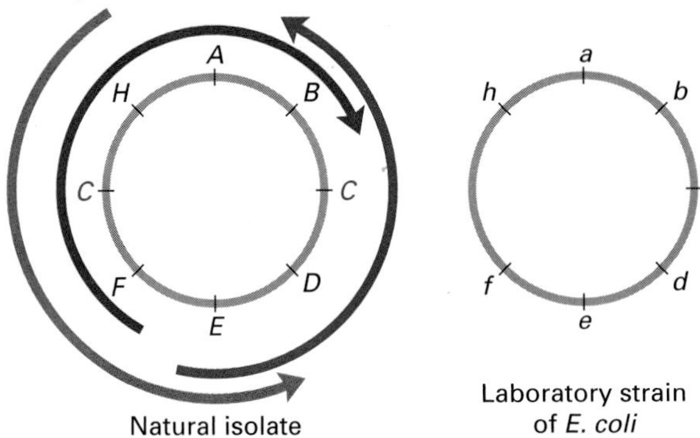


**Challenge Problem 3** The accompanying illustration shows a procedure known as *inverse PCR,* in which unknown sequences *flanking* a known sequence can be amplified. In the first step, genomic DNA is cleaved with a restriction enzyme (in this example, *Eco*RI) that does not cut within the known sequence. The resulting solution of restriction fragments is diluted and then treated with DNA ligase to join the ends together, forming circular DNA. (The solution is diluted to ensure that the ends of single fragments join together, rather than the ends of two or more different fragments.) Then PCR is carried out using oligonucleotide primers whose 3′ ends face *outward,* toward the unknown sequence (rather then inward, as in the usual PCR procedure). The result is amplification of the unknown fragment. Suppose that oligonucleotide primers of length 6 are adequate for the PCR (in reality, they are too short for most purposes), and suppose that known sequence is a transposable element whose ends are the inverted repeats

```
5′-GGTAAA . . . TTTACC-3′
3′-CCATTT . . . AAATGG-5′
```

**(a)** What primers would normally be used to amplify the known sequence?

**(b)** Are two different primers needed? Why or why not?

**(c)** What primers would be used to amplify the unknown sequence with inverse PCR?

**(d)** Would the PCR product be cleaved with *Eco*RI, and if so, into what fragments?

**(e)** If the PCR product were sequenced, how would one know which part originally flanked the known sequence on the left and which part originally flanked the known sequence on the right?

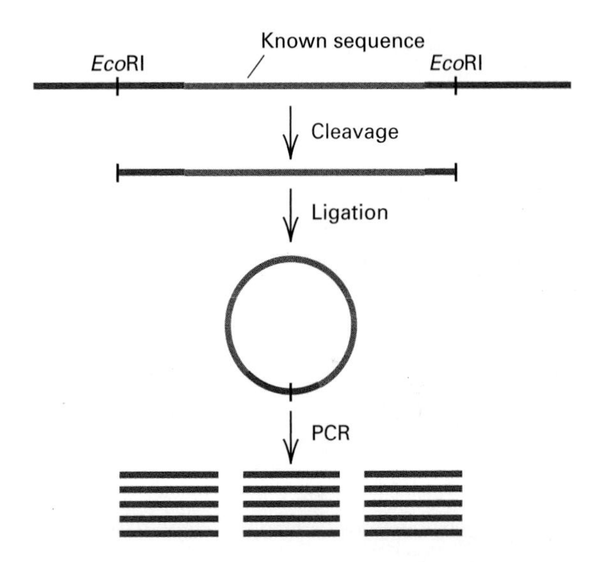


## GENETICS on the web

GeNETics on the Web will introduce you to some of the most important sites for finding genetics information on the Internet. To explore these sites, visit the Jones and Bartlett companion site to accompany *Genetics: Analysis of Genes and Genomes, Seventh Edition* at http://biology.jbpub.com/book/genetics.

There you will find a chapter-by-chapter list of highlighted keywords. When you select one of the keywords, you will be linked to a Web site containing information related to that keyword.

# CHAPTER 10

# Molecular Biology of Gene Expression

Masquerading as suspended seaweed, the leafy sea dragon *Phyllopteryx eques* is a species of fish that develops leafy outgrowths along its body. The genetic changes that underlie such morphological adaptations are largely unknown, but many of them are thought to be due largely to changes in patterns of gene expression rather than drastic changes in gene function.

## CHAPTER OUTLINE


**10.1** Amino Acids, Polypeptides, and Proteins

**10.2** Colinearity Between Coding Sequences and Polypeptides

**10.3** Transcription
- Overview of RNA Synthesis
- Types of RNA Polymerase
- Promoter Recognition
- Mechanism of Transcription
- Genetic Evidence for Promoters and Terminators

**10.4** Messenger RNA

**10.5** RNA Processing in Eukaryotes
- 5′ Capping and 3′ Polyadenylation
- Splicing of Intervening Sequences
- Characteristics of Human Transcripts
- Coupling of Transcription and RNA Processing
- Mechanism of RNA Splicing
- Effects of Intron Mutations
- Exon Shuffle in the Origin of New Genes

**10.6** Translation
- Nonsense-Mediated Decay
- Initiation by mRNA Scanning
- Elongation
- Release
- Protein Folding and Chaperones

**10.7** Complex Translation Units
- Polysomes
- Polycistronic mRNA

**10.8** The Standard Genetic Code
- Genetic Evidence for a Triplet Code
- How the Code Was Cracked
- Features of the Standard Code
- Transfer RNA and Aminoacyl-tRNA Synthetase Enzymes
- Redundancy and Wobble
- Nonsense Suppression

CONNECTION Messenger "Light"
Sydney Brenner, François Jacob, and Matthew Meselson 1961
*An Unstable Intermediate Carrying Information from Genes to Ribosomes for Protein Synthesis*

CONNECTION Poly-U
Marshall W. Nirenberg and J. Heinrich Matthaei 1961
*The Dependence of Cell-Free Protein Synthesis in* E. coli *upon Naturally Occurring or Synthetic Polyribonucleotides*

The term **gene expression** refers to the molecular processes by which the information contained in genes is converted into polypeptide chains that determine the metabolic and developmental capabilities of cells and organisms. As we saw in Chapter 1, the information transfer from DNA to protein is accomplished through the "central dogma" of gene expression:

$$DNA \rightarrow RNA \rightarrow Protein$$

In somewhat more detail, the principal steps in gene expression can be summarized as follows:

1. RNA molecules are synthesized enzymatically by an *RNA polymerase,* which uses the nucleotide sequence of a segment of a single strand of DNA as a template to create a complementary RNA strand in a stepwise manner. The overall process by which the segment corresponding to a particular gene is selected and an RNA molecule is made is *transcription.*
2. In eukaryotes, the RNA usually undergoes chemical modification in the nucleus called *RNA processing.*
3. The nucleotide sequence of the processed RNA molecule (the *messenger RNA*) is used in the process of *translation* to direct the sequential joining of amino acids in a particular order to create a polypeptide chain. The amino acid sequence in a polypeptide chain is therefore specified by the coding sequence in a gene.

## 10.1 Amino Acids, Polypeptides, and Proteins

Proteins are responsible for catalyzing most intracellular chemical reactions (enzyme proteins), for regulating gene expression (regulatory proteins), and for determining many features of the structures of cells, tissues, and organisms (structural proteins). A protein is composed of one or more **polypeptide chains**, each of which is a series of covalently joined amino acids. There are twenty different amino acids commonly found in polypeptides, and they can be joined in any order and in any number (typically 100–1000).

Each amino acid contains a carbon atom (the $\alpha$ carbon) to which is attached one carboxyl group ($-COOH$), one amino group ($-NH_2$), and a side chain commonly called an **R group** (**FIGURE 10.1**). The R groups are generally chains or rings of carbon atoms bearing various distinguishing atoms. The simplest R groups are those of glycine ($-H$) and of alanine

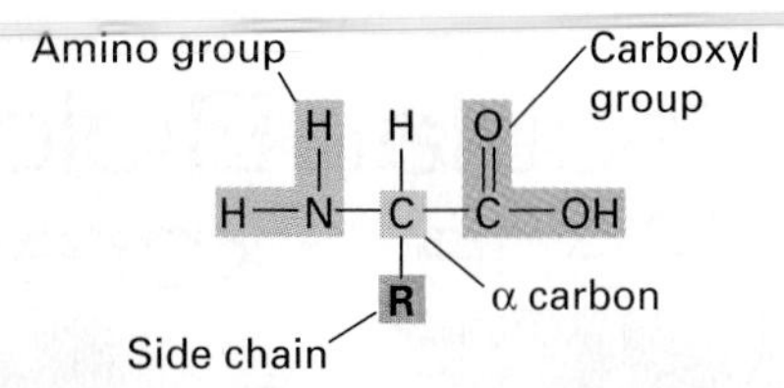


**FIGURE 10.1** The general structure of an amino acid.

($-CH_3$). For reference, the chemical structures of all twenty amino acids are shown in **FIGURE 10.2**. For each amino acid, the R group is indicated by a gold rectangle.

Each link in a polypeptide chain is formed when the carboxyl group of one amino acid becomes joined with the amino group of the next amino acid in line; the resulting chemical bond is a covalent bond called a **peptide bond** (**FIGURE 10.3**, part A). Thus, a polypeptide chain has an ordered array of side chains jutting off a backbone that consists of $\alpha$-carbon atoms alternating with peptide groups (Figure 10.3B).

The two ends of every polypeptide molecule are distinct. One end has a free amino ($-NH_2$) group and is called the **amino terminus**; the other end has a free carboxyl ($-COOH$) group and is the **carboxyl terminus**. Polypeptides are synthesized by adding successive amino acids to the carboxyl end of the growing chain. Conventionally, the amino acids of a polypeptide chain are numbered starting at the amino terminus.

Most polypeptide chains are highly folded, and a wide variety of three-dimensional shapes have been observed. The manner of folding is determined primarily by the sequence of amino acids—in particular, by noncovalent interactions between the side chains—so each polypeptide chain tends to fold into a unique three-dimensional shape. About 70–75 percent of polypeptide chains are sufficiently small that they fold correctly within milliseconds after release from the ribosome. Polypeptide chains that are large or that have a slow or very complex folding pathway recruit specialized proteins that facilitate the folding process. Folding is discussed further in Section 10.6.

Many protein molecules consist of more than one polypeptide chain. When this is the case, the protein is said to contain **protein subunits**. The subunits may be identical or different. For example, hemoglobin, the oxygen carrier of blood, consists of four subunits: two copies of a polypeptide designated the $\alpha$ chain, and two copies of a different polypeptide designated the $\beta$ chain.

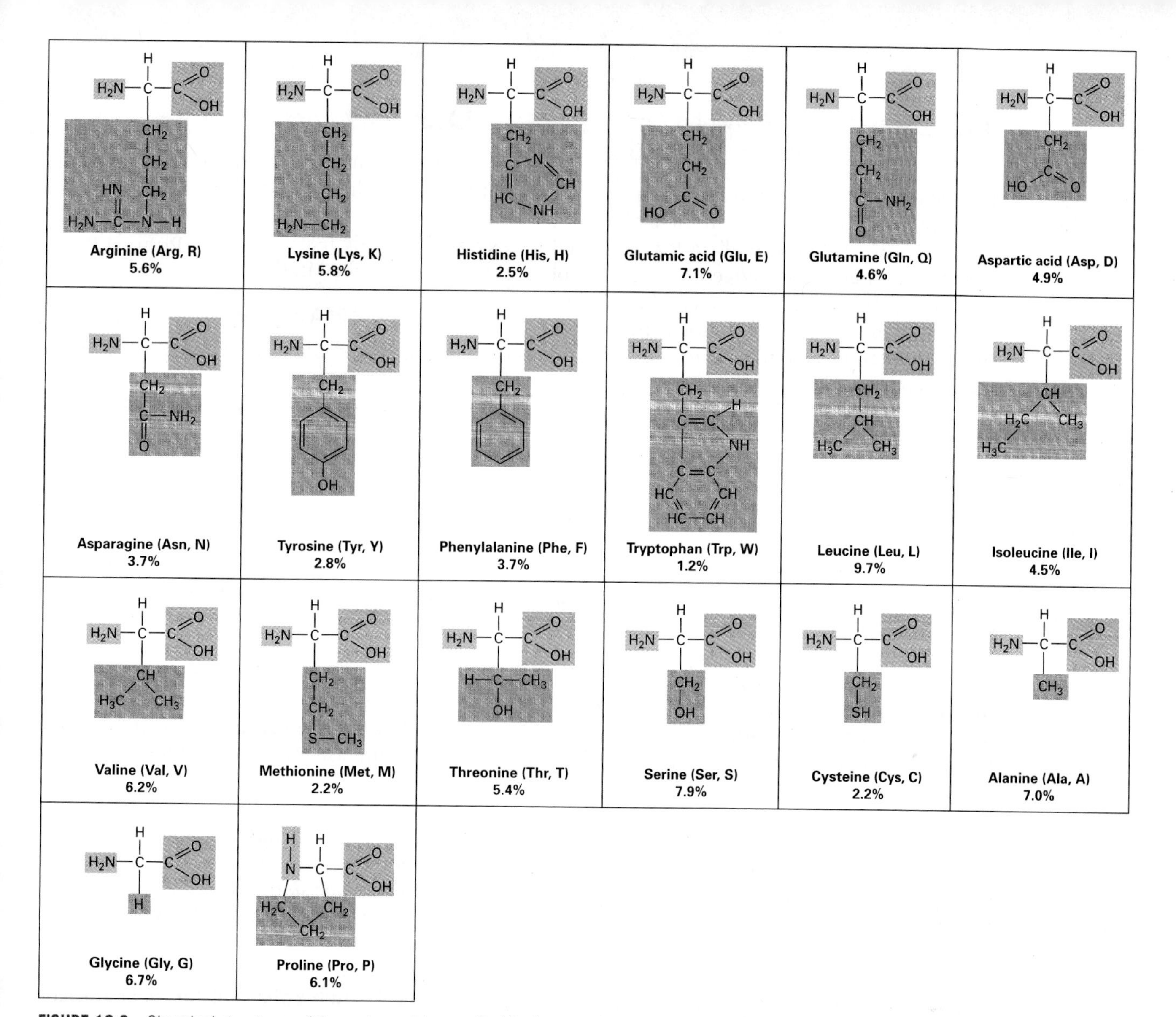


**FIGURE 10.2** Chemical structures of the amino acids specified in the genetic code, along with their conventional three-letter and one-letter abbreviations. Note that proline does not have the same general structure as the rest because it lacks a free amino group; proline is an *imino* acid. The percentage values give the relative abundance of each amino acid averaged over all proteins encoded in the human genome.

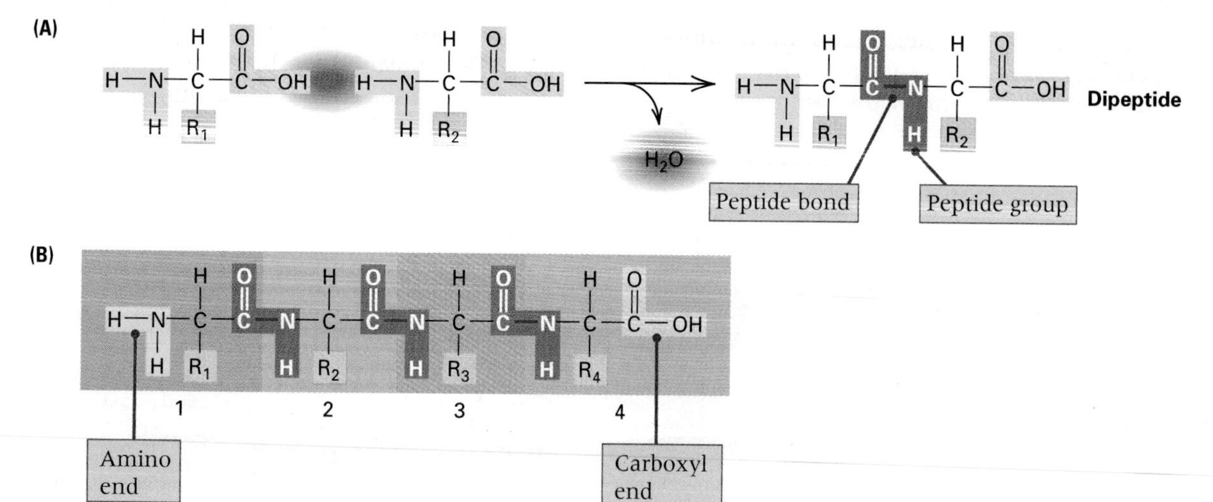


**FIGURE 10.3** Properties of a polypeptide chain. (A) Formation of a peptide bond by reaction of the carboxyl group of one amino acid (left) with the amino group of a second amino acid (right). A molecule of water (HOH) is eliminated to form the peptide bond (red line). (B) A tetrapeptide showing the alternation of $\alpha$-carbon atoms (black) and peptide groups (blue). The four amino acids are numbered below.

## 10.2 Colinearity Between Coding Sequences and Polypeptides

That genes specify the amino acid sequence in polypeptides was first shown in the early 1950s by the discovery that the allele for sickle-cell anemia brings about a change in the charge of the hemoglobin molecule by causing a substitution of an uncharged valine for a negatively charged glutamic acid at residue number 6 in the $\beta$-globin chain.

Most genes contain the information for the synthesis of only one polypeptide chain. Furthermore, the linear order of nucleotides in a gene determines the linear order of amino acids in a polypeptide. This point was first proved by studies of the tryptophan synthase gene *trpA* in *Escherichia coli,* a gene in which many mutations had been obtained and accurately mapped. The effects of numerous mutations on the amino acid sequence of the enzyme were determined by directly analyzing the amino acid sequences of the wildtype and mutant enzymes. Each mutation was found to result in a single amino acid substituting for the wildtype amino acid in the enzyme; more important, *the order of the mutations in the genetic map was the same as the order of the mutant amino acids in the polypeptide chain* (**FIGURE 10.4**). This attribute of genes and polypeptides is called **colinearity**, which means that the sequence of base pairs in DNA determines the sequence of amino acids in the polypeptide in a colinear, or point-to-point, manner. Colinearity is found almost universally in prokaryotes. However, we will see later that in eukaryotes, noncoding DNA sequences interrupt the coding sequences in most genes; in these genes, the order of mutations along a gene (but not their spacing) correlates with the respective amino acid substitutions.

## 10.3 Transcription

The first step in gene expression is **transcription**, the synthesis of an RNA molecule copied from the segment of DNA that constitutes the gene. The basic features of the production of RNA are described in this section.

### ■ Overview of RNA Synthesis

The essential chemical characteristics of the enzymatic synthesis of RNA resemble those of DNA synthesis discussed in Chapter 6.

1. The precursors in the synthesis of RNA are the four ribonucleoside 5′-triphosphates: adenosine triphosphate (ATP), guanosine triphosphate (GTP), cytidine triphosphate (CTP), and uridine triphosphate (UTP). They differ from the DNA precursors only in that the sugar is ribose rather than deoxyribose and in that the base uracil (U) replaces thymine (T) (**FIGURE 10.5**).
2. The enzyme used in transcription is **RNA polymerase** rather than DNA polymerase. The RNA polymerase binds to a DNA sequence of 20–200 nucleotides in length called a **promoter** and then initiates transcription at a nucleotide in or near the promoter called the *transcription start site.*
3. In the synthesis of RNA, a sugar–phosphate bond is formed between the 3′-hydroxyl group of one nucleotide and the 5′-triphosphate of the next nucleotide in line (**FIGURE 10.6**, parts A and B). This is the same chemical bond as in the synthesis of DNA, but the enzyme is different and the polymer being synthesized is RNA.
4. The linear order of nucleotides in an RNA molecule is determined by the

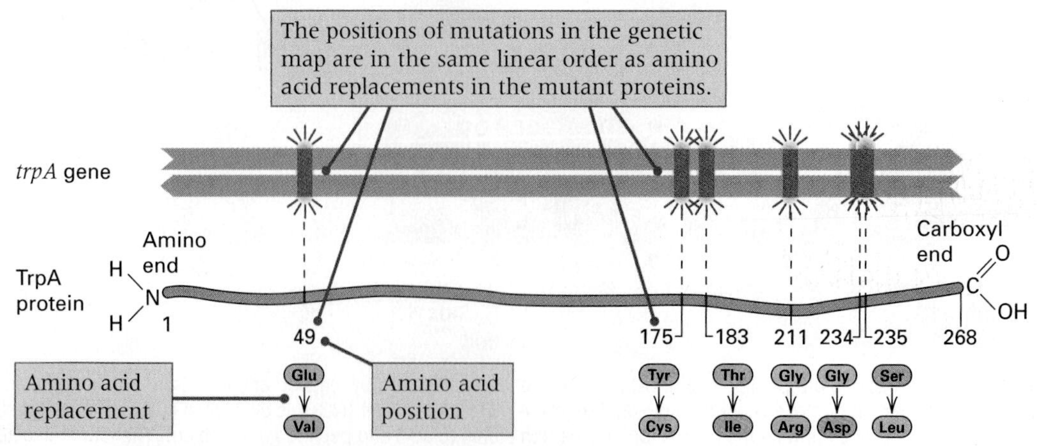


**FIGURE 10.4** Colinearity of DNA and protein in the *trpA* gene of *E. coli.*

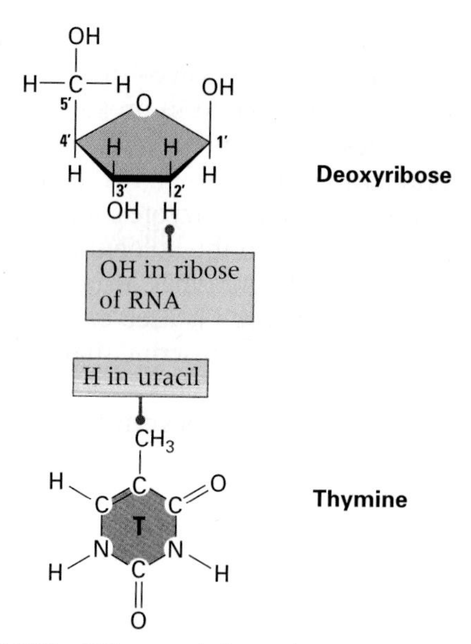


**FIGURE 10.5** Differences between the structures of ribose and deoxyribose and between those of uracil and thymine.

linear order of nucleotides in the DNA template. Each nucleotide added to the growing end of the RNA chain is chosen for its ability to base pair with the DNA template strand. Thus, the bases C, T, G, and A in a DNA strand cause G, A, C, and U, respectively, to be added to the growing end of an RNA molecule.

5. Nucleotides are added only to the 3′-OH end of the growing chain; as a result, the 5′ end of a growing RNA molecule bears a triphosphate group. Note that the 5′-to-3′ direction of RNA chain growth is the same as that in DNA synthesis.

A significant difference between DNA polymerase and RNA polymerase is that *RNA polymerase is able to initiate chain growth without a primer.* Furthermore:

Each RNA molecule produced in transcription derives from a single strand of DNA, because

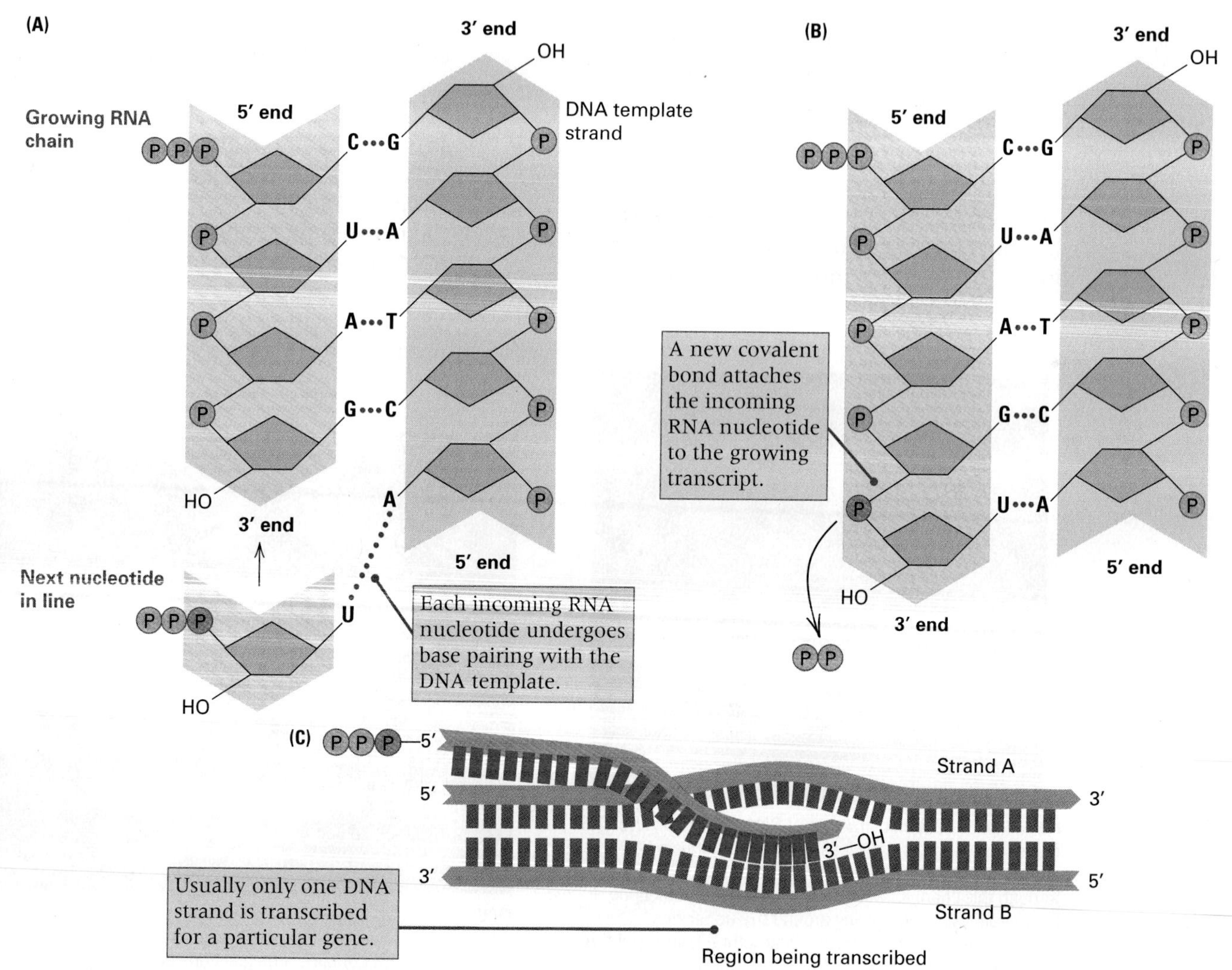


**FIGURE 10.6** RNA synthesis. (A) Base pairing with the template strand. (B) The polymerization step. (C) Geometry of RNA synthesis. RNA is copied from only one strand of a segment of a DNA molecule—in this example, strand B—without the need for a primer.

in any particular region of the DNA, only one strand serves as a template for RNA synthesis.

The implications of this statement are shown in Figure 10.6C.

## Types of RNA Polymerase

RNA polymerases are large, multisubunit complexes whose active form is called the **RNA polymerase holoenzyme**. Bacterial cells have only one RNA polymerase, the holoenzyme of which includes six polypeptide subunits. The molecular mass of the holoenzyme is about 400 kilodaltons (kD), and its widest dimension is 150 Å (**FIGURE 10.7**). In comparison, the mass of the hemoglobin tetramer is about 60 kD, and its widest dimension is about 60 Å. The long axis of the bacterial RNA holoenzyme is about the same length as a stretch of 45 nucleotides in duplex DNA. But in transcriptional initiation, the holoenzyme actually contacts 70–90 bp in the promoter region, which means that the DNA must wrap around the holoenzyme. Once transcription begins, the region of contact is reduced to about 35 nucleotides, centered on the nucleotide being added. The *processivity* of RNA polymerase (the number of nucleotides transcribed without dissociating from the template) is impressive: more than $10^4$ nucleotides in prokaryotes and more than $10^6$ nucleotides in eukaryotes. Processivity is important, because once the RNA polymerase separates from the template, it cannot resume synthesis. The rate of transcription is also impressive—approximately 70 nucleotides per second in prokaryotes and 40 nucleotides per second in eukaryotes.

Eukaryotic RNA polymerases are even larger than those in prokaryotes, and they include more subunits in the holoenzyme. There are also several different types, which include:

1. ***RNA polymerase I*** is used exclusively in producing the transcript that becomes processed into ribosomal RNA. The promoter region includes the transcription start site.
2. ***RNA polymerase II*** is the workhorse eukaryotic polymerase responsible for transcribing all protein-coding genes as well as the genes for a number of small nuclear RNAs used in RNA processing (U1, U2, U3, and so forth). The RNA polymerase II promoter is located near the transcription start site but upstream (on the 5′ side) of it. The mechanism of Pol II is the best understood of the eukaryotic polymerases. The holoenzyme contains twelve polypeptide subunits and has a molecular mass of about 500 kD. Its structure features a groove along the top (**FIGURE 10.8**), which helps guide the DNA template into the active site marked by the magnesium atom colored in pink.
3. ***RNA polymerase III*** is used in transcribing all transfer RNA genes as well as the 5S component of ribosomal RNA. The promoter for RNA polymerase III tran-

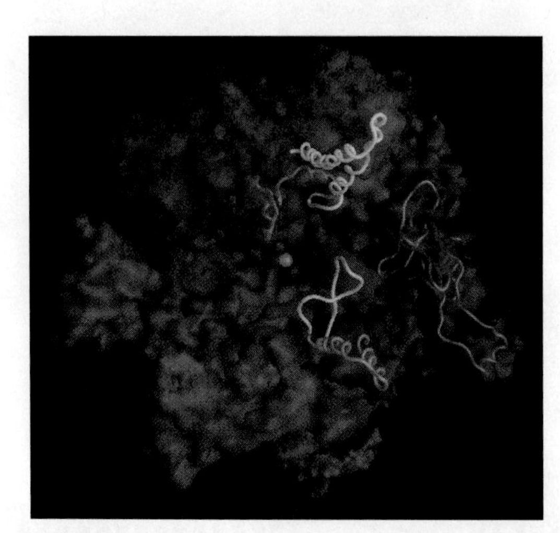

**FIGURE 10.7** Subunit structure of RNA polymerase from the bacterium *Thermus aquaticus.* The complex has a U-shaped channel running through it. DNA to be transcribed passes through this channel, with the promoter region emerging at the bottom right and curving around the upper right. The colored coils highlight some of the structural features of the complex; the pink sphere in the center is a $Mg^{2+}$ ion in the active site. [Reproduced from R. A. Mooney and R. Landick, "RNA polymerase unveiled," *Cell* 98 (1999): 687–690, © 1999, with permission from Elsevier.]

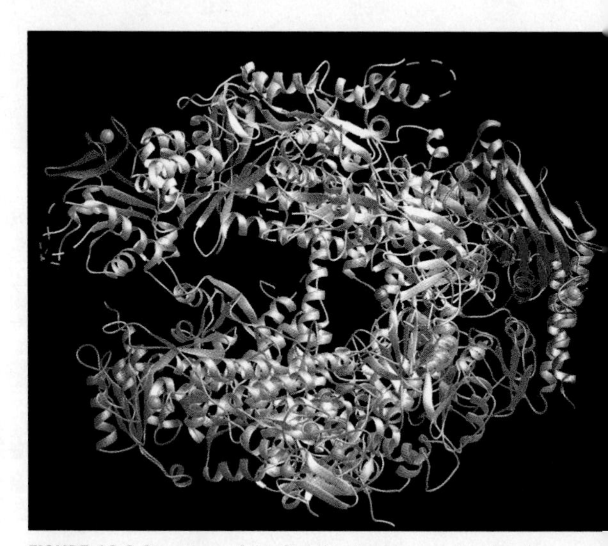

**FIGURE 10.8** Structure of the Pol II holoenzyme showing the groove that runs across the top. The groove terminates in the active site, which contains a magnesium ion (pink sphere). Different polypeptide components in the holoenzyme are shown in different colors. [Courtesy of David A. Bushnell and Roger Kornberg, Stanford University Medical School.]

scription is located near the transcription start site but downstream (on the 3′ side) of it.

## Promoter Recognition

Many promoter sequences have been isolated and their nucleotide sequences determined. The promoters for polymerases I, II, and III show no commonality, and within each class there is substantial sequence variation. The situation in bacteria is considerably simpler. Although bacterial promoters also differ in sequence, in part because they differ in their polymerase binding affinity, certain sequence patterns or *motifs* are quite frequent. Two such patterns that are often found in promoter regions in *E. coli* are illustrated in **FIGURE 10.9**. Each pattern is defined by a **consensus sequence** of nucleotides determined from the actual sequences by majority rule: Each nucleotide in the consensus sequence is the nucleotide most often observed at that position in actual sequences. Any particular sequence may resemble the consensus sequence very well or very poorly.

In the consensus sequences shown in Figure 10.9, the transcription start site is numbered as +1. One consensus motif is TTGACA, which is called the −35 motif because it is usually located approximately 35 base pairs upstream from the transcription start site. The other consensus motif is TATAAT, which is usually located near position −10. The −10 sequence is called the **TATA box**. The positions of the promoter sequences determine where the RNA polymerase begins synthesis.

The strength of the binding of RNA polymerase to different promoters varies greatly, which causes differences in the extent of expression from one gene to another. Most of the differences in promoter strength result from variations in the −35 and −10 promoter elements and in the spacing between them. Promoter strength among *E. coli* genes differs by a factor of $10^4$, and most of the variation can be attributed to the promoter sequences themselves. In general, the more closely the promoter elements resemble the consensus sequence, the stronger the promoter. Mutations that change the nucleotide sequence in a promoter can alter the strength of the promoter. Changes that result in less resemblance to the consensus sequence lower the promoter's strength, whereas those with greater resemblance to the consensus increase it. Furthermore, some promoters differ greatly from the consensus sequence in the −35 region.

All promoters typically require accessory proteins to activate transcription by RNA polymerase. In bacteria, among the most important accessory proteins for transcription are sigma factors. A **sigma factor** is a protein that allows RNA polymerase to bind properly to a promoter region. All bacteria produce sigma factors that allow transcription of genes needed for their normal growth and metabolism. There are also specialized sigma factors produced only under certain conditions that enable normally untranscribed genes to be transcribed. Various kinds of stress, including heat shock or starvation, induce the production of such specialized sigma factors.

Promoter sequences in eukaryotes are generally much longer and more complex than those in prokaryotes. Many promoters recognized by Pol II include a core region containing a TATA-box motif, which is analogous to that in prokaryotes but differs in its spacing relative to the transcriptional start site (**FIGURE 10.10**). Proper binding of Pol II to the promoter also requires a set of 26 **general transcription factors**, but even these proteins are not sufficient. They need to be recruited to the promoter by still other proteins that bind with other sequence motifs that are often located far upstream or sometimes even downstream from the core region contain-

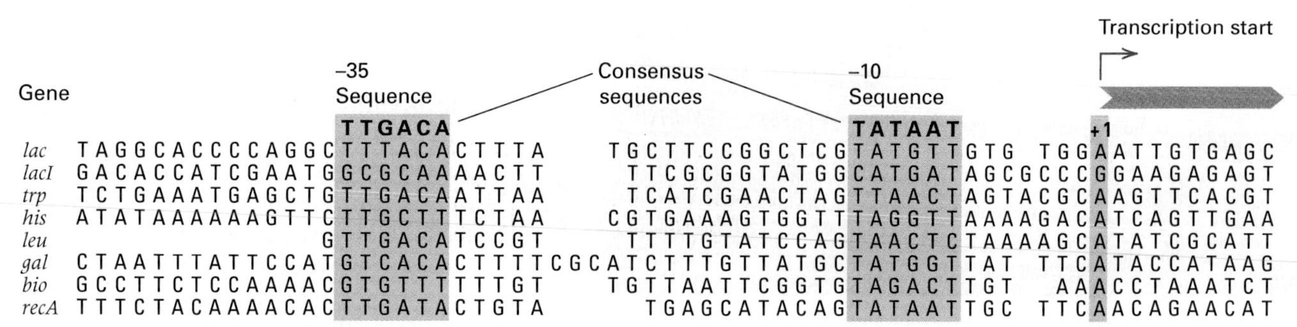


**FIGURE 10.9** Nucleotide sequences in promoter regions of several genes in *E. coli*. The consensus sequences located 10 and 35 nucleotides upstream from the transcription start site (+1) are indicated. Promoters vary tremendously in their ability to promote transcription. Much of the variation in promoter strength results from differences between the promoter elements and the consensus sequences at −10 and −35.

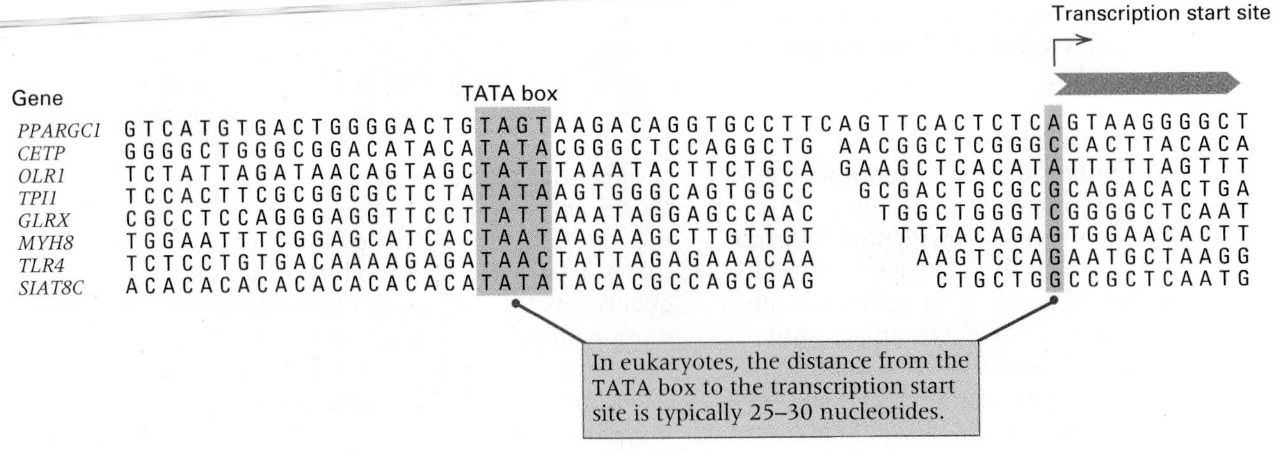


**FIGURE 10.10** Some human TATA-box genes showing sequences in the core promoter region near the transcription start site.

ing the TATA box. These sequence motifs, some of which act as *enhancers* and others as *silencers*, bind proteins that interact with the transcriptional machinery to regulate the level of transcription. This level of gene regulation is discussed in Chapter 11.

## Mechanism of Transcription

Once promoter recognition takes place, the mechanism of transcription can be described in terms of three discrete stages. These will be examined with regard to the mechanism of action of the eukaryotic Pol II polymerase, studies for which Roger D. Kornberg was awarded the 2006 Nobel Prize in Chemistry.

The Pol II complex does not act alone but in combination with five general transcription factors denoted TFIIB, TFIIE, TFIID, TFIIF, and TFIIH. Although these components can correctly transcribe naked DNA, they cannot transcribe native chromatin organized into nucleosomes (Chapter 7). In the cell, access to naked promoter DNA occurs by transient displacement of the nucleosomes through the action of *chromatin remodeling complexes* (Chapter 11). The specificity that causes transcription to begin only at the correct start site near a promoter is brought about by the general transcription factors associated with Pol II, each of which plays an essential role.

**FIGURE 10.11** Model showing how the TATA-box binding protein causes a right-angle bend in the promoter DNA, which flips the promoter DNA across the top of the Pol II polymerase. [Courtesy of Stephen Burley.]

1. ***Chain initiation*** In the first step in transcription, the **TATA-box binding protein** (TBP) binds the promoter DNA and bends it at almost a 90-degree angle (**FIGURE 10.11**). Physically, the TATA-box binding protein is closely associated with the polymerase, and the bend brings the promoter DNA into contact with the TFIIB component of the polymerase. The promoter DNA then follows a straight path across the top of the polymerase until, at a point 25–30 bp distant from the TATA box, the transcription start site is brought into position near the polymerase active site. At this point TFIIE joins the complex and recruits TFIIH, whose helicase activity destabilizes the DNA duplex. This makes the promoter duplex susceptible to thermal unwinding, which produces a transient unwound region or bubble. The unwound region is stabilized by TFIIF binding to the nontemplate strand, while the unbound template strand descends deeper into the groove in the polymerase

where the active site is situated. At a point three nucleotides deep into the groove, the template strand undergoes a sharp bend that flips the first base to be transcribed into the active site (**FIGURE 10.12**). At the same time, a large domain of the polymerase flips into position over the promoter DNA, forming a clamp that holds the transcription bubble in place.

Initiation of transcription now begins, a process stabilized by TFIIB. After the first nucleoside triphosphate of the RNA transcript is put in place, synthesis proceeds in the 5′-to-3′ direction. After about six nucleotides have been transcribed, the TFIIB is displaced and transcription continues.

2. ***Chain elongation*** As the RNA polymerase encounters new nucleotides along the template DNA strand, successive RNA nucleotides are added to the growing transcript. Only one DNA strand, the **template strand**, is transcribed. At steady state, the transcription bubble consists of fifteen nucleotides of unwound DNA duplex, of which eight to nine are paired with the 3′ end of the RNA transcript (Figure 10.6C). Each incoming nucleotide is added to the 3′ end of the transcript at a site three nucleotides from the point at which the DNA template strand unwinds from the nontemplate DNA.

The weak energy released by hydrogen bonding between complementary nucleotides is not adequate to account for the base-pairing specificity of transcription. The specificity is actually brought about by structural changes that take place around the active site of Pol II. The key movement is that of a domain near the active site, called the *trigger loop,* which moves into position only when correct base pairing occurs. The repositioning of the trigger loop brings a critical histidine residue into play, which promotes the flow of electrons that triggers polymerization. In this way, correct base pairing is physically coupled to formation of the phosphodiester bond in the growing RNA chain.

As each new bond is formed in the transcript, a helical segment of Pol II that is in contact with the single-stranded template lurches forward about 3 Å, which brings the next template nucleotide into the active site. About eight to nine nucleotides downstream from the active site, two segments of Pol II invade the RNA–DNA hybrid, breaking the hydrogen bonds, and a few nucleotides farther downstream from that point, the DNA template and nontemplate DNA strands are rejoined (Figure 10.6C).

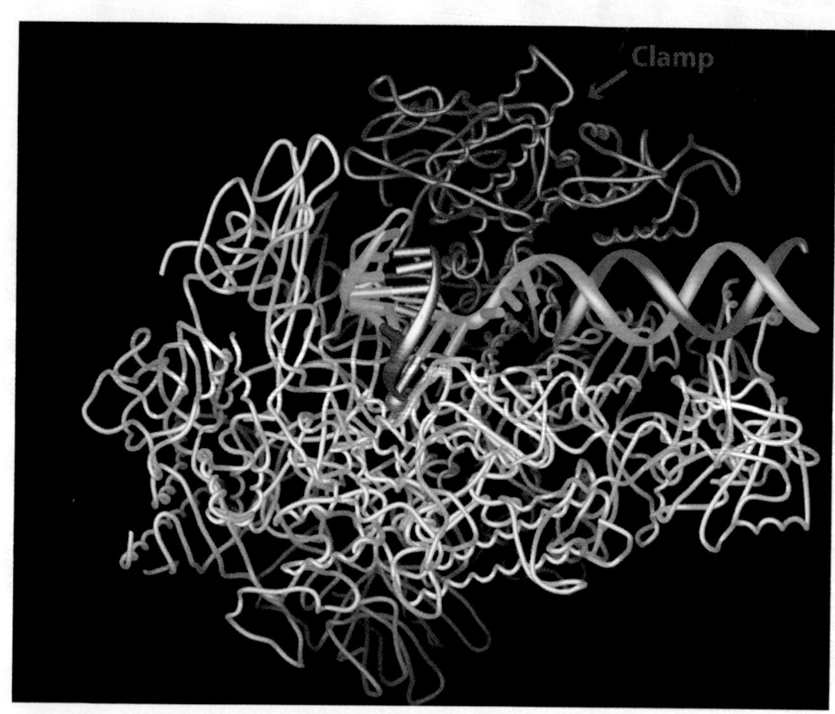


**FIGURE 10.12** Pol II in action. The template strand is shown in blue and the RNA transcript in red. The magnesium ion in the active site is indicated by the pink sphere. Part of the nontemplate strand is shown in green, paired with the template strand. The replication bubble is held in place by a mobile domain of Pol II (the clamp, orange), and the two largest subunits of Pol II are held together by a helix bridging between them (green helix). The other polypeptide chains in Pol II are shown in white. [Courtesy of David A. Bushnell and Roger Kornberg, Stanford University Medical School.]

3. ***Chain termination*** RNA polymerase reaches a chain-termination sequence, and both the newly synthesized RNA molecule and the polymerase are released. Two kinds of termination events are known: those that are self-terminating and depend only on the nucleotide sequence in the DNA template, and those that require the presence of a termination protein. In self-termination, which is the most common case in bacteria, transcription stops when the polymerase encounters a particular sequence of nucleotides in the transcribed DNA strand that is able to fold back upon itself to form a hairpin loop. An example of such a terminator found in *E. coli* is shown in **FIGURE 10.13**. The hairpin loop alone is not enough for termination of transcription;

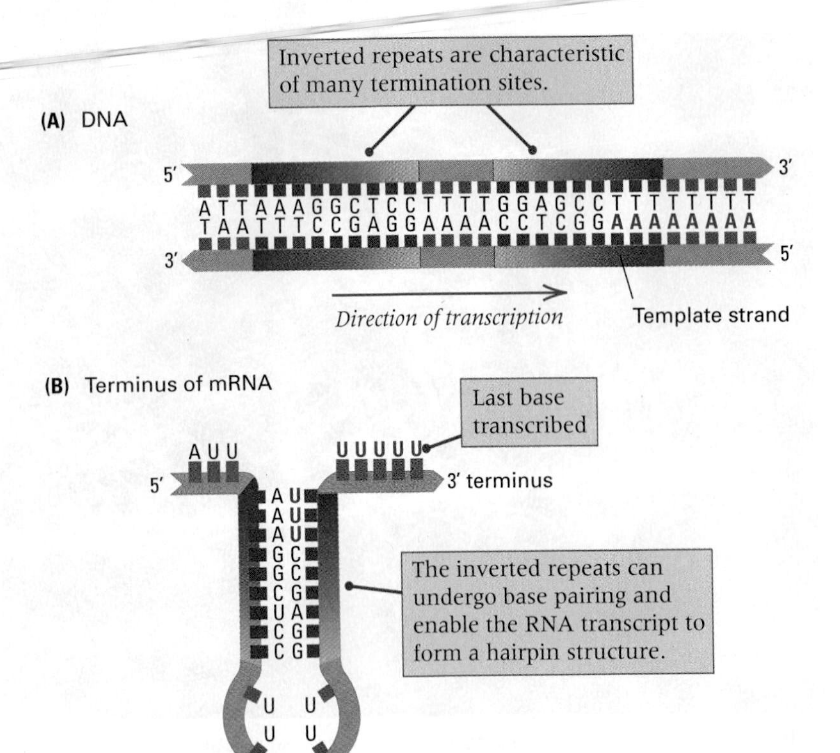
Inverted repeats are characteristic of many termination sites.
(A) DNA
5'
3'
ATTAAAGGCTCCTTTTGGAGCCTTTTTTTT
TAATTTCCGAGGAAAACCTCGGAAAAAAAA
3'
5'
Direction of transcription
Template strand
(B) Terminus of mRNA
Last base transcribed
AUU
UUUUU
5'
3' terminus
The inverted repeats can undergo base pairing and enable the RNA transcript to form a hairpin structure.

**FIGURE 10.13** (A) Nucleotide sequence of the transcription–termination region for the set of tryptophan-synthesizing genes in *E. coli*. (B) The 3′ terminus of the RNA transcript, folded to form a stem-and-loop structure. The sequence of *U*s found at the end of the transcript in this and many other prokaryotic genes is shown in red. The RNA polymerase, not shown here, terminates transcription when the loop forms in the transcript.

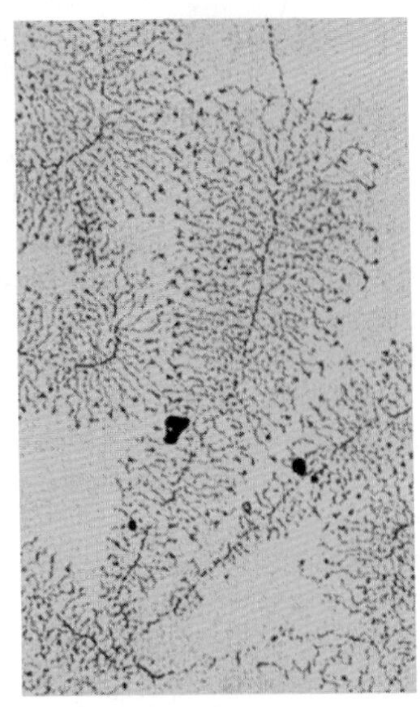

**FIGURE 10.14** Electron micrograph of part of the DNA of the newt *Triturus viridescens*, containing tandem repeats of genes being transcribed into ribosomal RNA. The thin strands forming each feather-like array are RNA molecules. A gradient of lengths can be seen for each rRNA gene. Regions in the DNA between the thin strands are spacer DNA sequences, which are not transcribed. [Courtesy of Oscar Miller.]

the run of *U*s at the end of the hairpin is also necessary. The requirements for transcription termination in eukaryotes are more complex.

Initiation of a second round of transcription need not await completion of the first, because the promoter becomes available once RNA polymerase has polymerized from fifty to sixty nucleotides. For a rapidly transcribed gene, such reinitiation occurs repeatedly, and a gene can be cloaked with numerous RNA molecules in various degrees of completion. The micrograph in **FIGURE 10.14** shows a region of the DNA of the newt *Triturus* that contains tandem repeats of a ribosomal RNA gene. Each repeat is associated with growing transcripts. The shortest transcripts are at the promoter end of the repeat; the longest are near the terminus.

### Genetic Evidence for Promoters and Terminators

The existence of promoters was first demonstrated in genetic experiments with *E. coli* by the isolation of a class of lactose-utilization mutations with the unique feature that the phenotype was $Lac^-$ only when the mutation was in the *cis* configuration (in the same DNA molecule) relative to the *lacZ* structural gene. These were denoted $p^-$ mutations. The relevance of the *cis* configuration can be seen by examining a cell with two copies of the gene *lacZ*—for example, a cell containing an F′ *lacZ* plasmid, which contains *lacZ* in the bacterial chromosome as well as *lacZ* in the F′ plasmid. Transcription of the *lacZ* gene enables the cell to synthesize the enzyme $\beta$-galactosidase. **TABLE 10.1** shows that a wildtype *lacZ* gene ($lacZ^+$) is inactive when it and a $p^-$ mutation are present in the same DNA molecule (either in the chromosome or in an F′ plasmid); this can be seen by comparing entries 4 and 5. Analysis of the RNA shows that in a cell with the genotype $p^-$ $lacZ^+$, the $lacZ^+$ gene is not transcribed, whereas in a cell with the genotype $p^+$ $lacZ^-$, a mutant RNA is produced. The $p^-$ mutations were called *promoter mutations*.

Mutations in *E. coli* have also been instrumental in defining the transcription–termination region. Mutations that create a new termination sequence upstream from the normal one have been isolated. When such a mutation is present, an RNA molecule shorter than the wildtype RNA is made. Other mutations eliminate the terminator, resulting in a longer transcript.

## 10.4 Messenger RNA

The RNA molecule produced from a DNA template is the **primary transcript**. Each gene has only one DNA strand that serves as the template strand, but which strand is the template strand can differ from gene to gene along

## connection
### Messenger "Light"

*Brenner and Jacob were guest investigators at the California Institute of Technology in 1961. At that time there was great interest in the mechanisms by which genes code for proteins. One possibility, which seemed reasonable at the time, was that each gene produced a different type of ribosome, differing in its RNA, which in turn produced a different type of protein. François Jacob and Jacques Monod had recently proposed an alternative, which was that the informational RNA ("messenger RNA") is actually an unstable molecule that breaks down rapidly. In this model, the ribosomes are nonspecific protein-synthesizing centers that synthesize different proteins according to specific instructions they receive from the genes through the messenger RNA. The key to the experiment is density-gradient centrifugation, which can separate macromolecules made "heavy" or "light" according to their content of $^{15}N$ or $^{14}N$, respectively. (This technique is described in Chapter 5.) The experiment is a purely biochemical proof of an issue absolutely critical for genetics—that genes code for proteins through the intermediary of a relatively short-lived messenger RNA.*

**Sydney Brenner,[1] François Jacob,[2] and Matthew Meselson[3] 1961**
[1]Cavendish Laboratory, Cambridge, England
[2]Institute Pasteur, Paris, France
[3]California Institute of Technology, Pasadena, California
*An Unstable Intermediate Carrying Information from Genes to Ribosomes for Protein Synthesis*

A large amount of evidence suggests that genetic information for protein structure is encoded in deoxyribonucleic acid (DNA) while the actual assembling of amino acids into proteins occurs in cytoplasmic ribonucleoprotein particles called ribosomes. The fact that proteins are not synthesized directly on genes demands the existence of an intermediate information carrier. . . . Jacob and Monod have put forward the hypothesis that ribosomes are non-specialized structures which receive genetic information from the gene in the form of an unstable intermediate or "messenger." We present here the results of experiments on phage-infected bacteria which give direct support to this hypothesis. . . . When growing bacteria are infected with T2 bacteriophage, synthesis of DNA stops immediately, to resume 7 minutes later, while protein synthesis continues at a constant rate; in all likelihood, the protein is genetically determined by the phage. . . . Phage-infected bacteria therefore provide a situation in which the synthesis of a protein is suddenly switched from bacterial to phage control. . . . It is possible to determine experimentally [whether an unstable messenger RNA is produced] in the following way: Bacteria are grown in heavy isotopes so that all cell constituents are uniformly labelled "heavy." They are infected with phage and transferred immediately to a medium containing light isotopes so that all constituents synthesized after infection are "light." The distribution of new RNA and new protein, labelled with radioactive isotopes, is then followed by density gradient centrifugation of purified ribosomes. . . . We may summarize our findings as follows: (1) After phage infection no new ribosomes can be detected. (2) A new RNA with a relatively rapid turnover is synthesized after phage infection. This RNA, which has a base composition corresponding to that of the phage DNA, is added to pre-existing ribosomes, from which it can be detached in a cesium chloride gradient by lowering the magnesium concentration. (3) Most, if not all, protein synthesis in the infected cell occurs in pre-existing ribosomes. . . . The results also suggest that the messenger RNA may be large enough to code for long polypeptide chains. . . . It is a prediction of the messenger RNA hypothesis that the messenger RNA should be a simple copy of the gene, and its nucleotide composition should therefore correspond to that of the DNA. This appears to be the case in phage-infected cells. . . . If this turns out to be universally true, interesting implications for the coding mechanisms will be raised.

***The results also suggest that the messenger RNA may be large enough to code for long polypeptide chains.***

Source: S. Brenner, F. Jacob, and M. Meselson, *Nature* 190 (1961): 576–581.

**Table 10.1 Effect of promoter mutations on transcription of the *lacZ* gene**

| Genotype | Transcription of *lacZ*$^+$ gene |
|---|---|
| 1. $p^+$ *lacZ*$^+$ | Yes |
| 2. $p^-$ *lacZ*$^+$ | No |
| 3. $p^+$ *lacZ*$^+$/$p^+$ *lacZ*$^-$ | Yes |
| 4. $p^-$ *lacZ*$^+$/$p^+$ *lacZ*$^-$ | No |
| 5. $p^+$ *lacZ*$^+$/$p^-$ *lacZ*$^-$ | Yes |

*Note: lacZ*$^+$ is the wildtype gene; *lacZ*$^-$ is a mutant that produces a nonfunctional enzyme.

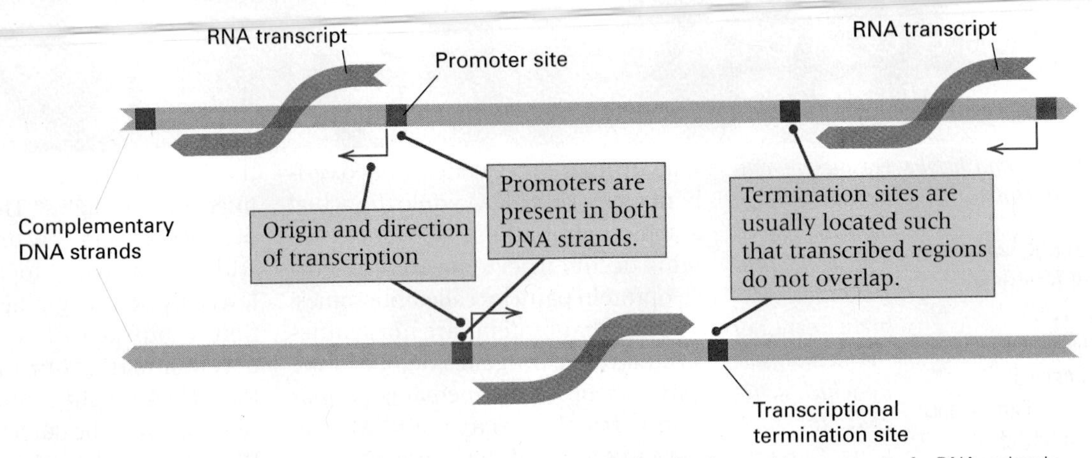


**FIGURE 10.15** A typical arrangement of promoters (green) and termination sites (orange) in a segment of a DNA molecule.

a DNA molecule. Therefore, in an extended segment of a DNA molecule, primary transcripts would be seen growing in either of two directions (**FIGURE 10.15**), depending on which DNA strand functions as a template in a particular gene. In prokaryotes, the primary transcript serves directly as t he **messenger RNA (mRNA)** used in polypeptide synthesis. In eukaryotes, the primary transcript is generally processed before it becomes mRNA.

Translation of an mRNA molecule never starts exactly at the 5′ end and proceeds straightaway to the other. Initiation of polypeptide synthesis usually begins many nucleotides downstream from the 5′ end of the mRNA. The 5′ untranslated region is followed by the **coding sequence**, also called the **open reading frame (ORF)**, which specifies the order in which the amino acids are present in the polypeptide chain. A typical coding sequence in an mRNA molecule is 500–3000 nucleotides in length, which is translated in nonoverlapping sets of three nucleotides (the *codons*). Beyond the coding region in the mRNA, following the termination codon, is a 3′ untranslated region.

In prokaryotes, most mRNA molecules are degraded within a few minutes after synthesis. In eukaryotes, a typical lifetime is several hours, although some last only minutes, and others persist for days. In both kinds of organisms, the degradation enables cells to dispose of molecules that are no longer needed. The short lifetime of prokaryotic mRNA is an important factor in regulating gene activity (Chapter 11).

## 10.5 RNA Processing in Eukaryotes

The process of transcription is very similar in prokaryotes and eukaryotes, but there are major differences in the relationship between the transcript and the mRNA used for polypeptide synthesis. In prokaryotes, the primary transcript is the mRNA. In contrast, the primary transcript in eukaryotes must be converted into mRNA. This conversion is called **RNA processing** and usually consists of three types of events: modification of the 5′ end, extension of the 3′ end, and excision of untranslated sequences embedded within the coding sequences. These events are illustrated diagrammatically in **FIGURE 10.16**.

### ■ 5′ Capping and 3′ Polyadenylation

Each end of a eukaryotic transcript is processed (Figure 10.16). The 5′ end is altered by the addition of a modified guanosine (7-methylguanosine) in an uncommon 5′-to-5′ (instead of 3′-to-5′) linkage; this terminal group is called a **cap**. The 5′ cap is necessary for the mRNA to bind with the ribosome to begin protein synthesis. The 3′ terminus of a eukaryotic mRNA molecule is usually modified by "trimming" (exonuclease digestion) back to a short sequence that serves as a substrate for the addition of a polyadenylate sequence (the **poly-A tail**) of as many as 200 nucleotides. The poly-A tail helps to determine mRNA stability.

### ■ Splicing of Intervening Sequences

The important feature of *splicing* of the primary transcript in eukaryotes is also shown in Figure 10.16. The segments that are excised from the primary transcript are called **introns** or **intervening sequences.** Accompanying the excision of introns is a rejoining of the segments that remain, which are called **exons**, to form the mRNA molecule. The excision of the introns and the joining of the exons constitute **RNA splicing**.

**(A) Transcription initiation and elongation**

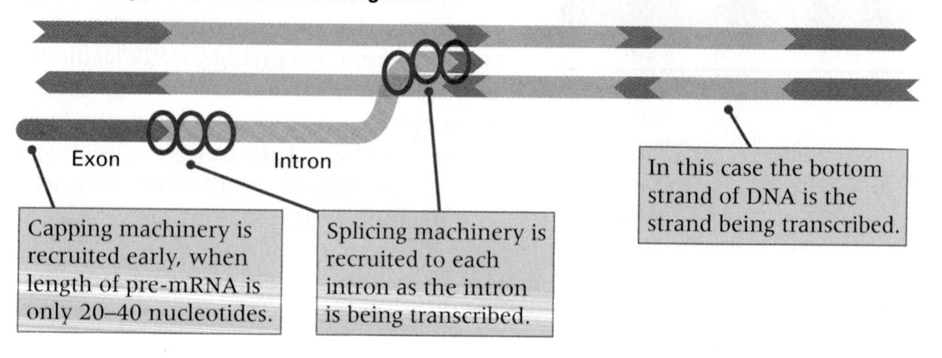

Exon
Intron
Capping machinery is recruited early, when length of pre-mRNA is only 20–40 nucleotides.
Splicing machinery is recruited to each intron as the intron is being transcribed.
In this case the bottom strand of DNA is the strand being transcribed.

**(B) Termination**

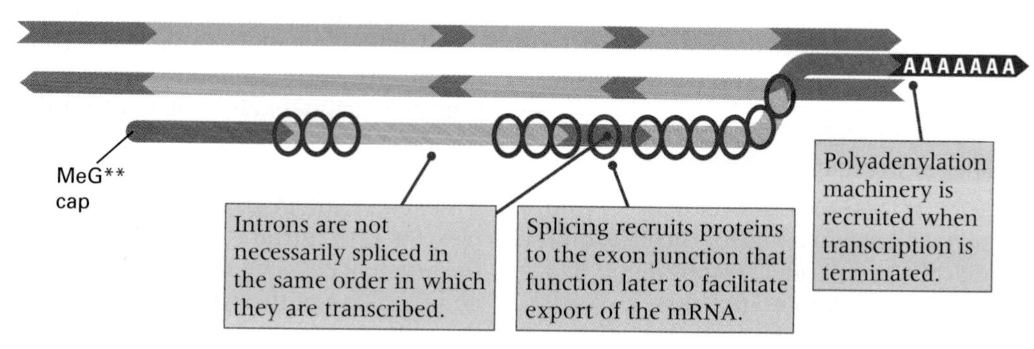

AAAAAAA
MeG**
cap
Introns are not necessarily spliced in the same order in which they are transcribed.
Splicing recruits proteins to the exon junction that function later to facilitate export of the mRNA.
Polyadenylation machinery is recruited when transcription is terminated.

**(C) Release and export**

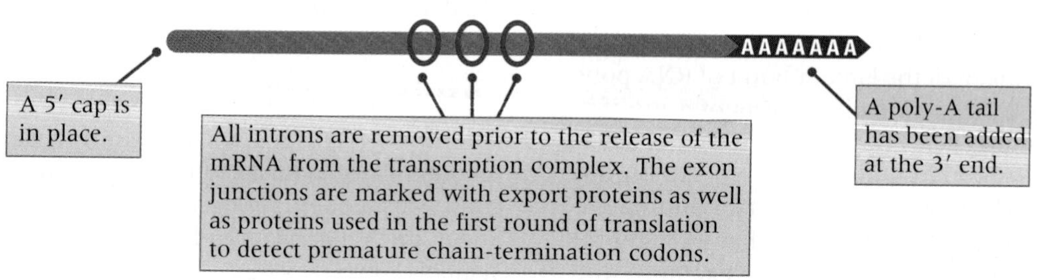


**FIGURE 10.16** In eukaryotes, transcription and RNA processing are coupled. Each step triggers the next in line. MeG denotes 7-methylguanosine (a modified form of guanosine), and the two asterisks indicate two nucleotides whose riboses are methylated.

## ■ Characteristics of Human Transcripts

**TABLE 10.2** summarizes features of the typical human gene. Both the mean and the median values are given because many of the size distributions have a very long right-hand tail, rendering the mean potentially misleading. For example, whereas the mean number of exons is 8.8, this average is unduly influenced by some genes that have a very large number of exons, such as the gene for the muscle protein titin, which includes 178 exons (the largest number for any human gene). Similarly, the distribution of intron sizes is strongly skewed. The most common intron length is 87 nucleotides, but the right-hand tail of the distribution is so stretched out that the mean is 3365 nucleotides. The median is the value that splits the distribution in the middle; half the values are above the median and half below.

One noteworthy feature of Table 10.2 is that human genes tend to be larger than those in nematode worms or fruit flies. Most human genes consist of small exons separated by long introns, and many genes are over 100 kb in length. The longest human gene encodes the muscle protein dystrophin; the gene is 2.4 Mb in length, and mutations in this gene are associated with X-linked Duchenne muscular dystrophy. The average human gene occupies 27 kb of genomic DNA, yet only 1.3 kb (about 5 percent) is used to encode amino acids. The picture is not much different for the medians. The median gene length is 14 kb, of which only 1.1 kb (about 8 percent) is used to encode amino

**Table 10.2 Characteristics of human transcripts**

| Gene feature | Median | Mean |
|---|---|---|
| Size of internal exon | 122 bp | 145 bp |
| Number of exons | 7 | 8.8 |
| Size of introns | 1023 bp | 3356 bp |
| 5′ untranslated region | 240 bp | 300 bp |
| 3′ untranslated region | 400 bp | 770 bp |
| Length of coding sequence | 1101 bp | 1341 bp |
| Number of amino acids (aa) | 367 | 447 |
| Extent of genome occupied | 14 kb | 27 kb |

Source: Data from International Human Genome Sequencing Consortium, *Nature* 409 (2001): 860–921.

acids. Most of the length of human genes is due to the long introns. Some introns are greater than 300 kb and require more than 2.8 hours to be transcribed.

## Coupling of Transcription and RNA Processing

An RNA transcript is not allowed to flop around in the nucleus waiting for the RNA processing machinery to find it. Transcription and RNA processing are **coupled processes**, which means that they are linked by physical or functional interactions so that the occurrence of one process initiates or regulates the next. Some of the interconnections are shown in Figure 10.16. A key player in the coupling is the carboxyl terminal domain of the large subunit of RNA polymerase II. This domain consists of a motif of seven amino acids that is repeated 27 times in the yeast protein and 52 times in the human protein. The motif includes sites for phosphorylation and dephosphorylation that switches the coupling functions on or off. For example, during the initiation of transcription, serine-5 in the repeating motif is phosphorylated, which causes release of the initiation factors and recruitment of the capping machinery that modifies the 5′ end of the transcript and protects it from degradation. Serine-5 is then dephosphorylated, triggering release of the capping machinery. At this point serine-2 is phosphorylated, which helps recruit the machinery for splicing and subsequent steps in RNA processing. The same repeating motif is also involved in recruiting the polyadenylation machinery when transcription is complete but before the transcript is released.

The effect of coupling, in which each successive step recruits the factors necessary for the next step, is to greatly increase the speed and specificity of RNA processing. Without the coupling, RNA processing would be dependent on diffusion and many mistakes would be made, especially in splicing the often large introns that separate relatively small exons. The recruitment of the splicing machinery while transcription is occurring greatly facilitates correct splicing. This is not to say that introns are spliced in exactly the same order in which they are transcribed. The order in which splicing takes place depends on the size and nucleotide composition of the introns, as well as on the overall rate of transcription.

There are numerous interconnecting links that couple the various steps in transcription with those in RNA processing. For example, proteins that bind with RNA polymerase to promote elongation also help recruit the splicing machinery, and the splicing machinery in turn stimulates elongation so that genes containing introns are more efficiently transcribed. The splicing machinery also helps recruit the polyadenylation machinery. The principal steps in RNA processing are all completed prior to release of the mRNA from the transcription complex. As each intron is spliced, proteins bind to the junction between the exons (Figure 10.16B and C). Some of these function after release of the mRNA to facilitate its export from the nucleus to the cytoplasm. Other of these proteins function in the first round of translation to identify defective mRNA molecules that are subsequently destroyed.

## Mechanism of RNA Splicing

The mechanism of RNA splicing is illustrated schematically in **FIGURE 10.17**. Part A shows the consensus sequence found at the 5′ (*donor*) end and at the 3′ (*acceptor*) end of most introns. The symbols are: N, any nucleotide; R, any purine (A or G); Y, any pyrimidine (C or U); and S, either A or C. In the first step of splicing, the 2′−OH of the adenylate (A) at the branch site, which is located a short distance upstream from a run of prymidines (Y) near the acceptor site, attacks the phosphodiester bond at the donor splice site junction. The attack results in cleavage at the donor splice site and formation of a branched molecule (Figure 10.17B) known as a **lariat** because it has a loop and a tail. The A−G linkage at the "knot" of the lariat is unusual in being 2′-to-5′ (instead of the usual 3′-to-5′). In the final step of splicing (Figure 10.17C), the 3′−OH of the guanylate of the donor exon attacks the phosphodiester bond at the acceptor splice site, freeing the lariat intron and joining the donor and acceptor exons together. The lariat intron is rapidly degraded into individual nucleotides by nucleases.

RNA splicing takes place in nuclear particles known as **spliceosomes**. These abundant

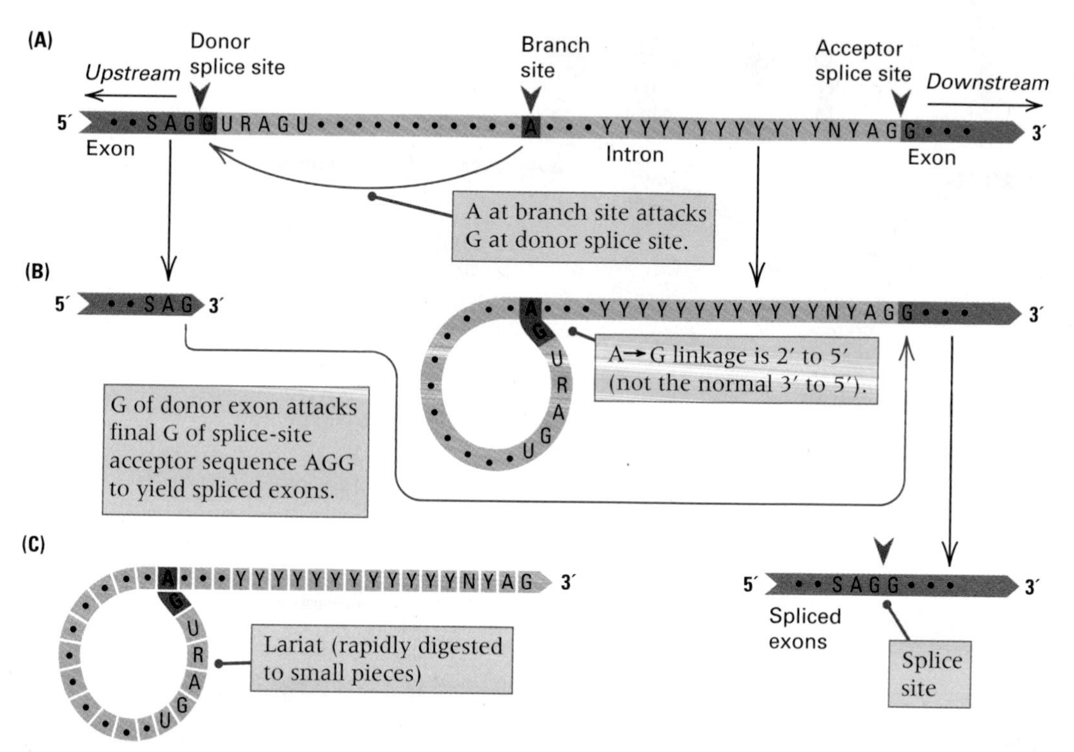


**FIGURE 10.17** Diagram showing removal of one intron from a primary transcript. (A) The A nucleotide at the branch site attacks the terminus of the 5′ exon, cleaving the exon–intron junction and forming a loop connected back to the branch site. (B) The 5′ exon is later brought to the site of cleavage of the 3′ exon, a second cut is made, and the exon termini are joined and sealed. (C) The loop is released as a lariat-shaped structure that is degraded. Because the loop includes most of the intron, the loop of the lariat is usually very much longer than the tail.

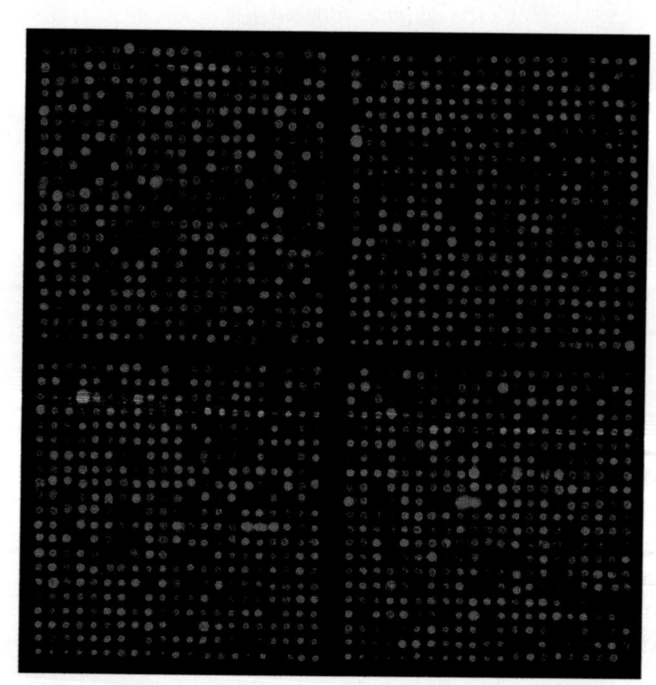

**ONE GENOMICS** technology for assaying the levels of gene expression of many genes simultaneously makes use of DNA probes that have been immobilized on a glass slide in an array of microdots, with $10^4$–$10^6$ different probes arrayed in an area about the size of a postage stamp. This is part of such an array containing 1764 probes for genes in budding yeast. The red, yellow, and green spots correspond to higher, equal, or lower levels of mRNA expression for genes from one yeast strain compared with those from another strain. This technology is discussed further in Chapter 12. [Courtesy of Jeffrey P. Townsend and Duccio Cavalieri, Yale University.]

particles are composed of protein and several specialized small RNA molecules, which are present in the cell as small nuclear ribonucleoprotein particles; the underlined letters give the abbreviation for these particles: **snRNPs** (pronounced *snerps*). The specificity of splicing comes from the five types of short RNA molecules (U1, U2, U4, U5, and U6) that are present in snRNPs. These RNAs contain sequences complementary to the splice junctions, to the branchpoint region of the intron, and/or to one another (**FIGURE 10.18**); as many as 100 spliceosome proteins may also be required for splicing. The ends of the intron are brought together by U1 RNA, which forms base pairs with nucleotides in the intron at both the 5′ and the 3′ ends, and U2 binds to the branchpoint region. As illustrated in Figure 10.18C, base pairing in the spliceosome includes U2 and U6, both of which are normally stabilized either by foldback pairing, in the case of U2 (part B) or pairing with U4, in the case of U6 (part A). These interactions bring the branchpoint region near to the donor splice site and allow the A in the branchpoint to attack the G of the donor splice site, freeing the upstream exon and forming the lariat intermediate (Figure 10.17). U5 RNA helps align the two

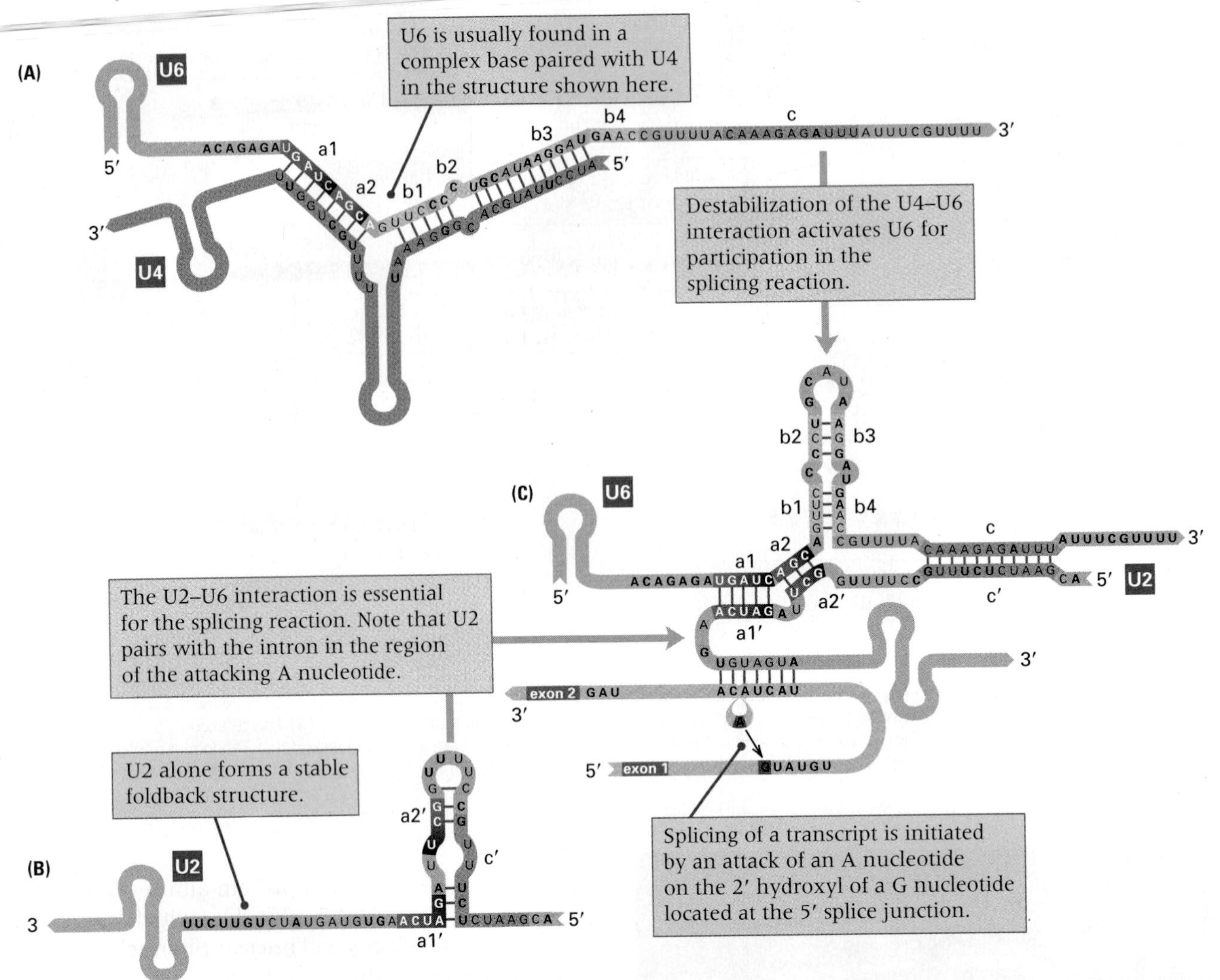

U6 is usually found in a complex base paired with U4 in the structure shown here.
Destabilization of the U4–U6 interaction activates U6 for participation in the splicing reaction.
The U2–U6 interaction is essential for the splicing reaction. Note that U2 pairs with the intron in the region of the attacking A nucleotide.
U2 alone forms a stable foldback structure.
Splicing of a transcript is initiated by an attack of an A nucleotide on the 2′ hydroxyl of a G nucleotide located at the 5′ splice junction.


**FIGURE 10.18** Dynamic interactions between some small nuclear RNAs present in snRNPs that are involved in splicing. (A) U6 snRNA is usually found complexed with U4 snRNA. (B) U2 snRNA forms a stable foldback structure on its own. (C) Essential to the splicing reaction is destabilization of the U4–U6 structure and formation of a U2–U6 structure in which U2 is base paired with part of the intron. An A in the paired region attacks the G at the 5′ splice junction, initiating the splicing reaction. The nucleotides in boldface are critical to the structures, judging by their having been conserved in very diverse species. Note that G–U base pairs are allowed in double-stranded RNA. [Adapted from H. D. Madhani and C. Guthrie, *Annu. Rev. Genet.* 1 (1994): 1–26.]

exons and somehow facilitates the final step in splicing, which results in scission of the intron from the downstream exon and ligation of the upstream and downstream exons.

Introns are also present in some genes in organelles, such as mitochondria, but the mechanisms of their excision differ from those of introns in nuclear genes because organelles do not contain spliceosomes. In one class of organelle introns, the intron contains a sequence coding for a protein that participates in removing the intron that codes for it. The situation is even more remarkable in the splicing of a ribosomal RNA precursor in the ciliate *Tetrahymena*. In this case, the splicing reaction is intrinsic to the folding of the RNA precursor; that is, the RNA precursor is *self-splicing* because the folded precursor RNA creates its own RNA-splicing activity. The self-splicing *Tetrahymena* RNA was the first example found of an RNA molecule that could function as an enzyme in catalyzing a chemical reaction; such enzymatic RNA molecules are called *ribozymes.*

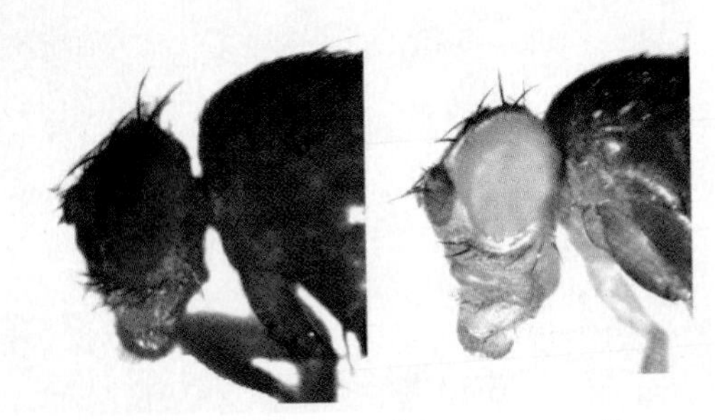

Mutations that prevent proper splicing are among the many known mutations in the *white* gene of *Drosophila*. [Courtesy of E. R. Lozovsky]

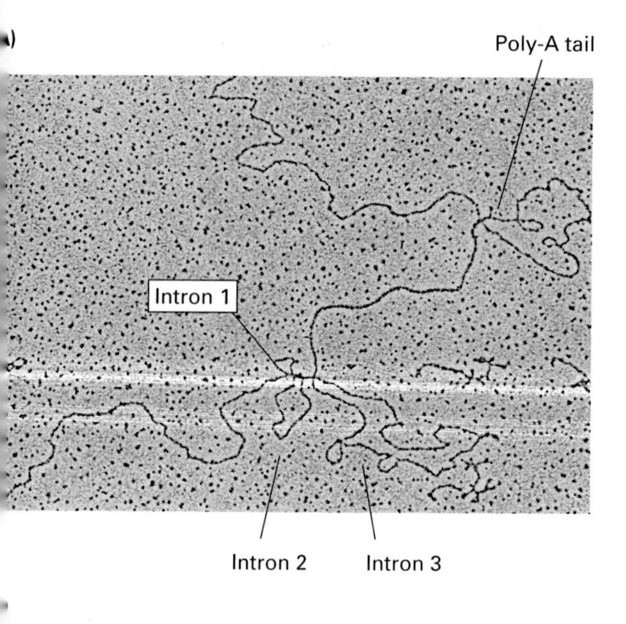


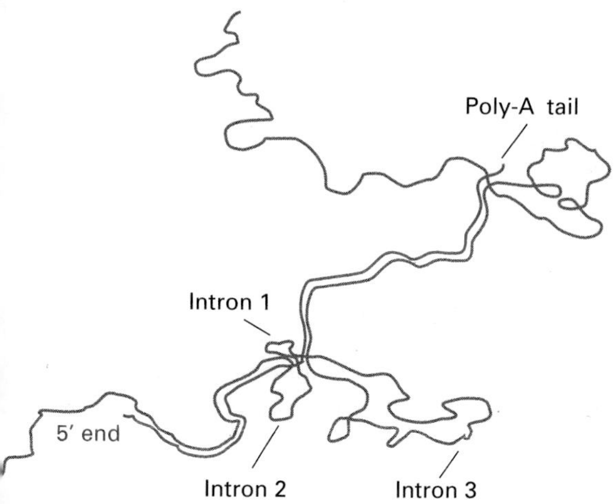


**URE 10.19** (A) An electron micrograph of a DNA–RNA hybrid ained by annealing a single-stranded segment of adenovirus DNA n one of its mRNA molecules. The loops are single-stranded DNA. An interpretive drawing. RNA and DNA strands are shown in red d blue, respectively. Four regions do not anneal, creating three single-nded DNA segments that correspond to the introns and the poly-A of the mRNA molecule. [Part A courtesy of Thomas R. Broker and ise T. Chow, University of Alabama at Birmingham. Original research npleted in 1977 at the Cold Spring Harbor Laboratory, New York.]

The existence and the positions of introns in a particular primary transcript are readily demonstrated by renaturing the transcribed DNA with the fully processed mRNA molecule. The DNA–RNA hybrid can then be examined by electron microscopy. An example of adenovirus mRNA (fully processed) and the corresponding DNA are shown in **FIGURE 10.19**. The DNA copies of the introns appear as single-stranded loops in the hybrid molecule, because no corresponding RNA sequence is available for hybridization.

## ■ Effects of Intron Mutations

Two observations suggest that there are few constraints on the length or nucleotide sequence of introns. First, many mutations in introns have no detectable effects on gene function. These include small deletions and insertions as well as nucleotide substitutions. Second, the nucleotide sequence of an intron undergoes very rapid change in the course of evolution. On the other hand, some sequences in introns are essential. These include the sequences near the ends that are necessary for correct splicing (Figure 10.17). Mutations that affect any of these sequences have important consequences because they interfere with the splicing reaction.

Two possible outcomes of splice-site mutations are illustrated in **FIGURE 10.20**. In part A, the intron with the mutated splice site fails to be removed, and it remains in the processed mRNA. The result is the production of a mutant protein with a normal sequence of amino acids up to the splice site but an abnormal sequence afterward. Most introns are long enough that, by chance, they contain a stop sequence that terminates protein synthesis, and once a stop is encountered, the protein grows no further. A second kind of outcome is shown in Figure 10.20B. In this case, splicing does occur, but at an alternative splice site. (The example shows the alternative site downstream from the mutation, but alternative sites can also be upstream.) The alternative site is called a **cryptic splice site** because it is not normally used. The cryptic splice site is usually a poorer match with the consensus sequence and is ignored when the normal splice site is available. The result of using the alternative splice site is again an incorrectly processed mRNA and a mutant protein. In some splice-site mutations, both outcomes can occur. Some transcripts leave the intron unspliced, whereas others are spliced at cryptic splice sites.

## ■ Exon Shuffle in the Origin of New Genes

Introns may play a role in gene evolution by facilitating the creation of new genes with novel combinations of exons. The evidence is that, in some cases, the exons in a gene code for segments of the completed protein that are relatively independent in their folding characteristics. For example, the central exon of the $\beta$-globin gene codes for the segment of the protein that folds around an iron-containing molecule of heme. Relatively autonomous folding units in proteins are known as **folding domains**, and the correlation between exons and domains found in some genes suggests that the genes were originally assembled from smaller pieces. In some cases, the ancestry of the exons

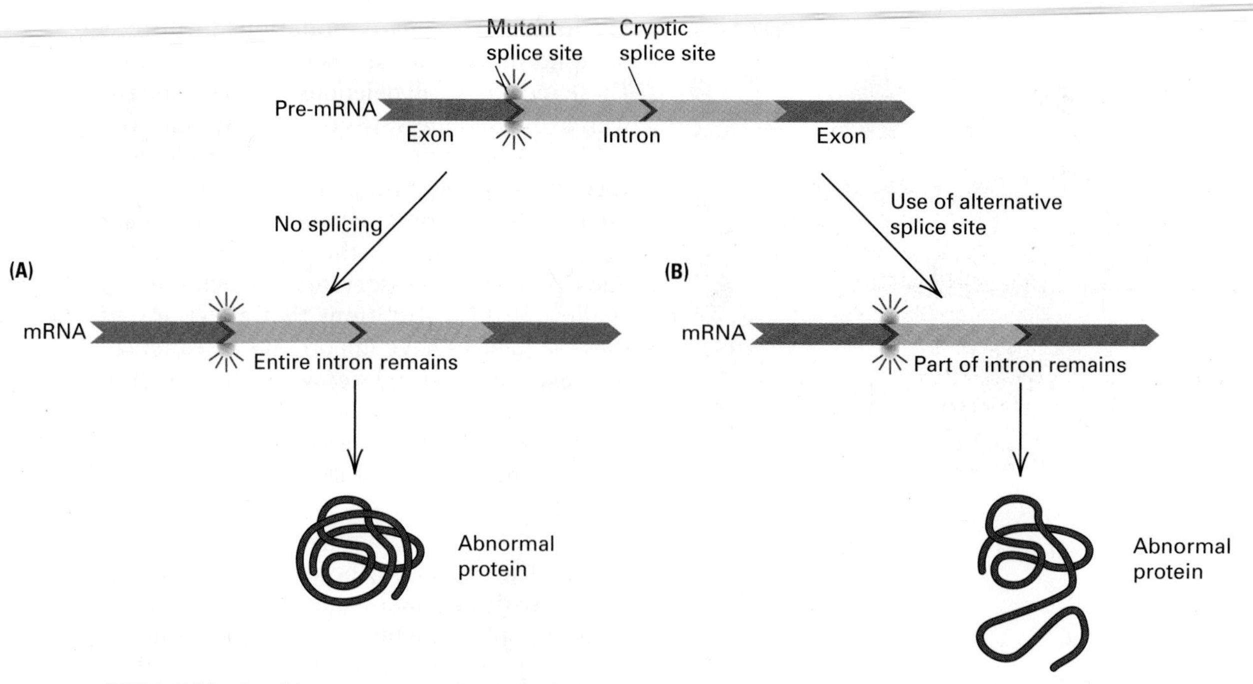


**FIGURE 10.20** Possible consequences of mutation in the donor splice site of an intron. (A) No splicing occurs, and the entire intron remains in the processed transcript. (B) Splicing occurs at a downstream cryptic splice site, and only the upstream part of the original intron remains in the processed transcript. Neither outcome results in a normal protein product.

can be traced. For example, the human gene for the low-density lipoprotein receptor that participates in cholesterol regulation shares exons with certain blood-clotting factors as well as epidermal growth factor. The model of protein evolution through the combination of different exons is called the **exon-shuffle model**. The mechanism for combining exons from different genes is not known. Although some genes support the model, in other genes the boundaries of the folding domains do not coincide with exons.

## 10.6 Translation

The synthesis of every protein molecule in a cell is directed by an mRNA originally copied from DNA. Polypeptide synthesis includes two kinds of processes: (1) the information transfer by which the RNA nucleotide sequence determines the amino acid sequence, and (2) the chemical processes by which the amino acids are linked together. The complete series of events constitutes **translation**.

The principal components of the translation apparatus are as follows:

- ***Messenger RNA,*** or ***mRNA*** Messenger RNA is needed to bring the ribosomal subunits together (described later) and to provide the coding sequence of nucleotides that determines the amino acid sequence in the resulting polypeptide chain.
- ***Ribosomes*** These components are particles on which protein synthesis takes place. They move along an mRNA molecule and align successive transfer RNA molecules; the amino acids are attached one by one to the growing polypeptide chain by means of peptide bonds. Ribosomes consist of two subunit particles. In *E. coli,* their sizes are 30S (the small subunit) and 50S (the large subunit). The counterparts in eukaryotes are 40S and 60S. (The S stands for *Svedberg unit,* which measures the rate of sedimentation of a particle in a centrifuge and so is an indicator of size.) Together, the small and large particles form a functional ribosome.
- ***Transfer RNA,*** or ***tRNA*** The sequence of amino acids in a polypeptide is determined by the nucleotide sequence in the mRNA by means of a set of *adaptor molecules,* the tRNA molecules, each of which is attached to a particular amino acid. Each group of three adjacent nucleotides in the mRNA forms a **codon** that binds to a particular group of three adjacent nucleotides in the tRNA (an **anticodon**), bringing the attached amino acid into line for

addition to the growing polypeptide chain.

- ***Aminoacyl tRNA synthetases*** This set of enzymes catalyzes the attachment of each amino acid to its corresponding tRNA molecule. A tRNA attached to its amino acid is called a **charged tRNA** or an **aminoacylated tRNA**.
- ***Initiation, elongation, and release factors*** Polypeptide synthesis can be divided into three stages: (1) initiation, (2) elongation, and (3) release. Each stage requires specialized proteins that are examined below.

In prokaryotes, all of the components for translation are present throughout the cell; in eukaryotes, they are located in the cytoplasm, as well as in mitochondria and chloroplasts.

In overview, the process of translation begins with an mRNA molecule binding to a ribosome. The aminoacylated tRNAs are brought along sequentially, one by one, to the ribosome that is translating the mRNA molecule. Peptide bonds are made between successively aligned amino acids, each time joining the amino group of the incoming amino acid to the carboxyl group of the amino acid at the growing end. Finally, the chemical bond between the last tRNA and its attached amino acid is broken, and the completed polypeptide is removed.

## Nonsense-Mediated Decay

In eukaryotes, the first round of translation of an mRNA is special because the mRNA has proteins bound to each site where an intron was present in the original transcript (Figure 10.16C). These bound proteins serve as quality-control markers to ensure that the introns have been removed. Most introns that failed to be removed from the mRNA are long enough that they are likely to contain at least one stop codon that terminates translation and triggers release of the polypeptide chain. (On average, a random RNA sequence will have a chain-termination codon about every 21 nucleotides.) In the first round of translation, as the codons across each exon-exon junction are translated, the proteins that marked the junction are removed. If all of the introns have been spliced out of the mRNA, then all of these protein markers will be removed prior to the moment when translation reaches the correct chain-termination codon at the end of the coding sequence. The mRNA is thereby freed of all the protein markers and becomes available for subsequent rounds of translation.

On the other hand, if an intron containing a chain-termination codon remains in the mRNA, then this termination codon will be encountered before all of the proteins marking the exon–exon junctions have been stripped away. When this happens, nucleases are recruited and the defective mRNA is destroyed. This quality-control process is known as **nonsense-mediated decay**. (The term *nonsense* derives from the original designation of chain-termination codons as nonsense codons.) Nonsense-mediated decay not only results in destruction of mRNA molecules that contain unspliced introns, but also in the destruction of mutant mRNA molecules in which a nucleotide substitution in the DNA results in a premature termination codon.

## Initiation by mRNA Scanning

Next we examine the details of translation as they occur in eukaryotes, pointing out differences in the prokaryotic mechanism that are significant. In the predominant mode of translation **initiation** in eukaryotes, the 5′ cap on the mRNA is instrumental (**FIGURE 10.21**). First the elongation factor eIF4F binds to the cap and then recruits eIF4A and eIF4B (Figure 10.21A). This creates a binding site for the other components of the initiation complex, which consist of a charged $tRNA^{Met}$ (that serves as an initiator tRNA), bound with elongation factor eIF2, and a small 40S ribosomal subunit together with elongation factors eIF3 and eIF5. These components all come together at the 5′ cap and form the 48S initiation complex (Figure 10.21B).

Once the initiation complex has formed, it moves along the mRNA in the 3′ direction, scanning for the first occurrence of the nucleotide sequence AUG, which signals the start of polypeptide synthesis. When this motif is encountered, the AUG is recognized as the initial methionine codon, and polypeptide synthesis begins. At this point eIF5 causes the release of all the initiation factors and the recruitment of a large 60S ribosomal subunit (Figure 10.21C). This subunit includes three binding sites for tRNA molecules. These sites are called the **E (exit)** site, the **P (peptidyl)** site, and the **A (aminoacyl)** site. Note that at the beginning of polypeptide synthesis, the initiator methionine tRNA is located in the P site and that the A site is the next site in line to be occupied. The tRNA binding is accomplished by hydrogen bonding between bases in the AUG codon in the mRNA and the three-base *anticodon* in the tRNA.

Prokaryotic mRNA molecules have no cap, and there is no scanning mechanism to locate

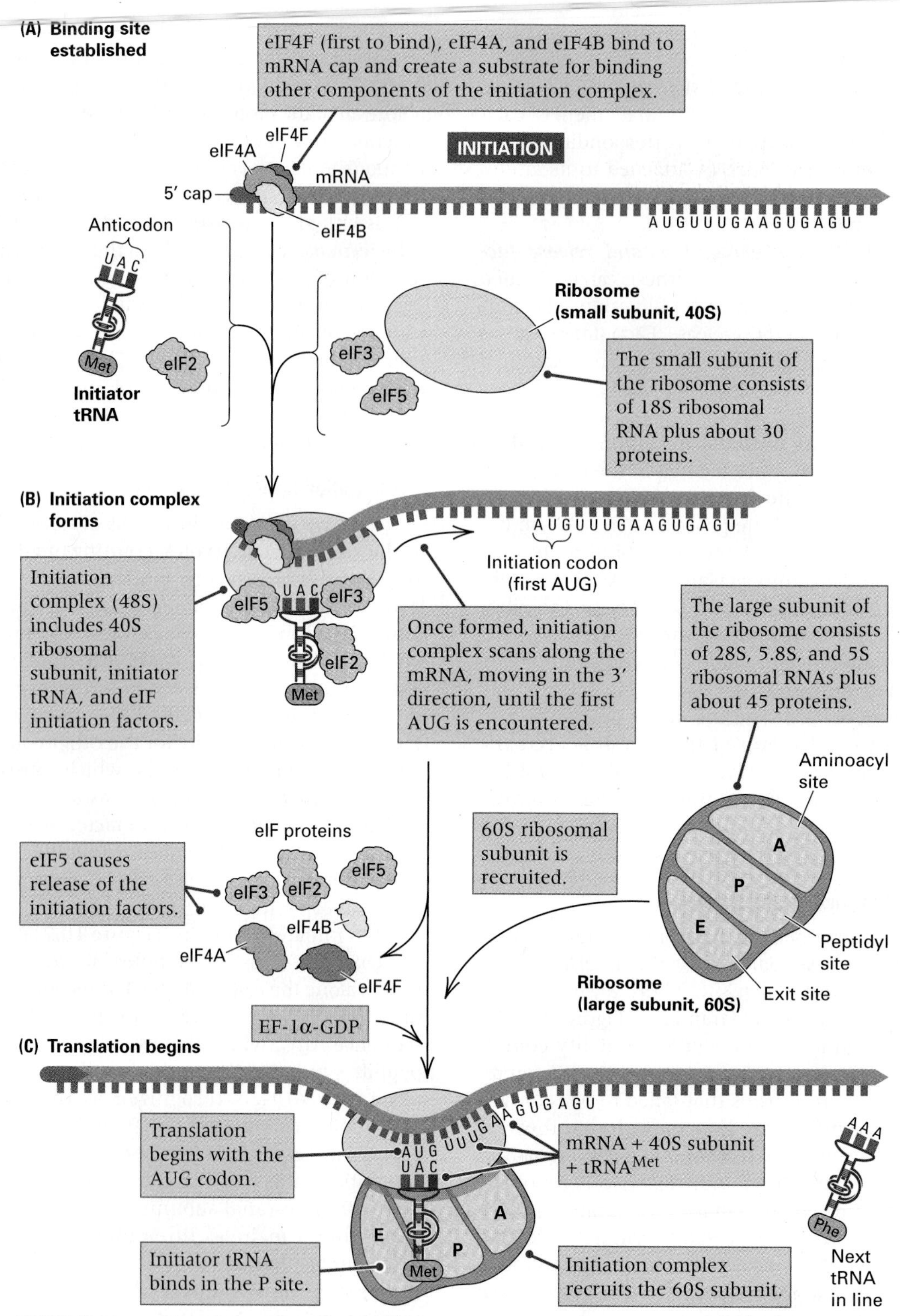

(A) Binding site established
eIF4F (first to bind), eIF4A, and eIF4B bind to mRNA cap and create a substrate for binding other components of the initiation complex.
INITIATION
eIF4F
eIF4A
mRNA
5′ cap
Anticodon
UAC
eIF4B
AUGUUUGAAGUGAGU
Ribosome (small subunit, 40S)
Met
eIF2
eIF3
eIF5
The small subunit of the ribosome consists of 18S ribosomal RNA plus about 30 proteins.
Initiator tRNA
(B) Initiation complex forms
AUGUUUGAAGUGAGU
Initiation codon (first AUG)
Initiation complex (48S) includes 40S ribosomal subunit, initiator tRNA, and eIF initiation factors.
Once formed, initiation complex scans along the mRNA, moving in the 3′ direction, until the first AUG is encountered.
The large subunit of the ribosome consists of 28S, 5.8S, and 5S ribosomal RNAs plus about 45 proteins.
Aminoacyl site
eIF proteins
60S ribosomal subunit is recruited.
eIF5 causes release of the initiation factors.
eIF3
eIF2
eIF5
eIF4B
eIF4A
eIF4F
A
P
E
Peptidyl site
Ribosome (large subunit, 60S)
Exit site
EF-1α-GDP
(C) Translation begins
Translation begins with the AUG codon.
mRNA + 40S subunit + tRNA^Met
Initiator tRNA binds in the P site.
Initiation complex recruits the 60S subunit.
Phe
Next tRNA in line


**FIGURE 10.21** Initiation of protein synthesis. (A) The initiation complex forms at the 59 end of the mRNA. (B) This consists of one 40S ribosomal subunit, the initiator tRNAMet, and the eIF initiation factors. (C) The initiation complex recruits a 60S ribosomal subunit in which the tRNAMet occupies the P (peptidyl) site of the ribosome. This complex travels along the mRNA until the first AUG is encountered, at which codon translation begins.

the first AUG. In the initiation of translation in *E. coli,* two initiation factors (IF-1 and IF-3) interact with the 30S subunit at the same time that another initiation factor (IF-2) binds with a special initiator tRNA charged with formylmethionine, symbolized $tRNA^{fMet}$. These components come together and combine with an mRNA, but not at the end. The attachment occurs by hydrogen bonding between the 3′ end of the 16S RNA present in the 30S subunit and a special sequence, the **ribosome-binding site**, in the mRNA (also called the *Shine-Dalgarno sequence*). Together, the 30S + $tRNA^{fMet}$ + mRNA complex recruits a 50S subunit, in which the $tRNA^{fMet}$ is positioned in the P site and aligned with the AUG initiation codon, just as in Figure 10.21C. In the assembly of the completed ribosome, the initiation factors dissociate from the complex.

One major functional difference between translational initiation in eukaryotes and prokaryotes is that in eukaryotes, because of the scanning mechanism of initiation, a single mRNA can usually encode only one polypeptide chain. In prokaryotic mRNA, in contrast, the ribosome-binding site can be present anywhere near an AUG, so polypeptide synthesis can begin at any AUG that is closely preceded by a ribosome-binding site. Prokaryotes put this feature to good use because, as we shall see later, many of their mRNAs contain more than one ribosome-binding site and initiation codon and hence code for more than one polypeptide, because translation can be initiated at any of the sites.

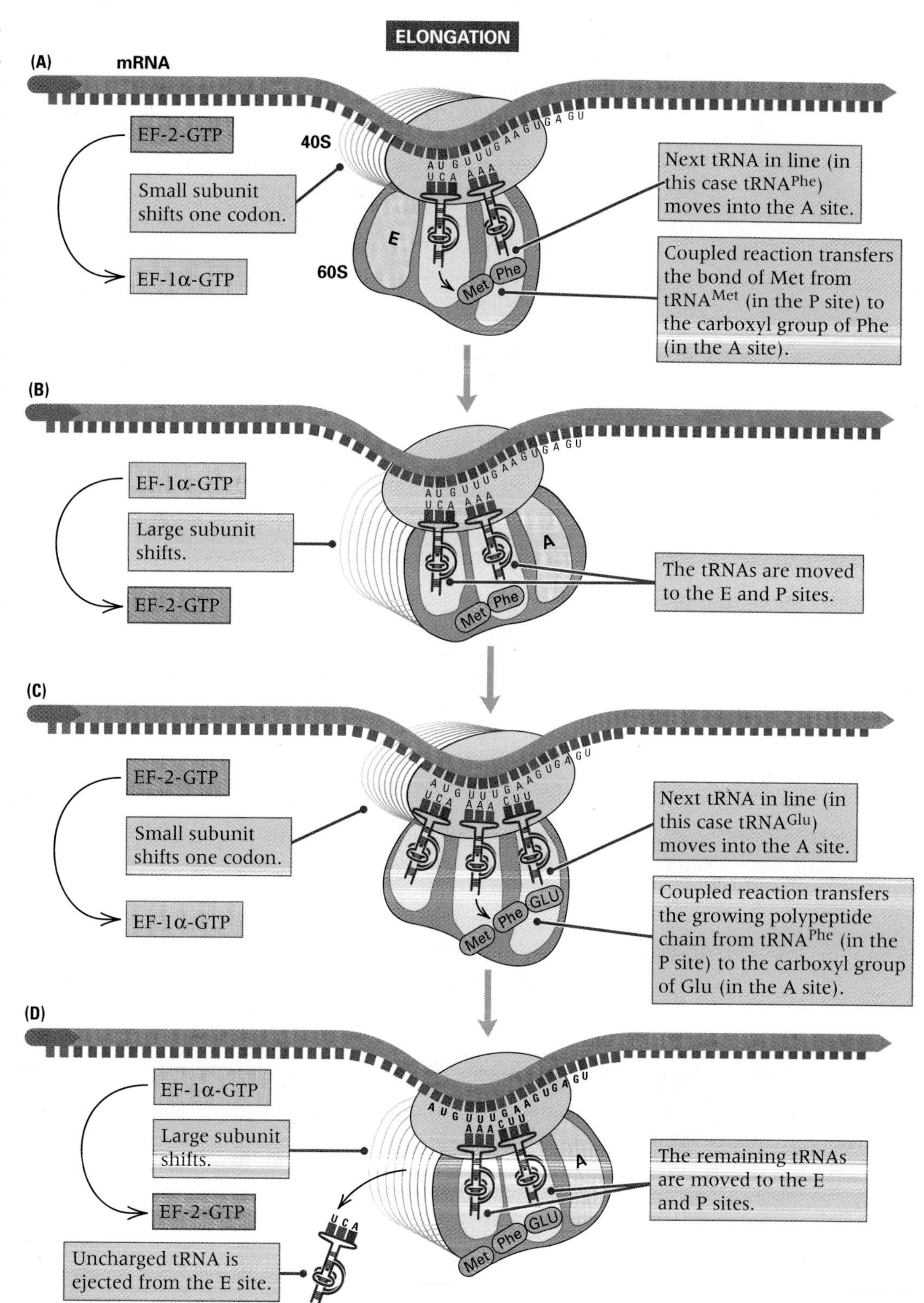


**FIGURE 10.22** The elongation cycle in protein synthesis.

## Elongation

Recruitment of the elongation factor EF-1α-GTP into the initiation complex begins the **elongation** phase of polypeptide synthesis (Figure 10.21C). Elongation consists of three processes executed iteratively:

1. Bringing each new aminoacylated tRNA into line
2. Forming the new peptide bond to elongate the polypeptide
3. Moving the ribosome to the next codon along the mRNA

The process of elongation is illustrated in **FIGURE 10.22**. The key players in providing the energy for translation are the elongation factors EF-2 and EF-1α, which alternately occupy the same ribosomal binding site. In their active forms (EF-2-GTP and EF-1α-GTP) the molecules are bound with guanosine triphosphate (GTP). Hydrolysis of the GTP to GDP releases the energy to move the ribosomal subunits along the messenger RNA, as well as to carry out the reactions needed to grow the polypeptide chain. Conversion of either elongation factor from its GTP-bound form into its GDP-bound form lowers its affinity for the ribosome, and the GDP-bound form diffuses away and is replaced by the GTP-bound form of the alternate elongation factor.

In the first step of elongation, the 40S ribosomal subunit moves one codon farther along the messenger RNA, and the charged tRNA corresponding to the new codon (in this case, $tRNA^{Phe}$) is brought into the A site on the 60S subunit (Figure 10.22A). The charged tRNA comes to the ribosome in a complex that also contains EF-1α-GTP, and at this step an important form of *kinetic proofreading* in translation takes place. When the charged tRNA does not match the codon as it should, GTP hydrolysis is delayed and the incorrect charged tRNA can diffuse away and be replaced. But if the codon-anticodon interaction is correct, then the rate of GTP hydrolysis by EF-1α increases by a factor of $5 \times 10^4$. The resulting EF-1α-GDP has a reduced affinity for the charged tRNA, which allows the charged tRNA to fit tightly within the active site of the 60S subunit where the peptide bond is formed by a **peptidyl transferase** activity. Peptide bond formation is a coupled reaction in which, in the example in Figure 10.22A, breakage of the bond connecting the methionine to the $tRNA^{Met}$ is coupled to formation of the peptide bond connecting the methionine to phenylalanine. Peptidyl transferase activity is not due to a single molecule but requires multiple components of the 60S subunit, including several proteins and the 28S ribosomal RNA in the 60S subunit. Some evidence indicates that the actual catalysis is carried out by the 28S RNA, which would suggest that 28S is an example of a ribozyme at work.

In the next step in chain elongation (part B), the 60S subunit swings forward to catch up with the 40S subunit, and at the same time the tRNAs in the P and A sites of the large subunit are shifted to the E and P sites, respectively.

One cycle of elongation is now completed, and the entire procedure is repeated for the next codon (Part C). The 40S subunit shifts one codon to the right, the next aminoacylated tRNA (in this case, $tRNA^{Glu}$) is brought into the A site, and after kinetic proofreading a new peptide bond is formed between the carboxyl group of Phe and the amino group of Glu. As shown in part D, the large subunit swings forward while at the same time the tRNAs in the P and A sites are shifted into the E and P sites. At this point the tRNA that formerly occupied the E site is ejected from the ribosome.

Polypeptide elongation consists of the steps C → D → C → D carried out repeatedly until a termination codon is encountered. The elongation cycle happens relatively rapidly. Under optimal conditions, eukaryotes synthesize a polypeptide chain at the rate of about 15 amino acids per second. Elongation in prokaryotes is a little faster (about 40 amino acids per second), but the essential processes are very similar. Owing to kinetic proofreading, the error rate is approximately one incorrect amino acid inserted per 2000 residues. In *E. coli*, the sizes of the ribosomal subunits are 30S (small) and 50S (large), and the complete ribosome is 70S. **FIGURE 10.23** shows a large ribosomal subunit from *E. coli*, reconstructed from the x-ray diffraction structure, depicting the locations of the tRNA molecules in their binding pockets: E in gold, P in blue, and A in green. Above the tRNAs is a channel through which the mRNA is moved along as translation progresses. The small subunit fits on top of the large subunit, leaving enough space for the tRNA molecules to bind. In prokaryotes the source of energy for elongation is also GTP hydrolysis. The *E. coli* analogs of EF-1α and EF-2 are EF-Tu and EF-G, respectively.

## ■ Release

Compared to initiation and elongation, the termination of polypeptide synthesis—the

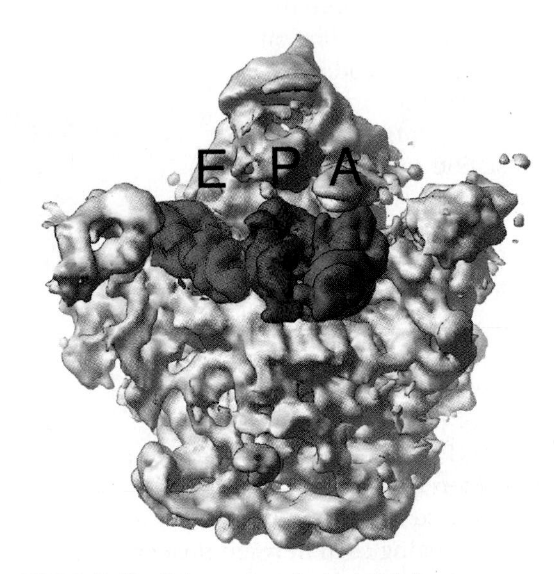


**FIGURE 10.23** Cutaway view of a bacterial ribosome, showing the groove in which the tRNAs are bound and the positions of tRNAs when present in the E (exit) site in gold, the P (peptidyl) site in blue, and the A (aminoacyl) site in green. [Reproduced from J. H. Cate, et al., *Science* 285 (1999): 2095–2104. Reprinted with permission from AAAS.]

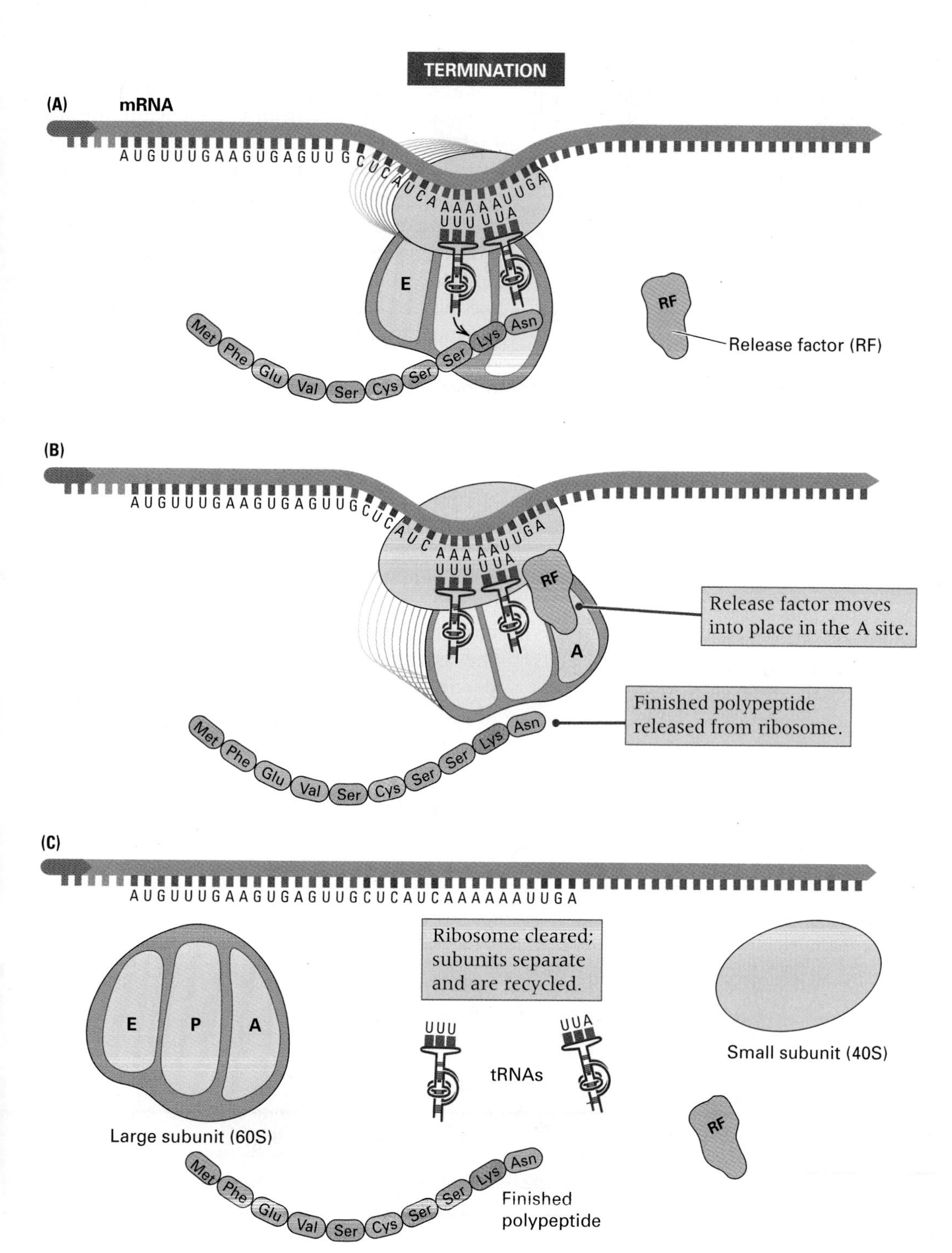


**FIGURE 10.24** Termination of protein synthesis. When a stop codon is reached (A), no tRNA can bind to that site (B), which causes the release of the newly formed polypeptide and the remaining bound tRNA (C).

**release** phase—is simple (**FIGURE 10.24**). When a stop codon is encountered, the tRNA holding the polypeptide remains in the P site, and a *release factor (RF)* binds with the ribosome. GTP hydrolysis provides the energy to cleave the polypeptide from the tRNA to which it is attached, as well as to eject the release factor and dissociate the 80S ribosome from the mRNA. At this point the 40S and 60S subunits are recycled to carry out translation of another mRNA. Eukaryotes have only one release factor that recognizes all three stop codons: UAA, UAG, and UGA. The situation differs in prokaryotes. In *E. coli*, the release factor RF-1 recognizes the stop codons UAA and UAG, whereas release factor RF-2 recognizes UAA and UGA. A third release factor, RF-3, is also required, but its function is uncertain.

## ■ Protein Folding and Chaperones

Each polypeptide chain tends to fold into a unique three-dimensional shape determined primarily by its sequence of amino acids. Generally speaking, polypeptide molecules fold so that amino acids with charged, hydrophilic side chains tend to be on the surface of the protein (in contact with water) and those with uncharged, hydrophobic side chains tend to be internal (hidden from water). Specific folded configurations also result from hydrogen bonding between peptide groups. Two fundamental polypeptide structures are the alpha ($\alpha$) helix and the beta ($\beta$) sheet (**FIGURE 10.25**). An $\alpha$ helix is formed by hydrogen bonding between peptide groups that are close together in the polypeptide backbone. In an $\alpha$ helix, often represented as a coiled ribbon, the backbone is twisted so that the N–H in each peptide group is hydrogen-bonded with the C=O in the peptide group located four amino acids farther along the helix. The helical twist may be right-handed or left-handed, but right-handed $\alpha$ helices are more common. Both $\alpha$ helices in Figure 10.25 are right-handed.

In contrast, a $\beta$ sheet is formed by hydrogen bonding between peptide groups in distant parts of the polypeptide chain, or even in different polypeptide chains. In a $\beta$ sheet, often represented as parallel "flat" ribbons, the backbones of the interacting polypeptide chains are held flat and relatively rigid (forming a "sheet"), because alternate N–H groups in one backbone are hydrogen-bonded with alternate C=O groups in the backbone of the adjacent chain. In each polypeptide chain, alternate C=O and N–H groups are free to form hydrogen bonds with their counterparts in a different chain on the opposite side, so a $\beta$ sheet can consist of multiple aligned segments in the same or different polypeptides. The orientation of the backbones in a $\beta$ sheet may be antiparallel (adjacent backbones reversed in orientation relative to their amino and carboxyl ends) or parallel, but antiparallel is more common. In Figure 10.25, both $\beta$ sheets are antiparallel.

Other types of interactions also are important in protein folding. For example, covalent bonds may form between the sulfur atoms of pairs of cysteines in different parts of the polypeptide. However, the rules of folding are so complex that, except for the simplest proteins, the final shape of a protein cannot usually be predicted from the amino acid sequence alone.

Although the folding of a polypeptide chain is determined by its amino acid sequence, some polypeptide chains need help to fold properly. An estimated 70–75 percent of polypeptides fold properly upon release from the

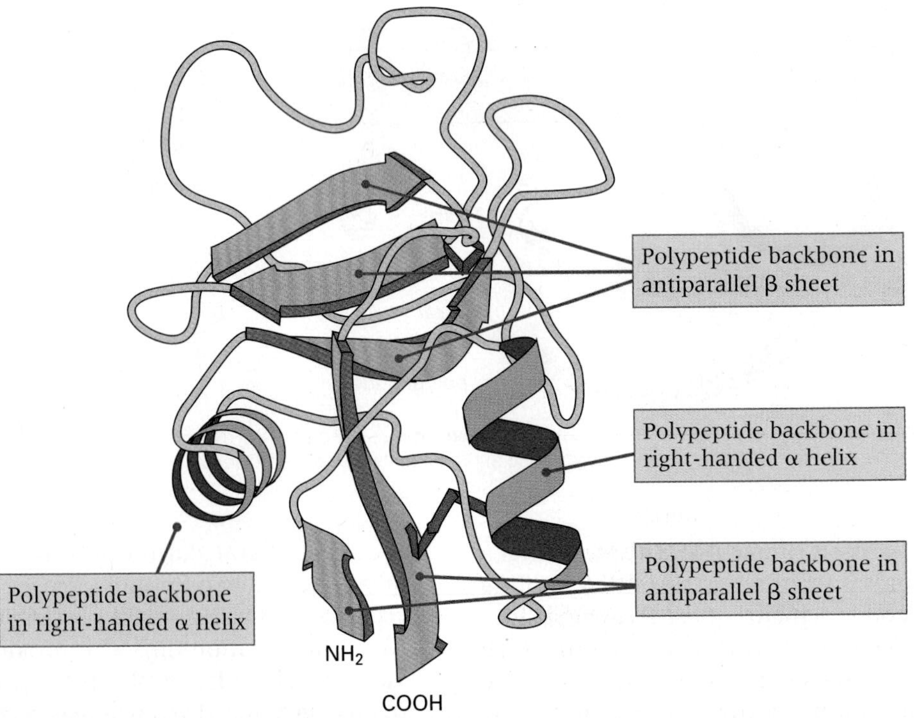


**FIGURE 10.25** A "ribbon" diagram of the path of the backbone of a polypeptide, showing the ways in which the polypeptide is folded. The flat arrows represent $\beta$ sheets, each of which is held to its neighboring $\beta$ sheet by hydrogen bonds. Helical regions are shown as coiled ribbons. The polypeptide chain in this example is a mannose-binding protein. [Adapted from W. I. Weiss, K. Drickamer, and W. A. Hendrickson, *Nature* 360 (1992): 127–134.]

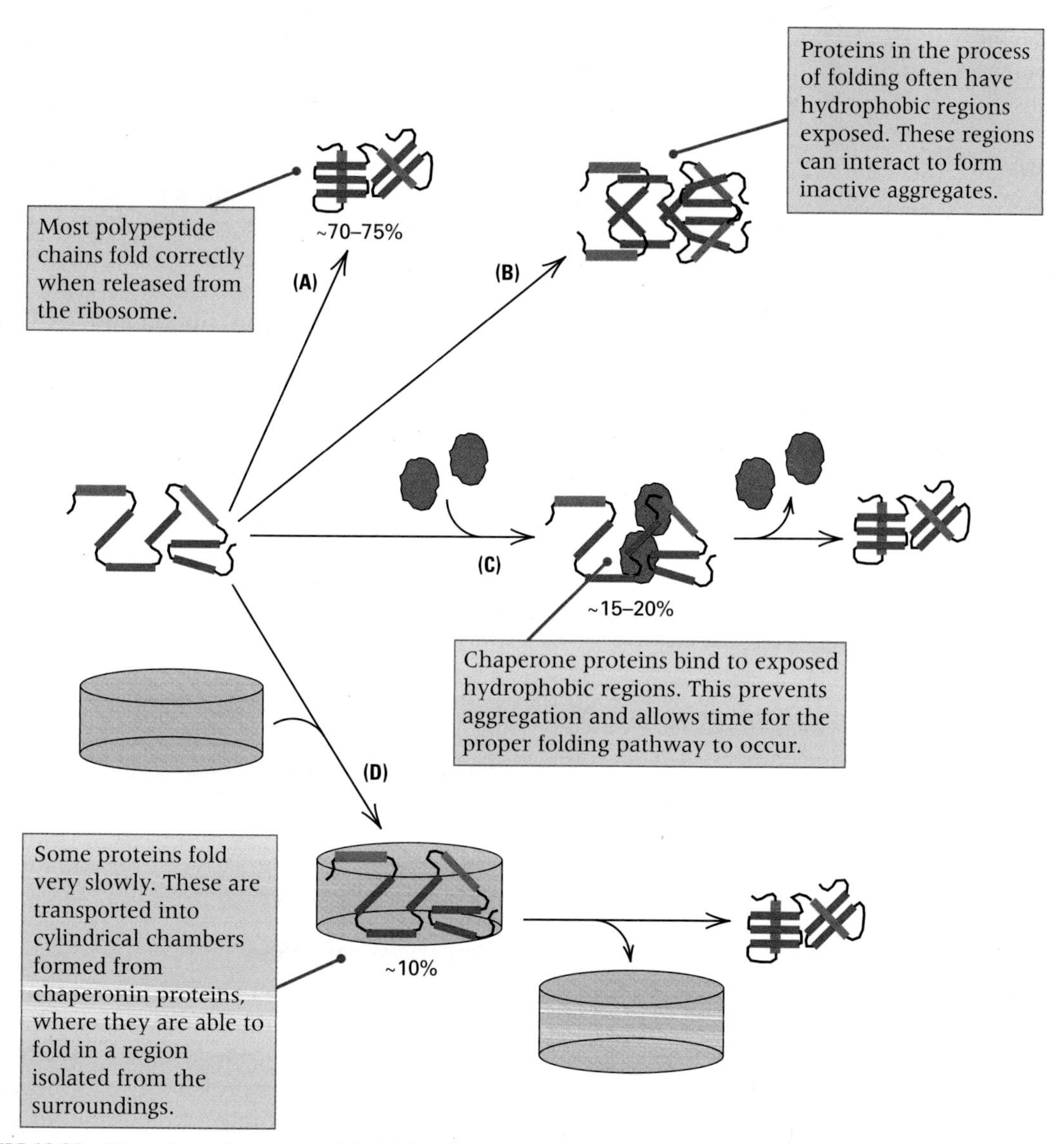

Proteins in the process of folding often have hydrophobic regions exposed. These regions can interact to form inactive aggregates.
Most polypeptide chains fold correctly when released from the ribosome.
~70–75%
(A)
(B)
(C)
~15–20%
Chaperone proteins bind to exposed hydrophobic regions. This prevents aggregation and allows time for the proper folding pathway to occur.
(D)
Some proteins fold very slowly. These are transported into cylindrical chambers formed from chaperonin proteins, where they are able to fold in a region isolated from the surroundings.
~10%


**FIGURE 10.26** Alternative pathways in protein folding. The green regions represent $\alpha$ helices and the red regions $\beta$ sheets.

ribosome (**FIGURE 10.26**, part A). These tend to be relatively small proteins or ones with relatively few folding domains. Their folding may be assisted by proteins associated with the ribosome, in prokaryotes called TF (*trigger factor*) and in eukaryotes NAC (*nascent chain-associated complex*). But larger proteins composed of multiple folding domains fold more slowly, during which hydrophobic residues are exposed to the high concentration of macromolecules in the cytoplasm. (In *E. coli*, the concentration of proteins and other macromolecules is 300–400 grams per liter.) Under such crowded conditions the exposed hydrophobic groups often attract each other and bind together, forming inactive protein aggregates (Figure 10.26B).

The proper folding of more complex polypeptides is aided by proteins called **chaperones** (Figure 10.26C). These proteins bind to hydrophobic groups and unstructured regions to shield them from aggregation, and by repeated cycles of binding and release they give the polypeptide time to find its proper folding pathway. In bacteria, the major chaperone proteins include DnaJ and DnaK, and in eukaryotes they include Hsp40, Hsp70 and Hsp90. The symbol Hsp stands for *heat-shock protein* and reflects the finding that these proteins are more highly expressed at high temperatures. The reason for their increased production is that high temperatures cause many proteins to begin to unfold (denature), and the presence of the chaperones stabilizes the partially unfolded states to enable proper refolding.

The most complex proteins with very slow and inefficient folding pathways are sequestered by a special class of proteins known as *chaperonins*. These form large, hollow cylindrical structures that trap the unstable intermediates inside and allow them to fold in a protected environment (Figure 10.26D). In bacteria the principal chaperonins are the GroEL and GroES proteins, and in eukaryotes they are the TRiC proteins. In eukaryotes, the most abundant polypeptides that make use of the chaperonin cylinders for folding are the cytoskeletal proteins actin and tubulin.

## 10.7 Complex Translation Units

The unit of translation is rarely one ribosome traversing one mRNA molecule. A single mRNA can be translated simultaneously by several ribosomes moving along it in tandem, and in prokaryotes a single mRNA can code for multiple polypeptide chains. These situations are examined in the following sections.

### ■ Polysomes

After about 25 amino acids have been joined together in a polypeptide chain, an AUG initiation codon is completely free of the ribosome, and a second initiation complex can form. The overall configuration is that of two ribosomes moving along the mRNA at the same speed. When the second ribosome has moved along a distance similar to that traversed by the first, a third ribosome can attach to the initiation site. The process of movement and reinitiation continues until the mRNA is covered with ribosomes at a density of about one ribosome per 80 nucleotides (**FIGURE 10.27**) This large translation unit is called a **polysome**, which is the usual form of the translation unit.

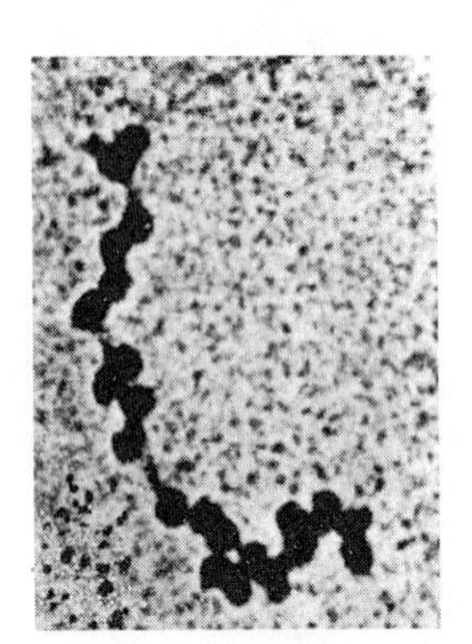

**FIGURE 10.27** Electron micrograph of *E. coli* polysomes. [Courtesy of Barbara A. Hamkalo, University of California, Irvine.]

In eukaryotes, transcription and RNA processing take place in the nucleus, whereas translation takes place in the cytoplasm. The processes are spatially and temporally uncoupled. In prokaryotes, because no nuclear envelope separates the DNA from the ribosome, the translational initiation complex can form even before the mRNA is released from the DNA. This allows the simultaneous occurrence of transcription and translation. **FIGURE 10.28** shows an electron micrograph of a DNA molecule with a number of attached mRNA molecules, each associated with ribosomes. Transcription of DNA is beginning at the upper left of the micrograph. The lengths of the polysomes increase with distance from the transcription initiation site; the mRNA is farther from that site and hence of greater length because the process of transcription has been going on longer.

### ■ Polycistronic mRNA

As we have noted, the eukaryotic scanning process for selecting the initiation codon implies that one mRNA can code for one and only one polypeptide chain. In prokaryotes, as shown in **FIGURE 10.29**, mRNA molecules commonly contain coding sequences for several different polypeptide chains; such a molecule is called a **polycistronic mRNA**. (The word *cistron* means a nucleotide sequence that encodes a single polypeptide chain). Eukaryotic mRNAs are always monocistronic. In a polycistronic mRNA,

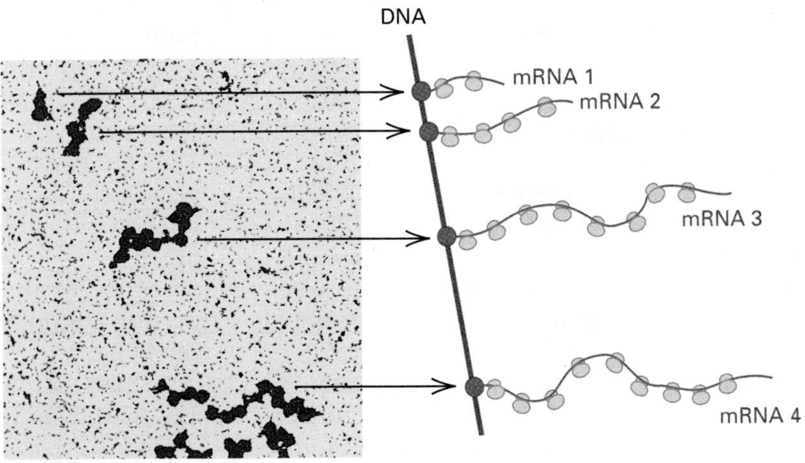


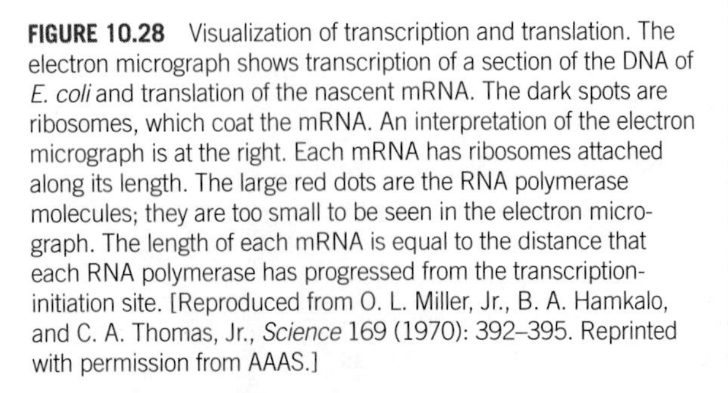

**FIGURE 10.28** Visualization of transcription and translation. The electron micrograph shows transcription of a section of the DNA of *E. coli* and translation of the nascent mRNA. The dark spots are ribosomes, which coat the mRNA. An interpretation of the electron micrograph is at the right. Each mRNA has ribosomes attached along its length. The large red dots are the RNA polymerase molecules; they are too small to be seen in the electron micrograph. The length of each mRNA is equal to the distance that each RNA polymerase has progressed from the transcription-initiation site. [Reproduced from O. L. Miller, Jr., B. A. Hamkalo, and C. A. Thomas, Jr., *Science* 169 (1970): 392–395. Reprinted with permission from AAAS.]

## connection

## Poly-U

**Marshall W. Nirenberg and J. Heinrich Matthaei 1961**
National Institutes of Health, Bethesda, Maryland
*The Dependence of Cell-Free Protein Synthesis in* E. Coli *upon Naturally Occurring or Synthetic Polyribonucleotides*

*In the years following the discovery of DNA structure by Watson and Crick in 1953, the biological implications of the discovery were largely ignored. A principal reason was that most biochemists still held strongly to the conviction that DNA had nothing to do with protein synthesis. The prevailing view was that proteins were made from small preexisting peptides by enzymes that joined the peptides together step by step in a specific order. It had been suggested that proteins might be made by amino acids being laid down in sequence upon an RNA template, but this hypothesis also was largely ignored. Not until this important paper appeared in 1961 was it shown that proteins are made by stepwise joining of individual amino acids in a sequence specified by a molecule of template RNA. The key finding was that in a cell-free mixture capable of supporting protein synthesis, the artificial polynucleotide poly-U (polyuridylic acid) resulted in the synthesis of a protein consisting only of the amino acid phenylalanine. The requirements for protein synthesis also included ribosomes (necessary for translation) and small RNA molecules (which include the charged transfer RNAs). After this paper appeared, the race was on to decipher the genetic code by which RNA specifies the amino acids in a protein.*

A stable cell-free system has been obtained from *E. coli* which incorporates [radioactive] valine into protein at a rapid rate. . . . The present communication describes a requirement for template RNA, needed for amino acid incorporation even in the presence of soluble [small] RNA molecules and ribosomes. The amino acid incorporation stimulated by the addition of template RNA has many properties expected of protein synthesis. Naturally occurring RNA as well as synthetic polynucleotides were active. The synthetic polynucleotide appears to contain a code for the synthesis of a "protein" containing only one amino acid. . . . [Specifically,] the addition of polyuridylic acid resulted in a remarkable stimulation of [radioactive] phenylalanine incorporation. Phenyalanine incorporation was almost completely dependent upon the addition of polyuridylic acid, and incorporation proceeded at a linear rate for approximately 30 minutes. No other polynucleotide tested could replace polyuridylic acid. . . . The product of the reaction had the same apparent solubility as authentic polyphenylalanine...[and contained] phenylalanine and no other amino acids. . . . The results indicate the polyuridylic acid contains the information for the synthesis of a protein having the characteristics of polyphenylalanine. . . . One or more uridylic acid residues therefore appears to be the code for phenylalanine. Whether the code is of the singlet, triplet, etc., type has not yet been determined. Polyuridylic acid seemingly functions as a synthetic template or messenger RNA, and this stable, cell-free system may well synthesize any protein corresponding to meaningful information contained in added RNA.

***The results indicate the polyuridylic acid contains the information for the synthesis of a protein having the characteristics of polyphenylalanine.***

Source: M. W. Nirenberg and J. H. Matthaei, *Proc. Natl. Acad. Sci. USA* 47 (1961): 1588–1602.

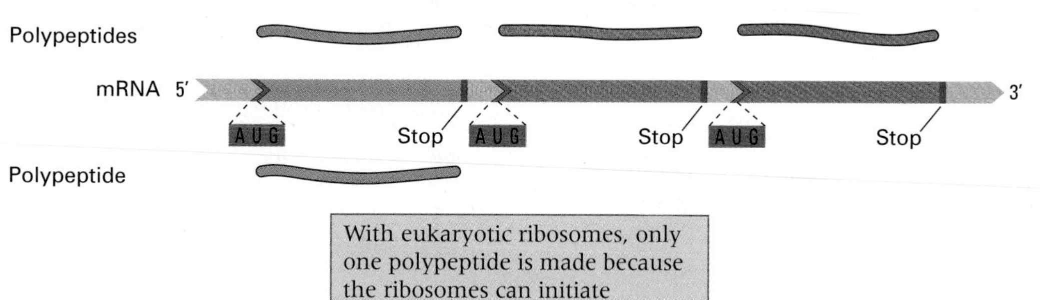

With prokaryotic ribosomes, three polypeptides are made because the ribosomes can initiate translation within an mRNA.
Polypeptides
mRNA 5′
3′
AUG
Stop
AUG
Stop
AUG
Stop
Polypeptide
With eukaryotic ribosomes, only one polypeptide is made because the ribosomes can initiate translation only at the 5′ end.

**FIGURE 10.29** Different products are translated from a three-cistron mRNA molecule by the ribosomes of prokaryotes and eukaryotes. The prokaryotic ribosome translates all of the open reading frames, but the eukaryotic ribosome translates only the open reading frame nearest the 5′ terminus of the mRNA. Translated sequences are shown in purple, yellow, and orange; stop codons in red; the ribosome binding sites in green; and the spacer sequences in light green.

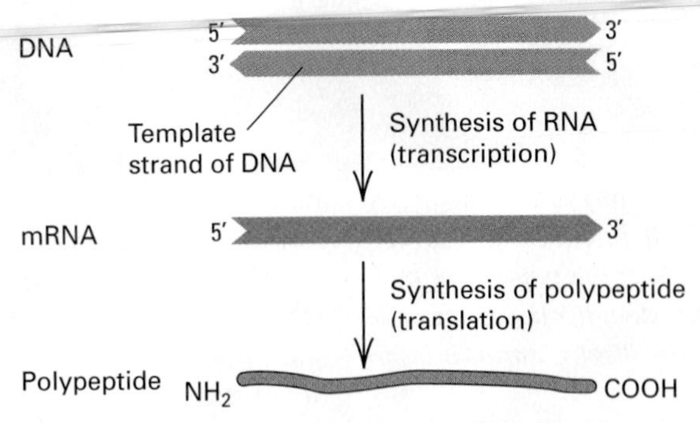


**FIGURE 10.30** Direction of synthesis of RNA with respect to the coding strand of DNA, and of synthesis of protein with respect to mRNA.

each polypeptide coding region is preceded by its own ribosome-binding site and AUG initiation codon. After the synthesis of one polypeptide is finished, the next along the way is translated. The coding sequences contained in a polycistronic mRNA molecule often encode the different proteins of a metabolic pathway. For example, in *E. coli*, the ten enzymes needed to synthesize histidine are encoded by one polycistronic mRNA molecule. The use of polycistronic mRNA is an economical way for a cell to regulate the synthesis of related proteins in a coordinated manner.

The definitive feature of translation is that it proceeds in a particular direction along the mRNA and the polypeptide:

> The mRNA is translated from an initiation codon to a stop codon in the 5′-to-3′ direction. The polypeptide is synthesized from the amino end toward the carboxyl end by the successive addition of amino acids to the carboxyl end.

These directions of synthesis are illustrated schematically in **FIGURE 10.30**. The convention in writing nucleotide sequences is to place the 5′ end at the left, and, in writing amino acid sequences, to place the amino end at the left. This means that an RNA sequence is written from left to right in the same order in which it is transcribed, and an amino acid sequence is written in the same order in which it is translated.

## 10.8 The Standard Genetic Code

The **genetic code** is the list of all codons and the amino acid that each one specifies. Before the nature of the genetic code was known, scientists reasoned that if all codons were assumed to have the same number of nucleotides, then each codon would have to contain at least three nucleotides. Codons consisting of pairs of nucleotides would be insufficient, because four nucleotides can form only $4^2 = 16$ pairs; triplets of nucleotides would suffice, because four nucleotides can form $4^3 = 64$ triplets. In fact, the genetic code is a triplet code, and all 64 possible codons carry information of some sort. Because only 20 amino acids need be specified, most amino acids are encoded by more than one codon. Furthermore, in the translation of mRNA molecules, the codons do not overlap but are used sequentially.

### ■ Genetic Evidence for a Triplet Code

Although theoretical considerations suggested that each codon must contain at least three nucleotides, codons of greater length could not be ruled out. The first widely accepted proof for a triplet code came from genetic experiments using *rII* mutants of bacteriophage T4 that had been induced by replication in the presence of the chemical *proflavin*. These experiments were carried out in 1961 by Francis Crick and collaborators. Proflavin-induced mutations typically resulted in total loss of function. The investigators suspected that the mutations were single-nucleotide insertions or deletions. Analysis of the properties of these mutations led directly to the deduction that the code is read in nonoverlapping groups of three nucleotides from a fixed point that establishes the **reading frame** of the mRNA. Mutations that delete or add a nucleotide pair shift the reading frame and are called **frameshift mutations. FIGURE 10.31** illustrates the profound effect of a frameshift mutation on the amino acid sequence of the polypeptide produced from a mutant mRNA.

The genetic analysis of the structure of the code began with an *rII* (rapid lysis) mutation called FC0, which was arbitrarily designated (+), as though it had an inserted nucleotide

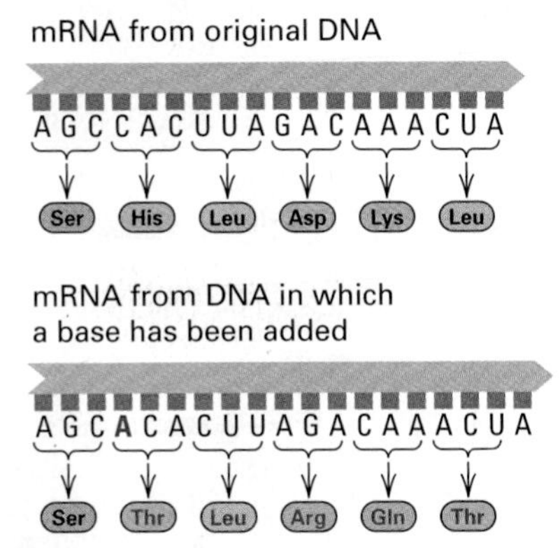


**FIGURE 10.31** The change in the amino acid sequence of a protein caused by the addition of an extra base, which shifts the reading frame. A deleted base also shifts the reading frame.

pair. [It could just as easily have been designated (−), as though it had a deleted nucleotide pair. Calling it (+) was a lucky guess, however, because when FC0 was sequenced, it did turn out to have a single-base insertion.] If FC0 has a (+) insertion, then it should be possible to revert the FC0 allele to "wildtype" by deletion of a nearby nucleotide. Selection for $r^+$ revertants was carried out by isolating plaques formed on a lawn of an *E. coli* strain K12 that was lysogenic for phage λ. The basis of the selection is that *rII* mutants are unable to propagate in K12(λ). Analysis of the revertants revealed that each still carried the original FC0 mutation and, in addition, another (intragenic suppressor) mutation that reversed the effects of the FC0 mutation. The suppressor mutations could be separated by recombination from the original mutation by crossing each revertant to wildtype; each suppressor mutation proved to be an *rII* mutation that, by itself, would cause the rapid lysis phenotype. If FC0 had an inserted nucleotide, then the suppressors should all result in deletion of a nucleotide pair; therefore, each suppressor of FC0 was inferred to have a single-nucleotide deletion (−). Three such revertants and their consequences for the translational reading frame, depicted using ordinary three-letter words for simplicity, are illustrated in **FIGURE 10.32**. The (−) mutations are designated $(-)_1$, $(-)_2$, and $(-)_3$, and those parts of the mRNA translated in the correct reading frame are indicated in green.

Each of the individual (−) suppressor mutations could, in turn, be used to select other "wildtype" revertants, with the expectation that these revertants would carry new suppressor mutations of the (+) variety, because the (−)(+) combination should yield a phage able to form plaques on K12(λ).

Once a large set of (+) and (−) mutations had been isolated, various double-mutant combinations could be brought together by phage recombination. Usually any (+) (−) combination and any (−)(+) combination resulted in a wildtype phenotype, whereas (+)(+) and (−)(−) combinations always resulted in the mutant phenotype. The key result came from the triple mutants (+)(+)(+) and (−)(−)(−), which often had a wildtype phenotype!

The phenotypes of the various (+) and (−) combinations were interpreted in terms of offsetting shifts in the reading frame, as shown in

| Phage type | Insertion/deletion | Translational reading frame of mRNA |
|---|---|---|
| Wildtype sequence | | THE BIG BOY SAW THE NEW CAT EAT THE HOT DOG · · |
| +1 insertion | (+) | THE BIG BOY SAW TTH ENE WCA TEA TTH EHO TDO G |
| Revertant 1 | $(-)_1$ (+) | THE BIG OYS AWT THE NEW CAT EAT THE HOT DOG · · |
| Revertant 2 | (+) $(-)_2$ | THE BIG BOY SAW TTH ENE WCA TEA THE HOT DOG · · |
| Revertant 3 | (+) $(-)_3$ | THE BIG BOY SAW TTH ENE WAT EAT THE HOT DOG · · |
| (–) deletion number 1 | $(-)_1$ | THE BIG OYS AWT HEN EWC ATE ATT HEH OTD OG · · · |
| (–) deletion number 2 | $(-)_2$ | THE BIG BOY SAW THE NEW CAT EAT HEH OTD OG · · · |
| (–) deletion number 3 | $(-)_3$ | THE BIG BOY SAW THE NEW ATE ATT HEH OTD OG · · · |
| Double (–) mutant | $(-)_1$ $(-)_2$ | THE BIG OYS AWT HEN EWC ATE ATH EHO TDO G · · · |
| Triple (–) mutant | $(-)_1$ $(-)_2$ $(-)_3$ | THE BIG OYS AWT HEN EWA TEA THE HOT DOG · · · · |

**FIGURE 10.32** Interpretation of the *rII* frameshift mutations, showing that combinations of appropriately positioned single-base insertions (+) and single-base deletions (−) can restore the correct reading frame (green). The key finding was that a combination of three single-base deletions, as shown in the bottom line, also restores the correct reading frame (green). Two single-base deletions do not restore the reading frame. These classic experiments gave strong genetic evidence that the genetic code is a triplet code.

Figure 10.32. The initial FC0 mutation shifts the reading frame from the point of the +1 insertion onward, resulting in an incorrect amino acid sequence and a nonfunctional protein. Deletion of a nearby nucleotide pair will restore the reading frame, although the amino acid sequence encoded between the two mutations will be different. In (+)(+) or (−)(−) double mutants, the reading frame is shifted by two nucleotides, and the resulting protein is still nonfunctional. However, in the (+)(+)(+) and (−)(−)(−) triple mutants, the reading frame is restored, though all amino acids encoded between the outside mutations are incorrect. The resulting protein is one amino acid longer for (+)(+)(+) and one amino acid shorter for (−)(−)(−) (Figure 10.32).

The genetic analysis of the insertion and deletion mutations strongly supported the following conclusions:

- Translation of an mRNA starts from a fixed point.
- There is a single reading frame maintained throughout the process of translation.
- Each codon consists of three adjacent nucleotides.

Crick and his colleagues also drew other inferences from these experiments. First, in the genetic code, most codons must function in the specification of an amino acid. Second, most amino acids must be specified by more than one codon. They reasoned that if each amino acid had only one codon, then only 20 of the 64 possible codons could be used for coding amino acids. In this case, most frameshift mutations should have affected one of the remaining 44 "noncoding" codons in the reading frame, and hence a nearby frameshift of the opposite-polarity mutation should not have suppressed the original mutation. Consequently, the code was inferred to contain synonymous codons (a property originally called *degeneracy*).

## How the Code Was Cracked

Polypeptide synthesis can be carried out in *E. coli* cell extracts obtained by breaking cells open. Various components can be isolated, and a functioning protein-synthesizing system can be reconstituted by mixing ribosomes, tRNA molecules, mRNA molecules, and the protein factors needed for translation. If radioactive amino acids are added to the extract, then radioactive polypeptides are made. Synthesis continues for only a few minutes because the mRNA is rapidly degraded by nucleases in the mixture. The experimental elucidation of the genetic code began with the observation that when the degradation of native mRNA was allowed to go to completion, and then the synthetic polynucleotide polyuridylic acid (poly-U) was added to the mixture to serve as an mRNA molecule, a polypeptide consisting only of phenylalanine (Phe−Phe−Phe. . .) was synthesized. From this simple result, and knowledge that the code is a triplet code, it follows that UUU must be a codon for the amino acid phenylalanine. Variations on this basic experiment identified other codons. For example, when a long sequence of guanines was added at the terminus of the poly-U, the polyphenylalanine was terminated by a sequence of glycines, indicating that GGG is a glycine codon (**FIGURE 10.33**). A trace of leucine or tryptophan was also present in the glycine-terminated polyphenylalanine. Incorporation of these amino acids was directed by the codons UUG and UGG at the transition point between U and G. When a single guanine was added to the terminus of a poly-U chain, the polyphenylalanine was terminated by leucine. Thus, UUG is a leucine codon, and UGG must be a codon for tryptophan. Similar experiments were carried out with poly-A, which yielded polylysine, and with poly-C, which produced polyproline.

Other experiments led to a complete elucidation of the code. The three codons

UAA UAG UGA

were found to be stop signals for translation, and the codon AUG, which encodes methio-

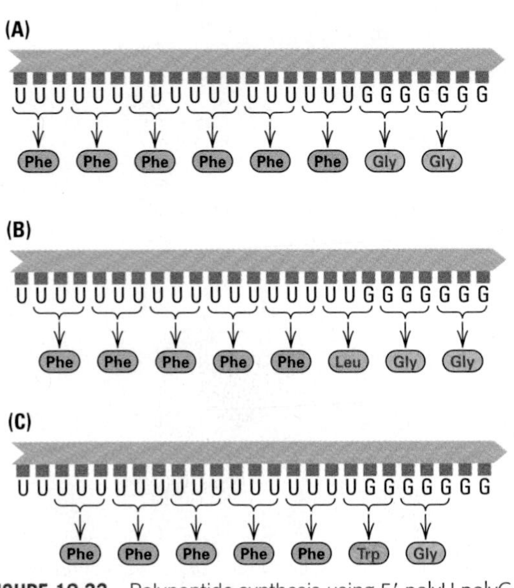


**FIGURE 10.33** Polypeptide synthesis using 5′-polyU-polyG-3′ as an mRNA in three different reading frames, showing the reasons for the incorporation of glycine, leucine, and tryptophan.

nine, was shown to be the initiation codon. AUG also codes for internal methionines.

## Features of the Standard Code

The *in vitro* translation experiments, which originally used components isolated from *E. coli*, have been repeated with components from many species of bacteria, yeast, plants, and animals. The standard genetic code deduced from these experiments is considered to be nearly universal because the same codon assignments can be made for nuclear genes in nearly all organisms that have been examined. However, some minor differences in codon assignments are found in certain protozoa and in the genetic codes of organelles.

The standard code is shown in **TABLE 10.3**. Altogether, 61 codons specify amino acids. In many cases, several codons specify the same amino acid. This feature confirms the inference from the *rII* frameshift mutations that the genetic code contains synonymous codons. All amino acids except tryptophan and methionine are specified by more than one codon. The pattern of synonymy is not random. For example, with the exception of serine, leucine, and arginine, all codons that correspond to the same amino acid are in the same box in Table 10.3. This means that *synonymous codons usually differ only in the third nucleotide position;* for example, GGU, GGC, GGA, and GGG all code for glycine. Furthermore, in all cases in which two codons code for the same amino acid, the third nucleotide is either A or G (both purines) or T or C (both pyrimidines).

Table 10.3 The standard genetic code

| First position (5′ end) | Second position | | | | Third position (3′ end) |
|---|---|---|---|---|---|
| | U | C | A | G | |
| U | UUU Phe F<br>UUC Phe F<br>UUA Leu L<br>UUG Leu L | UCU Ser S<br>UCC Ser S<br>UCA Ser S<br>UCG Ser S | UAU Tyr Y<br>UAC Tyr Y<br>**UAA Stop**<br>**UAG Stop** | UGU Cys C<br>UGC Cys C<br>**UGA Stop**<br>UGG Trp W | U<br>C<br>A<br>G |
| C | CUU Leu L<br>CUC Leu L<br>CUA Leu L<br>CUG Leu L | CCU Pro P<br>CCC Pro P<br>CCA Pro P<br>CCG Pro P | CAU His H<br>CAC His H<br>CAA Gln Q<br>CAG Gln Q | CGU Arg R<br>CGC Arg R<br>CGA Arg R<br>CGG Arg R | U<br>C<br>A<br>G |
| A | AUU Ile I<br>AUC Ile I<br>AUA Ile I<br>**AUG Met** M | ACU Thr T<br>ACC Thr T<br>ACA Thr T<br>ACG Thr T | AAU Asn N<br>AAC Asn N<br>AAA Lys K<br>AAG Lys K | AGU Ser S<br>AGC Ser S<br>AGA Arg R<br>AGG Arg R | U<br>C<br>A<br>G |
| G | GUU Val V<br>GUC Val V<br>GUA Val V<br>GUG Val V | GCU Ala A<br>GCC Ala A<br>GCA Ala A<br>GCG Ala A | GAU Asp D<br>GAC Asp D<br>GAA Glu E<br>GAG Glu E | GGU Gly G<br>GGC Gly G<br>GGA Gly G<br>GGG Gly G | U<br>C<br>A<br>G |

Note: Each amino acid is given its conventional abbreviation in both the single-letter and the three-letter format. The codon AUG, which codes for methionine (boxed) is generally used for initiation. The codons are conventionally written with the 5′ base on the left and the 3′ base on the right.

## Transfer RNA and Aminoacyl-tRNA Synthetase Enzymes

The decoding operation by which the nucleotide sequence within an mRNA molecule becomes translated into the amino acid sequence of a protein is accomplished by aminoacylated, or charged, tRNA molecules, each of which is linked to the correct amino acid by an **aminoacyl-tRNA synthetase**.

The tRNA molecules are small, single-stranded nucleic acids ranging in size from about 70 to 90 nucleotides. Like all RNA molecules, they have a 3′-OH terminus, but the opposite end terminates with a 5′-monophosphate rather than a 5′-triphosphate, because tRNA molecules are cut from a larger primary transcript. Internal complementary base sequences form short double-stranded regions, causing the molecule to fold into a structure in which open loops are connected to one another by double-stranded stems (**FIGURE 10.34**, part A). In two dimensions, a tRNA molecule is often drawn as a planar cloverleaf. Its three-dimensional structure is more complex, as is shown in Figure 10.34B. Note that the T$\psi$C loop and the DHU loop are in close proximity.

Particular regions of each tRNA molecule are used in the decoding operation. One region is the anticodon sequence, which consists of three nucleotides that can form base pairs with a codon sequence in the mRNA. No normal tRNA molecule has an anticodon complementary to any of the stop codons (UAG, UAA, and UGA), which is why these codons are stop signals. A second critical site is at the 3′ terminus of the tRNA molecule, where the amino acid attaches. A specific aminoacyl-tRNA synthetase matches the amino acid with the anticodon. At least one, and usually only one, aminoacyl-synthetase exists for each amino acid. To make the correct attachment, the synthetase must be able to distinguish one tRNA molecule from another. The necessary distinction is provided by recognition regions that encompass many parts of the tRNA molecule.

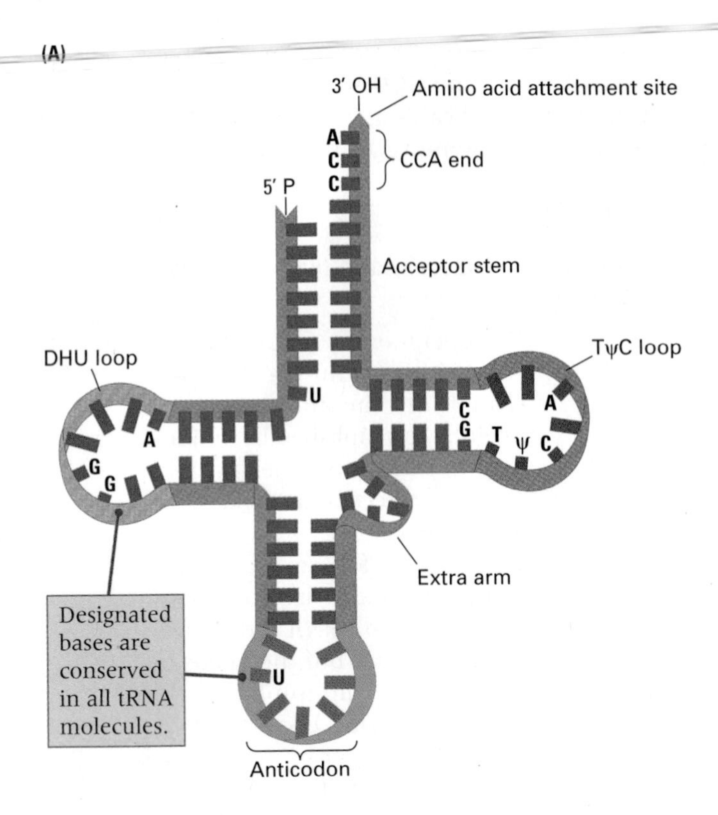


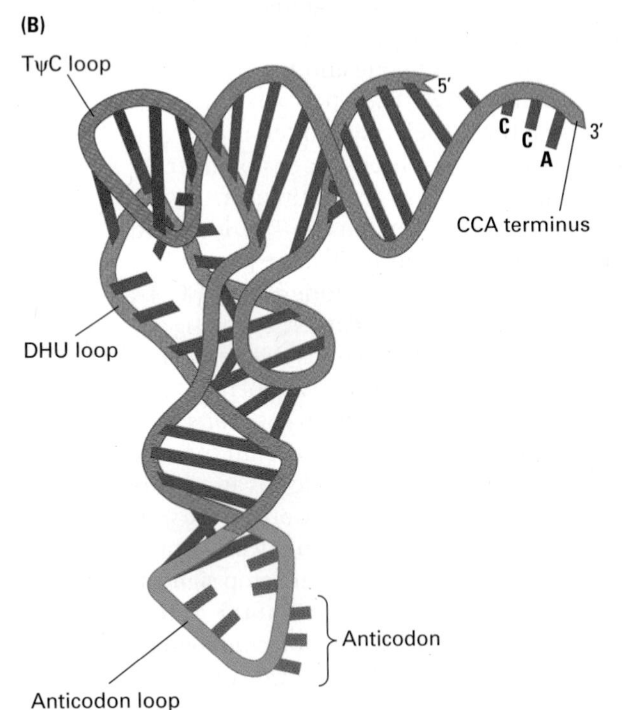


**FIGURE 10.34** (A) A tRNA "cloverleaf" configuration. The heavy black letters indicate a few bases that are conserved in the sequence of all tRNA molecules. The labeled loop regions are those found in all tRNA molecules. DHU refers to a base, dihydrouracil, found in one loop; the Greek letter Ψ is a symbol for the unusual base pseudouridine. (B) A schematic diagram of the three-dimensional structure of yeast $tRNA^{Phe}$. [Part B courtesy of Sung-Hou Kim, University of California, Berkeley.]

**FIGURE 10.35** shows the three-dimensional structure of the seryl-tRNA synthetase complexed with its tRNA. On binding with the tRNA, a part of the protein makes contact with the variable part of the TψC loop of the tRNA and guides the acceptor stem into the active pocket of the enzyme. These interactions depend primarily on recognition of the shape of the $tRNA^{Ser}$ through contacts with the backbone; they depend only secondarily on interactions that are specific to the anticodon.

## ■ Redundancy and Wobble

Several features of the genetic code suggest that normal Watson–Crick base pairing is not the only feature of codon–anticodon binding. First, the code is highly redundant. Second, the identity of the third nucleotide of a codon is often unimportant. For some amino acids—for example, proline (Pro), threonine (Thr), and glycine (Gly)—any nucleotide in the third position will do; and for other amino acids—for example, histidine (His), glutamine (Gln), and tyrosine (Tyr)—either purine (A or G) or either pyrimidine (U or C) in the third position will do. Third, the number of distinct tRNA molecules that have been isolated from a single organism is less than the number of codons; because all codons are used, the anticodons of some tRNA molecules must be able to pair with more than one codon. Experiments with several purified tRNA molecules showed this to be the case.

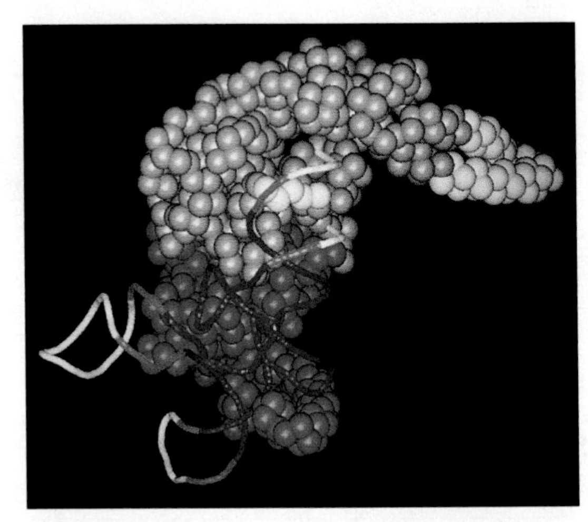

**FIGURE 10.35** Three-dimensional structure of seryl-tRNA synthetase (solid spheres) complexed with its tRNA. Note that there are many points of contact between the enzyme and the tRNA. The molecules are from *Thermus thermophilus*. [Reproduced from V. Biou, et al., *Science* 263 (1994): 1404–1410. Reprinted with permission from AAAS.]

To account for these observations, the **wobble** concept was advanced in 1966 by Francis Crick. He proposed that the first two nucleotides in a codon form base pairs with the tRNA anticodon according to the usual rules (A–U and G–C), but that the base at the 5′ end of the anticodon is less spatially constrained than the first two and can form hydrogen bonds with more than one base at the 3′ end of the codon. His suggestion was essentially correct, but the allowed base pairing differs somewhat among organisms (**TABLE 10.4**).

**Table 10.4 Wobble rules for tRNAs of *E. coli* and *S. cerevisiae***

| First base in anticodon (5′ position) | Allowed base in third codon position (3′ position) | |
|---|---|---|
| | ***E. coli*** | ***S. cerevisiae*** |
| A | U | — |
| C | G | G |
| U | A or G | A |
| G | C or U | C or U |
| I | A, C, or U | C or U |

Notes: In *S. cerevisiae*, an A at the 5′ position in the anticodon is always modified to I, which indicates inosine; inosine is structurally similar to adenosine except that the $-NH_2$ is replaced with $-OH$. Likewise, a U at the first anticodon position is often modified in this organism.

## Nonsense Suppression

A nucleotide substitution that creates a stop codon results in premature chain termination during translation. Such a mutation is called a **nonsense mutation**. For whimsical historical reasons, the stop codons are sometimes referred to as the *amber* (UAG), *ochre* (UAA), and *umber* (UGA). A remarkable observation was that some T4 phage mutants bearing nonsense mutations in the gene encoding the major head protein were able to propagate in some bacterial strains but not in others. How could this happen? On further analysis, it turned out that the strains able to support growth of the phage nonsense mutants carry suppressor mutations that act by changing the way the mRNA is read, not by changing the nucleotide sequence of the phage gene. The suppressor mutations proved to be mutant tRNA genes. These suppressor tRNA genes act by reading a stop codon as though it were a signal for a specific amino acid. The amino acid is inserted at that position, and translation continues. So long as the inserted amino acid is compatible with the function of the protein, the effects of the original mutations are suppressed, and plaques can be produced.

We will illustrate nonsense suppression by examining a chain-termination codon formed by mutation of the tyrosine codon UAC to the stop codon UAG (**FIGURE 10.36**, part A versus B). Part C shows how this mutation can be suppressed by a mutant leucine tRNA molecule. In *E. coli*, $tRNA^{Leu}$ has the anticodon 3′-AAC-5′, which pairs with the codon 5′-UUG-3′. A suppressor mutation in the $tRNA^{Leu}$ gene produces an altered tRNA with the anticodon AUC; this tRNA molecule is still charged with leucine but corresponds to the stop codon UAG rather than to the normal leucine codon UUG. Thus, in a cell that contains this suppressor tRNA, the mutant protein is completed, and the resulting protein will be functional provided that it can tolerate leucine instead of tyrosine. Many suppressor tRNA molecules of this type have been observed. Each suppressor is effective against only some nonsense mutations, because the resulting amino acid replacement may not yield a functional protein.

In *E. coli*, there are three classes of tRNA suppressors: those that suppress only UAG (amber suppressors), those that suppress both UAA and UAG (ochre suppressors), and those that suppress only UGA. They share the following properties:

1. The original mutant gene still contains the mutant nucleotide sequence (UAG in Figure 10.36).
2. The suppressor tRNA suppresses all chain-termination mutations with the same stop codon, provided the amino acid inserted is an acceptable amino acid at the site.
3. A cell can survive the presence of a tRNA suppressor *only* if the cell contains two or more copies of the same tRNA gene. Taking the example in Figure 10.36, if only one $tRNA^{Leu}$ gene were present in the genome and if it were mutated, then the normal leucine codon UUG would no longer be read as a sense codon, and all polypeptide chains would terminate wherever a UUG codon occurred. However, *multiple copies of most tRNA genes exist,* so if one copy is mutated to yield a suppressor tRNA, a normal copy nearly always remains.
4. Any chain-termination codon can be translated by a suppressor tRNA mutation that recognizes that codon. For example, translation of UAG by insertion of an amino acid would prevent termination of all wildtype mRNA reading frames terminating in UAG. However, the anticodon of the suppressor tRNA usually binds

**FIGURE 10.36** The mechanism of suppression by a nonsense suppressor tRNA molecule. (A) The wildtype gene. (B) A UAC → UAG chain-termination ("nonsense") mutation leads to an inactive, prematurely terminated protein. (C) A mutation in the tRNA$^{Leu}$ gene produces an altered tRNA molecule, which has an anticodon complementary to a UAG stop codon but can still be charged with leucine. This tRNA molecule allows the protein to be completed, but with a leucine at the site of the original tyrosine. Suppression will be achieved if the substitution restores activity to the protein.

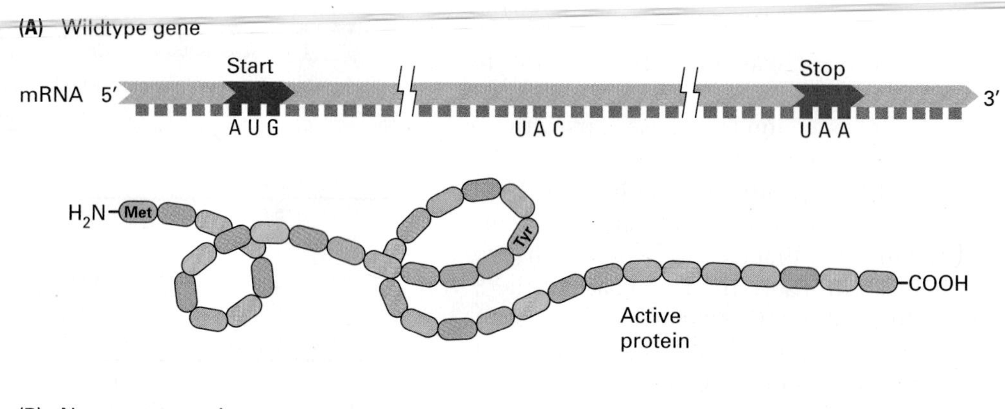


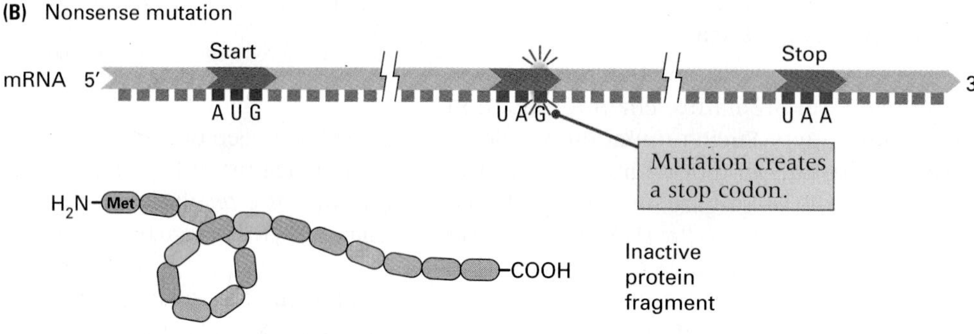


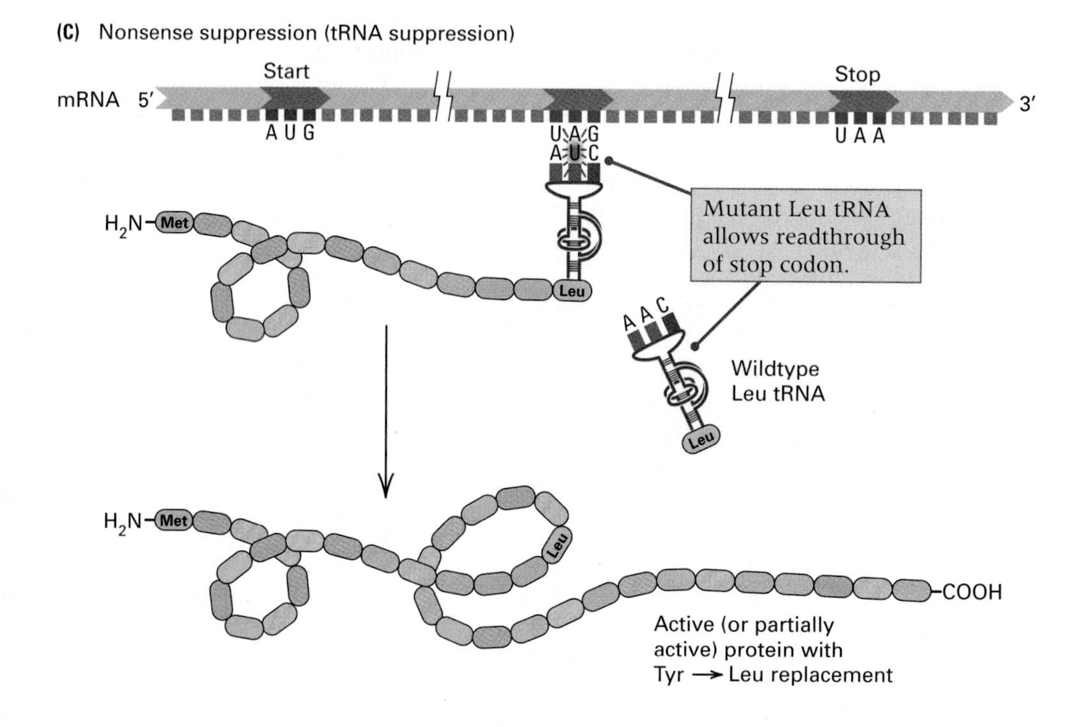


rather weakly to the stop codon, so the stop codon often results in termination anyway.

Suppressor tRNA mutations are very useful in genetic analysis, because they allow nonsense mutations to be identified through their ability to be suppressed. This is important because nonsense alleles usually result in a truncated and completely inactive protein and so are considered true loss-of-function alleles. Suppressor tRNAs have been widely used for genetic analysis in prokaryotes and yeast, because their other harmful effects are tolerated by the organism. In most metazoans, suppressor tRNAs have such severe harmful effects that they are of limited usefulness.

## CHAPTER SUMMARY

- In gene expression, information in the nucleotide sequence of DNA is used to dictate the linear order of amino acids in a polypeptide by means of an RNA intermediate.
- Transcription of an RNA from one strand of the DNA is the first step in gene expression.
- In eukaryotes, the RNA transcript is modified and may undergo splicing to make the messenger RNA (mRNA).
- The messenger RNA is translated on ribosomes in groups of three nucleotides (codons), each specifying an amino acid through base pairing with molecules of transfer RNA, each "charged" (covalently bonded) with one amino acid.
- Ribosomes are particles consisting of ribosomal RNA and numerous proteins combined in small and large ribosomal subunits. The subunits come together with other factors to initiate polypeptide synthesis, to translocate the ribosome progressively along the mRNA, and to release the finished polypeptide when a stop codon is encountered.
- Nearly all organisms use the standard genetic code with 61 codons for 20 amino acids and 3 stop codons, but exceptions are found in certain protozoa and in the genetic codes of mitochondrial and other organellar DNA.

## REVIEW THE BASICS

- What is meant by the term *gene expression*? Would you make a distinction between gene expression and gene regulation? Why or why not?
- Would you regard an original text and its translation into another language as "colinear"? Explain your answer.
- Suppose that a duplex DNA molecule undergoes two double-stranded breaks that tightly flank the promoter of a gene and that the promoter region is inverted prior to enzymatic repair of the breaks. Would you expect the inverted promoter to be able to recruit the transcription complex? What, if anything, would be wrong with the transcript of the gene?
- Is the DNA strand that serves as the template for RNA polymerase transcribed in the 5′-to-3′ or the 3′-to-5′ direction? Which end of the mRNA molecule is translated first? Which end of the polypeptide encoded in the mRNA is synthesized first?
- In a eukaryotic cell, four general types of RNA molecules are used in gene expression. What are these types of RNA called? Which type is not involved in gene expression in prokaryotic cells, and why not?
- What is a primary transcript? How does a primary transcript in eukaryotes differ from that in prokaryotes?
- How do prokaryotes and eukaryotes differ in the mechanism for selecting an AUG codon as a start for polypeptide synthesis?
- What role do release factors play in translation?
- What does it mean to say that the standard genetic code is redundant? Which (if any) amino acids are encoded by one codon? By two? By three? By four? By five? By six?
- Give an example of a genetic system that does not use the standard genetic code.
- What is a frameshift mutation? Explain how *rII* recombinants containing multiple, single-nucleotide frameshift mutations were used to show that the messenger RNA is translated in consecutive groups of three nucleotides (codons).

## GUIDE TO PROBLEM SOLVING

**Problem 1** The International Union of Biochemistry and Molecular Biology (IUBMB) has designated a single-letter code for abbreviating the nucleotide bases that allows for ambiguous assignments. The code is shown in the diagram below. The same code is used for DNA as for RNA. For ambiguous nucleotides, T and U are regarded as equivalent. Assuming standard Watson–Crick pairing in the unlabeled strand paired with that containing ambiguous bases, complete the sequence of the unlabeled strand, using the appropriate symbol from the standard ambiguity code.

**Answer** The ambiguity codes are very useful not only for designating uncertain nucleotides in DNA sequences, but also for summarizing the redundancies in the genetic code. The pairing relationships are straightforward for A, T, G, C, and U, but for ambiguous nucleotides one has to enumerate the possibilities and then select the symbol that expresses these ambiguities. A peculiar feature is that some symbols pair with themselves. For example, W (A or T) in one strand must also have a W (T or A) in the other strand, where the convention is that the paired nucleotides, though ambiguous, must obey the Watson–Crick pairing

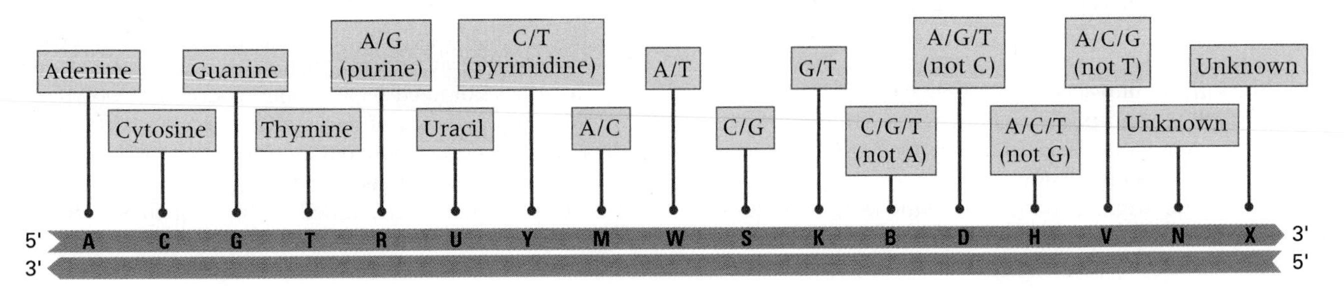

rules. All of the pairings can be worked out in this way, and the results are shown in the answer diagram in the middle of this page. There are two symbols—namely N and X—in use for "any nucleotide," so these can be paired however it is convenient.

**Problem 2** Rewrite the genetic code table using as many as possible of the single-letter codes for ambiguous bases established by the International Union of Biochemistry and Molecular Biology (IUBMB), as shown in Worked Problem 1.

**Answer** This problem requires that you examine the standard genetic code and select the proper symbol for ambiguous nucleotides. The version of the genetic code that results is shown here. It has considerably fewer entries than the standard format, and it shows the general structure of the code at a glance.

| First nucleotide in codon | Second nucleotide in codon: T | C | A | G |
|---|---|---|---|---|
| T | TTY Phe F<br>TTR Leu L | TCN Ser S | TAY Tyr Y<br>**TAR Stop** | TGY Cys C<br>**TGA Stop**<br>TGG Trp W |
| C | CTN Leu L | CCN Pro P | CAY His H<br>CAR Gln Q | CGN Arg R |
| A | ATH Ile I<br>**ATG Met** M | ACN Thr T | AAY Asn N<br>AAR Lys K | AGY Ser S<br>AGR Arg R |
| G | GTN Val V | GCN Ala A | GAY Asp D<br>GAR Glu E | GGN Gly G |

| 5' | A | C | G | T | R | U | Y | M | W | S | K | B | D | H | V | N | X | 3' |
|---|---|---|---|---|---|---|---|---|---|---|---|---|---|---|---|---|---|---|
| 3' | T | G | C | A | Y | A | R | K | W | S | M | V | H | D | B | N/X | X/N | 5' |

**Problem 3** The nontranscribed strand at the beginning of a coding region reads

5′-ATGCATCCGGGCTCATTAGTCT . . . -3′

Two mutations are studied. Mutation X has an insertion of another G immediately after the red G, and mutation Y has a deletion of the red A. What is the amino acid sequence of each of the following?

**(a)** the wildtype polypeptide
**(b)** the polypeptide in mutant X
**(c)** the polypeptide in mutant Y
**(d)** the polypeptide in a recombinant organism containing both mutations

**Answer**

**(a)** The sequence of the nontranscribed strand is given, and so the wildtype reading frame of the mRNA is AUG CAU CCG GGC UCA UUA GUC U . . . , which codes for Met–His–Pro–Gly–Ser–Leu–Val– . . ..
**(b)** The mutant X mRNA reading frame is AUG GCA UCC GGG CUC AUU AGU CU . . ., which codes for Met–Ala–Ser–Gly–Leu–Ile–Ser–Leu– . . ..
**(c)** The mutant Y mRNA reading frame is AUG CAU CCG GGC UCU UAG UCU . . ., which codes for Met–His–Pro–Gly–Ser (the UAG is a termination codon).
**(d)** The double-mutant mRNA reading frame is AUG GCA UCC GGG CUC UUA GUC U . . ., which codes for Met–Ala–Ser–Gly–Leu–Leu–Val– . . .. Note that the double mutant differs from wildtype only in the region between the insertion and deletion mutations, in which the reading frame is shifted.

## ANALYSIS AND APPLICATIONS

**10.1** In a random sequence of four ribonucleotides, what is the probability that any three adjacent nucleotides will be a start codon? A stop codon? In an mRNA molecule of random sequence, what is the average distance between stop codons?

**10.2** What possible amino acids are specified by a codon that consists of
**(a)** All pyrimidines?
**(b)** All purines?

**10.3** Each of the 61 codons that specify amino acids can have three possible nucleotide substitutions in the third position of the codon, yielding $61 \times 3 = 183$ possible third-position mutations. How many of these are synonymous mutations, which means that they specify the same amino acid as the original codon? What proportion of third-position changes are synonymous?

**10.4** A single codon in a double-stranded DNA molecule undergoes an inversion.
**(a)** If the original codon in the mRNA is 5′-AUG-3′, what is the codon in the transcript from the inversion?
**(b)** Would the inversion of any codon conserve the amino acid that is specified?
**(c)** What codons would, upon inversion, yield chain-termination codons?
**(d)** Which amino acids would inversion of each of the chain-termination codons specify?

**10.5** Poly–A codes for polylysine. If a C is added to the 5′ end of the molecule, the polylysine has a different amino acid at the amino terminus, and if a C is added to the 3′ end, there is a different amino acid at the carboxyl terminus. What are the amino acids?

**10.6** The synthetic polymer, poly–U, is used as an mRNA molecule in an *in vitro* protein-synthesizing system that does not need a special start codon. The polypeptide polyphenylalanine is synthesized. A single cytosine nucleotide is added to one end of the poly–U. The resulting polyphenylalanine has a valine at the amino terminus. Was the C added to the 3′ or the 5′ end of the poly–U?

**10.7** A novel animal virus is found with the unusual property that all of its G–C base pairs are oriented with the C in the transcribed strand of the DNA.
**(a)** What are the implications for the codons present in the viral DNA?
**(b)** What are the implications for the amino acids present in the viral proteins?

**10.8** What polypeptide products are made when the alternating polymer AAGAAGAAG ... is used in an *in vitro* protein-synthesizing system that does not need a start codon?

**10.9** What polypeptide products are made when the alternating polymer CACGCACGCACG... is used in an *in vitro* protein-synthesizing system that does not need a start codon?

**10.10** What polypeptide products are made when the alternating polymer AUAUAU... is used in an *in vitro* protein-synthesizing system that does not need a start codon?

**10.11** When a random polymer consisting of 1/3 U and 2/3 A is used to direct protein synthesis in a cell-free system, the resulting polypeptides are very short. Can you suggest an explanation.

**10.12** Some codons in the genetic code were determined experimentally by the translation of random polymers. If a ribonucleotide polymer is synthesized that contains 1/4 G and 3/4 U in random order, which amino acids would the resulting polypeptide chain contain, and in what frequencies?

**10.13** How many different sequences of nine ribonucleotides would code for the amino acids Trp–Tyr–Val? For Pro–Arg–Lys? Using the symbol Y for any pyrimidine, R for any purine, and N for any nucleotide, what are these sequences?

**10.14** Prior to the demonstration that the messenger RNA is translated in consecutive, nonoverlapping groups of three nucleotides (codons), the possibility of an *overlapping code* had to be considered. Such overlapping codes could be rejected because they impose constraints on which consecutive pairs of amino acids could be found in proteins. To understand why, suppose that the standard triplet codons were translated with an overlap of two. (In other words, the last two nucleotides of any codon are also the first two nucleotides of the next codon in line.) Which amino acids could follow:
**(a)** Cys
**(b)** Phe
**(c)** Trp
**(d)** Met

**10.15** If the standard genetic code were translated with an overlap of two nucleotides (i. e., the last two nucleotides of any codon are also the first two nucleotides of the next codon in line), which amino acids could be at the carboxyl end of a polypeptide chain?

**10.16** If the standard genetic code were translated with the codons overlapping, what amino acids could follow Met at the amino end of a polypeptide chain if:
**(a)** The overlap consisted of two nucleotides?
**(b)** The overlap consisted of one nucleotide?

**10.17** What codons could pair with the anticodon 5′–ICA–3′? (I stands for inosine, which can pair with A or U or C.) What amino acid would be incorporated?

**10.18** Two possible anticodons could pair with the codon UGG, but only one is actually used. Identify the possible anticodons, and explain why one of them is not used.

**10.19** The following is the nucleotide sequence of a strand of DNA.

TACGTCTCCAGCGGAGATCTTTTCCGGTCGCAACTGAGGTTGATC

The strand is transcribed from left to right and codes for a small polypeptide.
**(a)** Which end is the 3′ end and which the 5′ end?
**(b)** What is the sequence of the complementary DNA strand?
**(c)** What is the sequence of the transcript?
**(d)** What is the amino acid sequence of the peptide?
**(e)** In which direction along the transcript does translation occur?
**(f)** Which is the amino ($-NH_2$) and which the carboxyl ($-COOH$) end of the peptide?

**10.20** Protein synthesis occurs with relatively high fidelity. In prokaryotes, incorrect amino acids are inserted at the rate of approximately $5 \times 10^{-4}$ (that is, one incorrect amino acid per 2000 translated). What is the probability that a polypeptide of 300 amino acids has exactly the amino acid sequence specified in the mRNA?

**10.21** A DNA fragment containing a particular gene is isolated from a eukaryotic organism. This DNA fragment is mixed with the corresponding mRNA isolated from the organism, denatured, renatured, and observed by electron microscopy. Heteroduplexes of the type shown in the accompanying figure are observed. How many introns does this gene contain?

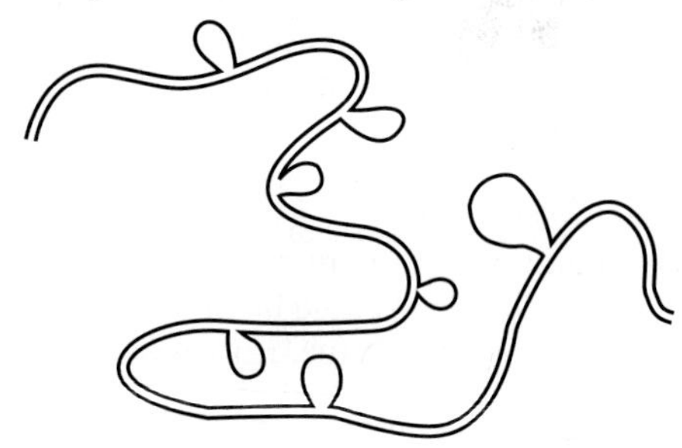

**10.22** If the DNA molecule shown below is transcribed from left to right, what are the sequence of the mRNA and the sequence of amino acids? What are the sequence of the mRNA and the sequence of amino acids if the segment in red is inverted?

5′-AGACTTAGCGCTAAACGTGGT-3′
3′-TCTGAATCGCGATTTGCACCA-5′

**10.23** Two *E. coli* genes, *A* and *B*, are known from mapping experiments to be very close to each other. A deletion mutation is isolated that eliminates the activity of both *A* and *B*. Neither the A nor the B protein can be found in the mutant, but a novel protein is isolated in which the amino-terminal 30 amino acids are identical to those of the *B* gene product and the carboxyl-terminal 30 amino acids are identical to those of the *A* gene product.

**(a)** With regard to the 5′-to-3′ orientation of the nontranscribed DNA strand, is the order of the genes *AB* or *BA*?

**(b)** Can you make any inference about the number of bases deleted in the coding region?

**10.24** Some RNA molecules contain sequences that are complementary to one another and can fold back upon themselves to form regions of double-stranded RNA in which the complementary bases are paired. These "stem-and-loop" or "hairpin" secondary structures are quite stable. If the RNA molecule shown here forms a stem-and-loop structure, with the "loop" consisting of the five bases immediately preceding the dashes, what nucleotides are needed at the 3′ end to form the stem?

5′-UACGGCUUCGAUGACAUGCG_ _ _ _ _ _ _ _ _ _ _ _ _ _ _ _-3′

**10.25** Researchers occasionally find a gene whose coding sequence fails match its protein product. In some instances, the cause has been traced to *RNA editing*, in which certain bases in the RNA are chemically modified after transcription. The modification is carried out by enzymes. One of these editing enzymes is a *cytosine deaminase*, which converts a C in the RNA into U. This type of editing occurs at some C residues in RNA transcripts from plant mitochondria and chloroplasts, and also in the apolipoprotein B RNA transcript in mammals. Consider the mRNA sequence shown here, which is translated in a reading frame starting with the first codon at the left.

5′-GUACCACGCUCGUCUCAU-3′

In this sequence, the accessible cytosines recognized by the cytosine deaminase are shown in red.

**(a)** What is the amino acid sequence if none of the accessible cytosines is modified?

**(b)** What is the amino acid sequence if all of the accessible cytosines are modified?

**(c)** Assume that any number from 0–4 of the accessible cytosines could be modified in any particular mRNA molecule. How many polypeptide sequences can this part of the gene encode?

**10.26** Another type of RNA editing is catalyzed by an adenosine deaminase that converts A into inosine (I), which is translated by the ribosome as G. This type of A–to–I editing is found in hepatitis delta virus, the mammalian transcript for the glutamate receptor, and several other systems. Consider the mRNA in Problem 10.25. If all of the A's in this transcript underwent A–to–I editing, how would the polypeptide sequence be altered?

**10.27** A molecular biologist discovers a gene in which there seems to be an intron with unusual 5′ and 3′ splice sites. The DNA sequence in the relevant region is shown here, where the dots indicate part of the putative intron whose splice sites are unknown.

5′-CTATACAGCGG...CATTCTGTGGGT-3′
3′-GATATGTCGCC...GTAAGACACCCA-5′

This region of DNA encodes the amino acid sequence Ser–Tyr–Met–Trp–Val, although the reading frame is not known. What are the likely splice sites?

**10.28** A DNA sequence acquired from an organism is thought to be part of an open reading frame. Neither the transcribed strand nor the correct reading frame is known. Can you deduce from the DNA sequence which strand is transcribed and what is the correct reading frame?

5′-CTAGGTGACCTAGCTTAA-3′
3′-GATCCACTGGATCGAATT-5′

**10.29** If the C–G base pair on the left in Problem 10.28 were an A–T base pair, how would it change the answer?

**10.30** An organism is discovered in which the promoter sequence in the template strand is 3′-TTTTT-5′, the transcript begins with the first nucleotide following the promoter sequence, and transcription terminates immediately prior to the sequence 3′-GGGGGG-5′ in the template strand. The primary transcript is capped and used directly for the mRNA, and translation is initiated by scanning from the 5′ end.

**(a)** What mRNA sequence would be transcribed from the DNA sequence below?

**(b)** What polypeptide chain would result?

3′-TTTTTATGGTACAGTTTGTCGCATACCATCGTCACGGGGGG-5′

## CHALLENGE PROBLEMS

**Challenge Problem 1** For two different frameshift mutations in the second codon of a gene, the amino terminal sequences of the mutant proteins are

Mutant 1: Met–Lys–UAG
Mutant 2: Met–Ile–Val–UAA

Mutant 1 has a single-nucleotide addition, and mutant 2 has a single-nucleotide deletion. Furthermore, the first five amino acids of the wildtype protein are known to be Met–(Asn, Val, Ser, Lys), where the parentheses mean that the order of the amino acids is unknown. Using the information provided by the frameshift mutations, determine the first five codons in the wildtype gene as well as the nature of each frameshift mutation.

**Challenge Problem 2** An organism is discovered whose RNA polymerase II has the following properties:

1. The DNA sequence in the template strand 3′–TATAATA–5′ serves as the promoter.
2. Transcription begins at the nucleotide immediately following the promoter.
3. Transcription continues until the transcript includes the sequence 5′–GGGGG–3′, at which point

the polymerase *switches to the other DNA strand* and continues transcribing (still adding nucleotides only to the 3′ end of the growing chain)

**4.** Transcription terminates immediately after transcribing 5′–TATATA–3′ from the template strand.

If the primary transcript is immediately capped and used as the mRNA, and translation is initiated by scanning from the 5′ end:

**(a)** What mRNA would be produced from the DNA sequence shown here?

**(b)** What polypeptide chain results?

5′-CCGTATATATTATGATCAATATGCATGCTCTCGGGGGTCACACT-3′

3′-GGCATATATAATACTAGTTATACGTACGAGAGCCCCCAGTGTGA-5′

**Challenge Problem 3** A prokaryote is discovered with the unusual property that some of its mRNA molecules are circular, but they are translated according to the standard genetic code using any 5′-AUG-3′ for initiation. For the circular mRNA shown here, deduce:

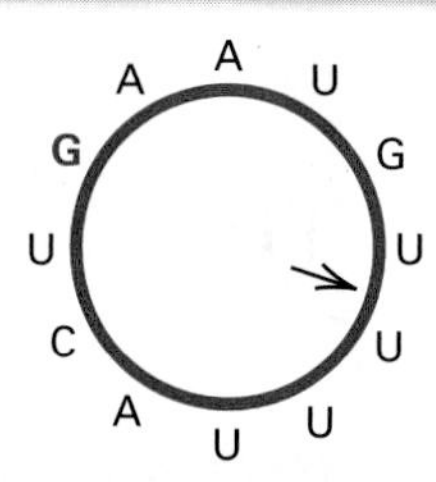


**(a)** The amino acid sequence of the wildtype polypeptide chain.

**(b)** The amino acid sequence of a mutant in which the red G were substituted with a U.

**(c)** The amino acid sequence of a mutant in which a C was inserted at the position indicated by the arrow.

**(d)** Explain why the polypeptide in part **b** is longer than that in part **c.**

**(e)** The amino acid sequence of the double mutant

## GENETICS *on the web*

GeNETics on the Web will introduce you to some of the most important sites for finding genetics information on the Internet. To explore these sites, visit the Jones and Bartlett companion site to accompany *Genetics: Analysis of Genes and Genomes, Seventh Edition* at http://biology.jbpub.com/book/genetics.

There you will find a chapter-by-chapter list of highlighted keywords. When you select one of the keywords, you will be linked to a Web site containing information related to that keyword.

# CHAPTER 11

# Molecular Mechanisms of Gene Regulation

The overriding principle of gene regulation is that some genes control the expression of other genes. For many years the regulation was thought to be mediated primarily by DNA-binding proteins. But it turns out that gene expression can also be regulated by RNA, especially small single-stranded RNA molecules processed from double-stranded RNA precursors. This molecular model shows how an RNA molecule containing complementary base sequences can form a stem-loop or hairpin structure, and how the seam of the hairpin is processed into microRNA able to regulate the expression of other genes.



## CHAPTER OUTLINE


**11.1 Transcriptional Regulation in Prokaryotes**
- Inducible and Repressible Systems of Negative Regulation
- Positive Regulation

**11.2 The Operon System of Gene Regulation**
- $Lac^-$ Mutants
- Inducible and Constitutive Synthesis and Repression
- The Repressor
- The Operator Region
- The Promoter Region
- The Operon System of Transcriptional Regulation
- Positive Regulation of the Lactose Operon
- Regulation of the Tryptophan Operon

**11.3 Regulation Through Transcription Termination**
- Attenuation
- Riboswitches

**11.4 Regulation in Bacteriophage Lambda**

**11.5 Transcriptional Regulation in Eukaryotes**
- Galactose Metabolism in Yeast
- Transcriptional Activator Proteins
- Transcriptional Enhancers and Transcriptional Silencers
- Deletion Scanning
- The Eukaryotic Transcription Complex
- Chromatin-Remodeling Complexes
- Alternative Promoters

**11.6 Epigenetic Mechanims of Transcriptional Regulation**
- Cytosine Methylation
- Methylation and Transcriptional Inactivation
- Genomic Imprinting in the Female and Male Germ Lines

**11.7 Regulation Through RNA Processing and Decay**
- Alternative Splicing
- Messenger RNA Stability

**11.8 RNA Interference**

**11.9 Translational Control**
- Small Regulatory RNAs Controlling Translation

**11.10 Programmed DNA Rearrangements**
- Gene Amplification
- Antibody and T-Cell Receptor Variability
- Mating-Type Interconversion
- Transcriptional Control of Mating Type

CONNECTION **Operator? Operator?**
François Jacob, David Perrin, Carmen Sanchez, and Jacques Monod 1960
*The Operon: A Group of Genes Whose Expression Is Coordinated by an Operator* [original in French]

CONNECTION **Double Trouble**
Andrew Fire, SiQun Xu, Mary K. Montgomery, Steven A. Kostas, Samuel E. Driver, and Craig C. Mell. 1998
*Potent and Specific Genetic Interference by Double-Stranded RNA in* Caenorhabditis Elegans

Humans and other vertebrate animals contain approximately 220 different cell types that are specialized in their functions. These differences are correlated with patterns of gene expression, because most genes differ in their level of expression according to cell type or stage in the cell cycle. The activity of genes is also keyed to the functions of the cell; for instance, the genes for hemoglobin are expressed at high levels only in precursors of the red blood cells. The control of synthesis of particular gene products is called *gene regulation.*

In many cases, gene activity is regulated at the level of transcription, either through signals originating within the cell itself or in response to external conditions. For example, many gene products are needed only on occasion, and transcription can be regulated in an on–off manner that enables such products to be present only when external conditions demand them. The flow of genetic information is regulated in other ways also. Control points for gene expression include:

1. ***Transcriptional regulation*** of the synthesis of RNA transcripts by controlling initiation or termination.
2. ***RNA processing,*** or regulation through RNA splicing or alternative patterns of splicing.
3. ***Translational control*** of polypeptide synthesis.
4. ***Stability of mRNA,*** because mRNAs that persist in the cell have longer-lasting effects than those that are degraded rapidly.
5. ***Posttranslational control,*** which includes a great variety of mechanisms that affect enzyme activity, activation, stability, and so on.
6. ***DNA rearrangements,*** in which gene expression changes depending on the position of DNA sequences in the genome.

The regulatory systems of prokaryotes and eukaryotes differ from each other in many details. Prokaryotes are generally free-living unicellular organisms that grow and divide indefinitely as long as environmental conditions are suitable and the supply of nutrients is adequate. Their regulatory systems are often geared to provide the maximum growth rate in a particular environment. In contrast, the cells in a developing multicellular organism modulate their growth rate as they undergo dramatic, coordinated differentiation in morphology and metabolism. In an adult animal, growth and division of most cell types has ceased, and each type of cell needs to maintain its identity through time.

## 11.1 Transcriptional Regulation in Prokaryotes

In bacteria and phages, on–off gene activity is often controlled through transcription. Under conditions when a gene product is needed, transcription of the gene is turned "on"; under other conditions, transcription is turned "off." The term *off* should not be taken literally. In bacteria, few examples are known of a system being switched off completely. When transcription is in the "off" state, a basal level of gene expression nearly always remains, often averaging one transcriptional event or fewer per cell generation; hence, "off" really means that there is very little synthesis of the gene product. Extremely low levels of expression are also found in certain classes of genes in eukaryotes, including many genes that participate in embryonic development. Regulatory mechanisms other than the on–off type also are known in both prokaryotes and eukaryotes; in these examples, the level of expression of a gene may be modulated in gradations from high to low according to conditions in the cell.

In bacterial systems, when several enzymes act in sequence in a single metabolic pathway, usually either all or none of these enzymes are produced. This **coordinate regulation** results from control of the synthesis of one or more polycistronic mRNA molecules encoding all of the gene products that function in the same metabolic pathway. This type of regulation is not found in eukaryotes because, as we saw in Chapter 10, eukaryotic mRNA is monocistronic.

### Inducible and Repressible Systems of Negative Regulation

The molecular mechanisms of regulation usually fall into either of two broad categories: *negative regulation* and *positive regulation*. In a system subject to **negative regulation** (**FIGURE 11.1**, part A), the default state is "on," and transcription takes place until it is turned "off" by a **repressor** protein that binds to the DNA upstream from the transcriptional start site. A negatively regulated system may be either *inducible* (Figure 11.1B) or *repressible* (Figure 11.1C), depending on how the active repressor is formed. In **inducible transcription,** a repressor DNA-binding protein normally keeps transcription in the "off" state. In the presence of a small molecule called the **inducer**, the

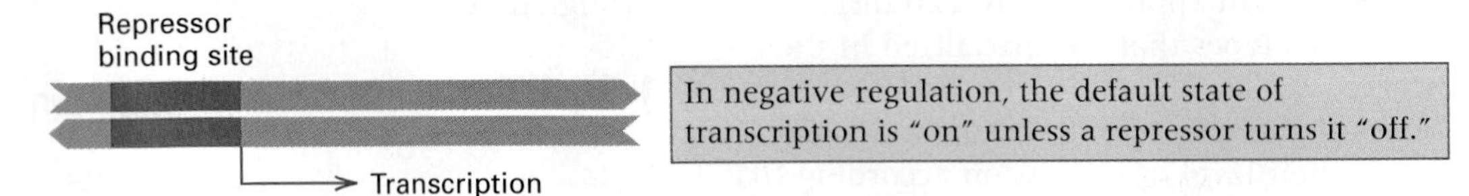


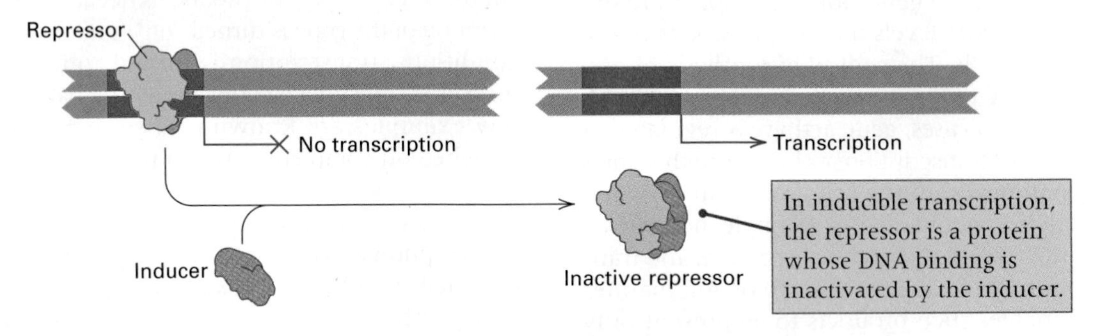


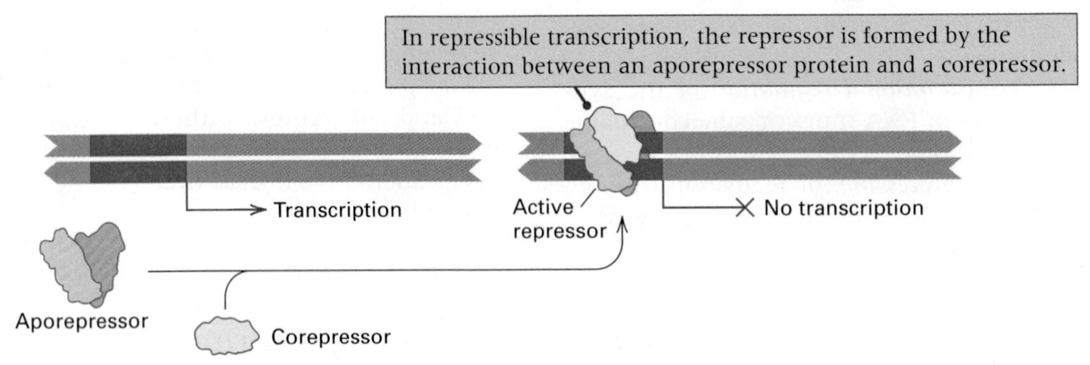


**FIGURE 11.1** Negative regulation (A) includes both inducible (B) and repressible (C) mechanisms of transcriptional control.

repressor binds preferentially with the inducer and loses its DNA-binding capability, allowing transcription to occur. Many degradative (catabolic) pathways are inducible and use the initial substrate of the degradative pathway as the inducer. In this way, the enzymes used for degradation are not synthesized unless the substrate is present in the cell.

In **repressible transcription** (Figure 11.1C), the default state is "on" until an active repressor is formed to turn it "off." In this case the regulatory protein is called an **aporepressor**, and it has no DNA-binding activity on its own. The active repressor that can bind to the DNA is formed by the combination of the aporepressor and a small molecule known as the **corepressor**. Presence of the corepressor thereby results in the cessation of transcription. Repressible regulation is often found in the control of the synthesis of enzymes that participate in biosynthetic (anabolic) pathways; in these cases the final product of the pathway is frequently the corepressor. In this way, the enzymes of the biosynthetic pathway are not synthesized until the concentration of the final product becomes too low to cause repression.

## Positive Regulation

Note that in negative regulation (Figure 11.1A), the default state of transcription is "on," and a repressor is required to turn it "off." Contrast this with a positively regulated system (**FIGURE 11.2**), in which the default state is "off," and binding with a regulatory protein is required to turn it "on." Such a regulatory protein is called a **transcriptional activator protein**. Negative and positive regulation are not mutually exclusive, and some systems are both positively and negatively regulated, utilizing two regulators to respond to different conditions in the cell. Negative regulation is more common in prokaryotes, positive regulation in eukaryotes.

Some genes exhibit **autoregulation**, which means that the protein product of a gene regu-

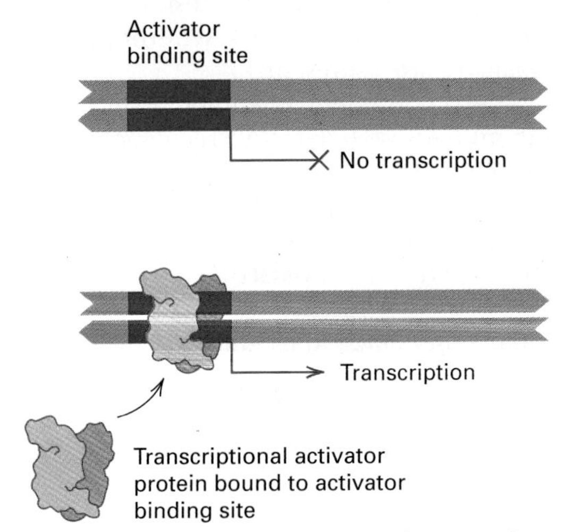


**FIGURE 11.2** In positive regulation, the default state of transcription is "off." Transcription is stimulated by the binding of a transcriptional activator protein.

lates its own transcription. In negative autoregulation, the protein inhibits transcription, and high concentrations of the protein result in less transcription of the mRNA. This mechanism automatically adjusts the steady-state level of the protein in the cell. In positive autoregulation, the protein stimulates transcription; as more protein is made, transcription increases to the maximum rate. Positive autoregulation is a common way for weak induction to be amplified. Only a weak signal is necessary to get production of the protein started, but then the positive autoregulation takes over and stimulates further production to the maximum level.

Next we examine two classical systems of regulation found in the bacterium *Escherichia coli*. These serve as specific examples of the general concepts introduced in Figures 11.1 and 11.2. We shall see that in the real world, most genes have overlapping mechanisms of control that include both positive and negative regulatory elements.

## 11.2 The Operon System of Gene Regulation

Gene regulation was first studied in detail in the system in *E. coli* that is responsible for degradation of the sugar lactose, and some of the key features of gene regulation were originally discovered in this system. It is a classic example of negative regulation in which transcription is inducible (Figure 11.1B).

Genomic sequencing has recently revealed that the genes for lactose utilization in *E. coli* are actually present in a *genomic island* of DNA of the type discussed in Section 9.1. Based on what we now know about the DNA sequence of these genes and their absence in closely related bacterial species, it is clear that the lactose genes reside in a genomic island of about 5 kb of DNA that was transferred into the *E. coli* genome from an unknown source at least 50 million years ago. This is one of at least 234 such islands that have become inserted into the *E. coli* genome since the time of its divergence from its nearest bacterial relative about 100 million years ago.

The first insights into the regulatory mechanisms came from a genetic analysis of mutations affecting lactose metabolism. The following sections describe the properties of these mutations and the interpretations that emerged from their analysis. The mechanisms of regulation inferred from the mutants has been abundantly confirmed by direct molecular studies.

### ■ Lac$^-$ Mutants

In *E. coli*, two proteins are necessary for the metabolism of lactose. They are the enzyme **$\beta$-galactosidase**, which cleaves lactose (a $\beta$-galactoside) to yield galactose and glucose; and a transporter molecule, **lactose permease**, which is required for the entry of lactose into the cell. The existence of two different proteins in the lactose-utilization system was first shown by a combination of genetic experiments and biochemical analysis.

First, hundreds of mutants unable to use lactose as a carbon source, designated Lac$^-$ mutants, were isolated. Some of the mutations were in the *E. coli* chromosome, and others were in an F′ *lac*, a plasmid carrying the genes for lactose utilization. By performing F′ × F$^-$ matings, investigators constructed partial diploids with the genotypes F′ *lac*$^-$/*lac*$^+$ and F′ *lac*$^+$/*lac*$^-$. (The genotype of the plasmid is given to the left of the slash and that of the chromosome to the right.) It was observed that all of these diploids always had a Lac$^+$ phenotype (that is, they made $\beta$-galactosidase and permease); thus, none produced an inhibitor that prevented functioning of the *lac* genes. Other partial diploids were then constructed in which both the F′ *lac* plasmid and the chromosome carried a *lac*$^-$ allele. These were tested for the Lac$^+$ phenotype, with the result that all of the mutants initially isolated could be placed into two complementation groups, called *lacZ* and *lacY*, a result that implies that the *lac* system consists of at least two genes. In particular, the partial diploids

$$F'\ lacZ^-lacY^+/lacZ^+lacY^- \text{ and}$$
$$F'\ lacZ^+lacY^-/lacZ^-lacY^+$$

had a Lac$^+$ phenotype (complementation), producing both β-galactosidase and permease. However, the genotypes

F′ *lacZ*$^-$*lacY*$^+$/*lacZ*$^-$*lacY*$^+$ and
F′ *lacZ*$^+$*lacY*$^-$/*lacZ*$^+$*lacY*$^-$

had a Lac$^-$ phenotype (noncomplementation). The F′ *lacZ*$^-$*lacY*$^+$/*lacZ*$^-$*lacY*$^+$ cells were unable to synthesize β-galactosidase, and hence *lacZ* encodes β-galactosidase; and the F′ *lacZ*$^+$*lacY*$^-$/*lacZ*$^+$ *lacY*$^-$ cells were unable to synthesize the permease, and hence *lacY* encodes the permease. (A third gene encoding a β-galactoside transacetylase was later discovered. It was not included among the early mutants because it is not essential for growth on lactose.) A final important result—that the *lacZ* and *lacY* genes are adjacent—was deduced from the high frequency of cotransduction observed in genetic mapping experiments.

## Inducible and Constitutive Synthesis and Repression

The on–off nature of the lactose-utilization system is shown by the following observations:

1. If a culture of Lac$^+$ *E. coli* is growing in a medium that does not include lactose or any other β-galactoside, then the intracellular concentrations of β-galactosidase and permease are exceedingly low, roughly 1–2 molecules of each per bacterial cell. However, if lactose is present in the growth medium, then the number of each of these molecules is about a thousandfold higher.
2. If lactose is added to a Lac$^+$ culture growing in a lactose-free medium (also lacking glucose, a point that will be discussed shortly), then both β-galactosidase and permease are synthesized nearly simultaneously, as shown in **FIGURE 11.3**. Analysis of the total mRNA present in the cells before and after the addition of lactose shows that almost no *lac* mRNA (the polycistronic mRNA that codes for β-galactosidase and permease) is present before lactose is added and that the addition of lactose triggers synthesis of *lac* mRNA.

These two observations led to the conclusions that transcription of the lactose genes is inducible and that lactose (or a derivative of lactose) is the inducer. Some analogs of lactose are also inducers, such as a sulfur-containing analog denoted IPTG (isopropylthiogalactoside); this synthetic inducer is convenient for experiments because it induces but is not cleaved by β-galactosidase, so IPTG levels are stable in cells whether or not β-galactosidase is present.

The key to understanding induction came from mutants with defective regulation. In one class of regulatory mutants, the *lac* mRNA was synthesized even in the absence of an inducer. Because of their constant synthesis, these mutants were termed **constitutive**. The constitutive mutants fell into two classes, *lacI*$^-$ and *lacO*$^c$. Mutants that failed to produce *lac* mRNA even when the inducer was present were also obtained. These uninducible mutants fell into two classes, *lacI*$^s$ and *lacP*$^-$. The characteristics of the mutants are shown in **TABLE 11.1** and discussed in the following sections.

## The Repressor

In Table 11.1, genotypes 3 and 4 show that *lacI*$^-$ mutations are recessive. In the absence of inducer, a *lacI*$^+$ cell does not synthesize *lac* mRNA, whereas the mRNA is produced in a *lacI*$^-$ mutant. These results suggest that:

> The *lacI* gene is a regulatory gene whose product is the repressor protein that keeps the system turned off. Because the repressor is necessary to shut off mRNA synthesis, regulation by the repressor is negative regulation.

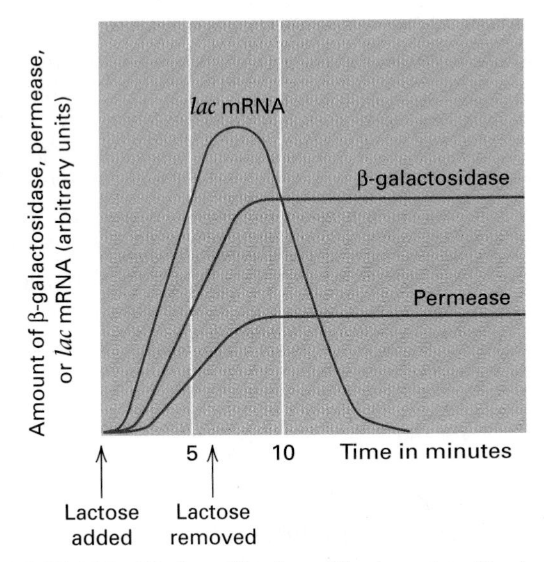


**FIGURE 11.3** The "on–off" nature of the *lac* system. The *lac* mRNA appears soon after lactose or another inducer is added; β-galactosidase and permease appear at nearly the same time but are delayed with respect to mRNA synthesis because of the time required for translation. When lactose is removed, no more *lac* mRNA is made, and the amount of *lac* mRNA decreases because of the degradation of mRNA already present. Both β-galactosidase and permease are stable proteins. Their amounts remain constant even when synthesis ceases. However, their concentration per cell gradually decreases as a result of repeated cell divisions.

**Table 11.1** Characteristics of partial diploids containing several combinations of *lacI*, *lacO* and *lacP* alleles

| Genotype | Synthesis of $lacZ^+$ mRNA | Lac phenotype |
|---|---|---|
| 1. F′ $lacO^c$ $lacZ^+$/$lacO^+$ $lacZ^+$ | Constitutive | + |
| 2. F′ $lacO^+$ $lacZ^+$/$lacO^c$ $lacZ^+$ | Constitutive | + |
| 3. F′ $lacI^-$ $lacZ^+$/$lacI^+$ $lacZ^+$ | Inducible | + |
| 4. F′ $lacI^+$ $lacZ^+$/$lacI^-$ $lacZ^+$ | Inducible | + |
| 5. F′ $lacO^c$ $lacZ^-$/$lacO^+$ $lacZ^+$ | Inducible | + |
| 6. F′ $lacO^c$ $lacZ^+$/$lacO^+$ $lacZ^-$ | Constitutive | + |
| 7. F′ $lacI^s$ $lacZ^+$/$lacI^+$ $lacZ^+$ | Uninducible | − |
| 8. F′ $lacI^+$ $lacZ^+$/$lacI^s$ $lacZ^+$ | Uninducible | − |
| 9. F′ $lacP^-$ $lacZ^+$/$lacP^+$ $lacZ^+$ | Inducible | + |
| 10. F′ $lacP^+$ $lacZ^+$/$lacP^-$ $lacZ^+$ | Inducible | + |
| 11. F′ $lacP^+$ $lacZ^-$/$lacP^-$ $lacZ^+$ | Uninducible | − |
| 12. F′ $lacP^+$ $lacZ^+$/$lacP^-$ $lacZ^-$ | Inducible | + |

How the *lacI* repressor protein binds to the DNA and prevents synthesis of *lac* mRNA will be explained shortly. A $lacI^-$ mutant lacks the repressor protein and hence transcription is constitutive. Wildtype copies of the repressor protein are present in a $lacI^+/lacI^-$ partial diploid, so transcription is repressed. It is important to note that the single $lacI^+$ gene prevents synthesis of *lac* mRNA from both the F′ plasmid and the chromosome. Therefore, the repressor protein must be diffusible within the cell, because it can shut off mRNA synthesis from both DNA molecules present in the partial diploid. Genetic mapping experiments placed the *lacI* gene very close to the *lacZ* gene and established the gene order *lacI lacZ lacY*.

Referring to Table 11.1 again, genotypes 7 and 8 indicate that the $lacI^s$ mutations are dominant and act to shut off mRNA synthesis from both the F′ plasmid and the chromosome whether or not the inducer is present (the superscript in $lacI^s$ signifies *super-repressor*). The $lacI^s$ mutations result in repressor molecules that fail to recognize and bind the inducer, and thus repress *lac* mRNA synthesis even in the presence of an inducer.

## ■ The Operator Region

Entries 1 and 2 in Table 11.1 show that $lacO^c$ mutants are dominant. However, the dominance is evident only in certain combinations of *lac* mutations, as can be seen by examining the partial diploids shown in entries 5 and 6. Both combinations are $Lac^+$ because a functional *lacZ* gene is present. However, in the combination shown in entry 5, synthesis of β-galactosidase is inducible even though a $lacO^c$ mutation is present. The difference is that in entry 5, the $lacO^c$ mutation is present in the same DNA molecule as the $lacZ^-$ mutation, whereas in entry 6, $lacO^c$ is contained in the same DNA molecule as $lacZ^+$. The key feature of these results is that:

> A $lacO^c$ mutation causes constitutive synthesis of β-galactosidase only when the $lacO^c$ and $lacZ^+$ alleles are contained in the same DNA molecule.

The $lacO^c$ mutation is said to be ***cis*-dominant**, because only genes in the *cis* configuration (in the same DNA molecule as that containing the mutation) are expressed in dominant fashion. Confirmation of this conclusion comes from an important biochemical observation: The mutant enzyme from the $lacZ^-$ sequence is synthesized constitutively in a $lacO^c$ $lacZ^-$/$lacO^+$ $lacZ^+$ partial diploid (entry 5), whereas the wildtype enzyme (coded by the $lacZ^+$ sequence) is synthesized only if an inducer is added. All $lacO^c$ mutations are located between the *lacI* and *lacZ* genes; hence the gene order of the four genetic elements of the *lac* system is:

*lacI lacO lacZ lacY*

An important feature of all $lacO^c$ mutations is that they cannot be complemented (a characteristic feature of all *cis*-dominant mutations); that is, a $lacO^+$ allele cannot alter the constitutive activity of a $lacO^c$ mutation. This observation implies that the *lacO* region does not encode a diffusible product and must instead define a site in the DNA that determines whether synthesis of the product of the adjacent *lacZ* gene is inducible or constitutive. The *lacO* region is called the **operator**. In the next section, we will see that the operator is in fact a binding site in the DNA for the repressor protein.

## ■ The Promoter Region

Entries 11 and 12 in Table 11.1 show that $lacP^-$ mutations, like $lacO^c$ mutations, are *cis*-dominant.

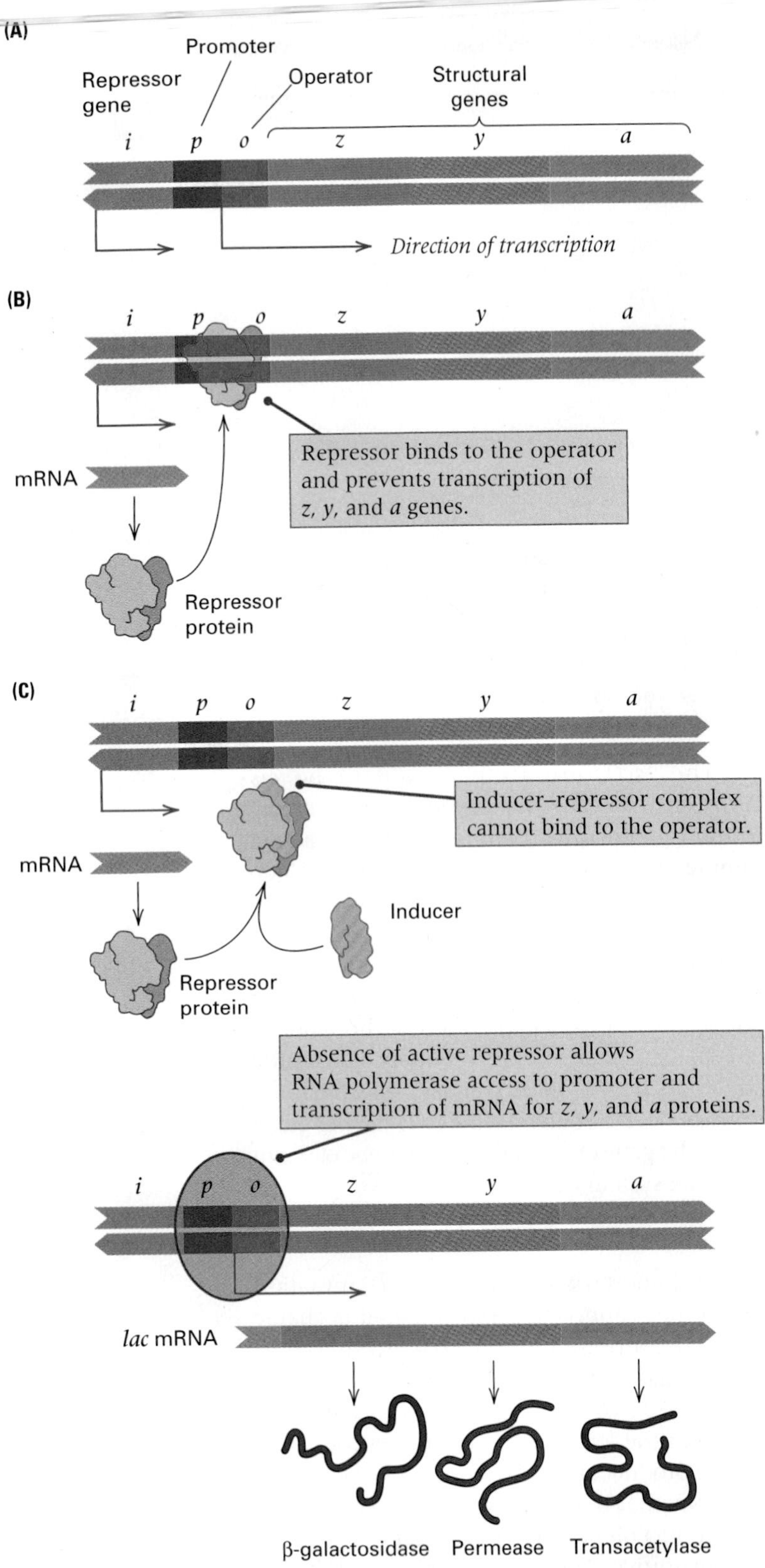
(A)
Promoter
Repressor gene
Operator
Structural genes
i p o z y a
Direction of transcription
(B)
Repressor binds to the operator and prevents transcription of z, y, and a genes.
mRNA
Repressor protein
(C)
Inducer–repressor complex cannot bind to the operator.
mRNA
Inducer
Repressor protein
Absence of active repressor allows RNA polymerase access to promoter and transcription of mRNA for z, y, and a proteins.
lac mRNA
β-galactosidase
Permease
Transacetylase

**FIGURE 11.4** (A) A map of the *lac* operon, not drawn to scale. The *p* and *o* sites are actually much smaller than the other regions and together comprise only 83 base pairs. (B) The *lac* operon in the repressed state. (C) The *lac* operon in the induced state. The inducer alters the shape of the repressor so that the repressor can no longer bind to the operator. The common abbreviations *i, p, o, z, y,* and *a* are used instead of *lacI, lacO,* and so on. The *lacA* gene is not essential for lactose utilization.

The *cis*-dominance can be seen in the partial diploid in entry 12. The genotype in entry 11 is uninducible, in contrast to the partial diploid of entry 12, which is inducible. The difference between the two genotypes is that in entry 11, the $lacP^-$ mutation is in the same DNA molecule with $lacZ^+$, whereas in entry 12, the $lacP^-$ mutation is combined with $lacZ^-$. This observation means that the presence of a $lacP^-$ mutation makes the adjacent genes uninducible. A $lacP^-$ mutation prevents a *cis*-located wildtype $lacZ^+$ gene from being transcribed. The $lacP^-$ mutations map between *lacI* and *lacO,* and the order of the five genetic elements of the *lac* system is:

*lacI lacP lacO lacZ lacY*

As expected because of the *cis*-dominance of $lacP^-$ mutations, they cannot be complemented; that is, a $lacP^+$ allele on another DNA molecule cannot supply the missing function to a DNA molecule carrying a $lacP^-$ mutation. Thus, *lacP,* like *lacO,* must define a site that determines whether synthesis of *lac* mRNA will take place. Because synthesis does not occur if the site is defective or missing, *lacP* defines an essential site for mRNA synthesis. The *lacP* region is called the **promoter**. It is, in fact, the site in the DNA at which RNA polymerase binds to initiate transcription.

## The Operon System of Transcriptional Regulation

The genetic regulatory mechanism of the *lac* system was first explained by the *operon model* of François Jacob and Jacques Monod, who carried out the pioneering studies discussed previously. The **operon model** is illustrated in **FIGURE 11.4**. (The figure uses the abbreviations *i, o, p, z, y,* and *a* for *lacI, lacO, lacP, lacZ, lacY,* and *lacA.*) The operon model has the following features:

1. The lactose-utilization system consists of two kinds of components—*structural genes* (*lacZ* and *lacY*), which encode proteins needed for the cleavage and transport of lactose, and *regulatory elements* (the repressor gene *lacI,* the promoter *lacP,* and the operator *lacO*).
2. The products of the *lacZ* and *lacY* genes are encoded by a single polycistronic mRNA molecule. (A third protein, encoded by *lacA,* is also translated from the mRNA. This protein is the enzyme transacetylase; it is used in the metabolism of certain $\beta$-galactosides other than lactose and will not be of further concern here.) The linked structural genes, together with *lacP* and *lacO,* constitute the *lac* **operon**.

3. The promoter mutations (*lacP*⁻) eliminate the ability to synthesize *lac* mRNA.
4. The product of the *lacI* gene is a repressor, which binds to a sequence of nucleotides in the DNA constituting the repressor-binding site, which is called the *operator*.
5. When the repressor is bound to the operator, initiation of transcription of *lac* mRNA by RNA polymerase is prevented.
6. Inducers stimulate mRNA synthesis by binding to, and inactivating, the repressor. In the presence of an inducer, the operator is not bound by the repressor, and the promoter is available for the initiation of mRNA synthesis.

Note that regulation of the operon requires that the *lacO* operator either overlap or be adjacent to the promoter of the structural genes, because binding with the repressor prevents transcription. Proximity of *lacI* to *lacO* is not strictly necessary, because the *lacI* repressor is a soluble protein and is therefore diffusible throughout the cell. The presence of inducer has a profound effect on the DNA-binding properties of the repressor; the inducer–repressor complex has an affinity for the operator that is approximately $10^3$ smaller than that of the repressor alone.

The ratio of the numbers of copies of β-galactosidase, permease, and transacetylase is 1.0 : 0.5 : 0.2 when the operon is induced. These differences are partly due to the order of the genes in the mRNA; downstream cistrons are less likely to be translated owing to failure of reinitiation when an upstream cistron has finished translation.

The operon model is supported by a wealth of experimental data and explains many of the features of the *lac* system, as well as numerous other negatively regulated genetic systems in prokaryotes. One aspect of the regulation of the *lac* operon—the effect of glucose—has not yet been discussed. Examination of this feature indicates that the *lac* operon is also subject to positive regulation, as we will see in the next section.

## Positive Regulation of the Lactose Operon

The function of β-galactosidase in lactose metabolism is to form glucose by cleaving lactose. (The other cleavage product, galactose, also is ultimately converted into glucose by the enzymes of the galactose operon.) Glucose is the preferred source of carbon and energy, and when both glucose and lactose are present in the growth medium, transcription of the *lac* operon does not take place, even in *lacI*⁻ mutants, until virtually all of the glucose in the medium has been consumed. This observation reveals a second level of regulation of the lac operon, which is responsive to the concentration of glucose.

The inhibitory effect of glucose on expression of the *lac* operon is indirect, mediated by the small molecule *cyclic adenosine monophosphate* (cAMP) (**FIGURE 11.5**). Cyclic AMP is synthesized by the enzyme *adenyl cyclase,* and the concentration of cAMP is regulated indirectly by glucose metabolism. When bacteria are growing in a medium containing glucose, the cAMP concentration in the cells is quite low. In a medium containing glycerol (or any other carbon source that does not use the glycolytic pathway for its metabolism), or when the bacteria are starved of an energy source, the cAMP concentration is high (**TABLE 11.2**). Glucose levels help regulate the cAMP concentration in the cell, and cAMP regulates the activity of the *lac* operon as well as several other operons that control degradative metabolic pathways.

*E. coli* and many other bacterial species contain a protein, the *cyclic AMP receptor protein* (CRP), encoded by the gene *crp*. Mutations in either *crp* or the adenyl cyclase gene drastically reduce synthesis of *lac* mRNA, which indicates that both CRP protein and cAMP are required for *lac* mRNA synthesis. CRP and cAMP bind to one another, forming a complex denoted **cAMP–CRP** that serves as a transcriptional activator that must bind DNA in the region of the *lac* promoter in order for normal levels of

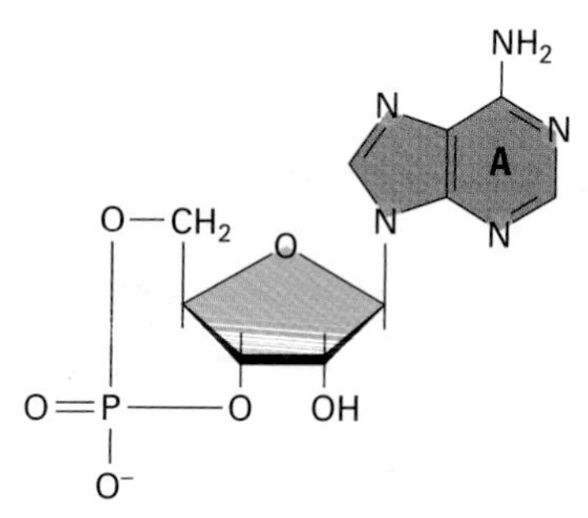


**FIGURE 11.5** Structure of cyclic AMP.

**Table 11.2 Concentration of cyclic AMP in cells growing in media with the indicated carbon sources**

| Carbon source | cAMP concentration |
|---|---|
| Glucose | Low |
| Glycerol | High |
| Lactose | High |
| Lactose + glucose | Low |
| Lactose + glycerol | High |

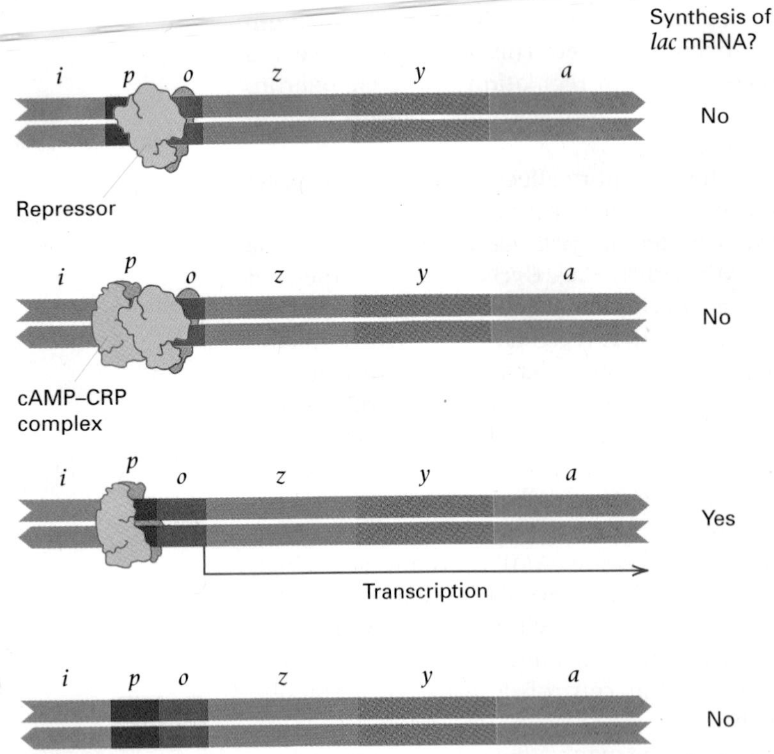


**FIGURE 11.6** Four regulatory states of the *lac* operon. The *lac* mRNA is synthesized only if cAMP–CRP is present and the repressor is absent.

transcription to occur (**FIGURE 11.6**). The requirement for cAMP–CRP is independent of the *lacI* repression system, because $crp^-$ and adenyl cyclase mutants are unable to make normal amounts of *lac* mRNA, even in $lacI^-$ or $lacO^c$ mutants. Unlike the repressor, which negatively regulates the *lac* operon, the cAMP–CRP complex positively regulates the operon.

Experiments carried out *in vitro* with purified *lac* DNA, *lac* repressor, cAMP–CRP, and RNA polymerase have established two further points:

1. In the absence of the cAMP–CRP complex, RNA polymerase binds only weakly to the promoter, but its binding is stimulated when cAMP–CRP is also bound to the DNA. The weak binding rarely leads to initiation of transcription, because the correct interaction between RNA polymerase and the promoter does not occur.
2. If the repressor is bound to the operator, then RNA polymerase cannot stably bind to the promoter.

These results explain how lactose and glucose function together to regulate transcription of the *lac* operon. The relationship of these elements to one another, to the start of transcription, and to the base sequence in the region is depicted in **FIGURE 11.7**.

A great deal is also known about the three-dimensional structure of the regulatory states of the *lac* operon. **FIGURE 11.8** shows how the repressor protein (violet) binds with two operator regions to form a loop that includes the site at which the CRP protein (dark blue) binds. The region of DNA shown in Figure 11.8 corresponds to the region in Figure 11.7 that extends from Operator 3 through Operator 1, and the tabs representing the bases in Figure 11.7 are color coded to match the regions in Figure 11.8. In Figure 11.8, the DNA region in red corresponds, on the right-hand side, to Operator 1 and, on the

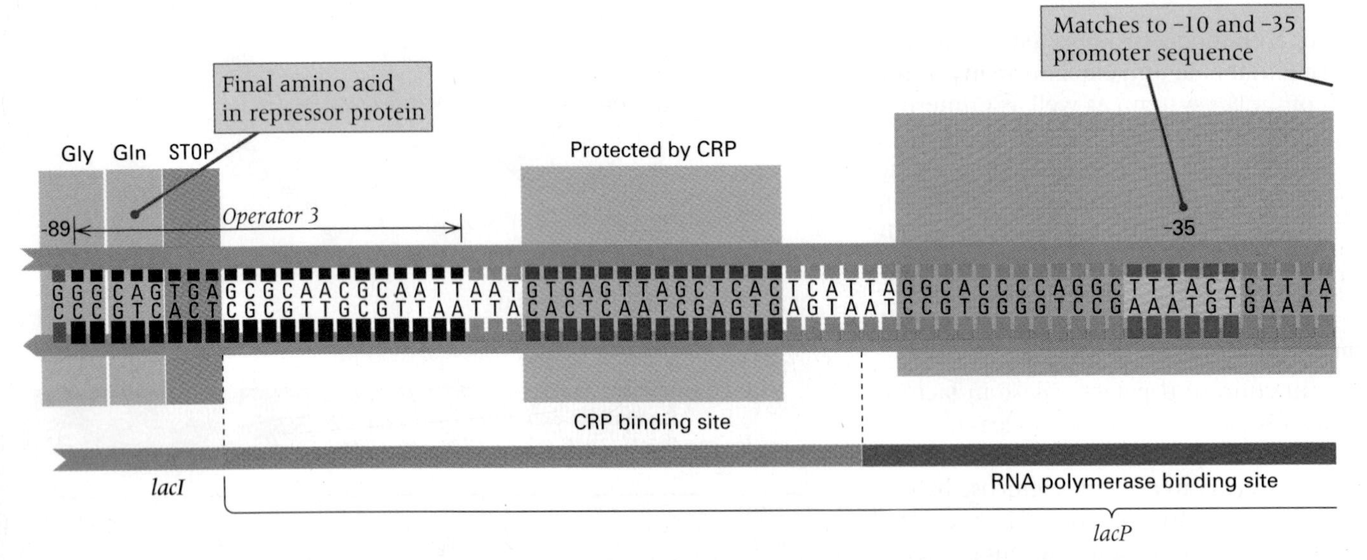


**FIGURE 11.7** (Above and facing page) The nucleotide sequence of the regulatory region of the *lac* operon, showing regions protected from DNase digestion by the binding of various proteins. The end of the *lacI* gene is shown at the extreme left; the ribosome binding site is the site at which the ribosome binds to the *lac* mRNA. The sites for CRP binding and for RNA polymerase binding are indicated along the bottom. The tabs representing the bases are color coded to match those in Figure 11.8.

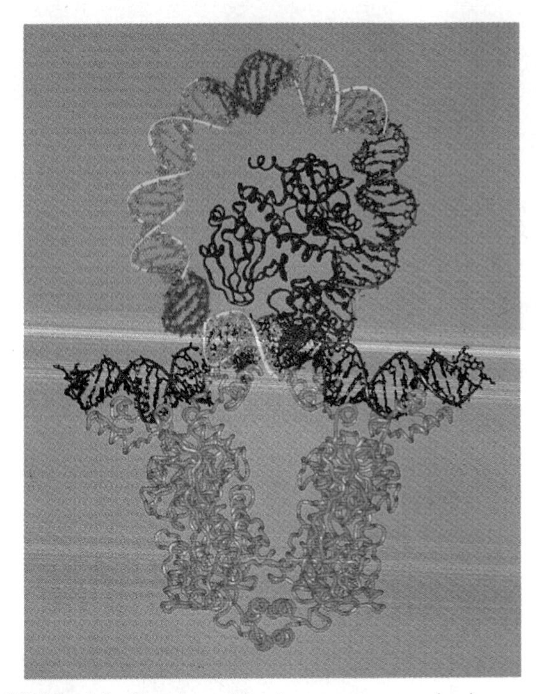

**FIGURE 11.8** Structure of the *lac* operon repression loop. The *lac* repressor, shown in violet, binds to two DNA regions (red) consisting of the symmetrical operator region indicated in Figure 11.7 and a second region immediately upstream from the CRP binding site. Within the loop is the CRP binding site (medium blue), shown bound with CRP protein (dark blue). The −10 and −35 promoter regions are in green. [Reproduced from M. Lewis, et al., *Science* 271 (1996): 1247–1254. Reprinted with permission from AAAS.]

left-hand side, to Operator 3. The *lac* repressor tetramer (violet) binds to these sites. The DNA loop is formed by the region between the repressor-binding sites and includes, in medium blue, the CRP binding site to which the CRP protein (dark blue) is bound. The DNA regions in green are the −10 and −35 sites in the *lacP* promoter. In the configuration in Figure 11.8, the *lac* operon is not transcribed. Removal of the repressor opens up the loop and allows transcription to occur.

DNA sequencing of the *lac* operon revealed the presence of three operator sequences. Operator sequences 1 and 3 are shown in Figure 11.7, whereas Operator 2 is located about 400 nucleotides into the *lacZ* coding region. The three operators have very different efficiencies of repressor binding. Operator 1 and Operator 2 (the one within the *lacZ* gene) bind repressor with high affinity, whereas Operator 3 binds with only about half the affinity of the other two. The most common repressed state of the operon is therefore one in which Operators 1 and 2 are bound with repressor in a conformation similar to that in Figure 11.8, but with a much longer upper loop. Full repression of transcription requires all three operators. The classic genetic experiments of Jacob and Monod identified only Operator 1, which actually accounts for only about 3 percent of the total repression. Nevertheless, that 3 percent was enough for them to carry out their pioneering genetic analysis, leading ultimately to their Nobel Prize in Physiology or Medicine in 1965.

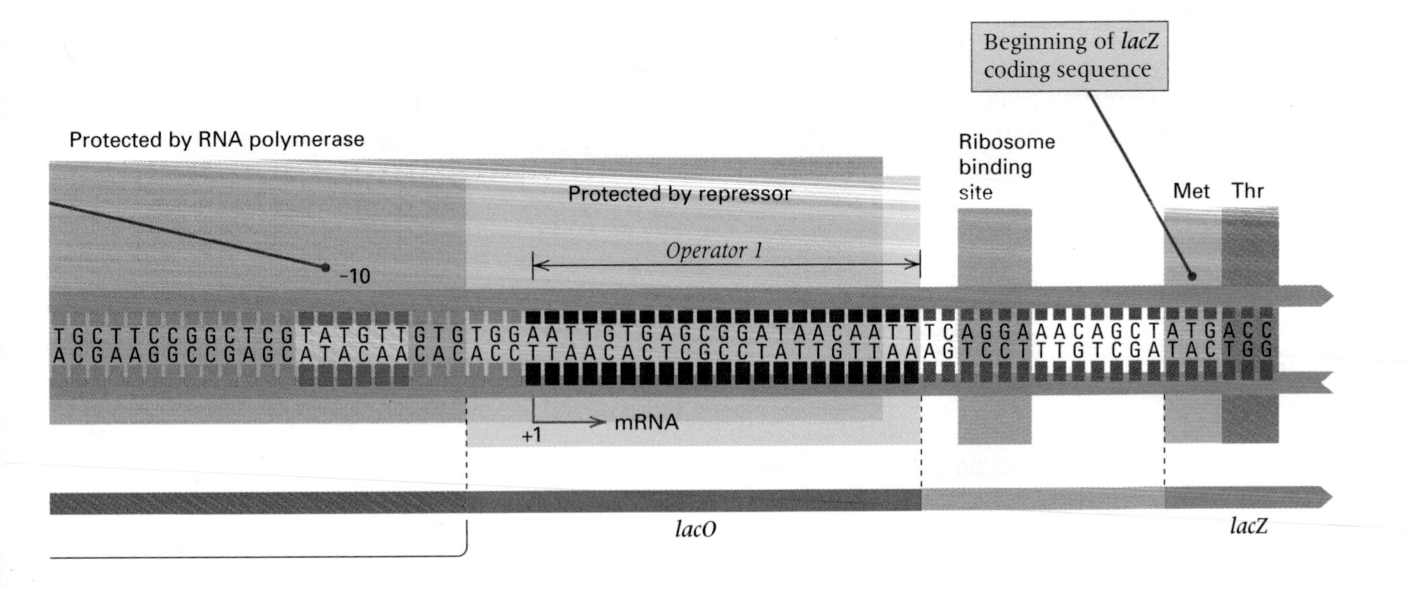

connection

## Operator? Operator?

*How is gene expression controlled? Before Jacob and Monod and their collaborators addressed this question experimentally, it was all a matter of speculation. Prior to this report, the researchers had previously discovered the* i (lacI) *gene that controls expression of the β-galactosidase* (z) *and permease* (y) *genes needed for lactose utilization. They also had strong reason to believe that lacI produces a regulatory protein. How does the regulatory protein work? Here they give evidence that it works by directly binding to a DNA "operator" adjacent to the genes it regulates. Furthermore, the* z *and* y *genes are adjacent and are controlled coordinately by the same "operator" upstream from* z*. The discovery was immediately recognized as fundamental. Jacob and Monod, along with André Lwoff, were awarded the Nobel Prize in 1965. We now know that coordinate regulation via operons is restricted to bacteria. However, the underlying principle—that regulatory genes often control their target genes by direct binding to DNA—is valid for all organisms.*

**François Jacob, David Perrin, Carmen Sanchez, and Jacques Monod 1960**
Institute Pasteur, Paris, France
*The Operon: A Group of Genes Whose Expression Is Coordinated by an Operator* [original in French]

The analysis of different bacterial systems leads to the conclusion that, in the synthesis of certain proteins, there is a dual genetic determination involving two types of genes with distinct functions: one (the gene for structure) is responsible for the structure of the protein molecule, and the other (the regulatory gene) governs the expression of the former through the intermediary action of a repressor. The regulatory genes that have so far been identified show the remarkable property of exercising a *coordinated effect,* each governing the expression of several genes for structure, closely linked together, and corresponding to enzyme proteins belonging to the *same biochemical pathway.* To explain this effect, it seems necessary to invoke a new type of genetic entity, called an "operator," which would be (a) adjacent to the group of genes and would control their activity; and (b) would be sensitive to the repressor produced by a particular regulatory gene. In the presence of the repressor, the expression of the group of genes would be inhibited through the mediation of the repressor. This hypothesis leads to some distinctive predictions concerning mutations that could affect the structure of the operator. (1) Certain mutations affecting an operator would be manifested by the loss of the capacity to synthesize the proteins determined by the group of linked genes "coordinated" by that operator. . . . (2) Other mutations, for example involving a loss of sensitivity (affinity) of the operator for the corresponding repressor, would be manifested by the constitutive synthesis of the proteins determined by the coordinated genes. . . . We have studied certain mutations affecting the metabolism of lactose in *Escherichia coli* that act simultaneously on the synthesis of β-galactosidase [the product of the *z* gene] and galactoside permease [the product of the *y* gene]. . . . The *i* gene is the regulatory gene synthesizing a repressor specific for the system. The genes *i, z* and *y* are closely linked. . . . Constitutive mutants ($o^c$) have now been isolated. [In partially diploid genotypes] only the allele of *z* or *y* that is *cis* with respect to $o^c$ is constitutively expressed. . . . Other mutants have been isolated that have lost the ability to synthesize both the permease and the β-galactosidase. . . . These mutants are recessive. . . . Genetic analysis shows that these mutations ($o^o$) are extremely closely linked to the $o^c$ mutations and that the order of the *lac* region is *i–o–z–y*. . . . The remarkable properties of the $o^c$ and $o^o$ mutations are inexplicable according to the "classical" concept of the genes for structure [*z* and *y*] and distinguish them equally from mutations affecting the regulatory gene *i*. On the other hand, they conform to the predictions arising from the hypothesis of the operator.

***It seems necessary to invoke a new type of genetic entity, called an "operator," which would be (a) adjacent to the group of genes and would control their activity; and (b) would be sensitive to the repressor produced by a particular regulatory gene.***

Source: F. Jacob, et al., *Comptes Rendus Hebd. Seances Acad. Sci.* 250 (1960): 1727–1729. Translated in E. A. Adelberg, *Papers in Bacterial Genetics.* Little, Brown and Company, Boston, MA (1966).

### ■ Regulation of the Tryptophan Operon

The tryptophan (*trp*) operon of *E. coli* serves as an example of negative regulation in which transcription is repressible (Figure 11.1C). The *trp* operon contains structural genes for enzymes that synthesize the amino acid tryptophan. When adequate tryptophan is present in the growth medium, transcription of the operon is repressed, and when the supply of tryptophan is insufficient, transcription takes place.

Tryptophan is synthesized in five steps, each requiring a particular enzyme. The genes coding for these enzymes are adjacent and are in the same linear order in the *E. coli* chromosome as the order in which the enzymes function in the biosynthetic pathway. The genes are called *trpE, trpD, trpC, trpB,* and *trpA,* and the enzymes are translated from a single polycistronic mRNA molecule. The *trpE* coding region is the first one translated. Upstream (on the 5′ side) of *trpE* are the promoter, the operator, and two regions called the *leader* and the *attenuator,* which are designated *trpL* and *trp a* (not *trpA*), respectively (**FIGURE 11.9**). The repressor gene, *trpR,* is located quite far from this operon.

The product of the *trpR* gene is an aporepressor protein, which requires a corepressor to become active. Mutations in either *trpR* or the operator cause constitutive initiation of transcription of *trp* mRNA. The role of tryptophan is that of a corepressor, which binds with the *trp* aporepessor to form the active repressor. Binding of the repressor to the *trp* operator shuts off synthesis. This basic on–off regulatory mechanism is outlined in **FIGURE 11.10**. When there is not enough tryptophan, the aporepressor adopts a

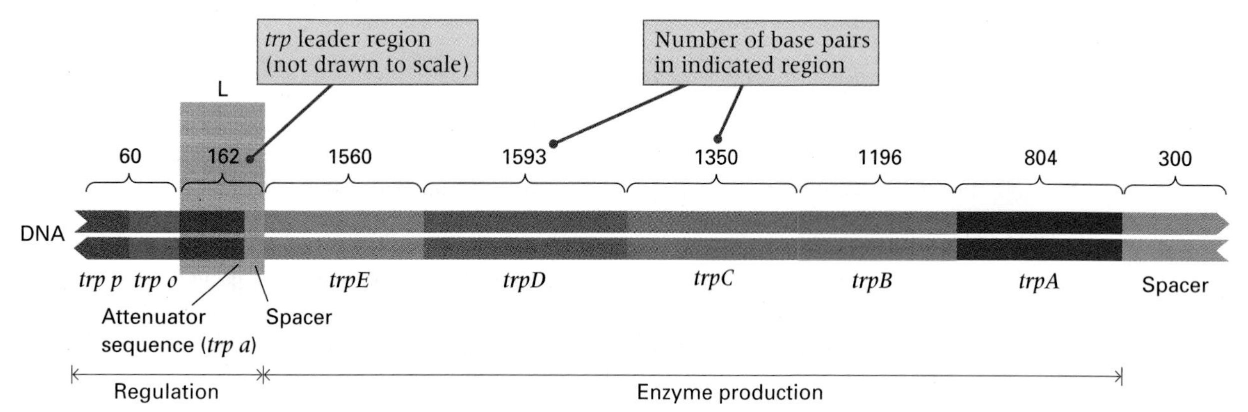


**FIGURE 11.9** The *E. coli trp* operon. For clarity, the regulatory region is enlarged with respect to the coding region. The actual sizes of the regions are indicated by the numbers of base pairs. Region *L* is the leader.

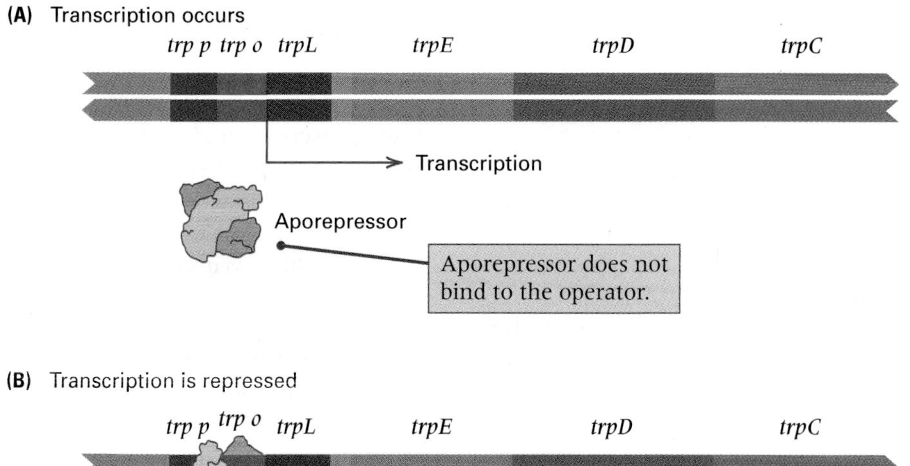


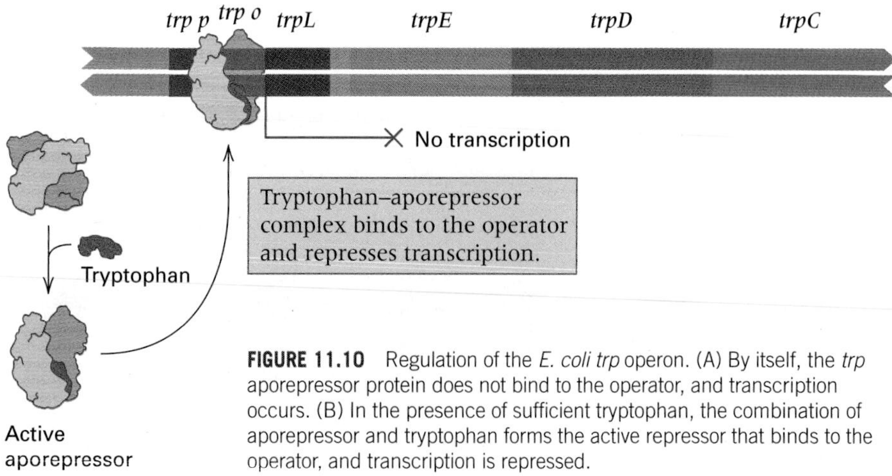


**FIGURE 11.10** Regulation of the *E. coli trp* operon. (A) By itself, the *trp* aporepressor protein does not bind to the operator, and transcription occurs. (B) In the presence of sufficient tryptophan, the combination of aporepressor and tryptophan forms the active repressor that binds to the operator, and transcription is repressed.

three-dimensional conformation unable to bind with the *trp* operator, and the operon is transcribed (Figure 11.10A). On the other hand, when tryptophan is present at high enough concentration, some molecules bind with the aporepressor and cause it to change conformation into the active repressor. The active repressor binds with the *trp* operator and prevents transcription (Figure 11.10B). Thus only when tryptophan is present in sufficient amounts is the active repressor molecule formed.

## 11.3 Regulation Through Transcription Termination

The mechanism of repressible transcription in Figure 11.10 provides a coarse level of regulation of the *trp* operon in that transcription can be either "off" or "on." There is another level of regulation that provides for fine tuning, which is employed when some tryptophan is present in the growth medium, but not enough to sustain optimal growth. Under these conditions, it is advantageous for cells to synthesize tryptophan, but at less than the maximum possible rate. The *trp* operon responds to this situation by means of a regulatory mechanism in which the amount of transcription in the derepressed state is controlled quantitatively by the concentration of tryptophan in the cell. This regulatory mechanism is called *attenuation,* and it is found in many operons responsible for amino acid biosynthesis.

### ■ Attenuation

The **attenuation** mechanism controls whether transcription, once started, will continue through the operon or be terminated prematurely. In wildtype cells, transcription of the *trp* operon is often initiated. However, in the presence of even small amounts of tryptophan, most of the mRNA molecules terminate in a specific 28-base region within the leader sequence (Figure 11.9). The result of termination is an RNA molecule containing only 140 nucleotides that stops short of the genes coding for

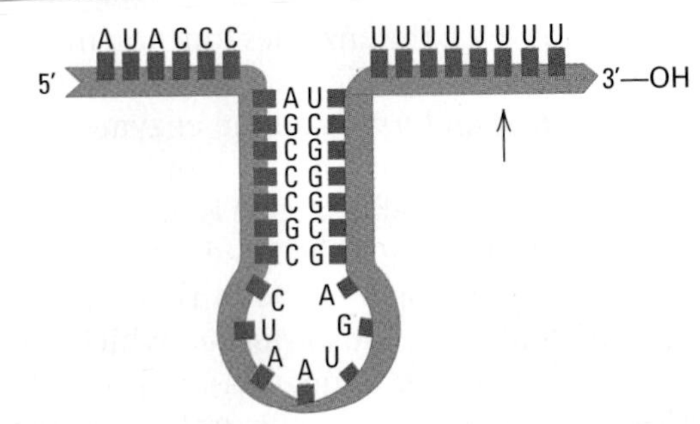


**FIGURE 11.11** The terminal region of the *trp* attenuator sequence. The arrow indicates the final uridylate in attenuated RNA. Nonattenuated RNA continues past this point. The nucleotides in red letters form a stem by base-pairing within the RNA.

the *trp* enzymes. The 28-base region in which termination occurs is called the **attenuator.** The nucleotide sequence of the attenuator region contains the usual features of a termination site, including a potential stem-and-loop configuration in the mRNA followed by a sequence of eight uridylates (**FIGURE 11.11**).

In *E. coli,* termination of transcription is determined by whether a small peptide encoded in the leader sequence can be translated. This coding sequence, shown in **FIGURE 11.12**, specifies a **leader polypeptide** 14 amino acids in length, and it includes two adjacent tryptophan codons at positions 10 and 11. When there is sufficient charged tryptophan tRNA to allow translation of these codons, the nascent transcript adopts a conformation in which the attenuator is exposed, and transcription is terminated. On the other hand, when there is insufficient charged tryptophan tRNA to allow translation of the leader polypeptide, the ribosome stalls at the tryptophan codons, which creates a situation in which the attenuator is hidden, and transcription continues throughout the entire operon.

The mechanism of attenuation is diagrammed in **FIGURE 11.13.** Part A shows the leader RNA molecule, including the two tryptophan codons in the leader polypeptide. Region 2 has a nucleotide sequence that enables it to pair either with region 1 or with re-

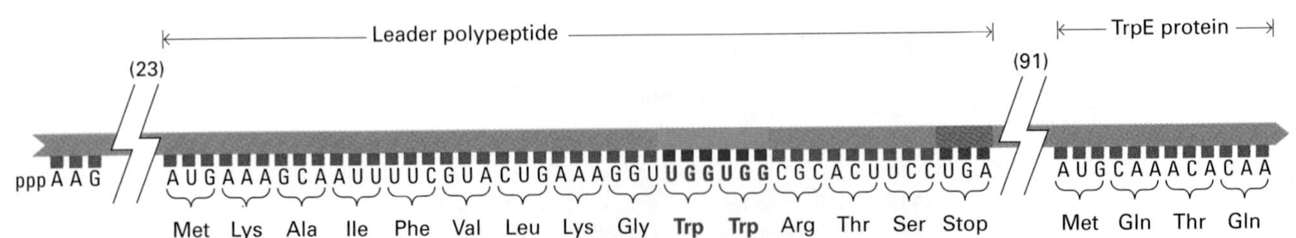


**FIGURE 11.12** The sequence of nucleotides in the *trp* leader mRNA, showing the leader polypeptide, the two tryptophan codons (red), and the beginning of the TrpE protein. The numbers 23 and 91 are the numbers of nucleotides in the sequence that, for clarity, are not shown.

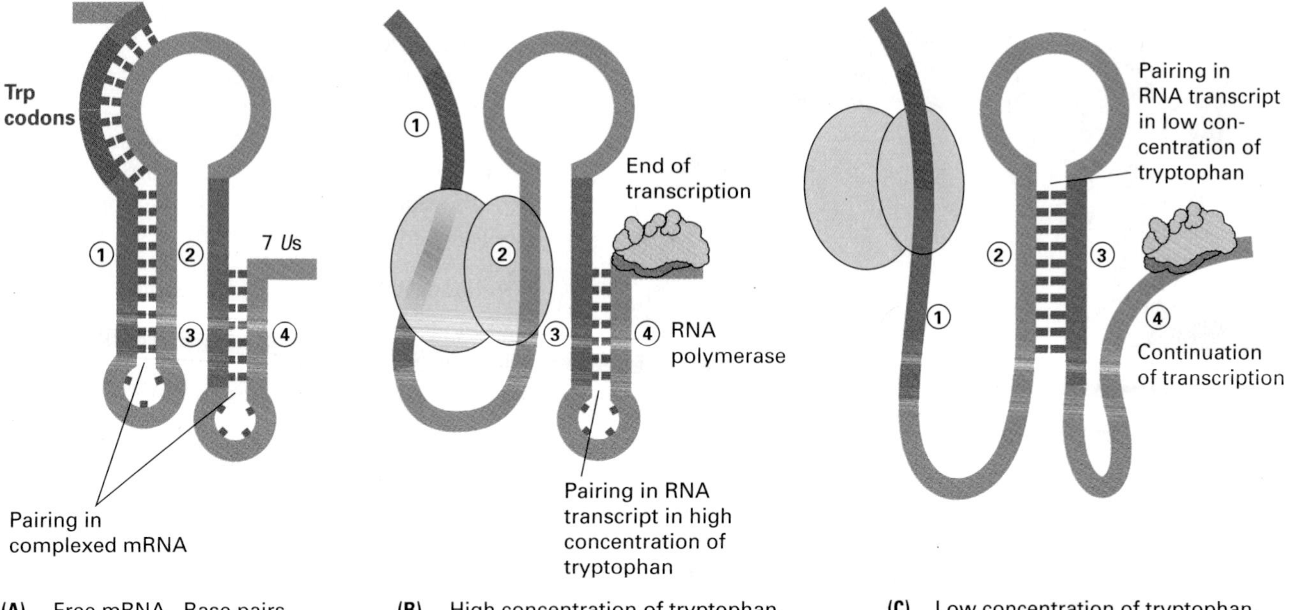


**(A)** Free mRNA. Base pairs between 1 and 2 and between 3 and 4.

**(B)** High concentration of tryptophan. Ribosome reaches region 2 and pairing of 3–4 causes termination of transcription.

**(C)** Low concentration of tryptophan. Ribosome stalled in region 1 at Trp codons permits pairing of 2–3 and transcription is not terminated.

**FIGURE 11.13** The explanation for attenuation in the *E. coli trp* operon. The tryptophan codons are highlighted in red.

gion 3. In the purified RNA, region 1 pairs with region 2, and region 3 pairs with region 4. Part B shows the configuration in a cell in which there is sufficient $tRNA^{Trp}$ to allow translation of the leader polypeptide. The ribosome moves beyond the Trp codons and blocks region 2, so the pairing that forms is between region 3 and region 4; this creates the transcriptional terminator, with termination occurring at the run of uridylates that follows region 4. Part C shows what happens when the ribosome stalls at the Trp codons as a result of insufficient $tRNA^{Trp}$. In this case, region 2 preferentially pairs with region 3, which disrupts the conformation of the terminator, allowing transcription to continue through the rest of the operon. The fine tuning of this system takes place at intermediate concentrations of tryptophan, when the fraction of nascent transcripts that are completed depends on how frequently translation is stalled, which in turn depends on the intracellular concentration of charged tryptophan tRNA.

Some bacteria have evolved a mechanism of regulation of the *trp* operon in which the intracellular concentration of tryptophan is detected in a manner that does not depend on translation. A particularly elegant example is found in *Bacillus subtilis.* In this organism, the *trp* operon is not repressible. There is no aporepressor, nor is there a short polypeptide encoded in the leader region. Regulation of transcription is controlled solely through attenuation, which in this case is mediated by a protein denoted TRAP (*trp* RNA-binding attenuation protein). Each monomer of the TRAP protein binds a molecule of tryptophan and, in so doing, undergoes a conformational change that enables it to bind with other monomers. The active form of TRAP is a radially symmetric 11-mer, which binds specifically with nucleotide sequences in the nascent mRNA in the regions corresponding to 1 and 2 in Figure 11.13B. The resulting structure, with regions 1 and 2 wrapped around the outside of the TRAP 11-mer, is shown in **FIGURE 11.14**. When regions 1 and 2 are trapped in this manner, region 3 pairs with region 4, much as is shown in Figure 11.13B, forming the transcriptional terminator. In the absence of the TRAP multimer, the terminator is not formed and transcription continues. The system is sensitive to the intracellular concentration of tryptophan; if the level is high enough to allow TRAP multimers to form, transcription is terminated, but if it is too low for TRAP multimers to form, the complete *trp* operon is transcribed.

In *E. coli,* many operons responsible for amino acid biosynthesis (for example, the leucine, isoleucine, phenylalanine, and histidine operons) are regulated by attenuators that function by forming alternative paired regions in the transcript. In the histidine operon, the coding region for the leader polypeptide contains seven adjacent histidine

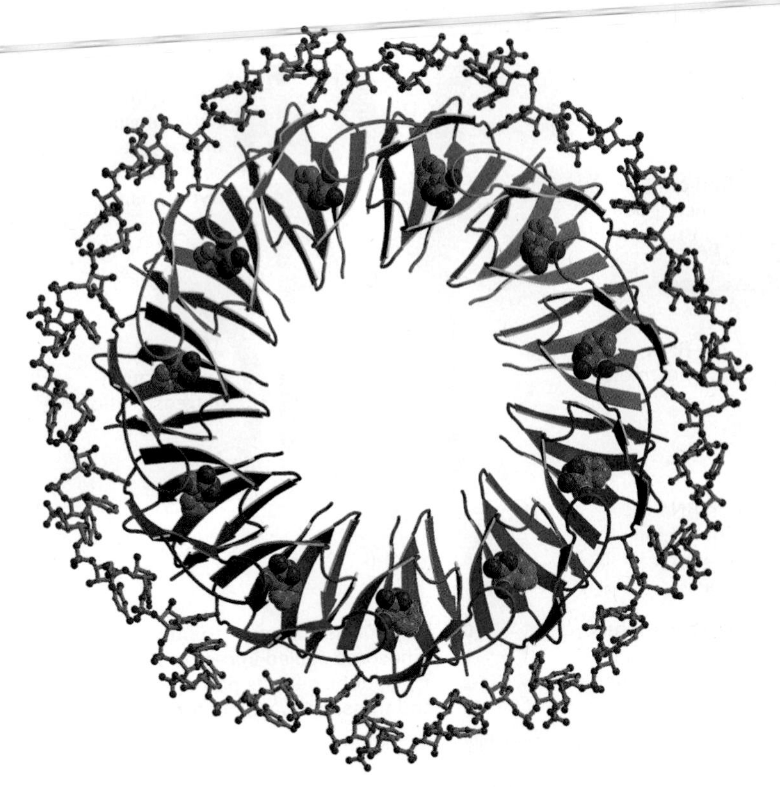

**FIGURE 11.14** Structure of TRAP protein (the tryptophan RNA-binding attenuation protein from *B. subtilis*) bound to trp RNA. The protein 11-mer is shown as a ribbon with each subunit in a different color, and the RNA is shown as a ball-and-stick model. The solid spheres in the inside of the TRAP subunits correspond to atoms of the tryptophan molecules. [Courtesy of Paul Gollnick, State University of New York at Buffalo, USA, and Alfred Antson, University of York, UK.]

codons (**FIGURE 11.15**, part A). In the phenylalanine operon, the coding region for the leader polypeptide contains seven phenylalanine codons divided into three groups (Figure 11.15B). This pattern, in which codons for the amino acid produced by enzymes of the operon are present at high density in the leader peptide mRNA, is characteristic of operons in which attenuation is coupled with translation. Through these codons, the cell monitors the level of aminoacylated tRNA charged with the amino acid that is the end product of each amino acid biosynthetic pathway.

Regulation of the *lac* and *trp* operons exemplifies some of the important mechanisms that control transcription of genes in prokaryotes. In the following section, we will see an example of how the leader sequence in a transcript can control transcriptional termination.

## ■ Riboswitches

Regulation by means of transcription termination is a relatively common mechanism for controlling gene expression in prokaryotes. Examples in which transcription termination is triggered by direct binding of a small molecule to a 5′ untranslated leader mRNA have emerged. The 5′ leader is able to adopt either of two conformations according to whether it binds with the small molecule. In the *antiterminator* conformation, transcription of the gene continues past the leader and through the remaining part of the gene. In the *terminator* conformation, which is triggered by binding with the small molecule, transcription is terminated. An RNA leader sequence able to switch between an antiterminator conformation and a terminator conformation is known as a **riboswitch**.

**FIGURE 11.16**, part A depicts the leader mRNA of the *yitJ* gene, which is involved in methionine biosynthesis in *Bacillus subtilis*. Nucleotides shown in red and blue can undergo two pairing configurations. In one, they form stem-and-loop structures of an anti-antiterminator (AAT) and a terminator (T), the latter followed by a string of uridylate residues. In this conformation, the RNA leader terminates transcription. The red and blue nucleotides can also pair with each other to form an antiterminator (AT) as shown at the upper right. In this conformation, transcription continues through the *yitJ* gene. The presence of S-adenosylmethionine (SAM), a modified form of methionine, results in conversion of the leader RNA from the read-through form (AT loop) to the terminator form (AAT and T loops) (Figure 11.16B). The 5′ leader RNA that binds S-adenosylmethionine is referred to as the *S box* RNA. S box RNA structures are found upstream of 11 transcriptional units that encode 26 proteins for sulfur metabolism in *B. subtilis*.

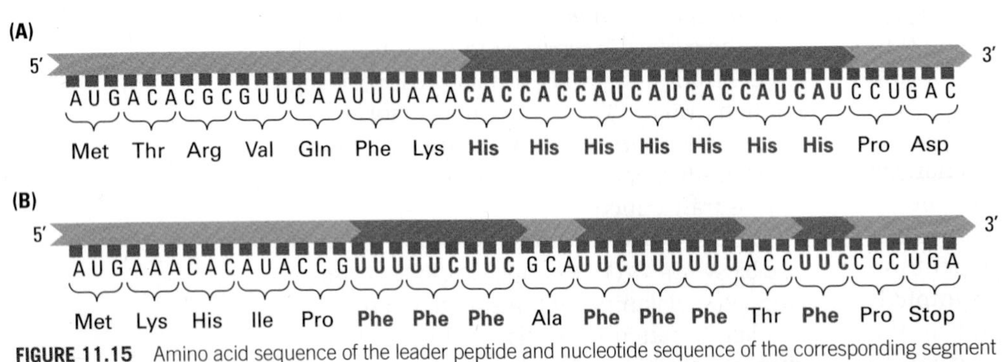


**FIGURE 11.15** Amino acid sequence of the leader peptide and nucleotide sequence of the corresponding segment of mRNA from the histidine operon (A) and the phenylalanine operon (B). The repetition of these amino acids is emphasized in red.

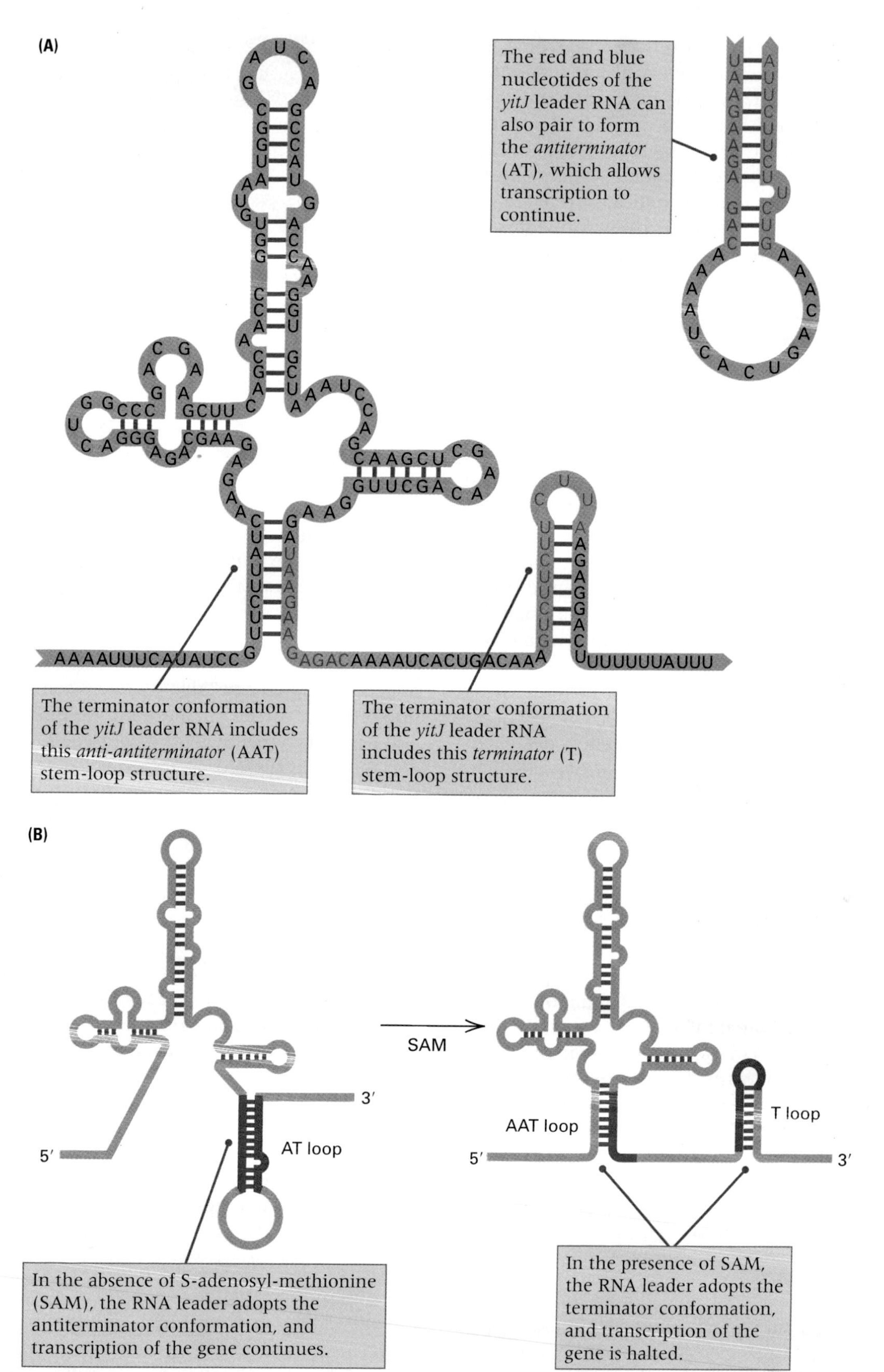

(A)
The red and blue nucleotides of the yitJ leader RNA can also pair to form the antiterminator (AT), which allows transcription to continue.
The terminator conformation of the yitJ leader RNA includes this anti-antiterminator (AAT) stem-loop structure.
The terminator conformation of the yitJ leader RNA includes this terminator (T) stem-loop structure.
(B)
SAM
AT loop
AAT loop
T loop
5′
3′
In the absence of S-adenosyl-methionine (SAM), the RNA leader adopts the antiterminator conformation, and transcription of the gene continues.
In the presence of SAM, the RNA leader adopts the terminator conformation, and transcription of the gene is halted.


**FIGURE 11.16** Riboswitch regulation of transcription termination by the *yitJ* leader RNA in *Bacillus subtilis*. (A) Structural model of the untranslated leader RNA of the *yitJ* gene is depicted in the terminator (T) conformation, which includes an anti-antiterminator (AAT) stem-loop structure. The red and blue nucleotides in the AAT and T stems can also pair to form an antiterminator (AT). (B) The presence of S-adenosylmethionine (SAM) results in conversion from the read-through (AT) form to the terminator (AAT +T) form. [After B. A. M. McDaniel et al. 2003. *Proceedings of the National Academy of Sciences USA* 100: 3083.]

Riboswitches have also been discovered in which the leader RNA binds flavin mononucleotide, thiamine pyrophosphate, vitamin $B_{12}$, guanine, adenine, uncharged tRNA, S-adenosylmethionine, or lysine to regulate synthesis or transport of the corresponding metabolites. DNA sequence analysis indicates that riboswitches are present in all three kingdoms—Archaea, Eubacteria, and Eukarya.

**Table 11.3** Characteristics of mixed infections containing several combinations of $\lambda$ *vir*, *cI*$^-$, *cII*$^-$, and *cIII*$^-$ mutants

| Infecting phages | Clear or turbid plaques |
|---|---|
| 1. $\lambda$ *vir* + $\lambda^+$ | Clear |
| 2. *cI*$^-$ + *cI*$^+$ | Turbid |
| 3. *cII*$^-$ + *cII*$^+$ | Turbid |
| 4. *cIII*$^-$ + *cIII*$^+$ | Turbid |

## 11.4 Regulation in Bacteriophage Lambda

When Jacob and Monod first proposed the operon model featuring negative regulation by a repressor protein, they suggested that the model could account not only for regulation in inducible and repressible operons for metabolic enzymes, but also for the lysogenic cycle of temperate bacteriophages. They proposed that $\lambda$ bacteriophage was kept quiescent and prevented from replicating within bacterial lysogens by a repressor. This explanation ultimately proved to be correct, although the biochemical mechanism of attaining the repressed, lysogenic state is more complicated than they imagined.

When $\lambda$ bacteriophage infects *E. coli*, each infected cell can undergo one of two possible outcomes:

1. A lytic infection, resulting in lysis and production of phage particles, or
2. A lysogenic infection, resulting in integration of the $\lambda$ molecule into the *E. coli* chromosome and formation of a lysogen.

Because of this dichotomy, $\lambda$ normally produces turbid (not completely clear) plaques on a lawn of *E. coli*. The initial infection and lysis produce a cleared region in the bacterial lawn, but a few lysogens grow within the cleared region, partially repopulating the cleared zone and producing a turbid plaque.

Mutations in regulatory genes in $\lambda$ were first identified in phage mutants that give clear rather than turbid plaques. The mutants proved to fall into four classes: $\lambda$*vir*, *cI*$^-$, *cII*$^-$, and *cIII*$^-$. The characteristics of these mutants are shown in **TABLE 11.3**. The genetic positions of the *cI*, *cII*, and *cIII* regions are shown in the simplified genetic map of $\lambda$ bacteriophage in **FIGURE 11.17**, in which the genes are grouped by functional categories. Recall from Chapter 9 that upon infection, the $\lambda$ DNA molecule circularizes, bringing the *R* and *A* genes adjacent to one another.

Among the mutants in Table 11.3, the "clear" mutants proved to be analogous to *lacI* and *lacO* mutants in *E. coli*. The $\lambda$*vir* mutant is dominant to the wildtype $\lambda^+$ in mixed infection, as indicated by the combination of infecting phage designated 1 in Table 11.3; this combination of phage carries out a productive infection and prevents lysogeny by the wildtype $\lambda^+$. The $\lambda$*vir* mutant is therefore analogous to the *lacO*$^c$ mutation. However, $\lambda$*vir* proves to be a *double* mutant, bearing mutations in two different operators, $O_L$ and $O_R$, as depicted in **FIGURE 11.18**. The *cI*$^-$ mutations are recessive, as can be seen in entry 2 in Table 11.3. These *cI*$^-$ mutations are analogous to *lacI*$^-$ mutations in that the *cI*$^+$ gene encodes the $\lambda$ repressor, which is diffusible. The *cII*$^+$ and *cIII*$^+$ genes encode not for repressor but, rather, for proteins needed in establishing lysogeny. The *cII*$^-$ and *cIII*$^-$ mutations are also recessive in mixed infections (entries 3 and 4 in Table 11.3).

The molecular basis on which the decision between the lysogenic and the lytic cycle is determined can be explained with reference to Figure 11.18. Upon infection of *E. coli* by $\lambda$, the $\lambda$ molecule circularizes, and RNA polymerase binds at $P_L$ and $P_R$ and initiates transcription of

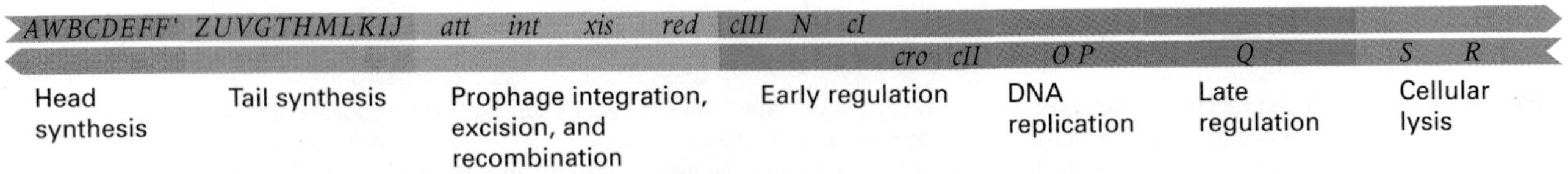

**FIGURE 11.17** Genetic map of $\lambda$ bacteriophage. The map is drawn to emphasize the functional organization of genes within the phage genome and to call attention to the regulatory features.

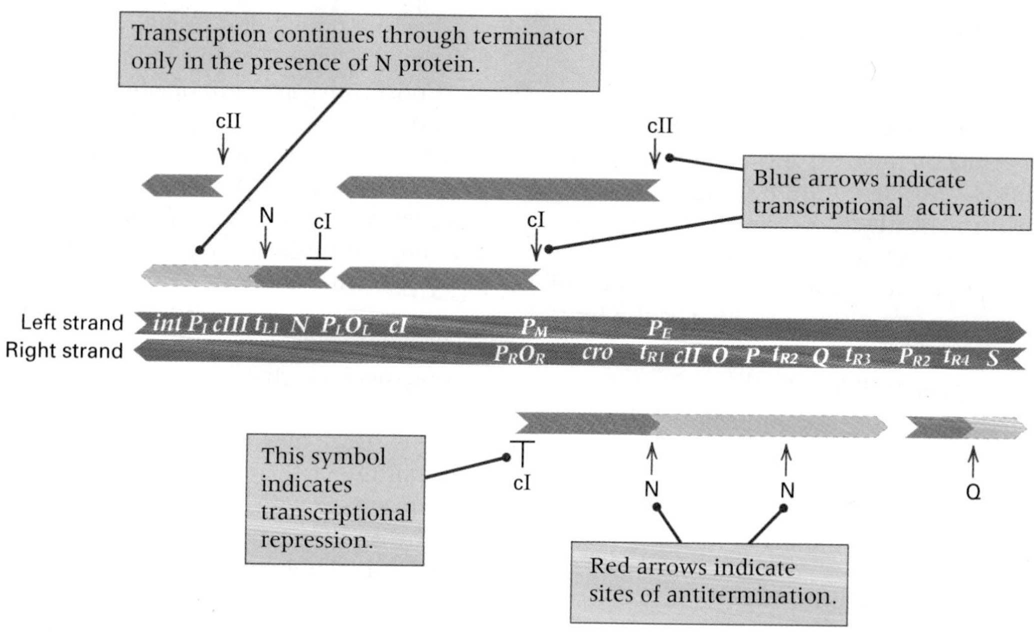


**FIGURE 11.18** Genetic and transcriptional map of the control region of bacteriophage λ as expressed in the early stages of a lysogenic infection. The green arrows show the origin, direction, and extent of transcription. Light green arrows indicate portions of transcripts that are synthesized as a result of antitermination activity of the N or Q proteins. The sites of antitermination activity are indicated with red arrows pointing to the interior of a transcript. Blue arrows pointing to the origin of a transcript indicate transcriptional activation by cI or cII proteins, and the sites of transcriptional repression by cI protein are indicated.

the *N* and *cro* genes. The N protein acts to prevent termination of the transcripts from $P_L$ and $P_R$, thus allowing production of cII protein; cII protein activates transcription at $P_E$ and $P_I$, thus allowing production of the cI and int (integration) proteins. The cI protein shuts down further transcription from $P_L$ and $P_R$ and stimulates transcription at $P_M$, increasing its own synthesis in a positive autoregulatory loop. Lysogeny is achieved if the concentration of cI protein reaches a level high enough to prevent transcription from $P_L$ and $P_R$ and to allow the int protein to catalyze site-specific recombination between the circular λ molecule and the *E. coli* chromosome at their respective attachment (*att*) sites. The result is the formation of a λ lysogen.

The alternative pathway is lytic development, which takes place when the cro protein dominates. The cro protein also can bind to $O_R$ and, in doing so, blocks transcription from $P_M$. If this occurs, then the concentration of repressor cannot rise to the level required to block transcription from $P_L$ and $P_R$. Transcription continues from $P_R$ and $P_{R2}$, and N and Q proteins prevent termination in the rightward (and also leftward) transcripts. Because the λ DNA molecule is in a circular configuration, rightward transcription moves through genes *S* and *R* and thence through the head and tail genes (*A* through *J* in Figure 11.17). The production of proteins needed for cellular lysis and formation of phage particles ensues, followed by phage assembly and cellular lysis to release phage.

In effect, the regulation of lysogeny in λ involves a sort of "race" between the production of the cro and cI proteins: The "winner" is whichever protein first achieves a concentration high enough to dominate the outcome of infection. The underlying molecular mechanism is that the cro and cI proteins compete for binding to $O_L$ and $O_R$ (each operator has three protein-binding sites that participate in the competition). If cro wins the contest, the lytic cycle results in that particular cell; if cI wins, a lysogen is formed. Hence,

> The cI and cro proteins function as a *genetic switch*: cI turns on lysogeny, and cro turns on the lytic cycle.

Additional factors that determine whether cI or cro controls the outcome of a particular infection include the nutritional state of the infected cell. This in turn influences the stabilities and levels of the regulatory proteins. Thus, the regulation of λ, an apparently "simple" system, is actually quite complex.

# 11.5 Transcriptional Regulation in Eukaryotes

Many eukaryotic genes are **housekeeping genes** that encode essential metabolic enzymes or cellular components and are expressed constitutively at relatively low levels in all cells. Other genes, which differ in their expression according to cell type or stage of the cell cycle, are often regulated at the level of transcription. Typically, levels of expression of eukaryotic genes may differ twofold to tenfold between the uninduced and induced levels, in contrast to the more dramatic differences seen in prokaryotes, in which the ratio between the uninduced and induced levels may be as great as a thousandfold.

## Galactose Metabolism in Yeast

We introduce transcriptional regulation in eukaryotes by examining the control of galactose metabolism in yeast and comparing it with the *lac* operon in *E. coli*. The first steps in the biochemical pathway for galactose degradation are illustrated in **FIGURE 11.19**, part A. Three enzymes, encoded by the genes *GAL1*, *GAL7*, and *GAL10*, are required for conversion of galactose to glucose-1-phosphate. These three genes are tightly linked, as shown in Figure 11.19B. Despite the tight linkage of the three genes, the genes are not part of an operon; the mRNAs are monocistronic. The *GAL1* and *GAL10* mRNAs are synthesized from divergent promoters lying between the genes, and *GAL7* mRNA is synthesized from its own promoter. The mRNAs are synthesized only when galactose is present as inducer; the genes are thus inducible.

Constitutive and uninducible mutants have been observed. In two types of mutants, *gal80* and *GAL81*$^c$, the mutants synthesize *GAL1*, *GAL7*, and *GAL10* mRNAs constitutively. Another type of mutant, *gal4*, is uninducible; it does not synthesize the mRNAs whether or not galactose is present. The characteristics of the mutants are shown in **TABLE 11.4**. The terms *cis* and *trans* are of no help in interpreting these results, because the regulatory genes are unlinked to the genes they regulate—*GAL1*, *GAL7*, and *GAL10* are on chromosome II, *GAL80* is on chromosome XIII, and *GAL4* and *GAL81* are on chromosome XVI.

The *gal80* mutation is recessive (entry 1 of Table 11.4). Thus, superficially, it behaves like a *lacI*$^-$ mutation. The wildtype *GAL80* allele does indeed encode a protein that negatively regulates transcription, and although the GAL80 protein is called a "repressor," it acts not by binding to an operator but by binding to a transcriptional activator protein and rendering it inactive. The activator that GAL80 binds to is the product of the *GAL4* gene, denoted GAL4.

The GAL4 protein is required for transcription of all three *GAL* genes. The *gal4* mutation is recessive, and in homozygous mutants the *GAL* genes are all uninducible. The GAL4 protein is a positive regulatory protein that activates transcription of the three *GAL* genes individually, starting at a different site upstream from each gene. The GAL4 protein bound with

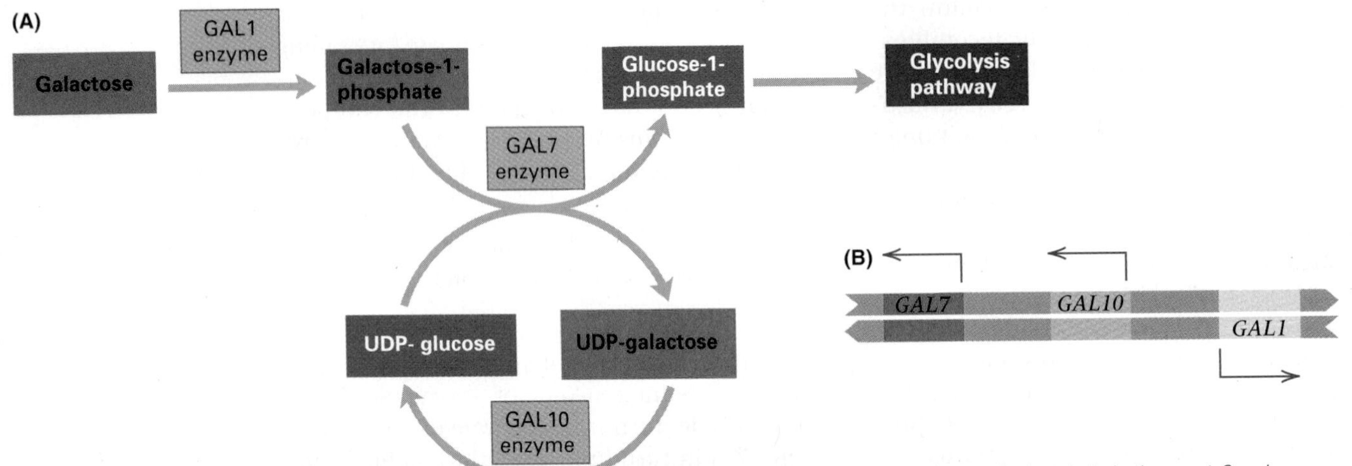


**FIGURE 11.19** (A) Metabolic pathway by which galactose is converted to glucose-1-phosphate in the yeast *Saccharomyces cerevisiae*. (B) Location of the *GAL* genes along chromosome II. Arrows indicate the direction of transcription. The *GAL1* and *GAL10* transcripts come from divergent promoters. *GAL7* has its own promoter. The enzymes corresponding to the GAL1, GAL7, and GAL10 proteins are, respectively, galactokinase, galactose-1-phosphate uridyl transferase, and UDP-glucose-4-epimerase.

**Table 11.4 Characteristics of diploids containing various combinations of *gal80*, *gal4*, and *GAL81*ᶜ mutations**

| Genotype | Synthesis of *GAL1*, *GAL7*, and *GAL10* mRNAs | Gal phenotype |
|---|---|---|
| 1. *gal80 GAL1*/*GAL80 GAL1* | Inducible | + |
| 2. *gal4 GAL1*/*GAL4 GAL1* | Inducible | + |
| 3. *GAL81ᶜ GAL1*/*GAL81 GAL1* | Constitutive | + |

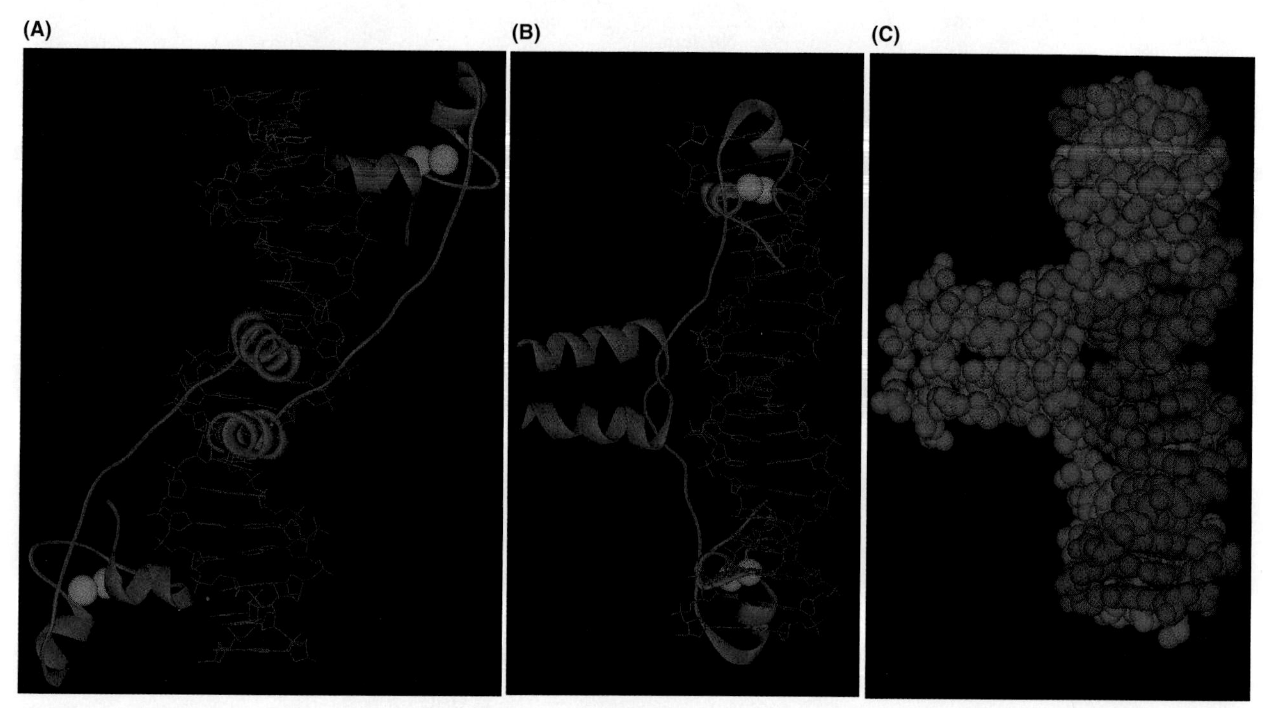


**FIGURE 11.20** Three-dimensional structure of the GAL4 protein (blue) bound to DNA (red). The protein is composed of two polypeptide subunits held together by the coiled regions in the middle. The DNA-binding domains are at the extreme ends, and each physically contacts three base pairs in the major groove of the DNA. The zinc ions in the DNA-binding domains are shown in yellow. The views in (A) and (B) are at right angles; (C) is a space-filling model. [Protein Data Bank 1D66. R. Marmorstein, et al., *Nature* 356 (1992): 408–414.]

its target site in the DNA is shown in **FIGURE 11.20**, in which the GAL4 protein (a dimer) is shown in blue and the DNA molecule in red. The small yellow spheres represent ions of zinc, which are essential components in the DNA binding.

The mechanism of *GAL* regulation is outlined in Figure 11.21. The symbols correspond to conventions in yeast genetics, in which the product of a gene, for example the gene *GAL3*, is denoted by the gene name with the suffix *p* added, in this case GAL3p. (Note that the gene symbol *GAL3* is printed in italics because it represents a gene, whereas the gene-product symbol GAL3p is not in italics.)

In *GAL* gene regulation, the key players are the products of the genes *GAL3*, *GAL80*, and *GAL4*. Although GAL80p superficially resembles a repressor protein because *gal80*/*gal80* homozygous mutants produce the GAL enzymes constitutively, the GAL80p protein does not bind with DNA. Rather, it has two binding sites, one for GAL4p (the transcriptional activator) and one for GAL3p.

In the presence of galactose (**FIGURE 11.21**, part A), GAL3p binds with galactose and ATP, and in this state GAL3p can bind with GAL80p and hold it in the cytoplasm. Inside the nucleus, the GAL4p protein attaches through one of its binding sites with an upstream activator sequence (UAS) located near each of the *GAL* genes. Another binding site on GAL4p recruits the transcriptional machinery, and the *GAL* genes are transcribed.

In the absence of galactose (part B), the GAL3p protein cannot bind with GAL80p. The GAL80p protein is therefore free to enter the nucleus. Inside the nucleus, GAL80p binds with the transcriptional activator site in GAL4p and thereby prevents recruitment of the

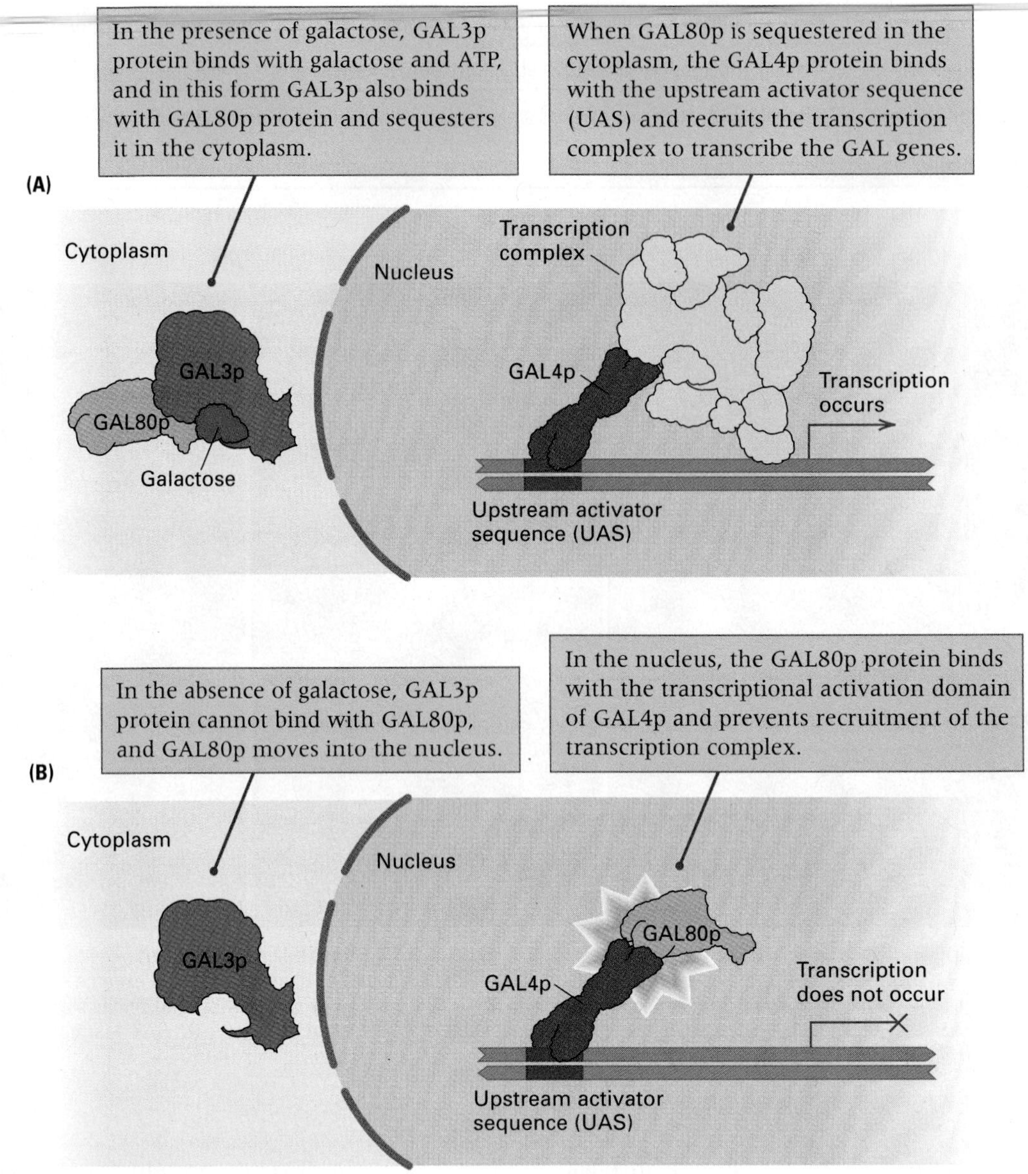

**(A)**

In the presence of galactose, GAL3p protein binds with galactose and ATP, and in this form GAL3p also binds with GAL80p protein and sequesters it in the cytoplasm.

When GAL80p is sequestered in the cytoplasm, the GAL4p protein binds with the upstream activator sequence (UAS) and recruits the transcription complex to transcribe the GAL genes.

**(B)**

In the absence of galactose, GAL3p protein cannot bind with GAL80p, and GAL80p moves into the nucleus.

In the nucleus, the GAL80p protein binds with the transcriptional activation domain of GAL4p and prevents recruitment of the transcription complex.

**FIGURE 11.21** Regulation of transcription of the *GAL* genes by the proteins encoded in *GAL3* (GAL3p), *GAL80* (GAL80p), and *GAL4* (GAL4p).

transcription complex. The *GAL* genes are therefore not transcribed in the absence of galactose.

The mysterious constitutive mutation *GAL81*$^{c}$ in Table 11.4 is dominant and results in constitutive synthesis the *GAL* genes. This mutant behaves as might be expected of an operator mutation, but it cannot be comparable to *lacO*$^{c}$ because it does not map near *GAL1*, *GAL7*, or *GAL10*. The mysterious allele was explained by the discovery that "*GAL81*$^{c}$" is actually a mutation in the *GAL4* gene in which GAL4p is unable to bind with GAL80p but can still recruit the transcription complex. Hence, the mutant GAL4p protein cannot bind with GAL80p and is free to activate transcription even in the absence of galactose. The transcriptional activation takes place whether or not wildtype GAL4p protein is also present, which also explains why "*GAL81*$^{c}$" is dominant.

*GAL* regulation in yeast also affords an example of how regulatory systems can evolve. In another species of yeast, *Kluyveromyces lactis*, the GAL3p (regulatory) and GAL1p (galactokinase) activities are part of a bifunctional protein encoded in a single gene. In this species, when the *GAL* genes are induced the induction is modest, in the range of three- to fivefold. The *Saccharomyces cerevisiae* genome had the opportunity to separate these functions owing to a whole-genome duplication that took place about 100 million years ago. Although many of the duplicated genes have been lost through deletion, genome sequencing clearly shows that *S. cerevisiae* is an ancient tetraploid.

Among the tetraploid genes that were not lost in *S. cerevisiae* were those corresponding to the bifunctional gene in *K. lactis*. In the ancient *S. cerevisiae*, one gene copy retained only the GAL3p function and the other gene copy retained only the GAL1p function. This specialization allowed a much greater transcriptional response of the *GAL* genes, because in *S. cerevisiae*, when *GAL1* is induced, it is induced about a thousandfold above its basal level of transcription. This type of gene specialization through duplication is known as *subfunctionalization*, and the process is discussed in greater detail in Chapter 17.

The main point of these comparisons is that the superficial similarity between the constitutive and uninducible mutations in the *lac* operon of *E. coli* and the *GAL* genes of yeast is not indicative of similar molecular regulatory mechanisms. Some physiological similarities remain in that, in both prokaryotes and eukaryotes, the genes for a particular metabolic (or developmental) pathway are expressed in a coordinated manner in response to a signal. The principle at work is that alternative molecular mechanisms are often employed to achieve similar ends.

## Transcriptional Activator Proteins

The GAL4 protein is an example of a *transcriptional activator protein*, which must bind with an upstream DNA sequence in order to prepare a gene for transcription. Some transcriptional activator proteins work by direct interaction with one or more components in the transcription complex, and in this way they recruit the transcription complex to the promoter of the gene to be activated. Other transcriptional activator proteins may initiate transcription by an already assembled transcription complex. In either case, the activator proteins are essential for the transcription of genes that are positively regulated.

Many transcriptional activator proteins can be grouped into categories on the basis of characteristics shared by their amino acid sequences. For example, one category has a *helix–turn–helix* motif, which consists of a sequence of amino acids forming a pair of $\alpha$-helices separated by a bend; the helices are so situated that they can fit neatly into the grooves of a double-stranded DNA molecule. The helix–turn–helix motif is the basis of the DNA-binding ability, although the sequence specificity of the binding results from other parts of the protein.

A second large category of transcriptional activator proteins includes a DNA-binding motif that is called a *zinc finger* because the folded structure incorporates a zinc ion. An already familiar example is the GAL4 transcriptional activator protein in yeast. The protein functions as a dimer composed of two identical GAL4 polypeptides oriented with their zinc-binding domains at the extreme ends (Figure 11.20 shows the zinc ions in yellow). The DNA sequence recognized by the protein is a symmetrical sequence, 17 base pairs in length, which includes a CCG triplet at each end that makes direct contact with the zinc-containing domains.

**FIGURE 11.22** DNA-binding domain of the GAL4 transcriptional activator protein in yeast. The four cysteine residues bind a zinc ion and form a peptide loop called a zinc finger. The zinc finger is a motif commonly found in DNA-binding proteins. The amino acids marked by an asterisk have been identified via mutations as sites at which amino acid replacements can abolish the DNA-binding activity of the protein. The result is that the target genes cannot be activated.

A more detailed illustration of the DNA-binding domain of the GAL4 protein is shown in **FIGURE 11.22**. Each of two zinc ions ($Zn^{2+}$) is chelated by bonds with four cysteine residues in characteristic positions at the base of a loop that extends for an additional 841 amino acids beyond those shown. The amino acids marked by a red asterisk are the sites of mutations that result in mutant proteins unable to activate transcription. Replacements at amino acid positions 15 (Arg $\rightarrow$ Gln), 26 (Pro $\rightarrow$ Ser), and 57 (Val $\rightarrow$ Met) are particularly interesting because they provide genetic evidence that zinc is necessary for DNA binding. In particular, the mutant phenotypes can be rescued by extra zinc in the growth medium, because the molecular defect reduces the ability of the zinc finger part of the molecule to chelate zinc. Extra zinc in the medium overcomes the defect and restores the ability of the mutant activator protein to attach to its particular binding sites in the DNA.

## Transcriptional Enhancers and Transcriptional Silencers

Some transcriptional activator proteins bind with particular DNA sequences known as **enhancers**. Enhancer sequences are typically rather short (usually fewer than 20 base pairs) and are found at a variety of locations around the gene they regulate. Most enhancers are upstream of the transcriptional start site (sometimes many kilobases away), others are in introns within the coding region, and a few are

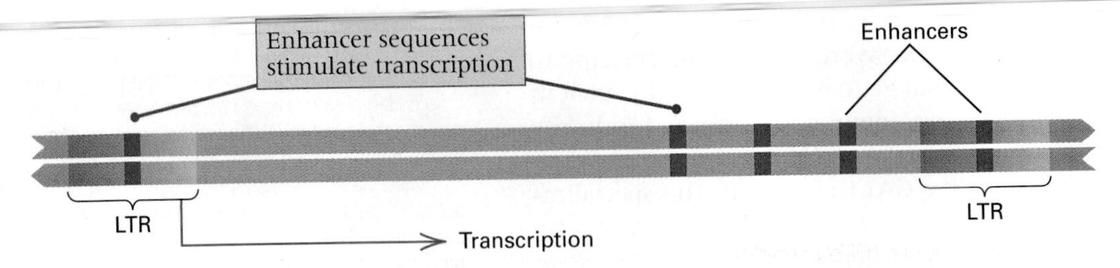


**FIGURE 11.23** Positions, in the mouse mammary tumor virus, of enhancers (orange) that allow transcription of the viral sequence to be induced by glucocorticoid steroid hormone. LTR stands for the long terminal repeated sequences found at the extreme ends of the virus.

even located at the 3′ end of the gene. One of the most thoroughly studied enhancers is in the mouse mammary tumor virus and determines transcriptional activation by the glucocorticoid steroid hormone. The consensus sequence of the enhancer is AGAQCAGQ, in which Q stands for either A or T. The virus contains five copies of the enhancer positioned throughout its genome (**FIGURE 11.23**), providing five binding sites for the hormone–receptor complex that activates transcription.

Enhancers are essential components of gene organization in eukaryotes because they enable genes to be transcribed only when proper transcriptional activators are present. Some enhancers respond to molecules from outside the cell—for example, steroid hormones that form receptor–hormone complexes. Other enhancers respond to molecules that are produced inside the cell (for example, during development), and these enhancers enable the genes under their control to participate in cellular differentiation or to be expressed in a tissue-specific manner. Many genes are under the control of several different enhancers, so they can respond to a variety of different molecular signals, both external and internal.

Some genes are also subjected to regulation by transcriptional **silencers**, which are short nucleotide sequences that are targets for DNA-binding proteins that, once recruited to the site, promote the assembly of large protein complexes that prevent transcription of the silenced genes. Examples of such silencing complexes are the *Drosophila* PcG (Polycomb group) proteins, which silence certain genes during development, and the complex in yeast that includes the RAP1, SIR, ORC, and histone 4 proteins, which is involved in silencing the unexpressed mating-type cassettes (described in Section 11.10).

## Deletion Scanning

Identifying the nucleotide sequences responsible for enhancer or silencer activity is usually difficult and labor intensive because the regulatory sequences typically are short, repeated, and often degenerate (that is, they can differ somewhat in nucleotide sequence from one copy to the next); their function is independent of orientation (an enhancer or silencer functions equivalently when inverted), and they are located at variable distances from the genes they control. Unlike protein-coding regions, they have no regular repeating pattern like the three-nucleotide phasing of codons, and so the genetic code provides no guidance in their identification. The study of enhancers and silencers was made much simpler by the use of reporter-gene constructs. A gene such as the *lacZ* gene from *E. coli*, which encodes the enzyme β-galactosidase, is substituted for the gene under study, and positioned downstream of the regulatory sequences to be analyzed. The regulatory sequence can be dissected genetically by making nucleotide substitutions or deletion mutations and examining the consequences on the expression of β-galactosidase activity. In this way, expression of the lacZ gene "reports" on the functional status of the upstream regulatory elements. Similarly, small segments of upstream sequence or a candidate regulatory motif can be placed upstream of *lacZ* and tested for the ability to confer authentic regulation on *lacZ*, in which case the expression of β-galactosidase will show the expected inducibility or repressibility, or in some cases the normal temporal or spatial regulation.

The advantages that *lacZ* provide as a reporter gene are several: the enzyme β-galactosidase is very stable, can be assayed through a wide range of enzyme activities, and has soluble substrates that yield either soluble or insoluble products. Soluble products allow accurate quantitation, while insoluble products allow direct visualization in colonies (**FIGURE 11.24**).

Figure 11.24A shows an example using quantitation. In these experiments, *lacZ* was fused to the yeast *GAL1* gene including 520 base pairs of upstream regulatory sequence. A series of deletions of the upstream region were made, introduced into yeast cells, and analyzed. The wildtype gene is induced by galactose by a factor of several thousand, whereas it is repressed by

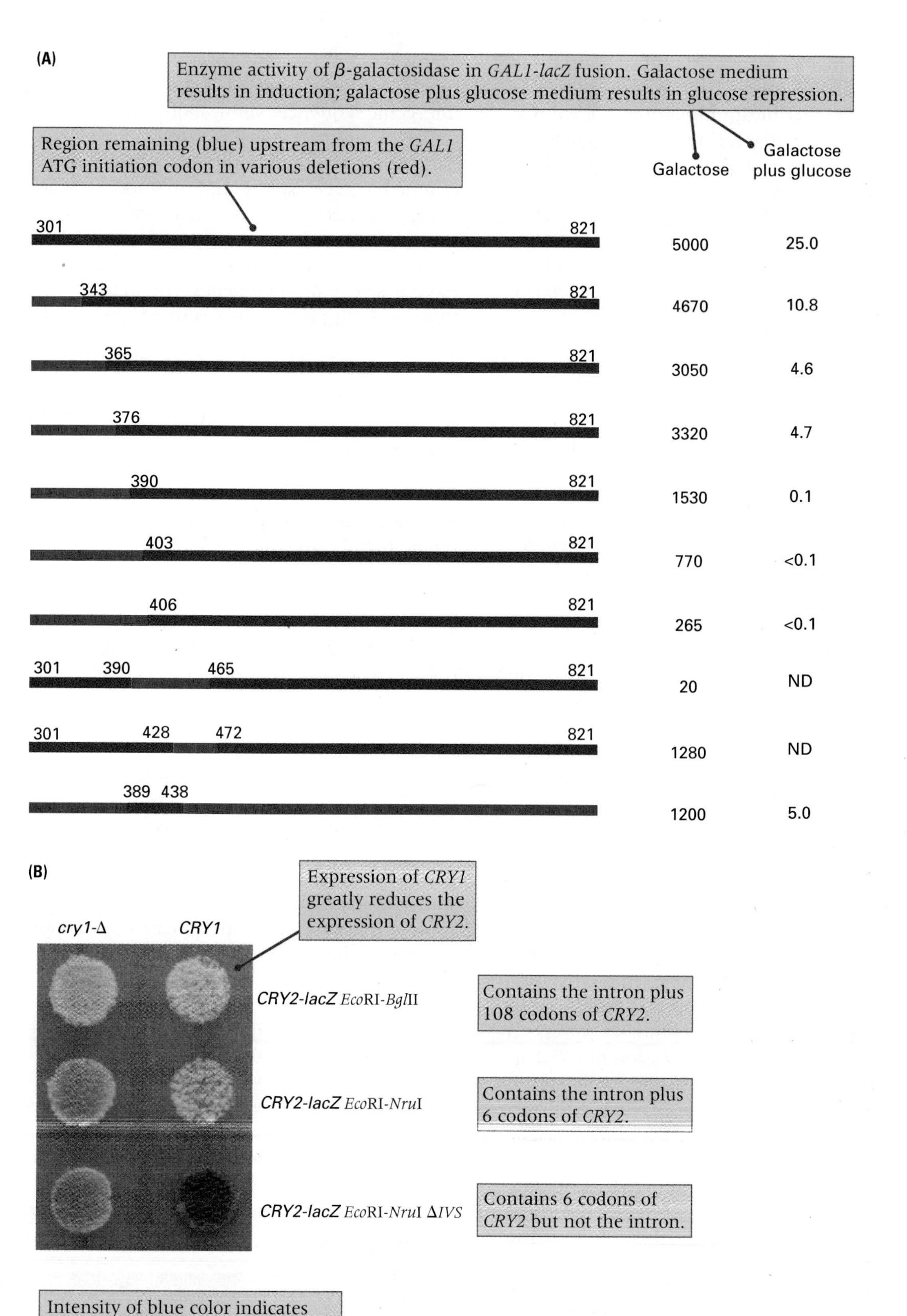


**FIGURE 11.24** Analyses of genetic regulation using reporter gene constructs. (A) Upstream regulatory sequences and about 30 codons of the *GAL1* gene of yeast were fused to the *lacZ* gene, and deletion mutations were made and transformed into yeast. Levels of $\beta$-galactosidase are given under inducing (galactose) or repressing (galactose plus glucose) conditions. (B) Expression of the *CRY1* ribosomal-protein gene greatly reduces expression of its *CRY2* duplicate copy. The deletions identify the intron of *CRY2* as containing the *cis*-acting component responsible for the reduced expression. [Part A adapted from R. W. West, Jr., R. R. Yocum, and M. Ptashne, *Mol. Cell Biol.* 4 (1984): 2467–2478; Part B courtesy of John L. Woolford, Jr. and Z. Li, Carnegie Mellon University.]

glucose. This behavior is recapitulated by the construct that begins at base pair 301 shown in the top line. Significant impairment of expression is seen when the region 301–343 is deleted, and deletion of the region 301–406 essentially extinguishes expression. Furthermore, deletion of the region 390–465 also eliminates expression.

The key result is shown at the bottom of the deletion series. This is a construct in which the segment from base pair 389 through 438 has been inserted upstream of *lacZ*. In this construct, *lacZ* becomes inducible by galactose and repressible by glucose, recapitulating the wildtype regulation (although not quantitatively, so other regions in the upstream region also contribute to GAL1 regulation). What is in the region 389–438? Further analysis indicated that the region 376–451 includes two repeats of a 17 base pair sequence in which each repeat itself includes two copies of the consensus sequence 5′-GGARGRC-3′ (R indicates either purine, A or G). The 17 base pair sequence is able to elicit authentic control, because this sequence is in fact the binding site for a dimer of GAL4, the transcriptional activator protein discussed in Section 11.5 and illustrated in Figure 11.20. The region 376–451 that includes two 17 base pair repeats is called the *Upstream Activating Sequence (UAS)* of *GAL1*.

The utility of *lacZ* as a reporter gene using insoluble products of β-galactosidase activity is shown in Figure 11.24B. A substrate called X-Gal (5-bromo-4-chloro-3-indolyl-β-D-galactopyranoside) yields an insoluble blue dye after cleavage. The intensity of the blue color provides a qualitative estimate of β-galactosidase activity. The genes *CRY1* and *CRY2* are duplicate genes for a ribosomal protein. Expression of *CRY1* greatly reduces expression of *CRY2*, as evidenced by the top pair of colonies in which a deletion (left) is compared with wildtype (right). The series of deletions indicates that the sequence accounting for the repression of the duplicate gene *CRY2* by the product of the gene *CRY1* is located within the intron *CRY2*. The key result is shown in the pair of colonies at the bottom, where the blue color resulting from X-Gal cleavage shows that deletion of the intron from *CRY2* eliminates the repression normally caused by the expression of *CRY1*.

Computational and genomic approaches are presently being used to identify regulatory sequences. An interesting example compared the complete genome sequences of six relatively closely related species of the yeast *Saccharomyces*. Many genes in these six species serve the same function and are regulated in similar ways. Their intergenic sequences show about 60% identity to the counterpart genes in *Saccharomyces cerevisiae*. On the assumption that sequences that are retained between species have regulatory functions, whereas those that have diverged do not, several dozen new regulatory motifs were identified for yeast species. Since these organisms have fewer than 6000 genes, the new motifs, together with previously identified ones, likely comprise a large fraction of the regulatory sequences of these species. Several other lines of evidence suggest that the comparative approach is very powerful. Previously known regulatory sequences have been rediscovered; novel regulatory motifs often are found upstream of genes with related functions or cellular processes such as meiosis, the cell cycle, or DNA damage and repair; novel regulatory motifs have been found in short DNA sequences located in chromatin precipitated by antibodies to transcriptional activators. A few of these novel motifs are shown in **TABLE 11.5**. The validity of the sequences is being tested using the reporter gene constructs such as those with *lacZ*.

## The Eukaryotic Transcription Complex

Many enhancers activate transcription by means of **DNA looping**, which refers to physical interactions between relatively distant regions along the DNA. The mechanism is illustrated in **FIGURE 11.25**. The factors necessary for transcription include a transcriptional activator protein that interacts with at least one protein subunit of the transcription complex to recruit the transcription complex to the gene.

The **basal transcription factors** are proteins in the transcription complex that are used widely in the transcription of many different genes. The basal transcription factors in

**Table 11.5 Novel regulatory sequence motifs found upstream of genes with similar expression patterns in budding yeast**

| Regulatory motif | Process governed |
|---|---|
| AATGTA | DNA damage |
| ACATAC | DNA damage, stress |
| TTTTCAT | Stress |
| TAGAAA | Cell cycle |
| TTCTTTC | Cell cycle |
| ACAAAA | Meiosis |
| CCCTTTT | Meiosis |

Source: P. Cliften et al. 2003. *Science* 301: 71.

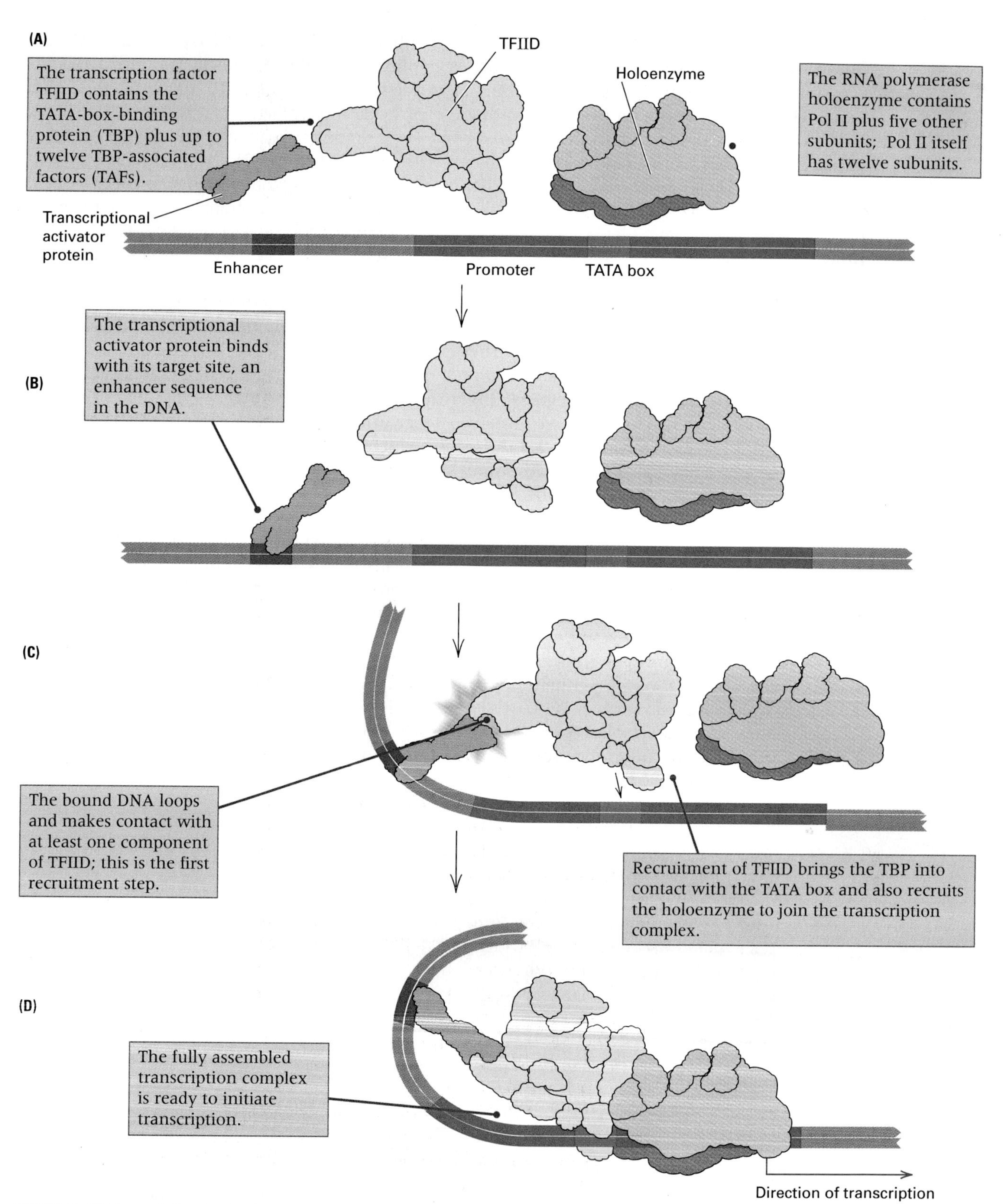


**FIGURE 11.25** Transcriptional activation by recruitment. (A) Relationship between enhancer and promoter and the protein factors that bind to them. (B) Binding of the transcriptional activator protein to the enhancer. (C) Bound transcriptional activator protein makes physical contact with a subunit in the TFIID complex, which contains the TATA-box-binding protein, and attracts ("recruits") the complex to the promoter region. (D) The Pol II holoenzyme and any remaining general transcription factors are recruited by TFIID, and the transcription complex is fully assembled and ready for transcription. In the cell, not all of the Pol II is found in the holoenzyme, and not all of the TBP is found in TFIID. In this illustration, transcription factors other than those associated with TFIID and the holoenzyme are not shown.

eukaryotes have been highly conserved in evolution. A minimal set necessary for accurate transcription *in vitro* includes TFIIB, TFIID, TFIIE, TFIIF, TFIIH, Pol II, and (less critically) TFIIA. (TF in these designations stands for *transcription factor*.) These components can assemble *in vitro* in stepwise fashion on a promoter. The first step is recruitment of TFIID, itself a complex of proteins that includes a **TATA-box-binding protein (TBP)**, which binds with the promoter in the region of the TATA box, and about 10 other proteins, called **TBP-associated factors (TAFs)**, which are the components that respond specifically to activator proteins. The TBP binds to the DNA in the minor groove and then bends the DNA by about 80 degrees. The Pol II RNA polymerase includes twelve protein subunits and is found in a complex called the **Pol II holoenzyme**, which also includes some of the basal transcription factors (TFIIB, TFIIF, and TFIIH) as well as other proteins.

It is not yet clear whether the transcription complex is recruited to the promoter and assembled stepwise, as *in vitro* studies suggest, or recruited in the form of one or more large preassembled complexes, which the composition of the Pol II holoenzyme suggests is the case. For simplicity, Figure 11.25 shows recruitment of one preassembled complex that includes TFIID, which in turn recruits the preassembled Pol II holoenzyme. To activate transcription (Figure 11.25B), the transcriptional activator protein binds to an enhancer in the DNA and to one of the TAF subunits in the TFIID complex. This interaction attracts ("recruits") the TFIID complex to the region of the promoter (Figure 11.25C). Attraction of the TFIID to the promoter also recruits the Pol II holoenzyme (Figure 11.25D), as well as any remaining general transcription factors, and once brought together, the transcriptional complex is ready for transcription to begin.

As Figure 11.25 suggests, the fully assembled transcription complex in eukaryotes is a very large structure. A real example, taken from early development in *Drosophila*, is shown in **FIGURE 11.26**. In this case, the enhancers, located a considerable distance upstream from the gene to be activated, are bound by the transcriptional activator proteins BCD and HB, which are products of the genes *bicoid (bcd)* and *hunchback (hb)*, respectively; these transcriptional activators function in establishing the anterior–posterior axis in the embryo. (Early *Drosophila* development is discussed in Chapter 13.) Note the position of the TATA box in the promoter of the gene. The TATA box binding is the function of the TBP. The functions of a number of other components of the transcription complex have also been identified. For example, the TFIIH contains both helicase and kinase activity to melt the DNA and to phosphorylate RNA polymerase II.

Phosphorylation allows the polymerase to leave the promoter and elongate mRNA. The

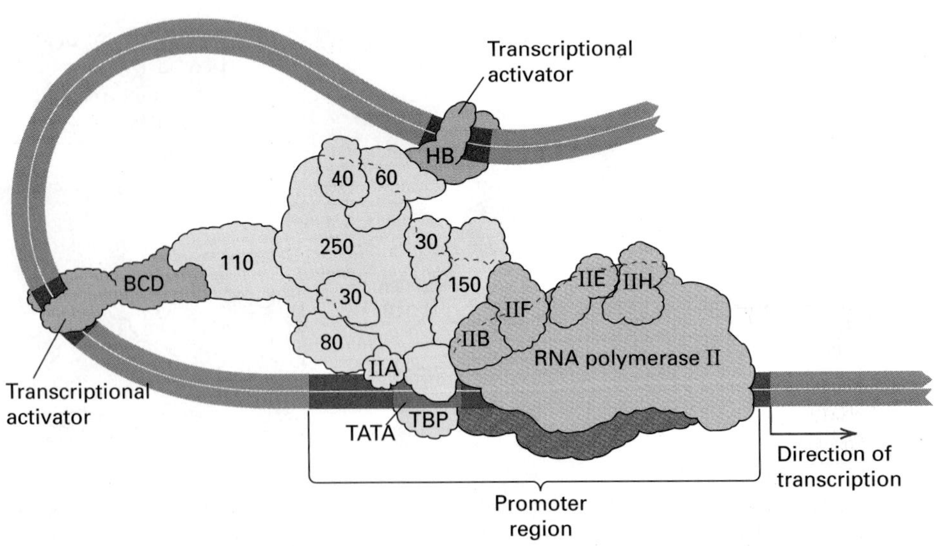


**FIGURE 11.26** An example of transcriptional activation during *Drosophila* development. The transcriptional activators in this example are bicoid protein (BCD) and hunchback protein (HB). The numbered subunits are TAFs (TBP-associated factors) that, together with TBP (TATA-box-binding protein) correspond to TFIID. BCD acts through a 110-kilodalton TAF, and HB through a 60-kilodalton TAF. The transcriptional activators act via enhancers to cause recruitment of the transcriptional apparatus. The fully assembled transcription complex includes TBP and TAFs, RNA polymerase II, and general transcription factors TFIIA, TFIIB, TFIIE, TFIIF, and TFIIH.

looping of the DNA effected by the transcriptional activators is an essential feature of the activation process. Transcriptional activation in eukaryotes is a complex process, especially when compared to the prokaryotic RNA polymerase, which consists of only six polypeptide chains.

### Chromatin-Remodeling Complexes

Eukaryotic DNA is typically found in the form of chromatin packaged with nucleosomes (Chapter 7). Special mechanisms are required for transcriptional activator proteins and the transcription complex to acquire access to the DNA, including the chemical modification of histone tails discussed in Chapter 7. The existence of such mechanisms is implied by the observation that the components of transcription sufficient to transcribe purified DNA *in vitro* are unable to initiate transcription of purified chromatin. The nucleosomes in chromatin must normally prevent the transcription complex from either binding to DNA or using it as a template.

Several different multiprotein complexes that can restructure chromatin and that enable it to be transcribed have been identified. These are known as **chromatin-remodeling complexes (CRCs)**. Among them are the SWI–SNF and RSC complexes in yeast and the NURF, CHRAC, and ACF complexes in *Drosophila*. SWI–SNF is named after the mutations, *swi* and *snf*, that led to its discovery, and the other abbreviations stand for *remodels the structure of chromatin* (RSC), *nucleosome remodeling factor* (NURF), *chromatin accessibility complex* (CHRAC), and *ATP-dependent chromatin assembly and remodeling factor* (ACF). All of these complexes use energy derived from ATP to restructure chromatin, and although the detailed mechanisms of chromatin remodeling are unknown, the complexes do differ in their activities. For SWI–SNF, for example, secondary mutations that suppress *swi* and *snf* map to the histone genes, the presence of the SWI–SNF complex renders the DNA in mononucleosomes susceptible to cleavage by DNAse, and purified SWI–SNF can restructure chromatin *in vitro*. RSC has similar, but not identical, properties. Furthermore, SWI–SNF is found as an integral component of the Pol II holoenzyme, and it enables the holoenzyme to disrupt nucleosomes. In contrast, NURF facilitates the binding of transcription factors to chromatin, CHRAC acts on nucleosomal spacing and increases the sensitivity of the DNA in chromatin to endonuclease activity, and ACF helps assemble properly spaced nucleosomal arrays and facilitates the binding of transcriptional activator proteins. Similar kinds of chromatin-remodeling complexes have been identified in higher eukaryotes, and they are probably present in all eukaryotes.

The molecular mechanism of chromatin remodeling is unknown, and because there are several distinct types of CRCs, there may be several mechanisms. In one general class of models, the CRC disrupts nucleosome structure without displacing the nucleosomes, rendering the DNA accessible to transcriptional activator proteins, the TATA-box-binding protein, and other components of the transcription complex. In another general class of models, the CRC repositions the nucleosomes along the DNA, making key DNA-binding sites accessible. An example is illustrated in **FIGURE 11.27**. Part A shows a transcriptionally inactive chromatin conformation, with the DNA-binding sites for a transcriptional activator protein (TAP) and TATA-box-binding protein (TBP) sequestered in nucleosomes and unavailable. Recruitment of a CRC to the site results in repositioning of the nucleosomes (part B), which renders the binding sites accessible (part C). In this chromatin configuration, TAP and TBP can bind with the DNA and recruit the rest of the transcription complex.

### Alternative Promoters

Some eukaryotic genes have two or more promoters that are active in different cell types. The different promoters result in different primary transcripts that contain the same protein-coding regions. An example from *Drosophila* is shown in **FIGURE 11.28**. The gene codes for alcohol dehydrogenase, and its organization in the genome, shown in part A, includes three protein-coding regions interrupted by two introns. Transcription in larvae (Figure 11.28B) uses a different promoter from that used in transcription in adults (Figure 11.28C). The adult transcript has a longer 5′ leader sequence, but most of this sequence is eliminated in RNA splicing. Alternative promoters make possible the independent regulation of transcription in larvae and adults.

The versatility of some enhancers results from their ability to interact with two different promoters in a competitive fashion; that is, at any one time, the enhancer can stimulate one promoter or the other, but not both. An example of this mechanism is illustrated in

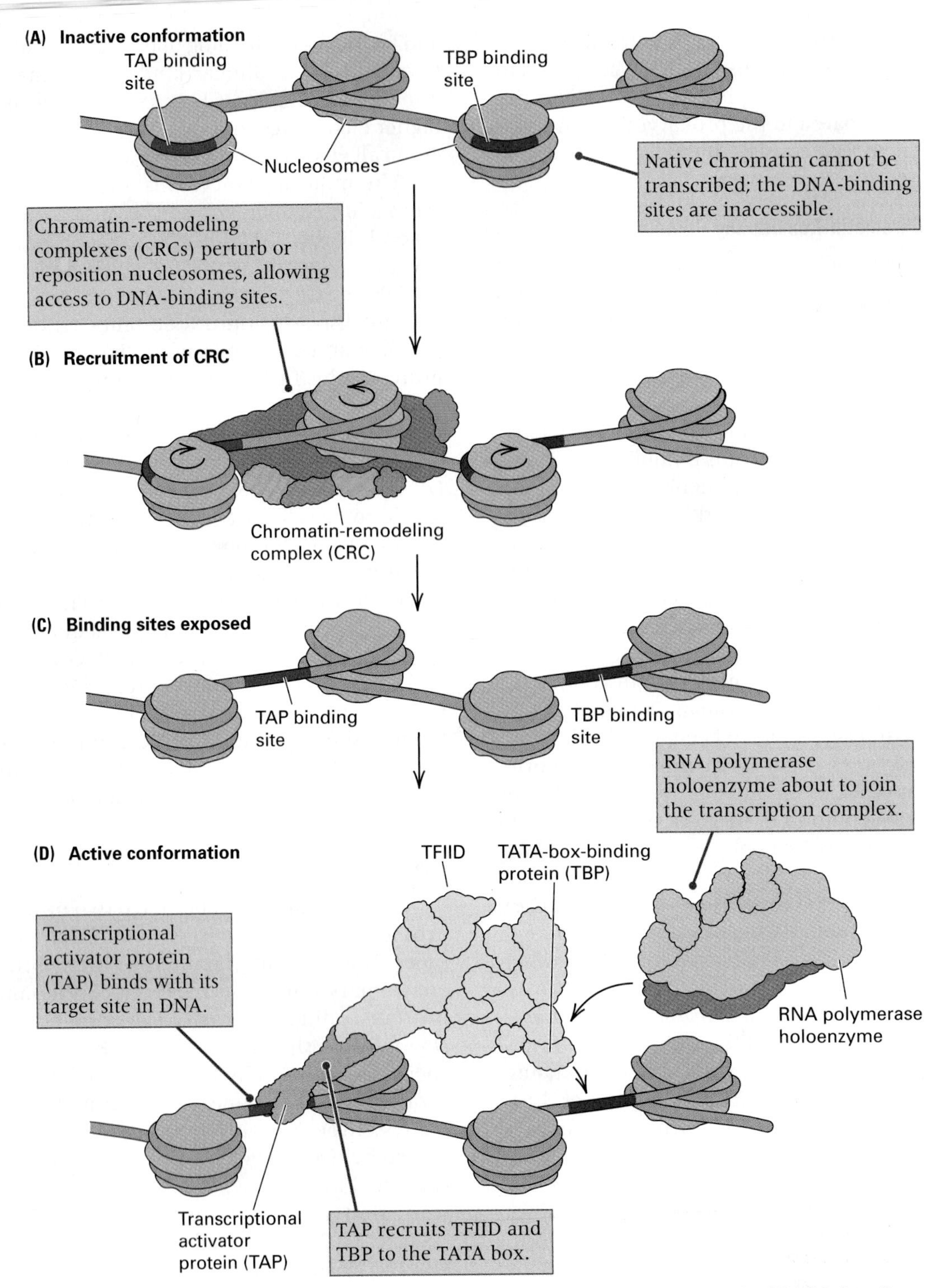


**FIGURE 11.27** Function of chromatin-remodeling complexes. (A) Native chromatin may conceal key DNA-binding sites. (B) A chromatin-remodeling complex either repositions the nucleosomes along the DNA or chemically modifies the histones. (C) DNA-binding sites become accessible. (D) The transcription complex is recruited to the site.

**FIGURE 11.29**, in which *P1* and *P2* are alternative promoters that compete for an enhancer located between them. When the enhancer is complexed with a transcriptional activator protein specific for promoter P1, the transcription complex is recruited to promoter P1 and transcription takes place (Figure 11.29A). When the enhancer is complexed with a different transcriptional activator specific for promoter P2, the transcription complex is recruited to promoter P2 (Figure 11.29B). In this way, competition for the enhancer serves as a switch mechanism for expres-

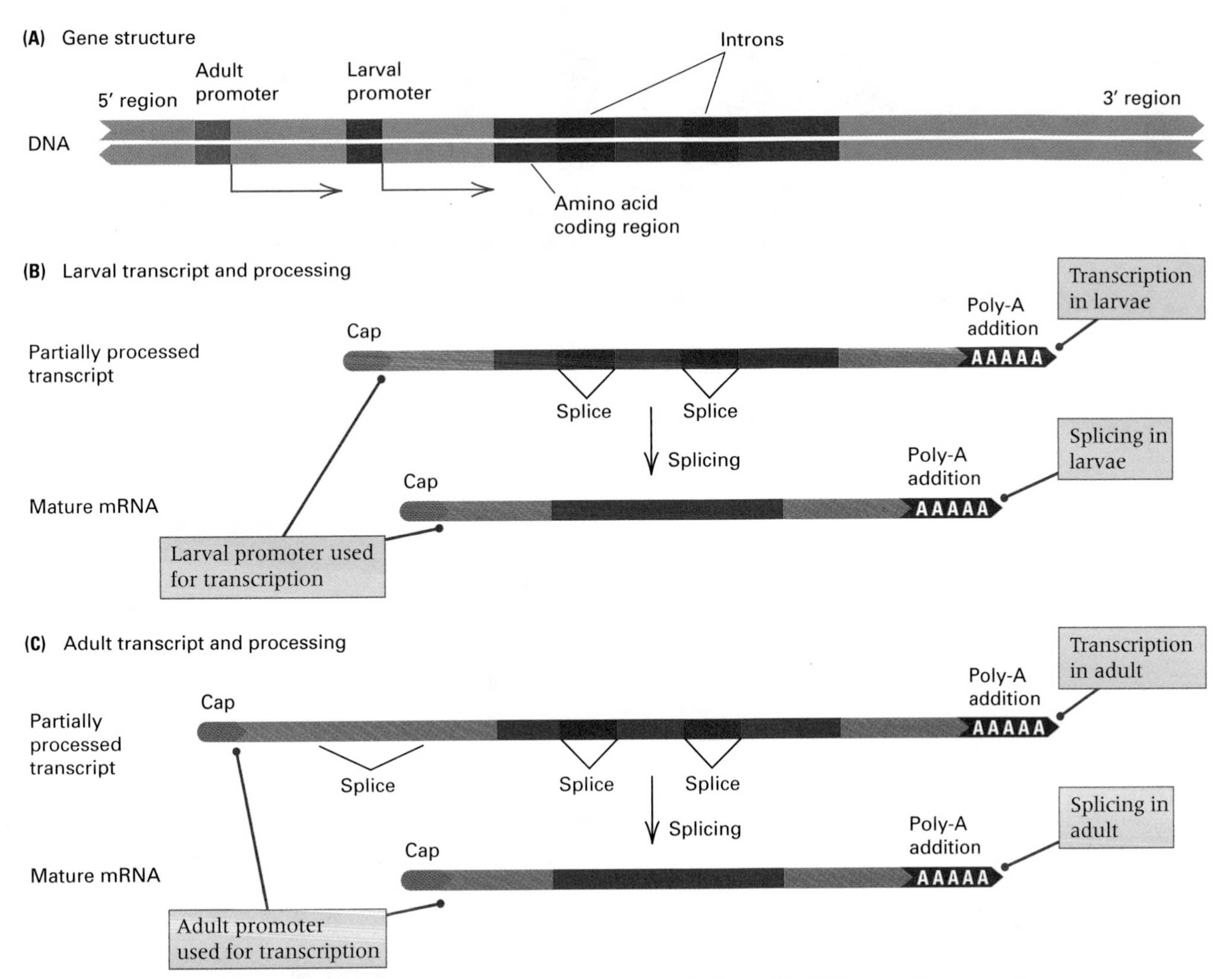


**FIGURE 11.28** Use of alternative promoters in the gene for alcohol dehydrogenase in *Drosophila*. (A) The overall gene organization includes two introns within the amino acid coding region. (B) Transcription in larvae uses the promoter nearest the 5′ end of the coding region. (C) Transcription in adults uses a promoter farther upstream, and much of the adult leader sequence is removed by splicing.

sion of the *P1* or *P2* promoter. This regulatory mechanism is present in chickens and results in a change from the production of embryonic β-globin to that of adult β-globin in development. In this case, the embryonic globin gene and the adult gene compete for a single enhancer, which in the course of development changes its preferred binding from the embryonic promoter to the adult promoter. In human beings, enhancer competition appears to control the developmental switch from the fetal γ-globin to that of the adult β-globin polypeptide chains. In persons in whom the β-globin promoter is deleted or altered in sequence and unable to bind with the enhancer, there is no competition for the enhancer molecules, and the γ-globin genes continue to be expressed in adult life when normally they would not be transcribed. Adults with these types of mutations have fetal hemoglobin instead of the adult forms. The condition is called *high-F disease* because of the persistence of fetal hemoglobin, but the clinical manifestations are very mild.

## 11.6 Epigenetic Mechanisms of Transcriptional Regulation

The prefix *epi* means "besides" or "in addition to"; **epigenetic** therefore refers to heritable changes in gene expression that are due not to changes the DNA sequence itself, but to something in addition to the DNA sequence, usually either chemical modification of the bases or protein factors bound with the DNA. In this section we discuss some examples of epigenetic regulation. We shall see that there is a great deal yet to be learned about the molecular mechanisms by which epigenetic modifications are established and maintained.

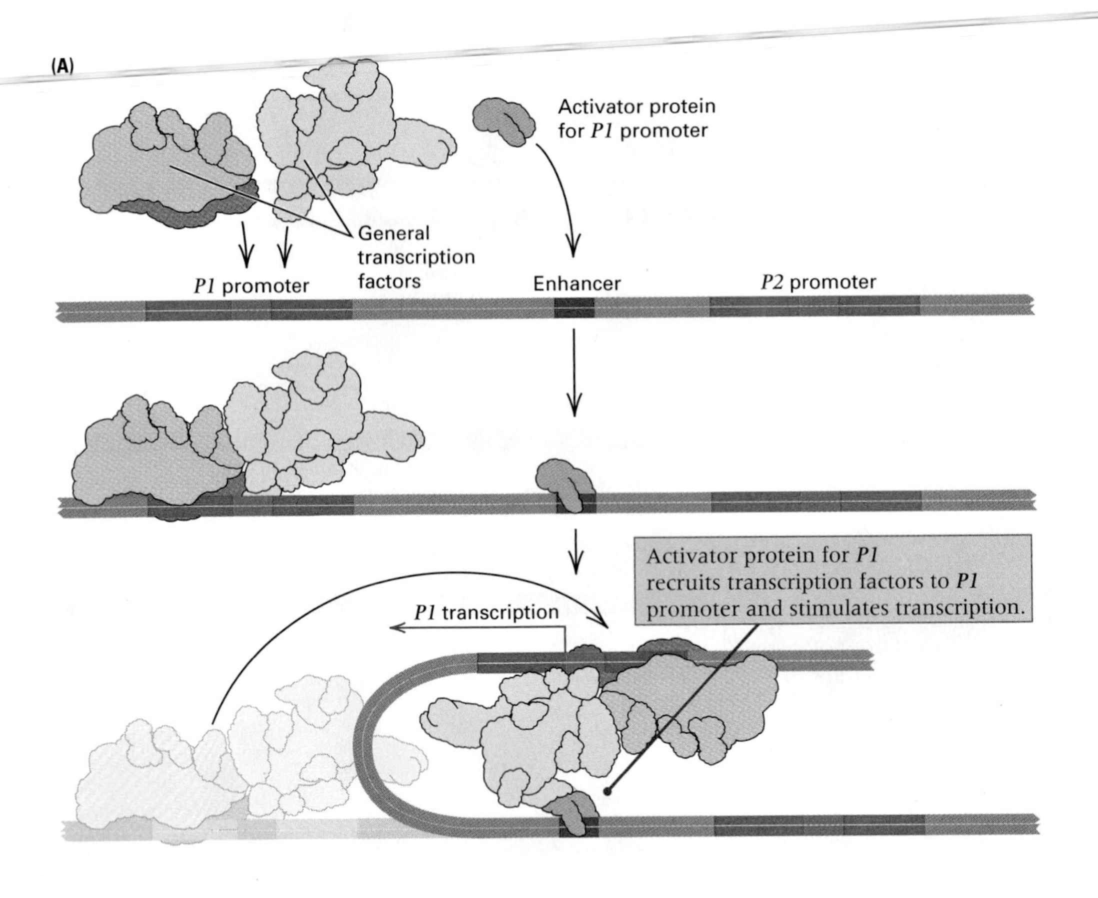


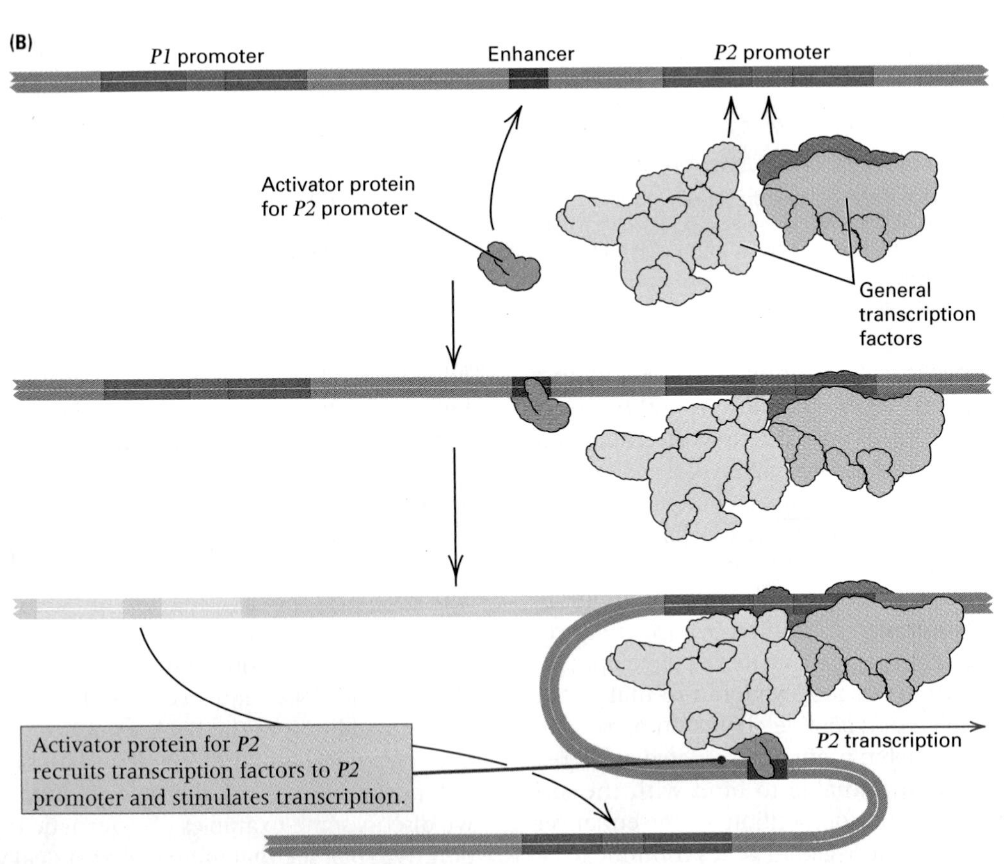


**FIGURE 11.29** Genetic switching regulated by competition for an enhancer. Promoters *P1* and *P2* compete for a single enhancer located between them. When complexed with an appropriate transcriptional activator protein, the enhancer binds preferentially with promoter *P1* (A) or promoter *P2* (B). Binding to the promoter recruits the transcription complex. If either promoter is mutated or deleted, then the enhancer binds with the alternative promoter. The location of the enhancer relative to the promoters is not critical.

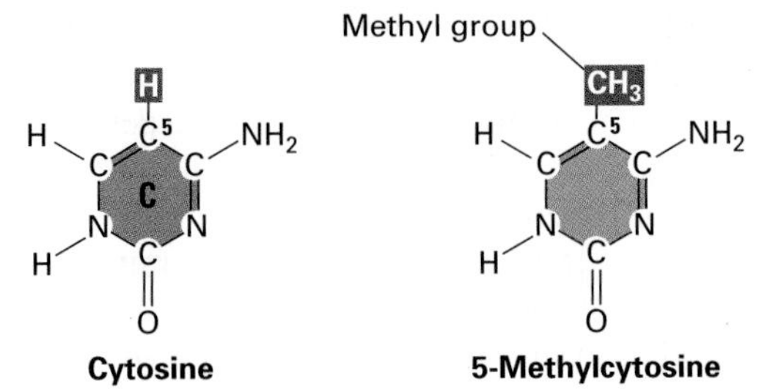

**FIGURE 11.30** Structures of cytosine and 5-methylcytosine.

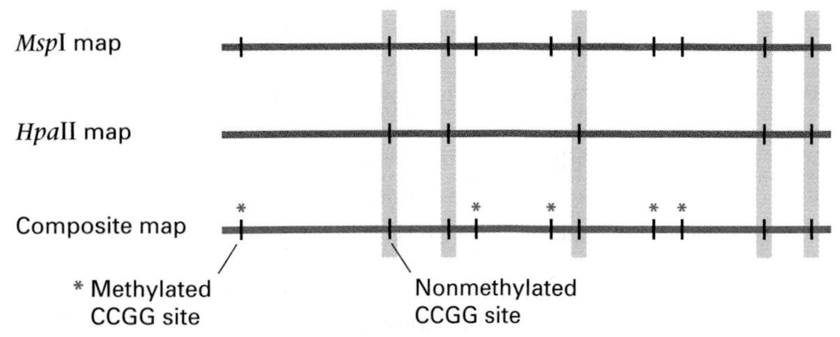

**FIGURE 11.31** Detection of methylated cytosines in CCGG sequences by means of restriction enzymes. The enzyme *Msp*I cleaves all CCGG sites regardless of methylation, whereas *Hpa*II cleaves only nonmethylated sites. The positions of the methylated sites are determined by comparing the restriction maps.

## Cytosine Methylation

In most higher eukaryotes, a proportion of the cytosine bases are modified by the addition of a methyl ($CH_3$) group to the number-5 carbon atom (**FIGURE 11.30**). The cytosines are incorporated in their normal, unmodified form in the course of DNA replication, and then the methyl group is added by an enzyme called **DNA methylase**. In mammals, cytosines are modified preferentially in 5′-CG-3′ dinucleotides. Many mammalian genes have CG-rich regions upstream of the coding region that provide multiple sites for methylation; these are called **CpG islands**, where the *p* represents the phosphate group in the polynucleotide backbone.

When a CG dinucleotide that is methylated in both strands undergoes DNA replication, each of the resulting daughter molecules contains one parental strand with a methylated CG and one newly synthesized strand with an unmethylated CG. The DNA methylase recognizes the half-methylation in these molecules and methylates the cytosines in the newly synthesized strands. Methylation of CG dinucleotides in the sequence CCGG can be detected by the use of the restriction enzymes *Msp*I and *Hpa*II. Both enzymes cleave the sequence CCGG. However, *Msp*I cleaves regardless of whether the interior C is methylated, whereas *Hpa*II cleaves only unmethylated DNA. Therefore, *Msp*I restriction sites that are not cleaved by *Hpa*II are sites at which the interior C is methylated (**FIGURE 11.31**).

## Methylation and Transcriptional Inactivation

A number of observations suggest that heavy methylation is associated with genes for which the rate of transcription is low. One example is the inactive X chromosome in mammalian cells, which is extensively methylated. In fact, in adult mammals, 60–70 percent of CpG dinucleotides in all chromosomes are methylated in somatic cells. The unmethylated CpGs are usually associated with the promoters of active housekeeping genes. The widespread methylation of inactive genes in adult somatic cells is thought to minimize accidental, low-level transcription from them.

Methylation is also widespread in the genomes of higher plants, where it is also correlated with transcriptional silencing. One example is found in the *Ac* transposable element in maize. Certain *Ac* elements lose activity of the transposase gene without any change in DNA sequence. These elements prove to have heavy methylation in a region particularly rich in the CG dinucleotides. Return to normal activity of the methylated *Ac* elements coincides with loss of methylation through the action of demethylating enzymes in the nucleus.

Methylation is also associated with transcriptional silencing in fungi. For example, in the phenomenon of *RIP (repeat-induced point mutation)* in *Neurospora*, discussed further in Chapter 14, repeated sequences are subjected to hypermutation, and many of the C-bearing nucleotides in the repeats become methylated, thereby inhibiting transcription. A related process affecting repeated sequences occurs in the sexual cycle in the fungi *Ascobolus immersus* and *Coprinus cinereus*. This process is called **methylation induced premeiotically (MIP)**, and it results in heavy methylation of cytosines in repeated sequences, leading to transcriptional silencing. In all of these examples, the methylation can be maintained through cell division; however, the mechanism of maintenance is not well understood.

Although there is a very strong correlation between heavy methylation and transcriptional silencing, heavy methylation may result from an earlier epigenetic signal that marks a gene for silencing and that recruits the methylase. If there is such an earlier signal, then it implies that methylation is the result of gene inactivity as well as its mechanism. In any case, treatment of cells with the cytosine analog *azacytidine* reverses methylation and can restore transcriptional activity. For example, in cell culture, some lineages of rat pituitary tumor cells express the gene for prolactin, whereas other related lineages do not. The gene is methylated in the nonproducing cells but is not methylated in the producers. Reversal of methylation in the nonproducing cells via azacytidine results in prolactin expression.

## ■ Genomic Imprinting in the Female and Male Germ Lines

Mammals feature an unusual type of epigenetic silencing known as **genomic imprinting**, a process with the following characteristics:

- Imprinting occurs in the germ line.
- It affects at most a few hundred genes (many of them located in clusters).
- It is accompanied by heavy methylation (though the primary signal for imprinting is unknown).
- Imprinted genes are differentially methylated in the female and male germ lines.
- Once imprinted and methylated, a silenced gene remains transcriptionally inactive during embryogenesis.
- Imprints are erased early in germ-line development, then later reestablished according to sex-specific patterns.

Although mammalian gametes are extensively methylated, most of the DNA is demethylated in preimplantation development, except for imprinted genes that retain their sex-specific patterns of methylation. The embryonic DNA is remethylated beginning after implantation, gradually attaining the heavy methylation levels found in adult somatic cells. In the germ line, the original imprints are erased when the DNA is globally demethylated, and remethylation takes place later in germ-line development. All remethylated genes acquire identical patterns of methylation in the germ line of both sexes, except for those few that have sex-specific patterns of imprinting and differential methylation. The imprinted genes undergo methylation during oocyte growth prior to ovulation in females, and in males probably around the time of birth. Because the methylation associated with imprinting is retained throughout embryonic development, any gene that is imprinted in either the female or the male germ line has, effectively, only one active copy in the embryo.

The epigenetic, sex-specific gene silencing associated with imprinting is dramatically evident in a pair of syndromes characterized by neuromuscular defects, mental retardation, and other abnormalities. These are *Prader–Willi syndrome* and *Angelman syndrome.* Both conditions are associated with rare, spontaneous deletions that include chromosomal region *15q11*. If the deletion takes place in the father, the result is Prader–Willi syndrome, whereas if it takes place in the mother, the result is Angelman syndrome. The reason is that *15q11* includes at least three genes (*SNRPN, necdin,* and *UBE3A*) that are imprinted and differentially methylated in the gametes. **FIGURE 11.32**, part A shows the pattern of

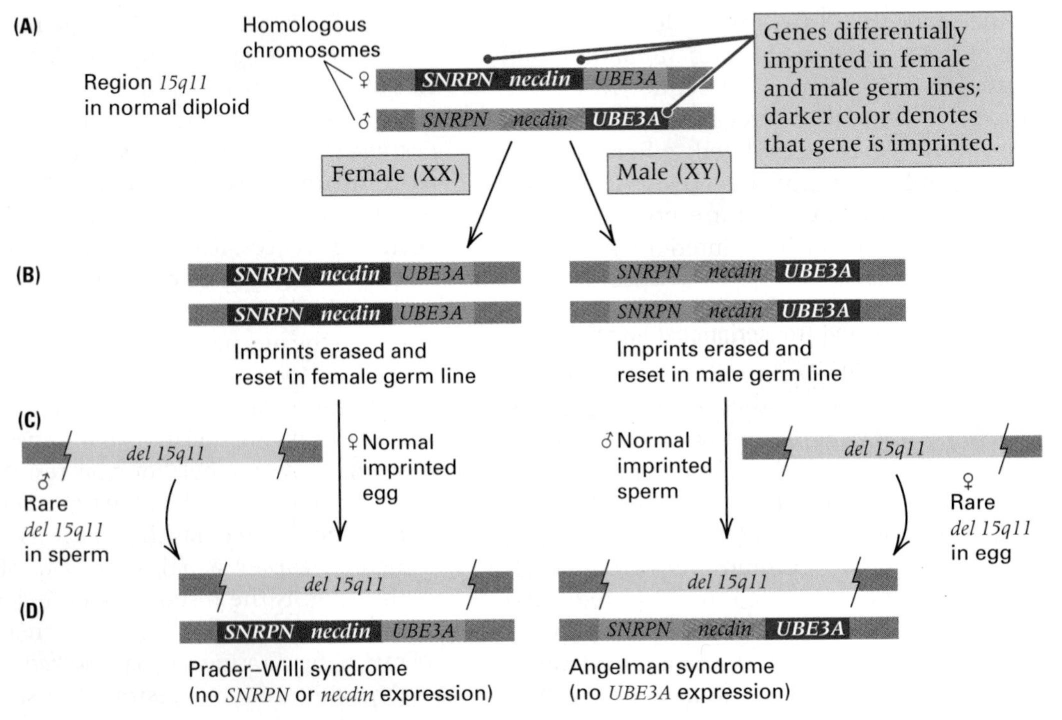


**FIGURE 11.32** Imprinting of genes in chromosomal region *15q11* results in different neuromuscular syndromes, depending on which parent contributes a *15q11* deletion and which an imprinted chromosome. (A) Pattern of imprinting in a normal diploid. The maternal chromosome is at the top, the paternal chromosome at the bottom. Imprinted and transcriptionally inactive genes are indicated. (B) In the germ line, the imprints are erased and reset in either female-specific or male-specific patterns. (C) An individual who inherits a maternally imprinted chromosome along with a *15q11* deletion has Prader–Willi syndrome, whereas one who inherits a paternally imprinted chromosome along with a *15q11* deletion has Angelman syndrome. Other genes in the region, not shown, may also be imprinted.

imprinting of these three genes in a normal embryo. *SNRPN* and *necdin* are imprinted in the egg, and *UBE3A* is imprinted in the sperm. In the embryo, therefore, *UBE3A* is transcriptionally active in the maternal chromosome, and *SNRPN* and *necdin* in the paternal chromosome. In the germ line of female and male embryos, shown in Figure 11.32B, the imprints are erased and reset according to sex; in the female both homologs have *SNRPN* and *necdin* imprinted, whereas in the male both homologs have *UBE3A* imprinted. If a normal, imprinted female gamete is fertilized by a sperm with a *15q11* deletion, the embryo has no transcriptionally active copy of either *SNRPN* or *necdin* and develops Prader–Willi syndrome. On the other hand, if a normal, imprinted male gamete fertilizes an egg with a *15q11* deletion, the embryo has no transcriptionally active copy of *UBE3A* and develops Angelman syndrome. These syndromes demonstrate not only the epigenetic control of gene expression by imprinting, but also differential imprinting in the sexes and the clustering of imprinted genes in the genome.

Why is there imprinting? One suggestion is that it evolved in early mammals with polyandry (each female mating with a series of males). In such a situation, it is to a male's benefit to silence genes that conserve maternal resources at the expense of the fetus, because this strategy maximizes the father's immediate reproduction. But it is to a female's benefit to silence genes that allocate resources to the fetus at the expense of the mother, because this strategy maximizes the female's long-term reproduction. This hypothesis is supported by the fact that some imprinted genes do affect the allocation of resources between mother and fetus in the direction that would be predicted. On the other hand, many genes that are imprinted have no obvious connection to maternal–fetal conflict.

## 11.7 Regulation Through RNA Processing and Decay

Although transcriptional control of gene expression is of cardinal importance, transcription is by no means the only level at which gene activity can be regulated. In this section we consider some mechanisms that act at the level of primary-transcript splicing or at the level of mRNA stability.

### ■ Alternative Splicing

Even when the same promoter is used to transcribe a gene, different cell types can produce different quantities of the protein (or even different proteins) because of differences in the mRNA produced in processing. The reason is that the same transcript can be spliced differently from one cell type to the next. The different splicing patterns may include exactly the same protein-coding exons, in which case the protein is identical, but the rate of synthesis differs because the mRNA molecules are not translated with the same efficiency. In other cases, the protein-coding part of the transcript has a different splicing pattern in each cell type, and the resulting mRNA molecules code for proteins that are not identical even though they share certain exons.

The insulin receptor gene in humans and other mammals provides an example of alternative splicing that results in the inclusion or exclusion of exon 11 in the messenger RNA. The resulting forms of the polypeptide chain differ in length by 12 amino acids. The relevant part of the primary transcript is shown in **FIGURE 11.33**. In the liver, all 20 exons are found in the mRNA for the long form of the receptor protein (Figure 11.33A), whereas in skeletal muscle

**(A)**

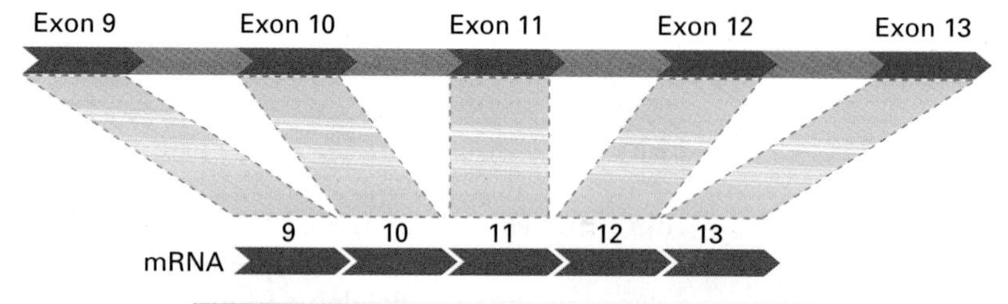
Primary transcript
Exon 9
Exon 10
Exon 11
Exon 12
Exon 13
9
10
11
12
13
mRNA
In RNA processing in the liver, the exons 9–13 are all included in the messenger RNA and the resulting protein has low affinity for insulin.

**(B)**

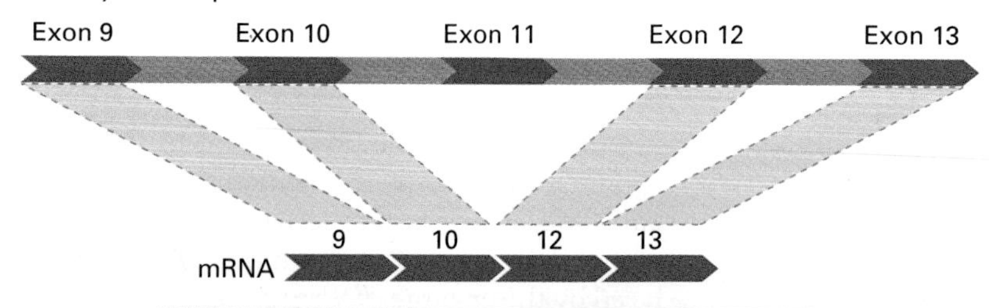
Primary transcript
Exon 9
Exon 10
Exon 11
Exon 12
Exon 13
9
10
12
13
mRNA
In RNA processing in skeletal muscle, the codons in exon 11 are excluded from the messenger RNA and the resulting protein has high affinity for insulin.

**FIGURE 11.33** Alternative splicing of the primary transcript of the gene encoding the $\alpha$ chain of the insulin receptor in humans and other mammals. (A) Splicing in the liver results in the low-affinity long form. (B) Splicing in skeletal muscle results in the high-affinity long form.

exon 11 is eliminated along with the flanking introns and excluded from the mRNA for the short form (Figure 11.33B). The long form of the receptor shows low affinity for insulin and is expressed in tissues such as the liver that are exposed to relatively high concentrations of insulin. The short form of the protein has a high affinity for insulin and is expressed preferentially in tissues such as skeletal muscle that are normally exposed to lower levels of insulin. Alternative splicing thus offers the possibility of generating proteins with different properties from the same gene. Although the human genome is estimated to contain fewer than 30,000 genes, the number of different proteins produced is several to many times greater because of alternative splicing.

### ■ Messenger RNA Stability

A short-lived mRNA produces fewer protein molecules than a long-lived mRNA, so features that affect the rate of mRNA stability affect the level of gene expression. One route of degradation is the *deadenylation-dependent pathway,* which begins with enzymatic trimming of length of the poly-A tail on the mRNA. When the poly-A tail is trimmed to a length of 25–60 nucleotides, the mRNA becomes susceptible to a decapping enzyme that removes the 5′ cap and renders the molecule unable to initiate translation; from this state the mRNA is rapidly degraded by exonucleases. An alternative pathway is the *deadenylation-independent pathway,* which is initiated either with decapping or with endonuclease cleavage of the mRNA, after which digestion goes to completion by exonuclease activity. The deadenylation-independent pathway is particularly active for mRNAs that contain early chain-termination codons or unspliced introns, and it prevents the accumulation of truncated polypeptides in the cell.

## 11.8 RNA Interference

In 1990 a group of plant geneticists reported experiments in which they manipulated genes for flower color in petunia (*Petunia hybrida*). The normal red or purple flower color in this plant results from a flavonoid pigment known as *anthocyanin,* which is synthesized via a metabolic pathway in which the rate-limiting step is catalyzed by the enzyme *chalcone synthase*. The investigators reasoned that an extra copy of this gene would increase the level of the enzyme and thus the amount of pigment, thereby yielding a darker flower color. In the actual experiment (Figure 11.34), the flower color of the genetically engineered plants was white! The total level of chalcone-synthase mRNA was about fifty times lower in the engineered plants than that in control plants, and in crosses the reduced pigmentation segregated along with an extra copy of the gene. Not only was the extra copy of the gene itself silent, but its presence caused the silencing of the wildtype copies of the gene in the same plant. The mechanism of this unexpected gene silencing remained a mystery until 1998 when researchers led by Andrew Fire and Craig C. Mello discovered that the presence of double-stranded RNA (dsRNA) produced such silencing in the nematode worm *Caenorhabditis elegans*, a finding for which they were awarded the 2006 Nobel Prize in Physiology or Medicine.

Gene silencing by dsRNA is an example of **RNA interference (RNAi)**. The ability to mount an RNAi response is widespread among eukaryotes and was probably present in the common ancestor, although this ability was lost in certain lineages including some fungi and parasitic protozoa. The molecular machinery of RNAi probably evolved originally as a defense against viruses and transposable elements that pass through a stage in which their genetic information is in the form of dsRNA. The silencing effect is highly specific and very potent, requiring only a few molecules of dsRNA per cell to be effective.

As might be expected of an RNAi response that evolved prior to the evolutionary diversification of eukaryotes, the mechanisms have been elaborated into several pathways that act somewhat differently, and organisms have made use of their components in multiple ways. Much is yet to be discovered and understood about the RNAi response, but some of the main outlines are clear. Two of the major pathways mediating an RNAi response are illustrated in Figure 11.35. Part A shows two major sources of dsRNA. That on the left is derived from transcription of the same duplex DNA molecule from both strands, as would happen if each strand had an upstream promoter. This is the type of dsDNA that would be produced by RNA viruses or certain transposable elements, and also that produced by the introduced chalcone-synthase gene in the petunias in **FIGURE 11.34**, in which the gene was inserted near a promoter that produced an *antisense transcript*, that is, a transcript from DNA strand that is not normally transcribed. When the antisense transcript formed an RNA duplex with the sense transcript, the resulting dsRNA set off the petunia RNAi response. In these types of dsRNA the two strands are exactly (or almost exactly) matching, and the dsRNA is the source of **small interfering RNA (siRNA)**.

**FIGURE 11.34** A bed of petunias with plants having wildtype flowers and those having white flowers. The white flowers are on plants with an extra copy of a gene for synthesis of the pigment anthocyanin, which was expected to make the flowers darker. Unexpectedly, the extra copy eliminated pigment altogether! [© Ivaschenko Roman/ Shutterstock, Inc.]

Both siRNA and miRNA are produced in the cytoplasm, and their pathways make use of similar components. One of these is an enzyme known as *dicer* (Figure 11.35B), which does what its name implies; it cleaves dsRNA into small double-stranded pieces about 25 base pairs in length with a short, single-stranded overhang at each end. Although the figure shows a single type of dicer enzyme, in many organisms the siRNA pathway and the miRNA pathway each uses its own version of dicer encoded in a different gene.

The short pieces of dsRNA produced by dicer are then incorporated into a RISC complex (Figure 11.35C), where RISC stands for *RNA-induced silencing complex*. Both strands are incorporated, but only one will serve as the *guide RNA* that identifies the target RNA by means of complementary base pairing; the other strand is a *passenger strand,* which in the activated RISC complex is degraded. As with dicer, in

On the right in **FIGURE 11.35**, part A is another source of dsRNA, in this case the stem of a stem-loop secondary structure formed in the transcript of a DNA duplex containing a duplicated sequence present in inverted orientation. (Such inverted repeat sequences are discussed in greater detail in Chapter 14.) Structures like the one shown are normally produced from longer transcripts by specialized enzymes present in the nucleus. The paired stem usually contains one or more base-pair mismatches, which accumulate in the inverted repeats as the genome evolves. These mismatches are the hallmark of another RNAi pathway mediated by **microRNA (miRNA)**.

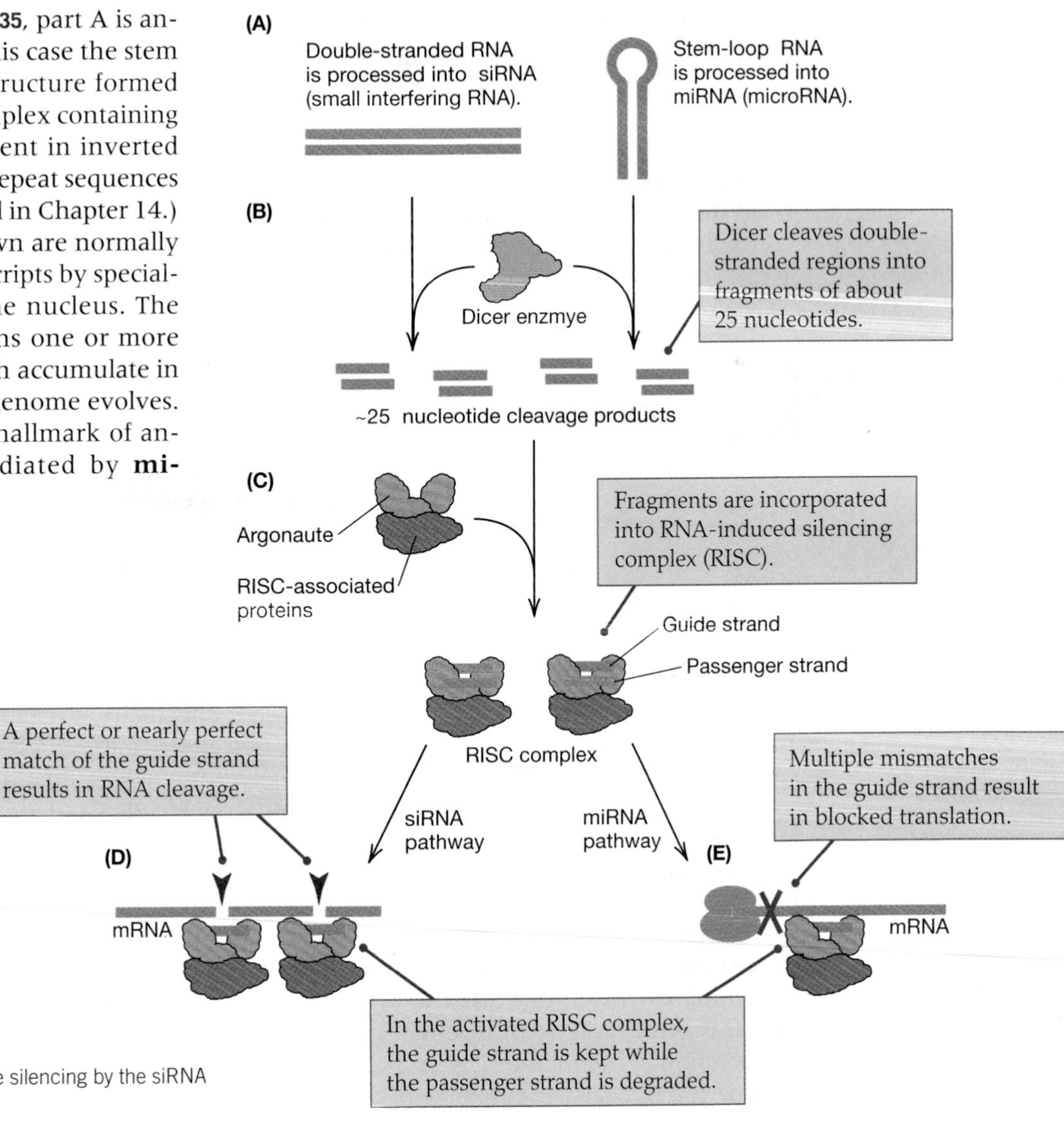


**FIGURE 11.35** Mechanisms of gene silencing by the siRNA and miRNA pathways.

## connection

# Double Trouble

*Weird and unexpected results began to be reported as soon as it became possible to introduce engineered RNA molecules into organisms. In extreme cases the engineered RNA prevented the expression of endogenous host genes with sequence homology. At first it seemed possible that the engineered RNA acted as an antisense inhibitor, in which the introduced RNA undergoes base pairing with the endogenous transcripts and interferes with their function. If this were true, the inhibitory effect of the introduced RNA should be strongly concentration-dependent. In this path-breaking paper, the authors show that introduced double-stranded RNA (dsRNA) mediates the inhibitory effects, and that only a few molecules per cell are required. The nematode worm* C. elegans *proved to be ideal for these experiments because, in contrast to some other organisms, dsRNA can be transported from cell to cell and from parent to offspring.*

**Andrew Fire[1], SiQun Xu[1], Mary K. Montgomery[1], Steven A. Kostas[1], Samuel E. Driver[2], and Craig C. Mello[2]**
**1998**
1. Carnegie Institution of Washington, Baltimore, Maryland
2. University of Massachusetts Medical School, Worcester, Massachusetts
*Potent and Specific Genetic Interference by Double-Stranded RNA in* Caenorhabditis Elegans

Experimental introduction of RNA into cells can be used in certain biological systems to interfere with the function of an endogenous gene. . . . Here we investigate the requirements for structure and delivery of the interfering RNA. To our surprise, we found that double-stranded RNA was substantially more effective at producing interference than was either strand individually. . . . Double-stranded RNA caused potent and specific interference. Only a few molecules of injected double-stranded RNA were required per affected cell, . . . suggesting that there could be a catalytic or amplification component of the interference process. . . . The *unc*-22 [*uncoordinated*-22] gene was chosen for initial comparisons of activity. [The] *unc*-22 [gene] encodes an abundant but nonessential myofilament protein. Several thousand copies of *unc*-22 mRNA are present in each striated muscle cell. . . . Decreases in *unc*-22 activity produce a severe twitching phenotype. . . . Purified antisense and sense RNAs covering a 742-nucleotide segment of *unc*-22 had only marginal interference activity, requiring very high doses of injected RNA to produce any observable effect. In contrast, a sense-antisense mixture produced highly effective interference with endogenous gene activity. The mixture was at least two orders of magnitude more effective than either strand alone. . . . The potent interfering activity of the sense-antisense mixture could reflect the formation of double-stranded RNA (dsRNA) or, conceivably, some other synergy between the strands. Electrophoretic analysis indicated that the injected material was predominantly double-stranded. . . . The phenotype produced by interference using *unc*-22 dsRNA was extremely specific. Progeny of injected animals exhibited behavior that precisely mimics loss-of-function mutations in *unc*-22. . . . Double-stranded RNA could conceivably mediate interference more generally in other nematodes, in other invertebrates, and, potentially, in vertebrates. RNA interference might also operate in plants. . . . Genetic interference by dsRNA could be used by the organism for physiological gene silencing.

***To our surprise, we found that double-stranded RNA was substantially more effective at producing interference than was either strand individually.***

Source: A. Fire, et al., *Nature* 391(1998): 806–810.

many organisms the components of the RISC complex—especially a key component known as *argonaute*—differ between the siRNA pathway and the miRNA pathway.

After RISC-complex formation, the siRNA and miRNA pathways function quite differently. In the siRNA pathway, the guide RNA matches the target RNA strand perfectly or almost perfectly, because the guide and the target are transcribed from opposite strands of the same duplex DNA. In this case the RISC complex cleaves the target RNA through the action of its version of argonaute (Figure 11.35D). It is the perfect or near-perfect match of the guide RNA that gives the siRNA pathway its great specificity. Normally only one target RNA is destroyed.

In the miRNA pathway (Figure 11.35E), the guide RNA and the target RNA generally include several mismatches because the guide RNA and the target RNA are transcripts from different regions in the genome. In this case the activated RISC complex attaches to the target RNA and, through its own version of argonaute, blocks translation. Because some mismatches are tolerated, a miRNA typically targets multiple transcripts, each from a different gene. In this way a single miRNA can regulate the expression of entire networks of genes. In humans, the miRNA pathway has been implicated in regulatory abnormalities in the cell cycle and the formation of tumors.

The RNAi response functions not only in regulating targets recognized through dsRNA, but components of the response are also implicated in genome structure and organization. In fission yeast, for example, argonaute protein is required for the induction and spread of heterochromatin, and dicer and other components of the siRNA pathway are used in maintaining the transcriptionally silent state of genes in heterochromatin.

The discovery of RNAi generated a great deal of excitement in genetics because of its potential in research as well as practical applications. The ability of introduced dsRNA to reduce the level of expression of genes having homologous transcripts is the basis of *genetic knockdowns* of activity. (They are called *knockdowns* rather than *knockouts* because the silencing is often incomplete.) In effect, the RNAi response affords a method of producing the equivalent of mutations that drastically reduce gene expression in organisms that do not have well-developed systems of mutagenesis and genetic manipulation.

RNAi also has important applications in biotechnology and medicine. In plant biotechnology, for example, RNAi has been used to reduce the production of toxins, much as chalcone synthase RNAi reduces the amount of anthocyanin pigment in petunia (Figure 11.34). In medicine, applications include novel therapies for controlling HIV, hepatitis, influenza, measles, and other viruses as well as new approaches to treat cancer and neurodegenerative diseases.

## 11.9 Translational Control

Because transcription and translation are uncoupled in eukaryotes, gene expression can be regulated at the level of translation separately from transcription. The principal types of translational control are:

- Inability of an mRNA molecule to be translated except under certain conditions
- Regulation of the overall rate of protein synthesis
- Aborted translation of the principal open reading frame because of the presence of a smaller open reading frame upstream in the mRNA
- Inhibition of translation by microRNAs that undergo base pairing with the mRNA (Figure 11.35E).

An important example of translational regulation is that of activating previously untranslated cytoplasmic mRNAs. This mechanism is prominent in early development, when newly fertilized eggs synthesize at a rapid rate many new proteins, virtually all of which derive from preexisting cytoplasmic mRNAs. In a few cases the molecular mechanism of mRNA activation is known. For example, in *Drosophila*, the mRNAs for the genes *bicoid*, *Toll*, and *torso* become activated because of the cytoplasmic elongation of their poly-A tail.

A dramatic example of translational control is the extension of the lifetime of silk fibroin mRNA in the silkworm. During cocoon formation, the silk gland predominantly synthesizes a single type of protein, silk fibroin, in very large amounts. The amount of fibroin is increased by three different mechanisms. First, the silk-gland cells become highly polyploid, accumulating thousands of copies of each chromosome. Second, transcription of the fibroin gene is initiated at a strong promoter, which results in the creation of about $10^4$ fibroin mRNA molecules per gene copy in a period of a few days. Third, the fibroin mRNA molecule has a very long lifetime. In contrast to a typical eukaryotic mRNA molecule, which has a lifetime of about 3 hours, fibroin mRNA survives for several days, during which each mRNA molecule is translated repeatedly to yield $10^5$ fibroin molecules. Thus each fibroin gene copy yields about $10^9$ protein molecules in the few days during which the cocoon is formed.

### ■ Small Regulatory RNAs Controlling Translation

Small regulatory RNAs that control translation have been described in both prokaryotes and eukaryotes, and analyses of genome sequences suggest that there will be many more examples. Although only a handful of cases have been studied in detail, most seem to involve regulatory RNAs that are complementary in sequence to part of the mRNA whose translation they control. We have already seen one example of translational inhibition by microRNA (Figure 11.35E).

An RNA sequence complementary to part of an mRNA is called an **antisense RNA**. The antisense regulatory RNAs act by pairing with the mRNA to either inhibit or activate translation (**FIGURE 11.36**). Although regulatory RNAs that control plasmid functions such as copy number target a single RNA, the bacterial RNAs discussed next often control translation of several mRNAs and serve as global regulators of cellular processes.

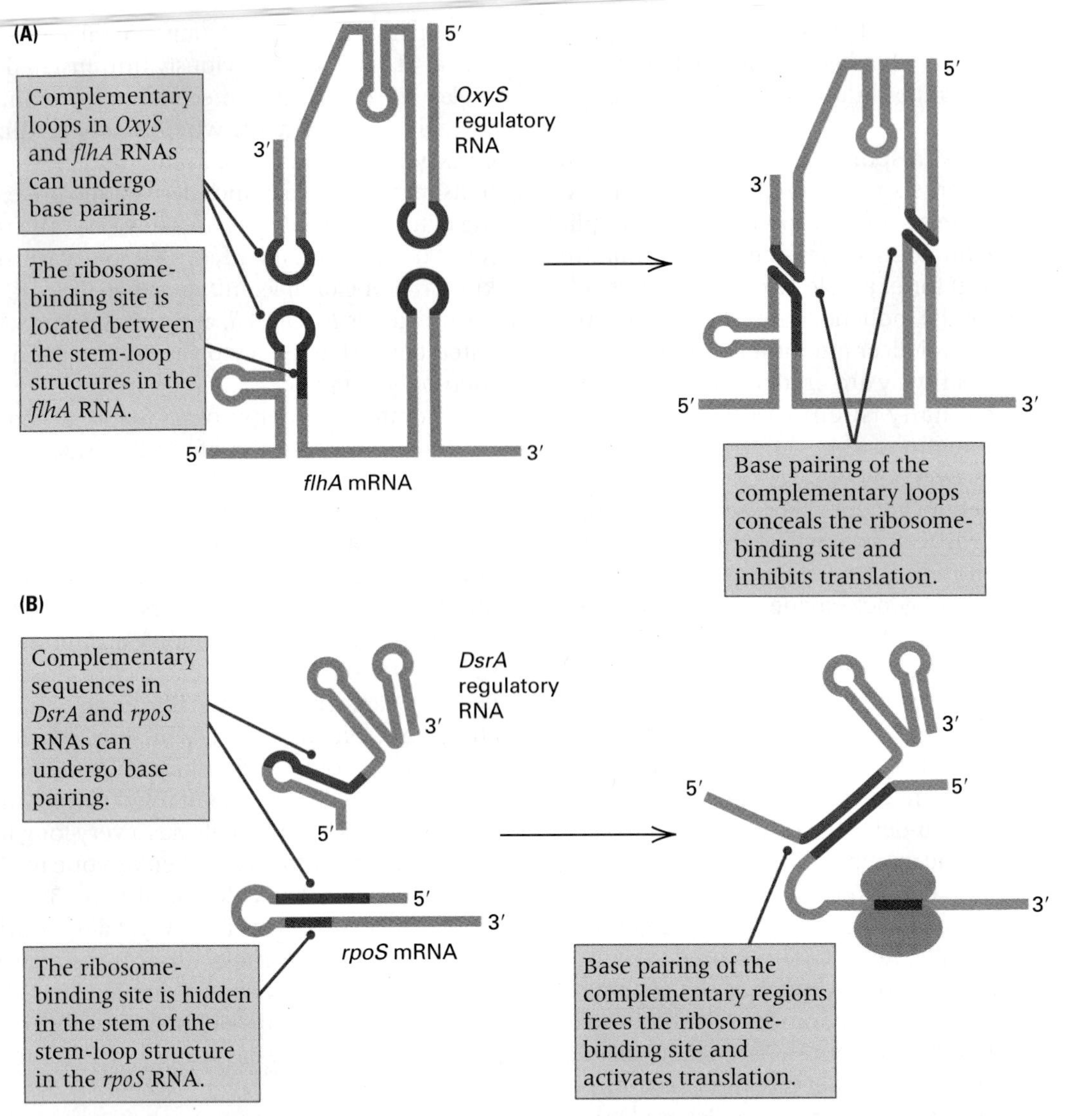
(A)
Complementary loops in OxyS and flhA RNAs can undergo base pairing.
The ribosome-binding site is located between the stem-loop structures in the flhA RNA.
5′
3′
OxyS regulatory RNA
5′
3′
flhA mRNA
5′
3′
5′
3′
Base pairing of the complementary loops conceals the ribosome-binding site and inhibits translation.
(B)
Complementary sequences in DsrA and rpoS RNAs can undergo base pairing.
DsrA regulatory RNA
3′
5′
5′
3′
rpoS mRNA
The ribosome-binding site is hidden in the stem of the stem-loop structure in the rpoS RNA.
3′
5′
5′
3′
Base pairing of the complementary regions frees the ribosome-binding site and activates translation.

**FIGURE 11.36** Regulation of translation of target mRNAs by (A) OxyS and (B) DsrA regulatory RNAs. [After S. Altuvia and E. G. H. Wagner. 2000. *Proc. Natl. Acad. Sci. USA* 97: 9824.]

Figure 11.36A shows an example of a small regulatory RNA that relieves oxidative stress in *E. coli*. One of the genes derepressed in the presence of hydrogen peroxide is *oxyS*, which encodes a regulatory RNA called OxyS. This RNA binds to several mRNAs. For any given interaction, only short stretches of OxyS are complementary to the target RNA and able to anneal with it. For example, two separate regions of OxyS bind to the mRNA of the gene *flhA*, which encodes a transcriptional activator protein (Figure 11.36A). The complementary regions are very short, in this example only seven nucleotides. One of the complementary regions is near the AUG translational start, and the other is more than forty nucleotides upstream. Base pairing between OxyS and the *flhA* mRNA conceals the ribosome-binding site and prevents translation. Such a bipartite complex composed of a small regulatory RNA and an mRNA is called a **kissing complex**.

Small regulatory RNAs can also activate translation. An example is the DsrA regulatory RNA from *E. coli* shown in Figure 11.36B. In this case the mRNA whose translation is controlled is from a gene *rpoS*, which encodes a sigma factor for RNA polymerase that allows transcription of a new set of RNAs from a special set of promoters at stationary phase in cell cultures when the cell density is high and the cells begin to slow their growth and division. The 5′ end of the *rpoS* mRNA is self-complementary and can form a hairpin that hides the ribosome-binding site and the translational start site. Two virtually contiguous regions of DsrA bind to the *rpoS* mRNA, and when binding occurs the ribosome-binding site is freed and translation can occur.

## 11.10 Programmed DNA Rearrangements

The introduction of changes in DNA sequence is an uncommon mechanism of gene regulation, but there are a number of important examples. In some cases the changes are reversible, but in others they are permanent. A permanent change in DNA sequence implies that in the cell lineage in which it takes place, the genotype becomes permanently altered. Such irreversible changes take place only in somatic cells, not in the germ line, so they are not genetically transmitted.

**NORMAN T-CELL** function requires the enzyme adenosine deaminase (ADA). This three-year-old boy had suffered from severe combined immunodeficiency disease (SCID), resulting from a mutant form of ADA. He spent the first months of his life inside a plastic bubble to protect him from fatal infections, and was among the first to benefit from gene therapy for SCID. After successful treatment, he is able to enjoy a day at the zoo like any other child. [© Peter Dejong/ AP Photos.]

### ■ Gene Amplification

In some cases, the number of gene copies increases temporarily by a process of **gene amplification**. An example is found in the development of the oocytes of the toad *Xenopus laevis*. The formation of an egg from its oocyte precursor is a complex process that requires a high rate of protein synthesis and consequently a great many (about $10^{12}$) ribosomes. Although they are already tandemly duplicated to about 600 copies, the number of rRNA genes is insufficient to produce so many ribosomes. In the development of the oocyte, the number of rRNA gene copies increases to about 2.4 million (an amplification by a factor of 4000). This amplification takes place over a 3-week period in oocyte development and comes about from the formation of a large number of extrachromosomal, replicating rolling circles of the rDNA genes. The molecular mechanism of excision of the circles from the chromosome and formation of the rolling circles is not known. When the oocyte is mature, no more rRNA needs to be synthesized, and the extrachromosomal copies of the rRNA cease to replicate and begin to be degraded. Amplification of rRNA genes during oogenesis occurs in many organisms, including insects, amphibians, and fish.

### ■ Antibody and T-Cell Receptor Variability

Perhaps the most impressive examples of programmed DNA rearrangement take place in the bone marrow–derived (B) cells and thymus–derived (T) cells that play key roles in the vertebrate immune system. B cells produce and secrete circulating antibodies that combine with foreign antigens and mark them for destruction, whereas T cells have specific receptors that recognize foreign antigens at the cell surface. Both antibodies and T-cell receptors have highly variable amino acid sequences from one B cell or T cell to the next, and much of this variability is generated by DNA rearrangements that assemble a relatively small number of coding sequences into a very large number of possible combinations, one in each B cell or T cell. We will examine the process of DNA splicing that generates diversity in antibodies. A similar process accounts for the variability of T-cell receptors.

A normal mammal is capable of producing more than $10^8$ different antibodies, only a fraction of which are produced at any one time. Each B cell can produce a single type of antibody, but the antibody is not secreted until the cell has been stimulated by the appropriate antigen. Once stimulated, the B cell undergoes successive mitoses and eventually produces a lineage of genetically identical cells that secrete the antibody. Antibody secretion may continue even if the antigen is no longer present.

There are five distinct classes of antibodies known as IgG, IgM, IgA, IgD, and IgE (Ig stands for **immunoglobulin**). These classes perform specialized functions in the immune response and exhibit certain structural differences. However, each contains two types of polypeptide chains differing in size: a large subunit called the *heavy (H) chain* and a small subunit called the *light (L) chain*. Immunoglobulin G (IgG) is the most abundant class of antibodies and has the simplest molecular structure (**FIGURE 11.37**). An IgG molecule consists of two heavy and two light chains held together by disulfide bridges (two covalently joined sulfur atoms) and has the overall shape of the letter Y. The sites for antibody specificity that combine with the antigen are located in the upper half of the arms above the fork of the Y. Each

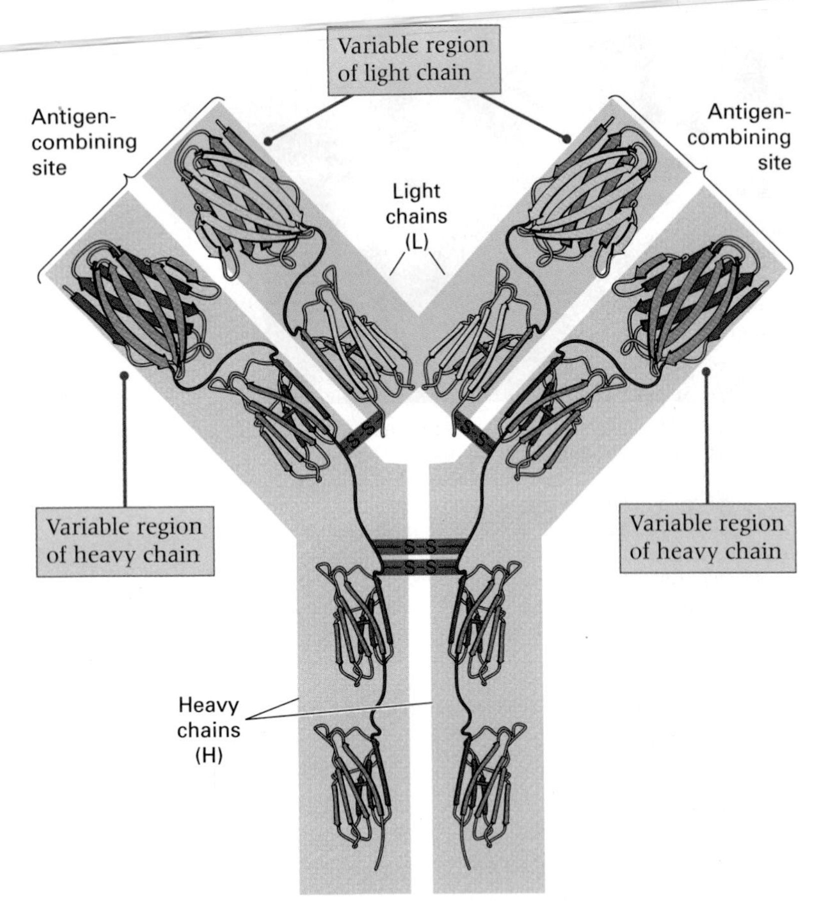


**FIGURE 11.37** Structure of the immunoglobulin G (IgG) molecule, showing the light chains (L, shaded blue) and heavy chains (H, shaded yellow). Variable and constant regions are indicated.

IgG molecule with a different specificity has a different amino acid sequence for the heavy and light chains in this part of the molecule. These specificity regions are called the **variable antibody regions** of the heavy and light chains. The remaining regions of the polypeptide are the **constant antibody regions**, which are called constant because they have virtually the same amino acid sequence in all IgG molecules.

Initial understanding of the genetic mechanisms responsible for variability in the amino acid sequences of antibody polypeptide chains came from cloning a gene for the light chain of IgG. The critical observation was made by comparing the nucleotide sequence of the gene in embryonic cells with that in mature antibody-producing B cells. In the genome of a B cell that was actively producing the antibody, the DNA segments corresponding to the constant and variable regions of the light chain were found to be very close together. However, in embryonic cells, these same DNA sequences were located far apart. Similar results were obtained for the variable and constant regions of the heavy chains. Segments encoding these regions were close together in B cells but widely separated in embryonic cells.

Complete DNA sequencing of the genomic region that codes for antibody proteins revealed not only the reason for the different gene locations in B cells and embryonic cells but also the mechanism for the origin of antibody variability. Cells in the germ line contain a small number of genes corresponding to the constant region of the light chain, and these are close together along the DNA. Separated from them, but on the same chromosome, is another cluster consisting of a much larger number of genes that correspond to the variable region of the light chain. In the differentiation of a B cell, one gene for the constant region is spliced (cut and joined) to one gene for the variable region, and this splicing produces a complete light-chain antibody gene. A similar splicing mechanism yields the constant and variable regions of the heavy chains.

The formation of a finished antibody gene is somewhat more complicated than this overview implies, because light-chain genes consist of three parts and heavy-chain genes consist of four parts. DNA splicing in the origin of a light chain is illustrated in **FIGURE 11.38**. For each of two parts of the variable region, the germ line contains multiple coding sequences called the **variable (V)** and **joining (J) antibody regions**. In the differentiation of a B cell, a deletion makes possible the joining of one of the V regions with one of the J regions. The DNA joining process is called **V–J joining**, and it can create many combinations of the V and J regions. Since the transcriptional enhancer lies in the region between the rightmost J region and the C region, the rearranged gene always contains the transcriptional enhancer. This also ensures that only the rearranged gene will be transcribed. When transcribed, the joined V–J sequence forms the 5′ end of the light-chain RNA transcript. Transcription continues on through the DNA region coding for the **constant (C) antibody region** of the gene. RNA splicing subsequently joins the V–J region with the C region, creating the light-chain mRNA. Combinatorial joining also takes place in the genes for the antibody heavy chains, but in this case the DNA splicing joins the heavy-chain counterparts of V and J with a third set of sequences, called D (for diversity), located between the V and J clusters.

The level of antibody variability created by combinatorial joining can be calculated. In mice, the light chains are formed from combinations of about 250 V regions and 4 J regions, giving $250 \times 4 = 1000$ different possible chains. For

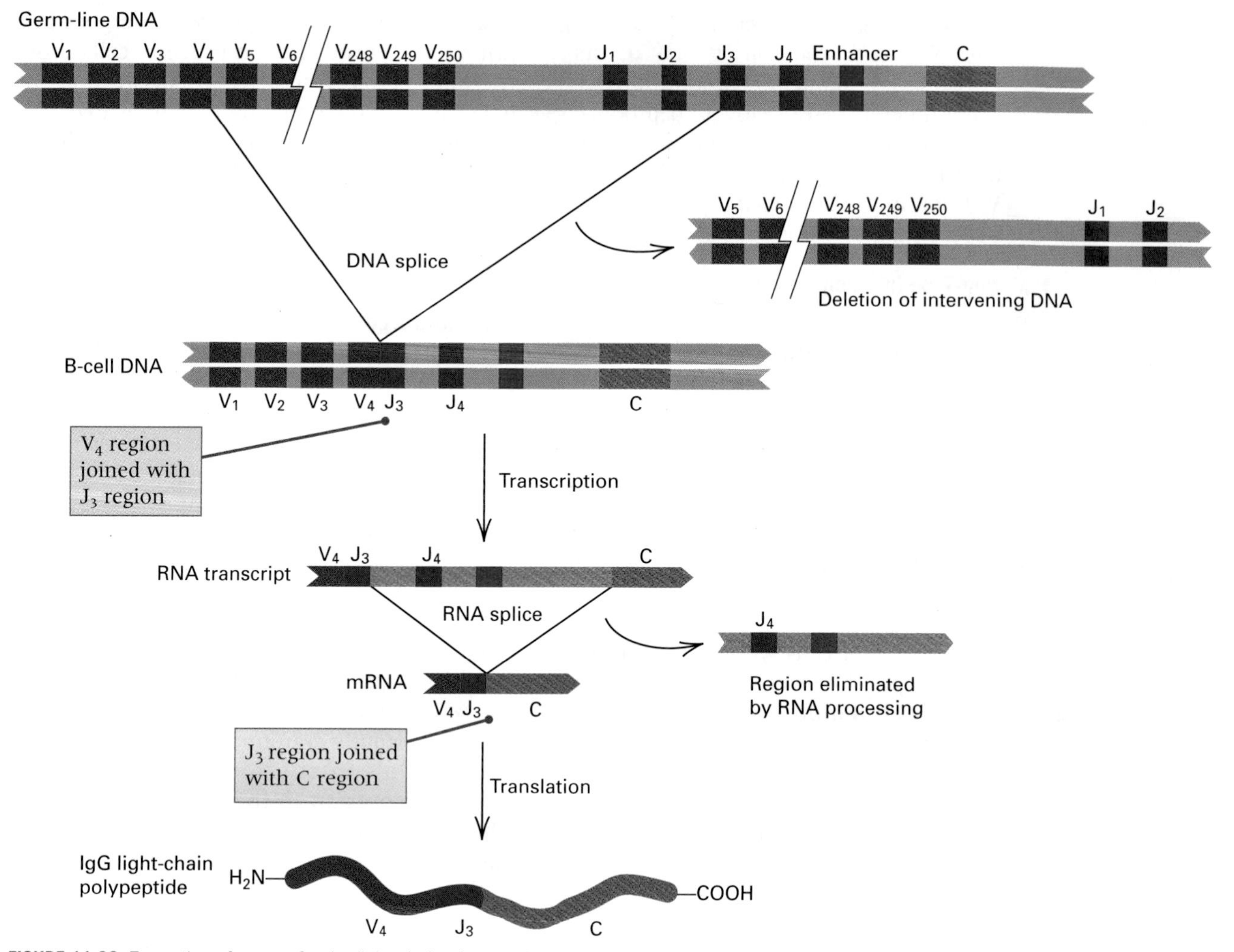


**FIGURE 11.38** Formation of a gene for the light chain of an antibody molecule. One variable (V) region is joined with one randomly chosen J region by deletion of the intervening DNA. The remaining J regions are eliminated from the RNA transcript during RNA processing.

the heavy chains, there are approximately 250 V, 10 D, and 4 J regions, producing $250 \times 10 \times 4 = 10{,}000$ possible combinations. Because any light chain can combine with any heavy chain, there are at least $1000 \times 10{,}000 = 10^7$ possible types of antibodies. The number of DNA sequences used for antibody production is quite small (about 500), but the number of possible antibodies is very large.

The value of $10^7$ different antibody types is an underestimate, because there are two additional sources of antibody variability:

1. The junction for V–J (or V–D–J) splicing in combinatorial joining can be between different nucleotides of a particular V–J combination in light chains (or a particular V–D or D–J combination in heavy chains). The different splice junctions can generate different codons in the spliced gene. For example, a particular combination of V and J sequences can be spliced in five different ways. At the splice junction, the V sequence contains the nucleotides CATTTC, and the J sequence contains CTGGGTG. The splicing event determines the codons for amino acids 97, 98, and 99 in the completed antibody light chain, and it can occur in any of the following ways, the last two of which result in altered amino acid sequences.

| **Spliced sequence** | **97** | **98** | **99** |
|---|---|---|---|
| CACTGGGTG | His | Trp | Val |
| CATTGGGTG | His | Trp | Val |
| CATTGGGTG | His | Trp | Val |
| CATTTGGTG | His | Leu | Val |
| CATTTCGTG | His | Phe | Val |

In this manner, variability in the junction of V–J joining can result in polypeptides that differ in amino acid sequence.

2. The V regions are susceptible to a high rate of *somatic mutation,* which occurs in B-cell development. These mutations allow different B-cell clones to produce different polypeptide sequences, even if they have undergone exactly the same V–J joining. The mechanism for the hypermutation is unknown.

## ■ Mating-Type Interconversion

Rearrangement of DNA sequences underlies the phenomenon of **mating-type interconversion** in yeast. As we saw in Chapter 5, *S. cerevisiae* has two mating types, denoted **a** and $\alpha$. Mating between haploid **a** and haploid $\alpha$ cells produces the **a**$\alpha$ diploid, which can undergo meiosis to produce four-spored asci containing haploid **a** and $\alpha$ spores in the ratio 2 : 2. If a single yeast spore of either the **a** or the $\alpha$ genotype is cultured in isolation from other spores, then mating between progeny cells would not be expected because the daughter cells would all have the same mating type. However, *S. cerevisiae* has a mating system called **homothallism**, in which some cells undergo a conversion into the opposite mating type that allows matings between cells in what would otherwise be a pure culture.

The outlines of mating-type interconversion are shown in **FIGURE 11.39**. An original haploid spore (in this example, $\alpha$) undergoes germination to produce two progeny cells. Both the mother cell (the original parent) and the daughter cell have mating type $\alpha$, as expected from a normal mitotic division. However, in the next cell division, a switching (interconversion) of mating type takes place in both the mother cell and its *new* progeny cell, in which the original $\alpha$ mating type is replaced with the **a** mating type. After this second cell division is complete, the $\alpha$ and **a** cells are able to undergo mating because they now are of opposite mating types. Fusion of the nuclei produces the **a**$\alpha$ diploid, which undergoes mitotic divisions and later sporulation to produce **a** and $\alpha$ haploid spores again.

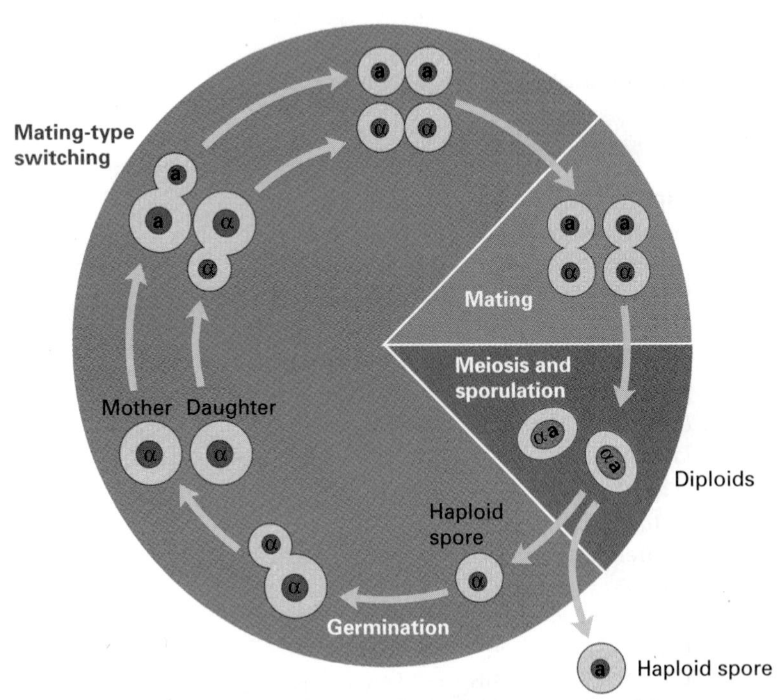


**FIGURE 11.39** Mating-type switching in the yeast *Saccharomyces cerevisiae.* Germination of a spore (in this example, one of mating type $\alpha$) forms a mother cell and a bud that grows into a daughter cell. In the next division, the mother cell and its new daughter cell switch to the opposite mating type (in this case, **a**). The result is two $\alpha$ and two **a** cells. Cells of opposite mating type can fuse to form **a**$\alpha$ diploid zygotes. In a similar fashion, germination of an **a** spore will result in switching to the $\alpha$ mating type.

The genetic basis of mating-type interconversion is DNA rearrangement, as outlined in **FIGURE 11.40**. The gene that controls mating type is the *MAT* gene in chromosome III, which can have either of two allelic forms, **a** or $\alpha$. If the allele in a haploid cell is *MAT***a**, then the cell has mating type **a**; if the allele is *MAT*$\alpha$, then the cell has mating type $\alpha$. However, both genotypes normally contain both **a** and $\alpha$ genetic information in the form of unexpressed **cassettes** present in the same chromosome. The *HML*$\alpha$ cassette, about 200 kb distant from the *MAT* gene, contains the $\alpha$ DNA sequence; and the *HMR***a** cassette, about 150 kb distant from *MAT* on the other side, contains the **a** DNA sequence. When mating-type interconversion occurs, a specific endonuclease, encoded by the *HO* gene elsewhere in the genome, is produced and cuts both strands of the DNA in the *MAT* region. The double-stranded break initiates a process in which genetic information in the unexpressed cassette containing the opposite mating type becomes inserted into *MAT*. In this process, the DNA sequence in the donor cassette is duplicated, so the mating type becomes converted, but the same genetic information is retained in unexpressed form in the cassette. The terminal regions of *HML, MAT,* and *HMR* are identical (they are shown in light blue and dark blue in Figure 11.40), and these regions are critical in pairing of the regions and interconversion. The unique part of the $\alpha$ region is 747 base pairs in length; that of the **a** region is 642 base pairs long. The molecular details of the conversion process are similar to those of the double-strand gap mechanism of recombination discussed in Chapter 6.

Figure 11.40 illustrates two sequential mating-type interconversions. In the first, an $\alpha$ cell (containing the *MAT*$\alpha$ allele) undergoes conversion into **a**, using the DNA sequence contained in the *HMR***a** cassette. The converted cell has the genotype *MAT***a**. In a later generation, a descendant **a** cell may become converted into mating type $\alpha$, using the unexpressed DNA sequence contained in *HML*$\alpha$. This cell has the genotype *MAT*$\alpha$. Mating-type switches can occur repeatedly in the lineage of any particular cell.

## Transcriptional Control of Mating Type

Although mating-type *interconversion* is controlled by DNA rearrangements, mating type itself is controlled at the level of transcription. Both *MAT***a** (mating type **a**) and *MAT*$\alpha$ (mating type $\alpha$) express a set of haploid-specific genes. They differ in that *MAT***a** expresses a set of **a**-specific genes and *MAT*$\alpha$ expresses a set of $\alpha$-specific genes. The haploid-specific genes expressed in cells of both mating types include *HO*, which encodes the HO endonuclease used in mating-type interconversion, and *RME1*, which encodes a repressor of meiosis-specific genes. The functions of the expressed genes that differ in the mating types include (1) secretion of a mating peptide that arrests cells of the opposite mating type before DNA synthesis and prepares them for cell fusion, and (2) production of a receptor for the mating peptide secreted by the opposite mating type. Therefore, when **a** and $\alpha$ cells are in proximity, they prepare each other for mating and undergo fusion.

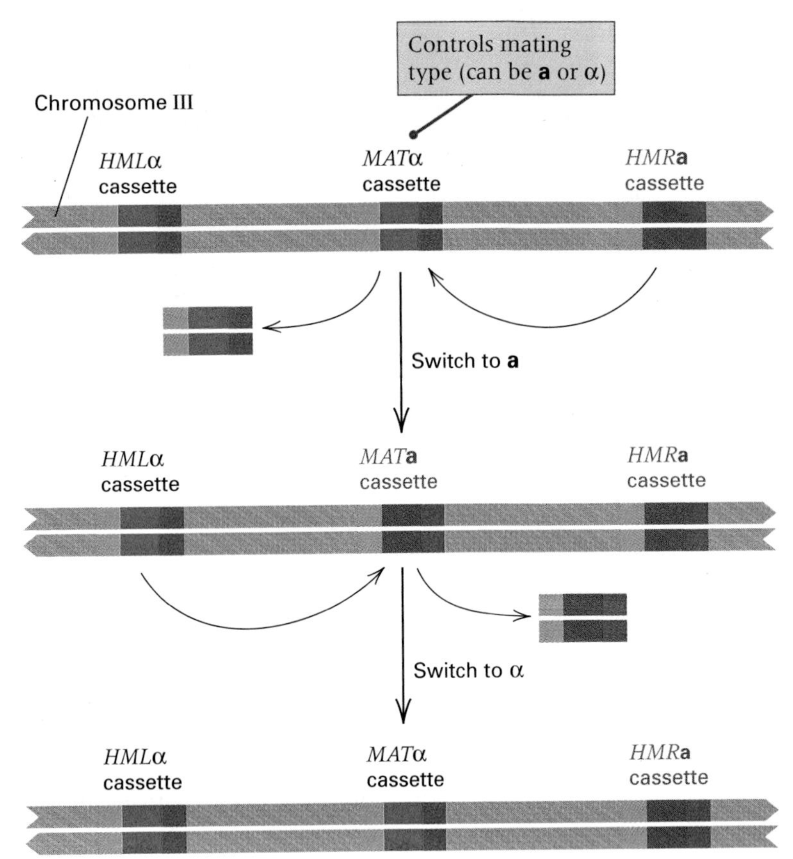


**FIGURE 11.40** Genetic basis of mating-type interconversion. The mating type is determined by the DNA sequence present at the *MAT* locus. The *HML* and *HMR* loci are cassettes that contain unexpressed mating-type genes, either $\alpha$ or **a**. In the interconversion from $\alpha$ to **a**, the $\alpha$ genetic information present at *MAT* is replaced with the **a** genetic information from *HMR***a**. In the switch from **a** to $\alpha$, the **a** genetic information at *MAT* is replaced with the $\alpha$ genetic information from *HML*$\alpha$.

Regulation of mating type at the level of transcription takes place according to the regulatory interactions diagrammed in **FIGURE 11.41**. These regulatory interactions were originally proposed on the basis of the phenotypes of various types of mutants, and most of the details have been confirmed by direct molecular studies. The symbols ***asg***, *$\alpha$sg*, and *hsg* represent the **a**-specific genes, the $\alpha$-specific genes, and the haploid-specific genes, respectively; each set of genes is represented as a single segment, and lack of a "sunburst" indicates that transcription does not take place. In a cell of mating type **a** (Figure 11.41A), the *MAT***a** region is transcribed and produces a polypeptide called **a**1. By itself, **a**1 has no regulatory activity, and in the absence of any regulatory signal, ***asg*** and *hsg*, but not *$\alpha$sg*, are transcribed. In a cell of mating type $\alpha$ (Figure 11.41B), the *MAT*$\alpha$ region is transcribed, and two regulatory proteins denoted $\alpha$1 and $\alpha$2 are produced: $\alpha$1 is a *positive regulator* of the $\alpha$-specific genes, and $\alpha$2 is a *negative regulator* of the **a**-specific genes. The result is that *$\alpha$sg* and *hsg* are transcribed, but transcription of ***asg*** is turned off. Both $\alpha$1 and $\alpha$2 bind with particular DNA sequences upstream from the genes that they control.

In the diploid (Figure 11.41C), both *MAT***a** and *MAT*$\alpha$ are transcribed, but the only polypeptides produced are **a**1 and $\alpha$2. The reason is that the **a**1 and $\alpha$2 polypeptides combine to form a negative regulatory protein that represses transcription of the $\alpha$1 gene in *MAT*$\alpha$ and of the haploid-specific genes. The $\alpha$2 polypeptide acting alone is a negative regulatory protein that turns off ***asg***. Because $\alpha$1 is not produced, transcription of *$\alpha$sg* is not turned on. The overall result is that the *$\alpha$sg* are not turned on because $\alpha$1 is absent, the ***asg*** are turned off because $\alpha$2 is present, and the *hsg* are turned off by the **a**1/$\alpha$2 complex. This ensures that meiosis can occur (because expression of *RME1* is turned off) and that mating-type switching ceases (because the HO endonuclease is absent). Thus, the homothallic **a**$\alpha$ diploid is stable and can undergo meiosis. The result is that the **a**$\alpha$ diploid does not transcribe either the mating-type-specific set of genes or the haploid-specific genes.

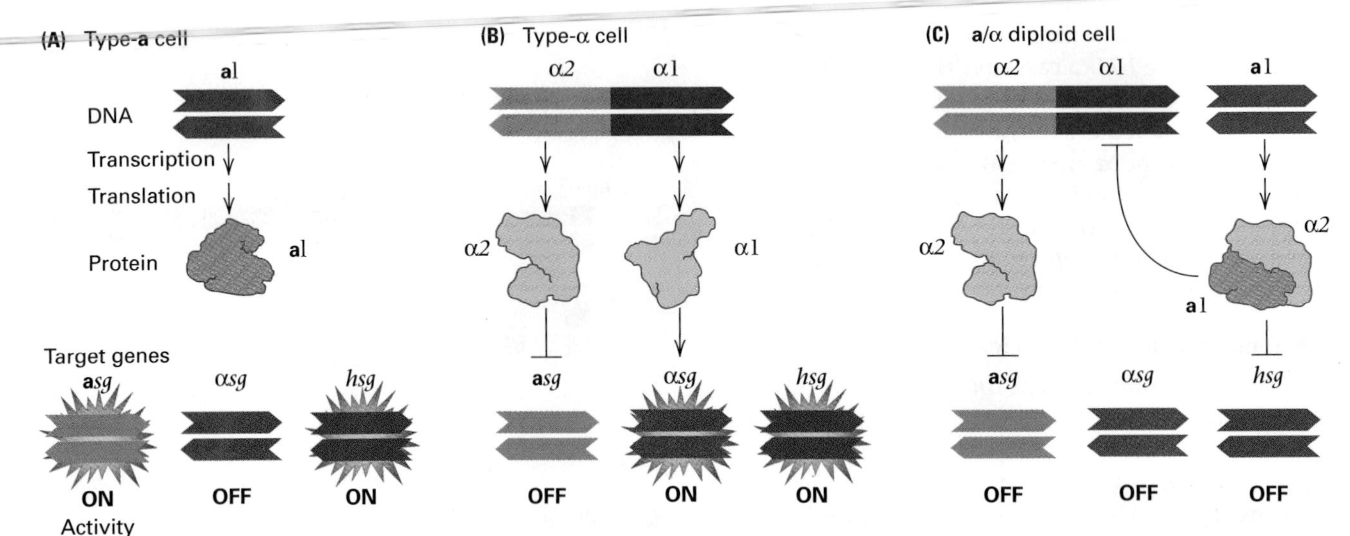


**FIGURE 11.41** Transcriptional regulation of mating type in yeast. The symbols **a***sg*, $\alpha$*sg*, and *hsg* denote sets of **a**-specific genes, $\alpha$-specific genes, and haploid-specific genes, respectively. Sets of genes represented with a "sunburst" are "on," and those without are "off." (A) In an **a** cell, the **a**1 peptide is inactive, and the sets of genes manifest their default states of activity (**a***sg* and *hsg* on and $\alpha$*sg* off), so the cell is an **a** haploid. (B) In an $\alpha$ cell, the $\alpha 2$ peptide turns the **a***sg* off and the $\alpha 1$ peptide turns the $\alpha$*sg* on, so the cell is an $\alpha$ haploid. (C) In an **a**/$\alpha$ diploid, the $\alpha 2$ and **a**1 peptides form a complex that turns $\alpha 1$ and the *hsg* off, the $\alpha 2$ peptide turns the **a***sg* off, and the $\alpha$*sg* manifest their default activity of off, so physiologically the cell is non-**a**, non-$\alpha$, and nonhaploid (that is, it is a normal diploid).

The repression of transcription of the haploid-specific genes mediated by the **a**1/$\alpha 2$ protein is an example of negative control of the type already familiar from the *lac* and *trp* systems in *E. coli*. The interesting twist in the yeast example is that the $\alpha 2$ protein has a regulatory role of its own in repressing transcription of the **a**-specific genes. Why does the $\alpha 2$ protein, on its own, not repress the haploid-specific genes as well? The answer lies in the specificity of its DNA binding. By itself, the $\alpha 2$ protein has low affinity for the target sequences in the haploid-specific genes. However, the **a**1/$\alpha 2$ heterodimer has both high affinity and high specificity for the target DNA sequences in the haploid-specific genes.

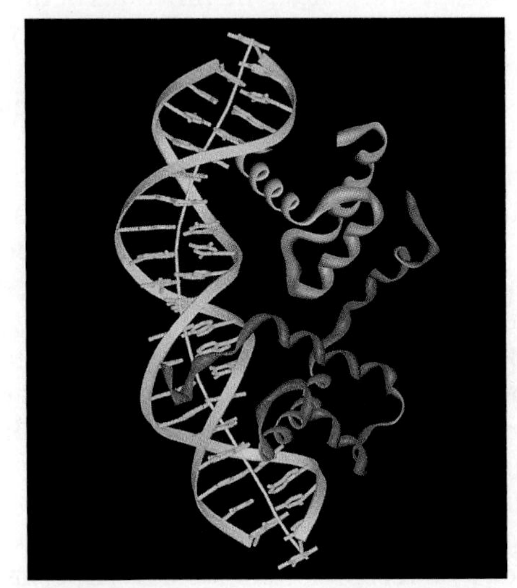

**STRUCTURE OF THE** a1/$\alpha 2$ protein bound with DNA. The **a**1 subunit is shown in blue, the $\alpha 2$ subunit in red. Contact with the DNA target results in a sharp bend in the DNA. [Reproduced from T. Li, et al., *Science* 270 (1995): 262–269. Reprinted with permission from AAAS.]

## CHAPTER SUMMARY

- Control of transcription is an important level of gene regulation, with mechanisms ranging from complete negative regulation ("on unless turned off") to complete positive regulation ("off unless turned on").
- Most genes are subject to multiple, overlapping regulatory mechanisms.
- In prokaryotes, the genes coding for the enzymes in a metabolic pathway are often clustered in the genome and controlled coordinately by a regulatory protein that binds with an operator region at the 5′ end of the cluster. This type of gene organization is known as an operon.
- In eukaryotes, most genes are not organized into operons. Genes at dispersed locations in the genome are coordinately controlled by one or more enhancer DNA sequences located near each gene that interact with transcriptional activator proteins.
- The transcription complex in eukaryotes consists of numerous protein components that are recruited to the promoter of a gene whose chromatin has been suitably reconfigured.
- Epigenetic mechanisms of transcriptional control are hereditary changes in gene expression mediated by modification of the DNA bases (usually cytosine methylation) or by binding with regulatory proteins.
- Gene expression can also be regulated at the level of RNA processing, alternative patterns of splicing, or transcript or mRNA stability.
- Double-stranded RNA molecules can be cleaved into short fragments that are used to target homologous RNA transcripts for cleavage or for blocking translation.
- Programmed DNA rearrangements underlie some genetic processes, such as the origin of antibody variability.

## REVIEW THE BASICS

- What is positive regulation of transcription? What is negative regulation of transcription? What is the role of the repressor in each case? Give an example of each type of regulation.
- What is autoregulation? Distinguish between positive and negative autoregulation. Which would be used to amplify a weak induction signal? Which would be used to prevent overproduction?
- What classes of *lac* mutants demonstrated that the presence of lactose in the growth medium was not necessary for expression of the genes for lactose utilization?
- How does an operon result in coordinate control of the genes included? Are operons usually found in eukaryotic organisms?
- In what sense does attentuation provide a "fine-tuning" mechanism for operons that control amino acid biosynthesis?
- An operon containing genes that encode the enzymes in a metabolic pathway for the synthesis of an amino acid includes a short, open reading frame in the leader sequence. This short, open reading frame contains multiple codons for the amino acid synthesized by the pathway. What does this observation suggest about regulation of the operon?
- What is a transcriptional activator protein? A transcriptional enhancer? A chromatin remodeling complex? What role do these elements play in eukaryotic gene regulation?
- How does the possibility of alternative splicing affect the generality of the statement that one gene encodes one polypeptide chain?
- What is meant by the term *epigenetic regulation*? Explain how epigenetic regulation can be mediated through cytosine methylation.
- What is the phenomenon of RNA interference (RNAi)? How is RNAi used in genetic analysis?
- Would you expect DNA splicing of the type observed in antibody formation to be a reversible process or an irreversible process? Why?

## GUIDE TO PROBLEM SOLVING

**Problem 1** The accompanying table shows genotypes for a number of *lac* operons in which the *z* gene has been genetically engineered. The symbol *zgfp* represents an engineered genes that encodes a green fluorescent protein; cells that produce this protein fluoresce green. The symbol *zrfp* represents an engineered gene that encodes a red fluorescent protein; cells that produce this protein fluoresce red. Cells that produce neither protein do not fluoresce, and those that produce both proteins fluoresce yellow (resulting from the mixture of green and red). Complete the table by entering red, green, yellow, or none corresponding to the fluorescence of cells of each genotype when grown in the absence (–) or presence (+) of inducer.

| | Genotype | + Inducer | – Inducer |
|---|---|---|---|
| **a.** | $i^+ o^c$ *zgfp* | | |
| **b.** | $i^- o^+$ *zgfp* | | |
| **c.** | $i^s o^+$ *zgfp* | | |
| **d.** | F' $i^+ o^c$ *zrfp* / $i^+ o^+$ *zgfp* | | |
| **e.** | F' $i^+ o^+$ *zrfp* / $i^+ o^c$ *zgfp* | | |
| **f.** | F' $i^- o^c$ *zrfp* / $i^+ o^+$ *zgfp* | | |
| **g.** | F' $i^- o^+$ *zrfp* / $i^+ o^c$ *zgfp* | | |
| **h.** | F' $i^s o^c$ *zrfp* / $i^+ o^+$ *zgfp* | | |
| **i.** | F' $i^s o^+$ *zrfp* / $i^+ o^c$ *zgfp* | | |

**Answer** The phenotypes can be deduced from the following principles: (1) the wildtype repressor is diffusible; (2) the wildtype repressor product can bind with $o^+$ but not with $o^c$; (3) the wildtype repressor binds with inducer; (4) the $i^-$ mutant repressor cannot bind with either $o^+$ or with $o^c$; (5) the $i^s$ mutant repressor binds with $o^+$ but cannot bind with $o^c$; (6) the $i^s$ mutant repressor does not bind with inducer. In the absence of inducer, the fluorescence phenotypes are (**a**) green, (**b**) green, (**c**) none, (**d**) red, (**e**) green, (**f**) red, (**g**) green, (**h**) red, (**i**) green. In the presence of inducer, the fluorescence phenotypes are (**a**) green, (**b**) green, (**c**) none, (**d**) yellow, (**e**) yellow, (**f**) yellow, (**g**) yellow, (**h**) red, (**i**) green.

**Problem** Consider genetically engineered strains of yeast in which red fluorescent protein (rfp) or green fluorescent protein (gfp) is fused to either GAL80p or GAL4p without changing any of the binding properties of GAP80p or GAL4p that are relevant to regulation of the *GAL* genes. Any cellular compartment in which the red fluorescent protein is concentrated fluoresces red, and any cellular compartment in which the green fluorescent protein is concentrated fluoresces green. If both are together in the same compartment, the compartment fluoresces yellow owing to the combination of red and green. Complete the table by entering red, green, yellow, or none corresponding to the fluorescence of the cytoplasm and nucleus of each genotype when grown in the absence (–) or presence (+) of galactose. For proteins that are transported into the nucleus, you may ignore the small amount of fluorescent signal that results from their translation in the cytoplasm. The *gal3[ko]* allele is a knockout mutation that eliminates GAL3p, and the *gal4[nls]* allele is a mutant *GAL4* in which the nuclear localization signal is deleted, causing the protein to remain in the cytoplasm.

| | | – Galactose | | + Galactose | |
|---|---|---|---|---|---|
| | Genotype | Cytoplasm | Nucleus | Cytoplasm | Nucleus |
| a. | *GAL3 GAL80–rfp GAL4–gfp* | | | | |
| b. | *GAL3 GAL80–gfp GAL4–rfp* | | | | |
| c. | *gal3[ko] GAL80–gfp GAL4–rfp* | | | | |
| d. | *GAL3 GAL80–rfp GAL4[nls]–gfp* | | | | |
| e. | *gal3[ko] GAL80–gfp GAL4[nls]–rfp* | | | | |

**Answer** The phenotypes can be deduced from the following principles: (1) in the presence of galactose, GAL3p binds GAL80p and keeps it in the cytoplasm; (2) in the absence of galactose, GAL3p does not bind GAL80p, which is transported into the nucleus; (3) GAL4p has a nuclear localization signal causing its transport into the nucleus. In the absence of galactose, the fluorescence phenotypes for the cytoplasm are (**a**) none, (**b**) none, (**c**) none, (**d**) green, (**e**) red; and for the nucleus they are (**a**) yellow, (**b**) yellow, (**c**) yellow, (**d**) red, (**e**) green. In the presence of galactose, the fluorescence phenotypes for the cytoplasm are (**a**) red, (**b**) green, (**c**) none, (**d**) yellow, (**e**) red; and for the nucleus they are (**a**)green, (**b**) red, (**c**) yellow, (**d**) none, (**e**) green.

**Problem** With regard to mating type in yeast, what phenotypes would each of the following haploid cells exhibit?

**(a)** A duplication of the mating-type gene giving the genotype *MAT***a**–*MAT*α

**(b)** A deletion of the *HML*α cassette in a *MAT***a** cell

**Answer**

**(a)** The haploid cell expresses both **a** and α, so the **a**-specific genes, the α-specific genes, and the haploid-specific genes are all inactive. The phenotype is similar to the **a**α diploid, but the cell is actually haploid. The cell will not mate, and if it attempts to sporulate, it will self-destruct. **b.** The phenotype is that of a normal **a** haploid, but mating-type switching to α cannot take place.

## ANALYSIS AND APPLICATIONS

**11.1** Is it necessary for the gene that codes for the repressor of a bacterial operon to be near the structural genes? Why or why not? Comparisons of the genomes of bacterial species related to *E. coli* clearly show that the *lac* operon is present in a genomic island introduced into the *E. coli* genome from another bacterial species at least 50 million years ago. In this particular case, how would the tight linkage of the *lacI* gene with the other *lac* genes have contributed to the maintenance of such a genomic island?

**11.2** The metabolic pathway for glycolysis, which is responsible for the degradation of glucose, is one of the fundamental energy-producing systems in living cells. Would you expect the enzymes in this pathway to be regulated? Why or why not?

**11.3** Among mammals, the reticulocyte cells in the bone marrow lose their nuclei in the process of differentiation into red blood cells. Yet the reticulocytes and red blood cells continue to synthesize hemoglobin. Suggest a mechanism by which hemoglobin synthesis can continue for a long period of time in the absence of the hemoglobin genes.

**11.4** Several eukaryotes are known in which a single effector molecule regulates the synthesis of different proteins encoded in distinct mRNA molecules, which for concreteness we may call X and Y. At what level in the process of gene expression does regulation occur in each of the following situations?

**(a)** Neither nuclear nor cytoplamic RNA can be found for either X or Y

**(b)** Nuclear but not cytoplasmic RNA can be found for both X and Y

**(c)** Both nuclear and cytoplasmic RNA can be found for both X and Y, but none of it is associated with polysomes

**11.5** Consider a eukaryotic transcriptional activator protein that binds to an enhancer sequence and promotes transcription. What change in regulation would you expect from a duplication in which several copies of the enhancer were present instead of just one?

**11.6** In studies of the operator region of an inducible operon in *E. coli*, the four constructs shown here were examined for level of transcription *in vitro*. The number associated with each construct is the relative level of transcription observed in the presence of the repressor protein. The symbols *E*, *B*, *H*, and *S* stand for the restriction sites *Eco*RI, *Bam*HI, *Hin*dIII, and *Sac*I. Construct **(a)** is the wildtype operator region, and in parts **(b–d)** the open boxes indicate restriction fragments that were deleted. What hypothesis about repressor–operator interactions can explain these results? How could this hypothesis be tested?

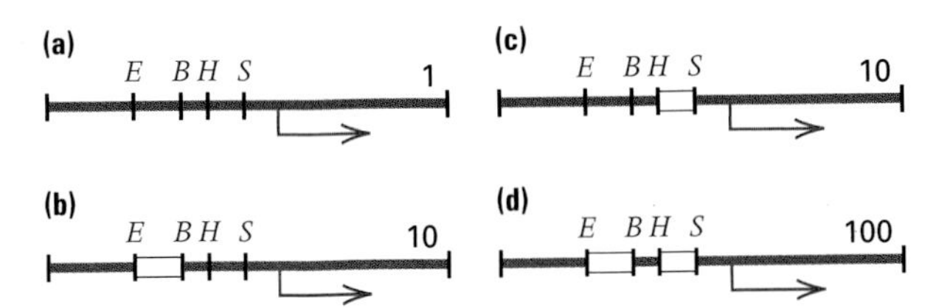


**11.7** The following questions pertain to the *lac* operon in *E. coli*.
- **(a)** Which proteins are regulated by the repressor?
- **(b)** How does binding of the *lac* repressor to the *lac* operator prevent transcription?
- **(c)** Is production of the *lac* repressor constitutive or inducible?

**11.8** Would you expect the regulation of a gene to be affected by an inversion of its promoter sequence? By an inversion of an enhancer sequence? Explain your answers.

**11.9** How do inducers enable transcription to occur in a bacterial operon that is under negative transcriptional control?

**11.10** The permease of *E. coli* that transports an alpha-galactoside known as melibiose can also transport lactose, but the melibiose permease is temperature sensitive: Lactose can be transported into the cell by the melibiose permease at 30°C but not at 37°C. In a strain that produces the melibiose permease constitutively, what are the phenotypes of *lacZ* and *lacY* mutants at 30°C and 37°C?

**11.11** The operon allows a type of coordinate regulation of gene activity in which a group of enzymes with related functions are synthesized from a single polycistronic mRNA. Does this imply that all proteins in the polycistronic mRNA are made in the same quantity? Explain.

**11.12** When glucose is present in an *E. coli* cell, is the concentration of cyclic AMP high or low? Can a mutant with either an inactive adenyl cyclase gene or an inactive *crp* gene synthesize beta-galactosidase? Does the binding of cAMP–CRP to DNA affect the binding of a repressor?

**11.13** Genes *A* and *B* are partially overlapping in the manner shown in the accompanying diagram, in which $P_A$ and $P_B$ are the promoters and the arrows show the origin, direction, and extent of transcription. The overlap includes a protein-coding exon. For mutations that take place in the protein-coding exon in the region of overlap, what are the consequences for reading frame B of a mutation that, in reading frame A, is:
- **(a)** A missense mutation?
- **(b)** A deletion?
- **(c)** A frameshift mutation?

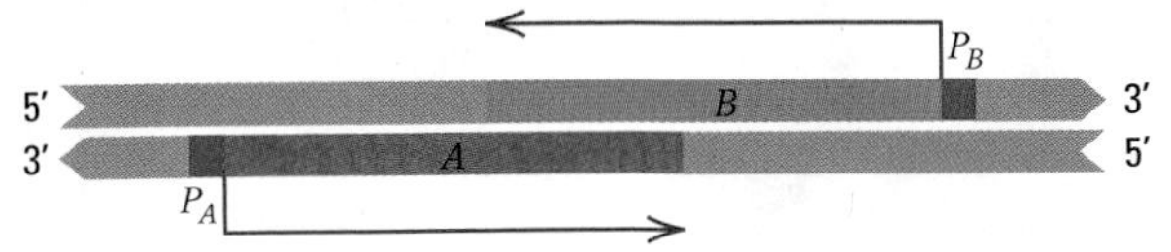


**11.14** Both repressors and aporepressors bind molecules that are substrates or products of the metabolic pathway encoded by the genes in an operon. Generally speaking, which binds the substrate of a metabolic pathway and which binds the product?

**11.15** In a bacterium related to *E. coli*, a biosynthetic operon containing genes for the synthesis of the amino acid proline exhibits the following features. The operon is found to have a regulatory system completely analogous to that of the *E. coli* tryptophan operon, with an aporepressor that binds to proline and a leader sequence in the mRNA that includes a sequence of five consecutive codons for proline.
- **(a)** Under what conditions would transcription of the proline operon be initiated?
- **(b)** Under what conditions would attenuation take place and transcription be halted?
- **(c)** Under what conditions would transcription continue through to the end of the operon?

**11.16** Is an attenuator a region of DNA that, like an operator, binds with a protein? Is RNA synthesis ever initiated at an attenuator?

**11.17** A mutant of *E. coli* is isolated that has a defective ribosomal protein with the property that translation stalls briefly whenever a codon for tryptophan is encountered, irrespective of the level of charged tryotophan tRNA. How would this mutation be expected to affect attenuation of the tryptophan operon?

**11.18** In order to study attenuation in the histidine operon, site-directed mutagenesis was used to convert the seven CAY codons for histidine in the *his* leader mRNA into UGG codons for tryptophan.
- **(a)** Under what conditions would the histidine operon initiate transcription?
- **(b)** Under what conditions would the attenuation take place, halting transcription?
- **(c)** Under what conditions would transcription continue through to the end of the operon?

**11.19** The methylation-sensitive restriction enzyme *Hpa*II, which cleaves only unmethylated 5′–CCGG–3′ sites, is used to study methylation in three species of mammals in a region of a gene with the restriction map shown in the diagram below, where the *Hpa*II (*H*) sites are numbered.

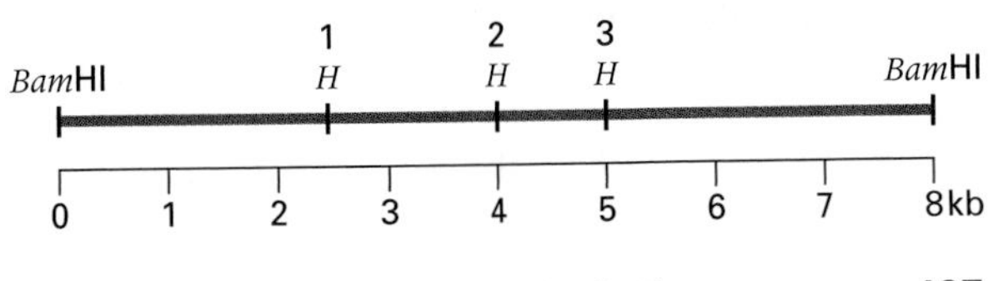

From each of the species, DNA is extracted from eggs and sperm, digested with *Bam*HI and *Hpa*II, and analyzed by Southern blotting using a probe that hybridizes with the entire region of interest. The results are shown in the gel diagram below.

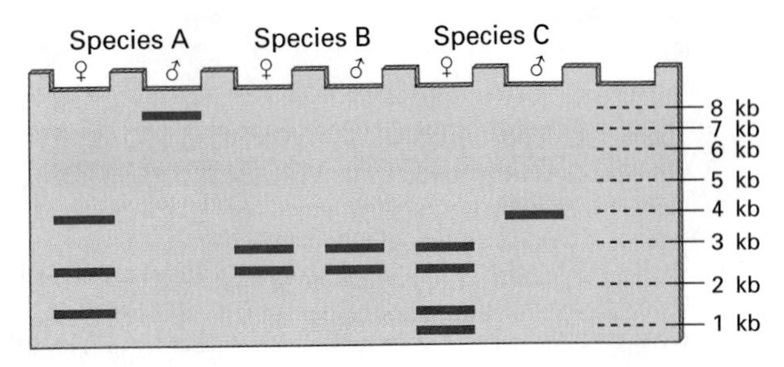


**(a)** In each type of gamete in each species, which *Hpa*II sites are methylated?

**(b)** Which species show differential methylation in the sexes?

**(c)** In which species do the male gametes show greater methylation than the female gametes?

**11.20** Among amino acid biosynthetic operons that have attenuation, some leader mRNA sequences contain very few codons for the amino acid that is synthesized (for example, in the *trp* operon), whereas others contain many more codons for the amino acid (for example, in the *phe* operon). Explain why this might be expected.

**11.21** What mating-type phenotype would you expect from a diploid cell of genotype *MAT***a**/*MAT*α with a mutation in *MAT*α in which the α2 gene product functions normally in turning off the **a**-specific genes but is unable to combine with the **a**1 product?

**11.22** What mating-type phenotype would you expect of a haploid cell of genotype *MAT***a**′, where the prime denotes a mutation that renders the **a**1 protein inactive? What mating type would you expect in the diploid *MAT***a**′ / *MAT*α?

**11.23** A mutant strain of *E. coli* is found that produces both beta-galactosidase and permease constitutively (that is, whether lactose is present or not).

**(a)** What are two possible genotypes for this mutant?

**(b)** A second mutant is isolated that produces no active beta-galactosidase at any time but produces permease if lactose is present in the medium. What is the genotype of this mutant?

**(c)** A partial diploid is created from the mutants in parts **(a)** and **(b)**: When lactose is absent, neither enzyme is made, and when lactose is present, both enzymes are made. What is the genotype of the mutant in part **(a)**?

**11.24** If a wildtype *E. coli* strain is grown in a medium without lactose or glucose, how many proteins are bound to the *lac* operon? How many are bound if the cells are grown in glucose?

**11.25** What amino acids can be inserted at the site of the UGA codon that is suppressed by a suppressor tRNA?

**11.26** A *lacI*$^+$ *lacO*$^+$ *lacZ*$^+$ *lacY*$^+$ Hfr strain is mated with an F$^-$ *lacI*$^-$ *lacO*$^+$ *lacZ*$^-$ *lacY*$^-$ strain. In the absence of any inducer in the medium, beta-galactosidase is made for a short time after the *lac* region has been transferred. Explain why it is made and why only for a short time.

**11.27** An *E. coli* mutant is isolated that is simultaneously unable to utilize a large number of sugars as sources of carbon. However, genetic analysis shows that the operons responsible for metabolism of each sugar are free of mutations. What genotypes of this mutant are possible?

**11.28** In the *Northern blot* technique, mRNA molecules are separated on an electrophoresis gel, transferred to a membrane filter, and hybridized with DNA that has been labeled with a radioactive or light-sensitive moiety. RNA molecules that contain sequences homologous to the probe become visible as discrete bands. A Northern blot experiment is carried out with mRNA from *E. coli*, using a DNA probe corresponding to the *lacA* gene, which is adjacent to *lacY*. Wildtype *lac* mRNA is approximately 5 kb in length. Two kinds of *lacZ* and *lacY* mutations are studied. The *lacZ*$^-$ and *lacY*$^-$ mutations are simple nucleotide substitution mutations that encode inactive proteins. The *lacZ** and *lacY** mutations are deletions. The *lacZ** deletion is missing 2 kb of *lacZ* coding sequence, *lacY** is missing 0.5 kb of *lacY* coding sequence, and neither allele produces a polypeptide product. The mRNA from the following six genotypes is analyzed via Northern blot after growth in either the presence (+) or the absence (–) of inducer. The RNA molecules are separated under conditions yielding the size scale shown by the dashed lines in the accompanying gel diagram. For each genotype, under both inducing (+) and the noninducing (–) conditions, indicate in the gel diagram where an mRNA band would be expected.

**(a)** *lacI*$^+$ $O^+$ $Z^+$ $Y^+$

**(b)** *lacI*$^-$ $O^+$ $Z^-$ $Y^-$

**(c)** *lacI*$^-$ $O^+$ $Z^*$ $Y^*$

**(d)** *lacI*$^+$ $O^c$ $Z^-$ $Y^*$

**(e)** F′ *lacI*$^+$ $O^c$ $Z^*$ $Y^-$ / *lacI*$^+$ $O^+$ $Z^-$ $Y^*$

**(f)** F′ *lacI*$^+$ $O^+$ $Z^*$ $Y^+$ / *lacI*$^+$ $O^c$ $Z^-$ $Y^*$

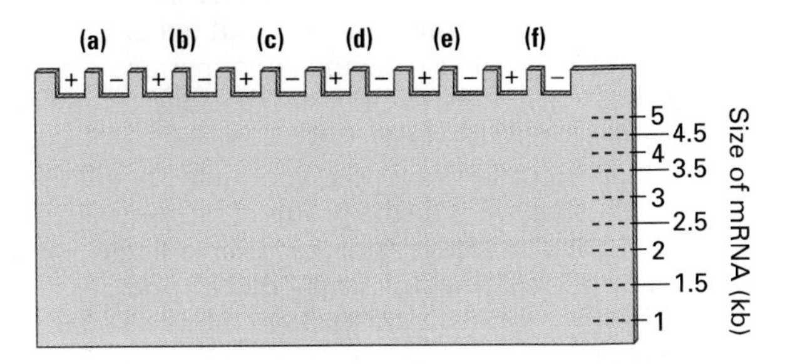


**11.29** In a *Western blot*, proteins are separated on an electrophoresis gel, transferred to a membrane filter, and mixed with a labeled antibody that combines specifically with one of the proteins of interest. The six strains of *E. coli* in the previous problem are analyzed using a Western blot stained with a mixture of two antibodies, one for beta-galactosidase and the other for the permease. The gel diagram below shows the positions to which these proteins would migrate, if they were present. Indicate in the gel diagram which bands, if any, would be found in each of the six strains grown under inducing (+) and under noninducing (–) conditions.

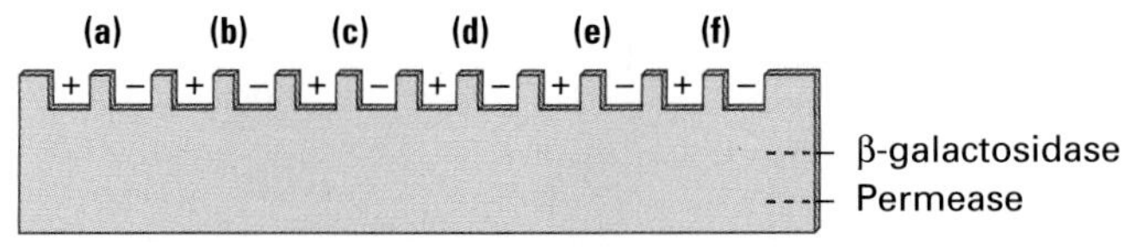


**11.30** The LacY permease is able to transport lactobionic acid as well as lactose, but LacZ cleaves lactobionic acid very inefficiently. Certain mutant forms of LacZ have an altered substrate specificity that allows the mutant enzyme to cleave lactobionic acid. These mutants are able to grow in medium containing lactobionic acid as the sole source of carbon and energy. However, they cannot grow unless they also have a *lacO*$^c$ mutation or unless IPTG (isopropylthiogalactoside) is present in the medium when the operator is wildtype. How can you explain these results?

## CHALLENGE PROBLEMS

**Challenge Problem 1** In a newly discovered organism the histidine operon is negatively regulated. It contains the structural genes for the enzymes needed to synthesize histidine, and the repressor protein is also coded within the operon—that is, within the polycistronic mRNA molecule that codes for the other proteins. Synthesis of this mRNA is controlled by a single operator regulating the activity of a single promoter. The corepressor of this operon is tRNA$^{His}$, to which histidine is attached. This tRNA is not coded by the operon itself. A collection of mutants with the following defects is isolated. Determine whether the histidine enzymes would be synthesized by each of the mutants and whether each mutant would be dominant, *cis*-dominant only, or recessive to its wildtype allele in a partial diploid.

**(a)** The promoter cannot bind with RNA polymerase.
**(b)** The operator cannot bind the repressor protein.
**(c)** The repressor protein cannot bind with DNA.
**(d)** The repressor protein cannot bind histidyl-tRNA$^{His}$.
**(e)** The uncharged tRNA$^{His}$ (that is, without histidine attached) can bind to the repressor protein.

**Challenge Problem 2** The attenuator of the histidine operon contains seven consecutive histidines. The relevant part of the attentuator coding sequence is

5′-AAACACCACCATCATCACCATCATCCTGAC-3′

A mutation occurs in which an additional A nucleotide is inserted immediately after the red A. What amino acid sequences are coded by the wildtype and mutant attenuators? What phenotype would you expect of the mutant?

**Challenge Problem 3** In a species of insect, the genes *cyclops* and *janus* have different tissue-specific patterns of expression; *cyclops* is expressed only in photoreceptor cells of the eye and *janus* only in the testes. Near its 5′ end *cyclops* contains an intron, indicated in the diagram by the hatching, within which there are two *Eco*RI restriction sites. When the *Eco*RI fragment is removed and inserted at an *Eco*RI site immediately upstream of the *janus* promoter, *janus* is transcribed in photoreceptor cells as well as in the testes. The same result occurs irrespective of the orientation in which the *Eco*RI fragment is inserted. What hypothesis can explain this result? How could the hypothesis be tested?

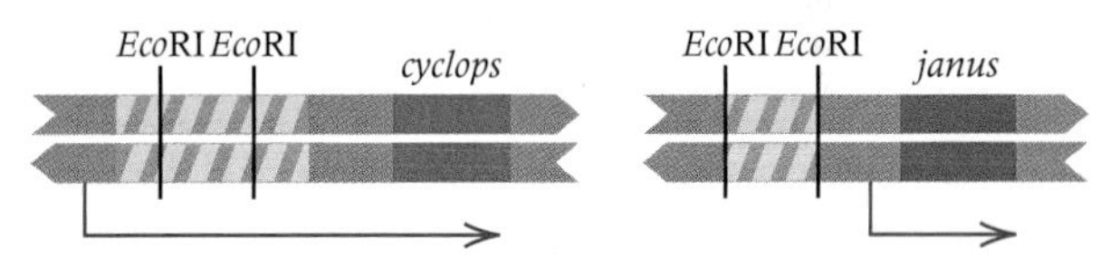


### GENETICS on the web

GeNETics on the Web will introduce you to some of the most important sites for finding genetics information on the Internet. To explore these sites, visit the Jones and Bartlett companion site to accompany *Genetics: Analysis of Genes and Genomes, Seventh Edition* at http://biology.jbpub.com/book/genetics.

There you will find a chapter-by-chapter list of highlighted keywords. When you select one of the keywords, you will be linked to a Web site containing information related to that keyword.

# CHAPTER 12

# Genomics, Proteomics, and Transgenics

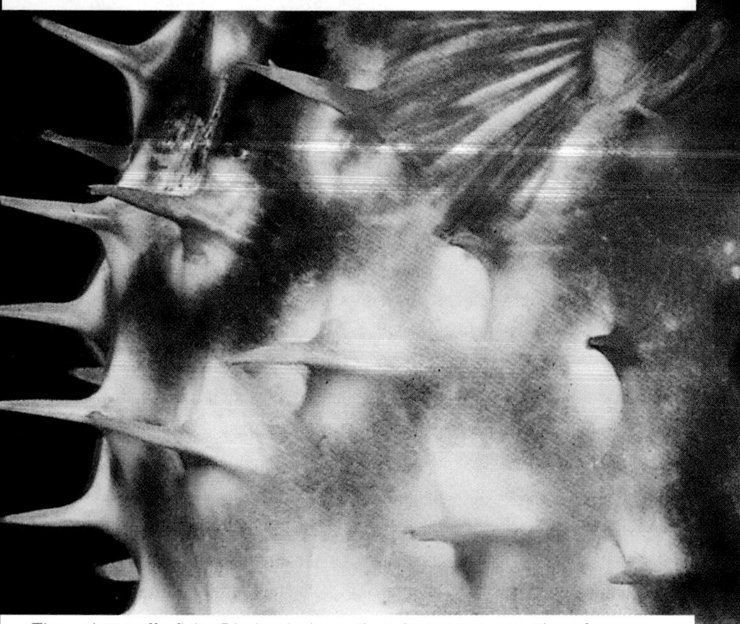

The spiny pufferfish, *Diodon holacanthus*, has a genome size of 800 million base pairs. The genome of its close relative, the Japanese pufferfish, *Takifugu rubripes*, is only 400 million base pairs—a mere 13 percent of the size of the human genome. Vertebrate organisms with such small genomes have been of great utility in comparative genomics to help identify the protein-coding genes and other important features of the human genome. [© AbleStock]

## CHAPTER OUTLINE


**12.1 Site-Specific DNA Cleavage and Cloning Vectors**
- Production of DNA Fragments with Defined Ends
- Recombinant DNA Molecules
- Plasmid, Lambda, and Cosmid Vectors

**12.2 Cloning Strategies**
- Joining DNA Fragments
- Insertion of a Particular DNA Molecule into a Vector
- The Use of Reverse Transcriptase: cDNA and RT-PCR

**12.3 Detection of Recombinant Molecules**
- Gene Inactivation in the Vector Molecule
- Cloning of Large DNA Fragments
- Screening for Particular Recombinants

**12.4 Genomics and Proteomics**
- Genomic Sequencing
- Genome Annotation
- Comparative Genomics
- Transcriptional Profiling
- Two-Hybrid Analysis of Protein Interactions

**12.5 Transgenic Organisms**
- Germ-Line Transformation in Animals
- Genetic Engineering in Plants
- Transformation Rescue
- Site-Directed Mutagenesis and Knockout Mutations

**12.6 Some Applications of Genetic Engineering**
- Giant Salmon with Engineered Growth Hormone
- Nutritionally Engineered Rice
- Production of Useful Proteins
- Genetic Engineering with Animal Viruses

CONNECTION **Hello, Dolly!**
Ian Wilmut, Angelika E. Schnieke, Jim McWhir, Alex J. Kind, and Keith H. S. Campbell 1997
*Viable Offspring Derived from Fetal and Adult Mammalian Cells*

CONNECTION **A Pinch of This and a Smidgen of That**
Oliver Smithies 2005
*Many Little Things: One Geneticist's View of Complex Diseases*

Human, chimpanzee, monkey, lemur, mouse, dog, cat, bat, squirrel, rabbit, guinea pig, armadillo, hedgehog, shrew, opposum, horse, elephant, pangolin, sloth, llama, dolphin—these are some of the mammalian genomes that have been sequenced. Add to this list many species of fruit flies, worms, and fungi, along with hundreds of bacteria, mitochondria, and chloroplasts as well as thousands of viruses, and you will begin to appreciate the vast amount of sequence data available for analysis and comparison. In addition to the genome sequences, methods are available for identifying which genes in the genome are transcribed in particular tissue types, at specific times in development, or at different stages of the cell cycle. These are the raw data of **genomics**, which deals with the DNA sequence, organization, function, and evolution of genomes. The counterpart at the level of proteins is **proteomics**, which aims to identify all the proteins in a cell or organism (including any posttranslationally modified forms) as well as their cellular localization, functions, and interactions. Proteomics makes use of methods discussed later in this chapter that identify which proteins in the cell undergo physical contact, thereby revealing *networks* of interacting proteins.

Genomics was made possible by the invention of techniques originally devised for the manipulation of genes and the creation of genetically engineered organisms with novel genotypes and phenotypes. We refer to this approach as **recombinant DNA**, but it also goes by the names *gene cloning* or *genetic engineering*. The basic technique is quite simple. DNA is isolated and cut into fragments by one or more restriction enzymes; then the fragments are joined together in a new combination and introduced back into a cell or organism to change its genotype in a directed, predetermined way. Such genetically engineered organisms are called **transgenic organisms**. Transgenics are often created for experimental studies, but an important application is the development of improved varieties of domesticated animals and crop plants, in which case a transgenic organism is often called a *genetically modified organism (GMO)*. Specific examples of genetically modified organisms are considered later in this chapter.

## 12.1 Site-Specific DNA Cleavage and Cloning Vectors

The technology of recombinant DNA relies heavily on *restriction enzymes*, which were examined in Section 2.3 with reference to their site-specific DNA cleavage in the generation of DNA fragments with defined ends. In recombinant DNA, after the DNA has been cleaved by a restriction enzyme, the restriction fragments are joined with a **vector** molecule, which is a (usually circular) DNA molecule into which a DNA fragment can be introduced and which can replicate in a suitable host organism (**FIGURE 12.1**). The bacterium *Escherichia coli* and the budding yeast *Saccharomyces cerevisiae* are typical host organisms for recombinant DNA. In this section, we take a closer look at the special features of restriction enzymes that make them so useful in DNA cloning. Some of the most popular types of vectors used in recombinant DNA are also examined.

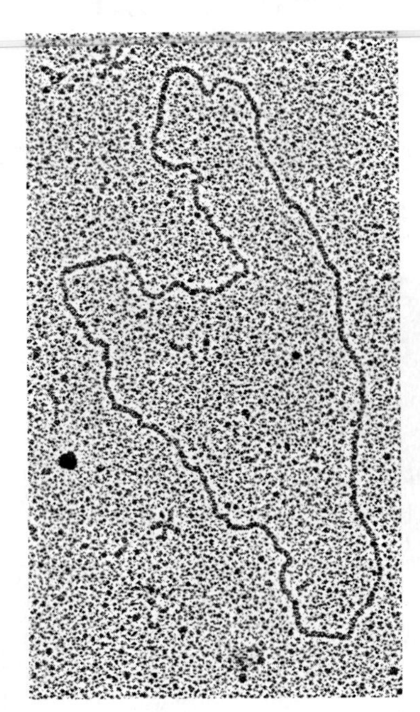

**FIGURE 12.1** Electron micrograph of a circular plasmid used as a vector for cloning in *E. coli*.

### Production of DNA Fragments with Defined Ends

A *restriction enzyme* is a nuclease that cleaves duplex DNA wherever the DNA molecule contains a particular short sequence of nucleotides matching the *restriction site* of the enzyme (Section 2.3). The backbone of each strand is cleaved, creating a free 3′ hydroxyl group (3′-OH) and a free 5′ phosphate group (5′-P). Most restriction sites consist of four or six nucleotides. About a thousand restriction enzymes, each with a different restriction site, have been isolated from microorganisms. Most restriction sites are symmetrical in the sense that the recognition sequence is identical in both strands of the DNA duplex. For example, the restriction enzyme *Eco*RI, isolated from *Escherichia coli,* has the restriction site 5′-GAATTC-3′. The sequence of the other strand is 3′-CTTAAG-5′, which is identical but is written with the 3′ end at the left. This kind of symmetry in a short sequence of DNA is called a *palindrome.* In double-stranded DNA, *Eco*RI cleaves each strand of the palindrome between the G and the A.

Observations with the electron microscope, made soon after restriction enzymes were discovered, indicated that the fragments produced by many restriction enzymes could spontaneously form circles. The circles could be made linear again by heating; however, if after circularization they were treated with the enzyme **DNA ligase**, which joins 3′-OH and 5′-P groups, then the ends became covalently joined. This observation was the first evidence for three important features of restriction enzymes:

1. Restriction enzymes make breaks in symmetrical sequences (palindromes).
2. The breaks need not be directly opposite one another in the two DNA strands.
3. Enzymes that cleave the DNA strands asymmetrically generate DNA fragments with single-stranded ends that have complementary base sequences.

These properties are illustrated for *Eco*RI in **FIGURE 12.2.**

Most restriction enzymes are like *Eco*RI in that the cuts in the DNA strands are staggered; the resulting single-stranded ends are called **sticky ends**, and they can adhere to each other because they have complementary nucleotide sequences. Some restriction enzymes, including *Eco*RI, leave a single-stranded overhang at the 5′ end (**FIGURE 12.3**, part A); others leave a 3′ overhang. A number of restriction enzymes cleave both DNA strands at the center of symmetry, forming **blunt ends**. Figure 12.3B shows the blunt ends produced by the enzyme *Bal*I. Blunt ends also can be joined via a ligase obtained from bacteriophage T4 (*T4 ligase*), which differs from most other ligases in that it can join blunt ends as well as single-stranded nicks in the backbone of duplex DNA. However, whereas ligation of sticky ends recreates the original restriction site, any blunt end can join with any other blunt end and not necessarily create a restriction site.

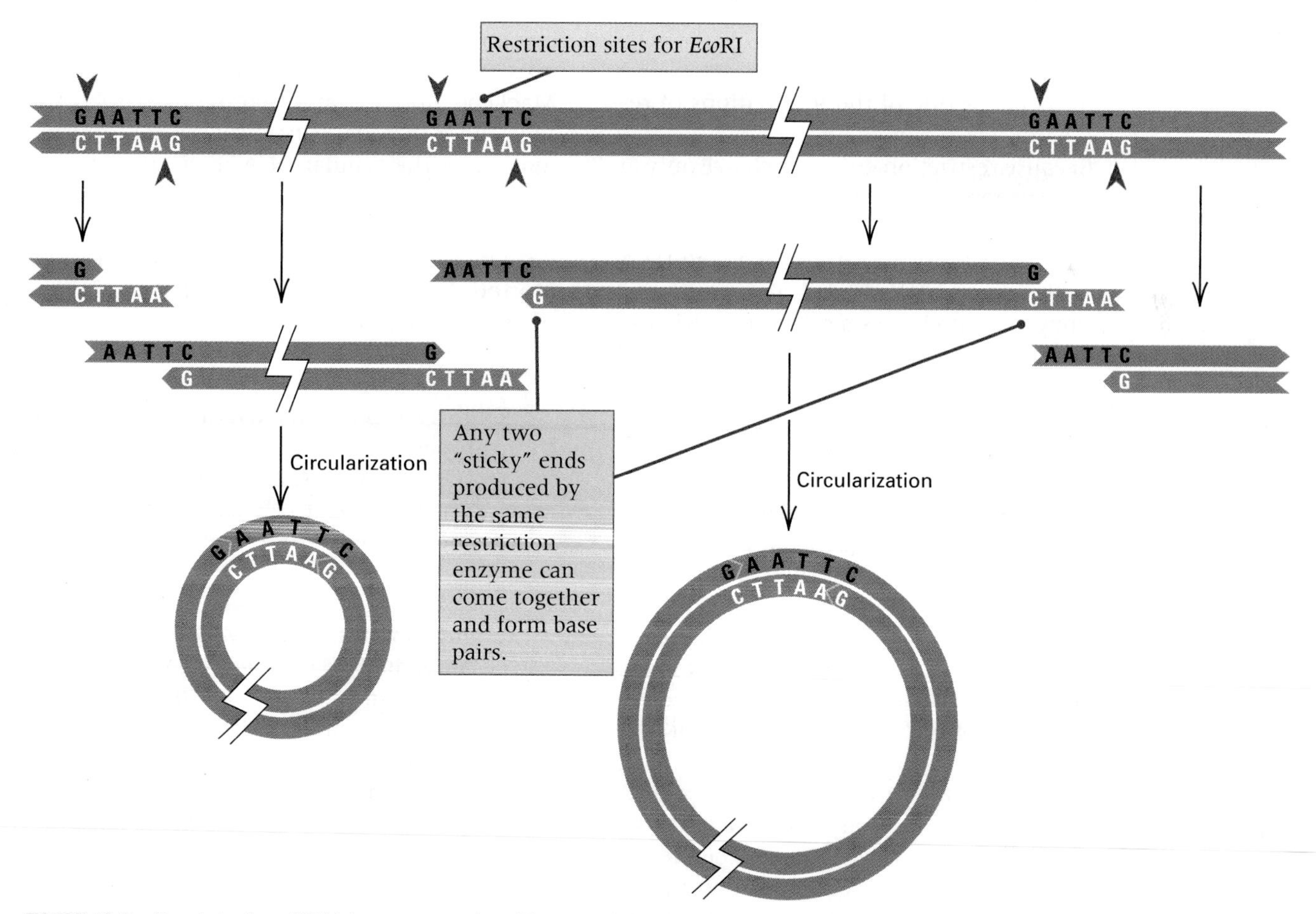


**FIGURE 12.2** Circularization of DNA fragments produced by a restriction enzyme. The red arrowheads indicate the *Eco*RI cleavage sites.

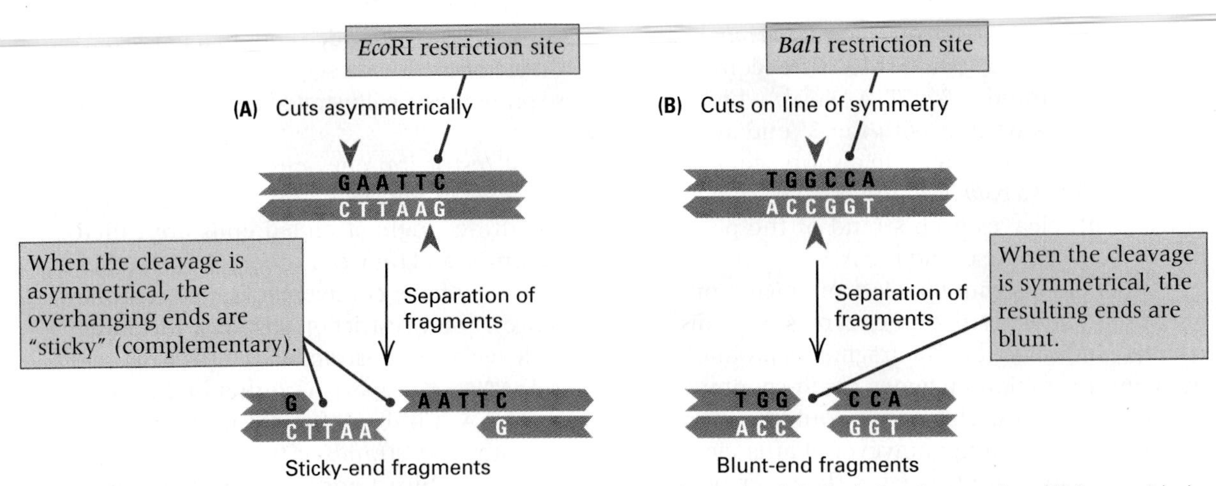


**FIGURE 12.3** Two types of cuts made by restriction enzymes. The red arrowheads indicate the cleavage sites. (A) Asymmetrical cuts made in each strand at an equal distance from the center of symmetry of the restriction site. (B) Symmetrical cuts made in each strand at the center of symmetry of the restriction site.

Most restriction enzymes recognize their restriction sequence without regard to the source of the DNA:

> If DNA fragments are produced by the same restriction enzyme, then the fragments obtained from one organism will have the same sticky ends as the fragments from another organism.

This principle is one of the foundations of recombinant DNA technology.

Because restriction enzymes cleave only at their restriction site, the number of cuts made in the DNA of an organism by a particular enzyme is limited. For example, an *E. coli* DNA molecule contains $4.6 \times 10^6$ base pairs, and any enzyme that cleaves a six-base restriction site will cut the molecule into about a thousand fragments; the reason is that, with equal and random frequencies of each of the four nucleotides, a six-base restriction site is expected to occur, on average, every $4^6 = 4096$ base pairs. Human DNA, with a genome size of $3 \times 10^9$ base pairs, would be cut into about a million fragments. These numbers, though large, are still small compared with the number of fragments that would be produced by random shearing of the DNA. Of special interest are cleavage products from smaller DNA molecules that may have only a few cleavage sites (or possibly none) for any particular enzyme, such as viral DNA or plasmid DNA. A **plasmid** is a DNA molecule present in a cell, usually a bacterial cell, that is capable of replicating independently of the bacterial chromosome. Different plasmids range in size from about 1 to 100 kb. As we shall see shortly, small circular plasmids containing a single cleavage site for a particular restriction enzyme are especially valuable as vectors in DNA cloning.

## Recombinant DNA Molecules

In recombinant DNA, a particular DNA segment of interest is joined to a vector DNA molecule. This recombinant molecule is introduced into a cell by means of a *transformation* procedure very similar to that used by Avery, MacLeod, and McCarty in proving that DNA is the genetic material (Chapter 1). After transformation, as the cellular DNA replicates, so does the recombinant DNA (**FIGURE 12.4**). When a stable transformant has been isolated, the DNA sequences linked to the vector are said to be **cloned**. In the following section, several types of vectors are described.

## Plasmid, Lambda, and Cosmid Vectors

The most generally useful vectors have three properties:

1. The vector DNA can be introduced into a host cell relatively easily.
2. The vector and any DNA it contains can be replicated inside the host cell.
3. Cells containing the vector can be identified in a straightforward manner, most conveniently through a novel phenotype conferred on the host by DNA sequences present in the vector (for example, antibiotic resistance).

The most commonly used vectors are *E. coli* plasmids and derivatives of the *E. coli* bacteriophages lambda (λ) and M13. Other plasmids and viruses also have been developed for cloning into cells of animals, plants, and bacteria. Plasmid and phage DNA can be introduced into cells by a transformation procedure in which

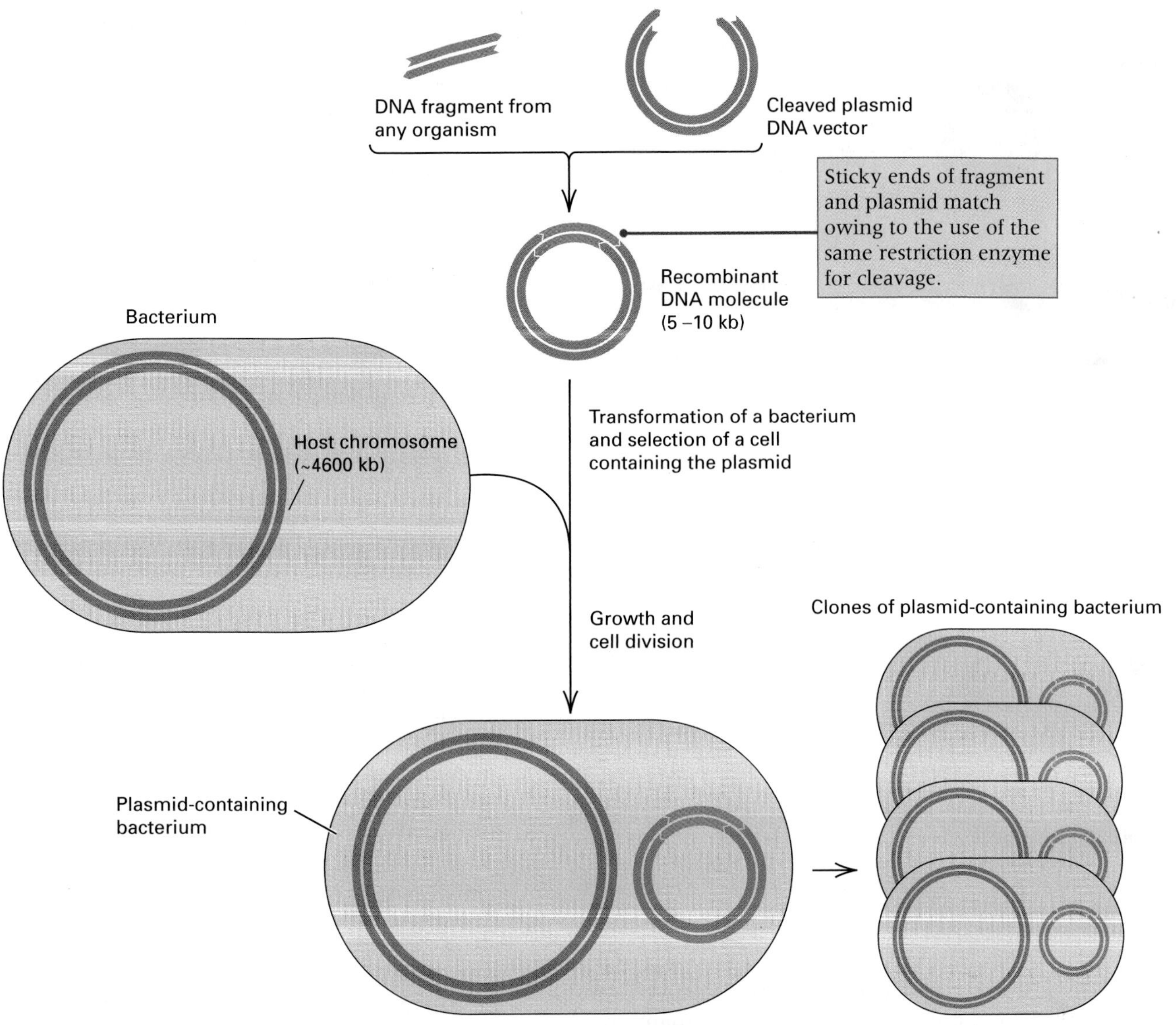


**FIGURE 12.4** An example of cloning. A fragment of DNA from any organism is joined to a cleaved plasmid. The recombinant plasmid is then used to transform a bacterial cell, where the recombinant plasmid is replicated and transmitted to the progeny bacteria. The bacterial host chromosome is not drawn to scale. It is typically about 1000 times larger than the plasmid.

cells are enabled to take up free DNA by exposure to a solution of calcium chloride. Recombinant DNA can also be introduced into cells by a kind of electrophoretic procedure called **electroporation**. After introduction of the DNA, the cells containing the recombinant DNA are detected by means of vector-encoded functions that become manifest during bacterial growth. For example, *E. coli* cells that have been successfully transformed with a vector that encodes resistance to an antibiotic, such as tetracycline, will grow on medium containing tetracycline, whereas untransformed cells will not grow. Variants of this procedure are used to transform animal or plant cells with suitable vectors (Section 12.5), but the technical details differ considerably.

Three types of vectors commonly used for cloning into *E. coli* are illustrated in **FIGURE 12.5**. Plasmids (Figure 12.5A) are most convenient for cloning relatively small DNA fragments (5–10 kb). Somewhat larger fragments can be cloned with bacteriophage λ (Figure 12.5B). The normal phage λ genome is approximately 50 kb in length, but because the central portion of the genome is not essential for infection and phage propagation, this portion can be removed and replaced with donor DNA. After the donor DNA has been ligated in place, the recombinant DNA is packaged into mature phage *in vitro*, and the recombinant phage are used to infect bacterial cells. However, to be packaged into a phage head, the recombinant DNA must be neither too large nor too small, which means that the donor

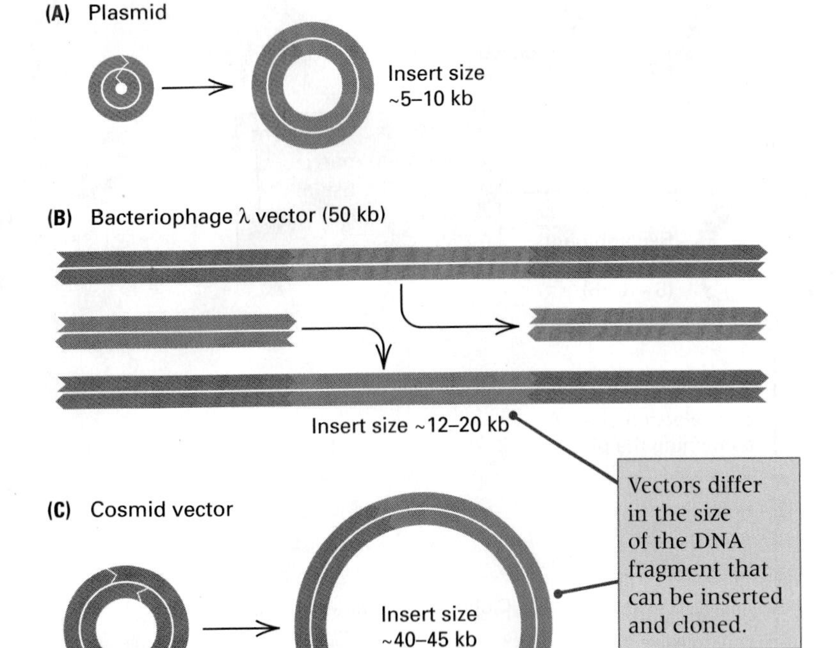


**FIGURE 12.5** Common cloning vectors for use with *E. coli*, not drawn to scale. (A) Plasmid vectors are ideal for cloning relatively small fragments of DNA. (B) Bacteriophage λ vectors contain convenient restriction sites for removing the middle section of the phage and replacing it with the DNA of interest. (C) Cosmid vectors are useful for cloning DNA fragments up to about 40 kb; they can replicate as plasmids but contain the cohesive ends of phage λ and so can be packaged in phage particles.

DNA must be roughly the same size as the portion of the λ genome that was removed. Most λ cloning vectors accept inserts ranging in size from 12 to 20 kb. Still larger DNA fragments can be inserted into cosmid vectors (Figure 12.5C). These vectors can exist as plasmids, but they also contain the phage single-stranded sticky ends of 12 bp known as *cohesive ends* (Section 9.7). Cosmid clones can be packaged into mature phage particles by virtue of the cohesive ends, which are cleaved to enable the phage DNA to be packaged into the phage heads. The size limitation on cosmid inserts usually ranges from 40 to 45 kb. Vectors for cloning even larger fragments of DNA are discussed in Section 12.3.

# 12.2 Cloning Strategies

In genetic engineering, the immediate goal of an experiment is usually to insert a *particular* fragment of chromosomal DNA into a plasmid or a viral DNA molecule. Any strategy for achieving this goal must include several steps:

1. Purification of donor DNA and vector DNA
2. Cleavage with one or more restriction enzymes
3. Joining donor DNA with vector DNA
4. Identification and isolation of desired recombinant clones

Steps 1 and 2 were considered in Section 12.1. Here we examine steps 3 and 4.

## ■ Joining DNA Fragments

The circularization of restriction fragments having terminal single-stranded, "sticky" ends with complementary bases was described in Section 12.1. Because a particular restriction enzyme produces fragments with *identical* sticky ends, without regard for the source of the DNA, fragments from DNA molecules isolated from two different organisms can be joined, as shown in **FIGURE 12.6**. In this example, the restriction enzyme *Eco*RI is used to digest DNA from any organism of interest and to cleave a bacterial plasmid that contains only one *Eco*RI restriction site. The donor DNA is cleaved into many fragments (one of which is shown), and the plasmid is cleaved into a single linear fragment. When the donor fragment and the linearized plasmid are mixed, recombinant molecules can form by base pairing between the complementary single-stranded ends. At this point, the DNA is treated with DNA ligase to seal the joints, and the donor fragment becomes covalently joined to the vector. The ability to join a donor DNA fragment of interest to a vector is the basis of the recombinant DNA technology.

Because any sticky end can base-pair with any other complementary sticky end, the joining of restriction fragments at random can produce a variety of products, including many that are useless. For example, consider a linear DNA molecule that is cleaved into four fragments—A, B, C, and D—present in the original molecule in the order A B C D (**FIGURE 12.7**). Reassembly of the fragments can occasionally yield the original molecule, but if B and C have the same pair of sticky ends, then molecules with different arrangements of the fragments are also formed, including arrangements in which one or more of the restriction fragments are inverted in orientation (indicated in Figure 12.7 by the reversed letters). Ligation of restriction fragments with blunt ends produces even more combinations, because any blunt end can be joined with any other blunt end.

Restriction fragments from the vector, if there are more than one, can also join together in the wrong order, but this potential problem can be eliminated by the use of a vector that has only

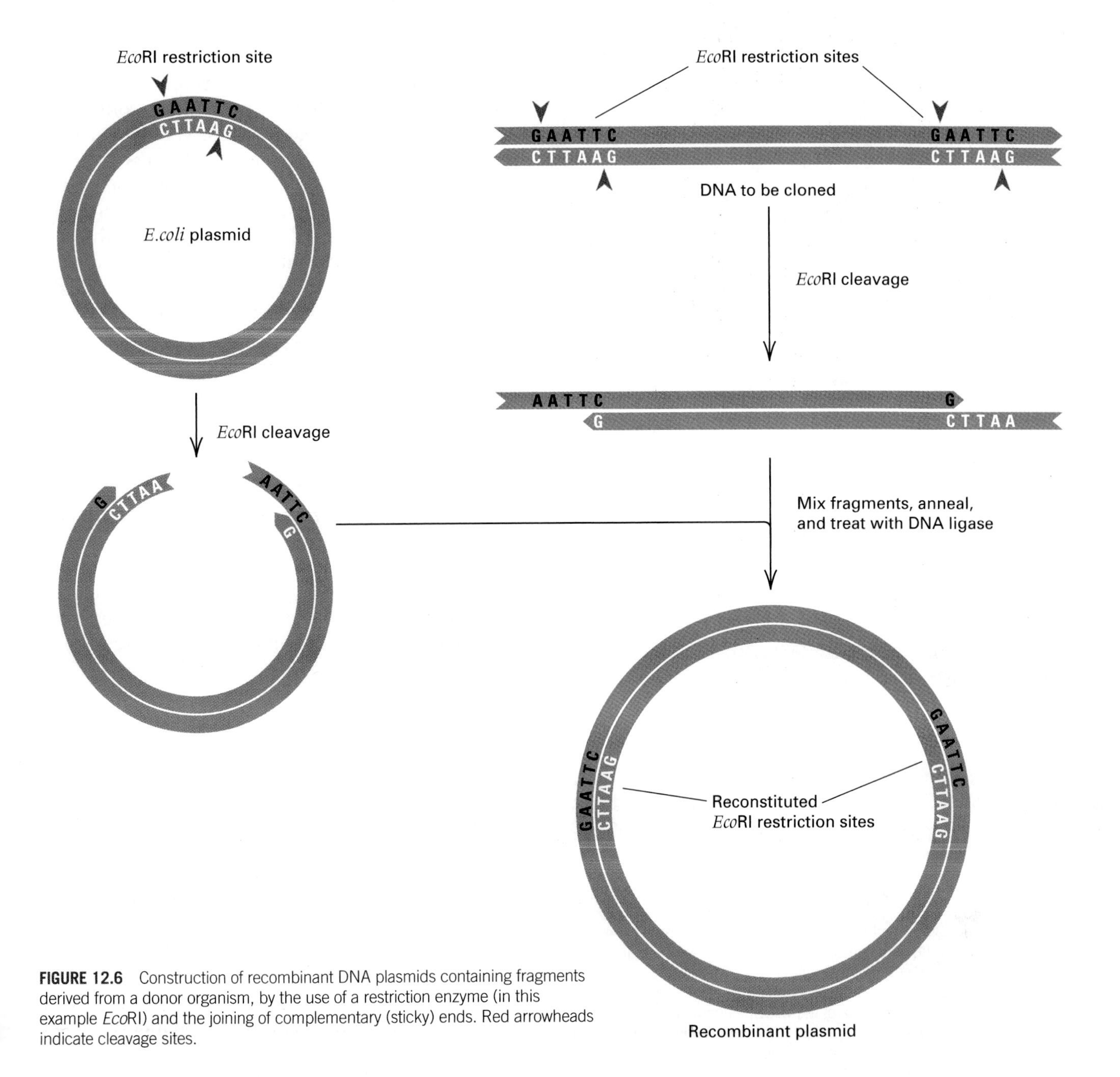


**FIGURE 12.6** Construction of recombinant DNA plasmids containing fragments derived from a donor organism, by the use of a restriction enzyme (in this example *Eco*RI) and the joining of complementary (sticky) ends. Red arrowheads indicate cleavage sites.

one cleavage site for a particular restriction enzyme. When a circular molecule has only one cleavage site for a restriction enzyme, as in Figure 12.6, cleavage occurs only at this one site, and any other DNA fragment of interest, if it has complementary ends, can be inserted into the gap. Many plasmids with single cleavage sites are available (most have been created by recombinant DNA). Many vectors contain unique sites for several different restriction enzymes, but generally only one or two enzymes are used at a time.

## Insertion of a Particular DNA Molecule into a Vector

In the recombinant DNA procedure described so far, a collection of fragments obtained by digestion with a restriction enzyme can be joined with a cleaved vector molecule, yielding a large number of recombinant molecules containing different fragments of donor DNA. If a particular cloned segment of DNA is desired, then the recombinant molecule containing that particular segment must be isolated from among what is often a large background of other recombinant molecules. The simplest method is direct selection of the desired clone on the basis of a phenotypic attribute that it confers.

As an example of direct selection for the recovery of recombinant molecules containing a particular gene, suppose that we wished to clone a gene used by *E. coli* for the biosynthesis

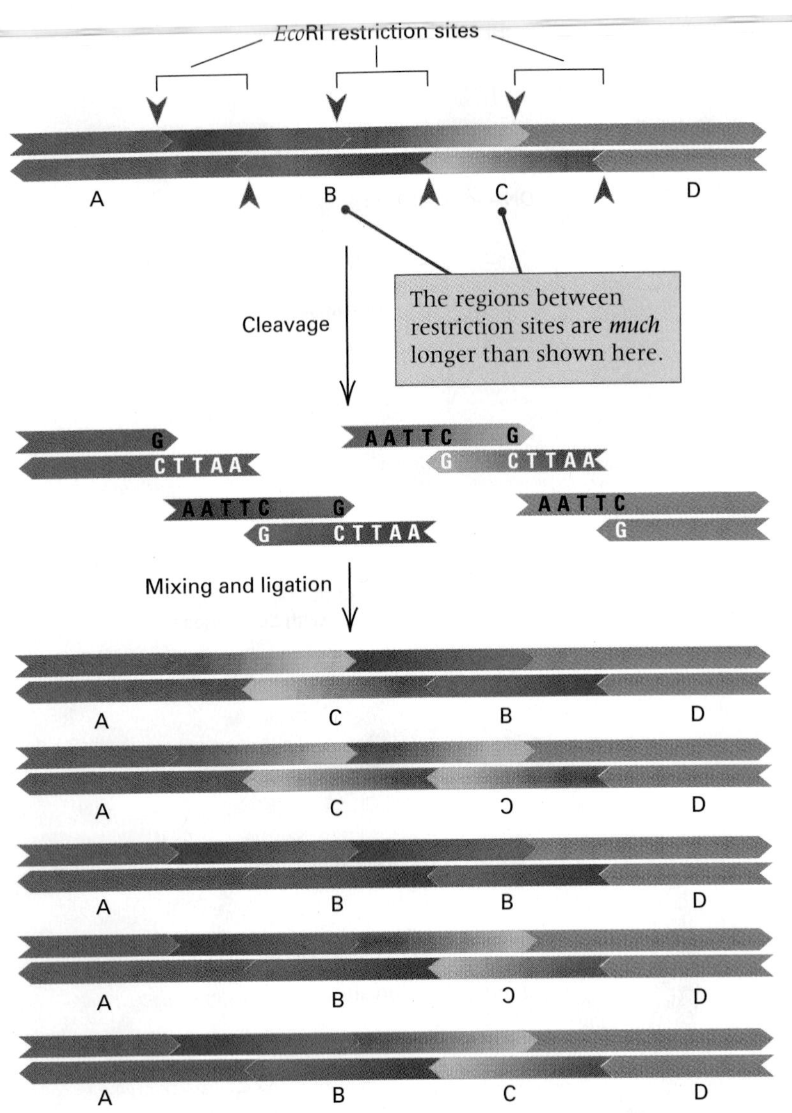


**FIGURE 12.7** Fragments produced by a restriction enzyme can be rejoined in arbitrary ways because the ends of the cleavage sites are compatible. In this example, a DNA molecule is cut into four different fragments (A, B, C, and D) by a restriction enzyme. A few examples of how these can be rejoined are shown. A backward letter means that the fragment has been ligated in reverse of the wildtype orientation. Even among the rejoined molecules that have A and D at the ends and two fragments in the middle, many have one fragment represented twice and the other not at all, or they have a fragment present in the reverse orientation. Only occasionally does a rejoining recreate the original order A–B–C–D.

of the amino acid leucine. Direct selection would make use of a *leu*$^-$ mutant strain having a mutation in the gene of interest. The *leu*$^-$ cells would be unable to form colonies on solid medium unless the growth medium contained leucine. To clone the *leu*$^+$ gene from a wildtype strain able to grow without leucine, DNA from the *leu*$^+$ strain is isolated and cleaved with a restriction enzyme, and the resulting fragments are ligated into a plasmid vector that is then used to transform mutant cells of genotype *leu*$^-$. The transformed cells are spread onto solid medium lacking leucine. Any untransformed cells, or cells transformed with DNA not containing the *leu*$^+$ gene, will be unable to grow and form colonies. The medium selects for *leu*$^+$ transformants because the few cells that do contain the *leu*$^+$ gene inserted into the vector will form colonies. This experiment will not work if the *leu*$^+$ gene contains a restriction site for the enzyme used, because cleavage within the gene will destroy its ability to function. For this reason, cloning experiments of this type are usually carried out under conditions of **partial digestion**, which means that the cleavage reaction is terminated when only a fraction of the restriction sites have been cleaved. In this way, even though the *leu*$^+$ gene may contain one or more restriction sites for the enzyme, some of the *leu*$^+$ genes will, by chance, still remain intact after partial digestion.

Direct selection for recombinant clones, such as in the *leu*$^+$ example, is usually not possible. If the clone of interest is either very rare or very difficult to detect, it is often preferable to purify the DNA fragment that contains the gene of interest before joining it with the vector. Two methods of fragment isolation are in common use:

1. Amplification by means of the *polymerase chain reaction (PCR),* discussed in Section 2.4, in which short oligonucleotide primers homologous to the ends of the fragment to be isolated are used to replicate the fragment of interest. After a sufficient number of rounds of replication, the resulting solution contains predominantly replicas of the amplified DNA fragment.
2. Purification of a restriction fragment known to contain the gene of interest. For example, if a gene is known to be contained within a particular restriction fragment, this fragment can be isolated by purifying the corresponding DNA band from a gel after electrophoresis (Section 2.3) and then joined to an appropriate vector.

Some genes in higher eukaryotes, including many human genes, are very long. They extend for hundreds of kilobases. Any DNA fragment containing the complete gene is too long to be amplified by PCR and is likely to contain multiple cleavage sites for any restriction enzyme. However, such large genes usually produce mRNA molecules of modest size, typically less than 3 kb. If the purpose of cloning is to produce the protein product of the gene, then all the necessary information is contained in the mRNA. How to clone a DNA molecule whose sequence corresponds to a particular mRNA is described next.

## The Use of Reverse Transcriptase: cDNA and RT-PCR

Some specialized animal cells contain a high abundance of a specific type of mRNA. For example, cells in the chicken oviduct contain so much mRNA for ovalbumin, an egg-white protein, that mRNA extracted from oviducts is predominantly ovalbumin mRNA. This purified mRNA can serve as the starting point for isolating recombinant clones that contain the coding sequence of ovalbumin.

Most RNA molecules are not as abundant as that of ovalbumin. Nevertheless, mRNA is often used for cloning, especially in organisms whose genes are interrupted by many long introns (as human genes are), because the mRNA includes the protein-coding sequences with the introns already removed.

In a typical eukaryotic cell, the levels of mRNA abundance can be separated, very roughly, into three categories:

1. About 1 percent of the expressed genes have an *abundant* mRNA, ranging from hundreds of copies to a few thousand copies per cell.
2. About 10 percent of the expressed genes have a *moderately abundant* mRNA, ranging from about 10 to 100 copies per cell.
3. The remaining almost 90 percent of the expressed genes have a *rare* mRNA, ranging from one to about ten copies per cell.

Cloning from mRNA molecules makes use of an unusual polymerase, **reverse transcriptase**, which can bind with a single-stranded RNA molecule and use it as a template for the synthesis of a complementary strand of DNA. Reverse transcriptase was originally discovered in RNA tumor viruses, in which single-stranded RNA is the genetic material, where its function is to make double-stranded DNA complementary to the viral RNA that can be integrated into the DNA of the host cell; reverse transcriptase converts the viral genetic information into DNA for replication and transmission to the daughter cells when the host cell divides. Like other DNA polymerases, reverse transcriptase requires a primer. When the enzyme is used in genetic engineering, the poly-A tail usually found at the 3′ end of eukaryotic mRNA serves as a convenient priming site, because the primer can be an oligonucleotide consisting of poly-T (**FIGURE 12.8**). Like any other single-stranded

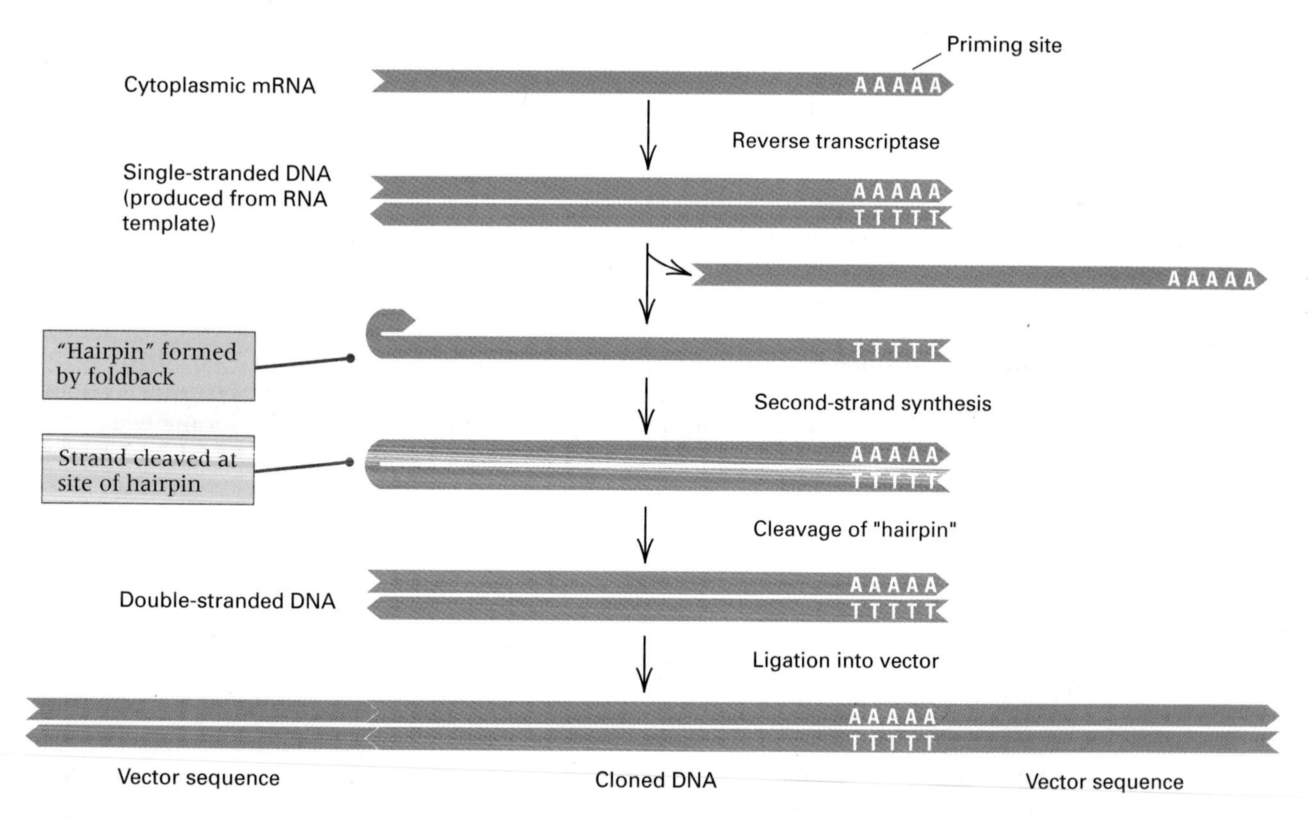


**FIGURE 12.8** Reverse transcriptase produces a single-stranded DNA complementary in sequence to a template RNA. In this example, a cytoplasmic mRNA is copied. As indicated here, most eukaryotic mRNA molecules have a tract of consecutive A nucleotides at the 3′ end, which serves as a convenient priming site. After the single-stranded DNA is produced, a foldback at the 3′ end forms a hairpin that serves as a primer for second-strand synthesis. After the hairpin is cleaved, the resulting double-stranded DNA can be ligated into an appropriate vector either immediately or after PCR amplification.

DNA molecule, the single strand of DNA produced from the RNA template can fold back upon itself at the extreme 3′ end to form a "hairpin" structure that includes a short double-stranded region consisting of a few base pairs. The 3′ end of the hairpin serves as a primer for synthesis of the second strand of DNA, forming a duplex DNA molecule. The second strand can be synthesized either by DNA polymerase or by reverse transcriptase. Reverse transcriptase itself produces the second strand in RNA viruses that encode a reverse transcriptase, such as the human immunodeficiency virus (HIV). Conversion into a conventional double-stranded DNA molecule is achieved by cleavage of the hairpin by a nuclease. The DNA complement of an RNA molecule is called **complementary DNA**, or **cDNA**.

In the reverse transcription of an mRNA molecule, the resulting full-length cDNA contains an uninterrupted coding sequence for the polypeptide of interest. Because many eukaryotic genes contain introns that are present in the primary transcript but removed in the production of the mature mRNA, the cDNA sequence is not identical to the genomic DNA. However, if the purpose of creating the recombinant DNA molecule is to identify the coding sequence or to synthesize the gene product in a genetically engineered organism, then cDNA formed from processed mRNA is the material of choice. The joining of cDNA to a vector can be accomplished by available procedures for joining blunt-ended molecules (Figure 12.8).

The cDNA molecules produced from rare RNAs will also be rare. However, the efficiency of cloning rare cDNA molecules can be markedly increased by PCR amplification prior to ligation into the vector. The only limitation on the procedure is the requirement that enough DNA sequence be known at both ends of the cDNA for appropriate oligonucleotide primers to be designed. PCR amplification of the cDNA produced by reverse transcriptase is called **reverse transcriptase PCR (RT-PCR)**. The resulting amplified molecules contain the coding sequence of the gene of interest with very little contaminating DNA.

## 12.3 Detection of Recombinant Molecules

When genomic restriction fragments produced by a restriction enzyme are mixed with a vector cleaved by the same enzyme, many types of recombinant molecules can result—including such examples as a self-joined circular vector that has not acquired any fragments, a vector containing one or more fragments, and a molecule consisting only of many joined fragments. To facilitate the isolation of a vector containing a particular DNA fragment, some means is needed to ensure (1) that the vector does indeed possess an inserted DNA fragment, and (2) that the fragment is in fact the DNA segment of interest. This section describes several useful procedures for detecting recombinant DNA molecules that have the desired characteristics.

### ■ Gene Inactivation in the Vector Molecule

When transformation is used to introduce recombinant plasmids into bacterial cells, the initial goal is to isolate bacteria that contain the plasmid from among the resulting mixture of plasmid-free and plasmid-containing cells. A common procedure is to use a plasmid that includes an antibiotic-resistance marker and to select the transformed bacteria on a medium containing the antibiotic. Only cells with the plasmid will be able to form colonies. An example of a widely used cloning vector is the pBluescript plasmid illustrated in **FIGURE 12.9**, part A. The entire plasmid is 2961 base pairs. Different regions contribute to its utility as a cloning vector:

- The plasmid *origin of replication* (the position at which DNA replication begins) is derived from the *E. coli* plasmid ColE1. The ColE1 is a high-copy-number plasmid, and its origin of replication enables pBluescript and its recombinant derivatives to exist in approximately 300 copies per cell.
- The ampicillin-resistance gene allows for selection of transformed cells in medium containing ampicillin.
- The cloning site, called a **multiple cloning site (MCS)** or **polylinker**, contains unique cleavage sites for many different restriction enzymes; it allows various types of restriction fragments to be inserted. In pBluescript, the MCS is a 108-bp sequence that contains cloning sites for 23 different restriction enzymes (Figure 12.9B).
- The detection of recombinant plasmids is achieved by means of a region containing the *lacZ* (β-galactosidase) gene from *E. coli*, shown in blue in Figure 12.9A. The basis of the selection is illustrated in **FIGURE 12.10**. When the *lacZ* region is interrupted by a fragment of DNA inserted into the MCS, the recombinant plasmids yield cells unable to produce the enzyme. Nonrecombinant plasmids lacking a DNA

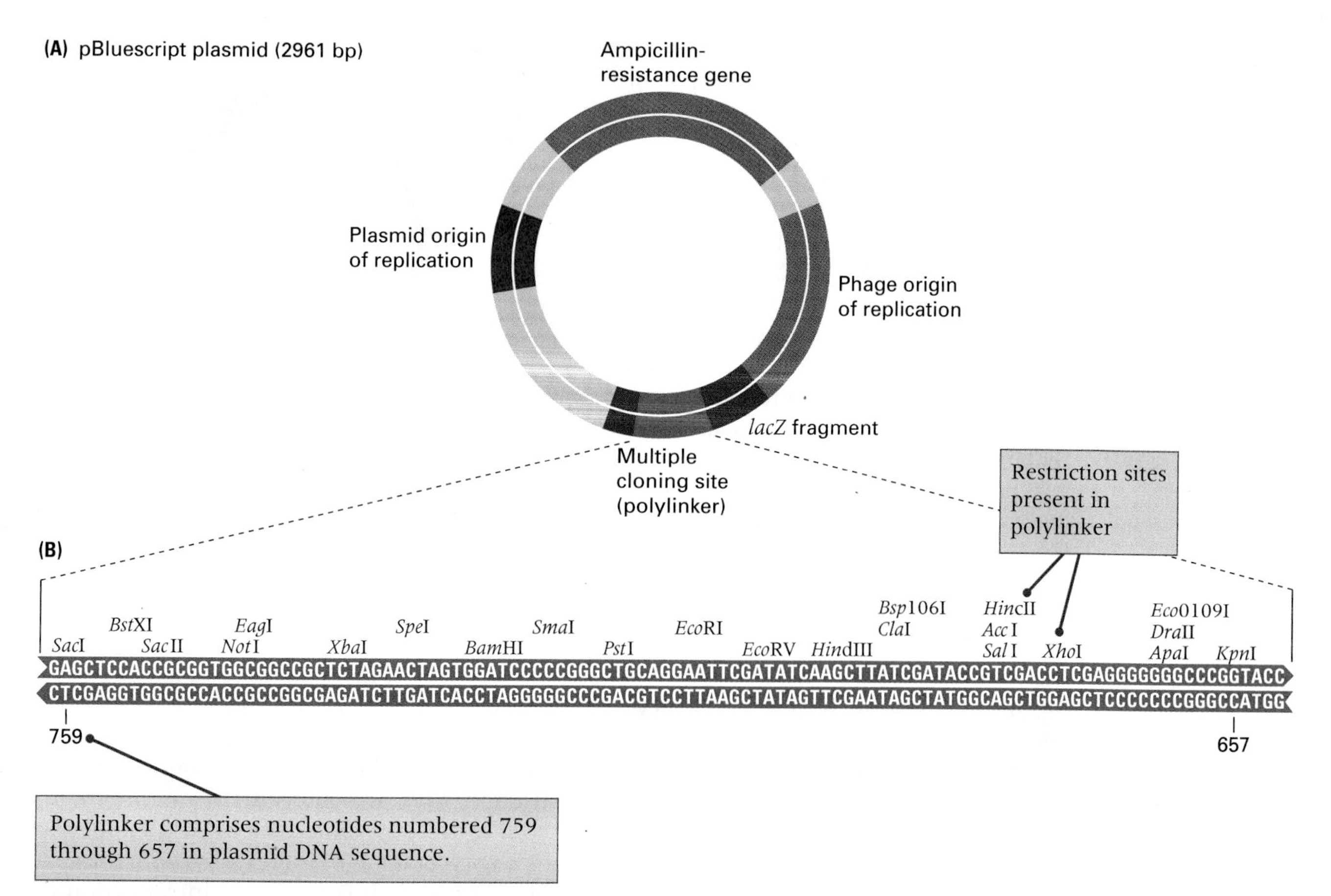


**FIGURE 12.9** (A) Diagram of the cloning vector pBluescript II. It contains a plasmid origin of replication, an ampicillin-resistance gene, a multiple cloning site (polylinker) within a fragment of the *lacZ* gene from *E. coli*, and a bacteriophage origin of replication. (B) Sequence of the multiple cloning site showing the unique restriction sites at which the vector can be opened for the insertion of DNA fragments. The numbers 657 and 759 refer to the position of the base pairs in the complete sequence of pBluescript. [Courtesy of Stratagene–An Agilent Technologies Company]

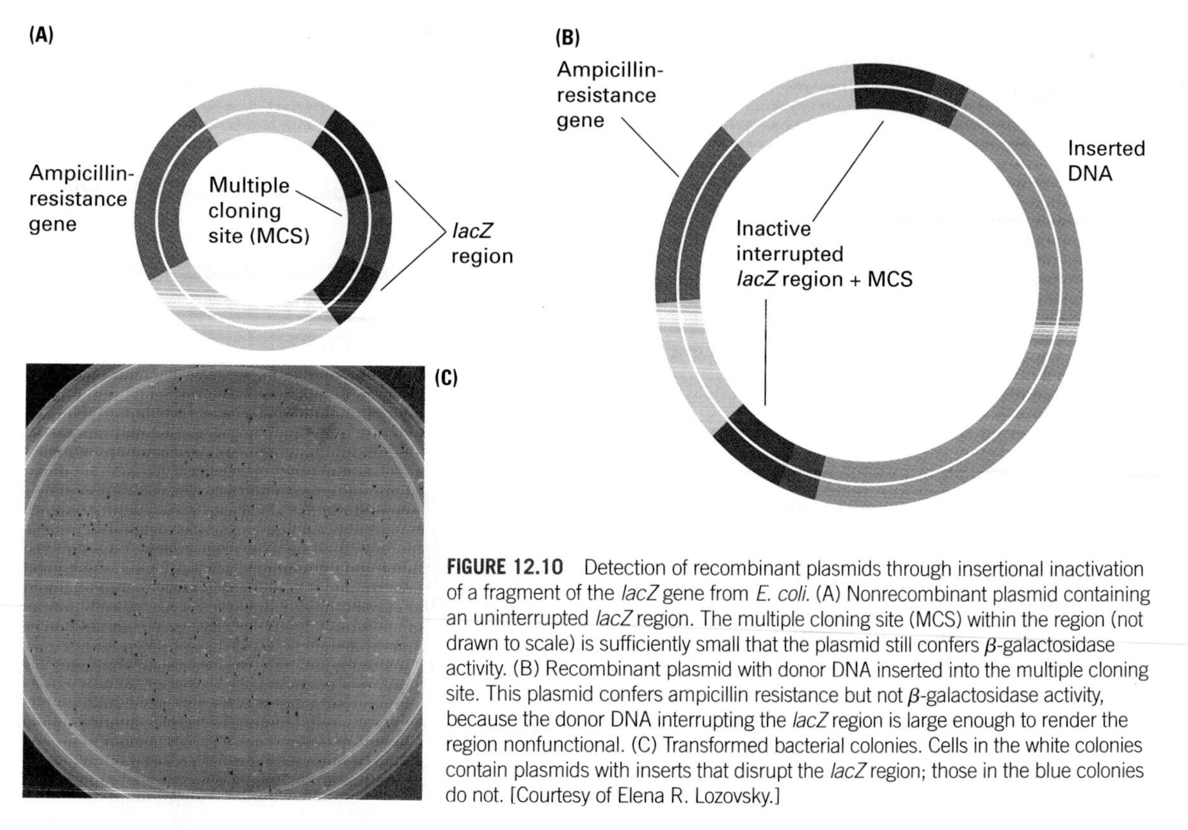


**FIGURE 12.10** Detection of recombinant plasmids through insertional inactivation of a fragment of the *lacZ* gene from *E. coli*. (A) Nonrecombinant plasmid containing an uninterrupted *lacZ* region. The multiple cloning site (MCS) within the region (not drawn to scale) is sufficiently small that the plasmid still confers $\beta$-galactosidase activity. (B) Recombinant plasmid with donor DNA inserted into the multiple cloning site. This plasmid confers ampicillin resistance but not $\beta$-galactosidase activity, because the donor DNA interrupting the *lacZ* region is large enough to render the region nonfunctional. (C) Transformed bacterial colonies. Cells in the white colonies contain plasmids with inserts that disrupt the *lacZ* region; those in the blue colonies do not. [Courtesy of Elena R. Lozovsky.]

fragment in the MCS yield cells that do produce the enzyme. The two types of cells can be distinguished by color when the growth medium contains a special $\beta$-galactoside compound called X-gal, which releases a deep blue dye when cleaved. On medium containing X-gal, as shown in Figure 12.10C, the colonies whose cells contain nonrecombinant plasmids produce $\beta$-galactosidase and turn a deep blue color, whereas colonies whose cells contain recombinant plasmids produce no $\beta$-galactosidase and remain the normal white color.

- The bacteriophage origin of replication derives from the single-stranded DNA phage f1. When cells containing a recombinant plasmid are infected with an f1 helper phage, the f1 origin enables a single strand of the inserted fragment, starting with *lacZ*, to be packaged in progeny phage. This feature is very convenient because it yields single-stranded DNA suitable as a template for DNA sequencing (Section 6.7). The plasmid shown in Figure 12.9A is the SK(+) variety. There is also an SK(−) variety in which the f1 origin is in the opposite orientation and packages the complementary DNA strand.

All good cloning vectors have an efficient origin of replication, at least one unique cloning site for the insertion of a DNA fragment, and a second gene whose interruption by inserted DNA yields a phenotype indicative of a recombinant plasmid.

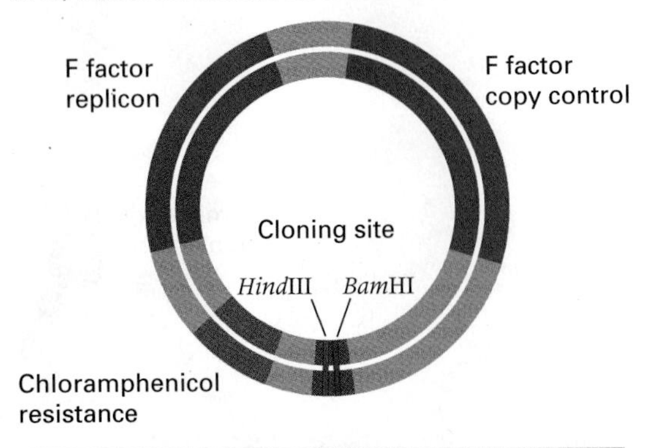


**FIGURE 12.11** Bacterial artificial chromosomes are based on a vector that contains the F factor replicon as well as genes for regulating copy number.

## ■ Cloning of Large DNA Fragments

Large DNA fragments can be cloned intact in bacterial cells with the use of specialized vectors that can accept large inserts. The vectors that can accept large DNA fragments are called *artificial chromosomes*. Among the most widely used are **bacterial artificial chromosomes (BACs)**. The BAC vector (**FIGURE 12.11**) is based on the F factor of *E. coli*, which was discussed in Chapter 9 in the context of its role in conjugation. The essential functions included in the 6.8-kb vector are genes for replication (*repE* and *oriS*), for regulating copy number (*parA* and *parB*), and for chloramphenicol resistance. BAC vectors with inserts greater than 300 kb can be maintained.

DNA fragments in the appropriate size range can be produced by breaking larger molecules into fragments of the desired size by physical means, by treatment with restriction enzymes that have infrequent cleavage sites (for example, enzymes such as *Not*I and *Sfi*I), or by treatment with ordinary restriction enzymes under conditions in which only a fraction of the restriction sites are cleaved (*partial digestion*). Cloning the large molecules consists of: mixing the large fragments of source DNA with the vector; ligation with DNA ligase; introduction of the recombinant molecules into cells of the host; and selection for the clones of interest. These methods are generally similar to those described in Section 12.2 for the production of recombinant molecules containing small inserts of cloned DNA. **FIGURE 12.12** shows a region of the *Drosophila* salivary gland chromosomes that hybridizes with a large-fragment clone containing an insert of 300 kb. The entire genome of *Drosophila* could be contained in only 550 clones of this size.

## ■ Screening for Particular Recombinants

Once a **library**, or large set of clones, has been obtained in a particular vector, the next problem is how to identify the particular recombinant clones that contain the gene of interest. In situations in which DNA or RNA molecules containing sequences complementary to the gene are available, either can be labeled with a fluorescent tag or radioactivity and used as a probe in hybridization experiments to identify the clones containing the gene. The hybridization procedure is known as **colony hybridization**, and it is outlined in **FIGURE 12.13**. Colonies to be tested are transferred (*lifted*) from a solid medium onto a nitrocellulose or nylon filter by gently pressing the filter onto the surface. A part of each colony remains on the agar medium,

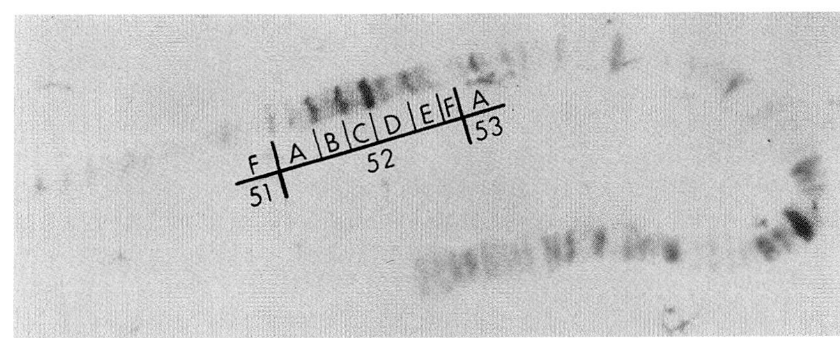
F
A
B
C
D
E
F
A
51
52
53

**FIGURE 12.12** Hybridization *in situ* between the giant salivary gland chromosomes and a recombinant DNA clone containing a large fragment of *Drosophila* DNA.

which constitutes the reference plate. The filter is treated with sodium hydroxide (NaOH), which simultaneously breaks open the cells and denatures the DNA. The filter is then saturated with a solution containing a probe consisting of labeled DNA or RNA, complementary to the gene being sought, and the cellular DNA is renatured. After washing to remove unbound probe, the positions of the bound probe identify the desired colonies. For example, with radioactively labeled probe, the desired colonies are located by means of autoradiography. A similar assay is done with bacterial cells that have been infected with phage vectors.

If transformed cells can synthesize the protein product of a cloned gene or cDNA, then immunological techniques may allow the protein-producing colony to be identified. In one method, the colonies are transferred as in colony hybridization, and the transferred copies are exposed to a labeled antibody directed against the particular protein. Colonies to which the antibody adheres are those that contain the gene of interest.

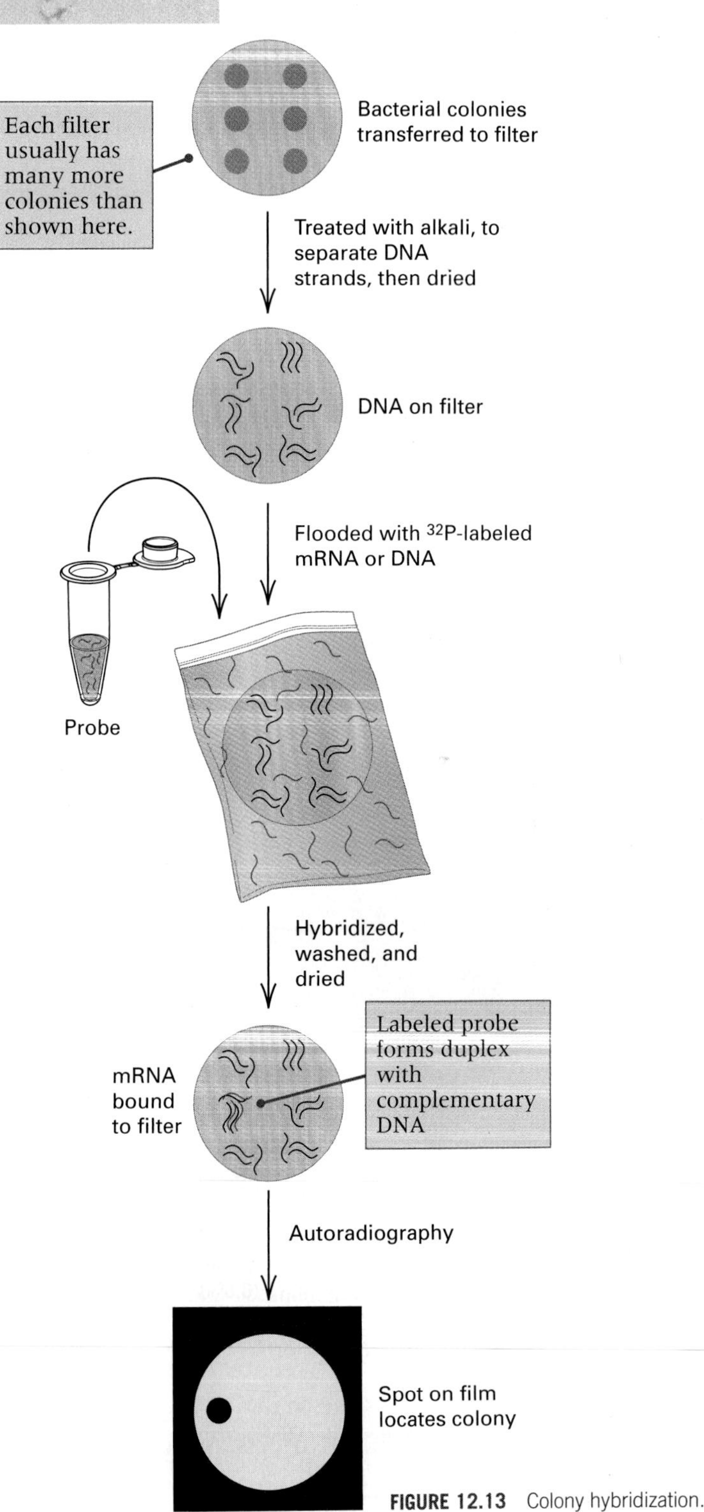
Each filter usually has many more colonies than shown here.
Bacterial colonies transferred to filter
Treated with alkali, to separate DNA strands, then dried
DNA on filter
Flooded with ³²P-labeled mRNA or DNA
Probe
Hybridized, washed, and dried
mRNA bound to filter
Labeled probe forms duplex with complementary DNA
Autoradiography
Spot on film locates colony

**FIGURE 12.13** Colony hybridization.

## 12.4 Genomics and Proteomics

In 1985 the idea that it would be useful to know the complete sequence of all three billion nucleotides in the human genome surfaced. This seemed an outlandish idea at the time, because the cost of DNA sequencing was about one dollar per base pair. But proponents of the idea argued that launching such a program would provide incentives for technology development, and sequencing costs would fall. The Human Genome Project was formally inaugurated in 1990, and by the time the human sequence was completed in 2003, sequencing costs had indeed fallen—to about one penny per base pair.

The goals of the Human Genome Project included sequencing not only the human genome but also the genomes of certain key model organisms used in genetic research because of their demonstrated utility in the discovery of gene function. Vast amounts of data would be generated, which would have to be stored and made accessible, and new methods for analyzing such sequences would have to be developed. The information would need to be made available for drug development and other purposes, and such ethical issues as the privacy of one's genetic information had to be dealt with.

The Human Genome Project was a great success, but it will require many years, probably decades, before genome function and regulation are understood in detail. Tools for understanding the human genome include methods for annotating its content, comparative genomics, transcriptional profiling, and studying protein expression, function, and interaction. These are some of the key approaches to genomics and proteomics, and they are discussed in the following sections.

## Genomic Sequencing

Among the first genomes to be sequenced were those of viruses and bacteria because they are quite small and relatively simple in their layout of regulatory and protein-coding sequences. Small, compact genomes are usually easier to interpret than more complex genomes. The genome of the bacterium *Mycoplasma genitalium* was one of the first to be sequenced. It has the smallest genome of any known free-living organism—a circular DNA molecule 580 kb in length that includes only 471 genes. The organism is a parasite associated with ciliated epithelial cells of the genital and respiratory tracts of primates, including human beings. It belongs to a large group of bacteria (the mycoplasmas) that lack a cell wall and that parasitize a wide range of plant and animal hosts.

Analysis of the *M. genitalium* genome enables us to identify what is probably a minimal set of genes necessary for a free-living cell. The cellular processes in which the gene products this organism participate are summarized in **FIGURE 12.14**. A substantial fraction of the genome is devoted to the synthesis of macromolecules such as DNA, RNA, and protein; another substantial fraction supports cellular processes and energy metabolism. There are very few genes for biosynthesis of small molecules. However, genes that encode proteins for salvaging and/or for transporting small molecules make up a significant fraction of the total, which underscores the fact that the bacterium is parasitic. The remaining genes are largely devoted to forming the cellular envelope and to helping the organism evade the immune system of the host.

Note in Figure 12.14 that one third of the genes have no identified function. This finding is typical of genomic sequences. In many genomic sequences, including the human genome, the proportion of genes with no identified function exceeds 50%. Hence, even when the genes in a genome are correctly identified, there are numerous additional issues. Genomic sequencing therefore should be thought of as only the initial stage in the quest to understand the higher and more integrated levels of biological organization and function.

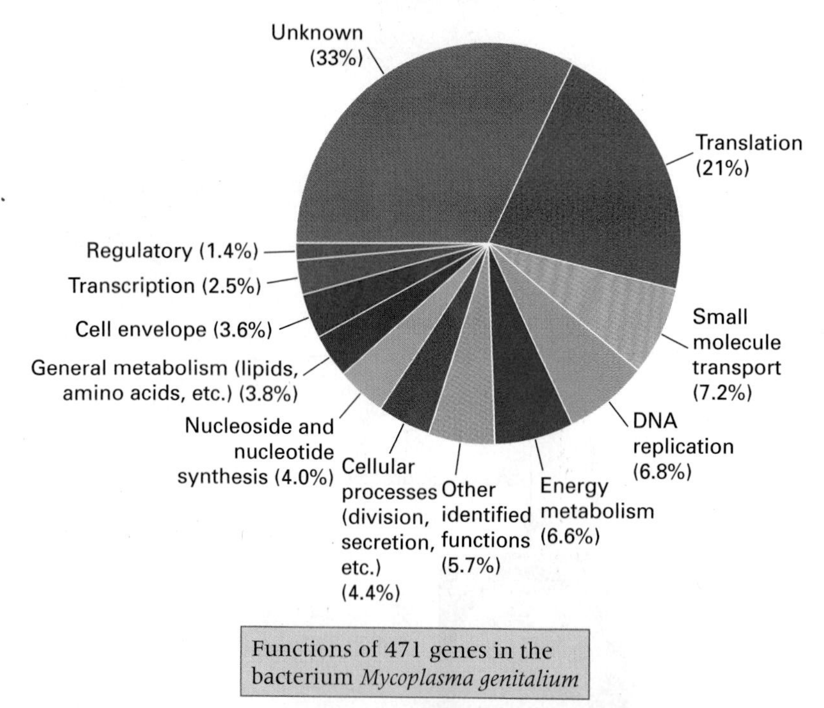


**FIGURE 12.14** Genes in the genome of *Mycoplasma genitalium* classified by function. [Data from C. M. Fraser, et al., *Science* 270 (1995): 397–404.]

## Genome Annotation

A genome sequence is not self-explanatory. It is like a book printed in an alphabet of only four letters, without spaces or punctuation, and lacking an index. To be useful, any genomic sequence must be accompanied by **genome annotation**, which refers to explanatory notes that accompany the sequence. A genome annotation specifies functional elements, notably sequences in or near coding regions that delineate protein-coding exons and introns as well as the upstream and downstream binding motifs that are targets of enhancer or silencer elements. Annotations also include sequences encoding functional RNAs such as tRNAs, small nuclear RNAs involved in splicing, and microRNAs. Annotations also include sequences corresponding to transposable elements and so forth.

Especially for large, complex genomes in which much of the DNA does not code for proteins, and in which most protein-coding exons are relatively small and interrupted by large introns, it is a daunting challenge to parse a genomic sequence into its protein-coding exons, to identify which protein-coding exons belong to the same gene, and to recognize the upstream and downstream regulatory regions that control gene expression. The annotation of genomic sequences at this level is one aspect of **computational genomics**, defined broadly as the use of computers in the interpretation and management of biological data.

Furthermore, especially in multicellular eukaryotes, even for genes whose functions can be assigned, it is not usually known when during the life cycle each gene is expressed, in which tissues it is expressed, or the presence, patterns, or tissue specificity of alternative splicing. Interactions among genes and gene products are also typically unknown. The greatest challenge is to understand how the genes in the genome function and are coordinately regulated to control development, metabolism, reproduction, behavior, and response to the environment.

One of the most informative sources of information in genome annotation consists of single-copy sequences that are transcribed and processed into mRNA. These can be studied by sequencing cDNA copies obtained as described in Section 12.2. Sequences identified from such cDNAs are called **expressed sequence tags (ESTs)**. One of the most ambitious EST projects to date included sequencing about 300,000 partial cDNAs, among them cDNAs obtained from cDNA libraries prepared from 37 distinct human organs and tissues. The total DNA sequence obtained was 83 million base pairs. Although the human genome contains an estimated 20,000—25,000 protein-coding genes, the ESTs suggest that there may be 100,000 or so alternatively spliced transcripts.

Computer matching among the ESTs revealed 87,983 distinct sequences, many of which could be assigned a function on the basis of similarity to known genes from human beings or from other organisms. **FIGURE 12.15** gives a breakdown of the ESTs by function. Approximately 40 percent of human genes are implicated in basic energy metabolism, cell structure, homeostasis, or cell division; a further 22 percent are concerned with RNA and protein synthesis and processing; and 12 percent are associated with signaling and communication between cells. **FIGURE 12.16** summarizes the results of examining the tissue-specific cDNA libraries.

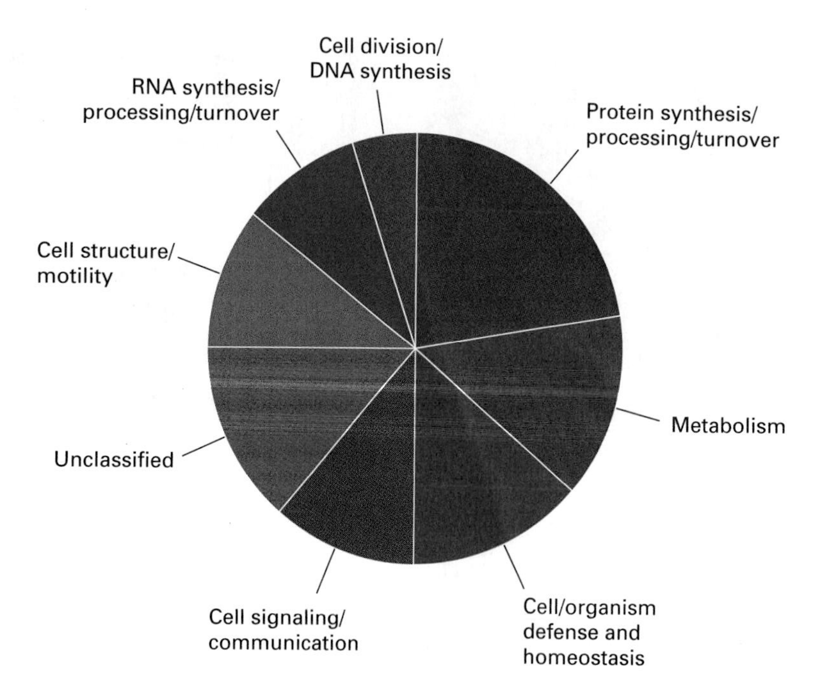


**FIGURE 12.15** Classification of cDNA sequences by function. The pie chart is based on over 13,000 distinct, randomly selected human cDNA sequences. [Data courtesy of Craig Venter and the Institute for Genomic Research.]

Protests were organized against genetically engineered corn when it appeared that the pollen could kill larvae of the monarch butterfly. [© J. Scott Applewhite/ AP Photos.]

For each organ or tissue type, the first number is the total number of cDNA clones sequenced, and the number in parentheses is the number of distinct cDNA sequences found among the total from that organ or tissue type.

## ■ Comparative Genomics

In many cases, useful information can be gained by identifying genes with similar sequences in other organisms, but if the organisms

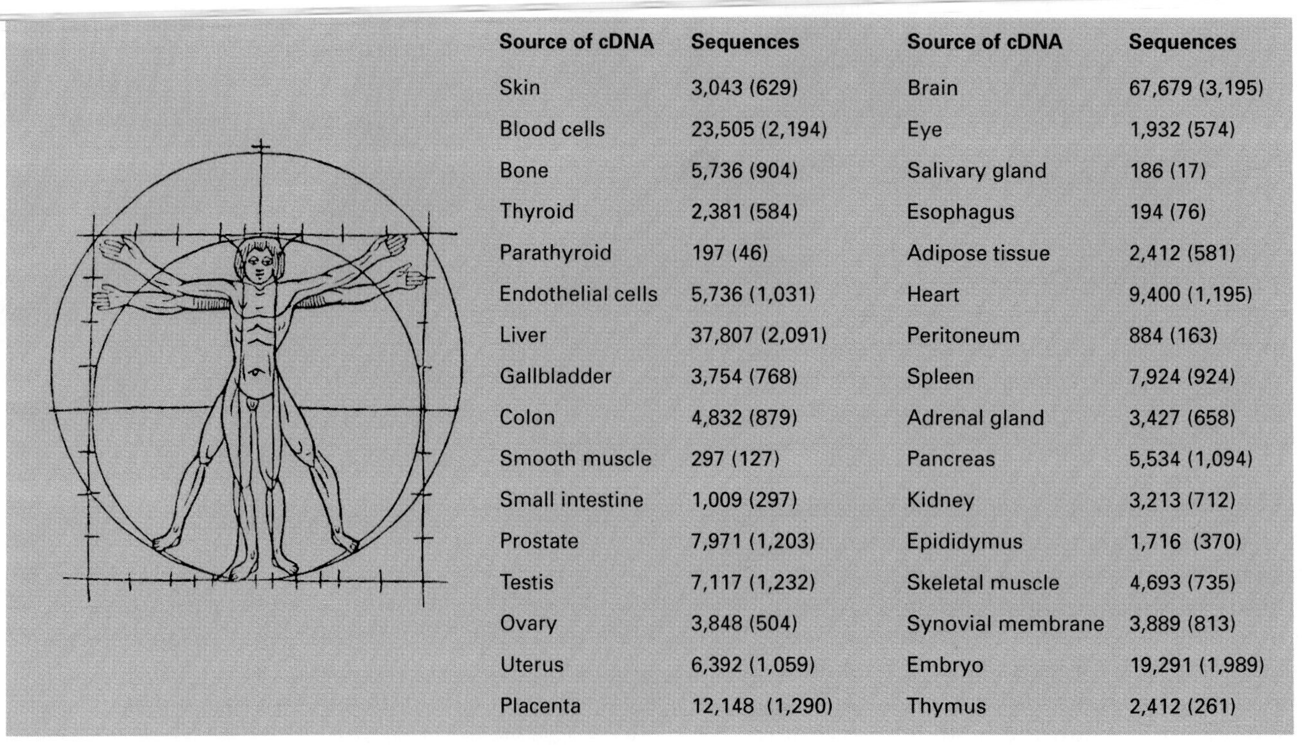

| Source of cDNA | Sequences | Source of cDNA | Sequences |
|---|---|---|---|
| Skin | 3,043 (629) | Brain | 67,679 (3,195) |
| Blood cells | 23,505 (2,194) | Eye | 1,932 (574) |
| Bone | 5,736 (904) | Salivary gland | 186 (17) |
| Thyroid | 2,381 (584) | Esophagus | 194 (76) |
| Parathyroid | 197 (46) | Adipose tissue | 2,412 (581) |
| Endothelial cells | 5,736 (1,031) | Heart | 9,400 (1,195) |
| Liver | 37,807 (2,091) | Peritoneum | 884 (163) |
| Gallbladder | 3,754 (768) | Spleen | 7,924 (924) |
| Colon | 4,832 (879) | Adrenal gland | 3,427 (658) |
| Smooth muscle | 297 (127) | Pancreas | 5,534 (1,094) |
| Small intestine | 1,009 (297) | Kidney | 3,213 (712) |
| Prostate | 7,971 (1,203) | Epididymus | 1,716 (370) |
| Testis | 7,117 (1,232) | Skeletal muscle | 4,693 (735) |
| Ovary | 3,848 (504) | Synovial membrane | 3,889 (813) |
| Uterus | 6,392 (1,059) | Embryo | 19,291 (1,989) |
| Placenta | 12,148 (1,290) | Thymus | 2,412 (261) |

**FIGURE 12.16** Classification of cDNA sequences by organ or tissue type. In each category, the initial number is the total number of cDNA clones examined. The number in parentheses is the number of distinct sequences found per organ or tissue type. [Data from M. D. Adams, et al., *Nature* (6547 Suppl.) 377 (1995): 3–174.]

diverged from a common ancestor too long ago, there is the problem of recognizing which sequences are sufficiently similar to be regarded as functionally equivalent. One way to get around his problem is to compare the genome sequences of groups of related species that have a graded series of divergence times. This approach is known as **comparative genomics**, which has become one of the most powerful strategies for identifying genetic elements in the human genome and those of model organisms.

The fruits of comparative genomics are exemplified in the genome sequences of 12 *Drosophila* species. **FIGURE 12.17** summarizes the evolutionary relationships among the species and their approximate divergence times. The species are very diverse in their geographical origins, global distribution, morphology, behavior, feeding habits, and other phenotypes, yet they share a similar cellular physiology, developmental program, and life cycle. Their genomes show substantial differences in sequence (5 million years in the scale of Figure 12.17 corresponds to about one nucleotide difference per 10 nucleotide sites), and they have also undergone multiple gene rearrangements primarily due to inversions. The 12-genome comparison therefore reveals how conserved gene functions are maintained in spite of extensive changes in genome structure and sequence.

Comparative genomics derives its power from the distinctive evolutionary patterns, called *evolutionary signatures*, whch different types of functional elements exhibit. Some examples from the 12 *Drosophila* species are illustrated in **FIGURE 12.18**. Part A shows characteristic evolutionary signatures of protein-coding sequences. Note the pronounced triplet periodicity uninterrupted by stop codons. Many of the nucleotide differences between species are in the third codon position, and the variant codons often encode the same amino acid (green). Deletions, when they occur, remove a number of nucleotides that is a multiple of three (gray), which conserves the proper reading frame. Contrast this pattern with that observed in noncoding regions (part B). Here the nucleotide differences between species are not concentrated in a particular triplet phase (red), triplets corresponding to chain-terminating (nonsense) codons come and go (blue), and deletions are not constrained to a multiple of three nucleotides (orange).

RNA transcripts that form foldback secondary structures, such as tRNAs, rRNAs, and some snRNAs, have distinctive evolutionary signature of their own. In Figure 12.18C, for

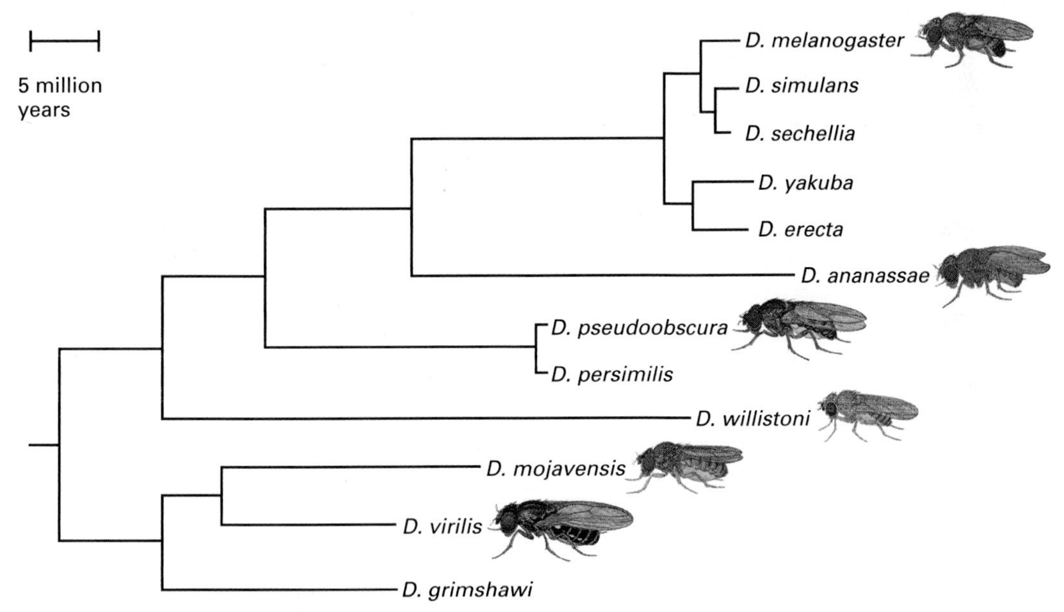


**FIGURE 12.17** Evolutionary relationships among twelve *Drosophila* species whose genomes were sequenced for comparative genomics, scaled by their approximate divergence times. [Reproduced from J. T. Patterson, *Studies in the Genetics of Drosophila, Part III and Part IV.* University of Texas Publications (1943 and 1944). Used with permission of the School of Biological Sciences, University of Texas at Austin.]

example, the matched parentheses show the conserved base pairs in the paired stem structures, but the identities of the paired bases often differ among species (paired nucleotides are color coded). A nice example involves the nucleotides at positions 29 and 38, which in *D. melanogaster* comprise a C-G pair but in *D. yakuba* comprise a U-G pair (U pairs with G as well as with A in double-stranded RNA).

MicroRNAs, important for their regulatory functions in the RNAi pathways, show yet another type of evolutionary signature (part D). In this case changes in the stem regions, even those that are complementary, are not well tolerated, but differences in the loop and other nonpaired regions are found. (The asterisks beneath the miRNA sequences indicate nucleotides that are identical to those in *D. melanogaster*.)

Finally, comparative genomics helps identify regulatory motifs that are the targets of enhancers and silencers. These are often difficult to recognize because they are relatively short, can be present on either DNA strand, and can change position within the gene promoter. The example in Figure 12.18E shows binding sites for the protein Mef-2. The consensus binding site has the sequence YTAWWWWTAR, where Y is any pyrimidine, R is any purine, and W means either A or T. The 12-species comparison shows the differing sequence and location of the Mef-2 biding site in one of its target genes. Some of the species have the binding site toward the 5′ end of the region shown, others have it near the 3′ end, and one species (*D. ananassae*) has a Mef-2 binding site at both locations. (The BLS scores are measures of closeness to the consensus sequence.)

The twelve-genomes comparisons were instrumental in correctly annotating hundreds of protein-coding genes in the *D. melanogaster* sequence, predicting the secondary structures of many noncoding RNAs with some likely to be involved in translational regulation, showing that some microRNA genes have multiple functional products that increase their regulatory repertoire, and revealing a network of pretranscriptional and posttranscriptional miRNA regulatory targets. The utility of a graded series of divergence times was also validated by the observation that the optimal divergence time for identifying evolutionary signatures depends on the length of the functional element. Longer functional elements are most easily recognized in closely related species, whereas shorter ones are most efficiently identified in more distantly related species.

### Transcriptional Profiling

A new approach to genetics called **functional genomics** focuses on genome-wide patterns of gene expression and the mechanisms by which gene expression is coordinated. As changes take place in the cellular environment—for example, through changes in the external conditions, development, or aging—the patterns of gene expression change also. But genes are usually deployed in sets, not individually. As the level

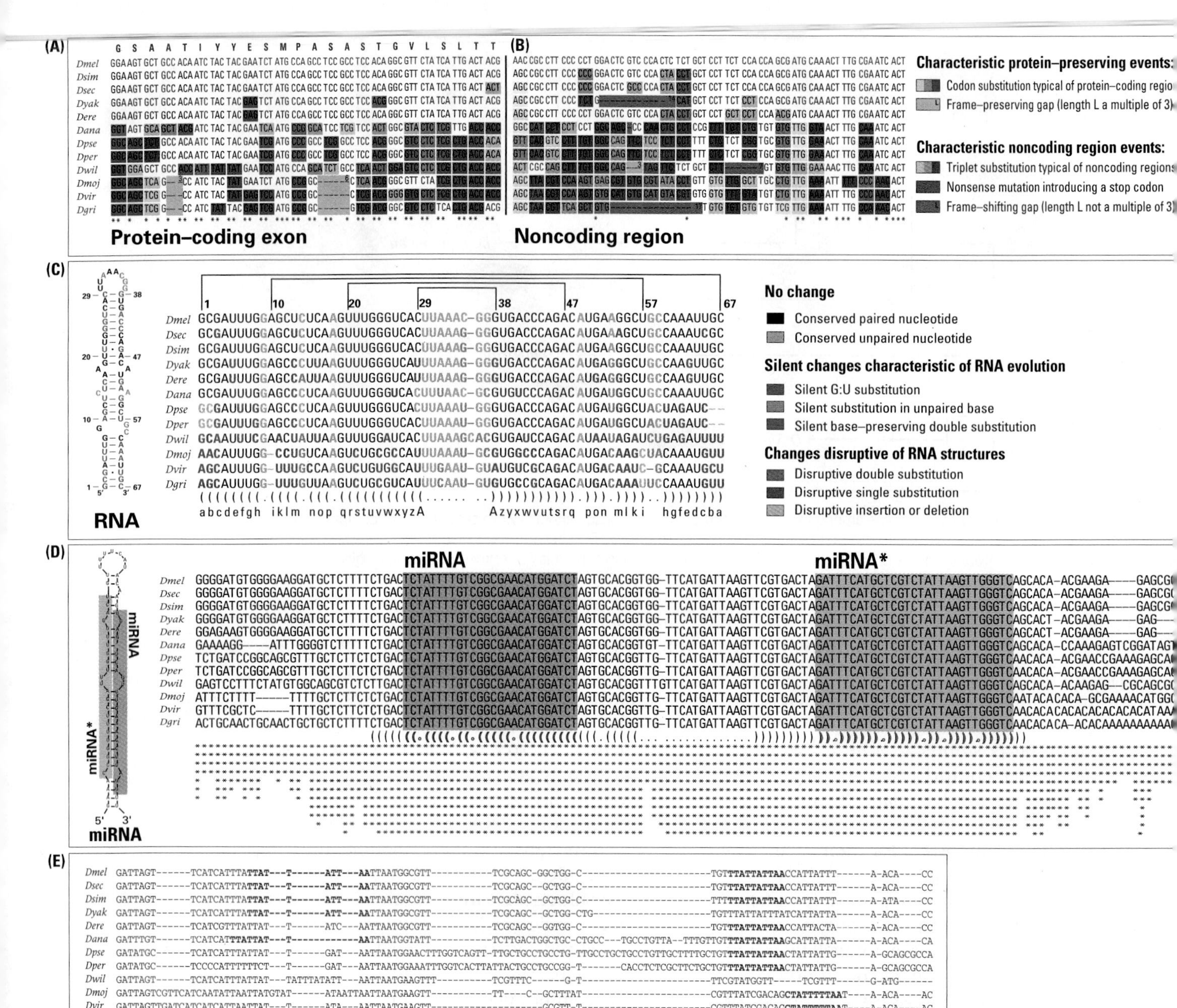


**FIGURE 12.18** Evolutionary signatures observed among the twelve *Drosophila* genomes in regions coding for (A) protein, (B) noncoding RNA, (C) a stem-loop secondary structure in RNA, (D) micoRNA, and (E) transcription-factor binding sites. [Reprinted by permission from Macmillan Publishers Ltd: A. Stark, et al., *Nature* 450 (2007): 219–232. http://www.nature.com/nature]

of expression of one coordinated set is decreased, the level of expression of a different coordinated set may be increased. How can one study tens of thousands of genes all at the same time?

The study of genome-wide patterns of gene expression became feasible with the development of the **DNA microarray** (or *chip*), a flat surface about the size of a postage stamp on which 10,000–100,000 distinct spots are present, each containing a different immobilized DNA sequence suitable for hybridization with DNA or RNA isolated from cells growing under different conditions, from cells not exposed or cells exposed to a drug or toxic chemical, from different stages of development, or from different types or stages of a disease such as cancer. Two types of DNA chips are presently in use:

1. A chip arrayed with oligonucleotides synthesized directly on the chip, one nucleotide at a time, by automated procedures; these chips typically have hundreds of thousands of spots per array.
2. A chip arrayed with denatured, double-stranded DNA sequences of 500–5000

bp, in which the spots, each about a millionth of a drop in volume, are deposited by capillary action from miniaturized fountain-pen-like devices mounted on the moveable head of a flatbed robotic workstation; these chips typically have tens of thousands of spots per array.

**FIGURE 12.19** shows one method by which DNA chips are used to assay the global levels of gene expression in an experimental sample relative to a control. At the upper right are shown six

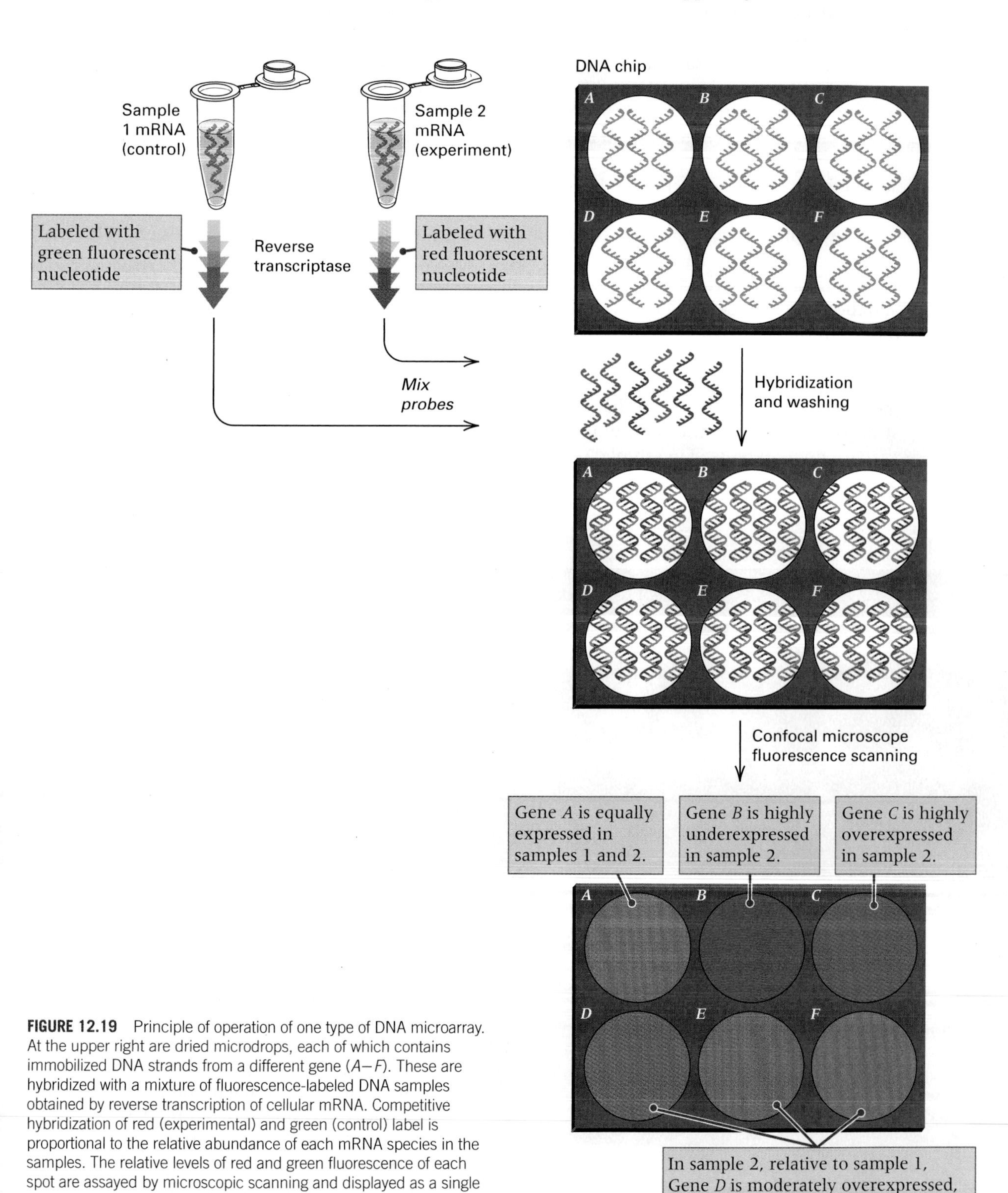


**FIGURE 12.19** Principle of operation of one type of DNA microarray. At the upper right are dried microdrops, each of which contains immobilized DNA strands from a different gene (*A–F*). These are hybridized with a mixture of fluorescence-labeled DNA samples obtained by reverse transcription of cellular mRNA. Competitive hybridization of red (experimental) and green (control) label is proportional to the relative abundance of each mRNA species in the samples. The relative levels of red and green fluorescence of each spot are assayed by microscopic scanning and displayed as a single color. Red or orange indicates overexpression in the experimental sample, green or yellow-green underexpression in the experimental sample, and yellow equal expression.

adjacent spots in the microarray, each of which contains a DNA sequence that serves as a probe for a different gene, *A* through *F*. At the left is shown the experimental protocol. Messenger RNA is first extracted from both the experimental and the control samples. This material is then subjected to one or more rounds of reverse transcription, as described in Section 12.2. In the experimental material (sample 2), the primer for reverse transcription includes a red fluorescent label; and in the control material (sample 1), the primer includes a green fluorescent label. When a sufficient quantity of labeled DNA strands have accumulated, the fluorescent samples are mixed and hybridized with the DNA chip.

The result of hybridization is shown in the middle part of Figure 12.19. Because the samples are mixed, the hybridization is competitive, and therefore the density of red or green strands bound to the DNA chip is proportional to the concentration of red or green molecules in the mixture. Genes that are overexpressed in sample 2 relative to sample 1 will have more red strands hybridized to the spot, whereas those that are underexpressed in sample 2 relative to sample 1 will have more green strands hybridized to the spot.

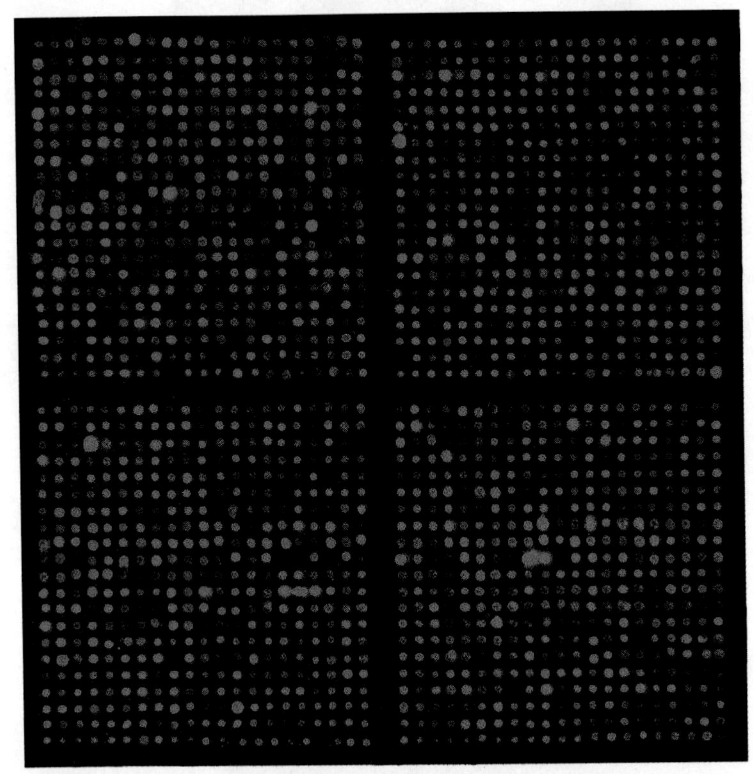

**FIGURE 12.20** Small part of a yeast DNA chip showing 1764 spots, each specific for hybridization with a different mRNA sequence. The color of each spot indicates the relative level of gene expression in experimental and control samples. The complete chip for all yeast open reading frames includes over 6200 spots. [Courtesy of Jeffrey P. Townsend and Duccio Cavalieri, Yale University.]

After hybridization, the DNA chip is placed in a confocal fluorescence scanner that scans each pixel (the smallest discrete element in a visual image) first to record the intensity of one fluorescent label and then again to record the intensity of the other fluorescent label. These signals are synthesized to produce the signal value for each spot in the microarray. The signals indicate the relative levels of gene expression by color, as shown in **FIGURE 12.20**. A spot that is red or orange indicates high or moderate overexpression of the gene in the experimental sample, a spot that is green or yellow-green indicates high or moderate underexpression of the gene in the experimental sample, and a spot that is perfectly yellow indicates equal levels of gene expression in the samples. In this manner, DNA chips can assay the relative levels of any mRNA species whose abundance in the sample is more than one molecule per $10^5$, and differences in expression as small as approximately twofold can be detected.

Gene-expression arrays have been used to identify groups of genes that are coordinately regulated in development. The example in **FIGURE 12.21** shows expression profiles for 20 groups of genes in the early stages of development in *Caenorhabditis elegans*. In these experiments, time in development was measured in minutes relative to the four-cell stage. Relative levels of gene expression are plotted on a logarithmic scale, and hence the changes in relative transcript abundance are often two or three orders of magnitude. Over the time period examined, the embryo undergoes a transition from control through maternal transcripts present in the egg to those transcribed in the embryo itself, and includes the times during which most of the major cell fates are specified. The microarrays used in these experiments allowed detection of transcripts from almost 9000 open reading frames, and the plots include traces for approximately 2500 genes, about 80% of all those that showed significant changes in transcript abundance over the time interval shown.

Up to the four-cell stage of development the patterns of transcription are all quite stable and then begin to change rapidly. Development in the earliest stages is supported largely by maternally derived transcripts. Many of these are cleared rapidly as development proceeds, for example, the transcripts plotted for clusters of 141, 244, and 568 genes in the lower-right panels. Production of transcripts from embryonic cells is clearly induced, as evidenced by those patterns for clusters of 431 and 153 genes in the panels at the upper right. The curves showing

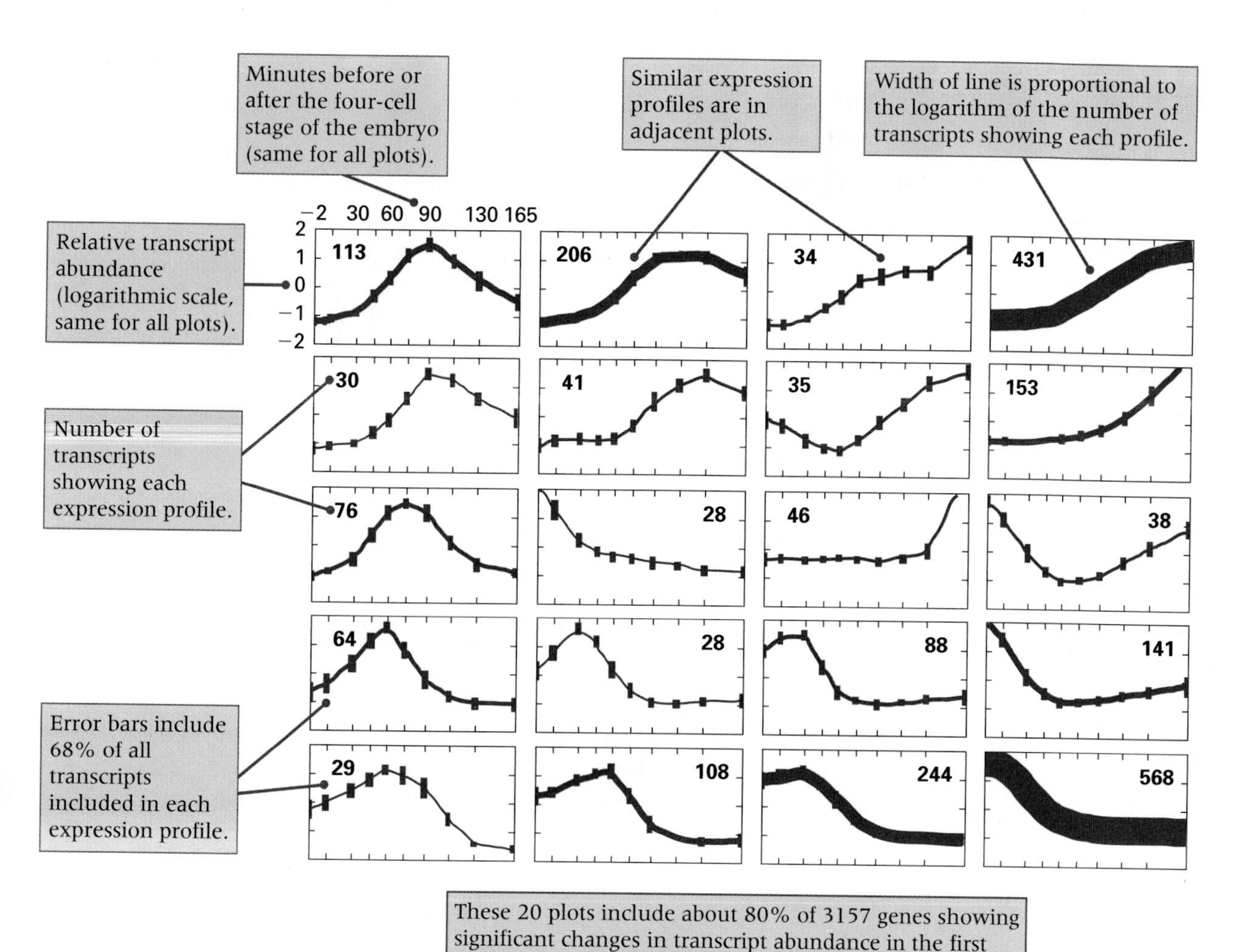

Minutes before or after the four-cell stage of the embryo (same for all plots).
Similar expression profiles are in adjacent plots.
Width of line is proportional to the logarithm of the number of transcripts showing each profile.
Relative transcript abundance (logarithmic scale, same for all plots).
Number of transcripts showing each expression profile.
Error bars include 68% of all transcripts included in each expression profile.
−2 30 60 90 130 165
2
1
0
−1
−2
113
206
34
431
30
41
35
153
76
28
46
38
64
28
88
141
29
108
244
568
These 20 plots include about 80% of 3157 genes showing significant changes in transcript abundance in the first 165 minutes after the four-cell stage in development.

**FIGURE 12.21** Patterns of transcriptional regulation of about 2500 genes during the first approximately 2.75 hours of development in *C. elegans*. Complete development requires about 14 hours. [Reproduced from L. R. Baugh, et al., *Development* 130 (2003): 889–900. Reproduced with permission of the Company of Biologists.]

the disappearance of the maternal transcripts and appearance of the embryonic transcripts intersect at about the time of gastrulation, indicating a somewhat earlier (mid-blastula) transition from maternal to embryonic control of development. Many of the gene transcription patterns are very complex, with a transient peak of expression suggesting that the transcript (though not necessarily the protein product) is needed for only a brief period in development. All five panels along the left-hand side of Figure 12.21 show this kind of pattern.

Although the transcriptional analysis in Figure 12.21 is a rather coarse, bird's-eye view of what takes place during development, the identification of groups of coordinately expressed genes is of considerable value in itself because it suggests that these genes may share common or overlapping *cis*-acting regulatory sequences that are controlled by common or overlapping sets of transcriptional activator proteins.

A different application of DNA microarrays is shown in **FIGURE 12.22**, which deals with the evolution of gene-expression differences among species. In this analysis, the relative abundance of 3809 transcripts was compared between adult males and females of two species of *Drosophila*, *D. melanogaster* and *D. simulans*, which became separate species approximately 2.5 million years ago. The horizontal axis of the graph is an index of the difference in relative level of expression of each transcript between the species, and the vertical axis is the number of genes. The three histograms correspond to male-biased genes (908 genes that show greater expression in males in both species, primarily genes expressed in the testes or other male reproductive organs), female-biased genes (1498 genes that show greater expression in females in both species, primarily genes expressed in the ovaries or other female reproductive organs), and 1403 non-sex-biased genes that show equal expression in the two sexes in both species.

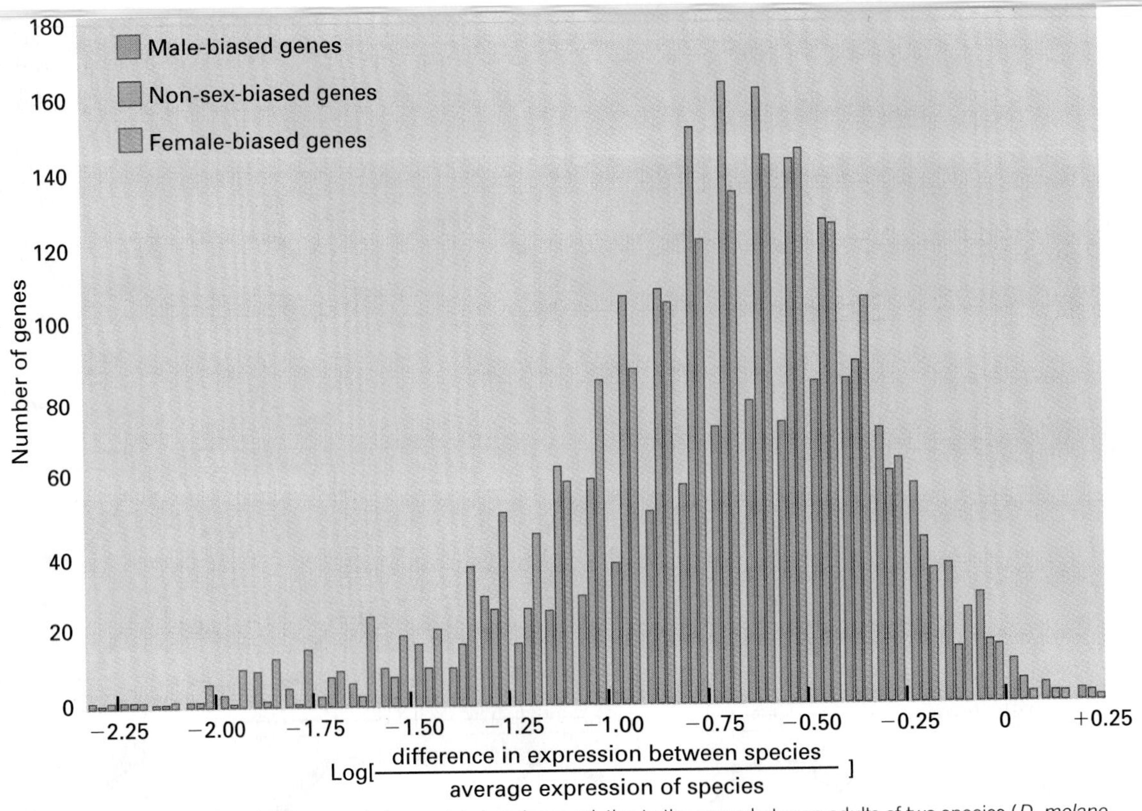


**FIGURE 12.22** Distribution of differences in transcript abundance relative to the mean between adults of two species (*D. melanogaster* and *D. simulans*) for genes expressed primarily in adult males (blue), expressed primarily in adult females (orange), or expressed approximately equally in the sexes (green). [Data from J. M. Ranz, et al., *Science* 300 (2003): 1742–1745.]

As the histograms show very clearly, genes associated with male reproduction (male-biased genes) show expression levels that are, on the average, more different between the species than the expression levels of genes associated with female reproduction (female-biased genes), and both of these show greater differences between species than genes that are expressed equally in both sexes. Interestingly, the same pattern of rapid evolution of male-biased and female-biased genes is observed for the rate of amino acid replacement in proteins as shown here for transcript abundance. The reasons for such dramatic differences between these classes of genes are not known for certain, but most evolutionary biologists believe they are related to *sexual selection*, the special type of natural selection associated with competition for mates and maximization of reproductive success.

## Two-Hybrid Analysis of Protein Interactions

Biological processes can also be explored from the standpoint of proteomics by examining protein–protein interactions. The rationale for studying such interactions is that proteins that participate in related cellular processes often interact with one another; hence, knowing which proteins interact can provide clues to the possible function of otherwise anonymous proteins.

One method for identifying protein–protein interactions makes use of the GAL4 transcriptional activator protein in budding yeast discussed in Section 11.5. The GAL4 protein includes two separate domains or regions, both of which are necessary for transcriptional activation. One domain is the zinc-finger DNA-binding domain that binds with the target site in the promoter of the *GAL* genes that are activated, and the other domain is the transcriptional activation domain that makes contact with the transcriptional complex and actually triggers transcription. In the wildtype GAL4 protein, these domains are tethered together because they are parts of the same polypeptide chain.

The key to identifying protein–protein interactions through the use of GAL4 is that the coding regions for the separate domains can be taken apart and each fused to a coding region for a different protein. The strategy is shown in **FIGURE 12.23**, part A, where the GAL4 DNA-binding domain and the transcriptional activation domain are depicted as separate entities, each fused to a different polypeptide chain,

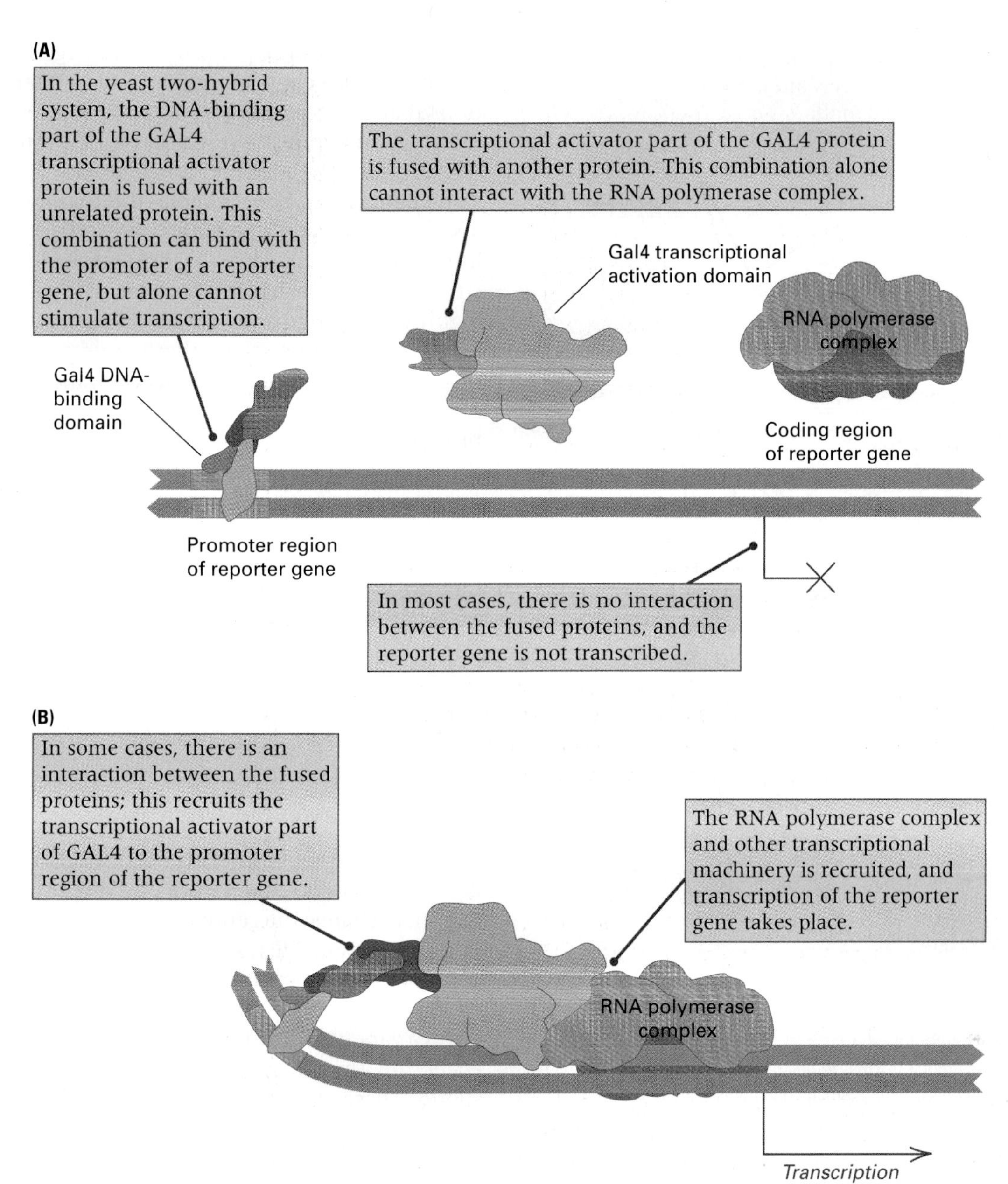

(A)
In the yeast two-hybrid system, the DNA-binding part of the GAL4 transcriptional activator protein is fused with an unrelated protein. This combination can bind with the promoter of a reporter gene, but alone cannot stimulate transcription.
The transcriptional activator part of the GAL4 protein is fused with another protein. This combination alone cannot interact with the RNA polymerase complex.
Gal4 transcriptional activation domain
RNA polymerase complex
Gal4 DNA-binding domain
Coding region of reporter gene
Promoter region of reporter gene
In most cases, there is no interaction between the fused proteins, and the reporter gene is not transcribed.
(B)
In some cases, there is an interaction between the fused proteins; this recruits the transcriptional activator part of GAL4 to the promoter region of the reporter gene.
The RNA polymerase complex and other transcriptional machinery is recruited, and transcription of the reporter gene takes place.
RNA polymerase complex
Transcription


**FIGURE 12.23** Two-hybrid analysis by means of the GAL4 protein. (A) When the proteins fused to the GAL4 domains do not interact, transcription of the reporter gene does not take place. (B) When the proteins do interact, the reporter gene is transcribed.

shown in the vicinity of a *GAL* promoter. The promoter is attached to a **reporter gene** whose transcription can be detected by means of, for example, a color change in the colony, the production of a fluorescent protein, or the ability of the cells to grow in the presence of an antibiotic. The fused DNA-binding domain and the fused transcriptional activation domain are both hybrid proteins, and for this reason the test system is called a **two-hybrid analysis**. In part A, the proteins fused to the GAL4 domains do not interact within the nucleus. The DNA-binding domain therefore remains separated from the transcriptional activation domain, and transcription of the reporter genes does not occur.

Figure 12.23B shows a case in which the protein fused to the GAL4 domains do interact. In this case, the DNA-binding domain and the transcriptional activation domain are brought into contact, and transcription of the reporter gene does take place. In this manner, transcription of the reporter gene in the two-hybrid analysis indicates that the proteins fused to the GAL4 domains undergo a physical interaction that brings the two hybrid proteins together.

An example of two-hybrid analysis is shown in **FIGURE 12.24**, which depicts a network of 318 protein–protein interactions among 329 nuclear proteins in yeast. The purpose of this analysis was to compare the observed network of

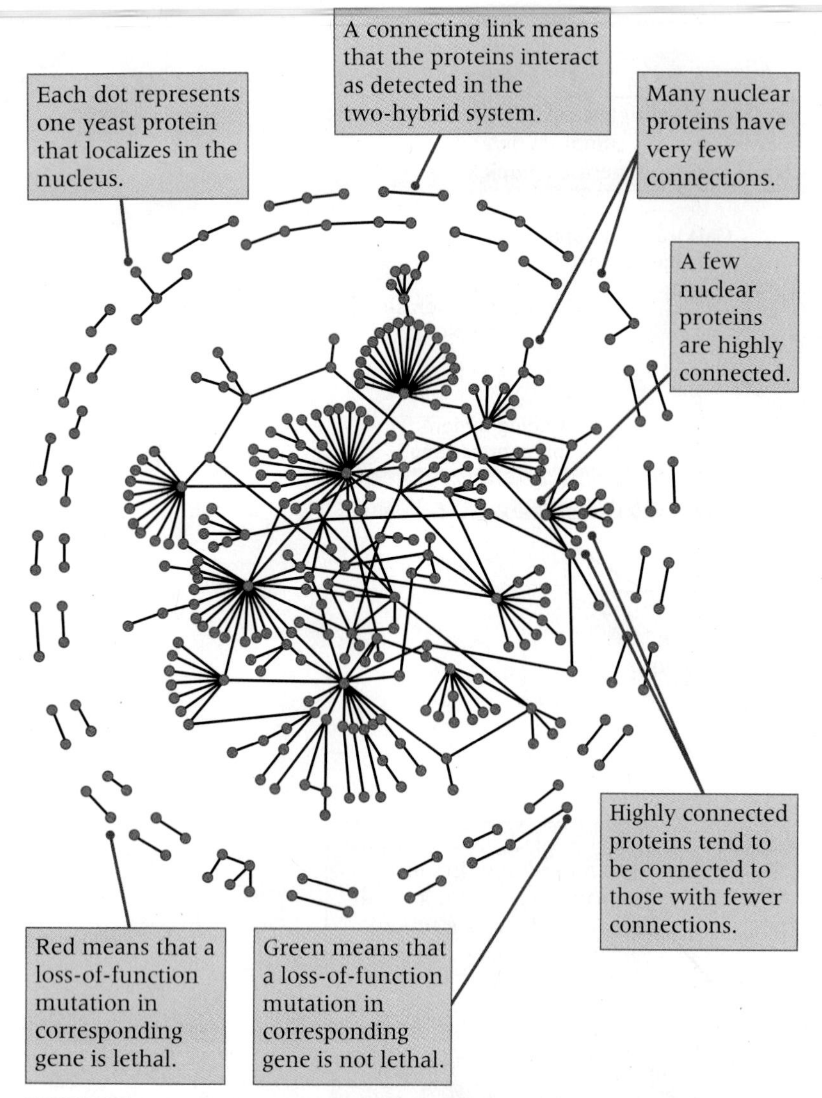

A connecting link means that the proteins interact as detected in the two-hybrid system.
Each dot represents one yeast protein that localizes in the nucleus.
Many nuclear proteins have very few connections.
A few nuclear proteins are highly connected.
Highly connected proteins tend to be connected to those with fewer connections.
Red means that a loss-of-function mutation in corresponding gene is lethal.
Green means that a loss-of-function mutation in corresponding gene is not lethal.

**FIGURE 12.24** Physical interactions among nuclear proteins in yeast. Not shown are nuclear proteins that exhibit no interactions. [Reproduced from S. Maslov and K. Sneppen, *Science* 296 (2002): 910–913. Reprinted with permission from AAAS. Illustration courtesy of Sergei Maslov, Brookhaven National Laboratory.]

interactions with random networks containing the same number of interactions, but with the interacting partners chosen at random. One interesting property of the network in Figure 12.24, as well as of other protein networks, is that there are fewer than expected interactions between proteins that are already highly connected. In other words, proteins that are highly connected to other proteins through many interactions tend to be connected not to other highly connected proteins, but to proteins with fewer connections. The systematic suppression of links between highly connected proteins has the effect of minimizing the extent to which random environmental or genetic perturbations in one part of the network spread to other parts of the network.

Two-hybrid analysis affords a powerful approach to discovering protein–protein interactions because it can be performed on a large scale, requires no protein purification, detects interactions that occur in living cells, and requires no information about the function of the proteins being tested. The method, however, does have some limitations. For example, the two-hybrid assay is qualitative, not quantitative, and so weak interactions cannot easily be distinguished from strong ones. The hybrid proteins are usually highly expressed to enhance the reliability of the assay, and so interactions that would not take place at normal concentrations can take place. The two-hybrid assay also requires that the protein–protein interactions take place in the nucleus, whereas some proteins may interact only in the environment of the cytoplasm. Finally, hybrid proteins may fold differently than native proteins, and the misfolded proteins may fail to interact when the native conformations do, or they may interact when the native conformations do not. The conclusion is that results from two-hybrid analyses need to be interpreted with care, but on the other hand the method has already yielded much valuable information.

## 12.5 Transgenic Organisms

One important application of genetic engineering is to employ genes taken from one species of organism to change the genotype of a different species of organism. The result of such genetic engineering is called a *transgenic organism.* To take a real example, the bacterium *Bacillus thuringiensis* produces a variety of insecticidal proteins called *delta-endotoxins* that form crystals in the bacterial spores. These proteins are toxic to insects of the orders *Lepidoptera* (moths and butterflies), *Diptera* (flies with a single pair of wings), and *Coleoptera* (beetles). When the bacterial spores are ingested by an insect, the crystals dissolve and are activated by proteases in the gut, where they bind to specific receptors in the intestinal epithelium and create "leakage channels" that ultimately kill the insect. Genetically engineered corn, cotton, and other crop plants whose genome contains a delta-endotoxin–coding region regulated by sequences taken from normal plant genes have been produced. Hence, the engineered plants produce the delta-endotoxin protein in their tissues, making them lethal when ingested by insects such as the European corn borer, a lepidopteran pest that was introduced into the United States in the early 1900s and that currently causes annual crop losses of field corn, popcorn, seed corn, and sweet corn in excess of $1 billion.

In this section, we shall examine some of the genetic procedures by which transgenic organisms are produced. We have already seen how recombinant DNA can be introduced into bacterial cells by means of transformation. Similar procedures work with yeast and many other unicellular microorganisms. Transformation in metazoan organisms, including domesticated animals and plants, requires different techniques.

## ■ Germ-Line Transformation in Animals

In the nematode *Caenorhabditis elegans,* the transformation procedure is based on the fact that DNA injected directly into the reproductive organs is sometimes spontaneously incorporated into the genome of germ-line cells. Therefore, if an animal is injected with recombinant DNA that includes a genetic marker along with any sequence of interest, transformation is manifested by presence of the marker phenotype in the progeny of the injected animals.

A somewhat more elaborate procedure is used for germ-line transformation in *Drosophila.* The usual method employs a transposable element called the ***P* element**, which is a DNA sequence 2.9 kb in length that includes a central region, encoding the transposase, flanked by a 31-base-pair sequence that is repeated in inverted orientation at each end (**FIGURE 12.25**, part A). The terminal inverted repeats are essential for recognition by the transposase, which transposes the *P* element by excision and insertion. (Details of cut-and-paste transposition will be examined in Section 14.3). For germ-line transformation, the vector is a plasmid containing a *P* element in which the transposase-coding region is replaced by the DNA sequence of interest, along with a suitable genetic marker (usually one affecting eye color). By itself, this *P* element cannot transpose because it encodes no transposase, but because it retains the inverted repeats it can be mobilized by transposase from another source. The source of transposase is usually a genetically engineered derivative of the *P* element, called *wings clipped,* that encodes functional transposase but cannot itself transpose because the ends of the inverted repeats are missing (Figure 12.25B).

In *Drosophila* transformation, any DNA sequence of interest can be placed between the ends of the deleted *P* element. The resulting vector, along with a different plasmid containing the wings-clipped element, are injected into the region of the early embryo containing the germ cells. The DNA is taken up by the germ cells, and the wings-clipped element makes possible the production of functional transposase (Figure 12.25B). This mobilizes the engineered *P* vector and results in its transposition into an essentially random location in the genome. Transformants are detected among the progeny of the injected flies because of the eye color or other genetic marker included in the *P* vector. Integration into the germ line is typically very

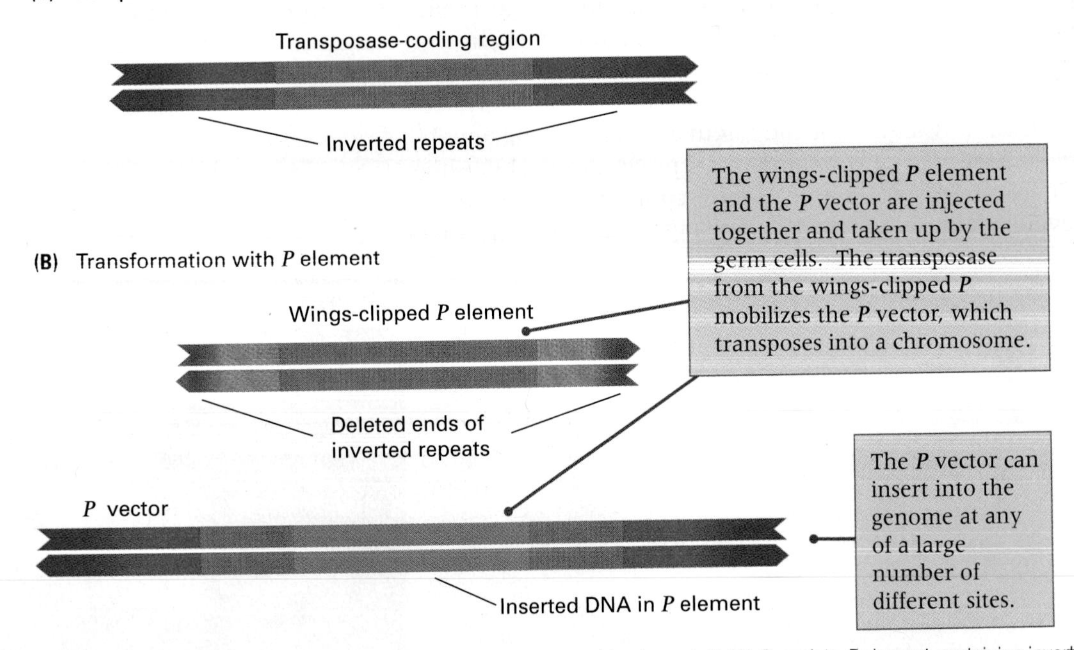
(A) Complete P element
Transposase-coding region
Inverted repeats
(B) Transformation with P element
Wings-clipped P element
Deleted ends of inverted repeats
P vector
Inserted DNA in P element
The wings-clipped P element and the P vector are injected together and taken up by the germ cells. The transposase from the wings-clipped P mobilizes the P vector, which transposes into a chromosome.
The P vector can insert into the genome at any of a large number of different sites.

**FIGURE 12.25** Transformation in *Drosophila* mediated by the transposable element *P*. (A) Complete *P* element containing inverted repeats at the ends and an internal transposase-coding region. (B) Two-component transformation system. The vector component contains the DNA of interest flanked by the recognition sequences needed for transposition. The wings-clipped component is a modified *P* element that codes for transposase but cannot transpose itself because critical recognition sequences are deleted.

connection

## Hello, Dolly!

*The Scottish Finn Dorset ewe known as "Dolly" is the first mammal to have been cloned from the nucleus of a cell taken from an adult mammal. The experiment created a press sensation. The President of the United States said that he would be against cloning people. A professor of public health said, "I don't think reasonable, rational people would want to clone themselves, but an eccentric millionaire might want to leave his money to a clone. There's no way of stopping the super-rich from going offshore to clone themselves." Some people were concerned for the animals. A spokesperson for People for the Ethical Treatment of Animals was quoted as saying, "It is time that society learned to respect our fellows, not exploit them for every fool thing." (Note in the excerpt that follows that the experiment was carried out with the prior approval of the appropriate Animal Welfare Committee.) Many important issues are raised by the cloning of animals. One is whether the cloned animals are truly normal genetically. Dolly, for example, was diagnosed with arthritis in 2002 and with progressive lung disease a year later. She was euthanized on February 14, 2003.*

**Ian Wilmut, Angelika E. Schnieke, Jim McWhir, Alex J. Kind, and Keith H. S. Campbell 1997**
Roslin Institute, Roslin, Midlothian, Scotland
*Viable Offspring Derived from Fetal and Adult Mammalian Cells*

Transfer of a single nucleus at a specific stage of development, to an enucleated unfertilized egg, provided an opportunity to investigate whether cellular differentiation to that stage involved irreversible genetic modification. The first offspring to develop from a differentiated cell were born after nuclear transfer from an embryo-derived cell that had been induced to become quiescent. Using the same procedure, we now report the birth of live lambs from adult mammary gland, fetus, and embryo. The fact that a lamb was derived from an adult cell confirms that differentiation of that cell did not involve the irreversible modification of genetic material required for development to term. . . . If the recipient cytoplasm is prepared by enucleation of an oocyte at metaphase II, it is only possible to avoid chromosomal damage and maintain normal ploidy by transfer of diploid nuclei. Our studies with cultured cells suggest that there is an advantage if cells are quiescent. . . . Together our results indicate that nuclei from a wide range of cell types should prove to be totipotent after enhancing opportunities for reprogramming by using appropriate combinations of these cell-cycle stages. In turn, the dissemination of the genetic improvement obtained within elite selection herds will be enhanced by limited replication of animals with proven performance by nuclear transfer of cells derived from adult animals. . . . The lamb born after nuclear transfer from a mammary gland cell is, to our knowledge, the first mammal to develop from a cell derived from an adult tissue. . . . This is consistent with the generally accepted view that mammalian differentiation is almost all achieved by systematic, sequential changes in gene expression brought about by interactions between the nucleus and the changing cytoplasmic environment. . . . These experiments were conducted under the Animals (Scientific Procedure) Act 1986 [United Kingdom] and with the approval of the Roslin Institute Animal Welfare Experiments Committee.

***The fact that a lamb was derived from an adult cell confirms that differentiation of that cell did not involve the irreversible modification of genetic material required for development to term.***

Source: I. Wilmut, et al., *Nature* 385 (1997): 810–813.

**"I LOVE EWE, MOM!"** Dolly and her natural-born lamb, whose fleece was white as snow. [Courtesy of the Roslin Institute, University of Edinburgh.]

efficient (10 to 20 percent of the injected embryos that survive and are fertile yield one or more transformed progeny). However, the efficiency decreases with the size of the DNA fragment in the *P* element, and the effective upper limit is approximately 20 kb.

In mammals, transformation of the germ line can be carried out in several ways. The most direct is the injection of vector DNA into the nucleus of fertilized eggs, which are then transferred to the uterus of foster mothers for development. The vector is usually a modified RNA virus, which uses a reverse transcriptase to convert the RNA genome into double-stranded DNA that becomes inserted into the genome in infected cells. Genetically engineered viruses containing inserted segments of RNA undergo the same integration process.

Another method of transforming mammals uses **embryonic stem cells** obtained from embryos a few days after fertilization (**FIGURE 12.26**). Although embryonic stem cells are not very hardy, they can be isolated and then grown in culture and transformed with recombinant DNA vectors. The genetically altered stem cells are transferred into a developing embryo that is transplanted into the uterus of a foster mother (Figure 12.26A), where the stem cells become incorporated into various tissues of the embryo and often participate in forming the germ line. If the embryonic stem cells carry a genetic marker, such as a gene for black coat color, then mosaic animals can be identified by their spotted coats (B). Some of these animals, when mated, produce black offspring (C), which indicates that the embryonic stem cells had become incorporated into the germ line. In this way,

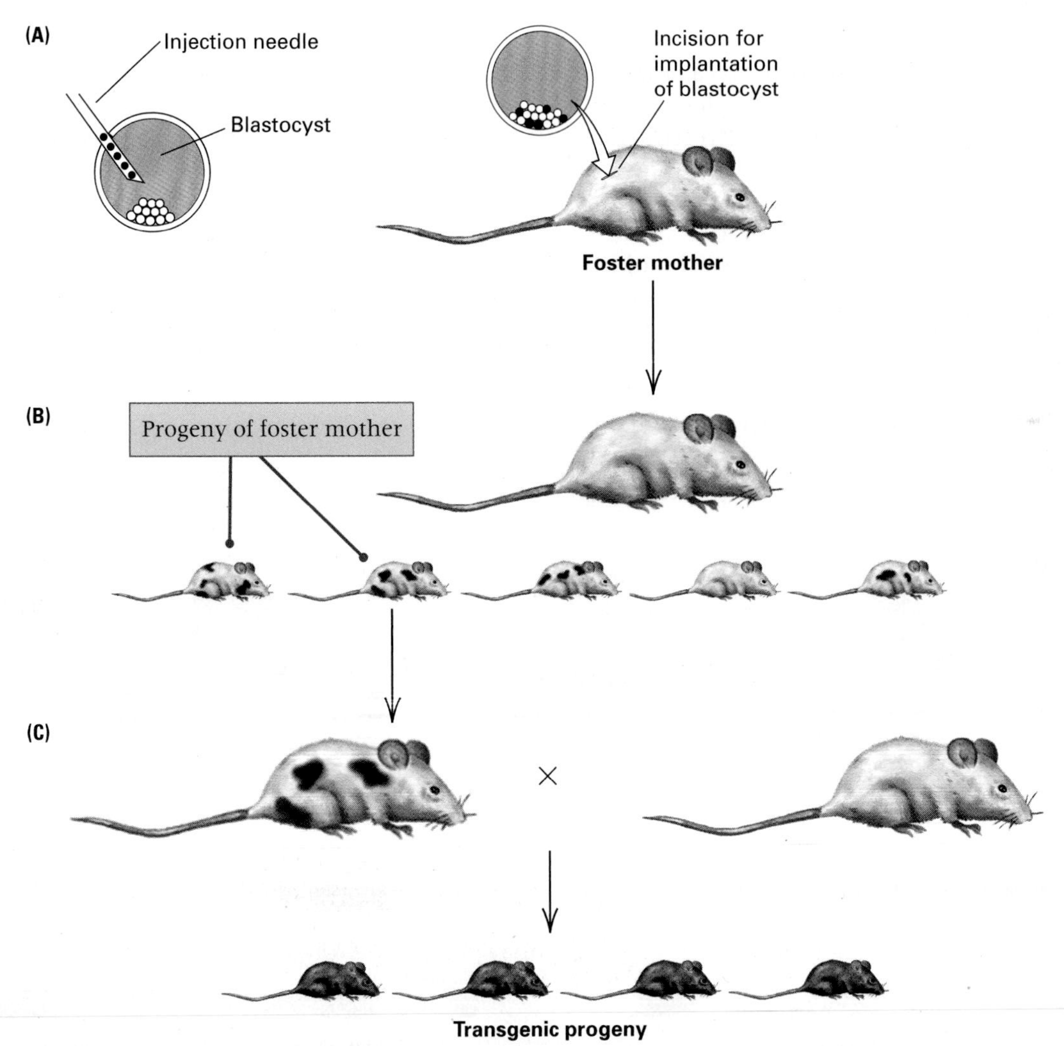


**FIGURE 12.26** Transformation of the germ line in the mouse via embryonic stem cells. (A) Stem cells obtained from an embryo of a black strain are isolated and, after genetic manipulation in culture, injected into the embryo of a white strain, which is then introduced into the uterus of a foster mother. (B) The resulting offspring are often mosaics that contain cells from both the black and the white strains. (C) If cells from the black strain colonize the germ line, then the offspring of the mosaic animal will be black. [Adapted from M. R. Capecchi, *Trends. Genet.* 5 (1989): 70–76.]

genetic changes introduced into the embryonic stem cells while they were in culture may become incorporated into the germ line of living animals. The method shown in Figure 12.27 has been used to create strains of mice with mutations in genes associated with such human genetic diseases as cystic fibrosis. These strains serve as mouse models for studying the disease and for testing new drugs and therapeutic methods.

The procedure for introducing changes into specific genes is called **gene targeting**, and for their groundbreaking discoveries that led to this technology, Mario R. Capecchi, Martin J. Evans, and Oliver Smithies were awarded the 2007 Nobel Prize in Physiology or Medicine. The specificity of gene targeting comes from DNA sequence homology between the introduced recombinant DNA and normal sequences already present in the chromosome. When the DNA sequences are similar enough, they can undergo pairing as the result of an exchange of complementary strands followed by a process of breakage and reunion. Two examples are illustrated in **FIGURE 12.27**, where the DNA sequences present in gene-targeting vectors are shown as looped configurations paired with homologous regions in the chromosome prior to recombination. The targeted gene is shown in pink. In Figure 12.28A, the vector contains the targeted gene interrupted by an insertion of a novel DNA sequence, and homologous recombination results in the novel sequence becoming inserted into the targeted gene in the chromosome. In Figure 12.27B, the vector contains only flanking sequences, not the targeted gene itself, so homologous recombination results in replacement of the targeted gene with an unrelated DNA sequence. In both cases, cells with targeted gene mutations can be selected by including an antibiotic-resistance gene, or other selectable genetic marker, in the sequences that are incorporated into the genome through homologous recombination. The molecular mechanism of homologous recombination is discussed in Section 6.8.

## Genetic Engineering in Plants

A procedure for the transformation of plant cells makes use of a plasmid found in the soil bacterium *Agrobacterium tumefaciens* and related species. Infection of susceptible plants with this bacterium results in the growth of tumors known as *crown gall tumors* at the entry site, usually a wound. Susceptible plants include about 160,000 species of flowering plants in the class *Dicotyledoneae*, which includes the great majority of common flowering plants.

The *Agrobacterium* contains a large plasmid of approximately 200 kb called the ***Ti* plasmid**, which includes a smaller (about 25-kb) region known as the **T DNA** terminated by 25-bp direct repeats (sequences repeated in the same orientation) at the extreme ends (**FIGURE 12.28**, part A). Infection with *Agrobacterium* causes a profound change in the metabolism of infected cells, because the T DNA encodes proteins that stimulate cell division, which causes the tumor,

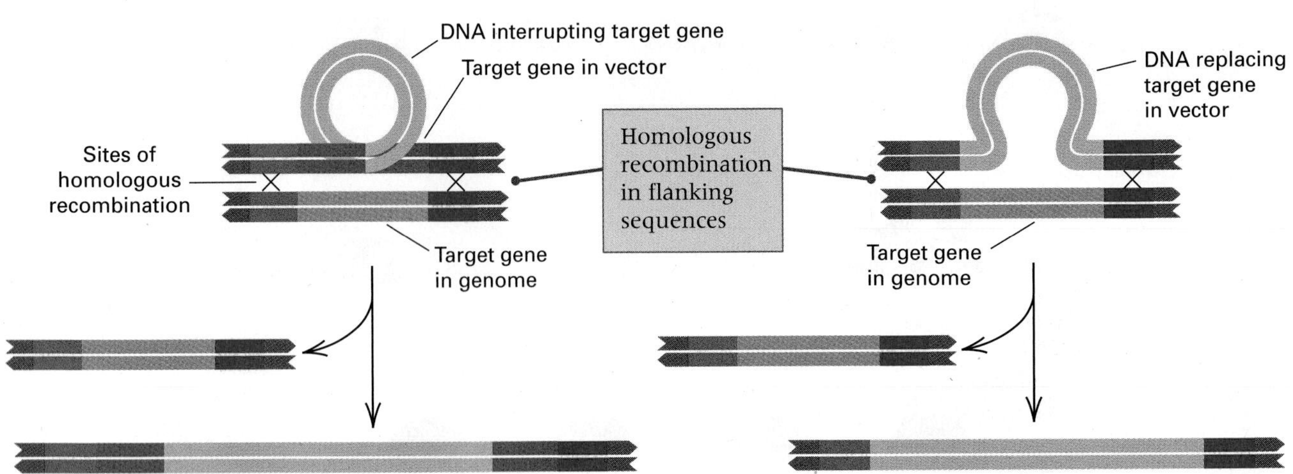


**FIGURE 12.27** Gene targeting in embryonic stem cells. (A) The vector (top) contains the target sequence (brick red) interrupted by an insertion. Homologous recombination introduces the insertion into the genome. (B) In this case the vector contains DNA sequences flanking the targeted gene. Homologous recombination results in replacement of the targeted gene with an unrelated DNA sequence. [Adapted from M. R. Capecchi, *Trends. Genet.* 5 (1989): 70–76.]

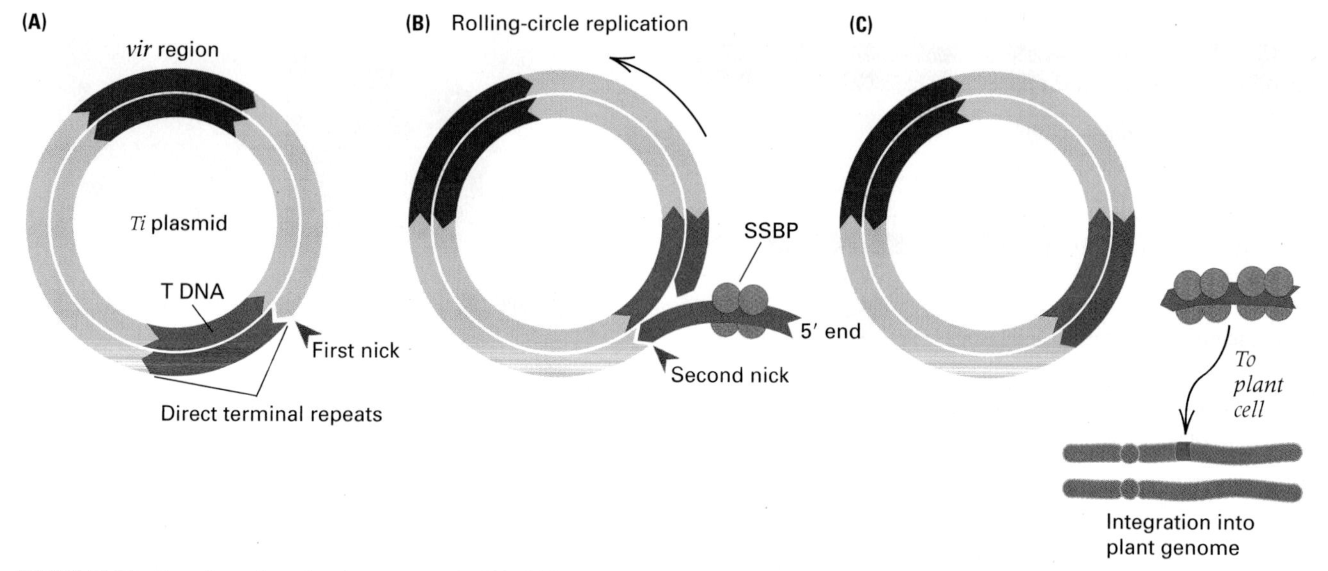


**FIGURE 12.28** Transformation of a plant genome by T DNA from the *Ti* plasmid. (A) A nick forms at the 5′ end of the T DNA. (B) Rolling-circle replication elongates the 3′ end and displaces the 5′ end, which is stabilized by single-stranded binding protein (SSBP). A second nick terminates replication. (C) The SSBP-bound T DNA is transferred to a plant cell and inserts into the genome.

and also encodes enzymes that convert the amino acid arginine into an unusual derivative (generally *nopaline* or *octopine*, depending on the particular type of *Ti* plasmid) that the bacterium needs in order to grow.

The change in plant cell metabolism is brought about by physical transfer and incorporation of the T DNA into the plant genome by the mechanism outlined in Figure 12.28. The transfer process is mediated by the products of six genes present in a 40-kb region of the *Ti* plasmid called the *virulence* (*vir*) region. The transfer of T DNA begins with the formation of a single-stranded nick (Figure 12.28A), from which the 5′ end of the T DNA peels off the plasmid; the resulting gap is closed by newly replicated DNA added onto the 3′ end of the nick, using the unnicked circle as the template strand (Figure 12.28B). The region of the plasmid that is transferred is delimited by a second nick at the other end of the T DNA, but the position of this nick is somewhat variable. The resulting single-stranded T DNA is bound with molecules of a single-stranded binding protein (SSBP) and is transferred into the plant cell and incorporated into the nucleus. There it is integrated into the chromosomal DNA by a mechanism that is still unclear (Figure 12.28C). Although the SSBP has certain similarities in amino acid sequence to the *recA* protein from *E. coli*, which plays a key role in homologous recombination, the integration of T DNA does not require sequence homology.

The use of T DNA in plant transformation is made possible by engineered plasmids in which the sequences normally present in the T DNA are removed and replaced with a selectable marker along with the DNA sequences to be incorporated into the plant genome. A second plasmid containing the *vir* genes permits mobilization of the engineered T DNA. In infected tissues, the *vir* functions mobilize the T DNA for transfer into the host cells and integration into the chromosome. Transformed cells are selected in plant cell culture by the use of the selectable marker and then grown into mature plants.

## ■ Transformation Rescue

One of the important applications of germ-line transformation is to define experimentally the limits of any particular gene along the DNA. Every gene includes upstream and downstream sequences that are necessary for its correct expression, as well as the coding sequences. As we saw in Chapter 11, these regulatory sequences control the time in development, the cell types, and the level at which transcription occurs. In defining the upstream and downstream limits of a gene, even knowing the complete nucleotide sequence of the coding region and the flanking DNA may be insufficient. The main reason is that there is no general method by which to identify regulatory sequences. Regulatory sequences are often composites of short, seemingly nondescript sequences, which are in fact the critical binding sites for regulatory proteins that control transcription.

To see how germ-line transformation is used to define the limits of a functional gene, consider the example of the *Drosophila* gene *white*, which

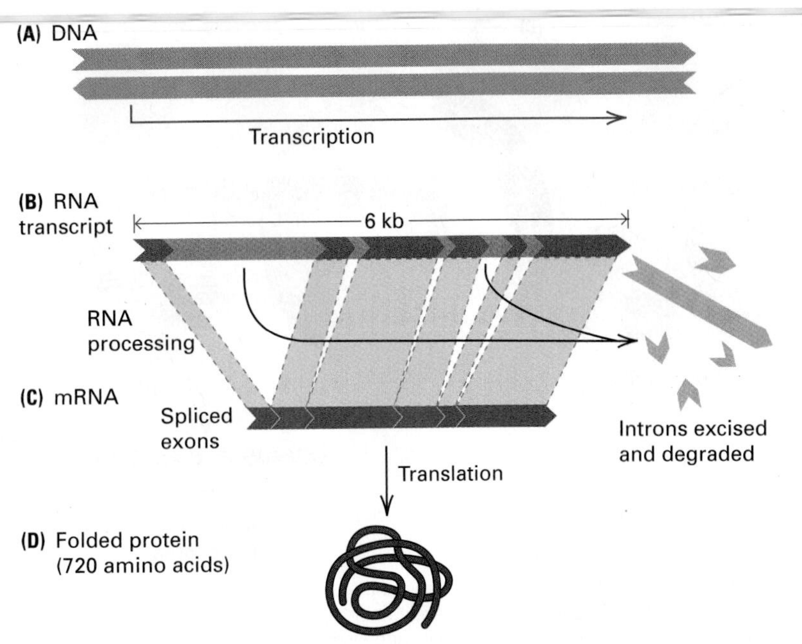


**FIGURE 12.29** Genetic organization of the *Drosophila white* gene.

when mutated, results in flies with white eyes instead of red. The genetic organization of *white* is illustrated in **FIGURE 12.29**. The primary RNA transcript (B) is a little more than 6 kb in length and includes five introns that are excised and degraded during RNA processing. The resulting mRNA (C) is translated into a protein of 720 amino acids (D), which is a member of a large family of related ATP-dependent transmembrane proteins known as *ABC transporters* that regulate the traffic of their target molecules or ions across the cell membrane. The white transporter is located in the pigment-producing cells in the eye, and its target molecule is one of the key precursors of the eye-color pigments; hence, flies with a mutant white transporter have white eyes.

The question addressed by germ-line transformation is this: How much DNA upstream and downstream of *white* is necessary for the fly to produce the wildtype transporter protein in the pigment cells? The experimental approach is first to clone a large fragment of DNA that includes the coding sequence for the wildtype white protein, and then to use germ-line transformation to introduce this fragment into the genome of a fly that contains a *white* mutation. If the introduced DNA includes all the 5′ and 3′ regulatory sequences necessary for correct gene expression, then the phenotype of the resulting flies will be wildtype, because the wildtype gene is dominant to the *white* mutation. The ability of an introduced DNA fragment to correct a genetic defect in a mutant organism is called **transformation rescue** (**FIGURE 12.30**), and it means that the fragment contains all the essential regulatory sequences. The

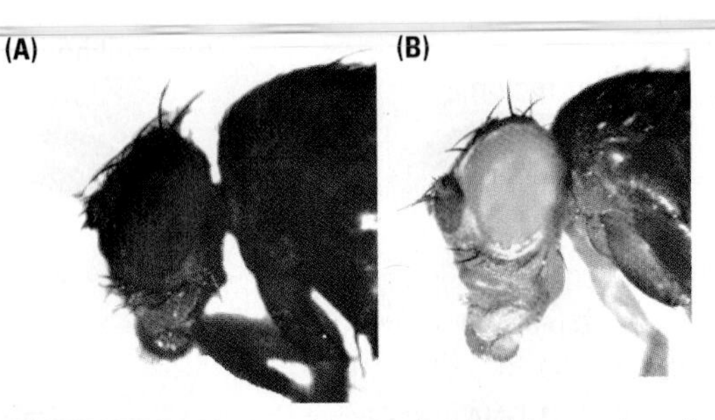


**FIGURE 12.30** Mutant white-eyed and wildtype red-eyed males of *Drosophila melanogaster*. [Courtesy of E. R. Lozovsky.]

fragment may also include some sequences that are nonessential, so finding the *minimal* 5′ and 3′ flanking regions requires that the smaller pieces of the original fragment also be assayed for transformation rescue. In the case of *white*, the minimal DNA fragment is about 8.5 kb in length, starting about 2 kb upstream from the transcription start site.

## Site-Directed Mutagenesis and Knockout Mutations

Once a gene has been cloned and sequenced, specific mutations can be introduced into the gene and transformed into the germ line in order to assay the resulting phenotype and make it possible to ascertain the function or functions of the gene. This approach is sometimes known as **reverse genetics** because it reverses the traditional flow of genetic analysis. Instead of starting with a mutant phenotype and trying to identify the mutant gene, reverse genetics starts with a known gene and determines how mutations affect the phenotype.

Several methods are available for creating a defined mutation at a specific site in cloned DNA. One method of **site-directed mutagenesis**, based on the polymerase chain reaction (PCR), is outlined in **FIGURE 12.31**. It starts with a unique restriction fragment isolated from a cloned wildtype gene, which in this example is assumed to be an *Eco*RI–*Bam*HI fragment. Two different PCR reactions are carried out. One PCR uses a primer pair designated 1 and 2, in which primer 1 extends beyond and overlaps the *Eco*RI site, allowing this site to be reconstituted upon amplification, and primer 2 is a *mutagenic primer* containing a single-base mismatch (in this example a G—A mismatch). The other PCR uses primer pairs 3 and 4, in which the 5′ end of primer 3 is adjacent to the 5′ end of primer 2, and primer 4 extends beyond and overlaps the *Bam*HI site, allowing this site to be reconstituted upon amplification. The amplification products are shown in part C. When these are ligated and then digested with *Eco*RI and *Bam*HI, the original *Eco*RI–*Bam*HI fragment is reconstituted, except that the new fragment has a specific A—T → G—C base-pair substitution at the site indicated. When the *Eco*RI–*Bam*HI fragment is substituted back into the original wildtype gene, the reconstituted gene has the A—T → G—C base-pair mutation that can be introduced into an organism by germ-line transformation.

Similar methods can be used to create *deletion mutations*, in which part of a gene is missing, or mutations in which part of a gene is replaced with a completely unrelated sequence. When such constructs are used in gene-targeting experiments to replace the wildtype gene in a genome by homologous recombination (see Figure 12.27), the result is a **knockout mutation** in which the wildtype gene has been rendered completely nonfunctional ("knocked out"). Two important applications of gene knockouts can be highlighted. The first is the generation of mouse strains with knockouts of the homologs of human disease genes in order to serve as models for the human disease. For example, the effectiveness of medication for different genetic causes of high blood pressure can be examined in mouse strains that have knockouts (or extra copies) of genes whose products are known to be important in regulating blood pressure. This application is highlighted in the Connection on page 463. A second important application of knockout mutagenesis is in determining gene function. For example, among the approximately 6000 genes identified from complete genomic sequencing of the genome of budding yeast (*Saccharomyces cerevisiae*), only about 25 percent are associated with genes previously identified through the isolation of mutants. Approximately another 35 percent have functions that may be inferred from homologies to genes that have been identified in other organisms. But the remaining approximately 40 percent have completely unknown functions. The functions of many of these mystery genes can be revealed by examining the phenotypes associated with knockout mutations under a variety of growth conditions.

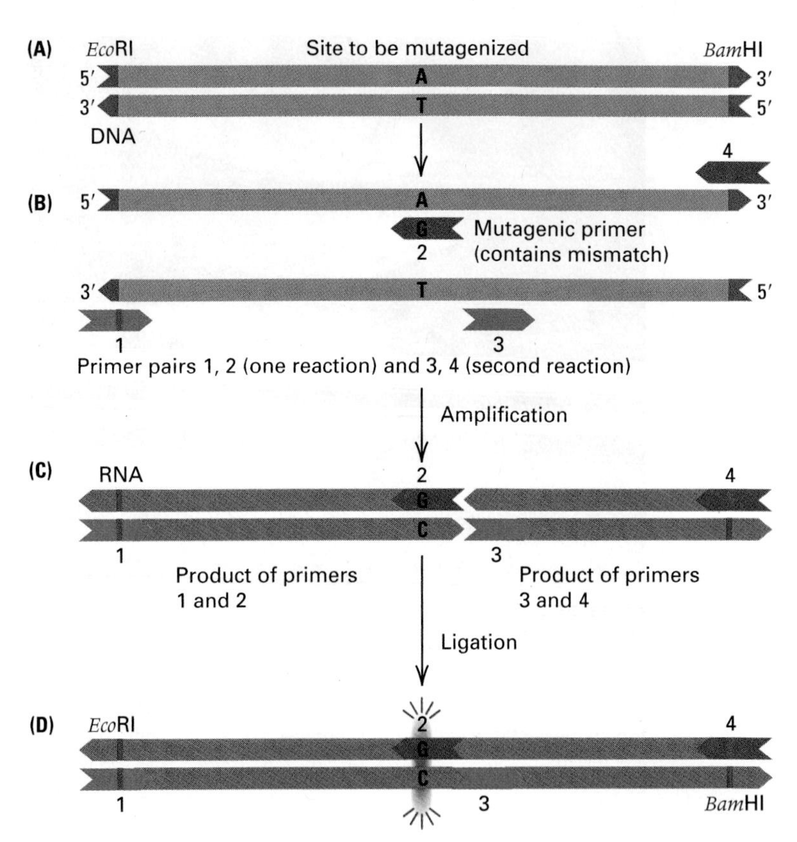


**FIGURE 12.31** Site-directed mutagenesis using PCR. (A) An A—T nucleotide pair is to be replaced with a G—C nucleotide pair. (B) The original fragment is amplified in two parts, using a primer (number 2 in this example) that contains a G instead of an A. (C) The amplified product on the left contains a G—C nucleotide pair at the targeted site. (D) Ligation of the amplified fragments reconstitutes a DNA duplex that contains the site-directed mutation.

## 12.6 Some Applications of Genetic Engineering

Recombinant DNA technology has revolutionized modern biology not only by opening up new approaches in basic research, but also by making possible the creation of organisms with novel genotypes for practical use in agriculture and industry. In this section we examine a few of many applications of recombinant DNA.

### ■ Giant Salmon with Engineered Growth Hormone

In many animals, the rate of growth is controlled by the amount of growth hormone produced. Transgenic animals with a growth-hormone gene under the control of a highly active promoter to drive transcription often grow larger than their normal counterparts. An example of a highly active promoter is found in the gene for **metallothionein**. The metallothioneins are proteins that bind heavy metals. They are ubiquitous in eukaryotic organisms and are encoded by members of a family of related genes. The human genome, for example, contains more than ten metallothionein genes that can be separated into two major groups according to their sequences. The promoter region of a metallothionein gene drives transcription of any gene to which it is attached, in response to heavy metals or steroid hormones. For example, when DNA constructs consisting of a rat growth-hormone gene under metallothionein control are used to produce transgenic mice, the resulting animals grow about twice as large as normal mice.

**FIGURE 12.32** Normal coho salmon (left) and genetically engineered coho salmon (right) containing a sockeye salmon growth-hormone gene driven by the regulatory region from a metallothionein gene. The transgenic salmon average 11 times the weight of the nontransgenic fish. The smallest fish on the left is about 10 cm. long, and the largest fish on the right is about 42 cm. [Reprinted by permission from Macmillan Publishers Ltd: R. H. Devlin, et al., *Nature* 371 (1994): 209–210.

The effect of another growth-hormone construct is shown in **FIGURE 12.32**. The fish are coho salmon at 14 months of age. Those on the left are normal, whereas those on the right are transgenic animals that contain a salmon growth-hormone gene driven by a metallothionein regulatory region. Both the growth-hormone gene and the metallothionein gene were cloned from the sockeye salmon. As an indicator of size, the largest transgenic fish on the right has a length of about 42 cm. On average, the transgenic fish are 11 times heavier than their normal counterparts, and the largest transgenic fish was 37 times the average weight of the nontransgenic animals. Not only do the transgenic salmon grow faster and become larger than normal salmon, they also mature faster.

## Nutritionally Engineered Rice

Although thus far we have focused on genetic engineering through the manipulation of single genes, it is also possible to create transgenic organisms that have entire new metabolic pathways introduced. A remarkable example is in the creation of a genetically engineered rice that contains an introduced biochemical pathway for the synthesis of β-carotene, a precursor of vitamin A found primarily in yellow vegetables and greens. (Deficiency of vitamin A affects some 400 million people throughout the world, predisposing them to skin disorders and night blindness.) The β-carotene pathway includes four enzymes, which in the engineered rice are encoded in genes from different organisms (**FIGURE 12.33**). Two of the genes come from the common daffodil (*Narcissus pseudonarcissus*), whereas the other two come from the bacterium *Erwinia uredovora*. Each pair of genes was cloned into T DNA and transformed into rice using *Agrobacterium tumefaciens* by the mechanism outlined in Figure 12.28. Transgenic plants were then crossed to produce progeny containing all four enzymes. The engineered rice seeds contain enough β-carotene to provide the daily requirement of vitamin A in 300 grams of cooked rice; they even have a yellow tinge (Figure 12.33B).

People on high-rice diets are also prone to iron deficiency because rice contains a small phosphorus-storage molecule called *phytate* (myo-inositol hexakis dihydrogen phosphate), which binds with iron and interferes with its absorption through the intestine. The transgenic β-carotene rice was also engineered to minimize this problem by introducing the fungal enzyme from *Aspergillus ficuum* that breaks down phytate, along with a gene encoding the iron-storage protein ferritin from the French bean, *Phaseolus vulgaris*, plus yet another gene from basmati rice that encodes a metallothionein-like gene that facilitates iron absorption in the human gut. Altogether, then, the transgenic rice strain rich in β-carotene and available iron contains six new genes taken from four unrelated species plus one gene from a totally different strain of rice!

## Production of Useful Proteins

Among the most important applications of genetic engineering is the production of large quantities of particular proteins that are otherwise difficult to obtain (for example, proteins that are present in only a few molecules per cell or that are produced in only a small number of cells or only in human cells). The method is simple in principle. A DNA sequence coding for the desired protein is cloned in a vector adjacent to an appropriate regulatory sequence. This step is usually done with cDNA, because cDNA has all the coding sequences spliced together in the right order. Using a vector with a high copy number ensures that many copies of the coding sequence will be present in each bacterial cell, which can result in synthesis of the gene product at concentrations ranging from 1 to 5 percent of the total cellular protein. In practice, the production of large quantities of a protein in bacterial cells is

## connection

### A Pinch of This and a Smidgen of That

Oliver Smithies 2005
University of North Carolina, Chapel Hill, NC
*Many Little Things: One Geneticist's View of Complex Diseases*

*This paper was written on the 25th anniversary of Oliver Smithies's first experimental success in introducing exogeneous DNA into a chosen site in a mammalian genome. The initial applications were to the genetic analysis of phenotypes in which single genes had a major effect. Here Smithies recounts how he came to realize that the approach also could be used to study complex diseases influenced by many genes, each with a relatively small effect. His idea was to create mouse chromosomes either lacking a gene or having an extra copy. By combining these chromosomes through crosses, he could create mice with 0 copies of the gene (if the mouse survived), or 1, 2, 3, or 4 copies. He could then examine the mice and look for associated changes in phenotype. His initial application of this approach was to regulation of blood pressure. This trait was of special interest to Smithies because his father had died from its complications, and because he himself required medication to control his own blood pressure. Smithies's insights and approach were a great success and well deserving of his share of the 2007 Nobel Prize in Physiology and Medicine.*

The common multifactorial diseases, such as atherosclerosis, hypertension and diabetes, are caused by complex interactions between multiple genetic and environmental factors. . . . Multifactorial diseases are challenging targets for research because of their complexity. They are attractive because a better understanding of their nature could affect the lives of many people, as more than half of those who live in affluent societies are likely to develop debilitating conditions that have multigenic causes. . . . It seemed likely, as going from two copies to one decreased the level of a gene product, that going from two copies to three copies would increase the level of the product. . . . I realized that this test of the effects of quantitative changes in gene expression could usefully be applied to all genes that potentially alter blood pressure. . . . My mindset had begun to change from the idea that a few major differences might determine this complex phenotype to the thought that it was more likely to be the result of 'many little things.' . . . I decided . . . [to set up] a computer simulation of the system. . . . Not being a computer buff, I spent a rather unenjoyable six months working off and on with the simulation. But the outcome was most revealing, and I have since, with ever-increasing enjoyment, used relatively simple computer simulations in much of my work. . . . I have found that the greatest value of these simulations is . . . that one is forced to identify crucial elements and define clearly the assumptions required to integrate them into a logical whole. In the case of our hypertension studies, the simulations showed that increases in the number of copies of the *AGT* [angiotensinogen] gene should cause . . . an increase in blood pressure. . . . My experiences during the course of this voyage into the field of multifactorial diseases led me to emphasize the importance of looking for quantitative differences as well as qualitative differences, when searching for genetic factors that determine individual risk of common disorders. Many quantitative differences have accumulated during human evolution—some easily seen as outward differences in our body proportions, some hidden as inward differences in gene expression—that are without sufficient effects to have been fixed or eliminated by selection. These 'many little things' are a joyful source of our individuality. But they are probably also a source of the poorly understood differences in individual susceptibility to [complex diseases].

***My mindset had begun to change . . . to the thought that [hypertension] was more likely to be the result of 'many little things.'***

Source: O. Smithies, *Nat. Rev. Genet.* 6 (2005): 419–425.

straightforward, but there are often problems that must be overcome, because in the bacterial cell, which is a prokaryotic cell, the eukaryotic protein may be unstable, may not fold properly, or may fail to undergo necessary chemical modification. Many important proteins are currently produced in bacterial cells, including human growth hormone, blood-clotting factors, and insulin. Patent offices in Europe and the United States have already issued tens of thousands of patents for the clinical use of the products of genetically engineered human genes. **FIGURE 12.34** gives a breakdown of the numbers of patents issued relative to clinical application.

#### Genetic Engineering with Animal Viruses

Genetic engineering in animal cells often exploits RNA viruses called **retroviruses** that use reverse transcriptase to make a double-stranded DNA copy of their RNA genome. The DNA copy then becomes inserted into the chromosomes of the cell. Ordinary transcription of DNA to RNA occurs only after the DNA copy is inserted. The infected host cell survives the

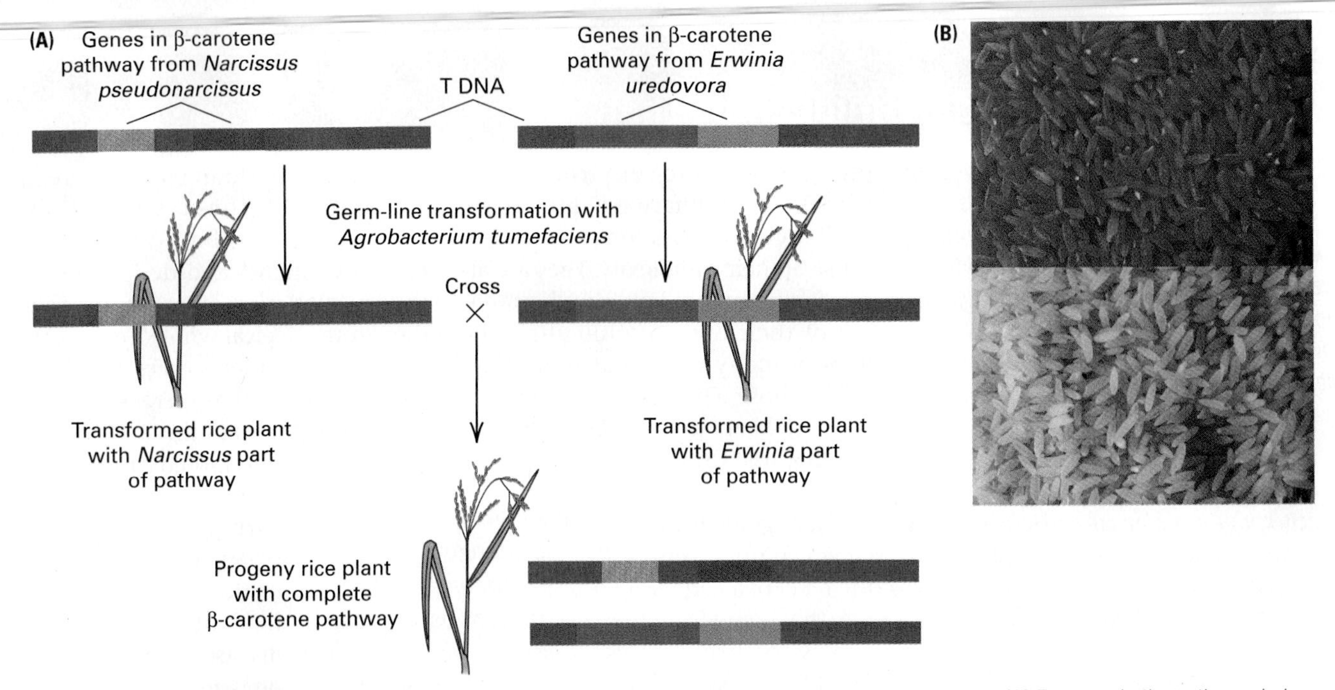


**FIGURE 12.33** Genetically engineered rice containing a biosynthetic pathway for $\beta$-carotene. (A) Enzymes in the pathway derive from genes in two different species. (B) Rice plants with both parts of the pathway produce grains with a yellowish cast (top) due to the $\beta$-carotene they contain, in contrast to the pure white grains (bottom) of normal plants. [Part B courtesy of Ingo Potrykus, Institute for Pflanzenuissenschaften, ETH Zurich.]

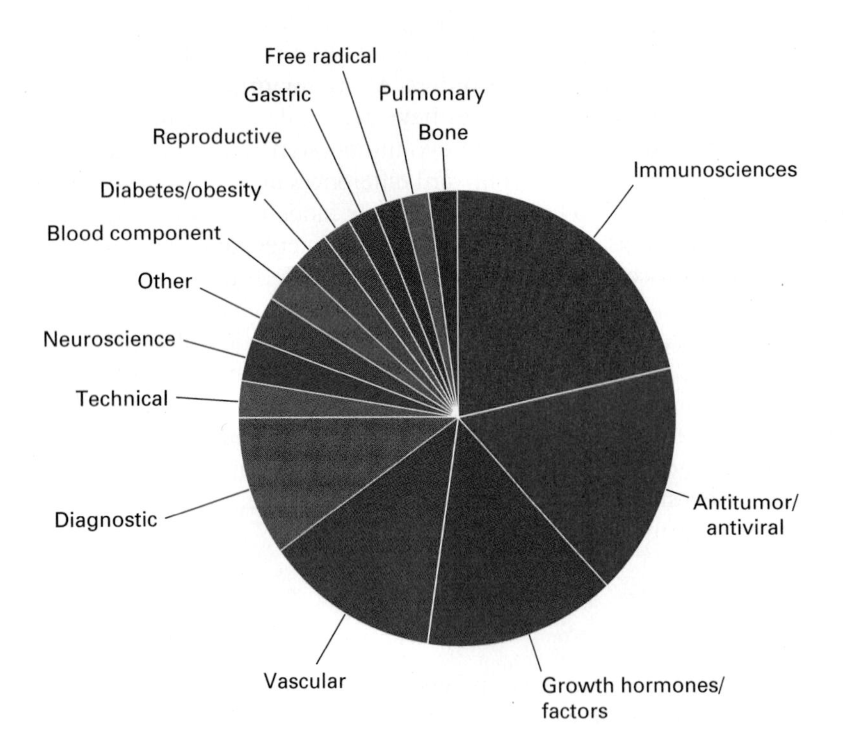


**FIGURE 12.34** Relative number of patents issued for various clinical applications of the products of genetically engineered human genes. [Data from S. M. Thomas, et al., *Nature* 380 (1996): 387–388.]

infection, retaining the DNA copy of the retroviral RNA in its genome. These features of retroviruses make them convenient vectors for the genetic manipulation of animal cells, including those of birds, rodents, monkeys, and human beings.

Genetic engineering with retroviruses allows the possibility of altering the genotypes of animal cells. Because a wide variety of retroviruses are known, including many that infect human cells, genetic defects may be corrected by these procedures in the future. The recombinant DNA procedure employed with retroviruses consists of the *in vitro* synthesis of double-stranded DNA from the viral RNA by means of reverse transcriptase. The DNA is then cleaved with a restriction enzyme and, using techniques already described, any DNA fragment of interest is inserted. Transformation yields cells with the recombinant retroviral DNA permanently inserted into the genome. However, many retroviruses contain genes that result in uncontrolled proliferation of the infected cell, thereby causing a tumor. When retrovirus vectors are used for genetic engineering, the cancer-causing genes are first deleted. The deletion also provides the space needed for incorporation of the desired DNA fragment.

Attempts are currently under way to assess the potential use of retroviral vectors in **gene therapy**, or the correction of genetic defects in somatic cells by genetic engineering. Examples might include correcting immunological deficiencies in patients with various kinds of inherited disorders. However, a number of major problems stand in the way of gene therapy becoming widely used. At this time, there is no completely reliable way to ensure that a gene will be inserted only into the appropriate target cell or target tissue. For example, in one of the earliest clinical trials with a group of four patients treated with retroviral vectors for severe combined immunodeficiency disease, one of the patients had a retroviral insertion into a site that caused aberrant expression of a gene, *LMO-2*, associated with acute lymphoblastic leukemia.

A major breakthrough in disease prevention would come through the development of synthetic vaccines produced by recombinant DNA. Production of natural vaccines is often unacceptable because of the extreme hazards of working with large quantities of the active virus—for example, the human immunodeficiency virus (HIV-1) that causes acquired immune deficiency syndrome (AIDS). The danger can be minimized by cloning and producing viral antigens in a nonpathogenic organism. Vaccinia virus, the agent used in smallpox vaccination, has attracted much attention as a candidate for this application. Viral antigens are often on the surface of virus particles, and some of these antigens can be engineered into the coat of vaccinia. For example, engineered vaccinia virus with certain surface antigens of hepatitis B virus, influenza virus, and vesicular stomatitis virus (which kills cattle, horses, and pigs) has already proved effective in animal tests. One of the great challenges is malaria, which affects approximately 300 million people in Africa, Asia, and Latin America, causing approximately 1–1.5 million deaths per year. Control of the disease by drugs is increasingly compromised by the spread of resistance mutations, but a recently developed vaccine has shown about 50 percent effectiveness in clinical trials.

## CHAPTER SUMMARY

- In recombinant DNA (gene cloning), DNA fragments are isolated, inserted into suitable vector molecules, and introduced into bacteria or yeast cells where they are replicated.
- Specialized methods for manipulating and cloning large fragments of DNA have resulted in integrated physical and genetic maps of the DNA in complex genomes.
- High-throughput automated DNA sequencing has resulted in the complete sequence of the genomes of many species of bacteria, archaeons, and eukaryotes including that of *Homo sapiens*.
- Comparisons among genomes of related species helps discover coding sequences and other functional genetic elements.
- Functional genomics using DNA microarrays enables global patterns and coordinated regulation of gene expression to be investigated. Proteomics methods such as two-hybrid analysis enable protein-protein interactions to be detected.
- Transgenic organisms carry DNA sequences that have been introduced by germ-line transformation or other methods.
- Recombinant DNA is widely used in research, medical diagnostics, and the manufacture of drugs and other commercial products.

## REVIEW THE BASICS

- What is meant by the term *recombinant DNA*? What are some of the practical uses of recombinant DNA?
- Why are restriction endonucleases essential in most types of recombinant DNA research?
- What features are essential in a bacterial cloning vector? How can a vector have more than one cloning site? What is a multiple cloning site (polylinker)?
- What is the reaction catalyzed by the enzyme reverse transcriptase? How is this enzyme used in recombinant DNA technology?
- Distinguish between a physical map and a genetic map. If recombination took place at uniform frequencies throughout a genome, how would the genetic map compare with the physical map?
- Explain the term *functional genomics*. What are DNA microarrays and how are they used in functional genomics? In the type of hybridization experiment described in the text, how would you interpret a spot on a DNA microarray that is red or orange? Yellow? Green or yellow-green?
- Describe the two-hybrid system that makes use of the yeast GAL4 protein, and explain how the two-hybrid system detects interactions between proteins.
- What is a transgenic organism? What are some of the practical uses of transgenic organisms?
- Explain the role of recombination in gene targeting in embryonic stem cells. How can gene targeting be used to create a "knockout" mutation?
- Where does T DNA come from, and how is it used in making transgenic plants?

**Problem 1** The restriction enzyme *Apa*I cleaves at the site GGGCC|C, where by convention the 5′ end of the sequence is written at the left and the 3′ end at the right. Cleavage takes place in double-stranded DNA at the position indicated by the vertical bar, and there is a symmetrical cut in the other strand. Using the same conventions, the restriction site of the restriction enzyme *Bsp*120I is G|GGCCC.

**(a)** What is the DNA structure at the ends of each *Apa*I restriction fragment?

**(b)** What is the DNA structure at the ends of each *Bsp*120I restriction fragment?

**(c)** Are the ends of *Apa*I fragments compatible with those of *Bsp*120I fragments in the sense they the single-stranded overhangs can undergo base pairing?

**Answer**

**(a)** In answering questions like this, it is best to write out the restriction site as a double-stranded molecule, showing the cleavage sites as arrows. In this case, the cleavage sites are

```
ApaI      ↓              ↓  Bsp120I
5'-GGGCCC-3'      5'-GGGCCC-3'
3'-CCCGGG-5'      3'-CCCGGG-5'
     ↑                   ↑
```

A typical *Apa*I restriction fragment would therefore have the structure shown below, where N indicates any nucleotide and the dots indicate an unspecified number of nuucleotides:

```
         5'-CNN...NNGGGCC-3'
ApaI  3'-CCGGGNN...NNC-5'
```

**(b)** A typical *Bsp*120I restriction fragment would have the following structure:

```
          5'-GGCCCNN...NNG-3'
Bsp120I        3'-GNN...NNCCCGG-5'
```

**(c)** The *Apa*I and *Bsp*120I restriction fragments are not compatible. The former has a 3′ overhang and the latter has a 5′ overhang, and these cannot undergo base pairing.

**Problem 2** In the diagrams shown here, the colored bars represent a gene you wish to clone (*yourgene*) in genomic DNA and two antibiotic-resistance genes (ampicillin resistance, *amp-r*, and tetracycline resistance, *tet-r*) in a plasmid. Also shown are the positions of restriction sites for the restriction enzymes *Eco*RI and *Bam*HI. The sites shown are the only sites present near *yourgene* and in the plasmid.

*Eco*RI *Bam*HI *Bam*HI *Eco*RI *Bam*HI

*yourgene*

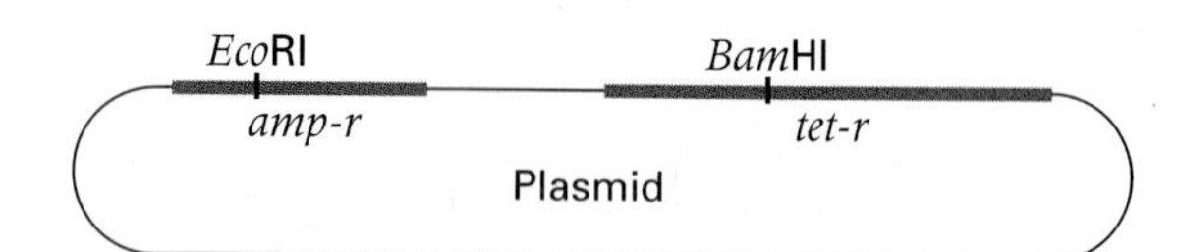


**(a)** Would you use *Eco*RI or *Bam*HI or both to digest genomic DNA in preparation for cloning *yourgene*?

**(b)** Would you use *Eco*RI or *Bam*HI or both to digest plasmid DNA in preparation for cloning *yourgene*?

**(c)** After digestion, mixing the DNAs, ligation, and transformation, which antibiotic would you use to select bacteria that contain the plasmid?

**(d)** Among the bacteria that contain the plasmid, how could you use medium containing an antibiotic to identify those transformed cells that potentially have inserted into the plasmid?

**Answer**

**(a)** You would not want to use *Bam*HI, since this enzyme cleaves within *yourgene* and would probably destroy its function. *Eco*RI would be the enzyme to use because *yourgene* is included in a single *Eco*RI fragment.

**(b)** You should digest the plasmid with *Eco*RI also, since you want the *Eco*RI sites flanking *yourgene* to be compatible with the ends of the cleaved plasmid.

**(c)** You should select transformants on tetracycline. Only those cells with the plasmid will grow.

**(d)** Among the cells that are tetracycline resistant, those that potentially have *yourgene* inserted into the *amp-r* gene should be sensitive to ampicillin, because the insertion interrupts the *amp-r* gene and destroys its function.

**Problem 3** The accompany graph shows the theoretical distribution of fragment sizes expected from digestion with a restriction enzyme. The curve is a so-called *exponential distribution*, symbolized here as $f(x)$, and the mean of the distribution is denoted $m$. Note that the curve is skewed to the left, so there are many more short fragments than long fragments. The distribution is divided by vertical lines into regions denoting fragments sizes smaller than half the mean ($m/2$), ranging in size from half the mean to the mean ($m/2$ to $m$), the mean to twice the mean ($m$ to $2m$), and so forth. Each region is labeled with the proportion of fragment sizes expected to be in that range.

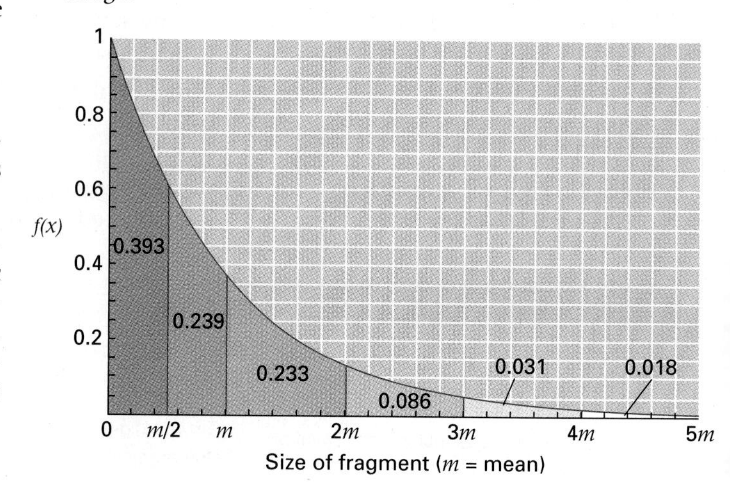


**(a)** What fraction of restriction fragments is expected to be smaller than the mean? To be smaller that two times the mean? Three times the mean? Four times the mean?

**(b)** What is the mean fragment size for a restriction enzyme whose restriction site consists of four nucleotides? For such an enzyme, what proportion of restriction fragments are expected to be smaller than 128 base pairs?

**(c)** What is the mean fragment size for a restriction enzyme whose restriction site consists of six nucleotides? For such an enzyme, what proportion of restriction fragments are expected to be larger than 16,384 base pairs?

**Answer**

**(a)** The expected proportion smaller than the mean is 0.393 + 0.239 = 0.632, smaller than two times the mean is 0.393 + 0.239 + 0.233 = 0.865, smaller than three times the mean is 0.393 + 0.239 + 0.233 + 0.086 = 0.951, and smaller than four times the mean is 0.393 + 0.239 + 0.233 + 0.086 + 0.031 = 0.982.

**(b)** The probability of a specific four-nucleotide sequence is 1/4 × 1/4 × 1/4 × 1/4 = 1/256, and the mean is the reciprocal of this, or 256 base pairs. A fragment length of 128 base pairs is half the mean, and 39.3 percent of the fragments are expected to be in this range.

**(c)** The probability of a specific six-nucleotide sequence is $(1/4)^6 = 1/4096$, and the mean is the reciprocal of this, or 4096 base pairs. A fragment length of 16,384 base pairs is four times the mean, and only 1 − 0.982 or 1.8 percent of the fragments are expected to be in this range.

## ANALYSIS AND APPLICATIONS

**12.1** Restriction enzymes generate one of three possible types of ends on the DNA molecules that they cleave. What are the three possibilities?

**12.2** In recombinant DNA, researchers typically prefer ligating restriction fragments that have sticky ends (single-stranded overhangs) rather than those that have blunt ends. Can you propose a reason why?

**12.3** In cloning a gene whose sequence is unknown, a researcher decides to use *partial digestion* with a restriction enzyme, in which the reaction is stopped when only a fraction of the restriction sites have been cleaved. What possible advantage is there is using a partial digest rather than a complete digest?

**12.4** What is the average distance between restriction sites for each of the following restriction enzymes? Assume that the DNA substrate has a random sequence with equal amounts of each nucleotide. The symbol N stands for any nucleotide, R for any purine (A or G), and Y for any pyrimidine (T or C).

**(a)** TCGA *Taq*I

**(b)** GGTACC *Kpn*I

**(c)** GTNAC *Mae*III

**(d)** GGNNCC *Nla*IV

**(e)** GRCGYC *Acy*I

**12.5** The restriction enzymes *Sep*I and *Ppu*10I have the restriction sites

ATGCA|T (*Sep*I)
A|TGCAT (*Ppu*10I)

where the 5′ end is written at the left and cleavage is at the site of the vertical bar. Are the sticky ends produced by these restriction enzymes compatible? Explain.

**12.6** The restriction enzyme *Aas*I has the restriction site 5′–GACNNNN|NNGTC–3′, where N means any nucleotide (matched by a complementary nucleotide in the other strand) and cleavage is at the site of the vertical bar. What is the probability that two random *Aas*I sites have compatible sticky ends?

**12.7** Do the ends of different restriction fragments produced by a particular restriction enzyme always have the same sequence? Must opposite ends of each restriction fragment be the same? Explain your answer?

**12.8** Will the sequences 5′-GGCC-3′ and 3′-GGCC-5′ in a double-stranded DNA molecule be cut by the same restriction enzyme?

**12.9** The restriction enzymes *Hae*I and *Aat*I have the restriction sites

WGGCCW (*Hae*I)
AGGCCT (*Aat*I)

where the 5′ end is written at the left and W stands for either A (paired with T in the other strand) or T (paired with A in the other strand). Assuming a random nucleotide sequence, what is the probability that a *Hae*I site is an *Aat*I site? What is the probability that an *Aat*I site is an *Hae*I site?

**12.10** The restriction enzyme *Fae*I has the restriction site CATG|, where the 5′ end is written at the left and the cleavage site is indicated by the vertical bar. If an mRNA has the sequence CAUG, where AUG is the start codon for translation, would *Fae*I be a good enzyme to use to cleave the DNA immediately after the initiation codon? Explain your answer.

**12.11** The restriction enzyme *Aoc*II has the restriction site GDGCH|C, where D denotes A or G or C (with a complementary base in the other strand) and H denotes A or C or T (with a complementary base in the other strand). How many *Aco*II sites are possible? Which of these are palindromes?

**12.12** The restriction enzyme *Aoc*II has the restriction site GDGCH|C, where D denotes A or G or C (with a complementary base in the other strand) and H denotes A or C or T (with a complementary base in the other strand). If two random *Aoc*II sites are cleaved, what is the probability that their sticky ends are compatible?

**12.13** The restriction enzyme *Acc*I has the restriction site GT|MKAC, where M denotes A or C (with a complementary base in the other strand) and K denotes G or T (with a complementary base in the other strand). If two random *Acc*I sites are cleaved, what is the probability that their sticky ends are compatible?

**12.14** The restriction enzyme *Hpy*8I has the restriction site GCNNAC, where N denotes any nucleotide, and the restriction enzyme *Acc*I has the restriction site GTMKAC, where M denotes A or C (with a complementary base in the other strand) and K denotes G or T (with a complementary base in the other strand). What is the probability that two random *Hpy*8I sites are identical? What is the probability that a random *Hpy*8I site is also an *Acc*I site?

**12.15** A circular plasmid has two restriction sites for the enzyme *Bgl*II, which cleaves the site A|GATCT. After digestion the fragments are ligated together, and a circular product is isolated that includes one copy of each of the fragments. Does this mean that the ligated plasmid is the same as the original? Explain.

**12.16** The 8-kb plasmid shown here has restriction sites for the enzymes *Pst*I, *Xba*I, and *Bam*HI. What size fragments would be produced by digestion with each of the enzymes individually, with each possible pair of enzymes, and with all three together?

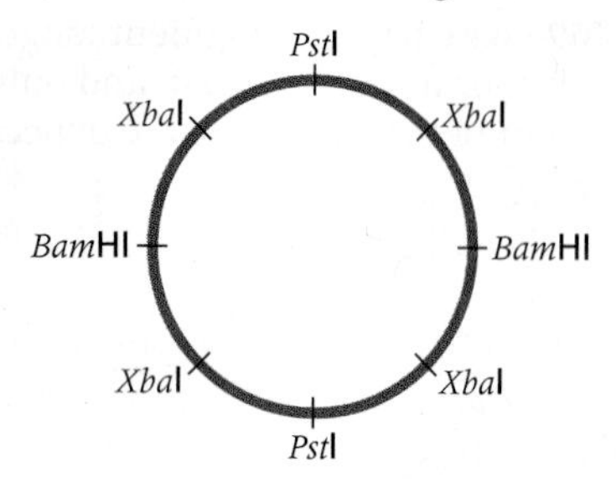


**12.17** In cloning into bacterial plasmid vectors, why is it useful to insert DNA fragments to be cloned into a restriction site inside an antibiotic-resistance gene? Why is another gene for resistance to a second antibiotic also convenient?

**12.18** How frequently would the restriction enzymes *Taq*I (restriction site TCGA) and *Mae*III (restriction site GTNAC, in which N is any nucleotide) cleave double-stranded DNA molecules containing random sequences of

**(a)** 20 percent A–T base pairs and 80 percent G–C base pairs?

**(b)** 50 percent A–T base pairs and 50 percent G–C base pairs?

**(c)** 80 percent A–T base pairs and 20 percent G–C base pairs?

**12.19** You want to introduce a human gene into a bacterial vector downstream of a bacterial promoter in hopes of producing a large amount of the human protein. Should you use the genomic DNA or the cDNA? Explain your reasoning.

**12.20** Shown here is a restriction map of a 12-kb linear plasmid isolated from cells of *Borrelia burgdorferi*, a spirochete bacterium transmitted by the bite of Ixodes ticks that causes Lyme disease. The symbols *C, P, K, A, E, B,* and *X* represent cleavage sites for the restriction enzymes *Cla*I, *Pst*I, *Kpn*I, *Ava*I, *Eco*RI, *Bal*I, and *Xba*I, respectively. In the accompanying gel diagram, show the positions at which bands would be found after digestion of the plasmid with the indicated restriction enzyme or enzymes.

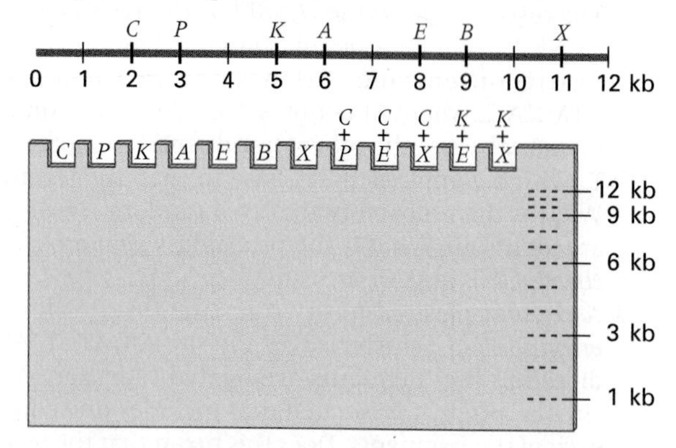


**12.21** A circular plasmid gives the illustrated pattern of bands when digested singly or in combination with three restriction enzymes, *Ava*I (*A*), *Bgl*II (*B*), and *Mbo*I (*M*). Draw a restriction map of the plasmid. (*Hint*: Some bands may conceal multiple fragments, because the total size of all fragments produced must be the same across all lanes in the gel.)

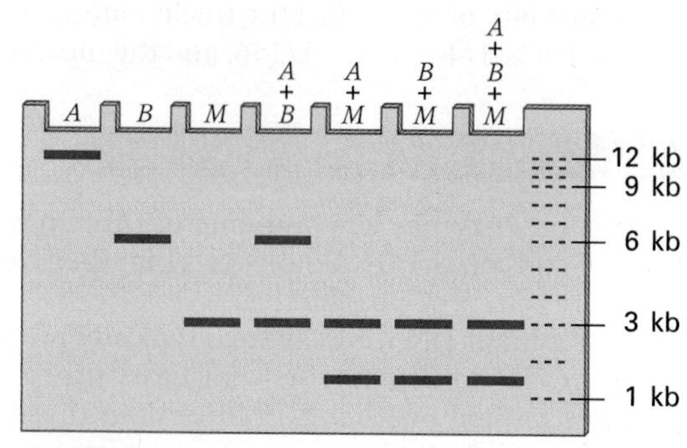


**12.22** A DNA microarray is competitively hybridized as described in the text with human genomic DNA from females and males. The female DNA is labeled with a moiety that fluoresces red, the male DNA with one that fluoresces green. What color would you expect a spot to fluoresce if

**(a)** It corresponds to an autosomal gene?

**(b)** It corresponds to an X-linked gene?

**(c)** It corresponds to a Y-linked gene?

**12.23** A functional genomics experiment is carried out using a DNA microarray to assay levels of gene expression in a species of bacteria. What genes would you expect to find overexpressed in cells grown in minimal medium compared to cells grown in complete medium?

**12.24** Many engineered vectors have a polylinker, or multiple cloning site (MCS), instead of a single restriction site available for insertion of DNA. A polylinker is a short DNA sequence containing multiple different restriction sites, each unique to the vector, any combination of which can be cleaved with the appropriate restriction enzymes to yield a linear vector molecule with defined restriction sites at the ends. The accompanying diagram illustrates five restriction sites in the polylinker of a vector; the vector sequences adjacent to the polylinker are denoted X and Y. Also shown is the restriction map of a genomic target sequence whose ends are denoted A and B. The restriction enzymes *Sal*I, *Kpn*I, *Pst*I, and *Xma*I all produce sticky ends; *Hpa*I produces blunt ends.

**(a)** If the vector and target were both digested with *Hpa*I and the resulting fragments ligated in such a way that each vector molecule was ligated with one and only one target fragment, in what orientation would the A fragment be ligated into the polylinker?

**(b)** What restriction enzyme (or enzymes) would you use to digest the vector and target so that after mixing of the fragments and ligation, the cloned DNA would have the sequence X B A Y?

**(c)** What restriction enzyme (or enzymes) would you use to digest the vector and target so that after mixing of the fragments and ligation, the cloned DNA would have the sequence X A B Y?

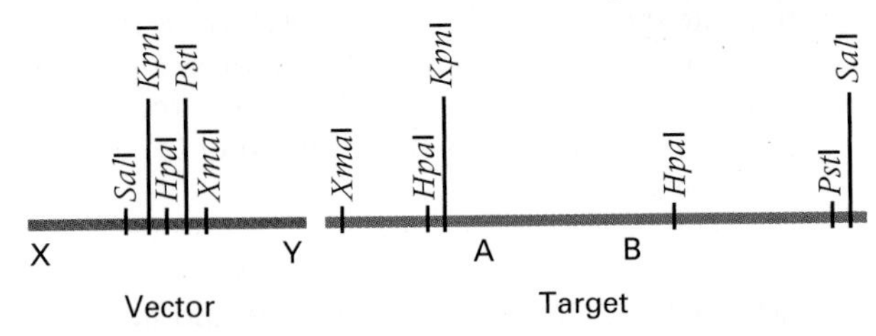


**12.25** A circular plasmid gives the illustrated pattern of bands when digested singly or in combination with three restriction enzymes, *Sal*I (*S*), *Eco*RI (*E*), and *Cla*I (*C*). Draw a restriction map of the plasmid. (*Hint*: Some bands may conceal multiple fragments, because the total size of all fragments produced must be the same across all lanes in the gel.)

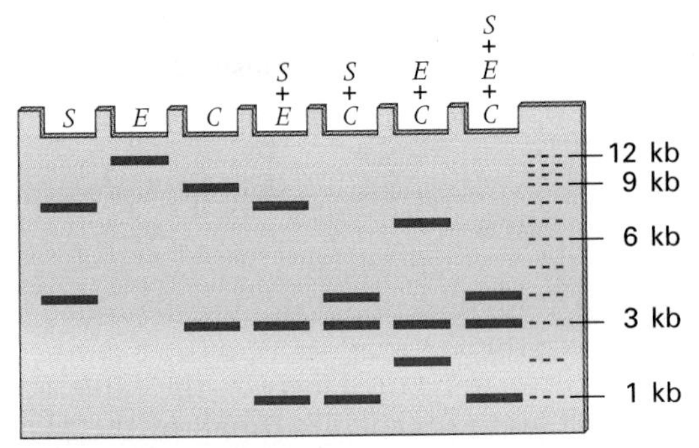


**12.26** Shown here is a gene with its transcription start site and direction of transcription indicated, along with the multiple cloning site (MCS) in a plasmid vector. The symbols *A, B, C, D, E, F, G.* and *H* represent cleavage sites for the restriction enzymes *Aat*I, *Bcl*I, *Cla*I, *Dra*II, *Eco*RI, *Fsi*I, *Gdi*I, and *Hin*dIII, respectively, each of which cleaves DNA asymmetrically and produces complementary sticky ends. A team of researchers wish to place the coding sequence of the gene under the control of a promoter located to the left of the MCS, as shown in the diagram. To do this, they isolate the *Aat*I fragment containing the coding sequence, add it to a solution containing a plasmid that has been cleaved with *Aat*I, ligate the sticky ends together, and transform bacterial cells. They pick a single transformant containing a plasmid that has the *Aat*I fragment inserted in the right place but find, to their disappointment, that the gene is not transcribed. Suggest an explanation.

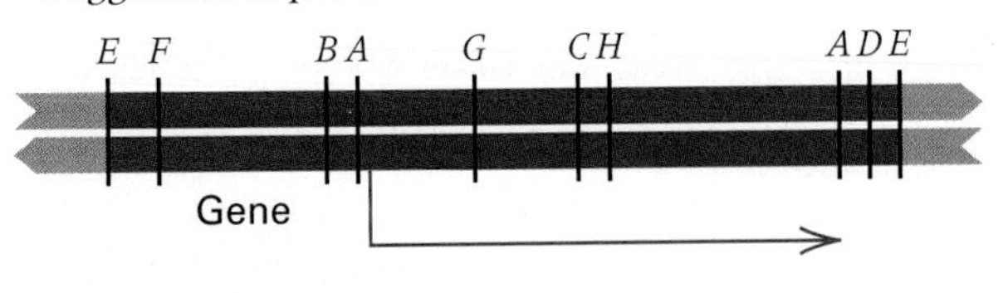


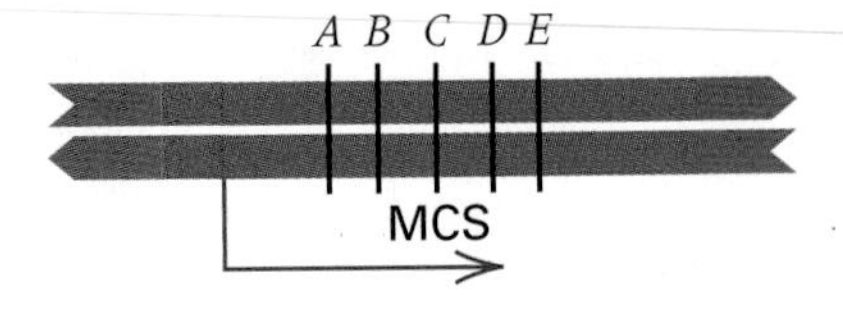


**12.27** The nonfunctional recombinant DNA clone described in the previous problem is analyzed to deduce a restriction map in the vicinity of the MCS. The result is shown in the accompanying illustration. Does this finding help explain the unexpected result?

**12.28** Analysis of additional clones from Problem 12.26 reveals that they fall into two classes with respect to the restriction map in the vicinity of the MCS. One of the restriction maps is shown in the previous problem. What does the other look like? Would this type of clone be expected to transcribe the gene of interest? Why or why not?

**12.29** Another geneticist examines the restriction maps in Problem 12.26 and suggests that *directional cloning* would be a more fruitful approach to making sure that the coding sequence is transcribed in the recombinant plasmid. In this approach, the gene of interest and the plasmid are both cleaved with the same two restriction enzymes, which are chosen so that their sticky ends will ensure that the fragment of interest is inserted in the correct orientation in the plasmid. For directional cloning of the fragment in Problem 12.26, which two restriction enzymes should be used? What is the restriction map of the resulting plasmid in the vicinity of the MCS?

**12.30** A geneticist is studying a double-stranded circular plasmid of 4 kb. This plasmid carries genes whose protein products confer resistance to tetracycline (*tet-r*) and ampicillin (*amp-r*) in host bacteria. The DNA has a single site for each of the following restriction enzymes: *Sst*I, *Kpn*I, *Bam*HI, *Eco*RI, and *Hin*dIII. Cloning DNA into the *Sst*I site does not affect resistance to either drug. Cloning DNA into the *Kpn*I, *Bam*HI, and HindIII sites abolishes tetracycline resistance. Cloning into the *Eco*RI site abolishes ampicillin resistance. Digestion with the following mixtures of restriction enzymes yields fragments with the sizes listed below. Indicate the positions of the *Eco*RI, *Kpn*I, *Bam*HI, and HindIII cleavage sites on a restriction map, relative to the *Sst*I cleavage site.

| Enzyme | Fragment size mixture (kb) |
|---|---|
| *Sst*I *Eco*RI | 0.70, 3.30 |
| *Sst*I *Kpn*I | 0.30, 3.70 |
| *Sst*I *Bam*HI | 0.08, 3.92 |
| *Sst*I *Hin*dIII | 0.85, 3.15 |
| *Sst*I *Kpn*I *Eco*RI | 0.30, 0.70, 3.00 |

**Challenge Problem 1** A geneticist decides to use yeast DNA microarrays to study the effects of deletion mutations in the yeast mating type locus *MAT*. To perform these experiments, RNA is extracted from two strains of yeast to be compared, and labeled cDNA probe is prepared from each strain by reverse transcription of the RNA in the presence of either green (Cy3) or red (Cy5) fluorescent label. The differently labeled cDNA solutions are mixed and applied to a microarray with spots of DNA sequence for each of the types of genes shown in the accompanying table. *MATα1*, *MATα2*, and *MAT***a***1* are DNAs for the possible coding sequences present at the *MAT* locus, *αsg* and ***a****sg* spots contain DNA of genes expressed specifically in *α* and **a** cells, respectively, and *hsg* spots contain DNA of genes expressed specifically in haploid cells. For each of the hybridizations in the accompanying table, indicate the expected color of the hybridization (red, green, yellow, or none). Yellow indicates equal amounts of Cy3 and Cy5, and "none" indicates lack of expression in both strains. The symbol Δ means a deletion mutation. (Before tackling this problem, you may wish to review the genetic control of yeast mating type in Section 11.10.)

**Challenge Problem 2** The accompanying illustration shows a gene with its transcription start site and direction of transcription, along with the multiple cloning site (MCS) in two plasmid vectors, A and B, which differ according to the position of a promoter adjacent to the MCS. The symbols *A, D, E, H, P, S,* and *X* represent cleavage sites for the restriction enzymes *Alu*I, *Dde*I, *Eco*RI, *Hin*dIII, *Pst*I, *Sac*I, and *Xho*II, respectively. In directional cloning, a gene of interest and a plasmid are both cleaved with the same two restriction enzymes, which are chosen so that their sticky ends will ensure that the fragment of interest is inserted in the correct orientation in the plasmid. A geneticist plans to use directional cloning to place the coding region of the gene under the control of the promoter in each of the two plasmids.

**(a)** Which restriction enzymes should be used for directional cloning in plasmid A?

**(b)** Which restriction enzymes should be used for directional cloning in plasmid B?

**(c)** Draw restriction maps of the resulting plasmids in the vicinity of the MCS.

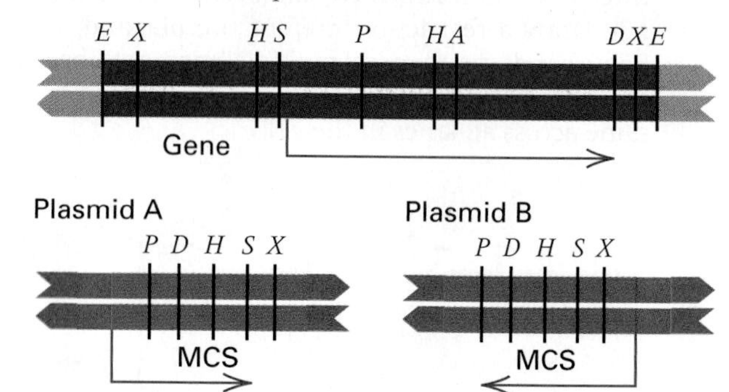


**Challenge Problem 3** The directional cloning described in the previous problem was successful in both cases, but unfortunately, the labels on the tubes became smudged and unreadable. Do all of the plasmids have to be restriction mapped in full, or is there a simpler tell-tale digest that could be done to distinguish between them?

| Red (Cy5) | Wildtype *α* cells | Wildtype *α* cells | Wildtype **a** cells | Δ*MAT***a***1* cells | Δ*MATα1* cells | Δ*MAT(α2α1)* Δ*MAT***a***1* |
|---|---|---|---|---|---|---|
| **Green (Cy3)** | **Wildtype a cells** | **Wildtype a/α cells** | **Wildtype a/α cells** | **Wildtype a cells** | **Wildtype α cells** | **Wildtype a cells** |
| *MATα1* | | | | | | |
| *MATα2* | | | | | | |
| *MAT***a***1* | | | | | | |
| *αsg* | | | | | | |
| **a***sg* | | | | | | |
| *hsg* | | | | | | |

## GENETICS on the web

GeNETics on the Web will introduce you to some of the most important sites for finding genetics information on the Internet. To explore these sites, visit the Jones and Bartlett companion site to accompany *Genetics: Analysis of Genes and Genomes, Seventh Edition* at http://biology.jbpub.com/book/genetics.

There you will find a chapter-by-chapter list of highlighted keywords. When you select one of the keywords, you will be linked to a Web site containing information related to that keyword.

# CHAPTER 13

# Genetic Control of Development

A Hawaiian "happy face" spider *Theridion grallator*. Spider populations on different islands show a variety of happy-face patterns and colors. Developmental geneticists try to identify the genetic mechanisms that regulate the differences in color, pattern, and shape among these and other organisms. [Courtesy of Geoff Oxford, Department of Biology, University of York.]

## CHAPTER OUTLINE


**13.1** Genetic Determinants of Development

**13.2** Early Embryonic Development in Animals
- Autonomous Development and Intercellular Signaling
- Composition and Organization of Oocytes
- Early Development and Activation of the Zygotic Genome

**13.3** Genetic Analysis of Development in the Nematode
- Analysis of Cell Lineages
- Mutations Affecting Cell Lineages
- Programmed Cell Death
- Loss-of-Function and Gain-of-Function Alleles
- Epistasis in the Analysis of Developmental Switches

**13.4** Genetic Control of Development in *Drosophila*
- Maternal-Effect Genes and Zygotic Genes
- Genetic Basis of Pattern Formation in Early Development
- Coordinate Genes
- Gap Genes
- Pair-Rule Genes
- Segment-Polarity Genes
- Interactions in the Regulatory Hierarchy
- Metamorphosis of the Adult Fly
- Homeotic Genes
- *HOX* Genes in Evolution

**13.5** Genetic Control of Development in Higher Plants
- Flower Development in *Arabidopsis*
- Combinatorial Determination of the Floral Organs

CONNECTION Distinguished Lineages
John E. Sulston, E. Schierenberg, J. G. White, and J. N. Thomson 1983
*The Embryonic Cell Lineage of the Nematode* Caenorhabditis Elegans

CONNECTION Embryo Genesis
Christiane Nüsslein-Volhard and Eric Wieschaus 1980
*Mutations Affecting Segment Number and Polarity in* Drosophila

Understanding gene regulation is central to understanding how an organism as complex as a human embryo develops from a fertilized egg. In development, genes are expressed according to a prescribed program to ensure that the fertilized egg divides repeatedly and that the resulting cells become specialized in an orderly way to give rise to the fully differentiated organism. The genotype determines not only the events that take place in development, but also the temporal order in which the events unfold.

Genetic approaches to the study of development often make use of mutations that alter developmental patterns. These mutations interrupt developmental processes and make it possible to identify factors that control development and to study the interactions among them. This chapter demonstrates how genetics is used in the study of development. Many of the examples come from the nematode *Caenorhabditis elegans*, the fruit fly *Drosophila melanogaster*, and the flowering plant *Arabidopsis thaliana*, because these organisms have been studied intensively from the standpoint of developmental genetics. The key process in development is **pattern formation**, which means the emergence of spatially organized and specialized cells that form the embryo from cell division and differentiation of the fertilized egg.

## 13.1 Genetic Determinants of Development

Conceptually, the relationship between genotype and development is straightforward:

> The genotype contains a developmental program that unfolds and results in the expression of different sets of genes in different types of cells.

In other words, development consists of a program that results in the specific expression of some genes in one cell type and not in another. As development unfolds, cell types progressively that differ qualitatively in the genes that are expressed appear. Whether a particular gene is expressed may depend on the presence or absence of a particular transcription factor, a change in chromatin structure, or the synthesis of a particular receptor molecule. These and other molecular mechanisms define the pattern of interactions by which one gene controls another. Collectively, these regulatory interactions ultimately determine the fate of the cell. Moreover, many interactions that are important in development make use of more general regulatory elements. For example, developmental events often depend on regional differences in the concentration of molecules within a cell or within an embryo, and the activity of enhancers may be modulated by the local concentration of these substances.

The formation of an embryo is also affected by the environment in which development takes place. The genotype determines the developmental potential of the organism. Genes determine whether a developing embryo will become a nematode, a fruit fly, a chicken, or a mouse. However, the expression of the genetically determined developmental potential is also influenced by the environment—in some cases, very dramatically. Fetal alcohol syndrome is one example in which an environmental agent, chronic alcohol poisoning, affects various aspects of fetal growth and neurological development.

Development includes many examples in which genes are selectively turned on or off by the action of regulatory proteins that respond to environmental signals. The identification of genetic regulatory interactions that operate during development is an important theme in developmental genetics. The implementation of different regulatory interactions means that genetically identical cells can become qualitatively different in the genes that are expressed.

Among the most intensively studied developmental genes are those that act early in development, before tissue differentiation, because these genes establish the overall pattern of development.

## 13.2 Early Embryonic Development in Animals

The early development of the animal embryo establishes the basic developmental plan for the whole organism. Fertilization initiates a series of mitotic **cleavage divisions** in which the embryo becomes multicellular. There is little or no increase in overall size compared with the egg, because the cleavage divisions are accompanied by little growth; they merely partition the fertilized egg into progressively smaller cells. The cleavage divisions form the **blastula**, which is essentially a ball of about $10^4$ cells containing a cavity (**FIGURE 13.1**). Completion of the cleavage divisions is followed by the formation of the **gastrula** through an infolding of part of the blastula wall and extensive cellular migration. In the reorganization of cells in the gastrula, the cells become arranged in several distinct layers. These layers establish the basic body plan of an animal. In higher plants, as we shall see later,

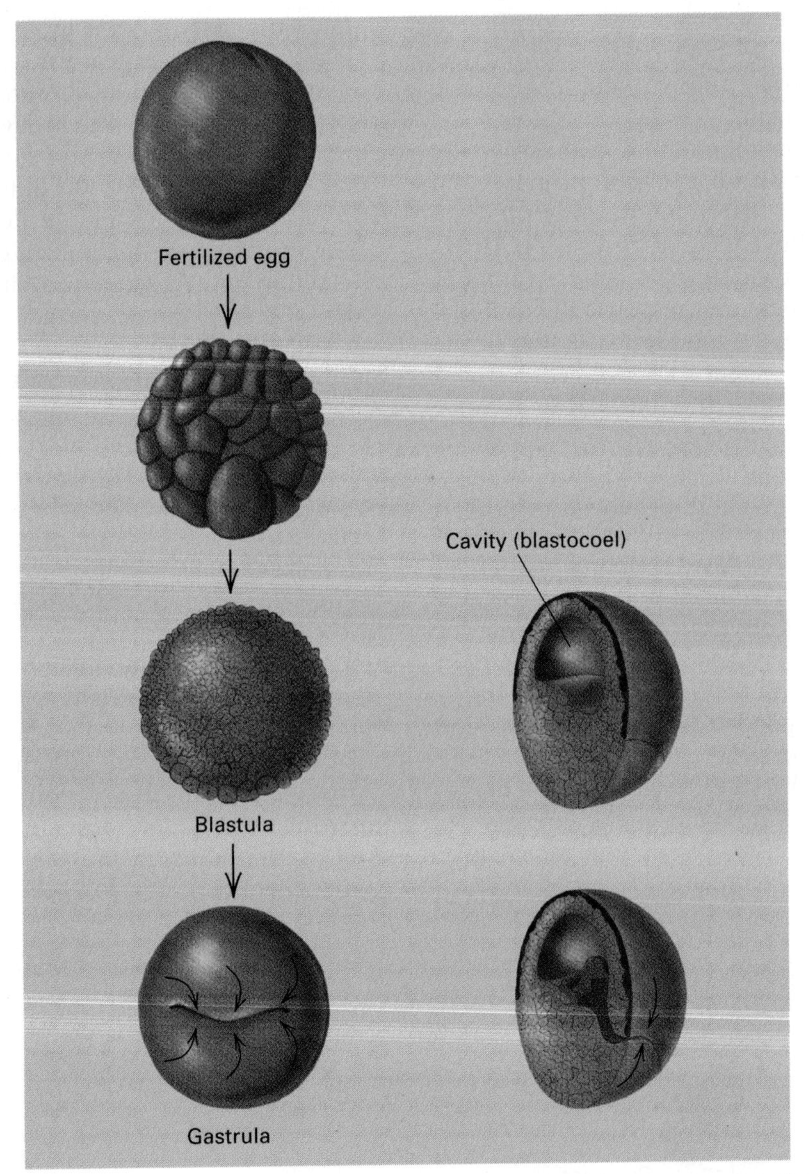


**FIGURE 13.1** Early development of the animal embryo. The cleavage divisions of the fertilized egg result in first a clump of cells and then a hollow ball of cells (the blastula). Extensive cell migrations form the gastrula and establish the basic body plan of the embryo.

the developmental processes differ substantially from those in animals.

A fertilized animal egg has full developmental potential because it contains the genetic program in its nucleus and macromolecules present in the oocyte cytoplasm that are necessary for giving rise to an entire organism. Full developmental potential is maintained in cells produced by the early cleavage divisions. However, the developmental potential is progressively channeled and limited in early embryonic development. Cells within the blastula usually become committed to particular developmental outcomes, or **cell fates**, that limit the differentiated states possible among descendant cells.

## Autonomous Development and Intercellular Signaling

Two principal mechanisms progressively restrict the developmental potential of cells within a lineage:

1. Developmental restriction may be **autonomous**, which means that it is determined by genetically programmed changes in the cells themselves.
2. Cells may respond to **positional information**, which means that developmental restrictions are imposed by the position of cells within the embryo. Positional information may be mediated by signaling interactions between neighboring cells or by gradients in the concen-

tration of morphogens. A **morphogen** is a molecule that participates directly in the control of growth and form during development.

Restriction of developmental fate is usually studied by transplanting cells of the embryo to new locations to determine whether they can substitute for the cells that they displace. An example showing the distinction between autonomous development and response to positional information is illustrated in **FIGURE 13.2**. In the normal embryo (Figure 13.2A), cells 1 and 2 have different fates either because of autonomous developmental programs or because they respond to positional information near the anterior and posterior ends. When the cells are transplanted (Figure 13.2B), autonomous development is indicated if the developmental fate of cells 1 and 2 is unchanged in spite of their new locations; in this case, the embryo has anterior and posterior reversed. When development depends on positional information, the transplanted cells respond to their new locations, and the embryo is normal.

In some eukaryotes, such as the soil nematode *Caenorhabditis elegans*, many lineages develop autonomously according to genetic programs that are induced by interactions with neighboring cells very early in embryonic development. Subsequent cell interactions also are important, and each stage in development is set in motion by the successful completion of the preceding stage. **FIGURE 13.3** illustrates the first three cell divisions in the development of *C. elegans*, which result in eight embryonic cells that differ in their genetic activity and developmental fate. The determination of cell fate in these early divisions is in part autonomous and in part a result of interactions between cells. Figure 13.3B shows the lineage relationships between the cells. Cell-autonomous mechanisms are illustrated by the transmission of cytoplasmic particles called *polar granules* from the cells P0 to P1 to P2 to P3. Normal segregation of the polar granules is a function of microfilaments in the cytoskeleton. Cell-signaling mechanisms are illustrated by the effects of P2 on EMS and on ABp. The EMS fate is determined by the activity of the *mom-2* gene in P2. The P2 cell also produces a signaling molecule, APX-1, which determines the fate of ABp through the cell-surface receptor GLP-1. In contrast to *C. elegans*, in which many developmental decisions are cell-autonomous, in *Drosophila* and *Mus* (the

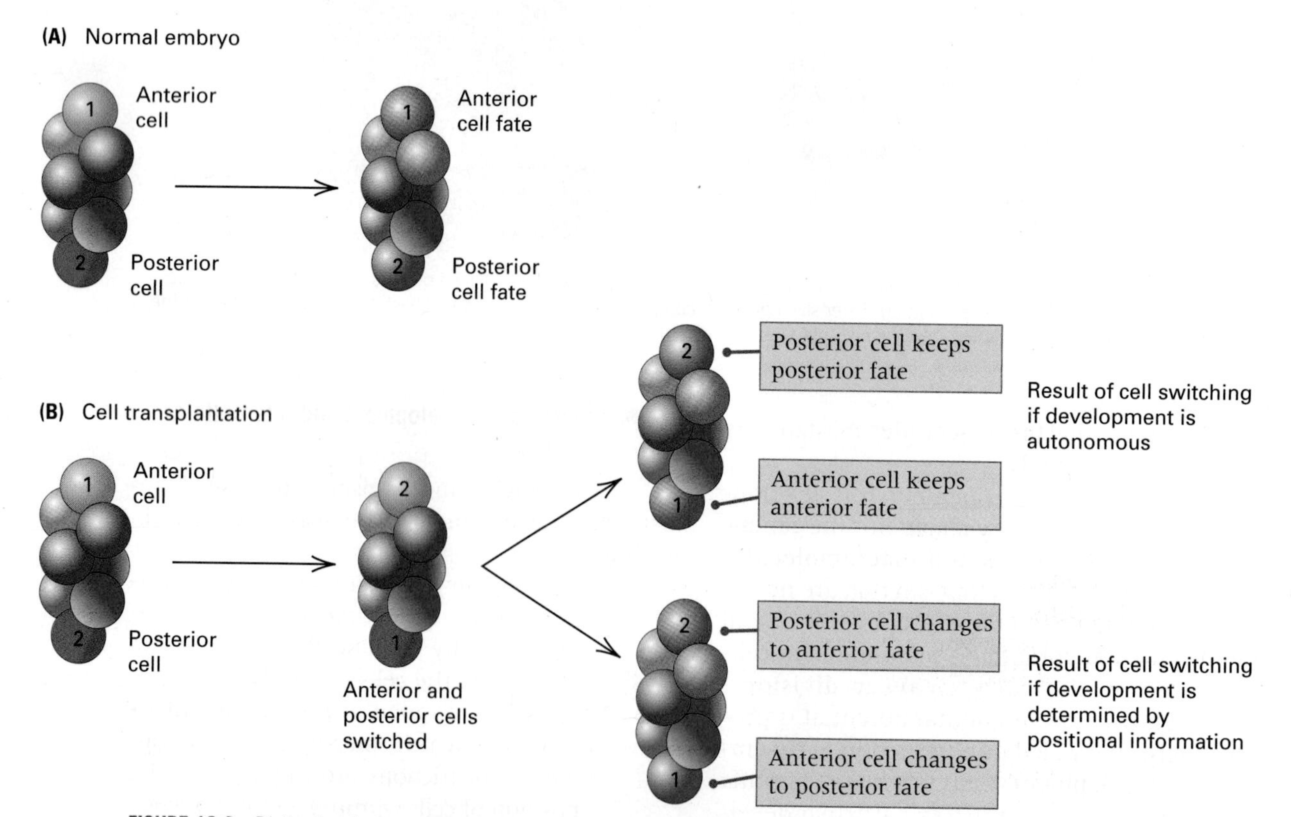


**FIGURE 13.2** Distinction between autonomous determination and positional information. (A) Cells 1 and 2 differentiate normally as shown. (B) Transplantation of the cells to reciprocal locations. If the transplanted cells differentiate autonomously, then they differentiate as they would in the normal embryo, but in their new locations. If they differentiate in response to positional information (signals from neighboring cells), then their new positions determine their fate.

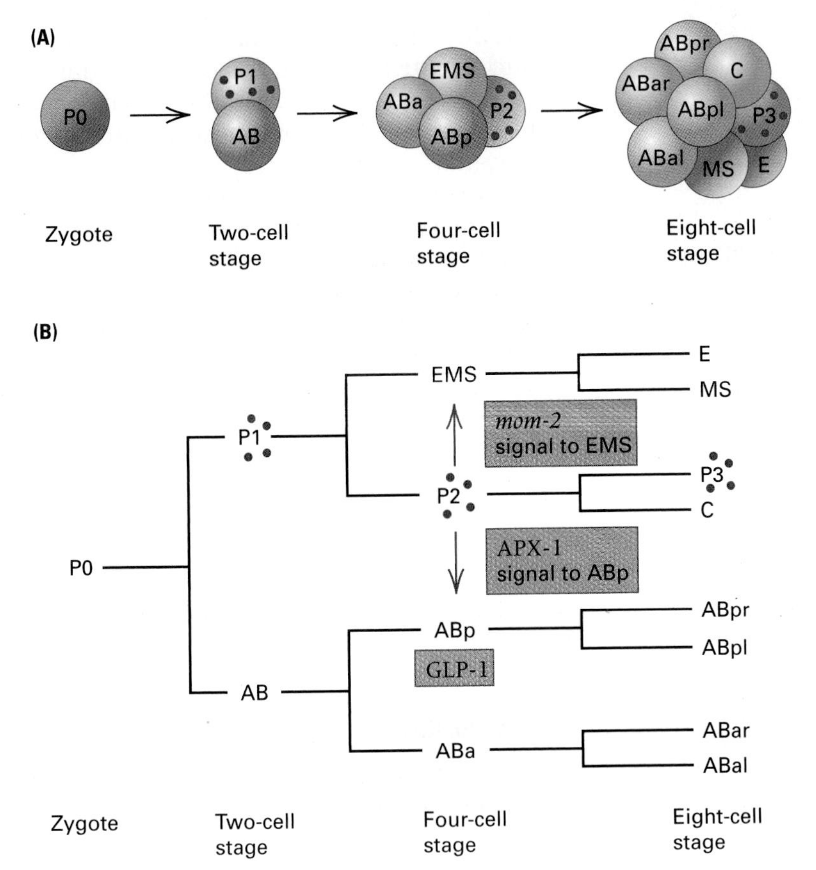


**FIGURE 13.3** Early cell divisions in *C. elegans* development. (A) Spatial organization of cells. (B) Lineage relationships of the cells. The transmission of the polar granules illustrates cell-autonomous development. The arrows denote cell-to-cell signaling mechanisms that determine developmental fate. [Adapted from W. Roush, *Science* 272 (1996): 1871.]

mouse), regulation by intercellular signaling is more the rule than the exception. The use of intercellular signaling to regulate development provides a sort of insurance that helps to overcome the death of individual cells that might happen by chance during development.

One special case of intercellular signaling is **embryonic induction**, in which the development of a major embryonic structure is determined by a signal sent from neighboring cells. An example is found in early development of the sea urchin, *Strongylocentrotus purpuratus*, in which, after the fourth cleavage division, specialized cells called *micromeres* are produced at the ventral pole of the embryo (**FIGURE 13.4**, parts A and B). After additional cell proliferation, the region of the embryo immediately above the micromeres folds inward to form a tube, the *archenteron*, that eventually develops into the stomach, intestine, and related structures. If micromere cells from a sixteen-cell embryo are removed and transplanted into the dorsal region of another embryo at the eight-cell stage (Figure 13.4C), then the next cleavage results in an embryo with two sets of micromeres (Figure 13.4D). In these embryos, as development proceeds, an archenteron forms in the normal position from cells lying above the ventral micromeres. But a second archenteron also forms beneath the transplanted micromeres (Figure 13.4E), which indicates that the transplanted micromeres are capable of inducing archenteron development in adjacent cells.

## ■ Composition and Organization of Oocytes

The oocyte is a diploid cell during most of oogenesis. The cytoplasm of the oocyte is extensively organized and regionally differentiated (**FIGURE 13.5**). This spatial differentiation ultimately determines the different developmental fates of cells in the blastula. The cytoplasm of the animal egg has two essential functions:

1. Storage of the molecules needed to support the cleavage divisions and the rapid RNA and protein synthesis that take place in early embryogenesis.
2. Organization of the molecules in the cytoplasm to provide the positional information that results in differences between cells in the early embryo.

To establish the proper composition and organization of the oocyte requires the participation of many genes within the oocyte itself and gene products supplied by adjacent helper cells of various types (**FIGURE 13.6**). Numerous maternal genes are transcribed in oocyte development, and the mature oocyte typically contains an abundance of transcripts that code for proteins needed in the early embryo. Some of the mater-

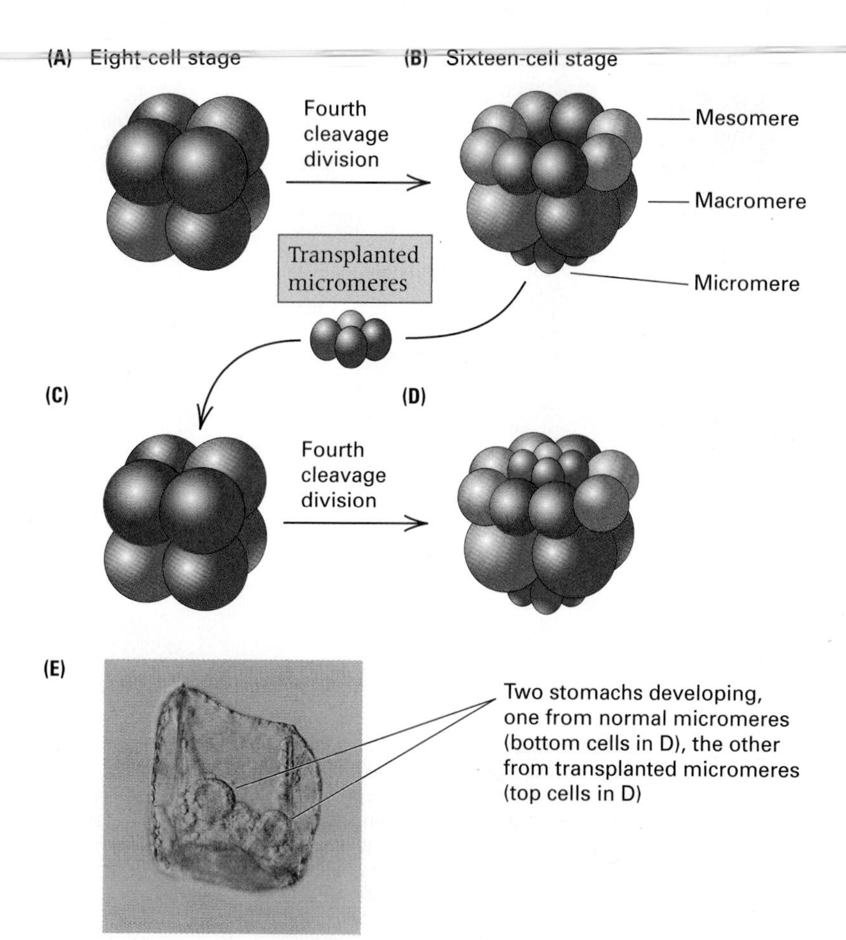


**FIGURE 13.4** Embryonic induction of the archenteron by micromere cells in the sea urchin. (A) Normal embryo at the 8-cell stage. (B) Normal embryo at the 16-cell stage, showing micromeres on the ventral side. (C) Transplanted 8-cell embryo with micromeres from another embryo placed at the dorsal end. (D) Transplanted embryo after the fourth cleavage division with two sets of micromeres. (E) The result of the transplant is an embryo with two archenterons. The doughnut-shaped objects in this embryo are the developing stomachs. [Part E reproduced from A. Ransick and E. H. Davidson, *Science* 259 (1993): 1134–1138. Reprinted with permission from AAAS.

**FIGURE 13.5** The animal oocyte is highly organized, which is revealed by the visible differences between the dorsal (dark) and ventral (light) parts of these *Xenopus* eggs. The regional organization of the oocyte determines many of the critical events in early development. [Courtesy of Michael Klymkowsky, MCDB, University of Colorado at Boulder.]

nal mRNA transcripts are masked in special ribonucleoprotein particles and cannot be translated, and the release of the mRNA, enabling it to be translated, does not take place until after fertilization.

Developmental instructions in oocytes are determined in part by the presence of distinct types of molecules at different positions within the cell and in part by gradients of morphogens that differ in concentration from one position to the next. Although the oocyte contains the products of many genes, only a small number of genes are expressed exclusively in the formation of the oocyte. Most genes expressed in oocyte formation are also important in the development of other tissues or at later times in development. Therefore, it is not only gene products, but also their spatial organization within the cell that give the oocyte its unique developmental potential.

## Early Development and Activation of the Zygotic Genome

Fertilized frog eggs with the nucleus removed are still able to carry out the cleavage divisions and form rudimentary blastulas. Similarly, when gene transcription in sea urchin or amphibian zygotes is blocked, there is no effect on the cleavage divisions or on blastula formation, but gastrulation does not take place. These observations imply that early development through blastula formation does not depend on genetic information in the zygotic cell nucleus. All the necessary macromolecules are in the oocyte cytoplasm.

Following this early period when zygotic genes are not needed, the embryo becomes dependent on the activity of its own genes. In mammals, the zygotic genes are needed much earlier than in lower vertebrates. Inhibitors of transcription stop development of the mouse embryo when they are applied at any time after the first cleavage division. This result implies that the shift from control by the maternal genome to control by the zygotic genome begins in the two-cell stage. This is also the stage when proteins encoded by the zygotic nucleus are first detected. However, even in mammals, the earliest stages of development are greatly influenced by the cytoplasm of the oocyte, which determines the planes of the initial cleavage divisions and other events that ultimately affect cell fate.

Early activation of the zygotic nucleus in mammals may be necessary because, in gamete formation, certain genes undergo the process of epigenetic inactivation known as *imprinting*

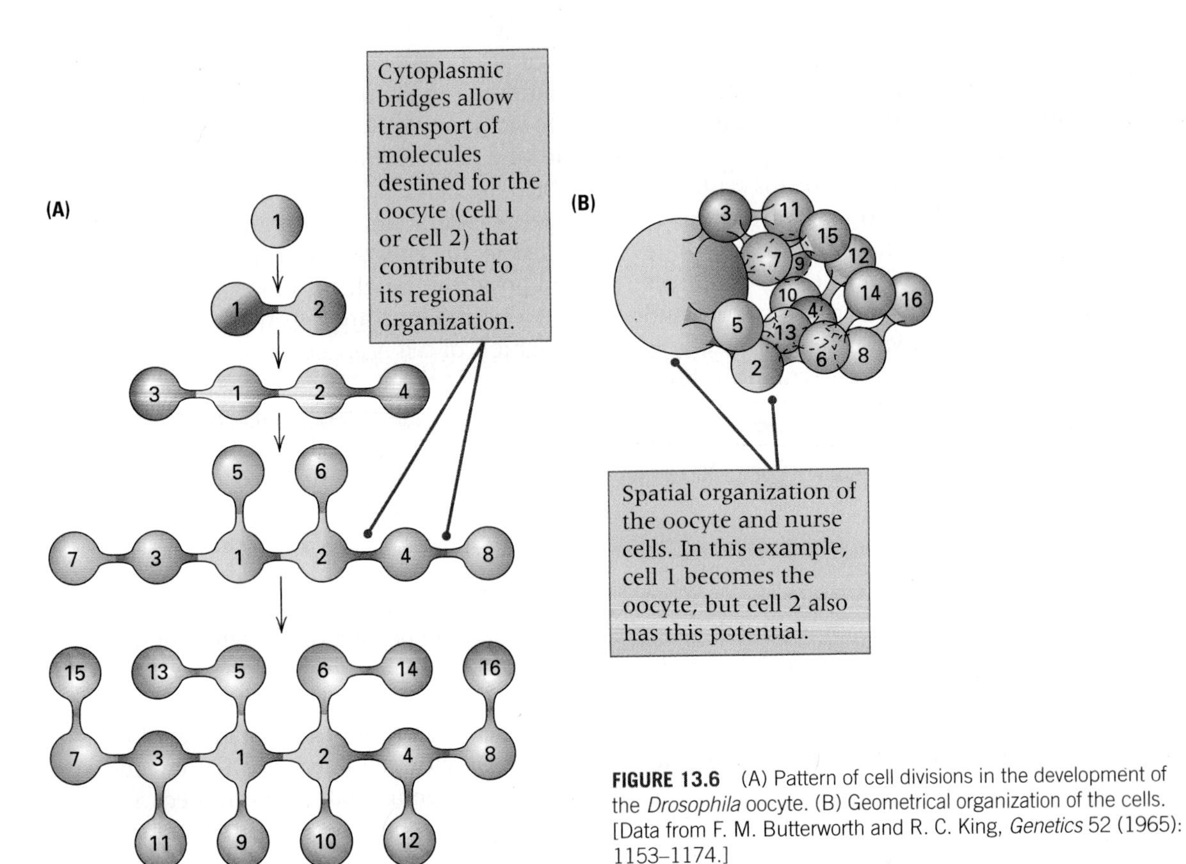


**FIGURE 13.6** (A) Pattern of cell divisions in the development of the *Drosophila* oocyte. (B) Geometrical organization of the cells. [Data from F. M. Butterworth and R. C. King, *Genetics* 52 (1965): 1153–1174.]

(discussed in Chapter 11). Not many genes are subject to imprinting. Among these are the gene for an insulin-like growth factor (*Igf2*), which is imprinted during oogenesis, and the gene for the *Igf2* transmembrane receptor (*Igf2-r*), which is imprinted during spermatogenesis. Therefore, expression of both *Igf2* and *Igf2-r* in the embryo requires activation of the sperm nucleus for *Igf2* and of the egg nucleus for *Igf2-r*. Imprinting also affects genes in the human genome and has been implicated in conditions such as the fragile-X syndrome (Chapter 14) and the Prader-Willi and Angelman syndromes (Chapter 11).

## 13.3 Genetic Analysis of Development in the Nematode

The soil nematode *Caenorhabditis elegans* (**FIGURE 13.7**) is popular for genetic studies because it is small, is easy to culture, and has a short generation time with a large number of offspring. The worms are grown on agar surfaces in petri dishes and feed on bacterial cells such as *E. coli*. Because they are microscopic in size, as many as $10^5$ animals can be contained in a single petri dish. Sexually mature adults of *C. elegans* are capable

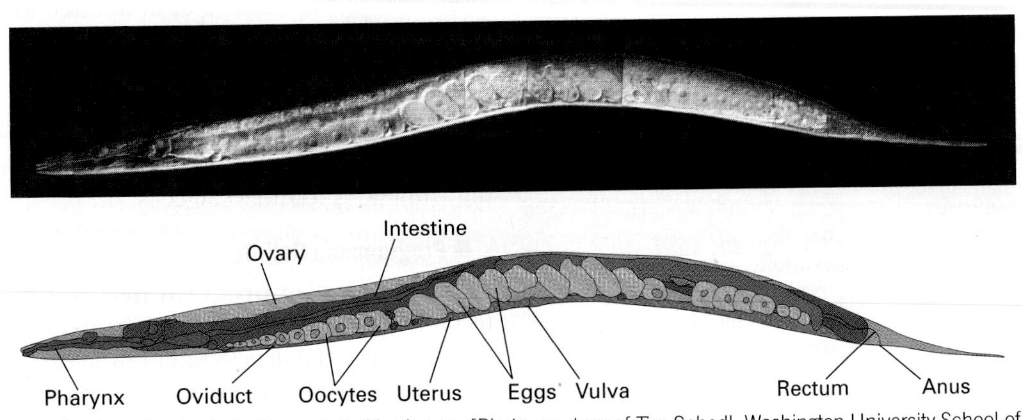


**FIGURE 13.7** The soil nematode *Caenorhabditis elegans*. [Photo courtesy of Tim Schedl, Washington University School of Medicine.]

of laying more than 300 eggs within a few days. At 20°C, it requires about 60 hours for the eggs to hatch, undergo four larval molts, and become sexually mature adults.

Nematodes are diploid organisms with two sexes. In *C. elegans*, the two sexes are the hermaphrodite and the male. The hermaphrodite contains two X chromosomes (XX), produces both functional eggs and functional sperm, and is capable of self-fertilization. The male produces only sperm and fertilizes the hermaphrodites. The sex-chromosome constitution of *C. elegans* consists of a single X chromosome; there is no Y chromosome, and the male karyotype is XO.

## Analysis of Cell Lineages

The transparent body wall of the worm has made possible the study of the division, migration, and differentiation or death of all cells present in the course of development. Nematode development is unusual in that the pattern of cell division and differentiation is highly stereotyped, with almost no variation from one individual to the next. The hermaphrodite contains exactly 959 somatic cells, the male exactly 1031. The complete developmental history of each somatic cell is known at the cellular and ultrastructural levels. The pedigree relationships among a group of cells is known as a cell **lineage**. A cell lineage can be illustrated with a **lineage diagram** that shows the ancestor–descendant relationships and the terminal differentiated state of each cell. **FIGURE 13.8** shows a generalized lineage diagram of a cell A whose descendants undergo either programmed cell death or terminal differentiation into cell type W, X, or Y. The letter symbols are the kind normally used for cells in nematodes, in which the name denotes the cell lineage according to ancestry and position in the embryo. For example, the cells A.a and A.p are the anterior and posterior daughters of cell A, respectively; and A.aa and A.ap are the anterior and posterior daughters of cell A.a.

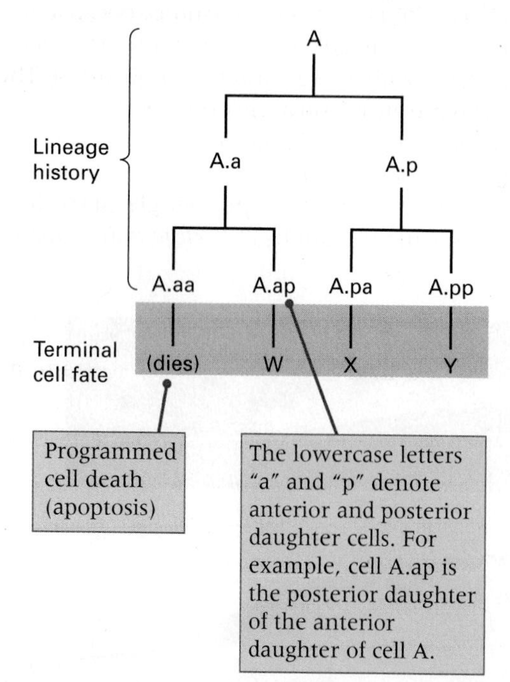


**FIGURE 13.8** Generalized cell-lineage diagram. Different terminally differentiated cell fates are denoted W, X, and Y. One cell in the lineage (A.aa) undergoes programmed cell death.

## Mutations Affecting Cell Lineages

Many mutations that affect cell lineages have been studied in nematodes. They reveal several general features by which genes control development:

- The division pattern and fate of a cell are generally affected by more than one gene and can be disrupted by mutations in any of them.
- Most genes that affect development are active in more than one type of cell.
- Complex cell lineages often include simpler, genetically determined lineages within them; these components are called *sublineages* because they are expressed as an integrated pattern of cell division and terminal differentiation.
- The lineage of a cell may be triggered autonomously within the cell itself or by signaling interactions with other cells.
- Regulation of development is controlled by genes that determine the different sublineages that cells can undergo and the individual steps within each sublineage.

Mutations can affect cell lineages in two major ways. One is through nonspecific metabolic blocks (for example, in DNA replication) that prevent the cells from undergoing division or differentiation. The other is through specific molecular defects that result in patterns of division or differentiation that are normally found elsewhere in the embryo or at a different time in development. From the standpoint of genetic analysis of development, the latter class of mutants is the more informative, because the mutant genes must be involved in developmental processes rather than in general "housekeeping" functions found in all cells.

## Programmed Cell Death

The process of **programmed cell death**, technically known as **apoptosis**, is an important feature of normal development in many organisms. Apoptosis is a process in which, at the appropriate time in development, a cell commits suicide. In many cases, the signaling molecules

## connection

### Distinguished Lineages

*The data produced in this landmark study form the basis for interpreting developmental mutants in the nematode worm. This long paper offers voluminous data, and it is presently available through the Internet. During embryogenesis, 1030 cells are generated; 131 of these, or 13%, undergo programmed cell death. What is the reason for such a high proportion of programmed cell deaths? The embryonic lineage is highly invariant—the same from one organism to the next. Why is there not more developmental flexibility, as is found in most other organisms? These issues are addressed in this excerpt, in which the emphasis is on the historical background and motivation of the study, the big picture of development, and interpretation of the lineage in terms of the evolution of the nematode. The technique of Nomarski microscopy is a modern invention also called differential interference contrast microscopy. When light passes through living material, it changes phase according to the refractive index of the material. Adjacent parts of a cell or organism that differ in refractive index cause different changes in phase. When two sets of waves combine after passing through an object, the difference in phase creates an interference pattern that yields an image of the object. The major advantage of Nomarski microscopy is that it can be used to observe living tissue.*


**John E. Sulston,[1] E. Schierenberg,[2] J. G. White,[1] and J. N. Thomson[1] 1983**
[1]Medical Research Council Laboratory for Molecular Biology, Cambridge, England, United Kingdom
[2]Max Planck Institute for Experimental Medicine, Gottingen, Germany

*The Embryonic Cell Lineage of the Nematode* Caenorhabditis Elegans

This report marks the completion of a project begun over one hundred years ago—namely the determination of the entire cell lineage of a nematode. Nematode embryos were attractive to nineteenth-century biologists because of their simplicity and the reproducibility of their development, and considerable progress was made in determining their lineages by the use of fixed specimens. By the technique of Nomarski microscopy, which is nondestructive and yet produces high resolution, cells can be followed in living larvae. The use of living material lends a previously unattainable continuity and certainty to the observations, and has permitted the origin and fate of every cell in one nematode species [*Caenorhabditis elegans*] to be determined. Thus, not only are the broad relationships between tissues now known unambiguously but also the detailed pattern of cell fates is clearly revealed. . . . The lineage is of significance both for what it can tell us immediately about relationships between cells and also as a framework into which future observations can be fitted. . . . The embryonic cell lineage is essentially invariant. The patterns of division, programmed cell death, and terminal differentiation are constant from one individual to another, and no great differences are seen in timing. . . . We shall use the term "sublineage" as an abbreviation for the more descriptive, but cumbersome, phrase "intrinsically determined sublineage"—namely, a fragment of the lineage which is thought to be generated by a program within its precursor cell. . . . Two of the available criteria for postulating the existence of a sublineage are: (1) the generation of the same lineage, giving rise to the same cell fates, from a series of precursors of diverse origin and position; (2) evidence for cell autonomy within the lineage, obtained from laser ablation experiments or the study of mutants. . . . The large number of programmed cell deaths, and their reproducibility, are evident from the lineage. The most likely reason for the occurrence of most of them is that, because of the existence of sublineages, unneeded cells are frequently generated along with needed ones. . . . Perhaps the most striking findings are firstly the complexity and secondly the cell autonomy of the lineages. . . .

***The nematode belongs to an ancient phylum, and its cell lineage is a piece of frozen evolution. In the course of time, new cell types were generated from precursors selected not so much for their intrinsic properties as for the accident of their position in the embryo.***

With hindsight, we can rationalize both this complexity and this rigidity. The nematode belongs to an ancient phylum, and its cell lineage is a piece of frozen evolution. In the course of time, new cell types were generated from precursors selected not so much for their intrinsic properties as for the accident of their position in the embryo. . . . Cell-cell interactions that were initially necessary for developmental decisions may have been gradually supplanted by autonomous programs that were fast, economical, and reliable, the loss of flexibility being outweighed by the gain in efficiency. On this view, the perverse assignments, the cell deaths, the long-range migration—all the features which could, it seems, be eliminated from a more efficient design—are so many developmental fossils. These are the places to look for clues both to the course of evolution and to the mechanisms by which the lineage is controlled today.

Source: J. E. Sulston, et al., *Dev. Biol.* 100 (1983): 64–119.

that determine this fate have been identified (a number are known to be transcription factors). Failure of programmed cell death often results in specific developmental abnormalities. When apoptosis fails and the surviving cells differentiate into recognizable cell types, the result is the presence of supernumerary cells of that type. For example, with mutations in the *ced-3* gene (*ced* = *cell death abnormal*) in *C. elegans*, a particular cell that normally undergoes programmed cell death survives and often differentiates into a supernumerary neuron. There are exactly 131 programmed cell deaths in the development of the *C. elegans* hermaphrodite. None of these deaths is essential. Mutants that cannot execute programmed cell death are viable and fertile but are slightly impaired in development and in some sensory capabilities. On the other hand, mutants in *Drosophila* that fail to execute apoptosis are lethal, and in mammals, including human beings, failure of programmed cell death results in severe developmental abnormalities or, in some instances, leukemia or other forms of cancer (Chapter 15).

## Loss-of-Function and Gain-of-Function Alleles

Genes that cause cells to differ in developmental fate are not always easy to recognize. For example, a mutation may identify a gene that is *necessary* for the expression of a particular developmental fate, but the gene may not be *sufficient* to determine that developmental fate in the cells in which it is expressed. This possibility complicates the search for genes that control major developmental decisions.

Genes that control decisions about cell fate can sometimes be identified by the unusual characteristic that dominant and recessive mutations have opposite effects. The reasoning is that, if different mutations in the same gene can alter cell fate in opposite directions, then the product of the gene must be both necessary and sufficient for expression of the fate. Recessive alleles in genes controlling development are often *loss-of-function mutations*, such as deletions, nonsense mutations, or inactivating amino acid replacements. Dominant alleles in developmental-control genes are often *gain-of-function mutations* in that the gene is overexpressed or is expressed ectopically in the wrong cells or at the wrong time.

In *C. elegans*, only a small number of genes have dominant and recessive alleles that affect cell fate in opposite ways. Among them is the *lin-12* gene, which controls developmental decisions in a number of cells. One example concerns the cells denoted Z1.ppp and Z4.aaa in **FIGURE 13.9**. These cells lie side by side in the embryo, but they derive from cell $P_0$ (the zygote) via completely different lineages. Normally, as shown in **FIGURE 13.10**, part A, one of the cells differentiates into an *anchor cell* (AC), which participates in development of the vulva, and the other differentiates into a *ventral uterine precursor cell* (VU). Z1.ppp and Z4.aaa are equally likely to become the anchor cell.

Direct cell–cell interaction between Z1.ppp and Z4.aaa controls the AC-VU decision. If either cell is ablated (burned away) by a laser microbeam, then the remaining cell differentiates into an anchor cell (Figure 13.10B). This result implies that the preprogrammed fate of both Z1.ppp and Z4.aaa is that of an anchor cell. In the normal situation, commitment of either Z1.ppp or Z4.aaa to the anchor-cell fate also commits the other cell to the ventral uterine precursor fate. As noted, recessive and dominant mutations of *lin-12* have opposite effects. Recessive mutations in which *lin-12* activity is lacking or is greatly reduced are denoted *lin-12(0)*, and in the mutants both Z1.aaa and Z4.aaa

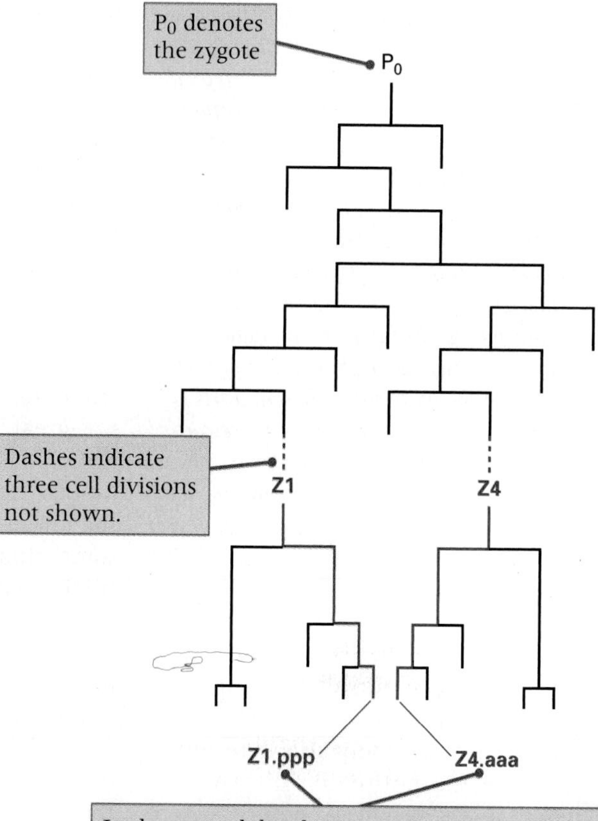
P0 denotes the zygote
P0
Dashes indicate three cell divisions not shown.
Z1
Z4
Z1.ppp
Z4.aaa
In the normal development of the vulva, Z1.ppp and Z4.aaa are equally likely to differentiate into the anchor cell. Whichever cell remains differentiates into a ventral uterine precursor cell.

**FIGURE 13.9** Complete lineage of Z1.ppp and Z4.aaa in *C. elegans*.

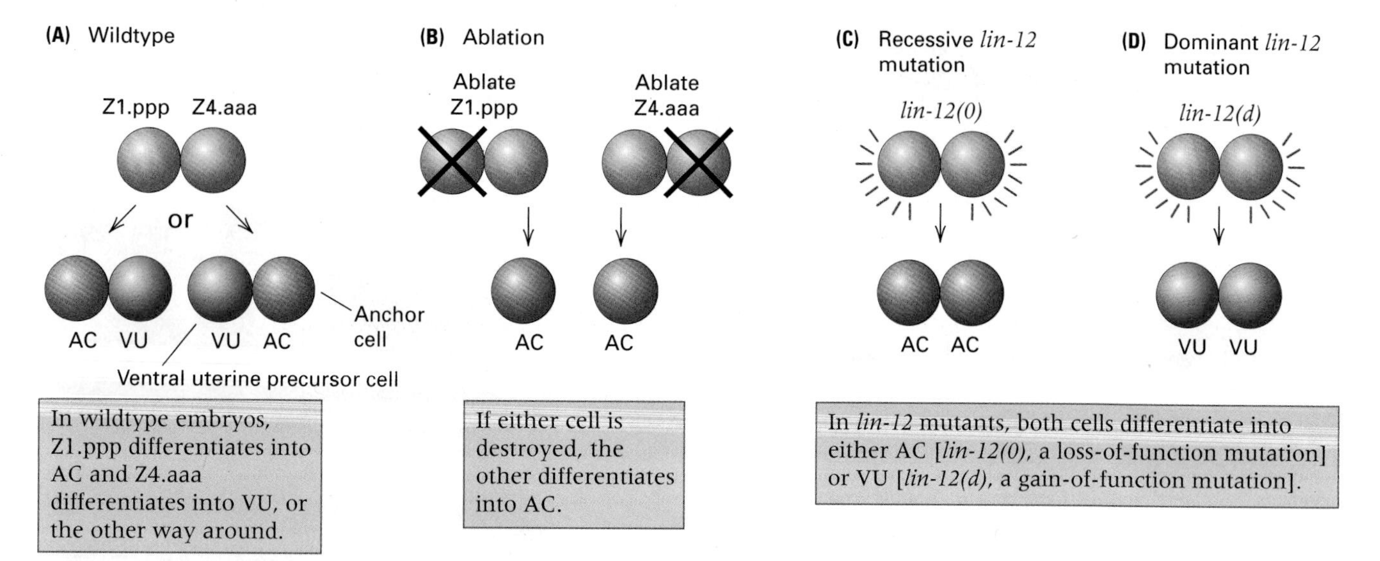


**FIGURE 13.10** Control of the fates of Z1.ppp and Z4.aaa in vulval development and genetic control of cell fate by the *lin-12* gene. With recessive loss-of-function mutations [*lin-12(0)*], both cells become anchor cells; with dominant gain-of-function mutations [*lin-12(d)*], both cells become ventral uterine precursor cells.

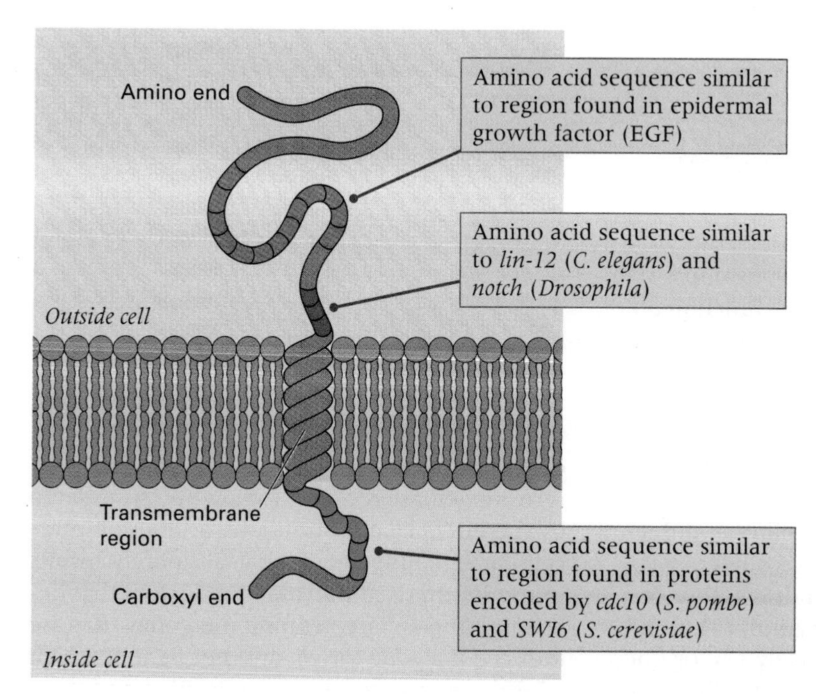


**FIGURE 13.11** The structure of the LIN-12 protein is that of a receptor protein containing a transmembrane region and various types of repeated units that resemble those in epidermal growth factor and other developmental-control genes.

become anchor cells (Figure 13.10C). In contrast, dominant mutations in which *lin-12* activity is overexpressed are denoted *lin-12(d)*, and in these mutants both Z1.aaa and Z4.ppp become ventral uterine precursor cells (Figure 13.10D).

Molecular analysis of *lin-12* revealed a gene encoding a typical transmembrane receptor protein, a function consistent with the opposite effects of *lin-12* mutations. Too little receptor renders cells insensitive to signal; too much receptor renders them super sensitive. The transmembrane region of the LIN-12 protein (**FIGURE 13.11**) divides the molecule into an extracellular part (the amino end) and an intracellular part (the carboxyl end). The extracellular part contains thirteen repeats of a domain found in a mammalian peptide hormone, epidermal growth factor (EGF), as well as in the product of the *notch* gene in *Drosophila*, which controls the decision between epidermal and neural cell fates. Nearer the transmembrane region, the amino end contains three repeats of a cysteine-rich domain also found in the *notch* protein. Inside the cell, the carboxyl part of the LIN-12 protein contains six repeats of a domain also found in the genes *SWI6* and *cdc10*, which control cell division in budding yeast and fission yeast.

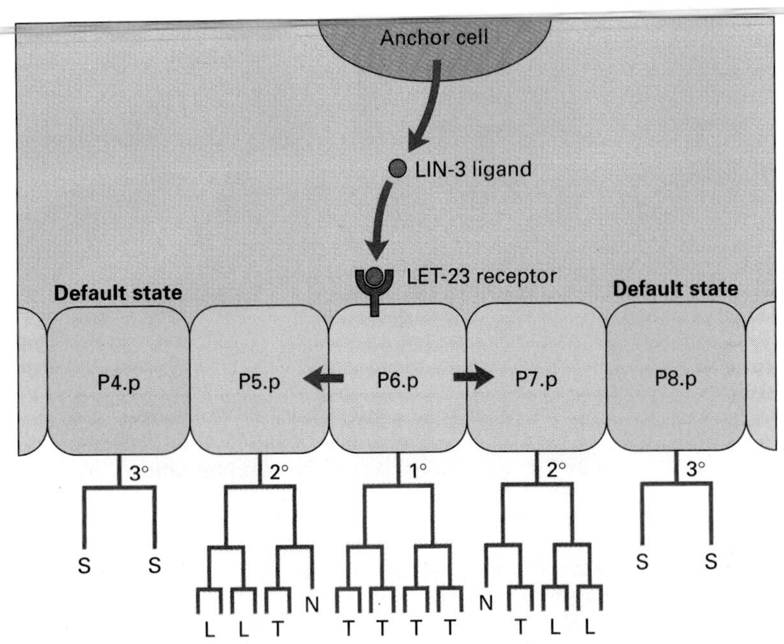


**FIGURE 13.12** Determination of vulval differentiation by means of intercellular signaling. Cells P4.p through P8.p in the hermaphrodite give rise to lineages in the development of the vulva. The three types of lineages are designated 1°, 2°, and 3°. The 1° lineage is induced in P6.p by the ligand LIN-3 produced in the anchor cell (AC), which stimulates the LET-23 receptor tyrosine kinase in P6.p. The P6.p cell, in turn, produces a ligand that stimulates receptors in P5.p and P7.p to induce the 2° fate. The 3° fate is the default condition, which P4.p and P8.p adopt normally and all cells adopt in the absence of AC.

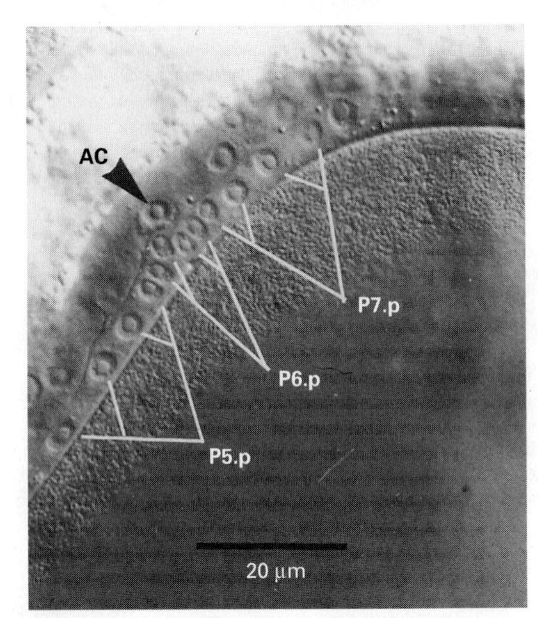


**FIGURE 13.13** Spatial organization of cells in the vulva, including the anchor cell (black arrowhead) and the daughter cells produced by the first two divisions of P5.p through P7.p (white tree diagrams). The length of the scale bar equals 20 micrometers. [Reproduced from G. D. Jongeward, T. R. Clandirin, and P. W. Sternberg, *Genetics* 139 (1995): 1553–1566.]

**FIGURE 13.12** illustrates another example in which loss-of-function and gain-of-function alleles have opposite effects on phenotype. The anchor cell expresses a signaling gene, called *lin-3*, whose product controls the fate of target cells in the development of the vulva. Figure 13.12 illustrates five precursor cells, P4.p through P8.p, that participate in the development of the vulva. Each precursor cell has the capability of differentiating into one of three fates, called the 1°, 2°, and 3° lineages, which differ in whether descendant cells remain in a syncytium (S) or divide longitudinally (L), transversely (T), or not at all (N). The precursor cells normally differentiate as shown, yielding five lineages in the order 3°–2°–1°–2°–3°. The vulva itself is formed from the 1° and 2° cell lineages. The spatial arrangement of some of the key cells is shown in the photograph in **FIGURE 13.13**. The black arrow indicates the anchor cell, and the white lines show the pedigrees of twelve cells. The four cells in the middle derive from P6.p, and the four on each side derive from P5.p, and P7.p.

The important role of the *lin-3* gene product (LIN-3) is suggested by the opposite phenotypes of loss-of-function and gain-of-function alleles. Loss of LIN-3 results in the complete absence of vulval development. Conversely, overexpression of LIN-3 results in excess vulval induction. LIN-3 is a typical example of an interacting molecule, or **ligand**, that interacts with an EGF-type transmembrane receptor. In this case, the receptor is located in cell P6.p and is the product of the gene *let-23*. The LET-23 protein is a tyrosine–kinase receptor that, when stimulated by the LIN-3 ligand, stimulates a series of intracellular signaling events that ultimately results in the synthesis of transcription factors that determine the 1° fate. Among the genes that are induced is a gene for yet another ligand, which stimulates receptors on the cells P5.p and P7.p, causing these cells to adopt the 2° fate (blue horizontal arrows in Figure 13.12). The evidence for horizontal signaling is found in genetic mosaic worms in which the LET-23 receptor is missing in some or all of P5.p through P7.p. If the receptor is missing in all three cell types, none of the cells adopts its normal fate. However, if the receptor is present in P6.p but absent in P5.p and P7.p, then all three cell types differentiate as they should, which implies that receipt of the LIN-3 signal is necessary for 1° determination and that a stimulated P6.p is necessary for 2° determination.

In vulval development, the adoption of the 3° lineages by the P4.p and P8.p cells is determined not by a positive signal, but by the lack of a signal because in the absence of the anchor cell, all of the cells P4.p through P8.p express the 3° lineage. Thus, development of the 3° lineage is the default, or uninduced, state, which means that the 3° fate is preprogrammed into the cell and must be overridden by another signal if the cell's fate is to be altered.

### Epistasis in the Analysis of Developmental Switches

Beadle and Tatum first showed how mutations could be used to discover the enzymatic steps in biochemical pathways for the synthesis of small molecules essential to growth (Chapter 1). A similar approach can be used for genetic dissection of developmental pathways, but the logic is somewhat different. Biochemical pathways are *substrate-dependent pathways*, in which each step must be completed successfully before the next step can proceed. Developmental pathways are *switch-regulation pathways*, in which each component in the pathway either inhibits or stimulates the activity of the next component in line. Components that inhibit the pathway are *negative regulators*, and those that stimulate the pathway are *positive regulators*.

The control of vulval development in *C. elegans* illustrates the logic behind the genetic analysis of switch-regulation pathways, but before proceeding we need to emphasize some caveats:

- The recessive mutant alleles that are analyzed must be complete loss-of-functional alleles, because in some cases even a residual activity will give misleading results.
- Each mutant gene must have a unique and nonredundant function in the pathway. Pathways with genetic redundancy resulting from duplicate genes or genes with overlapping functions are not suitable for this type of analysis.
- The mutant alleles should affect components in the same developmental pathway.
- The regulatory pathway should be linear, with each component interacting only with its nearest neighbors, without branching or parallel pathways.

Despite these caveats, many switch-regulation pathways do lend themselves to genetic analysis. The first step is to isolate a large number of recessive loss-of-functional alleles that encode components of the pathway, and then to assign these to complementation groups (genes) as described in Chapters 1 and 3. In switch-regulation pathways, two classes of genes are typically observed, in which loss-of-function alleles yield opposite phenotypes. For developmental determination of the vulva in *C. elegans*, mutant alleles in one class of genes result in no vulva (called vulvaless) whereas mutant alleles in the other class of genes results in the development of multiple vulvas (multivulva).

The genetic analysis of a switch-regulation pathway is based on the phenotypes of all pairs of double mutants of genes with opposite phenotypes. For vulva development, for example, we would make all possible pairs of vulvaless–multivulva double mutants. The order of the components in the switch-regulation pathway is determined by the type of epistasis observed in the double mutants. In Chapter 3 we defined *epistasis* as any interaction between mutant alleles that altered the 9 : 3 : 3 : 1 ratio expected from independent assortment of two genes. In the analysis of switch-regulation pathways the term is used in a somewhat different sense. For double mutants with alleles in genes with contrasting phenotypes, a gene is called an **epistatic gene** if its mutant phenotype masks the mutant phenotype of another gene. For example, if the phenotype of *aa bb* is the same as that of *aa* $b^+b^+$, then the gene *a* is said to be epistatic to the gene *b*. The gene whose mutant phenotype is concealed is called a **hypostatic gene**, and in the previous example the *b* gene is hypostatic to the *a* gene. To take a specific example, in a vulvaless–multivulva double mutant, if the double mutant is vulvaless, then the vulvaless gene is epistatic to the multivulva gene. Equivalently, we could say that the multivulva gene is hypostatic to the vulvaless gene. Epistasis helps to determine the order of components in a developmental pathway for the following reason:

> **Principle of epistasis:** In a linear switch-regulation pathway, the product of the epistatic gene acts *downstream* in the pathway relative to the product of the hypostatic gene; in other words, the product of the hypostatic gene acts *upstream* relative to that of the epistatic gene.

The logic behind this principle is illustrated in **FIGURE 13.14**. Two switch-regulation pathways in which mutants yield opposite phenotypes, either vulvaless or multivulva, are shown. In both pathways, A and B represent the wildtype gene products of the genes *a* and *b*, respectively. By convention, the arrowhead implies positive regulation (stimulation) and the bar implies negative regulation (inhibition). The gene products A and

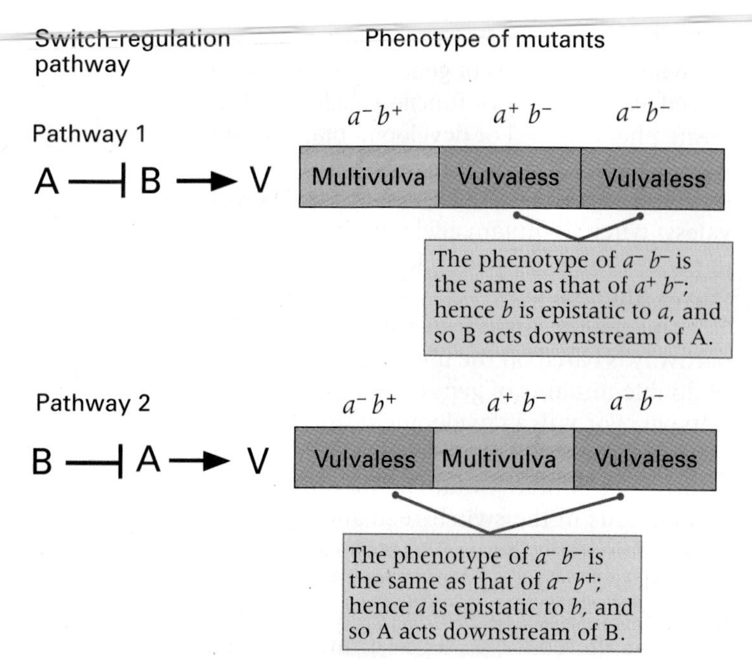


**FIGURE 13.14** Logic underlying the principle of epistasis: The gene product of the epistatic gene acts downstream of the product of the hypostatic gene.

B need not interact directly. If there are intervening components, then the arrows and bars represent the net effect on all the intermediates in the pathway.

The arrows and bars in Figure 13.14 were chosen so that $a^-a^-\ b^+b^+$ and $a^+a^+\ b^-b^-$ mutants would have different phenotypes. In pathway 1, A inhibits B and B stimulates vulva development (V). The $a^-a^-\ b^+b^+$ genotype therefore has less inhibition of B, which means greater activity of B, which implies a phenotype of multivulva. In contrast, the $a^+a^+\ b^-b^-$ genotype lacks B and hence no vulva development occurs, yielding a phenotype of vulvaless. Pathway 2 has the order of A and B interchanged, and in this case the $a^-a^-\ b^+b^+$ phenotype is vulvaless and that of $a^+a^+\ b^-b^-$ is multivulva.

The epistatic interactions of *a* and *b* are shown by the colons. In pathway 1 the $a^-a^-\ b^-b^-$ double mutant is vulvaless because the animal lacks B, whereas in pathway 2 the $a^-a^-\ b^-b^-$ double mutant is vulvaless because the animal lacks A. In both cases, in accord with the principle of epistasis, the gene that is downstream in the pathway is epistatic to the gene that is upstream in the pathway.

Now we are in a position to apply the principle of epistasis to a number of mutants that affect vulval development in *C. elegans*. Mutations in any of four different genes result in the vulvaless phenotype (*lin-3*, *let-23*, *lin-45*, and *let-60*), and mutations in either of two different genes result in the multivulva phenotype (*lin-1* and *lin-15*). All of these mutations are loss-of-function mutations, and the phenotypes of the homozygous mutants are shown color coded in **FIGURE 13.15**, part A, pink for vulvaless and green for multivulva. Also shown in the same color codes are the epistatic relations among all vulvaless–multivulva combinations. Note that *lin-1* is epistatic to all of the vulvaless mutants, which implies that the product of *lin-1* acts downstream of the products of the vulvaless genes. Also, *lin-15* is hypostatic to *let-23*, *lin-45*, and *let-60*, which implies that the product of *lin-15* acts upstream of the products of these three genes. Finally, *lin-15* is epistatic to *lin-3*, which implies that the product of *lin-15* acts downstream of that of *lin-3*. These inferences are all combined in the pathway in part B. The inference that the interactions are bars rather than arrows follows from the nature of the phenotypes. For example, because the phenotype of a *lin-1* mutant is multivulva, then the wildtype product of *lin-1* must inhibit vulva formation. Going back one step, because the phenotype of *let-23* (and also *lin-45* and *let-60*) is vulvaless, then these gene products must negatively regulate *lin-1*, because in the mutants failure to down-regulate *lin-1* would result in overproduction of the *lin-1* inhibitor, and hence suppress vulva formation.

As is apparent from Figure 13.15B, analysis of the loss-of-function mutants in part A does not reveal the order in which the products of *let-23*,

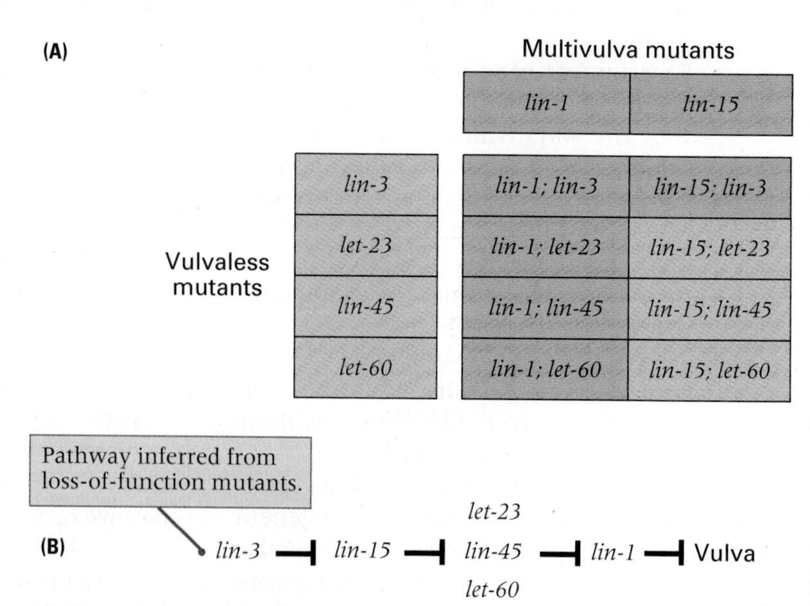


**FIGURE 13.15** Application of the principle of epistasis to vulval specification. (A) Experimental results. (B) Inferred switch-regulation pathway.

*lin-45*, and *let-60* function in the pathway because these mutants all yield the vulvaless phenotype. In such situations a gain-of-function mutation can be used to resolve the pathway. One such mutation is a *let-60* gain-of-function mutation, symbolized *let-60(gf)*, which results in a phenotype of multivulva (the opposite of a *let-60* loss-of-function mutation). **FIGURE 13.16**, part A shows the color-coded phenotypes for *let-60(gf)* in combination with either *let-23* or *lin-45*. The principle of epistasis still holds. Using this principle, the observation that *let-60(gf)* is epistatic to *let-23* means that *let-60* acts downstream of *let-23*, and the observation that *let-60(gf)* is hypostatic to *lin-45* means that *let-60* acts upstream of *lin-45*.

The complete pathway for vulva determination implied by all of the epistatic interactions is shown in Figure 13.16B. Based solely on genetic analysis, such a pathway should be regarded as a working hypothesis whose predictions must be confirmed by studies in molecular and cellular biology. In the case of vulva determination, the order of gene products in the pathway has been verified and the gene products identified (part C). The gene products are as follows:

- The wildtype *lin-3* gene encodes a ligand resembling vertebrate epidermal growth factor, whose role in vulval development is illustrated in Figure 13.12.
- The wildtype *lin-15* gene controls two distinct and novel genetic activities whose functions in the pathway are still unclear.
- The wildtype *let-23* gene encodes a receptor resembling the epidermal growth-factor receptor in vertebrates (Figure 13.11), which binds with the *lin-3* ligand (Figure 13.12).
- The wildtype *let-60* gene encodes a cellular growth factor resembling the vertebrate growth factor Ras.
- The wildtype *lin-45* gene encodes a tyrosine kinase resembling the vertebrate Ras tyrosine kinase.
- The wildtype *lin-1* gene encodes a transcription factor containing a domain rich in positively charged and aromatic amino acid residues known as the ETS domain.

Why so much emphasis on vulval development in worms, you might ask? One reason is that vulval development demonstrates how epistasis is used in the analysis of switch-regulation pathways. Another reason is that ligands for epidermal growth-factor receptors that activate Ras and downstream tyrosine kinases have been implicated in many human cancers. The role of the Ras protein in cancer is discussed in more detail in Chapter 15, but here we should emphasize that many of these cancers have mutations that are analogous to the gain-of-function mutation in *let-60(gf)*.

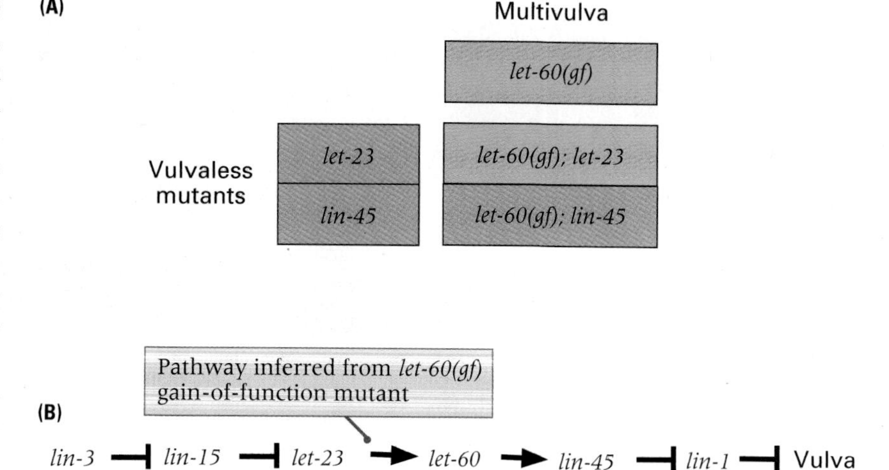


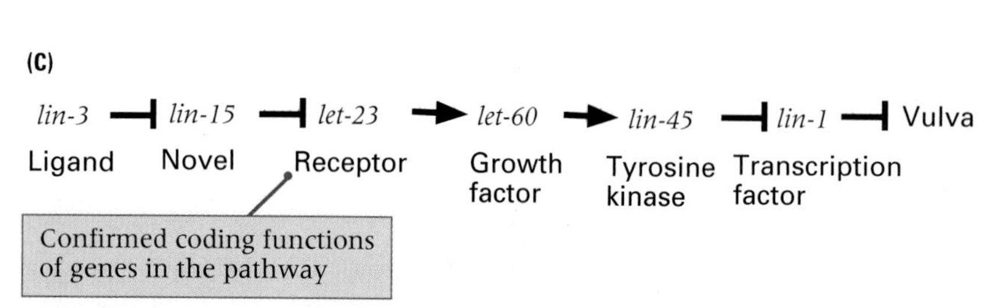


**FIGURE 13.16** The principle of epistasis also applies to gain-of-function *(gf)* mutants. In this case, analysis of *let-60(gf)* allows the order of action of the wildtype products of *let-23*, *let-60*, and *lin-45* to be determined.

## 13.4 Genetic Control of Development in *Drosophila*

Many important insights into developmental processes have been gained from genetic analysis in *Drosophila melanogaster*. **FIGURE 13.17** summarizes the developmental cycle through egg, larval, pupal, and adult stages. Early development includes a series of cell divisions, migrations, and infoldings that result in the gastrula. About 24 hours after fertilization, the first-stage larva, composed of about $10^4$ cells, emerges from the egg. Each larval stage is called an *instar*. Two successive larval molts that give rise to the second and third instar larvae are followed by pupation and a complex metamorphosis that gives rise to the adult fly composed of more than $10^6$ cells. In wildtype strains reared at 25°C, development requires from 10 to 12 days.

**FIGURE 13.18** shows side views of *Drosophila* embryos at various stages early in development, viewed through a scanning electron microscope to highlight surface detail. Each embryo is

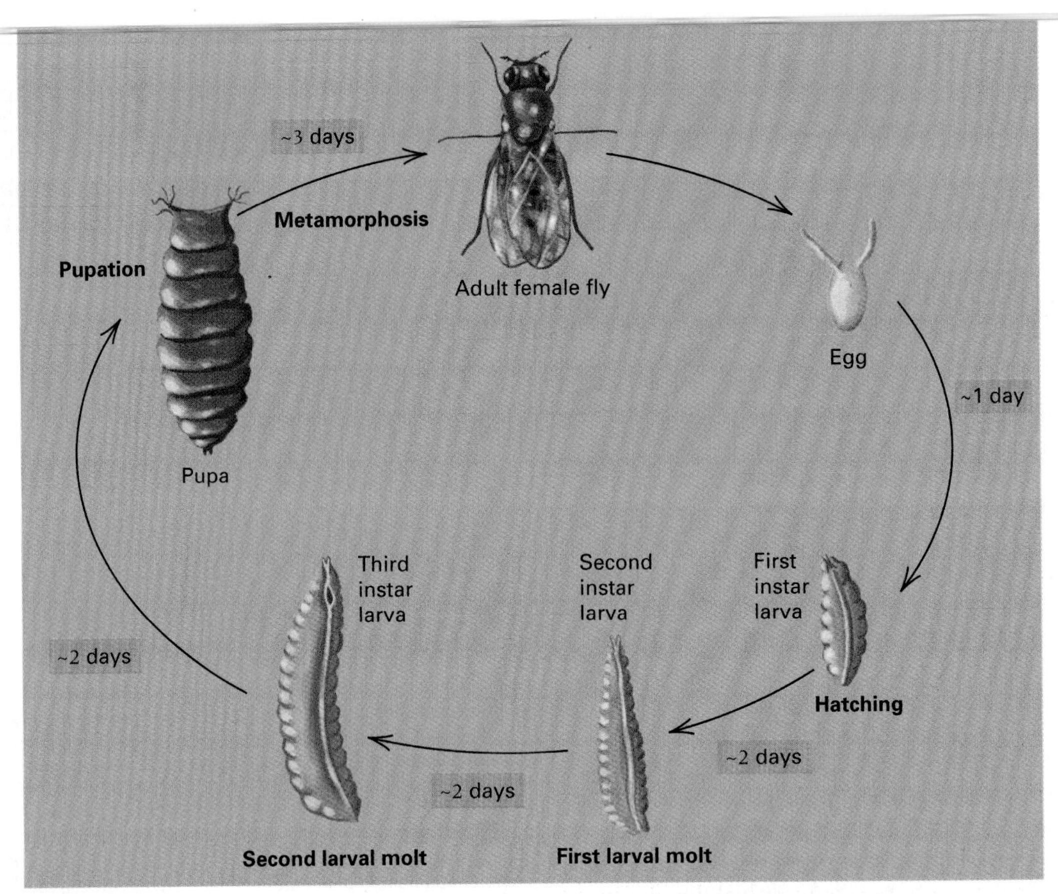


**FIGURE 13.17** Developmental program of *Drosophila melanogaster*. The durations of the stages are at 25°C.

arranged with the anterior end at the left and the ventral side down. The stages are arranged chronologically from 1 through 12. Stages 1–5 represent the early cleavage divisions, beginning with the newly laid fertilized egg (stage 1) and ending with the cellular blastoderm (stage 5) about 3.25 hours later. Gastrulation (stages 6–8) is followed by formation of a characteristic pattern of segments along the body, ending with stage 12 about 9 hours after fertilization. Early development in *Drosophila* takes place within the egg case, but in these photographs the egg case has been removed.

A key issue in developmental genetics is to identify the genes and the interactions among the genes that drive the pattern of differentiation in Figure 13.18. This is done through studies of mutants that show abnormalities in development. In order to discuss these mutants, we first need to examine the cellular events that take place in the formation of the cellular blastoderm, and then show how the geometry of the cellular blastoderm is related to the morphology of the adult animal.

The cellular events in stages 1–5 in Figure 13.18 are diagrammed in **FIGURE 13.19**. In these diagrams the egg case is shown. Part A corresponds to a newly laid fertilized egg in which the male and female nuclei have fused (black dot). The first nine mitotic divisions occur in rapid succession, without division of the cytoplasm, and produce a cluster of nuclei within the egg (stage 2). The nuclei migrate to the periphery, and the germ line is formed from about 10 **pole cells** set off at the posterior end (stage 3); the pole cells undergo two additional divisions and are reincorporated into the embryo by invagination. The nuclei within the embryo undergo four more mitotic divisions without division of the cytoplasm, forming the **syncytial blastoderm**, which contains about 6000 nuclei (stage 4). Cellularization of the blastoderm takes place from about 150 to 195 minutes after fertilization by the synthesis of membranes that separate the nuclei. The **blastoderm** formed by cellularization (stage 5) is a flattened hollow ball of cells that corresponds to the blastula in higher animals.

The ablation of patches of cells within a *Drosophila* blastoderm results in localized defects in the larva and adult. The spatial correlation between the position of the cells destroyed and the type of defects results in a **fate map** of the blastoderm, which specifies the cells in the blas-

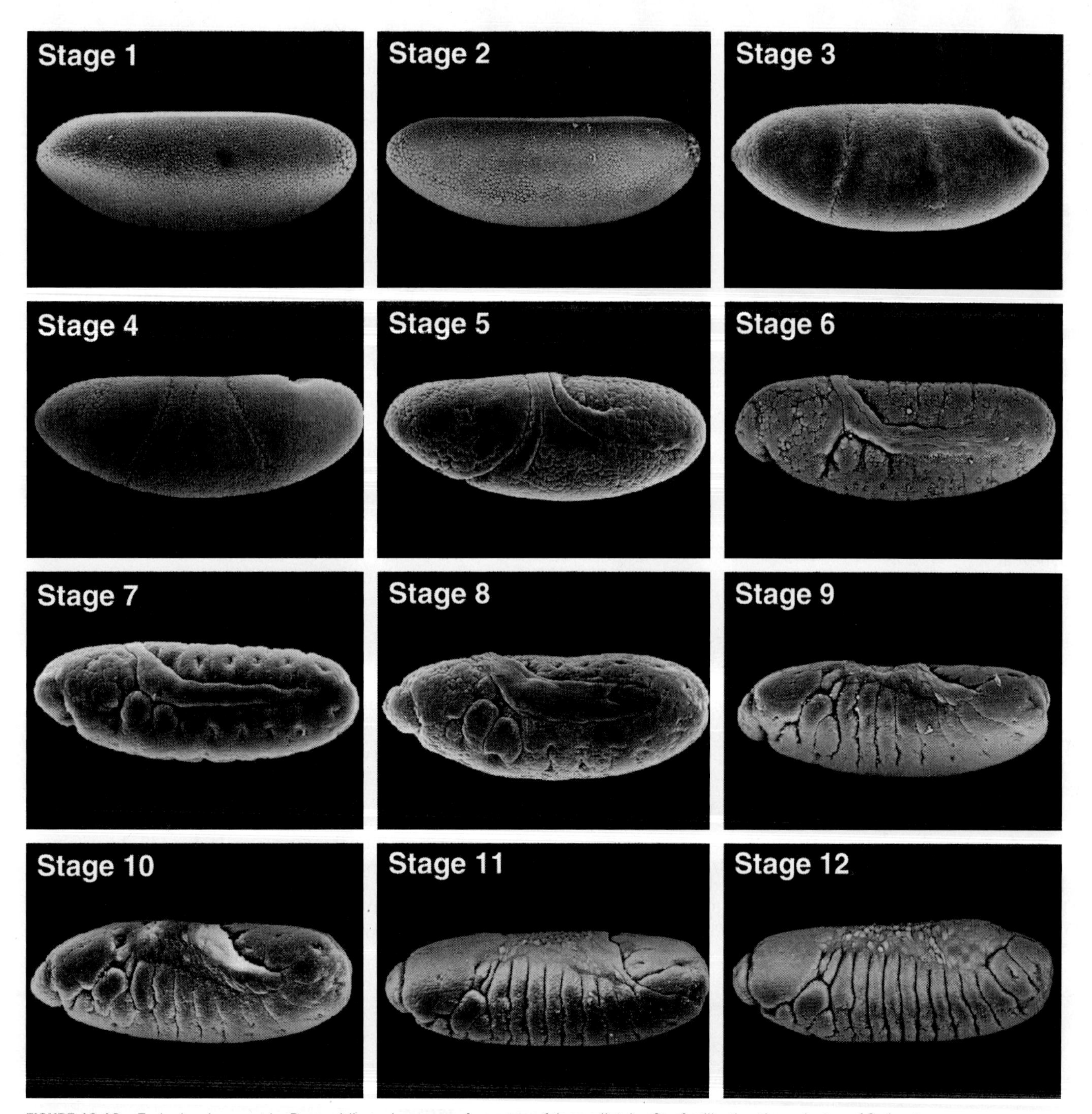


**FIGURE 13.18** Early development in *Drosophila melanogaster* from stage 1 immediately after fertilization through stage 12 about 9 hours after fertilization. [Courtesy of Thomas Kaufman and F. Rudolf Turner, Indiana University at Bloomington.]

toderm that give rise to the various larval and adult structures. Use of genetic markers in the blastoderm has made possible further refinement of the fate map (**FIGURE 13.20**). Cell lineages can be genetically marked during development by inducing recombination between homologous chromosomes in mitosis, which results in genetically different daughter cells (Chapter 5). Much like the cells in the early blastula of *Caenorhabditis*, cells in the blastoderm of *Drosophila* have predetermined developmental fates, with little ability to substitute in development for other, sometimes even adjacent, cells. Evidence for this conclusion comes from experiments in which cells from a genetically marked blastoderm are implanted into host blastoderms. Blastoderm cells implanted into the equivalent regions of the host become part of the normal adult structures. However, blastoderm cells implanted into different regions develop autonomously and are not integrated into host structures.

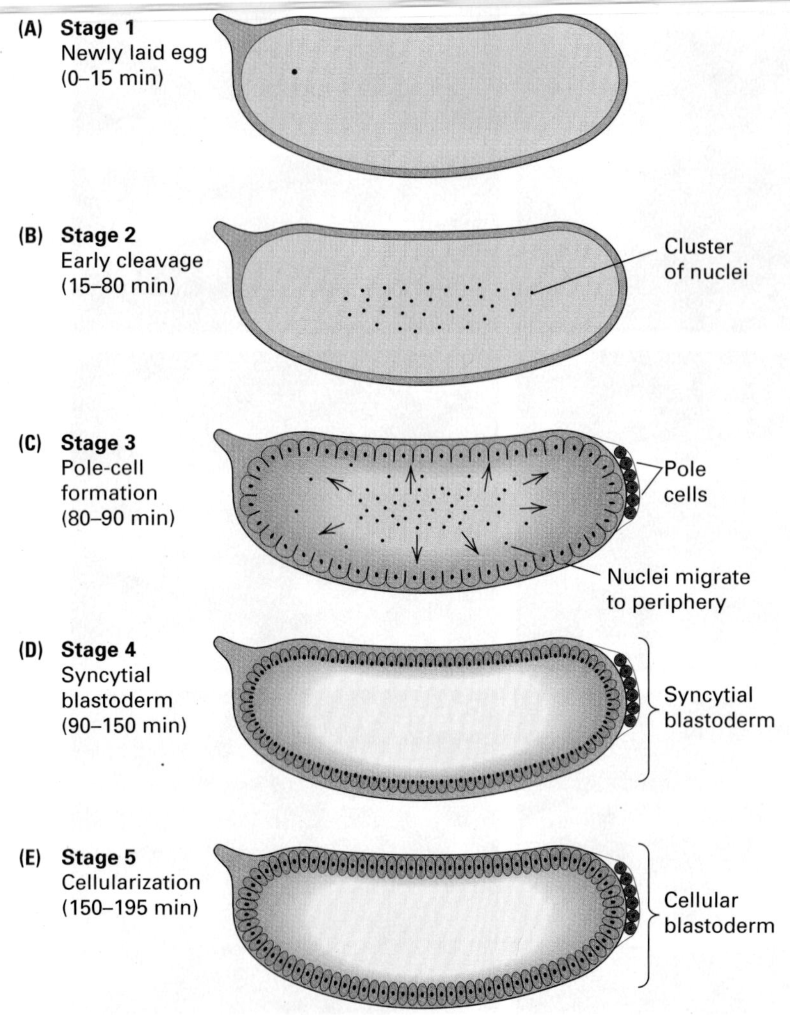


**FIGURE 13.19** Early development in *Drosophila.* (A) The nucleus in the fertilized egg. (B) Mitotic divisions take place synchronously within a syncytium. (C) Some nuclei migrate to the periphery of the embryo, and at the posterior end, the pole cells (which form the germ line) become cellularized. (D) Additional mitotic divisions occur within the syncytial blastoderm. (E) Membranes are formed around the nuclei, giving rise to the cellular blastoderm.

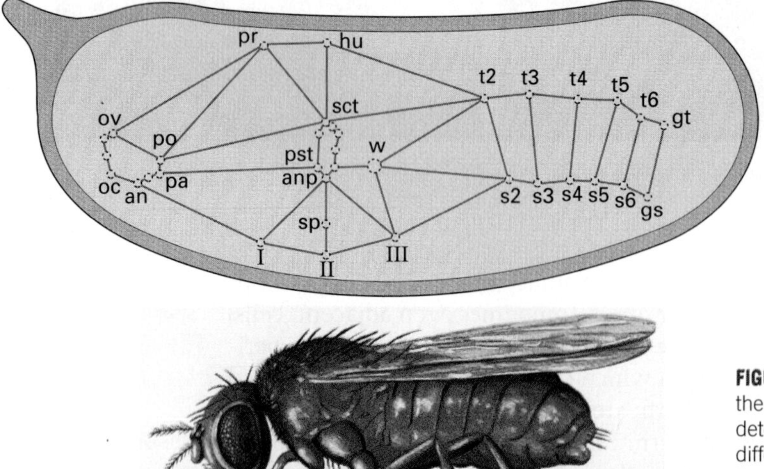


**FIGURE 13.20** Fate map of the *Drosophila* blastoderm, which shows the adult structures that derive from various parts. The map was determined by correlating the expression of genetic markers in different adult structures in genetic mosaics. The abbreviations stand for various body parts in the adult fly. For example, ov and oc are head structures; w is the wing; I, II, and III are the first, second, and third legs; and gs and gt are genital structures. [Adapted from J. C. Hall, W. M. Gelbart, and D. R. Kankel. (M. Ashburner and E. Novitski, eds.). *The Genetics and Biology of Drosophila, vol. 1a.* Academic Press (1976): 265–314.]

Because of the relatively high degree of determination in the blastoderm, genetic analysis of *Drosophila* development has tended to focus on the early stages of development, when the basic body plan of the embryo is established and key regulatory processes become activated. The following sections summarize the genetic control of these early events.

## ■ Maternal-Effect Genes and Zygotic Genes

Early development in *Drosophila* requires translation of maternal mRNA molecules present in the oocyte. Blockage of protein synthesis during this period arrests the early cleavage divisions. Expression of the zygote genome is also required, but the timing is different. Blockage of transcription of the zygote genome at any time after the ninth cleavage division prevents formation of the blastoderm.

Because the earliest stages of *Drosophila* development are programmed in the oocyte, mutations that affect oocyte composition or structure can upset development of the embryo. Genes that function in the mother and are needed for development of the embryo are called **maternal-effect genes**, and developmental genes that function in the embryo are called **zygotic genes**. The interplay between the two types of genes is as follows:

> The zygotic genes interpret and respond to the positional information laid out in the egg by the maternal-effect genes.

Mutations in maternal-effect genes result in a phenotype in which homozygous females produce eggs that are unable to support normal embryonic development, whereas homozygous males produce normal sperm. Therefore, reciprocal crosses give dramatically different results. For example, a recessive maternal-effect mutation, *m,* will yield the following results in reciprocal crosses:

$$m/m \text{ ♀} \times +/+ \text{ ♂} \rightarrow +/m \text{ progeny}$$
(abnormal development)

$+/+$ ♀ $\times$ $m/m$ ♂ $\rightarrow$ $+/m$ progeny
(normal development)

The $+/m$ progeny of the reciprocal crosses are genetically identical, but development is upset when the mother is homozygous $m/m$.

The reason why maternal-effect genes are needed in the mother is that the maternal-effect genes establish the polarity of the *Drosophila* oocyte even before fertilization takes place. They are active during the earliest stages of embryonic development, and they determine the basic body plan of the embryo. Maternal-effect mutations provide a valuable tool for investigating the genetic control of pattern formation and for identifying the molecules that are important in morphogenesis.

## Genetic Basis of Pattern Formation in Early Development

The *Drosophila* larva features fourteen superficially similar repeating units visible as a pattern of stripes along the main trunk (**FIGURE 13.21**). The

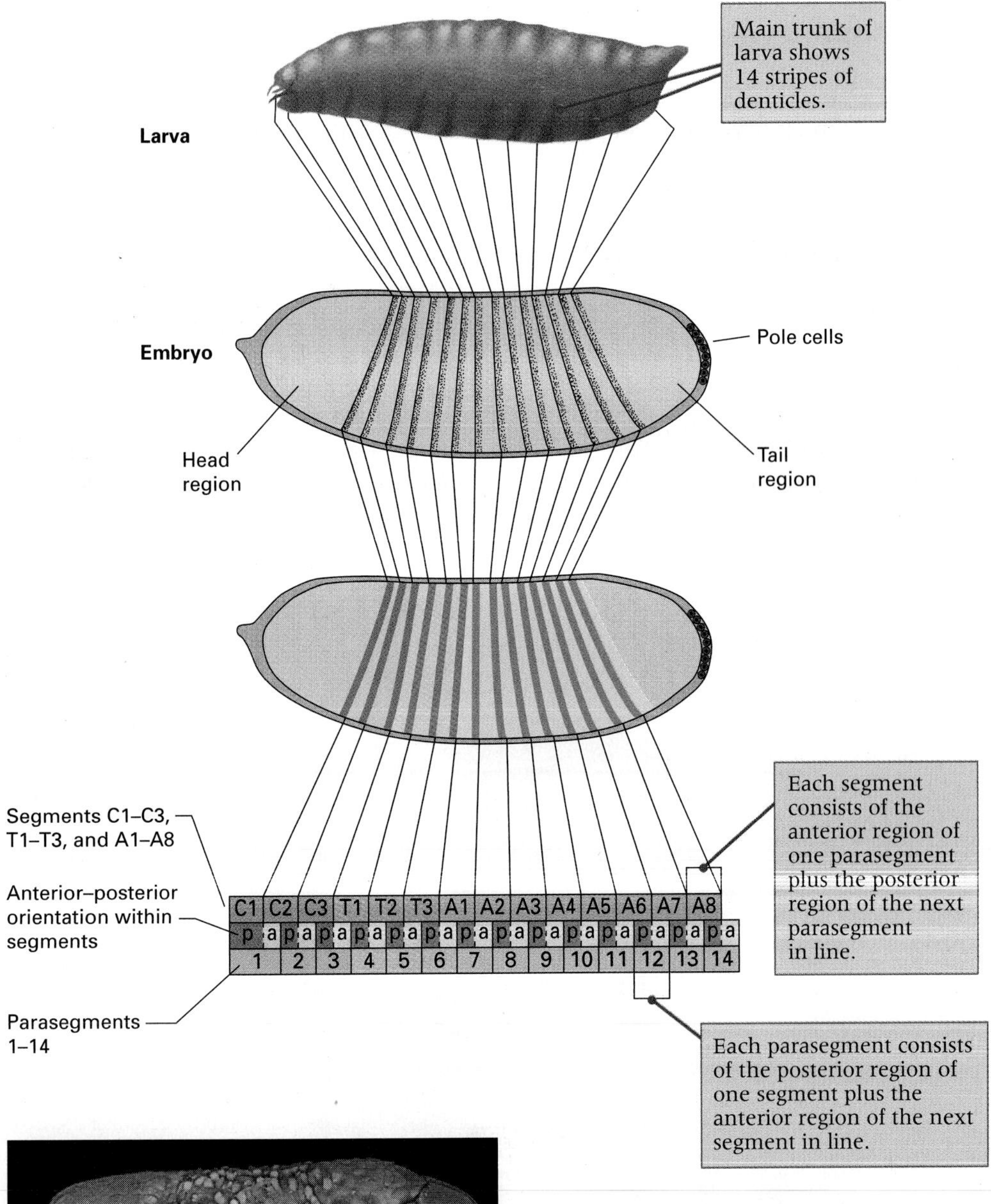


**FIGURE 13.21** Segmental organization of the *Drosophila* embryo and larva. The segments are defined by successive indentations formed by the sites of muscle attachment in the larval cuticle. The parasegments are not apparent morphologically but include the anterior and posterior regions of adjacent segments. The photograph shows the segments C1–C3, T1–T3, and A1–A8 in a stage 12 embryo about 9 hours after fertilization. [Photo courtesy of Thomas Kaufman and F. Rudolf Turner, Indiana University at Bloomington.]

stripes can be recognized externally by the bands of *denticles*, which are tiny, pigmented, tooth-like projections from the surface of the larva. The fourteen stripes in the embryo correspond to the segments in the larva that forms from the embryo. Each **segment** is defined morphologically as the region between successive indentations formed by the sites of muscle attachment in the larval cuticle. The designations of the segments are indicated in Figure 13.21. There are three head segments (C1–C3), three thoracic segments (T1–T3), and eight abdominal segments (A1–A8). In addition to the segments, another type of repeating unit is also important in development. These repeating units are called **parasegments**; each parasegment consists of the posterior region of one segment and the anterior region of the adjacent segment. Parasegments have a transient existence in embryonic development. Although they are not visible morphologically, they are important in gene expression because the patterns of expression of many genes coincide with the boundaries of the parasegments rather than with the boundaries of the segments.

The early stages of pattern formation are determined by genes that are often called **segmentation genes** because they determine the origin and fate of the segments and parasegments. There are four classes of segmentation genes, which differ in their times and patterns of expression in the embryo.

1. The ***coordinate genes*** determine the principal coordinate axes of the embryo: the anterior–posterior axis, which defines the front and rear, and the dorsal–ventral axis, which defines the top and bottom.
2. The ***gap genes*** are expressed in contiguous groups of segments along the embryo (**FIGURE 13.22**, part A), and they establish the next level of spatial organization. Mutations in gap genes result in the absence of contiguous body segments, so gaps appear in the normal pattern of structures in the embryo.
3. The ***pair-rule genes*** determine the separation of the embryo into discrete segments (Figure 13.22B). Mutations in pair-rule genes result in missing pattern elements in alternate segments. The reason for the two-segment periodicity of pair-rule genes is that the genes are expressed in a zebra-stripe pattern along the embryo.
4. The ***segment-polarity genes*** determine the pattern of anterior–posterior development within each segment of the embryo (Figure 13.22C). Mutations in segment-polarity genes affect all segments or parasegments in which the

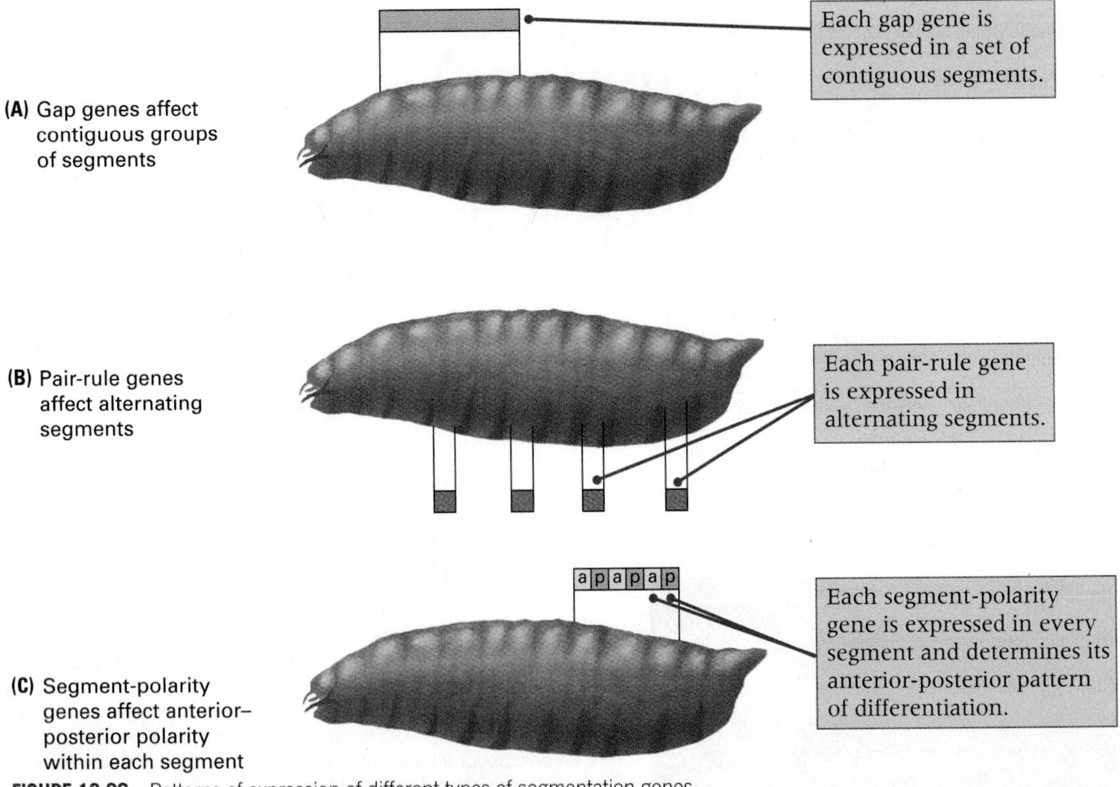

Each gap gene is expressed in a set of contiguous segments.
(A) Gap genes affect contiguous groups of segments
(B) Pair-rule genes affect alternating segments
Each pair-rule gene is expressed in alternating segments.
a p a p a p
Each segment-polarity gene is expressed in every segment and determines its anterior-posterior pattern of differentiation.
(C) Segment-polarity genes affect anterior–posterior polarity within each segment

**FIGURE 13.22** Patterns of expression of different types of segmentation genes.

normal gene is active. Many segment-polarity mutations have the normal number of segments, but part of each segment is deleted and the remainder is duplicated in mirror-image symmetry.

Evidence for the existence of these four classes of segmentation genes is presented in the following sections.

## Coordinate Genes

The **coordinate genes** are maternal-effect genes that establish early polarity through the presence of their products at defined positions within the oocyte or through gradients of concentration of their products. As illustrated in **FIGURE 13.23**, the genes that determine the anterior–posterior axis can be classified into three groups—*anterior, posterior,* and *terminal*—according to the morphology of mutant phenotypes.

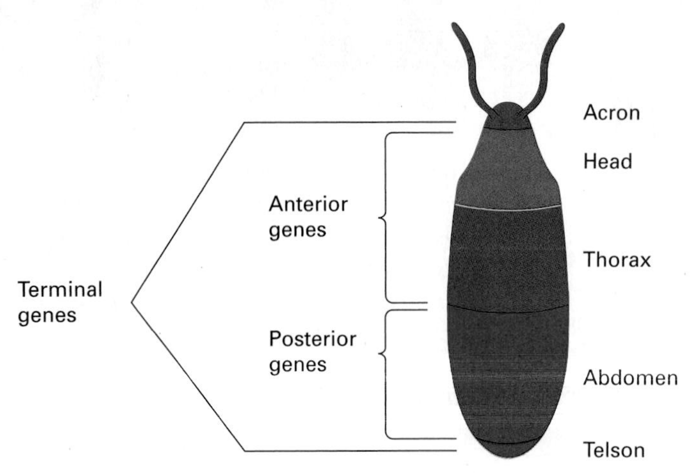


**FIGURE 13.23** Regional differentiation of the early *Drosophila* embryo along the anterior–posterior axis. Mutations in any of the classes of genes shown result in elimination of the corresponding region of the embryo.

- The *anterior genes* affect the head and thorax. The key gene in this class is *bicoid*. Mutations in *bicoid* produce embryos that lack the head and thorax and occasionally have abdominal segments in reverse polarity duplicated at the anterior end. The *bicoid* phenotype resembles that produced by certain kinds of surgical manipulations. For example, when *Drosophila* eggs are punctured and small amounts of cytoplasm allowed to escape, loss of cytoplasm from the anterior end results in embryos in which some posterior structures develop in place of the head. Similarly, replacement of anterior cytoplasm with posterior cytoplasm by injection yields embryos with two mirror-image abdomens and no head.

The *bicoid* gene product is a transcription factor for genes that determine anterior structures. Because the *bicoid* mRNA is localized in the anterior part of the early-cleavage embryo, these genes are activated primarily in the anterior region. The *bicoid* mRNA is produced in nurse cells (the cells surrounding the oocyte in Figure 13.7) and exported to a localized region at the anterior pole of the oocyte. The protein product is less localized and, during the syncytial cleavages, forms an anterior–posterior concentration gradient with the maximum at the anterior tip of the embryo (**FIGURE 13.24**). The bicoid protein is a principal morphogen in determining the blastoderm fate map. The protein is a transcriptional activator that contains a helix–turn–helix motif for DNA binding (Chapter 11). Genes affected by the bicoid protein contain multiple upstream binding domains that consist of nine nucleotides resembling the consensus sequence 5′-TCTAATCCC-3′. Binding sites that differ by as many as two base pairs from the consensus sequence can bind the bicoid protein with high affinity, and sites that contain four mismatches bind with low affinity. The combination of high- and low-affinity binding sites determines the concentration of bicoid protein needed for gene activation; genes with many high-affinity binding sites can be activated at low concentrations, but those with many low-affinity binding sites need higher concentrations. Such differences in binding affinity mean that the level of gene expression can differ from one regulated gene to the next along the bicoid concentration gradient. One of the important genes activated by *bicoid* is the gap gene *hunchback*. Other genes in the anterior class code for cellular components necessary for *bicoid* localization.

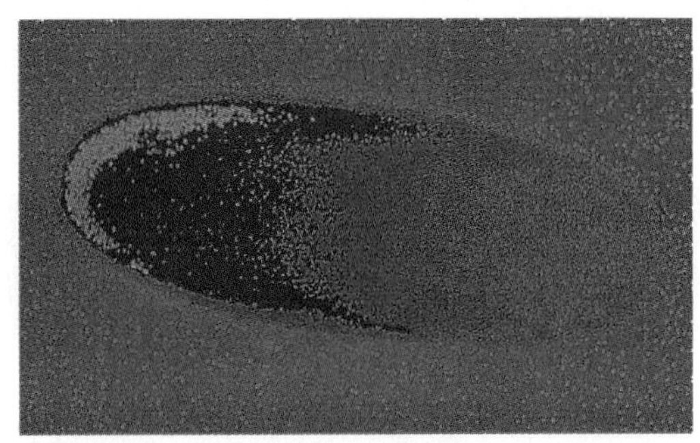

**FIGURE 13.24** A gradient of gene expression resembling that of *bicoid* in the early *Drosophila* embryo. In this photograph, the intensity of the fluorescent signal has been pseudocolored so that the region of highest concentration is pink and the region of lowest concentration is green. [Courtesy of James Langeland, Sean Carroll, and Stephen Paddock, University of Wisconsin at Madison.]

- The *posterior genes* affect the abdominal segments (Figure 13.23). Some of the mutants also lack pole cells. One of the posterior mutations, *nanos,* yields embryos with defective abdominal segmentation but normal pole cells, which resemble abnormalities produced by sur-

## connection
### Embryo Genesis

**Christiane Nüsslein-Volhard and Eric Wieschaus 1980**
European Molecular Biology Laboratory, Heidelberg, Germany
*Mutations Affecting Segment Number and Polarity in* Drosophila

*Nüsslein-Volhard and Wieschaus were exceptionally bold in supposing that the molecular mechanisms governing a process as complex as early embryonic development could be understood by the genetic and molecular analysis of mutations. The phenotype of such mutants is superficially identical—the embryo dies. The* Drosophila *genetic map was already littered with mutations classified collectively as "recessive lethals." These were generally considered as not amenable to further analysis because, in any particular case, the search for the specific defect was regarded as a needle-in-a-haystack problem. Nüsslein-Volhard and Wieschaus ignored most of the existing mutants. They set out to acquire systematically a new set of recessive-lethal mutants, each showing a specific and characteristic type of defect in the formation of organized patterns in the early embryo. Their first efforts, reported in this paper, yielded a number of mutations in each of three major classes of genes concerned with development. The paper sparked an enormous interest in* Drosophila *developmental genetics. Today, a typical Annual Drosophila Research Conference includes approximately 500 presentations (mainly posters) dealing with aspects of* Drosophila *development. Nüsslein-Volhard and Wieschaus were awarded a Nobel Prize in 1995. They shared it with Edward B. Lewis for his pioneering genetic studies of the homeotic genes.*

The construction of complex form from similar repeating units is a basic feature of spatial organization in all higher animals. Very little is known for any organism about the genes involved in this process. In *Drosophila*, the metameric [repeating] nature of the pattern is most obvious in the thoracic and abdominal segments of the larval epidermis and we are attempting to identify all loci required for the establishment of this pattern. . . . We have undertaken a systematic search for mutations that affect the segmental pattern. We describe here mutations at 15 loci which show one of three novel types of pattern alteration: pattern duplication (segment polarity mutants; six loci), pattern deletion in alternating segments (pair-rule mutants; six loci) and deletion of a group of adjacent segments (gap mutants; three loci). . . . Segment polarity mutants have the normal number of segments. However, in each segment a defined fraction of the normal pattern is deleted and the remainder is present as a mirror-image duplication. The duplicated part is posterior to the 'normal' part and has reversed polarity. . . . In pair-rule mutants homologous parts of the pattern are deleted in every other segment. Each of the six loci is characterized by its own pattern of deletions. . . . One of the striking features of the [segment-polarity and pair-rule] classes is that the alteration in the pattern is repeated at specific intervals along the antero-posterior axis of the embryo. No such repeated pattern is found in mutants of the gap class and instead a group of up to eight adjacent segments is deleted. . . . The lack of a repeated pattern suggests that the loci are involved in processes in which position along the antero-posterior axis of the embryo is defined by unique values. . . . The majority of mutants described here have been isolated in systematic searches for mutations affecting the segmentation pattern. These experiments are still incomplete. . . . In *Drosophila*, it would seem feasible to identify all genetic components involved in the complex process of embryonic pattern formation.

***In* Drosophila*, it would seem feasible to identify all genetic components involved in the complex process of embryonic pattern formation.***

Source: C. Nüsslein-Volhard and E. Wieschaus, *Nature* 287 (1980): 795–801.

gical removal of the posterior cytoplasm. The phenotype does not result merely from a generalized disruption of development at the posterior end, because the pole cells (as well as a posterior structure called the telson that normally develops between the pole cells and the abdomen) are not affected in either *nanos* or the surgically manipulated embryos. The *nanos* mRNA is localized tightly to the posterior pole of the oocyte, and the gene product is a repressor of translation. Among the mRNAs not translated in the presence of nanos protein is the *hunchback* mRNA. Hence *hunchback* expression is controlled jointly by the bicoid and nanos proteins; bicoid protein activates transcription in an anterior–posterior gradient, and nanos protein represses translation in the posterior region.

- The *terminal genes* affect both the most anterior structure (the acron) and the most posterior structure (the telson) (Figure 13.23). The key gene in this class is *torso*, which codes for a transmembrane receptor that is uniformly distributed throughout the embryo in the early developmental stages. The torso receptor is activated by a signal released only at the poles of the egg by the nurse cells in that location. The torso receptor is a tyrosine kinase that initiates cellular differentiation by means of phosphorylation of specific tyrosine residues in one or more target proteins.

Apart from the three sets of genes that determine the anterior–posterior axis of the embryo, a fourth set of genes determines the dorsal–ventral axis. The morphogen for dorsal–ventral determination is the product of the gene *dorsal*, which is present in a pronounced ventral-to-dorsal gradient in the late syncytial blastoderm. The dorsal protein is a transcription factor. More than a dozen other genes are known to affect dorsal–ventral determination. Mutations in these genes eliminate ventral and lateral pattern elements. In many cases, the mutant embryos can be rescued by the injection of wildtype cytoplasm, no matter where the wildtype cytoplasm is taken from or where it is injected. Examples include the genes *snake*, *gastrulation-defective*, and *easter*. All three genes code for proteins called *serine proteases*. Serine proteases are synthesized as inactive precursors that require a specific cleavage for activation. They often act in a temporal sequence, which means that activation of one enzyme in the pathway is necessary for activation of the next enzyme in line. The serial activation of the enzymes results in a *cascade effect* that greatly amplifies an initial signal. Each step in the cascade multiplies the signal produced in the preceding step, like a nuclear chain reaction in which each event triggers the occurrence of multiple new events at the next instant in time.

## Gap Genes

The main role of the coordinate genes is to regulate the expression of a small group of approximately six gap genes along the anterior–posterior axis. The genes are called **gap genes** because mutations in them result in the absence of pattern elements derived from a group of contiguous segments (Figure 13.22A). Gap genes are zygotic genes. The gene *hunchback* serves as an example of the class because *hunchback* expression is controlled by offsetting effects of *bicoid* and *nanos*. Transcription of *hunchback* is stimulated in an anterior-to-posterior gradient by the bicoid transcription factor, but posterior *hunchback* expression is prevented by translational repression because of the posteriorly localized nanos protein. In the early *Drosophila* embryo in **FIGURE 13.25**, the gradient of *hunchback* expression is indicated by the green fluorescence of an antibody specific to the *hunchback* gene product. The superimposed red

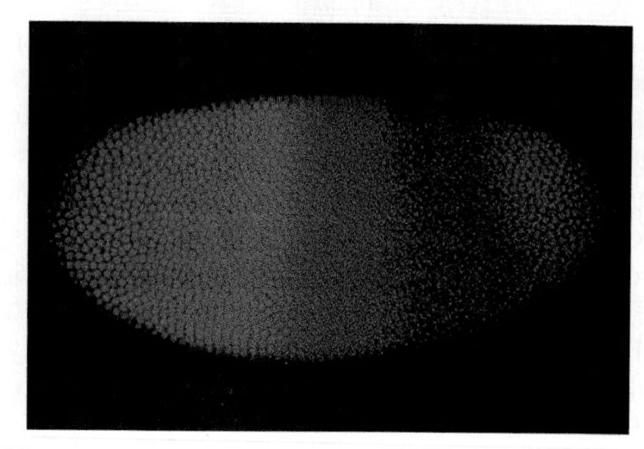

**FIGURE 13.25** An embryo of *Drosophila*, approximately 2.5 hours after fertilization, showing the regional localization of the *hunchback* gene product (green), the *Krüppel* gene product (red), and their overlap (yellow). [Courtesy of James Langeland, Sean Carroll, and Stephen Paddock, University of Wisconsin at Madison.]

fluorescence results from antibody specific to the product of *Krüppel*, another gap gene. The region of overlapping gene expression appears in yellow. The products of both *hunchback* and *Krüppel* are transcription factors of the zinc-finger type (Chapter 11). Other gap genes also encode transcription factors. Together, the gap genes have a pattern of regional specificity and partly overlapping domains of expression that enable them to act in combinatorial fashion to control the next set of genes in the segmentation hierarchy, the pair-rule genes.

## Pair-Rule Genes

The coordinate and gap genes determine the polarity of the embryo and establish broad regions within which subsequent development takes place. As development proceeds, the progressively more refined organization of the embryo is correlated with the patterns of expression of the segmentation genes. Among these are the **pair-rule genes**, in which the mutant phenotype has alternating segments absent or malformed (Figure 13.22B). Approximately eight pair-rule genes have been identified. For example, mutations of the pair-rule gene *even-skipped* affect even-numbered segments, and those of another pair-rule gene, *odd-skipped*, affect odd-numbered segments. The function of the pair-rule genes is to give the early larva a segmented body pattern with both repetitiveness and individuality of segments. For example, there are eight abdominal segments that are repetitive in that they are regularly spaced and share several common features, but they differ in the details of their differentiation.

One of the earliest pair-rule genes expressed is *hairy*, whose pattern of expression is under both positive and negative regulation by the products of *hunchback, Krüppel,* and other gap genes. Expression of *hairy* yields seven stripes (**FIGURE 13.26**). The striped pattern of pair-rule gene expression is typical, but the stripes of expression of one gene are usually slightly out of register with those of another. Together with the continued regional expression of the gap genes, the combinatorial patterns of gene expression in the embryo are already complex and linearly differentiated. **FIGURE 13.27** shows an embryo stained for the products of three genes: *hairy* (green), *Krüppel* (red), and *giant* (blue). The regions of overlapping expression appear as color mixtures—orange, yellow, light green, or blue. Even at the early stage in Figure 13.27, there is a unique combinatorial pattern of gene expression in every segment and parasegment. The complexity of combinatorial control can be appreciated by considering that the expression of the *hairy* gene in stripe 7 depends on a promoter element smaller than 1.5 kb that contains a series of binding sites for the protein products of the genes *caudal, hunchback, knirps, Krüppel, tailless, huckebein, bicoid,* and perhaps still other proteins yet to be identified. The combinatorial patterns of gene expression of the pair-rule genes define the boundaries of expression of the segment-polarity genes, which function next in the hierarchy.

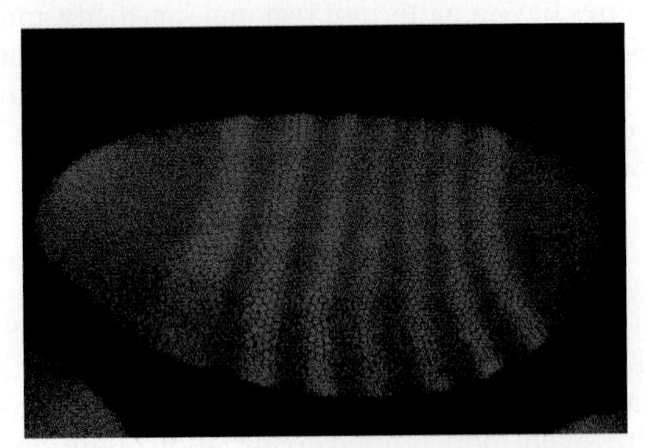

**FIGURE 13.26** Characteristic seven stripes of expression of the gene *hairy* in a *Drosophila* embryo approximately 3 hours after fertilization. [Courtesy of James Langeland, Sean Carroll, and Stephen Paddock, University of Wisconsin at Madison.]

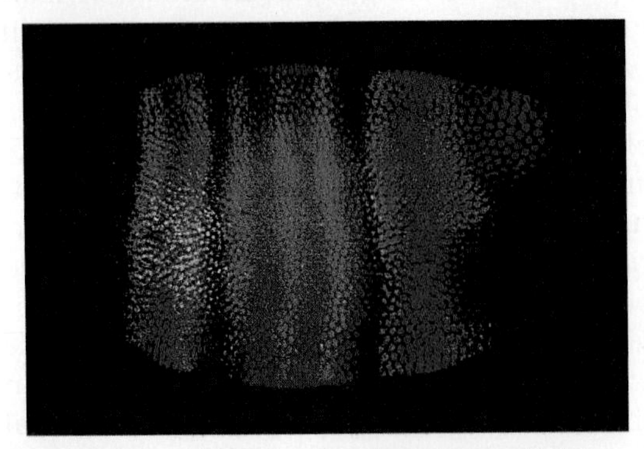

**FIGURE 13.27** Combined patterns of expression of *hairy* (green), *Krüppel* (red), and *giant* (blue) in a *Drosophila* embryo approximately 3 hours after fertilization. Already, considerable linear differentiation is apparent in the patterns of gene expression. [Courtesy of James Langeland, Sean Carroll, and Stephen Paddock, University of Wisconsin at Madison.]

## ■ Segment-Polarity Genes

Whereas the pair-rule genes determine the body plan at the level of segments and parasegments, the **segment-polarity genes** create a spatial differentiation within each segment. Approximately fourteen segment-polarity genes have been identified. The mutant phenotype has repetitive deletion of pattern along the embryo (Figure 13.22C) and usually a mirror-image duplication of the part that remains. Among the earliest segment-polarity genes expressed is *engrailed,* whose stripes of expression approximately coincide with the boundaries of the parasegments and so divide each segment into anterior and posterior domains (**FIGURE 13.28**).

The expression of the segment-polarity genes finally establishes the early polarity and linear differentiation of the embryo into segments and parasegments.

## ■ Interactions in the Regulatory Hierarchy

Genes in the regulatory hierarchy are controlled by a complex set of interactions that ensure an orderly progression through the molecular events of development. Interactions among some of the coordinate genes, gap genes, pair-rule genes, and segment-polarity genes are shown in **FIGURE 13.29**. The green connectors indicate stimulatory effects, and the red connectors indicate inhibitory effects. Many of these interactions were originally discovered from genetic analysis using the principle of epistasis discussed in Section 13.3. Most of the genes are controlled by a complex set of stimulatory and

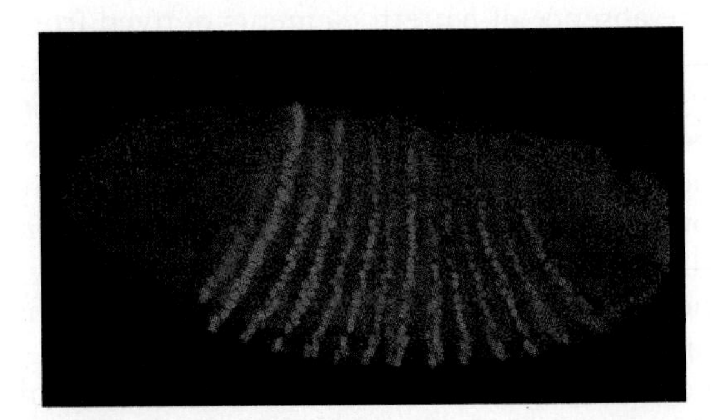

**FIGURE 13.28** Expression of the segment-polarity gene *engrailed* partitions the early *Drosophila* embryo into 14 regions. These eventually differentiate into three head segments, three thoracic segments, and eight abdominal segments. [Courtesy of James Langeland, Sean Carroll, and Stephen Paddock, University of Wisconsin at Madison.]

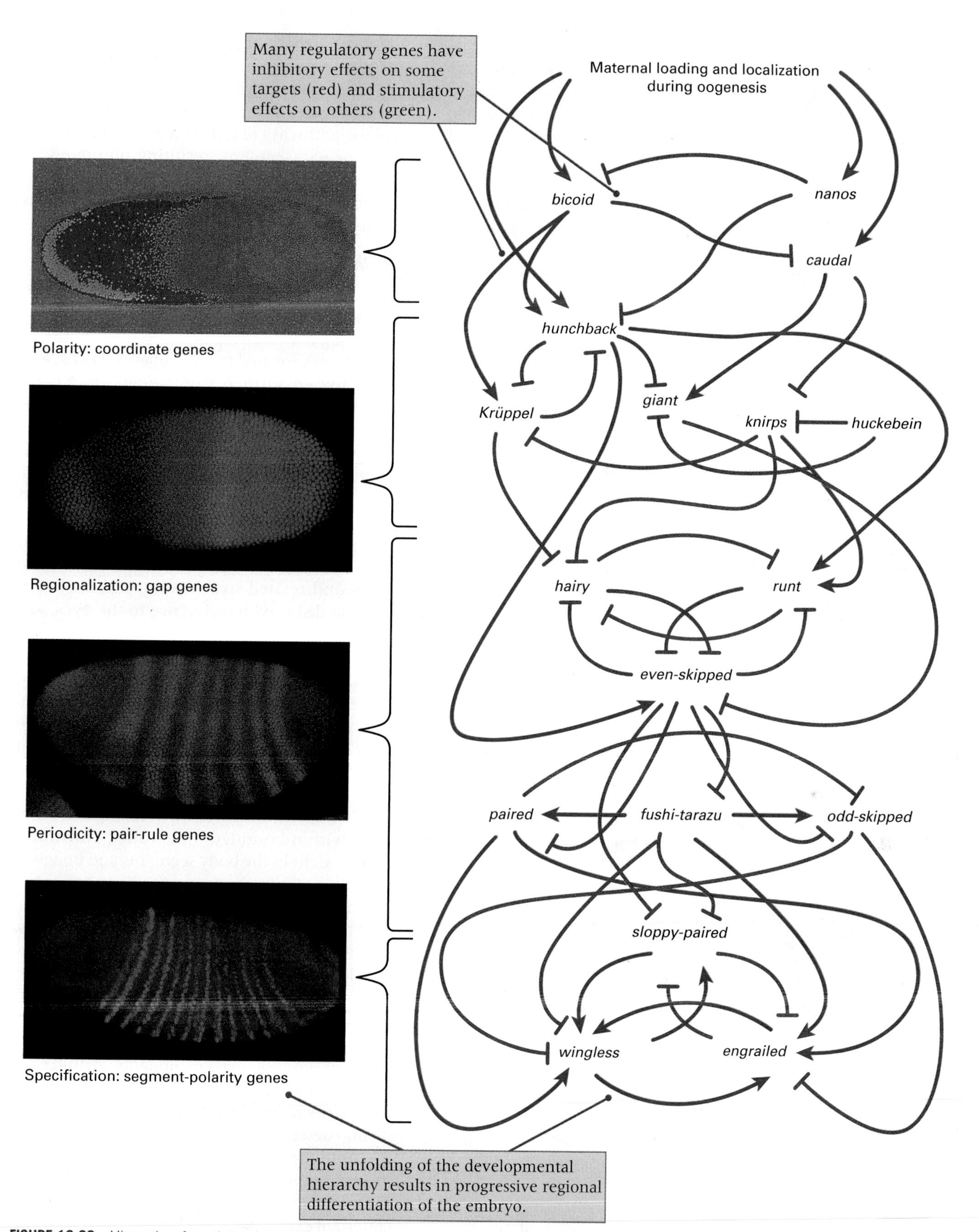


**FIGURE 13.29** Hierarchy of regulatory interactions among genes controlling early development in *Drosophila*. [Photos courtesy of James Langeland, Sean Carroll, and Stephen Paddock, University of Wisconsin at Madison; Illustration courtesy of George von Dassow, Center for Cell Dynamics, University of Washington.]

inhibitory effects acting together. The coordinate genes act first to establish the polarity of the embryo, then the gap genes to differentiate large regions, after which the pair-rule genes establish the periodicity of the embryo indicated by the zebra stripes, and finally the segment-polarity genes act in the specification of the developmental identity and fate of each of the body segments. At each level in the regulatory hierarchy, the genes act to regulate other genes expressed at the same level, and also act to regulate the activity of genes that are expressed in the next downstream level in the hierarchy.

The segment-polarity genes also act to regulate downstream developmental genes that control the pathways of differentiation in each segment or parasegment, resulting ultimately in the morphology of the adult fly. The metamorphosis of the adult fly and how it emerges is discussed next.

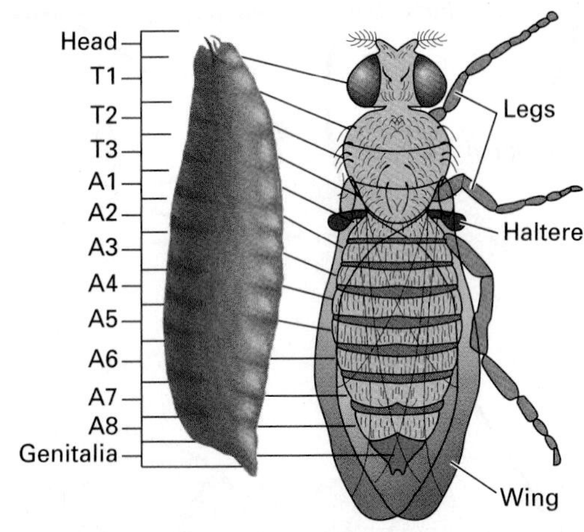


**FIGURE 13.30** Relationship between larval and adult segmentation in *Drosophila*. Each of the three thoracic segments in the adult carries a pair of legs. The wings develop on the second thoracic segment (T2) and the halteres (flight balancers) on the third thoracic segment (T3).

## ■ Metamorphosis of the Adult Fly

As with many other insects, the larvae and adults of *Drosophila* have a segmented body plan consisting of a head formed from segments C1–C3, a thorax formed from segments T1–T3, and an abdomen formed from segments A1–A8 (**FIGURE 13.30**). Metamorphosis makes use of about twenty structures called **imaginal disks** present inside the larvae (**FIGURE 13.31**). Formed early in development, the imaginal disks ultimately give rise to the principal structures and tissues in the adult organism. Examples of imaginal disks are the pair of wing disks (one on each side of the body), which give rise to the wings and related structures; the pair of eye-antenna disks, which give rise to the eyes, antennae, and related structures; and the genital disk, which gives rise to the reproductive apparatus. During the pupal stage, when many larval tissues and organs break down, the imaginal disks progressively unfold and differentiate into adult structures. The morphogenic events that take place in the pupa are initiated by the hormone *ecdysone*, secreted by the larval brain.

Cell determination in *Drosophila* also takes place within bounded units called **compartments**. Cells in the body segments and imaginal disks do not migrate across the boundaries between compartments. For example, the *Drosophila* wing disk includes five compartment boundaries, and most body segments include one boundary that divides the segment into anterior and posterior halves. The evidence for compartments comes from genetic marking of individual cells by means of mitotic recombination and observation of the positions of their descendants. Within each compartment, neighboring groups of cells not necessarily related by ancestry undergo developmental determination together.

As in the early embryo, overlapping patterns of gene expression and combinatorial control guide later events in *Drosophila* development. The pattern of expression of a key gene in wing development, *vestigial*, in a wing disk is shown in **FIGURE 13.32**, part A. The apparently uniform and approximately circular pattern of expression is actually the summation of *vestigial* response to two

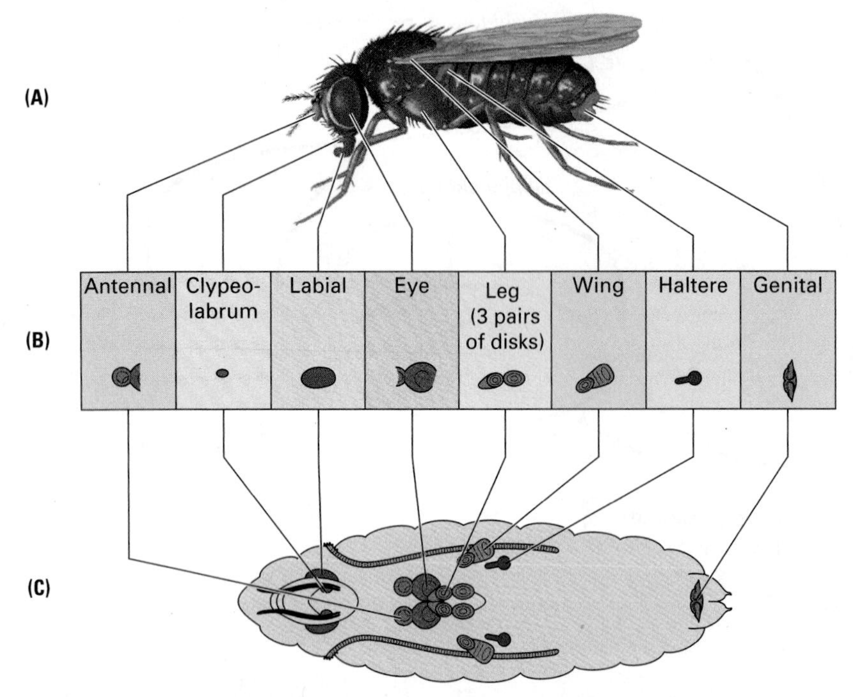


**FIGURE 13.31** Adult *Drosophila* structures (A) and the imaginal disks (B) from which they derive. (C) General morphology and positions of the disks in the late third instar larva.

separate signaling pathways shown in **FIGURE 13.33**, which result in the cross-shaped and four-part patterns of expression shown at the bottom. Separate visualization of these patterns in the wing disk is shown in Figure 13.32B. The signaling pathway A in Figure 13.33 consists of the products of the genes *Apterous, fringe, Serrate,* and so forth; and pathway B consists of the products of the genes *engrailed, hedgehog,* and so forth. The suppressor-of-hairy protein binds to a *boundary*

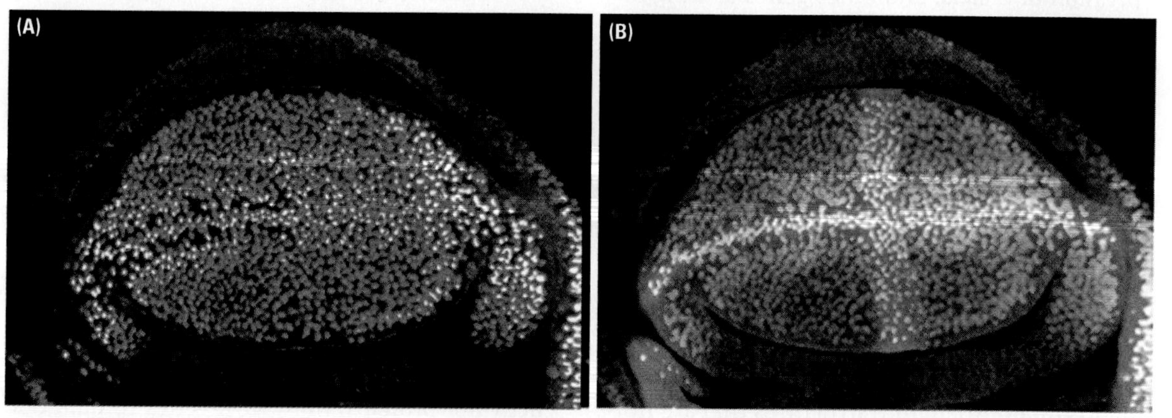


**FIGURE 13.32** (A) Expression of the *vestigial* gene (green) in the developing wing imaginal disk. The approximately circular area of expression gives rise to the wing proper. (B) Visualization of the underlying boundary and quadrant patterns of *vestigial* expression in the same disk. [Courtesy of Jim Williams, Sean Carroll, and Stephen Paddock, University of Wisconsin at Madison.]

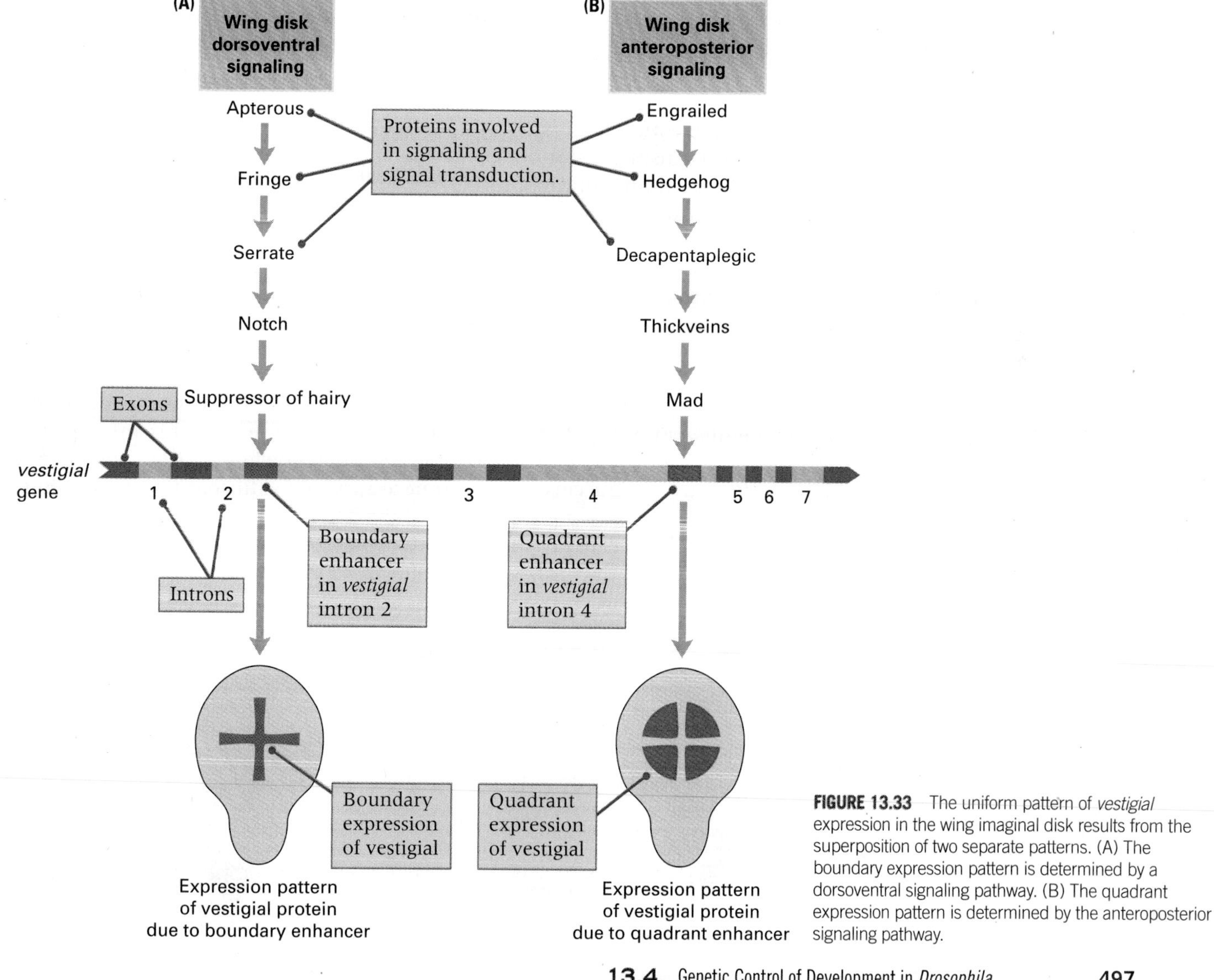


**FIGURE 13.33** The uniform pattern of *vestigial* expression in the wing imaginal disk results from the superposition of two separate patterns. (A) The boundary expression pattern is determined by a dorsoventral signaling pathway. (B) The quadrant expression pattern is determined by the anteroposterior signaling pathway.

*enhancer* in the vestigial gene, which induces gene expression in the cross-shaped pattern. The mad protein binds to a separate *quadrant enhancer*, which induces gene expression in the quadrant pattern. Such overlapping patterns of gene expression of *vestigial* and other genes in wing development ultimately yield the exquisitely fine level of cellular and morphological differentiation observed in the adult animal.

## Homeotic Genes

Among the genes that transform the periodicity of the *Drosophila* embryo into a body plan with linear differentiation are two small sets of **homeotic (*HOX*) genes**. Homeotic genes are defined by mutations that result in the transformation of one body segment into another; the phenotype is that the affected segments differentiate into structures that are normally present elsewhere in the organism. The normal functions of homeotic genes are to regulate target genes concerned with such characteristics as rates of cell division, orientation of mitotic spindles, and the capacity to differentiate bristles, legs, and other features. Homeotic genes are also important in restricting the activities of groups of structural genes to definite spatial patterns. One class of homeotic mutation is illustrated by *bithorax*, which causes transformation of the anterior part of the third thoracic segment into the anterior part of the second thoracic segment, with the result that the halteres normally formed from segment T3 are transformed into a pair of wings in addition to the pair normally formed from segment T2 (**FIGURE 13.34**).

The other class of homeotic mutation is illustrated by *Antennapedia*, which results in transformation of the antennae into legs. The *HOX* genes represented by *bithorax* and *Antennapedia* are in fact gene clusters. The cluster containing *bithorax* is designated BX-C (stands for *bithorax* complex), and that containing *Antennapedia* is called ANT-C (stands for *Antennapedia* complex). Both gene clusters were initially discovered through their homeotic effects in adults. Later they were shown to affect the identity of larval segments. The BX-C is primarily concerned with the development of larval segments T3 through A8 (Figure 13.30), with principal effects in T3 and A1. The BX-C region extends across approximately 300 kb of DNA yet contains only three essential coding regions. The rest of the region appears to consist of a complex series of enhancers and other regulatory elements that function to specify segment identity by activating the different coding regions to different degrees in particular parasegments. The ANT-C is primarily concerned with the development of the head (H) and thoracic segments T1 and T2.

The homeotic genes encode transcriptional activator proteins. Most *HOX* genes contain one or more copies of a characteristic sequence of about 180 nucleotides called a **homeobox**, which is also found in key genes concerned with the development of embryonic segmentation in organisms as diverse as segmented worms, frogs, chickens, mice, and human beings. Homeobox sequences are present in exons and code for a protein-folding domain that includes a helix–turn–helix DNA-binding motif (Chapter 11).

## *HOX* Genes in Evolution

In *Drosophila*, a *HOX* gene called *ultrabithorax* (*Ubx*) is normally active in segment T3, and only in segment T3, which gives rise to the halteres. Reduced *Ubx* expression makes the haltere more wing-like, and in the absence of *Ubx* expression, the haltere disk forms a structurally normal wing (Figure 13.34). Given this information, it is tempting to say that *Ubx* controls the development of the haltere. But this inference is wrong, because it implies that *Ubx* expression is both necessary and sufficient for the development of the haltere. The fact is that *Ubx* homologs are expressed in T3 in probably all insects, but segment T3 carries halteres only in the dip-

**FIGURE 13.34** (A) Wildtype *Drosophila* showing wings and halteres (the pair of knob-like structures protruding posterior to the wings). (B) A fly with four wings produced by mutations in the *bithorax* complex. The mutations convert the third thoracic segment into the second thoracic segment, and the halteres that are normally present on the third thoracic segment become converted into the posterior pair of wings. [Courtesy of Edward B. Lewis. Used with permission of Pamela Lewis.]

teran insects. The ancestral condition in insects is similar to that in today's dragonfly, in which the hind wings are virtually identical to the forewings; the dragonfly *Ubx* homolog clearly does not make a haltere. In the insect lineage leading to present-day lepidopterans, the hind wing became modified from the forewing, but not to such an extreme extent as in the dipterans; yet the lepidopteran *Ubx* homolog is expressed in the hind wing.

The resolution of the paradox of how *Ubx* can seemingly control the development of the haltere in dipterans without doing so in other insects is found in the gradual evolution of sets of genes that come under *Ubx* control. The situation is exemplified in **FIGURE 13.35**, in which the small spheres denote genes and the arrows gene activation in segment T3. The condition in part A is assumed to be the ancestral state, in which the gene *c1* has evolved in such a way that it is activated by the *Ubx* homolog. In this organism, the active genes in T3 will consist of the genes activated by the *Ubx* homolog (*c1*), plus genes further along the regulatory hierarchy activated by these genes (*c2* and *c3*), plus other genes (not shown) whose expression in T3 is independent of the *Ubx* pathway. The complete set of *Ubx*-induced genes, direct and indirect, is shown at the right of the regulatory hierarchy. These are the genes that account for the slight modifications of the hind wing as compared with the forewing. In the absence of the *Ubx* homolog, the hind wing would lose these slight modifications and become even more like a forewing.

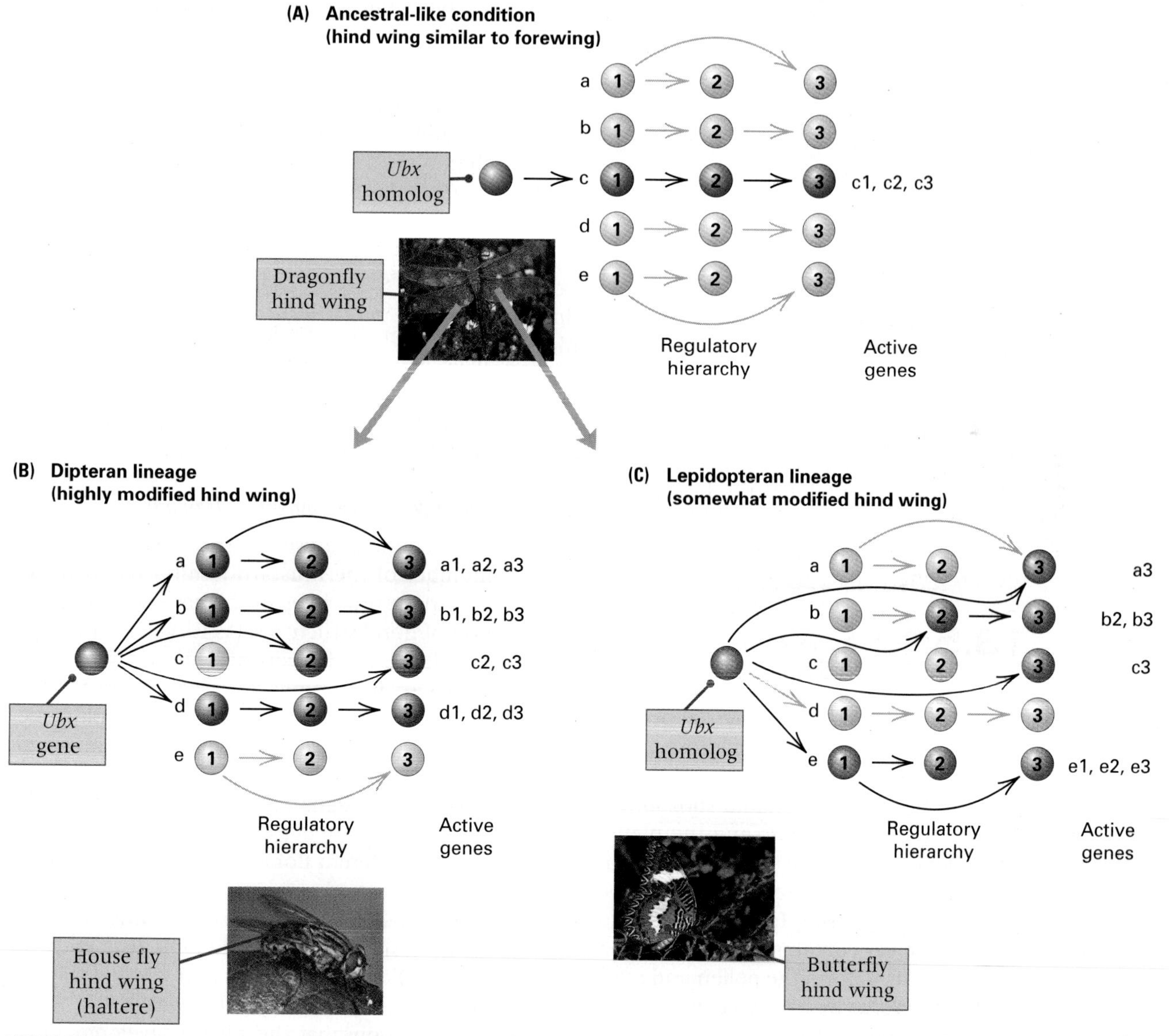


**FIGURE 13.35** Evolutionary scheme by which the *Ubx* homolog expressed in segment T3 produces different developmental pathways for the hind wing in different insect lineages. As each lineage evolves, mutations gradually accumulate, through natural selection, that bring different sets of genes under the control of the *Ubx* homolog. [Illustration based on concepts by Sean Carroll; photos © Photodisc.]

As evolution proceeds, mutations that, by chance, can either release a gene from *Ubx* control or bring a gene under *Ubx* control can occur. Such mutations would typically include the loss or gain of *Ubx*-responsive enhancers. A newly evolved *Ubx* responsiveness, or loss of *Ubx* responsiveness, would not affect the expression of the same genes in other tissues, because *Ubx* is expressed specifically in T3. If the novel pattern of T3 gene expression is favored by natural selection, then the mutation will become incorporated into the species. After a long period of evolution, numerous genes will have evolved in this way, progressively modifying the hind-wing pattern in a manner favored by natural selection. In dipterans, for example, the selection was for extreme modification of size and structure to form the haltere. The result of the evolution is shown diagrammatically in Figure 13.35B, where *c1* is depicted as having lost *Ubx* activation and *a1, b1, c2* and *c3*, and *d1* as having gained it. (Note that the *Ubx* gene can control genes anywhere in the regulatory hierarchy, not only at the beginning.) Similarly, in the lepidopteran lineage (part C), there was selection for less extreme modification, which is depicted as loss of *c1* activation but gain of *a3, b2, c3*, and *e1* activation. The point is that, independently in each lineage, the hind wing was modified gradually while a sequence of genes was brought under or released from *Ubx* control. The result is that in dipterans the *Ubx* gene controls haltere development, whereas in lepidopterans it controls a less extreme modification of the forewing. In each lineage, the *Ubx* homolog is critical in developmental control, but it regulates different genes. In each lineage also, mutations in *Ubx* will make the hind wing more similar to the forewing.

## 13.5 Genetic Control of Development in Higher Plants

Reproductive and developmental processes in plants differ significantly from those in animals. For example, plants have an alternation of generations between the diploid sporophyte and the haploid gametophyte, and the plant germ line is not established in a discrete location during embryogenesis but rather at many locations in the adult organism. In a corn plant, for example, each ear contains somatic cells that undergo meiosis to form the pollen and ovules.

In animals, as we have seen, most of the major developmental decisions are made early in life, in embryogenesis. In higher plants, differentiation takes place almost continuously throughout life in regions of actively dividing cells called **meristems** in both the vegetative organs (root, stem, and leaves) and the floral organs (sepal, petal, pistil, and stamen). The shoot and root meristems are formed during embryogenesis and consist of cells that divide in distinctive geometric planes and at different rates to produce the basic morphological pattern of each organ system. The floral meristems are established by a reorganization of the shoot meristem after embryogenesis and eventually differentiate into floral structures characteristic of each particular species. One important difference between animal and plant development is that:

> In higher plants, as groups of cells leave the proliferating region of the meristem and undergo further differentiation into vegetative or floral tissue, their developmental fate is determined almost entirely by their position relative to neighboring cells.

The critical role of positional information in the development of higher plants stands in contrast to animal development, in which cell lineage often plays a key role in determining cell fate.

The plastic or "indeterminate" growth patterns of higher plants are the result of continuous production of both vegetative and floral organ systems. These patterns are conditioned largely by day length and the quality and intensity of light. The plasticity of plant development gives plants a remarkable ability to adjust to environmental insults. **FIGURE 13.36** shows a tree that, over time, adjusts to the presence of a nearby fence by engulfing it into the trunk.

### Flower Development in *Arabidopsis*

Genetic analysis of *Arabidopsis thaliana* (a member of the mustard family) has revealed important principles in the genetic determination of floral structures. As is typical of flowering plants, the flowers of *Arabidopsis* are composed of four types of organs arranged in concentric rings, or whorls. **FIGURE 13.37** illustrates the geometry, looking down at a floral meristem from the top. From outermost to innermost, the whorls are designated 1, 2, 3, and 4 (part A). In the development of the flower, each whorl gives rise to a different floral organ (part B). Whorl 1 yields the sepals (the green, outermost floral leaves), whorl 2 the petals (the white, inner floral leaves), whorl 3 the stamens (the male organs, which form pollen), and whorl 4 the carpels (which fuse to form the ovary).

Mutations that affect floral development fall into three major classes, each with a characteristic phenotype (**FIGURE 13.38**). Compared

**FIGURE 13.36** The ability of plant development to adjust to perturbations is illustrated by this tree. Encountering a fence, it eventually incorporates the fence into the trunk. [Courtesy of Robert E. Pruitt, Purdue University.]

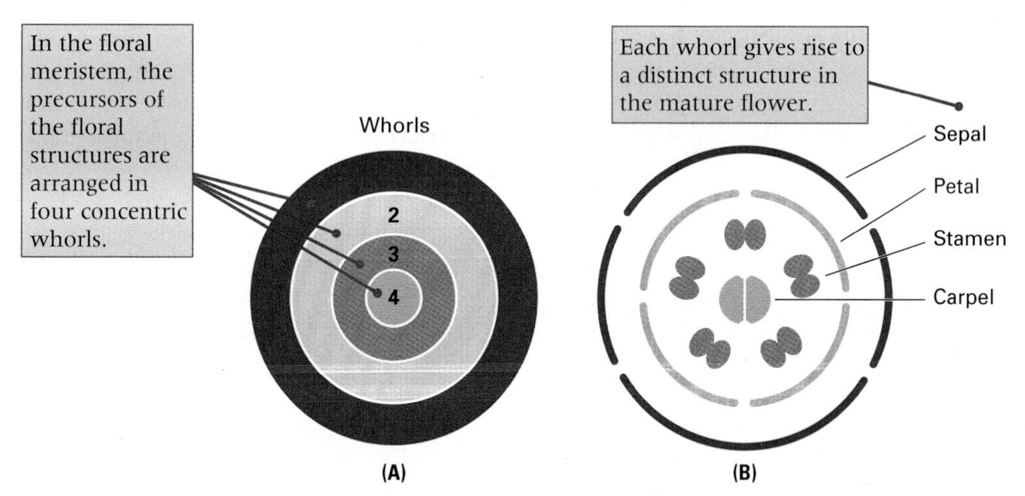


**FIGURE 13.37** Origin of distinct floral structures from concentric whorls in the floral meristem.

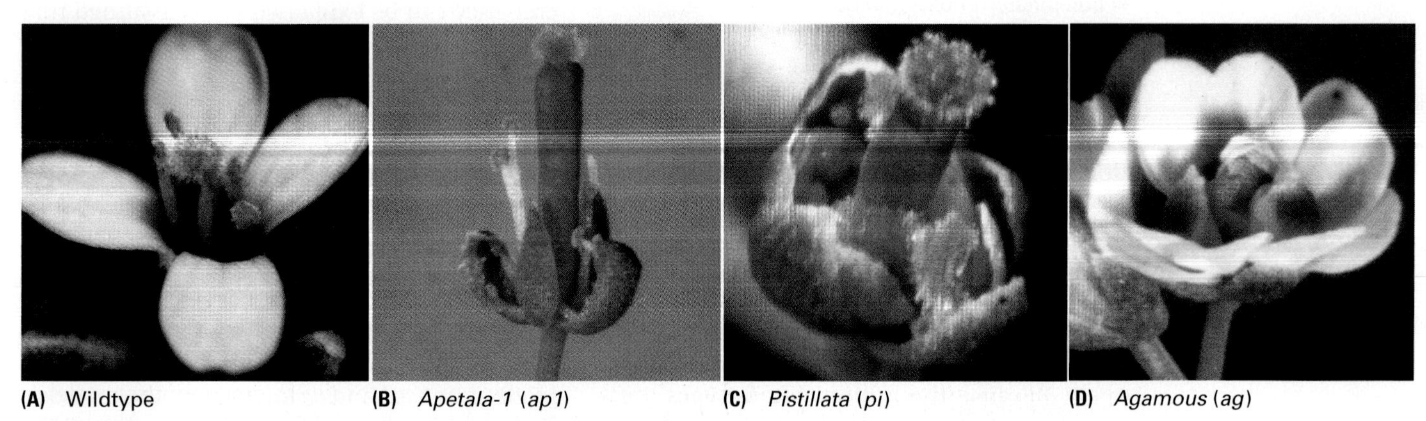


**FIGURE 13.38** Phenotypes of the major classes of floral mutations in *Arabidopsis.* (A) The wildtype floral pattern consists of concentric whorls of sepals, petals, stamens, and carpels. (B) The homozygous mutation *ap1* (*apetala-1*) results in flowers missing sepals and petals. (C) Genotypes that are homozygous for either *ap3* (*apetala-3*) or *pi* (*pistillata*) yield flowers that have sepals and carpels but lack petals and stamens. (D) The homozygous mutation *ag* (*agamous*) yields flowers that have sepals and petals but lack stamens and carpels. [Parts A, C, and D courtesy of Elliot M. Meyerowitz, California Institute of Technology; Part B reproduced from E. M. Meyerowitz and J. L. Bowman, *Sci. Am.* 271 (1994): 56–65. Used with permission of Elliot M. Meyerowitz, California Institute of Technology.]

**Table 13.1 Floral development in mutants of *Arabidopsis***

| | Whorl | | | |
|---|---|---|---|---|
| **Genotype** | **1** | **2** | **3** | **4** |
| wildtype | sepals | petals | stamens | carpels |
| *ap1/ap1* | carpels | stamens | stamens | carpels |
| *ap3/ap3* | sepals | sepals | carpels | carpels |
| *pi/pi* | sepals | sepals | carpels | carpels |
| *ag/ag* | sepals | petals | petals | sepals |

**Table 13.2 Domains of expression of genes determining floral development**

| Whorl | Genes expressed | Determination |
|---|---|---|
| 1 | *ap1* | sepal |
| 2 | *ap1* + *ap3* and *pi* | petal |
| 3 | *ap3* and *pi* + *ag* | stamen |
| 4 | *ag* | carpel |

with the wildtype flower (panel A), one class lacks sepals and petals (panel B), another class lacks petals and stamens (panel C), and the third class lacks stamens and carpels (panel D). On the basis of crosses between homozygous mutant organisms, these classes of mutants can be assigned to different complementation groups, each of which defines a different gene. Each gene and the phenotype of a plant homozygous for a recessive mutation in the gene are shown in **TABLE 13.1**. The phenotype lacking sepals and petals is caused by mutations in the gene *ap1 (apetala-1)*. The phenotype lacking stamens and petals is caused by a mutation in either of two genes, *ap3 (apetala-3)* or *pi (pistillata)*. The phenotype lacking stamens and carpels is caused by mutations in the gene *ag (agamous)*. Each of these genes encodes a transcription factor. The transcription factors encoded by *ap3, pi,* and *ag* are members of what is called the *MADS box* family of transcription factors; each member of this family contains a sequence motif of 58 amino acids. MADS box transcription factors are very common in plants and are also found, less frequently, in animals.

## Combinatorial Determination of the Floral Organs

The role of the *ap1, ap3, pi,* and *ag* transcription factors in the determination of floral organs can be inferred from the phenotypes of the mutations. The logic of the inference is based on the observation (Table 13.1) that mutation in any of the genes eliminates two floral organs that arise from adjacent whorls. This pattern suggests that *ap1* is necessary for sepals and petals, *ap3* and *pi* are both necessary for petals and stamens, and *ag* is necessary for stamens and carpels. Because the mutant phenotypes are caused by loss-of-function alleles of the genes, it may be inferred that *ap1* is expressed in whorls 1 and 2, that *ap3* and *pi* are expressed in whorls 2 and 3, and that *ag* is expressed in whorls 3 and 4. The overlapping patterns of expression are shown in **TABLE 13.2**.

The model of gene expression in Table 13.2 suggests that floral development is controlled in combinatorial fashion by the four genes. Sepals develop from tissue in which only *ap1* is active; petals are evoked by a combination of *ap1, ap3,* and *pi;* stamens are determined by a combination of *ap3, pi,* and *ag;* and carpels derive from tissue in which only *ag* is expressed. This model is illustrated graphically in **FIGURE 13.39**.

This model of floral determination is often called the **flower ABC model** because the wildtype activity of *ap1* was originally designated A, that of *ap3* and *pi* acting together as B, and that of *ag* as C. Therefore, the combination of activities present in each whorl would be represented as A in whorl 1, AB in whorl 2, BC in whorl 3, and C in whorl 4 (Figure 13.39).

You may have noted already that the model in Table 13.2 does not account for all of the phenotypic features of the *ap1* and *ag* mutations in Table 13.1. In particular, according to the combinatorial model in Table 13.2, the development of carpels and stamens from whorls 1 and 2 in homozygous *ap1* plants would require expression of *ag* in whorls 1 and 2. Similarly, the development of petals and sepals from whorls 3 and 4 in homozygous *ag* plants would require expression of *ap1* in whorls 3 and 4. This discrepancy can be explained if it is assumed that *ap1* expression and *ag* expression are mutually exclusive. In the presence of the AP1 transcription factor, *ag* is repressed; in the presence of the AG transcription factor, *ap1* is repressed. If this were the case, then in *ap1* mutants, *ag* expression would spread into whorls 1 and 2; in *ag* mutants, *ap1* expression would spread into whorls 3 and 4. This additional assumption enables us to explain the phenotypes of the single and even double mutants.

With the additional assumption we have made about *ap1* and *ag* interaction, the model in Table 13.2 fits the data, but is the model true? For these genes, the patterns of gene expression, assayed by *in situ* hybridization of RNA in floral cells with labeled probes for each of the genes, fit the patterns in Table 13.2. In particular, *ap1* is expressed in whorls 1 and 2, *ap3* and *pi* in whorls

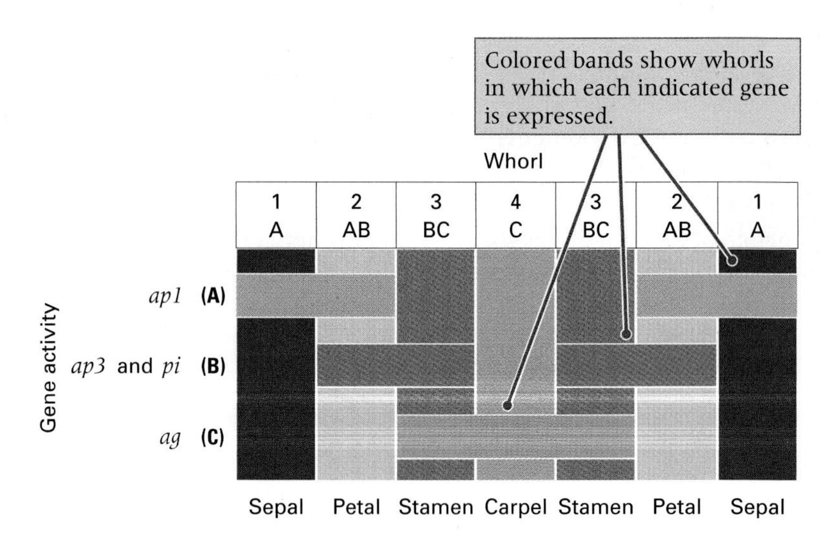


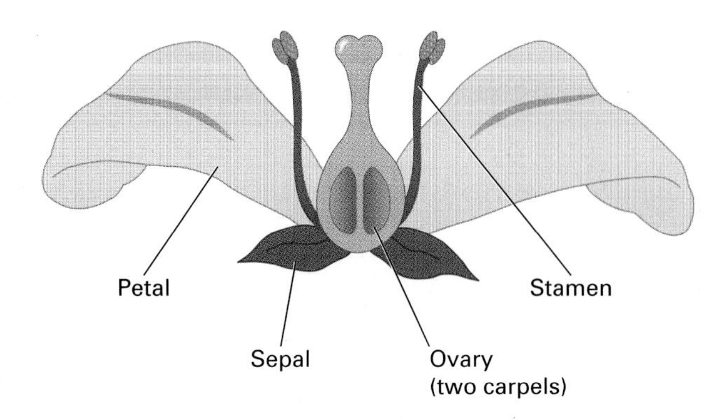


**FIGURE 13.39** Control of floral development in *Arabidopsis* by the overlapping expression of four genes. The sepals, petals, stamens, and carpels are floral organ systems that form in concentric rings, or whorls. The developmental identity of each concentric ring is determined by the genes *ap1*, *ap3* and *pi*, and *ag*, each of which is expressed in two adjacent rings. Therefore, each whorl has a unique combination of active genes indicated by the combinations of A, B, and C.

2 and 3, and *ag* in whorls 3 and 4. Furthermore, the seemingly arbitrary assumption about *ap1* and *ag* expression being mutually exclusive turns out to be true. In *ap1* mutants, *ag* is expressed in whorls 1 and 2; reciprocally, in *ag* mutants, *ap1* is expressed in whorls 3 and 4. It is also known how *ap3* and *pi* work together. The active transcription factor that corresponds to these genes is a dimeric protein composed of Ap3 and Pi polypeptides. Each component polypeptide, in the absence of the other, remains inactive in the cytoplasm. Together, they form an active dimeric transcription factor that migrates into the nucleus.

Given the critical role of the Ap1, Ap3/Pi, and Ag transcription factors in floral determination, it might be speculated that triple mutants lacking all three types of transcription factors would have very strange flowers. The phenotype of the *ap1 pi ag* triple mutant is shown in **FIGURE 13.40**. The flowers have none of the normal floral organs. They consist merely of leaves arranged in concentric whorls.

**FIGURE 13.40** The homozygous triple mutant *ap1 pi ag* lacks all of the transcription factors needed for floral development. Hence, the flowers lack all of the wildtype floral structures. They have no sepals, petals, stamens, or carpels. Without the floral genetic determinants, the flowers consist entirely of leaves arranged in concentric whorls. [Courtesy of Elliot M. Meyerowitz, California Institute of Technology.]

## CHAPTER SUMMARY

- In animal cells, maternal gene products in the oocyte control the earliest stages of development, including the establishment of the main body axes.
- Developmental genes are often controlled by gradients of gene products, either within cells or across parts of the embryo.
- Regulation of developmental genes is hierarchical—genes expressed early in development regulate the activities of genes expressed later.
- The principle of epistasis helps to determine the order in which genes act in a linear switch–regulation developmental pathway. This principle asserts that the epistatic gene acts downstream of the hypostatic gene.
- Regulation of developmental genes is combinatorial—each gene is controlled by a combination of other genes.
- Many of the fundamental processes of pattern formation appear to be similar in animals and plants.

## REVIEW THE BASICS

- What is meant by the term *positional information* in regard to development? How can positional information affect cell fate?
- How does knowledge of the complete cell lineage of nematode development demonstrate the importance of programmed cell death (apoptosis) in development?
- If a gene is both necessary and sufficient for determining a developmental pathway, explain why loss-of-function mutants would be expected to have a different phenotype than gain-of-function mutants.
- What is a receptor? What is a ligand? What role do these types of molecules play in signaling between cells?
- What is meant by the term *polarity* in reference to the mature oocyte?
- How was the study of maternal-effect lethal genes a key to deciphering the genetic control of early embryogenesis in *Drosophila*?
- Among genes that control embryonic development in *Drosophila*, distinguish among coordinate genes, gap genes, pair-rule genes, and segment-polarity genes. Generally speaking, what is the temporal order of expression of these classes of genes?
- What is a homeotic mutation? Give an example from *Drosophila*. Do homeotic mutations occur in organisms other than *Drosophila*?
- Do plants have a germ line in the same sense as animals? What does the difference in germ-cell origin imply about the potential role of "somatic" mutations in the evolution of each type of organism?
- What is the genetic basis of the developmental determination of sepals, petals, stamens, and carpels in floral development in *Arabidopsis*?

## GUIDE TO PROBLEM SOLVING

**Problem 1** In the diagram shown here, a substance B inhibits the development of red pigments in a flower. The wildtype color of the flower is pink, but mutants are known that are either white or red. Assuming that the substance B is the product of a gene *B*, what flower-color phenotype would you expect of a loss-of-function mutation in gene *B*? What flower-color phenotype would you expect of a gain-of-function mutation in gene *B*? What principle in developmental genetics does this situation exemplify?

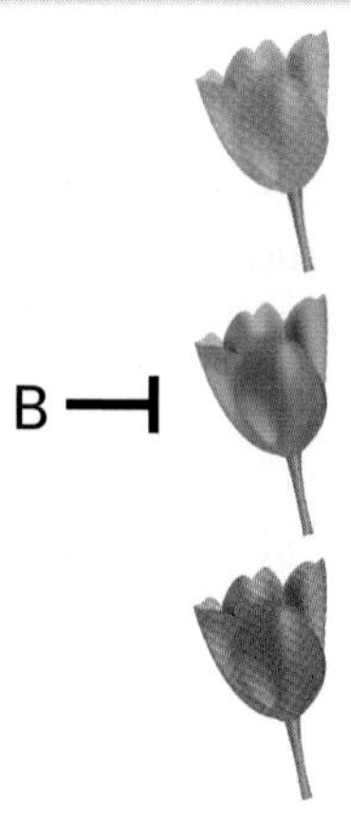


**Answer** Because B inhibits flower color, and the wildtype phenotype is pink, loss-of-function in gene *B* would eliminate B and hence reduce the amount of inhibition; the expected phenotype would therefore be red. Similarly, a gain-of-function mutation in gene *B* would increase the amount of B and therefore intensify the inhibition, and the expected flower-color phenotype would be white. This situation exemplifies the principle that, in developmental-control genes, a loss-of-function mutation often has the opposite phenotypic effect of a gain-of-function mutation.

**Problem 2** The illustration on the next page gives detail about the linear switch-regulation pathway controlling flower pigmentation. The substance A is an intensifier of B. If A is the product of a gene *A*, what flower-color phenotype would you expect from a loss-of-function mutation in gene *A*? What phenotype would you expect from a gain-of-function mutation in gene *A*?

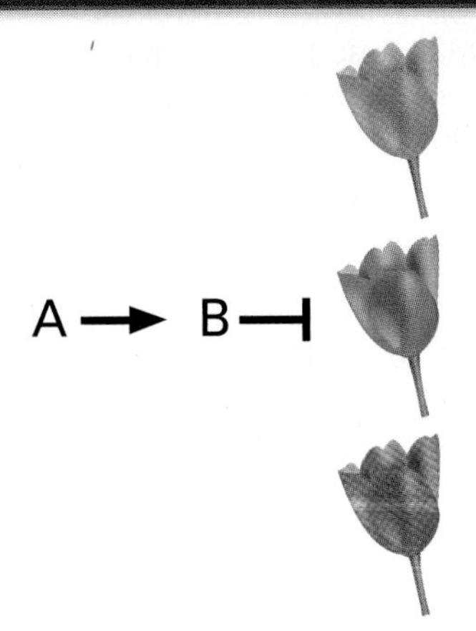


**Answer** Because A intensifies B, a loss-of-function mutation in gene *A* would eliminate A and thereby reduce the level of inhibition due to B; the expected flower-phenotype would therefore be red. Conversely, a gain-of-function mutation in gene *A* would increase the effect of B, thereby increasing the inhibition of flower color due to B, and the expected flower-color phenotype would be white.

**Problem** Shown here are three possible phenotypes observed on a distal appendage of a certain insect species. The wildtype phenotype consists of a single row of bristles. Some mutants yield a phenotype of no bristles, whereas others yield a phenotype in which the appendage has two rows of bristles. The development of these bristles is known to be due to a linear switch-regulation pathway involving the substances A, B, C, and D, which are encoded in the genes *A*, *B*, *C*, and *D*, respectively. The order of action of *A*, *B*, *C*, and *D* is unknown. The mutant alleles a and b result in missing bristles (M), whereas the mutant alleles c and d result in an extra row of bristles (E). The matrix shows the phenotype observed in all possible double mutants. What is the implied order of gene action of substances A, B, C, and D?

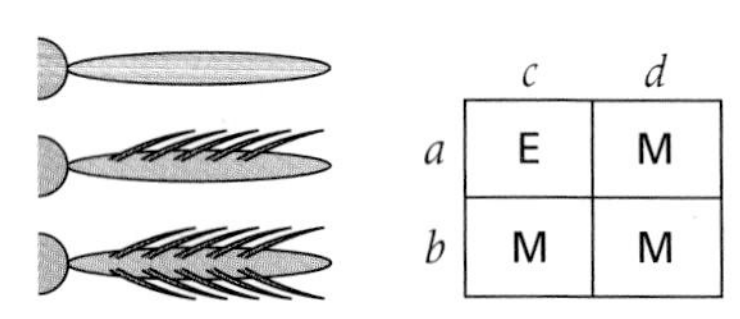

| | *c* | *d* |
|---|---|---|
| *a* | E | M |
| *b* | M | M |

**Answer** According to the principle of epistasis, the product of the epistatic gene in a double-mutant combination acts downstream of the product of the hypostatic gene. The first row therefore implies the order A C and D A, whereas the second row implies the order C B and D B. Putting all this information together, the order of action of the substances must be D–A–C–B.

## ANALYSIS AND APPLICATIONS

**13.1** A *heterochronic* mutation is one that alters the timing of developmental events relative to one another. A mutation is found in which the developmental pattern is normal but slow. Does this qualify as a heterochronic mutation? Explain.

**13.2** Distinguish between a loss-of-function mutation and a gain-of-function mutation. Can the same gene undergo both types of mutations? Can the same allele have both types of effects?

**13.3** What is the principle of epistasis, and how is it used in the genetic analysis of linear switch-regulation pathways in development?

**13.4** A double mutant of *Drosophila* is constructed. One of the mutants alone causes enlarged eyes, whereas the other mutant alone causes small eyes. The double mutant has small eyes. Which of the genes is epistatic? Which is hypostatic? Assuming a linear switch-regulation pathway for eye size, what do these results imply about the order of action of the gene products in the developmental pathway?

**13.5** What is the result of a maternal-effect lethal allele in *Drosophila*? If an allele is a maternal-effect lethal, how can a fly be homozygous for it?

**13.6** How would you determine whether two independently isolated maternal-effect lethal mutations were alleles of the same gene?

**13.7** Consider a particular gene that is necessary, but not sufficient, for the development of a certain morphological feature. What is the expected phenotype of a loss-of-function mutation in the gene? Is the allele expected to be dominant or recessive?

**13.8** Consider a particular gene that is sufficient for the development of a certain morphological feature. What is the expected phenotype of a gain-of-function mutation in the gene? Is the allele expected to be dominant or recessive?

**13.9** A linear switch-regulation pathway of the type shown here controls flower color in a certain species. The wildtype color is violet, but mutants can be either white or purple. Gene products A and B act in the pathway at the positions and in the manner shown (A is inhibitory, B is intensifying). A and B are the products of genes *A* and *B*, respectively. What are the expected phenotypes of:

**(a)** A loss-of-function mutation in *A*?
**(b)** A gain-of-function mutation in *A*?
**(c)** A loss-of-function mutation in *B*?
**(d)** A gain-of-function mutation in *B*?

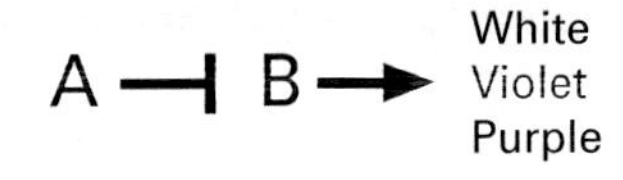


**13.10** In a species related to that in Problem 13.9, the linear switch-regulation pathway controlling flower color is shown here. The wildtype color is again violet, and mutants may be either white or purple. In this species both A and B are inhibitory. What are the expected phenotypes of:

**(a)** A loss-of-function mutation in *A*?
**(b)** A gain-of-function mutation in *A*?
**(c)** A loss-of-function mutation in *B*?
**(d)** A gain-of-function mutation in *B*?

A —| B —| White / Violet / Purple

**13.11** *Drosophila* normally has four bristles on the scutellum, a small triangular region of cuticle at the base of the thorax. Some mutants have no bristles on the scutellum, and some have extra bristles. Assume a linear switch-regulation pathway involving the gene products U, V, W, X, Y, and Z encoded in the genes *U, V, W, X, Y,* and *Z,* respectively. The mutant alleles *u, v,* and *w* result in extra bristles (E), whereas the mutant alleles *x, y,* and *z* result in missing bristles (M). From the double-mutant data shown here, what can you deduce about the order of action of *U, V, W, X, Y,* and *Z* in the bristle-number pathway?

| | *x* | *y* | *z* |
|---|---|---|---|
| *u* | E | M | E |
| *v* | M | M | E |
| *w* | E | E | E |

**13.12** A species of flowering plants has flowers that are either yellow (wildtype), white, or orange. The intensity of the pigment is determined by a linear switch-regulation pathway with components A, B, C, D, E, and F encoded in the genes *A, B, C, D, E,* and *F,* respectively. Mutant alleles *a, b,* and *c* result in orange flowers (O), whereas mutant alleles *d, e,* and *f* result in white flowers (W). The matrix shown here summarizes the phenotypes observed in the double mutants. What do these data imply about the order of action of the components in the pathway?

| | *d* | *e* | *f* |
|---|---|---|---|
| *a* | O | O | O |
| *b* | O | O | W |
| *c* | W | O | W |

**13.13** The *bicoid* mutation in *Drosophila* results in absence of head structures in the early embryo. A similar phenotype results when cytoplasm is removed from the anterior end of a *Drosophila* embryo. What developmental effect would you expect in each of the following circumstances?

**(a)** A *bicoid* embryo was injected in the anterior end with some cytoplasm taken from the anterior end of a wildtype embryo.

**(b)** A wildtype *Drosophila* embryo was injected in the middle with some cytoplasm taken from the anterior end of another wildtype embryo

**13.14** The mRNA from the *bicoid* gene includes a sequence at the 3′ end that results in its being localized at the anterior end of the egg. What phenotype would you search for in a mutant screen for genes necessary for this localization?

**13.15** A certain crustacean can have a medium sized claw (wildtype), and enlarged claw, or a reduced claw. Claw size is determined by a linear switch-regulation pathway with components A through G encoded in the genes *A* through *G*. The components operate in the order E – C – G – A – D – B – F. Mutant alleles *a, b,* and *c* result in an enlarged claw (E), whereas mutant alleles *d, e, f,* and *g* result in a reduced claw (R). Complete the matrix of double mutants by entering into each cell the expected phenotype of the double mutant (E or R).

| | *d* | *e* | *f* | *g* |
|---|---|---|---|---|
| *a* | | | | |
| *b* | | | | |
| *c* | | | | |

**13.16** The nuclei of brain cells in the adult frog normally do not synthesize DNA or undergo mitosis. However, when transplanted into developing oocytes, the brain cell nuclei behave as follows: (a) In rapidly growing premeiosis oocytes, they synthesize RNA. (b) In more mature oocytes, they do not synthesize DNA or RNA, but their chromosomes condense and they begin meiosis. How would you explain these results?

**13.17** Consider a hypothetical mutant protease that affects floral development in *Arabidopsis thaliana.* The protease has an altered substrate specificity that enables it to cleave and inactivate both Ap1 and Ag proteins (the products of *ap1* and *ag,* respectively). In view of the fact that tissue containing the Ap3/Pi dimeric protein, but neither Ap1 nor Ag alone, develops into floral organs intermediate between petals and stamens, what floral phenotype would be expected in the protease mutant? (*Hint:* The default state is "leaf.")

**13.18** What phenotype would be expected of a gain-of-function mutation in *Arabidopsis* that resulted in expression of *ap3* and *pi* in whorl 1?

**13.19** What is the expected floral structure arising from each whorl in an *Arabidopsis* mutant in which *ag* (*agamous*) is restricted to whorl 3? (*Hint:* The default state is "leaf.")

**13.20** What is the expected floral structure arising from each whorl in an *Arabidopsis* mutant in which *ap1* (*apetala-1*) is restricted to whorl 2? (*Hint:* The default state is "leaf.")

**13.21** Genetic analysis is being carried out on a specialized appendage of a marine invertebrate that is used in immobilizing prey. The wildtype appendage, shown at the top in the accompanying diagram, has three segments. The proximal segment, attached directly to the body, has a single row of sparse bristles; the medial segment has a single row of dense bristles; and the distal segment has two symmetrical rows of dense bristles. Three mutations that affect the appendage are isolated; they cause the phenotypes shown in the diagram. The presence of a segment lacking bristles may be regarded as the default state of development of this appendage. Suggest a hypothesis to explain how the genes *a, b,* and *c* are involved in the developmental determination of this appendage.

$a^+a^+\ b^+b^+\ c^+c^+$

$a^-a^-\ b^+b^+\ c^+c^+$

$a^+a^+\ b^-b^-\ c^+c^+$

$a^+a^+\ b^+b^+\ c^-c^-$

**13.22** To gain further insight into the genetic control of development of the appendage discussed in the previous problem, all possible double mutants are constructed. They have the phenotypes shown here.

**(a)** Are these phenotypes consistent with the inferences drawn from those of the single mutants?

**(b)** What additional information is provided by the double mutants?

**(c)** Propose a detailed model of which gene products are necessary for the development of each segment.

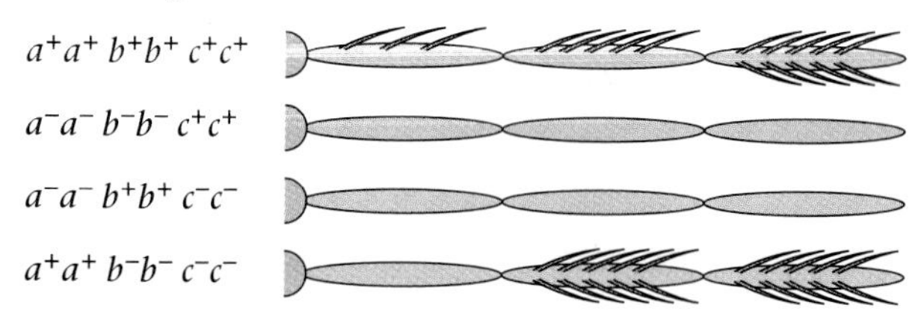


**13.23** From the model for the genetic control of appendage development that emerges from data in the previous two problems, sketch the phenotype associated with each of the following types of mutations.

**(a)** A mutation in which the domain of expression of $c^+$ expands into the distal segment.

**(b)** A $b^-$ mutation in which the domain of expression of $a^+$ expands into the proximal segment.

**(c)** An $a^-$ mutation in which the domain of expression of $b^+$ expands into the medial segment.

**13.24** The drug actinomycin D prevents RNA transcription but has little direct effect on protein synthesis. When fertilized sea urchin eggs are immersed in a solution of the drug, development proceeds to the blastula stage, but gastrulation does not take place. How would you interpret this finding?

**13.25** What floral phenotype would be expected of a loss-of-function mutation in *Arabidopsis* whose function was necessary for Ap3/Pi and Ag to exert their effects on development?

**13.26** A mutation in the axolotl designated *o* is a maternal-effect lethal because embryos from *oo* females die at gastrulation, irrespective of their own genotype. However, the embryos can be rescued by injecting oocytes from *oo* females with an extract of nuclei from either $o^+/o^+$ or $o^+/o$ eggs. Injection of cytoplasm is not as effective. Suggest an explanation for these results.

**13.27** What floral phenotype would be expected of a mutation resulting in a loss-of-function of *ap3* and *pi*? A gain-of-function mutation in which both *ap3* and *pi* were expressed in whorls 1 and 4 as well as whorls 2 and 3? What do these results imply about Ap3/Pi being necessary or sufficient for the developmental fate of the four whorls?

**13.28** Explain why, in accounting for the phenotypes of floral mutants in *Arabidopsis*, it was necessary to postulate that:

**(a)** In *agamous* mutants, the domain of expression of *apetala-1* expands to whorls 3 and 4.

**(b)** In *apetala-1* mutants, the domain of expression of *agamous* expands to whorls 1 and 2.

**13.29** If *agamous* expression did not expand into whorls 1 and 2 in *apetala-1* mutants, could the phenotype of whorls 1 and 2 be predicted? Why or why not?

**13.30** Mutants affecting a linear switch-regulation pathway result in either enlarged ears or reduced ears in a species of gerbil. The pathway involves the components A through D encoded in the genes *A* through *D*. Mutant alleles *a* and *b* result in enlarged ears (E), and mutant alleles *c* and *d* result in reduced ears (R). The double mutants have the phenotypes shown in the accompanying matrix. From these data, what can you infer about the order in which A, B, C, and D function in the pathway?

| | *c* | *d* |
|---|---|---|
| *a* | R | E |
| *b* | R | E |

## CHALLENGE PROBLEMS

**Challenge Problem 1** You wish to demonstrate that during segmentation of the *Drosophila* embryo, normal pair-rule patterns of expression require normal expression of the gap genes, whereas gap gene expression does not require pair-rule expression. You have the following four mutations available:

**(a)** A mutation in the zygotic-effect gap gene *knirps* (*kni*).

**(b)** A mutation in the zygotic-effect pair-rule gene *fushi tarazu* (*ftz*).

**(c)** A transgene consisting of a reporter gene (*lacZ*) fused to the enhancer elements of *kni*.

**(d)** A transgene consisting of a reporter gene (*lacZ*) fused to the enhancer elements of *ftz*.

Describe the strains you would need and how you would use them to show that wildtype expression of *kni* is needed for proper expression of *ftz*, but that wildtype expression of *ftz* is not needed for proper expression of *kni*. You do not need to give details of the crosses.

**Challenge Problem 2** A type of jellyfish has a structure composed of eight circularly arranged cells from which projections emerge, forming a sort of pore. Whereas some of the cells have identical projections, others are quite different, as shown in the accompanying diagram. The letters a through d indicate the products of four developmental-control genes, *a* through *d*, that are found in each cell around the circle.

In the following diagrams, sketch the expected phenotype of each of the single mutants, *a* through *d*, assuming that the default state of a cell, with none of the gene products present, is the absence of projections.

**Challenge Problem 3** For the developmental-control system described in the previous problem, sketch the expected phenotypes of each of the six possible types of double mutants:

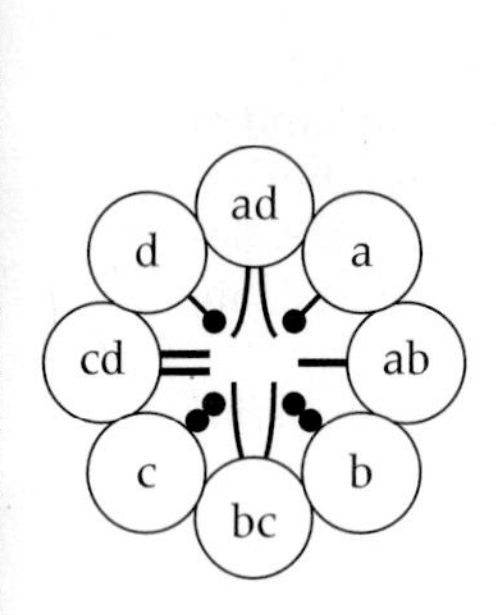


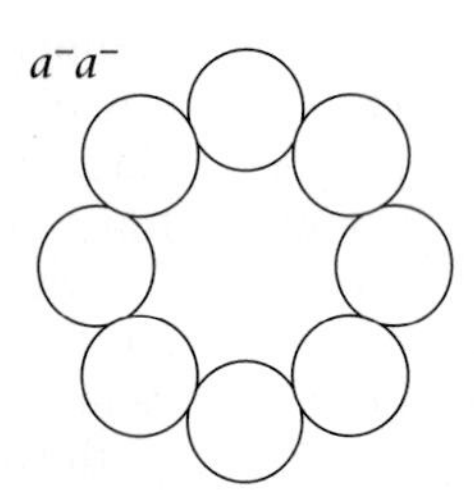


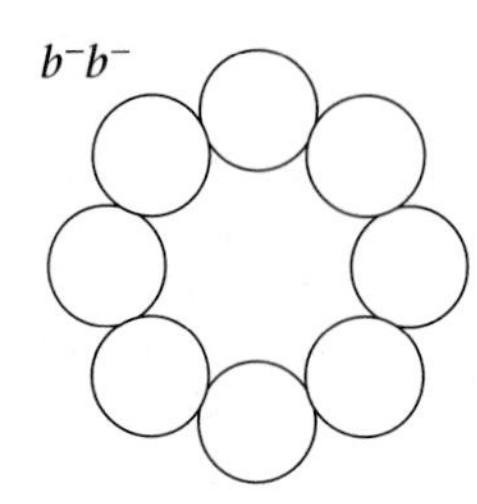


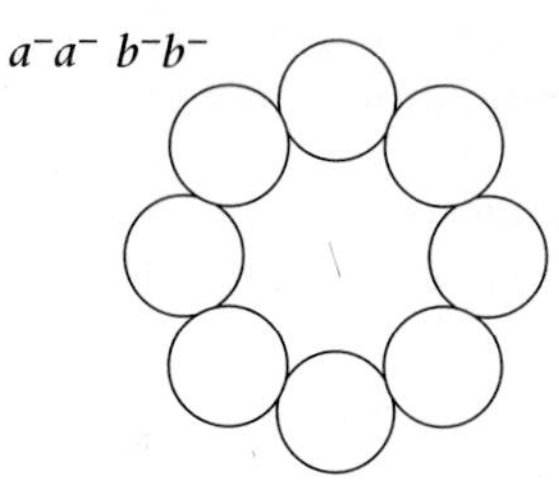


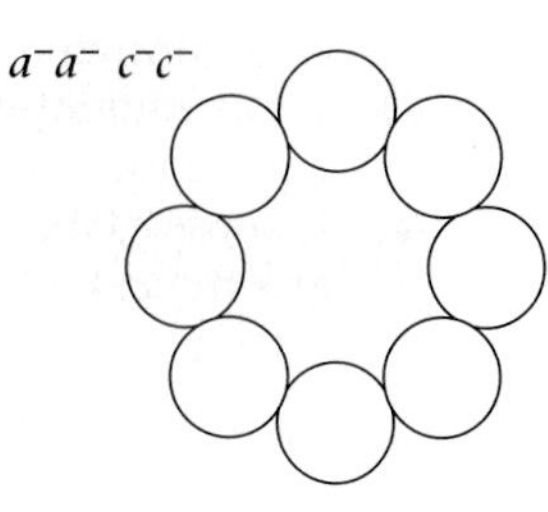


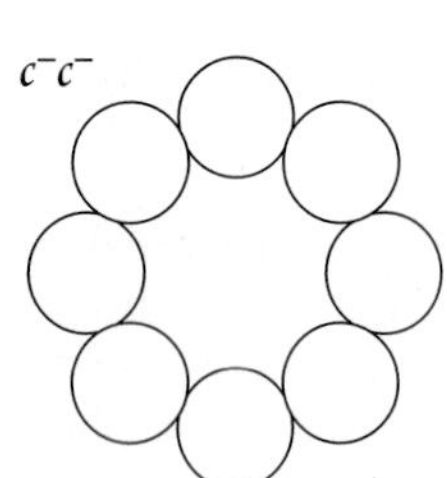


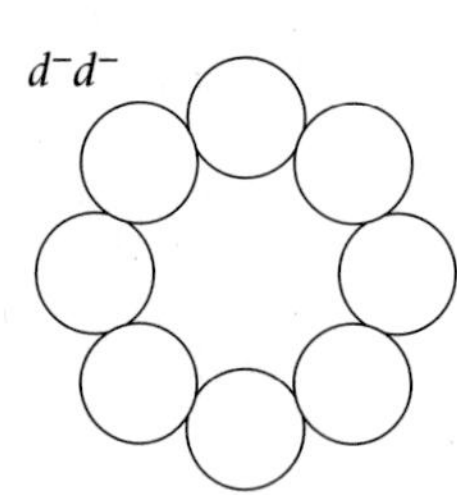


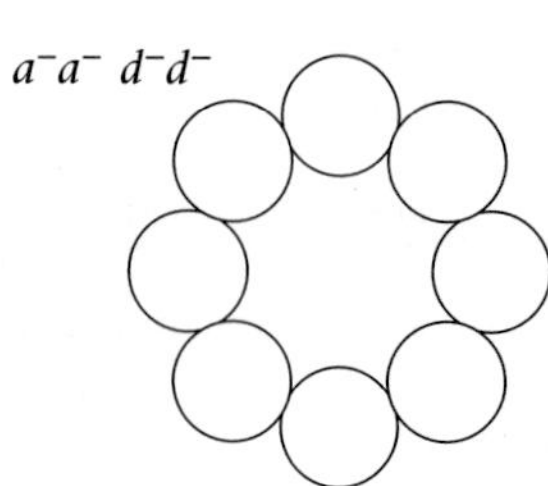


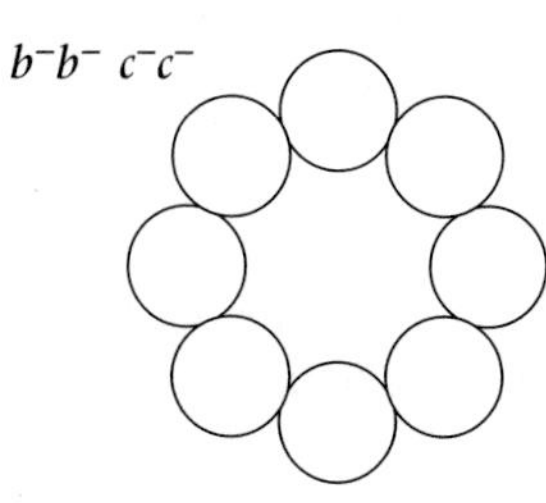


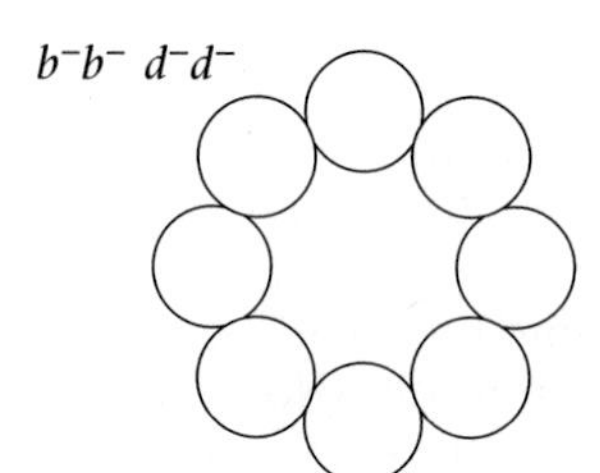


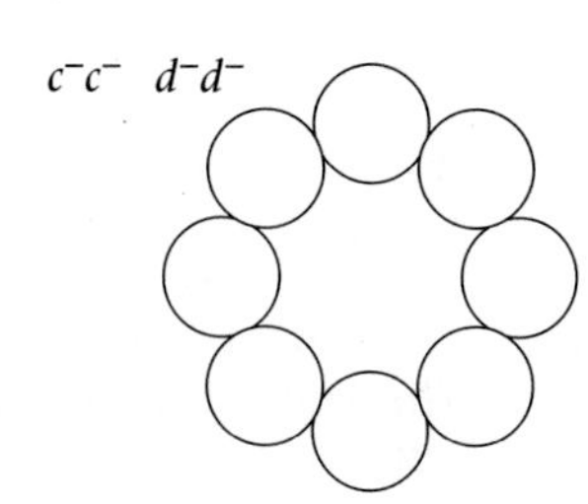


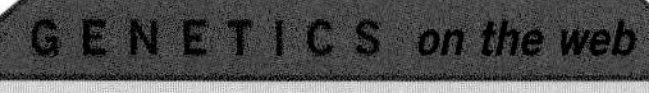


GeNETics on the Web will introduce you to some of the most important sites for finding genetics information on the Internet. To explore these sites, visit the Jones and Bartlett companion site to accompany *Genetics: Analysis of Genes and Genomes, Seventh Edition* at http://biology.jbpub.com/book/genetics.

There you will find a chapter-by-chapter list of highlighted keywords. When you select one of the keywords, you will be linked to a Web site containing information related to that keyword.

# CHAPTER 14

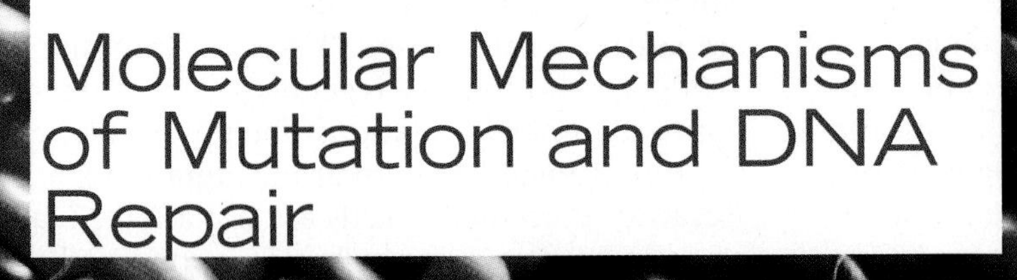

# Molecular Mechanisms of Mutation and DNA Repair

Maize *(Zea Mays)* was domesticated from the wild grass teosinte *(Zea mexicana)* about 5–10 thousand years ago by native American Indians living near what is now Mexico City. They soon developed hundreds of different varieties differing in kernel color and other characteristics. Some of the mutants they selected are now known to result from the action of transposable elements.



## CHAPTER OUTLINE


**14.1 Types of Mutations**
- Germ-Line and Somatic Mutations
- Conditional Mutations
- Classification by Function

**14.2 The Molecular Basis of Mutation**
- Nucleotide Substitutions
- Missense Mutations: The Example of Sickle-Cell Anemia
- Insertions, Deletions, and Frameshift Mutations
- Dynamic Mutation of Trinucleotide Repeats
- Cytosine Methylation and Gene Inactivation

**14.3 Transposable Elements**
- Molecular Mechanisms of Transposition
- Transposable Elements as Agents of Mutation
- Transposable Elements in the Human Genome
- RIP: A Defense Against Transposons

**14.4 Spontaneous Mutation**
- The Nonadaptive Nature of Mutation
- Estimation of Mutation Rates
- Hot Spots of Mutation

**14.5 Mutagens**
- Depurination
- Oxidation
- Base-Analog Mutagens
- Chemical Agents That Modify DNA
- Intercalating Agents
- Ultraviolet Irradiation
- Ionizing Radiation
- Genetic Effects of the Chernobyl Nuclear Accident

**14.6 Mechanisms of DNA Repair**
- Mismatch Repair
- Base Excision Repair
- AP Repair
- Nucleotide Excision Repair
- Photoreactivation
- DNA Damage Bypass
- The SOS Repair System

**14.7 Reverse Mutations and Suppressor Mutations**
- Intragenic Suppression
- Intergenic Suppression
- The Ames Test for Mutagen/Carcinogen Detection

CONNECTION **Her Feeling for the Organism**
Barbara McClintock 1950
*The Origin and Behavior of Mutable Loci in Maize*

CONNECTION **X-Ray Daze**
Hermann J. Muller 1927
*Artificial Transmutation of the Gene*

CONNECTION **Replication Slippage in Unstable Repeats**
Micheline Strand, Tomas A. Prolla, R. Michael Liskay, and Thomas D. Petes 1993
*Destabilization of Tracts of Simple Repetitive DNA in Yeast by Mutations Affecting DNA Mismatch Repair*

Any heritable change in a gene is a **mutation.** In this chapter, we examine the nature of mutations at the molecular level. We examine how mutations occur and what effects they have on genes and gene expression. Although mutations take place spontaneously, they can also be induced by radiation and a variety of chemical agents that damage nucleotides or that produce breaks in polynucleotide chains. Most types of DNA damage are corrected by special repair enzymes almost immediately after they occur. The new mutations that are transmitted in any generation therefore represent the minority of changes in DNA that have escaped repair.

## 14.1 Types of Mutations

Mutations can happen at any time and in any cell. The phenotypic effects can range from minor alterations that are detectable only by molecular methods (for example, DNA polymorphisms) to drastic changes in essential processes that cause death of the cell or organism.

### ■ Germ-Line and Somatic Mutations

Mutations can be classified in a variety of ways (**TABLE 14.1**). Most mutations are *spontaneous*, which means that they occur in the absence of any known cause. Mutations can also be *induced* through damage to DNA caused by chemical agents or radiation. In multicellular organisms, one important distinction is based on the type of cell in which the mutation first occurs. Those that arise in cells that ultimately form gametes are **germ-line mutations**, all others are **somatic mutations**. A somatic mutation yields an organism that is genotypically, and for many dominant mutations phenotypically, a mixture (**mosaic**) of normal and mutant tissue. Because reproductive cells are not affected, a somatic mutation cannot be transmitted to the progeny and may not be amenable to genetic analysis. In higher plants, however, somatic mutations can often be propagated by vegetative means without going through seed production; these means include grafting and the rooting of stem cuttings. Vegetative propagation is typical of many commercially important fruits, such as the Delicious apple and the Florida navel orange.

### ■ Conditional Mutations

Among the mutations that are most useful for genetic analysis are those whose effects can be turned on or off by the experimenter. These are called **conditional mutations** because they produce changes in phenotype in one set of environmental conditions (called the **restrictive conditions**) but not in another (called the **permissive conditions**). For example, a **temperature-sensitive mutation** is a conditional mutation whose expression depends on temperature. Usually, the restrictive temperature is high (in *Drosophila*, 29°C), and the organism exhibits a mutant phenotype above this critical temperature; the permissive temperature is lower (in *Drosophila*, 18°C), and under permissive conditions the phenotype is wildtype or nearly wildtype. Proteins containing amino acid replacements are often

**Table 14.1 Major types of mutations and their distinguishing features**

| Basis of classification | Major types of mutations | Major features |
|---|---|---|
| Origin | Spontaneous<br>Induced | Occurs in absence of known mutagen<br>Occurs in presence of known mutagen |
| Cell type | Somatic<br>Germ-line | Occurs in nonreproductive cells<br>Occurs in reproductive cells |
| Expression | Conditional<br>Unconditional | Expressed only under restrictive conditions (such as high temperature)<br>Expressed under permissive conditions as well as restrictive conditions |
| Effect on function | Loss-of-function (knockout, null)<br>Hypomorphic (leaky)<br>Hypermorphic<br>Gain-of-function (ectopic expression) | Eliminates normal function<br>Reduces normal function<br>Increases normal function<br>Expressed at incorrect time or in inappropriate cell types |
| Molecular change | Nucleotide substitution<br>Transition<br>Transversion<br>Insertion<br>Deletion | One base pair in duplex DNA replaced with a different base pair<br>Pyrimidine (T or C) to pyrimidine, or purine (A or G) to purine<br>Pyrimidine (T or C) to purine, or purine (A or G) to pyrimidine<br>One or more extra nucleotides present<br>One or more missing nucleotides |
| Effect on translation | Synonymous (silent)<br>Missense (nonsynonymous)<br>Nonsense (termination)<br>Frameshift | No change in amino acid encoded<br>Change in amino acid encoded<br>Creates translational termination codon (UAA, UAG, or UGA)<br>Shifts triplet reading of codons out of correct phase |

temperature sensitive; the protein folds properly and functions nearly normally under permissive conditions, but it is unstable and denatures under restrictive conditions. Temperature-sensitive amino acid replacements are frequently used to block particular biochemical pathways under restrictive conditions in order to test the importance of the pathways in various cellular processes, such as DNA replication.

An example of temperature sensitivity is found in the Siamese cat, with its black-tipped paws, ears, and tail (**FIGURE 14.1**). In this breed, an enzyme in the pathway for deposition of the black pigment melanin is temperature-sensitive. The pathway is blocked at normal body temperature, and consequently the pigment is absent in the hair over most of the body where the temperature is warmest. The tips of the legs, ears, snout, and tail are cooler than the rest of the body, so the pigment is deposited in these areas.

**FIGURE 14.1** Siamese cat showing the characteristic pattern of pigment deposition. [© Oleg V. Ivanov/ShutterStock, Inc.]

### Classification by Function

Mutations can also be classified according to the manner in which they affect gene expression. The major categories are described below. Classical terms for the categories were introduced by Herman J. Muller in 1931, but most of them have been superseded by modern equivalents. We will use the modern terms.

- A mutation that results in complete gene inactivation or in a completely nonfunctional gene product is a **loss-of-function mutation**. Examples of loss-of-function mutations include a deletion of all or part of a gene, and an amino acid replacement that inactivates the protein. Loss-of-function mutations are also called **null mutations** or **knockout mutations**. In Muller's terminology they are known as *amorphic mutations*.
- A mutation that reduces, but does not eliminate, the level of expression of a gene or the activity of the gene product is called a **hypomorphic mutation.** Hypomorphic mutations typically result from nucleotide substitutions that prevent a protein from being produced at the normal level, or from an amino acid replacement that impairs protein function. This class of mutations is sometimes referred to as *leaky*. The basis of the term is that, because the level of expression or activity differs from individual to individual by chance, a few individuals have enough enzyme activity to "leak through" to produce a quasi-normal phenotype. The opposite of a hypomorphic mutation is a **hypermorphic mutation**. As the prefix *hyper* implies, a hypermorphic mutant produces a greater-than-normal level of gene expression, typically because the mutation changes the regulation of the gene so that the gene product is overproduced. Less commonly, an amino acid replacement may increase the activity of an enzyme or other protein.
- A **gain-of-function mutation** is one that qualitatively alters the action of a gene. For example, a gain-of-function mutation may cause a gene to become active in a type of cell or tissue in which the gene is not normally active. Or it may result in the expression of a gene in development at a time during which the wildtype gene is not normally expressed. Whereas most loss-of-function and hypomorphic mutations are recessive, many gain-of-function mutations are dominant. Muller's term for a gain-of-function mutation is *neomorphic mutation*.

Expression of a wildtype gene in an abnormal location is also called **ectopic expression**. For example, expression of the wildtype gene product of the *Drosophila* gene *eyeless* in tissues that do not normally form eyes results in the development of parts of compound eyes, complete with eye pigments, in abnormal locations (**FIGURE 14.2**). The locations can be in any tissues in which the gain-of-function is expressed, including on the legs or mouthparts, in the abdomen, or on the wings.

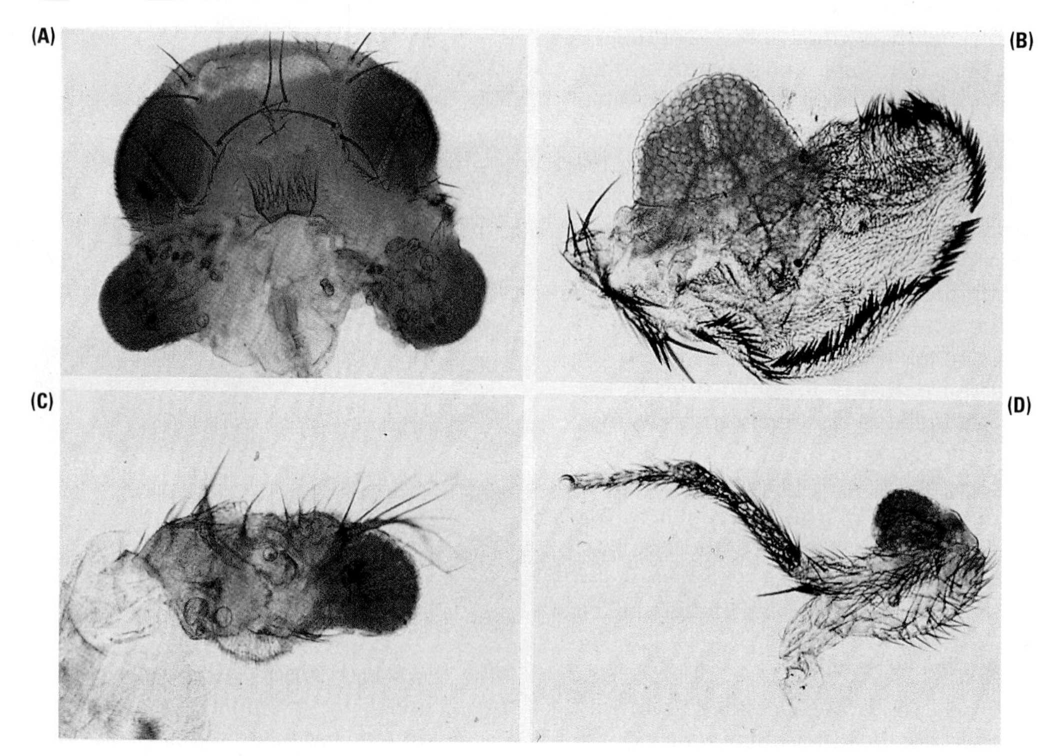


**FIGURE 14.2** Ecoptic expression of the wildtype allele of the *eyeless* gene in *Drosophila* results in misplaced eye tissue. (A) An adult head in which both antennae form eye structures. (B) A wing with eye tissue growing out from it. (C) A single antenna in which most of the third segment consists of eye tissue. (D) Middle leg with an eye outgrowth at the base of the tibia. [Reproduced from G. Halder, P. Callaerts, and W. J. Gehring, *Science* 267 (1995): 1788–1792. Reprinted with permission from AAAS.

## 14.2 The Molecular Basis of Mutation

All mutations result from changes in the nucleotide sequence of DNA or from deletions, insertions, or rearrangement of DNA sequences in the genome. The major types of chromosomal rearrangements are discussed in Chapter 8. Here we examine mutations at the molecular level.

### Nucleotide Substitutions

At the molecular level, the simplest type of mutation is a **nucleotide substitution**, in which a nucleotide pair in a DNA duplex is replaced with a different nucleotide pair. For example, in an A → G substitution, an A is replaced with a G in one of the DNA strands. This substitution temporarily creates a mismatched G–T base pair; at the very next replication, the mismatch is resolved as a proper G–C base pair in one daughter molecule and as a proper A–T base pair in the other daughter molecule. In this case, the G–C base pair is the mutant and the A–T base pair is nonmutant. Similarly, in an A → T substitution, an A is replaced with a T in one strand, creating a temporary T–T mismatch, which is resolved by replication as T–A in one daughter molecule and A–T in the other. In this example, the T–A base pair is mutant and the A–T base pair is nonmutant. The T–A and the A–T are not equivalent, as can be seen by considering the nucleotide polarity of the strands: 5′-T-3′/3′-A-5′ is the base pairing in one case, 5′-A-3′/3′-T-5′ in the other.

Some nucleotide substitutions replace one pyrimidine base with the other or one purine base with the other. These are called **transition mutations**. The possible transition mutations are:

$$\text{T} \rightarrow \text{C} \quad \text{or} \quad \text{C} \rightarrow \text{T}$$
$$(\text{pyrimidine} \rightarrow \text{pyrimidine})$$

$$\text{A} \rightarrow \text{G} \quad \text{or} \quad \text{G} \rightarrow \text{A}$$
$$(\text{purine} \rightarrow \text{purine})$$

Other nucleotide substitutions replace a pyrimidine with a purine or the other way around. These are called **transversion mutations**. The possible transversion mutations are:

$$\text{T} \rightarrow \text{A},\ \text{T} \rightarrow \text{G},\ \text{C} \rightarrow \text{A}, \quad \text{or} \quad \text{C} \rightarrow \text{G}$$
$$(\text{pyrimidine} \rightarrow \text{purine})$$

$$\text{A} \rightarrow \text{T},\ \text{A} \rightarrow \text{C},\ \text{G} \rightarrow \text{T}, \quad \text{or} \quad \text{G} \rightarrow \text{C}$$
$$(\text{purine} \rightarrow \text{pyrimidine})$$

Altogether, there are four possible transitions and eight possible transversions. Therefore, if nucleotide substitutions were strictly random, then one would expect a 1 : 2 ratio of transitions to transversions. However:

> Spontaneous nucleotide substitutions are often biased in favor of transitions. Among spontaneous nucleotide substitutions in the human genome, the ratio of transitions to transversions is approximately 2 : 1.

Examination of the genetic code (Table 1.1 on page 25) shows that the bias toward transitions has an important implication for nucleotide substitutions in the third position of codons. In all codons with a pyrimidine in the third position, it does not matter which pyrimidine is present; likewise, in most codons that end in a purine, either purine will do. This means that most transition mutations in the third codon position do not change the amino acid in the encoded protein. Mutations that change the nucleotide sequence without changing the amino acid sequence are called **synonymous mutations** or **silent mutations**, because they change one codon into a "synonym" for the same amino acid.

Mutational changes in nucleotides that are outside of coding regions can also be silent. Most DNA polymorphisms are in this category. In noncoding regions, which include regulatory regions and the DNA between genes, the precise nucleotide sequence is often not critical. These sequences can undergo base substitutions, small deletions or additions, insertions of transposable elements, and other rearrangements, and yet the mutations may have no detectable effect on phenotype. On the other hand, some noncoding sequences do have essential functions, and mutations in these sequences often do have phenotypic effects.

This cat is an eye-color mosaic. Such mosaicism usually results from a somatic mutation in a gene affecting eye color that occurs in a cell lineage giving rise to the pigmented iris. [© Sara Robinson/ShutterStock, Inc.]

### ■ Missense Mutations: The Example of Sickle-Cell Anemia

Most nucleotide substitutions in coding regions do result in changed amino acids; these are called **missense mutations** or **nonsynonymous mutations**. A change in the amino acid sequence of a protein may alter the biological properties of the protein. The classic example of a phenotypic effect of a single amino acid change is the change responsible for the human hereditary disease *sickle-cell anemia*. The molecular basis of the sickle-cell mutation is shown in **FIGURE 14.3**. It is an A–T → T–A transversion in the second codon position for the sixth amino acid in the $\beta$-globin chain of hemoglobin, causing the normal glutamic acid (Glu or E) to be replaced with valine (Val or V). The amino acid replacement is therefore denoted as either Glu6Val or E6V. When the sickle-cell allele is homozygous, the defective $\beta$-polypeptide chain gives the hemoglobin protein a tendency to form long, needle-like polymers. Red blood cells in which polymerization takes place become deformed into crescent, sickle-like shapes. Some of the deformed red blood cells are destroyed immediately (reducing the oxygen-carrying capacity of the blood and causing the anemia), whereas others become trapped in capillary vessels and cause partial or complete blocks in capillary circulation.

The consequences of the Glu6Val replacement are a profound set of pleiotropic effects. All of these effects are related to the breakdown of red blood cells, to the decreased oxygen-carrying capacity of the blood, or to physiological adjustments the body makes to try to compensate for the disease (such as enlargement of the spleen). Patients with sickle-cell anemia suffer bouts of severe pain. The anemia causes impaired growth, weakness, and jaundice. Affected people are so generally weakened that they are susceptible to bacterial infections, which are the most common cause of death in children with the disease.

Although sickle-cell anemia is a severe genetic disease that often results in premature death, it is relatively frequent in areas of Africa and the Middle East in which a type of malaria

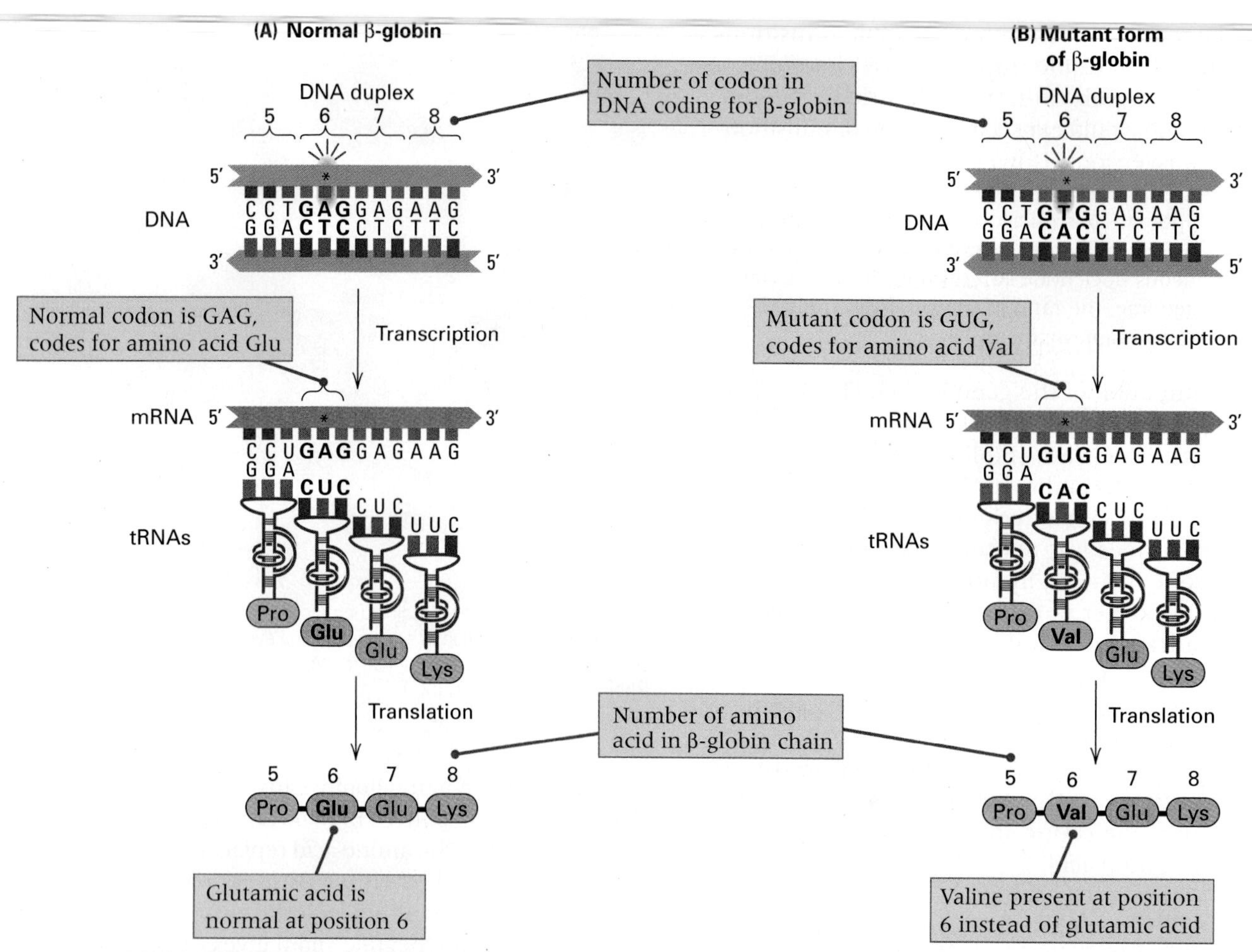


**FIGURE 14.3** Molecular basis of sickle-cell anemia. (A) Part of the DNA in the normal $\beta$-globin gene is transcribed into a messenger RNA coding for the amino acid sequence Pro–Glu–Glu–Lys. The T in the marked A–T base pair is transcribed as the A in the GAG codon for Glu (glutamic acid). (B) Mutation of the normal A–T base pair to a T–A base pair results in the codon GUG instead of GAG. The codon GUG codes for Val (valine), so the polypeptide sequence in this part of the molecule is Pro–Val–Glu–Lys. The resulting hemoglobin is defective and tends to polymerize at low oxygen concentration.

caused by the protozoan parasite *Plasmodium falciparum* is widespread. The association between sickle-cell anemia and malaria is not coincidental. The association results from the ability of the mutant $\beta$-hemoglobin to afford some protection against malarial infection. In the life cycle of the parasite, it passes from a mosquito to a human through the mosquito's bite. The initial stages of infection take place in cells in the liver, where specialized forms of the parasite that are able to infect and multiply in red blood cells are produced. Widespread infection of red blood cells impairs the ability of the blood to carry oxygen, causing the weakness, anemia, and jaundice characteristic of malaria. However, in people who are heterozygous for the sickle-cell allele, infection with malaria is less likely and also less severe. The partial resistance of heterozygous sickle-cell genotypes creates a genetic balancing act between the prevalence of the genetic disease sickle-cell anemia and that of the parasitic disease malaria. If the mutant sickle-cell hemoglobin becomes too frequent, more lives are lost from sickle-cell anemia than are gained by the protection against malaria; on the other hand, if the mutant sickle-cell hemoglobin becomes too rare, fewer lives are lost from sickle-cell anemia but the gain is offset by more deaths from malaria. The end result of this kind of genetic balancing act is discussed in quantitative terms in Chapter 17.

In contrast to the Glu6Val replacement, some amino acid replacements in hemoglobin have negligible effects on phenotype. Amino acid replacements are often tolerated in other proteins as well. The key issue is whether a particular amino acid replacement affects protein folding, function, or stability, and this depends on the protein, the amino acid and its precise role in the structure and function of the protein, and the nature of the amino acid replacement. For example, the replacement of one hydrophilic amino acid with another is expected to have a lesser effect than the replacement of a hydrophilic amino acid with a hydrophobic amino acid. Any change in the active site of an enzyme usually decreases enzymatic activity.

**FIGURE 14.4** illustrates the nine possible codons that can result from a single nucleotide substitution in the UAU codon for tyrosine, organized according to whether the first, second, or third nucleotide in the codon is mutant. One mutation is silent (box), and six are missense mutations that change the amino acid inserted in the polypeptide chain. The other two mutations create a stop codon that results in the premature termination of translation and the production of a truncated polypeptide. A nucleotide substitution that creates a new stop codon is called a **nonsense mutation**. Because nonsense mutations cause premature chain termination, the remaining polypeptide fragment is nearly always nonfunctional.

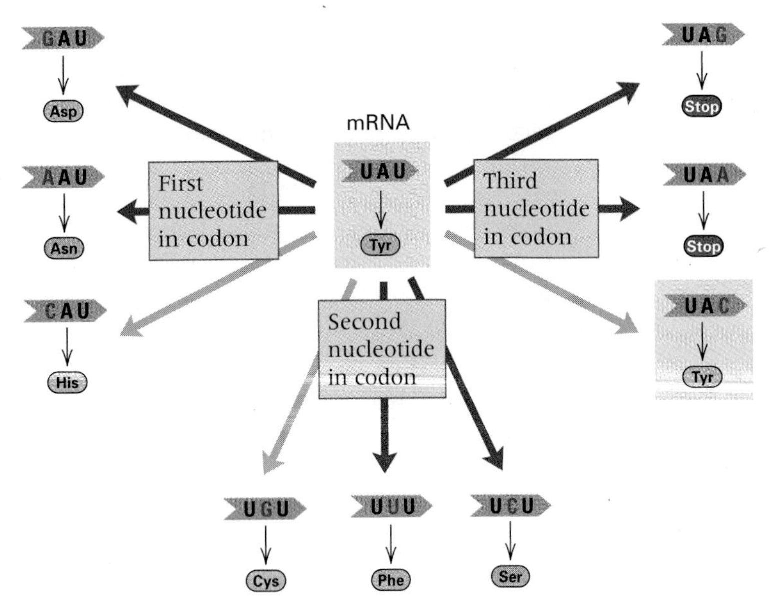


**FIGURE 14.4** The nine codons that can result from a single base change in the tyrosine codon UAU. Blue arrows indicate transversions, gray arrows transitions. Tyrosine codons are in boxes. Two possible stop (nonsense) codons are shown in red. Altogether, the codon UAU allows for six possible missense mutations, two possible nonsense mutations, and one silent mutation.

## Insertions, Deletions, and Frameshift Mutations

Insertions or deletions of nucleotides can also occur in DNA, but at a rate considerably lower than that of nucleotide substitution. Most insertions and deletions in mammalian genomes are small (typically 1–6 base pairs), but some are large enough to eliminate entire genes or part of a chromosome (Chapter 8). If they occur outside of coding regions, small insertions or deletions may have negligible effects on phenotype unless they affect the expression of nearby genes.

Within a coding region, the insertion or deletion of exactly three nucleotides results in the addition or elimination of one amino acid from the encoded polypeptide chain, because three nucleotides is the length of a codon. An insertion or deletion of six nucleotides adds or eliminates two amino acids, and that of nine nucleotides adds or eliminates three amino acids. In general, provided that the number of nucleotides inserted or deleted is an exact multiple of three, an insertion or deletion adds or eliminates some number of amino acids from the polypeptide chain.

If the length of an insertion or deletion is not an exact multiple of three nucleotides, the mutation shifts the phase in which the ribosome reads the triplet codons and, consequently, alters all of the amino acids downstream from the site of the mutation. Such mutations are called **frameshift mutations** because they "shift" the reading frame of the codons in the mRNA. A common type of frameshift mutation is a single-nucleotide insertion or deletion. An example of a single-nucleotide insertion in the human $\beta$-globin gene is illustrated in **FIGURE 14.5**, where

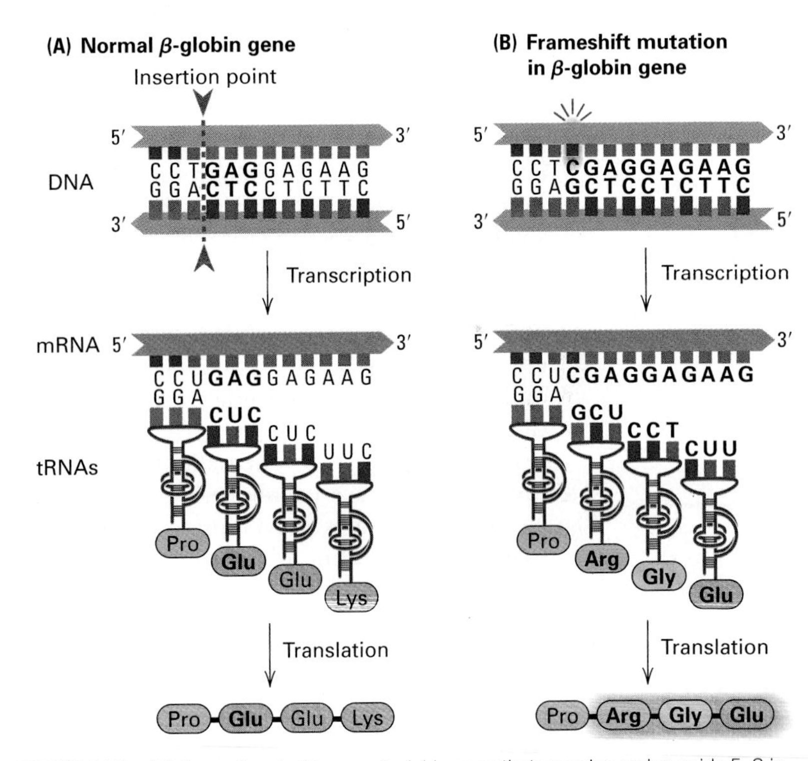


**FIGURE 14.5** (A) A small part of human $\beta$-globin gene that encodes amino acids 5–8 in the polypeptide chain. (B) Result of a frameshift mutation (in this example, one that inserts a C–G base pair at the position indicated by the "burst"). All codons downstream from the site of the mutation are changed because of the shifted reading frame, so the entire amino acid sequence differs from that point on.

(A) is the wildtype sequence and (B) is the mutant sequence with a C–G nucleotide pair inserted at the position of the sunburst. The result of the insertion is that, starting with the codon in which the insertion occurs, the entire amino acid sequence of the polypeptide is altered because the reading frame is shifted, in this example by +1. Translation continues in the shifted reading frame until a termination codon is encountered. (Translation of a completely random nucleotide sequence would be expected to continue for only about twenty amino acids before running into a termination codon.) Any insertion or deletion that is not a multiple of three nucleotides results in a shifted reading frame, and all amino acids downstream from the insertion are different from the original. Unless it is very near the carboxyl terminus of a protein, a frameshift mutation almost always results in the synthesis of a nonfunctional protein.

## ■ Dynamic Mutation of Trinucleotide Repeats

Genetic studies of an X-linked form of mental retardation revealed an unexpected class of mutations called **dynamic mutations** because of the extraordinary genetic instability of the region of DNA involved. The X-linked condition, one of at least twelve genetic disorders associated with dynamic mutation, is associated with a class of X chromosomes that tends to fracture in cultured cells that are starved for DNA precursors, such as nucleotides. The position of the fracture is in region Xq27.3, near the end of the long arm. The X chromosomes containing this site are called *fragile-X* chromosomes, and the associated form of mental retardation is the *fragile-X syndrome*. The fragile-X syndrome affects about 1 in 2500 males and 1 in 5000 females. It accounts for about half of all cases of X-linked mental retardation and is second only to Down syndrome (Chapter 8) as a cause of inherited mental impairment.

The fragile-X syndrome is highly variable in severity. Males are usually more severely affected than females. Developmental delays in speech and communication skills are common, as well as delays in such gross motor skills as sitting up and learning to walk. Physical symptoms may include a long face with protruding ears, weakness in connective tissues resulting in poor muscle tone and extremely flexible joints, and enlarged testicles in males past puberty. Mental retardation is usually moderate in males and mild in females. Behavioral effects may include anxiety, poor concentration, trouble coping with sensory stimuli, avoidance of eye contact, and tantrums or emotional outbursts. These symptoms are nonspecific and overlap with such conditions as autism and attention deficit-hyperactivity disorder.

A hint of something unusual about the fragile-X syndrome was the paradoxical pattern of its inheritance, key features of which are illustrated in **FIGURE 14.6**. Approximately one in five males who carry the fragile X chromosome are themselves phenotypically normal and also have phenotypically normal children. The oddity is that the heterozygous daughters of such a "transmitting male" often have affected children of both sexes. In Figure 14.6, the transmitting male denoted I-2 is not affected, but the X chromosome that he transmits to his daughters (II-2 and II-5) somehow becomes altered in the females in such a way that sons and daughters in the next generation (III) are affected. Both affected and normal granddaughters of the transmitting male may have affected progeny (generation IV).

The molecular basis of the fragile-X chromosome has been traced to a **trinucleotide repeat** of the form CGG (or, equivalently, CCG on the other strand) present in the DNA at the site where the breakage takes place (**FIGURE 14.7**). Normal X chromosomes have 6–54 tandem copies of the repeating unit, with an average of about thirty, whereas affected persons have 230–2300 or more copies of the repeat. The

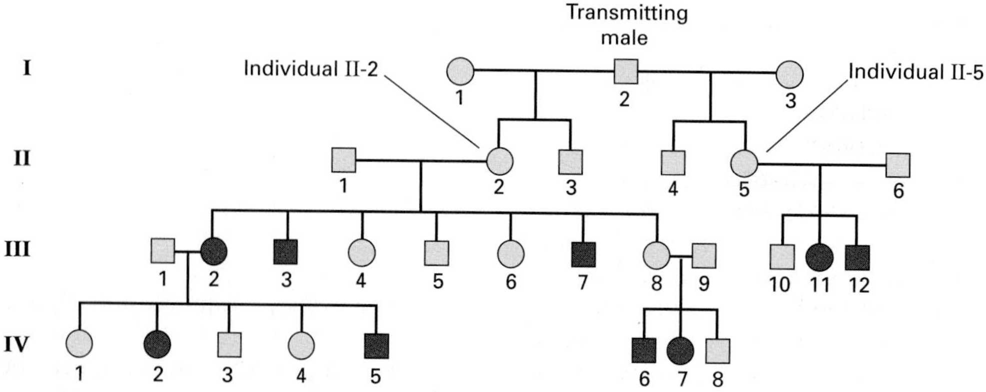


**FIGURE 14.6** Pedigree showing transmission of the fragile-X syndrome. Male I-2 is not affected, but his daughters (II-2 and II-5) have affected children and grandchildren. [Adapted from C. D. Laird, *Genetics* 117 (1987): 587–599.]

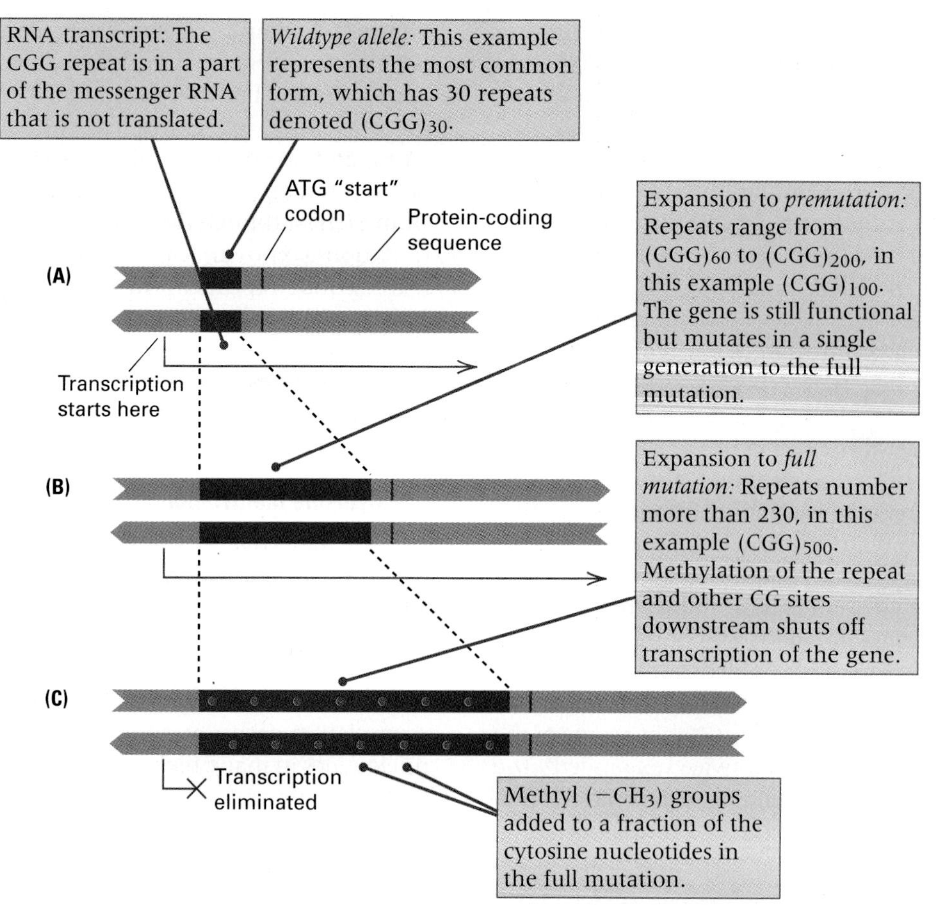
RNA transcript: The CGG repeat is in a part of the messenger RNA that is not translated.
Wildtype allele: This example represents the most common form, which has 30 repeats denoted $(CGG)_{30}$.
ATG "start" codon
Protein-coding sequence
Expansion to premutation: Repeats range from $(CGG)_{60}$ to $(CGG)_{200}$, in this example $(CGG)_{100}$. The gene is still functional but mutates in a single generation to the full mutation.
(A)
Transcription starts here
(B)
Expansion to full mutation: Repeats number more than 230, in this example $(CGG)_{500}$. Methylation of the repeat and other CG sites downstream shuts off transcription of the gene.
(C)
Transcription eliminated
Methyl ($-CH_3$) groups added to a fraction of the cytosine nucleotides in the full mutation.

**FIGURE 14.7** Dynamic mutation in the CGG repeat present in the *FMR1* gene implicated in the fragile-X syndrome. (A) The wildtype allele typically has 30 copies of the repeat. (B) The premutation has 60–200 copies, which predisposes to further amplification when transmitted through a female. (C) The full mutation, containing >230 copies. In the full mutation, absence of transcription of the gene is associated with methylation of certain CG dinucleotides in the region.

trinucleotide repeat in the X chromosome in transmitting males is called the *premutation* and has an intermediate number of copies, ranging from 52 to 230. Approximately 1 in 250 females and 1 in 800 males carries an X chromosome with the premutation. The unprecedented feature of the trinucleotide premutation is that when transmitted by females (and only by females), it often increases in copy number (called *trinucleotide expansion*) to a level of 230 copies or more, at which stage the chromosome causes the fragile-X syndrome. The amplification does not take place in transmission through a male.

The functional basis of the disorder is that an excessive number of copies of the CGG repeat cause loss of function of a gene designated *FMR1* (fragile-site mental retardation-1) in which the CGG repeat is present. Most fragile-X patients exhibit no *FMR1* messenger RNA, whereas normal persons and carriers do show expression. The *FMR1* gene is expressed primarily in brain and testes, which explains the strange association between mental and testes abnormalities in affected males.

There is about an 80 percent chance that a premutation transmitted by a female will undergo amplification. Surprisingly, the amplification does not take place in the mother's germ line but in somatic cells of the early embryo. Amplification occurs to a different extent in different somatic cells, and so individuals with the fragile-X syndrome are somatic mosaics for cells with different numbers of copies of the CGG repeat in the X chromosome. This accounts for the great variation in severity of the fragile-X syndrome from one affected individual to the next.

Other genetic diseases associated with dynamic mutation include the neurological disorders myotonic dystrophy (chromosome 19, with an unstable repeat of CTG), Kennedy disease (X chromosome, repeat AGC), Friedreich ataxia (chromosome 9, AAG), spinocerebellar ataxia type 1 (chromosome 6, AGC), and Huntington disease (chromosome 4, AGC). For the fragile-X syndrome and myotonic dystrophy, the trinucleotide expansions occur primarily or exclusively when transmitted by females, but for spinocerebellar ataxia type I and Huntington disease, they occur primarily or exclusively when transmitted by males. Some trinucleotide repeats can undergo amplification when transmitted by either sex.

Some dynamic mutations also show a phenomenon called **anticipation**, in which the number of repeats gradually increases as the trinucleotide repeat is transmitted through several generations. This results in a more severe disease with a progressively earlier age of onset through time.

The molecular mechanism of trinucleotide expansion is illustrated in **FIGURE 14.8**. The process is called **replication slippage** (also called *slipped-strand mispairing)*. As replication is proceeding along a template strand containing the repeats (A), the replication complex momentarily dissociates from the template strand. In reassociating with the template, the 3′ end of the new strand backtracks along the template and pairs with an upstream set of repeats (B). Replication continues normally from this point (C), but some of the repeats will be replicated twice (expanded), the level of expansion depending on how far the replication complex backtracked in reassociating.

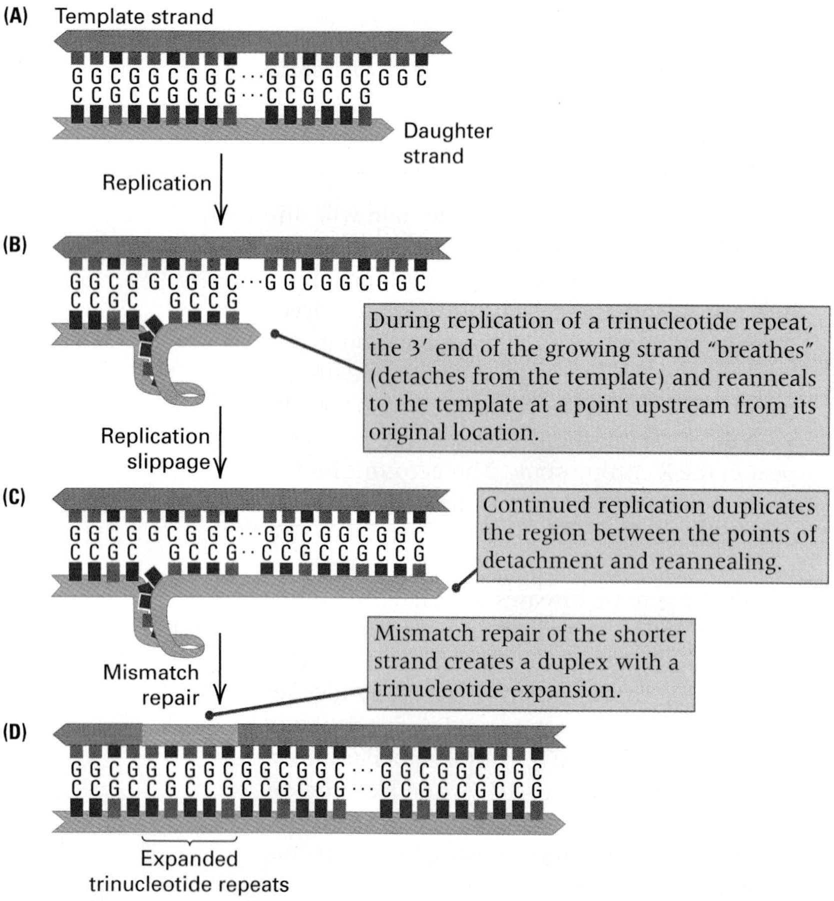

**FIGURE 14.8** Model of replication slippage.

The template and the daughter strand cannot pair properly because they have a different number of repeats, but this situation is corrected by the nucleotide-excision repair system (Section 14.6), and one outcome of repair is that the expanded region is introduced into the template strand (D). Although the mechanism of dynamic mutation is known, it is not known why some trinucleotide repeats in the genome are genetically unstable whereas others are stable, or why the trinucleotide premutation state is uniquely prone to expansion whereas chromosomes that may have only somewhat fewer copies are genetically stable.

## Cytosine Methylation and Gene Inactivation

The molecular mechanism of *FMR1* inactivation is associated with the enzymatic addition of a methyl ($-CH_3$) group to each of certain of the cytosine nucleotides in the 5′ region of the *FMR1* gene (Figure 14.7). As we saw in Chapter 11 when we discussed gene regulation, cytosine methylation occurs at a fraction of the cytosine nucleotides in many higher eukaryotes. In mammals it occurs preferentially at CG dinucleotides, and each CGG repeat in the amplified region of *FMR1* includes a potential methylation site. A high density of methylated CG dinucleotides is usually associated with repression of transcription of the affected gene. In the case of *FMR1*, the lack of transcription of the gene in affected individuals is associated with the methylation of the expanded CGG repeat as well as increased methylation of other CG dinucleotides nearby.

What does the *FMR1* protein do? The protein, called FMRP (for fragile-X mental retardation protein), is an RNA-binding protein that binds with messenger RNAs and regulates either their translation into protein or their localization in the cytoplasm or both. FMRP does not bind all mRNA molecules, but only a specific subset that contain strategically placed guanine nucleotides that allow the guanines to undergo base pairing in groups of four when the RNA molecules form a loop and fold back upon themselves (**FIGURE 14.9**, part A). The configuration formed by four guanine nucleotides pairing in this fashion is called a **G quartet,** or less commonly *Hoogsteen base pairing*. Figure 14.9B shows a simplified version of a G quartet as viewed from the top, and the details of the hydrogen bonding between the guanines are shown in part C.

In the brain, mRNA molecules containing G quartets often encode proteins that function in the development of the facial bones and the nervous system or that function in learning and memory. Many are synaptic proteins, which

## connection
## Her Feeling for the Organism

**Barbara McClintock 1950**
Cold Spring Harbor Laboratory, Cold Spring Harbor, New York
*The Origin and Behavior of Mutable Loci in Maize*

*Many geneticists regard McClintock's papers on transposable elements as difficult. Her discoveries were completely novel. Genes that could move from one place to another in the genome were unheard of. She had no terminology with which to discuss such things, so she had to adapt the conventional terminology to describe a unique situation. What we now call a transposable element, and believe to be universal among organisms, McClintock calls a "chromatin element." McClintock was a superb geneticist and cytogeneticist, perhaps the best of her generation. In tracking down transposable elements by genetic means, McClintock had to use considerable ingenuity in designing crosses that were sometimes quite complex. Her writing style is also uniquely McClintock. The excerpt that follows is a good example. It is a relentless marshaling of observation and hypothesis, experiment and result, interpretation and deduction. McClintock was awarded a Nobel Prize in 1983. The following passage deals with the discovery of* Ds (Dissociation) *and* Ac (Activator). *In modern terminology, we call* Ac *an autonomous transposable element because it codes for all the proteins it needs for its own transposition;* Ds *is a nonautonomous transposable element because it requires* Ac *to provide the proteins necessary for it to move.*

In the course of an experiment designed to reveal the genic composition of the short arm of chromosome 9, a phenomenon of rare occurrence (or recognition) in maize began to appear with remarkably high frequencies in the cultures. The terms mutable genes, unstable genes, variegation, mosaicism, [and] mutable loci have been applied to this phenomenon. Its occurrence in a wide variety of organisms has been recognized. . . . A fortunate discovery was made early in the study of the mutable loci which proved to be of singular importance in showing the kinds of events that are associated with their origin and behavior. A locus was found in the short arm of chromosome 9 at which breaks were occurring in somatic cells. The time and frequency of the breakage events occurring at this *Ds* (*Dissociation*) locus appeared to be the same as the time and frequency of the mutation-producing mutable loci. An extensive study of the *Ds* locus has indicated the reason for this relationship and has produced the information required to interpret the events occurring at mutable loci. It has been concluded that the changed phenotypic expression of such loci is related to changes in a chromatin element other than that composing the genes themselves, and that mutable loci arise when such chromatin is inserted adjacent to the genes that are affected. . . . The *Ds* locus is composed of this kind of material. Various types of alterations are observed as a consequence of events occurring at the *Ds* locus. . . . They involve chromosome breakage and fusion. The breaks are related, however, to events occurring at this one specific locus in the chromosome—the *Ds* locus. . . . [Among the known events is] transposition of *Ds* activity from one position to another in the chromosome complement with or without an associated gross chromosomal rearrangement. . . . It is from the transpositions of *Ds* that some of the new mutable loci may arise. . . . In one case the transposed *Ds* locus appeared in a single gamete of a plant carrying chromosome 9 with a dominant *C* (colored aleurone) allele. . . . The new position of *Ds* corresponded to the known location of *C*. . . . Significantly, the appearance of *Ds* activity at this new position was correlated with the disappearance of the normal action of the *C* locus. The resulting phenotype was the same as that produced by the known recessive allele *c*. . . . That the *c* phenotype in this case was associated with the appearance of *Ds* at the *C* locus was made evident because mutations from *c* to *C* occurred [in which] *Ds* action concomitantly disappeared. . . . The origin and behavior of this mutation at the *c* locus have been interpreted as follows: Insertion of the chromatin composed of *Ds* adjacent to the *C* locus is responsible for complete inhibition of the action of *C*. Removal of this foreign chromatin can occur. In many cases, the mechanism associated with the removal results in restoration of former genic organization and action. . . . The mutation-producing mechanisms involve only *Ds*. No gene mutations occur at the *C* locus; the restoration of its action is due to the removal of the inhibiting *Ds* chromatin. . . . [The movement] of *Ds* requires an activator. This activator has been designated *Ac*. . . . *Ac* shows a very important characteristic not exhibited in studies of the usual genetic factors. This characteristic is the same as that shown by *Ds*. Transposition of *Ac* takes place from one position in the chromosomal complement to another—very often from one chromosome to another. Again, as in *Ds*, changes in state may occur at the *Ac* locus. . . . It should be emphasized that when no *Ac* is present in a nucleus, no mutation-producing events occur at *Ds*.

***Transposition of* Ac *takes place from one position in the chromosomal complement to another—very often from one chromosome to another.***

---

Source: B. McClintock, *Proc. Natl. Acad. Sci. USA* 36 (1950): 344–355.

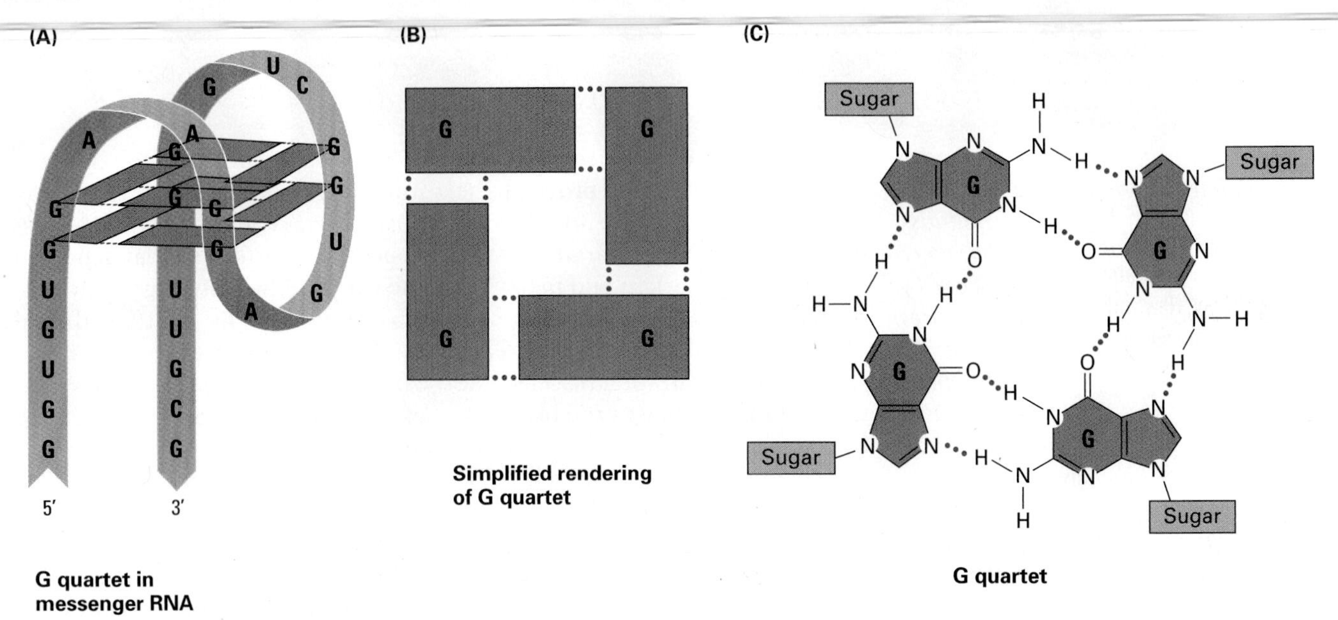


**FIGURE 14.9** Part of a messenger DNA containing a G quartet, which serves as a target for binding and regulation by the FMRP protein. (A) The G quartet forms when the mRNA folds back upon itself. (B) G quartet viewed from the top. (C) Hydrogen bonding in a G quartet.

work at the point of contact between the axon (for outgoing signals) of one neuron and the dendrite (for incoming signals) of a neighboring neuron. The symptoms of the fragile-X syndrome are consistent with these functions, and mRNAs containing G quartets have been shown to be abnormally regulated in fragile-X patients.

## 14.3 Transposable Elements

In a 1940s study of the genetics of kernel mottling in maize (**FIGURE 14.10**), Barbara McClintock discovered a genetic element that not only regulated the mottling, but also caused chromosome breakage. She called this element *Dissociation (Ds)*. Genetic mapping showed that the chromosome breakage always occurs at or very near the location of *Ds*. McClintock's critical observation was that *Ds* does not have a constant location but occasionally moves to a new position (**transposition**), causing chromosome breakage at the new site. Furthermore, *Ds* moves only if a second element, called *Activator (Ac)*, is also present in the same genome. In addition, *Ac* itself moves within the genome and can also cause modification in the expression of genes at or near its insertion site. Since McClintock's original discovery, many other **transposable elements** have been discovered. They can be grouped into "families" based on similarity in DNA sequence. The genomes of most organisms contain multiple copies of each of several distinct families of transposable elements. Once situated in the genome, transposable elements can persist for long periods and undergo multiple mutational changes. Approximately 50 percent of the human genome consists of transposable elements; as we shall see later, most of these are evolutionary remnants no longer able to transpose.

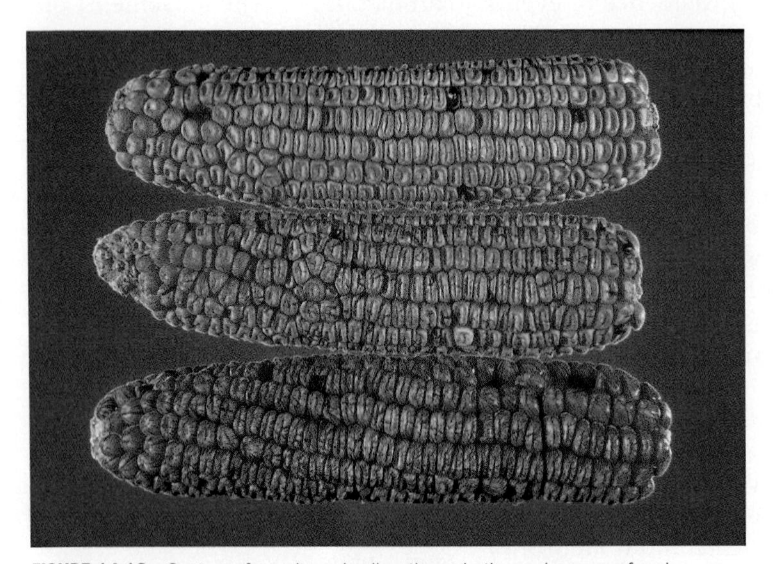

**FIGURE 14.10** Sectors of purple and yellow tissue in the endosperm of maize kernels resulting from the presence of the transposable elements *Ds* and *Ac*. The different level of sectoring in some ears results from dosage effects of *Ac*. [Courtesy of Jerry L. Kermicle, Professor Emeritus, University of Wisconsin at Madison.]

### ■ Molecular Mechanisms of Transposition

The molecular mechanism of *Ds* transposition is illustrated in **FIGURE 14.11**. In this example the insertion goes into the wildtype *shrunken* gene in maize chromosome 9, causing a knockout mutation. To initiate the process, the target

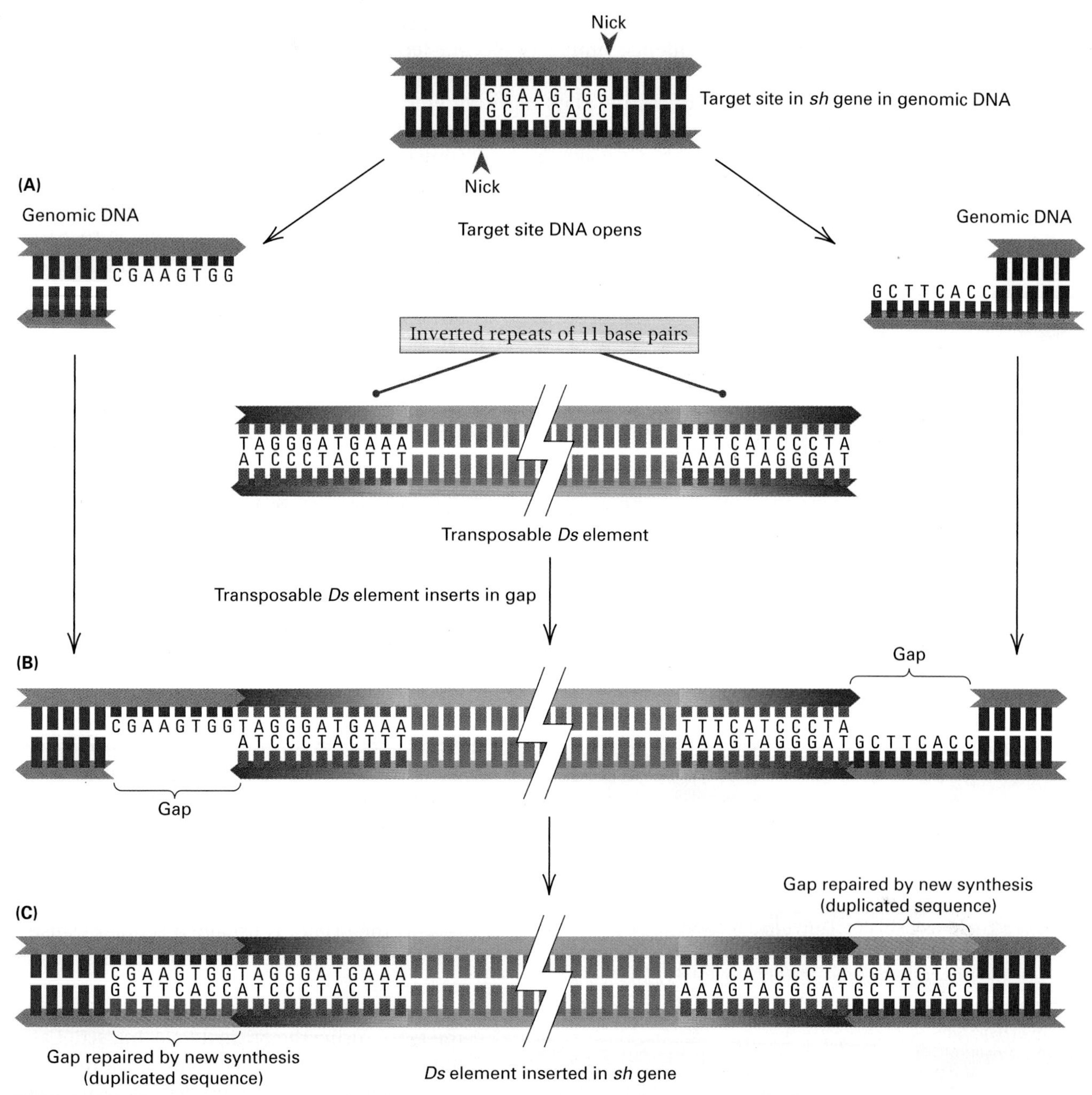


**FIGURE 14.11** The sequence arrangement of a cut-and-paste transposable element (in this case the *Ds* of maize) and the changes that take place when it inserts into the genome. *Ds* is inserted into the maize *sh* gene at the position indicated. In the insertion process, a sequence of eight base pairs next to the site of insertion is duplicated and flanks the *Ds* element.

site for insertion is cleaved with a staggered cut, leaving a 3′ overhang of eight nucleotides on each strand (part A). The overhanging 3′ ends are ligated with the 5′ ends of the *Ds* element to be inserted, leaving an eight-nucleotide gap in each strand (B). When the gap is filled by repair enzymes, the result is a new insertion of *Ds* flanked by an eight-base-pair duplication of the target sequence (C). The *Ds* element can insert in either orientation.

The presence of a target-site duplication is characteristic of most transposable element insertions, and it results from asymmetrical cleavage of the target sequence. For elements like *Ds*, target-site cleavage is a function of a transposase protein that catalyzes transposition. Each family of transposable elements has its own **transposase** that determines the distance between the cuts made in the target DNA strands. Depending on the particular transposable element, the distance may be 1–12 base pairs, and this determines the length of the target-site duplication. Most transposable elements have many potential target sites scattered throughout the genome, and they usually show little or no sequence similarity from one site to the next.

The *Ds* element is one of a large class of transposable elements called **cut-and-paste elements** that insert via a mechanism like that in Figure 14.11. Characteristic of these elements is the presence of *terminal inverted repeats,* a sequence repeated in inverted orientation at each end of the element. In the case of *Ds,* the terminal inverted repeats are 11 bp in length (Figure 14.11), but in other families of transposable elements, they can be up to a few hundred base pairs long. The terminal repeats are usually essential for efficient transposition, because they contain binding sites for the transposase that allow the element to be recognized and ligated into the cleaved target site. Many transposable elements encode their own transposase in sequences located in the central region between the terminal inverted repeats, so these elements are able to promote their own transposition. Elements in which the transposase gene has been deleted or inactivated by mutation are transposable only if another member of the family, encoding a functional transposase, is present in the genome to provide this activity. The inability of the maize *Ds* element to transpose without *Ac* results from the absence of a functional transposase gene in *Ds.* The presence of an *Ac* element provides *trans-activation* that enables a *Ds* element to transpose.

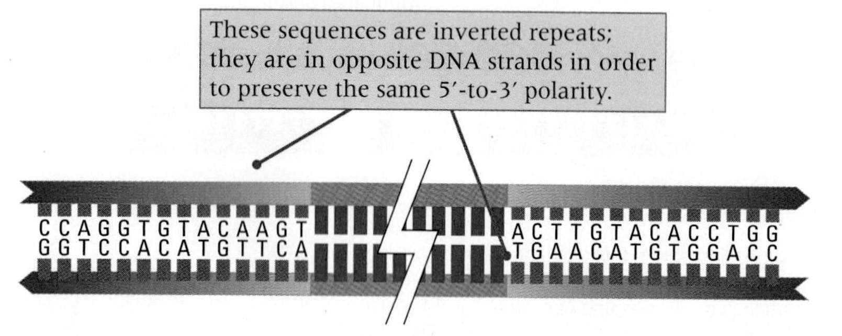

(A) Direct repeat
These sequences are direct repeats; they have the same 5′-to-3′ polarity and are in the same DNA strand.
CCAGGTGTACAAGT
GGTCCACATGTTCA
CCAGGTGTACAAGT
GGTCCACATGTTCA
(B) Inverted repeat
These sequences are inverted repeats; they are in opposite DNA strands in order to preserve the same 5′-to-3′ polarity.
CCAGGTGTACAAGT
GGTCCACATGTTCA
ACTTGTACACCTGG
TGAACATGTGGACC

**FIGURE 14.12** (A) In a direct repeat, a DNA sequence is repeated in the same left-to-right orientation. (B) In an inverted repeat, the sequence is repeated in the reverse left-to-right orientation *in the opposite strand,* in order to maintain the correct 5′-to-3′ polarity.

Another large class of transposable elements possesses terminal direct repeats, typically 200–500 bp in length, called *long terminal repeats,* or *LTRs.* As the name implies, terminal direct repeats are present in the same orientation at both ends of the element (**FIGURE 14.12**, part A), whereas terminal inverted repeats are present in reverse orientation (Figure 14.12B). These elements are called **LTR retrotransposons** because they transpose using an RNA transcript as an intermediate. A typical LTR retrotransposon is the *copia* element from *Drosophila* illustrated in **FIGURE 14.13**; in this case each LTR is itself flanked by short terminal inverted repeats. Transposition of a retrotransposon begins with transcription of the element into an RNA copy. Among the encoded proteins is an enzyme known as **reverse transcriptase**, which can "reverse-transcribe," using the RNA transcript as a template for making a complementary DNA daughter strand. A primer is needed for reverse transcription. For retrotransposons the primer is usually a cellular transfer RNA molecule whose 3′ end is complementary to part of the LTR. The reverse transcriptase adds successive deoxyribonucleotides to the 3′ end of the tRNA, using the original RNA transcript as a template. Single-stranded cleavage of the RNA template by an element-encoded RNase provides a primer for second-strand DNA synthesis using the first DNA strand as a template. In this way a double-stranded DNA copy is made of the RNA transcript, and this is inserted into the target site in a fashion analogous to the mechanism shown

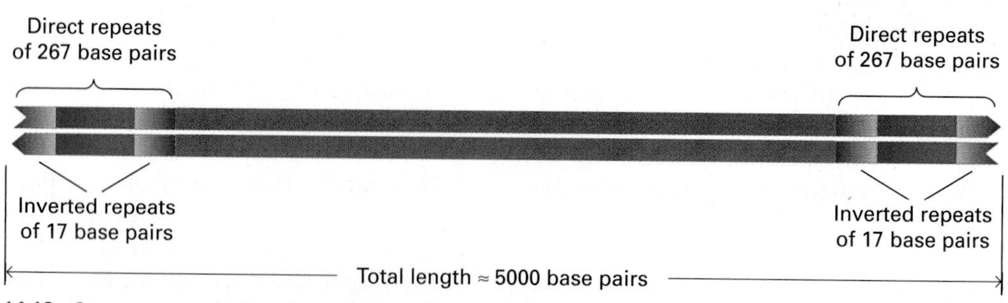


**FIGURE 14.13** Sequence organization of a *copia* retrotransposable element of *Drosophila melanogaster.*

in Figure 14.11, using yet another element-encoded protein. (We have glossed over many details, such as how the primers are removed and replaced with DNA.)

Some retrotransposable elements have no terminal repeats and are called **non-LTR retrotransposons.** This class includes elements denoted *LINE* (long interspersed elements) and *SINE* (short interspersed elements). LINE and SINE elements are the most abundant types of transposable elements in mammalian genomes, although cut-and-paste and LTR retrotransposons are also found. An example of a SINE in the human genome is a set of related sequences called the *Alu* family because its members contain a characteristic restriction site for the restriction endonuclease *Alu*I. The *Alu* sequences are about 300 base pairs in length and are present in approximately 500,000 copies in the human genome. The *Alu* family alone accounts for about 11 percent of human DNA. In many organisms, transposable elements of various families constitute a significant part of the total genome size.

## Transposable Elements as Agents of Mutation

Transposable elements can cause mutations. For example, in some genes in *Drosophila*, approximately half of all spontaneous mutations that have visible phenotypic effects result from insertions of transposable elements (**FIGURE 14.14**). The wrinkled-seed mutation in Mendel's peas is another good example of mutation due to a transposable element. In this case, the transposable element is a cut-and-paste element related to the maize *Ac* element that also produces an 8-bp target-site duplication. The insertion site is in the gene for starch-branching enzyme I (SBEI), and the insertion creates a loss-of-function allele. Most transposable elements are present in nonessential regions of the genome and usually cause no detectable phenotypic change. But when an element transposes, it can insert into an essential region that disrupts the region's function. If transposition inserts an element into a coding region of DNA, then the inserted element interrupts the coding region. Because most transposable elements contain coding regions of their own, either transcription of the transposable element interferes with transcription of the gene into which it is inserted, or transcription of the gene terminates within the transposable element. The insertion therefore causes a knockout mutation. Even if transcription proceeds through the element, the phenotype will be mutant because the coding region then contains incorrect sequences.

Transposable elements can also cause chromosomal rearrangements indirectly through homologous recombination between copies present at different positions in the genome.

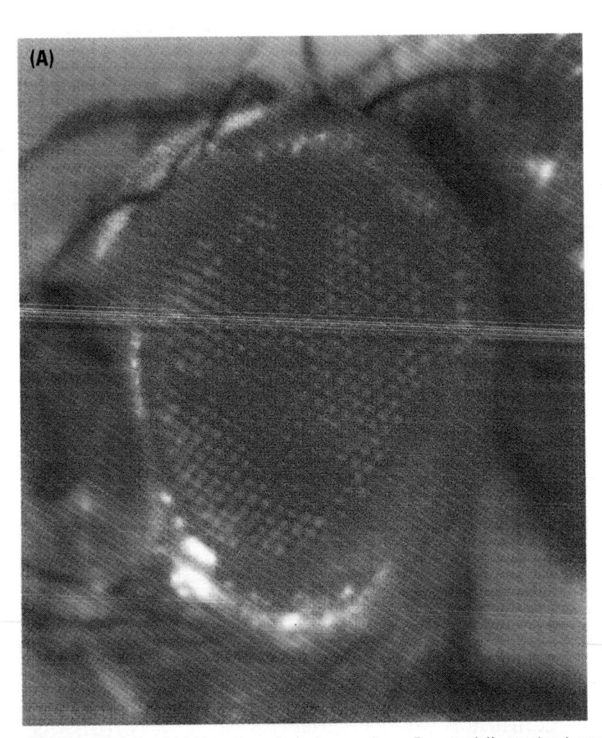


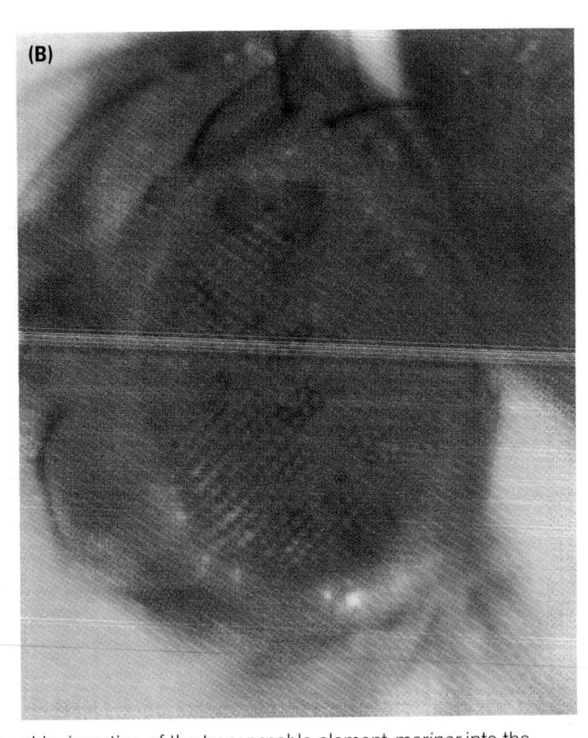


**FIGURE 14.14** (A) Peach-colored eyes in a *Drosophila* mutant caused by insertion of the transposable element *mariner* into the upstream regulatory region of the *white* gene, which alters the level of expression of the gene. (B) Mosaic eye color. The patches of red are caused by excision of the transposable element, restoring normal *white*-gene expression to the cells in which excision took place.

This process is known as *ectopic recombination*, and its consequences have been examined in Chapter 8. Ectopic recombination between copies present in the same orientation in the same chromosome results in a deletion of the material between the copies (Figure 8.14), and ectopic recombination between copies present in inverted orientation in the same chromosome results in an inversion of the material between the copies (Figure 8.20). Similarly, ectopic recombination between copies present in different chromosomes can result in reciprocal translocations.

## ■ Transposable Elements in the Human Genome

Part of the reason why the human genome is so relatively large yet contains fewer than 30,000 genes is that it includes a high proportion of transposable elements. The principal categories and their abundances are shown in **TABLE 14.2**. The general categories are shown, along with specific examples. The human genome consists of almost 50 percent transposable elements. The largest single category consists of SINE elements, of which the *AluI* family is the most abundant. Although transposable elements were long regarded as "selfish DNA"—a sort of genomic parasite—evidence is beginning to suggest that at least the *AluI* family may benefit the human genome. First, *AluI* elements are disproportionately represented in gene-rich regions of the genome that are high in G + C content, which suggests that they play some functional role. Second, in human beings, as in some other organisms, SINE elements are transcribed when the organism is under stress. The resulting transcripts can bind to a particular protein kinase that normally blocks translation under stress. In this way SINE elements may be able to promote translation under organismic stress.

The second major class of human transposable elements consists of LINE elements, of which *LINE1* is the most abundant. Third on the list are the LTR retrotransposable elements, followed a distant fourth by DNA elements, of which the *mariner* transposon is an example. The *mariner* transposon is of some interest because it is widespread among eukaryotic genomes. About 14 percent of all insect species carry *mariner*, for example. One reason for its wide distribution is that *mariner* is relatively efficient in being transferred from one species to another, even unrelated, species, but the mechanisms by which this horizontal transmission takes place are largely matters of speculation.

The human genomic DNA sequence implies that most transposons in the genome are no longer capable of transposition. One type of evidence derives from comparing sequences of different copies of the same element throughout the genome. Because a transpositionally active element will give rise to new copies that are identical or nearly identical in sequence from one to the next, close sequence similarity among copies suggests active transposition. On the other hand, copies of transposons that can no longer move are free to change in sequence as successive mutations take place and are incorporated into the population, and so large sequence differences among copies suggest a low rate of transposition. Because the average rate of nucleotide substitution per base pair in the human genome is roughly constant through time, the amount of sequence divergence between copies can be used to estimate the time since transposition.

The analysis of sequence differences among human transposable elements suggests that the overall activity of transposable elements in the human genome has decreased substantially, and quite steadily, over the past 35 to 50 million years. The ancient times mean that the decrease in transposition was taking place in the hominid lineage long before human beings existed as a species. Other mammals that have been studied show greater and more typical rates of transposition. In the mouse, for example, the rate of transposition of SINE and LINE elements, relative to that in the human genome, has increased from 1.7 times higher in the past 100 million years to 2.6 times higher in the past 25 million years. This comparison is consistent with the finding that about 1 in 10 new mutations in the mouse is due to transposition, whereas only about 1 in 600 new mutations in the human genome is due to transposition. LTR retrotransposons exhibit no convincing evidence of ongoing transposition in the human genome, and DNA transposons seem to have lost their ability

**Table 14.2 Transposable elements in the human genome**

| Type | Number of copies | Percentage of total genome |
|---|---|---|
| SINEs | 1,558,000 | 13.1 |
| *AluI* | 1,090,000 | 10.6 |
| LINEs | 868,000 | 20.4 |
| *LINE1* | 516,000 | 16.9 |
| LTR elements | 443,000 | 8.3 |
| DNA elements | 294,000 | 2.8 |
| *mariner* | 14,000 | 0.1 |
| Unclassified | 3,000 | 0.1 |
| **Total of all types** | | **44.7** |

Source: Data from International Human Genome Sequencing Consortium, *Nature* 409 (2001): 860–921.

to transpose about 50 million years ago. Hence, human beings stand in contrast to many other organisms, including other mammals, in which transposition is a major source of mutation as well as evolutionary innovation.

### RIP: A Defense Against Transposons

Transposable elements are like molecular parasites in that they invade and multiply within the genome. Organisms have evolved various defense mechanisms to attack sequences that are present in multiple copies in the genome. Most of these are mechanisms of "silencing," in which the sequences remain intact but their transcription is prevented or their transcripts are destroyed. These mechanisms are discussed in the context of gene regulation in Chapter 11. However, some organisms have evolved defense mechanisms that attack repeated sequences by means of mutation. Notable among these is a process known as **RIP (repeat-induced point mutation)**, which occurs in the filamentous fungus *Neurospora crassa* and related species.

RIP takes place in haploid cells undergoing the sexual cycle. Because the cells are haploid, each gene should be present in only a single copy. Sequences that are longer than a few hundred base pairs and that are repeated in the haploid genome are subjected to RIP. The mechanisms of repeat recognition and mutagenesis are poorly understood, but the result of RIP is that the repeated sequences undergo multiple transition mutations in which G—C nucleotide pairs are substituted with A—T nucleotide pairs. The hypermutation creates multiple missense and nonsense mutations in coding sequences that may be present in the repeated sequences, effectively eliminating a functional gene product. In addition, many of the remaining C-bearing nucleotides in the repeats become methylated, which decreases transcription as noted in Section 14.2 in the context of the fragile-X syndrome. RIP is so efficient at detecting and mutating repeats that it affords a convenient approach to obtaining new mutations in single-copy genes. The procedure is to introduce extra partial or complete copies of the gene of interest into the genome by DNA transformation, after which RIP rips into the repeats, including the original single-copy gene that is present. (The *Neurospora* genome does, however, include some sequences that are tandemly repeated, as they are in other organisms, such as those transcribed into ribosomal RNA and those encoding the histone chromosomal proteins; these sequences are protected from RIP by an unknown mechanism.)

## 14.4 Spontaneous Mutation

Mutations are statistically random events. There is no way of predicting when, or in which cell, a mutation will take place. However, every gene mutates spontaneously at a characteristic rate, so it is possible to assign probabilities to particular mutational events. The various kinds of mutational alterations in DNA differ substantially in complexity, so their probabilities of realization are quite different. A fundamental principle concerning mutation is that:

> The mutational process is random in the sense that whether a particular mutation happens is unrelated to any adaptive advantage it may confer on the organism in its environment. A potentially favorable mutation does not arise *because* the organism has a need for it.

The experimental basis for this conclusion is presented in the next section.

### The Nonadaptive Nature of Mutation

The concept that mutations are spontaneous, statistically random events unrelated to adaptation was not widely accepted until the late 1940s. Before that time, it was believed that mutations occurred in bacterial populations *in response to* particular selective conditions. The basis for this belief was the observation that when antibiotic-sensitive bacteria are spread on a solid growth medium containing the antibiotic, some colonies that consist of cells that have an inherited resistance to the drug form. The initial interpretation of this observation (and similar ones) was that these adaptive variations were *induced* by the selective agent itself.

Several types of experiments showed that adaptive mutations take place spontaneously and hence were present at low frequency in the bacterial population even *before* it was exposed to the antibiotic. One experiment utilized a technique developed by Joshua and Esther Lederberg called **replica plating** (**FIGURE 14.15**). In this procedure, a suspension of bacterial cells is spread on a solid medium. After colonies have formed, a piece of sterile velvet mounted on a solid support is pressed onto the surface of the plate. Some bacteria from each colony stick to the fibers, as shown in Figure 14.15A. Then the velvet is pressed onto the surface of fresh medium, transferring some of the cells from each colony, which give rise to new colonies that have positions identical to those on the first plate. Figure 14.15B shows how this method was used to demonstrate the spontaneous origin of mutant *E. coli* that are

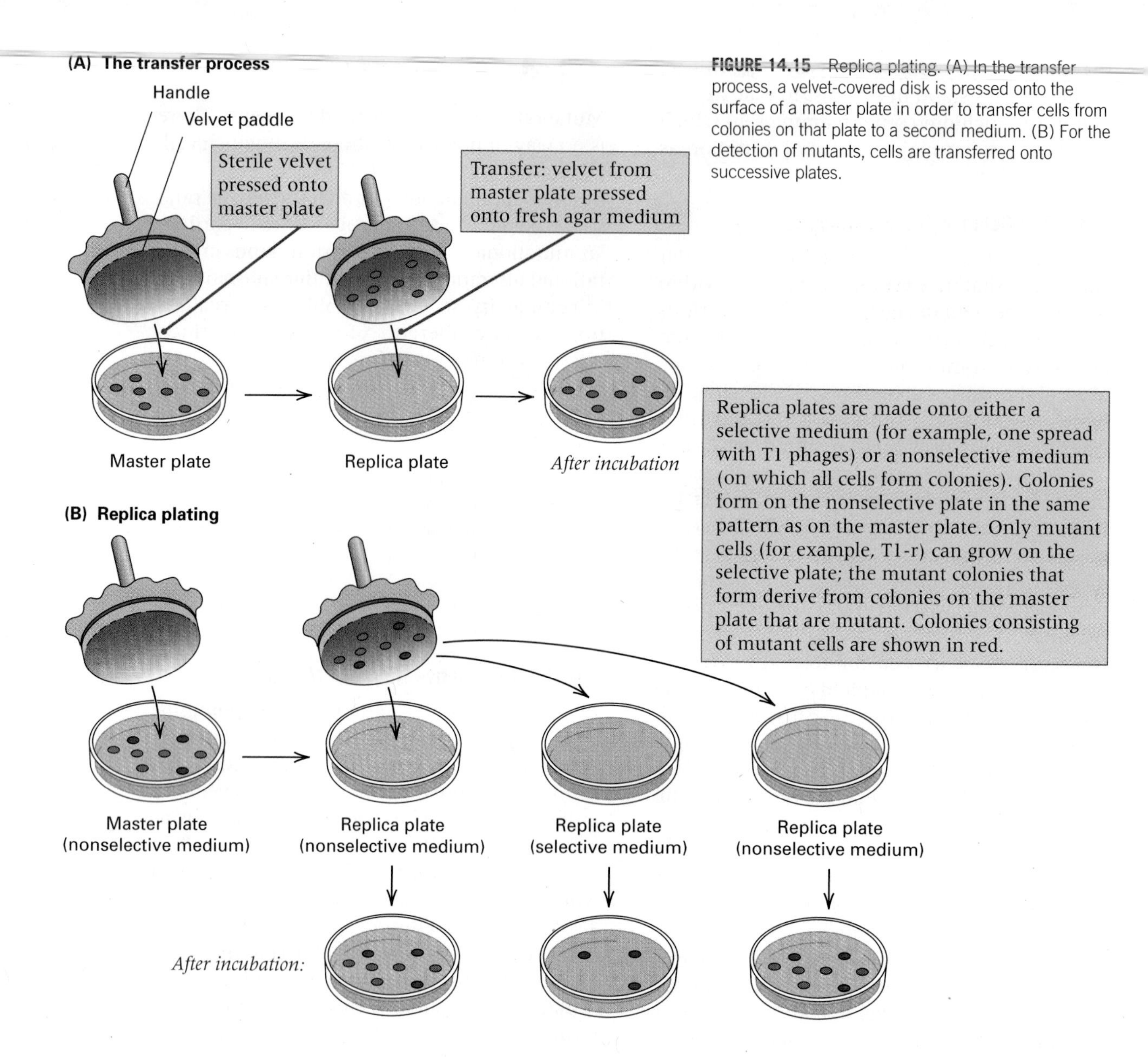
(A) The transfer process
Handle
Velvet paddle
Sterile velvet pressed onto master plate
Transfer: velvet from master plate pressed onto fresh agar medium
Master plate
Replica plate
After incubation
(B) Replica plating
Replica plates are made onto either a selective medium (for example, one spread with T1 phages) or a nonselective medium (on which all cells form colonies). Colonies form on the nonselective plate in the same pattern as on the master plate. Only mutant cells (for example, T1-r) can grow on the selective plate; the mutant colonies that form derive from colonies on the master plate that are mutant. Colonies consisting of mutant cells are shown in red.
Master plate (nonselective medium)
Replica plate (nonselective medium)
Replica plate (selective medium)
Replica plate (nonselective medium)
After incubation:

**FIGURE 14.15** Replica plating. (A) In the transfer process, a velvet-covered disk is pressed onto the surface of a master plate in order to transfer cells from colonies on that plate to a second medium. (B) For the detection of mutants, cells are transferred onto successive plates.

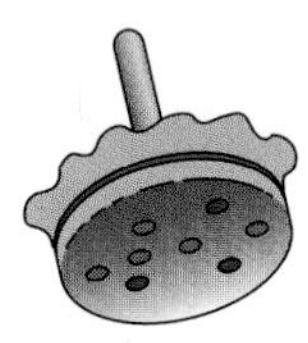
Velvet paddle

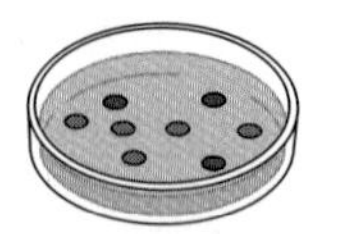
Replica plate (nonselective medium)

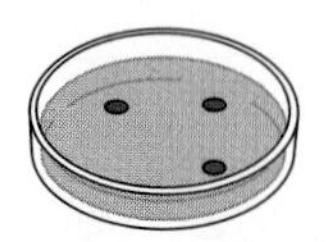
Replica plate (selective medium)

resistant to the bacteriophage T1 (T1-r mutants). A master plate containing about $10^7$ cells growing on nonselective medium (lacking T1 phage) was replica-plated onto a series of plates that had been spread with about $10^9$ T1 phages. After incubation for a time sufficient for colony formation, a few colonies of phage-resistant bacteria appeared in the same positions on each of the selective replica plates. This meant that the T1-r cells that formed the colonies must have been transferred from corresponding positions on the master plate. Because the colonies on the master plate had never been exposed to the phage, the mutations to resistance must have been present, by chance, in a few original cells not exposed to the phage.

The replica-plating experiment illustrates that:

> Selective techniques merely select mutants that preexist in a population.

This principle is the basis for understanding how natural populations of rodents, insects, and disease-causing bacteria become resistant to the chemical substances used to control them. A familiar example is the high level of resistance to insecticides, such as DDT, that now exists in many insect populations. The resistance is due to selection for spontaneous mutations affecting behavioral, anatomical, and enzymatic traits that enable the insect to become tolerant to the chemical. Similar problems are encountered in controlling plant pathogens. For example, the introduction of a new variety of a crop plant that is resistant to a particular strain of disease-causing fungus results in only temporary protection against the disease. The resistance inevitably breaks down because of the occurrence of spontaneous mutations in the fungus that enable it to attack the new plant genotype. Such mutations have a clear selective advantage, and the mutant alleles rapidly become widespread in the fungal population.

## Estimation of Mutation Rates

Spontaneous mutations tend to be rare (except for unusual phenomena such as dynamic mutation), and the methods used to estimate the frequency with which they arise require large populations and (often) special techniques. The **mutation rate** is the probability that a gene undergoes mutation in a single generation or in forming a single gamete. Measurement of mutation rates is important in population genetics, in studies of evolution, and in analysis of the effects of environmental mutagens.

One of the earliest techniques to be developed for measuring mutation rates is the ***ClB* method**. *ClB* refers to a special X chromosome of *Drosophila melanogaster.* This X chromosome has a large inversion (*C*) that (for reasons described in Chapter 8) prevents the recovery of recombinant chromosomes in the progeny from a female heterozygous for the chromosome; a recessive lethal (*l*); and the dominant marker bar (*B*), which reduces the normal round eye to a bar shape. The presence of a recessive lethal in the X chromosome means that males with that chromosome and females homozygous for it cannot survive. The technique is designed to detect mutations arising in a normal X chromosome.

In the *ClB* procedure, females heterozygous for the *ClB* chromosome are mated with males that carry a normal X chromosome (**FIGURE 14.16**). Females with the bar phenotype are chosen from the progeny and then individually mated with normal males. (The presence of the bar phenotype indicates that the females are heterozygous for the *ClB* chromosome and the normal X chromosome from the male parent.) A ratio of two females to one male is expected among the progeny from such a cross, as shown in the lower part of the illustration. The critical observation in determining the mutation rate is the fraction of males produced in the second cross. Because the *ClB* males die, all of the males in this generation must contain an X chromosome derived from the X chromosome of the initial normal male (top row of Figure 14.16). Furthermore, it must be a nonrecombinant X chromosome because of the inversion (*C*) in the *ClB* chromosome. Occasionally, the progeny include no males, and this means that the X chromosome present in the original sperm underwent a mutation somewhere along its length to yield a new recessive lethal. The method provides a quantitative estimate of the rate at which mutation to a recessive lethal allele occurs *at any of the large number of genes* in the X chromosome. About 0.15 percent of the X chromosomes acquire new recessive lethals in spermatogenesis; this means that the mutation rate is $1.5 \times 10^{-3}$ recessive lethals per X chromosome per generation. Note that the *ClB* method tells us nothing about the mutation rate for a particular gene, because the method does not reveal how many genes on the X chromosome would cause lethality if they were mutant. Since the time that the *ClB* method was devised, a variety of other methods have been developed for determining mutation rates in *Drosophila* and other organisms. Of significance is the fact that mutation rates vary widely from one gene to another. For example, the yellow-body mutation in *Drosophila* occurs at a frequency of $10^{-4}$ per gamete per generation, whereas mutations to streptomycin resistance in *E. coli* occur at a frequency of $10^{-9}$ per cell per generation. Furthermore, within a single organism, the frequency can vary enormously; in *E. coli* it ranges from $10^{-5}$ for some genes to $10^{-9}$ for others.

## Hot Spots of Mutation

Certain DNA sequences are called mutational **hot spots** because they are more likely to undergo mutation than others. Mutational hot spots may include monotonous runs of a single nucleotide or tandem repeats of a short sequence, such as an unstable trinucleotide repeat, that may expand or contract by replication slippage or other mechanisms. Hot spots are found at many sites throughout the genome and within genes. For genetic studies of mutation, the existence of hot spots means that a relatively small number of sites accounts for a disproportionately large fraction of all mutations. For example, in the analysis of the *rII* gene in bacteriophage T4 (Chapter 9), one extreme hot spot accounted for approximately 20 percent of all the mutations obtained.

In many organisms, including bacteria, maize, and mammals (but not yeast, fruit flies, or nematodes), at least 1 percent of the cytosine bases are methylated at the carbon-5 position, yielding 5-methylcytosine instead of ordinary cytosine (**FIGURE 14.17**). The methylation is carried out in duplex DNA and catalyzed by a *cytosine methylase* that uses S-adenosylmethionine as the donor of the methyl group ($-CH_3$). The methylation is not random but, rather, occurs preferentially at certain target sequences that contain cytosine—in mammals at cytosines in CG dinucleotides. In

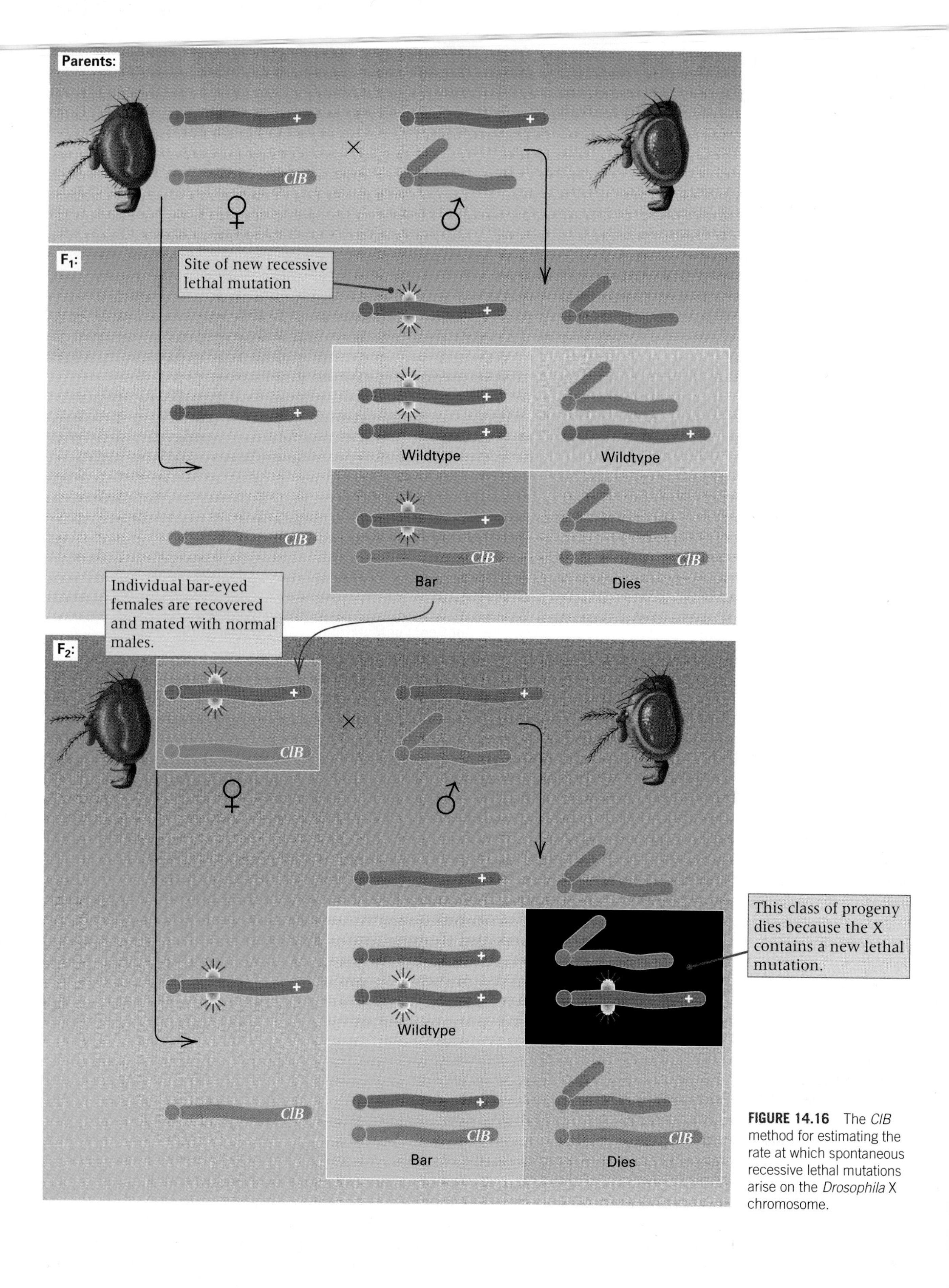


**FIGURE 14.16** The *ClB* method for estimating the rate at which spontaneous recessive lethal mutations arise on the *Drosophila* X chromosome.

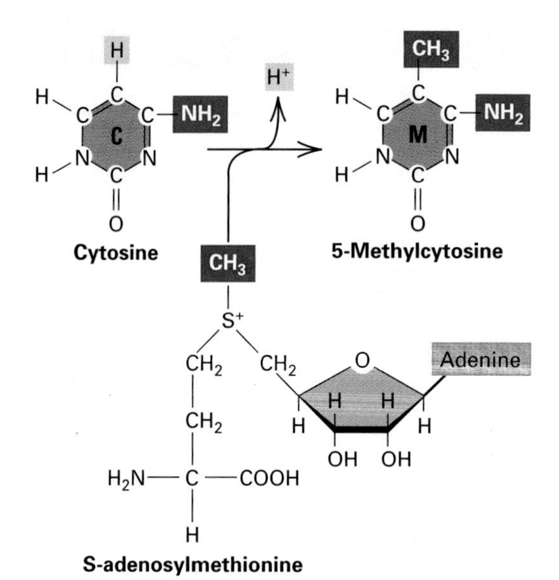


**FIGURE 14.17** Methylation of cytosine at the number-5 position in the base. The methyl donor is S-adenosylmethionine.

(A)

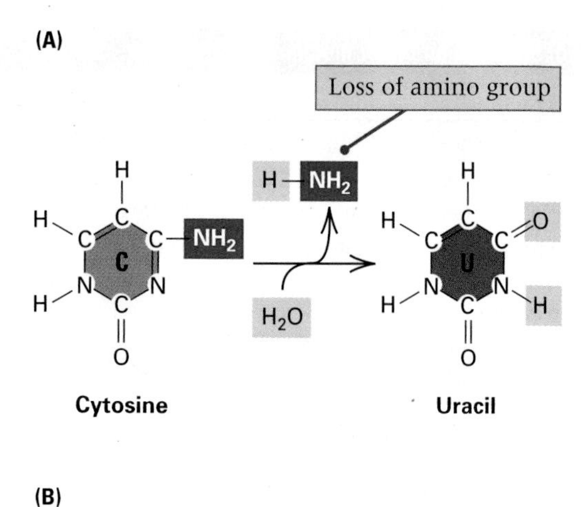


(B)

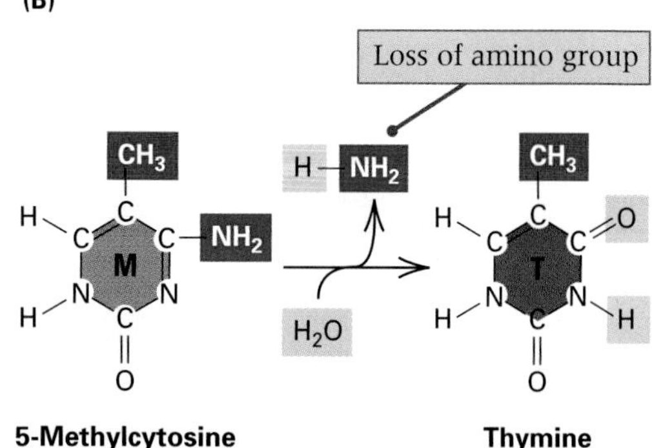


**FIGURE 14.18** (A) Spontaneous loss of the amino group of cytosine to yield uracil. (B) Spontaneous loss of the amino group of 5-methylcytosine to yield thymine.

DNA replication, the 5-methylcytosine pairs with guanine and replicates normally.

As we have seen in the discussion of the fragile-X syndrome, high levels of methylation in or around certain genes are associated with a dramatic reduction in the level of transcription (Figure 14.7). In addition to its role in repressing gene transcription, cytosine methylation is an important contributor to mutational hot spots because the methylated cytosines are usually highly mutable. The mutations are usually G−C → A−T transitions, for the reason illustrated in **FIGURE 14.18.** Both cytosine and 5-methylcytosine are subject to occasional loss of an amino group, a process known as **deamination**. For cytosine, this loss yields *uracil*. Because uracil pairs with adenine instead of guanine, replication of a molecule containing a G−U base pair would ultimately lead to substitution of an A−T pair for the original G−C pair, because in successive rounds of replication the affected base pair becomes G−U → A−U → A−T.

Fortunately, as we shall see in Section 14.6, uracil is one of the bases recognized by a DNA repair system called *base-excision repair*. In this repair process, the uracil is removed and replaced with normal cytosine. Hence, deamination of normal cytosine resulting in uracil is easily detected and repaired.

Not so for 5-methylcytosine, however. Unlike normal cytosine, deamination of 5-methylcytosine yields *thymine* (Figure 14.18B), which is a normal DNA base and hence is not removed by the DNA uracil glycosylase. In this case, deamination changes the original G−C pair into a G−T pair, which in the next round of replication gives G−C (wildtype) and A−T (mutant) daughter molecules.

## 14.5 Mutagens

A **mutagen** is any agent that causes an increase in the rate of mutation above the spontaneous background. The first evidence that external agents could increase the mutation rate was presented in 1927 by Hermann Muller, who showed that x-rays are mutagenic in *Drosophila*. Since then, a large number of physical agents and chemicals have been shown to increase the mutation rate. Some of the principal agents that damage DNA are listed in **TABLE 14.3**, along with the major types of damage they produce. Because some environmental contaminants are mutagenic, as are numerous chemicals found in tobacco products, mutagens are of great importance in public health.

### Depurination

Heading the list in Table 14.3 is water. In purine nucleotides, the sugar–purine bonds are relatively labile and subject to hydrolysis. The loss of the purine base, called **depurination,** is illustrated in **FIGURE 14.19.** Loss of a pyrimidine base (*depyrimidation*) can also occur through

| Table 14.3 Major agents of mutation and their mechanisms of action | | |
|---|---|---|
| **Agent of mutation** | **Examples** | **Principal mechanism of mutagenesis** |
| Water | Hydrolysis | Depurination (A or G detached from its deoxyribose sugar) |
| Oxidizing agent | Nitrous acid | Deamination ($-NH_2 \rightarrow =O$): C → U, 5-MeC → T, A → Hypoxanthine |
| Base analog | 5-Bromodeoxyuridine | Increased rate of base mispairing |
| Alkylating agent | Ethylmethane sulfonate<br>Nitrogen mustard | Bulky attachments made to side groups on bases |
| Intercalating agent | Proflavin | Causes topoisomerase II to leave a nick in DNA strand; misrepair results in the insertion or deletion of one or a few nucleotides |
| Ultraviolet light | Natural sunlight<br>UV lamps | Forms pyrimidine dimers (covalent bonds between adjacent pyrimidines, primarily T) present in the same DNA strand |
| Ionizing radiation | X-rays<br>Radon gas<br>Radioactive materials | Single- and double-stranded breaks in DNA; damage to nucleotides |

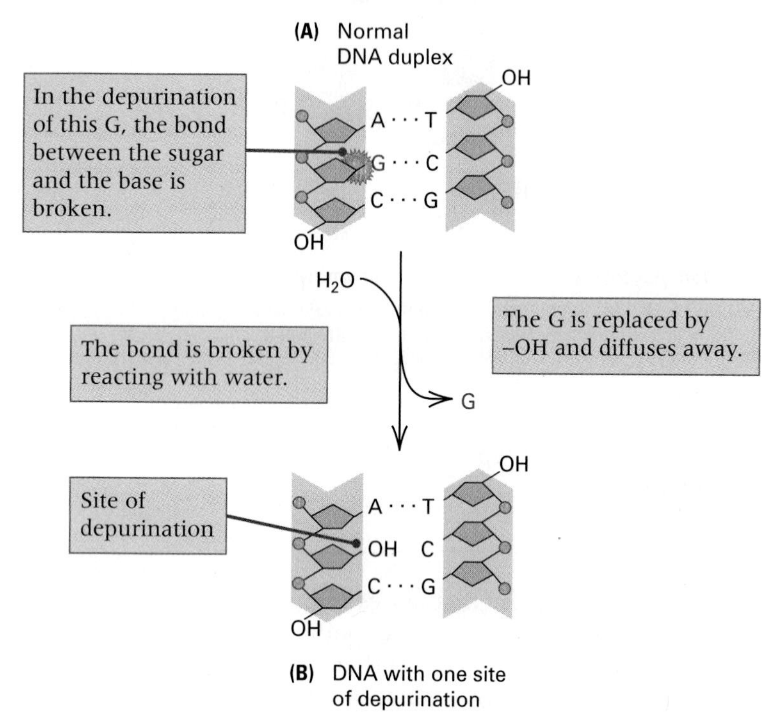


**FIGURE 14.19** Depurination. (A) Part of a DNA molecule prior to depurination. The bond between the labeled G and the deoxyribose to which it is attached is about to be hydrolyzed. (B) Hydrolysis of the bond releases the G purine, which diffuses away from the molecule and leaves a hydroxyl (–OH) in its place in the depurinated DNA.

hydrolysis, but to a lesser extent than depurination. Loss of the base from a deoxyribose sugar is not always mutagenic, because the site lacking the base can be corrected by the AP repair system (Section 14.6). If, however, the replication fork reaches the apurinic site before repair has taken place, then replication almost always inserts an adenine nucleotide in the daughter strand opposite the apurinic site. After another round of replication, what was originally a G–C pair becomes a T–A pair, which is an example of a transversion mutation.

In air, the rate of spontaneous depurination is approximately $3 \times 10^{-9}$ depurinations per purine nucleotide per minute. This rate is at least tenfold greater than any other single source of spontaneous DNA degradation. At this rate, the half-life of a purine nucleotide exposed to air is about 300 years. This sets a practical limit to how long DNA can persist in the environment before losing its biological activity.

## Oxidation

Oxidizing agents are also prominent in their ability to damage DNA. An *oxidizing agent* is any chemical that readily transfers oxygen atoms. Common oxidizing agents include bleach (NaOCl), hydrogen peroxide ($H_2O_2$), and nitrous acid ($HNO_2$). Oxidation of the bases changes their properties of hydrogen bonding, which causes incorrect nucleotides to be incorporated into duplex DNA. The example in **FIGURE 14.20** shows one of the most common types of oxidative damage to DNA, the oxidation of guanine to produce 8-oxoguanine. Whereas guanine pairs with cytosine, 8-oxoguanine pairs with adenine. Hence, 8-oxoguanine results in C to A transversions. As we will discuss in Section 14.6, however, 8-oxoguanine is one of the modified bases recognized and repaired by the base-excision repair system.

Nitrous acid acts as a mutagen by deamination of the bases adenine, cytosine, and gua-

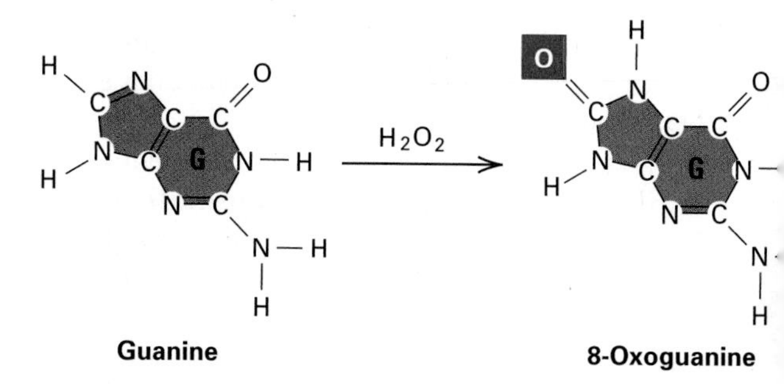


**FIGURE 14.20** Oxidation of guanine results in 8-oxoguanine.

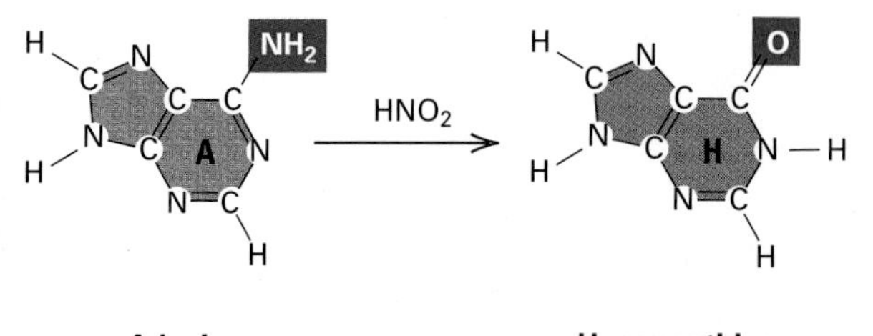


**Adenine** **Hypoxanthine**

**FIGURE 14.21** Deamination of adenine results in hypoxanthine.

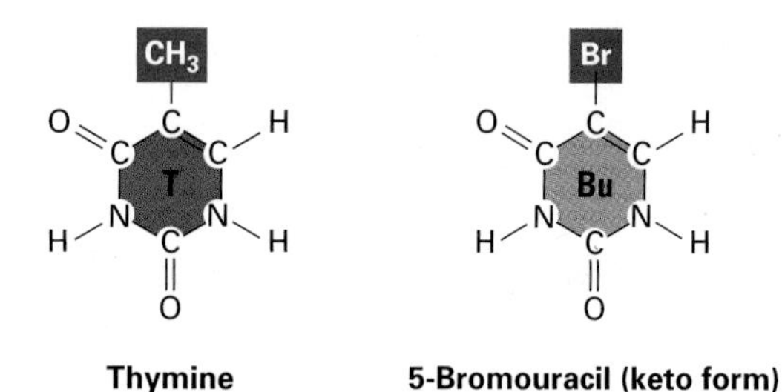


**Thymine** **5-Bromouracil (keto form)**

**FIGURE 14.22** Structures of thymine and 5-bromouracil.

nine. Deamination alters the hydrogen-bonding specificity of each base. As we have seen in Figure 14.18, deamination of 5-methylcytosine results in thymine, and deamination of cytosine results in uracil. The result of deamination of adenine is illustrated in **FIGURE 14.21**. The product is a base called *hypoxanthine*, which pairs with cytosine rather than thymine, so the result of deamination of A is an A–T to G–C transition.

## Base-Analog Mutagens

A **base analog** is a molecule sufficiently similar to one of the four DNA bases that it can be incorporated into a DNA duplex in the course of normal replication. Such a substance must be able to pair with a normal base in the template strand. Some base analogs are mutagenic because they are more prone to mispairing than are the normal bases. This mechanism of mutagenesis can be illustrated with 5-bromouracil (Bu), a commonly used base analog that is efficiently incorporated into the DNA of bacteria and viruses.

The base 5-bromouracil is an analog of thymine, and the bromine atom is about the same size as the methyl group of thymine (**FIGURE 14.22**). Normally, 5-bromouracil is in the *keto* form, in which it pairs with adenine (**FIGURE 14.23**, part A), but it occasionally shifts its configuration to the *enol* form, in which it pairs with guanine (part B). The shift is influenced by the bromine atom and takes place in 5-bromouracil more frequently than in thymine.

There are two pathways by which 5-bromouracil can be mutagenic. These are illustrated in **FIGURE 14.24**. In pathway A, the 5-bromouracil is incorporated in its enol form, paired with G. This mode of incorporation is rare, but the mutagenic base pair is created in the first round of replication. In the next round of replication, the Bu will usually pair with A, which leads to a G–C to A–T transition. In pathway B, the 5-bromouracil is incorporated in its keto form, paired with A. This is by far the more frequent mode of incorporation, but the mutagenic base pair is not formed until a later round of replication when Bu pairs with G. In this case the result is an A–T to G–C transition.

## Chemical Agents That Modify DNA

Some mutagens react with DNA in a variety of different ways and produce a broad spectrum of effects. Among these are the **alkylating agents**, which are highly reactive chemicals that act as potent mutagens in both prokaryotes and eukaryotes. Examples of alkylating agents are ethyl methanesulfonate (EMS) and nitrogen mustard, the structures of which are shown in **FIGURE 14.25**. Nitrogen mustard is a gas causing extreme pain and extensive lung damage when inhaled, and was used for chemical warfare in Europe in the First World War (1914–1918).

**(A) Adenine paired with 5-bromouracil**

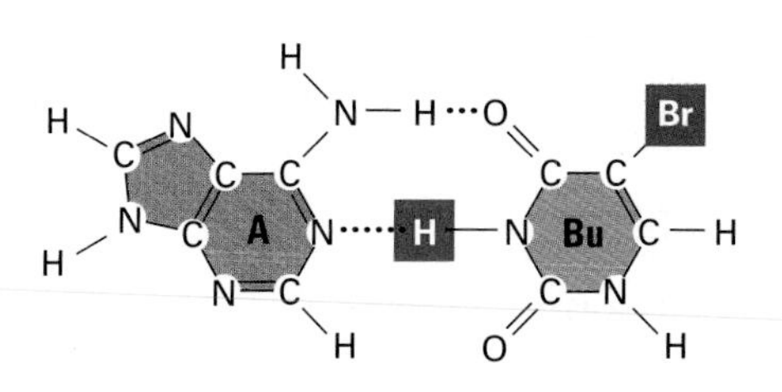


**(B) Guanine paired with 5-bromouracil**

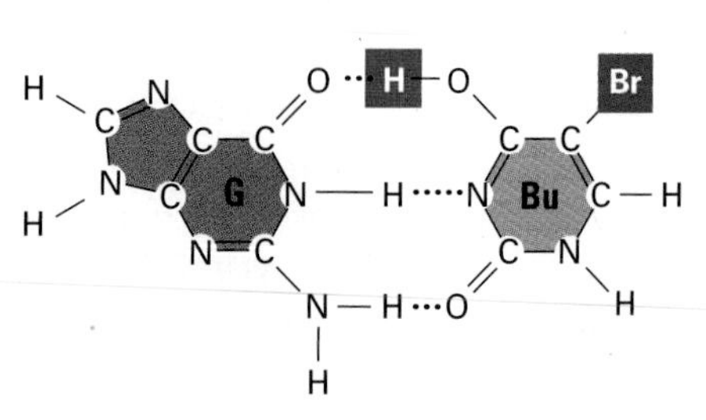


**Adenine** **5-Bromouracil (keto form)** **Guanine** **5-Bromouracil (enol form)**

**FIGURE 14.23** Mispairing mutagenesis by 5-bromouracil. (A) A base pair between adenine and the keto form of 5-bromouracil. (B) A base pair between guanine and the rare enol form of 5-bromouracil. One of 5-bromouracil's hydrogen atoms changes position to create the keto form.

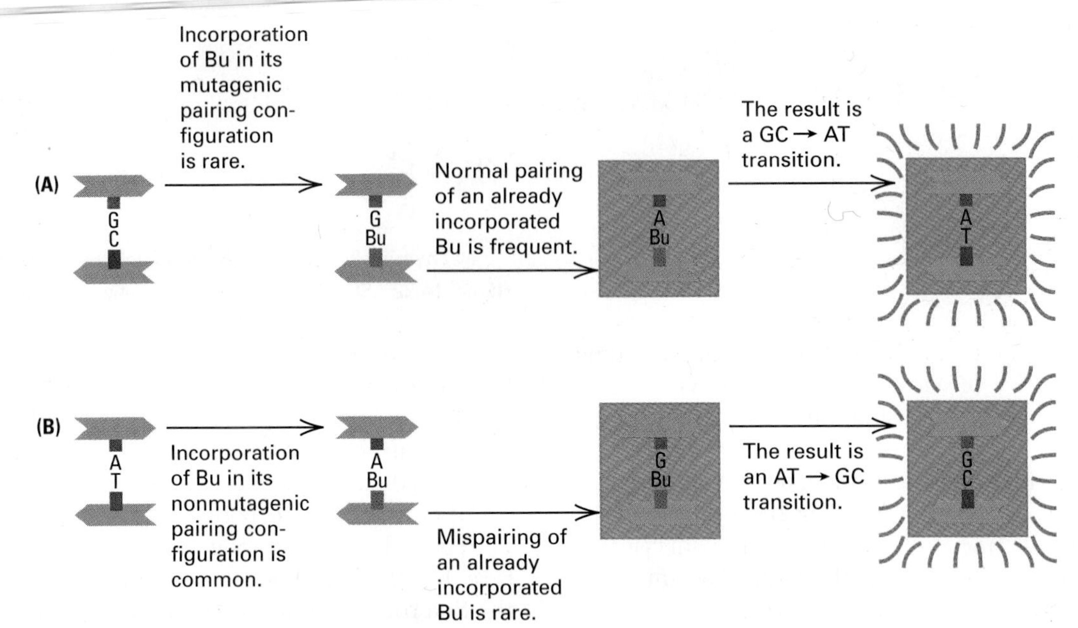


**FIGURE 14.24** Shown are two pathways for mutagenesis by 5-bromouracil (Bu). The position of the arrow shows which strand of each DNA duplex is being followed through the next round of replication. (A) Incorporation paired with G is rare. In this case the mutagenic base pair is formed in the first round of replication. (B) Incorporation paired with A is frequent. In this case the mutagenic base pair is formed after the first round of replication.

$CH_3-CH_2-O-S(=O)_2-CH_3$

**Ethyl methanesulfonate**

$CH_3-N(CH_2-CH_2-Cl)_2$

**Nitrogen mustard**

**FIGURE 14.25** The chemical structures of two highly mutagenic alkylating agents; the alkyl groups are shown in red.

Ethyl methanesulfonate is a soluble solid and has been used widely to induce mutations for genetic research. The alkylating agents add bulky side groups to the DNA bases that either alter their base-pairing properties or cause structure distortion of the DNA molecule. For example, the reaction of EMS with guanine results in $O^6$-ethylguanine (**FIGURE 14.26**). Alkylation of either guanine or thymine causes mispairing, leading to the transitions A–T to G–C or G–C to A–T. EMS reacts less readily with adenine and cytosine.

## Intercalating Agents

The **acridine** molecules are planar three-ringed molecules whose dimensions are roughly the same as those of a purine–pyrimidine pair. Acridine orange is an example whose structure is shown below.

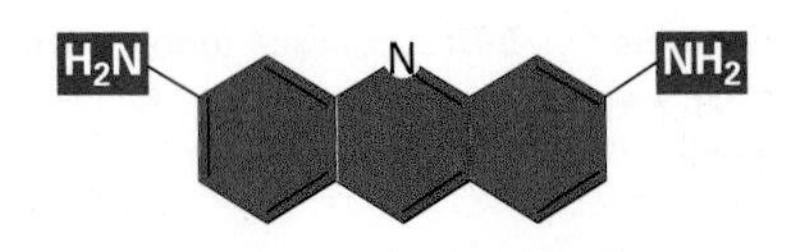


The acridines cause short insertions or deletions in a DNA molecule, primarily consisting of a single nucleotide. Because of their structure, the acridines were once thought to act by inserting themselves between the base pairs in DNA (a process called **intercalation**) causing a distortion resulting in a small insertion or deletion. Current data, however, suggest that acridines actually do their damage by interfering with topoisomerase II. This enzyme relieves torsional stress in DNA by making a double-stranded break, rotating the free ends, and then sealing the break. In the presence of acridine, the enzyme leaves the DNA nicked, and failure of prompt repair results in the addition or deletion of one or occasionally a few base pairs at the site. The result of a single-base addition or deletion in a coding region is a frameshift mutation (Figure 14.5).

## Ultraviolet Irradiation

Ultraviolet (UV) light is a mutagen in all viruses and cells. The effects are caused by chemical changes in the bases resulting from absorption of the energy of the light. The major products formed in DNA after UV irradiation are covalently joined pyrimidines (**pyrimidine dimers**), primarily thymine (**FIGURE 14.27**, part A), that are adjacent in the same polynucleotide strand. This chemical linkage brings the bases closer together, causing a distortion of the helix (Figure 14.27B), which blocks transcription and transiently blocks DNA replication. Pyrimidine dimers can be repaired in ways discussed later in this chapter. Nevertheless, excessive exposure of the skin to the UV rays in sunlight increases the risk of skin cancer.

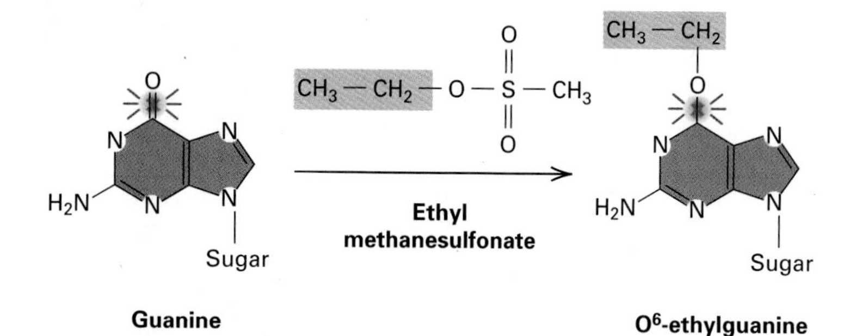


**FIGURE 14.26** Mutagenesis of guanine by ethyl methanesulfonate (EMS).

**(A)**

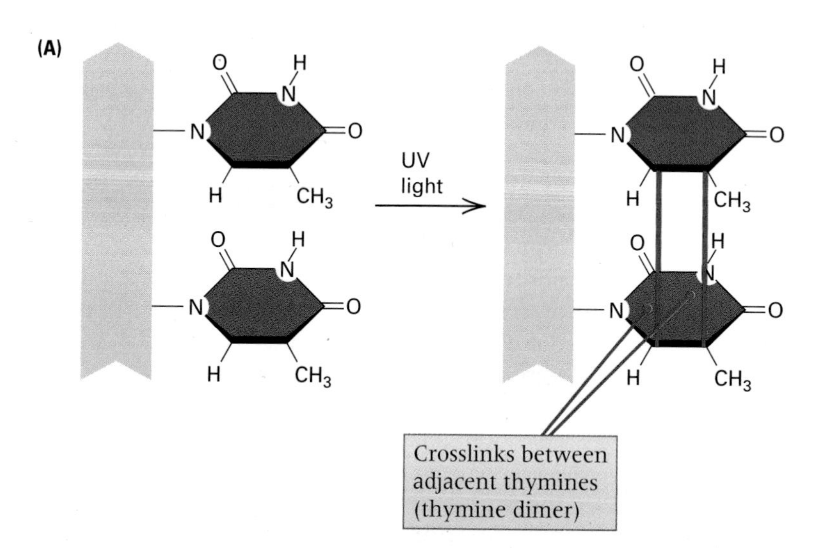


**(B)**

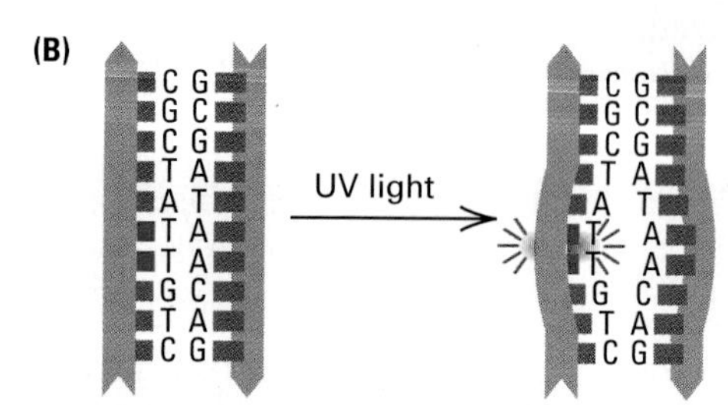


**FIGURE 14.27** (A) Structural view of the formation of a thymine dimer. Adjacent thymines in a DNA strand that have been subjected to ultraviolet (UV) irradiation are joined by formation of the bonds shown in red. Other types of bonds between the thymine rings also are possible. Although not drawn to scale, these bonds are considerably shorter than the spacing between the planes of adjacent thymines, so the double-stranded structure becomes distorted. The shape of each thymine ring also changes. (B) The distortion of the DNA helix caused by two thymines moving closer together when joined in a dimer.

## ■ Ionizing Radiation

**Ionizing radiation** includes x-rays and the particles and radiation released by radioactive elements ($\alpha$ and $\beta$ particles and $\gamma$-rays). When x-rays were first discovered late in the nineteenth century, their power to pass through solid materials was regarded as a harmless entertainment and a source of great amusement:

> By 1898, personal x rays had become a popular status symbol in New York. The *New York Times* reported that "there is quite as much difference in the appearance of the hand of a washerwoman and the hand of a fine lady in an x-ray picture as in reality." The hit of the exhibition season was Dr. W. J. Morton's full-length portrait of *"the x-ray lady,"* a "fashionable woman who had evidently a scientific desire to see her bones." The portrait was said to be a "fascinating and coquettish" picture, the lady having agreed to be photographed without her stays and corset, the better to satisfy the "longing to have a portrait of well-developed ribs." Dr. Morton said women were not afraid of x rays: "After being assured that there is *no danger* they take the rays without fear."
>
> The titillating possibility of using x rays to see through clothing or to invade the privacy of locked rooms was a familiar theme in popular discussions of x rays and in cartoons and jokes. Newspapers carried advertisements for *"x-ray proof underclothing"* for those seeking to protect themselves from x-ray inspection.
>
> The luminous properties of radium soon produced a full-fledged radium craze. A famous woman dancer performed *radium dances* using veils dipped in fluorescent salts containing radium. *Radium roulette* was popular at New York casinos, featuring a "roulette wheel washed with a radium solution, such that it glowed brightly in the darkness; an unseen hand cast the ball on the turning wheel and sparks marked its course as it bounded from pocket to glimmery pocket." A patent was issued for a process for making women's gowns luminous with radium, and Broadway producer Florenz Ziegfeld snapped up the rights for his stage extravaganzas.
>
> Even while the unrestrained use of x rays and radium was growing, evidence was accumulating that the new forces might not be so benign after all. Hailed as tools for fighting cancer, they could also cause cancer. Doctors using x rays were the first to learn this bitter lesson. (Quoted from S. Hilgartner, R. C. Bell, and R. O'Connor, *Nukespeak* (San Francisco: Sierra Club Books, 1982).

Doctors were indeed the first to learn the lesson. Many suffered severe x-ray burns or required amputation of overexposed hands or arms. Many others died from radiation poisoning or from radiation-induced cancer. By the mid-1930s, the number of x-ray deaths had grown so large that a monument to the "x-ray martyrs" was erected in a hospital courtyard in Germany. Yet the full hazards of x-ray exposure were not widely appreciated until the 1950s.

When ionizing radiation interacts with water or with living tissue, highly reactive ions called *free radicals* are formed. The free radicals react with other molecules, including DNA, which results in the carcinogenic and mutagenic effects. The intensity of a beam of ionizing radiation can be described quantitatively in several ways. There are, in fact, a bewildering variety of units in common use (**TABLE 14.4**). Some of the units (becquerel, curie) deal with the number of disintegrations emanating from a material, others (roentgen) with the number of ionizations the radiation produces in air, still others (gray, rad) with the amount of energy imparted to material exposed to the radiation, and some (rem, sievert) with the effects of radiation on living tissue. The types of units have proliferated through the years in attempts to encompass different types of radiation, including nonionizing radiation, in a common frame of reference. The units in Table 14.4 are presented only for ease of interpreting the multitude of units found in the literature on radiation genetics. They need not be memorized.

Genetic studies of ionizing radiation support the following general principle:

> Over a wide range of x-ray doses, the frequency of mutations induced by x-rays is proportional to the radiation dose.

One type of evidence supporting this principle is the frequency with which X-chromosome recessive lethals are induced in *Drosophila* (**FIGURE 14.28**). The mutation rate increases linearly with

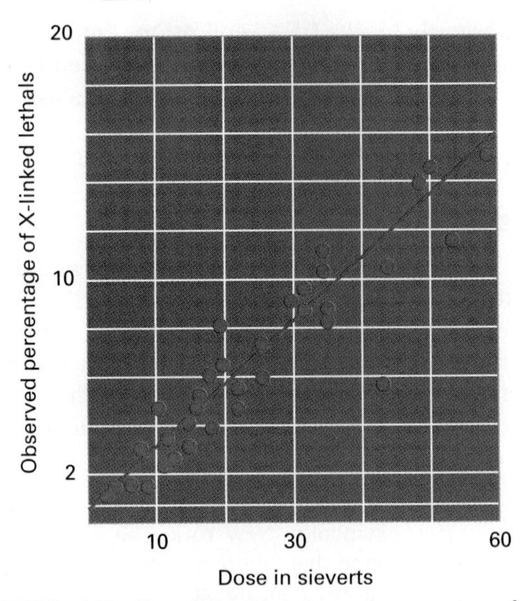


**FIGURE 14.28** The relationship between the percentage of X-linked recessive lethals in *D. melanogaster* and x-ray dose, obtained from several experiments. The frequency of spontaneous X-linked lethal mutations is 0.15 percent per X chromosome per generation.

increasing x-ray dose. For example, an exposure of 10 sieverts increases the frequency from the spontaneous value of 0.15 percent to about 3 percent. The mutagenic and lethal effects of ionizing radiation at low to moderate doses result primarily from damage to DNA. Three types of damage in DNA are produced by ionizing radiation: single-strand breakage (in the sugar–phosphate backbone), double-strand breakage, and alterations in nucleotide bases. The single-strand breaks are usually efficiently repaired, but the other damage is responsible for mutation and lethality. In eukaryotes, ionizing radiation also results in chromosome breaks. Although mechanisms exist for repairing the breaks, the repair often leads to chromosome rearrangements such as translocations (Chapter 8). In human cells in culture, a dose of 0.2 sievert results in an average of one visible chromosome break per cell.

Ionizing radiation is widely used in tumor therapy. The basis for the treatment is the increased frequency of chromosomal breakage (and the consequent lethality) in cells undergoing mitosis compared with cells in interphase. Tumors usually contain many more mitotic cells than most normal tissues, so more tumor cells than normal cells are destroyed. Because not all tumor cells are in mitosis at the same time, irradiation is carried out at intervals of several days to allow interphase tumor cells to enter mitosis. Over a period of time, most tumor cells are destroyed.

**Table 14.4 Units of radiation**

| Unit (abbreviation) | Magnitude |
|---|---|
| becquerel (Bq)* | 1 disintegration/second = $2.7 \times 10^{-11}$ Ci |
| curie (Ci) | $3.7 \times 10^{10}$ disintegrations/second = $3.7 \times 10^{10}$ Bq |
| gray (Gy)* | 1 joule/kilogram = 100 rad |
| rad (rad) | 100 ergs/gram = 0.01 Gy |
| rem (rem) | damage to living tissue done by 1 rad = 0.01 Sv |
| roentgen (R) | produces 1 electrostatic unit of charge per cubic centimeter of dry air under normal conditions of pressure and temperature. (By definition, 1 electrostatic unit repels with a force of 1 dyne at a distance of 1 centimeter.) |
| sievert (Sv) | 100 rem |

*Units officially recognized by the International System of Units as defined by the General Conference on Weights and Measures.

## connection
### X-Ray Daze

Hermann J. Muller 1927
University of Texas, Austin, Texas
*Artificial Transmutation of the Gene*

*Mutagenesis, which means the deliberate induction of mutations, plays an important role in genetics because it makes possible the identification of genes that control biological processes, such as development and behavior. For the demonstration that x-rays could be used to induce new mutations, Muller was awarded a Nobel Prize in 1946. His discovery also had important practical implications in regard to x-rays (and other mutagenic agents) present in the environment. After World War II ended with the nuclear bombing of Hiroshima and Nagasaki, Muller became a leader in the fight against the above-ground testing of nuclear bombs because of the release of radioactivity into the atmosphere. He also condemned the casual use of x-rays, such as for the fitting of children's shoes, and he campaigned for the engineering of safer x-ray machines for medical use. Muller's prose could be turgid at times, thanks to his use of subordinate clauses. This excerpt captures the style.*

Most modern geneticists will agree that gene mutations form the chief basis of organic evolution, and therefore of most of the complexities of living things. Unfortunately for the geneticists, however, the study of these mutations, and, through them, of the genes themselves, has heretofore been very seriously hampered by the extreme infrequency of their occurrence under ordinary conditions, and by the general unsuccessfulness of attempts to modify decidedly, and in a sure and detectable way, this sluggish "natural" mutation rate. . . . On theoretical grounds, it has appeared to the present writer that radiations of short wave length should be especially promising for the production of mutational changes, and for this and other reasons a series of experiments concerned with this problem has been undertaken during the past year on the fruit fly, *Drosophila melanogaster*. . . . It has been found quite conclusively that treatment of the sperm with relatively heavy doses of x-rays induces the occurrence of true "gene mutations" in a high proportion of the treated germ cells. Several hundred mutants have been obtained in this way in a short time and considerably more than a hundred have been followed through three, four or more generations. They are (nearly all of them, at any rate) stable in their inheritance. . . . Regarding the types of mutations produced, it was found that the recessive lethals greatly outnumbered the nonlethals producing a visible morphological abnormality. . . . In addition to gene mutations, it was found that there is also caused by x-ray treatment a high proportion of rearrangements in the linear order of the genes. . . . The transmuting action of x-rays on the genes is not confined to the sperm cells, for treatment of the unfertilized females causes mutations about as readily as treatment of the males. . . . In conclusion, the attention of those working along classical genetic lines may be drawn to the opportunity, afforded them by the use of x-rays, of creating in their chosen organisms a series of artificial races for use in the study of genetics. If, as seems likely on genetic considerations, the effect is common to most organisms, it should be possible to produce, "to order," enough mutations to furnish respectable genetic maps.

***Treatment of the sperm with relatively heavy doses of x-rays induces the occurrence of true "gene mutations" in a high proportion of the treated germ cells.***

Source: H. J. Muller, *Science* 66 (1927): 84–87.

**FIGURE 14.29** gives representative values of doses of ionizing radiation received by human beings in the United States in the course of a year. The unit of measure is the millisievert, which equals 0.1 rem. The exposures in Figure 14.29 are on a yearly basis, so over the course of a generation, the total exposure is approximately 100 millisieverts. Note that, with the exception of diagnostic x-rays, which yield important compensating benefits, most of the total radiation exposure comes from natural sources, particularly radon gas. Less than 20 percent of the average radiation exposure comes from artificial sources. Nevertheless, there are dangers inherent in any exposure to ionizing radiation, particularly in an increased risk of leukemia and certain other cancers in the exposed persons. With regard to increased genetic diseases in future generations, the risk that a small amount of additional radiation will have mutagenic effects is low enough that most geneticists are currently more concerned about the effects of the many chemical mutagens (as well as carcinogens) that are introduced into the environment from a variety of sources.

The National Research Council of the National Academy of Sciences (USA) regularly updates the estimated risks of radiation exposure. The latest estimates are summarized in **TABLE 14.5**. The message is that an additional 10 millisieverts of radiation per generation

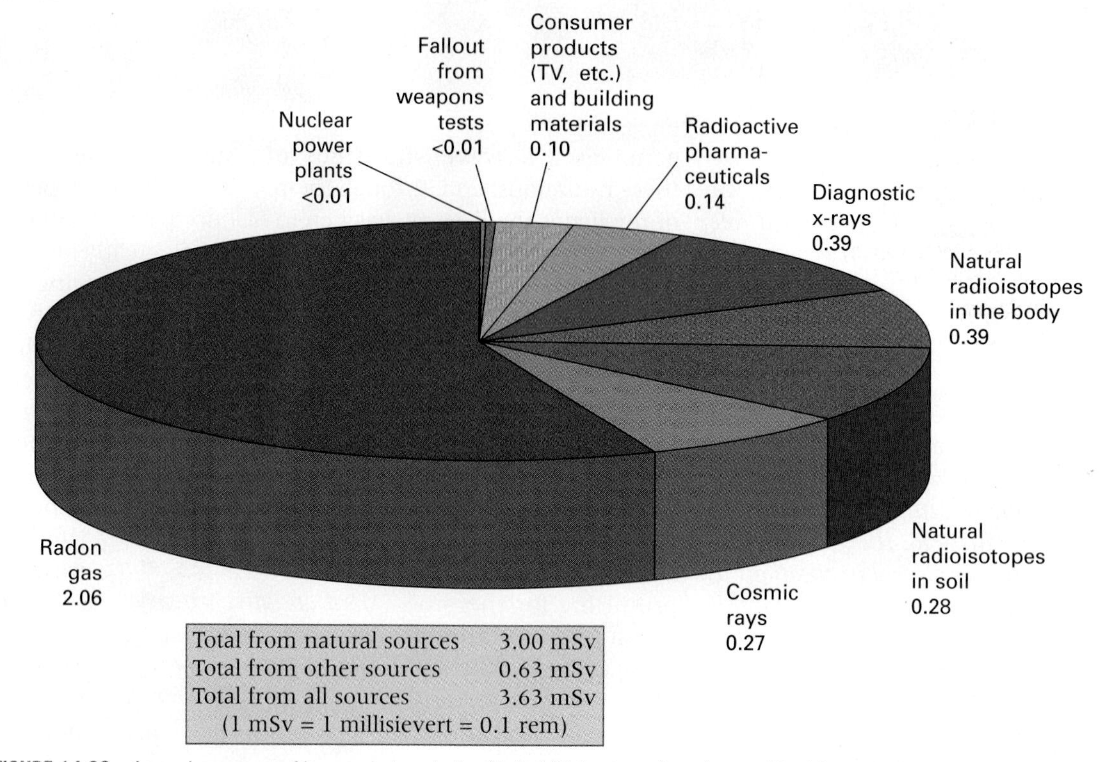


**FIGURE 14.29** Annual exposure of human beings in the United States to various forms of ionizing radiation. [Data from the National Research Council.]

**Table 14.5 Estimated genetic effects of an additional 10 millisieverts per generation**

| Type of disorder | Current incidence per million liveborn | Additional cases per million liveborn per 10 mSv per generation: First generation | Additional cases per million liveborn per 10 mSv per generation: At equilibrium |
|---|---|---|---|
| Autosomal dominant | | | |
| Clinically severe | 2500 | 5–20 | 25 |
| Clinically mild | 7500 | 1–15 | 75 |
| X-linked | 400 | <1 | <5 |
| Autosomal recessive | 2500 | <1 | Very slow increase |
| Chromosomal | | | |
| Unbalanced translocations | 600 | <5 | Very little increase |
| Trisomy | 3800 | <1 | <1 |
| Congenital abnormalities (multifactorial) | 20,000–30,000 | 10 | 10–100 |
| Other multifactorial disorders | | | |
| Heart disease | 600,000 | Unknown | Unknown |
| Cancer | 300,000 | Unknown | Unknown |
| Others | 300,000 | Unknown | Unknown |

Source: Reproduced from The Committee on Biological Effects of Ionizing Radiations. *Health Effects of Exposure to Low Levels of Ionizing Radiation.* National Academies Press, Washington, D.C. (1990). Reprinted with permission from the National Academies Press, Copyright 1990, National Academy of Sciences.]

(about a 10 percent increase in the annual exposure) is expected to cause a relatively modest increase in diseases that are wholly or partly due to genetic factors. The most common conditions in the table are heart disease and cancer. No estimate for the radiation-induced increase is given for either of these traits because the genetic contribution to the total is still very uncertain.

## Genetic Effects of the Chernobyl Nuclear Accident

On April 26, 1986, a nuclear power plant exploded near the city of Chernobyl (also spelled

Chornobyl) in the Ukraine. The Soviet President, Mikhail Gorbachev, addressed the nation on television:

> Good evening, comrades. All of you know that there has been an incredible misfortune. . . . For the first time we confront the real force of nuclear energy out of control.

The immediate area was heavily contaminated, and clouds of radioactive debris disseminated over long distances. It was the largest publicly acknowledged nuclear accident in history. It is estimated to have released 50–200 million curies of radiation, ten times greater than that released by the atomic bomb over Hiroshima in 1945. Iodine-131 and cesium-137 were the principal radioactive contaminants. More than 200,000 people living in the area were evacuated almost immediately, but many were heavily exposed to radiation. Some children absorbed an amount of radiation equal to that of a thousand chest x-rays. Within a short time there was a notable increase in the frequency of thyroid cancer in children, and other health effects of radiation exposure were also detected.

At first, little attention was given to possible genetic effects of the Chernobyl disaster, because relatively few people were acutely exposed and because the radiation dose to people outside the immediate area was considered too small to worry about. In the district of Belarus, some 200 kilometers north of Chernobyl, the average exposure to iodine-131 was estimated at approximately 0.185 sievert per person. This is a fairly high dose, but the exposure was brief because the half-life of iodine-131 is only 8 days. Radiation from the much longer-lived cesium-137 (half-life of 30 years) was estimated at less than 5 millisieverts per year. On the basis of data from laboratory animals and studies of the survivors of the Hiroshima and Nagasaki atomic bombs, little detectable genetic damage was expected from these exposures.

Nevertheless, 10 years after the meltdown, studies of people living in Belarus did indicate a remarkable increase in the mutation rate. The observations focused on simple sequence repeat (SSR) polymorphisms because of their intrinsically high mutation rate due to replication slippage and other factors. Each SSR was examined in parents and their offspring to detect any DNA fragments that increased or decreased in size, indicative of an increase or decrease in the number of repeating units contained in the DNA fragment. Five SSR polymorphisms were studied, with the results summarized in **FIGURE 14.30**. Two of the loci (blue dots) showed no evidence for an increase in the mutation rate. This is the expected result. The unexpected finding was that three of the loci (red dots) did show a significant increase, and the level of increase was consistent with an approximate doubling of the mutation rate. It is still not known whether the increase was detected because replication slippage is more sensitive to radiation than other types of mutations or because the effective radiation dose sustained by the Belarus population was much higher than originally thought.

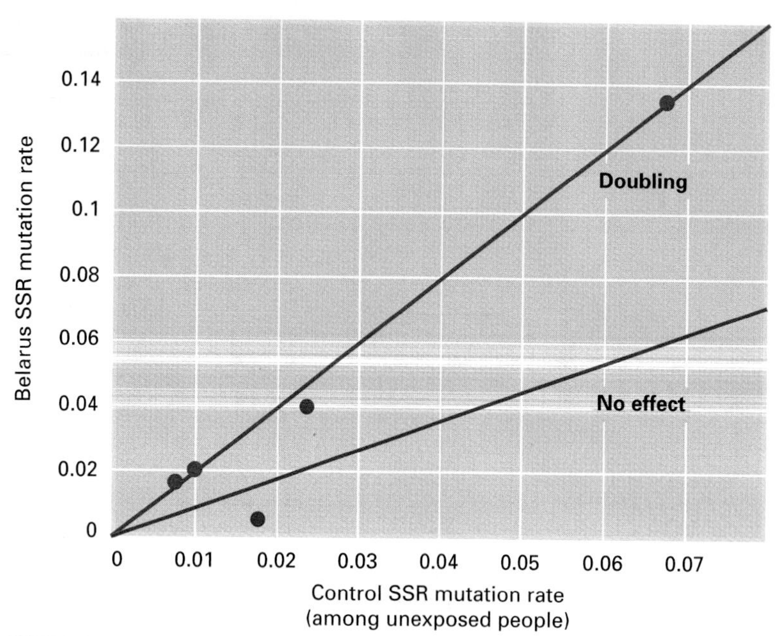


**FIGURE 14.30** Mutation rates of five SSR polymorphisms among people of Belarus exposed to radiation from Chernobyl and among unexposed British people. Three of the loci (red dots) show evidence of an approximately twofold increase in the mutation rate. [Data from Y. E. Dubrova, et al., *Nature* 380 (1996): 683–686.]

## 14.6 Mechanisms of DNA Repair

Spontaneous damage to DNA in human cells takes place at a rate of approximately 1 event per billion nucleotide pairs per minute (or, per nucleotide pair, at a rate of $1 \times 10^{-9}$ per minute). This may seem like quite a low rate, but it implies that every 24 hours, in every human cell, the DNA is damaged at approximately 10,000 different sites. Fortunately for us, under ordinary circumstances much of the damage done to DNA by spontaneous chemical reactions in the nucleus, as well as that caused by chemical mutagens and radiation, can be repaired.

**TABLE 14.6** summarizes some of the most important mechanisms of DNA repair. The first on the list is straightforward. A nick in a DNA strand is a site at which one of the phosphodiester bonds

**Table 14.6 Types of DNA damage and mechanism of repair**

| Type of damage | Major mechanism of repair |
|---|---|
| Nicks in DNA strand | Repaired by DNA ligase |
| Chemically modified base | Base excision (removal by base-specific DNA glycosylase) |
| Mismatched bases | Corrected by mismatch repair (excision and resynthesis) |
| Apurinic or apyrimidinic site | Fixed by AP endonuclease repair system |
| Damaged region of DNA | Nucleotide excision repair (excision and resynthesis across partner strand) |
| Pyrimidine dimers (from UV light) | Enzymatically reversed |
| Damaged region of DNA | DNA damage bypass (sequence in damaged region recovered via recombination) |

along the backbone is broken. Nicks are repaired by the enzyme DNA ligase, which restores the covalent bond. In the following sections, we examine other key molecular mechanisms for the repair of aberrant or damaged DNA.

## Mismatch Repair

We have already encountered the mismatch repair system in the context of gene conversion and genetic recombination (Section 6.8). However, the most important role of mismatch repair is as a "last chance" error-correcting mechanism in replication. During DNA replication, mismatched nucleotides are incorporated at the rate of about $10^{-5}$ per template nucleotide per round of replication. As shown in **FIGURE 14.31**, approximately 99 percent of these are immediately corrected by the proofreading function (3′-to-5′ exonuclease activity) of the major replication polymerase. This leaves a mismatch rate of $10^{-7}$ per template nucleotide per round of replication; 99.9 percent of the remaining mismatches are corrected by the mismatch repair system, yielding an overall mismatch rate of

$$10^{-5} \times 10^{-2} \times 10^{-3} = 10^{-10}$$

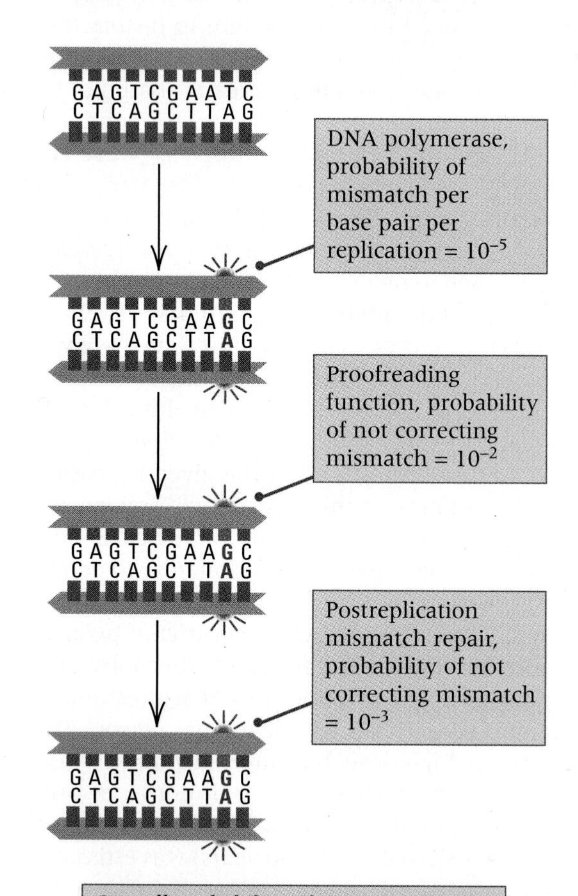


**FIGURE 14.31** Summary of rates of error in DNA polymerization, proofreading, and postreplication mismatch repair. The initial rate of nucleotide misincorporation is $10^{-5}$ per base pair per replication. The proofreading function of DNA polymerase corrects 99 percent of these errors, and of those that remain, postreplication mismatch repair corrects 99.9 percent. The overall rate of misincorporated nucleotides that are not repaired is $10^{-10}$ per base pair per replication.

The operation of **mismatch repair** is illustrated in **FIGURE 14.32**. When a mismatched base is detected, one of the strands is cut in two places and a region around the mismatch is removed. The excised region is variable in size. In *E. coli*, an enzyme first cleaves the DNA at the nearest GATC sequence on the unmethylated (newly synthesized) strand, provided that the site is within about 1 kb of the mismatch. Then an exonuclease degrades the cleaved DNA strand to a position on the other side of the mismatch (the excision step), creating a single-stranded gap. After excision, DNA polymerase fills in the gap by using the remaining strand as a template, thereby eliminating the mismatch. The mismatch-repair system also corrects most small insertions or deletions.

If the DNA strand that is removed in mismatch repair is chosen at random, then the repair process sometimes creates a mutant molecule by cutting the strand that contains the correct base and using the mutant strand as a template. However, this is prevented from happening in newly synthesized DNA because the daughter strand is less methylated than the

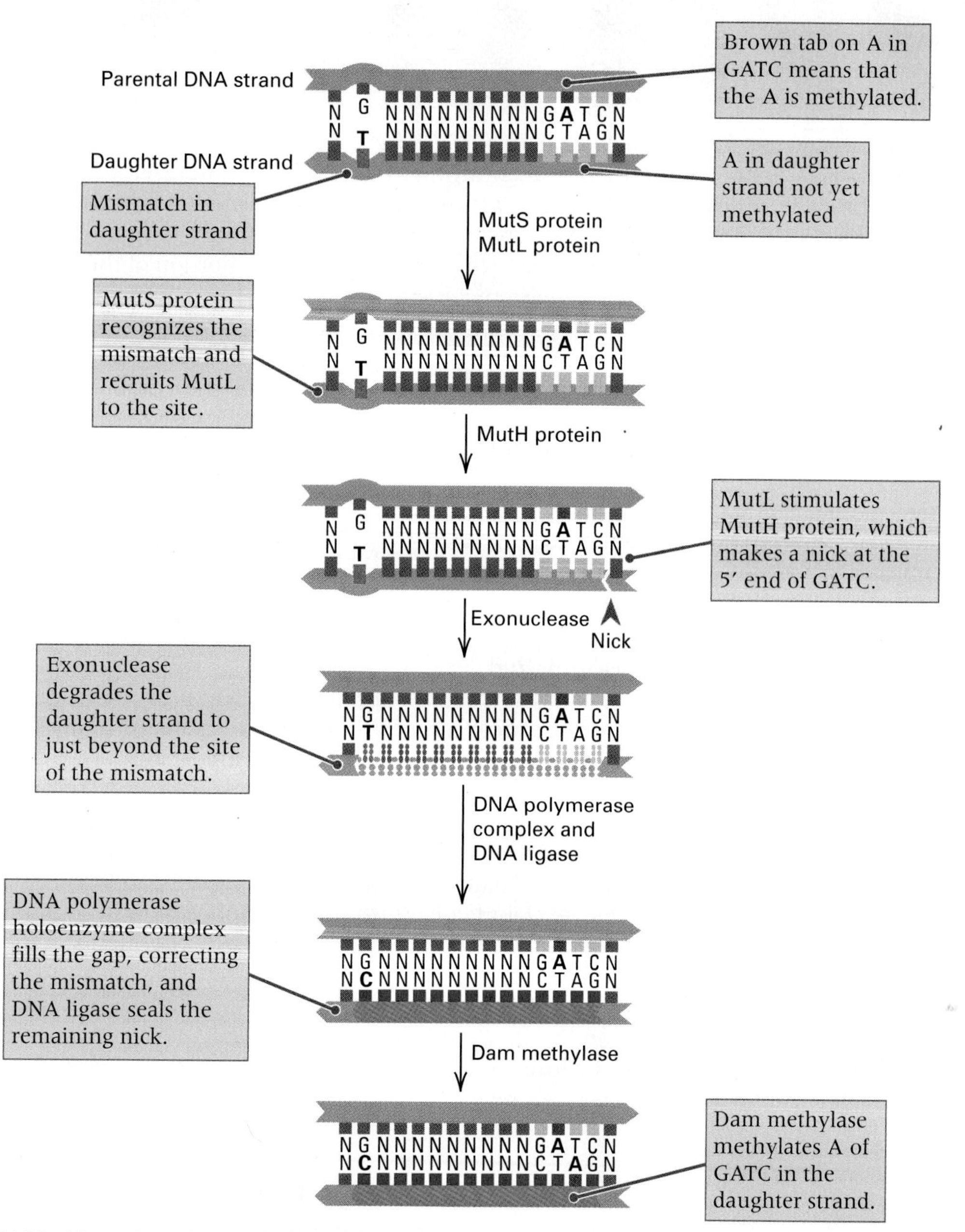

Parental DNA strand
Daughter DNA strand
Brown tab on A in GATC means that the A is methylated.
A in daughter strand not yet methylated
Mismatch in daughter strand
MutS protein
MutL protein
MutS protein recognizes the mismatch and recruits MutL to the site.
MutH protein
MutL stimulates MutH protein, which makes a nick at the 5′ end of GATC.
Exonuclease
Nick
Exonuclease degrades the daughter strand to just beyond the site of the mismatch.
DNA polymerase complex and DNA ligase
DNA polymerase holoenzyme complex fills the gap, correcting the mismatch, and DNA ligase seals the remaining nick.
Dam methylase
Dam methylase methylates A of GATC in the daughter strand.

**FIGURE 14.32** Mismatch repair consists of the excision of a segment of a DNA strand that contains a base mismatch, followed by repair synthesis. In *E. coli*, cleavage takes place at the nearest methylated GATC sequence in the unmethylated strand. An exonuclease removes successive nucleotides until just past the mismatch, and the resulting gap is repaired. Either strand can be excised and corrected, but in newly synthesized DNA, methylated bases in the template strand often direct the excision mechanism to the newly synthesized strand that contains the incorrect nucleotide.

parental strand. The mismatch-repair system recognizes the degree of methylation of a strand and *preferentially excises nucleotides from the undermethylated strand.* This helps ensure that incorrect nucleotides incorporated into the daughter strand in replication will be removed and repaired. The daughter strand is always the undermethylated strand because its methylation lags somewhat behind the moving replication fork, whereas the parental strand was fully methylated in the preceding round of replication.

The mechanism of mismatch repair has been studied extensively in bacteria. Mutants defective in the process were identified as having high rates of spontaneous mutation. The products of two genes, *mutL* and *mutS,* recognize and bind to a mismatched base pair. This triggers the excision of a tract of nucleotides from the newly synthesized strand. Experiments

## connection

# Replication Slippage in Unstable Repeats

**Micheline Strand,[1] Tomas A. Prolla,[2] R. Michael Liskay,[2] and Thomas D. Petes[1] 1993**
[1]University of North Carolina, Chapel Hill, North Carolina
[2]Yale University, New Haven, Connecticut
*Destabilization of Tracts of Simple Repetitive DNA in Yeast by Mutations Affecting DNA Mismatch Repair*

*One of the most important principles of genetics is that many fundamental processes are carried out in much the same way in many different organisms. Basic cellular processes are similar in all eukaryotic cells, and the basic mechanisms of DNA replication and repair are fundamentally the same in eukaryotes and prokaryotes. The similarity of mismatch repair among organisms motivated this paper. The authors report that yeast cells deficient in mismatch repair are very deficient in their ability to replicate faithfully tracts of short repeating sequences. Because this type of instability also characterizes certain human hereditary colon cancers, the authors correctly predicted that the mutant genes causing the colon cancer would be genes involved in mismatch repair.*

The genomes of all eukaryotes contain tracts of DNA in which a single base or a small number of bases is repeated. Expansions of such tracts have been associated with several human disorders including the fragile-X syndrome. In addition, simple repeats are unstable in certain forms of colorectal cancer, suggesting a defect in DNA replication of repair. . . . One mechanism [for changes in repeat number] is DNA polymerase slippage: during replication of the tract, the primer and template strand transiently dissociate and then reassociate in a misaligned configuration. If the unpaired bases are in the primer strand, continued synthesis results in an elongation of the tract whereas unpaired bases in the template strand result in a deletion. . . . The most common simple repeat in most eukaryotes is poly(GT)$_{10\text{-}30}$. We examined the genetic control of the stability of poly(GT) tracts. . . . Three mutations affecting mismatch repair in yeast were examined, *pms1* and *mlh1* (homologs of *MutL*) and *msh2* (a homolog of *MutS*). . . . In two different genetic backgrounds using two different instability assay systems, all three mutations led to 100- to 700-fold elevated levels of tract instability. . . . Yeast contains two DNA polymerases that have 3′–5′ exonuclease ("proofreading") activities. In strains with mutations in the nuclease portion of these genes, neither affects tract stability as dramatically as mutations that reduce mismatch repair. . . . The results show that the instability of poly(GT) tracts in yeast is increased by mutations in the mismatch repair genes but is relatively unaffected by mutations affecting the proofreading functions of DNA polymerase. These results indicate that DNA polymerase *in vivo* has a very high rate of slippage on templates containing simple repeats, but most of the errors are corrected by cellular mismatch repair systems. Thus, the instability of simple repeats observed for some human diseases may be a consequence of either an increased rate of DNA polymerase slippage or a decreased efficiency of mismatch repair.

**The instability of poly(GT) tracts in yeast is increased by mutations in the mismatch repair genes.**

Source: M. Strand, et al., *Nature* 365 (1993): 274–276.

carried out in yeast revealed that tandem repeats of short nucleotide sequences are a thousandfold less stable in mutants deficient in mismatch repair than in wildtype yeast. By that time it was already known that some forms of human hereditary colorectal cancer result in decreased stability of simple repeats, and the yeast researchers suggested that these high-risk families might be segregating for an allele causing a mismatch-repair deficiency. Within two years, scientists in several laboratories had identified four human genes homologous to *mutL* or *mutS*, any one of which, when mutated, results in hereditary nonpolyposis colorectal cancer (HNPCC). Most cases of this type of cancer may be caused by mutations in one of these four mismatch-repair genes.

### ■ Base Excision Repair

We have already seen that DNA duplexes will often contain uracil formed from the deamination of cytosine and 8-oxoguanine formed from the oxidation of guanine. These and other incorrect bases in DNA also commonly occur at the rate of many thousands per day per cell. The vast majority of these are removed almost immediately by a process of base excision repair using any of about eight enzymes that recognize specific incorrect bases. For example, the enzyme *DNA uracil glycosylase* recognizes an incorrect G–U base pair in duplex DNA, and the enzyme *DNA 8-oxoguanine glycosylase* recognizes an incorrect 8-oxoG–C base pair.

The base excision part of the repair process is shown in **FIGURE 14.33**. Part A shows deamina-

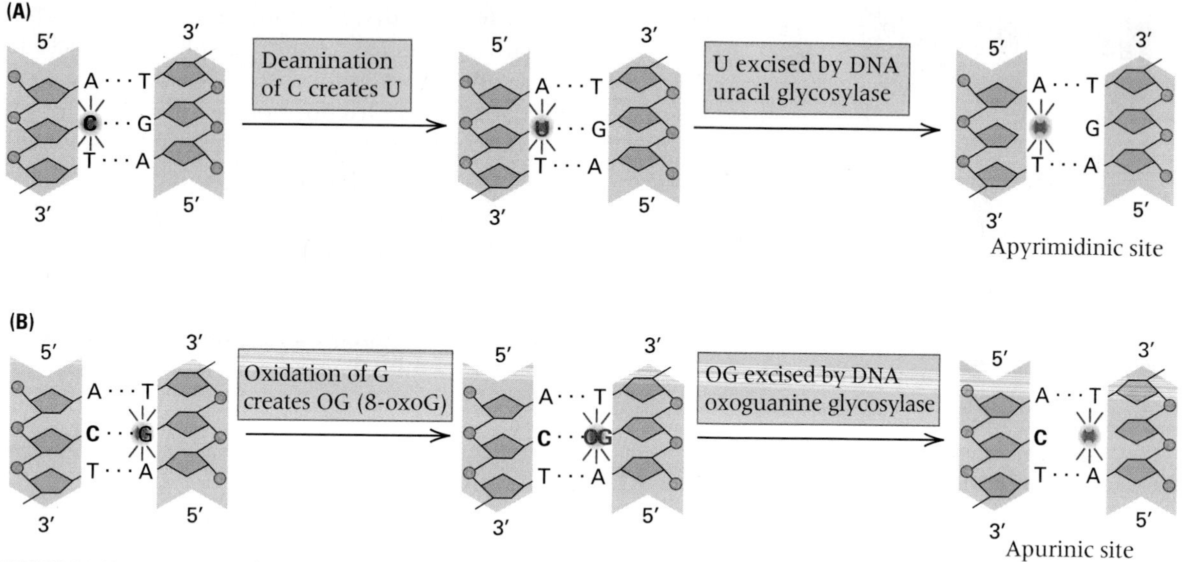


**FIGURE 14.33** Base excision. (A) Excision of uracil results in an apyrimidinic site. (B) Excision of 8-oxoguanine results in an apurinic site.

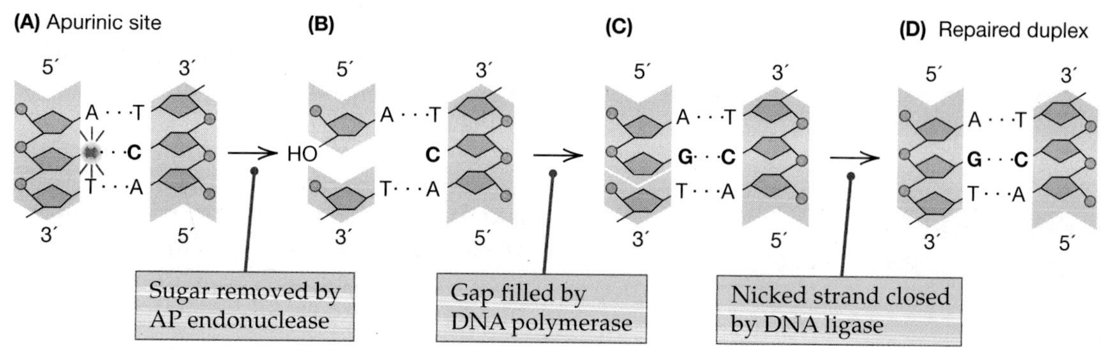


**FIGURE 14.34** Action of AP endonuclease. (A) An apurinic site in a DNA duplex. (B) AP endonuclease excises the empty deoxyribose from the DNA strand. (C) A specialized DNA polymerase fills the gap using the continuous strand as a template. (D) The remaining nick is closed by DNA ligase, restoring the original sequence. The AP endonuclease acts similarly to repair apyrimidinic sites.

tion of cytosine leading to the presence of a uracil-containing base. The incorrect G–U base pair is recognized by its specific glycosylase, which cleaves the offending uracil from the deoxyribose sugar to which it is attached. The enzyme works by scanning along duplex DNA until a uracil nucleotide is encountered, at which point a specific arginine residue in the enzyme intrudes into the DNA through the minor groove. The intrusion compresses the DNA backbone flanking the uracil and results in the uracil flipping out so that it becomes accessible to cleavage by the uracil glycosylase. This leaves behind a deoxyribose sugar in the duplex DNA that lacks a pyrimidine base, which is known as an *apyrimidinic site*. Figure 14.33B shows how 8-oxyguanine is removed by its specific glycosylase, resulting in this case in an *apurinic site*. Both apyrimidinic sites and apurinic sites are repaired by the mechanism discussed next.

## AP Repair

Restoration of apurinic and apyrimidinic sites occurs by means of AP repair. The process is illustrated for an apurinic site in **FIGURE 14.34**. The key enzyme is the *AP endonuclease*, which cleaves the baseless sugar from the DNA, leaving a single-stranded gap that is repaired by a specialized DNA polymerase (polymerase beta). The gap filling leaves one nick remaining in the repaired strand, which is closed by DNA ligase, completing the repair.

## Nucleotide Excision Repair

**Nucleotide excision repair** is a ubiquitous multistep enzymatic process by which a stretch of a damaged DNA strand is removed from a duplex molecule and replaced by resynthesis using the undamaged strand as a template (**FIGURE 14.35**). The substrate for repair can be any distortion in the duplex molecule. The distortion recruits repair

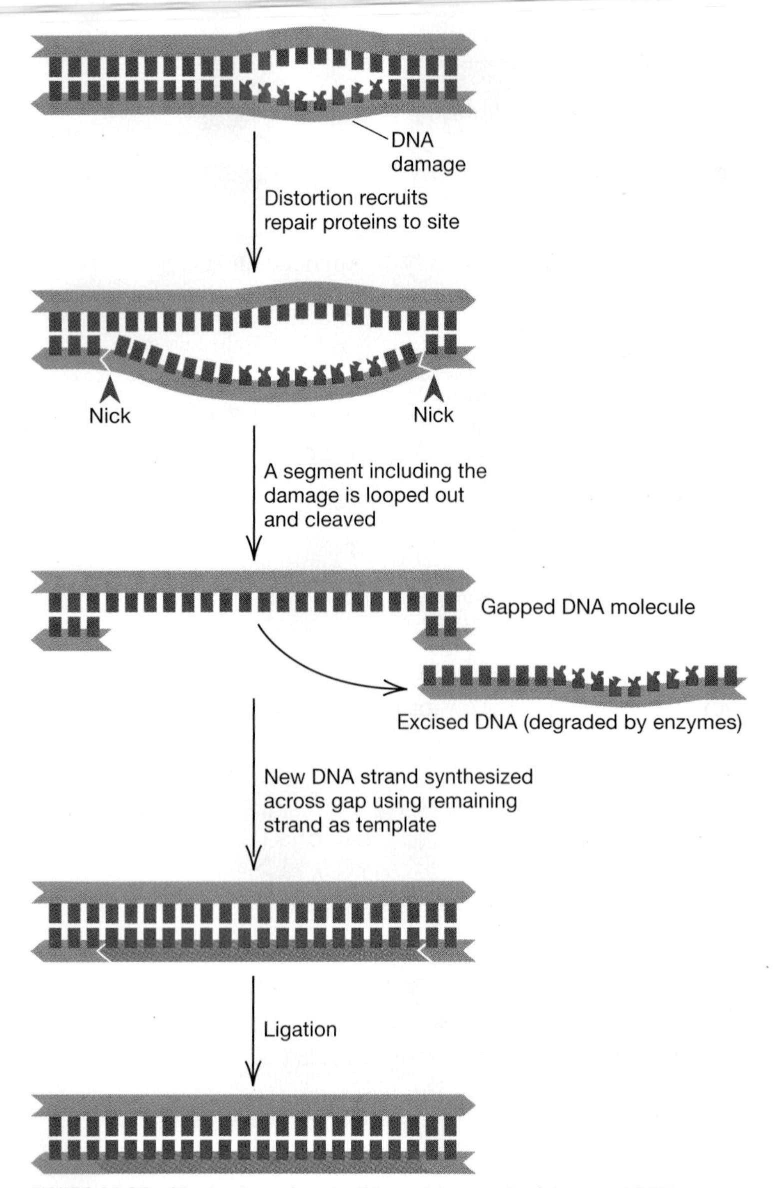
DNA damage
Distortion recruits repair proteins to site
Nick
Nick
A segment including the damage is looped out and cleaved
Gapped DNA molecule
Excised DNA (degraded by enzymes)
New DNA strand synthesized across gap using remaining strand as template
Ligation

**FIGURE 14.35** Mechanism of nucleotide excision repair of damaged DNA.

proteins to the site, and the DNA duplex is unwound. Interestingly, the enzyme system responsible for the unwinding in eukaryotes is one of the normal transcription factors (TFIIH). After the unwinding, repair endonucleases make two cuts in the sugar–phosphate backbone. The excised region is usually quite precise. In prokaryotes, the cleavage sites are eight nucleotides from the 5′ end and five nucleotides from the 3′ end of the damage; and in eukaryotes, the cleavage sites are 24 nucleotides from the 5′ end and five nucleotides from the 3′ end of the damage. The free 3′ hydroxyl group at one end of the gap serves as a primer for synthesis of a new strand by specialized DNA polymerases (polymerases delta and epsilon). The final step of the repair process is the joining of the newly synthesized segment to the contiguous strand by DNA ligase.

Several disease syndromes are associated with defects in excision repair. These include xeroderma pigmentosum, Cockayne's syndrome, and trichothiodystrophy. Although these syndromes share such symptoms as skin abnormalities or neurological defects, they differ dramatically other respects. For example, xeroderma pigmentosum patients have severe light sensitivity and a high incidence of early-onset skin cancer, those with Cockayne syndrome have pronounced dwarfism, and patients with trichothiodystrophy have sulfur-deficient, brittle hair and often characteristic facial abnormalities.

## ■ Photoreactivation

Various enzymes can recognize and catalyze the direct reversal of specific DNA damage. A classic example found in some organisms is the reversal of UV-induced pyrimidine dimers by **photoreactivation**, in which an enzyme breaks the bonds that join the pyrimidines in the dimer and thereby restores the original bases (Figure 14. 27). The enzyme binds to the dimers in the dark but then utilizes the energy of blue light to cleave the bonds. Another important example of enzymatic reversal of DNA damage is an enzyme that removes the ethyl group from $O^6$-ethylguanine (Figure 14.26), which restores the original guanine base. Without the enzymatic repair, the $O^6$-ethylguanine would pair with thymine instead of cytosine.

## ■ DNA Damage Bypass

Sometimes DNA damage persists rather than being reversed or removed, but its harmful effects can be minimized. This often requires that the damaged area be skipped over during replication, so the process is called **DNA damage bypass**. For example, when DNA polymerase reaches a damaged site (such as a pyrimidine dimer), it stops synthesis of the strand. However, after a brief time, synthesis is reinitiated beyond the damage and chain growth continues, producing a gapped strand with the damaged spot in the gap.

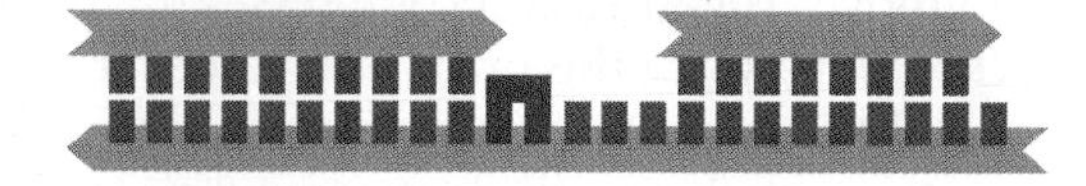

The gap can be filled by strand exchange with the parental strand that has the same polarity, and the secondary gap produced in that strand can be filled by repair synthesis (**FIGURE 14.36**). The products of this exchange and resynthesis are two intact single strands, each of which can then serve in the next round of replication

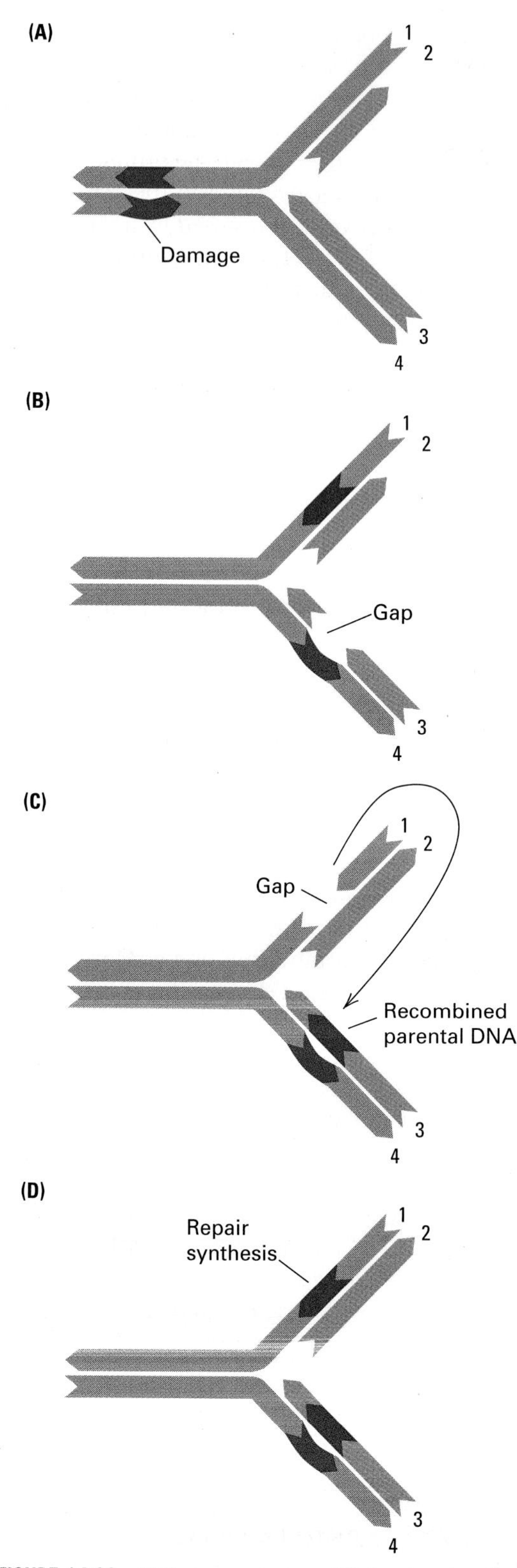


**FIGURE 14.36** DNA damage bypass. (A) A molecule with DNA damage in strand 4 is being replicated. (B) By reinitiation of synthesis beyond the damage, a gap is formed in strand 3. (C) A segment of parental strand 1 is excised and inserted in strand 3. (D) The gap in strand 1 is next filled in by repair synthesis.

as a template for the synthesis of an undamaged DNA molecule.

### ■ The SOS Repair System

**SOS repair**, which is found in *E. coli* and related bacteria, is a complex set of processes that includes an alternative bypass system that allows DNA replication to take place across pyrimidine dimers or other DNA distortions, but at the cost of the fidelity of replication. Even though intact DNA strands are formed, the strands are often defective. SOS repair is said to be *error-prone*. A significant feature of the SOS repair system is that it is not always active but is induced by DNA damage. Once activated, the SOS system allows the growing fork to advance across the damaged region, adding nucleotides that are often incorrect because the damaged template strand cannot be replicated properly. The proofreading system of DNA polymerase is relaxed to allow polymerization to proceed across the damage, despite the distortion of the helix.

## 14.7 Reverse Mutations and Suppressor Mutations

In most of the mutations we have considered so far, a wildtype (normal) gene is changed into a form that results in a mutant phenotype, an event called a **forward mutation**. Mutations are frequently reversible, and an event that restores the wildtype phenotype is called a **reversion**. A reversion may result from a **reverse mutation**, an exact reversal of the alteration in base sequence that occurred in the original forward mutation, restoring the wildtype DNA sequence. A reversion may also result from the occurrence of a second mutation at some other site in the genome, which compensates for the effect of the original mutation. Reversion by the exact reversal mechanism is infrequent. The second-site mechanism is much more common, and a mutation of this kind is called a **suppressor mutation**. A suppressor mutation can occur at a different site in the gene containing the mutation that it suppresses (*intragenic* suppression) or in a different gene (*intergenic* suppression). Most suppressor mutations do not fully restore the wildtype phenotype, for reasons that will be apparent in the following discussion of the two kinds of suppression.

### ■ Intragenic Suppression

Reversion of frameshift mutations usually results from **intragenic suppression**; the mutational effect of the addition (or deletion) of a nucleotide pair in changing the reading frame of the mRNA is rectified by a compensating deletion (or addition) of a second nucleotide pair at a nearby site in the gene. In an ingenious experiment discussed in Chapter 10, intragenic frameshift suppressors were used to demonstrate that the genetic code is a triplet code.

A second type of intragenic suppression is observed when loss of activity of a protein, caused by one amino acid change, is at least partly restored by a change in a second amino acid in the same protein. An example of this type of suppression in the protein product of the *trpA* gene in *E. coli* is shown in **FIGURE 14.37**. The TrpA polypeptide, composed of 268 amino acids, is one of two polypeptides that make up the enzyme tryptophan synthase in *E. coli*. The mutation shown, one of many that inactivate the enzyme, causes a change of amino acid number 210 from glycine in the wildtype protein to glutamic acid. This glycine is not in the active site; rather, the inactivation is caused by a change in the folding of the protein, which indirectly affects the active site. The activity of the mutant protein is partly restored by a second mutation that results in a change in amino acid number 174 from tyrosine to cysteine. A protein in which only the tyrosine has been replaced by cysteine is inactive, again because of a change in the shape of the protein. This change at the second site, found repeatedly in different reversions of the original mutation, restores activity because the two regions that contain amino acids 174 and 210 interact fortuitously to produce a protein with the correct folding. This type of reversion can usually be taken to mean that the two regions of the protein interact, and studies of the amino acid sequences of mutants and revertants are often informative in elucidating the structure of proteins.

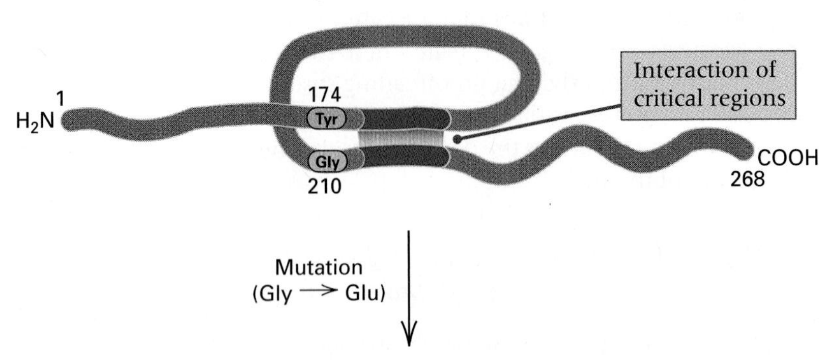


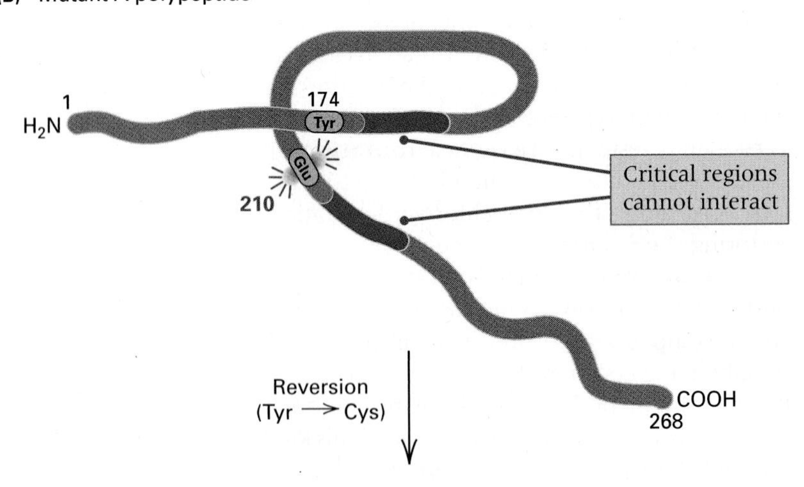


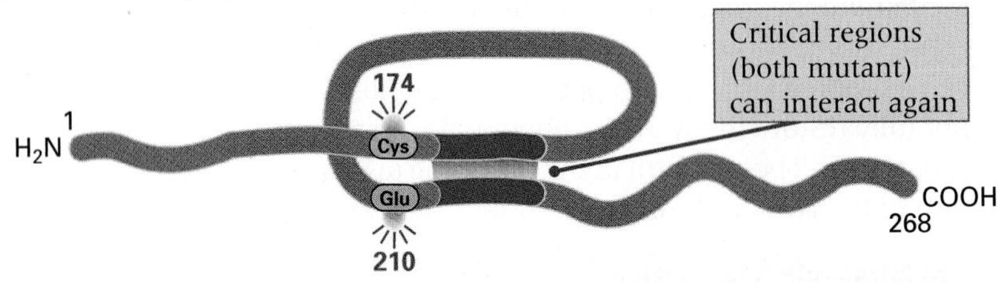


**FIGURE 14.37** Model for the effect of mutation and intragenic suppression on the folding and activity of the A protein of tryptophan synthase in *E. coli*. (A) The wildtype protein in which two critical regions (orange segments) of the polypeptide chain interact. (B) Disruption of proper folding of the polypeptide chain by a substitution of amino acid 210 prevents the critical regions from interacting. (C) Suppression of the effect of the original forward mutation by a subsequent change in amino acid 174. The structure of the region that contains this amino acid is altered, bringing the critical regions together again.

## Intergenic Suppression

**Intergenic suppression** refers to a mutational change *in a different gene* that eliminates or suppresses the mutant phenotype. Many intergenic suppressors are very specific in being able to suppress the effects of mutations in only one or a few other genes. For example, phenotypes resulting from the accumulation of intermediates in a metabolic pathway as a consequence of mutational inactivation of one step in the pathway can sometimes be suppressed by mutations in other genes that function earlier in the pathway, which reduces the concentration of the intermediate. These suppressors act only to suppress mutations that affect steps further along the pathway, and they suppress all alleles of these genes.

In prokaryotes, the best-understood type of intergenic suppression is found in some tRNA genes, mutant forms of which can suppress the effects of particular mutant alleles of many other genes. These suppressor mutations change the anticodon sequence in the tRNA and thus the specificity of mRNA codon recognition by the tRNA molecule. Mutations of this type were first detected in certain strains of *E. coli*, which were able to suppress particular phage-T4 mutants that failed to form plaques on standard bacterial strains. These strains were also able to suppress particular alleles of numerous other genes in the bacterial genome. The suppressed mutations were, in each case, nonsense mutations—those in which a stop codon (UAA, UAG, or UGA) had been introduced within the coding sequence of a gene, with the result that polypeptide synthesis was prematurely terminated and only an amino-terminal fragment of the polypeptide was synthesized. The suppressors are therefore called **nonsense suppressors**. The mechanism of suppression is discussed in detail in Chapter 10, but an overview is that the suppressor mutation occurs in the anticodon of one of the normal tRNA molecules, allowing it to pair with a nonsense codon and insert an amino acid so that translation can continue. Such mutations are not lethal because most tRNA genes are present in several copies, so nonmutated copies of the tRNA are also present to allow correct translation of the codon.

## The Ames Test for Mutagen/Carcinogen Detection

In view of the increased number of chemicals present as environmental contaminants, tests for the mutagenicity of these substances have become important. Furthermore, most carcinogens are also mutagens, so mutagenicity provides an initial screening for potential hazardous agents. One simple method of screening large numbers of substances for mutagenicity is a reversion test that uses nutritional mutants of bacteria. In the simplest type of reversion test, a compound that is a potential mutagen is added to solid growth medium, a known number of a mutant bacterium are plated, and the revertant colonies are counted. An increase in the reversion frequency significantly greater than that obtained in the absence of the test compound identifies the substance as a mutagen. However, simple tests of this type fail to demonstrate the mutagenicity of a large number of potent carcinogens. The explanation for this failure is that many substances are not directly mutagenic (or carcinogenic); rather, they require a conversion into mutagens by enzymatic reactions that take place in the livers of animals and that have no counterpart in bacteria. The normal function of these enzymes is to protect the organism from various naturally occurring harmful substances by converting them into soluble nontoxic substances that can be disposed of in the urine. However, when the enzymes encounter certain human-made and natural compounds, they convert these substances, which may not be harmful in themselves, into mutagens or carcinogens. The enzymes of liver cells, when added to the bacterial growth medium, activate the compounds and allow their mutagenicity to be recognized. The addition of liver extract is one step in the Ames test for carcinogens and mutagens.

In the **Ames test**, histidine-requiring ($His^-$) mutants of the bacterium *Salmonella typhimurium* containing either a base substitution or a frameshift mutation are tested for reversion to $His^+$. In addition, the bacterial strains have been made more sensitive to mutagenesis by the incorporation of several mutant alleles that inactivate the excision-repair system and that make the cells more permeable to foreign molecules. Because some mutagens act only on replicating DNA, the solid medium used contains enough histidine to support a few rounds of replication but not enough to permit the formation of a visible colony. The medium also contains the potential mutagen to be tested and an extract of rat liver. If the test substance is a mutagen or is converted into a mutagen, some colonies are formed. A quantitative analysis of reversion frequency can also be carried out by incorporating various amounts of the potential mutagen in the medium. The reversion frequency generally depends on the concentration of the substance being tested and, for a known carcinogen or mutagen, correlates roughly with its carcinogenic potency in animals. The Ames test is simple, rapid, inexpensive, and exquisitely sensitive. Some chemicals can be detected to be mutagenic in amounts as small as $10^{-9}$ g, and a condensate of as little as 1/100 of a cigarette can be shown to be mutagenic. The test is also highly quantitative (**FIGURE 14.38**). Chemicals need not be classified simplistically as "mutagenic" or "nonmutagenic." They can be classified according to their potency as mutagens, because more than a millionfold range in potency can be detected.

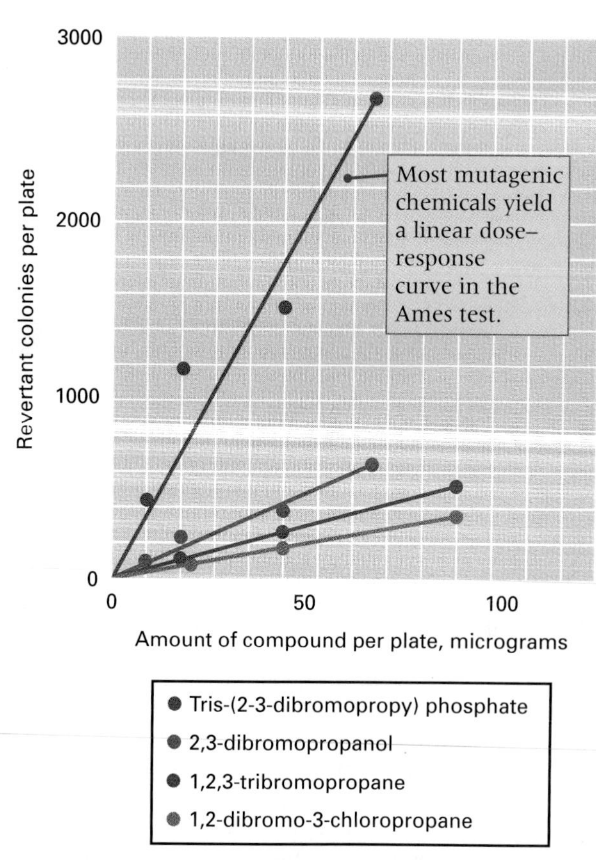


**FIGURE 14.38** Linear dose–response relationships obtained with various chemical mutagens in the Ames test. [Data from B. N. Ames, *Science* 204 (1979): 587–593.]

## CHAPTER SUMMARY

- Mutations can be classified by type of change at the molecular level, by conditions of expression, by effects on gene function, and in other ways.
- Many mutations occur in noncoding DNA sequences. Among those that do occur in coding sequences, missense (nonsynonymous) mutations are nucleotide substitutions resulting in an amino acid replacement, synonymous (silent) substitutions change a codon into a synonymous codon, and nonsense mutations create a translation-terminating codon. Single-base insertions or deletions result in frameshift mutations, in which the translational reading frame is offset.
- Expansion (growth in copy number) of trinucleotide repeats is an important mechanism of mutation associated with several human genetic diseases, including the fragile-X syndrome and Huntington disease.
- Transposable elements are DNA sequences that are able to change their location within a chromosome or to move between chromosomes. Insertion of transposable elements is an important mechanism of spontaneous mutation.
- Mutations can be induced by various agents, including base analogs, intercalating agents, highly reactive chemicals, and x-rays; most mutagens are also carcinogens.
- Cells contain enzymatic pathways for the repair of different types of damage to DNA. Among the most important repair mechanisms are mismatch repair of mispaired nucleotides, AP repair of apurinic or apyrimidinic sites, base excision repair, nucleotide excision repair, and repair following replication bypass of severe DNA damage.
- Suppressor mutations diminish or eliminate the phenotypic effects of a known mutation.

## REVIEW THE BASICS

- What is a spontaneous mutation? An induced mutation? A conditional mutation?
- Define and give a hypothetical example of each of the following types of nucleotide substitutions: transition, transversion, missense (nonsynonymous), synonymous (silent), nonsense.
- For a conditional mutation, what determines whether the mutant phenotype will be expressed?
- How is it possible for an organism to be homozygous for a mutation that is a temperature-sensitive lethal?
- What is a trinucleotide expansion disease and how is the "expansion" related to replication slippage? Name one human disease associated with trinucleotide expansions.
- What is meant by the term *mutation rate?* In practice, what does it mean to say that a particular allele has a mutation rate of $10^{-6}$ per gene per generation?
- What are transposable elements? Explain how a transposable element can cause a new mutation.
- How was the technique of replica plating used to demonstrate that antibiotic-resistance mutations are present at low frequency in populations of bacteria even in the absence of the antibiotic?
- What is a hot spot of mutation? Why might a mutational hot spot coincide with the site of a methylated cytosine in the DNA?
- What is mismatch repair and how does it happen? Name one human disease associated with a defective mismatch-repair system.
- Spontaneous depurination is the most frequent cause of damage to DNA, but the overwhelming majority of such lesions are repaired. How does the repair take place?

## GUIDE TO PROBLEM SOLVING

**Problem 1** The nontemplate DNA strand in a protein-coding region of the *rIIB* gene in bacteriophage T4 is shown here:

5′- · · · AGACTAGACAAA · · · -3′

This particular region of the polypeptide is quite tolerant of amino acid replacements, so many missense mutations in the region do not destroy function. Two frameshift mutations are isolated: −C is a deletion of the red C, and +T is an insertion of T between the red A and the red G. Recombination is used to combine the mutations into a single gene.

**(a)** Assuming that the wildtype reading frame includes the AGA codon in the mRNA, what is the wildtype amino acid sequence across this region?
**(b)** What is the effect of the −C mutation?
**(c)** What is the effect of the +T mutation?
**(d)** What happens in the double mutant?

**Answer**

**(a)** Because it is the sequence of the nontemplate DNA strand that is shown, the opposite strand is the strand transcribed. Transcription takes place in a right-to-left direction (because the RNA transcript is elongated by adding nucleotides to the 3′ end), and so the mRNA across the region has the same base sequence as the nontemplate strand (with U in the mRNA replacing T in the DNA). The amino acid sequence across this region is therefore Arg–Leu–Asp–Lys.
**(b)** In the −C mutant, the mRNA sequence across the region would be 5′–AGAUAGACAAA–3′, which is translated as Arg–End, where End indicates a UAG termination codon.
**(c)** In the +T mutant, the mRNA sequence across the region would be 5′–AGACUAUGACAAA–3′, which is translated as Arg–Leu–End, where End indicates a UGA termination codon.

**(d)** In the double mutant, the mRNA has the sequence 5′–AGAUAUGACAAA–3′, which is translated as Arg–Tyr–Asp–Lys. The frameshift mutations compensate and wildtype reading frame is restored. The only difference in the protein is that the Leu at position 2 in wildtype is replaced with Tyr in the double mutant.

**Problem 2** The base hypoxanthine (H) is an analog of guanine that pairs with cytosine but occasionally pairs with cytosine. What are two major mechanisms of mutagenesis due to hypoxanthine, and what types of mutations are produced?

**Answer** In problems of this sort, first note the base that is replaced, and then deduce what can happen in subsequent rounds of DNA replication. In the more common mechanism of mutagenesis, the H will enter a DNA strand paired with C, and in subsequent rounds of replication occasionally pair with T. The result will be a GC to AT transition mutation. In the less common mechanism of mutagenesis, the H will enter a DNA strand paired with T. In subsequent rounds of DNA replication the H will usually pair with C. In this mechanism, the result is an AT to GC transition mutation.

**Problem 3** Spontaneous depurination of DNA occurs at the rate of approximately $3 \times 10^{-9}$ depurinations per purine nucleotide per minute. A diploid human cell has a genome size of approximately $6 \times 10^9$ base pairs. Approximately how many spontaneous depurinations must be repaired per day in each cell?

**Answer** Because each base pair contains one purine, the number of spontaneous depurinations per diploid cell per minute is $(3 \times 10^{-9})(6 \times 10^9) = 18$ depurinations per cell per minute, or $18 \times 60 \times 24 = 26{,}000$ depurinations per diploid cell per day.

## ANALYSIS AND APPLICATIONS

**14.1** There are 12 possible substitutions of one nucleotide pair for another (for example, AT → GC).
- **(a)** Which changes are transitions and which transversions?
- **(b)** Assuming random nucleotide substitutions, what is the expected ratio of transitions : transversions?

**14.2** What is the minimum number of single-nucleotide substitutions that would be necessary for each of the following amino acid replacements?
- **(a)** Trp → Lys
- **(b)** Met → His
- **(c)** Tyr → Gly
- **(d)** Ala → Asp

**14.3** Weedy plants that are resistant to the herbicide atrazine have a single nucleotide substitution in the gene *psbA* that results in the replacement of a serine with an alanine in the polypeptide. Is the base change in the *psbA* gene that results in this amino acid replacement a transition or a transversion?

**14.4** The herbicide glyphosate acts by inhibiting the enzyme 5-enolpyruvylshikimate-3-phosphate synthase (EPSPS). Resistance to the herbicide has evolved in a few plant species. In some case the changes are in the EPSPS enzyme itself, and have been traced to the combined effects of two amino acid substitutions, T102I and P106S. (In this symbolism, the number is the amino acid residue in the protein: the wildtype amino acid is written at its left, and the mutant amino acid at its right.)
- **(a)** What nucleotide substitution can results in T102I?
- **(b)** What nucleotide substitution can results in P106S?
- **(c)** Suggest an explanation of why naturally occurring glyphosate mutants are so rare?

**14.5** A mutation is isolated that cannot be induced to revert. What types of molecular changes might be responsible?

**14.6** Occasionally, a person is found who has one blue eye and one brown eye or who has a sector of one eye a different color from the rest. Can these phenotypes be explained by new mutations? If so, in what types of cells must the mutations occur?

**14.7** An organism is discovered deep in an ocean vent whose genetic mechanisms are much like those of other organisms except that its DNA contains three base pairs instead of two. These are G paired with C, A paired with T, and M paired with N. The base M is a purine and N is a pyrimidine.
- **(a)** What types of nucleotide substitution can occur in this organism?
- **(b)** Among all possible nucleotide substitutions, what is the ratio of transitions : transversions?
- **(c)** How does the transition : transversion ratio compare with that expected in normal DNA.

**14.8** In the vent organism described in Problem 7, translation treats M as equivalent to G and N as equivalent to N.
- **(a)** What are the start codons in translation?
- **(b)** What are the stop codons?

**14.9** In an mRNA with a random sequence of nucleotides, what is the average distance between stop codons?

**14.10** In a coding sequence made up of equal proportions of A, U, G, and C, what is the probability that a random nucleotide substitution will result in a chain-termination codon?

**14.11** Does the nucleotide sequence of a mutation tell you anything about its dominance or recessiveness?

**14.12** You are studying a bacteriophage with a genome consisting of double-stranded DNA, and you discover a mutant that yields exceptionally large plaques. The gel diagram shows the bands obtained from the wildtype (Wt) and mutant (Mt) phage when their DNA is digested with the restriction enzyme *Sal*I. Suggest a molecular basis for the large-plaque mutation.

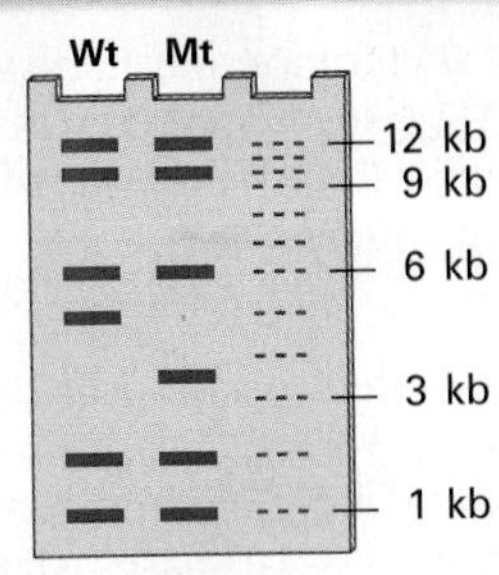


**14.13** A small plasmid contains a gene whose nontemplate strand includes the protein-coding sequence

5′-ATGCATCCCCCCCCCATGCAT-3′

The translational reading frame is ATG–CAT–CCC and so forth. The plasmid contains only two ATGCAT restriction sites, which the enzyme Ppu10I cleaves at the position of the vertical bar in each strand: A|TGCAT. You digest the plasmid with Ppu10I and then ligate the fragments together again. You isolate plasmids that contain one and only one copy of the sequence above in the correct position. Half the plasmids result in the amino acid sequence Met–His–Pro–Pro–Pro–Met–His. What do the other half encode, and what is their origin?

**14.14** The accompanying diagrams show two double-stranded DNA molecules, each of which contains two copies of a transposable element (shaded) in either (a) direct or (b) inverted orientation. In each case, diagram the result of homologous recombination between the transposable elements.

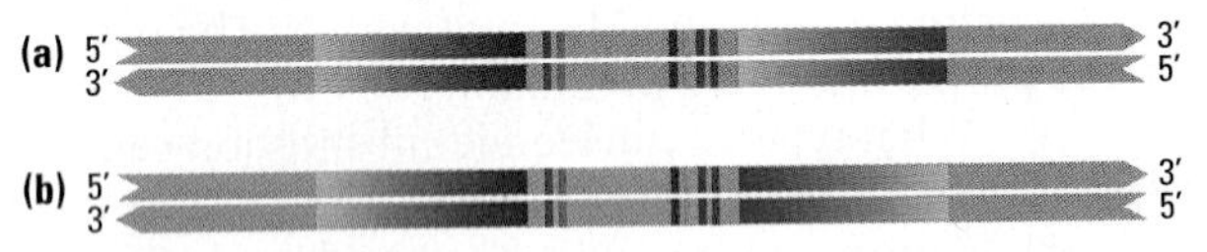


**14.15** A *lac*$^+$ *tet-r* plasmid is cleaved in the *lac*$^+$ gene with a restriction enzyme. The enzyme has a four-base restriction site and generates fragments with a two-base single-stranded overhang. The cutting site in the *lac* gene is in the codon for the second amino acid in the chain, a site that can tolerate any amino acid without loss of function. After cleavage, the single-stranded ends are converted to blunt ends with DNA polymerase I, and then the ends are joined by blunt-end ligation to recreate a circle. A *lac*$^-$ *tet-s* bacterial strain is transformed with the DNA, and tetracycline-resistant bacteria are selected. What is the Lac phenotype of the colonies? Explain.

**14.16** Human hemoglobin C is a variant in which a lysine in the beta-hemoglobin chain is substituted for a particular glutamic acid. What single-base substitution can account for the hemoglobin-C mutation?

**14.17** How many amino acids can substitute for tyrosine by a single-base substitution? (Do not assume that you know which tyrosine codon is being used.)

**14.18** Dyes are often incorporated into a solid medium to determine whether bacterial cells can utilize a particular sugar as a carbon source. For instance, on eosin-methylene blue (EMB) medium containing lactose, a Lac$^+$ cell yields a purple colony and a Lac$^-$ cell yields a pink colony. If a population of Lac$^+$ cells is treated with a mutagen that produces Lac$^-$ mutations and the population is allowed to grow for many generations before the cells are placed on EMBlactose medium, a few pink colonies are found among a large number of purple ones. However, if the mutagenized cells are plated on the medium immediately after exposure to the mutagen, some colonies appear that are *sectored* (half purple and half pink). Explain how the sectored colonies arise.

**14.19** Which of the following amino acid replacements would be expected with the highest frequency among mutations induced by 5-bromouracil? (1) Met → Leu, (2) Met → Lys, (3) Leu → Pro, (4) Pro → Thr, (5) Thr → Arg.

**14.20** Among several hundred missense mutations in the gene for the A protein of tryptophan synthase in *E. coli*, fewer than 30 of the 268 amino acid positions are affected by one or more mutations. Explain why the number of positions affected by amino acid replacements is so low.

**14.21** The molecule 2-aminopurine (Ap) is an analog of adenine that pairs with thymine. It also occasionally pairs with cytosine. What types of mutations will be induced by 2-aminopurine?

**14.22** For mutations that result from spontaneous missense substitutions, the rate of forward mutation (from wildtype allele to mutant allele) is always much greater than the rate of reverse mutation (from a mutant allele back to wildtype). Explain why.

**14.23** The X chromosome of *Drosophila* contains about 2700 genes. In each generation, about 1 percent of all X chromosomes undergo spontaneous mutation to a new recessive lethal. What proportion of all X-linked genes are capable of mutating to a lethal if the rate of spontaneous mutation is $10^{-5}$ per gene per generation.

**14.24** A *Drosophila* male carries an X-linked temperature-sensitive recessive allele that is viable at 18°C but lethal at 29°C. What sex ratio would be expected among the progeny if the progeny were reared at 29°C and the male were mated to:

**(a)** A normal XX female?

**(b)** An attached-X female?

**14.25** Among genomes as diverse as the bacteriophages lambda and T4, the bacterium *Escherichia coli*, and the fungi *Neurospora crassa* and *Saccharomyces cerevisiae*, the spontaneous mutation rate is approximately 0.003 mutation per genome per DNA replication. The genome sizes lambda, *E. coli*, and *S. cerevisiae* are approximately 50 thousand, 4.6 million, and 13.5 million base pairs, respectively.

**(a)** What proportion of genomes escape spontaneous mutation in each replication?

**(b)** What is the mutation rate per base pair per DNA replication in each of these genomes?

**(c)** What might the differences in mutation rate per base pair imply about the evolution of mutation rates?

**14.26** If a gene in a particular chromosome has a probability of mutation of $5 \times 10^{-5}$ per generation, and if the allele in a particular chromosome is followed through successive generations:

**(a)** What is the probability that the allele does not undergo a mutation in 10,000 consecutive generations?

**(b)** What is the average number of generations before the allele undergoes a mutation?

**14.27** In the mouse, a dose of approximately 1 gray (Gy) of x-rays produces a rate of induced mutation equal to the rate of spontaneous mutation.

**(a)** Taking the spontaneous mutation rate into account, what is the total mutation rate at 1 Gy?

**(b)** What dose of x-rays is expected to increase the mutation rate by 50 percent?

**(c)** What dose is expected to increase the mutation rate by 10 percent?

**14.28** The accompanying diagrams show two nonhomologous chromosomes, each containing a copy of a transposable element (shaded), which can be oriented either (a) in the same direction or (b) in opposite directions with respect to the centromere. For clarity, the length of the transposable element is greatly exaggerated relative to the length of the chromosome. (In reality, the average transposable element in *Drosophila* is about 0.01 percent of the length of a chromosome.) Draw diagrams illustrating the consequences of ectopic recombination between the transposable elements.

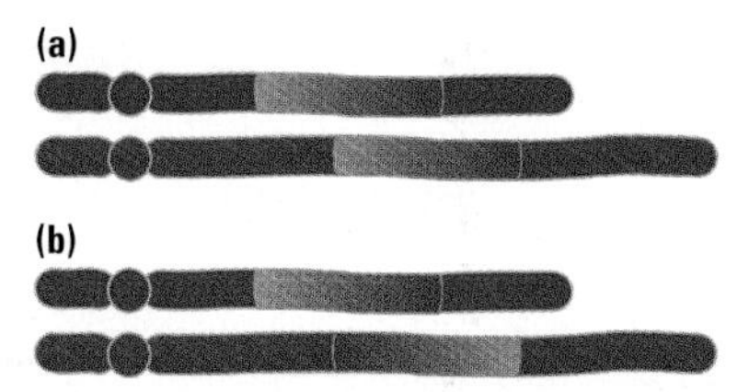


**14.29** This problem illustrates how conditional mutations can be used to determine the order of genetically controlled steps in a developmental pathway. A certain organ undergoes development in the sequence of stages A → B → C, and both gene *X* and gene *Y* are necessary for the sequence to proceed. A conditional mutation *X[hs]* is sensitive to heat (the gene product is inactivated at high temperatures), and a conditional mutation *Y[cs]* is sensitive to cold (the gene product is inactivated at cold temperatures). The double mutant *X[hs]/X[hs]; Y[cs]/Y[cs]* is created and reared at either high or low temperatures. How far would development proceed in each of the following cases at the high temperature and at the low temperature?

**(a)** Both *X* and *Y* functions are necessary for the A → B step.

**(b)** Both *X* and *Y* functions are necessary for the B → C step.

**(c)** *X* function is necessary for the A → B step, and *Y* function for the B → C step.

**(d)** *Y* function is necessary for the A → B step, and *X* function for the B → C step.

**14.30** The nontemplate DNA strand in a protein-coding region of a gene is shown here:

5′-· · · AGAGTAACGCTCAGA · · · -3′

The translational reading frame across this region is Arg–Val–Thr–Leu–Arg and so forth. This particular region of the polypeptide is quite tolerant of amino acid replacements, so many missense mutations in the region do not destroy function. Four frameshift mutations are isolated: −G is a deletion of the red G, +G is a duplication of the red G, −C is a deletion of the red C, and +C is a duplication of the red C. As expected, all four mutations lead to a nonfunctional protein. Recombination is used to combine the mutations, in the expectation that a deletion (or duplication) of one nucleotide will be compensated for by a duplication (or deletion) slightly farther along, because the second mutation will shift the reading frame back to the normal reading frame, and changes in the amino acids between the mutations should not destroy function. An unexpected result was obtained. Although the combination +G with −C did, indeed, restore protein function, the combination −G with +C was still mutant. How can you account for this result?

**14.31** A *Neurospora* strain that is unable to synthesize arginine (and that therefore requires this amino acid in the growth medium) produces a revertant colony able to grow in the absence of arginine. A cross is made between the revertant and a wildtype strain. What proportion of the progeny from this cross would be arginine-independent if the reversion occurred by each of the following mechanisms?

**(a)** A precise reversal of the nucleotide change that produced the original *arg* mutant allele.

**(b)** A mutation to a suppressor of the *arg* allele occurring in another gene located in a different chromosome.

**(c)** A suppressor mutation occurring in another gene located 10 map units away from the *arg* allele in the same chromosome.

## CHALLENGE PROBLEMS

Challenge Problem 1 Which of the following 8-spore asci might arise from gene conversion at the *m* locus? Describe how these asci might arise.

| Ascus type | A | B | C | D | E | F |
|---|---|---|---|---|---|---|
| Spore 1 | + | + | + | + | + | + |
| Spore 2 | + | + | + | + | + | + |
| Spore 3 | + | *m* | + | + | *m* | *m* |
| Spore 4 | + | *m* | *m* | *m* | *m* | *m* |
| Spore 5 | *m* | *m* | *m* | + | + | *m* |
| Spore 6 | *m* | *m* | *m* | *m* | + | *m* |
| Spore 7 | *m* | + | *m* | *m* | *m* | *m* |
| Spore 8 | *m* | + | *m* | *m* | *m* | *m* |

Challenge Problem 2 A strain of *E. coli* contains a mutant $tRNA^{Leu}$. Instead of recognizing the codon 5′–CUG–3′, as it does in nonmutant cells, the mutant tRNA recognizes the codon 5′–GUG–3′. A missense mutation of another gene, affecting amino acid number 28 along the chain, is suppressed in cells with the mutant $tRNA^{Leu}$.

**(a)** Assuming normal Watson–Crick base pairs, determine the anticodons of the wildtype and mutant $tRNA^{Leu}$ molecules.

**(b)** What kind of mutation is present in the mutant tRNALeu gene?

**(c)** What amino acid would be inserted at position 28 of the mutant polypeptide chain if the missense mutation were not suppressed?

**(d)** What amino acid is inserted at position 28 when the missense mutation is suppressed?

### GENETICS *on the web*

GeNETics on the Web will introduce you to some of the most important sites for finding genetics information on the Internet. To explore these sites, visit the Jones and Bartlett companion site to accompany *Genetics: Analysis of Genes and Genomes, Seventh Edition* at http://biology.jbpub.com/book/genetics.

There you will find a chapter-by-chapter list of highlighted keywords. When you select one of the keywords, you will be linked to a Web site containing information related to that keyword.

# CHAPTER 15

# Molecular Genetics of the Cell Cycle and Cancer

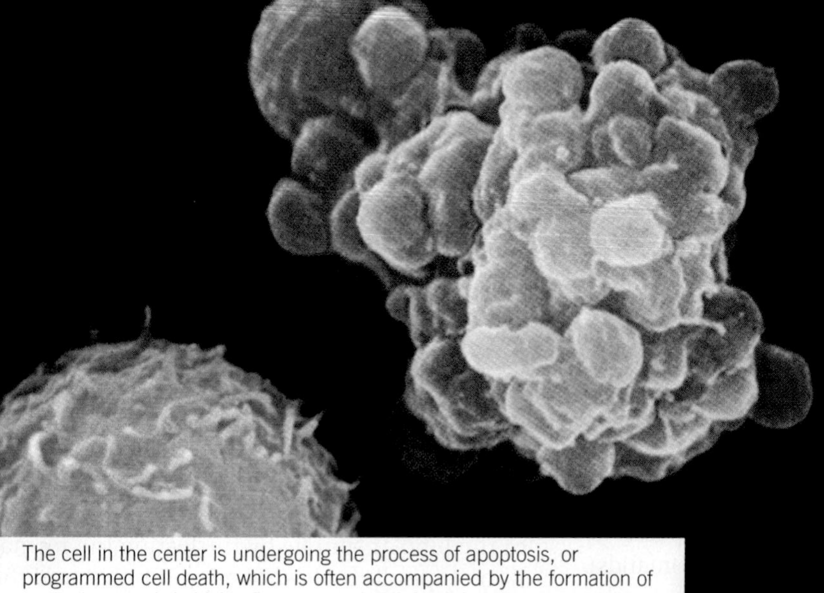

The cell in the center is undergoing the process of apoptosis, or programmed cell death, which is often accompanied by the formation of such characteristic blebs. Programmed cell death helps to maintain the integrity of the genome from generation to generation. Cancer cells are cancerous in part because they undergo mutations that circumvent programmed cell death. [© Phototake, Inc./Alamy Images]

## CHAPTER OUTLINE


**15.1** The Cell Cycle
- Key Events in the Cell Cycle
- Transcriptional Program of the Cell Cycle

**15.2** Genetic Analysis of the Cell Cycle
- Mutations Affecting Progression Through the Cell Cycle

**15.3** Progression Through the Cell Cycle
- Cyclins and Cyclin-Dependent Protein Kinases
- Targets of the Cyclin–CDK Complexes
- Triggers for the $G_1$/S and $G_2$/M Transitions
- Protein Degradation Helps Regulate the Cell Cycle

**15.4** Checkpoints in the Cell Cycle
- The DNA Damage Checkpoint
- The Centrosome Duplication Checkpoint
- The Spindle Assembly Checkpoint
- The Spindle Position Checkpoint

**15.5** Cancer Cells
- Oncogenes and Proto-Oncogenes
- Tumor-Suppressor Genes

**15.6** Hereditary Cancer Syndromes
- Defects in Cell-Cycle Regulation and Checkpoints
- Defects in DNA Repair

**15.7** Genetics of the Acute Leukemias

CONNECTION Cycle-Ops
Leland H. Hartwell, Joseph Culotti, John R. Pringle, and Brian J. Reid 1974
*Genetic Control of the Cell Division Cycle in Yeast*

CONNECTION Sick of Telomeres
William C. Hahn, Christopher M. Counter, Ante S. Lundberg, Roderick L. Beijersbergen, Mary W. Brooks, and Robert A. Weinberg 1999
*Creation of Human Tumor Cells with Defined Genetic Elements*

CONNECTION Two Hits, Two Errors
Alfred G. Knudson 1971
*Mutation and Cancer: Statistical Study of Retinoblastoma*

Cancer is a disease characterized by the uncontrolled proliferation of cells. The normal mechanisms that regulate cellular growth and division break down. Cancer is a genetic disease. It results from mutations that overcome the normal limits to the number of cell divisions that can take place before a cell dies. These mutations usually occur in somatic cells, and full-blown malignant cancer usually requires a number of sequential mutations to get started. But occasionally a mutation affecting cell-cycle regulation is inherited through the germ line, and persons who inherit the mutation have a greatly increased risk (sometimes approaching 100 percent) of developing malignancies due to additional somatic mutations. To understand cancer, therefore, one must understand the normal mechanisms that govern cellular proliferation. This is where we begin.

## 15.1 The Cell Cycle

The essential functions of the mitotic cell cycle are:

1. To ensure that each chromosomal DNA molecule is replicated once and only once per cycle.
2. To ensure that the identical replicas of each chromosome (the sister chromatids) are distributed equally to the two daughter cells.

As we saw in Chapter 4, the cell cycle is divided into four phases: the *M (mitosis) phase*, in which the sister chromatids are physically separated into the two daughter nuclei, and a three-part *interphase* composed of *$G_1$ (gap 1)*, *S (DNA synthesis)*, and *$G_2$ (gap 2)*, which occur in that order. The cell cycle may thus be illustrated as shown in **FIGURE 15.1** for a mammalian cell. In this representation, *cytokinesis* (the division of the cytoplasm into two approximately equal parts containing the daughter nuclei) is included in the M period.

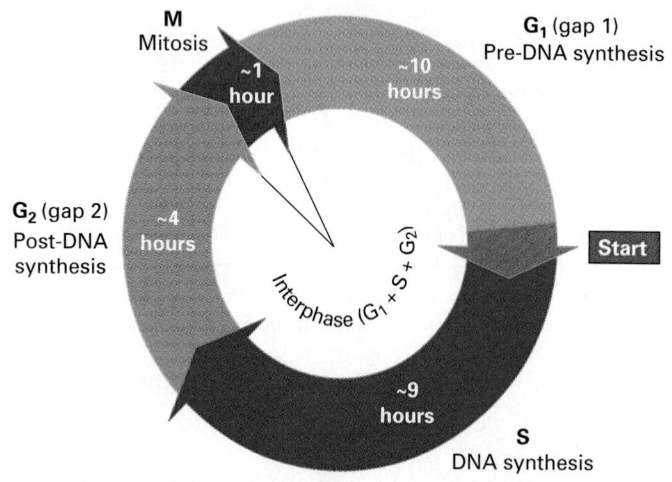


**FIGURE 15.1** The cell cycle of a typical mammalian cell growing in tissue culture with a generation time of 24 hours.

### Key Events in the Cell Cycle

**FIGURE 15.2** shows a number of key events that must occur to ensure the proper duplication and distribution of the chromosomes. The spindle pole is organized around a small region of clear cytoplasm near the interphase nucleus called the **centrosome**; in many organisms this role is played by a pair of **centrioles**, which are more particulate in appearance. Both are microtubule organizing centers that must be duplicated and positioned. The duplication begins late in $G_1$ and is completed during S phase (in some cells, duplication may occur somewhat later). The duplicated poles then slowly begin to migrate to positions on opposite sides of the nucleus. Meantime, within the nucleus during the S phase, DNA replication takes place. Completion of DNA replication marks the beginning of $G_2$. Soon after the M phase commences, the centrosomes reach their final destinations and the chromosomes begin to condense. Each centrosome organizes one pole of the bipolar spindle by nucleating the formation of spindle and astral microtubules, and the condensed chromosomes become attached to spindle microtubules on both sides of the centromere (kinetochore) as the nuclear envelope breaks down. Each chromosome is thus physically attached to both spindle poles and is maneuvered to a position approximately halfway between them. As anaphase begins the spindle elongates, the centromeres separate, and the sister chromatids migrate toward opposite poles. Once the daughter chromosomes reach the poles, the spindle disassembles, the chromosomes decondense, and the nuclear membrane is formed again. These events return the cell to the $G_1$ phase.

The mechanisms by which cell growth is regulated and the cell cycle controlled have been approached via the methods of biochemistry, cell biology, and genetics. The most extensive genetic studies have focused on two yeasts, the budding yeast (*Saccharomyces cerevisiae*) and the fission yeast (*Schizosaccharomyces pombe*). As its common name implies, *S. cerevisiae* multiplies by budding. The position of a cell within the cell cycle can be monitored visually by the size of the bud. This convenient feature of the cell cycle is depicted diagrammatically in **FIGURE 15.3**. The bud first emerges shortly after entry into S phase and grows throughout the cell cycle. Mitosis takes place

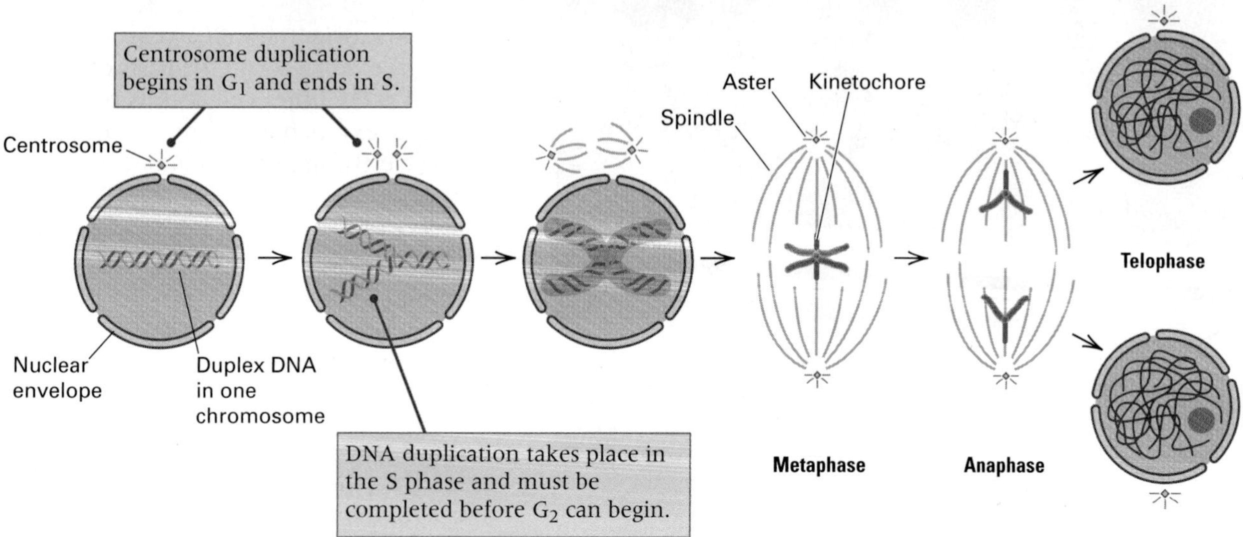


**FIGURE 15.2** Major events in the cell cycle. In yeast, the spindle pole body serves the same function as the centrosome in many other organisms; both are microtubule organizing centers from which the spindle emerges. In most other organisms, nuclear division takes place and the pinching off of cells (cytokinesis) follows. In yeast, the "shell" of the daughter cell forms and enlarges before nuclear division takes place. The nucleus (its membrane never breaks down) moves into the bridge between the mother and daughter cell, and nuclear division occurs there. After the daughter nuclei move into the two cell bodies, a septum is laid down between the two cells. [Adapted from L. H. Hartwell and M. B. Kastan, *Science* 266 (1994): 1821–1828.]

*within* the nucleus without breakdown of the nuclear envelope. In this case the spindle poles, which are known as *spindle pole bodies,* are embedded within the nuclear envelope. To position the spindle, one spindle pole body moves while still within the nuclear envelope to a position exactly opposite the second spindle pole body across the way. By this time the nucleus already contains the duplicated chromosomes. Capture of the tips of the cytoplasmic microtubules at the bud tip connects one spindle pole body to the bud cortex, which results in movement of the nucleus to the neck of cytoplasm connecting the mother and daughter cell at $G_2$. Mitosis occurs when the bud has grown to a size nearly as large as the mother cell. Shortly after chromosome separation, a barrier or *septum* is laid down between the mother and daughter cells. At cell separation, the daughter cell is only slightly smaller than the mother cell, but it typically must grow this extra bit to achieve the optimal size before it can start its own cell cycle. The various stages of bud growth from beginning to end can be identified in the electron micrograph in **FIGURE 15.4**.

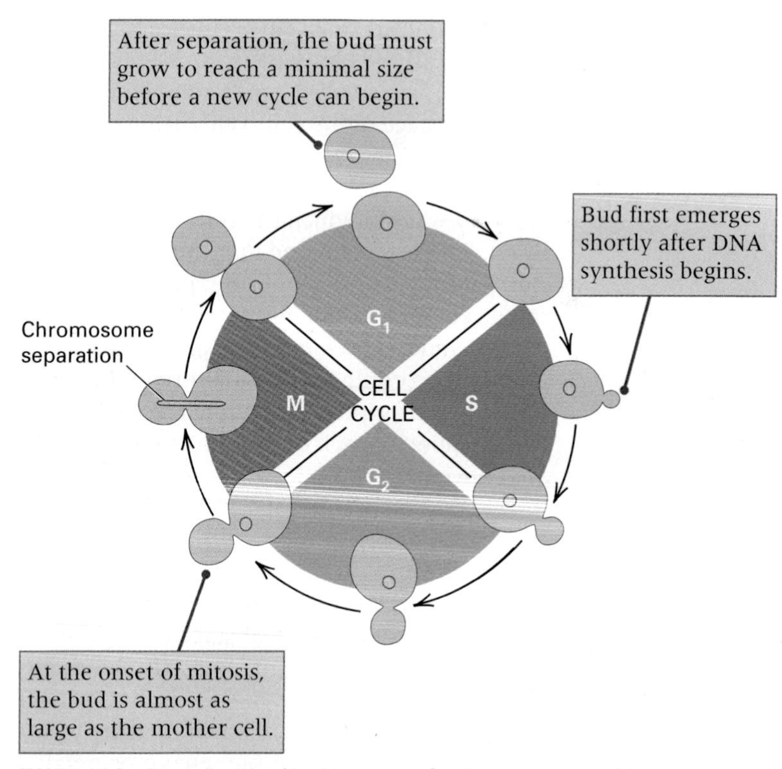


**FIGURE 15.3** The cell cycle of budding yeast, *Saccharomyces cerevisiae.*

## Transcriptional Program of the Cell Cycle

In budding yeast, because the DNA sequence of the entire genome is known, it is possible to analyze the transcription patterns of all of the 5538 genes in the cell in a single experiment using the type of high-density gene arrays (DNA chips) discussed in Section 12.4 and illustrated in Figure 12.20 on page 450. Such experiments have shown that the transcript levels of about 800 genes vary in a periodic or cyclic pattern through the cell cycle. This result is

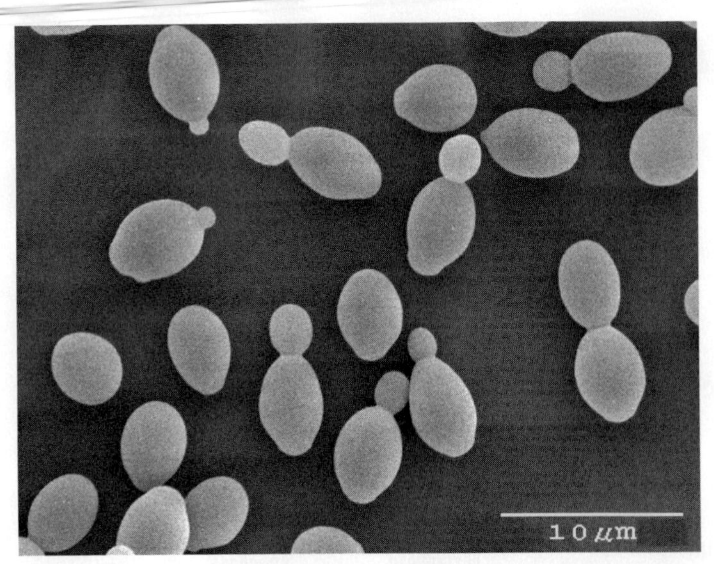


**FIGURE 15.4** Scanning electron micrograph of cells of diploid budding yeast in various stages of the cell cycle. Bud size is correlated with the position of the cell within the cell cycle. Here the buds range in size from quite small to nearly as large as the mother cell. [© Biphoto/Photo Researches, Inc.]

dramatically illustrated in **FIGURE 15.5**. Each horizontal stripe represents the expression pattern of a single gene through two or more cell cycles in synchronized cells. (In a culture of *synchronized cells,* all cells are at the same stage in the cell cycle.) The colored bands across the top indicate the stage of the cell cycle (using the same colors for the stages as labeled on the right), and the designations Alpha, *cdc15*, and *cdc28* refer to different ways in which the cells were synchronized. For each gene, red indicates overexpression relative to the level of expression of the same gene in a nondividing cell, and green indicates underexpression. The genes have been grouped according to the similarity in their pattern of transcription through the cell cycle. Transcription of each of these 800 genes is initiated once per cell cycle, an event indicated by the red portion of the stripe as the cell cycle progresses. Some genes are transcribed in $G_1$, some in $G_2$, a few in S, and some in M. The genes are arranged in the order in which they are transcribed, and their principal stage of expression is indicated at the far right. Typically, genes encoding proteins that are needed in one part of the cycle are transcribed in the immediately preceding period. For example, enzymes needed for synthesis of the trinucleotide precursors of DNA and for DNA replication are made in $G_1$ immediately prior to their use in S phase. Similarly, the histone proteins are synthesized during S phase immediately prior to their incorporation into chromatin and their use in chromosome condensation.

## 15.2 Genetic Analysis of the Cell Cycle

The ready assignment of a yeast cell to a position in the cell cycle on the basis of the relative sizes of the mother cell and its bud has made possible extensive genetic analyses of the cell cycle through the isolation and study of temperature-sensitive mutants. Temperature-sensitive **cell division cycle (*cdc*) mutants** are typically wildtype at 23°C (the *permissive* temperature) but unable to complete the cell cycle at 36°C (the *restrictive* temperature). At the higher temperature, mutant cells accumulate at a characteristic stage in the cell cycle. This is the stage at which their progression in the cell cycle is halted. The stage-specific stop is exceptional among mutations affecting cellular functions. For example, temperature-sensitive mutants with defects in protein synthesis do not cease growing abruptly when the temperature is raised. Each cell continues along until it runs out of functional proteins, and new proteins need to be synthesized for continued progression in the cell cycle. The stopping point differs from cell to cell, so the mutant cells stop growing at different stages in the cell cycle. In the microscope one sees unbudded cells, cells with small buds, and cells with large buds—essentially the same distribution one sees in an asynchronous cell population. But for each *cdc* mutant the stop is at a specific stage, which differs from one mutant to the next and is related to the function of the mutant gene product. Once it has been established where in the cell cycle a *cdc* mutant is blocked, further analysis can be used to determine whether specific processes can occur in the mutants, such as DNA synthesis, spindle formation, or formation of the division septum. In the next section we consider the analysis of two *cdc* mutants that stop, or **arrest**, at the $G_2$/M boundary.

### Mutations Affecting Progression Through the Cell Cycle

The micrograph in **FIGURE 15.6**, part A, shows cells of a mutant designated *cdc13* grown in an unsynchronized culture at the permissive temperature of 23°C. The morphology and variation in size are normal; some unbudded cells are present, as well as dividing cells with buds ranging in size from very small to nearly as large as the mother cell. When this culture was placed at the restrictive temperature of 36°C for 6 hours, the cells took on the appearance shown in Figure 15.6B. Two types of configurations are seen. One consists of a pair of large cells. These

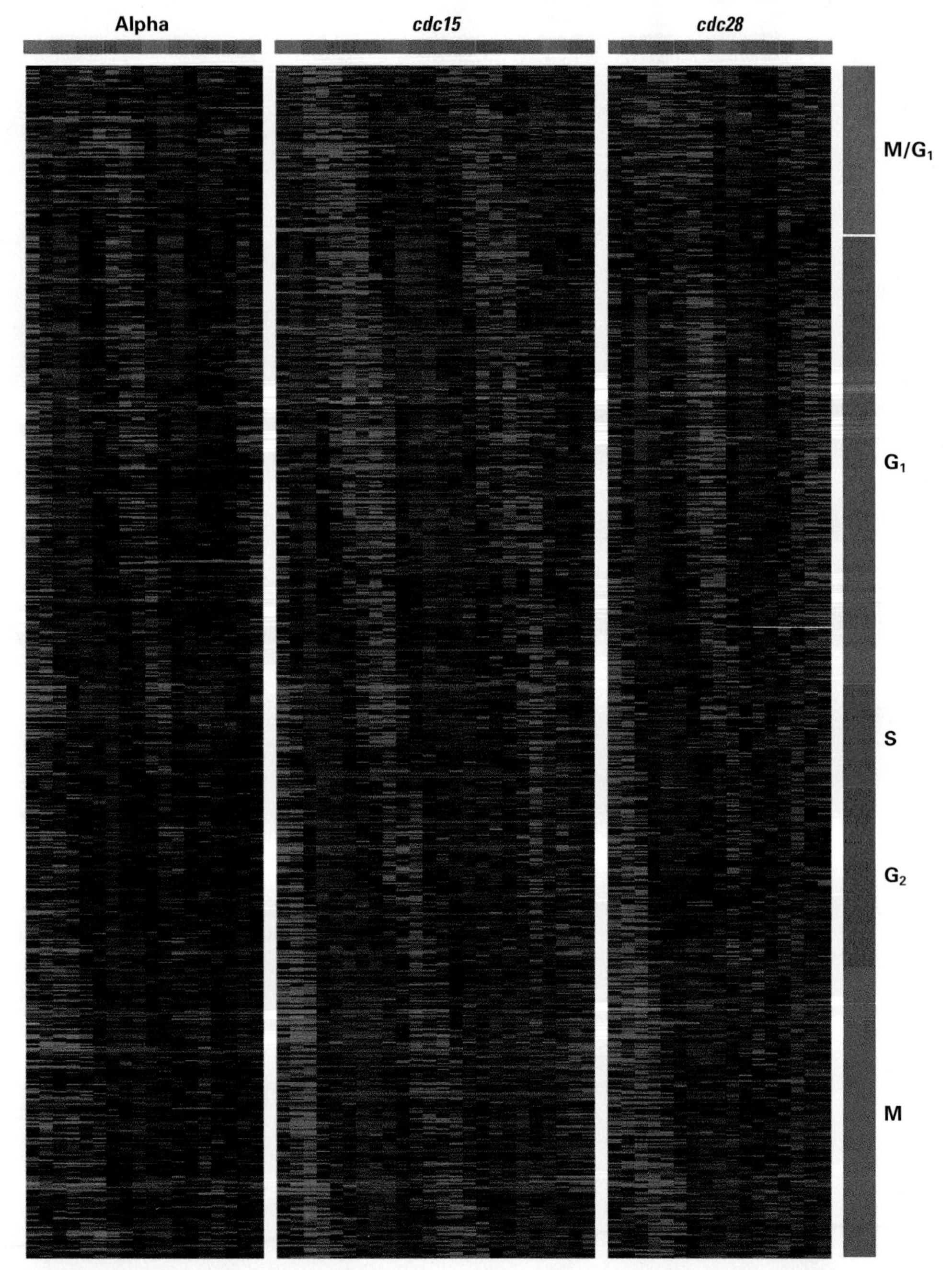


**FIGURE 15.5** Expression patterns of about 800 yeast genes whose level of transcription varies systematically through the cell cycle. Each horizontal stripe indicates the transcription pattern of a different gene through the cell cycle. Green indicates underexpression relative to a nondividing cell, red overexpression. The color bars on the right indicate the phase in which each gene is maximally expressed: $M/G_1$ transition, $G_1$, S, $G_2$, or M. These same colors indicate the cell-cycle phase along the top. Alpha, *cdc15,* and *cdc28* indicate different ways in which the cell cycle has been synchronized. [Reproduced from P. T. Spellman, et al., *Mol. Biol. Cell* 9 (1998): 3273–3297.]

are cells that were not dividing or had small buds at 23°C; after the temperature shift they continued in the cell cycle until they arrested at the large-bud stage. The other configuration consists of four large cells (*quartets*). These derive from cells that, prior to the temperature shift, were nearing the end of division and had large buds; after the temperature shift they completed their current cycle, and both daughter cells, still side-by-side, initiated one more cell cycle and arrested at the large-bud stage, yielding the quartet.

**FIGURE 15.7** illustrates how these results can be interpreted. Single mutant cells that are transferred to 36°C prior to the time that the affected protein is needed arrest as large budded cells, because in the absence of functional protein they cannot proceed beyond this point.

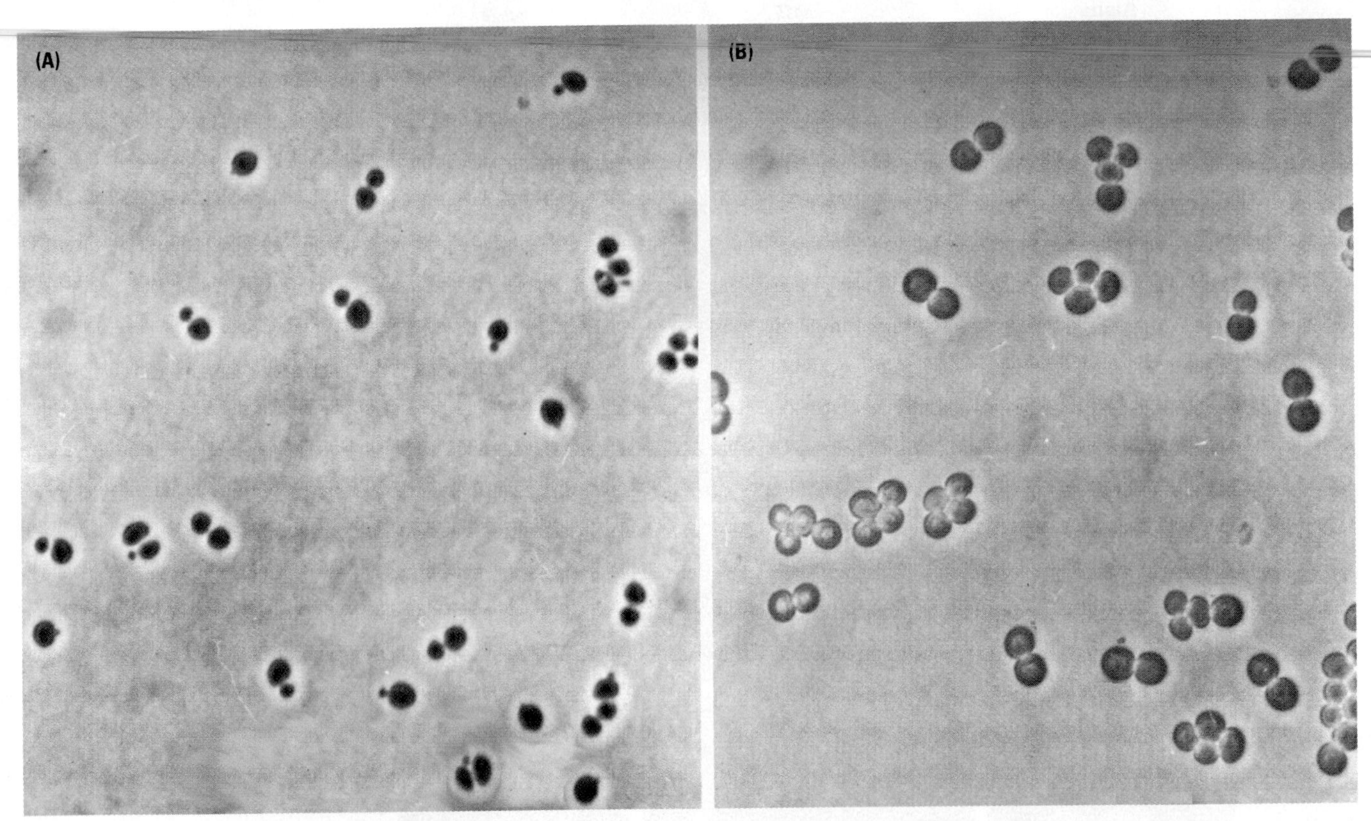


**FIGURE 15.6** Cell-cycle arrest in temperature-sensitive *cdc13* mutants. (A) An asynchronous culture of a *cdc13* mutant grown at 23°C (the permissive temperature). The cells either are unbudded or have buds ranging in size from very small to nearly the size of the mother cell. (B) The same cells after incubation for 6 hours at 36°C (the restrictive temperature). Some cells are arrested with very large buds. Others have completed their current division and started another. These are the cells arrested in quartets. [Reprinted from *Exp. Cell Res.*, vol. 67, J. Culotti and L. H. Hartwell, "Genetic control of the cell division cycle in yeast . . .," pp. 389–401, copyright (1971), with permission from Elsevier. (http://www.sciencedirect.com/science/journal/00144827)]

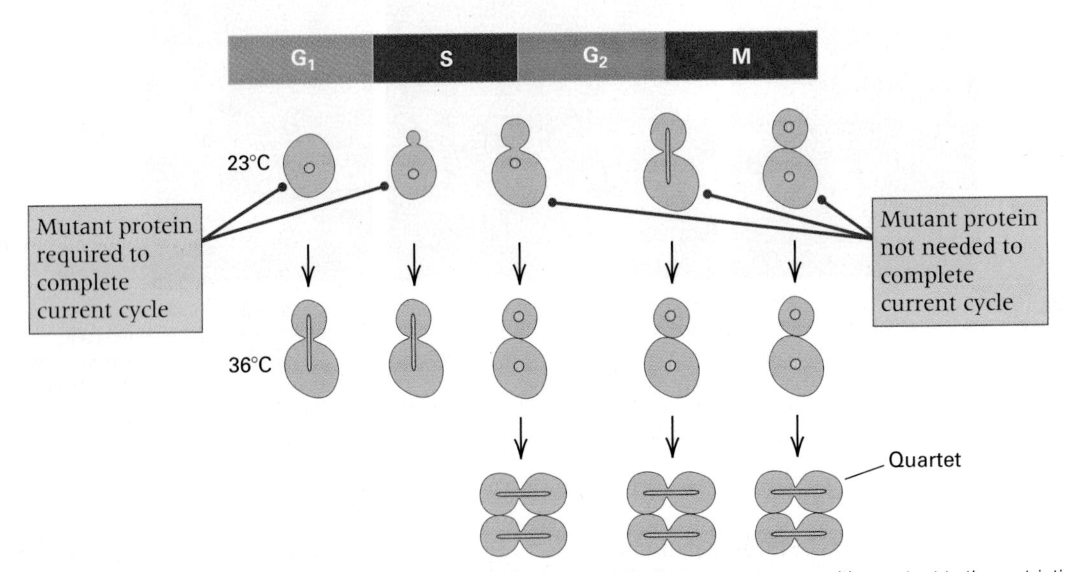


**FIGURE 15.7** Determining the time of function of a cell-cycle protein by shift of a temperature-sensitive mutant to the restrictive temperature. Cells in a stage prior to when the defective protein is needed undergo arrest in the current division cycle at the stage when the defective protein is needed. These yield the doublet configurations. Cells that have passed the time when the protein is needed in the current cycle complete the current cycle, begin another, and arrest in this second cycle at the stage when the defective protein is needed. These yield the quartet configurations. [Adapted from L. E. Hartwell (J. G. Fortner and J. E. Rhodes, eds.). *Accomplishments in Cancer Research.* J. B. Lippincott Company, Philadelphia (1992).]

Cells that are transferred to 36°C when the protein is no longer needed in the current cycle (some time late in the S phase in this example) complete the current cycle and initiate a new cycle that is terminated when both daughter cells arrest with large buds because the affected protein is now unavailable. The *cdc13* mutant showing the pattern in Figure 15.6 has a defect in telomere processing, which occurs late in the S phase; when *cdc13* cells are grown at the restrictive temperature, abnormally extended single-stranded DNA is found at the telomeres.

**FIGURE 15.8** shows the results of an experiment with another cell-division-cycle mutant; this one is designated *mob1-77*, which also arrests at the large-bud stage like the cells in Figure 15.6B. These data indicate that *mob1-77* cells, although they cannot complete the division process, can replicate their DNA at the restrictive temperature. At the permissive temperature (the upper graphs in Figure 15.8), asynchronous cultures of both wildtype and mutant cells include some cells with 1*C* (haploid) and some cells with 2*C* (diploid) DNA content, depending on whether they have completed DNA replication. At 36°C, the *mob1-77* mutant cells all become arrested at the large-bud stage with the 2*C* DNA content characteristic of cells at the $G_2$/M interface. These and other experiments allow the mechanism of a mutation-induced block in the cell cycle to be progressively narrowed down. Ultimately, the protein affected and the nature of the mutation are identified via cloning and sequencing of the mutant allele. On the basis of this information, further experiments are designed to zero in on the protein's exact function.

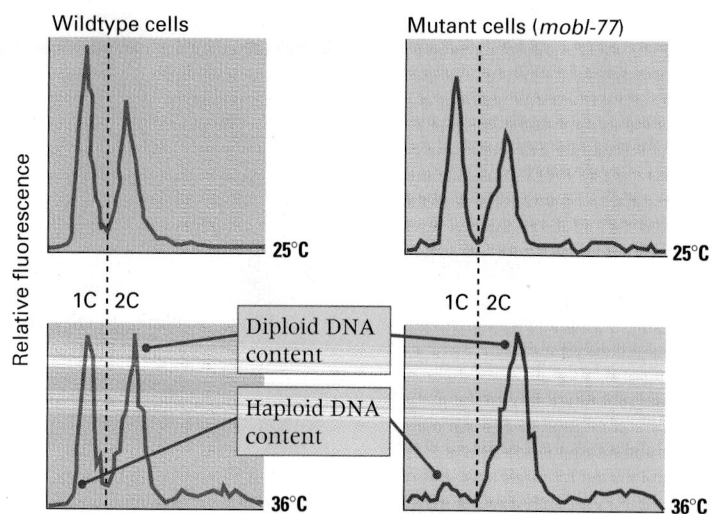


**FIGURE 15.8** Replication of DNA in *cdc* mutant cells, starting with a population of haploid (1*C*) cells. At the permissive temperature (top panels), both mutant and wildtype cells have about 67 percent of cells with a haploid DNA content and 33 percent with a diploid DNA content; the latter have completed S phase but have not yet completed M. At the restrictive temperature (bottom panels), the mutant cells (right) accumulate with a diploid DNA content (2*C*), indicating that they are unable to complete the cell cycle. [Adapted from F. C. Luca and M. Winey, *Mol. Biol. Cell* 9 (1998): 29–46.]

## 15.3 Progression Through the Cell Cycle

Much of the molecular machinery that regulates the cell cycle has been conserved in evolution. For this reason, our current understanding of the way the cell cycle works includes inferences made from studies of cells of organisms as diverse as yeasts, frogs, and human beings.

### Cyclins and Cyclin-Dependent Protein Kinases

In the early stages of the cell cycle, progression from one phase to the next is controlled by characteristic protein complexes that are called **cyclin–CDK complexes** because they are composed of **cyclin** subunits combined with **cyclin-dependent protein kinase (CDK)** subunits (**FIGURE 15.9**). All eukaryotes regulate progression through the cell cycle by means of cyclin–CDK complexes, although the details of their structure and their mechanisms of action may differ slightly from one organism to another. Typically, a eukaryotic cell has a small *family* of genes whose individual members encode cyclins that function in specific portions of the cell cycle, or

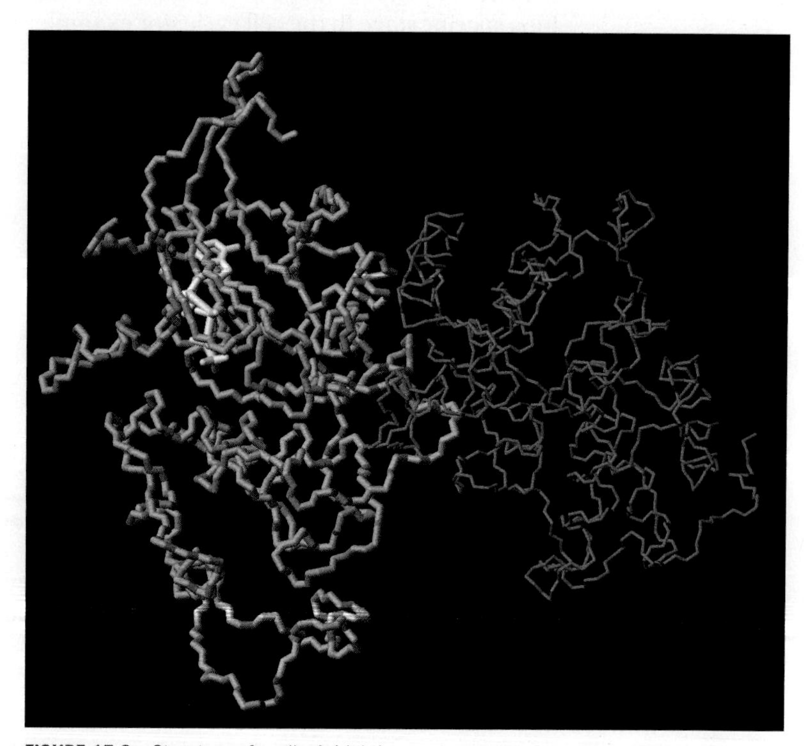

**FIGURE 15.9** Structure of cyclin A (right) complexed with Cdk2 (left). Cyclin A binds to one side of the catalytic cleft in Cdk2, inducing a large conformational change that opens the active site and activates the kinase domain. The ATP (white) bound to Cdk2 is the phosphate donor in the kinase reaction. [Courtesy of Carlos Bustamante, Cornell University. Coordinate data from P. D. Jeffrey, et al., *Nature* 376 (1995): 313–321.]

**Table 15.1 Cyclin–CDK interactions in eukaryotic cells**

| | Stage of cell cycle | | | | | |
|---|---|---|---|---|---|---|
| | $G_1$ phase | | S phase | | $G_2$/M phase | |
| **Organism** | **cyclin** | **CDK** | **cyclin** | **CDK** | **cyclin** | **CDK** |
| Budding yeast | Cln1, Cln2, Cln3 | Cdc28 | Clb5, Clb6 | Cdc28 | Clb2 | Cdc28 |
| Fission yeast | Cig1 | Cdc2 | Cig2 | Cdc2 | Cdc13 | Cdc2 |
| Higher eukaryotes | Cyclin D1, D2, D3<br>Cyclin E | Cdk4, Cdk6<br>Cdk2 | Cyclin A | Cdk2 | Cyclin B<br>Cyclin B3 | Cdc2 (Cdk1) |
| | **These regulate growth and morphogenesis** | | **These regulate DNA replication** | | **These regulate mitosis** | |

subprograms of differentiation, as shown in **TABLE 15.1**. In unicellular eukaryotes, a single CDK can interact with any of several different cyclins at different times in the cell cycle; this allows a single CDK to play multiple roles, because interaction with the cyclin determines the substrate specificity of the CDK. Such multiple roles are played by Cdc28 in budding yeast and by Cdc2 in fission yeast. On the other hand, even unicellular eukaryotes can have more than one cyclin for a specific stage in the cell cycle; for example, cyclins Cln1, Cln2, and Cln3 all function in the $G_1$ stage in budding yeast. Mammalian cells express different CDKs through the cell cycle, and some CDKs interact with different cyclins in different stages. For example, Cdk2 interacts with cyclin E in $G_1$ but with cyclin A in S. Despite these differences, the function of some CDKs has been remarkably conserved in evolution. A striking illustration of this principle is the finding that coding sequences for the human proteins Cdk2 and Cdc2 are able to substitute for the *CDC28* gene in budding yeast, and the human proteins allow the yeast cell to grow and mate essentially normally. Among all the enzyme complexes listed, the mammalian cyclin B–Cdc2 complex that forms in $G_2$/M is especially noteworthy, because it controls the passage from $G_2$ into M. Even though the cyclin B–Cdc2 complex is assembled prior to mitosis, its catalytic activity is restricted to mitosis, because until then, it is inhibited by phosphorylation.

The term *cyclin* is apt because the abundance of these proteins changes cyclically in the cell cycle, and some are present only at specific times. Most cyclins appear abruptly and disappear a short time later (**FIGURE 15.10**). Each cyclin appears at its characteristic time in the cell cycle because its transcription is linked to the cell cycle via a previously expressed cyclin. Typically, the presence of an active cyclin–CDK complex results in the activation of a transcription factor that leads to transcription of the next cyclin needed in the cell cycle. The cyclin disappears when its gene is no longer transcribed and the previously made mRNA and protein are degraded. Active cyclin–CDK complexes also cause

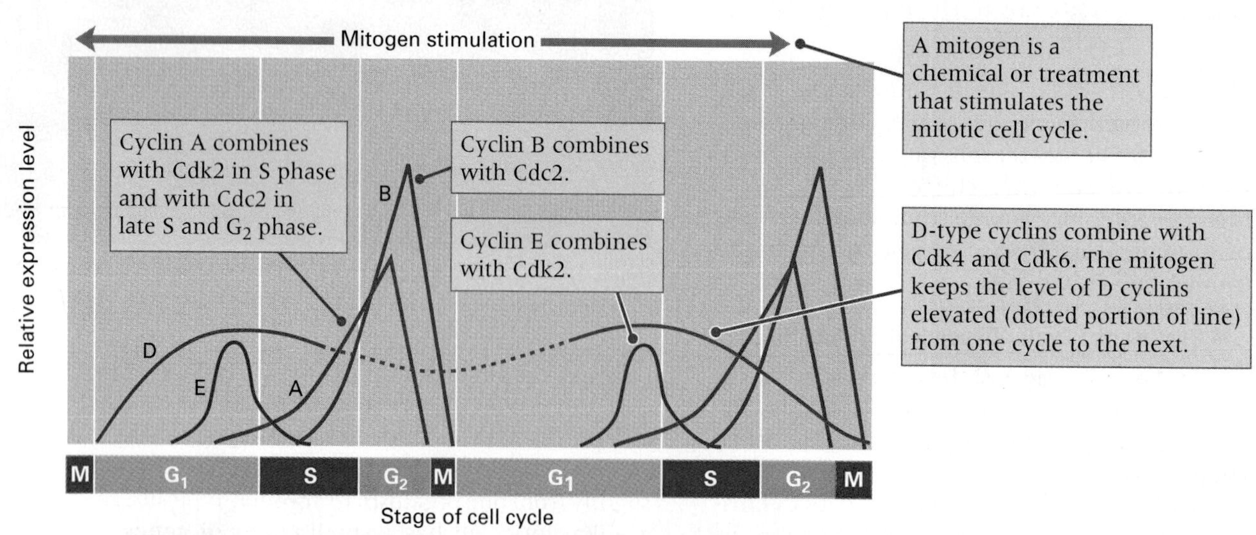


**FIGURE 15.10** Fluctuations of cyclin levels during the cell cycle. Expression of cyclins E, A, and B (mitotic cyclins) are periodic, whereas cyclin D is expressed throughout the cell cycle in response to mitosis-stimulating drugs (mitogens). [Adapted from C. J. Sherr, *Science* 274 (1996): 1672–1677.]

## connection
### Cycle-Ops

*This classic paper is one of the high points of genetic analysis. It shows how the phenotypes of mutants can be used to dissect a complex temporal process, in this case the cell cycle. The results demonstrate that the yeast cell cycle includes two parallel pathways emerging from a start that defines the point at which a cell becomes irreversibly committed to mitosis rather than to mating.*

**Leland H. Hartwell, Joseph Culotti, John R. Pringle, and Brian J. Reid**
**1974**
University of Washington, Seattle, Washington
*Genetic Control of the Cell Division Cycle in Yeast*

Two features which distinguish the cell cycle of *Saccharomyces cerevisiae* from most other eukaryotes are particularly useful for an analysis of the gene functions that control the cell division cycle. First, the fact that both haploid and diploid cells undergo mitosis permits the isolation of recessive mutations in haploids and their analysis by complementation in diploids. Second, the daughter cell is recognizable at an early stage of the cell cycle as a bud on the surface of the parent cell. Since the ratio of bud size to parent cell size increases progressively during the cycle, this ratio provides a visual marker of the position of the cell cycle. We have taken advantage of these features in the isolation and characterization of 150 temperature-sensitive mutants of the cell division cycle (*cdc* mutants). These mutants are temperature-sensitive in the sense that they are unable to reproduce at 36°C (the restrictive temperature) but do grow normally at 23°C (the permissive temperature). . . . These mutations define 32 genes, each of whose products plays an essential role in the completion of one event in the mitotic cycle. . . . The phenotypes of the *cdc* mutants suggest that the following events are ordered in a single dependent pathway: "start," initiation of DNA synthesis, DNA synthesis, medial nuclear division, late nuclear division, cytokinesis, and cell separation. Hence, the temporal sequence of these events is easily accounted for by the fact that no event in this pathway can occur without the prior occurrence of all preceding events. A second dependent pathway is comprised of the events "start," bud emergence, nuclear migration, cytokinesis, and cell separation. Thus, the temporal sequence of these five events is also assured. Furthermore, the integration of the two pathways is accomplished by the facts that both diverge from a common event, "start," and that both converge on a common event, cytokinesis.

***These mutations define 32 genes, each of whose products plays an essential role in the completion of one event in the mitotic cycle.***

Source: L. H. Hartwell, et al., *Science* 183 (1974): 46–51.

the transcriptional activation of genes other than cyclins.

### Targets of the Cyclin–CDK Complexes

How do the cyclin–CDK complexes control progression through the cell cycle? Through their protein kinase activity, they add phosphate groups to the hydroxyl groups present in the amino acids serine, threonine, and tyrosine found in proteins. The cyclin component of the complex binds to the protein substrate and tethers it, allowing the CDK component to phosphorylate the tethered substrate. Once the targeted protein is phosphorylated, it dissociates from the cyclin–CDK complex. The activities of the phosphorylated form of a protein often differ dramatically from those of the unphosphorylated form. Phosphorylation may activate one enzyme but inactivate another. Even in a single enzyme molecule, phosphorylation at different sites may have opposite effects. Another route of cell-cycle regulation is mediated through *phosphatase* enzymes that dephosphorylate proteins that the cyclin-dependent kinases have phosphorylated. Reversing the effects of CDKs, the phosphatases activate enzymes that are inactive or deactivate ones that are active.

In mammalian cells, specific cyclin–CDK complexes can be detected at different times in the cell cycle, as illustrated in **FIGURE 15.11**. The cyclin D–Cdk4 and cyclin D–Cdk6 complexes appear in the early or middle part of $G_1$, whereas cyclin E–Cdk2 and cyclin A–Cdk2 appear later in $G_1$. The cyclin A–Cdk2 complex is present throughout the S phase and into M (mitosis), and the cyclin B–Cdc2 complex carries the cell through the $G_2$/M transition. The functional activity of the cyclin–CDK complexes is controlled in part by their state of phosphorylation, but the activity of cyclin D–Cdk4 and cyclin D–Cdk6 are also controlled by a protein designated the p16 inhibitor.

### Triggers for the $G_1$/S and $G_2$/M Transitions

In the cell cycle, the cells monitor their internal and external conditions. Until the conditions are suitable to initiate a division cycle, the

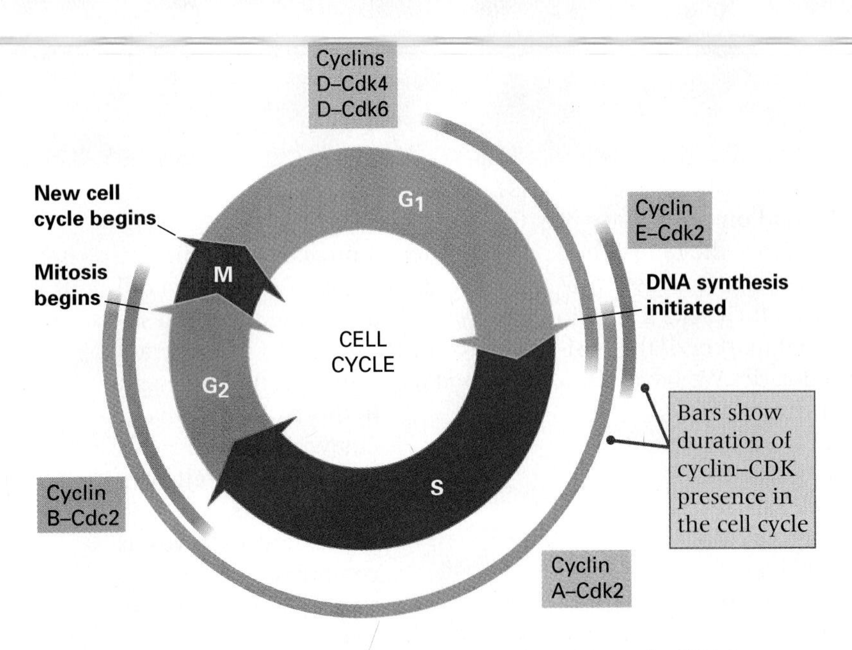


**FIGURE 15.11** The temporal expression pattern of activities of the cyclin–CDKs in mammalian cells.

cells accumulate at a point in $G_1$ called the **$G_1$ restriction point** or *start*. In animal cells, a protein called the **retinoblastoma (RB) protein** holds cells at the restriction point by binding to and sequestering the transcription factor E2F, which is needed for further progression (**FIGURE 15.12**, part A). The RB protein was first identified in human pedigrees because mutant forms of the protein are associated with the formation of malignant tumors in the retina, which can require surgical removal of the eyes.

Growth is triggered by a growth factor combining with its receptor. This event activates a signal-transduction pathway that culminates in the production of cyclin D. In the middle of $G_1$ phase, the RB protein begins to be phosphorylated by both the cyclin D–Cdk4 kinase and the cyclin D–Cdk6 kinase; late in $G_1$ the RB phosphorylation is completed by the cyclin E–Cdk2 kinase as cells approach the **$G_1$/S transition** when they become committed to DNA replica-

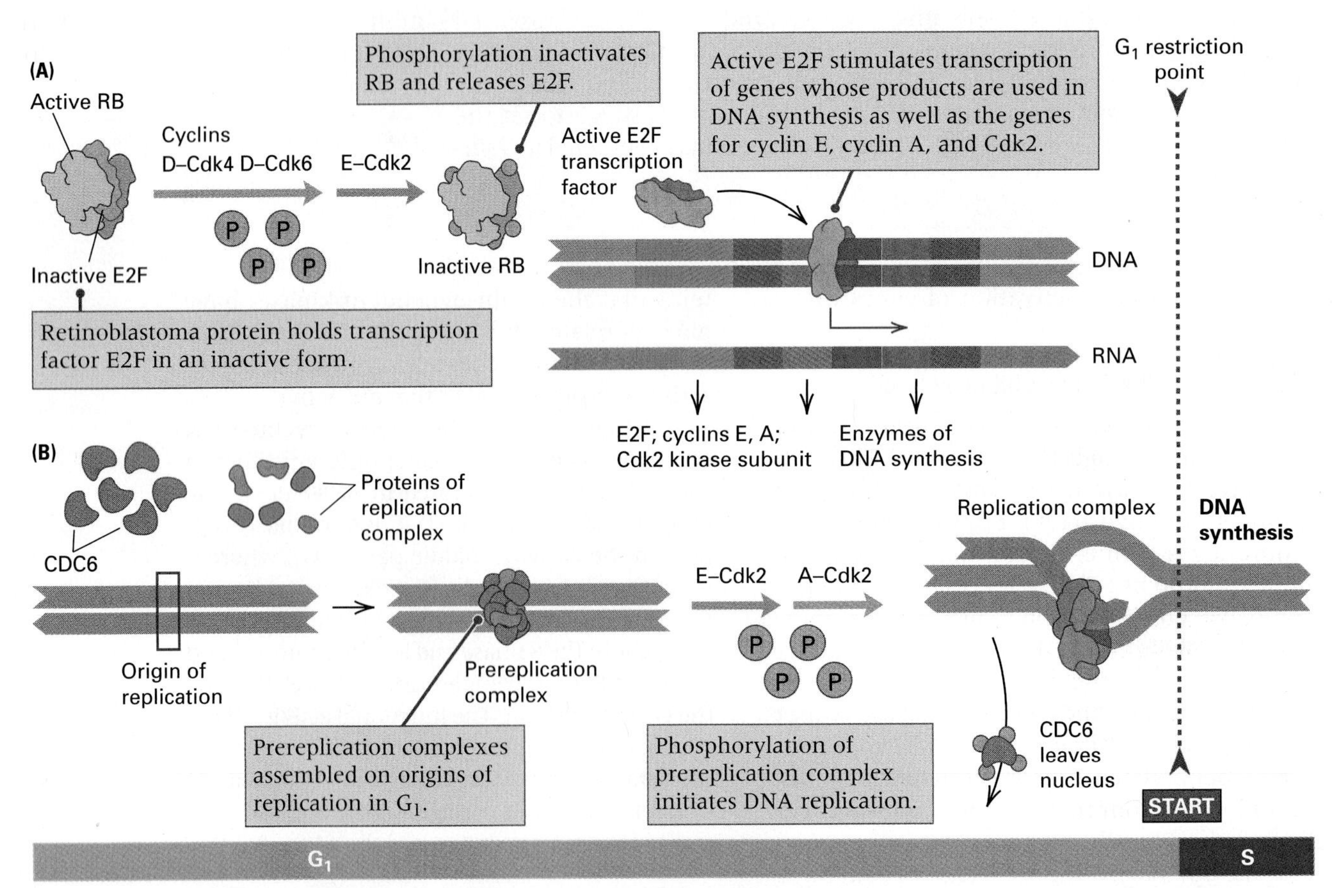


**FIGURE 15.12** (A) Role of the retinoblastoma protein RB in controlling the transition from $G_1$ phase to S phase. The cyclin D-dependent kinases Cdk4 and Cdk6 initiate phosphorylation of RB in mid-$G_1$ phase, a process completed by cyclin E–Cdk2; this frees the transcription factor E2F, which activates transcription of enzymes for DNA synthesis. (B) The free E2F also activates transcription of the genes for cyclin A, cyclin E, and Cdc2, which help convert prereplication complexes to replication complexes for transition to S phase.

tion. Phosphorylation of RB inactivates the protein and frees the bound E2F transcription factor (Figure 15.12A). Release of the E2F results in transcription of the genes and translation of the enzymes responsible for DNA replication, including DNA polymerase. The cells therefore begin to accumulate the precursors and enzymes required for DNA synthesis. E2F also activates transcription of the gene for E2F itself (an example of positive autoregulation), as well as those for the cyclins E and A and the Cdk2 kinase subunit.

Cells in $G_1$ also prepare for DNA synthesis by assembling **prereplication complexes** at origins of replication. Assembly of each of the complexes is initiated by association of six proteins constituting the **origin recognition complex (ORC)** with an origin of replication in the DNA. CDC6 protein (named after a protein in yeast) attaches to each ORC and recruits other components, which completes the assembly of a prereplication complex poised to initiate DNA synthesis. Cyclin E–Cdk2 kinase and cyclin A–Cdk2 kinase (Figure 15.12B) phosphorylate components of the complexes, including the CDC6 protein, which in turn trigger the initiation of DNA synthesis and thus the $G_1/S$ transition. Once phosphorylated, CDC6 leaves the initiation complex (this event correlates with the actual initiation of DNA synthesis) and moves from the nucleus to the cytoplasm. The exit of CDC6 appears to ensure that DNA synthesis will not reinitiate at the same origin of replication. Once cells have entered S phase, the cyclin A–Cdk2 phosphorylates E2F and inhibits its binding to DNA, thus inactivating its function as a transcription factor. The cyclin A–Cdk2 activity is required throughout S phase, apparently to keep the RB protein heavily phosphorylated.

The progression from $G_2$ to M (the **$G_2/M$ transition**) is controlled by a cyclin B–Cdc2 complex also known as *maturation-promoting factor.* After synthesis and assembly in the cytoplasm, the cyclin B–Cdc2 complex remains in the cytoplasm in an inactive form, because its rate of import into the nucleus is smaller than its rate of export from the nucleus. The balance is tipped in favor of import when cyclin B is phosphorylated just prior to the $G_2/M$ transition, which masks the nuclear export signal. Reducing nuclear export ensures that cyclin B–Cdc2 accumulates in the nucleus. The cyclin B–Cdc2 complex is activated by dephosphorylation at different sites, and it then carries out phosphorylation of its substrates to bring about events that complete the $G_2/M$ transition. These include (1) phosphorylation of a kinesin-related motor protein associated with duplication of the spindle pole and formation of the bipolar spindle and (2) phosphorylation of nuclear lamins, causing their disassembly and ultimately the breakdown of the nuclear envelope. At this time the chromosomes condense and assemble onto the spindle, and the chromosome segregation machinery becomes active.

## Protein Degradation Helps Regulate the Cell Cycle

A fundamental feature of the cell cycle is that it is a true cycle; it is not reversible. The cycle is propelled forward by a process of protein degradation that complements the periodic activation of cyclin–CDK complexes. Protein degradation (*proteolysis*) eliminates proteins that were used in the preceding phase as well as proteins that would inhibit progression into the next phase. In the early stages of the cell cycle, progression requires the sequential activation of cyclin–CDKs. Entry into each new phase also requires destruction of the cyclins used in the preceding phase. In the later stages of the cell cycle, progression is propelled by proteolysis alone. This process is best understood in yeast, and we describe it here using the yeast terminology.

In the completion of mitosis and the return to $G_1$ phase, two key regulatory events must occur:

1. The sister chromatids must separate (marking the onset of anaphase).
2. The cells must exit from mitosis, which entails chromosome decondensation, spindle disassembly, inactivation of the chromosome segregation machinery, and cytokinesis.

Both of these key events are triggered by protein degradation, as indicated in **FIGURE 15.13** for yeast. Exit from mitosis requires the destruction of cyclin B (yeast Clb). Cyclin B is most abundant when cells enter mitosis, but it disappears after chromosome disjunction has occurred at the transition from metaphase to anaphase. Cyclin B is marked for destruction by the **anaphase-promoting complex (APC/C)**, which is a ubiquitin–protein ligase responsible for ubiquitinating its target proteins. Ubiquitin is a 76-residue protein that is transferred to certain abnormal proteins to mark them for degradation into amino acids through digestion in a *proteasome*—a large, multifunctional, multisubunit complex responsible for most of the cytoplasmic proteolysis in the cell beyond that which takes place in lysosomes. The same process is used to mark certain proteins for destruction when their function is no longer needed. Among the

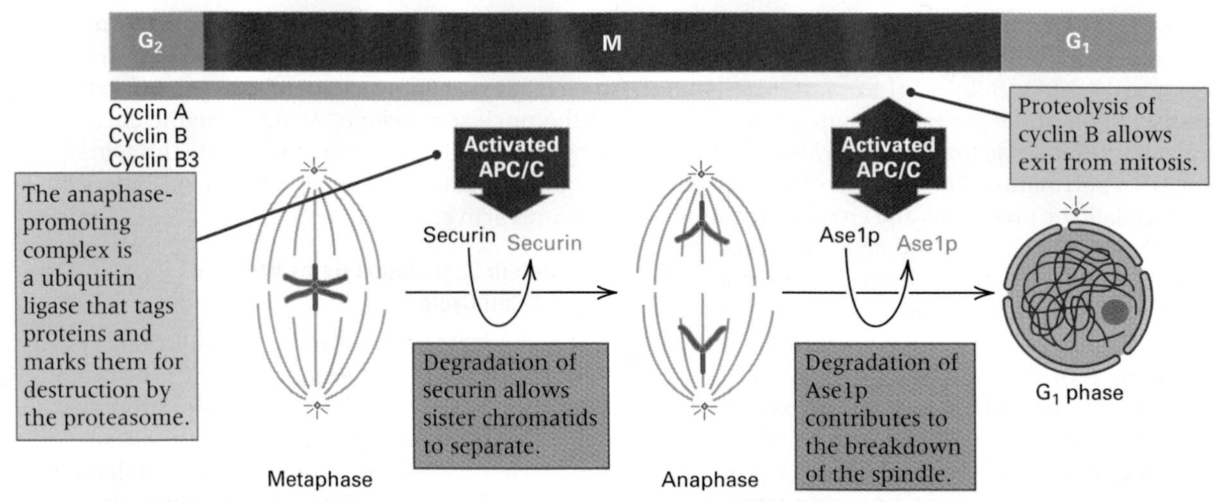


**FIGURE 15.13** Role of activated anaphase-promoting complex (APC/C) in controlling the proteolysis necessary for the transition to anaphase and the exit from mitosis. APC/C is a ubiquitin ligase that marks proteins for proteasome destruction. Its substrates include securin, which unless destroyed inhibits breakdown of the protein "glue" (Scc1p) that holds the sister chromatids together, and Ase1p, a microtubule-associated protein that binds antiparallel microtubules from the spindle pole bodies in the spindle midzone. Destruction of cyclin B somewhat later allows the exit from mitosis.

substrates for anaphase-promoting complex is cyclin B, which when ubiquitinated is marked for degradation via the proteasome.

Other substrates for the APC/C are the protein securin, which inhibits a protease called separase, and Ase1p, a microtubule-associated protein that binds antiparallel microtubules from the spindle pole bodies in the midzone of the spindle. As the securin is degraded, the separase protein becomes free to degrade Scc1p, a component of the cohesin complex that condenses chromosomes and also holds sister chromatids together. Cleavage of Scc1p results in its dissociation from the chromatin, which allows the sister chromatids to come apart and be pulled toward opposite poles of the spindle. Degradation of Ase1p is thought to release the microtubule ends and allow disassembly of the spindle. Thus, the metaphase/anaphase and mitosis/$G_1$ transitions do not require cyclin-dependent protein kinases, but rather require the activity of the anaphase-promoting complex to target specific proteins for degradation.

## 15.4 Checkpoints in the Cell Cycle

Cells monitor their external environment as well as their internal physiological state and functions. In the absence of needed nutrients or growth factors, animal cells may exit from the cell cycle and become quiescent in a state called $G_0$. Upon growth stimulation, they reenter the cell cycle in a process that requires cyclin D–CDK activity. Cells also have mechanisms that respond to symptoms of stress, including DNA damage, oxygen depletion, inadequate pools of nucleoside triphosphates, and (in the case of animal cells) loss of intercellular adhesion. Inside the cell, several key events in the cell cycle are monitored. When defects are identified, progression through the cell cycle is halted at a **checkpoint**. By allowing time for correction and repair, the checkpoints serve to maintain the correct order of steps with respect to each other in the cell cycle. Three principal checkpoints are:

1. A DNA damage checkpoint
2. A centrosome duplication checkpoint
3. Checkpoint for spindle assembly and position

These checkpoints are summarized in **FIGURE 15.14**. All three types of checkpoints are important in maintaining the stability of the chromosome complement. When there is an error in any of the three processes monitored by the checkpoints, failure to stop at the checkpoint may lead to aneuploidy, polyploidy, or an increase in mutation rate. In the following sections, we shall examine how some of these checkpoints work.

### The DNA Damage Checkpoint

A **DNA damage checkpoint** arrests the cell cycle when DNA is damaged or replication is not completed. DNA damage includes either modification of nitrogenous bases or breakage of the phosphodiester backbone. When modified nucleotides are repaired by an excision repair pathway (Chapter 14), the repair entails removal of the affected nucleotides, followed by resynthesis with a repair polymerase and finally ligation. DNA molecules broken in the back-

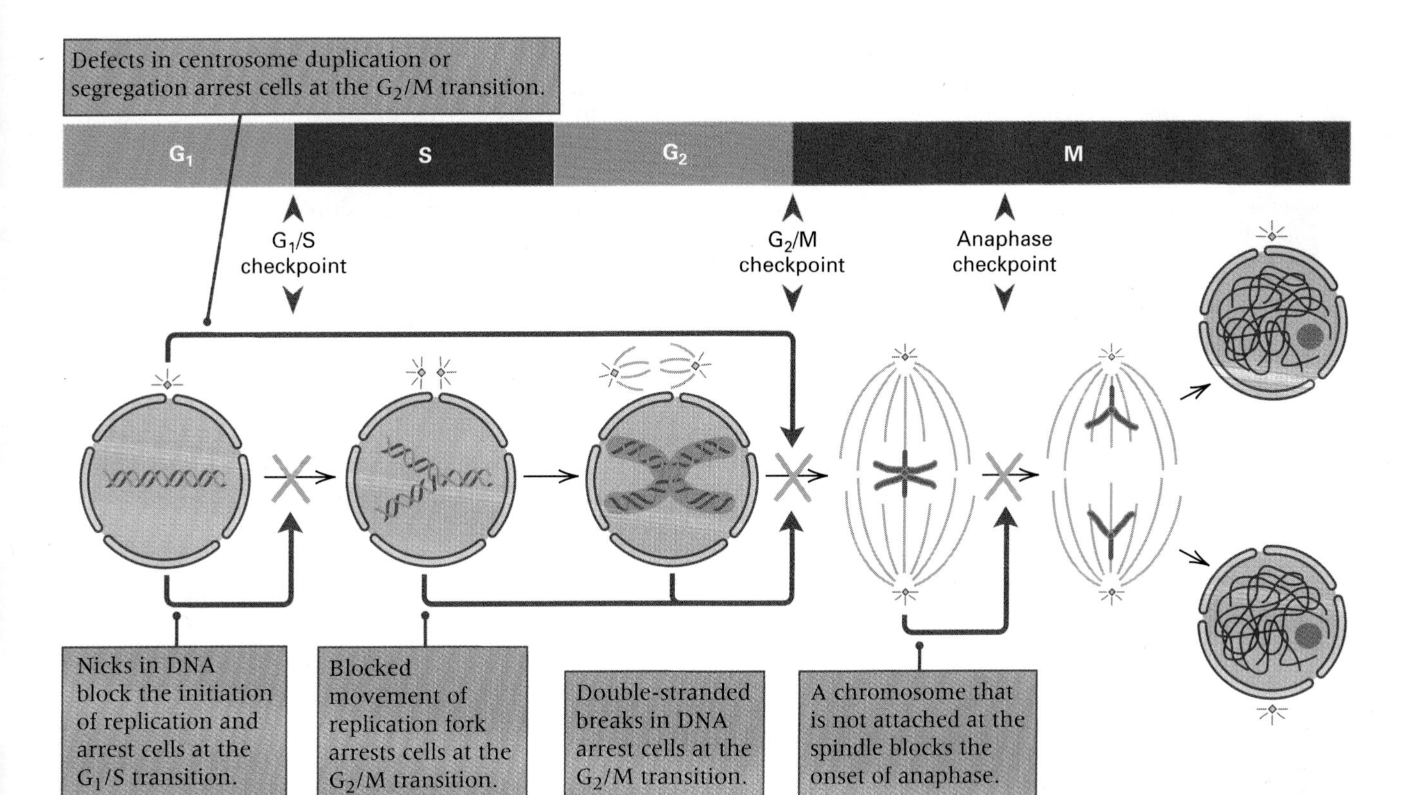
Defects in centrosome duplication or segregation arrest cells at the G2/M transition.
G1
S
G2
M
G1/S checkpoint
G2/M checkpoint
Anaphase checkpoint
Nicks in DNA block the initiation of replication and arrest cells at the G1/S transition.
Blocked movement of replication fork arrests cells at the G2/M transition.
Double-stranded breaks in DNA arrest cells at the G2/M transition.
A chromosome that is not attached at the spindle blocks the onset of anaphase.

**FIGURE 15.14** Cell-cycle checkpoints. The text in the boxes explains the events monitored and the steps affected. [Adapted from L. H. Hartwell and M. B. Kastan, *Science* 266 (1994): 1821–1828.]

bone are repaired by homology-based recombination, nonhomologous end joining, or addition of new telomeres. In animal cells, a DNA damage checkpoint acts at three stages in the cell cycle: at the $G_1/S$ transition, in S phase, and at the $G_2/M$ boundary. The S-phase checkpoint continuously monitors the progression of DNA synthesis throughout S. Although there are three DNA damage checkpoints, each monitors DNA damage or incomplete replication. If either type of problem is detected, the DNA checkpoint acts to block the cell cycle at multiple points.

Key proteins in the mammalian cell's response to stress in general and to DNA damage in particular are several slightly different forms of a protein called the **p53 transcription factor**. In normal cells, the level of activated p53 protein is very low, even though substantial amounts of p53 mRNA and protein are present. The activity of p53 is kept low by another protein called Mdm2. To function as a transcription factor, p53 must be activated first by phosphorylation and then by acetylation. Mdm2 binds to p53 and prevents phosphorylation and subsequent steps in the activation of p53 as a transcription factor. In addition, Mdm2 continuously shuttles between the nucleus and the cytoplasm, and in this process it continuously exports p53 from the nucleus for degradation by the proteasome in the cytoplasm. The p53–Mdm2 export cycle is illustrated in **FIGURE 15.15**.

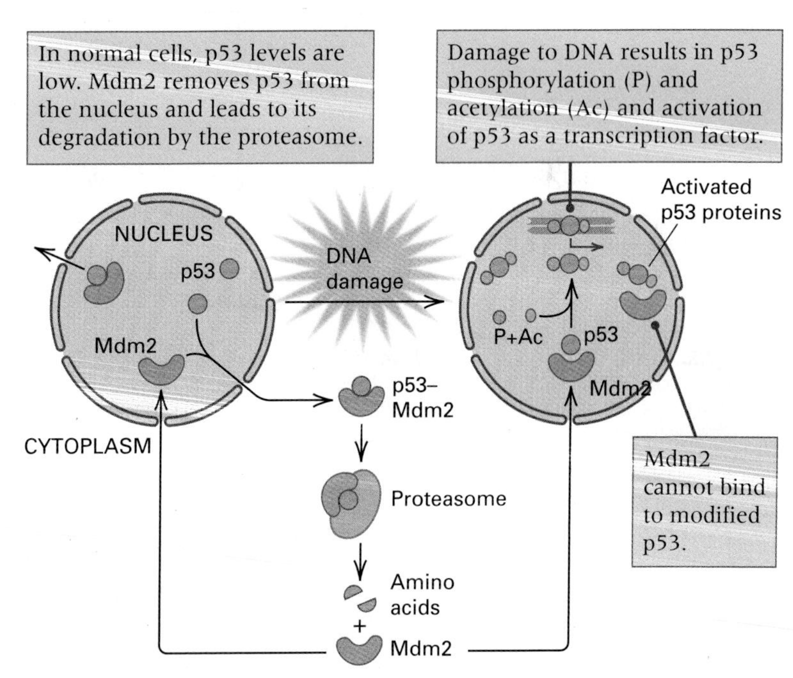
In normal cells, p53 levels are low. Mdm2 removes p53 from the nucleus and leads to its degradation by the proteasome.
Damage to DNA results in p53 phosphorylation (P) and acetylation (Ac) and activation of p53 as a transcription factor.
NUCLEUS
p53
Mdm2
DNA damage
Activated p53 proteins
P+Ac
p53
Mdm2
p53–Mdm2
CYTOPLASM
Proteasome
Amino acids
+
Mdm2
Mdm2 cannot bind to modified p53.

**FIGURE 15.15** The effect of DNA damage on transcription factor p53. In normal cells the level of p53 is low, in part because Mdm2 shuttles it to the cytoplasm where it is destroyed by the proteasome. DNA damage results in phosphorylation and acetylation of p53, rendering it unable to bind Mdm2. Hence p53 levels rise.

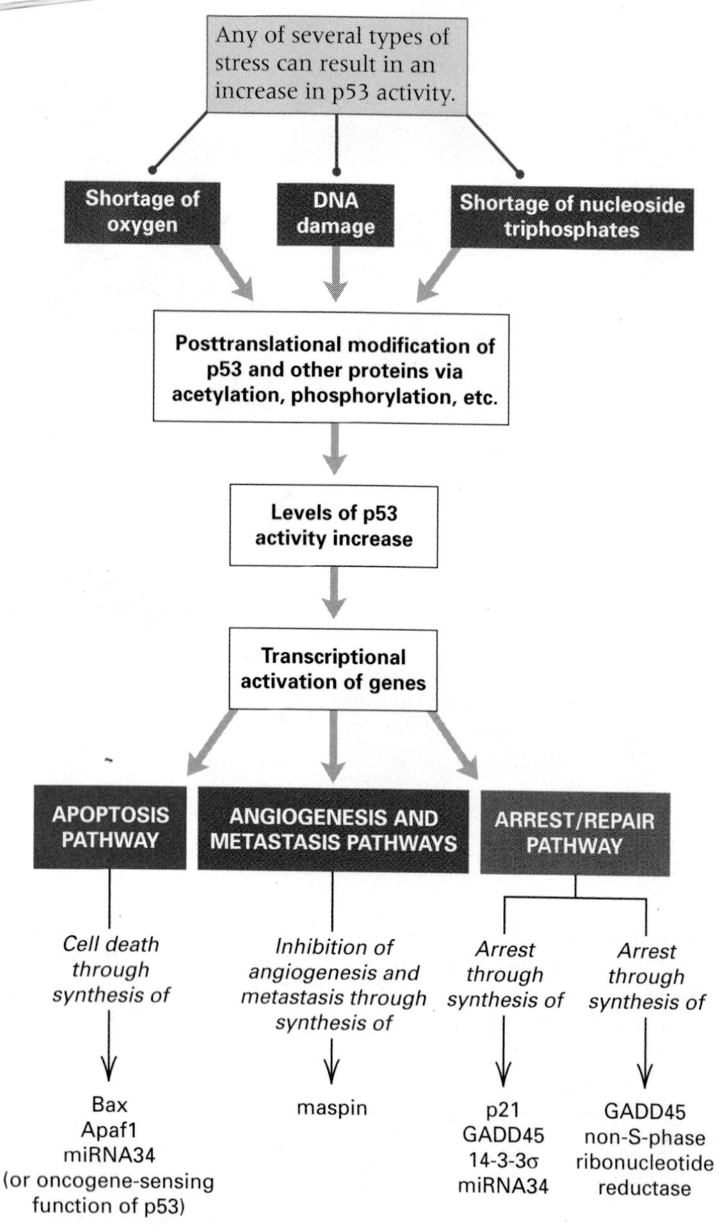


**FIGURE 15.16** Downstream events triggered by p53 include transcriptional activation of the genes for p21, GADD45, 14-3-3$\sigma$, Bax, maspin, Apaf1, and miRNA34a and b/c. Activation of the apoptosis pathway by p53 follows transcriptional activation of Bax and, possibly, direct sensing of activated oncogenes (cancer-causing genes) of either cellular or viral origin.

When cells are treated with agents that damage DNA, some cells arrest in $G_1$ and others arrest in $G_2$. The DNA damage signal is sensed and transmitted, and this causes p53 to become activated by protein kinases and acetylases, overriding the inhibitory effect of Mdm2. Activation of p53 causes its release from Mdm2, which stabilizes the active p53 transcription factor and results in increased levels of the p53 protein. The activated p53 then triggers the transcription of a number of genes and the repression of others. Some of the key proteins whose genes are transcriptionally activated by p53 are listed in **TABLE 15.2**, and a flow chart of how these proteins affect processes in the cell cycle and other cellular properties is given in **FIGURE 15.16**.

Increased transcription of the genes for p21, GADD45, 14-3-3$\sigma$, and miRNA34a, as well as decreased transcription of the gene for cyclin B, all serve to block the cell cycle at particular points, as illustrated in **FIGURE 15.17**. The $G_1/S$ transition checkpoint is mediated by the increased levels of p21, which result in inhibition of the $G_1$ cyclin–CDKs and in this way block the $G_1/S$ transition. Hence, if DNA damage is detected in $G_1$, the cell is blocked in the $G_1/S$ transition. The microRNAs miRNA34a and b/c inhibit translation of mRNAs for Cdk416 through interaction with Dicer protein. The S-phase response to DNA damage is mediated by the p21 protein and GADD45, which form a complex with proliferating cell nuclear antigen (PCNA) and reduce the processivity of DNA polymerase. The **processivity** of a DNA polymerase is the number of consecutive nucleotides in the template strand that are replicated before the polymerase detaches from the template. Decreasing the processivity of DNA polymerase therefore slows down DNA synthesis, in effect buying time for the cell to repair DNA damage. The $G_2/M$ checkpoint is mediated by the 14-3-3$\sigma$ protein, which hinders activation of cyclin

**Table 15.2 Products of genes transcriptionally activated by p53**

| Gene product | Function |
|---|---|
| p21 | Inhibits several cyclin-dependent kinases; arrests cells at $G_1/S$ boundary. |
| 14-3-3$\sigma$ | Predicted to bind to and sequester phosphorylated Cdc25C phosphatase in the cytoplasm, which prevents Cdc25C from activating the cyclin B–Cdc2 kinase; arrests cells at the $G_2/M$ boundary. |
| GADD45 | Binds to proliferating cell nuclear antigen (PCNA), blocking its role as a processivity factor for DNA polymerase and hence blocking DNA replication; functions directly in DNA repair. |
| Bax | Acts as a positive regulator of apoptosis (programmed cell death). |
| Apaf1 | Scaffold protein that, when activated by cytochrome c, oligomerizes caspases into the complex that promotes apoptosis. |
| Maspin | Acts as an inhibitor of serine proteases and is an inhibitor of angiogenesis (formation of blood vessels) and metastasis. |
| miRNA34a, b/c | MicroRNAs that induce senescence and apoptosis. |

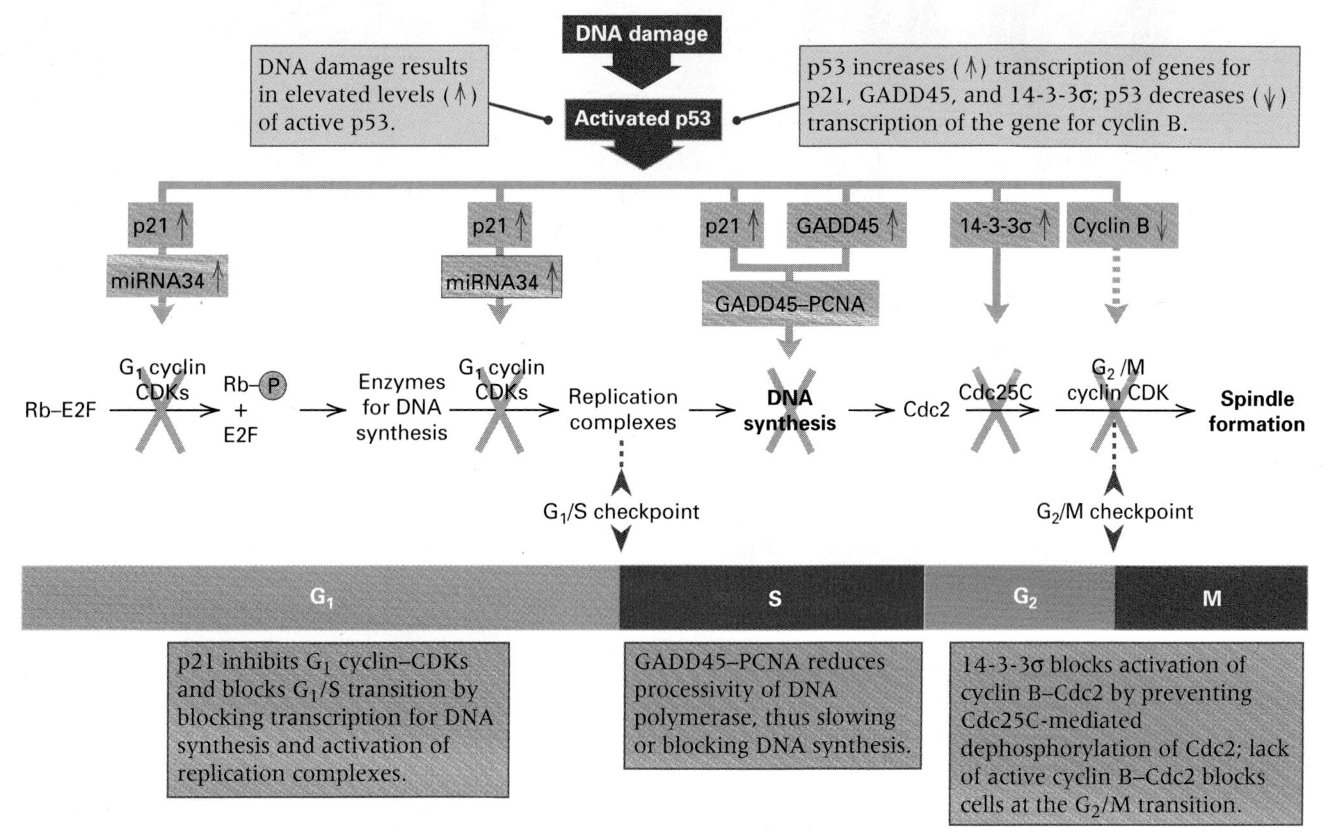


**FIGURE 15.17** Role of p53 in the DNA damage checkpoint. DNA damage results in an increase in the level of active p53, which increases transcription of the genes for p21, GADD45, and 14-3-3σ and decreases transcription of the genes for cyclin B.

B–Cdc2 by preventing dephosphorylation of Cdc2 by the phosphatase Cdc25C, thus blocking the $G_2$/M transition. At the same time, the decrease in the level of cyclin B reduces the level of the active cyclin B–Cdc2 complex, which also ensures that the cell remains in $G_2$. Hence, if DNA damage is detected in the S or the $G_2$ phase, the cell cycle arrests at the $G_2$/M checkpoint.

DNA damage also triggers activation of another pathway, a pathway for **apoptosis**, or **programmed cell death**. When the apoptotic pathway is activated, a cascade of proteolysis that culminates in cell suicide is initiated. The proteases involved are called *caspases*. Their activation ultimately results in destruction of the cellular DNA, internal organelles, and the actin cytoskeleton, and it is accompanied by nuclear condensation and usually followed by engulfment of the cellular remnants by phagocytes. There is also evidence that p53 acts directly to increase the permeability of mitochondria, resulting in the release of cytochrome c, which triggers apoptosis. The p53 transcription factor also activates the apoptotic pathway by activating transcription of *Bax* and *Apaf1* and inhibiting synthesis of Bcl2. As illustrated in **FIGURE 15.18**, Bax normally exists in a heterodimer with an inhibitor of apoptosis called Bcl2. When p53 activates transcription of *Bax* and reduces expression of Bcl2, the balance is tilted in favor of Bax homodimers and against the Bcl2 heterodimer, which promotes apoptosis and self-destruction of the cell. On the other hand, activation of **oncogenes**, which are genes associated with cancers, can increase the level of activated (phosphorylated) Bcl2, which prevents apoptosis and allows the affected cells to grow and divide indefinitely. Cellular immortality does not always follow oncogene activation, because in some cases it activates the apoptotic pathway, possibly through an oncogene-sensing function of p53 illustrated in Figure 15.16. In all of its functions, p53 acts to protect the integrity of the genome with respect to nucleotide sequence and strand integrity and with respect to euploidy. Instability in the genome, unbalanced genomes, and damaged DNA pose a hazard to the organism. Apoptosis is a mechanism for killing such damaged—and thus dangerous—cells.

To gain an appreciation of the importance of the DNA damage checkpoint, look again at Figure 15.17 and consider what would happen if part or all of this fail-safe mechanism should malfunction. Suppose, for example, that cells lacked p21 protein but retained the rest of the mechanism. If DNA damage were properly sensed, and p53 functioned as usual to increase levels

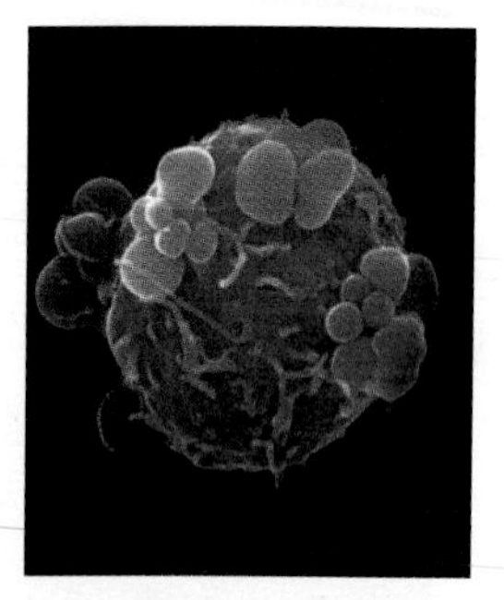

**THE FORMING** of large blebs ("blisters") from the cell membrane is a characteristic of apoptosis. [© Dr. Andrejs Liepins/Photo Researchers, Inc.]

## connection
### Sick of Telomeres

**William C. Hahn,[1,2] Christopher M. Counter,[3] Ante S. Lundberg,[1,2] Roderick L. Beijersbergen,[1] Mary W. Brooks,[1] and Robert A. Weinberg[1] 1999**
[1] Massachusetts Institute of Technology, Cambridge, Massachusetts
[2] Harvard Medical School, Boston, Massachusetts
[3] Duke University Medical Center, Durham, North Carolina
*Creation of Human Tumor Cells with Defined Genetic Elements*

*One day, about 20 years ago, at a well-known midwestern medical school, the head of the Department of Medicine confronted the head of the Department of Genetics and said, "Your Professor Blank up the hall studies telomeres. Who cares about telomeres? Nobody ever gets sick because of their telomeres!" We recount this story to emphasize that the directions of basic research that prove most important in the long run cannot usually be predicted, even by experts. As this important paper shows,* lots *of people get sick—very sick—because of their telomeres.*

During malignant transformation, cancer cells acquire genetic mutations that override the normal mechanisms controlling cellular proliferation. Primary rodent cells are efficiently converted into tumorigenic cells by the coexpression of cooperating oncogenes [genes that cause cancer when mutated]. However, similar experiments with human cells have consistently failed to yield tumorigenic transformants, indicating a fundamental difference in the biology of human and rodent cells. . . . One ostensibly important difference between rodent and human cells derives from their telomere biology. Murine somatic cells express telomerase activity and have much longer telomeres than their human counterparts, which lack telomerase activity. Because normal human cells progressively lose telomeric DNA with passage in culture, telomeric erosion is thought to limit cellular life span. Ectopic expression of the *hTERT* gene, which encodes the catalytic subunit of the telomerase, enables some . . . pre-senescent primary human cells to multiply indefinitely. . . . To determine whether *hTERT* human cells were tumorigenic, . . . we also introduced large-T antigen [which inhibits both the retinoblastoma and p53 tumor-suppressor proteins] and an oncogenic *ras* [signal transduction] mutation. . . . We observed efficient colony formation in soft agar [an assay for the tumorigenic state] only with cells expressing the combination of large-T, *ras* and *hTERT*. . . . When these cells were introduced into [immunologically deficient] mice, rapidly growing tumors were repeatedly observed with high efficiency. . . . We conclude that ectopic expression of a defined set of genes . . . suffices to convert normal human cells into tumorigenic cells. . . . It is now highly likely that telomere maintenance contributes directly to oncogenesis by allowing precancerous cells to proliferate beyond the number of replicative doublings allotted to their normal precursors.

***It is now highly likely that telomere maintenance contributes directly to oncogenesis***

Source: W. C. Hahn, et al., *Nature* 400 (1999): 464–468.

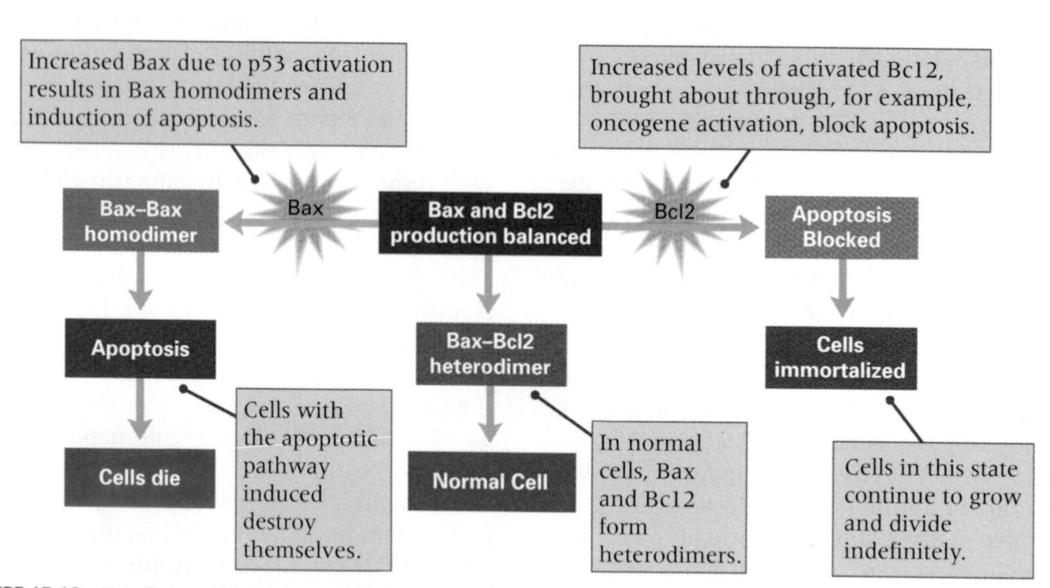


**FIGURE 15.18** How Bax and Bcl2 interact to regulate apoptosis.

of 14-3-3σ and GADD45, cells would accumulate in $G_2$ and be unable to undergo mitosis. However, DNA would continue to be synthesized. In addition, lacking p21 protein, the cells would have a defective $G_1$/S checkpoint and so could embark on additional rounds of DNA synthesis. The expected result of p21 malfunction would therefore be polyploid cells, and this is the result observed.

To take another example, consider the consequence of loss of p53 function. Even if DNA damage were detected, the cell would be unable to mount a response—it would be unable to buy time to repair the DNA damage. The cells could initiate new rounds of synthesis with damaged chromosomes. Such synthesis would not only result in an increased frequency of mutation but would also permit gene amplification. Amplification of genes encoding cyclin D or Cdk4 would allow cells to escape the normal controls on DNA synthesis and proliferation. Cells already in S or $G_2$ would enter mitosis with damaged chromosomes, because not enough time for repair of lesions would have elapsed. In addition, the organism would have lost its ultimate protection against such damaged cells: apoptosis. The absence of p53 means that transcription of *Bax* and *Apaf1* would not be increased and that the apoptotic pathway would not be turned on; the normal balance between Bax and Bcl2 would be maintained, ensuring the survival of these damaged cells and thus putting the organism at risk. If the description of these runaway cells reminds you of the unchecked proliferation of cancer cells, this is because cancer cells *become* cancer cells by subverting the checkpoint mechanisms.

## The Centrosome Duplication Checkpoint

A **centrosome duplication checkpoint** monitors formation of a bipolar spindle. It seems to be coordinated with entry into mitosis, because activation of cyclin B–Cdc2 kinase is correlated with centrosome duplication and formation of the spindle. This duplication checkpoint may also be coordinated, in some organisms, with the spindle checkpoint and the exit from mitosis.

## The Spindle Assembly Checkpoint

The **spindle assembly checkpoint** monitors assembly of the spindle and attachment of the kinetochores to the spindle. (The *kinetochore* is the spindle fiber attachment site on the chromosome.) Improper or incomplete assembly triggers a block to anaphase by preventing activation of the anaphase-promoting complex (Figure 15.13) and thus separation of the sister chromatids.

Cells can detect a single unattached or misattached chromosome and delay anaphase. Studies on insect and mammalian chromosomes suggest that the absence of tension at the kinetochore is the initiating signal for cell-cycle arrest. Tension on the kinetochore is related to the level of phosphorylation at kinetochores; unattached kinetochores have relatively more of a phosphorylated entity than do attached kinetochores. If all chromosomes form stable, bipolar attachments to spindle microtubules at their kinetochores, the Cdc20 protein (initially identified and named in yeast) is able to activate the anaphase-promoting complex (Figure 15.13). The anaphase-promoting complex not only degrades cyclin B, allowing exit from mitosis, but also degrades proteins that inhibit the separation of sister chromatids and disassembly of the spindle. Thus, the anaphase-promoting complex promotes anaphase by allowing the separation of chromosomes. As illustrated in **FIGURE 15.19**, when an unattached kinetochore is detected, the Bub, Mad, and Mps1 proteins act

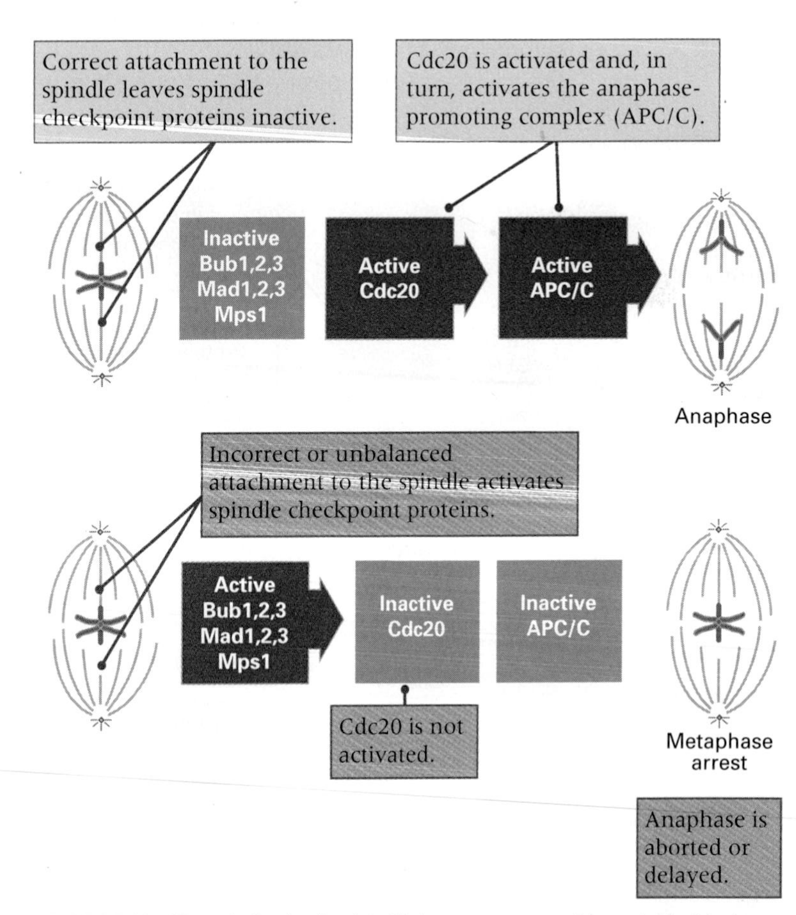


**FIGURE 15.19** The spindle checkpoint. All chromosomes must form stable, bipolar attachments to spindle microtubules to activate the anaphase-promoting complex, which promotes the onset of anaphase and separation of the sister chromatids.

to block the onset of anaphase, apparently by inhibiting the protein Cdc20. Because Cdc20 is not activated, the anaphase-promoting complex is not activated, and the cells remain at metaphase until all chromosomes form stable, bipolar attachments to spindle microtubules.

If any one of the Mad, Bub, or Mps1 proteins is defective, Cdc20 is still activated even if a microtubule has failed to attach to a kinetochore. Hence, the spindle checkpoint is not activated. The Cdc20 is activated normally, and in turn the anaphase-promoting complex is activated, and anaphase takes place. The result is two aneuploid daughter cells, one lacking a chromosome and one with an extra copy.

### The Spindle Position Checkpoint

In budding yeast, the mitotic spindle is aligned along an axis connecting the mother cell and the daughter cell (the bud). Spindle elongation during anaphase delivers one spindle pole body into the daughter cell, and the elongated spindle occupies the neck of cytoplasm between the mother cell and the daughter cell. It is within the neck region that septation occurs to form a barrier separating the mother cell from the daughter cell. If the spindle is misaligned or a spindle pole body does not pass into the daughter cell, the spindle position checkpoint is activated and prevents further progression of the cell cycle. Evidence suggests that spindle position is also monitored in animal cells and that this checkpoint is active until the spindle is aligned with the polarized axis, at least in cultured cells. However, most of what we know about the spindle position checkpoint comes from studies of budding yeast.

The characteristic feature of mutants defective in the spindle position checkpoint is that they exit mitosis without having a spindle pole body in the daughter cell. In these mutants, the spindle pole enters the bud, but the spindle returns to the mother cell. At this point the spindle is disassembled and cytoplasmic division (cytokinesis) takes place, giving rise to a mother cell with two nuclei and a daughter cell with none. Exit from mitosis depends on activation of a GTPase called Tem1p to a form in which it is bound with GTP (**FIGURE 15.20**). Tem1p is normally kept inactive by the protein Bub2p, which is a GTPase activating protein (GAP) that maintains Tem1p in its inactive GDP-bound form. Activation of Tem1p is catalyzed by Lte1p, a guanine nucleotide exchange factor (GEF). Activated Tem1p in turn activates a protein kinase and phosphatase cascade that ultimately results in activation of Cdh1p–APC/C, the anaphase promoting complex, which triggers degradation of cyclin B and thus an exit from mitosis.

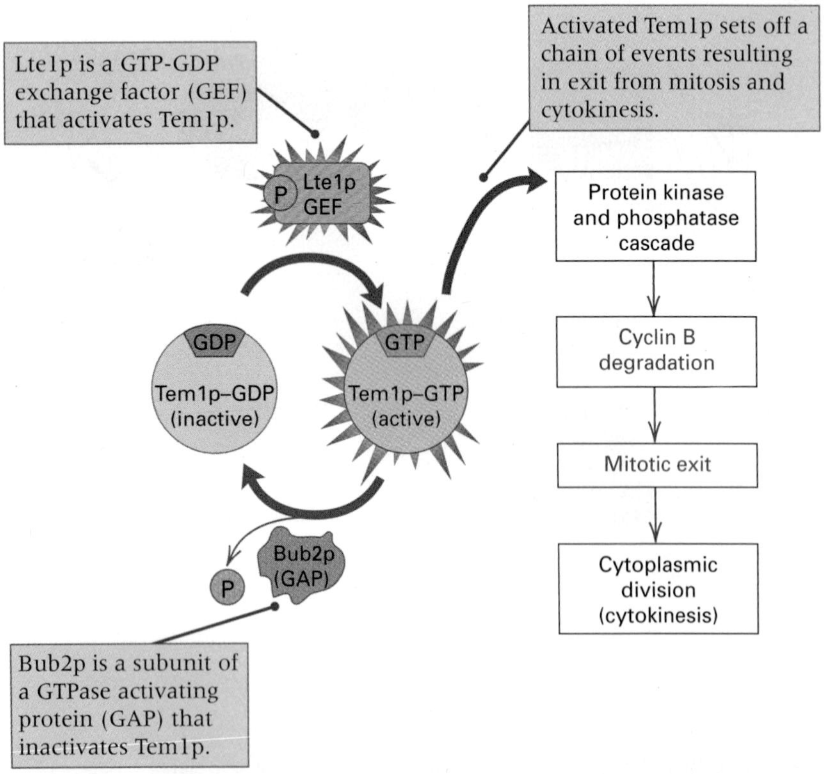

Spindle position controls the mitotic exit network as illustrated in **FIGURE 15.21**. The key feature is that the components of the mitotic exit network are kept apart owing to their cellular locations, until they come together when needed to bring mitosis to an end. In particular, Tem1p and Bub2p are localized to the cytoplasmic face of the spindle pole body in the daughter cell (Figure 15.21). In this location, Tem1p is inactive because Bub2p acts to keep Tem1p in the inactive form bound with GDP. The protein Lte1p necessary for Tem1p activation is localized in the bud in the cell cortex. Thus, if the spindle pole body is brought to the cortex at the bud tip, then cytoplasmic microtubules can mediate the interaction between Lte1p and Tem1p to activate Tem1p. The mitotic exit network will therefore be activated just when the spindle is correctly positioned as the spindle pole body in the daughter cell approaches the cell cortex at the bud tip. Two proteins are responsible for capturing the astral (cytoplasmic) microtubules and drawing them toward the cell cortex: Bim1p and Kar9p. Bim1p is located on the free (plus) ends of astral microtubules. It becomes associated with Kar9p, which is located in the cell cortex of the bud tip, bringing it into contact with the ends of the

Lte1p is a GTP-GDP exchange factor (GEF) that activates Tem1p.

Activated Tem1p sets off a chain of events resulting in exit from mitosis and cytokinesis.

Bub2p is a subunit of a GTPase activating protein (GAP) that inactivates Tem1p.

**FIGURE 15.20** Control of the mitotic exit network by Tem1p–GTP.

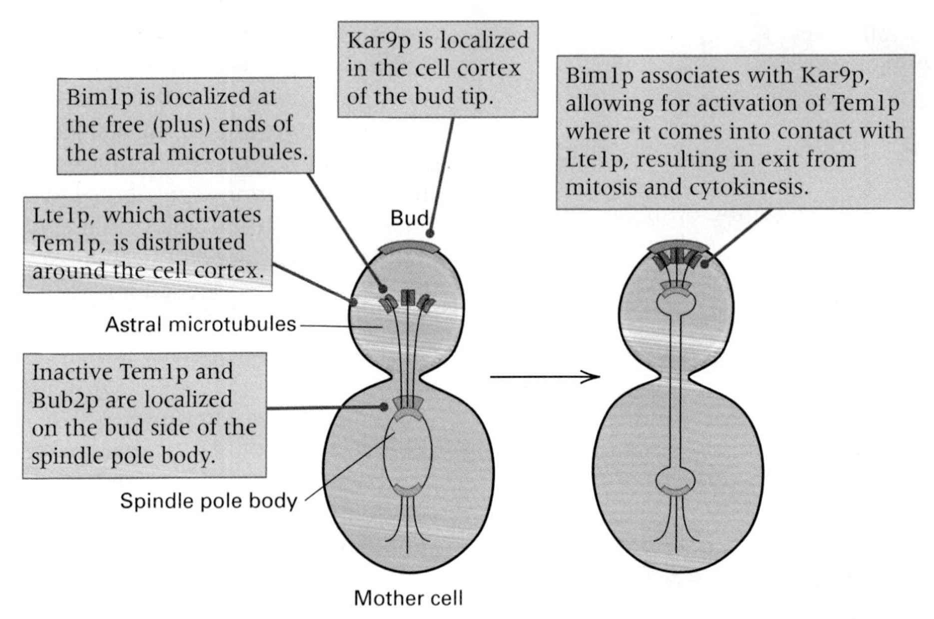
Kar9p is localized in the cell cortex of the bud tip.
Bim1p is localized at the free (plus) ends of the astral microtubules.
Bim1p associates with Kar9p, allowing for activation of Tem1p where it comes into contact with Lte1p, resulting in exit from mitosis and cytokinesis.
Lte1p, which activates Tem1p, is distributed around the cell cortex.
Bud
Astral microtubules
Inactive Tem1p and Bub2p are localized on the bud side of the spindle pole body.
Spindle pole body
Mother cell

**FIGURE 15.21** Cellular localization of components of the spindle positioning machinery and the mitotic exit network. Bim1p and Kar9p, together with other components and motor proteins, draw the nucleus into the bud. Once the nucleus enters the bud, Tem1p becomes active via Lte1p and activates the mitotic exit network.

microtubules. It is thought that depolymerization of the microtubules from the plus end occurs, drawing the nucleus into the daughter cell and toward the cell cortex, which in turn allows for activation of Tem1p and the mitotic exit network.

Failure of any checkpoint results in genetic instability. The particular kinds of genomic instability associated with defects in the three checkpoints discussed are summarized in **FIGURE 15.22**. Malfunctioning of the spindle itself can lead to aneuploidy, whereas a failure to duplicate a spindle pole can lead to polyploidy (Chapter 8). Defects in the DNA damage checkpoints can result in chromosomal aberrations of various kinds, including translocations, deletions, and amplification of genes or subregions of chromosomes. The amplified genes may be found as tandem repeats within a chromosome or as extrachromosomal circles lacking a centromere and telomeres.

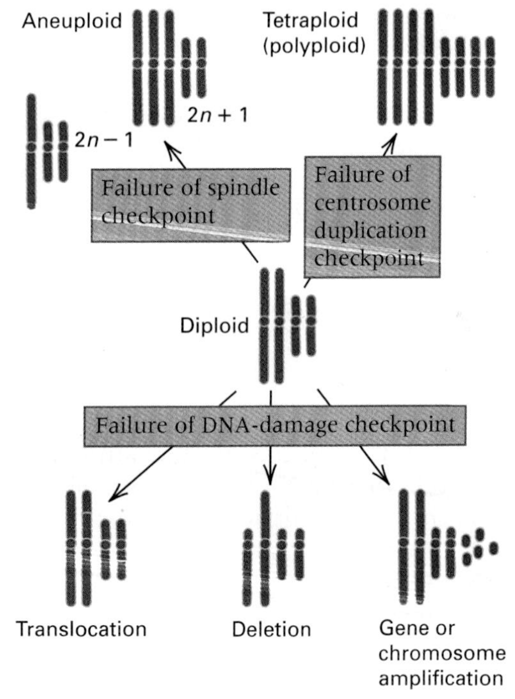
Aneuploid
Tetraploid (polyploid)
$2n + 1$
$2n - 1$
Failure of spindle checkpoint
Failure of centrosome duplication checkpoint
Diploid
Failure of DNA-damage checkpoint
Translocation
Deletion
Gene or chromosome amplification

**FIGURE 15.22** Contributions of checkpoint failures to genetic instability. [Adapted from L. H. Hartwell and M. B. Kastan, *Science* 266 (1994): 1821–1828.]

## 15.5 Cancer Cells

**Cancer** is not one disease but rather many diseases that share similar attributes; the cells show uncontrolled growth as a result of mutations that affect a limited number of genes. It is a disease of somatic cells. About 1 percent of cancer cases are **familial**, which means there is clear evidence for segregation of a gene in the pedigree that predisposes cells in affected individuals to progress to the cancerous state. The other 99 percent of cases are called **sporadic**, which in this context means *not familial*, and are the result of genetic changes only in somatic cells. Cancer cells share several properties not found in normal cells. One of these is the effect of cell-to-cell contact. In normal cells, cell-to-cell contact inhibits further growth and division, a process called **contact inhibition**.

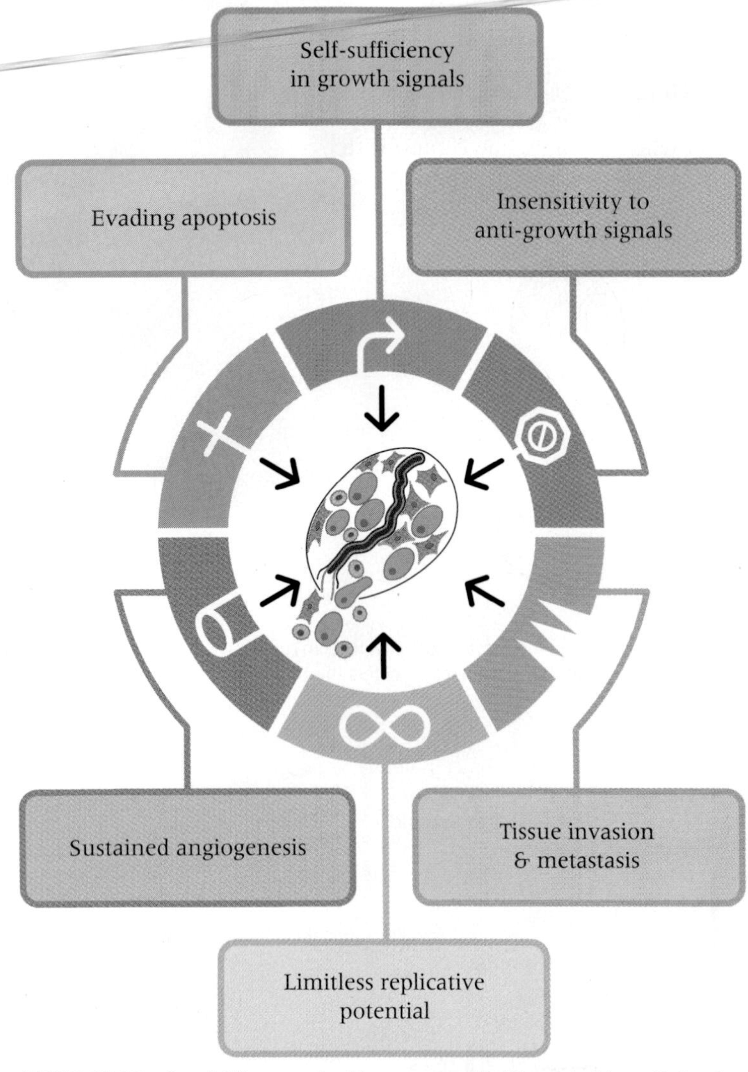


**FIGURE 15.23** Capabilities acquired by cancer cells. [Reprinted from *Cell*, vol. 100, D. Hanahan and R. A. Weinberg, "The hallmarks of cancer," pp. 57–70, copyright (2000), with permission from Elsevier, http://www.sciencedirect.com/science/journal/00928674.]

**FIGURE 15.23** highlights six other attributes of cancer cells that are not found in normal cells:

1. Loss of growth-factor dependence.
2. Insensitivity to anti-growth signals.
3. Evasion of apoptosis.
4. Immortality (no cell senescence).
5. Ability to metastasize and invade other tissues.
6. Sustained angiogenesis (formation of blood vessels).

Within an organism, tumor cells are clonal. They descend from a single ancestral cell that became cancerous. This conclusion is based in part on the observation that tumor cells in a female express the genes in only one of the X chromosomes, as a result of the normal inactivation of the other X chromosome (Chapter 8). If the tumor cells were not clonal, gene expression from both X chromosomes would be expected, reflecting the random inactivation of one X chromosome in each cell lineage. The finding of expression from a single X chromosome must therefore reflect clonality.

The conversion of normal somatic cells to cancer cells is a process that requires multiple steps. An important contributor to cancer conversion is genetic instability in the precursor cell population. The genetic instability may occur at the level of nucleotide sequences in the DNA or at the level of chromosome structure or number. The genetic instability is manifested as mutations, gene amplification, chromosomal rearrangements, or aneuploidy. The mutations result in a cell population that is genetically heterogeneous. In such a mixed population, any cell that has a proliferative advantage will contribute a greater fraction of descendants to the future cell population than will its neighbors, and because of this advantage, its clone expands at the expense of others. Subsequent mutations in the descendants of this cell and further clonal expansions of the derivative cells can give rise to a clone of cells with the proliferative capacity typical of cancer cells.

Many cancers are the result of alterations in cell-cycle control, particularly in control of the $G_1$-to-S transition, and in the $G_1$/S checkpoint associated with this transition. These alterations also affect apoptosis through their interactions with p53. **FIGURE 15.24** summarizes the key elements of the regulatory circuitry that governs the $G_1$/S transition, including the proteins that promote cell-cycle progression, that activate the checkpoint, and that govern the apoptotic pathway. Tumor cells commonly show altered expression or inactivation of function of one or more of these genes. Some genes for which alterations have been detected in cancer cells are listed in **TABLE 15.3.** The major mutational targets for the multistep cancer progression are of two types: *proto-oncogenes* and *tumor-suppressor genes*. These are discussed in the following sections.

## Oncogenes and Proto-Oncogenes

**Oncogenes** are gain-of-function mutations associated with cancer progression. They are derived from normal cellular genes called **proto-oncogenes**. Oncogenes are gain-of-function mutations because they improperly enhance the expression of genes that promote cell proliferation or inhibit apoptosis. In Figure 15.24, the proteins enclosed in rectangles are the products of proto-oncogenes, which when mutated can give rise to oncogenes. In this section, some of the oncogenes identified in Figure 15.24 and Table 15.3 are discussed in more detail. These

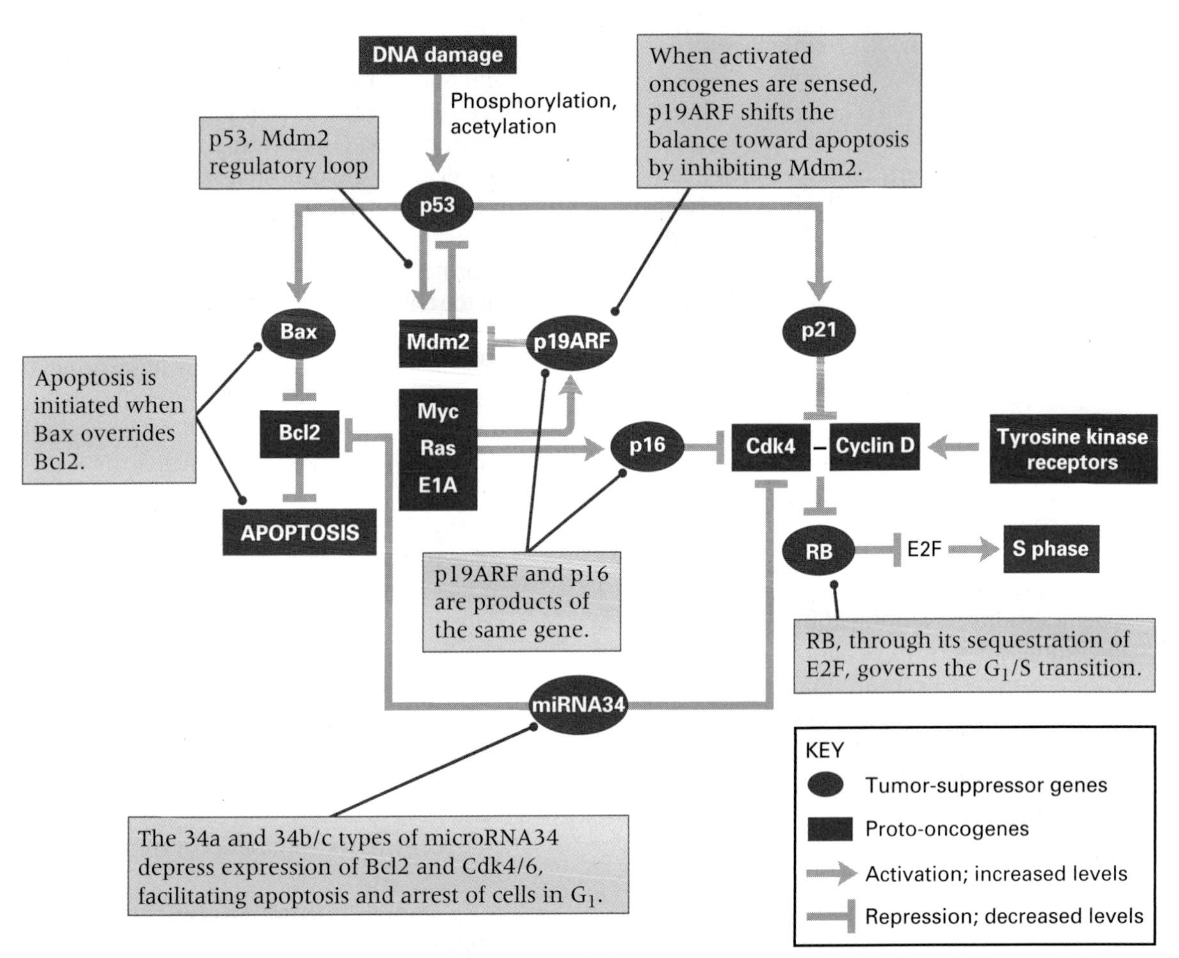


**FIGURE 15.24** Interactions between the p53 pathway and the RB (retinoblastoma protein) pathway in controlling apoptosis or DNA synthesis.

**Table 15.3 Cell cycle regulatory genes affected in tumors**

| Protein oncogenes | Alteration | Consequence |
|---|---|---|
| *Cyclin D* | Amplification or overexpression | Promotes entry into S phase. |
| *Cdk4* | Amplification | Promotes entry into S phase. |
| *Cdk4* | Mutation | Cdk4 resistant to inhibition by p16; promotes entry into S phase. |
| *EGFR* (epidermal growth-factor receptor, a tyrosine kinase receptor) | Amplification | Promotes proliferation by constitutive activation of pathway from growth-factor receptor. |
| *FGFR* (fibroblast growth-factor receptor) | Amplification | Promotes proliferation by constitutive activation of pathway from growth-factor receptor. |
| *Ras* | Amplification | Promotes proliferation by constitutively transmitting growth signal. |
| *Ras* | Mutation | Inactivates GTPase activity; constitutive activation of pathway from growth-factor receptor. |
| *Bcl2* | Overexpression by translocation next to strong enhancer | Blocks apoptosis. |
| *Mdm2* | Amplification | Mimics loss of p53 with loss of $G_1$/S, S, and $G_2$/M checkpoint functions. |
| Telomerase | Overexpression | Cells no longer undergo senescence. |
| **Tumor-suppressor genes** | **Alteration** | **Consequence** |
| *p53* | Mutation | Loss of $G_1$/S,S, and $G_2$/M checkpoint functions. |
| *p21* | Mutation | Loss of $G_1$/S and S checkpoint functions. |
| *RB* | Mutation | Promotes proliferation; E2F uninhibited. |
| *Bax* | Mutation | Failure to promote apoptosis of damaged cells. |
| *Bub1p* | Mutation | Loss of spindle assembly checkpoint function. |
| *E-cadherin* | Mutation | Loss of contact inhibition; tissue invasion and metastasis. |

include cyclins and CDKs, growth-factor receptors, Ras, and Mdm2.

**Cyclin D and Cdk4** Overexpression of cyclin D promotes unscheduled entry into S phase. Overexpression is often the result of gene amplification. Amplification of the cyclin D gene is found in about 35 percent of esophageal carcinomas, 15 percent of bladder cancers, and about 15 percent of breast cancers. On the other hand, although only 15 percent of breast cancers show *amplification* of the cyclin D gene, more than 50 percent of breast cancers show *overexpression* of cyclin D; this discrepancy implies that other mechanisms of overexpression also occur. Paralleling the increase in cyclin D expression, the gene for cyclin D's CDK partner, Cdk4, is amplified in some tumors, including 12 percent of gliomas (brain tumors) and 11 percent of sarcomas (muscle or connective tissue cancer).

**Growth-factor receptors and Ras** Growth factors stimulate growth by binding to a growth-factor receptor at the membrane. The binding activates a signal transduction pathway that acts through Ras (discussed in the next column), cyclin D, and its partner CDKs. The gene that encodes the receptor for epidermal growth factor (EGFR) is amplified in 45 percent of malignant astrocytomas (a kind of brain tumor), 35 to 50 percent of glioblastomas (another kind of brain tumor), 20 percent of breast cancers, 15 to 30 percent of ovarian cancers, 10 percent of head and neck cancers, and melanomas. The genes for fibroblast growth-factor receptor (FGFR) are amplified in about 20 percent of breast cancers.

Activated tyrosine kinase receptors such as EGFR and FGFR activate a signal transduction pathway that activates a small signaling protein called a **G protein**, as illustrated in **FIGURE 15.25** for the G protein Ras. Ras is usually bound to GDP, and in this state it is inactive. Receptor activation results in activation of a protein that stimulates the exchange of GTP to replace GDP in the Ras protein, and the Ras protein is active when it is bound to GTP. The Ras–GTP, in turn, activates a downstream protein, propagating a growth signal. The activation of Ras as Ras–GTP is transient; yet another protein (GAP) activates the intrinsic GTPase activity of Ras, so Ras hydrolyzes the GTP and returns itself to the inactive state of Ras–GDP. Mutations in the *RAS* gene that inactivate its GTPase function occur; its GTPase activity cannot be activated by GAP, and it remains in its active form of Ras–GTP, whether or not the cell is receiving signals from growth-factor receptors. A signal for growth is therefore transmitted constitutively, and unrestrained growth and division take place. This loss of growth-factor dependence is one of the hallmarks of cancer cells.

**EGFR and FGFR** Amplification or overexpression of EGFR or FGFR results in autoactivation of the receptor and transmission of a constitutive growth signal acting through the Ras pathway. Overexpression of Ras also leads to enhanced signal transduction and renewed cycles of proliferation.

**Mdm2** Amplification of the gene that encodes Mdm2 acts to promote cell proliferation, but it does so indirectly. Overexpression of Mdm2 effectively overwhelms any increase in production of p53, because the

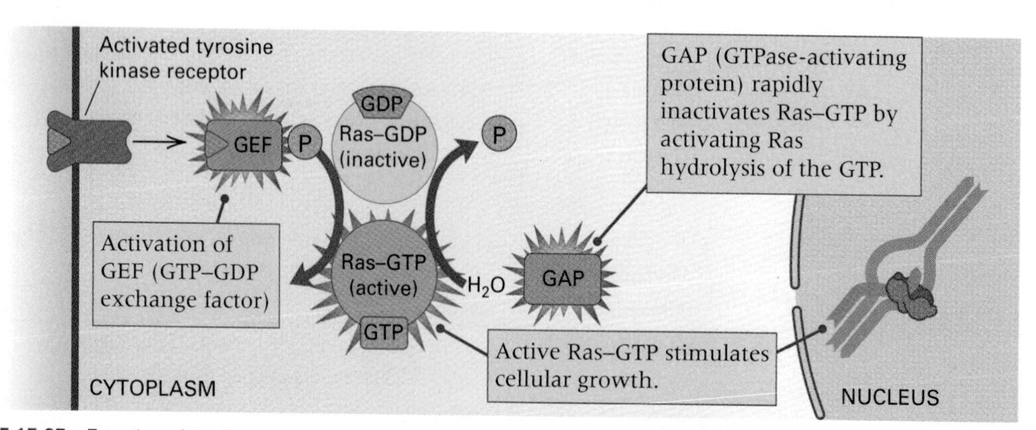


**FIGURE 15.25** Function of the Ras protein, which acts as a switch in stimulating cellular growth in the presence of growth factors. Ras is the product of a proto-oncogene. Certain mutant Ras proteins, such as a G19V (valine substitution for glycine at position 19), lack GTPase activity and remain in the form of Ras–GTP; hence a growth-promoting signal is present even in the absence of growth factors.

Mdm2 shuttles the extra p53 from the nucleus to the cytoplasm, where it is destroyed by the proteasome (Figure 15.15). This process effectively prevents functioning of the $G_1$/S checkpoint. Hence, amplification of the *Mdm2* gene and the resultant overexpression of Mdm2 protein is equivalent in its effects to inactivation of the gene for p53. Amplification of the *Mdm2* gene has been found in 19 of 28 tumor types examined, most commonly in tumors of adipose tissue (42 percent), soft-tissue sarcomas (20 percent), osteosarcoma (16 percent), and esophageal carcinoma (13 percent).

**Telomerase** Normal cells in culture cease to divide after about fifty doublings (a process called *senescence*), whereas cancer cells in culture divide indefinitely—they are immortal. The senescent behavior of normal cells is associated with a loss of telomerase activity. The telomeres are no longer elongated, which contributes to the onset of senescence and cell death. Cancer cells have high levels of telomerase, which help to protect them from senescence, making them immortal.

### Tumor-Suppressor Genes

**Tumor-suppressor genes** are genes that normally negatively control cell proliferation or that activate the apoptotic pathway, in which loss-of-function mutations contribute to cancer progression. Some of the key functions of tumor-suppressor genes are illustrated in Figure 15.24 and listed in Table 15.3. They are examined in more detail in this section.

**p53** Loss of function of p53 results in acquisition of two characteristics of cancer cells: insensitivity to antigrowth signals and evasion of apoptosis. Loss of function of p53 eliminates the DNA checkpoint that monitors DNA damage in $G_1$ and S, as we saw earlier in Figures 15.14 and 15.17. In the absence of functional p53, the proteins and microRNAs responsible for arresting cells in $G_1$ or $G_2$ (p21, GADD45, 14-3-3$\sigma$, and miRNAs 34a and 34b/c) are not synthesized in response to DNA damage. There is therefore no block to a cell's proceeding into S phase or into M phase even if it has damaged chromosomes, altered ploidy, or amplified genes. In addition, loss of p53 function costs the organism its ultimate defense against aberrant cells, because DNA damage is no longer able to trigger enhanced expression of Bax and Apaf1 and reduce expression of Bcl2, and hence the self-destruction of aberrant cells; the damaged cells survive and proliferate. Furthermore, their genetic instability increases the probability of additional genetic changes and thus progression toward the cancerous state. Given this scenario and the key role of p53 in protecting the cell against the consequences of DNA damage it is not surprising that p53 proves to be nonfunctional in 55 percent of all cancers. In addition, as mentioned previously, Mdm2 overexpression—equivalent to loss of p53 function—occurs in other tumors. Because loss of p53 and overexpression of Mdm2 rarely occur in the same tumor, the actual loss of p53 function by one mechanism or another is probably substantially greater than 55 percent. Mutant p53 proteins are found frequently in melanomas, several kinds of lung cancers, colorectal tumors, bladder and prostate cancers, and astrocytomas.

Testicular embryonal carcinoma is one of the most curable of cancers, with a cure rate of 90 to 95 percent. A striking observation is that these carcinomas never include mutations in the p53 gene. The curability and the lack of p53 involvement may be causally related. The p53 gene is not expressed in embryonal testicular cells, possibly because the process of recombination with breakage and rejoining of DNA strands would be sensed as DNA damage, and imposition of the checkpoints would block meiosis. Because the p53 gene is not expressed, there can be no selective advantage conferred on cells by mutations in the p53 gene in embryonal testicular cells. The current interpretation of the high success rate with treatment using certain drugs (for example, *cisplatin*) is that the drug induces the expression of p53, which activates the apoptotic pathway and leads to suicide of the cancer cells.

**p21** Loss of p21 function results in renewed rounds of DNA synthesis without accompanying mitosis, and the level of ploidy of the cell increases. Mutations in the gene that encodes p21 occur in some prostate cancers.

**p16/p19ARF** The p16 and p19ARF proteins are products of the same gene; they derive from two distinct transcripts from different promoters, each with a different 5′ exon. The p16 product can inhibit the cyclin-D–Cdk4 complex and help control entry into S phase; the gene encoding p16 is a very frequent target for inactivation during tumor progression. The gene is deleted in 55 percent of gliomas (a form of brain tumor), 55 percent of mesotheliomas, more than 50 percent of melanomas, 40 percent of nasopharyngeal carcinomas, 50 percent of biliary-tract carcinomas, and 30 percent of esophageal carcinomas. Figure 15.24 shows the functional interactions of p19ARF as a tumor suppressor. If cellular or viral oncogenes are being overexpressed, the synthesis of p19ARF is elevated, and at this higher concentration p19ARF binds to Mdm2 and prevents it from binding with p53 and shuttling p53 from the nucleus. Given this mode of action, one might expect that some tumors might lack p19ARF. In fact, no mutation has yet been reported that inactivates p19ARF without also inactivating p16, although the opposite kind of mutation has been observed.

**RB** The retinoblastoma protein controls the transition from $G_1$ to S phase by controlling the activity of the transcription factor E2F, as illustrated in Figure 15.12. Loss of RB function frees E2F to initiate transcription of the enzymes for DNA synthesis at all times in the cell cycle; hence excessive rounds of DNA synthesis are continuously being initiated. Loss of RB function is found in melanomas, in small-cell lung carcinoma, in osteosarcoma, and in liposarcomas. Just as loss of p53 and overexpression of Mdm2 appear to be equivalent in their effects, so do loss of either of the tumor-suppressor proteins RB or p16 and overexpression of either cyclin D or Cdk4. One basis for this conclusion is the observation that tumor cells are nearly always altered for RB, p16, cyclin D, or Cdk4, but rarely for more than one. This finding suggests that loss of function of either p16 or RB is equivalent to overexpression of either cyclin D or Cdk4. Any one of these defects leads to the same result: The cells lose control of the $G_1/S$ transition and embark upon unscheduled rounds of DNA synthesis; they become insensitive to antigrowth signals.

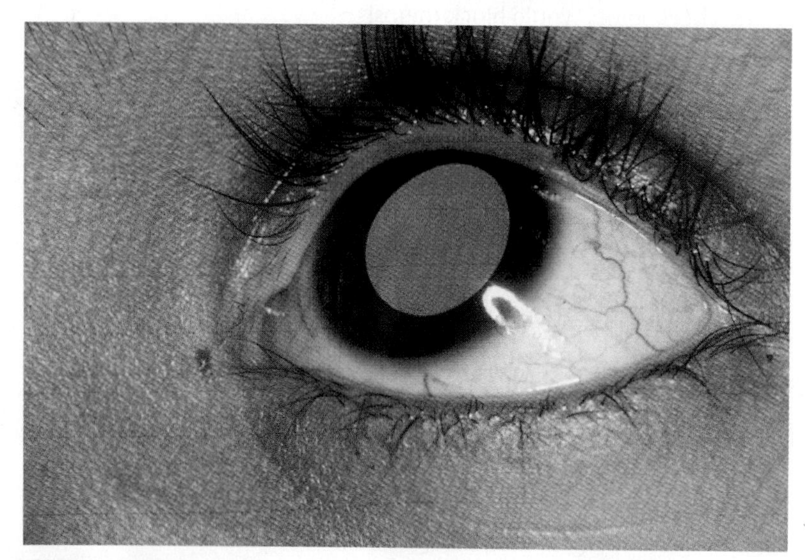

**A RETINOBLASTOMA** tumor first appears as a white mass in the retina but may grow to fill the entire eye, as in this case, giving the pupil a white or yellow cast. [© Custom Medical Stock Photo.]

**Bax** The Bax tumor-suppressor protein promotes apoptosis (Figure 15.18). Loss of Bax function is found particularly in gastric adenocarcinomas and in colorectal carcinomas, particularly those associated with microsatellite instability because of defective mismatch repair. Cells that are defective in mismatch repair are especially prone to undergo replication slippage leading to deletions or additions of nucleotides in runs of short tandem repeats (discussed in Chapter 14). One of the sequences prone to replication slippage is a run of eight consecutive G nucleotides within the *Bax* gene, which makes this gene especially susceptible to frameshift mutation by the addition or deletion of a base pair in this region.

**Bub1** Bub1 is a protein that is primarily involved in the spindle checkpoint (Figure 15.19). A subset of colon cancers show chromosomal instability; they do not maintain a constant karyotype but, rather, are aneuploid. Some of these unstable lines are defective in Bub1.

**E-cadherin** In normal cells, cell-to-cell contact inhibits further growth and division, a process called *contact inhibition*. Cancer cells have lost contact inhibition. They continue to grow and divide and even pile up on one another. E-cadherin is uniquitously expressed on epithelial cells. Interaction between molecules on adjacent cells allows transmission of anti-growth factors and other signals via the cytoplasmic contacts. E-cadherin acts to suppress invasion and metastasis by epithelial cells, and inactivation (either by mutation in the gene itself or in the gene for its interacting partner β-catenin) is a key step in acquisition of the capability for metastasis.

## 15.6 Hereditary Cancer Syndromes

As we have noted, approximately 99 percent of cancers are sporadic; all of the genetic changes occur in somatic cells. Furthermore, the change from a normal cell into a tumor cell is progressive. It goes step by step as each new somatic mutation along the way compromises a different mechanism of cell-cycle control.

### ■ Defects in Cell-Cycle Regulation and Checkpoints

Approximately 1 percent of cancers are familial. In these cases, one of the cancer-progression mutations that may occur in sporadic tumors is inherited through the germ line. The presence of this mutation predisposes the individual to cancer, because it makes fewer somatic mutations necessary for a precancerous cell to progress to malignancy. **FIGURE 15.26** illustrates some of the genetic mutations and changes in cell morphology that take place in the progression of a type of familial colon cancer called adenomatous polyposis. In this case the inherited mutation is in the gene *APC* in chromosome 5, which is a tumor-suppressor gene whose normal function is to transduce the signal of contact inhibition into the cell to inhibit further growth and division. Subsequent mutations in the progression to malignancy include an oncogenic mutation in a *Ras* gene and loss or mutation of the gene encoding p53. This is only one possible route of progression. Different tumors may progress by different pathways, depending on what mutations occur and in what order.

Familial adenomatous polyposis of the colon and some other important familial cancer syndromes are identified in **TABLE 15.4**. The Li-Fraumeni syndrome, familial retinoblastoma, and familial melanoma are all associated with germ-line mutations in genes that are often found to be mutant in sporadic tumors. The genes affected in these syndromes are the *p53* gene, the *RB1* gene, and the *p16* gene, respectively. A pedigree of a cancer-prone family segregating for a mutation in the *p53* gene (Li-Fraumeni syndrome) is given in **FIGURE 15.27**. This syndrome shows clear autosomal dominant

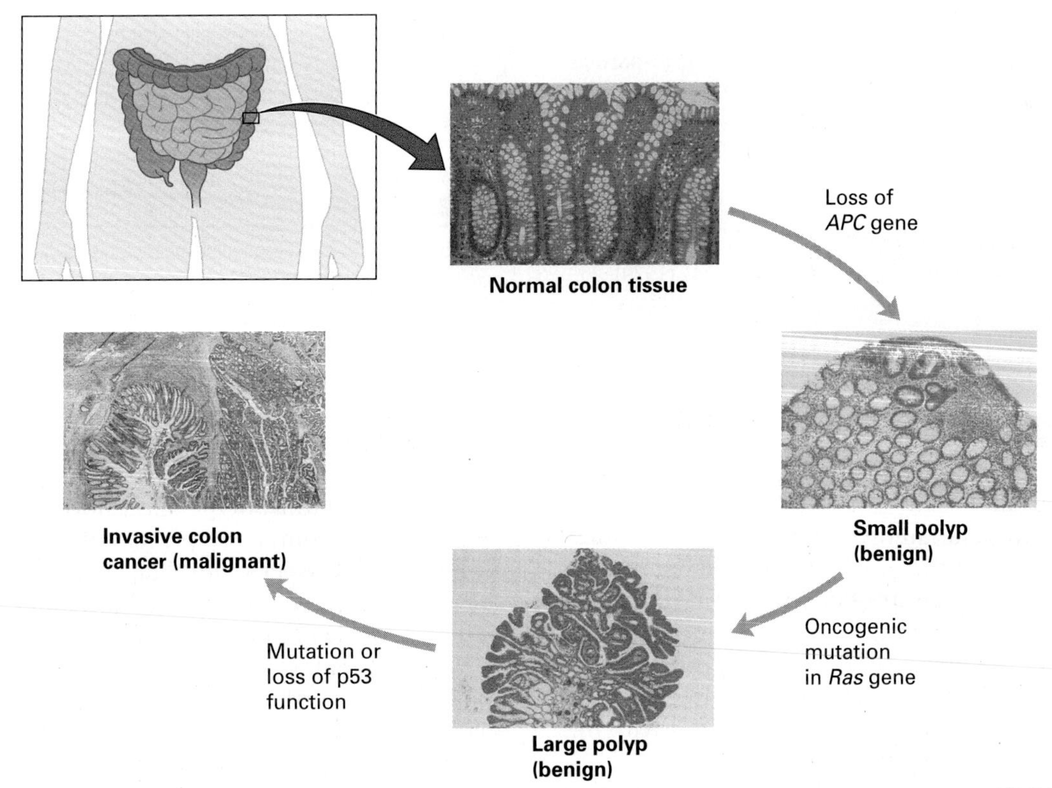


**FIGURE 15.26** Somatic mutations and cell morphology in the progression of invasive colon cancer. [Photos courtesy of Kathleen R. Cho, University of Michigan Medical School.]

**Table 15.4 Inherited cancer syndromes**

| Syndrome | Primary tumors | Associated tumors | Chromosome | Gene | Proposed function |
|---|---|---|---|---|---|
| Li-Fraumeni syndrome | Sarcomas, breast cancer | Brain tumors, leukemia | *17p13* | *p53* | Transcription factor |
| Familial retinoblastoma | Retinoblastoma | Osteosarcoma | *13q14* | *RB1* | Cell-cycle regulator |
| Familial melanoma | Melanoma | Pancreatic cancer | *9p21* | *p16* | Inhibitor of Cdk4 and Cdk6 |
| Hereditary nonpolyposis colorectal cancer (HNPCC) | Colorectal cancer | Ovarian, glioblastoma | *2p22*<br>*3p21*<br>*2q32*<br>*7p22* | *MSH2*<br>*MLH1*<br>*PMS1*<br>*PMS2* | DNA mismatch repair |
| Familial breast cancer | Breast cancer | Ovarian cancer | *17q21* | *BRCA1* | Repair of DNA double-strand breaks |
| Familial adenomatous polyposis of the colon | Colorectal cancer | Other gastrointestinal tumors | *5q21* | *APC* | Regulation of β-catenin |
| Xeroderma pigmentosum | Skin cancer | | Several complementation groups | *XPB*<br>*XPD*<br>*XPA* | DNA-repair helicases, nucleotide excision repair |

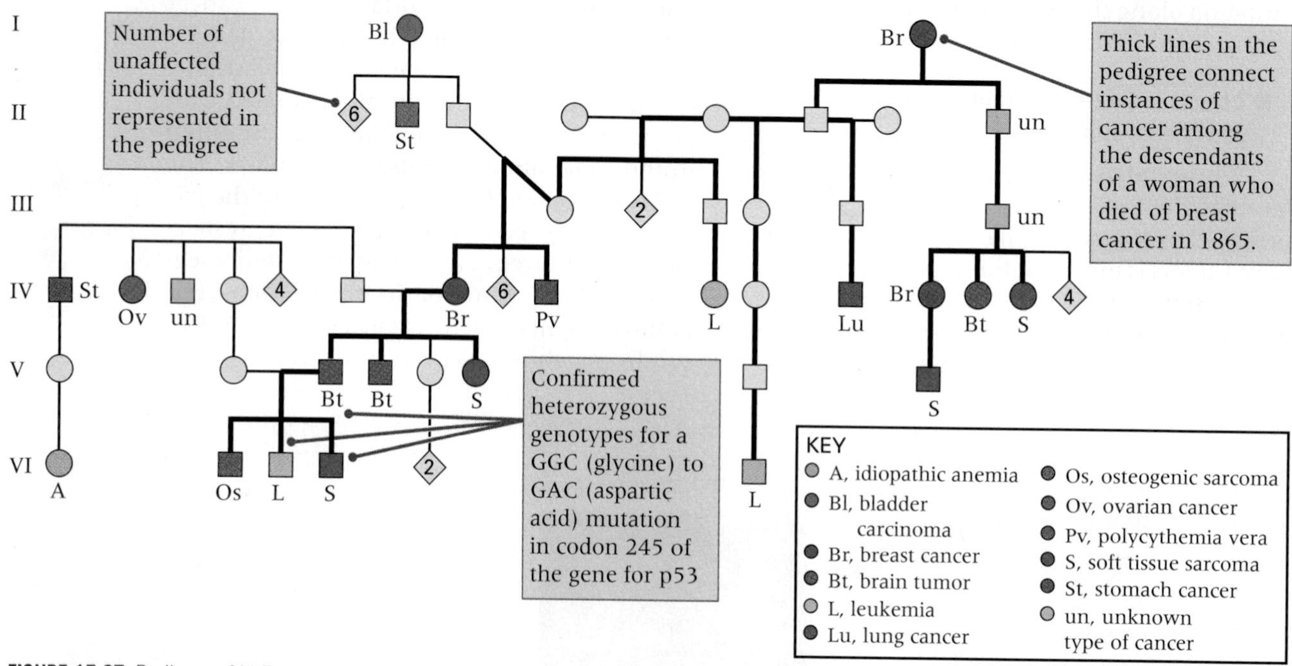


**FIGURE 15.27** Pedigree of Li-Fraumeni syndrome associated with mutation in the gene for p53. Individuals are prone to develop any of a variety of cancers. [Data from W. A. Blattner, et al., *J. Am. Med. Assoc.* 241 (1979): 259–261.]

inheritance. However, the affected individuals show a range of different tumors and often have more than one, including osteosarcoma, leukemia, breast cancer, lung cancer, soft tissue sarcoma, and brain tumors. A large fraction of Li-Fraumeni families show segregation for a mutation in the *p53* gene. For this family, affected members are heterozygous for a single nucleotide pair in the *p53* gene, which changes codon 245 from GGC (glycine) to GAC (aspartic acid). The observation that the affected individuals are heterozygous is consistent with the dominant autosomal inheritance.

A situation analogous to the human Li-Fraumeni syndrome has been created in mice by experimental knockout (loss of function) of the *p53* gene via the germ-line transformation methods discussed in Chapter 12. Animals heterozygous or homozygous for the p53 knockouts were compared to wildtype mice. Normal mice do not develop tumors; 50 percent of them live to the age of 120 weeks, and about 5 percent live for at least 160 weeks. In contrast, half of the heterozygous p53 knockout mice are dead by 55 weeks, and almost none of them live as long as 120 weeks. Furthermore, 50 percent of these mice develop tumors (mostly osteosarcomas) by 18 months of age. The homozygous knockout mice fare even worse—50 percent are dead by 20 weeks, and all are dead by 40 weeks; among the homozygous animals, 75 percent develop tumors by 6 months, and by 10 months all have developed tumors or are dead. In the homozygous genotype, the tumors are lymphomas rather than osteosarcomas, possibly because normal development of the immune system requires massive apoptosis within the thymus gland, and in the p53 homozygous knockouts, apoptosis is abolished because Bax is not upregulated as in normal mice and p53 knockout heterozygotes.

The fact that animals heterozygous for the p53 knockout are severely affected for both longevity and incidence of tumors does not necessarily imply that the *p53* mutation is dominant at the level of the individual cell. In the case of the *p53* knockout mutation, the effects of the mutation are manifested only in somatic cells that have become homozygous for the knockout mutation or in those that have lost (or undergone somatic mutation at) the wildtype allele in the homologous chromosome. In a heterozygous animal, only one copy of the wildtype *p53* gene is present to protect the cell; inactivation of the lone wildtype allele disables the checkpoints that depend on the p53 protein. Because there are so many somatic cells in which such a rare aberration can occur, it is nearly certain that inactivation or loss of *p53* will take place somewhere in the heterozygous organism, initiating the sequence of mutations that results in cancer.

On the other hand, most observed mutations in the p53 gene are dominant-negative missense mutations. They are dominant at both the organismal and cellular level. The p53 protein functions as a tetramer. In the heterozygous genotype, fifteen of the sixteen possible types of tetramers contain at least one mutant subunit. If the mutant subunit poisons the activity of the tetramer, the remaining 1/16 of the normal activity is not sufficient for normal function, and the mutation will be dominant (**FIGURE 15.28**).

The idea that loss of the wildtype allele of a tumor-suppressor gene might be the triggering event at the cellular level for tumors in heterozygous genotypes was first suggested from studies on retinoblastoma by Alfred Knudson in 1971, long before the function of the *RB* gene was identified. As the excerpts in the Connection on page 580 entitled *Two Hits, Two Errors* indicate, Knudson noted that sporadic cases of retinoblastoma usually had a single tumor, whereas familial retinoblastoma cases usually had bilateral tumors and more than one tumor. On the basis of a statistical analysis of the time of tumor diagnosis, he suggested that genesis of a tumor in familial cases required a single hit in a somatic cell, whereas genesis of a tumor in sporadic cases required two hits, which would happen only rarely. The two-hit model would also explain why sporadic cases rarely have a second tumor; each individual hit is a rare event.

Retinoblastoma, like the p53 deficiency, is inherited in pedigrees as a simple Mendelian dominant. But Knudson's hypothesis implied that even in familial cases, there must be another mutational event that triggers tumor development. Once the gene itself, called *RB1*, was located, analysis of genetic markers around the gene in tumor cells revealed that the triggering event is the loss of the wildtype *RB1* allele. Any of several mechanisms can uncover the mutant allele, including chromosome loss, mitotic recombination, deletion, and inactivating nucleotide substitutions. These can be distinguished from one another on the basis of the genetic markers that are lost or retained in the tumor cells; several of the mechanisms known to occur are diagrammed in **FIGURE 15.29**. At the organismic level, expression of the mutant gene is dominant. However, at the cellular level, expression of the mutant gene is recessive. Uncovering of the recessive allele by various mechanisms is called **loss of heterozygosity**.

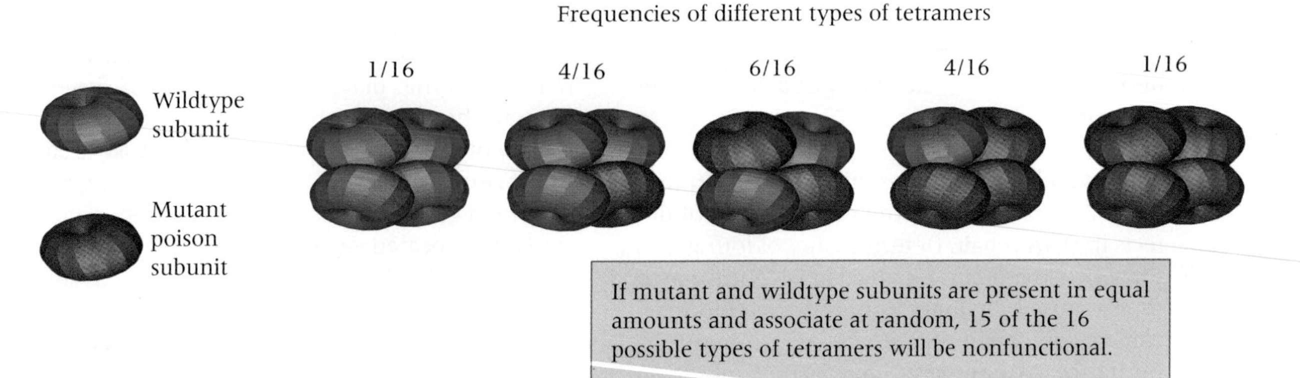
Frequencies of different types of tetramers
1/16
4/16
6/16
4/16
1/16
Wildtype subunit
Mutant poison subunit
If mutant and wildtype subunits are present in equal amounts and associate at random, 15 of the 16 possible types of tetramers will be nonfunctional.

**FIGURE 15.28** Consequences for p53 function of a dominant-negative mutation in the *p53* gene.

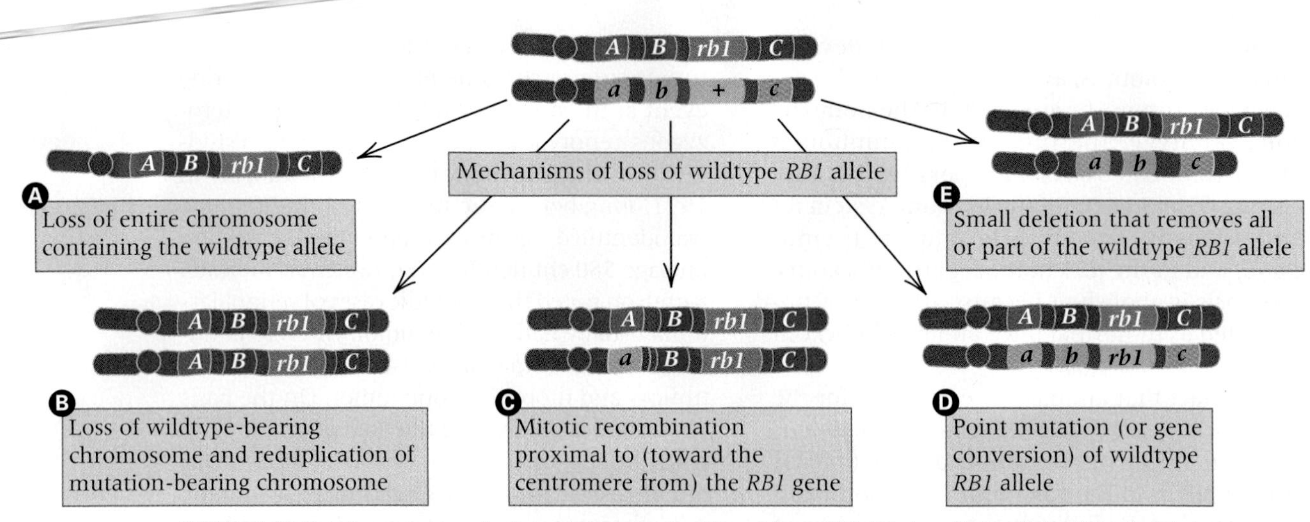


**FIGURE 15.29** Genetic mechanisms for loss of heterozygosity of the wildtype *RB1* allele in patients with familial retinoblastoma.

The inherited cancer syndromes listed in Table 15.4 result from mutations in tumor-suppressor genes. All show autosomal dominant inheritance except for xeroderma pigmentosum, which is inherited as a recessive. In all cancer syndromes that show dominant inheritance, loss of heterozygosity is required for manifestation of the tumor phenotype. Expression of the mutant allele *predisposes* the cell to become cancerous but is not in itself sufficient for generation of a cancer cell, with the notable exception of the p53 poison subunits discussed earlier. Tumor progression still occurs only when additional somatic mutations and clonal expansions take place. The germ-line mutations do not themselves cause cancer; they merely make it much more likely that the progression will occur, because the tumor, in effect, has been given a head start.

### Defects in DNA Repair

Genetic instability clearly contributes to the genesis of tumor cells. We know from studies on yeast and bacteria that cells have extensive mechanisms for repairing DNA lesions (see Chapter 14). Defects in these processes result in greatly elevated mutation rates and genetic instability. It was therefore not surprising when hereditary cancer syndromes that result from inherited defects in DNA repair were discovered.

Several of the inherited cancer syndromes listed in Table 15.4 are the consequence of defects in DNA repair. Defects in any of four genes that encode proteins involved in DNA mismatch repair cause hereditary nonpolyposis colorectal cancer, which shows autosomal dominant inheritance. Mutant cells have higher mutation rates, which promote progression toward the cancerous state. Inherited breast cancer syndromes prove to be associated with mutations in either of two genes, *BRCA1* or *BRCA2*. Less is known about the function of these genes. Current evidence suggests that *BRCA1* is somehow involved in repair of double-strand breaks. Inherited skin cancer syndromes are called xeroderma pigmentosum. Xeroderma pigmentosum cells are defective in nucleotide excision repair; they are unable to repair defects such as thymine dimers that are induced by ultraviolet light. Individuals with this syndrome are very sensitive to the ultraviolet light that is present in sunlight and emitted by fluorescent lights.

## 15.7 Genetics of the Acute Leukemias

The human acute leukemias arise in progenitors of the white blood cells. The initial genetic events that result in the acute leukemias are quite different from those indicated in the cancers examined previously. They do not arise as a consequence of alterations in cell-cycle regulation or checkpoints, nor are they familial.

Up to 65 percent of the acute leukemias arise as a consequence of chromosomal translocations involving genes that encode transcription factors that play a role in blood-cell development (hematopoiesis). The translocations are mainly of the two types illustrated in **FIGURE 15.30**. In a **promoter fusion** (part A), the coding region for a gene that encodes a transcription factor is translocated near an enhancer for an immunoglobulin heavy-chain gene or a T-cell receptor gene. The result is overexpression of the transcription factor, derangement of normal hematopoiesis, and overproduction of lymphocytes. One important aspect of normal hemato-

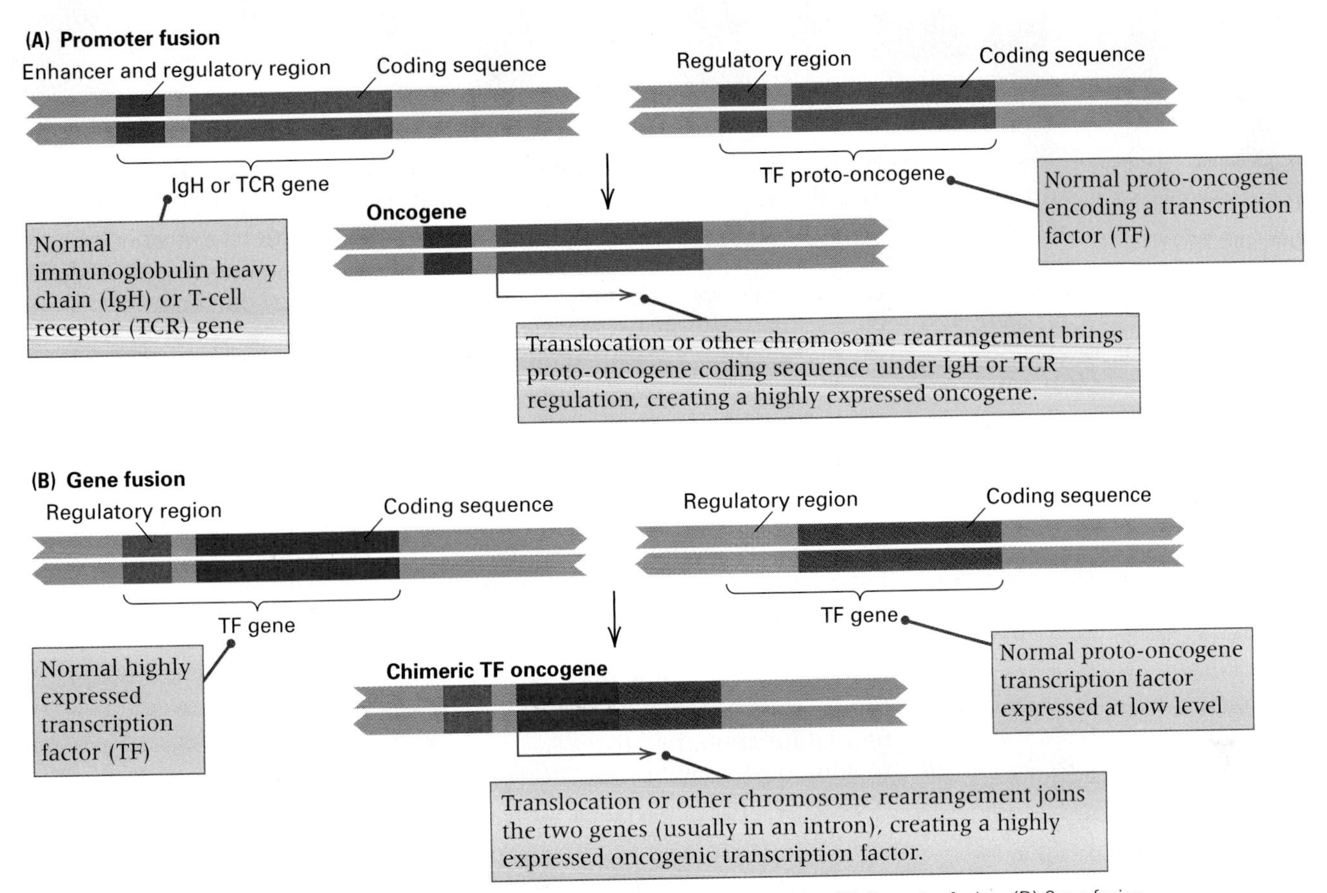
(A) Promoter fusion
Enhancer and regulatory region
Coding sequence
Regulatory region
Coding sequence
IgH or TCR gene
TF proto-oncogene
Oncogene
Normal immunoglobulin heavy chain (IgH) or T-cell receptor (TCR) gene
Normal proto-oncogene encoding a transcription factor (TF)
Translocation or other chromosome rearrangement brings proto-oncogene coding sequence under IgH or TCR regulation, creating a highly expressed oncogene.
(B) Gene fusion
Regulatory region
Coding sequence
Regulatory region
Coding sequence
TF gene
TF gene
Normal highly expressed transcription factor (TF)
Chimeric TF oncogene
Normal proto-oncogene transcription factor expressed at low level
Translocation or other chromosome rearrangement joins the two genes (usually in an intron), creating a highly expressed oncogenic transcription factor.

**FIGURE 15.30** Translocations that aberrantly activate transcription factors in acute leukemias. (A) Promoter fusion. (B) Gene fusion.

poiesis is the apoptotic destruction of progenitor cells that fail to rearrange their antigen-receptor genes productively (Chapter 11). Estimates are that 75 percent of B-cell and 95 percent of T-cell precursors self-destruct during normal development. *Bcl2* (*Bcl* stands for B-cell lymphoma) is an example of an oncogene in Table 15.4 that was originally discovered in a promoter fusion that placed *Bcl2* next to an immunoglobulin heavy-chain gene enhancer. Overexpression of Bcl2 blocks apoptosis by preventing formation of Bax homodimers (Figure 15.18). These undeservedly alive lymphocytes proliferate and come to dominate the population of white blood cells, but they are useless in fighting infection because they have not rearranged their immunoglobulin genes properly. In such cases the bone marrow becomes almost totally occupied by the cancerous leukocytes, leading to severe anemia and bleeding. Chemotherapy can sometimes be quite effective in treating acute leukemias in young children by activating the *p53* gene, provided that the aberrant leukocytes still carry the wildtype alleles encoding p53.

The second type of translocation, a **gene fusion** (part B of Figure 15.30), is more common than a promoter fusion. Most commonly, breakpoints occur in introns of two different transcription factor-encoding genes on different chromosomes. The result is a fusion gene that encodes a chimeric transcription factor with altered function, which may have the capacity to interfere with normal hematopoiesis. The uniqueness of these chimeric proteins, and the fact that they are present only in cancer cells and not in normal cells, makes them inviting targets for drugs or chemotherapy. If one could successfully attack cells expressing the chimeric protein, one could selectively kill the cancer cells.

Chronic myeloid leukemia (CML) accounts for 15 to 20 percent of all cases of leukemia. The hallmark of CML is the presence of the *Philadelphia chromosome,* which results from a reciprocal translocation between chromosomes 9 and 22 in hematopoietic stem cells of the bone marrow. These are the cells that differentiate into various specialized types of cells that become part of the blood and immune system. The molecular result of the *t(9, 22)* translocation is replacement of the first exon of *c-abl* with sequences from the *bcr* gene, resulting in a Bcr-Abl fusion protein. Because the N-terminal region of Abl normally inhibits function of the catalytic domain, the fusion protein, having lost the

## connection
### Two Hits, Two Errors

*This landmark paper was a turning point in thinking about genetics and cancer. Pedigree studies had already shown that there were hereditary predispositions to certain cancers, such as retinoblastoma. But sporadic cases also occurred, in which only one member of a kindred was affected. What is the relationship between these two forms? By an ingenious statistical analysis, Knudson showed that the sporadic cases exhibit two-hit kinetics as a function of age, indicating that two independent mutations in the same retinal cell are involved, whereas familial cases exhibit one-hit kinetics. The simplest interpretation is that in sporadic cases, the first mutation knocks out a key gene and the second knocks out its allele; in familial cases, the initial mutant allele is inherited, and consequently only one mutation (in the remaining functional allele) is needed to cause the disease. At the level of the individual cell, therefore, a mutation in the retinoblastoma gene (RB1) is recessive, whereas at the familial level, the disease is inherited as a dominant, because almost every person who inherits one mutant allele will undergo the second mutation in at least one retinal cell.*


**Alfred G. Knudson 1971**
M. D. Anderson Hospital
The University of Texas, Houston, Texas
*Mutation and Cancer: Statistical Study of Retinoblastoma*


The hypothesis is developed that retinoblastoma is a cancer caused by two mutational events. In the dominantly inherited form, one mutation is inherited via the germinal cells and the second occurs in somatic cells. In the nonhereditary form, both mutations occur in somatic cells. . . . Several authors have concluded that retinoblastoma may be caused by either a germinal or a somatic mutation. . . . All bilateral cases [25–30 percent] should be counted as hereditary because the proportion of affected offspring closely approximates the 50 percent expected with dominant inheritance. . . . If a second, single event is involved [in the hereditary cases], the distribution of bilateral cases with time should be an exponential function, i.e., the fraction of the total cases that develop should be constant, as expressed in the relationship $dS/dt = -kS$, and $\ln(S) = -kt$, where $S$ is the fraction of survivors not yet diagnosed at time $t$, and $dS$ is the change in this fraction in the interval $dt$. As shown in the figure, this is indeed the case. By contrast, the fractional decrease in unilateral cases per unit time does not show this relationship. . . . The exponential decline in new hereditary cases with time reflects the occurrence of a second event at a constant rate . . . of the order of $2 \times 10^{-7}$ per year. . . . The two-mutation hypothesis is consistent with current thought . . . that the common cancers are produced by about 3–7 mutations. Interestingly, one of the lowest estimates [is] for brain tumors, which are, like retinoblastoma, derived from neural elements.

Source: A. G. Knudson, *Proc. Natl. Acad. Sci. USA* 68 (1971): 820–823.

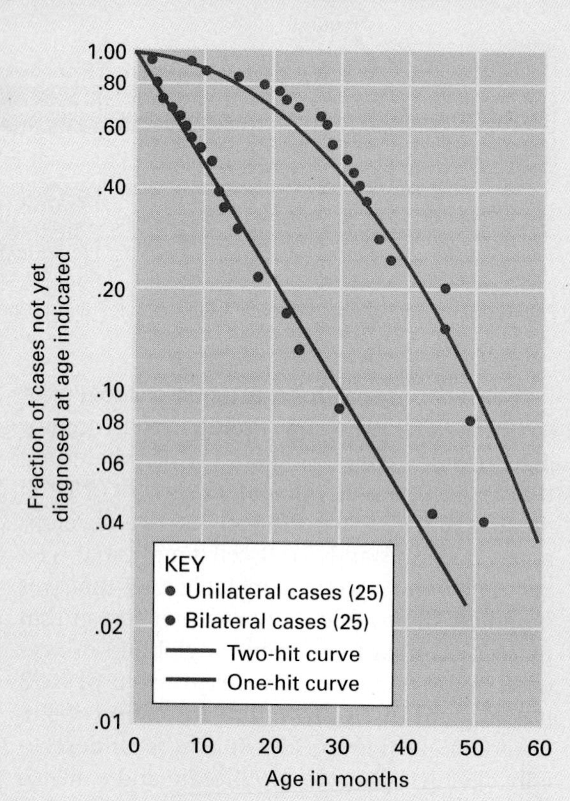


inhibitory domain, is a constitutively active protein kinase.

The tyrosine kinase inhibitor STI571 (imatinib mesylate) is a therapeutic agent specifically directed toward the Bcr-Abl tyrosine kinase. By inhibiting the Bcr-Abl kinase, the drug not only inhibits the kinase activity, but also inhibits proliferation of cells expressing the kinase *in vivo* and *in vitro*. Continuous exposure to the drug eradicates most cancer cells. It is thought to be very effective because Bcr-Abl may be the sole molecular abnormality early in the course of chronic myeloid leukemia. STI571 also inhibits the Kit tyrosine kinase and has proven effective against gastrointestinal stromal tumors, in which mutations that constitutively activate the *c-kit* gene encoding Kit are involved.

In one study of colorectal cancers, 182 cancers were tested for mutations in 138 genes encoding protein kinases, out of a total of 518 known protein-kinase genes in the human genome. About 30 percent of the colorectal cancers proved to have a mutation in one of the genes encoding tyrosine kinases. These results, along with the positive outcomes of treatment with drugs like STI571, have led to an increased focus on protein kinases as therapeutic possibilities.

## CHAPTER SUMMARY

- About 10–15 percent of the genes in a yeast cell function in cell division and in the transition from one phase of the cell cycle to the next; their transcription is initiated once per cell cycle.
- Progression through the cell cycle is controlled by cyclin-dependent protein kinases and by protein degradation.
- Checkpoints monitor a dividing cell for DNA damage, cellular defects, other abnormalities in the cell cycle, and cell size. Detection of abnormalities elicits a response that arrests the cell cycle, allowing time for repair of defects (or cell death—apoptosis), and ensuring that the phases remain in correct order.
- Cancer cells show uncontrolled growth and proliferation and loss of contact inhibition.
- Progression from a normal to a cancerous state requires several genetic changes. Most cancers are sporadic (not inherited).
- A small proportion of cancers are associated with mutations transmitted through the germ line, which predispose the somatic cells of people carrying them to undergo cancer progression.
- The genetic changes that take place in cancer progression often involve defects or overexpression of genes that function in cell cycle regulation or checkpoint control.

## REVIEW THE BASICS

- What role does the centriole (in some organisms, the centrosome) play in cell division?
- What molecular process that takes place in the nucleus defines the S period?
- In yeast, what phenotype defines a *cdc* mutant? For a temperature-sensitive *cdc* mutant grown at the restrictive temperature, why do cells arrest at a particular stage in the cell cycle?
- What does the abbreviation CDK stand for? What is a cyclin–CDK complex? What determines the substrate specificity of a cyclin–CDK complex?
- What is the role of proteolysis in regulating the cell division cycle?
- What is a cell-cycle checkpoint? Which checkpoints are emphasized in this chapter, and what does each check for?
- What is apoptosis and what role does it play in preserving the integrity of the genome of a multicellular organism?
- Mutations in genes whose products are involved in DNA repair are often associated with an increased risk of cancer. What does this observation imply about the role of spontaneous mutations in the development of cancer?
- Reconcile the following statements: "Cancer is a genetic disease" and "Most cancers are sporadic (not familial)."
- Distinguish between proto-oncogenes and tumor-suppressor genes, and give one example of each. In which class of genes does a loss-of-function mutation predispose to cancer? A gain-of-function mutation? Explain your answer.
- What is loss of heterozygosity, and how is this phenomenon related to the progression of some types of cancer?

## GUIDE TO PROBLEM SOLVING

**Problem 1** Distinguish between an oncogene and a tumor-suppressor gene. Why are oncogene mutations associated with cancer gain-of-function mutations, whereas tumor-suppressor mutations associated with cancer are loss-of-function mutations?

**Answer** Oncogenes are mutant genes whose normal counterparts, called proto-oncogenes, encode products that promote cell proliferation or inhibit apoptosis. The products of tumor-suppressor genes inhibit cell proliferation or activate the apoptotic pathway. Oncogenes are gain-of-function mutations because their enhanced expression makes uncontrolled cell division possible or prevents the apoptosis pathway. Tumor-suppressor mutations associated with cancer are loss-of-function mutations because absence of a functional gene product fails in inhibiting cell division or in activating the apoptosis pathway.

**Problem 2** DNA from cells of a patient with retinoblastoma was analyzed using a Southern blot with a probe for a particular restriction fragment in the *RB1* gene. The result from nontumor cells and the results from cells taken from a tumor in each eye are shown in the accompanying diagram.

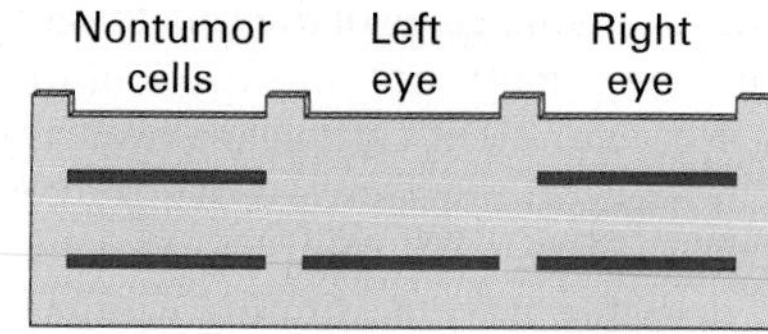


**(a)** Which band is associated with the mutant allele and which with the nonmutant allele?

**(b)** How is it possible for the bands from the tumor in the left eye to be different from those from the tumor in the right eye?

**(c)** Explain how cells in the tumor in the right eye can have a "loss of heterozygosity" even though the bands are indistinguishable from those observed from nontumor cells.

**Answer**

**(a)** Because both tumors contain the smaller band (the one nearer the bottom of the gel), this band is likely the one that is associated with the mutant *RB1* allele; the larger band is associated with the nonmutant allele.

**(b)** The loss of heterozygosity is independent in different tumors; hence the mechanism need not be the same in both eyes.

**(c)** This result can be explained if the wildtype allele has undergone a new mutation (or perhaps a small deletion or insertion), which inactivates the gene but does not detectably change the size of the band.

**Problem** In chronic myelogenous leukemia, white blood cells proliferate ceaselessly. In affected white blood cells, the *BCR-ABL* oncogene, the result of a gene fusion, transmits a constitutive growth signal. The severity of the disease correlates with the fraction of white blood cells that carry the *BCR-ABL* oncogene. One treatment for the disease is a drug that inhibits the BCR-ABL protein and prevents it from transmitting the growth signal. This can result in remission of the disease by preventing proliferation of the cancer cells. During treatment of the disease, the effectiveness of treatment is monitored by assessing the level of the *BCR-ABL* oncogene present in a fixed quantity of white blood cells. Suppose that the structure of the DNA in the nonmutant *BCR* and *ABL* genes are as shown here, and that of the *BCR-ABL* oncogene is as depicted.

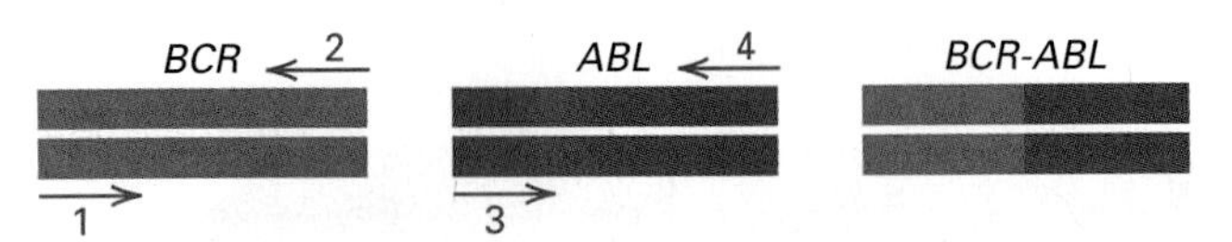


Short oligonucleotides with the polarities shown (with the 3′ end at the tip of each arrow) are labeled 1, 2, 3, and 4.

**(a)** How would you detect the presence of the *BCR-ABL* oncogene?

**(b)** How would you monitor the level of the cancer cells bearing the *BCR-ABL* oncogene relative to levels of the normal *BCR* and *ABL* proto-oncogenes?

**Answer**

**(a)** The presence of the oncogene could be detected by the polymerase chain reaction using primers 1 and 4.

**(b)** The relative levels of the cancer cells could be monitored by carrying out three PCR reactions using primers 1 + 2, 3 + 4, and 1 + 4. If done carefully, the amount of product in a PCR reaction is proportional to the amount of template in the original mixture. Primers 1 + 2 and 3 + 4 assay for the number of copies of *BCR* and *ABL*, respectively, among the white blood cells, whereas primers 1 + 4 assay for the number of copies of the *BCR-ABL* oncogene. The number of copies of *BCR-ABL* relative to *BCR* and *ABL* provides an estimate of the relative abundance of white blood cells that carry the oncogene.

## ANALYSIS AND APPLICATIONS

**15.1** What does it mean to say that a mutant allele in a protein-coding gene encodes a "poison subunit"? In a heterozygous genotype for such a mutation, if the protein subunits in a cell are synthesized in equal numbers and associate at random, what fraction of active protein molecules is expected if the functional form of the protein is a dimer? A trimer? A tetramer?

**15.2** What phenotype would you expect of a yeast strain carrying a temperature-sensitive mutation in a gene required for formation of the septum between mother cell and daughter cell:
**(a)** At the permissive temperature.
**(b)** At the restrictive temperature.

**15.3** A yeast cell carries a temperature-sensitive mutation in a gene necessary to repair double-stranded breaks in DNA. If cells were irradiated at the restrictive temperature, at what stage of the cell division cycle would you expect the mutant cells to accumulate?

**15.4** How do cyclin–CDK complexes exert their effects on cells? What counterpart class of enzymes reverses the effects of cyclin–CDK complexes?

**15.5** How does the normal retinoblastoma protein function to hold mammalian cells at the $G_1$ restriction point ("start")?

**15.6** Name two proteins that are targets of the anaphase-promoting complex (APC/C), and explain how this complex acts on these proteins.

**15.7** What cell-cycle checkpoints were highlighted in this chapter? What event or events cause each checkpoint to be activated to stop the cell division cycle?

**15.8** What protein is the major player in activating a DNA damage checkpoint? How is this protein normally kept from triggering the checkpoint? What happens to the protein when there is DNA damage?

**15.9** The protein GADD45, for which p53 is a transcription factor, reduces the processivity of the DNA polymerase. What does *processivity* mean? How does a reduction in processivity help in the repair of damaged DNA?

**15.10** What are the roles of Bax and Bcl2 proteins in programmed cell death? How is the balance in the amounts of these proteins affected by activated p53? By certain cellular oncogenes?

**15.11** If cancer is a "genetic disease," how can it be true that most cases are sporadic (that is, not familial)?

**15.12** Li-Fraumeni syndrome is a familial cancer syndrome in which affected relatives show early onset of a diverse set of tumors, including brain, breast, lung cancers, and sarcomas. Nonsense mutations, missense mutations, and mutations altering splice donor sites of the *p53* gene have been found segregating in diverse Li-Fraumeni families. In about half of the families showing the syndrome, no mutation in the *p53* coding regions or in splice donor or acceptor sites has been detected. The question is whether, in these families, the syndrome is due to changes in *p53* or to some other cause.

**(a)** How would you determine whether the mutations in such families are actually associated with the *p53* gene in any way?

**(b)** If the disease gene is *p53*-associated, what kinds of changes would you look for?

**(c)** Suppose that the evidence indicates that the segregating mutation is not linked to the *p53* gene. What other possible causes of the syndrome might you suggest? (Remember that inheritance of the syndrome is dominant.)

**15.13** You use a cloned cDNA encoding p53 to try to determine what kinds of *p53* mutations are present in several cancer patients by using Southern blot analysis. None of the patients has relatives with cancer, and the tumors are classified as sporadic. DNA is isolated from normal and tumor cells from the four patients, cleaved with *Not*I (an eight-base cutter), subjected to agarose gel electrophoresis, transferred to a nitrocellulose filter, and probed with radioactive *p53* cDNA. The result is depicted here.

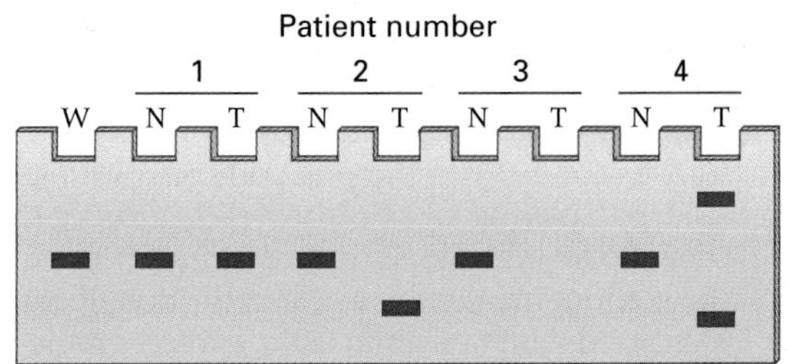


The lane labeled W is the pattern obtained from the wildtype *p53* gene; the other lanes contain DNA from nontumor (N) and tumor (T) tissue from four patients. Assuming that all four patients do carry *p53* mutations in their tumor cells, what kinds of mutations have occurred in the *p53* genes of these patients?

**15.14** Many types of cancer cells have defects in the $G_1/S$ checkpoint. These also tend to have abnormalities in chromosome number or structure. Why would chromosome abnormalities be expected in such cases?

**15.15** The p53 protein is defective in more than half of all cancers. Why might this be expected?

**15.16** What role does Ras–GTP play in intracellular signalling that makes it a proto-oncogene?

**15.17** Draw a diagram showing how recombination between homologous chromosomes during mitosis can result in a cell lineage with loss of heterozygosity for a *p53* mutation. What other genes also lose heterozygosity in this process?

**15.18** What does it mean to say that mutations in the retinoblastoma gene *RB1* are "dominant at the organismic level but recessive at the cellular level"?

**15.19** In familial retinoblastoma, there is an average of three retinal tumors per heterozygous carrier of the mutation. Assuming that the number of retinal cells at risk is $2 \times 10^6$ in each eye, and that each tumor results from an independent loss of heterozygosity, what is the estimated rate of loss of heterozygosity per cell? Should this be regarded as a "mutation rate"? Why or why not?

**15.20** Considering the data on familial retinoblastoma in the previous problem, what is the probability that an individual who inherits an *RB1* mutation will not develop any retinoblastomas? What is the "penetrance" of familial retinoblastoma?

**15.21** In patients with bilateral retinoblastoma, would the mechanism of loss of heterozygosity in tumors in different eyes be expected to be the same or different? Explain your answer.

**15.22** The p53 protein subunits contain a tetramerization domain and a DNA-binding domain. In order to serve as a functional transcription factor, all four subunits in the tetramer must be able to bind their target sites in the DNA.

**(a)** Would you expect a mutation that inactivates the tetramerization domain to be dominant or recessive at the organismal level? At the cellular level? Why?

**(b)** Would you expect a mutation that inactivates the DNA-binding domain to be dominant or recessive at the organismal level? At the cellular level? Why?

**15.23** The gel diagram here shows the pattern of bands observed in a Southern blot for a 12-kb *Eco*RI restriction fragment from a nonmutant *RB1*$^+$ allele and a mutant allele *RB1*$^-$ in which there is an internal 10- kb deletion in the *Eco*RI fragment. A scale showing the electrophoretic mobility of DNA fragments of various sizes appears at the right. Show what pattern of bands would be expected in:

**(a)** Normal retinal cells from a heterozygous *RB1*$^+$/*RB1*$^-$ individual.

**(b)** Cells of retinoblastoma tumors caused by loss of the wildtype *RB1*$^+$-bearing chromosome.

**(c)** Cells of retinoblastoma tumors caused by mitotic recombination.

**(d)** Cells of retinoblastoma tumors caused by a new missense substitution in *RB1*$^+$.

**(e)** Cells of retinoblastoma tumors caused by a 6-kb deletion in the *Eco*RI fragment (not including the region that hybridizes with the probe).

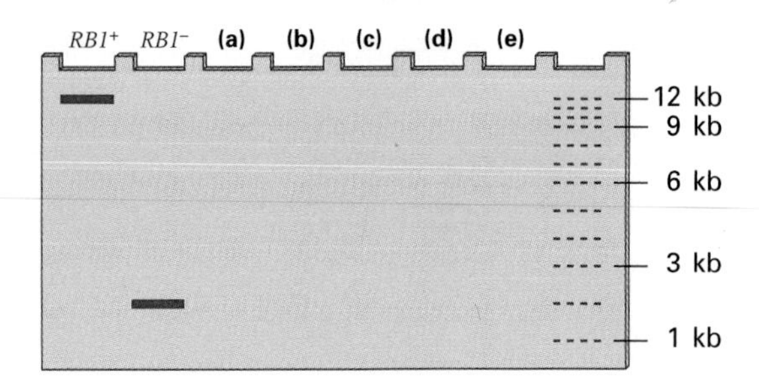

**15.24** The pedigree shown below includes individuals affected with adenomatous polyposis, and the diagram of the gel shows a restriction fragment from a number of alleles of *APC*, mutant forms of which are associated with this cancer. Four sizes of restriction fragments (a through d) are observed. Individuals in generations I and II are old enough that they will have developed the cancer if they carry a mutant *APC* allele, but the individuals in generation III are all too young to have developed the disease. Identify the high-risk individuals in generation III and those who are not at risk. (Note that a mutant allele and a nonmutant allele can yield the same size restriction fragment.)

**15.25** Mutagenesis of a *RAS* gene of budding yeast yields a temperature-sensitive conditional mutation.

- **(a)** Would you expect a cell that is carrying a mutation that prevents Ras from exchanging GTP for GDP at 36°C to continue to divide at the restrictive temperature?
- **(b)** Would this mutation be dominant or recessive?
- **(c)** Would you expect a cell that is carrying a mutation that inactivates the GTPase activity of Ras at 36°C to continue to divide at the restrictive temperature?
- **(d)** Would this mutation be dominant or recessive?

**15.26** You are studying a particular leukemia in humans that is associated with a translocation between chromosomes 8 and 14. How would you determine experimentally whether the translocation is likely to be of the promoter-fusion or the gene-fusion type?

**15.27** Human papilloma virus (HPV) is present in greater than 90 percent of cervical cancers. HPV encodes two proteins, E6 and E7, that are potent contributors to its ability to induce tumors. E7 is known to disable RB; E6 binds to p53 and targets it for degradation. Discuss how these activities might contribute to the development of the cancerous state in infected cells.

**15.28** Among 24 kindreds that manifested a syndrome of sarcoma, breast cancer, and other cancers in young patients, and that showed autosomal dominant inheritance, one family had the following pedigree. If inheritance is due to an autosomal dominant gene, which individuals must be heterozygous for the disease allele but nevertheless be cancer-free?

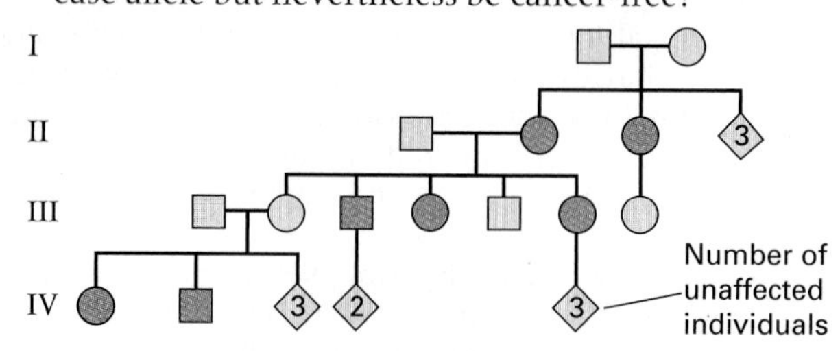


**15.29** Mutant *cdc9* alleles of budding yeast are defective for DNA ligase activity at 36°C but have normal activity at 23°C. If *cdc9* cells grown at 23°C are transferred to 36°C, they arrest at the $G_2$/M boundary. Upon return to 23°C, they resume the cell cycle after a delay during which they ligate all the nicks in the newly synthesized DNA. The cells are viable. Mutant *rad9* alleles are defective in the $G_2$/M DNA checkpoint. The *rad9 cdc9* double mutants grow normally at 23°C. If transferred to 36°C, the cells continue through mitosis. Upon plating at 23°C, however, most of the cells prove to be inviable. If, prior to transfer to 36°C and throughout incubation at 36°C, the double mutant is incubated with benzimidazole, a microtubule poison akin to colchicine, then upon return to 23°C accompanied by washout of the benzimidazole, the cells behave like the *cdc9* mutant cells. Propose an explanation for these observations.

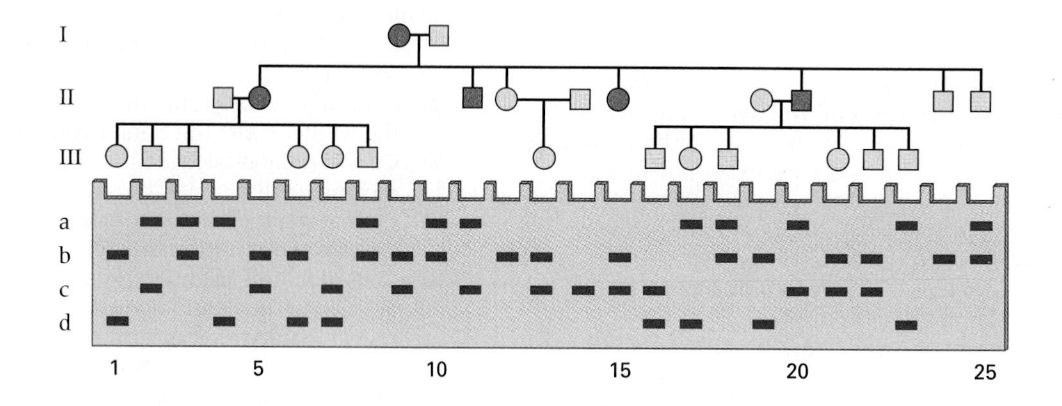

**15.30** The properties of the *rad9 cdc9* double mutants led Weinert and Hartwell to investigate whether other *cdc rad9* double mutants continued to cycle and lose viability upon transfer to 36°C, in order to determine the specificity of the *RAD9* checkpoint. The results of these analyses are summarized in the table below:

| Double mutant | Function of the CDC gene | Dies rapidly at 36°C? |
|---|---|---|
| *cdc28 rad9* | Cyclin-dependent kinase | No |
| *cdc7 rad9* | Spindle pole body duplication (centrosome) | No |
| *cdc8 rad9* | Thymidylate kinase (TTP synthesis) | No |
| *cdc6 rad9* | ORC component | No |
| *cdc2 rad9* | DNA polymerase subunit | Yes |
| *cdc13 rad9* | Telomere-binding protein | Yes |
| *cdc17 rad9* | DNA polymerase subunit | Yes |
| *cdc16 rad9* | Anaphase-promoting complex (APC/C) component | No |
| *cdc20 rad9* | Required for exit from mitosis | No |
| *cdc23 rad9* | APC/C component | No |
| *cdc15 rad9* | Required for degradation of B cyclins | No |

What would you deduce about *RAD9*'s role in the several checkpoints in the yeast cell cycle?

## CHALLENGE PROBLEMS

**Challenge Problem 1** The human oncogene *MYC* in chromosome 8 (near 8q24) was first discovered in association with Burkitt lymphoma. The diagram illustrates the result of a Southern blot obtained when DNA from noncancer cells (the control) was compared with various dilutions (up to 1000-fold) of DNA isolated from an equivalent number of cancer cells from the same patient. The blot was hybridized with a labeled probe for *MYC*. The dilutions clearly differ in the intensity (darkness) of the hybridization signal. Propose a hypothesis to account for these results.

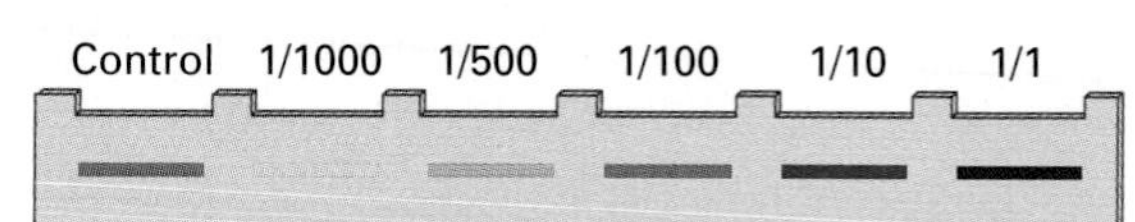


**Challenge Problem 2** The cDNAs for mammalian Cdk2 and Cdc2 can substitute for the *CDC28* gene in budding yeast. Design an experiment to test whether the cDNAs for human cyclin B and cyclin D can substitute for Clb2 in budding yeast.

**Challenge Problem 3** You are given a diploid strain of budding yeast and are told that it is homozygous for a mutation that inactivates a checkpoint. For chromosome V, the strain has the genotype

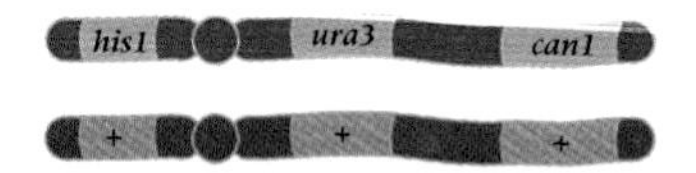


The *can1* mutation results in recessive resistance to canavanine, *ura3* results in a recessive requirement for uracil and resistance to 5-fluoro-orotic acid, and *his1* results in a recessive requirement for histidine.

**(a)** What property would you expect the strain to exhibit (for these genetic markers) if it were homozygous for a mutation that inactivates the spindle checkpoint?

**(b)** What property would you expect the strain to show if it were homozygous for a mutation affecting a DNA-damage checkpoint?

### GENETICS on the web

GeNETics on the Web will introduce you to some of the most important sites for finding genetics information on the Internet. To explore these sites, visit the Jones and Bartlett companion site to accompany *Genetics: Analysis of Genes and Genomes, Seventh Edition* at http://biology.jbpub.com/book/genetics.

There you will find a chapter-by-chapter list of highlighted keywords. When you select one of the keywords, you will be linked to a Web site containing information related to that keyword.

# CHAPTER 16

# Mitochondrial DNA and Extranuclear Inheritance

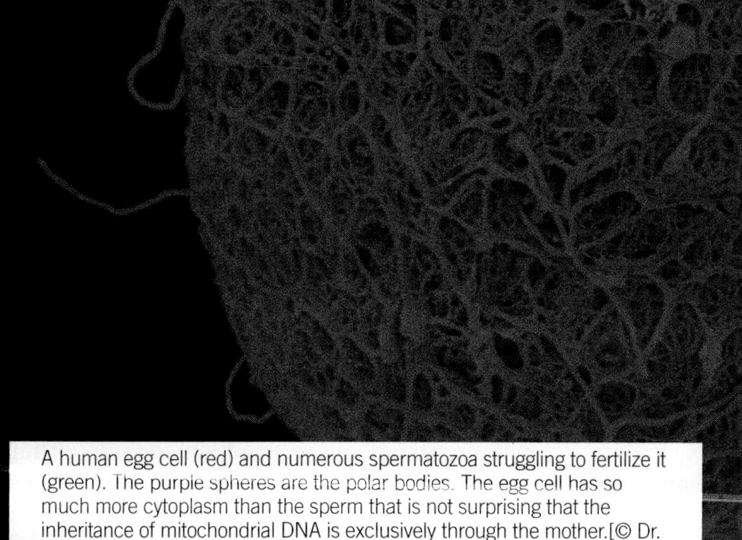

A human egg cell (red) and numerous spermatozoa struggling to fertilize it (green). The purple spheres are the polar bodies. The egg cell has so much more cytoplasm than the sperm that is not surprising that the inheritance of mitochondrial DNA is exclusively through the mother.[© Dr. Y. Nikas/Phototake, Inc./Alamy Images.]

## CHAPTER OUTLINE


**16.1** Patterns of Extranuclear Inheritance
- Mitochondrial Genetic Diseases
- Heteroplasmy
- Maternal Inheritance and Maternal Effects
- Tracing Population History Through Mitochondrial DNA

**16.2** Organelle Heredity
- RNA Editing
- The Genetic Codes of Organelles
- Leaf Variegation in Four-O'Clock Plants
- Drug Resistance in *Chlamydomonas*
- Respiration-Defective Mitochondrial Mutants
- Cytoplasmic Male Sterility in Plants

**16.3** The Evolutionary Origin of Organelles

**16.4** Cytoplasmic Transmission of Symbionts

**16.5** Maternal Effect in Snail Shell Coiling

CONNECTION *Chlamydomonas* Moment
Ruth Sager and Zenta Ramanis 1965
*Recombination of Nonchromosomal Genes in* Chlamydomonas

CONNECTION A Coming Together
Lynn Margulis (formerly Lynn Sagan) 1967
*The Origin of Mitosing Cells*

Although most hereditary traits in eukaryotes are determined by genes in the nucleus, some are not. The exceptions include traits determined by the DNA in mitochondria and in chloroplasts. These are self-replicating cytoplasmic **organelles** specialized for respiration (mitochondria) or for photosynthesis (chloroplasts). Each of these organelles contains its own DNA that codes for numerous proteins and RNA molecules. The DNA is replicated within the organelles and transmitted to daughter organelles as they form. The organelle genetic system is separate from that of the nucleus, and traits determined by organelle genes show patterns of inheritance quite distinct from those of nuclear genes.

Many organisms also contain intracellular parasites or symbionts, including cytoplasmic bacteria, viruses, and other elements. In some cases, such cytoplasmic entities confer inherited phenotypic traits on the infected cell or organism. Organelle heredity and other examples of the diverse phenomena that make up **extranuclear inheritance** (also called **cytoplasmic inheritance**) are presented in this chapter.

## 16.1 Patterns of Extranuclear Inheritance

No criterion other than a non-Mendelian pattern of inheritance is universally applicable to distinguish extranuclear from nuclear inheritance. In higher organisms, extranuclear inheritance is usually indicated by **uniparental inheritance**, which means transmission through only one parent. Genetic transmission of cytoplasmic factors, such as mitochondria or chloroplasts, is determined by maternal and paternal contributions at the time of fertilization, by mechanisms of elimination from the zygote, and by the irregular sorting of cytoplasmic elements during cell division. Uniparental transmission through the mother constitutes **maternal inheritance**; uniparental transmission through the father is **paternal inheritance**.

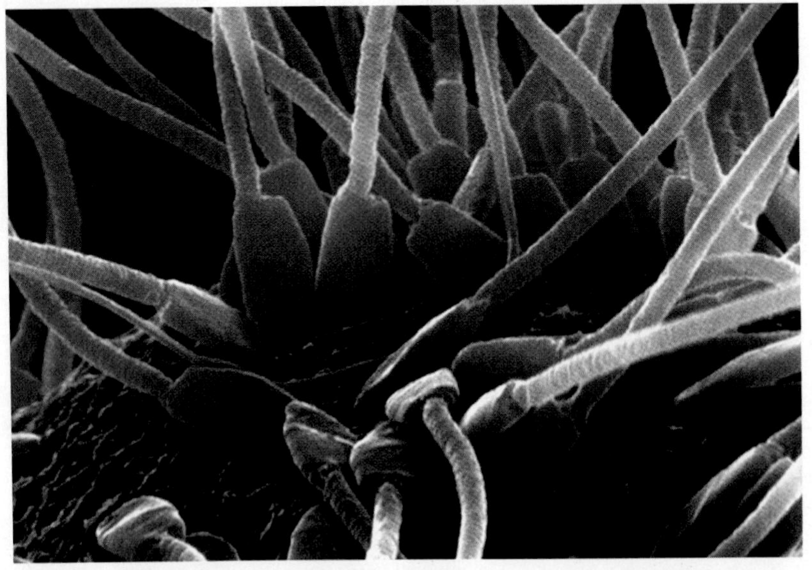

**A HUMAN SPERM** and egg. The volume of the egg cell is about 5000 times the volume of the sperm head and contributes virtually all of the cytoplasm to the zygote, including the mitochondria. [© David M. Phillips/Science Source/Photo Researchers, Inc.]

### Mitochondrial Genetic Diseases

In higher animals, mitochondria are typically inherited through the mother because the egg is the major contributor of cytoplasm to the zygote. Therefore, mitochondria usually show maternal inheritance; the progeny of a mutant mother and a normal father are mutant, whereas the progeny of a normal mother and a mutant father are normal.

The inheritance of mitochondrial DNA (mtDNA) can be tracked through the use of DNA polymorphisms of the type illustrated in **FIGURE 16.1**. Part A shows the positions of cleavage of human mtDNA by a restriction enzyme, in this example *Hae*II. When mtDNA from different individuals is cleaved with this restriction enzyme, the cleavage products include either one fragment of 8.6 kb or two smaller fragments of 4.5 kb and 4.1 kb. The pattern with two smaller fragments is more common than the pattern with one larger fragment. A number of other *Hae*II fragments are also observed (part B), but these are the same size in the mtDNA molecules from different individuals.

Maternal inheritance mtDNA is indicated in **FIGURE 16.2**. In the gel diagram in part A, the polymorphism of one large (red) versus two small *Hae*II fragments (blue) is indicated, whereas the other bands that do not differ in size among individuals are not shown. The pedigree shown in part B has the pedigree symbols color coded in red or blue according to the absence or presence of the polymorphic *Hae*II site. The maternal inheritance of the mtDNA type is clear. The characteristic feature is that females (I-3, I-5, and II-8) transmit their mtDNA to *all* of their progeny, whereas males (I-2, II-7, and II-10) transmit their mtDNA to *none* of their progeny.

Although the mutation in the *Hae*II site yielding the 8.6-kb fragment is not associated with any disease, a number of other mutations in mitochondrial DNA do cause diseases and exhibit similar patterns of mitochondrial inheritance. Most of these conditions decrease

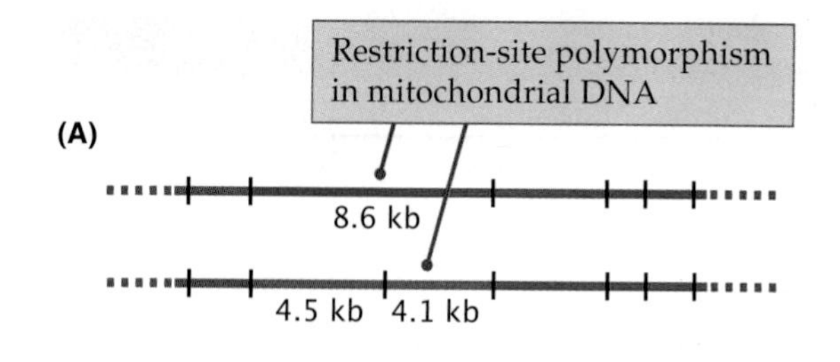


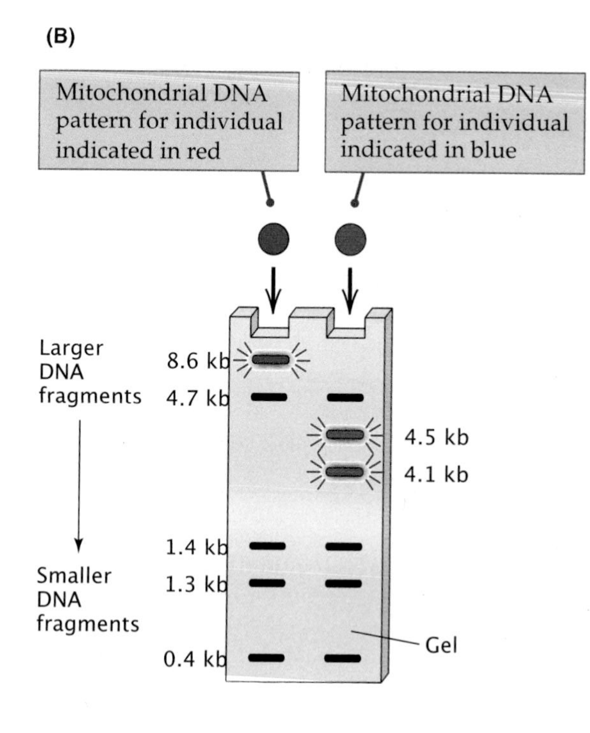


**FIGURE 16.1** Gel patterns resulting from presence or absence of a *Hae*II restriction site in human mitochondrial DNA. (A) Diagram of mtDNA showing the locations of the *Hae*II sites. The mtDNA molecule is diagrammed as a linear fragment, but the dots at the ends indicate that the molecule is actually circular. In one type of mtDNA molecule, a particular HaeII site is missing, and cleavage with *Hae*II results in an 8.6-kb DNA fragment (red). In another type of mtDNA the site is present, and cleavage results in a 4.5-kb fragment and a 4.1-kb fragment (blue). (B) Separation of *Hae*II fragments from mtDNA in a gel, highlighting the 8.6-band (red) versus the 4.5-kb plus 4.1-kb band (blue) polymorphism. The bands resulting from cleavage at other *Hae*II sites are indicated in black.

the ATP-generating capacity of the mitochondria and affect the function of muscle and nerve cells, particularly in the central nervous system, leading to blindness, deafness, or stroke. **TABLE 16.1** lists some of the human diseases caused by mitochondrial mutations. Many of these conditions are lethal unless some of the mitochondria are normal, and there is variable expressivity because the proportions of normal and mutant mitochondria differ among affected persons.

An example of a human pedigree for inheritance of a mitochondrially inherited disease is shown in **FIGURE 16.3**. This disease, which is known as MERRF (myoclonic epilepsy associated with ragged-red muscle fibers), affects the central nervous system and skeletal muscle. The cause is a mutation in a $tRNA^{Lys}$ gene encoded in

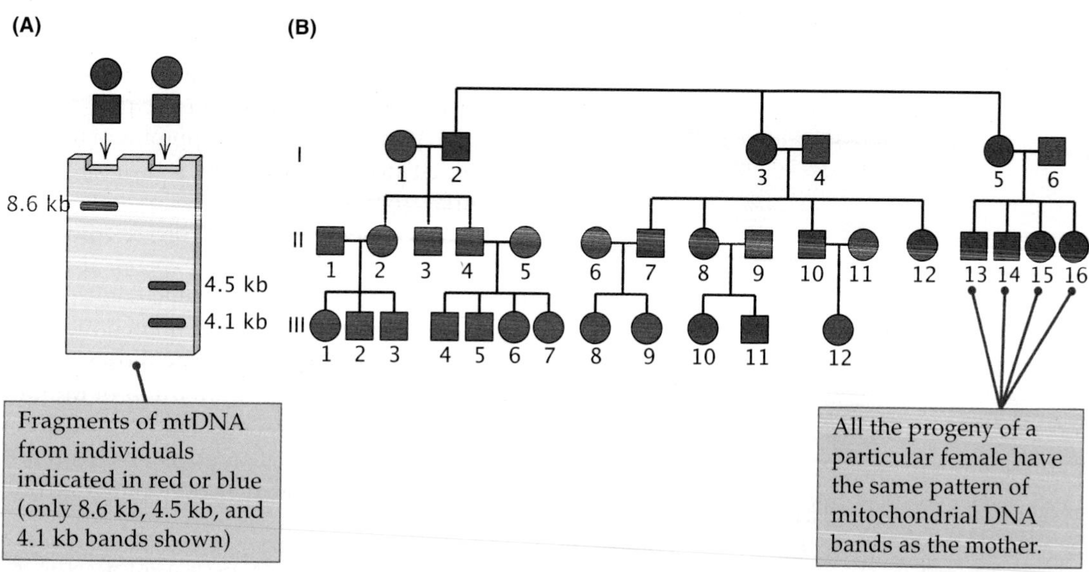


**FIGURE 16.2** Maternal inheritance of human mitochondrial DNA. (A) Bands in a gel highlighting the *Hae*II restriction-site polymorphism illustrated in Figure 16.1. Absence of the *Hae*II site results in a single band of 8.6-kb, whereas presence of the site results in bands of 4.5 kb and 4.1 kb. Other bands resulting from HaeII digestion are not shown. (B) Pedigree showing maternal inheritance of the DNA pattern with the 8.6-kb fragment (red pedigree symbols) versus the 4.5 kb plus 4.1 kb fragments (blue pedigree symbols). The mitochondrial DNA type is transmitted only through the mother. [Adapted from D. C. Wallace, *Trends Genet.* 5 (1989): 9–13.]

**Table 16.1 Phenotypes associated with some mitochondrial mutations**

| Nucleotide changed | Mitochondrial component affected | Phenotype[a] |
|---|---|---|
| 3460 | ND1 of Complex I[b] | LHON |
| 11778 | ND4 of Complex I | LHON |
| 14484 | ND6 of Complex I | LHON |
| 8993 | ATP6 of Complex V[b] | NARP |
| 3243 | $tRNA^{Leu(UUR)}$[c] | MELAS, PEO |
| 3271 | $tRNA^{Leu(UUR)}$ | MELAS |
| 3291 | $tRNA^{Leu(UUR)}$ | MELAS |
| 3251 | $tRNA^{Leu(UUR)}$ | PEO |
| 3256 | $tRNA^{Leu(UUR)}$ | PEO |
| 5692 | $tRNA^{Asn}$ | PEO |
| 5703 | $tRNA^{Asn}$ | PEO, myopathy |
| 5814 | $tRNA^{Cys}$ | Encephalopathy |
| 8344 | $tRNA^{Lys}$ | MERRF |
| 8356 | $tRNA^{Lys}$ | MERRF |
| 9997 | $tRNA^{Gly}$ | Cardiomyopathy |
| 10006 | $tRNA^{Gly}$ | PEO |
| 12246 | $tRNA^{Ser(AGY)}$[c] | PEO |
| 14709 | $tRNA^{Glu}$ | Myopathy |
| 15923 | $tRNA^{Thr}$ | Fatal infantile multisystem disorder |
| 15990 | $tRNA^{Pro}$ | Myopathy |

[a]LHON Leber's hereditary optic neuropathy; NARP Neurogenic muscle weakness, ataxia, retinitis pigmentosa; MERRF Myoclonic epilepsy and ragged-red fiber syndrome; MELAS Mitochondrial myopathy, encephalopathy, lactic acidosis, stroke-like episodes; PEO Progressive external ophthalmoplegia.
[b]Complex I is NADH dehydrogenase. Complex V is ATP synthase.
[c]In $tRNA^{Leu(UUR)}$, the R stands for either A or G; in $tRNA^{Ser(AGY)}$, the Y stands for either T or C.

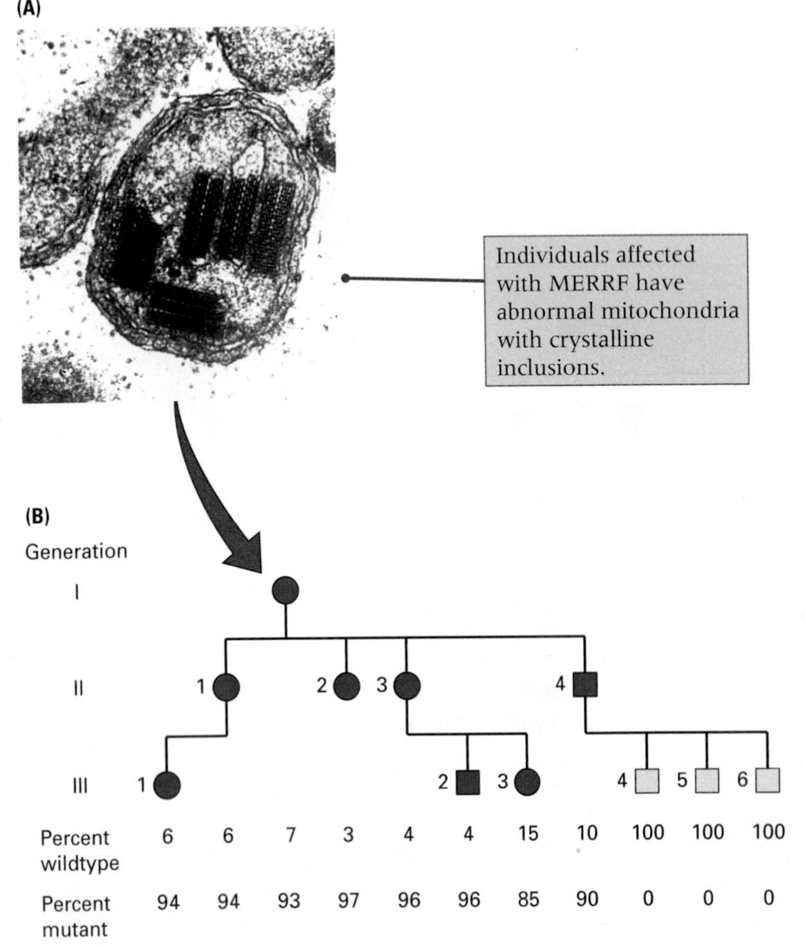


mitochondrial DNA (mtDNA). **FIGURE 16.4** shows the organization of genes in human mtDNA. The $tRNA^{Lys}$ mutation is highly pleiotropic because it affects the synthesis of all proteins encoded in the mitochondrial DNA in all cells. Like many mitochondrial mutations, MERRF is lethal in the absence of normal mitochondria. All of the individuals indicated by red symbols in the pedigree in Figure 16.3B manifest some symptoms of the disease, which include hearing loss, seizures, fatigue, dementia, tremors, and jerkiness of movement. The affected individuals in the pedigree differ considerably in the symptoms they express. The variable expressivity is due to differences in the proportions of normal and mutant mitochondria in the cells of affected persons. Those most severely affected have the highest proportion of mutant mitochondria; those least affected have the lowest, as indicated in Figure 16.3B. An electron micrograph of an abnormal mitochondrion from one

**FIGURE 16.3** Inheritance of myoclonic epilepsy with ragged-red fiber disease (MERRF) in humans. (A) Electron micrograph of an abnormal MERRF mitochondrion containing paracrystalline inclusions. (B) The pedigree shows inheritance of MERRF in one family and the percentage of the mitochondria in each person found to be wildtype or mutant. [Reproduced from J. M. Shoffner, et al., "Myoclonic epilepsy and ragged–red fiber disease. . .," *Cell* 61 (1990): 931–937, © 1990, with permission from Elsevier, http://www.sciencedirect.com/science/journal/00928674]

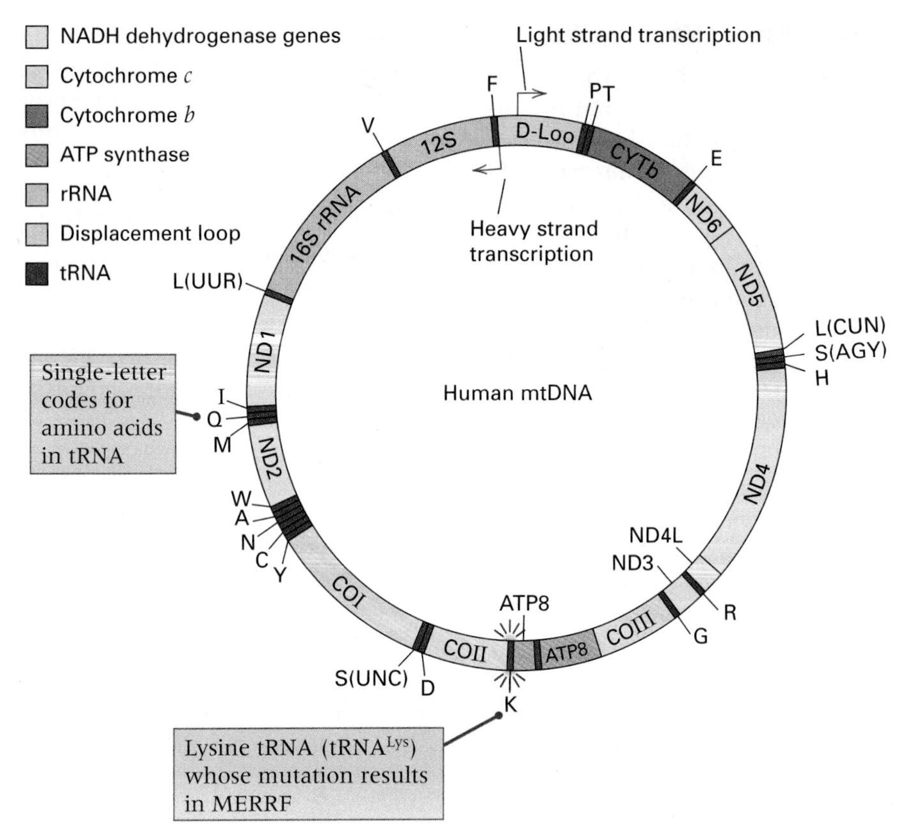


**FIGURE 16.4** Genes in human mitochondrial DNA. The tRNA genes are indicated by the one-letter amino acid symbols; hence $tRNA^{Lys}$ is denoted K. The positions of these and other genes in the mitochondrial DNA are indicated by color according to the key at the upper left. The arrows indicate the promoters for transcription of the heavy and light strands. [Reprinted with permission from the *Annual Review of Genetics*, Volume 29, © 1995 by Annual Reviews [www.AnnualReviews.org.]. Illustration courtesy of David A. Clayton, Howard Hughes Medical Institute.]

of the affected patients is shown in Figure 16.3A. It contains numerous paracrystalline inclusion bodies formed from abnormal protein aggregates.

Consideration of the phenotypes caused by the mutations listed in Table 16.1 and the variable expressivity seen in pedigrees of mitochondrial diseases indicates that mutations in different mitochondrial genes can result in similar phenotypes and that mutations in the same mitochondrial gene can result in different phenotypes. Some of the variation appears to arise from differences among affected persons in the ratio of mutant to wildtype mitochondria, as seen in Figure 16.3B. Another source of variation results from differences in the structural and metabolic consequences of each individual mutation.

## Heteroplasmy

The condition in which two or more genetically different types of mitochondria (or other organelle) are present in the same cell is known as **heteroplasmy**. Heteroplasmy of mitochondria is unusual among animals. For example, in a typical human cell containing 1000–10,000 mitochondria, only one mtDNA sequence is usually represented.

The rarity of heteroplasmy made it possible to identify the remains of the last Russian tsar, Nicholas II, who, together with his family and servants, was secretly executed by a Bolshevik firing squad on the night of July 16, 1918. Their bodies were hidden in a shallow grave. DNA analysis of bone samples exhumed in 1992 and thought to be from members of the murdered family resulted in identification of nine bodies, five of whom were related. The mother, Tsarina Alexandra, and three of her daughters were identified because their mitochondrial DNA was identical to that of living relatives. The father, Tsar Nicholas, proved to be heteroplasmic for mitochondrial DNA. To establish this identification, the body of the tsar's brother, Georgij, who had died of tuberculosis in 1899, was exhumed. The mitochondrial DNA sequence of Georgij Romanov not only matched that of the tsar, but was also heteroplasmic at the identical base pair. The two brothers differed in the ratios of the two mitochondrial types: Georgij Romanov had about 62 percent T and 38 percent C at the heteroplasmic base pair, whereas Nicholas had 28 percent T and 72 percent C.

## ■ Maternal Inheritance and Maternal Effects

A maternal pattern of inheritance usually indicates extranuclear inheritance, but not always. The difficulty is in distinguishing *maternal inheritance* from *maternal effects.* A **maternal effect** is an influence of the mother's nuclear genotype on the phenotype of the progeny. Such effects may be mediated either through substances present in the egg that affect early development or through nurturing. Examples of developmental and nurturing effects are, respectively, the intrauterine environment of female mammals and their ability to produce milk. The distinction between maternal inheritance and maternal effects is as follows:

1. In *maternal inheritance,* the hereditary determinants of a trait are extranuclear, and genetic transmission is only through the maternal cytoplasm; Mendelian segregation is not observed.
2. In a *maternal effect,* the nuclear genotype of the mother determines the phenotype of the progeny. The hereditary determinants are nuclear genes transmitted by both sexes and, in suitable crosses, undergo Mendelian segregation.

In higher plants, organelle inheritance is often uniparental, but the particular type of uniparental inheritance depends on the organelle and the organism. For example, among angiosperms, mitochondria typically show maternal inheritance, but chloroplasts may be inherited maternally (the predominant mode), paternally, or from both parents, depending on the species. In conifers, chloroplast DNA is usually inherited paternally, but a few offspring result from maternal transmission. The redwood *Sequoia sempervirens* shows paternal transmission of both mitochondrial and chloroplast DNA. Whatever the pattern of cytoplasmic transmission, most species have a few exceptional progeny, which indicates that the predominant mode of transmission is not absolute.

**FIGURE 16.5** A small sample of approximately 150 registered breeds of dogs. [© Carolyn A. McKeone/Photo Researchers, Inc.]

## ■ Tracing Population History Through Mitochondrial DNA

Mitochondrial DNA has a number of features that make it useful for studying the genetic relationships among organisms. One important feature of mtDNA is that, in many organisms including humans, the molecule does not undergo genetic recombination. The absence of recombination implies that the DNA molecule in any mitochondrion derives from a single mtDNA molecule present in an ancestor. In humans and other organisms with maternal transmission, this ancestor would have to be female. The absence of recombination also means that mtDNA contains a great deal of genetic information about ancestry, because any mutations that occur in a mtDNA lineage are inherited together.

In Chapter 17 we shall see how variation in nucleotide sequences in human mtDNA has revealed the ancestral history and patterns of migration among modern human populations. This approach is also useful for tracing the ancestry or many other organisms, including our best friend, the dog (**FIGURE 16.5**). About one third of all U.S. households have at least one dog, totaling more than 50 million animals. About half of these are mixed breeds, and the rest are any of about 150 pure breeds currently recognized by the American Kennel Club. Each breed has its own combination of physical attributes and a characteristic temperament. Perennial favorites are Labrador Retrievers, Golden Retrievers, German Shepherds, Beagles, Dachshunds, Yorkshire Terriers, Boxers, Poodles, Chihuahuas, and Shih Tzus. Studies of mitochondrial DNA sequences from 67 breeds and 162 wild wolves demonstrate a very close genetic relationship of dogs to wolves. The results suggest that domestication began at least 100,000 years ago. The data imply at least two independent events of domestication as well as repeated episodes of genetic admixture resulting from mating between dogs and wolves during the period when domestication was taking place.

# 16.2 Organelle Heredity

As noted earlier, mitochondria are respiratory organelles, and chloroplasts are photosynthetic organelles. The DNA of these organelles is usually in the form of supercoiled circles of double-

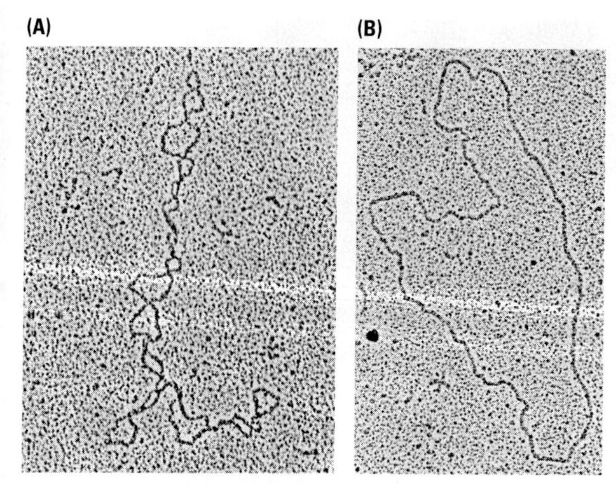


**FIGURE 16.6** Mitochondrial DNA from rat liver. (A) Native supercoiled configuration. (B) Relaxed configuration produced by a nick in the DNA strand in the process of isolation. The size of each molecule is 16,298 nucleotide pairs. [Courtesy of David R. Wolstenholme, University of Utah.]

stranded molecules (**FIGURE 16.6**). The chloroplast DNA in most plants ranges in size from 120 to 160 kb. Mitochondrial genomes are usually smaller and have a greater size range. For example, the mitochondrial genome in mammals is about 16.5 kb, that in *D. melanogaster* about 18.5 kb, and that in some higher plants more than 100 kb. The mitochondrial genomes of higher plants are exceptional in consisting of two or more circular DNA molecules of different sizes. Several copies of the genome are present in each organelle. There are typically from 2 to 10 copies in mitochondria and from 20 to 100 copies in chloroplasts. In addition, most cells contain multiple copies of each organelle.

Organelle genes code for DNA polymerases that replicate the organelle DNA. They also code for other components essential to their function or to replication, but many organelle components are determined by nuclear genes. These components are synthesized in the cytoplasm and imported into the organelle. Mitochondrial DNA contains a relatively small number of genes. For example, both the human mitochondrial genome and that of yeast contain approximately 40 genes. Chloroplast genomes are generally larger than those of mitochondria and contain more genes. Many genes in mitochondria and chloroplasts code for the ribosomal RNA and transfer RNA components used in protein synthesis. **FIGURE 16.7** shows the genetic functions of the genes in the 120-kb chloroplast genome of the liverwort *Marchantia polymorpha*.

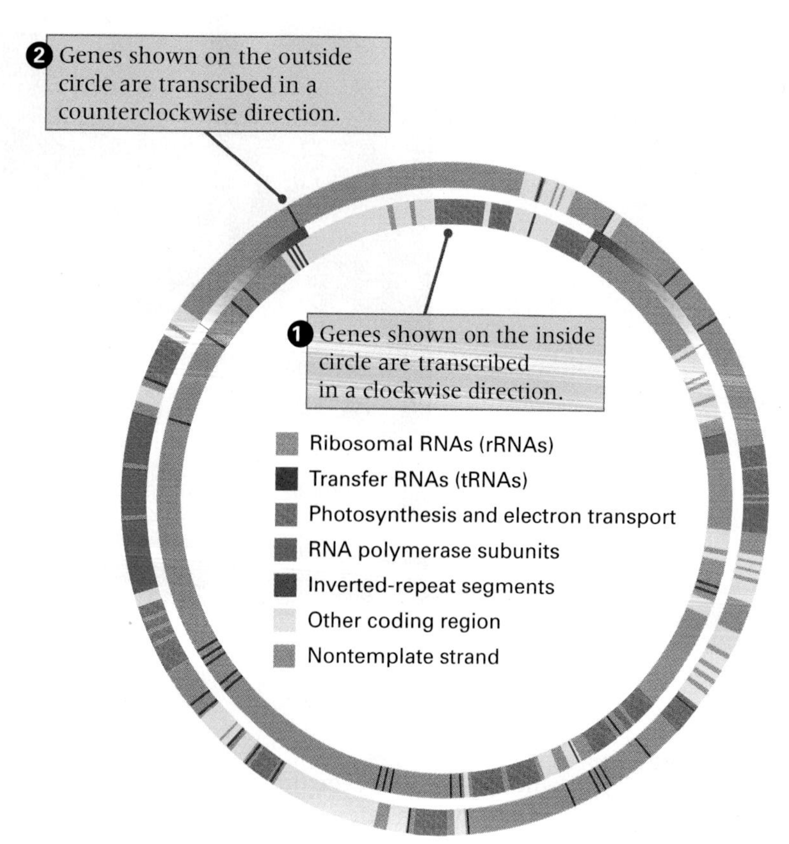


**FIGURE 16.7** Organization of genes in the chloroplast genome of the liverwort *Marchantia polymorpha*. The large inverted-repeat segments (green) that include the genes coding for ribosomal RNA (light brown) are a feature of many other chloroplast genomes. The length of the entire molecule is 121,024 nucleotide pairs. [Data from K. Ohyama, et al., *Nature* 322 (1986): 572–574.]

## RNA Editing

The mRNAs of mitochondria and chloroplasts often are modified posttranscriptionally in processes called **RNA editing**. The mechanisms of modification as well as the predominant patterns of modification differ between plants and animals. In most land plants, except algae and mosses, mRNAs may be extensively modified by conversion of cytosine residues to uracils. As a result of RNA editing, a CGG codon for arginine present in an organelle gene is changed into a UGG codon for tryptophan in the mRNA. The modification appears to be accomplished by deamination of selected cytosine residues in the mRNA, leaving uracil residues in their places. In the mitochondria of the protozoans that cause sleeping sickness (*Trypanosoma brucei*), the insertion or deletion of uridine residues is directed by a guide RNA in a process that resembles mRNA splicing (Chapter 10). Whatever the mechanism, the outcome is an mRNA whose sequence may differ substantially from that of the DNA from which it was transcribed.

## The Genetic Codes of Organelles

Translation in organelles follows the standard decoding process in which the organelle tRNA molecules translate codon by codon along

**FIGURE 16.8** Leaf variegation in *Mirabilis jalapa*. The sectors result from cytoplasmic segregation of normal (green) and defective (white) chloroplasts.

the mRNA. However, the genetic code in animal mitochondria often differs somewhat from that in nuclear genes, prokaryotes, and archaeans. Human mitochondria depart from the standard code in three principal ways:

1. The UGA codon is not a stop codon but encodes tryptophan instead.
2. The AGA and AGG codons are stop codons rather than arginine codons.
3. AUA encodes methionine rather than isoleucine.

The genetic code in yeast mitochondria is identical to the standard code except that UGA is a tryptophan codon and all four codons of the form CUN code for threonine rather than leucine. In animals, the mitochondrial genetic code can differ slightly from one species to the next, and some such codes are identical to the standard code. In plants, the standard genetic code is used in both mitochondria and chloroplasts.

Because the structure and function of mitochondria and chloroplasts depend on both nuclear and organelle genes, it is sometimes difficult to distinguish these contributions. Another complication is that numerous mitochondria or chloroplasts are present in a zygote, but a variable number of them may contain a mutation. The result is that traits determined by organelle genes exhibit a pattern of transmission very different from simple Mendelian traits determined entirely by nuclear genes. Some examples of traits determined by chloroplast and mitochondrial genes are examined in the following sections.

### Leaf Variegation in Four-O'Clock Plants

Chloroplast transmission accounts for the unusual pattern of inheritance observed with leaf variegation in a strain of the four-o'clock plant, *Mirabilis jalapa*. Variegation refers to the appearance of white regions in the leaves and stems that result from the lack of green chlorophyll (**FIGURE 16.8**). On the variegated plants, some branches are completely green, some completely white, and others variegated. Branches of all three types produce flowers, and these flowers can be used to perform nine possible crosses (**TABLE 16.2**). From each cross, the seeds are collected and planted, and the phenotypes of the progeny are examined. The results of the crosses are summarized in Table 16.2. Two significant observations are that:

1. Reciprocal crosses yield different results. For example, the cross

   green ♀ × white ♂

   yields green plants, whereas the cross

   white ♀ × green ♂

   yields white plants. (These usually die shortly after germination because they lack chlorophyll and hence photosynthetic activity.)
2. The phenotype of the branch bearing the female parent in each case determines the phenotype of the progeny (compare columns 1 and 3 in Table 16.2). The male makes no contribution to the phenotype of the progeny.

Furthermore, when flowers present on either the green or the variegated progeny plants are used in subsequent crosses, the patterns of transmission are identical with those of the original crosses. These observations suggest direct transmission of the trait through the mother, or maternal inheritance.

The genetic explanation for the variegation is as follows:

- Green color depends on the presence of chloroplasts, and pollen contains no chloroplasts.
- Segregation of chloroplasts into daughter cells is determined by cytoplasmic division and is somewhat irregular.
- The cells of white tissue contain mutant chloroplasts that lack chlorophyll.

These points are depicted in the model in **FIGURE 16.9**. Variegated plants germinate from embryos that are heteroplasmic for green and white (mutant) chloroplasts. Random segregation of the chloroplasts in cell division results in some branches derived from cells that contain only green chloroplasts and some branches derived from cells that contain only mutant chloroplasts. The flowers on green branches produce ovules with normal chloroplasts, so all progeny are green. The flowers on white branches produce ovules with only mutant chloroplasts, so only white progeny are formed. The ovules formed by flowers on variegated branches are of three types: those with only chlorophyll-containing chloroplasts, those with only mutant chloroplasts, and those that are heteroplasmic. The heteroplasmic class again yields variegated plants.

**Table 16.2 Crosses and progeny phenotypes in variegated four-o'clock plants**

| Phenotype of branch bearing egg parent | Phenotype of branch bearing pollen parent | Phenotype of progeny |
|---|---|---|
| white | white | white |
| white | green | white |
| white | variegated | white |
| green | white | green |
| green | green | green |
| green | variegated | green |
| variegated | white | variegated, green, or white |
| variegated | green | variegated, green, or white |
| variegated | variegated | variegated, green, or white |

## Drug Resistance in *Chlamydomonas*

Cytoplasmic inheritance has been studied extensively in the unicellular green alga

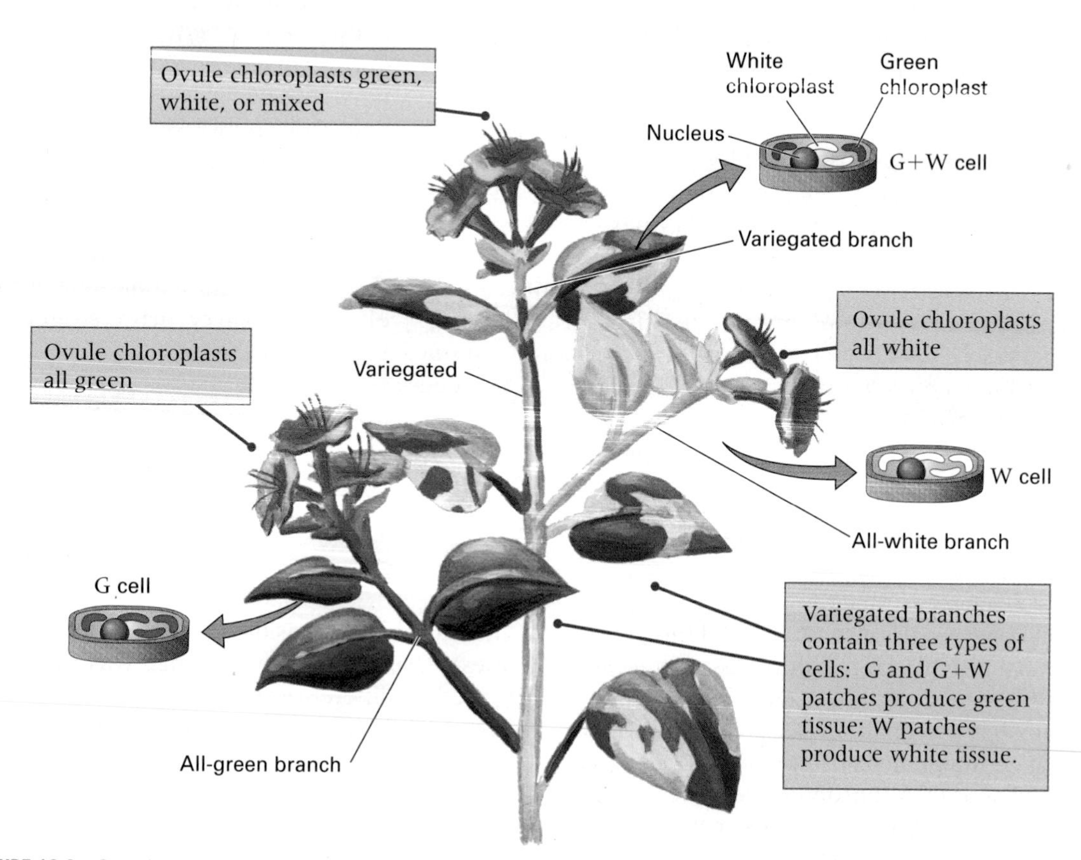


**FIGURE 16.9** Genetic model for leaf variegation. Branches that are all green, all white, and variegated form on the same plant. Flowers form on all three kinds of branches. The insets show the chloroplast composition of cells in each type of branch. Cells in all-green or all-white branches contain only green or white chloroplasts, respectively, whereas cells in variegated branches are heteroplasmic.

## connection

### Chlamydomonas Moment

**Ruth Sager and Zenta Ramanis 1965**
Columbia University, New York, New York
*Recombination of Nonchromosomal Genes in* Chlamydomonas

*Before the widespread use of polymorphic DNA markers in mitochondria and chloroplasts, the study of extranuclear inheritance was hindered in most organisms not only by uniparental inheritance of cellular organelles, but also by the lack of mutant phenotypes showing extranuclear transmission. The alga* Chlamydomonas *proved to be suitable for studies of extranuclear inheritance. Mutants showing nonchromosomal inheritance could be obtained, and strains in which the inheritance of nonchromosomal genes was biparental were discovered, making it possible to examine segregation of nonchromosomal alleles. The compelling evidence for the existence of genetic systems apart from genes in the nucleus was the discovery of recombination in nonchromosomal genes. In the matings described here, the parental genotypes for the nonchromosomal genes were* $ac_1$ sd *(slow growth on acetate, streptomycin requiring) and* $ac_2$ sr *(acetate-requiring, streptomycin-resistant). The intragenic recombinant products were either wildtype* $ac^+$ *(acetate-independent) versus the double mutant* $ac_1$ $ac_2$ *or they were wildtype* ss *(streptomycin-sensitive) versus the double mutant* sd sr. *The authors leave no doubt about their opinion (which turned out to be correct) that the nonchromosomal genetic systems are due to DNA in the cellular organelles.*

The analysis of nonchromosomal heredity has made slow progress, often under severe attack, [in part because of] the difficulty in obtaining mutations of nonchromosomal genes. . . . Within the past few years, a concerted attack has been made on this problem with the alga *Chlamydomonas*. . . . Two findings have been of key importance in our investigation. First, streptomycin was developed as [an agent for selecting] nonchromosomal (NC) genes. . . . Second, the existence of an exceptional class of zygotes was discovered that transmits NC genes from both parents to the progeny, in contrast to the standard pattern of maternal inheritance. With this material it was established that the NC genes are particulate in nature, genetically autonomous, and stable. . . . In this paper we describe what is to our knowledge the first evidence of recombination of nonchromosomal genes. . . . The occurrence of linked recombination demonstrates that NC genes, like the chromosomal ones, are capable of close pairing and precise reciprocal exchange. . . . The salient features of our results are the following: (1) Segregation of NC genes did not occur during meiosis. . . . (2) NC gene segregation began at the first or second mitotic division after meiosis, with pure clones arising so long as any heterozygous cells remained. (3) The average segregation ratio is 1 : 1 although the progeny of individual zygotes gave ratios significantly different from 1 : 1. (4) The acetate alleles segregated independently of the streptomycin alleles. (5) Novel classes of progeny were recovered that were phenotypically indistinguishable from wildtype $ac^+$ and from wildtype *ss*. New mutant phenotypes were also observed with new levels of acetate requirement and new levels of streptomycin resistance and dependence. [These were the reciprocal, double-mutant types.] . . . From existing knowledge, we infer that intragenic recombination, in which the wildtype gene is reconstituted, requires a close intermolecular pairing and exchange of the sort only known to occur between nucleic acids. Consequently, we view our results as providing evidence of the nucleic acid nature of a nonchromosomal genetic system. These genetic findings have been paralleled by the discovery of DNA of high molecular weight in chloroplasts and mitochondria. It is a reasonable surmise that these organelle DNAs carry primary genetic information.

**We view our results as providing evidence of the nucleic acid nature of a nonchromosomal genetic system.**

Source: R. Sager and Z. Ramanis, *Proc. Natl. Acad. Sci. USA* 53 (1965): 1053–1061.

*Chlamydomonas* (**FIGURE 16.10**). Cells of this organism have a single large chloroplast containing 75–80 copies of a duplex DNA molecule approximately 195 kb in size. Cells of the two mating types, designated $mt^+$ and $mt^-$, are of equal size and appear to contribute the same amount of cytoplasm to the zygote. After cells of opposite mating type fuse, the diploid zygote undergoes meiosis to form a tetrad of four haploid cells. The cells can be grown on solid growth medium to form visible clusters of cells (colonies). Cytoplasmic inheritance was first suggested by differences in the result of reciprocal crosses between mutant, antibiotic-resistant strains and wildtype, antibiotic-sensitive strains. For example, for streptomycin resistance (*str-r*) versus sensitivity (*str-s*),

$$str\text{-}r\ mt^+ \times str\text{-}s\ mt^-$$
yields only *str-r* progeny

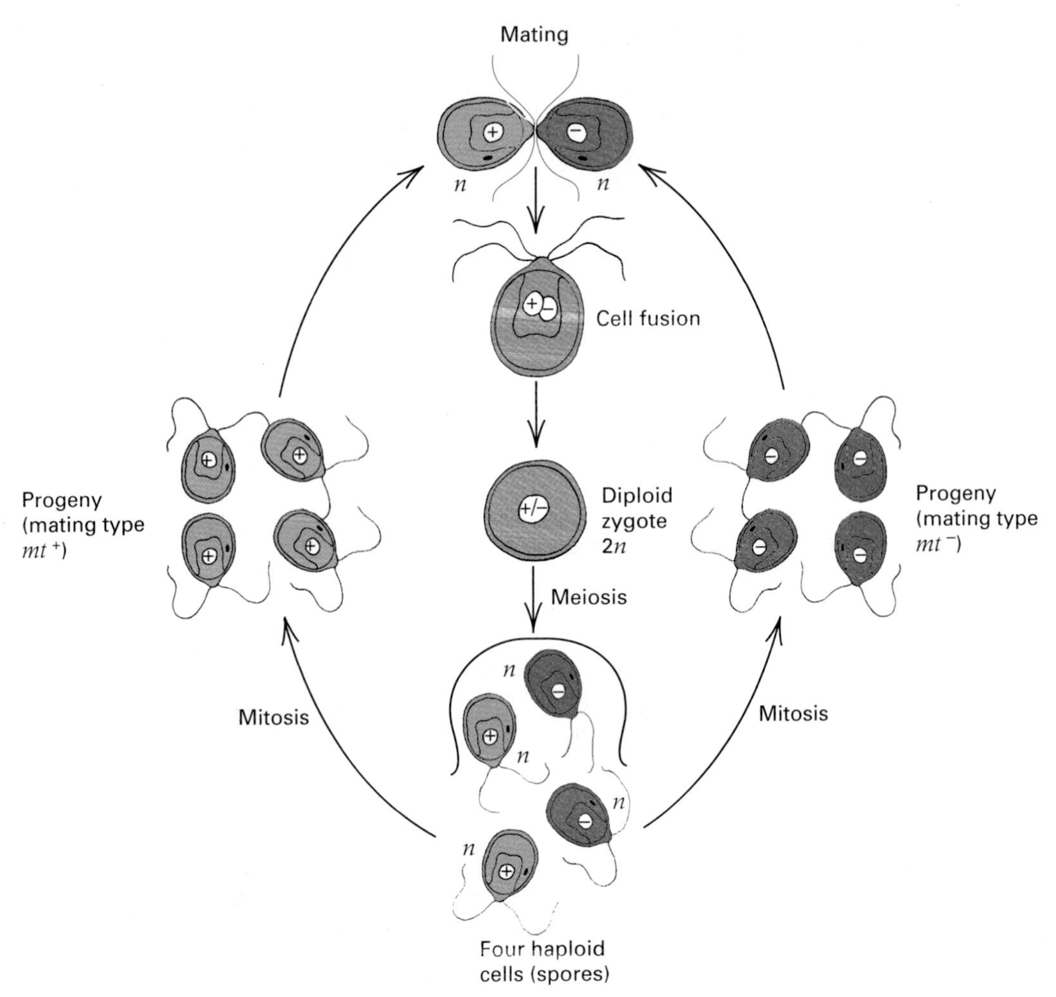


**FIGURE 16.10** Life cycle of *Chlamydomonas reinhardtii.*

*str-s* $mt^+$ × *str-r* $mt^-$
yields only *str-s* progeny

This is a clear example of uniparental inheritance, because the resistance or sensitivity comes from the $mt^+$ parent. In contrast, the *mt* alleles in the nucleus behave in strictly Mendelian fashion, yielding the 1 : 1 progeny ratio expected of nuclear genes.

A large number of antibiotic-resistance markers with uniparental inheritance have been examined in *Chlamydomonas.* In each case, the antibiotic resistance has been traced to the chloroplast. Haploid *Chlamydomonas* contains only one chloroplast. If the chloroplast could be derived from the $mt^+$ or $mt^-$ cell with equal probability, then uniparental inheritance would not be observed because the chloroplast could come from either parent. But direct tracking of chloroplast DNA via molecular markers indicates that the chloroplast of the $mt^-$ parent is preferentially lost after mating. This accounts for the fact that the antibiotic-resistance phenotype of the $mt^+$ parent is transmitted to the progeny.

About 5 percent of the progeny from *Chlamydomonas* crosses are heteroplasmic. Both parental chloroplasts are retained, although they ultimately segregate in later cell divisions. The heteroplasmic cells have been valuable for genetic analysis, because recombination can occasionally take place between the DNA in the two chloroplasts. Analysis of the recombination frequencies for the various phenotypes has made possible the construction of fairly detailed genetic maps of the chloroplast genome.

The inheritance of mitochondria in *Chlamydomonas* is also uniparental, but it is through the $mt^-$ parent. This pattern is observed for *MUD2* mutations, which occur in the mitochondrial cytochrome *b* gene and confer resistance to the antibiotic myxothiazol. When crosses are made between mutant, antibiotic-resistant (*MUD2*) and wildtype, antibiotic-sensitive (*mud2*) strains, the results are as follows:

*MUD2 mt*$^+$ × *mud2 mt*$^-$
yields only *mud2* (sensitive) progeny

*mud2 mt*$^+$ × *MUD2 mt*$^-$
yields only *MUD2* (resistant) progeny

In each cross, the tetrads yield 2 *mt*$^+$ : 2 *mt*$^-$ spores. However, all spore progeny have the resistance phenotype of the *mt*$^-$ parent. Inheritance of resistance to myxothiazol is therefore uniparental, because the phenotype of the progeny is the same as that of the *mt*$^-$ parent. Because resistance to myxothiazol results from a change in the mitochondrial DNA, we can also infer that the progeny must receive their mitochondrial DNA from the *mt*$^-$ parent. In contrast, as indicated by the streptomycin example above, they receive their chloroplast DNA from the *mt*$^+$ parent. It is not presently known whether having mitochondria come from one parent and chloroplasts from the other has any biological significance, or whether the patterns are merely a matter of evolutionary chance.

## ■ Respiration-Defective Mitochondrial Mutants

In budding yeast, *Saccharomyces cerevisiae*, very small colonies are occasionally observed when the cells are grown on solid medium containing glucose. These are called **petite mutants**. Microscopic examination indicates that the cells are of normal size even though the colonies are small. Physiological studies show that the colonies grow normally during the early stages when they are fermenting glucose and making ethanol, but they soon stop growing because of a defect in the oxygen-requiring respiration that is needed for further metabolism of the ethanol. The colonies are small because the cells obtain ATP only via fermentation and are unable to obtain the much larger yields of ATP via metabolism of the ethanol.

Among several types of petite mutants, two major types can be distinguished by their behavior in crosses with wildtype nonpetites. The results are illustrated in **FIGURE 16.11**. One type, called **segregational petites**, exhibits the typical segregation of a simple Mendelian recessive. In a cross between a segregational petite and wildtype, the diploid zygotes are normal; and when the diploids undergo meiosis, half of the ascospores in an ascus produce petite colonies and half form wildtype colonies. The 1 : 1 ratio (actually 2 : 2) indicates that these petites are the result of a nuclear mutation and that the allele for petite is recessive.

For the second type, **neutral petites**, the result is completely different. In a cross with wildtype, all diploid zygotes produce wildtype ascospores (4 : 0 segregation). The same pattern is found when the progeny from such a cross are backcrossed with neutral petites. In both cases, the phenotype that is inherited is from the normal parent, so the inheritance is uniparental. The explanation for these results is that the majority of neutral petites have deletions of most of the mitochondrial DNA genes involved in oxidative respiration. When a neutral-petite cell mates with a wildtype cell, the cytoplasm of the wildtype cell contributes normal mitochondrial DNA, which is transmitted to the progeny spores.

Petites arise at a frequency of roughly $10^{-5}$ per generation. The neutral petite phenotype is the result of large deletions in mitochondrial DNA. For unknown reasons, yeast mitochondria often fuse and fragment, which may cause occasional deletions of DNA. The production of neutral petites is apparently a result of the segregation of aberrant mitochondrial DNA from normal DNA, and the further sorting out of mutant mitochondria, in the course of cell division.

## ■ Cytoplasmic Male Sterility in Plants

An example of extranuclear inheritance that is important in agriculture is **cytoplasmic male sterility**, a condition in which a plant does not produce functional pollen, but the female reproductive organs and fertility are normal. This type of sterility is used extensively in the production of hybrid corn seed, because it circumvents the need for manual detasseling of the plants to be used as females to produce the hybrid. The tassel is the pollen-bearing organ, and in male-fertile varieties, detasseling is necessary to prevent self-fertilization of female plants.

The genetically transmitted male sterility is not controlled by nuclear genes but is transmitted through the egg cytoplasm from generation to generation. The pattern of inheritance of cytoplasmic male sterility in corn is summarized in **FIGURE 16.12**. The key observation is that repeated backcrossing of a male-sterile variety, using pollen from a male-fertile variety, does not restore male fertility. The resulting plants, in which nearly all nuclear genes from the male-sterile variety have been replaced with those from the male-fertile variety, remain male-sterile. Because a small amount of functional pollen is produced by the male-sterile variety, it is also possible to carry out the reciprocal cross:

male-fertile ♀ × male-sterile ♂

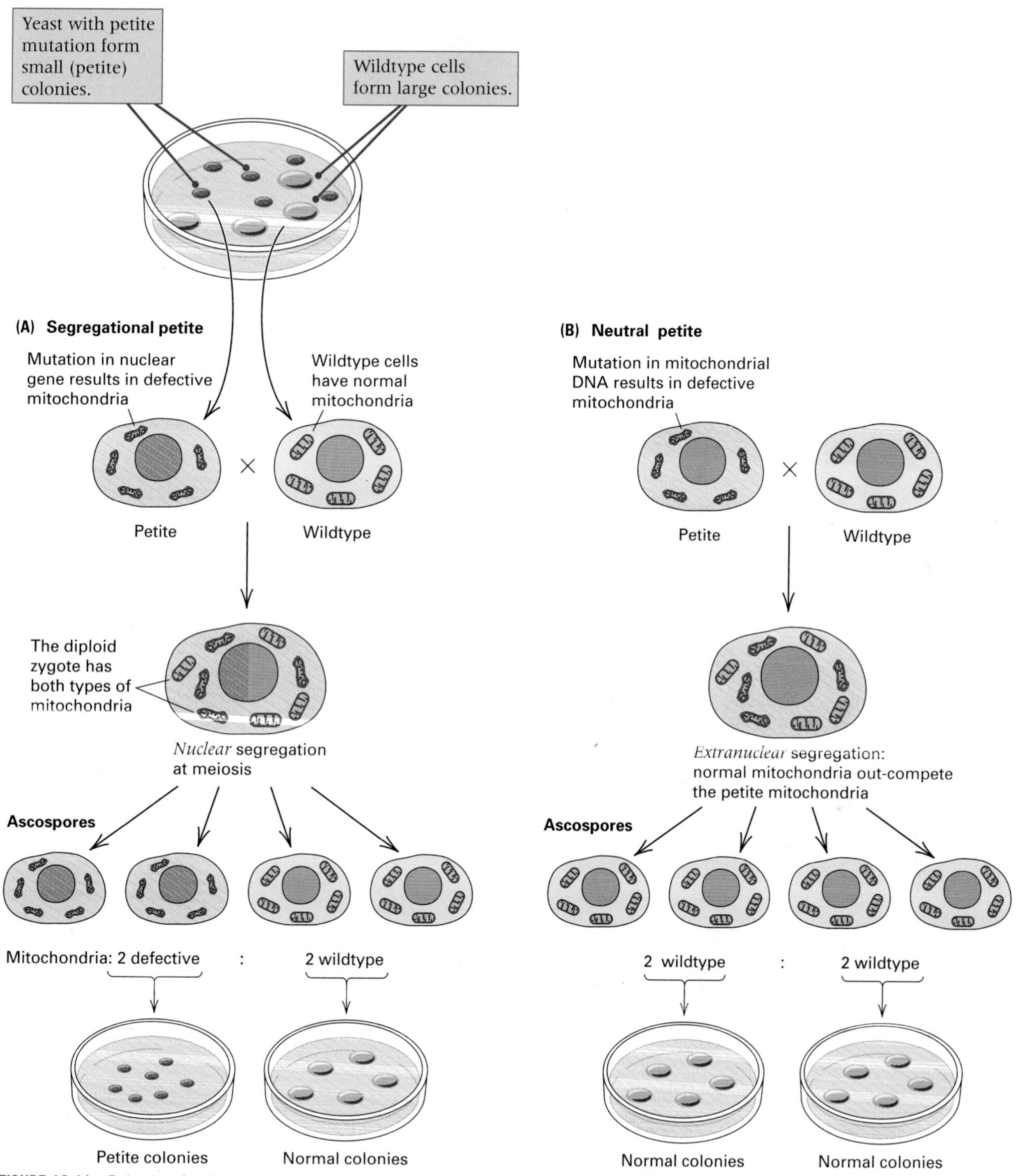

Yeast with petite mutation form small (petite) colonies.
Wildtype cells form large colonies.
(A) Segregational petite
Mutation in nuclear gene results in defective mitochondria
Wildtype cells have normal mitochondria
Petite
Wildtype
The diploid zygote has both types of mitochondria
Nuclear segregation at meiosis
Ascospores
Mitochondria: 2 defective : 2 wildtype
Petite colonies
Normal colonies
(B) Neutral petite
Mutation in mitochondrial DNA results in defective mitochondria
Petite
Wildtype
Extranuclear segregation: normal mitochondria out-compete the petite mitochondria
Ascospores
2 wildtype : 2 wildtype
Normal colonies
Normal colonies

**FIGURE 16.11** Behavior of petite mutations in genetic crosses. The brown circles represent petite cells. The brown nucleus in part A contains an allele resulting in petite colonies. The blue cells have normal mitochondria.

In this case, the progeny are fully male fertile. The difference between the reciprocal crosses means that male sterility in these varieties is maternally inherited.

Cytoplasmic male sterility in maize results from rearrangements in mitochondrial DNA. For example, one DNA rearrangement fuses two mitochondrial genes and creates a novel protein that causes male sterility. Although cytoplasmic male sterility is maternally inherited, certain nuclear genes called **fertility restorers** can suppress the male-sterilizing effect of the

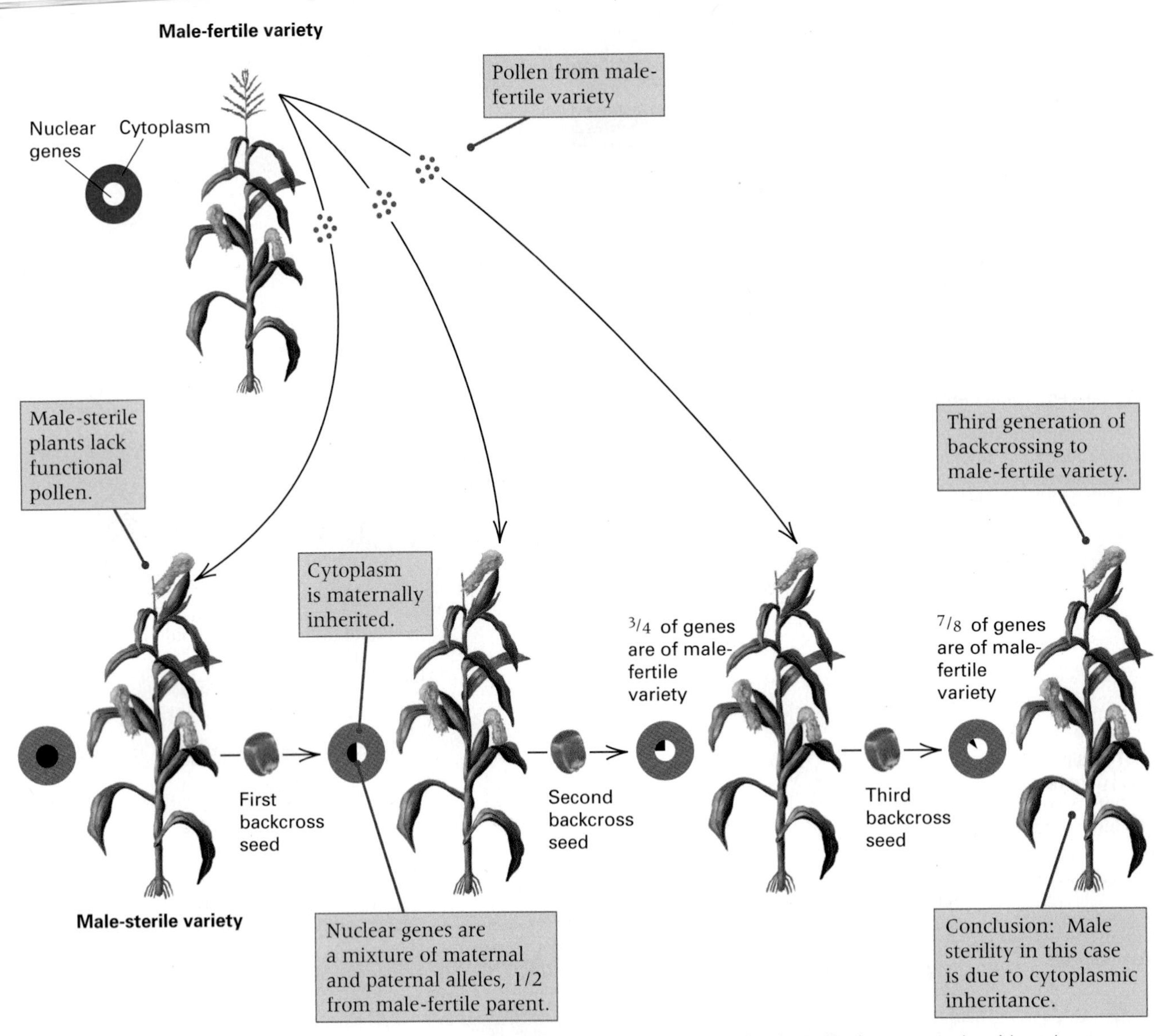
Male-fertile variety
Pollen from male-fertile variety
Nuclear genes
Cytoplasm
Male-sterile plants lack functional pollen.
Third generation of backcrossing to male-fertile variety.
Cytoplasm is maternally inherited.
3/4 of genes are of male-fertile variety
7/8 of genes are of male-fertile variety
First backcross seed
Second backcross seed
Third backcross seed
Male-sterile variety
Nuclear genes are a mixture of maternal and paternal alleles, 1/2 from male-fertile parent.
Conclusion: Male sterility in this case is due to cytoplasmic inheritance.

**FIGURE 16.12** Backcross experiment demonstrating maternal inheritance of male sterility (orange cytoplasm) in maize.

cytoplasm. Different types of cytoplasmic male sterility can be classified on the basis of their response to restorer genes. For example, in one case, restoration of fertility requires the presence of dominant alleles of two different nuclear genes, and plants with male-sterile cytoplasm and both restorer alleles produce normal pollen. In another case, suppression requires the dominant allele of a different restorer gene, but this gene acts only in the gametophyte; hence plants with male-sterile cytoplasm that are heterozygous for the restorer allele produce a 1 : 1 ratio of normal to aborted pollen grains.

## 16.3 The Evolutionary Origin of Organelles

A widely accepted theory holds that eukaryotic organelles evolved more than half a billion years ago from prokaryotes that lived inside primitive cells in **symbiosis**, a mutually beneficial interaction. This is the **endosymbiont theory** (**FIGURE 16.13**). Mitochondria are thought to have originated from aerobic bacteria. Their closest living relatives based on ribosomal RNA comparisons are the α proteobacteria. The bacterial symbiont provided the molecular machinery for respiration, in which carbohydrates are broken down in stepwise fashion using oxygen to yield energy in the form of ATP and waste products in the form of carbon dioxide and water. Chloroplasts are thought to have originated from cyanobacteria much like those that exist today. The cyanobacterial symbiont provided the molecular machinery for photosynthesis, in which carbohydrates are synthesized from carbon dioxide and water using the energy absorbed from sunlight by chlorophyll.

If an aerobic bacterium and a cyanobacterium invaded a host cell and lived in symbiosis, what were the characteristics of the host cell? An ancestor resembling present-day archaeans is suspected. The archaeans share some features with eubacteria, including their prokaryotic cellular organization, but their mechanisms of macromolecular syntheses—replication, transcription, and translation—resemble those of eukaryotes. In any event, according to the endosymbiont theory, the mutually beneficial association of cells eventually evolved into the diverse assemblage of modern eukaryotes depicted in Figure 1.25 on page 32.

One line of evidence for the endosymbiont theory is that mitochondrial and chloroplast organelles of eukaryotes share with prokaryotes numerous features that distinguish them from eukaryotic cells. Among these common features:

- The genomes are composed of circular DNA that is not extensively complexed with histone-like proteins.
- The genomes are organized with functionally related genes close together and often expressed as a single unit.
- The ribosome particles on which protein synthesis takes place have major subunits whose size is similar in organelles and in prokaryotes but differs from that in cytoplasmic ribosomes in eukaryotic cells.
- The nucleotide sequences of the key RNA constituents of ribosomes are similar in chloroplasts, cyanobacteria such as *Anacystis nidulans,* and even bacteria such as *E. coli.*

The genomes of today's organelles are small compared with those of bacteria and cyanobacteria. The chloroplast genome is only 3 to 5 percent of that of cyanobacteria. Organelle evolution was probably accompanied by major genome rearrangements, some genes being transferred into the nuclear genome of the host and others being eliminated. The transfer of organelle genes into the nucleus differed from one lineage to the next. For example, the gene for one subunit of the mitochondrial ATP synthase is located in the mitochondrial genome in yeast but in the nuclear genome in *Neurospora.*

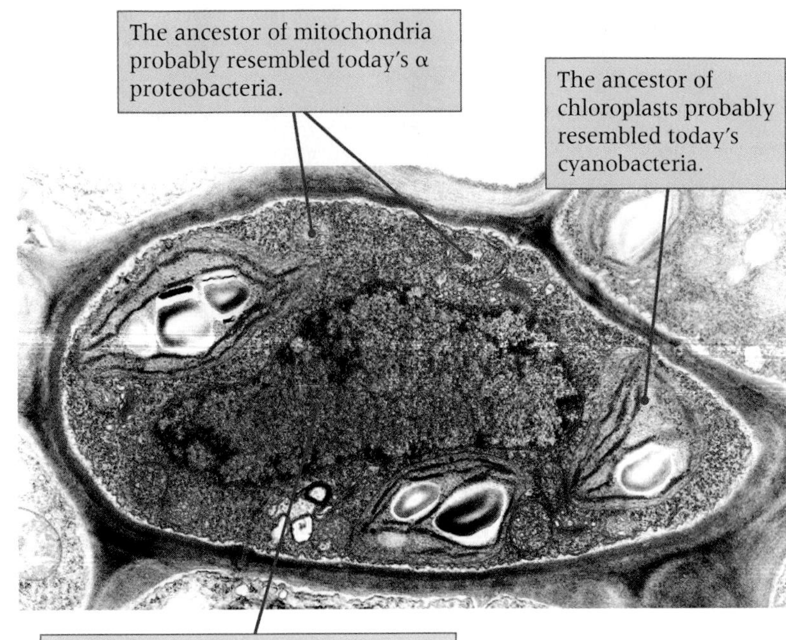

The ancestor of mitochondria probably resembled today's α proteobacteria.
The ancestor of chloroplasts probably resembled today's cyanobacteria.
In the course of evolution, most genes originally present in mitochondria and chloroplasts became incorporated into the nucleus. Many proteins that are encoded in nuclear genes contain sequences that result in their being transported into either the mitochondria or the chloroplasts.

**FIGURE 16.13** Mitochondria (red) and chloroplasts (dark green) inside a leaf cell of lambsquarters (*Chenopodium album*). The nucleus is orange. [© Martha J. Powell/Visuals Unlimited.]

## 16.4 Cytoplasmic Transmission of Symbionts

In eukaryotes, a variety of cytoplasmically transmitted traits result from the presence of bacteria and viruses living in the cytoplasm of certain cells. One of the classical examples is the killer phenomenon found in certain strains of the protozoan *Paramecium aurelia.* Killer strains of *Paramecium* release to the surrounding medium a substance that is lethal to many other strains of the protozoan. The killer phenotype requires the presence of cytoplasmic bacteria referred to as **kappa particles**, whose maintenance is dependent on a dominant nuclear gene, *K.* The kappa particles are cells of *Caedibacter taeniospiralis,* which produce the killer substance. Why killer strains themselves are immune to the substance has not yet been determined.

*Paramecium* is a diploid protozoan that undergoes sexual exchange through a mating process called **conjugation** (part A of **FIGURE 16.14**). Initially, each cell has two diploid micronuclei. Many protozoans contain micronuclei (small nuclei) and a macronucleus (a large nucleus with specialized functions), but only the micronuclei are relevant to the genetic processes described here. When two cells come into contact for conjugation, the two micronuclei in each cell undergo meiosis, forming eight micronuclei in each cell. Seven of the micronuclei and the

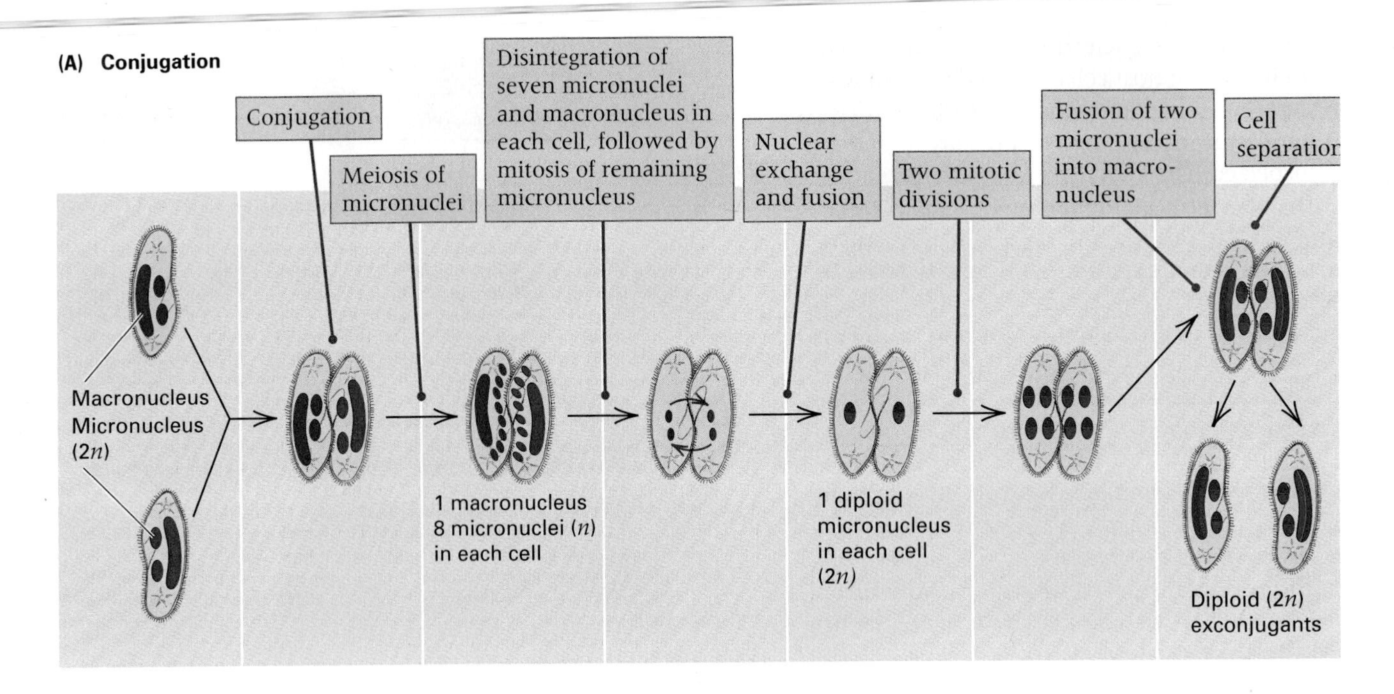


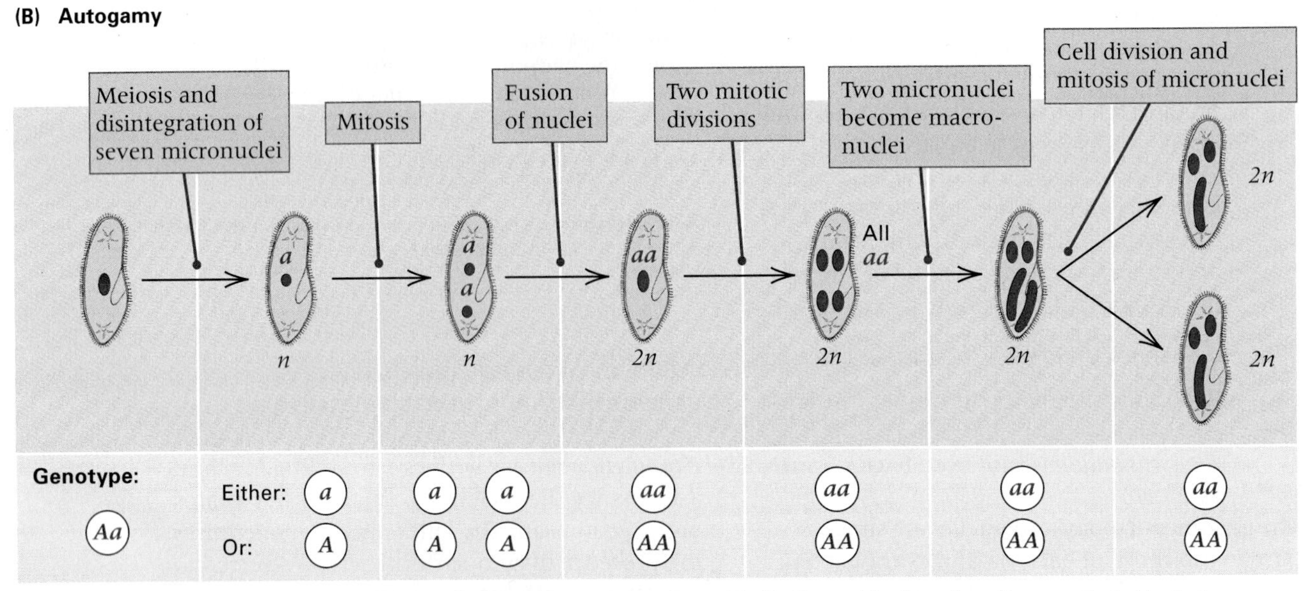


**FIGURE 16.14** (A) Conjugation in *Paramecium* results in reciprocal fertilization and the formation of two genetically identical exconjugant cells. (B) Autogamy in *Paramecium*. The alleles in the heterozygous *Aa* cell segregate in meiosis, with the result that the progeny cells become homozygous for either *A* or *a*. The production of *aa* progeny cells is shown in the diagram. The initial steps resemble those in conjugation in part A: The micronuclei undergo meiosis, after which seven of the products and the macronucleus degenerate.

macronucleus disintegrate, leaving each cell with one micronucleus, which then undergoes a mitotic division. The cell membrane between the two cells breaks down slightly, and the cells exchange a single micronucleus; then the nuclei fuse to form a diploid nucleus. After nuclear exchange, the two new cells (the exconjugants) are genotypically identical. In this sequence of events, only the micronuclei are exchanged, without any mixing of cytoplasm.

Single cells of *Paramecium* also occasionally undergo an unusual nuclear phenomenon called **autogamy** (Figure 16.14B). Meiosis takes place, and again seven of the micronuclei and the macronucleus disintegrate. The surviving nucleus then undergoes mitosis and nuclear fusion to recreate the diploid state, and a new macronucleus is formed. The genetic importance of autogamy is that, even if the initial cell is heterozygous, the newly formed diploid nucleus becomes homozygous because it is derived from a single haploid meiotic product. In a population of cells undergoing autogamy, the surviving micronucleus is selected randomly, so for any pair of alleles, half of the new cells contain one allele and half contain the other.

In conjugation, the exchange of micronuclei is occasionally accompanied by small amounts of cytoplasmic mixing. A comparison of the phenotypes following conjugation of killer (*KK*) cells and sensitive (*kk*) cells, with and without cytoplasmic exchange, yields evidence for cytoplasmic inheritance of the killer phenotype (**FIGURE 16.15**). Without cytoplasmic mixing (part A), the expected 1 : 1 ratio of killer to sensitive exconjugant cells is observed; with cytoplasmic mixing (part B), both exconjugants are killer cells because each cell has kappa particles derived from the cytoplasm of the killer parent. Conjugation is eventually followed by autogamy of each of the exconjugants (Figure 16.15). Autogamy of *Kk* killer cells results in an equal number of *KK* and *kk* cells, but the *kk* cells cannot maintain the kappa particle and so become sensitive.

Another example of a symbiont-related cytoplasmic effect is a condition in *Drosophila* called **maternal sex ratio**, which is characterized by the production of almost no male progeny. The daughters of sex-ratio females pass on the trait, whereas the sons that are occasionally produced do not. Cytoplasm taken from the eggs of sex-ratio females transmits the condition when injected into the female embryos from unaffected cultures of the same or other *Drosophila* species. A species of bacteria has been isolated from the cytoplasm of sex-ratio females; when such bacteria are allowed to infect other *Drosophila* females, these acquire the sex-ratio trait. The causative agent of the sex-ratio condition is not the bacteria themselves, but rather a virus that multiplies in them. When released by the bacterial cells, this virus kills most male *Drosophila* embryos. Why female embryos are not killed by the virus is still a mystery.

## 16.5 Maternal Effect in Snail Shell Coiling

Maternal transmission is not always mediated by cytoplasmic organelles or symbionts; in some cases, it is mediated through the cytoplasmic transmission of the products of nuclear genes. An example is determination of the direction of coiling of the shell in the snail *Limnaea peregra*. The direction of coiling, as viewed by looking into the opening of the shell, may be either to the right (dextral coiling) or to the left (sinistral coiling). Reciprocal crosses between homozygous strains give the following results:

dextral ♀ × sinistral ♂ → all $F_1$ dextral

sinistral ♀ × dextral ♂ → all $F_1$ sinistral

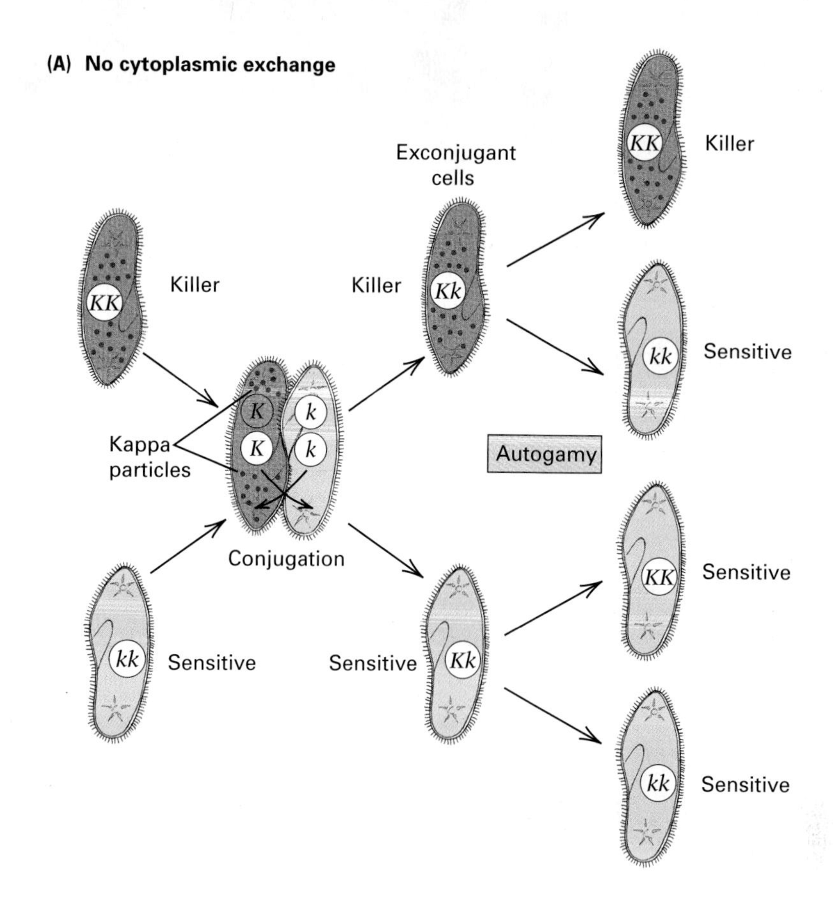


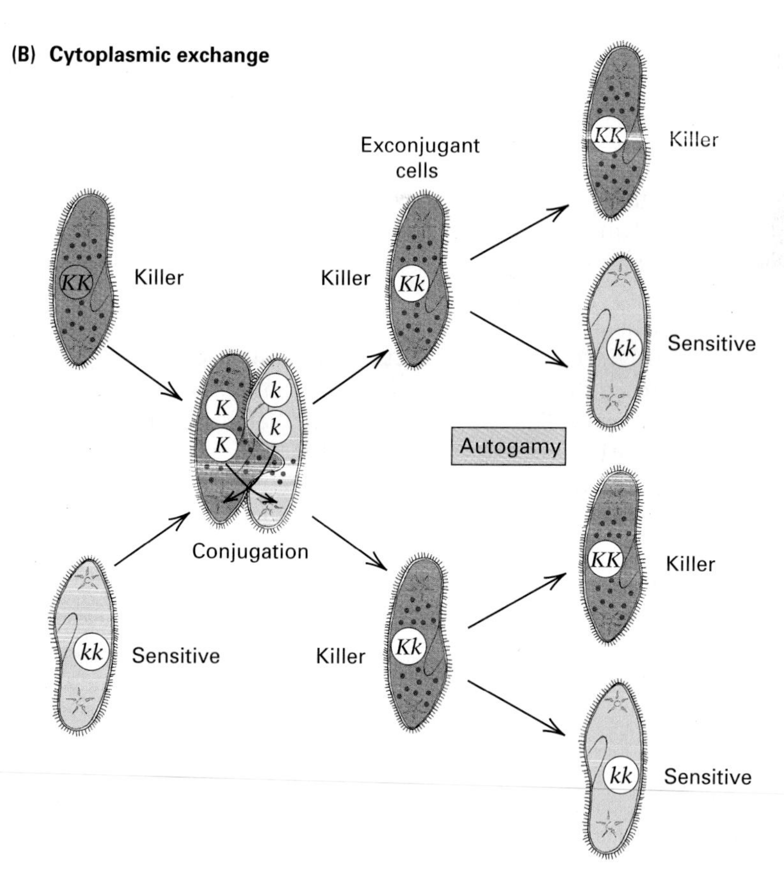


**FIGURE 16.15** Crosses between killer (*KK*) and sensitive (*kk*) *Paramecium.* (A) The result when no cytoplasmic exchange takes place in conjugation. (B) The result when cytoplasmic exchange occurs. Cells containing kappa are shown in yellow. The genotype and phenotype of each cell are indicated.

## connection
### A Coming Together

**Lynn Margulis (formerly Lynn Sagan) 1967**
Boston University, Boston, Massachusetts
*The Origin of Mitosing Cells*

*By the 1960s, certain features of the genetic apparatus of cellular organelles were already known to resemble those in prokaryotes. A number of scientists had suggested that the similarities may reflect an ancient ancestral relationship. Lynn Margulis took these ideas a giant step further by proposing a comprehensive hypothesis of the origin of the eukaryotic cell as a series of evolved symbiotic relationships. The main players are the mitochondria, the kinetosomes (basal bodies) of the flagella and cilia, and the photosynthetic plastids. All are hypothesized to have originated as prokaryotic symbionts. When the paper was submitted for publication, the anonymous reviewers were so strongly negative that it was almost rejected. Through the years, specific predictions of the hypothesis have been experimentally verified. A few revisions have been necessary; for example, current evidence suggests that the earliest symbiont was kinetosome bearing, not a protomitochondrion. However, no major inconsistencies have been uncovered. Today the serial endosymbiont theory is widely accepted.*

This paper presents a theory of the origin of the discontinuity between eukaryotic and prokaryotic cells. Specifically, the mitochondria, the basal bodies [kinetosomes] of the flagella, and the photosynthetic plastids can be considered to have derived from free-living cells, and the eukaryotic cell is the result of the evolution of ancient symbioses. Although these ideas are not new, in this paper they have been synthesized in such a way as to be consistent with recent data on the biochemistry and cytology of subcellular organelles. . . . Many aspects of this theory are verifiable by modern techniques of molecular biology. . . . Prokaryotic cells containing DNA, synthesizing protein on ribosomes, and using messenger RNA as intermediate between DNA and protein are ancestral to all extant cellular life. Such cells arose under reducing conditions of the primitive terrestrial atmosphere. . . . Eventually, a population of cells arose using photoproduced ATP with water as the source of hydrogen atoms in the reduction of $CO_2$ for the production of cell material. This led to the formation of gaseous oxygen as a by-product of photosynthesis. . . . The continued production of free oxygen resulted in a crisis. . . . It is suggested that the first step in the origin of eukaryotes from prokaryotes was related to survival in the new oxygen-containing atmosphere: an aerobic prokaryotic microbe (the protomitochondrion) was ingested into the cytoplasm of a heterotrophic anaerobe. This symbiosis became obligate and resulted in the first aerobic amoeboid [without, at this stage, mitosis]. . . . Some of the amoeboids ingested certain motile prokaryotes. The genes of the parasite coded for its characteristic morphology, (9 + 2) fibrils in cross section. This parasite also became symbiotic, forming primitive amoeboflagellates . . . that could actively pursue their own food. . . . The replicating nucleic acid of the endosymbiont genes (which determines its characteristic 9 + 2 structure of sets of microtubules) was eventually utilized to form the chromosomal centromeres and centrioles of eukaryotic mitosis and to distribute newly synthesized host nuclear chromatin to host daughters. . . . Eukaryotic plant cells acquired photosynthesis by symbiosis with photosynthetic prokaryotes (protoplastids), which themselves evolved from organisms related to cyanobacteria. . . . In order to document this theory, the following, at the very least, are required: (1) The theory must be consistent with geological and fossil records. (2) Each of the three symbiotic organelles (the mitochondria, the [9 + 2] homologues, and the plastids) must demonstrate general features characteristic of cells originating in hosts as symbionts. None may have features that conflict with such an origin. (3) Predictions based on this account of the origin of eukaryotes must be verified. The rest of this paper [24 pages] discusses the evidence for this theory in terms of assumptions based on molecular biology which can be made concerning evolutionary mechanisms.

**The eukaryotic cell is the result of the evolution of ancient symbioses.**

Source: L. Sagan, *J. Theor. Biol.* 14 (1967): 225–274.

In these crosses, the $F_1$ snails have the same genotype, but the reciprocal crosses give different results; the direction of progeny coiling is the same as that of the mother. This result is typical of traits with maternal inheritance. However, in this case, all $F_2$ progeny from both crosses exhibit dextral coiling, a result that is inconsistent with cytoplasmic inheritance. The $F_3$ generation provides the explanation.

The $F_3$ generation, obtained by self-fertilization of the $F_2$ snails (the snail is hermaphroditic), indicates that the inheritance of coiling direction depends on nuclear genes

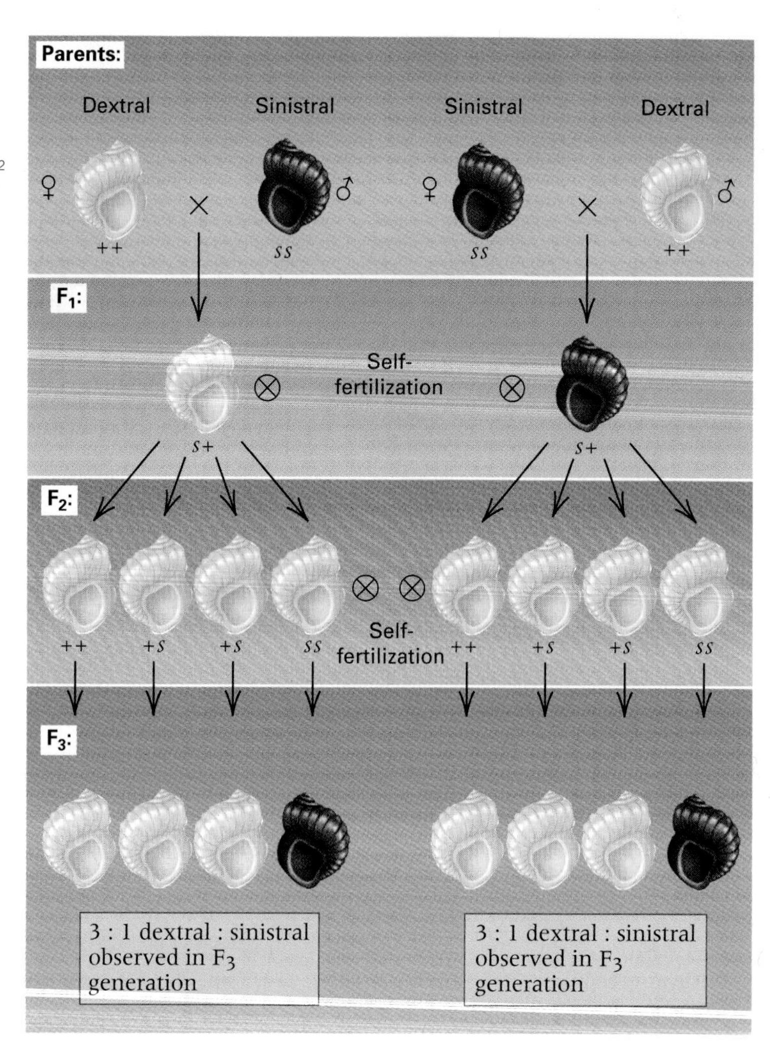


**FIGURE 16.16** Inheritance of the direction of shell coiling in the snail *Limnaea.* Sinistral coiling is determined by the recessive allele *s*, dextral coiling by the wildtype + allele. However, the direction of coiling is determined by the nuclear genotype of the mother, not by the genotype of the snail itself or by extranuclear inheritance. The $F_2$ and $F_3$ generations can be obtained by self-fertilization because the snail is hermaphroditic and can undergo either self-fertilization or cross-fertilization.

rather than on extranuclear factors. Three fourths of the $F_3$ progeny have dextrally coiled shells and one fourth have sinistrally coiled shells (**FIGURE 16.16**). This is a typical 3 : 1 ratio with dextral coiling (+/+ and *s*/+) dominant over sinistral coiling (*s*/*s*). The ratio indicates Mendelian segregation, but it is observed in the $F_3$ generation, not in the $F_2$ generation. The Mendelian ratio is delayed for a generation because:

> The coiling phenotype of an individual snail is determined by the genotype of its mother.

This means that the direction of shell coiling is not a case of maternal inheritance, but rather a case of maternal effect (Section 16.1). Cytological analysis of developing eggs has provided this explanation: The genotype of the mother determines the orientation of the spindle in the initial mitotic division after fertilization, and this in turn controls the direction of shell coiling of the offspring (**FIGURE 16.17**). This classic example of maternal effect indicates that more than a single generation of crosses is needed to provide conclusive evidence of extranuclear inheritance of a trait.

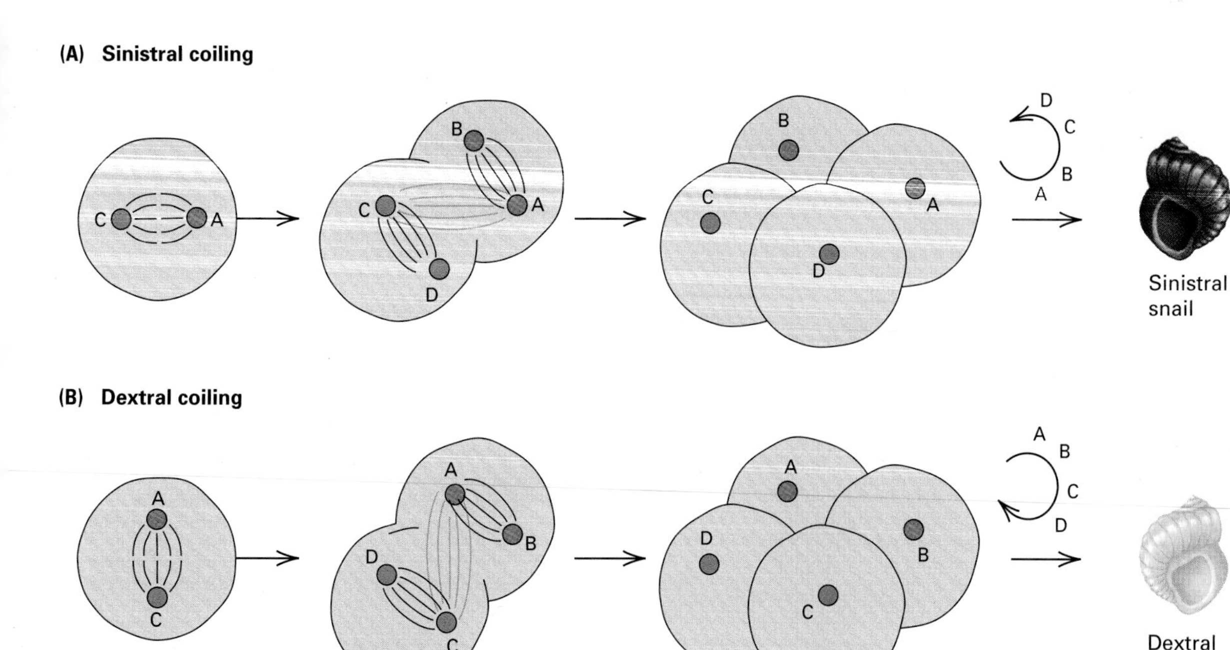


**FIGURE 16.17** The direction of shell coiling in *Limnaea* is determined by the orientation of the mitotic spindles during the second cleavage division in the zygote, which is predetermined in the egg cytoplasm as a result of the mother's genotype. The spiral patterns established by the cleavage result in (A) sinistral (leftward) or (B) dextral (rightward) coiling.

**FIGURE 16.18** Mating between two snails of the species *Euhadra congenita*. In this case both partners have dextrally coiled shells. Each is about 30 mm in width. [Courtesy of Takahiro Asami, Shinshu University.]

The direction of shell coiling may be more than a mere biological curiosity. It affects how easily two snails can mate. In some species of snails, mating partners undergo mating face-to-face (**FIGURE 16.18**). The genital apparatus is positioned in such a way that the genital apparatus of one partner will most easily match that of the other partner if their shells are coiled in the same direction. (In Figure 16.18 both partners are dextral.) For this reason dextral snails can mate most easily mate with other dextral snails. Likewise, sinistral snails can mate most easily with other sinistral snails. Face-to-face mating of dextral with sinistral is possible, but in this case one partner has to contort the genital apparatus to fit that of its mate. A rare sinistral snail in a largely dextral species may therefore be a very lonely snail, and likewise for a rare dextral snail in a largely sinistral species.

## CHAPTER SUMMARY

- Some inherited traits are determined by DNA molecules present in organelles, principally mitochondria and chloroplasts. A hallmark of such traits is uniparental inheritance, either maternal or paternal.
- Progeny resembling the mother is characteristic of both maternal inheritance and maternal effects. Although maternal inheritance results from the transmission of nonnuclear genetic factors, maternal effects result from nuclear genes expressed in the mother that determine the phenotype of the progeny.
- Mitochondria, photosynthetic plastids, and certain other cellular organelles are widely believed to have originated through independent episodes of symbiosis in which a prokaryotic cell, the progenitor of the present organelle, became an intracellular symbiont of the proto-eukaryotic cell.

## REVIEW THE BASICS

- What are the normal functions of mitochondria and chloroplasts?
- Can eukaryotic cells survive without mitochondria? Can plant cells survive without chloroplasts?
- Give an example of a human disease caused by a mutation in mitochondrial DNA.
- Distinguish between maternal inheritance and maternal effect.
- What is the endosymbiont theory of the origin of certain cellular organelles?
- What is variegation?
- How does leaf variegation in four-o'clock plants demonstrate heteroplasmy of chloroplasts in certain cell lineages and homoplasmy in others?
- Of what practical use is cytoplasmic male sterility in maize breeding?
- What does the occurrence of segregational petites in yeast imply about the relationship between the nucleus and mitochondria?

## GUIDE TO PROBLEM SOLVING

**Problem 1** The following mRNA is part of a coding sequence isolated from a human cell. Is it likely to be from a nuclear gene or a mitochondrial gene? Explain your answer.

5'-AUGAGACAUAAUGAUAGAGGGAGGCCCAAUCGC-3'

**Answer** Most likely nuclear; the corresponding amino acid sequence is Met–Arg–His–Asn–Asp–Arg–Gly–Arg–Pro–Asn–Arg. In mitochondria, the translation would be Met–stop–His–Asn–Asp–stop–Gly–stop–Pro–Asn–Arg. Three stop codons in the first 11 codons is incompatible with an open reading frame.

**Problem 2** Two strains of maize exhibit male sterility. One is cytoplasmically inherited; the other is Mendelian. What crosses could you make to identify which is which?

**Answer** Cytoplasmic male sterility is transmitted through the female, so outcrosses (crosses to unrelated strains) using the cytoplasmic male-sterile strain as the female parent will yield male-sterile progeny. Repeated outcrosses using the strain with Mendelian male sterility will result in either no male sterility (if the sterility gene is recessive) or 50 percent sterility (if it is dominant).

**Problem 3** Analysis of the mitochondrial DNA of a marine snail consistently yields two different restriction maps in the mtDNA present in a single animal. These are shown in the accompanying diagram. Suggest a hypothesis for the difference in the restriction maps, and explain why both types of mtDNA are found in the same individual.

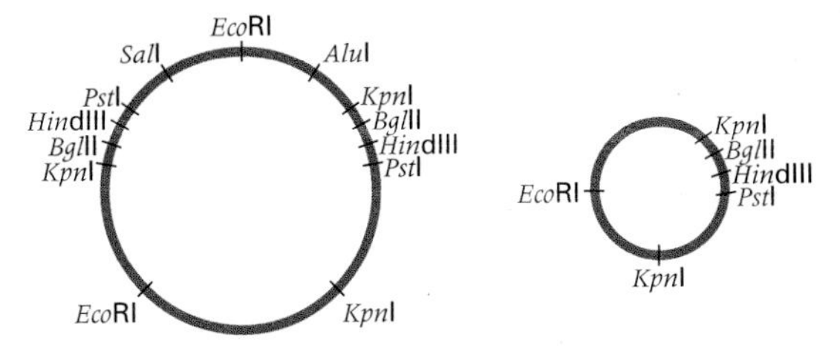


**Answer** The cluster of four restriction sites suggests a repeated sequence. Recombination between the direct repeats deletes the segment between them. The nonrecombinant mtDNA and the recombinant DNA therefore yield two different restriction maps. They are both present in single individuals because the mtDNA with the deletion is derived from the other by recombination. The recombination event must take place at a sufficiently high rate that each individual has both types of mtDNA. Since the deletion is irreversible, the failure to find any individuals with only the deletion suggests that this condition is lethal.

## ANALYSIS AND APPLICATIONS

**16.1** Why are so many inherited mitochondrial diseases characterized by muscle insufficiency or weakness?

**16.2** Why are diseases that result from mitochondrial mutations highly variable in severity from one patient to the next?

**16.3** In the course of evolution, many genes originally present in mitochondrial DNA have been transferred to the nucleus. If a present-day human mitochondrial coding sequence were to be transferred to the nucleus and translated successfully, what constraints on codon usage would be necessary?

**16.4** Is the genetic code used in human mitochondrial DNA completely redundant in the sense that, at any third-codon position, either purine or either pyrimidine is equivalent? How does this situation differ from the standard genetic code?

**16.5** The accompanying pedigree and gel diagram shows the pattern of inheritance of an organelle genome present in a parasitic microorganism related to the malaria parasite. What mode of genetic transmission do these data suggest?

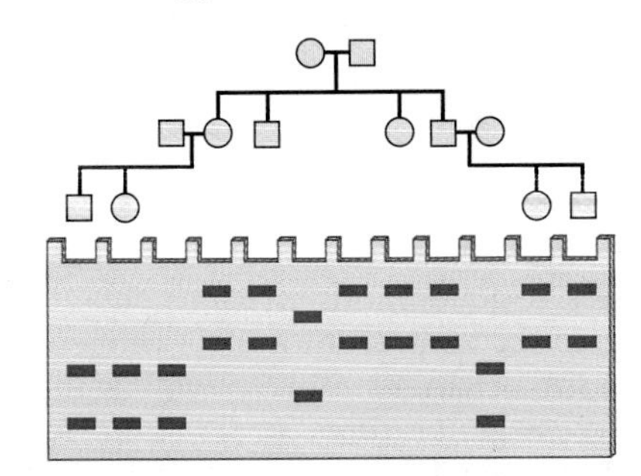

**16.6** A mutant plant is found with yellow instead of green leaves. Microscopic and biochemical analyses show that the cells contain very few chloroplasts and that the plant manages to grow by making use of other pigments. The plant does not grow very well, but it does breed true. The inheritance of yellowness is studied in the following crosses:

green male × yellow female, which yields all yellow progeny

yellow male × green female, which yields all green progeny

What do these results suggest about the type of inheritance?

**16.7** For the phenotype in the preceding problem, suppose that variegated progeny are found in the crosses at a frequency of about 1 per 1000 progeny. How might this be explained?

**16.8** In the human mitochondrial genetic code, what fraction of transition mutations in the third-codon position are synonymous? What fraction of transversion mutations in the third position are synonymous?

**16.9** Some plant species have distinct sexes determined by a mechanism analogous to the XX–XY chromosomal system in animals. In these species, how could paternal transmission of the Y chromosome be distinguished from paternal transmission of an organelle?

**16.10** In the maternally inherited leaf variegation in *Mirabilis*, how does the $F_2$ generation of the cross green female × white male differ from that which would be expected if white leaves were due to a conventional X-linked recessive gene in animals?

**16.11** What kinds of progeny would be expected to result from the following crosses with the four o'clock plant *Mirabilis*?

**(a)** green female × white male
**(b)** white female × green male
**(c)** variegated female × green male
**(d)** green female × variegated male

**16.12** Analysis of the mitochondrial DNA of a marine snail consistently yields two different restriction maps in the mtDNA present in a single animal. These are shown in the diagram on the next page. Suggest a hypothesis for the difference in the restriction maps, and explain why both types of mtDNA are found in the same individual.

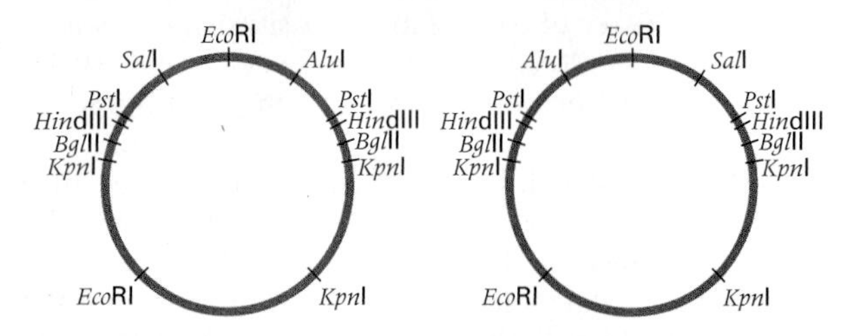


**16.13** Assuming that chloroplasts duplicate and segregate randomly in cell division, what is the probability that both replicas of a single mutant chloroplast will segregate to the same daughter cell during mitosis? After two divisions, what is the probability that at least one daughter cell will lack the mutant chloroplast?

**16.14** If *a* and *b* are chloroplast markers in *Chlamydomonas,* what genotypes of progeny are expected from the cross $a^+\ b^-\ mt^+ \times a^-\ b^+\ mt^-$?

**16.15** The pedigree illustrated here shows the pattern of bands observed in several generations of crosses of a certain plant species. Are the patterns consistent with Mendelian inheritance? Explain why or why not. If not, suggest an explanation for the patterns.

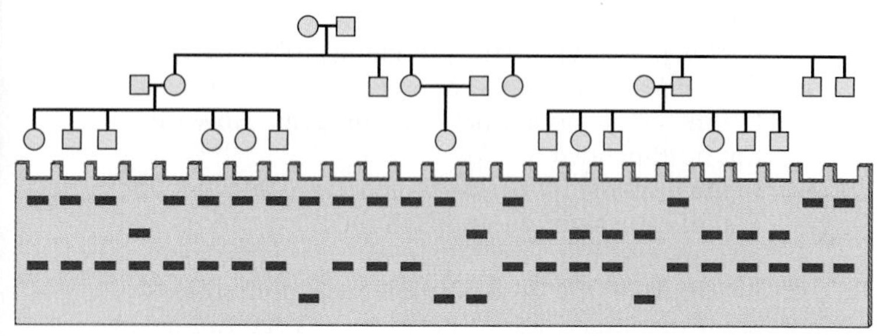

**16.16** Consider three types of antibiotic resistance (*r*) versus sensitivity (*s*) in *Chlamydomonas:*

**(a)** *cpl-r* results from a mutation in chloroplast DNA

**(b)** *mit-r* results from a mutation in mitochondrial DNA

**(c)** *nuc-r* results from a mutation in nuclear chromosomal DNA

If a cross of resistant $mt^+ \times$ sensitive $mt^-$ is carried out for each of these three types of resistances individually, what types of progeny are expected, and in what proportions? Explain your answers.

**16.17** For the antibiotic-resistance markers described in the previous problem, what phenotypes would be expected, and in what ratio, from the following crosses?

**(a)** *cpl-r mit-s nuc-r* $mt^+ \times$ *cpl-s mit-r nuc-s* $mt^-$

**(b)** *cpl-s mit-r nuc-s* $mt^+ \times$ *cpl-r mit-s nuc-r* $mt^-$

**16.18** Why are segregational petites in yeast called segregational?

**16.19** A new antibiotic-resistance, *klv-r,* is found in *Chlamydomonas*. Random spores chosen from the cross *klv-r* $mt^+ \times$ *klv-s* $mt^-$ yield the following progeny:

92 *klv-r* $mt^+$
88 *klv-s* $mt^-$
11 *klv-r* $mt^-$
9 *klv-s* $mt^+$

What type of inheritance does the antibiotic resistance show? How can the association with mating type be explained?

**16.20** An antibiotic-resistant haploid strain of yeast is isolated. Mating with a wildtype (antibiotic-sensitive) strain produces a diploid that, when grown for some generations and induced to undergo sporulation, produces tetrads containing either four antibiotic-resistant spores or four antibiotic-sensitive spores. What can you conclude about the inheritance of the antibiotic resistance?

**16.21** What is the phenotype of a diploid yeast produced by crossing a segregational petite with a neutral petite?

**16.22** Equal numbers of *AA* and *aa Paramecium* undergo conjugation in random pairs. What are the expected matings? What genotypes and frequencies are expected after the conjugating pairs separate? What genotypes and frequencies are expected after the cells undergo autogamy?

**16.23** What are the possible genotypes of the progeny of an *Aa Paramecium* that undergoes autogamy?

**16.24** A cross is carried out in maize using a female parent from a variety with cytoplasmically inherited male sterility and a male parent from a variety homozygous for a dominant allele, *Rsf,* that overrides the cytoplasmic sterility and restores fertility. The genotype of the female parent is homozygous *rsf.* From this cross, $F_1$ plants are grown and self-fertilized. What are the expected genotypes and fertilities of the female and male parts of the plants in the next generation?

**16.25** In conjugation between a *KK* killer strain of *Paramecium* and a *kk* nonkiller strain, why is cytoplasmic exchange necessary to obtain a 1 : 1 ratio of killer : nonkiller after autogamy of the exconjugants has taken place? What is the ratio of killer : nonkiller in the absence of cytoplasmic exchange?

**16.26** A strain of *Drosophila* recovered from nature has the following properties. In every generation, about 90 percent of the progeny are females. When the females are crossed to males from a normal strain in successive generations, in each generation they produce about 90 percent female progeny. The males from the original strain, as well as those from the backcross generations, produce normal sex ratios when mated with normal females. Suggest an explanation for these results.

**16.27** Corn plants of genotype *AA bb* with cytoplasmic male sterility and normal *aa BB* plants are planted in alternate rows, and pollination is allowed to occur at random. What progeny genotypes would be expected from the two types of plants?

**16.28** The illustration below shows the locations of restriction sites A through N in the mitochondrial DNA molecules of five species of crabs. The presence of a restriction site is represented by a short vertical line. Assuming that the proportion of shared restriction sites between two molecules is an indication of their closeness of relationship, draw a diagram showing the inferred relationships among the species.

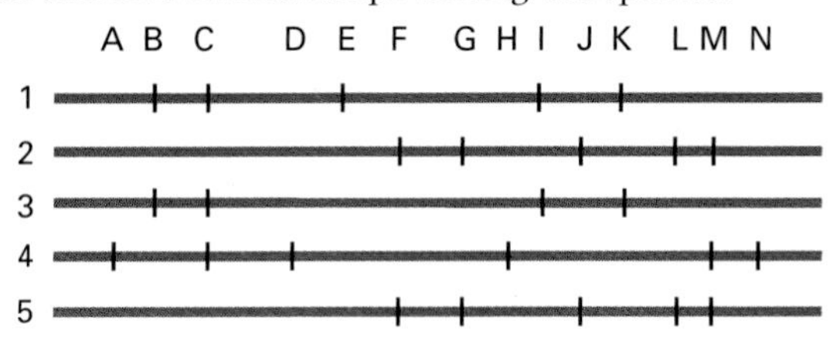


**16.29** A snail of the species *Limnaea peregra* in which the shell coils to the left undergoes self-fertilization, and all of the progeny coil to the right. What is the genotype of the parent? If the progeny are individually self-fertilized, what coiling phenotypes are expected?

**16.30** Females of a certain strain of *Drosophila* are mated with wildtype males. All progeny are female. There are three possible explanations: (1) The females are homozygous for an autosomal recessive allele that produces lethal male embryos (that is, a maternal effect). (2) The females are homozygous for an X-linked allele that is lethal in males. (3) The females carry a cytoplasmically inherited factor that kills male embryos. A cross is carried out between $F_1$ females and wildtype males, and only female progeny are produced. Does this result help to decide among these hypotheses?

## CHALLENGE PROBLEMS

**Challenge Problem 1** Genetic suppression of nonsense (chain-terminating) mutations is often due to mutant tRNA molecules whose anticodon binds with the chain-terminating codon, allowing a fraction of the mRNA molecules to translate through the termination codon and continue polypeptide synthesis. Would you expect nonsense suppression through tRNA mutations to be possible in human mitochondrial genes? Explain your answer.

**Challenge Problem 2** In many lineages of organisms, nucleotide sequences in mitochondrial DNA change at a relatively constant rate through evolutionary time. In the canid family, which includes dogs, wolves, foxes, coyotes, and jackels, the rate is approximately 2.86 percent per million years. This estimate is based on a comparison of mtDNA in wolves and coyotes. Their mtDNA differ at 5.72 percent of the nucleotide sites, and fossil evidence suggests that these species diverged from a common ancestor about a million years ago. Since the common ancestor, there have been a total of 2 million years of evolution (a million years from the common ancestor to today's wolf plus a million years from the common ancestor to today's coyote). The estimated rate of evolution of mtDNA is therefore $5.72/(2 \times 10^6) = 2.86 \times 10^{-6}$ percent per year, or 2.86 percent per million years. On the other hand, the mtDNA of domesticated dogs and wolves differs by about 1 percent. What does this finding imply about the date of the most recent common ancestor of dogs and wolves? Is this estimate consistent with archeological evidence of the earliest remains of domesticated dogs dating to 10,000–15,000 years ago?

**Challenge Problem 3** In yeast, an *ade6*$^-$ mutation results in a requirement for adenine. An *ade6*$^-$ strain with normal colony size was crossed to each of three *ade6*$^+$ petite strains. Among 100 colonies arising from random spores examined from each cross, the nutritional status and sizes were as follows:

**(a)** 26 Ade$^-$ petite : 23 Ade$^-$ normal : 24 Ade$^+$ petite : 27 Ade$^+$ normal

**(b)** 46 Ade$^-$ normal : 54 Ade$^+$ normal

**(c)** 7 Ade$^-$ petite : 45 Ade$^-$ normal : 40 Ade$^+$ petite : 8 Ade$^+$ normal

State what type of petite is involved in each cross, and indicate whether anything further can be said about the genetic basis of the petite phenotype.

### GENETICS on the web

GeNETics on the Web will introduce you to some of the most important sites for finding genetics information on the Internet. To explore these sites, visit the Jones and Bartlett companion site to accompany *Genetics: Analysis of Genes and Genomes, Seventh Edition* at http://biology.jbpub.com/book/genetics.

There you will find a chapter-by-chapter list of highlighted keywords. When you select one of the keywords, you will be linked to a Web site containing information related to that keyword.

# CHAPTER 17

# Molecular Evolution and Population Genetics

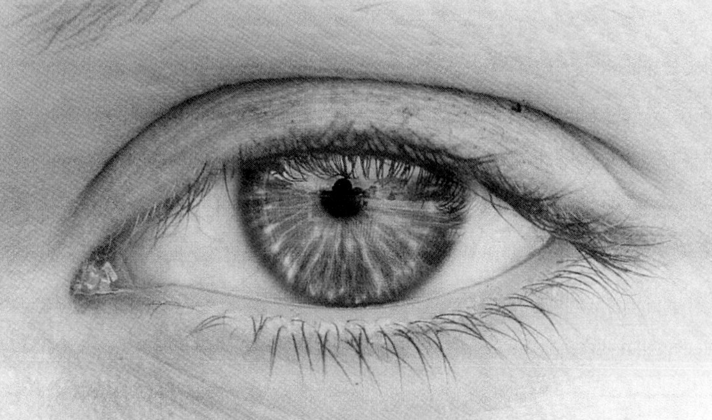

In many pedigrees blue eyes is inherited as a simple Mendelian recessive trait. Genome-wide association studies have pinpointed a likely molecular mechanism. A single-nucleotide polymorphism in an intron of a gene in chromosome 15 appears to reduce the expression of an adjacent gene designated *OCA2*. Mutations in *OCA2* are known to cause one type of albinism (lack of pigment in skin, hair, and eyes), and the gene is the human counterpart of a mouse gene that is critical for the function of melanocyte pigment cells. [© Stephen Mcsweeny/ShutterStock, Inc.]

## CHAPTER OUTLINE


**17.1 Molecular Evolution**
- Gene Trees
- Bootstrapping
- Gene Trees and Species Trees
- Rates of Protein Evolution
- Rates of DNA Evolution
- Origins of New Genes: Orthologs and Paralogs

**17.2 Population Genetics**
- Allele Frequencies and Genotype Frequencies
- Random Mating and the Hardy–Weinberg Principle
- Implications of the Hardy–Weinberg Principle
- A Test for Random Mating
- Frequency of Heterozygous Genotypes
- Multiple Alleles
- DNA Typing
- X-Linked Genes

**17.3 Inbreeding**
- The Inbreeding Coefficient
- Allelic Identity by Descent
- Calculation of the Inbreeding Coefficient from Pedigrees
- Effects of Inbreeding

**17.4 Genetics and Evolution**

**17.5 Mutation and Migration**
- Irreversible Mutation
- Reversible Mutation

**17.6 Natural Selection**
- Selection in a Laboratory Experiment
- Selection in Diploid Organisms
- Components of Fitness
- Selection–Mutation Balance
- Heterozygote Superiority

**17.7 Random Genetic Drift**

**17.8 Tracing Human History Through Mitochondrial DNA**

CONNECTION A Yule Message from Dr. Hardy
Godfrey H. Hardy 1908
*Mendelian Proportions in a Mixed Population*

CONNECTION Resistance in the Blood
Anthony C. Allison 1954
*Protection Afforded by Sickle-Cell Trait Against Subtertian Malarial Infection*

## 17.1 Molecular Evolution

Macromolecules such as DNA, RNA, and protein are linear polymers of subunits. The specific sequence of subunits along each molecule determines its information content or function. With the vast outpouring of data from large-scale genomic sequencing, there is great interest in comparing the sequences of related molecules among species, motivated in part by the hope of correlating differences in sequence with differences in function, especially in proteins.

Although the sequences of macromolecules contain information about function, they also contain information about evolutionary history. Sequences change through time even among macromolecules whose function remains identical. In fact, it is often difficult to distinguish which differences in sequence between species are important to the function of a molecule, and which differences have such small effects that they simply reflect changes that take place by chance over evolutionary time.

The study of how (and why) the sequences of macromolecules change through time constitutes *molecular evolution.* In this section we consider several aspects of molecular evolution, beginning with reconstructing the evolutionary history of a set of sequences and ending with the origin of new genes.

### Gene Trees

Because sequences change through time, it follows that sequence differences accumulate through time. The accumulation of differences is the basis of **molecular phylogenetics**, which is the analysis of molecular sequences in order to infer their evolutionary relationships.

The principles of reconstructing evolutionary history from molecular sequences can best be illustrated by example. We will use the data in **FIGURE 17.1**, part A, which depicts the sequence of the first fifty amino acids of the beta globin chain of adult hemoglobin from each of seven organisms. The technical term used for the source of each sequence is **taxon** (plural **taxa**), and in this case the taxa are species of vertebrates. In evolutionary studies, the sequences to be compared are normally much longer, and there can be many more species, but this small data set will serve to illustrate the methods. Each raised dot means that the amino acid at the site is identical to that in the molecule listed along the top (in this case human). The cow and sheep sequences can be aligned with the others only if gaps are introduced at positions 2 and 3.

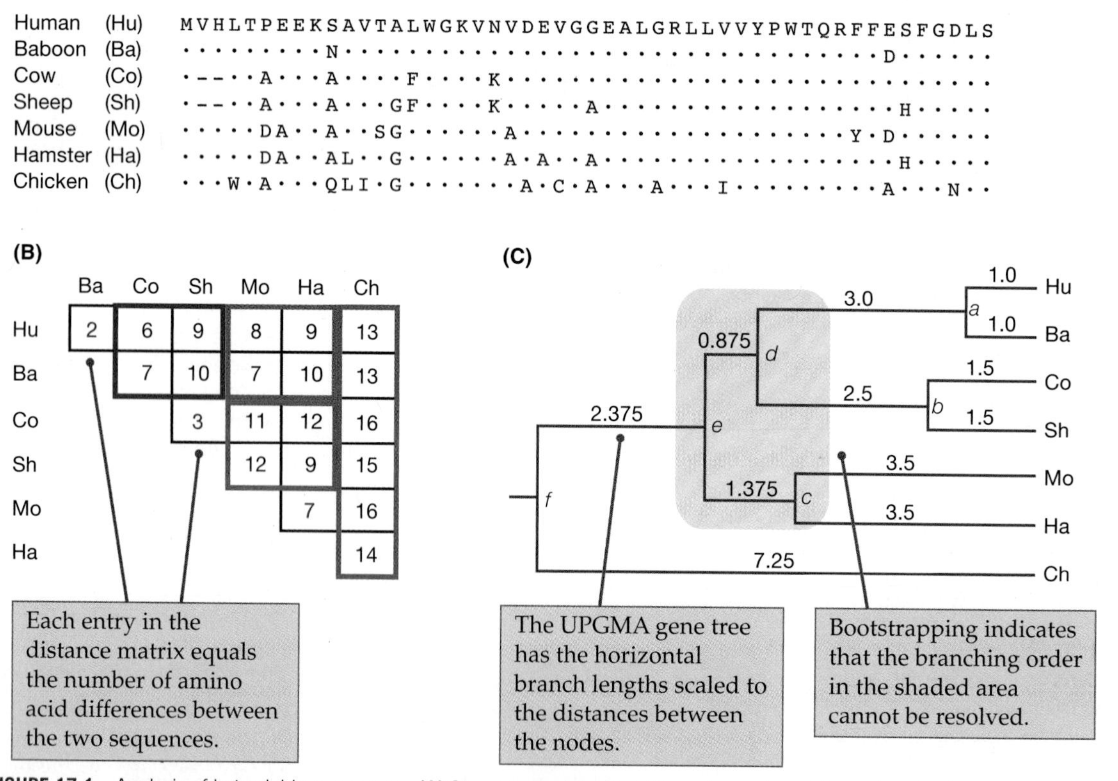


| | Ba | Co | Sh | Mo | Ha | Ch |
|---|---|---|---|---|---|---|
| Hu | 2 | 6 | 9 | 8 | 9 | 13 |
| Ba | | 7 | 10 | 7 | 10 | 13 |
| Co | | | 3 | 11 | 12 | 16 |
| Sh | | | | 12 | 9 | 15 |
| Mo | | | | | 7 | 16 |
| Ha | | | | | | 14 |

**FIGURE 17.1** Analysis of beta globin sequences. (A) Sequences of the first fifty amino acid residues of each of seven taxa. (B) Distance matrix showing the number of differences between each pair of sequences. (C) Gene tree inferred from the distance matrix using UPGMA.

These are indicated by dashes and correspond to deletions of 6 bp in the cow and sheep genes relative to the others. (Alternatively, there could be a 6-bp insertion in each of the other sequences, but this interpretation is less likely.)

Because the sequences of biological macromolecules change through time, the amount of difference between any two sequences can be taken as a measure of how long they have been evolving along separate evolutionary paths. There are two problems in assuming such an equivalence. The first problem is that changes in sequence are a matter of chance depending on which mutations take place and their likelihood of being fixed in the population. (A mutant allele is said to be **fixed** if it replaces all other alleles in the population.) Because new mutations occur at random times, any two molecules separated by the same length of time may differ at more or fewer sites purely by chance. In practice, this problem can be minimized by avoiding sequences that are so closely related that their expected number of differences is of the same magnitude as the random variation. The second problem is that some mutations may increase the ability of the organism to survive and reproduce, and these mutations have a much better chance of being fixed in the population. Thus, problems can be minimized by studying regions of DNA, RNA, or protein that are less likely to include favorable mutation, but it is often difficult to know where these regions are located.

A frequent goal of molecular evolution is to estimate the pattern of evolutionary relationships among a set of sequences, which is called a **gene tree**. It is a gene tree because it is based on a single gene (or, in this example, part of a gene). As will be explained, the pattern of evolutionary relationships among the sequences in a gene tree is not necessarily the same as the pattern of evolutionary relationships among the taxa. A number of methods can be used to estimate a gene tree, and standard software packages are available for doing the calculations. Each method has its own strengths and limitations, and many evolutionary biologists analyze their data using multiple methods with the hope that the resulting trees will differ only in unimportant details. The most commonly used methods can be classified under four broad headings.

1. *Distance methods,* which are based on the number of differences between each pair of sequences.
2. *Parsimony methods,* which systematically search among all possible gene trees to find those that minimize the number of fixed mutations needed to account for the data.
3. *Maximum likelihood methods,* which invoke a theoretical model of how nucleotide or amino acid substitutions occur, and then identify the gene tree that maximizes the probability of observing the actual data based on this model.
4. *Bayesian methods,* which use a generalization of Bayes theorem (Chapter 3) to infer the relative probability of any gene tree based on prior assumptions about the distribution of possible trees (for example, in the simplest case, that all possible gene trees are equally likely).

Among the simplest methods to implement are distance methods, and we will illustrate one of these distance methods using the data in Figure 17.1A. In the first step, all possible pairs or sequences are compared. For each pairwise comparison, the *distance* between the sequences is estimated based on the number of differences between them. In cases in which the sequences are sufficiently different that multiple changes may have taken place at any given site during the course of evolution, then a correction for such *multiple hits* needs to be made. In Figure 17.1A, the sequences are similar enough that a multiple-hit correction is not needed, and we can take the distance between each pair of sequences as the number of amino acid sites that differ between the sequences. (For convenience in this example, we will also count any deleted amino acids as differences.) These pairwise differences are then displayed in the form of a **distance matrix** (Figure 17.1B), which provides the data for the analysis.

One way to estimate a gene tree from a distance matrix is known as **neighbor joining**. At each step, the sequences separated by the shortest distances (the neighbors) are joined together by branches in the gene tree leading to their inferred common ancestor. Neighbor joining is conceptually simple and often yields a gene tree that is identical or very similar to the best trees produced by other methods. We will illustrate a method that defines the distance between groups of sequences according to the unweighted arithmetic average of their pairwise distances. This neighbor-joining method is known as *UPGMA* (an acronym for *unweighted pair group method with arithmetic mean*). This method assumes that rates of evolution are approximately constant among the different branches, and it is very sensitive to departures from this assumption. Nevertheless, its simplicity makes it useful for purposes of illustration.

The gene tree in Figure 17.1C is based on the distance matrix in part B. The tips at the far right are the sequences from the existing taxa (Hu, Ba, Co, etc.), and each node in the tree represents the sequence present in an inferred common ancestor (*a, b, c,* etc.) To see how this tree was derived, note first that the closest pairs of sequences are those whose distances are denoted in red in the distance matrix. In the first step, therefore, we combine these neighbors, namely, human with baboon (Hu/Ba), cow with sheep (Co/Sh), and mouse with hamster (Mo/Ha). The branches leading to *a* depict the joining of Hu with Ba, those leading to *b* depict the joining of Co with Sh, and those leading to *c* the joining of Mo with Ha. The pairwise differences between Hu/Ba, Co/Sh, and Mo/Ha are 2, 3, and 7, respectively (red numbers in the distance matrix), and note that the branch lengths are scaled according to half of each of these distances. The reason is that the sum of the two branches connecting to each node should equal the total distance, for example, in the case of Mo/Ha to *c*, for $3.5 + 3.5 = 7$.

The next step in estimating the gene tree is to decide whether to join *a* (the common ancestor of Hu/Ba) with *b* (that of Co/Sh) or *c* (that of Mo/Ha). In the distance matrix, the distance between the group Hu/Ba and the group Co/Sh is the average of the numbers in the red square, or $(6 + 9 + 7 + 10)/4 = 8$; the distance between the group Hu/Ba and the group Mo/Ha is the average of the numbers in the blue square, or $(8 + 9 + 7 + 10)/4 = 8.5$. These are close, but the rule is to combine the groups with the shortest distance, so we join Hu/Ba with Co/Sh and connect them to the common ancestor *d*. Again the distances are halved, and the branch lengths in the gene tree are scaled so that each sequence from an existing taxon has the same distance back to the common ancestor. Hence, we make the distance *a–d* equal to 3.0 and the distance *b–d* equal to 2.5 so that Hu, Ba, Co, and Sh are all equally distant from *d*.

Now we join *c* with *d*. The reason is that the distance between the group Mo/Ha and the group Hu/Ba/Co/Sh is given by the average of the numbers in the blue and green boxes, or $(8 + 9 + 7 + 10 + 11 + 12 + 12 + 9)/8 = 9.75$, whereas that between the group Mo/Ha and Ch is given by $(16 + 14)/2 = 15$ (these are the two entries at the bottom right in the distance matrix). The branch length *d–e* is set equal to 0.875 and that of *c–e* is set equal to 1.375, so that the total branch length from *e* to any existing taxon equals $9.75/2 = 4.875$.

Finally, we join *e* with Ch. The total distance between Ch and all the other taxa is given by the average of the numbers in the purple rectangle, or $(13 + 13 + 16 + 15 + 16 + 14)/6 = 14.5$. This distance is split, apportioning a branch length of 7.15 to the *f*–Ch lineage and 2.375 to the branch *f–e*. In this way, the total branch length between *f* and the sequence in any existing taxon is 7.25.

In estimating a gene tree, the calculations are normally not done by hand as we have done here for illustration. The calculations are tedious, even for such a small set of sequences as we have been considering. Many methods are impossible to implement without high-speed computers, and some are challenging even for the fastest computers. Software for implementing any of the gene-tree methods discussed above is readily available from sources that can be identified by searching on the Internet, using key words such as *molecular phylogenetics software*.

## Bootstrapping

A gene tree is an estimate of the true pattern of evolutionary relationships among a set of sequences. It is an estimate because there is random variation in the number of substitutions, and the true gene tree is unknown. As might be expected, shorter branches in a gene tree are less reliable than longer branches. But what criterion can be used to assess the reliability of a particular branching order? For example, in Figure 17.1C, do the data really imply that the Mo/Ha lineage splits off from the common ancestor prior to the Hu/Ba/Co/Sh lineage?

A common technique for assessing the reliability of a node in a gene tree is called **bootstrapping**. In this procedure, 1000 or more different data sets are constructed from the actual data by choosing sites at random. Bootstrap sampling is carried out with replacement, which means that the same site can, by chance, be chosen two or more times. A bootstrap sample from the sequence in Figure 17.1A would therefore consist of a sample of fifty sites chosen at random with replacement. In a particular bootstrap sample of fifty, eighteen sites are expected to be present once, nine twice, and five three or more times—and eighteen sites not at all. Hence, if a branching in the gene tree is supported by many of the sites in the sequences, then gene trees from most bootstrap samples will include the same branching, but if a branching is supported by relatively few sites, the gene trees from many bootstrap samples will not include it. In the gene tree in Figure 17.1C, fewer than 50 percent of 1000 bootstrap samples support the branching order included in the shaded area. In practical terms, what this means is that, insofar as this small region of the protein is concerned, the groups Hu/Ba, Co/Sh, and Mo/Ha became separated so closely in time that the question of which taxon split off first cannot be resolved.

## Gene Trees and Species Trees

It seems reasonable to suppose that the evolutionary relationships among a set of genes from different species (the gene tree) must be the same as the evolutionary relationships among the species themselves (the species tree), because the genes are present in the genomes of the species. But this is not the case. The gene tree is not necessarily the same as the species tree.

One way in which the gene tree can differ from the species tree is shown in **FIGURE 17.2**. The ancestral population is originally fixed for the $A_1$ allele (top), but then an $A_2$ mutation occurs and the population becomes polymorphic. The polymorphism is maintained even as species 1 splits off and as species 2 and 3 become separated. However, loss of one allele (and fixation of the other) eventually takes place, and in this example species 1 and 2 become fixed for allele $A_2$ whereas species 3 becomes fixed for allele $A_1$. This means that the gene tree would group the alleles from species 1 and 2 as being the most closely related, whereas the species tree shows that species 2 and 3 are actually the most closely related.

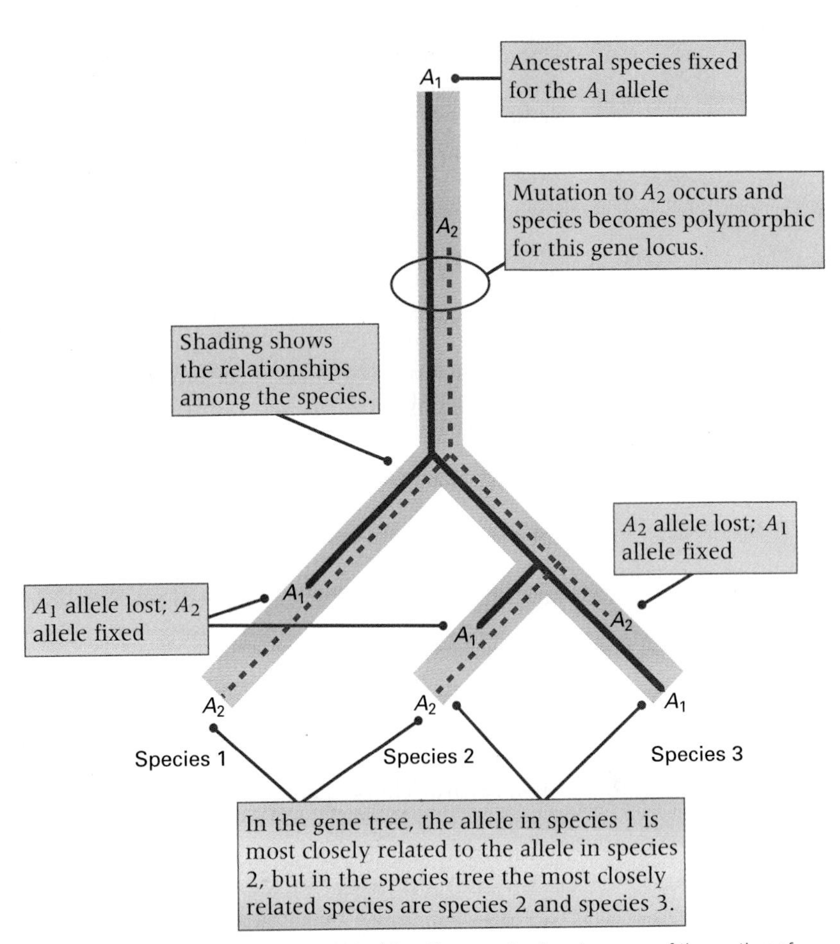


**FIGURE 17.2** A gene tree may not coincide with a species tree because of the sorting of polymorphic alleles in the different lineages.

The situation becomes even more complicated because of recombination. When genes can undergo recombination, it implies that different genes can have different evolutionary histories. Because recombination within a gene can occur, it is even possible for different parts of the same gene to have different evolutionary histories. This is one reason why the nonrecombining part of the Y chromosome (Chapter 8) and mitochondrial DNA (Section 17.8) are of such utility in tracing the recent history of human populations. However, these studies yield estimates of either the Y chromosome tree or the mitochondrial DNA tree, and it is important to bear in mind that the gene trees for nuclear genes will not necessarily be the same, and in fact may differ from one nuclear gene to the next.

On the other hand, for genes with polymorphisms that persist for relatively short times or for taxa that are sufficiently old, gene trees often do coincide with species trees. Discordance between gene trees and species trees is a potential problem primarily for genes that can maintain polymorphisms for long periods of time or for taxa that are closely related. The ancient polymorphism problem is exemplified by many genes that function in the immune system that are highly polymorphic and that can maintain these polymorphisms for periods that are as long or longer than the time required for the formation of new species. In such cases, the allele sorting process in Figure 17.2 can result in discrepancies between the gene tree and the species tree. The young taxa problem is exemplified by the branching order between human, chimpanzee, and gorilla. The species are still so relatively young that polymorphisms in the ancestral populations have been sorted so that some genes support one branching order and other genes support another. (As data have accumulated, however, the majority of gene trees have supported a branching order in which the gorilla splits off first.)

## Rates of Protein Evolution

By the *rate* of sequence evolution of a molecule, we mean the fraction of sites that undergo a change in some designated interval of time. For example, for the entire beta globin molecule between mouse and human, the rate of sequence evolution is 1.23 amino acid replacements per amino acid site per billion years, or $1.23 \times 10^{-9}$ amino acid replacements per amino acid site per year.

Different proteins (and sometimes different parts of proteins) evolve at very different rates. Molecules such as the histones H3 and H4 evolve very slowly. For example, among the 103 amino acids in histone H4, the molecules in rice and humans differ at only two sites. Other molecules evolve relatively rapidly. For example, between mouse and humans the antiviral protein gamma interferon has evolved at a rate of about $5 \times 10^{-9}$ amino acid replacements per amino acid site per year, a rate approximately four times faster than beta globin and very much faster than histone H4 (there are no differences in histone H4 between mouse and human).

Although protein sequences evolve at very different rates, some proteins in some taxa show a rough constancy in their rate of amino acid replacement over long periods of evolutionary time. The apparently constant rate of sequence change has been called a **molecular clock** and affords a basis for attaching a time scale to a gene tree and therefore a time scale for the branching of species independent of the fossil record. There is an elegant theoretical argument that explains why a constant rate of sequence evolution might be expected, as least in the simplest cases. Consider a gene in a population of $N$ diploid individuals, so that in the entire population there are $2N$ copies of the gene. Suppose that some new mutations are **selectively neutral**, which means that they have no effects on the ability of the organisms to survive and reproduce, and that the rate of mutation to selectively neutral alleles is $\mu$ per gene per generation. Then in any one generation, the expected number of new selectively neutral alleles is $2N\mu$. As time goes on, because the population is finite in size, some of the lineages of these genes will become extinct by chance and they will be replaced with other gene lineages. Eventually a time will come when all the gene lineages will have become extinct except one. The probability that any particular gene lineage replace all others is $1/(2N)$, and because there are $2N\mu$ new mutations, the expected number of new mutations (sequence changes) that become fixed in each generation is:

$$\text{Rate of neutral evolution} = \frac{2N\mu}{2N} = \mu$$

The expected rate of neutral evolution is therefore equal to the rate of neutral mutation, which constitutes a sort of molecular clock whose ticks are mutations that become fixed.

In applying this model there are some important caveats. One warning is that the neutral mutation rate can differ from one gene to the next according to what fraction of new mutations is neutral or nearly neutral. For genes that evolve extremely slowly, like histone H4, most amino acid replacements are probably very deleterious, and so the neutral mutation rate will be very low. On the other hand, for genes that evolve extremely rapidly, such as gamma interferon, a significant fraction of the amino acid replacements that become fixed may be favorable mutations, which violates the assumption of neutrality. A second caveat is that the molecular clock is not like a timepiece that ticks at reproducible intervals. It is a random or stochastic clock, in which only the average interval between ticks is predictable. An analogy may be made with radioactive decay, which is a random but clocklike process, but even this analogy has the shortcoming that the random variation in a molecular clock is much larger, relative to the mean, than the random variation in an atomic clock. A third warning is that in many cases molecular clocklike behavior is not observed, because there are different rates of sequence evolution along different branches of the gene tree. For example, in the lineages of humans and mouse, the rate of sequence evolution along the branch leading to mouse is about two times faster than that along the branch leading to human, which is thought to be due to the shorter generation time of organisms in the mouse lineage, resulting in more generations along that lineage. Methods for reconstructing gene trees need to take this source of variation into account, which the UPGMA method discussed earlier does not.

## Rates of DNA Evolution

The general principles of the molecular evolution of protein sequences also apply to DNA sequences, but there are some important differences. Proteins consist of twenty amino acids, but DNA consists of only four nucleotides. The smaller number of subunits means that independent but identical changes at nucleotide sites are much more likely than at amino acid sites, and also that a DNA site that undergoes two sequential mutations is more likely to return to the original state than is an amino acid site. In analyzing DNA sequences, these kinds of events often require that multiple-hit corrections be made to the distance matrix.

There are also different kinds of nucleotide sites depending on their position and function in the genome, and these evolve at different rates (**FIGURE 17.3**). Note that the vertical scale in Figure 17.3 is logarithmic and covers three orders of magnitude, so the differences are very large. On

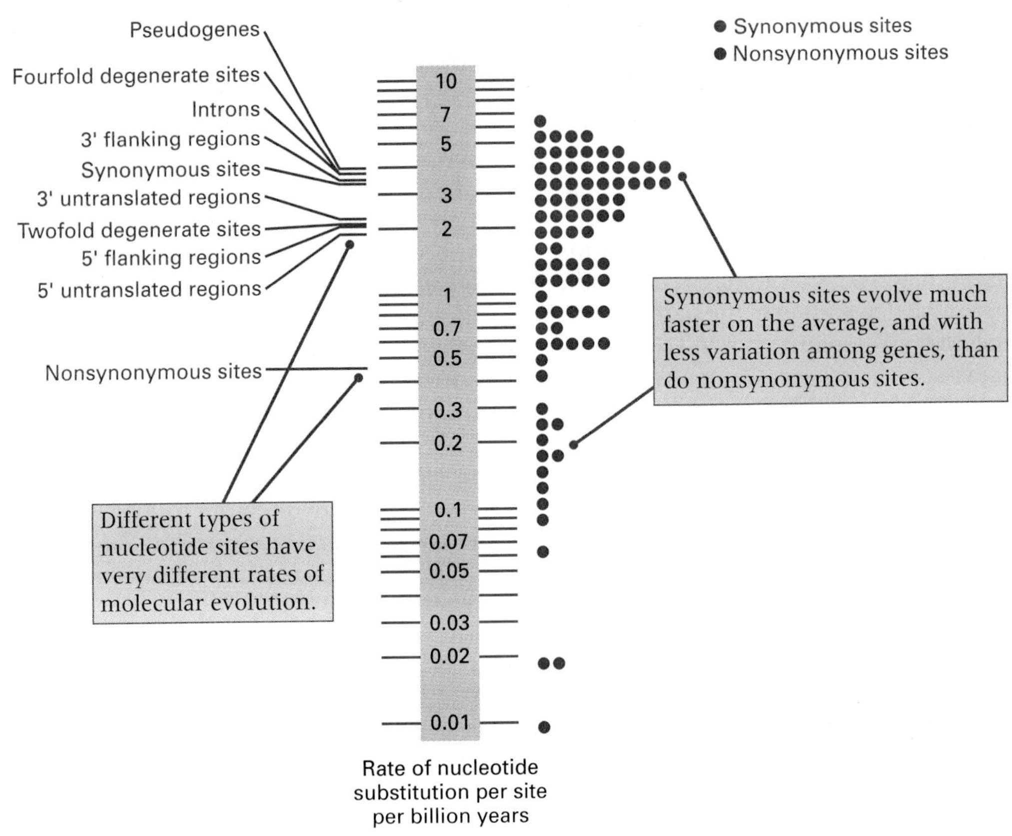


**FIGURE 17.3** Rates of nucleotide substitution in mammalian genes. The left side shows the average rates for different classes of sequence; the right side shows the rates for synonymous and nonsynonymous sites in a sample of 43 genes.

the right are shown the rates of synonymous substitution (red) and nonsynonymous substitution (blue) in each of 43 genes. A *synonymous substitution* in a coding sequence does not result in an amino acid replacement. Synonymous sites are sites at which synonymous substitutions can occur, primarily at the third codon position. A *nonsynonymous substitution* does result in an amino acid replacement. Nonsynonymous nucleotide sites occur primarily at first and second codon positions. Reflecting the great variation in rate of amino acid replacement among different proteins, the rates of nonsynonymous substitution in Figure 17.3 are highly variable among genes. The rates of synonymous substitution in the same genes are much less variable and also much faster.

Plotted on the left in Figure 17.3 are the average rates of nucleotide substitution for different classes of DNA sequence. The fastest evolving DNA sequences are those of **pseudogenes**, which are duplicate genes that have lost their function because of mutation. Introns and fourfold degenerate sites also evolve very rapidly. (A *fourfold degenerate site* is a synonymous nucleotide site at which the same amino acid is specified whatever the identity of the nucleotide; a *twofold degenerate site* is one at which the encoded amino acid depends only on whether the nucleotide at the site is a pyrimidine or a purine.) The differences in the average rate of nucleotide substitution among the different classes of DNA sequence are thought to reflect differing tolerance for nucleotide substitutions. DNA sequences in which most nucleotide substitutions are deleterious are relatively intolerant to nucleotide substitutions, and the rate of nucleotide substitution is relatively low. Conversely, in sequences in which many nucleotide substitutions are equivalent or nearly equivalent in their effects on survival and reproduction, the rate of nucleotide substitution is relatively high. The high rate of nucleotide substitution in pseudogenes is understandable from this point of view, as are the high rates in fourfold degenerate sites and introns.

## Origins of New Genes: Orthologs and Paralogs

In the course of evolution, new genes usually come from preexisting genes, and new gene functions evolve from previous gene functions. The raw material for new genes comes from duplications of regions of the genome, which may include one or more genes. Duplications take place relatively frequently. Analysis of the genomic sequences of a wide variety of eukaryotes suggests that a eukaryotic genome containing

30,000 genes may be expected to undergo roughly 60–600 duplications per million years.

From an evolutionary standpoint, two types of duplications need to be distinguished. The first is typified by beta globin in the gene tree in Figure 17.1. Each time a speciation event took place, which is represented by a branching of the tree, the beta globin gene became duplicated in the sense that each derived species has a copy of the beta globin gene that existed in the parental species. Genes that are duplicated as an accompaniment to speciation and that retain the same function are known as **orthologous genes**.

New gene functions can arise from duplications that take place in the genome of a single species. Duplications within a genome result in **paralogous genes**. An example of paralogous genes are the beta globin gene and the alpha globin gene. Although the genes are distinct now, they are sufficiently similar in sequence to show without ambiguity that they originated from a single gene in a remote common ancestor. Hence, the beta globin and alpha globin genes are paralogs. (To take this discussion one step further, the beta genes in any two mammalian species are orthologs, as are the alpha genes. However, the beta gene in one species and the alpha gene in another are paralogs.)

When a gene duplication has taken place, the paralogs are redundant and one is free to evolve along any path. Probably the most common event is that one of the paralogs undergoes a mutation that destroys its function, or a deletion that eliminates it. But occasionally mutations take place that cause the functions of the paralogs to diverge. They may evolve different pH optima, for example, so that one gene product performs optimally in compartments of the cell that are relatively basic and the other in compartments that are relatively acidic. Or genetic rearrangements can fuse two unrelated genes and yield a new activity. We have already seen examples in Chapter 15 of how translocations that fuse a transcription factor with an oncogene can lead to acute leukemia. These particular rearrangements are extremely deleterious, but the creation of chimeric genes affords an example of how new functions can be acquired.

Gene duplications also allow the paralogous copies to evolve more specialized functions. An example is shown in **FIGURE 17.4**. At the top is a gene in a multicellular eukaryote that has enhancers for expression in tissue types 1 (red) and 2 (green). The other panels show a gene duplication followed by sequential mutations that knock out either the type-2 enhancer elements (left) or the type-1 enhancer elements (right). The result is that the paralog on the left is expressed only in tissue type 1 and that on the right only in tissue type 2. Specialization of paralogs accompanying loss of functional capabilities is known as **subfunctionalization**. It may often be advantageous because each of the specialized genes is free to evolve toward optimal function in its own domain of expression.

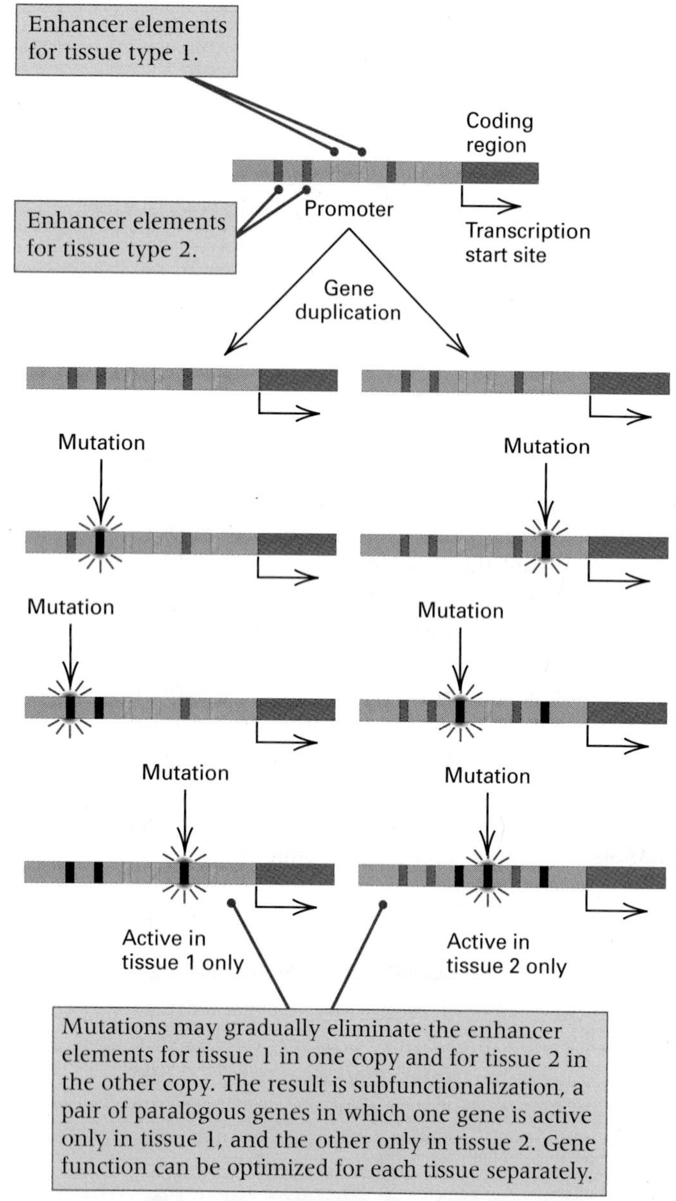


**FIGURE 17.4** Specialization of paralogous genes by subfunctionalization. In this example the paralogs become specialized for tissue-specific expression.

## 17.2 Population Genetics

Our discussion of molecular evolution focused on new mutations that become fixed in lineages because the rate of sequence evolution depends on how rapidly such fixations can take place.

Between the time that a new mutation occurs and the time that it ultimately becomes fixed or lost, it is subject to important processes that determine what the ultimate fate of the mutation will be and how rapidly the fate will be realized. The study of the processes that determine the fate of alleles in populations constitutes *population genetics.*

## ■ Allele Frequencies and Genotype Frequencies

The term *population* refers to a group of organisms of the same species living within a prescribed geographical area. Many geographically widespread species are subdivided into more or less distinct breeding populations, called **subpopulations**, that live within limited geographical areas. Each subpopulation is a *local population.* Although matings occur primarily within each subpopulation, because of occasional migration between subpopulations they all share a common pool of genetic information called the **gene pool**.

We will illustrate the basic concepts of population genetics with an example of an AIDS-resistance phenotype determined by a relatively common mutant allele. The gene in question is a chemokine receptor gene, *CCR5*, found in the human population. (*Chemokines* are molecules that white blood cells of the immune system use to attract one another.) The *CCR5* receptor enables the HIV virus to combine with the plasma membrane and infect the CD4(+) class of T cells of the immune system, which is necessary for an HIV infection to progress. Most human subpopulations contain a *CCR5* allele known as Δ*32* because it has a 32-bp deletion within the coding region. The molecular consequences of the deletion are shown in **FIGURE 17.5**. The relevant part of the normal *CCR5* DNA sequence is shown, grouped into codons, along with the amino acid sequence of the polypeptide, given in the single-letter abbreviations. The nucleotides missing in the Δ*32* deletion are highlighted in color. Above the DNA sequence is the normal CCR5 polypeptide; below is the Δ*32* mutant polypeptide. The deletion creates a frameshift in translation following codon 184, which results in the insertion of 31 incorrect amino acids until a termination codon is encountered after amino acid 215. The truncated protein is nonfunctional and does not support HIV entry into CD4(+) cells. The Δ*32* mutation was originally discovered among persons at high risk of HIV-1 infection who had remained AIDS-free for at least 10 years. The usual frequency of heterozygous genotypes, as we shall see, is about 20 percent, but among AIDS nonprogressors the frequency is approximately 40 percent. The homozygous Δ*32* genotypes, which are much less frequent, seem to have even greater protection.

Why is the Δ*32* homozygous genotype much less frequent than the Δ*32* heterozygous genotype? And what is the relationship between the homozygous and heterozygous frequencies?

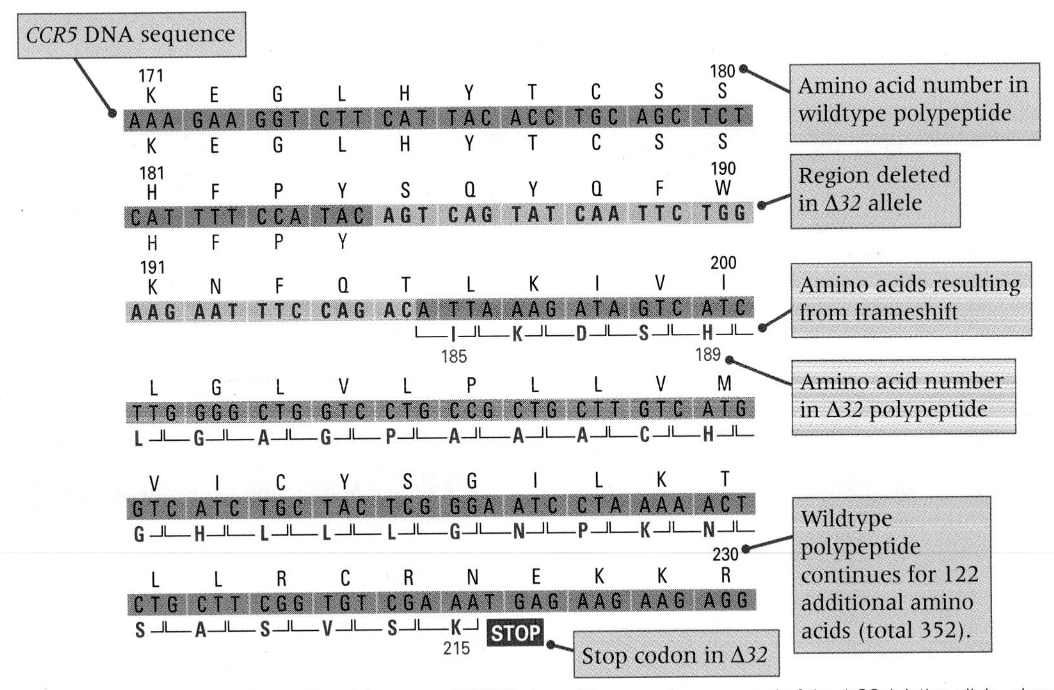


**FIGURE 17.5** DNA sequence of a portion of the normal *CCR5* chemokine receptor gene and of the Δ*32* deletion allele, along with their products of translation.

To begin to answer these questions, we should look at some data. For convenience, we will symbolize the normal *CCR5* allele as *A* and the Δ*32* deletion allele as *a*. In one study of 1000 French people whose DNA was genotyped for *CCR5*, the numbers of homozygous normal (*AA*), heterozygous Δ*32* (*Aa*), and homozygous Δ*32* (*aa*) individuals were as follows:

795 *AA* 190 *Aa* 15 *aa*

These numbers contain a great deal of information about the population, such as whether it is a single homogeneous population or a mixture of genetically somewhat different subpopulations. To interpret this information, first note that the sample contains two types of data: (1) the number of each of the three genotypes, and (2) the number of individual *CCR5* normal (*A*) and Δ*32* (*a*) alleles. Furthermore, the 1000 genotypes represent 2000 alleles of the *CCR5* gene, because each human genome is diploid. These alleles break down as shown in **FIGURE 17.6**. Each homozygous *AA* genotype represents two *A* alleles, each homozygous *aa* genotype two *a* alleles, and each heterozygous *Aa* genotype one allele of each type. By the kind of allele counting in Figure 17.6, the sample of 1000 people therefore represents 1780 *CCR5* normal (*A*) alleles and 220 Δ*32* (*a*) alleles.

Usually, it is more convenient to analyze the data in terms of relative frequency than in terms of the observed numbers. For genotypes, the **genotype frequency** in a population is the proportion of organisms that have the particular genotype. For each allele, the **allele frequency** is the proportion of all alleles that are of the specified type. In the *CCR5* example, the genotype frequencies are obtained by dividing the observed numbers by the total sample size—in this case, 1000. Therefore, the genotype frequencies are:

0.795 *AA* 0.190 *Aa* 0.015 *aa*

Similarly, the allele frequencies are obtained by dividing the observed number of each allele by the total number of alleles (in this case, 2000):

$$\text{Allele frequency of } A = \frac{1780}{2000} = 0.89$$

$$\text{Allele frequency of } a = \frac{220}{2000} = 0.11$$

Note that the genotype frequencies add up to 1.0, as do the allele frequencies. This is a consequence of their definition in terms of proportions; they must add up to 1.0 when all of the possibilities are taken into account. An allele with a frequency of 1.0 is *fixed,* and an allele whose frequency has reached a value of 0 is *lost.*

It is also noteworthy that the allele frequencies among adults, as we have calculated them in Figure 17.6, are the same as those among the gametes produced by the adults. This is true because Mendelian segregation ensures that each heterozygous genotype will produce equal numbers of each type of gamete (**FIGURE 17.7**). Thus, when we calculate the gametic frequencies, the heterozygous genotypes contribute equally to both classes of gametes, whereas the homozygous *AA* and homozygous *aa* genotypes contribute only *A*-bearing or only *a*-bearing gametes, respectively. Consequently, when the genotype frequencies among the parents are taken into account, as well as Mendelian segregation in the heterozygous *Aa* genotypes, the gametes produced in the population have the composition deduced in Figure 17.7. The equality between the allele frequencies among the adults and those among the gametes produced by the adults, namely 0.89 *A* and 0.11 *a*, must be true whenever each adult in the population produces the same number of functional gametes. The apparent random pairing of alleles in the adult genotypes is also important, because it is a clue as to how the gametes are united in fertilization. This issue is examined in the next section.

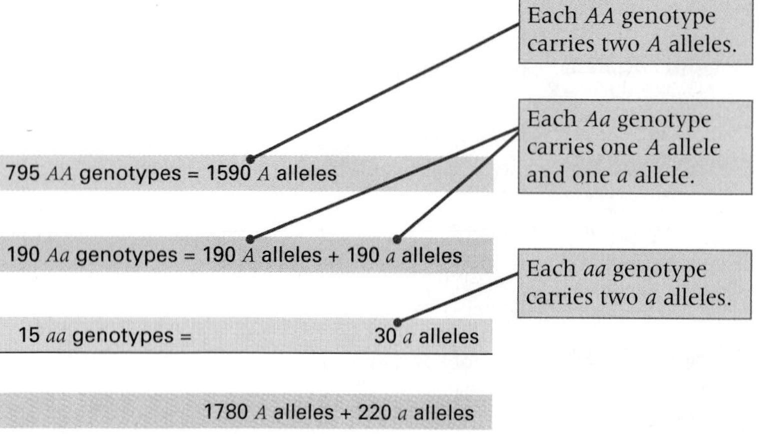
Each AA genotype carries two A alleles.
Each Aa genotype carries one A allele and one a allele.
Each aa genotype carries two a alleles.
795 AA genotypes = 1590 A alleles
190 Aa genotypes = 190 A alleles + 190 a alleles
15 aa genotypes = 30 a alleles
1780 A alleles + 220 a alleles

**FIGURE 17.6** Analysis of the alleles present among a sample of 1000 people genotyped for the *CCR5* receptor gene. The symbol *a* refers to the Δ*32* allele.

## Random Mating and the Hardy-Weinberg Principle

The transmission of genetic material from one generation to the next must be analyzed in terms of alleles rather than genotypes, because genotypes are disrupted in each generation by the processes of segregation and recombination. When gametes unite in fertilization, the geno-

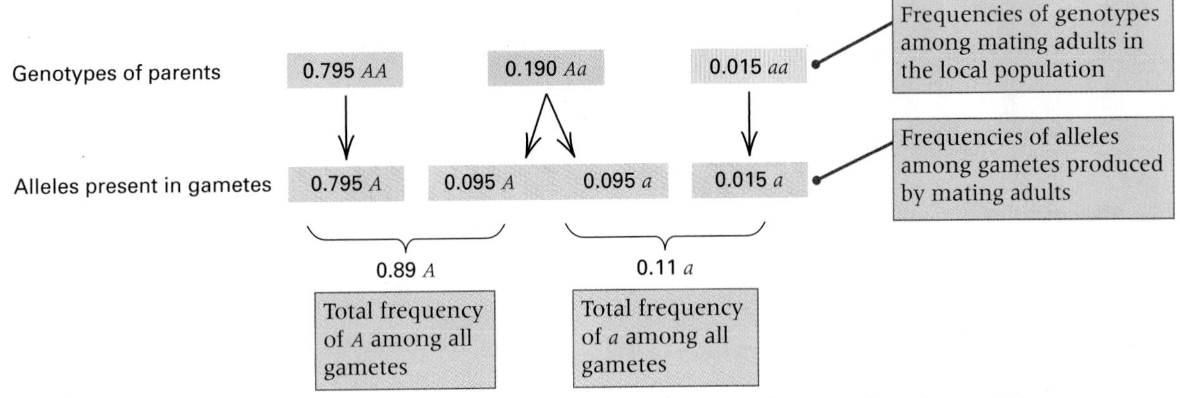


**FIGURE 17.7** Calculation of the allele frequencies in gametes. In any population, the frequency of gametes containing any particular allele equals the frequency of the genotypes that are homozygous for the allele, plus one half the frequency of all genotypes that are heterozygous for the allele.

types of the next generation are formed. The genotype frequencies in the zygotes of the progeny generation are determined by the frequencies with which the parental gametes come together, and these frequencies are, in turn, determined by the manner in which the genotypes pair as mates. In **random mating**, the mating pairs are formed without regard to genotype. Random mating is by far the most prevalent mating system for most species of animals and plants, except for plants that regularly reproduce through self-fertilization. When mating is random, each type of mating pair is formed as often as would be expected through chance encounters between the genotypes.

Random mating implies a pleasingly simple relationship between allele frequency in one generation and genotype frequency in the next generation, because random mating of individuals is equivalent to random union of gametes. Conceptually, we may imagine all the gametes of a population as present in a large container. To form zygote genotypes, pairs of gametes are withdrawn from the container at random. To be specific, consider two alleles $A$ and $a$ with allele frequencies $p$ and $q$, respectively, where $p + q = 1$. In the sample of *CCR5* receptor genotypes considered earlier, we calculated:

$$p = 0.89 \text{ for the } A \text{ allele}$$
$$q = 0.11 \text{ for the } a \text{ allele}$$

The genotype frequencies expected with random mating can be deduced from the tree diagram in **FIGURE 17.8**. The gametes at the left represent the eggs and those in the middle the sperm. The genotypes that can be formed with two alleles are shown at the right, and with random mating, the frequency of each genotype is calculated by multiplying the allele frequencies of the corresponding gametes. However, the genotype

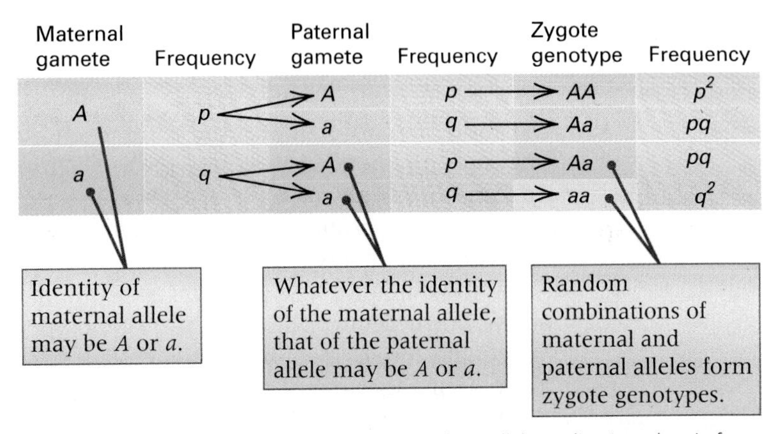


**FIGURE 17.8** When gametes containing either of two alleles unite at random to form the next generation, the genotype frequencies among the zygotes are given by the ratio $p^2 : 2pq : q^2$. This constitutes the Hardy–Weinberg principle.

$Aa$ can be formed in the following two ways: The $A$ allele could have come from the mother (top part of diagram) or from the father (bottom part of diagram). In each case, the frequency of the $Aa$ genotype is $pq$; considering both possibilities, we find that the frequency of $Aa$ is $pq + pq = 2pq$. Consequently, the overall genotype frequencies expected with random mating are:

$$AA: p^2 \quad Aa: 2pq \quad aa: q^2 \qquad (1)$$

The frequencies $p^2$, $2pq$, and $q^2$ result from random mating for a gene with two alleles; they constitute the **Hardy–Weinberg principle**, after Godfrey Hardy and Wilhelm Weinberg, who derived it independently of each other in 1908. Sometimes the Hardy–Weinberg principle is demonstrated by the type of Punnett square illustrated in **FIGURE 17.9**. Such a square is completely equivalent to the tree diagram in Figure 17.8. Although the Hardy–Weinberg principle is exceedingly simple, it has a number of important implications that are not obvious. These are described in the following sections.

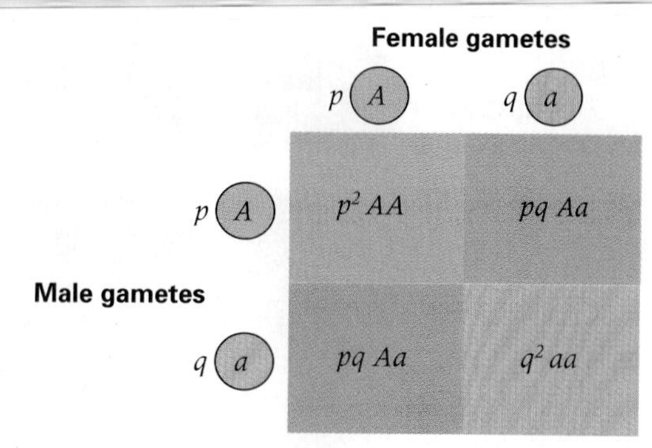


**FIGURE 17.9** A Punnett square showing, in a cross-multiplication format, the ratio $p^2 : 2pq : q^2$ characteristic of random mating.

## Implications of the Hardy–Weinberg Principle

One important implication of the Hardy–Weinberg principle is that *the allele frequencies remain constant from generation to generation.* To understand this point, note that with random mating and gametic frequencies $p$ and $q$, the frequencies of zygotes with the genotypes *AA, Aa,* and *aa* are $p^2$, $2pq$, and $q^2$, respectively. Assuming equal *viability* (ability to survive) among the zygotes, these frequencies will equal those among adults. Furthermore, if all of the adult genotypes are equally fertile, then the frequency, $p'$, of the allele *A* among gametes of the next generation will be,

$$p' = p^2 + 2pq/2 = p(p + q) = p$$

because $p + q = 1$. This argument shows that the frequency of allele *A* remains constant at the value of $p$ through the passage of one (or any number of) complete generation(s). This principle depends on certain assumptions, of which the most important are the following:

- Mating is random; there are no subpopulations that differ in allele frequency.
- Allele frequencies are the same in males and females.
- All the genotypes are equal in viability and fertility (*selection* does not operate).
- Mutation does not occur.
- Migration into the population is absent.
- The population is sufficiently large that the frequencies of alleles do not change from generation to generation merely because of chance.

## A Test for Random Mating

As a real application of the Hardy–Weinberg principle, consider again the *CCR5* receptor. The frequencies of the *CCR5* normal allele (*A*) and *Δ32* allele (*a*) among 1000 adults were 0.89 and 0.11, respectively; assuming random mating, the expected genotype frequencies can be calculated from Equation (1) as:

$$AA: \ (0.89)^2 = 0.7921$$
$$Aa: \ 2(0.89)(0.11) = 0.1958$$
$$aa: \ (0.4575)^2 = 0.0121$$

Note that these are the same frequencies calculated earlier on the basis of the supposition that the alleles were joined in pairs at random. Now we know what this means. It means genotype frequencies according to the Hardy–Weinberg principle. Because the total number of people in the sample is 1000, the expected numbers of the genotypes are 792.1 *AA*, 195.8 *Aa*, and 12.1 *aa*. The *observed* numbers were 795 *AA*, 190 *Aa*, and 15 *aa*. Goodness of fit between the observed and expected numbers can be determined by means of the $\chi^2$ test described in Chapter 4. In this case,

$$\chi^2 = \frac{(795 - 792.1)^2}{792.1} + \frac{(190 - 195.8)^2}{195.8} + \frac{(15 - 12.1)^2}{12.1}$$
$$= 0.877$$

The fit is obviously satisfactory, but to evaluate it quantitatively we need to use the chart in Figure 4.22 on page 141. There is also an adjustment needed in the degrees of freedom. Normally, with three classes of data, we would have 2 degrees of freedom. But when any quantity, such as an allele frequency, is estimated from the data, 1 degree of freedom must be deducted for each quantity estimated. In this case we estimated the gene frequency $p$ ($q = 1 - p$ follows automatically), so we lose 1 of the 2 degrees of freedom we would otherwise have, and thus the appropriate number of degrees of freedom for the test is 1. For this we obtain a $P$ value of 0.35, which means that the hypothesis of random mating can account for the data. On the other hand, the $\chi^2$ test detects only deviations that are rather large, so a good fit to Hardy–Weinberg frequencies should not be overinterpreted, because:

> It is entirely possible for one or more assumptions of the Hardy–Weinberg principle to be violated, including the assumption of random matings, and still not produce deviations from the expected genotype frequencies that are large enough to be detected by the $\chi^2$ test.

The use of a $\chi^2$ test to determine departures from random mating may seem somewhat circular, inasmuch as the data themselves are used in calculating the expected genotype frequencies, but the use of the data is precisely compensated for by deducting 1 degree of freedom for every quantity estimated from the data.

## Frequency of Heterozygous Genotypes

Another important implication of the Hardy–Weinberg principle is that *for a rare allele, the frequency of heterozygotes far exceeds the frequency of the rare homozygote.* For example, when the frequency of the rarer allele is $q = 0.1$, the ratio of heterozygous genotypes to homozygous genotypes is

$$2pq/q^2 = 2p/q = 2(0.9)/0.1$$

or approximately 20; when $q = 0.01$, this ratio is about 200; and when $q = 0.001$, it is about 2000. In other words:

> When an allele is rare, there are many more heterozygotes than there are homozygotes for the rare allele.

The reason for this perhaps unexpected relationship is shown in **FIGURE 17.10**, which plots the frequencies of homozygous and heterozygous genotypes with random mating. Note that at allele frequencies near 0 or 1, the frequency of the heterozygous genotype goes to 0 as a linear function, whereas that of the rarer, homozygous genotype goes to 0 more rapidly.

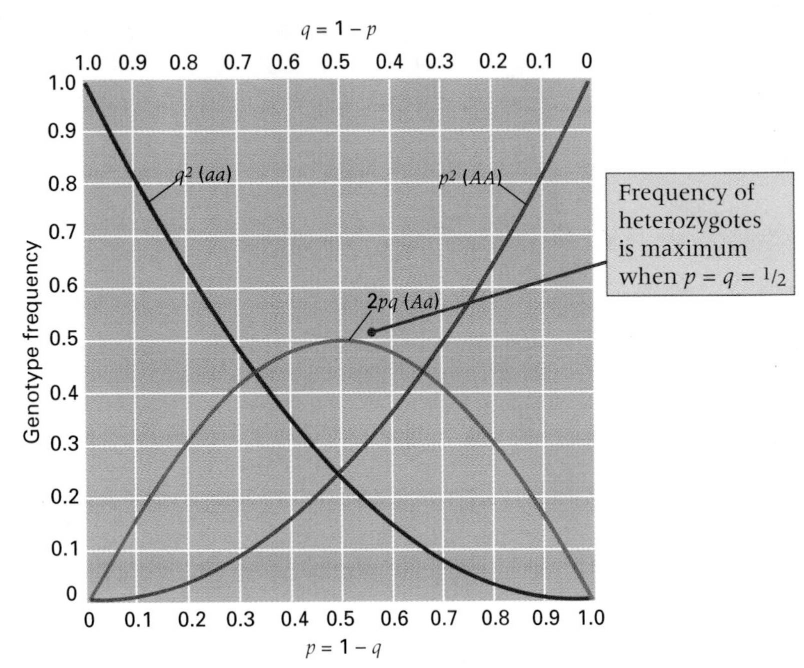


**FIGURE 17.10** Graphs of $p^2$, $2pq$, and $q^2$. If the allele frequencies are between 1/3 and 2/3, then the heterozygous genotype has the greatest frequency.

One practical implication of this principle is seen in the example of cystic fibrosis, an inherited secretory disorder of the pancreas and lungs. Cystic fibrosis is one of the most common recessive severe genetic disorders among Caucasians; it affects about 1 in 1700 newborns. This is the frequency of homozygous recessives, but what are the allele frequencies? The calculation is straightforward, because with random mating, the frequency of homozygous recessives must correspond to $q^2$. In the case of cystic fibrosis,

$$q^2 = 1/1700 = 0.00059 \quad \text{or}$$
$$q = (0.00059)^{1/2} = 0.024$$

and consequently,

$$p = 1 - q = 1 - 0.024 = 0.976$$

The frequency of heterozygous carriers of the allele for cystic fibrosis is therefore,

$$2pq = 2(0.976)(0.024) = 0.047 = 1/21$$

This calculation implies that, for cystic fibrosis, although only 1 person in 1700 is affected with the disease (homozygous), about 1 person in 21 is a carrier (heterozygous). The calculation should be regarded as approximate because it is based on the assumption of Hardy–Weinberg genotype frequencies. Nevertheless, considerations like these are important in predicting the outcome of population screening for the detection of carriers of harmful recessive alleles, which is essential in evaluating the potential benefits of such screening programs.

## Multiple Alleles

Extension of the Hardy–Weinberg principle to multiple alleles of a single autosomal gene can be illustrated by a three-allele case. **FIGURE 17.11** shows the situation with three alleles, designated $A_1$, $A_2$, and $A_3$, with allele frequencies $p_1$,

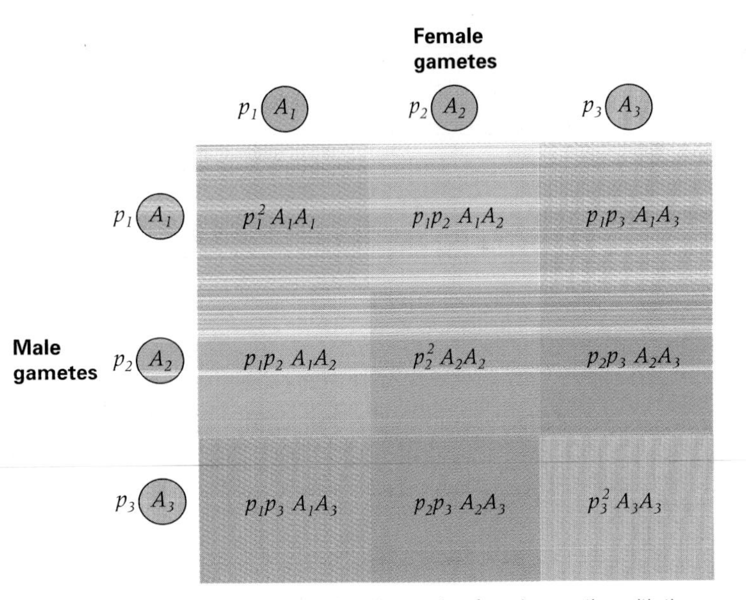


**FIGURE 17.11** Punnett square showing the results of random mating with three alleles.

$p_2$, and $p_3$. As always, the allele frequencies must sum to 1, so $p_1 + p_2 + p_3 = 1.0$. As in Figure 17.9, the entry in each square is obtained by multiplying the frequencies of the alleles at the corresponding margins; any homozygote (such as $A_1A_1$) has a random-mating frequency equal to the square of the corresponding allele frequency (in this case, $p_1^2$). Any heterozygote (such as $A_1A_2$) has a random-mating frequency equal to twice the product of the corresponding allele frequencies (in this case, $2p_1p_2$). The extension to any number of alleles is straightforward:

Frequency of any homozygote =
square of allele frequency
Frequency of any heterozygote =
2 × product of allele frequencies (2)

Multiple alleles determine the human ABO blood groups (Chapter 3), and there are three principal alleles, designated $I^A$, $I^B$, and $I^O$. In one study of 3977 Swiss people, the allele frequencies were found to be 0.27 $I^A$, 0.06 $I^B$, and 0.67 $I^O$. Applying the rules for multiple alleles, we can expect the genotype frequencies resulting from random mating to be as shown in **FIGURE 17.12**. Because both $I^A$ and $I^B$ are dominant to $I^O$, the expected frequency of blood-group *phenotypes* is that shown at the right in the illustration. Note that the majority of A and B phenotypes are heterozygous for the $I^O$ allele; this is because the $I^O$ allele has such a high frequency in the population.

## DNA Typing

Genetic variation in the form of DNA polymorphisms is common in the human genome as well as in genomes of many other organisms. This much is clear from the application of the variety of experimental methods by which DNA polymorphisms can be detected, which we examined in Chapter 2. In the human genome there is so much genetic variation that, except for identical twins and other multiple births that arise from a single zygote, no two human beings are (or ever have been) genetically identical. Each human genotype is unique. One practical application of the principle of genetic uniqueness has come through **DNA typing**, a procedure in which biological samples of unknown human origin are matched with their source through the use of polymorphic DNA markers. Even small samples of human material often contain enough DNA that the genotype can be determined for a number of markers and matched against those present among a group of suspects. Typical examples of crime-scene evidence include blood, semen, hair roots, and skin cells. Even a small number of cells is sufficient, because predetermined regions of DNA can be amplified by the polymerase chain reaction (Chapter 2).

If a suspect's DNA contains sequences that are clearly not present in a sample collected at the crime scene, then the sample must have originated from a different person. On the other hand, if a suspect's DNA *does* match that from the crime scene, then the suspect could be the source. The strength of the DNA evidence depends on the number of polymorphisms examined and on the number of alleles present in the population. The greater the number of polymorphisms that match, especially if they are highly polymorphic, the stronger the evidence linking the suspect to the sample taken from the scene of the crime. With a sufficient number and quality of matching DNA markers, DNA typing can be as reliable a method of individual identification as fingerprints.

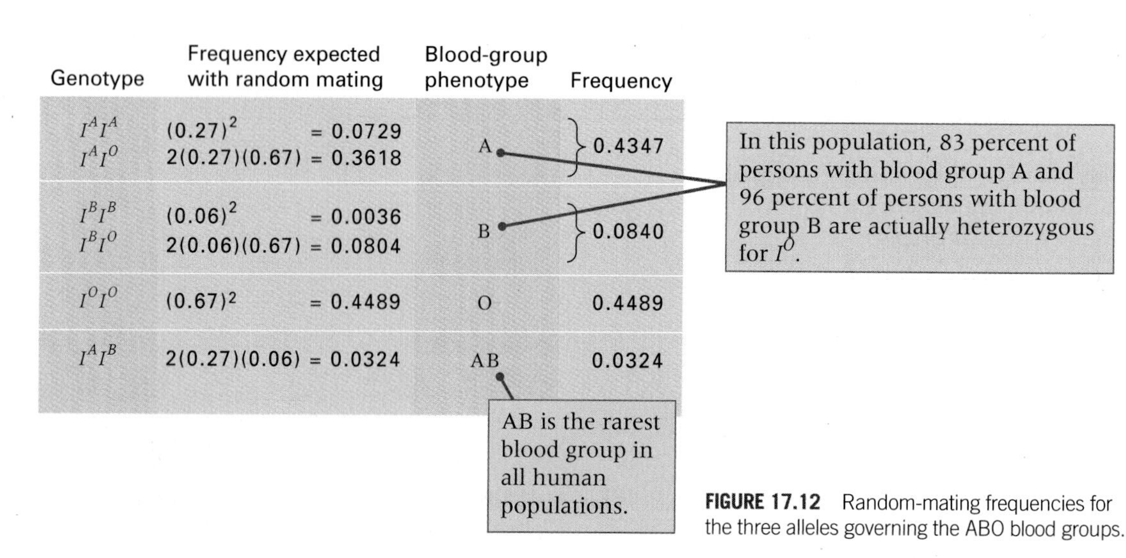

| Genotype | Frequency expected with random mating | Blood-group phenotype | Frequency |
|---|---|---|---|
| $I^AI^A$ | $(0.27)^2 = 0.0729$ | A | 0.4347 |
| $I^AI^O$ | $2(0.27)(0.67) = 0.3618$ | | |
| $I^BI^B$ | $(0.06)^2 = 0.0036$ | B | 0.0840 |
| $I^BI^O$ | $2(0.06)(0.67) = 0.0804$ | | |
| $I^OI^O$ | $(0.67)^2 = 0.4489$ | O | 0.4489 |
| $I^AI^B$ | $2(0.27)(0.06) = 0.0324$ | AB | 0.0324 |

In this population, 83 percent of persons with blood group A and 96 percent of persons with blood group B are actually heterozygous for $I^O$.
AB is the rarest blood group in all human populations.

**FIGURE 17.12** Random-mating frequencies for the three alleles governing the ABO blood groups.

## connection

### A Yule Message from Dr. Hardy

**Godfrey H. Hardy 1908**
Trinity College, Cambridge, England
*Mendelian Proportions in a Mixed Population*

*An early argument against Mendelian inheritance was that dominant traits, even harmful ones, should eventually come to have a 3:1 ratio in any population. This is obviously not the case. There are many dominant traits in human beings that are rare now and have always been rare, such as brachydactyly [short fingers], which is the specific issue in this paper. Hardy, a mathematician, realized that, with random mating, and assuming no differences in survival or fertility, the frequencies of dominant and recessive genotypes would have no innate tendency to change frequency of their own accord. The form of the principle as written by Hardy is that, if the genotype frequencies of* AA, Aa, *and* aa *are* P, 2Q, *and* R *in any generation, then, with random mating, the frequencies will be* $(P + Q)^2 : 2(P + Q)(Q + R) : (Q + R)^2$ *in the next and subsequent generations. The allele frequencies* p *and* q *of* A *and* a *are not stated explicitly, and so the familiar form of the principle as stated today—* $p^2 : 2pq : q^2$*—does not appear as such in Hardy's paper. But it is there implicitly, because* $p = P + Q$ *and* $q = Q + R$.

I am reluctant to intrude in a discussion concerning matters of which I have no expert knowledge, and I should have expected the very simple point which I wish to make to have been familiar to biologists. However, some remarks of Mr. Udny Yule suggest that it may still be worth making. . . . Mr. Yule is reported to have suggested, as a criticism of the Mendelian position, that if brachydactyly is dominant "in the course of time one would expect, in the absence of counteracting forces, to get three brachydactylous persons to one normal." . . . It is not difficult to prove, however, that such an expectation would be quite groundless. Suppose that *A*, *a* is a pair of Mendelian characters, *A* being dominant, and that in any given generation the numbers of pure dominants (*AA*), heterozygotes (*Aa*), and pure recessives (*aa*) are as $P:2Q:R$. Finally, suppose that the numbers are fairly large, so that the mating may be regarded as random, that the sexes are evenly distributed among the three varieties, and that they are all equally fertile. A little mathematics of the multiplication-table type is enough to show that in the next generation the numbers will be as

$$(P + Q)^2 : 2(P + Q)(Q + R) : (Q + R)^2$$

or as $P_1 : 2Q_1 : R_1$, say. The interesting question is—in what circumstances will this distribution be the same as that in the generation before? It is easy to see that the condition for this is $Q^2 = PR$. And since $Q_1^2 = P_1R_1$, whatever the values of *P*, *Q*, and *R* may be, the distribution will in any case continue unchanged after the second generation. . . . In a word, there is not the slightest foundation for the idea that a dominant character should show a tendency to spread, or that a recessive should tend to die out. . . . I have, of course, considered only the very simplest hypothesis possible. Hypotheses other than that of purely random mating will give different results, and, of course, if, as appears to be the case sometimes, the character is not independent of sex, or has an influence on fertility, the whole question may be greatly complicated. But such complications seem to be irrelevant to the simple issue raised by Mr. Yule's remarks.

***There is not the slightest foundation for the idea that a dominant character should show a tendency to spread, or that a recessive should tend to die out.***

Source: G. H. Hardy, *Science* 28 (1908): 49–50.

**FIGURE 17.13** shows one type of polymorphism used in DNA typing. The restriction fragments that correspond to each allele differ in length because they contain different numbers of units repeated in tandem; the polymorphism is said to be a *variable number of tandem repeats (VNTR) polymorphism* (see Figure 2.25 on page 68 for an illustration). Such markers are of value in DNA typing because many alleles are possible, owing to the variable number of repeating units. In Figure 17.13, the lanes in the gel labeled M contain multiple DNA fragments of known size to serve as molecular-weight markers. Each of lanes 1–9 contains DNA from a single person. Two typical features of VNTRs are to be noted:

1. Most people are heterozygous for VNTR alleles that produce restriction fragments of different sizes. Heterozygosity is indicated by the presence of two distinct bands. In Figure 17.13, only the person numbered 1 appears to be homozygous for a particular allele.
2. The restriction fragments from different people cover a wide range of sizes. The variability in size indicates that the population as a whole contains many VNTR alleles. Although many alleles

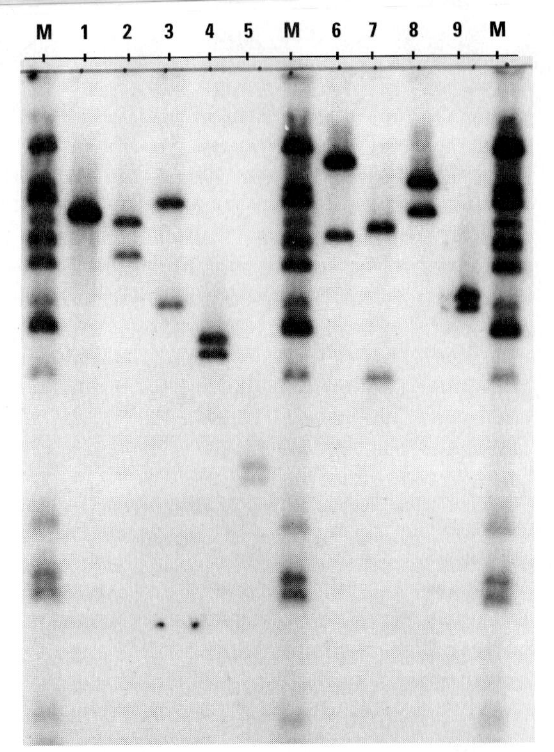


**FIGURE 17.13** Genetic variation in a VNTR used in DNA typing. Each numbered lane contains DNA from a single person; the DNA has been cleaved with a restriction enzyme, separated by electrophoresis, and hybridized with radioactive probe DNA. The lanes labeled M contain molecular-weight markers. [Courtesy of Robert W. Allen, Human Identity Testing Laboratory, Oklahoma State University.]

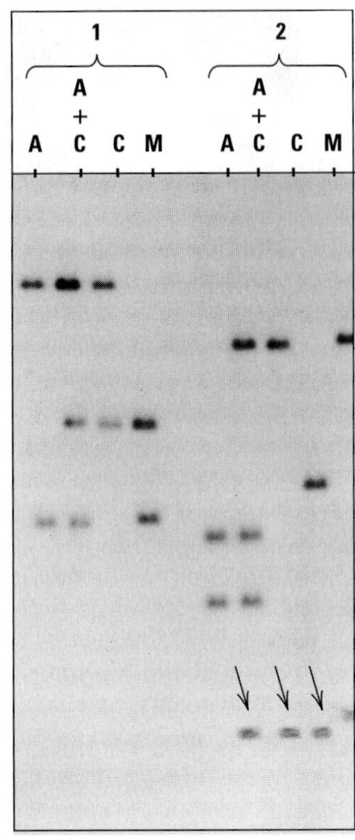


**FIGURE 17.14** Use of DNA typing in paternity testing. The sets of lanes numbered 1 and 2 contain DNA samples from two different paternity cases. In each case, the lanes contain DNA fragments from the following sources: M, the mother; C, the child; A, the accused father. The lanes labeled A + C contain a mixture of DNA fragments from the accused father and the child. The arrows in case 2 point to bands of the same size that are present in lanes M, C, and A + C. Note that the male accused in case 2 could not be the father because neither of his bands is shared with the child. [Courtesy of Robert W. Allen, Human Identity Testing Laboratory, Oklahoma State University.]

can be present in the population as a whole, each person can have no more than two alleles.

The reason why VNTRs are useful in DNA typing is also evident in Figure 17.13; each of the nine people tested has a different pattern of bands and thus could be identified uniquely by means of this VNTR. On the other hand, the uniqueness of each person in Figure 17.13 is due in part to the high degree of polymorphism of the VNTR and in part to the small sample size. If more people were examined, then pairs that matched in their VNTR types by chance would certainly be found. This is why matches at multiple, independent loci are needed for reliable DNA typing. The greater the number of loci, the less likely it is that a match of DNA types is due to chance. With enough matching markers, it becomes virtually certain that the two DNA samples came from the same person.

DNA typing not only can implicate the guilty, it also can exonerate the innocent whose DNA types do *not* match the evidence. An example that illustrates the use of DNA typing in paternity testing is shown in **FIGURE 17.14**. The numbers 1 and 2 designate different cases, in each of which a man was accused of fathering a child. In each case, DNA was obtained from the mother (M), the child (C), and the accused man (A). The pattern of bands obtained for one VNTR is shown. The lanes labeled A + C contain a mixture of DNA from the accused man and the child. In case 1, the lower band in the child was inherited from the mother and the upper band from the father; because the upper band is the same size as one of those in the accused, the accused man could have contributed this allele to the child. This finding does not *prove* that the accused man is the father; it says only that he cannot be excluded on the basis of this particular gene. (On the other hand, if a large enough number of genes are studied, and the man cannot be excluded by any of the genes, then it makes it more likely that he really is the father.) Case 2 in Figure 17.14 is an exclusion. The band denoted by the arrows is the band inherited by the child from the mother. The other band in the child does not match either of the bands in the accused, so the accused man could not be the biological father. (In theory, mutation could be invoked to explain the result, but this possibility is extremely unlikely.)

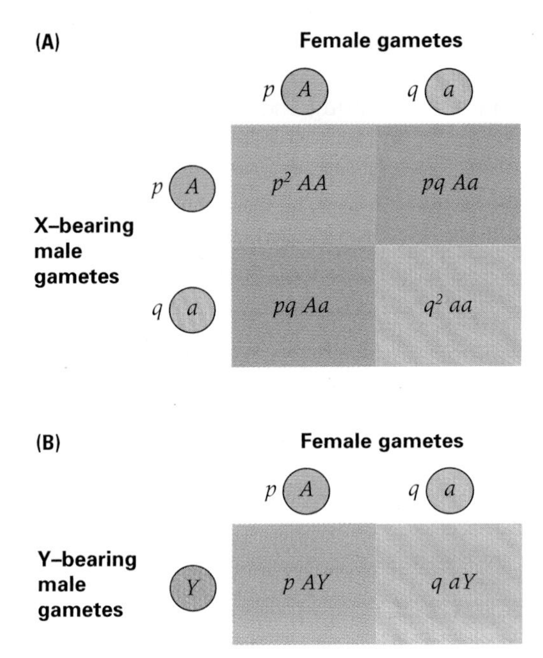


**FIGURE 17.15** The results of random mating for an X-linked gene. (A) Genotype frequencies in females. (B) Genotype frequencies in males.

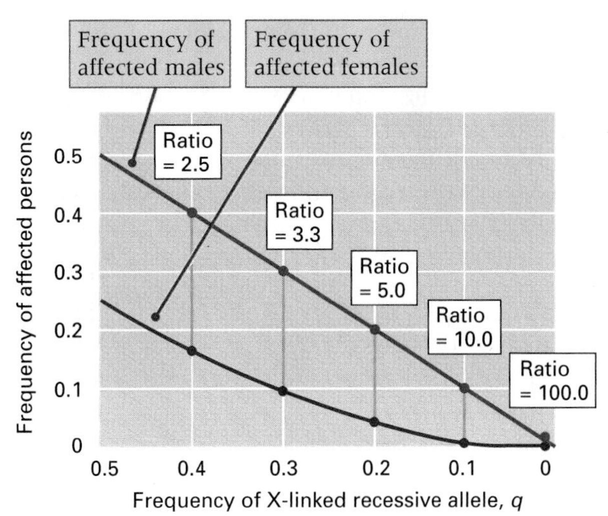


**FIGURE 17.16** For a recessive X-linked allele, the upper curve gives the frequency of affected males ($q$) and the lower curve the frequency of affected females ($q^2$), for values of $q \leq 0.5$. Although both frequencies decrease as the recessive allele becomes rare, the frequency of affected females decreases faster. The result is that the ratio of affected males to affected females increases as the allele frequency decreases.

## ■ X-Linked Genes

If a gene with alleles $A$ and $a$ is X-linked, the result of random mating is illustrated in **FIGURE 17.15**. The principles are the same as those considered earlier, but male gametes carrying the X chromosome (part A) must be distinguished from those carrying the Y chromosome (part B). When the male gamete carries an X chromosome, the Punnett square is exactly the same as that for the two-allele autosomal gene in Figure 17.9, but all the offspring are female. Consequently, among female progeny, the genotype frequencies are:

$$AA : p^2 \quad Aa : 2pq \quad aa : q^2$$

When the male gamete carries a Y chromosome, the outcome is quite different (Figure 17.15B). All the offspring are male, and each has only one X chromosome, which is inherited from the mother. Therefore, each male offspring receives only one copy of each X-linked gene, and the genotype frequencies among males are the same as the allele frequencies: $A$ males with frequency $p$, and $a$ males with frequency $q$.

An important implication of Figure 17.15 is that if $a$ is a rare recessive allele, then there will be many more males exhibiting the trait than females, because the frequency of affected females ($q^2$) will be much smaller than the frequency of affected males ($q$). This principle is illustrated in **FIGURE 17.16**. As the allele frequency of the recessive decreases toward 0, the frequencies of affected males and females both decrease, but the frequency of affected females decreases faster. The result is that the ratio of affected males to affected females increases. At an allele frequency of $q = 0.3$, for example, the ratio of affected males to affected females is 3.3, but for an allele frequency of $q = 0.1$, the ratio of affected males to affected females is 10.0. In general, the ratio of affected males to affected females is $q/q^2$, or $1/q$.

For an X-linked recessive trait, the frequency of affected males provides an estimate of the frequency of the recessive allele. A specific example is found in the common form of X-linked color blindness in human beings. This trait affects about 1 in 20 males among Caucasians, so $q = 1/20 = 0.05$. The expected frequency of color-blind females is therefore estimated as $q^2 = (0.05)^2 = 0.0025$, or about 1 in 400.

# 17.3 Inbreeding

The term **inbreeding** means mating between relatives, such as first cousins. In human pedigrees, a mating between relatives is often called a **consanguineous mating**. The principal consequence of inbreeding is that the frequency of heterozygous offspring is smaller than it is with random mating. This effect is seen most dramatically in repeated self-fertilization, which happens naturally in certain plants. Consider a hypothetical population consisting exclusively of $Aa$ heterozygotes. With self-fertilization, each plant would produce offspring in the proportions 1/4 *AA*, 1/2 *Aa*, and 1/4 *aa*, which means that one generation of self-fertilization reduces the proportion of heterozygotes from 1 to 1/2. In the second generation, only the heterozygous plants can again produce heterozygous offspring, and only half of their offspring will again be heterozygous. Heterozygosity is therefore reduced to 1/4 of what it was originally. Three generations of self-fertilization reduces the heterozygosity to $1/4 \times 1/2 = 1/8$, and so on. The remainder of this section demonstrates how the reduction in heterozygosity because of inbreeding can be expressed in quantitative terms.

## ■ The Inbreeding Coefficient

Repeated self-fertilization is a particularly intense form of inbreeding, but weaker forms of inbreeding are qualitatively similar in that they also lead to a reduction in heterozygosity. A convenient measure of the effect of inbreeding is based on the reduction in heterozygosity. To put this into symbols, suppose that $H_I$ is the frequency of heterozygous genotypes in a population of inbred individuals. The most widely used measure of inbreeding is called the **inbreeding coefficient**, symbolized *F*, which is defined as the proportionate reduction in $H_I$ compared with the value of $2pq$ that would be expected with random mating. Expressed as an equation, this definition implies that:

$$F = (2pq - H_I)/2pq \qquad (3)$$

Equation (3) can be rearranged as:

$$H_I = 2pq(1 - F)$$

The homozygous genotypes increase in frequency as the heterozygotes decrease in frequency, and overall, the genotype frequencies in the inbred population are given by:

$$AA\!: \; p^2(1 - F) + pF$$
$$Aa\!: \; 2pq(1 - F)$$
$$aa\!: \; q^2(1 - F) + qF \qquad (4)$$

These expressions are modifications of the Hardy–Weinberg principle that take inbreeding into account. When $F = 0$ (no inbreeding), the genotype frequencies are the same as those given in the Hardy–Weinberg principle: $p^2$, $2pq$, and $q^2$. At the other extreme, when $F = 1$ (complete inbreeding), the inbred population consists entirely of *AA* and *aa* genotypes in the frequencies *p* and *q*, respectively. Inbreeding alone does not change the allele frequencies in a population. This you can verify for yourself by calculating that, for any value of the inbreeding coefficient *F*, the allele frequencies of *A* and *a* remain at *p* and *q*, respectively.

A graphical representation of the genotype frequencies in Equation (4) is shown in **FIGURE 17.17**. For any gene in a population of organisms, a proportion of the population equaling $1 - F$ may be considered as having, by chance, escaped the effects of inbreeding, which means that the genotype frequencies are those expected with random mating: $p^2$, $2pq$, and $q^2$. In another proportion, equaling *F*, the alleles may be considered to have become homozygous solely because of inbreeding, with the *AA* and *aa* genotype frequencies equal to *p* and *q*, respectively.

**FIGURE 17.18** shows graphically the reduction in heterozygosity due to inbreeding. Each curve is given by an equation of the form $H_I = 2pq(1 - F)$, where the case $F = 0$ corresponds to random mating. The curves show that as the inbreeding coefficient increases, the frequency of heterozygous genotypes decreases proportionally until, when $F = 1$, there are no heterozygotes remaining in the inbred population. In other words, in a highly inbred population with $F = 1$, the genotypes consist either of *AA* or of *aa* in the relative proportions *p* and *q*, respectively.

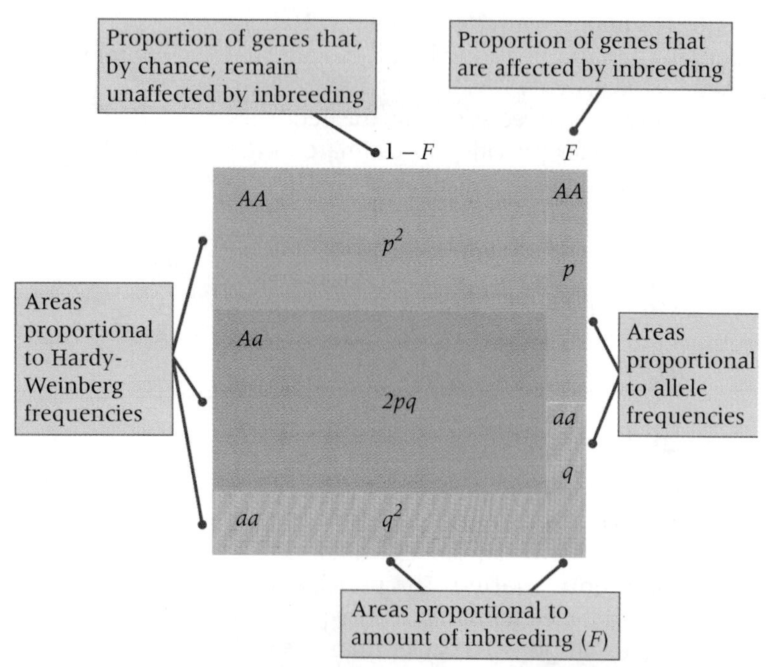


**FIGURE 17.17** Effect of inbreeding on genotype frequencies. The large rectangles on the left pertain to alleles whose ancestries, by chance, are not affected by inbreeding and for which the genotype frequencies remain in Hardy–Weinberg proportions. The narrow rectangles on the right pertain to alleles whose ancestries are affected by the inbreeding, and in this case the genotype frequencies of *AA* and *aa* are related as $p : q$. (Note that there are no heterozygous genotypes in the latter case.)

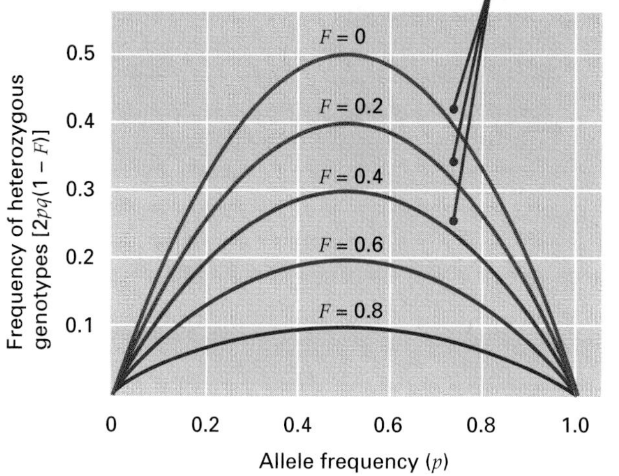


**FIGURE 17.18** Frequency of heterozygous genotypes in an inbred population ($y$-axis) against allele frequency ($x$-axis). As the inbreeding becomes more intense (greater inbreeding coefficient $F$), the proportion of heterozygous genotypes decreases.

A naturally self-fertilizing plant, *Phlox cuspidata,* provides a specific example of the effect of inbreeding. In this species, about 78 percent of the plants undergo self-fertilization, and it can be shown that the inbreeding coefficient corresponding to this level of self-fertilization is $F = 0.64$. In one sample of three genotypes the observed numbers were 15 $Pgm^a\ Pgm^a$, 6 $Pgm^a\ Pgm^b$, and 14 $Pgm^b\ Pgm^b$. These are not at all in agreement with the Hardy–Weinberg expectations for the allele frequencies of $Pgm^a$ ($p = 0.51$) and $Pgm^b$ ($q = 0.49$), because the numbers expected with random mating are 9.1, 17.5, and 8.4. Note that the largest discrepancy is that there are too few observed heterozygous genotypes.

Equation (4) tells us how to adjust the expected numbers to take the inbreeding into account. For the allele frequencies 0.51 and 0.49, and an inbreeding coefficient $F = 0.64$, the expected genotype frequencies are:

$$Pgm^a\ Pgm^a\text{: } (0.51)^2(1 - 0.64) + (0.51)(0.64) = 0.420$$

$$Pgm^a\ Pgm^b\text{: } 2(0.51)(0.49) \times (1 - 0.64) = 0.180$$

$$Pgm^b\ Pgm^b\text{: } (0.49)^2(1 - 0.64) + (0.49)(0.64) = 0.400$$

In a sample size of 35, the expected numbers of the three genotypes are therefore $0.420 \times 35 = 14.7$, $0.180 \times 35 = 6.3$, and $0.400 \times 35 = 14$, compared with the observed numbers 15, 6, and 14. In this example, the agreement with the inbreeding model is excellent.

## Allelic Identity by Descent

In human genetics, as well as in animal and plant breeding, a geneticist may wish to calculate the inbreeding coefficient of a particular individual whose pedigree is known in order to determine the probability of each possible genotype for the individual. This section introduces a concept that is particularly suited to pedigree calculations.

The diagnostic criterion of inbreeding is a closed loop in the pedigree. Such a closed loop is often called a *path.* **FIGURE 17.19**, part A is an inbreeding pedigree in which individual I is the product of a mating between half siblings, and the path (closed loop) from I through the common ancestor A is evident. A streamlined version of this pedigree useful for inbreeding calculations is depicted in Figure 17.19B; individuals of both sexes are shown as circles, diagonal lines trace alleles transmitted from parent to

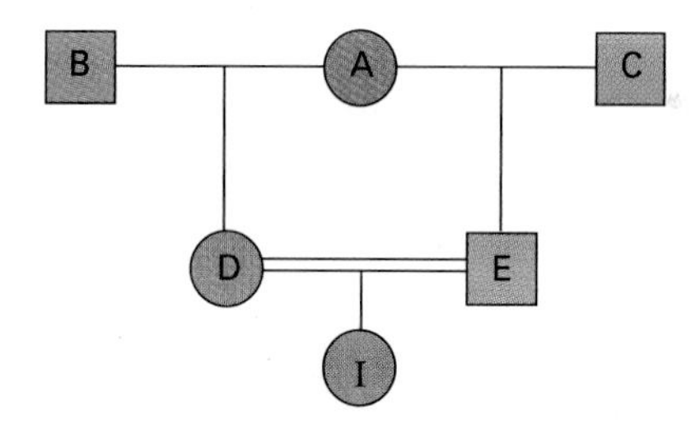


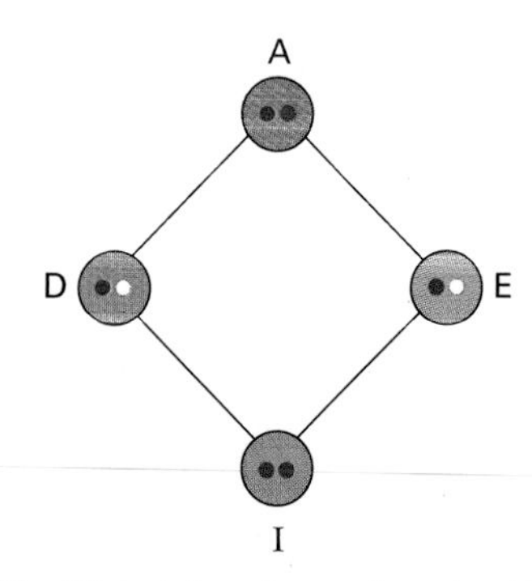


**FIGURE 17.19** (A) An inbreeding pedigree in which individual I results from a mating between half siblings. (B) The same pedigree redrawn to show gametes in the closed loop leading to individual I. The colored dots represent alleles of a particular gene.

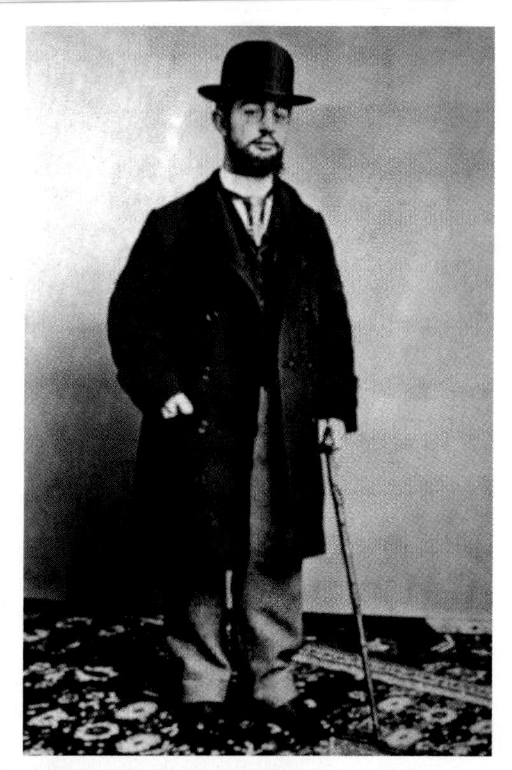

**THE FAMOUS** 19th century French artist Henri Toulouse-Lautrec probably suffered from the rare bone disease pyknodystosis. The condition results from a recessive mutation in chromosome 1, and his parents were first cousins. The principal symptoms are short stature (his adult height was 4 feet 11 inches), disproportional facial features, and bone fractures (he suffered two femur fractures). [© Pictorial Press Ltd/Alamy Images]

offspring, and ancestors who did not contribute to the inbreeding of I (ancestors B and C) are left out altogether. The contrasting dots inside the circles represent particular alleles present in the corresponding individuals, and the depicted pattern of inheritance of the alleles represents the principal consequence of inbreeding and illustrates why closed loops are important. Specifically, individuals D and E both received the blue allele from A and received a white allele from B and C, respectively. The blue alleles in D and E have a very special relationship; they both originate by replication of the blue allele in A and therefore are said to show **identity by descent**. In the interval of the relatively few generations involved in most pedigrees, alleles that are identical by descent can be assumed to be identical in nucleotide sequence, because mutation can usually be ignored in so few generations. Individual I is shown as having received the blue allele from both D and E, and because the alleles in question are identical by descent, this individual must be homozygous for the allele inherited from the common ancestor. Without the closed loop in the pedigree, identity by descent of the alleles in I would be much less probable. The concept of identity by descent provides an alternative definition of the inbreeding coefficient, which can be shown to be completely equivalent to the definition in terms of heterozygosity given earlier. Specifically:

> The inbreeding coefficient of an individual is the probability that the individual carries alleles that are identical by descent.

## Calculation of the Inbreeding Coefficient from Pedigrees

**FIGURE 17.20** illustrates how the probability of identity by descent can be calculated in a pedigree, and the alleles are drawn as dots on the diagonal lines to emphasize their pattern of transmission. The double-headed arrows indicate which pairs of alleles must be identical by descent in order for I to have alleles that are identical by descent. The values of 1/2 represent the probability that a diploid genotype will transmit the allele inherited from a particular parent; this value must be 1/2 because of Mendelian segregation. The term $(1/2)(1 + F_A)$ can be explained by looking at the situation in **FIGURE 17.21**, which shows four possible pairs of gametes that A can transmit, each equally likely. In cases A and B, the transmitted alleles are certainly identical by descent because they are products of DNA replication of the same allele (blue or red). In cases C and D, the transmitted alleles have contrasting colors, but they can still be identical by descent if

The loop around the individual at the top is the probability that each of two offspring receives an allele that is identical by descent.
$(1/2)(1 + F_A)$
1/2
1/2
The inbreeding coefficient of this individual is the product of the probabilities in all three loops:
$1/2 \times 1/2(1 + F_A) \times 1/2$
The loop around this individual is the probability that the allele (•) transmitted to an offspring is identical by descent to that inherited from its parent.

**FIGURE 17.20** Calculation of the probability that the alleles indicated by the double-headed arrows are identical by descent.

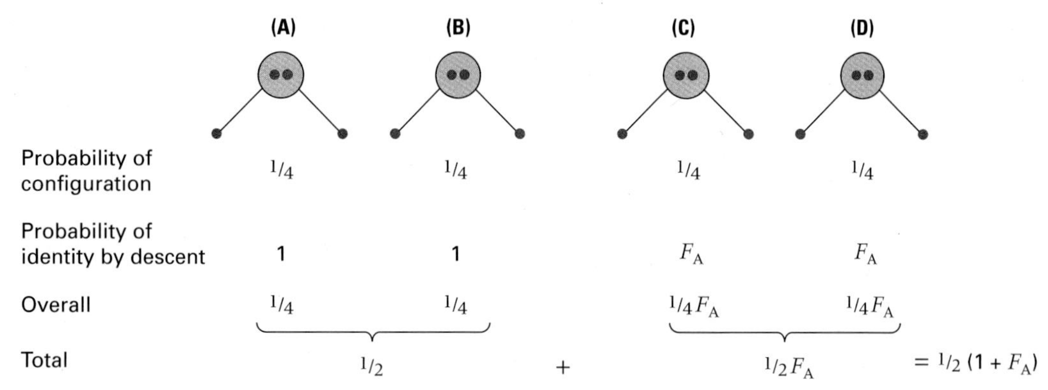


**FIGURE 17.21** A diagram illustrating the reason why $(1/2)(1 + F_A)$ is the probability that two different gametes from the same individual, A, carry alleles that are identical by descent.

A is already inbred. However, the probability that A carries alleles that are identical by descent is, by definition, the inbreeding coefficient of A, and this is designated $F_A$. Altogether, the probability that A transmits two alleles that are identical by descent is the sum of the probabilities of the four eventualities in Figure 17.21, and this sum is equal to $(1/2)(1 + F_A)$. Matters are somewhat simpler when A is not itself inbred, because in this case $F_A = 0$, and the corresponding probability is simply 1/2.

Returning now to Figure 17.20, the alleles in I can be identical by descent only if all three identities indicated by the loops are true, and this probability is the product of the three numbers shown. That is, the inbreeding coefficient of I, symbolized $F_I$, is:

$$F_I = (1/2) \times (1/2)(1 + F_A) \times (1/2)$$
$$= (1/2)^3(1 + F_A)$$

For a more remote common ancestor, the formula becomes,

$$F_I = (1/2)^n(1 + F_A) \qquad (5)$$

where $n$ is the number of individuals in the path traced from one of the parents of I back through the common ancestor and forward again to the other parent of I.

Most pedigrees of interest are more complex than the one in Figure 17.20, but the inbreeding coefficient of any individual can still be calculated by using Equation (5). Application to complex pedigrees requires several additional rules:

1. If there are several distinct paths that can be traced through a common ancestor, then each path makes a contribution to the inbreeding coefficient calculated as in Equation (5).
2. If there are several common ancestors, then each is considered in turn in order to calculate its separate contribution to the inbreeding.
3. The overall inbreeding coefficient is then calculated as the sum of all these separate contributions of the common ancestors.

The extension of Equation (5) to complex pedigrees may be written as,

$$F_I = \Sigma(1/2)^n(1 + F_A) \qquad (6)$$

where $\Sigma$ denotes the summation of all paths through all common ancestors.

Application of Equation (6) can be illustrated in the pedigree in **FIGURE 17.22**. Each path and its contribution to the inbreeding coefficient of I is shown below (the red letter indicates the common ancestor).

EBD $(1/2)^3(1 + F_B)$
ECD $(1/2)^3(1 + F_C)$
ECABD $(1/2)^5(1 + F_A)$
EBACD $(1/2)^5(1 + F_A)$

The inbreeding coefficient of I is the sum of these contributions. If none of the common ancestors is inbred (that is, if $F_A = F_B = F_C = 0$), then

$$F_I = (1/2)^3 + (1/2)^3 + (1/2)^5 + (1/2)^5$$
$$= 5/16$$

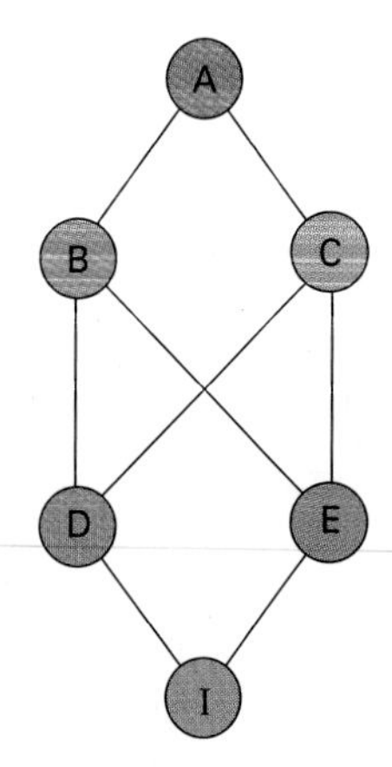


**FIGURE 17.22** A complex pedigree in which the inbreeding of individual I results from several independent paths through three different common ancestors.

The value 5/16 is to be interpreted to mean that the probability that individual I will be heterozygous at any particular locus is smaller, by the proportion 5/16, than the probability that a noninbred individual will be heterozygous at the same locus. Equivalently, the 5/16 could be interpreted to mean that individual I is heterozygous at 5/16 fewer loci in the whole genome than a noninbred individual.

### Effects of Inbreeding

The effects of inbreeding differ according to the normal mating system of an organism. At one extreme, in regularly self-fertilizing plants, inbreeding is already so intense and the organisms are so highly homozygous that additional inbreeding has virtually no effect. However:

> In most species, inbreeding is harmful, and much of the effect is due to rare recessive alleles that would not otherwise become homozygous.

Among human beings, inbreeding is usually uncommon because of social conventions and laws, although in small isolated populations (isolated villages, religious communities, aboriginal groups) it does occur, mainly through matings between relatives more distant than second cousins (remote relatives). The most common type of close inbreeding is between first cousins. The effect is always an increase in the frequency of genotypes that are homozygous for rare, usually harmful recessives. For example, among American whites, the frequency of albinism among offspring of matings between nonrelatives is approximately 1 in 20,000; among offspring of first-cousin matings, the frequency is approximately 1 in 2000. The reason for the tenfold greater risk may be understood by comparing the genotype frequencies of homozygous recessives in Equation (4) for inbreeding and Equation (1) for random mating. Among the offspring of a mating between first cousins, $F = 1/16$, or 0.062. Therefore, the frequency of homozygous recessives produced by first-cousin mating will be,

$$q^2(1 - 0.062) + q(0.062)$$

whereas among the offspring of nonrelatives, the frequency of homozygous recessives is $q^2$. For albinism, $q = 0.007$ (approximately), and the calculated frequencies are $5 \times 10^{-4}$ for the offspring of first cousins and $5 \times 10^{-5}$ for the offspring of nonrelatives. The increased frequency of rare homozygous recessive genotypes is the principal consequence of first-cousin mating (or any degree of inbreeding). The effect of first-

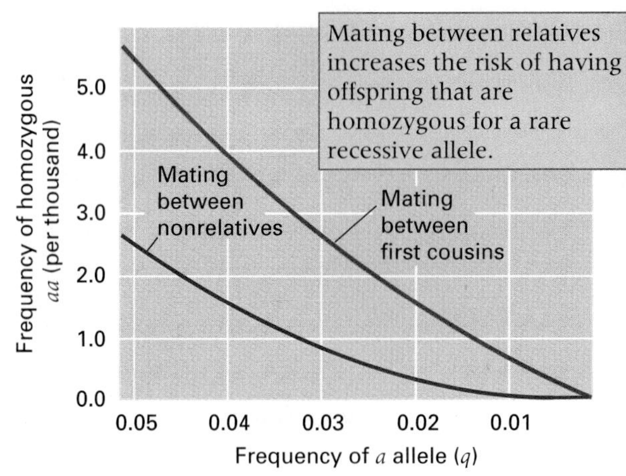


**FIGURE 17.23** Effect of first-cousin mating on the frequency of offspring genotypes that are homozygous for a rare autosomal recessive allele. Although both curves decrease as the allele becomes rare, the curve for mating between nonrelatives decreases faster. As a consequence, the more rare the recessive allele, the greater the proportion of all affected individuals that result from first-cousin matings.

cousin mating is illustrated in **FIGURE 17.23**, which shows a greater effect for rare alleles. For example, with an allele frequency $q = 0.05$, the relative risk of a homozygous offspring is $0.0335938/0.0025 = 13.4$, whereas with an allele frequency $q = 0.01$, the relative risk of a homozygous offspring is $0.00634375/0.0001 = 63.4$.

## 17.4 Genetics and Evolution

**Evolution** consists of progressive changes in the gene pool. Some authors prefer a stricter requirement that the changes must be associated with the progressive adaptation of a population to its environment. Whichever definition is adopted, evolution occurs because genetic variation exists in populations and because there is a natural selection favoring organisms that are best adapted to the environment. Genetic variation and natural selection are population phenomena, so they are conveniently discussed in terms of allele frequencies.

Four processes account for most of the changes in allele frequency in populations. They form the basis of cumulative change in the genetic characteristics of populations, leading to the descent with modification that characterizes the process of evolution. These process are:

1. **Mutation**, the origin of new genetic capabilities in populations by means of spontaneous heritable changes in genes.
2. **Migration**, the movement of individuals among subpopulations within a larger population.

3. **Natural selection**, which results from the differing abilities of individuals to survive and reproduce in their environment.
4. **Random genetic drift**, the random, undirected changes in allele frequency that occur by chance in all populations, but particularly in small ones.

Some of the implications of these processes for population genetics are considered in the following sections.

## 17.5 Mutation and Migration

Mutation is an essential process in evolution because it is the ultimate source of genetic variation. But it is a relatively weak force for changing allele frequencies, primarily because typical mutation rates are so low. Moreover, most newly arising mutations are harmful to the organism and are eliminated from the population in a few generations.

Migration is similar to mutation in that new alleles can be introduced into a local population, although the alleles derive from another subpopulation rather than from new mutations. In the absence of migration, the allele frequencies in each local population can change independently, so local populations can undergo considerable genetic differentiation. Genetic differentiation among subpopulations means that there are differing frequencies of common alleles among the local populations or that some local populations possess certain rare alleles not found in others. The accumulation of genetic differences among subpopulations can be minimized if the subpopulations exchange individuals (undergo migration). In fact, only a relatively small amount of migration among subpopulations (on the order of just a few immigrant individuals in each local population in each generation) is usually sufficient to prevent the accumulation of high levels of genetic differentiation. On the other hand, genetic differentiation can occur in spite of migration if other evolutionary forces, such as natural selection for adaptation to the local environment, are sufficiently strong.

### Irreversible Mutation

To appreciate the magnitude of mutation as an evolutionary force, consider an allele *A* and a selectively neutral mutant allele *a*. (The mutational change in *a* could be, for example, a nucleotide substitution at an irrelevant site in an intron, or a change from one synonymous codon to another.) Suppose allele *A* undergoes mutation to *a* at a rate of $\mu$ per *A* allele per generation. What this rate means is that in any generation, the probability that a particular *A* allele mutates to *a* is $\mu$. Equivalently, we could say that in any generation, the proportion of *A* alleles that do *not* mutate to *a* equals $1 - \mu$. If we let $p$ and $q$ represent the allele frequencies of *A* and *a*, respectively, then the allele frequency of *A* changes according to the rule that the allele frequency of *A* in any generation equals the allele frequency of *A* in the previous generation times the proportion of *A* alleles that failed to mutate in that generation. (Here we are assuming that reverse mutation of *a* to *A* can be neglected.) In symbols, the rule implies that,

$$p_n = p_{n-1}(1 - \mu)$$

where the subscript $n$ denotes any generation and $n - 1$ denotes the previous generation. Hence the allele frequency of *A* in generations 0, 1, 2, 3, and so forth changes according to:

$$p_1 = p_0(1 - \mu)$$
$$p_2 = p_1(1 - \mu)$$
$$p_3 = p_2(1 - \mu)$$

Substituting $p_1$ from the first equation into the second, we obtain:

$$p_2 = p_1(1 - \mu) = p_0(1 - \mu)^2$$

Substituting this into the equation for $p_3$ yields:

$$p_3 = p_2(1 - \mu) = p_0(1 - \mu)^3$$

The general rule follows by successive substitutions:

$$p_n = p_0(1 - \mu)^n \qquad (7)$$

where $p_n$ represents the allele frequency of *A* in the *n*th generation.

If we assume that allele *A* is fixed in the population at time $n = 0$, then $p_0 = 1$, and Equation (7) becomes:

$$p_n = (1 - \mu)^n \qquad (8)$$

For realistic values of the mutation rate, Equation (8) implies a very slow rate of change in allele frequency. The pattern of change is illustrated by the lower curve in **FIGURE 17.24**, which assumes a mutation rate of $\mu = 1 \times 10^{-5}$ per generation. With this value of $\mu$, the allele frequency of *A* decreases by half its present value every 69,314 generations. To put this rate in human terms, assuming 20 years per generation, it would require $69{,}314 \times 20 = 1{,}368{,}280$ years, or approximately 1.4 million years, to reduce the allele frequency of *A* by half. For example, if a human allele *A* were fixed at the time of *Homo erectus*, 1.4 million years ago, and *A* underwent mutation to *a* at a rate of $1 \times 10^{-5}$

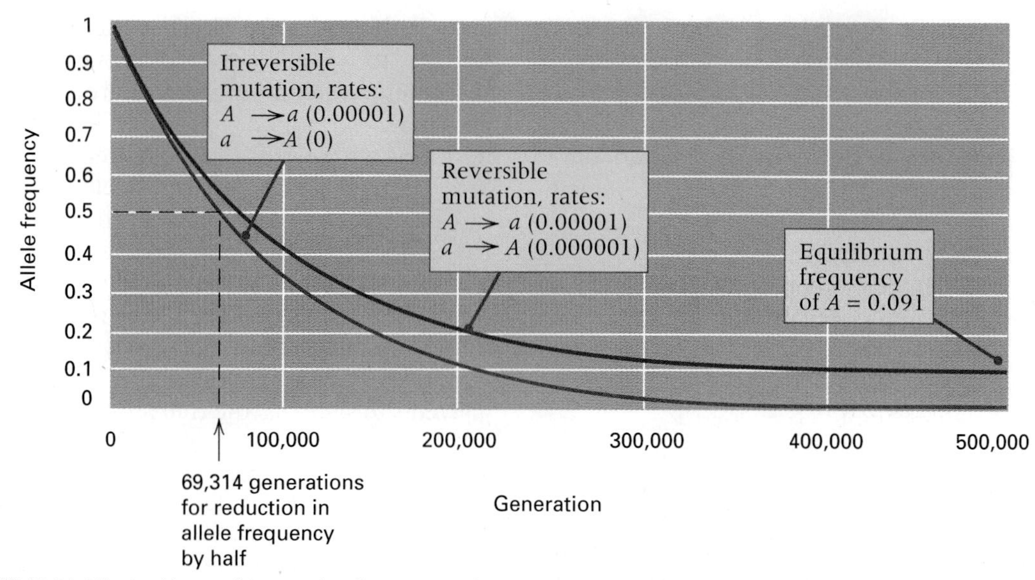


**FIGURE 17.24** Effects of irreversible mutation (bottom curve) and reversible mutation (top curve) on allele frequency. In this example, the forward mutation rate is $\mu = 1 \times 10^{-5}$ and the reverse mutation rate is $\nu = 1 \times 10^{-6}$.

per generation, then 50 percent of the alleles present in the modern human population would still be of type *A*.

### Reversible Mutation

When the possibility of reverse mutation is taken into account, the **forward mutation** of *A* to *a* is offset by **reverse mutation** of *a* to *A*. Eventually, an **equilibrium** of allele frequency is reached at which the allele frequencies become stable and no longer change, even though there is continuous mutation. The reason for the equilibrium is that at this point, each new *A* allele that is lost by forward mutation is replaced by a new *A* allele created by reverse mutation.

It is quite straightforward to deduce the equilibrium frequencies when mutation is reversible. As before, let *A* and *a* have allele frequencies *p* and *q*, respectively. Suppose that the rate of forward mutation ($A \rightarrow a$) is $\mu$ and that of reverse mutation ($a \rightarrow A$) is $\nu$. In any generation, the number of *A* alleles lost to forward mutation is proportional to $p\mu$, and the number of new *A* alleles created by reverse mutation is proportional to $q\nu$. At equilibrium, the loss of *A* must equal the gain of *a*; hence $p\mu = q\nu$. Substituting $q = 1 - p$, we obtain,

$$p\mu = (1 - p)\nu = \nu - p\nu$$

and therefore, at equilibrium,

$$\hat{p} = \frac{\nu}{\mu + \nu} \qquad (9)$$

where the symbol $\hat{p}$ is used, by convention, to mean the equilibrium value of *p*. As a check of the correctness of Equation (9), note that $\hat{q} = \mu/(\mu + \nu)$ and so $\hat{p}\mu = \hat{q}\nu$ as required.

In Figure 17.24, the upper curve gives the change in allele frequencies for reversible mutation with $\mu = 1 - 10^{-5}$ and $\nu = 1 \times 10^{-6}$. The equilibrium in this example is:

$$\hat{p} = \frac{1 \times 10^{-6}}{1 \times 10^{-5} + 1 \times 10^{-6}} = \frac{1}{11}$$

The time required to reach equilibrium is again very long. When $\mu = 1 \times 10^{-5}$ and $\nu = 1 \times 10^{-6}$, it requires 63,013 generations for any specified values of *p* and *q* to go halfway to their equilibrium values. (The half-time of 69,314 generations derived earlier is different owing to the effect of reverse mutation.)

## 17.6 Natural Selection

The driving force of adaptive evolution is *natural selection,* which is a consequence of hereditary differences among organisms in their ability to survive and reproduce in the prevailing environment. Since it was first proposed by Charles Darwin in 1859, the concept of natural selection has been revised and extended, most notably by the incorporation of genetic concepts. In its modern formulation, the occurrence of natural selection rests on two premises, both of which are universally acknowledged to be true, and one deduction:

**Premise 1**: In all organisms, more offspring are produced than survive and reproduce.

(Darwin borrowed this part of the theory from Thomas Malthus.)

**Premise 2**: Organisms differ in their ability to survive and reproduce, and some of these differences are due to genotype.

**Deduction**: In every generation, the genotypes that promote survival in the prevailing environment (favored genotypes) must be present in excess among individuals of reproductive age, and hence the favored genotypes will contribute disproportionately to the offspring of the next generation.

By means of this process, the alleles that enhance survival and reproduction increase in frequency from generation to generation, and the population becomes progressively better able to survive and reproduce in the prevailing environment. This progressive genetic improvement in populations constitutes the process of evolutionary **adaptation**.

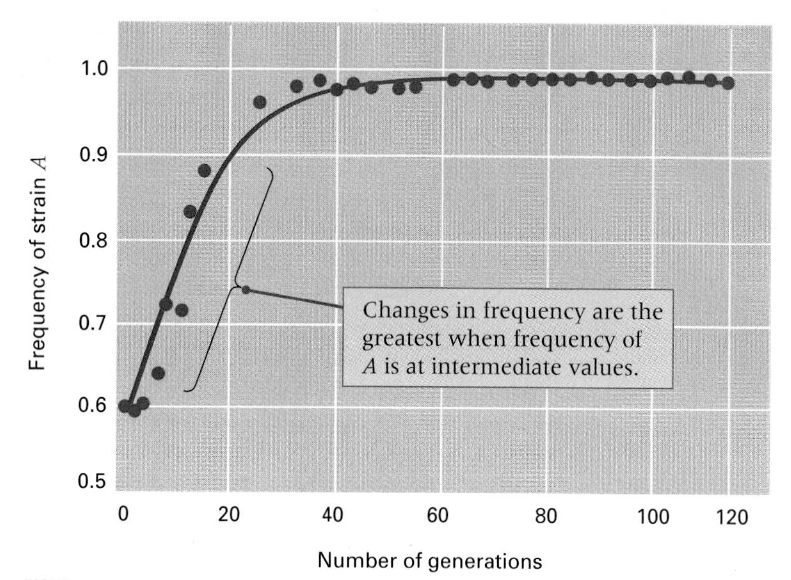


**FIGURE 17.25** Increase in frequency of a favored strain of *E. coli* resulting from selection in a continuously growing population. The *y*-axis is the number of *A* cells at any time divided by the total number of cells ($A + B$). Note that the changes in frequency are the greatest when the frequency of the favored strain, *A*, is at intermediate values.

## Selection in a Laboratory Experiment

Selection is easily studied experimentally in bacterial populations because of their short generation time (about 30 minutes). **FIGURE 17.25** shows the result of competition between two bacterial genotypes, *A* and *B*, under conditions in which *A* is the superior competitor. In the experiment, the competition was allowed to continue for 120 generations, in which time the proportion of *A* genotypes ($p$) increased from 0.60 to 0.995 and that of *B* genotypes decreased from 0.40 to 0.005. The data points give a satisfactory fit to an equation of the form,

$$\frac{p_n}{q_n} = \left(\frac{p_0}{q_0}\right)\left(\frac{1}{w}\right)^n \qquad (10)$$

where $p_0$ and $q_0$ are the initial frequencies of *A* and *B* (in this case 0.6 and 0.4, respectively), $p_n$ and $q_n$ are the frequencies after $n$ generations of competition (here, $n = 120$, $p_n = 0.995$, and $q_n = 0.005$), and $w$ is a measure of the competitive ability of *B* when competing against *A* under the conditions of the experiment. Equation (10) can be derived step by step as follows: Because the relative rates of survival and reproduction of strains *A* and *B* are in the ratio $1 : w$, the ratio of frequencies in any generation, $p_n/q_n$, must equal the ratio of frequencies in the previous generation, $p_{n-1}/q_{n-1}$, times $1/w$. Therefore,

$$(p_1/q_1) = (p_0/q_0) \times (1/w)$$
$$(p_2/q_2) = (p_1/q_1) \times (1/w)$$
$$(p_3/q_3) = (p_2/q_2) \times (1/w)$$

and so on. Substituting successively from the first equation into the second, the second into the third, and so forth shows how Equation (10) is ultimately obtained.

The value of $w$ is called the **relative fitness** of the *B* genotype relative to the *A* genotype under these particular conditions. The data in Figure 17.25 fit Equation (10) for a value of $w = 0.96$, which generates the smooth curve. Relative fitness measures the comparative contribution of each parental genotype to the pool of offspring genotypes produced in each generation. In this example, for each offspring cell produced by an *A* genotype, a *B* genotype produces an average of 0.96 offspring cell.

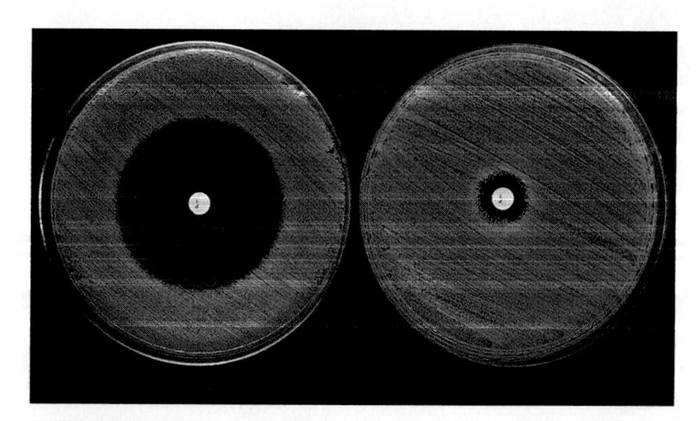

**WIDESPREAD USE** of antibiotics has resulted in the natural selection of antibiotic resistance. These petri dishes show the growth of the bacterium *Staphylococcus aureus*. The small white disk contains a solution of penicillin, which diffuses out into the medium forming the halo around the disk. The bacterial strain on the left is sensitive. That on the right has a gene for a penicillinase protein and is resistant except for the highest doses nearest the disk. [John Durham/Photo Researchers, Inc.]

In population genetics, relative fitnesses are usually calculated by taking the most favored genotype ($A$ in this case) as the standard with a fitness of 1.0. However, the selective disadvantage of a genotype is often of greater interest than its relative fitness, because some of the equations are simplified. The selective disadvantage of a disfavored genotype is called the **selection coefficient** against the genotype, and it is calculated as the difference between the fitness of the standard (taken as 1.0) and the relative fitness of the genotype in question. In the case at hand, the selection coefficient against $B$, denoted $s$, is:

$$s = 1.000 - 0.96 = 0.04 \qquad (11)$$

In words, the selective disadvantage of strain $B$ is 4 percent per generation. When the fitnesses are known, Equation (10) also enables us to predict the allele frequencies in any future generation, given the original frequencies. Alternatively, it can be used to calculate the number of generations required for selection to change the allele frequencies from any specified initial values to any later ones. For example, from the relative fitnesses of $A$ and $B$ just estimated, we can calculate the number of generations required to change the frequency of $A$ from 0.1 to 0.8. In this example, $p_0/q_0 = 0.1/0.9$, $p_n/q_n = 0.8/0.2$, and $w = 0.96$. A little manipulation of Equation (10) gives:

$$n = \frac{\log(\frac{0.1}{0.9}) - \log(\frac{0.8}{0.2})}{\log(0.96)}$$

$$= 87.8 \text{ generations}$$

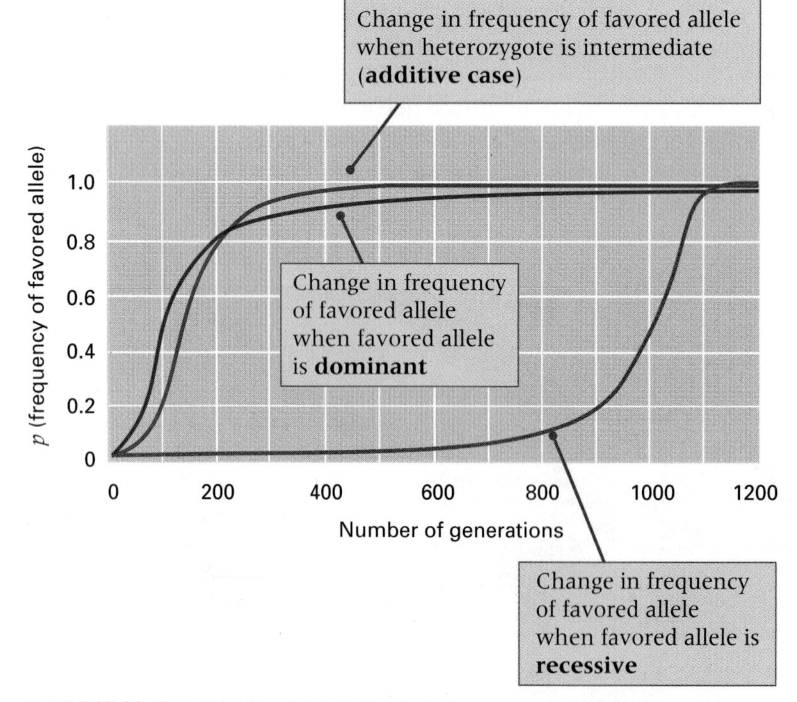


**FIGURE 17.26** Theoretically expected change in frequency of an allele favored by selection in a diploid organism undergoing random mating when the favored allele is dominant, when the favored allele is recessive, and when the heterozygote is exactly intermediate in fitness (additive). In each case, the selection coefficient against the least fit homozygous genotype is 5 percent.

## Selection in Diploid Organisms

Selection in diploids is analogous to that in haploids, but dominance and recessiveness create additional complications. **FIGURE 17.26** shows the change in allele frequencies for both a favored dominant and a favored recessive in a random-mating diploid population. The striking feature of the figure is that the frequency of the favored dominant allele changes very slowly when the allele is common, and the frequency of the favored recessive allele changes very slowly when the allele is rare. The reason is that with random mating, rare alleles are found much more frequently in heterozygotes than in homozygotes. With a favored dominant at high frequency, most of the recessive alleles are present in heterozygotes, and the heterozygotes are not exposed to selection and hence do not contribute to change in allele frequency. Conversely, with a favored recessive at low frequency, most of the favored alleles are in heterozygotes, and again the heterozygotes are not exposed to selection and do not contribute to change in allele frequency. The principle is quite general:

> Selection for or against a recessive allele is very inefficient when the recessive allele is rare.

The inefficiency of selection against rare recessive alleles has an important practical implication. There is a widely held belief that medical treatment to save the lives of persons with rare recessive disorders will cause a deterioration of the human gene pool because of the reproduction of persons who carry the harmful genes. This belief is unfounded. With rare alleles, the proportion of homozygous genotypes is so small that reproduction of the homozygous genotypes contributes a negligible amount to change in allele frequency. Considering their low frequency, it matters little whether homozygous genotypes reproduce or not. Similar reasoning applies to eugenic proposals to cleanse the human gene pool of harmful recessives by preventing the reproduction of affected persons. People with severe genetic disorders rarely reproduce anyway, and even when they do, they have essentially no effect on allele frequency. The underlying principle is that:

> The main reservoir of harmful recessive alleles in the human population is in the genomes of phenotypically normal carriers who are heterozygous.

## Components of Fitness

Data like those in Figure 17.25 are essentially unobtainable in higher organisms because of their long generation times. Observations extending over 120 generations would take 10 years with *Drosophila* and 2400 years with humans. How, then, is one to estimate fitness? The usual approach is to divide fitness into its component parts and estimate these separately. The two major components of the fitness of a genotype are usually considered to be:

- The **viability**, which means the probability that a newly formed zygote of the specified genotype survives to reproductive age
- The **fertility**, which means the average number of offspring produced by an individual of the specified genotype during the reproductive period

Together, viability and fertility measure the contribution of each genotype to the offspring of the next generation.

**TABLE 17.1** is an example of experiments carried out to estimate the viability of two genotypes coding for different electrophoretic forms of the enzyme alcohol dehydrogenase in *Drosophila*. The genotypes are homozygous for either the *F* or the *S* allele. The tabulated numbers are averages obtained in seven separate experiments in which larvae were placed together in competition in culture vials. The ratio of surviving larvae to those introduced originally is the probability of survival of each genotype. As noted at the bottom of the table, these numbers can be converted into relative viabilities in two ways, using either *SS* or *FF* as the standard. Because of the convention that the larger of the two relative values should be 1.0, the situation can be summarized either by saying that the viability of *FF* relative to *SS* is 0.94 or by saying that the selective disadvantage of *FF* relative to *SS* with respect to viability is 0.06.

Estimates of overall fitness based on viability and fertility components may be used for predicting changes in allele frequency that would be expected in experimental or natural populations. However, the fitness of a genotype may depend on the environment and on other factors, such as population density and the frequencies of other genotypes. Consequently, fitnesses estimated in a particular laboratory situation may have little relevance to predicting changes in allele frequency in natural populations unless other factors are comparable.

## Selection–Mutation Balance

Natural selection usually acts to minimize the frequency of harmful alleles in a population. However, the harmful alleles can never be totally eliminated, because mutation of wildtype alleles continually creates new harmful mutations. With the forces of selection and mutation acting in opposite directions, the population eventually attains a state of equilibrium, or **selection–mutation balance**, at which new mutations exactly offset selective eliminations. In determining the allele frequencies at equilibrium, it makes a great deal of difference whether the harmful allele is completely recessive or has, instead, a small effect (partial dominance) on the fitness of the heterozygous carriers. To deduce the equilibrium frequencies, we will symbolize the wildtype allele as *A* and the harmful allele as *a* and represent the allele frequencies as $p$ and $q$, respectively.

In the case of a complete recessive, the fitnesses of *AA*, *Aa*, and *aa* genotypes can be written as $1 : 1 : 1 - s$, where $s$ represents the selection coefficient against the *aa* homozygote and

**Table 17.1 Relative viability of alcohol dehydrogenase genotypes in *Drosophila***

| | Genotype | |
|---|---|---|
| **Experimental data and relative viability** | ***FF*** | ***SS*** |
| Experimental data | | |
| Number of larvae introduced per vial | 200 | 100 |
| Average number of survivors per vial | 132.72 | 70.43 |
| Probability of survival | 0.6636 | 0.7043 |
| Relative viability | | |
| With *SS* as standard | 0.6636/0.7043 = 0.94 | 0.7043/0.7043 = 1.0 |
| With *FF* as standard | 0.6636/0.6636 = 1.0 | 0.7043/0.6636 = 1.06 |

measures the fraction of *aa* genotypes that fail to survive or reproduce. For a recessive lethal, for example, $s = 1$. When $q$ is very small (as it will be near equilibrium), the number of *a* alleles eliminated by selection will be proportional to $q^2s$, assuming Hardy–Weinberg proportions. At the same time, the number of new *a* alleles introduced by mutation will be proportional to $\mu$. At equilibrium, the selective eliminations must balance the new mutations, and so $q^2s = \mu$, or:

$$\hat{q} = \sqrt{\frac{\mu}{s}} \qquad (12)$$

As before, $\hat{q}$ means the equilibrium value. To apply Equation (12) to a specific example, consider the recessive allele for Tay–Sachs disease, which in most non-Jewish populations has a frequency of $q = 0.001$. Because the condition is lethal, $s = 1$. Assuming that $q = 0.001$ represents the equilibrium frequency and that the allele has no effect on the fitness of the heterozygous carriers, the mutation rate required to account for the observed frequency would be:

$$\mu = q^2s = (0.001)^2 \times 1.0 = 1 \times 10^{-6}$$

This example shows how considerations of selection–mutation balance can be used in estimating human mutation rates.

If a harmful allele shows partial dominance in having a small detrimental effect on the fitness of the heterozygous carriers, then the fitnesses of *AA*, *Aa*, and *aa* can be written as $1 : 1 - hs : 1 - s$, in which $hs$ is the selection coefficient against the heterozygous carriers. The parameter $h$ is called the **degree of dominance**. When $h = 1$, the harmful allele is completely dominant; when $h = 0$, it is completely recessive; and when $h = 1/2$, the fitness of the heterozygous genotype is exactly the average of those of the homozygous genotypes. For most harmful alleles that show partial dominance, the value of $h$ is considerably smaller than $1/2$.

With random mating, when an allele is rare (as a harmful allele will be at equilibrium), the frequency of heterozygous genotypes greatly exceeds the frequency of those that are homozygous for the harmful recessive. This means that most of the alleles that are eliminated by selection are eliminated because of the selection against the heterozygous genotypes, because there are so many of them. For each *Aa* genotype that fails to survive or reproduce, half of the alleles that are lost are *a*, so the number of alleles lost to selection in any generation is proportional to:

$$2pq \times hs \times 1/2$$

On the other hand, the number of new *a* alleles created by $A \rightarrow a$ mutation equals $\mu$. At equilibrium, the loss due to selection must balance the gain from mutation, so

$$2pq \times hs \times 1/2 = \mu,$$

or

$$\hat{p}\,\hat{q}hs = \mu$$

Because a harmful allele will be rare at equilibrium, especially if it shows partial dominance, $\hat{p}$ will be so close to 1 that it can be replaced with 1 in the above expression. Therefore, the equilibrium value for the frequency of a partially dominant harmful allele is:

$$\hat{q} = \frac{\mu}{hs} \qquad (13)$$

Note that Equation (13) does not include a square root, which makes a substantial difference in the allele frequency. This effect can be seen by comparing the curves in **FIGURE 17.27** for a harmful allele that is lethal when homozygous ($s = 1$). The top curve is for a complete recessive ($h = 0$), and the other curves refer to various degrees of partial dominance; for example, $h = 1/2$ means that the heterozygous genotype has a fitness exactly intermediate between the homozygous genotypes. Even a small amount of partial dominance has a substantial effect on reducing the equilibrium allele frequency.

## Heterozygote Superiority

So far we have considered examples of selection in which the fitness of the heterozygote is intermediate between those of the homozygotes (or possibly equal to that of one homozygote). In these cases, the allele associated with the most fit homozygote eventually becomes fixed, unless the selection is opposed by mutation. In this section we consider the possibility that the fitness of the heterozygote is greater than that of both homozygotes; this case is called **overdominance** or **heterozygote superiority**.

When there is overdominance, neither allele can be eliminated by selection, because in each generation, the heterozygotes produce more offspring than the homozygotes. The selection in favor of heterozygous genotypes keeps both alleles in the population. Eventually there is an equilibrium in which the allele frequency no longer changes. With overdominance, the fitnesses of *AA*, *Aa*, and *aa* may be written as $1 - s : 1 : 1 - t$, where $s$ and $t$ are the selection coefficients against *AA* and *aa*, respectively, relative to *Aa*. Assuming random mating, in any generation, the proportion of *A* alleles eliminated by selection is $p^2s/p = ps$, and the proportion of *a* alleles eliminated by selection is $q^2t/q = qt$. At equilibrium, the selective eliminations of *A* must balance the selective eliminations of *a*, and hence $\hat{p}s = \hat{q}t$, or:

$$\hat{p} = \frac{t}{s + t} \qquad (14)$$

With overdominance, the allele frequencies always go to equilibrium, but the rate of approach depends on the magnitudes of *s* and *t*. The equilibrium is attained most rapidly when there is strong selection against the homozygotes.

Overdominance does not appear to be a particularly common form of selection in natural populations. However, there are several well-established cases, the best known of which involves the sickle-cell hemoglobin mutation. The initial hint of a connection between sickle-cell anemia and malaria caused by the protozoan *Plasmodium falciparum* was the substantial overlap of the geographical distribution of sickle-cell anemia and that of malaria (**FIGURE 17.28**).

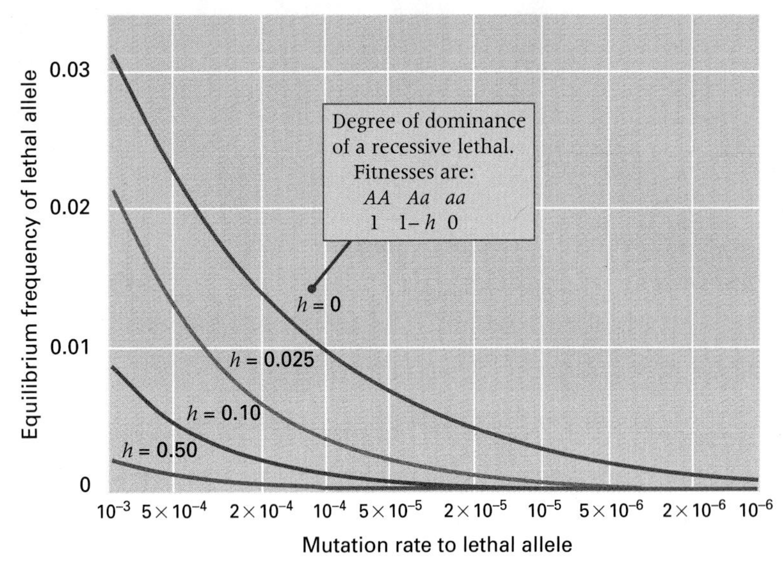


**FIGURE 17.27** Effect of partial dominance on the equilibrium frequency of a harmful allele that is lethal when homozygous. The degree of dominance is measured by *h*. For example, a value of $h = 0.025$ means that the relative fitnesses of *AA* and *Aa* are 1 : 0.975.

On the basis of the rates at which people of different genotypes are infected with malaria, and the mortality of infected people, the relative fitnesses of the genotypes *AA*, *AS*, and *SS* have been estimated as 0.85 : 1 : 0. (*S* represents the sickle-cell mutation.) In the absence of intensive medical treatment, the homozygous *SS* genotype is lethal as a result of severe anemia. On the other hand, there is overdominance because *AS* heterozygotes are more resistant to malarial infection than either of the homozygous genotypes.

From these fitnesses, we have $s = 0.15$ and $t = 1$, so the predicted equilibrium frequency of *A* from Equation (14) equals $1/(0.15 + 1) = 0.87$. The predicted equilibrium frequency of *S* is therefore $1 - 0.87 = 0.13$, which is a value reasonably close to that observed in many parts of West Africa.

## 17.7 Random Genetic Drift

The process of *random genetic drift* comes about because populations are not infinitely large (as we

**(A)**

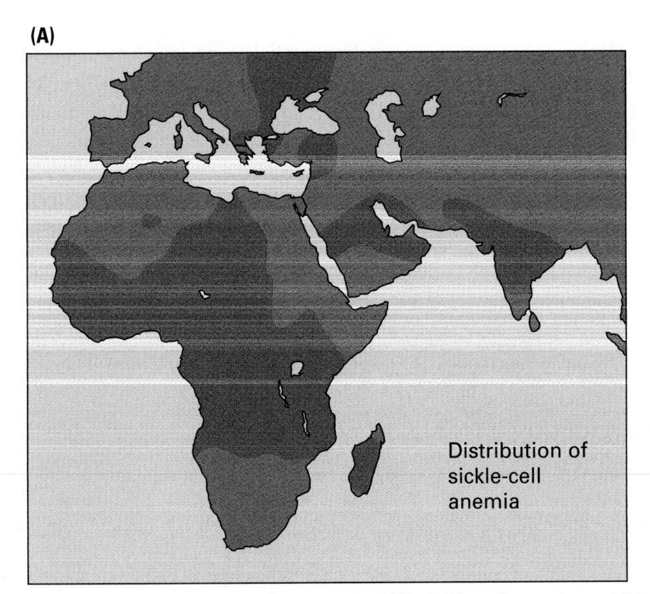


**(B)**

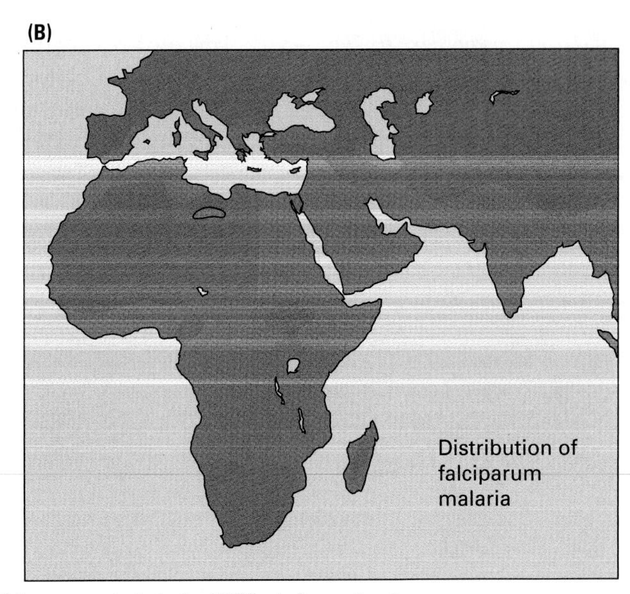


**FIGURE 17.28** Geographic distribution of (A) sickle-cell anemia and (B) falciparum malaria in the 1920s, before extensive malaria-control programs were launched. Shades of brown indicate the areas in question.

# connection

## Resistance in the Blood

**Anthony C. Allison 1954**
Radcliffe Infirmary,
Oxford, England
*Protection Afforded by Sickle-Cell Trait Against Subtertian Malarial Infection*

*Malaria is the most prevalent infectious disease in tropical and subtropical regions of the world, infecting up to 250 million people each year and causing as many as 2 million deaths. The disease is characterized by recurrent episodes of fever with alternating shivering and sweating. Patients suffer anemia due to destruction of red blood cells, as well as enlargement of the spleen, inflammation of the digestive system, bronchitis, and many other severe complications. The type of malaria caused by the protozoan parasite* Plasmodium falciparum *is called "subtertian" malaria because there is less than a 3-day interval between bouts of fever. This parasite is spread through bites by the mosquito* Anopheles gambiae. *Upon transmission, the parasites multiply in the liver for about a week and then begin to infect and multiply in red blood cells (parasitemia), which are destroyed after a few days. In regions of Africa, the Middle East, the Mediterranean region, and India where falciparum malaria is endemic, there is also a relatively high frequency of the sickle-cell mutation affecting the amino acid sequence of the beta chain of hemoglobin. Heterozygous carriers have no severe clinical symptoms, but they have the so-called "sickle-cell trait," in which the red blood cells collapse into half-moon, or sickle, shapes after 1 to 3 days when sealed under a cover slip on a microscope slide. Homozygous affected persons have "sickle-cell anemia," in which many red blood cells sickle spontaneously while still in the bloodstream, causing severe complications and often death. Why would a genetic disease that is effectively lethal when homozygous be maintained at a frequency of 10 percent or more in a population? Allison noted the correlation between the sickle-cell mutation and malaria and speculated that the sickle-cell trait gives heterozygous carriers some protection against malaria. Key evidence supporting this hypothesis is presented here. Later work showed that the heterozygous carriers have an approximately 15 percent selective advantage as a result of their malaria resistance.*

During the course of field work in Africa in 1949 I was led to question the view that the sickle-cell trait is neutral from the point of view of natural selection and to reconsider the possibility that it is associated with a selective advantage. I noted that the incidence of the sickle-cell trait was higher in regions where malaria was prevalent than elsewhere. . . . It became imperative, then, to ascertain whether sickle cells can afford some degree of protection against malarial infection. . . . Children were chosen rather than adults as subjects for the observations so as to minimize the effect of acquired immunity to malaria. The recorded incidence of parasitemia in a group of 290 Ganda children [living near Kampala, Uganda] is presented in the accompanying Table. . . .

**It became imperative . . . to ascertain whether sickle cells can afford some degree of protection against malarial infection.**

| | With parasitemia | Without parasitemia | Total |
|---|---|---|---|
| Sicklers | 12 (27.9%) | 31 (72.1%) | 43 |
| Non-sicklers | 113 (45.7%) | 134 (54.3%) | 247 |

It is apparent that the incidence of parasitemia is lower in the sickle-cell group than in the group without sickle cells. The difference is statistically significant ($\chi^2 = 4.8$ for 1 degree of freedom). . . . The parasite density in the two groups also differed: of 12 sicklers with malaria, 66.7% had only slight parasitemia while 33.3% had a moderate parasitemia. Of the 113 non-sicklers with malaria, 34% had slight parasitemia, the parasite density in the remainder being moderate or severe. . . . [Among a group of adult males who volunteered to be bitten by heavily infected *Anopheles gambiae* mosquitoes], an infection with *Plasmodium falciparum* was established in 14 out of 15 men without the sickle-cell trait, whereas in a comparable group of 15 men with the trait only 2 developed parasites. It is concluded that the abnormal erythrocytes of individuals with the sickle-cell trait are less easily parasitized by *P. falciparum* than are normal erythrocytes. Hence, those who are heterozygous for the sickle-cell gene will have a selective advantage in regions where malaria is hyperendemic. This fact may explain why the sickle-cell gene remains common in these areas in spite of the elimination of genes in patients dying of sickle-cell anemia.

Source: A. C. Allison, *Br. Med. J.* 1 (1954): 290–294.

have been assuming all along) but rather finite, or limited, in size. The breeding individuals of any one generation produce a potentially infinite pool of gametes. In the absence of fertility differences, the allele frequencies among gametes will equal the allele frequencies among adults. However, because of the limited size of the population, only a few of these gametes will participate in fertilization and be represented among zygotes of the next generation. In other words, a process of *random sampling* takes place in going from one generation to the next. Because there is variation among samples as a result of chance, the allele frequencies among gametes and zygotes may differ as a result of chance. Changes in allele frequency that come about because of the generation-to-generation sampling in finite populations are the cause of random genetic drift.

To look at this process more generally, consider a population consisting of exactly $N$ diploid individuals in each generation. At each autosomal locus, there will be exactly $2N$ alleles. Suppose that $i$ of these are of type $A$ and the remaining $2N - i$ are of type $a$. The allele frequency of $A$, designated $p$, therefore equals $i/2N$. The number of $A$ alleles in the next generation cannot be specified with certainty, because it is governed by the chance sampling process of random genetic drift. On the other hand, it is possible to specify a *probability* for each possible number of $A$ alleles in the next generation. This probability is given by the terms of the binomial distribution (Chapter 4). In particular, the probability $P(k|i)$ that there will be exactly $k$ copies of $A$ in the next generation, given that there are exactly $i$ copies among the parents, is given by:

$$P(k|i) = \frac{2N!}{k!\,(2N - k)!}\, p^k q^{2N-k} \qquad (15)$$

In Equation (15), $p = i/2N$, $q = 1 - p$, and $k$ can equal 0, 1, 2, and so forth, up to and including $2N$. The part of the equation with factorials is the number of possible orders in which the $k$ copies of allele $A$ can be chosen. (The $A$ alleles could be the first $k$ chosen, the last $k$ chosen, alternating with $a$, and so on.) The powers of $p$ and $q$ represent the probability that any particular order will be realized. For any specified values of $N$ and $i$, one can use Equation (15) along with a table of random numbers (or, better yet, a personal computer) to calculate an actual value for $k$.

Let's consider a concrete example to illustrate the essential features of random genetic drift. **FIGURE 17.29**, part A shows the result of random genetic drift in twelve subpopulations, each consisting of eight diploid individuals, and

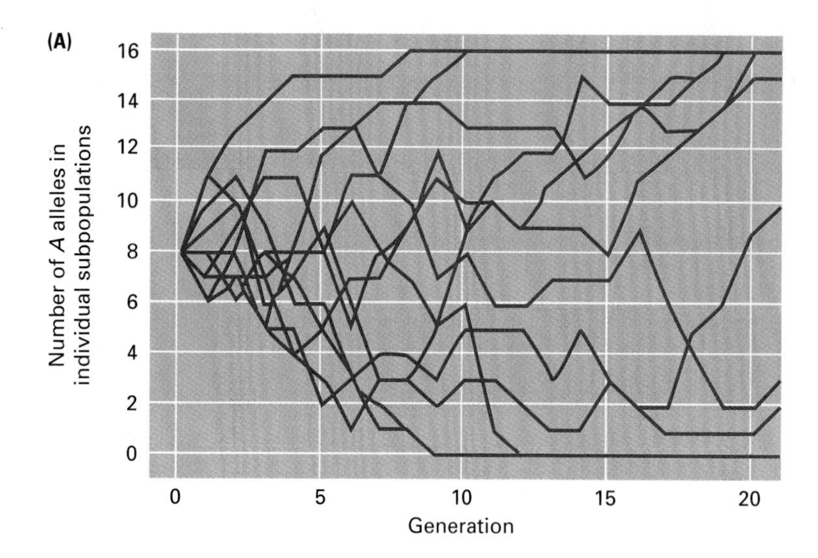


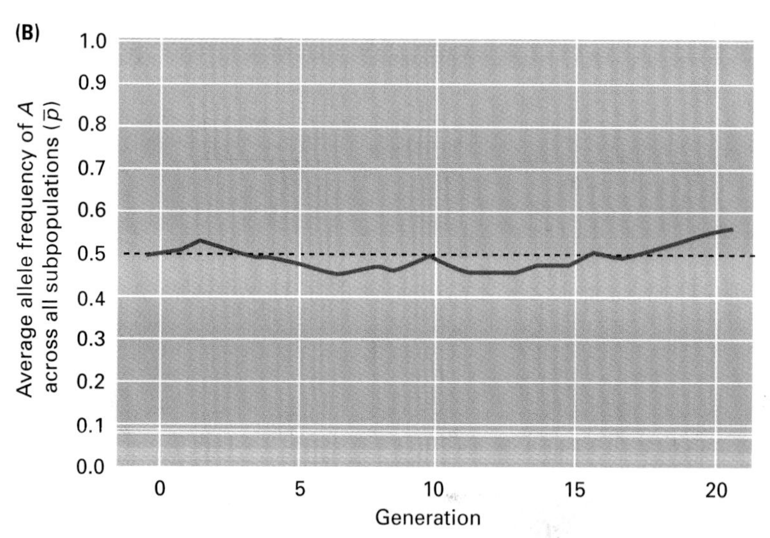


**FIGURE 17.29** (A) Random genetic drift in twelve hypothetical subpopulations of eight diploid oganisms. (B) Average allele frequency among the subpopulations in part A.

initially containing equal numbers of $A$ and $a$ alleles. Within each subpopulation, mating is random. A computer program generating values of $k$ with Equation (15) was used to calculate the number of $A$ alleles in each subpopulation in each successive generation. These numbers are shown on the $y$ axis, and the dispersion of allele frequencies resulting from random genetic drift is apparent. These changes in allele frequency would be less pronounced and would require more generations in larger populations than in the very small populations illustrated here, but the overall effect would be the same. Because of the prominent role of $2N$ in Equation (15), the dispersion of allele frequency resulting from random genetic drift depends on population size; the smaller the population, the greater the dispersion and the more rapidly it takes place.

In Figure 17.29A, the principal effect of random genetic drift is evident even in the first few

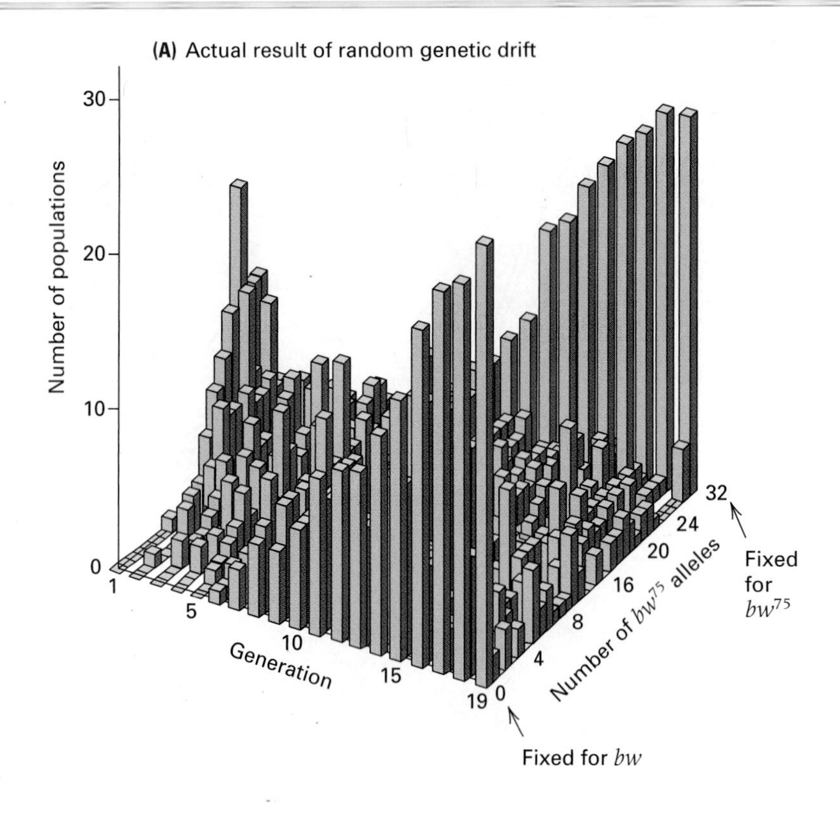


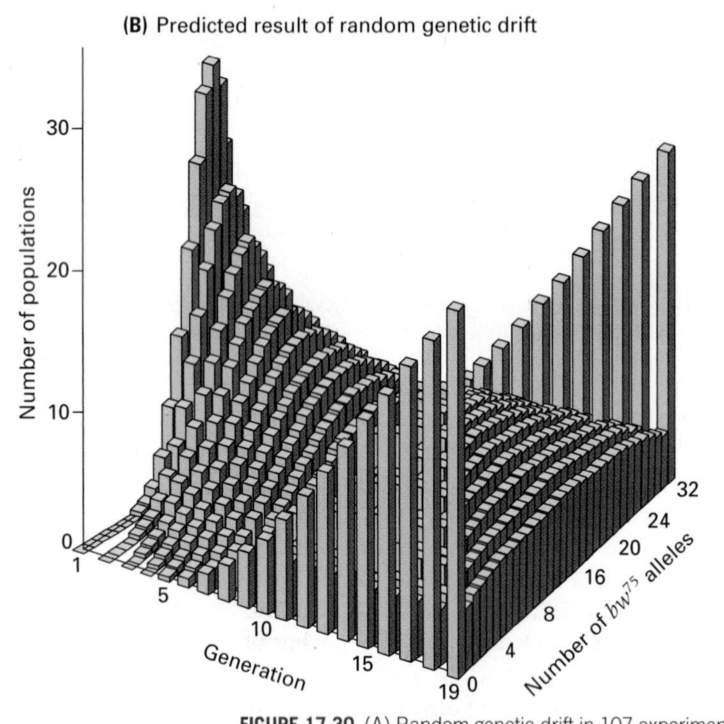


**FIGURE 17.30** (A) Random genetic drift in 107 experimental populations of *Drosophila melanogaster*, each consisting of eight females and eight males. (B) Theoretical expectation of the same situation, calculated from the binomial distribution. [Part A data from P. Buri, *Evolution* 10 (1956): 367–402. Graphs reproduced from D. L. Hartl and A. G. Clark, *Principles of Population Genetics, Third edition.* Sinauer Associates (1997).]

generations. The allele frequencies begin to spread out over a wider range. By the seventh generation, the spreading is extreme, and the number of *A* alleles ranges from 1 to 15. This spreading out means that the allele frequencies among the subpopulations become progressively more different. In general:

> Random genetic drift causes differences in allele frequency among subpopulations; it can be a major cause of genetic differentiation among subpopulations.

Although allele frequencies among subpopulations spread out over a wide range because of random genetic drift, the *average* allele frequency among subpopulations remains approximately constant. This point is illustrated in Figure 17.29B. The average allele frequency stays close to 0.5, its initial value. If an infinite number of subpopulations were being considered instead of only twelve subpopulatons, then the average allele frequency would be exactly 0.5 in every generation. This principle implies that in spite of the random drift of allele frequency in individual subpopulations, the average allele frequency among a large number of subpopulations remains constant and equal to the average allele frequency among the original subpopulations.

After a sufficient number of generations of random genetic drift, some of the subpopulations become fixed for *A* and others for *a*. Because we are excluding the occurrence of mutation, a population that becomes fixed for an allele remains fixed thereafter. After 21 generations in Figure 17.29A, only four of the populations are still segregating; eventually, these, too, will become fixed. Because the average allele frequency of *A* remains constant, it follows that a fraction $p_0$ of the populations will ultimately become fixed for *A* and a fraction $1 - p_0$ will become fixed for *a*. (The symbol $p_0$ represents the allele frequency of *A* in the initial generation.) Therefore:

> With random genetic drift, the probability of ultimate fixation of a selectively neutral allele is equal to the frequency of the allele in the original population.

In Figure 17.29A, five of the fixed populations become fixed for *A* and three for *a*, which is not very different from the equal numbers expected theoretically with an infinite number of subpopulations.

**FIGURE 17.30**, part A shows an actual example of random genetic drift in small experimental populations of *Drosophila*. The figure is based on 107 subpopulations, each initiated with eight $bw^{75}/bw$ females ($bw$ = brown eyes) and eight $bw^{75}/bw$ males and maintained at a constant size of sixteen by randomly choosing eight males and eight females from among the progeny of each generation. Note how the allele frequencies among subpopulations spread out because of random genetic drift and how subpopulations soon begin to be fixed for either $bw^{75}$ or $bw$. Although the data are somewhat rough because there are only 107 subpopulations, the overall pattern of genetic differentiation has a reasonable resemblance to that expected from the theory based on the binomial distribution, which is shown in Figure 17.30B.

If random genetic drift were the only force at work, then all alleles would eventually become either fixed or lost and there would be no polymorphism. On the other hand, many factors can act to retard or prevent the effects of random genetic drift, of which the following are the most important: (1) large population size; (2) mutation and migration, which impede fixation because alleles lost by random genetic drift can be reintroduced by either process; and (3) natural selection, particularly those modes of selection that tend to maintain genetic diversity, such as heterozygote superiority.

## 17.8 Tracing Human History Through Mitochondrial DNA

Random genetic drift is sometimes referred to as *random failure to breed*. This description captures a key feature of random genetic drift, which is that as the lineages of genes are traced forward in time, in every generation some of these lineages go extinct whereas other lineages expand. Eventually a time is reached in which all lineages have gone extinct except one. Looking backward in time, this means that all alleles present in the current population can trace their ancestry back to just one allele present in some ancestral population. The convergence of separate gene genealogies onto a common ancestral gene is known as **coalescence**. For a selectively neutral allele, the average time required for coalescence to a single common ancestor is $4N$ generations, where $N$ is roughly the number of breeding individuals present in each generation.

The coalescence of gene genealogies is easiest to track for DNA molecules that do not undergo recombination, because then the ancestral history of a set of sequences can be estimated using the methods of molecular phylogenetics discussed in Section 17.1. One type of DNA molecule that does not undergo recombination is mitochondrial DNA (mtDNA). Coalescence of mtDNA implies that, if one traces mtDNA sequences far enough back in time, the DNA molecule in any mitochondrion present in the current population derives from a single mtDNA molecule present in a common ancestor. In humans and other organisms with maternal transmission of mtDNA, this common ancestor would have to be a female.

In addition to the lack of recombination, mtDNA is a good genetic marker for tracing ancestry because in many species it evolves considerably more rapidly than nuclear genes. In humans, differences in mitochondrial DNA sequences accumulate among lineages at an average rate of approximately 1 change per mitochondrial lineage every 3800 years. For example, the people of Papua New Guinea have been relatively isolated genetically from the aboriginal people of Australia ever since these areas were colonized approximately 40,000 years ago and 30,000 years ago, respectively. The total time for evolution between the populations is therefore 70,000 years (40,000 years in Papua New Guinea and 30,000 years in Australia). If one nucleotide change accumulates every 3800 years on the average, the expected number of differences in the mitochondrial DNA of modern-day Papua New Guineans and Australian aborigines is expected to be 18.4 nucleotides (calculated as 70,000/3800), with of course some statistical variation from one pair of persons to the next. This example shows how the rate of mtDNA evolution can be used to predict the number of differences between populations. In practice, the calculation is usually done the other way around, and the observed number of differences between pairs of populations is used to estimate the number of years since the populations have been geographically separated.

Archaeological evidence and sequence comparisons of nuclear genes suggest that modern human populations originated in sub-Saharan Africa approximately 100,000 years ago. The

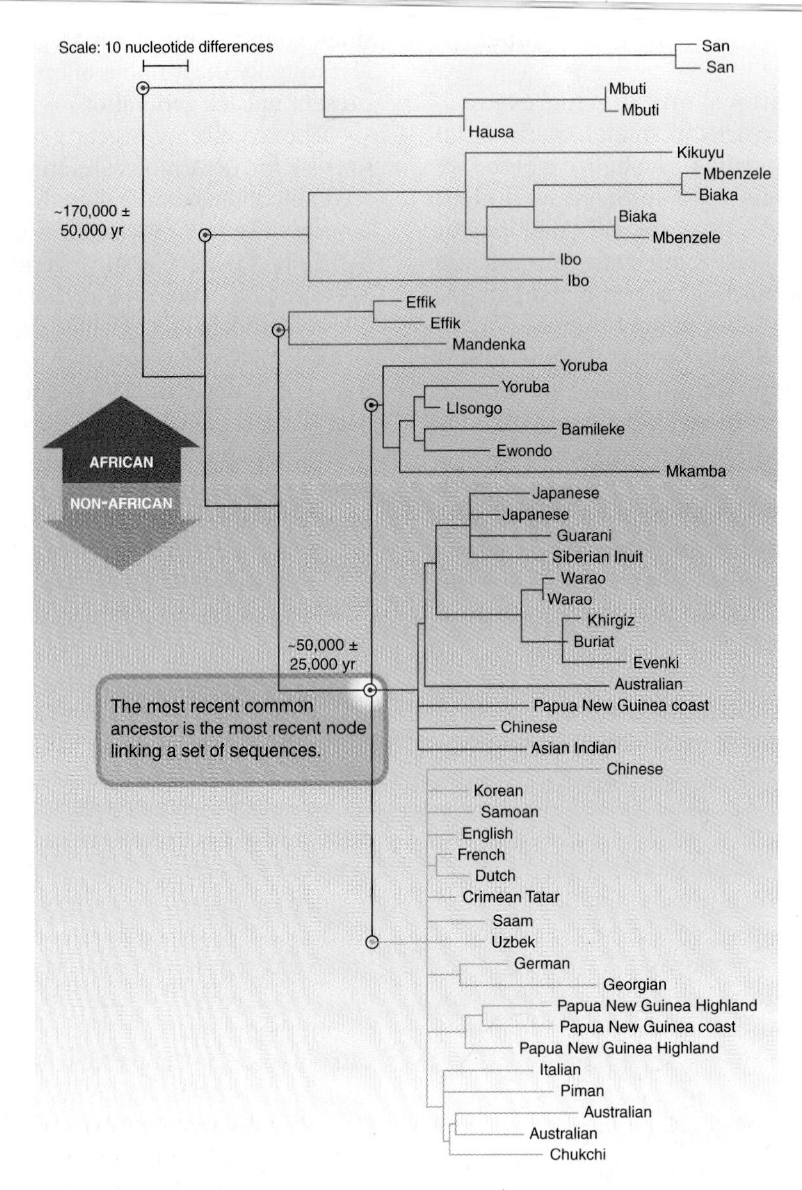


**FIGURE 17.31** Inferred evolutionary relationships among human mitochondrial DNA molecules based on analysis of the complete nucleotide sequence of mtDNA from 53 persons. [Adapted from M. Ingman, et al., *Nature* 408 (2000): 708–713.]

probable historical relationships among human populations can be reconstructed based on differences in the nucleotide sequences of mtDNA. **FIGURE 17.31** shows the inferred ancestral relationships of mtDNA based on the complete mtDNA sequences from 53 persons representing human populations from throughout the world. Because mtDNA undergoes no recombination, the tree is the genetic history of only a single DNA molecule, and because mtDNA is maternally inherited, it is the genetic history of females. Nevertheless, the mtDNA tree shows several remarkable features:

- Much of the mtDNA diversity in African populations is not found among non-Africans; on the average, the mtDNA of Africans shows about twice as much genetic variation as the mtDNA among non-Africans.
- Three of the four major lineages of mtDNA are found only in sub-Saharan

Africans (green); the age of the most recent common ancestor of these sequences (green circle) is approximately 170,000 ± 50,000 years.
- A restricted subset of mtDNA lineages is found among non-Africans (purple and red); these sequences share a more recent common ancestor with five African mtDNAs (orange) than these African mtDNAs share with other Africans.
- The age of the most recent common ancestor of the mtDNA lineage joining African and non-African populations (blue circle) is approximately 50,000 ± 25,000 years.

These features of the mtDNA tree are consistent with a history in which all modern human populations are derived from a migration out of Africa that took place approximately 100,000 years ago (**FIGURE 17.32**). There were also earlier migrations out of Africa, but the descendants of these earlier migrants were apparently displaced by people who came later. Figure 17.32 shows the dates of the earliest fossil and archaeological finds on each continent. The age of the oldest fossils of the modern subspecies of human beings (*Homo sapiens sapiens*) found in Africa is 130,000 years. This time is consistent with 170,000 ± 50,000 years for the time of the most recent common ancestor of all of the mtDNA lineages in Figure 17.31. The estimated time of the Great Migration out Africa is 50,000 ± 25,000 years (blue circle in Figure 17.31). At first sight, this estimate seems inconsistent with the 100,000 years based on archeological evidence and that from nuclear genes. However, the genetic history of mtDNA is the genetic history of females, and it reasonable to suppose that some of the mtDNA diversity among the original migrants was lost in the first few tens of thousands of years in the subsequent expansion. It is not known whether the original migrants came from a restricted geographical locality, but it is worth noting that the contemporary mtDNA lineages in orange in Figure 17.31 are present in geographically dispersed individuals.

The Great Migration 100,000 years ago went initially to the Middle East and Northern Europe (first fossil and archaeological evidence 40,000 years), then east and south to Asia ( 67,000 years) and Australasia (40,000–60,000 years), and finally from East Asia across the Bering Strait to North America (20,000 years) and South America (13,000 years). At a rate of sequence evolution averaging one nucleotide substitution every 3800 years, the average number of differences between any two mtDNA sequences descended from the Great Migration is about fifty in a molecule of total length 16.5 kb. In other words, our mtDNA molecules are 99.7 percent identical.

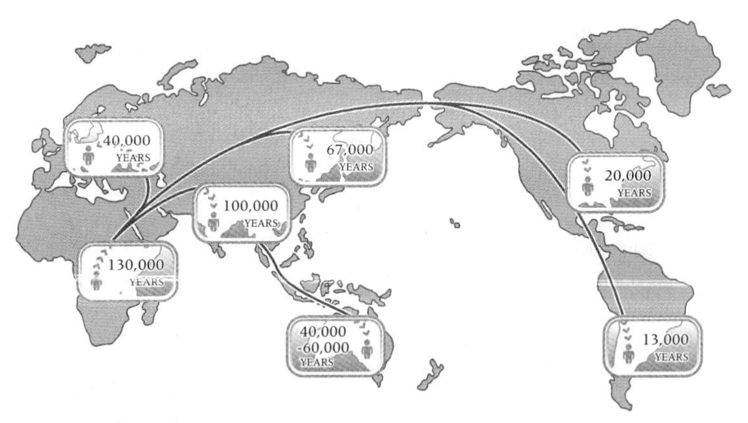


**FIGURE 17.32** The dispersal of modern human populations from sub-Saharan Africa began approximately 100,000 years ago. The dates on the map are based on the earliest fossil and archaeological evidence from each continent. [Adapted from S. B. Hedges, *Nature* 408 (2000): 652–653.]

## CHAPTER SUMMARY

- The ancestral history of protein and nucleic acid molecules can be inferred because the sequences accumulate changes through evolutionary time.
- Many genes in natural populations are polymorphic—they have two or more common alleles. Genetic polymorphisms can be used as genetic markers in pedigree studies and for individual identification (DNA typing).
- The Hardy–Weinberg principle asserts that, with random mating, the frequencies of the genotypes *AA, Aa,* and *aa* in a population are expected to be $p^2$, $2pq$, and $q^2$, respectively, where $p$ and $q$ are the allele frequencies of *A* and *a*. One implication of this principle is that the great majority of rare, harmful recessive alleles are present in heterozygous carriers.
- Inbreeding means mating between relatives. Through inbreeding, replicas of an allele present in a common ancestor may be passed down both sides of a pedigree and come together in an inbred organism. Such alleles

are said to show identity by descent. Because inbreeding can result in alleles that are identical by descent, inbreeding results in an excess of homozygous genotypes relative to the frequencies of genotypes expected with random mating.

- Mutation and migration both introduce new alleles into populations. Because mutation rates are typically small, mutation is a weak force for changing allele frequencies.
- Natural selection can change allele frequency in a systematic direction because of the differential reproductive success of certain genotypes. Selection can result in a stable genetic polymorphism if the heterozygous genotype has the highest fitness. Selection against recurrent harmful mutations leads to a selection–mutation balance in which the allele frequency of the harmful recessive remains constant generation after generation.
- Random genetic drift consists of random changes in allele frequency that occur in finite populations as a result of chance excess reproduction (or lack of reproduction) among certain genotypes, because all breeding adults cannot be expected to contribute equally to the next generation.
- Evidence from human mitochondrial DNA is consistent with a scenario in which all non-African human populations were founded from a relatively small group of people who migrated out of Africa approximately 100,000 years ago.

## REVIEW THE BASICS

- What is a gene tree? A species tree? Must they always agree? Explain why or why not.
- Distinguish between orthologous genes and paralogous genes. Which type of duplication provides the raw material for the evolution of new gene functions?
- What does the Hardy–Weinberg principle imply about the relative frequencies of heterozygous carriers and homozygous affected organisms for a rare, harmful recessive allele?
- Traits due to recessive alleles in the X chromosome are usually much more prevalent in males than in females. Explain why this discrepancy is expected with random mating.
- Why are the effects of inbreeding more easily observed with a rare recessive allele than with a rare dominant allele?
- Name four evolutionary processes that can change allele frequencies in natural populations. How do allele frequencies change in the absence of these processes?
- What is the role of mutation in the evolutionary process? Why is mutation a weak force for changing allele frequencies?
- What do population geneticists mean by *fitness*? What is the fitness of an organism that dies before the age of reproduction? What is the fitness of an organism that is sterile?
- Why, in a random-mating population, does the allele frequency of a rare recessive allele change slowly under selection?
- Many recessive alleles are extremely harmful when homozygous, so there is selection in every generation that tends to reduce the allele frequency. Yet harmful recessive alleles are maintained at a low frequency for almost every gene. What process prevents harmful recessive alleles from being completely eliminated?
- Heterozygote superiority of the type observed with sickle-cell anemia is sometimes called balancing selection. Do you think this is an appropriate term? Explain why or why not.
- What is random genetic drift and why does it occur? Explain why this process implies that, in the absence of other forces, the ancestry of all alleles present at a locus in a population can eventually be traced back to a single allele present in some ancestral population.

## GUIDE TO PROBLEM SOLVING

**Problem 1** The gel diagram shown here summarizes the results of a study of RFLP genotypes among 300 individuals samples from a natural population of the plant *Arabidopsis lyrata*. The RFLP alleles are denoted $A_1$ and $A_2$, and above each lane is the number of individuals observed whose genomic DNA yielded that pattern of bands. What are the allele frequencies of $A_1$ and $A_2$? What are the expected numbers of the three genotypes, assuming random mating?

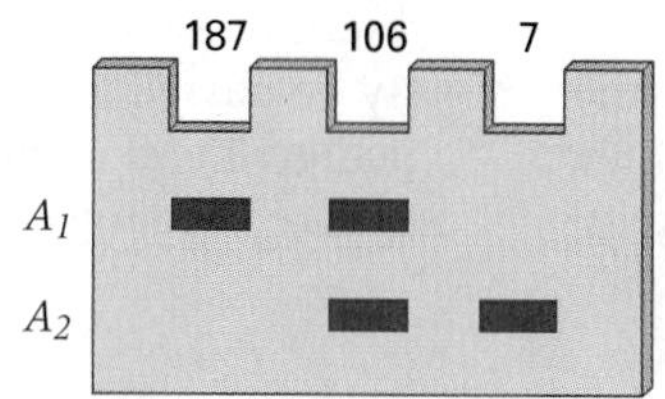


**Answer** The allele frequencies are determined by counting the alleles. The 187 $A_1A_1$ plants represent 374 $A_1$ alleles, the 106 $A_1A_2$ plants represent 106 $A_2$ and 106 $A_1$ alleles, and the 7 $A_2A_2$ plants represent 14 $A_2$ alleles, for a total of $300 \times 2 = 600$ alleles altogether. The allele frequency $p$ of $A_1$ is $(374 + 106)/600 = 0.8$, and the allele frequency $q$ of $A_2$ is $(106 + 14)/600 = 0.2$. As a check on the calculations, note that the allele frequencies sum to unity, as they should. For the second part of the question, the expected genotype frequencies with random mating are $p^2$ $A_1A_1$, $2pq$ $A_1A_2$, and $q^2$ $A_2A_2$, so the expected numbers are $A_1A_1$: $(0.8)^2 \times 300 = 192$, $A_1A_2$: $2(0.2)(0.8) \times 300 = 96$, $A_2A_2$: $(0.2)^2 \times 300 = 12$. As a check on these calculations, note that $192 + 96 + 12 = 300$.

**Problem 2** The gel patterns shown here correspond to a VNTR (variable number of tandem repeat) polymorphism in genomic DNA among four pairs of parents (1–8) and four children (A–D). One child comes from each pair of parents.

**(a)** Why does the DNA from each person exhibit two bands?

**(b)** Would it be possible for a person's DNA to exhibit only one band? How?

**(c)** What type of dominance is illustrated by the VNTR alleles?

**(d)** Which pairs of people who have the DNA in lanes 1–8 are the parents of each child?

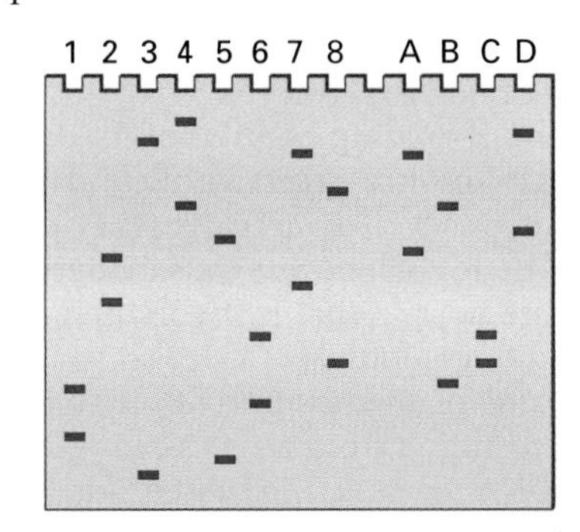


**Answer**

**(a)** The two bands in genomic DNA from each individual indicate heterozygosity for two different VNTR alleles.

**(b)** Genomic DNA from a person who is homozygous for a VNTR allele will exhibit a single VNTR band.

**(c)** VNTR alleles are codominant because bands from both alleles are detected in heterozygous genotypes.

**(d)** Because none of the bands in the four pairs of parents have the same electrophoretic mobility, the parents of each child can be identified as the persons who share a single VNTR band with the child. Child A has parents 7 and 2; child B has parents 4 and 1; child C has parents 6 and 8; and child D has parents 3 and 5.

**Problem 3** Autosomal recessive polycystic kidney disease (AR-PKD) is a severe genetic disorder characterized by the growth of numerous cysts in the kidneys. Children with AR–PKD suffer from high blood pressure, urinary tract infections, and frequent urination. They often develop kidney failure and require dialysis before adulthood. The frequency of affected children is about 1 per 10,000.

**(a)** Assuming Hardy-Weinberg proportions, what is the frequency of heterozygous carriers of AR-PKD?

**(b)** What is the expected frequency of AR-PKD among the children of first cousins?

**(c)** What is the ratio of AR-PKD among the offspring of first-cousin matings to that among the offspring of nonrelatives?

**Answer** First we must calculate the frequency of the recessive allele, $q$, using the information that the frequency of homozygous recessives is 1 in 10,000. Assuming random mating frequencies, $q = \sqrt{(1/10,000)} = 0.01$.

**(a)** Assuming Hardy-Weinberg proportions, the frequency of heterozygous genotypes equals $2(0.01)(0.99) = 0.0198$, or about 1 in 50 individuals.

**(b)** Among the offspring of first-cousin matings, the expected frequency of AR-PKD equals $q^2(1 - F) + qF$, where $F$ is the inbreeding coefficient and equals 1/16 for the offspring of first cousins. In this case, the formula yields $(0.01)^2(15/16) + (0.01)(1/16) = 0.000718$, or about 1 in 1400 births.

**(c)** The ratio is given in the following formula, where $q = 0.01$ and $F = 1/16$.

$$\text{Ratio} = \frac{q^2(1-F)+qF}{q^2} = (1-F)+F/q$$

In this case the ratio is $(15/16) + (1/16) \times (1/0.01) = 7.19$. In other words, the offspring of first cousins have more than a 7-fold greater risk of being homozygous for a recessive mutant allele causing AR-PKD.

## ANALYSIS AND APPLICATIONS

**17.1** If the genotype $AA$ is an embryonic lethal and the genotype $aa$ is fully viable but sterile, what genotype frequencies would be found in adults in an equilibrium population containing the $A$ and $a$ alleles? Is it necessary to assume random mating?

**17.2** Three codominant alleles of a single gene determine different forms of alcohol dehydrogenase in the flowering plant *Phlox drummondii*. In one sample of 35 plants, the following data were obtained:

| Genotype | $A_1A_1$ | $A_1A_2$ | $A_2A_2$ | $A_2A_3$ | $A_3A_3$ | $A_1A_s$ |
|---|---|---|---|---|---|---|
| Number | 2 | 5 | 12 | 10 | 5 | 1 |

What are the frequencies of the alleles $A_1$, $A_2$, and $A_3$ in this sample?

**17.3** The illustration at the top of the next page shows a human pedigree, along with the pattern of restriction fragments observed in a DNA sample from each of the individuals when hybridized with a labeled probe. Examine the pedigree and the band patterns, and suggest a mode of inheritance. Is the Hardy–Weinberg principle relevant to this situation? Explain why or why not.

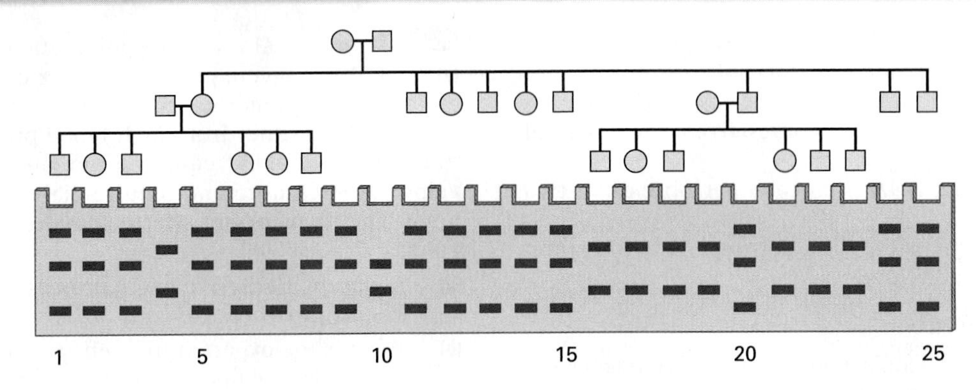


**17.4** DNA from 100 unrelated people was digested with the restriction enzyme *Hin*dIII, and the resulting fragments were separated and probed with a sequence for a particular gene. Four fragment lengths that hybridized with the probe were observed—namely 5.7, 6.0, 6.2, and 6.5 kb—where each fragment defines a different restriction-fragment allele. The accompanying illustration shows the gel patterns observed; the number of individuals with each gel pattern is shown across the top. Estimate the allele frequencies of the four restriction-fragment alleles.

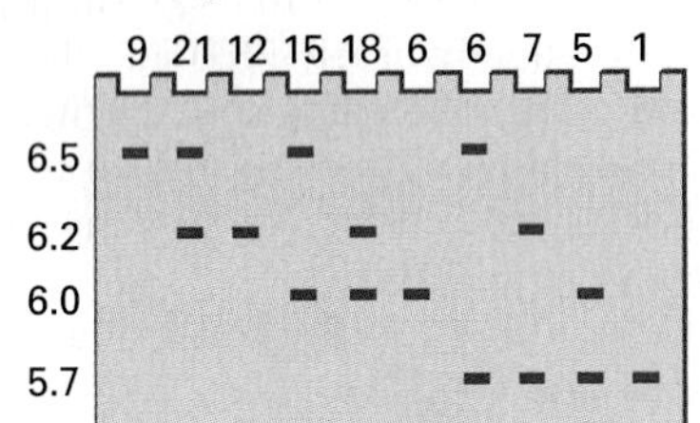


**17.5** If the frequencies of genotypes *AA*, *Aa*, and *aa* are denoted *P*, *Q*, and *R*, respectively, show that the Hardy–Weinberg principle implies that $Q^2/4 = PR$.

**17.6** Which of the following genotype frequencies of *AA*, *Aa*, and *aa*, respectively, satisfy the Hardy–Weinberg principle?

**(a)** 0.25, 0.50, 0.25
**(b)** 0.36, 0.55, 0.09
**(c)** 0.49, 0.42, 0.09
**(d)** 0.64, 0.27, 0.09
**(e)** 0.29, 0.42, 0.29

**17.7** A population geneticist observed the RFLP phenotypes shown in the accompanying gel diagram, where the number above each well is the number of individuals exhibiting the corresponding phenotype. Test whether the observed phenotype frequencies are consistent with those expected from a single locus with three alleles in a random-mating population, using the chi-square criterion. Explain why, in this test, the appropriate number of degrees of freedom is 3.

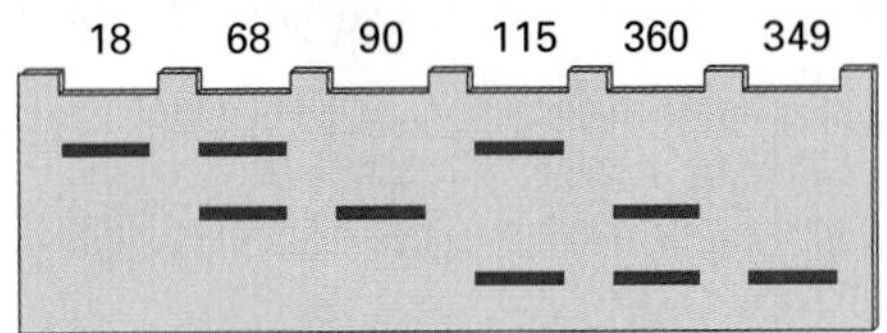


**17.8** Hartnup disease is an autosomal-recessive disorder of intestinal and renal transport of amino acids. The frequency of affected newborn infants is about 1 in 14,000. Assuming random mating, what is the frequency of heterozygotes?

**17.9** A randomly mating population of dairy cattle contains an autosomal-recessive allele causing dwarfism. If the frequency of dwarf calves is 10 percent, what is the frequency of heterozygous carriers of the allele in the entire herd? What is the frequency of heterozygotes among nondwarf cattle?

**17.10** Xeroderma pigmentosum (XP) is an often-fatal skin cancer resulting from a recessive mutant allele that affects DNA repair. In the United States, the frequency of homozygous-recessive affected people is approximately 1 in 250,000. (The mutant allele can be in any one of about eight different genes, but for the purposes of this problem it makes no difference.)

**(a)** What is the expected frequency of XP among the offspring of first-cousin matings?
**(b)** What is the ratio of XP among the offspring of first-cousin matings to that among the offspring of nonrelatives?

**17.11** In a Pygmy group in Central Africa, the frequencies of alleles determining the ABO blood groups were estimated as 0.74 for $I^O$, 0.16 for $I^A$, and 0.10 for $I^B$. Assuming random mating, what are the expected frequencies of ABO genotypes and phenotypes?

**17.12** In certain grasses, the ability to grow in soil contaminated with the toxic metal nickel is determined by a dominant allele.

**(a)** If 60 percent of the seeds in a randomly mating population are able to germinate in contaminated soil, what is the frequency of the resistance allele?
**(b)** Among plants that germinate, what proportion are homozygous?

**17.13** If an X-linked recessive trait is present in 2 percent of the males in a population with random mating, what is the frequency of the trait in females? What is the frequency of carrier females?

**17.14** Among 35 individuals of the flowering plant *Phlox roemariana*, the following genotypes were observed for the alleles of a gene determining different forms of the enzyme phosphoglucose isomerase: 2 $A_1A_1$, 13 $A_1A_2$, 20 $A_2A_2$.

**(a)** What are the frequencies of the alleles $A_1$ and $A_2$?
**(b)** Assuming random mating, what are the expected numbers of the genotypes?

**17.15** How does the frequency of heterozygotes in an inbred population compare with that in a randomly mating population with the same allele frequencies?

**17.16** In a population of *Drosophila*, an X-linked recessive allele causing yellow body color is present in genotypes at frequencies typical of random mating; the frequency of the recessive allele is 0.20. Among 1000 females and 1000 males, what are the expected numbers of the yellow and wildtype phenotypes in each sex?

**17.17** Galactosemia is an autosomal-recessive condition associated with liver enlargement, cataracts, and mental retardation. Among the offspring of unrelated individuals, the frequency of galactosemia is $8.5 \times 10^{-6}$. What is the expected frequency among the offspring of first cousins ($F = 1/16$) and among the offspring of second cousins ($F = 1/64$)?

**17.18** Which of the following genotype frequencies of *AA*, *Aa*, and *aa*, respectively, are suggestive of inbreeding?

**(a)** 0.25, 0.50, 0.25
**(b)** 0.36, 0.55, 0.09
**(c)** 0.49, 0.42, 0.09
**(d)** 0.64, 0.27, 0.09
**(e)** 0.29, 0.42, 0.29

**17.19** Consider the accompanying pedigree.

**(a)** Identify all of the common ancestors.
**(b)** Identify all paths through each common ancestor, underlining the common ancestor in each path.
**(c)** Calculate the inbreeding coefficient of individual L, assuming that $F_B$ is unknown.
**(d)** Calculate the inbreeding coefficient of individual L, assuming that $F_B = 0$.

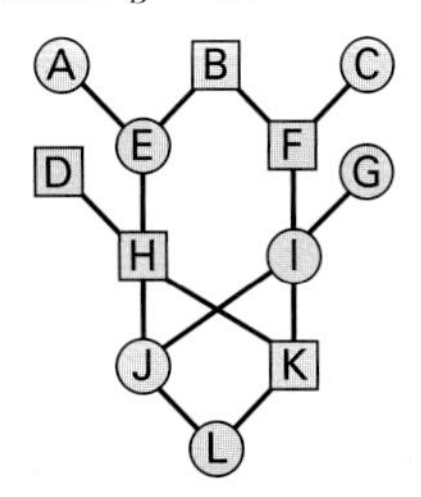


**17.20** Suppose the frequency of a recessive genetic disorder among the children of nonrelatives in a population is 1 in 2.6 million.

**(a)** What is the estimated frequency of the harmful recessive allele?
**(b)** What is the expected frequency of homozygous recessives among the offspring of first cousins?
**(c)** If first-cousin matings make up 1 percent of all matings in the population, and all other matings are random, what is the total frequency of homozygous recessive genotypes?
**(d)** What proportion of all homozygous recessive genotypes come from the first-cousin matings?

**17.21** Self-fertilization in the annual plant *Phlox cuspidata* results in an average inbreeding coefficient of $F = 0.66$.

**(a)** What frequencies of the genotypes for the enzyme phosphoglucose isomerase would be expected in a population with alleles $A_1$ and $A_2$ at frequencies 0.43 and 0.57, respectively?
**(b)** What frequencies of the genotypes would be expected with random mating?

**17.22** What is the UPGMA distance matrix for the gene tree shown here?

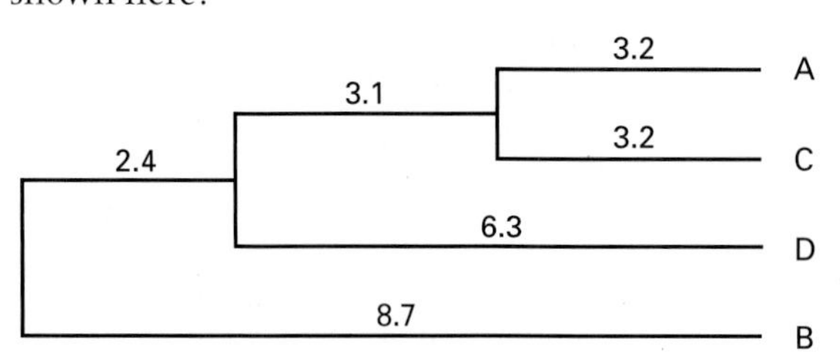


**17.23** A strain of *Drosophila* has an insertion of the transposable element *mariner* in the wildtype $w^+$ allele of the gene *white* in the X chromosome. The insertion mutation results in peach-colored eyes instead of white. Suppose this transposable element undergoes spontaneous excision from its site of insertion with a frequency of 5 percent per generation:

**(a)** What is the frequency of the mutant allele in X chromosomes after 1 generation?
**(b)** After 10 generations?
**(c)** After 20 generations?
**(d)** After how many generations will the frequency of mutant X chromosomes be below 0.01?

**17.24** Two strains of *Escherichia coli*, A and B, are inoculated into a chemostat in equal frequencies and undergo competition. After 40 generations, the frequency of the B strain is 35 percent. What is the fitness of strain B relative to strain A, under the particular experimental conditions, and what is the selection coefficient against strain B?

**17.25** An allele *A* mutates to a nonfunctional allele *a* at a rate $10^{-6}$ per generation. Reverse mutation from *a* to *A* can be ignored. Determine the expected equilibrium frequency of allele *a* if:

**(a)** The allele *a* is a recessive lethal with no effect in heterozygous genotypes.
**(b)** The allele *a* is a "recessive" lethal that results in a 1 percent selective disadvantage in heterozygous genotypes.
**(c)** By what percentage does the small amount of dominance reduce the equilibrium frequency of *a*?

**17.26** An allele *A* is a hotspot of mutation and undergoes mutation to *a* at a frequency of $10^{-4}$ per generation. If the frequency of reverse mutation from *a* to *A* is $10^{-7}$ per generation, what is the expected equilibrium allele frequency of *a*?

**17.27** In the distance matrix shown here, which pair of taxa should be joined first, and what is the resulting UPGMA distance matrix?

| | B | C | D |
|---|---|---|---|
| A | 8 | 15 | 19 |
| B | | 18 | 16 |
| C | | | 5 |

**17.28** Huntington disease is a form of neuromuscular degeneration, usually beginning in middle age, that results from a dominant mutation in a gene encoding a protein called huntingtin. The early onset reduces the reproductive fitness of affected individuals by about 20 percent. A large study in Michigan estimated that the frequency of the dominant allele is about 0.00005. Assume that this population undergoes random mating and is in equilibrium.

**(a)** What is the expected frequency of individuals who carry the dominant allele?

**(b)** What is the estimated mutation rate to the dominant deleterious allele?

**17.29** An overdominant locus has relative fitnesses of *AA*, *Aa*, and *aa* given by 0.90 : 1.00 : 0.50.

**(a)** What is the expected equilibrium frequency of the *a* allele?

**(b)** What is the expected frequency of heterozygous genotypes at equilibrium?

**(c)** If homozygous *aa* were lethal, how would this change the equilibrium allele frequency of *a*?

**(d)** What is the expected frequency of heterozygous genotypes at the new equilibrium?

**17.30** In the following situations, which allele frequency will change faster? Explain why.

**(a)** A rare beneficial allele that is dominant or a rare beneficial allele that is recessive?

**(b)** A common deleterious allele that is dominant or a common deleterious allele that is recessive?

## CHALLENGE PROBLEMS

**Challenge Problem 1** Human genetic polymorphisms in which there are a variable number of tandem repeats (VNTR) are often used in DNA typing. The table below shows the alleles of four VNTR loci and their estimated average frequencies in genetically heterogeneous Caucasian populations. (The allele 9.3 of the *HUMTHO1* locus has a partial copy of one of the tandem repeats.) Assuming random mating, estimate:

**(a)** The probability that an individual will be of genotype (6, 9.3) for *HUMTHO1*, (10, 11) for *HUMFES*, (6, 6) for *D12S67*, and (18, 24) for *D1S80*.

**(b)** The probability that an individual will be of genotype (8, 10) for *HUMTHO1*, (8, 13) for *HUMFES*, (1, 2) for *D12S67*, and (19, 19) for *D1S80*.

| ***HUMTHO1*** | | ***HUMFES*** | | ***D12S67*** | | ***D1S80*** | |
|---|---|---|---|---|---|---|---|
| **Repeats** | **Freq.** | **Repeats** | **Freq.** | **Repeats** | **Freq.** | **Repeats** | **Freq.** |
| 6 | 0.230 | 8 | 0.007 | 1 | 0.015 | 18 | 0.293 |
| 7 | 0.160 | 10 | 0.321 | 2 | 0.005 | 19 | 0.011 |
| 8 | 0.105 | 11 | 0.373 | 3 | 0.058 | 20 | 0.021 |
| 9 | 0.193 | 12 | 0.233 | 4 | 0.078 | 21 | 0.032 |
| 9.3 | 0.310 | 13 | 0.066 | 5 | 0.118 | 22 | 0.043 |
| 10 | 0.002 | | | 6 | 0.324 | 23 | 0.016 |
| | | | | 7 | 0.196 | 24 | 0.335 |
| | | | | 8 | 0.127 | 25 | 0.037 |
| | | | | 9 | 0.059 | 26 | 0.016 |
| | | | | 10 | 0.020 | 28 | 0.078 |
| | | | | | | 29 | 0.059 |
| | | | | | | 30 | 0.016 |
| | | | | | | 31 | 0.043 |

**Challenge Problem 2** In the pedigree shown here, individual I is the result of two successive generations of brother-sister mating. Assume that $F_A = F_B = 0$.

**(a)** Calculate the inbreeding coefficients of individuals E and F (who result from one generation of brother–sister mating).

**(b)** Calculate the inbreeding coefficient of I.

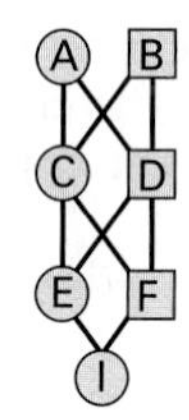


**Challenge Problem 3** There are an average of 78 nucleotide differences in mitochondrial DNA between two Africans, an average of 58 differences between two Asians, and an average of 40 differences between two Caucasians. Assuming a rate of evolution of mitochondrial DNA of 1 nucleotide difference per 1500 to 3000 years, what is the estimated average age of the common ancestor of the mitochondrial DNA among Africans? Among Asians? Among Caucasians?

### GENETICS on the web

GeNETics on the Web will introduce you to some of the most important sites for finding genetics information on the Internet. To explore these sites, visit the Jones and Bartlett companion site to accompany *Genetics: Analysis of Genes and Genomes, Seventh Edition* at http://biology.jbpub.com/book/genetics.

There you will find a chapter-by-chapter list of highlighted keywords. When you select one of the keywords, you will be linked to a Web site containing information related to that keyword.

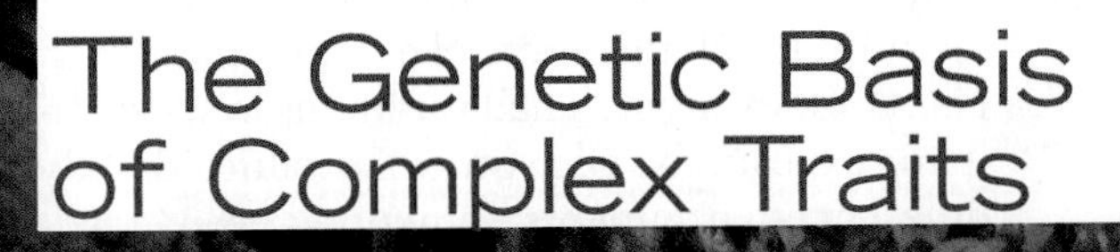

# CHAPTER 18

# The Genetic Basis of Complex Traits

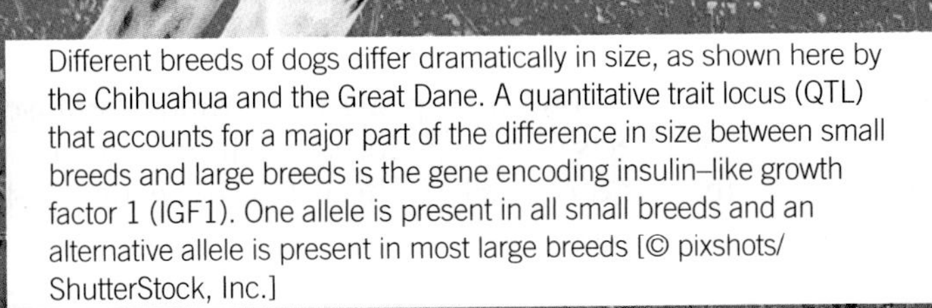

Different breeds of dogs differ dramatically in size, as shown here by the Chihuahua and the Great Dane. A quantitative trait locus (QTL) that accounts for a major part of the difference in size between small breeds and large breeds is the gene encoding insulin–like growth factor 1 (IGF1). One allele is present in all small breeds and an alternative allele is present in most large breeds [© pixshots/ ShutterStock, Inc.]

## CHAPTER OUTLINE


**18.1 Complex Traits**
- Continuous, Categorical, and Threshold Traits
- The Normal Distribution

**18.2 Causes of Variation**
- Genotypic Variation
- Environmental Variation
- Genetics and Environment Combined
- Genotype-by-Environment Interaction and Association

**18.3 Genetic Analysis of Complex Traits**
- The Number of Genes Affecting Complex Traits
- Broad-Sense Heritability
- Twin Studies

**18.4 Artificial Selection**
- Narrow-Sense Heritability
- Phenotypic Change with Individual Selection: A Prediction Equation
- Long-Term Artificial Selection
- Inbreeding Depression and Heterosis

**18.5 Correlation Between Relatives**
- Covariance and Correlation
- The Geometrical Meaning of a Correlation
- Estimation of Narrow-Sense Heritability

**18.6 Heritabilities of Threshold Traits**

**18.7 Identification of Genes Affecting Complex Traits**
- Linkage Analysis in the Genetic Mapping of Quantitative Trait Loci
- The Number and Nature of QTLs
- Candidate Genes for Complex Traits

CONNECTION A Maize'n Grass
George W. Beadle 1972
*The Mystery of Maize*

CONNECTION Win, Place, or Show?
Hugh E. Montgomery and eighteen other investigators 1998
*Human Gene for Physical Performance*

Earlier chapters have emphasized traits in which differences in phenotype result from alternative genotypes of a single gene. These traits are particularly suited for genetic analysis through the study of pedigrees, not only because of the small number of genotypes and phenotypes, but also because of the simple correspondence between genotype and phenotype. However, many traits that are important in medical genetics, plant breeding, and animal breeding are influenced by *multiple* genes as well as by the environment. These are known as **complex traits** or **multifactorial traits** because of the multiple genetic and environmental factors implicated in their causation, and they are said to show *complex inheritance.* The inheritance is complex because a single genotype can have many possible phenotypes (depending on the environment), and a single phenotype can include many possible genotypes. Genetic analysis of complex inheritance requires special concepts and methods, which are introduced in this chapter.

## 18.1 Complex Traits

Complex traits are typically influenced not only by the alleles of two or more genes, but also by the environment. The phenotype of an organism is therefore potentially influenced by:

- ***Genetic factors*** in the form of alternative genotypes of one or more genes.
- ***Environmental factors*** in the form of conditions that are favorable or unfavorable for the development of the trait.

Examples of environmental factors include the effect of nutrition on the growth rate of animals and the effects of fertilizer, rainfall, and planting density on the yield of crop plants.

With some complex traits, differences in phenotype result largely from differences in genotype, and the environment plays a minor role. With others, differences in phenotype result largely from the effects of environment, and genetic factors play a minor role. However, most complex traits fall between these extremes, and both genotype and environment must be taken into account in their analysis.

In a genetically heterogeneous population, many genotypes are formed by the processes of segregation and recombination. Variation in genotype can be eliminated by studying inbred lines, which are homozygous for most genes, or the $F_1$ progeny from a cross of inbred lines, which are uniformly heterozygous for all genes in which the parental inbreds differ. In contrast, complete elimination of environmental variation is impossible, no matter how hard the experimenter may try to render the environment identical for all members of a population. With plants, for example, small variations in soil quality or exposure to the sun will produce slightly different environments, sometimes even for adjacent plants. Similarly, highly inbred *Drosophila* still show variation in phenotype (for example, in body size) brought about by environmental differences among animals within the same culture bottle. Therefore, traits that are susceptible to small environmental effects will never be uniform, even in inbred lines.

Most traits that are important in plant and animal breeding are complex traits. In agricultural production, one economically important complex trait is yield—for example, the size of the harvest of corn, tomatoes, soybeans, or grapes per unit area. In domestic animals, important complex traits include meat quality, milk production per cow, egg laying per hen, fleece weight per sheep, and litter size per sow. Important complex traits in human genetics include infant growth rate, adult weight, blood pressure, serum cholesterol, and length of life. In evolutionary studies, fitness is the preeminent complex trait.

Most complex traits cannot be studied by means of the usual pedigree methods, because the effects of segregation of alleles of one gene may be concealed by effects of other genes, and environmental effects may cause identical genotypes to have different phenotypes. Therefore, individual pedigrees do not fit any simple pattern of dominance, recessiveness, or X linkage. Nevertheless, genetic effects on the traits can be assessed by comparing the phenotypes of relatives who, because of their familial relationship, must have a certain proportion of their genes in common. Such studies utilize many of the concepts of population genetics discussed in Chapter 17.

Three categories of traits are frequently found to have complex inheritance. They are described in the following section.

### Continuous, Categorical, and Threshold Traits

Most phenotypic variation in populations is not manifested in a few easily distinguished categories. Instead, the traits vary continuously from one phenotypic extreme to the other, with no clear-cut breaks in between. Height and weight are prime examples. Other examples include milk production in cattle, growth rate in poultry, yield in corn, and blood pressure in human beings. Such traits are called **continuous traits** or **quantitative traits**, because there is a continuous or quantitative gradation from one phenotype to the next.

Two other types of complex traits are not continuous:

**Categorical traits** are traits in which the phenotype corresponds to any one of a number of discrete categories. Typically the phenotype corresponds to a count of, for example, number of skin ridges forming the fingerprints, number of kernels on an ear of corn, number of eggs laid by a hen, number of bristles on the abdomen of a fly, and number of puppies in a litter.

**Threshold traits** are traits that have only two or a few phenotypic classes, but their inheritance is determined by the effects of multiple genes acting together with the environment. Examples of threshold traits are twinning in cattle and parthenogenesis (development of unfertilized eggs) in turkeys. In a threshold trait, each organism has an underlying risk or *liability* to express the trait. Even though the underlying liability is not directly observable, if it is high enough (above a threshold), the trait will be expressed (for example, a pregnancy results in twins); otherwise, the trait is not expressed (the pregnancy results in a single birth). In many threshold-trait disorders, the phenotypic classes are affected and not affected. Examples of threshold-trait disorders in human beings include adult-onset diabetes, schizophrenia, and many congenital abnormalities, such as spina bifida. Threshold traits can be interpreted as continuous traits by imagining that the underlying *liability* is a continuous variable.

## The Normal Distribution

The **distribution** of a trait in a population is a description of the population in terms of the proportion of individuals who have each of the possible phenotypes. To create such a description with continuous traits, it is usually convenient to begin by grouping similar phenotypes into classes. Data for an example pertaining to the distribution of height among 4995 British women are given in **TABLE 18.1** and in **FIGURE 18.1**. One can imagine each bar in the graph in Figure 18.1 being built, step by step, as each of the women's height is measured, by placing a small square along the *x*-axis at the location corresponding to the height of that woman. As sampling proceeds, the squares begin to pile up in certain places, leading ultimately to the histogram shown.

**Table 18.1 Distribution of height among British women**

| Interval number ($i$) | Height interval (inches) | Midpoint ($x_i$) | Number of women ($f_i$) |
|---|---|---|---|
| 1 | 53–55 | 54 | 5 |
| 2 | 55–57 | 56 | 33 |
| 3 | 57–59 | 58 | 254 |
| 4 | 59–61 | 60 | 813 |
| 5 | 61–63 | 62 | 1340 |
| 6 | 63–65 | 64 | 1454 |
| 7 | 65–67 | 66 | 750 |
| 8 | 67–69 | 68 | 275 |
| 9 | 69–71 | 70 | 56 |
| 10 | 71–73 | 72 | 11 |
| 11 | 73–75 | 74 | 4 |
| | | | Total $N$ = 4995 |

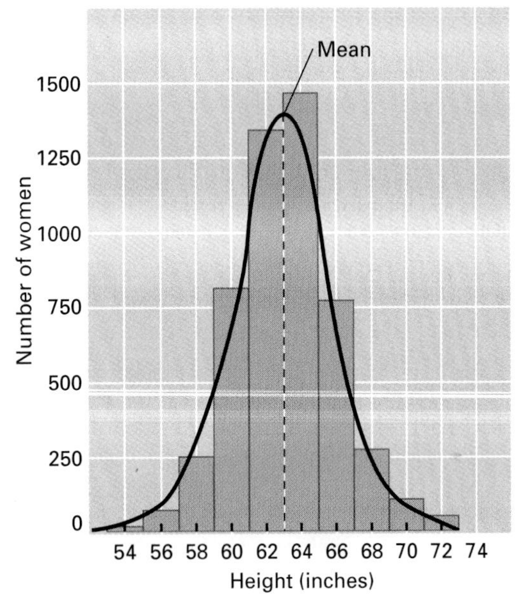


**FIGURE 18.1** Distribution of height among 4995 British women and the smooth normal distribution that approximates the data.

Displaying a distribution completely, in either tabular form (Table 18.1) or graphical form (Figure 18.1), is always helpful but often unnecessary; it often happens that a description of the distribution in terms of two major features—the *mean* and the *variance*—is sufficient. To discuss the mean and the variance in quantitative terms, we will need some symbols. In Table 18.1, the height intervals are numbered from 1 (53–55 inches) to 11 (73–75 inches). The symbol $x_i$ designates the midpoint of the height interval numbered $i$; for example, $x_1 = 54$ inches, $x_2 = 56$ inches, and so on. The number of women in height interval $i$ is designated $f_i$; for example, $f_1 = 5$ women, $f_2 = 33$ women, and so forth. The total size of the sample, in this case 4995, is denoted by $N$. The mean and variance serve to characterize the distribution of height among these women as well as the distribution of many other quantitative traits.

The **mean**, or average, is the peak of the distribution. The mean of a population is estimated from a sample of individuals from the population, as follows:

$$\bar{x} = \frac{\Sigma f_i x_i}{N} \qquad (1)$$

where $\bar{x}$ is the estimate of the mean and $\Sigma$ symbolizes summation over all classes of data (in this example, summation over all 11 height intervals). In Table 18.1, the mean height in the sample of women is 63.1 inches.

The **variance** is a measure of the spread of the distribution and is estimated in terms of the squared *deviation* (difference) of each observation from the mean. The variance is estimated from a sample of individuals as follows:

$$s^2 = \frac{\Sigma f_i(x_i - \bar{x})^2}{N - 1} \qquad (2)$$

where $s^2$ is the estimated variance and $x_i$, $f_i$, and $N$ are as in Table 18.1. The difference $(x_i - \bar{x})$ is the deviation of each height category from the mean height in the entire sample. The variance describes the extent to which the phenotypes are clustered around the mean, as shown in **FIGURE 18.2**. A large value implies that the distribution is spread out, and a small value implies that it is clustered near the mean. From the data in Table 18.1, the variance of the population of British women is estimated as $s^2 = 7.24$ in$^2$. (Note that the units of the variance are squared.)

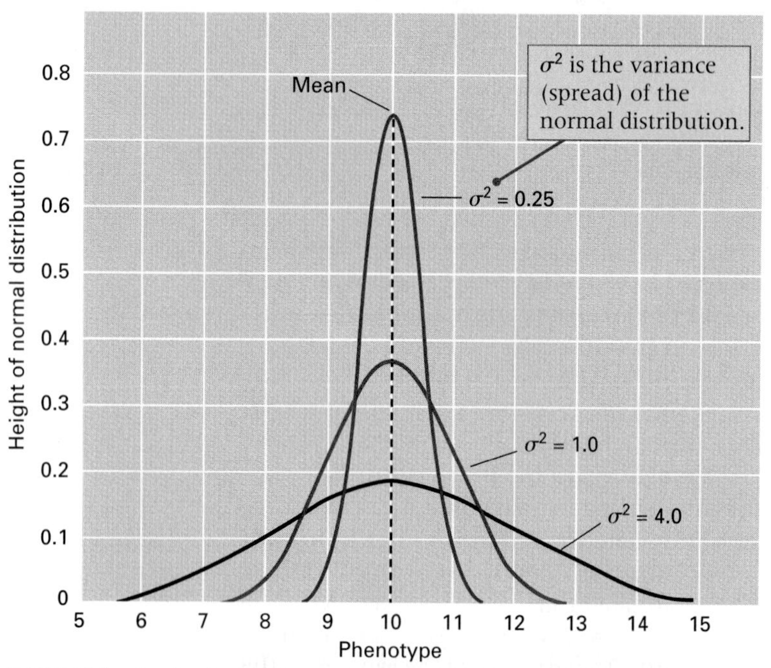


**FIGURE 18.2** Graphs showing that the variance of a distribution measures the spread of the distribution around the mean. The area under each curve covering any range of phenotypes equals the proportion of individuals having phenotypes within that range.

A quantity closely related to the variance—the **standard deviation** of the distribution—is defined as the square root of the variance. For the data in Table 18.1, the estimated standard deviation $s$ is obtained from Equation (2) as follows:

$$s = \sqrt{s^2} = \sqrt{7.24 \text{ in}^2} = 2.69 \text{ inches}$$

The standard deviation has the useful feature of having the same units of dimension as the mean—in this example, inches. Just as the mean of the sample, $\bar{x}$, is used to estimate the unknown mean of the population, $\mu$, the standard deviation of the sample, $s$, is used to estimate the unknown standard deviation of the population, $\sigma$, and $s^2$ used to estimate $\sigma^2$.

When the data are symmetrical, or approximately symmetrical, the distribution of a trait can often be approximated by the smooth, arching curve known as the **normal distribution**, which is shown for the height example in Figure 18.1. Because the normal distribution is symmetrical, half of its area is determined by points that have values greater than the mean and half by points with values less than the mean, and thus the proportion of phenotypes that exceed the mean is 1/2. One important characteristic of the normal distribution is that it is completely determined by the value of the mean and the variance.

The mean and the standard deviation (square root of the variance) of a normal distribution provide a great deal of information about the distribution of phenotypes in a population, as is illustrated in **FIGURE 18.3**. Specifically, for a normal distribution:

1. Approximately 68 percent of the population have a phenotype within one standard deviation of the mean (in the symbols of Figure 18.3, between $\mu - \sigma$ and $\mu + \sigma$).
2. Approximately 95 percent lie within two standard deviations of the mean (between $\mu - 2\sigma$ and $\mu + 2\sigma$).
3. Approximately 99.7 percent lie within three standard deviations of the mean (between $\mu - 3\sigma$ and $\mu + 3\sigma$).

Applying these rules to the data in Figure 18.1, in which the mean and standard deviation are 63.1 and 2.69 inches, reveals that approximately 68 percent of the women are expected to have heights in the range from 63.1 − 2.69 inches to 63.1 + 2.69 inches (that is, 60.4–65.8), and approximately 95 percent are expected to have heights in the range from 63.1 − 2(2.69) inches to 63.1 + 2(2.69) inches (that is, 57.7–68.5).

Real data frequently conform to the normal distribution. Normal distributions are usually the rule when the phenotype is determined by the cumulative effect of many individually small independent factors. This is the case for many multifactorial traits.

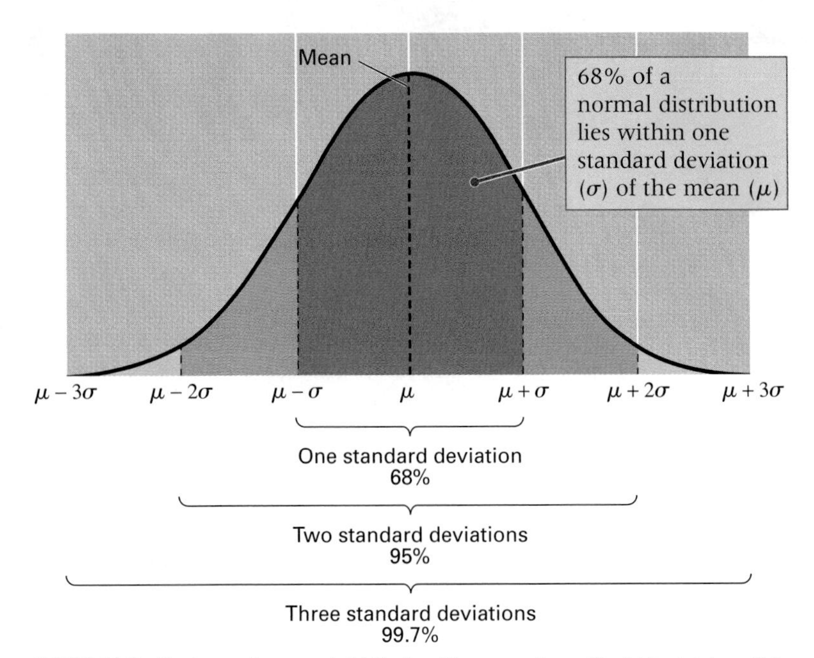


**FIGURE 18.3** Features of a normal distribution. The proportions of individuals lying within 1, 2, and 3 standard deviations of the mean are approximately 68 percent, 95 percent, and 99.7 percent, respectively. In this normal distribution, $\mu$ represents the mean and $\sigma$ the standard deviation.

## 18.2 Causes of Variation

In considering the genetics of complex traits, an important objective is to assess the relative importance of genotype and environment. In some cases in experimental organisms, it is possible to separate genotype and environment with respect to their effects on the mean phenotype. For example, a plant breeder may study the yield of a series of inbred lines grown in a group of environments that differ in planting density or amount of fertilizer. It would then be possible:

- To compare yields of the same genotype grown in different environments, and thereby rank the *environments* relative to their effects on yield.
- To compare yields of different genotypes grown in the same environment and thereby rank the *genotypes* relative to their effects on yield.

Such a fine discrimination between genetic and environmental effects is not usually possible, particularly in human genetics. For example, with regard to the height of the women in Figure 18.1, environment could be considered favorable or unfavorable for tall stature only in comparison with the mean height of a genetically identical population reared in a different environment. This reference population does not exist. Likewise, the genetic composition of the population could be judged as favorable or unfavorable for tall stature only in comparison with the mean of a genetically different population reared in an identical environment. This reference population does not exist either.

Without such standards of comparison, it is impossible to distinguish genetic from environmental effects on the mean phenotype. However, it is still possible to assess genetic versus environmental contributions to the *variance* in phenotype, because instead of comparing the means of two or more populations, we can compare the phenotypes of individuals within the *same* population. Some of the differences in phenotype result from differences in genotype and others from differences in environment, and it is often possible to separate these effects.

In any distribution of phenotypes, such as the one in Figure 18.1, four sources contribute to phenotypic variation:

1. Genotypic variation
2. Environmental variation
3. Variation due to genotype-by-environment interaction
4. Variation due to genotype-by-environment association

These sources of variation are discussed in the following sections.

**AN EAR OF WILD TEOSINTE** (left) looks dramatically different from an ear of modern cultivated maize (right), but the organisms are very closely related, producing fertile hybrids in both reciprocal crosses (middle). Early Native Americans first domesticated teosinte and, through many generations, selected the mutant phenotypes most favorable for cultivation, leading ultimately to modern maize. [Courtesy of John Doebley, University of Wisconsin—Madison.]

## connection
### A Maize'n Grass

**George W. Beadle 1972**
University of Chicago
*The Mystery of Maize*

*Cultivated corn (*Zea mays*) is a wonder crop. It can be boiled, baked, popped, and chopped; it can be ground, fermented, fried, and dried. It is fodder for livestock and raw material for industry. Domesticated 5 to 10 thousand years ago by Native American Indians in the Balsas River drainage, about 100 miles southwest of Mexico City, maize became a staple for many people in the Middle Americas. The Native Americans developed 200 to 300 varieties, essentially all that we have today. Accomplished without knowledge of genetics, their achievement in plant breeding has been called the most extraordinary in all of human existence. But the ancestral source of maize was for a long time very obscure. A few early geneticists suspected teosinte because they could imagine a handful of major gene differences affecting growth and development that might convert teosinte into something resembling maize. Genetic evidence for this hypothesis is presented in this piece by George Beadle. Best known for his work in* Neurospora *and for the "one gene, one enzyme" concept (see the Connection on page 402), Beadle had begun his career studying corn, and when he retired as president of the University of Chicago in 1968, he immediately resumed his maize experiments in a half-acre lot off Ellis Avenue near 55th Street in the middle of the city, carrying out crosses that he had first conceived as a graduate student at Cornell 40 years earlier. Sure enough, the major differences between the species could be accounted for by mutations in a few key genes. A number of these have since been cloned and sequenced and their functions identified. One is a gene that in maize is known as* teosinte branched, *whose mutant phenotype is a highly branched plant similar in growth habit to teosinte.*

For most cultivated plants, wild relatives are known from which they could reasonably have been derived. . . . In the case of corn, early plant explorers had great difficulty finding any wild relative that could conceivably have been its ancestor. . . . Finally they found what looked as if it might be a relative. It was a plant that natives in parts of Mexico called *teocentli*, from the Aztec language, meaning "support of the gods." This is now anglicized to "teosinte." . . . I well remember discussions with Rollins A. Emerson [at Cornell University in about 1930] in which he pointed out that just two mutations could make teosinte an easily usable food plant—one to a non-shattering rachis [the elongated axis of the ear], so the fruits would not be scattered and lost as food; the other to a soft fruitcase, so the kernels could be threshed free of them. . . . To jump ahead now forty years, . . . I decided that on retirement [as president of the University of Chicago] I would return to a study of the relation of teosinte to corn. . . . Mendel's laws say that if original parents differ from each other by only one gene, in the second generation of descendants [from the hybrid cross] each original parental type will appear with a statistical frequency of one in four. With a difference of two genes, the frequency with each parental type will be reproduced in one in sixteen; and so on for more genes. . . . I resolved to grow up to 50,000 second-generation plants, if necessary. . . . With this year's planting, I estimate that we will have grown and examined just about 50,000. . . . What have been the results? In the second-generation cross of Chapalote corn and Chalco teosinte we found good parent corn types and good parent teosinte types with a frequency of about 1 in each 500 plants. These frequencies are intermediate between those expected with four and those expected with five gene differences. Of course, the genetic complexities are greater than this kind of arithmetic implies, but I will not go into them here. . . . It seems clear that the genetic differences between corn and teosinte cannot be so great as to [exclude] an ancestral relationship of teosinte to corn. And it seems reasonable to assume that [Native American Indians] could have selected and preserved the relatively few mutations required to produce a useful plant from teosinte—early corn.

**It was a plant that natives in parts of Mexico called teocentli, from the Aztec language, meaning "support of the gods."**

Source: G. W. Beadle, *Field Museum Natural History Bulletin* 43 (1972): 2–11.

### Genotypic Variation

The variation in phenotype caused by differences in genotype among individuals is termed **genotypic variance**. **FIGURE 18.4** illustrates the genetic variation expected among the $F_2$ generation from a cross of two inbred lines that differ in genotype for three unlinked genes. The alleles of the three genes are represented as *Aa*, *Bb*, and *Cc*, and the genetic variation in the $F_2$ caused by segregation and recombination is evident in the differences in color. Relative to a categorical trait (a trait whose phenotype is determined by counting, such as ears per stalk in corn), if we assume that each uppercase allele is favorable for the ex-

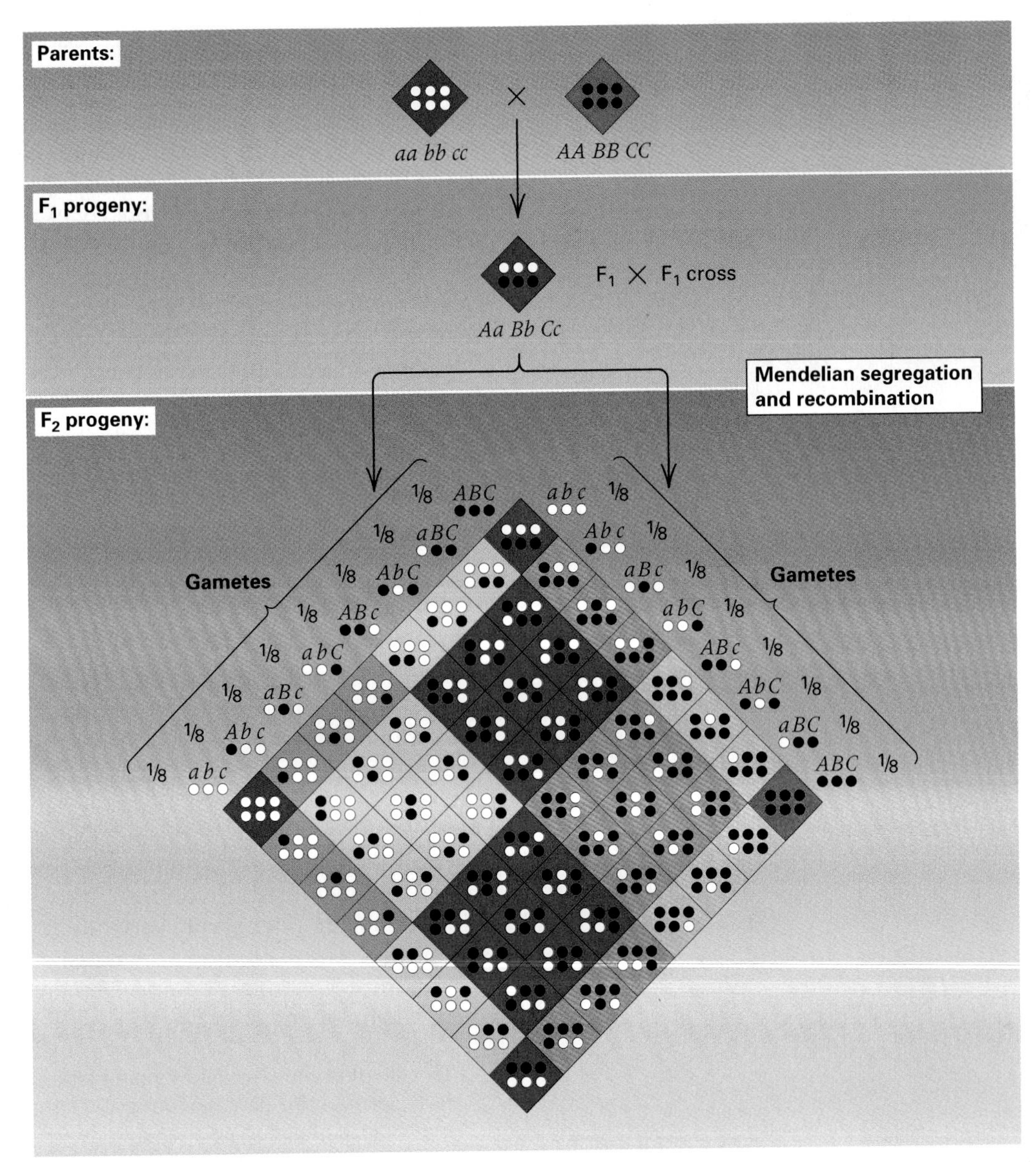


**FIGURE 18.4** Segregation of three independent genes affecting a quantitative trait. Each uppercase allele in a genotype contributes one unit to the phenotype.

pression of the trait and adds one unit to the phenotype, whereas each lowercase allele is without effect, then the *aa bb cc* genotype has a phenotype of 0, and the *AA BB CC* genotype has a phenotype of 6. There are seven possible phenotypes (0–6) in the $F_2$ generation. The distribution of phenotypes in the $F_2$ generation is shown in the colored bar graph in **FIGURE 18.5**. The normal distribution approximating the data has a mean of 3 and a variance of 1.5. In this case, we are assuming that *all* of the variation in phenotype in the population results from differences in genotype among the individuals.

Figure 18.5 also includes a bar graph with diagonal lines that represent the theoretical distribution when the trait is determined by 30 unlinked genes segregating in a randomly mating population, grouped into the same number of phenotypic classes as the three-gene case. We assume that fifteen of the genes are nearly fixed for the favorable allele and fifteen for the unfavorable allele. The contribution of each favorable allele to the phenotype has been chosen to make the mean of the distribution equal to three and the variance equal to 1.5. Note that the distribution with thirty genes is virtually identical to that with three genes and that both are approximated by the same normal distribution. If such distributions were encountered in actual research, the

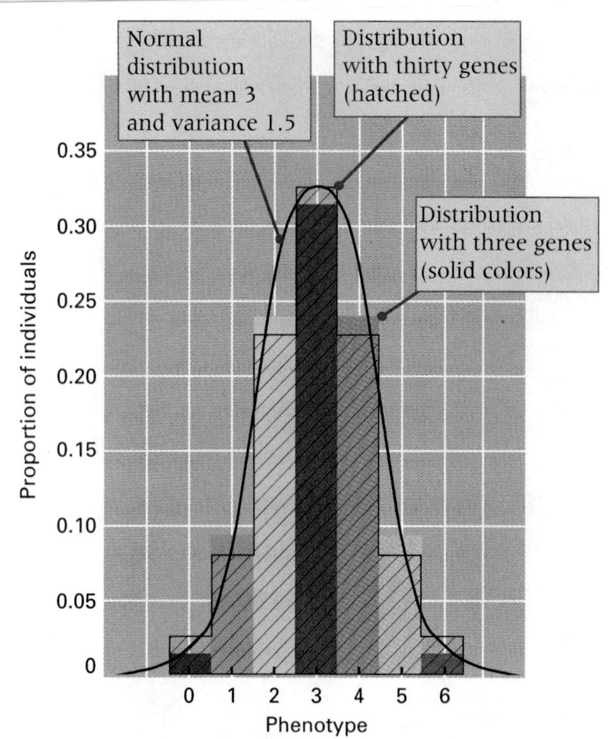


**FIGURE 18.5** The distributions of phenotypes determined by the segregation of three genes and thirty independent genes. Both distributions are approximated by the same normal distribution (black curve).

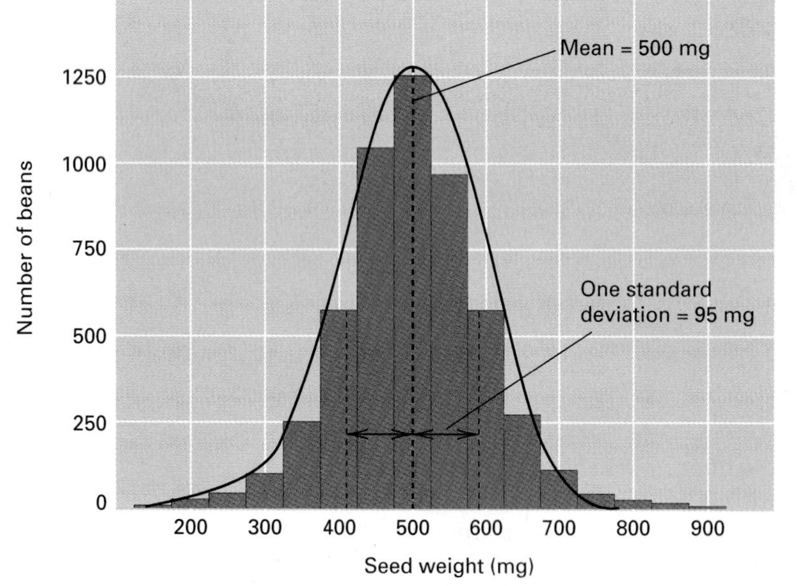


**FIGURE 18.6** Distribution of seed weight in a homozygous line of edible beans. All variation in phenotype among individuals results from environmental differences.

experimenter would not be able to distinguish between them. The key point is that:

> Even in the absence of environmental variation, the distribution of phenotypes, by itself, provides no information about the number of genes influencing a trait and no information about the dominance relations of the alleles.

However, the number of genes influencing a complex trait is important in determining the potential for long-term genetic improvement of a population by means of artificial selection. For example, in the three-gene case in Figure 18.5, the best possible genotype would have a phenotype of 6, but in the thirty-gene case, the best possible genotype (homozygous for the favorable allele of all thirty genes) would have a phenotype of 60.

Later in this chapter, some methods for estimating how many genes affect a quantitative trait will be presented. All the methods depend on comparing the phenotypic distributions in the $F_1$ and $F_2$ generations of crosses between nearly or completely homozygous lines.

## Environmental Variation

The variation in phenotype caused by differences in environment among individuals is termed **environmental variance**. **FIGURE 18.6** is an example showing the distribution of seed weight in edible beans. The mean of the distribution is 500 mg, and the standard deviation is 95 mg. All of the beans in this population are genetically identical and homozygous because they are highly inbred. Therefore, in this population, *all* of the phenotypic variation in seed weight results from environmental variance. A comparison of Figures 18.5 and 18.6 demonstrates the following principle:

> The distribution of a trait in a population provides no information about the relative importance of genotype and environment. Variation in the trait can be entirely genetic, it can be entirely environmental, or it can reflect a combination of both influences.

## Genetics and Environment Combined

Genotypic and environmental variance are seldom separated as clearly as in Figures 18.5 and 18.6, because usually they work together. Their combined effects are illustrated for a simple hypothetical case in **FIGURE 18.7**. Figure 18.7A is the distribution of phenotypes for three genotypes, ignoring (for the moment) environmental effects. As depicted, the trait can have one of three distinct and nonoverlapping phenotypes determined by the effects of two additive alleles. The genotypes are in random-mating proportions for an allele frequency of 1/2, and the distribution of phenotypes has mean 5 and variance 2. Because it results solely from differences in genotype, this variance is *genotypic variance*, which is symbolized $\sigma_g^2$. Figure 18.7B is the distribution

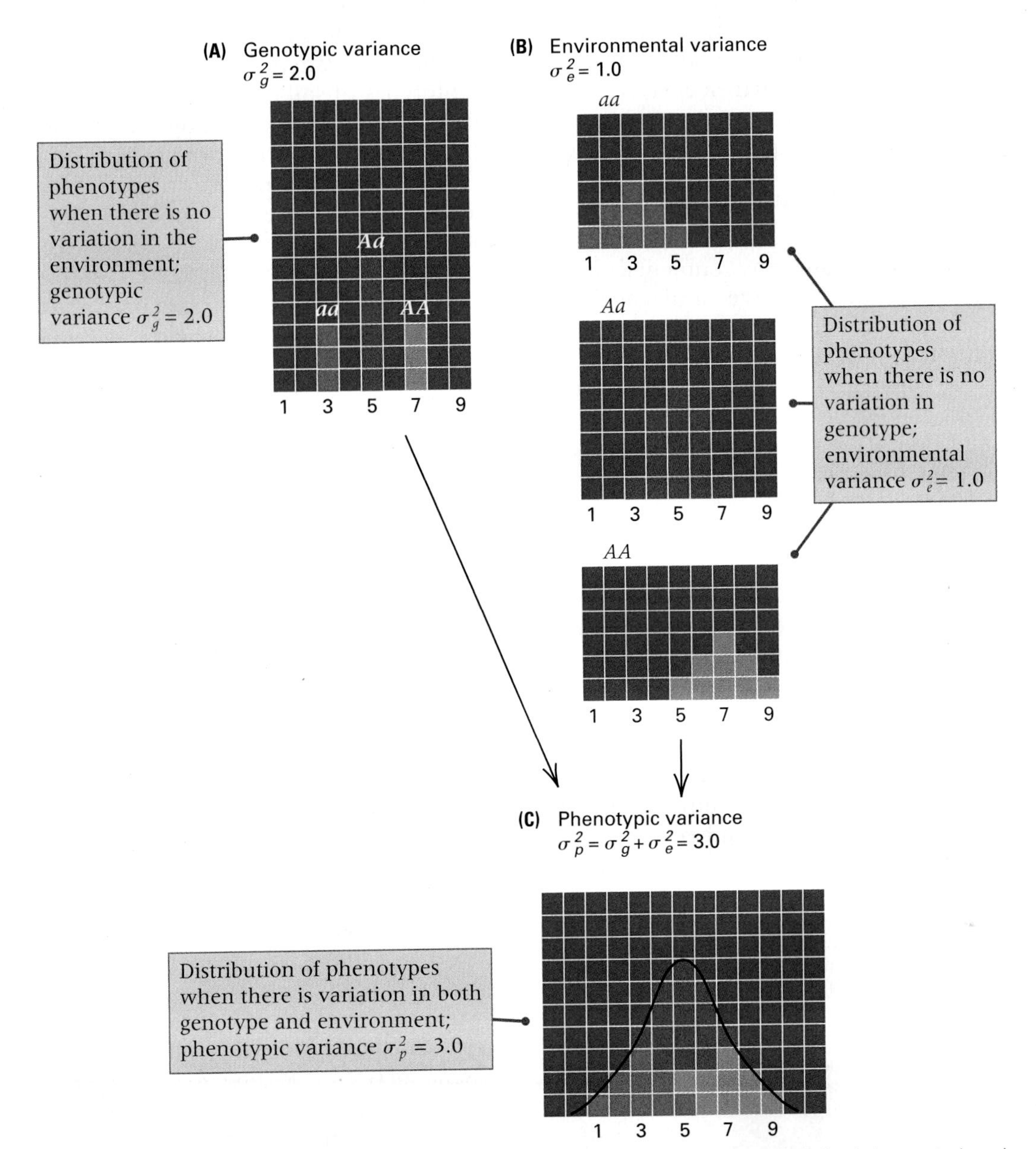


**FIGURE 18.7** The combined effects of genotypic and environmental variance. (A) Population affected only by genotypic variance, $\sigma_g^2$. (B) Populations of each genotype affected only by environmental variance, $\sigma_e^2$. (C) Population affected by both genotypic and environmental variance, in which the phenotypic variance, $\sigma_p^2$, equals the sum of $\sigma_g^2$ and $\sigma_e^2$.

of phenotypes in the presence of environmental variation, illustrated for each genotype separately. Each distribution of environmental effects has a mean of 0 and a variance of 1. Because this variance results solely from differences in environment, it is *environmental variance,* which is symbolized $\sigma_e^2$. When the effects of genotype and environment are combined in the same population, each genotype is affected by environmental variation, and the distribution shown in Figure18.7C of the figure results. The variance of this distribution is the **phenotypic variance**, which is symbolized $\sigma_p^2$. Because we are assuming that genotype and environment have separate and independent effects on phenotype, we expect $\sigma_p^2$ to be greater than either $\sigma_g^2$ or $\sigma_e^2$ alone. In fact,

$$\sigma_p^2 = \sigma_g^2 + \sigma_e^2 \qquad (3)$$

In words, Equation (3) states that:

> When genetic and environmental effects contribute independently to phenotype, the total variance equals the sum of the genotypic variance and the environmental variance.

Equation (3) is one of the most important relations in the analysis of complex inheritance. How it can be used to analyze data will be explained shortly. Although the equation serves as an excellent approximation in very many cases, it is valid in an exact sense only when genotype and environment are independent in their effects on phenotype. The two most important departures from independence are discussed in the next section.

### ■ Genotype-by-Environment Interaction and Association

In the simplest cases, environmental effects on phenotype are additive, and each environment adds (or detracts) the same amount to (or from) the phenotype, independent of the genotype. When this is not true, environmental effects on phenotype differ according to genotype, and a **genotype-by-environment interaction (G–E interaction)** is said to be present. In some cases, G–E interaction can even change the relative ranking of the genotypes, so a genotype that is superior in one environment may become inferior in another.

An example of genotype-by-environment interaction in maize is illustrated in **FIGURE 18.8**. The two strains of corn are hybrids formed by crossing different pairs of inbred lines, and their overall means, averaged across all of the environments, are approximately the same. However, the strain designated A clearly outperforms B in the stressful (negative) environments, whereas the performance is reversed when the environment is of high quality. (Environmental quality is judged on the basis of soil fertility, moisture, and other factors.) Because of G–E interaction, no one plant variety will outperform all others in all types of soil and climate, and therefore plant breeders must develop special varieties that are suited to each growing area.

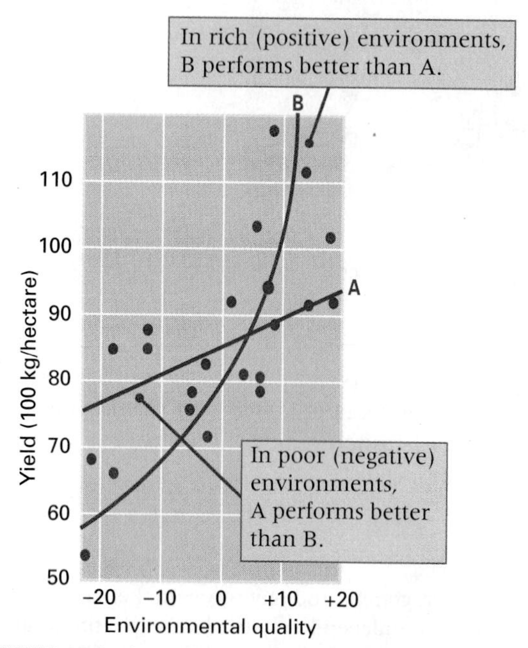


**FIGURE 18.8** Genotype-by-environment interaction in maize. Strain A is superior when environmental quality is low (negative numbers), but strain B is superior when environmental quality is high. [Data from W. A. Russell, *Annual Corn & Sorghum Research Conference* 29 (1974): p. 81.]

In some organisms, particularly plants, experiments like those illustrated in Figure 18.8 can be carried out to estimate the relative contribution of G–E interaction to the total observed variation in phenotype. In other organisms, particularly human beings, the effect of G–E interaction cannot usually be evaluated separately. One particularly well documented form of G–E interaction in animals is **genotype-by-sex interaction**, in which the same genes have different phenotypic effects in females and males. (It may seem odd to regard sex as an environment, but for purposes of explaining genotype-by-sex interaction, this is a convenient way to think about it.) A familiar example of genotype-by-sex interaction in humans is adult height. Although the distributions of adult stature in females and males are strongly overlapping, females are on the average shorter than males. Yet the genes affecting adult stature are shared between the sexes. It is their phenotypic effects that differ according to the hormonal environment.

When the different genotypes are not distributed at random in all the possible environments, there is **genotype-by-environment association (G–E association)**. In these circumstances, certain genotypes are preferentially associated with certain environments, which may either increase or decrease the phenotype of these genotypes compared with what would result in the absence of G–E association. An example of deliberate genotype-by-environment association can be found in dairy husbandry, in which some farmers feed each of their cows in proportion to its level of milk production. Because of this practice, cows with superior genotypes with respect to milk production also have a superior environment in that they receive more feed. In plant breeding, genotype-by-environment association can often be eliminated or minimized by appropriate randomization of genotypes within the experimental plots. In other cases, human genetics again being a prime example, the possibility of G–E association cannot usually be controlled.

## 18.3 Genetic Analysis of Complex Traits

Equation (3) can be used to separate the effects of genotype and environment on the total phenotypic variance. Two types of data are required:

1. The phenotypic variance of a genetically heterogeneous population, which provides an estimate of $\sigma_p^2 = \sigma_g^2 + \sigma_e^2$
2. The phenotypic variance of a genetically uniform population, which provides an estimate of $\sigma_e^2$ because a genetically uniform population has a value of $\sigma_g^2 = 0$

An example of a genetically uniform population is the $F_1$ generation from a cross between two highly homozygous strains, such as inbred lines. An example of a genetically heterogeneous population is the $F_2$ generation from the same cross. If the environments of both populations are the same, and if there is no G–E interaction, then the estimates may be combined to deduce the value of $\sigma_g^2$.

As an illustration of this approach, we will consider the Mexican cave-dwelling fish *Astyanax fasciatus*, one of about 80 species of cave fishes that are known worldwide. Among other genetic changes accompanying evolution in perpetual darkness, these species have all evolved reduced eyes and body pigment (**FIGURE 18.9**). To estimate the number of genes contributing to reduced eye size, an inbred line from a cave population was mated with an inbred line from a surface population, and the offspring were reared in the same environment. The variances in eye diameter in the $F_1$ and $F_2$ generations were estimated as 0.057 and 0.563, respectively. Written in terms of the components of variance, these are:

$$F_2 \text{ variance: } \sigma_p^2 = \sigma_g^2 + \sigma_e^2 = 0.563$$

$$F_1 \text{ variance: } \sigma_e^2 = 0.057$$

The estimate of genotypic variance, $\sigma_g^2$, is obtained by subtracting the second equation from the first; that is,

$$(\sigma_g^2 + \sigma_e^2) - \sigma_e^2 = \sigma_g^2$$

and therefore,

$$0.563 - 0.057 = 0.506 = \sigma_g^2$$

Hence, the estimate of $\sigma_g^2$ is 0.506, whereas that of $\sigma_e^2$ is 0.057. In this example, the genotypic variance is much greater than the environmental variance, but this is not always the case.

The next section shows what other information can be obtained from an estimate of the genotypic variance.

## The Number of Genes Affecting Complex Traits

When the number of genes influencing a complex trait is not too large, knowledge of the genotypic variance can be used to estimate the number of genes. The required information consists of the means and variances of two phenotypically divergent strains, as well as the variances of the $F_1$, $F_2$, and backcross generations. In ideal cases, the data appear as in **FIGURE 18.10**, in which $P_1$ and $P_2$ represent the divergent parental strains (for example, inbred lines). The points lie on a triangle, with increasing variance according to the increasing genetic heterogeneity (genotypic variance) of the populations. If the $F_1$ and backcross means lie exactly between their parental means, then these means will lie at the midpoints along the sides of the triangle, as shown in Figure 18.10. This finding implies that the alleles affecting the trait are *additive;* that is, for each gene, the phenotype of the heterozygote is the average of the phenotypes

**FIGURE 18.9** Reduced eye size and pigmentation in a cave-dwelling *Astyanax* (above) and a surface-dwelling relative (below). [Courtesy of Richard L. Borowsky and Horst Wilkens, New York University.]

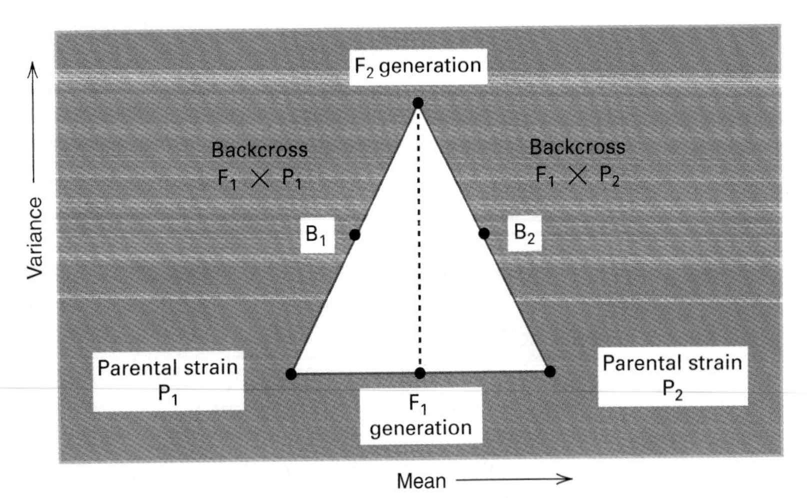


**FIGURE 18.10** Means and variances of parents (P), backcross ($B_1$ and $B_2$), and hybrid ($F_1$ and $F_2$) progeny of inbred lines for an ideal quantitative trait affected by unlinked and completely additive genes. [Adapted from R. Lande, *Genetics* 99 (1981): 541–553.]

of the corresponding homozygotes. In such a simple situation, it can be shown that the number, $n$, of genes contributing to the trait is,

$$n = \frac{D^2}{8\sigma_g^2} \qquad (4)$$

where $D$ represents the difference between the means of the original parental strains, $P_1$ and $P_2$. This equation can be verified by applying it to the ideal case of 3 genes in Figure 18.4. The parental strains are the homozygous genotypes with mean phenotypes of 0 and 6. Consequently, $D = 6 - 0 = 6$. The genotypic variance is given in Figure 18.5 as $\sigma_g^2 = 1.5$. Substituting $D$ and $\sigma_g^2$ into Equation (4), we obtain $n = (6)^2/(8 \times 1.5) = 3$, which is correct because there are three independent and equivalent genes that affect the trait.

Applied to actual data, Equation (4) requires several assumptions that are not necessarily valid. In addition to the assumption that all generations are reared in the same environment, the theory also makes the genetic assumptions that: (1) the alleles of each gene are additive, (2) the genes contribute equally to the trait, (3) the genes are unlinked, and (4) the original parental strains are homozygous for alternative alleles of each gene. However, when the assumptions are invalid, the calculated $n$ is smaller than the actual number of genes affecting the trait, and usually much smaller. The estimated number is a minimum because almost any departure from the genetic assumptions leads to a smaller genotypic variance in the $F_2$ generation and so, for the same value of $D$, would yield a larger value of $n$ in Equation (4). This is why the estimated $n$ is the *minimum* number of genes that can account for the data.

For the cave-dwelling *Astyanax* fish discussed in the preceding section, the parental strains had average phenotypes of 7.05 and 2.10, giving $D = 4.95$. The estimated value of $\sigma_g^2 = 0.506$, so the minimum number of genes affecting eye diameter is $n = (4.95)^2/(8 \times 0.506) = 6.0$. Therefore, at least six different genes affect the diameter of the eye of the fish. Three of these genes have been identified using the genetic mapping methods discussed in Section 18.7.

The number of genes that affect a quantitative trait is important because it influences the amount by which a population can be genetically improved by selective breeding. With traits determined by a small number of genes, the potential for change in a trait is relatively small, and a population consisting of the best possible genotypes may have a mean value that is only 2 or 3 standard deviations above the mean of the original population. However, traits determined by a large number of genes have a large potential for improvement. For example, after a population of the flour beetle *Tribolium* was selected for increased pupa weight for many generations, the mean value for pupa weight was found to be 17 standard deviations above the mean of the original population. Determination of traits by a large number of genes implies that:

> Selective breeding can create an improved population in which the value of *every* individual greatly exceeds that of the *best* individuals that existed in the original population.

This principle at first seems paradoxical, because in a large enough population, every possible genotype should be created at some low frequency. The resolution of the paradox is that real populations subjected to selective breeding typically consist of a few hundred organisms (at most), and hence in the real world many of the theoretically possible genotypes are never formed because their expected frequencies are much too low. However, as selection takes place, and the allele frequencies change, these genotypes become more common and allow the selection of superior organisms in future generations.

## Broad-Sense Heritability

Estimates of the number of genes that determine quantitative traits are frequently unavailable because the necessary experiments are impractical or have not been carried out. Another attribute of quantitative traits, which requires less data to evaluate, makes use of the ratio of the genotypic variance to the total phenotypic variance. This ratio of $\sigma_g^2$ to $\sigma_p^2$ is called the **broad-sense heritability**, symbolized as $H^2$, and it measures the importance of genetic variation, relative to environmental variation, in causing variation in the phenotype of a trait of interest. Broad-sense heritability is a ratio of variances, specifically:

$$H^2 = \frac{\sigma_g^2}{\sigma_p^2} = \frac{\sigma_g^2}{(\sigma_g^2 + \sigma_e^2)} \qquad (5)$$

Substitution of the data for eye diameter in *Astyanax*, in which $\sigma_g^2 = 0.506$ and $\sigma_g^2 + \sigma_e^2 = 0.563$, into Equation (5) yields $H^2 = 0.506/0.563 = 0.90$ for the estimate of broad-sense heritability. This value implies that 90 percent of the variation in eye diameter in this population results from differences in genotype among the fish.

Knowledge of heritability is useful in the context of plant and animal breeding, because heritability can be used to predict the magnitude and speed of improvement in the population. The broad-sense heritability defined in Equation (5) is used in predicting the outcome of selection practiced among clones, inbred lines, or varieties. Analogous predictions for randomly bred populations utilize another type of heritability, different from $H^2$, which we will discuss shortly. Broad-sense heritability measures how much of the total variance in phenotype results from differences in genotype. For this reason, $H^2$ is often of interest in regard to human quantitative traits.

### Twin Studies

In human beings, twins would seem to be ideal subjects for separating genotypic and environmental variance, because **identical twins**, which arise from the splitting of a single fertilized egg, are genetically identical and are often strikingly similar in such traits as facial features and body build. **Fraternal twins**, which arise from two fertilized eggs, have the same genetic relationship as ordinary siblings, and therefore only half of the genes in either twin are identical with those in the other. Theoretically, the variance between members of an identical-twin pair would be equivalent to $\sigma_e^2$, because the twins are genetically identical, whereas the variance between members of a fraternal-twin pair would include not only $\sigma_e^2$ but also part of the genotypic variance (approximately $\sigma_g^2/2$, because of the identity of half of the genes in fraternal twins). Consequently, both $\sigma_g^2$ and $\sigma_e^2$ could be estimated from twin data and combined as in Equation (5) to estimate $H^2$. TABLE 18.2 summarizes estimates of $H^2$ based on twin studies of several traits.

Unfortunately, twin studies are subject to several important sources of error, most of which increase the similarity of identical twins, so the numbers in Table 18.2 should be considered very approximate and probably too high. Four of the potential sources of error are:

1. Genotype-by-environment interaction, which increases the variance in fraternal twins but not in identical twins
2. Frequent sharing of embryonic membranes between identical twins, resulting in more similar intrauterine environments than those of fraternal twins
3. Greater similarity in the treatment of identical twins by parents, teachers, and peers, resulting in a decreased environmental variance in identical twins
4. Different sexes in half of the pairs of fraternal twins, in contrast with the same sex of identical twins

These pitfalls and others imply that data from human twin studies should be interpreted with caution and reservation.

**Table 18.2 Broad-sense heritability, in percent, based on twin studies**

| Trait | Heritability ($H^2$) |
|---|---|
| Longevity | 29 |
| Height | 85 |
| Weight | 63 |
| Amino acid excretion | 72 |
| Serum lipid levels | 44 |
| Maximum blood lactate | 34 |
| Maximum heart rate | 84 |
| Verbal ability | 63 |
| Numerical ability | 76 |
| Memory | 47 |
| Sociability index | 66 |
| Masculinity index | 12 |
| Temperament index | 58 |

## 18.4 Artificial Selection

**Artificial selection** is the breeders' practice of choosing a select group of organisms from a population to become the parents of the next generation. When artificial selection is carried out either by choosing the best organisms in a species that reproduces asexually or by choosing the best subpopulation among a series of inbred subpopulations, the broad-sense heritability is used to predict how rapidly progress can be achieved. Broad-sense heritability is important in this context, because with clones or inbred lines, superior genotypes can be perpetuated without disruption of favorable gene combinations by segregation and recombination. An example is the selection of superior varieties of plants that are propagated asexually by means of cuttings or grafts. Because there is no sexual reproduction, each plant has exactly the same genotype as its parent.

In sexually reproducing populations that are genetically heterogeneous, broad-sense heritability is not relevant in predicting progress resulting from artificial selection, because superior genotypes must necessarily be broken up by the processes of segregation and recombination.

For example, if the best genotype is heterozygous for each of two unlinked loci, *Aa Bb*, then because of segregation and independent assortment, among the progeny of a cross between parents with the best genotypes—*Aa Bb* × *Aa Bb*—only 1/4 will have the same favorable *Aa Bb* genotype as the parents. The rest of the progeny will be genetically inferior to the parents. For this reason, to the extent that high genetic merit may depend on particular combinations of alleles, each generation of artificial selection results in a slight setback in that the offspring of superior parents are generally not quite so good as the parents themselves. Progress under selection can still be predicted, but the prediction must make use of another type of heritability, the narrow-sense heritability, which is discussed in the next section.

## Narrow-Sense Heritability

**FIGURE 18.11** illustrates a typical form of artificial selection and its result. The organism is *Nicotiana longiflora* (tobacco), and the trait is the length of the corolla tube (the corolla is a collective term for all the petals of a flower). Figure 18.11A shows the distribution of phenotypes in the parental generation, and Figure 18.11B shows the distribution of phenotypes in the off-

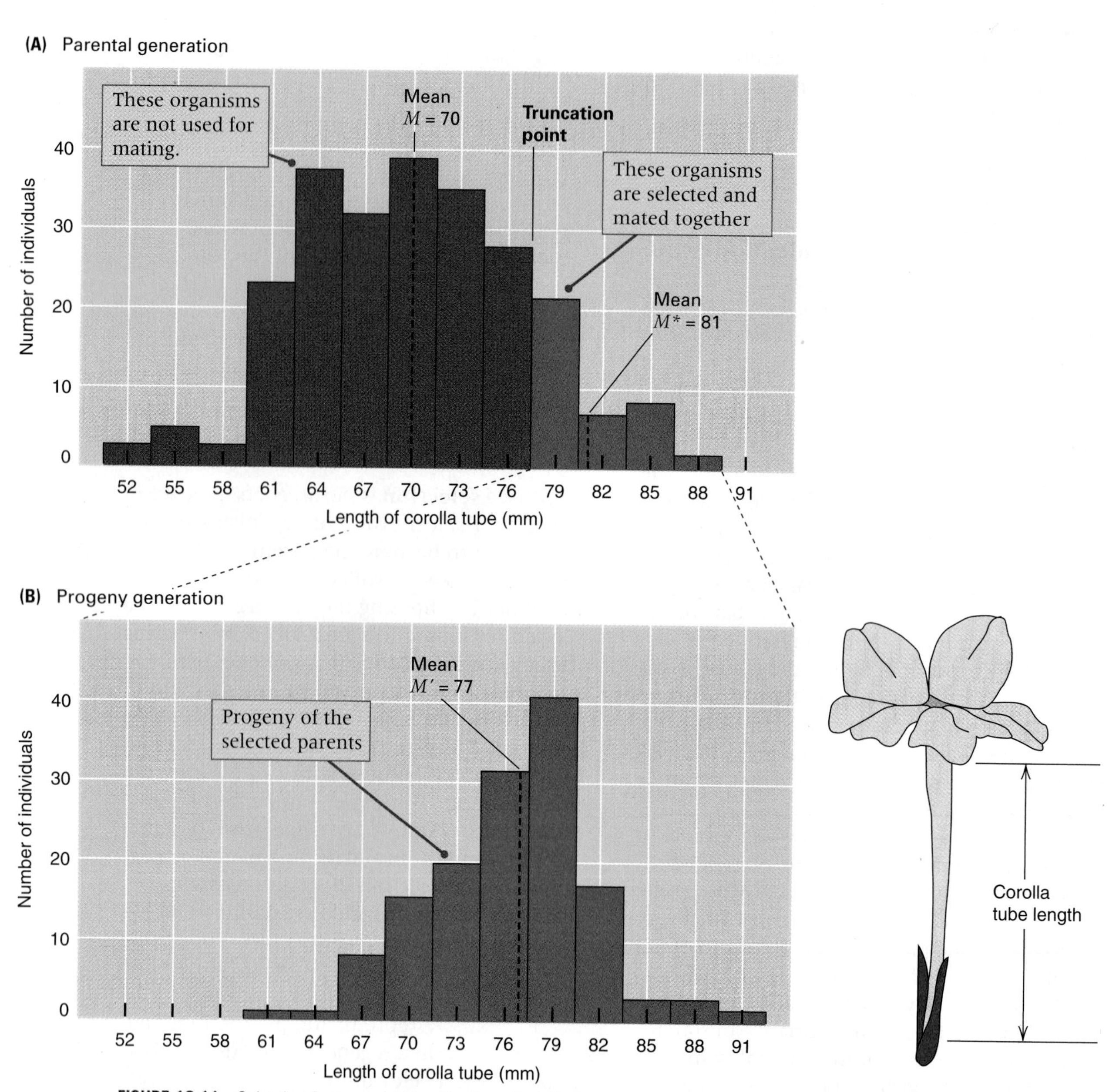


**FIGURE 18.11** Selection for increased length of corolla tube in tobacco. (A) Distribution of phenotypes in the parental generation. The symbol *M* denotes the mean phenotype of the entire population, and *M** denotes the mean phenotype of the organisms chosen for breeding (organisms with a phenotype that exceeds the truncation point). (B) Distribution of phenotypes among the offspring bred from the selected parents. The symbol *M′* denotes the mean.

spring generation. The parental generation is the population from which the parents were chosen for breeding. The type of selection is called **individual selection**, because each member of the population to be selected is evaluated according to its own individual phenotype. The selection is practiced by choosing some arbitrary level of phenotype, called the **truncation point**, that determines which individuals will be saved for breeding purposes. All individuals with a phenotype above the threshold are randomly mated among themselves to produce the next generation.

### Phenotypic Change with Individual Selection: A Prediction Equation

In evaluating progress through individual selection, three distinct phenotypic means are important. These means are symbolized in Figure 18.11 as $M$, $M^*$ and $M'$ and are defined as follows:

1. $M$ is the mean phenotype of the entire population in the parental generation, including both the selected and the nonselected individuals.
2. $M^*$ is the mean phenotype among those individuals selected as parents (those with a phenotype above the truncation point).
3. $M'$ is the mean phenotype among the progeny of selected parents.

The relationship among these three means is given by:

$$M' = M + h^2(M^* - M) \qquad (6)$$

where the symbol $h^2$ is the **narrow-sense heritability** of the trait in question.

Later in this chapter, a method for estimating narrow-sense heritability from the similarity in phenotype among relatives will be explained. In Figure 18.11, $h^2$ is the only unknown quantity, so it can be estimated from the data themselves. Rearranging Equation (6) and substituting the values for the means from Figure 18.11, we find that:

$$h^2 = \frac{M' - M}{M^* - M} = \frac{77 - 70}{81 - 70}$$

$$= 0.636$$

In a manner analogous to the way in which total phenotypic variance can be split into the sum of the genotypic variance and the environmental variance (Equation 3), the genotypic variance can be split into parts resulting from the additive effects of alleles, dominance effects, and effects of interaction between alleles of different genes. The difference between the broad-sense heritability, $H^2$, and the narrow-sense heritability, $h^2$, is that $H^2$ includes all of these genetic contributions to variation, whereas $h^2$ includes only the additive effects of alleles. From the standpoint of animal or plant improvement, $h^2$ is the heritability of interest, because:

> The narrow-sense heritability, $h^2$, is the proportion of the variance in phenotype that can be used to predict changes in population mean with individual selection according to Equation (6).

The distinction between the broad-sense heritability and the narrow-sense heritability can be appreciated intuitively by considering a population in which there is a rare recessive gene. In such a case, most homozygous recessive genotypes come from matings between heterozygous carriers. Some matings have more than one affected offspring. Hence, affected siblings can resemble each other more than they resemble their parents. For example, if *a* is a recessive allele, the mating $Aa \times Aa$, in which both parents show the dominant phenotype, may yield two offspring that are *aa*, which both show the recessive phenotype; in this case, the offspring are more similar to each other than to either of the parents. It is the dominance of the wildtype allele that causes this paradox, contributing to the broad-sense heritability of the trait but not to the narrow-sense heritability. The narrow-sense heritability includes only those genetic effects that contribute to the resemblance between parents and their offspring, because narrow-sense heritability measures how similar offspring are to their parents.

In general, the narrow-sense heritability of a trait is smaller than the broad-sense heritability. For example, in the parental generation in Figure 18.11, the broad-sense heritability of corolla tube length is $H^2 = 0.82$. The two types of heritability are equal only when the alleles affecting the trait are additive in their effects; with additive effects, each heterozygous genotype shows a phenotype that is exactly intermediate between the phenotypes of the respective homozygous genotypes.

Equation (6) is of fundamental importance in quantitative genetics because of its predictive value. This can be seen in the following example. The selection in Figure 18.11 was carried out for several generations. After two generations, the mean of the population was 83, and parents having a mean of 90 were selected. By use of the estimate $h^2 = 0.636$, the mean in the next generation can be predicted. The information provided is that $M = 83$ and $M^* = 90$. Therefore, Equation (6) implies that the predicted mean is:

$$M' = 83 + (0.636)(90 - 83) = 87.4$$

This value is in good agreement with the observed value of 87.9.

### ■ Long-Term Artificial Selection

Artificial selection is analogous to natural selection in that both types of selection cause an increase in the frequency of alleles that improve the selected trait (or traits). The principles of natural selection discussed in Chapter 17 also apply to artificial selection. For example, artificial selection is most effective in changing the frequency of alleles that are in an intermediate range of frequency ($0.2 < p < 0.8$). Selection is less effective for alleles with frequencies outside this range and is least effective for rare recessive alleles. For complex traits, including fitness, the total selection is shared among all the genes that influence the trait, and the selection coefficient affecting each allele is determined by: (1) the magnitude of the effect of the allele; (2) the frequency of the allele; (3) the total number of genes affecting the trait; (4) the narrow-sense heritability of the trait; and (5) the proportion of the population that is selected for breeding.

The value of heritability is determined both by the magnitude of effects and by the frequency of alleles. If all favorable alleles were fixed ($p = 1$) or lost ($p = 0$), then the heritability of the trait would be 0. Therefore, the heritability of a quantitative trait is expected to decrease over many generations of artificial selection as a result of favorable alleles becoming nearly fixed. For example, ten generations of selection for less fat in a population of Duroc pigs decreased the heritability of fatness from 73 to 30 percent because of changes in allele frequency that resulted from the selection.

Population improvement by means of artificial selection cannot continue indefinitely. A population may respond to selection until its mean is many standard deviations different from the mean of the original population, but eventually the population reaches a **selection limit** at which successive generations show no further improvement. Progress may stop because all alleles affecting the trait are either fixed or lost, and so the narrow-sense heritability of the trait becomes 0. However, it is more common for a selection limit to be reached because natural selection counteracts artificial selection. Many genes that respond to artificial selection as a result of their favorable effect on a selected trait also have indirect harmful effects on fitness. For example, selection for increased size of eggs in poultry results in a decrease in the number of eggs, and selection for extreme body size (large or small) in most animals results in a decrease in fertility. When one trait (for example, number of eggs) changes in the course of selection for a different trait (for example, size of eggs), then the unselected trait is said to have undergone a **correlated response** to selection. Correlated response of fitness is typical in long-term artificial selection. Each increment of progress in the selected trait is partially offset by a decrease in fitness because of a correlated response; eventually, artificial selection for the trait of interest is exactly balanced by natural selection against the trait. Thus, a selection limit is reached, and no further progress is possible without a change in the strategy of selection.

### ■ Inbreeding Depression and Heterosis

Inbreeding can have harmful effects on economically important traits such as yield of grain and egg production. This decline in performance is called **inbreeding depression**, and it results principally from rare harmful recessive alleles becoming homozygous because of inbreeding. The degree of inbreeding is measured by the inbreeding coefficient $F$ discussed in Chapter 17. **FIGURE 18.12** is an example of inbreeding depression of yield in corn, in which the yield decreases linearly as the inbreeding coefficient increases.

Most highly inbred strains suffer from many genetic defects, as might be expected from the uncovering of deleterious recessive alleles. One would also expect that if two different inbred strains were crossed, then the $F_1$

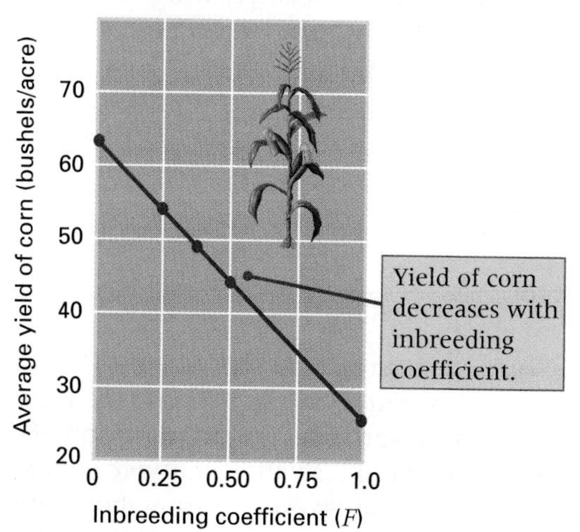


**FIGURE 18.12** Inbreeding depression for yield in corn. [Data from N. P. Neal, *Agron. J.* 27 (1935): 666–670.]

would show improved features, because a harmful recessive allele inherited from one parent would be likely to be covered up by a normal dominant allele from the other parent. In many organisms, the $F_1$ generation of a cross between inbred lines is superior to the parental lines, and the phenomenon is called **hybrid vigor** or **heterosis**. The phenomenon, which is widely used in the production of corn and other agricultural products, yields genetically identical hybrid plants with traits that are sometimes more favorable than those of the ancestral plants from which the inbreds were derived. The features that most commonly distinguish hybrid plants from their inbred parents are their rapid growth, larger size, and greater yield. Furthermore, the $F_1$ plants have a fairly uniform phenotype (because $\sigma_g^2 = 0$). Genetically heterogeneous crops with high yields or certain other desirable features can also be produced by traditional plant-breeding programs, but growers often prefer hybrid plants because of their uniformity. For example, uniform height and time of maturity facilitate machine harvesting, and plants that all bear fruit at the same time accommodate picking and shipping schedules.

Hybrid varieties of corn are used almost exclusively in the United States for commercial crops. A farmer cannot plant the seeds from his crop because the $F_2$ generation consists of a variety of genotypes, most of which do not show hybrid vigor. The production of hybrid seeds is a major industry in corn-growing regions of the United States.

## 18.5 Correlation Between Relatives

Studies of complex inheritance rely extensively on similarity among relatives to assess the importance of genetic factors. Particularly in the study of complex behavioral traits in human beings, genetic interpretation of familial resemblance is not always straightforward because of the possibility of nongenetic, but nevertheless familial, sources of resemblance. However, the situation is usually less complex in plant and animal breeding, because genotypes and environments are under experimental control.

### Covariance and Correlation

Genetic data about families are frequently pairs of numbers: pairs of parents, pairs of twins, or pairs consisting of a single parent and a single offspring. An important issue in quantitative genetics is the degree to which the numbers in each pair are associated. The usual way to measure this association is to calculate a statistical quantity called the *correlation coefficient* between the members of each pair.

The correlation coefficient among relatives is based on the covariance in phenotype among them. Much as the variance describes the tendency of a set of measurements to vary (Equation 2), the covariance describes the tendency of pairs of numbers to vary together (co-vary). Calculation of the covariance is similar to calculation of the variance in Equation (2) except that the squared deviation term $(x_i - \bar{x})^2$ is replaced with the product of the deviations of the pairs of measurements from their respective means—that is, $(x_i - \bar{x})(y_i - \bar{y})$.

For example, $(x_i - \bar{x})$ could be the deviation of a particular father's height from the overall father mean, and $(y_i - \bar{y})$ could be the deviation of his son's height from the overall son mean. In symbols, let $f_i$ be the number of pairs of relatives with phenotypic measurements $x_i$ and $y_i$. Then the estimated **covariance** (*Cov*) of the trait among the relatives is,

$$Cov(x,y) = \frac{\Sigma f_i (x_i - \bar{x})(y_i - \bar{y})}{N - 1} \quad (7)$$

where $N$ is the total number of pairs of relatives studied.

The **correlation coefficient** ($r$) of the trait between the relatives is calculated from the covariance as follows,

$$r = \frac{Cov(x,y)}{s_x s_y} \quad (8)$$

where $s_x$ and $s_y$ are the standard deviations of the measurements, estimated from Equation (2). The correlation coefficient can range from $-1.0$ to $+1.0$. A value of $+1.0$ means perfect association. When $r = 0$, $x$ and $y$ are not associated.

### The Geometrical Meaning of a Correlation

**FIGURE 18.13** shows a geometrical interpretation of the correlation coefficient. Just as the distribution of a single variable can be built up, step by step, by placing small squares corresponding to each individual along the $x$-axis, so too can the joint distribution of two variables (for example, height of fathers and height of sons) be built up, step by step, by placing small cubes on the $x$-$y$ plane at a position determined by each pair of measurements. As these cubes pile up, the joint distribution assumes the form of an inverted bowl, cross sections of which are themselves normal distributions. When there is no correlation between the variables, the surface is completely

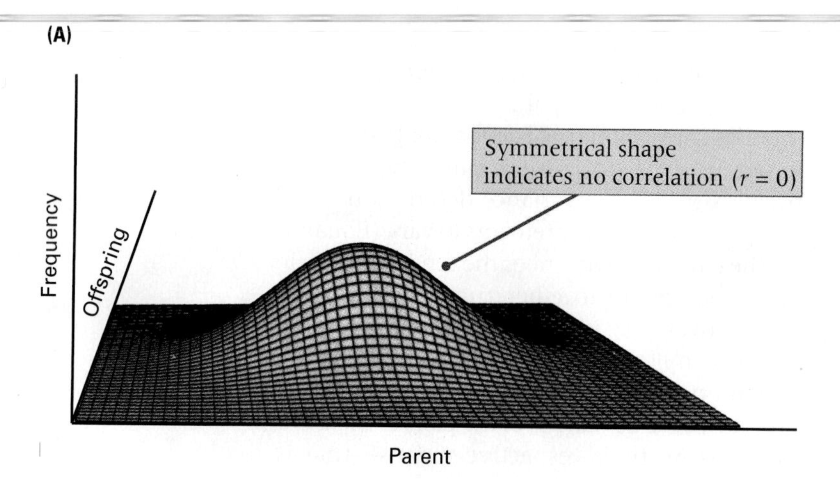


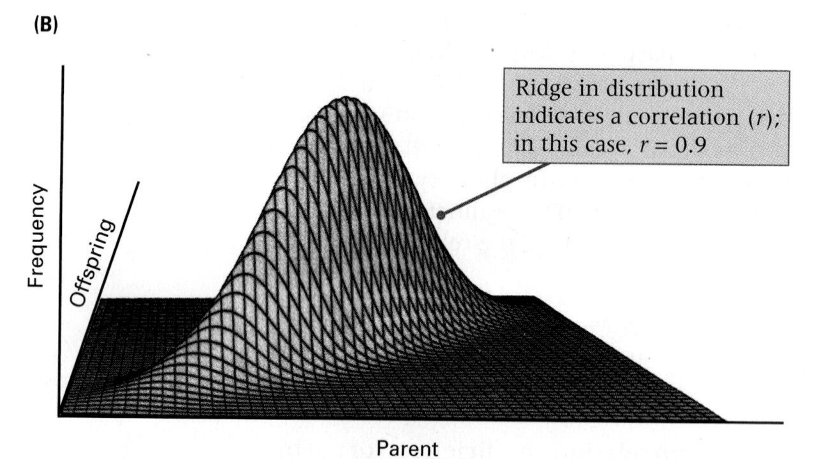


**FIGURE 18.13** Distribution of a quantitative trait in parents and offspring when there is no correlation (A) or a high correlation (B) between them. [From BIOMETRY by Robert R. Sokal and F. James Rohlf. © 1969, 1981, 1995 by W. H. Freeman and Company. Used with permission.]

symmetrical (Figure 18.13A), which means that the distribution of $y$ is completely independent of that of $x$. However, when the distributions are correlated, the distribution of $y$ varies according to the particular value of $x$, and the joint distribution becomes asymmetrical with a ridge built up in the $x$-$y$ plane (Figure 18.13B).

There are usually not enough data to form a full three-dimensional joint distribution of $x$ and $y$ as depicted in Figure 18.13. Instead, the data are in the form of individual points, sparse enough to be plotted on a scatter diagram like those in **FIGURE 18.14**. In these diagrams, the clustering of points around each line corresponds to the ridge of points in Figure 18.13. The higher the ridge, the more clustered the points are around the line.

## Estimation of Narrow-Sense Heritability

Covariance and correlation are important in quantitative genetics, because the correlation coefficient of a trait between individuals with various degrees of genetic relationship is related fairly simply to the narrow-sense or broad-sense heritability, as shown in **TABLE 18.3**. The table gives the theoretical values of the correlation coefficient for various pairs of relatives; $h^2$ represents the narrow-sense heritability and $H^2$ the broad-sense heritability. For parent–offspring, half-sibling, or first-cousin pairs, narrow-sense heritability can be estimated directly by multiplication. Specifically, $h^2$ can be estimated as twice the parent–offspring correlation, four times the half-sibling correlation, or eight times the first-cousin correlation.

**Table 18.3 Theoretical correlation coefficient in phenotype between relatives**

| Degree of relationship | Correlation coefficient* |
|---|---|
| Offspring and one parent | $h^2/2$ |
| Offspring and average of parents | $h^2/2$ |
| Half siblings | $h^2/4$ |
| First cousins | $h^2/8$ |
| Monozygotic twins | $H^2$ |
| Full siblings | $\sim H^2/2$ |

*Contributions from interactions among alleles of different genes have been ignored. For this and other reasons, $H^2$ correlations are approximate.

With full siblings and identical twins, the correlation coefficient is related to the broad-sense heritability, $H^2$, because phenotypic resemblance depends not only on additive effects, but also on dominance. In these relatives, dominance contributes to resemblance because the relatives can share *both* of their alleles as a result of their common ancestry, whereas parents and offspring, half siblings, and first cousins can share at most a single allele of any gene because of common ancestry. Therefore, to the extent that phenotype depends on dominance effects, full siblings can resemble each other more than they resemble their parents.

The potentially greater resemblance between siblings than between parents and offspring may again be understood by considering an autosomal recessive trait caused by a rare recessive allele. In a randomly mating population, the probability that an offspring will be affected, if the mother is affected, is $q$ (the allele frequency), which corresponds to the random-mating probability that the sperm giving rise to the offspring carries the recessive allele. In contrast, the probability that both of two siblings will be affected, if one of them is affected, is 1/4, because most of the matings that produce affected individuals will be between heterozygous parents. When the trait is rare, the parent–offspring resemblance will be very small (as

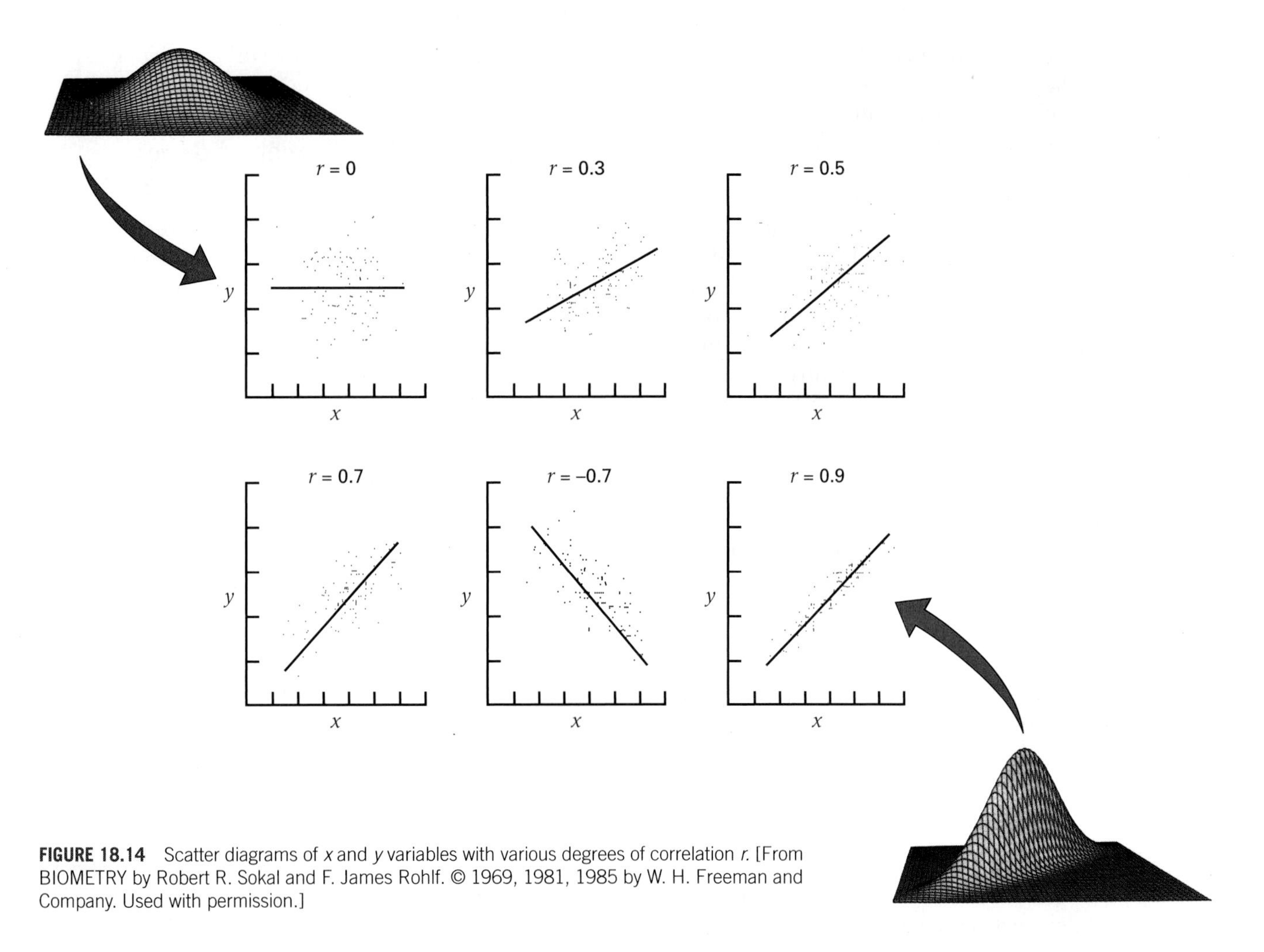


**FIGURE 18.14** Scatter diagrams of *x* and *y* variables with various degrees of correlation *r*. [From BIOMETRY by Robert R. Sokal and F. James Rohlf. © 1969, 1981, 1985 by W. H. Freeman and Company. Used with permission.]

$q$ approaches 0), whereas the sibling–sibling resemblance will remain large. This discrepancy is entirely a result of dominance, and it arises only because, at any particular locus, full siblings may have the same genotype.

## 18.6 Heritabilities of Threshold Traits

Application of the concepts of individual selection can be used to understand the meaning of heritability in the context of threshold traits. The analogy is illustrated in **FIGURE 18.15**. Threshold traits depend on an underlying quantitative trait, called the **liability**, which refers to the degree of genetic risk or predisposition to the trait. Only those individuals with a liability greater than a certain threshold develop the trait. Figure 18.15A shows the distribution of liability in a population, where liability is

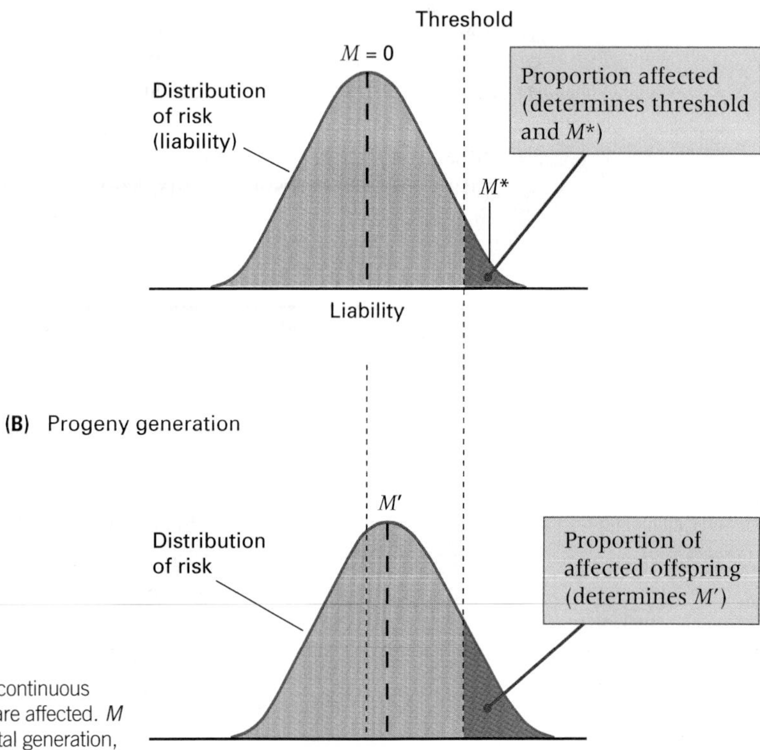


**FIGURE 18.15** Interpretation of a threshold trait in terms of an underlying continuous distribution of liability. Individuals with a liability greater than the threshold are affected. *M* (arbitrarily set equal to 0), $M^*$, and $M'$ are the mean liabilities of the parental generation, of affected individuals, and of the offspring of affected individuals, respectively.

measured on an arbitrary scale so that the mean liability ($M$) equals 0 and the variance in liability equals 1. Affected individuals are in the right-hand area of the distribution. Using various characteristics of the normal distribution, we can use the observed proportion of affected individuals to calculate the position of the threshold and the value of $M^*$, which in this context is the mean liability of affected individuals.

The distribution of liability among offspring of affected individuals is shown in Figure 18.15B. It is shifted slightly to the right, which implies that, relative to the population as a whole, a greater proportion of individuals with affected parents have liabilities above the threshold and so will be affected. The observed proportion of affected individuals in the offspring generation can be used to calculate $M'$. The important point is that $M^*$ and $M'$ can be obtained from the observed proportions corresponding to the shaded red areas in Figure 18.15. With $M = 0$, Equation (6) can be used to calculate the narrow-sense heritability of liability of the trait in question.

Examples of narrow-sense heritability of the liability to important congenital abnormalities are given in **TABLE 18.4**. How these heritabilities translate into risk is illustrated in **FIGURE 18.16**, along with observed data for the most common congenital abnormalities in Caucasians. The theoretical curves correspond to the relationships among the incidence of the trait in the general population, the type of inheritance, and the risk among first-degree relatives of affected individuals. (First-degree relatives of an affected individual include parents, offspring, and siblings.) Simple Mendelian inheritance produces the highest risks, but these traits, as a group, tend to be rare. The most common traits are threshold traits, and with a few exceptions, their risks tend to be moderate or low, corresponding to the heritabilities indicated.

## 18.7 Identification of Genes Affecting Complex Traits

A gene that affects a complex trait is a **quantitative-trait locus (QTL)**. The term derives from the fact that many complex traits of medical or commercial importance are quantitative traits. QTLs cannot usually be identified in pedigrees, because their individual effects are obscured by the segregation of other genes and by environmental variation. Even so, genes that affect the traits can be localized if they are genetically linked with polymorphic DNA markers, such as those discussed in Chapter 2, because the effects of the genotype affecting the complex trait will be correlated with the genotype of a linked genetic marker. Locating QTLs in the genome is important to the manipulation of genes in breeding programs and to the identification and study of the genes in order to identify their functions. Another approach to the identification of genes that influence complex traits is through the study of candidate genes that are known to be involved in physiological or developmental processes that affect the trait. These complementary approaches to complex inheritance are examined next.

### Linkage Analysis in the Genetic Mapping of Quantitative Trait Loci

Polymorphic DNA markers are abundant, are distributed throughout the genome, and often have multiple codominant alleles; thus, they are ideally suited for linkage studies of quantitative traits. In QTL studies, as many widely scattered markers as possible are monitored, along

**Table 18.4 Narrow-sense heritability of liability for congenital abnormalities**

| Trait | Description | Population frequency (%) | Narrow-sense heritability (%) |
|---|---|---|---|
| Cleft lip | Upper lip not completely formed | 0.08 | 65 |
| Spina bifida | Exposed spinal cord | 0.11 | 65 |
| Club foot | Abnormally formed foot | 0.08 | 80 |
| Pyloric stenosis | Obstructed stomach | 0.40 | 85 |
| Dislocation of hip | Hip joint mispositioned | 0.06 | 95 |
| Hydrocephalus | Fluid accumulation around brain | 0.12 | 40 |
| Celiac disease | Inability to digest fats, starches, and sugars | 0.04 | 80 |
| Hypospadias | Abnormally formed penis | 0.30 | 75 |
| Atrial septal defect | Hole between upper chambers of heart | 0.07 | 70 |
| Patent ductus arteriosus | Hole between aorta and pulmonary artery; blood bypasses lungs | 0.08 | 75 |

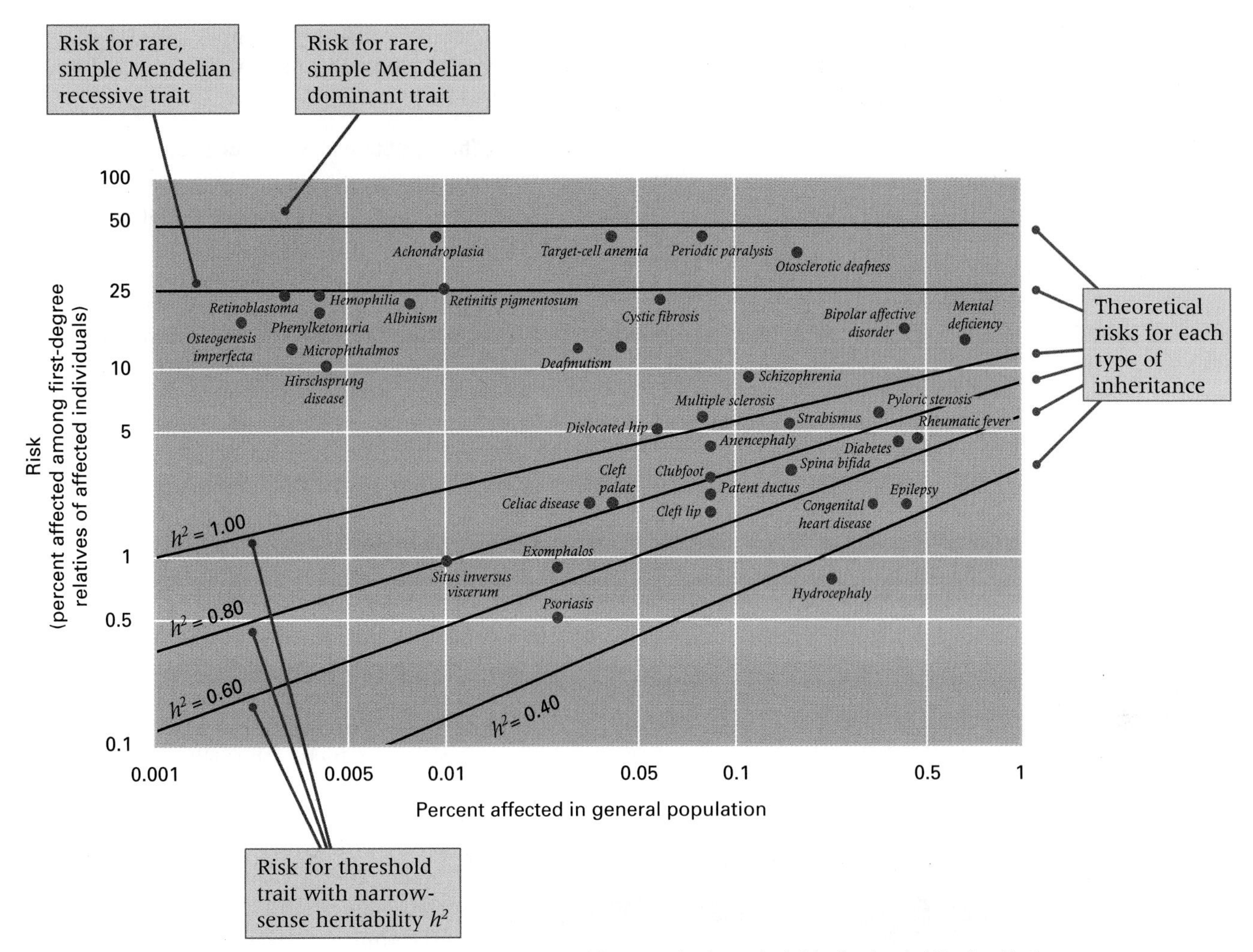


**FIGURE 18.16** Recurrence risks of common abnormalities. Diagonal lines are the theoretical risks for threshold traits with the indicated values of the narrow-sense heritability of liability. Horizontal lines are the theoretical risks for simple dominant or recessive traits.

with the complex trait, in successive generations of a genetically heterogeneous population. Statistical studies are then carried out to identify which DNA markers are associated with the complex trait in the sense that their presence in a genotype is consistently accompanied by the phenotype of interest. These DNA markers identify regions of the genome that contain one or more QTLs that have important effects on the trait, and the DNA markers can be used to trace the segregation of the important regions in breeding programs and even as starting points for identifying genes with particularly large effects.

An example of genetic mapping of QTLs for several quantitative traits in tomatoes is illustrated in **FIGURE 18.17**. More than 300 highly polymorphic genetic markers have been mapped in the tomato genome, with an average spacing between markers of 5 map units. The chromosome maps in Figure 18.17 show a subset of 67 markers that were segregating in crosses between the domestic tomato and a wild South American relative. The average spacing between the markers is 20 map units. Backcross progeny of the cross $F_1 \times$ domestic tomato were tested for the genetic markers, and the fruits of the backcross progeny were assayed for three quantitative traits—fruit weight, content of soluble solids, and acidity. Statistical analysis of the data was carried out in order to detect marker alleles that were associated with phenotypic differences in any of the traits; a significant association indicates genetic linkage between the marker gene and one or more QTLs affecting the trait. A total of six QTLs affecting fruit weight were detected (green bars), as well as four QTLs affecting soluble solids (blue bars) and five QTLs affecting acidity (dark red bars). Although additional QTLs with smaller effects undoubtedly remained undetected in these types of experiments, the effects of the mapped QTLs are substantial; the mapped QTLs account for 58 percent of the total phenotypic variance in fruit weight, 44 percent of the phenotypic variance in

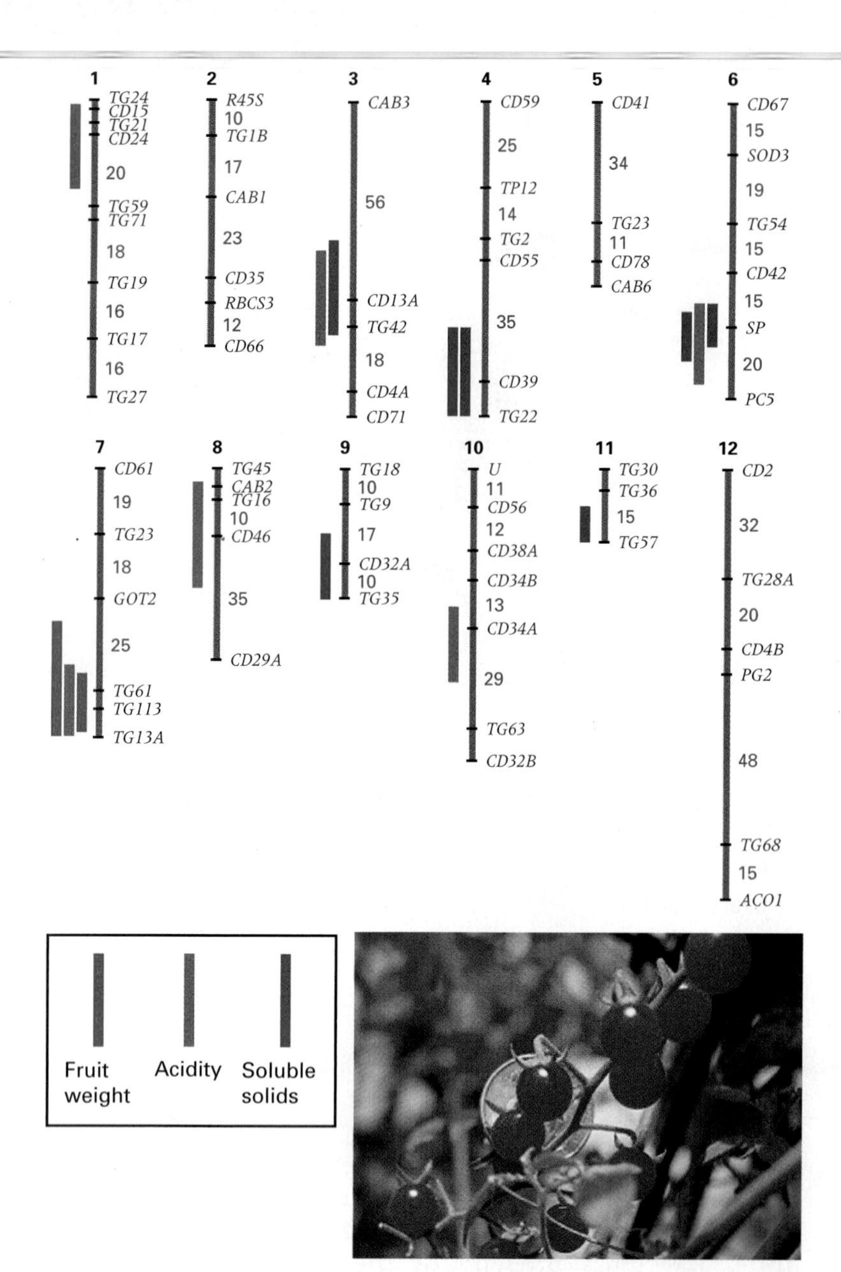


**FIGURE 18.17** Location of QTLs for several quantitative traits in the tomato genome. The genetic markers are shown for each of the 12 chromosomes. The numbers in red are distances in map units between adjacent markers, but only map distances of 10 or greater are indicated. The regions in which the QTLs are located are indicated by the bars: green bars, QTLs for fruit weight; blue bars, QTLs for content of soluble solids; and dark red bars, QTLs for acidity (pH). The data are from crosses between the domestic tomato (*Solanum lycopersicon*) and a wild South American relative with small fruit (*Solanum chmielewskii*). The photograph is of fruits of wild tomato. Note the small size in comparison with the coin (a U.S. quarter). The $F_1$ generation was backcrossed with the domestic tomato, and fruits from the progeny were assayed for the genetic markers and each of the quantitative traits. [Data from A. H. Paterson et al., *Nature* 335 (1988): 721–725. Photo courtesy of Steven D. Tanksley, Cornell University.]

soluble solids, and 48 percent of the phenotypic variance in acidity. The genetic markers linked to the QTLs with substantial effects make it possible to trace the transmission of the QTLs in pedigrees and to manipulate the QTLs in breeding programs by following the transmission of the linked marker genes. Figure 18.17 also indicates a number of chromosomal regions containing QTLs for two or more of the traits—for example, the QTL regions on chromosomes 6 and 7, which affect all three traits.

## The Number and Nature of QTLs

The fine-scale genetic localization and identification of QTLs has become possible only recently, and it is currently a very active field of genetic research. Most studies in human populations relay on *whole-genome association* studies, which are sometimes denoted by the acronym WGA. A typical WGA study involves some number of patients with a disease and a comparable number of controls without the disease, matched with the patients for age, sex, ethnic group, place of residence, and so forth. The sample sizes need to be large, minimally several thousand of both patients and control subjects, and often tens of thousands. Each patient and control subject is genotyped, typically using a single-nucleotide polymorphism (SNP) chip of the type described in Chapter 2, in which the genotype at 500 thousand or more SNPs throughout the genome are determined simultaneously. Then statistical analysis is carried out to ascertain which of the SNP polymorphisms are associated with the disease, and which polymorphic nucleotides increase risk of the disease and which are protective. The SNP itself is highly unlikely to be causally related to the disease, but the association implies that there is a genetically linked QTL that is causally related. Although there is much more to be learned from WGA studies, enough have been carried out to be able to make some cautious generalizations about the nature and number of QTLs and the magnitude of their effects, such as:

- Almost all complex traits are affected by many QTLs. For example, at least 35 QTLs affect the level of serum cholesterol.
- Many QTLs affecting complex traits show pleiotropy; they affect several complex traits simultaneously.
- Epistasis (gene interaction) among QTLs affecting complex traits is common. Because of epistasis, the effects of any combination of QTLs are not necessarily predictable from knowledge of the individual effects. In some instances, the entire effect of a QTL is due to its interaction with another QTL somewhere else in the genome.
- The effects of QTLs are often highly dependent on the environment. *Genotype-by-sex* interaction is particularly important; that is, the effect of a QTL can differ dramatically between females and males. Some QTLs have effects that are popula-

## connection

## Win, Place, or Show?

**Hugh E. Montgomery and eighteen other investigators 1998**
University College, London, and six other research institutions
*Human Gene for Physical Performance*

*A professional basketball executive once said of superstar Michael Jordan, "Gene for gene, DNA for DNA, nobody can replace him." The implication that Jordan's phenomenal success was not due exclusively to his work ethic, practice, commitment, consistency, and character is interesting. But nobody doubts that Jordan's success in basketball was associated in part with his being tall. And height is a multifactorial trait partly under genetic control. About 50 percent of the variation in height among people is due to genetic differences. But might invisible, metabolic differences also influence physical performance? Some people find this possibility unsettling, because it potentially raises all sorts of issues about the value of effort and practice, the fairness of competition, genetic testing for entry into elite physical training programs, and so forth. The extent to which genotype influences physical performance is not known, but some genetic differences may be important. In this paper, a polymorphism in the angiotensin-converting enzyme (ACE) is reported to have a major effect on physical endurance. It is not difficult to see how ACE could play a role. The enzyme degrades a class of vasodilator proteins and also converts angiotensin I into a vasoconstrictor. Note, however, that the authors are cautious in their conclusions and are careful not to speculate that the gene for ACE is a QTL (quantitative-trait locus) affecting physical performance. It is still too early to reach firm conclusions.*

A specific genetic factor that strongly influences human physical performance has not so far been reported, but here we show that a polymorphism in the gene encoding angiotensin-converting enzyme does just that. An "insertion" allele of the gene is associated with elite endurance performance among high-altitude mountaineers. Also, after physical training, repetitive weight-lifting is improved elevenfold in individuals homozygous for the "insertion" allele compared with those homozygous for the "deletion" allele. The endocrine renin angiotensin system is important in controlling the circulation. . . . A polymorphism in the human *ACE* gene for angiotensin-converting enzyme has been described in which the deletion (*D*) rather than insertion (*I*) allele is associated with higher activity by tissue ACE. . . . Our initial studies suggested that the *I* allele was associated with improved endurance performance. . . . High-altitude mountaineers perform extreme-endurance exercise. [The *ACE* genotypes of 25] elite unrelated male British mountaineers . . . were compared with those of 1906 British males. . . . Both the genotype distribution and allele frequency differed significantly between climbers and controls [see figure].

***We show that a polymorphism in the gene encoding angiotensin-converting enzyme [influences human physical performance].***

In a second study, *ACE* genotype was determined in 123 Caucasian males recruited into the United Kingdom army consecutively. . . . The maximum duration (in seconds) for which they could perform repetitive elbow flexion while holding a 15-kg barbell was assessed both before and after a 10-week physical training period. . . . Duration of exercise improved significantly for those of *II* and *ID* genotypes, but not for those of *DD* genotype (Figure B). . . . Increased performance is likely to be due to an improvement in the endurance characteristics of the tested muscles. . . . Further work will be needed to determine whether this correlation holds beyond the limited group studied here.

Source: H. E. Montgomery, et al., *Nature* 393 (1998): 221–222.

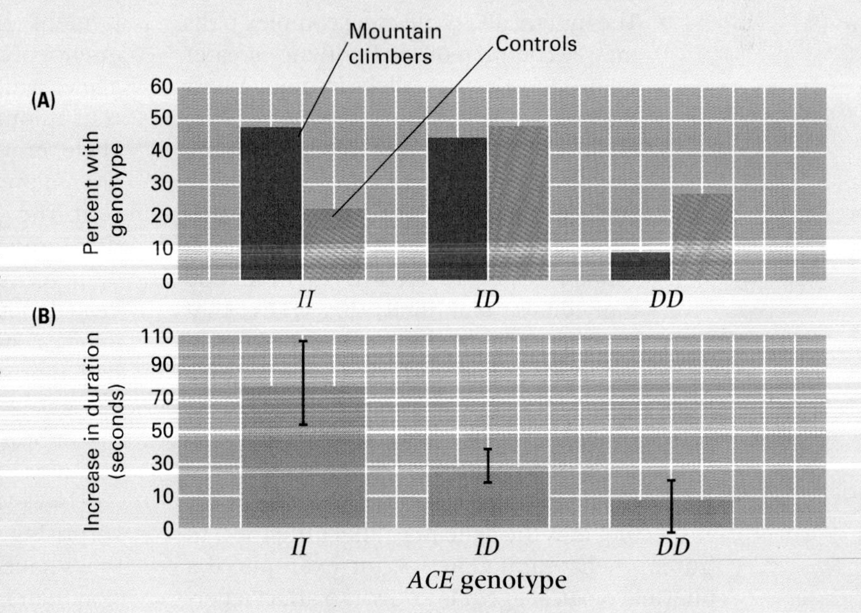

tion specific. For example, a QTL in human chromosome 9 leads to a 30–40 percent increase in the risk of coronary artery disease, but only in Caucasians, not in African Americans.
- The distribution of effects of QTLs is exponential; that is, most QTLs have very small effects, many fewer have moderate effects, and very few have large effects. Specifically, most QTLs for complex human diseases increase the risk by 10–30 percent, and some have effects in the range 30–50 percent. QTLs with larger effects are uncommon.
- Many QTLs are not due to single genes but rather to the combined effects of several closely linked genes. This finding reflects the process of duplication and divergence by which functionally related genes evolve (Chapter 17), and it makes sense that functionally related genes would affect the same complex traits.
- Although some genes affecting complex traits might have been predicted based on the biology of the trait and the inferred function of annotated genes in the genomic sequence, many more genes could not have been predicted. Those with known functions have no obvious connection to the trait, and many are genes whose function is unknown.
- The mutations in genes affecting complex traits may be in protein-coding regions or in regions that affect gene regulation.
- The mutant alleles affecting complex traits may be common (allele frequency greater than 0.05) or rare (allele frequency less than 0.05).
- The majority of common alleles affecting complex traits accounts for a relatively small fraction of the total phenotypic variance in the trait. An extreme example is human height, where a study of 40,000 individuals identified 20 QTLs that together accounted for only 3 percent of the total variance in height.

## Candidate Genes for Complex Traits

A **candidate gene** for a complex trait is a gene for which there is some *a priori* basis for suspecting that the gene may affect the trait. In human behavioral genetics, for example, if a pharmacological agent is known to affect a personality trait, and the molecular target of the drug is known, then the gene that codes for the target molecule and any genes whose products interact with the drug or with the target molecule are all candidate genes for affecting the personality trait.

One example of the use of candidate genes in the study of human behavioral genetics is in the identification of a naturally occurring genetic polymorphism associated with serious depression in response to stressful life experiences. The neurotransmitter substance serotonin (5-hydroxytryptamine) is known to influence a variety of psychiatric conditions, such as anxiety and depression. Among the important components in serotonin action is the serotonin transporter protein SLC6A4 (SLC stands for *solute carrier*), encoded in a gene in chromosomal region *17q11.1–q12*. Neurons that release serotonin to stimulate other neurons also take it up again through the SLC6A4 transporter. The uptake terminates the stimulation and also recycles the molecule for later use.

The serotonin transporter became a strong candidate gene for depression when it was discovered that the transporter is the target of a class of antidepressants known as selective serotonin reuptake inhibitors. The widely prescribed antidepressant Prozac is an example of such a drug.

Motivated by the strong suggestion that the serotonin transporter might be involved in depression, researchers looked for evidence of genetic polymorphisms affecting the SLC6A4 transporter gene in human populations. Such a polymorphism was found in the promoter region of the transporter. About 1 kb upstream from the transcription start site is a series of 16 tandem repeats of a nearly identical DNA sequence of about 15 base pairs. This is the *L* (long) allele, which has an allele frequency of 57 percent among Caucasians. There is also an *S* (*short*) allele, in which three of the repeated sequences are not present. The *S* allele frequency is 43 percent. The genotypes *LL*, *LS*, and *SS* are found in the Hardy–Weinberg proportions of 32 percent, 49 percent, and 19 percent. Further studies revealed that the polymorphism does have a physiological effect. In cells grown in culture, *LL* cells had approximately 50 percent more mRNA for the transporter than *LS* or *SS* cells, and *LL* cells had approximately 35 percent more membrane-bound transporter protein than *LS* or *SS* cells.

In view of these differences, it is perhaps not surprising that individuals with *SS* or *SL* genotypes have a much higher risk of severe depression than *LL* genotypes. Interestingly, the higher risk shows a genotype-by-environment interaction in that it occurs only among individuals who have experienced three or more stressful life events (threat, defeat, humiliation, or death

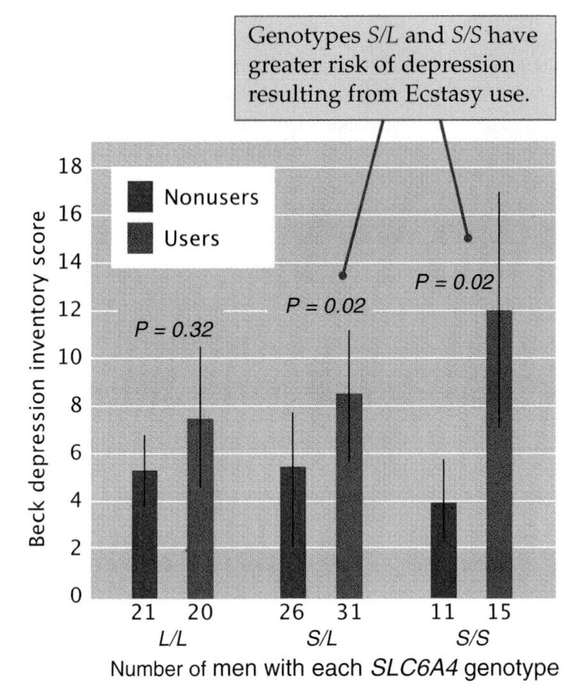


**FIGURE 18.18** Depression score of male users of the illicit drug Ecstasy as a function of genotype for the serotonin uptake transporter SLC6A4. The *P* values are significance levels for each genotype as compared with male nonusers. [Data from J. P. Roiser, et al., *Am. J. Psychiatry* 162 (2005): 609–612.]

of a loved one) occurring within the previous five years.

The *S* and *L* forms of the serotonin transporter also affect drug response. One of the illicit drugs in widespread use is the club drug MDMA (3,4-methylenedioxymethamphetamine), also called Ecstasy, which binds to the serotonin transporter and inhibits reuptake. Habitual use of Ecstasy is associated with long-term changes in the release and reuptake of serotonin. This is potentially troubling, as in 2006 some 12.3 million Americans aged 12 or older had tried the drug at least once, representing about 5 percent of the U.S. population in that age group.

The effects of the *SLC6A4* polymorphism in habitual Ecstasy users are shown in **FIGURE 18.18.** In this study, male Ecstasy users were classified according to their *SLC6A4* genotype (*LL*, *SL*, or *SS*) and compared with male nonusers with respect to the Beck Depression Inventory score, a standardized rating scale for depression based on a questionnaire covering a variety of different symptoms of depression. A score greater than 9 is regarded as evidence of depression.

Although the Ecstasy users were required to abstain from the drug for at least 3 weeks prior to testing, nevertheless the long-term effects of use are evident in Figure 18.18. Although the difference in the depression score is not statistically significant in the *LL* genotype ($P = 0.32$), both the *SL* and *SS* genotypes have a significantly greater depression score among Ecstasy users than among nonusers ($P = 0.02$). Roughly speaking, there is a twofold greater risk of depression among the *SL* genotypes and a tenfold greater risk of depression among the *SS* genotypes. This type of genotype-by-environment interaction is an example of what has been called *pharmacogenomics,* a genotype-specific response to pharmaceutical agents.

## CHAPTER SUMMARY

- Complex traits are determined by multiple genetic and environmental factors acting together.
- The relative contributions of genotype and environment to a trait are measured by the variance due to genotype (genotypic variance) and the variance due to environment (environmental variance).
- Correlations between relatives are used to estimate various components of variation, such as genotypic variance, additive variance, and dominance variance.
- Additive variance accounts for the parent–offspring correlation; dominance variance accounts for the sib–sib correlation over and above that expected from the additive variance.
- Narrow-sense heritability is the ratio of additive (transmissible) variance to the total phenotypic variance; it is widely used in plant and animal breeding to predict the response to artificial selection.
- Genes that affect quantitative traits can be identified by genetic mapping using highly polymorphic DNA markers, or by specifying candidate genes on the basis of knowledge of the physiology and development of the trait.

## REVIEW THE BASICS

- Give an example of a trait in human beings that is affected both by environmental factors and by genetic factors. Specify some of the environmental factors that affect the trait.
- Does the distribution of phenotypes of a trait in a population tell one anything about the relative importance of genes and environment in causing differences in phenotype among individuals? Does it tell one anything about the number of genes that may affect the trait? Explain.
- What is the genotypic variance of a quantitative trait? What is the environmental variance?
- Distinguish between the variance due to genotype-by-enviroment interaction and the variance due to genotype-by-environment association. Which type of variance is more easily controlled by the experimenter?
- In regard to the genotypic variance, what is special about an inbred line or about the $F_1$ progeny of a cross between two inbred lines?
- Distinguish between broad-sense heritability and narrow-sense heritability. Which type of heritability is relevant to individual truncation selection?
- The distribution of bristle number on one of the abdominal segments in a population of *Drosophila* ranges from 12 to 26. The narrow-sense heritability of the trait is approximately 50 percent. Do you think that it would be possible, by practicing artificial selection over a number of generations, to obtain a population in which the *mean* bristle number was greater than 26? a population in which the *minimum* bristle number was greater than 26? Explain why or why not.
- Why are correlations between relatives of interest to the quantitative geneticist?
- In the context of multifactorial inheritance, what is a QTL? Does a QTL differ from any other kind of gene? How can QTLs be detected?
- In the context of multifactorial inheritance, what is a candidate gene? Would you regard the human *PAH* gene for phenylalanine hydroxylase as a natural candidate gene for mental retardation? Why or why not?

## GUIDE TO PROBLEM SOLVING

**Problem 1** Two inbred lines of *Drosophila* are crossed, and the $F_1$ generation has a mean number of abdominal bristles of 20 and a standard deviation of 2. The $F_2$ generation has a mean of 20 and a standard deviation of 3. What are the environmental variance, the genetic variance, and the broad-sense heritability of bristle number in this population?

**Answer** The phenotypic variance in the $F_1$ generation equals $2^2 = 4$, and the phenotypic variance in the $F_2$ generation equals $3^2 = 9$. Because the $F_1$ generation results from a cross of inbred lines, it is genetically homogeneous and therefore the phenotypic variance is equal to the environmental variance $\sigma_e^2$, owing to the fact that the genetic variance $\sigma_g^2$ equals 0 in a genetically homogeneous population. Hence $\sigma_e^2 = 4$. The phenotypic variance in the genetically heterogeneous $F_2$ generation is the sum $\sigma_g^2 + \sigma_e^2 = 9$. Therefore, $\sigma_g^2$ in the $F_2$ generation equals $9 - 4 = 5$. The broad-sense heritability $H^2$ equals $\sigma_g^2/(\sigma_g^2 + \sigma_e^2) = 5/9 =$ 55.6 percent.

**Problem 2** A flock of broiler chickens has a mean weight gain of 700 grams between ages 5 and 9 weeks, and the narrow-sense heritability of weight gain in this flock is 0.80. Selection for increased weight gain is carried out for five consecutive generations, and in each generation the average of the parents is 50 grams greater than the average of the population from which the parents were derived. Assuming that the heritability of the trait remains constant at 80 percent:

**(a)** What is the expected mean weight gain among the broilers after one generation of selection?

**(b)** What is the expected mean weight gain among the broilers after five consecutive generations of selection?

**Answer**

**(a)** In the first generation of selection, the mean of the population is $M = 700$ grams, that of the selected parents is $M^* = 750$ grams, and the narrow-sense heritability $h^2$ is given as 0.80. The formula below gives the prediction equation for this case, and the predicted average weight gain is 740 grams per individual.

$$M' = M + h^2(M^* - M)$$
$$= 700 + (0.80)(750 - 700)$$
$$= 740 \text{ grams}$$

**(b)** It is clear from the formula that if $M^* - M = 50$ in each generation, and $h^2$ remains at 0.80, then the expected increase in weight gain per generation is 40 grams. After five generations the expected gain is 200 grams, yielding an expected average weight gain after five generations of 900 grams per individual.

**Problem 3** A restriction fragment polymorphism (RFLP) in *Drosophila* is linked to a QTL (quantitative trait locus) affecting the number of abdominal bristles. A probe for the RFLP hybridizes with a 6.5-kb restriction fragment, with a 4.0-kb restriction fragment, or with both (in heterozygotes). The QTL genotypes may be designated *QQ*, *Qq*, and *qq*. The average abdominal bristle numbers in these genotypes are 20, 18, and 16, respectively. The typical standard deviation in bristle number is about 2, so the effect of the favorable allele is equal to one phenotypic standard deviation. This would usually be regarded as a rather large effect. A mating is carried out as shown below, and each of the progeny is examined for bristle number and then subjected to electrophoresis and DNA hybridization to classify the RFLP genotype.

$$\frac{6.5\ Q}{4.0\ q}\ ♀♀ \times \frac{4.0\ q}{4.0\ q}\ ♂♂$$

As expected, the progeny phenotypes with respect to the RFLP fall into two classes: those with both bands (4.0/6.5 heterozygotes) and those with the 4.0 band only (4.0/4.0 homozygotes). In the boxes under the lanes in the gel shown below, write the average bristle number for each of the two RFLP phenotypes, assuming that the frequency of recombination $r$ between the RFLP locus and the QTL is

**(a)** $r = 0$

**(b)** $r = 10$ percent

**(c)** $r = 50$ percent

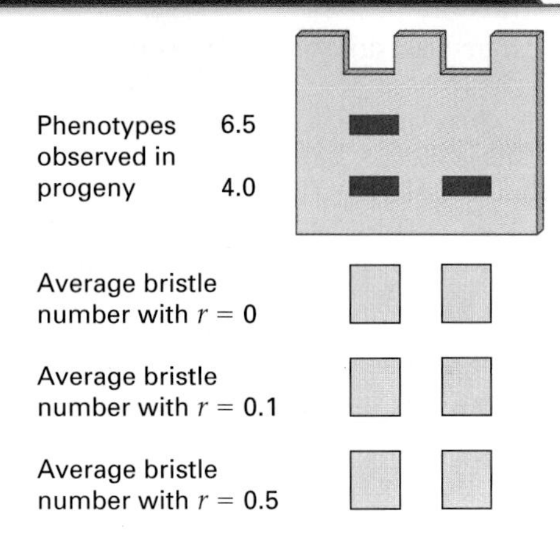


**Answer** The female gametes are of four types: 6.5 $Q$ [frequency $(1 - r)/2$], 6.5 $q$ [frequency $r/2$], 4.0 $Q$ [frequency $r/2$], and 4.0 $q$ [frequency $(1 - r)/2$]. The male gametes are only 4.0 $q$. Therefore, the four types of female gametes, listed in the same order as above, produce progeny of genotypes 6.5 $Q$/4.0 $q$ (average bristle number 18), 6.5 $q$/4.0 $q$ (average bristle number 16), 4.0 $Q$/4.0 $q$ (average bristle number 18), and 4.0 $q$/4.0 $q$ (average bristle number 16). Considering only the 6.5/4.0 RFLP genotypes, the average bristle number is $(1 - r)(18) + (r)(16) = 18 - 2r$; and considering only the 4.0/4.0 RFLP genotypes, the average bristle number is $(r)(18) + (1 - r)(16) = 16 + 2r$. The answers are therefore **(a)** for $r = 0$, RFLP type 6.5/4.0 averages 18 bristles and RFLP type 4.0/4.0 averages 16 bristles; **(b)** for $r = 0.10$, RFLP type 6.5/4.0 averages 17.8 bristles and RFLP type 4.0/4.0 averages 16.2 bristles; **(c)** for $r = 0.10$, RFLP type 6.5/4.0 averages 17.0 bristles and RFLP type 4.0/4.0 averages 17.0 bristles. The point is that the detection of QTL by means of linked genetic markers is very sensitive to the tightness of the linkage. For very tight linkage, the ability to detect the QTL is maximal, but then the power to detect the QTL decreases linearly with the frequency of recombination until, at $r = 0.50$, there is no ability to detect the QTL.

## ANALYSIS AND APPLICATIONS

**18.1** Two varieties of corn, A and B, are field-tested in Indiana and North Carolina. Strain A is more productive in Indiana, but strain B is more productive in North Carolina. What phenomenon in quantitative genetics does this example illustrate?

**18.2** A distribution has the feature that the standard deviation is equal to the variance. What are the possible values for the variance?

**18.3** The following questions pertain to a normal distribution.

- **(a)** What term applies to the value along the $x$-axis that corresponds to the peak of the distribution?
- **(b)** If two normal distributions have the same mean but different variances, which is the broader?
- **(c)** What proportion of the population is expected to lie within one standard deviation of the mean? Within two standard deviations?

**18.4** Distinguish between the broad-sense heritability of a quantitative trait and the narrow-sense heritability. If a population is fixed for all genes that affect a particular quantitative trait, what are the values of the broad-sense and narrow-sense heritabilities?

**18.5** When we compare a quantitative trait in the $F_1$ and $F_2$ generations obtained by crossing two highly inbred strains, which set of progeny provides an estimate of the environmental variance? What determines the variance of the other set of progeny?

**18.6** For the difference between the domestic tomato, *Solanum lycopersicum*, and the wild South American relative, *Solanum chmielewskii*, the environmental variance $\sigma_e^2$ accounts for 13 percent of the total phenotypic variance $\sigma_p^2$ of fruit weight, 9 percent of $\sigma_p^2$ of soluble-solid content, and 11 percent of $\sigma_p^2$ of acidity. What are the broad-sense heritabilities of these traits?

**18.7** Ten female mice had the following numbers of live-born offspring in their first litters: 11, 9, 13, 10, 9, 8, 10, 11, 10, 13. Considering these females as representative of the total population from which they came, estimate the mean, variance, and standard deviation of size of the first litter in the entire population.

**18.8** Suppose a quantitative trait is determined by three independently segregating genes in a randomly mating population. Alleles $A$, $B$, and $C$ are dominant to their recessive counterparts $a$, $b$, and $c$. Genotype $aa\ bb\ cc$ has a phenotype of 0, and two units in phenotype are added for each locus that is heterozygous or homozygous for the favorable allele.

- **(a)** What is the expected distribution of phenotypes in a population in which each favorable allele has a frequency of 1/2?
- **(b)** Calculate the mean, variance, and standard deviation of the phenotype in this population. (Note: In problems such as this, the variance is most easily calculated as the mean of the squares of the phenotypic values minus the square of the mean phenotypic value: in symbols, $s^2 = Mean(x^2) - [Mean(x)]^2$.)
- **(c)** How do the answers to parts **(a)** and **(b)** compare to those for a trait affected by additive alleles (in which each favorable allele present in a genotype adds one unit to the phenotype)?

**18.9** Values of IQ score are distributed approximately according to a normal distribution with mean 100 and standard deviation 15. What proportion of the population has a value above 130? Below 85? Above 85?

**18.10** The data in the table below pertain to milk production over an 8-month lactation among 300 two-year-old Jersey cows.

| Pounds of milk produced | Number of cows |
|---|---|
| 2500–3000 | 23 |
| 3000–3500 | 60 |
| 3500–4000 | 58 |
| 4000–4500 | 67 |
| 4500–5000 | 54 |
| 5000–5500 | 23 |
| 5500–6000 | 15 |

The data have been grouped by intervals, but for purposes of computation, each cow may be treated as though the milk production were equal to the midpoint of the interval (for example, 2750 for cows in the 2500–3000 interval).

**(a)** Estimate the mean, variance, and standard deviation in milk yield.

**(b)** What range of yield would be expected to include 68 percent of the cows?

**(c)** What range of yield would be expected to include 95 percent of the animals.

**18.11** In the $F_2$ generation of a cross of two cultivated varieties of tobacco, the number of leaves per plant was distributed according to a normal distribution with mean 18 and standard deviation 3. What proportion of the population is expected to have the following phenotypes?

**(a)** between 15 and 21 leaves

**(b)** between 12 and 24 leaves

**(c)** fewer than 15 leaves

**(d)** more than 24 leaves

**(e)** between 21 and 24 leaves

**18.12** Two highly homozygous inbred strains of mice are crossed and the 6-week weight of the $F_1$ progeny determined.

**(a)** What is the magnitude of the genotypic variance in the $F_1$ generation?

**(b)** If all alleles affecting 6-week weight were additive, what would be the expected mean phenotype of the $F_1$ generation compared with the average 6-week weight of the original inbred strains?

**18.13** In a cross between two cultivated inbred varieties of tobacco, the variance in leaf number per plant is 1.46 in the $F_1$ generation and 5.97 in the $F_2$ generation. What are the genotypic and environmental variances? What is the broad-sense heritability in leaf number?

**18.14** A genetically heterogeneous population of wheat has a variance in the number of days to maturation of 40, whereas two inbred populations derived from it have a variance in the number of days to maturation of 10.

**(a)** What is the genotypic variance, $\sigma_g^2$, the environmental variance, $\sigma_e^2$, and the broad-sense heritability, $H^2$, of days to maturation in this population?

**(b)** If the inbred lines were crossed, what would be the predicted variance in days to maturation of the $F_1$ generation?

**18.15** Maternal effects are nongenetic influences on offspring phenotype that derive from the phenotype of the mother. For example, in many mammals, larger mothers have larger offspring, in part because of a maternal effect on birth weight. What result would a maternal effect have on the correlation in birth weight between mothers and their offspring compared with that between fathers and their offspring? How would such a maternal effect influence the estimate of narrow-sense heritability?

**18.16** In terms of the narrow-sense heritability, what is the theoretical correlation coefficient in phenotype between first cousins who are the offspring of monozygotic twins?

**18.17** If the correlation coefficient of a trait between first cousins is 0.09, what is the estimated narrow-sense heritability of the trait?

**18.18** The asexual unicellular protozoan *Difflugia* has a number of well-formed and easily counted teeth encircling the region of the cell that functions as the mouth. The cells also have a variable number of spiny projections. The correlation coefficients between a parent and its offspring in a genetically heterogeneous population are 0.956 and 0.287 for tooth number and length of longest spine, respectively.

**(a)** Estimate the broad-sense heritabilities of these traits.

**(b)** In view of the fact that parents and offspring in asexual organisms are related as identical twins in sexual organisms, how can the large difference in broad-sense heritability be explained?

**18.19** In an experiment focusing on weight gain between ages 3 and 6 weeks in mice, the difference in mean phenotype between two strains was 17.6 grams and the genotypic variance was estimated as 0.88. Estimate the minimum number of genes affecting this trait.

**18.20** Estimate the minimum number of genes affecting fruit weight in a population of the domestic tomato produced by crossing two inbred strains. When average fruit weight is expressed as the logarithm of fruit weight in grams, the inbred lines have average fruit weights of –0.137 and 1.689. The $F_1$ generation had a variance of 0.0144, and the $F_2$ generation had a variance of 0.0570.

**18.21** Artificial selection for fruit weight is carried out in a tomato population. From each plant, all the fruits are weighed, and the average weight per tomato is regarded as the phenotype of the plant. In the population as a whole, the average fruit weight per plant is 75 grams. Plants whose fruit weight averages 100 grams per tomato are selected and mated at random to produce the next generation. The narrow-sense heritability of average fruit weight in this population is estimated to be 20 percent.

**(a)** What is the expected average fruit weight among plants after one generation of selection?

**(b)** What is the expected average fruit weight among plants after five consecutive generations of selection when, in each generation, the chosen parents have an average fruit weight 25 grams above the population average in that generation? Assume that the narrow-sense heritability remains constant during this time.

**18.22** In a selection experiment for increased plasma cholesterol levels in mice, parents with a mean level of 2.37 units were selected from a population with a mean of 2.26 units, and the progeny of the selected parents had an average level of 2.33 units. Estimate the narrow-sense heritability of this trait from these data.

**18.23** A mouse population has an average weight gain between ages 3 and 6 weeks of 12 grams, and the narrow-sense heritability of the weight gain between 3 and 6 weeks is 20 percent.

**(a)** What average weight gain would be expected among the offspring of parents whose average weight gain was 16 grams?

**(b)** What average weight gain would be expected among the offspring of parents whose average weight gain was 8 grams?

**18.24** To estimate the heritability of maze-learning ability in rats, a selection experiment was carried out. From a population in which the average number of trials necessary to learn the maze was 10.8, with a variance of 4.0, animals were selected that managed to learn the maze in an average of 5.8 trials. Their offspring required an average of 8.8 trials to learn the maze. What is the estimated narrow-sense heritability of maze-learning ability in this population?

**18.25** A replicate of the population in the preceding problem was reared in another laboratory under rather different conditions of handling and other stimulation. The mean number of trials required to learn the maze was still 10.8, but the variance was increased to 9.0. Animals with a mean learning time of 5.8 trials were again selected, and the mean learning time of the offspring was 9.9.

**(a)** What is the heritability of the trait under these conditions?

**(b)** Is this result consistent with that in Problem 18.24 and how can the difference be explained?

**18.26** A representative sample of lamb weights (in pounds) at the time of weaning in a large flock is shown below. If the narrow-sense heritability of weaning weight is 20 percent, and the half of the flock consisting of the heaviest lambs is saved for breeding for the next generation, what is the best estimate of the average weaning weight of the progeny? (*Note*: If a normal distribution has mean $\mu$ and standard deviation $\sigma$, then the mean of the upper half of the distribution is given by $\mu + 0.8\sigma$.)

| | | | | |
|---|---|---|---|---|
| 81 | 81 | 83 | 101 | 86 |
| 65 | 68 | 77 | 66 | 92 |
| 94 | 85 | 105 | 60 | 90 |
| 94 | 90 | 81 | 63 | 58 |

**18.27** A quantitative trait is affected by 20 genetically unlinked, additive loci at which, for each locus, the presence of 1 or 2 favorable alleles adds 1 or 2 units to the phenotype, respectively. A population is segregating for all 20 loci with allele frequencies equal to 1/2 for both the favorable allele and the unfavorable allele at each locus.

**(a)** What is the mean phenotype in this population?

**(b)** If artificial selection were carried out until the favorable allele at each locus was fixed, what would the phenotype of the selected population be?

**(c)** In the original population, what is the expected frequency of the genotype that is homozygous for the favorable allele at every locus?

**(d)** How many individuals would have to be present in the original population so that one individual would be expected that is homozygous for all favorable alleles? (This calculation shows why artificial selection can result in a selected population whose mean phenotypic value is much greater than that of any genotype found in the original population.)

**18.28** For a phenotype determined by a single, completely penetrant recessive allele at frequency $q$ in a random-mating population, it can be shown that the narrow-sense heritability is equal to $2q/(1+q)$.

**(a)** Calculate the narrow-sense heritability for $q =$ 1.0, 0.5, 0.1, 0.05, 0.01, 0.005, and 0.001.

**(b)** Note that the narrow-sense heritability goes to 0 as $q$ goes to 0, yet the phenotype is completely determined by heredity. How can this be explained?

**18.29** In artificial selection, the deviation of the mean of the selected parents from the mean of the population ($M^* - M$) is often called the *selection differential* and is symbolized $S$. If artificial selection is carried out with a constant selection differential of $S$ per generation, and $h^2$ remains constant, show that after $t$ generations of selection, the mean of the selected population is given by $M_t - M_0 = th^2S$, where $M_0$ is the mean of the original population.

**18.30** With reference to the selection differential $S$ defined in the previous problem, what is the corresponding formula for $M_t$ if the selection differential can change and in generation $i$ is symbolized as $S_i$?

## CHALLENGE PROBLEMS

**Challenge Problem 1** A quantitative trait is determined by three independently segregating, completely additive genes in a randomly mating population. Alleles *A*, *B*, and *C* are favorable for the trait. Genotype *aa bb cc* has a phenotype of 0, and in every other genotype one unit in phenotype is added for each favorable allele that is present. Consider a random-mating population in which the allele frequencies are 0.5 for the favorable allele and the unfavorable allele at each locus.

**(a)** What is the mean phenotype in the population?

**(b)** If artificial selection is carried out such that the selected parents have a phenotype of either 5 or 6, what is the mean phenotype among the selected parents?

**(c)** What are the allele frequencies among the selected parents?

**(d)** If the selected parents are mated at random, what is the expected mean phenotype among the offspring?

**(e)** Do these results conform to the formula $M' = M + h^2(M^* - M)$?

**(f)** What is the value of $h^2$? Is this value unexpected?

**Challenge Problem 2** Two varieties of chickens differ in the average number of eggs laid per hen per year. One variety yields an average of 300 eggs per year, the other an average of 270 eggs per year. The difference is due to a single quantitative-trait locus (QTL) located between two restriction fragment length polymorphisms. The genetic map is as shown here.

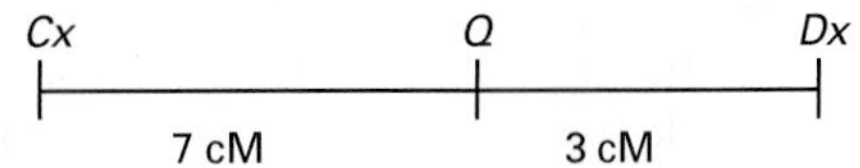


In this diagram, *Cx* and *Dx* are the locations of the restriction fragment length polymorphisms, and *Q* is the location of the QTL affecting egg production. The map distances are given in centimorgans (cM). The *Cx* alleles are denoted *Cx4.3* and *Cx2.1* according to the size of the restriction fragment produced by digestion with a particular restriction enzyme, and the *Dx* alleles are denoted *Dx3.2* and *Dx1.1*. The allele *Q* is associated with the higher egg production and *q* with the lower egg production, with the *Qq* genotype having an annual egg production equal to the average of the genotypes *QQ* (300 eggs) and *qq* (270 eggs). A cross of the following type is made:

*Cx4.3 Q Dx1.1* / *Cx2.1 q Dx3.2* × *Cx4.3 q Dx1.1* / *Cx2.1 q Dx3.2*

Female offspring are classified into the following four types:

**(a)** *Cx4.3 Dx1.1 / Cx2.1 Dx3.2*
**(b)** *Cx4.3 Dx3.2 / Cx2.1 Dx3.2*
**(c)** *Cx2.1 Dx1.1 / Cx2.1 Dx3.2*
**(d)** *Cx2.1 Dx3.2 / Cx2.1 Dx3.2*

What is the expected average egg production of each of these genotypes? You may assume complete chromosome interference across the region.

**Challenge Problem 3** The pedigree shown on the next page includes information about the alleles of three genes that are segregating.

• Gene 1 has four alleles denoted *A, B, C,* and *D*. Each allele is identified according to the electrophoretic mobility of a DNA restriction fragment that hybridizes with a probe for gene 1. The pattern of bands observed in each individual in the pedigree is shown in the diagram of the gel below the pedigree. When an allele is homozygous, the band is twice as thick as otherwise, because homozygous genotypes yield twice as many restriction fragments corresponding to the allele than do homozygotes. The small table at the upper right of the pedigree gives the frequencies of the gene 1 alleles in the population as a whole.

• Gene 2 has four alleles denoted *E, F, G,* and *H*. These also are identified by the electrophoretic mobility of restriction fragments. For each person in the pedigree, the genotype with respect to gene 2 is indicated. The population frequencies of the gene-2 alleles are also given in the table at the upper right.

• Gene 3 is associated with a rare genetic disorder. Affected people are indicated by the red symbols in the pedigree. The genotype associated with the disease has complete penetrance. There are two alleles of gene 3, which are denoted *Q* and *q*.

**(a)** What genetic hypothesis can explain the pattern of affected and nonaffected individuals in the pedigree?
**(b)** If the male IV-2 fathers a daughter, what is the probability that the daughter will be affected?
**(c)** Assuming random mating, if the male IV-2 fathers a daughter, what is the probability that the daughter will have the genotype *HF* for gene 2?
**(d)** Identify the genotype of each person in the pedigree with respect to genes 1, 2, and 3. Wherever possible, specify the linkage phase (coupling or repulsion) of the alleles present in each individual.
**(e)** Examine the data for evidence of linkage. Are any of the pairs of genes linked? On the basis of this pedigree, what is the estimated recombination frequency between gene 1 and gene 2? between gene 2 and gene 3?

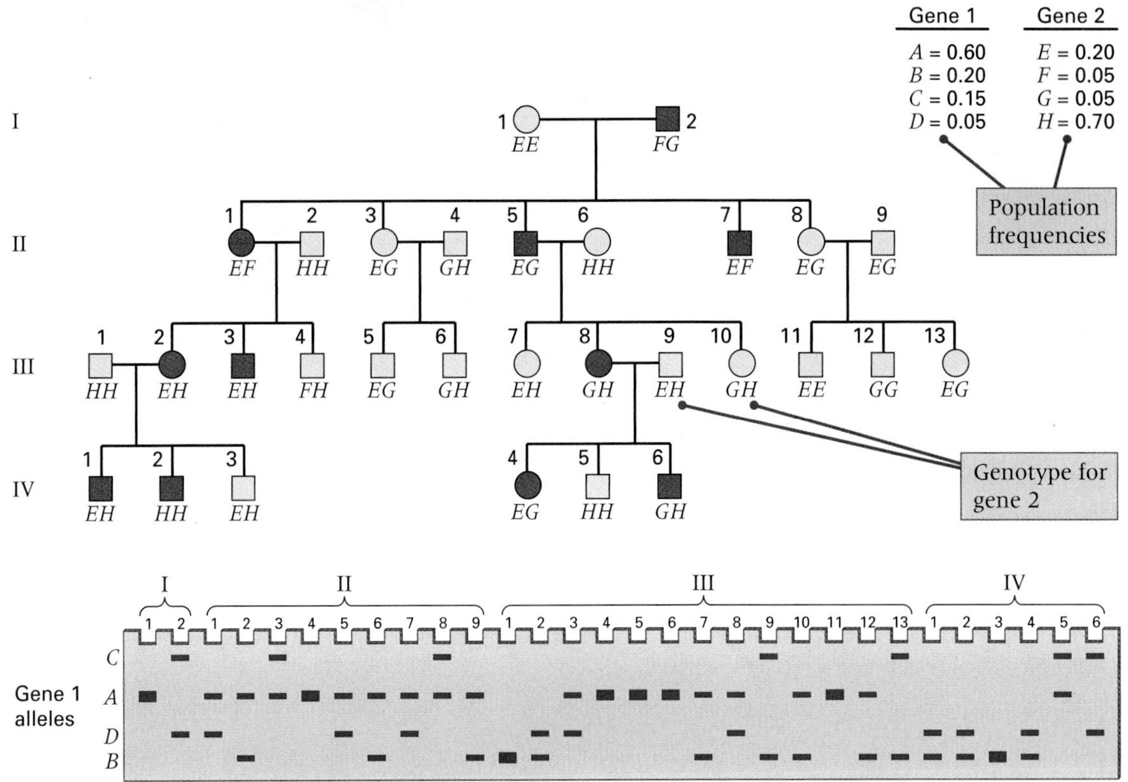


Electrophoresis pattern when DNA from each individual is probed for gene 1

## GENETICS *on the web*

GeNETics on the Web will introduce you to some of the most important sites for finding genetics information on the Internet. To explore these sites, visit the Jones and Bartlett companion site to accompany *Genetics: Analysis of Genes and Genomes, Seventh Edition* at http://biology.jbpub.com/book/genetics.

There you will find a chapter-by-chapter list of highlighted keywords. When you select one of the keywords, you will be linked to a Web site containing information related to that keyword.

# Answers to Even-Numbered Problems

## CHAPTER 1

**1.2** In regard to replication, they inferred that each strand of the double helix was used as a template for the formation of a new daughter strand having a complementary sequence of bases. In regard to coding capability, they noted that genetic information could be coded by the sequence of bases along the DNA molecule, analogous to letters of the alphabet printed on a strip of paper. Finally, in regard to mutation, they noted that changes in genetic information could result from errors in replication, and the altered nucleotide sequence could be then perpetuated.

**1.4** A mixture of heat-killed S cells and living R cells was able to cause pneumonia in mice, but neither heat-killed S cells alone nor living R cells alone could do so.

**1.6** Phenol would not destroy the transforming activity, strong alkali would.

**1.8** Because the mature T2 phage contains only DNA and protein; the labeled RNA was left behind in the material released by the burst cells.

**1.10** In this bacteriophage DNA the amount of A does not equal that of T, and the amount of G does not equal that of C. You can therefore deduce that the DNA molecule is single stranded, not double stranded.

**1.12** 5′-CTGAT-3′

**1.14** The complementary strand has the sequence 5′-TACTACTAC. . .-3′.

**1.16** 5′-AUCAG-3′

**1.18** The complementary sequence is 5′-AUACGAUA-3′

**1.20** The codon for leucine must be 5′-UUA-3′.

**1.22** Six possible reading frames would need to be examined. Either DNA strand could be transcribed, and each transcript could be translated in any one of three reading frames

**1.24** The result means that an mRNA is translated in non-overlapping groups of three nucleotides: the genetic code is a triplet code.

**1.26** The codon 5′–UGG–3′ codes for Trp, and in this random polymer the Trp codon is expected with a frequency of $1/4 \times 3/4 \times 3/4 = 9/64$. The amino acid Val could be specified by either 5′–GUU–3′ or 5′–GUG–3′; the former has an expected frequency of $3/4 \times 1/4 \times 1/4 = 3/64$ and the latter of $3/4 \times 1/4 \times 3/4 = 9/64$, totaling 12/64 or 3/16. The amino acid Phe could be specified only by 5′–UUU–3′ in this random polymer, and so Phe would have an expected frequency of $(1/4)^3 = 1/64$

**1.28**

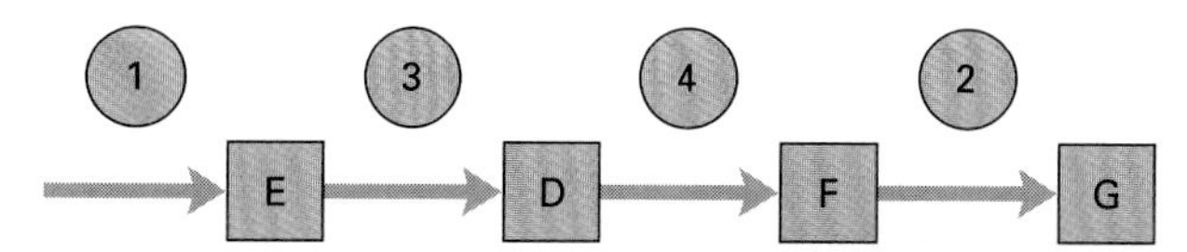


## CHAPTER 2

**2.2** (e) is the 5′ carbon and carries the phosphate group; (c) is the 3′ carbon and carries the hydroxyl group

**2.4** A DNA palindrome reads the same forward and backward, but along opposite strands. **a.** This is a palindrome because the partner strand reads 3′-GGCC-5′; **b.** No, because the partner strand reads 3′-AAAA-5′; **c.** Yes, because the partner strand reads 3′-CGATCG-5′, **d.** No, because the partner strand reads 3′-GGCGAG-5′; **e.** No, because the partner strand reads 3′-TTCCAA-5′.

**2.6** **a.** *Sca*I produces blunt ends; **b.** *Nhe*I produces a 5′ overhang; **c.** *Cfo*I produces a 3′ overhang.

**2.8**

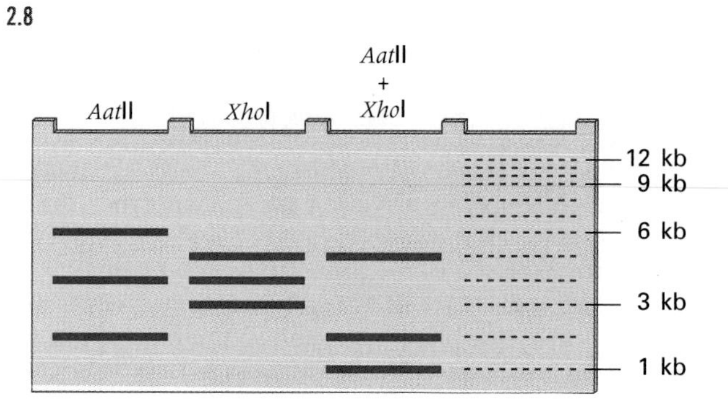


**2.10** **a.** $B_5$ = T; **b.** $B_6$ = G; **c.** $B_7$ = the complementary pyrimidine; **d.** $B_8$ = T or A.

**2.12** The probability of a restriction site is $1/4^n$, where $n$ is the number of nucleotides in the site. **a.** $1/4^4 = 0.0039$; **b.** $1/4^6 = 0.00024$; **c.** $1/4^8 = 0.000015$.

**2.14** **a.** $4.6 \times 10^6/256 \approx 1.8 \times 10^4$; **b.** $4.6 \times 10^6/4096 \approx 1.1 \times 10^3$; **c.** $4.6 \times 10^6/65536 \approx 70$.

**2.16** There are two *Kas*I sites in the original molecule (shown in red).

```
5'-CTGGGGCGCCCTCGTCAGCGAGGGGGCGCCGAT-3'
3'-GACCCCGCGGGAGCAGTCGCTCCCCCGCGGCTA-5'
```

Cleavage produces the fragments:

```
5'-CTGGG       -3'
3'-GACCCCGCG -5'

5'-GCGCCCTCGTCAGCGAGGG          -3'
3'-      GGAGCAGTCGCTCCCCCGCGG -5'

5'-GCGCCGAT-3'
3'-      GCTA-5'
```

If the fragments come back together in the same order as originally and the breaks are sealed, the middle fragment could be present in either the original orientation or reversed, since the ends are complementary. Hence, the possible products are

5′-CTGGGGCGCCCTCGTCAGCGAGGGGGCGCCGAT-3′
3′-GACCCCGCGGGAGCAGTCGCTCCCCGCGGGCTA-5′

or

5′-CTGGGGGCGCCCCTCGCTGACGAGGGCGCCGAT-3′
3′-GACCCCCGCGGGGAGCGACTGCTCCCGCGGCTA-5′

**2.18** Cleavage with *Pst*I destroys the *Pvu*II site, but cleavage with *Pvu*II leaves the *Pst*I site intact.

**2.20** Probe A yields bands of 4 and 12 kb, probe B yields bands of 8 and 12 kb.

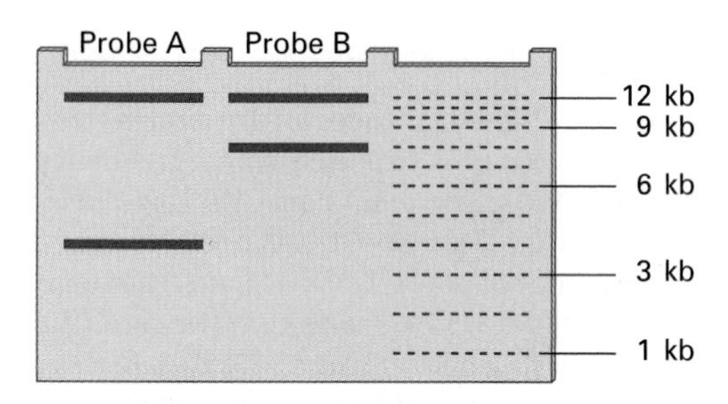


**2.22** This restriction enzyme and probe would yield an RFLP with two alleles, one (call it $A_1$) identified by a 6-kb band and the other (call it $A_2$) identified by a 10-kb band. The possible genotypes are $A_1A_1$ (homozygous), $A_1A_2$ (heterozygous), and $A_2A_2$ (homozygous), whose phenotypes are, for $A_1A_1$, one band of 6 kb; for $A_1A_2$, bands of 6 kb and 10 kb; and, for $A_2A_2$, one band of 10 kb.

**2.24** In this case three alleles would be detected. Let $A_1$ represent the allele that yields the DNA molecule at the top, $A_2$ that in the middle, and $A_3$ that at the bottom. The alleles differ in the size of the restriction fragment that hybridizes with the probe. $A_1$ yields a 3 kb fragment, $A_2$ a 5 kb fragment, and $A_3$ a 6 kb fragment. With three alleles there are 6 possible genotypes. The genotypes and phenotypes (bands observed) are as follows: $A_1A_1$ (3 kb band only), $A_1A_2$ (3 kb + 5 kb), $A_2A_2$ (5 kb only), $A_1A_3$ (3 kb + 6 kb), $A_2A_3$ (5 kb + 6 kb), and $A_3A_3$ (6 kb only).

**2.26** No, they would not amplify the same fragment because these primers have the wrong polarity. These primers would not pair with any regions in the target DNA strands.

**2.28** A RAPD polymorphism is an amplified band that appears in some, but not all, samples of individuals, hence in this example bands 3, 5, 7, and 10 are RAPD polymorphisms.

**2.30** For each locus, the child must share one band with the mother and one with the father. For locus 1, the larger band is shared with the mother, and the allele yielding the smaller band must have come from A. Locus 1 therefore rules out male B as the father. For locus 2, the smaller band in the child is shared with the mother, and the allele for the larger band could have come from either A or B. Hence locus 2 does not rule out either male. Altogether, because of locus 1, the evidence rules out B but does not rule out A as the possible father.

## CHAPTER 3

**3.2** 3/4

**3.4** $3 \times 2 \times 2 \times 1 \times 3 = 36$

**3.6** 2/3

**3.8** Evidently the 7 kb band is a marker for the recessive allele. **a.** The expected phenotype of II-1 is a single band of 7 kb. **b.** Because II-2 is not affected, the possible molecular phenotypes are a single band of 3 kb or two bands of 3 kb and 7 kb.

**3.10** **a.** Because the trait is rare, it is reasonable to assume that the affected father is heterozygous *HD*/*hd*, where *hd* represents the normal allele. Half of his gametes contain the *HD* allele, so the probability is 1/2 that his son received the allele and will later develop the disorder. **b.** We do not know whether the young man is heterozygous *HD*/*hd*, but the probability is 1/2 that he is; if he is heterozygous, half of his gametes will carry the *Hd* allele. Therefore, the overall probability that his child has the *HD* allele is $(1/2) \times (1/2) = (1/4)$.

**3.12** **a.** The trait is most likely to be due to a recessive allele because there is consanguinity (mating between relatives) in the pedigree. **b.** The double line indicates consanguineous mating. **c.** III-1 and III-2 are first cousins. **d.** I-1, I-2, II-2, II-3, III-1, and III-2 are most likely *Aa*, II-1 and II-4 are most likely *AA*.

**3.14** The probability that the parent has the *AA BB* genotype is Pr{*AA BB*} = 1/2. The probability of producing an *A b* gamete, given that a parent has the *AA BB* genotype, is Pr{*A b*|*AA BB*} $= (1/2) \times 0 = 0$. The probability that the parent has the *AA Bb* genotype is Pr{*AA Bb*} = 1/2. The probability of producing an *A b* gamete, given that the parent has the *AA Bb* genotype, is Pr{*A b*|*AA Bb*} $= (1/2) \times (1/2) = 1/4$. Therefore, the overall probability of producing an *A b* gamete is Pr{*A b*}= $(1/2)(0) + (1/2)(1/4) = 1/8$.

**3.16** $(0.002)(1) + (1 - 0.002)(0.002) = 0.004$.

**3.18** Parents have genotypes *WW* and *ww*. All $F_1$ progeny from this cross must be heterozygous *Ww*. The ratio of different phenotypes in the $F_2$ generation is 3/4 round seeds (*WW* + *Ww*) and 1/4 wrinkled (*ww*). Among the $F_2$ round seeds, 2/3 are *Ww* and 1/3 are *WW*. Therefore, among the progeny of the cross *en masse* of (2/3 *Ww* + 1/3 *WW*) × *ww*, the expected proportion with wrinkled seeds will be $(2/3) \times (1/2) \times 1 = 1/3$.

**3.20** The 9 : 3 : 3 : 1 ratio is that of the genotypes *A– B–*, *A– bb*, *aa B–*, and *aa bb*. The modified 9 : 7 implies that all of the last three genotypes have the same phenotype. The $F_2$ testcross is between the genotypes *Aa Bb* and *aa bb*, and the progeny are expected in the proportions 1/4 *Aa Bb*, 1/4 *Aa bb*, 1/4 *aa Bb* and 1/4 *a abb*. The last three genotypes would again have the same phenotype, and so the ratio of phenotypes among the progeny of the testcross is 1 : 3.

**3.22** Since the woman is affected with the trait, she has a probability of 1/2 of passing the dominant allele to her child, and the child will have a probability of 1/2 of actually exhibiting the trait (because the trait shows 50% penetrance). Thus, the probability that the child will be affected is therefore $1/2 \times 1/2 = 1/4$.

**3.24** Use Bayes theorem with event *A* as the event that the $F_1$ individual with the dominant phenotype has the genotype *Nn* (probability 2/3). Event *A′* is that the individual is homozygous dominant *NN* (probability 1/3). Let *B* be the event that the four offspring from the testcross have the dominant phenotype. Then,

$$\Pr\{A \mid B\} = \frac{(1/2)^4(2/3)}{(1/2)^4(2/3) + (1)^4(1/3)} = 1/9 = 0.11$$

**3.26** To avoid ambiguity, we will list the bands in the order to which the alleles are written rather than in order of their size. The $F_1$ progeny ($A_1A_2\ B_1B_2$) each exhibit four bands of 4, 8, 2, and 6 kb. The $F_2$ progeny consist of 1/16 $A_1A_1\ B_1B_1$ (bands of 4 and 6 kb), 2/16 $A_1A_1\ B_1B_2$ (bands of 4, 6, and 2 kb), 1/16 $A_1A_1\ B_2B_2$ (bands of 4 and 2 kb), 2/16 $A_1A_2\ B_1B_1$ (bands of 4, 8, and 6 kb), 4/16 $A_1A_2\ B_1B_2$ (bands of 4, 8, 6, and 2 kb), 2/16 $A_1A_2\ B_2B_2$ (bands of 4, 8, and 2 kb), 1/16 $A_2A_2\ B_1B_1$ (bands of 8 and 6 kb), 2/16 $A_2A_2\ B_1B_2$ (bands of 8, 6, and 2 kb), and 1/16 $A_2A_2\ B_2B_2$ (bands of 8 and 2 kb).

**3.28**

| Paternal/Maternal | 1/2 *A* | 1/2 *a* |
|---|---|---|
| 1/3 *A* | 1/6 *AA* | 1/6 *Aa* |
| 2/3 *a* | 2/6 *Aa* | 2/6 *aa* |

The expected ratio is 1/6 *AA* : 3/6 *Aa* : 2/6 *aa*

## CHAPTER 4

**4.2** Illustration (A) is anaphase II of meiosis, because the chromosome number has been reduced by half. Illustration (B) is anaphase of mitosis, because the homologous chromosomes are not paired. Illustration (C) is anaphase I of meiosis, because the homologous chromosomes are paired.

**4.4** Since the two genes are in different chromosomes (the mutation for phenylkentonuria is autosomal, that for hemophilia X linked), they will assort independently, and so we can consider them separately. Because both parents are heterozygous for an autosomal recessive phenylketonuria mutation, there is a 1/4 chance that their child will be homozyous recessive. For X-linked hemophilia, only a boy could be affected; a girl would have inherited a normal X chromosome from her father, so she could not have hemophilia. So the probability of a girl being affected with both diseases is 0 and the probability of a boy being affected with both diseases is $1/4 \times 1/2 = 1/8$. Overall, the probability is $(1/2)(0) + (1/2)(1/8) = 1/16$.

**4.6** Because the child has two Y chromosomes, the nondisjunction took place in the father, during the second meiotic division.

**4.8** **a.** For each gene, half of the offspring will be heterozygous. Since the three genes undergo independent assortment, the proportions for each can be multiplied to get the expected frequency of offspring heterozygous for all three genes: $(1/2)^3 = 1/8$. **b.** For each gene, half of the offspring will be homozygous, so you can multiply the proportions for each gene to get the expected frequency of offspring homozygous for all three genes: $(1/2)^3 = 1/8$. (Note that the problem does not say that all three genes must be recessive or dominant. So *AABBCC, aabbcc, AAbbcc, AAbbCC.* etc. are all equivalent.)

**4.10** **a.** $(1/2)^6 = 1/64$. **b.** $6(1/6)^6 = 6/64$. **c.** $15(1/6)^6 = 15/64$. **d.** $20(1/6)^6 = 20/64$. **e.** $1 - (1/64) - (6/64) - (15/64) = 42/64$.

**4.12** The observed $F_2$ ratio suggests that two different genes are involved in the determination of bulb color. Moreover, since the ratio departs from the classic Mendelian ratio of 9 : 3 : 3 : 1, it means that the genes interact with each other. (They show epistasis.) Let us use the symbol *R* to denote a dominant allele necessary to develop the red color, and *r* be a recessive allele of the same gene, so that onion bulbs with the genotype *rr* are white. Let us also use the symbol *W* to denote a dominant allele of another gene that suppresses the expression of *R*. On this hypothesis, bulbs with the genotypes *R− W−* are also white. Based on this hypothesis, the initial cross was between an *RR ww* red plant and an *rr WW* white plant. The $F_1$ progeny of this cross are genotypically *Rr Ww* and white. The $F_2$ progeny still have an underlying genotypic ratio of 9 *R− W−* : 3 *R− ww* : 3 *rr W−* : 1 *rr ww*. If onion bulbs with the genotype *rr ww* are yellow, those with the genotype *R− ww* are red, and those with the genotypes *R−W−* and *rr W−* are white, then we can explain the observations because 9 (*R− W−*) + 3 (*rr W−*) = 12.

**4.14** The chi-square value is 4.00.

**4.16** There are 21 possible genotypes for an autosomal gene: 6 homozygous and 15 heterozygous. For an X-linked gene the number of possible genotypes is 27: 6 homozygous and 15 heterozygous genotypes in females, and 6 hemizygous genotypes in males

**4.18** **a.** The probability that their first child will have brown eyes is 3/4 because both parents must be heterozygous. **b.** 2/3. **c.** The probability that this couple will have three children with blue eyes is $(1/4)^3 = 1/64$. The probability that none of the three children will have blue eyes is $(3/4)^3 = 27/64$.

**4.20** For convenience we will use the symbols *W* and *w* to denote the wildtype and mutant alleles, respectively, of the X-linked gene. Let event *A* be that III-1 is heterozygous *Ww*. Let event *B* be that none of IV-1, 2, 3 are affected. The complement of *A*, the event *A′*, is the event that III-1 is homozygous *WW*. Then, $\Pr\{A\} = 1/4$ and $\Pr\{A'\} = 3/4$, since these do not depend on *B*. Furthermore, $\Pr\{B|A'\} = 0$ and $\Pr\{B|A\} = (1/2)^2$. The reason the 1/2 is squared and not cubed is that the female IV-2 is not informative with regard to III-1; in particular, because the male III-2 has genotype *WY*, the female IV-2 will be unaffected no matter what the genotype of III-1. With these expressions in place we can use Bayes theorem as follows:

$$\Pr\{A \mid B\} = \frac{\Pr\{B \mid A\}\Pr\{A\}}{\Pr\{B \mid A\}\Pr\{A\} + \Pr\{B \mid A'\}\Pr\{A'\}}$$

$$= \frac{(1/2)^2(1/4)}{(1/2)^2(1/4) + (1)^2(3/4)} = \frac{1}{13} = 0.077$$

**4.22** For convenience we will use the symbols $W$ and $w$ to denote the wildtype and mutant alleles, respectively, of the X-linked gene. Let event A be that III-1 is heterozygous $Ww$ and let event B be that the individuals IV-1, 2, 3 are all not affected. Since II-2 must be heterozygous $Ww$, then the probability that III-1 is heterozygous is Pr{A} = 1/2. The complement of A is the event A′ that III-1 is homozygous $WW$, and so Pr{A′} = 1/2. Since the male III-2 is affected, and none of IV-1, IV-2, and IV-3 is affected, all of these offspring must have inherited the $W$ allele from III-1. Hence $\Pr\{B|A\} = (1/2)^3$ and Pr{B|A′} = 1. Applying Bayes theorem to this situation we have

$$\Pr\{A \mid B\} = \frac{\Pr\{B \mid A\}\Pr\{A\}}{\Pr\{B \mid A\}\Pr\{A\} + \Pr\{B \mid A'\}\Pr\{A'\}}$$

$$= \frac{(1/2)^3(1/2)}{(1/2)^3(1/2) + (1)(1/2)} = \frac{1}{9} = 0.11$$

**4.24** X-linked inheritance is suggested by the fact that each male has only one band, which is transmitted to his daughters but not his sons. The deduced genotypes are: I-1 ($A_1A_1$), I-2 ($A_2$), II-1 ($A_1$), II-2 ($A_1A_2$), II-3 ($A_1$), II-4 ($A_1$), II-5 ($A_1A_2$), II-6 ($A_2$), III-1 ($A_2$), II-2 ($A_1A_2$), III-3 ($A_1A_2$), and III-4 ($A_1$).

**4.26** The accompanying Punnet square shows the outcome of the cross. The X- and Y-chromosomes from the male are denoted in boldface, the attached X-chromosomes are yoked by a ^ sign.

| Eggs | Sperm X | Sperm Y |
|---|---|---|
| $\hat{XX}$ | $\hat{XX}$ **X** | $\hat{XX}$ **Y** |
| Y | **X** Y | **Y** Y |

The surviving progeny consists of yellow males and wildtype females in equal proportions. Attached-X inheritance differs from the typical situation in that the sons, rather than the daughters, receive their fathers' X-chromosome.

**4.28** The results of the crosses reveal how the three alleles control plumage pattern. If we let $P^r$ be the allele controlling the restricted pattern, then this allele must be dominant to the $P^m$ and $P^d$ alleles responsible for the mallard and dusky patterns, respectively. Also the $P^m$ allele is dominant to $P^d$. **a.** an $F_1$ male from cross 1 has the genotype $P^rP^m$; an $F_1$ female from cross 2 has the genotype $P^mP^d$. Half of their progeny would have the restricted and the other half the mallard plumage patterns. **b.** In this case 3/4 of the progeny would have restricted and 1/4 would have mallard plumage pattern.

**4.30** **a.** All affected individuals, except one (individual 9), are homozygous for both FMF and VNTR allele number 1. This implies that the FMF mutation and VNTR1 are close together in the same chromosome. **b.** Affected individual 9 is heterozygous for the VNTR locus, which can be explained by recombination between the VNTR locus and the FMF locus in one of the parents.

## CHAPTER 5

**5.2** **a, b, c,** and **d** are correct; **e, f,** and **g** are false.

**5.4** If the frequency of recombination is 0.05, the frequency of choosing a nonrecombinant gamete at random is 1 − 0.05 = 0.95.

**5.6** 1 : 2 : 1

**5.8** 100 percent. In real organisms, the maximum frequency of recombination is 50 percent.

**5.10** Three: parental ditype, nonparental ditype, and tetratype.

**5.12** Since we know the frequencies of recombination (0.10 and 0.15) and also that the coefficient of coincidence equals 0.40, we can write the frequency of double crossovers that should be observed as 0.40 × 0.10 × 0.15 = 0.006. These are of two types, *A b C* and *a B c*, and so the expected number among 1000 gametes is 3 of each. In the *a–b* interval, the expected proportion of single recombinants equals 0.10 − 0.006 = 0.094. (The trick here is to remember to subtract the double crossovers.) There are two types of single crossovers in this interval, *A b c* and *a B C*, and among 1000 gametes the expected number of each is 94/2 = 47. Similarly, in the *b–c* interval, the expected proportion of single recombinants equals 0.15 − 0.006 = 0.144. In this case the two types are *A B c* and *a B C*, and each has an expected number of 144/2 = 72. The remaining gametes, of which there are 1000 − 6 − 94 − 144 = 756, are expected to be divided equally between the nonrecombinant *A B C* and *a b c* types, for an expected number of 756/2 = 378 of each.

**5.14** **a.** This is a 3 : 1 ratio, which means that the $F_1$ parent is homozygous for one gene, in this case *R*, and heterozygous for the other. The genotype of the $F_1$ parental plant is therefore *R T / R t*; **b.** This is also a 3 : 1 ratio, but in this case the genotype of the $F_1$ parental plant is *r T / r t*; **c.** Here again is a 3 : 1 ratio, and the genotype of the $F_1$ parental plant is *R T / r T* **d.** In this case there are four classes of progeny, and so we can assume that $F_1$ parent was doubly heterozygous. We know that the genes are linked, which explains why the ratio is different from 9 : 3 : 3 : 1. The observed distribution of phenotypes in the $F_2$ generation can be explain by the $F_1$ plant having a genotype *R T / r t*. Although it is not obvious from the data, the frequency of recombination is 20%. You can verify this for yourself with a Punnett square.

**5.16** For each pair of genes, classify the tetrads as PD, NPD and TT, and tabulate the results as follows:

*leu2 – trp1* PD 230 + 235 = 465
NPD 215 + 220 = 435
TT 54 + 46 = 100

For this pair of genes, PD = NPD, therefore, *leu2* and *trp1* are unlinked.

*leu2 – met14* PD 230 + 235 = 465
NPD 235 + 220 = 455
TT 54 + 46 = 100

For this pair of genes, PD = NPD, therefore, *leu2* and *met14* are unlinked.

*trp1 – met14* PD 235 + 215 + 46 = 496
NPD 230 + 220 + 54 = 504
TT = 0

For this pair of genes, PD = NPD and therefore, *trp1* and *met14* are also unlinked. However, there are no TT for *trp1* and *met14*, which means that both genes are closely linked to their respective centromeres. Because they are so close to their centomeres, the segregation of *trp1* and *met14* can be used as genetic markers of centromere segregation, which allows us to map the *leu2* gene with respect to its centromere. Because the other two genes do not recombine with their centromeres, all TT for *leu2* will result from recombination of *leu2* with its centromere. The map distance between *leu2* and its centromere is therefore (1/2) × (100/1000) × 100 = 5 cM

**5.18** **a.** The most frequent classes of the progeny imply that the genotype of the heterozygous parent is *F g H* / *f G h*, with the gene order still to be determined. **b.** Comparison of the parental classes with the double-recombinants (least common) identifies the gene in the middle, which is *G* in this case. Thus, the correct linear order of the genes is *F G H* or *H G F*. **c.** The map distance between *F* and *G* is (74 + 6)/1000 = 0.08. The map distance between *G* and *H* is (114 + 6)/1000 = 0.12. Hence the map distance between the middle gene *G* and its nearest neighbor *F* is 0.08 = 8% = 8 map units = 8 cM. **d.** The map distance between the middle gene *G* and its farthest neighbor *H* is 0.12 = 12% = 12 map units = 12 cM. **e.** The expected number of double crossovers is the product of the frequencies of recombination in each region and the total number of progeny: 0.08 × 0.12 × 1000 = 9.6. **f.** The coincidence *c* is the observed number of double crossovers over the expected or 6/9.6 = 0.625. The interference is *i* = (1 – *c*) = 1 – 0.625 = 0.375

**5.20** The frequency of *dpy*-21 *unc*-34 recombinants is expected to be 0.24/2 = 0.12 among both eggs and sperm, so the expected frequency of *dpy*-21 *unc*-34/ *dpy*-21 *unc*-34 zygotes is 0.12 × 0.12 = 1.44 percent.

**5.22** Mutations 1, 2 and 7 form one complementation group, mutation 2 is the only representative of another complementation group, mutations 4 and 6 make up a third complementation group, mutations 5 and 9 form a fourth complementation group, and a fifth complementation group is represented by mutation 8.

**5.24** For determining linkage to the centromere, consider each gene from the standpoint of first- or second-division segregation. Gene *c* gives (159 + 116) = 275 asci with first-division segregation and (83 + 41 + 1) = 125 with second-division segregation. The distance between *c* and its centromere is therefore (1/2) × (125/400) × 100 = 15.6 map units. Gene *v* gives (159 + 83 + 116 + 1) = 359 asci with first-division segregation and 41 with second-division segregation. The distance between *v* and its centromere is therefore (1/2) × (41/400) × 100 = 5.1 map units. Gene *a* shows (159 + 83 + 41) = 283 asci with first-division segregation and 116 with second-division segregation, hence the distance between *a* and its centromere is (1/2) × (116/400) × 100 = 14.5 map units.

**5.26** **a.** The coat-color gene cannot be linked to the gene with alleles $A_1$ and $A_2$, because this gene is autosomal and the coat-color gene is X-linked. (One indication that the $A_1$, $A_2$ allele pair is autosomal is that some males exhibit two both bands and so are heterozygous. **b.** The coat-color gene is linked to the gene with alleles $B_3$, and $B_4$, because most progeny from a doubly heterozygous mother carry a nonrecombinant X chromosome. **c.** There are six recombinant genotypes, namely, numbers 2, 6, 11, 16, 32, and 34. **d.** There are a total of 30 offspring in whom recombination in the mother could be detected, and so the estimated frequency of recombination is 6/30 = 20 percent.

**5.28** **a.** Individual I-2 is homozygous for both the recessive allele *a* and the a restriction-fragment length polymorphism (RFLP) allele $M_2$. If individual I-1 has the coupling configuration (*C*), then *a* is in the same chromosome as $M_2$. Then for II-1 to be homozygous for $M_2$ and for *a* (we know her genotype is *aa* because she is affected), there would have to be no recombination between the two loci. If the recombination frequency is *r*, the frequency of no recombination equals 1 – *r*. The similar argument is valid for II-2. Since he is heterozygous for the RFLP marker and heterozygous *Aa* (we know his genotype is *Aa* because he is not affected), he must have inherited an $M_1$ *A* chromosome from I-1, which is also nonrecombinant and therefore a frequency of 1 – *r* as well. Altogether, the probability of the pedigree given that I-1 has the coupling configuration is $(1 - r)^2$. On the other hand, if I-1 has the repulsion configuration (*R*), then the gametes inherited by II-1 and II-2 would both have to be recombinant, and so the probability of the pedigree given that I-1 has the repulsion configuration is $r^2$.

Now let's apply Bayes theorem:

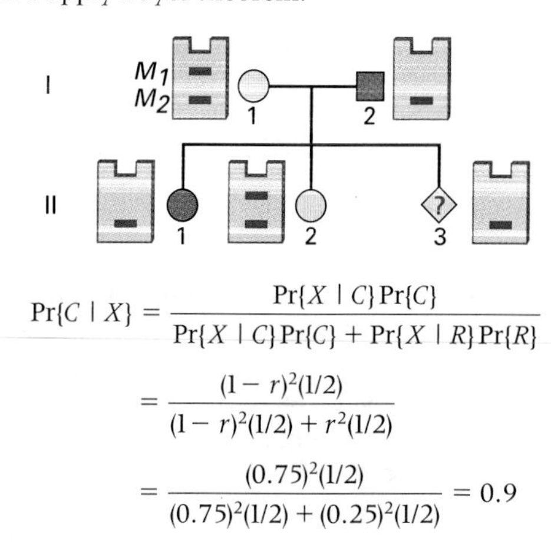


$$\Pr\{C \mid X\} = \frac{\Pr\{X \mid C\}\Pr\{C\}}{\Pr\{X \mid C\}\Pr\{C\} + \Pr\{X \mid R\}\Pr\{R\}}$$

$$= \frac{(1-r)^2(1/2)}{(1-r)^2(1/2) + r^2(1/2)}$$

$$= \frac{(0.75)^2(1/2)}{(0.75)^2(1/2) + (0.25)^2(1/2)} = 0.9$$

**b.** Because the coupling and repulsion configurations are mutually exclusive,

$$\Pr\{R \mid X\} = 1 - \Pr\{C \mid X\} = 1 - 0.9 = 0.1.$$

**c.** From the gel pattern, we see that II-3 inherited the $M_2$ allele. So we need to determine the probability that II-3 also inherited the *a* allele. This would occur without recombination for coupling and with recombination for repulsion. Using the law of total probability and the answers from parts **a** and **b**:

$$\Pr\{Y\} = \Pr\{Y \mid C\}\Pr\{C\} + \Pr\{Y \mid R\}\Pr\{R\} = (0.75)(0.9) + (0.25)(0.1) = 0.7.$$

## CHAPTER 6

**6.2** The 3′-to-5′ exonuclease activity of DNA polymerase has the critically important proofreading function. If this error-correcting mechanism is incapacitated, the frequency of mutations resulting from the incorporation of wrong nucleotides will increase by a factor of about hundred (see Chapter 14 for more detail).

**6.4** This protein is called a helicase.

**6.6** Rolling circle replication begins with a single-strand break, which produces 5′-P and 3′-OH groups. The polymerase extends the 3′-OH terminus and displaces the 5′-P end. Since the DNA strands are antiparallel, the strand complementary to the displaced strand must be terminated by a 3′-OH group.

**6.8** **c.** density-gradient centrifugation

**6.10** The 5′ carbon carries the group *a*, which is $-CH_2OH$. In a nucleotide, *a* corresponds to one or more phosphate groups; *b* is –H; *c* is –OH; *d* is –H, and *e* corresponds to one of the bases, A, T, G, or C.

**6.12** DNA polymerase requires a single-stranded template, a primer, and all four trinucleotides; it adds nucleotides only to an existing 3′ end. In the molecule on the left, the DNA polymerase could elongate each 3′ end using the other strand as a template. The resulting product would have the sequence

5′-ATGGATCCTTATAACTA-3′
3′-TACCTAGGAATATTGAT-5′

In the molecule on the right, on the other hand, the 3′ ends cannot be elongated, as there is no template. In other words, a 5′ overhang (as in the molecule on the left) can be filled in by DNA polymerase, but a 3′ overhang (as in the molecule on the right) cannot.

**6.14** For the semiconservative mode of replication, after one round of replication all DNA molecules would be $^{15}N$–$^{14}N$; after two rounds half would be $^{15}N$–$^{14}N$ and the other half $^{14}N$–$^{14}N$; and after three rounds of replication, 1/4 would be $^{15}N$–$^{14}N$ and 3/4 would be $^{14}N$–$^{14}N$. For the conservative mode, after one round of replication, half of DNA molecules would be $^{15}N$–$^{15}N$ and another half $^{14}N$–$^{14}N$; after two rounds, 1/4 would be $^{15}N$–$^{15}N$ and 3/4 $^{14}N$–$^{14}N$; and after three rounds, 1/8 $^{15}N$–$^{15}N$ and 7/8 $^{14}N$–$^{14}N$.

**6.16** **a.** In each complete 360 degree turn of the double helix, there are 10 bp of DNA. Hence 800 bp of DNA corresponds to 80 complete turns, so each replication fork makes 80 revolutions per second, and hence $80 \times 60 = 4800$ revolutions per minute. **b.** Two replication forks move away from the origin of replication in opposite directions, which meet when each has moved half-way around the genome. The time required is $(0.5 \times 4.3 \times 10^6)/800$ sec = 2687.5 sec. or 44.79 min.

**6.18** The percentage of adenine in the DNA would have increased by 5 percent.

**6.20** DNA replication could occur nearly normally at 37°C but not at 42°C. At the higher temperature, the mutant would not be able to grow. It would be a lethal.

**6.22** The DNA would not necessarily have the same nearest neighbors. If the species both had random sequences, similar base ratios would indeed result in similar nearest-neighbor frequencies. But consider the following counterexample: One of the species has long stretches of T's alternating with long stretches of G's, whereas the other species has a random sequence. In this case, the base ratios would be the same, but the nearest-neighbor frequencies would be very different.

**6.24** Rolling circle replication begins with a single-stranded nick, which produces a 5′–P and 3′–OH group. DNA polymerase adds nucleotides to 3′–OH end using the intact (unnicked) strand as a template; this growing strand is the leading strand. The displaced strand with the free 5′–P is the lagging strand.

**6.26** If the Holliday junction is resolved with no recombination of the outside markers, one molecule should be "light" and the other "heavy." If the Holliday junction is resolved with recombination of the outside markers, one molecule should be "light" on the left-hand side and "heavy" on the right-hand side, or the other way around; in this case, the density of both molecules would, on the average, be intermediate.

**6.28** In the chromosome on the left, the DNA in each chromatid should have one solid line (labeled strand) and one dashed line (unlabeled strand). In the chromosomes on the right, the darker chromatids should have DNA with both lines solid (both strands labeled), and the lighter chromatid should have one line solid (labeled strand) and one dashed (unlabeled strand).

**6.30** The first (5′) nucleotide is G in sequence (A) and T in sequence (B). In addition, Sequence (B) has an insertion of 5′-AG-3′ between nucleotides 10 and 11 (counting from the 5′ end, at the bottom). Alternatively, you could also say that sequence (A) has a deletion of 5′-AG-3′ between nucleotides 10 and 11. Without knowing the ancestral DNA sequences from which these originated, it is not possible to say whether (B) had an addition or (A) a deletion. In either case, if the insertion/deletion happens to be in a coding region, it would result in a shift in the translational reading frame, and therefore an incorrect amino acid sequence.

**7.2** DNA structure is a long, thin thread, and its total length in most cells greatly exceeds the diameter of a nucleus, even allowing for a large number of fragments per cell. This forces you to conclude that each DNA molecule must be folded back on itself repeatedly.

**7.4** Each nucleosome includes 145 bp, and with a linker size of 55 bp, the total DNA length per nucleosome is 200 bp. This value corresponds to 5 nucleosomes per kb and 250 nucleosomes per 50 kb.

**7.6** No, because the *E. coli* chromosome is circular. There are no termini.

**7.8** The telomers would become longer in each successive generation.

**7.10** In *Drosophila*, the heterochromatin occupies about 25 percent of each chromosome arm nearest the centromere; these regions are not highly replicated in polytene nuclei but clump together to form the chromocenter. The outer 75% of each is chromosome arm consists of euchromatin that is highly replicated and that forms the banded regions.

**7.12** The molecules are of the same length, so the molecule with the greater proportion of G–C base pairs will be more stable, and so will require a higher temperature for denaturation, because each G–C pair has three hydrogen bonds whereas each A–T base pair has only two. Hence the duplex with 8 G–C base pairs will be more stable than the duplex with only 4.

**7.14** 42,000 meters is 42,000,000 millimeters. Scaling this length back down by a factor of 500,000 yields 84 millimeters. A length of 84 millimeters is about the actual length of the fully extended DNA duplex in human chromosome 2.

**7.16** The chromosome becomes successively shorter with each round of DNA replication. The new mutations to *w* result from the successive shortening of the chromosome, when DNA replication can no longer replicate the sequences needed for wildtype expression of the gene.

**7.18** The junction has the sequence shown here

```
5'-TTAGGGTTAGGGTTAGGG-3'
3'-AATCCCAATCCCAATCCC-5'
```

**7.20** **a.** There are 10 base pairs for every turn of the double helix. With four turns of unwinding, $4 \times 10 = 40$ base pairs are broken. **b.** Each twist compensates for the underwinding of one full turn of the helix; hence, there will be four twists.

**7.22** A change of 51 in the percentage of G + C base pairs results in a change of 20.9°C in $T_m$, hence the relationship is $T_m = 78 + (20.9/51)(g - 22)$, or $T_m = 69.0 + 0.41g$ °C.

**7.24** Because $g = (T_m - 69.0)/0.41$, the answers are: **a.** $g =$ 38.3 percent G + C. **b.** g = 43.9 percent G + C.

**7.26** If you double the concentration, renaturation will take place twice as fast because the complementary strands will find each other more rapidly. Hence the $C_0t_{1/2}$ value will remain the same. This result implies that the estimated complexity of a DNA sample does not depend on the initial concentration, since any change in initial concentration is balanced by a corresponding change in the time of renaturation.

**7.28** **a.** The helix is overwound one complete turn. **b.** Supercoiling does take place. **c.** Positive supercoiling relieves overwinding.

**7.30** A schematic $C_0t$ curve showing the characteristics described in the problem is shown in the accompanying illustration.

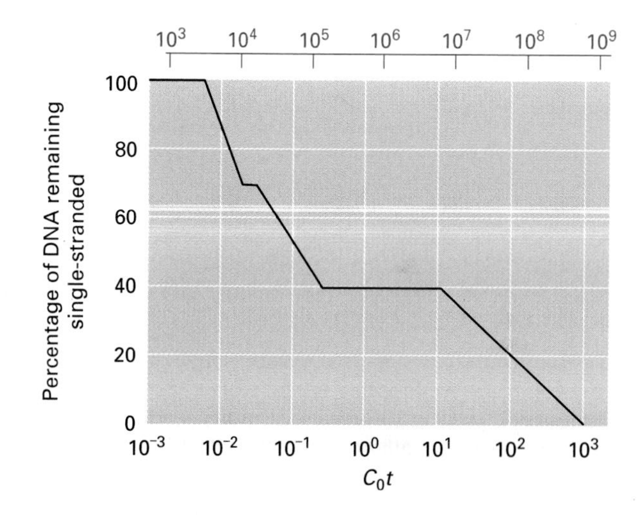




**8.2** The genetic consequence of the obligatory crossover is that alleles in the pseudoautosomal region in the X chromosome can be interchanged with their homologous alleles in the pseudoautosomal region in the Y chromosome, yielding a pattern of inheritance in pedigrees that is indistinguishable from that of ordinary autosomal inheritance.

**8.4** **a.** The Klinefelter karyotype is 47, XXY, hence one Barr body; **b.** the Turner karyotype is 45, X, hence no Barr bodies; **c.** the Down syndrome karyotype is 47, +21; people with this condition have one or zero Barr bodies depending on whether they are female (XX) or male (XY); **d.** males with the karyotype 47, XYY have no Barr bodies; **e.** females with the karyotype 47, XXX have two Barr bodies.

**8.6** The only 45-chromosome karyotype found at appreciable frequencies in spontaneous abortions in 45, X, and so 45, X is the probable karyotype. Had the fetus survived, it would have been a 45, X female with Turner syndrome.

**8.8** Because the X chromosome in the 45, X daughter contains the color-blindness allele, the 45, X daughter must have received the X chromosome from her father through a normal X-bearing sperm. The nondisjunction must therefore have occurred in the mother, resulting in an egg cell lacking an X chromosome.

**8.10** The inversion has the sequence *A B E D C F G*, the deletion *A B F G*. The possible translocated chromosomes are (a) *A B C D E T U V* and *M N O P Q R S F G* or (b) *A B C D E S R Q P O N M* and *V U T F G*. One of these possibilities includes two monocentric chromosomes and the other includes a dicentric and an acentric. Only the translocation with two monocentrics is genetically stable.

**8.12** The mother has a Robertsonian translocation that includes chromosome 21. The child with Down syndrome has 46 chromosomes, including two copies of the normal chromosome 21 plus an additional copy attached to another chromosome (the Robertsonian translocation). This differs from the usual situation, in which Down syndrome children have trisomy 21 (karyotype 47, +21).

**8.14** The genes uncovered by a single deletion must be contiguous. The gene order is deduced from the overlaps between the deletions. The overlaps are *a* and *d* between the first and second deletions and *e* between the second and third. The gene order (as far as can be determined from these data) is diagrammed in part A of the accompanying illustration. The deletions are shown in red. Gene *b* is at the far left, then *a* and *d* (in unknown order), then *c*, *e*, and *f*. (Gene *c* must be to the left of *e*, because otherwise *c* would be uncovered by deletion 3.) Part B is a completely equivalent map with gene *b* at the right. The ordering can be completed with a three-point cross between *b*, *a*, and *d* or between *a*, *d*, and *c* or by examining additional deletions. Any deletion that uncovers either *a* or *d* (but not both) plus at least one other marker on either side would provide the information to complete the ordering.

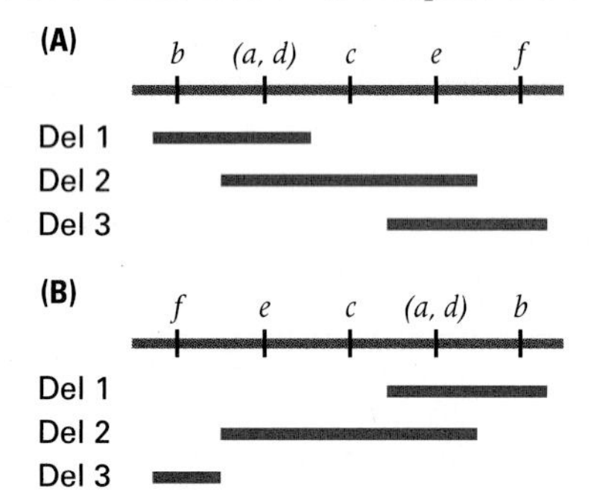


**8.16** The order is (*a e*) *d f c b*, where the parentheses mean that the order of *a* and *e* cannot be determined from these data. The answer *b c f d* (*a e*) is also correct.

**8.18** The group of 4 synapsed chromosomes implies that a reciprocal translocation has taken place. The group of 4 includes both parts of the reciprocal translocation and their nontranslocated homologues.

**8.20** The observation that the cross between these species yields a hybrid in which the chromosomes do not pair indicates that the chromosomes of the two parental species are quite different genetically. One possibility for the origin of *G. hirsutum* is that it arose by polyploidization in such a hybrid, which would account for the 26 chromosome pairs in *G. hirsutum*. The backcross results are also consistent with this hypothesis, because each yields a hybrid with 13 bivalents and 13 univalents. The data therefore suggest that present-day American cultivated cotton originally arose as an allotetraploid from a hybrid between a wild species related to *G. arboreum* and another wild species related to *G. thurberi*.

**8.22** An *AABBBBCC* type of polyploid could be formed by 9 possible types of hybridization and endoreduplication: [(*AA* × *BB*) × (*BB*)] × *CC*)], [(*AA* × *BB*) × (*CC*)] × *BB*)], [(*BB* × *CC*) × (*BB*)] × *AA*)], [(*BB* × *CC* × (*AA*)] × *BB*)], [(*BB* × *BB*) × (*AA*)] × *CC*)], [(*BB* × *BB*) × (*CC*)] × *AA*)], [(*AA* × *CC*) × (*BB*)] × *BB*)], [(*AA* × *BB*) × (*BB* × *CC*)], [(*AA* × *CC*) × (*BB* × *BB*)].

**8.24** In this male, the tip of the X chromosome containing $y^+$ became attached to the Y chromosome, yielding the genotype *y* X / $y^+$ Y. The gametes are *y* X and $y^+$ Y, and so all of the offspring are either yellow females or wildtype males.

**8.26** **a.** The frequency of recombination between the gene for virescent leaves and the translocation breakpoint in chromosome 2 equals (36 + 42)/(36 + 240 + 282 + 42) = 13.0 map units. **b.** If there were no translocation, the expected ratio would be 1 : 1 : 1 : 1.

**8.28** Viable progeny are expected to result only from adjacent segregation, in which the Y chromosme segregates with the normal chromosome 2, and the two parts of the X–2 translocation segregate together. The expected result is 50 percent ruby, dumpy males and 50 percent nonruby, nondumpy females.

**8.30** The unusual yeast strain is disomic for chromosome 2, and the result of the cross is a trisomic. Segregation of markers on chromosome 2, including *his7*, is aberrant, but segregation of markers on the other chromosomes is completely normal.

## CHAPTER 9

**9.2** The final dilution should contain 700 viable cells per ml, which is a $10^6$-fold dilution of the original suspension. You could do any combination of 100-fold and 10-fold dilutions that multiply to $10^6$, but the simplest would be three successive 100-fold dilutions.

**9.4** The T2*h* plaques will be clear because both strains are sensitive; the T2 plaques will be turbid because of the resistance of the B/2 cells.

**9.6** The *E. coli* strain contains 5,000 kilobase pairs, which implies approximately 50 kb per minute. Because the l genome is 50 kb, the genetic length of the prophage is one minute.

**9.8** Plate on minimal medium that lacks leucine (selects for Leu$^+$) but contains streptomycin (selects for Str-r). The *leu* allele is the selected marker, and *str-r* the counterselected marker.

**9.10** **a.** Use $m = 2 - 2d^{1/3}$, in which $m$ is map distance in minutes and $d$ cotransduction frequency. With the values of $d$ given, the map distances are 0.74, 0.41, and 0.18 minutes, respectively. **b.** Use $d = [1 - (m/2)]^3$. With the values of $m$ given, cotransduction frequencies are 0.42, 0.12, and 0, respectively. For markers greater than two minutes apart, the frequency of cotransduction equals zero. (c) Map distance *a–b* equals 0.66 minutes, *b–c* equals 1.07 minutes, and these are additive, giving 1.73 minutes for the distance *a–c*. The predicted cotransduction frequency for this map distance is 0.002.

**9.12** The only gene order consistent with the data is *arg–tau–suc–top–mot*.

**9.14** **a.** Medium containing lactose, tryotophan, histidine, and streptomycin. **b.** Medium containing galactose, tryotophan, histidine, and streptomycin. **c.** Medium containing glucose (or any sugar other than lactose or galactose), histidine, and streptomycin. **d.** Medium containing glucose (or any sugar other than lactose or galactose), tryptophan, and streptomycin.

**9.16** **a** is true because many mating pairs will break apart after the entry of $a^+$ but before the entry of $c^+$; **b** is false because $b^+$ has to enter before $c^+$; **c** is true because many of the $b^+$ *str-r* colonies will be $a^-$ $b^+$ *str-r*; **d** is false because many of the $b^+$ *str-r* colonies will be $a^-$ $b^+$ *str-r*; **e** is true because more recombination events are needed to result in $a^+$ $b^-$ $c^+$ *str-r* than in $a^+$ $b^+$ $c^+$ *str-r* ; **f** is true because more recombination events are needed to result in $a^+$ $b^-$ $c^+$ *str-r* than in $a^-$ $b^+$ $c^+$ *str-r* ; **g** is true because many mating pairs will break apart after the entry of $a^+$ but before the entry of $b^+$.

**9.18** Apparently, *h* and *tet* are closely linked, and so recombinants containing the $h^+$ allele of the Hfr tend to also to contain the *tet-s* allele of the Hfr, and these recombinants are eliminated by the counterselection for *tet-r.*

**9.20** Depending on the size of the F′ plasmid, it could be F′ *g*, F′ *g h*, F′ *g h i*, and so forth, or F′ *f*, F′ *f e*, F′ *f e d*, and so forth.

**9.22** The order 500 $his^+$, 250 $leu^+$, 50 $trp^+$ implies that the genes are transferred in the order $his^+$ $leu^+$ $trp^+$. The *met* mutation is the counterselected marker that prevents growth of the Hfr parent. The medium containing histidine selects for $leu^+$ $trp^+$, and the number is small because both genes must be incorporated by recombination.

**9.24** The time of 15 minutes is too short to allow the transfer of a terminal gene. Thus, one explanation for a $z^+$ *str-r* colony is that aberrant excision of the F factor occurred in an Hfr cell, yielding an F′ plasmid carrying terminal markers. When the F′ $z^+$ plasmid is transferred into an $F^-$ recipient, the genotype of the resulting cell would be F′ $z^+/z^-$ *str-r*. Alternatively, a $z^+$ *str-r* colony can originate by reverse mutation of $z^-$ to $z^+$ in the $F^-$ recipient. The possibilities can be distinguished because the F′ $z^+/z^-$ *str-r* cell would able to transfer the $z^+$ marker to other cells.

**9.26** **a.** Consider each Hfr in turn and, starting with the earliest-entering gene, write down the name of each gene as it enters. The difference in time of entry between adjacent genes is the distance in minutes between the genes. When a partial genetic map of gene transfer from each Hfr has been made, arrange the maps so that any shared markers between two or more Hfr strains coincide. This process yields the genetic maps shown below, where the arrows indicate direction of transfer and the numbers are times of transfer in minutes.

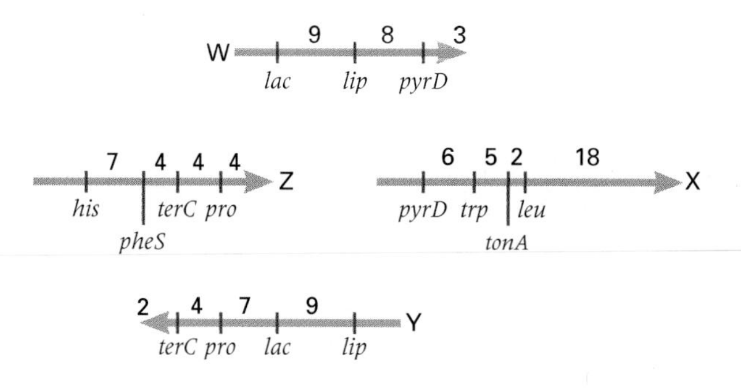


Now arrange the composite genetic map in the manner required, in the form of a circle of 100 minutes, with 0 minutes at the top and *leu* at about 2 minutes. The map is as shown here.

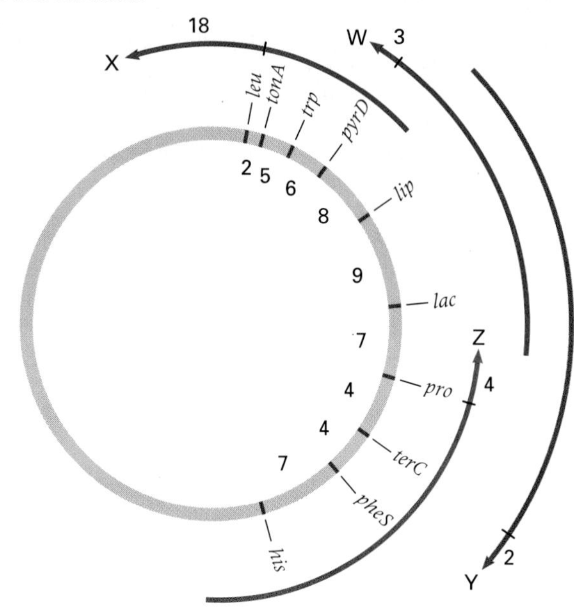


**b.** Comparison of this map with that in Figure 9.13 indicates that a group of markers present in a contiguous segment of chromosome is in the opposite order as compared with the standard map. The simplest explanation is that the strain in question has undergone an inversion of part of the chromosome, with breakpoints in the *tonA–pro* and *trp–terC* intervals in the standard map.

**9.28** The accompanying map shows the genetic intervals defined by the endpoints of the deletions described in Problem 9.27 and the locations of the mutations with respect to these intervals. Also shown are the boundaries of the rIIA and rIIB cistrons as defined by the complementation test in Problem 9.29.

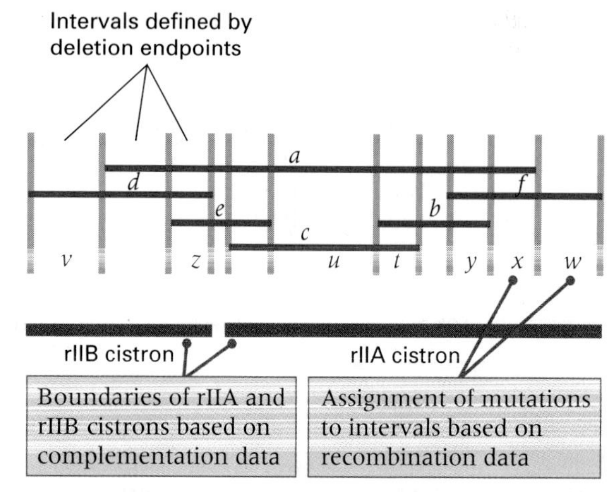


**9.30** The mean number of P2 phage per cell is 3, and the probability of a cell remaining uninfected by P2 is $e^{-3} = 0.050$; the mean number of P4 phage per cell is 2, and the probability of a cell remaining uninfected by P4 is $e^{-2} = 0.135$. **a.** Not be infected with either phage: $0.050 \times 0.135 \times 10^8 = 6.75 \times 10^5$; **b.** Survivors: $6.75 \times 10^5 + 0.050 \times (1 - 0.135) \times 10^8 = 5.00 \times 10^6$. **c.** Producers of P2 progeny: $(1 - 0.050) \times 0.135 \times 10^8 = 1.28 \times 10^7$. **d.** Producers of P1 progeny: $(1 - 0.050) \times (1 - 0.135) \times 10^8 = 8.22 \times 10^7$.

**10.2** **a.** All pyrimidines: Phe, Ser, Leu, and Pro. **b.** All purines: Lys, Arg, Glu, and Gly.

**10.4** **a.** The original DNA must have the sequence of bases 5′–ATG–3′/3′–TAC–5′, where the bottom strand (written to the right of the slash) is the one transcribed; the inversion has the sequence 5′–CAT–3′/3′–GTA–5′, where again the bottom strand is transcribed. The mRNA therefore contains 5′–CAU–3′ (the "reverse complement" of the original codon), which specifies His. **b.** None of the reverse-complement codons are synonymous. This answer can be seen by checking the reverse complement of each codon in turn, or more easily by noting that there are no synonymous codons with a complementary base in the second position. **c.** The chain-termination codons are 5′–UAA–3′, 5′–UAG–3′, and 5′–UGA–3′, which have reverse complements 5′–UUA–3′ (Leu), 5′–CUA–3′ (Leu), and 5′–UCA–3′ (Ser), respectively. **d.** Same answer as in part **c.**

**10.6** Codons are read from the 5′ end of the mRNA. The first amino acid in the growing polypeptide is at amino end. Therefore, if the C were added to the 3′ end, the last codon would be UUC, which also codes for the phenylalanine owing to the degeneracy of the genetic code. On the other hand, adding the C to the 5′ end would make the first codon CUU, which specifies valine (Val) at the amino end of the polyphenylalanine chain.

**10.8** Translation could be in one of three reading frames: AAG AAG AAG …., AGA AGA AGA …., or GAA GAA GAA ….. The first codes for polylysine, the second for polyarginine, and the third for polyglutamic acid.

**10.10** The RNA would be translated as an alternating sequence of the codons AUA–UAU–AUA–UAU– …., which corresponds to a polypeptide with the repeating sequence Ile–Tyr–Ile–Tyr ….

**10.12** In a random sequence there are $2^3 = 8$ possible codons. Their identities, coding capacities, and frequencies are

| | |
|---|---|
| GGG (Gly) | $(1/4)^3 = 1/64$ |
| GGU (Gly) | $(1/4)^2(3/4) = 3/64$ |
| GUG (Val) | $(1/4)^2(3/4) = 3/64$ |
| UGG (Trp) | $(3/4)(1/4)^2 = 3/64$ |
| GUU (Val) | $(1/4)(3/4)^2 = 9/64$ |
| UUG (Leu) | $(3/4)^2(1/4) = 9/64$ |
| UGU (Cys) | $(3/4)^2(1/4) = 9/64$ |
| UUU (Phe) | $(3/4)^3 = 27/64$ |

Therefore, the amino acids expected in a random polymer and their frequencies are glycine (4/64 = 6.25%), valine (12/64 = 18.75%), tryptophan (3/64 = 4.69%), leucine (9/64 = 14.06%), cysteine (9/64 = 14.06%) and phenylalanine (27/64 = 42.19%).

**10.14** **a.** UGYN would be translated as UGY–GYN, which corresponds to Cys–Val (if Y is a T) or Cys–Ala (if Y is a C). Hence, cystein could be followed only by valine or alanine. **b.** UUYN would be translated as Phe–Phe, Phe–Leu, or Phe–Ser. **c.** UGGN would be translated as Trp–Arg. **d.** AUGN would be translated as Met–Cys, Met–Trp (unless N is A, in which case Met is the final amino acid in the polypeptide chain).

**10.16** **a.** AUGN would be translated as AUG–UGN, so the second amino acid could be Cys or Trp. **b.** In this case translation would be as AUG–GNN, so the second amino acid could be Val, Ala, Asp, Glu, or Gly.

**10.18** 5′–UGG–3′ codes for tryptophan. This sequence can pair with either of the anticodons 3′–ACC–5′ or 3′–ACU–5′ anticodons. However, 3′–ACU–5′ can also pair with 5′–UGA–3′, which is a chain-termination codon. If this anticodon were used for tryptophan , the amino acid would be inserted instead of terminating the chain. Therefore, only 3′–ACC–5′ is used.

**10.20** The probability of a correct amino acid at a particular site is $1 - 5 \times 10^{-4} = 0.9995$, and so the probability of all 300 amino acids being correctly translated is $(0.9995)^{300} = 0.86$, or 86%.

**10.22** If transcribed from left to right, the transcribed strand is the bottom one, and the mRNA sequence is AGA CUU AGC GCU AAA CGU GGU, which codes for Arg–Leu–Ser–Ala–Lys–Arg–Gly. To invert the molecule, the strands must be interchanged in order to preserve the correct polarity, and so the sequence of the inverted molecule is:

5′-AGAACGTTTAGCGCTAAGGGT-3′
3′-TCTTGCAAATCGCGATTCCCA-5

This codes for an mRNA with sequence AGA ACG UUU AGC GCU AAG GGU, which codes for Arg–Thr–Phe–Ser–Ala–Lys–Gly.

**10.24** The missing part of the stem is

5′-GUCAUCGAAGCCGUA-3′.

The stem and loop structure is shown here.

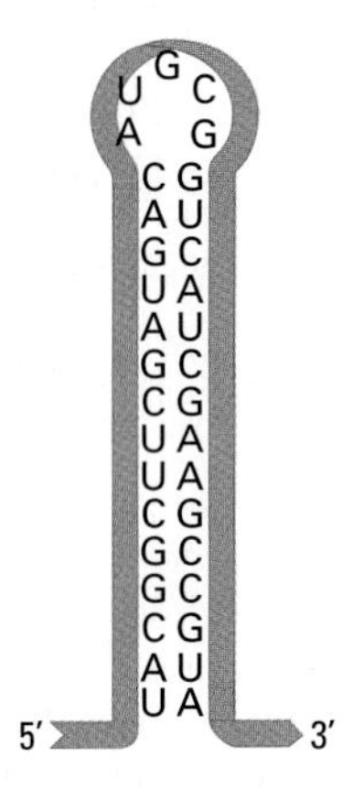


**10.26** With no A–to–I modification, the polypeptide sequence is Val–Pro–Arg–Ser–Ser–His; with A–to–I modification, the mRNA (separated into codons) reads 5′–GUI CCI CGC UCG UCU CIU–3′, which is translated as Val–Pro–Arg–Ser–Ser–Arg.

**10.28** Because either strand could be transcribed, and in the transcript there are three possible reading frames, there are six ways in which the DNA could be transcribed and translated. The mRNA will have the same sequence as the non-transcribed strand (with U replacing T), and so the RNA sequences could be as follows:

5′-CUA GGU GAC CUA GCU UAA-3′
3′-GAU CCA CUG GAU CGA AUU-5′

The red triplets are stop codons. In this case you are lucky because the mRNA at the top has stop codons in all three reading frames and the bottom mRNA has stop codons in two of the three. The reading frame of the mRNA that could be a coding sequence is

5′-UU AAG CUA GGU CAC CUA G-3′

in which the complete codons specify Lys–Leu–Gly–His. Hence, the transcribed strand is the top strand and the reading frame is the one specified here.

**10.30** **a.** 5′-UACCAUGUCAAACAGCGUAUGGUAGCAGUG-3′. **b.** Met–Ser–Asn–Ser–Val–Trp.

## CHAPTER 11

**11.2** The enzymes are expressed constitutively because glucose is metabolized in virtually all cells. However, the levels of enzyme are regulated to prevent runaway synthesis or inadequate synthesis.

**11.4** The most likely level of regulation is **(a)** transcription, **(b)** RNA processing, **(c)** translation.

**11.6** There are two possibilities. The first is that there are two repressor-binding sites (one contained in the *Eco*RI–*Bam*HI fragment and the other in the *Hin*dIII–*Sac*I fragment), which act synergistically. Each alone allows about a 10-fold repression, but together the repression is 100 fold. This hypothesis could be tested by sequencing the restriction fragments; if each contains a repressor-binding site, a region of similar sequence should be found in each fragment. The other hypothesis is that the repressor-binding site is very large, extending across the entire *Eco*RI–*Sac*I region. In this case the *Bam*HI–*Hin*dIII fragment should also be part of the repressor-binding site, so deletion of this fragment should have a major effect, whereas, with two separate binding sites, the deletion would have at most a minor effect.

**11.8** An inversion not only inverts the sequence, but because the DNA strands in a duplex are antiparallel, it exchanges the strands themselves. The inversion of a promoter sequence would therefore have a profound effect because the inverted promoter is in the opposite strand pointing in the wrong direction. On the other hand, one would expect little or no effect of the inversion of an enhancer element, because enhancers are targets for DNA binding proteins irrespective of their strand or orientation in the DNA.

**11.10** At 37°C, both $lacZ^-$ and $lacY^-$ would be phenotypically $Lac^-$; at 30°C, the $lacY^-$ strain would be $Lac^+$ and the $lacZ^-$ strains would be $Lac^-$.

**11.12** The cAMP concentration is low in the presence of glucose. Both types of mutation are phenotypically $Lac^-$. The cAMP-CRP binding does not interfere with repressor binding because the DNA binding sites of cAMP–CRP and lacI are distinct.

**11.14** With a repressor, the inducer is usually an early (often the first) substrate in the pathway, and the repressor is inactivated by combining with the inducer. With an aporepressor, the effector molecule is usually the product of the pathway, and the aporepressor is activated by the binding.

**11.16** The attenuator is not a binding site for any protein, and RNA synthesis does not begin at an attenuator. An attenuator is strictly a potential termination site for transcription.

**11.18** **a.** The histidine operon would initiate transcription when the level of histidine is low enough that the His aporepressor is no longer bound, derepressing transcription. **b.** Attenuation would take place (halting transcription) when the level of charged tryptophan tRNA is high enough that translation of the leader polypeptide can take place. **c.** Transcription would continue through to the end of the operon when the level of charged tryptophan tRNA is so low that translation stalls at the Trp codons in the leader sequence.

**11.20** Amino acids that normally have among the lowest levels of charged tRNA in the cell should have few codons in the leader mRNA, otherwise stalling would take place even at the normal low levels. On the other hand, amino acids that have among the highest levels of charged tRNA in the cell should include multiple codons in the leader mRNA, otherwise stalling might not take place often enough.

**11.22** The **a**1 protein functions only in combination with the α2 protein in diploids; hence, the *MAT***a**′ haploid cell would appear normal. However, the diploid *MAT***a**′/*MAT*α has the phenotype of an α haploid. The reason is that the genotype lacks an active **a**1/α2 complex to repress the haploid-specific genes as well as repressing α1, and the α1 gene product activates the α-specific genes.

**11.24** In the absence of lactose or glucose, two proteins are bound—the *lac* repressor and CRP-cAMP. In the presence of glucose, only the repressor is bound.

**11.26** The *lac* genes are transferred by the Hfr cell and enter the recipient. No repressor is present in the recipient initially, and so *lac* mRNA is made. However, the *lacI* gene is also transferred to the recipient, and soon afterward the repressor is made and *lac* transcription is repressed.

**11.28** Wildtype *lac* mRNA, or that containing either of the point mutations, should yield a band at 5 kb; mRNA containing *lacZ** will be shorter by 2 kb, and that containing lacY* shorter by 0.5 kb. The expected pattern of bands is shown in the accompanying diagram.

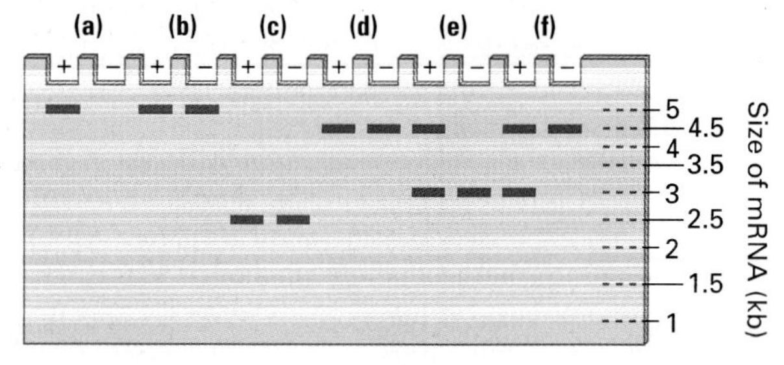


**11.30** In order to grow on lactobionic acid, even the mutant LacZ strain must have both *lacZ* and *lacY* expressed. Evidently lactobionic acid is not an inducer of the *lac* operon, and therefore a $lacO^+$ mutant cannot grow in lactobionic acid unless IPTG, an inducer of the *lac* operon, is also present to induce transcription of the operon. The finding with $lacO^c$ mutants is consistent with this explanation, because when LacY and the mutant LacZ are produced constitutively, the strains can grow in lactobionic acid.

**12.2** Sticky ends are preferred because a sticky end will join only with other sticky ends that are complementary in sequence, whereas blunt ended fragments will join with any other blunt ended fragments.

**12.4** The average distance between restriction sites equals the reciprocal of the probability of occurrence of the restriction site. You must therefore calculate the probability of occurrence of each restriction site in a random DNA sequence. The answers are **(a)** 256, **(b)** 4096, **(c)** 256 (because probability of N, any nucleotide at a site, is 1), **(d)** 256, **(e)** 1024 (because the probability of R is 1/2 and the probability of Y is 1/2).

**12.6** The two N's in the overhangs would have to match perfectly, and the probability of this event is $(1/4)^2 = 1/16$.

**12.8** No, because the 5′–to–3′ polarity of the sequences is different.

**12.10** *Fae*I would be a good choice. Since the mRNA has the sequence 5′–CAUG–3′, the transcribed DNA strand has the sequence 3′–GTAC–5′ and the nontranscribed strand has the sequence 5′–CATG–3′. Hence *Fae*I cleaves the coding sequence immediately after the initiation codon.

**12.12** There are $3 \times 3 = 9$ possible *Aoc*II sites, and so the probability that any two random sites have compatible ends is 1/9.

**12.14** $(1/4)^2 = 1/16$; $(1/2)^2 = 1/4$.

**12.16** *Pst*I 4 kb, *Xba*I 2 kb, *Bam*HI 4 kb; *Pst*I + *Xba*I 1 kb + 2 kb; *Pst*I + *Bam*HI 2 kb; *Xba*I + *Bam*HI 1 kb + 2 kb; *Pst*I + *Xba*I, + *Bam*HI 1 kb.

**12.18** The frequency of sites would be the same for both enzyme for any base ratio so long as the sequence is random. The average number of nucleotide pairs between sites are: **a.** 625; **b.** 256, **c.** 625.

**12.20** *C*, 2 kb + 10 kb; *P*, 3 kb + 9 kb; *K*, 5 kb + 7 kb; *A*, 6 kb; *E*, 4 kb + 8 kb; B, 3 kb + 9 kb; *X*, 1 kb + 11 kb; *C* + *P*, 1 kb + 2 kb + 9 kb; *C* + *E*, 2 kb + 4 kb + 6 kb; *C* + *X*, 1 kb + 2 kb + 9 kb; *K* + *E*, 3 kb + 4 kb + 5 kb; *K* + *X*, 1 kb + 5 kb + 6 kb.

**12.22** **a.** For autosomal genes, the ratio of copy number in females to that in males is 1 : 1, and so a yellow spot is expected. **b.** For an X-linked gene, the ratio of copy number in females to males is 2 : 1, and so a red spot is expected. **c.** For a Y-linked gene, the ratio of copy number in females to males is 0 : 1, and so a green spot is expected.

**12.24** **a.** *Hpa*I produces blunt ends, and any blunt end can be ligated onto any other. Hence, the *Hpa*I fragment A B can be ligated into the polylinker in either of two orientations. The resulting clones are expected to be X A B Y and X B A Y in equal frequency. **b.** A restriction enzyme produces fragments whose ends are identical, so either end can be ligated onto a complementary sticky end. To force the orientation X A B Y, one needs to cleave the vector and the target with two restriction enzymes that produce different sticky ends. The site nearest X in the vector must match the site nearest A in the target, and the site nearest Y in the vector must match the site nearest B in the vector. In this case, if the vector and target are both cleaved with *Kpn*I and *Pst*I, the resulting clone is expected to have the orientation X A B Y. **c.** Following the logic of part **(b)**, digestion of the vector and target with *Xma*I and *Sal*I will yield clones with the orientation X B A Y.

**12.26** Because any two compatible ends of restriction fragments can be ligated, the gene is probably present in the plasmid in an inverted orientation, and therefore cannot function.

**12.28** In the clones that transcribe the gene of interest, the configuration of the gene relative to the position of the promoter is in the correct orientation. The restriction map of these clones is shown here. In this orientation the gene would be expected to be transcribed.

**12.30** The restriction map is shown in the accompanying diagram.

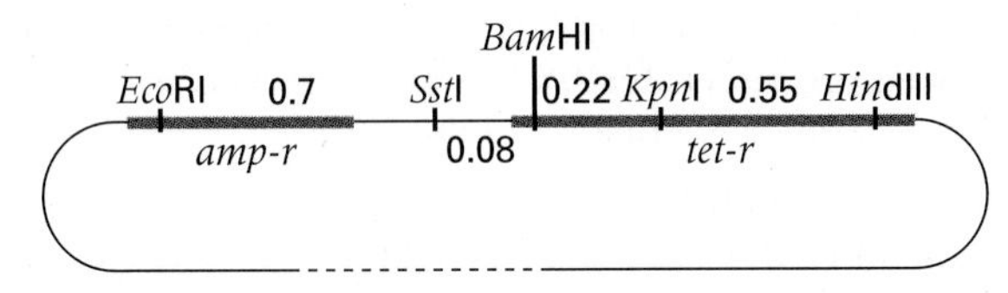


Rest of plasmid between *Eco*RI and *Hin*dIII = 2.45 kb



**13.2** In a loss-of-function mutation, the genetic information in a gene is not expressed in some or all cells. In a gain-of-function mutation, the genetic information is expressed in inappropriate amounts, at inappropriate times, or in inappropriate cells. Both mutations can occur in the same gene: Some alleles may be loss-of-function and others gain-of-function. However, a particular allele must be one or the other (or neither).

**13.4** The small-eye mutation is epistatic to the large-eye mutation, and the large-eye mutation is hypostatic to the small-eye mutation. The result implies that the gene product of the small-eye gene acts downstream in the pathway from that of the large-eye gene.

**13.6** Carry out a complementation test. Cross two heterozygous genotypes and determine whether the female progeny have the maternal-effect lethal phenotype. If they are alleles, 1/4 of the progeny would be expected to have this phenotype. If they are not alleles, all of the progeny should be normal.

**13.8** Because the gene is sufficient, a gain-of-function mutation will induce the developmental fate and the morphological feature will be expressed in whichever cell types the gain-of-function allele is expressed. The allele will be dominant because a normal allele in the homologous chromosome will not prevent the developmental pathway from occurring.

**13.10** **a.** white; **b.** purple; **c.** purple; **d.** white.

**13.12** The first row implies the order D A, E A, and F A; the second D B, E B, and B F; and the third C D, E C, and C F. Together these imply that the components act in the order E − C − D − B − F − A.

**13.14** Look for embryos with a phenotype similar to *bicoid* mutants, lacking the head and thorax.

**13.16** The transplanted nuclei respond to the cytoplasm of the oocyte and behave like the oocyte nucleus at each stage of development.

**13.18** Petal–petal–stamen–carpel.

**13.20** Whorls 1, 2, 3, and 4 yield leaf, petal, stamen, and carpel, respectively.

**13.22** **a.** The results are consistent with the single mutants. **b.** The phenotypes of the double mutants indicate that $a^+$ alone is sufficient to produce the distal segment, that $a^+$ and $c^+$ are necessary to produce the medial segment, and that $b^+$ and $c^+$ are necessary to produce the proximal segment. **c.** A detailed model is shown in the accompanying diagram.

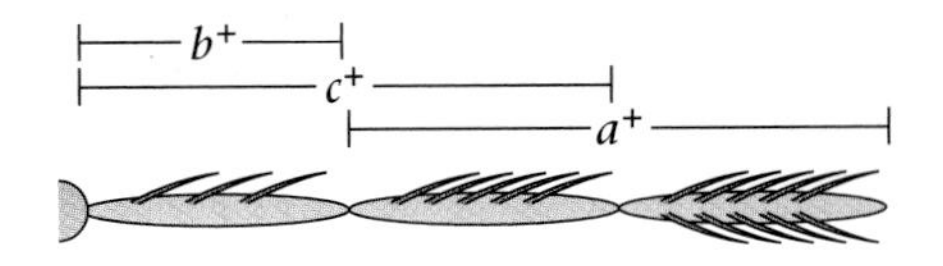


**13.24** The proteins required for cleavage and blastula formation are translated from mRNAs present in the mature oocyte, but transcription of zygotic genes is required for gastrulation.

**13.26** The material required for rescue is either mRNA transcribed from the maternal *o* gene or a protein product of the gene that is localized in the nucleus.

**13.28** **a.** Because, in *agamous* mutants, whorls 1–4 yield sepals, petals, petals, and sepals, respectively, but petals and sepals require *apetala-1* expression. **b.** Because, in *apetala-1* mutants, whorls 1–4 yield carpels, stamens, stamens, and carpels, respectively, but carpels and stamens require *agamous* expression.

**13.30** The first row implies the order C A and C B, whereas the second row implies the order A D and B D. It is clear that C acts first in the pathway and that D acts last, but the order of action of A and B is ambiguous. The ambiguity can be represented symbolically by the pathway C − (A, B) − D.

## CHAPTER 14

**14.2** **a.** 2; **b.** 3; **c.** 2; **d.** 1.

**14.4** **a.** T102I results from a C → T transition; **b.** P106S results from a C → T transition; **c.** Glyphosate-resistant mutants are so rare because they are double mutants, and the occurrence of two mutations in the same gene is a much rarer event than that of either mutation alone.

**14.6** Such people are somatic mosaics and can be explained by somatic mutations in the pigmented cells of the iris of the eye or in their precursor cells.

**14.8** **a.** The start codons are AUG and AUM. **b.** The stop codons are UAA, UAG, UAM, UGA, and UMA.

**14.10** Any single nucleotide substitution creates a new codon with a random nucleotide sequence, so the probability that it is a chain-termination codon (UAA, UAG, or UGA) is 3/64, or approximately 5 percent.

**14.12** The mutation is likely a deletion, because while all of the other restriction fragments match in size, one fragment that is 5 kb in the wildtype phage is only about 3.5 kb in the mutant phage. This difference suggests a deletion.

**14.14** The results of homologous recombination are shown in the diagrams.

**14.16** The possible lysine codons are AAA and AAG, and the possible glutamic acid codons are GAA and GAG. Therefore, the mutation results from an AT to GC transition in the first position of the codon.

**14.18** The sectors are explained by a nucleotide substitution in only one strand of the DNA duplex. Replication of this heteroduplex yields one daughter DNA molecule with the wildtype sequence and another with the mutant sequence. Cell division produces one Lac$^+$ and one Lac$^-$ cell, located side by side. Since the cells do not move on the agar surface, the Lac$^+$ cells form a purple half-colony and the Lac$^-$ cells form a pink half-colony.

**14.20** Amino acid replacements at many positions in the A protein do not eliminate its ability to function; only amino acid replacements at critical positions that do eliminate tryptophan synthase activity are detectable as Trp mutants.

**14.22** Because there are many ways in which an amino acid replacement can impair the activity of a wildtype protein, but very few ways in which mutant protein can regain wildtype activity other than restoration of the original amino acid.

**14.24** **a.** 1 : 1 because the mutation is recessive and because males get their X chromosome from the mother. **b.** All females,

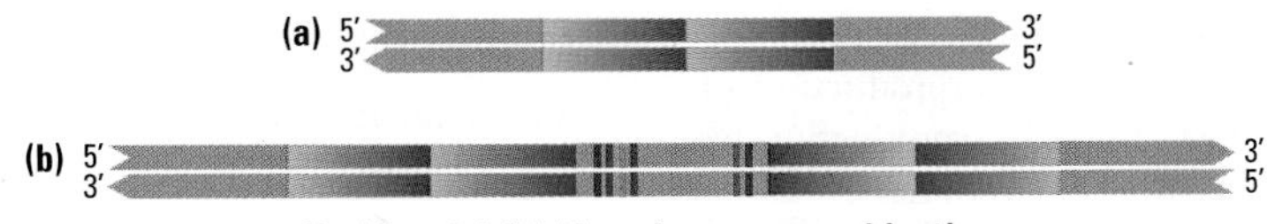


**Problem 14.14 Homologous recombination**

because at 29°C the males (who in this case get their X chromosome from their father) die.

**14.26** **a.** $5 \times 10^{-5}$ is the probability of a mutation per generation, and so $1 - (5 \times 10^{-5}) = 0.99995$ is the probability of no mutation per generation. The probability of no mutations in 10,000 generations equals $(0.99995)^{10,000} = 0.61$. **b.** The average number of generations until a mutation occurs is $1/(5 \times 10^{-5}) = 20,000$ generations.

**14.28** A crossover between the transposable elements in (a) results in a reciprocal translocation, whereas a crossover between the transposable elements in (b) results in a dicectric chromosome and an acentric chromosome. The products of recombination are shown in the accompanying illustration.

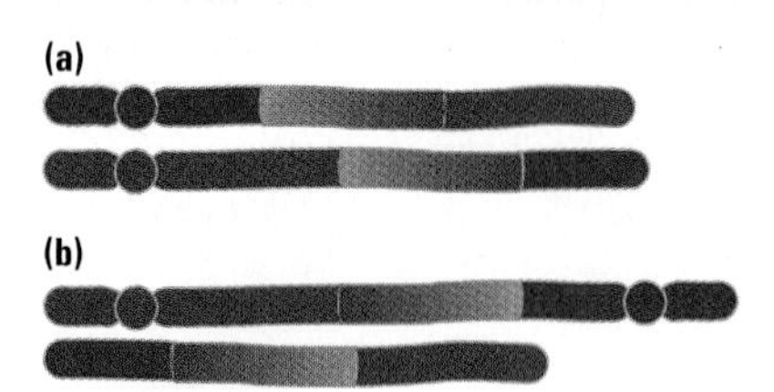


**14.30** Deduce the sequence of the reading frame of both of the double mutants. The +G with −C combination has the reading frame AGAGGUAAGTCTAGA, which translates as Arg–Gly–Lys–Ser–Arg. This polypeptide has three adjacent amino acid replacements but could function if the replacements could be tolerated. On the other hand, the −G with +C mutant combination has the reading frame AGATAACGCCTCAGA, which translates as Arg–End, where End represents the UAA stop codon. In this double mutant, the downstream duplication does not correct the upstream termination codon.

## CHAPTER 15

**15.2** **a.** At the permissive temperature the phenotype is normal. **b.** At the restrictive temperature the mother cell and daughter cell are unable to separate; they remain attached through a small cytoplasmic bridge, and successive divisions form clumps of interconnected cells, each cell attached to its mother and its own daughters through such cytoplasmic bridges.

**15.4** Cyclin–CDK complexes transfer phosphate groups to their target protein; the counterparts that reverse phosphorylation are phosphatases.

**15.6** Two important targets of the anaphase-promoting complex are Pds1p and cyclin B. This complex is a ubiquitin-protein ligase that ubiquitinates the proteins and tags them for destruction in the proteasome. Destruction of Pds1p allows disjunction of the sister chromatids and is associated with the metaphase/anaphase transition; destruction of cyclin B is associated with the mitosis/$G_1$ transition.

**15.8** The p53 protein is the major player. Normally p53 does not accumulate in the nucleus because it is continuously exported by Mdm2. In the presence of DNA damage, p53 is phosphorylated and acetylated, which reduces its binding by Mdm2; the protein accumulates in the nucleus and activates transcription of the genes for p21, GADD45, 14-3-3$\sigma$, and miRNA34a and miRNA34b/c. The gene products are associated with a block in the $G_1$/S transition, prolongation of S phase, and a block in the $G_2$/M transition, respectively.

**15.10** Bax and Bcl2 are normally present as heterodimers. Overproduction of Bax results in Bax homodimers that activate the apoptosis pathway. The p53 protein plays a role in apoptosis because p53 is a transcription factor for the gene encoding Bax. Certain cellular oncogenes result in overproduction of Bcl2, which counteracts the effect of Bax overproduction by sequestering Bax subunits in Bax/Bcl2 heterodimers.

**15.12** **a.** Test for genetic linkage of the segregating cancer gene to DNA markers very near the *p53* gene. If there is tight linkage between the disease gene and a DNA marker at *p53*, then it is likely that the disease is *p53*-related. If the disease gene is not genetically linked to *p53*, another gene is probably involved. **b.** Possible sites for change include: promoter changes that reduce expression of the *p53* gene, changes in introns that correspond to essential splice sites, or a deletion that removes the whole gene (so that the sequence of *p53* from a heterozygous individual is that of the wildtype $p53^+$ homolog). **c.** Anything that can mimic loss of *p53* function could underlie the phenotype. A mutation that causes overexpression of Mdm2 is one possibility. Other possibilities are mutations that abrogate the ability of a protein kinase that phosphorylates p53 (if there is a crucial one) or a protein acetylase that acetylates p53, so long as lack of either modification results in failure of p53 levels to rise in response to DNA damage (presumably because Mdm2 can still bind to p53). For all three suggestions, the syndrome should show dominant inheritance but loss of heterozygosity should be required for expression at the cellular level.

**15.14** Chromosome abnormalities are expected because the normal $G_1$/S checkpoint serves to allow DNA damage (such as broken chromosomes) to be repaired prior to entering the S period. In the absence of a normal $G_1$/S checkpoint, the cell transitions into S even when there are broken chromosomes and other abnormalities.

**15.16** Ras–GTP is a central step in transmitting a growth signal in response to activation of certain tyrosine kinase receptors. Constitutive Ras–GTP transmits the growth signal continuously.

**15.18** It means that most individuals who inherit a mutant *RB1* allele from either parent develop retinoblastoma, so in families the mutant allele behaves as a dominant with respect to the retinoblastoma phenotype. However, somatic cells in heterozygous individuals do not become cancerous unless they undergo a loss of heterozygosity, making them effectively homozygous for the mutation, so at the cellular level the mutation is recessive.

**15.20** Since the probability of loss of heterozygosity per cell is $7.5 \times 10^{-7}$ and there are $4 \times 10^{6}$ cells at risk in both eyes together, the probability of no loss of heterozygosity in any cell is $(1 - 7.5 \times 10^{-7})$ raised to the power of $4 \times 10^{6}$, which equals 0.05, or 5 percent. The penetrance of familial retinoblastoma is therefore 95 percent. (Students familiar with the Poisson distribution can use the fact that the mean number of tumors per heterozygous individual is 3, from which the Poisson distribution implies that the number of individuals with no tumors is $e^{-3} = 0.05$.)

**15.22** **a.** A mutation affecting the tetramerization domain would be dominant at the organismal level but recessive at the cellular level, because mixed tetramers cannot be formed. In order for a heterozygous cell to become mutant, there must be a loss of heterozygosity. **b.** A mutation affecting the DNA-binding domain would be dominant at both the organismal and cellular level. The heterozygous cells make very little functional transcription factor because of the poison subunits.

**15.24** Inspection of the pedigree indicates that the band c from the mother in generation I (individual 9) derives from a mutant allele, since all of her affected offspring, and none of her normal offspring, inherit this allele. Thus individual 5 carries the mutant allele associated with band c, and among her offspring individuals 2 and 7 are at high risk, the others not. The individual 20 also carries the mutant allele yielding band c, hence among his offspring individuals 16, 21, and 22 are at high risk, the others not. The remaining individual 13 in generation III is not at risk, even though her DNA yields a band that comigrates with c. The reason is that this band was derived from an allele in her father, who is unrelated to any of the affected individuals in the pedigree; hence the band c in this case marks a nonmutant allele.

**15.26** One approach is first to determine whether genes for T cell receptor or the heavy chain of immunoglobulins map near the translocation site. If either is near the breakpoint, one would expect a promoter fusion to be present. As confirmation, it should be determined whether the mRNA of the gene next to the breakpoint is the normal size, as it would be expected to be in a promoter fusion. If there is no suggestion that the T cell receptor or immunoglobulin genes are at the site of the translocation, one would expect a gene fusion. To test this, make DNA probes from DNA flanking the site of the translocation and use them to probe RNA blots from normal and leukemic cells (the probes have to be long enough to extend into exons or they will not hybridize to mRNAs). For a gene fusion, the mRNA in the leukemic cell should be a chimera, and one would expect a single mRNA from the leukemic cells to hybridize with both flanking DNAs; for wildtype cells, the two probes should hybridize to mRNAs of different sizes. A more refined analysis would tell whether the gene fusion took place in introns.

**15.28** The individuals that must be heterozygous for the disease allele but nevertheless be cancer-free are either I-1 or I-2 and also III-3.

**15.30** *RAD9* is probably not involved in the $G_1/S$ DNA checkpoint, because the *cdc28, cdc6,* and *cdc8 rad9* double mutants survive incubation at 36°C, implying that the $G_1/S$ checkpoint is intact in the *rad9* mutant. *RAD9* is also probably not involved in the centrosome duplication checkpoint, because the *cdc7 rad9* double mutant survives incubation at 36°C. *RAD9* is involved in the S phase DNA checkpoint, because the *cdc2, cdc13,* and *cdc17 rad9* (as well as the *cdc9 rad9*) double mutants die at 36°C. *RAD9* is probably not involved in the spindle or in the exit from mitosis checkpoints, because the *cdc16, cdc20, cdc23,* and *cdc15 rad9* double mutants survive incubation at 36°C.

## CHAPTER 16

**16.2** The severity is variable because the severity depends on the proportion of defective mitochondria that are present in the affected individual.

**16.4** Yes, it is true in the human mitochondrial code. In the standard code the exceptions to the rule are AUA and AUG, which code for isoleucine and methionine, respectively, and UGA and UGG, which code for "stop" and tryptophan, respectively.

**16.6** Inheritance is strictly maternal, and so yellow is probably inherited cytoplasmically. A reasonable hypothesis is that the gene for yellow is contained in chloroplast DNA.

**16.8** All transition mutations in the third position are synonymous. Among 128 possible transversion mutations, 56, or 56/128 = 43.8 percent, are synonymous.

**16.10** With X-linked inheritance, half of the $F_2$ males (1/4 of the total progeny) would be white. The observed result is that all of the $F_2$ progeny are green.

**16.12** The cluster of four restriction sites suggests an inverted-repeat sequence. Recombination between the inverted repeats inverts the segment between them, yielding the two different restriction maps. They are both present in single individuals because one is derived from the other by recombination, which must take place at a sufficiently high rate that each individual has both types of mtDNA.

**16.14** The chloroplast from the $mt^-$ strain is preferentially lost, and the *mt* alleles segregate 1 : 1. The expected result is 1/2 $a^+\ b^-\ mt^+$ and 1/2 $a^+\ b^-\ mt^-$.

**16.16** **a.** All *cpl-r* because the chloroplast comes from the $mt^+$ parent. **b.** All *mit-s* because the mitochondria come from the $mt^-$ parent. **c.** Resistant : sensitive in a ratio of 2 : 2 because a nuclear gene undergoes Mendelian segregation.

**16.18** Segregational petites are called segregational because the petite phenotype results from a mutant nuclear gene that undergoes normal 2 : 2 segregation in meiosis, yielding 2 petite : 2 nonpetite progeny from a cross.

**16.20** Because every tetrad shows uniparental inheritance, it is likely that the trait is determined by factors transmitted through the cytoplasm, such as mitochondria. The 4 : 0 tetrads result from diploids containing only mutant factors, and the 0 : 4 tetrads result from diploids containing only wildtype factors.

**16.22** The mating pairs are 1/4 *AA* × *AA*, 1/2 *AA* × *aa*, and 1/4 *aa* × *aa*. These produce only *AA*, *Aa* and *aa* progeny, respectively, and so the progeny are 1/4 *AA*, 1/2 *Aa*, and 1/4 *aa*. In autogamy, the *Aa* genotypes become homozygous for either *A* or *a*, and so the expected result of autogamy is 1/2 *AA* and 1/2 *aa*.

**16.24** The $F_1$ nuclear genotype is *Rsf*/*rsf*, and all the progeny receive male-sterile cytoplasm (restored to male fertility by the *Rsf* allele). In the next generation, 3/4 of the plants are *Rsf*/− and are both female and male fertile, but 1/4 of the plants are *rsf*/*rsf*, which are female fertile but male sterile.

**16.26** The most likely possibility is maternal transmission of a factor, possibly a virus or bacteria, that results in a high pre-adult mortality among the male offspring.

**16.28** The accompanying table shows the percent of restriction sites shared by each pair of mitochondrial DNA molecules.

| | 1 | 2 | 3 | 4 | 5 |
|---|---|---|---|---|---|
| 1 | 100 | 0 | 80 | 10 | 0 |
| 2 | | 100 | 0 | 10 | 100 |
| 3 | | | 100 | 11 | 0 |
| 4 | | | | 100 | 10 |

The entries in the table may be explained by example. Molecules 3 and 4 together have 9 restriction sites, of which 1 is shared, so the entry in the table is 1/9 = 11 percent. The diagonal entries of 100 mean that each molecule shares 100 percent of the restriction sites with itself. The two most closely related molecules are 2 and 5 (in fact, they are identical), so these must be grouped together. The two next closest are 1 and 3, so these also must be grouped together. The 2–5 group shares an average of 10 percent restriction sites with molecule 4, and the 1–3 group shares an average of 10.5 percent restriction sites [that is, (10 + 11)/2] with molecule 4. Therefore, the 2–5 group and the 1–3 group are about equally distant from molecule 4. The diagram shows the inferred relationships among the molecules. Data of this type do not indicate where the "root" should be positioned in the tree diagram.

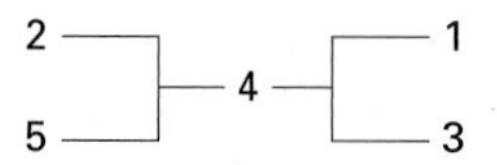


**16.30** The result excludes hypothesis (1) but does not distinguish between hypotheses (2) and (3).

## CHAPTER 17

**17.2** The numbers of alleles of each type are $A_1$ = 2 + 2 + 5 + 1 = 10; $A_2$ = 5 + 12 + 12 + 10 = 39; $A_3$ = 10 + 5 + 5 + 1 = 21. The total number of alleles is 70, so the allele frequencies are $A_1$, 10/70 = 0.14; $A_2$, 39/70 = 0.56; and $A_3$, 21/70 = 0.30.

**17.4** Allele 5.7 frequency = (2 × 9 + 21 + 15 + 6)/200 = 0.30; allele 6.0 frequency = (2 × 12 + 21 + 18 + 7)/200 = 0.35; allele 6.2 frequency = (2 × 6 + 15 + 18 + 5)/200 = 0.25; allele 6.5 frequency = (2 × 1 + 6 + 7 + 5)/200 = 0.10.

**17.6** If the genotype frequencies satisfy the Hardy-Weinberg principle, they should be in the proportions $p^2$, $2pq$, and $q^2$, respectively. In each case, $p$ equals the frequency of *AA* plus one-half the frequency of *Aa*, and $q = 1 - p$. The allele frequencies and expected genotype frequencies with random mating are **a**, $p$ = 0.50, expected: 0.25, 0.50, 0.25; **b**, $p$ = 0.635, expected: 0.403, 0.464, 0.133; **c**, $p$ = 0.7, expected: 0.49, 0.42, 0.09; **d**, $p$ = 0.775, expected: 0.601, 0.349, 0.051; **e**, $p$ = 0.5, expected: 0.25, 0.50, 0.25. Therefore, only **a** and **c** fit the Hardy-Weinberg principle.

**17.8** The frequency of the recessive allele is $q = \sqrt{(1/14{,}000)}$ = 0.0085. Hence, $p$ = 1 − 0.0085 = 0.9915, and the frequency of heterozygotes is $2pq$ = 0.017 (about 1 in 60).

**17.10** Assuming random mating, the frequency of the recessive allele is $q = \sqrt{(1/250{,}000)}$ = 0.002. **a.** Among the offspring of first-cousin matings, the expected frequency of XP equals $q^2(1 - F) + qF$, where $F$ = 1/16 for the offspring of first cousins. In this case the formula yields $(0.002)^2(15/16) + (0.002)(1/16)$ = 0.00012875, or about 1 in 7767. **b.** The ratio is given by (1/7767)/(1/250,000) = 32.2. That is, the offspring of first cousins have more than a 32-fold greater risk of being homozygous for a recessive mutant allele causing XP.

**17.12** **a.** Because the proportion 0.60 can germinate, the proportion 0.40 that cannot are the homozygous recessives. Thus, the allele frequency $q$ of the recessive is $\sqrt{(0.4)}$ = 0.63. This implies that $p$ = 1 − 0.63 = 0.37 is the frequency of the resistance allele. **b.** The frequency of heterozygotes is $2pq$ = 2 × 0.37 × 0.63 = 0.46. The frequency of homozygous dominants is $(0.37)^2$ = 0.14, and the proportion of the surviving genotypes that are homozygous is 0.14/0.60 = 0.23.

**17.14** **a.** There are 17 $A_1$ and 53 $A_2$ alleles, and so the allele frequencies are: $A_1$, 17/70 = 0.24; $A_2$, 53/70 = 0.76. **b.** The expected genotype frequencies are: $A_1A_1$, $(0.24)^2 = 0.058$; $A_1A_2$, $2(0.24)(0.76) = 0.365$; $A_2A_2$, $(0.76)^2 = 0.578$. The expected numbers are 2.0, 12.8, and 20.2, respectively.

**17.16** The frequency of the yellow allele *y* is 0.2. In females, the expected frequencies of wildtype (+/+ and *y*/+) and yellow (*y* / *y*) are $(0.8)^2 + 2(0.8)(0.2) = 0.96$ and $(0.2)^2 = 0.04$, respectively. Among 1000 females, there would be 960 wildtype and 40 yellow. The phenotype frequencies are very different in males. In males, the expected frequencies of wildtype (+/Y) and yellow (*y*/Y) are 0.8 and 0.2, respectively, where Y represents the Y chromosome. Among 1000 males, there would be 800 wildtype and 200 yellow.

**17.18** Inbreeding is suggested by a deficiency of heterozygotes relative to $2pq$ expected with random mating. The suggestive populations are d with expected heterozygote frequency 0.349 and e with expected heterozygote frequency 0.5.

**17.20** **a.** Note that $1/(2.6 \times 10^6) = 3.846 \times 10^{-7}$; set $q^2 = \sqrt{(3.846 \times 10^{-7})}$, hence $q = 0.000620$. **b.** $q^2(15/16) + q(1/16) = 3.912 \times 10^{-5}$, or 1 in 25,561. **c.** $(0.99 \times 3.846 \times 10^{-7}) + (0.01 \times 3.912 \times 10^{-5}) = 7.720 \times 10^{-7}$. **d.** $(0.01 \times 3.912 \times 10^{-5})/(7.720 \times 10^{-7}) = 50.7$ percent. Although only 1 percent of the matings are between first-cousins, they account for more than half of the homozygous recessives for this gene.

**17.22** For each pair of taxa, the UPGMA distance is the sum of the branch lengths connecting them. The distance matrix is therefore

| | B | C | D |
|---|---|---|---|
| A | 17.4 | 6.4 | 12.6 |
| B | | 17.4 | 17.4 |
| C | | | 12.6 |

**17.24** Use Equation 17-10 with $n = 40$, $p_0/q_0 = 1$ (equal amounts inoculated), and $p_{40}/q_{40} = 0.35/0.65$. The solution for *w* is $w = 0.9846$, which is the fitness of strain B relative to strain A, and the selection coefficient against B is $s = 1 - 0.9846 = 0.0154$.

**17.26** $10^{-4}/(10^{-4} + 10^{-7}) = 0.999$.

**17.28** **a.** Expected frequency of heterozygous genotypes, $2 \times 0.00005 \times 0.99995 = 10^{-4}$, or 1 in 10,000. **b.** Because $q = \mu/hs$, where $q = 0.00005$ and $hs = 0.20$ are given, the estimated mutation rate is $\mu = 0.00005 \times 0.20 = 1 \times 10^{-5}$.

**17.30** **a.** A rare beneficial dominant will change faster in frequency because the heterozygous genotypes are exposed to the selection favoring it. **b.** Note that this is the same case as in part **a**, because the allele of a rare beneficial dominant is a common deleterious recessive, and the other way around. Hence, the frequency of the common deleterious recessive changes faster, because the beneficial selection acts on the rare heterozygous genotypes.

## CHAPTER 18

**18.2** Since the variance equals the square of the standard deviation, the only possibilities are that the variance equals either 1 or 0.

**18.4** Broad-sense heritability is the proportion of the phenotypic variance attributable to all differences in genotype, or $H^2 = \sigma_g^2/\sigma_p^2$, which includes dominance effects and interactions between alleles. Narrow-sense heritability is the proportion of the phenotypic variance due only to additive effects, or $h^2 = \sigma_a^2/\sigma_p^2$, which is used to predict the resemblance between parents and offspring in artificial selection. If all allele frequencies equal 1 or 0, both heritabilities equal zero.

**18.6** For fruit weight, $\sigma_e^2/\sigma_p^2 = 0.13$ is given. Because $\sigma_p^2 = \sigma_g^2 + \sigma_e^2$, then $H^2 = \sigma_g^2/\sigma_p^2 = 1 - \sigma_e^2/\sigma_p^2 = 1 - 0.13 = 0.87$. Therefore, 87 percent is the broad-sense heritability of fruit weight. The values of $H^2$ for soluble-solid content and acidity are 91 percent and 89 percent, respectively.

**18.8** **a.** Phenotypes 0, 2, 4, and 6 with frequencies 0.015625, 0.140625, 0.421875, and 0.421875, respectively. **b.** Mean 4.50, variance 2.25, standard deviation 1.50. **c.** Phenotypes 0, 1, 2, 3, 4, 5, and 6 with frequencies 0.0156, 0.0938, 0.2344, 0.3125, 0.2344, 0.0938, and 0.0156, respectively. Mean 3.00, variance 1.50, standard deviation 1.22.

**18.10** **a.** To determine the mean, multiply each value for the pounds of milk produced (using the midpoint value) by the number of cows, and divide by 300 (the total number of cows) to obtain a value of the mean of 4080 pounds. The estimated variance is 626,100 pounds$^2$, and the estimated standard deviation, which is the square root of the variance, is 791 pounds. **b.** The range that includes 68 percent of the cows is the mean minus the standard deviation to the mean plus the standard deviation, or 4080 − 791 = 3289 pounds to 4080 + 791 = 5662 pounds. **c.** For 95 percent of the animals, the range is within two standard deviations, or 2498 − 5698 pounds.

**18.12** **a.** The genotypic variance is 0 because all $F_1$ mice have the same genotype. **b.** With additive alleles, the mean of the $F_1$ generation will equal the average of the parental strains.

**18.14** The phenotypic variance, $\sigma_p^2$, in the genetically heterogeneous population is the sum $\sigma_g^2 + \sigma_e^2 = 40$. The variance in the inbred lines equals the environmental variance (because $\sigma_g^2 = 0$ in a genetically homogeneous population); hence $\sigma_e^2 = 10$. Therefore, $\sigma_g^2$ in the heterogeneous population equals 40 − 10 = 30. The broad-sense heritability $H^2$ equals $\sigma_g^2/(\sigma_g^2 + \sigma_e^2) = 30/40 = 75$ percent. **b.** If the inbred lines were crossed, the $F_1$ generation would be genetically uniform, and consequently the predicted variance would be $\sigma_e^2 = 10$.

**18.16** The offspring are genetically related as half-siblings, and Table 18.3 says that the theoretical correlation coefficient is $h^2/4$.

**18.18** **a.** Because *Difflugia* is an asexual organism, parent and offspring have the same genetic relationship as monozygotic twins; hence, the broad-sense heritabilities are 0.956 and

0.287, respectively. **b.** The environmental component of the variance for length of longest spine is larger than that for tooth number.

**18.20** $\sigma_g^2 = 0.0570 - 0.0144 = 0.0426$ and $D = 1.689 - (-0.137) = 1.826$. The minimum number of genes is estimated as $n = (1.826)^2/(8 \times 0.0426) = 9.8$.

**18.22** $h^2 = (2.33 - 2.26)/(2.37 - 2.26) = 63.6$ percent.

**18.24** Use Equation 18-6 with $M = 10.8$, $M^* = 5.8$, and $M = 8.8$. The narrow-sense heritability is estimated as $h^2 = (8.8 - 10.8)/(5.8 - 10.8) = 0.40$.

**18.26** The mean weaning weight of the 20 individuals is 81 pounds, and the standard deviation is 13.8 pounds. The mean of the upper half of the distribution can be calculated by using the relationship given in the problem, which says that the mean of the upper half of the population is $81 + (0.8)(13.8) = 92$ pounds. Using Equation 18-6, we find that the expected weaning weight of the offspring is $81 + 0.2(92 - 81) = 83.2$ pounds.

**18.28** **a.** The narrow-sense heritabilities are 1.0, 0.67, 0.18, 0.10, 0.02, 0.01, and 0.002, respectively. **b.** The magnitude of the narrow-sense heritability depends on the parent–offspring correlation in phenotype. With rare recessive alleles, most homozygous offspring come from matings between heterozygous genotypes, so there is essentially no parent–offspring correlation in phenotype.

**18.30** Successive substitutions in this case yield $M_t = M_0 + h^2(S_1 + S_2 + S_3 + \ldots + S_{t-1})$.

# Word Roots prefixes, suffixes and combining forms

| Roots and Prefixes | Meaning | Example |
|---|---|---|
| **a-, an-** | *absence or lack* | acentric, lack of a centromere |
| **ab-** | *departing from, away from* | abnormal, departing from normal |
| **ac-, acro-** | *extreme or extremity, peak* | acrocentric, centromere near the end of a chromosome |
| **allel-** | *of one another* | alleles, alternative forms of a gene |
| **amphi-** | *on both sides, of both kinds* | amphidiploid, an organism containing diploid genomes from two different organisms (also called an allotetraploid) |
| **ana-** | *apart, up, again* | anaphase of mitosis, when the chromosomes separate (move apart) |
| **ant-, anti-** | *opposed to, preventing or inhibiting* | antibiotic, preventing or inhibiting life |
| **ante-** | *preceding, before* | antedate, precedes a date |
| **apo-** | *former, from* | aporepressor, precursor to repressor |
| **aut-, auto-** | *self* | autogenous, self-generated |
| **bi-** | *two* | bidirectional, going in two directions |
| **bio-** | *life* | biology, the study of life and living organisms |
| **blast-** | *bud or germ* | blastoderm, structure formed in early development |
| **carcin-** | *cancer* | carcinogen, a cancer-causing agent |
| **cata-** | *down* | catabolism, chemical breakdown |
| **caud-** | *tail* | caudal (directional term) |
| **chiasm-** | *crossing* | chiasma, the cross-shaped figure that occurs at the site of crossing over between homologous chromosomes |
| **chrom-** | *colored* | chromosome (staining body), so named because they stain darkly |
| **circum-** | *around* | circumnuclear, surrounding the nucleus |
| **co-, con-, com-** | *together* | codominant, expression of both alleles in a heterozygote |
| **cyt-** | *cell* | cytology, the study of cells |
| **de-** | *undoing, reversal, loss, removal* | deoxy, lacking an oxygen atom |
| **di-** | *twice, double* | dicentric, having two centromeres |
| **dys-** | *difficult, faulty, painful* | dysfunctional, disturbed function |
| **ec-, ex-, ecto-** | *out, outside, away from* | excise, to cut away |
| **ectop-** | *displaced* | ectopic, expression that occurs in the wrong tissue or cell type |
| **endo-** | *within, inner* | endonuclease, cleaving a nucleic acid in the interior, not the end, of a molecule |
| **epi-** | *over, above* | episome, a genetic element over (beyond) the core genome |
| **eu-** | *good, well* | eukaryote, a cell with a good or true nucleus |
| **exo-** | *outside, outer layer* | exonuclease, an enzyme that digests nucleic acids beginning at the ends of the molecule |
| **extra-** | *outside, beyond* | extracellular, outside the body cells of an organism |
| **flagell-** | *whip* | flagellum, the tail of a sperm cell |
| **gam-, gamet-** | *married, spouse* | gametes, the sex cells |
| **gene** | *beginning, origin* | genetics |
| **gon-, gono-** | *seed, offspring* | gonads, the sex organs |
| **haplo-** | *single* | haploid, gametic chromosome number |
| **hema-, hemato-, hemo-** | *blood* | hemoglobin, blood protein |
| **hemi-** | *half* | hemimethylated, methylated on one DNA strand only |
| **hetero-** | *different or other* | heterosexuality, sexual desire for a person of the opposite sex |
| **holo-** | *whole* | holoenzyme, form of an enzyme containing all subunits |
| **hom-, homo-** | *same* | homozygous, having the same allele of a gene in homologous chromosomes |
| **homeo-** | *similar* | homeotic, related structurally |
| **hyper-** | *excess* | hypermorphic, state in which a gene is expressed at levels greater than normal |
| **hypo-** | *below, deficient* | hypomorphic, state in which a gene is expressed at levels less than normal |
| **in-** | *in, into* | induce, to lead into (a new state) |
| **inter-** | *between* | intercellular, between the cells |
| **intercal-** | *insert* | intercalated dyes, dyes that insert between adjacent base pairs in DNA |
| **intra-** | *within, inside* | intracellular, inside the cell |

| Roots and Prefixes | Meaning | Example |
|---|---|---|
| **iso-** | *equal, same* | isothermal, equal, or same, temperature |
| **juxta-** | *near, close to* | juxtapose, place near or next to |
| **karyo-** | *kernel, nucleus* | karyotype, the set of the nuclear chromosomes |
| **kin-, kines-** | *move* | kinetic energy, the energy of motion |
| **lact-** | *milk* | lactose, milk sugar |
| **lumen** | *light* | lumen, center of a hollow structure |
| **lys-** | *dissolution or loosening* | lysis, disruption by dissolution |
| **macro-** | *large* | macromolecule, large molecule |
| **mal-** | *bad, abnormal* | malfunction, abnormal functioning of an organ |
| **mater-** | *mother* | maternal, pertaining to the mother |
| **mega-** | *large* | megabase, a million base pairs |
| **meio-** | *less* | meiosis, nuclear division that halves the chromosome number |
| **mero-** | *partial* | merodiploid, partial diploid (in bacteria) |
| **meta-** | *beyond, between, transition* | metaphase, the stage of mitosis or meiosis in which chromosomes are located between the poles of the spindle |
| **micro-** | *small* | microscope, an instrument used to make small objects appear larger |
| **mito-** | *thread, filament* | mitochondria, small, filament like structures located in cells |
| **mono-** | *single* | monohybrid, heterozygous for one allelic pair |
| **morpho-** | *form* | morphology, the study of form and structure of organisms |
| **multi-** | *many* | multinuclear, having several nuclei |
| **muta-** | *change* | mutation, change in the base sequence of DNA |
| **nano-** | *dwarf* | nanometer, one billionth of a meter |
| **nucle-** | *pit, kernel, little nut* | nucleus |
| **nulli-, nullo-** | *none* | nullisomic, having no copies of a particular chromosome |
| **oligo-** | *few* | oligonucleotide, a nucleic acid molecule containing a few nucleotides |
| **onco-** | *a mass* | oncology, study of cancer |
| **oo-** | *egg* | oocyte, precursor of female gamete |
| **org-** | *living* | organism |
| **ov-, ovi-** | *egg* | ovum, oviduct |
| **oxy-** | *oxygen* | oxygenation, the saturation of a substance with oxygen |
| **para-** | *near, beside* | paracentric, beside the centromere |
| **pater-** | *father* | paternal, pertaining to the father |
| **patho-** | *pathogen* | disease-causing |
| **pent-** | *five* | pentose, a 5-carbon sugar |
| **per-** | *through* | permease, a protein that carries a small molecule through the cell membrane |
| **peri-** | *around* | pericentric, around the centromere |
| **phago-** | *eat* | bacteriophage, a virus that infects bacteria |
| **pheno-** | *show, appear* | phenotype, the physical appearance of an individual |
| **pili** | *hair* | arrector pili muscles of the skin, which make the hairs stand erect |
| **poly-** | *multiple* | polymorphism, multiple forms |
| **post-** | *after, behind* | posterior, places behind (a specific) part |
| **pre-, pro-** | *before, ahead of* | prenatal, before birth |
| **proto-** | *first or original* | prototroph, having the nutritional requirements of the wildtype |
| **pseudo-** | *false* | pseudodominant, appears but isn't dominant |
| **re-** | *back, again* | reinfect |
| **retro-** | *backward, behind* | retrovirus, to move "backward" from RNA into DNA |
| **sanguin-** | *blood* | consanguineous, indicative of a genetic relationship between individuals |
| **se-** | *apart* | segregate, to move apart |
| **semi-** | *half* | semicircular, having the form of half a circle |
| **septum** | *fence* | septum, membrane or wall between cells |
| **soma-** | *body* | somatic cell |
| **sub-** | *under* | subunit |
| **super-** | *above, over* | supercoil |
| **telo-** | *end* | telomere, the end of a chromosome arm |
| **tetra-** | *four* | tetraploid, having four sets of chromosomes |
| **thermo-** | *heat* | thermophile, heat-loving, able to grow at high temperatures |
| **topo-** | *locale, local* | topoisomerase, altering the local state |
| **toti-** | *wholly* | totipotent, having the ability to generate all cell types |

| Roots and Prefixes | Meaning | Example |
| --- | --- | --- |
| **tra-, trans-** | *across, through* | transgenic, placement of novel DNA into an organism |
| **tri-** | *three* | triploid, three complete sets of chromosomes in a cell |
| **ultra-** | *beyond* | ultraviolet radiation, beyond the band of visible light |
| **vita-** | *life* | vital, alive |
| **vitre-** | *glass* | in vitro, in "glass"—the test tube |
| **viv-** | *live* | in vivo |
| **zyg-** | *a yoke, twin* | zygote |

| Suffixes | Meaning | Example |
| --- | --- | --- |
| **-able** | *able to, capable of* | viable, ability to live or exist |
| **-ac** | *referring to* | cardiac, referring to the heart |
| **-age** | *action, process* | cleavage, process of cleaving |
| **-al** | *relating to, pertaining to* | chromosomal |
| **-ary** | *associated with, relating to* | coronary, associated with the heart |
| **-bryo** | *swollen* | embryo |
| **-cide** | *destroy or kill* | germicide, an agent that kills germs |
| **-ell, -elle** | *small* | organelle |
| **-emia** | *condition of the blood* | anemia, deficiency of red blood cells |
| **-gen** | *an agent that initiates* | pathogen, any agent that produces disease |
| **-gram** | *data that are systematically recorded, a record* | autoradiogram, a record of where radioactive atoms decayed |
| **-ic** | *having the character of* | acidic |
| **-logy** | *the study of* | pathology, the study of changes in structure and function brought on by disease |
| **-lysis** | *loosening or breaking down* | hydrolysis, chemical decomposition of a compound into other compounds as a result of taking up water |
| **-oid** | *like, resembling* | cuboid, shaped as a cube |
| **-oma** | *tumor* | lymphoma, a tumor of the lymphatic tissues |
| **-ory** | *referring to, of* | auditory, referring to hearing |
| **-pathy** | *disease* | osteopathy, any disease of the bone |
| **-phil, -philo** | *like, love* | hydrophilic, water-attracting (e.g., molecules) |
| **-phobia** | *fear* | acrophobia, fear of heights |
| **-phragm** | *partition* | diaphragm, which separates the thoracic and abdominal cavities |
| **-plasm** | *form, shape* | cytoplasm |
| **-scope** | *instrument used for examination* | microscope, instrument used to examine small things |
| **-some** | *body* | chromosome |
| **-stasis** | *arrest, fixation, stand* | epistasis, "stand upon", the genotype at one locus affects the phenotypic expression of the genotype at a second locus |
| **-troph** | *nutrition* | prototroph, nutritional requirements of wildtype organism |
| **-ula, -ule** | *diminutive* | blastula, little "bud", sphere of cells in early development |
| **-zyme** | *ferment* | enzyme |

# Concise Dictionary of Genetics and Genomics

## A

**aberrant 4 : 4 segregation** The presence of equal numbers of alleles among the spores in a single ascus, yet with a spore pair in which the two spores have different genotypes.

**acentric chromosome** A chromosome with no centromere.

**acridine** A chemical mutagen that causes single-base insertions or deletions.

**acrocentric chromosome** A chromosome with the centromere near one end.

**active site** The part of an enzyme at which substrate molecules bind and are converted into their reaction products.

**acylated tRNA** A tRNA molecule to which an amino acid is linked.

**adaptation** Any characteristic of an organism that improves its chance of survival and reproduction in its environment; the evolutionary process by which a species undergoes progressive modification favoring its survival and reproduction in a given environment.

**addition rule** The principle that the probability that any one of a set of mutually exclusive events is realized equals the sum of the probabilities of the separate events.

**additive variance** The magnitude of the genetic variance that results from the additive action of genes and that accounts for the genetic component of the parent-offspring correlation; the value that the genetic variance would assume if there were no dominance or interaction of alleles affecting the trait.

**adenine (A)** A nitrogenous purine base found in DNA and RNA.

**adenyl cyclase** The enzyme that catalyzes the synthesis of cyclic AMP.

**adjacent segregation** Segregation of a heterozygous reciprocal translocation in which a translocated chromosome and a normal chromosome segregate together, producing an aneuploid gamete.

**adjacent-1 segregation** Segregation from a heterozygous reciprocal translocation in which homologous centromeres go to opposite poles of the first-division spindle.

**adjacent-2 segregation** Segregation from a heterozygous reciprocal translocation in which homologous centromeres go to the same pole of the first-division spindle.

**agarose** A component of agar used as a gelling agent in gel electrophoresis; its value is that few molecules bind to it, so it does not interfere with electrophoretic movement.

**agglutination** The clumping or aggregation, as of viruses or blood cells, caused by an antigen-antibody interaction.

**albinism** Absence of melanin pigment in the iris, skin, and hair of an animal; absence of chlorophyll in plants.

**alkaptonuria** A recessively inherited metabolic disorder in which a defect in the breakdown of tyrosine leads to excretion of homogentisic acid (alkapton) in the urine.

**alkylating agent** An organic compound capable of transferring an alkyl group to other molecules.

**allele** Any of the alternative forms of a given gene.

**allele frequency** The relative proportion of all alleles of a gene that are of a designated type.

**allopolyploid** A polyploid formed by hybridization between two different species.

**allosteric protein** Any protein whose activity is altered by a change of shape induced by binding a small molecule.

**allozyme** Any of the alternative electrophoretic forms of a protein coded by different alleles of a single gene.

**α helix** A fundamental unit of protein folding in which successive amino acids form a right-handed helical structure held together by hydrogen bonding between the amino and carboxyl components of the peptide bonds in successive loops of the helix.

**alpha satellite** *See* **alphoid DNA.**

**alphoid DNA** Highly repetitive DNA sequences associated with mammalian centromeres.

**alternate segregation** Segregation from a heterozygous reciprocal translocation in which both parts of the reciprocal translocation separate from both nontranslocated chromosomes in the first meiotic division.

**amber codon** Common jargon for the UAG stop codon; an amber mutation is a mutation in which a sense codon has been altered to UAG.

**Ames test** A bacterial test for mutagenicity; also used to screen for potential carcinogens.

**amino acid** Any one of a class of organic molecules that have an amino group and a carboxyl group; 20 different amino acids are the usual components of proteins.

**amino acid attachment site** The 3′ terminus of a tRNA molecule at which an amino acid is attached.

**aminoacylated tRNA** A tRNA covalently attached to its amino acid; charged tRNA.

**aminoacyl site** The tRNA-binding site on the ribosome to which each incoming charged tRNA is initially bound.

**aminoacyl tRNA synthetase** The enzyme that attaches the correct amino acid to a tRNA molecule.

**amino terminus** The end of a polypeptide chain at which the amino acid bears a free amino group ($-NH_2$).

**amniocentesis** A procedure for obtaining fetal cells from the amniotic fluid for the diagnosis of genetic abnormalities.

**amorph** A mutation in which the function of the gene product is completely abolished.

**amplification** Process by which large quantities of a DNA fragment can be obtained by selective replication.

**amplified fragment length polymorphism (AFLP)** A type of DNA marker assayed by PCR amplification of genomic restriction fragments onto whose ends specific oligonucleotide sequences have been ligated to serve as primer-binding sites.

**analog** *See* **base analog.**

**anaphase** The stage of mitosis or meiosis in which chromosomes move to opposite ends of the spindle. In anaphase I of meiosis, homologous centromeres separate; in anaphase II, sister centromeres separate.

**anaphase-promoting complex (APC/C)** A ubiquitin-protein ligase that targets proteins whose destruction enables a cell to transition from metaphase into anaphase.

**aneuploid** A cell or organism in which the chromosome number is not an exact multiple of the haploid number; more generally, aneuploidy refers to a condition in which particular genes or chromosomal regions are present in extra or fewer copies compared with wildtype.

**annealing** The coming together of two strands of nucleic acid that have complementary base sequences to form a duplex molecule.

**antibiotic-resistant mutant** A cell or organism that carries a mutation conferring resistance to an antibiotic.

**antibiotic-resistant mutation** A mutation conferring resistance to one or more antibiotics.

**antibody** A blood protein produced in response to a specific antigen and capable of binding with the antigen.

**anticipation** A phenomenon observed in certain trinucleotide-repeat diseases in which, in each successive generation, the mean age of onset of the disease is less than in the previous generation.

**anticodon** The three bases in a tRNA molecule that are complementary to the three-base codon in mRNA.

**antigen** A substance able to stimulate the production of antibodies.

**antigen presentation** An immune process in which a cell carrying an antigen bound to an MHC protein stimulates a responsive cell.

**antigen-presenting cell** Any cell, such as a macrophage, expressing certain products of the major histocompatibility complex and capable of stimulating T cells that have appropriate antigen receptors.

**antimorph** A mutation whose phenotypic effects are opposite to those of the wildtype allele.

**antiparallel** The chemical orientation of the two strands of a double-stranded nucleic acid molecule; the 5′-to-3′ orientations of the two strands are opposite one another.

**antisense RNA** An RNA molecule complementary in nucleotide sequence to all or part of a messenger RNA.

**antiterminator** A sequence in RNA that allows transcription to continue through the gene.

**AP endonuclease** An endonuclease that cleaves a DNA strand at any site at which the deoxyribose lacks a base.

**apoptosis** Genetically programmed cell death, especially in embryonic development.

**aporepressor** A protein converted into a repressor by binding with a particular molecule.

**AP repair** Repair of duplex DNA in which one or more deoxyribose sugars has lost its base and become an apurinic or apyrimidinic nucleotide.

**Archaea** One of the three major classes of organisms; previously called archaebacteria, they are unicellular microorganisms, usually found in extreme environments, that differ as much from bacteria as either group differs from eukaryotes. An archaeon is a single cell or species of archaea. *See also* **Bacteria.**

**artificial selection** Selection, imposed by a breeder, in which organisms of only certain phenotypes are allowed to breed.

**ascospore** *See* **ascus.**

**ascus** A sac containing the spores (ascospores) produced by meiosis in certain groups of fungi, including *Neurospora* and yeast.

**asexual polyploidization** Formation of a polyploid through the fusion of normal gametes followed by endoreduplication of the chromosome sets in the hybrid.

**A site** *See* **aminoacyl site.**

**assortative mating** Nonrandom selection of mating partners with respect to one or more traits; it is positive when like phenotypes mate more frequently than would be expected by chance and is negative when the reverse occurs.

**ATP** Adenosine triphosphate, the primary molecule for storing chemical energy in a living cell.

**attached-X chromosome** A chromosome in which two X chromosomes are joined to a common centromere; also called a compound-X chromosome.

**attachment site** The base sequence in a bacterial chromosome at which bacteriophage DNA can integrate to form a prophage; either of the two attachment sites that flank an integrated prophage.

**attenuation** *See* **attenuator.**

**attenuator** A regulatory base sequence near the beginning of an mRNA molecule at which transcription can be terminated; when an attenuator is present, it precedes the coding sequences.

**attractant** A substance that draws organisms to it through chemotaxis.

**AUG** The usual initiation codon for polypeptide synthesis as well as the internal codon for methionine.

**autogamy** Reproductive process in ciliates in which the genotype is reconstituted from the fusion of two identical haploid nuclei resulting from mitosis of a single product of meiosis; the result is homozygosity for all genes.

**autonomous determination** Cellular differentiation determined intrinsically and not dependent on external signals or interactions with other cells.

**autopolyploidy** Type of polyploidy in which there are more than two sets of homologous chromosomes.

**autoradiography** A process for the production of a photographic image of the distribution of a radioactive substance in a cell or large cellular molecule; the image is produced on a photographic emulsion by decay emission from the radioactive material.

**autoregulation** Regulation of gene expression by the product of the gene itself.

**autosomes** All chromosomes other than the sex chromosomes.

**auxotroph** A mutant microorganism unable to synthesize a compound required for its growth but able to grow if the compound is provided.

## B

**backcross** The cross of an $F_1$ heterozygote with a partner that has the same genotype as one of its parents.

**Bacteria** One of the major kingdoms of living things; includes most bacteria. *See also* **Archaea.**

**bacterial artificial chromosome (BAC)** A recombinant DNA plasmid in which a large fragment of DNA has been ligated into a suitable vector, making possible its replication and segregation in bacterial cells.

**bacterial attachment site** *See* **attachment site.**

**bacterial transformation** *See* **transformation.**

**bacteriophage** A virus that infects bacterial cells; commonly called a phage.

**band** In gel electrophoresis, a compact region of heavy staining due to the accumulation of molecules of a given size; in *Drosophila* genetics, any of the horizontal striations found in the giant polytene chromosomes in the salivary glands of third in-star larvae; in metaphase chromosomes, any of the horizontal striations revealed by special staining procedures.

**Barr body** A darkly staining body found in the interphase nucleus of certain cells of female mammals; consists of the condensed, inactivated X chromosome.

**basal transcription factors** Transcription factors that are associated with transcription of a wide variety of genes.

**base** Single-ring (pyrimidine) or double-ring (purine) component of a nucleic acid.

**base analog** A chemical so similar to one of the normal bases that it can be incorporated into DNA.

**base composition** The relative proportions of the bases in a nucleic acid or of A + T versus G + C in duplex DNA.

**base excision repair** A mechanism of DNA repair initiated by removal of a mismatched or damaged base from its associated deoxyribonucleotide sugar.

**base pair** A pair of nitrogenous bases, most commonly one purine and one pyrimidine, held together by hydrogen bonds in a double-stranded region of a nucleic acid molecule; commonly abbreviated bp, the term is often used interchangeably with the term *nucleotide pair.* The normal base pairs in DNA are AT and GC.

**base pairing** Specific hydrogen bonding of A with T or of G with C in duplex DNA.

**base stacking** The tendency of the hydrophobic faces of nucleotides to come together in solution to minimize contact with the water molecules.

**base-substitution mutation** Incorporation of an incorrect base into a DNA duplex.

**Bayes theorem** A classical equation in probability theory used for calculating conditional probabilities.

**B cells** A class of white blood cells, derived from bone marrow, with the potential for producing antibodies.

**becquerel** A unit of radiation equal to 1 disintegration/second, abbreviated Bq; equal to $2.7 \times 10^{-11}$ curie.

**behavior** The observed response of an organism to stimulation or environmental change.

**B-form DNA** The right-handed helical structure of DNA proposed by Watson and Crick.

**β-galactosidase** An enzyme that cleaves lactose into its glucose and galactose constituents; produced by a gene in the *lac* operon.

**β sheet** A fundamental structure formed in protein folding in which two polypeptide segments are held together in a parallel or antiparallel array by hydrogen bonds.

**bidirectional replication** DNA replication proceeding in opposite directions from an origin of replication. *See also* **θ replication.**

**binding site** A DNA or RNA base sequence that serves as the target for binding by a protein; the site on the protein that does the binding.

**biochemical pathway** A diagram showing the order in which intermediate molecules are produced in the synthesis or degradation of a metabolite in a cell.

**bioinformatics** The use of computers in the interpretation and management of biological data.

**bivalent** A pair of homologous chromosomes, each consisting of two chromatids, associated in meiosis I.

**blastoderm** Structure formed in the early development of an insect larva; the syncytial blastoderm is formed from repeated cleavage of the zygote nucleus without cytoplasmic division; the cellular blastoderm is formed by migration of the nuclei to the surface and their inclusion in separate cell membranes.

**blastula** A hollow sphere of cells formed early in development.

**block** In a biochemical pathway, a defective step at which the normal biochemical reaction cannot be carried out, often because of a defective enzyme encoded by a mutant gene.

**blood-group system** A set of antigens on red blood cells resulting from the action of a series of alleles of a single gene, such as the ABO, MN, or Rh blood-group systems.

**blunt ends** Ends of a DNA molecule in which all terminal bases are paired; the term usually refers to termini formed by a restriction enzyme that does not produce single-stranded ends.

**bootstrapping** Analysis of multiple data sets formed by random sampling with replacement from an actual data set in order to estimate a degree of confidence in a particular branch or branching pattern in a gene tree.

**branched pathway** A metabolic pathway in which an intermediate serves as a precursor for more than one product.

**branch migration** In a DNA duplex invaded by a single strand from another molecule, the process in which the size of the heteroduplex region increases by progressive displacement of the original strand.

**broad-sense heritability** The ratio of genotypic variance to total phenotypic variance.

## C

**cAMP-CRP complex** A regulatory complex consisting of cyclic AMP (cAMP) and the CRP protein; the complex is needed for transcription of certain operons.

**cancer** Any of a large number of diseases characterized by the uncontrolled proliferation of cells.

**candidate gene** A gene proposed to be involved in the genetic determination of a trait because of the role of the gene product in the cell or organism.

**cap** A complex structure at the 5′ termini of most eukaryotic mRNA molecules, having a 5′-5′ linkage instead of the usual 3′-5′ linkage.

**CAP protein** Acronym for catabolite activator protein; also called CRP, for cAMP receptor protein.

**carbon-source mutant** A cell or organism that carries a mutation preventing the use of a particular molecule or class of molecules as a source of carbon.

**carboxyl terminus** The end of a polypeptide chain at which the amino acid has a free carboxyl group (—COOH).

**carcinogen** A physical agent or chemical reagent that causes cancer.

**carrier** A heterozygote for a recessive allele.

**cascade** A series of enzyme activations serving to amplify a weak chemical signal.

**cassette** In bacterial genetics, a circular molecule of duplex DNA containing a target sequence for a site-specific recombinase (integrase) and usually one or more protein-coding sequences. In yeast, either of two sets of inactive mating-type genes that can become active by relocating to the *MAT* locus.

**catabolite activator protein** *See* **CAP protein.**

**categorical trait** A complex trait in which each possible phenotype can be classified into one of a number of discrete categories. Also called a meristic trait.

***cdc* mutants** Mutants in which the cell division cycle is defective.

**cDNA** *See* **complementary DNA.**

**cell cycle** The growth cycle of a cell; in eukaryotes, it is subdivided into $G_1$ (gap 1), S (DNA synthesis), $G_2$ (gap 2), and M (mitosis).

**cell cycle arrest** A blockage in the progression of the cell cycle.

**cell fate** The pathway of differentiation that a cell normally undergoes.

**cell hybrid** Product of fusion of two cells in culture, often from different species.

**cell lineage** The ancestor-descendant relationships of a group of cells in development.

**cell senescence** A normal process in which mammalian cells in culture cease dividing after about 50 doublings.

**cellular oncogene** A gene coding for a cellular growth factor whose abnormal expression predisposes to malignancy. *See also* **oncogene.**

**centimorgan** A unit of distance in the genetic map equal to 1 percent recombination; also called a map unit.

**central dogma** The concept that genetic information is transferred from the nucleotide sequence in DNA to the nucleotide sequence in an RNA transcript to the amino acid sequence of a polypeptide chain.

**centriole** In animal cells, one of a pair of particulate structures composed of an array of microtubules around which the spindle is organized.

**centromere** The region of the chromosome that is associated with spindle fibers and that participates in normal chromosome movement during mitosis and meiosis.

**centrosome** A localized region of clear cytoplasm found near the nucleus of nondividing cells, which in dividing cells duplicates to form the centers around which the spindle is organized. *See also* **centriole.**

**centrosome duplication checkpoint** A mechanism that arrests the cell cycle while the centrosome remains undivided.

**chain elongation** The process of addition of successive amino acids to the growing end of a polypeptide chain.

**chain initiation** The process by which polypeptide synthesis is begun.

**chain termination** The process of ending polypeptide synthesis and releasing the polypeptide from the ribosome; a chain-termination mutation creates a new stop codon, resulting in premature termination of synthesis of the polypeptide chain.

**chaperone** A protein that assists in the three-dimensional folding of another protein.

**Chargaff's rules** In double-stranded DNA, the amount of A equals that of T, and the amount of G equals that of C.

**charged tRNA** A tRNA molecule to which an amino acid is linked; aminoacylated tRNA.

**checkpoint** Any mechanism that arrests the cell cycle until one or more essential processes are completed.

**chemoreceptor** Any molecule, usually a protein, that binds with other molecules and stimulates a behavioral response, such as chemotaxis, sensation, etc. *See also* **chemosensor.**

**chemosensor** In bacteria, a class of proteins that bind with particular attractants or repellents. Chemosensors interact with chemoreceptors to stimulate chemotaxis. *See also* **chemotaxis.**

**chemotaxis** Behavioral response in which organisms move toward chemicals that attract them and away from chemicals that repel them.

**chiasma** The cytological manifestation of crossing-over; the cross-shaped exchange configuration between nonsister chromatids of homologous chromosomes that is visible in prophase I of meiosis. The plural can be either *chiasmata* or *chiasmas.*

**chimeric gene** A gene produced by recombination, or by genetic engineering, that is a mosaic of DNA sequences from two or more different genes.

**chi-square ($\chi^2$)** A statistical quantity calculated to assess the goodness of fit between a set of observed numbers and the theoretically expected numbers.

**chromatid** Either of the longitudinal subunits produced by chromosome replication.

**chromatid interference** In meiosis, the effect that crossing-over between one pair of nonsister chromatids may have on the probability that a second crossing-over in the same chromosome will involve the same or different chromatids; chromatid interference does not generally occur.

**chromatin** The aggregate of DNA and histone proteins that makes up a eukaryotic chromosome.

**chromatin-remodeling complex** Any of a number of complex protein aggregates that reorganizes the nucleosomes of chromatin in preparation for transcription.

**chromocenter** The aggregate of centromeres and adjacent heterochromatin in nuclei of *Drosophila* larval salivary gland cells.

**chromomere** A tightly coiled, bead-like region of a chromosome most readily seen during pachytene of meiosis; the beads are in register in the polytene chromosomes of noncycling cells in which they occur, resulting in the banded appearance of the chromosome.

**chromosome** In eukaryotes, a DNA molecule that contains genes in linear order to which numerous proteins are bound and that has a telomere at each end and a centromere; in prokaryotes, the DNA is associated with fewer proteins, lacks telomeres and a centromere, and is often circular; in viruses, the chromosome is DNA or RNA, single-stranded or double-stranded, linear or circular, and often free of bound proteins.

**chromosome complement** The set of chromosomes in a cell or organism.

**chromosome interference** In meiosis, the phenomenon by which a crossover at one position inhibits the occurrence of another crossover at a nearby position.

**chromosome map** A diagram showing the locations and relative spacing of genes along a chromosome.

**chromosome painting** Use of differentially labeled, chromosome-specific DNA strands for hybridization with chromosomes to label each chromosome with a different color.

**chromosome territory** The three-dimensional region occupied by a particular chromosome in the nucleus of an interphase or noncycling cell.

**chromosome theory of heredity** The theory that chromosomes are the cellular objects that contain the genes.

**circadian rhythm** A biological cycle with an approximate 24-hour period synchronized with the cycle of day and night.

**circular permutation** A permutation of a group of elements in which the elements are in the same order but the beginning of the sequence differs.

***cis* configuration** The arrangement of linked genes in a double heterozygote in which both mutations are present in the same chromosome—for example, $a_1\ a_2/++$; also called coupling.

***cis*-dominant mutation** A mutation that affects the expression of only those genes on the same DNA molecule as that containing the mutation.

***cis* heterozygote** *See* ***cis* configuration.**

***cis–trans* test** *See* **complementation test.**

**cistron** A DNA sequence specifying a single genetic function as defined by a complementation test; a nucleotide sequence coding for a single polypeptide.

***ClB* method** A genetic procedure used to detect X-linked recessive lethal mutations in *Drosophila melanogaster;* so named because one X chromosome in the female parent is marked with an inversion (*C*), a recessive lethal allele (*l*), and the dominant allele for Bar eyes (*B*).

**cleavage division** Mitosis in the early embryo.

**clone** A collection of organisms derived from a single parent and, except for new mutations, genetically identical to that parent; in genetic engineering, the linking of a specific gene or DNA fragment to a replicable DNA molecule, such as a plasmid or phage DNA.

**cloned DNA** A DNA sequence incorporated into a vector molecule capable of replication in the same or a different organism.

**cloning** The process of producing cloned genes.

**coalescence** The convergence of gene lineages onto a common ancestor as the gene genealogies are traced backward in time.

**coding region** The part of a DNA sequence that codes for the amino acids in a protein.

**coding sequence** A region of a DNA strand with the same sequence as is found in the coding region of a messenger RNA, except that T is present in DNA instead of U.

**coding strand** In a transcribed region of a DNA duplex, the strand that is not transcribed.

**codominance** The expression of both alleles in a heterozygote.

**codon** A sequence of three adjacent nucleotides in an mRNA molecule, specifying either an amino acid or a stop signal in protein synthesis.

**coefficient of coincidence** An experimental value obtained by dividing the observed number of double recombinants by the expected number calculated under the assumption that the two events take place independently.

**cohesive end** A single-stranded region at the end of a double-stranded DNA molecule that can adhere to a complementary single-stranded sequence at the other end or in another molecule.

**cointegrate** A DNA molecule, usually circular and formed by recombination, that joins two replicons.

**colchicine** A chemical that prevents formation of the spindle in nuclear division.

**colinearity** The linear correspondence between the order of amino acids in a polypeptide chain and the corresponding sequence of nucleotides in the DNA molecule.

**colony** A visible cluster of cells formed on a solid growth medium by repeated division of a single parental cell and its daughter cells.

**colony hybridization assay** A technique for identifying colonies containing a particular cloned gene; many colonies are transferred to a filter, lysed, and exposed to radioactive DNA or RNA complementary to the DNA sequence of interest, after which colonies that contain a sequence complementary to the probe are located by autoradiography.

**color blindness** In human beings, the usual form of color blindness is X-linked red-green color blindness. Unequal crossing-over between the adjacent red and green opsin pigment genes results in chimeric opsin genes causing mild or severe green-vision defects (deuteranomaly or deuteranopia, respectively) and mild or severe red-vision defects (protanomaly or protanopia, respectively).

**combinatorial control** Strategy of gene regulation in which a relatively small number of time- and tissue-specific positive and negative regulatory elements are used in various combinations to control the expression of a much larger number of genes.

**combinatorial joining** The DNA splicing mechanism by which antibody variability is produced, resulting in many possible combinations of V and J regions in the formation of a light-chain gene and many possible combinations of V, D, and J regions in the formation of a heavy-chain gene.

**common ancestor** Any ancestor of both one's father and one's mother.

**comparative genomics** The analysis of genomic sequences among multiple related species to determine their similarities and differences.

**compartment** In *Drosophila* development, a group of descendants of a small number of founder cells with a determined pattern of development.

**compatible** Blood or tissue that can be transfused or transplanted without rejection.

**complementary DNA (cDNA)** A DNA molecule made by copying RNA with reverse transcriptase; usually abbreviated cDNA.

**complementary pairing** Where each base along one strand of DNA is matched with a base in the opposition position on the other strand.

**complementation** The phenomenon in which two recessive mutations with similar phenotypes result in a wildtype phenotype when both are heterozygous in the same genotype; complementation means that the mutations are in different genes.

**complementation group** A group of mutations that fail to complement one another.

**complementation test** A genetic test to determine whether two mutations are alleles (are present in the same functional gene).

**complete medium** Culture medium containing all required nutrients to support growth and cell division.

**complex** A term used to refer to an ordered aggregate of molecules, as in an enzyme-substrate complex or a DNA-histone complex.

**complex trait** A trait whose expression is determined by multiple interacting genes and usually also by environmental factors.

**compound-X chromosome** *See* **attached-X chromosome.**

**computational genomics** The use of computers in the analysis of genomic sequences.

**condensation** The coiling process by which chromosomes become shorter and thicker during prophase of mitosis or meiosis.

**condensins** A class of proteins that participates in chromosome condensation.

**conditional mutation** A mutation that results in a mutant phenotype under certain (restrictive) environmental conditions but results in a wildtype phenotype under other (permissive) conditions.

**conditional probability** The probability that an event A is realized, given that another event B is known to have occurred; symbolized $Pr\{A \mid B\}$.

**congenital** Present at birth.

**conjugation** A process of DNA transfer in sexual reproduction in certain bacteria; in *E. coli*, the transfer is unidirectional, from donor cell to recipient cell. Also, a mating between cells of *Paramecium*.

**conjugative plasmid** A plasmid encoding proteins and other factors that make possible its transmission between cells.

**consanguineous mating** A mating between relatives.

**consensus sequence** A generalized base sequence derived from closely related sequences found in many locations in a genome or in many organisms; each position in the consensus sequence consists of the base found in the majority of sequences at that position.

**conserved sequence** A base or amino acid sequence that changes very slowly in the course of evolution.

**constant region** The part of the heavy and light chains of an antibody molecule that has the same amino acid sequence among all antibodies derived from the same heavy-chain and light-chain genes.

**constitutive mutation** A mutation that causes synthesis of a particular mRNA molecule (and the protein that it encodes) to take place at a constant rate, independent of the presence or absence of any inducer or repressor molecule.

**contact inhibition** A phenomenon in normal mammalian cells in culture whereby cells cease to grow and divide when they are in close physical proximity.

**contig** A set of cloned DNA fragments overlapping in such a way as to provide unbroken coverage of a contiguous region of the genome; a contig contains no gaps.

**continuous trait** A trait in which the possible phenotypes have a continuous range from one extreme to the other rather than falling into discrete classes.

**coordinate gene** Any of a group of genes that establish the basic anterior-posterior and dorsal-ventral axes of the early embryo.

**coordinate regulation** Control of synthesis of several proteins by a single regulatory element; in prokaryotes, the proteins are usually translated from a single mRNA molecule.

**copy number polymorphism** Process by which a substantial portion of the human genome can be duplicated or deleted in microscopic chunks ranging from 1 kb to 1 Mb.

**core particle** The aggregate of histones and DNA in a nucleosome, without the linking DNA.

**co-repressor** A small molecule that binds with an aporepressor to create a functional repressor molecule.

**correlated response** Change of the mean in one trait in a population accompanying selection for another trait.

**correlation coefficient** A measure of association between pairs of numbers, equaling the covariance divided by the product of the standard deviations.

**cosuppression** A phenomenon, originally discovered in transgenic plants, in which extra copies of a gene result in a shutdown of expression of all copies that are present in the genome.

**Cot** Mathematical product of the initial concentration of nucleic acid ($C_0$) and the time ($t$).

**Cot curve** A graph of percent nucleic acid renaturation as a function of the initial concentration ($C_0$) multiplied by time.

**cotransduction** Transduction of two or more linked genetic markers by one transducing particle.

**cotransformation** Transformation in bacteria of two genetic markers carried on a single DNA fragment.

**counterselected marker** A mutation used to prevent growth of a donor cell in an Hfr $\times$ $F^-$ bacterial mating.

**coupled processes** Processes that are biochemically or functionally connected such that the occurrence of the earlier process initiates or regulates the later process.

**coupled transcription-translation** In prokaryotes, the translation of an mRNA molecule before its transcription is completed.

**coupling** *See* ***cis*** **configuration.**

**courtship song** In *Drosophila*, the regular oscillation in the time interval between pulses in successive episodes of wing vibration that the male directs toward the female during mating behavior.

**covalent bond** A chemical bond in which electrons are shared.

**covalent circle** *See* **covalently closed circle.**

**covalently closed circle** A ring-shaped duplex DNA molecule whose ends are joined by covalent bonds.

**covariance** A measure of association between pairs of numbers that is defined as the average product of the deviations from the respective means.

**CpG island** A region of DNA with a high density of CpG dinucleotides; in mammalian cells it is often associated with the promoter regions of genes.

**C region** *See* **constant region.**

**Crick strand** The lower strand of duplex DNA when the strands are conventionally represented as parallel horizontal lines with the 5′ end of the top strand at the extreme left.

**crossing-over** A process of exchange between nonsister chromatids of a pair of homologous chromosomes that results in the recombination of linked genes.

**crown gall tumor** Plant tumor caused by infection with *Agrobacterium tumefaciens*.

**CRP protein** Acronym for cyclic AMP repressor protein, which binds with cAMP and regulates the activity of inducible operons in prokaryotes.

**cryptic splice site** A potential splice site not normally used in RNA processing unless a normal splice site is blocked or mutated.

**curie** A unit of radiation equal to $3.7 \times 10^{10}$ disintegrations/second, abbreviated Ci; equal to $3.7 \times 10^{10}$ becquerels.

**cut-and-paste elements** Transposable elements that are not replicated in the process of transposition but are cleaved ("cut") from an existing site in the genome and inserted ("pasted") at a new target site.

**C-value paradox** The often massive, counterintuitive, and seemingly arbitrary differences in genome size observed among eukaryotic organisms.

**cyclic adenosine monophosphate (cAMP)** *See* **cyclic AMP.**

**cyclic AMP** Cyclic adenosine monophosphate, usually abbreviated cAMP, used in the regulation of cellular processes; cAMP synthesis is regulated by glucose metabolism; its action is mediated by the CRP protein in prokaryotes and by protein kinases (enzymes that phosphorylate proteins) in eukaryotes.

**cyclic AMP receptor protein (CRP)** *See* **CRP protein.**

**cyclin** Any of a number of proteins that help regulate the cell cycle and whose abundance rises and falls rhythmically during the cell cycle.

**cyclin–CDK complex** Protein complex formed by the interaction between a cyclin and a cyclin-dependent protein kinase (CDK).

**cyclin-dependent protein kinase (CDK)** Any of a number of proteins that are activated by combining with a cyclin and that regulate the cell cycle by phosphorylation of other proteins.

**cytogenetics** The study of the genetic implications of chromosome structure and behavior.

**cytokinesis** Division of the cytoplasm.

**cytological map** Diagrammatic representation of a chromosome.

**cytoplasmic inheritance** Transmission of hereditary traits through self-replicating factors in the cytoplasm—for example, mitochondria and chloroplasts.

**cytoplasmic male sterility** Type of pollen abortion in maize with cytoplasmic inheritance through the mitochondria.

**cytosine (C)** A nitrogenous pyrimidine base found in DNA and RNA.

**cytotoxic T cell** Type of white blood cell that participates directly in attacking cells marked for destruction by the immune system.

## D

**daughter strand** A newly synthesized DNA or chromosome strand.

**deamination** Removal of an amino group ($-NH_2$) from a molecule.

**decaploid** An organism with ten sets of chromosomes.

**deficiency** *See* **deletion.**

**degeneracy** The feature of the genetic code in which an amino acid corresponds to more than one codon; also called redundancy.

**degenerate code** *See* **degeneracy.**

**degree of dominance** The extent to which the phenotype of a heterozygous genotype resembles one of the homozygous genotypes.

**degrees of freedom** An integer that determines the significance level of a particular statistical test. In the goodness-of-fit type of chi-square test in which the expected numbers are not based on any quantities estimated from the data themselves, the number of degrees of freedom is one less than the number of classes of data.

**deletion** Loss of a segment of the genetic material from a chromosome; also called deficiency.

**deletion mapping** The use of overlapping deletions to locate a gene on a chromosome or a genetic map.

**deme** *See* **local population.**

**denaturation** Loss of the normal three-dimensional shape of a macromolecule without breaking covalent bonds, usually accompanied by loss of its biological activity; conversion of DNA from the double-stranded into the single-stranded form; unfolding of a polypeptide chain.

**denaturation mapping** An electron-microscopic technique for localizing regions of a double-stranded DNA molecule by noting the positions at which the individual strands separate in the early stages of denaturation.

**denatured DNA** Duplex DNA whose strands have become partially or completely disassociated.

**deoxyribonuclease** An enzyme that breaks sugar-phosphate bonds in DNA, forming either fragments or the component nucleotides; abbreviated DNase.

**deoxyribonucleic acid** *See* **DNA.**

**deoxyribose** The five-carbon sugar present in DNA.

**dependent** In probability, two events are dependent if the occurrence of one provides information regarding the occurrence of the other.

**depurination** Removal of purine bases from DNA.

**derepression** Activation of a gene or set of genes by inactivation of a repressor.

**deuteranomaly** *See* **color blindness.**

**deuteranopia** *See* **color blindness.**

**deviation** In statistics, a difference from an expected value.

**diakinesis** The substage of meiotic prophase I, immediately preceding metaphase I, in which the bivalents attain maximum shortening and condensation.

**dicentric chromosome** A chromosome with two centromeres.

**dideoxyribose** A deoxyribose sugar lacking the 3′ hydroxyl group; when incorporated into a polynucleotide chain, it blocks further chain elongation.

**dideoxy sequencing method** Procedure for DNA sequencing in which a template strand is replicated from a particular primer sequence and terminated by the incorporation of a nucleotide that contains dideoxyribose instead of deoxyribose; the resulting fragments are separated by size via electrophoresis.

**differentiation** The complex changes of progressive diversification in cellular structure and function that take place in the development of an organism; for a particular line of cells, this results in a continual restriction in the types of transcription and protein synthesis of which each cell is capable.

**diffuse centromere** A centromere that is dispersed throughout the chromosome, as indicated by the dispersed binding of spindle fibers, rather than being concentrated at a single point.

**dihybrid** Heterozygous at each of two loci; progeny of a cross between true-breeding or homozygous strains that differ genetically at two loci.

**dimer** A protein formed by the association of two polypeptide subunits.

**diploid** A cell or organism with two complete sets of homologous chromosomes.

**diplotene** The substage of meiotic prophase I, immediately following pachytene and preceding diakinesis, in which pairs of sister chromatids that make up a bivalent (tetrad) begin to separate from each other and chiasmata become visible.

**direct repeat** Copies of an identical or very similar DNA or RNA base sequence in the same molecule and in the same orientation.

**discontinuous variation** Variation in which the phenotypic differences for a trait fall into two or more discrete classes.

**disease gene** Any gene with one or more alleles that increase the risk of occurrence of a disease phenotype in an individual carrying such an allele.

**disjunction** Separation of homologous chromosomes to opposite poles of a division spindle during anaphase of a mitotic or meiotic nuclear division.

**displacement loop** Loop formed when part of a DNA strand is dislodged from a duplex molecule because of the partner strand's pairing with another molecule.

**distance matrix** A matrix showing the amount of sequence divergence between all possible pairs of a set of protein or nucleic acid sequences.

**distribution** In quantitative genetics, the mathematical relation that gives the proportion of members in a population that have each possible phenotype.

**disulfide bond** Two sulfur atoms covalently linked; found in proteins when the sulfur atoms in two different cysteines are joined.

**dizygotic twins** Twins that result from the fertilization of separate ova and that are genetically related as siblings; also called fraternal twins.

**D loop** *See* **displacement loop.**

**DNA** Deoxyribonucleic acid, the macromolecule, usually composed of two polynucleotide chains in a double helix, that is the carrier of the genetic information in all cells and many viruses.

**DNA chip** An array of tiny dots ("microarray") of DNA molecules immobilized on glass or on another solid support used for hybridization with a probe of fluorescently labeled nucleic acid.

**DNA cloning** *See* **cloned DNA.**

**DNA damage bypass** A mechanism of DNA repair in which a damaged region of DNA is skipped over during replication and replaced later by a region of DNA excised from the equivalent strand in the daughter duplex.

**DNA damage checkpoint** A mechanism that arrests the cell cycle while damaged DNA remains unrepaired.

**DNA fingerprinting** *See* **DNA typing.**

**DNA gyrase** One of a class of enzymes called topoisomerases, which function during DNA replication to relax positive supercoiling of the DNA molecule and which introduce negative supercoiling into nonsupercoiled molecules early in the life cycle of many phages.

**DNA ligase** An enzyme that catalyzes formation of a covalent bond between adjacent 5′-P and 3′-OH termini in a broken polynucleotide strand of double-stranded DNA.

**DNA looping** A mechanism by which enhancers that are distant from the immediate proximity of a promoter can still regulate transcription; the enhancer and promoter, both bound with suitable protein factors, come into indirect physical contact by means of the looping out of the DNA between them. The physical interaction stimulates transcription.

**DNA marker** Any feature of genomic DNA that differs among individuals and that can be used to disinguish homologous DNA molecules among the individuals in a population or in a pedigree.

**DNA methylase** An enzyme that adds methyl groups ($-CH_3$) to certain bases, particularly cytosine.

**DNA microarray** Tiny dots of each of a large number of DNA fragments or oligonucleotides immobilized on a solid surface about the size of a postage stamp; typically used for hybridizing DNA or RNA samples to identify genotypes, copy-number polymorphisms, or gene-expression levels.

**DNA polymerase** Any enzyme that catalyzes the synthesis of DNA from deoxynucleoside 5′-triphosphates, using a template strand.

**DNA polymerase complex** Aggregate of proteins including DNA polymerase that functions in DNA replication.

**DNA polymerase I (Pol I)** In *E. coli* DNA replication, the enzyme that fills in gaps in the lagging strand and replaces the RNA primers with DNA.

**DNA polymerase III (Pol III)** The major DNA replication enzyme in *E. coli*.

**DNA polymorphism** Any feature of genomic DNA that frequently differs among individuals; rare mutations are not usually considered polymorphisms. *See also* **DNA marker.**

**DNA repair** Any of several different processes for restoration of the correct base sequence of a DNA molecule into which incorrect bases have been incorporated or whose bases have been chemically modified.

**DNA replication** The copying of a DNA molecule.

**DNase** *See* **deoxyribonuclease.**

**DNA typing** Electrophoretic identification of individual persons by the use of DNA probes for highly polymorphic regions of the genome, such that the genome of virtually every person exhibits a unique pattern of bands; sometimes called DNA fingerprinting.

**DNA uracyl glycosylase** An enzyme that removes uracil bases when they occur in double-stranded DNA.

**domain** A folded region of a polypeptide chain that is spatially somewhat isolated from other folded regions.

**dominance** Condition in which a heterozygote expresses a trait in the same manner as the homozygote for one of the alleles. The allele or the corresponding phenotypic trait expressed in the heterozygote is said to be dominant.

**dominance variance** The part of the genotypic variance that results from the dominance effects of alleles affecting the trait.

**dominant allele** An allele associated with a phenotype that is expressed when the allele is either heterozygous or homozygous.

**dominant-negative mutation** A mutation whose presence knocks out the function of the homologous wildtype allele; the usual mechanism is one in which mutant protein subunits combine with wildtype subunits to produce inactive multimers.

**dosage compensation** A mechanism regulating X-linked genes such that their activities are equal in males and females; in mammals, random inactivation of one X chromosome in females results in equal amounts of the products of X-linked genes in males and females.

**double-strand break-repair model** A model of recombination in which the exchange process is initiated by a duplex DNA molecule containing a double-stranded break.

**double-stranded DNA** A DNA molecule consisting of two antiparallel strands that are complementary in nucleotide sequence.

**double-Y syndrome** The clinical features of the karyotype 47,XYY.

**Down syndrome** The clinical features of the karyotype 47,+21 (trisomy 21).

**drug-resistance plasmid** A plasmid that contains genes whose products inactivate certain antibiotics.

**duplex DNA** A double-stranded DNA molecule.

**duplication** A chromosome aberration in which a chromosome segment is present more than once in the haploid genome; if the two segments are adjacent, the duplication is a tandem duplication.

**dynamic mutation** Term applied to certain sequences composed of tandem repeats that show genetic instability owing to a relatively high frequency of change in the number of copies of the repeat from one generation to the next.

## E

**ectopic expression** Gene expression that occurs in the wrong tissues or the wrong cell types.

**ectopic rerecombination** Recombination between homologous DNA sequences present at nonhomologous sites in the genome, such as between two copies of a transposable element.

**editing function** The activity of DNA polymerases that removes incorrectly incorporated nucleotides; also called the proofreading function.

**effector** A molecule that brings about a regulatory change in a cell, as by induction.

**electrophoresis** A technique used to separate molecules on the basis of their different rates of movement in response to an applied electric field, typically through a gel.

**electroporation** Introduction of DNA fragments into a cell by means of an electric field.

**elongation** Addition of amino acids to a growing polypeptide chain or of nucleotides to a growing nucleic acid chain.

**embryo** An organism in the early stages of development (the second through seventh weeks in human beings).

**embryoid** A small mass of dividing cells formed from haploid cells in anthers that can give rise to a mature haploid plant.

**embryonic induction** Developmental process in which the fate of a group of cells is determined by interactions with nearby cells in the embryo.

**embryonic stem cells** Cells in the blastocyst that give rise to the body of the embryo.

**endonuclease** An enzyme that breaks internal phosphodiester bonds in a single- or double-stranded nucleic acid molecule; usually specific for either DNA or RNA.

**endoreduplication** Doubling of the chromosome complement because of chromosome replication and centromere division without nuclear or cytoplasmic division.

**endosperm** Nutritive tissues formed adjacent to the embryo in most flowering plants; in most diploid plants, the endosperm is triploid.

**endosymbiont theory** Theory that mitochondria and chloroplasts were originally free-living organisms that invaded ancestral eukaryotes, first perhaps as parasites, later becoming symbionts.

**enhancer** A base sequence in eukaryotes and eukaryotic viruses that increases the rate of transcription of nearby genes; the defining characteristics are that it need not be adjacent to the transcribed gene and that the enhancing activity is independent of orientation with respect to the gene.

**enhancer trap** Mutagenesis strategy in which a reporter gene with a weak promoter is introduced at many random locations in the genome; tissue-specific expression of the reporter gene identifies an insertion near a tissue-specific enhancer.

**environmental variance** The part of the phenotypic variance that is attributable to differences in environment.

**enzyme** A protein that catalyzes a specific biochemical reaction and is not itself altered in the process.

**epigenetic** Inherited changes in gene expression resulting from altered chromatin structure or DNA modification (usually methylation) rather than changes in DNA sequence.

**episome** A DNA element that can exist in the cell either as an autonomously replicating entity or become incorporated into the genome.

**epistasis** A term referring to an interaction between nonallelic genes in their effects on a trait. Generally, *epistasis* means any type of interaction in which the genotype at one locus affects the phenotypic expression of the genotype at another locus. *See also* **hypostasis.**

**epistatic gene** In developmental genetics, any gene whose mutant phenotype masks that of another gene. *See* **hypostatic gene.**

**equational division** Term applied to the second meiotic division because the haploid chromosome complement is retained throughout.

**equilibrium** *See* **genetic equilibrium.**

**equilibrium centrifugation in a density gradient** A method for separating macromolecules of differing density by high-speed centrifugation in a solution of approximately the same density as the molecules to be separated; the centrifugation produces a gradient of density inside the centrifuge tube, and each macromolecule migrates to a position in the gradient that matches its own density.

**equilibrium density-gradient centrifugation** *See* **equilibrium centrifugation in a density gradient.**

**erythroblastosis fetalis** Hemolytic disease of the newborn; blood cell destruction that occurs when anti-$Rh^+$ antibodies in a mother cross the placenta and attack $Rh^+$ cells in a fetus.

**E site** *See* **exit site.**

**estrogen** A female sex hormone; of great interest in the study of cellular regulation in eukaryotes because the number of types of target cells is large.

**ethidium bromide** A fluorescent molecule that binds to DNA and changes its density; used to purify supercoiled DNA molecules and to localize DNA in gel electrophoresis.

**euchromatin** A region of a chromosome that has normal staining properties and undergoes the normal cycle of condensation; relatively uncoiled in the interphase nucleus (compared with condensed chromosomes), it apparently contains most of the genes.

**Eukarya** One of the major kingdoms of living organisms, in which the cells have a true nucleus and divide by mitosis or meiosis.

**eukaryote** A cell with a true nucleus (DNA enclosed in a membranous envelope) in which cell division takes place by mitosis or meiosis; an organism composed of eukaryotic cells.

**euploid** A cell or an organism having a chromosome number that is an exact multiple of the haploid number.

**event** Any combination of elementary outcomes.

**evolution** Cumulative change in the genetic characteristics of a species through time, resulting in greater adaptation.

**excision** Removal of a DNA fragment from a molecule.

**excisionase** An enzyme that is needed for prophage excision; works together with an integrase.

**excision repair** Type of DNA repair in which segments of a DNA strand that are chemically damaged are removed enzymatically and then resynthesized, using the other strand as a template.

**exit site** The tRNA-binding site on the ribosome that binds each uncharged tRNA just prior to its release.

**exon** The sequences in a gene that are retained in the messenger RNA after the introns are removed from the primary transcript.

**exon shuffle** The theory that new genes can evolve by the assembly of separate exons from preexisting genes, each coding for a discrete functional domain in the new protein.

**exonuclease** An enzyme that removes a terminal nucleotide in a polynucleotide chain by cleavage of the terminal phosphodiester bond; nucleotides are removed successively, one by one; usually specific for either DNA or RNA and for either single-stranded or double-stranded nucleic acids. A 5′-to-3′ exonuclease cleaves successive nucleotides from the 5′ end of the molecule; a 3′-to-5′ exonuclease cleaves successive nucleotides from the 3′ end.

**expressed sequence tags (ESTs)** Nucleotide sequences of cDNA molecules that have been reverse-transcribed from messenger RNAs.

**expressivity** The degree of phenotypic expression of a gene.

**extranuclear inheritance** Inheritance mediated by self-replicating cellular factors located outside the nucleus—for example, in mitochondria or chloroplasts.

## F

**familial** Tending to be present in multiple generations of a pedigree.

**fate** In development, the final product of differentiation of a cell or of a group of cells.

**fate map** A diagram of the insect blastoderm identifying the regions from which particular adult structures derive.

**favism** Destruction of red blood cells (anemia) triggered by eating raw fava beans or by coming into contact with certain other environmental agents; associated with a deficiency of the enzyme glucose-6-phosphate dehydrogenase.

**feedback inhibition** Inhibition of an enzyme by the product of the enzyme or, in a metabolic pathway, by a product of the pathway.

**fertility** The ability to mate and produce offspring.

**fertility factor** *See* **F plasmid.**

**fertility restorer** A suppressor of cytoplasmic male sterility.

**F factor** *See* **F plasmid.**

**$F_1$ generation** The first generation of descent from a given mating.

**$F_2$ generation** The second generation of descent, produced by intercrossing or self-fertilizing $F_1$ organisms.

**first-division segregation** Separation of a pair of alleles into different nuclei in the first meiotic division; happens when there is no crossing-over between the gene and the centromere in a particular cell.

**first meiotic division** The meiotic division that reduces the chromosome number; sometimes called the reduction division.

**fitness** A measure of the average ability of organisms with a given genotype to survive and reproduce.

**5′-P group** The end of a DNA or RNA strand that terminates in a free phosphate group not connected to a sugar farther along.

**five prime end (5′ end)** The end of a polynucleotide chain that terminates with a 5′ carbon unattached to another nucleotide.

**fixed allele** An allele whose allele frequency equals 1.0.

**flagella** The whip-like projections from the surface of bacterial cells that serve as the organs of locomotion.

**flower ABC model** A model of floral determination in which a unique combination of gene activities present in each whorl of the floral meristem results in the differentiation of a distinct organ in the mature flower.

**fluctuation test** A statistical test used to determine whether bacterial mutations occur at random or are produced in response to selective agents.

**folded chromosome** The form of DNA in a bacterial cell in which the circular DNA is folded to have a compact structure; contains protein.

**folding domain** A short region of a polypeptide chain within which interactions between amino acids result in a three-dimensional conformation that is attained relatively independently of the folding of the rest of the molecule.

**forward mutation** A change from a wildtype allele to a mutant allele.

**founder effect** Random genetic drift that results when a group of founders of a population are not genetically representative of the population from which they were derived.

**F plasmid** A bacterial plasmid—often called the F factor, fertility factor, or sex plasmid—that is capable of transferring itself from a host ($F^+$) cell to a cell not carrying an F factor ($F^-$ cell); when an F factor is integrated into the bacterial chromosome (in an Hfr cell), the chromosome becomes transferrable to an $F^-$ cell during conjugation.

**F′ plasmid** An F plasmid that contains genes obtained from the bacterial chromosome in addition to plasmid genes; formed by aberrant excision of an integrated F factor, taking along adjacent bacterial DNA.

**fragile-X chromosome** A type of X chromosome containing a site toward the end of the long arm that tends to break in cultured cells that are starved for DNA precursors; causes fragile-X syndrome.

**fragile-X syndrome** A common form of inherited mental retardation associated with a type of X chromosome prone to trinucleotide expansion.

**frameshift mutation** A mutational event caused by the insertion or deletion of one or more nucleotide pairs in a gene, resulting in a shift in the reading frame of all codons following the mutational site.

**fraternal twins** Twins that result from the fertilization of separate ova and are genetically related as siblings; also called dizygotic twins.

**free radical** A highly reactive molecule produced when ionizing radiation interacts with water; free radicals are potent oxidizing agents.

**frequency of recombination** The proportion of gametes carrying combinations of alleles that are not present in either parental chromosome.

**functional genomics** Use of DNA microarrays and other methods to study the coordinated expression of many genes simultaneously.

## G

**gain-of-function mutation** Mutation in which a gene is overexpressed or inappropriately expressed.

**galactosidase** *See* **β-galactosidase.**

**gamete** A mature reproductive cell, such as sperm or egg in animals.

**gametophyte** In plants, the haploid part of the life cycle that produces the gametes by mitosis.

**gap gene** Any of a group of genes that control the development of contiguous segments or parasegments in *Drosophila* such that mutations result in gaps in the pattern of segmentation.

**gastrula** Stage in early animal development marked by extensive cell migration.

**G bands** Bands in mammalian or other chromosomes resulting from treatment with the Giemsa staining reagent.

**gel electrophoresis** *See* **electrophoresis.**

**gene** The hereditary unit containing genetic information transcribed into an RNA molecule that is processed and either functions directly or is translated into a polypeptide chain; a gene can mutate to various forms called alleles.

**gene amplification** A process in which certain genes undergo differential replication either within the chromosome or extrachromosomally, increasing the number of copies of the gene.

**gene cloning** *See* **cloned DNA.**

**gene conversion** The phenomenon in which the products of a meiotic division in an *Aa* heterozygous genotype are in some ratio other than the expected 1*A* : 1*a*—for example, 3*A* : 1*a*, 1*A* : 3*a*, 5*A* : 3*a*, or 3*A* : 5*a*.

**gene dosage** Number of gene copies.

**gene expression** The multistep process by which a gene is regulated and its product synthesized.

**gene flow** Exchange of genes among populations resulting from either dispersal of gametes or migration of individuals; also called migration.

**gene fusion** Juxtaposition of parts of two or more genes to create a new genetic unit capable of expression.

**gene library** A large collection of cloning vectors containing a complete (or nearly complete) set of fragments of the genome of an organism.

**gene pool** The totality of genetic information in a population of organisms.

**gene product** A term used for the polypeptide chain translated from an mRNA molecule transcribed from a gene; if the RNA is not translated (for example, ribosomal RNA), the RNA molecule is the gene product.

**gene regulation** Processes by which gene expression is controlled in response to external or internal signals.

**gene silencing** Any mechanism that results in the failure of a wildtype gene to be expressed.

**gene targeting** Procedure for genetic engineering of genes *in vivo*, usually through homologous recombination.

**gene therapy** Method for treatment of disease by means of DNA rather than proteins or other pharmaceuticals.

**gene tree** A graph depicting the genealogical relationships among a set of alleles of the same gene in one or more species.

**generalized transduction** *See* **transducing phage.**

**general transcription factor** A protein molecule needed to bind with a promoter before transcription can proceed; transcription factors are necessary, but not sufficient, for transcription, and they are shared among many different promoters.

**genetic analysis** Use of genetic methods to identify the genes whose products participate in pathways of metabolism, development, behavior, or other biological processes.

**genetic code** The set of 64 triplets of bases (codons) corresponding to the twenty amino acids in proteins and the signals for initiation and termination of polypeptide synthesis.

**genetic differentiation** Accumulation of differences in allele frequency between isolated or partially isolated populations.

**genetic divergence** *See* **genetic differentiation.**

**genetic engineering** The linking of two DNA molecules by *in vitro* manipulations for the purpose of generating a novel organism with desired characteristics.

**genetic equilibrium** In population genetics, a situation in which the allele frequencies remain constant from one generation to the next.

**genetic linkage** *See* **linkage.**

**genetic map** *See* **linkage map.**

**genetic mapping** Procedure in which mutant alleles or DNA markers are assigned relative positions along a chromosome on the basis of the frequencies of recombination between them.

**genetic marker** Any pair of alleles whose inheritance can be traced through a mating or through a pedigree.

**genetics** The study of biological heredity.

**genome** The total complement of genes contained in a cell or virus; in eukaryotes, commonly used to refer to all genes present in one complete haploid set of chromosomes.

**genome annotation** Explanatory notes added to a genomic sequence that specify the locations of known or hypothesized functional elements, such as protein-coding genes or microRNAs.

**genome equivalent** The number of clones of a given size that contain the same aggregate amount of DNA as the haploid genome size of an organism.

**genomic DNA** The totality of DNA in a cell or organism.

**genomic imprinting** *See* **imprinting.**

**genomics** Study of the DNA sequence, organization, function, and evolution of genomes.

**genotype** The genetic constitution of an organism or virus, typically with respect to one or a few genes of interest, as distinguished from its appearance, or phenotype.

**genotype-by-environment association** The condition in which genotypes and environments are not in random combinations.

**genotype-by-environment interaction** The condition in which genetic and environmental effects on a trait are not additive.

**genotype-by-sex interaction** A type of genotype-by-environment interaction in which the phenotypic effects of genes differ according to the sex of the individual.

**genotype frequency** The proportion of members of a population that are of a prescribed genotype.

**genotypic variance** The part of the phenotypic variance that is attributable to differences in genotype.

**germ cell** A cell that gives rise to reproductive cells.

**germinal mutation** A mutation in a cell from which gametes are derived, as distinguished from a somatic mutation.

**germ line** Cell lineage consisting of germ cells.

**germ-line mutation** *See* **germinal mutation.**

**glucose-6-phosphate dehydrogenase (G6PD)** An enzyme used in one biochemical pathway for the metabolism of glucose. In human beings, a defective G6PD enzyme (G6PD deficiency) is associated with a form of anemia in which the red blood cells are prone to burst.

**glyphosate** A widely used herbicide.

**goodness of fit** The extent to which observed numbers agree with the numbers expected on the basis of some specified genetic hypothesis.

**G protein** One of a family of signaling proteins that is activated by binding to a molecule of guanosine triphosphate (GTP).

**G quartet** A fold-back structure formed in a single strand of DNA or RNA as a result of unusual base pairing between four guanine nucleotides.

**$G_1$ period** *See* **cell cycle.**

**G6PD** *See* **glucose-6-phosphate dehydrogenase.**

**G6PD deficiency** *See* **glucose-6-phosphate dehydrogenase.**

**$G_1$ restriction point** The "start" point at which cells are arrested in the $G_1$ phase until they become committed to division.

**$G_1$/S transition** Transition between the first "growth" phase of the cell cycle to DNA synthesis.

**$G_2$/M transition** Transition between the second "growth" phase of the cell cycle to mitosis.

**$G_2$ period** *See* **cell cycle.**

**gray** A unit of absorbed radiation equal to 1 joule of radiation energy absorbed per kilogram of tissue, abbreviated Gy; equal to 100 rad.

**guanine (G)** A nitrogenous purine base found in DNA and RNA.

**guide RNA** The RNA template present in telomerase.

**gynandromorph** A sexual mosaic; an individual organism that exhibits both male and female sexual differentiation.

**gyrase** A type of topoisomerase II that cleaves and rejoins both strands of a DNA duplex.

## H

**Haldane's mapping function** A mapping function based on the assumption of no chromosome interference.

**haploid** A cell or organism of a species containing the set of chromosomes normally found in gametes.

**haplotype** The allelic form of each of a set of linked genes present in a single chromosome.

**Hardy–Weinberg principle** The genotype frequencies expected with random mating.

**H chain** *See* **heavy chain.**

**heavy (H) chain** One of the large polypeptide chains in an antibody molecule.

**helicase** A protein that separates the strands of double-stranded DNA.

**helix** *See* **$\alpha$ helix.**

**helix–loop–helix** A DNA-binding motif found in many regulatory proteins.

**helper T cell** A type of white blood cell involved in activating other cells in the immune system.

**hemizygous gene** A gene present in only one dose, such as the genes on the X chromosome in XY males.

**hemoglobin** The oxygen-carrying protein in red blood cells.

**hemophilia A** One of two X-linked forms of hemophilia; patients are deficient in blood-clotting factor VIII.

**heritability** A measure of the degree to which a phenotypic trait can be modified by selection. *See also* **broad-sense heritability** and **narrow-sense heritability.**

**heterochromatin** Chromatin that remains condensed and heavily stained during interphase; commonly present adjacent to the centromere and in the telomeres of chromosomes. Some chromosomes are composed primarily of heterochromatin.

**heterochromatin compartment** The physical region occupied by the heterochromatin in the nucleus of an interphase or noncycling cell.

**heterochronic mutation** A mutation in which an otherwise normal gene is expressed at the wrong time.

**heteroduplex** A duplex molecule of nucleic acid whose strands are derived from different sources, such as from different homologous chromosomes or even from different organisms.

**heteroduplex region** Region of a double-stranded nucleic acid molecule in which the two strands have different hereditary origins; produced either as an intermediate in recombination or by the *in vitro* annealing of single-stranded complementary molecules.

**heterogametic** Refers to the production of dissimilar gametes with respect to the sex chromosomes; in most animals, the male is the heterogametic sex, but in birds, moths, butterflies, and some reptiles, it is the female. *See also* **homogametic.**

**heterogeneous nuclear RNA (hnRNA)** The collection of primary RNA transcripts and incompletely processed products found in the nucleus of a eukaryotic cell.

**heterokaryon** A cell or individual having nuclei from genetically different sources, the result of cell fusion not accompanied by nuclear fusion.

**heteroplasmy** The presence of two or more genetically different types of the same organelle in a single organism.

**heterosis** The superiority of hybrids over either inbred parent with respect to one or more traits; also called hybrid vigor.

**heterozygote superiority** The condition in which a heterozygous genotype has greater fitness than either of the homozygotes.

**heterozygous** Carrying dissimilar alleles of one or more genes; not homozygous.

**hexaploid** A cell or organism with six complete sets of chromosomes.

**H4 histone** *See* **histone.**

**Hfr** An *E. coli* cell in which an F plasmid is integrated into the chromosome, making possible the transfer of part or all of the chromosome to an $F^-$ cell.

**high-copy-number plasmid** A plasmid for which there are usually considerably more than 2 (and often more than 20) copies per cell.

**highly repetitive sequence** A DNA sequence present in thousands of copies in a genome.

**highly significant** *See* **statistically significant.**

**histocompatability** Acceptance by a recipient of transplanted tissue from a donor.

**histocompatibility antigens** Tissue antigens that determine transplant compatibility or incompatibility.

**histocompatibility genes** Genes coding for histocompatibility antigens.

**histone** Any of the small basic proteins bound to DNA in chromatin; the five major histones are designated H1, H2A, H2B, H3, and H4. Each nucleosome core particle contains two molecules each of H2A, H2B, H3, and H4. The H1 histone forms connecting links between nucleosome core particles.

**histone tail** Region at the amino end of histone subunits that is susceptible to chemical modification affecting gene activity.

**Holliday junction** A cross-shaped configuration of two DNA duplexes formed as an intermediate in recombination.

**Holliday junction resolving enzyme** An enzyme that catalyzes the breakage and rejoining of two DNA strands in a Holliday junction to generate two independent duplex molecules.

**Holliday model** A molecular model of genetic recombination in which the participating duplexes contain heteroduplex regions of the same length.

**Holliday structure** A cross-shaped configuration of two DNA duplexes formed as an intermediate in recombination.

**holocentric chromosome** A chromosome with a diffuse centromere. *See also* **diffuse centromere.**

**holoenzyme** *See* **RNA polymerase holoenzyme.**

**homeobox** A DNA sequence motif found in the coding region of many regulatory genes; the amino acid sequence corresponding to the homeobox has a helix-loop-helix structure.

**homeotic gene** Any of a group of genes that determine the fundamental patterns of development. *See also* **homeotic mutation.**

**homeotic mutation** A developmental mutation that results in the replacement of one body structure by another body structure.

**homeotic selector gene** *See* **homeotic gene.**

**homogametic** Producing only one kind of gamete with respect to the sex chromosomes. *See also* **heterogametic.**

**homogentisic acid** The molecule, formerly called alkapton, that is excreted in the urine of alkaptonurics and turns black upon oxidation.

**homologous** In reference to DNA, having the same or nearly the same nucleotide sequence.

**homologous chromosomes** Chromosomes that pair in meiosis and have the same genetic loci and structure; also called homologs.

**homologous recombination** Genetic exchange between identical or nearly identical DNA sequences.

**homothallism** The capacity of cells in certain fungi to undergo a conversion in mating type to make possible mating between cells produced by the same parental organism.

**homozygous** Having the same allele of a gene in homologous chromosomes.

**H1 histone** *See* **histone.**

**Hoogsteen base pairing** An unusual type of base pairing in which four guanine bases form a hydrogen-bonded quartet.

**horizontal transmission** Transfer of genes from one species to another, as by transmissible bacterial plasmids.

**hormone** A small molecule in higher eukaryotes, synthesized in specialized tissue, that regulates the activity of other specialized cells. In animals, hormones are transported from their source to a target tissue in the blood.

**hot spot** A site in a DNA molecule at which the mutation rate is much higher than the rate for most other sites.

**housekeeping gene** A gene that is expressed at the same level in virtually all cells and whose product participates in basic metabolic processes.

**H substance** The carbohydrate precursor of the A and B red-blood-cell antigens.

**H3 histone** *See* **histone.**

**H2A histone** *See* **histone.**

**H2B histone** *See* **histone.**

**human genome project** A worldwide project to map genetically and sequence the human genome and those of selected model organisms.

**Huntington disease** Dominantly inherited degeneration of the neuromuscular system, with onset in middle age.

**hybrid** An organism produced by the mating of genetically unlike parents; also, a duplex nucleic acid molecule produced of strands derived from different sources.

**hybrid vigor** *See* **heterosis.**

**hydrogen bond** A weak noncovalent linkage between two negatively charged atoms in which a hydrogen atom is shared.

**hydrophobic interaction** A noncovalent interaction between nonpolar molecules or nonpolar groups, causing the molecules or groups to cluster when water is present.

**hypermorphic mutation** A mutation in which the gene product is produced in greater abundance than wildtype.

**hypomorphic mutation** A mutation in which the gene product is produced in less abundance than wildtype.

**hypostasis** The condition in which the characteristic phenotype of a mutant allele is masked by the presence of a mutant allele of a different gene. For example, in *Drosophila,* the *white* (*w*) mutation causes white eyes, and its presence thereby masks the effects of any other genotype affecting eye color, such as homozygosity for *vermilion* (*v*), *brown* (*bw*), *garnet* (*g*), *scarlet* (*st*), and so forth. In classical terminology, *w* is said to be epistatic to *v, bw, g,* and *st;* and the latter are said to be hypostatic to *w.*

**hypostatic gene** In developmental genetics, any gene whose mutant phenotype is masked by that of another gene. *See* **epistatic gene.**

## I

**identical twins** *See* **monozygotic twins.**

**identity by descent** The condition in which two alleles are both replicas of a single ancestral allele in the recent past.

**imaginal disk** Structures present in the body of insect larvae from which the adult structures develop during pupation.

**immune response** Tendency to develop resistance to a disease-causing agent after an initial infection.

**immunity** A general term for resistance of an organism to specific substances, particularly agents of disease.

**immunoglobulin** One of several classes of antibody proteins.

**immunoglobulin class** The category in which an immunoglobulin is placed on the basis of its chemical characteristics, including the identity of its heavy chain.

**imprinting** A process of DNA modification in gametogenesis that affects gene expression in the zygote; one probable mechanism is the methylation of certain bases in the DNA.

**inborn error of metabolism** A genetically determined biochemical disorder, usually in the form of an enzyme defect that produces a metabolic block.

**inbreeding** Mating between relatives.

**inbreeding coefficient** A measure of the genetic effects of inbreeding in terms of the proportionate reduction in heterozygosity in an inbred organism compared with the heterozygosity expected with random mating.

**inbreeding depression** The deterioration in a population's fitness or performance that accompanies inbreeding.

**incompatible** Refers to blood or tissue whose transfusion or transplantation results in rejection.

**incomplete dominance** A situation in which the heterozygous genotype has a phenotype that falls somewhere in the range between the two homozygous genotypes.

**incomplete penetrance** Condition in which a mutant phenotype is not expressed in all organisms with the mutant genotype.

**independent** In probability, two events are independent if the occurrence of one provides no information regarding the occurrence of the other.

**independent assortment** Random distribution of unlinked genes into gametes, as with genes in different (nonhomologous) chromosomes or genes that are so far apart on a single chromosome that the recombination frequency between them is 1/2.

**individual selection** Selection based on each organism's own phenotype.

**induced mutation** A mutation formed under the influence of a chemical mutagen or radiation.

**inducer** A small molecule that inactivates a repressor, usually by binding to it and thereby altering the ability of the repressor to bind to an operator.

**inducible** A gene that is expressed, or an enzyme that is synthesized, only in the presence of an inducer molecule. *See also* **constitutive.**

**inducible transcription** *See* **inducible.**

**induction** Activation of an inducible gene; prophage induction is the derepression of a prophage that initiates a lytic cycle of phage development.

**initiation** The beginning of protein synthesis.

**inosine (I)** One of a number of unusual bases found in transfer RNA.

**insertional inactivation** Inactivation of a gene by interruption of its coding sequence; used in genetic engineering as a means of detecting insertion of a foreign DNA sequence into the coding region of a gene.

**insertion sequence** A DNA sequence capable of transposition in a prokaryotic genome; such sequences usually code for their own transposase.

***in situ* hybridization** Renaturation of a labeled probe nucleic acid to the DNA or RNA in a cell; used to localize particular DNA molecules to chromosomes or parts of chromosomes and to identify the time and tissue distribution of RNA transcripts.

**integrase** Any site-specific recombinase that can join two DNA molecules containing the target sequence.

**integration** The process by which one DNA molecule is inserted intact into another replicable DNA molecule, as in prophage integration and the integration of plasmid or tumor viral DNA into a chromosome.

**integron** In bacteria, a DNA structure consisting of a promoter, a flanking target site for a site-specific recombinase (integrase), a coding sequence for the integrase, and usually one or more protein-coding cassettes that have been "captured" by site-specific recombination.

**intercalation** Insertion of a planar molecule between the stacked bases in duplex DNA.

**interference** The tendency for crossing-over to inhibit the formation of another crossover nearby.

**intergenic complementation** Complementation between mutations in different genes. *See also* **complementation test.**

**intergenic suppression** Suppression of a mutant phenotype because of a mutation in a different gene; often refers specifically to a tRNA molecule able to recognize a stop codon or two different sense codons.

**interphase** The interval between nuclear divisions in the cell cycle, extending from the end of telophase of one division to the beginning of prophase of the next division.

**interrupted-mating technique** In an Hfr × $F^-$ cross, a technique by which donor and recipient cells are broken apart at specific times, allowing only a particular length of DNA to be transferred.

**intervening sequence** *See* **intron.**

**intragenic suppression** Suppression of a mutant phenotype because of a compensatory mutation in the same gene.

**intron** A noncoding DNA sequence in a gene that is transcribed but is then excised from the primary transcript in forming a mature mRNA molecule; found primarily in eukaryotic cells. *See also* **exon.**

**inversion** A structural aberration in a chromosome in which the order of several genes is reversed from the normal order. A pericentric inversion includes the centromere within the inverted region, and a paracentric inversion does not include the centromere.

**inversion loop** Loop structure formed by synapsis of homologous genes in a pair of chromosomes, one of which contains an inversion.

**inverted repeat** Either of a pair of base sequences present in the same molecule that are identical or nearly identical but are oriented in opposite directions; often found at the ends of transposable elements.

***in vitro* experiment** An experiment carried out with components isolated from cells.

***in vivo* experiment** An experiment performed with intact cells.

**ionizing radiation** Electromagnetic or particulate radiation that produces ion pairs when dissipating its energy in matter.

**IS element** *See* **insertion sequence.**

**isochromosome** A chromosome with two identical arms containing homologous genes.

**isotopes** The forms of a chemical element that have the same number of electrons and the same number of protons but differ in the number of neutrons in the atomic nucleus; unstable isotopes undergo transitions to a more stable state and, in so doing, emit radioactivity.

## J

**J (joining) region** Any of multiple DNA sequences that code for alternative amino acid sequences of part of the variable region of an antibody molecule; the J regions of heavy and light chains are different.

**just-so story** Made-up stories, not supported by hard evidence, about the presumed adaptive significance of a trait; named after the 1902 book *Just So Stories* by Rudyard Kipling.

## K

**kappa particle** An intracellular parasite in *Paramecium* that releases a substance capable of killing sensitive cells.

**karyotype** The chromosome complement of a cell or organism; often represented by an arrangement of metaphase chromosomes according to their lengths and the positions of their centromeres.

**kilobase pair (kb)** Unit of length of a duplex DNA molecule; equal to 1000 base pairs.

**kindred** A group of relatives.

**kinetochore** The cellular structure, formed in association with the centromere, to which the spindle fibers become attached in cell division.

**kissing complex** The bipartite structure formed by base pairing between a small regulatory RNA and a messenger RNA.

**Klinefelter syndrome** The clinical features of human males with the karyotype 47,XXY.

**knockout mutation** Any mutation that completely eliminates the function of the gene; synonyms include *null mutation, loss-of-function mutation,* and *amorphic mutation.*

**Kosambi's mapping function** A mapping function based on the assumption that the coincidence between crossovers is proportional to the distance between them.

**kuru** A neurological disease caused by a prion protein.

## L

***lac* operon** The set of genes required to metabolize lactose in bacteria.

**Lac repressor** A protein coded by the *lacI* gene that, in the absence of lactose, binds with the operator and prevents transcription.

**lactose** Milk sugar; each molecule consists of a joined glucose and galactose.

**lactose permease** An enzyme responsible for transport of lactose from the environment into bacteria.

**lagging strand** The DNA strand whose complement is synthesized in short fragments that are ultimately joined together.

**lariat structure** Structure of an intron immediately after excision in which the 5′ end loops back and forms a 5′-2′ linkage with another nucleotide.

**L chain** *See* **light chain.**

**leader** *See* **leader polypeptide, leader sequence.**

**leader polypeptide** A short polypeptide encoded in the leader sequence of some operons coding for enzymes in amino acid biosynthesis; translation of the leader polypeptide participates in regulation of the operon through attenuation.

**leader sequence** The region of an mRNA molecule from the 5′ end to the beginning of the coding sequence, sometimes containing regulatory sequences; in prokaryotic mRNA, it contains the ribosomal binding site.

**leading strand** The DNA strand whose complement is synthesized as a continuous unit.

**leptotene** The initial substage of meiotic prophase I during which the chromosomes become visible in the light microscope as unpaired thread-like structures.

**lethal mutation** A mutation that results in the death of an affected organism.

**liability** Risk, particularly toward a threshold type of quantitative trait.

**library** *See* **gene library.**

**ligand** The molecule that binds to a specific receptor.

**light (L) chain** One of the small polypeptide chains in an antibody molecule.

**likelihood ratio** The ratio of the probabilities of obtaining a set of observed data under two different hypotheses—for example, linkage at some specified value of recombination versus no linkage.

**lineage** The ancestral history of a cell in development or of a species in evolution.

**lineage diagram** A diagram of cell lineages and their developmental fates.

**linkage** The tendency of genes located in the same chromosome to be associated in inheritance more frequently than expected from their independent assortment in meiosis.

**linkage equilibrium** The condition in which the alleles of different genes in a population are present in gametes in proportion to the product of the frequencies of the alleles.

**linkage group** The set of genes present together in a chromosome.

**linkage map** A diagram of the order of genes in a chromosome in which the distance between adjacent genes is proportional to the rate of recombination between them; also called a genetic map.

**linker** In genetic engineering, synthetic DNA fragments that contain restriction-enzyme cleavage sites that are used to join two DNA molecules.

**local population** A group of organisms of the same species occupying an area within which most individual members find their mates; synonomous terms are *deme* and *Mendelian population.*

**localized centromere** The type of centromere organization found in most eukaryotic cells in which the centromere is located in one small region of the chromosome.

**locomotor activity** Movement of organisms from one place to another under their own power.

**locus** The site or position of a particular gene on a chromosome.

**lod score** The log-odds, or logarithm of the likelihood ratio; in genetic linkage studies, the logarithm is usually taken to the base 10.

**loss-of-function mutation** A mutation that eliminates gene function; also called a null mutation.

**loss of heterozygosity** Loss of the presence of the wildtype allele, or loss of its function, in a heterozygous cell, enabling the phenotype of a recessive mutant allele to be expressed; mechanisms for loss of heterozygosity include chromosome loss, gene conversion, and mutation.

**lost allele** An allele no longer present in a population; its frequency is 0.

**LTR retrotransposon** A type of transposable element that transposes via an RNA intermediate and that contains long terminal repeats (LTRs) in direct orientation at its ends.

**lysis** Breakage of a cell caused by rupture of its cell membrane and cell wall.

**lysogen** Clone of bacterial cells that have acquired a prophage.

**lysogenic cycle** In temperate bacteriophage, the phenomenon in which the DNA of an infecting phage becomes part of the genetic material of the cell.

**lysozyme** One of a class of enzymes that dissolves the cell wall of bacteria; found in chicken egg white, human tears, and coded by many phages.

**lytic cycle** The life cycle of a phage, in which progeny phage are produced and the host bacterial cell is lysed.

## M

**macrophage** One of a class of white blood cells that processes antigens for presentation as a necessary step in stimulation of the immune response.

**major groove** In B-form DNA, the larger of two continuous indentations running along the outside of the double helix.

**major histocompatibility complex (MHC)** The group of closely linked genes coding for antigens that play a major role in tissue incompatibility and that function in regulation and other aspects of the immune response.

**map-based cloning** A strategy of gene cloning based on the position of a gene in the genetic map; also called positional cloning.

**map distance** The genetic distance between two marker genes expressed as the sum of the length in map units across a number of small, nonoverlapping intervals between the marker genes.

**mapping function** The mathematical relation between the genetic map distance across an interval and the observed percentage of recombination in the interval.

**map unit** A unit of distance in a linkage map that corresponds to a recombination frequency of 1 percent. Technically, the map distance across an interval in map units equals one-half the average number of crossovers in the interval expressed as a percentage. Map units are sometimes denoted centimorgans (cM).

**masked mRNA** Messenger RNA that cannot be translated until specific regulatory substances are available; present in eukaryotic cells, particularly eggs; storage mRNA.

**maternal effect** A phenomenon in which the genotype of a mother affects the phenotype of the offspring through substances present in the cytoplasm of the egg.

**maternal-effect gene** A gene that influences early development through its expression in the mother and the presence of the gene product in the oocyte.

**maternal inheritance** Extranuclear inheritance of a trait through cytoplasmic factors or organelles contributed by the female gamete.

**maternal sex ratio** Aberrant sex ratio in certain *Drosophila* species caused by cytoplasmic transmission of a parasite that is lethal to male embryos.

**mating system** The norms by which members of a population choose their mates; important systems of mating include random mating, assortative mating, and inbreeding.

**mating-type interconversion** Phenomenon in homothallic yeast in which cells switch mating type as a result of the transposition of genetic information from an unexpressed cassette into the active mating-type locus.

**Maxam–Gilbert method** A technique for determining the nucleotide sequence of DNA by means of strand cleavage at the positions of particular nucleotides.

**MCS** Multiple cloning site. *See also* **polylinker.**

**MCP** *See* **methyl-accepting chemotaxis protein.**

**mean** The arithmetic average.

**megabase pair** Unit of length of a duplex nucleic acid molecule; equal to 1 million base pairs.

**meiocyte** A germ cell that undergoes meiosis to yield gametes in animals or spores in plants.

**meiosis** The process of nuclear division in gametogenesis or sporogenesis in which one replication of the chromosomes is followed by two successive divisions of the nucleus to produce four haploid nuclei.

**melting curve** A graph of the amount of denatured DNA present in a solution (measured by UV light adsorption) as a function of increasing temperature.

**melting temperature** The midpoint of the narrow temperature range at which the strands of duplex DNA denature.

**Mendelian genetics** The mechanism of inheritance in which the statistical relations between the distribution of traits in successive generations result from (1) particulate hereditary determinants (genes), (2) random union of gametes, and (3) segregation of unchanged hereditary determinants in the reproductive cells.

**Mendelian population** *See* **local population.**

**meristem** The mitotically active growing point of plant tissue.

**meristic trait** A trait in which the phenotype is determined by counting, such as number of ears on a stalk of corn or number of eggs laid by a hen.

**merodiploid** A bacterial cell carrying two copies of a region of the genome and therefore diploid for the genes in this region.

**messenger RNA (mRNA)** An RNA molecule transcribed from a DNA sequence and translated into the amino acid sequence of a polypeptide. In eukaryotes, the primary transcript undergoes elaborate processing to become the mRNA.

**metabolic pathway** A set of chemical reactions that take place in a definite order to convert a particular starting molecule into one or more specific products.

**metabolite** Any small molecule that participates in metabolism as a substrate, product, or cofactor.

**metacentric chromosome** A chromosome with its centromere about in the middle so that the arms are equal or almost equal in length.

**metallothionein** Any of a class of proteins that bind heavy metals.

**metaphase** In mitosis, meiosis I, or meiosis II, the stage of nuclear division in which the centromeres of the condensed chromosomes are arranged in a plane between the two poles of the spindle.

**metaphase plate** Imaginary plane, equidistant from the spindle poles in a metaphase cell, on which the centromeres of the chromosomes are aligned by the spindle fibers.

**methyl-accepting chemotaxis protein** Alternative designation of chemoreceptors in bacterial chemotaxis; abbreviated MCP.

**methylation** The modification of a protein or a DNA or RNA base by the addition of a methyl group ($-CH_3$).

**methylation induced premeiotically (MIP)** An epigenetic mechanism of gene silencing found in certain filamentous fungi in which genes present in multiple copies are heavily methylated prior to the onset of meiosis.

**MHC** *See* **major histocompatibility complex.**

**micrococcal nuclease** A nuclease that is isolated from a bacterium and that cleaves double-stranded DNA without regard to sequence.

**microRNA** Small double-stranded RNA molecules that repress translation of mRNAs containing complementary sequences.

**microsatellite** A type of DNA marker based on a short sequence (2–9 base pairs) that is present at one or more sites in the genome and is repeated in tandem at each site.

**middle repetitive DNA sequence** A DNA sequence present tens to hundreds of times per genome.

**migration** Movement of organisms among subpopulations; also, the movement of molecules in electrophoresis.

**minimal medium** A growth medium consisting of simple inorganic salts, a carbohydrate, vitamins, organic bases, essential amino acids, and other essential compounds; its composition is precisely known. Minimal medium contrasts with complex medium or broth, which is an extract of biological material (vegetables, milk, meat) that contains a large number of compounds, the precise composition of which is unknown.

**minisatellite** A type of DNA marker based on a sequence of 10–60 base pairs that is present at one or more sites in the genome and is repeated in tandem at each site.

**minor groove** In B-form DNA, the smaller of two continuous indentations running along the outside of the double helix.

**mismatch** An arrangement in which two nucleotides opposite one another in double-stranded DNA are unable to form hydrogen bonds.

**mismatch repair** Removal of one nucleotide from a pair that cannot properly hydrogen-bond, followed by replacement with a nucleotide that can hydrogen-bond.

**missense mutation** An alteration in a coding sequence of DNA that results in an amino acid replacement in the polypeptide.

**mitosis** The process of nuclear division in which the replicated chromosomes divide and the daughter nuclei have the same chromosome number and genetic composition as the parent nucleus.

**mitotic spindle** The set of fibers that arches through the cell during mitosis and to which the chromosomes are attached. *See also* **spindle.**

**mobile DNA** Any DNA sequence capable of moving from one position to another in the genome; usually refers to transposable elements.

**molecular clock** A condition in which a protein or nucleic acid molecule has the same probability of change per unit time in every branch of a gene tree.

**molecular systematics** A group of statistical methods for estimating gene trees and often, by inference, the evolutionary relationships among the taxa of which the genes are representative.

**monocistronic mRNA** A mRNA molecule that codes for a single polypeptide chain.

**monoclonal antibody** Antibody directed against a single antigen and produced by a single clone of B cells or a single cell line of hybridoma cells.

**monohybrid** A genotype that is heterozygous for one pair of alleles; the offspring of a cross between genotypes that are homozygous for different alleles of a gene.

**monomer** Any of the individual polypeptide chains (subunits) that make up a protein composed of multiple polypeptide chains.

**monomorphic gene** A gene for which the most common allele is virtually fixed.

**monoploid** The basic chromosome set that is reduplicated to form the genomes of the species in a polyploid series; the smallest haploid chromosome number in a polyploid series.

**monosomic** Condition of an otherwise diploid organism in which one member of a pair of chromosomes is missing.

**monozygotic twins** Twins developed from a single fertilized egg that splits into two embryos at an early division; also called identical twins.

**morphogen** Substance that induces differentiation.

**mosaic** An organism composed of two or more genetically different types of cells.

**M period** *See* **cell cycle.**

**mRNA** *See* **messenger RNA.**

**multifactorial trait** A trait determined by the combined action of many factors, typically some genetic and some environmental.

**multiple alleles** The presence in a population of more than two alleles of a gene.

**multiple cloning site** *See* **polylinker.**

**multiplication rule** The principle that the probability that all of a set of independent events are realized simultaneously equals the product of the probabilities of the separate events.

**multivalent** An association of more than two homologous chromosomes resulting from synapsis during meiosis in a polysomic or polyploid individual.

**mutagen** An agent that is capable of increasing the rate of mutation.

**mutagenesis** The process by which a gene undergoes a heritable alteration; also, the treatment of a cell or organism with a known mutagen.

**mutant** Any heritable biological entity that differs from wildtype, such as a mutant DNA molecule, mutant allele, mutant gene, mutant chromosome, mutant cell, mutant organism, or mutant heritable phenotype; also, a cell or organism in which a mutant allele is expressed.

**mutant screen** A type of genetic experiment in which the geneticist seeks to isolate multiple new mutations that affect a particular trait.

**mutation** A heritable alteration in a gene or chromosome; also, the process by which such an alteration happens. Used incorrectly, but with increasing frequency, as a synonym for mutant, even in some excellent textbooks.

**mutation pressure** The generally very weak tendency of mutation to change allele frequency.

**mutation rate** The probability of a new mutation in a particular gene, either per gamete or per generation.

**mutually exclusive (disjoint) events** In probability, two events are mutually exclusive if the occurrence of one precludes the occurrence of the other.

## N

**narrow-sense heritability** The fraction of the phenotypic variance revealed as resemblance between parents and offspring; technically, the ratio of the additive genetic variance to the total phenotypic variance.

**native conformation** The three-dimensional shape that a molecule is thought to acquire inside the cell.

**natural selection** The process of evolutionary adaptation in which the genotypes genetically best suited to survive and reproduce in a particular environment give rise to a disproportionate share of the offspring and so gradually increase the overall ability of the population to survive and reproduce in that environment.

**nature versus nurture** Archaic term for genetics versus environment; usually a false dichotomy because most traits are influenced by both genetic and environmental factors.

**negative assortative mating** *See* **assortative mating.**

**negative chromatid interference** In meiosis, when two or more crossovers occur in a chromosome, a tendency for the nonsister chromatids that participate in one event to also participate in another event.

**negative regulation** Regulation of gene expression in which mRNA is not synthesized until a repressor is removed from the DNA of the gene.

**negative supercoiling** Supercoiling of DNA that counteracts underwinding of the DNA.

**neighbor joining** A method for estimating a gene tree in which pairs of taxa are joined sequentially according to which pair are separated by the shortest distance.

**neomorph** A mutation in which the mutant allele expresses either a novel gene product or a mutant gene product that has a novel phenotypic effect.

**neutral allele** An allele that has a negligible effect on the ability of the organism to survive and reproduce.

**neutral petite** A strain of yeast with impaired respiration and small colonies resulting from a mutation in mitochondrial DNA.

**nick** A single-strand break in a DNA molecule.

**nicked circle** A circular DNA molecule containing one or more single-strand breaks.

**nitrous acid** $HNO_2$, a chemical mutagen.

**noncomplementation** Failure of two mutant alleles to show complementation; noncomplementation implies that the alleles are mutant forms of the same gene.

**nondisjunction** Failure of chromosomes to separate (disjoin) and move to opposite poles of the division spindle; the result is loss or gain of a chromosome.

**nonhistone chromosomal proteins** A large class of proteins, not of the histone class, found in isolated chromosomes.

**non-LTR retrotransposon** A type of transposable element that transposes via an RNA intermediate and that does not contain long terminal repeats (LTRs) at its ends.

**nonparental ditype** An ascus containing two pairs of recombinant spores.

**nonpermissive conditions** Environmental conditions that result in expression of the phenotype of a conditional mutation.

**nonselective medium** A growth medium that allows growth of wildtype and of one or more mutant genotypes.

**nonsense-mediated decay** In eukaryotes, a process in which a messenger RNA containing a premature chain-terminating codon is destroyed in the first round of translation owing to the presence of proteins that are bound to the exon-exon junctions.

**nonsense mutation** A mutation that changes a codon specifying an amino acid into a stop codon, resulting in premature polypeptide chain termination; also called a chain-termination mutation.

**nonsense suppression** Suppression of the phenotype of a polypeptide chain-termination mutation mediated by a mutant tRNA that inserts an amino acid at the site of the termination codon.

**nonsense suppressor** Any mutation that suppresses the phenotype of a chain-terminating ("nonsense") mutation; nonsense suppressors are usually mutations in tRNA genes.

**nonsynonymous mutation** A nucleotide substitution in a codon that changes the amino acid specified by the codon.

**nontaster** Individual with the inherited inability to taste the substance phenylthiocarbamide (PTC); phenotype of the homozygous recessive.

**normal distribution** A symmetrical bell-shaped distribution characterized by the mean and the variance; in a normal distribution, approximately 68 percent of the observations are within 1 standard deviation from the mean, and approximately 95 percent are within 2 standard deviations.

**nuclease** An enzyme that breaks phosphodiester bonds in nucleic acid molecules.

**nucleic acid** A polymer composed of repeating units of phosphate-linked five-carbon sugars to which nitrogenous bases are attached. *See also* **DNA** and **RNA.**

**nucleic acid hybridization** The formation of duplex nucleic acid from complementary single strands.

**nucleoid** A DNA mass, not bounded by a membrane, within the cytoplasm of a prokaryotic cell, chloroplast, or mitochondrion; often refers to the major DNA unit in a bacterium.

**nucleolar organizer region** A chromosome region containing the genes for ribosomal RNA; abbreviated NOR.

**nucleolus (*plural* nucleoli)** Nuclear organelle in which ribosomal RNA is made and ribosomes are partially synthesized; usually associated with the nucleolar organizer region. A nucleus may contain several nucleoli.

**nucleoside** A purine or pyrimidine base covalently linked to a sugar.

**nucleosome** The basic repeating subunit of chromatin, consisting of a core particle composed of two molecules each of four different histones around which a length of DNA containing about 145 nucleotide pairs is wound, joined to an adjacent core particle by about 55 nucleotide pairs of linker DNA associated with a fifth type of histone.

**nucleosome core particle** A unit of chromatin structure composed of two molecules each of histones H2A, H2B, H3, and H4, around which is wound a segment of duplex DNA of about 145 nucleotide pairs.

**nucleotide** A nucleoside phosphate.

**nucleotide analog** A molecule that is structurally similar to a normal nucleotide and that is incorporated into DNA.

**nucleotide excision repair** A mechanism of DNA repair in which nucleotides with mismatched or damaged bases are cleaved from a DNA strand and replaced using the complementary DNA strand as a template.

**nucleotide substitution** A mutation in which one nucleotide is replaced with a different nucleotide; in duplex DNA the partner nucleotide in the complementary strand is also changed to preserve the Watson–Crick base pairing.

**nucleus** The organelle, bounded by a membranous envelope, that contains the chromosomes in a eukaryotic cell.

**nullisomic** A cell or individual that contains no copies of a particular chromosome.

**null mutation** Any mutation that completely eliminates the function of the gene; synonyms include *knockout mutation, loss-of-function mutation,* and *amorphic mutation.*

**nutritional mutation** A mutation in a metabolic pathway that creates a need for a substance to be present in the growth medium or that eliminates the ability to utilize a substance present in the growth medium.

## O

**ochre codon** Jargon for the UAA stop codon; an ochre mutation is a UAA codon formed from a sense codon.

**octoploid** A cell or organism with eight complete sets of chromosomes.

**Okazaki fragment** Any of the short strands of DNA produced during discontinuous replication of the lagging strand; also called a precursor fragment.

**oligomerization domain** A region of a polypeptide chain that binds to a region of another polypeptide chain to form a protein with two or more subunits.

**oligonucleotide** A short, single-stranded nucleic acid, usually synthesized for use in DNA sequencing, as a primer in the polymerase chain reaction, as a probe, or for oligonucleotide site-directed mutagenesis.

**oligonucleotide primer** A short, single-stranded nucleic acid synthesized for use in DNA sequencing or as a primer in the polymerase chain reaction.

**oligonucleotide site-directed mutagenesis** Replacement of a small region in a gene with an oligonucleotide that contains a specified mutation.

**oncogene** A gain-of-function mutation in a cellular gene, called a proto-oncogene, whose normal function is to promote cellular proliferation or inhibit apoptosis; oncogenes are often associated with tumor progression.

**open reading frame** In the coding strand of DNA or in mRNA, a region containing a series of codons uninterrupted by stop codons and therefore capable of coding for a polypeptide chain; abbreviated ORF.

**operator** A regulatory region in DNA that interacts with a specific repressor protein in controlling the transcription of adjacent structural genes.

**operon** A collection of adjacent structural genes regulated by an operator and a repressor.

**operon model** A model depicting the coordinate regulation of a set of closely linked genes, such as those for lactose utilization in E. coli; many genes for metabolic pathways in bacteria are organized into operons.

**ORF** *See* **open reading frame.**

**organelle** A membrane-bounded cytoplasmic structure that has a specialized function, such as a chloroplast or mitochondrion.

**origin of replication** A DNA base sequence at which replication of a molecule is initiated.

**origin recognition complex (ORC)** A protein complex that initiates the formation of pre-replication complexes at origins of replication in eukaryotic cells.

**orthologous genes** Genes present in two or more related species that descend from a single gene present in a common ancestor of the species.

**outcrossing** A mating system in which mating pairs are not genetically closely related; avoidance of close inbreeding.

**overdominance** A condition in which the fitness of a heterozygote is greater than the fitness of both homozygotes.

**overlapping genes** Genes that share part of their coding sequences.

**ovule** The structure in seed plants that contains the embryo sac (female gametophyte) and that develops into a seed after fertilization of the egg.

## P

**pachytene** The middle substage of meiotic prophase I in which the homologous chromosomes are closely synapsed.

**pair-rule gene** Any of a group of genes active early in *Drosophila* development that specifies the fates of alternating segments or parasegments. Mutations in pair-rule genes result in loss of even-numbered or odd-numbered segments or parasegments.

**palindrome** In nucleic acids, a segment of DNA in which the sequence of bases on complementary strands reads the same from a central point of symmetry—for example 5′-GAATTC-3′; the sites of recognition and cleavage by restriction endonucleases are frequently palindromic.

**paracentric inversion** An inversion that does not include the centromere.

**paralogous genes** Two or more genes that derive from a common ancestral gene through one or more gene duplications followed by sequence divergence.

**parasegment** Developmental unit in *Drosophila* consisting of the posterior part of one segment and the anterior part of the next segment in line.

**parental combination** Alleles present in an offspring chromosome in the same combination as that found in one of the parental chromosomes.

**parental ditype** An ascus containing two pairs of nonrecombinant spores.

**parental strand** In DNA replication, the strand that served as the template in a newly formed duplex.

**parental types** In a genetic cross, any configuration of alleles in an offspring that is present in either of the parents.

**partial digestion** Condition in which a restriction enzyme cleaves some, but not all, of the restriction sites present in a DNA molecule.

**partial diploid** A cell in which a segment of the genome is duplicated, usually in a plasmid.

**partial dominance** A condition in which the phenotype of the heterozygote is intermediate between the corresponding homozygotes but resembles one more closely than the other.

**Pascal's triangle** Triangular configuration of integers in which the $n$th row gives the binomial coefficients in the expansion of $(x + y)^{n-1}$. The first and last numbers in each row equal 1, and the others equal the sum of the adjacent numbers in the row immediately above.

**paternal inheritance** Extranuclear inheritance of a trait through cytoplasmic factors or organelles contributed by the male gamete.

**path in a pedigree** A closed loop in a pedigree, with only one change of direction, that passes through a common ancestor of the parents of an individual.

**pathogen** An organism that causes disease.

**pathogenicity island** A horizontally transferred region of a bacterial DNA molecule containing genes causing disease.

**pattern formation** The creation of a spatially ordered and differentiated embryo from a seemingly homogeneous egg cell.

**PCR** *See* **polymerase chain reaction.**

**pedigree** A diagram representing the familial relationships among relatives.

***P* element** A cut-and-paste type of transposable element in *Drosophila* used for germ-line transformation and other types of *in vivo* genetic engineering.

**penetrance** The proportion of organisms having a particular genotype that actually express the corresponding phenotype. If the phenotype is always expressed, penetrance is complete; otherwise, it is incomplete.

**peptide bond** A covalent bond between the amino group ($-NH_2$) of one amino acid and the carboxyl group ($-COOH$) of another.

**peptidyl site** The tRNA-binding site on the ribosome to which the tRNA bearing the nascent polypeptide becomes bound immediately after formation of the peptide bond.

**peptidyl transferase** The enzymatic activity of ribosomes responsible for forming a peptide bond.

**percent G + C** The proportion of base pairs in duplex DNA that consist of GC pairs.

**pericentric inversion** An inversion that includes the centromere.

**permease** A membrane-associated protein that allows a specific small molecule to enter the cell.

**permissive condition** An environmental condition in which the phenotype of a conditional mutation is not expressed; contrasts with nonpermissive or restrictive condition.

**petite mutant** A mutation in yeast resulting in slow growth and small colonies.

**PEV** *See* **position effect variegation.**

**PFGE** *See* **pulse-field gel electrophoresis.**

**p53 transcription factor** An important protein that helps regulate a mammalian cell's response to stress, especially to DNA damage.

**$P_1$ generation** The parents used in a cross, or the original parents in a series of generations; also called the P generation if there is no chance of confusion with the grandparents or more remote ancestors.

**P group** *See* **5′-P group.**

**phage** *See* **bacteriophage.**

**phage-attachment site** The base sequence in a bacterial chromosome at which bacteriophage DNA can integrate to form a prophage.

**phage repressor** Regulatory protein that prevents transcription of genes in a prophage.

**phenotype** The observable properties of a cell or an organism, which result from the interaction of the genotype and the environment.

**phenotypic variance** The total variance in a phenotypic trait in a population.

**phenylalanine hydroxylase (PAH)** The enzyme that converts phenylalanine to tyrosine and that is defective in phenylketonuria.

**phenylketonuria (PKU)** A hereditary human condition resulting from inability to convert phenylalanine into tyrosine; causes severe mental retardation unless treated in infancy and childhood by a low-phenylalanine diet; abbreviated PKU.

**phenylthiocarbamide** *See* **PTC tasting.**

**Philadelphia chromosome** Abnormal human chromosome 22, resulting from reciprocal translocation and often associated with a type of leukemia.

**phosphodiester bond** In nucleic acids, the covalent bond formed between the 5′-phosphate group (5′-P) of one nucleotide and the 3′-hydroxyl group (3′-OH) of the next nucleotide in line; these bonds form the backbone of a nucleic acid molecule.

**photoreactivation** The enzymatic splitting of pyrimidine dimers produced in DNA by ultraviolet light; requires visible light and the photoreactivation enzyme.

**phylogenetics** Inference of the ancestral relationships among species or among the sequences of macromolecules.

**phylogenetic tree** A diagram showing the genealogical relationships among a set of genes or species.

**physical map** A diagram showing the relative positions of physical landmarks in a DNA molecule; common landmarks include the positions of restriction sites and particular DNA sequences.

**plaque** A clear area in an otherwise turbid layer of bacteria growing on a solid medium, caused by the infection and killing of the cells by a phage; because each plaque results from the growth of one phage, plaque counting is a way of counting viable phage particles. The term is also used for animal viruses that cause clear areas in layers of animal cells grown in culture.

**plasmid** An extrachromosomal genetic element that replicates independently of the host chromosome; it may exist in one or many copies per cell and may segregate in cell division to daughter cells in either a controlled or a random fashion. Some plasmids, such as the F factor, may become integrated into the host chromosome.

**pleiotropic effect** Any phenotypic effect that is a secondary manifestation of a mutant gene.

**pleiotropy** The condition in which a single mutant gene affects two or more distinct and seemingly unrelated traits.

**point mutation** A mutation caused by the substitution, deletion, or addition of a single nucleotide pair.

**polarity** The 5′-to-3′ orientation of a strand of nucleic acid.

**pole cell** Any of a group of cells, set off at the posterior end of the *Drosophila* embryo, from which the germ cells are derived.

**Pol III holoenzyme** A large protein complex containing the major DNA polymerase that elongates each growing DNA strand during replication.

**Pol II holoenzyme** A large protein complex containing the type of RNA polymerase used in transcribing most protein-coding genes.

**poly-A tail** The sequence of adenines added to the 3′ end of many eukaryotic mRNA molecules in processing.

**polycistronic mRNA** An mRNA molecule from which two or more polypeptides are translated; found primarily in prokaryotes.

**polygenic inheritance** The determination of a trait by alleles of two or more genes.

**polylinker** A short DNA sequence that is present in a vector and that contains a number of unique restriction sites suitable for gene cloning.

**polymer** A regular, covalently bonded arrangement of basic subunits or monomers into a large molecule, such as a polynucleotide or polypeptide chain.

**polymerase** An enzyme that catalyzes the covalent joining of nucleotides—for example, DNA polymerase and RNA polymerase.

**polymerase $\alpha$** The major DNA replication enzyme in eukaryotic cells.

**polymerase chain reaction** Repeated cycles of DNA denaturation, renaturation with primer oligonucleotide sequences, and replication, resulting in exponential growth in the number of copies of the DNA sequence located between the primers.

**polymerase $\gamma$** The enzyme responsible for replication of mitochondrial DNA.

**polymorphic gene** A gene for which there is more than one relatively common allele in a population.

**polymorphism** The presence in a population of two or more relatively common forms of a gene, chromosome, or genetically determined trait.

**polynucleotide chain** A polymer of covalently linked nucleotides.

**polypeptide** *See* **polypeptide chain.**

**polypeptide chain** A polymer of amino acids linked together by peptide bonds.

**polyploidy** The condition of a cell or organism with more than two complete sets of chromosomes.

**polyprotein** A protein molecule that can be cleaved to form two or more finished protein molecules.

**polyribosome** *See* **polysome.**

**polysome** A complex of two or more ribosomes associated with an mRNA molecule and actively engaged in polypeptide synthesis; a polyribosome.

**polysomy** The condition of a diploid cell or organism that has three or more copies of a particular chromosome.

**polytene chromosome** A giant chromosome consisting of many identical strands laterally apposed and in register, exhibiting a characteristic pattern of transverse banding.

**P1 artificial chromosome** A recombinant DNA plasmid in which a large fragment of DNA has been ligated into a vector derived from the *E. coli* bacteriophage P1.

**P1 bacteriophage** Temperate virus that infects *E. coli*; in cells lysogenic for P1, the phage DNA is not integrated into the chromosome and replicates as a plasmid.

**population** A group of organisms of the same species.

**population genetics** Application of Mendel's laws and other principles of genetics to entire populations of organisms.

**population structure** *See* **population substructure.**

**population substructure** Organization of a population into smaller breeding groups between which migration is restricted. Also called population subdivision.

**positional cloning** A strategy of gene cloning based on the position of a gene in the genetic map; also called map-based cloning.

**positional information** Developmental signals transmitted to a cell by virtue of its position in the embryo.

**position effect** A change in the expression of a gene depending on its position within the genome.

**position-effect variegation** Mosaic phenotype (variegation) due to variation in the level of expression of a gene in different cell lineages owing to its position in the genome.

**positive assortative mating** *See* **assortative mating.**

**positive chromatid interference** In meiosis, when two or more crossovers occur in a chromosome, a tendency for the nonsister chromatids that participate in one event not to participate in another event.

**positive regulation** Mechanism of gene regulation in which an element must be bound to DNA in an active form to allow transcription. Positive regulation contrasts with negative regulation, in which a regulatory element must be removed from DNA.

**postmeiotic segregation** Segregation of genetically different products in a mitotic division following meiosis, as in the formation in *Neurospora* of a pair of ascospores that have different genotypes.

**postreplication repair** DNA repair that takes place in nonreplicating DNA or after the replication fork is some distance beyond a damaged region.

**post-transcriptional cosuppression** A type of gene silencing in which transcripts of the silenced genes are produced but the gene products are not.

**posttranslocation state** The state of the ribosome immediately following the movement of the small subunit one codon farther along the mRNA; the aminoacyl (A) site is unoccupied.

**precursor fragment** *See* **Okazaki fragment.**

**prediction equation** In quantitative genetics, an equation used to predict the improvement in mean performance of a population brought about by artifical selection; it always includes heritability as one component.

**pre-replication complex** Protein complexes assembled at origins of replication that are activated by certain cyclin–CDKs, triggering the initiation of DNA synthesis.

**pretranslocation state** The state of the ribosome immediately prior to the movement of the small subunit one codon farther along the mRNA; the exit (E) site is unoccupied.

**Pribnow box** A base sequence in prokaryotic promoters to which RNA polymerase binds in an early step of initiating transcription.

**primary transcript** An RNA copy of a gene; in eukaryotes, the transcript must be processed to form a translatable mRNA molecule.

**primase** The enzyme responsible for synthesizing the RNA primer needed for initiating DNA synthesis.

**primer** In nucleic acids, a short RNA or single-stranded DNA segment that functions as a growing point in polymerization.

**primer adapter** A short, double-stranded oligonucleotide ligated to the end of a DNA duplex for use as a priming site for amplification by the polymerase chain reaction.

**primer oligonucleotide** A single-stranded DNA molecule, typically from 18 to 22 nucleotides in length, that can hybridize with a longer DNA strand and serve as a primer for replication. In the polymerase chain reaction, two primers anneal to opposite strands of a DNA duplex with their 3′-OH ends facing. *See also* **polymerase chain reaction.**

**primosome** The enzyme complex that forms the RNA primer for DNA replication in eukaryotic cells.

**principle of epistasis** In the genetic analysis of linear developmental-switch pathways, the principle that the epistatic gene in a double mutant genotype acts downstream in the pathway from the hypostatic gene.

**prion protein** Protein infectious agent; a protein that can transmit disease.

**probability** A mathematical expression of the degree of confidence that certain events will or will not be realized.

**probe** A labeled DNA or RNA molecule used in DNA-RNA or DNA-DNA hybridization assays.

**processing** A series of chemical reactions in which primary RNA transcripts are converted into mature mRNA, rRNA, or tRNA molecules or in which polypeptide chains are converted into finished proteins.

**processivity** Refers to the number of consecutive nucleotides in a template strand of nucleic acid that are traversed before a DNA polymerase or an RNA polymerase detaches from the template.

**product molecule** The molecule resulting from an enzyme-catalyzed change in a substrate molecule.

**programmed cell death** Cell death that happens as part of the normal developmental process. *See also* **apoptosis.**

**prokaryote** An organism that lacks a nucleus; prokaryotic cells divide by fission.

**promoter** A DNA sequence at which RNA polymerase binds and initiates transcription.

**promoter fusion** Joining of the promoter region of one gene with the protein-coding region of another.

**promoter mutation** A mutation that increases or decreases the ability of a promoter to initiate transcription.

**promoter recognition** The first step in transcription.

**proofreading function** The activity of some DNA polymerases that removes incorrectly incorporated nucleotides; also called the editing function.

**prophage** The form of phage DNA in a lysogenic bacterium; the phage DNA is repressed and usually integrated into the bacterial chromosome, but some prophages are in plasmid form.

**prophage induction** Activation of a prophage to undergo the lytic cycle.

**prophase** The initial stage of mitosis or meiosis, beginning after interphase and terminating with the alignment of the chromosomes at metaphase; often absent or abbreviated between meiosis I and meiosis II.

**protanomaly** *See* **color blindness.**

**protanopia** *See* **color blindness.**

**protein** A molecule composed of one or more polypeptide chains.

**protein folding** The process by which a polypeptide chain conforms itself in three dimensions to attain its native conformation.

**protein subunit** Any of the polypeptide chains in a protein molecule made up of multiple polypeptide chains.

**proteome** The complete set of proteins encoded in the genome.

**proteomics** Study of the complement of proteins present in a cell or organism in order to identify their localization, functions, and interactions.

**proto-oncogene** A eukaryotic gene that functions to promote cellular proliferation or inhibit apoptosis, in which gain-of-function mutations (oncogenes) are associated with tumor progression.

**prototroph** Microbial strain capable of growth in a defined minimal medium that ideally contains only a carbon source and inorganic compounds. The wildtype genotype is usually regarded as a prototroph.

**pseudoautosomal region** In mammals, a small region of the X and Y chromosome containing homologous genes.

**pseudogene** A DNA sequence that is not functional because of one or more mutations but that has a functional counterpart in the same organism; pseudogenes are regarded as mutated forms of ancient gene duplications.

**P site** *See* **peptidyl site.**

**PTC tasting** Genetic polymorphism in the ability to taste phenylthiocarbamide.

***P* transposable element** A *Drosophila* transposable element used for the induction of mutations, germ-line transformation, and other types of genetic engineering.

**Punnett square** A sort of cross-multiplication matrix used in the prediction of the outcome of a genetic cross, in which male and female gametes and their frequencies are arrayed along the edges.

**pulse-field gel electrophoresis** A type of electrophoresis in which the electric field is manipulated to make possible the separation of large DNA fragments.

**purine** An organic base found in nucleic acids; the predominant purines are adenine and guanine.

**pyrimidine** An organic base found in nucleic acids; the predominant pyrimidines are cytosine, uracil (in RNA only), and thymine (in DNA only).

**pyrimidine dimer** Two adjacent pyrimidine bases, typically a pair of thymines, in the same polynucleotide strand, between which chemical bonds have formed; the most common lesion formed in DNA by exposure to ultraviolet light.

## Q

**QTL** *See* **quantitative trait locus.**

**quantitative trait** A trait—typically measured on a continuous scale, such as height or weight—that results from the combined action of several or many genes in conjunction with environmental factors.

**quantitative trait locus** The locus of a gene whose alleles have significant differential effects on the phenotype of a quantitative trait among individuals differing in genotype for the alleles. *See* **QTL.**

## R

**race** A genetically or geographically distinct subgroup of a species.

**rad** Radiation absorbed dose; an amount of ionizing radiation resulting in the dissipation of 100 ergs of energy in 1 gram of matter; equal to 0.01 gray.

**random amplified polymorphic DNA (RAPD)** A type of DNA marker identified by a band in a gel after amplification of genomic DNA by the polymerase chain reaction using one or more short oligonucleotide primers of random sequence.

**random genetic drift** Fluctuation in allele frequency from generation to generation resulting from restricted population size.

**random mating** System of mating in which mating pairs are formed independently of genotype and phenotype.

**random spore analysis** In fungi, the genetic analysis of spores collected at random rather than from individual tetrads.

**reading frame** The phase in which successive triplets of nucleotides in mRNA form codons; depending on the reading frame, a particular nucleotide in an mRNA could be in the first, second, or third position of a codon. The reading frame actually used is defined by the AUG codon that is selected for chain initiation.

**reannealing** Reassociation of dissociated single strands of DNA to form a duplex molecule.

**recessive** Refers to an allele, or the corresponding phenotypic trait, expressed only in homozygotes.

**reciprocal cross** A cross in which the sexes of the parents are the reverse of another cross.

**reciprocal translocation** Interchange of parts between nonhomologous chromosomes.

**recombinant** A chromosome that results from crossing-over and that carries a combination of alleles differing from that of either chromosome participating in the crossover; the cell or organism containing a recombinant chromosome.

**recombinant DNA** A DNA molecule produced by the joining of two or more DNA fragments that are not normally in physical proximity and are often from different organisms.

**recombinant DNA technology** Procedures for creating DNA molecules composed of one or more segments from other DNA molecules.

**recombinant types** In a genetic cross, any configuration of alleles in an offspring that is not present in either of the parents.

**recombination** Exchange of parts between DNA molecules or chromosomes; recombination in eukaryotes usually entails a reciprocal exchange of parts.

**recombination repair** Repair of damaged DNA by exchange of good for bad segments between two damaged molecules.

**recruitment** The process in which a transcriptional activator protein interacts with one or more components of the transcription complex and attracts it to the promoter.

**red–green color blindness** *See* **color blindness.**

**reductional division** Term applied to the first meiotic division because the chromosome number (counted as the number of centromeres) is reduced from diploid to haploid.

**redundancy** The feature of the genetic code in which an amino acid corresponds to more than one codon; also called degeneracy.

**regulatory gene** A gene with the primary function of controlling the expression of one or more other genes.

**regulatory RNA** An RNA molecule that plays a role in gene regulation, such as small RNAs that regulate translation.

**rejection** An immune response against transfused blood or transplanted tissue.

**relative fitness** The fitness of a genotype expressed as a proportion of the fitness of another genotype.

**relative risk** The ratio of the probabilities of observing a particular phenotype, usually a disease, among individuals that differ in genotype.

**relaxed DNA** A DNA circle whose supercoiling has been removed either by introduction of a single-strand break or by the activity of a topoisomerase.

**release factor** A protein that functions in translation in the release of a completed polypeptide chain from the ribosome.

**rem** The quantity of any kind of ionizing radiation that has the same biological effect as one rad of high-energy gamma rays; rem stands for roentgen equivalent man.

**renaturation** Restoration of the normal three-dimensional structure of a macromolecule; in reference to nucleic acids, the term means the formation of a double-stranded molecule by complementary base pairing between two single-stranded molecules.

**repair** *See* **DNA repair.**

**repair synthesis** The enzymatic filling of a gap in a DNA molecule at the site of excision of a damaged DNA segment.

**repellent** Any substance that elicits chemotaxis of organisms in a direction away from itself.

**repetitive sequence** A DNA sequence present more than once per haploid genome.

**replica plating** Procedure in which a particular spatial pattern of colonies on an agar surface is reproduced on a series of agar surfaces by stamping them with a template that contains an image of the pattern; the template is often produced by pressing a piece of sterile velvet upon the original surface, which transfers cells from each colony to the cloth.

**replication** *See* **DNA replication; $\theta$ replication.**

**replication fork** In a replicating DNA molecule, the region in which nucleotides are added to growing strands.

**replication origin** The base sequence at which DNA synthesis begins.

**replication slippage** The process in which the number of copies of a small tandem repeat can increase or decrease during replication.

**replicon** A DNA molecule that has at least one replication origin.

**reporter gene** A gene whose expression can readily be monitored.

**repressible transcription** A mode of regulation in which the default state of transcription is "on" unless a protein (the repressor) binds to the DNA and turns it "off."

**repression** A regulatory process in which a gene is temporarily rendered unable to be expressed.

**repressor** A protein that binds specifically to a regulatory sequence adjacent to a gene and blocks transcription of the gene.

**repulsion** *See* ***trans*** **configuration.**

**restriction endonuclease** A nuclease that recognizes a short nucleotide sequence (restriction site) in a DNA molecule and cleaves the molecule at that site; also called a restriction enzyme.

**restriction enzyme** *See* **restriction endonuclease.**

**restriction fragment** A segment of duplex DNA produced by cleavage of a larger molecule by a restriction enzyme.

**restriction fragment length polymorphism (RFLP)** Genetic variation in a population associated with the size of restriction fragments that contain sequences homologous to a particular probe DNA; the polymorphism results from the positions of restriction sites flanking the probe, and each variant is essentially a different allele.

**restriction map** A diagram of a DNA molecule showing the positions of cleavage by one or more restriction endonucleases.

**restriction site** The base sequence at which a particular restriction endonuclease makes a cut.

**restrictive condition** A growth condition in which the phenotype of a conditional mutation is expressed.

**retinoblastoma** An inherited cancer caused by a mutation in the tumor-suppressor gene located in chromosome band *13q14.* Inheritance of one copy of the mutation results in multiple malignancies in retinal cells of the eyes in which the mutation becomes homozygous—for example, through a new mutation or mitotic recombination.

**retinoblastoma protein** Any of a family of proteins found in animal cells that functions to hold cells at the $G_1/S$ restriction point ("start") by binding to and sequestering a transcription factor that initiates the cell cycle.

**retrovirus** One of a class of RNA animal viruses that cause the synthesis of DNA complementary to their RNA genomes on infection.

**reverse genetics** Procedure in which mutations are deliberately produced in cloned genes and introduced back into cells or the germ line of an organism.

**reverse mutation** A mutation that undoes the effect of a preceding mutation.

**reverse transcriptase** An enzyme that makes complementary DNA from a single-stranded RNA template.

**reverse transcriptase PCR (RT-PCR)** Amplification of a duplex DNA molecule originally produced by reverse transcriptase using an RNA template.

**reversion** Restoration of a mutant phenotype to the wildtype phenotype by a second mutation.

**RFLP** *See* **restriction fragment length polymorphism.**

**R group** *See* **side chain.**

**Rh** Rhesus blood-group system in human beings; maternal-fetal incompatibility of this system may result in hemolytic disease of the newborn.

**Rh-negative** Refers to the phenotype in which the red blood cells lack the D antigen of the Rh blood-group system.

**rhodopsin** One of a class of proteins that function as visual pigments.

**Rh-positive** Refers to the phenotype in which the red blood cells possess the D antigen of the Rh blood-group system.

**ribonuclease** Any enzyme that cleaves phosphodiester bonds in RNA; abbreviated RNase.

**ribonucleic acid** *See* **RNA.**

**ribonucleoprotein particle** Nuclear particle containing a short RNA molecule and several proteins; involved in intron excision and splicing and other aspects of RNA processing.

**ribose** The five-carbon sugar in RNA.

**ribosomal RNA (rRNA)** RNA molecules that are components of the ribosomal subunits; in eukaryotes, there are four rRNA molecules—5S, 5.8S, 18S, and 28S; in prokaryotes, there are three—5S, 16S, and 23S.

**ribosome** The cellular organelle on which the codons of mRNA are translated into amino acids in protein synthesis. Ribosomes consist of two subunits, each composed of RNA and proteins. In prokaryotes, the subunits are 30S and 50S particles; in eukaryotes, they are 40S and 60S particles.

**ribosome-binding site** The base sequence in a prokaryotic mRNA molecule to which a ribosome can bind to initiate protein synthesis; also called the Shine–Dalgarno sequence.

**riboswitch** A 5′ RNA leader sequence that, according to whether it is bound with a small molecule, can adopt either of two configurations, one of which permits transcription and the other of which terminates transcription.

**ribozyme** An RNA molecule able to catalyze one or more biochemical reactions.

**ring chromosome** A chromosome whose ends are joined; one that lacks telomeres.

**RIP (repeat-induced point mutation)** A mutational mechanism of gene silencing found in certain filamentous fungi in which genes present in multiple copies undergo multiple nucleotide substitutions.

**risk factor** Any genetic or environmental agent that increases the risk of a particular phenotype, usually a disease.

**RNA** Ribonucleic acid, a nucleic acid in which the sugar constituent is ribose; typically, RNA is single-stranded and contains the four bases adenine, cytosine, guanine, and uracil.

**RNA editing** Process in which certain nucleotides in RNA are chemically changed to other nucleotides after transcription.

**RNA interference (RNAi)** The ability of small fragments of double-stranded RNA to silence genes whose transcripts contain homologous sequences.

**RNA polymerase** An enzyme that makes RNA by copying the base sequence of a DNA strand.

**RNA polymerase holoenzyme** A large protein complex in eukaryotic nuclei consisting of Pol II in combination with at least 9 other subunits; Pol II itself includes 12 subunits.

**RNA processing** The conversion of a primary transcript into an mRNA, rRNA, or tRNA molecule; includes splicing, cleavage, modification of termini, and (in tRNA) modification of internal bases.

**RNase** Any enzyme that cleaves RNA.

**RNA splicing** Excision of introns and joining of exons.

**Robertsonian translocation** A chromosomal aberration in which the long arms of two acrocentric chromosomes become joined to a common centromere.

**roentgen** A unit of ionizing radiation, defined as the amount of radiation resulting in $2.083 \times 10^9$ ion pairs per $cm^3$ of dry air at 0°C and 1 atm pressure; abbreviated R; 1 R produces 1 electrostatic unit of charge per $cm^3$ of dry air at 0°C and 1 atm pressure.

**rolling-circle replication** A mode of replication in which a circular parent molecule produces a linear branch of newly formed DNA.

**Rous sarcoma virus** A type of retrovirus that infects the chicken.

**R plasmid** A bacterial plasmid that carries drug-resistance genes; commonly used in genetic engineering.

**rRNA** *See* **ribosomal RNA.**

**RT-PCR** *See* **reverse transcriptase PCR.**

## S

**S** *See* **cell cycle.**

**salvage pathway** A minor source of deoxythymidine triphosphate (dTTP) for DNA synthesis that uses the enzyme thymidine kinase (TK).

**Sanger sequencing** A method of sequencing DNA in which elongation of a daughter strand along a template strand is interrupted by the incorporation of a dideoxyribonucleotide lacking a free 3′ hydroxyl group.

**satellite DNA** Eukaryotic DNA that forms a minor band at a different density from that of most of the cellular DNA in equilibrium density gradient centrifugation; consists of short sequences repeated many times in the genome (highly repetitive DNA) or of mitochondrial or chloroplast DNA.

**scaffold** A protein-containing material in chromosomes, believed to be responsible in part for the compaction of chromatin.

**second-division segregation** Segregation of a pair of alleles into different nuclei in the second meiotic division, the result of crossing-over between the gene and the centromere of the pair of homologous chromosomes.

**second meiotic division** The meiotic division in which the centromeres split and the chromosome number is not reduced; also called the equational division.

**segment** Any of a series of repeating morphological units in a body plan.

**segmentation gene** Any of a group of genes that determines the spatial pattern of segments and parasegments in *Drosophila* development.

**segment-polarity gene** Any of a group of genes that determines the spatial pattern of development within the segments of *Drosophila* larvae.

**segregation** Separation of the members of a pair of alleles into different gametes in meiosis.

**segregational petites** Slow-growing yeast colonies deficient in respiration because of mutation in a chromosomal gene; inheritance is Mendelian.

**segregation mutation** Developmental mutation in which sister cells express the same fate when they should express different fates.

**selected marker** A genetic mutation that allows growth in selective medium.

**selection** In evolution, intrinsic differences in the ability of genotypes to survive and reproduce; in plant and animal breeding, the choosing of organisms with certain phenotypes to be parents of the next generation; in mutation studies, a procedure designed in such a way that only a desired type of cell can survive, as in selection for resistance to an antibiotic.

**selection coefficient** The amount by which relative fitness is reduced or increased.

**selection differential** In artificial selection, the difference between the mean of the selected organisms and the mean of the population from which they were chosen.

**selection limit** The condition in which a population no longer responds to artificial selection for a trait.

**selection-mutation balance** Equilibrium determined by the opposing effects of mutation tending to increase the frequency of a deleterious allele and selection tending to decrease its frequency.

**selection pressure** The tendency of natural or artificial selection to change allele frequency.

**selectively neutral mutation** A mutation that has no (or negligible) effects on fitness.

**selective medium** A medium that allows growth only of cells with particular genotypes.

**self-fertilization** The union of male and female gametes produced by the same organism.

**selfish DNA** DNA sequences that do not contribute to the fitness of an organism but are maintained in the genome through their ability to replicate and transpose.

**semiconservative replication** The usual mode of DNA replication, in which each strand of a duplex molecule serves as a template for the synthesis of a new complementary strand, and the daughter molecules are composed of one old (parental) and one newly synthesized strand.

**semisterility** A condition in which a significant proportion of the gametophytes produced by a plant or of the zygotes produced by an animal are inviable, as in the case of a translocation heterozygote.

**senaptonemal complex** A ladder-like protein structure that forms between homologous chromosomes as they undergo synapsis early in meiosis.

**sense strand** Originally defined as the nontemplate strand in duplex DNA, which has the same nucleotide sequence as the transcript (although U replaces T in the RNA). Some later authors used the term in the opposite sense to mean the template DNA strand. Because both usages persist, the term is ambiguous and is best avoided. Use *template strand* or *transcribed strand* instead.

**sequence-tagged site (STS)** A DNA sequence, present once per haploid genome, that can be amplified by the use of suitable oligonucleotide primers in the polymerase chain reaction in order to identify clones that contain the sequence.

**sex chromosome** A chromosome, such as the human X or Y, that has a role in the determination of sex.

**sex-influenced trait** A trait whose expression depends on the sex of the individual.

**sex-limited trait** A trait expressed in one sex and not in the other.

**sex-linked** Refers to a trait determined by a gene on a sex chromosome, usually the X.

**sexual polyploidization** Formation of a polyploid through the fusion of unreduced gametes.

**Shine–Dalgarno sequence** *See* **ribosome-binding site.**

**shotgun sequencing** Method of genomic sequencing in which sequenced clones are chosen at random and the sequences are assembled by computer according to their overlaps.

**shuttle vector** A vector capable of replication in two or more organisms—for example, yeast and *E. coli*.

**sib** *See* **sibling.**

**sibling** A brother or sister, each having the same parents.

**sibship** A group of brothers and sisters.

**sickle-cell anemia** A severe anemia in human beings inherited as an autosomal recessive and caused by an amino acid replacement in the $\beta$-globin chain; heterozygotes tend to be more resistant to falciparum malaria than are normal homozygotes.

**side chain** In protein structure, the chemical group attached to the $\alpha$-carbon atom of an amino acid by which the chemical properties of each amino acid are determined; also called R group.

**sievert** A unit of radiation exposure of biological tissue equal to the dose in gray multiplied by a quality factor that measures the relative biological effectiveness of the type of radiation; abbreviated Sv. Roughly speaking, 1 sievert is the amount of radiation equivalent in its biological effects to one gray of gamma rays; 1 Sv equals 100 rem.

**sigma (σ) subunit** The subunit of RNA polymerase needed for promoter recognition.

**signal transduction** The process by which a regulatory signal ultimately determines cell fate by activation of a series of regulatory proteins.

**significant** *See* **statistically significant.**

**silencer** A nucleotide sequence that binds with certain proteins whose presence prevents or modulates gene expression.

**silent mutation** A mutation that has no phenotypic effect.

**simple sequence length polymorphism (SSLP)** Tandem repeats of a short DNA sequence whose repeat number differs among homologous chromosomes.

**simple tandem repeat polymorphism (STRP)** A DNA polymorphism in a population in which the alleles differ in the number of copies of a short, tandemly repeated nucleotide sequence.

**single-active-X principle** In mammals, the genetic inactivation of all X chromosomes except one in each cell lineage, except in the very early embryo.

**single-copy sequence** A DNA sequence present only once in each haploid genome.

**single nucleotide polymorphism (SNP)** A DNA marker in which a single nucleotide pair differs in the DNA sequence of homologous chromosomes and in which each of the alternative sequences occurs relatively frequently.

**single-stranded DNA** A DNA molecule that consists of a single polynucleotide chain.

**single-stranded DNA binding protein (SSBP)** A protein able to bind single-stranded DNA.

**sister chromatid exchange** Recombination between duplex DNA molecules present in sister chromatids; it normally has no genetic consequence and does not occur in some organisms.

**sister chromatids** Chromatids produced by replication of a single chromosome.

**site-directed mutagenesis** Method of engineering mutations at specific nucleotide sites or in specific regions of DNA.

**site-specific exchange** *See* **site-specific recombination.**

**site-specific recombinase** An enzyme that catalyzes intermolecule recombination between two duplex DNA molecules at the site of a target sequence that they have in common.

**site-specific recombination** Recombination that is catalyzed by a particular enzyme and that always takes place at the same site between the same DNA sequences.

**small interfering RNA (siRNA)** Small cleavage products of double-stranded RNA used to target RNAs containing complementary sequences for destruction or for inhibition of their function.

**small nuclear ribonucleoprotein particle (snRNP)** Small nuclear particle that contains short RNA molecules and several proteins. They are involved in intron excision and splicing and other aspects of RNA processing.

**snRNP** Abbreviation for *small nuclear ribonucleoprotein particle.*

**SNP chips** A type of microarray that enables the SNP genotype of an individual to be determined with 100% accuracy.

**somatic cell** Any cell of a multicellular organism other than the gametes and the germ cells from which gametes develop.

**somatic-cell genetics** Study of somatic cells in culture.

**somatic mosaic** An organism that contains two genetically different types of cells.

**somatic mutation** A mutation arising in a somatic cell.

**SOS repair** An inducible, error-prone system for repair of DNA damage in *E. coli.*

**Southern blot** A nucleic acid hybridization method in which, after electrophoretic separation, denatured DNA is transferred from a gel to a membrane filter and then exposed to labeled DNA or RNA under conditions of renaturation; the labeled regions locate the homologous DNA fragments on the filter.

**spacer sequence** A noncoding base sequence between the coding segments of polycistronic mRNA or between genes in DNA.

**specialized transduction** *See* **transducing phage.**

**species** Genetically, a group of actually or potentially inbreeding organisms that is reproductively isolated from other such groups.

**S period** *See* **cell cycle.**

**spindle** A structure composed of fibrous proteins on which chromosomes align during metaphase and move during anaphase.

**spindle assembly checkpoint** A mechanism that arrests the cell division cycle until the spindle is properly deployed.

**splice acceptor** The 5′ end of an exon.

**splice donor** The 3′ end of an exon.

**spliceosome** An RNA-protein particle in the nucleus in which introns are removed from RNA transcripts.

**spontaneous mutation** A mutation that happens in the absence of any known mutagenic agent.

**sporadic** An instance of a disease that is solitary, lacking other affected members in the same pedigree.

**spore** A unicellular reproductive entity that becomes detached from the parent and can develop into a new organism upon germination; in plants, spores are the haploid products of meiosis.

**sporophyte** The diploid, spore-forming generation in plants, which alternates with the haploid, gamete-producing generation (the gametophyte).

**SSR** Simple sequence repeat; tandem repeat of a short DNA sequence.

**SSRP** Simple sequence repeat polymorphism, a polymorphism in which the alleles differ in the number of copies of a simple sequence repeat.

**standard deviation** The square root of the variance.

**start codon** An mRNA codon, usually AUG, at which polypeptide synthesis begins.

**statistically significant** Said of the result of an experiment or study that has only a small probability of happening by chance on the assumption that some hypothesis is true. Conventionally, if results as bad or worse would be expected less than 5 percent of the time, the result is said to be statistically significant; if less than 1 percent of the time, the result is called statistically highly significant; both outcomes cast the hypothesis into serious doubt.

**sticky end** A single-stranded end of a DNA fragment produced by certain restriction enzymes capable of reannealing with a complementary sequence in another such strand.

**stop codon** One of three mRNA codons—UAG, UAA, and UGA—at which polypeptide synthesis stops.

**structural gene** A gene that encodes the amino acid sequence of a polypeptide chain.

**STRP** *See* **simple tandem repeat polymorphism.**

**STS** *See* **sequence-tagged site.**

**subfunctionalization** The loss of a subset of tissue-specific regulatory elements in one or both members of a pair of duplicated genes.

**submetacentric chromosome** A chromosome whose centromere divides it into arms of unequal length.

**subpopulation** Any of the breeding groups within a larger population between which migration is restricted.

**subspecies** A relatively isolated population or group of populations distinguishable from other populations in the same species by allele frequencies or chromosomal arrangements and sometimes exhibiting incipient reproductive isolation.

**substrate** A substance acted on by an enzyme.

**subunit** A component of an ordered aggregrate of macromolecules—for example, a single polypeptide chain in a protein containing several chains.

**supercoiled DNA** *See* **supercoiling.**

**supercoiling** Coiling of double-stranded DNA in which strain caused by overwinding or underwinding of the duplex makes the double helix itself twist; a supercoiled circle is also called a twisted circle or a superhelix.

**suppressive petites** A class of cytoplasmically inherited, small-colony mutants in yeast that are not corrected by the addition of normal mitochondria.

**suppressor mutation** A mutation that partially or completely restores the function impaired by another mutation at a different site in the same gene (intragenic suppression) or in a different gene (intergenic suppression).

**suppressor tRNA** Usually, a tRNA molecule capable of translating a stop codon by inserting an amino acid in its place, but a few suppressor tRNAs replace one amino acid with another.

**symbiosis** The living together in intimate association of organisms that are mutually interdependent for their survival and reproduction.

**synapsis** The pairing of homologous chromosomes or chromosome regions in zygotene of the first meiotic prophase.

**synaptonemal complex** A complex protein structure that forms between synapsed homologous chromosomes in the pachytene substage of the first meiotic prophase.

**syncytial blastoderm** Stage in early *Drosophila* development formed by successive nuclear divisions without division of the cytoplasm.

**syndrome** A group of symptoms that appear together with sufficient regularity to warrant designation by a special name; also, a disorder, disease, or anomaly.

**synonymous mutation** A nucleotide substitution in a codon that does not alter the amino acid specified by the codon.

**syntenic genes** Genes present in the same chromosome; not equivalent to linked genes because syntenic genes can show independent assortment if they are far enough apart.

**synteny** The presence of two different genes on the same chromosome.

**synteny group** A group of genes present in a single chromosome in two or more species.

## T

**tandem duplication** A pair of identical or closely related DNA sequences that are adjacent and in the same orientation.

**taster** One who has the inherited ability to taste the substance phenylthiocarbamide (PTC); the phenotype of the homozygous dominant and heterozygote.

**TATA box** The base sequence 5′-TATA-3′ in the DNA of a promoter.

**TATA-box binding protein (TBP)** A protein that binds to the TATA motif in the promoter region of a gene.

**tautomeric shift** A reversible change in the location of a hydrogen atom in a molecule, altering the molecule from one isomeric form to another. In nucleic acids, the shift is typically between a keto group (keto form) and a hydroxyl group (enol form).

**taxon** A population, species, or other group of organisms of which a protein or nucleic acid sequence, or a set of such sequences, is regarded as representative.

**TBP** TATA-box binding protein. *See also* **TATA-box binding protein.**

**TBP-associated factor (TAF)** Any protein found in close association with TATA-box binding protein.

**T cells** A class of white blood cells instrumental in various aspects of the immune reponse.

**T DNA** Transposable element found in *Agrobacterium tumefaciens,* which produces crown gall tumors in a wide variety of dicotyledonous plants.

**telomerase** An enzyme that adds specific nucleotides to the tips of the chromosomes to form the telomeres.

**telomere** The tip of a chromosome, containing a DNA sequence required for stability of the chromosome end.

**telophase** The final stage of mitotic or meiotic nuclear division.

**temperate phage** A phage capable of both a lysogenic and a lytic cycle.

**temperature-sensitive mutation** A conditional mutation that causes a phenotypic change only at certain temperatures.

**template strand** A nucleic acid strand whose base sequence is copied in a polymerization reaction to produce either a complementary DNA or RNA strand.

**terminal redundancy** In bacteriophage T4, a short duplication found at opposite ends of the linear molecule; the duplicated region differs from one phage to the next.

**termination** Final stage in polypeptide synthesis when the ribosome encounters a stop codon and the finished polypeptide is released from the ribosome.

**terminator** A sequence in RNA that halts transcription.

**terminus of replication** In a circular DNA molecule that undergoes bidirectional $\theta$ replication, the point at which the replication forks meet.

**testcross** A cross between a heterozygote and a recessive homozygote, resulting in progeny in which each phenotypic class represents a different genotype.

**testis-determining factor (TDF)** Genetic element on the mammalian Y chromosome that determines maleness.

**tetrad** The four chromatids that make up a pair of homologous chromosomes in meiotic prophase I and metaphase I; also, the four haploid products of a single meiosis.

**tetrad analysis** A method for the analysis of linkage and recombination using the four haploid products of single meiotic divisions.

**tetramer** A protein composed of four polypeptide subunits, which may or may not be identical.

**tetraploid** A cell or organism with four complete sets of chromosomes; in an autotetraploid, the chromosome sets are homologous; in an allotetraploid, the chromosome sets consist of a complete diploid complement from each of two distinct ancestral species.

**tetratype** An ascus containing spores of four different genotypes—one each of the four genotypes possible with two alleles of each of two genes.

**thermophile** An organism that normally lives at an unusually high temperature.

**$\theta$ replication** Bidirectional replication of a circular DNA molecule, starting from a single origin of replication.

**30-nm chromatin fiber** The level of compaction of eukaryotic chromatin resulting from coiling of the extended, nucleosome-bound DNA fiber.

**three-point cross** Cross in which three genes are segregating; used to obtain unambiguous evidence of gene order.

**3′-OH group** The end of a DNA or RNA strand that terminates in a sugar and so has a free hydroxyl group on the number-3′ carbon.

**threshold trait** A trait with a continuously distributed liability or risk; organisms with a liability greater than a critical value (the threshold) exhibit the phenotype of interest, such as a disorder.

**thymine (T)** A nitrogenous pyrimidine base found in DNA.

**thymine dimer** *See* **pyrimidine dimer.**

**time of entry** In an Hfr $\times$ $F^-$ bacterial mating, the earliest time that a particular gene in the Hfr parent is transferred to the $F^-$ recipient.

***Ti* plasmid** A plasmid present in *Agrobacterium tumefaciens* and used in genetic engineering in plants.

**topoisomerase** An enzyme that introduces or eliminates either underwinding or overwinding of double-stranded DNA. It acts by introducing a single-strand break, changing the relative positions of the strands, and sealing the break.

**topoisomerase I** A nick-closing type of topoisomerase enzyme in which one supercoil is removed from DNA per reaction cycle.

**topoisomerase II** A topoisomerase enzyme able to remove both positive and negative supercoils from DNA. These enzymes can also cleave duplex DNA and pass another duplex DNA molecule through the gap.

**total variance** Summation of all sources of genetic and environmental variation.

**totipotent cell** A cell capable of differentiation into a complete organism; the zygote is totipotent.

**trait** Any aspect of the appearance, behavior, development, biochemistry, or other feature of an organism.

***trans* configuration** The arrangement in linked inheritance in which a genotype heterozygous for two mutant sites has received one of the mutant sites from each parent—that is, $a_1+/+a_2$.

**transcript** An RNA strand that is produced from, and is complementary in base sequence to, a DNA template strand.

**transcription** The process by which the information contained in a template strand of DNA is copied into a single-stranded RNA molecule of complementary base sequence.

**transcriptional activator protein** Positive control element that stimulates transcription by binding with particular sites in DNA.

**transcriptional cosuppression** A mechanism of gene silencing in which the silenced genes fail to be transcribed.

**transcription complex** An aggregate of RNA polymerase (consisting of its own subunits) along with other polypeptide subunits that makes transcription possible.

**transducing phage** A phage type capable of producing particles that contain bacterial DNA (transducing particles). A specialized transducing phage produces particles that carry only specific regions of chromosomal DNA; a generalized transducing phage produces particles that may carry any region of the genome.

**transduction** The carrying of genetic information from one bacterium to another by a phage.

**transfer RNA (tRNA)** A small RNA molecule that translates a codon into an amino acid in protein synthesis; it has a three-base sequence, called the anticodon, complementary to a specific codon in mRNA, and a site to which a specific amino acid is bound.

**transformation** Change in the genotype of a cell or organism resulting from exposure of the cell or organism to DNA isolated from a different genotype; also, the conversion of an animal cell, whose growth is limited in culture, into a tumor-like cell whose pattern of growth is different from that of a normal cell.

**transformation mutation** A mutation in which affected cells undergo a developmental fate that is characteristic of other cells.

**transformation rescue** The ability of a wildtype gene to complement a recessive mutation when introduced into a mutant organism by germ-line transformation.

**transgenic** Refers to animals or plants in which novel DNA has been incorporated into the germ line.

***trans* heterozygote** *See* ***trans* configuration.**

**transition mutation** A mutation resulting from the substitution of one purine for another purine or that of one pyrimidine for another pyrimidine.

**translation** The process by which the amino acid sequence of a polypeptide is synthesized on a ribosome according to the nucleotide sequence of an mRNA molecule.

**translocation** Interchange of parts between nonhomologous chromosomes; also, the movement of mRNA with respect to a ribosome during protein synthesis. *See also* **reciprocal translocation.**

**transmembrane receptor** A receptor protein containing amino acid sequences that span the cell membrane.

**transmission genetics** The study of the pattern of inheritance of genes and of other genetic elements from generation to generation.

**transposable element** A DNA sequence capable of moving (transposing) from one location to another in a genome.

**transposase** Protein necessary for transposition.

**transposition** The movement of a transposable element.

**transposon** A transposable element that contains bacterial genes—for example, for antibiotic resistance; also loosely used as a synonym for transposable element.

**transposon tagging** Insertion of a transposable element that contains a genetic marker into a gene of interest.

**transversion mutation** A mutation resulting from the substitution of a purine for a pyrimidine or that of a pyrimidine for a purine.

**trinucleotide expansion** Increase in the number of tandem copies of a trinucleotide repeat during DNA replication.

**trinucleotide repeat** A tandemly repeated sequence of three nucleotides; genetic instability in trinucleotide repeats is the cause of a number of human hereditary diseases.

**triplet code** A code in which each codon consists of three bases.

**triplication** The presence of three copies of a DNA sequence that is ordinarily present only once.

**triploid** A cell or organism with three complete sets of chromosomes.

**trisomic** A diploid organism with an extra copy of one of the chromosomes.

**trisomy-X syndrome** The clinical features of the karyotype 47,XXX.

**trivalent** Structure formed by three homologous chromosomes in meiosis I in a triploid or trisomic chromosome when each homolog is paired along part of its length with first one and then the other of the homologs.

**tRNA** *See* **transfer RNA.**

**true-breeding** Refers to a strain, breed, or variety of organism that yields progeny like itself; homozygous.

**truncation point** In artificial selection, the value of the phenotype that determines which organisms will be retained for breeding and which will be culled.

**tumor-suppressor gene** A gene that normally controls cell proliferation or that activates the apoptotic pathway, in which loss-of-function mutations are associated with cancer progression.

**Turner syndrome** The clinical features of human females with the karyotype 45,X.

**twin spot** A pair of adjacent, genetically different regions of tissue, each derived from one daughter cell of a mitosis in which mitotic recombination took place.

**two-hybrid analysis** A method for detecting protein-protein interactions that makes use of two fused (hybrid) proteins, one including the DNA-binding domain and the other the transcriptional activation domain of a trancriptional activator protein. If the polypeptide chains attached to these components interact within the nucleus, then the interaction brings the domains together and transcription of a reporter gene takes place.

## U

**ultracentrifuge** A centrifuge capable of speeds of rotation sufficiently high to separate molecules of different density in a suitable density gradient.

**uncharged tRNA** A tRNA molecule that lacks an amino acid.

**uncovering** The expression of a recessive allele present in a region of a structurally normal chromosome in which the homologous chromosome has a deletion.

**underwound DNA** A DNA molecule whose strands are untwisted somewhat; hence, some of its bases are unpaired.

**unequal crossing-over** Crossing-over between nonallelic copies of duplicated or other repetitive sequences—for example, in a tandem duplication, between the upstream copy in one chromosome and the downstream copy in the homologous chromosome.

**union of A + B (or A ∪ B)** An event composed of the elementary outcomes present in A only, present in B only, or present in both.

**uniparental inheritance** Extranuclear inheritance of a trait through cytoplasmic factors or organelles contributed by only one parent. *See also* **maternal inheritance.**

**unique sequence** A DNA sequence that is present in only one copy per haploid genome, in contrast with repetitive sequences.

**univalent** Structure formed in meiosis I in a monoploid or a monosomic when a chromosome has no pairing partner.

**unlinked genes** Genes whose frequency of recombination is 50 percent (independent assortment); unlinked genes can be in the same chromosome if they are sufficiently far apart.

**uracil (U)** A nitrogenous pyrimidine base found in RNA.

**uridylate** A nucleotide in RNA in which the base is uridine.

**UTR** The untranslated region of a messenger RNA; the 5′ UTR is upstream of the protein-coding region and the 3′ UTR is downstream.

## V

**variable expressivity** Differences in the severity of expression of a particular genotype.

**variable number of tandem repeats (VNTR)** Any region of the genome that, among the members of a population, is highly poplymorphic in the number of tandem repeats of a short nucleotide sequence; the result is polymorphism in the size of the restriction fragment that contains the repeated sequence.

**variable region** The portion of an immunoglobulin molecule that varies greatly in amino acid sequence among antibodies in the same subclass. *See also* **V region.**

**variance** A measure of the spread of a statistical distribution; the mean of the squares of the deviations from the mean.

**variegation** Mottled or mosaic expression.

**vector** A DNA molecule, capable of replication, into which a gene or DNA segment is inserted by recombinant DNA techniques; a cloning vehicle.

**viability** The probability of survival to reproductive age.

**viral oncogene** A class of genes found in certain viruses that predispose to cancer. Viral oncogenes are the viral counterparts of cellular oncogenes. *See also* **cellular oncogene, oncogene.**

**virulent phage** A phage or virus capable only of a lytic cycle; contrasts with temperate phage.

**virus** An infectious intracellular parasite able to reproduce only inside living cells.

**VNTR** *See* **variable number of tandem repeats.**

**Von Hippel–Lindau disease** Hereditary disease marked by tumors in the retina, brain, other parts of the central nervous system, and various organs throughout the body. The disease is caused by an autosomal dominant mutation in chromosome 3.

**V region** One of multiple DNA sequences coding for alternative amino acid sequences of part of the variable region of an antibody molecule. *See also* **variable region.**

**V-type position effect** A type of position effect in *Drosophila* that is characterized by discontinuous expression of one or more genes during development; usually results from chromosome breakage and rejoining such that euchromatic genes are repositioned in or near centromeric heterochromatin.

## W

**Watson–Crick pairing** Base pairing in DNA or RNA in which A pairs with T (or with U in RNA) and G pairs with C.

**Watson strand** The upper strand of duplex DNA when the strands are conventionally represented as parallel horizontal lines with the 5′ end of the top strand at the extreme left.

**wildtype** The most common phenotype or genotype in a natural population; also, a phenotype or genotype arbitrarily designated as a standard for comparison.

**wobble** The acceptable pairing of several possible bases in an anticodon with the base present in the third position of a codon.

## X

$\chi^2$ *See* **chi-square.**

**X chromosome** A chromosome that plays a role in sex determination and that is present in two copies in the homogametic sex and in one copy in the heterogametic sex.

**xeroderma pigmentosum** An inherited defect in the repair of ultraviolet-light damage to DNA, associated with extreme sensitivity to sunlight and multiple skin cancers.

**X inactivation** *See* **single-active-X principle.**

**X-linked inheritance** The pattern of hereditary transmission of genes located in the X chromosome; usually evident from the production of nonidentical classes of progeny from reciprocal crosses.

## Y

**YAC** *See* **yeast artificial chromosome.**

**Y chromosome** The sex chromosome present only in the heterogametic sex; in mammals, the male-determining sex chromosome.

**yeast artificial chromosome (YAC)** In yeast, a cloning vector that can accept very large fragments of DNA; a chromosome introduced into yeast derived from such a vector and containing DNA from another organism.

## Z

**Z-form DNA** One of the unusual three-dimensional structures of duplex DNA in which the helix is left-handed.

**zinc finger** A structural motif, found in many DNA-binding proteins, in which finger-like projections entrap a zinc ion.

**zygote** The product of the fusion of a female and a male gamete in sexual reproduction; a fertilized egg.

**zygotene** The substage of meiotic prophase I in which homologous chromosomes synapse.

**zygotic gene** Any of a group of genes that control early development through their expression in the zygote.

# Index

## A


A base. *See* adenine
A (aminoacyl) site, 359
ABC transporters, 460
abnormalities in human chromosomes. *See also* mutation(s); *specific condition*
  structural, 269–282
    deletions and duplications, 269–275
    inversions, 275–277
    translocations, 277–282
  trisomy and monosomy, 264–269
    environmental effects, 269
ABO blood groups (human), 103–105
abundant mRNA, 439
Ac (Activator), 519, 520
acceptor end of DNA strands. *See* 3′ (acceptor) end of DNA strands
accessory DNA molecules. *See* plasmids
accessory proteins, for transcription, 347
ACE (angiotensin-converting enzyme), 673
acentric chromosomes, 255–256
  chromosomal inversions, 276
ACF complex, 407
acridine, 532
acrocentric chromosomes, 255–256
Activator (Ac), 519, 520–523
acute leukemias, 465, 578–581, 582, 618
adaptive evolution, 634–639
adaptor molecules, 358
addition rule (probability), 93
  for predicting progeny distribution, 136–138
additive alleles, 661–662
adenine (A), 8, 41
  base pair stability, 264
  bonding to thymine. *See* pairing of DNA bases
  deamination of, 529, 531
  mismatching in base pairs. *See* incorporation error
  molar concentration of, 42–43
adenomatous polyposis, 575
adenosine triphosphate (ATP), 344
adenovirus, 357
adenyl cyclase, 387
adjacent-1 segregation, 278
adjacent-2 segregation, 278
Africa, Great Migration out of, 645
agarose concentration in gel electrophoresis, 51
*Agrobacterium tumefaciens*, plasmid vector in, 458–459, 462
Ahmed, Shawn, 237
albinism, pedigree for, 98
alignment of chromosomes during cell cycle, 119
alkaptonuria, 10–13
alkylating agents, 531
allele(s), 78
  additive, 661–662
  codominant alleles, 80
    common in SSRPs, 100
    defined, 66
  defined, 60
  dominant. *See* dominant alleles
  fixed, 613
  during gamete formation. *See* segregation
  harmful. *See* risk factors (genetic)
  meiosis and, 119–128
  multiple, Hardy-Weinberg principle and, 623–624
  multiple (tandem repeat polymorphisms), 102–103
  pedigree analysis, 97–99
  recessive. *See* recessive alleles
  standard notation of, 61
  *trans* and *cis* configurations, 153
allele frequency, 619–620, 625
  equilibrium of, 634, 638–639
  generational constancy of, 622
  overdominance, 638–639
  population changes in, 632–633, 642
  and selection in diploid organisms, 636–637
  in X-linked genes, 627
allelic identity by descent, 629–630
Allison, Anthony C., 640
allopolyploidy, 286–288
  segmental, 292
alpha (α) helix, 364
alpha globin sequences, 618
alphoid DNA, 240
alternate segregation, 278
alternative promoters, 407–410
alternative splicing, 32
*Alu* sequences, 523, 524
amber codon. *See* UAG codon
Ames test, 545
amino acids
  defined, 11–14
  structure of, 342, 343
  twenty different, 21–22
amino terminus, 342
aminoacyl tRNA synthetases, 359, 371–372
aminoacylated tRNA, 359
amniocentesis, 267
amoebozoa, 32
amorphic mutations. *See* loss-of-function mutations
amplification, 55–60, 419. *See also* PCR amplification
amylopectin, 78
anaphase I stage (meiosis), 125–127
anaphase II stage (meiosis), 128
anaphase (mitosis), 118, 119
  mitotic recombination, 181–182
anaphase-promoting complex (APC/C), 561–562, 567
ancestral commonalities. *See* unity of life
ancestry among organisms. *See* unity of life
anchor cell (AC), 480–482
anemia, sickle-cell, 513–515, 639, 640
aneuploid conditions, 264
  environmental effects in humans, 269
aneuploidy, 569
Angelman syndrome, 412–413, 477
angiotensin-converting enzyme (ACE), 673
anhidrotic ectodermal dysplasia, 292
animal(s). *See also* mammals; *specific species*
  early embryonic development in, 472–477
    versus plants, 500
  meiosis, 120–121. *See also* meiosis
  mitochondrial genetic code, 594

animal breeding. *See also* artificial selection
complex inheritance in, 652, 660, 662–663
DNA typing in, 69
animal models, knockout mutations in, 461, 463, 576–577
animal viruses, genetic engineering with, 463–465
Annual Drosophila Research Conference, 492
*Antennapedia* (ANT-C) gene, 498
anterior genes, 491
anthocyanin, 414
anti-terminator conformation, 394
antibiotic resistance, 296, 299, 303, 526, 597, 635
cassettes, 300–301
multiple-antibiotic-resistant bacteria, 300, 302
antibodies, defined, 104
antibody variability, DNA rearrangements in, 419–422
anticipation, 518
anticodon, 358, 359, 371
antigends, defined, 104
antiparallelism of DNA strands, 44–45
strand elongation and, 191
antisense RNA, 417–418
antisense transcript, 414
AP endonuclease, 541
AP repair, 541
APC/C (anaphase-promoting complex), 561–562, 567
apoptosis, 478–480, 565–566, 571
aporepressor, 382, 391–392
apurinic site, 541
apyrimidine site, 541
AR-PKD (autosomal recessive polycystic kidney disease), 647
*Arabidopsis thaliana*, developmental genetics in, 472, 500–503
Arber, Werner, 48
archaeans, 601
archeans, ancestry of, 30–31
archenteron, 475, 476
arginine-requiring mutants (Beadle–Tatum experiments), 17
argonaute, 416–417
ARS (autonomously replicating sequences), 191
artificial chromosomes, 442
artificial selection, 635–636, 658, 663–667
defined, 663–667
long-term, 666
selection differential, 679
*Ascobolus immersus*, transcriptional silencing in, 411
ascospore, 173
ascus (asci), 173
ordered, 177–180
ASDs (autism spectrum disorders), 270
asexual polyploidization, 286
*Aspergillus ficuum*, 462
mitotic recombination, 181–182
*Astyanax fasciatus* (Mexican cave-dwelling fish), 661–662
ATP (adenosine triphosphate), 344
attached-X chromosomes, 161–163
attenuation, 392–394
attenuator, 392
AUG codon, 25–26
discovery of, 370–371
translation initiated at, 359–360
Australian Aborigines, 643
autism spectrum disorders (ASDs), 270
autogamy, 602–603
autonomous development, 473–475
autonomously replicating sequences (ARS), 191
autopolyploidy, 286–288
autoregulation, 382–383
autosomal recessive polycystic kidney disease (AR-PKD), 647
autosomes
defined, 128
trisomy of. *See* trisomy (human)
auxotrophs, 303
average (mean), 653, 654
phenotypic, 665
Avery, Oswald, 4, 434
azacytidine, 411

## B

B (bone marrow-derived) cells, DNA rearrangements in, 419–422
B form of DNA, 43–45
*Bacillus subtilis*, gene regulation in, 393, 394–395
*Bacillus thuringiensis*, endotoxins produced by, 454
backcross experiments, 87. *See also* ratios (from genetic crosses)
BACs (bacterial artificial chromosomes), 442
bacteria. *See also specific species*
ancestry of, 30–31
chaperones in, 365–366
chromosomal structure, 226
discover of conventional sexual cycle, 313
drug resistance in, 526
genome size, 222–223
holoenzymes in, 346
mismatch repair in, 539–540
promoter recognition in, 347
selection studied in, 635
SOS repair in, 543
bacterial artificial chromosomes (BACs), 442
bacterial genetics, 296–316
conjugation, 297, 305–312
DNA-mediated transformation, 303–305
mobile DNA, 296–302
transduction, 312–316
mechanism in bacteria, 303
phage lytic cycle, 312–313
specialized, 332–333
bacteriophage experiments
messenger RNA and, 351
triplet code and, 368–370
bacteriophage genetics, 316–325
generalized transduction, 313–316
genetic and physical mapping, 319–321
lysogeny, 326–332, 396–397
phage lytic cycle, 312–313
plaque formation, 317–318
recombination in lytic cycle, 318–319
specialized transduction, 332–334
bacteriophage infection, types of, 396–397
bacteriophage λ
gene regulation in, 396–397
lysogeny and specialized transduction, 325–334
as recombinant DNA vector, 434–436, 442
bacteriophage lytic cycle, 312–313
recombination in, 318–319
bacteriophage mutants, 396
bacteriophage(s), role of DNA in, 4–6
balanced translocation, 264
balancing selection. *See* heterozygote superiority
*Bal*I enzyme, 433
*BamHI* restriction enzyme, 48–50, 52
*Bam*HI site, 460–461
bands (regions of DNA), 50–51
deletion mapping, 271–273
electrophoresis. *See* gel electrophoresis
equilibrium density-gradient centrifugation, 192–193
nomenclature in human karyotypes, 253
transverse banding, 232

bar-shaped eyes (*Drosophila* mutation), 145, 273
Barr bodies, 292
basal transcription factors, 404–407
base analog, 531
base-analog mutagens, 531
base composition of DNA, 43
base-excision repair, 529, 540–541
base pairing. *See* pairing of DNA bases
base stacking, 44–45
  denaturation and, 53
bases in DNA, 8, 41. *See also specific base*
  mismatching in base pairs. *See* incorporation error
  molar concentration of, 42–43
Bax protein, 574
Bayes, Thomas, 96
Bayes theorem, 96–97
Bayesian methods, 613
*Bcl2 gene*, 579
Beadle, George W., 14, 15, 21, 105–106, 483, 656
Beadle–Tatum experiments, 14–21, 105–106
behavioral genetics, 674–675
Benzer, Seymour, 321, 329
beta (β) sheet, 364
β-carotene, 462, 464
β-galactosidase, 383–389. *See also lacZ* gene
beta globin sequences, 612–614, 618
*bicoid* gene, 491, 493, 506
bidirectional DNA replication, 198
binding sites, tRNA, 359–361
binomial distribution, 136–139
  significance of binomial coefficient, 138–139
biochemical expression of genotype. *See* phenotypes
biochemical pathways, 11–13
bisphenol A, 269
*bithorax* (BX-C) gene, 498
bivalent chromosomes, 125
  double-strand break and repair model, 212–215
  polyploidy in plant evolution, 284–289
    allopolyploidy, 286–288, 292
    autopolyploids, 286–288
    monoploids, 288–289
    sexual vs. asexual polyploidization, 285–286
black urine disease (alkaptonuria), 10–13
Blackburn, Elizabeth H., 241
blastoderm
  formation of, 486–488
  genetically marked, 487
blastula
  cell fates in, 475–476
  formation of, 472, 476
bleb formation, 565
blender experiment (Hershey–Chase), 6–7
block, metabolic pathway, 12
blood– brain barrier, genes affecting, 29
blood types, human, 103–105
blunt ends, 433–434, 436, 438
blunt ends of cleaved DNA strands, 50
bonding to DNA bases. *See* pairing of DNA bases
bone marrow-derived (B) cells, DNA rearrangements in, 419–422
bootstrapping, 614–615
Botstein, David, 64
boundary enhancers, 497–498
*BRCA* genes, 578
*BRCA1* gene, 68
breaking of DNA, intentional. *See* separation of DNA fragments
breaking of DNA, unintentional
  chromosomal inversions, 275–277
    human X and Y chromosomes, 260–262
  deletions from, 270
breeding animals, 69. *See also* artificial selection
  complex inheritance in, 652, 660, 662–663
  DNA typing in, 69
breeding failure, random. *See* random genetic drift
breeding plants
  DNA typing in, 69
  with monoploid organisms, 288–289
Brenner, Sydney, 351
Bridges, Calvin, 134
broad-sense heritability, 662–663, 668
broccoli, 100
bromodeoxyuridine (BUdR), 194–196
5-bromouracil (Bu), mutagenesis by, 531–532
Bub1 protein, 574
Burkitt lymphoma, 585

## C

C base. *See* cytosine
C-value paradox, 223
*Caedibacter taeniospiralis*, 601
*Caenorhabditis elegans* (soil nematode)
  anatomy of, 477
  developmental genetics in, 472, 474–475, 477–485
  developmental mutations in, 480–481
  DNA microarrays of, 450–451
  frequency of recombination, 186
  gene silencing in, 414, 416
  germ-line transformation in, 455
calico cat, 187, 259
cAMP-CRP complex, 387–388
cAMP (cyclic adenosine monophosphate), 387
Campbell, Allan, 331
Campbell, Keith, 456
*can1* mutation, 585
cancer, 569–575. *See also specific type*
  defined, 569
  familial, 569, 575–578
  genetic research on, 566
  pedigree studies of, 580
  radiotherapy for, 534
  sporadic, 569
cancer cells, capabilities acquired by, 570
candidate genes, 674–675
canid family (dogs), mitochondrial DNA studies in, 592, 609
cap, 352, 359
Capecchi, Mario R., 458
carbon atoms, numbering of, 41
carbon-source mutants (bacterial genetics), 303
carboxyl terminus, 342
carcinogens, Ames test for, 545
Carothers, E. Eleanor, 127
carriers
  of mutual alleles, 98
  sickle-cell trait, 640
  X-linked conditions, 516
cascade effect, 493
caspases, 565
cassettes, 300–301, 422
categorical traits, 653
cats, mutations in, 187, 259, 511, 513
*CCR5* receptor gene, 619–620
*cdc* mutants, 554–557, 584–585
CDK (cyclin-dependent protein kinase), 557–559

*Cdk4* gene, 572
cDNA. *See* complementary DNA (cDNA)
*ced-3* gene, 480
cell cycle, 116, 552–554
  checkpoints in. *See* checkpoints
  genetic analysis of, 554–557
  key events in, 552–553
  progression through, 557–562
    mutations affecting, 554–557, 559, 570–571
  regulation of
    defects in, 575–578
    by proteolysis, 561–562, 565
  transcriptional program of, 553–554
cell death, programmed (apoptosis), 478–480, 565–566, 571
cell division, 115. *See also* mitosis
cell division cycle (*cdc*) mutants, 554–557, 584–585
cell fate, 473
  determination of, 474, 480–483
cell fate map, 486–488
cell lineage. *See* lineage
cells, metabolic activities of, 10
centimorgam, defined, 156
central dogma of molecular genetics, 22–23, 342
centrioles, 552
centromeres, 117, 127
  chromosomal inversions that include, 277
  chromosome stability and, 255–256
  molecular structure of, 239–241
  ordered tetrads, analysis of, 177–180
centromeric core particle, 228–229, 239
centrosome duplication checkpoint, 567
centrosomes, 117, 552
CEPH (Centre d'Etude du Polymorphisme Humain), 170, 171
cercozoa, 32
cervical cancer, 584
chain elongation, 191, 202–204
  dideoxy sequencing method and, 207
  telomeres, 242–243
  in transcription, 349
  in translation, 359, 361–362
chain initiation, 191
  multiple, in eukaryotes, 199
  by primosome complexes, 201–202
  in transcription, 348–349
  in translation, 359–361
chain termination
  in transcription, 349–350
    regulation through, 392–396
  in translation, 359, 362–363
chain termination codon, 359
chalcone synthase, 414
chance. *See* probability
chaperones, 365–366
chaperonins, 366
Chargaff, Erwin, 42–43
Chargaff's rules, 43, 44
charged tRNA, 359, 371–372
Chase, Martha, 4–6
checkpoints in cell cycle, 116, 562–569
  centrosome duplication, 567
  defects in, 575–578
  DNA damage, 562–567
  S-phase, 563
  spindle assembly, 567–568
  spindle position, 568–569
chemical isolation of DNA, 47–55
  gel electrophoresis, 50–52, 438
  nucleic acid hybridization, 52–54
    backcross experiments, 87
    CNP detection, 68
    SNPs (single-nucleotide polymorphisms), 62–63
  restriction enzymes, 48–50
    defined, 432
    features of, 433–434
    use of, 436–437
  Southern blot, 54–55, 65
chemokines, 619–620
Chernobyl nuclear accident, 536–537
chi-square ($\chi 2$) method, 139–143
  to test for linkage, 153–154
chiasma (chiasmata), 125, 155–156
  double-strand break and repair model, 212–215
chimeric genes, 274
*Chlamydomonas* (green alga), drug resistance in, 595–598
chloroplast DNA, 588, 592–593
  evolutionary origin of, 600–601
  transmission of, 594–595
chorionic villus sampling (CVS), 267
CHRAC complex, 407
chromalveolates, 32
chromatid interference, 164
chromatids, 117
  multiple crossovers and, 164–165. *See also* recombination of genes
  separation into sister chromatids, 119
chromatin, 226–229
  heterochromatin and euchromatin, 239
  30-nm chromatin fiber, 229
chromatin domains, 229
chromatin element, 519
chromatin loops, 229
chromatin remodeling, 229
chromatin-remodeling complexes (CRCs), 348, 407–408
chromocenters, 232
chromosomal scaffold, 232
chromosome(s), 39, 114–149
  artificial, 442
  defined, 61
  discovery of, 2–3, 367
  gene densities and
    chromosome territories and, 230
    human chromosomes, 255–256
  genes on, 39, 61, 115
  genetic markers, 39
    applications of, 68–70
    cotransduction studies, 315
    developmental studies using, 487
    in DNA typing, 625–626, 670–671
    harmless vs. harmful polymorphism, 99–100
    in human pedigree analysis, 100–101
    human Y chromosome, 262–264
    physical distance between, 169
    simple sequence repeats (SSRs), 67, 262
    for tracing ancestry, 643
  human. *See* human genome
  linkage and recombination of genes in, 151–155
  maps. *See* genetic mapping
  meiosis. *See* meiosis
  mitosis. *See* mitosis
  molecular organization of. *See* molecular organization of chromosomes
  number of, in various species, 115
  polyploidy in plant evolution, 284–289
    allopolyploidy, 286–288, 292
    autopolyploids, 286–288

monoploids, 288–289
sexual vs. asexual polyploidization, 285–286
progeny distributions, predicting, 136–139
hypothesis testing (goodness of fit), 139–144
sex-chromosome inheritance, 128–136. *See also* X-linked inheritance
tandem repeat polymorphisms, 66–67, 102–103. *See also* DNA markers
Hardy-Weinberg principle and, 623–624
transposable elements, 80, 520–525
as agents of mutation, 523–524
defined, 520
discovery of, 519, 520
in human genome, 520, 523, 524–525
chromosome 21 (human), translocation of. *See* Down syndrome
chromosome breakage. *See entries at* breaking of DNA
chromosome complements. *See also* chromosomes
human. *See* human genome; human karyotypes
stability of, 115–116
chromosome condensation, 230–232
chromosome ends. *See* telomeres
chromosome interference in double crossovers, 167–168
chromosome maps. *See* genetic mapping
chromosome painting, 252–255
allopolyploid analysis, 286–287
chromosome territories, 229–230
chromosome theory of heredity, 134
chronic myeloid leukemia (CML), 579–580
*Chrysanthemum* polyploidy, 284–285, 286
cI protein, 397
circular chromosomes
bacterial. *See* bacterial genetics
supercoiling of, 223
circular DNA molecules
rolling-circle replication of, 199
theta replication of, 197–198
circular permutations, 321
*cis* configuration of alleles, 153
recombination frequency, 154–155
*cis*-dominant, 385–386
cisplatin, 573
cistrons, 325
classical genetics, 2
*ClB* method, 527–528
cleavage division, 472, 476
cleaving DNA molecules with restriction enzymes, 48–50
defined, 432
features of, 433–434
use of, 436–437
clonal tumor cells, 570
clone(s), 456
of genetic risk factors, 170
cloned sequences, 434
cloning vectors, 432–436. *See also* recombinant DNA technology
gene inactivation in, 440–442
insertion of particular DNA molecules into, 437–438
origin of replication in, 441–442
CML (chronic myeloid leukemia), 579–580
CNPs (copy-number polymorphisms), 68
in human genome, 270. *See also* deletions in human chromosomes
coalescence, 643
Cockayne's syndrome, 542
coding sequence, 342, 352
colinearity between polypeptides and, 344
overlapping, 377
coding sequence motifs, 347–348
codominant alleles, 80
common in SSRPs, 100
defined, 66
codon(s), 352, 358. *See also specific codon*
genetic code for, 25–26
initiation. *See* AUG codon
number of nucleotides on. *See* triplet code
overlapping, 377
stop, 359, 363, 370
synonymous, 370, 371
coefficient of coincidence, 167
Cohanim, 263
cohesive ends, 326, 436
cointegrates, 299–300
colchicine, 289
colinearity, 344
colonies, bacterial, 302
colony hybridization, 442–443, 449–450
color blindness (human), 273–275
X-linked, 627
colorectal cancers, 575, 580–581
common-ancestor inheritance. *See* inbreeding
common ancestry among organisms. *See* unity of life
comparative genomics, 445–447, 612
compartments, 496
complement, chromosomal. *See also* chromosomes
human. *See* human genome; human karyotypes
stability of, 115–116
complement of an event, defined, 92
complementary DNA (cDNA), 440
sequencing of, 445
complementary pairing of DNA bases. *See* pairing of DNA bases
complementary strands of DNA. *See* pairing of DNA bases; structure of DNA
complementation, 18–20
epistasis and, 106
complementation groups, 20
complementation matrix, 20
complementation test for gene mutations, 17–20, 106
complete medium, defined, 16
complete penetrance of genetic disorder, 98
complex inheritance, 652
correlation between relatives, 667–669
complex traits, 651–681
candidate genes for, 674–675
categorical, 653
continuous (quantitative), 652, 662
defined, 652
genes affecting, 660–663
identification of, 652, 670–675
mutations in, 674
number of, 661–662
normal distribution of, 653–655
threshold, 653, 669–670
variation in, causes of, 655–660
complex translation units, 366–368
complexity, genetic, 233–234
alternative splicing and. *See* alternative splicing
compound-X chromosomes. *See* attached-X chromosomes
computational genomics, 445
concatemeric molecules, 320–321
concentration of DNA bases, 42–43
renaturation and, 233
condensation, chromosomal, 230–232
condensins, 117, 230
conditional mutations, 510–511

conditional probability, 94–96
congenital abnormalities, narrow-sense heritability of liability for, 670
conjugation, 297, 305–312, 601–603
F′ plasmids, 311–312
mechanism in bacteria, 303
time-of-entry mapping, 306–311
frequency of cotransduction and, 315–316
conjugative junction, 297
conjugative plasmids, 297, 299
consanguineous individuals, defined, 99
consanguineous mating, 627, 632. *See also* inbreeding
consensus sequence, 347, 354–357
conservation of DNA sequences, defined, 227
constant (C) antibody regions, 420
constitutive mutants, 384–385
contact inhibition, 569, 574–575
continuous traits, 652, 662
coordinate genes, 490, 491–493
in regulatory hierarchy, 494–496
coordinate regulation, 381
*Coprinus cinereus*, transcriptional silencing in, 411
copy-number polymorphisms (CNPs), 68
in human genome, 270. *See also* deletions in human chromosomes
core particles of nucleosomes, 228–229, 239
corepressor, 382, 391–392
corn. *See* maize
correlated response, 666
correlation, 667–669
geometrical meaning of, 667–668
correlation coefficient, 667–668
cosmid vectors, 434–436
Cot analysis (Cot curves), 235–237
cotransduction, 315
cotransformation, 305
counterselected marker, defined, 306
coupled processes, 354, 362
coupling configuration of alleles, 153
recombination frequency, 154–155
covalent bonding, in peptide groups, 364
covariance, 667
CpG islands, 411
CRCs (chromatin-remodeling complexes), 348, 407–408
Cre recombinase, 300
Crick, Francis H. C., 6, 367
bacteriophage experiments, 368–370
landmark double-helix paper (excerpt), 47
wobble concept, 373
*cro* gene, 397
cross-fertilization (outcrossing), 80–81. *See also* backcross experiments; testcross experiments
polyploids, 286
crown gall tumors, 458
CRP (cyclic AMP receptor protein), 387–388
*crp* gene, 387–388
cryptic splice site, 357
CTP (cytidine triphosphate), 344
cut-and-paste elements, 522
CVS (chorionic villus sampling), 267
cyanobacteria, 600–601
cyclic adenosine monophosphate (cAMP), 387
cyclic AMP receptor protein (CRP), 387–388
cyclin, 557–559
cyclin B, 561
cyclin-CDK complexes, 557–559
cyclin D, 572
cyclin-dependent protein kinase (CDK), 557–559
cystic fibrosis, 623
cytidine triphosphate (CTP), 344
cytokinesis, 116, 552
cytological maps of chromosomes, 232
cytoplasm
oocyte, functions of, 475, 476
symbionts in. *See* symbionts
cytoplasmic inheritance. *See* extranuclear inheritance
cytoplasmic male sterility, 598–600, 606
cytosine (C), 8, 41
base pair stability, 246
bonding to guanine. *See* pairing of DNA bases
deamination of, 529, 531
mismatching in base pairs. *See* incorporation error
molar concentration of, 42–43
cytosine deaminase, 378
cytosine methylation, 411–412, 518–520, 529
mutational hot spots and, 527–529

## D

D loops, 213
Darwin, Charles, 634
dATP (deoxyadenosine 5′-triphosphate), 42, 56
Davis, Ronald W., 64
dCTP (deoxycytidine 5′-triphosphate), 42, 56
deadenylation-dependent pathway, 414
deadenylation-independent pathway, 414
deamination, 529, 531
decaploids, 284
defective proteins, 15–20
defects, enzyme. *See* inborn errors of metabolism
deficiencies in human chromosomes. *See* abnormalities in human chromosomes; deletions in human chromosomes
degeneracy, 370, 371
degree of dominance, 638–639
degrees of freedom (df), 140–142
deletion mapping, 322–323
deletion mutations, 461, 515–516
deletion scanning, 402–404
deletions in human chromosomes, 270–273
mapping, 271–273
delta-endotoxins, 454
denaturation (unwinding), 52–53, 200–201
melting temperature, 247
density-gradient centrifugation, 351
denticles, 490
deoxyadenosine, 42
deoxyadenosine 5′-triphosphate (dATP), 42, 56
deoxycytidine, 42
deoxycytidine 5′-triphosphate (dCTP), 42, 56
deoxyguanosine, 42
deoxyguanosine 5′-triphosphate (dGTP), 42, 56
deoxyribose, 8
structure of, 345
deoxythymidine, 42
deoxythymidine 5′-triphosphate (dTTP), 42, 56
dephosphorylation, 354
depolymerization of tubulin, 117
depression, candidate gene for, 674–675
depurination, 529–530
depyrimidation, 529–530
deutranopia and deuteranomaly, 274
development, autonomous, 473–475
developmental genetics, 471–508
basics of, 472
in *C. elegans*, 472, 474–475, 477–485

in *Drosophila*, 474–477, 485–500
early embryonic, 472–477
in plants, 500–503
developmental mutations, 472, 479, 480–485, 488, 491–492, 500–502
developmental switches, epistasis in, 483–485
deviation, standard, 654
dGTP (deoxyguanosine 5′-triphosphate), 42, 56
diakinesis, 125
dicentric chromosomes, 255–256
chromosomal inversions, 276
dicer, 415
dideoxy sequencing method, 207
dideoxyribose, 207
diploid cells, 115
diploid organisms
extra chromosome copies. *See* trisomy (human)
missing copy of chromosome, 269
selection in, 636–637
diplotene, 125
direct repeats, 270
directional cloning, 469
directionality of DNA strands. *See* polarity of DNA strands
discontinuous replication of lagging DNA strands, 204–207
disease genes, 68. *See also specific disease or gene*
cancer, 569–575
mitochondrial, 588–591
X-linked, 516–518
disjoint events, 93
predicting progeny distribution, 136–138
disjunctional segregation, 278
mitotic recombination, 181–182
Dissociation (Ds), 519, 520–523
distance between genetic markers, 169. *See also* map distance
distance matrix, 613
distance methods, 613
distribution of complex traits, 653–655
DNA, 2–10
chloroplast. *See* chloroplast DNA
complementary. *See* complementary DNA (cDNA)
function of, 2–6, 46
as genetic blueprint, 2
genomic. *See* genomes
mitochondrial. *See* mitochondria; mitochondrial DNA (mtDNA)
structure and replication, 6–10
totality in single cell. *See* genomes
transfer of genetic information to RNA. *See* transcription
DNA 8-oxoguanine glycosylase, 540
DNA bases. *See* bases in DNA
DNA-binding motif, 401
DNA breakage. *See entries at* breaking of DNA
DNA cleavage, site-specific, 432–436
DNA cloning, 51–52
DNA concentration. *See* concentration of DNA bases
DNA damage, types of, 538. *See also* DNA repair
DNA damage bypass, 542–543
DNA damage checkpoint, 562–567
DNA evolution, rates of, 616–617. *See also* molecular evolution
DNA fragments
with defined ends, production of, 432–434
isolation methods, 438
joining, 436–437
large, cloning of, 442
in replication forks (precursor fragments), 204–207
selective replication of, 55–60, 419. *See also* PCR amplification
separation and identification of, 47–55
gel electrophoresis, 50–52, 438
nucleic acid hybridization. *See* nucleic acid hybridization
restriction enzymes. *See* restriction enzymes
Southern blot, 54–55, 65
DNA ligase, 433
DNA looping, 404–407
DNA markers. *See* genetic markers
DNA-mediated transformation, 303–305
DNA methylase, 411
DNA microarrays, 355, 448–452, 553–554
CNP detection, 68
defined, 62
DNA modification, chemical agents that cause, 531–532
DNA polymerase complex, 202
DNA polymerases, 56
PCR with. *See* PCR amplification
processivity of, 564
DNA polymorphisms, 39–40. *See also* genetic markers
in DNA typing, 624–626, 670–671
extranuclear inheritance and, 588
in gene trees, 615
as silent mutations, 513
types of, 61–62
DNA rearrangements, programmed, 381, 419–424
DNA repair, 510, 537–543
AP, 541
base excision, 540–541
damage bypass, 542–543
defects in, 578
mismatch, 538–540
nucleotide excision, 541–542
photoreactivation, 542
SOS system, 543
DNA replication, 9–10
chromosome territories and, 230
discontinuous, of lagging DNA strands, 204–207
DNA structure and, 46
elongation. *See* elongation of DNA strands
initiation of. *See* chain initiation
meiosis and, 119
during mitosis, 116, 552
problems with, 191–192
proofreading. *See* proofreading
selective, of fragments, 55–60, 419. *See also* PCR amplification
semiconservative replication, 192–200
bidirectional replication in eukaryotes, 198–200
in chromosomes, 194–196
rolling-circle DNA replication, 199
theta replication of circular DNA molecules, 197–198
stabilization and stress relief, 200–201
unidirectional and bidirectional, 198
DNA-RNA hybrid, introns demonstrated using, 357
DNA sequencing, 8. *See also* coding sequence
complexity of. *See* complexity, genetic
conservation of, defined, 227
flanking sequences, 339
genome determination, 32–33
mapping. *See* genetic mapping; mapping of human genome
polyploidy, 288
repetitive, 233, 237
eukaryotic genomes, 237–239
telomeric DNA, for various organisms, 244
terminator sequencing, 207–210
transposable elements, 80, 520–525

DNA sequencing *(Continued)*
as agents of mutation, 523–524
defined, 520
discovery of, 519, 520
in human genome, 520, 523, 524–525
DNA strands. *See* DNA structure
DNA structure, 6–9, 40–47
antiparallelism of strands, 44–45
base pairing. *See* pairing of DNA bases
base stacking, 44–45, 53
cohesive ends, 326, 436
double helix, 7–8
antiparallelism of, 45–46
characteristics of, 43
major and minor grooves, 44–45
problems with Watson–Crick model, 191–192
unwinding of (denaturation), 52–53, 200–201, 247
5′ (donor) end of DNA strands, 8, 354–357
discontinuous replication of lagging strand, 204–207
strand elongation and, 191, 204
polynucleotide chains. *See* polynucleotide chains
as related to function, 46
3′ (acceptor) end of DNA strands, 8, 354–357, 440
restoring ends of DNA molecules, 242
strand elongation and, 191, 203–204
DNA transfer. *See* transcription
DNA typing (DNA fingerprinting), 67, 69, 624–626, 650, 670–671
DNA uracil glycosylase, 540
dogs, mitochondrial DNA studies in, 592, 609
"Dolly" (cloned sheep), 456
domestication science, DNA typing in, 69
dominance
degree of, 638–639
over-, 638–639, 646
dominant alleles, 78
codominant alleles, 80
common in SSRPs, 100
defined, 66
epistasis, 105–107
among quantitative-trait loci, 672
defined, 483
in developmental switches, 483–485
principle of, 483–484, 505
in genes controlling development. *See* gain-of-function mutations
incomplete dominance, 101–107
pedigree analysis, 97–99
phenotypic ratios ($F_2$ generation), 82. *See also* segregation
selection and, 636–637
testcross and backcross experiments, 86, 87. *See also* ratios (from genetic crosses)
genetic mapping in three-point testcrosses, 166–169
with unlinked genes, 88–89
donor end of DNA strands. *See* 5′ (donor) end of DNA strands
*dorsal* gene, 493
dosage compensation, 256–259
double crossovers, 164–165, 166
chromosome interference in, 167–168
double helix structure of DNA, 7–8. *See also* DNA structure
antiparallelism of, 45–46
characteristics of, 43
major and minor grooves, 44–45
problems with Watson–Crick model, 191–192
unwinding of (denaturation), 52–53, 200–201
melting temperature, 247
double mutants, 396
double-strand break and repair model, 212–215
double-stranded DNA, 8
double-stranded RNA (dsRNA), 414–416
double-Y karyotype (human), 269
Down syndrome, 266–267
chromosomal translocations and, 280–282
landmark paper on (excerpt), 265
*Drosophila* (fruit fly)
alternative promoters in, 407–409
attached-X chromosomes, 148
bar-shaped eyes (mutation), 145, 273
base composition of DNA, 43
brown eyes (mutation), 110
chromosomes without ends, 241, 242
*ClB* method using, 527–528
curly wings (mutation), 293
developmental genetics in, 474–477, 485–500
developmental mutations in, 480
DNA microarrays of, 451–452
dosage compensation, 257
ectopic expression in, 511–512
gene cloning in, 442–443
genome of, 33, 76
genomic sequencing of, 446–448
genotype fitness in, 637
germ-line transformation in, 455–457, 459–460
inbred lines, 652, 676
linkage and recombination, 159–162
attached-X chromosomes, 161, 162, 163
experiments [Morgan], 151–155
map distance and physical distance, 169
recombination in females only, 155
recombination within genes, 180–181
maternal sex ratio in, 603
multiple initiation of DNA strands, 199–200
polytene chromosomes, 232
position effect, 283
random genetic drift in, 642–643
repetitive and unique nucleotide sequences, 238–239
ruby eye color, 294
sex determination in, 135–136
transcriptional complexes in, 406, 407
transposable elements in, 522, 523
X-linked inheritance, 129–130
x-ray-induced mutations in, 529, 534, 535
yellow body color, 147
drug resistance, 526, 595–598, 635
drug response, DNA polymorphism affecting, 675
Ds (Dissociation), 519, 520
dsRNA (double-stranded RNA), 414–416
dTTP (deoxythymidine 5′-triphosphate), 42, 56
Duchenne muscular dystrophy, X-linked, 353
duplex DNA, 8. *See also* double helix structure of DNA
equilibrium density-gradient centrifugation, 192–193
duplications
in human chromosomes, 273–275
in origin of new genes, 617–618
Duroc-Jersey pigs, 109
dynamic mutations, 516–518
dystrophin gene, 353

## E

E-cadherin, 574–575
E (exit) site, 359
ecdysone, 496
ecological indicators, DNA polymorphisms as, 69
*Eco*RI site, 433, 436–437, 460
Ecstasy (MDMA), 675

ectopic expression, 511–512
ectopic recombination, 270, 524
  chromosomal inversions, 275–277
    human X and Y chromosomes, 260–262
  unequal crossing over, 273
    red–green color blindness, 273–275, 627
EGFR, 572
electrophoresis, 50, 438
electroporation, 435
elementary outcomes, 91, 92–93
elongation cycle
  in transcription, 349
  in translation, 361–362
elongation factors, 359–362
elongation of DNA strands, 191, 202–204
  dideoxy sequencing method and, 207
  telomeres, 242–243
  in transcription, 349
  in translation, 359, 361–362
embryonic development, early, 472–477
embryonic induction, 475
embryonic stem cells, germ-line transformations using, 457–458
EMS (ethyl methanesulfonate), 531–532
endoreduplication, 286
endosymbiont theory, 600–601
ends of DNA strands. *See* 5' (donor) end of DNA strands; 3' (acceptor) end of DNA strands
*engrailed* gene, 494
enhancers, 348, 401–402, 408–410
  evolutionary signature, 447
  pattern formation, 497–498
environmental effects on genes, 28–29
  development and, 472
  nondisjunction in humans, 269
  phenotypes vs. genotypes, 61–62, 84
environmental variance, 652, 655, 658–660
enzyme defects. *See* inborn errors of metabolism
enzymes. *See also specific enzyme*
  connection between genes and, 14–21
  defined, 10
  DNA polymerases, 56
    PCR with. *See* PCR amplification
    processivity of, 564
  proofreading, 192, 204, 205–207
  restriction enzymes, 48–50
    defined, 432
    features of, 433–434
    use of, 436–437
  topoisomerase enzymes, 201, 224–225
epidemiology, DNA typing in, 69
epigenetic regulation, 409–413
episomes, 305, 326
epistasis, 105–107
  among quantitative-trait loci, 672
  defined, 483
  in developmental switches, 483–485
  principle of, 483–484, 505
epistatic gene, 483
equational division. *See* second meiotic division
equilibrium density-gradient centrifugation, 192–193
equilibrium, of allele frequency, 634, 638–639
*Escherichia coli*, 4–6. *See also* bacterial genetics
  bacteriophage infection of, 396–397
  circular genetic map of, 310
  cultures and colonies, 302
  DNA repair in, 543, 544
  generalized transduction, 313–316
  *lacZ* gene. *See lacZ* gene
  pathogenicity of, 302
  plasmids, 296–297
  promoter regions, 347, 350
  protein folding in, 365
  as recombinant DNA host, 432, 434–438, 440–442
  rII gene structure, 321–325, 329
  spontaneous mutations in, 525–526
  transcription in, 349–350
    regulation of, 383–394, 398–401, 424
  translation in, 360–362, 368
    control of, 418
  tRNA suppressors in, 373–374
  *trpA* gene, 344, 544
estrogen and nondisjunction, 269
ESTs (expressed sequence tags), 445
ethyl methanesulfonate (EMS), 531–532
euchromatin, 169, 238–239
  PEV (position-effect variegation), 283–284
Eukarya group, 30
eukaryotes. *See also* animal(s); fungi; plant(s); *specific class or species*
  ancestry of, 30–31
  bidirectional DNA replication, 198–200
  cell cycle, 116, 557–560
  chromosomal structure, 226–232
  evolution of, 600–601, 604
  evolutionary relationships among, 31–32
  gene expression in, 342
  genomes of
    genome size, 222–223
    repetitive and unique nucleotide sequences, 232–239
  germ-line transformation in, 455–458
  messenger RNA in, 352, 366, 439
  molecular evolution in, 617–618
  origins of replication, 191, 198–200
  primer molecules for initiation, 202
  primer removal and replacment, 206
  promoter sequences in, 347
  protein folding in, 365–366
  recombination, 210
    bacteria vs., 316
  ribosomes in, 358, 367
  RNA interference in, 414–417
  RNA polymerases in, 346
  RNA processing in, 352–358
  semiconservative replication in chromosomes of, 194–196
  supercoiling of DNA molecules, 224–225
  symbionts in, 588, 601–603, 604
  transcription in
    regulation of, 381, 398–409
    silencing of, 411–412, 518–520
  translation in, 359, 361–363, 366
    control of, 417–418
European Gypsies, 264
Evans, Martin J., 458
*even-skipped* genes, 493
events, probability of, 92–97
evolution. *See also* molecular evolution
  adaptive, 634–639
  defined, 632
  genetics and, 632–633
  grass family genome, 289–290
  human, tracing of, 592, 615, 643–645
  of human Y chromosome, 260–264
  of organelles, 600–601, 604
  plants, polyploidy in, 284–289
  process of, 30–32

evolutionary complexity, 222–223. *See also* genome size
evolutionary genetics, DNA markers for, 70
evolutionary relationships among species, studying, 70
evolutionary signatures, 446–448
excavates, 32
exchange, genetic. *See* recombination of genes
excision, 332
  base-excision repair, 529, 540–541
  F plasmids from Hfr cells, 311–312
excisionase, 332
! notation. *See* factorials
exon(s), 352–353
exon-shuffle model, 357–358
exonuclease, 204
exploration and stabilization principle, 117–119
exponential distribution, 466
expressed sequence tags (ESTs), 445
expression. *See* gene expression
extranuclear inheritance, 587–609. *See also* chloroplast DNA; mitochondrial DNA (mtDNA)
  defined, 588
  discovery of, 596
  patterns of, 588–592
*eyeless* gene, 511–512

## F

F factor, 297
F plasmids, 297
F′ plasmids, 311–312
F plasmids
  excision from Hfr cells, 311–312
  matings with Hfr cells, 305–311
  R plasmids and, 302
$F_1$ generation, defined, 81
$F_2$ generation, defined, 82
factorials, 138
failure to breed, random. *See* random genetic drift
familial adenomatous polyposis, 575
familial cancer, 569, 575–578
families of genes, identifying with nucleic acid hybridization, 52
families of shared proteins, 32–33
fate map, 486–488. *See also* cell fate
female germ line, genomic imprinting in, 412–413
fertility, 637
fertility restorers, 599
fertilization, 472
  polyploids, 285, 286
  semisterility from translocation, 278, 282
fetal alcohol syndrome, 472
FGFR, 572
fingerprinting. *See* DNA typing
Fire, Andrew, 414, 416
first-division segregation, 177–180
first meiotic division, 120, 122, 124–127
  segregation in, 177–180
Fischer, Emil, 10
Fisher, Ronald Aylmer, 142–143
fitness. *See* genotype fitness
5′ capping, 352, 359
5′ (donor) end of DNA strands, 8, 354–357
  discontinuous replication of lagging strand, 204–207
  strand elongation and, 191, 204
5′ leader, 394
5′-phosphate (5′-P) group, 42
fixed alleles, 613
flanking sequences, 339
*flhA* gene, 418
flower ABC model, 502–503
flower development, developmental genetics of, 500–503
FMF (familial Mediterranean fever), 148
*FMR1* gene, 517, 518
folding domains, 357–358
food safety science, DNA typing in, 69
forced heterozygosity of Y chromosome, 260–261
forensics, 624–626
forward mutation, 543, 634
four-strand double crossover, 164
four-strand stage of meiosis, 161–163
fourfold degenerate site, 617
fragile-X syndrome, 477, 516–518, 529
frameshift mutations, 36, 368–370, 376, 378, 515–516
  defined, 515–516
  reversion of, 543–544
fraternal twins, 663
free radicals, 534
frequency of crossing over, 156. *See also* map distance
frequency of recombination, 152, 154–155, 156–159
  in bacteriophage mapping, 319–321
  lod scores and, 172
  map distance and, 156, 168–169
  minimum and maximum, 164
fruit flies. *See Drosophila*
functional genomics, 447–452
fungi. *See specific species*
fungi, transcriptional silencing in, 411
fusion genes, 579

## G

G-bands, 253, 256
G base. *See* guanine
G protein, 572
G quartets, 243, 518–520
$G_1$ and $G_2$ periods of interphase, 116, 552, 559–561, 570
$G_1$ restriction point, 560
$G_1$/S transition, 560–561
$G_2$/M transition, 561
gain-of-function mutations, 480–483, 504–505, 511
*GAL* genes
  proteomic studies of, 452–454
  regulation of, 398–401, 404
galactose, 387, 398–401
galactosemia, 649
gametes. *See also* oocytes
  defined, 83
  formation of, alleles during. *See* segregation
  homogametic and homogametic sexes, 128, 133–134
  meiosis and, 120. *See also* meiosis
  methylation of, 412–413
  nuclear division, 115
  polyploidy in plant evolution, 284–289
    allopolyploidy, 286–288, 292
    autopolyploids, 286–288
    monoploids, 288–289
    sexual vs. asexual polyploidization, 285–286
  random union of. *See* random mating
  trisomy, mechanics of. *See* trisomy (human)
  viability of, 622, 637
gap genes, 490, 493
  in regulatory hierarchy, 494–496
Garrod, Archibald, 10–13
gastrula, 472
Gautier, Marthe, 265

gel electrophoresis, 50–52, 438
Sanger sequencing with, 208
gene amplification, 419. *See also* PCR amplification
gene cloning. *See* cloning vectors; recombinant DNA technology
gene conversation, 211–212
gene densities
chromosome territories and, 230
human chromosomes, 255–256
gene discovery, genetic analysis for, 20–21
gene dosage, trisomy and, 264
gene duplication
in human chromosomes, 273–275
in origin of new genes, 617–618
gene evolution. *See* molecular evolution
gene expression, 21–26, 341–379. *See also* phenotypes
control of. *See* gene regulation
defined, 342
overview of, 342
position effect, 282–284
telomeric silencing, 242
transcription (DNA to RNA). *See* transcription
translation (synthesis of polypeptides). *See* translation
gene fusion, 579
gene interaction. *See* epistasis
gene libraries, 442, 445–446
gene mutation. *See* mutation(s)
gene order, 305
gene pool, 619
gene regulation, 380–429
control points of, 381
defined, 381
in development. *See* developmental genetics
discovery of, 390
evolution of, 400–401
motif identification, 402–404
through DNA rearrangements, 381, 419–424
through RNA interference, 414–417
through RNA processing and decay, 381, 413–414
transcriptional. *See* transcriptional regulation
translational, 381, 417–418
gene segregation. *See* segregation
gene silencers, 348, 401–402, 447
gene silencing
as defense against transposons, 525
discovery of, 416
epigenetic mechanisms of, 411–413
by RNA interference, 414–417
in X-chromosome inactivation, 259
gene targeting, 458
gene therapy, 464–465
gene transition. *See* transmission genetics
gene trees, 612–615
genealogical studies. *See* human(s), population history
general transcription factors, 347–349
generalized transduction, 313–316
genes. *See also specific gene by name*
alleles. *See* alleles
chimeric genes, 274
on chromosomes, 39, 61, 115. *See also* chromosome(s)
common among ancestors. *See* unity of life
connection between enzymes and, 14–21
defined, 2, 60, 325
disease genes, 68. *See also specific disease or gene*
cancer, 569–575
mitochondrial, 588–591
X-linked, 516–518
DNA structure and, 46
interaction with environment. *See* environmental effects on genes
linkage between. *See* linkage between genes
multiple alleles of. *See* multiple alleles (tandem repeat polymorphisms)
mutant. *See* mutant genes; mutation(s)
position effect, 282–284
recombination *between*, 180–182
recombination *of*. *See* recombination of genes
standard notation of, 61
X-linked genes, 130–131
genetic analysis, 14–21
complementation test for gene mutations, 17–20, 106
conjugation in bacterial genetics, 306
for gene discovery, 20–21
mapping. *See* genetic mapping
phage mutants, 318
probability in, 90–97
renaturation kinetics, 235–237
terminology of, 60–62
tetrad analysis, 173–180
ordered tetrads, 177–180
unordered tetrads, 173–177
genetic code, 25–26, 368–374
for codons, 25–26
defined, 368
degeneracy of, 370, 371
discovery of, 370–371
features of, 371
of organelles, 593–594
redundancy of, 372
standard abbreviations, 375–376
wobble in, 373
genetic complexity, 233–234
alternative splicing and. *See* alternative splicing
genetic differences among individuals, 39–40
genetic distance, 151
genetic drift, random, 633, 639–643
genetic engineering. *See* recombinant DNA technology; transgenic organisms
genetic exchange. *See* recombination of genes
genetic factors affecting phenotype, 652, 655
genetic inheritance, 2. *See also* transmission genetics
pseudoautosomal inheritance, 259
with sex chromosomes, 128–136. *See also* X-linked inheritance
genetic knockdown, 416–417
genetic linkage, 68–69
genetic map distance. *See* map distance
genetic mapping, 69, 151, 155–165. *See also* mapping of human genome
deletion mapping, 322–323
*Escherichia coli* T4 mutants, 319–321
frequency of recombination, 156–159
in human pedigrees, 170–173
multiple crossovers, 164–165
of quantitative trait loci, 670–672
tetrad analysis, 173–180
in three-point testcrosses, 166–169
time-of-entry (transfer-order) mapping, 306–311
frequency of cotransduction and, 315–316
timing of crossing over, 161–163
translocation breakpoints, 278–280
using DNA-mediated transformation, 304–305
genetic mapping functions, 168–169
genetic markers, 39
applications of, 68–70
cotransduction studies, 315

genetic markers *(Continued)*
developmental studies using, 487
in DNA typing, 625–626, 670–671
harmless vs. harmful polymorphism, 99–100
in human pedigree analysis, 100–101
human Y chromosome, 262–264
physical distance between, 169
simple sequence repeats (SSRs), 67
Y-chromosome genealogical studies, 262
for tracing ancestry, 643
genetic recombination, 159. *See also* transcription
bacteria vs. eukaryotes, 316
in bacteriophage lytic cycle, 318–319
characteristic frequency of, 154–155
chi-square test for linkage, 154
defined, 151
deletions created by, 270
ectopic, 524
evolutionary history and, 615
exchange mechanism in bacteria, 303
frequency of crossing over, 156. *See also* map distance
human X and Y chromosomes, 260–262
mitotic, 181–182
molecular mechanisms of, 210–215
multiple crossovers, 164–165
probability of crossing over. *See* genetic distance
site-specific, 300, 326–332
tetrad analysis, 174
timing of crossing over, 161–163
unequal crossing over, 273
red–green color blindness, 273–275, 627
genetic risk factors. *See* risk factors (genetic)
genetic switching, 397, 408–410
genetic transmission. *See* transmission genetics
genetic uniqueness, principle of, 624–626
genetic variation. *See* polymorphism
genetically modified organism (GMO). *See* transgenic organisms
genetically uniform population, example of, 661
genetics
behavioral, 674–675
defined, 2
developmental. *See* developmental genetics
evolution and, 632–633
population. *See* population genetics
reverse, 460
genome annotation, 444–445
genome size, 222–223
analysis by renaturation kinetics, 235–237
genomes (genomic DNA)
about, 32–33
complete set of proteins in. *See* proteins
detection of DNA markers, 39–40
determination of, 32–33
grass family, evolution in, 289–290
human. *See* human genome
mitochondrial, 593, 594
polyploidy in plant evolution, 284–289
allopolyploidy, 286–288, 292
autopolyploids, 286–288
monoploids, 288–289
sexual vs. asexual polyploidization, 285–286
zygotic, early development and activation of, 476–477
genomic imprinting, 476–477
in germ lines, 412–413
genomic islands, 302, 383
genomic sequencing, 432, 443, 444. *See also* genetic mapping; mapping of human genome
genomics, 443–452
comparative, 445–447, 612
computational, 445
defined, 2, 432
functional, 447–452
pharmaco-, 675
techniques used in, 432
genotype(s)
of ABO blood group system, 104
bacterial genetics, 302–303
phage mutants, 318
defined, 61
expression of. *See* phenotypes
with multiple alleles, 67
phenotypes vs., 61–62, 84
incomplete dominance, 101–107
genotype-by-environment interaction (G-E interaction), 660, 663, 675
genotype-by-sex interaction, 660
quantitative-trait loci and, 672, 674
genotype fitness
as complex trait, 652
components of, 637
relative, 635–636
genotype frequencies, 82–85, 89–90, 619–620, 625. *See also* segregation
goodness-of-fit testing (example), 141–142
heterozygous, 623
inbreeding and, 628–629
Mendel's data "too good to be true", 142–144
in X-linked genes, 627
genotypic variance, 656–660
germ cells. *See* gametes
germ line, genomic imprinting in, 412–413
germ-line mutations, 510
germ-line transformation, 455–458
rescue, 459–460
*giant* gene, 494
Giemsa (G-bands), 253, 256
gigabases, defined, 223
glucose, *lac* operon regulated by, 387–389
glucosinolates, 100
glycine codon, discovery of, 370
GMO (genetically modified organism). *See* transgenic organisms
goodness-of-fit testing, 139–144
Mendel's data "too good to be true", 142–144
Gorbachev, Mikhail, 537
grass family genome, evolution of, 289–290
great apes, species tree for, 615
Greider, Carol W., 241
Griffith, Frederick, 3
grooves in DNA double helix, 44–45
grosshoppers, meiosis of, 127
growth-factor receptors, 572
growth hormone, engineered, 461–462
GTP (guanosine triphosphate), 344, 362
guanine (G), 8, 41
base pair stability, 246
bonding to cytosine. *See* pairing of DNA bases
elongated strands of telomeres, 243
mismatching in base pairs. *See* incorporation error
molar concentration of, 42–43
guanosine triphosphate (GTP), 344, 362
guide RNA, 242, 415–416
Gypsies, European, 264
gyrase, 201

## H

*Hae*II site, 588
Hahn, William C., 566
*hairy* gene, 493–494
Haldane's mapping function, 168
haploid nuclei, 115
  creation of. *See* meiosis
  polyploidy and, 285
haplotypes, 262
  Cohanim, 263
  European Gypsies, 264
  Genghis Khan, 262
Hardy, Godfrey, 621, 625
Hardy-Weinberg principle, 621–622, 625
  inbreeding and, 628
  and multiple alleles, 623–624
Hartnup disease, 648
Hartwell, Leland H., 559
Hazara ethnicity, 262–263
heat-shock protein (Hsp), 365
heavy (H) chain, 419–420
helical structure of DNA. *See* double helix structure of DNA
helicase proteins, 200
helix-turn-helix motif, 401
hemophilia A, 133, 145
hereditary cancer syndromes, 569, 575–578
hereditary nonpolyposis colorectal cancer (HNPCC), 540
heritability
  broad-sense, 662–663, 668
  narrow-sense, 664–665, 670
    estimation of, 666, 668–669
  threshold traits and, 669–670
Hershey, Alfred, 4–6
Hershey–Chase blender experiment, 6
heterochromatin, 169, 238–239
  PEV (position-effect variegation), 283–284
heterochronic mutation, 505
heteroduplex regions, 212, 214
heterogametic sex, 128, 133–134
heterokaryon, 18–20
heteroplasmy, 591
heterosis, 667
heterozygosity
  example of, 661
  inbreeding and, 628–629
  loss of, 577
  selection and, 636–637
heterozygote superiority, 638–639, 646
heterozygous carriers of mutual alleles, 98
heterozygous genotypes
  defined, 61
  forced heterozygosity of Y chromosome, 260–261
  frequency of, 623
  inversions, causes of, 275
  with multiple alleles, 67
  principle of segreation and, 80–83. *See also* segregation
  testcross and backcross experiments, 86, 87. *See also* ratios (from genetic crosses)
    genetic mapping in three-point testcrosses, 166–169
    with unlinked genes, 88–89
hexaploids, 284
Hfr cells, 305–306. *See also* conjugation
  excision of F plasmids, 311–312
  time-of-entry mapping, 306–311
high-copy-number plasmids, 296
high-F disease, 409
highly repetitive DNA sequences, 237–239
highly statistically significant, defined, 140
histidine operon, 393, 394, 429
histone proteins, 227
histone tails, 229
history of domestication, DNA typing for, 69
HIV (human immunodeficiency virus), 440, 465, 619–620
HNPCC (hereditary nonpolyposis colorectal cancer), 540
Hodgkin, Jonathan, 237
Holliday, Robin, 214
Holliday junction-resolving enzymes, 214–215
Holliday junctions, 214–215
holocentric chromosomes, 240
holoenzyme, 346, 406
homeobox, 498
homeotic (*HOX*) genes, 498–500
homogametic sex, 128, 133–134
  dosage compensation, 256–259
homogentistic acid, 10–13
homologous chromosomes
  defined, 61
  meiosis and, 119–120
  synapsis of, 124
homothallism, 422
homozygotes, selection and, 636–637
homozygous genotypes
  defined, 61
  with multiple alleles, 67
  principle of segreation and, 80–83. *See also* segregation
  testcross and backcross experiments, 86, 87. *See also* ratios (from genetic crosses)
    genetic mapping in three-point testcrosses, 166–169
    with unlinked genes, 88–89
Hoogsteen base pairing. *See* G quartet
hotspots of mutation, 325, 527–529
housekeeping genes, 398
*HOX* (homeotic) genes, 498–500
*Hpa*II enzyme, 427–428
HPV (human papillomavirus), 584
Hsp (heat-shock protein), 365
human chroionic gonadotropin (HCG), 266
human genome, 39. *See also* genomes
  compared to other species, 446
  DNA polymorphisms in, 624–626
  mapping. *See* mapping of human genome
  mitochondrial, 593
  structural abnormalities, 269–282
    deletions and duplications, 268–275
    inversions, 275–277
    translocations, 277–282
  transposable elements in, 520, 523, 524–525
  trisomy and monosomy, 264–269
    environmental effects, 269
  X and Y chromosomes. *See* sex chromosomes
Human Genome Project, 443–444
human(s) (*Homo sapiens sapiens*)
  ABO blood groups, 103–105
  alternative promoters in, 409
  complex traits in, 652, 660, 663
  inbreeding in, 627, 632
  mitochondrial genome, 590–591, 594
  pedigree analysis, 97–101
    genetic mapping, 170–173
    molecular markers, 100–101
    X-linked inheritance, 133
  population history
    DNA typing for, 69
    tracing through Y chromosome, 260–264
  size of gene, 353

human(s) *(Continued)*
tracing ancestry of, 592, 615, 643–645
transcripts, characteristics of, 353–354
human immunodeficiency virus (HIV), 440, 465, 619–620
human karyotypes, 251–264. *See also* human genome
centromeres and chromosomal stability, 255–256
conventional symbols in, 255
dosage composition (X chromosome), 256–259
pseudoautosomal inheritance, 259–260
sex-chromosome abnormalities, 268–269
standard, 252–255
Y chromosome evolution, 260–264
human papillomavirus (HPV), 584
human red–green color blindness, 273–275
X-linked, 627
*hunchback* gene, 491–492, 493
huntingtin gene (human), 34, 650
huntingtin protein (human), 101
Huntington's disease, 185, 649–650
pedigree for, 97–98
paper on (excerpt), 101
hybrid vigor, 667
hybridization. *See* nucleic acid hybridization
hybrids, defined, 81
hydrogen bonding, 364, 530
hydrogen bonds between paired bases, 43–44
hydrophobic, DNA bases as, 45
hypermorphic mutations, 511
hypomorphic mutations, 511
hypostatic gene, 483
hypothesis testing (goodness of fit), 139–144
hypoxanthine, 531

## I

identical twins, 663
identification of DNA fragments, 47–55
gel electrophoresis, 50–52, 438
nucleic acid hybridization, 52–54
backcross experiments, 87
CNP detection, 68
restriction enzymes, 48–50
defined, 432
features of, 433–434
use of, 436–437
SNPs (single-nucleotide polymorphisms), 62–63. *See also* DNA markers
Southern blot, 54–55, 65
identification of individual organisms. *See* DNA typing
identity by descent, 629–630
imaginal disks, 496
imatinib mesylate (tyrosine kinase inhibitor STI571), 580
immune response, DNA rearrangements and, 419–422
immunity (from phage infection), 332
immunoglobulins, DNA rearrangements and,, 419–422
inactivation of X chromosome, 257–259
active genes, 260
gene silencing, 259
inborn errors of metabolism, 10–14, 26
inbreeding, 99, 627–632
complex inheritance studies, 652, 676
defined, 627
effects of, 632, 666–667
inbreeding coefficient, 628–629
calculation of, from pedigrees, 630–632
inbreeding depression, 666
inbreeding pedigree, 629–632
incomplete dominance, 101–107
incomplete linkage between genes, 151
incorporation error, 191–192
independent assortment, 88, 152
independent events
defined, 93
predicting progeny distribution, 137–138
independent fertilizations, 94
independent segregation, 94
indirect effects of mutant genes, 28–29
individual selection, 665–666
individuals, identifying. *See* DNA typing
induced mutations, 510, 525
inducer, 381–382
example of, 384, 387
inducible transcription, 381–382
example of, 383–392
inheritance, genetic, 2. *See also* transmission genetics
pseudoautosomal inheritance, 259
with sex chromosomes, 128–136. *See also* X-linked inheritance
initiation codon. *See* AUG codon
initiation factors, 359–360
initiation of DNA strands. *See* chain initiation
insecticide resistance, mutations causing, 526–527
insertion mutations, 515–516
insertion sequences (IS elements), 298
instar, 485
insulin-like growth factor (*Igf*) gene, 477
insulin receptor gene, 413–414
integrase, 300
integrons, 300–301
interaction between genes. *See* epistasis
intercalating agents, 532
intercalation, 532
intercellular signaling, 473–475
cell fate determination by, 480–483
interference
chromatid interference, 164
chromosome interference in double crossovers, 167–168
intergenic suppression, 544–545
intermediary RNA
central dogma of molecular genetics, 22–23
evolutionary perspective on, 30
synthesis of polypeptides. *See* translation
International Unio of Biochemistry and Molecular Biology (IUBMB) code abbreviations, 375–376
interphase periods, 116, 552, 559–561, 570
chromatin organization during, 230
interrupted-mating technique, 308
intersection of events (probability), 93–94
intervening sequences. *See* intron(s)
intragenic suppression, 543–544
introns, 352–357
effects of mutations, 357–358
in gene evolution, 357–358
in organelles, 356
in translation, 359
inverse PCR, 339
inversion loops, 275–277
inversions, chromosomal, 275–277
human X and Y chromosomes, 260–262
inverted repeats, 270
ionizing radiation, 529, 533–536
iron deficiency, 462
irreversible mutations, 633–634
IS elements (insertion sequences), 298

## J


Jacob, François, 351
  operon model, 386, 389, 390, 396
joining (J) antibody regions, 420
Jordan, Michael, 673
Judaism, common ancestries of, 263


## K


kappa particles, 601
karyotypes, human, 251–264. *See also* human genome
  centromeres and chromosomal stability, 255–256
  conventional symbols in, 255
  dosage composition (X chromosome), 256–259
  pseudoautosomal inheritance, 259–260
  sex-chromosome abnormalities, 268–269
  standard, 252–255
  Y chromosome evolution, 260–264
Khan, Ghengis, 262–263
kilobases, defined, 48, 223
kinetic proofreading, 362
kinetochore, 239, 552, 567
kinetochores, 117, 119
kinetosome, 604
kissing complex, 418
Klinefelter syndrome, 269
*Kluyveromyces lactis*, gene regulation in, 400–401
knockout mutations, 461, 463, 576–577. *See also* loss-of-function mutations
Knudson, Alfred, 577, 580
Kornberg, Roger D., 348
Kosambi's mapping function, 168
*Krüppel* gene, 493–494


## L


*lac* operon, 386–389, 425, 427
  versus galactose metabolism, 398–401
  regulatory states of, 388–389
lactose permease, 383–389
lactose-utilization mutations, 350–351, 383–384, 390
*lacZ* gene, 350–351, 383–392
  in recombinant DNA, 440–442
  as reporter gene, 402–404
lagging DNA strand
  defined, 191, 205
  discontinuous replication of, 204–207
λ bacteriophage
  gene regulation in, 396–397
  lysogeny and specialized transduction, 325–334
  as recombinant DNA vector, 434–436, 442
lariat, 354–355
leader mRNA, 394–396
leader polypeptide, 392
leading DNA strand, defined, 191, 205
leaf variegation, 594–595
leaky mutations. *See* hypomorphic mutations
Lederberg, Esther, 525
Lederberg, Joshua, 305, 313, 525
Lejeune, Jerome, 265
Lemba people, 263
leptotene, 124
leucine codon, discovery of, 370
leukemias, acute, 465, 578–581, 582, 618
Lewis, Edward B., 492
Li-Fraumeni syndrome, 576, 582
liability, threshold traits and, 653, 669–670
libraries, gene, 442, 445–446
life cycle of organism, 121
ligand, 482
light (L) chain, 419–420
likelihood ratio, 171–172
*Limnaea peregra*, maternal effect in, 603–606
*lin* genes, 480–482, 484–485
LINE (long interspersed elements), 523, 524
lineage, 478–479
  genetically marked, 487
lineage diagram, 478
lineage mutations, 478
linkage between genes, 151–154
  analysis of, in mapping, 670–672. *See also* genetic mapping
    in human pedigrees, 170–173
  characteristic frequency of recombination, 154–155
  chi-square test for, 153–154
  defined, 151
  tetrad analysis, 173–180
    ordered tetrads, 177–180
    unordered tetrads, 173–177
linkage groups, 159
local population, 619
localized centromeres, 240
locus of a gene, 61
lod scores, 171–172
long terminal repeats (LTRs), 522
loss-of-function mutations, 480–485, 504–505, 511. *See also* knockout mutations
loss of heterozygosity, 577
*lozenge* (*lz*) gene, 180–181
LTR retrotransposons, 522–523
LTRs (long terminal repeats), 522
Lwoff, André, 390
lymphoma, Burkitt, 585
Lyon, Mary F., 258
lysis, 317
lysogen, 326
lysogenic infection, 396–397
lysogeny, 326–332
lytic cycle, bacteriophage, 312–313
  recombination in, 318–319
lytic infection, 396–397


## M


M period of interphase, 116, 552, 561. *See also* mitosis
M1V mutant (PAH gene), 27
MacLeod, Colin, 4, 434
macromolecules, 612
*MADS box* transcription factors, 502
maize
  cytoplasmic male sterility in, 589–600
  hybrid varieties of, 667
  phenotypic variation in, 655, 656, 660
  transposable elements in, 519–523
major grooves in DNA double helix, 44–45
malaria, 465, 514, 639, 640
male germ line, genomic imprinting in, 412–413
male sterility, cytoplasmic, 598–600, 606
Malthus, Thomas, 635
mammals. *See also entries at* human; *specific species*
  cell cycle checkpoints in, 563
  cell cycle progression in, 557–560
  cloning of, 456
  developmental mutations in, 480
  gene-line transformation in, 456–457
  genomic imprinting in, 412–413
  methylation in, 411

mammals (Continued)
mitochondrial genome of, 593, 594
molecular evolution in, 616–617
zygotic genome activation in, 476–477
map distance, 156–159
defined, 158
frequency of cotransduction, 315–316
physical distance and, 169
recombination frequency and, 156, 168–169
tetrad analysis, 175–177
map unit, defined, 156
mapping functions, 168–169
mapping of human genome, 130–131, 171, 443–444. *See also* genetic mapping
deletions in human chromosomes, 271–273
landmark paper on (excerpt), 64
translocation breakpoints, 278–280
*Marchantia polymorpha*, chloroplast genome of, 593
Margulis, Lynn (Sagan), 604
*mariner* transposon, 524, 649
massively parallel sequencing, 209–210
*MAT* gene, 422–423
maternal-age effect, 269
maternal effect, 592, 678
in snail shell coiling, 603–606
maternal-effect genes, 476–477, 488–489. *See also* coordinate genes
maternal-effect mutations, 488–489, 491–492
maternal inheritance, 588
versus maternal effect, 592, 605
of mitochondrial DNA, 588–591
maternal sex ratio, 603
mating
consanguineous, 627, 632. *See also* inbreeding
random. *See* random mating
snail shell coiling and, 606
mating-type interconversion, 422–423
mating type, transcriptional control of, 423–424
Matthaei, J. Heinrich, 367
maturation-promoting factor, 561
maximum likelihood methods, 170–173, 613
Mb (megabase pairs), defined, 68
McCarty, Maclyn, 4, 434
McClintock, Barbara, 241, 519, 520
MCS (multiple cloning site), 440–441, 468
*Mdm2* gene, 572–573
MDMA (Ecstasy), 675
mean (average), 653, 654
phenotypic, 665
megabase pairs, defined, 68
megabases, defined, 223
meiocytes, 119
meiosis, 119–128
function of tetrads, 172–174. *See also* tetrad analysis
sexual polyploidization, 285–286
timing of crossing over, 161–163
in translocation heterozygotes, 278
trisomic segregation, 267–268
meiotic drive, 112
melibiose, 427
Mello, Craig C., 414, 416
melting temperature, 247
Mendel, Gregor, 2
landmark paper (excerpt), 84
results of, "too good to be true", 142–144
single-gene segration experiments, 80–87
Mendelian genetics. *See* transmission genetics
Mendel's first law. *See* segregation
Mendel's second law. *See* principle of independent assortment
mental retardation, X-linked, 516–517
meristems, 500
merodiploids, 312
MERRF (myoclonic epilepsy associated with ragged-red muscle fibers), 589–590
Meselson, Matthew, 192, 195
Meselson–Stahl experiment, 192–194, 195
messenger RNA (mRNA), 342, 350–352
alternative splicing, 32
central dogma of molecular genetics, 22–23, 342
discovery of, 351
evolutionary perspective on, 30
leader, 394–396
levels of abundance, 439
of mitochondria and chloroplasts, 593
polycistronic, 366–368, 386
reverse transcription of, 439–440
role in translation, 358–361
stability of, 352, 381, 414
synthesis of polypeptides. *See* translation
metabolic activities of cells, 10
metabolic pathways, 11–13
metabolism, inborn errors of, 10–14, 26
metabolites, 13
metacentric chromosomes, 255–256
metallothionein, 461
metamorphosis, developmental genetics of,, 496–498
metaphase I stage (meiosis), 125
trisomic segregation, 267–268
metaphase II stage (meiosis), 128
trisomic segregation, 267–268
metaphase (mitosis), 117–119
metaphase plate, 119
metaphase spread, 252
methionine codon, 370–371
methylation, 411
mutational hot spots and, 527–529
restriction enzyme studies of, 427–428
transcriptional inactivation and, 411–412, 518–520, 529
methylation induced premeiotically (MIP), 411
5-methylcytosine, deamination of, 529, 531
microarrays. *See* DNA microarrays
micromeres, 475, 476
microRNA (miRNA), 415–417, 447
microscopy, Nomarski, 479
microtubules (mitotic spindle), 117, 119
in localized centromeres, 240
middle-repetitive DNA sequences, 237, 239
Miescher, Friedrich, 2
migration, as evolutionary force, 632–634, 645
minimal medium
defined, 14
nutritional mutants (bacterial genetics), 303
minor grooves in DNA double helix, 44–45
MIP (methylation induced premeiotically), 411
*Mirabilis jalapa* (four o'clock plants), leaf variegation in, 594–595
mismatch repair, 538–540
mismatching in base pairs
in DNA renaturation, 233
mismatch repair, 211–212
mismatching of base pairs. *See* incorporation error
missense (nonsynonymous) mutations, 513–515
mitochondria
eukaryotes lacking, evolution and, 31–32
evolutionary origin of, 600–601
genetic code in, 594

mitochondrial DNA (mtDNA), 588
coalescence of, 643
evolution of, 609
heteroplasmy of, 591
human, genes in, 590–591
inheritance of, 588
tracing population history through, 592, 615, 643–645
mitochondrial genetic diseases, 588–591
mitochondrial mutants, respiration-defective, 598
mitosis, 116–119, 552
asexual polyploidization, 286
interphase chromatin organization, 230
recombination between genes during, 181–182
mitotic exit network, 568–569
mitotic recombination, 181–182
mitotic spindle, 117, 239
mobile DNA, 296–302
moderately abundant mRNA, 439
modern (molecular) genetics. *See* molecular genetics
molar concentration of DNA bases, 42–43
molecular clock, 616
molecular evolution, 612–618
comparative genomics and, 446–448, 612
defined, 612
DNA microarray studies of, 451–452
exon-shuffle model of, 357–358
homeotic genes in, 498–500
mitochondrial, 609
origin of new genes in, 617–628
rate of, 615–617
in regulatory systems, 400–401
study techniques, 612–615
transposition in, 524–525
molecular genetics, 2
central dogma of, 22–23, 342
molecular organization of chromosomes, 221–249
bacteria, 226
centromere structure, 239–241
eukaryotes, 226–232
repetitive and unique nucleotide sequences, 232–239
genome size and evolutionary complexity. *See* genome size
polytene chromosomes, 232
supercoiling of DNA, 223–226
telomere structure, 241–244
molecular phenotypres, 78–80
molecular phylogenetics, 70, 612
molecular structure of DNA, 6–9, 40–47
antiparallelism of strands, 44–45
base pairing. *See* pairing of DNA bases
base stacking, 44–45, 53
cohesive ends, 326, 436
double helix structure of DNA, 7–8
antiparallelism of, 45–46
characteristics of, 43
major and minor grooves, 44–45
problems with Watson–Crick model, 191–192
unwinding of (denaturation), 52–53, 200–201, 247
5′ (donor) end of DNA strands, 8, 354–357
discontinuous replication of lagging strand, 204–207
strand elongation and, 191, 204
polynucleotide chains. *See* polynucleotide chains
as related to function, 46
3′ (acceptor) end of DNA strands, 8, 354–357, 440
restoring ends of DNA molecules, 242
strand elongation and, 191, 203–204
molecular unity of life, 29–33
ancestral commonalities for all humans, 262
Monod, Jacques, 351
operon model, 386, 389, 390, 396
monoploid organisms, 288–289
monoploid sets of chromosomes, 284–285
monosomy of human X chromosome, 269
Montgomery, Hugh E., 673
Morgan, Thomas Hunt, 129–131, 151, 152, 153
morphogen, 474
morphological genetics, 2
morphological phenotypes, 78–80
Morton, W. J., 533
mosaicism, 510, 513
defined, 255
somatic mosaics, 182
in X-chromosome inactivation, 258–259
mRNA. *See* messenger RNA (mRNA)
mtDNA. *See* mitochondrial DNA (mtDNA)
Muller, Herman J., 241, 511, 529, 535
Mullis, Kary B., 57
multifactorial traits. *See* complex traits
multiple alleles (tandem repeat polymorphisms), 67, 102–103
Hardy-Weinberg principle and, 623–624
multiple-antibiotic-resistant bacteria, 300, 302
multiple cloning site (MCS), 440–441, 468
multiple hits, 613
multiplication rule (probability), 94
for predicting progeny distribution, 137–138
multiplicity of genes, 24
Murry, Jeffrey C., 171
*Mus* (mouse)
developmental genetics in, 474–476
knockout mutations in, 461, 463, 576–577
transposable elements in, 524
mutagen(s), 529–537. *See also specific agent*
Ames test for, 545
base-analog, 531
defined, 529
mutagenic primer, 460
mutant genes, 15–20. *See also* mutation(s)
CNPs located near, 68
complementation test for, 17–20, 106
heterozygous carriers, 98
trait effects from, 28–29
*trans* vs. *cis* configurations, 153
mutant organism, defined, 26
mutant screens, defined, 15–16
mutation(s), 26–27
in bacterial genetics, 302–303
bacteriophage, 396
cancer. *See* cancer
cell cycle progression, 554–557, 559, 570–571
checkpoint failure, 568–569
complex traits, 674
conditional, 510–511
defined, 510
deletion, 461, 515–516
developmental, 472, 479, 480–485, 488, 491–492, 500–502
DNA structure and, 46–47
dynamic, 516–518
as evolutionary force, 632–634
forward, 543, 634
frameshift. *See* frameshift mutations
functional classification of, 511
gain-of-function, 480–483, 504–505, 511
gene linkage. *See* linkage between genes
germ-line, 510

mutation(s) *(Continued)*
heterochronic, 505
homeotic, 498–500
hot spots of, 527–529
hypermorphic, 511
hypomorphic, 511
induced, 510, 525
insertion, 515–516
intron, 357–358
irreversible, 633–634
knockout, 461, 463, 576–577
lactose-utilization, 350–351, 383–384, 390
lineage, 478
loss-of-function, 480–485, 504–505, 511. *See also* knockout mutations
maternal-effect, 488–489, 491–492
missense (nonsynonymous), 513–515
mitochondrial DNA, 588–591, 598
molecular basis of, 512–520
multiple alleles (tandem repeat polymorphisms), 67, 102–103
Hardy-Weinberg principle and, 623–624
nonadapative nature of, 525–527
nonsense, 373–374, 515
phase mutants. *See* bacteriophage genetics
promoter, 387
promoter binding strength, 347
reducing with proofreading. *See* proofreading
reverse, 543–545, 634
reversible, 634
RIP (repeat-induced point), 411, 525
selectively neutral, 616
somatic, 422, 510
spontaneous, 510, 513, 525–529
suppressor, 369, 373–374, 543–545
synonymous (silent), 513
temperature-sensitive, 510–511, 554, 556
transcription-termination region, 350–351
transition, 512
transposable elements as agents of, 523–524
transversion, 512–513
types of, 510–511
Y chromosome (human), 260–262
mutation rate
defined, 527
estimation of, 527–528
mutation-selection balance, 637–638
mutually exclusive events, 93
predicting progeny distribution, 136–138
*Mycoplasma genitalium*, genomic sequencing of, 444

## N

NAC (nascent chain-associated complex), 365
*nanos* gene, 491–492, 493
narrow-sense heritability, 664–665, 670
estimation of, 666, 668–669
nascent chain-associated complex (NAC), 365
natural selection, 633, 634–639, 666
negative autoregulation, 383
negative chromatid interference, 164
negative regulation, 483
examples of, 383–392, 424
inducible, 381–382
repressible, 381–382
examples of, 384–385, 390–392
negatively supercoiled DNA molecules, 224
neighbor joining, 613
neomorphic mutations. *See* gain-of-function mutations
*Neurospora crassa*, 15–20, 106
recombination, 210–211
RIP mutations in, 525
transcriptional silencing in, 411
neutral evolution, rate of, 616
neutral petites, 598
new genes, origins of, 617–618
Nicholas, Tsar, 591
*Nicotiana longiflora* (tobacco), narrow-sense heritability in, 664–665
9:4:3 ratio, 107
9:6:1 ratio, 107
9:7 ratio, 106
Nirenberg, Marshall W., 367
nitrogen-containing bases in DNA. *See* bases in DNA
nitrogen mustard, 531–532
Nomarski microscopy, 479
non-LTR retrotransposons, 523
noncomplementation, 18. *See also entries at* complementation
noncomputation, 20
nonconjugative plasmids, 297, 299
nondisjunction, 134
double-strand break and repair model, 212–215
estrogen and (human), 269
of human chromosome 21. *See* Down syndrome
nondisjunctional segregation, 278
nonparental ditype (NPD) tetrad, 174
nonreciprocal (Robertsonian) translocations, 280–281
nonselective medium (bacterial genetics), 303
nonsense-mediated decay, 359
nonsense mutations, 373–374, 515
nonsense suppressors, 545, 609
nonsynonymous (missense)mutations, 513–515
nonsynonymous substitution, 617
nopaline, 459
normal distribution of complex traits, 653–655
Northern blot technique, 428
NPD (nonparental ditype), 174
nuclear accidents, 536–537
nuclear division, 115. *See also* mitosis
nucleic acid hybridization, 52–54
backcross experiments, 87
CNP detection, 68
SNPs (single-nucleotide polymorphisms), 62–63
nucleoids, 226
nucleosides, 41
nucleosomes, 227–228
in chromatin, 407
nucleotide deletions, 461, 515–516
nucleotide excision repair, 541–542
nucleotide insertions, 515–516
nucleotide substitutions, 512–513
nucleotides, 41
nucleus, chromosome territories in, 229–230
null mutations. *See* loss-of-function mutations
numbering of carbon atoms, 41
NURF complex, 407
nurse cells, 491
Nüsslein-Volhard, Christiane, 492
nutritional mutants (bacterial genetics), 303

## O

observed vs. expected events. *See* goodness-of-fit testing
ochre codon. *See* UAA codon
octopine, 459
octoploids, 284, 286

*odd-skipped* genes, 493
*Oenothera,* translocation complexes in, 282
Okazaki fragments. *See* precursor fragments
oligonucleotides, 56, 57. *See also* PCR amplification
CNP detection, 68
SNPs (single-nucleotide polymorphisms), 62–63
oncogenes, 570–573, 581. *See also specific gene*
activation of, 565
defined, 570
1′ carbon, defined, 41
1:2:1 ratio (Mendelian), 82–85
oocytes, 120. *See also* gametes
composition and organization of, 475–476, 488
open reading frame (ORF). *See* coding sequence
operator, defined, 385
operator region, example of, 385–386
operator sequences, *lac* operon, 389
operon model, 386–387, 389, 390, 396
operon system, 383–392. *See also specific operon*
examples of, 387–392
opsithokonts, 32
ORC (origin recognition complex), 561
ordered tetrads, analysis of, 177–180
ORF (open reading frame). *See* coding sequence
organelle(s), 588
evolutionary origin of, 600–601, 604
genetic code in, 593–594
organelle heredity, 588, 592–600. *See also* chloroplast DNA; extranuclear inheritance; mitochondrial DNA (mtDNA)
organelle introns, 356
organismic genetics, 2
origin of replication, 191
in cloning vectors, 440–442
origin recognition complex (ORC), 561
origins of replication, multiple, in eukaryotes, 198–200
orthologous genes, 618
orthologs, 617–618
outcrossing, 80–81. *See also* backcross experiments; testcross experiments
polyploids, 286
ovalbumin mRNA, 439
overdominance, 638–639, 646
overlapping code, 377
oxidation, 530–531
oxidizing agent, 530–531
*oxyS* gene, 418

## P

*P* element, 455–457
P (peptidyl) site, 359
$P_1$ generation, defined, 81
p16/p19ARF proteins, 573–574
*p21* function, loss of, 573
*p53* function, loss of, 567, 573, 576–577, 583
*p53* transcription factor, 563–565, 571
loss of function, 567, 573, 576–577, 583
pachytene, 125
PAH (phenylalanine hydroxylase), 13
mutants of, 26–27
pair-rule genes, 490, 493–494
in regulatory hierarchy, 494–496
paired chromosomes, 115
pairing of DNA bases, 8, 40
code abbreviations, 375–376
between DNA and RNA, 23–24
incorporation error (mismatching), 191–192
molar concentrations of bases and, 43
renaturation
kinetics of, 233–235
nucleic acid hybridization, 52–54, 62–63, 68, 87
in RNA splicing, 355
stability of, 246
during transcription, 349
palindromes (DNA sequences), 50, 433
Papua New Guineans, 643
paracentric inversion, 275–277
paralogous genes, 618
paralogs, 617–618
*Paramecium aurelia,* killer strains of, 601–602
parasegments, 490
parental ditype (PD) parental ditype, 174
parental types of chromosomes, 152
parsimony methods, 613
partial digestion, 438, 442, 467
partial diploids, 312
Pascal's triangle, 137–138
passenger strand, 415
paternal inheritance, 588
paternity testing, 626
path, pedigree, 629–630
pathgenicity islands, 301–302
pattern formation, 472, 489–491, 497–498
pBluescript plasmid, 440–441
PCR adaptors, 210
PCR amplification, 57–60, 438
inverse PCR procedure, 339
massively parallel sequencing, 209–210
reverse transcriptase, 440
site-directed mutagenesis, 460–461
tandem repeats and, 67
PD (parental ditype), 174
pedigree, 97–101
human
genetic mapping in, 170–173
X-linked inheritance, 133
molecular markers, 100–101
pseudoautosomal inheritance and, 259
pedigree studies
cancer, 580
inbreeding, 629–632
mitochondrial genetic disease, 589–590
penetrance of genetic disorder, 98
peptide bond, 342
peptidyl transferase, 362
percent recombination. *See* map distance
permissive conditions, 510, 554
petite mutants, 598–599
*Petunia hybrida,* gene regulation in, 414–415
PEV (position-effect variegation), 283–284
phage. *See entries at* bacteriophage
phage lytic cycle, 312–313
recombination in, 318–319
phage repressor, 331–332
pharmacogenomics, 675
*Phaseolus vulgaris,* 462
phenotype frequencies, 82–85, 89–90. *See also* segregation
goodness-of-fit testing (example), 141–142
Mendel's data "too good to be true", 142–144
phenotypes
of ABO blood group system, 104
bacterial genetics, 302–303
defined, 61–62
epistasis, 105–107
among quantitative-trait loci, 672

phenotypes *(Continued)*
defined, 483
in developmental switches, 483–485
principle of, 483–484, 505
genotypes vs., 61–62, 84
incomplete dominance, 101–107
morphological vs. molecular, 78–80
position effect, 282–284
telomeric silencing, 242
testcross and backcross experiments, 86, 87. *See also* ratios (from genetic crosses)
genetic mapping in three-point testcrosses, 166–169
with unlinked genes, 88–89
phenotypic means, 665
phenotypic ratios, 82. *See also* segregation
phenotypic variance, 659
phenotypic variation
causes of, 652, 655–660
with individual selection, 665–666
physical performance and, 673
phenylalanine, 11–14, 367, 370
breakdown and excretion of, 29
phenylalanine hydroxylase (PAH), 13
mutants of, 26–27
phenylalanine operon, 394
phenylketonuria (PKU), 13
traits affected by, 28–29
phenylthiocarbamide (PTC), 99–100
Philadelphia chromosome, 579–580
*Phlox cuspidata*, inbreeding in, 629
phosphatase enzymes, 559
phosphodiester bonds, 41–42
phosphorylation, 354
photoreactivation, 542
phylogenetics, molecular, 612
physical distance between genetic markers, 169
*Escherichia coli* T4 mutants, 319–321
physical expression of genotype. *See* phenotypes
physical performance, phenotypic variation and, 673
phytate, 462
PKU (phenylketonuria), 13
traits affected by, 28–29
plant(s). *See also specific species*
breeding, complex inheritance in, 652, 656, 660, 662–663. *See also* artificial selection
cytoplasmic male sterility in, 598–600, 606
developmental genetics of, 500–503
drug resistance in, 526–527
extranuclear inheritance in, 592. *See also* chloroplast DNA
genetic engineering in, 445, 454, 458–459, 464
hybrid, 667
methylation in, 411
mitochondrial genome of, 593, 594
repeated self-fertilization in, 627–628, 632
somatic mutations in, 510
plantae, 32
plants
breeding
DNA typing in, 69
with monoploid organisms, 288–289
grass family genome, evolution of, 289–290
meiosis in, 120–121. *See also* meiosis
polyploidy in evolution of, 284–289, 569
allopolyploidy, 286–288, 292
autopolyploids, 286–288
monoploids, 288–289
sexual vs. asexual polyploidization, 285–286
plaque formation (bacteriophage genetics), 317–318
plasmid vectors, 434–436, 437, 440–441
in plants, 458–459
plasmids, 222, 296
defined, 32, 434
F plasmids, 297, 302
F′ plasmids, 311–312
F plasmids
matings with Hfr cells, 305–311
origins of replication, 191
R plasmids, 302
*Plasmodium falciparum*, 639, 640
pleiotropy, 29
pneumonia (Griffith's experiment with mice), 3–4
point centromeres, 240
Pol II holoenzyme, 406
polar granules, 474
polarity of DNA strands, 8, 42
pole cells, 486, 488
poly-A tail, 352
as priming site in recombinant DNA, 439
trimming of, 414
polyadenylation, 352, 354
polycistronic mRNA, 366–368, 386
polylinker, 440–441, 468
polylysine, 370
polymerase α (alpha), 202
polymerase chain reaction. *See* PCR amplification
polymerase δ (delta), 202
polymerization of tubulin, 117
polymorphism. *See* genetic markers
polynucleotide chains, 41–43
as antiparallel in double helix, 45–46
cleaving with restriction enzymes, 48–50
defined, 432
features of, 433–434
use of, 436–437
rewinding of (renaturation)
kinetics of, 233–235
nucleic acid hybridization, 52–54, 62–63, 68, 87
supercoiling of, 223–226
twisting of. *See* double helix structure of DNA
unwinding of (denaturation), 52–53, 200–201
melting temperature, 247
polypeptide chains, 21. *See also* central dogma of molecular genetics
colinearity between coding sequences and, 344
structure of, 342, 343, 364
synthesis of. *See* protein synthesis; translation
polyploidy in plant evolution, 284–289, 569
allopolyploidy, 286–288, 292
autopolyploids, 286–288
monoploids, 288–289
sexual vs. asexual polyploidization, 285–286
polyproline, 370
polysomes, 366
polytene chromosomes, 232
polyuridylic acid (poly-U), 367, 370
population
defined, 619
genetically uniform, example of, 661
population genetics, 618–627
basic concepts of, 619–620
defined, 619
evolution and, 632–633
migration in, 633–634
mitochondrial DNA used in, 592, 615, 643–645
mutations in, 633–634
population studies, DNA markers for, 70

position effect, 282–284
positional information, 473–474, 500
positive autoregulation, 383
positive chromatid interference, 164
positive regulation, 381–383
  of lactose operon, 387–389
positive regulators, 483
positively supercoiled DNA molecules, 224
posterior genes, 491–492
posttranslational control, 381
Prader-Willi syndrome, 412–413, 477
precursor fragments, 204–207
premutation, 517
prereplication complexes, 561
primary transcript, 350–352
  alternative splicing of, 413–414
primase, 202
primer
  mutagenic, 460
  in recombinant DNA, 439–440
  for retrotransposons, 522–523
primer molecules, 56, 57
  initiation of DNA strands, 191, 201–202
  removal and replacement of, 205–207
primosome complex, 201–202
principle of independent assortment, 88, 152
principle of segregation. *See* segregation
prior information, in probability, 94
probability
  basic, in genetics, 91–97
  goodness-of-fit testing, 139–144
  progeny distributions, 136–144
  in random genetic drift, 641–642
probe DNA, 53
processivity, of DNA polymerase, 564
product molecules, 11–13
proflavin, 368–370
progeny distributions, predicting, 136–139
  hypothesis testing (goodness of fit), 139–144
programmed cell death (apoptosis), 478–480, 565–566, 571
prokaryotes. *See also* bacteria; *specific class or species*
  ancestry of, 30–31
  chromosomal structure, 226
  colinearity in, 344
  DNA replication in, 200
    initiation of DNA strands, 202
  genome size, 222–223
  messenger RNA in, 352, 359–360, 366–368
  organelle genetic code in, 594
  protein folding in, 365–366
  repetitive and unique nucleotide sequences, 238
  ribosomes in, 358, 362, 367
  RNA polymerases in, 346
  transcriptional regulation in, 381–383
  translation in, 359–363, 366
promoter(s), 344
  alternative, 407–410
  binding strength of, 347
  example of, 386
  genetic evidence for, 350
promoter fusion, 578
promoter mutations, 387
promoter recognition, 347
promoter region, 346
  example of, 385–386
  of metallothionein gene, 461
proofreading (3′-to-5′ exonuclease activity), 192, 204, 205–207
prophage, 326–332
prophage induction, 332
prophase I stage (meiosis), 124–125
prophase II stage (meiosis), 128
prophase (mitosis), 117, 118
protanopia and protanomaly, 274
proteasome, 561
protein(s), 10. *See also specific protein; specific protein*
  antibodies, defined, 104
  common among ancestors. *See* unity of life
  for compacting DNA, 226
  complete set in genomes, 32–33
  defective proteins, 15–20
  degradation of, 561–562, 565
  functions of, 342
  polypeptide chains, 21. *See also* central dogma of molecular genetics
    colinearity between coding sequences and, 344
    structure of, 342, 343, 364
    synthesis of. *See* protein synthesis; translation
  shared, families of, 32–33
  structure of, 342, 343
  synthesis of, 358–359. *See also* translation
    discovery of, 367
    by transgenic organisms, 462–463
protein evolution, rates of, 615–616
protein folding, 364–366
protein kinases, cyclin-dependent, 557–559
protein markers, in translation, 359
proteobacteria, 600
proteolysis, cell cycle regulation by, 561–562, 565
proteomes, about, 32–33
proteomics, 452–454
  defined, 432
proto-oncogenes, 570–573, 581
prototrophs, 303
pseudoautosomal inheritance, 259
pseudogenes, 39, 617
PTC (phenylthiocarbamide), 99–100
Punnett, Reginald C., 85
Punnett square, 85, 621–622, 627
  algebraic alternative to, 89–90
  example of, 83
  X-linked inheritance, 132–133
purine bases, 41–43. *See also* adenine (A); guanine (G)
  molar concentration of, 43
  pairing. *See* pairing of DNA bases
pyknodystosis, 630
pyrimidine bases, 41–43. *See also* cytosine (C); thymine (T)
  pairing. *See* pairing of DNA bases
pyrimidine dimers, 532–533

## Q

quadrant enhancer, 498
quantitative-trait loci (QTLs), 670
  genetic mapping of, 670–672
  number and nature of, 672–674
quantitative traits, 652, 662
quartets, 555

## R

R-bands, 253
R cells, injected into mice, 3–4
R group, 342
R plasmids, 302
R408W mutant (PAH gene), 27–28

radiation
ionizing, 529, 533–536
ultraviolet, 532–533
units of, 534, 535
radioactivity, 536–537
radiotherapy, 534
Ramanis, Zenta, 596
random failure to breed. *See* random genetic drift
random genetic drift, 633, 639–643
random mating, 620–621, 625
selection and, 636–637
test for, 622–623
random sampling, 641
random shearing, 48
random-spore analysis, 177
RAPD (randomly amplified polymorphic DNA), 75, 111
RapI protein, 242
rare mRNA, 439, 440
rare recessive inheritance, 98–99
Ras protein, 572
ratios (from genetic crosses)
basic Mendelian, 82–85
hypothesis testing (goodness of fit), 139–144
incomplete dominance and epistasis, 106–109
RB (retinoblastoma) protein, 560, 571, 574, 577, 580, 583
reading frame, 368–370
recessive alleles, 78
in genes controlling development. *See* loss-of-function mutations
incomplete dominance, 101–107
narrow-sense heritability and, 665, 668–669
pedigree analysis, 97–99
phenotypic ratios ($F_2$ generation), 82. *See also* segregation
rare recessive inheritance, 98–99
selection against, 636–637
testcross and backcross experiments, 86, 87. *See also* ratios (from genetic crosses)
genetic mapping in three-point testcrosses, 166–169
with unlinked genes, 88–89
uncovering of, 271
reciprocal crosses, 81–82
X-linked inheritance, 129–133
reciprocal translocations, 278–280
without semisterility, 282
recombinant chromosomes, 152
recombinant DNA technology, 432–443
applications of, 461–465
cloning strategies, 436–440, 442, 469. *See also* cloning vectors
defined, 432
detection of recombinant molecules, 440–443
physical exchange between chromosomes, 159–161
principles of, 434
reverse transcriptase used in, 439–440
screening for particular recombinants, 442–443
recombination frequency, 152, 154–155, 156–159
in bacteriophage mapping, 319–321
lod scores and, 172
map distance and, 156, 168–169
minimum and maximum, 164
recombination of genes, 159, 180–182. *See also* transcription
bacteria vs. eukaryotes, 316
in bacteriophage lytic cycle, 318–319
characteristic frequency of, 154–155
chi-square test for linkage, 154
defined, 151
deletions created by, 270
ectopic, 524
evolutionary history and, 615
exchange mechanism in bacteria, 303
frequency of crossing over, 156. *See also* map distance
human X and Y chromosomes, 260–262
mitotic, 181–182
molecular mechanisms of, 210–215
multiple crossovers, 164–165
probability of crossing over. *See* genetic distance
site-specific, 300, 326–332
tetrad analysis, 174
timing of crossing over, 161–163
unequal crossing over, 273
red–green color blindness, 273–275, 627
recombination within genes, 180–181
recruitment, transcriptional activation by, 404–405
red–green color blindness, 273–275
X-linked, 627
reductional division. *See* first meiotic division
redundancy of genetic code, 372
regional centromeres, 240
regulatory elements, 386
regulatory hierachy, interactions in, 494–496
relative fitness, 635–636
relative risk, defined, 68
relaxase, 297
relaxed DNA molecules, 223, 226
release factor (RF), 363
release phase, translation, 362–363
removal of primer molecules, 205–207
renaturation
kinetics of, 233–235
nucleic acid hybridization, 52–54
backcross experiments, 87
CNP detection, 68
SNPs (single-nucleotide polymorphisms), 62–63
repair of mismatched base pairs, 211–212
repetitive sequences of DNA, 233
analysis by renaturation kinetics, 235–237
repica DNA strands, 10
replica plating, 525–526
replication bubbles, 199
replication forks, 198
fragments in, 204–207
replication of DNA, 9–10
chromosome territories and, 230
discontinuous, of lagging DNA strands, 204–207
DNA structure and, 46
elongation. *See* elongation of DNA strands
initiation of. *See* chain initiation
meiosis and, 119
during mitosis, 116, 552
problems with, 191–192
proofreading. *See* proofreading
selective, of fragments, 55–60, 419. *See also* PCR amplification
semiconservative replication, 192–200
bidirectional replication in eukaryotes, 198–200
in chromosomes, 194–196
rolling-circle DNA replication, 199
theta replication of circular DNA molecules, 197–198
stabilization and stress relief, 200–201
unidirectional and bidirectional, 198
replication slippage, 518
replicons, defined, 191
reporter-gene constructs, 402–404
reporter genes, in proteomics, 453
repressible transcription, 381–382
examples of, 384–385, 390–392

repressor, 381–382
examples of, 384–385, 387, 391–392
repulsion configuration of alleles, 153
recombination frequency, 154–155
restriction conditions, 510, 554, 556
restriction enzymes, 48–50
defined, 432
features of, 433–434
use of, 436–437
restriction fragment polymorphism (RFLP), 63–66, 646, 648, 676. *See also* DNA markers
Mendel's monohybrid exeriments, 82
restriction fragments, 50
purification of, 438
restriction maps, 51
restriction site, 432
restriction sites of enzymes, 50
retinoblastoma (RB) protein, 560, 571, 574, 577, 580, 583
retrotransposons, 522–523
retroviruses, 444–445
reverse genetics, 460
reverse mutations, 543–545, 634
reverse transcriptase, 439–440, 522–523
reverse transcriptase PCR (RT-PCR), 440
reversible mutations, 634
reversible terminators, 210
reversion, 543
reversion test, 545
rewinding of double helix. *See* renaturation
RF (release factor), 363
RFLP (restriction fragment polymorphism), 63–66, 646, 648, 676. *See also* DNA markers
Mendel's monohybrid exeriments, 82
rhodopsin pigment. *See* red–green color blindness
ribonucleic acid (RNA)
antisense, 417–418
central dogma of molecular genetics, 22–23, 342
double-stranded, 414–416
guide, 242, 415–416
messenger. *See* messenger RNA (mRNA)
micro-, 415–417, 447
S box, 394
small interfering, 414–415
transfer. *See* transfer RNA (tRNA)
ribose, 201–202
structure of, 345
ribosomal RNA (rRNA)
central dogma of molecular genetics, 22–23, 342
evolutionary perspective on, 30
ribosome-binding site, 361
ribosomes, 22, 25, 351
role in translation, 358, 367
structure of, 358, 362
riboswitches, 394–396
ribozymes, 356
rice, transgenic, 462, 464
rII gene structure, 321–325, 329
RIP (repeat-induced point) mutation, 411, 525
RISC (RNA-induced silencing) complex, 415–416
risk factors (genetic), 68
cloning, 170
RNA (ribonucleic acid)
antisense, 417–418
central dogma of molecular genetics, 22–23, 342
double-stranded, 414–416
evolutionary perspective on, 30
guide RNA, 242, 415–416
messenger. *See* messenger RNA (mRNA)
micro-, 415–417, 447
primer molecules, 56, 57
initiation of DNA strands, 191, 201–202
removal and replacement of, 205–207
processing, chromosome territories and, 230
S box, 394
small interfering, 414–415
synthesis of polypeptides. *See* translation
transfer. *See* transfer RNA (tRNA)
transfer of genetic information from DNA. *See* transcription
RNA editing, 378, 593
RNA-induced silencing complex (RISC), 415–416
RNA interference (RNAi), 414–417
RNA polymerase, 342, 344–345
processivity of, 346
promoter binding strength, 347
types of, 346–347
RNA polymerase holoenzyme, 346
RNA polymerase I, 346
RNA polymerase II, 346
carboxyl terminal domain, 354
transcription factors associated with, 348–349
RNA polymerase III, 346–347
RNA processing, 342
coupling of transcription and, 354
defined, 352
in eukaryotes, 352–358
gene regulation through, 381, 413–414
speed and specificity of, 354
RNA splicing, 352–353
gene regulation through, 381, 413–414
mechanism of, 354–357
mutations at site of, 357–358
order of, 354
self-, 356
specificity of, 355
RNA synthesis, overview of, 344–346
RNA viruses, 440, 457, 463–465
Robertsonian translocations, 280–281
rolling-circle DNA replication, 199, 303, 307
Romanov, Georgij, 591
RPA (replication protein A), 206
*rpoS* gene, 418
rRNA. *See* ribosomal RNA (rRNA)
RSC complex, 407
RT-PCR (reverse transcriptase PCR), 440

## S

S-adenosylmethionine (SAM), 394, 528–529
S box RNA, 394
S cells, injected into mice, 3–4
S period of interphase, 116, 552, 560–561, 570
S-phase checkpoint, 563
*Saccharomyces cerevisiae*
cell cycle in, 552–553, 559
gene regulation in, 400–401, 404
knockout mutations in, 461
mating-type interconversion in, 422–424
mitochondrial mutants in, 598
as recombinant DNA host, 432
Sagan, Lynn (Margulis), 604
Sager, Ruth, 596
salmon, transgenic, 461–462
*Salmonella typhimurium*
Ames test using, 545
F plasmid infection, 336
sample space, defined, 91–92

Sanger, Frederik, 208
Sanger sequencing, 208
satellite DNA, 233, 239
*Schizosaccharomyces pombe*, cell cycle in, 552–553
second-division segregation, 177–180
second meiotic division, 120, 123, 128
  segregation in, 177–180
secondary effects of mutant genes, 28–29
segment, 490
segment-polarity genes, 490–491, 494
  in regulatory hierarchy, 494–496
segmental allopolyploidy, 292
segmentation genes, 490–491
  regulatory hierarchy of, 494–496
segregation
  chi-square test for linkage, 154
  defined, 83
  first-division and second-division, 177–180
  independent, 94
  independent assortment, 88, 152
  of single gene, 80–87
  tetrad formation in meiosis, 174
  of three or more genes, 89–90
  of two genes, 87–89
  verification of, 85–87
segregational petites, 598
selected marker, defined, 306
selection
  artificial. *See* artificial selection
  balancing. *See* heterozygote superiority
  in diploid organisms, 636–637
  individual, 665–666
  natural, 633, 634–639, 666
selection coefficient, 636
selection differential, 679
selection limit, 666
selection-mutation balance, 637–638
selective breeding, 662
selective medium (bacterial genetics), 303
selectively neutral mutations, 616
self-fertilization
  polyploids, 285, 286
  repeated, 627–628, 632
self-pollination (self-fertilization), 80–81
self-splicing, 356
self-sterility, 103
self-termination, 349–350
semiconservative DNA replication, 192–200
  bidirectional replication in eukaryotes, 198–200
  in chromosomes, 194–196
  rolling-circle DNA replication, 199
  theta replication of circular DNA molecules, 197–198
semisterility from translocation, 278
  reciprocal translocations without semisterility, 282
senescence, 573
separation of DNA fragments, 47–55
  gel electrophoresis, 50–52, 438
  nucleic acid hybridization, 52–54
    backcross experiments, 87
    CNP detection, 68
    SNPs (single-nucleotide polymorphisms), 62–63
  restriction enzymes, 48–50
    defined, 432
    features of, 433–434
    use of, 436–437
  Southern blot, 54–55, 65
septum, 553
sequence of DNA bases, 8. *See also* coding sequence
  complexity of. *See* complexity, genetic
  conservation of, defined, 227
  flanking sequences, 339
  genome determination, 32–33
  mapping. *See* genetic mapping; mapping of human genome
  polyploidy, 288
  repetitive, 233, 237
    eukaryotic genomes, 237–239
  telomeric DNA, for various organisms, 244
  terminator sequencing, 207–210
serial dilutions, 335
serine-5, 354
serine proteases, 493
serotonin, 674–675
seryl-tRNA synthetase, 372
severe combined immunodeficiency disease, 465
sex chromosomes. *See also entries at* X; Y chromosome
  dosage compensation, 256–259
  in human karyotypes, 252–253
  human trisomies and monosomies, 268–269
  inheritance with, 128–136. *See also* X-linked inheritance
sex determination (chromosomal), 128–129, 133–134
  *SRY* gene (human), 260
sex ratio, maternal, 603
sex-specific gene silencing, 412–413
sexual polyploidization, 285–286
sexual selection, 452
shared protein families, 32–33
sheep, cloning of, 456
Shine-Dalgarno sequence. *See* ribosome-binding site
shotgun sequencing, 208
*shrunken* gene, 520
Siamese cat, 511
siblings, defined, 97
sickle-cell anemia, 513–515, 639, 640
sigma factor, 347
significant (in probability), defined, 140
silencers, 348, 401–402, 447. *See also* gene silencing
silent (synonymous) mutations, 513
simple sequence repeat polymorphisms (SSRPs), 100–101
simple sequence repeats (SSRs), 67
  Y-chromosome genealogical studies, 262
simultaneous transformation (cotransformation), 305
SINE (short interspersed elements), 523, 524
single-active-X principle, 258, 268
single-copy DNA sequences, 237–238
single-nucleotide polymorphism (SNP), 61–63. *See also* DNA markers
  that eliminate restriction sites. *See* RFLP
single-nucleotide polymorphisms (SNPs), 672
single-stranded binding protein (SSBP), 200–201, 459
single-stranded DNA, 8
siRNA (small interfering RNA), 414–415
sister chromatids, 119
  exchange of, 196
site-directed mutagenesis, 460–461
site-specific DNA cleavage, 48–50
  defined, 432
  features of, 433–434
  use of, 436–437
site-specific recombination, 300, 326–332
size, genome. *See* genome size
Skolnick, Mark, 64
sleeping sickness, 593
slipped-strand mispairing. *See* replication slippage
small interfering RNA (siRNA), 414–415
small regulatory RNAs, translational control by, 417–418
Smithies, Oliver, 458, 463

snail shell coiling, maternal effect in, 603–606
SNP chips, 61
SNPs (single-nucleotide polymorphisms), 61–63, 672. *See also* DNA markers
  that eliminate restriction sites. *See* RFLP
snRNPs, 355, 356
software, gene-tree estimation, 613–614
somatic cells
  conversion to cancer cells, 570
  nuclei, 115
somatic mosaics, 182
somatic mutations, 422, 510
SOS repair system, 543
Southern, Edwin, 54–55
Southern blot, 54–55
  with RFLPs, 65
specialized transduction, 332–333
species trees, 615
sperm. *See* gametes
spermatocytes, 120
spindle assembly checkpoint, 567–568
spindle fiber, 117, 239
spindle, malfunctioning of, 569
spindle pole, 552
spindle pole bodies, 553
spindle position checkpoint, 568–569
spliceosomes, 354–355
splicing. *See* RNA splicing
spontaneous mutations, 510, 513, 525–529
sporadic cancer, 569
spores, 120
sporophytes, 120
*SRY* gene (human), 260
SSB (single-stranded DNA-binding protein), 200–201
SSBP (single-stranded binding protein), 459
SSRPs (simple sequence repeat polymorphisms), 100–101
SSRs (simple sequence repeats), 67
  Y-chromosome genealogical studies, 262
stability of chromosomes
  base pairing differences, 246
  centromeres and, 255–256
  chromosome complements, 115–116
stabilization of DNA template strands, 201
Stahl, Franklin W., 192, 195
standard deviation, 654
standard genetic code, 25–26, 368–374. *See also* genetic code
  for codons, 25–26
  defined, 368
  degeneracy of, 370, 371
  discovery of, 370–371
  features of, 371
  of organelles, 593–594
  redundancy of, 372
  standard abbreviations, 375–376
  wobble in, 373
statistical significance
  defined, 140
  for linkage between genes, 153–154
Stein, Curt, 159
stem cells, embryonic, germ-line transformations using, 457–458
sterility. *See* fertilization; semisterility from translocation
sticky ends of cleaved DNA strands, 50, 433–434, 436, 438
stop codons, 359, 363, 370, 373
  UAA codon, 26, 370, 373
  UGA codon, 26, 370, 373
stop sequence, in introns, 357
strand invasion, 213–214
Strand, Micheline, 540
strands of DNA. *See* DNA structure
*Streptoccus pneumoniae*, 3
stress on DNA strands during replication, 201
*Strongylocentrotus purpuratus*, developmental genetics in, 475
structural genes, 386
structure of DNA, 6–9, 40–47
  antiparallelism of strands, 44–45
  base pairing. *See* pairing of DNA bases
  base stacking, 44–45, 53
  cohesive ends, 326, 436
  double helix structure of DNA, 7–8
    antiparallelism of, 45–46
    characteristics of, 43
    major and minor grooves, 44–45
    problems with Watson–Crick model, 191–192
    unwinding of (denaturation), 52–53, 200–201, 247
  5′ (donor) end of DNA strands, 8, 354–357
    discontinuous replication of lagging strand, 204–207
    strand elongation and, 191, 204
  polynucleotide chains. *See* polynucleotide chains
  as related to function, 46
  3′ (acceptor) end of DNA strands, 8, 354–357, 440
    restoring ends of DNA molecules, 242
    strand elongation and, 191, 203–204
structure of human chromosomes, abnormal, 269–282
  deletions and duplications, 268–275
  inversions, 275–277
  translocations, 277–282
Sturtevant, Alfred H., 155
subfunctionalization, 401, 618
sublineages, 478
submetacentric chromosomes, 255–256
subpopulation, 619, 633
substrate-dependent pathways, 483
substrate molecules, 11–13
Sulston, John E., 479
supercoiling of DNA, 223–226
suppression
  intergenic, 544–545
  intragenic, 543–544
suppressor mutations, 369, 373–374, 543–545
suppressors, nonsense, 545, 609
SWI-SNF complex, 407
switch-regulation pathways, 483–485, 504
*Sxl* gene (*Drosophila*), 136
symbionts, 588, 604
  cytoplasmic transmission of, 601–603
symbiosis, 600
synapsis of homologous chromosomes, 124
synaptonemal complex, 124
synchronized cells, 554
syncytial blastoderm, 486, 488
synonymous codons, 370, 371
synonymous (silent) mutations, 513
synonymous substitution, 617
syntenic genes, 152
  coupling vs. repulsion, 153
synteny groups, defined, 290
synthesis of DNA. *See* elongation of DNA strands
synthesis of polypeptides. *See* translation

## T

T base. *See* thymine
T DNA, 458–459
T (thymus-derived) cells, DNA rearrangements in, 419–422

T4 bacteriophage
  genetic recombination, 318–319
  mapping of, 319–321
  rII gene structure, 321–325, 329
T4 ligase, 433
TAFs (TBP-associated factors), 406
tandem duplication, 273–275
tandem repeat polymorphisms, 66–67, 102–103. *See also* DNA markers
  Hardy-Weinberg principle and, 623–624
TAP (transcriptional activator protein), 382, 401, 407, 452
  homeotic genes encoding, 498
*Taq* polymerase, 60
taster polymorphism with PTC, 99–100
TATA box, 347–348
TATA-box-binding protein (TBP), 348, 406, 407–408
Tatum, Edward L., 14, 15, 21, 105–106, 313, 483
taxon, 612
TBP-associated factors (TAFs), 406
TBP (TATA-box-binding protein), 348, 406, 407–408
TDF (testis-determining factor), 260
telomerase, 242, 573
telomeres, 566
  molecular structure of, 241–244
  sequences of, for various organisms, 244
telomeric silencing, 242
telophase I stage (meiosis), 127
telophase II stage (meiosis), 128
telophase (mitosis), 118, 119
temperate phage, 326–332
temperature-sensitive mutations, 510–511, 554, 556
template DNA, 10
  massively parallel sequencing of, 209–210
  PCR amplification with, 60
  Sanger sequencing of, 208
template strand, 349, 350–352
Temujin. *See* Khan, Ghengis
teosinte, 656
terminal genes, 492–493
terminal inverted repeats, 522
terminal redundancy, 320
termination proteins, 349–350
terminator conformation, 394
terminator sequencing of DNA, 207–210
  Sanger and shotgun sequencing, 208
terminators, genetic evidence for, 350
terminus of replication, 198
testcross experiments, 86. *See also* ratios (from genetic crosses)
  genetic mapping in three-point testcrosses, 166–169
  with unlinked genes, 88–89
testicular embryonal carcinoma, 573
testing segregation principle, 85–87
testis-determining factor (TDF), 260
tetrad analysis, 173–180
  ordered tetrads, 177–180
  unordered tetrads, 173–177
*Tetrahymena*, RNA splicing in, 356
tetraploids, 264, 284
tetratype (TT) tetrad, 174
TF. *See* transcription factors (TF); trigger factor (TF)
thermophiles, defined, 60
theta replication of circular DNA molecules, 197–198
13:3 ratio, 107, 109
30-nm chromatin fiber, 229
three-point testcrosses, 165, 166–169
three-strand double crossover, 164
3′ (acceptor) end of DNA strands, 8, 354–357, 440
  restoring ends of DNA molecules, 242
  strand elongation and, 191, 203–204
3′-hydroxyl (3′-OH) group, 42
3′ polyadenylation, 352
3′-to-5′ exonuclease activity, 192, 204
3:1 ratio (Mendelian), 82–85
  goodness-of-fit testing (example), 141–142
  Mendel's data "too good to be true", 142–144
threshold traits, 653
  heritability and, 669–670
thymine dimer, formation of, 533
thymine (T), 8, 41
  base pair stability, 264
  bonding to adenine. *See* pairing of DNA bases
  mismatching in base pairs. *See* incorporation error
  molar concentration of, 42–43
  structure of, 345
thymus-derived (T) cells, DNA rearrangements in, 419–422
*Ti* plasmid, 458–459
time-of-entry mapping, 306–311
  frequency of cotransduction and, 315–316
tissue antigens, 189
titin gene, 353
tolerance mechanism, 104
tomatoes, quantitative trait mapping in, 671–672
topoisomerase enzymes, 201, 224–225
torsional stress during DNA replication, 201
*torso* gene, 492–493
Toulouse-Lautrec, Henri, 630
traits, relationships with genes, 24. *See also* phenotypes
tranformation of *S. penumoniae* (experiment), 4–5
*trans* configuration of alleles, 153
  recombination frequency, 154–155
transacetylase, 386, 387
transcript(s)
  antisense, 414
  defined, 23
  human, characteristics of, 353–354
  primary, 350–352
    alternative splicing of, 413–414
transcription (DNA to RNA), 22–24, 342, 344–350. *See also* recombination of genes
  accessory proteins for, 347
  alternative splicing, 32
  base-pairing specificity of, 349
  chromosome territories and, 230
  coupling of RNA processing and, 354
  defined, 344
  elongation cycle in, 349
  in eukaryotes, 352–358
  evolutionary perspective on, 30
  inducible, 381–382
    example of, 383–392
  initiation of, 348–349
  mechanism in bacteria, 303
  mechanism of, 348–350
  plasmid replication, 297
  rate of, 346, 354, 411–412
  reinitiation of, 350
  repressible, 381–382
    examples of, 384–385, 390–392
  reverse, 439–440, 522–523
  termination of, 349–350
    regulation through, 392–396
  X-chromosome inactivation and, 259–260
  of *Xist* (in dosage compensation), 257
transcription complex, eukaryotic, 404–407

transcription factors (TF), 406
basal, 404–407
in floral development, 502–503
gap genes encoding, 493
general, 347–349
p53, 563–565, 571
loss of function, 567, 573, 576–577, 583
transcription start site, 344, 346, 347
transcriptional activator proteins (TAP), 382, 401, 407, 452
homeotic genes encoding, 498
transcriptional enhancers, 348, 401–402
evolutionary signature, 447
transcriptional inactivation
methylation and, 411–412, 518–520, 529
in vector molecules, 440–442
transcriptional profiling, 447–452
transcriptional program, of cell cycle, 553–554
transcriptional regulation, 381–383
in bacteriophages, 396–397
epigenetic, 409–413
in eukaryotes, 381, 398–409
of mating type, 423–424
operon system of, 383–392
in prokaryotes, 381–383
through termination, 392–396
transcriptional silencers, 348, 401–402, 447. *See also* gene silencing
transducing phage, defined, 312
transduction, 312–316
mechanism in bacteria, 303
phage lytic cycle, 312–313
specialized, 332–333
transfer-order maps, 306–311
frequency of cotransduction and, 315–316
transfer RNA (tRNA)
aminoacylated, 359, 371–372
binding sites for, 359–361
central dogma of molecular genetics, 22, 342
charged, 359, 371–372
evolutionary perspective on, 30
intergenic suppression in, 544
mutations, 373–374
role in translation, 24–25, 358–359
structure of, 371, 372
transformation, DNA-mediated, 303
transformation procedure, in recombinant DNA technology, 434–435
transformation rescue, 459–460
transfusions, blood (human), 104–105
transgenic organisms, 454–461
defined, 432, 454
examples of, 461–465
transition mutations, 512
translation (synthesis of polypeptides), 22, 24–25, 342, 358–366
defined, 358
direction of, 368
elongation cycle in, 361–362
initiation of, 359–361
nonsense-mediated decay, 359
in organelles, 593–594
overview of, 359
release phase of, 362–363
translational control, 381, 417–418
translocation complexes in *Oenothera*, 282
translocations, chromosomal, 277–282
genetic mapping of breakpoints, 278–280
reciprocal, 278–280, 282
Robertsonian, 280–281
transmission genetics, 77–113
human pedigree analysis, 97–101
molecular markers, 100–101
incomplete dominance and epistasis, 101–107
morphological phenotypes, 78–80
probability in genetic analysis, 90–97
progeny distributions, predicting, 136–139
hypothesis testing (goodness of fit), 139–144
pseudoautosomal inheritance, 259
segregation of single gene, 80–87
segregation of three or more genes, 87–89
segregation of two genes, 87–89
sex-chromosome inheritance, 128–136. *See also* X-linked inheritance
transmission of genes together. *See* linkage between genes
transportable element (*P* element), 455–457
transposable elements, 80, 520–525
as agents of mutation, 523–524
defined, 520
discovery of, 519, 520
in human genome, 520, 523, 524–525
transposase, 455, 521–523
transposition, 520–523
in bacterial genetics, 298–299
transposon, 299
transverse banding, 232
transversion mutations, 512–513
TRAP (*trp* RNA-binding attenuation protein), 393, 394
*Tribolium* (flour beetle), complex inheritance in, 662
trichothiodystrophy, 542
trigger factor (TF), 365
trigger loop, 349
trinucleotide expansion, 517
trinucleotide repeats, dynamic mutation of, 516–518
triplet code, 368
genetic evidence for, 368–370
triploids, 264, 285
trisomic zygotes, 269
trisomy (human), 264
sex-chromosome trisomies, 268–269
trisomies 18 and 13, 266
trisomy 21 (Down syndrome), 266–267
chromosomal translocations and, 280–282
landmark paper on (excerpt), 265
*Triturus*, transcription in, 350
trivalent chromosomes, 267–268
tRNA. *See* transfer RNA (tRNA)
*trp* operon, 390–394
*trp* RNA-binding attenuation protein (TRAP), 393, 394
*trpA* gene, 344, 544
true-breeding, defined, 81
truncation point, 665
*Trypanosoma brucei*, 593
tryptophan codon, discovery of, 370
tryptophan operon, regulation of, 390–394
TT (tetratype), 174
tubulin polymerization and depolymerization, 117
tumor-suppressor genes, 573–575
Turner syndrome, 269
Turpin, Raymond, 265
12:3:1 ratio, 107
twin spot (mitotic recombination), 181–182
twin studies, 663
twisting of polynucleotide chains. *See* double helix structure of DNA
two-gene testcrosses, 87–89
two-hybrid analysis, 452–454

two-strand double crossover, 164
twofold degenerate site, 617
type II restriction endonucleases, 48
typing. *See* DNA typing
tyrosine, 11–13
tyrosine kinase inhibitor STI571 (imatinib mesylate), 580

## U

U base. *See* uracil
UAA codon, 26, 370, 373
UAG codon, 26, 370, 373
UAS (upstream activator sequence), 399, 404
UGA codon, 26, 370, 373
*ultrabithorax* (*Ubx*) gene, 498–500
ultraviolet radiation, 532–533
umber codon. *See* UGA codon
unbalanced translocation, 264
*unc*-22 gene, 416
uncovering of recessive alleles, 271
underwinding of DNA molecules, 223–224
unequal crossing over, 273
  red–green color blindness, 273–275
    X-linked, 627
unidirectional DNA replication, 198
union of events (probability), 93
uniparental inheritance
  defined, 588
  example of, 597–598
unique DNA sequences, 237–238
unity of life, 29–33
  ancestral commonalities for all humans, 262
univalent chromosomes, 267–268
universal recipients and donors (ABO), 105–107
unlinked genes, testcrosses with, 88–89
unordered tetrads, analysis of, 173–177
unpaired chromosomes. *See* univalent chromosomes
unreduced gametes, 285–286
unwinding of double helix. *See* denaturation
UPGMA (unweighted pair group method with arithmetic mean), 613, 616
upstream activator sequence (UAS), 399, 404
uracil (U), 22, 23, 202
  pairing. *See* pairing of DNA bases
  structure of, 345
uridine triphosphate (UTP), 344
UTP (uridine triphosphate), 344
UV (ultraviolet) light, 532–533

## V

V-J joining, 420–421
V-type position effect, 283–284
vaccines, recombinant DNA, 465
vaccinia virus, 465
variable (V) antibody regions, 420
variable number of tandem repeats (VNTRs), 67, 625–626, 647, 650
variance, 653–655
  co-, 667
  environmental, 652, 655, 658–660
  genotypic, 656–660
  phenotypic, 659. *See also* phenotypic variation
variegated-type position effect, 283–284
vectors. *See* cloning vectors
vegetative propagation, somatic mutations in, 510
ventral uterine precursor cell (VU), 480–481
verification of segregation, 85–87
*vestigial* gene, 496–498
viability of gametes, 622, 637
viroids, defined, 32
virulence (vir) region, 459
viruses. *See also specific virus*
  animal, genetic engineering with, 463–465
  genetically engineered, 457
  genome size, 222–223
  repetitive and unique nucleotide sequences, 238
  RNA, 440, 457, 463–465
vitamin A deficiency, 462
VNTR (variable number of tandem repeats) polymorphism, 67, 625–626, 647, 650
vulval differentiation, determination of, 482–485

## W

W chromosome, 134
water, DNA base repellance to, 45
Watson, James D., 6, 367
  landmark double-helix paper (excerpt), 47
Watson–Crick double helix. *See* double helix structure of DNA
Watson–Crick pairing. *See* pairing of DNA bases
Weinberg, Wilhelm, 621
Western blot technique, 428
WGA (whole-genome association), 672
White, Raymond L., 64
White Leghorn and While Wyandotte chickens, 107, 109
whole-genome association (WGA), 672
Wieschaus, Eric, 492
Wilmut, Ian, 456
wings clipped, 455
wobble concept, 372–373
Wright, Sewall, 144

## X

X chromosome, 128. *See also entries at* X-linked; sex chromosomes
  cancer and, 570
  *Drosophila* linkage experiments, 152
  human trisomies and monosomies, 268–269
  inactivation of (X inactivation), 257–259
    active genes, 260
    gene silencing, 259
  inactive, 411
  inversion with Y chromosome, 260–262
  X-linked genes, 130–131
X-Gal, 404, 442
X-linked conditions, dynamic mutations in, 516–518
X-linked Duchenne muscular dystrophy, 353
X-linked genes, allele frequency in, 627
X-linked inheritance, 129–133
  dosage compensation, 256–259
  human pedigree characteristics, 133
  red–green color blindness, 274
x-rays, mutagenicity of, 529, 533–536
*Xenopus laevis*
  gene amplification in, 419
  oocytes, 476
xeroderma pigmentosum, 542, 578, 648
*XIC* site, 257
*Xist* region, 257

## Y

Y chromosome, 128, 268–269. *See also* sex chromosomes
  in *Drosophila*, 135–136
  ectopic recombination in, 270
  evolutionary history and, 615
  gene content and evolution of, 260–264
yeast. *See also specific species*
  cell cycle in, 552–553, 559
  DNA microarray analysis of, 355, 450, 553–554
  galactose metabolism in, 398–401
  mating-type interconversion, 422–424
  mismatch repair in, 540
  mitochondrial genetic code, 594
  mitochondrial mutants in, 598
  proteomic studies of, 452–454
  spindle position checkpoint in, 568
  transcription complexes in, 407

## Z

Z chromosome, 134
Z form of DNA, 46
Z-linked inheritance, 134
zinc finger, 401, 493
zygotene, 124
zygotes (human), monosomic and trisomic, 266
zygotic genes, 476–477, 488–489